ASTM DICTIONARY OF ENGINEERING SCIENCE & TECHNOLOGY

Sponsored by ASTM Committee E02 on Terminology

Ninth Edition
2000

ASTM Stock Number: DEF00

ASTM
100 Barr Harbor Drive
PO Box C700
West Conshohocken, PA 19428-2959
e-mail: service@astm.org
website: http://www.astm.org

ORDER INFORMATION: Additional copies of this book in print, on CD-ROM, or reprints (single or multiple copies) of individual standards may be obtained by contacting ASTM at the above address or at 610-832-9585 (phone), 610-832-9555 (fax), service@astm.org (e-mail), or through www. astm.org (website).

Printed in Baltimore, MD
September 2000

FOREWORD

Committee E02 on Terminology is very proud to sponsor this edition of the *ASTM Dictionary of Engineering Science & Technology*, (formerly the *Compilation of ASTM Standard Definitions*). This document enables its users to reference terminology, compare definitions across disciplines, and take advantage of existing terms and definitions rather than reinvent terms and develop definitions unnecessarily.

Ron Buchan, former lexicographer for the National Aeronautics and Space Administration, guided ASTM in developing the structure of the compilation. ASTM was recommended to me because of its world class reputation in terminology and definitions. I use the compilation in the standardization of data elements for computer applications for the federal government. ASTM's definitions are developed by subject matter experts in specific disciplines such as Healthcare, Metrics, and Steel. The definitions are developed using the structure proposed in the *Form and Style for ASTM Standards,* which makes comparison of definitions across the various disciplines easier. As technology becomes more prevalent in our industries, the compilation will become an even more valuable asset.

Angela D. Booker
Chairman
ASTM Committee E02 on Terminology

ASTM TERMINOLOGY STANDARDS CITED IN THIS COMPILATION

The definitions contained in this compilation are standard definitions that are published in the terminology standards listed below. The designation of the terminology standard and the committee having jurisdiction appear at the end of each definition. Further discussion of terms may be included in the text of the terminology standard.

Designation	Title	Committee
A 340	Magnetic Testing, Symbols	A06
A 644	Iron Castings	A04
A 902	Metallic Coated Steel Products	A05
A 941	Steel, Stainless Steel, Related Alloys, and Ferroalloys	A01
B 243	Powder Metallurgy	B09
B 354	Uninsulated Metallic Electrical Conductors	B01
B 374	Electroplating	B08
B 542	Electrical Contacts and Their Use	B02
B 713	Superconductors	B01
B 846	Copper and Copper Alloys	B05
B 881	Aluminum- and Magnesium-Alloy Products	B07
B 899	Non-Ferrous Metals and Alloys	B02
C 11	Gypsum and Related Building Materials and Systems	C11
C 43	Structural Clay Products	C15
C 51	Lime and Limestone as used by the Industry	C07
C 71	Refractories	C08
C 119	Dimension Stone	C18
C 125	Concrete and Concrete Aggregates	C09
C 162	Glass and Glass Products	C14
C 168	Thermal Insulating Materials	C16
C 219	Hydraulic Cement	C01
C 242	Ceramic Whitewares and Related Products	C21
C 274	Structural Sandwich Constructions	D30
C 286	Porcelain Enamel and Ceramic-Metal Systems	B08
C 460	Asbestos-Cement	C17
C 634	Environmental Acoustics	E33
C 709	Manufactured Carbon and Graphite	D02
C 717	Building Seals and Sealants	C24
C 822	Concrete Pipe and Related Products	C13
C 859	Nuclear Materials	C26
C 896	Clay Products	C04
C 904	Chemical-Resistant Nonmetallic Materials	C03
C 1145	Advanced Ceramics	C28
C 1154	Non-Asbestos Fiber-Reinforced Cement Products	C17
C 1180	Mortar and Grout for Unit Masonry	C12
C 1209	Concrete Masonry Units and Related Units	C15
C 1213	Precast Concrete Products	C27
C 1232	Masonry	C15
D 8	Materials for Roads and Pavements	D04
D 9	Wood	D07

D 16	Paint, Related Coatings, Materials and Applications	D01
D 121	Coal and Coke	D05
D 123	Textiles	D13
D 459	Soaps and Other Detergents	D12
D 653	Soil, Rock, and Contained Fluids	D18
D 804	Naval Stores, Including Tall Oil and Related Products	D01
D 883	Plastics	D20
D 907	Adhesives	D14
D 996	Packaging and Distribution Environments	D10
D 1038	Veneer and Plywood	D07
D 1079	Roofing, Waterproofing, and Bituminous Materials	D08
D 1129	Water	D19
D 1356	Sampling and Analysis of Atmospheres	D22
D 1517	Leather	D31
D 1554	Wood-Base Fiber and Particle Panel Materials	D07
D 1566	Rubber	D11
D 1695	Cellulose and Cellulose Derivatives	D01
D 1711	Electrical Insulation	D09
D 1968	Paper and Paper Products	D06
D 2050	Zippers	D13
D 2507	Rheological Properties of Gelled Rocket Propellants	F07
D 2652	Activated Carbon	D28
D 2825	Polishes and Related Materials	D21
D 2864	Electrical Insulating Liquids and Gases	D27
D 2946	Asbestos	C17
D 3053	Carbon Black	D24
D 3064	Aerosol Products	D10
D 3743	Bridge Deck and Substructure Protection	D04
D 3766	Catalysts and Catalysis	D32
D 3878	Composite Materials	D30
D 3888	Open-End Spinning	D13
D 3918	Reinforced Plastic Pultruded Products	D20
D 3990	Fabric Defects	D13
D 4023	Humidity Measurements	D22
D 4092	Dynamic Mechanical Measurements on Plastics	D20
D 4127	Ion-Selective Electrodes	D19
D 4150	Gaseous Fuels	D03
D 4175	Petroleum, Petroleum Products, and Lubricants	D02
D 4391	Burning Behavior of Textiles	D13
D 4410	Fluvial Sediment	D19
D 4439	Geosynthetics	D35
D 4466	Multicomponent Textile Fabrics	D13
D 4523	Feather-Filled and Down Filled Products	D13
D 4538	Protective Coating and Lining Work for Power Generation Facilities	D33
D 4725	Engine Coolants	D15
D 4790	Aromatic Hydrocarbons and Related Chemicals	D16
D 4848	Force, Deformation, and Related Properties of Textiles	D13
D 4850	Fabric	D13
D 4920	Moisture in Textiles	D13
D 4965	Seams and Seam Finishes Used in Home Sewing	D13
D 5077	Electrostatic Discharge (ESD) Packaging Materials	D10
D 5219	Body Dimensions for Apparel Sizing	D13
D 5253	Writing Care Instructions and General Refurbishing Procedures for Textile Floor Coverings and Textile	D13
D 5488	Environmental Labeling of Packaging Materials and Packages	D10
D 5497	Buttons	D13
D 5535	Formed-in-Place Sealants for Joints and Cracks in Pavement	D04
D 5646	Basic Sewing Machine Stitches for Home Use	D13
D 5681	Waste and Waste Management	D34
D 5684	Pile Floor Coverings	D13
D 6161	Crossflow Microfiltration, Ultrafiltration, Nanofiltration and Reverse Osmosis Membrane Processes	D19
D 6384	Biodegradability and Ecotoxicity of Lubricants and Lubricant Components	D02

D 6440	Hydrocarbon Resins	D01
D 6488	Print Problems	D01
E 6	Methods of Mechanical Testing	E28
E 7	Metallography	E04
E 41	Conditioning	G03
E 131	Molecular Spectroscopy	E13
E 135	Analytical Chemistry for Metals, Ores, and Related Materials	E01
E 170	Radiation Measurements and Dosimetry	E10
E 175	Microscopy	E41
E 176	Fire Standards	E05
E 253	Sensory Evaluation of Materials and Products	E18
E 284	Appearance	E12
E 344	Thermometry and Hydrometry	E20
E 349	Space Simulation	E21
E 456	Quality and Statistics	E11
E 473	Thermal Analysis	E37
E 609	Pesticides	E35
E 631	Building Constructions	E06
E 673	Surface Analysis	E42
E 772	Solar Energy Conversion	E44
E 833	Building Economics	E06
E 856	Physical and Chemical Characteristics of Refuse Derived Fuel	D34
E 867	Vehicle-Pavement Systems	E17
E 943	Biological Effects and Environmental Fate	E47
E 957	Geothermal Energy	E44
E 1102	Agricultural Chemical Application	E35
E 1126	Biomass Fuels	E48
E 1142	Thermophysical Properties	E37
E 1187	Conformity Assessment	E36
E 1316	Nondestructive Examinations	E07
E 1325	Design Of Experiments	E11
E 1445	Hazardous Potential of Chemicals	E27
E 1480	Facility Management Building-Related	E06
E 1481	Railing Systems and Rails for Buildings	E06
E 1519	Agricultural Tank Mix Adjuvants	E35
E 1542	Occupational Health and Safety	E34
E 1547	Industrial and Specialty Chemicals	E15
E 1605	Abatement of Hazards from Lead-Based Paint in Buildings and Related Structures	E06
E 1620	Liquid Particles and Atomization	E29
E 1638	Sieve Analysis	E29
E 1658	Expressing Conclusions of Forensic Document Examiners	E30
E 1705	Biotechnology	E48
E 1732	Forensic Science	E30
E 1749	Rigid Wall Relocatable Shelters	E06
E 1778	Pavement Distress	E17
E 1807	Metal Connector Plates	E06
E 1823	Fatigue and Fracture	E08
E 1992	Terminology Management	E02
E 2035	Forensic Psychophysiology	E52
F 17	Flexible Barrier Materials	F02
F 109	Surface Imperfections on Ceramics	C21
F 141	Resilient Floor Coverings	F06
F 149	Optical Character Recognition	F05
F 221	Carbon Paper and Inked Ribbon Products and Images Made Therefrom	F05
F 335	Electrostatic Copying	F05
F 395	Vacuum Cleaners	F11
F 412	Plastic Piping Systems	F17
F 425	Lithographic Copy Products	F05
F 471	Combination Locks	F12
F 472	Geometry of Alpine Skis	F27
F 538	Characteristics and Performance of Tires	F09

F 547	Nails for Use with Wood and Wood-Base Materials	F16
F 549	Carbonless Copy Products	F05
F 552	Chain Link Fencing	F14
F 592	Collated and Cohered Fasteners and Their Application Tools	F16
F 740	Filtration	F21
F 747	Amusement Rides and Devices	F24
F 818	Spill Response Barriers	F20
F 819	Electrical Protective Equipment for Workers	F18
F 869	Athletic Shoes and Biomechanics	F08
F 909	Printers	F05
F 1078	Surgical Scissors-Inserted and Non-Inserted Blades	F04
F 1107	Snowboarding	F27
F 1156	Product Counterfeit Protection Systems	F12
F 1177	Emergency Medical Services	F30
F 1241	Silicon Technology	F01
F 1251	Polymeric Biomaterials in Medical and Surgical Devices	F04
F 1294	Automatic Identification Hard Copy Printing Systems	F05
F 1327	Barrier Materials for Medical Packaging	F02
F 1379	Barbed Tape	F14
F 1457	Laser Printers	F05
F 1479	Microwave Food Packaging	F02
F 1490	Search and Rescue	F32
F 1494	Protective Clothing	F23
F 1549	Underwater Search, Rescue, and Recovery Activities	F32
F 1582	Spinal Implants	F04
F 1600	Bioremediation	F20
F 1611	Sizing of IMFDs for Application to Reamed Diaphyseal Bone Associated Instrumentation	F04
F 1623	Thermal Imaging Products	F05
F 1638	Surgical Tissue/Dressing/Pick-Up Forceps (Thumb-Type)	F44
F 1646	Safety and Traction for Footwear	F13
F 1773	Climbing And Mountaineering Equipment And Practices	F08
F 1789	F16 Mechanical Fasteners	F16
F 1827	Food Service Equipment	F26
F 1840	Surgical Suture Needles	F04
F 1857	Ink Jet Printers and Images Made Therefrom	F05
F 1972	Candles and Associated Accessory Items	F15
F 2002	Anesthesia and Respiratory Equipment	F29
G 15	Corrosion and Corrosion Testing	G01
G 40	Wear and Erosion	G02
G 113	Natural and Artificial Weathering Tests of Nonmetallic Materials	G03
G 126	Compatibility and Sensitivity of Materials In Oxygen Enriched Environments	G04

A

A4A—symbol for ISO standard metric cut-size paper which is nominally 20 by 30 cm or exactly 210 by 297 mm. **F 1457,** F05

AASHTO compaction—see **compaction test.** **D 653,** D18

AATCC blue wool lightfastness standards, *n*—standard dyed-wool samples of seven grades, each step in the series representing a doubling of lightfastness. **E 284,** E12

abandonment—see **decommissioning.** **D 653,** D18

Abbe condenser—see **condenser, Abbe.** **E 175,** E41

Abbé value—the reciprocal dispersive power, a value used in optical design, expressed mathematically as:

$$Abbé\ value = (n_d - 1)/(n_F - n_C)$$

where n_d is the refractive index for the helium line at 587.6 nm and n_F and n_C are the refractive indices for the hydrogen lines at 486.1 and 656.3 nm, respectively. See synonymous term **nu-value** and related term **dispersion.** **C 162,** C14

abbreviation, *n*—a shortened form of a single-element term created by the omission of some of its letters. **E 1992,** E02

abduction—the movement of a body part from the longitudinal (midline) of the body or in reference to fingers and toes, movement away from the midline of hand or foot. **F 869,** F08

aberration—any error that results in image degradation. Such errors may be chromatic, spherical, astigmatic, comatic, distortion, or curvature of field; and can result from design or execution, or both. **E 175,** E41

abietic acid, commercial grade, *n*—a product consisting chiefly of rosin acids in substantially pure form, separated either from rosin or tall oil commercially for specific purposes and in which abietic acid and its isomers are the principal components. **D 804,** D01

ablation, *n*—a self-regulating heat and mass transfer process in which incident thermal energy is expended by sacrificial loss of material. **E 349,** E21

above-grade, *adj*—above the surface of the ground, *as related to floor location,* above a well-ventilated space with at least 18 in. between the bottom of the lowest horizontal structural member and any point of the ground. **F 141,** F06

abrasion, *n*—the wearing away of any part of a material by rubbing against another surface. **D 123,** D13

abrasion—a rubbing and wearing away. (ISRM) **D 653,** D18

abrasion—the mechanical wearing, grinding, scraping or rubbing away (or down) of rock surfaces by friction or impact, or both. **D 653,** D18

abrasion, *n*—the surface loss of a material due to frictional forces. **D 1566,** D11

abrasion, *n*—wearing, grinding, or rubbing away by friction. **F 141,** F06

abrasion-corrosion, *n*—a synergistic process involving both abrasive wear and corrosion in which each of these processes is affected by the simultaneous action of the other and, in many cases, is thereby accelerated. **G 40,** G02

abrasion mark, *n*—an area damaged by friction. **D 123,** D13

abrasion of refractories, *n*—wearing away of refractory surfaces by the scouring action of moving solids. **C 71,** C08

abrasion resistance—the degree to which a porcelain enamel will resist attack by abrasive materials. **C 286,** B08

abrasion resistance—the property of a particle to resist attrition or wearing away by friction. **D 2652,** D28

abrasion resistance—the ability of the polished shoe finish to withstand scuff marks versus the nonpolished shoe finish. **D 2825,** D21

abrasion resistance (*coatings*)—the ability of a coating to resist being worn away and to maintain its original appearance and structure when subjected to rubbing, scraping, or wear. **D 16,** D01

abrasion resistance (coatings)—ability of a coating to resist being worn away and to maintain its original appearance, integrity, and structure when subjected to rubbing, scraping, or wear. **E 1605,** E06

abrasion resistance index, *n*—a measure of the abrasion resistance of a rubber relative to that of a standard rubber under the same specified conditions, expressed as a percentage. **D 1566,** D11

abrasion tester—a machine for determining the quantity of material lost by friction wear under specified conditions. **F 869,** F08

abrasive—any rock, mineral, or other substance that, owing to its superior hardness, toughness, consistency, or other properties, is suitable for grinding, cutting, polishing, scouring, or similar use. **D 653,** D18

abrasive blast cleaning—the removal from a concrete surface of all laitance, oil, grease, curing compounds, or coatings by the action of a high-velocity stream of abrasive in air or water followed by a blast of air to provide a concrete surface free of abrasive dust and other particles. The abrasive may be clean silica sand, mineral grit, steel shot, steel grit, or mineral slag. **D 3743,** D04

abrasive blasting—a process for cleaning or finishing by means of an abrasive directed at high velocity against the work piece. **B 374,** B08

abrasiveness—the property of a material to remove matter when scratching and grinding another material. (ISRM) **D 653,** D18

abrasive wear, *n*—wear due to hard particles or hard protuberances forced against and moving along a solid surface. **G 40,** G02

abrasivity, *n*—the ability of a material or substance to cause abrasive wear. **G 40,** G02

abridged spectrophotometry, *n*—the measurement of reflectance factor or transmittance factor in a number of wavelength bands rather than as continuous functions of wavelength. **E 284,** E12

ABS—an abbreviation for alkyl benzene sulfonate. Although strictly speaking this might apply to any such compound, present practice is to use it for those containing branched chains. (See **LAS**). **D 459,** D12

absolute coil—a coil (or coils) that respond(s) to the total detected electric or magnetic properties, or both, of a part or section of the test part without comparison to another section of the part or to another part. **E 1316,** E07

absolute filter—See **filter.** **D 1356,** D22

absolute humidity, *n*—the mass of water vapor present in a unit volume of air. (See **humidity** and **relative humidity**) **D 123,** D13

absolute humidity, vapor concentration, vapor density (d_v), *n*—the ratio of the mass of water vapor, m_v, to the total volume of the moist air, v:

$$d_v = \frac{m_v}{v}$$

relative humidity, n—the ratio of the actual water vapor pressure to the saturation pressure.

relative humidity with respect to ice (U_i), *n*—the ratio in percent of the mole fraction of water vapor, x_v, in moist air to the mole fraction of water vapor, x_{vi}, that the moist air would have if it were saturated with respect to ice at the same pressure, p, and temperature, t.

$$U_i = \frac{X_v}{X_{vi}} \times 100$$

relative humidity with respect to water *(U_w)*—the ratio in percent of mole fraction of water vapor, x_v, in moist air to the mole fraction of water vapor, x_{vw}, that the moist air would have if it were saturated with respect to water at the same pressure, p, and temperature, t

$$U_w = \frac{X_v}{X_{vw}} \times 100$$

D 1356, D22

absolute impact velocity— See **impact velocity**. **G 40,** G02

absolute judgment, *n*—an evaluation of a stimulus made without direct comparison. **E 253,** E18

absolute manometer—a manometer whose calibration can be calculated from the measurable physical constants of the instrument and for which calibration is the same for all ideal gases. **E 1316,** E07

absolute measurements—in electromagnetic testing, measurements made without a direct reference using an absolute coil in contrast to differential and comparative measurements. (See also **absolute coil**). **E 1316,** E07

absolute or true density—See *absolute or true density* under **density.** **C 242,** C21

absolute pressure—pressure above the absolute zero corresponding to empty space, that is, local atmospheric pressure plus gage pressure. **E 1316,** E07

absolute readout—in electromagnetic testing, the signal output of an absolute coil. (See also **absolute coil**). **E 1316,** E07

absolute sealing—a level of sealing that requires all seams, slots, holes, and fasteners passing through the seal plane to be sealed. **E 1749,** E06

absolute system—a system that uses a coil assembly and associated electronics to measure the total electromagnetic properties of a test part without direct comparison to another section of the part or to another part (see **absolute coil**). (E 566) **E 1316,** E07

absolute temperature—See **temperature.** **D 1356,** D22

absorbance—the logarithm of that fraction of an incident light beam that is dissipated in the sample, being neither transmitted nor reflected. **C 242,** C21

absorbance, *n*—the logarithm to the base of 10 of the reciprocal of *transmittance.* **D 1356,** D22

absorbance—the logarithm to the base 10 of the reciprocal of the relative transmittance, *T*, expressed as:

$$A = \log_{10}(1/T) = -\log_{10} T$$

D 4790, D16

absorbance, *n*—*in spectrometry,* the logarithm to the base 10 of the reciprocal of the transmittance, *T*.

$$A = \log_{10}(1/T) = -\log_{10} T$$

E 135, E01

absorbance, *A*—*in spectrochemical analysis,* the logarithm to the base 10 of the reciprocal of the transmittance.

$$A = \log\ (1/T)$$

C 859, C26

absorbance, *A*— the logarithm to the base 10 of the reciprocal of the transmittance, (*T*).

$$A = \log_{10}(1/T) = -\log_{10}T$$

E 131, E13

absorbance, *A, n*—logarithm to the base 10 of the reciprocal of the internal transmittance T_I. $A = \log_{10}(1/T_I) = -\log_{10} T_I$. **E 284,** E12

absorbate, *n*—material that has been retained by the process of absorption. **D 1356,** D22

absorbed dose—the amount of energy imparted by ionizing radiation per unit mass of irradiated matter. Denoted by "rad;" 1 rad = 0.01 j/kg. SI unit is "gray;" 1 gray = 1 j/kg. **E 1316,** E07

absorbed dose (D)—Quantity of ionizing radiation energy imparted per unit mass of a specified material. The SI unit of absorbed dose is the gray (Gy), where 1 gray is equivalent to the absorption of 1 joule per kilogram of the specified material (1 Gy = 1 J/kg). The mathematical relationship is the quotient of $d\bar{\varepsilon}$ by dm, where $d\bar{\varepsilon}$ is the mean incremental energy imparted by ionizing radiation to matter of incremental mass dm (see ICRU 33).

$$D = d\bar{\varepsilon} \,/\, dm$$

E 170, E10

absorbed dose rate—the absorbed dose per unit of time; rads/s. SI unit, grays/s. **E 1316,** E07

absorbed dose rate, $\dot{D}$—the absorbed dose in a material per incremental time interval, that is, the quotient of d *D* by d *t* (see ICRU Report 33).

$$\dot{D} = \mathrm{d}D/\mathrm{d}t$$

SI unit: $Gy{\cdot}s^{-1}$. **E 170,** E10

absorbed moisture—water held mechanically in the material and having physical properties not substantially different from ordinary water at the same temperature and pressure. **C 242,** C21

absorbed moisture, *n*—water held within the materials and having physical properties not substantially different from ordinary water at the same temperature and pressure. **C 1145,** C28

absorbed water—water held mechanically in a soil or rock mass and having physical properties not substantially different from ordinary water at the same temperature and pressure. **D 653,** D18

absorbent, *n*—material in which absorption occurs. **D 1356,** D22

absorbent compound, *n*—*in textile cleaning,* sponge-like particles which, when saturated with water or dry solvent and detergent and brushed into the textile, remove soil from the textile. See **powder cleaner**. **D 123,** D13

absorbent packing— See **packing.** **D 996,** D10

absorbent pad, *n*—*for cleaning textile floor coverings,* a damp textile material (fabric, felt, sponge, or mop) used to agitate and wipe the pile and, in the process, absorb soil. **D 123,** D13

absorber—that part of a solar collector whose primary function is to absorb radiant energy and transform it into another form of energy.

NOTE—A thermal absorber usually possesses a solid surface through which energy is transmitted by thermal conduction to the transfer fluid; however, the transfer fluid itself can be the absorber in the case of an optically transparent container and a "black liquid". A photovoltaic absorber converts part of the incident solar flux into electrical energy, and part to thermal energy. **E 772,** E44

absorptance, *n*—the ratio of the radiant flux absorbed by a body to that incident upon it. **C 168,** C16

absorptance α, *n*—the ratio of the absorbed radiant or luminous flux to the incident flux. **E 284,** E12

absorptance, *n*—ratio of the absorbed radiant or luminous flux to the incident flux. Symbol: α_e, α_v, α.

NOTE—In general, the value of the absorptance depends upon the mode of irradiation, the spectral composition, and the state of polarization of the incident radiation. **E 349,** E21

absorptance, α— the ratio of the absorbed radiant or luminous flux to the incident flux. (Practice E 491, for Solar Simulation for Thermal Balance Testing of Spacecraft). See **radiometric properties and quantities**. **E 772,** E44

absorptance, α—the ratio of radiant flux absorbed by a surface to that incident upon it. **E 1316,** E07

absorption, *n*—weight of water picked up by a clay masonry unit during immersion at prescribed conditions expressed in relation to the dry weight of the unit.

Two conditions of immersion are designated in standards relating to brick: 24 h in room temperature (60–86°F (15.5–30°C)) water or 5 h in boiling water. (Different time intervals are specified for structural tile and other products.) The resulting absorptions are termed *cold absorption* and *boiling absorption*.

Absorption values are used in brick and tile standards as one factor in classifying these products into durability grades. Absorptions are indicators of the extent of firing during manufacture as well as being indicators of durability. **C 43,** C15

absorption, *n*—the process by which a liquid is drawn into and tends to fill permeable pores in a porous solid body; also, the increase in mass of a porous solid body resulting from the penetration of a liquid into its permeable pores. **C 125,** C09

absorption, *n*—transformation of radiant energy to a different form of energy by interaction with matter. **C 168,** C16

absorption—(*1*) the relationship of the weight of the water absorbed by a ceramic specimen, subjected to prescribed immersion procedure, to the weight of the dry specimen.

(*2*) the capacity of a substance to take up a substance, usually a liquid or gas, with the formation of an apparently homogeneous mixture. **C 242,** C21

absorption—the increase in weight of concrete resulting from the penetration of water into the concrete. **C 822,** C13

absorption—difference in the amount of water contained within a concrete masonry unit or related unit between a saturated and oven-dry condition in accordance with the requirements of Test Methods C 140. It is expressed as weight of water per volume of concrete. **C 1209,** C15

absorption, *n*—a process in which one material (the absorbent) takes in or absorbs another (the absorbate); as the absorption of moisture by fibers. (See also **adsorption,** and **moisture equilibrium for testing**. Compare **desorption** and **resorption**.) **D 123,** D13

absorption—the assimilation of fluids into interstices. **D 653,** D18

absorption, *n*—a process in which one material (the absorbent) takes up and retains another (the absorbate) with the formation of an homogeneous mixture having the attributes of a solution. **D 1356,** D22

absorption—a process in which fluid molecules are taken up by a liquid or solid and distributed throughout the body of that liquid or solid. **D 2652,** D28

absorption, *n*—the process by which a liquid is drawn into and tends to fill permeable pores in a porous solid body, also, the increase in mass of a porous solid body resulting from penetration of a liquid into its permeable pores. **D 4439,** D35

absorption—the holding of a substance within a solid by cohesive or capillary forces. **D 6161,** D19

absorption—the decrease in intensity which radiation undergoes during its passage through matter when the ratio of transmitted or reflected luminous flux to incident is less than 1. **E 7,** E04

absorption, *n*—the transformation of radiant energy to a different form of energy by interaction with matter. **E 284,** E12

absorption, *n*—transformation of radiant energy to a different form of energy by interaction with matter. **E 349,** E21

absorption—a process in which one material (the absorbent) takes in and retains another (the absorbate). **E 609,** E35

absorption—the process by which incident radiant energy is transformed into another form of energy by interaction with matter. **E 772,** E44

absorption—the process whereby the incident particles or photons of radiation are reduced in number or energy as they pass through matter. **E 1316,** E07

absorption—in leak testing, the binding or incorporation of gas in the interior of a solid (or liquid). **E 1316,** E07

absorption band—a region of the absorption spectrum in which the absorbance passes through a maximum. **E 131,** E13

absorption coefficient—specific factor characteristic of a substance on which its absorption radiation depends. The rate of decrease of the natural logarithm of the intensity of a parallel beam per unit distance traversed in a substance. For X-rays, the linear absorption coefficient is the natural logarithm of the ratio of the incident intensity of an X-ray beam incident on unit thickness of an absorbing material to the intensity of the beam transmitted. If $I_e$is the incident intensity of a beam of X-rays, $I_t$the transmitted intensity, and X the thickness of the absorbing material, then:

$$I_t = I_e exp(-\mu X)$$

Here μ is the linear absorption coefficient. The mass absorption coefficient is given by μ/ρ where ρ is the density. **E 7,** E04

absorption coefficient, α—a measure of absorption of radiant energy from an incident beam as it traverses an absorbing medium according to Bouguer's law, $P/P_o = e^{-\alpha b}$. **E 131,** E13

absorption coefficient, α, *n*—measure of the absorption of radiant energy from an incident beam (P_o) as it traverses an absorbing medium according to Bouguer's law, $P = P_o\ e^{-\alpha b}$, where b is the sample optical pathlength. **E 284,** E12

absorption edge—an abrupt change in absorption coefficient at a particular wavelength. The absorption coefficient is always larger on the short wavelength side of the absorption edge. **E 7,** E04

absorption limit—See **absorption edge.** **E 7,** E04

absorption loss—that part of transmitted energy (mechanical) lost due to dissipation or conversion into other forms (heat, etc.). **D 653,** D18

absorption (of electromagnetic radiation), *n*—a decrease in radiant energy when passing through matter, resulting in a corresponding increase in the energy of the absorbing system. **E 135,** E01

absorption parameter, *a*—the relative reflection loss per reflection that results from the absorption of radiant energy at a reflecting surface: a = 1 – *R*, and *R* = the reflected fraction of incident radiant power. **E 131,** E13

absorption spectrometry, *n*—the branch of spectroscopy pertaining to the absorption of electromagnetic radiation by atoms, ions, radicals, and molecules. **E 135,** E01

absorption spectroscopy—See **emission spectroscopy**. **F 1156,** F12

absorption spectrum—a plot, or other representation, of absorbance, or any function of absorbance, against wavelength, or any function of wavelength. **E 131,** E13

absorption test—a test made to determine the absorption of concrete. **C 822,** C13

absorption tinting strength, *n*—relative change in the absorption properties of a standard white material when a specified amount of an absorbing colorant, black or chromatic, is added to it. **E 284,** E12

absorptivity, *a—in spectrochemical analysis*, the absorbance (*A*) divided by the product of the cell path length (*b*) and the concentration (*c*).

$$a = A/bc$$

C 859, C26

absorptivity, *a*— the absorbance divided by the product of the concentration of the substance and the sample pathlength, $a = A/bc$. The units of *b* and *c* shall be specified. **E 131,** E13

absorptivity, *a, n*—the absorbance divided by the product of the concentration, *c*, of the substance and the sample optical pathlength, *b*, $a = A/bc$. The units of *b* and *c* shall be specified. **E 284,** E12

absorptivity, molar, ϵ—the product of the absorptivity, *a*, and the molecular weight of the substance. **E 131,** E13

absorptivity of an absorbing material, *n*—internal absorptance of a layer of the material such that the path of the radiation is of unit length. **E 349,** E21

abundance sensitivity (isotopic measurement)—the ratio of the ion beam intensity of the major isotope, *M*, to the background current at the adjacent mass positions.

$$\text{Abundance sensitivity} = \frac{\text{ion current at mass } M}{\text{ion current at } M \pm 1}$$

C 859, C26

ac—symbol used to designate an electric voltage or current whose amplitude varies periodically as a function of time, its average value over one complete period being zero. One complete repetition of the wave pattern is referred to as a CYCLE, and the number of cycles occurring in one second is called the FREQUENCY, measured in hertz (Hz). For example, the electricity supplied by commercial utility companies in the United States is, in most localities, 60 Hz, although other frequencies may be encountered. **D 2864,** D27

accelerant, *n*—any material used to initiate or promote the spread of a fire. The most common accelerants are flammable or combustible liquids. Whether a substance is an accelerant depends not on its chemical structure but on its use (source: IAAI Forensic Science Committee, *Glossary of Terms Related to Chemical and Instrumental Analysis of Fire Debris*) (use: Test Method E 1387) CRIM. **E 1732,** E30

accelerated adsorption tests—adsorption tests in which the end point is hastened by testing at conditions more severe than those anticipated in service. **D 2652,** D28

accelerated aging, *n—in textile processing and testing*, the use of controlled environmental conditions to promote rapid physical or chemical change in a textile material. **D 123,** D13

accelerated aging—the deterioration of a material faster than normal by subjecting the material to conditions specified by the test method being followed. **F 869,** F08

accelerated aging—a technique to simulate the effects of package aging by subjecting the product/package system to environmental stress; generally accomplished using temperature and humidity extremes, depending on application; other environmental elements, for example, ultraviolet light, may have to be considered. **F 1327,** F02

accelerated erosion—erosion at a rate greater than geologic or natural erosion. **D 4410,** D19

accelerated life test, *n*—a test method designed to approximate in a short time, the deteriorating effect of normal long-term service conditions. **D 1566,** D11

accelerated life test—method designed to approximate, in a short time, the deteriorating effect of normal, long-term service conditions. (D 1566, D-11) **F 869,** F08

accelerated outdoor weathering, *n*—outdoor weathering using the sun as the source of irradiance, and where the rate of deterioration is accelerated over that of the in-service exposure position increasing one or more of the influencing parameters. **G 113,** G03

accelerated service life—the elapsed time until the end point is reached in an accelerated adsorption test. **D 2652,** D28

accelerated test—See **test, accelerated**. **E 1749,** E06

accelerated weathering, *v*—the exposure of a specimen to a specified test environment for a specified time with the intent of producing, in a shorter time period, effects similar to actual weathering. **D 1079,** D08

accelerating potential—a relatively high voltage applied between the cathode and anode of an electron gun to accelerate electrons. **E 7,** E04

accelerating potential—the difference in electric potential between the cathode and anode in an X-ray tube through which a charged particle is accelerated; usually expressed in units of kV or MV. **E 1316,** E07

acceleration period, *n—in cavitation and liquid impingement erosion*, the stage following the incubation period during which the erosion rate increases from near zero to a maximum value. (See also **erosion rate-time pattern.**) **G 40,** G02

accelerator, *n*—a material that will shorten the setting time of gypsum plasters. **C 11,** C11

accelerator—*as used in a furnace*, a material used to promote the rapid and complete reaction of a sample. **C 859,** C26

accelerator—a material that increases the rate at which chemical reactions would otherwise occur. **D 653,** D18

accelerator, delayed action, *n*—an accelerator that, in conjunction with other curing agent(s), produces, at vulcanizing temperatures, a period of no significant cross-linking, followed by a period of rapid cross-link formation. **D 1566,** D11

accelerator (for pyrohydrolysis)—a substance that speeds a reaction, in this case, a pyrohydrolysis. **C 859,** C26

accelerator (rubber), *n*—a compounding material used in small amounts with a vulcanizing agent to increase the speed of vulcanization. **D 1566,** D11

accelerometer, *n*—an instrument that senses inertial reaction to measure linear or angular acceleration. **F 538,** F09

acceptability/unacceptability, *n*—degree to which a stimulus is judged to be favorable or unfavorable. **F 253,** E18

acceptable quality level—the maximum percent defective or the maximum number of units defective per hundred units that, for the purpose of sampling test, can be considered satisfactory as a process average. **E 1316,** E07

acceptable quality level (AQL), *n*—the maximum percent nonconforming which, for purposes of sampling inspection, can be considered satisfactory as a process average. **D 1711,** D09

acceptable quality level (*AQL or* p_1), *n—in acceptance sampling*, the maximum fraction of nonconforming items at which the process average can be considered satisfactory; the process average at which the risk of rejection is called the producer's risk. **D 123,** D13

acceptance angle, *n—for an optical fiber*, the maximum angle, measured from the longitudinal axis or centerline of the fiber to an incident ray, within which the ray will be accepted for transmission along the fiber by total internal reflection. **E 131,** E13

acceptance angle (±α, deg), *n*—the angular distance, centered on the array axis of symmetry, over which the following conditions are met: (*a*) wind components are unambiguously defined, and (*b*) flow across the transducers is unobstructed or remains within the angular range for which transducer shadow corrections are defined. **D 1356,** D22

acceptance (control chart or acceptance control chart usage, *n*), *n*—a decision that the process is operating in a satisfactory manner with respect to the statistical measures being plotted: action limits: *control limits*. **E 456,** E11

acceptance level—a test level above or below which test specimens are acceptable in contrast to rejection level. **E 1316,** E07

acceptance limit (AL), *n*—a numerical value that defines the point between acceptable and unacceptable quality. **D 4175,** D02

acceptance limits—test levels used in electromagnetic sorting which establish the group into which the material under test belongs. (E 566) **E 1316,** E07

acceptance number, *n*—the maximum allowable number of nonconformities for a given AQL and sample size (lot-sample size). **D 1711,** D09

acceptance number (*c*), *n*—*in acceptance sampling*, the maximum number of nonconforming items in a sample that allows the conclusion that the lot conforms to the specification. **D 123,** D13

acceptance of a lot—the approval of a lot as conforming to contract or specification. **D 1517,** D31

acceptance range—a relative range of variation in impact pressure and release characteristic over which the image acceptor media will receive an image of specified quality. **F 221,** F05

acceptance sampling, *n*—sampling done to provide specimens for acceptance testing. **D 123,** D13

acceptance standard—in tubing inspection, a tube used to establish the acceptance level with artificial discontinuities as specified in the applicable product standard. **E 1316,** E07

acceptance standard—a tube with artificial discontinuities specified in the applicable product standard used to establish the acceptance level. (E 215) **E 1316,** E07

acceptance test—a test, or series of tests, conducted by the procuring agency, or an agent thereof, to determine whether an individual lot of materials conforms to the purchase order or contract or to determine the degree of uniformity of the material supplied by the vendor, or both. **C 859,** C26

acceptance test, *n*—a test, or series of tests conducted by the procuring agency, or an agent thereof, upon receipt to determine whether an individual lot of materials conforms to the purchase order or contract or to determine the degree of uniformity of the material supplied by the vendor, or both. (Compare **preproduction test** and **qualification test.**) **D 907,** D14

acceptance testing, *n*—testing performed to decide if a material meets acceptance criteria. **D 123,** D13

acceptance testing—testing performed on a product to determine whether or not an individual lot of the product conforms with specified requirements. **F 412,** F17

accepted engineering practice—that which conforms to accepted principles, tests, or standards of nationally recognized technical or scientific authorities. **F 747,** F24

accepted reference value, *n*—a value that serves as an agreed-upon reference for comparison, and which is derived as: (*1*) a theoretical or established value, based on scientific principles, (*2*) an assigned or certified value, based on experimental work of some national or international organization, or (*3*) a consensus or certified value, based on collaborative experimental work under the auspices of a scientific or engineering group. **E 456,** E11

acceptor—an impurity in a semiconductor that accepts electrons excited from the valence band, leading to hole conduction. **F 1241,** F01

accepts, *n*—the output stream from a materials separation device that contains the highest concentration (purity) of the components that the device is designed to separate. **D 5681,** D34

access, *n*—the process of reaching the patients or subjects and establishing physical contact with them. **F 1490,** F32

accessibility—the fraction of total cellulose present that is accessible to certain reagents under certain specified conditions. The conditions of determination should always be indicated. **D 1695,** D01

accessible—permitting close approach or contact that could include requiring removal or opening of an access panel or door. **E 1705,** E48

accessible bag, *n*—*in sampling*, a bag (in a pile of bags of asbestos fiber) of which at least one side or one end is fully exposed. **D 2946,** C17

accessible surface, *n*—*in sampling,* in a pile of bags of asbestos fiber, a side (of the pile) in which all the bags are accessible bags. **D 2946,** C17

accessible surface—interior or exterior surface (usually up to 5 ft (1.5 m) from floor or ground) that is accessible for young child to mouth or chew. See also **chewable surface**. **E 1605,** E06

accessories, *n*—linear formed metal, metal and paper, or plastic members fabricated for the purpose of forming corners, edges, control joints, or decorative effects in conjunction with gypsum board and plaster assemblies. **C 11,** C11

accessories—subordinate material such as fasteners, backer strips, closure strips, ridge and corner rolls, roofing starters and finishing pieces, couplings, gaskets, pipe fittings or other supplementary material necessary for the proper application of primary asbestos-cement products. **C 460,** C17

accessories, *n*—subordinate material such as fasteners, backer strips, closure strips, ridge and corner rolls, roofing starters and finishing pieces, couplings, gaskets, pipe fittings or other supplementary material necessary for the proper application of primary fiber-reinforced cement products. **C 1154,** C17

accessories—subordinate material such as fasteners, backer strips, closure strips, ridge and corner rolls, roofing starters and finishing pieces, couplings, gaskets, pipe fittings or other supplementary material necessary for the proper application of primary asbestos-cement products. **D 2946,** C17

accessories—*optional* mechanical devices used on or in conjunction with a boom system but not included with the basic boom and end connector; for example, lights, paravanes, drogues, buoys, anchor systems, storage bags, boxes or reels, bulkhead connectors or repair kits, and so forth. **F 818,** F20

acclimation period—the time necessary for an animal to adjust to the laboratory and specific test environments. **E 609,** E35

Ac_{cm}, Ac_1, Ac_3, Ac_4—See **transformation temperature**. **A 941,** A01

accommodation, visual—adjustment of the eye, either the focus or the iris opening, to optimize its performance under the specific viewing conditions prevailing. **E 1316,** E07

accountability—a term used to designate the internal control system used for nuclear materials, but does not include physical protection. **C 859,** C26

accreditation, *n*—official authorization, approval, or recognition accorded an individual or organization based upon specific qualification. **E 631,** E06

accreditation—procedure by which an authoritative body gives formal recognition that a body or person is competent to carry out specific tasks, (ISO Guide 2). **E 1187,** E36

accreditation criteria, *for laboratory accreditation, n*—set of requirements used by an accrediting body which a testing laboratory must meet to be accredited, (ISO Guide 2, *A*). **E 1187,** E36

accredited laboratory—testing laboratory to which accreditation has been granted, (ISO Guide 2, *A*). **E 1187,** E36

accredited laboratory test report—test report that includes a statement by the testing laboratory that it is accredited for the test reported and that the test has been performed in accordance with the conditions prescribed by the accrediting body, (ISO Guide 2, *A*). **E 1187,** E36

accrediting authority, *n*—a body that evaluates the capability of a testing agency or an inspection agency, or both, in certain specific fields of activity. **D 1356,** D22

accrediting body—governmental or non-governmental body that conducts and administers a laboratory accreditation system and grants accreditation, (ISO Guide 2, *A*). **E 1187,** E36

accretion, *n*—a phenomenon consisting of the increase in size of particles by the process of external additions. **D 1356,** D22

accretion—a process of sediment accumulation. **D 4410,** D19

accumulation condition—the region of the C–V curve for which a 5-V increment toward a more negative voltage for *p*–type material, or toward a more positive voltage for *n*–type material, results in less than a 1 % change in the maximum capacitance, C_{max}. **F 1241,** F01

accumulation period, *n—in cavitation and liquid impingement erosion,* a less-preferred term for **acceleration period**. **G 40,** G02

accumulation test—a leak test used to detect very small leaks in which gas contained in a component being tested will, if a leak is present, collect for a specified period of time in a closed evacuated chamber into which the component has been placed. At the end of the test period the chamber is opened to a leak detector which is sensitive to the gas. **E 1316,** E07

accumulator—a pulsation dampener installed on the suction and/or discharge lines of pumps, generally plunger type, to minimize pressure surges and provide uniformity of flow. **D 6161,** D19

accuracy, *n—of a test method,* the degree of agreement between the true value of the property being tested (or an accepted standard value) and the average of many observations made according to the test method, preferably by many observers. (See also **bias** and **precision**.) **D 123,** D13

accuracy—a measure of the degree of conformity of a value generated by a specific procedure to the assumed or accepted true value, and includes both precision and bias. **D 1129,** D19

accuracy, *n*—the degree of conformity of a value generated by a specific procedure to the assumed or accepted true value and includes both precision and bias. **D 1356,** D22

accuracy—the closeness of agreement between an observed value and an accepted reference value. Where an accepted reference value is not available, a measure of the degree of conformity of a value generated by a specific procedure to the assumed or accepted true value, and includes both precision and bias. **D 6161,** D19

accuracy—the closeness of agreement between an observed value and an accepted reference value. **E 131,** E13

accuracy—the closeness of agreement between a measurement result and an accepted reference value (see Terminology E 456). **E 170,** E10

accuracy, *n*—the closeness of agreement between a test result and an accepted reference value. **E 284,** E12

accuracy, *n—of a temperature measurement,* closeness of agreement between the result of a temperature measurement and a true value of the temperature. **E 344,** E20

accuracy, *n*—the closeness of agreement between a test result and an accepted reference value.

NOTE—The term accuracy, when applied to a set of test results, involves a combination of a random component and of a common systematic error or bias component. **E 456,** E11

accuracy—generally, a term used to indicate the reliability of a measurement, or an observation. Specifically, a measure of closeness of agreement between a test result and a recognized standard value. For example, the difference between the observed and a standard value for the sulfur content of a sample of refuse-derived fuel. This difference includes both random and systematic errors. **E 856,** D34

accuracy—the closeness or degree of agreement (within a stated tolerance and probability of conformity) between a quantity measured or estimated by a WIM system and an accepted reference value. **E 867,** E17

accuracy—the agreement between an experimentally determined value and the accepted reference value. In chemical work, this term is frequently used to express freedom from bias, but in other fields it assumes a broader meaning as a joint index of precision and bias. To avoid confusion, the term *bias* will be used in appraising the systematic error of test methods for industrial chemicals. See also **bias**. **E 1547,** E15

accuracy, *n*—degree of conformity of a measured or calculated value to some recognized standard or specified value. **E 1605,** E06

accuracy, *n*—a measurement concept that describes the degree of correspondence between an average measured value and an accepted reference or standard value for the object, material or phenomenon under test. **F 538,** F09

accuracy (as distinguished from precision)—the degree of conformity of a measured or calculated value to some recognized standard or specified value. This concept involves the systematic error of an operation, which is seldom negligible. (E 380) **F 221,** F05

acetal plastic, *n*—a plastic based on polymers having a predominance of acetal linkages in the main chain. (See also polyoxymethylene.) **F 1251,** F04

acetal plastics, *n*—plastics based on polymers having a predominance of acetal linkages in the main chain. (See also **polyoxymethylene.**) **D 883,** D20

acetal plastics, *n*—highly crystalline linear thermoplastic homopolymers or copolymers characterized by repeating oxymethylene units. **F 412,** F17

acetylation—substitution of an acetyl radical for an active hydrogen. Specifically, formation of cellulose acetate from cellulose. **D 1695,** D01

acetylation—substitution of an acetyl radical for an active hydrogen. Specifically, formation of cellulose acetate from cellulose. **D 6161,** D19

acetylation pulps—pulps used in the manufacture of cellulose acetate or other esters, and subject to various specifications by the manufacturers, including those of purity, moisture content, sheet properties, and viscosity. **D 1695,** D01

acetyl groups—the characteristic groups of acetic acid; CH_3CO—. **D 1695,** D01

ac excitation, N_1I/ℓ_1—the ratio of the rms ampere-turns of exciting current in the primary winding of an inductor to the effective flux path length of the inductor. **A 340,** A06

achieved width, *n*—the calculated bar or other element width based on measurements. **F 1294,** F05

achromatic—literally, color-free. A lens or prism is said to be achromatic when corrected for two colors. The remaining color seen in an image formed by such a lens is said to be secondary chromatic aberration. See **apochromatic objective** **E 7,** E04

achromatic—literally, color-free. A lens or prism is said to be achromatic when corrected for two colors. The remaining color seen in an image formed by such a lens is said to be secondary chromatic aberration. **E 175,** E41

achromatic, *adj*—(*1*) *for primary light sources,* the computed chromaticity of the equal-energy spectrum.

(*2*) *for surface colors,* the color of a whitish light, serving as the illuminant, to which adaptation has taken place in the visual system of the observer.

(*3*) perceived as having no hue, that is, as white, gray, or black. **E 284,** E12

achromatic objective—an objective that is corrected chromatically for two colors, and spherically for one, usually in the yellow-green part of the spectrum. **E 7,** E04

achromatic objective—an objective that is corrected chromatically for two colors, and spherically for one, usually in the yellow-green part of the spectrum. **E 175,** E41

achromatic objective lens—an objective lens with longitudinal chromatic correction for green and blue, and spherical chromatic correction for green. Note—Lens should be used with a green filter. **E 7,** E04

acicular powder—needle shaped particles. **B 243,** B09

acid annealing—an annealing process in which ferrous metal shapes are coated with acid before and in conjunction with the annealing. **C 286,** B08

acid content, *n*—*of felt,* the number of milliequivalents of acid present per unit weight of felt, measured under prescribed conditions. **D 123,** D13

acid etch, *v*—to clean or alter a surface by the application of acid. **C 904,** C03

acid-etched—treated in an acid bath, usually phosphoric, to provide a rough surface. **F 547,** F16

acid-extractable material—substances dissolved by an acid under specified conditions. **D 2652,** D28

acid extraction—See **extraction.** **E 7,** E04

acid foods—foods that have a natural pH of 4.6 or below. **F 17,** F02

acid gas—natural gas containing high concentrations of hydrogen sulfide or carbon dioxide, or both, which is acidic when in contact with water or water vapor. **D 4150,** D03

acid groups—functional groups having the properties of acids. In cellulose and its derivatives, these are usually carboxyl groups. **D 1695,** D01

acidifier—a material that can be added to spray mixtures to lower the pH. **E 1519,** E35

acidity—the quantitative capacity of aqueous media to react with hydroxyl ions. **D 1129,** D19

acidity—the number of milligrams of sodium hydroxide consumed when 100 mL of the sample are titrated under the conditions prescribed in this method. **D 4790,** D16

acidity—the quantitative capacity of aqueous media to react with hydroxyl ions. **D 6161,** D19

acidity, free mineral—the quantitative capacity of aqueous media to react with hydroxyl ions to pH 4.3. **D 1129,** D19

acidity, theoretical free mineral—the free mineral acidity that would result from the conversion of the anions of strong acids in solution to their respective free acids. **D 1129,** D19

acid number, *n*—the quantity of base, expressed in milligrams of potassium hydroxide, that is required to titrate acidic constituents present in 1 g of sample. **D 907,** D14

acid number (*coatings*)—the number of milligrams of potassium hydroxide (KOH) required to neutralize the free acids in 1 g of an oil, resin, varnish, or other substance; generally reported on the nonvolatile content. **D 16,** D01

acid polishing—the polishing of a glass surface by acid treatment. **C 162,** C14

acid rain—rain having a pH of less than 5.65. **E 631,** E06

acid reaction—a characteristic of materials producing the acid-color of the indicator used under the conditions prescribed in this method. **D 4790,** D16

acid refractories, n— see **refractories, acid.** **C 71,** C08

acid resistance—the degree to which a porcelain enamel will resist attack by acids. **C 286,** B08

acid-resistant brick, *n*—brick suitable for use in contact with chemicals, usually in conjunction with acid-resistant mortars. See Specification C 279. **C 43,** C15

acid-sized paper, *n*—paper that has been manufactured using a procedure or process at pH values below 7 (usually 4.0 to 6.5) that results in paper that has resistance to water penetration. See **sizing.** **D 1968,** D06

acid treating—a refining process in which an unfinished petroleum insulating oil is contacted with sulfuric acid to improve its color, odor, stability, and other properties. **D 2864,** D27

acid value (coatings)— see **acid number.** **D 16,** D01

acid-wash color—the color developed in the separated acid when a sample of detergent alkylate is agitated with sulfuric acid under the conditions prescribed by the method. **D 459,** D12

acid wash color—the color developed in the separated acid when a sample is agitated with sulfuric acid under the condition prescribed in this method. **D 4790,** D16

AC impedance—See **electrochemical impedance**. **G 15,** G01

ACK, *n*—a communication control character transmitted by a receiver as an affirmative response to a sender; a signal that the printer receiver sends to the host indicating that the printer has received a message and is ready to receive the next message. **F 1457,** F05

ac loss characteristics—those properties of a dielectric or insulation system (such as dissipation factor, power factor, and loss index) that may be used as a measure of the power or energy losses that would result from the use of such material in an ac electric field. **D 2864,** D27

acoustical board—a low-density, sound absorbing cellulosic fiberboard having a factory-applied finish and a fissured, felted-fiber, slotted or perforated surface pattern provided to reduce sound reflection. Usually supplied for use in the form of tiles. **D 1554,** D07

acoustical material—any material considered in terms of its acoustical properties. *Commonly and especially,* a material designed to absorb sound. **C 634,** E33

acoustic emission (AE)—the class of phenomena whereby transient elastic waves are generated by the rapid release of energy from localized sources within a material, or the transient waves so generated. Acoustic emission is the recommended term for general use. Other terms that have been used in AE literature include (1) stress wave emission, (2) microseismic activity, and (3) emission or acoustic emission with other qualifying modifiers. **E 1316,** E07

acoustic emission channel—see **channel, acoustic emission.** **E 1316,** E07

acoustic emission count (emission count) (N)—see **count, acoustic emission.** **E 1316,** E07

acoustic emission count rate—see **count rate, acoustic emission (emission rate or count rate) ($\dot{N}$).** **E 1316,** E07 *acoustic emission event*—see **event, acoustic emission.** **E 1316,** E07

acoustic emission event energy—see **energy, acoustic event.** **E 1316,** E07

acoustic emission sensor—see **sensor, acoustic emission.** **E 1316,** E07

acoustic emission signal amplitude—see **signal amplitude, acoustic emission.** **E 1316,** E07

acoustic emission signal (emission signal)—see **signal, acoustic emission.** **E 1316,** E07

acoustic emission signature (signature)—see **signature, acoustic emission.** **E 1316,** E07

acoustic emission transducer—see **sensor, acoustic emission.** **E 1316,** E07

acoustic emission waveguide—see **waveguide, acoustic emission.** **E 1316,** E07

acoustic impedance, $Z \equiv R + jX$; $[ML^{-4}T^{-1}]$; mks acoustic ohm ($Pa{\cdot}s/m^3$)—*of a surface, for a given frequency,* the complex quotient obtained when the sound pressure averaged over the surface is divided by the volume velocity through the surface. The real and imaginary components are called, respectively, **acoustic resistance** and **acoustic reactance.** **C 634,** E33

acoustic pathlength (*d,* (m)), *n*—the physical distance between transducer transmitter-receiver pairs. **D 1356,** D22

acousto-ultrasonics (*AU*)—a nondestructive examination method that uses induced stress waves to detect and assess diffuse defect states, damage conditions, and variations of mechanical properties of a test structure. The AU method combines aspects of acoustic emission (AE) signal analysis with ultrasonic materials characterization techniques. **E 1316,** E07

acromion, *n*—*in anatomy,* that part of the shoulder blade located at the end of the spine which articulates with the collar bone. (See also **shoulder joint**.) **D 123,** D13

across grain, *n*—Synonym for against grain. **C 709,** D02

across machine direction, *n*—the direction perpendicular to which a product moves through the manufacturing process. **F 141,** F06

across shoulder, *n*—*in body measurements,* the distance from shoulder joint to shoulder joint across the back. **D 123,** D13

acrylic plastic, *n*—a plastic based on polymers made with acrylic acid or a structural derivative of acrylic acid. **F 1251,** F04

acrylic plastics—plastics based on polymers made with acrylic acid or a structural derivative of acrylic acid. **D 883,** D20

acrylic resin—under **resin, synthetic,** see *acrylic resin.* **D 16,** D01

acrylonitrile-butadiene-styrene (ABS) pipe and fitting plastics—plastics containing polymers or blends of polymers, or both, in which the minimum butadiene content is 6%, the minimum acrylonitrile content is 15 %, the minimum styrene or substituted styrene content, or both, is 15 %, and the maximum content of all other monomers is not more than 5 %; plus lubricants, stabilizers, and colorants. **F 412,** F17

actinic radiation, *n*—the spectral region(s) of a light source responsible for the photodegradation of a particular material. **G 113,** G03

actinide series—the series of elements beginning with actinium, Element No. 89, and continuing through lawrencium, Element No. 103. **C 859,** C26

actinolite asbestos—asbestiform variety of the monoclinic amphibole silicate minerals of the tremolite-actinolite series. Its empirical formula is $Ca_2(Mg, Fe^{+2})_5Si_8O_{22}(OH)_2$. Its Chemical Abstracts number is 77536-66-4. **D 2946,** C17

activated carbon—a family of carbonaceous substances manufactured by processes that develop adsorptive properties. **D 2652,** D28

activated carbon—granulated or powdered activated carbon used to remove tastes, odor, chlorine, chloramines, and some organics from water. A family of carbonaceous substances manufactured by processes that develop adsorptive properties. **D 6161,** D19

activated charcoal, *n*—activated charcoal refers to properly conditioned coconut-shell charcoal. **D 1356,** D22

activated sintering—a sintering process during which the rate of sintering is increased, for example by addition of a substance to the powder or by changing sintering conditions. **B 243,** B09

activation—elimination of a passive condition on a surface. **B 374,** B08

activation—a process in which contamination of the surface of contacts causes arcing at lower than usual voltage or arcing persists at lower than usual current, or both. For example, palladium contacts operated in an organic vapor produce arcs at voltage and current less than the minimum arcing voltage and current because of the presence of carbon on the contact surfaces. **B 542,** B02

activation—any process whereby a substance is treated to develop adsorptive properties. **D 2652,** D28

activation—in neutron radiography, the process of causing a substance to become artificially radioactive by subjecting it to bombardment by neutrons or other particles. **E 1316,** E07

activation—the process of causing a substance to become artificially radioactive by subjecting it to bombardment by neutrons or other particles. **E 1316,** E07

activation analysis—method of chemical analysis involving creation of a radio-active isotope that can be identified by determining the nature of its radiative emission. **F 1156,** F12

activation cross section—the cross section for processes in which the product nucleus is radioactive (see **cross section**). **E 170,** E10

activation temperature—temperature which generates an optical density of 0.20. **F 1623,** F05

activation time—time between the application of a stimulus and the onset of the response to it. **F 1156,** F12

activator—*in diffusion coatings,* a chemical, usually a halide salt, that enters into a reaction with the source or master alloy, depositing the source on the substrate. **B 374,** B08

activator—a material that causes a catalyst to begin its function. **D 653,** D18

activator, *n*—compounding material used in small proportions to increase the effectiveness of an accelerator. **D 1566,** D11

activator—a material that increases the biological efficacy of agrichemicals. **E 1519,** E35

active—the negative direction of electrode potential. (Also used to describe corrosion and its associated potential range when an electrode potential is more negative than an adjacent depressed corrosion rate [passive] range.) **G 15,** G01

active chemical deposit rate—the amount of active ingredient deposited per unit area. **E 1102,** E35

active chemical rate—the amount of active ingredient (a.i.) applied per unit treated, expressed in terms of mass per relevant unit treated. (For area treatment, kg a.i./ha, lb a.i./A, or oz a.i./1000 ft. of row; for space application, mg a.i./m^3, or oz a.i./1000 ft^3; for individual units, kg a.i./plant or animal). **E 1102,** E35

active coat, *n*—a coating layer which contains the primary image (color) forming ingredients. **F 1623,** F05

active earth pressure—see **earth pressure**. **D 653,** D18

active fiber optic chemical sensor, *n*— a fiber optic chemical sensor in which a transduction mechanism other than the intrinsic spectroscopic properties of the analyte is used to modulate the optical signal. **E 131,** E13

active hours—See **hours of operation.** **E 1480,** E06

active ingredient—component of an aerosol formulation that produces the specific effect for which the formulation is designed. **D 3064,** D10

active ingredient—a component of the formulation that produces a specific effect for which the formulation is designed. **E 1519,** E35

active ingredient of a synthetic detergent—the organic surface-active material present in the detergent. **D 459,** D12

active oxygen—*in cleaning compounds,* the oxidizing power of oxygen present as peroxide or other active oxygen-containing moieties in solution expressed as oxygen (equivalent weight 8.00). **D 459,** D12

active (real) power, *P*—the product of the rms current, *I*, in an electrical circuit, the rms voltage, *E*, across the circuit, and the cosine of the angular phase difference, θ between the current and the voltage.

$$P = EI \cos\theta$$

NOTE—The portion of the active power that is expended in a magnetic core is the total core loss, P_c. **A 340,** A06

active solar energy system—See **building subsystem.** **E 631,** E06

active state of plastic equilibrium—see **plastic equilibrium.** **D 653,** D18

activity—a generic term used to describe the capacity to adsorb in general; also, the adsorptive capacity of an adsorbent as measured by a standard test. **D 2652,** D28

activity—*of a catalyst,* the measure of the rate of a specific catalytic reaction conducted in the presence of a catalyst. **D 3766,** D32

activity—the thermodynamically effective concentration of a free ion in solution. In dilute solutions, ionic activity, and concentration are practically identical, but in solutions of high ionic strength, or in the presence of complexing agents, activity may differ significantly from concentration. Ionic activity, not concentration, determines both the rate and the extent of chemical reactions. **D 4127,** D19

activity, ***A***—of an amount of radioactive nuclide in a particular energy state at a given time, the quotient of d*N* by d*t*, where d*N* is the expectation value of the number of spontaneous nuclear transitions from that energy state in the time interval d*t* (ICRU).

$$A = dN/dt$$

Unit: s^{-1}

The special name for the unit of activity is the becquerel (Bq).

$$1\ Bq = 1\ s^{-1}$$

E 170, E10

activity coefficient—a factor, γ, that relates activity, *A*, to the concentration, *C* of a species in solution:

$$A = \gamma C$$

The activity coefficient is dependent on the ionic strength of the solution. Ions of similar size and charge have similar activity coefficients. **D 4127,** D19

activity (ion)—the ion concentration corrected for deviations from ideal behavior. Concentration multiplied by activity coefficient. **B 374,** B08

activity limit—the lowest activity that can be measured in a solution that is well buffered with respect to the ion being measured. The activity limit will vary as a function of solution pH. **D 4127,** D19

activity standard—a standardizing solution whose value is reported in terms of ionic activity. If the electrode is calibrated using activity standards, the activity of the free, unbound ion in the sample is determined. **D 4127,** D19

acuity, *n*—the ability to repeatedly detect or discriminate sensory stimuli. **E 253,** E18

acute ecotoxicity, *n*—the propensity of a test material to produce adverse behavioral, biochemical, or physiological effects in non-human organisms or populations in a short period, usually not constituting a substantial portion of the life span. **D 6384,** D02

acute ecotoxicity test, *n*—a comparative ecotoxicity test in which a representative subpopulation of organisms is exposed to different treat rates of a test material and is observed for a short period, usually not constituting a substantial portion of their life span. **D 6384,** D02

acute radiation syndrome—the immediate effects of a short term, whole body overexposure of a person to ionizing radiation. These effects include nausea and vomiting, malaise, increased temperature, and blood changes. **E 1316,** E07

acute test—a comparative study in which organisms, that are subjected to different treatments, are observed for a short period usually not constituting a substantial portion of their life span. **E 943,** E47

acyl groups—radicals derived from carboxylic acids by removal of the hydroxyl group. **D 1695,** D01

Adams color difference, *n*—color difference calculated by using the Adams-Nickerson opponent-color equations, based on applying the Munsell Value function to CIE 1931 tristimulus values *X, Y, Z*. **E 284,** E12

adapt—See **building modification.** **E 631,** E06

adaptation, sensory, *n*—a decrease in sensitivity to a given stimulus which occurs as a result of exposure to that stimulus. **E 253,** E18

adaptive location—source location by iterative use of simulated sources in combination with computed location. **E 1316,** E07

ADC—abbreviation for **automatic density control.** **F 1457,** F05

add—See **building modification.** **E 631,** E06

addition, *n*—a material that is interground or blended in limited amounts into a hydraulic cement during manufacture either as a "processing addition" to aid in manufacturing and handling the cement or as a "functional addition" to modify the use properties of the finished product.

functional addition, n—an addition introduced to modify one or more properties of a hydraulic cement.

air-entraining addition, n—an addition that will entrain air in mortar or concrete.

processing addition, n—an addition introduced to aid in the manufacture or handling, or both, of a hydraulic cement.

C 219, C01

addition agent—a material added in small quantities to a solution to modify its characteristics. It is usually added to a plating solution for the purpose of modifying the character of a deposit. **B 374,** B08

addition polymerization—polymerization in which monomers are linked together without the splitting off of water or other simple molecules. **D 883,** D20

addition polymerization, *n*—polymerization in which monomers are linked together without the splitting off of water or other simple molecules and involves the opening of a double bond. **F 1251,** F04

additive—any substance added in small quantities to another substance, usually to improve properties; sometimes called a modifier. **D 16,** D01

additive—any material other than the basic components of a grout system. **D 653,** D18

additive—a chemical compound or compounds added to an insulating fluid for the purpose of imparting new properties or altering those properties which the fluid already has. **D 2864,** D27

additive, *n*—a material added to another, usually in small amounts, to impart or enhance desirable properties or to suppress undesirable properties. **D 4175,** D02

additive color mixture, *n*—superposition or other nondestructive combination of lights of different perceived colors. **E 284,** E12

additive color stimulus mixture, *n*—method of simulation that combines on the retina the actions of various color stimuli in such a manner that they cannot be perceived individually. **E 284,** E12

additive color theory—the three primary colors of this theory are red, green, and yellow; when mixed together in the proper proportions, they create white by the addition of the color elements, not black as would happen with the mixing of the three subtractive primaries: magenta, cyan, and yellow. **F 1156,** F12

additive primaries, n— same as **primary color stimuli**. **E 284,** E12

adhere, *v*—to cause two surfaces to be held together by adhesion. **D 907,** D14

adherence—(*1*) the degree of adhesion of a porcelain enamel or other ceramic coating to a metal substrate.

(*2*) Stress necessary to cause separation of one material from another at their interface. **C 286,** B08

adherend, *n*—a body held to another body by an adhesive. (See also **substrate**.) **D 907,** D14

adherend preparation, n—See **surface preparation**. **D 907,** D14

adhesion—the attractive force that exists between an electrodeposit and its substrate that can be measured as the force required to separate an electrodeposit and its substrate. **B 374,** B08

adhesion—the physical attraction of two substances, especially the macroscopically observable attraction of dissimilar substances. **C 904,** C03

adhesion, *n*—the property denoting the ability of a material to resist delamination or separation into two or more layers. **D 123,** D13

adhesion, *n—in tire fabrics,* the force required to separate a textile material from rubber or other elastomer by a definite prescribed method. **D 123,** D13

adhesion—shearing resistance between soil and another material under zero externally applied pressure.

	Symbol	Unit
Unit Adhesion	c_a	FL^{-2}
Total Adhesion	C_a	F or FL^{-1}

D 653, D18

adhesion—shearing resistance between two unlike materials under zero externally applied pressure. **D 653,** D18

adhesion, *n*—the state in which two surfaces are held together by interphase forces which may consist of chemical forces or interlocking action, or both. (R)

mechanical adhesion, n—adhesion between surfaces in which the adhesive holds the parts together by interlocking action.

specific adhesion, n—adhesion between surfaces which are held together by valence forces of the same type as those which give rise to cohesion. **D 907,** D14

adhesion, *n—in asbestos,* see **fiber adhesion**. **D 2946,** C17

adhesion, *n*—the bonding of the ink to the substrate. **D 6488,** D01

adhesion failure, n—in characterizing the locus of an adhesive-joint-failure, the fracture occurring within the interphase region. (Compare *cohesion failure*.) **D 907,** D14

adhesion failure, *n*—the loss of structural integrity due to the separation of two bonded surfaces at the bond interface. **D 1566,** D11

adhesion, practical, *n*—the force or work required to detach or remove a coating from the underlayer or substrate; it can be measured in terms of peel, pull, or shear strength as an experimentally determined quantity. **B 374,** B08

adhesion promoter, *n*—a substance used to improve bonding of the adhesive to the substrate. (Compare coupling agent and primer.) **D 907,** D14

adhesion promoters—materials built into the binder or added to the paint to form primary bonds to either the substrate or the previously applied coating with the specific aim of improving the dry or wet adhesion, or both. **D 16,** D01

adhesion value, *n*—force required to cause adhesion failure. **D 1566,** D11

adhesive, *n*—a substance capable of holding materials together by surface attachment. **C 11,** C11

adhesive, *n*—a substance capable of holding materials together by surface attachment. (See also **glue, gum, mucilage, paste, resin,** and **sizing**.)

adhesive, assembly, n— See **assembly adhesive**.
adhesive, bulk, n— See **bulk adhesive**.
adhesive, casein, n— See **casein adhesive.**
adhesive, cellular, n— See **cellular adhesive**.
adhesive, cold-setting, n—See **cold-setting adhesive**.
adhesive, contact, n— See **contact adhesive**.
adhesive, foamed, n— See **foamed adhesive**.
adhesive, gap-filling, n—See **gap-filling adhesive.**
adhesive, heat activated, n—See **heat activated adhesive**.
adhesive, hot-melt, n— See **hot-melt adhesive**.
adhesive, hot-setting, n—See **hot-setting adhesive**.
adhesive, intermediate-temperature-setting, n—See **intermediate-temperature-setting adhesive**.
adhesive, mastic, n— See **mastic adhesive**.
adhesive, multiple-layer, n—See **multiple-layer adhesive**.
adhesive, polyvinyl acetate emulsion, n—See **polyvinyl acetate emulsion adhesive**.
adhesive, pressure-sensitive, n—See **pressure-sensitive adhesive**.
adhesive, room-temperature-setting, n—See **room-temperature-setting adhesive**.
adhesive, separate application, n—See **separate-application adhesive**.
adhesive, solvent, n— See **solvent adhesive.**
adhesive, solvent-activated, n—See **solvent-activated adhesive**.
adhesive, structural, n—See **structural adhesive**.
adhesive, warm-setting, n—See **warm-setting adhesive**.
adhesive, urea-formaldehyde, n—See **urea-formaldehyde adhesive.** **D 907,** D14

adhesive, *n*—a substance capable of holding materials together by surface attachment. **D 996,** D10

adhesive—a substance capable of holding materials together by surface attachment.

NOTE—adhesive is the general term and includes, among others, cement, glue, mucilage, and paste. All of these terms are loosely used interchangeably. Various descriptive adjectives are applied to the term adhesive to indicate certain characteristics as follows:

(1) Physical form, that is, liquid adhesive, tape adhesive
(2) Chemical type, that is, silicate adhesive, resin adhesive
(3) Materials bonded, that is, paper adhesive, metal-plastic adhesive, can label adhesive
(4) Conditions of use, that is, hot-setting adhesive **D 1038,** D07

adhesive—a substance capable of holding materials together by means of surface attachment.

cold setting adhesive—an adhesive which sets at temperatures below 20°C (68°F).

contact pressure adhesive—a resinous adhesive which is aggressively and permanently tacky at room temperature and adheres to a variety of surfaces upon contact with a minimum of pressure required. (Syn. **pressure-sensitive adhesives.**)

core splice adhesive—a film adhesive, capable of expansion of at least 175 % of its original thickness, used primarily to join or splice together two or more separate sections of core material in sandwich constructions.

foamed adhesive—an adhesive, the apparent density of which has been decreased substantially by the presence of numerous gaseous cells dispersed throughout its mass.

supported film adhesive—an adhesive material incorporating a carrier that remains in the bond when the adhesive is employed; carrier support material is usually composed of organic/inorganic fibers which may be in woven (knit) or nonwoven (mat) form.

unsupported film adhesive—an adhesive material in film form without a carrier support. **E 1749,** E06

adhesive—a substance capable of holding materials together by surface attachment. **F 412,** F17

adhesive, anaerobic, *n*—See **anaerobic adhesive.** **D 907,** D14

adhesive assembly, *n*—a group of materials or parts, including adhesive, placed together for bonding or which has been bonded together. (See **assembly adhesive.**) **D 907,** D14

adhesive bonded joint—see **joint, adhesive bonded.** **F 412,** F17

adhesive, contact—an adhesive that is apparently dry to the touch and that will adhere to itself instantaneously upon contact. **E 1749,** E06

adhesive dispersion, *n*—a two-phase system in which one phase is suspended in a liquid. (Compare to **emulsion.**) **D 907,** D14

adhesive failure, *n*—*in building construction,* failure of the bond between the sealant, adhesive, or coating and the substrate surface. **C 717,** C24

adhesive failure, *n*—a separation of two bonded surfaces that occurs within the bonding material. **D 1566,** D11

adhesive joint, *n*—See *adhesive joint* under **joint.** **D 907,** D14

adhesive-joint failure, *n*—the locus of fracture occurring in an adhesively-bonded joint resulting in a loss of load-carrying capability. **D 907,** D14

adhesive, n—a substance capable of holding materials together by surface attachment. **C 274,** D30

adhesive, solvent—see **solvent cement.** **F 412,** F17

adhesive transfer—a condition occurring when an adhesive-coated material is peeled away from an opposing material to which it has been sealed and shows visible evidence of the adhesive being left on the opposing material. This evidence is in the form of an adhesive layer that remains with the opposing material, the adhesive having separated either adhesively from the coated web or cohesively within the adhesive itself. **F 1327,** F02

adhesive treated-tire cord, *n*—a tire cord whose adhesion to rubber or other elastomer has been improved by the application of a dip followed by rapid drying and (normally) additional heat treatment. **D 123,** D13

adhesive wear, *n*—wear due to localized bonding between contacting solid surfaces leading to material transfer between the two surfaces or loss from either surface. **G 40,** G02

adiabatic calorimeter, *n*—a calorimeter that has a jacket temperature adjusted to follow the calorimeter temperature as closely as possible so as to maintain zero thermal head. **D 5681,** D34

adiabatic decomposition temperature rise, $(T)_d$, *n*—an estimation of the computed temperature which a specimen would attain if all of the enthalpy (heat) of decomposition reaction were to be absorbed by the sample itself. High values represent high hazard potential. **E 1445,** E27

adiabatic extrusion—a method of extrusion in which, after the extrusion apparatus has been heated sufficiently by conventional means to plastify the material, the extrusion process can be continued with the sole source of heat being the conversion of the drive energy, through viscous resistance of the plastic mass in the extruder. **D 883,** D20

adiabatic extrusion—a method of extrusion in which the sole source of heat is the conversion of the drive energy, through viscous resistance of the plastic mass in the extruder. (D-20) **F 412,** F17

adjacency—two OCR characters printed on the same line with character spacing reference lines separated by the proper space for the font and system. **F 149,** F05

adjunct instructor—an individual with specialized subject matter expertise, who, on occasion, instructs a specific topic of a curriculum under the direction of the course instructor/coordinator. **F 1177,** F30

adjustable pressure limiting valve (APL), *n*—operator-adjustable device intended to limit the maximum pressure in an anesthesia breathing system when it is part of the flow pathway of the breathing system, by controlling the rate of escape of gases from the system. **F 2002,** F29

adjusted internal rate-of-return (AIRR)—the compound rate of interest that, when used to discount the terminal values of costs and benefits of a project over a given study period, will make the costs equal the benefits when cash flows are reinvested at a specified rate. (Syn. *financial management rate of return (FMRR)*) **E 833,** E06

adjusted serviceability score—See **serviceability score.** **E 1480,** E06

adjustment factors, *n*—traffic summary statistic corrections for multiple axle vehicles, weekday traffic variation, seasonal traffic variation, and regional traffic characteristics; and, the estimate of annual growth or decline for traffic summary statistics on road segments not recounted during a year. **E 867,** E17

adjuvant—a material added to a tank mix to aid or modify the action of an agrichemical, or the physical characteristics of the mixture. **E 1519,** E35

administrative removal—(*of workers*), temporary removal of workers from a job site prior to blood-lead levels reaching values requiring medical removal. **E 1605,** E06

admittance ratio, $y\rho c \equiv g\rho c - jb\rho c$; **[dimensionless]**—the reciprocal of the impedance ratio. The real and imaginary components are called, respectively, **conductance ratio** and **susceptance ratio.** **C 634,** E33

admittance, Y— the reciprocal of impedance. **E 1142,** E37

admixture, *n*—a material other than water, aggregate, or inorganic cementitious material that is used as an ingredient in gypsum plaster or cement plaster and is added to the batch immediately before or during job mixing. **C 11,** C11

admixture, *n*—a material other than water, aggregates, hydraulic cement, and fiber reinforcement used as an ingredient of concrete or mortar and added to the batch immediately before or during its mixing.

accelerating admixture, n—admixture that accelerates the setting and early strength development of concrete. (C 494/ 494M)

air-entraining admixture, n—admixture that causes the development of a system of microscopic air bubbles in concrete or mortar during mixing.

retarding admixture, n—admixture that retards the setting of concrete. (C 494/ 494M)

water-reducing admixture, n—admixture that either increases the slump of freshly mixed mortar or concrete without increasing the water content or that maintains the slump with a reduced amount of water due to factors other than air entrainment. **C 125,** C09

admixture—a material other than water, aggregates, cement and fibre reinforcing used as an ingredient of concrete and added to the batch immediately before or during its mixture. **C 822,** C13

admixture—a material other than water, aggregates, or cementitious material, used as a grout ingredient for cement-based grouts. **D 653,** D18

adsorbate, *n*—material that has been retained by the process of adsorption. **D 1356,** D22

adsorbate—any substance that is or can be adsorbed. **D 2652,** D28

adsorbed water—water in a soil or rock mass attracted to the particle surfaces by physiochemical forces, having properties that may differ from those of pore water at the same temperature and pressure due to altered molecular ar-rangement; adsorbed water does not include water that is chemically combined within the clay minerals. **D 653,** D18

adsorbent, *n*—solid material on the surface of which adsorption takes place. **D 1356,** D22

adsorbent—any solid having the ability to concentrate significant quantities of other substances on its surface. **D 2652,** D28

adsorption—the capacity of a substance to accept and retain on its surface a layer of another substance, usually a gas or a liquid. **C 242,** C21

adsorption, *n*—a process in which the surface of a solid takes on or absorbs in an extremely thin layer molecules of gases, of dissolved substances, or of liquids with which it is in contact. (See also **absorption, moisture equilibrium of testing**. Compare **desorption** and **resorption**.) **D 123,** D13

adsorption—the attachment of water molecules or ions to the surfaces of soil particles. **D 653,** D18

adsorption, *n*—a physical process in which molecules of gas, of dissolved substances, or of liquids, adhere in an extremely thin layer to the surfaces of solid bodies with which they are in contact. **D 1356,** D22

adsorption—a process in which fluid molecules are concentrated on a surface by chemical or physical forces, or both. **D 2652,** D28

adsorption—the holding of a substance onto the surface of a solid by chemical surface forces, without forming new chemical bonds. **D 6161,** D19

adsorption wave—see **mass transfer zone.** **D 2652,** D28

adsorption zone—see **mass transfer zone.** **D 2652,** D28

advanced ceramic, *n*—a highly engineered, high performance, predominately non-metallic, inorganic, ceramic material having specific functional attributes. **C 1145,** C28

advanced life support—medically accepted life sustaining, invasive or non-invasive procedures; provided at the detection of a physician or other authorized health care provider. **F 1177,** F30

advancing slope grouting—a method of grouting by which the front of a mass of grout is caused to move horizontally by use of a suitable grout injection sequence. **D 653,** D18

adventitious carbon referencing—*XPS*, a method of determining the charging potential of a particular specimen by comparing the experimentally determined binding energy of the *C* 1s peak maximum from contaminating hydrocarbon or hydrocarbon groups on the specimen to a standard binding energy value. **E 673,** E42

AE activity, *n*—the presence of acoustic emission during a test. **E 1316,** E07

Ae_{cm}, Ae_1, Ae_3, Ae_4—See **transformation temperature**. **A 941,** A01

aeolian deposits—wind-deposited material such as dune sands and loess deposits. **D 653,** D18

aeolian vibration—resonant vibration of a conductor caused by the alternate shedding of wind-induced vortices from opposite sides of the conductor. (Frequency is usually less than 200 Hz and amplitude rarely exceeds one conductor diameter.) **B 354,** B01

aerating nozzle—a device to atomize liquid for the purpose of aeration.

NOTE—Although this term is occasionally used to designate certain types of airblast or internal mixing pneumatic atomizers, it is ambiguous and is not recommended for describing the latter devices. **E 1620,** E29

aeration system—a type of circulation system that causes the beverage to cascade across the top and down the sides of the bowl interior incorporating sir into the beverage. **F 1827,** F26

AE rms, *n*—the rectified, time averaged AE signal, measured on a linear scale and reported in volts. **E 1316,** E07

aerobes—organisms that require air or free oxygen for growth. **F 1600,** F20

aerobic, *n*—a condition in which a measurable volume of air is present in the incubation chamber or system. **D 4439,** D35

aerobic—able to live, grow, or take place only where free oxygen is present. **E 1126,** E48

aerobic—able to live, grow, or take place only where free oxygen is present. **E 1705,** E48

aerobic bacteria—bacteria that require oxygen for growth. See **bacteria, aerobes**. **D 6161,** D19

aerobic fermentation—fermentation processes that require the presence of air. **E 1126,** E48

aerobic fermentation—fermentation processes that require the presence of air. **E 1705,** E48

aerodynamic diameter—the diameter of a hypothetical sphere having a specific gravity of unity and the same settling velocity as the actual particle. **E 1620,** E29

aerosol, *n*—a dispersion of solid or liquid particles in a gaseous medium. **D 1356,** D22

aerosol, *n*—a dispersion of solid particles or liquid particles, or both, in gaseous media. **E 1620,** E29

aerosol can side seam—cans which are formed from rectangular sheets have a soldered, bonded or welded strip which joins two corresponding or matching side gilt edges to form a cylinder. **D 3064,** D10

aerosol development—development in which the toner is carried to the field of the electrostatic image by means of a suspending gas. **F 335,** F05

aerosol generator—any mechanical or thermal device that produces a liquid dispersion having a volume median less than 50 μm. **E 1102,** E35

aerosol package—See **package.** **D 996,** D10

aerosol packaging—pressurizing sealed containers with liquefied or compressed gases, enabling the product to self-dispense. The term "aerosol" as used here is not confined to the scientific definition (that is, a suspension of fine solid or liquid particles in air or gas). **D 3064,** D10

aerosols—distribution of droplets with $D_{V.5} \leq 50\ \mu m$. **E 1102,** E35

AE signal duration—the time between AE signal start and AE signal end. **E 1316,** E07

AE signal end—the recognized termination of an AE signal, usually defined as the last crossing of the threshold by that signal. **E 1316,** E07

AE signal generator—a device which can repeatedly induce a specified transient signal into an AE instrument. **E 1316,** E07

AE signal rise time—the time between AE signal start and the peak amplitude of that AE signal. **E 1316,** E07

AE signal start—the beginning of an AE signal as recognized by the system processor, usually defined by an amplitude excursion exceeding threshold. **E 1316,** E07

affective test, *n*—any method to assess acceptance, attitudes, emotions, of preference for a stimulus(i). **E 253,** E18

aftercure, *n*—a continuation of the process of vulcanization after the energy source has been removed. **D 1566,** D11

after effects, *n*—total array of sensations that occur after elimination of the physical stimulus from the sensing field. **E 253,** E18

afterflame, *n*—persistent flaming of a material after the ignition source has been removed. **D 123,** D13

after-flame time, *n*—the length of time for which a material continues to flame after the ignition source has been removed. **D 123,** D13

after-flame time, *n*—the length of time for which a material continues to flame after the ignition source has been removed. **F 1494,** F23

afterglow, *n*—glow in a material after the removal of an external ignition source or after the cessation (natural or induced) of flaming of the material. (See also **flame, glow,** and **smoldering.**) **D 123,** D13

afterglow, *n*—emission of light, usually subsiding, from a material undergoing combustion, but occurring after flaming has ceased. **E 176,** E05

afterglow, *n*—a glow in a material after the removal of an external ignition source or after the cessation (natural or induced) of flaming of the material (see also **glow**). **F 1494,** F23

afterglow time, *n*—the time afterglow continues after the cessation of flaming or after removal of the ignition source. **D 123,** D13

aftersensation, *n*—sense impression after direct stimulation of the sense organ has ceased. **E 253,** E18

aftertaste, *n*—the oral or nasal sensations that occur after the source of the stimulus has been removed from the oral cavity. See **after effects.** **E 253,** E18

against grain, *n*—*in manufactured carbon and graphite product technology*, the direction in a carbon or graphite body with preferred orientation that has the maximum c-axis alignment (syn. *across grain*). **C 709,** D02

agar, *n*—a semisolid culture medium used to support the growth of bacteria and other micro-organisms. **F 1494,** F23

aged insulation value—thermal resistance (R-value) of a thermal insulation material as determined after standard conditioning to simulate service exposure. **E 631,** E06

age hardening, *n*—hardening by **aging**, usually after rapid cooling or **cold working**. **A 941,** A01

ageing, accelerated, *n*—exposure of rubber to a test environment with the intent of producing, in a shorter time period, effects similar to normal ageing. **D 1566,** D11

ageing (act of), *n*—exposure of materials to a deteriorating environment for a specified time interval. **D 1566,** D11

ageing, air bomb, *n*—the process of exposing materials to the action of air at an elevated temperature and pressure. **D 1566,** D11

ageing, air oven, *n*—The process of exposing materials to the action of air at an elevated temperature at atmospheric pressure. **D 1566,** D11

ageing, natural, *n*—ageing under normal service conditions. **D 1566,** D11

ageing, oxygen bomb, *n*—the process of exposing materials to the action of oxygen at an elevated temperature and pressure. **D 1566,** D11

ageing (rubber), *n*—the irreversible change of material properties during exposure to a deteriorating environment, for a specified time interval. **D 1566,** D11

ageing, shelf, *n*—ageing during storage. **D 1566,** D11

agency, *n*—an organization or part of an organization engaged in the activities of testing or inspection, or both. **D 1356,** D22

age resistance, *n*—the ability of a material to resist deterioration caused by ageing. **D 1566,** D11

ageusia, *n*—lack of sensitivity to taste stimuli. **E 253,** E18

agglomerate—several particles adhering together. **B 243,** B09

agglomerate—a jumbled mass or collection of two or more particles or aggregates, or a combination thereof, held together by relatively weak cohesive forces caused by weak chemical bonding or an electrostatic surface charge generated by handling or processing. **C 242,** C21

agglomerate, *n*—*in manufactured carbon and graphite product technology*, a composite particle containing a number of grains. **C 709,** D02

agglomerate, *n*—*as used in fractography,* a cluster of grains, particles, platelets, or whiskers, or a combination thereof, present in a larger solid mass. **C 1145,** C28

agglomerate, *n*—two or more particles adhering together. **E 1638,** E29

agglomerate, compounding material, *n*—a cluster of particles of one or more compounding materials loosely held together. **D 1566,** D11

agglomerate, latex, *n*—a cluster of rubber particles in a colloidal aqueous suspension of such particles. **D 1566,** D11

agglomerates, *n*—clusters of particles of compounding materials contained in a continuous rubber phase. **D 1566,** D11

agglomerating, *adj*—*as applied to coal*, the property of softening when it is heated to above about 400°C in a nonoxidizing atmosphere, and then appearing as a coherent mass after cooling to room temperature. **D 121,** D05

agglomeration, *n*—a process of contact and adhesion whereby the particles of a dispersion form clusters of increasing size. **D 1356,** D22

agglomeration (latex), *n*—reversible or irreversible joining together of latex particles. **D 1566,** D11

agglomeration or flocculation—the coalescence of dispersed suspended matter into large flocs or particles which settle rapidly. **D 4410,** D19

aggradation—the geologic process by which stream beds, flood plains, and the bottoms of other water bodies are raised in elevation by the deposition of material eroded and transported by water from other areas. **D 4410,** D19

aggregate, *n*—an inert granular material which may be added to gypsum plasters. **C 11,** C11

aggregate, *n*—granular material, such as sand, gravel, crushed stone, or iron blast-furnace slag, used with a cementing medium to form hydraulic-cement concrete or mortar.

coarse aggregate, n—(1) aggregate predominantly retained on the 4.75-mm (No. 4) sieve; or *(2)* that portion of an aggregate retained on the 4.75-mm (No. 4) sieve.

fine aggregate, n—(1) aggregate passing the ⅜-in. (9.5-mm) sieve and almost entirely passing the 4.75-mm (No. 4) sieve and predominantly retained on the 75-μm (No. 200) sieve; or *(2)* that portion of an aggregate passing the 4.75-mm (No. 4) sieve and retained on the 75-μm (No. 200) sieve.

heavyweight aggregate, n—aggregate of high density; such as barite, magnetite, limonite, ilmenite, iron, or steel.

lightweight aggregate, n—aggregate of low density used to produce lightweight concrete, including: pumice, scoria, volcanic cinders, tuff, and diatomite; expanded or sintered clay, shale, slate, diatomaceous shale, perlite, vermiculite, or slag; and end products of coal or coke combustion. **C 125,** C09

aggregate—a dense mass of particles held together by strong intermolecular or atomic cohesive forces that is stable to normal mixing techniques, including high-speed stirring and ultrasonics. **C 242,** C21

aggregate—any inert material such as sand, gravel, slag, carbon, etc., usually consisting of various-sized particles and used with chemical-resistant binders to form chemical-resistant mortars. **C 904,** C03

aggregate, *n*—a dense mass of particles held together by strong intermolecular or atomic cohesive forces. It is stable to normal handling and ordinary mixing techniques including high-speed stirring and ultrasonics.**(C-21, C242)** **C 1145,** C28

aggregate, *n*—a granular material of mineral composition such as sand, gravel, shell, slag, or crushed stone, used with a cementing medium to form mortars or concrete, or alone as in base courses, railroad ballasts, etc. **D 8,** D04

aggregate—*as a grouting material,* relatively inert granular mineral material, such as sand, gravel, slag, crushed stone, etc. "Fine aggregate" is material that will pass a No. 4 (6.4-mm) screen,

"Coarse aggregate" is material that will not pass a No. 4 (6.4-mm) screen. Aggregate is mixed with a cementing agent (such as Portland cement and water) to form a grout material. **D 653,** D18

aggregate—*(1)* crushed stone, crushed slag, or water-worn gravel used for surfacing a built-up roof;

(2) any granular mineral material. **D 1079,** D08

aggregate—granular material such as sand, gravel, crushed stone. **D 6161,** D19

aggregates, *n*—a granular mineral material such as natural sand, manufactured sand, gravel, crushed stone, and air cooled blast furnace slag. **C 1180,** C12

aggressive tack, n— See *aggressive tack* under **tack.** **D 907,** D14

aggressivity index, *n*—a measure of the corrosivity of water toward asbestos-cement defined as pH + log(*AH*) where pH = negative of the logarithm of the hydronium molarity (index of acidity of the water), pH units. *A* = total alkalinity, ppm as $CaCO_3$, as determined by Test Methods D 1067 and *H* = calcium hardness, ppm as $CaCO_3$, as determined by Test Method D 1126. **C 460,** C17

aggressivity index, *n*—measure of the corrosiveness of water toward asbestos-cement defined as:

$$\text{pH} + \log (AH)$$

where:

pH = negative of the logarithm of the hydronium molarity (index of acidity of the water) pH units.

A = total alkalinity, ppm as $CaCO_3$, as determined by Test Methods D 1067, and

H = calcium hardness, ppm as $CaCO_3$, as determined by Test Method D 1126.

D 2946, C17

aging, *n*—a change in the properties of certain **steels** that occurs at ambient or moderately elevated temperatures after hot working or a heat treatment (**quench aging, natural aging,** or **artificial aging**) or after a cold-working operation (**strain aging**). **A 941,** A01

aging—the storing of porcelain enamel slips or powders before use. The change occurring in slips or powders with the lapse of time. **C 286,** B08

aging, *n—(1)* the effect on materials of exposure to an environment for an interval of time. *(2)* the process of exposing materials to an environment for an interval of time. **D 883,** D20

aging—in general, the change of properties with the passage of time. Specifically, the changes occurring in shredded alkali cellulose when allowed to stand exposed to air. **D 1695,** D01

aging, *n—(1)* the effect on materials of exposure to an environment for an interval of time.

(*2*) the process of exposing materials to an environment for an interval of time. (D-20) **F 412,** F17

aging—*(1)* the effect on materials of exposure to an environment for an interval of time. *(2)* the process of exposing materials to an environment for an interval of time. (D 883, D-20) **F 869,** F08

aging, *n*—the process of exposing materials to an environment for an interval of time. **F 1251,** F04

aging—the exposure of a material to stress, such stress of which may include time, pressure, contact with materials or chemicals, temperature, abrasion, ionizing radiation, light, impact with gas or particles, tensile or compressive force (either static or cyclic), or any other feature that may be present during a material's service life. These stressors may be present individually or in combination. **G 126,** G04

aging coefficient—the percentage change in a specific magnetic property resulting from a specific aging treatment.

NOTE—The aging treatments usually specified are:
(*a*) 100 h at 150°C or
(*b*) 600 h at 100°C. **A 340,** A06

aging effect, *n*—a change in a material brought about by exposure of the material to an environment for an interval of time. **F 1251,** F04

aging, magnetic—the change in the magnetic properties of a material resulting from metallurgic change due to a normal or specified aging condition.

NOTE—This term implies a deterioration of the magnetic properties of magnetic materials for electronic and electrical applications, unless otherwise specified. **A 340,** A06

aging time, *n*—See **joint conditioning time.** **D 907,** D14

agitator, *n*—a device that is in contact with the surface to be cleaned and assists in dirt removal by mechanical action, rotary and otherwise. **F 395,** F11

agitators—a removable device attached to the power output shaft that extends downward into the mixer bowl and converts or imparts the desired action on the contents of the bowl. **F 1827,** F26

agitator tank—a tank, usually vertical and with open top, with rotation paddles used to prevent segregation of grout after mixing.

D 653, D18

agricultural lime—either ground quicklime or hydrated lime whose calcium and magnesium content is capable of neutralizing soil acidity. **C 51,** C07

agricultural limestone—ground or pulverized limestone whose calcium and magnesium content is capable of neutralizing soil acidity. **C 51,** C07

agricultural spray oil—See **crop oil, dormant oil, modified vegetable oil, naphtha-based oil, oil, paraffinic oil, petroleum oil, phytobland oil, horticultural mineral oil, and vegetable oil**. **E 1519,** E35

"A" Horizon—see **horizon**. **D 653,** D18

AIAG, *n*—an abbreviation for the Automotive Industry Action Group; a standards body composed of representatives from the U.S. automobile industry responsible for the development of a common automotive industry standard relating to symbol technologies and Electronic Data Interchange (EDI). **F 1294,** F05

Aiken code—two-phase self-clocking code based upon the assumption of a predetermined flux reversal for every bit space such that a reversal between two consecutive clock cycles is treated as one, and no reversal is a zero. **F 1156,** F12

AIM, *n*—an acronym for the Automatic Identification Manufacturers; a trade association of automatic identification suppliers. **F 1294,** F05

air assist nozzle—a pneumatic atomizer in which pressurized air is utilized to enhance the atomization produced by pressurized liquid. The air may be required only for part of the operating range (for example, low liquid flow rates). **E 1620,** E29

air at normal conditions (standard air), *n*—air at 50 % relative humidity, 25°C and 101.3 kPa (77°F and 760 mm Hg). See also **atmosphere.** **D 1356,** D22

air atomizing—air used to atomize powder and to control powder/air mix and powder cloud density. **C 286,** B08

air atomizing nozzle—see **pneumatic atomizer**. **E 1620,** E29

air bells—bubbles of irregular shape formed generally during the pressing or molding operations in the manufacture of optical glass. **C 162,** C14

air blast nozzle—see **airblast nozzle**. **E 1620,** E29

airblast nozzle—a pneumatic atomizer that utilizes a relatively large volume of low-pressure air.

NOTE—The term is occasionally used to designate the entire class of pneumatic atomizers. **E 1620,** E29

airborne drift—the dispersion of chemical particles to the atmosphere outside the intended target. **E 1102,** E35

airborne sound—sound that arrives at the point of interest, such as one side of a partition, by propagation through air. **C 634,** E33

air bubble barrier—special-purpose barrier created by rising stream of air bubbles and entrained water, produced by injecting air at some depth below water surface. **F 818,** F20

air carrier sprayer—an apparatus consisting of a pressure source and controls for the spray liquid and a blower with suitable ducts to produce an air jet in which spray nozzles are located. Air from the blower carries the spray for a distance for deposition on the target being treated. **E 1102,** E35

air chain, *n—in mica*, a series of air inclusions in the form of a chain or streak. **D 1711,** D09

air classification—the separation of powder into particle size fractions by means of an air stream of controlled velocity. **B 243,** B09

air conditioning—the simultaneous control of all, or at least the first three, of those factors affecting both the physical and chemical conditions of the atmosphere within any structure. These factors include temperature, humidity, motion, distribution, dust, bacteria, odor, and toxic gases. **E 41,** G03

air content, *n*—the volume of air voids in cement paste, mortar, or concrete, exclusive of pore space in aggregate particles, usually expressed as a percentage of total volume of the paste, mortar, or concrete. **C 125,** C09

air content, *n*—of freshly mixed mortar the volume of air (and other gases) in mortar, expressed as a percentage of total volume of mortar. **C 219,** C01

air-cooled blast-furnace slag, *n*—the material resulting from solidification of molten blast-furnace slag under atmospheric conditions; subsequent cooling may be accelerated by application of water to the solidified surface. **C 125,** C09

air-dried moisture—this term has been used inappropriately to refer to both residual moisture and air-dry loss. Because of the potential for confusion, this term shall not be used. **D 121,** D05

air drop, *n*—lengths of open run conductors or cables supported only at each end. **E 176,** E05

air-dry—a condition applied to paper or pulp whereby its moisture content is in equilibrium with the atmospheric conditions to which it is exposed. According to trade custom, pulps are generally understood to be air dry when they contain 10 % of moisture, for example, a pound of air-dry pulp contains 0.9 lb of oven-dry pulp and 0.1 lb of moisture (see *STP 60-B*). **D 1695,** D01

air, dry—air containing no water vapor. **E 41,** G03

air drying, *n*—a process of partial drying of coal to bring its moisture near to equilibrium with the atmosphere in the room in which further reduction and division of the sample will take place. **D 121,** D05

air-drying—a process of partial drying of refuse-derived fuel (RDF) to bring its moisture content near to equilibrium with the atmosphere in which further reduction, division, and characterization of the sample are to take place. In order to bring about this equilibrium, the RDF is usually subjected to drying under controlled temperature conditions ranging from 30 to 40°C. **E 856,** D34

air-dry loss, *n*—in coal, the loss in mass, expressed as a percentage, resulting from each air-drying stage or the sum of all sequential air-drying stages in which the results from each stage are adjusted to the as-received basis. **D 121,** D05

air-dry loss—the decrease in sample mass due to air-drying. This decrease is presumed to be moisture. **E 856,** D34

air-entraining hydraulic cement, *n*—a hydraulic cement containing an air-entraining addition in such amount as to cause air to be entrained in mortar within specified limits when measured by the prescribed method. **C 219,** C01

air entry value—the applied suction at which water menisci of the porous segment of a suction sampler break down, and air enters. **D 653,** D18

air-felting—forming of a fibrous-felted board from an air suspension of damp or dry fibers on a batch or continuous forming machine (sometimes referred to as the dry or semi-dry process). **D 1554,** D07

air flow rate—the flow rate of air, expressed in volume per relevant unit (ft^3/min, m^3/s, m^3/tree). **E 1102,** E35

airflow resistance, *R;* [$ML^{-4}T^{-1}$]; mks acoustic ohm (Pa·s/m^3)—the quotient of the air pressure difference across a specimen divided by the volume velocity of airflow through the specimen. The pressure difference and the volume velocity may be either steady or alternating. **C 634,** E33

airflow resistivity, r_o; [$ML^{-3}T^{-1}$]; mks rayl/m (Pa·s/m^2)—*of a homogeneous material,* the quotient of its specific airflow resistance divided by its thickness. **C 634,** E33

air fluidizer—air used to impart fluid-like properties to powder via a fluid bed. **C 286,** B08

air-gap magnetic field strength, H_g—the magnetic field strength required to produce the induction existing at some point in a nonmagnetic gap in a magnetic circuit.

NOTE—In the cgs-emu system of units, H_g is numerically equal to the induction existing at such a point and exceeds the magnetic field strength in the magnetic material. **A 340,** A06

air handling unit—a device used for distributing conditioned air supply to a room, space, or area. **E 772,** E44

air leakage, *n*—in buildings, the passage of uncontrolled air through **cracks** or openings in the building envelope or its **components,** such as ducts, because of air pressure or temperature difference. **E 631,** E06

air-lock—an intermediate enclosed chamber of a vacuum or pressure system through which an object may be passed without materially changing the vacuum or pressure of the system. **E 7,** E04

air mass, AM—the ratio of the mass of atmosphere in the actual observer-sun path to the mass that would exist if the observer was at sea level, at standard barometric pressure, and the sun was directly overhead.

NOTE—(Sometimes called air mass ratio.) Air mass varies with the zenith angle of the sun and the local barometric pressure, which changes with altitude. For sun zenith angle, *Z*, of 62° or less and local atmospheric pressure, *P*, where P_o is standard atmospheric pressure, AM ≃ sec Z (P/P_o). **E 772,** E44

air permeability, *n*—the rate of air flow through a material under a differential pressure between the two fabric surfaces. (See **permeability, porosity**) **D 123,** D13

air pollution, *n*—the presence of unwanted material in the air. **D 1356,** D22

air, saturated—a mixture of dry air and water vapor in which the latter is at its maximum concentration for the prevailing temperature and pressure. **E 41,** G03

air scour—distributing air over the entire filter area at the bottom of a filter media flowing upward to improve the effectiveness of backwashing or to permit the use of lower backwash water flow rate, or both. **D 6161,** D19

air-slaked lime—the product containing various proportions of the oxides, hydroxides, and carbonates of calcium and magnesium which results from the exposure of quicklime to the air in sufficient quantity to show physical signs of hydration (difficult to determine visually in pulverized quicklime). **C 51,** C07

air-space ratio, G_a **(D)**—ratio of: (*1*) volume of water that can be drained from a saturated soil or rock under the action of force of gravity, to (*2*) total volume of voids. **D 653,** D18

air-supported roof, *n*—a fabric roof-system that is properly secured and primarily supported and held in place by air pressure. **D 123,** D13

air-supported structure—an amusement device that incorporates a structural and mechanical system and employs a high-strength fabric or film that achieves its strength, shape, and stability by pretensioning with internal air pressure, all of which are intended to provide an enclosed area for the self enjoyment of those so confined within. **F 747,** F24

air void, *n*—a space in cement paste, mortar, or concrete filled with air; an entrapped air void is characteristically 1 mm or more in width and irregular in shape; an entrained air void is typically between 10 and 1000 μm in diameter and spherical or nearly so. **C 125,** C09

air-void ratio, G_v **(D)**—the ratio of: (*1*) the volume of air space, to (*2*) the total volume of voids in a soil or rock mass. **D 653,** D18

air wicking, *n—in tires*, the passage of air longitudinally along or through yarns in a fabric that has been encased and cured in rubber or other elastomer, that is, air permeability in the plane of the fabric. **D 123,** D13

Airy disk—the image of a bright point object, as focused by a lens system. With monochromatic light, it consists of a central point of maximum intensity surrounded by alternate circles of light and darkness caused by the reinforcement and interference of diffracted rays. The light areas are called maxima and the dark areas minima. The distribution of light from the center to the outer areas of the figure was investigated mathematically by Sir George Airy. The diffraction disk forms a basis for determining the resolving power of an ideal lens system. The diameter of the disk depends largely on the aperture of the lens. The diffraction of light causing the Airy disk is a factor limiting the resolution of a well corrected optical system. **E 175,** E41

alabaster glass—a milky-white glass that diffuses light without fiery color. **C 162,** C14

albedo—the use of the term *albedo* is discouraged in favor of the preferred term, **reflectance**. **E 772,** E44

ALC—apparent lead concentration. **E 1605,** E06

Alclad, , *adj*—having an aluminum or aluminum-alloy coating that is metallurgically bonded to either one side or both surfaces of an aluminum alloy product, and that is anodic to the alloy to which it is bonded, thus electrolytically protecting the core alloy against corrosion. (See also individual product type such as *Alclad plate, Alclad sheet*, etc.). **B 881,** B07

alclad sheet and plate—composite sheet (and plate) having on both surfaces a metallurgically bonded aluminum or aluminum alloy coating that is anodic to the core alloy to which it is bonded, thus electrolytically protecting the core alloy against corrosion. **E 1749,** E06

alcohol-benzene solubility—solubility of a cellulosic pulp in a mixture of ethanol and benzene. The term is without precise meaning unless complete specification of an analytical procedure is attached explicitly or implicitly. **D 1695,** D01

alcohols—the family name of a group of organic compounds that vary in chain length and consist of a hydrocarbon plus one or more hydroxyl groups; for example, the straight chain series CH $_3$-(CH_2)n-OH. Includes methanol, ethanol, butanol, propanol, and higher alcohols. If the hydrocarbon is aromatic, the compounds are called phenols. **E 1126,** E48

alcohols—series of liquid products composed of a hydrocarbon plus a hydroxyl group, such as ethanol (C_2H_5OH). **E 1705,** E48

alcove—a narrow channel to convey molten glass from refiner to forehearth or to the revolving pot where it is gathered by the Owens machine. **C 162,** C14

aldehyde groups—carbonyl groups to which a hydrogen atom is attached; the first oxidation stage of an alcohol;— CHO. **D 1695,** D01

aldehydes—a broad class of organic compounds having a generic formula RCHO, and characterized by a carbonyl group. **D 4790,** D16

algae—any of a group of chiefly aquatic mono cellular plants with chlorophyll often masked by a brown or red pigment. **D 6161,** D19

alginite—See *alginite* under **maceral**. **D 121,** D05

Algorithm—a computational methodology or rule, typically a mathematical formula that is applied to a given data set, in order to produce a related, secure data set, that is, encryption.

F 1156, F12

aliases, *n—in a fractional factorial design,* two or more effects which are estimated by the same contrast and which, therefore, cannot be estimated separately. **E 456,** E11

aliases, *n—in a fractional factorial design,* two or more effects which are estimated by the same contrast and which, therefore, cannot be estimated separately. **E 1325,** E11

aliasing—the appearance of features at wavenumbers other than their true value caused by using a sampling frequency less than twice the highest modulation frequency in the interferogram; also known as "folding." **E 131,** E13

aliasing, *n*—the spectrum of a digitized data record exists over the range of frequencies from zero to one half the sampling frequency. If the spectrum of the original signal extends beyond one half the sampling frequency, then those components of the signal at frequencies higher than one half the sampling frequency will, when digitized, be folded back into the spectrum of the digitized signal. The excessive high frequency components will thus be "aliased" into low frequency components. **E 867,** E17

aligning stiffness coefficient, *n—of a tire,* the ratio of tire aligning stiffness to the tire normal force. **F 538,** F09

aligning stiffness, [*FL/γ*], *n—of a tire,* the rate of change of tire aligning torque with respect to change in tire slip angle, usually evaluated at zero slip angle. **F 538,** F09

aligning torque, [*FL*], *n—of a tire,* the component of a tire moment vector tending to rotate a tire about the Z'-axis, positive clockwise when looking in the positive direction of the Z'-axis. **F 538,** F09

alignment—a mechanical or electrical adjustment of the components of an optical device in such a way that the path of the radiating beam coincides with the optical axis or other predetermined path in the system. In electron optics there are three general types:

(*1*) *magnetic alignment*—an alignment of the electron optical axis of the electron microscope such that the image rotates about a point in the center of the viewing screen when the current flowing through a lens is varied.

(*2*) *mechanical alignment*—a method of aligning the geometrical axis of the electron microscope by relative physical movement of the components, usually as a step preceding either magnetic or voltage alignment.

(*3*) *voltage alignment*—a condition of alignment of an electron microscope such that the image expands or contracts symmetrically about the center of the viewing screen when the accelerating voltage is changed. **E 7,** E04

alignment, *n*—the adjustment of various parts of the vehicle's suspension system to ensure proper handling stability and to minimize abnormal tire treadwear. **F 538,** F09

aliquant—a representative portion of a whole that divides the whole leaving a remainder. **C 859,** C26

aliquant (solution)—a measured quantity of a homogeneous, liquid mixture that, when taken by either weight or volume, is not an evenly divisible portion of the total mixture. **C 859,** C26

aliquot—a representative portion of a whole that divides the whole exactly without leaving a remainder. **C 859,** C26

aliquot, *n*—a representative portion of the whole that can be expressed as the inverse of an integer. **D 1356,** D22

alkali—an industrial term for the oxide of sodium or potassium; less frequently of lithium. **C 162,** C14

alkali aggregate reaction—a chemical reaction between Na_2O and K_2O in the cement and certain silicate minerals in the cement and certain silicate minerals in the aggregate, which causes expansion resulting in weakening and cracking of Portland cement grout. See **reactive aggregate**. **D 653,** D18

alkali equivalent, n—deprecated term; see **equivalent alkalies**. **C 219,** C01

alkali ion diode—a sensor for halogen gases. See also **halogen leak detector** (2). **E 1316,** E07

alkaline detergent—under **detergent,** see *inorganic alkaline detergent.* **D 459,** D12

alkaline earth solutions (AES)—an aqueous solution of the oxide or hydroxide of an element of group IIa in the periodic table, such as calcium or magnesium. These solutions may be strongly alkaline. See **pH**. **C 51,** C07

alkaline-filled paper, *n*—a paper containing an alkaline filler such as calcium carbonate; having a pH value in excess of 7 (extract pH usually in the range from 7.5 to 10.0), and containing a reserve buffering capacity that can neutralize acidic materials formed in the paper or acidic gases sorbed from the atmosphere. **D 1968,** D06

alkaline or basic reaction—a characteristic of the materials producing the alkalicolor of the indicator used under the conditions prescribed in this method. **D 4790,** D16

alkaline reserve, *n*—the level, expressed as moles per kilogram or percent by weight of paper, of alkaline materials (such as calcium carbonate) capable of neutralizing either acidic degradation products formed in paper during its use and storage, or acidic gases sorbed by the paper from the atmosphere. **D 1968,** D06

alkaline-sized paper, *n*—paper that has been manufactured using a procedure or process at a pH value above 7 (usually 7.5 to 10.0) that results in paper that has resistance to liquid penetration. See **sizing**. **D 1968,** D06

alkalinity—the quantitative capacity of aqueous media to react with hydrogen ions. **D 1129,** D19

alkalinity—the quantitative capacity of aqueous media to react with hydrogen ions. "M" alkalinity is that which will react with acid as the pH of the sample is reduced to the methylorange endpoint of about 4.5. "P" alkalinity is that which reacts with acid as the pH of the sample is reduced to the phenolphthalein end point of 8.3." M" is the total alkalinity which is the sum of hydroxide plus carbonate plus bicarbonate contents, "P" includes all the hydroxyl and half the carbonate content. **D 6161,** D19

alkali resistance—*for porcelain enamels,* the degree to which a porcelain enamel will resist attack by aqueous alkaline solutions. **C 286,** B08

alkali resistance—for a cellulosic pulp, the fraction insoluble in alkali, usually sodium hydroxide, of a fixed concentration under specified conditions. The term is without precise meaning unless complete specification of an analytical procedure is attached explicitly or implicitly. (see also **alkali solubility**) **D 1695,** D01

alkali solubility, *n—in wool,* the percent of clean wool that is soluble in a specified alkaline solution under controlled conditions of temperature and time. **D 123,** D13

alkali solubility—for a cellulosic pulp, the fraction in alkali of a fixed concentration under specified conditions. This term is without precise meaning unless complete specification of an analytical procedure is attached explicitly or implicitly. Some related terms imply at least a partial specification, for example, "ten percent potassium hydroxide solubility" or "nondilution alkali solubility." **D 1695,** D01

alkali staining—discoloration caused by the presence of an alkali. **D 1695,** D01

alkyd plastics—plastics based on alkyd resins. **D 883,** D20

alkyd resin—under **resin, synthetic,** see *alkyd resin.* **D 16,** D01

alkyd resin—a polyester convertible into a crosslinked form; requiring a reactant of functionality higher than two, or having double bonds. **D 883,** D20

alkyd resin, *n*—a polyester convertible into a crosslinked form; requiring a reactant of functionality higher than two, or having double bonds. **F 1251,** F04

alkyl benzene sulfonate (in the context of soaps and detergents)—the detergent produced by sulfonating detergent alkylate; any surface-active substance having the molecular structure of a benzene sulfonic acid having as a ring substituent(s) an alkyl group(s) sufficiently large to confer detergent properties. **D 459,** D12

alkyl groups—monovalent aliphatic radicals derived from aliphatic hydrocarbons by removal of a hydrogen. **D 1695,** D01

alligator—a general term used for leather made from the skins of all aquatic species with a grain similar to the american alligator which cannot be killed legally and, therefore, is not used by the United States industry. **D 1517,** D31

alligator (crocodile) cracking, *n*—interconnected or interlaced cracks forming a pattern which resembles an alligator's hide. **E 867,** E17

alligator (crocodile) cracking, *n*— interconnected or interlaced cracks forming a pattern which resembles an alligator's hide. **E 1778,** E17

alligator-grained leather—leather of various types, such as calf, sheep or cattlehide embossed to resemble the grain of alligator hide. **D 1517,** D31

alligator hide—a defect characterized by an extreme roughness of the porcelain enamel surface: a severe case of orange peel. **C 286,** B08

alligatoring—the cracking of the surfacing bitumen on a built-up roof, producing a pattern of cracks similar to an alligator's hide; the cracks may not extend through the surfacing bitumen. **D 1079,** D08

alligatoring—*in protective coatings*, surface cracking of coating film having an appearance similar to alligator hide. **D 4538,** D33

allotriomorphic crystal—a crystal whose lattice structure is normal, but whose outward shape is imperfect since it is determined to some extent by the surroundings; the grains in a metallic aggregate are allotriomorphic crystals. **E 7,** E04

allowable bearing value (allowable soil pressure), q_a, p_a **(FL^{-2})**—the maximum pressure that can be permitted on foundation soil, giving consideration to all pertinent factors, with adequate safety against rupture of the soil mass or movement of the foundation of such magnitude that the structure is impaired. **D 653,** D18

allowable pile bearing load, Q_a, P_a **(F)**—the maximum load that can be permitted on a pile with adequate safety against movement of such magnitude that the structure is endangered. **D 653,** D18

allowable properties—mechanical properties of materials as prepared for design use. Allocwable properties of wood are identified with stress-grade descriptions and reflect the orthotropic structure of wood. Often considered synonomous with allowable unit stresses, working stresses, and design stresses. **D 9,** D07

alloy, *n*—a unique composition of two or more polymers that has one or more of the polymers treated or processed in a special way to confer enhanced performance characteristics on the resulting material. **D 1566,** D11

alloy, *n—in plastics,* two or more immiscible polymers united, usually by another component, to form a plastic vein having enhanced properties. (D-20) **F 412,** F17

alloy, *n* **(in plastics)**—two or more immiscible polymers united, usually by another component, to form a plastic resin having enhanced performance properties. **D 883,** D20

alloy steel, *n*—a **steel**, other than a **stainless steel**, that conforms to a specification that requires one or more of the following elements, by mass percent, to have a minimum content equal to or greater than: 0.30 for aluminum; 0.0008 for boron; 0.30 for chromium; 0.30 for cobalt; 0.06 for columbium (niobium); 0.40 for copper; 0.40 for lead; 1.65 for manganese; 0.08 for molybdenum; 0.30 for nickel; 0.60 for silicon; 0.05 for titanium; 0.30 for tungsten (wolfram); 0.10 for vanadium; 0.05 for zirconium; or 0.10 for any other alloying element, except sulphur, phosphorus, carbon, and nitrogen. **A 941,** A01

alloy steel—steel is considered to be alloy when the maximum range given for manganese exceeds 1.65 % or a definite minimum quantity for any of the following elements is specified or required within the limits of the recognized field of constructional alloy steels: chromium, molybdenum, nickel, or any other alloying element added to obtain a desired alloying effect. **F 1789,** F16

alloy system—a complete series of compositions produced by mixing in all proportions any group of two, or more, components, at least one of which is a metal. **E 7,** E04

all purpose compound, *n*—a compound formulated and manufactured to serve as a taping or finishing compound, or both. **C 11,** C11

alluvial channel—See **alluvial stream**. **D 4410,** D19

alluvial deposit—sediment deposited by the action of moving water. **D 4410,** D19

alluvial fans—sediment deposited in the shape of a segment of a cone formed because of a sudden flattening of a stream gradient especially at debouchures of tributaries on main stream flood plains. **D 4410,** D19

alluvial stream—a stream whose boundary is composed of appreciable quantities of the sediments transported by the flow and which generally changes its bed forms as the rate of flow changes. **D 4410,** D19

alluviation—the process of accumulating sediment deposits at places where the flow is retarded. **D 4410,** D19

alluvium—soil, the constituents of which have been transported in suspension by flowing water and subsequently deposited by sedimentation. **D 653,** D18

alluvium—a general term for all fluvial deposits resulting directly or indirectly from the sediment transport of (modern) streams, thus including the sediments laid down in riverbeds, flood plains, lakes, fans, and estuaries. **D 4410,** D19

allyl plastics—plastics based on allyl resins. **D 883,** D20

allyl resin—a resin made by polymerization of chemical compounds containing the allyl group. **D 883,** D20

alpaca, *n*—the fleece and fiber produced by the alpaca, an animal of the genus Llama (*Lama glama pacus*). **D 123,** D13

alpha activity—the spontaneous emission of doubly charged helium ions from the nucleus. **C 859,** C26

alpha-amylase—enzyme that acts specifically to accelerate the hydrolysis of starch to dextrins. **E 1705,** E48

alpha brass—a solid solution phase of one or more alloying elements in copper and having the same crystal lattice as copper. **E 7,** E04

alpha-cellulose—(*1*) Historically, a term used to indicate the pure, relative undegraded cellulose found in pulps. (*2*) Alpha-cellulose content is often measured by TAPPI Method T 203 om-93 where alpha-cellulose is that portion of the pulp which does not dissolve under the test conditions. **D 1695,** D01

alpha complementation—the ability of a short amino-terminal fragment (alpha fragment) of β-galactosidase to form a functional complex with the carboxyl terminal fragment (omega fragment). **E 1705,** E48

alpha iron (Fe)—solid phase of pure iron which is stable at temperatures below 910°C and possesses the body-centered cubic lattice. It is ferro-magnetic below 768°C. **E 7,** E04

alpha (α) loss peak—*in dynamic mechanical measurement,* first peak in the damping curve below the melt, in order of decreasing temperature or increasing frequency, (D 4092, D-20). **E 1142,** E37

alpha (α) loss peak (*in dynamic mechanical measurement*)—the first peak in the damping curve below the melt, in order of decreasing temperature or increasing frequency. **D 4092,** D20

alphameric—See **alphanumeric.** **F 149,** F05

alpha-methylstyrene—2-phenylpropene (C_9H_{10}) mol weight 119.16; colorless liquid; subject to polymerization by heat or catalysts; freezing point, −23.21°C; boiling point, 165.38°C. **D 4790,** D16

alphanumeric—term pertaining to both numbers and alphabetical characters, typically used to designate a device capable of handling both types of characters. **E 1316,** E07

alphanumeric—pertaining to a character set that contains letters, digits, and usually other characters such as punctuation marks. *Syn.* **alphameric.** **F 149,** F05

alphanumeric—print set containing both alphabetic and numeric characters. **F 1156,** F12

alphanumeric character set—a character set that contains both letters and digits and may contain control characters, special characters, and the space character. **F 149,** F05

alphanumeric character subset—a character subset that contains both letters and digits and may contain control characters, special characters, and the space character. **F 149,** F05

alpha particle—a positively charged particle emitted by certain radio-nuclides. It consists of two protons and two neutrons, and is identical to the nucleus of a helium atom. **E 1316,** E07

alpha pulps—see **chemical cellulose.** **D 1695,** D01

altazimuthal mount—a supporting device that facilitates tracking of the sun and allows rotation about horizontal and vertical axes. It can be used to aim equipment such as heliostats, concentrating collectors, exposure specimens, or radiometers. **E 772,** E44

alter—See **building modification.** **E 631,** E06

alter—to change fastener properties such as hardness, tensile strength, surface finish, length, or other characteristics of the fastener through such processes as heat treatment, plating, and machining. **F 1789,** F16

alteration—the modification of a document or article with the intent that it will pass as genuine with minimum risk of detection in circumstances of ordinary use. **F 1156,** F12

alternating load—See **loading amplitude.** **E 1823,** E08

alum—aluminum sulfate, $AL_2(SO_4)_3XH_2O$ (X = 14-18), a coagulant. **D 6161,** D19

alumina-chromia brick, *n*—a refractory brick, which may be burned or unburned, manufactured predominantly of a mixture of a high-alumina raw material and chromic oxide (Cr_2O_3), with the alumina (Al_2O_3) content being 50 % or greater by weight and with chromic oxide (Cr_2O_3) predominating by weight among the other constituent oxides. **C 71,** C08

alumina porcelain—See *alumina porcelain* under **porcelain.** **C 242,** C21

alumina whiteware—See *alumina whiteware* under **ceramic whiteware.** **C 242,** C21

aluminized—dipped in molten aluminum for coating purposes resulting in smooth, continuous, and adherent aluminum coating. **F 547,** F16

aluminized—*in chain link fabric,* describes fabric woven from steel wire that is coated with aluminum before weaving (see Specification A 491). **F 552,** F14

aluminized coating, *n*—a coating on steel of aluminum alloy, typically 89 to 99 % aluminum; generally applied to steel sheet and wire by the hot-dip process. **A 902,** A05

aluminizing—forming of an aluminum or aluminum alloy coating on a metal by hot dipping, hot spraying, or diffusion. **B 374,** B08

aluminous cement, n—deprecated term. **C 219,** C01

aluminum common nail—plain-shank, aluminum-alloy, 1 by 0.072 to 6 by 0.262-in. nails with flat 5⁄32 to 17⁄32-in. head and medium diamond point. **F 547,** F16

aluminum enamel—a porcelain enamel specifically designed for application to aluminum. **C 286,** B08

alum leather—leather produced by alum tannage, usually in combination with other substances. Before the invention of chrome-tanning this was the principal method of tanning with mineral agents. **D 1517,** D31

ambient, *adj*—surrounding on all sides. **D 1356,** D22

ambient field, *n*—when an object or light source is viewed, the complete area beyond the surround from which light might reach the observer's eyes and influence the object's appearance. See **surround.** **E 284,** E12

ambient light—light not provided by the visual testing system. **E 1316,** E07

ambient light—See **lighting.** **E 1480,** E06

ambient noise—the composite of airborne sound from many sources near and far associated with a given environment. No particular sound is singled out for interest. **C 634,** E33

ambient temperature—the temperature of the surrounding atmosphere as determined by an instrument shielded from direct or reflected rays of the sun. **D 2864,** D27

ambient temperature—the temperature of the surroundings, typically 20°– 25°C. **D 6161,** D19

ambient temperature, *n*—the existing temperature of the air or of an object in thermal equilibrium with the surrounding atmosphere. **G 113,** G03

ambulance—a vehicle for transportation of the sick and injured, equipped and staffed to provide emergency medical care during transit. **F 1177,** F30

ambulance service—a qualified provider of medical transportation for patients requiring treatment and/or monitoring due to illness or injury. **F 1177,** F30

ambulance service provider—a person or organization, either public or private, responsible for operation, maintenance, and administation of an ambulance service. **F 1177,** F30

American grain count, *n*—a direct yarn numbering system for expressing linear density, equal to the mass in grains per 120 yards of sliver or roving. **D 123,** D13

American Method—a method of application for roofing shingles, generally rectangular in shape, to provide double coverage with head lap and no side lap. **C 460,** C17

American method, *n*— *in shingles,* a method of application for roofing shingles, generally rectangular in shape, to provide double coverage with head lap and no side lap. **C 1154,** C17

American method—method of application for roofing shingles, generally rectangular in shape, to provide double coverage with head lap and no side lap. **D 2946,** C17

American Public Health Association (APHA) color, n—see **platinum cobalt color scale.** **E 284,** E12

American softwood lumber standard—a voluntary product standard of the U. S. Department of Commerce which serves as a basis for the preparation and revision of industry-sponsored grading rules for lumber. **D 9,** D07

American Standard Code for Information Interchange—a data communications code set consisting of a 7-bit-plus-parity code that can be translated with a leading ''0" as an 8-bit set. (See **ASCII**.) **F 1457,** F05

American standard lumber—lumber conforming to the basic provisions of the American Softwood Lumber Standard. **D 9,** D07

Ames bar code—See **bar code**. **F 1156,** F12

amino plastics, *n*—plastics based on amino resins. **D 883,** D20

amino resin, *n*—a resin made by polycondensation of a compound containing amino groups, such as urea or melamine, with an aldehyde, such as formaldehyde, or an aldehyde-yielding material. **D 883,** D20

ammeter shunt—a low-resistance precision resistor with high current carrying capacity connected in parallel with an ammeter. **E 1316,** E07

amorphous—noncrystalline, or devoid of regular structure. **B 374,** B08

amorphous—non crystalline, devoid of regular cohesive structure. **D 6161,** D19

amorphous alloy—a semiprocessed alloy produced by a rapid quenching, direct casting process resulting in metals with noncrystalline structure. **A 340,** A06

amorphous peat—see **sapric peat**. **D 653,** D18

amorphous regions—those regions within a cellulosic material which, on the basis of X-ray diffraction or other suitable technique, do not show any evidence of crystalline structure. The technique should be specified. **D 1695,** D01

amorphous silicon (α-Si) X-ray detector, *n*—an amorphous silicon (α-Si) X-ray detector consists of a glass substrate with a matrix of photodiodes fabricated from amorphous silicon and switches arranged in rows and columns upon it; the photodiodes are activated by light photons emitted from a scintillator which is activated by X rays and is usually in close contact with the diode matrix. **E 1316,** E07

amosite—the acronym assigned to grunerite asbestos, and derived from the name of the first developers of a major deposit of this mineral. **C 460,** C17

amosite, *n*—the acronym assigned to **grunerite asbestos,** and derived from the name of the first developers of a major deposit of this mineral. **D 2946,** C17

amount, *n*—*in cotton fiber testing with the Length Analyzer*, a measure of the thickness or pneumatic density of the test beard, proportional to the number of fibers present at various distances from the base of the specimen clamp jaws. **D 123,** D13

amount, *n*—*in cotton length testing with the Fibrograph*, a measure of the thickness, or optical density, of the test beard, proportional to the number of fibers present at various distances from the comb(s). **D 123,** D13

ampere—the current that will deposit silver at the rate of 0.0011180 g/s. Current flowing at the rate of 1 C/s. **B 374,** B08

ampere per metre, A/m—the unit of magnetic field strength in the SI system of units. **A 340,** A06

ampere (turn), A—the unit of magnetomotive force in the SI system of units. The symbol A represents the unit of electric current, ampere in the SI system of units. **A 340,** A06

ampere turns—the product of the number of turns of a coil and the current in amperes flowing through the coil. **E 1316,** E07

amphibole asbestos, *n*—asbestiform amphibole silicate minerals including the orthorhombic anthophyllite series and the monoclinic cummingtonite [grunerite asbestos (amosite)] series, the tremolite-actinolite series, and the alkali amphibole [riebeckite asbestos (crocidolite)] series among others. The amphiboles contain essential (OH) groups in the structure, and the Si:O ratio is 4:11. A considerable amount of elemental substitution can take place in these varieties of asbestos. The crystal structures are composed of strips or ribbons of linked polyhedra that join to form fibrils. The individual strips are made up from three components; these are two double chains of linked (Si, Al)O_4 tetrahedra and a strip of linked MgO_6, FeO_6, or $A1O_6$ octahedra. **C 460,** C17

amphibole asbestos, *n*—asbestiform amphibole silicate minerals including the orthorhombic anthophyllite series and the monoclinic cummingtonite [grunerite asbestos (amosite)] series, the tremolite-actinolite series, and the alkali amphibole [riebeckite asbestos (crocidolite)] series, among others. The amphiboles contain essential (OH) groups in the structure, and the Si:O ratio is 4:11. A considerable amount of elemental substitution can take place in these varieties of asbestos. The crystal structures are composed of strips or ribbons of linked polyhedra that join to form fibrils. The individual strips are made up from three components; these are two double chains of linked (Si, Al)O_4 tetrahedra and a strip of linked MgO_6, FeO_6, or AlO_6 octahedra. **D 2946,** C17

ampholytic or amphoteric (zwitterionic) surfactant—a chemical compound capable of producing either (or both) positively or negatively charged surface active ions in an aqueous solution. The charge carried on the *ampholytic*surface active ions depends upon the pH of the solution but that on *amphoteric (zwitterionic)* surfactants does not. **D 459,** D12

ampholytic surfactant or amphoteric surfactant—see **surface-active agent.** **D 459,** D12

amphoteric—capable of acting as an acid or a base. **D 6161,** D19

amphoteric surfactant—a surface-active agent capable of forming, in aqueous solution, either surface-active anions or surface-active cations depending on the pH. **E 1519,** E35

amplification factor—ratio of dynamic to static displacement. **D 653,** D18

amplifier—a negative lens, used in lieu of an eyepiece, to project under magnification the image formed by an objective. The amplifier is especially designed for flatness of field and should be used with an apochromatic objective. **E 7,** E04

ampliphan eyepiece— See **amplifier.** **E 7,** E04

amplitude—the vertical pulse height of a signal, usually base to peak, when indicated by an A-scan presentation. **E 1316,** E07

amplitude distortion— same as **harmonic distortion.** **E 1316,** E07

amplitude hologram—a recording of the variation of light intensity caused by the interference between the reference beam and the object beam, as light or dark areas on the recording medium. The light and dark interference lines in the recording medium diffract laser light to produce the reconstruction. **E 1316,** E07

amplitude response—that property of a test system whereby the amplitude of the detected signal is measured without regard to phase. (See also **harmonic analysis and phase analysis**.) **E 1316,** E07

ampoule—a glass container designed to be filled and sealed by fusion of the glass neck. **C 162,** C14

ampoule, *n*—a hermetically sealed, small bulbous glass or plastic vessel. Opening is achieved by breaking the stem. (Also *ampule* or *ampul*.) **D 996,** D10

amusement ride or amusement device—a device or combination of devices or elements that carry, convey, or direct a person(s) over or through a fixed or restricted course or within a defined area, for the primary purpose of amusement or entertainment. **F 747,** F24

amylaceous, *adj*—pertaining to, or of the nature of, starch; starchy. **D 907,** D14

anaerobes—organisms that grow in the absence of air or oxygen and do not use molecular oxygen in respiration. **F 1600,** F20

anaerobic, *n*—a condition in which no measurable volume of air is present in the incubation chamber or system. **D 4439,** D35

anaerobic—living or active in an airless environment. **E 1126,** E48

anaerobic—living or active in an airless environment. **E 1705,** E48

anaerobic adhesive, *n*—an adhesive that is kept in the uncured state by oxygen, as in air, and that cures in the absence of oxygen when exposed to metal ions, especially copper or iron. **D 907,** D14

anaerobic bacteria—bacteria that do not use oxygen. Oxygen is toxic to them. See **bacteria, anaerobes**. **D 6161,** D19

anaerobic bacteria—microbes whose metabolisms require the absence of free oxygen. **E 1126,** E48

anaerobic bacteria—microbes whose metabolisms require the absence of free oxygen. **E 1705,** E48

anaerobic digester—a chemical reactor in which anaerobic bacteria are used to decompose biomass or organic wastes to produce methane and carbon dioxide. **E 1126,** E48

anaerobic digester—a chemical reactor in which anaerobic bacteria are used to decompose biomass or organic wastes to produce methane and carbon dioxide. **E 1705,** E48

anaerobic digestion—degradation of organic matter by microbes in the absence of air (oxygen) to produce methane and carbon dioxide (biogas). **E 1126,** E48

anaerobic digestion—degradation of organic matter by microbes in the absence of air (oxygen) to produce methane and carbon dioxide (biogas). **E 1705,** E48

anaerobic fermentation—fermentation processes conducted in the absence of air. The following anaerobic fermentation processes are significant in obtaining useful forms of energy from biomass: (*1*) alcoholic fermentation, fermentation processes whereby certain microorganisms convert glucose and other substrates with alcohol as an end product, (*2*) methane fermentation, generally termed anaerobic digestion (See also **anaerobic digestion**). **E 1126,** E48

anaerobic fermentation—fermentation processes conducted in the absence of air. The following anaerobic fermentation processes are significant in obtaining useful forms of energy from biomass: (*1*) alcoholic fermentation, fermentation processes whereby certain microorganisms convert glucose and other substrates with alcohol as an end product, (*2*) methane fermentation, generally termed anaerobic digestion (See also **anaerobic digestion**). **E 1705,** E48

analate addition—a variation of the known addition measurement technique in which the sample (analate) is added to a reagent containing the ion being measured. The electrode is placed in the reagent, and the sample concentration is calculated from the change in electrode potential after the addition of the sample. **D 4127,** D19

analate subtraction—a variation of the known subtraction measurement technique in which the sample (analate) is added to a reagent containing an ion that reacts with the species being determined. The electrode is placed in the reagent, the change in electrode potential is observed when the sample is added, and the sample concentration calculated. **D 4127,** D19

analog image—an image produced by a continuously variable physical process (for example, exposure of film). **E 1316,** E07

analog pH/mV meter—an instrument that displays voltages, pH units, or other concentration units by means of a needle pointer and scale. **D 4127,** D19

analog to digital converter (a/d)—a device that changes an analog signal to a digital representation of the signal. **E 1316,** E07

analysis:

area—a two-dimensional region of a specimen surface, usually measured in the plane of the specimen surface, from which a specified percentage of the signal is detected.

volume—a three dimensional region of a specimen from which a specified percentage of the measured signal is generated. Also see **information depth**. **E 673,** E42

analysis, *n*—an act of inspecting the S/H image and associating this image with a known calibration reference. **F 538,** F09

analysis bandwidth—spectral band used in a photometric instrument, such as a densitometer, for the measurement of optical absorbance or reflectance. **E 170,** E10

analysis of variance (ANOVA), *n*—a procedure for dividing the total variation of a set of data into two or more parts, one of which estimates the error due to selecting and testing specimens and the other part(s) possible sources of additional variation. **D 123,** D13

analysis (physical or chemical)—the determination of physical or chemical properties or composition of a material. **C 859,** C26

analysis sample—See *analysis sample* under **sample**. **D 121,** D05

analysis sample, *n*—the final subsample prepared from the air-dried laboratory sample but reduced in particle size by passing through a mill with a 0.5 mm (0.02–in.) size or smaller final screen. **D 5681,** D34

analysis sample—final subsample prepared from the air-dried laboratory sample but reduced by passing through a mill with a 0.5-mm (0.02-in.) size or smaller final screen. **E 856,** D34

analysis wavelength—wavelength used in a spectrophotometric instrument for the measurement of optical absorbance or reflectance. **E 170,** E10

analyte—a possible sample component whose presence and concentration is of interest. **D 1129,** D19

analyte, *n*—*in methods of chemical analysis*, the component determined by a method. **E 135,** E01

analytical curve—the graphical representation of a relation between some function of radiant power and the concentration or mass of the substance emitting or absorbing it. **E 131,** E13

analytical curve—see **calibration curve.** **E 135,** E01

analytical detection limit, *n*—a number, expressed in units of concentration (or amount), that describes the lowest concentration level (or amount) that an analyst can determine to be different from an analytical blank (background level). **F 1494,** F23

analytical gap, *n*—*in optical emission spectrometry*, the region between two electrodes in which the specimen is excited and from which radiant energy is used for analysis. **E 135,** E01

analytical line, *n*—the particular wavelength of an element used in determining the presence or concentration of that element. **E 135,** E01

analytical reagent grade—reagents of high purity that meet ACS specifications, as published by the American Chemical Society. **C 859,** C26

analytical technique, *n*—a procedure whereby the concentration of the test chemical in a collection medium is quantitatively determined. **F 1494,** F23

analytical wavelength—any wavelength at which an absorbance measurement is made for the purpose of the determination of a constituent of a sample. **E 131,** E13

analyzer, *n*—the instrumental equipment necessary to perform automatic analysis of ambient air through the use of physical and chemical properties and giving either cyclic or continuous output signal.

analyzer system, n—all sampling, analyzing, and readout instrumentation required to perform ambient air quality analysis automatically.

full scale, n—the maximum measuring limit for a given range of an analyzer.

lag time, n—the time interval from a step change in the input concentration at the analyzer inlet to the first corresponding change in the analyzer signal readout.

linearity, n—the maximum deviation between an actual analyzer reading and the reading predicted by a straight line drawn between upper and lower calibration points.

minimum detection limit, n—the smallest input concentration that can be determined as the concentration approaches zero.

noise, n—random deviations from a mean output not caused by sample concentration changes.

open path analyzer, n—an analytical system that measures the average atmospheric or emission compound concentration along one or more monitoring paths open to the atmosphere. See **monitoring path.**

operating humidity range of analyzer, n—the range of ambient relative humidity of air surrounding the analyzer, over which the analyzer will meet all performance specifications.

operating temperature range of analyzer, n—the range of ambient temperatures of air surrounding the analyzer, over which the monitor will meet all performance specifications.

operational period, n—the period of time over which the analyzer can be expected to operate unattended within specifications.

output, n—a signal that is related to the measurement, and intended for connection to a readout or data acquisition device.

range, n—the concentration region between the minimum and maximum measurable limits.

readout instrumentation, n—output meters, recorder, or data acquisition system for monitoring analytical results.

response time, n—the time interval from a step change in the input concentration at the analyzer inlet to an output reading of 90 % of the ultimate reading.

rise time, n—response time minus lag time.

sample system, n—equipment necessary to provide the analyzer with a continuous representative sample.

span drift, n—the change in analyzer output over a stated time period, usually 24 h of unadjusted continuous operation, when the input concentration is at a constant, stated upscale value.

zero drift, n—the change in analyzer output over a stated time period of unadjusted continuous operation when the input concentration is zero; usually expressed as a percentage change of full scale over a 24-h operational period.

See also **point analyzer.** **D 1356,** D22

analyzer—an optical device, capable of producing plane polarized light, used for detecting the state of polarization. **E 7,** E04

analyzer—an optical device, capable of producing plane polarized light, used for detecting the state of polarization. **E 175,** E41

analyzer lag, *n*—Deprecated term. Use the preferred term *instrument response time.* **D 4175,** D02

analyzer system—See **analyzer.** **D 1356,** D22

analyzer transmission—see **spectrometer transmission.** **E 673,** E42

anchor—in general, a metal shape inserted into a slot or hole in the stone that provides for the transfer of loads from the stone to the building structure, either directly or through an intermediate structure. **C 119,** C18

anchor, *v*—to secure firmly (*Webster*). **D 996,** D10

anchor, *n*—a device used to connect securely a **building component** to adjoining construction, to a supporting member, or to the ground. **E 631,** E06

anchor, *n*—components that are directly attached to the bony elements of the spine (sacrum, lamina, pedicle, vertebral body, spinous process, transverse process, the pelvis, or ribs). **F 1582,** F04

anchorage—the system consisting of stone, anchor and primary structure, secondary structure or back-up preventing lateral movement of the stone. **C 119,** C18

anchorage, *n*—a means of connecting securely, by using an **anchor,** a **building component** to adjoining construction, supporting member(s), or to the ground. **E 631,** E06

anchorage system—a group of interacting elements, components, and structures. **E 631,** E06

anchor bolt—a steel rod or bar, one end of which is intended to be cast in concrete while the opposite end is threaded and projects from the concrete for anchoring other material to the concrete. The end cast in concrete may be either straight or provided with an anchor, such as a bent hook, forged head, or a tapped or welded attachment to resist forces imposed on the anchor bolt as required. **F 1789,** F16

anchoring point, *n*—a reference point against which other items are judged . **E 253,** E18

anchoring system—a group of interacting anchors and elements. **E 631,** E06

anchor point—structural point on the end connector or along the length of a boom section designed for the attachment of anchor or mooring lines. **F 818,** F20

ancillary equipment—mechanical devices *essential* to the operation of a given boom system; for example, air pumps, hydraulic power supplies, control manifolds, and so forth. **F 818,** F20

andalusite—a polymorph, along with sillimanite and kyanite, of composition $Al_2O_3{\cdot}SiO_2$ which on firing dissociates to yield principally mullite. **C 242,** C21

anesthetic-gas scavenging systems, *n*—complete systems that collect and remove the excess gases released from equipment used in administering anesthesia, or exhaled by the patient during anesthesia for the purpose of conveying these gases to an appropriate place of discharge. **F 2002,** F29

angle, n—see **aperture angle, aperture solid angle, azimuthal angle, entrance angle, observation angle, rotation angle, specular angle.** **E 284,** E12

angle:

collection—SIMS, the angle between the normal to the original specimen surface and the axis of the secondary ion collection optics.

of detector—EIA, SIMS, the angle between the incident beam direction and the direction pointing from the beam spot to the center of the detector.

of emission—AES, XPS, the angle of emission or ejection of electrons from a solid measured relative to the normal to the surface.

of incidence—the angle between the incident beam and the normal to the surface.

of scattering—EIA, the angle between the incident beam direction and the direction in which a particle is traveling after it is scattered. If the particle is incident on the detector, this angle will be the same as **angle** *of detector.*

solid, of detector—EIA, the solid angle intercepted by the detector, with the radius originating at the beam spot.

takeoff—AES, XPS the angle at which particles leave a specimen measured relative to the plane of the specimen surface. (see **angle** *of emission*). **E 673,** E42

angle—on countersunk head, total included angle formed by conical underportion or bearing surface. **F 547,** F16

angle—See **lead angle.** **F 547,** F16

angle beam—a term used to describe an angle of incidence or refraction other than normal to the surface of the test object, as in angle beam examination, angle beam search unit, angle beam longitudinal waves, and angle beam shear waves. **E 1316,** E07

angle lapping—a method of specimen preparation in which a specimen is mechanically polished at an angle so that changes in composition as a function of position, usually depth, can be more easily determined. **E 673,** E42

angle (normal)—perpendicular to the surface described. **F 1156,** F12

angle of acceptance—widest angle off the normal at which a light ray, etc., may enter an aperture and still be detected by the sensor. **F 1156,** F12

angle of attack, *n— in impingement erosion,* the angle between the direction of motion of an impinging liquid or solid particle and the tangent to the surface at the point of impact. **G 40,** G02

angle of bend, *n*—the change in the angle between the two legs of the specimen during a bend test, measured before release of the bending forces. **E 6,** E28

angle of external friction (angle of wall friction), δ (degrees) —angle between the abscissa and the tangent of the curve representing the relationship of shearing resistance to normal stress acting between soil and surface of another material. **D 653,** D18

angle of friction (angle of friction between solid bodies), $\phi\ s$ **(degrees)**—angle whose tangent is the ratio between the maximum value of shear stress that resists slippage between two solid bodies at rest with respect to each other, and the normal stress across the contact surfaces. **D 653,** D18

angle of illumination, *n*—angle between the specimen normal and the illuminator axis. **E 284,** E12

angle of incidence, θ—the angle between an incident radiant beam and a perpendicular to the interface between two media. **E 131,** E13

angle of incidence, *n*—the angle between a ray impinging on a surface at a point and the perpendicular to the surface at that point. In *the description of a beam,* the angle of incidence of the ray at the center of the beam. **E 284,** E12

angle of incidence—the angle between a ray and the normal to the plane on which it is incident. (The plane of incidence may be the aperture plane, the collector, or any other plane of interest.) **E 772,** E44

angle of incidence, *n*—the angle between an incident ray and the normal to a surface at the point of contact. When referring to a scanner reading a printed symbol, the angle between the light source scanner beam and the symbol position. **F 1294,** F05

angle of incidence, *n— in impingement erosion,* the angle between the direction of motion of an impinging liquid or solid particle and the normal to the surface at the point of impact. **G 40,** G02

angle of inclination—an angle at which the test sheet and supporting base are inclined from the horizontal. (F 254) **F 221,** F05

angle of internal friction (angle of shear resistance), ϕ (degrees) —angle between the axis of normal stress and the tangent to the Mohr envelope at a point representing a given failure-stress condition for solid material. **D 653,** D18

angle of obliquity, α, β, ϕ, Ψ (degrees)—the angle between the direction of the resultant stress or force acting on a given plane and the normal to that plane. **D 653,** D18

angle of placement of metal connector plate—angle of inclination of lengthwise axis of metal connector plate parallel to longitudinal axis of coiled metal strip, that is, main direction of metal connector plate to direction of test-load application to wood member of connection; with *zero-degree angle* defined as that of lengthwise plate axis being parallel to load direction; and *angle greater than zero* defined as that of lengthwise plate axis being rotated clockwise away from the loading axis when facing the plated connection. **E 1807,** E06

angle of reflection: (*1*) *reflection*—the angle between the reflected beam and the normal to the reflecting surface.
(*2*) *diffraction*—the angle between the diffracted beam and the diffracting planes. **E 7,** E04

angle of reflection, *n*—the angle between a ray reflected from a surface at a point and the perpendicular to the surface at that point. **E 284,** E12

angle of reflection—the angle between the direction of propagation of a reflected ray and the normal to the surface at the point of reflection. **E 772,** E44

angle of refraction—the angle between the direction of propagation of a refracted ray and the normal to the interface at the point of refraction. **E 772,** E44

angle of repose, *n*—the greatest angle, measured from horizontal, attained by a coal so that material on the sides of top of the inverted cone or trench created in car top sampling remains stable, that is, will not shift or slide. Coal size and moisture content are contributing factors to this angle. In all cases, the coal shall be dug so that the physical angle is less than the angle of repose. **D 121,** D05

angle of repose, α (degrees)—angle between the horizontal and the maximum slope that a soil assumes through natural processes. For dry granular soils the effect of the height of slope is negligible; for cohesive soils the effect of height of slope is so great that the angle of repose is meaningless. **D 653,** D18

angle of shear resistance—see **angle of internal friction.** **D 653,** D18

angle of view, *n*—angle between the normal to the surface of the specimen and the axis of the receiver. **E 284,** E12

angle of wall friction—see **angle of external friction.** **D 653,** D18

angle ply—any filamentary lamina orientated in a direction other than that specified as 0° (that is, the reference axis) within a composite assembly. **E 1749,** E06

angleply laminate—any balanced laminate consisting of plus and minus theta plies where theta is an acute angle with respect to a reference direction. **D 3878,** D30

angle resolved AES—the recording of Auger electron spectra as a function of angle emission. **E 673,** E42

angle, rotation, n— see **rotation angle.** **E 284,** E12

angstrom (A)—a unit of length equaling 10^{-10} metres, $10^{-4}\mu$ metres, 10^{-8} centimetres and 3.937×10^{-9} in. The symbol is Å, A or A.U. **D 6161,** D19

ångström unit—a unit of linear measure named after A. J. Ångström. It is 1×10^{-10} metres; 1 μm = 10,000 Å. It is generally abbreviated as A. in the United States; elsewhere, it is variously abbreviatedÅ , A., A.U., Å., or ÅU. **E 175,** E41

angström unit (Å)—10^{-8} cm. **B 374,** B08

angstrom unit (Å)—a unit of length which may be used to express the wavelength of electromagnetic radiation, that is, light. One angstrom unit is equal to 0.1 nanometres. (1 nm = 10^{-9}m).

E 1316, E07

Angstrom unit (abbreviation) = A,Å , or *A. U*—a unit of length equal to 10^{-8} cm. This is the standard unit of measurement in X-ray crystallography. **E 7,** E04

angular aggregate—aggregate, the particles of which possess well-defined edges formed at the intersection of roughly planar faces. **D 653,** D18

angular aperture—See **aperture, optical.** **E 7,** E04

angular aperture—see **aperture, angular.** **E 175,** E41

angular distribution of secondary ions—see **secondary ions.** **E 673,** E42

angular frequency, ω—the number of radians per second traversed by a rotating vector that represents any periodically varying quantity. **E 1142,** E37

angular strain, n—use *shear strain.* **E 6,** E28

angular subtense, *n— visual,* the angle subtended (by an object) at the first nodal point of the eye. **E 284,** E12

anhydrite, *n*—the mineral consisting primarily of anhydrous calcium sulfate, $CaSO_4$. **C 11,** C11

anhydrite, n—see **calcium sulfate**. **C 219,** C01

anhydroglucose units—the repeating unit of many polysaccharides, including cellulose; since the glucose molecules have combined with elimination of water, the unit is called "anhydroglucose" rather than "glucose." **D 1695,** D01

anhydrous—a material that does not contain water either absorbed on its surface or as water of crystallization; a water-free product. **E 1126,** E48

anhydrous—a material that does not contain water either absorbed on its surface or as water of crystallization; a water-free product. **E 1705,** E48

anhydrous ethanol—100 % ethanol, neat ethanol, 199 + proof ethanol. **E 1705,** E48

anhydrous soap—under **soap,** see *anhydrous soap.* **D 459,** D12

anhydrous, without water—term used in chemistry to denote absence of water. 199 + proof ethanol is considered anhydrous ethanol. **E 1705,** E48

aniline dyed leather—leather which has been colored by transparent dyes as distinguished from leather treated by pigments or other opaque material. **D 1517,** D31

aniline finish—a clear finish with little or no pigmentation. **D 1517,** D31

aniline point—the minimum temperature for complete miscibility of equal volumes of aniline and the sample under test. See ASTM Test Methods D 611, for Aniline Point and Mixed Aniline Point of Petroleum Products and Hydrocarbon Solvents. Aniline point is a function of both molecular weight and composition. In comparing two samples of similar molecular weight, the aniline point can be used as a means of comparing aromatic content of the two samples. A product of high aniline point will be low in aromatics and naphthenes, and therefore high in paraffins. **D 2864,** D27

aniline point, *n*—the minimum equilibrium solution temperature for equal volumes of aniline (aminobenzene) and sample. **D 4175,** D02

animal fiber, *n*—any natural protein-base fiber. **D 123,** D13

anion—a negatively-charged ion. **B 374,** B08

anion—negatively charged ion. **D 6161,** D19

anion—a negatively charged ion. **G 15,** G01

anion-exchange material—a material capable of the reversible exchange of negatively charged ions. **D 1129,** D19

anion exchange material—a material capable of the reversible exchange of negatively charged ions. **D 6161,** D19

anionic detergent—under **detergent,** see *anionic detergent.* **D 459,** D12

anionic emulsion, *n*—a type of emulsion such that a particular emulsifying agent establishes a predominance of negative charges on the discontinuous phase. **D 8,** D04

anionic emulsion—an emulsion in which the emulsifying system establishes a predominance of negative charges on the discontinuous phase. **D 1079,** D08

anionic polyelectrolyte—usually acrylamide and acrylamide and acrylic copolymers, negatively charged, used for coagulation/flocculation, see Polyelectrolytes. **D 6161,** D19

anionic surfactant—a surface-active agent in which the active portion of the molecule containing the lipophilic segment forms exclusively a negative ion (anion) when placed in aqueous solution. **E 1519,** E35

anisotropic, *adj—as used in Test Method D 5061,* exhibiting optical properties of different values when viewed with an optical microscope having mutually exclusive polarized light, for example, crossed nicols. **D 121,** D05

anisotropic—not isotropic; having mechanical or physical properties, or both, that vary with direction relative to natural reference axes in a material. **E 1749,** E06

anisotropic, *adj.*—exhibiting different physical properties in different directions. **F 1241,** F01

anisotropic, *adj*—having different values for a specific property in different directions. **F 1494,** F23

anisotropic etch—a selective etch that exhibits an accelerated etch rate along specific crystallographic planes. **F 1241,** F01

anisotropic mass—a mass having different properties in different directions at any given point. **D 653,** D18

anisotropic material—a material in which the magnetic properties differ in various directions. **A 340,** A06

anisotropic membrane—a nonuniform structure in cross-section; typically the support substructure has pores much larger than the barrier layer. See **asymmetric membranes**. **D 6161,** D19

anisotropic (replaces anisotropy)—having different values for a property, in different directions. **E 7,** E04

anisotropy—having different properties in different directions. (ISRM) **D 653,** D18

ankle, *n—in anatomy,* the joint between the foot and the lower leg. **D 123,** D13

ankle girth, *n—in body measurements,* the circumference of the leg over the greatest prominence of the ankle. **D 123,** D13

ankle height, *n—in body measurements,* with the subject standing barefoot, the distance from the center of the prominent outside ankle bone to the floor. **D 5219,** D13

anneal—to attain acceptably low stresses, or desired structure, or both, in glass by controlled cooling from a suitable temperature. **C 162,** C14

anneal, *vt—in fire assay,* to heat and then gradually cool a metal to remove internal stresses and make the material less brittle. **E 135,** E01

anneal (annealing)—a thermal treatment to change the properties or grain structure of the product. **B 846,** B05

annealed—heated and subsequently cooled to provide increased ductility. **F 547,** F16

annealed wire—see **soft wire.** **B 354,** B01

annealing, *n*—a generic term covering any of several **heat treatments**. **A 941,** A01

annealing—a controlled cooling process for glass designed to reduce residual stress to a commercially acceptable level and modify structure. **C 162,** C14

annealing—a general term applied to a variety of thermal treatments applied to fasteners for the purpose of softening or homogenizing material properties. The specific types of annealing are: **F 1789,** F16

annealing acid—see **acid annealing** and **annealing.** **C 286,** B08

annealing point (A.P.)—that temperature corresponding either to a specific rate of elongation of a glass fiber when measured by Test Method C 336, or a specific rate of midpoint deflection of a glass beam when measured by Test Method C 598. At the annealing point of glass, internal stresses are substantially relieved in a matter of minutes. **C 162,** C14

annealing range—the range of glass temperature in which stress in glass can be relieved at a commercially practical rate. For purposes of comparing glasses, the annealing range is assumed to correspond with the temperature between the annealing point (A.P.) and the strain point (St.P.). **C 162,** C14

annealing-twin bands— See **twin bands.** **E 7,** E04

annihilation radiation—gamma radiation produced by the annihilation of a positron and an electron. For particles at rest, two photons are produced, each having an energy corresponding to the rest mass of an electron (511 keV). **E 170,** E10

annual average daily traffic, (AADT), *n*—the estimate of typical traffic on a road segment for all days of the week, Sunday through Saturday, over the period of one year and calculated from permanent counter data as the sum of Monthly Average Daily Traffic (MADTs) divided by the number of MADTs, and for short term counts by using adjustment factors. **E 867,** E17

annual average days of the week, (AADW), *n*—a traffic volume mean statistic for each day of the week, over the period of one year, and calculated from included permanent counter data as the sum of Monthly Average Days of the Week (MADWs) for a year divided by the number of MADWs. **E 867,** E17

annual average weekday traffic, (AAWDT), *n*—the estimate of typical traffic over the period of one year, for the days Monday through Thursday, calculated from permanent counter data as the sum of Monthly Average Weekday Traffic (MAWDTs) divided by the number of MAWDTs. **E 867,** E17

annual average weekend traffic, (AAWET), *n*—the estimate of typical traffic over the period of one year, for the days Saturday through Sunday, calculated from permanent counter data as the sum of Monthly Average Weekend Traffic (MAWETs) divided by the number of MAWETs. **E 867,** E17

annually recurring costs—those costs that are incurred in a regular pattern each year throughout the study period. **E 833,** E06

annual ring—the growth layer produced by the tree in a single growth year, including earlywood and latewood. **D 9,** D07

annual space; annulus—the space between two concentric tubes or casings, or between the casing and the borehole wall. This would include the space(s) between multiple strings of tubing/casings in a borehole installed either concentrically or multi-cased adjacent to each other. **D 653,** D18

annual summary statistics, *n*—yearly estimates of the central tendency of traffic volume, vehicle classification, and Equivalent Single Axle Load (ESALs) by vehicle classification, based on traffic measurements during the weekday combinations of Sunday through Saturday, Monday through Thursday, and Friday through Sunday. **E 867,** E17

annual value—a uniform annual amount equivalent to the project costs or benefits taking into account the time value of money throughout the study period (Syn. *annual worth, equivalent uniform annual value*). **E 833,** E06

annual vehicle miles traveled (AVMT), *n*—average annual traffic on a road segment, expressed as AADT multiplied by the number of days in the year, multiplied by the length of the road segment. **E 867,** E17

annual worth—See **annual value**. **E 833,** E06

annular, *adj*—descriptor for directional illuminating (or viewing) geometry in which the illuminator provides radiation (or the receiver possesses responsivity) that is distributed continuously and uniformly throughout the 360° of azimuth of the measurement. (See also *circumferential.*) **E 284,** E12

annular coil clearance—the mean radial distance between adjacent coil assembly and test part surface in electromagnetic encircling coil examination. **E 1316,** E07

annular coils—see **encircling coils.** **E 1316,** E07

annular space—the space between the inner surface of the female end and the outer surface of the male end of an assembled pipe joint. **C 822,** C13

annular thread—multiple ring-like threads rolled completely around shank in planes perpendicular to nail axis; having a lead angle of zero degree. **F 547,** F16

anode—the electrode in electrolysis, at which negative ions are discharged, positive ions are formed, or other oxidizing reactions occur. **B 374,** B08

anode—positive electrode. **D 6161,** D19

anode—the positive electrode of a discharge tube. In an X-ray tube, the anode carries the target. **E 1316,** E07

anode—the electrode of an electrolytic cell at which oxidation is the principal reaction. (Electrons flow away from the anode in the external circuit. It is usually the electrode where corrosion occurs and metal ions enter solution.) **G 15,** G01

anode aperture—See **aperture.** **E 7,** E04

anode corrosion—dissolution of anode metal by the electrochemical action in an electrolytic cell. **B 374,** B08

anode corrosion efficiency—the ratio of the actual corrosion (weight loss) of an anode to the theoretical corrosion (weight loss) calculated by Faraday's law from the quantity of electricity that has passed. **G 15,** G01

anode current—the electrons passing from the cathode to the anode in an X-ray tube, minus the small loss incurred by the back scattered fraction. **E 1316,** E07

anode efficiency—current efficiency of a specified anodic process. **B 374,** B08

anode fall—the potential difference between the anode and the electrical discharge plasma. **B 542,** B02

anode film—(*1*) the layer of solution in contact with the anode that differs in composition from that of the bulk of the solution.

(*2*) The outer layer of the anode itself consisting of oxidation or reaction products of the anode metal. **B 374,** B08

anode polarization—See **polarization.** **B 374,** B08

anodes, impressed current—inactive materials such as high-silicon cast iron and graphite. These anodes provide the electrical connection to the conductive overlay, but require use of an outside source for protective energy, that is, an impressed current. At the current densities recommended by the manufacturers, these anodes are highly resistant to electrolytic attack but deteriorate at known rates. **D 3743,** D04

anodes, sacrificial—chemically active metals such as zinc, aluminum, and magnesium which, when electrically connected to the reinforcing bar, will provide the energy needed to cathodically protect the reinforcing bar. Sacrificial anodes deteriorate in service at a rate proportional to the energy needed to protect the reinforcing bar plus whatever may deteriorate by local-action corrosion. **D 3743,** D04 *anodic (anode) material transfer*—see **material transfer.** **B 542,** B02

anodic coating—a protective, decorative, or functional coating, formed by conversion of the surface of a metal in an electrolytic oxidation process. **B 374,** B08

anodic inhibitor—an inhibitor that reduces the corrosion rate by acting on the anodic (oxidation) reaction. **G 15,** G01

anodic polarization—the change of the electrode potential in the noble (positive) direction due to current flow. (See **polarization.**) **G 15,** G01

anodic protection—a technique to reduce the corrosion rate of a metal by polarizing it into its passive region where dissolution rates are low. **G 15,** G01

anodized aluminum—natural-colored or surface-colored aluminum having increased anodic corrosion resistance. **F 547,** F16

anodizing—an electrolytic oxidation process in which the surface of a metal, when anodic, is converted to a coating having desirable protective, decorative, or functional properties. **B 374,** B08

anolyte—the portion of electrolyte in the vicinity of the anode; in a divided cell, the portion of electrolyte on the anode side of the diaphragm. **B 374,** B08

anolyte—the electrolyte adjacent to the anode of an electrolytic cell. **G 15,** G01

anomaly, *n*—a change in the strain pattern of the rubber surface of a straining block as a result of applied stress brought about through a change in atmospheric pressure on the rubber surface. **F 538,** F09

anosmia, *n*—lack of sensitivity to odor stimuli. **E 253,** E18

ANSI—American National Standards Institute. **F 1156,** F12

antagonism, *n*—joint action of two or more stimuli whose combination elicits a level of sensation lower than that expected from combining the effects of each stimulus taken separately. **E 253,** E18

antelope finish suede—applied to lambskin, goatskin, or calfskin, sueded and finished to resemble antelope. **D 1517,** D31

antelope leather—a fine, soft leather made from an antelope skin, velvety in texture and sheen, sueded on the flesh side. This leather is so rare, that for practical purposes, it is virtually nonexistent. **D 1517,** D31

anteroposterior—extending from the front to the rear. **F 869,** F08

anthophyllite asbestos, *n*—asbestiform variety of the orthorhombic amphibole silicate minerals of the anthophyllite series. Its empirical formula is $Mg_7Si_8O_{22}(OH)_2$. Its Chemical Abstracts number is 77536-67-5. **D 2946,** C17

anthracite—See *anthracite* under **rank**. **D 121,** D05

anthracite—a granular hard coal used as a filtration media, commonly used as the coarser layer in dual and multimedia filters. **D 6161,** D19

anthracite-coal-base carbon refractory, n—see **carbon refractory, anthracite-coal-base.** **C 71,** C08

anthracitic class—See *anthracitic class* under **rank**. **D 121,** D05

anthropometry—the science of the measurement of the human body and its parts. **F 869,** F08

anti-aliasing filter, *n*—a low-pass analog filter applied to the original analog profile signal to suppress those components of the signal at frequencies higher than one half the intended digital sampling frequency. **E 867,** E17

anticoagulant (natural rubber latex), *n*—a substance added to field latex to retard bacterial action which would otherwise cause rapid coagulation of the latex. **D 1566,** D11

antidegradant, *n*—compounding material used to retard deterioration caused by oxidation, ozone, light, and combinations of these. **D 1566,** D11

antidunes—bed forms that occur at a velocity higher than that velocity which forms dunes and plane beds. Antidunes commonly move upstream, and are accompanied by, and in phase with, waves on the water surface. **D 4410,** D19

antiferromagnetic material—a feebly magnetic material in which almost equal magnetic moments are lined up antiparallel to each other. Its susceptibility increases as the temperature is raised until a critical (Neél) temperature is reached; above this temperature the material becomes paramagnetic. **A 340,** A06

antiflex cracking agent, *n*—a compounding material used to retard cracking caused by cyclic deformations. **D 1566,** D11

antifoam, *n*—a substance added to engine coolant concentrate, corrosion inhibitor packages, or supplemental coolant additives to prevent or suppress foam. **D 4725,** D15

antifoaming agent—a material used to inhibit or prevent the formation of foam. **E 1519,** E35

antifoulant—see **antiscalant**. **D 6161,** D19

antifreeze, *n*—a term frequently used in the marketplace for engine coolant concentrate. (See **engine coolant concentrate**.) **D 4725,** D15

anti-knock index, *n*—the arithmetic average of the Research octane number (RON) and Motor octane number (MON). **D 4175,** D02

anti-lock braking system (ABS), *n*—a collection of sensing and control hardware installed on a vehicle to prevent wheel lock-up during brake application. **F 538,** F09

antimony—an industrial term for an oxide of antimony. **C 162,** C14

antioxidant, *n*—compounding material used to retard deterioration caused by oxidation. **D 1566,** D11

antioxidant—compounding ingredient used to retard deterioration caused by oxidation. (D-11) **F 412,** F17

antioxidant—see **stabilizer**. **F 1623,** F05

antiozonant, *n*—compounding material used to retard deterioration caused by ozone. **D 1566,** D11

anti-pitting agent—an addition agent for the specific purpose of preventing gas pits in a deposit. **B 374,** B08

antiscalant—a compound added to a water which inhibits the precipitation of sparingly soluble inorganic salts. **D 6161,** D19

anti-scale compound—a preparation that is applied to burning tools to protect them from scaling in service. **C 286,** B08

anti-skid plate—See **loading.** **D 996,** D10

antistat,—a chemical additive, generally added to the back side or printhead side of thermal products, which enhances the conductivity of the coating, allowing static electric charges to be bled off during the production processes or during the operation of the product in a machine. **F 1623,** F05

antistatic agent, *n*—a material which reduces the tendency for accumulation of electric charge on the surface of an article. **D 1566,** D11

antistatic agent—a chemical compound which, when impregnated or formulated into or topically applied to a primary material or substrate, gives the primary material antistatic properties. See **antistatic property**. **D 5077,** D10

antistatic property—the prevention of triboelectric charge generation by effectively minimizing the production of a static charge when materials are separated from another surface. **D 5077,** D10

anti-Stokes line (band)—a Raman line (band) that has a frequency higher than that of the incident monochromatic beam. **E 131,** E13

anti-telescoping device—a plastic or metal device attached to the ends of a spiral wound cartridge to prevent movement of the cartridge leaves in the feed flow direction, due to high feed flows. **D 6161,** D19

anvil—the base on which objects for hardness test are placed. **E 7,** E04

anvil, *n*—the smooth, hardened surface upon which the test sample or cup containing the sample rests. **E 1445,** E27

anvil—tool arm against which staple legs are driven to form clinch. **F 592,** F16

anvil effect—the effect caused by use of too high a load or when testing the hardness of too thin a specimen, resulting in a bulge or shiny spot on the under side of the specimen. **E 7,** E04

AOC—assimilable organic carbon. **D 6161,** D19

apartment—See **dwelling unit.** **E 631,** E06

apartment building—See **building.** **E 631,** E06

aperture—an opening in material, space, or time over which an element is considered to be active. **E 1316,** E07

aperture angle, 2κ, *n*—angle subtended at a point on a specimen by the maximum dimension of the illuminator or receiver, within which the flux in a directional beam is contained. **E 284,** E12

aperture, angular—the angle between the most divergent rays that can pass through a lens to form the image of an object. **E 175,** E41

aperture area—see **area, aperture**. **E 772,** E44

aperture diaphragm—a device to define the aperture. **E 7,** E04

aperture, effective—the diameter of the entrance pupil; it is the apparent diameter of the limiting aperture measured from the front. **E 175,** E41

aperture, electron:

anode aperture— the opening in the accelerating voltage anode shield of the electron gun through which the electrons must pass to illuminate or irradiate the specimen.

condenser aperture—an opening in the condenser lens controlling the number of electrons entering the lens and the angular aperture of the illuminating beam. The angular aperture can also be controlled by the condenser lens current.

physical objective aperture—a metal diaphragm, centrally pierced with a small hole, used to limit the cone of electrons accepted by the objective lens. This improves image contrast since highly scattered electrons are prevented from arriving at the Gaussian image plane and therefore can not contribute to background fog. **E 7,** E04

aperture leak—a leak of such geometric configuration that the length of the leakage path is much smaller than the shortest diameter of the path, so that the leak may be considered the equivalent of an opening in an infinitesimally thin wall. **E 1316,** E07

aperture mode, *n*—color seen through an aperture which prevents its association with a specific object or source. **E 284,** E12

aperture of an IRE, *A'*—that portion of the IRE surface that can be utilized to conduct light into the IRE at the desired angle of incidence. **E 131,** E13

aperture, optical—the working diameter of a lens or a mirror.

angular aperture— the angle between the most divergent rays which can pass through a lens to form the image of an object. **E 7,** E04

aperture size, *n*—the dimension defining an opening in a sieving medium. **E 1638,** E29

aperture solid angle, ω, *n*—solid angle subtended at a point on the specimen, defined by the sum of rays from the illuminator or the sum of directions in which the receiver is sensitive to incoming radiation. **E 284,** E12

aperture stop, *n*—the physical diameter that limits the size of the cone of radiation that an optical system will accept from an axial point on the object. **E 284,** E12

apex (aka pocket)—pocket formed at the downstream end of a U, V, J, or W shaped configuration. **F 818,** F20

API gravity—an arbitrary scale developed by the American Petroleum Institute and frequently used in reference to petroleum insulating oil. The relationship between API gravity and specific gravity 60/60°F is defined by the following:

$$\text{Deg API Gravity at } 60°\text{F} = 141.5/(\text{sp gr } 60/60°\text{F}) - 131.5$$

D 2864, D27

aplanatic—corrected for spherical aberration and coma. **E 175,** E41

apochromatic objective—an objective with longitudinal chromatic correction for red, green and blue, and spherical chromatic correction for green and blue. This is the best choice for high resolution or color photomicrography. **E 7,** E04

apochromatic objective—a lens system whose secondary chromatic aberrations have been substantially reduced. (See *achromatic*). **E 175,** E41

apodization—modification of the ILS function by multiplying the interferogram by a weighting function the magnitude of which varies with retardation. **E 131,** E13

apparent area of contact, *n*—*in tribology*, the area of contact between two solid surfaces defined by the boundaries of their macroscopic interface. (Contrast with **real area of contact.**) **G 40,** G02

apparent attenuation—the observed ultrasound energy loss. In addition to the true loss, the apparent attenuation may also include losses attributable to instrumentation, specimen configuration, beam divergence, interface reflections, and measurement procedure. (E 664) **E 1316,** E07

apparent coercive field strength, H'_c—the value of apparent magnetic field strength on a flux-current loop coincident, with zero induction. **A 340,** A06

apparent cohesion—see **cohesion**. **D 653,** D18

apparent density, *n*—the mass of a unit volume of powder, usually expressed as grams per cubic centimetre, determined by a specified method. **B 243,** B09

apparent density—the mass of a material divided by its apparent volume including closed pores. **C 859,** C26

apparent density—See **density, apparent.** **D 883,** D20

apparent density, *n*—See **density, bulk,** the preferred term. **D 1566,** D11

apparent density—the density calculated when the pycnometer is calibrated with water, weighed in air, and when the sample is weighed in air and no air buoyancy correction is used for either weighing, even though the density in vacuum of water is used in calculating the apparent volume of the pycnometer. **D 4790,** D16

apparent density—the weight per unit volume of a material including voids inherent in the material as tested. **F 412,** F17

apparent density—see **density, apparent.** **F 1251,** F04

apparent density at 60°F—the weight in air of a unit volume of sample at 60°F; in this method, the weight is in pounds and the volume in U.S. liquid gallons. Average air in this method is assumed to have a density of 0.0012 g/cm^3. **D 4790,** D16

apparent hardness—the hardness, including the effects of porosity, of a P/M material measured under specified conditions, using standard macro-indentation hardness equipment. **B 243,** B09

apparent lead concentration (ALC)—the average of at least three XRF analyzer readings on a coated surface. **E 1605,** E06

apparent magnetic field strength, H'—the magnetic field strength value expressed in terms of the instantaneous exciting current, i:

$$\text{(In cgs−emu system of units) } H' = 0.4\pi N_1 I/\ell_1$$

A 340, A06

apparent magnetic field strength, incremental, H'_Δ—the magnetic field strength value developed for incremental flux-current loop. **A 340,** A06

apparent magnetic field strength, maximum, H'_{max}—the maximum value of H' attained in a flux-current loop. **A 340,** A06

apparent opening size (AOS), O_{95}, *n—for a geotextile,* a property which indicates the approximate largest particle that would effectively pass through the geotextile. **D 4439,** D35

apparent or pycnometric density—See *apparent or pycnometric density* under **density.** **C 242,** C21

apparent porosity, *n*—in cemented carbides, microstructural features that appear to be pores in a properly prepared, unetched surface; these features may result from uncombined carbon or nonmetallic inclusions as well as actual porosity. **B 243,** B09

apparent porosity—See *apparent porosity* under **porosity.** **C 242,** C21

apparent power, P_a—the product (volt-amperes) of the rms exciting current and the applied rms *terminal* voltage in an *electric* circuit containing inductive impedance. The components of this impedance as a result of the winding will be linear, while the components as a result of the magnetic core will be nonlinear. The unit of apparent power is the volt-ampere, VA. **A 340,** A06

apparent power, specific, $P_{a(B;f)}$—the value of the apparent power divided by the active mass of the specimen, that is, volt-amperes per unit mass. The values of voltage and current are those developed at a maximum value of cyclically varying induction B and specified frequency f. **A 340,** A06

apparent rank, *n—of coal,* the rank designation obtained on samples other than channel samples, but otherwise conforming to procedures of Classification D 388. **D 121,** D05

apparent residual induction, B'_r—the value of induction on a flux-current loop when the apparent magnetic field strength has a zero value. **A 340,** A06

apparent solar time, apt—the hours of the day as computed from the position of the sun using the equation of time. (See ASHRAE *Handbook of Applications,* 1982, Chapter 57.) **E 772,** E44

apparent surface resistivity—the surface resistance between two electrodes forming opposite sides of a square on the surface of the copy substrate or the electrical equivalent. **F 335,** F05

apparent temperature—the temperature of an object as determined solely from the measured radiance, assuming an emissivity of unity. **E 1316,** E07

apparent thermal conductivity, λ_a, k_a, *n*—a thermal conductivity assigned to a material that exhibits thermal transmission by several modes of heat transfer resulting in property variation with specimen thickness, or surface emittance. See **conductivity, thermal.** **C 168,** C16

apparent thermal resistivity, r_a, *n*—a thermal resistivity assigned to a material that exhibits thermal transmission by several modes of heat transfer resulting in property variation with specimen thickness, or surface emittance. See **resistivity, thermal.** **C 168,** C16

apparent viscosity, *n*—resistance to shear at a given rate of shear, expressed as viscosity in absolute units. (See **viscosity.**) **D 907,** D14

apparent viscosity (of a non-Newtonian fluid)—the viscosity of a Newtonian fluid that produces the same reading in the same apparatus under identical conditions. **D 2507,** F07

appearance, *n*—(*1*) the aspect of visual experience by which things are recognized.

(*2*) *in psychophysical studies,* visual perception in which the spectral and geometric aspects of a visual stimulus are integrated with its illuminating and viewing environment. **E 284,** E12

apple-box nail—coated, regular-stock-steel, 1⅝ and 1¾ by 0.080-in. nails with flat 15⁄64-in. head and medium diamond point. **F 547,** F16

applicable standards—those having the capability of being applied in some fashion to the host standard. **F 1789,** F16

application rate—the quantity (mass, volume or thickness) of material applied per unit area. **D 1079,** D08

application rate—the amount of any material applied per unit treated: **E 1102,** E35

application temperature, *n*—the range of material temperature, as recommended by the manufacturer, that is to be used when installing sealant; for hot-applied sealants, the application temperature is any temperature between the minimum application temperature and safe heating temperature. **D 5535,** D04

applicator—any device used to mechanically apply molten sealants that is capable of delivering the required volume at the specified application temperature. **C 717,** C24

applied-K curve—a curve (a fixed-load or fixed-displacement crack-extension-force curve) obtained from a fracture mechanics analysis for a specific configuration. The curve relates the stress-intensity factor to crack size and either applied load or displacement. **E 1823,** E08

applique—decoration laid on and applied to another surface. **F 869,** F08

approach, *n*—a set of lanes accommodating all left turn, through, and right turn movements arriving at an intersection from a given direction. **E 867,** E17

approving authority—the individual official, board, department, or agency established and authorized by a state, county, city, or other political subdivision, created by law to administer and enforce specified requirements. **C 896,** C04

approving authority—the individual official, board, department, or agency established and authorized by a state, county, city, or other political subdivision, created by law to administer and enforce specified requirements. **F 412,** F17

apron leathers—any one of several varieties of leather used in connection with textile machinery and blacksmith aprons. Comber and Gill Box apron leather is soft, mellow, tough leather, tanned from steerhides, heavily stuffed and boarded or otherwise softened. Rub Roll apron leather is a flexible but firm, dry, strong leather. **D 1517,** D31

aqueous extract, *n—in wool testing,* the solution obtained by digesting a material with water or with a sodium chloride solution to dissolve soluble materials. **D 123,** D13

aquiclude—a relatively impervious formation capable of absorbing water slowly but will not transmit it fast enough to furnish an appreciable supply for a well or spring. **D 653,** D18

aquifer—a geologic formation, group of formations, of part of a formation that is saturated and is capable of providing a significant quantity of water. **D 653,** D18

aquifer—a water-bearing geological formation that provides a ground water reservoir. **D 6161,** D19

aquifer, *n*—a water-bearing, permeable body of rock or granular material below the surface of the earth. **E 957,** E44

aquifer, confined—an aquifer bounded above and below by confining beds and in which the static head is above the top of the aquifer. **D 653,** D18

aquifer, unconfined—an aquifer that has a water table. **D 653,** D18

aquitard—a confining bed that retards but does not prevent the flow of water to or from an adjacent aquifer; a leaky confining bed. **D 653,** D18

araban—a pentosan yielding essentially only arabinose on hydrolysis. **D 1695,** D01

arabinogalactan—a polysaccharide consisting of arabinose and galactose units, like the water-soluble polysaccharide of larch. **D 1695,** D01

arabinomethylglucuronoxylan—a hemicellulose containing arabinose, 4-O-methylglucuronic acid, and xylose groups in its structure. **D 1695,** D01

arabinose—a pentose that occurs as one of the sugar units in some hemicelluloses. **D 1695,** D01

aramid—a fully aromatic polyamide. **D 6161,** D19

arc, anode—the arc that occurs at less than a critical electrode spacing (see arc, cathode), and results in anode material loss. **B 542,** B02

arc, cathode—the arc that occurs at greater than a critical electrode spacing (see arc, anode), and results in cathode material loss. **B 542,** B02

arc, condensed—see **triggered capacitor discharge.** **E 135,** E01

arc, continuous dc, *n*—a self-maintaining dc discharge. **E 135,** E01

arc discharge—a self-sustaining, high current density, high temperature discharge, uniquely characterized by a cathode fall nearly equal to the ionization potential of the gas or vapor in which it exists. **B 542,** B02

arc gap—distance between the arc electrodes (inch). **F 819,** F18

arch, *n*—a part of a melter; a crown. **C 162,** C14

arch, *v*—to heat a pot in a pot arch. **C 162,** C14

arch, *n*—an inner portion of a shoe resembling an arch in form or function. **F 1646,** F13

arch-footwear—the bottom curve of a shoe last from heel to ball. **F 869,** F08

arching—the transfer of stress from a yielding part of a soil or rock mass to adjoining less-yielding or restrained parts of the mass. **D 653,** D18

architectural coatings—coatings intended for on-site application to interior or exterior surfaces of residential, commercial, institutional or industrial buildings. **D 16,** D01

architectural program— See **facility program**. **E 1480,** E06

architectural-use, *n*—*in the building trade*, a descriptive term for fabrics used in fabric roof-systems or similar industrial applications. (See also **fabric roof-system.)** **D 123,** D13

archival quality—the ability of a print to retain specified characteristics and legibility over a period of use and storage under specified conditions. **F 221,** F05

archival quality—the properties of a copy necessary to retain information of permanent value under specified conditions of storage and use. **F 335,** F05

archive sample—any sample retained for record or for future use. **C 859,** C26

arch support—a device of leather, metal, or other material shaped to the contour of the longitudinal arch of the foot and inserted or built in a shoe. **F 869,** F08

arcing—in electron diffraction, the production of segments of circular patterns, indicating a departure from completely random orientation of the crystals of the specimen. **E 7,** E04

arc line—not recommended, see **atom line.** **E 135,** E01

Ar_{cm}, Ar_1, Ar_3, Ar_4—See **transformation temperature**. **A 941,** A01

arc, noncapacitive ac, *n*—*in optical emission spectrometry*, a series of separate electrical discharges, individually self-initiating or initiated separately by another means, in which each current pulse has a polarity that is reversed from the previous one. **E 135,** E01

arc, noncondensed, intermittent dc—see **arc, noncapacitive, intermittent dc.** **E 135,** E01

arc propagation, *n*—the movement of an electric arc from its point of inception to another location. **D 1711,** D09

arc, shortest—a limiting state of an arc in which the total arc voltage approaches the sum of the cathode and anode falls. **B 542,** B02

arc strikes—localized burn damage to a part from an arc caused by making or breaking an energized electrical circuit. **E 1316,** E07

arctic—the biogeographic realm that comprises the tundra and treeless grounds lying north of the timberline in the Northern hemisphere or south of the timberline in the Southern hemisphere. **F 1600,** F20

arc tracking, *n*—the process producing tracks when arcs occur on or close to the insulation surface. **D 1711,** D09

arc welding—a group of welding processes wherein coalescence is produced by heating with an arc or arcs, with or without the application of pressure and with or without the use of filler metal. **B 846,** B05

area—See **space categories**. **E 1480,** E06

area, *A*—the geometric cross-sectional area of a magnetic path which is perpendicular to the direction of the induction. **A 340,** A06

area, absorber—the total uninsulated heat transfer surface area of the absorber, including unirradiated as well as irradiated portions. **E 772,** E44

area amplitude response curve—a curve showing the changes in amplitude at normal incidence from planar reflectors of different areas located at equal distances from the search unit in an ultrasonic-conducting medium. **E 1316,** E07

area, aperture— of a flat plate collector, (*1*) the maximum projected area of a solar collector through which the unconcentrated solar radiant energy may be admitted to the absorber. (*2*) *effective aperture area*—the area as defined above projected normal to the sun's rays and corrected for any shading. Units: square metres (m^2) [square feet (ft^2)]. **E 772,** E44

area, collector panel—the total area of the panel assembly (with its containing box, if present), projected on the aperture plane. **E 772,** E44

area contamination—*on a semiconductor wafer*, matter, unintentionally added to the surface of the wafer, of extent greater than a single localized light scatterer. **F 1241,** F01

area defect—*not preferred*, see **extended light scatterer**. **F 1241,** F01

area density—mass per unit area. **B 354,** B01

area, gross aperture— *of a concentrating collector*, the maximum projected area through which the unconcentrated solar radiant energy is admitted, including any area of the reflector or refractor shaded by the receiver and its supports, and including gaps between reflector segments within a collector module. **E 772,** E44

area, gross collector—the maximum area of the complete collector module, including integral mounting means, projected on the aperture plane. **E 772,** E44

area grouting—grouting a shallow zone in a particular area utilizing holes arranged in a pattern or grid. **D 653,** D18

area, net aperture—*of a concentrating collector,* the maximum projected area through which the unconcentrated solar radiant energy is admitted, excluding any area of the reflector or refractor shaded by the receiver and its supports, and excluding gaps between reflector segments within a collector module. **E 772,** E44

area of influence of a well, α (L^2)—area surrounding a well within which the piezometric surface has been lowered when pumping has produced the maximum steady rate of flow. **D 653,** D18

area of interest—the specific portion of the object image on the radiograph that is to be evaluated. **E 1316,** E07

area ratio of a sampling spoon, sampler, or sampling tube, A_r (D)—the area ratio is an indication of the volume of soil displaced by the sampling spoon (tube), calculated as follows:

$$A_r = [(D_e^2 - D_i^2/D_i^2] \times 100$$

where:

D_e = maximum external diameter of the sampling spoon, and

D_i = minimum internal diameter of the sampling spoon at the cutting edge.

D 653, D18

area reflector, *n*—reflector subtending a relatively large solid angle at the observer's eye, so that the observer can clearly distinguish its size and shape. **E 284,** E12

arithmetic mean sound pressure level—*of several related sound pressure levels measured at different positions or different times, or both, in a specified frequency band,* the sum of the sound pressure levels divided by the number of levels. **C 634,** E33

arm—cantilever part of tool to hold the clinching mechanism, permitting placement of fastener away from edges of the work. **F 592,** F16

armhole, *n*—*in garment construction,* the area of a garment through which the arm passes or into which a sleeve is fitted. (Compare **armscye.**) **D 123,** D13

armhole, *n*—*in garment construction,* the area of a garment through which the arm passes or into which a sleeve is fitted. (See **armscye.**) **F 1494,** F23

arm length, *n*—*in body measurements,* with the arm bent at 90 degrees and the clenched fist placed on the hip, the distance from the shoulder joint along the outside of the arm over the elbow to the greatest prominence on the outside of the wrist. **D 123,** D13

armor—the artificial surfacing of bed, banks, shore, or embankment to resist erosion or scour. **D 653,** D18

armoring—the formation of a resistant layer of relatively large particles by erosion of the finer particles. **D 4410,** D19

armor stone—(generally one ton to three tons in weight) stone resulting from blasting, cutting, or by other methods to obtain rock heavy enough to require handling two individual pieces by mechanical means. **D 653,** D18

armpit, *n*—*in anatomy,* the hollow under the junction of the arm and the shoulder. **D 123,** D13

armscye, *n*—*in garment construction,* the opening in a garment for the attachment of a fitted sleeve. (Compare **armhole.**) **D 123,** D13

armscye, *n*—*in garment construction,* the opening in a garment for the attachment of a fitted sleeve. (See **armhole.**) **F 1494,** F23

armscye circumference, *n*—*in body measurements,* with the arm hanging down, the distance from the shoulder joint through the front-break point, the armpit, the back-break point and to the starting point. **D 123,** D13

aroma, *n*—perception resulting from stimulating the olfactory receptors; in a broader sense, the term is sometimes used to refer to the combination of sensations resulting from stimulation of the entire nasal cavity. **E 253,** E18

aromatic oil, *n*—a hydrocarbon process oil containing at least 35 %, by mass, of aromatic hydrocarbons. **D 1566,** D11

aromatic polyester, *n*—a polyester derived from monomers in which all the hydroxyl and carboxyl groups are linked directly to aromatic nuclei. **D 883,** D20

aromatics—that class of organic compounds which behave chemically like benzene. They are cyclic unsaturated organic compounds that can sustain an induced electronic ring current due to delocalization of electrons around the ring. **D 2864,** D27

array—an arrangement of devices connected to common feed, product and reject headers; that is, a 2:1 array. **D 6161,** D19

array, *n*—a group of two or more AE sensors positioned on a structure for the purposes of detecting and locating sources. The sources would normally be within the array. **E 1316,** E07

array processor—a special purpose logical processing device that performs extremely fast mathematical operations on digital arrays. **E 1316,** E07

arrest—that portion of a cooling curve in which temperature is invariant with time (for example, thermal or eutectic arrest). **E 7,** E04

arrester, *n*—a term for an air cleaning device. **D 1356,** D22

arrest mark—See **dwell mark.** **C 162,** C14

arrhenius equation—a mathematical relationship between the specific reaction rate and the temperature given as

$$k = Ae^{-E/RT}$$

where:

k is the reaction rate constant, A is the frequency factor, E is the energy of activation, R is the gas constant, and T is the absolute temperature. **E 1142,** E37

Arrhenius equation—$k = Ze^{-E/RT}$ where k is the specific reaction rate constant in reciprocal minutes for first order, Z is the pre-exponential factor in reciprocal minutes, E is the Arrhenius activation energy in J/mol, R is the gas constant, 8.32 J/mol K, and T is the temperature in kelvin. **E 1445,** E27

Arrhenius plot, *n*—a graph of the logarithm of thermal life as a function of the reciprocal of absolute temperature. **D 1711,** D09

arris—the junction of two planes of the same stone forming an external corner. **C 119,** C18

arris (of an arch), *n*—the outside corner or angle formed by the meeting of a wall surface with the curved surface of an arch. **C 11,** C11

arrival time interval (Δt_{ij})—see **interval, arrival time.** **E 1316,** E07

arsenic—an industrial term for an oxide of arsenic. **C 162,** C14

artifact—a false microstructural feature that is not an actual characteristic of the specimen; it may be present as a result of improper or inadequate preparation, handling methods, or optical conditions for viewing. **E 7,** E04

artifact, *n*—an object (as a tool, ornament, or element of a structure) showing human workmanship or modification. **E 631,** E06

artifact—spurious indication on a radiograph arising from, but not limited to, faulty manufacture, storage, handling, exposure, or processing. **E 1316,** E07

artificial aging, *n*—**aging** above room temperature. **A 941,** A01

artificial aging—aging in which a stress variable is outside the domain of exposure that the material might see in a component for oxygen service or in which an alternative mechanism is used to produce an effect that simulates the results of natural aging. The degree of artificiality may vary on a large scale. An example of mild artificiality might be exposure of a material to a greater pressure than it experiences in the use conditions. An example of extreme artificiality would be the use of sand paper to increase a material's surface roughness to simulate particle-impact abrasion that occurs in the use condition. A high degree of artificiality affects the strength of conclusion that can be drawn, because it may be difficult to relate the results to the use condition. Artificial aging that accelerates natural aging but does not alter it is preferred. **G 126,** G04

artificial daylight, *n*—an artificial light that has a spectral power distribution approximating that of a phase of natural daylight. **E 284,** E12

artificial discontinuity—reference discontinuities, such as holes, grooves, or notches that are introduced into a reference standard to provide accurately reproducible sensitivity levels for electromagnetic test equipment. **E 1316,** E07

artificially soiled cloth (sometimes called "standard soiled cloth")—cloth soiled with one or more materials and used to evaluate the effectiveness of detergents or washing equipment. **D 459,** D12

artificial weathering—exposure to laboratory conditions, which may be cyclic, involving changes in temperature, relative humidity, radiant energy, and any other elements found in the atmosphere in various geographical areas. **D 883,** D20

artificial weathering, *v*—exposure to conditions, which may be cyclic, involving temperature, relative humidity, radiant energy, and/or any other conditions or pollutants found in the atmosphere in various geographical areas, which may accelerate changes in properties of materials over those of natural weather conditions. **D 1079,** D08

artificial weathering—exposure to laboratory conditions, which may be cyclic, involving temperature, relative humidity, radiant energy, or any other conditions or pollutants found in the atmosphere in various geographical areas; or both. **F 412,** F17

artificial weathering, *n*—exposure of a material to laboratory conditions that simulate outdoor weathering. **F 1251,** F04

as-analyzed moisture—synonym for *as-determined moisture.* **D 121,** D05

asbestiform—having an inherent fine-textured morphology, resulting from unequal relative development of the principal crystal axes in a silicate mineral, that predetermines subdivision into strong flexible fibers having microscopic to submicroscopic thickness and a high length to width ratio when the mineral is subjected to comminution. Term derived from "asbestos." **D 2946,** C17

asbestos—the naturally occurring inorganic hydrated silicates, occurring in layered structures composed of chains of silicon and oxygen tetrahedra, that can subdivide into flexible fibers. **C 460,** C17

asbestos—see **industrial talc** and **industrial talc, nonasbestos type.** **D 16,** D01

asbestos—a group of natural fibrous impure silicate materials. **D 1079,** D08

asbestos, *n*—the generic term for naturally occurring inorganic hydrated silicates, occurring in layered structures composed of chains of silicon and oxygen tetrahedra, that can subdivide into flexible fibers. Refer to STP 834. **D 2946,** C17

asbestosboard nail—galvanized, hardened-steel, helically threaded, 1¼ and 1½ by 0.083-in. screwnails with flat, slightly countersunk, 3⁄16-in. head and blunt diamond point. **F 547,** F16

asbestos fiber, *n*—acicular silicate mineral, with a structure based upon silicon-oxygen tetrahedra, that fits the definition of a fiber, and is composed of single crystals in predominantly parallel orientation. Common usage also designates a collectivity of asbestos fibers as asbestos fiber. **D 2946,** C17

asbestos-shingle nail— See **shingle nail.** **F 547,** F16

as-built, *adj*—pertaining to the as-constructed, **as-fabricated,** as-manufactured, or as-furnished state of a finished product relating to size, shape, materials, and finish regardless of drawings or specifications. **E 631,** E06

as-built drawing—See **drawing.** **E 1480,** E06

A-scan—a method of data presentation utilizing a horizontal base line that indicates distance, or time, and a vertical deflection from the base line which indicates amplitude. **E 1316,** E07

ascending fork point—in a ternary phase diagram, the configuration at the convergence of the three bivariant curves upon each of the four phases associated in Class II univariant equilibrium; for example, the union of two ascending liquidus surface valleys to form one ascending liquidus surface valley. **E 7,** E04

ascent rate, *n*—the speed of vertical movement toward the water's surface. **F 1549,** F32

ASCII—American Standard Code for Information Interchange. **F 1457,** F05

as-determined basis—See *as-determined basis* under **reporting bases**. **D 121,** D05

as-determined basis, *n*—analytical data obtained from an analysis sample after conditioning and preparation which represent the numerical values obtained at the particular moisture and ash level in the sample at the time of analysis. **D 5681,** D34

as-determined basis—analytical data obtained from the analysis sample after conditioning and preparation in accordance with Method E 829. As-determined basis data represent the numerical values obtained at the particular moisture and ash level in the sample at the time of analysis. **E 856,** D34

aseptic—as applied to aseptic packaging, synonymous with commercially sterile. **F 17,** F02

aseptic, *adj*—sterile, free from viable microbiological contamination. **F 1494,** F23

aseptic packaging—filling of a commercially sterilized product into presterilized containers, followed by hermetic sealing in a commercially sterile atmosphere. **F 17,** F02

aseptic packaging—See Terminology F 17. **F 1327,** F02

aseptic sampling—sampling process in which no extraneous microorganisms or substances are introduced into the sample or its original bulk material as a result of the sampling system and activity. **E 1705,** E48

as-fabricated, *adj*—(*1*) *of a milled metal product,* pertaining to the surface appearance and texture or temper produced by the original forming process. (*2*) *of a formed metal product,* pertaining to the surface appearance of the product to removal of disfigurations caused by the forming process. **E 631,** E06

ash, *n*—*in carbon and graphite technology,* the residue remaining after oxidation of a carbon or graphite. **C 709,** D02

ash, *n*—inorganic residue remaining after ignition of combustible substances, determined by definite prescribed methods. **D 121,** D05

ash, *n*—the residue from incineration of a material under specified conditions. **D 1566,** D11

ash—the inorganic residue obtained by igniting a specimen of pulp, paper, or other cellulosic material in such a way that all combustible and volatile compounds are removed. Conditions of ashing should be specified. **D 1695,** D01

ash, *n—in wood, pulp, or paper; general term,* the residue after the ignition of a specimen of wood, pulp, or paper at a specified temperature for a specified time so as to remove combutable and volatile components. **D 1968,** D06

ash—residue after the combustion of a substance under specified conditions. **D 2652,** D28

ash—the residue remaining after ignition of refuse-derived fuel determined by definite prescribed methods. The ash may not be identical in composition or quantity with the inorganic substances present in the refuse-derived fuel before ignition. **E 856,** D34

ash—inorganic residue remaining after combustion, determined by definite prescribed methods. **E 1126,** E48

ash—the residual inorganic matter obtained on ignition of a sample in air at a specified temperature. **E 1547,** E15

ash—inorganic residue remaining after combustion, determined by definite prescribed methods. **E 1705,** E48

ash content—the percentage by dry weight of material remaining after an oven dry organic soil or peat is burned by a prescribed method. **D 653,** D18

ash content, *n*—the residue from an engine coolant concentrate, antirust, or engine coolant that remains after evaporation, charring, and ignition at strong heat. **D 4725,** D15

ash content of paper, *n*—the solid residue remaining after combustion of the paper under specified conditions, expressed as a percentage of the dry mass of the original paper. **D 1711,** D09

ash fusion temperature—melting point of ash, usually expressed in degrees Fahrenheit. Variations include oxidizing atmosphere or reducing atmosphere, initial softening, or final fluid temperature. Some specifications include two intermediate points between initial softening and final fluid. **E 1126,** E48

ash fusion temperature—melting point of ash, usually expressed in degrees Fahrenheit. Variations include oxidizing atmosphere or reducing atmosphere, initial softening, or final fluid temperature. Some specifications include two intermediate points between initial softening and final fluid. **E 1705,** E48

ashlar—(1) a squared block of building stone; (2) a masonry of such stones; (3) a thin-dressed rectangle of stone for facing of walls (often called ashlar veneer). **C 119,** C18

ash *n*—the residue remaining after ignition of a substance as determined by definite prescribed methods. **D 5681,** D34

as is basis—as received. **D 2652,** D28

askarel—a generic term for a group of synthetic, fire-resistant, chlorinated aromatic hydrocarbons used as electrical insulating liquids. They have a property under arcing conditions such that any gases produced will consist predominantly of noncombustible hydrogen chloride with lesser amounts of combustible gases. **D 2864,** D27

as-mined coal—*for the purpose of Test Method D 4749,* same as **run-of-mine (ROM) coal**. **D 121,** D05

aspect ratio—ratio of the longer to the shorter transverse dimensions of a rectangular composite superconductor. **B 713,** B01

aspect ratio—the length-to-width ratio of a microstructural feature in a two-dimensional plane. **E 7,** E04

aspect ratio, *n—in a bar code symbol,* the ratio of bar height to symbol length. **F 1294,** F05

aspecular, *adj*—away from the specular direction. **E 284,** E12

aspecular angle, *n*—viewing angle measured from the specular direction, in the illuminator plane unless otherwise specified. **E 284,** E12

aspergillus niger—one of the most common mold growths found on vegetable tanning vats and on leather, usually greenish or blackish in color. **D 1517,** D31

asperity, *n—in tribology,* a protuberance in the small-scale topographical irregularities of a solid surface. **G 40,** G02

asphalt, *n*—a dark brown to black cementitious material in which the predominating constituents are bitumens which occur in nature or are obtained in petroleum processing. **D 8,** D04

asphalt—a dark brown to black cementitious material in which the predominating constituents are bitumens which occur in nature or are obtained in petroleum processing. **D 1079,** D08

asphalt, air blown—an asphalt produced by blowing air through molten asphalt at an elevated temperature to raise its softening point and modify other properties. **D 1079,** D08

asphalt cement, *n*—a fluxed or unfluxed asphalt specially prepared as to quality and consistency for direct use in the manufacture of bituminous pavements, and having a penetration at 25°C (77°F) of between 5 and 300, under a load of 100 g applied for 5 s. **D 8,** D04

asphaltene—a high molecular weight hydrocarbon fraction precipitated from asphalt by a designated paraffinic naphtha solvent at a specified temperature and solvent-asphalt ratio. **D 1079,** D08

asphaltenes, *n*—the high molecular weight hydrocarbon fraction precipitated from asphalt by a designated paraffinic naphtha solvent at a specified solvent-asphalt ratio. **D 8,** D04

asphalt felt—an asphalt-saturated felt. **D 1079,** D08

asphalt felt, breather type—an underlayment sheet material saturated with asphalt for use with asbestos-cement products, which allows the transmission of water vapor. **C 460,** C17

asphalt felt, breather type, *n*—an underlayment sheet material saturated with asphalt, which allows the transmission of water vapor. **C 1154,** C17

asphalt felt, breather type—underlayment sheet material saturated with asphalt for use with asbestos-cement products, which allows the transmission of water vapor. **D 2946,** C17

asphaltite—a natural asphalt found below ground level. **D 1079,** D08

asphalt mastic—a mixture of asphaltic material and graded mineral aggregate that can be poured when heated, but requires mechanical manipulation to apply. **D 1079,** D08

asphalt rock (rock asphalt), *n*—a naturally occurring rock formation, usually limestone or sandstone, impregnated throughout its mass with a minor amount of bitumen. **D 8,** D04

asphalt rock (rock asphalt)—a naturally occurring rock formation, usually limestone or sandstone, containing throughout its mass a minor amount of asphalt. **D 1079,** D08

asphalt roof cement, *n*—a trowelable mixture of asphalt, solvents, mineral stabilizers, fillers or fibers, or both. See Specifications D 2822 and D 4586. **D 1079,** D08

asphalt roof cement, *n*—see *flashing cement.* **D 1079,** D08

asphalt-rubber, *n*—a blend of asphalt cement, reclaimed tire rubber, and certain additives in which the rubber component is at least 15 % by weight of the total blend and has reacted in the hot asphalt cement sufficiently to cause swelling of the rubber particles. **D 8,** D04

asphalt-shingle nail— See **roofing nail.** **F 547,** F16

asphalt, steam blown—an asphalt produced by blowing steam through molten asphalt to modify its properties. **D 1079,** D08

asphalt tile, *n*—an obsolete floor surfacing unit composed of asphalt or hydrocarbon resins, or both, crysotile asbestos fibers, mineral fillers, and pigments. **F 141,** F06

aspirated psychrometer—See **psychrometer.** **D 1356,** D22

aspirating nozzle—see **siphon nozzle**. **E 1620,** E29

aspirator, *n*—any apparatus such as a squeeze bulb, fan, pump, or venturi that produces a movement of a fluid by suction.

D 1356, D22

aspirator valve—a valve in which the propellant vapor is aspirated through an orifice in the valve chamber, causing a suction effect that draws the product up the dip tube and into the valve. **D 3064,** D10

a-spot—the areas of two mating contacts through which current flows from one contact to the other. **B 542,** B02

as-received basis—See *as-received basis* under **reporting bases**. **D 121,** D05

as-received basis, *n*—test data calculated to the condition of the sample as it arrived in the laboratory and before any laboratory processing or conditioning. **D 5681,** D34

as-received basis—the test data calculated to the moisture condition of the sample as it arrived in the laboratory and before any laboratory processing or conditioning. **E 856,** D34

assay, *n*—analysis of a mixture to determine the presence or concentration of a particular component. **F 1494,** F23

assay fluid, *n*—a sterile liquid used to wash the test specimen surface to determine microbiological penetration. **F 1494,** F23

assembled button, *n*—a decorative button consisting of combinations of similar or dissimilar materials, such as plastic and metal or metal and metal, which have been joined together by such processes as gluing, swedging or metal stamping. **D 5497,** D13

assembly, n—See **adhesive assembly**. **D 907,** D14

assembly, *n*—a unit or structure composed of a combination of materials or products, or both. **E 176,** E05

assembly, *n*— a complete implant configuration (not including spine, pelvis, ribs, or substitute material) as intended for surgical use. **F 1582,** F04

assembly adhesive, *n*—an adhesive that can be used for bonding parts together, such as in the manufacture of a boat, airplane, furniture, and the like. **D 907,** D14

assembly lot—a combination of different products which respectively meet the definition of lot. For example, a bolt, nut, and washer assembly having each component meet the lot definition is an assembly lot. **F 1789,** F16

assembly time, *n*—the time interval between applying the adhesive on the substrate and the application of pressure, or heat, or both, to the assembly.

closed assembly time, n—the time interval between the closing together of substrates after the adhesive has been applied, and the application of pressure, heat, or both, to the assembly.

open assembly time, n—the time interval between applying adhesive on the substrates and closing them together before bonding. **D 907,** D14

assessment monitoring—an investigative monitoring program that is initiated after the presence of a contaminant in ground water has been detected. The objective of this program is to determine the concentration of constituents that have contaminated the ground water and to quantify the rate and extent of migration of these constituents. **D 653,** D18

assessment, *of a laboratory*—the activity of evaluating a laboratory's compliance with accreditation criteria. **E 1187,** E36

assessor, *n*—a general term for any individual responding to stimuli in a sensory test. **E 253,** E18

assessor, *of a laboratory*—an individual who carries out some or all functions related to laboratory assessment, (ISO Guide 2, *B*). **E 1187,** E36

as-shipped or produced coal—*for the purpose of Test Method D 4749,* raw or prepared coal in any state or condition at which it leaves the mine property or loading facility. **D 121,** D05

assignable area—See **space categories**. **E 1480,** E06

assignable cause, *n*—a factor which contributes to variation and is feasible to detect and identify. **D 123,** D13

assignable cause, *n*—a factor that contributes to variation, and which is feasible to detect and identify.

NOTE—Many factors will contribute to variation but it may not be feasible (economically or otherwise) to identify some of them. **E 456,** E11

assigned test value (ATV), *n*—the average of all results obtained in several laboratories which are considered acceptable based on the reproducibility of the test method. **D 4175,** D02

associated gas—natural gas, also known as gas-cap gas or dome gas, that overlies and is in immediate contact, but not in solution, with crude oil in a reservoir. **D 4150,** D03

associate instructor—an individual who possesses the qualifications and education/training of a course instructor/coordinator, but, in a specific course, assumes a supportive or assisting role to the course instructor/coordinator. This individual may substitute for the course instructor/coordinator in case of necessity or, in other courses, serves as a course instructor/coordinator. **F 1177,** F30

associative counterfeit—illegal use of a name or product shape that differs from the original product but that the consumer will associate with the original. **F 1156,** F12

associative evidence, *n*—that evidence which tends to link a person, place, or thing with another person, place, or thing. **E 1732,** E30

associative thickener—water soluble polymers containing hydrophobic groups that are capable of nonspecific hydrophobic association similar to surfactants that elevate viscosity presumably by association between thickener particles or thickener and dispersed particles that may be present in the aqueous system such as latex particles rather than through high molecular weight or chain stiffness of the thickener molecules themselves. **D 16,** D01

assurance of conformity—procedure resulting in a statement giving confidence that a product, process or service fulfills specified requirements, (ISO Guide 2). **E 1187,** E36

A-stage, *n*—an early stage in the preparation of certain thermosetting resins in which the material is still soluble in certain liquids, and may be liquid or capable of becoming liquid upon heating. **D 883,** D20

A-stage, *n*—*in polymer chemistry,* an early stage in a thermosetting resin reaction in which the material melts when heated and dissolves in certain liquids. (Compare **B-stage** and **C-stage**.) **D 907,** D14

A-stage—an early stage in the reaction of certain thermosetting resins in which the material is fusible and still soluble in certain liquids. (Syn. *resol.*) (Compare with **B-stage** and **C-stage**.) **E 1749,** E06

asterism—a lengthening of diffraction spots usually in the radial direction. **E 7,** E04

astigmatism—a defect in a lens or optical system which causes rays in one plane parallel to the optical axis to focus at a distance different from those in the plane at right angles to it. **E 7,** E04

ASTM cement types—Portland cements meeting the requirements of Specifications C 150. Cement types have slightly different formulations that result in various characteristics which address different construction conditions and different physical and chemical environments. They are as follows:

Type I (Portland)—a general-purpose construction cement with no special properties.

Type II (Portland)—a construction cement that is moderately resistant to sulfates and generates a lower head of hydration at a slower rate than Type I

Type III (Portland: high early strength)—a construction cement that produces a high early strength. This cement reduces the curing time required when used in cold environments, and produces a higher head of hydration than Type I.

Type IV (Portland)—a construction cement that produces a low head of hydration (lower than Types I and II) and develops strength at a slower rate.

Type V (Portland)—a construction cement that is a high sulfate resistant formulation. Used when there is severe sulfate action from soils and ground water. **D 653,** D18

ASTM grain size number— See **grain size.** **E 7,** E04

astringency, *n*—the complex of sensations due to shrinking, drawing, or puckering of the epithelium as a result of exposure to substances such as alums or tannins. **E 253,** E18

as-welded condition—a condition created as a result of forming annealed sheet or plate into tubular form and welding without subsequent heat treatment or cold work. **B 846,** B05

asymmetrical—this refers to a snowboard ski shape that does not have a longitudinal line of symmetry. Heel-side and toe-side sidecuts shaped and offset differently from each other; they are not mirror images of each other. This typically requires that a different snowboard ski be utilized for regular-foot (left foot forward) and goofy-foot (right foot forward) snowboard binding mounting positions. **F 1107,** F27

asymmetrical offset, O_s, O_t—the distance along the longitudinal axis that each side of an asymmetrical shape is offset from the other side. Offset may be different at the shoulder and heel. **F 1107,** F27

asymmetric membrane—membrane which has a change in pore structure with depth. See **anisotropic membranes.** **D 6161,** D19

asymmetry potential—the potential across a glass pH electrode membrane when the inside and outside of the membrane are in contact with solutions of identical pH. This term has also been used to define the observed potential differences between identical electrode pairs placed in identical solutions. **D 4127,** D19

asynchronous, *adj*—a data transmission in which the time between transmitted characters can vary, it is controlled by start and stop codes at the beginning and end of data sets. (See **synchronous.**) **F 1457,** F05

ATD—see **anti-telescoping device.** **D 6161,** D19

athermal—not isothermal, with changing rather than constant temperature conditions. **E 7,** E04

at-line analysis—analytical procedure performed in a process environment using manually entered samples. **D 4790,** D16

atmosphere, *n*—the gaseous envelope which surrounds the earth and includes ambient air, indoor air, and workplace air. See also **air at normal conditions.**

synthetic atmosphere, n—a specific gaseous mass containing any number of constituents and in any proportion produced for a special purpose. **D 1356,** D22

atmosphere for testing, *n*—air at ambient conditions of relative humidity and temperature in which tests or experiments are conducted. (See also **standard atmosphere for testing.**) **D 123,** D13

atmosphere for testing geosynthetics, *n*—air maintained at a relative humidity between 50 to 70 % and a temperature of 21 ± 2°C (70° ± 4°F). **D 4439,** D35

atmosphere for testing textiles, *n*—*for glass*, air maintained at a relative humidity of at least 48 % and no greater than 67 % and at a temperature of at least 20°C (68°F) and no greater than 25°C (77°F). **D 123,** D13

atmosphere, standard—air maintained at a specified temperature, relative humidity, and standard atmospheric pressure. **E 41,** G03

atmosphere (standard)—the pressure exerted by a mercury column 760 mm in height at 0°C under standard acceleration of gravity; equivalent to 101 325 Pa. **E 1316,** E07

atmospheric pressure—the pressure due to the weight of the atmosphere. It is the pressure indicated by a barometer that registers actual atmospheric pressure which is not corrected to sea level equivalence. Standard atmospheric pressure is a pressure of 76 cm Hg (101325 Pa) having a density of 13.5951 g/cm^3, under standard gravity of 980.665 cm/s^2. **E 41,** G03

atmospheric pressure—the pressure of the atmosphere at a specified place and time. **E 1316,** E07

atomic absorption—the absorption of radiant energy by ground state atoms. Substances when dispersed as an atomic vapor will absorb characteristic radiations identical to those which the same substances can emit. This property is the basis for analysis by atomic absorption spectroscopy. **D 2864,** D27

atomic absorption—absorption of radiant energy by ground-state atoms. **E 1605,** E06

atomic absorption spectrometry—the measurement of the light absorbed at the wavelength of resonance lines by the unexcited atoms of an element. **C 859,** C26

atomic absorption spectrometry, *n*—the branch of spectroscopy pertaining to the absorption of electromagnetic radiation by free atoms. **E 135,** E01

atomic mass unit (amu)—the unit of measure of the mass of a particle (atom, molecule, ion, etc.), defined as 1/12 of the mass of carbon-12. The numerical value of the mass of a particle in terms of amu is identical with the older atomic weight. **E 1316,** E07

atomic replica—See **replica.** **E 7,** E04

atomic scattering factor—the ratio of the amplitude of the wave scattered by an atom to that scattered by a single electron. Symbol = *f.* **E 7,** E04

atomic weight—the mean weight of the atom of an element in relation to $^{12}C = 12.000$. **C 859,** C26

atomic weight—the relative mass of an atom based on a scale in which a specific carbon atom (carbon 12) is assigned a mass value of 12. **D 6161,** D19

atomization—the dispersion of a molten metal into particles by a rapidly moving gas or liquid stream, or by mechanical means. **B 243,** B09

atomization, *n*—the process of atomizing. **E 1620,** E29

atomize, *v*—to transform bulk liquid or slurry into particles. **E 1620,** E29

atomized metal powder—metal powder produced by the dispersion of a molten metal by a rapidly moving gas, or liquid stream, or by mechanical dispersion. **B 243,** B09

atomizer, *n*—a device for atomizing. **E 1620,** E29

atom line, *n*—a spectral line resulting from radiation emitted during electron transition as an excited atom returns to a lower energy level. **E 135,** E01

ATP—adenosine triphosphate. **D 6161,** D19

atraumatic—teeth that would interdigitate except for being spaced apart a predesigned distance so they will not stress, crush, or otherwise traumatize the tissue being grasped. **F 1638,** F44

attached cushion, *n*—*for pile yarn floor covering*, a material, bonded to the backing fabric side of a pile yarn floor covering to provide additional dimensional stability, thickness, and padding. **D 5684,** D13

attached upholstery fabric, *n*—the exterior fabric covering secured to a furniture unit by the furniture manufacturer or custom upholsterer. (See **furniture coverings**) **D 123,** D13

attachment area, *n*—portion of the needle where the attachment of the suture takes place. For example, eyed, drilled, and channel. **F 1840,** F04

attachment hub—a device featuring an industry-standard square drive, that supports and powers ancillary devices such as a vegetable slicer or chopping attachment. **F 1827,** F26

attachment points, *n*—points where alternate pairs of coil loops are joined around the circumference creating the concertina effect. **F 1379,** F14

attapulgite clay—a chain-lattice clay mineral. The term also applies to a group of clay materials that are lightweight, tough, matted, and fibrous. **D 653,** D18

attenuated total reflection (ATR)—reflection that occurs when an absorbing coupling mechanism acts in the process of total internal reflection to make the reflectance less than unity. **E 131,** E13

attenuation—reduction of amplitude with time or distance. **D 653,** D18

attenuation, *n*—the decrease in AE amplitude per unit distance, normally expressed in dB per unit length. **E 1316,** E07

attenuation—a factor that describes the decrease in ultrasound intensity with distance. Normally expressed in decibel per unit length.

NOTE—The attenuation parameter is sometimes expressed in nepers (Np) per unit length. The value in decibels (dB) is 8.68 times the value in nepers. If the loss over a path is 1 Np, then the amplitude has fallen to 1/e of its initial value (e = 2.7183...). (E 664) **E 1316,** E07

attenuation coefficient—for a parallel beam of specified particles or radiation, the quantity μ in the expression $\mu \Delta x$ for the fraction removed in passing through a thin layer Δx of a substance in the limit as Δx approaches zero, where Δx is measured in the direction of the beam. **E 673,** E42

attenuation coefficient—related to the rate of change in the intensity of a beam of radiation as it passes through matter. See **linear and mass attenuation coefficient.** **E 1316,** E07

attenuation cross section—the probability, expressed in barns, that a neutron will be totally absorbed by the atomic nucleus. **E 1316,** E07

attenuation index, κ—a measure of the absorption of radiant energy by an absorbing material. κ is related to the absorption coefficient by: $n\kappa = \alpha c_o / 4\pi\nu$, where c_o = the speed of light in vacuo, ν = the frequency of radiant energy, and n = the refractive index of the absorbing medium. **E 131,** E13

attenuation period, *n*— *in cavitation and liquid impingement erosion,* a less-preferred term for **deceleration period**. **G 40,** G02

attenuator—a device for altering the amplitude of an ultrasonic indication in known increments, usually decibels. **E 1316,** E07

attic, *n*—an accessible enclosed space in a building immediately below the roof and wholly or partly within the roof framing. **E 176,** E05

attic—See **building space.** **E 631,** E06

attitude, *n*—a predisposition to respond in a characteristic way toward a class of objects, concepts, or stimuli. **E 253,** E18

attitude scale, *n*—a means for eliciting indications of the attitudes or opinions held, usually on a measuring system using marks or value designations. **E 253,** E18

attractant—an agent that increases the attentive frequency of an organism. **E 609,** E35

attractant—a material that attracts specific pests. **E 1519,** E35

attraction—a response towards or to facilitate contact with a material or condition. **E 943,** E47

attribute, *n*—a specific characteristic of a thing. (See **attribute data.**) **D 123,** D13

attribute, *n*—a perceived characteristic. **E 253,** E18

attribute data, *n*—observed values or determinations which indicate the presence or absence of specific characteristics. **D 123,** D13

attribute data, *n*—observed values or determinations which indicate the presence or absence of specific characteristics. **E 456,** E11

attributes, method of, *n*—measurement of quality by the method of attributes consists of noting the presence (or absence) of some characteristic or attribute in each of the units in the group under consideration, and counting how many units do (or do not) possess the quality attribute, or how many such events occur in the unit, group, or area. **E 456,** E11

attributes of color—(*1*) *for the object mode of appearance*, hue, lightness, and saturation. In the Munsell system, Munsell Hue, Munsell Value, and Munsell Chroma.

(*2*) *for the illuminant or aperture mode*, hue, brightness, and saturation. **E 284,** E12

attrital coal—See *attrital coal* under **coal**. **D 121,** D05

A-tuft, *n*—a single-pass process for aligning hook free fibers on the Fibroliner FL-101. **D 123,** D13

audible leak indicator—an accessory to a leak detector which converts the output signal to an audible note whose frequency is a function of the leakage rate. **E 1316,** E07

audition, *n*—the sense of hearing. **E 253,** E18

Auger:

analysis volume—see *volume* under **analysis**.

chemical effects—*AES,* see **chemical.**

chemical shift—*AES,* see **chemical.**

current—the electron current due to the emission of Auger electrons.

electron—an electron emitted as the result of an Auger process.

electron yield—the probability that an atom with a vacancy in a particular inner shell will relax by an Auger process.

line scan—a plot of Auger signal strength as a function of displacement along a designated line on the specimen surface. Normally, the abscissa is the line along which the signal is measured and the ordinate is directly proportional to signal strength.

line shape—the energy distribution in an Auger spectrum for a particular Auger transition.

map—two dimensional image of the specimen surface showing the location of emission of Auger electrons from a particular element. A map is normally produced by rastering the incident electron beam over the specimen surface and simultaneously recording the Auger signal strength for a particular transition as a function of position.

matrix effects—see **matrix effects,** *Auger*.

parameter—*XPS,* the kinetic energy of the sharpest Auger peak in the spectrum minus the kinetic energy of the most intense photoelectron peak from the same element; the energy of the ionizing photons must be specified.

peak energy for dN(E)/dE, N(E)—the designation of the energy of the Auger electron distribution. In *dN/dE* spectra, peak energies should be measured at the most negative excursions of the Auger features. In *N(E)* spectra, peak energies are measured at peak maxima. (Peak energies in *dN/dE* spectra are not the same as those in *N(E)* spectra.)

process—the relaxation, by electron emission, of an atom with a vacancy in an inner electron shell.

signal strengths—*AES, XPS,* in *dN/dE* spectra, signal strengths are measured as the peak-to-peak heights of the Auger features. In *N(E)* spectra, signal strengths are measured as the heights of the

Auger peaks above background. In *I(E)*, signal strengths are measured as the areas under the electron energy distribution, *N(E)*.

spectrum, dN(E)/dE, N(E), I(E)—AES, the display of Auger signal strength as a function of electron energy. Auger spectra from solids may be measured as the first derivative of the electron energy distribution and may be designated by *dN/dE.* The Auger electron energy distribution may be designated as *N(E).* With certain type analyzers (for example, the CMA) the displays are *dEN(E)/*dE and *EN(E).* The area under Auger peaks may be designated as *I(E)* with background subtraction method, and integration limits specified.

transition—transitions involved in electron emission by an Auger process are designated by indicating the electron shells. The first letter designates the shell containing the initial vacancy and the last two letters designate the shells containing electron vacancies created by Auger emission (for example, KLL, and LMN). When a bonding electron is involved the letter V is used (for example, LMV and KVV). When a particular subshell involved is known this can also be indicated (for example, KL_1L_2). Coupling terms may also be added where known ($L_3M_{4,5}M_{4,5};^1D$). More complicated Auger processes (such as, multiple initial ionizations and additional electronic excitations) can be designated by separating the initial and final states by a dash (for example, LL-VV and K-VVV). When an Auger process involves an electron from the same principal shell as the initial vacancy (for example, L_1L_2M) it is referred to as a Coster-Kronig transition. If both electrons are from the same principal shell as the initial vacancy (for example, $M_1M_2M_3$) it is called a super Coster-Kronig transition.

transition rate—the probability per unit time for two bound electrons to undergo energy state transitions such that one will fill an initial core hole vacancy and the other will go to a final state in the positive energy continuum. **E 673,** E42

Auger electron spectroscopy—the measurement of Auger signal strength as a function of electron energy. **E 673,** E42

auger increment, *n*—the retained portion of one extraction operation of the auger. **D 121,** D05

ausferrite, *n*—a cast iron matrix microstructure, produced by a controlled thermal process, which consists of predominantly acicular ferrite and high carbon austenite. (See **austempered ductile iron.**) **A 644,** A04

austempered ductile iron, *n*—a ductile cast iron that has been produced by a controlled thermal process which results in a matrix microstructure consisting of predominately acicular ferrite and high carbon austenite. **A 644,** A04

austempering, *n*—**heat treatment** involving **quenching** a steel object from a temperature above the **transformation range** in a medium maintained at a temperature above the **martensite range** sufficiently fast to avoid the formation of high temperature transformation products, and then holding it at that temperature until transformation is complete. **A 941,** A01

austenite—a face-centered cubic solid solution of carbon or other elements in gamma iron. **E 7,** E04

austenite grain size—the grain size which exists or existed in austenite at a given temperature. See Test Methods E 112. **E 7,** E04

austenitic stainless alloys—steel alloys that contain a minimum of 15 % chromium and from a residual to 20 % nickel. Some alloys may contain as much as 18 % manganese. The metal is predominantly face centered cubic in structure and hardenable only by cold working. Essentially nonmagnetic in its wire form, it may become slightly magnetic from cold working. Austenitic stainless steels can be grouped into three categories: 300 series alloy, Cr-Ni-Mn alloys, and Cr-Ni-Mo-Ti. **F 1789,** F16

austenitize, *vt*—to convert the matrix of a ferrous alloy to austenite by heating above the transformation temperature. **A 644,** A04

austenitizing, *n*—forming austenite by heating a steel object above the **transformation range**. **A 941,** A01

authenticate—to prove a document or article to be genuine or as represented. **F 1156,** F12

authority—the person (organization) authorizing the testing. **C 717,** C24

authority—body that has legal powers and rights, (EN 45020). **E 1187,** E36

authority having jurisdiction, *n*— the organization, office, or individual responsible for approving any equipment, installation, or procedure. **F 1494,** F23

autocatalytic plating—deposition of a metal coating by a controlled chemical reduction, catalyzed by the metal or alloy being deposited. **B 374,** B08

autoclave—a closed vessel for producing an environment of fluid pressure, with or without heat, to an enclosed object undergoing a chemical reaction or other operation. **E 1749,** E06

autoclaved products, *n—in asbestos-cement*, those that have been treated in a saturated steam atmosphere at between 689 and 1517 MPa (100 and 220 psi) for at least 8 h, and that contain portland cement as defined in Specifications C 150 and C 618, together with silica in the ratio of 3:2, that can react to form calcium silicate reaction products. **C 460,** C17

autoclaved products, *n*—those that have been treated in a saturated steam atmosphere at between 620 and 1517 kPa (90 and 220 psi,) for at least 6 h, and that contain portland cement as defined in Specifications C 150 and C 618 together with silica in the ratio of 3:2 that can react to form calcium silicate reaction products. **C 1154,** C17

autoclaved products, *n—in asbestos-cement*, those that have been treated in a saturated steam atmosphere at between 689 and 1517 MPa (100 and 200 psi) for at least 8 h, and that contain portland cement as defined in Specifications C 150 and C 618, together with silica in the ratio of 3:2, that can react to form calcium silicate reaction products. **D 2946,** C17

autoclave molding—a process where the lay-up or other assembly is covered by a vacuum bag and placed in an autoclave capable of providing heat and pressure for curing the part. **E 1749,** E06

autoclave (rubber), *n*—a pressure vessel used for vulcanizing rubber products by means of steam under pressure. **D 1566,** D11

autodeposition—a single-step immersion metal finishing process in which an organic coating is applied by means of unique surface chemical reactions carried out in an aqueous latex dispersion, also referred to as chemiphoresis. Components within the bath give rise to chemical reactions that slightly solubilize the metallic surface and lead to destabilization, deposition, and coalescence of the dispersed latex particles at that surface. **D 16,** D01

autodiscrimination, *n*—the ability of a symbol reader to automatically recognize and decode multiple symbologies. **F 1294,** F05

autodoping—*of an epitaxial layer*, incorporation of dopant originating from the substrate into epitaxial layer. Also called **self-doping.** **F 1241,** F01

auto-fill—a water height sensor devise that activates a fresh water fill solenoid if the water level in the cooking vessel drops below predetermined height. **F 1827,** F26

autographic dilatometer—a dilatometer that automatically records instantaneous and continuous changes in dimensions and some other controlled variable such as temperature or time.

E 7, E04

autographic pyrometer— See **pyrometer.** E 7, E04

autohesion, *n*—adhesion developed by interdiffusion of the molecules of two surfaces of the same material and consequent obliteration of the interface between them. (Sometimes called "autoadhesion.") (Compare **blocking.**) **D 907,** D14

autoignition, *n*—the ignition of a material caused by the application of pressure, heat, or radiation, rather than by an external ignition source, such as a spark, flame, or incandescent surface. **D 4175,** D02

autoignition, *n*—the ignition of a material commonly in air as the result of heat liberation due to an exothermic oxidation reaction in the absence of an external ignition source such as a spark or flame. **E 1445,** E27

autoignition temperature, *n*—the minimum temperature at which autoignition occurs> **D 4175,** D02

autoignition temperature, *n*—the minimum temperature at which autoignition occurs under the specified conditions of test. **E 1445,** E27

autoignition temperature—the lowest temperature at which a material will spontaneously ignite in oxygen under specific test conditions. **G 126,** G04

automatic density control , *n*—the process whereby the toner layer deposited on the substrate is controlled by the printer. (See **ADC.**) **F 1457,** F05

automatic image analysis—the separation and quantitative evaluation of an image into its elements with or without operator interaction. It includes the enhancement, detection, and quantification of the features contained in an image through the use of optical, geometrical, and stereological parameters and a computer program. Image analysis data output can provide individual measurements on each separate feature (feature specific) or totals for all features of a particular type in the field (field specific). **E 7,** E04

automatic image analyzer—a device which can be programmed to detect and measure features of interest in an image. It may include accessories such as automatic focus and an automatic traversing stage to permit unattended operation. **E 7,** E04

automatic lock slider, *n—in zippers,* a slider that provides involuntary, positive locking action on the chain when the pull is released. **D 123,** D13

automatic machine (or conveyor)—a machine for mechanically processing parts through treatment cycles, such as cleaning, anodizing, or plating. **B 374,** B08

automatic plating: (*1*) *full*—plating in which the cathodes are automatically conveyed through successive cleaning and plating tanks.

(*2*) *semi*—plating in which the cathodes are conveyed automatically through only one plating tank. **B 374,** B08

automatic trip—machine-activated tool mechanism providing continuous cycling while trip is in contact with the work. **F 592,** F16

automobile leather—see **upholstery leather.** **D 1517,** D31

automotive, *adj*—descriptive of equipment associated with self-propelled machinery, usually vehicles driven by internal combustion engines. **D 4175,** D02

auto nail—nail of ¼ to 5¼-in. length, sheared off bright, smooth, knurled, or helically fluted, regular-stock-steel or stiff-stock, 0.032-in. (21-gage) to 0.162-in. (8-gage) wire and driven subsequently by the same machine at a rapid rate. This nail has a sheared-bevel or sheared-square point. **F 547,** F16

autopsy—the dissection of a membrane module or element to investigate causes of unsatisfactory performance. **D 6161,** D19

autoradiograph—the image of an object containing a radioelement obtained, on a recording medium, by means of its own radiation. **E 1316,** E07

auxiliary anode—a supplementary anode employed during electrodeposition to achieve a desired thickness distribution of the deposit. **B 374,** B08

auxiliary cathode—See **thief.** **B 374,** B08

auxiliary electrode—see **counter electrode.** **G 15,** G01

auxiliary energy subsystem—in solar energy applications, equipment using nonsolar energy sources to supplement or backup the output provided by a solar energy system. **E 772,** E44

auxiliary solvent—liquid material used in addition to the primary solvent. It is generally used to replace part of the primary solvent to produce a specific effect, or, as a matter of economics. **D 3064,** D10

availability—the on-stream time or rated operating capacity of a water treatment system. **D 6161,** D19

available chlorine in cleaning compounds—the oxidizing power of chlorine present as hypochlorite or other oxidizing chlorine moieties in solution, expressed as chlorine of equivalent weight 35.45, and as determined by thiosulfate titration. **D 459,** D12

available lime index—those constituents of a lime which enter into a desired reaction under the conditions of a specific method or process. **C 51,** C07

available program, *n*—a qualifying term which can be used in the definition of recyclable, reusable, refillable, returnable, compostable, establishing limits; for example by population and access within geographic area. **D 5488,** D10

a-value—membrane water permeability coefficient. The coefficient is defined as the amount of water produced per unit area of membrane when net driving pressure (NDP) is unity, a unit of measurement is $m^3/hr/m^2/kPa$. **D 6161,** D19

aventurine—glass containing colored, opaque spangles of nonglassy material. **C 162,** C14

average, *n—for a series of observations,* the total divided by the number of observations. **D 123,** D13

average background reflectance—expressed as a percent, is the simple arithmetic average of the background reflection readings from at least five different points on a sheet. **F 149,** F05

average coating thickness—determined as either the value obtained by analytical methods or the mean value of a specified number of local thickness measurements that are evenly distributed over the significant surface. **F 1789,** F16

average coefficient of cubical expansion— average change in unit volume of a substance per unit change in temperature over a specified range of temperature. **E 7,** E04

average coefficient of linear expansion— average change in unit length of a body per unit change in temperature over a specified range of temperature. **E 7,** E04

average coefficient of thermal expansion— general term. (See also **average coefficient of cubical expansion** and **average coefficient of linear expansion.**) **E 7,** E04

average diameter, *n*—the average of the maximum and minimum outside the diameters, as determined at any one section of the pipe or tube. **B 899,** B02

average diameter (for round tubes only)— the average of the maximum and minimum outside diameters or the maximum and minimum inside diameters, whichever is applicable, as determined at any one cross section of the tube. **B 846,** B05

average discharge (corona) current (I_t), *n*—the sum of the absolute magnitudes of the individual discharges during a certain time interval divided by that time interval. **D 1711,** D09

average edge—an imaginary line bisecting the irregularities of the character edge. **F 149,** F05

average emission function decay length—the negative reciprocal slope of the logarithm of a specified exponential approximation to the emission depth distribution function over a specified range of depths, as determined by a straightline fit to the emission depth distribution function plotted on a logarithmic scale versus depth on a linear scale. **E 673,** E42

average erosion rate, *n*—a less preferred term for **cumulative erosion rate.** **G 40,** G02

average fiber diameter, *n*—*in wool and other animal fibers,* the average width of a group of fibers when measured on a projected image. **D 123,** D13

average grade—See **grade.** **E 631,** E06

average grain diameter— See **grain size.** **E 7,** E04

average injection velocity, *n*—the mean value of the velocity of the molten plastic flow front within a cavity during the injection time that is calculated from the shot volume and injection time. **D 883,** D20

average interstitial velocity—see **velocity, average interstitial.** **D 653,** D18

average particle size—a single value representing the entire particle-size distribution. **C 242,** C21

average run length (ARL)—*(1)* **sample sense,** *n*—the average number of times that a process will have been sampled and evaluated before a shift in process level is signaled, and (*2*) **unit sense,** *n*—the average number of units that will have been produced before a shift in level is signaled. **E 456,** E11

average signal level, *n*—the rectified, time averaged AE logarithmic signal, measured on the AE amplitude logarithmic scale and reported in dB_{ae} units (where 0 dB_{ae} refers to 1 μV at the preamplifier input). **E 1316,** E07

average sound pressure level—*of several related sound pressure levels measured at different positions or different times, or both, in a specified frequency band,* ten times the common logarithm of the arithmetic mean of the squared pressure ratios from which the individual levels were derived. **C 634,** E33

average tire tread depth, [*L*], *n*—the average of all tire groove (void) depth measurements. **F 538,** F09

aversion, *n*—feeling of dislike provoking avoidance of a stimulus. **E 253,** E18

aviation gasoline, *n*— *designed for spark-ignition aircraft engines,* fuel that has specific properties including knock-free performance, low volatility, and good stability. **D 4175,** D02

avicide—a chemical used to kill, control, or cause other adverse effects on birds. **E 609,** E35

avoidance—a response away from or to limit contact with a material or condition. **E 943,** E47

avulsion—a sudden, natural change of a stream channel, so that the water flows elsewhere than in its previous course. **D 4410,** D19

AWWA—American Water Works Association. **D 6161,** D19

AWWARF—American Water Works Association Research Foundation. **D 6161,** D19

axial ratio—the ratio of the length of one axis to that of another (for example, *c/a*) or the continued ratio of three axes (for example, *a:b:c*). **E 7,** E04

axial strain, **n**—linear strain in a plane parallel to the longitudinal axis of the specimen. **E 6,** E28

axis (crystal)—the edge of the unit cell of a space lattice. Any one axis of any one lattice is defined, in length and direction, with respect to the other axes of that lattice. **E 7,** E04

axis, optic—the direction, or directions in an anisotropic crystal along which light is not doubly refracted. **E 175,** E41

axis, optical—the line formed by the coinciding principal axes of a series of optical elements comprising an optical system. It is the line passing through the centers of curvature of the optical surfaces. **E 175,** E41

axle configuration code, *n*—alphanumeric code, up to nine characters long, defining the number and type of units in a vehicle, number of axle groups in each unit, and number of axles in each axle group. **E 867,** E17

axle correction factor, *n*—an adjustment of vehicle axle sensor summarized base data for incidence of multiple axle vehicles. **E 867,** E17

axle count, *n*—the total number of a vehicle's axles in contact with the pavement. **E 867,** E17

axle efficiency, [*nd*], *n*—*in a vehicle,* the ratio of the sum of the wheel torques at the driven wheels to the product of driveshaft torque and axle ratio. **F 538,** F09

axle group, *n*—one or more adjacent axles, usually sharing a common connection to the body of a vehicle, which jointly support a portion of the vehicle's weight. **E 867,** E17

axle-group load—the sum of all tire loads on a group of adjacent axles. **E 867,** E17

axle load—the sum of all tire loads on an axle. An axle is comprised of two or more wheel assemblies lying approximately on a common axis oriented transversely to the nominal direction of motion of the vehicle. **E 867,** E17

axle spacing, *n*—for each axle, the horizontal distance between the center of that axle and that of the preceding axle; the axle spacing for the vehicle's front axle is assumed to be zero. **E 867,** E17

azeotrope—constant boiling mixture, for ethanol-water, the azeotrope of 95.6 % ethanol and 4.4 % water (both percentages by volume) boils at one atmosphere pressure. **E 1705,** E48

azeotropic distillation—the use of an organic solvent to create a new constant boiling point mixture, a method used to produce anhydrous ethanol from the ethanol water azeotrope. **E 1705,** E48

azimuth—*in ellipsometry,* the angle between the major axis of the ellipse and the plane of incidence. **F 1241,** F01

azimuthal angle, η, *n*—angle between the plane containing the axis of the illuminator (or the path of illumination) and the specimen normal and the plane containing the axis of the receiver (or the path of reception) and the specimen normal. The origin and direction of measure of the angle should be specified when required. **E 284,** E12

azimuthal viewing, *n*—deprecated term; do not use. Replace *azimuthal* by *annular* or *circumferential.* **E 284,** E12

azimuth angle, *n*—an angle of a plane to the horizon measured clockwise to the object. **G 113,** G03

B

Babo's law—the vapor pressure over a liquid solvent is lowered approximately in proportion to the quantity of a nonvolatile solute dissolved in the liquid. **E 7,** E04

back—the side reverse to the face of a panel, or the poorer side of a panel in any grade of plywood calling for a face and back. **D 1038,** D07

back—a crop with the head trimmed off behind the horn holes. **D 1517,** D31

back bedding, n—See **bedding**. **C 717,** C24

back bedding—See **windows and doors. E 631,** E06

back breakpoint, *n—in anatomy*, the location on the back of the body where the arm separates from the body. **D 123,** D13

back coating, *n—in textile floor covering*, an adhesive compound applied for such purpose as locking pile yarn tufts into a carpet backing, bonding a secondary backing to a primary backing, increasing fabric body or stiffness and increasing dimensional stability. **D 5684,** D13

backdrafting, *n*—the reversal of the normal (upward) direction of air flow in a vent for a vented combustion appliance (boiler, fireplace, furnace, or water heater), when the vented appliance is operating. **D 1356,** D22

backed exposure, *n*—a technique of weathering in which the test specimens being exposed are mounted onto a solid backing material, of sufficient strength to hold the specimen. When the specimen and the backing are in direct contact the backing material must be of a type that will not contaminate the specimen. When two materials are intimately joined together to form one composite, the materials below the top surface are not considered as a backing. **G 113,** G03

back emf (electromotive force)—the potential set up in an electrolytic cell that opposes the flow of current, caused by such factors as concentration polarization and electrode films. See **emf (electromotive force). B 374,** B08

back emission—the electrical breakdown of air due to excessive charge build-up in the porcelain enamel powder film during powder application. This is due to the self-limiting characteristic of electrostatic powders. **C 286,** B08

backer material, *n*—a compressible material that is placed in joints or cracks before applying sealant to prevent bonding of the sealant on the bottom of the joint, control sealant depth, and prevent sagging of the sealant. **D 5535,** D04

backer printing—printing on the reverse side of the sheet. For OCR forms, the paper should have sufficient opacity so that printing on the back can't be seen on the front by the optical scanner. **F 149,** F05

backer strips—water-repellent strips of asphalt-coated felt applied behind each joint where the vertical edges of two shingles meet. **C 460,** C17

backer strips, *n— in shingles*, water-repellent strips of asphalt-coated felt applied behind each joint where the vertical edges of two shingles meet. **C 1154,** C17

backer strips—water-repellent strips of asphalt-coated felt applied behind each joint where the vertical edges of two shingles meet. **D 2946,** C17

backfill—all the material used to fill the trench from bedding to finished surface. **C 896,** C04

backfill—all material used to fill the trench from bedding to finished surface. **F 412,** F17

backfill, final—material used to fill the trench from initial backfill to finished surface. **C 896,** C04

backfill, final—material used to fill the trench from initial backfill to finished surface. **F 412,** F17

backfill, initial—material used to fill the trench from top of bedding to a designated height over the pipe. **C 896,** C04

backfill, initial—material used to fill the trench from top of bedding to a designated height over the pipe. **F 412,** F17

backfill, unconsolidated—non-compacted material in place in the trench. **C 896,** C04

backfill, unconsolidated—noncompacted material in place in trench. **F 412,** F17

back flushing, *n*—a process by which liquid is forced in the reverse direction to the flow direction. **D 4439,** D35

background—apparent absorption caused by anything other than the substance for which the analysis is being made. **E 131,** E13

background:

inelastic—ISS, the response of the energy filtering and detection system to probe ions that have undergone inelastic scattering events at the specimen surface.

instrumental—ISS, the response of the energy filtering and detection system to events other than those induced by bombardment of the specimen surface by a beam of probe ions.

secondary ion—ISS, the response of the energy filtering and detection system to secondary ions produced by bombardment of the target material with probe ions.

signal—for a specific measurement, any signal present at a particular position due to processes or sources other than those of primary interest. **E 673,** E42

background—the surface of the test part against which the indication is viewed. It may be the natural surface of the test part or the developer coating on the surface. **E 1316,** E07

background—in magnetic particle examination, the appearance of the surface of the test part against which indications are viewed. **E 1316,** E07

background—the reflectance of the receptor virgin surface. **F 221,** F05

background—toner particles in the non-image areas of an electrostatic print. **F 335,** F05

background, *n*—a measure of the reflectance or density of the unimaged portion of a thermal paper; typically measured using a densitometer, reflectometer, or opacimeter. **F 1623,** F05

background blackening—a continuous, slowly varying blackening of photographic film which has been exposed to diffracted X-rays, on which the blackening due to diffracted spots or lines is superimposed. **E 7,** E04

background intensity—in spectrochemical analyses, the optical density or darkening of the photographic plate or film near the spectrographic line being measured, or the intensity of light measured with a phototube near the spectrographic line being measured. **C 859,** C26

background noise—noise from all sources unrelated to a particular sound that is the object of interest. Background noise may include airborne, structureborne, and instrument noise. **C 634,** E33

background potential—see **residual potential. F 335,** F05

background radiation—all radiation received by the infrared sensing device that was not emitted by the specified area of the surface being examined. **E 1316,** E07

background reflectance—a measurement of the brightness of paper referring to the amount of light reflected back from the paper at a particular point when that point is flooded with light, as compared with the known value representing absolute white (such as $BaSO_4$).

F 149, F05

background signal—in leak testing, the steady or fluctuating output signal of the leak detector caused by the presence of residual tracer gas or other substance to which the detecting element responds. **E 1316,** E07

background, target—that portion of the background which is confined to the field of view. **E 1316,** E07

backing, *n—for pile yarn floor covering,* all materials in a pile yarn floor covering other than pile yarn. **D 123,** D13

backing—laminating, by the use of adhesives, a piece of fabric to leather or other materials used for shoe uppers. Its purpose is to make leather or shoe stronger and possibly firmer. **F 869,** F08

backing board, *n*—a noncombustible insulating board, mounted behind the specimen during actual testing to satisfy the theoretical analysis assumption of no heat loss through the specimen. It shall be roughly 25 ± 5 mm thick with a density no greater than 200 ± 50 kg/m^3. **E 176,** E05

backing fabric, *n—in textiles,* (*1*) a fabric into which a pile yarn is inserted, or (*2*) a reinforcing layer adhered to the reverse side of a fabric. **D 123,** D13

backing fabric, *n—in textile conservation,* a support textile fabric located behind the textile artifact. **D 123,** D13

backing film—a film used as auxiliary support for the thin replica or specimen-supporting film. **E 7,** E04

backing pump—Same as **fore pump**. **E 1316,** E07

backing space—the space between a backing pump (fore pump) and the associated diffusion pump (or other type of pump requiring a fore pump). See also **ballast.** **E 1316,** E07

backing space technique—a method of testing for leaks in which the leak detector is connected to the backing space to take advantage of the compression of gas that occurs between the vacuum system and the backing pump, due to the action of the diffusion pump (or other type of pump of high speed relative to its backing pump). **E 1316,** E07

back ionization—see **back emission**. **C 286,** B08

back joint—a vertical mortar joint, parallel to the vertical substrate, between a chemical-resistant construction unit and a substrate or another chemical-resistant construction unit. **C 904,** C03

backnailing—the practice of blind-nailing roofing felts to a substrate in addition to hot-mopping to prevent slippage. **D 1079,** D08

backpack grouting—the filling with grout of the annular space between a permanent tunnel lining and the surrounding formation. **D 653,** D18

back-packing—any material (usually granular) that is used to fill the empty space between the lagging and the rock surface. (ISRM) **D 653,** D18

backpart molding—a preparatory shaping operation to heel seat prior to lasting, usually performed on the thermoplastic counters where backpart components are heated on a metal heel form and molded into the backpart heel shape combining counter, upper and lining before the actual lasting operation is performed on the backpart of the shoe. **F 869,** F08

back pressure—Same as **forepressure**. **E 1316,** E07

back pressure test—Same as **pressure-evacuation test**. **E 1316,** E07

back putty, n—See **bedding**. **C 717,** C24

back putty—See **windows and doors.** **E 631,** E06

back reflection—the diffraction of X-rays at a Bragg angle approaching 90°. **E 7,** E04

back reflection—indication of the echo from the far boundary of the material under test. **E 1316,** E07

backrinding, *n*—a molding defect in which the rubber adjacent to the spew line shrinks below the surface of the molded product, with the spew line often being ragged and torn. **D 1566,** D11

backscattered electrons—*AES,* electrons originating in the incident beam which are emitted after interaction with the target. By convention, electrons with energies greater than 50 eV are considered as backscattered electrons. **E 673,** E42

back scattered radiation—radiation which is scattered more than 90° with respect to the incident beam, that is, backward in the general direction of the radiation source. **E 1316,** E07

backscattering:

energy—EIA, energy of a particle from the analyzing beam after it has undergone a backscattering collision and escaped the specimen.

factor—AES, the fractional increase in the Auger current due to backscattered electrons.

spectrum—EIA, a plot of backscattering yield (ordinate) versus backscattering energy (abscissa).

yield—EIA, the number of particles detected (counts) per unit backscattering energy per incident ion. **E 673,** E42

backscatter peak—a peak in the observed photon spectrum (normally below about 0.25 MeV) resulting from large-angle (>110°) Compton scattering of gamma rays from materials near the detector. This peak will not have the same shape as the full-energy peaks (being wider and skewed toward lower energy). **E 170,** E10

backset—the liquid portion of the thin stillage that is recycled as part of the process liquid in mash preparation. **E 1705,** E48

backshelf—used for counter-height equipment typically located against a wall, but could be freestanding. **F 1827,** F26

back side, *n—in textile materials,* the side of the material that faces inward in the completed object (Ant. **face side**). **D 123,** D13

backside—Not preferred, use **back surface.** **F 1241,** F01

backside density, *n*—the optical density of ink image on topside of a substrate as measured from the backside of the media on which it was printed. **F 1857,** F05

back surface—the end of a reference block that is opposite the entry surface. (E 127) **E 1316,** E07

back surface—*of a semiconductor wafer,* the exposed surface opposite to that upon which active semiconductor devices have been or will be fabricated. **F 1241,** F01

back titration—a titration in which an excess of reagent is added to the sample to complex or precipitate the species being determined. The excess reagent is then determined by titration, usually with the species that was originally in the sample. Back titrations are useful for increasing accuracy in the analysis of very dilute solutions. **D 4127,** D19

back-up material—See **sealant backing**. **C 717,** C24

back waist length, *n—in body measurements,* the vertical distance along the spine from the cervicale to the waist. **D 123,** D13

back waist length, *n— in body measurements,* the vertical distance along the spine from the cervical to the waist. **F 1494,** F23

back wall—the wall at the charging end of a melter. **C 162,** C14

backwash—reverse the flow of water with/without air either across or through a medium or membrane designed to remove the collected foreign material from the bed or membranes. **D 6161,** D19

back width, *n—in body measurements,* the distance from back-breakpoint to back-breakpoint. **D 123,** D13

bacteria—any of a class of microscopic single-celled organisms reproducing by fission or by spores. Characterized by round, rod-like spiral or filamentous bodies, often aggregated into colonies or mobile by means of flagella. Widely dispersed in soil, water, organic matter, and the bodies of plants and animals. Either autotrophic (self-sustaining, self-generative), saprophytic (derives nutrition from non-living organic material already present in the environment), or parasitic (deriving nutrition from another living organism). Often symbiotic (advantageous) in man, but sometimes pathogenic. **D 6161,** D19

bactericide—agent capable of killing bacteria. **D 6161,** D19

bacteriophage—a virus that infects bacteria. **E 1705,** E48

bacteriophage, *n*—a type of virus which infects bacteria. **F 1494,** F23

bacteriostat—substance that prevents bacterial growth and metabolism but does not necessarily kill them. **D 6161,** D19

BAF (bioaccumulation factor), *n*—the quotient obtained by dividing the concentration of a substance in an organism (or specified tissue) by its concentration in a specified exposure medium, for example, air, food, sediment, soil, water, when several media are possible sources (see **bioaccumalation**). **E 943,** E47

baffle—a mold part used to close the delivery or baffle hole in a blank mold. **C 162,** C14

baffle—a pier, weir, sill, fence, wall, or mound built on the bed of a stream to parry, deflect, check, or regulate the flow or to float on the surface to dampen the wave action. **D 653,** D18

baffle filters—a series of vertical baffles designed to capture grease and drain it away to a container. The filters are arranged in a channel or bracket for easy insertion into, and removal from, the hood for cleaning, and are usually constructed of aluminum, steel, or stainless steel and they come in various standard sizes. See **exhaust hood**. **F 1827,** F26

baffle mark—a mark or seam on a bottle resulting from a mold joint between blank mold and baffle. **C 162,** C14

baffle wall—a wall used to deflect gases or flames in a melter. See **shadow wall**. **C 162,** C14

bag, *n*—a preformed **container** of tubular construction made of flexible material, generally enclosed on all sides except one forming an opening that may or may not be sealed after filling. (See also **pouch.**)

paper multiwall-sack—a flexible **container** made of several plies, usually of kraft paper. The various plies may be specially treated, such as waxed paper, glassine, greaseproof, polyethylene, **wet strength paper,** or other specialty sheets. The particular nature of the sack depends upon the material to be packed and the type of transportation to be employed.

bag liner—See **liner.** **D 996,** D10

bag, *n*—*in sampling,* any quantity of asbestos fiber corresponding to one particular grade that is packed in a suitable container. **D 2946,** C17

bagasse—residue remaining after extraction of a sugar-containing juice from plants like sugar cane. **E 1126,** E48

bagasse—residue remaining after extraction of a sugar-containing juice from plants like sugar cane. **E 1705,** E48

bag cure, *n*—a method of vulcanization in which an inflated flexible bag is used to impart positive internal pressure on the article being vulcanized. **D 1566,** D11

bagging, *n*—a very heavy, loosely woven fabric used for bale covering, etc., woven from heavy, tightly twisted roving, consisting largely of reworked and waste fibers (principally cotton and jute). **D 123,** D13

baghouse—a chamber containing bag filters for the removal of particles from a process exhaust stream. **C 162,** C14

bag leather—a general term for leathers used in traveling bags and suitcases. It does not include the light leathers employed for women's fancy handbags. The staple material for bag and case leather at present is leather made from the hides of animals of the bovine species, but heavy sealskins and goatskins are also used. **D 1517,** D31

bag modeling—a method of molding or laminating which involves the application of fluid pressure, usually by means of air, steam, water or vacuum, to a flexible barrier material which transmits the pressure to the material being molded or bonded. **D 883,** D20

bag molding, *n*—a method of molding or bonding involving the application of fluid pressure, usually by means of air, steam, water, or vacuum, to a flexible cover which, sometimes in conjunction with the rigid die, completely encloses the material to be bonded. **D 907,** D14

bag molding—a method of molding or bonding involving the application of fluid pressure, usually by means of air, steam, water, or vacuum, to a flexible cover which, sometimes in conjunction with a rigid die, completely encloses the material to be bonded. (Compare with **vacuum bag molding**.) **E 1749,** E06

bagout port—*in the nuclear industry*, an opening in a glovebox to which a bag is attached and in which material and equipment can be encased, then removed without possible contamination of the surrounding environment. **C 859,** C26

bag sampler—a sampler that utilizes a collapsible bag as the sample collection container. **D 4410,** D19

bail, *n*—*in zippers*, a portion or portions of the slider to which the pull or pulls are attached. **D 123,** D13

bail, *n*—the usually arched handle of a pail or can. **D 996,** D10

bailer—a hollow tubular receptacle used to facilitate withdrawal of fluid from a well or borehole. **D 653,** D18

bail-out bottle, *n*—A diver-carried supply of air or mixed gas (as appropriate) sufficient under standard operating conditions to allow the diver to reach the surface or another source of breathing gas, or to be reached by a safety diver. **F 1549,** F32

bainite—upper, lower, intermediate— metastable microstructure or microstructures resulting from the transformation of austenite at temperatures between those which produce pearlite and martensite. These structures may be formed on continuous (slow) cooling if the transformation rate of austenite to pearlite is much slower than that of austenite to bainite. Ordinarily, these structures may be formed isothermally at temperatures within the above range by quenching austenite to a desired temperature and holding for a period of time necessary for transformation to occur. If the transformation temperature is just below that at which the finest pearlite is formed, the bainite (upper bainite) has a feathery appearance. If the transformation temperature is just above that at which martensite is produced, the bainite (lower bainite) is acicular, resembling slightly tempered martensite. At the higher resolution of the electron microscope, upper bainite is observed to consist of plates of cementite in a matrix of ferrite. These discontinuous carbide plates tend to have parallel orientation in the direction of the longer dimension of the bainite areas. Lower bainite consists of ferrite needles containing carbide platelets in parallel array cross-striating each needle axis at an angle of about 60°. Intermediate bainite resembles upper bainite; however, the carbides are smaller and more randomly oriented. **E 7,** E04

bait—the tool dipped into molten glass to start any drawing operation. **C 162,** C14

bake-out—in leak testing, the degassing of a vacuum system by heating during the pumping process. **E 1316,** E07

baking, *n*—heating to a low temperature in order to remove gases. **A 941,** A01

baking duration—the time measured from when the plated product reaches a specified temperature in the baking furnace or oven until it is removed. **F 1789,** F16

baking finish—a paint or varnish that requires baking at temperatures above 150°F (65°C) for the development of desired properties. **D 16,** D01

baking temperature—a temperature above 150°F (65°C). **D 16,** D01

bal—a laced shoe in which the quarters meet and the vamp is stitched over the quarters at the front of the throat. Bal is an abbreviation of Balmoral, the Scottish castle in which this style was first worn. **F 869,** F08

balance—See **windows and doors.** **E 631,** E06

balanced construction—See under **construction.** **D 1038,** D07

balanced filters (X-rays)—a pair of filters used to eliminate all but a narrow range of wavelengths. The filter materials and thicknesses are chosen so that their absorption edges lie very close together and so that they have the same absorption except for wavelengths lying in the range between their absorption edges. When these filters are used alternately, the difference in effect, if any, is due to X-rays that have wavelengths in this range. Balanced filters thus can be made to serve as a crude monochromator. **E 7,** E04

balanced incomplete block design (BIB), *n*—an incomplete block design in which each block contains the same number *k* of different versions from the *t* versions of a single principal factor arranged so that every pair of versions occurs together in the same number, λ, of blocks from the *b* blocks. **E 456,** E11

balanced incomplete block design (BIB), *n*—an incomplete block design in which each block contains the same number *k* of different versions from the *t* versions of a single principal factor arranged so that every pair of versions occurs together in the same number, λ, of blocks from the *b* blocks. **E 1325,** E11

balanced laminate—any laminate that contains one ply of minus theta orientation with respect to the laminate principal axis for every identical ply with a plus theta orientation. **D 3878,** D30

balanced laminate—a composite laminate in which all laminae occur in pairs symmetric about the midplane (but not necessarily adjacent to each other). See **symmetrical laminate**. **E 1749,** E06

balancing, *n*—a process to correct for heavy or light areas of a tire and wheel assembly. **F 538,** F09

balata, *n*—a hard thermoplastic consisting of approximately equal proportions of trans-polyisoprene and resin, obtained from trees of the *Sapotaceae* family. **D 1566,** D11

bald roof—See **smooth-surfaced roof**. **D 1079,** D08

bale, *n*—in packaging a shaped unit, bound with cord or metal ties under tension, and containing compressed articles or materials. It may be wrapped. **D 996,** D10

bale coating, *n*—a coating applied to surfaces of rubber bales that inhibits adhesion to other surfaces. **D 1566,** D11

bale (jute), *n*—a bag, sack, etc., containing packages of yarn. **D 123,** D13

ballast—materials used to provide stability to a buoyant object (such as casing within a borehole filled with water). **D 653,** D18

ballast—in leak testing, a backing space large enough to maintain a low forepressure when the fore pump is temporarily stopped. **E 1316,** E07

ballast—weight applied to the skirt to improve boom performance. **F 818,** F20

ball burnishing—See **barrel burnishing.** **B 374,** B08

ball clay—a secondary clay, commonly characterized by the presence of organic matter, high plasticity, high dry strength, long vitrification range, and a light color when fired. **C 242,** C21

ball cratering—a method of specimen preparation in which a specimen is polished by a sphere in order to expose compositional changes below the original surface of a specimen with the intent that the depth of these layers can be related to the position on the surface created by the ball. **E 673,** E42

ball element—an element used in an impact printing device in which the fully formed characters are located on the outer surface of a sphere-like device. Frequently called *golf-ball element.* **F 909,** F05

ball head—spherical head. **F 547,** F16

ballistic point—bullet-shaped point. **F 547,** F16

ball measurement—the line running completely around the foot or last coplanar to the joints of large and small toes. **F 869,** F08

ball mill—*in porcelain enamels,* a dense, ceramic-lined rotating cylinder in which ceramic materials are wet or dry ground, generally using pebbles or porcelain balls as grinding media. **C 286,** B08

ball mill, *n*—a closed rotating cylinder containing hard balls (or other shaped members) that serves to grind coarse materials to a finer particle size. **D 1566,** D11

ball milling—a method of grinding and mixing material, with or without liquid, in a rotating cylinder or conical mill partially filled with grinding media such as balls or pebbles. **C 242,** C21

ball point—spherical point on shank of nail, having a diameter equal to that of shank. **F 547,** F16

ball printer—an impact printing device utilizing a ball element. The paper on the carriage remains stationary while the element is positioned and caused to strike the paper to produce characters in their proper positions. **F 909,** F05

balsam, Canada—a resin from the balsam fir *Abies balsamea.* Dissolved in xylene, toluene, or benzene it is used as a mountant for permanent microscopical preparations. Its refractive index may vary from 1.530 to 1.545 and its softening point from room temperature to 100°C, these properties varying with age and solvent content. If impure it discolors with age. **E 175,** E41

baluster—See **railing systems.** **E 631,** E06

baluster, baluster bar, *n*—one of a series of closely spaced, upright, and parallel infill members of a balustrade, located between top rail or handrail and bottom rail or tread or floor beneath balustrade. Synonym for **picket.** **E 1481,** E06

baluster casting—an ornamental cast element attached to a baluster. Also, cast element designed to attach baluster to top and bottom rails. **E 1481,** E06

baluster (picket), *n*—one of a series of closely-spaced upright members that support the handrail in a railing system. **E 1605,** E06

baluster railing system—a system consisting of posts, balusters, top rail, and bottom rail. **E 1481,** E06

balustrade—See **railing systems.** **E 631,** E06

balustrade, *n*—a railing system consisting of a row of balusters capped by a rail or handrail. **E 1481,** E06

band—in electron diffraction, a broad intensity maximum with sharp edges. **E 7,** E04

band—the light frequency spectrum between two defined limits; also light band. **F 149,** F05

band, *n*—a flexible anchor component with a noncircular cross section that connects the bony elements of the spine, pelvis, or ribs to each other or to other implant components using a knot or similar tying mechanism, forming a locked, closed loop. **F 1582,** F04

band aid sealant joint, *n*—Use **bridge sealant joint** (preferred term). **C 717,** C24

band application—an application of a pesticide to a continuous restricted area such as in or along a crop row rather than over the entire field area. **E 609,** E35

band application—an application of a pesticide to a continuous restricted area such as in or along a crop row rather than over the entire field area (see Definitions E 609). **E 1102,** E35

banded coal—See *banded coal* under **coal.** **D 121,** D05

banded structure (banding)—alternate bands parallel with the direction of working resulting from the elongation of segregated areas. **E 7,** E04

banding—Use **strapping.** **D 996,** D10

banding, *n*—uniform density variations or voids in a given color which appear in the direction that the printhead travels. **F 1857,** F05

banding (railing)—a portion of wood or other material extending around one or more edges of a plywood panel. **D 1038,** D07

band pass filter—a wave filter having a single transmission band; neither of the cut-off frequencies being zero or infinity. **E 1316,** E07

band printer—a type of printer that employs an endless steel band containing fully formed characters. The band transverses the width of the paper form. Hammers strike the back of the form, synchronized in such a way as to produce the desired characters in their proper position. **F 909,** F05

bandwidth, [1/ *T*], *n*—the range of frequencies within which certain performance characteristics occur; specific limits normally apply. **F 538,** F09

bank—a grouping of devices. See **array, block, train.** **D 6161,** D19

bank gravel, *n*—gravel found in natural deposits, usually more or less intermixed with fine material, such as sand or clay, or combinations thereof; gravelly clay, gravelly sand, clayey gravel, and sandy gravel indicate the varying proportions of the materials in the mixture. **D 8,** D04

banking—the alignment of the first graphic shape in a line with respect to the left (right) margin, by certain devices (that is, typewriters, line printers, etc.). **F 149,** F05

bank (mill, calender, or spreader), *n*— a reservoir of material at the opening between rolls (mill or calender), or at the spreader bar. **D 1566,** D11

bar, *n*—solid wrought product that is long in relation to its cross section, is square or rectangular with sharp or rounded corners/edges (excluding plate and flattened wire), or is a regular hexagon or octagon, and in which at least one perpendicular distance between parallel faces is 0.375 in. or greater [over 10.00 mm].

bus bar, n—rigid electric conductor in the form of a bar.

extruded bar, n—bar brought to final dimensions by hot extruding.

rolled bar, n—bar brought to final dimensions by hot rolling.

saw-plate bar, n—bar brought to final thickness by hot or cold rolling and to final width by sawing. **B 881,** B07

bar, *n*—an elongated, forged or rolled metal product with uniform strength, length and section (such as rectangular, square, round, oval or hexagonal). **B 899,** B02

bar—unit of pressure; 14.50 lbs/in. ^{2}, 1.020 kg/cm^2, 0.987 atm, 0.1 MPa. **D 6161,** D19

bar, *n*—a round, square, rectangular, or other polygonal solid member having a length greater than its width or thickness; and usually of rolled, drawn, or extruded metal (if of steel, having dimensions of 0.204 in. (5.2 mm) or more in thickness, and 8.0 in. (20.3 mm) or less in width). **E 631,** E06

bar—See **type bar.** **F 909,** F05

bar, *n*—one of two types of elements comprising a bar code symbol. An element of a bar code symbol whose reflectance is less than the Global Threshold. **F 1294,** F05

bar—a solid rolled or forged section that is long in relationship to its cross-sectional dimensions with a relatively constant cross-section throughout its length. Carbon and alloy steel bars are produced from hot rolled or cast billets, or from blooms rolled single strand into coils. **F 1789,** F16

barb, *n—as related to barbed wire*, a short length of wire, with exposed ends cut on a bias to produce sharp points. **A 902,** A05

barb, *n—in down*, the primary structure of plumage emanating from a quill point of a down cluster. **D 123,** D13

barb, *n—in feathers*, the primary structure of plumage emanating from the quill of a feather. **D 123,** D13

barbed—shank provided with repetitive, shallow or deep, symmetrical or nonsymmetrical, cross-wise or oblique, diagonal or perpendicular indentations and ridges, excluding grip marks. **F 547,** F16

barbed tape—strip of metal, machined to produce clusters of sharp points (see Terminology F 1379 for further information). **F 552,** F14

barbed tape, *n*—strip of metal, machined to produce clusters of sharp points. **F 1379,** F14

barbed wire, *n*—a fabricated wire product consisting of two line wires twisted to form a two-wire strand, into which 2-point or 4-point barbs are tightly wrapped and locked into place at specified intervals. **A 902,** A05

barbed wire arm—a post fitting used to hold barbed wire in place. May be vertical or angled to hold three or more strands of barbed wire, or "V" shaped to hold six or more strands of barbed wire. **F 552,** F14

barbed wire arm base—a casting of aluminum, malleable iron, or steel, used in combination with one or two barbed wire extension arms held in place by a bolt and nut. **F 552,** F14

barbed wire extension arm—pressed aluminum or pressed galvanized steel used in combination with barbed wire arm base to hold three strands of barbed wire either vertically or angled in or out from the fence line. Also used in pairs in a "V" configuration to hold 6 strands of barbed wire. **F 552,** F14

barb length, *n*—measured from the barb point to the center line of the barb cluster. **F 1379,** F14

barb length classification, *n*—commonly used barbed types describing length and shape; such as long barb, medium barb, and short barb. **F 1379,** F14

barb spacing, *n*—the circumferential or linear distance between the center lines of the barb clusters. **F 1379,** F14

bar code—a binary coding system consisting of vertical marks or bars that, when read by an optical scanner, can be converted to machine language. **F 149,** F05

bar code—an array of rectagular lines and spaces that are arranged in a predetermined pattern following unambiguous rules and representing data that are referred to as characters. **F 1156,** F12

bar code:

Ames—discrete, self-checking numeric bar code symbology similar to ***Codabar***

Codabar—bar code symbology with 4 bars and 3 spaces per numeric (or control) character, 16 in all; self-checking, very good accuracy rate.

Code 2 of 5—bar code, numeric with control symbols; 2 wide bars and 3 narrow ones; binary display, bars only.

Code 2 of 5 matrix—bar code with 10 digits and start/stop; similar to **Code 11**; 1-modulo, 10 check digit; subject to substitution errors.

Code 3 of 9—also known as **Code 39**; bar code symbology using 3 wide bars or spaces (2 bars, 1 space) interspersed with 6 narrow bars or spaces in an interleaved grouping of 44 alphanumeric and control characters; both bars and spaces are used to convey information; self-checking system offers substitution error rates calculated to be less than 1 in (1)0 × 10^6.

Code 11—bar code symbology for high density numeric uses with 11 characters, no self-checking; uses 3 bars and 2 spaces—one of each is wide.

Code 39—See **Code 3 of 9**

Code 93—very high density alphanumeric bar code symbology, similar to **Code 39**, continuous, not self-checking.

Code 128—128-character bar code symbology, high density, similar to UPC (**see UPC**); not self-checking; continuous; 3 bars, 3 spaces in 11-module character.

Delta Distance—bar code symbology of discrete extended 16-character numeric code using 6 narrow bars with variable width spaces between them.

EAN—bar code symbology, European Article Numbering; fully compatible with UPC (Universal Product Code), using 7 modules of 2 bars and 2 spaces each; self-checking.

Interleavened 2 of 5—bar code symbology based on **Code 2 of 5**, except that the spaces also contribute to the information content.

Nixdorf—a computer manufacturer Nixdorf's bar code symbology using 3 bars of 3 different widths and 2 spaces; likely to be superceded by the EAN code.

Plessey—proprietary bar code symbology using 4 bars and the adjacent spaces; wide bar followed by narrow spaces are 1 bits and narrow bars followed by wide spaces are 0 bits; also known as a pulse-width modulated code, or an Anker code **F 1156,** F12

bar code, *n*—an array of parallel rectangular marks and spaces in a predetermined pattern. **F 1294,** F05

bar code density, *n*—the number of characters that can be represented in a lineal inch. (See **symbol density**). **F 1294,** F05

bar code label, *n*—an adhesive-backed carrier bearing printed bar code information suitable for a affixing to an apparatus, equipment, or container surface. **F 1294,** F05

bar code reader—see **bar code scanner. F 1294,** F05

bar code scanner, *n*—a device used to identify and decode a bar code symbol. **F 1294,** F05

bar code symbol, *n*—an array of rectangular bars and spaces, arranged in a predetermined pattern, following specific rules that represent elements of data referred to as characters. **F 1294,** F05

bare conductor—a conductor having no nonmetallic covering. **B 354,** B01

barefoot, *adv or adj*—with the feet uncovered or unclothed; without shoes or stockings. **F 1646,** F13

bar height, *n*—the dimension of a symbol element perpendicular to its width. **F 1294,** F05

barium sulfate ($BaSO_4$)—a standard reflecting agent used to calibrate instruments for measuring the whiteness and reflectance of papers. **F 149,** F05

bark—the layer of a tree outside the cambium comprising the inner bark and the outer bark.

inner bark—the layer of living bark (phloem) that separates the outer bark from the cambium and which in the living tree generally is moist and soft.

outer bark—the layer of dead bark outside the inner bark, forming the exterior surface of the tree stem. The outer bark frequently is corky and dry.

bark product—an opening between annual growth rings that contains bark. Bark pockets appear as dark streaks on radial surfaces and as rounded areas on tangential surfaces. **D 9,** D07

barkometer—a hydrometer used for determining the specific gravity of tanning solutions. A specific gravity of 1.000 is equivalent of 0 deg barkometer (Bk), and each additional degree Bk is equivalent to an increase of 0.001 in specific gravity. **D 1517,** D31

bark tannage—leather tanned by use of vegetable tannins found in barks, wood, and other plant derivatives, as distinguished from mineral tannages. **D 1517,** D31

barn—a unit of area equal to 10^{-24} cm^2, used to express the probability of the occurrence of a specific nuclear interaction. **C 859,** C26

barn—a unit of area used for expressing the area of nuclear cross sections

$$1 \text{ barn} = 10^{-24} \text{ cm}^2.$$

E 1316, E07

barometric efficiency—the ratio of the change in depth to water in a well to the inverse of water-level change in barometric pressure, expressed in length of water. **D 653,** D18

bar printer—an impact printer in which the type slugs are carried on a **type bar.** *Syn.* print bar. **F 909,** F05

barré, *n*—an unintentional, repetitive visual pattern of continuous bars and stripes usually parallel to the filling of woven fabric or to the courses of circular knit fabric. (Compare **warp streaks** and **mixed filling**, see **filling blend**.) **D 123,** D13

bar reflectance (Rb), *n*—the smallest reflectance value in a bar. **F 1294,** F05

barrel—the cylindrical portion of a vitrified clay pipe exclusive of branches, spurs, joints, and handling rings or lugs. **C 896,** C04

barrel, *n*—a bulged cylindrical **container** of greater length than breadth, made of wooden **staves** bound together with hoops and having two flat ends of equal diameter. (Compare **drum.**)

cask, n— a term used synonymously with **barrel** but usually of large size or capacity.

keg, n—a small slack or tight **barrel** of 30-gal capacity or less.

barrier material:

grease-resistant barrier—a material that prevents or retards the transmission of grease or oils.

water-resistant barrier—a material that retards the transmission of liquid water.

water-vapor-resistant barrier—a material that retards the transmission of water vapor. **D 996,** D10

barrel burnishing—the smoothing of surfaces by means of tumbling the work in rotating barrels in the presence of metallic or ceramic shot, and in the absence of abrasive. In ball burnishing, the shot consists of hardened steel balls. **B 374,** B08

barrel distortion— See **distortion. E 7,** E04

barrel electroplating—an electroplating process in which electrodeposits are applied to articles in bulk in a rotating, oscillating, or otherwise moving container. **B 374,** B08

barrel finishing (or tumbling)—bulk processing in barrels, in either the presence or absence of abrasives or burnishing shot, for the purpose of improving the surface finish. **B 374,** B08

barrel-galvanized— See **hot-galvanized. F 547,** F16

barrel, glass container—{archaic} See **sidewall, glass container**. **C 162,** C14

barrel nail—bright or coated, regular-stock-steel, ⅝ by 0.067 to 1½ by 0.092-in. nails with flat 0.148 to 0.219-in. head and medium diamond point. **F 547,** F16

barrel plating (or cleaning)—plating or cleaning in which the work is processed in bulk in a rotating container. **B 374,** B08

barrel processing—mechanical, chemical, cleaning, or electrolytic treatment of articles in bulk or in a rotating, oscillating, or otherwise moving container. **B 374,** B08

barrier—any material limiting passage through itself of solids, liquids, semisolids, gases, or forms of energy such as ultraviolet light. **B 713,** B01

barrier—any material limiting passage through itself of solids, liquids, semisolids, gases, vapors, or forms of energy such as ultraviolet light. **F 17,** F02

barrier—See Terminology F 17. **F 1327,** F02

barrier layer—*in anodizing aluminum*, the thin, pore-free, semiconducting aluminum oxide region nearest the metal surface and distinct from the main anodic oxide coating which has a pore structure. **B 374,** B08

barrier materials—specialized porous or nonporous packaging materials that provide environmental protection to the package contents as well as protection to the environment from the package contents: (*1*) gas, vapor, humidity, liquid, microbial, or light resistant materials that control or eliminate the amount of those environmental constituents that pass into or out of a package; (*2*) a porous material preventing the passage of microorganisms that might contaminate the contents of the package. **F 1327,** F02

bar-size section—a hot-rolled steel angle, channel, tee, or zee having a maximum cross-section dimension of less than (76 mm) (3.0 in.) **E 631,** E06

bartack, *n—in garment construction*, a reinforcement at points of strain. **F 1494,** F23

bar width, *n*—the thickness of a bar measured from the edge closest to the symbol start character to the trailing edge of the same bar. **F 1294,** F05

bar width ratio, *n—in symbologies having two widths*, the ratio of the widest to the narrowest width within the bar code symbol. (See also **wide to narrow ratio.**) **F 1294,** F05

bar width reduction, *n*—a diminution of the nominal bar width dimension on film masters or printing plates to compensate for systematic errors in some printing processes. **F 1294,** F05

basal application—application of a chemical to the base of a plant. **E 1102,** E35

basal plane—that plane of a hexagonal or tetragonal crystal which is perpendicular to the axis of highest symmetry. Its Miller indices are (0001) or (001), respectively. **E 7,** E04

basalt ware—a black unglazed vitreous ceramic ware having the appearance of basalt rock. **C 242,** C21

base—{archaic} the bottom of a bottle. **C 162,** C14

base—*in grouting*, main component in a grout system. **D 653,** D18

baseball leather—leather used for covers of baseballs. The better grades of balls have covers of alum-tanned horsehide front leather. Some cheaper grades are made of kip and sheepskins. **D 1517,** D31

baseboard, *n*—a molding covering the juncture of a wall and the adjoining floor. **E 1605,** E06

baseboard application—application to a building on the lower portion of the inside walls. **E 1102,** E35

base building, *n* (immeuble de base)—a general-purpose office building intended, but not yet adapted, to suit the operational requirements of a specific tenant. **E 1480,** E06

base coat, *n*—the first layer or layers of plaster applied over a lath or other substrate. The first application is normally called a scratch coat and the second application is referred to as a brown coat. **C 11,** C11

base coat—for two coat-one fire application, the thin layer of bonding frit applied first and used to promote adherence after firing to the metal substrate. **C 286,** B08

basecoat, *n*—a coating applied to a base substrate to produce a level surface for the application of the active coat; may also function to improve the heat/energy transfer of the active coat layer. **F 1623,** F05

base coat floating, *n*—the finishing act of spreading, compacting, and smoothing of the base coat plaster to a reasonably true plane. **C 11,** C11

base conditions—temperature and pressure conditions at which natural gas volumes are determined for purposes of custody transfer. In natural gas measurement the properties of interest are temperature, pressure, and composition. Assuming ideal gas properties, for simplicity, tables of pure compounds can be prepared for use in calculating gas properties for any composition at "base conditions." These "base conditions" are chosen near ambient. **D 4150,** D03

base course (base)—a layer of specified or selected material of planned thickness constructed on the subgrade or subbase for the purpose of serving one or more functions such as distributing load, providing drainage, minimizing frost action, etc. **D 653,** D18

base data, *n*—the unedited and unadjusted measurements of traffic volume, vehicle classification, and vehicle or axle weight. **E 867,** E17

base date—See **base time.** **E 833,** E06

base exchange, *n*—a surface property exhibited by collodial inorganic materials, usually clays, whereby absorbed surface cations are replaced by other cations. **C 1145,** C28

base exchange—the physicochemical process whereby one species of ions adsorbed on soil particles is replaced by another species. **D 653,** D18

base flow—stream flow that is sustained by ground water and other delayed sources. **D 4410,** D19

baselength, *n*—the length of a segment of a pavement macrotexture profile being analyzed required to be 100 mm. **E 867,** E17

baseline—any line drawn on an absorption spectrum to establish a reference point representing a function of the radiant power incident on a sample at a given wavelength. **E 131,** E13

base line—the time of flight or distance trace (horizontal) across the A-scan CRT display (for no signal condition). **E 1316,** E07

base line—a reference line used to specify the nominal relative vertical position of OCR characters printed on the same line. **F 149,** F05

basement—See **building space.** **E 631,** E06

basement—See **space categories.** **E 1480,** E06

base metal, *n—as related to metallic-coated steel*, the steel to which the coating is applied, as distinguished from the coating metal. **A 902,** A05

base metal—(*1*) See **basis metal**; (*2*) *in diffusion coatings*, the metal present in the largest proportion in an alloy. **B 374,** B08

base metal—the sheet or plate from which the pipe is formed. **B 846,** B05

base metal—the metal to which porcelain enamel is applied. **C 286,** B08

base paper, *n*—the fiber network existent prior to the application of any material onto the surface of that fiber network. **D 1968,** D06

base ply—the bottom or first ply in a built-up roofing membrane when additional plies are to be subsequently installed. **D 1079,** D08

base sheet—a product intended to be used as a base ply in a built-up roofing system. **D 1079,** D08

base side, *adj—when referring to a film master*, the nonimage bearing side (See also the reverse, **emulsion side.**) **F 1294,** F05

base stock—see **conductive base stock.** **F 335,** F05

base time—the date to which all future and past benefits and costs are converted when a present value method is used (usually the beginning of the study period) (Syn. *base date*). **E 833,** E06

base weight—the weight of a specified area of paper stock expressed in grams per square metre or in pounds per ream, where the size and number of sheets per ream may vary for each particular grade of paper. Common ream sizes used in electrostatic papers are:

25 by 38 in. by 500 sheets (3300 ft^2)
24 by 36 in. by 500 sheets (3000 ft^2)
17 by 22 in. by 500 sheets (1300 ft^2) **F 335,** F05

basic color terms, *n*—a group of eleven color names found in anthropological surveys to be in wide use in fully developed languages: white, black, red, green, yellow, blue, brown, gray, orange, purple, pink. **E 284,** E12

basic fiber—unprocessed glass fibers directly from the forming equipment. **C 162,** C14

basic hydrolysis—the chemical addition of water to a compound. **E 1705,** E48

basic life support—medically accepted non-invasive procedures used to sustain life. **F 1177,** F30

basic life support/cardiopulmonary resuscitation (BLS/CPR)—a set of skills that includes airway management, chest compressions, and others as defined by the American Heart Association. **F 1177,** F30

basic NMR frequency, ν_0—the frequency, measured in hertz (Hz), of the oscillating magnetic field applied to induce transitions between nuclear magnetic energy levels. **E 131,** E13

basic refractories, n— see **refractories, basic.** **C 71,** C08

basic stress *(archaic)*—the term once used for an allowable property for clear, straight-grained lumber. **D 9,** D07

basil—uncolored vegetable-tanned sheepskin and lambskin. **D 1517,** D31

basis—See **reporting bases**. **D 121,** D05

basis metal—see **base metal.** **C 286,** B08

basis metal (or material)—material upon which coatings are deposited. **B 374,** B08

basis weight—deprecated term (do not use in the sense of mass per unit area). **D 4439,** D35

basis weight—the weight in pounds of a ream cut to a specified basic size. The number of sheets in a ream is usually 500. The basic size for writing papers commonly used in OCR applications is 17 by 22 in. Also measured metrically in grams per square metre (g/m^2) and referred to as grammage. **F 149,** F05

basket, *n*—a semirigid **container** usually open at the top and provided with one or two handles for carrying. (Compare **hamper.**) **D 996,** D10

basket nail—bright, regular-stock-steel, ⅝and ¾ by 0.048-in. nails with large flat head and medium needle point. **F 547,** F16

basket, pickle—see **pickle basket** and **pickling.** **C 286,** B08

bast and leaf fiber, *n*—fiber derived from the inner fibrous bark and the hard coarse leaves of dicotyledonous plants such as flax, hemp, jute, and abaca. **D 123,** D13

bat, *v*—to reject or discard a brick or shape. **C 71,** C08

BAT—best available technology. **D 6161,** D19

batch, *n*—the component raw materials properly weighed, proportioned, and mixed for delivery to a processing unit. Also, the product output from a processing unit in which there is essentially no product output until all component materials are charged and processed. **A 644,** A04

batch—(*1*) the recipe of batch ingredients.
(*2*) the raw materials weighed but unmixed.
(*3*) the raw materials, properly proportioned and mixed, for delivery to the furnace. **C 162,** C14

batch—a quantity of material made in a production sequence that is expected to be uniform in isotopic, physical, and chemical characteristics. **C 859,** C26

batch—the total quantity of a material produced in a single final mixing operation after all production processes are complete, or just prior to filling. **D 16,** D01

batch—*in grouting*, quantity of grout mixed at one time. **D 653,** D18

batch, *n*—the manufactured unit or a blend of two or more units of the same formulation and processing. (Compare **manufactured unit.**) **D 907,** D14

batch, *n*—the product of one mixing operation. **D 1566,** D11

batch, *n*—a definite quantity of some product or material produced under conditions that are considered uniform.

NOTE—A batch is usually smaller than a lot. **E 456,** E11

batch—the quantity of material that has been formulated in a single continuous operation and subjected to chemical processing or physical mixing to produce a homogeneous material. **E 1749,** E06

batch average thickness—the calculated average thickness of a coating if it were uniformly distributed on the surfaces of the items. **F 1789,** F16

batch charger—a mechanical device for introducing batch to the melter. **C 162,** C14

batch coating, *n*— *of metallic coated steel products,* the process of discontinuous-sequential passage of steel articles through the various steps of the coating process, such as, cleaning, pickling, fluxing, and coating. **A 902,** A05

batch feeder—See **batch charger**. **C 162,** C14

batch fermentation—batch of nutrient mixture and microorganisms mixed in a vessel and allowed to ferment. **E 1705,** E48

batch house—the place where batch materials are received, handled, weighed, and mixed. **C 162,** C14

batch method—*in grouting*, a quantity of grout materials are mixed or catalyzed at one time prior to injection. **D 653,** D18

batch mixer—*in grouting*, a machine that mixes batches of grout, in contrast to a continuous mixer. **D 653,** D18

batch sampling—sampling over some time period in such a way as to produce a single test sample for analysis. **E 176,** E05

batch smelter—any smelter that operates as a periodic unit, being charged, fired, and discharged according to a predetermined cycle. **C 286,** B08

bate—to treat unhaired hides or skins with a warm aqueous solution of an enzyme in order to remove certain undesirable nitrogenous constituents. **D 1517,** D31

bath—synonymous with **float bath**. **C 162,** C14

bath—see **suspension.** **E 1316,** E07

bath mat, *n*—an absorbent textile floor covering normally used in the bathroom as a pad on which to step when getting out of the tub or shower. **D 123,** D13

bathochromic shift, *n*—change of a spectral band to longer wavelength (lower frequency) because of structural modifications or environmental influence; also known as "red shift." **E 131,** E13

bathroom—See **building space.** **E 631,** E06

bathroom soil—the soil composed of water insoluble, or practically insoluble, materials or a mixture of these materials, present on typical bathroom surfaces other than those of floors and toilets. A major component of this soil is the insoluble precipitate, commonly referred to as "soap scum," that is deposited when soap is used in hard water. **D 459,** D12

bath rug, *n*—a scatter rug used in the bathroom. **D 123,** D13
bath sheet, *n*—a textile terry product with end hems or fringes and side hems or selvages that is generally much larger than a bath towel. **D 123,** D13
bath towel, *n*—a textile terry product with end hems or fringes and side hems or selvages, which is used to dry a person's body after bathing or swimming. **D 123,** D13
batten—a long narrow strip of asbestos-cement, either flat or corrugated, used to conceal the joints in butt joint application of flat or corrugated sheets. **C 460,** C17
batten, *n*—a long narrow strip, either flat or corrugated, used to conceal the joints in butt joint application of flat or corrugated sheets. **C 1154,** C17
batten—a narrow strip of lumber for covering the adjoining edges of roofing or siding, often designated as "batts." **D 9,** D07
batten—See **box.** **D 996,** D10
batten—long narrow strip of asbestos-cement, either flat or corrugated, used to conceal the joints in butt joint application of flat or corrugated sheets. **D 2946,** C17
batten lugs, *n*—protrusions on the underside of tile designed to engage over the upper edge of tiling battens. **C 43,** C15
batting, *n*—a textile filling material consisting of a continuous web of fibers formed by carding, garnetting, air laying, or other means. **D 123,** D13
batting integrity, *n*—the ability of a textile filling material to resist distortion or change when subjected to multiple home launderings or drycleanings. **D 123,** D13
baud, *n*—a measure of data transmission speed; ideally one baud equals one bit per second. **F 1457,** F05
Baumé gravity—a unit of density based on specific gravity and defined by the following equation:

$$\text{Baumé gravity} = 145 - [145/\text{sp gr}] \text{ at } 15.5/15.5°C\ (60/60°F)$$

See also **density** and **specific gravity**. **E 1547,** E15
baume scale,°Be—a measure of the density of a solution relative to water.

$$°BE = 145 - \frac{145}{\text{specific gravity*}}$$

United States for densities greater than unity.

$$°BE = \frac{140}{\text{specific gravity*}} - 130$$

For densities less than unity.
*60°F/60°F **D 6161,** D19
Bayard-Alpert ionization gage—see **ionization vacuum gage.** **E 1316,** E07
BCD—binary coded decimal. **F 1156,** F12
BCD—abbreviation for **block circuit diagram**. **F 1457,** F05
bead—(*1*) an enlarged, rounded raised section on a glass article.
(*2*) a small piece of glass tubing bonded around a wire lead.
(*3*) in fiber glass, a tear drop-shaped glass mass which forms as a result of the interruption of the fiber forming process below an orifice. **C 162,** C14
bead, *n*—*in building construction*, in sealants and glazing, a strip of applied sealant, glazing compound, or putty. **C 717,** C24
bead, *n*—*in a separate element zipper*, an enlarged section on inner edge of each tape to which interlockable elements are affixed. **D 123,** D13
bead, *n*—*in continuous element zippers*, an optional enlarged section of the tape located at the outer edge of the continuous interlockable elements and against which the slider flanges bear. **D 123,** D13
bead—See **windows and doors.** **E 631,** E06
bead, *n*—*of a tire*, the part of a tire that comes in contact with the rim and is shaped to secure the tire to the rim. **F 538,** F09
beadboard, *n*—molded **expanded polystyrene thermal insulation board;** also called **MEPS.** **E 631,** E06
bead carryover—beads of developer appearing on a print or copy, or both; a deletion around each particle is common. **F 1457,** F05
beading—(*1*) the application of porcelain enamel, usually of a contrasting color, to the edge or rim of porcelain enameled articles.

(*2*) Removal of excess slip from the edge of dipped ware.

(*3*) In dry processing enameling, a bead of porcelain enamel along the edge of ware. **C 286,** B08
beading enamel—any of the special porcelain enamels used for beading. **C 286,** B08
bead separation, *n*—a breakdown of bond between tire components in the bead area. **F 538,** F09
beam, *n*—*in textiles*, a large spool containing many ends of yarn wound parallel, and used for such purposes as weaving or warp knitting. **D 123,** D13
beam—a convex wooden slab sloping downward from about waist height over which a hide is placed for trimming off excess flesh and ragged edges and scudding by hand. **D 1517,** D31
beam, *n*—*in optics*, a concentrated unidirectional flow of radiant energy. **E 284,** E12
beam, *n*—a structural member intended primarily to resist transverse forces, and subject to bending by these forces. **E 631,** E06
beam:
analyzing—same as *incident.*
current—the total current incident on the specimen by the primary particle source.
current density—the current incident on the specimen per unit area.
diameter—*in surface analysis,* the full width of the incident beam at half maximum intensity measured in a plane normal to the beam direction. This plane must be specified and is often taken at the intersection of the beam center with the specimen.
divergence, convergence—angles spanned by the directions of all particles of the incident beam.
energy—the energy of the particles incident on the specimen surface, expressed in electron volts (eV).
energy, primary—the kinetic energy of the primary beam, usually expressed in kiloelectronvolts (keV).
incident—the energetic particles incident on the specimen.
particle—atomic or molecular species contained in the incident beam, regardless of state of ionization.
primary—a directed flux of particles (ions or neutrals) incident on the specimen.
profile, primary ion—the spatial distribution of the primary ion current in a plane perpendicular to the primary ion beam axis.
size—the full width at half-maximum of the beam at a given point in space that must be defined.
spot—the area on the specimen surface illuminated by the incident beam. **E 673,** E42
beam axis—the acoustic centerline of a search unit's beam pattern ac defined by the locus of points of maximum sound pressure in the far field, and its extension into the near field. **E 1316,** E07
beam loading—the application of a load to a pipe between two points of support, usually expressed in newtons (or pounds-force) and the distance between the centers of the supports. **F 412,** F17
beam, radiant energy—a collection of rays confined to a specific path. **E 772,** E44
beam ratio—the measured intensity of the reference beam divided by the measured intensity of the object beam in the plane of the

recording medium. **E 1316,** E07

beam set, *n—in textiles,* one or more beams of yarn in a single shipment to be further processed together for a specific end use. **D 123,** D13

beam shear—a term describing the stresses developed in planes parallel to facing planes of flat sandwich constructions when subjected to flatwise flexure in such a manner that the applied moments produce curvature of the plane of a sheet of the sandwich construction (see Test Method C 393). **E 1749,** E06

beamsplitter—a semireflecting device used to create, and often to recombine, spatially separate beams. **E 131,** E13

beamsplitter—an optical device for dividing a beam into two or more separate beams. **E 1316,** E07

beamsplitter efficiency—the product $4RT$, where R is the reflectance and T is the transmittance of the beamsplitter. **E 131,** E13

beam spread—a divergence of the ultrasonic beam as the sound travels through a medium. **E 1316,** E07

bearer arch—See **rider arch**. **C 162,** C14

bearer bars, *n*—solid lines that are located at the perimeter of some bar code symbols whose purpose is to reduce partial, incorrect scans and support the printing plate. Bearer bars do not contain information. **F 1294,** F05

bearing area—for chemical-resistant polymer machinery grout, the portion of the grout surface in contact with the underside of a supported surface. **C 904,** C03

bearing area [L^2], *n*—the product of the pin diameter and specimen thickness. **E 6,** E28

bearing capacity—see **ultimate bearing capacity**. **D 653,** D18

bearing capacity (of a pile), Q_p, P_p **(F)**—the load per pile required to produce a condition of failure. **D 653,** D18

bearing load [F], *n*—a compressive load on an interface. **E 6,** E28

bearing strain, *n*—the ratio of the bearing deformation of the bearing hole, in the direction of the applied force, to the pin diameter. **E 6,** E28

bearing strength—the non-destructive limit of pipe load, as determined by 3-edge bearing test method, used to determine field supporting strength. **C 896,** C04

bearing strength [FL^{-2}], *n*—the maximum bearing stress which a material is capable of sustaining. **E 6,** E28

bearing stress [FL^{-2}], *n*—the force per unit of bearing area. **E 6,** E28

bearing surface—the outside surface of an item of glassware on which it rests when in its normal upright position. **C 162,** C14

bearing surface—underside of head in contact with nailed member. **F 547,** F16

bearing wall—See **wall.** **E 631,** E06

bearing yield strength [FL^{-2}], *n*—the bearing stress at which a material exhibits a specified limiting deviation from the proportionality of bearing stress to bearing strain. **E 6,** E28

bed, *v—in building construction in glazing,* to apply a bead of sealant between a lite of glass or a panel and the stationary stops or sight bars of the sash or frame. **C 717,** C24

bed, n—See **bedding.** **C 717,** C24

bedcovering, *n*—a textile product used on a bed over the sheets for warmth or decoration. **D 123,** D13

bed depth—the depth of the filter medium or ion exchange resin in a vessel. **D 6161,** D19

bedding, *n—in building construction in glazing,* a bead of sealant applied between a lite of glass or a panel and the stationary stop or sight bar of the sash or frame. It is usually the first bead to be applied when setting glass or panels. **C 717,** C24

bedding—the materials, their placement, consolidation, and configuration, as designed to support, and to develop field supporting strength of vitrified clay pipe. **C 896,** C04

bedding—applies to rocks resulting from consolidation of sediments and exhibiting surfaces of separation (bedding planes) between layers of the same or different materials, that is, shale, siltstone, sandstone, limestone, etc. (ISRM) **D 653,** D18

bedding—collective term signifying the existence of layers of beds. Planes or other surfaces dividing sedimentary rocks of the same or different lithology. **D 653,** D18

bedding, *n*—materials that provide support for buried pipe. **F 412,** F17

bedding, *v*—placement of support materials for buried pipe. **F 412,** F17

bedding coat, *n*—that coat of plaster to receive aggregate or other decorative material of any size, impinged or embedded into its surface, before it sets. **C 11,** C11

bed expansion—the depth increase of filter medium or ion exchange resin that occurs during backwashing. **D 6161,** D19

bed joint—a horizontal mortar joint between a chemical-resistant construction unit and a substrate or another chemical-resistant construction unit. **C 904,** C03

bed-load—material moving on or near the stream bed by rolling, sliding, and skipping. **D 4410,** D19

bed-load discharge—the quantity of bed-load passing a cross section of a stream in a unit of time. **D 4410,** D19

bed-load sampler—a device for sampling the bed-load. **D 4410,** D19

bed material—the sediment mixture of which the stream bed is composed. **D 4410,** D19

bed-material discharge—that part of the total sediment discharge composed of grain sizes occurring in appreciable quantities in the bed material. **D 4410,** D19

bed-material load—that part of the total load which is composed of particle sizes present in appreciable quantities in the shifting portions of the stream bed. **D 4410,** D19

bed moisture—synonym for *inherent moisture.* **D 121,** D05

bedrock—the more or less continuous body of rock which underlies the overburden soils. (ISRM) **D 653,** D18

bedrock (ledge)—rock of relatively great thickness and extent in its native location. **D 653,** D18

bedspread, *n*—a type of bedcovering that is placed over the blankets and sheets for appearance and warmth. **D 123,** D13

bed surface, *n*—the bed surface of the brick is the in situ non-vertical surface intended to be joined by mortar. **C 43,** C15

beehive coke—See *beehive coke* under **coke**. **D 121,** D05

beer—term used to describe the product of ethanol fermentation by microorganisms. **E 1705,** E48

beer-case cleat nail—bright, regular-stock-steel, 1 by 0.080 to 1⅞ by 0.106-in. nails with 3⁄16 to ¼-in. oval head and duckbill or clinch point. **F 547,** F16

beer-case hinge, lock and latch nail—bright, regular-stock-steel, ⅝ by 0.080 to 1 5⁄16 by 0.092-in. nails with oval 3⁄16 to 7⁄32-in. head and duckbill or clinch point. **F 547,** F16

beer-case strap nail—bright, regular-stock-steel, 1¼ by 0.092 to 2 by 0.113-in. nails with oval 15⁄64 to -in. head and medium diamond point. **F 547,** F16

Beer's law—the absorbance of a homogeneous sample containing an absorbing substance is directly proportional to the concentration of the absorbing substance. See also **absorptivity.** **E 131,** E13

Beer's law, *n*—the absorbance of a homogeneous sample containing an absorbing substance is directly proportional to the concentration of the absorbing substance. See also **absorptivity**. **E 284,** E12

behavior, *n*—observable, recordable, or measurable actions or activity of an organism. **E 943,** E47

Beilby layer, *n*—an altered surface layer of supposedly amorphous material formed on a crystalline solid during mechanical polishing, whose existence was proposed in Sir George Beilby's writings. The existence of such a layer is not supported by recent research, and the use of this term is therefore considered archaic and is strongly discouraged. **G 40,** G02

belayer, *n*—the one who performs the belay. **F 1490,** F32

bell—see **female end of pipe.** **C 822,** C13

bell—the flared-end portion of a vitrified clay pipe or fitting, designed to function in the joining of other such pipe. **C 896,** C04

bell-and-spigot joint—see **joint, bell-and-spigot gasket.** **F 412,** F17

Belleek china—a highly translucent whiteware composed of a body containing a significant amount of frit and normally having a luster glaze. **C 242,** C21

bell end—the enlarged portion of a pipe that resembles the socket portion of a fitting and that is intended to be used to make a joint. **F 412,** F17

bell jar—a container, open at one end (usually the bottom), which is used as a vacuum chamber or test vessel. **E 1316,** E07

bell jar testing—a test used for detecting leakage from an object completely or partially filled with a tracer gas and placed in a vacuum chamber or bell jar. **E 1316,** E07

bellows length—the distance from the eyepiece to the photosensitive material or viewing screen in a photomicrographic apparatus. **E 7,** E04

belly—that part of the hide below the belly line. For steerhide leather, the belly line (RU) passes through a point at or above the top of the rear break. For cowhide leather, the belly line passes through a point at or above the top of the front break and a point not more than 2½ in. (64 mm) below the top of the rear break. **D 1517,** D31

below-grade, *adj*—below the surface of the ground, *as related to floor location*, part or all of the floor is below the ground. **F 141,** F06

belt, *n*—*in a tire*, a breaker that substantially restricts the carcass in a circumferential direction. **F 538,** F09

belting butt—a double back with the tail cut off at the butt line. **D 1517,** D31

belting butt bend—a double bend with the tail cut off at the butt line. **D 1517,** D31

beltmarks—See **chain marks**. **C 162,** C14

belt printer—a type of printer that employs an endless belt, the outer surface of which contains fully formed characters. The belt is caused to transverse the width of the paper form. Hammers are caused to strike the back of the form, synchronized in such a way as to produce the desired characters in their proper positions. **F 909,** F05

belt separation, *n*—a breakdown of bonding between the belts or plies or tread, or combination thereof. **F 538,** F09

bench—See **siege**. **C 162,** C14

bench—(*1*) the unexcavated rock having a nearly horizontal surface which remains after a top heading has been excavated, or (*2*) step in a slope; formed by a horizontal surface and a surface inclined at a steeper angle than that of the entire slope. (ISRM) **D 653,** D18

benchmark neutron field—a well-characterized neutron field which will provide a fluence of neutrons for validation or calibration of experimental techniques and methods and for validation of cross sections and other nuclear data. The following classification of benchmark neutron fields for reactor dosimetry has been made:

controlled neutron field—a neutron field physically well-defined, and with some spectrum definition, employed for a restricted set of validation experiments.

reference neutron field—a permanent and reproducible neutron field less well characterized than a standard field but accepted as a measurement reference by a community of users.

standard neutron field—a permanent and reproducible neutron field with neutron fluence rate and energy spectra, and their associated spatial and angular distributions characterized to state-of-the-art accuracy. Important field quantities must be verified by interlaboratory measurements and calculations. **E 170,** E10

bench marks, *n*—marks placed on a specimen to define gage length, that is, the portion of the specimen that will be evaluated in a specific test. **D 123,** D13

bench marks, *n*—two marks of known separation applied to a specimen to measure the strain of the specimen during extension. **D 1566,** D11

bench test—a modified service test in which the service conditions are approximated, but the equipment is laboratory equipment and not necessarily identical with that in which the product will be employed. **F 869,** F08

bend—a back with the shoulder cut off at right angles to the backbone line at the break of the fore flank. **D 1517,** D31

bend, *vt*—*in mechanics*, to force an object from its natural or manufactured shape into a curve or into increased curvature. **D 4439,** D35

bend—a fitting either molded separately or formed from pipe for the purpose of accommodating a directional change. **F 412,** F17

bendable bolts—bolts furnished with an altered section at some location at which the bolt will bend. **F 1789,** F16

bend angle—180° minus the internal angle created by applying a load on the weld joining two sheets of plastic. **C 904,** C03

bending—process of deformation normal to the axis of an elongated structural member when a moment is applied normal to its long axis. (ISRM) **D 653,** D18

bending length, *n*—*(1) general*—a measure of the interaction between fabric weight and fabric stiffness as shown by the way in which a fabric bends under its own weight. It reflects the stiffness of a fabric when bent in one plane under the force of gravity, and is one component of drape.

(2) specific—the cube root of the ratio of the flexural rigidity to the weight per unit area. **D 123,** D13

bending moment, *n*— *of paper*, the work (force multiplied by the distance over which it is applied) required to deflect the test piece under specified conditions. **D 1968,** D06

bending stiffness, n—the sandwich property which resists bending deflections. D=EI; the facing modulus times the panel moment of inertia. **C 274,** D30

bending stress—a stress system that simultaneously imposes a compressive component at one surface, graduating to an imposed tensile component at the opposite surface of a glass section. **C 162,** C14

bend test, *n*—a test for ductility (Note 1) performed by bending or folding a specimen, usually by steadily applied forces but in some instances by blows. The bending may be interrupted to examine the bent surface for cracks. **E 6,** E28

bend test—various tests in which a fastener is bent through its axis or on a round mandrel to determine the toughness and ductility of the fastener. **F 1789,** F16

benefit-cost analysis—a method of evaluating projects or investments by comparing the present value or annual value of expected benefits to the present value or annual value of expected cost. **E 631,** E06

benefit-to-cost ratio (BCR)—benefits divided by costs, where both are discounted to a present value or equivalent uniform annual value (Syn. *benefit-cost ratio*). **E 833,** E06

bent glass—flat glass that has been shaped while hot into a body having curved surfaces. **C 162,** C14

bent head—wire bent and upset to form head. (See **hook head**.) **F 547,** F16

bentonite—a distinct type of fine-grained clay containing not less than 85 % montmorillionite clay having the formula $(OH)_4Si_8Al_4O_{20}nH_2O$ and composed of units made up of two silica tetrahedral sheets with a central alumina octahedral sheet. **C 242,** C21

bentonitic clay—a clay with a high content of the mineral montmorillonite, usually characterized by high swelling on wetting. **D 653,** D18

benzene—cyclohexatriene, benzol (obsolete) (C_6H_6) mol weight 78.11; clear, colorless, highly flammable liquid; characteristic odor; solidification point +5.5°C; boiling point 80.1°C. **D 4790,** D16

benzene, carbon disulfide-free—benzene treated with alcoholic sodium hydroxide and used as a spectrophotometric reference standard. **D 4790,** D16

benzene-535, refined—benzene with impurities limited to trace amounts by a solidification point of 5.35°C and having a total distillation range of no more than 1.0°C. Refer to Specification D 2359 for complete specifications. **D 4790,** D16

benzene-545, refined—benzene with impurities limited to trace amounts by a solidification point of 5.45°C and having a total distillation range of no more than 1°C. Refer to Specification D 4734. **D 4790,** D16

benzene-485, refined (nitration grade)—benzene with impurities limited by a solidification point of 4.85°C and having a total distillation range of no more than 1.0°C. Refer to Specification D 835 for complete specifications. **D 4790,** D16

benzene, thiophene-free—benzene refined by special treatment and used as a reagent in ASTM standards. **D 4790,** D16

Berglund-Liu atomizer—a vibratory atomizer in which a piezoelectric transducer transmits high-frequency oscillations to a liquid stream discharged through an orifice, creating relatively uniform drops whose size is a function of the frequency of oscillation and the flow rate of the liquid through the orifice. **E 1620,** E29

berm—a shelf that breaks the continuity of a slope. **D 653,** D18

berry-box nail—bright, regular-stock-steel, ¾ to 1¼ by 0.054 and 0.062-in. nails with flat and medium diamond or needle point. **F 547,** F16

Bertrand lens—an auxiliary removable lens in the body of a microscope, used to examine images in the back focal plane of the objective, for example, interference figures with polarized light. **E 7,** E04

Bertrand lens—see **lens, Bertrand.** **E 175,** E41

beryllium oxide (beryllia) (BeO)—an inorganic material of exceptionally high thermal conductivity which is toxic in the powder form. **C 242,** C21

beta activity—the spontaneous emission of electrons from a nucleus. **C 859,** C26

beta-cellulose—(*1*) Historically, a term used to indicate impurities of moderate chain lengths found in pulps, predominately degraded cellulose. (*2*) Beta-cellulose content, as measured by TAPPI Method T 203 om-93, is the pulp fraction soluble in caustic, which precipitates upon acidification. **D 1695,** D01

beta (β) loss peak—*in dynamic mechanical measurement*, second discrete peak in damping curve below the melt, in order of decreasing temperature or increasing frequency, (D 4092, D-20). **E 1142,** E37

beta (β) loss peak (*in dynamic mechanical measurement*)—the second peak in the damping curve below the melt, in order of decreasing temperature or increasing frequency. **D 4092,** D20

beta structure—structurally analogous body-centered cubic phases (similar to beta brass), or electron compounds, that have ratios of 3 valence electrons to 2 atoms. **E 7,** E04

betatron—an electron accelerator in which acceleration is provided by a special magnetic field constraining the electrons to a circular orbit. This type of equipment usually operates at energies between 10 and 31 MEV. **E 1316,** E07

bevel angle, α—the smaller of the angles between the wafer surface and the section plane. **F 1241,** F01

beveled bead—a bead of glazing compound having a sloped surface to drain water away from the glass or panel. **C 717,** C24

beveled pipe—a pipe with an end angled to mate with a complimentary pipe end or adjust to another surface. **C 896,** C04

beveled pipe—a pipe with an end chamfered to mate or adjust to another surface or to assist in assembly. **F 412,** F17

beveled square point—extremity of nail shank opposite head sheared obliquely to shank axis. (See **sheared-bevel point.)** **F 547,** F16

beveling—the process of edge finishing flat glass to a bevel angle. **C 162,** C14

bevel point—point sheared obliquely to staple-leg axis, with beveled face across staple-leg end; used to produce an outward clinch or to provide additional penetration, or both, in thin stapling member. **F 592,** F16

"B" horizon—see **horizon**. **D 653,** D18

bias—a constant or systematic error, as opposed to a random error, manifesting itself as a persistent positive or negative deviation of the method average from the accepted reference value. **C 242,** C21

bias—a persistent positive or negative deviation of the method average from the correct value or accepted reference value. (Also referred to as "constant" or "systematic error.") **C 859,** C26

bias, *n*—the difference between the population mean of the test results and an accepted reference value. **D 121,** D05

bias, *n*—*in statistics*, a constant or systematic error in test results. **D 123,** D13

bias—the persistent positive or negative deviation of the method average value from the assumed or accepted true value. **D 1129,** D19

bias, *n*—a systematic (nonrandom) deviation of the method average value or the measured value from an accepted value.

laboratory bias, n—systematic differences between the true value and a value reported by a laboratory due to errors of application such as losses, contamination, miscalibration, and faulty manipulations, for example.

method bias, n—systematic departures of the limiting mean from the true value of the parameter measured caused by physical or chemical phenomena inherent in the methodology. **D 1356,** D22

bias—a persistent positive or negative deviation of a test method average value from the assumed or accepted true value. **D 4790,** D16

bias—a systematic error that contributes to the difference between a population mean of the measurements or test results and an accepted or reference value. **E 131,** E13

bias, *n*—systematic error manifested as a persistent positive or negative deviation of the method average from its accepted true value. **E 253,** E18

bias, *n*—a systematic difference between the sample mean of the measurements or test results and an accepted reference value. **E 284,** E12

bias, *n*—the scatter between the mean values of subsets of data, from each other or from the accepted value. **E 344,** E20

bias, *n*—the difference between the expectation of the test results and an accepted reference value.

NOTE—Bias is the total systematic error as contrasted to random error. There may be one or more systematic error components contributing to the bias. A larger systematic difference from the accepted reference value is reflected by a larger bias value. **E 456,** E11

bias—a systematic error that is consistently negative or consistently positive. The mean of errors resulting from a series of observations which does not tend toward zero. **E 856,** D34

bias—a constant or systematic error as opposed to a random error. It manifests itself as a persistent positive or negative deviation of the method average from the accepted reference value. See also **accuracy**. **E 1547,** E15

bias, *n*—systematic error that contributes to the difference between a population mean of the measurements or test results and an accepted reference or true value. **E 1605,** E06

bias, *n*—the difference between the average measured test result and the accepted reference value; it measures in an inverse manner the accuracy of a test. **F 538,** F09

biased gun—an electron gun in which there is a bias voltage on the cathode cap. (See also **self-biased gun.**) **E 7,** E04

biaxial compression—compression caused by the application of normal stresses in two perpendicular directions. (ISRM) **D 653,** D18

biaxial state of stress—state of stress in which one of the three principal stresses is zero. (ISRM) **D 653,** D18

bicellular sealant backing, *n*—a flexible and very compressible extruded shape (usually cylindrical), with a surface skin, that is composed of both open and closed cell material. **C 717,** C24

bicheroux process—{archaic} an intermittent process for making plate glass, in which the glass is cast between rolls, onto driven conveyer rolls, or a flat moving table. **C 162,** C14

bicomponent fiber, *n*—a fiber consisting of two polymers which are chemically different, physically different, or both. **D 123,** D13

biconical, adj— see the preferred but not equivalent term, **bidirectional,** as in bidirectional optical measuring system. **E 284,** E12

biconstituent fiber, *n*—deprecated term. Use the preferred term *bicomponent bigeneric fiber.* **D 4466,** D13

bidirectional, adj— see **bidirectional optical measuring system.** **E 284,** E12

bi-directional, *adj*—*in reference to bar codes,* symbologies that may be decoded whether scanned in one direction or the reverse direction. **F 1294,** F05

bidirectional optical measuring system, *n*—an optical system for measuring the reflecting or transmitting properties of specimens, wherein the illuminator and receiver each subtend small angles at the specimen surface. **E 284,** E12

biff—to beat a salted hide that has been placed on a rack, in order to shake loose salt from the hair. **D 1517,** D31

bifilar eyepiece—a Filar eyepiece with motion in two mutually perpendicular directions. **E 7,** E04

bilateral agreement—recognition arrangement that covers acceptance of each other's results by two parties, (ISO Guide 2). **E 1187,** E36

billet—refinery shape used for piercing or extrusion into tubular products or for extrusion into rods, bars, and shapes. Circular in cross section, usually 3 to 16 in. (76 to 406 mm) in diameter, normally ranging in weight from 100 to 4200 lb (63 to 28 200 kg). **B 846,** B05

billet, *n*—a formed shape that may be further worked, or a solid, semifinished, round, or rectangular product that has been hot-worked by forging, rolling, or extrusion. **B 899,** B02

binary alloy—any specific composition in a binary system. **E 7,** E04

binary cycle plant, *n*—a facility that generates electric power by transferring heat from produced geothermal fluids to a non-aqueous working fluid that vaporizes and causes a turbine to rotate the shaft of a generator. **E 957,** E44

binary digit, *n*—a unit of electronic data. (See **bit.**) **F 1457,** F05

binary elastic scattering event—*ISS,* the collision between an incident probe ion and a single surface atom in which the total kinetic energy and momentum are conserved. **E 673,** E42

binary elastic scattering peak—*ISS,* an increase in the spectrometer detection system response above the background level which can be attributed to binary elastic scattering of the probe ion from a surface atom of a particular mass. **E 673,** E42

binary synchronous transmission, *n*—a form of data sending. (See **binary digit, synchronous,** and **asynchronous.**) **F 1457,** F05

binary system—the complete series of compositions produced by mixing a pair of components in all proportions. **E 7,** E04

binder—a cementing medium; either a material added to the powder to increase the green strength of the compact, and which is expelled during sintering; or a material (usually of relatively lower melting point) added to a powder mixture for the specific purpose of cementing together powder particles which alone would not sinter into a strong body. **B 243,** B09

binder, *n*—a substance added to a granular material to give it workability and green or dry strength. **C 71,** C08

binder—(*1*) for a continuous filament process, a constituent of a fiber glass sizing that couples the fiber to the composite matrix.

(*2*) for insulation, material applied to glass fibers to hold them in a desired arrangement. **C 162,** C14

binder—a cementing medium; either a material added to the mixture to increase the green or dry strength as compacted, and which may be expelled during sintering or calcining, or a material added to a mixture for the purpose of cementing together particles. **C 242,** C21

binder—a substance, usually an organic material such as coal tar pitch or petroleum pitch, used to bond the coke or other filler material prior to baking. **C 709,** D02

binder—a cementing medium; either a material added to the powder to increase the green strength of the compact, and that is expelled during sintering; or a material (usually of relatively low melting point) added to a powder mixture for the specific purpose of cementing together powder particles that alone would not sinter into a strong body. **C 859,** C26

binder—a substance used to bond aggregates or fillers, or both, into a solid mass. **C 904,** C03

binder—anything that causes cohesion in loosely assembled substances, such as clay or cement. **D 653,** D18

binder, *n*—in a reinforced plastic, the continuous phase which holds together the reinforcement. **D 883,** D20

binder, *n*—a component of an adhesive composition that is primarily responsible for the adhesive forces which hold two bodies together. (See also **extender** and **filler**.) **D 907,** D14

binder—an extraneous bonding agent, either organic or inorganic, used to bind particles together to produce a particle board. **D 1554,** D07

binder—the resinous adhesive component of a pigmented coating composition. **F 335,** F05

binder—in a reinforced plastic, the continuous phase that holds together the reinforcement. **F 412,** F17

binder, *n*—a material employed to bind the image-forming materials to the substrate. **F 1623,** F05

binder phase, *n—as used in Test Method D 5061*, a continuous solid carbon matrix formed during the thermoplastic deformation of those coal macerals that become plastic during carbonization. **D 121,** D05

binder removal, *n*—the chemical or thermal extraction of binder from a compact. Synonymous with **debinding.** **B 243,** B09

binders—in reference to cartridge filters, chemicals used to hold, or 'bind', short fibers together in a filter. **D 6161,** D19

binder (soil binder)—portion of soil passing No. 40 (425-μm) U.S. standard sieve, **D 653,** D18

binder tape—see **core wrap (binder tape).** **D 1711,** D09

binding—in surface filtration, a build-up of particulates on the filter, restricting fluid flow through the filter at normal pressures. **D 6161,** D19

binding—a narrow strip of material attached to the edge of shoe uppers to change appearance and protect the cut edges against fraying. **F 869,** F08

binding energy—the work that must be expended in removing an electron from a given electronic level to a reference level, such as the vacuum level or the Fermi level. **E 673,** E42

binding site, *n—for pile yarn floor covering*, a place at which the pile yarn is, or can be, bound to the backing fabric. **D 123,** D13

binodal curve—in a two-dimensional phase diagram, a continuous line consisting of both of the pair of conjugate boundaries of a two-phase equilibrium and which join, without inflection, at a critical point. See **miscibility gap.** **E 7,** E04

binomial distribution, *n*—the frequency distribution which has the probability function:

$$P(r) = (n!/[r!(n - r)!])p^r q^{n-r}$$

where:

$P(r)$ = probability of obtaining exactly r "successes" in n independent trials,
p = probability, constant from trial to trial, of obtaining a "success" in a single trial, and
q = $1 - p$.

D 123, D13

bioaccumulation—the net accumulation of a substance by an organism as a result of uptake from all environmental sources. **E 943,** E47

bioassay—an experiment that uses living whole organisims, tissues or cells to measure the presence, the concentration or the relative potency of one or more chemicals. **E 943,** E47

bioaugmentation—the addition of microorganisms (predominantly bacteria) to increase the biodegradation rate of target pollutants. **F 1600,** F20

biocide, *n*—a chemical used to kill bacteria and other microorganisms. **D 4439,** D35

biocide—a substance that kills all living organisms. **D 6161,** D19

bioconcentration—the net accumulation of a substance by an aquatic organism as a result of uptake directly from aqueous solution. **E 943,** E47

bioconversion—a general term describing the use of biological systems to transform one compound into another. Examples are digestion of organic wastes or sewage by microorganisms to produce methane. **E 1126,** E48

bioconversion—a general term describing the use of biological systems to transform one compound into another. Examples are digestion of organic wastes or sewage by microorganisms to produce methane. **E 1705,** E48

biodegradable, *adj*—capable of undergoing decomposition into carbon dioxide, methane, water, inorganic compounds, or biomass in which the predominant mechanism is the enzymatic action of micro-organisms, that can be measured by standardized tests, in a specified period of time, reflecting available disposal conditions. **D 5488,** D10

biodegradable plastic, *n*—a degradable plastic in which the degradation results from the action of naturally-occurring microorganisms such as bacteria, fungi, and algae. **D 883,** D20

biodegradable plastic, n—See **degradable plastic.** **D 883,** D20

biodegradation—the chemical alteration and breakdown of a substance, usually to smaller products caused by microorganisms or their enzymes. **F 1600,** F20

biofuel—biomass-derived fuel. **E 1126,** E48

biofuel—biomass-derived fuel. **E 1705,** E48

biogas—a gaseous mixture of mainly carbon dioxide and methane produced by the anaerobic digestion of organic matter. **E 1126,** E48

biogas—a composition of methane and carbon dioxide and minor constituents produced by the digestion of organic substrates in the absence of oxygen. **E 1705,** E48

biological deposits—deposits of organisms or the products of their life processes. **D 1129,** D19

biological deposits—the debris left by organisms as a result of their life processes. **D 6161,** D19

biological evaluation test (biotest)—a test which involves exposure of sealed packages to biological indicators and is designed to determine the microbiological integrity of a package under the specific conditions of the test. **F 17,** F02

biological evaluation test (biotest)—See Terminology F 17. **F 1327,** F02

biological monitoring—analysis of a person's blood or urine, or both, to determine the level of lead contamination in the body. **E 1605,** E06

biological monitoring, *n*—the chemical analysis of chemicals or metabolites, or both, from a worker's blood, urine, fingernails, sweat, breath, and so forth. **F 1494,** F23

biomagnification—the increase in tissue concentration of poorly depurated materials in organisms along a series of predator-prey associations, primarily through the mechanism of dietary accumulation. **E 943,** E47

biomarker, *n*—a biological measure (within organisms) of exposure to, effects of or susceptibility to, environmental stress using molecular, genetic, biochemical, histological, or physiological techniques. **E 943,** E47

biomarker assay—an experiment that uses a molecular, genetic, biochemical, histological, anatomical or physiological technique to assess exposure, response or susceptibility of organisims, tissue or cells to environmental stress. **E 943,** E47

biomass—any material which is or was a living organism or excreted from a micro-organism. **D 6161,** D19

biomass, *n*—any material, excluding fossil fuels, which is or was, a living organism or component of a living organism. **D 6384,** D02

biomass—any material, excluding fossil fuels, which is or was a living organism that can be used as a fuel peanut hulls, agricultural waste, corn and other grains, sugar, and bagasse are all examples of biomass. **E 772,** E44

biomass—total weight of living matter in a given volume. When considered as an energy source, biomass is further subdivided into: (*1*) primary biomass, rapidly growing plant material that may be used directly or after a conversion process for the production of energy, and (*2*) secondary biomass, biomass residues remaining after the production of fiber, food, or other products of agriculture, or biomass by-products from animal husbandry or food preparation that are modified physically rather than chemically. Examples include waste materials from agriculture and forestry industries (manure, sewage, etc.) from which energy may be produced. The above distinction noted between primary and secondary biomass is based on economic factors; these are defined differently in ecological science. **E 1126,** E48

biomass—any material, excluding fossil fuels, which is or was a living organism that can be used as a fuel directly or after a conversion process. Peat is not a biomass. **E 1126,** E48

biomass—total weight of living matter in a given volume. When considered as an energy source, biomass is further subdivided into: (*1*) primary biomass, rapidly growing plant material that may be used directly or after a conversion process for the production of energy, and (*2*) secondary biomass, biomass residues remaining after the production of fiber, food, or other products of agriculture, or biomass by-products from animal husbandry or food preparation that are modified physically rather than chemically. Examples include waste materials from agriculture and forestry industries (manure, sewage, etc.) from which energy may be produced. The above distinction noted between primary and secondary biomass is based on economic factors; these are defined differently in ecological science. **E 1705,** E48

biomass—any material, excluding fossil fuels, which is or was a living organism that can be used as a fuel directly or after a conversion process. Peat is not a biomass. **E 1705,** E48

biomass fuel—fuel derived from biomass. **E 772,** E44

biomass fuel—fuel derived from biomass. **E 1126,** E48

biomass fuel—fuel derived from biomass. **E 1705,** E48

biomass pellet—see **fuel pellet**. **E 1126,** E48

biometric—use of biologic or physiologic data to identify someone; for example, fingerprints or facial characteristics. **F 1156,** F12

bioremediation—the enhancement of biodegradation. **F 1600,** F20

bioremediation agents—inorganic and organic compounds and microorganisms that enhance biological degradation processes, predominantly microbial. **F 1600,** F20

biostat—a substance that inhibits biological growth. **D 6161,** D19

biostimulation—the addition of microbial nutrients, oxygen, heat, or water, or some combination thereof, to enhance the rate of biodegradation of target pollutants by indigenous species (predominantly bacteria and fungi). **F 1600,** F20

bipolar—semiconductor device fabrication technology that produces transistors that use both holes and electrons as charge carriers. **F 1241,** F01

bipolar electrode—an electrode that is not directly connected to the power supply but is so placed in the solution between the anode and the cathode that the part nearest the anode becomes cathodic and the part nearest the cathode becomes anodic. **B 374,** B08

bipolar field—see **field, bipolar.** **E 1316,** E07

birdseye, *n*—*in knitted fabrics*, an unintentional tuck stitch. **D 123,** D13

birdseyes—small localized areas in wood with the fibers indented and otherwise contorted to form small circular or elliptical figures remotely resembling birds' eyes on the tangential surface. Common in sugar maple; rare in other hardwood species. **D 9,** D07

birefringence, *n*—(*double refraction*) a property of anisotropic materials which manifests itself as a splitting of a light ray into components having different vibration directions which are transmitted at different velocities. **D 123,** D13

birefringence—a double-refraction phenomenon in which an unpolarized beam of light is divided into two beams with different directions and relative velocities. **F 1241,** F01

birefringent—having more than one refractive index. Such materials exhibit alternately bright and dark reflections at 45° intervals during a 360° rotation with plane-polarized light. (See also **anisotropic**.) **E 7,** E04

bisectrix, acute—in biaxial crystals, that principal axis of the ellipsoid of indexes which bisects the smaller angle between the optic axes. **E 175,** E41

bisectrix, obtuse—in biaxial crystals, that principal axis of the ellipsoid of indexes which bisects the larger angle between the optic axes. **E 175,** E41

bisque—a coating of wet-process porcelain enamel that has been dried, but not fired. **C 286,** B08

bisque fire—See *bisque fire* under **firing.** **C 242,** C21

bisulfiting—the treatment of hot solutions of vegetable tanning extracts with sodium bisulfite in order to increase their solubility and rate of take-up by hides. **D 1517,** D31

bisynic (BSC), *n*—an IBM-developed method of binary synchronous transmission. **F 1457,** F05

bit—any device that may be attached to or is an integral part of a drill string and is used as a cutting tool to bore into or penetrate rock or other materials. **D 653,** D18

bit—binary digit. **F 1457,** F05

bite, *n*—*in building construction in glazing*, the dimensional depth by which the framing system (such as frames, stops, sash, or lock-strip gaskets) overlaps the edge of the glass or panel. **C 717,** C24

bite—See **windows and doors.** **E 631,** E06

bite, chemical, *n*—stinging experienced primarily in the oral cavity as a result of exposure to substances such as highly carbonated beverage. **E 253,** E18

bit map—a pattern of bits representing the dots in a printed image. **F 1457,** F05

bitter, (taste), *adj*—taste produced by substances such as quinine or caffeine when in solution. **E 253,** E18

bitumen, *n*—a class of black or dark-colored (solid, semisolid, or viscous) cementitious substances, natural or manufactured, composed principally of high molecular weight hydrocarbons, of which asphalts, tars, pitches, and asphaltites are typical. **D 8,** D04

bitumen—(*1*) a class of amorphous, black or dark-colored, (solid, semi-solid, or viscous) cementitious substances, natural or manufactured, composed principally of high molecular weight hydrocarbons, soluble in carbon disulfide, and found in asphalts, tars, pitches, and asphaltites;

(*2*) a generic term used to denote any material composed principally of bitumen. **D 1079,** D08

bitumen, polymer-modified—a black or dark-colored cementitious, high-molecular weight hydrocarbon such as asphalt, tar, or pitch in which polymer has been dispersed to modify the properties of the bitumen. **D 3743,** D04

bitumen, rubberized—a special form of polymer-modified bitumen where an asphalt, tar, or pitch has been combined with natural or synthetic rubber to modify its properties. **D 3743,** D04

bitumen trap—See **pitch pocket.** **D 1079,** D08

bituminized, *adj*— impregnated with bitumen. Example: bituminized fiber pipe. **D 1079,** D08

bituminous, *adj*—containing or treated with bitumen (also *bituminized*). Examples: bituminous concrete, bituminized felts and fabrics, bituminous pavement. **D 8,** D04

bituminous, *adj*— containing or treated with bitumen. Examples: bituminous concrete, bituminous felts and fabrics, bituminous pavement. **D 1079,** D08

bituminous bleeding, *n*—excess bitumen on the surface of the pavement, usually found in the wheel paths. **E 867,** E17

bituminous bleeding, *n*— excess bitumen on the surface of the pavement, usually found in the wheel paths. **E 1778,** E17

bituminous class—See *bituminous class* under **rank**. **D 121,** D05

bituminous emulsion, *n*—(*1*) a suspension of minute globules of bituminous material in water or in an aqueous solution,
(*2*) a suspension of minute globules of water or of an aqueous solution in a liquid bituminous material. **D 8,** D04

bituminous emulsion—(*1*) a suspension of minute globules of bituminous material in water or in an aqueous solution;
(*2*) a suspension of minute globules of water or of an aqueous solution in a liquid bituminous material (invert emulsion). **D 1079,** D08

bituminous grout—a mixture of bituminous material and fine sand that will flow into place without mechanical manipulation when heated. **D 1079,** D08

bituminous pavement, *n*—a pavement comprising an upper layer or layers of aggregate mixed with a bituminous binder, (such as asphalt, coal tars and natural tars) and surface treatments such as chip seals, slurry seals, sand seals, and cape seals are also included. **E 867,** E17

bituminous pavement, *n*—a pavement comprising an upper layer or layers of aggregate mixed with a bituminous binder, such as asphalt, coal tars, and natural tars for purposes of this terminology; surface treatments such as chip seals, slurry seals, sand seals, and cape seals are also included. **E 1778,** E17

bituminous varnish— under **varnish,** see *bituminous varnish.* **D 16,** D01

bivariant equilibrium—a stable state among a number of phases equal to the number of components in a system and in which any two of the external variables (temperature, pressure, or concentrations) may be varied, at will, without necessarily causing a change in the number of phases; sometimes called divariant equilibrium. **E 7,** E04

black, *adj*—color description applied to opaque objects that are highly light absorbing throughout the visible spectrum. **D 2946,** C17

black annealing, *n*—**box annealing** steel sheet, strip, or wire. **A 941,** A01

blackboard enamel— see **chalkboard enamel.** **C 286,** B08

blackbody, *n*—the ideal, perfect emitter and absorber of thermal radiation. It emits radiant energy at each wavelength at the maximum rate possible as a consequence of its temperature, and absorbs all incident radiance. **C 168,** C16

blackbody—the ideal, perfect emitter and absorber of thermal radiation which emits radiant energy at the maximum rate possible, as a consequence of its temperature, and absorbs all incident radiation. **C 242,** C21

blackbody, n— see the preferred term, **full radiator.** **E 284,** E12

blackbody, *n*—the perfect or ideal source of thermal radiant power having a spectral distribution described by the Planck equation. **E 344,** E20

blackbody—a hypothetical "body" that completely absorbs all incident radiant energy, independent of wavelength and direction; that is, one which neither reflects nor transmits any of the incident radiant energy. It is the emitter of electromagnetic radiant energy which, at a given temperature, presents the maximum spectral density of radiant exitance or radiance at all wavelengths.

NOTE—No real material is a blackbody. A completely enclosed cavity with opaque walls at a uniform temperature contains blackbody radiation. A blackbody radiator can be approximated in the laboratory to any desired degree of approximation by a furnace containing a cavity with opaque walls at a uniform temperature, that contains an aperture through which the blackbody radiation is observed. The degree of approximation to a true blackbody radiator is inversely related to the ratio of the area of the aperture to the area of the interior wall of the cavity. **E 772,** E44

blackbody—an ideal thermal radiator (emissivity = 1.0) that emits and absorbs all of the available thermal radiation at a given temperature. **E 1316,** E07

blackbody equivalent temperature—the apparent temperature of an object as determined from the measured radiance and the assumption that it is an ideal blackbody with emissivity of 1.0. **E 1316,** E07

blackbody radiant energy— see **radiant energy, blackbody**. **E 772,** E44

blackbody temperature, *n*—the temperature of a perfect radiator—a surface with an emissivity of unity and, therefore, a reflectivity of zero. **E 176,** E05

black box, *n*—a black painted aluminum box with an open top where the flat test specimens to be exposed constitute the top surface of the box. The box is equipped with mounting strips to hold the test specimens firmly in place. The top surface of the box must be completely filled at all times; any blank spaces on the top surface must be occupied by black "dummy" panels to maintain correct operating condition. **G 113,** G03

black box under glass, *n*—a glass covered enclosure or cabinet of any convenient size. It shall be constructed of corrosion resistant metal and be enclosed to prevent ambient air from circulating over the samples. Exterior non-glass surfaces shall be painted black. The interior shall remain unpainted. **G 113,** G03

black density—see **image density.** **F 335,** F05

black edging—a black porcelain enamel applied over the ground coat and exposed in specified areas by brushing the cover coat bisque prior to firing (see also **edging**). **C 286,** B08

black felt, *n*—those classifications of felt manufactured to various shades of the color black. **D 123,** D13

black light—electromagnetic radiation in the near-ultraviolet range of wavelength. (330–390 nm) (3300–3900Å). **E 1316,** E07

black light—electromagnetic radiation in the near ultraviolet range of wavelength (330 to 390 nm) (3300 to 3900 Å). **E 1316,** E07

black light filter—a filter that transmits near-ultraviolet radiation while absorbing other wavelengths. **E 1316,** E07

black light filter—a filter that transmits near ultraviolet radiation while absorbing other wavelengths. **E 1316,** E07

black market—the distribution (sale) of legitimate goods through illicit commerce. (See **fence.**) **F 1156,** F12

black marking—black marks on a flooring surface usually caused by the impact of the soles and heels of footwear. **D 2825,** D21

black marking—black smudges on the surface of the pultruded product that cannot be removed by cleaning or scrubbing or wiping with solvent.

NOTE—Black marking results from excessive pressures in the die when the pultrusion is rubbing against soft or unchromed die surfaces.
D 3918, D20

black oxide—a finish on metal produced by immersing a metal in hot oxidizing salts or salt solutions. **B 374,** B08

black panel thermometer, *n*—a temperature measuring device consisting of a metal panel, having a black coating which absorbs all wavelengths uniformly, with a thermal sensitive element firmly attached to the center of the exposed surface. The black panel thermometer is used to control an artificial weathering device and to provide an estimate of the maximum temperature of samples exposed to a radiant energy source. **G 113,** G03

black sidewall, *n*—a sidewall on which only black compounds comprise the outer visible surface of the tire. **F 538,** F09

black speck—a defect that appears in the fired cover coat as a small dark spot. **C 286,** B08

black write—a process in electrostatic printing where the photoconductive element is charged with a charge of the same sign as that of the toner. A light beam, used like a "stylus" is used to discharge only those areas that are to receive toner to form the image. In the development process, the charged background areas repel the like charged toner to the discharged areas on the photoconductor.
F 909, F05

blade—the segment that contains the cutting edge which may be with or without serrations. **F 1078,** F04

blade alignment—the positioning of the blades with respect to tip match-up and blade setting. **F 1078,** F04

blade clincher—thin clinching arm; usually designed for insertion between layers of corrugated boards. **F 592,** F16

blaine fineness—the fineness of powdered materials, such as cement and pozzolans, expressed as surface area usually in square centimetres per gram. **D 653,** D18

blank—a pressed, presintered or fully sintered compact, usually in the unfinished condition, requiring cutting, machining, or some other operation to give it its final shape. **B 243,** B09

blank—a piece of flat product intended for subsequent fabrication by forming, bending, cupping, drawing, hot pressing, and so forth.
B 846, B05

blank—(*1*) a preliminary shape from which a finished article is further formed, molded, or cut.

(*2*) a semi-finished piece of glass for making an optical element, such as a lens or prism. Also known as a pressing.
C 162, C14

blank—the piece cut from metal sheet that is to be used in forming the finished article. **C 286,** B08

blank—matrix carried through all or part of the analytical process, where the analyte is not present, or where the analyte response is suppressed.

NOTE—A blank must be appropriate to the analytical process it is being used with.

NOTE—A blank is typically used to monitor contamination or to establish a baseline for quantitation. **D 1129,** D19

blank, *n*—a portion of a rubber compound of suitable volume to fill the cavity of a mold. **D 1566,** D11

blank, *n*—*in biodegradability testing,* a test system containing all system components with the exception of the test material.
D 6384, D02

blank, *n*—the contamination level of the fluid when the test coupon is omitted. **G 126,** G04

blank carburizing, *n*—simulating the **carburizing** operation without introducing carbon. **A 941,** A01

blanket, *n*—*for bedding,* an unquilted fabric covering designed primarily to provide thermal insulation.
D 123, D13

blanket—the sheet of elastomer-coated fabric or equivalent placed on the blanket cylinder to receive ink from the plate and offset it to the sheet or web on the impression cylinder. **F 425,** F05

blanket cylinder—a cylinder to which the blanket is attached.
F 425, F05

blanket feed—a method for charging batch designed to produce an even distribution of batch across the width of the melter.
C 162, C14

blanket grouting—a method in which relatively closely spaced shallow holes are drilled and grouted on a grid pattern over an area, for the purpose of making the upper portions of the bedrock stronger and less pervious. **D 653,** D18

blanket insulation, *n*—a relatively flat and flexible insulation in coherent sheet form furnished in units of substantial area.
C 168, C16

blanket insulation, metal mesh, *n*—blanket insulation covered by flexible metal-mesh facings attached on one or both sides.
C 168, C16

blanket roll-up—a container made from fabric or similar material and specifically designed to protect the blanket from damage during storage or transportation. **F 819,** F18

blanket wash—the solvent used to clean the blanket.
F 425, F05

blanking band—a band of uniform height with its longitudinal center positioned optimally between the highs and lows of the surface record depicting at least 100 ft (30 m) of pavement.
E 867, E17

blank mold—the metal mold that first shapes the glass in the manufacture of hollow ware. **C 162,** C14

blank nitriding, *n*—simulating the nitriding operation without introducing nitrogen. **A 941,** A01

blank sample—unexposed specimen of the *medium* used in testing, such as a wipe or a filter, which is analyzed with other samples to determine whether samples are either (*1*) contaminated before collection (for example, in the field, or at the testing site), or are (*2*) contaminated after collection (for example, during transportation to the laboratory or in the laboratory), or both. Also called a *media blank*, or a *dummy specimen*. **E 1605,** E06

blast-furnace slag, *n*—the nonmetallic product, consisting essentially of silicates and aluminosilicates of calcium and other bases, that is developed in a molten condition simultaneously with iron in a blast furnace. **C 125,** C09

blast-furnace slag, *n*—the nonmetallic product, consisting essentially of silicates and aluminosilicates of calcium and other bases, that is developed in a molten condition simultaneously with iron in a blast furnace. **C 219,** C01

blast-furnace slag, *n*—the nonmetallic product, consisting essentially of silicates and alumino-silicates of lime and of other bases, that is developed simultaneously with iron in a blast furnace.
D 8, D04

blast-furnace slag—the nonmetallic product, consisting essentially of silicates and alumino-silicates of calcium and other bases, that is developed in a molten condition simultaneously with iron in a blast furnace. **D 1079,** D08

blastibility—index value of the resistance of a rock formation to blasting. (ISRM) **D 653,** D18

blasting—See **sand blasting; grit blasting; wet blasting.** **B 374,** B08

blasting—see **feathering.** **F 335,** F05

blasting cap (detonator, initiator)—a small tube containing a flashing mixture for firing explosives. (ISRM) **D 653,** D18

blast nozzle—see **pneumatic atomizer**. **E 1620,** E29

bleach, *n—in care of textiles*, a product for brightening and aiding in the removal of soils and stains from textile materials by oxidation that is inclusive of both chlorine and non-chlorine products. **D 123,** D13

bleachability—the capacity of a pulp to bleach to a given whiteness. This is approximately and indirectly related to lignin content. **D 1695,** D01

bleached, *adj*—having been subjected to the process of bleaching. **D 1968,** D06

bleached specimen, *n*—specimen whose absorptance has been decreased by chemical or radiant means. **E 284,** E12

bleaching—*(1)* the process of removing oxidized tannins and insoluble materials from the surface layers of leather, in order to prevent crackiness of the grain. It is performed by dipping the leather in a weak alkaline solution to render the tannin readily soluble, dipping in water, neutralizing in weak acid solution and washing.

(*2*) the process of lightening the color of chrome leather by treating with synthetic tannins or precipitating white pigment in the surface of the leather. **D 1517,** D31

bleaching, *n*—a process involving a chemical or biological treatment of pulp, primarily to increase whiteness and brightness; such a process may alter or remove noncellulosic materials, such as but not limited to lignin, resin, and colorants. **D 1968,** D06

bleed, *n*—the spreading or running of a dye or pigment color by the action of a solvent. **D 6488,** D01

bleed, *n*—the spreading of ink into an unwanted area on the printed sheet. **D 6488,** D01

bleed, *n*—ink feathering of one color into an adjacent color over time. **F 1857,** F05

bleeder cloth—a nonstructural layer of material used in the manufacture of composite assemblies to allow the escape of excess gas and resin during cure. **E 1749,** E06

bleeders—holes or pipes through bridge decks to provide for drainage. **D 3743,** D04

bleeding, *n*—the autogenous flow of mixing water within, or its emergence from, newly placed concrete or mortar caused by the settlement of the solid materials within the mass, also called water gain. **C 125,** C09

bleeding—the diffusion of coloring matter through a coating from the substrate; also, the discoloration arising from such diffusion. In the case of printing ink, the spreading or running of a pigment color by the action of a solvent such as water or alcohol. **D 16,** D01

bleeding—*in grouting*, the autogeneous flow of mixing water within, or its emergence from, newly placed grout caused by the settlement of the solid materials within the mass. **D 653,** D18

bleeding—the transfer of materials exuded from leather to other materials that come in contact with it. **D 1517,** D31

bleeding, *n*—the exuding of a liquid compounding material from the surface of a vulcanized or unvulcanized rubber. **D 1566,** D11

bleeding, *n*—the unintentional transfer of coloring matter from one medium to or through another. **E 284,** E12

bleeding of coating—the diffusion of coloring matter through a substrate from the coating, or to copy paper from the coating. **F 221,** F05

bleeding rate—*in grouting*, the rate at which water is released from grout by bleeding. **D 653,** D18

bleed of image—spread or diffusion of an image with time. **F 221,** F05

bleedout—the action of an entrapped liquid penetrant in surfacing from discontinuities to form indications. **E 1316,** E07

bleed through—glue or components of glue that have seeped through the outer layer or ply of a glued wood product and that show as a blemish or discoloration on the surface. **D 1038,** D07

bleed through, *n*—the diffusion of ink through the substrate. **F 1857,** F05

blemish—in dry process enameling, an insignificant imperfection in the porcelain enamel surface. **C 286,** B08

blemish—in grading lumber, anything marring the appearance. May not be classified as a defect. **D 9,** D07

blemish—anything marring the appearance of the veneer that is not classifiable as a defect. **D 1038,** D07

blemish—strained or discolored area attributable to normal composition or forming, or both. (See also **inclusion.**)

F 109, C21

blend—a combining of various cementitious materials. **C 822,** C13

blended hydraulic cement, *n*—a hydraulic cement consisting of two or more inorganic constituents (at least one of which is not portland cement or portland cement clinker) which separately or in combination contribute to the strength-gaining properties of the cement, (made with or without other constituents, processing additions and functional additions, by intergrinding or other blending). **C 219,** C01

blended soap—under **soap,** see *blended soap.* **D 459,** D12

blending—the thorough intermingling of powders of the same nominal composition (not to be confused with mixing). **B 243,** B09

blending-batch—{archaic} stepwise changes in batch composition to arrive at the final change in finished glass. **C 162,** C14

blending plan, *n*—the instructions for mixing fibers during specimen preparation. **D 123,** D13

blind—an image area on a plate that will not take ink. This term should not be used where no image is present. **F 425,** F05

blind clinch—clinch between the layers of corrugated boards, usually buried with wide-crown retractable anvil tools. **F 592,** F16

blind hem stitch, *n—in home sewing*, a complex machine stitch pattern consisting of small groups of straight or narrow zigzag stitches separated by a wide zigzag stitch unit at consistent intervals. **D 123,** D13

blinding, *n—for geotextiles*, the condition where soil particles block the surface openings of the fabric, thereby reducing the hydraulic conductivity of the system. **D 4439,** D35

blinding—placing selected material to completely cover the pipe or tubing. **F 412,** F17

blind ink—See **reflective ink**. **F 149,** F05

blind nailing—the use of nails that are not exposed to the weather in the finished roofing. **D 1079,** D08

blister—a dome-shaped imperfection or defect, resulting from loss of adhesion between a metallic deposit and the substrate. **B 374,** B08

blister—a relatively large gaseous inclusion in glass. **C 162,** C14

blister—a defect caused by gas evolution consisting of a bubble that forms during fusion and remains when the porcelain enamel solidifies. **C 286,** B08

blister—a convex, raised area on the pipe surface indicating an internal separation. **C 896,** C04

blister, *n*—*in bonded, fused, or laminated fabrics,* a bulge, swelling, or similar surface condition on either the face fabric or the backing fabric characterized by the fabric being raised from the plane of the underlying component over a limited area to give a puffy appearance. **D 123,** D13

blister, *n*—an imperfection, a rounded elevation of the surface of a plastic, with boundaries that may be more or less sharply defined, somewhat resembling in shape a blister on the human skin. **D 883,** D20

blister, *n*—an elevation of the surface of an adherend, somewhat resembling in shape a blister on the human skin; its boundaries may be indefinitely outlined and it may have burst and become flattened. **D 907,** D14

blister—*in plywood,* an elevation of the surface of an adherend (separation between plies), somewhat resembling in shape a blister on the human skin; its boundaries may be indefinitely outlined and it may have burst or become flattened. **D 1038,** D07

blister—(*1*) a raised portion of a roofing membrane resulting from local internal pressure;

(*2*) the similarly formed protuberances in coated prepared roofing. **D 1079,** D08

blister—a rounded elevation of the pultruded surface with boundaries that may be more or less sharply defined.

NOTE—The rounded elevation somewhat resembles in shape a blister on the surface of human skin. Blisters may exist within the pultrusion as a hollow delaminated area (usually gas-filled) under a raised portion of the surface. **D 3918,** D20

blister—bubble or gaseous inclusion at the surface which if broken could form a pit, pock, or hole.

F 109, C21

blister, *n*—rounded elevation of the surface of a plastic, with boundaries that may be more or less sharply defined, somewhat resembling in shape a blister on the human skin. (D-20) **F 412,** F17

blister, *n*—in sheet plastics, an imperfection, a rounded elevation of the surface, with boundaries that may be more or less sharply defined, somewhat resembling in shape a blister on the human skin. **F 1251,** F04

blistered compact, *n*—a sintered object characterized by having blisters or eruptions on the surface. **B 243,** B09

blistering—the development during firing of enclosed or broken macroscopic vesicles or bubbles in a body, or in a glaze or other coating. **C 242,** C21

blistering, *n*—the formation of small air pockets under the printed paper surface. **D 6488,** D01

blistering, *n*—an enclosed raised spot or air bubble on the paper surface that can be caused by moisture trapped during the drying process. **D 6488,** D01

blistering resistance (*coatings*)—the ability of a coating to resist the formation in the film of dome-shaped, liquid- or gas-filled projections resulting from local loss of adhesion and lifting of the film from the previously applied coating or the substrate. **D 16,** D01

blister pack—See **pack.** **D 996,** D10

blister(s), *n*—surface or internal imperfection(s), produced by entrapped gases or other volatiles, during the manufacture of rubber articles. **D 1566,** D11

blisters—gas- or water vapor-filled bubbles or cavities in the waterproofing membrane caused by entrapped air, the volatilization of liquids applied to the concrete, or the vaporization of water on or in the concrete. **D 3743,** D04

bloach—{archaic} an imperfection resulting from incompletely grinding plate glass, caused by a low place in the plate which retains part of the original rough surface. **C 162,** C14

bloating—substantial swelling produced by a heat treatment that causes the formation of a vesicular structure. **C 242,** C21

bloating of refractories, *v*—substantial swelling produced by a heat treatment that causes the formation of a vesicular structure. **C 71,** C08

Bloch wall—a domain wall in which the magnetic moment at any point is substantially parallel to the wall surface. See also **domain wall.** **A 340,** A06

block, *n*—*in experimenting,* a group of units that is relatively homogeneous within itself, but may differ from other similar groups. **D 123,** D13

block, *v*—to align warp and weft yarns at right angles, by some form of manipulation. **D 123,** D13

block—a grouping of devices in a single unit having common control. See **array, bank, train**. **D 6161,** D19

block—*in a honeycomb core material,* a single production unit of honeycomb before slicing. **E 1749,** E06

block—*in fatigue loading,* a specified number of constant amplitude loading cycles applied consecutively, or a spectrum loading sequence of finite length that is repeated identically. **E 1823,** E08

block, *n*—synonym for *element.* **F 538,** F09

block, *n*—a group of data transmitted as a unit. **F 1457,** F05

block circuit diagram, *n*—an illustration of the interconnection of the major elements of a system, each element being presented by a block. (See **BCD**.) **F 1457,** F05

block copolymer—an essentially linear copolymer in which there are repeated sequences of polymeric segments of different chemical structure. **D 883,** D20

block copolymer, *n*—an essentially linear copolymer in which there are repeated sequences of polymeric segments of different chemical structure. **F 1251,** F04

block cracking, *n*—a pattern of cracks that divide the pavement into approximately rectangular pieces, ranging in size from approximately 0.1 m^2 to 10 m^2 (1 to 100 ft^2). **E 867,** E17

block cracking, *n*—a pattern of cracks that divide the pavement into approximately rectangular pieces, ranging in size from approximately 0.1 m^2 to 1.0 m^2(1 ft ^{2} to 100 ft^2). **E 1778,** E17

blocked curing-agent, *n*—a curing agent or hardener rendered unreactive, which can be reactivated as desired by physical or chemical means. (Compare **hardener**.) **D 907,** D14

block flow—the distance an adhesive, sealant, or coating will sag on a vertical surface in a given period of time. Also referred to as *slump*. **E 1749,** E06

blocking—(*1*) shaping a gather of glass in a cavity of wood or metal.

(2) *archaic,* promoting mixing and fining of glass by immersion of a wooden block or other object to create bubbles.

(3) reprocessing to remove surface imperfections.

(4) *archaic,* mounting of glass for grinding and polishing.

(5) *archaic,* idling a furnace at reduced temperatures.

See **hot hold**. **C 162,** C14

blocking, *n*—sticking of a coated surface to an adjacent surface when the two surfaces have been in contact with each other for an extended period of time. **D 16,** D01

blocking, *n—of coated fiber glass yarn solar screening,* an undesired adhesion between touching layers of a material, such as occurs under moderate pressure, during storage or use. **D 123,** D13

blocking, *n*—the measurement of the development of surface tack and the thermal softening point of the material. **D 123,** D13

blocking—wood blocks placed between the excavated surface of a tunnel or shaft and the main bracing system. (ISRM) **D 653,** D18

blocking, *n*—unintentional adhesion between plastic films or between a film and another surface. **D 883,** D20

blocking, *n*—an undesired adhesion between touching layers of a material, such as occurs under moderate pressure during storage or use. **D 907,** D14

blocking—See **loading.** **D 996,** D10

blocking—(*1*) wood built into a roofing system above the deck and below the membrane and flashing to (*a*) stiffen the deck around an opening, (*b*) act as a stop for insulation, (*c*) serve as a nailer for attachment of the membrane or flashing.

(*2*) wood cross-members installed between rafters or joists to provide support at cross-joints between deck panels.

(*3*) cohesion or adhesion between similar or dissimilar materials in roll or sheet form that may interfere with the satisfactory and efficient use of the material. **D 1079,** D08

blocking—the adhesion between touching layers of leather such as occurs under moderate pressures during storage or use. **D 1517,** D31

blocking, *n—of paper or paperboard,* undesired cohesion or adhesion that interferes with the satisfactory and efficient use of the material. **D 1968,** D06

blocking, *adv*—an undesired adhesion between sheets of printed material that might occur under moderate pressure or increased temperature, or both, while in storage or in use. **D 6488,** D01

blocking—the tendency for coated sheets to stick together under pressure. (F 335) **F 221,** F05

blocking—the tendency for coated sheets to stick together under pressure. **F 335,** F05

blocking, *n*—image is transferred upon contact from one substrate to another substrate which may cause the images or substrate coatings, or both, to stick to one another. See also **transference, offset.** **F 1857,** F05

blocking geometry—*EIA,* experimental situation wherein the atom rows or planes of a single crystal target are aligned parallel to a vector from the specimen to the detector. **E 673,** E42

blocking or masking—surrounding specimens or covering their sections with absorptive material. **E 1316,** E07

blocking resistance, *n— of paper or paperboard,* the capacity of a given paper or paperboard to resist blocking. See **blocking.** **D 1968,** D06

block insulation, *n*—rigid insulation preformed into rectangular units. **C 168,** C16

block mold—a one-piece mold. **C 162,** C14

block reek, rake—{archaic} a scratch imperfection caused by cullet lodged in the felt in the polishing operation. **C 162,** C14

block structure—See **cell structure.** **F 1241,** F01

blood-lead level (blood level)—concentration of lead in the blood, 1 μmole/L = 20.72 μg/mL. **E 1605,** E06

blood-lead testing—testing by laboratories to determine the blood-lead level. **E 1605,** E06

bloom—a visible exudation or efflorescence on a surface. **B 374,** B08

bloom—(*1*) a visible surface film resulting from attack by the atmosphere or from the deposition of particulate or vapor condensates. (See also **smoked.**)

(*2*) a blemish in float glass appearing on the bottom (tin contact) surface after reheating as a result of the presence of tin diffused into the surface. **C 162,** C14

bloom—see **efflorescence.** **C 460,** C17

bloom, *n—in building construction,* a substance formed by blooming. **C 717,** C24

bloom, *n*—a visible exudation or efflorescence on the surface of a material. **D 883,** D20

bloom—a light-colored deposit of ellagic acid appearing on the grain surface of leather tanned with certain pyrogallo tannins, such as myrabolans, valonia, and dividivi. The appearance may be objectionable for some purposes, but bloom does not significantly affect the other physical properties of the leather. **D 1517,** D31

bloom—a condition in which moisture has condensed upon and is being trapped by a polish film, rendering a haze over the surface. **D 2825,** D21

bloom, *n*—in *asbestos-cement,* see **efflorescence (bloom).** **D 2946,** C17

bloom, *n*—the scattering of light in directions near the specular angle of reflection by a deposit on or exudation from a specimen. **E 284,** E12

bloom, *n*—a visible exudation or efflorescence on the surface of a material. (D-20) **F 412,** F17

bloom—coating that forms on rubber blankets when they are left standing. The bloom should be washed off before the blanket is used again. **F 425,** F05

bloom—a visible exudation of efflorescence on the surface of a material. (D 883, D-20) **F 869,** F08

bloom, *n*—a visible exudation or efflorescence of a performance additive on the surface of a material. **F 1251,** F04

blooming, *v—in building construction,* movement or diffusion of a component such as a plasticizer, monomer, unreacted polymer or other formulation ingredient to a sealant, coating, or membrane surface. **C 717,** C24

blooming—in radiologic real-time imaging, an undesirable condition exhibited by some image conversion devices and television pickup tubes brought about by exceeding the allowable input brightness for the device, causing the image to go into saturation, producing a fuzzy image of degraded spatial resolution and grey scale rendition. **E 1316,** E07

blooming/blushing, *n*—a foggy appearance in a printed ink film commonly caused by incompatibility of the ingredients in the ink. **D 6488,** D01

blooming, fiber—a pultrusion surface condition exhibiting a fiber prominence or fiber show that usually has a white or bleached color and a sparkling appearance.

NOTE—The surface generally feels rough when touched by the fingers and is of superficial thickness easily removed by buffing or light sanding. **D 3918,** D20

blooming, undercure—a dull and bleached surface color that is evident in pultruded material not exposed to the weather.

NOTE—This condition is usually the result of insufficient surface cure. **D 3918,** D20

bloom (rubber), *n*—a liquid or solid material that has migrated to the surface of a rubber and generally changes the surface appearance. **D 1566,** D11

bloom (rubber)—a liquid or solid material that has migrated to the surface of a rubber. (D 1566, D-11) **F 869,** F08

blotch, *n*—an offcolored area of any shape caused by grease or oil. (*Syn.* **oil spot**) **D 3990,** D13

blotting—the action of the developer in soaking up the penetrant from the discontinuity to accelerate bleedout. **E 1316,** E07

blow, *n*—the volume expansion that occurs during the production of cellular or sponge rubber. **D 1566,** D11

blow-and-blow process—the process of forming hollow ware in which both the preliminary and final shapes are formed by air pressure. **C 162,** C14

blow back—the enlargement of a minified radiograph to its original size by use of an optical direct reader. **E 1316,** E07

blowby, *n*—*in internal combustion engines,* that portion of the combustion products and unburned air/fuel mixture which leaks past piston rings into the engine crankcase during operation. **D 4175,** D02

blower—one who forms glass by blowing. (See also **gaffer**.) **C 162,** C14

blow head—part of a forming machine serving to introduce air under pressure to blow any hollow glass article. **C 162,** C14

blow hole—an unintended hole or void in a metal casting resulting from entrained gases. **E 631,** E06

blowholes—a hole produced in a casting by gas which was trapped during solidification. **E 7,** E04

blow-in—the inflow of ground water and unconsolidated material into a borehole or casing caused by differential hydraulic heads; that is, caused by the presence of a greater hydraulic head outside of a borehole/casing than inside. **D 653,** D18

blowing agent—a compounding ingredient used to produce gas by chemical or thermal action, or both, in manufacture of hollow or cellular articles. **D 883,** D20

blowing agent, *n*—a compounding material used to produce gas by chemical or physical action, or both, in the manufacture of hollow or cellular articles. **D 1566,** D11

blowing agent—a compounding ingredient used to produce gas by chemical or thermal action, or both, in manufacture of hollow or cellular articles. (D-20). **F 412,** F17

blowing iron—See **blowpipe**. **C 162,** C14

blow mold—the metal mold in which a blown glass article is finally shaped. **C 162,** C14

blow molding—a method of fabrication in which a heated parison is forced into the shape of a mold cavity by internal gas pressure. **D 883,** D20

blow molding—a method of fabrication in which a parison (hollow tube) is forced into the shape of the mold cavity by internal gas pressure. (D-20) **F 412,** F17

blown glass—glassware shaped by air pressure, as by compressed air or by mouth blowing. **C 162,** C14

blow-off resistance—the degree to which a deposited layer of powder resists being blown off by a standard jet of air. **C 286,** B08

blowout—the displacement and lengthening of an arc to facilitate its extinction. The blowout effect can be achieved by a magnetic field, air blast, etc. **B 542,** B02

blowout—a sudden or violent uncontrolled escape of fluids or gas, or both, from a borehole. **D 653,** D18

blow-over—the thin-walled bubble formed above a blow mold in hand-shop operation to facilitate bursting-off. **C 162,** C14

blowpipe—the pipe used by a glassmaker for gathering and blowing by mouth. **C 162,** C14

blowups, *n*—localized upward movement of the pavement surface at transverse joints or cracks, often accompanied with shattering of the concrete in that area. **E 867,** E17

blowups, *n*—localized upward movement of the pavement surface at transverse joints or cracks, often accompanied with shattering of the concrete in that area. **E 1778,** E17

blucher—a laced shoe in which the quarters extend forward over the vamp and are left loose at the inner edge. **F 869,** F08

blue—usually in the phrase "in the blue," applied to hides or skins that have been chrome-tanned but not dyed nor fat-liquored. **D 1517,** D31

blueberry—See **strawberry**. **D 1079,** D08

blued—heated to result in oxidized bluish surface of steel nail. **F 547,** F16

blue dip—a solution, once widely used, containing a mercury compound used to deposit mercury upon a metal by immersion, usually prior to silver plating. **B 374,** B08

blue enamel—(*1*) in dry-process porcelain enameling, an area of enamel coating so thin that it appears blue in color.

(*2*) In wet-process enameling, a cover coat applied too thin to hide the substrate. **C 286,** B08

blueing—the formation of a thin oxide film on steel, either by heating in air, or by immersion in oxidizing solutions. **B 374,** B08

bluing, *n*—subjecting the scale-free surface of a steel object to the action of air, steam, or other agents at a suitable temperature, thereby forming a thin blue film of oxide and improving the object's appearance and corrosion resistance. **A 941,** A01

blunging—the wet process of blending, or suspending ceramic material in liquid by agitation. **C 242,** C21

blunt chisel point, short chisel point— chisel point with large included angle. **F 547,** F16

blunt diamond point, short diamond point— diamond point with large included angle. **F 547,** F16

blunted-point—point end of nail purposely dulled by nail user prior to driving of nail. (See **dull point.**) **F 547,** F16

blunting line—*in fracture testing*, a line that approximates apparent crack advance due to crack-tip blunting in the absence of slow stable crack tearing. The line is defined based on the assumption that the crack advance is equal to one half of the crack-tip opening displacement. This estimate of pseudo-crack advance, Δa_B , is based on the effective yield strength of the material tested.

$$\Delta a_B = J/2\,\sigma_Y$$

E 1823, E08

blunt point, short point—point with large included angle; designed to punch and not to split. (See **dull point.**) **F 547,** F16

blush—a premature coloration of a CFB sheet caused by component migration from the CB to the CF within the sheet. **F 549,** F05

blushing—dulling or mottling of the finish of the leather resulting from condensed moisture during the drying of the finish. Also referred to as lacquer bloom. **D 1517,** D31

boarded leather—leather on which a false or accentuated grain has been produced by folding the grain side and working the leather back and forth. Hand boarding is done with a curved cork board attached to the worker's arm and rolled over the folded skin. **D 1517,** D31

board foot—a unit of measurement represented by a board 1 ft long, 1 ft wide, and 1 in. thick. Abbreviation ft. b.m.; bd. ft., fmb. In finished or surfaced lumber, the board-foot measure is based on the nominal size. In practice, the working unit is 1000 board feet. Abbreviation M bd. ft.; M B.M.; M B.F. **D 9,** D07

board insulation, *n*—semirigid insulation preformed into rectangular units having a degree of suppleness particularly related to their geometrical dimensions. **C 168,** C16

board measure—a unit of measurement of the volume in board feet of logs or lumber. Abbreviation B.M. **D 9,** D07

boards—See **lumber.** **D 9,** D07

boardy—adjective applied to stiff, inflexible leather. **D 1517,** D31

boat nail—light-duty or heavy-duty, bright or galvanized, regular-stock-steel, 1½ by 3⁄16 to 4 by 3⁄8-in., round-wire nails with oval, countersunk, 13⁄32 to 3⁄4-in. head and medium chisel point. Also, bright, annularly or helically threaded, nonferrous or stainless-steel, 1 by 0.062 to 3 by 0.165-in., round-wire nails with flat or oval, slightly countersunk, 1⁄8 to 3⁄8-in. head and medium diamond point. Also, galvanized, regular-stock-steel, 2 to 2 ½-in. square, cut nails with oval head. **F 547,** F16

bobbin coil—see **ID coil.** **E 1316,** E07

BOD (biochemical oxygen demand)—the amount of dissolved oxygen utilized by natural agencies in water in stabilizing organic matter at specified test conditions. **D 6161,** D19

body—the attribute of molten glass, associated with viscosity and homogeneity, which is conducive to workability. **C 162,** C14

body—the structural portion of a ceramic article, or the material or mixture from which it is made. **C 242,** C21

body—See **pipe body.** **C 896,** C04

body, *n*—the structural portion of a ceramic article, or the material or mixture from which it is made.**(C21, C 242)** **C 1145,** C28

body—See **container.** **D 996,** D10

body, *n*—central portion of the needle intended to be grasped by the needle holder. **F 1840,** F04

body-centered—having an atom (or group of atoms) separated by a translation of ½, ½, ½ from a similar atom (or group of atoms). The number of atoms in a body-centered cell must be a multiple of two. **E 7,** E04

body color, *n*—color produced by absorption and scattering of light by colorants within a colored material. **E 284,** E12

body dimension, *n*—*in garment construction*, a body measurement which can be used to build a sizing system or to select an appropriately sized garment. (See also **sizing system**.) **D 123,** D13

body dimension, *n*— *in garment construction*, a body measurement which can be used to build a sizing system or to select an appropriately sized garment. **F 1494,** F23

body feed—the continuous addition of filter medium (for example, diatomaceous earth) to sustain the efficacy of the filter. **D 6161,** D19

body fluid simulant, *n*—a liquid which is used to act as a model for human body liquids. **F 1494,** F23

body (food), *n*—the quality of a food or beverage relating either to its consistency, compactness of texture, fullness, flavor, or combination thereof. **E 253,** E18

body force—a force such as gravity whose effect is distributed throughout a material body by direct action on each elementary part of the body independent of the others. (ISRM) **D 653,** D18

body measurement, *n*— *in anthropometry*, a standardized distance between two specified points on the human anatomy. **F 1494,** F23

body measurements, *n*—*in anthropometry*, a standardized distance between two specified points on the human anatomy. **D 123,** D13

bog—a peat covered area with a high water table and a surface dominated by a carpet of mosses, chiefly sphagnum. It is generally nutrient poor and acidic. It may be treed or treeless. **D 653,** D18

boghead coal—See *boghead coal* under **coal**. **D 121,** D05

bogus—false; fake or simulation of the genuine article; counterfeit in that it pretends or is purported to be genuine; slang for counterfeit. (See **look-alike**.) **F 1156,** F12

Bohr magneton—a constant that is equal to the magnetic moment of an electron because of its spin. The value of the constant is ($9\ 274\ 078 \times 10^{-21}$ erg/gauss or $9\ 274\ 078 \times 10^{-24}$ J/T). **A 340,** A06

boil—(*1*) {archaic} an imperfection; a gaseous inclusion larger in size than a seed.

(*2*) turbulence caused by gases escaping from the melting batch. **C 162,** C14

boiler—self-contained electric, gas, or steam coil-powered vessel wherein water is boiled to produce steam for the steam cooker. **F 1827,** F26

boiling—a defect visible in the fired porcelain enamel caused by gas evolution which results in the formation of blisters, pinholes, black specks, dimples, or spongy surface. **C 286,** B08

boiling point, *n*—the temperature at which the vapor pressure of an engine coolant reaches atmospheric pressure under equilibrium boiling conditions. **D 4725,** D15

boiling pressure—at a specified temperature, the pressure at which a liquid and its vapor are in equilibrium. **E 7,** E04

boiling pressure—at a specific temperature, the value of the vapor pressure of the liquid at which it is equal to the external pressure. **E 1142,** E37

boiling temperature—at a specified pressure, the temperature at which a liquid and its vapor are in equilibrium. **E 7,** E04

boiling temperature—at a specific pressure, the temperature at which the vapor pressure of the liquid is equal to the external pressure. **E 1142,** E37

boilup rate, *n*— *in column distillation*, the quantity of vapor entering the column per unit of time. **D 4175,** D02

bole—the stem or trunk of a tree of size sufficient to yield lumber, veneer, or poles. **D 9,** D07

bolometer—instrument for measuring irradiance. Its principle is based on the variation of electrical resistance, with the incoming radiation, as a result of temperature change, of one or both of the resistance elements which comprise the instrument. **E 772,** E44

bolster, *n*—pillow or similarly shaped unit containing upholstery material covered by upholstery cover material that may or may not be attached to the upholstered furniture item but is sold and delivered with it. **E 176,** E05

bolt—(*1*) a short section of a tree trunk or limb;

(*2*) a short log of a length suitable for peeling in a lathe for veneer; and

(*3*) a short portion of a log prepared for production of shingles, staves, etc. **D 9,** D07

bolt—the component or part of a combination lock that locks or blocks another mechanism from operating until it is retracted. **F 471,** F12

bolt, *n*—an anchor component that connects to the bony elements of the spine, pelvis, or ribs by means of threads with the lead threads accomodating a nut thus sandwiching the bony element or implant component between the nut or washer and bolt head or other fixed stop. **F 1582,** F04

bolt—a headed and externally threaded fastener designed to be assembled with a nut. **F 1789,** F16

bolt-hole brush—a special round brush used to remove porcelain enamel bisque from in and around small openings in the ware. **C 286,** B08

bolt interconnection, *n*—an interconnection having an implant component sandwiched between two nuts or between a nut and fixed stop. **F 1582,** F04

bolt load - elongation behavior— when tensile loaded, a bolt will elongate elastically until stressed beyond its proportional limit where it will behave plastically. **F 1789,** F16

bolt (veneer)—a short log cut to length suitable for peeling in a lathe; also block. **D 1038,** D07

bomb test—see **pressure-evacuation test.** **E 1316,** E07

bond—see **adherence.** **C 286,** B08

bond, *n*—the union of materials by adhesives.

edge joint, n—in wood bonding, a type of laminate joint made by bonding adherends edge-to-edge with grain directions parallel to form wider stock.

face joint, n—in wood bonding, a type of laminate joint made by bonding adherends face-to-face with grain directions parallel to form thicker stock.

laminate joint, n—in wood bonding, a joint made by bonding layers of adherends face-to-face or edge-to-edge to form thicker or wider stock. **D 907,** D14

bond, *v*—to unite materials by means of an adhesive. (Synonym for **glue**, v.) (See also **adhere**.) (Compare **laminate,** v.) **D 907,** D14

bond, *n*—the attachment at an interface between an adhesive and an adherend. **D 1038,** D07

bond, *v*—to attach materials together by means of an adhesive. **D 1038,** D07

bond—the adhesive and cohesive forces holding two roofing components in intimate contact. **D 1079,** D08

bond breaker—*in building construction,* a material to prevent adhesion at a designated interface. **C 717,** C24

bond breaker, *n*—a material installed in pavement cracks or joints that is intended to prevent adherence of the sealant to the bottom of the crack or joint. **D 5535,** D04

bonded fabric, *n*—a layered fabric structure wherein a face or shell fabric is joined to a backing fabric, such as tricot, with an adhesive that does not significantly add to the thickness of the combined fabrics. (See also **laminated fabric, coated fabric**.) **D 123,** D13

bonderized—phosphate coated. (See **parkerized.**) **F 547,** F16

bond fireclay, n—see **fireclay, plastic or bond.** **C 71,** C08

bondline, *n*—the layer of adhesive which attaches two adherends. (Synonym for *glue line.*) **D 907,** D14

bond paper, *n*—one of many grades of paper covering a wide range of quality, from grades requiring superior permanence, strength and durability to applications where permanence and durability are less important, but in all cases requiring good printing properties, color fidelity, erasability, and cleanliness. **D 1968,** D06

bond plaster, *n*—a calcined gypsum plaster specially formulated for application over rough monolithic concrete as a bonding coat for a subsequent gypsum plaster layer. **C 11,** C11

bond strength, *n—of bonded, fused, or laminated fabrics*, the tensile force expressed in ounces per 25 mm (1 in.) of width, required to separate the component layers under specified conditions. **D 123,** D13

bond strength—*in grouting*, resistance to separation of set grout from other materials with which it is in contact; a collective expression for all forces such as adhesion, friction, and longitudinal shear. **D 653,** D18

bond strength, *n*—the unit load applied to tension, compression, flexure, peel, impact, cleavage, or shear, required to break an adhesive assembly with failure occurring in or near the plane of the bond. (See also **adhesion** and **bond**.)

dry strength, n—the strength of an adhesive joint determined immediately after drying under specified conditions or after a period of conditioning in a standard laboratory atmosphere.

wet strength, n—the strength of an adhesive joint determined immediately after removal from a liquid in which it has been immersed under specified conditions of time, temperature, and pressure. **D 907,** D14

bond strength, *n*—a measure of the force required to separate surfaces which have been bonded together. **D 1711,** D09

bone ash—calcined bone consisting essentially of calcium phosphate. **C 242,** C21

bone china—a translucent china made from a ceramic whiteware body composition containing a minimum of 25 % bone ash. **C 242,** C21

bone coal—See *bone coal* under **coal**. **D 121,** D05

bone-dry—see **oven-dry.** **D 1695,** D01

bonnet, *n*—*in cleaning pile floor coverings*, an absorbent pad which can be mounted under a rotary shampoo machine. **D 123,** D13

BOO—build, own, operate. **D 6161,** D19

bookbinder's wire—wire used in stitchers to fasten paper; measured according to AWG sizes. **F 592,** F16

book bulk, *n*—the overall thickness of a given number of sheets. See **thickness**. **D 1968,** D06

book fold, *n*—a fabric doubled selvage to selvage, then folded back and forth upon itself in predetermined lengths. (See also **shoe fold**.) **D 123,** D13

book paper, *n*—a general term for a group of uncoated or coated papers (exclusive of newsprint) suitable for the graphic arts. **D 1968,** D06

boom—floating mechanical barrier used to control the movement of substances that float. **F 818,** F20

boom planing—heeling over of a boom and loss of draft. **F 818,** F20

boom section—length of boom between two end connectors. **F 818,** F20

boom segment—repetitive identical portion of the boom section. **F 818,** F20

boom sprayer—a sprayer apparatus consisting of a pressure source and controls, and employing a boom (vertical or hotizontal) with atomizers (hydraulic, rotary, or other) arranged to provide uniform coverage of the treated surfaces. **E 1102,** E35

boom submergence (aka submarining)—containment failure due to loss of freeboard. **F 818,** F20

boom weight—dry weight of a fully assembled boom section including end connectors. **F 818,** F20

booster, n— synonym for **secondary accelerator**, which is the preferred term. **D 1566,** D11

booster heater—the water heater responsible for heating and maintaining the final sanitizing rinse water (to a minimum of 180°F) to dishwater, may be separate from dishwater or integral. **F 1827,** F26

booster heater inlet temperature—the temperature of water being supplied to the booster heater (see **dishwashing machine, commercial:** *booster heater*) measured at the booster heater inlet. **F 1827,** F26

boost melting—See **electric boosting**. **C 162,** C14

boot—a suspended enclosure in the nose of a melter protecting a portion of the surface and serving as a gathering opening. **C 162,** C14

BOOT—build, own, operate and transfer. **D 6161,** D19

borax glass—vitreous anhydrous sodium tetraborate ($Na_2B_4O_7$). **C 162,** C14

borderline pumping temperature, *n*— the lowest temperature at which the critical yield stress or critical viscosity that allows a fluid to be pumped occurs. **D 4175,** D02

borehole, *n*—the circular hole through soil and rock strata made by boring. **D 121,** D05

borehole—a hole of circular cross-section made in soil or rock. **D 653,** D18

borehole log—the record of geologic units penetrated, drilling progress, depth, water level, sample recovery, volumes and types of materials used, and other significant facts regarding the drilling of an exploratory borehole or well. **D 653,** D18

borehole television log—a borehole or well video record produced by lowering a television camera into the borehole or well. This record is useful in visually observing downhole conditions such as collapsed casing or a blocked screen. **D 653,** D18

borer holes—voids made by wood-boring insects. **D 1038,** D07

borescope—A flexible or rigid tube-like instrument used for a remote direct viewing visual aid. The instrument may consist of mirrors, prisms, lenses, optic-fibers, or a miniature CCD camera to transmit images to the viewing or recording medium. **E 1316,** E07

boron equivalent—*in the nuclear industry*, the absorptive capacity for thermal neutrons of weights of various elements expressed in terms of the weight of natural boron. **C 859,** C26

borosilicate glass—any silicate glass having at least 5 % of boron oxide (B_2O_3). **C 162,** C14

bottle—See **container.** **D 996,** D10

bottom—See **box.** **D 996,** D10

bottom assembly, *n*—the components of the lowermost part of a slide fastener which determine whether the slide fastener will be non-separable or separable. (See also **non-separable zipper** and **separable zipper**.) **D 123,** D13

bottom charge—concentrated explosive charge at the bottom of a blast hole. (ISRM) **D 653,** D18

bottom echo—see **back reflection.** **E 1316,** E07

bottom filler—the material used for filling the cavity between the inner and outer soles of the shoe; or, a material, usually felt or a mixture of adhesive and cork, or sawdust, that is used to fill the space between the shank and the lasted over edges of the uppers. **F 869,** F08

bottom out, *v*—to deform a tire by radial load on the tread until radial movement of the inside surface is stopped by the rim or other tire inside surface. **F 538,** F09

bottom rail—See **railing systems.** **E 631,** E06

bottom rail—the lowest member of a railing system, supporting balusters or panels, if any. **E 1481,** E06

bottom rail—horizontal member of the framework running continuously along the bottom edge of the fence. **F 552,** F14

bottoms—the underface of the shoe sole which extends from the toe to the heel breast. The heel is not a part of the bottom. **F 869,** F08

bottom scissor half—the component which contains the threaded end of the screw. **F 1078,** F04

bottomset bed—fine-grained material (usually silts and clays) slowly deposited on the bed of a quiescent body of water which may in time be buried by foreset beds and topset beds. **D 4410,** D19

bottomsize, nominal—*for the purpose of Test Method D 4749*, the sieve designating the lower limit or bottomsize shall be that sieve of the series given in the Standard Series of Sieves section with the largest openings through which passes a total of less than 15 % of the sample. This defined bottomsize is not to be confused with the size of the smallest particles in the lot. **D 121,** D05

bottom stop, *n*—a part affixed to both stringers immediately below, or over, the chain, holding the two stringers together at the bottom and preventing the slider from leaving the chain. **D 123,** D13

bottom-tension boom—boom with tension member located along the bottom of the skirt. **F 818,** F20

bottom time, *n*—the total elapsed time measured in minutes from the time that the diver leaves the surface in descent to the time that the diver begins ascent. **F 1549,** F32

bottom trip—tool activation by tool nose touching the work, while at the same time activating trigger trip. **F 592,** F16

Bouguer's law—the absorbance of a homogeneous sample is directly proportional to the thickness of the sample in the optical path. **E 131,** E13

Bouguer's law, *n*—the absorbance of a homogeneous sample is directly proportional to the thickness of the sample in the optical path. (Also known as **Lambert's [thickness] law.**) **E 284,** E12

boulder clay—a geological term used to designate glacial drift that has not been subjected to the sorting action of water and therefore contains particles from boulders to clay sizes. **D 653,** D18

boulders—a rock fragment, usually rounded by weathering or abrasion, with an average dimension of 12 in. (305 mm) or more. **D 653,** D18

boulder size (fluvial sediment)—larger than 256 mm in diameter. **D 4410,** D19

boulevard clamp—a two piece clamp with carriage bolts and nuts designed for 180° horizontal rail connections to a line post (also called *line rail clamp*). **F 552,** F14

boundary friction, *n*—friction at low sliding speeds (0.02 m/min or less) where lubrication occurs under thin-film lubricant conditions. **D 123,** D13

boundary grain—in the Jeffries' method for grain size measurement, a grain that is intersected by the boundary of the standard area and is, therefore, counted only as one-half grain. (See also **Jeffries' Method.**) **E 7,** E04

boundary layer—a thin layer at the membrane surface where water velocities deviate significantly less than those in the bulk flow. **D 6161,** D19

bound monomer, *n*—monomer that is combined or reacted with itself or other types of monomers in a polymerization reaction to form a polymer. **D 1566,** D11

bound seam-finish, *n*—*in home sewing,*a finish for the raw edges of a plain seam, in which another fabric encloses the raw edges of one or more seam allowances. (Compare **Hong Kong seam-finish.**) **D 123,** D13

bow—the distortion of a piece of lumber in which there is a deviation in a direction perpendicular to the flat face from a straight line from end to end of the piece. **D 9,** D07

bow, *n*—a fabric condition resulting when filling yarns or knitted courses are displaced from a line perpendicular to the selvages and form one or more arcs across the width of the fabric. (See also **double bow.**) **D 123,** D13

bow—a condition of longitudinal curvature in pultruded parts.

D 3918, D20

bow—*of a semiconductor wafer,* the deviation of the center point of the median surface of a free, unclamped wafer from a median-surface reference plane established by three points equally spaced on a circle with diameter a specified amount less than the nominal diameter of the wafer. **F 1241,** F01

bowl—See **spout**. **C 162,** C14

bowl guard—the barrier, assembled over the bowl area, intended to reduce the user's access to the hazards that exist in the bowl during machine operation. **F 1827,** F26

box, *n*—a rigid **container** having closed faces and completely enclosing the contents. When this term is used in connection with fiberboard boxes, such fiber boxes must comply with all the requirements of the carrier rules. (See **carton.**)

bottom, n— the **face** of a **box** on which it usually rests while filling.

box batten, n—a reinforcing member, (*1*) for a **wood box** internally or externally applied to the sides, top and bottoms. When applied externally it should be applied in pairs; (*2*) in a **wirebound box,** a batten is a reinforcement used on the ends of the container only.

cleated fiberboard box—a rigid **container** having five or six **panel faces** with wood strips fastened to them, the panels being made of **solid** or **corrugated fiberboard.**

cleated plywood box—a rigid **container** having five or six **panel faces** with wood strips fastened to them, the panels being made of plywood.

flange, n— in **fiberboard boxes,** an extension to a panel similar to a short flap that may be folded in or out, usually at angles of 90 or 180° to the panel.

nailed wood box— a rigid **container** constructed of wood in several standard styles, assembled by fastening sides, top and bottom to the ends with nails or other suitable fasteners.

skid box—a metal, wooden, or fiber **box** fastened to a platform raised on skid members or legs; it may or may not be collapsible. (See also **skid.**)

wirebound box— a rigid **container** whose sides, top, and bottom are of rotary-cut lumber, sliced lumber, **resawn lumber, fiberboard,** or combinations thereof, usually ⅜ in. (9.5 mm) or less in thickness, fastened to **cleats** and to each other by means of binding wires and staples; and ends of similar material, plain or stapled to **battens** or **liners,** fastened in place by means of nails or staples or wires stapled thereto.

D 996, D10

box annealing, *n*—**annealing** in a sealed container under conditions that minimize oxidation. **A 941,** A01

boxboard—See **paperboard.** **D 996,** D10

box calf or sides—sides or skins finished by folding with the grain side in and rubbing the flesh side with a cork-surfaced instrument known as a hand board. Machinery is now also used. The effect is sometimes imitated by embossing. Also called "box" or "willow" finish. **D 1517,** D31

boxcar truncation—identical effective weighting of all points in the measured interferogram prior to the Fourier transform; all points outside of the range of the measured interferogram take a value of zero. **E 131,** E13

boxed heart—the term used when the pith falls entirely within the four faces of a piece of wood throughout its length. **D 9,** D07

boxed pith—See **boxed heart.** **D 9,** D07

box furnace—a furnace in which, periodically, a load of ware is introduced; fired, and removed. **C 286,** B08

box nail—bright, coated or galvanized, regular-stock-steel, 1 by 0.058 to 5 by 0.162-in. nails, made of lighter-gage wire than common nails and sinkers, with flat 11⁄64 to 13⁄32-in. head and medium diamond point. **F 547,** F16

box section—a concrete pipe with a rectangular cross section. **C 822,** C13

box section bottom slab—lower horizontal portion of a box section in the installed condition. **C 822,** C13

box section top slab—upper horizontal portion of a box section in the installed condition. **C 822,** C13

box section wall—vertical sides of a box section in the installed condition. **C 822,** C13

box stay wire—wire used in stitchers for assembly of containers; with dimensions measured in thousandths of inches. **F 592,** F16

BP—an abbreviation for base paper, referring to an uncoated paper that is coated to form a carbonless product. Also known as base stock or rawstock. **F 549,** F05

brace band—a symmetrically formed strip of metal shaped to fit around a post and used with a carriage bolt and nut to attach the rail end or brace rail end truss rod tightener to the post. Also used for attaching barbed wire, tension wire, and other items to a terminal post. **F 552,** F14

brace rail—a compression member at terminal posts or corner posts. **F 552,** F14

brace rail end—a cup-shaped fitting similar to a rail end with provision for attaching a truss rod. **F 552,** F14

bracing—See **loading.** **D 996,** D10

bracket, *n*—projecting element or hardware attached to the surface of a member to support other members. **E 631,** E06

brackish water—water that contains dissolved matter at an approximate concentration range from 1000 to 30 000 mg/L. **D 1129,** D19

brackish water—water with an approximate concentration of total dissolved solids ranging from 1000 to 10 000 mg/L. See **high brackish water, sea water.** **D 6161,** D19

brad—small nail with small head.

brad, common wire— slender, regular-stock-steel, ⅜ by 0.035 to 6 by 0.262-in. wire nails with brad 0.050 to 0.331-in. head and medium diamond point.

brad, cut—slender, usually small, regular-stock-steel nails of same thickness throughout, but tapering in width; with slight projection on one side serving as head. Also, tapering, square-bodied, finishing nail with countersunk head. **F 547,** F16

brad-clinched point—pointed end of nail, having been driven through member, flattened and bent sideways to a limited extent when striking flat anvil plate. **F 547,** F16

brad head—small-diameter, deep, circular, barrel-shaped head with flat or concavely cupped top surface, as found on finishing nails and common brads for countersinking where concealment is important. **F 547,** F16

Bragg angle—the angle between the incident beam and the lattice planes considered. **E 7,** E04

Bragg equation:—

$$n\lambda = 2d \sin \theta$$

where:
n = order of reflection,
λ = wavelength of X-rays,
d = distance between lattice planes, and
θ = Bragg angle.

E 7, E04

Bragg method—a method of X-ray diffraction in which a single crystal is mounted on a spectrometer with a crystal face parallel to the axis of the instrument. **E 7,** E04

Bragg's rule—an empirical rule formulated by W. H. Bragg and R. Kleeman that states that the stopping cross section of a compound specimen is equal to the sum of the products of the elemental stopping cross sections for each constituent and its atomic fraction, that is,

$$\epsilon(A_xB_y) = x\epsilon_A + y\epsilon_B$$

where:

$\epsilon(A_xB_y)$	=	the stopping cross section of the compound, and
A_xB_y and ϵ_A and ϵ_B	=	the stopping cross section of elements *A* and *B* respectively.

E 673, E42

braid, *n*—a narrow tubular or flat fabric produced by intertwining strands of materials according to a definite pattern. **B 713,** B01

braid, *n*—a narrow tubular or flat fabric produced by intertwining a single set of yarns according to a definite pattern (Maypole process). **D 123,** D13

braid, *n*—(*1*) woven metallic wire used as a shield for insulated conductors and cables.

(*2*) A woven fibrous protective outer covering over an insulated conductor or cable. **D 1711,** D09

braided fabric, *n*—a structure produced by interlacing several ends of yarns in a manner such that the paths of the yarns are not parallel to the fabric axis. **D 123,** D13

braided fabric—see *braided fabric* under **fabric.** **D 3878,** D30

braided river—a wide- and shallow-river where the flow passes through a number of small interlaced channels separated by bars or shoals. **D 4410,** D19

braided rope, *n*—a cylindrically produced rope made by intertwining, maypole fashion, several to many strands according to a definite pattern with adjacent strands normally containing yarns of the opposite twist. **D 123,** D13

braising pan—equipment suitable for the preparation of foods by several methods, such as frying, braising, boiling and simmering. **F 1827,** F26

brake bar rack, *n*—a variable friction device consisting of a u-shaped metal bar to which are attached several metal bars that create friction on the rope. Also known as a rappel rack. **F 1490,** F32

braking coefficient, [*nd*], *n*—the ratio of the braking force to the normal force on a tire. **F 538,** F09

braking force coefficient, *n*—*of a tire,* the ratio of braking force to normal force. **F 538,** F09

braking force coefficient, peak, *n*—*of a tire,* the maximum value of tire braking force coefficient that occurs prior to wheel lockup as the braking torque is progressively increased. **F 538,** F09

braking force coefficient, slide, *n*—*of a tire,* the value of braking force coefficient obtained on a locked wheel. **F 538,** F09

braking force coefficient, tire—the ratio of braking force to vertical load. **E 867,** E17

braking force coefficient, tire, peak—the maximum value of tire braking force coefficient that occurs prior to wheel lockup as the braking torque is progressively increased. **E 867,** E17

braking force coefficient, tire, slide—the value of the braking force coefficient obtained on a locked wheel. **E 867,** E17

braking force, [*F*], *n*—*of a tire,* the negative longitudinal force resulting from braking torque application. **F 538,** F09

braking force, tire—the negative longitudinal force resulting from braking torque application. **E 867,** E17

braking torque—the negatively directed wheel torque. (See **torque (T), wheel**). **E 867,** E17

braking torque, [ML^2/T^2], *n*—*of a vehicle,* the negative wheel torque. **F 538,** F09

branched polyethylene plastics, *n*—those containing significant amounts of both short-chain and long-chain branching and having densities in the 0.910 to 0.940 g/cm^3 range. **D 883,** D20

branching—*in the nuclear industry,* the existence of two or more modes by which a radionuclide can undergo radioactive decay. **C 859,** C26

brand, *n*—*of a tire,* the name under which one or more tire lines are marketed. **F 538,** F09

brand, tire, private, *n*—a brand name used by a tire seller or group of sellers who are not manufacturers. **F 538,** F09

brashness—a condition that causes some pieces of wood to be relatively low in shock resistance for the species and, when broken in bending, to fail abruptly without splintering at comparatively small deflections. **D 9,** D07

Braun's law—the ratio of the solubility change with pressure, temperature being constant, $(\delta X/\delta P)T$, to the solubility change with temperature, pressure being constant, $(\delta X/\delta T)P$, is equal to the negative of the product of the absolute temperature, *T*, and the (fictitious) volume change (Δv) which accompanies the solution of 1 g·molecular weight of the solute in an infinitely large quantity of the saturated solution at *T* degrees, divided by the amount of heat, *Q*, developed in the process:

$$(\delta X/\delta P)T/(\delta X\ \delta T)P = -T{\cdot}\Delta v/Q$$

E 7, E04

brazing—the joining of ends of two wires, rods, or groups of wires with a nonferrous filler metal at temperatures above 800°F [427°C]. **B 354,** B01

break—(*1*) *heavy leather*—the places, in the areas where the fore shank and hind shank join the body of the hide, where the texture of the leather changes quite sharply from the firm, close weave of the bend to a loose, open texture.

(*2*) *shoe upper leather*—the superficial wrinkling formed when the leather is bent, grain inward. Adjectives commonly used to describe this characteristic are "tight," "fine," "loose," "coarse," and "pipy," or "flanky." **D 1517,** D31

break, *n*—a crack or tear extending into or through the reinforcing material. **F 538,** F09

breakaway staple—staple with its crown designed to break off if removal is attempted; used to discourage pilfering and shop-lifting. **F 592,** F16

breakaway torque—torque necessary to start a fastener in motion after the axial load of the mating components has been reduced to zero. **F 1789,** F16

breakdown, electrical—the electrical discharge or arc occurring between the electrodes and through the equipment being tested. **F 819,** F18

breakdown potential—the least noble potential where pitting or crevice corrosion, or both, will initiate and propagate. **G 15,** G01

breakdown voltage—see **dielectric breakdown voltage.** **D 1711,** D09

breaker, *n*—*in a tire,* one or more plies under the tread region of a tire that are additional to those which extend from bead to bead. **F 538,** F09

breakeven analysis—a technique for determining that value of a variable which results in benefits (savings) just equal to costs.

E 833, E06

break factor, *n—in yarn testing*, the comparative breaking load of a skein of yarn adjusted for the linear density of the yarn expressed in an indirect system. **D 123,** D13

break-in, n—See **run-in**. **G 40,** G02

break in, v—See **run in**. **G 40,** G02

breaking force, *n*—the maximum force applied to a material carried to rupture. (Compare **breaking point.**. See **breaking strength.**) **D 123,** D13

breaking force, *(F), J, n*—the force at failure. **D 4439,** D35

breaking length, *n—of pulp and paper*, a calculated value expressed as the length of a strip of paper, usually stated in metres, which would break of its own weight when suspended vertically; calculated from the tensile strength and the basis weight of the sheet. **D 1968,** D06

breaking load, *n*—the maximum force applied to a specimen in a tensile test carried to rupture. **D 4439,** D35

breaking load [F], *n*—the load at which fracture occurs. **E 6,** E28

breaking point, *n—on a force-elongation curve, or stress-strain curve*, the point corresponding with the breaking force or the breaking stress in a tensile test. (Compare **breaking force**.) **D 4848,** D13

breaking strength, *n— as related to wire*, the maximum force developed prior to fracture during tension testing of wire and wire products. **A 902,** A05

breaking strength—the maximum load that a conductor attains when tested in tension to rupture. **B 354,** B01

breaking strength, *n*—the ability or capacity of a specific material to withstand the ultimate tensile load or force required for rupture. (See also **tensile strength**) **D 123,** D13

breaking strength, *n*—strength expressed in terms of breaking force. (See also **breaking force** and **strength**. *Syn*., strength at break) **D 4848,** D13

breaking tenacity, *n*—the tenacity at the breaking force. **D 123,** D13

breaking tenacity, *n*—the tenacity corresponding to the breaking force. **D 123,** D13

breaking toughness, *n*—the actual work per unit volume or per unit mass of material that is required to rupture the material. **D 123,** D13

breaking toughness, *n*—toughness up to the breaking force of a material. **D 4848,** D13

breaking toughness, *T, (FL⁻), Jm^{-2}, n—for geotextiles*, the actual work-to-break per unit surface area of material. **D 4439,** D35

break-in, [*L*], *n*—one or more periods of initial standardized tire operation during which tire is brought to the state which will lead to more consistent test results. **F 538,** F09

break loose torque—torque applied in a removal direction necessary to start the fastener in motion from its fully preloaded installed position. **F 1789,** F16

break-open—in testing thermal protective material, a response evidenced by the formation of a hole in the material which allows the molten substance to pass through the material. **F 1494,** F23

break out—in dry process enameling, a defect characterized by an area of blisters with well defined boundaries. **C 286,** B08

breakout—fiber separation or break on surface plies at drilled, machined, etc., edges. **E 1749,** E06

breakout pressure, *n—for inflatable restraints*, the pressure level during deployment which ruptures the module cover. **D 123,** D13

breakpoint—the appearance in the effluent of a specified concentration of an adsorbate. **D 2652,** D28

breakpoint chlorination—the point at which the water chlorine demand is satisfied and any further chlorine is the chlorine residual, the "free" chlorine species. **D 6161,** D19

break tank—a storage device used for hydraulic isolation and surge protection. **D 6161,** D19

breakthrough—the first appearance in the effluent of an adsorbate of interest under specified conditions. **D 2652,** D28

breakthrough detection time, *n*—the elapsed time measured from the start of the test to the sampling time that immediately precedes the sampling time at which the test chemical is first detected. **F 1494,** F23

breakup, *n*—liquid disintegration that occurs during atomization. **E 1620,** E29

breakup length, *n*—the distance between the liquid discharge point of an atomizing device and the point where liquid breakup commences. **E 1620,** E29

breakwater stone—(generally three tons to twenty tons in weight) stone resulting from blasting, cutting, or other means to obtain rock heavy enough to require handling individual pieces by mechanical means. **D 653,** D18

breast, *n*—the fore part or the heel of a shoe next to the shank; the front face of a shoe heel. **F 1646,** F13

breast wall—(*1*) the entire side wall of a melter between the flux block and the crown, but not including the ends.

(*2*) refractory wall between pillars of a pot furnace and in front of or surrounding the front of a pot. **C 162,** C14

breather—a loosely woven cloth (such as glass fabric) which serves as a continuous vacuum path over a part but does not come in contact with the resin. **E 1749,** E06

breather finish—coating system allowing the passage of water vapor. **E 631,** E06

breathing zone, *n*—that location in the atmosphere at which persons breathe. **D 1356,** D22

breezing—{archaic} buckwheat anthracite coal or coarse sand spread on the siege before setting of pots. **C 162,** C14

bremsstrahlung—*XPS*, photon radiation, continuously distributed in energy up to the energy of the incident electrons, emitted from an anode due to deceleration of incident electrons within the anode. The bremsstrahlung from a conventional X-ray source contributes to the background and the Auger signal strengths in an XPS spectrum. **E 673,** E42

brick, *n*—a solid masonry unit of clay or shale, usually formed into a rectangular prism while plastic and burned or fired in a kiln. Brick is a ceramic product. **C 43,** C15

brick face, *n*—the face of brick is any surface intended for use to form the exposed surface of the masonry structure. **C 43,** C15

brick-siding nail—galvanized, colored (baked-lacquer finished), plain-shank barbed or annularly threaded, regular-stock-steel, ⅞ by 0.092 to 2 ½ by 0.099-in. nails with flat checkered 3⁄16 or ¼-in. head and medium diamond point. **F 547,** F16

bridge—the structure formed by the end walls of the adjacent melter and refiner compartments of a melting furnace and the covers spanning the gap between the end walls. **C 162,** C14

bridge, *n—in buttons*, the area between the holes partially covered by the sewing threads with dimensions varying upon design and end use. **D 123,** D13

bridge—an obstruction within the annulus which may prevent circulation or proper emplacement of annular materials. **D 653,** D18

bridge, *n*—the area of a button between the holes partially covered by the sewing threads with dimensions varying upon design and end use. **D 5497,** D13 *bridge cover*—See **bridge wall cover.** **C 162,** C14

bridge sealant joint, *n—in building construction,* a joint where sealant is applied over the joint to the face of substrates that are in approximately the same plane. **C 717,** C24

bridge top stop, *n—in zippers,* a part affixed immediately above the chain, holding the tops of the two stringers together and preventing the slider from leaving the chain. **D 123,** D13

bridgewall—that part of a melting furnace forming a bridge or separation between melter and refiner. **C 162,** C14

bridge wall cover—refractory blocks spanning the space between the bridge walls. **C 162,** C14

bridging—the formation of arched cavities in a powder mass. **B 243,** B09

bridging, *n—in framing,* sections of wood or metal pieces used between framing members to stiffen, give lateral support, and minimize rotation. **C 11,** C11

bridging—spanning a feature without full contact, such as tape or fabric spanning a radius, step, core edge, etc., or vacuum bagging material spanning tool or part surfaces. **E 1749,** E06

bridging—enlargement of a graphic shape beyond the COL, which produces undesired character fill in. **F 149,** F05

bridging—combination of peaks and smudges that may close or partially close the loop of a character (F 149). **F 335,** F05

bridging, *v*—the clumping of toner which causes a hollow area in the toner supply that prevents the free flow of toner to the dispenser auger. **F 1457,** F05

bridging failure—portions of a boom emerging from the water due to poor wave conformance, with resulting containment failure. **F 818,** F20

bridle—device attached to a boom to distribute the load exerted by towing or anchoring the boom. **F 818,** F20

bridle leather—a harness-finished strap leather. **D 1517,** D31

brief (of a facility)— See **facility program.** **E 1480,** E06

bright—term applied to nails with natural bare finish resulting from cleaning of nails which have not undergone treatments affecting finish, such as hardening, bluing, coating, plating, etching, painting, etc. Also applied to polished appearance after plating. **F 547,** F16

bright annealing, *n*—**annealing** in a protective medium to prevent discoloration of the bright surface. **A 941,** A01

bright dip (nonelectrolytic)—a solution used to produce a bright surface on a metal. **B 374,** B08

bright electroplating—a process that produces an electrodeposit having a high degree of specular reflectance in the as-plated condition. **B 374,** B08

bright electroplating range—the range of current densities within which an electroplating solution produces a bright deposit under a given set of operating conditions. **B 374,** B08

brightener—an addition agent that leads to the formation of a bright plate, or that improves the brightness of the deposit. **B 374,** B08

brightener—see **fluorescent whitening agent.** **D 459,** D12

brightening agent—see **fluorescent whitening agent.** **D 459,** D12

brightfield illumination—*for reflected light,* the illumination which causes specularly reflected surfaces normal to the axis of a microscope to appear bright. *For transmission electron microscopy,* the illumination of an object so that it appears on a bright background. **E 7,** E04

bright finish—non-coated wire finish. **F 547,** F16

bright glaze—See *bright glaze* under **glaze.** **C 242,** C21

brightness—as commonly used in the paper industry, the reflectivity of a sheet of pulp or paper for blue light measured under standardized conditions on a particular instrument designed and calibrated specifically for the purpose. **D 1695,** D01

brightness, *n—in paper and paperboard,* reflectance of an infinitely thick stack of material measured for blue light with centroid wavelength of 457 nm under specified spectral and geometric conditions. **D 1968,** D06

brightness, *n—(1)* aspect of visual perception whereby an area appears to emit more or less light. **[CIE] (F 923)**

(*2*) *of an object color,* combination of lightness and saturation.

(*3*) *in the textile industry,* perceived as saturated, vivid, deep, or clean.

(*4*) *of paper,* reflectance of an infinitely thick specimen (reflectivity) measured for blue light with a centroid wave-length of 457 nm under specified spectral and geometric conditions of measurement.

(*5*) *dyer's,* the color quality, combining lightness and saturation, that would be decreased by adding black, gray, or a complementary color to a chromatic dye. **E 284,** E12

brightness—*in paper,* a characteristic of white paper measured in terms of reflectance in the blue and violet portions of the spectrum. **F 149,** F05

bright or mirror finish—highly reflective surface. **F 1840,** F04

bright plating—a process that produces an electrodeposit having a high degree of specular reflectance in the as-plated condition. **B 374,** B08

bright plating range—the range of current densities within which a given plating solution produces a bright plate. **B 374,** B08

bright-throwing power—the measure of the ability of a plating solution or a specified set of plating conditions to deposit uniformly bright electroplate upon an irregularly shaped cathode. **B 374,** B08

bright-wood—wood free of stain or any discoloration. **D 9,** D07

brine—water that contains dissolved matter at an approximate concentration of more than 30 000 mg/L. **D 1129,** D19

brine—the concentrate (reject) stream from a crossflow membrane device performing desalination. Portion of the feed stream which does not pass through the membrane. **D 6161,** D19

brine, *n*—in geothermal, fluids in a liquid phase that have been produced from geothermal wells or from hot springs and that contain appreciable amounts of sodium chloride and other salts. **E 957,** E44

brine (concentrate) seal—a rubber lip seal on the outside of a spiral wound cartridge which prevents feed by-pass between the cartridge and the inside pressure vessel wall. **D 6161,** D19

Brinell hardness number, HB [FL^{-2}], *n*—a number related to the applied load and to the surface area of the permanent impression made by a ball indenter computed from the equation:

$$HB = 2P/\beta D(D - \sqrt{D^2 - d^2}$$

where:

P = applied load, kgf,
D = diameter of ball, mm, and
d = mean diameter of the impression, mm.

E 6, E28

Brinell hardness test, *n*—an indentation hardness test using calibrated machines to force a hard ball, under specified conditions, into the surface of the material under test and to measure the diameter of the resulting impression after removal of the load. **E 6,** E28

brinelling, *n*—damage to a solid bearing surface characterized by one or more plastically formed indentations caused by static or impulsive overloads, especially as found in rolling contact bearings. (See also **false brinelling**.) **G 40,** G02

brine seal carrier—see ATD. **D 6161,** D19

brine system staging—a process in which the concentrate, under pressure, of a group of membrane devices is fed directly to another set of membrane devices to improve the efficiency of the water separation. **D 6161,** D19

brining—a process of curing hides by soaking with salt solution (sodium chloride). **D 1517,** D31

briquet—see **compact.** **B 243,** B09

briquette, *n*—a cylindrical block composed of granulated coal or coke particles compressed and embedded with an epoxy binder. **D 121,** D05

brittle erosion behavior, *n*—erosion behavior having characteristic properties that can be associated with brittle fracture of the exposed surface; that is, little or no plastic flow occurs, but cracks form that eventually intersect to create erosion fragments. (See also **ductile erosion behavior**.) **G 40,** G02

brittle failure—a pipe failure mode which exhibits no visible (to the naked eye) material deformation (stretching, elongation, or necking down) in the area of the break. **F 412,** F17

brittle fracture, *n*—fracture that occurs without appreciable plastic deformation of the material. **A 644,** A04

brittle fracture area, *n*—The fraction or percent of the fracture surface that formed by brittle fracture. (When observed with no or low magnification, brittle fracture appears whiter and shinier than ductile fracture.) **A 644,** A04

brittleness—the tendency of a material to break at a very low strain, elongation, or deflection, and to exhibit a clean fracture surface with no indications of plastic deformation. **E 1749,** E06

broadcast application—the distribution of a material uniformly over the entire area to be treated and not just to portions of the area. **E 609,** E35

broadcast application—an application of a material over the entire area of a field. **E 1102,** E35

broadcast resin monolithic floor surfacing—a flooring system whereby a film of catalyzed resin binder is applied on a prepared concrete substrate followed immediately by the seeding to excess, into this wet film, of a dry inert filler. Upon hardening, any unbonded filler is removed. The application may be repeated once or several times to yield a greater thickness of surfacing. A top coat may be applied. **C 904,** C03

broadgoods—non-preimpregnated or uncured preimpregnated materials wider than 12 in. (300 mm). **E 1749,** E06

broad head—See **large head.** **F 547,** F16

broiler, conveyor—equipment that carries the food product on a wire rack through a tunnel that heats using high temperature radiant heat sources above and or below the rack, for cooking on one or both sides of the food product at once. See **broiler, overfired** and **broiler, underfired**. **F 1827,** F26

broiler, overfired—equipment with a high temperature radiant heat source above a grate for cooking food. **F 1827,** F26

broiler, underfired—equipment with a high temperature radiant heat source below a grate for cooking food, including charbroiler, radiant-broilers, smokeless broilers, etc. **F 1827,** F26

broken end, *n—in woven fabrics,* a void in the warp direction due to yarn breakage. **D 123,** D13

broken filament, *n—in multifilament yarn,* breaks in one or more filaments. **D 123,** D13

broken grain (shelling, leafing, grain separation)—a separation on veneer surface between annual rings. **D 1038,** D07

broken pick, *n—in woven fabrics,* a discontinuity in the filling direction caused by a break or cut in the filling yarn. **D 123,** D13

bromine index, *n*—the number of milligrams of bromine that will react with 100 g of sample under the conditions of the test. **D 4175,** D02

bromine index—the number of milligrams of bromine consumed by 100 g of sample under given conditions. **D 4790,** D16

bromine number, *n*—the number of grams of bromine that will react with 100 g of the sample under the conditions of the test. **D 4175,** D02

bromine number, total—number of centigrams of bromine equivalent to the total unsaturation present in 1 g of sample. It is a measure of the total ethylenic unsaturation present in the designated aliphatic compound. **E 1547,** E15

bronzing—the application of a chemical finish to copper or copper alloy surfaces to alter the color. **B 374,** B08

bronzing—excessive concentration of crystallized dyestuff on the surface of the leather tending to give a metallic sheen. **D 1517,** D31

bronzing—reflection of reddish-brown highlights from surface, characteristic of nonuniformly deposited film. **D 2825,** D21

bronzing, *n*—an image which has a bronze sheen. **F 1857,** F05

bronzy color (or bronzing), *n*—a metallic coloration observed when viewing the light reflected at angles near the angle of specular reflection, the color usually being quite different from that observed for other directions. **E 284,** E12

brooming—embedding a ply by using a broom to smooth it out and ensure contact with the adhesive under the ply. **D 1079,** D08

broom nail—bright, regular-stock-steel, ⅝ to ¾ by 0.072 or 0.080-in. nails with flat 13⁄64 or 7⁄32 in. head and medium diamond point. **F 547,** F16

brown coat, *n*—the second coat in three-coat gypsum plaster application. **C 11,** C11

bruise—synonymous with **percussion cone.** **C 162,** C14

bruise, *n—in fabrics,* an area that has been subjected to impact or pressure, which differs from the adjacent normal fabric. **D 123,** D13

brush—a sliding contact member consisting of one or more sliders (see sliders). **B 542,** B02**brush**—to remove bisque in a definite pattern by means of a brush. **C 286,** B08

brush, *n*—a hand-held cleaning tool consisting of a base into which bristles are inserted. **D 123,** D13

brush, *vt—in refurbishing textiles,* (1) to use a brush to remove surface particulate soils, (2) work a cleaning solution or spot cleaning agent into carpet pile or upholstery fabric, or (3) restore the appearance of pile fabrics. **D 123,** D13

brush, bolt-hole—see **bolt-hole brush.** **C 286,** B08

brush coat—*in sealants,* a thin layer of Class A curing type sealant used alone or in conjunction with a Type B sealant. **E 1749,** E06

brush coloring—the application of dye-stuff to leather with a brush or swab, the leather being laid on a table. Also called table dyeing. **D 1517,** D31

brush-drag, *n*—resistance encountered when applying a coating by brush, directly related to the high-shear viscosity of the coating. **D 16,** D01

brushing—see **brush.** **C 286,** B08

brush plating—a method of plating in which the plating solution is applied with a pad or brush, within which is an anode and which is moved over the cathode to be plated. **B 374,** B08

brush polishing (electrolytic)—a method of electropolishing (q.v.) in which the electrolyte is applied with a pad or brush in contact with the part to be polished. **B 374,** B08

BS—abbreviation for base stock. (see **BP**) **F 549,** F05

B-scan presentation—a means of ultrasonic data presentation which displays a cross section of the specimen indicating the approximate length (as detected per scan) of reflectors and their relative positions. **E 1316,** E07

B-stage, *n*—an intermediate stage in the reaction of certain thermosetting resins in which the material swells when in contact with certain liquids and softens when heated, but may not entirely dissolve or fuse. **D 883,** D20

B-stage, *n*—*in polymer chemistry,* an intermediate stage in a thermosetting resin reaction in which the material softens when heated, and swells but does not dissolve in certain liquids. (Compare **A-stage** and **C-stage.**) **D 907,** D14

B-stage—an intermediate stage, in the reaction of certain thermosetting resins in which the material softens when heated and swells in contact with certain liquids, but may not entirely fuse or dissolve. The resin in an uncured thermosetting adhesive is usually in this stage. Sometimes referred to as *resitol.* **E 1749,** E06

BTU—one British Thermal Unit is the amount of heat required to raise 1 lb of water 1°F. **E 1705,** E48

B-tuft, *n*—a two-pass process for aligning hooked fibers on the Fibroliner FL-101. **D 123,** D13

bubble immersion test—a form of leak test of gas-containing enclosures in which a leak is indicated by the formation of a bubble at the site of a leak. **E 1316,** E07

bubble packaging material—a material consisting of a flexible plastic film having uniformly spaced bubbles integrally molded therein. **D 996,** D10

bubble point pressure—the pressure necessary to displace a liquid held by surface tension forces from the largest equivalent capillaries in a membrane filter. **D 6161,** D19

bubble point test—a nondestructive membrane filter test used to assess filter integrity and proper installation. **D 6161,** D19

bubbler, *n*—a sampling device consisting of a gas disperser immersed in an absorbing liquid.
fritted bubbler, n—a bubbler having a frit as the gas disperser. **D 1356,** D22

bubbler—a device using a liquid stream to couple an ultrasonic beam to the test piece. **E 1316,** E07

bubble structure—size and spatial distribution of voids within the fired porcelain enamel. **C 286,** B08

bubbling pressure—the applied air pressure at which water menisci of the porous segment of a suction sampler break down, and air exists. **D 653,** D18

buck—a special support for ware during the firing of porcelain enamel on heavy ware. **C 286,** B08

bucket pump sprayer—a sprayer apparatus consisting of a manually operated pump that may be held or mounted in a bucket containing the spray solution. The pump is connected to an atomizing device that forms and distributes the spray. **E 1102,** E35

bucking coils—same as **differential coils.** **E 1316,** E07

buckle (centre bulge or oil can)—alternate bulges and hollows recurring along the length of a strip with the edges remaining relatively flat. **B 846,** B05

buckling—a bulge, bend, bow, kink, or wavy condition produced in sheets, plates, columns, or beams by compressive stresses. **D 653,** D18

buck sides—cattlehide shoe upper leather finished to resemble buckskin. **D 1517,** D31

buckskin—leather from deer and elk skins, used for shoes, gloves, and clothing. Only the outer cut of the skin from which the surface grain has been removed may be correctly defined as "genuine buckskin." Leather finished from the split or under-cut of deerskin must be described as "split buckskin." **D 1517,** D31

buddy system, *n*—a means of organizing employee work groups whereby each participant is matched with another so that prompt assistance can be rendered in the case of any emergency. **F 1494,** F23

buffable—capable of improvement in gloss or general appearance, or both, of a polish film by a mechanical action. **D 2825,** D21

buffalo leather—leather made from the hides of domesticated water buffalo of the Far East, not the American bison. **D 1517,** D31

buffer—a compound or mixture that, when contained in solution, causes the solution to resist change in pH. Each buffer has a characteristic limited range of pH over which it is effective. **B 374,** B08

buffer, *n*—a material or device, such as folded up **corrugated fiberboard,** placed in a container to position and protect the contents from the forces of impact. **D 996,** D10

buffer—*in fiber optics,* see **fiber optic buffer**. **E 131,** E13

buffer, *n*—*in spectrometric analysis,* a substance that tends to minimize the effects of one or more elements on the emission of other elements. **E 135,** E01

buffer—a solution of partially ionized acids or bases capable of reducing pH changes in the presence of added alkalis or acids. **F 1600,** F20

buffer action—the resistance of a solution to change in pH. **D 459,** D12

buffer or buffering agent—a compound or mixture that, when contained in solution, causes the solution to resist change in pH. Each buffer has a characteristic limited range of pH over which it is effective. **E 1519,** E35

buffing—the smoothing of a surface by means of a rotating flexible wheel to the surface of which fine, abrasive particles are applied in liquid suspension, paste, or grease stick form. **B 374,** B08

buffing—the mechanical operation of removing a superficial portion of the grain of flesh surface with sand paper or knife in order to produce a velvet surface or to minimize natural surface defects on the grain or flesh. **D 1517,** D31

buffing dust—the fine particulate leather fiber that is sanded off during buffing. **D 1517,** D31

buffing marks—the characteristic surface condition resulting from a buffing operation. **F 869,** F08

buffing paper—rolls or continuous paper coated with specially graded and treated abrasive particulates that is used to buff leather on a belt sanding machine. **D 1517,** D31

buffing rubber, *n*—particulate rubber produced as a byproduct of the buffing operation in the carcass preparation stage of a tire retreading; characterized by a wide range of particle sizes that are predominately elongated or acicular in shape. See **particulate rubber**. **D 1566,** D11

buffing-type of floor polish—a floor polish that requires buffing to maintain or enhance appearance, or both. **D 2825,** D21

builder—a material added to a soap or synthetic detergent formulation that enhances or maintains the cleaning efficiency of the surfactant, principally by inactivating water hardness either by sequestration, precipitation, or ion exchange. Other functions, depending on the performance capability of the builder compound used, include supplying alkalinity, buffering to maintain alkalinity at effective cleaning levels, helping to keep removed soil in suspension, and emulsifying oily soils. **D 459,** D12

builder's model, *n*—a reference standard of quality for specific building **components,** denoting by example, the level of quality adopted by a builder. **E 631,** E06

building, *n*—(*1*) a shelter comprising a partially or totally enclosed space, erected by means of a planned process of forming and combining materials. (*2*) the act or process of constructing.

apartment building—a **building** containing more than two **dwelling units** not intended for individual unit ownership.

condominium, n—an **apartment building,** group of townhouses, or single dwellings in which each **dwelling unit** is individually owned and each owner holds an interest in common areas. Also commonly used to denote an individual unit.

house, n—a **building** intended in its entirety as a **dwelling.**

split-level house—one divided vertically so that the floor level of rooms in one part is approximately midway between the levels of two successive stories in an adjoining part.

industrialized building—a **manufactured building** (preferred term).

manufactured building—a structure wholly or substantially made in a manufacturing plant for installation or assembly at the building site.

manufactured home—a **manufactured building** intended to be used as a **dwelling.**

packaged building—Use **manufactured building** or **precut building.**

precut building—a **manufactured building** produced largely of elements cut to size in a factory and transported for assembly at the erection site.

prefabricated building—Use **manufactured building.**

E 631, E06

building, *n*—a structure comprising a partially or totally enclosed space, erected by means of a planned process of forming and combining materials. **E 1481,** E06

building, *n* (bâtiment)—a shelter comprising a partially or totally enclosed space, erected by means of a planned process of forming and combining materials. Compare **building construction**.

E 1480, E06

building board—a natural finished multi-purpose cellulosic fiberboard. **D 1554,** D07

building brick, *n*—brick for building purposes not especially produced for texture or color (formerly called common brick). See Specification C 62. **C 43,** C15

building code—See **code.** **E 631,** E06

building component, *n*—a building element using industrial products that are manufactured as independent units capable of being joined with other elements. **E 631,** E06

building component, *n* (composant de bâtiment)—a building element using industrial products that are manufactured as independent units capable of being joined with other elements. **E 1480,** E06

building construction, *n*—(*1*) the act or process of making or forming a **building** by assembling or combining elements, **components,** or systems. (*2*) the structure or part thereof so formed.

closed construction—a method by which a **building, system,** assembly, or **component** is manufactured, in such a manner that portions cannot be readily inspected at the installation site without disassembly or destruction.

industrialized building process—the process of constructing manufactured **buildings.**

open construction—a method by which a **building, component,** assembly or system is manufactured in such a manner that all portions can be readily inspected on site without disassembly or destruction.

panelized construction—a building method using **panels** as major elements. **E 631,** E06

building construction, *n* (construire de bâtiment)—(*1*) the act or process of making or forming a building by assembly or combining elements, components or systems. (*2*) the structure or part thereof so formed. Compare **building.** **E 1480,** E06

building construction joint, *n*—a designed division of a building that allows movement of all component parts of the building, in any plane, which may be caused by thermal, seismic, wind loading or any other force. The construction of the separation is accomplished by one of the following methods: (*1*) manufactured devices suitable for this application, or (*2*) by field fabrication of suitable materials.

C 11, C11

building core and service area—See **space categories.**

E 1480, E06

building decision—a decision regarding the design, financing, engineering, construction, management, or operation of a building.

E 833, E06

building drain—that part of the lowest horizontal piping of a drainage system that receives the discharge from soil, waste, and other drainage pipes inside the walls of the building and conveys it to the building sewer beyond the foundation walls of the building or structure. **F 412,** F17

building drain (sanitary)—a building drain that conveys gray water or sewage, or both. **F 412,** F17

building drain (storm)—a building drain that conveys storm water only. **F 412,** F17

building economics—the application of economic analysis to the design, financing, engineering, construction, management, operation, or ownership of buildings. **E 833,** E06

building enclosure—Use **building envelope.** **E 631,** E06

building envelope—the outer elements of a **building,** both above and below ground, that divide the external from the internal environments. **E 631,** E06

building envelope, *n* (enveloppe d'un immeuble)—perimeter elements of a **building**, both above and below ground, that divide the external from the internal environment. **E 1480,** E06

building environmental survey—*for lead* systematic inspection of a building and associated on-site paint, dust, soil, and water for the presence of lead compounds. **E 1605,** E06

building fabric—(*1*) elements, components, parts, materials, or systems of a building separately or in combination; (*2*) loadbearing part of a structure without windows, doors, interior or exterior finishes. **E 631,** E06

building gross area— See **space categories**. **E 1480,** E06

building heat loss factor—a measure of the heat loss rate of a building expressed in joules per degree day (or Btu per degree day). This factor is multiplied by the number of degree days in a given period to estimate the energy required to heat the building during that period. **E 772,** E44

building modification—change or activity affecting the materials, structure, operations, or appearance of a building or its systems.

adapt, v—in building, to make suitable for a particular purpose by means of change or modification.

add, v—in building, to extend by means of new construction, or by enclosing an existing structure.

alter, v—in building, to make different, or to rearrange the layout.

improve, v—to enhance the quality or value of land or property.

maintain, v—to keep in working order, or to preserve from decline or failure.

modernize, v—in building, to adapt to current needs, tastes, or usage by **remodeling** or **repair.**

rebuild, v—to return a **building** to its previous state or condition.

reconstruct, v—to reproduce in the exact form and detail a

building, structure, or **artifact** as it appeared at a specific period in time.

reconstruction, n—the act or process of reproducing by new construction the exact form and detail of a vanished **building,** other structure, or **artifact** as it appeared at a specific period in time.

remodel, v—to replace or **improve** a **building** or its parts.

repair, v—to replace or correct damaged or faulty **components** or **subsystems** of a **building** to **maintain** operating capability.

retrofit, v—in building, to add new materials or equipment not provided at the time of original construction. **E 631,** E06

building occupant— See **occupant**. **E 1480,** E06

building or construction lime—a lime whose chemical and physical characteristics and method of processing make it suitable for the ordinary or special construction uses of the product. **C 51,** C07

building performance, *n*—the behavior in service of a construction as a whole, or of the **building components.**

durability, n—the capability of a **building,** assembly, **component,** product, or construction to maintain **serviceability** over at least a specified time.

serviceability, n—the capability of a **building,** assembly, **component,** product, or construction to perform the function(s) for which it is designed and used. **E 631,** E06

building performance, *n* (rendement d'édifice)—the behavior in service of a construction as a whole, or of the building components. Compare **facility performance**.

durability, n (durée)—the capability of a building, assembly, component, product, or construction to maintain serviceability for at least a specified period of time. Compare **facility durability**.

serviceability, n (fonctionnalité)—the capability of a **building**, assembly, component, product, or construction to perform the function(s) for which it is designed or used, or both. Compare **facility serviceability**. **E 1480,** E06

building permit, *n*—an authorization granted by the agency having jurisdiction to an applicant to proceed with construction on a specific project. **E 631,** E06

building preservation, *n*—measures taken to conserve, protect, rehabilitate, restore, or stabilize a building. See **preservation.** **E 631,** E06

building projection, *n* (saillie d'un immeuble)—pilaster, convector, baseboard heating unit, radiator, or other building element located in the interior of a building wall that prevents the use of that space for furniture, equipment, circulation, or other functions. **E 1480,** E06

building sanitary sewer—that part of the horizontal piping of a sanitary drainage system which extends from the building sanitary drain, receives the discharge of the building sanitary drain, and conveys it to a public sewer, private sewer, individual sewage disposal system, or other point of disposal. **F 412,** F17

building seal—See **seal.** **C 717,** C24

building service area—Preferred term is **building core and service area**. See **space categories**. **E 1480,** E06

building space:

attic, n—an accessible enclosed space immediately below the roof and wholly or partly within the roof framing.

basement—a space partly below **average grade** having less than one half of its clear height (measured from floor level to ceiling level) below **average grade.**

bathroom—a room containing a bathtub or shower, or both, and usually a lavatory (wash basin) and toilet (water closet).

cellar—a space wholly or partly below **average grade** having more than one half of its clear height (measured from floor level to ceiling level) below **average grade.**

environmental chamber, n—an enclosed space, used for testing designed and constructed to provide control of interior atmosphere to specified conditions.

habitable space—**occupiable space** normally used for living, including such activities as sleeping, eating, and cooking.

half bath—a room containing a lavatory (wash basin) and a toilet (water closet).

kitchen—a space containing facilities primarily for the preparation of food.

occupiable space—space normally used by people.

office, n—a place, such as a room, suite, or building, in which business, clerical, or professional activities are conducted.

open-plan workstation—office workspace for one person, not enclosed by full-height walls.

primary circulation area—portion of building area dedicated to public corridor, lobby, or atrium; or required for access to stairs, elevators, restroom facilities, or building exits.

secondary circulation area—portion of building area not defined as **primary circulation area,** but required for access to some subdivision of space, whether or not bounded by **walls.**

story, n—a space excluding **attics, basements,** and **cellars,** between successive floor levels or between **floor** and roof.

first story—the lowermost **story** of a **building** entirely above the **average grade** (also used as a synonym for **ground floor**).

top story—the uppermost **story** of a building. **E 631,** E06

building space—See **space categories**. **E 1480,** E06

building stone—natural rock of adequate quality to be quarried and cut as dimension stone as it exists in nature, as used in the construction industry. **C 119,** C18

building storm sewer—that part of the horizontal piping of a storm drainage system which extends from the building storm drain, receives the discharge of the building storm drain, and conveys it to a public storm sewer, private storm sewer, or other point of disposal. **F 412,** F17

building subsystem—a complete, integrated set of parts that functions as a unit within the finished **building.** See also **cladding system, hard-coat system, railing systems.**

solar energy system—a **building subsystem** to convert solar energy into thermal energy for space heating or cooling, water heating, or process energy.

active solar energy system—a **building subsystem** in which solar energy is collected and transferred predominantly by mechanical power not derived from solar radiation.

passive solar energy system—a **building subsystem** in which solar energy is collected and transferred predominantly by natural means, namely, conduction, convection, radiation, or evaporation. **E 631,** E06

building subsystem, *n* (sous-système d'immeuble)—complete, integrated set of parts that functions as a unit within the finished **building**. **E 1480,** E06

building supply—See **water service.** **F 412,** F17

building system—(*1*) group of structural or non-structural components or assemblies, or both, of a building interacting to serve a common purpose; (*2*) method for fabricating or erecting an entire structure. See also **anchorage system, anchoring system, hard-coat system, structural system, exterior installation, finish system.**

closed system—a building system having interchangeability of only its own **subsystems,** subassemblies, and **components.**

industrialized building system—the integration of **subsystems** and **components** into an overall process, utilizing factors of production, transportation, and on-site assembly techniques.

open system—a building system, designed to have interchangeability of its **subsystems,** subassemblies, **components,** or building elements with like **subsystems,** subassemblies, **components,** or elements of other systems.

prefabricated panel system—building-panel system fabricated away from its ultimate position on a building. **E 631,** E06

building system—an aggregation or assemblage of items joined in regular interaction or interdependence in buildings or building construction. **E 833,** E06

building system, *n* (système d'immeuble)—collection of equipment, facilities, and software designated to perform a specific function. **E 1480,** E06

building unit, *n*—a unit, the specifications for which include measures of durability, strength, and other structural properties, but not requirements affecting appearance. **C 43,** C15

building up—electroplating for the purpose of increasing the dimensions of an article. **B 374,** B08

build-up—condition resulting from lack of self-sensitivity in a polish, whereby new film deposits over old, with little or no self-cleaning action. **D 2825,** D21

buildup factor—for radiation passing through a medium, the ratio of the total value of a specified radiation quantity (such as absorbed dose) at any point in that medium to the contribution to that quantity from the incident uncollided radiation reaching that point. **E 170,** E10

built soap—under **soap,** see *built soap.* **D 459,** D12

built-up laminated wood, *n*—See *built-up laminated wood* under **wood laminate**. **D 907,** D14

built-up roofing—a continuous, semiflexible membrane consisting of plies of saturated felts, coated felts, fabrics or mats assembled in place with alternate layers of bitumen, and surfaced with mineral aggregate, bituminous materials, or a granule surfaced sheet (abbreviation, BUR). **D 1079,** D08

built-up timbers—an assembly made by joining layers of lumber together with mechanical fastenings so the grain of all laminations is essentially parallel. **D 9,** D07

bulb, *n—of a liquid-in-glass thermometer*, reservoir for the thermometric liquid. **E 344,** E20

bulb edge—the heavy rounded edge or bead of flat glass. **C 162,** C14

bulb of pressure—see **pressure bulb**. **D 653,** D18

bulge—doughnut-like protrusion perpendicular to nail shank below nail head. **F 547,** F16

bulk adherend, *n— as related to interphase,* the adherend, unaltered by the adhesive. (Compare to **bulk adhesive**.) **D 907,** D14

bulk adhesive, *n— as related to interphase*, the adhesive, unaltered by the adherend. (Compare to **bulk adherend**.) **D 907,** D14

bulk density, *n*—the mass per unit volume of a powder under nonstandard conditions, for example in a shipping container (not to be confused with apparent density). **B 243,** B09

bulk density, *n—of aggregate,* the mass of a unit volume of bulk aggregate material (the unit volume includes the volume of the individual particles and the volume of the voids between the particles). **C 125,** C09

bulk density—the mass of a material divided by its exterior volume. **C 859,** C26

bulk density, *n*—apparent mass per unit volume. **D 123,** D13

bulk density, ρ —the mass of a quantity of a bulk solid divided by its total volume. **D 653,** D18

bulk density, *n*—the weight per unit volume of a loosely packed material, such as a molding powder or pellets. **D 883,** D20

bulk density, *n*—the mass per unit volume of a material, including voids inherent in the material as tested. **E 1638,** E29

bulk density—the weight per unit volume of a material including voids inherent in the material as tested. **F 1251,** F04

bulk factor, *n*—the ratio of the volume of a given mass of molding material to its volume in the molded form. **D 883,** D20

bulk factor, *n*—the ratio of the volume of a given mass of molding material to its volume in the molded form. **F 412,** F17

bulk factor, *n*—the ratio of the volume of a given mass of molding material to its volume in the molded form. **F 1251,** F04

bulkhead—a steep or vertical structure supporting natural or artificial embankment. **D 653,** D18

bulking—the increase in volume of a material due to manipulation. Rock bulks upon being excavated; damp sand bulks if loosely deposited, as by dumping, because the apparent cohesion prevents movement of the soil particles to form a reduced volume. **D 653,** D18

bulking number, *n— of paper,* the number of sheets required to produce a stack of 25 mm thickness (approximately 1 in.). See **thickness**. **D 1968,** D06

bulking value—solid volume of a unit weight of material, usually expressed as gallons per pound. For practical purposes this is 0.120 divided by the specific gravity. **D 16,** D01

bulk molding compound (BMC), *n*—a putty-like mixture of any thermosetting resin containing fillers, fiber reinforcements, catalysts and thickening agents, or thermoplastic polymers, often extruded into logs or ropes. **D 883,** D20

bulk packaging—see **packaging.** **D 996,** D10

bulk reflection—reflection in which radiant energy is returned exclusively from within the specimen. **E 131,** E13

bulk sample, *n—in the sampling of bulk material,* one or more portions which (*1*) are taken from material that does not consist of separately identifiable units and (*2*) can be identified after sampling as separate or composited units. (Compare to **discrete sample, lot sample.**) **D 123,** D13

bulk shrinkage, *n*—a measure of potential stretch and power of stretch yarns or a measure of bulk of textured-set yarns. **D 123,** D13

bulk solid —an assembly of solid particles handled in sufficient quantities that its characteristics can be described by the properties of the mass of particles rather than the characteristics of each individual particle. May also be referred to as granular material, particulate solid or powder. Examples are sugar, flour, ore, and coal. **D 653,** D18

bulk specific gravity, *n*—the ratio of the mass of a volume of a material (including the permeable and impermeable voids in the material, but excluding the voids between particles of the material) at a stated temperature to the mass of an equal volume of distilled water at a stated temperature. **C 125,** C09

bulk specific gravity *(saturated surface dry)*, *n*—the ratio of the mass of a volume of a material including the mass of water within the pores in the material (but excluding the voids between particles) at a stated temperature, to the mass of an equal volume of distilled water at a stated temperature. **C 125,** C09

bulk storage—the storage of hose or covers together with one or more layers piled neatly, but without the benefit of spacers, supports, or special protective containers. **F 819,** F18

bullhides—hides from bulls are characterized by thick and rough head, neck and shoulders, and coarse flanks. They are often poor in quality and heavy, ranging from 60 lb (27 kg) up. **D 1517,** D31

bull's eye—(*1*) a tempered solid cylindrical sight glass.
(*2*) the glass left by the punty in the center of a flat disk of glass made by the hand blown crown process.
(*3*) in flat glass, an optical distortion that arises from a polishing depression or a solid inclusion trapped between layers of laminated glass. **C 162,** C14

bump check—See **percussion cone.** **C 162,** C14

bumping, molding process, *n*—the application, release, and reapplication of pressure prior to the start of vulcanization to vent entrapped gases, thereby facilitating complete filling of the mold. **D 1566,** D11

bunch, *n*—a defect in a yarn characterized by a segment not over 6 mm (¼ in.) in length that shows an abrupt increase in diameter caused by more fibers matted in this particular place. (See **slug, slub.**) **D 123,** D13

bunch-stranded conductor—a conductor composed of wires twisted together with a given length and direction of lay in such manner that the respective wires at successive cross sections along the length of the conductor do not necessarily form a symmetrical geometric pattern, nor necessarily occupy the same positions relative to each other. **B 354,** B01

bundle, *n*—two or more articles held together with rope, wire, or **strapping** so as to form a shipping unit; it may be wrapped. **D 996,** D10

bundle—an assemblage of asbestos in which the fibers remain entirely in their original close-packed parallel configuration (or not appreciably displaced therefrom) and having a transverse dimension typically between 2 and 8 mm. **D 2946,** C17

bundle—a general term for a collection of essentially parallel filaments. **D 3878,** D30

bundle—a general term for a collection of parallel filaments or fibres. **D 6161,** D19

bung hole—*in packaging,* an opening in a **barrel** or **drum** through which material can be poured to fill, empty or vent. **D 996,** D10

bunker — synonym for **bin**, but sometimes understood as being a bin without any or only a samll vertical part at the top of the hopper. **D 653,** D18

Bunsen coefficient, *n*—the solubility of a gas expressed as the volume, reduced to 273 K (32°F) and 101.3 kPa (1 atm), dissolved by one volume of liquid at the specified temperature and 101.3 kPa. **D 4175,** D02

buoyancy chamber (aka flotation chamber)—enclosed compartment of air or other buoyant material providing flotation for the boom. **F 818,** F20

buoyant unit weight (*submerged unit weight*)—see **unit weight.** **D 653,** D18

burden—*in an explosive blasting,* the distance between the charge and the free face of the material to be blasted. **D 653,** D18

burden—distance between charge and free surface in direction of throw. (ISRM) **D 653,** D18

burglar-resistant spindle—a tapered, shouldered, or other specially designed and treated spindle intended to resist punching, pulling, and drilling. **F 471,** F12

buried layer—a diffused region in a substrate that is, or is intended to be, covered with an epitaxial layer. **F 1241,** F01

buried pile yarn, *n*—*for concrete pile yarn floor covering*, the part of pile yarn tufts which is in or under the primary backing. **D 5684,** D13

burl—(*1*) a hard, woody outgrowth on a tree, more or less rounded in form, usually resulting from the entwined growth of a cluster of adventitious buds.
(*2*) in wood or veneer, a localized severe distortion of the grain generally rounded in outline, usually resulting from overgrowth of dead branch stubs, varying from 1/2 in. to several inches in diameter; frequently includes one or more clusters of several small contiguous conical protuberances, each usually having a core of pith but no appreciable amount of end grain (in tangential view) surrounding it. **D 9,** D07

burlap, *n*—a coarse, heavy, plainweave fabric of coarse single jute yarn. (See also **hessians.**) **D 123,** D13

burn, *v*—the heat treatment to which refractory materials are subjected in the firing process. **C 71,** C08

burn—a discoloration, distortion, or destruction of the pultruded surface as a result of thermal decomposition. **D 3918,** D20

burn, *n*—*in emission spectrometry,* (*1*) that portion of a solid specimen from which atoms were volatilized;
(*2*) the act of burning. **E 135,** E01

burn, *vt*—*in emission spectrometry,* to vaporize and excite a specimen with sufficient energy to generate spectral radiation. **E 135,** E01

burn, *v*—to undergo combustion. **E 176,** E05

burnable neutron absorber—a neutron absorber, such as boron, purposely included in a reactor to help control long-term reactivity changes by its progressive burnup. (Also referred to as burnable poison.) **C 859,** C26

burnable poison—See **burnable neutron absorber.** **C 859,** C26

burn chemical, *n*—perception of increased temperature and irritation resulting from exposure to such substances as ethyl alcohol, or high concentrations of NaCl or acids. The sensation lingers a short time after the stimulus is removed. **E 253,** E18

burn distance, *n*—the measurement from the bottom edge of the specimen to the farthest point that shows evidence of damage due to combustion. **F 1494,** F23

burner block—a refractory block with one or more orifices through which fuel is admitted to a furnace. **C 162,** C14

burner fuel oil, *n*—any petroleum liquid suitable for the generation of heat by combustion in a furnace or firebox as a vapor or a spray, or a combination of both. **D 4175,** D02

burning—see **firing.** **C 286,** B08

burning bars, points, or tools—equipment used to suspend or support ware during the firing operations. **C 286,** B08

burning behavior, *n*—all the changes that take place when materials or products are exposed to a specified ignition source. **D 123,** D13

burning behavior, *n*—all the changes that take place when materials or products are exposed to a specified ignition source. **F 1494,** F23

burning (burnt, burned)—a term applied to metal which has been permanently damaged by having been heated to a temperature close to or within the melting range. This results in a structure exhibiting incipient melting or intergranular oxidation. **E 7,** E04

burning (firing) of refractories, *v*—the final heat treatment in a kiln to which refractory brick and shapes are subjected in the process of manufacture for the purpose of developing bond and other necessary physical and chemical properties. **C 71,** C08

burning tool mark—a defect in the porcelain enamel appearing on the surface opposite to the point of contact with the supporting burning tool. **C 286,** B08

burnishing—the smoothing of surfaces by rubbing, accomplished chiefly by the movement rather than the removal of the surface layer. **B 374,** B08

burnishing—similar to buffing but using polishing paper/cloths with no or very fine abrasives that produces a very smooth polished appearance on the leather surface. **D 1517,** D31

burnishing—enhancement of the appearance of a polish accomplished by dry mechanical abrasion with a suitable machine and accessories. **D 2825,** D21

burn off—the unintentional removal of an autocatalytic deposit from a nonconducting substrate, during subsequent electroplating operations, owing to the application of excess current or a poor contact area. **B 374,** B08

burn-off—the process of severing an unwanted portion of a glass article by fusing the glass. **C 162,** C14

burn-off papers—materials sensitive to and that will produce an image by electric current. **F 221,** F05

burn rate—the rate at which a material burns after removal of the ignition heat source. **E 1749,** E06

burnt deposit—a rough, noncoherent or otherwise unsatisfactory deposit produced by the application of an excessive current density and usually containing oxides or other inclusions. **B 374,** B08

burnt lime—calcined limestone (CaO · MgO, dolomitic), or CaO (calcitic), or a mixture of these. **C 162,** C14

burr—fragment of excess material or foreign particle adhering to the surface.

F 109, C21

burr—thin, wing-like ridge protruding from side or edge of point or underside of top of head; defect formed during pointing or heading process and intended to be removed during manufacturing process. (See **whisker.**) **F 547,** F16

burr-wool waste, *n*—waste removed by the burr guard of cards or burr pickers having a very short fiber and full of burrs or seeds. **D 123,** D13

burst—an open break in the metal during forging located on the flats or corners of bolt and screw heads, or at the periphery of a flanged or circular headed bolt or screw, or on the flats or corners of the nut. **F 1789,** F16

burst emission—see **emission, burst.** **E 1316,** E07

bursting-off—the breaking of the blowover. **C 162,** C14

bursting strength, *n*—the force or pressure required to rupture a textile by distending it with a force, applied at right angles to the plane of the fabric, under specified conditions. **D 123,** D13

bursting strength— See **package testing.** **D 996,** D10

bursting strength—*of paper or paperboard*, the maximum liquid pressure required to produce rupture of the material when the pressure is increased at a controlled rate through a rubber diaphragm to a constrained circular area of the material. **D 1968,** D06

bursting strength of paper, *n*—the hydrostatic pressure required to produce rupture of a circular area of the material under specified test conditions. **D 1711,** D09

bursting strength "points", *n—in paper*, a unit of measure for bursting strength, measured in pounds per square inch, that should be considered colloquial and directly interchangeable with "pounds per square inch". See **bursting strength**. **D 1968,** D06

bursting strength test— See **package testing.** **D 996,** D10

burst strength—the internal pressure required to cause a pipe or fitting to fail. **F 412,** F17

burst strength—a measure of the internal pressure necessary to rupture a package or seal. **F 1327,** F02

bus bar—includes material of solid rectangular or square cross section or a solid section with two plane parallel surfaces and round or other simple regular-shaped edges. **B 846,** B05

bus (bus bar)—a rigid conducting section, for carrying current to the anode and cathode bars. **B 374,** B08

bus conductor, *n*—rigid electric conductor of any cross section. **B 881,** B07

bus conductor stock—a bar, rod, or shape of high-conductivity copper used to make bus conductors. **B 846,** B05

bushing—(*1*) a liner in the feeder orifice for molten glass.

(*2*) a precious metal or refractory/metal structure with single or multiple hole(s) through which glass flows and is attenuated into fiber(s). **C 162,** C14

bust girth, *n—in body measurements*, the circumference of the body over the fullest part of the breasts and parallel to the floor. (Compare **chest girth.**) **D 123,** D13

bust girth, *n—in body measurement*, the circumference of the body over the fullest part of the breasts and parallel to the floor. (See **chest girth.**) **F 1494,** F23

bust point to bust point, *n—in body measurements*, the distance across the front from the apex of one breast to the apex of the other. **D 123,** D13

butcher cuts—damage to hides caused by improper removal from the animal, usually in the form of cuts or furrows on the flesh side. **D 1517,** D31

butler finish—a finish composed of fine, uniformly distributed parallel lines, having a characteristic luster usually produced with rotating wire brushes or cloth wheels with applied abrasives. **B 374,** B08

butt—that part of the hide or skin covering the rump or hind part of the animal. **D 1517,** D31

butted wood member—wood member with its squared end or side placed adjoining the squared end or side of another wood member; with both abutting members of same thickness and in a single plane. **E 1807,** E06

butterfly bruise—See **percussion cone.** **C 162,** C14

butterfly point—a defective point with one or two thin fins projecting. (See **burr.**) **F 547,** F16

buttering—the application of a glazing compound to a flat contact surface of a member before placing that member in position, such as the buttering of a removable stop before fastening it in place. **C 717,** C24

butt-fused joint—see **joint, butt-fused.** **F 412,** F17

butt glazing, *n*—in building construction, a glazing system wherein butt joints are formed between lites and filled with a glazing sealant. **C 717,** C24

butt joint, n—See **butt sealant joint**. **C 717,** C24

butt joint—See **joint.** **E 631,** E06

button, *n*—a knob, disc, or similar object which when forced through a narrow opening or buttonhole, fastens one part of a garment or other flexible substrate to another. (See also **sew-through flange button.**) **D 123,** D13

button clincher—circular clincher. **F 592,** F16

button head—cylindrical head with flat upper and bearing surfaces, smaller in diameter and thicker than flat head for same size nail shank. **F 547,** F16

buttonhole stitch, *n—in home sewing*, a complex machine stitch pattern made by coordinated motions of needle and feed, appearing as very close stitches forming a narrow rectangle of stitching that is usually composed of four stitch segments, one on each side and on each of the two ends of the rectangle. **D 123,** D13

button sample—*in sealants*, an identified small amount of sealant extruded from a mixed sealant cartridge. **E 1749,** E06

button test—a test designed to determine relative fusibility of porcelain enamel frit or powder and so called because the completed specimens resemble buttons.

NOTE—See **flow button** . **C 286,** B08

buttress-type thread—thread with flank on head side of crest almost perpendicular to nail axis; while flank on point side of crest is noticeably inclined. **F 547,** F16

butt sealant joint, *n*—*in building construction*, a joint where sealant is applied within the joint between approximately parallel substrate surfaces that are face-to-edge or edge-to-edge. **C 717,** C24

butt welding (machine)—the fusing together of two pieces of plastic which are aligned in the same plane, with the same mating thickness, by application of heat and pressure. Also called *hot-plate welding*. **C 904,** C03

butylene plastic—plastic based on resins made by the polymerization of butene or copolymerization of butene with one or more unsaturated compounds, the butene being in greatest amount by weight. **F 1251,** F04

butylene plastics—plastics based on resins made by the polymerization of butene or copolymerization of butene with one or more unsaturated compounds, the butene being in greatest amount by weight. **D 883,** D20

butylene plastics—plastics based on resins made by the polymerization of butene or copolymerization of butene with one or more unsaturated compounds, the butene being in greatest amount by weight. (D-20) **F 412,** F17

B-value—salt diffusion coefficient— The coefficient is defined as the amount of salt transferred per unit area of membrane when the difference in salt concentration across the membrane is unity. A unit of measurement is m/h. **D 6161,** D19

BWRO—brackish water reverse osmosis. **D 6161,** D19

by mechanical treatment—without causing a significant change in size. **B 846,** B05

by-pass clinch—clinch with legs paralleling and adjacent to each other. **F 592,** F16

bypass nozzle—see **by-pass nozzle**. **E 1620,** E29

by-pass nozzle—a swirl chamber atomizer containing by-pass orifice(s) or port(s) through which part of the inlet liquid may be withdrawn from the swirl chamber and returned to the supply tank or pressure pump suction. The discharge flow is modulated by controlling the pressure in the bypass line, using a valve in the line. **E 1620,** E29

by-product coke—See *by-product coke* under **coke**. **D 121,** D05

byte, *n*—a set of seven or eight bits used to represent a character or control function. **F 1457,** F05

by thermal treatment—without causing recrystallization. **B 846,** B05

C

C—a computer language. **F 1156,** F12

cabal glass—a glass consisting primarily of the oxides of *ca*lcium, *b*oron, and *al*uminum. **C 162,** C14

cable, *n*—a multi-strand, flexible longitudinal element designed primarily to resist axial tension loading. **F 1582,** F04

cable (concentric lay conductor)—conductor constructed with a central core surrounded by one or more layers of helically laid wires. **B 713,** B01

cable twist, *n*—the construction of cabled yarn, cord, or rope in which each successive twist is in the opposite direction to the preceding twist; an S/Z/S or Z/S/Z construction. **D 123,** D13

cable wrap, *n*—paper used for mechanical protection or for space-filling (rather than as electrical insulation) in low-voltage cables with nonmetallic sheaths. **D 1711,** D09

cabretta—skin of hair sheep, originally Brazilian, used principally for glove and garment leathers. (See also **cape.**) **D 1517,** D31

CAC—combined available chlorine. **D 6161,** D19

cadmium ratio—the ratio of the neutron reaction rate measured with a given bare neutron detector to the neutron reaction rate measured with an identical neutron detector enclosed by a particular cadmium cover and exposed in the same neutron field at the same or an equivalent spatial location. **E 170,** E10

cadmium ratio—the ratio of the neutron reaction rate measured with a given bare neutron detector to the reaction rate measured with an identical neutron detector enclosed by a particular cadmium cover and exposed in the same neutron field at the same or an equivalent spatial location.

NOTE—In practice, meaningful experimental values can be obtained in an isotropic neutron field by using a cadmium filter approximately 1-mm thick. **E 1316,** E07

CADS—acronym for Counterfeiting, Altering, Duplicating, and Simulating; also for Currency Authenticating and Denominating System, and Computer Aided Design System. **F 1156,** F12

Cady test—See **package testing.** **D 996,** D10

cage—an assembled unit of steel reinforcement consisting of circumferential and longitudinal bars or wires. **C 822,** C13

cake, *n*—a bonded mass of unpressed metal powder. **B 243,** B09

cake—refinery shape used for rolling into plate, sheet, strip, or shape. Rectangular in cross section and of various sizes, normally ranging in weight from 140 to 62 000 lb (63 to 28 200 kg). **B 846,** B05

caking, *n*—a settling out of pigments in a fluid ink during storage (sedimentation). **D 6488,** D01

calcarenite—a limestone composed predominantly of clastic sand-size grains of calcite, or rarely aragonite, commonly as tiny fossils, shell fragments, or other fossil debris. **C 119,** C18

calcareous—originating from predominately calcium carbonate or one of its derivative forms. **C 51,** C07

calcia—the chemical compound calcium oxide (CaO). **C 51,** C07

calcine—a ceramic material or mixture fired to less than fusion for use as a constituent in a ceramic composition. **C 242,** C21

calcine, *v* **(calcination,** *n***)**—firing or heating a granular or particulate solid at less than fusion temperature, but sufficient to remove most of its chemically combined volatile matter (that is, H_2O, CO_2) and otherwise to develop the desired properties for use. **C 1145,** C28

calcine, calcines, *n*—refractory material, often fireclay, that has been heated to eliminate volatile constituents and to produce desired physical changes. **C 71,** C08

calcined coke, *n*—petroleum coke that has been thermally treated to drive off the volatile matter and to develop crystalline structure. **D 4175,** D02

calcined gypsum, *n*—a dry powder; primarily calcium sulfate hemihydrate, resulting from calcination of gypsum; cementitious base for production of most gypsum plasters: also called plaster of paris; sometimes called stucco. **C 11,** C11

calcined refractory dolomite, n—see **dolomite, calcined refractory.** **C 71,** C08

calcining of refractory materials, *v*—the heat treatment to which raw refractory materials are subjected, preparatory to further processing or use, for the purpose of eliminating volatile chemically combined constituents and producing volume changes. **C 71,** C08

calcite—a doubly refracting mineral used in the manufacture of polarizing prisms. It is uniaxial negative and in the trigonal diversion of the hexagonal system of crystals. Its indexes are $\epsilon = 1.486$, $\omega = 1.658$; its hardness is 3 on the Mohr scale and specific gravity 2.711. **E 175,** E41

calcium carbonate equivalents (mg/L as $CaCO_3$)—a method for expressing mg/L as ion in terms of calcium carbonate. Concentration in calcium carbonate equivalents is calculated by multiplying concentration in mg/L of the ion by the equivalent weight of calcium carbonate (50) and dividing by the equivalent weight of the ion. **D 6161,** D19

calcium hypochlorite—Ca $(HClO)_2$, a disinfection agent. **D 6161,** D19

calcium silicate, *n*—insulation composed principally of hydrous calcium silicate, and which usually contains reinforcing fibers. **C 168,** C16

calcium-silicate brick, *n*—a pressed and subsequently autoclaved unit that consists of sand and lime, with or without the inclusion of other materials. **C 1209,** C15

calcium sulfate, *n*—the chemical compound $CaSO_4$. **C 11,** C11

calcium sulfate, *n*—in cement manufacture, a material composed essentially of calcium sulfate in one or more of its hydration states: **anhydrite** ($CaSO_4$), **gypsum** ($CaSO_4 \cdot 2H_2O$), or **calcium sulfate hemihydrate** ($CaSO_4 \cdot \frac{1}{2}H_2O$). **C 219,** C01

calendar staple—staple formed to provide a hanger for use with calendars or booklets. **F 592,** F16

calender, *n*—a machine with two or more parallel, counter-rotating rolls with a controllable, roll-to-roll spacing, rotating at selected surface speeds and controlled temperatures, used for sheeting, laminating, skim coating (topping) and friction coating, to a controlled thickness and/or surface condition. **D 1566,** D11

calf girth, *n*—*in body measurements*, the maximum circumference around the leg between the knee and ankle, parallel to the floor. **D 123,** D13

calf leather—leather made from the skins of young cattle from a few days up to a few months old, the skins weighing up to 25 lb (11.3 kg). It is finer grained, lighter in mass and more supple than cowhide, or kip leathers. **D 1517,** D31

calf leather—leather made from the skins of young cattle from a few days up to a few months old, the skins weighing up to 15 lb (6.8 kg). It is finer grained, lighter in mass and more supple than cowhide or kip leathers. (D 1517, D-31) **F 869,** F08

calibrant, *n*—a reference material used for a calibration. **E 135,** E01

calibrate, *v*—to determine and record the relationship between a set of standard units of measure and the output of an instrument or test procedure. **D 123,** D13

calibrate, *vt*—(*1*) to establish the relationship between the response of an instrument and the concentration or mass of the analyte; (*2*) to establish a table of corrections to improve the accuracy of equipment used to measure physical properties such as mass, volume, temperature, etc. **E 135,** E01

calibrate, *v*—to find and eliminate systematic errors of an instrument scale or method of measurement by use of material standards and techniques traceable to an authorized national or international measurement system. **E 284,** E12

calibrated instrument—an instrument for which the response has been documented upon being directly compared with the response of a standard instrument, both having been exposed to the same radiation field under the same conditions; or one for which the response has been documented upon being exposed to a standard radiation field under well-defined conditions. **E 170,** E10

calibration—the determination of the values of the significant parameters by comparison with values indicated by a reference instrument or by a set of reference standards. **C 859,** C26

calibration, *n*—the act or process of calibrating; the recorded relationship resulting from calibrating. **D 123,** D13

calibration—1) the act or process of determining the relationship between a set of standard units of measure and the output of an instrument or test procedure,
2) the graphical or mathematical relationship relating the desired property (expressed in a standard unit of measure such as micrometers or Kg/mm^2) to the instrument output (instrument units such as filar divisions or pixels). **E 7,** E04

calibration, *n*—the act, process, or result of establishing: (*1*) the relationship between the response of an instrument and the concentration or mass of the substance determined; (*2*) a table of corrections to improve the accuracy of equipment used to measure physical properties such as mass, volume, temperature, etc. **E 135,** E01

calibration, *n—of a thermometer or thermometric system,* the set of operations that establish, under specified conditions, the relationship between the values of a thermometric quantity indicated by a thermometer or thermometric system and the corresponding values of temperature realized by standards. **E 344,** E20

calibration—the set of operations that establishes, under specified conditions, the relationship between values indicated by a measuring instrument or measuring system, or values represented by a material, and the corresponding known values of measurement, (ISO Guide 25). **E 1187,** E36

calibration cotton standards, *n*—cotton samples taken from blended bulk source on which fiber properties have been determined under the International Calibration Cotton Standards Program. **D 123,** D13

calibration curve—the graphical or mathematical representation of a relationship between the assigned (known) values of standards and the measured responses from the measurement system. **C 859,** C26

calibration curve—a plot of the potential (emf) of a given ion-selective electrode cell assembly (ion-selective electrode combined with an identified reference electrode) versus the logarithm of the ionic activity (concentration) of a given species. For uniformity, it is recommended that the potential be plotted on the ordinate (vertical axis) with the more positive potentials at the top of the graph and that pa_A (−log activity of the species measured, *A*) or pc_A(−log concentration of species measured, *A*) be plotted on the abscissa (horizontal axis) with increasing activity to the right. **D 4127,** D19

calibration curve, *n*—the graphical or mathematical representation of the relationship between the response of an instrument and the concentration or mass of the analyte. **E 135,** E01

calibration curve—graphical or mathematical representation of a relation between a measured parameter and a property of the standard for the substance under consideration. **E 1605,** E06

calibration, instrument, *n*—the comparison of an instrument with, or the adjustment of an instrument to, a known reference(s) often traceable to the National Institute of Standards and Technology (NIST). (See also **standardization, instrument.)** **E 1316,** E07

calibration laboratory—laboratory that performs calibration, (ISO Guide 25). **E 1187,** E36

calibration method—defined technical procedure for performing a calibration, (ISO Guide 25). **E 1187,** E36

calibration platform—a moving platform for applying a force in the contact plane of a tire, and associated means for measuring the applied force. The calibration platform consists of a rigid plate with a high friction surface,,in contact with the tire footprint, supported on a frictionless, preferably air, bearing. It may also be instrumented to measure vertical forces (loads). **E 867,** E17

calibration point, *n*—a specific value, established by a reference, at which the indication or output of a measuring device is determined. **E 344,** E20

calibration reference signals—repeatable signals in the range of expected wheel-force transducer system loading. These signals could either be constant voltages or preferably produced by a strain-gage calibration shunt resistor. **E 867,** E17

calibration source or field—see **electron standard field, γ-ray standard field,** and **X-ray standard field.** **E 170,** E10

calibration standard—a standard having an accepted value (reference value) for use in calibrating a measurement instrument or system. **C 859,** C26

calibration tire, *n*—a witness tire designed to provide a fixed or known test value for selected properties. **F 538,** F09

California bearing ratio, *CBR* **(D)**—the ratio of: (*1*) the force per unit area required to penetrate a soil mass with a $3\text{-in.}^2(19\text{-cm})^2$ circular piston (approximately 2-in. (51-mm) diameter) at the rate of 0.05 in. (1.3 mm)/min, to (*2*) that required for corresponding penetration of a standard material. The ratio is usually determined at 0.1-in. (2.5-mm) penetration, although other penetrations are sometimes used. Original California procedures required determination of the ratio at 0.1-in. intervals to 0.5 in. (12.7 mm). Corps of Engineers' procedures require determination of the ratio at 0.1 in. and 0.2 in. (5.1 mm). Where the ratio at 0.2 in. is consistently higher than at 0.1 in., the ratio at 0.2 in. is used. **D 653,** D18

california process—this process requires the accurate drafting of patterns, cutting, and stitching of the upper, sock lining, and platform cover. The upper and sock lining are stitched together. The platform cover is stitched to the other two parts in a separate operation. The last is then inserted into the upper. After the last is inserted, the platform is accurately pressed into place. **F 869,** F08

caliper—See **package testing.** **D 996,** D10

caliper, *n—of paper and paperboard,* see **thickness.** **D 1968,** D06

caliper—the thickness of a sheet of paper measured under specified conditions and usually expressed in thousandths of an inch (mils). **F 149,** F05

caliper—the thickness of a sheet expressed in millimetres or in thousandths of an inch. **F 335,** F05

caliper diameter (Feret's diameter)—the length of a line normal to two parallel lines, tangent to opposite edges of a phase or object. **E 7,** E04

caliper log—a geophysical borehole log that shows to scale the variations with depth in the mean diameter of a cased or uncased borehole. **D 653,** D18

C_5 aliphatic resin, *n*—a resin produced by the cationic polymerization of a low-boiling aliphatic steam-cracker fraction. **D 6440,** D01

call rotation—a system in which emergency medical responses are allocated sequentially to multiple providers. **F 1177,** F30

calm water boom—boom intended for use in calm waters (see Practice F 625 for environmental descripters). **F 818,** F20

calomel half cell (calomel electrode)—a half cell containing a mercury electrode in contact with a solution of potassium chloride of specified concentration that is saturated with mercurous chloride (calomel). **B 374,** B08

calomel reference electrode—a reference electrode with an internal reference element of mercury/mercurous chloride. The filling solution is usually saturated KCl, saturated with mercurous chloride. **D 4127,** D19

calorific value, *n*—the heat of combustion of a unit quantity of a substance. **D 121,** D05

calorific value, *n*—the heat produced by combustion of a unit quantity of a specimen under specified conditions. **D 5681,** D34

calorific value—the energy of combustion of a unit quantity of refuse-derived fuel. It may be expressed in kilojoules per kilogram (kJ/kg), British thermal units per pound (Btu/lb), or calories per gram (cal/g) (obsolete). **E 856,** D34

calorimeter, *n—as used in Test Method D 3286*, the bomb and its contents, the calorimeter vessel with stirrer, the water in which the bomb is immersed, and the portions of the thermometer and the ignition leads within the calorimeter vessel. **D 121,** D05

calorimeter—an instrument capable of making absolute measurements of energy deposition (or absorbed dose) in a material through measurement of its change in temperature and a knowledge of the characteristics of its material construction. **E 170,** E10

calorimeter jacket, *n*—the insulating medium surrounding the calorimeter. **D 121,** D05

calorimeter jacket, *n*—the insulating medium surrounding a calorimeter. **D 5681,** D34

calorizing—imparting resistance to oxidation to an iron or steel surface by heating in aluminum powder at 800 to 1000°C (1470 to 1830°F). **B 374,** B08

camber—a single arch of curvature. (See also **waviness.**)

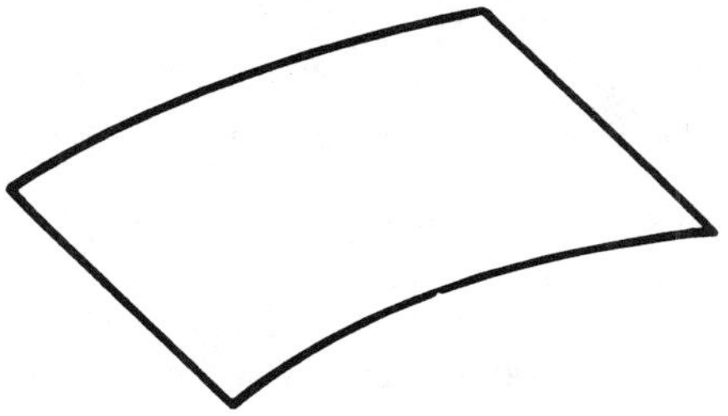

F 109, C21

camber height, *h*—distance between the running surface of the ski and a plane surface, measured with the ski resting freely under its own mass. X_h indicates the location of camber height from the tail of the ski. **F 472,** F27

cambium—the layer of cells that lies between the inner bark and the wood of a tree, that repeatedly subdivides to form new wood and bark cells. **D 9,** D07

cam lock slider, *n—in zippers*, a slider that incorporates a curled projection or projections on the pull that extends through a window or windows to effect a locking action by pressing against the interlocking elements when the cam lock slider is in the locked position. **D 123,** D13

camouflet—the underground cavity created by a fully contained explosive. (ISRM) **D 653,** D18

campaign—the working life of a melting furnace between major cold repairs. **C 162,** C14

camping tentage, *n*—any portable temporary shelter or structure designed to protect persons from the elements, all or a portion of the covering which is made of fabric or other pliable materials. **D 123,** D13

can, *n—in packaging,* a receptacle generally of 10-gal capacity or less, normally not used as a **shipping container.** **D 996,** D10

Canada balsam—see **balsam, Canada.** **E 175,** E41

canal—that part of a melting furnace leading from the fining area to the forming area. See **channel** and **forehearth.** **C 162,** C14

candela—a unit of luminous intensity (formerly candle). One candela is the luminous intensity in the perpendicular direction of a surface of 1/600 000 m2 of a blackbody radiator at the temperature of freezing platinum under a pressure of 101 325 Pa. One candela produces 1 lumen of luminous flux per steradian of a solid angle measured from the source. **E 1316,** E07

candela, *cd, n*—the SI unit of luminous intensity; the luminous intensity, in a given direction, of a source that emits monochromatic radiation of frequency 540×10^{12} hertz and that has a radiant intensity in that direction of (1/683) watt per steradian. **E 284,** E12

candidate method, *n*—an analytical method or measurement process being considered for standardization. **D 1356,** D22

candidate tire, *n*—a test tire that is part of a test program. **F 538,** F09

candidate tire set, *n*—a set of candidate tires. **F 538,** F09

cane—solid glass rods. **C 162,** C14

canister vacuum cleaner, *n*—a portable floor-supported vacuum cleaner, having a nozzle separated from the cleaner housing by a hose and designed for normal-duty cleaning of household dirt. In use, only the nozzle is guided over the surface area to be cleaned. The cleaner may have detachable nozzles, attachments, and wands for both floor and above-the-floor cleaning. The nozzle may employ a driven agitator to assist in cleaning. **F 395,** F11

cannel coal—See *cannel coal* under **coal.** **D 121,** D05

cannot be associated/cannot be connected—these terms are too vague and may be interpreted as reflecting bias as they have no counterpart suggesting that the writer cannot be eliminated either. **E 1658,** E30

canonical-form definition, *n*—see **intensional definition** **E 1992,** E02

canopy—a covering fixed above cooking equipment and overhanging on all sides of its unclosed sides, whose lower edge is a minimum height of 6 ft 6 in. from the finished floor. The purpose of the canopy is to contain and capture the unwanted by-products resulting from cooking activities and may be Type I or Type II. See *Type I, Type II,* and **exhaust hood.** **F 1827,** F26

canopy penetrating agent—an adjuvant that increases the penetration of the spray material into the crop canopy. See **deposition aid.** **E 1519,** E35

cant—a log that has been slabbed on one or more sides, usually with the intention of resawing at right angles to the widest sawn face.

D 9, D07

cantilever, *n*—an overhanging portion or a member or slab projecting beyond support(s) sufficiently to induce bending and shear stresses in projecting part(s) when subjected to transverse loading including uniform, concentrated, or other load types. **E 631,** E06

cantilever slide gate—any horizontal slide gate spanning an opening lacking a top or bottom support within that opening (see Specification F 1184). **F 552,** F14

cant strip—a beveled strip used under flashings to modify the angle at the point where the roofing or waterproofing membrane meets any vertical element. **D 1079,** D08

cap, *n*—(*1*) another name for crown.
(*2*) a type of bottle closure. **C 162,** C14

cap, *v*—{archaic} to cut off the ends of a glass cylinder. **C 162,** C14

cap—a horizontal wood member, generally square or near square; used to crown and bind the top ends of upright posts and piles together as a unit and to support upper construction. **D 9,** D07

cap—See **container (cover).** **D 996,** D10

cap—See **railing systems.** **E 631,** E06

cap, *n*—a fitting or plug used to close the end of a pipe, tubular post, newel, or rail. **E 1481,** E06

capability index (C_p), *n*—for a stable process, the specification range divided by six times the standard deviation.

$$C_p = \frac{(USL - LSL)}{6 \times s}$$

A 644, A04

capability index (C_{pk}), *n*—for a stable process, the smaller of the upper capability index (CPU) or the lower capability index (CPL). **A 644,** A04

capacitance—that property of a system of conductors and dielectrics that permits the storage of electrical charge when a potential difference exists between the conductors. **E 1142,** E37

capacitance, *C*, *n*—that property of a system of conductors and dielectrics which permits the storage of electrically separated charges when potential differences exist between the conductors. **D 1711,** D09

capacitivity—the same as **permittivity, relative.** **D 2864,** D27

capacitor tissue, *n*—very thin (5 to 50 μm) pure, nonporous paper used as the dielectric in capacitors, usually in conjunction with an insulating liquid. **D 1711,** D09

capacity, *n*—*for tensile testing machines*, the maximum force for which the machine is designed. **D 123,** D13

capacity, *n*—the power which a component of a geothermal facility (e.g., a well, a reservoir, a power plant, or a direct-use facility) is capable of supplying at a point in time, assuming that other required components of the geothermal facility are available. Capacity is expressed in units of power (e.g., Megawatts, kilowatts). **E 957,** E44

cape, skin or leather—skin of South African hair sheep. Fine-grained leather, superior to wool sheep for gloves and garments. Loosely applied to all hair sheep, but should be qualified to show origin, if other than South African. **D 1517,** D31

cap flashing—See **flashing**. **D 1079,** D08

cap head—extra large, flat or dished, round or squared sheet-metal head applied to nail shank during nail manufacture. **F 547,** F16

capillary action—the phenomenon of intrusion of a liquid into interconnected small voids, pores, and channels in a solid, resulting from surface tension. **C 242,** C21

capillary action, *n*—the phenomenon of intrusion of a liquid into interconnected small voids, pores, and channels in a solid, resulting from surface tension. **C 1145,** C28

capillary action (capillarity)—the rise or movement of water in the interstices of a soil or rock due to capillary forces. **D 653,** D18

capillary flow—see **capillary action.** **D 653,** D18

capillary fringe—the basal region of the vadose zone comprising sediments that are saturated, or nearly saturated, near the water table, gradually decreasing in water content with increasing elevation above the water table. **D 653,** D18

capillary fringe zone—the zone above the free water elevation in which water is held by capillary action. **D 653,** D18

capillary head, *h* **(L)**—the potential, expressed in head of water, that causes the water to flow by capillary action. **D 653,** D18

capillary migration—see **capillary action.** **D 653,** D18

capillary migration—*of water*, movement of water induced by the force of molecular attraction (surface tension) between the water and the material it contacts. Compare **rising damp.** **E 631,** E06

capillary rise (height of capillary rise), h_c **(L)**—the height above a free water elevation to which water will rise by capillary action. **D 653,** D18

capillary water—water subject to the influence of capillary action. **D 653,** D18

capital cost—the costs of acquiring, substantially improving, expanding, changing the functional use of, or replacing a building or building system. **E 833,** E06

capital cost, *n* (coût en capital (frais d'immobilisations))—costs of acquiring, substantially improving, expanding, changing the functional use of, or replacing, a **building** or **building system**. **E 1480,** E06

cap nail—bright, galvanized or electroplated, plain-shank or threaded, regular-stock-steel, ¾ by 0.105 to 6 by 0.135-in. nails with integral nominal 1-in. cap head and medium diamond point. **F 547,** F16

capped steel, *n*—a **rimmed steel** in which, during ingot solidification, the rimming action was limited by mechanical or chemical means. **A 941,** A01

cap rail—See **railing systems.** **E 631,** E06

cap rail—a secondary railing element, often a handrail, fastened to the top rail of a railing system. (Syn. *rail cap.*) **E 1481,** E06

cap sheet—a granule-surfaced coated felt used as the top ply of a built-up roofing membrane. **D 1079,** D08

capsomere—a structural subunit of the outer protein shell (capsid) of a virus consisting of protein monomers. **E 1705,** E48

capstan head— See **dual head.** **F 547,** F16

caravan, *n*—*for purposes of a tire test*, two or more vehicles running in the same time frame, over the same test course(s), under similar but independent conditions. **F 538,** F09

carbide—a compound of carbon with one or more elements, which, in customary formulation, are considered as being more positive than carbon. **E 7,** E04

carbide precipitation "sensitization"—a condition which affects some austenitic stainless steels which have been subjected to temperatures in the critical range, typically 800° to 1400°F. Complex chromium carbides precipitate and reside primarily at the grain boundaries, causing deterioration of its corrosion resistance by depleting its adjacent areas of chromium. **F 1789,** F16

carbide, primary, *n*—carbide precipitated in cast iron during solidification. **A 644,** A04

carbohydrates—molecules consisting of carbon, hydrogen and oxygen that include celluloses, starches and sugars. **E 1705,** E48

carbohydrates not cellulose—the noncellulosic carbohydrates of a cellulosic material. **D 1695,** D01

carbon—an element, number 6 of the periodic table of elements, electronic ground state $1s\ ^2 2s^2\ 2p^2$. **C 709,** D02

carbon—*in carbon and graphite technology*, an artifact consisting predominantly of the element carbon and possessing limited long range order. **C 709,** D02

carbonate carbon, *n*—the carbon content present in the solid products derived from the combustion or reaction of coal, coal by-products, or coke as carbonates and which is noncombustible in standard industry practice. **D 121,** D05

carbonate hardness—the hardness in a water caused by carbonates and bicarbonates of calcium and magnesium. The amount of hardness equivalent to the alkalinity formed and deposited when water is boiled. In boilers, carbonate hardness is readily removed by blowdown. **D 6161,** D19

carbonation, *n*—building(*s*), a process of chemical weathering whereby minerals that contain sodium oxide, calcium oxide, potassium oxide, or other basic oxides are changed to carbonates by the action of carbonic acid derived from atmospheric carbon dioxide and water. **E 631,** E06

carbon black, *n*—a material consisting essentially of elemental carbon in the form of near-spherical colloidal particles and coalesced particle aggregates of colloidal size, obtained by partial combustion or thermal decomposition of hydrocarbons. **D 1566,** D11

carbon black, *n*—material consisting essentially of elemental carbon in the form of near-spherical colloidal particles and coalesced particle aggregates of colloidal size, obtained by partial combustion or thermal decomposition of hydrocarbons. **D 3053,** D24

carbon black agglomerate, *n*—a cluster of physically bound and entangled aggregates. **D 3053,** D24

carbon black aggregate, *n*—a discrete, rigid, colloidal mass of extensively coalesced particles; it is the smallest dispersible unit. **D 3053,** D24

carbon black, carcass grade, *n*—a type of furnace carbon black having an average particle size in the range from 31 to 200 nm. **D 3053,** D24

carbon black, furnace, *n*—a type of carbon black produced by the decomposition reaction of hydrocarbons when injected into a high-velocity stream of combustion gases under controlled conditions. **D 3053,** D24

carbon black, hard, *n*—See **carbon black, tread grade**, the preferred term. **D 3053,** D24

carbon black microstructure, *n*—arrangement of carbon atoms within a carbon black particle. **D 3053,** D24

carbon black particle, *n*—a small spheroidally shaped (paracrystalline, non-discrete) component of a carbon black aggregate; it is separable from the aggregate only by fracturing. **D 3053,** D24

carbon black particle diameter, *n*—arithmetic average of the diameters of particles within a carbon black aggregate as measured by electron microscopy. **D 3053,** D24

carbon black pellet, *n*—a relatively large agglomerate mass that has been densified in spheroidal form to facilitate handling and processing. **D 3053,** D24

carbon black reinforcing, *n*—See **carbon black, tread grade**, the preferred term. **D 3053,** D24

carbon black, semi-reinforcing, *n*—See **carbon black, carcass grade**, the preferred term. **D 3053,** D24

carbon black, soft, *n*—See **carbon black, carcass grade**, the preferred term. **D 3053,** D24

carbon black structure, *n*—the quality of irregularity and deviation from sphericity of the shape of a carbon black aggregate. **D 3053,** D24

carbon black, surface activity, *n*—the inherent ability of the carbon black surface to interact physically or chemically, or both, with rubber or other molecules. **D 3053,** D24

carbon black, target value, *n*—a consensus value for selected primary properties on which producers center their manufacturing process and users center their specification. **D 3053,** D24

carbon black, thermal, *n*—a type of carbon black produced under controlled conditions by the thermal decomposition of hydrocarbons in the absence of air or flames. **D 1566,** D11

carbon black, thermal, *n*—a type of carbon black produced under controlled conditions by the thermal decomposition of hydrocarbons in the absence of air or flames. **D 3053,** D24

carbon black, thermal, acetylenic, *n*— a thermal black produced from acetylene gas. **D 1566,** D11

carbon black, thermal, acetylenic, *n*—a thermal black produced from acetylene gas. **D 3053,** D24

carbon black, tread grade, *n*—a type of furnace carbon black having an average particle size in the range from 1 to 30 nm. **D 3053,** D24

carbon black, typical value, *n*—a consensus value for those carbon black properties which are not specifically targeted for control in the manufacturing process and which are somewhat dependent upon the targeted properties. **D 3053,** D24

carbon black weight mean particle size, *n*—ratio equal to the sum of individual particle diameters, each raised to the fourth power, divided by the sum of the individual particle diameters, raised to the third power, **D 3053,** D24

carbon-ceramic refractory, *n*—a manufactured refractory comprised of carbon (including graphite) and one or more ceramic materials such as fireclay and silicon carbide. **C 71,** C08

carbon fiber precursor—a material from which carbon fiber is made by pyrolysis. Polyacrylonitrile, rayon or pitch fibers are commonly used precursors. **D 3878,** D30

carbon form, *n*—*as used in Test Method D 5061*, microscopically distinguishable carbonaceous textural components of coke, but excluding mineral carbonates. **D 121,** D05

carbon gel (carbon-rubber-gel), *n*— the portion of rubber that is not leached by a solvent under specified conditions, from an intimate mixture of carbon black and unvulcanized rubber. **D 1566,** D11

carbonitriding, *n*—**case hardening** in which a suitable steel object is heated above Ac_1 in a gaseous atmosphere of such composition as to cause simultaneous absorption of carbon and nitrogen by the surface and, by diffusion, to create a concentration gradient. **A 941,** A01

carbonized and neutralized wool, *n*—a term descriptive of scoured wool processed to destroy cellulosic impurities by treating with a mineral acid or an acid salt, drying and baking, crushing, and dusting out the embrittled cellulosic matter followed by neutralization of the acidified wool. **D 123,** D13

carbonless manifold set—a set that contains multiple plies of carbonless paper consisting of one CB and one CF, or one CB, one or more CFBs and one CF. **F 549,** F05

carbonless paper—a sheet coated or treated to provide manifolding capability without the use of carbon paper interleaves. **F 549,** F05

carbon or carbon sheet—see **carbon paper.** **F 221,** F05

carbon paper—a sheet composed of a supporting substrate on one or both sides of which is a coating containing a transferable (usually colored) material. The coating is of such nature that it will transfer in part or entirely to a copy sheet at the point of pressure contact. **F 149,** F05

carbon paper—a sheet composed of a supporting substrate on one or both sides of which is a coating containing a transferable (usually colored) material. The coating is of such nature that it will transfer in part or entirely to a copy sheet at the point of pressure contact. (F 129) **F 221,** F05

carbon potential, *n*—the carbon content at the surface of a specimen of pure iron in equilibrium with the carburizing medium considered, and under the conditions specified. **A 941,** A01

carbon refractory, *n*—a manufactured refractory comprised substantially or entirely of carbon (including graphite). **C 71,** C08

carbon refractory, anthracite-coal base, *n*— a manufactured refractory comprised substantially of calcined anthracite coal. **C 71,** C08

carbon refractory, graphite-base, *n*—a manufactured refractory comprised substantially of graphite. **C 71,** C08

carbon refractory, metallurgical-coke-base, *n*— a manufactured refractory comprised substantially of metallurgical coke. **C 71,** C08

carbon refractory, petroleum-coke-base, *n*— a manufactured refractory comprised substantially of calcined petroleum coke. **C 71,** C08

carbon residue, *n*—the residue formed by evaporation and thermal degradation of a carbon-containing material. **D 4175,** D02

carbon restoration, *n*—replacing the carbon lost from the surface layer in previous processing by carburizing this layer to substantially the original carbon level. **A 941,** A01

carbon steel, *n*—steel for which (*1*) no minimum content is specified for chromium, cobalt, molybdenum, nickel, titanium, tungsten, or zirconium, or any other element added to obtain a desired alloying effect; (*2*) for which the specified minimum for copper does not exceed 0.40 %; (*3*) for which the specified maximum for any of the following elements does not exceed these percentages: manganese 1.65 %, silicon 0.60 %, or copper 0.60 %; and (*4*) in which the incidental content of the following elements does not exceed these percentages: nickel, 0.25 %; chromium, 0.20 %; and molybdenum, 0.06 %. **A 902,** A05

carbon steel, *n*—a **steel** that conforms to a specification that prescribes a maximum limit, by **heat analysis** in mass percent, of not more than: 2.00 for carbon and 1.65 for manganese, but does not prescribe a minimum limit for chromium, cobalt, columbium (niobium), molybdenum, nickel, tungsten (wolfram), vanadium, or zirconium. **A 941,** A01

carbon steel—steel for which no minimum content is specified or required for chromium, molybdenum, nickel, or any other element added to obtain a desired alloying effect; or steel for which maximum content specified for manganese does not exceed 1.65 %. When specified, boron may be added to killed carbon steel with a maximum allowable of 0.003 %. **F 1789,** F16

carbon tissue—a substrate manufactured for use in the production of carbon paper. **F 221,** F05

carbonyl group—the bivalent radical,— CO—, especially as it occurs in aldehydes or ketones. **D 1695,** D01

carbonyl powder—a metal powder prepared by the thermal decomposition of a metal carbonyl. **B 243,** B09

carboxyhemoglobin saturation, *n*—the percent of blood hemoglobin converted to carboxyhemoglobin from reaction with inhaled carbon monoxide. **E 176,** E05

carboxyl group—the radical —COOH characteristic of most organic acids. **D 1695,** D01

carboxymethylcellulose, CMC—the common name for a cellulose ether of glycolic acid. It is usually marketed as a water-soluble sodium salt, more properly called sodium carboxymethylcellulose. In the early literature, it is sometimes called cellulose glycolate or cellulose glycolic acid. **D 1695,** D01

carboy, *n*—a **container** made of glass, ceramic, plastic, or metal, having a capacity of 5 to 15 gal (19 to 57 L) with the pouring and filling opening at the top. **D 996,** D10

carburizing, *n*—a process in which an austenitized steel object is brought into contact with a carbonaceous environment of sufficient carbon potential to cause absorption of carbon at the surface and, by diffusion, to create a concentration gradient. **A 941,** A01

carcass, *n*—the fabric, cord, or metal reinforced section, or all three, of a rubber product as distinguished from the rubber tube, cover, or tread. **D 1566,** D11

carcass, *n—of a tire*, the part of a tire structure that does not include the tread and sidewall rubber. **F 538,** F09

carcinogen—an agent producing or inciting cancerous growth. **E 609,** E35

carcinogen—cancer-causing agent. **F 1600,** F20

carded wool, *n*—scoured wool which has been processed through a carding machine. **D 123,** D13

carding leather—a type of tannage of side leather used on the cards of textile machinery. **D 1517,** D31

cardiovascular tracing, *n*—a display of physiological patterns of the subject=s relative blood volume and pulse rate. The cardiograph component records this activity. **E 2035,** E52

care and maintenance, *n*—effective cleaning to remove soil and maximize use life of garments while maintaining (not removing) protective properties. **F 1494,** F23

career apparel, *n*—garments, the styling and performance of which are designed for various end uses so as to be suitable for on-the-job wear in a variety of businesses and professions. (See also **dress career apparel** and **vocational career apparel.**) **D 123,** D13

care instructions, *n—in textiles*, a series of directions that describes practices which should refurbish a product without adverse effects and warn against any part of the directions which one could reasonably be expected to use that my harm the item. **D 123,** D13

care label, *n—in textiles,*a label or other affixed instructions that report how a product should be refurbished **D 123,** D13

care procedure, *n—in textiles*, one or more refurbishing methods to which products may be subjected for soil and stain removal and aesthetic improvement such as appearance restoration or hand. **D 123,** D13

care symbol, *n*—a pictorial symbol that gives directions for refurbishing a consumer textile product. **D 123,** D13

car nail—light-duty or heavy-duty, bright or coated, barbed or helically threaded, regular-stock-steel, 1½ by 0.105-in. (light) or 0.135-in. (heavy) to 6 by 0.225-in. (light) or 0.244-in. (heavy) nails and screwnails with flat or oval countersunk head and medium diamond point. Also, light-duty or heavy-duty, coated, barbed or helically threaded, regular-stock-steel, 1½ by 0.092 or 0.106 to 6 by 0.207 or 0.225-in. nails with oval countersunk 15⁄64 to 15⁄32-in. head and medium diamond point. **F 547,** F16

carnival—a mobile enterprise principally devoted to offering amusement or entertainment to the public in, upon, or by means of portable amusement rides or devices or temporary structures in any number or combination, whether or not associated with other structures or forms of public attraction. **F 747,** F24

carnival glass—glass having an iridescent coloration obtained by firing metallic salts applied onto a colored glass body. See **lusters.** **C 162,** C14

C_9 aromatic resin, *n*—a resin produced by the cationic polymerization of heavy aromatic steam-cracker or coal-tar fractions.

D 6440, D01

carpet, *n*—all textile floor coverings not designated as rugs. **D 123,** D13

carpet-embedded dust—See **dust.** **D 1356,** D22

carpet module, *n*—textile floor covering sections usually having dimensions of less than 1 m^2. **D 123,** D13

carpet sweeper, *n*—a manually powered machine which has rotary brushes and which is used for light surface cleaning of carpets and rugs. **D 123,** D13

carpincho leather—leather from the skin of the carpincho, a large South American rodent. The skin is used in making glove leather, usually chrome-tanned and washable. In the glove-leather trade, it is classified as a pigskin. It resembles pigskin in appearance, a characteristic being the occurrence of bristle holes in straight-line groups, usually five holes in a group. **D 1517,** D31

carriage—the device for holding the food product that is manually or automatically reciprocated to bring the product in contact with the knife. **F 1827,** F26

carriage return, *n*—an operation which advances the cursor to the beginning of a new line. (See **CR.**) **F 1457,** F05

carrier, *n*—*in braiding machinery,* that part of a braiding machine that holds the package of yarn, thread, or cord, and carries the yarn when the machine is operated. **D 123,** D13

carrier—(*1*) a gas, liquid, or solid used to propel or transport a pesticide;(*2*) an organism that bears an infectious agent, but that shows no marked symptoms of the disease caused by that agent. **E 609,** E35

carrier—a liquid, either aqueous or nonaqueous, in which liquid penetrant examination materials are dissolved or suspended. **E 1316,** E07

carrier—See **scrim.** **E 1749,** E06

carrier—that substance in a developer that conveys a toner, but does not itself become a part of the image in the finished print. In a liquid developer the carrier may be called a dispersant. **F 335,** F05

carrier—*in semiconductor materials,* an entity capable of carrying electric charge through a solid, for example, valence holes and conduction electrons in semiconductors. Also called **charge carrier.** **F 1241,** F01

carrier density, *n* (electrons); *p* (holes)—the number of majority carriers per unit volume in an extrinsic semiconductor, usually given in number/cm^3 although the SI unit is number/m^3. **F 1241,** F01

carrier detect, *n*—a communication control character used in an RS232 system to signal the sender that the receiver is on-line. (See **CD.**) **F 1457,** F05

carrier fluid—the fluid in which fluorescent and nonfluorescent magnetic particles are suspended to facilitate their application. **E 1316,** E07

carrier fluid, *n*— *in impingement or slurry erosion,* fluid medium that transports impinging solid or liquid particles and that gives the particles their momentum relative to the solid surface on which they are impinging. **G 40,** G02

carrier gas—a gas that is inert in the analysis that is used to sweep gaseous products through the analysis system. **C 859,** C26

carrier return, *n*—a code which instructs the printer to begin a new line. **F 1457,** F05

carry-in—manual lehr loading. **C 162,** C14

carton, *n*—a folding **box,** generally made from **boxboard** for merchandising consumer quantities of products (for example, shelf packages or prime packages). **D 996,** D10

carton—See **packaging.** **F 547,** F16

cartridge—see **spiral-wound cartridge**. **D 6161,** D19

cartridge—a case containing an inked ribbon correction ribbon, or dry ink toner for insertion into an image producing device. **F 221,** F05

car vacuum cleaner, *n*—a relatively small, lightweight, portable cleaner that is designed for operation from a d-c power source, generally a 12-V automotive battery (see also **hand-held vacuum cleaner**). **F 395,** F11

cascade development—development in which dry toner and carrier are drawn over an electrostatic image by gravity. **F 335,** F05

cascade impactor—See **impactor.** **D 1356,** D22

cascade mixing—the rearrangement of the constituents of a solid, within the penetration depth of an incident particle, caused by collisions between the incident particles and the atoms of the solid. **E 673,** E42

cascading, *n*—a condition where ink is absorbed into the substrate before it has time to spread, causing negative stitch lines or low density bands between print swaths in solid image areas. **F 1857,** F05

cascading booms—booming configuration formed by positioning two or more booms in a deflection mode such that successive booms progressively move oil to the desired area. **F 818,** F20

cascading water—perched ground water that enters a well casing via cracks or uncovered perforations, trickling, or pouring down the inside of the casing. **D 653,** D18

case, *n*—*in case hardening,* the outer portion that has been made harder than the **core** as a result of altered composition or microstructure, or both, from treatments such as **carburizing, nitriding,** and **induction hardening**. **A 941,** A01

case, *n*—*in textiles,* a shipping unit, usually a carton, box, bale, or other container holding a number of yarn packages. **D 123,** D13

case—See **container.** **D 996,** D10

case—*in a ferrous alloy,* the outer portion that has been made harder than the inner portion (see **core**) as a result of altered composition, or structure, or both, from treatments such as carburizing, nitriding, and induction hardening. **E 7,** E04

cased glass—glassware whose surface layer has a different composition from that of the main glass body. **C 162,** C14

casehardened—a term sometimes used for tempered glass. (See **tempered glass.**) **C 162,** C14

case-hardened, surface-hardened—surface of steel nail carburized and subsequently hardened, by suitable heat treatment, leaving a soft core. **F 547,** F16

case hardening, *n*—a generic term covering any of several processes applicable to **steel** that change the chemical composition or microstructure, or both, of the surface layer. **A 941,** A01

casehardening—a condition of stress and set in wood due to drying in which outer fibers are under compressive stress and inner fibers under tensile stress, a condition that often exists during subsequent processing. In commerce, the term often connotes a degree of severity regarded as detrimental in use. **D 9,** D07

casein adhesive, *n*—an aqueous colloidal dispersion of casein that may be prepared with or without heat, may contain modifiers, inhibitors, and secondary binders to provide specific adhesive properties, and includes a subclass, usually identified as *casein glue,* that is based on a dry blend of casein, lime, and sodium salts, mixed with water and prepared without heat. **D 907,** D14

case leather—see **bag leather.** **D 1517,** D31

case liner—See **liner.** **D 996,** D10

cash flow—the stream of monetary (dollar) values—costs and benefits—resulting from a project investment. **E 833,** E06

cashmere, *n*—*in roving, yarn, or fabrics,* cashmere hair or products made therewith having a cashmere coarse-hair content not exceeding a specified maximum percentage by length. **D 123,** D13

cashmere coarse-hair, *n*—those coarse fibers in cashmere hair having widths greater than 30 μm. **D 123,** D13

cashmere coarse-hair content, *n*—the total length of the cashmere coarse-hair fibers that are present, expressed as a percentage of the total length of all the cashmere hair fibers; that is, the percentage by length of cashmere coarse-hair in cashmere hair. **D 123,** D13

cashmere down, *n*—those fibers in cashmere hair widths of 30 μm or less. **D 123,** D13

cashmere hair, *n*—the fibers produced by a form of goat (*Capra hircus*) indigenous to Asia and known as the cashmere goat. **D 123,** D13

casing—pipe, finished in sections with either threaded connections or bevelled edges to be field welded which is installed temporarily or permanently to counteract caving, to advance the borehole, or to isolate the zone being monitored, or combination thereof. **D 653,** D18

casing, *n—of a tire*, a used or treadless tire to which additional tread rubber may be attached for the purpose of retreading. **F 538,** F09

casing head—deep countersunk (32 deg) head with flat or cupped, circular top surface as found on casing nails to facilitate flush driving of nail. **F 547,** F16

casing nail—bright or galvanized, slender, regular-stock-steel, 1 by 0.067 to 3½ by 0.135-in. nails with flat or cupped 0.099 to 0.177-in. casing head and medium diamond point for countersinking where concealment is important. **F 547,** F16

casing, protective—a section of larger diameter pipe that is emplaced over the upper end of a smaller diameter monitoring well riser or casing to provide structural protection to the well and restrict unauthorized access into the well. **D 653,** D18

casing, surface—pipe used to stabilize a borehole near the surface during the drilling of a borehole that may be left in place or removed once drilling is completed. **D 653,** D18

cask—See **barrel.** **D 996,** D10

cassette—a light-tight film or plate holder. **E 7,** E04

cassette—a light-tight container for holding radiographic recording media during exposure, for example, film, with or without intensifying or conversion screens. **E 1316,** E07

cassette—a light-tight device for holding film or conversion screens and film in close contact during exposure. **E 1316,** E07

cassette—see **cartridge.** **F 221,** F05

cassiterite (SnO_2)—an inorganic mineral of the tetragonal form used as a source of tin and tin oxide. **C 242,** C21

CASS test (copper accelerated salt spray)—an accelerated corrosion test for some electrodeposits and for anodic coatings on aluminum. **B 374,** B08

cast—an impression or mold taken from a person or thing. **F 869,** F08

castable, *n*—a combination of refractory grain and a suitable amount of bonding agent that, after the addition of a proper liquid, is generally poured into place to form a refractory shape or structure which becomes rigid because of chemical action (see Classification C 401). **C 71,** C08

castable, *n*—a combination of filler and suitable binder that is generally poured or compacted into place and which hardens. **C 904,** C03

cast analysis, *n*—Deprecated term. Use preferred term *heat analysis.* **A 902,** A05

cast analysis—Deprecated term. Use the preferred term **heat analysis.** **A 941,** A01

cast film—a film made by depositing a layer of plastic, either molten, in solution, or in a dispersion, onto a surface, solidifying and removing the film from the surface. **D 883,** D20

cast film, *n*—a film made by depositing a layer of plastic, either molten, in solution, or in a dispersion, onto a surface, solidifying the deposit and removing the film from the surface. **F 1251,** F04

casting—a general term for a metal object produced at or near-finished shape by pouring or otherwise introducing molten metal into a mold and allowing it to solidify. **B 846,** B05

casting, *n*—metal object, at or near dimensions shape, produced by introducing molten metal into a mold or a die and allowing it to solidify.

die casting, n—casting produced by introducing molten metal under substantial pressure into a metal die and characterized by a high degree of fidelity to the die cavity.

investment casting, n—casting produced by surrounding (investing) an expendable pattern (usually wax or plastic) with a refractory slurry that sets at room temperature, after which the pattern is removed through the use of heat, and the resultant cavity is filled with molten metal and allowed to solidify.

permanent mold casting, n—casting produced by introducing molten metal by gravity or low pressure into a mold constructed of durable material, usually iron or steel, and allowing it to solidify.

sand casting, n—casting produced by pouring molten metal into a sand mold and allowing it to solidify.

semi-permanent mold casting, n—permanent mold casting which is made using an expendable core such as sand. **B 881,** B07

casting—a process of shaping glass by pouring molten glass into molds, onto tables, or between rollers. **C 162,** C14

casting—a process for forming ceramic ware by introducing a body slip into a porous mold which absorbs sufficient water (or other liquid) from the slip to produce a semirigid article.

drain casting (hollow casting)—forming ceramic ware by introducing a body slip into an open porous mold, and then draining off the remaining slip when the case has reached the desired thickness.

solid casting—forming ceramic ware by introducing a body slip into a porous mold which usually consists of two major sections, one section forming the contour of the outside and the other forming the contour of the inside of the ware and allowing a solid cast to form between the two mold faces. **C 242,** C21

casting, centrifugal—a casting produced in a cylindrical mold rotating on its axis with the major axis of the product coinciding with the axis of rotation. The axis of rotation may be horizontal, vertical, or any angle in between. **B 846,** B05

casting, centrifuged—a casting produced in a mold, a number of which may be mounted around a central sprue. The molds are rotated, in a vertical position, about a central axis concentric with the central sprue. **B 846,** B05

casting, continuous—a casting produced by the continuous pouring and solidification of molten metal through a water-cooled mold which determines the cross-sectional shape. The length of the product is not restricted by mold dimensions. **B 846,** B05

casting, drain (hollow casting), *v*—forming ceramic ware by introducing a body slip into an open, porous mold, and then draining off the remaining slip when the cast piece has reached the desired thickness.**(C21, C 242)** **C 1145,** C28

casting, permanent mold—a product produced in a reusable mold constructed of a durable material, usually iron or steel, with the molten metal being introduced by gravity, low-pressure or vacuum. **B 846,** B05

casting, sand—a casting produced in a sand mold. **B 846,** B05

casting, semicentrifugal—a casting produced in a manner similar to the centrifugal casting except that a central core is used to allow the formation of other than a cylindrical inside surface. The axis of rotation is always vertical. **B 846,** B05

cast iron, *n*—a generic term for a series of alloys primarily of iron, carbon, and silicon in which the carbon is in excess of the amount which can be retained in solid solution in austenite at the eutectic temperature. **A 644,** A04

cast iron enamel—a porcelain enamel specifically designed for application to cast iron. **C 286,** B08

cast lead head—half-round or bell-shaped lead head serving as head and washer requiring approximately 25 % more lead than compressed lead head. (See **lead head.**) **F 547,** F16

cast replica—See **replica.** **E 7,** E04

cast stone—an architectural precast concrete building unit intended to simulate natural cut stone. **C 1213,** C27

cast structure—the structure, on a macroscopic or microscopic scale, of a casting. **E 7,** E04

catalyst—a substance whose presence initiates or changes the rate of a chemical reaction, but does not itself enter into the reaction. **C 904,** C03

catalyst—a material that causes chemical reactions to begin. **D 653,** D18

catalyst, *n*—a substance that initiates or changes the rate of chemical reaction, but is not consumed or changed by the reaction. (See also **hardener**.) (Compare **inhibitor**.) **D 907,** D14

catalyst—a substance whose presence initiates or changes the rate of a chemical reaction, but does not itself enter into the reaction. **D 6161,** D19

catalyst—a substance that increases the rate of a chemical reaction; used extensively in polymerization reactions. **E 1749,** E06

catalyst bed support—an essentially inert plate, grid, particulate bed, or other structural component designed to hold up or bear a quantity of catalyst in a catalytic reactor. **D 3766,** D32

catalyst carrier—a solid, generally porous material upon the surface or into the voids of which catalytic materials are placed to create the desired catalyst. **D 3766,** D32

catalyst substrate—less preferred alternative term for **catalyst carrier.** **D 3766,** D32

catalyst support—less preferred alternative term for **catalyst carrier.** **D 3766,** D32

catalyst system—those materials that, in combination, cause chemical reactions to begin; catalyst systems normally consist of an initiator (catalyst) and an activator. **D 653,** D18

cataphoresis—See **electrophoresis.** **B 374,** B08

catastrophic period, *n*— *in cavitation or liquid impingement erosion,* a stage during which the erosion rate increases so drastically that continued exposure threatens or causes gross disintegration of the exposed surface. This stage is not inevitable; it is observed most commonly with some brittle materials. When it does occur, it may begin during any stage of the more common erosion rate-time pattern. **G 40,** G02

catastrophic wear, *n*—rapidly occurring or accelerating surface damage, deterioration, or change of shape caused by wear to such a degree that the service life of a part is appreciably shortened or its function is destroyed. **G 40,** G02

catch up, *n*—a condition that occurs when the non-image area of the plate becomes ink receptive. See **scum**. **[GATF].** **D 6488,** D01

catenary:

filament catenary—the difference in length of the filaments in a specified length of tow, end, or strand as a result of unequal tension; the tendency of some filaments in a taut horizontal tow, end, or strand to sag lower than others.

roving catenary—the difference in length of the ends, tows, or strands in a specified length of roving as a result of unequal tension; the tendency of some ends, tows, or strands in a taut horizontal roving to sag lower than others. **D 3878,** D30

catenary configuration (aka "U," "J" configuration)—booming configuration formed by towing or anchoring each end of a length of boom, resulting in a characteristic" U" or "J" shape. **F 818,** F20

catenary drag force—load imposed on a boom, deployed in a catenary configuration, resulting from towing, current, and/or wind forces. **F 818,** F20

cat eye—an imperfection; an elongated bubble containing a piece of foreign matter. **C 162,** C14

cathode—the electrode in electrolysis at which positive ions are discharged, negative ions are formed, or other reducing actions occur. **B 374,** B08

cathode—unmelted, electrodeposited, and somewhat rough flat plate normally used for melting. The customary size is about 3 ft (0.914 m) square, about ½ to ⅞ in. (12.7 to 22.2 mm) thick, weighing up to about 300 lb (136 kg) and may have hanging loops attached. Cathodes may also be cut to smaller dimensions. **B 846,** B05

cathode—negative electrode. **D 6161,** D19

cathode—the electrode of an electrolytic cell at which reduction is the principal reaction. (Electrons flow toward the cathode in the external circuit.) **G 15,** G01

cathode efficiency—the current efficiency of a specified cathodic process. **B 374,** B08

cathode fall—the potential difference between the cathode and the electric discharge plasma. **B 542,** B02

cathode film—the layer of solution in contact with the cathode that differs in composition from that of the bulk of the solution. **B 374,** B08

cathode lens—a lens field terminated on one side by a surface at zero potential (cathode) normal to the optic axis. A cathode lens occurs in any system in which a cathode is imaged by its own electron emission, be it thermionic, photoelectric, secondary, or field emission. **E 7,** E04

cathode polarization—See **polarization.** **B 374,** B08

cathodic (cathode) material transfer—see **material transfer.** **B 542,** B02

cathodic corrosion—corrosion of a metal when it is a cathode. (It usually happens to metals because of a rise in pH at the cathode or as a result of the formation of hydrides.) **G 15,** G01

cathodic inhibitor—an inhibitor that reduces the corrosion rate by acting on the cathodic (reduction) reaction. **G 15,** G01

cathodic polarization—the change of the electrode potential in the active (negative) direction due to current flow. (See **polarization.**) **G 15,** G01

cathodic protection, *n*—protection of a metal from corrosion by making it a cathode through the galvanic sacrifice of a less noble metal or through an impressed electric current. **B 899,** B02

cathodic protection—the control of corrosion by the application of a cathodic current to the reinforcing steel by means of impressed current or a sacrificial anode. **D 3743,** D04

cathodic protection—a technique to reduce the corrosion rate of a metal surface by making it the cathode of an electrochemical cell. **G 15,** G01

catholyte—the portion of the electrolyte in the vicinity of the cathode; in a divided cell the portion of the cathode side of the diaphragm. **B 374,** B08

catholyte—the electrolyte adjacent to the cathode of an electrolytic cell. **G 15,** G01
cation—a positively-charged ion. **B 374,** B08
cation—an ion that moves, or would move toward a cathode; thus nearly always synonymous with positive ion. **D 653,** D18
cation—positively charged ion. **D 6161,** D19
cation—a positively charged ion. **G 15,** G01
cation conductivity—a conductivity measurement performed on water after cations have been exchanged for protons using cation exchange media. **D 1129,** D19
cation exchange—see **base exchange**. **D 653,** D18
cation exchange capacity (CEC)—the total capacity of a porous system to absorb cations from a solution. **D 653,** D18
cation-exchange material—a material capable of the reversible exchange of positively charged ions. **D 1129,** D19
cation exchange material—a material capable of the reversible exchange of positively charged ions. **D 6161,** D19
cationic detergent—under **detergent,** see *cationic detergent.* **D 459,** D12
cationic emulsion, *n*—a type of emulsion such that a particular emulsifying agent establishes a predominance of positive charges on the discontinuous phase. **D 8,** D04
cationic emulsion—an emulsion in which the emulsifying system establishes a predominance of positive charges on the discontinuous phase. **D 1079,** D08
cationic polyelectrolyte—a polymer containing positively charged groups used for coagulation/flocculation, usually dimethyl - aminoethyl methacrylate or dimethyl-aminoethyl acrylate. See **polyelectrolyte**. **D 6161,** D19
cationic surfactant—a surface-active agent in which the active portion of the molecule containing the lipophilic segment forms exclusively a positive ion (cation) when placed in aqueous solution. **E 1519,** E35
cat scratch—an imperfection; surface irregularities on glassware resembling the marks of a cat's claws. **C 162,** C14
caul, *n*—a sheet of material employed singly or in pairs in hot or cold pressing of assemblies being bonded. **D 907,** D14
caul—a sheet of material employed singly or in pairs in hot or cold pressing of assemblies being bonded. **E 1749,** E06
caulk, *v*—*in building construction,* to install or apply a sealant across or into a joint, crack, or crevice. **C 717,** C24
caulk, n—See **sealant.** **C 717,** C24
caulk, *v*—to fill joints, **cracks,** or crevices in order to prevent the passage of air or water. **E 631,** E06
caulking—a material ranging in physical characteristics from plastic, to solid, to preformed, used to seal and waterproof joints and overlaps in structures, other assemblies or portions thereof where movement may occur. **C 460,** C17
caulking, n—See **sealant.** **C 717,** C24
caulking, *n*—a material ranging in physical characteristics from plastic, to solid, to preformed, used to seal and waterproof joints and overlaps in structures, other assemblies, or portions thereof where movement may occur. **C 1154,** C17
caulking—a composition of vehicle and pigment, used at ambient temperatures for filling joints, that remains plastic for an extended time after application. **D 1079,** D08
caulking—material ranging in physical characteristics from plastic, to solid, to preformed, used to seal and waterproof joints and overlaps in structures, other assemblies or portions thereof where movement may occur. **D 2946,** C17
caulking compound, n—See **sealant.** **C 717,** C24
caulking compound—a soft, plastic material, consisting of pigment and vehicle, used for sealing joints in buildings and other structures where normal structural movement may occur. Caulking compound retains its plasticity for an extended period after application. It is available in forms suitable for application by gun and knife and in extruded preformed shapes. **D 16,** D01
cauls, *n*—sheets of material employed singly or in pairs in hot or cold pressing of assemblies being bonded. Cauls are employed usually to protect either the faces or the press platen or both against marring and staining, to prevent sticking, to facilitate press loading, or to impart a desired surface texture or finish, and to provide uniform pressure distribution.

Note—Cauls may be made of any material such as aluminum, stainless steel, hardboard, fiberboard, or plastic, with the length and width generally equal to the platen size of the press in which they are employed. **D 1038,** D07

caustic cracking—stress corrosion cracking of metals in caustic solutions. (See also **stress corrosion cracking.**) **G 15,** G01
caustic dip—*in diffusion coatings,* a strongly alkaline treatment applied by dip or spray for neutralizing acid residues. **B 374,** B08
caustic embrittlement—a form of metal failure that occurs in steam boilers at riveted joints and at tube ends, the cracking being predominantly intercrystalline. **D 1129,** D19
caustic embrittlement— see **caustic cracking.** **G 15,** G01
caves or washouts, *n*—zones of increased hole diameter caused by rock fragments that fall from the walls of a borehole and can block the hole or contaminate the cuttings and which erode or abrade the sidewall of the borehole by the action of the drilling. These zones can affect the accuracy of certain geophysical logs (especially density). Corrections to other geophysical logs can be made if a caliper log is available. The most common causes of caves or washouts include soft or fractured lithologies, the presence of water-producing zones, and the downhole pressure of the drilling medium (fluid or air) that often causes differential erosion of various strata within the borehole. **D 121,** D05
caving; sloughing—the inflow of unconsolidated material into a borehole which occurs when the borehole walls lose their cohesive strength. **D 653,** D18
cavitating disk device (or apparatus), *n*—a flow cavitation test device in which cavitating wakes are produced by holes in, or protuberances on, a disk rotating within a liquid-filled chamber. Erosion test specimens are attached flush with the surface of the disk, at the location where the bubbles are presumed to collapse. **G 40,** G02
cavitating jet, *n*—a continuous liquid jet (sometimes submerged) in which cavitation is induced by the nozzle design, or sometimes by an obstruction placed in the center of the flow passage. **G 40,** G02
cavitating wake, n— See **flow cavitation**. **G 40,** G02
cavitation, *n*—the formation of vapor-filled cavities in the interior or on the solid boundaries of liquids in motion where the pressure is reduced to a critical value without a change in ambient temperature. **E 1620,** E29
cavitation—the formation and rapid collapse within a liquid of cavities or bubbles that contain vapor or gas or both. **G 15,** G01
cavitation, *n, n*—the formation and subsequent collapse, within a liquid, of cavities or bubbles that contain vapor or gas or both. **G 40,** G02
cavitation cloud, *n*—a collection of a large number of cavitation bubbles. The bubbles in a cloud are small, typically less than 1 mm

(0.04 in.) in cross section. A surface that is being eroded by cavitation is usually obscured by a cavitation cloud. **G 40,** G02

cavitation corrosion, *n*—a form of localized, accelerated corrosion characterized by deep pitting and caused by high mechanical forces resulting from coolant vapor bubble collapse at the surface of the metal. **D 4725,** D15

cavitation corrosion—the conjoint action of cavitation-erosion and corrosion. **G 15,** G01

cavitation damage—the degradation of a solid body resulting from its exposure to cavitation. (This may include loss of material, surface deformation, or changes in properties or appearance.) **G 15,** G01

cavitation damage, n— See **damage.** **G 40,** G02

cavitation-erosion—loss of material from a solid surface due to mechanical action of continuing exposure to cavitation. **G 15,** G01

cavitation erosion, *n*—progressive loss of original material from a solid surface due to continued exposure to cavitation. **G 40,** G02

cavitation erosion corrosion, *n*—the mechanical removal of protective films on metal by the formation and collapse of vapor bubbles in a liquid,and the abrasive action of a liquid, which may contain suspended solids, moving at high velocity. **D 4725,** D15

cavitation erosion test, *n*—a procedure whereby the surface of a solid is subjected to cavitation attack under specified, or measurable, or at least repeatable conditions. **G 40,** G02

cavitation number, σ, *n*—a dimensionless number that measures the tendency for cavitation to occur in a flowing stream of liquid, and that is computed from the equation:

$$\sigma = \left(P_o - P_v\right)/\frac{1}{2}\rho V_o^2$$

where:
P_v = vapor pressure,
P_o = static pressure in the stream in an undisturbed state,
V_o = undisturbed stream velocity, and
ρ = liquid density.

G 40, G02

cavitation tunnel, *n*—a flow cavitation test facility in which liquid is pumped through a pipe or tunnel, and cavitation is induced in a test section by conducting the flow through a constriction, or around an obstacle, or a combination of these. **G 40,** G02

cavity—a natural underground opening that may be small or large. **D 653,** D18

cavity—underground opening created by a fully contained explosive. (ISRM) **D 653,** D18

cavity, *n*—in specimen preparation, the part of the hollow space of a mold that forms one specimen. **D 883,** D20

cavity—a void portion of a mold made by casting, machining, bobbing, or a combination of these methods. **F 869,** F08

cavity—that portion of the oven in which food may be heated, cooked, or defrosted. **F 1827,** F26

CB—an abbreviation for coated back, referring to a carbonless paper of the two-coat, transfer type. **F 549,** F05

"C" blade—blade clincher resembling the letter "C"; designed to fasten partially overlapping container panels. **F 592,** F16

CBR—an abbreviation for *chemical, biological, radiological.* **E 1749,** E06

CCITT—abbreviation for Comite Consultatif International Telegraphique et Telephonique (Consultive Committee for International Telephone and Telegraph). **F 1457,** F05

CD—compact disk. **F 1156,** F12

CD—abbreviation for **carrier detect.** **F 1457,** F05

CED (cupriethylenediamine) viscosity—see **viscosity, cupriethylenediamine.** **D 1695,** D01

ceiling limit—an exposure which shall not be exceeded during any part of the workday. If instantaneous monitoring is not feasible, then the ceiling limit shall be assessed as a 15-min time-weighted average exposure which shall not be exceeded at any time over a working day except for substances which cause immediate irritation upon short exposure. **E 1542,** E34

cell, *n*—a single small cavity surrounded partially or completely by walls. **C 717,** C24

cell—a general term for the minute units of wood structure having distinct cell walls and cell cavities including wood fibers, vessel segments, and other elements of diverse structure and function. **D 9,** D07

cell, *n*—a small cavity surrounded partially or completely by walls. **D 883,** D20

cell, *n*—a single small cavity surrounded partially or completely by walls. **D 1566,** D11

cell, *n*—a small partially or completely enclosed cavity. (D-20) **F 412,** F17

cell, *n*—a small partially or completely enclosed cavity. **F 1251,** F04

cellar—See **building space.** **E 631,** E06

cellar—See **space categories.** **E 1480,** E06

cell, closed—a cell totally enclosed by its walls and hence not interconnecting with other cells. (ISO) (See **cell** and **cell, open.**) **D 883,** D20

cell, closed, *n*—a cell totally enclosed by its walls, hence not interconnecting with other cells. **D 1566,** D11

cell, closed—a cell totally enclosed by its walls and hence not interconnecting with other cells. (ISO, D-20)(See also **cell** and **cell, open.**) **F 412,** F17

cell, closed—see **closed cell.** **F 1251,** F04

cell, open—a cell not totally enclosed by its walls and hence interconnecting with other cells. (See **cell** and **cell, closed.**) **D 883,** D20

cell, open, *n*—a cell not totally enclosed by its walls and hence interconnecting with other cells. **D 1566,** D11

cell, open—a cell not totally enclosed by its walls and hence interconnecting with other cells. (D-20) (See also **cell** and **cell, closed.**) **F 412,** F17

cell, open—see **open cell.** **F 1251,** F04

cellophane pulps—pulps used in the manufacture of cellophane, and subject to various specifications by the manufacturers, including those of purity, moisture content, sheet properties, and viscosity. **D 1695,** D01

cells /core holes, *n*—continuous openings or perforations within extruded clay products. The extent of permissible openings is specified for each product as the percentage of gross area in the normal bedding surface plane that must be net (solid) area. Core hole is generally used for brick while cell is used for structural tile. Cells are distinguished from core holes by being larger in size. As an illustration, cells must be larger than 1 in.^{2}(645 mm^2) under Specification C 34, and 1½ in.^{2}(968 mm^2) under Specification C 652. **C 43,** C15

cell structure—malformations attributable to crystal inhomogeneities and that have their origins in the crystal growth process. Also known as **block structure.** **F 1241,** F01

cellular, *adj*—having a composition of plastic or rubber with relative density decreased by the presence of cells disposed throughout its mass. In closed-cell materials, the cells are predominantly separate from each other. In open-cell materials, the cells are predominantly interconnected. **C 717,** C24

cellular adhesive, n—Synonym for **foamed adhesive.** **D 907,** D14

cellular concrete, *n*—a lightweight hydraulic-cement concrete having a homogeneous void or cell structure attained using gas-forming chemicals or foaming agents. **C 125,** C09

cellular elastomeric, *n*—insulation composed principally of natural or synthetic elastomers, or both, processed to form a flexible, semirigid, or rigid foam which has a predominantly closed-cell structure. **C 168,** C16

cellular glass, *n*—insulation composed of glass processed to form a rigid foam having a predominantly closed-cell structure. **C 168,** C16

cellular material, *n*—a generic term for materials containing many cells (either open, closed, or both) dispersed through the mass. **C 717,** C24

cellular material, *n*—a generic term for materials containing many cells (either open, closed, or both) dispersed throughout the mass. **D 1566,** D11

cellular material, collapse, *n*—undesirable densification of a cellular material resulting from the breakdown of its cellular structure. **D 1566,** D11

cellular material, cored, *n*—cellular material containing a multiplicity of holes (usually, but not necessarily, cylindrical in shape), molded or cut into the material in some pattern normally perpendicular to the largest surface, and extending part or all the way through the piece. **D 1566,** D11

cellular material, flexible, *n*—a cellular organic polymeric material that will not rupture when a specimen 200 by 25 by 25 mm (8 by 1 by 1 in) is bent around a 25-mm (1-in) diameter mandrel at a uniform rate of one lap in 5 s at a temperature between 18 and 29°C. **D 1566,** D11

cellular plastic—a plastic containing numerous cells, intentionally introduced, interconnecting or not, distributed throughout the mass. (See also **syntactic cellular plastics.**) **D 883,** D20

cellular plastic—a plastic containing numerous cells, intentionally introduced, interconnecting or not, distributed throughout the mass. (D-20) **F 412,** F17

cellular plastic—a plastic containing numerous cells, intentionally introduced, interconnecting or not, distributed throughout the mass. (D 883, D-20) **F 869,** F08

cellular polystyrene, *n*—insulation composed principally of polymerized styrene resin processed to form a rigid foam having a predominantly closed-cell structure. **C 168,** C16

cellular polystyrene, *n*—polymerized styrene resin processed to form a rigid foam having a predominately closed-cell structure making it suitable as thermal insulation. **E 631,** E06

cellular polyurethane, *n*—insulation composed principally of the catalyzed reaction product of polyisocyanate and polyhydroxy compounds, processed usually with fluorocarbon gas to form a rigid foam having a predominantly closed-cell structure. **C 168,** C16

cellular striation, *n*—a condition characterized by a layer within a cellular material that differs greatly from the characteristic cell structure. **D 883,** D20

cellular striation, *n*—a condition characterized by a layer within a cellular material that differs greatly from the characteristic cell structure. **D 1566,** D11

cellulose—the carbohydrate that is the principal constituent of wood and forms the structural framework of the wood cells. **D 9,** D07

cellulose—(*1*) the main solid constituent of woody plants; it occurs widely elsewhere in the vegetable kingdom, and to a small extent in the animal kingdom. (*2*) chemically, cellulose is β-1-4 glucan of high degree of polymerization. It is desirable to apply "cellulose" to this material only and to designate the predominantly cellulosic residue obtained by subjecting woody tissues to various pulping processes as "cellulosic residues," "cellulosic pulps," or the like. **D 1695,** D01

cellulose—an amorphous carbohydrate ($C_6H_{10}O_5$) that is the principal constituent of wood and plants. **D 6161,** D19

cellulose acetate—in the broad sense, any of several esters of cellulose and acetic acid. (see also **cellulose triacetate**) **D 1695,** D01

cellulose acetate butyrate—a mixed ester of cellulose containing both acetate and butyrate groups. **D 1695,** D01

cellulose acetate butyrate (CAB) plastics—plastic made by compounding a cellulose acetate butyrate ester with plasticizers and other ingredients. Cellulose acetate butyrate ester is a derivative of cellulose (obtained from cotton or wool pulp, or both) made by converting some of the hydroxyl groups in cellulose to acetate and butyrate groups with chemicals. **F 412,** F17

cellulose acetate (CA)—in the broad sense, any of several esters of cellulose and acetic acid. **D 6161,** D19

cellulose acetate phthalate—a mixed ester of cellulose containing both acetate and phthalate groups. **D 1695,** D01

cellulose acetate propionate—a mixed ester of cellulose containing both acetate and propionate groups. **D 1695,** D01

cellulose derivative—a substance derived from cellulose by substitution of one or more of the hydroxyl groups with some other radical. Most derivatives are ethers or esters. **D 1695,** D01

cellulose esters—derivatives of cellulose in which one or more of the hydroxyl hydrogens have been replaced acyl groups. **D 1695,** D01

cellulose ethers—derivatives of cellulose in which one or more of the hydroxyl hydrogens have been replaced by alkyl groups. **D 1695,** D01

cellulose I—the crystalline modification of cellulose that normally occurs in nature. **D 1695,** D01

cellulose II—the crystalline modification of cellulose that is found in mercerized cellulose, in regenerated cellulose, and in cellulose produced by the hydrolysis of various cellulose derivatives. **D 1695,** D01

cellulose III—a crystalline modification of cellulose produced by treatment, under certain conditions, with ammonia or sometimes by amines. The method of removing the reagent determines the modification produced. **D 1695,** D01

cellulose IV—a crystalline modification of cellulose produced by heat treatment of cellulose II. **D 1695,** D01

cellulose lacquer— see **lacquer.** **D 16,** D01

cellulose lacquer—a liquid coating composition containing as the basic film-forming ingredient a cellulose ester or ether and plasticizers with or without resins or pigments. **D 1695,** D01

cellulose mixed ester—a cellulose ester containing more than one type of acyl group. **D 1695,** D01

cellulose nitrate—any of various nitrate esters of cellulose. **D 1695,** D01

cellulose plastics—plastics based on cellulose compounds, such as esters (cellulose acetate) and ethers (ethylcellulose). **D 1695,** D01

cellulose propionate—any ester of cellulose with propionic acid. **D 1695,** D01

cellulose, purified cotton—see **cotton cellulose, purified.** **D 1695,** D01

cellulose, purified wood—see **wood cellulose, purified.**

D 1695, D01

cellulose sodium glycolate—see **carboxymethyl cellulose.** D 1695, D01

cellulose triacetate—that form of cellulose acetate in which the degree of substitution approaches 3 sufficiently that the product is not soluble in acetone. **D 1695,** D01

cellulose X—a crystalline modification of cellulose produced by treatment of cellulose with strong hydrochloric acid or phosphoric acid. **D 1695,** D01

cellulose xanthates—the salts of cellulose xanthic acid. Commonly, cellulose xanthate refers to sodium cellulose xanthate, the essential constituent of the viscose solution, from which viscose rayon is spun. **D 1695,** D01

cellulosic fiber, *n*—insulation composed principally of cellulose fibers usually derived from paper, paperboard stock, or wood, with or without binders. **C 168,** C16

cellulosic fiberboard—a generic term for a homogeneous panel made from lignocellulosic fibers (usually wood or cane) characterized by an integral bond produced by inter-felting of the fibers, to which other materials may have been added during manufacture to improve certain properties, but which has not been consolidated under heat and pressure as a separate stage in manufacture, said board having a density of less than 31 lb/ft^3 (specific gravity 0.50) but having a density of more than 10 lb/ft^3 (specific gravity 0.16). **D 1554,** D07

cellulosic plastics, *n*—plastics based on cellulose compounds, such as esters (cellulose acetate) and ethers (ethyl cellulose). **D 883,** D20

Celsius, *adj*—pertaining to or denoting something related to the expression of temperature in degrees Celsius. **E 344,** E20

Celsius—designation of the degree on the International Practical Temperature Scale; also used for the name of the scale, as "Celsius Temperature Scale." Formerly (prior to 1948) called" Centigrade." The Celsius temperature scale is related to the International Kelvin Temperature Scale by the equation $T_c = T - 273.16K$. **E 1142,** E37

celsius (°C)—the designation of the degree on the International Practical Temperature Scale. Formerly called centigrade,° C = K minus 273.15. K = Kelvin. **D 6161,** D19

cement; Portland cement—commonly known as Portland cement. A mixture that consists of a calcareous argillaceous, or other silica-, alumina,- and iron-oxide bearing materials that is manufactured and formulated to produce various types which are defined in Specification C 150. Portland cement is also considered a hydraulic cement because it must be mixed with water to form a cement-water paste that has the ability to harden and develop strength even if cured under water (see **ASTM cement types**). **D 653,** D18

cement, n—See Discussion under **adhesive.** **D 907,** D14

cement, v—See **bond.** **D 907,** D14

cement, *n*—a general term for a binding element. See specific terms such as Portland cement, Keene's cement, and adhesive cement. **E 631,** E06

cement—a term broadly applied to a number of adhesives used commonly in shoemaking. **F 869,** F08

cement, API, Class A—a cement intended for use from the surface to a depth of 6000 ft (1828 m). This cement is similar to ASTM Type I cement. **D 653,** D18

cement, API, Class B—a cement intended for use from the surface to a depth of 6000 ft (1828 m) when conditions require moderate- to high-sulfate resistance. This cement is similar to ASTM Type II cement. **D 653,** D18

cement, API, Class C—this cement is intended for use from the surface to a depth of 6000 ft (1828 m) when conditions require high early strength. This cement is similar to ASTM Type III cement. Also available as a high sulfate resistant type. **D 653,** D18

cement, API, Class G—this cement is intended for use from the surface to a depth of 8000 ft (2438 m). It can be used with accelerators or retarders to cover a wide range of well depths and temperatures. No additions other than calcium sulfate or water, or both, can be intergound or blended with the clinker during manufacture of the cement. Also available as several sulfate-resistant types. **D 653,** D18

cement, API, Class H—this cement is intended for use from the surface to a depth of 8000 ft (2438 m). It can be used with accelerators or retarders to cover a wide range of well depths and temperatures. No additions other than calcium sulfate or water, or both, can be interground or blended with the clinker during manufacture of the cement. Also available as a sulfate-resistant type. **D 653,** D18

cement, API, Class J—this cement is intended for use from depths of 12 000 to 16 000 ft (3658 to 4877 m) under conditions of extremely high temperatures and pressures. It can be used with accelerators and retarders to cover a range of well depths and temperatures. No additions of retarders other than calcium sulfate, or water, or both, can be interground or blended with the clinker during manufacture of the cement. **D 653,** D18

cementation, *n*—the introduction of one or more elements into the outer portion of a steel object by means of diffusion at high temperature. **A 941,** A01

cementation—See **pack cementation.** **B 374,** B08

cement-bonded particle board, *n*—manufactured flat sheets of hydraulic cementitious matrices and fibrous wood particles. **C 1154,** C17

cement bond (sonic) log—a borehole geophysical log that can be used to determine the effectiveness of a cement seal of the annular space of a well. **D 653,** D18

cement-coated—surface coated by tumbling or immersion in natural resin or shellac to produce a limited temporary bond between driven nail and surrounding wood, provided coating is not removed during driving, and to reduce rusting during storage. **F 547,** F16

cement construction—a process in which the outsole is attached to the upper by cementing instead of sewing or by other methods. Also known as the compo process, after Compo Industries, Inc., which introduced this method commercially into the U.S. about 1930; also known as cement process and as cemented process. **F 869,** F08

cement factor—quantity of cement contained in a unit volume of concrete or grout, expressed as weight, or volume (specify which). **D 653,** D18

cement, finishing, *n*—a mixture of dry fibrous or powdery materials, or both, that when mixed with water develops a plastic consistency, and when dried in place forms a relatively hard, protective surface. **C 168,** C16

cement grout—a grout in which the primary cementing agent is Portland cement. **D 653,** D18

cement, insulating, *n*—a mixture of dry granular, flaky, fibrous, or powdery materials that when mixed with water develops a plastic consistency, and when dried in place forms a coherent covering that affords substantial resistance to heat transmission. **C 168,** C16

cementite, *n*—a very hard and brittle compound of iron and carbon corresponding to the empirical formula Fe_3C, commonly known as iron carbide. **A 644,** A04

cementite—a very hard and brittle compound of iron and carbon corresponding to the empirical formula Fe_3C. It is commonly known as iron carbide and possesses an orthorhombic lattice. In "plain-carbon steels" some of the iron atoms in the cementite lattice are replaced by manganese, and in "alloy steels" by other elements such as chromium or tungsten. Cementite will often appear as distinct lamellae or as spheroids or globules of varying size in hypo-eutectoid steels. Cementite is in metastable equilibrium and has a tendency to decompose into iron and graphite, although the reaction rate is very slow. **E 7,** E04

cementite, primary, *n*—cementite precipitated in cast iron during solidification. Also known as primary carbide. (See **cementite.**) **A 644,** A04

cementitious factor—quantity of cement and other cementitious materials contained in a unit volume of concrete or grout, expressed as weight or volume (specify which). **D 653,** D18

cementitious material, *n*—a material that, when mixed with water, with or without aggregate, provides the plasticity and the cohesive and adhesive properties necessary for placement, and the formation of a rigid mass. **C 11,** C11

cementitious material, *n*—when proportioning masonry mortars, the following are considered cementitious material: portland cement, blended hydraulic cement, masonry cement, lime putty, and hydrated lime. **C 1180,** C12

cementitious material ***(hydraulic),*** *n*—an inorganic material or a mixture of inorganic materials that sets and develops strength by chemical reaction with water by formation of hydrates and is capable of doing so under water. **C 125,** C09

cementitious material, hydraulic, *n*—an inorganic material or a mixture of inorganic materials which sets and develops strength by chemical reaction with water by formation of hydrates, and which is capable of doing so underwater. **C 219,** C01

cementitious material, hydraulic, *n*— an inorganic material or a mixture of inorganic materials, that sets and develops strength by chemical reaction with water forming hydrates and is capable of doing so under water. **C 1180,** C12

cementitious mixture, *n*—a mixture (mortar, concrete, or grout) containing hydraulic cement. **C 125,** C09

cement paste, *n*—an unhardened or hardened mixture of finely divided hydraulic cementious material and water. **C 219,** C01

cement, portland—hydraulic cement produced by pulverizing clinker consisting essentially of hydraulic calcium silicates, and usually containing one or more of the forms of calcium sulfate as an interground addition and conforming to Specification C 150. **D 2946,** C17

cement, portland blast-furnace slag—essentially an intimately interground mixture of portland cement clinker and granulated blast-furnace slag or an intimate and uniform blend of portland cement and fine granulated blast-furnace slag in which the amount of the slag constituent is within specified limits conforming to Specification C 150. **D 2946,** C17

cement, portland pozzolan—intimate and uniform blend of portland cement or portland blast-furnace slag cement and fine pozzolan produced by intergrinding portland cement and pozzolan, by blending portland cement or portland blast-furnace slag cement and finely divided pozzolan, or a combination of intergrinding and blending, in which the amount of the pozzolan constituent is within specified limits. **D 2946,** C17

cement, rubber, *n*—an adhesive that is either a liquid dispersion or solution of raw or compounded rubber, or both. **D 1566,** D11

center—inner layers whose grain direction runs parallel to that of the outer plies; may be of parallel laminated plies. (See also **core.**) **D 1038,** D07

center back waist length, *n*—*in body measurements*, the vertical distance along the spine from the cervicale to the waist. **D 123,** D13

centerburst—the region of greatest amplitude in an interferogram. **E 131,** E13

center front waist length, *n*—*in body measurements*, the vertical distance from the neck baseline at the center front to the waist level. **D 123,** D13

centering—an operation on lens elements wherein the element is optically aligned with the axis of rotation and the edges ground concentric with the optical axis. **C 162,** C14

centerline—the vertical axis around which character elements are located for letters, numerals, or symbols of an OCR font. **F 149,** F05

centerline, *n*—*in foot protective devices*, a line which extends from the toe of the footwear horizontally along the sole to the heel vertically to the top of the footwear, and diagonally to the point of intersection at the toe. **F 1494,** F23

center match—matched lumber with tongues and grooves precisely centered on the edges. **D 9,** D07

center of tire contact, *n*—the intersection of the wheel plane and the vertical projection of the spin axis of a wheel onto the road plane. **F 538,** F09

center rib, *n*—a rib at or near the circumferential centerline of the tread band. **F 538,** F09

center row, *n*—a row located at or near the circumferential centerline. **F 538,** F09

center stop—a device to receive and hold the drop bar on a double gate. **F 552,** F14

center wavelength, *n*—a wavelength, usually near the middle of the band of radiant power over which a radiation thermometer responds, that is used to characterize its performance. **E 344,** E20

center wear, *n*—a type of irregular wear characterized by a wear rate continuously increasing from shoulder to center of the tread band. **F 538,** F09

centigrade—since 1948, now called Celsius, a temperature scale. **D 6161,** D19

Centigrade—see **Celsius.** **E 1142,** E37

central conductor—a conductor passed through a hollow part and used to produce circular magnetization within the part. **E 1316,** E07

centralizer—a device that assists in the centering of a casing or riser within a borehole or another casing. **D 653,** D18

central pencil—a bundle of rays originating in the axis with an angular aperture equal to the effective aperture of the lens. These rays pass through the lens aperture and contribute to the formation of the image. **E 7,** E04

central vacuum cleaning system, *n*—a cleaning system consisting of a stationary vacuum producer and dust collector that incorporates the use of a tubing system internal to a building structure and a flexible hose, or both, for conveying dust from the area being cleaned to the dust collector. The system is designed for all-purpose cleaning including various types of larger debris and may be designed for liquid pickup. The system is used by inserting one end of a hose into a wall vacuum inlet valve and attaching a cleaning nozzle to the other end. In use, the nozzle is guided over the surface area to be cleaned. The system may contain a driven agitator to assist in dirt removal and it normally has detachable nozzles, attachments, and wands for both floor and above-the-floor cleaning. **F 395,** F11

centrifugal atomizer—a device wherein a rotating solid surface is the primary source of energy utilized to produce a spray.

NOTE—Alternatively, an atomizer that rotates to distribute the liquid. **E 1620,** E29

centrifugal cast button, *n*—see *rotation cast button.* **D 5497,** D13

centrifugal pressure nozzle—see **swirl chamber atomizer**. **E 1620,** E29

centrifuge—machine that separates a mixture of solids and liquids by centrifugal force. **E 1705,** E48

centrifuge—device for separating items of various masses by inducing gravitational forces/actions through the use of high rotational velocities. **F 1156,** F12

centrifuge moisture equivalent—see **moisture equivalent**. **D 653,** D18

centripetal acceleration, $[ML^2/T^2]$, *n—of a vehicle*, the component of the vector acceleration (of a point in a vehicle) perpendicular to the tangent of the path of the point and parallel to the road plane. **F 538,** F09

centroid wavelength, *n*—wavelength marking the center, in terms of area under a curve, of a function of wavelength weighted by multiplication with a specified response function. **E 284,** E12

centronics interface, *n*—a parallel interface standard that connects elements of a communications system. **F 1457,** F05

ceramic article—an article having a glazed or unglazed body of crystalline or partly crystalline structure, or of glass, which body is produced from essentially inorganic, nonmetallic substances and either is formed from a molten mass which solidifies on cooling, or is formed and simultaneously or subsequently matured by the action of the heat. **C 242,** C21

ceramic coating—an inorganic, essentially nonmetallic coating, on metal. **C 286,** B08

ceramic colorant—see **color oxide.** **C 286,** B08

ceramic glass enamel (also ceramic enamel or glass enamel)—a decorative, usually colored, vitreous inorganic coating for bonding to glass at temperatures above 425°C (800°F). **C 162,** C14

ceramic ink—an ink containing a ceramic pigment that develops its color on firing. Also known as stamping, screening, or printing ink. **C 286,** B08

ceramic junction reference electrode—a reference electrode in which the filling solution contacts the sample solution by means of a small porous ceramic junction. **D 4127,** D19

ceramic membrane—generally a glass, silica, alumina, or carbon based membrane. Generally used in micro and ultrafiltration. They tend to withstand high temperatures and wide pH ranges and be more chemically inert than polymeric membranes. **D 6161,** D19

ceramic-metal coating—a mixture of one or more ceramic materials in combination with a metallic phase applied to a metallic substrate which may or may not require heat treatment prior to service. This term may also be used for coatings applied to nonmetallic substrates, for example, graphite. **C 286,** B08

ceramic mosaic tile—an unglazed tile formed by either the dust-pressed or plastic method, usually ¼ to ⅜ in. (6.4 to 9.5 mm) thick, and having a facial area of less than 6 in.^{2} (39 cm^2) and which is usually mounted on sheets approximately 1 by 2 ft (0.3 by 0.6 m) to facilitate setting. **C 242,** C21

ceramic paste—a French term synonymous with "ceramic body." **C 242,** C21

ceramic process—the production of articles or coatings from essentially inorganic, nonmetallic materials, the article or coating being made permanent and suitable for utilitarian and decorative purposes by the action of heat at temperatures sufficient to cause sintering, solid-state reactions, bonding, or conversion partially or wholly to the glassy state. **C 242,** C21

ceramics—a general term applied to the art or technique of producing articles by a ceramic process, or to the articles so produced. **C 242,** C21

ceramic whiteware—a fired ware consisting of a glazed or unglazed ceramic body which is commonly white and of fine texture, designating such product classifications as tile, china, porcelain, semivitreous ware and earthenware.

alumina whiteware— any ceramic whiteware in which alumina (Al_2O_3) is an essential crystalline phase.

cordierite whiteware—any ceramic whiteware in which cordierite ($2MgO{\cdot}2Al_2O_3{\cdot}5SiO_2$) is the essential crystalline phase.

forsterite whiteware—any ceramic whiteware in which forsterite ($2MgO{\cdot}SiO_2$) is the essential crystalline phase.

steatite whiteware—any ceramic whiteware in which magnesium metasilicate ($MgO{\cdot}SiO_2$) is the essential crystalline phase.

titania whiteware—any ceramic whiteware in which titania (TiO_2) is the essential crystalline phase.

zircon whiteware—any ceramic whiteware in which zircon ($ZrO_2{\cdot}SiO_2$) is the essential crystalline phase. **C 242,** C21

ceric-cerous sulfate dosimeter—a liquid chemical radiation dosimetry system composed of water with ceric sulfate or ceric ammonium sulfate in aqueous sulfuric acid solution, and whose response is based quantitively on the amount of reduction of ceric to cerous ions by ionizing radiation, as analyzed by spectrophotometry or electrochemical potentiometry. It is generally considered to be a reference-standard dosimetry system. **E 170,** E10

cermet, *n*—a composite material or article comprised of a ceramic and a metal or metal alloy, interdistributed in any of various geometrical forms but intimately bonded together. **C 1145,** C28

cermet coating—see **ceramic-metal coating.** **C 286,** B08

certainty equivalent technique—a technique used to adjust economic measures of project worth to reflect risk exposure and risk attitude. **E 833,** E06

certificate of competence—document issued under the rules of a certification system indicating that adequate confidence is provided that the named person is competent in performing specific services, (EN 45020). **E 1187,** E36

certificate of compliance, *n—in manufactured products*, a document that states that the product was manufactured, sampled, tested, and inspected in accordance with the requirements of the specification (including year of issue) and any other requirements specified in the purchase order or contract, and has been found to meet such requirements. **A 941,** A01

certificate of compliance—a document or electronic record, signed by an authorized party, affirming that the supplier of the fastener and/or related service has met the requirements of the relevant specifications, contract, or regulation. **F 1789,** F16

certificate of conformance—a document or electronic record affirming that the fastener has met the requirements of the relevant specifications, contract, or regulation. **F 1789,** F16

certificate of conformity—document issued under the rules of a certification system, indicating that adequate confidence is provided that a duty identified product, process or service is in conformity with a specific standard or other normative document, (ISO Guide 2). **E 1187,** E36

certificate of test—a written, printed, or signed document attesting to the validity of the test performed. **C 859,** C26

certification—*in protective coatings*, the written documentation of the qualification of personnel or material. **D 4538,** D33

certification, *n*—a written declaration that a particular product or service complies with stated criteria. **E 631,** E06

certification, *n*—a system whereby an organization determines that a manufacturer has demonstrated the ability to make a product that complies with the requirements of the specification, authorizes the manufacturer to use a label on products that comply with the requirements of the specification, and conducts a follow-up program to verify the methods the manufacturer uses to determine compliance with the requirements of this specification. **F 1494,** F23

certification—the procedure and action by a duly authorized body of determining, verifying, and attesting in writing to the qualifications of personnel, processes, procedures, or items in accordance with applicable requirements. **F 1789,** F16

certification body—body that conducts certification of conformity, (ISO Guide 2). **E 1187,** E36

certification organization, *n*—an independent, third party organization that determines product compliance with the requirements of the specification with a labeling and listing follow-up program. **F 1494,** F23

certification scheme—certification system as related to specified products, processes or services to which the same particular standards and rules, and the same procedure, apply, (ISO Guide 2). **E 1187,** E36

certification system—system that has its own rules of procedure and management for carrying out certification of conformity, (ISO Guide 2). **E 1187,** E36

certified quality assurance system—a system so designated officially by a recognized accrediting body as having met all of the criteria within a national or an international third party quality system standard. **F 1789,** F16

certified reference material—a reference material with one or more property values that are certified by a technically valid procedure, accompanied by or traceable to a certificate or other documentation that is issued by a certifying body. A certifying body is a technically competent body (organization or firm, public or private) that issues a reference material certificate. A reference material certificate is a document certifying one or more property values for a certified reference material, stating that the necessary procedures have been carried out to establish their validity. (ISO definition: ISO Guide 30, *Terms and Definitions Used in Connection with Reference Materials*.) **C 859,** C26

certified reference material—a reference material, the composition or properties of which are certified by a recognized standardizing agency or group. Typically such a material is accompanied by documentation (certificate). **E 7,** E04

certified reference material—a material that has been characterized by a recognized standard or testing laboratory, for some of its chemical or physical properties, and that is generally used for calibration of a measurement system, or for development or evaluation of a measurement method. **E 170,** E10

certified reference material (CRM), *n*—a reference material, the composition or properties of which are certified by a recognized standardizing agency or group. **E 135,** E01

cervical, *n*—*in body measurements*, the most prominent bone at the base of the neck. **F 1494,** F23

cervicale, *n*—*in anatomy*, the prominent point of the seventh or lowest neck vertebra at the back of the body. **D 123,** D13

cervicale to bust point, *n*—*in body measurements*, the distance from the cervicale around the base of the neck and down to the bust point. **D 123,** D13

cervicale to wrist, *n*—*in body measurements*, with the arm bent, the distance from the cervical to the shoulder joints, along the outside of the arm, over the elbow to the greater prominence on the outside of the wrist. **D 123,** D13

cetane index, *n*—an approximation of the cetane number (the ignition performance) of distillate diesel fuel, which does not contain a cetane improver additive, calculated from the density and the mid-boiling point temperature (see also **diesel index**). **D 4175,** D02

cetane number, cn *n*—a measure of the ignition performance of a diesel fuel obtained by comparing it to reference fuels in a standardized engine test. **D 4175,** D02

CF—an abbreviation for coated front, referring to a carbonless paper of the two-coat, transfer type. **F 549,** F05

CFB—an abbreviation for coated front and backed, referring to a carbonless paper of the two-coat, transfer type. **F 549,** F05

CFR—Code of Federal Regulations. **E 1605,** E06

CFU—colony forming unit; unit used in the measure of total bacteria count (TBC). **D 6161,** D19

cGMP—abbreviation for current good manufacturing practices as defined in CFR Title 21, Parts 210 and 211. **E 1705,** E48

cgs-emu system of units—the system for measuring physical quantities in which the base units are the centimetre, gram, and second, and the numerical value of the magnetic constant, Γ_m, is unity. **A 340,** A06

chafer fabric, *n*—*in tire fabrics*, a woven fabric, usually coated with unvulcanized rubber, which is laid around the bead of a tire before vulcanization. **D 123,** D13

chain, *n*—*in zippers*, the assemblage formed by interlocking several elements of two stringers. **D 123,** D13

chain length distribution—in a linear polymer like cellulose, the frequency distribution of molecular size, usually expressed in units of degree of polymerization. **D 1695,** D01

chain marks—marks made on the bottoms of glass articles as they ride through a lehr on a chain belt slightly overheated. **C 162,** C14

chain printer—a type of printer that employs an endless chain, each link of which contains a fully formed character. The chain is caused to transverse the width of the paper form. Hammers are caused to strike the back of the form, synchronized in such a way as to produce the desired characters in their proper position. **F 909,** F05

chain sampling, *n*—*in acceptance sampling*, a sampling plan for which the decision to accept or reject a lot is based in part on the results of inspection of the lot and in part on the results of inspection of the immediately preceding lots. **D 123,** D13

chain saw, *n*—a portable power operated tool used for cutting wood which has cutters linked in a chain. **F 1494,** F23

chain speed, *n*—the velocity of synchronized movement of linked cutters around a bar and sprocket. **F 1494,** F23

chainstitch—a single thread stitch, characterized by the fact that the entire thread may be pulled out when one stitch is cut or broken. **F 869,** F08

chainstop, *n*—*for chainsaw cut resistance*, the resulting action when a material clogs (jams) the drive sprocket or slows the speed sufficiently to prevent advancement of the chain saw. **F 1494,** F23

chain thickness, *n*—*in zippers*, the measurement from front to back of the chain. **D 123,** D13

chain width, *n*—the measurement between the shoulders of the interlocked elements or between the outermost edges of the bead if the bead extends beyond the elements. **D 123,** D13

chalk, *n*—*in building construction*, a powder formed by chalking **C 717,** C24

chalkboard enamel—a special type of mat porcelain enamel used to provide a writing surface for chalk. **C 286,** B08

chalking, *v—in building construction,* formation of a powder on the surface of a sealant, coating, or membrane that is caused by the disintegration of the polymer or binding medium due to weathering. **C 717,** C24

chalking, *n*—(plastics) a powdery residue on the surface of a material resulting from degradation or migration of an ingredient, or both. **D 883,** D20

chalking , *n*—formation on a pigmented coating of a friable powder evolved from the film itself at or just beneath the surface. **E 1605,** E06

chalking, *n—in plastics,* a powdery residue on the surface of a material resulting from degradation or migration of an ingredient, or both. (D-20) **F 412,** F17

chalking resistance (*coatings*)—the ability of a pigmented coating to resist the formation of a friable powder on its surface caused by the disintegration of the binding medium by degradative weather factors. **D 16,** D01

chalking (rubber), *n*—the formation of a powdery residue on the surface of a rubber, commonly resulting from surface degradation. **D 1566,** D11

chalky or chalked—the condition of a porcelain enameled surface that has lost its natural gloss and become powdery. **C 286,** B08

challenge chemical—a chemical used to contact a protective clothing material sample to determine chemical/protective clothing material interactions or compatibility. **F 1494,** F23

challenge suspension, *n*—a liquid containing an agent that is used to test the penetration resistance of materials. **F 1494,** F23

chamber—a large room excavated underground, for example, for a powerhouse, pump station, or for storage. (ISRM) **D 653,** D18

chamber blasting (coyotehole blasting)—a method of quarry blasting in which large explosive charges are confined in small tunnel chambers inside the quarry face. (ISRM) **D 653,** D18

chamfer—the broken external edges of the instrument. **F 1078,** F04

chamois leather—a soft, pliable absorbent leather which is recognized in this country and abroad as being made from the inner side of a sheepskin, known technically as flesher, from which the outer or grain side has been split prior to tanning. **D 1517,** D31

champion, *n*—an advocate, or small group of advocates, in an agency that recognizes the need for a pavement management system and works to get it adopted and implemented. **E 867,** E17

chance error, *n*—error that has equal probability of being positive or negative. The mean of the chance errors resulting from a series of observations tends toward zero as the number of observations approaches infinity. **D 121,** D05

channel—(*1*) in container glass, that part of a forehearth which carries the glass from the melter to the flow spout and in which temperature adjustments are made.

(*2*) in fiber glass, the structure to deliver glass from the melter to the refiner or forehearth. **C 162,** C14

channel—a three-sided, U-shaped recess in a sash or frame to receive a light of glass or a panel; may be formed between an integral and a removable stop or in "pocket" glazing, with no removable stop. **C 717,** C24

channel—a natural or artificial waterway that periodically or continuously contains moving water. **D 4410,** D19

channel—*EIA,* an interval of the measured energy of backscattered particles defined by adjacent energy thresholds in the analog-to-digital converter used for spectrum production. **E 673,** E42

channel—a slanting cut made around the edge of an outsole or insole to provide a groove for the stitching and to keep the line of thread below the surface of the leather. Soles so stitched are called "channel stitched." **F 869,** F08

channel—any unimpaired pathway across the entire width of the intended seal. **F 1327,** F02

channel, acoustic emission—an assembly of a sensor, preamplifier or impedance matching transformer, filters secondary amplifier or other instrumentation as needed, connecting cables, and dedector or processor.

NOTE—A channel for examining fiberglass reinforced plastic (FRP) may utilize more than one sensor with associated electronics. Channels may be processed independently or in predetermined groups having similar sensitivity and frequency characteristics. **E 1316,** E07

channel-fill deposits—deposits of sediment within a channel, partly or completely filling the channel. Such materials accumulate where the transporting capacity has been insufficient to remove it as rapidly as it has been delivered. **D 4410,** D19

channel glazing—the method of glazing in which lights or panels are set in a channel formed by two stops. **C 717,** C24

channeling—the process of forming a vertical cavity resulting from a faulty cement job in the annular space. **D 653,** D18

channeling—the greater flow of fluid through passages of lower resistance which can occur in fixed beds or columns of particles due to nonuniform packing, irregular sizes and shapes of the particles, gas pockets, wall effects, and other causes. **D 2652,** D28

channeling—unequal flow distribution in the desalination bundle or filter bed. **D 6161,** D19

channeling—motion of energetic particles along certain axial or planar directions of a crystalline solid as the particles penetrate the specimen. The potentials of the individual atoms of the solid combine to reduce scattering with those atoms. **E 673,** E42

channeling—*SIMS,* the process by which particles preferentially penetrate crystalline specimens in certain crystallographic directions because of the relatively open arrangement of atoms presented to the impinging particle beam. **E 673,** E42

channel mopping—See **mopping,** *(3) strip.* **D 1079,** D08

char, *v*—to form carbonaceous residue during pyrolysis or during incomplete combustion. **E 176,** E05

char, *n*—a carbonaceous residue formed by pyrolysis or incomplete combustion. **E 176,** E05

character—(*1*) a member of a set of elements upon which agreement has been reached and that is used for the organization, control, or representation of information. Characters may be letters, digits, punctuation marks, or other symbols, often represented in the form of a spatial arrangement of adjacent or connected strokes or in the form of other physical conditions in data media.

(*2*) a letter, digit, or other symbol that is used as part of the organization, control, or representation of data. A character is often in the form of a spatial arrangement of adjacent or connected strokes. **F 149,** F05

character alignment—the vertical or horizontal position of characters with respect to a given reference line. **F 149,** F05

character boundary—*in character recognition,* the largest rectangle with a side parallel to the document reference edge, each of whose sides is tangential to a given character outline. **F 149,** F05

character cel, *n*—the rectangular-shaped areas on a page containing a single character with its portion of the space which separates it from adjacent characters. **F 1457,** F05

character erase—an OCR graphic shape that will cover a single character or a single space and will be read by the interpreter as a deletion. **F 149,** F05

characteristic—a distinguishing feature or trait. In grading lumber and other wood products, a feature in or on the wood which by its nature, extent, and frequency of occurrence determines the grade or level of quality. **D 9,** D07

characteristic, *n*—a property of items in a sample or population which, when measured, counted, or otherwise observed, helps to distinguish between the items. **D 123,** D13

characteristic, *n*—a property of items in a sample or population which, when measured, counted or otherwise observed, helps to distinguish between the items. **E 456,** E11

characteristic—a directly measurable distinguishing property of the pavement surface. Examples are pavement longitudinal profile, transverse profile, and separations in the continuity of a pavement surface. **E 867,** E17

characteristic, *n—in terminology*, any of the properties that constitute or distinguish a concept. **E 1992,** E02

characteristic curve—the curve showing the relationship between exposure and resulting density in a photographic image. It is usually plotted as the density against the log exposure. Called also the "H and D curve" and the "sensitometric curve." **E 7,** E04

characteristic curve—the plot of density versus log of exposure or of relative exposure. (Also called the D-log E curve or the *H* and D curve.) **E 1316,** E07

characteristic electron loss phenomena—*AES,* the inelastic scattering of electrons in solids that produces a discrete energy loss determined by the characteristics of the material. The most probable form is due to excitation of valence electrons. For some solids (for example, nontransition metals), inelastic scattering is dominated by plasmon excitations (a collective excitation of valence electrons). For other solids, the inelastic scattering may be due to a combination of plasmon excitation and single valence electron excitations. Inelastic scattering can also occur through the excitation of core level electrons when this is energetically possible. **E 673,** E42

characteristic emulsion curve—see **emulsion calibration curve.** **E 135,** E01

characteristic impedance of the medium, ρc; [$ML^{-2}T^{-1}$]; mks rayl (Pa·s/m)—the specific normal acoustic impedance at a point in a plane wave in a free field. It is a pure specific resistance since the sound pressure and the particle velocity are in phase and it is equal in magnitude to the product of the density of the medium, ρ, and the speed of sound in the medium, c. Its value when the medium is air at 20°C and 101.325 kPa is 413 mks rayl (Pa·s/m). **C 634,** E33

characteristic product size, *n*—the screen size corresponding to 63.2 % cumulative passing by mass. **D 5681,** D34

characteristic radiation—X-radiation of a particular set of wavelengths, produced by and characteristic of a particular element used as a target whenever its excitation potential is exceeded. **E 7,** E04

characteristic radiation, *n—of X rays*, a unique set of X rays emitted by an element. **E 135,** E01

characteristics—see **hazardous waste characteristics**. **E 1605,** E06

characteristic X-rays—photons emitted by ionized atoms and having a particular distribution in energy and intensity characteristic of the atomic number and chemical environment of the atom; in XPS, the term is ordinarily used in reference to the X-ray source of the spectrometer. **E 673,** E42

characterize, *v*—to specify the parameters or performance of an instrument or method of measurement. **E 284,** E12

character outline limit (COL)—the minimum, nominal, and maximum limits of a given graphic shape. **F 149,** F05

character pitch, *n*—the number of characters that can be printed in a horizontal 1 in. (25.4 mm). **F 1457,** F05

character reader—an input unit that performs character recognition. **F 149,** F05

character reading—machine reading of alpha or numeric characters, or symbols, or both, by optical means (as opposed to optical mark reading). **F 149,** F05

character recognition—(*1*) The identification of characters by automatic means.

(*2*) See **magnetic ink character recognition; optical character recognition**. **F 149,** F05

characters, *n—in bar code symbols*, the smallest group of elements that represent one or more numbers, letters, punctuation marks or other information. **F 1294,** F05

character set—(*1*) a finite set of different characters upon which agreement has been reached and that is considered complete for some purpose, for example, each of the character sets contained in ANSI X3.17-81 Character Set for Optical Character Recognition (OCR-A) and ANSI X3.49-75 (R 1982) Character Set for Optical Character Recognition (OCR-B).

(*2*) an ordered set of unique representations called characters, for example, the 26 letters of the English alphabet, Boolean 0 and 1, the set of symbols in the Morse code, and the 128 ASCII characters. **F 149,** F05

character set, *n*—those characters available for encodation in a particular automatic identification technology. **F 1294,** F05

character set—the collection of all the characters available in a given font. **F 1457,** F05

character skew—the rotational deviation of the printed image from its intended orientation relative to a document reference edge. **F 149,** F05

character spacing—the pitch distance between adjacent characters. **F 149,** F05

character stroke width—the distance between the average edges of a character element. **F 149,** F05

character subset—a selection of characters from a character set, comprising all characters that have a specified common feature, for example, in each of the character sets contained in ANSI X3.17-81 Character Set for Optical Character Recognition (OCR-A) and ANSI X3.49-75, (R 1982) Character Set for Optical Character Recognition (OCR-B), the digits 0 to 9 may constitute a character subset. **F 149,** F05

charge, *n*—See batch (*3*). **C 162,** C14

charge, *v*—to add batch to a melter. **C 162,** C14

charge:

charge modification—any method used to alter the amount or the distribution of charge on a specimen surface.

charge neutralization—*ISS, SIMS,* a technique in which a surface under ion bombardment is maintained at a known potential by compensating for the accumulated charge.

charge referencing—any method used to adjust the energy scale calibration of a spectrometer to accommodate the effects of steady-state charging of a specimen surface. **E 673,** E42

charge acceptance—the apparent surface voltage present on an electrostatic recording medium immediately after charging. **F 335,** F05

charge capacity—see **thermal capacity**. **E 772,** E44

charge coupled device, CCD, *n*—an array (linear or matrix) of transductive elements wherein packets of electrons are set in each element as a result of the quantity of light received during an exposure interval, and where these packets are recovered from the array in the form of a pulse height-modulated electric signal. **F 1294,** F05

charged area development—(**CAD**), a process in electrostatic copying where the photoconductive element is charged with a charge of the opposite sign as that of the toner. A light source is used to discharge all areas on the photoconductor that are NOT to receive toner to form the image. The toner is attracted to the non-discharged areas of the photoconductor when the latent electrostatic image is developed. **F 335,** F05

charge decay—loss of charge on the deposited powder due to electrical leakage. **C 286,** B08

charge decay rate—loss of charge per unit of time. **C 286,** B08

charged particle equilibrium—a condition that exists in an incremental volume within a material under irradiation if the kinetic energies and number of charged particles (of each type) entering that volume are equal to those leaving that volume. **E 170,** E10

charge neutralizer gun—an electron gun used to dissipate the charges which tend to build up on specimen surfaces, within an electron-diffraction camera, which would introduce undesired electrostatic fields. **E 7,** E04

charge retention—the ability of an electrically charged layer to retain its initial charge. **C 286,** B08

charge to mass ratio—ratio of the charge on a powder expressed in coulombs to the mass of the powder expressed in kilograms. **C 286,** B08

charging—the process of establishing an electrostatic surface charge of uniform density on an insulating medium. **F 335,** F05

charging potential—*in surface analysis*, the electrical potential of the surface of an insulating specimen caused by irradiation. If the specimen is heterogeneous, there may be different charging potentials on different areas of the surface. **E 673,** E42

charring, *n*—the formation of carbonaceous residue as the result of pyrolysis or incomplete combustion. **D 123,** D13

charring—the formation of a carbonaceous residue as the result of pyrolysis or incomplete combustion. **F 1494,** F23

chatter, *n*—an undesirable pattern created on the surface of a work piece, usually at regularly spaced intervals, due to an out-of-round, out-of-balance condition or due to an induced natural frequency, or its harmonics, or both, in a grinding machine. **C 1145,** C28

chatter marks, *n*—defect on calendered sheeting consisting of transverse narrow bands of alternately thicker and thinner material. **D 1566,** D11

chatter sleek—{archaic} See **frictive track**. **C 162,** C14

check—an imperfection; a surface crack in a glass article. **C 162,** C14

check—a separation of the wood along the fiber direction that usually extends across the rings of annual growth, commonly resulting from stresses set up in wood during seasoning.

end check—a seasoning check occurring on the end of a board or other piece of wood.

heart check—a check that extends across the growth layers in one or more directions from the pith toward, but not to, the surface of a piece of wood. A synonym is *pith check*.

pith check—see *heart check*.

roller check—a crack in the wood structure caused by a piece of cupped lumber being flattened between machine rollers.

star check—a heart check in which the separation extends in more than one direction from the pith.

surface check—a check occurring on the surface of a piece of wood, usually on the tangential face not extending through the piece.

through check—a check that extends through a piece of wood, or from a surface to the opposite or to an adjoining surface. **D 9,** D07

check analysis, *n*—Deprecated term. Use preferred term **product analysis**. **A 902,** A05

check analysis—(See *Product Analysis*). **F 1789,** F16

check character, *n*—a calculated character included within a symbol whose value is used to perform a mathematical check of the accuracy of the decoded data. **F 1294,** F05

checkered head—flat head with raised or depressed, checker or grid marks of various designs. **F 547,** F16

checkers—(*1*) an open structure of firebrick serving as a heat exchanger.

(*2*) slang for regenerators of this type.

(*3*) slang for refractory brick used in such a construction. **C 162,** C14

checking—*in protective coatings*, the formation of slight breaks in a coating film that do not penetrate to the underlying surface. **D 4538,** D33

checking (coatings), *n*—phenomenon manifested in paint films by slight breaks in the film that do not penetrate to the underlying surface. **E 1605,** E06

checking resistance (*coatings*)—the ability of a coating to resist slight breaks in the film that do not penetrate to the previously applied coating or to the substrate. The breaks should be called cracks if penetration extends to the previously applied coating or to the substrate. See **cracking resistance.** **D 16,** D01

cheesemelter—a low input unit, designed to melt cheese on top of specialty foods, but usually incapable of fully cooking a food item such as steak or chicken. See **broiler, overfired.** **F 1827,** F26

chelate compound—a compound in which the metal is contained as an integral part of a ring structure and is not readily ionized. **B 374,** B08

chelating agent—a compound capable of forming a chelate compound with a metal ion. See **chelate compound.** **B 374,** B08

chelating agent—a sequestering or complexing agent that, in aqueous solution, renders a metallic ion inactive through the formation of an inner ring structure with the ion. **D 459,** D12

chelating agents— a sequestering or complexing agent that, in aqueous solution, renders a metallic ion inactive through the formation of an inner ring structure with the ion. **D 6161,** D19

chemical:

effects—AES, any change in the shape of an Auger spectrum or in the Auger peak energy for an element which is due to chemical bonding.

shift—AES, XPS, a change in peak energy because of a change in the chemical environment of the atom. **E 673,** E42

chemical adsorption— see **chemisorption.** **D 2652,** D28

chemical assay—a chemical measurement of the quantity of one or more components of a material. **C 859,** C26

chemical attack—chemical reaction or solvent effect, causing failure or deterioration of plastic and rubber parts, organic coating, metals, or lithography involved in the completed package. **D 3064,** D10

chemical carbonless transfer paper—a transfer paper that produces a visible image by transferring a relatively colorless material from one sheet to another where it reacts to form a visible image. **F 549,** F05

chemical cellulose—a chemically purified cellulosic material that is intended for chemical treatment to produce derivatives. **D 1695,** D01

chemical cleaner—see **cleaner, chemical.** **F 412,** F17

chemical conversion coating—a protective or decorative nonmetallic coating produced *in situ* by chemical reaction of a metal with a chosen environment. (It is often used to prepare the surface prior to the application of an organic coating.) **G 15,** G01

chemical cotton—chemical cellulose prepared from cotton; generally, but not necessarily, cotton linters. **D 1695,** D01

chemical deposition—the precipitation of one metal from a solution of its salts by the addition of another metal or reagent to the solution. **B 243,** B09

chemical durability—the lasting quality (both physical and chemical) of a glass surface. It is frequently evaluated, after prolonged weathering or storing, in terms of chemical and physical changes in the glass surface, or in terms of changes in the contents of a vessel. **C 162,** C14

chemical feed pump—a pump used to meter chemicals, such as chlorine of polyphosphate, into a feed water supply. **D 6161,** D19

chemical grout—any grouting material characterized by being a true solution; no particles in suspension. See also **particulate grout.** **D 653,** D18

chemical grout system—any mixture of materials used for grouting purposes in which all elements of the system are true solutions (no particles in suspension). **D 653,** D18

chemical lime—a quicklime or hydrated lime whose chemical and physical characteristics and method of processing make it suitable for one or more of the many and varied chemical and industrial uses of the product. **C 51,** C07

chemical luminescence, *n*—luminescence resulting from a chemical reaction. (See also **luminescence**.) **E 284,** E12

chemically bonded brick, n— see **unburned brick.** **C 71,** C08

chemically curing sealant, *n*—a compound that cures primarily through chemical reaction. **C 717,** C24

chemically curing sealant, *n*—a sealant that cures primarily through chemical reaction. **D 5535,** D04

chemically foamed polymeric material—a cellular material in which the cells are formed by gases generated from thermal decomposition or other chemical reaction. **D 883,** D20

chemically foamed polymeric material, *n*—a cellular material in which the cells are formed by gases generated by thermal decomposition or other chemical reaction. **F 1251,** F04

chemically formed polymeric material—a cellular material in which the cells are formed by gases generated from thermal decomposition or other chemical reaction. (D-20) **F 412,** F17

chemically precipitated metal powder—powder produced by the reduction of a metal from a solution of its salts either by the addition of another metal higher in the electromotive series, or by other reducing agent. **B 243,** B09

chemically refined copper—copper recovered from an aqueous solution by other than electrolytic means. Usually when this term is used alone it refers to chemically refined tough pitch copper. This designation applies to the following:
—copper cast in refinery shapes suitable for hot or cold working, or both, and by extension, to fabricators products made therefrom,
—ingots or ingot bars suitable for remelting. **B 846,** B05

chemically strengthened—glass that has been ion-exchanged to produce a compressive stress layer at the treated surface. **C 162,** C14

chemical milling—the shaping of a work piece by immersion in an etchant employing a resist for selective removal of material. **B 374,** B08

chemical plating—deposition of a metal coating by chemical, nonelectrolytic methods. See also **immersion plate**, **contact plating**. **B 374,** B08

chemical polishing—the improvement in surface smoothing of a metal by simple immersion in a suitable solution. See **bright dip (nonelectrolytic**). **B 374,** B08

chemical porcelain—See *chemical porcelain* under **porcelain**. **C 242,** C21

chemical potential—(μ_i or $\bar{G}_i$) the partial molar free energy of component *i*, that is, the change in free energy of a solution upon adding one mole of component *i* to an infinite amount of solution of given composition,

$$(\delta G/\delta n_i)_{T,P,n_{1...}} = \bar{G}_i = \mu_i$$

where:
G = Gibbs free energy, and
n_i = number of moles of the i^{th}component. **E 7,** E04

chemical protective clothing (CPC), *n*— any material or combination of materials used in an item of clothing for the purpose of isolating parts of the body from direct contact with a potentially hazardous chemical. **F 1494,** F23

chemical-protective suit ensemble—the combination of a chemical-protective suit (totally encapsulating, splash-protective) with the wearer's respiratory protective equipment, gloves, boots, communications system, and cooling device, or some combination of those. **F 1494,** F23

chemical pulp, *n*—fibrous material obtained by a predominantly chemical treatment of wood or other plant material; principal processes are sulfate (also known as kraft), sulfite, and soda. (see **mechanical pulp** and **semichemical pulp**) **D 1968,** D06

chemical pulps—in the paper industry, pulps produced by chemical processes, as contrasted to those produced by mechanical processes. (see also **chemical wood pulp**). **D 1695,** D01

chemical reaction papers—papers that produce a visible image by the reaction of relatively colorless materials which come into contact under pressure. **F 221,** F05

chemical repellent—any substance whose odor, taste, appearance, tactile sensation, or combination thereof produces an aversion response in the target animal. **E 609,** E35

chemical resistance—*in asbestos cement products*, the ability of the product to resist chemical attack, dissolution, decomposition or other chemical changes when in contact with liquid, gaseous, or solid media normally encountered in its service environment. **D 2946,** C17

chemical resistance, *n*—the ability to resist chemical attack. **D 4439,** D35

chemical resistance—the ability to resist chemical attack. **E 1749,** E06

chemical resistance—the ability to resist chemical attack. **F 412,** F17

chemical-resistant—the ability of a material to resist degradation by reaction with, dissolution by, or reduction of physical continuity from contact with a chemical agent or agents, thereby retaining its capacity to perform as a structural or aesthetic entity. **C 904,** C03

chemical-resistant carbon brick—a brick comprised of carbon (including graphite) that is carbon bonded and fired to a minimum temperature of 1850°F (1010°C). This brick exhibits good resistance to thermal shock and resists exposure to a wide range of alkalies and acids, particularly hydrofluoric acid. When used in oxidizing conditions, temperature exposure should be less than 600°F (315°C). **C 904,** C03

chemical-resistant construction unit—a modular nonmetallic material, either vitreous or nonvitreous, used in industrial processes primarily for applications where chemical, thermal, and mechanical resistance is required. **C 904,** C03

chemical-resistant fireclay or shale brick—a kiln fired brick from clay, shale, or mixtures thereof that exhibits low absorption and high resistance to a wide range of chemical environments. This brick should not be used in hot caustic or where hydrofluoric acid or other fluoride chemical compounds are found. Commonly referred to as acid-resistant brick or acid-proof brick. **C 904,** C03

chemical-resistant polymer concrete—a construction material composed of a continuous phase (binder) of a polymer and a discontinuous phase (aggregate) generally used in applications where chemical resistance is required. **C 904,** C03

chemical-resistant polymer machinery grout—an intimate mixture of liquid resinous material, selected filler materials, and setting agent, which harden by chemical reaction to provide support for machine bases. **C 904,** C03

chemical-resistant resin grout—a mixture of liquid resin, filler and setting agent. The components form a mixture that subsequently hardens by chemical reaction. The setting agent may be separate or incorporated in the filler. **C 904,** C03

chemical-resistant resin monolithic surfacing—an intimate mixture of liquid resinous materials, selected filler materials, and setting agent. These components are mixed together, placed to a minimum thickness of 60 mils (1.5 mm), then hardened to form a bonded overlay. **C 904,** C03

chemical-resistant resin mortar—an intimate mixture of liquid resinous material, selected filler materials, and setting agent. These components form a trowelable mortar that subsequently hardens by chemical reaction. **C 904,** C03

chemical-resistant tile grout—an intimate mixture of liquid resinous material, selected filler materials, and setting agent. These components form a flowable mixture that subsequently hardens by chemical reaction. **C 904,** C03

chemical sanitizing—solution to destroy or kill any residual bacteria from multiple-use tableware. **F 1827,** F26

chemical sanitizing, recirculated wash, fresh water rinse type —machines with a final rinse using fresh water from an outside source combined with chemical sanitizing solution; additional parts: chemical sanitizing equipment. See **dishwashing machine, commercial**. **F 1827,** F26

chemical sanitizing, stationary rack, dump type—machines with chemical sanitizing solution added to the rinse cycle; additional parts: chemical sanitizing equipment. See **dishwashing machine, commercial**. **F 1827,** F26

chemical separation—the removal, isolation, or separation of the desired substance from the remainder of the sample by methods that involve difference(s) in the chemical properties of the substances involved. **C 859,** C26

chemical setting silicate and silica chemical-resistant mortar—an intimate mixture of a silicate or silica binder, a chemically inert solid filler, and a setting agent. The binder may be a liquid silicate or silica, or a powder to which water is added. These components are subsequently hardened by the chemical reaction between the setting agent and the binder. **C 904,** C03

chemical shift (NMR), δ—the defining equation for δ is the following:

$$\delta = \frac{\Delta\nu}{\nu_R} \times 10^6$$

where ν_R is the frequency with which the reference substance is in resonance at the magnetic field used in the experiment and $\Delta\nu$ is the frequency difference between the reference substance and the substance whose chemical shift is being determined, at constant field. The sign of $\Delta\nu$ is to be chosen such that shifts to the high frequency side of the reference shall be positive. **E 131,** E13

chemical treatment, *n*—a passivating treatment normally applied to metallic coatings to retard the formation of corrosion products (storage stain) during shipment and storage. **A 902,** A05

chemical vapor deposition—*in semiconductor technology*, a process in which a controlled chemical reaction produces a thin surface film. **F 1241,** F01

chemical wash, *n*—*in rug cleaning*, a specialized professional process used on oriental rugs. **D 123,** D13

chemical wood pulp—a term used in the paper industry for pulps obtained by digestion of wood with solutions of various chemicals. **D 1695,** D01

chemisorption, *n*—adsorption, especially when irreversible, by means of chemical forces in contrast with physical forces. **D 1356,** D22

chemisorption (chemical adsorption)—the binding of an adsorbate to the surface of a solid by forces whose energy levels approximate those of a chemical bond. **D 2652,** D28

chem-mechanical polish—*in semiconductor technology*, a process for the removal of surface material from the wafer that utilizes chemical and mechanical actions to achieve a mirror-like surface for subsequent processing. **F 1241,** F01

chest, *n*—*in garment construction*, a measurement taken from below each armhole seam straight across the garment while it is laid flat. **F 1494,** F23

chest girth, *n*—*in body measurements*, the circumference of the body over the shoulder blades, under the arms and across the upper chest. (Compare **bust girth**.) **D 123,** D13

chest girth, *n*— *in body measurements*, the circumference of the body over the shoulder blades, under the arms and across the upper chest. (See **bust girth.**) **F 1494,** F23

chest harness, *n*—a type of harness worn around the chest for upper body support. **F 1490,** F32

chestnut extract—a vegetable tanning agent extracted from the wood of the chestnut tree and used in tanning/re-tanning heavy leathers. **D 1517,** D31

chevron configuration—booming configuration used in narrow watercourses, formed by positioning two lengths of boom in a deflection mode, the leading end of each length is positioned in the middle of the watercourse and the trailing ends lead to opposite shores. **F 818,** F20

chewable surface—surface easily accessible to children (usually up to five feet from the floor or ground), and likely to be chewed-on, such as window sills, balusters, and handrails. See **accessible surface**. **E 1605,** E06

child-resistant packaging— See **packaging.** **D 996,** D10

chill, *n*—an object, usually metal, imbedded in a portion of the mold to accelerate the local rate of heat removal from the metal being cast. **A 644,** A04

chill, *v*—to accelerate the freezing rate of cast iron, usually in a localized region, to refine the graphite structure or cause formation of primary carbides. **A 644,** A04

chilled iron, *n*—a cast iron that would normally solidify free of primary carbide which is purposely caused to solidify as white cast iron, locally or entirely, by accelerated cooling. **A 644,** A04

chill mark—a wrinkled surface condition on glassware resulting from uneven cooling in the forming process. **C 162,** C14

chill, microstructural, *n*—a localized region of primary carbides in a casting made from a cast iron that would normally solidify free of primary carbides. **A 644,** A04

chime (chine), *n*— *in packaging,* the rim of a **container,** such as a **drum, barrel,** or **can.** **D 996,** D10

chimney effect, *n*—a phenomenon consisting of a vertical movement of a localized mass of air or other gases due to temperature differences. **D 1356,** D22

chimney effect—upward thrust of smoke and hot gases by convection currents confined in a vertical enclosure. **E 176,** E05

china—a glazed or unglazed vitreous ceramic whiteware made by the china process and used for nontechnical purposes, designating such products as dinnerware, sanitary ware, and artware when they are vitreous. (See also **bone china.**) **C 242,** C21

china clay—See **kaolin.** **C 242,** C21

china process—the method of producing glazed ware by which the ceramic body is fired to maturity, following which the glaze is applied and matured by firing at a lower temperature. **C 242,** C21

china sanitary ware (sanitary plumbing fixtures)—glazed, vitrified whiteware fixtures having a sanitary service function. **C 242,** C21

Chinese script eutectic—a configuration of eutectic constituents, found particularly in some cast alloys of aluminum containing iron and silicon and in magnesium alloys containing silicon, which resembles in appearance the characters in Chinese script. **E 7,** E04

Chinese wok range—equipment characterized as having one or more "wells" or chamber and designed to use a wok as the cooking utensil. See **range**. **F 1827,** F26

chip—an irregularly-shaped piece of stone, usually with one convex surface, that has been dislodged from a snip. **C 119,** C18

chip—an imperfection due to breakage of a small fragment out of an otherwise regular surface. **C 162,** C14

chip—a small piece of broken-off material, or the location where a small piece of the unit material has been broken off. **C 896,** C04

chip—crushed angular rock fragment of a size smaller than a few centimetres. (ISRM) **D 653,** D18

chip—area along an edge or corner where the material has broken off.

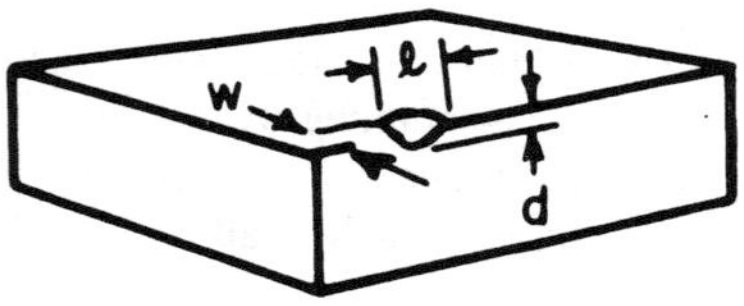

where:
w = width
l = length, and
d = depth.

F 109, C21

chip—*in semiconductor wafers*, region where material has been unintentionally removed from the surface or edge of the wafer. **F 1241,** F01

chip and tear, *n*—a special type of irregular wear characterized by a rough tread surface which may contain cracks, abrasion pits or surface ruptures. **F 538,** F09

chipboard—See **paperboard.** **D 996,** D10

chip marks—shallow depressions or indentations in or on the surface of dressed lumber caused by chips or shavings getting embedded in the surface during dressing. **D 9,** D07

chipped glass—a glass article with chipped surface produced intentionally. **C 162,** C14

chipping—the process of removing thin extra glass prior to grinding. **C 162,** C14

chipping—fracturing and breaking away of fragments of a porcelain enameled surface. **C 286,** B08

chipping resistance (*coatings*)—the ability of a coating or layers of coatings to resist removal, usually in small pieces, resulting from impact by hard objects or from wear during service. **D 16,** D01

chipping resistance (coatings)—ability of a coating or layers of coatings to resist removal, usually in small pieces, resulting from impact by hard objects or from wear during service. **E 1605,** E06

chips—small fragments of wood chopped or broken by cuts such as by a planer, chipper, mechanical hog, hammermill. **D 9,** D07

chips—small pieces of wood chopped off a block by ax-like cuts as in a chipper of the paper industry, or produced by mechanical hogs, hammermills, etc. **D 1554,** D07

chips—minor damage to the pultruded surface that removes material but does not cause a crack or craze. **D 3918,** D20

chirping—the process of dispersing the zero phase difference points for different wavelengths across the interferogram, so that the magnitude of the signal is reduced in the short region of the interferogram where all wavelengths would otherwise constructively interfere. **E 131,** E13

chirp test, *n*—the progressive application of brake torque required to produce the maximum value of longitudinal braking force that will occur prior to wheel lockup, with subsequent brake release to prevent any wheel lockup (tire slide). **E 867,** E17

chisel—the steel cutting tool used in percussion drilling. (ISRM) **D 653,** D18

chisel point—point with two major planes forming "V" and pair of minor planes on each flank; forming hexagonal cross section. **F 547,** F16

chisel point—point with two symmetrically beveled planes forming "V" at end of staple leg, resulting in straight penetration. (See **cross-cut chisel point.**) **F 592,** F16

chloramine—a combination of chlorine and ammonia in water which has bactericidal qualities for a longer time than does free chlorine. **D 6161,** D19

chlorinated poly(vinyl chloride) plastics—plastics based on chlorinated poly(vinyl chloride) in which the chlorinated poly(vinyl chloride) is in the greatest amount by weight. **D 883,** D20

chlorinated poly(vinyl chloride) plastics—plastics based on chlorinated poly(vinyl chloride) in which the chlorinated poly(vinyl chloride) is in the greatest amount by weight. (D-20) **F 412,** F17

chlorine—chemical used for its qualities as a bleaching or oxidizing agent and disinfectant in water purification. **D 6161,** D19

chlorine bleach, *n*—a bleach that releases the hypochlorite ion in solution, for example, sodium hypochlorite. **D 123,** D13

chlorine, combined available—residual chlorine combined with ammonia nitrogen or nitrogenous compounds. **D 1129,** D19

chlorine demand—the amount of chlorine used up by reacting with oxidizable substances in water before chlorine residual can be measured. **D 6161,** D19

chlorine, free available—the hypochlorite ions (OCl^-), hypochlorous acid (HOCl) or the combination thereof present in water. **D 1129,** D19

chlorine, free available—the chlorine (Cl_2), hypochlorite ions (OCl^-), hypochlorous acid (HOCl) or the combination thereof present in water. **D 6161,** D19

chlorine number—the number of grams of chlorine gas or of bleaching powder (expressed as its equivalent in chlorine) that is consumed by 100 g of ovendry pulp in a definite time under certain specified conditions. The chlorine number is an indication of the bleach requirement of the pulp and an indirect estimate of the lignin content. **D 1695,** D01

chlorine or volatile halide extraction—See **extraction.** **E 7,** E04

chlorine requirement—the amount of chlorine required to achieve, under specified conditions, the objectives of chlorination. **D 1129,** D19

chlorine residual—the amount of available chlorine present in water at any specified time. **D 1129,** D19

chlorine, residual—the amount of available chlorine present in water at any specified time. **D 6161,** D19

chlorine, total available—the sum of free available chlorine plus chloramines present in water. **D 6161,** D19

chlorofluorocarbon plastic, *n*—a plastic based on polymers made with monomers composed of chlorine, fluorine, and carbon only. **F 1251,** F04

chlorofluorocarbon plastics—plastics based on polymers made with monomers composed of chlorine, fluorine, and carbon only. (ISO) **D 883,** D20

chlorofluorocarbon plastics—plastics based on polymers made with monomers composed of chlorine, fluorine, and carbon only. (ISO, D-20) **F 412,** F17

chlorofluorohydrocarbon plastic, *n*—a plastic based on polymers made with monomers composed of chlorine, fluorine, hydrogen, and carbon only. **F 1251,** F04

chlorofluorohydrocarbon plastics, *n*—plastics based on polymers made with monomers composed of chlorine, fluorine, hydrogen, and carbon only. (ISO) **D 883,** D20

chlorofluorohydrocarbon plastics, *n*—plastics based on polymers made with monomers composed of chlorine, fluorine, hydrogen, and carbon only. (ISO, D-20) **F 412,** F17

chlorosity—the concentration of the dissolved chloride equivalent in water at 20°C. **D 1129,** D19

choke—an imperfection consisting of an insufficient opening in the finish and neck of a container. **C 162,** C14

chopped fiber—fiber glass strand which has been chopped to specified lengths. **C 162,** C14

chopped strand, *n*—*in glass textiles,* a strand made from short predetermined lengths of cut continuous filament and used as a reinforcing material. (See also **strand.**) **D 123,** D13

chord length—(LTS) the straight-line distance between the snowboard ski tail and the snowboard ski tip with the snowboard ski pressed flat to a plane surface to take out the camber. **F 1107,** F27

chord length, *n*—the straight line distance between the two ends of a curved needle. **F 1840,** F04

chord length, $\mathbf{L}_{TS}$—straight line distance measured between the ski tail and ski tip with the ski pressed against a plane surface.

NOTE—Either method at the manufacturer's discretion may be used to indicate nominal ski length or ski size when rounded to common increment. **F 472,** F27

chord modulus, *n*—in a *stress-strain curve,* the ratio of the change in stress to the change in strain between two specified points on the curve. **D 123,** D13

"C" Horizon—see **horizon.** **D 653,** D18

chroma, *n*—one of three terms used in the Munsell notations to denote color, referring to the saturation or purity dimension. **E 253,** E18

chroma, *n*—(*1*) attribute of color used to indicate the degree of departure of the color from a gray of the same lightness. See also **Munsell chroma.**

(*2*) C^*, (in the CIE 1976 L^*, a^*, b^* or L^*, u^*, v^* system) the quantity $C^*_{ab} = (a^{*2} + b^{*2})^{1/2}$ or $C^*_{uv} = (u^{*2} + v^{*2})^{1/2}$..

(*3*) attribute of a visual perception, produced by an object color, that permits a judgment to be made of the amount of pure chromatic color present, irrespective of the amount of achromatic color. **E 284,** E12

chromatic, *adj*—perceived as having a hue; not white, gray, or black. **E 284,** E12

chromatic aberration—a defect in a lens or lens system as a result of which the lens possesses different focal lengths for radiation of different wavelengths. **E 7,** E04

chromatic aberration—a defect in a lens or lens system as a result of which the lens possesses different focal lengths for radiation of different wavelengths. **E 175,** E41

chromatic adaptation, *n*—changes in the visual system's sensitivities due to changes in the spectral quality of illuminating and viewing conditions. **E 284,** E12

chromaticity, *n*—the color quality of a color stimulus definable by its chromaticity coordinates, or by its dominant (or complementary) wavelength and its purity taken together. **E 284,** E12

chromaticity coordinates, *n*—the ratio of each of the tristimulus values of any viewed light to the sum of the three. **E 284,** E12

chromaticity diagram, *n*—a plane diagram in which points specified by chromaticity coordinates represent the chromaticities of lights (color stimuli). **E 284,** E12

chromaticness, *n*—(*1*) attribute of visual perception combining the hue and saturation.

(*2*) attribute of a visual perception according to which the color of an area appears to be more or less chromatic. **E 284,** E12

chromating—the process for producing a conversion coating (q.v.) containing chromium compounds. **B 374,** B08

chrome brick, *n*—a refractory brick, which may be burned or unburned, manufactured predominantly or entirely of refractory-grade chrome ore, and having a Cr_2O_3 content of 30 % or greater (see also chromic oxide brick). **C 71,** C08

chrome-finished griddle—equipment with a polished chrome-finished cooking surface that is easier to clean and radiates less heat towards the chef and the kitchen. See **griddle.** **F 1827,** F26

chrome-magnesia brick, *n*—a refractory brick, which may be burned or unburned, manufactured predominantly of a mixture of refractory-grade chrome ore and refractory-grade magnesia in which the refractory-grade chrome ore predominates by weight; the refractory-grade magnesia may be either seawater magnesia, dead-burned magnesite, fused magnesia, or a combination of these materials. **C 71,** C08

chrome ore, refractory-grade, *n*—a refractory ore consisting essentially of chrome-bearing spinel with only minor amounts of accessory minerals and having physical and chemical properties suitable for making refractory products. **C 71,** C08

chrome retannage—the process of applying trivalent chromium salts as a secondary adjunct to the main tannage to impart or enhance certain leather properties; usually applied after splitting and shaving. **D 1517,** D31

chrome tannage—conversion of raw hides/skin into leather with chromium compounds, thereby substantially increasing strength properties and resistance to various biological and physical agents. **D 1517,** D31

chromia-alumina brick, *n*—a refractory brick, which may be burned or unburned, manufactured predominantly of a mixture of chromic oxide (Cr_2O_3) and a high-alumina raw material, with the chromic oxide content being 50 % or greater by weight and with alumina (Al_2O_3) predominating by weight among the other constituent oxides. **C 71,** C08

chromic oxide brick—a refractory brick, which may be burned or unburned, manufactured predominantly of chromic oxide (Cr_2O_3) and having a Cr_2O_3 content of 90 % or greater (may also be called chromia brick). **C 71,** C08

chromizing—a surface treatment at elevated temperatures, generally carried out in pack, vapor, or salt baths, in which an alloy is formed by the inward diffusion of chromium into the base metal. **B 374,** B08

chromophore—color forming element. **F 1156,** F12

chronic ecotoxicity test, *n*—a comparative ecotoxicity test in which a representative subpopulation of organisms is exposed to different treat rates of a test material and is observed for a period of time which constitutes a major portion of their life span. **D 6384,** D02

chronic test—a comparative study in which organisms, that are subjected to different treatments, are observed for a long period or a substantial portion of their life span. **E 943,** E47

chrysotile—an asbestos mineral belonging to the serpentine group, having a chemical composition close to$Mg_3Si_2O_5(OH)_4$. Moderate amounts of aluminum may substitute for silicon and moderate amounts of iron may substitute for magnesium. Small amounts of MnO, CaO, K_2O, and Na_2O are also reported in the chemical analyses. The crystal structure of crysotile asbestos consists of double layers, each consisting of a layer of SiO_4 tetrahedra that is coordinated to a second layer of $MgO_2(OH)_4$ octahedra linked through the sharing of oxygen atoms; the composite double layer rolls up, like a scroll, to form long hollow tubes. The outer diameters of the individual tubes are in the order of 25 nm; the length-to-diameter ratio can vary from 20 to well over 10 000. Chrysotile is characterized by a combination of a distinctive morphology, a chemical composition close to $Mg_3Si_2O_5(OH)_4$, and the characteristic X-ray and electron diffraction patterns. Its Chemical Abstracts number is 12001-29-5 (see Practice E 849). **C 460,** C17

chrysotile, *n*—an asbestos mineral belonging to the serpentine group, having a chemical composition close to $Mg_3SiO_5(OH)_4$. Moderate amounts of aluminum may substitute for silicon and moderate amounts of iron may substitute for magnesium. Small amounts of MnO, CaO, K_2O, and Na_2O are also reported in the chemical analyses. The crystal structure of chrysotile asbestos consists of double layers, each consisting of a layer of linked SiO_4 tetrahedra that is coordinated to a second layer of linked $MgO_2(OH)_4$ octahedra linked through the sharing of oxygen atoms; the composite double layer rolls up, like a scroll to form long hollow tubes. The outer diameters of the individual tubes are in the order of 25 nm; the length-to-diameter ratio can vary from 20 to well over 10 000. Chrysotile is characterized by a combination of a distinctive morphology, a chemical composition close to $Mg_3Si_2O_5(OH)_4$, and characteristic X-ray and electron diffraction patterns. Its Chemical Abstracts number is 12001-29-5. **D 2946,** C17

chrysotile asbestos—see **chrysotile.** **D 2946,** C17

chuck mark—any physical mark on either surface of a wafer caused by robotic end effector, a chuck, or wand. **F 1241,** F01

chunk glass—{archaic} optical glass obtained in breaking open a pot of transfer glass. **C 162,** C14

CIAP—an abbreviation for *corrosion inhibiting adhesive primer.* **E 1749,** E06

cidaphobia—sensitivity of target animals that increases their aversion to the high mortality within the population. **E 609,** E35

CIE, *n*—abbreviation for International Commission on Illumination, which in French is Commission Internationale de l'Éclairage. **D 2946,** C17

CIE, *n*—the abbreviation for the French title of the International Commission on Illumination, Commission Internationale de l'Éclairage. **E 284,** E12

CIE—Commission Internationale de l'Eclairage. **F 1156,** F12

CIE color-rendering index, *R, n*—measure of the degree to which the computed chromaticity of a CIE test color sample illuminated by a test illuminant conforms to that of the same sample illuminated by a reference illuminant. **E 284,** E12

CIELAB color difference, *n*—color difference calculated by using the CIE 1976 L^* a^* b^* opponent-color scales, based on applying a cube-root transformation to CIE 1931 tristimulus values X, Y, Z or CIE 1964 tristimulus values X_{10}, Y_{10}, Z_{10}. **E 284,** E12

CIELUV color difference, *n*—color difference calculated by using the CIE 1976 L^* u^* v^* opponent-color scales, based on a linear transformation of CIE chromaticity coordinates x, y, or x_{10}, y_{10} and a cube-root transformation of Y or Y_{10} to L^*, applied to CIE 1931 tristimulus values X, Y, Z or CIE 1964 tristimulus values X_{10}, Y_{10}, Z_{10}. **E 284,** E12

CIE observer, *n—in color determination,* see **observer, standard, CIE 1931**; and **observer, supplementary, CIE 1964.** **D 2946,** C17

CIE primaries, *n*—the primary color stimuli used in the CIE system of colorimetry. **E 284,** E12

CIE source C, *n*—see **standard source.** **D 2946,** C17

CIE spectral tristimulus values, *n*— tristimulus values or color-matching functions of the spectral components of an equal-energy spectrum in the CIE (*XYZ*) system. **E 284,** E12

CIE 1931 standard colorimetric system, *n*—a system for determining the tristimulus values of any spectral power distribution using the set of reference color stimuli X, Y, Z and the three CIE color-matching functions $\bar{x}(\lambda)$, $\bar{y}(\lambda)$, $\bar{z}(\lambda)$ adopted by the CIE in 1931. **E 284,** E12

CIE standard illuminant *A*, *n*—colorimetric illuminant, representing the full radiator at 2855.6 K, defined by the CIE in terms of a relative spectral power distribution. **E 284,** E12

CIE standard illuminant *B*, *n*—colorimetric illuminant, representing direct sunlight with a correlated color temperature of 4874 K, defined by the CIE in terms of a relative spectral power distribution. Declared obsolete by the CIE in 1983. **E 284,** E12

CIE standard illuminant *C*, *n*—colorimetric illuminant, representing daylight with a correlated color temperature of 6774 K, defined by the CIE in terms of a relative spectral power distribution. **E 284,** E12

CIE standard illuminant D_{65}, *n*—colorimetric illuminant, representing daylight with a correlated color temperature of 6504 K, defined by the CIE in terms of a relative spectral power distribution. **E 284,** E12

CIE 1931 standard observer, *n*—ideal colorimetric observer with color matching functions $\bar{x}(\lambda)$, $\bar{y}(\lambda)$, $\bar{z}(\lambda)$ corresponding to a field of view subtending a 2° angle on the retina; commonly called the "2° standard observer." **E 284,** E12

CIE standard source *A*, *n*—a gas-filled tungsten-filament lamp operated at a correlated color temperature of 2855.6 K. **E 284,** E12

CIE standard source *B*, *n*—standard source A combined with a specified liquid filter, to provide radiant flux with a correlated color temperature of 4874 K. Declared obsolete by the CIE in 1983. **E 284,** E12

CIE standard source *C*, *n*—standard source A combined with a specified liquid filter, to provide radiant flux with a correlated color temperature of 6774 K. **E 284,** E12

CIE 1964 supplementary standard colorimetric system, *n*—a system for determining the tristimulus values of any spectral power distribution using the set of reference color stimuli X_{10}, Y_{10}, Z_{10}, and the three CIE color-matching functions $\bar{x}_{10}(\lambda)$, $\bar{y}_{10}(\lambda)$, $\bar{z}_{10}(\lambda)$ adopted by the CIE in 1964. **E 284,** E12

CIE 1964 supplementary standard observer, *n*—ideal colorimetric observer with color matching functions $\bar{x}_{10}(\lambda)$, $\bar{y}_{10}(\lambda)$, $\bar{z}_{10}(\lambda)$ corresponding to a field of view subtending a 10° angle on the retina; commonly called the "10° standard observer." **E 284,** E12

CIE 1976 uniform-chromaticity-scale diagram, *n*—the uniform-chromaticity-scale diagram produced by plotting in rectangular coordinates v' against u', quantities defined as follows:

$$u' = 4X/(X + 15Y + 3Z) = 4x/(-2x + 12y + 3)$$

$$v' = 9Y/(X + 15Y + 3Z) = 9y/(-2x + 12y + 3)$$

for the CIE 1931 standard colorimetric system, or v'_{10} against u'_{10} for the CIE 1964 supplementary standard colorimetric system, in which case in the above formulae X_{10}, Y_{10}, Z_{10} are used instead of *X, Y, Z* and x_{10}, y_{10} instead of *x, y*. **E 284,** E12

CIE 1976 (u', v') or (u'_{10}, v'_{10}) chromaticity diagram, *n* —chromaticity diagram in which the CIE 1976 L^* u^* v^* (CIELUV) chromaticity coordinates are plotted, with u' (or u'_{10}) as abscissa and v' (or v'_{10}) as ordinate. **E 284,** E12

CIE 1931 (*x*, *y*) chromaticity diagram, *n*—chromaticity diagram for the CIE 1931 standard observer, in which the CIE 1931 chromaticity coordinates are plotted, with *x* as abscissa and *y* as ordinate. **E 284,** E12

CIE 1964 (x_{10}, y_{10}) chromaticity diagram, *n*—chromaticity diagram for the CIE 1964 supplementary standard observer, in which the CIE 1964 chromaticity coordinates are plotted, with x_{10} as abscissa and y_{10} as ordinate. **E 284,** E12

cigar-box nail—bright, smooth, or barbed, regular-stock-steel, ½, ⅝¾, and by 0.044-in. nails with flat slightly countersunk 0.099-in. head and medium diamond point. **F 547,** F16

cinematography—an instrumentation system for filming, measuring, and analyzing movements of the athlete, usually including a computerized digitizor for data analysis. **F 869,** F08

cine-radiography—the production of a series of radiographs that can be viewed rapidly in sequence, thus creating an illusion of continuity. **E 1316,** E07

CIP—cleaning-in-place. **D 6161,** D19

circle, *n*—circular blank fabricated from plate, sheet, or foil. **B 881,** B07

circle breathing system, *n*—breathing system in which the direction of gas flow is through separate inspiratory and expiratory pathways that form a circle. **F 2002,** F29

circuit—in filament winding, the winding produced by a single revolution of mandrel or form. **D 883,** D20

circuit grouting—a grouting method by which grout is circulated through a pipe extending to the bottom of the hole and back up the hole via the annular space outside the pipe. Then the excess grout is diverted back over a screen to the agitator tank by means of a packing gland at the top of the hole. The method is used where holes tend to cave and sloughing material might otherwise clog openings to be grouted. **D 653,** D18

circular anisotropic phase, *n—as used in Test Method D 5061*, a group of binder-phase anisotropic carbon textures that are distinguished by approximately circular domains (that is length equals width) and composed of fine circular (0.5- to 1.0-μm), medium circular (1.0- to 1.5-μm), and coarse circular (1.5- to 2.0-μm) size categories. **D 121,** D05

circular bend, *n*—simultaneous, multidirectional deformation of a fabric in which one face of a flat specimen becomes concave and the other becomes convex. **D 123,** D13

circular field—see **field, circular.** **E 1316,** E07

circularity, *n*—ratio of a single ink dot height divided by its width with 1.0 being a perfect circle. **F 1857,** F05

circular magnetization—the magnetization in a part resulting from current passed directly through the part or through a central conductor. **E 1316,** E07

circular mil (cmil)—a unit of area equal to the area of a circle one mil (0.001 in., 0.0254 mm) in diameter. The area of a circle in circular mils is equal to the square of the diameter in mils [area(cmils) = diameter(mils)^2], 1 cmil = 7.854×10^{-7} in.^{2} [5.067×10^{-10} m^2]. **B 354,** B01

circular reinforcement—a circular-shaped line of reinforcement. **C 822,** C13

circular ride—an amusement ride whose motion is primarily rotary in a fixed or variable plane from horizontal to 45° above horizontal. **F 747,** F24

circulation—applies to the fluid rotary drilling method; drilling fluid movement from the mud pit, through the pump, hose and swivel, drill pipe, annular space in the hole and returning to the mud pit. **D 653,** D18

circulation rate [L^3 T^{-1}]—*in fatigue testing*, the volume rate of change of the environment chamber volume. **E 1823,** E08

circulation space— See **space categories**. **E 1480,** E06

circulation system—the system that moves the beverage within the bowl to ensure proper cooling and mixing. **F 1827,** F26

circumduction—the movement of a body part about an axis so as to describe a cone or the conical movement of a body part about an axis. **F 869,** F08

circumferential, *adj*—descriptor for directional illuminating (or viewing) geometry in which the illuminator provides radiation (or the receiver possesses responsivity) in many beams (or directions), normally distributed at uniform intervals throughout the 360° of azimuth of the measurement. The number and angular distribution of the beams (or directions) should be specified. (See also *annular*.) **E 284,** E12

circumferential coils— see **encircling coils.** **E 1316,** E07

circumferential line, *n—on a tire*, any real or imaginary circle on the surface of a tire, lying in a plane that is perpendicular to the spin axis. **F 538,** F09

circumferential patternation, *n*—measurements taken in a circumferential direction, showing the variation in liquid flux about the nozzle axis. **E 1620,** E29

circumferential reinforcement—reinforcement that is approximately perpendicular to the longitudinal axis of the concrete pipe. **C 822,** C13

ciromg accelerator—a material added to cement to decrease the time for curing. Examples are sodium chloride, calcium sulfate (gypsum), and aluminum powder. **D 653,** D18

citric acid—$C_3H_4(OH)(CO_2H)_3$, membrane cleaning chemical. **D 6161,** D19

clad—See **cladding**. **E 131,** E13

clad—surface sheathed. **F 547,** F16

cladding—nonload-bearing stone used as the facing material in wall construction that contains other materials. **C 119,** C18

cladding, *n—of an optical fiber*, a layer of a optically transparent lower refractive index material in intimate contact with a core of higher refractive index material used to achieve total internal reflection. **E 131,** E13

cladding ratio—ratio by percent thickness of the component layers, for example. **B 846,** B05

cladding system, *n*—material assembly applied to a building as a non-load-bearing wall, or attached to a wall surface as a protective and ornamental covering. **E 631,** E06

clad wire—wire comprised of a given metal covered with a relatively thick application of a different metal. **B 354,** B01

clamp, *n*—that part of a testing machine used to grip the specimen by means of suitable jaws or jaw liners, or both. **D 123,** D13

clamp, *n*—an interconnection component whose mechanism to secure the longitudinal element is through a squeezing action. **F 1582,** F04

clamping pressure—the pressure exerted by the lip of the lock-strip gasket on material installed in the channel, when the lock-strip is in place. **C 717,** C24

clamp load—sometimes called preload or initial load. It is a tension on a bolt or screw, which results in equal and opposite forces which exist at the interface between two members generated through the cumulative effect of tightening one or more fasteners. **F 1789,** F16

clarifier—a tank in which precipitate settles and supernatant overflows, a liquid-solids separation unit using gravity to remove solids by sedimentation. **D 6161,** D19

clarity, *n*—the characteristic of a transparent body whereby distinct high-contrast images or high-contrast objects (separated by some distance from the body) are observable through the body. **E 284,** E12

clark degree—number of grains of substance per one British imperial gallon of water expressed $CaCO_3$. Concentration in Clark or English degree is calculated by dividing concentration in calcium carbonate equivalents by 14.3. One grain weighs 1/7000 lb and one imperial gallon of water weighs 10 lbs at 25°C. **D 6161,** D19

class, *n*—a group, set, or kind marked by common attributes or a common attribute (source: *Webster's Unabridged Dictionary*, 1967) (use: Test Method E 1387) CRIM. **E 1732,** E30

class characteristic(s), *n*—the attribute(s) that establish membership in a class. **E 1732,** E30

classes of buildings, *adj* (catégories d'immeubles)—buildings categorized by selected attributes concerning **facility serviceability** and performance. **E 1480,** E06

classic gully—a channel that is formed by gully erosion and is not interrupted by mechanical tillage operations to fill the resulting void. Gully depth can exceed 30 m. (See **gully erosion**.) **D 4410,** D19

classification—separation of a powder into fractions according to particle size. **B 243,** B09

classification, *n*— *in engine oils*, the systematic arrangement into categories according to differing levels of performance in specified engine tests. **D 4175,** D02

classification, *n*—a method of sorting stimuli into predefined categories . **E 253,** E18

classification, *n*—see **vehicle classification.** **E 867,** E17

classification, *n*—the systematic arrangement of persons or objects into categories (groups or classes) based on shared traits or characteristics (source: Osterburg and Ward, *Criminal Investigation*, 1992, p. 835) (use: Test Method E 1387). **E 1732,** E30

Class III—Non-Newtonian Shear-Thickening Fluid—a fluid in which the shear stress is not directly proportional to the shear rates, and in which the shear stress-shear rate ratio increases as the shear stress increases.

(*a*) *Type A—Dilatant Fluid*—a Class III fluid that exhibits a shear stress-shear rate ratio that is independent of the duration of application of shear stress.

(*b*) *Type B—Rheopectic Fluid*—A Class III fluid that exhibits time-dependent, reversible changes of the shear stress-shear rate ratio. **D 2507,** F07

Class III quaternary equilibrium—in a four-component system, the stable univariant coexistence of five phases, three of which appear in both of the associated bivariant equilibria at temperatures and pressures above, while the other two phases occur in all three of the associated bivariant equilibria below; for example, $L + \alpha + \beta = \gamma + \delta$. **E 7,** E04

Class III quinary equilibrium—in a five-component system, the stable univariant coexistence of six phases, three of which appear in all three of the associated bivariant equilibria at temperatures and pressures above, while the other three occur in all three of the associated bivariant equilibria below; for example, $L + \alpha + \beta = \gamma + \delta + \epsilon$. **E 7,** E04

Class III ternary equilibrium—in a three-component system, the stable univariant coexistence of four phases, one of which must disappear at higher temperature or pressure; for example, the ternary peritectic equilibrium, $L + \alpha + \beta = \gamma$. **E 7,** E04

Class II—Non-Newtonian Shear-Thinning Fluid—a fluid in which the shear stress is not directly proportional to the shear rate and in which the shear stress-shear rate ratio decreases as the shear stress increases.

(*a*) *Type A—Plastic Fluid*—a Class II fluid that exhibits a change in shear rate directly proportional to the change in shear stress above the yield stress.

(*b*) *Type B—Pseudoplastic Fluid*—a Class II fluid that exhibits a shear stress-shear rate ratio that is independent of the duration of application of shear stress.

(*c*) *Type C—Thixotropic Fluid*—a Class II fluid that exhibits time-dependent, reversible changes of the shear stress-shear rate ratio. **D 2507,** F07

Class II quaternary equilibrium—in a four-component system, the stable univariant coexistence of five phases, two of which appear in each of the three associated bivariant equilibria at temperatures and pressures above, while the other three phases all occur in both of the associated bivariant equilibria below; for example, $L + \alpha = \beta + \gamma + \delta$. **E 7,** E04

Class II quinary equilibrium—in a five-component system, the stable univariant coexistence of six phases, two of which appear in each of the four associated bivariant equilibria at temperatures and pressures above, while the other four phases occur in both of the associated bivariant equilibria below; for example, $L + \alpha = \beta + \gamma + \delta + \epsilon$. **E 7,** E04

Class II ternary equilibrium—in a three-component system, the stable univariant coexistence of four phases, two of which appear in both of the associated bivariant equilibria at higher temperature and pressure, while the other two phases occur in both bivariant equilibria below; for example, $L + \alpha = \beta + \gamma$. **E 7,** E04

Class I—Newtonian Fluid— a fluid that exhibits a direct proportionality between shear stress and shear rate in the region of laminar flow. **D 2507,** F07

Class I quaternary equilibrium—in a four-component system, the stable univariant coexistence of five phases, one of which must disappear upon lowering the temperature or pressure; for example, the quaternary eutectic equilibrium, $L = \alpha + \beta + \gamma + \delta$. **E 7,** E04

Class I quinary equilibrium—in a five-component system, the stable univariant coexistence of six phases, one of which must disappear upon lowering the temperature or pressure; for example, the quinary eutectic equilibrium, $L = \alpha + \beta + \gamma + \sigma + \epsilon$. **E 7,** E04

Class I ternary equilibrium—in a three-component system, the stable univariant coexistence of four phases, one of which must disappear upon lowering the temperature or pressure; for example, the ternary eutectic equilibrium, $L = \alpha + \beta + \delta$. **E 7,** E04

Class IV quaternary equilibrium—in a four-component system, the stable univariant coexistence of five phases, one of which must disappear at higher temperature or pressure; for example, the quaternary peritectic equilibrium, $L + \alpha + \beta + \gamma = \delta$. **E 7,** E04

Class IV quinary equilibrium—in a five-component system, the stable univariant coexistence of six phases, four of which appear in both associated bivariant equilibria at temperatures and pressures above, while the other two occur in all four associated bivariant equilibria below; for example, $L + \alpha + \beta + \gamma = \delta + \epsilon$. **E 7,** E04

Class V quinary equilibrium—in a five-component system, the stable univariant coexistence of six phases, one of which must disappear upon increasing the temperature or pressure; for example, the quinary peritectic equilibrium, $L + \alpha + \beta + \gamma + \delta = \epsilon$. **E 7,** E04

Clausius-Clapeyron equation—the rate of change of the pressure of a heterogeneous equilibrium with change of temperature dP/dT is equal to the heat of transformation from the low to the high temperature state $\Delta\ Hv$ divided by the product of the absolute temperature of the equilibrium and the volume change of the transformation $T\Delta V$:

$$dP/dT = \Delta Hv/T\Delta V$$

or,

$$d \ln P/dT \approx Hv/RT$$

where R is the gas constant and the equilibrium is not near a critical point. **E 7,** E04

clay, *n*—an earthy or stony mineral aggregate consisting essentially of hydrous silicates of alumina, plastic when sufficiently pulverized and wetted, rigid when dry, and vitreous when fired to a sufficiently high temperature. **C 43,** C15

clay—a natural mineral agglomerate, consisting essentially of hydrous aluminum silicates; plastic when sufficiently wetted, rigid when dried en masse, and vitrified when fired to a sufficiently high temperature. **C 242,** C21

clay—an earthy or stony mineral aggregate consisting essentially of hydrous silicates of alumina, plastic when sufficiently pulverized and wetted, rigid when dry, and vitreous when fired to a sufficiently high temperature. **C 896,** C04

clay (clay soil)—fine-grained soil or the fine-grained portion of soil that can be made to exhibit plasticity (putty-like properties) within a range of water contents, and that exhibits considerable strength when air-dry. The term has been used to designate the percentage finer than 0.002 mm (0.005 mm in some cases), but it is strongly recommended that this usage be discontinued, since there is ample evidence from an engineering standpoint that the properties described in the above definition are many times more important. **D 653,** D18

clay roof tile, *n*—a solid unit of clay or shale, or both, formed into any of a range of generally rectangular planar shapes while plastic and fired in kiln. Clay roof tile is a ceramic product. See Specification C 1167. **C 43,** C15

clay size—that portion of the soil finer than 0.002 mm (0.005 mm in some cases) (see also **clay**). **D 653,** D18

clay size (fluvial sediment)—0.00024 to 0.004 mm in diameter. **D 4410,** D19

clay soil—see **clay**. **D 653,** D18

CLC—corrected lead concentration. **E 1605,** E06

clean—free of contaminants. **E 1316,** E07

clean—free from visible or tactile contamination. **F 1646,** F13

cleanability—the relative ease with which soils or stains can be removed from a material. **C 286,** B08

cleaned—oil and other foreign matter removed from surface by tumbling with sawdust or by chemical process. **F 547,** F16

cleaner—a solution, usually alkaline, used to remove oil, grease, drawing compounds, and loose dirt from metal as a step in preparing the surface for porcelain enameling. **C 286,** B08

cleaner, chemical—an organic solvent used to remove foreign matter from the surface of plastic pipe and fittings. **F 412,** F17

cleaner, mechanical—an abrasive material or device used to remove foreign matter and gloss from the surface of plastic pipe and fittings. **F 412,** F17

cleaner sheets—sheets of blotter-like stock which are sometimes used as an aid in washing up the inked rollers. **F 425,** F05

clean-finish seam finish, *n*—*in home sewing*, a seam finish in which the cut edge is folded under and the fold line is edge stitched. (Syn. **turned and stitched seam finish.**) **D 4965,** D13

cleaning—the removal of grease, oxides, or other foreign material from a surface.

alkaline cleaning—cleaning by means of alkaline solutions.

anodic or reverse cleaning—electrolytic cleaning in which the work is the anode.

cathodic or direct cleaning—electrolytic cleaning in which the work is the cathode.

diphase cleaning—cleaning by means of solutions that contain a solvent layer and an aqueous layer. Cleaning is effected both by solvent and emulsifying action.

direct current cleaning—See *cathodic or direct cleaning.*

electrolytic cleaning—alkaline cleaning in which a current is passed through the solution, the work being one of the electrodes.

emulsion cleaning—cleaning by means of solutions containing organic solvents, water, and emulsifying agents.

immersion—See *soak cleaning.*

reverse current cleaning—See *anodic or reverse cleaning.*

soak cleaning—cleaning by immersion without the use of current, usually in alkaline solution.

solvent cleaning—cleaning by means of organic solvents.

spray cleaning—cleaning by means of spraying.

ultrasonic cleaning—cleaning by any chemical means aided by ultrasonic energy. **B 374,** B08

cleaning—the removal of unwanted materials, such as mold release agents. **B 374,** B08

cleaning—a process of removing undesirable matter. **D 459,** D12

dry cleaning—cleaning fabrics in a substantially nonaqueous liquid medium. **D 459,** D12

wet cleaning—a term used in the dry cleaning industry to denote cleaning in an aqueous medium. **D 459,** D12

complexing agent—see **sequestering agent.** **D 459,** D12

cleaning—removal of marks, dust, and other extraneous materials from the surface. **D 2825,** D21

cleaning ability, dry, *n*—the potential of a vacuum cleaner to remove dirt from a surface (sometimes referred to in the industry as *cleanability, dry*). **F 395,** F11

cleaning agent, *n*—a chemical compound or formulation of several compounds which loosens, disperses, dissolves, or emulsifies soil to facilitate removal by mechanical action. **D 123,** D13

cleaning agent, *n*—a chemical compound or formulation of several compounds which loosens, disperses, dissolves or emulsifies, or causes a chemical change in the soil to facilitate removal by mechanical action. **D 5253,** D13

cleaning effectiveness, dry, *n*—the ratio of the quantity of dirt removed to the quantity of dirt distributed on a test area. **F 395,** F11

cleaning effectiveness factor (CEF), *n*—the fraction of contaminant removed from an initially contaminated test coupon and determined by gravimetric techniques. **G 126,** G04

cleaning-emulsifiable solvent—two-stage cleaning system wherein a concentrate containing organic solvents and surface-active agents is applied to a surface, subsequently emulsified, and removed along with the soil, by water rinsing. **B 374,** B08

cleaning tool, *n*—a customer-installed device for a vacuum cleaner that is applied to the surface to be cleaned and is attached to the hose or the nozzle, for specialty cleaning functions. **F 395,** F11

cleanliness—the degree of freedom from undesirable and unintentional ink transfer from carbon paper and inked ribbons. **F 221,** F05

clean room—an area in which the temperature, humidity, and airborne contamination particulate are controlled as required. **C 859,** C26

clean-up—in leak testing, the time required for a leak testing system to reduce its signal output to 37 % of the signal indicated at the time the tracer gas ceases to enter the leak system. Also called clean-up time. **E 1316,** E07

cleanup, *n*—wet-sweeping, HEPA-vacuuming, and washing down of surfaces within the work area at the end of each day. Compare **final cleanup**. **E 1605,** E06

clean wool fiber present, *n*—*in raw wool*, the mass of wool base present in the raw wool, adjusted to a moisture content of 12 %, an alcohol-extractable content of 1.5 %, and a mineral matter content of 0.5 %. **D 123,** D13

clear—free of defects or imperfections. **D 9,** D07

clear—free of turbidity. **D 4790,** D16

clearance—plain section of shank between head and shank deformation (disregarding gripmarks). **F 547,** F16

clear area—that region of a document reserved for OCR characters and the required clear space around these characters. **F 149,** F05

clear area, *n*—a space containing no dark marks, which precedes the start character of a symbol and which follows the stop character. Also, the region of a document reserved for OCR characters and the required unmarked space around these characters. (See also **quiet zone**.) **F 1294,** F05

clear cross test—an experimental method for determining which of two conceivable two-phase equilibria is real; at that composition at which the two conceivable two-phase fields cross, an alloy is brought to equilibrium and the phases identified; the same principle may be applied to higher-order equilibria in higher-order systems. **E 7,** E04

clear frit—a frit that remains essentially transparent or nonopaque when processed into a porcelain enamel. **C 286,** B08

clear glass focusing screen—a glass screen polished on both sides and mounted for use in a camera, in lieu of photo-sensitive material, for the purpose of establishing a plane on which to focus an image prior to recording it. **E 7,** E04

clear glaze—See *clear glaze* under **glaze.** **C 242,** C21

clear to send, *n*—a control circuit that indicates to the data terminal equipment that data can or cannot be transmitted. (See **CTS**.) **F 1457,** F05

cleat, *n*—the joint system of coal seams, usually oriented normal or nearly normal to the bedding. **D 121,** D05

cleat, *n*—a wood or metal strip attached along the edge of a **panel** of a **container** for the attaching of an adjacent panel, or fastened to the panel between the edges, or to barrel heads, for reinforcement and stiffening. **D 996,** D10

cleat—L-shaped nail. **F 547,** F16

cleated fiberboard box— See **box.** **D 996,** D10

cleated plywood box— See **box.** **D 996,** D10

cleat nail—bright, regular-stock-steel, $1\frac{1}{8}$ by 0.080 to 1 $\frac{7}{8}$ by 0.106-in. nails with oval $\frac{3}{16}$ to $\frac{1}{4}$-in. head and duckbill or clinch point. (See **clinch nail.**) **F 547,** F16

cleavage—*in crystallography*, the splitting, or tendency to split, along planes determined by the crystal structure. *In petrology*, a tendency to cleave or split along definite, parallel, closely spaced planes. It is a secondary structure, commonly confined to bedded rocks. **D 653,** D18

cleavage—the tendency to cleave or split along definite parallel planes, which may be highly inclined to the bedding. It is a secondary structure and is ordinarily accompanied by at least some recrystallization of the rock. (ISRM) **D 653,** D18

cleavage, *n*—*in an adhesively bonded joint*, a separation in the joint caused by wedge or other crack-opening type action. **D 907,** D14

cleavage crack—damage produced by the translation of a hard, sharp object across a glass surface. This fracture system typically includes a plastically deformed groove on the damaged surface, together with median and lateral cracks emanating from this groove. **C 162,** C14

cleavage-peel strength, *n*—the force per unit width of bondline required to produce progressive separation by wedge or other crack-opening type action of two adherends where one or both undergo significant bending. **D 907,** D14

cleavage plane—a crystallographically preferred fracture plane. **F 1241,** F01

cleavage planes—the parallel surfaces along which a rock or mineral cleaves or separates; the planes of least cohesion, usually parallel to a certain face of the mineral or crystal. **D 653,** D18

cleavage planes—that family of planes of a crystal along which the crystal is easily split. **E 7,** E04

cleavage strength, *n*—the force per unit width of bondline required to produce progressive separation by wedge or other crack-opening type action of two adherends of sufficient thickness to produce no significant bending. **D 907,** D14

cleft water—water that exists in or circulates along the geological discontinuities in a rock mass. **D 653,** D18

climatological types, *n*—major regions of significantly different recurring weather patterns. In weathering, several distinct climatological types are used to evaluate the atmospheric durability of materials. Within any single climatological variation, at a specific geographic location, short term weather patterns may fluctuate significantly. This may cause variability in short term exposures. Major climatic variations that are used for the study of weathering are: **G 113,** G03

climbing, *n*—a rope of kernmantle construction, specifically designed to provide support, facilitate movement, and protect a climber by cushioning the impact of a fall. Climbing ropes sometimes are referred to as dynamic ropes. **F 1773,** F08

climbing drum peel test—See **test, climbing drum peel**. **E 1749,** E06

clinch—point end bent sideways to provide increased holding power or to eliminate protrusion of point end of excessively long fastener, or both. **F 547,** F16

clinch—protruding point end turned over or flattened when driven or driven against clinching plate. **F 592,** F16

clinched point—pointed end of nail, having been driven through member bent sideways. (See **brad-clinched point, J-clinched point, plate-clinched point.**) **F 547,** F16

clincher—part of tool that folds fastener legs to form clinch. **F 592,** F16

clinching plate—(See **anvil.**) Also, hardened flat metal plate; used in clinching nails and staples designed for clinching. **F 592,** F16

clinch nail—any nail designed for clinching after driving. Bright, regular-stock-steel, 1 by 0.080 to 4 by 0.177-in. nails with oval 3/16 to 3/8-in. head and duckbill or clinch point. (See **cleat nail.**) **F 547,** F16

clinch point—end of wire flattened or notched or both for easy clinching when striking anvil plate or "nail buck". (See **self-clinching.**) **F 547,** F16

clinch point—point designed to facilitate clinching when driven against clinching plate. (See **step point.**) **F 592,** F16

clinch strip, *n*—high-modulus or high-hardness compound applied between the carcass and the sidewall in the bead area to reinforce the bead. **F 538,** F09

clinical certification—a standardized process for evaluation and recognition of an acceptable level of competence in a specific aspect of patient care. **F 1177,** F30

clinical experience—exposure to and practice in an area of patient care. **F 1177,** F30

clinical/field preceptor—an individual who supervises and evaluates the students during clinical or field experiences under the direction of the course instructor/coordinator. **F 1177,** F30

clinical medical practice—patient diagnosis and treatment, including treatment protocols, which are the purview of qualified professionals (as determined by the state or other appropriate authority). **F 1177,** F30

clinical thermometer, *n*—thermometer of any type designed to measure human body temperature. **E 344,** E20

clinker, *n*—generally a fused or partly fused by-product of the combustion of coal, but also including lava and portland-cement clinker, and partly vitrified slag and brick. **D 8,** D04

clip, *n—in shingles*, see **storm anchor**. **C 460,** C17

clip—a wire spring device used to hold glass in a rabbeted metal sash for face-glazing. **C 717,** C24

clip, n—in shingles—See **storm anchor.** **C 1154,** C17

clip, *n—in shingles*, see **storm anchor (clip)**. **D 2946,** C17

clip, *n*—a small fastening device, usually of metal, designed to hold an element or **component** in place. **E 631,** E06

clip—See **strip.** **F 592,** F16

clip, *n*—a mechanical means of joining two loops. **F 1379,** F14

clip mark, *n*—a visible deformation near the edge of a fabric parallel with the lengthwise direction caused by pressure exerted by a clasping device on a clip tenter frame. (See also **pin mark.**) **D 123,** D13

clipped head—misnomer for D head. (See **notched head.**) **F 592,** F16

clipper—the shearing machine used to dimension dry or green veneers to width. **D 1038,** D07

clipping—*in fatigue spectrum loading*, the process of decreasing or increasing the magnitude of all loads (strains) that are, respectively, above or below a specified level, referred to as clipping level; the loads (strains) are decreased or increased to the clipping level. **E 1823,** E08

clo, *n*—unit of thermal resistance equal to 0.155 Km^2/W. **D 123,** D13

clo, *n*—a unit of thermal resistance (insulation) equal to 0.155 K-m$_2$/W. **F 1494,** F23

clogging, *n—for geotextiles,* the condition where soil particles move into and are retained in the openings of the fabric, thereby reducing the hydraulic conductivity. **D 4439,** D35

clogging potential, *n—in geotextiles,* the tendency for a given geotextile to decrease permeability due to soil particles that have either lodged in the geotextile openings or have built up a restrictive layer on the surface of the geotextile. **D 4439,** D35

closed assembly time, *n*—See *closed assembly time* under **assembly time.** **D 907,** D14

closed cell, *n*—a cell totally enclosed by its walls and hence not interconnecting with other cells. **C 717,** C24

closed cell, *n*—a cell totally enclosed by its walls and hence not interconnecting with other cells. (See also **cell** and **open cell.**) **F 1251,** F04

closed-cell cellular plastics—cellular plastics in which almost all the cells are noninterconnecting. **D 883,** D20

closed-cell cellular plastics—cellular plastics in which almost all the cells are noninterconnecting. (D-20) **F 412,** F17

closed-cell foamed plastic, *n*—a plastic in which almost all the cells are noninterconnecting. **F 1251,** F04

closed-cell foamed plastics—See **closed-cell cellular plastics.** **F 412,** F17

closed cell material, *n*—a cellular material in which substantially all cells in the mass are closed cells. **C 717,** C24

closed cell sealant backing, *n—in building construction,* a flexible and somewhat compressible extruded shape (usually cylindrical), with a surface skin, that is composed of closed cell material. **C 717,** C24

closed chip—fractured area on the edge or corner when the material has not broken off (Syn. *potential chip).*

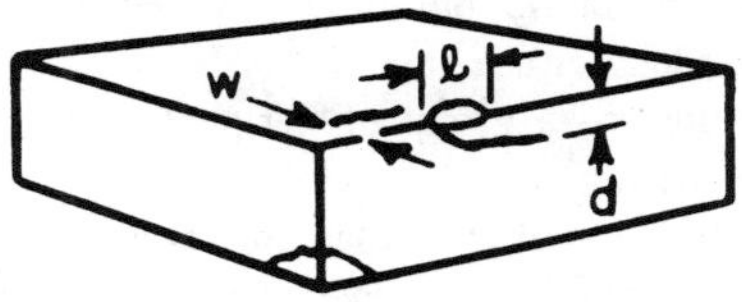

where:
W = width
l = length, and
d = depth.

F 109, C21

closed construction—See **building construction.** **E 631,** E06

closed-face fabric, *n*—a face or shell fabric of closed construction so that no open-face areas appear. **D 123,** D13

closed-loop, *adj*—refers to a testing mode in which the collection medium volume is fixed. **F 1494,** F23

closed pore—a pore not communicating or connected with an exterior surface. **B 243,** B09

closed porosity—See *closed porosity* under **porosity.** **C 242,** C21

closed pot—a pot having a crown protecting the glass from the furnace atmosphere. **C 162,** C14

closed system—See **building system.** **E 631,** E06

close nipple—a nipple with no shoulder or unthreaded portion between two threads; the shortest possible pipe nipple with complete threads. **B 846,** B05

close out—enclosure of honeycomb or other core material within a structure that may contain hard edges or attachment points, or both. **E 1749,** E06

close packed—a geometric arrangement whereby a collection of equally sized spheres (atoms) may be packed together in a minimum total volume. **E 7,** E04

closure—See **compression joint.** **C 896,** C04

closure—the opening is reduced in dimension to the extent that it cannot be used for its intended purpose. (ISRM) **D 653,** D18

closure—*in grouting*, closure refers to achieving the desired reduction in grout take by splitting the hole spacing. If closure is being achieved, there will be a progressive decrease in grout take as primary, secondary, tertiary, and quanternary holes are grouted. **D 653,** D18

closure, *n—in packaging,* a means of closing a **container** to retain the contents.

plug, n—in packaging, a type of **closure** that is designed to be inserted into a **container** opening. It may be held by friction or by screw threads. (See **cap.**) **D 996,** D10

closure—process by which a person cognitively completes patterns or shapes that are incompletely perceived. **E 1316,** E07

closure strip—an asphalt or rubber preformed filler strip having the same shape and pitch as the asbestos-cement corrugated product and used to close openings in the corrugated sheets at window beads, eaves, lower edge of siding, and similar places. **C 460,** C17

closure strip, *n*—an asphalt or rubber preformed filler strip having the same shape and pitch as the corrugated product, used to close openings in the corrugated sheets at window beads, eaves, lower edge of siding, and similar places. **C 1154,** C17

closure strip—asphalt or rubber preformed filler strip having the same shape and pitch as the asbestos-cement corrugated product and used to close openings in the corrugated sheets at window beads, eaves, lower edge of siding, and similar places. **D 2946,** C17

cloth, *n*—any textile fabric, but especially one designed for apparel, domestic, or industrial use. **D 123,** D13

clothing ensemble, *n*—a group of garments worn together on the body at the same time. **F 1494,** F23

cloud, *n*—any collection of particulate matter in the atmosphere dense enough to be perceptible to the eye, especially a collection of water drops.

cloud water, n—an aggregate of condensed water vapor or ice crystals that are suspended in the atmosphere. **D 1356,** D22

cloud, *n*—any collection of particulate matter in the atmosphere dense enough to be perceptible to the eye, especially a collection of water drops. **E 1620,** E29

cloud cover—that portion of the sky which is covered by clouds, usually expressed in tenths of sky covered. **E 772,** E44

cloud development—see **aerosol development.** **F 335,** F05

cloud point,, *n*—the temperature at which a defined liquid mixture, under controlled cooling, produces perceptible haze or cloudiness due to the formation of fine particles of an incompatible material. **D 6440,** D01

cloud point of phenol—the temperature at which a separate phase forms when a homogeneous solution of phenol in water is allowed to cool at a prescribed rate from a temperature above that at which phase separation occurs. It may precisely be defined as follows: when a homogeneous solution of phenol and water is allowed to cool at a prescribed rate with stirring, the solution will show a slight cloudiness or turbidity as the cloud point is approached. On further cooling, the cloudiness will increase rapidly and the thermometer bulb, which is centrally located in the test tube, will suddenly become invisible. The temperature at which the thermometer bulb becomes invisible is taken as the cloud point. **D 4790,** D16

cloud water—See **cloud.** **D 1356,** D22

cloudy—qualitative expression of turbidity. **D 4790,** D16

clout nail—bright, regular-stock-steel, ¾ by 0.072 to 1 ½ by 0.092-in. nails with large flat 0.225 to 0.262-in. head and long side point or duckbill point. **F 547,** F16

clusec—an obsolete unit of flow rate equal to 10-2 lusecs. **E 1316,** E07

cluster controller, *n*—a device in an IBM Model 3274, 3276, or equivalent that controls the flow of information in a local area network. **F 1457,** F05

cluster sampling, *n*—when the primary sampling unit comprises a bundle of elementary units or a group of subunits, the term cluster sampling may be applied. **E 456,** E11

cmc (*l:c*) color difference, *n*—color difference calculated by use of the formula developed by the Colour Measurement Committee of the Society of Dyers and Colourists of Great Britain. **E 284,** E12

coadsorption—the adsorption of two or more components on an adsorbent, each affecting the adsorbability of the other. **D 2652,** D28

coagent, *n*—a compounding ingredient used in small amounts to increase the crosslinking efficiency of certain non-sulfur vulcanizing systems or to modify the properties given by such systems. **D 1566,** D11

coagulant—chemical added in water and wastewater applications to cause destalization of suspended particles and subsequent formation of flocs that adsorb, entrap, or otherwise bring together suspended matter that is so fine, it is defined as colloidal. Compounds of iron and aluminum are generally used to form flocs to allow removal of turbidity, bacteria, color, and other finely divided matter from water and waste water. **D 6161,** D19

coagulant dipping (latex), *n*—a dipping process in which the product form is first immersed in a coagulant solution, withdrawn, dried, and then immersed in the latex. **D 1566,** D11

coagulation—the agglomeration of colloidal or finely divided suspended matter caused by the addition to the liquid of an appropriate chemical coagulant, by biological processes, or by other means (see also **agglomeration**). **D 4410,** D19

coagulation (rubber latex), *n*—irreversible agglomeration of particles originally dispersed in a rubber latex. **D 1566,** D11

coagulum—an agglomerate of particles. **D 2825,** D21

coagulum (latex), *n*—an undersirable, irreversible, agglomeration of rubber particles retained on a 180 μm mesh screen. **D 1566,** D11

coal, *n*—a brown to black combustible sedimentary rock (in the geological sense) composed principally of consolidated and chemically altered plant remains.

lithotype, n—any of the constituents of banded coal: vitrain, fusain, clarain, durain, or attrital coal or a specific mixture of two or more of these.

banded coal, n—coal that is visibly heterogeneous in composition, being composed of layers of vitrain and attrital coal, and, commonly, fusain.

attrital coal, n—the ground mass or matrix of banded coal in which vitrain and, commonly, fusain layers as well, are embedded or enclosed.

fusain, n—coal layers composed of chips and other fragments in which the original form of plant tissue structure is preserved; commonly has fibrous texture with a very dull luster.

vitrain, n—shiny black bands, thicker than 0.5 mm, of subbituminous and higher rank banded coal.

nonbanded coal, n—consistently fine-granular coal essentially devoid of megascopic layers.

cannel coal, n—nonbanded coal in which the liptinite is predominantly sporinite.

boghead coal, n—nonbanded coal in which the liptinite (the waxy component) is predominantly alginite.

impure coal, n—coal having 25 weight % or more, but less than 50 weight %, of ash on the dry basis.

bone coal, n—impure coal that contains much clay or other fine-grained detrital mineral matter.

mineralized coal, n—impure coal that is heavily impregnated with mineral matter, either dispersed or discretely localized along cleat joints or other fissures. Pyritic or calcareous mineralized coal is most common. **D 121,** D05

coalescence, *n*—a process by which the particles of a dispersion combine into one body. **D 1356,** D22

coalescence—growth of grains at the expense of the remainder by absorption or the growth of a phase or particle at the expense of the remainder by absorption or by reprecipitation. **E 7,** E04

coalescence, *n*—the merging of two or more liquid particles to form a single liquid particle. **E 1620,** E29

coalescence, *n*—the puddling or pooling of adjacent ink drops on the substrate before they can be dried or absorbed resulting in nonuniformity of color density. **F 1857,** F05

coalescing—the separation of mixtures of immiscible fluids (such as oil and water) based on different specific gravities and surface tensions. Can occur whenever two or more droplets collide and remain in contact and then become larger by passing through a coalescer. The enlarged drops then separate out of solution more rapidly. **D 6161,** D19

coal seam, *n*—the stratum, layer, or bed of coal that lies between two other rock layers whose compositions differ significantly from that of coal. **D 121,** D05

coal tar, *n*—a dark brown to black cementitious material produced by the destructive distillation of bituminous coal. **D 8,** D04

coal tar—a dark brown to black cementitious material produced by the destructive distillation of coal. **D 1079,** D08

coal-tar felt—a felt that has been saturated with refined coal tar. **D 1079,** D08

coal-tar pitch—a dark brown to black, solid cementitious material obtained as residue in the partial evaporation or distillation of coal tar. **D 1079,** D08

coal tar roof cement, *n*—a trowelable mixture of processed coal tar base, solvents, mineral fillers or fibers, or both. See Specification D 4022. **D 1079,** D08

coal washability, *n*—the determination of the theoretical limits for the removal of mineral impurities from coal by beneficiation processes that rely on specific gravity separations. **D 121,** D05

coarse aggregate, *n*—(*1*) aggregate predominantly retained on the 4.75-mm (No. 4) sieve: or (*2*) that portion of an aggregate retained on the 4.75-mm (No. 4) sieve. **D 8,** D04

coarse coal—See *coarse coal* under **sample**. **D 121,** D05

coarse end, *n*—a larger than normal diameter warp end. (Compare **fine end.**) **D 123,** D13

coarse grain—See **grain.** **D 9,** D07

coarse-grained—containing predominately large grains that are greater than 0.1 mm in size. **C 709,** D02

coarse grains—grains either larger than normal for the particular wrought metal or alloy, or grains of such a size that a surface roughening, popularly known as "orange peel" or" alligator skin," is produced. **E 7,** E04

coarse pick, *n*—*in woven fabrics*, one or more picks of larger diameter than the normal filling yarn in the fabric. **D 123,** D13

coarse rough fiber—fibers of flesh surfaces of leather or splits which are frayed, separated, and present a shaggy appearance. **D 1517,** D31

coarse sprays—distribution of droplets with $D_{V.5} > 1200\ \mu m$. **E 1102,** E35

coat, *n*—an outer garment which covers at least the upper half of the body, has sleeves and a front opening, and is usually worn over another garment, such as a shirt or dress. **D 123,** D13

coated—covered fully or partially with natural resin or conversion coating to provide ease of driving, increased holding power, or corrosion resistance, or a combination of these. **F 547,** F16

coated fabric, *n*—a flexible material composed of a textile fabric and an adherent polymeric material applied to one or both surfaces. **D 123,** D13

coated fabric, *n*—a flexible product composed of a textile fabric and an adherent polymeric material applied to one or both surfaces. **D 1566,** D11

coated fabric—a flexible product composed of a textile fabric and an adherent polymeric material applied to one or both surfaces. (D 1566, D-11) **F 869,** F08

coated fastener—a fastener with appropriate material applied to its surface to increase the fastener-withdrawal resistance. **F 592,** F16

coated offset master—a planographic surface of proper specifications on which the image to be reproduced by the lithographic process is placed. **F 425,** F05

coated paper, *n*—paper which has been coated on one or both sides with a minimum coat weight of 2.5 lb/3300-ft ^{2}(3.7 g/m^2) of coating material per side. see **coating.** **D 1968,** D06

coated paper—paper which has one or both surfaces covered with a substance to produce certain desired properties for use in specific electrophotographic processes. **F 335,** F05

coated paper copying—a form of direct electrophotographic copying. **F 335,** F05

coated sheet (or felt)—(*1*) an asphalt felt that has been coated on both sides with harder, more viscous asphalt;
(*2*) a glass fiber felt that has been simultaneously impregnated and coated with asphalt on both sides. **D 1079,** D08

coated wire—wire comprised of a given metal covered with a relatively thin application of a different metal. **B 354,** B01

coating, *n*—a liquid or semiliquid that dries or cures to form a protective finish, suitable for application to thermal insulation or other surfaces in thickness of 30 mils (0.76 mm) or less, per coat. **C 168,** C16

coating—see **ceramic coating** and **ceramic-metal coating.** **C 286,** B08

coating—a liquid, liquefiable or mastic composition that is converted to a solid protective, decorative, or functional adherent film after application as a thin layer. **D 16,** D01

coating, *n*—*of paper,* the layer of pigment and adhesive applied to the surface of paper or paperboard to create a new surface. **D 1968,** D06

coating, *n*—a liquid, liquifiable, or mastic composition that, after application as a thin layer, is converted to a solid protective, or decorative, or functional adherent film. **E 631,** E06

coating, *n*—liquid, liquefiable, or mastic composition that is converted to a solid protective, or decorative, or functional adherent film after application as a thin layer. **E 1605,** E06

coating—a layer of any substance intentionally applied to a surface to modify its functional or decorative characteristics. **F 1646,** F13

coating applicator—*in protective coatings*, an organization or individual responsible for applying a protective or decorative coating. **D 4538,** D33

coating cycle—specific time and temperature to achieve desired depth of diffusion. **B 374,** B08

coating, diffusion—an alloy coating produced by applying heat to one or more coatings deposited on a metal substrate. **B 374,** B08

coating pick, *n*—the removal of coating particles from the sheet surface that occurs when the tack of the ink exceeds the surface strength of the substrate. **D 6488,** D01

coating powder, *n*—a heat-fusible, finely-divided solid resinous material used to form electrical insulating coatings. **D 1711,** D09

coating system—*in protective coatings*, a protective film consisting of one or more coats, applied in a predetermined order by prescribed methods. **D 4538,** D33

coating weight—the wet or dry weight of coating material applied to a paper or other substrate expressed as grams per square metre or pounds per specified ream. **F 335,** F05

coating work—*in protective coatings*, an all-inclusive term to define all operations required to accomplish a complete coating job; construed to include materials, equipment, labor, preparation of surfaces, control of ambient conditions, application of coating systems, and inspection. **D 4538,** D33

coaxial thermocouple—a thermocouple consisting of a thermoelement in wire form within a thermoelement in tube form with the wire being electrically insulated from the tube except at the measuring junction. **E 344,** E20

cobble (cobblestone)—a rock fragment, usually rounded or semirounded, with an average dimension between 3 and 12 in. (75 and 305 mm). **D 653,** D18

cobble size (fluvial sediment)—64 to 256 mm in diameter. **D 4410,** D19

cobwebbing—*in protective coatings*, the formation of fine filaments (cobwebs) or partly dried coating, during spray application. **D 4538,** D33

cobwebbing, *n*—a filmy buildup of dried ink or coating on the doctor blade, ends of the impression roll, or printing cylinder. **D 6488,** D01

cockle—hard, firm nodules appearing on the necks and bellies of sheepskins. **D 1517,** D31

cockle, *n—of paper*, a defective, puckered condition of a paper sheet as a result of non-uniform hygro-expansion which can be related to any non-uniformity in the sheet, including mass distribution and drying stresses. **D 1968,** D06

cockle finish, *n— of paper*, an intentional rough, puckered surface, typically obtained by rewetting and drying of a paper sheet without physical restraint. **D 1968,** D06

cockles, *n—in yarns*, irregular, thick, uneven lumps. **D 123,** D13

cocoon, *v—in packaging,* to employ strippable, usually plastic, sometimes multi-layered films to encapsulate an item. **D 996,** D10

cocuring—the act of curing a composite laminate and simultaneously bonding it to some other hard detail during the same cure cycle (for example, curing a skin laminate and bonding it to honeycomb core simultaneously). **E 1749,** E06

co-curing, n—curing a composite laminate and simultaneously bonding it to the sandwich core. **C 274,** D30

Codabar bar code—See **bar code.** **F 1156,** F12

COD—chemical oxygen demand—the amount of oxygen required under specified test conditions for the oxidation of water borne organic and inorganic matter. **D 6161,** D19

code, *v*—to assign numbers, letters, words, or symbols as identifying marks to **containers,** packaged materials, or articles to convey information concerning the qualities of the container or its contents, date, place of manufacture, or other significant identification. (Compare **marking.**) **D 996,** D10

code—(*1*) a system of symbols, letters or numbers, used to convey a message requiring brevity; (*2*) a set of rules established by a legal or quasi-legal body. **F 412,** F17

code:

Hamming—any of a class of forward error correcting codes based upon the use of Galois Fields (extended, commutative, finite, sometimes imbedded, algebraic rings) utilizing 2^n, and represented as: GF(2^n), where n is the degree of an arbitrary polynomial. **F 1156,** F12

Code 11 bar code—See **bar code.** **F 1156,** F12

Code 39 bar code—See **bar code.** **F 1156,** F12

Code 93 bar code—See **bar code.** **F 1156,** F12

Code 128 bar code—See **bar code.** **F 1156,** F12

code, classification—a code that identifies a plastic material by its properties in accordance with the pertinent ASTM specification. **F 412,** F17

code density, *n*—the number of characters that can appear per unit of length, usually expressed as characters per inch (cpi). **F 1294,** F05

code, *n* (in the Law)—a collection of laws (regulations, ordinances, or statutory requirements) adopted by governmental (legislative) authority.

building code, n—a **code** applicable to **buildings,** adopted and administered with the primary intent of protecting public health, safety, and welfare.

model code, n—a proposed **code** that is established within the procedural framework of a group of knowledgeable people, and is designed for adoption by governmental authority. **E 631,** E06

code, manufacturer's—a code that provides manufacturing identity for a piping product. **F 412,** F17

code medium, *n*—the material used to construct a machine-readable code; such materials may be retroreflective, luminescent, magnetic, opaque, transponder or conductive. **F 1294,** F05

Code 2 of 5 bar code—See **bar code.** **F 1156,** F12

Code 3 of 9 bar code—See **bar code.** **F 1156,** F12

Code of Federal Regulations (CFR)—basic component of the *Federal Register* publication system. The CFR is a codification of the regulations of the various Federal agencies. **E 1605,** E06

Code 2 of 5 matrix bar code—See **bar code.** **F 1156,** F12

code set, *n*—the specific assignment of data characters to symbol characters. **F 1294,** F05

code, thermoplastic pipe materials designation—ciphers and letters for the designation of stress-rated thermoplastic compound, which consists of two or three letters to indicate the abbreviation as listed in Terminology D 1600, for the type of thermoplastic resin—followed by four arabic numerals—two to describe the short-term properties, in accordance with the ASTM standard being referenced, and two to designate the hydrostatic design stress when tested in water at 73°F (23°C) in units of 100 psi, with any decimal figures dropped. **F 412,** F17

code word, *n—in reference to bar codes*, an arrangement of bars and spaces used to communicate one or more data characters. **F 1294,** F05

co-dispensing valve—an arrangement whereby two components of a product are separated inside the container and mixed at the time of use, when ejected through dual channels into the valve. **D 3064,** D10

coefficient of absolute viscosity—see **coefficient of viscosity.**

D 653, D18

coefficient of active earth pressure—see **coefficient of earth pressure.** D 653, D18

coefficient of compressibility (coefficient of compression), α_v **(L^2F^{-1})**—the secant slope, for a given pressure increment, of the pressure-void ratio curve. Where a stress-strain curve is used, the slope of this curve is equal to $\alpha_v/(1 + e)$. **D 653,** D18

coefficient of consolidation, c_v **(L^2T^{-1})**—a coefficient utilized in the theory of consolidation, containing the physical constants of a soil affecting its rate of volume change.

$$c_v = k(1 + e)/\alpha_v\gamma_w$$

where:
k = coefficient of permeability, LT^{-1},
e = void ratio, D,
α_v = coefficient of compressibility, L^2F^{-1}, and
γ_w = unit weight of water, FL^{-3}.

D 653, D18

coefficient of earth pressure, K **(D)**—the principal stress ratio at a point in a soil mass.

coefficient of earth pressure, active, K_A (D)—the minimum ratio of (*1*) the minor principal stress, to (*2*) the major principal stress. This is applicable where the soil has yielded sufficiently to develop a lower limiting value of the minor principal stress.

coefficient of earth pressure, at rest, K_O (D)—the ratio of: (*1*) the minor principal stress, to (*2*) the major principal stress. This is applicable where the soil mass is in its natural state without having been permitted to yield or without having been compressed.

coefficient of earth pressure, passive, K_P (D)—the maximum ratio of: (*1*) the major principal stress, to (*2*) the minor principal stress. This is applicable where the soil has been compressed sufficiently to develop an upper limiting value of the major principal stress. **D 653,** D18

coefficient of expansion— see **coefficient of linear thermal expansion.** **E 1142,** E37

coefficient of friction—the ratio of the parallel component of force required to overcome or have a tendancy to overcome the resistance to relative motion of two surfaces in physical contact one with another, but otherwise unconstrained, to the normal component of the force—usually the force as a result of gravity—applied through the object which tends to cause the friction. **C 242,** C21

coefficient of friction, *n*—the ratio of the tangential force that is needed to maintain uniform relative motion between two contacting surfaces to the perpendicular force holding them in contact. **D 123,** D13

coefficient of friction (coefficient of friction between solid bodies), f **(D)**—the ratio between the maximum value of shear stress that resists slippage between two solid bodies with respect to each other, and the normal stress across the contact surfaces. The tangent of the angle of friction is ϕs. **D 653,** D18

coefficient-of-friction (COF or μ**)**—a dimensionless number: the ratio of two forces acting at the interface of two contacting solid bodies. The force used in the numerator is parallel to the surfaces and the force used in the denominator is perpendicular (normal) to the surfaces. See also *dynamic coefficient of friction.* **F 1646,** F13

coefficient of friction, f—a constant proportionality factor, μ, relating normal stress and the corresponding critical shear stress at which sliding starts between two surfaces: $T = \mu \cdot \sigma$. (ISRM) **D 653,** D18

coefficient of friction μ *or* f, *n—in tribology*, the dimensionless ratio of the friction force (F) between two bodies to the normal force (N) pressing these bodies together. (See also **static coefficient of friction** and **kinetic coefficient of friction.**)

$$\mu = (F/N)$$

G 40, G02

coefficient of image removal—the change in reflectance of the printed pattern divided by one hundred. (F 362) **F 221,** F05

coefficient of internal friction, μ (D)—the tangent of the angle of internal friction (angle of shear resistance) (see **internal friction**). **D 653,** D18

coefficient of kinetic or sliding friction, *n—of paper,* the ratio of the force required to sustain the uniform relative movement of the surfaces, to the normal force. **D 1968,** D06

coefficient of length variation, *n*—a measure of fiber length distribution. **D 123,** D13

coefficient of linear thermal expansion, α'—change in length, relative to the length of the specimen, accompanying a unit change of temperature, at a specified temperature. **E 1142,** E37

coefficient of line retroreflection, R_M, *n—of a reflecting stripe*, the ratio of the coefficient of luminous intensity (R_I) of a retroreflecting stripe to its length (l), expressed in candelas per lux per metre ($cd \cdot lx^{-1} \cdot m^{-1}$). $R_M = (R_I/l)$. **E 284,** E12

coefficient of luminous intensity R_I, *n—of a retroreflector*, ratio of the luminous intensity (I) of the retroreflector in the direction of observation to the illuminance ($E_\perp$) at the retroreflector on a plane perpendicular to the direction of the incident light, expressed in candelas per lux ($cd \cdot lx^{-1}$). $R_I = (I/E_\perp)$. **E 284,** E12

coefficient of permeability (permeability), k **(LT^{-1})**—the rate of discharge of water under laminar flow conditions through a unit cross-sectional area of a porous medium under a unit hydraulic gradient and standard temperature conditions (usually 20°C). **D 653,** D18

coefficient of retroreflected luminance, R_L, *n*— ratio of the luminance, L, of a projected surface to the normal illuminance, $E_\perp$, at the surface on a plane normal to the incident light, expressed in candelas per square metre per lux ($cd \cdot m^{-2} \cdot lx^{-1}$). $R_L = (L/E_\perp)$. **E 284,** E12

coefficient of retroreflection, R_A, *n—of a plane reflecting surface*, the ratio of the coefficient of luminous intensity (R_I) of a plane retroreflecting surface to its area (A), expressed in candelas per lux per square metre ($cd \cdot lx^{-1} \cdot m^{-2}$). $R_A = (R_I/A)$. **E 284,** E12

coefficient of scatter—the rate of increase of reflectance with thickness at infinitesimal thickness of porcelain enamel over an ideally black backing. **C 286,** B08

coefficient of shear resistance—see **coefficient of internal friction,** μ (D). **D 653,** D18

coefficient of static or starting friction, *n—of paper,* the ratio of the force resisting initial motion of the surfaces, to the normal force. **D 1968,** D06

coefficient of subgrade reaction (modulus of subgrade reaction), k, k_s **(FL^{-3})**—ratio of: (*1*) load per unit area of horizontal surface of a mass of soil, to (*2*) corresponding settlement of the surface. It is determined as the slope of the secant, drawn between the point corresponding to zero settlement and the point of 0.05-in. (1.3-mm) settlement, of a load-settlement curve obtained from a plate load test on a soil using a 30-in. (762-mm) or greater diameter loading plate. It is used in the design of concrete pavements by the Westergaard method. **D 653,** D18

coefficient of thermal expansion—change in unit of length (or volume) accompanying a unit change of temperature, at a specified temperature. **E 7,** E04

coefficient of transmissibility—the rate of flow of water in gallons per day through a vertical strip of the aquifer 1 ft (0.3 m) wide, under a unit hydraulic gradient. **D 653,** D18

coefficient of uniformity, C_u (D)—the ratio D_{60}/D_{10}, where D_{60} is the particle diameter corresponding to 60 % finer on the cumulative particle-size distribution curve, and D_{10} is the particle diameter corresponding to 10 % finer on the cumulative particle-size distribution curve. **D 653,** D18

coefficient of variation—a measure of relative precision calculated as the standard deviation of a series of values divided by their average. It is often multiplied by 100 and expressed as a percentage. **E 1547,** E15

coefficient of variation, CV, *n*—a measure of the dispersion of observed values equal to the standard deviation for the values divided by the average of the values; may be expressed as a percentage of the average (CV %). **D 123,** D13

coefficient of variation unevenness, *n*—*in textiles,* the standard deviation of the linear densities over which unevenness is measured expressed as a percentage of the average linear density for the total length within which unevenness is measured. (See also **unevenness** and **mean deviation unevenness.**) **D 123,** D13

coefficient of viscosity (coefficient of absolute viscosity), η (FTL^{-2})—the shearing force per unit area required to maintain a unit difference in velocity between two parallel layers of a fluid a unit distance apart. **D 653,** D18

coefficient of volume compressibility (modulus of volume change), m_v **(L^2F^{-1})**—the compression of a soil layer per unit of original thickness due to a given unit increase in pressure. It is numerically equal to the coefficient of compressibility divided by one plus the original void ratio, or $a_v/(1 + e)$. **D 653,** D18

coefficient of volume thermal expansion α_v—for a solid or liquid, the change in volume, relative to the volume of the specimen, accompanying a change of temperature at a specified temperature. **E 1142,** E37

coercive field strength, H_c—the (dc) magnetic field strength required to restore the magnetic induction to zero after the material has been symmetrically cyclically magnetized. **A 340,** A06

coercive field strength, intrinsic, H_{ci}—the (dc) magnetic field strength required to restore the instrinsic magnetic induction to zero after the material has been symmetrically cyclically magnetized. **A 340,** A06

coercive force—the magnetizing force at which the magnetic flux density is equal to zero. The corresponding field intensity value is indicative of the ease of difficulty or demagnetization. **E 1316,** E07

coercivity, H_{cs}—the maximum value of coercive field strength that can be attained when the magnetic material is symmetrically cyclically magnetized to saturation induction, B_S. **A 340,** A06

coextrusion—*in flexible barrier materials,* (*1*) a process whereby two or more plastic streams are forced simultaneously through one or more shaping orifices to become one continuously-formed multilayered structure. (*2*) Also, the product resulting from such a process. **F 17,** F02

coextrusion—a process whereby two or more heated or unheated plastic material streams forced through one or more shaping orifice(s) become one continuously formed piece. **F 412,** F17

coextrusion—See Terminology F 17. **F 1327,** F02

COF—coefficient of friction **F 1646,** F13

co-fab, n—fabrication process where close-outs and inserts are bonded into the panel the same time the facings are bonded to the core. **C 274,** D30

cohered—fastened together in strip form, usually with adhesive. (See **collated.**) **F 547,** F16

cohered—assembled in strip, coil, or other predetermined form as defined in Terminology F 547. **F 592,** F16

coherence—a property of a beam of electromagnetic radiation in which the phase relationship between any two points across the beam or in time remains essentially constant (see **coherence length**). **E 1316,** E07

coherence—a measure of the ability a source of electromagnetic radiation has to interfere. **F 1241,** F01

coherence length—the path difference between the object beam and the reference beam at which interference fringes reduce in contrast by a factor of $\sqrt{2}/2$ (0.707) from the point of maximum contrast. The coherence length is related to the width of the spectral line emitted from the laser: $L_c = c/\Delta\nu$, where c is the speed of light and$\Delta\nu$ is the bandwidth of the spectral emission line. **E 1316,** E07

coherent precipitate—a precipitated particle of a second phase, the lattice of which still maintains registry with the matrix lattice. Because the lattice spacings are usually different, strains usually exist at the interface. **E 7,** E04

coherent scattering—a kind of X-ray electron scattering in which the phase of the scattered beam has a definite (not random) relation to the phase of the incident beam. Also called unmodified scattering. **E 7,** E04

cohesion—the mutual attraction by which elements of a substance are held together. **C 904,** C03

cohesion—shear resistance at zero normal stress (an equivalent term in rock mechanics is intrinsic shear strength). (ISRM) **D 653,** D18

cohesion, *n*—the state in which the constituents of a mass of material are held together by chemical and physical forces. **D 907,** D14

cohesion, *n*—*in asbestos,* lateral force resisting separation of adjacent fibrous elements. **D 2946,** C17

cohesion, c **(FL^{-2})**—the portion of the shear strength of a soil indicated by the term c, in Coulomb's equation, $s = c + p \tan \phi$. See **intrinsic shear strength.**

apparent cohesion—cohesion in granular soils due to capillary forces. **D 653,** D18

cohesion failure, n—rupture of an adhesive bond, such that the separation appears to be within the adhesive. (Compare *adhesive failure.*) **D 907,** D14

cohesionless soil—a soil that when unconfined has little or no strength when air-dried and that has little or no cohesion when submerged. **D 653,** D18

cohesive blocking, n— See **blocking.** **D 907,** D14

cohesive failure, *n*—*in building construction,* failure characterized by rupture within the sealant, adhesive, or coating. **C 717,** C24

cohesive failure (bonded assembly), *n*— a rupture occurring entirely within any single uniform layer of the assembly. **D 1566,** D11

cohesive force, *n*—*in a textile strand,* the force required to overcome fiber cohesion as the strand is being reduced in linear density. **D 123,** D13

cohesive sediments—that material whose resistance to initial movement or erosion depends upon the strength of the bond between particles. **D 4410,** D19

cohesive soil—a soil that when unconfined has considerable strength when air-dried and that has significant cohesion when submerged. **D 653,** D18

coil, absolute—see **absolute coil.** **E 1316,** E07

coiled—assembled in coil form. **F 592,** F16

coiled sheet, *n*—sheet in coils with slit edges. **B 899,** B02

coil-fed tool—tool utilizing a coil of collated fasteners or a coil or wire. **F 592,** F16

coil loop, *n*—one complete rotation (360°) of barbed tape. **F 1379,** F14

coil method—a method of magnetization in which part, or whole, of the component is encircled by a current-carrying coil. **E 1316,** E07

coil, reference—see **reference coil.** **E 1316,** E07

coil rotation, *n*—the spiraling effect that occurs as the concertina coils are deployed. **F 1379,** F14

coil size—the dimension of a coil, for example, length or diameter. **E 1316,** E07

coil spacing—in electromagnetic testing, the axial distance between two encircling coils of a differential system. **E 1316,** E07

coil technique—a technique of magnetization in which all, or a portion, of the part is encircled by a current-carrying coil. **E 1316,** E07

coil, test—in electromagnetic testing, the section of the probe or coil assemby that excites and /or detects the electromagnetic field in the material under test. **E 1316,** E07

coin, *v*—to re-press a sintered compact to obtain a definite surface configuration. Synonymous with emboss (not to be confused with **re-strike** or **size**). **B 243,** B09

coincidence sum peak—a peak in the observed photon spectrum produced at an energy corresponding to the sum of the energies of two or more gamma- or x-rays from a single nuclear event when the emitted photons interact with the detector within the resolving time of the detector. **E 170,** E10

coining—the pressing of a sintered compact to obtain a definite surface configuration (not to be confused with repressing or sizing). **B 243,** B09

coke—a carbonaceous solid produced from coal, petroleum, or other materials by thermal decomposition with passage through a plastic state. **C 709,** D02

coke—a carbonaceous solid produced from coal, petroleum, or other materials by thermal decomposition with passage through a plastic state.

beehive coke, n—coke manufactured in beehive, rectangular, or similar forms of ovens in a horizontal bed, where heat for the coking process is secured by combustion within the oven chamber.

by-product coke, n—coke manufactured with attendant recovery of by-products, in ovens that are heated externally.

coke breeze, n—the fine screenings from crushed coke or from coke as taken from the ovens, of a size varied in local practice but usually passing a 12.7-mm (½-in.) or 19.0-mm (¾-in.) screen opening.

dry coke, n—a laboratory term applied to coke that has been dried to constant weight in accordance with definite prescribed methods. **D 121,** D05

coke-oven tar, *n*—coal tar produced in by-product coke ovens in the manufacture of coke from bituminous coal. **D 8,** D04

coke-oven tar—See **coal tar.** **D 1079,** D08

coke pore, *n*—*as used in Test Method D 5061*, a microscopically distinguishable void that is a structural element of coke. **D 121,** D05

coke reactivity, *n*—*as used in Test Method D 5061*, a measure of the mass loss when coke, held at a designated temperature, is contacted with gaseous carbon dioxide over a specific time interval. **D 121,** D05

coke wall, *n*—*as used in Test Method D 5061*, a predominantly carbonaceous layer that encloses a coke pore and which is a structural element and essence of coke. **D 121,** D05

COL—See **character outline limit.** **F 149,** F05

Colburn sheet process—{archaic} manufacture of sheet glass by bending the vertically drawn sheet over a roll which establishes the definition of draw. **C 162,** C14

cold-applied—capable of being applied without heating as contrasted to hot-applied; cold-applied products are furnished in a liquid state, whereas hot-applied products are furnished as solids that must be heated to liquefy them. **C 717,** C24

cold-cathode gun—an electron gun in which electrons produced in a gas discharge are accelerated through a small aperture in the anode. **E 7,** E04

cold-cathode ionization gage—see **ionization vacuum gage.** **E 1316,** E07

cold checks, *n*—defect on calendered sheeting consisting of surface roughness. **D 1566,** D11

cold-drawing—reducing the cross section by pulling through a die or dies, at a temperature lower than the recrystallization temperature. **B 354,** B01

cold filling—the pressurizing of a container by cooling the propellant (and sometimes the product) below its boiling point and transferring it into the container before the valve is put in place. The operation is usually carried out at atmospheric pressure (that is, high pressure equipment is not needed). **D 3064,** D10

cold flow—See **creep.** **D 883,** D20

cold flow, n— See **creep**. **D 907,** D14

cold flow—See **creep.** **F 412,** F17

cold flow—see preferred term **creep.** **F 1251,** F04

cold flow (unvulcanized rubber), *n*— slow deformation, under gravitational force, at or below room temperature. **D 1566,** D11

cold forming—the process of forming material below the recrystallization tempertaure by forcing or pressing metal into various dies. **F 1789,** F16

cold isostatic pressing, *n*—the pressing of a powder, compact or sintered object by subjecting it, at ambient temperature, to nominally equal presure from every direction. **B 243,** B09

cold joint, *n*—a plane of weakness in concrete caused by an interruption or delay in the placing operation, which permits the first batch to start setting before the next batch is placed, resulting in little or no bond between the two batches. **C 717,** C24

cold joint—See **joint.** **E 631,** E06

cold junction—See **reference junction.** **E 7,** E04

cold junction correction— See **reference junction correction.** **E 7,** E04

cold molding—a special process of compression molding in which the molding is formed at room temperature and subsequently baked at elevated temperatures. **D 883,** D20

cold molding—a special process of compression molding in which the molding is formed at room temperature and subsequently baked at elevated temperatures. (D-20) **F 412,** F17

cold pressing—the forming of a compact at room temperature. **B 243,** B09

cold pressing, *n*—a bonding operation in which an assembly is subjected to pressure without the application of heat. **D 907,** D14

cold-process roofing—a continuous, semiflexible membrane consisting of plies of felts, mats, or fabrics laminated on a roof with alternate layers of roof cement and surfaced with a cold-applied coating. **D 1079,** D08

cold resistant—the ability to withstand the effect of a specified cold or low temperature without loss of serviceability. **F 869,** F08

cold-rolled steel—a low-carbon, cold-reduced and annealed sheet steel. **C 286,** B08

cold-setting adhesive, *n*—an adhesive that sets at temperatures below 20°C (68°F). (See also **hot-setting adhesive, intermediate-temperature-setting adhesive**, and **room-temperature adhesive**.) **D 907,** D14

cold setting adhesive—See **adhesive, cold setting**. **E 1749,** E06

cold top melter—an all electric melting furnace in which a thermally insulating layer of batch is maintained on top of the molten glass. **C 162,** C14

cold treatment, *n*—exposing a steel object to temperatures below room temperature for the purpose of obtaining desired conditions or properties, such as dimensional or structural stability. **A 941,** A01

cold welding—cohesion between two surfaces of metal, generally under the influence of externally applied pressure, at room temperature. **B 243,** B09

cold work—controlled mechanical operations for changing the form or cross section of a product and for producing a strain-hardened product at temperatures below the recrystallization temperature. **B 846,** B05

cold worked structure—a microstructure resulting from plastic deformation of a metal or alloy below its recrystallization temperature. **E 7,** E04

cold working, *n*—mechanical deformation of a metal at temperatures below its **recrystallization temperature**. **A 941,** A01

cold zone—the volume in the fryer below the heating element or heat exchanger surface designed to remain cooler than the cook zone. See **fryer**. **F 1827,** F26

collaborative test, *n*—an interlaboratory study of a test method wherein the participants analyze or make measurements on subsamples of the same test material. **D 1356,** D22

collagen—the principal fibrous protein in the corium or derma layer of a hide or skin. **D 1517,** D31

collapse—the flattening of single cells or rows of cells during the drying or pressure treatment of wood, characterized by a caved-in or corrugated appearance. **D 9,** D07

collapse, *n*—inadvertent densification of cellular material during manufacture resulting from breakdown of cell structure. **D 883,** D20

collapse, *n*—(*1*) inadvertent densification of cellular material during manufacture resulting from breakdown of cell structure; (*2*) the buckling of the inner liner of composite piping; (*3*) the buckling or flattening of a plastic rehabilitation liner; (*4*) the buckling or crushing of a plastic pipe from external forces, such as earth loads or external hydrostatic load. **F 412,** F17

collapsible tube—See **tube.** **D 996,** D10

collar—*in grouting*, the surface opening of a borehole. **D 653,** D18

collar—See **railing systems.** **E 631,** E06

collar, n— Synonym for **escutcheon**. **E 1481,** E06

collar head—intermediate flat head with shanks protruding in both directions of nail axis. **F 547,** F16

collar leather—a subdivision of harness leather, made from very light cattlehides in full thickness, or of cattlehide splits, and used for covering horse collars. **D 1517,** D31

collar nail—See **collar head, gudgeon, projection head.** **F 547,** F16

collated—fastened next to each other in strip form. (See **cohered.**) **F 547,** F16

collated—assembled in strip or other predetermined form. **F 592,** F16

collecting surface, *n*—*in the rotor of an open-end spinning machine*, that portion of the internal surface of the rotor, often in the form of a groove, in which the fibers are condensed for assembly into yarn. **D 123,** D13

collection angle—See **angle.** **E 673,** E42

collection efficiency—See **efficiency.** **D 1356,** D22

collection efficiency, *n*— *in impingement erosion and particulate flows*, the cross-sectional area of undisturbed fluid containing particles that will all ultimately impinge on a given solid surface, divided by the projected area of the solid surface, where these two areas are perpendicular to the direction of relative motion between the solid surface and the particles in the undisturbed fluid. **G 40,** G02

collection medium, *n*—a liquid or gas that does not affect the measured permeation and in which the test chemical is freely soluble or adsorbed to a saturation concentration greater than 0.5 weight or volume percent. **F 1494,** F23

collector, *n*—a reagent used in froth flotation to promote contact and adhesion between particles and air bubbles. **D 121,** D05

collector, *n*—a device for removing and retaining contaminants from air or other gases. **D 1356,** D22

collector, *n*—optical components, such as the cornea and lens of the eye, which guide radiant flux from a specimen being observed or measured to a sensor. **E 284,** E12

collector, concentrating—a solar collector that uses reflectors, lenses, or other optical elements to redirect and concentrate the solar irradiance on the collector aperture onto an absorber of which the surface area is smaller than the collector aperture area. **E 772,** E44

collector cover (glazings)— see **cover plate, collector.** **E 772,** E44

collector efficiency— see **efficiency, collector**. **E 772,** E44

collector, evacuated tube—a solar collector made from transparent tubing (usually glass) with an evacuated space between the tube and the absorber. The absorber may consist of an inner tube or another shape, with means for removal of thermal energy and may be specially coated. **E 772,** E44

collector, flat plate—a nonconcentrating solar collector in which the absorbing surface is essentially planar. **E 772,** E44

collector, line-focus—a concentrating solar collector that concentrates the solar flux in one dimension only. **E 772,** E44

collector, point focus—a concentrating collector that focuses the solar flux to a point. **E 772,** E44

collector, solar thermal—a device designed to absorb solar irradiance and to transfer the thermal energy to a fluid passing through it. **E 772,** E44

collector subsystem—that portion of the solar system which includes the solar collectors and related piping or ducts. **E 772,** E44

collector, tracking—a solar collector that moves so as to follow the apparent motion of the sun during the day, rotating about one axis or two orthogonal axes. **E 772,** E44

collector, trickle—a flat plate solar collector in which unpressurized liquid flows or "trickles" over the absorber. **E 772,** E44

collimate—to render fibers parallel. **E 1749,** E06

collimated light—light with all of the rays going in exactly the same direction; parallel light rays (do not spread apart as they move away from their source.) **F 1156,** F12

collimation—the operation of controlling a beam of radiation so that its rays are as nearly parallel as possible. **E 7,** E04

collimation—the operation of controlling a beam of radiation so that if the light source were a point, the light rays would become parallel. The total bundle of rays diverge as the source size increases. **E 175,** E41

collimator—a device of radiation absorbent material intended for defining the direction and angular divergence of the radiation beam. **E 1316,** E07

collimator—a device for controlling the size and direction of the ultrasonic beam. **E 1316,** E07

collision cascade—a sequential energy transfer between atoms in a solid as a result of bombardment by an energetic species. **E 673,** E42

collodion replica— See **replica.** **E 7,** E04

colloid—a substance of very fine particle size, typically between 0.1 and 0.001 μm in diametre suspended in liquid or dispersed in gas. A system of at least two phases, including a continuous liquid plus solid, liquid or gaseous particles so small that they remain in dispersion for a practicable time. **D 6161,** D19

colloidal grout—*in grouting*, a grout in which the dispersed solid particles remain in suspension (colloids). **D 653,** D18

colloidal mixer—*in grouting*, a mixer designed to produce colloidal grout. **D 653,** D18

colloidal particle—an electrically-charged particle, generally smaller in size than 200 mμ, dispersed in a second continuous phase. **B 374,** B08

colloidal particle, *n*—a dispersed particle with a linear dimension of 5 to 100 nm. **C 1145,** C28

colloidal particles—particles that are so small that the surface activity has an appreciable influence on the properties of the aggregate. **D 653,** D18

colloids (fluvial sediment)—smaller than 0.00024 mm in diameter. **D 4410,** D19

colluvial deposits—that material accumulated along valley margins by mass movements from the adjacent hillsides. **D 4410,** D19

colony forming unit (CFU)—unit used in the measure of total bacterial count (TBC). **D 6161,** D19

colophony, *n*—a term denoting medium and high grades of rosin. **D 804,** D01

colophony, n— See **rosin**. **D 907,** D14

color—a quality of visible phenomena of insulating fluids, the numerical value for which is derived by comparing this quality using transmitted light with that of a series of numbered reference standards. **D 2864,** D27

color, *n*—(*1*) *of an object*, aspect of object appearance distinct from form, shape, size, position, or gloss that depends upon the spectral composition of the incident light, the spectral reflectance or transmittance of the object, and the spectral response of the observer, as well as the illuminating and viewing geometry.

(*2*) *perceived*, attribute of visual perception that can be described by color names such as white, gray, black, yellow, brown, vivid red, deep reddish purple, or by combinations of such names.

(*3*) *colorimetric*, characteristics of a color stimulus denoted by a colorimetric specification with three values, such as tristimulus values. **E 284,** E12

Colorado steer—a side-branded steerhide, not necessarily from Colorado. **D 1517,** D31

color anodizing—*in anodizing aluminum*, formation of a colored coating on aluminum where the colored compound, pigment, or dye is incorporated after the coating has been formed. **B 374,** B08

colorant, *n*—dye, pigment, or other agent used to impart a color to a material. **E 284,** E12

colorant—a material used to alter the color of the tank mix. **E 1519,** E35

color atlas, *n*—a collection of color samples arranged according to a color order system. **E 284,** E12

color bleeding, *n*—the loss of color from a dyed fabric when immersed in water, drycleaning solvent, or similar liquid medium, with consequent coloring of the liquid medium. (Compare **color staining, crocking.**) **D 123,** D13

color bleed resistance—the freedom from intermixing of the inks on multiple-colored ribbons. **F 221,** F05

color blindness, *n*—total or partial inability to differentiate certain hues. **E 253,** E18

color constancy, *n*—the general tendency of the colors of an object to remain constant when the color of the illumination is changed. **E 284,** E12

color contrast, *n*—*in textiles*, a general term for a visible color difference between two adjacent areas. **D 123,** D13

color difference—(*1*) the magnitude and character of the difference between two colors, described by such terms as redder, bluer, lighter, darker, grayer, or cleaner. (*2*) the magnitude and direction of the difference between a sample and a standard, computed from tristimulus values, or chromaticity coordinates and luminance factor, by means of a specified set of color difference equations. **C 242,** C21

color difference, *n*—(*1*) *perceived*, the magnitude and character of the difference between two colors described by such terms as redder, bluer, lighter, darker, grayer, or cleaner.

(*2*) *computed*, the magnitude and direction of the difference between two psychophysical color stimuli and their components computed from tristimulus values, or chromaticity coordinates and luminance factor, by means of a specified set of color-difference equations. **E 284,** E12

color-difference units, *n*—units of size of the color differences calculated according to various equations. Such color differences *cannot* be accurately converted between different equations by the use of average factors. **E 284,** E12

color donor film—film support, usually thin plastic, coated or impregnated with a volatile dyestuff that can be sublimed off onto a target substrate. **F 1156,** F12

colored—See **anodized, blued, enameled, lacquered, painted.** **F 547,** F16

colored fiber, *n*—*in wool top*, any fiber the color or shade of which differs from the normal color or shade of the fiber mass of the sample. **D 123,** D13

colored frit—a frit containing a colorant in order to produce a strong color in the porcelain enamel. **C 286,** B08

colorfastness, *n*—the resistance of a material to change in any of its color characteristics, to transfer its colorant(s) to adjacent materials, or both, as the result of exposure of the material to any real or simulated environment that might be encountered during processing, storage, use or testing of the material. **D 123,** D13

color film—a photographic film consisting of several emulsion layers, where the individual layers selectively record various wavelengths of light. **E 7,** E04

color former—see **leuco dye**. **F 1623,** F05

colorfulness, n— see **chromaticness (2)**. **E 284,** E12

color grading, *n*—the act of identifying a specimen by a color grade or color score that is specific to the color and the material graded. **D 123,** D13

color grading, *n*—the act of identifying a specimen by a color grade or color score, which is specific to the color and the material graded. **E 284,** E12

colorimeter, *n*—an instrument used for color measurement based on optical comparison with standard colors. **D 1356,** D22

colorimeter, n— see **tristimulus colorimeter, visual colorimeter**. **E 284,** E12

colorimetric—See the preferred term, **spectrophotometry.** **C 859,** C26

colorimetric purity, p_c, *n*—the fraction of spectrally pure light in an additive mixture with reference achromatic (white) light to produce a color that matches that of the color stimulus considered. (As a reference achromatic light, the CIE recommends an equal energy source for self-luminous bodies and illuminant D_{65} (daylight) for nonself-luminous bodies.) **E 284,** E12

colorimetry—See the preferred term,**spectrophotometry.** **C 859,** C26

colorimetry, *n*—the science of color measurement. **E 284,** E12

coloring—(*1*) the production of desired colors on metal surfaces by appropriate chemical or electrochemical action.

(*2*) light buffing of metal surfaces for the purpose of producing a high luster. Called "color buffing." **B 374,** B08

color lamp, *n*—*in color determination of cotton with a Color Meter*, a lamp with a specific energy output function used in conjunction with special tristimulus filters to obtain a desired response function. **D 123,** D13

color match, *n*—(*1*) condition existing when colors match within a specified or agreed tolerance. Sometimes called *commercial color match.*

(*2*) condition existing when colors are indistinguishable; a normal observer is usually implied. Sometimes called an *exact color match.* **E 284,** E12

color matching, *n*—procedure for providing, by selection, formulation, adjustment, or other means, a trial color that is indistinguishable from, or within specified tolerances of, a specified standard color under specified conditions. **E 284,** E12

color-matching functions, *n*—the amounts, in any trichromatic system, of the three reference color stimuli needed to match by an additive mixture monochromatic components of an equal energy spectrum. **E 284,** E12

color measurement, *n*—process of deriving, by visual or instrumental means, a set of three numbers that describe the attributes of a color, in the form of a color notation or a colorimetric specification. **E 284,** E12

color meter, *n*—an instrument which measures the fiber sample color as presented in the viewing window, in terms of the tristimulus values *Y* and *Z* and transmits these values to the IC/TC for further processing. **D 123,** D13

color mixture, n— see **additive color mixture, additive color stimulus mixture,** or **subtractive color mixture**. **E 284,** E12

color notation, *n*—the symbols used in a systematic way to designate colors. **E 284,** E12

color of an object—the aspect of the appearance of an object dependent upon the spectral composition of the incident light, the spectral reflectance or transmittance of the object, and the spectral response of the observer.

hue, n—The attribute of color perception by means of which a color is judged to be red, orange, yellow, green, blue, purple, or imtermediate between adjacent pairs of these, considered in a close ring, red and purple being an adjacent pair. (White, gray and black colors possess no hue).

lightness, n—(*1*) The attribute by which a perceived color is judged to be equivalent to a member of a series of grays ranging from black to white. (*2*) The attribute of color perception by which a non-self-luminous body is judged to reflect more or less light.

saturation— attribute of a visual sensation that permits a judgment to be made of the proportion of pure chromatic color in the total sensation. **D 16,** D01

color order system, *n*—a rational method or plan of ordering and specifying all producible object or display colors, or all within a limited domain, by means of a set of physical standards selected and displayed so as to represent adequately the whole set of such colors under consideration. **E 284,** E12

color oxide—a material used to impart color to a porcelain enamel. **C 286,** B08

color perception, *n*—subjective impression of color, as modified by the conditions of observation and by mental interpretation of the stimulus object. **E 284,** E12

color preference, *n*—preference, within a specific application, for one color over other related colors. **E 284,** E12

color, psychophysical, *n*—characteristics of a color stimulus (that is, light producing a visible sensation that is a function of its wave length) denoted by three dimension values such as three tristimulus values. **D 2946,** C17

color rendering, *n*—effect of a light source on the color appearances of objects compared to their color appearances under a reference light source. (See also **CIE color rendering index**.) **E 284,** E12

color reversion—a process common to almost all bleached cellulose pulps in which the color darkens to a greater or lesser extent on standing. **D 1695,** D01

color scale, n— see **Gardner color scale, petroleum color scale, platinum-cobalt color scale, Saybolt color.** **E 284,** E12

color separations—process of (preparing printing plates by) photographing the subject image through color filters to produce (printing) plates that can be used to print the sequence of discrete colors used to create a multi-color printed image. **F 1156,** F12

color solid, n— see **color space**. **E 284,** E12

color space—a three dimensional arrangement for representing all possible colors; for example, in the color space defined by the color scales *L, a,* and *b* used to describe the color of opaque specimens, scale *L* is a measure of lightness, *a* is a measure of redness (plus) or greenness (minus), and *b* is a measure of yellowness (plus) or blueness (minus). **C 242,** C21

color space, *n*—*specific to this standard*, the daylight color of opaque specimens are represented by points in a space in terms of three color scales: reflectance, R_d, and the chromaticity coordinates for redness or greenness, $\pm a$, and yellowness or blueness, $\pm b$. **D 123,** D13

color space, *n*—a geometric space, usually of three dimensions, in which colors are arranged systematically. **E 284,** E12

color specification, *n*—notation or set of three color-scale values used to designate a color in a specified color system. Practical color specifications may include color tolerances as well as target color designation. **E 284,** E12

color stability, *n*—*in coated glass textiles*, the ability of the applied coating to resist fading from exposure to sunlight and water. **D 123,** D13

color stability—the ability of the applied coating color to resist fading from exposure to sunlight and water. **D 2825,** D21

color staining, *n*—the undesired pickup of color by a fabric: (*1*) when immersed in water, drycleaning solvent, or similar liquid medium, that contains dyestuffs or coloring material not intended for coloring the fabric, or (*2*) by direct contact with other dyed material from which color is transferred by bleeding or sublimation. (Compare **crocking** and **color bleeding.**) **D 123,** D13

color staining, *n*—the discoloration of a material by transfer of colorant from another material. **E 284,** E12

color standard—a plaque or other physical standard of established color value, against which standardization of an instrument is made. **C 242,** C21

color stimulus, *n*—a radiant flux capable of producing a color perception. **E 284,** E12

color stimulus function, $\phi(\lambda)$, *n*—description of a color stimulus by the spectral concentration of a radiometric quantity, such as radiance or radiant power, as a function of wavelength. **E 284,** E12

color temperature—the temperature of a blackbody in degrees Kelvin (K). In photography, the apparent temperature in K of a luminous source which may be measured by its emission ratio of blue to red light. **E 7,** E04

color temperature, *n*— *of a source*, the temperature, usually expressed in kelvins, of a full radiator that would emit light of the same chromaticity as the source. (See also **correlated color temperature, distribution temperature**.) **E 284,** E12

color temperature—temperature in degrees Kelvin (K) at which a black body must be operated to give a color equal to that of the source in question. **E 1142,** E37

color tolerance, *n*—the permissible color difference between sample and specified color. **E 284,** E12

color tolerance set, *n*—a group of colored standards, usually seven painted chips, arranged on a single card, one exhibiting a desired color, and two each exhibiting the limits of the permissible range of color variation in each of the color attributes. **E 284,** E12

color variation—property of nonuniform color exhibited by asbestos-cement products before or after weathering. **D 2946,** C17

column, *n*—a building member, usually structural and vertical, subjected to longitudinal (axial) compression and also to lateral forces such as bending. **E 631,** E06

columnar structure—a macro- or microstructure characterized by elongated grains whose long axes are parallel, for example, to solidification direction, electroplated direction, etc. **E 7,** E04

column, electron microscope—the assembly of gun, lenses, and specimen, viewing and plate chambers. **E 7,** E04

coma—a lens aberration occurring in that part of the image field that is some distance from the principal axis of the system. It results from different magnification in the various lens zones. Extra-axial object points appear as short comet-like images with the brighter small head toward the center of the field (positive coma) or away from the center (negative coma). **E 175,** E41

combed finish, *n*—units whose face surfaces are altered by more or less parallel scratches or scarfs in manufacture. **C 43,** C15

comber/brusher, *n*—an instrument which prepares the test beard of fibers for length, length uniformity, strength, and elongation measurements by combing the test specimen to remove loose or unclamped fibers and paralleling the individually clamped fibers, and by brushing the clamped fibers to remove fiber crimp and smooth the test beard of cotton. **D 123,** D13

comber leather—a steerhide leather, heavily stuffed and usually boarded, used in textile combing machines. **D 1517,** D31

combination change key—a key used in resetting a key change combination. **F 471,** F12

combination electrode—an electrochemical apparatus that incorporates an ion-selective electrode and a reference electrode in a single assembly thereby avoiding the need for a separate reference electrode. **D 4127,** D19

combination last (or shoe)—designed to provide a proper fit for the individual with thinner than normal instep or heel. Length and width will be of standard measurements, but narrower fitting qualities will prevail through instep, waist, and heel. **F 869,** F08

combination of features, n—See **feature**—*of a facility*. **E 1480,** E06

combination oven—equipment that combines the function of hot air convection or super-heated steam heating, or both, to perform steaming, baking, roasting, rethermalizing, and proofing of various food products. See **oven.** **F 1827,** F26

combination oven/steamer—See *combination oven.* **F 1827,** F26

combination stranded conductor—a conductor constructed with wires of different diameters with such differences occurring either within a single layer or from layer to layer. **B 354,** B01

combination tanned—formerly, tanned with a blend of vegetable extracts. Today, tanned with two or more types of tanning materials, such as chromium compounds and vegetable extracts, or chromium compounds and synthetic tannins. **D 1517,** D31

combination vacuum cleaner, *n*—a canister vacuum cleaner having a motorized nozzle separated from the cleaner housing but connected to it by means of a hose or hose and wand. **F 395,** F11

combine, *v*—*in sampling*, to join together two or more increments without prior division or comminution. **E 135,** E01

combined, *adj*—the application of two or more techniques to different samples at the same time. (ICTAC) **E 473,** E37

combined sewer—a pipeline intended to convey sewage and storm water. **C 822,** C13

combined water, *n*—the water chemically held, as water of crystallization, by the calcium sulfate dihydrate or hemihydrate crystal. **C 11,** C11

combing wool, *n*—wool that is strong and strictly of combing length, that is, 2 in. (50 mm) or more. **D 123,** D13

comb-rack—(*1*) a burning tool shaped like a comb used for supporting ware during firing.

(*2*) A comb-like tool for supporting ware during the metal pickling operation. **C 286,** B08

combustible, *adj*—capable of undergoing combustion. **E 176,** E05

combustible—that portion of the refuse-derived fuel sample which is consumed upon ignition exclusive of the moisture present in the sample. **E 856,** D34

combustible carbon, *n*—carbon content remaining in the solid products derived from the combustion or reaction of coal, coal by-products, or coke, exclusive of carbonate in any form. **D 121,** D05

combustible gases—flammable gases formed from breakdown (partial or complete) of some insulating materials subjected to electrical or thermal stress, or both. **D 2864,** D27

combustible liquid—a liquid having a flash point at or above 37.8°C (100°F). The flash point of a liquid having a viscosity less than 45 SUS at 37.8°C (100°F) and a flash point below 93.4°C (200°F) shall be determined in accordance with the Test Methods D 93, for Flash Point by the Pensky-Martens Closed Tester. **E 772,** E44

combustibles, *n*—the value obtained by subtracting the dry weight (in percent) of the ash (as determined in Test Method D 3174) from 100 % representing original weight of analyzed sample. **D 121,** D05

combustibles, n—the portion of a sample which is consumed by oxidation upon ignition and exclusive of the moisture present in the sample. **D 5681,** D34

combustible textile, *n*—a textile that will ignite and burn or that will give off vapors that will ignite and burn when subjected to external sources of ignition. (See **noncombustible textile**) **D 123,** D13

combustion, *n*—a chemical process of oxidation that occurs at a rate fast enough to produce heat and usually light either as glow or flames. **D 123,** D13

combustion, *n*—a chemical process of oxidation that occurs at a rate fast enough to produce temperature rise and usually light either as a glow or flame. (See also **glow** and **smoldering.**) **E 176,** E05

combustion, *n*—a chemical process of oxidation that occurs at a rate fast enough to produce heat and usually light either as glow or flames. **F 1494,** F23

combustion products—airborne effluent from a material undergoing combustion; this may also include pyrolysates. **E 176,** E05

combustion system downdrafting, *n*—the reversal of the ordinary (upward) direction of air flow in a combustion system when vented combustion appliances are not operating. **D 1356,** D22

combustion system spillage, *n*—entry of combustion products into a building, caused by backdrafting, vent blockage, or a leaky heat exchanger. **D 1356,** D22

comeback—the time required for a box furnace to return to temperature after the introduction of a load of ware. **C 286,** B08

comet, *n*—extraneous ink deposited in the shape of a round dot with a trailing tail. **D 6488,** D01

comforter, *n*—a bedcovering assembly, consisting of an insulating filler secured between two layers of fabric, used primarily to reduce heat loss. **D 123,** D13

Comite Consultatif International Telegraphique et Telephonique (Consultive Committee for International Telephone and Telegraph), *n*—a committee established under the United Nations to recommend worldwide communication usage standards. (See **CCITT**.) **F 1457,** F05

command, *n*—a byte or sequence of bytes from the host computer which invokes options available with the laser printer. **F 1457,** F05

commercial allowance, *n*—an arbitrary value equal to the commercial moisture regain plus a specified allowance for finish, used with the mass of scoured, oven-dried yarn, to compute (1) yarn linear density, (2) the commercial or legal mass of a shipment or delivery of any specific textile material (see also **commercial moisture regain**) or (3) the mass of a specific component in the analysis of fiber blends. **D 123,** D13

commercial composition, *n*—*in wool*, the percentages by weight of wool base, moisture, and other non-wool base components in wool to which a specific commercial designation is applied. **D 123,** D13

commercial designation, *n*—*in wool*, a term applied to a lot of wool in a stated form, and having a specified commercial composition. **D 123,** D13

commercial laundering—a process by which textile products or specimens may be washed, bleached, rinsed, dried, and pressed typically at higher temperatures, higher pH and longer times than used for home laundering. **D 123,** D13

commercial mass, *n*—billed mass as determined by a generally accepted method or as agreed upon between the purchaser and seller. **D 123,** D13

commercial moisture content, *n*—*for wool*, the moisture calculated as a percentage of the weight of the wool, top, noils, yarn, fabric, etc., in the "as-is" condition; that is, containing whatever moisture, oil, grease, or other extraneous matter that may be present. **D 123,** D13

commercial moisture regain, *n*—a formally adopted arbitrary value, to be used with the oven-dried mass of textile fibers, when calculating the commercial mass of a shipment or delivery. **D 123,** D13

commercial moisture regain—see **moisture regain, commercial.** **D 1695,** D01

commercial sterility—*of thermally processed food*, the condition achieved by application of heat, sufficient, alone or in combination with other appropriate treatments, to render the food free of microorganisms capable of growing in the food at normal nonrefrigerated conditions at which the food is likely to be held during distribution and storage. **F 17,** F02

commercial vacuum cleaner, *n*—a vacuum cleaner suitable for the heavy-duty and sometimes continuous cleaning tasks encountered in establishments such as hotels, motels, office buildings, churches, clubs, etc. **F 395,** F11

commercial vehicle, *n*—a vehicle with heavy duty chassis and suspension designed for commercial freight haulage. **E 867,** E17

commingling—mixing of fasteners from different lots that are of the same grade and size in the same container. **F 1789,** F16

comminution—the act or process of reduction of particle size with attendant increase in surface area and population of particles, usually but not necessarily by grinding, milling, or pulverizing. **C 242,** C21

comminution, *n*—the act or process of reduction in particle size, usually but not necessarily by grinding or milling. **C 1145,** C28

comminution, *n*—*in sample preparation,* processes which improve the microscale homogeneity of a sample by reducing the particle size. **E 135,** E01

common area—room or area that is accessible to all tenants in a building or development (for example, hallway, vestibule, laundry area). **E 1605,** E06

common brad—See **brad.** **F 547,** F16

common cause—a source of variation that affects all the individual values of the process output being studied; in control chart, it appears as part of the random process variation. **F 1789,** F16

common nail, common wire nail—bright, plain-shank, regular-stock-steel, ½ by 0.035 to 6 by 0.262-in. nails with flat 11⁄64 to 17⁄32-in. head and medium diamond point. Diameter is larger than that for sinkers, coolers, corkers, and box nails of same length. **F 547,** F16

communicating pores— see **interconnected porosity.** **B 243,** B09

communication—*in grouting*, subsurface movement of grout from an injection hole to another hole or opening. **D 653,** D18

communication resource—an entity responsible for implementation of direct medical direction and/or entities responsible for response and scene two-way communication (a. k. a. medical control resource). **F 1177,** F30

comonomer, *n*—one of the two or more monomer species that polymerize to form a copolymer. **D 1566,** D11

compact—an object produced by the compression of metal powder, generally while confined in a die, with or without the inclusion of nonmetallic constituents. Synonymous with **briquet.** **B 243,** B09

compact—to treat glass in a manner, such as by heat treatment, to approach maximum density. **C 162,** C14

compacted backfill, *n*— *in pipe laying*, backfill material which has been compacted to the density specified by the engineer. **C 1154,** C17

compacted backfill, *n*— *in pipe laying*, backfill material that has been compacted to the density specified by the engineer. **D 2946,** C17

compacted graphite iron, *n*—a cast iron that has been treated in the liquid state so as to cause its graphitic carbon to occur in the compacted graphite shape in the as-cast condition. (See **graphite, compacted** and **graphite, spheroidal**.) **A 644,** A04

compactibility, *n*—a conceptual term, encompassing the *powder* characteristics of compressibility, green strength, edge retention, and lamination tendency, that relates to the ability of a powder to be consolidated into a usable green compact. **B 243,** B09

compactibility—measure of a boom's storage volume per unit length

(m^3/m). **F 818,** F20

compacting—See **debulking**. **E 1749,** E06

compacting tool set— see **die.** **B 243,** B09

compaction—in ceramics or powder metallurgy, the preparation of a compact or object produced by the compression of a powder, generally while confined in a die, with or without the inclusion of lubricants, binders, etc., and with or without the concurrent applications of heat. **C 859,** C26

compaction—mechanical or hydraulic consolidation of backfill to achieve stability. **C 896,** C04

compaction—the densification of a soil by means of mechanical manipulation. **D 653,** D18

compaction—in crossflow filtration, the result of applied pressure and temperature compressing a polymeric membrane which may result in a decline in flux. **D 6161,** D19

compaction curve (Proctor curve) (moisture-density curve)—the curve showing the relationship between the dry unit weight (density) and the water content of a soil for a given compactive effort. **D 653,** D18

compaction ratio, *n*—the measurement of the relationship of volume displacement of a package before and after simulated landfill conditions as determined in standardized tests. **D 5488,** D10

compaction, soil—act of packing soil with mechanical force to increase its density. **F 412,** F17

compaction test (moisture-density test)—a laboratory compacting procedure whereby a soil at a known water content is placed in a specified manner into a mold of given dimensions, subjected to a compactive effort of controlled magnitude, and the resulting unit weight determined. The procedure is repeated for various water contents sufficient to establish a relation between water content and unit weight. **D 653,** D18

compact stranded conductor—a unidirectional or unilay or conventional concentric conductor, constructed with a central core surrounded by one or more layers of helically laid wires and formed into final shape by rolling, drawing, or other means; and manufactured to a specified nominal diameter, approximately 8 to 10 % below the nominal diameter of a conventional noncompact conductor of the same cross-sectional area. **B 354,** B01

comparative measurements—in electromagnetic testing, measurements made in which the unbalance in the system is measured using comparator coils in contrast to differential and absolute measurements. (See also **comparator coils**). **E 1316,** E07

comparative readout—in electromagnetic testing, the signal output of comparator coils. (See also **comparator coils**). **E 1316,** E07

comparative system—a system that uses coil assemblies and associated electronics to detect any electric or magnetic condition, or both, that is not common to the test specimen and the standard (see **comparator coils**). (E 566) **E 1316,** E07

comparator coils—in electromagnetic testing, two or more coils electrically connected in series opposition but arranged so that there is no mutual induction (coupling) between them such that any electric or magnetic condition, or both, that is not common to the test specimen and the standard, will produce an unbalance in the system and thereby yield an indication. **E 1316,** E07

comparator, spectral—an instrument for the inspection and wavelength measurement of spectrograms. (E 135, E-2) **C 859,** C26

comparison sample, *n*—(fire debris) 1) a sample of material collected from a fire scene which is, to the best of the investigator's knowledge, identical in every respect to a sample suspected of containing ignitable substance, but which does not contain ignitable substance. 2) a sample of suspected ignitable substance submitted for the purpose of comparing with any ignitable substance separated from a debris sample. **E 1732,** E30

comparison standard—a standard micrograph or a series of micrographs, usually taken at 75 or 100 diameters, or a suitable equivalent built into the eyepiece and used to determine grain size by direct comparison with the image. **E 7,** E04

compatibility, *n—in building construction*, the capability of two or more materials to be placed in contact or close proximity with one another and each material maintaining its usual physical or chemical properties, or both. **C 717,** C24

compatibility—the ability of various components or an aerosol formulation to be used together without undesirable physical or chemical results. **D 3064,** D10

compatibility agent—a surface-active agent that allows simultaneous application of liquid fertilizer and agrichemical, or two or more agrichemical formulations, as a uniform tank mix, or improves the homogeneity of the mixture and the uniformity of the application. **E 1519,** E35

compatible—nonstaining to substrate and sealant and nondetrimental to sealant performance. **C 717,** C24

compatible—(*1*) a condition wherein components of a plastic piping system or different specific plastic materials, or both, can be joined together for satisfactory joints. (*2*) in relation to elastomeric seal joints, a condition wherein the elastomer does not adversely affect the pertinent properties of the plastic pipe or fittings, or both, when the sealing gasket is in intimate contact with the plastic for a prolonged period. **F 412,** F17

compatible—not injurious to or changing the physical or electrical characteristics of the blankets or affecting their application, use, or acceptability. **F 819,** F18

compatible materials, *n—in building construction*, compounds or substances that can exist in close proximity to one another without detrimental effects on either. **C 717,** C24

compensating extension wires, *n*—those extension wires fabricated from materials basically different in composition from the thermocouple. **E 344,** E20

compensating eyepiece—one designed for use with objectives such as apochromats, the lateral chromatic aberration of which is undercorrected. **E 7,** E04

compensating eyepieces—those designed for use with objectives such as apochromats in order to correct chromatic aberration. **E 175,** E41

compensating lead wires—wires leading from a thermocouple to the voltage-measuring instrument. These wires must be of such compositions that they will generate an *emf* equivalent to the *emf* generated by the reference junction of the couple. **E 7,** E04

compensating thermocouple, *n*—a thermocouple for the purpose of generating an electrical signal representing long-term changes in the stack metal temperatures wherein a fraction of the signal generated is subtracted from the signal developed by the stack-gas thermocouples. **E 176,** E05

compensation—reduction in number of free carriers resulting from the presence of impurities other than the majority dopant density impurity. **F 1241,** F01

complementary color pair—in color theory, two discrete colors opposite one another on the color chart, such as blue and orange or red and green (additive theory). **F 1156,** F12

complementary color percepts, *n*—(*1*) pairs of color percepts, one of which is induced by the other through simultaneous contrast; (*2*) pairs of color percepts, one of which is the negative after-image of the other. **E 284,** E12

complementary colors, *n*—color stimuli that produce a specified achromatic stimulus when they are suitably mixed in an additive manner. **E 284,** E12

complementary color stimuli, *n*—pairs of color stimuli that, by additive mixture produce an achromatic stimulus. **E 284,** E12

complementary wavelength, *n*—the wavelength of a spectrally pure light that when added to the light reflected or transmitted by the specimen will produce a combination that color matches a reference achromatic (white) light. **E 284,** E12

complete immersion thermometer, *n*—a liquid-in-glass thermometer designed to indicate temperatures correctly when the entire thermometer is exposed to the temperature being measured. (Compare **total immersion thermometer** and **partial immersion thermometer.**) **E 344,** E20

completely alloyed powder, *n*—see **pre-alloyed powder**. **B 243,** B09

completely randomized design, *n*—a design in which the treatments are assigned at random to the full set of experimental units. **E 456,** E11

completely randomized design, *n*—a design in which the treatments are assigned at random to the full set of experimental units. **E 1325,** E11

completely randomized factorial design, *n*—a factorial experiment (including all replications) run in a completely randomized design. **E 456,** E11

completely randomized factorial design, *n*—a factorial experiment (including all replications) run in a completely randomized design. **E 1325,** E11

complexing agent—a compound that will combine with metallic ions to form complex ions. See **complex ion.** **B 374,** B08

complex ion—an ion composed of two or more ions or radicals, both of which are capable of independent existence, for example, cuprocyanide $(Cu(CN)_3)^{=}$. **B 374,** B08

complex machine stitch pattern, *n—in home sewing,* a machine stitch pattern formed when two or more simple machine stitch patterns are combined in one repeating unit. (Compare simple machine stitch pattern.) **D 5646,** D13

complex modulus, *E, *G**, or *K****—ratio of the stress to strain where each is a factor that may be represented by a complex number as follows: $E^* = E' + iE''$, $G^* = G' + iG''$, and $K^* = K' + iK''$.

where:
E* = complex modulus, measured in tension or flexure,
E′ = storage modulus, measured in tension or flexure,
E″ = loss modulus, measured in tension or flexure,

G* = complex modulus, measured in shear,
G′ = storage modulus, measured in shear,
G″ = loss modulus, measured in shear,

K* = complex modulus, measured in compression,
K′ = storage modulus, measured in compression
K″ = loss modulus, measured in compression, and
i = $\sqrt{-1}$, measured in compression.

The complex modulus may be measured in tension or flexure, (*E**), compression, (*K**), or in shear, (*G**), (D 4092, D-20). **E 1142,** E37

complex modulus, *E or *G****—the ratio of the stress to strain where each is a vector that may be represented by a complex number.

$$E^* = E' + iE''$$
$$G^* = G' + iG''$$
$$K^* = K' + iK''$$

where:
E* = complex modulus, measured in tension or flexure,
E′ = storage modulus, measured in tension or flexure,
E″ = loss modulus, measured in tension or flexure,

G* = complex modulus, measured in shear,
G′ = storage modulus, measured in shear,
G″ = loss modulus, measured in shear,

K* = complex modulus, measured in compression,
K′ = storage modulus, measured in compression
K″ = loss modulus, measured in compression, and
i = $\sqrt{-1}$, measured in compression.

D 4092, D20

complexometric titration—a titration in which the titrant and sample ion form a complex. Many classic complexometric titrations can conveniently be followed by electrode. The success of the titration depends on the lower limit of detection of the electrode, the strength of the complex, and the possible presence of ions that compete for the complexing agent or which interfere with the electrode. Examples of complexometric titrations are the determination of calcium with EDTA, and acid-base titrations. **D 4127,** D19

complex radiation, *n*—radiation composed of a number of monochromatic radiations. **E 349,** E21

complex seam, *n—, in home sewing,*a seam made in two or more steps. (Ant. **plain seam.**) **D 123,** D13

complex shear compliance, *J**—the reciprocal of complex shear modulus.

$$J^* = \frac{1}{G^*}$$

D 4092, D20

complex shear compliance, *J**—reciprocal of complex shear modulus, where $J^* = 1/G^*$, (D 4092, D-20). **E 1142,** E37

complex silicate inclusions—a general term describing silicate inclusions containing visible constituents besides the silicate matrix. An example would be corundum or spinel crystals occurring in a silicate matrix in steel. **E 7,** E04

complex tensile compliance, *D**—the reciprocal of complex tensile modulus.

$$D^* = \frac{1}{E^*}$$

D 4092, D20

complex tensile compliance, *D**—reciprocal of complex tensile modulus, where $D^* = 1/E^*$, (D 4092, D-20). **E 1142,** E37

compliance, *C* (Pa $^{-1}$)—the quotient of strain and stress.

$$C = \frac{c}{\sigma} = \frac{1}{M}$$

D is the tensile compliance; *J* is the shear compliance; *B* is the bulk compliance; *O* is the longitudinal compression compliance. **D 4092,** D20

compliance indicator, *n*—a specified string of three characters indicating that the message which follows conforms to the requirements of a particular standard. **F 1294,** F05

compliance (LF^{-1}], *n*— the ratio of displacement increment to load increment. **E 1823,** E08

component—an individual piece or a complete assembly of individual pieces. **B 713,** B01

component, *n—as used with textile fiber polymers,* a polymer with distinguishable properties. **D 123,** D13

component—one of the independently variable substances by means of which the composition of each phase of a system of heterogeneous equilibrium may be described completely; usually an element, or a compound that remains undissociated throughout the range of temperature and pressure concerned. **E 7,** E04

component—See **building component.** **E 631,** E06

component—See **building component.** **E 1480,** E06

component, *n*— any single element used in an assembly. **F 1582,** F04

component of variance, *n*—a part of a total variance identified with a specified source of variability. **D 123,** D13

component of variance, *n*—a part of a total variance identified with a specified source of variability. **E 456,** E11

components, *n—for pile yarn floor covering*, the individual yarn or fabric elements into which a pile yarn floor covering can be dissected. **D 123,** D13

components pattern—process of identifying specific building components containing LBP at a hazardous level within a building or group of buildings. **E 1605,** E06

composite, *n*—a solid product consisting of two or more distinct phases, including a binding material (matrix) and a particulate or fibrous material. **D 883,** D20

composite:

composite material—a substance consisting of two or more materials, insoluble in one another, which are combined to form a useful engineering material possessing certain properties not possessed by the constituents. **D 3878,** D30

composite coating—a coating consisting of deposits incorporating particles of another material. See also **dispersion coating**. **B 374,** B08

composite compact—a metal powder compact consisting of two or more adhering layers, rings, or other shapes of different metals or alloys with each material retaining its original identity. **B 243,** B09

composite conductor—a conductor consisting of two or more types of wire, each type of wire being plain, clad, or coated; stranded together to operate mechanically and electrically as a single conductor. **B 354,** B01

composite conductor—a conductor consisting of two or more types of material, each type of material being plain, clad, or coated, and assembled together to operate mechanically and electrically as a single conductor. **B 713,** B01

composite design, *n*—a design developed specifically for fitting second order response surfaces to study curvature, constructed by adding further selected treatments to those obtained from a 2^n factorial (or its fraction). **E 456,** E11

composite design, *n*—a design developed specifically for fitting second order response surfaces to study curvature, constructed by adding further selected treatments to those obtained from a 2^n factorial (or its fraction). **E 1325,** E11

composite, filamentary—a major form of advanced composites in which the fiber constituent consists of continuous filaments. **E 1749,** E06

composite lot sample—a single sample prepared by combining samples from several containers in the same ratio as the quantities in the containers sampled. **C 859,** C26

composite material, *n*—structured combination of two or more discrete materials. **E 176,** E05

composite material—a material consisting of any combination of high-strength, high-modulus fibers, whiskers, or particles in a homogeneous matrix. **E 1749,** E06

composite membrane—a membrane having two or more layers with different physical or chemical properties. Membrane manufactured by forming a thin desalinating barrier layer on a porous carrier membrane. **D 6161,** D19

composite pipe—pipe consisting of two or more different materials arranged with specific functional purpose to serve as pipe. **F 412,** F17

composite plate—an electrodeposit consisting of two or more layers of metal deposited successively. **B 374,** B08

composite powder—a powder in which each particle consists of two or more distinct constituents. **B 243,** B09

composite sample—a combination of two or more samples. **D 1129,** D19

composite sample—a portion of leather, which may be the scraps from the cuttings of physical test specimens, that has been taken from each of the sample units constituting the sample. The leather is composited as specified for the purpose of testing a lot for chemical properties. **D 1517,** D31

composite sample, *n— in asbestos,* a set of unit samples of asbestos fiber (drawn systematically or at random) taken from a lot, comprising not less than two and not more than 200 bags, for use in the laboratory as a test sample, that is, as a source of test specimens. **D 2946,** C17

composite sample—a sample formed by combining two or more individual samples or representative portions of the samples. **D 4410,** D19

composite superconductor—a conductor incorporating superconductive material. **B 713,** B01

composite tube—See **tube.** **D 996,** D10

composite viewing—the viewing of two or more superimposed radiographs from a multiple film exposure. **E 1316,** E07

composition—the quantity of each of the components of a mixture; usually expressed in terms of the weight percentage, or the atomic percentage of each of the components in the mixture. **B 713,** B01

composition—the quantity of each of the components of a mixture; usually expressed in terms of the weight percentage, or the atomic percentage of each of the components in the mixture. **E 7,** E04

composition—quantity of the components of a mixture; usually expressed in terms of the weight percentage, or the atomic percentage of each of the components in the mixture, (E 7, E-4). **E 1142,** E37

composition—materials composed of granulated fillers, such as cork, leather, fibers, minerals, in a resinous matrix, usually an elastomer. Compressed and molded into sheet materials, compositions are used for insoles, heel bases, etc. **F 869,** F08

compositional depth profile—the chemical composition and the atomic concentration measured as a function of distance from the surface. **E 673,** E42

compositional inhomogeneity, *n—as used in fractography,* a microstructural irregularity related to the nonuniform distribution of an additive, a different crystalline or glass phase or in a multiphase material the nonuniform distribution of a second phase. **C 1145,** C28

compost—the product of composting. **D 883,** D20

compostable, *adj*—capable of undergoing biological decomposition in a compost site as part of an available program, such that the material (that is, feedstock) is not visually distinquishable and breaks down to carbon dioxide, water, inorganic compounds, and biomass, at a rate consistent with known compostable materials. **D 5488,** D10

compostable plastic—a plastic that undergoes biological degradation during composting to yield carbon dioxide, water, inorganic compounds, and biomass at a rate consistent with other known compostable materials and leaves no visually distinguishable or toxic residues. **D 883,** D20

compound, *n*—an intimate mixture of all the ingredients necessary for a finished material or product. **C 717,** C24

compound, *n*—an intimate admixture of (a) polymer(s) with all the materials necessary for the finished product. **D 883,** D20

compound, *n*—an intimate admixture of a polymer(s) with all the materials necessary for the finished article. **D 1566,** D11

compound, *n*—a mixture of a polymer with other ingredients such as fillers, stabilizers, catalysts, processing aids, lubricants, modifiers, pigments, or curing agents. **F 412,** F17

compound—a mixture of a polymer with other ingredients such as fillers, stabilizers, catalysts, processing aids, lubricants, modifiers, pigments, or curing agents. (F 412, F-17) **F 869,** F08

compound compact—a metal powder compact consisting of mixed metals, the particles of which are joined by pressing or sintering or both, with each metal particle retaining substantially its original composition. **B 243,** B09

compounding ingredient, n—See **compounding material**, the preferred term. **D 1566,** D11

compounding material, *n*—a substance used as part of a rubber mix. **D 1566,** D11

compound, standard, *n*—a control or reference compound prepared according to a prescribed formula and mixing procedure. **D 1566,** D11

compound term, *n*—a term that contains more than one word in the natural spoken word order. **E 1992,** E02

compreg—synthetic resin-treated, compressed wood with reduced swelling and shrinking characteristics and increased density and strength properties. **D 1038,** D07

compressed air sprayer—a spraying apparatus that uses air pressure to move liquid from a container through an atomizing device. **E 1102,** E35

compressed natural gas (CNG)—natural gas that is typically pressurized to 3600 psi. CNG is primarily used as a vehicular fuel. **D 4150,** D03

compressed stranded conductor—a unidirectional or unilay or conventional concentric conductor manufactured to a specified nominal diameter 3 % less than the calculated diameter of noncompressed conductor of the same construction and cross-sectional area. **B 354,** B01

compressed thickness (t, (L), mm), *n*—thickness under a specified stress applied normal to the material. **D 4439,** D35

compressibility, *n*—the capacity of a metal powder to be densified under a uniaxially applied pressure in a closed die. **B 243,** B09

compressibility—property of a soil or rock pertaining to its susceptibility to decrease in volume when subjected to load. **D 653,** D18

compressible-washer-type direct tension indicator—a direct tension indicator having the capability of indicating the achievement of a required minimum bolt tension by the degree of its plastic deformation. **F 1789,** F16

compression, *n*—the act, process, or result of compacting, condensing, or concentrating. **D 4848,** D13

compressional wave—see **longitudinal wave.** **E 1316,** E07

compression coupling—See **compression joint.** **C 896,** C04

compression curve—see **pressure-void ratio curve**. **D 653,** D18

compression deflection—the amount of deformation of material when being compressed. **F 869,** F08

compression disk—a disk of compressible material placed between the ends of adjacent pipe for the purpose of distributing the jacking force. **C 896,** C04

compression failure—deformation of the wood fibers resulting from excessive compression along the grain either in direct end compression or in bending. It may develop in standing trees due to bending by wind or snow or to internal longitudinal stresses developed in growth, or it may result from stresses imposed after the tree is cut. In surfaced lumber, compression failures appear as fine wrinkles across the face of the piece. **D 9,** D07

compression fitting joint—see **joint, compression fitting.** **F 412,** F17

compression gasket, *n*—a gasket designed to be used under compression. **C 717,** C24

compression gasket joint—see **joint, compression gasket.** **F 412,** F17

compression glazing, n—see **compression glazing system**. **C 717,** C24

compression glazing, *v*—*in building construction*, the act of installing components into a compression glazing system. **C 717,** C24

compression glazing system, *n*—*in building construction*, gasket(s) or a combination of sealant(s) and gasket(s) used to provide weather-tightness by way of continuous pressure on the components in the joint. **C 717,** C24

compression index, C_c (D)—the slope of the linear portion of the pressure-void ratio curve on a semi-log plot. **D 653,** D18

compression joint—a joint designed so that a sealing action is obtained by compressing elastomeric components. **C 896,** C04

compression load—a load which tends to compress or shorten the member. The value for compressive strength may depend upon the degree of distortion. **F 1789,** F16

compression mold—a mold which is open when the material is introduced and which shapes the material by heat and by the pressure of closing. **F 869,** F08

compression molded button, *n*—a button or button blank which is produced by compression molding. **D 123,** D13

compression molded button, *n*—a button or button blank which is produced by compression molding.thermoset-molding compounds such as urea-formaldehyde, melamine-formaldehyde, styrene-modified polyester, or any combination thereof. This method using styrene modified polyester resin, and having in its formulation pearlescent pigments which are oriented in the molding process, form a button or button blank which resembles natural shell. **D 5497,** D13

compression molding, *n*—the method of molding a material already in a confined cavity by applying pressure and usually heat. **D 123,** D13

compression molding—the method of molding a material already in a confined cavity by applying pressure and usually heat. **D 883,** D20

compression molding, *n*—molding process in which the material is placed directly in the mold cavity and compressed to shape by closure of the mold. **D 1566,** D11

compression molding, *n*—the method of molding a material already in a confined cavity by applying pressure and usually heat. **D 5497,** D13

compression molding—the method of molding a material in a confined cavity by applying pressure and usually heat. (D-20) **F 412,** F17

compression molding, *n*—a process for molding a material in a confined cavity by applying pressure and usually heat. **F 1251,** F04

compression parallel to grain—compression, endwise (parallel to the grain). The imposition of a compressive stress that acts in a direction parallel to the grain of the wood, as in a column. **D 9,** D07

compression perpendicular to grain—compression, sidewise (perpendicular to the grain). The imposition of a compressive stress that acts in a direction approximately perpendicular to the grain of the wood. **D 9,** D07

compression ratio—the ratio of the volume of the loose powder to the volume of the compact made from it. **B 243,** B09

compression recovery, *n*—the degree to which a material returns to its original dimension(s) after removal of a compressive force. **D 4848,** D13

compression seal, *n*—a seal which is attained by a compressive force on the sealing material. **C 717,** C24

compression set, *n*—the residual deformation of a material after removal of the compressive stress. **D 1566,** D11

compression test—a test made on a concrete specimen to determine the compressive strength. **C 822,** C13

compression wave (irrotational)—wave in which element of medium changes volume without rotation. **D 653,** D18

compression wood—abnormal wood formed typically on the lower side of branches and inclined trunks of softwood trees. Compression wood is identified by its relatively wide annual rings, usually eccentric, relatively large amount of latewood, sometimes more than 50 % of the width of the annual rings in which it occurs, and its lack of demarcation between earlywood and latewood in the same annual rings. Compression wood shrinks excessively lengthwise, as compared with normal wood. **D 9,** D07

compressive force, *n*—the perpendicular force applied to surface(s) of a material in compaction. **D 4848,** D13

compressive strength, *n*—the maximum load sustained by a standard specimen of a material when subjected to a crushing force. **C 11,** C11

compressive strength—a property of solid material that indicates its ability to withstand a uniaxial compressive load. **C 709,** D02

compressive strength—the maximum resistance of a concrete specimen to axial compressive loading; or the specified resistance used in design calculations. **C 822,** C13

compressive strength—the maximum stress that a specimen or material will support when subjected to a crushing force applied at a specified rate. **C 904,** C03

compressive strength, *n*—the maximum compressive load which a specimen will support divided by the cross sectional area of the specimen. **C 1180,** C12

compressive strength—See **strength, compressive.** **E 1749,** E06

compressive strength [FL^{-2}], *n*—the maximum compressive stress which a material is capable of sustaining. Compressive strength is calculated from the maximum load during a compression test and the original cross-sectional area of the specimen. **E 6,** E28

compressive strength (unconfined or uniaxial compressive strength), p_c, q_u, C_o **(FL^{-2})**—the load per unit area at which an unconfined cylindrical specimen of soil or rock will fail in a simple compression test. Commonly the failure load is the maximum that the specimen can withstand in the test. **D 653,** D18

compressive stress—normal stress tending to shorten the body in the direction in which it acts. (ISRM) **D 653,** D18

compressive stress [FL^{-2}], *n*—normal stress due to forces directed toward the plane on which they act. **E 6,** E28

Compton edge (E_c)—the maximum energy value of electrons of the Compton scattering continuum. The energy value of the Compton edge is

$$E_c = E_\gamma - \frac{E_\gamma}{1 + \dfrac{2E_\gamma}{0.511}}$$

which corresponds to 180° scattering of the primary photon of energy E_γ (MeV). For a 1 MeV photon, the Compton edge is about 0.8 MeV. **E 170,** E10

Compton scattering—X-ray scattering by atoms in which the scattered beam has, relative to the incident beam, a longer wavelength and a random phase relationship. Also called incoherent or modified scattering. **E 7,** E04

Compton scattering—elastic scattering of a photon by an atomic electron, under the condition of conservation of momentum, that is, the vector sum of the momenta of the outgoing electron and photon is equal to the momentum of the incident photon. The scattered photon energy, E'_γ, is given by

$$E'_\gamma = \frac{E_\gamma \; E_\gamma (1 - \cos\theta)}{1 + \quad 0.511}$$

where E_γ is the incident photon energy in MeV and θ is the angle between the direction of the primary and scattered photon. The electron energy, E_e, is equal to $E_\gamma - E'_\gamma$. **E 170,** E10

compton scatter radiation—the scattered X-ray or gamma ray which results from the inelastic scattering of an incident X-ray or gamma ray on an electron. Since the ejected electron has short range in most materials, it is not considered part of the scattered radiation. **E 1316,** E07

computed radiology (photo stimulated luminescence method)—a two-step radiological imaging process; first, a storage phosphor imaging plate is exposed to penetrating radiation; second, the luminescence from the plate's photostimulable luminescent phosphor is detected, digitized, and presented via hard copy or a CRT. **E 1316,** E07

computer generated seed—random number arbitrarily generated by computer and used to create a string of random digits; especially useful in cryptography. **F 1156,** F12

Conbur test—See **package testing.** **D 996,** D10

concatenation, *n*—the linking or chaining together of either separate items of data in a bar code symbol, or of the data contained in two or more separate bar code symbols (also referred to as "message append"). **F 1294,** F05

concave surface—a local concave curvature in the flat surfaces of pultruded plastic parts as measured transversely to their length. **D 3918,** D20

concealed membrane waterproofing, *n*—also referred to as **structural slab waterproofing**; *(1)* for below grade: refers to a form of waterproofing where the membrane is applied to the mud mat and later covered with a topping, usually concrete, to act as a wearing layer or course, and *(2)* for elevated structural concrete deck: applied over the structural surface and covered/concealed by other components such as a topping slab, pavers, ballast, pavement, and plantings. **D 1079,** D08

concentrate, *n*—the froth product recovered in coal froth flotation. **D 121,** D05

concentrate—the product mix to which the propellant is added. **D 3064,** D10

concentrate—the stream exiting a crossflow membrane device which has increased concentration of solutes and particles over the feed stream. Portion of the feed stream which does not pass through the membrane. The stream in which dissolved solids or particulates, or both, are concentrated in a membrane separation process.

D 6161, D19

concentrate recycle—a technique for improving recovery in which a fraction of the concentrate is recycled through the membrane system. **D 6161,** D19

concentrate (reverse osmosis)—the residual portion of an aqueous solution applied to a membrane. **D 1129,** D19

concentrating collector— see **collector, concentrating.** **E 772,** E44

concentration, *n*—the quantity of a substance contained in a total unit quantity of sample.

mass concentration, n—concentration expressed in terms of mass of substance per unit volume of gas or liquid.

ppb(v), n—a unit of measure of the concentration of gases in air expressed as parts of the gas per billion (10^9) parts of the air-gas mixture, both by volume.

ppm(v), n—a unit of measure of the concentration of gases in air expressed as parts of the gas per million parts of the air-gas mixture, both by volume.

vapor concentration, n—concentration expressed in terms of gaseous volume of substance per unit volume of air or other gas usually expressed in percent or parts per million by volume. See also **absolute humidity.**

volume concentration, n—concentration expressed in terms of gaseous volume of substance per unit volume of air or other gas usually expressed in percent or parts per million. **D 1356,** D22

concentration—the actual amount of a substance in a given volume of solution. When measuring ionic concentrations by electrode, a distinction is made between the concentration of the free, unbound ion, and total concentration that includes ions bound to complexing agents. **D 4127,** D19

concentration—amount of the active ingredient contained in the chemical formulation expressed as a percent or mass per relevant unit basis. **E 1102,** E35

concentration, *n*—quantity of substance in a unit quantity of sample. **E 1605,** E06

concentration—see **number density.** **E 1620,** E29

concentration—relative amount of a minority constituent of a mixture to the majority constituent (for example parts per million, parts per billion, or percent) by either volume or weight. **F 1241,** F01

concentration, *c*— the quantity of the substance contained in a unit quantity of sample. **E 131,** E13

concentration, (*c*), *n*—the quantity of a substance contained in a unit quantity of sample. **E 135,** E01

concentration cell—an electrolytic cell, the emf of which is caused by a difference in concentration of some component in the electrolyte. (This difference leads to the formation of discrete cathode and anode regions.) **G 15,** G01

concentration factor, CF—

$$CF = \frac{C_B \text{ (brinewater concentration)}}{C_F \text{ (feedwater concentration)}}$$

$$= \frac{1}{1 - \text{conversion}} \text{ (approximation)}$$

D 6161, D19

concentration factor, *n* **(D)**—a parameter used in modifying the Boussinesq equations to describe various distributions of vertical stress. **D 653,** D18

concentration index, *n*—the concentration of an element at which the intensities of the analytical and internal standard lines are equal. **E 135,** E01

concentration of sediment (by mass)—the ratio of the mass of dry sediment in a water-sediment mixture to the mass of the mixture. **D 4410,** D19

concentration polarization—that part of the total polarization that is caused by changes in the activity of the potential-determining components of the electrolyte. **B 374,** B08

concentration polarization—the increase of the solute concentration over the bulk feed solution which occurs in a thin boundary layer at the feed side of the membrane surface, resulting from the removal of the solvent. **D 6161,** D19

concentration range, *n—in an analytical method,* the concentrations within which a method has been tested and found suitable for use. **E 135,** E01

concentration ratio—in leak testing, the ratio of the number of atoms (molecules) of a given constituent of a (gas) mixture to the total number of atoms (molecules) in the mixture. For ideal gases the concentration ratio has the same value as the volume fraction or the partial pressure of the constituent. **E 1316,** E07

concentration ratio, geometric—the ratio of the collector aperture area to the absorber area. **E 772,** E44

concentration standard—a standardizing solution whose value is reported in terms of total concentration of the ion of interest. If the electrode is calibrated using pure-concentration standards and measurements made on untreated samples, results must be corrected for the sample ionic strength. More commonly, a reagent is added to all standards and samples before measurement in order to fix the ionic strength, thus avoiding the need for correction. **D 4127,** D19

concentration-time curve, *n*—a plot of the concentration of a gaseous toxicant as a function of time. **E 176,** E05

concentration (volume)—the ratio of the volume of dry sediment to the volume of the water-sediment mixture. **D 4410,** D19

concentrator—an optical device (lenses or mirrors) that, as part of a solar collector, receives the unconcentrated solar irradiance and redirects (concentrates) it to a smaller area (the receiver). **E 772,** E44

concentricity, *n*—the ratio, expressed in percent, of the minimum wall thickness to the maximum wall thickness. **D 1711,** D09

concentric-lay conductor—conductor constructed with a central core surrounded by one or more layers of helically laid wires. Several types are as follows:

compact round conductor—see **compact stranded conductor**.

conventional concentric conductor—conductor constructed with a round central core surrounded by one or more layers of helically laid round wires. The direction of lay is reversed in successive layers, and generally with an increase in length of lay for successive layers.

equilay conductor—conductor constructed with a central core surrounded by more than one layer of helically laid wires, all layers having a common length of lay, direction of lay being reversed in successive layers.

parallel core conductor—conductor constructed with a central core of parallel-laid wires surrounded by one layer of helically laid wires.

rope-lay conductor—conductor constructed of a bunch-stranded or a concentric-stranded member or members, as a central core, around which are laid one or more helical layers of such members.

unidirectional conductor—conductor constructed with a central core surrounded by more than one layer of helically laid wires, all layers having a common direction of lay, with increase in length of lay for each successive layer.

unilay conductor—conductor constructed with a central core surrounded by more than one layer of helically laid wires. All layers having a common direction of lay and essentially similar length of lay, which is a multiple of the outer diameter of the finished conductor, for each successive layer. The lay length of the inner layers shall not vary by more than ±10 % of the lay length of the outer layer. **B 354,** B01

concentric-lay conductor, *n*—a conductor composed of a central core surrounded by one or more layers of helically laid strands. **D 1711,** D09

conceptual model—a simplified representation of the hydrogeologic setting and the response of the flow system to stress. **D 653,** D18

concertina, *n*—a pattern formed by attaching adjacent loops of helical coils to one another at specified points on the circumference, resulting in an accordion-like configuration. **F 1379,** F14

concertina spacing, *n*—distance between adjacent concertina attachments in a concertina configuration, as measured down the length of the extended coil (also called *concertina attachment spacing*). **F 1379,** F14

concrete, *n*—a composite material that consists essentially of a binding medium within which are embedded particles or fragments of aggregate; in hydraulic-cement concrete, the binder is formed from a mixture of hydraulic cement and water. **C 125,** C09

concrete—a homogeneous mixture of portland cement, aggregates, and water which may contain admixtures. **C 822,** C13

concrete, *n*—a hard, strong material made by mixing a cementing material (commonly portland cement) and a mineral aggregate (as washed sand and gravel or broken rock) with sufficient water to cause the cement to set and bind the entire mass. **F 141,** F06

concrete brick, *n*—a concrete masonry unit made from portland cement, water, and suitable aggregates, with or without the inclusion of other materials. See Specification C 55. **C 1209,** C15

concrete nail, concrete stub nail—hardened-steel, ½ by 0.135 to 3 ½ by 0.207-in. nails with flat countersunk 5⁄16 to ½-in. head and medium diamond point. **F 547,** F16

concrete, refractory, *n*—a hardened castable. **C 71,** C08

concretion, *n*—*in a geological sense*, a mass of mineral matter found in rock of a composition different from its own and produced by deposition from aqueous solution in the rock. **D 121,** D05

concurrent analysis—an analysis performed at the same or nearly the same time, or in immediate sequence, often in relationship or association to another analysis. In analytical chemistry, a term used to describe a series of measurements in which reference material samples are analyzed in tandem with samples whose chemical composition or particular chemical property(ies) are being determined. **C 859,** C26

condensate, *n*—liquid or solid matter formed by condensation from the vapor phase. **D 1356,** D22

condensate—a mixture of condensed steam and cooling water, exiting cooking equipment and if applicable, directed to the floor drain. **F 1827,** F26

condensation, *n*—a chemical reaction in which two or more molecules combine with the separation of water or some other simple substance. (See also **polymerization.**) **D 907,** D14

condensation—the conversion of water vapor or other gas to liquid as the temperature drops or atmospheric pressures rises. (See also **dew point**). **D 1079,** D08

condensation, *n*—the process of converting a material in the gaseous phase to a liquid or solid state by decreasing temperature or by increasing pressure, or both. **D 1356,** D22

condensation polymer—a polymer made by condensation polymerization. **D 883,** D20

condensation polymer, *n*—polymerization in which during an acid/base reaction a small molecule is often split out. **F 1251,** F04

condensation polymerization—polymerization in which monomers are linked together with the splitting off of water or other simple molecules. **D 883,** D20

condensation sampling—See **sampling.** **D 1356,** D22

condensed system—a pure substance, or mixture, at a pressure and temperature such that the vapor phase does not exist. **E 7,** E04

condenser—a term applied to lenses or mirrors designed to collect, control, and concentrate radiation in an illumination system. **E 7,** E04

condenser, Abbe—originally a two-lens substage condenser combination designed by Ernst Abbe. It lacks chromatic correction though designed for a minimum of spherical aberration and has only a very low-angle aplanatic cone. It may be rated with a numerical aperture as high as 1.3. **E 175,** E41

condenser aperture— See **aperture.** **E 7,** E04

condenser, darkfield—a condenser forming a hollow cone of light with its apex (or focal point) in the plane of the specimen. When used with an objective having a numerical aperture lower than the minimum numerical aperture of the hollow cone, only light deviated by the specimen enters the objective. Objects are seen as bright images against a dark background. **E 175,** E41

condenser, darkfield, bispheric—a darkfield condenser consisting of a convex spherical reflector mounted concentric with a larger concave reflector. The rays are formed into a diverging cone by the convex reflector. The annular concave reflector then forms a hollow converging cone which is focused on the subject. See **condenser, darkfield.** **E 175,** E41

condenser, darkfield, paraboloid—a darkfield condenser consisting of a reflecting surface in the form of a segment of a paraboloid of revolution. Parallel rays entering the condenser around the periphery of the central stop are reflected from the curved surfaces and converge at the focus of the paraboloid. See **condenser, darkfield.** **E 175,** E41

condenser lens—a device used to focus radiation in or near the plane of the object. **E 7,** E04

condenser or condenser lens—a term applied to lenses or mirrors designed to collect, control, and concentrate radiation in an illumination system. **E 175,** E41

condenser, variable-focus—essentially an Abbe condenser in which the upper lens element is fixed and the lower movable. The lower lens may be used to focus the illumination between the elements so that it emerges from the stationary lens as a large diameter parallel bundle. The field of low-power objectives may thus be filled without removing the top element. At the opposite extreme it can be adjusted to have a numerical aperture as high as 1.3. **E 175,** E41

condensoid, *n*—the particles of a dispersion formed by condensation.

constant flow high-volume sampler—See **sampler.** **D 1356,** D22

condition, *v*—to bring a material to moisture equilibrium with a specified atmosphere. **D 123,** D13

conditioning—the conversion of a surface to a suitable state for successful treatment in succeeding steps. **B 374,** B08

conditioning, *v*— the storage of a specimen under specified temperature, humidity, etc., for a specified time prior to testing. **D 1079,** D08

conditioning, *n*—*in sampling,* the process by which the fiber is put into a consistent state to be tested. **D 2946,** C17

conditioning—the exposure of a material to the influence of a prescribed atmosphere for a stipulated period of time or until a stipulated relation is reached between material and atmosphere. **E 41,** G03

conditioning agent—an additive to a water suspension that will impart specific properties such as proper wetting, particle dispersion, or corrosion resistance. **E 1316,** E07

conditioning agents, *n*—all chemicals that enhance the performance of the collectors or frothers. Conditioning agents change the characteristics of the surface of the minerals or the environment. There are many subgroups according to their function: activators, depressants, emulsifiers, dispersants, flocculants, chelating reagents, froth depressants, pH modifiers, etc. **D 121,** D05

conditioning (environmental), *n*—the storage of a rubber, under specified conditions (time, temperature, humidity) prior to testing. **D 1566,** D11

conditioning heat treatment, *n*—a preliminary **heat treatment** used to prepare a steel object for a desired reaction to a subsequent **heat treatment**. **A 941,** A01

conditioning (mechanical), *n*—the prescribed program of deformation of a specimen prior to testing. **D 1566,** D11

conditioning time, n— See **joint conditioning time**. (See also **curing time** and **setting time**.) **D 907,** D14

condition, standard—the condition reached by a specimen when it is in temperature and moisture equilibrium with a standard atmosphere. **E 41,** G03

condominium—See **building.** **E 631,** E06

conductance—the capacity of a medium, usually expressed in mhos, for transmitting electric current. The reciprocal of resistance. **B 374,** B08

conductance—the ratio of the current carried through a material to the difference in potential applied across the material. It is the reciprocal of *resistance*. The unit is: $(\text{ohm})^{-1}$ or siemens. **D 2864,** D27

conductance—in leak testing, the ratio of the throughput (under steady state, conservative conditions) of a gas flowing through a conduit or an orifice to the difference in the partial pressures of the gas at the two ends of the conduit or on the two sides of the orifice, expressed in volume units per unit time, such as cubic metres per second. **E 1316,** E07

conductance, apparent dc—the dc conductance measured at the end of a specific electrification time. The "apparent dc conductance" is the reciprocal of the "apparent dc resistance." The unit is: $(\text{ohm})^{-1}$ or siemens. **D 2864,** D27

conductance, dc—the ratio of the total current (in amperes) passing through a material to the dc voltage (in volts) applied between two electrodes that are in contact with, or immersed in a specimen. The "dc conductance" is the reciprocal of the "dc resistance." The unit is: $(\text{ohm})^{-1}$ or siemens. **D 2864,** D27

conductance, film, *n*—the time rate of heat flow from a unit area of a surface to its surroundings, induced by a unit temperature difference between the surface and the environment. **C 168,** C16

conductance, insulation, *n*—the ratio of the total volume and surface current between two electrodes (on or in a specimen) to the dc voltage applied to the two electrodes. **D 1711,** D09

conductance (specific)—a measure of the ability of the water to conduct an electric current at 77°F (25°C). It is related to the total concentration of ionizable solids in the water. It is inversely proportional to electrical resistance. **D 653,** D18

conductance, surface, *n*—the ratio of the current between two electrodes (on the surface of a specimen) to the dc voltage applied to the electrodes. **D 1711,** D09

conductance, thermal—the thermal transmission in unit time through unit area of a particular body or assembly having defined surfaces, when unit average temperature difference is established between the surfaces. $C = \text{W/m}^2\cdot\text{K}$ ($C = \text{Btu/h}\cdot\text{ft}^2\cdot{}^\circ\text{F}$). **D 1079,** D08

conductance, thermal, *C, n*—the time rate of steady state heat flow through a unit area of a material or construction induced by a unit temperature difference between the body surfaces.

$$C = q/\Delta T$$

A conductance (C) associated with a material shall be specified as a material C. A conductance (C) associated with a system or construction of materials shall be specified as a system C. (C in SI units: $\text{W/m}^2\text{K}$.) (C in inch-pound units: $(\text{Btu/h})/\text{ft}^2/\text{F} = \text{Btu/h ft}^2\text{F}$.) **C 168,** C16

conductance, volume, *n*—the ratio of the current in the volume of a specimen between two electrodes (on or in the specimen) to the dc voltage applied to the two electrodes. **D 1711,** D09

conducting material (conductor), *n*—a material within which an electric current is produced by application of a voltage between points on, or within, the material. **D 1711,** D09

conducting salt—a salt added to the solution in order to increase its conductivity. **B 374,** B08

conductive base stock—paper stock intended as a substrate for electrostatic coating with surface or volume conductivity over a range of relative humidities controlled by special treatment. **F 335,** F05

conductive bituminous overlay—a first course bituminous overlay using aggregate or high-crystalline structure carbon, or equivalent, which is capable of distributing impressed current from the anodes to all parts of the bridge deck roadway surface. **D 3743,** D04

conductive ceramic tile—tile made from special body compositions or by methods that result in specific properties of electrical conductivity while retaining other normal physical properties of ceramic tile. **C 242,** C21

conductive polymer mortar—a rigid material formulated by polymerization of a select resin system and conductive calcined petroleum coke which is capable of distributing impressed anodic current. The material is used to fill cut slots, as strips or ribbons in grid-fashion on a bridge deck or structure, or applied as a thin overlay to substructure concrete members for stopping the corrosion of reinforcing steel. **D 3743,** D04

conductivity—the ratio of the current density carried through a specimen to the potential gradient paralleling the current. This is numerically equal to the conductance between opposite faces of a unit cube of liquid. It is the reciprocal of **resistivity.** **D 2864,** D27

conductivity—the property of a substance's (in this case, water and dissolved ions) ability to transmit electricity. The inverse of resistivity. Measured by a conductivity meter, and described in microsiemens/cm or micromhos/cm, μS/cm. **D 6161,** D19

conductivity—the intrinsic property of a particular material to carry electric current; it is commonly expressed in percent IACS (International Annealed Copper Standard) or MS/m (MegaSiemens/meter). **E 1316,** E07

conductivity, apparent dc volume—the "dc volume conductivity" measured at the end of a specified electrification time. It is the reciprocal of the apparent dc volume resistivity. The unit most commonly used is: $(\text{ohm-centimetre})^{-}$ or siemens per centimetre. The SI unit is $(\text{ohm-metre})^{-}$. **D 2864,** D27

conductivity, dc—the ratio of the current density passing through a specimen at a given instant of time and under prescribed conditions, to the dc potential gradient paralleling the current. It is the reciprocal of the dc resistivity. In common practice the "dc conductivity" is numerically equal to the "dc conductance" between opposite faces of a centimetre cube of liquid. The unit is: (ohm-centimetre) $^{-}$ or siemens per centimetre. The SI unit is: (ohm-metre) $^{-1}$. **D 2864,** D27

conductivity, dc volume—the property of a material that permits the flow of electricity through its volume. It is numerically equal to the ratio of the steady-state current density to the steady direct voltage gradient parallel with the current in the material. The dc volume conductivity is the reciprocal of the dc volume resistivity. The unit commonly used is: (ohm-centimetre)$^{-1}$ or siemens per centimetre. The SI unit is (ohm-metre) $^{-1}$. **D 2864,** D27

conductivity, electrical—the ratio of the current density carried through a specimen to the potential gradient paralleling the current. This is numerically equal to the conductance between opposite faces of a unit cube of material. It is the reciprocal of resistivity. **B 713,** B01

conductivity (electrical)—a measure of the ease with which charge carriers flow in a material; the reciprocal of resistivity. **F 1241,** F01

conductivity, electrical (volume), σ— the ratio of the current density ($A{\cdot}cm^{-2}$) through a specimen to the potential gradient (V/cm) in the same direction as the current. **E 1142,** E37

conductivity, specific conductance—the current transferred across unit area per unit potential gradient. In the metric system, K = amperes per square centimetre divided by volts per centimetre. The reciprocal of resistivity. **B 374,** B08

conductivity, surface, *n*—the surface conductance multiplied by that ratio of specimen surface dimensions (distance between electrodes divided by the width of electrodes defining the current path) which transforms the measured conductance to that obtained if the electrodes had formed the opposite sides of a square. **D 1711,** D09

conductivity, thermal—the thermal transmission, by conduction only, in unit time through unit area between two isothermal surfaces of an infinite slab of a homogeneous material of unit thickness, in a direction perpendicular to the surface, when unit temperature difference is established between the surfaces. $k = W/m{\cdot}K$ ($k = Btu{\cdot}in./h{\cdot}ft^{2}{\cdot}°F$). **D 1079,** D08

conductivity, thermal, λ or k, *n*—the time rate of steady state heat flow through a unit area of a homogeneous material induced by a unit temperature gradient in a direction perpendicular to that unit area. (λ or k in SI units: $(W/m^2)/(K/m) = W/m\ K$.) (λ or k in inch-pound units: $(Btu/h)/ft^{2}/(F/ft) = Btu/h\ ft\ F$) or $(Btu/h)/ft^2/(F/in.) = Btu\ in./h\ ft^2\ F$.) **C 168,** C16

conductivity type—*of a semiconductor crystal or wafer*, a property that identifies the majority charge carrier in the semiconductor. See also ***n*-type, *p*-type.** **F 1241,** F01

conductivity, volume, *n*—the volume conductance multiplied by that ratio of specimen volume dimensions (distance between electrodes divided by the cross-sectional area of the electrodes) which transforms the measured conductance to that conductance obtained if the electrodes had formed the opposite sides of a unit cube. **D 1711,** D09

conductor—a wire or combination of wires not insulated from one another, suitable for carrying an electric current. **B 354,** B01

conductor—a wire or combination of wires not insulated from one another, suitable for carrying an electrical current. **B 713,** B01

conductor, *n*—a wire, or combination of wires not insulated from each other, suitable for carrying electric current. **D 1711,** D09

conductor—a substance or body that allows a flow of electric current to pass continuously along it or through it when a sufficient voltage is applied across any two points. **D 5077,** D10

conductor, compact round—a conductor constructed with a central core surrounded by one or more layers of helically laid wires and formed into final shape by rolling, drawing, or other means. **B 713,** B01

conductor, composite—see **composite conductor.** **B 713,** B01

conductor, conventional concentric—conductor constructed with a round central core surrounded by one or more layers of helically laid round wires. The direction of lay is reversed in successive layers, and generally with an increase in length of lay for successive layers. **B 713,** B01

conductor core—the center strand or member about which one or more layers of wires or members are laid helically to form a concentric-lay. **B 354,** B01

conductor, coreless—a conductor constructed of one or more layers of helically laid wires and formed into final shape by rolling, drawing, or other means. **B 713,** B01

conductor, equilay—conductor constructed with a central core surrounded by more than one layer of helically laid wires, all layers having a common length of lay, direction of lay being reversed in successive layers. **B 713,** B01

conductor, hollow (also tubular conductor)—a conductor in which the individual elements are disposed about one or more hollow passages, the direction of which is along the axial length of the conductor. **B 713,** B01

conductor, parallel core—conductor constructed with a central core of parallel-laid wires surrounded by one layer of helically laid wires. **B 713,** B01

conductor, rope-lay—conductor constructed of a bunch-stranded or a concentric-stranded member or members, as a central core, around which are laid one or more helical layers of such members. **B 713,** B01

conductor, tape—a conductor constructed in the form of a flat ribbon or strip. **B 713,** B01

conductor, tubular—a conductor constructed in the form of a tube. **B 713,** B01

conductor, unidirectional—conductor constructed with a central core surrounded by more than one layer of helically laid wires, all layers having a common direction of lay, with increase in length of lay for each successive layer. **B 713,** B01

conductor, unilay—conductor constructed with a central core surrounded by more than one layer of helically laid wires, all layers having a common length and direction of lay. **B 713,** B01

conduit, *n*—*in asbestos-cement*, asbestos-cement pipe used to protect wires for electric-power or communication systems, for both undergound and exposed situations. **C 460,** C17

conduit—a pipe for conveying fluid. **C 896,** C04

conduit, *n*—pipe used to protect wires for electric-power or communication systems, for both underground and exposed situations. **C 1154,** C17

conduit, *n*—*in asbestos-cement*, asbestos-cement pipe used to protect wires for electric-power or communication systems, for both underground and exposed situations. **D 2946,** C17

conduit—a solid or flexible tube, pipe, or channel through which insulated electrical wires are run or through which water or some other fluid flows. **E 1749,** E06

conduit—a tubular raceway for carrying electric wires, cables, or other conductors. **F 412,** F17

conduit nail—bright or copper-plated, regular-stock-steel or hardened-steel, 1¾ to 3 ½ by 0.161-in. nails with annularly threaded shank; bent, curved hook head, and medium needle point. Different types of head designed for fastening ½, ¾, or 1-in. conduit, tubing, pipe, cable, etc. **F 547,** F16

cone, *n*—in *textiles*, a yarn holder or bobbin of conical shape used as a core for a yarn package of conical form, also called a cone core. **D 123,** D13

cone, *n*—the yarn package obtained when yarn is wound upon a cone core. **D 123,** D13

"cone and quartering" method—a form of sampling in which a powder is piled into a conical heap, spread into a circular cake, and divided into quarters, of which a diagonally opposite pair are taken as samples. **C 859,** C26

cone atomizer—an atomizer that produces a conical spray pattern. **E 1620,** E29

cone head—head consisting of truncated cone with large diameter usually nearest shank of nail. **F 547,** F16

cone of impression, *n*—a rise of the potentiometric surface in the approximate shape of a cone that develops around an injection well. **D 653,** D18

cone pattern, *n*—a diverging spray pattern that is nominally symmetric about the nozzle axis and whose apex is located at or near the nozzle discharge orifice. **E 1620,** E29

cone penetration—See **penetration**. **D 1079,** D08

cone-screen test—a method for testing fineness of enamel with a cone-shaped sieve. (see also **screen test**) **C 286,** B08

cone spray nozzle—see **cone atomizer**. **E 1620,** E29

confidence interval, *n*—an interval estimate of a population parameter computed so that the statement "the population parameter lies in this interval" will be true, on the average, in a stated proportion of the times such statements are made. **D 123,** D13

confidence interval, *n*—the range of values that may be expected to encompass the true value, generally stated at some probability, the confidence level. **E 135,** E01

confidence interval—an interval estimate of a population parameter computed so that the statement ″the population parameter included in this interval" will be true, on the average, in a stated proportion of the times such computations are made based on different samples from the population. **E 1823,** E08

confidence level, *n*—the probability, or expected percent of the times, that the selected percent (P %) of the actual population lies within the tolerance interval calculated from the data sample. **A 644,** A04

confidence level, *n*—the stated proportion of times the confidence interval is expected to include the population parameter. **D 123,** D13

confidence level, *n*—the probability that the true value lies within a stated range (the confidence interval). **E 135,** E01

confidence level (or coefficient)—the stated proportion of the times the confidence interval is expected to include the population parameter. **E 1823,** E08

confidence limits, *n*—the two statistics that define the ends of a confidence interval. **D 123,** D13

confidence limits—the limits on either side of the mean value of a group of observations which will, in a stated fraction or percent of the cases include the expected value. Thus the 95 % confidence limits are the values between which the population mean will be situated in 95 out of 100 cases. **D 4790,** D16

confidence limits—the two statistics that define a confidence interval. **E 1823,** E08

configuration, *n*—the state of the various interface and printing options that are set for the host and printer. **F 1457,** F05

confining bed—a hydrogeologic unit of less permeable material bounding one or more aquifers. **D 653,** D18

confining unit—a term that is synonymous with "aquiclude," "aquitard," and "aquifuge": defined as a body of relatively low permeable material stratigraphically adjacent to one or more aquifers. **D 653,** D18

conformance—ability of a boom to maintain freeboard and draft when subjected to a given set of environmental conditions. **F 818,** F20

conformity—fulfillment by a product, process or service of specified requirements, (ISO Guide 2). **E 1187,** E36

conformity surveillance—evaluation for conformity to determine the continuing conformity with specified requirements, (ISO Guide 2). **E 1187,** E36

conformity testing—evaluation for conformity by means of testing, (EN 45020). **E 1187,** E36

confounded factorial design, *n*—a factorial experiment in which only a fraction of the treatment combinations are run in each block and where the selection of the treatment combinations assigned to each block is arranged so that one or more prescribed effects is(are) confounded with the block effect(s), while the other effects remain free from confounding.

NOTE—All factor level combinations are included in the experiment. **E 456,** E11

confounded factorial design, *n*—a factorial experiment in which only a fraction of the treatment combinations are run in each block and where the selection of the treatment combinations assigned to each block is arranged so that one or more prescribed effects is(are) confounded with the block effect(s), while the other effects remain free from confounding.

NOTE—All factor level combinations are included in the experiment. **E 1325,** E11

confounding, *n*—combining indistinguishably the main effect of a factor or a differential effect between factors (interactions) with the effect of other factor(s), block factor(s) or interactions(s).

NOTE—Confounding is a useful technique that permits the effective use of specified blocks in some experiment designs. This is accomplished by deliberately preselecting certain effects or differential effects as being of little interest, and arranging the design so that they are confounded with block effects or other preselected principal factor or differential effects, while keeping the other more important effects free from such complications. Sometimes, however, confounding results from inadvertent changes to a design during the running of an experiment or from incomplete planning of the design, and it serves to diminish, or even to invalidate, the effectiveness of an experiment. **E 456,** E11

confounding, *n*—combining indistinguishably the main effect of a factor or a differential effect between factors (interactions) with the effect of other factor(s), block factor(s) or interaction(s).

NOTE—Confounding is a useful technique that permits the effective use of specified blocks in some experiment designs. This is accomplished by deliberately preselecting certain effects or differential effects as being of little interest, and arranging the design so that they are confounded with block effects or other preselected principal factor or differential effects, while keeping the other more important effects free from such complications. Sometimes, however, confounding results from inadvertent changes to a design during the running of an experiment or from incomplete planning of the design, and it serves to diminish, or even to invalidate, the effectiveness of an experiment. **E 1325,** E11

congealing point, *n*—that temperature at which molten petroleum wax ceases to flow, when allowed to cool under prescribed conditions. **D 4175,** D02

congruent phases—those states of matter of unique composition that co-exist at equilibrium at a single point in temperature and pressure;

for example, the two coexisting phases of a two-phase equilibrium (E 7, E-4). **E 1142,** E37

congruent transformation—an isothermal, or isobaric, phase change in which both of the phases concerned have the same composition throughout the process; the order of a system becomes unary at a composition of congruency. **E 7,** E04

congruent transformation—an isothermal, or isobaric, phase change in which both of the phases concerned have the same composition throughout the process; the order of a system becomes unary at a composition of congruency, (E 7, E-4). **E 1142,** E37

conical, ω— over a solid angle larger than an infinitesimal element of solid angle and less than a hemisphere. The geometry of the solid angle must be described in the text. For incident beams it is assumed that the radiance is constant over the entire solid angle. (See **Radiometric properties and quantities)** **E 772,** E44

conical head— See **countersunk head.** **F 547,** F16

conifer—a tree belonging to the order *Coniferae,* usually evergreen, with cones and needle-shaped or scalelike leaves, and producing wood known commerically as "softwood." **D 9,** D07

conjugate joints (faults)—two sets of joints (faults) that formed under the same stress conditions (usually shear pairs). (ISRM) **D 653,** D18

conjugate phases—those states of matter of unique composition which coexist at equilibrium at a single point in temperature and pressure; for example, the two coexisting phases of a two-phase equilibrium. **E 7,** E04

conjugate planes—two planes of an optical system such that one is the image of the other. **E 7,** E04

connate water, *n*—water entrapped in the voids of a sedimentary or extrusive igneous rock at the time of its deposition or emplacement. **D 653,** D18

connected porosity—See *connected porosity* under **porosity.** **C 242,** C21

connecting ring, *n—in zippers*, a device shaped like the letter "D" used to secure a pull, having more than one component in its design, to the bail of the slider. **D 123,** D13

connection—device or method used to fasten together two or more components of a structural system using mechanical means, welding, adhesives, or a combination of them. **E 631,** E06

connection, *n*—structural junction of two or more wood members, components, or assemblies, designed to be connected with mechanical fasteners, adhesives, welds, or a combination thereof, to transmit structural forces safely. Colloquially, the term *joint* is used in place of the term *connection.* **E 1807,** E06

connection head, *n*—a housing enclosing a terminal block for an electrical temperature-sensing device and usually provided with threaded openings for attachment to a protecting tube and for attachment of conduit. **E 344,** E20

connector, *n*—within the restrictions of this terminology, abbreviation for **metal connector plate**. **E 1807,** E06

connector hole—opening in metal connector plate, resulting from punching integral tooth from, or nail hole in, connector plate during its fabrication. Also called *slot* when opening is not round. **E 1807,** E06

consensus, *n*—substantial agreement achieved through a **consensus process,** but not necessarily unanimity. **E 631,** E06

consensus—general agreement, characterized by the absence of substantial issues by any important part of the concerned interests and by a process that involves seeking to take into account the views of all parties concerned and to reconcile any conflicting arguments, (ISO Guide 2). **E 1187,** E36

consensus process, *n*—a formal procedure for reaching **consensus** that includes the elements of due process. **E 631,** E06

consensus standard—a widely available standard developed by ASTM, ASME, SAE, ISO, or any other standards-setting organization which has under its structure those parties which include users, producers, and other interested persons. **F 1789,** F16

conservation, *n*—the examination, preservation, and restoration of cultural objects with minimal sacrifice of their aesthetic and historic integrity. **D 123,** D13

conservation—See **preservation.** **E 631,** E06

conservator, *n*—a person whose activity involves the science, technology, and documentation associated with the conservation of cultural objects. **D 123,** D13

consistency, *n*—a property of a material determined by the complete flow force relation. **C 11,** C11

consistency, *n*—the degree of mobility of a refractory castable as determined by Practices C 860. **C 71,** C08

consistency, *n—of fresh concrete, mortar, or grout,* the relative mobility or ability to flow. **C 125,** C09

consistency—the properties of a slip that control its draining, flowing, and spraying behavior. **C 286,** B08

consistency—the relative ease with which a soil can be deformed. **D 653,** D18

consistency—*in grouting*, the relative mobility or ability of freshly mixed mortar or grout to flow; the usual measurements are slump for stiff mixtures and flow for more fluid grouts. **D 653,** D18

consistency, *n*—that property of a liquid adhesive by virtue of which it tends to resist deformation. (See also **viscosity** and **viscosity coefficient.**) **D 907,** D14

consistency index—see **relative consistency**. **D 653,** D18

consistency (normal), *n*—the number of millilitres of water per 100 g of gypsum plaster or gypsum concrete required to produce a mortar or a slurry of specified fluidity. **C 11,** C11

consistent with—there are times when this expression is perfectly appropriate, such as when "evidence consistent with disguise is present" or "evidence consistent with a simulation or tracing is present, but "the known writing is consistent with the questioned writing" has no intelligible meaning. **E 1658,** E30

console, *n*—an instrument package that contains one or more instruments used to monitor depth, bottom time, surface intervals, air pressure, compass direction, decompression status, or some combination thereof. **F 1549,** F32

consolidated-drained test (slow test)—a soil test in which essentially complete consolidation under the confining pressure is followed by additional axial (or shearing) stress applied in such a manner that even a fully saturated soil of low permeability can adapt itself completely (fully consolidate) to the changes in stress due to the additional axial (or shearing) stress. **D 653,** D18

consolidated-undrained test (consolidated quick test)—a soil test in which essentially complete consolidation under the vertical load (in a direct shear test) or under the confining pressure (in a triaxial test) is followed by a shear at constant water content. **D 653,** D18

consolidation—the gradual reduction in volume of backfill matter to achieve stability. **C 896,** C04

consolidation—the gradual reduction in volume of a soil mass resulting from an increase in compressive stress

initial consolidation (initial compression)—a comparatively sudden reduction in volume of a soil mass under an applied load due principally to expulsion and compression of gas in the soil voids preceding primary consolidation.

primary consolidation (primary compression) (primary time effect) —the reduction in volume of a soil mass caused by the application of a sustained load to the mass and due principally to a squeezing out of water from the void spaces of the mass and accompanied by a transfer of the load from the soil water to the soil solids.

secondary consolidation (secondary compression) (secondary time effect)—the reduction in volume of a soil mass caused by the application of a sustained load to the mass and due principally to the adjustment of the internal structure of the soil mass after most of the load has been transferred from the soil water to the soil solids. **D 653,** D18

consolidation—reduction in volume of soil as a result of gravitational forces. **F 412,** F17

consolidation curve—see **consolidation time curve.** **D 653,** D18

consolidation grouting—injection of a fluid grout, usually sand and Portland cement, into a compressible soil mass in order to displace it and form a lenticular grout structure for support. **D 653,** D18

consolidation ratio, U_s (D)—the ratio of: (*1*) the amount of consolidation at a given distance from a drainage surface and at a given time, to (*2*) the total amount of consolidation obtainable at that point under a given stress increment. **D 653,** D18

consolidation test—a test in which the specimen is laterally confined in a ring and is compressed between porous plates. **D 653,** D18

consolidation-time curve (time curve) (consolidation curve) (theoretical time curve)—a curve that shows the relation between: (*1*) the degree of consolidation, and (*2*) the elapsed time after the application of a given increment of load. **D 653,** D18

conspicuity, *n*—the characteristics of an object that determine the likelihood that it will come to the attention of an observer. **E 284,** E12

constant amplitude loading— *in fatigue loading*, a loading (straining) in which all of the peak loads (strains) are equal and all of the valley loads (strains) are equal. **E 1823,** E08

constant dollars—dollars of uniform purchasing power exclusive of general inflation or deflation. **E 833,** E06

constant energy resolution—*AES, XPS,* a mode of operation in which the instrumental resolution is constant over the spectrum. See **fixed analyzer transmission.** **E 673,** E42

constant-head boundary—the conceptual representation of a natural feature such as a lake or river that effectively fully penetrates the aquifer and prevents water-level change in the aquifer at that location. **D 653,** D18

constant life diagram—*in fatigue*, a plot (usually on rectangular coordinates) of a family of curves each of which is for a single fatigue life, *N*, relating stress amplitude, S_a, to mean stress, S_m, or maximum stress, S_{max}, or both, to minimum stress, S_{min}. The constant life fatigue diagram is usually derived from a family of *S-N* curves each of which represents a different stress ratio (*A* or *R*) for a 50 % probability of survival. **E 1823,** E08

constant load—See **load.** **D 996,** D10

constant-rate-of-extension tensile type testing machine (*CRE*), *n*—a testing machine in which the rate of increase of specimen length is uniform with time. **D 123,** D13

constant-rate-of-extension type tensile testing machine (*CRE*), *n*—*in tensile testing*, an apparatus in which the pulling clamp moves at a uniform rate, and the force-measuring mechanism moves a negligible distance with increasing force, less than 0.13 mm (0.005 in.). **D 123,** D13

constant-rate-of-load tensile testing machine (*CRL*), *n*—a testing machine in which the rate of increase of the load being applied to the specimen is uniform with time after the first 3 s. **D 123,** D13

constant-rate-of-load tensile testing machine (CRL), *n*—a testing machine in which the rate of increase of the load being applied to the specimen is uniform with time after the first 3 s. **D 4439,** D35

constant-rate-of-traverse tensile testing machine (*CRT*), *n*—a testing machine in which the pulling clamp moves at a uniform rate and the load is applied through the other clamp which moves appreciably to actuate a weighing mechanism, so that the rate of increase of load or elongation is dependent upon the extension characteristics of the specimen. **D 123,** D13

constant-temperature stability (CTS) value—the maximum temperature at which a chemical compound or mixture may be held for a 2-h period under the conditions of the test without exhibiting a measurable exothermic reaction. **E 1445,** E27

constant velocity transport, *n*—a mechanism which feeds the paper through the printer at a steady rate. (See **CVT.**) **F 1457,** F05

constant weight—the condition of a substance in which all volatile components have been vaporized, and repeated exposure to a specified temperature, for any period of time, causes no change in weight. **C 896,** C04

constituent—a phase, or combination of phases, which occurs in a characteristic configuration in an alloy micro-structure. **E 7,** E04

constitutional diagram—graphical representation of the compositions, temperatures, pressures or combinations thereof at which the heterogeneous equilibria of an alloy system occur; also called "phase diagram," or "equilibrium diagram." **E 7,** E04

constitutional diagram—graphical representation of the compositions, temperatures, pressures, or combinations thereof at which the heterogeneous equilibria of a system occur. **E 1142,** E37

constitutive equation—force deformation function for a particular material. (ISRM) **D 653,** D18

constraint, *n*—any restriction to the deformation of a body. **E 6,** E28

constriction resistance—the increase in resistance arising from a change in current density distribution. In electric contacts it is that portion of contact resistance resulting from the convergence of current into the a-spots. **B 542,** B02

construct, *n*— a complete implant configuration attached to and including the spine, pelvis, ribs or substitute material as intended for surgical use. **F 1582,** F04

construction—details of arrangement or thickness or both, of veneers and other components used in the fabrication of plywood.

all-veneer construction—plywood in which all plies are veneer. Ordinarily no single ply of veneer will exceed 5⁄16 in. (7.9 mm) in thickness.

balanced construction—a construction such that the forces induced by uniformly distributed changes in moisture content will not cause warpage.

composite construction—a panel assembly consisting of veneers and other wood-based materials. Normally the non-veneer component is identified in describing the construction.

lumber core construction—plywood in which the center ply or core is of lumber rather than of veneer. Ordinarily cores that are 3⁄8 in. (9.5 mm) or greater in thickness will be of lumber.

symmetrical construction—plywood panels in which the plies on one side of the panel center line are essentially equal in thickness, grain direction, properties, and arrangement to those on the other side of the center line. **D 1038,** D07

constructional units, *n*—*in pile floor covering*, the needles, pitch, rows, shot, etc. into which the warp and filling yarns are commonly grouped. **D 123,** D13

construction contingency—the funds added to estimated and known costs in case of cost overruns during construction. **E 833,** E06

construction documents—materials that convey the physical, aesthetic, technical, performance, and administrative requirements necessary to initiate a contract for construction of the proposed project. **E 833,** E06

construction joint, *n—in building construction,* a formed or assembled joint at a predetermined location where two successive placements ("lifts") of concrete meet. **C 717,** C24

construction joint—See **joints.** **E 631,** E06

constructor, *n—in pipe laying,* the party that furnishes the work and materials for placement and installation. **C 460,** C17

constructor, *n— in pipe laying,* the party that furnishes the work and materials for placement and installation. **C 1154,** C17

constructor, *n—in pipe laying,* party that furnishes the work and materials for placement and installation. **D 2946,** C17

consumer care, *n—of consumer textile products,* cleaning and maintenance procedures as customarily undertaken by the ultimate user. **D 123,** D13

consumer panel, *n*—a group that is representative of the potential user population and that does not have technical knowledge of the products to be tested. **E 253,** E18

consumer's risk (β), *n*— the probability of accepting a lot when the process average is at the limiting quality level. **D 123,** D13

consumer textile product, *n*—a textile product intended to satisfy human wants and needs. **D 123,** D13

contact, *n*—a) a generic term that applies to a device or part of a device and that has the capability of completing or interrupting the flow of an electrical signal in a circuit. b) may also be used with modifiers such as: electrical contact, arcing contact, noble metal contact, separable contact, etc. **B 542,** B02

contact, *adj*—contact area, the part of an electrical device that is actually touching and where the electrical signal is expected to pass. Contact member, one of the electrical path parts that can make or break an electrical path. **B 542,** B02

contact adhesive, *n*—an adhesive having the property of autohesion. (Also called *contact bond adhesive* and *dry-bond adhesive.*) **D 907,** D14

contact adhesive, *n*—an adhesive that is apparently dry to the touch and that will adhere to itself instantaneously upon contact; also called contact bond adhesive or dry bond adhesive (D 907, D-14). **D 996,** D10

contact adhesive—see **adhesive.** **D 996,** D10

contact adhesive—See **adhesive, contact.** **E 1749,** E06

contact angle, *n— for paper wettability,* the angle formed by a paper substrate and the tangent to the surface of the liquid drop at the point of contact with the substrate when measured under specified conditions. **D 1968,** D06

contact, arcing—an electrical contact whose primary mode of wearout occurs on the contacting surfaces as a result of an arc formed between separating or closing contact pairs. **B 542,** B02

contact batch operation—an adsorption process in which an adsorbent is dispersed in a fluid to be treated and then separated when practical equilibrium is attained. **D 2652,** D28

contact bond adhesive, n—Synonym for **contact adhesive.** **D 907,** D14

contact bounce—the unwanted operation of contacts immediately following intentional operation. **B 542,** B02

contact chatter—the unwanted operation of contacts resulting from external forces operating on them. For example, vibration may cause contacts to open and close or "chatter." **B 542,** B02

contact-closing force—the transient force between contacts during closure. At the first instant of closure this force is zero. It then builds up to a maximum value dependent on the forces and inertia of the contact system and finally stabilizes at the static contact force. **B 542,** B02

contact, composite—a contact made of two or more distinct materials or alloys bonded to each other. For example, a contact with a facing of a precious metal bonded to a base-metal backing. **B 542,** B02

contact force—the force to close, maintain, or open contacts. See also **insertion force, withdrawal force,** and **normal force.** **B 542,** B02

contact grouting—see **backpack grouting.** **D 653,** D18

contact head—electrode assembly used to clamp and support a part to facilitate passage of electrical current through the part for circular magnetization. **E 1316,** E07

contact herbicide—a chemical that kills those plant parts with which it comes into contact. **E 609,** E35

contact length—the difference between the projected length, L_p, and the sum of $L_t + L_s$ or $L_c = L_p - (L_t + L_s)$. **F 1107,** F27

contact length, l_C—the difference between the projected length, L_P and the sum of l_T plus l_S or $l_C = L_P - (l_T + l_S)$. **F 472,** F27

contact noise—a varying voltage across a pair of electric contacts due to conditions at their interface. **B 542,** B02

contact, non-arcing—mating electrical contact surfaces that do not experience wearout due to arc erosion that is opposite of arcing contacts. **B 542,** B02

contact pad—replaceable metal pad, usually of copper braid, placed on electrodes to give good electrical contact, thereby preventing damage, such as arc strikes, to the part under test. **E 1316,** E07

contact, pitted—a contact that has numerous discrete hollows in its surface. **B 542,** B02

contact plating—deposition of a metal by the use of an internal source of current by immersion of the work in solution in contact with another metal. **B 374,** B08

contact potential—the potential difference at the junction of two dissimilar substances. **B 374,** B08

contact pressure—the force per unit area of physical contact between two contacts. This term is frequently but improperly used when contact force is meant. The area of physical contact is usually difficult to determine and quite different from the apparent area of contact. **B 542,** B02

contact pressure—an imprecise term denoting the minimum amount of pressure necessary to ensure an essentially void-free area between two mating surfaces. **E 1749,** E06

contact pressure molding, *n*—a method of molding or laminating in which the pressure, usually less than 70 kPa (10 psi), is only slightly more than necessary to hold the materials together during the molding operation. **D 883,** D20

contact pressure, *p* **(FL^{-2})**—the unit of pressure that acts at the surface of contact between a structure and the underlying soil or rock mass. **D 653,** D18

contact resistance—the resistance to current flow offered by the contact interface, comprising the sum of the constriction resistance plus the film resistance.

NOTE—In a practical measurement, correction must be made for bulk resistance consisting of contact material, lead wires, etc. **B 542,** B02

contact(s), butting—a type of contacts in which the direction of the motion of the moving contact is perpendicular to the contact faces. The contacts close and open with no appreciable sliding or rolling action. **B 542,** B02

contact scanner, *n*—a bar code reader that requires physical contact between the code medium and the scanner. **F 1294,** F05

contact, screw—a contact fabricated with an external thread for attachment to a support member or for adjustment. **B 542,** B02

contact, sliding—an electric contact which is expected to do its primary function during sliding. **B 542,** B02

contact, solid—a monolithic contact member. **B 542,** B02

contact, spring—a contact system in which one piece of material is used for both the driving spring and electric contact. **B 542,** B02

contact, static—an electric junction designed for infrequent separation and connection. **B 542,** B02

contact stress—the tensile stress component imposed at a glass surface immediately surrounding the contact area between the glass surface and an object generating a locally applied force. **C 162,** C14

contact surface area—the product of the average width times the contact length expressed quantitatively as follows:

$$A_C = \left[\frac{(b_H + 2\, b_M + b_V)}{4} \right] [l_C]$$

F 472, F27

contact surface area—the product of the average width times the contact length expressed quantitatively as follows:

$$A_c = \frac{b_h + 2b_m + b_v}{4} (L_c)$$

F 1107, F27

contacts, wiping—contacts that have some sliding motion during opening or closing. **B 542,** B02

contact testing—a technique in which the search unit makes contact directly with the test piece through a thin layer of couplant. **E 1316,** E07

contact time, *n— in an intermittent contact test,* the duration during each cycle that the challenge side chamber of the permeation cell is filled with the test chemical. **F 1494,** F23

contact trip—See **bottom trip.** **F 592,** F16

contact wipe—relative tangential motion between contacting surfaces that occurs during the normal course of contact closure. **B 542,** B02

container, *n*—a receptacle designed to hold a material, or to give integrity to the material. **D 123,** D13

container—a nonspecific term for a receptacle capable of **closure** (See also: **bag, barrel, basket, box, can, carton, crate, cylinder, drum, envelope, hamper, pail, tube.**)

body, n—in packaging, the principal part of a **container,** usually the largest part in one piece containing the sides.

bottle, n— a rigid or semirigid **container** typically of glass or plastic, having a comparatively narrow neck or mouth, and usually no handle (*Webster*).

case, n— a nonspecific term for a **shipping container.** In domestic commerce, case usually refers to a box made from **corrugated** or **solid fiberboard** wood, or metal.

cover, n—in packaging, the top or bottom, or both of a **container,** usually the part that closes the filling and dispensing opening. It is often called a cap when used with **fiberboard containers.** (See also **shroud.**)

cylinder, n— a rigid cylindrical metal **container** designed as a portable vessel for the storage and transportation of compressed gases. Generally equipped with protected valve closure and suitable pressure-relief safety device.

die-cut, adj— (*1*) a method of preparation in which a part or **container** has been cut, slotted, and scored or any combination of these by custom-made dies; (*2*) *n,* a part so made.

expendable container—a **container** for shipping or storage, or both, intended primarily for a single trip.

face, n—in packaging, any one of the plane surfaces of a **container.**

fast pack container—a standard size, **reusable container** with foam cushion **inserts.**

fiberboard container—a **box, package,** or **drum** made of **fiberboard.** When the term box is used for classification purposes, the structure must comply with all requirements of the carrier rules.

flap, n— one of the closing members of a **fiberboard container.**

glass container— any glass receptacle capable of holding a **seal** or **closure** for retention of contents.

intermodal container—a reusable **shipping container** manufactured to standard dimensions intended to unitize cargo or freight for shipping by one or more modes of transportation without the need for intermediate handling of the contents.

jar, n—a widemouthed container made typically of glass, plastic, or earthenware.

jug, n—a large, deep, usually glass, plastic or earthenware container with a narrow mouth and a handle.

manufacturer's joint—that part of a **fiberboard container** where the ends of the box blank are joined together in the manufacturing process by taping, stitching, or gluing.

modular container— a family of **containers** designed to be assembled into a **unit load.**

returnable container—a **shipping container** of any material designed to be used for more than one shipment.

reusable container—a shipping and storage **container** designed for reuse without impairment of its protective function.

seam, n (when referring to a **fiberboard container***)*—the lines of junction created by any free edge of a container flap or wall where it abuts or overlaps another portion of the **container** (except the **manufacturer's joint**).

shipping container—a container that is sufficiently strong to be used in commerce for **packing,** storing, and shipping commodities. (See also **barrel, crate, drum.**) **D 996,** D10

container, *n*—a usually portable device in which material is stored, transported, treated, disposed of, or otherwise handled. **E 1605,** E06

containerboard—any **paperboard** made specifically for the manufacture of **corrugated** and **solid fiberboard shipping containers.** Basis weight is expressed in pounds per 1000 ft^2 (or grams per square metre). It is customarily shipped in rolls.

cylinder kraft—**containerboard** made from kraft pulp on a cylinder machine.

Fourdrinier kraft—**containerboard** made from kraft pulp on a Fourdrinier machine, basically of single-ply formation, although possibly with supplementary second-ply, with less prominent grain direction. The sheet is formed on a traveling endless-wire screen which may also be vibrated to obtain more random orientation of fibers.

solid fiberboard— a solid board made by laminating two or more plies of **containerboard.** **D 996,** D10

containerization, *n*—(*1*) a shipping method in which material (such as merchandise) is packaged together in one **container.** (*2*) the use of transport **containers** to unitize cargo for transportation, supply, and storage. **D 996,** D10

containment material—in a solar energy system, a material that encloses the heat-transfer fluid or is in contact with the heat transfer or heat storage material, or both. **E 772,** E44

containment mode—placement of a boom to prevent free movement of a floating substance. **F 818,** F20

contaminant—an undesirable substance not normally present in water or soil. **D 653,** D18

contaminant, *n*—a material added by human or natural activities which may, in sufficient concentrations, render the atmosphere unacceptable. **D 1356,** D22

contaminant, *n*—a general term applicable to various extraneous and undesirable materials in pulp or other papermaking raw materials. **D 1968,** D06

contaminant—any foreign substance present which will adversely affect performance or quality. **D 6161,** D19

contaminant—any foreign substance present on the test surface or in the inspection materials which will adversely affect the performance of liquid penetrant materials. **E 1316,** E07

contaminant—any substance between and not a part of either of the two surfaces; *contaminant, contaminating, contamination.* **F 1646,** F13

contaminant—(*1*) a foreign or unwanted substance that can have deleterious effects on system operation, life or reliability. (*2*) unwanted molecular and particulate matter that could affect or degrade the performance of the components upon which they reside. **G 126,** G04

contaminants, *n—in asbestos,* any foreign matter (other than associated minerals and fines) in a sample of asbestos fiber. **D 2946,** C17

contaminants—all adventitious substances or microorganisms present in raw materials, bulk drugs, or final products. **E 1705,** E48

contamination—the presence of a substance not intentionally incorporated in a product. **F 412,** F17

contamination—*on a semiconductor wafer,* see **area contamination** or **particulate contamination**. **F 1241,** F01

contamination—any undesirable solid or liquid material that rests upon a surface, or exists between two mating surfaces; *contaminant, contaminating.* **F 1646,** F13

contamination, *n*—a process of contaminating. **G 126,** G04

context effect, *n*—effect upon the perception of a stimulus arising from its interrelationship with other stimuli in a presentation set. **E 253,** E18

contingency plan—document setting out an organized, planned, and coordinated course of action to be followed in case of an emergency, such as a fire or explosion, or a release of hazardous waste or hazardous waste constituents from a treatment, storage, or disposal facility that could threaten human health or the environment. **E 1605,** E06

continuity of coating—the degree to which a porcelain enamel or ceramic coating is free of defects, such as bare spots, boiling, blisters or copperheads, that could reduce its protective properties. **C 286,** B08

continuous bar code, *n*—a bar code symbology where all spaces within the symbol are parts of characters and no inter-character spaces are present. **F 1294,** F05

continuous cast drawing stock—the as-rolled section, normally round, produced by a continuous casting and rolling process. **B 354,** B01

continuous cleaning (coating)—a term describing a type of porcelain enamel designed to provide the continuous removal, at normal use temperatures, of food soils accumulated on the interior surfaces of ovens. **C 286,** B08

continuous coating, *n— of metallic coated steel products,* the process of uninterrupted passage of long lengths of steel products, usually steel sheet or wire, through the various processing steps such as cleaning, annealing, and coating. **A 902,** A05

continuous deionization—a deionization process that does not require regular interruptions in service to discharge ionic materials collected from the water being processed. **D 6161,** D19

continuous element, *n*—a configured element formed continuously along a length of monofilament. **D 123,** D13

continuous element zipper, *n—in zippers,* a zipper consisting of two continuously formed elements, each attached to one of the opposing edges of two tapes, which are engaged and disengaged by the movement of a slider. (Compare **separate element zipper**.) **D 123,** D13

continuous emission— see **emission, continuous.** **E 1316,** E07

continuous fermentation—nonstop flow of nutrients into a fermenting vessel, with the simultaneous outflow of products, organisms, and by-products. **E 1705,** E48

continuous fiber ceramic composite (CFCC), *n*—a ceramic matrix composite in which the reinforcing phase(s) consists of continuous filaments, fibers, yarn, or knitted or woven fabrics. **C 1145,** C28

continuous filament—a single glass fiber of sufficiently small diameter to be flexible enough for textile uses and of great or indefinite length. **C 162,** C14

continuous filament yarn, *n*—a yarn made of filaments that extend substantially throughout the length of the yarn. **D 123,** D13

continuous form, *n*—a quantity of paper made up of numerous connected individual perforated sheets, folded to form a pack. **D 1968,** D06

continuous furnace—synonymous with **melter**. **C 162,** C14

continuous furnace—a furnace into which ware is fed continuously and through which it progresses during firing. **C 286,** B08

continuous jet, n— See **liquid jet**. **G 40,** G02

continuously reinforced concrete pavement (CRCP), *n*—Portland cement concrete pavement with sufficient longitudinal steel reinforcement to control transverse crack spacings and openings in lieu of transverse contraction joints for accommodating concrete volume changes and load transfer. **E 867,** E17

continuously reinforced concrete pavement (CRCP), *n*—Portland cement concrete pavement with sufficient longitudinal steel reinforcement to control transverse crack spacings and openings in lieu of transverse contraction joints for accommodating concrete volume changes and load transfer. **E 1778,** E17

continuous method—with relation to magnetic particle inspection: a method wherein the indicating medium is applied while the magnetizing force is present. **E 1316,** E07

continuous mixer—a mixer into which the ingredients of the mixture are fed without stopping, and from which the mixed product is discharged in a continuous stream. **D 653,** D18

continuous moving bed—an adsorption process characterized by flow of a fluid through a continuously moving bed of granular adsorbent with continuous withdrawal of spent adsorbent and continuous addition of reprocessed or virgin adsorbent. **D 2652,** D28

continuous partial discharges (continuous corona), *n*—discharges that recur at rather regular intervals; for example on approximately every cycle of an alternating voltage or at least once per minute for an applied direct voltage. **D 1711,** D09

continuous phase—the phase forming the matrix or background in which other phases may be dispersed as isolated units. **E 7,** E04

continuous sampling—See **sampling.** **D 1356,** D22

continuous sintering—presintering, or sintering, in such manner that the objects are advanced through the furnace at a fixed rate by manual or mechanical means. Synonymous with **stoking.** **B 243,** B09

continuous smelter—a type of smelter into which the raw mix is fed continuously and from which the molten product is discharged continuously. **C 286,** B08

continuous spectrum (X-rays)—the polychromatic radiation given off by the target of an X-ray tube, containing all wavelengths above a certain minimum value called the short wave length limit. **E 7,** E04

continuous spun bonded polypropylene mat—a sheet material, resembling paper or felt, made from polypropylene plastic fibers, used in the manufacture of preformed membrane sheet for waterproofing. **D 3743,** D04

continuous tone—a positive or negative print or transparency composed of a range of optical densities from black through grey to white in which the greys are formed by varying the amounts of silver, dye, or pigment, or combination thereof. **F 1156,** F12

continuous variate, *n*—a variate that is a measurement based on a scale that is assumed to be continuous. **D 123,** D13

continuous waste—a drain connecting two or more plumbing fixtures or components of plumbing fixtures to a common trap. **F 412,** F17

continuous wave—a constant flow of ultrasonic waves, as opposed to pulsed. **E 1316,** E07

continuum—the smooth distribution of energy deposited in a gamma detector arising from partial energy absorption from Compton scattering or other processes (for example, Bremsstrahlung). See **Compton scattering.** **E 170,** E10

contract furniture, *n*—furniture manufactured for use in non-household applications. **D 123,** D13

contraction—linear strain associated with a decrease in length. (ISRM) **D 653,** D18

contractor—see **lead abatement contractor**. **E 1605,** E06

contralateral, *adj*—located, occurring, or acting on the opposite side of the body. **F 1646,** F13

contrast, *n*—*visual,* the degree of dissimilarity in appearance of two parts of a field of view seen simultaneously or successively. **E 253,** E18

contrast, *n*—*(1) objective,* the degree of dissimilarity of a measured quantity such as luminance of two areas, expressed as a number computed by a specified formula.

(*2*) *subjective*, the degree of dissimilarity in appearance of two parts of a field of view seen simultaneously or successively. **E 284,** E12

contrast, *n*—a linear function of the observations for which the sum of the coefficients is zero.

Note—With observations Y_1, Y_2,..., Y_n, the linear function $a_1Y_1 + a_2Y_2 + ... + a_n\ Y_n$ is a contrast if, and only if $\Sigma a_i = 0$, where the a_i values are called the contrast coefficients. **E 456,** E11

contrast—the difference in visibility (brightness or coloration) between an indication and the background. **E 1316,** E07

contrast—the difference between the amount of light reflected or transmitted by an object and by the background within the field of view. **E 1316,** E07

contrast, *n*—a linear function of the observations for which the sum of the coefficients is zero.

Note—With observations Y_1, Y_2, ..., Y_{n}, the linear function $a_1Y_1 + a\ _2Y_2 + ... + a_1Y_{n}$ is a contrast if, and only if $\Sigma a_{i} = 0$, where the $a_i$values are called the contrast coefficients. **E 1325,** E11

contrast—*(1)in optical character recognition*, the difference between color or shading of the printed material on a document and the background on which it is printed.

(*2*) See **print contrast ratio**. **F 149,** F05

contrast, *n*—the difference in reflectance between dark bars and light spaces of a bar code symbol. **F 1294,** F05

contrast agent—a material added to a component to enhance details by selective absorption of the incident radiation. **E 1316,** E07

contrast analysis, *n*—a technique for estimating the parameters of a model and making hypothesis tests on preselected linear combinations of the treatments (contrasts).

Note—Contrast analysis involves a systematic tabulation and analysis format usable for both simple and complex designs. When any set of orthogonal contrasts is used, the procedure, as in the example, is straightforward. When terms are not orthogonal, the orthogonalization process to adjust for the common element in nonorthogonal contrast is also systematic and can be programmed. **E 456,** E11

contrast analysis, *n*—a technique for estimating the parameters of a model and making hypothesis tests on preselected linear combinations of the treatments (contrasts).

Note—Contrast analysis involves a systematic tabulation and analysis format usable for both simple and complex designs. When any set of orthogonal contrasts is used, the procedure, as in the example, is straightforward. When terms are not orthogonal, the orthogonalization process to adjust for the common element in nonorthogonal contrast is also systematic and can be programmed. **E 1325,** E11

contrast effect, *n*—special case of context effect in which the perceived degree of difference between stimuli is exaggerated as a result of their interrelationship. **E 253,** E18

contrast enhancement (electron optics)— an improvement in electron image contrast by the use of an objective aperture diaphragm, shadow casting, or other means. **E 7,** E04

contrast gloss, **n**— see **luster**. **E 284,** E12

contrast perception—the ability to differentiate various components of the object structure by different intensity levels in the image. **E 7,** E04

contrast, photographic—the word contrast has been used in many different senses in connection with various photographic characteristics; there are different types of photographic contrast and different methods of measuring it. It is frequently used to designate the magnitude of the density difference resulting from a given exposure difference. (For another use, see **gamma**.) **E 7,** E04

contrast ratio—the ratio of the reflectance of a coating over black backing to its reflectance over a backing of reflectance of 0.80 (80 percent). **C 286,** B08

contrast ratio—ratio of the reflectance of a dry paint film over a black substrate of 5 % or less reflectance, to the reflectance of the same paint, equivalently applied and dried, over a substrate of 80 % reflectance. **D 16,** D01

contrast ratio, **n**— see the preferred terms **contrast** or **opacity**. **E 284,** E12

contrast ratio—optical measurement of the ratio of the intensity of the lightest and darkest elements in an image. **F 1156,** F12

contrast sensitivity—a measure of the minimum percentage change in an object which produces a perceptible density/brightness change in the radiological image. **E 1316,** E07

contrast stretch—a function that operates on the greyscale values in an image to increase or decrease image contrast. **E 1316,** E07

control—(evaluation), *n*—an evaluation to check, test, or verify; **(authority):** the act of guiding, directing, or managing; **(stability):** a state of process in which the variability is attributable to a constant system of chance causes. **E 456,** E11

control, *n*—material of established origin that is used to evaluate the performance of a test or comparison. **E 1732,** E30

control—a standard production-inked ribbon (film or fabric), carbon paper, manifold sets, etc. that has known values in normal use.

(F 153) **F 221,** F05

control, ***n*—in weathering**, the term control has three current widespread uses:

1. A material which is of similar composition and construction to the test material used for comparison, exposed at the same time. **G 113,** G03

control access system—a device, barrier or mechanism that prohibits unauthorized activation of a ride or device control system. **F 747,** F24

control analyses—the determination of specific parameters used as criteria for proper operation of a system. **D 1129,** D19

control block—a group of devices having a common piping and control system. **D 6161,** D19

control chart factor, *n*—a factor, usually varying with sample size, to convert specified statistics or parameters into a central line value or control limit appropriate to the control chart. **E 456,** E11

control chart method, *n*—the method of using control charts to determine whether or not processes are in a stable state. **E 456,** E11

control coupon (witness coupon), *n*—a coupon made from the same material and prepared in exactly the same way as the test coupons, which is used to verify the validity of the method or part thereof. **G 126,** G04

control echo—reference signal from a constant reflecting surface, such as a back reflection. **E 1316,** E07

control (expansion-contraction) joint, *n*—a designed separation in the system materials that allows for movement caused by expansion or contraction of the system. The construction of the separation is accomplished by one of the following methods: (1) manufactured devices suitable for this application, or (2) by field fabrication of suitable materials. **C 11,** C11

control joint, *n*—*in building construction*, a formed, sawed, tooled, or assembled joint acting to regulate the location and degree of cracking and separation resulting from the dimensional change of different elements of a structure. **C 717,** C24

control joint—See **joint.** **E 631,** E06

controlled blasting—includes all forms of blasting designed to preserve the integrity of the remaining rocks, that is, smooth blasting or pre-splitting. (ISRM) **D 653,** D18

controlled cooling, *n*—cooling a steel object from an elevated temperature in a predetermined manner to avoid hardening, cracking, or internal damage, or to produce a desired microstructure or mechanical properties. **A 941,** A01

controlled fission—*in the nuclear industry*, fission under conditions of continuous adjustment of control rods and of other control mechanisms in a reactor that compensates for the changes in reactivity. **C 859,** C26

controlled flow—a characteristic of a resin system with elevated viscosity during cure. **E 1749,** E06

controlled low strength material (CLSM)—flowable low compressive strength cementitious material used in the pipe zone as a bedding material. Also referred to as controlled density fill, flowable fill, slurry, or lean concrete. **C 896,** C04

controlled-pore filter—See **filter.** **D 1356,** D22

controlled potential coulometry—the measurement of the number of coulombs required for an electrochemical reaction occurring under conditions where the working electrode potential is precisely controlled. **C 859,** C26

controlled-rate thermal analysis, (*CRTA*), *n*—a family of techniques that monitors the temperature versus time profile needed to maintain a chosen, fixed rate of change of a property of a substance. (ICTAC)

NOTE—Compared to controlled-temperature experiments, where the reaction rate tends to increase exponentially and the rate can become limited by heat or mass transfer, CRTA experiments are more likely to involve the chemical reaction as the limiting step. This technique can also improve the resolution of multiple reactions. For example, in controlled rate experiments, power to the furnace is controlled to ensure a fixed rate of mass loss (or gain). **E 473,** E37

controlled-strain test—a test in which the load is so applied that a controlled rate of strain results. **D 653,** D18

controlled-stress test—a test in which the stress to which a specimen is subjected is applied at a controlled rate. **D 653,** D18

controlled-temperature program, *n*—the temperature history experienced by a sample during the course of a thermal analysis experiment.

NOTE—In contrast to controlled-rate experiments, power to the furnace is controlled to ensure a fixed rate of temperature change for controlled-temperature experiments. The program may include heating or cooling segments in which the temperature is changed at a fixed rate, isothermal segments in which time becomes the explicit independent variable, or any sequence of these individual segments. If the atmosphere (or vacuum) around the sample is changed by some external action (depending on the independent variable only—temperature or time) during the course of the experiment, that too becomes part of the controlled-temperature program. **E 473,** E37

control limit—limits on a control chart which are used as criteria for signaling the need for action, or for judging whether a set of data does or does not indicate a state of statistical control. **F 1789,** F16

control limits, *n*—predetermined ranges based on the variability of past observations between which the instrument data for a test must fall to be considered valid. **D 123,** D13

control limits, *n*—limits on a control chart which are used as criteria for signaling the need for action, or for judging whether a set of data does or does not indicate a state of statistical control. **E 456,** E11

control plan—a written description of a system for controlling fasteners and the processes used in their manufacture. Three distinct phases are used in a control plan, including prototype, pre-launch, and production. **F 1789,** F16

control plate—See **solid metal-coupon control specimen.** **E 1807,** E06

control rinse water—water used for equipment washing and rinsing having a known chemistry. **D 653,** D18

control rod—*in the nuclear industry*, a device used to control chain reactions, usually a neutron-absorbing material. **C 859,** C26

control sediment—a sediment that is essentially free of contaminants and is used routinely to assess the acceptability of a test. **E 943,** E47

control specimen—See **solid metal-coupon control specimen.** **E 1807,** E06

control tire, *n*—a reference tire, used in a specified manner throughout a test program **F 538,** F09

control undercover garment, *n*—a garment having a known history, the performance of which in a specific end-use has been established previously, and which is used as a standard of comparison. **D 123,** D13

control well—well by which the aquifer is stressed, for example, by pumping, injection, or change of head. **D 653,** D18

convection—the transport of heat by fluid flow. **E 772,** E44

convection, forced—convection caused by mechanical forces such as fans and injectors. **E 772,** E44

convection, natural—convection within a fluid, due to density differences caused by temperature differences. **E 772,** E44

convection oven, commercial—equipment for cooking food by forcing hot air over surface of the food using a fan in a closed cavity. See **oven**. **F 1827,** F26

conventional blanket, *n*—a blanket woven in either a plain or twill weave that is napped on both sides. **D 123,** D13

conventional sulfur vulcanizing system, *n*—as applied to natural and isoprene rubbers and butadiene-based synthetic rubbers, a vulcanizing system using a relatively high proportion of sulfur in which the combined sulfur exists predominantly in the polysulfidic cross links chain modifications. **D 1566,** D11

conventional true value of a quantity, *n*—value attributed to a particular quantity and accepted, sometimes by convention, as having an uncertainty appropriate for a given purpose.

NOTE—"Conventional true value" is sometimes called "assigned value", "best value", "conventional value", or "reference value". "Reference value", in this sense, should not be confused with "reference value" in the sense of an influence quantity affecting a measuring instrument.

NOTE—Frequently, a number of results of measurements of a quantity is used to establish a conventional true value. **E 456,** E11

convergence—generally refers to a shortening of the distance between the floor and roof of an opening, for example, in the bedded sedimentary rocks of the coal measures where the roof sags and the floor heaves. Can also apply to the convergence of the walls toward each other. (ISRM) **D 653,** D18

convergence, *n*—tendency of a stimulus to be perceived as similar to prior stimulus or stimuli. **E 253,** E18

convergence effect, *n*—special case of context effect in which the perceived degree of difference between stimuli is diminished as a result of their interrelationship. **E 253,** E18

conversion coating—a coating produced by chemical or electrochemical treatment of a metallic surface that gives a superficial layer containing a compound of the metal, for example, chromate coatings on zinc and cadmium, oxide coating on steel. **B 374,** B08

conversion efficiency—the ratio of the actual to theoretical fuel ethanol yield per unit mass of the feedstock. **E 1705,** E48

conversion, hardness—the exchange of a hardness number determined by one method for an equivalent hardness number of a different scale. **E 7,** E04

conversion screen—a device that converts the imaged neutron beam to radiation or light that exposes the radiographic film. **E 1316,** E07

conversion solution—an aqueous solution used to define the non-image areas of an electrostatic lithographic master by making them water receptive and ink repellent. **F 425,** F05

conversion treatment—a chemical or electrochemical process producing a superficial layer containing a compound of the metal. **B 374,** B08

conversion (Y)—product water flow rate divided by feed water flow rate. Also called recovery; given as fraction or decimal. See **recovery**. **D 6161,** D19

convex head—See **umbrella head.** **F 547,** F16

convex surface—a local convex curvature in the flat surfaces of pultruded plastic parts as measured transversely to their length. **D 3918,** D20

conveyor—a mechanism designed to move racked/unracked dishware through the treatment stages and out to the clean end of conveyor type dishwashing machines. **F 1827,** F26

conveyor oven—equipment that carries the food product on a moving conveyor into and through a heated chamber. The chamber may be heated by gas or electric forced convection, radiants, or quartz tubes. Top and bottom heat may be independently controlled. See **oven**. **F 1827,** F26

convolution—the combination of local measurements of drop size distribution and number density into equivalent line-of-sight values of drop size distribution and optical extinction. **E 1620,** E29

convoy, *n*—*in tire testing*, two or more vehicles running at the same time, over the same test course, under the same interdependent conditions. **F 538,** F09

cooking area—See **cooking surface**. **F 1827,** F26

cooking cavity—the volume within equipment. **F 1827,** F26

cooking container—a vessel used to hold the food product that is being heated by the equipment. **F 1827,** F26

cooking energy—energy consumed (Btu or kJ) by the equipment as it is used to cook a specified food product to a specified cooked condition under all cooking-load (see **load**) scenarios, $E_{equipment}$. **F 1827,** F26

cooking energy efficiency—quantity of energy imparted to the specified food product, expressed as a percentage of energy consumed by the equipment during the cooking event, η_{cook}.

$$\eta_{cook} = \frac{E_{food}}{E_{appliance}} \times 100$$

where:

$\eta cook$ = see *cooking energy efficiency*,
$E_{equipment}$ = energy into the equipment (see *cooking energy*),
E_{food} = see energy to food, and

$$= E_{sens} + E_{thaw} + E_{evap}$$

where:

$E_{s\text{-}ens}$ = quantity of heat added to food product, that causes their temperature to increase from the starting temperature to the average bulk temperature of a "done" food product,

$$= (W_i)\,(C_p)\,(T_f - T_i)$$

where:

W_i = initial weight of food product, lb (kg), as specified in the applicable standard test method,
C_p = specific heat of food product, Btu/lb, °F (kJ/kg, °C), as specified in the applicable standard test methods,
T_f = final cooked temperature of food product, °F (°C), as specified in the applicable standard test method,
T_i = initial internal temperature of food product, °F (°C), as specified in the applicable standard test method.

where:

E_{thaw} = latent heat (of fusion) added to the food product, that causes the moisture (in the form of ice) contained in the food product to melt when the temperature of the food product reaches 32°F (0°C) (the additional heat required to melt the ice is not reflected by a change in the temperature of the food product), Btu (kJ),

$$= W_{iw} \times Hf$$

where:

W_{iw} = initial weight of water in the food product, lb (kg),
H_f = heat of fusion, Btu/lb (kJ/kg), and
= 144 Btu/lb (36 kJ/kg) at 32°F (0°C).

where:

$E_{e\text{-}vap}$ = latent heat (of vaporization) added to the food product, that causes some of the moisture contained in the food product to evaporate. Similar to the heat fusion, the heat of vaporization cannot be perceived by a change in temperature and must be calculated after determining how much moisture was lost from a "done" food product.

$$= W_{loss} \times H_v$$

where:
W_{loss} = weight loss of water during cooking, lb (kg),
H_v = heat of vaporization, Btu/lb (kJ/Kg),
= 970 Btu/lb (2256 kJ/kg) at 212°F (100°C), and

where:
$E_{equipment}$ = energy into equipment, Btu (kJ).
F 1827, F26

cooking energy rate—average rate of energy consumption (Btu/h or kJ) during the cooking energy efficiency tests, refers to all loading scenarios, $E_{cook\ rate}$.

$$E_{cook\ rate} = \frac{E_{cook} \times 60}{t_{cook}}$$

where:
E_{cook} = equipment energy consumption, and
t_{cook} = cook time, min.
F 1827, F26

cooking medium—a substance within equipment that transfers heat to the food product, e.g., oil in a fryer, water in a pasta cooker, or air in an oven. **F 1827,** F26

cooking surface—an area of the equipment that receives a source of heat upon which a food product is cooked (or defrosted), for example, griddle surface, element or burner of a range, or broiler grate. **F 1827,** F26

cooking unit—a heating device located on the equipment that is powered by a single heat source comprised of either a gas burner or an electrical element, that is independently controlled. **F 1827,** F26

cook time—time required to cook the specified food product to a specified cooked condition during a cooking energy efficiency (see **cooking energy efficiency**) test. **F 1827,** F26

cook zone—the volume of oil in the fryer in which foods are cooker; typically the entire volume from the heating element(s), or heat exchange surface to the surface of the frying-medium. See **fryer**. **F 1827,** F26

coolant additive package, *n*—the combination of inhibitors added to an engine coolant to mitigate cooling system degradation, corrosion, scaling, and foaming, or to provide other desirable properties. **D 4725,** D15

cooler nail—usually coated, slender, regular-stock-steel, 1 by 0.062 to 2⅞ by 0.120-in. nails with flat $^{11}/_{64}$ to $^{19}/_{64}$-in. head and medium diamond point, with head diameter same as or smaller than that of common wire nail of same length. **F 547,** F16

cool-flame, *n*—a faint, pale blue luminescence or flame occurring below the autoignition temperature (AIT). **E 1445,** E27

cooling, chemical, *n*—sensation of reduced temperature experienced as a result of exposure to certain substances such as menthol or anise. The sensation usually persists after the stimulus is removed. **E 253,** E18

cooling curve—graphical representation of the course of temperature fall of a chemical mixture as a function of time commonly exhibiting more or less abrupt changes of rate at, or near, those temperatures at which phase changes begin; used in finding the temperatures at which phase changes occur. Occasionally, some property or function other than time may be used; for example, thermal expansion. **E 7,** E04

cooling curve—graphical representation of specimen temperature or temperature change as a function of time or decreasing environment temperature. **E 1142,** E37

cooling-down period—{archaic} (*1*) the time elapsing after a covered pot is opened before the glass is cool enough to work.

(*2*) period between fining stage and the removal of the glass from the furnace. **C 162,** C14

cooling, physical, *n*—sensation of reduced temperature experienced as a result of exposure to thermally cold substances, such as ice; to substances that evaporate rapidly, such as acetone or alcohol; or to substances that have a negative heat of solution, such as crystalline sorbitol. The duration of the sensation is usually limited to the time of direct contact with the stimulus. **E 253,** E18

cooling rate—See **setting rate.** **C 162,** C14

cooling rate—the average slope of the time-temperature curve taken over a specified time and temperature interval. **E 7,** E04

cooling rate—average slope of the time-temperature curve taken over a specific time and temperature interval as the temperature is decreased. **E 1142,** E37

cooling time, *n*—in molding, the time interval from the start of forward screw movement until the mold starts to open. **D 883,** D20

cooling zone—that part of the continuous furnace in which the ware is allowed to cool after firing. **C 286,** B08

coping—a covering on top of a wall exposed to the weather, usually sloped to carry off water. **D 1079,** D08

copolymer—See **polymer.** **D 883,** D20

copolymer, n— See **polymer.** **D 907,** D14

copolymer, *n*—a polymer formed from two different monomers. **D 1566,** D11

copolymer, *n*—a polymer consisting of molecules characterized by the repetition (neglecting ends, branch junctions and other irregularities) of two or more different types of monomeric units. See **polymer.** **F 1251,** F04

copolymerization—See **polymerization.** **D 883,** D20

copolymerization, n— See **polymerization.** **D 907,** D14

copolymerization—see **polymerization** and **copolymer.** **F 1251,** F04

copper common nail—bright, solid-copper, ⅜ by 0.065 to 6 by 0.284-in. nails with flat head and medium diamond point. **F 547,** F16

copper corrosion—a qualitative indication of reactive impurities in aromatic hydrocarbons. An iridescent, gray, or black discoloration of polished copper strip is considered cause for rejection. **D 4790,** D16

coppered, copper-washed—all surfaces chemically plated with copper, usually by chemical rather than electrolytic process. (See **electroplated.**) **F 547,** F16

copper enamel—a porcelain enamel specifically designed for application to copper. **C 286,** B08

copperhead—a defect occurring in sheet metal ground coat that appears as a small freckle or pimple-like spot, reddish brown in color. **C 286,** B08

copper number—the number of grams of copper in the cuprous oxide reduced from a cupric compound by 100 g of pulp or paper (after deduction of moisture, ash, and sizing materials) treated under specified conditions with an excess of cupric solution. The copper number is an indication of the relative number of reducing groups in the pulp or paper. **D 1695,** D01

copper number—the weight in grams of copper reduced from the cupric to the cuprous state by exposure to 100 g of paper, paperboard, or pulp as determined by a specified method; indicates the relative number of reducing groups in the pulp or paper and is used as a measure of its chemical quality and stability. **D 1968,** D06

copper spark method—*in emission spectroscopy,* a method using high voltage spark excitation for the determination of elemental constituents of material deposited, usually by evaporation, on a pair of copper electrodes. **C 859,** C26

copying—the act of producing an image on a document or other receptor media which is a duplication of the image of another document, such as by a photographic, xerographic, or facsimile process or by carbon or carbonless papers. **F 335,** F05

copying—the act of producing an image on a document or other receptor media which is a duplication of the image of another document, such as by a photographic, xerographic or facsimile process or by carbon or carbonless papers. **F 909,** F05

coquina—a limestone composed predominantly of unaltered shells or shell fragments loosely cemented by calcite. **C 119,** C18

cord—a generally attenuated glassy inclusion with properties differing from those of the surrounding glass. **C 162,** C14

cord, *n*— a twisted or formed structure composed of one or more single or plied filaments, strands, or yarns of organic polymer or inorganic materials. **D 123,** D13

cord, *n—in zippers,* a strand of multiple yarns for use in forming a bead. **D 123,** D13

cord, *n—of glass fiber,* a strand made by combining multiple ends of filament strands, including cabled yarns, primarily for structural application. **D 123,** D13

cord, *n—in a tire,* filament(s) or plied yarns used in making a tire ply. **F 538,** F09

cordierite porcelain—See *cordierite porcelain* under **porcelain.** **C 242,** C21

cordierite whiteware—See *cordierite whiteware* under **ceramic whiteware.** **C 242,** C21

cordovan—leather made from the tight, firm portion of horse butts. It has very fine pores and characteristic waxy finish, and is very durable. **D 1517,** D31

cordovan—leather made from the tight, firm portion of horse butts. It has very fine pores, a characteristic waxy finish and is very durable. **F 869,** F08

cord twist, *n*—the amount of twist in a cord made from two or more single or plied yarns. **D 123,** D13

corduroy, *n*—a filling cut-pile fabric in which the cut fibers form a surface of wales (rounded cords or ribs) which usually run warpwise. **D 123,** D13

corduroy flesh—a rough condition of the flesh side of leather caused by failure to remove the twitching muscles. **D 1517,** D31

core, *n—in case hardening,* the interior portion of unaltered composition or microstructure, or both, of a case hardened steel object. **A 941,** A01

core, *n—in clad products,* the central portion of a multilayer composite metallic material. **A 941,** A01

core—see **conductor core.** **B 354,** B01

core—a cylinder of concrete obtained from concrete by means of a core drill. **C 822,** C13

core, *n—in drilling,* a cylindrical section of rock (coal) that is usually 5 to 10 cm in diameter, taken as part of the interval penetrated by a core bit and brought to the surface for geologic examination, representative sampling and laboratory analyses. **D 121,** D05

core, *n—in sampling fiber packages,* the portion of wool or other fiber obtained using a sampling tube. **D 123,** D13

core, *n*—a filament or strand that serves as an extended axis about which other elements can be wound. **D 123,** D13

core—a cylindrical sample of hardened grout, concrete, rock, or grouted deposits, usually obtained by means of a core drill. **D 653,** D18

core, *n—in packaging,* a cylindrical structure used as a carrier of flexible material that is wound around it. **D 996,** D10

core—*(1) in decorative or hardwood plywood,* the center most ply. It may be of lumber (either edgeglued or closely assembled) or other wood-based panel material, or of one or more thicknesses of veneer.

(2) in construction plywood, all plies or layers between the face and back. **D 1038,** D07

core—*(1) case hardening*—interior portion of unaltered composition, or microstructure, or both, of a case-hardened steel article.

(*2*) *clad products*—the central portion of a multilayer composite metallic material. **E 7,** E04

core, *n—of an optical fiber,* the center region of an optical waveguide through which radiant energy is transmitted. **E 131,** E13

core—a generally centrally located layer or composite component of a sandwich construction, usually low density, which separates and stabilizes the facings and transmits shear between them and provides most of the shear rigidity of the construction. **E 1749,** E06

core—See **rail.** **F 592,** F16

core, *n*—a paper, plastic, or fiber core upon which the thermal product is wound. **F 1623,** F05

core barrels, *n*—two nested tubes above the bit of a core drill, the outer rotating with the bit, the inner receiving and preserving a continuous section or core of the material penetrated. The following two types of inner barrels are commonly used.

split-tube barrel, n—a type of inner barrel consisting of two longitudinal halves of pipe bound together by reinforced tape at intervals along the barrel length that allows easy access to a relatively intact core (by cutting the tape). (This is the preferred barrel type for coal exploration, where available.)

solid-tube barrel, n—a type of inner barrel consisting of a single solid-walled length of pipe in which removal of the core is accomplished by mechanical or hydraulic pressure at one end of the pipe thus extruding the core onto a core tray. (The core is likely to be less intact than when a split-tube barrel is used.) **D 121,** D05

core block—in cutting rotary veneer, the portion of the bolt remaining after available veneer has been removed. **D 1038,** D07

core compressive modulus—the ratio of the compressive load (below the proportional limit of the core) per unit of original area to the corresponding deformation per unit of original thickness. **E 1749,** E06

cored, annular, conductor—a conductor in which one or more layers of wires are laid helically around a central core of metallic or nonmetallic material. The core is used as a mandrel for the application of subsequent layers of wire to maintain consistent spacing of these members around the conductor axis. **B 354,** B01

cored bar—a compact of bar shape heated by its own electrical resistance to a temperature high enough to melt its interior. **B 243,** B09

cored braid, *n—in rope,* a hollow braid construction, either plain or twill, the center of which is filled with yarns which are not braided. **D 123,** D13

core drilling; diamond drilling—a rotary drilling technique, using diamonds in the cutting bit, that cuts out cylindrical rock samples. (ISRM) **D 653,** D18

cored tile or block, *n*—see **gypsum tile or block** **C 11,** C11

core, laminated—a magnetic component constructed by stacking suitably thin pieces of magnetic material which are stamped, sheared, or milled from sheet or strip material. Individual pieces usually have an insulating surface coating to minimize eddy current losses in the assembled core. **A 340,** A06

core loss, ac eddy current, incremental, $P_{\Delta e}$—the power loss caused by eddy currents in a magnetic material that is cyclically magnetized. **A 340,** A06

core loss, ac eddy current, normal, P_e—the power losses as a result of eddy currents in a magnetic material that is symetrically cyclically magnetized.

NOTE—The voltage is generally assumed to be across the parallel combination of core inductance, L_1, and core resistance, R_1. **A 340,** A06

core loss, ac, incremental, $P_{c\Delta}$—the core loss in a magnetic material when the material is subjected simultaneously to a dc biasing magnetizing force and an alternating magnetizing force. **A 340,** A06

core loss, ac, specific, $P_{c(B;f)}$—the active power (watts) expended per unit mass of magnetic material in which there is a cyclically varying induction of a specified maximum value, *B*, at a specified frequency, *f*. **A 340,** A06

core loss, ac, (total), P_c—the active power (watts) expended in a magnetic circuit in which there is a cyclically alternating induction.

NOTE—Measurements of core loss are normally made with sinusoidally alternating induction, or the results are corrected for deviations from the sinusoidal condition. **A 340,** A06

core loss current, ac - I_c—See **current, ac core loss.** **A 340,** A06

core loss, residual, P_r—the portion of the core loss power, P_c, which is not attributed to hysteresis or eddy current losses from classical assumptions. **A 340,** A06

core, mated—two or more magnetic core segments assembled with the magnetic flux path perpendicular to the mating surface. **A 340,** A06

core module—See **module.** **E 631,** E06

core, n—a centrally located layer of a sandwich construction, usually low density, which separates and stabilizes the facings and transmits shear between the facings and provides most of the shear rigidity of the construction. **C 274,** D30

core (of an electromagnetic inspection circuit)—that part of the magnetic circuit which is within the electrical winding. **E 1316,** E07

core (of gypsum board), *n*—the hardened material filling the space between the face and back papers consisting substantially of rehydrated gypsum with additives. **C 11,** C11

core plate—a generic term for any insulating material, formed metallurigically or applied externally as a thin surface coating, on sheet or strip stock used in the construction of laminated and tape wound cores. **A 340,** A06

core, powder (dust)—a magnetic core comprised of small particles of electrically insulated metallic ferromagnetic material. These cores are characterized by low hysteresis and eddy current losses. **A 340,** A06

core recovery—ratio of the length of core recovered to the length of hole drilled, usually expressed as a percentage. **D 653,** D18

core rod—a member of a die used to produce a hole in a compact. **B 243,** B09

core sample, *n*—that part of a core of rock or coal obtained so as to represent accurately a thickness of a unit penetrating by drilling. **D 121,** D05

core shear—the shear stress applied to the core material used in sandwich panel construction. **E 1749,** E06

core shear modulus—the ratio of the shear stress to the corresponding shear strain for stresses below the proportional limit in shear of the core. **E 1749,** E06

core splice adhesive—See **adhesive, core splice.** **E 1749,** E06

core-spun yarn, *n*—a compound structure in which a filament or strand serves as an axis around which a cover of either loose fiber or a yarn is wound. **D 123,** D13

core stabilization—a process to rigidize honeycomb core materials to prevent distortion during machining or curing. **E 1749,** E06

core, tape-wound—a magnetic component constructed by the spiral winding of strip material onto a suitable mandrel. The strip material usually has an insulating surface coating which reduces interlaminar eddy current losses in the finished core. **A 340,** A06

core wrap (binder tape), *n*—paper used to wrap groups of insulated wire into cable configuration prior to sheathing. **D 1711,** D09

coring, *v*—the process of perforating structural clay products, generally performed during extrusion by supporting cores (rods) within the shaping cap of the extruder. **C 43,** C15

coring—a variable composition between the center and outside of a unit of structure, (such as a dendrite, grain, carbide particle) resulting from non-equilibrium growth which occurs over a range of temperatures or compositions. **E 7,** E04

corker nail—usually coated, slender, regular-stock-steel, 1 by 0.062 to 5⅞ by 0.244-in. nails with 5⁄32 to ½-in. sinker head and medium diamond point. **F 547,** F16

cork-insulation nail—galvanized, regular-stock-steel, 3 to 9 by 0.148-in. nails with flat ½-in. head and medium diamond point. **F 547,** F16

cork tile, *n*—a floor surfacing unit made from natural cork shavings compressed and baked to be thoroughly and uniformly bonded together. **F 141,** F06

corner barbed wire arm—a combination post cap and fitting used to hold barbed wire in place on a corner post. May be angled in or out from the fence line to hold three or more strands of barbed wire, or "V" shaped to hold six or more strands of barbed wire. **F 552,** F14

corner breaks, *n—(JCP only)* a portion of the slab separated by a crack which intersects the adjacent transverse and longitudinal joints, describing approximately a 45° angle with the direction of traffic, where the length of the sides is from 0.3 m (1 ft) to one half the width of the slab. **E 867,** E17

corner breaks, *n*—(*JCP only*) a portion of the slab separated by a crack that intersects the adjacent transverse and longitudinal joints, describing approximately a 45 degree angle with the direction of traffic where the length of the sides is from 0.3 m (1 ft) to one half the width of the slab. **E 1778,** E17

corner effect—the reflection of an ultrasonic beam directed at normal incidence to the line of intersection of two perpendicular planes. **E 1316,** E07

cornering force coefficient, [*nd*], *n*—the ratio of cornering force to the normal force on a tire. **F 538,** F09

cornering force, [ML^2/T^2], *n—of a tire,* the horizontal force acting perpendicularly to the instantaneous motion vector of the center of contact for a tire operating at a slip angle. **F 538,** F09

cornering stiffness coefficient, *n—of a tire,* the ratio of tire cornering stiffness to tire normal force. **F 538,** F09

cornering stiffness, [*F/γ*], *n—of a tire,* the negative of the rate of change of tire lateral force with respect to change in tire slip angle, usually evaluated at zero slip angle. **F 538,** F09

cornering traction coefficient, *n*—the ratio of the cornering tractive force vector and normal force. **F 538,** F09

cornering traction vector angle, *n*—the angle between the resultant cornering traction force vector and the X' axis. **F 538,** F09

cornering tractive force, (F), *n*—the vectorial sum of lateral and longitudinal tractive force components. **F 538,** F09

corner post—a terminal post, at which the direction of the line of fence changes in two or more directions. **F 552,** F14

corner radius on square or rectangular wire—any configuration on the corner between a chamfer and a full radius. The measurement of a corner radius is the distance from the blend point on one surface to the extension of the other surface. **B 846,** B05

corner rail clamp—a two-piece clamp with carriage bolts and nuts designed for horizontal rail connections to a 90° corner post. **F 552,** F14

corner reinforcement, exterior, *n*—a preformed section of wire or expanded sheet steel, for the reinforcement of exterior stucco external corners (arrises). **C 11,** C11

corner rolls—half-round units of asbestos-cement used to trim and flash corners in asbestos-cement corrugated application. **C 460,** C17

corner rolls, *n*—half-round units used to trim and flash corners in corrugated application. **C 1154,** C17

corner rolls—half-round units of asbestos cement used to trim and flash corners in asbestos-cement corrugated applications. **D 2946,** C17

corn stover—the stalks of the maize plant. **E 1126,** E48

corn stover—the stalks of the maize plant. **E 1705,** E48

corona, *n*—visible partial discharges in gases adjacent to a conductor. **D 1711,** D09

corona—a luminous discharge due to ionization of the air surrounding an electrode, caused by the high electric field strength in the vicinity of the electrode, exceeding a certain critical (that is, threshold) value. **D 2864,** D27

corona—(*1*) the visible glow produced by ionized air in the path of high-voltage discharge.

(*2*) the type of discharge and the device from which the discharge emanates. **F 335,** F05

corona, *n*—a device used to place a uniform electrical charge on the surface of a xerographic photoreceptor. **F 1457,** F05

corona, *n*—chemical separation of ink dispersant and pigment caused by interaction with substrate which appears as an image with a dark center surrounded by a lighter shadow. **F 1857,** F05

corona charging—charging by means of gaseous ions generated by a corona device. **F 335,** F05

corona (discharge)—a self-sustaining discharge characterized by highly asymmetrical electric fields with the result that ionization predominantly occurs near the electrode with the higher potential gradient. **B 542,** B02

corona effect—light emitted in the UV range of the electromagnetic spectrum by electronically excited molecules that have reached a singlet state and have not consumed the absorbed energy by other physical process. **D 2864,** D27

corona (partial discharge) extinction voltage, *CEV*—the highest voltage at which partial discharge (or corona) no longer exceeds a specified intensity as the applied voltage is gradually decreased from a value above the corona inception voltage. Where the applied voltage is alternating the CEV is expressed as $1/\sqrt{2}$ of the peak voltage. **D 2864,** D27

corona (partial discharge) inception voltage, *CIV*—the lowest voltage at which continuous partial discharge (or corona) exceeding a specified intensity is observed as the applied voltage is gradually increased. Where the applied voltage is alternating, the *CIV* is expressed as $1/\sqrt{2}$ of the peak voltage. **D 2864,** D27

corona shield—a smooth, rounded metal covering placed around exposed high-voltage components to prevent electrical discharge. **E 7,** E04

corotron, *n*—a name for a specific type of corona. **F 1457,** F05

correctable (as related to typewriter ribbons)—an image that may be completely removed from the substrate by means of lift-off rather than erasure or cover-up. **F 221,** F05

corrected air flow, *n*—the volume of air movement per unit of time under standard atmospheric conditions. The flow is expressed in cubic feet per minute or litres per second. **F 395,** F11

corrected grain—portions of the grain surface lightly abraded with emery wheel or sandpaper, so as to lessen the effect of grain damage. **D 1517,** D31

corrected temperature rise, *n*—the temperature of the calorimeter, caused by the process that occurs inside the bomb; that is, the observed temperature change corrected for various effects. **D 121,** D05

corrected temperature rise, *n*—the increase in temperature of the calorimeter caused by the process that occurs inside the bomb; the observed temperature change corrected for various effects. **D 5681,** D34

correlated color temperature, *n*—*of a source*, the temperature, usually expressed in kelvins, of a full radiator that would emit light of the chromaticity most closely resembling that of the light from the source. **E 284,** E12

correlation coefficient (r)—a measure of the strength of the linear relationship between X and Y, calculated by the equation:

$$r_{xy} = \frac{(\sum_{i=1}^{n} X_i Y_i)}{(\sum_{i=1}^{n} X_i^2)^{1/2} (\sum_{i=1}^{n} Y_i^2)^{1/2}}$$

where:

n = the number of observations in X and Y.

E 131, E13

corresponding force, *n*—synonym for force at specified elongation. **D 4439,** D35

corridor, *n*—an enclosed space connecting a room or compartment with an exit. The corridor may include normal extensions, such as lobbies and other enlarged spaces. **E 176,** E05

Corrodkote test—an accelerated corrosion test for electrodeposits. **B 374,** B08

corrosion—(*1*) gradual solution or oxidation of a metal.

(*2*) solution of anode metal by the electrochemical action in the plating cell. **B 374,** B08

corrosion, *n*—the chemical or electrochemical reaction between a material, usually a metal, and its environment that produces a deterioration of the material and its properties (G 15, G-1).

corrosion, inhibitor, n—a chemical substance or combination of substances that, when present in the proper concentration and form in the environment, prevents or reduces corrosion.

volatile corrosion inhibitor (VCI)—a material that slowly releases vapor to inhibit corrosion within a **package** by neutralizing the effects of moisture-laden air. **D 996,** D10

corrosion—the formation of rust. **F 1078,** F04

corrosion—the chemical or electrochemical reaction between a material, usually a metal, and its environment that produces a deterioration of the material and its properties. **G 15,** G01

corrosion fatigue—the process by which fracture occurs prematurely under conditions of simultaneous corrosion and repeated cyclic loading at lower stress levels or fewer cycles than would be required in the absence of the corrosive environment. **E 1823,** E08

corrosion fatigue—the process in which a metal fractures prematurely under conditions of simultaneous corrosion and repeated cyclic loading at lower stress levels or fewer cycles than would be required in the absence of the corrosive environment. **G 15,** G01

corrosion fatigue strength—the maximum repeated stress that can be endured by a metal without failure under definite conditions of corrosion and fatigue and for a specific number of stress cycles and a specified period of time. **G 15,** G01

corrosion inhibitor—a chemical substance or combination of substances that, when present in the proper concentration and forms in the environment, prevents or reduces corrosion. **G 15,** G01

corrosion inhibitor package, *n*—the combination of inhibitors added to an engine coolant to mitigate cooling system corrosion. **D 4725,** D15

corrosion of refractories, *n*—destruction of refractory surfaces by the chemical action of external agencies. **C 71,** C08

corrosion potential—the potential of a corroding surface in an electrolyte relative to a reference electrode measured under open-circuit conditions. **G 15,** G01

corrosion products—products that result from chemical or electrochemical reaction between a metal and its environment. **D 1129,** D19

corrosion products—products that result from chemical or electrochemical reaction between a metal and its environment. **D 6161,** D19

corrosion rate—the amount of corrosion occurring in unit time. (For example, mass change per unit area per unit time; penetration per unit time) **G 15,** G01

corrosion-resistant—term used to describe a material or treatment designed to inhibit corrosion. (See **rustproof.**) **F 547,** F16

corrosive substance—*in Committee D-16 Standards*, material in industrial aromatic hydrocarbons that discolors or tarnishes polished copper. **D 4790,** D16

corrosive sulfur—*n*, elemental sulfur and thermally unstable sulfur compounds in electrical insulating oil that can cause corrosion of certain transformer metals such as copper and silver. **D 2864,** D27

corrosive water, *n*—a standard solution containing 100 ppm each of sulfate, chloride, and bicarbonate ions introduced as the sodium salts to distilled water. **D 4725,** D15

corrosive wear, *n*—wear in which chemical or electrochemical reaction with the environment is significant. **G 40,** G02

corrugated—a term denoting an asbestos-cement sheet product having a design of alternating ridges and valleys manufactured according to a specified pitch. **C 460,** C17

corrugated, *adj*—pertaining to a sheet product having a design of sinusoidal alternating ridges and valleys manufactured according to a specific pitch. **C 1154,** C17

corrugated—denotes an asbestos-cement sheet product having a design of alternating ridges and valleys manufactured according to a specified pitch. **D 2946,** C17

corrugated box—See **box.** **D 996,** D10

corrugated fiberboard:

(*1*) *single face*—the structure formed by one corrugated member glued to the flat facing;

(*2*) *single wall*—the structure formed by one corrugated inner member glued between two flat facings; also known as double face;

(*3*) *double wall*—the structure formed by three flat facings and two intermediate corrugated members;

(*4*) *triple wall*—the structure formed by four flat facings and three intermediate corrugated members. (See also **containerboard.**)

corrugating medium—**paperboard** used in forming the fluted portion of the **corrugated board.**

corrugation flute— one of the wave shapes formed in the inner member, that is, the corrugating medium, of corrugated fiberboard. Flutes most commonly used are:

	Number, per Linear ft	Span Between Adjacent Flutes, mm	Height, in. (mm)[A,B]
A-flute	36 ± 3	7.9 to 9.1	3 / 16 (4.7)
B-flute	50 ± 3	4.7 to 6.6	3 / 32 (2.4)
C-flute	42 ± 3	6.8 to 7.8	9 / 64 (3.6)
E-flute	94 ± 4	3.0 to 3.5	3 / 64 (1.2)

[A] The values are approximate.
[B] Height does not include thickness of facing.

V-board—a term adopted from the grade symbol of **corrugated** or **solid fiberboard** made to comply with the weather-resistant class as defined in Federal Specification PPP-F-320, made of wet strength **paperboard.** Components are especially made to exhibit high strength (against bursting, tearing, or rupturing) when wet.

W-board—same as V-board except the "W" grades are of lower test requirements and primarily for use as interior or intermediate containers. **D 996,** D10

corrugated glass—glass rolled to produce a corrugated contour. **C 162,** C14

corrugated roofing/siding nail—See **roofing nail.** **F 547,** F16

corrugation, *n*—transverse undulations at regular intervals in the surface of the pavement consisting of alternate valleys and crests not more than 1 m (3 ft) apart. **E 867,** E17

corrugation, *n*—transverse undulations at regular intervals in the surface of the pavement consisting of alternate valleys and crests not more than 1 m (3 ft) apart. **E 1778,** E17

cortex, *n*—*in mammalian hair fibers*, the principal body of the fiber made up of elongated cells. **D 123,** D13

corundum—a naturally occurring hexagonal mineral of the composition Al_2O_3, which can also be prepared synthetically to high purity; noted for its hardness (9 on Mohs scale) and refractoriness (M.P. = 2045°C). **C 242,** C21

cosolvent—solvent used to improve the mutual solubility of other ingredients. **D 3064,** D10

cost analysis—subdividing the project estimate into component parts to find and compare their relationship to previously established historical costs. **E 833,** E06

cost effective—the condition whereby the present value benefits (savings) of an investment exceeds its present value costs. **E 833,** E06

Coster-Kronig transition—*AES, XPS* Auger process involving an electron from the same principal shell as the initial vacancy (for example, $L_1\ L_2M$). **E 673,** E42

cost limitations—the budget boundaries for project elements. **E 833,** E06

cost model—the description of the project divided into discrete elements showing quantities and unit price for each element. **E 833,** E06

cost overruns—in project design or construction, or both, the unanticipated increase in cost due to factors such as unanticipated site conditions, changes in the cost of building materials, labor, weather, and labor disputes. **E 833,** E06

cost professional—in project design or construction, or both, a person engaged in the occupation of understanding, managing or estimating project costs, or a combination thereof. **E 833,** E06

cotton, *n*—a vegetable seed fiber consisting of unicellular hairs attached to the seed of several species of the genus *Gossypium* of the family Malvaceae. **D 123,** D13

cotton cellulose, purified—chemical cellulose from cotton fiber or linters. (see also **chemical cotton**) **D 1695,** D01

cotton color diagram, *n*—a diagram showing the color ranges of standards officially established by the U.S. Department of Agriculture for the various grades of cotton in relation to scales of reflectance. R_d, on the vertical axis, and yellowness, $+b$, on the horizontal axis. **D 123,** D13

cotton count, *n*—an indirect yarn numbering system generally used in the cotton system equal to the number of 840-yd lengths of yarn per pound. **D 123,** D13

cotton linters—see **linters.** **D 1695,** D01

cotton linters, *n*—the short fibers adhering to cottonseed after the operation of ginning (seed removal and cleaning); cut from the seed in a series of passes through cutting blades and referred to as "first-cut linters," "second-cut linters," "mill run," and so forth; used primarily in the manufacture of cotton fiber content paper and cellulose derivatives. **D 1968,** D06

cotton maturity, *n*—the degree of fiber wall development. **D 123,** D13

cotton system, *n*—a spinning system adapted to fibers less than 65 mm (2.5 in.) in length. **D 123,** D13

cotton waste, *n*—material removed from seed cotton, ginned lint, or stock in process by any cleaning or processing machinery and usually consisting of undesirable fibers or a mixture of cotton fibers with foreign matter. **D 123,** D13

could not be identified/cannot identify—these terms are objectionable not only because they are ambiguous but also because they are biased; they imply that the examiner's task is only to identify the suspect, not to decide whether or not the suspect is the writer. If one of these terms is used, it should always be followed by "or eliminate[d]". **E 1658,** E30

coulomb—the quantity of electricity that is transmitted through an electric circuit in 1 s when the current in the circuit is 1 A. The quantity of electricity that will deposit 0.0011180 g of silver. **B 374,** B08

coulometer—an electrolytic cell arranged to measure the quantity of electricity by the chemical action produced in accordance with Faraday's law. **B 374,** B08

coulometer—an electrolytic cell or electronic device arranged to measure the quantity of electricity by the chemical action produced in accordance with Faraday's law. **C 859,** C26

count, *n—in woven textiles*, the number of warp yarns (ends) and filling yarns (picks) per unit distance as counted while the fabric is held under zero tension, and is free of folds and wrinkles. **D 123,** D13

count, *n—in knitted fabrics*, the number (counted units) of wale loops and course loops per 25 mm (1 in.). **D 123,** D13

count, *n*—the activity of measuring and recording traffic characteristics such as vehicle volume, classification, speed, weight, or a combination of these characteristics. **E 867,** E17

count—approximate number of nails per pound influenced by sizing of individual elements. **F 547,** F16

count, acoustic emission (emission count) (N)—the number of times the acoustic emission signal exceeds a preset threshold during any selected portion of a test. **E 1316,** E07

counter—a device for the measurement of radiation intensity by means of an electrical triggering principle (Geiger-Müller). **E 7,** E04

counter—a piece of stiffening material inserted between the lining and the outside of a shoe upper at the back of the shoe. The purpose of the counter is to strengthen the back part and to prevent it from sagging and losing its shape. There are two types of counter: flat and molded. **F 869,** F08

countercurrent adsorption—an adsorption process in which the flow of fluid is in a direction opposite to the movement of the adsorbent. **D 2652,** D28

counter electrode, *n—in optical emission spectrometry*, the electrode in an analytical pair that does not contain the specimen being analyzed. **E 135,** E01

counter electrode—the electrode in an electrochemical cell that is used to transfer current to or from a test electrode. **G 15,** G01

counter electrode (electrochemical)—an electrode in an electrochemical cell that is used to transfer current to or from a working or a reference electrode, or both. **C 859,** C26

counterfeit—the reproduction of a document, article, or security feature with the intent to deceive the close scrutiny of a qualified examiner. **F 1156,** F12

counterflashing—formed metal or elastomeric sheeting secured on or into a wall, curb, pipe, roof-top unit, or other surface, to cover and protect the upper edge of a base flashing and its associated fasteners. **D 1079,** D08

counterfoil—the stub of a check, postal money order, receipt, etc., kept by the user as a record of the transaction, such as a rectangular shaped security label affixed so it overlaps the two center pages on the stitching of a United States Non-Immigrant Visa (NIV). **F 1156,** F12

counter pocket—a piece of lining material attached to inside quarters of unlined shoes to conceal counter. **F 869,** F08

countersunk head—head with bearing surface conically shaped: "deep" having 32 to 74-deg included angle; "medium" having 75 to 99-deg included angle; "slight" having 100 to 160-deg included angle. **F 547,** F16

count, event (N_e)—the number obtained by counting each discerned acoustic emission event once. **E 1316,** E07

counting method—*in fatigue spectrum loading*, a method of counting the occurrences and defining the magnitude of various loading parameters from a load-time history; (some of the counting methods are: level crossing count, peak count, mean crossing peak count, range count, range-pair count, rain-flow count, racetrack count). **E 1823,** E08

count median size, *n*—a measurement of particle size of samples of particulate matter, consisting of that diameter of particle such that one half of the number of particles is larger and half is smaller. **D 1356,** D22

count period, *n*—the beginning and ending time of traffic characteristic measurement. **E 867,** E17

count rate, acoustic emission (emission rate or count rate) ($\dot{N}$)—the time rate at which emission counts occur. **E 1316,** E07

count, ring-down—see **count, acoustic emission,** the preferred term. **E 1316,** E07

country hides—hides taken off by butchers and farmers; their quality is usually lower than that of packer hides because they are removed by less skilled hands and are not cured as well as packer hides. **D 1517,** D31

counts—*EIA,* events recorded by the detector and registered in a channel of a backscattering spectrum. Counts constitute the ordinate of a BS spectrum.

integrated—sum of all counts registered in a given set of channels or their corresponding energy ranges.

per channel—number of counts in a single channel; unit of the ordinate of a backscattering spectrum. **E 673,** E42

count type, *n*—the traffic characteristic being measured, the measurement device, and period. **E 867,** E17

couplant—a material used at the structure-to-sensor interface to improve the transmission of acoustic energy across the interface during acoustic emission monitoring. **E 1316,** E07

couplant—a substance used between the search unit and test surface to permit or improve transmission of ultrasonic energy. **E 1316,** E07

coupler, color—a substance capable of reacting with the oxidation product of a color-forming developer to produce a colored dye image. **E 7,** E04

coupling, *n—in asbestos-cement conduit, sewer, underdrain and storm drain pipe,* component made from a larger diameter pipe of the same type or type II and of the same class, or of a higher class, or produced otherwise to yield at least equal performance, for joining asbestos-cement pipe that when properly installed, forms a silt-tight joint, allows alignment corrections and slight changes in direction, and provides an assembled joint equivalent in serviceability and strength to the pipe sections. Alternatively, for storm drain couplings, plastic sleeves that, when properly installed develop sufficient tightness to prevent the surrounding soil from entering the drain, may be used as couplings. **C 460,** C17

coupling, *n—in asbestos-cement nonpressure sewer pipe,* a section for joining asbestos-cement nonpressure sewer pipe, that when properly installed with the proper accessories, develops an assembled joint equivalent in serviceability and strength to the pipe sections when tested in accordance with 8.3 of Specification C 428. **C 460,** C17

coupling, *n—in asbestos-cement pressure pipe,* section for joining asbestos-cement pipe that, when properly installed with the proper accessories, develops a joint equivalent in strength and serviceability to the pipe sections. **C 460,** C17

coupling, *n—in nonpressure sewer pipe,* a section for joining nonpressure sewer pipe, that when properly installed with the proper accessories, develops an assembled joint equivalent in serviceability and strength to the pipe sections, when tested in accordance with 8.3 of Specification C 428. **C 1154,** C17

coupling, *n—in asbestos-cement conduit, sewer, underdrain, and storm drain pipe,* component made from a larger diameter pipe of the same type or type II and of the same class, or of a higher class, or produced otherwise to yield at least equal performance, for joining asbestos-cement pipe that when properly installed, forms a silt-tight joint, allows alignment corrections and slight changes in direction, and provides an assembled joint equivalent in serviceability and strength to the pipe sections. Alternatively, for storm drain couplings, plastic sleeves that, when properly installed develop sufficient tightness to prevent the surrounding soil from entering the drain, may be used as couplings. **D 2946,** C17

coupling, *n—in asbestos-cement nonpressure sewer pipe,* section for joining asbestos-cement nonpressure sewer pipe, that when properly installed with the proper accessories, develops an assemble joint equivalent in serviceability and strength to the pipe sections when tested in accordance with 8.3 of Specification C 428. **D 2946,** C17

coupling, *n—in asbestos-cement pressure pipe,* section for joining asbestos-cement pipe that, when properly installed with the proper accessories, develops a joint equivalent in strength and serviceability to the pipe sections. **D 2946,** C17

coupling—two electric circuits are said to be coupled to each other when they have an impedance in common so that a current in one causes a voltage in the other. **E 1316,** E07

coupling agent, *n*—a molecule, having different or like functional groups, that is capable of reacting with surface molecules of two different substances, thereby chemically bridging the substances. **D 907,** D14

coupling coefficient, *k′*—the ratio of the mutual inductance between two windings and the geometric mean of the individual self-inductances of the windings. **A 340,** A06

coupon, *n*—a portion of a material or laboratory sample from which multiple specimens can be taken for testing. **D 4439,** D35

course, *n—in knitted fabrics,* the series of successive loops lying crosswise of a knitted fabric, that is, lying at right angles to a line passing through the open throat to the closed end of the loops. **D 123,** D13

course administrator—an individual responsible for managing administrative details of a course, separate from actual instruction of the course. **F 1177,** F30

course instructor/coordinator (I/C)—an individual who is authorized by the appropriate entity to present and assess competence in all of the subject matter contained in a curriculum. This person also oversees all instruction in the course and makes final evaluations concerning student competence. **F 1177,** F30

cover, *n—in yarns,* the outside layer of fibers that form the surface of a yarn. **D 123,** D13

cover—the perpendicular distance from any point in the roof of an underground opening to the ground surface. (ISRM) **D 653,** D18

cover—*in grouting,* the thickness of rock and soil material overlying the stage of the hole being grouted. **D 653,** D18

cover—See **container.** **D 996,** D10

cover—the section of the lock case designed to be removed for access to components within the case. **F 471,** F12

cover—an electrically insulated enclosure designed to be installed temporarily on various types of irregularly shaped electrical equipment to protect personnel and equipment working in close proximity. **F 819,** F18

coverage, *n*—the area to be covered per unit volume of coating to obtain specified dry thickness and desired performance. **C 168,** C16

coverage—the surface area to be continuously covered by a specific quantity of a particular material. **D 1079,** D08

coverage, coverage rate, covering power— ambiguous terms that are used in some instances to refer to hiding power and in others to mean spreading rate. The precise terms **hiding power** and **spreading rate** are preferred. **D 16,** D01

cover coat—a porcelain enamel finish applied and fused over a ground coat or direct to the metal substrate. **C 286,** B08

covered yarn, *n*—a compound structure which contains distinguishable inner and outer fibrous elements which can be different. **D 123,** D13

cover flange—See **railing systems.** **E 631,** E06

cover flange— Synonym for **escutcheon**. **E 1481,** E06

covering capacity, dry, *n*—the area covered to a dry thickness of 1 in. (25 mm) by 100 lb (45.4 kg) of dry cement when mixed with the recommended amount of water, molded and dried to constant weight. **C 168,** C16

covering capacity, wet, *n*—the area covered to a wet thickness of 1 in. (25 mm) by 100 lb (45.4 kg) of dry cement when mixed with the recommended amount of water, and molded. **C 168,** C16

covering power—the ability of a plating solution under a specified set of plating conditions to deposit metal on the surfaces of recesses or deep holes. (To be distinguished from throwing power.) **B 374,** B08

covering power—the ability of a glaze to cover the surface of the fired ware uniformly and completely. **C 242,** C21

covering power—the degree to which a porcelain enamel coating obscures the underlying surface. **C 286,** B08

cover plate—Synonym for **escutcheon**. **E 1481,** E06

cover plate, collector—a sheet of transparent (or translucent) glazing placed above the absorber in a solar collector, to provide thermal and environmental protection. **E 772,** E44

cover ring—Synonym for **escutcheon**. **E 1481,** E06

coverstrip, *n*—a thin layer of black compound which covers the unexposed white sidewall portion of a finished tire. **F 538,** F09

covert—discrete; not readily discernible; hidden. **F 1156,** F12

cover-up—the obliteration of one or more images by means of an opaque material similar in color to the substrate. **F 221,** F05

cowhide—term specifically applied to leather made from hides of cows, although it is sometimes loosely used to designate any leather tanned from hides of animals of the bovine species. **D 1517,** D31

CPU—chloroplatinate unit (color indicator). **D 6161,** D19

CR—abbreviation for **carriage return**. **F 1457,** F05

crack—a partial break in the stone (see fracture, microcrack, seam). **C 119,** C18

crack—an irregular separation with well-defined sharp edges visible on the surface of a pipe. **C 896,** C04

crack, *n—as used in fractography,* a plane of fracture without complete separation. **C 1145,** C28

crack—a small fracture, that is, small with respect to the scale of the feature in which it occurs. (ISRM) **D 653,** D18

crack—a visual separation that occurs internally or penetrates down from the pultruded surface to the equivalent of one full ply or more of reinforcement. **D 3918,** D20

crack, *n*—fissure or discontinuity of the pavement surface not necessarily extending through the entire thickness of the pavement. **E 867,** E17

crack, *n*—fissure or discontinuity of the pavement surface not necessarily extending through the entire thickness of the pavement. **E 1778,** E17

crack—line of fracture without complete separation.

F 109, C21

crack—any narrow opening or fissure in the surface that is visible to the naked eye. **F 412,** F17

crack—*on semiconductor wafers,* cleavage or fracture that extends to the surface of a wafer and which may or may not pass through the entire thickness of the wafer. **F 1241,** F01

crack—a crystalline fracture passing through or along the grain boundaries which is normally caused by overstressing the metal during manufacturing, such as forging, forming, or heat treating. **F 1789,** F16

crack and crevice application—application by a means that projects the material into cracks and crevices of a structure. **E 1102,** E35

crack-arrest fracture toughness, K_a **[$FL^{-3/2}$]**—the value of the stress-intensity factor shortly after crack arrest. **E 1823,** E08

crack (*building defect*), *n*—a flaw consisting of complete or incomplete separation within a single element or between contiguous elements of constructions. **E 631,** E06

crack deflection, *n*—a toughening mechanism in advanced ceramics or ceramic matrix composites characterized by fracture surface roughening and crack tilting/twisting during propagation around grains or a reinforcing component caused by stress fields around the grains or component developed through mismatches in thermal expansion or mechanical properties (such as elastic modulus), or both, between grains or between reinforcement and matrix. **C 1145,** C28

crack depth, a **[L]**—*in part-through surface-crack specimens (PS),* the distance from, and normal to, the cracked plate surface to the point of maximum penetration of the crack front into the material. Crack depth is less than the specimen thickness. **E 1823,** E08

crack displacement [L]—the load-induced separation vector between two points (on the facing surfaces of a crack) that were initially coincident. **E 1823,** E08

cracker, *n*—heavy-duty mill having two deeply corrugated or pyramid-cut rolls for breaking down a rubber or a mix, or for cutting rubber or a mix into pieces. **D 1566,** D11

crack extension, Δa **[L]**—an increase in crack size. **E 1823,** E08

crack-extension force, G **[FL^{-1} or FLL^{-2}]**—the elastic energy per unit of new separation area that is made available at the front of an ideal crack in an elastic solid during a virtual increment of forward crack extension. **E 1823,** E08

crack-extension resistance, K_R **[$FL^{-3/2}$],** G_R **[FL^{-1}] or** J_R **[FL^{-1}]**—a measure of the resistance of a material to crack extension expressed in terms of the stress-intensity factor, K; crack-extension force, G; or values of J derived using the J-integral concept. **E 1823,** E08

crack filler, *n*—bituminous material used to fill and seal cracks in existing pavements. **D 8,** D04

cracking—a defect in the bisque consisting of fractures or separations. **C 286,** B08

cracking—*in protective coatings,* the formation of breaks in a coating film that extend through to the underlying surface. **D 4538,** D33

cracking—in leak testing, same as **dissociation**. **E 1316,** E07

cracking, *n*—a condition where ink that has been absorbed into a substrate causes the coating to shrink to a state much smaller than the original coating dimension causing fractures in the image area. **F 1857,** F05

cracking, alligator (crocodile)—see **alligator (crocodile) cracking.** **E 867,** E17

cracking, block—see **block cracking.** **E 867,** E17

cracking (coatings), *n*—phenomenon manifested in paint films by a break extending through to the surface painted. **E 1605,** E06

cracking, durability "D"—see **durability "D" cracking.** **E 867,** E17

cracking, edge—see **edge cracking.** **E 867,** E17

cracking load—a load which creates a bending moment of enough magnitude to produce a tensile stress greater than the sum of the induced compression plus the tensile strength of the concrete resulting in tensile cracks on the tension face of the pole. **C 1213,** C27

cracking, longitudinal—see **longitudinal cracking.** **E 867,** E17

cracking, map—see **map cracking.** **E 867,** E17

cracking, reflection, at joints—see **reflection cracking at joints.**

E 867, E17

cracking resistance (*coatings*)—the ability of a coating to resist breaks of the film where the breaks extend through to the surface painted and the previously applied coating or the substrate is visible. The use of a minimum magnification of 10 diameters is recommended in cases where it is difficult to differentiate between cracking and checking. See **checking resistance.** **D 16,** D01

cracking, slippage—see **slippage cracking.** **E 867,** E17

cracking, transverse—see **transverse cracking.** **E 867,** E17

crack jump behavior— *in chevron-notch specimen tests*, that type of sporadic crack growth which is characterized primarily by periods during which the crack front is nearly stationary until a critical load is reached, whereupon the crack becomes unstable and suddenly advances at high speed to the next arrest point, where it remains nearly stationary until the load again reaches a critical value, and so forth. **E 1823,** E08

crackled—glassware, the surface of which has been intentionally cracked by water immersion and partially healed by reheating before final shaping. **C 162,** C14

crackled—a mottled textural effect in a wet process porcelain enamel resembling a wrinkled surface. **C 286,** B08

crack length, *a* [L]—See **crack size** and **surface crack length.**. **E 1823** **C 286,** B08

crack length, *a* [L]—in Test Method E 1457, the physical crack size is represented as a_p . The subscript *p* is everywhere implied. **E 1823,** E08

crack mark, *n*—an open place causing a streak of variable length approximately parallel to the length or width. **D 123,** D13

crack mark, *n—in bonded, fused, or laminated fabrics*, a sharp break or crease in the surface contour of either the face fabric or the backing fabric that becomes evident when the bonded, fused, or laminated composite is rolled, bent, draped, or folded. **D 123,** D13

crack-mouth opening displacement (*CMOD*), $2v_m$ [L]—the Mode 1 (also called opening-mode) component of crack displacement resulting from the total deformation (elastic plus plastic), measured under load at the location on a crack surface that has the greatest elastic displacement per unit load. **E 1823,** E08

crack-off—the process of severing a glass article by breaking, as by scratching and then heating. **C 162,** C14

crack-plane orientation—an identification of the plane and direction of a fracture in relation to product configuration. This identification is designated by a hyphenated code with the first letter(s) representing the direction normal to the crack plane and the second letter(s) designating the expected direction of crack propagation. **E 1823,** E08

cracks, *n*—fissure(s) originating in the surface of a rubber vulcanizate or product as a result of natural weathering. **D 1566,** D11

cracks, flex, *n*—fissures originating in the surface of a rubber vulcanizate, resulting from cyclic deformation (usually bending). **D 1566,** D11

crack size, *a* [L]—a lineal measure of a principal planar dimension of a crack. This measure is commonly used in the calculation of quantities descriptive of the stress and displacement fields and is often also termed crack length or depth. **E 1823,** E08

crack(s), ozone, *n*—fissure(s) originating in the surface of a rubber vulcanizate, caused by exposure to an ozone-containing environment; these fissures are perpendicular to the direction of strain, and usually occur in rubbers having main chain unsaturation. **D 1566,** D11

crack strength, σ_c [FL^{-2}]—the maximum value of the nominal stress that a cracked structure is capable of sustaining. **E 1823,** E08

crack-tip opening displacement (*CTOD*), δ, [L]—the crack displacement resulting from the total deformation (elastic plus plastic) at variously defined locations near the original (prior to load application) crack tip. **E 1823,** E08

crack-tip plane strain—a stress-strain field (near the crack tip) that approaches plane strain to the degree required by an empirical criterion. **E 1823,** E08

crack-tip plane stress—a stress-strain field (near the crack tip) that is not in plane strain. **E 1823,** E08

craftsman, *n*—an artisan who is skilled in creating new cultural objects. **D 123,** D13

crash towel, *n*—a plain weave nonterry product with hems or selvages which has a rough texture caused by uneven yarns. **D 123,** D13

crate, *n*—a rigid **shipping container** of framed construction joined together with nails, bolts or any equivalent method of fastening. The framework may or may not be enclosed with sheathing. It may be demountable (reusable) or nondemountable. (See also **rubbing strip, strut.**)

open crate—a **crate** with exposed frame members and not enclosed by **sheathing.**

sheathed crate— a **crate** that is enclosed by having the frame members completely covered with **sheathing** boards or material. **D 996,** D10

crater—excavation (generally of conical shape) generated by an explosive charge. (ISRM) **D 653,** D18

crater, *n*—a small, shallow surface imperfection. **D 883,** D20

crater—a small, shallow pultrusion surface imperfection. **D 3918,** D20

crater, *n*—a small, shallow surface imperfection. (D-20) **F 412,** F17

crater edge effect—*SIMS,* a signal caused by secondary ions that originate from depths shallower than the maximum depth of the crater formed by primary bombardment. **E 673,** E42

cratering—*in protective coatings*, the formation of round depressions in a coating film that do not expose the previous coat or the substrate. **D 4538,** D33

cratering—a surface texture of irregular closed ridges with smooth central regions. **F 1241,** F01

crawling—a parting and contraction of the glaze on the surface of ceramic ware during drying or firing, resulting in unglazed areas bordered by coalesced glaze. **C 242,** C21

crawling—a defect in the porcelain enamel appearing as agglomerates or irregularly shaped islands. **C 286,** B08

crawling, *n*—defect in which the wet film recedes from localized areas of the substrate (usually caused by insufficient wetting) leaving those areas uncoated. **D 16,** D01

crawling—*in protective coatings*, a defect in which a wet coating film recedes from a small area to form an uneven surface shortly after application. **D 4538,** D33

crawling, *n*—a nonuniform contraction of the printing ink on the substrate due to poor wetting. Also referred to as *reticulation or pearling*. **D 6488,** D01

craze—multiple fine cracks at or under the pultruded surface. **D 3918,** D20

craze—to produce many small cracks or fracture lines. **F 1156,** F12

craze, crazing—a defect appearing as one or more fine cracks in the porcelain enamel. **C 286,** B08

craze, hairline—multiple fine pultrusion surface separation cracks that exceed ¼ in. (6.4 mm) in length and do not penetrate in depth to the equivalent of full ply of reinforcement. **D 3918,** D20

craze, resin—multiple fine separation cracks at the pultruded surface not penetrating into the reinforcement.

NOTE—This condition is usually due to resin shrinkage during cure in resin-rich areas. **D 3918,** D20

craze, star—multiple fine pultrusion surface separation cracks exceeding ¼ in. (6.4 mm) in length but not penetrating in depth to the equivalent of a full ply of reinforcement, that appear to emanate from a central point.

NOTE—This condition is often caused by impact damage. **D 3918,** D20

crazing—a network of fine hairline cracks in a coating. **B 374,** B08

crazing—the cracking that occurs in fired glazes or other ceramic coatings as a result of tensile stresses. May also occur in the surface portion of uncoated (unglazed) whiteware bodies. **C 242,** C21

crazing, *n*—apparent fine cracks at or under the surface of a plastic. **D 883,** D20

crazing, *n*—fine cracks that may extend in a network on or under the surface of or through a layer of adhesive. **D 907,** D14

crazing, *n*—formation of a random pattern of shallow cracks on a rubber surface usually due to ageing by light. **D 1566,** D11

crazing—*in protective coatings,* the formation of a criss-cross pattern of minute cracks on the surface of a coating film. **D 4538,** D33

crazing, *n*—a network of apparent fine cracks on or beneath the surface of materials such as in transparent plastics, glazed ceramics, glass, or clear coatings. **E 284,** E12

crazing—the development of a multitude of very fine cracks in a material such as ceramic glaze, varnish, paint, etc., often the result of exposure to sunlight, weathering, or certain solvents. **E 1749,** E06

crazing, *n*—apparent fine cracks at or under the surface of a plastic. **F 412,** F17

crazing, *n*—minute, closely grouped, generally superficial cracks that usually results from light activated oxidation. **F 538,** F09

crazing, *n*—apparent fine cracks at or under the surface of a plastic. **F 1251,** F04

crazing—a network of checks or cracks appearing on a coated surface. **G 15,** G01

CRC—abbreviation for **cyclic redundancy check**. **F 1457,** F05

CRE—abbreviation for constant-rate-of-extension. **D 123,** D13

creaming—the separation of a layer of the dispersed phase of an emulsion polish to the surface of the liquid continuous phase. **D 2825,** D21

crease, *n*—a fabric defect evidenced by a break, line, or mark in a fabric generally caused by a sharp fold. **D 123,** D13

crease mark, *n*—a visible deformation left in a fabric after a crease has been incompletely removed during fabric processing. **D 123,** D13

crease retention, *n*—that property of a fabric which enables it to maintain an inserted crease. **D 123,** D13

creep, *n*—the time dependent part of a strain resulting from stress. **C 717,** C24

creep—time dependent deformation of a material under load. **C 904,** C03

creep—slow movement of rock debris or soil usually imperceptible except to observations of long duration. Time-dependent strain or deformation, for example, continuing strain with sustained stress. **D 653,** D18

creep, *n*—the time-dependent part of strain resulting from stress. **D 883,** D20

creep, *n*—*in an adhesive,* the time-dependent increase in strain resulting from a sustained stress. **D 907,** D14

creep—the time-dependent part of a strain resulting from stress. **D 1079,** D08

creep, *n*—the time-dependent part of a strain resulting from stress. **D 1566,** D11

creep, *n*—the time-dependent increase in accumulative strain in a material resulting from an applied constant force. **D 4439,** D35

creep, *n*—the time-dependent increase in strain in a solid resulting from force. **E 6,** E28

creep, *n*—the time-dependent part of strain resulting from stress. (D-20) **F 412,** F17

creep, *n*—the time-dependent part of strain resulting from stress. **F 1251,** F04

creep crack growth behavior—a plot of the time rate of crack growth, da/dt, as a function of C*(t). **E 1823,** E08

creep recovery, *n*—the time-dependent decrease in strain in a solid, following the removal of force. **E 6,** E28

creep rupture strength [FL^{-2}], *n*—the stress that will cause fracture in a creep test at a given time, in a specified constant environment. **E 6,** E28

creep strain, *n*—the total strain, at any given time, produced by the applied stress during a creep test. (A) **D 907,** D14

creep strength [FL^{-2}], *n*—the stress that causes a given creep in a creep test at a given time in a specified constant environment. **E 6,** E28

creep zone boundary—the locus of points ahead of the crack front where the equivalent strain caused by the creep deformation equals the equivalent strain caused by the elastic deformation. **E 1823,** E08

crenulations—multiple kinks that can increase the apparent fiber cohesion. **D 2946,** C17

creped duplex paper— See **paper.** **D 996,** D10

creped paper—See **paper.** **D 996,** D10

crepe rubber—originally unvulcanized natural rubber with light color and knobbly surface for soles and heels. Most crepe rubber for shoes is now of synthetic elastomers. **F 869,** F08

crescent crack—damage having the appearance of a crescent, produced in a glass surface by the frictive translation of a hard, blunt object across the glass surface. The crescent shape is concave toward the direction of translation on the damaged surface. **C 162,** C14

crescents—structures with parallel major axes, attributed to substrate defects either above or below the surface plane of silicon substrates after epitaxial deposition (see also **fishtails**). **F 1241,** F01

cresol(s)—methyl phenol, hydroxymethyl benzene (C_7H_8O) mol weight 108.13; colorless, yellowish, brownish, yellow, or pinkish liquid; phenolic odor. Three isomeric cresols exist. **D 4790,** D16

crest—outermost part of thread, joining flanks. (See **double crest.**) **F 547,** F16

crest diameter—twice the distance between nail axis and crest, measured perpendicular to nail axis. Thread crest diameters for given shank diameters:

Wire Diameter, in.	Helically Threaded Nails Stiff-stock, in.	Helically Threaded Nails Hardened, in.	Annularly Threaded Nails, Stiff-Stock and Hardened, in.
0.120	0.134 to 0.138	0.140 to 0.144	0.132 to 0.136
0.135	0.161 to 0.166	0.161 to 0.166	0.149 to 0.154
0.148	0.173 to 0.180	0.173 to 0.180	0.161 to 0.166
0.165	0.196 to 0.201	0.196 to 0.201	0.180 to 0.185
0.177	0.209 to 0.213	0.209 to 0.213	0.193 to 0.197

F 547, F16

crest factor, *cf*—the ratio of the maximum value of a periodically alternating quantity to its rms value.

NOTE—For a sinusoidal variation the crest factor is $\sqrt{2}$. **A 340,** A06

cresylic acids—commercial mixtures of phenolic materials which may include phenol, cresols, xylenols, and other alkylated phenols. **D 4790,** D16

crevice corrosion—localized corrosion of a metal surface at, or immediately adjacent to, an area that is shielded from full exposure to the environment because of close proximity between the metal and the surface of another material. **G 15,** G01

cricket, *n*—a construction to divert water around or away from a chimney, curb, wall, expansion joint, or other penetration. (See **saddle**). **D 1079,** D08

criminalistics, *n*—a brance of forensic science concerned with the examination and interpretation of physical evidence, for the purpose of aiding forensic investigation. **E 1732,** E30

crimp, *n—as applied to a continuous-element zipper,* the predetermined formation of the monofilament cross-section at the point where the continuous element is interlocked. **D 123,** D13

crimp—an operation that mechanically seals the valve to the container. **D 3064,** D10

crimp contraction, *n*—an indicator of crimp capacity or a characterization of a yarn's ability to contract under tension. **D 123,** D13

crimp development medium, *n—for testing of textured yarn,* an environment that allows the temporary set of fiber crimp to be overcome and that allows the filaments to assume their permanently set configuration. **D 123,** D13

crimped head—head with undulating or pronged rim, or both. **F 547,** F16

crimp frequency, *n—in man-made staple fibers,* the number of crimps or waves per unit length of extended or straightened fiber. **D 123,** D13

crimp index, *n*—an indirect measure of the amplitude of the crimp. **D 123,** D13

crimp recovery, *n*—a measure of the ability of a yarn to return to its original crimped state after being subjected to tension. **D 123,** D13

crinkled—a textural effect in a porcelain enamel surface having the appearance of fine wrinkles or ridges. **C 286,** B08

criterion, *n*—An established precedent, rule, measure, norm, or **code** upon which a decision may be based. **E 631,** E06

criterion—See **requirement statement**. **E 1480,** E06

criterion of failure—complete separation, or the presence of a crack of specified length visible at a specified magnification. Other criteria may be used but should be clearly defined. **E 1823,** E08

critical angle—the incident angle of the ultrasonic beam beyond which a specific refracted wave no longer exists. **E 1316,** E07

critical angle, θ_c—the angle whose sine is equal to the relative refractive index for light striking an interface from the greater to the lesser refractive medium: $\theta_c = \sin^{-1} n_{21}$, where n_{21} = the ratio of the refractive indices of the two media. **E 131,** E13

critical anodic current density—the maximum anodic current density observed in the active region for a metal or alloy electrode that exhibits active-passive behavior in an environment. **G 15,** G01

critical bed depth—the minimum depth of an adsorbent bed required to contain the mass transfer zone. **D 2652,** D28

critical circle (critical surface)—the sliding surface assumed in a theoretical analysis of a soil mass for which the factor of safety is a minimum. **D 653,** D18

critical cooling rate, *n*—the slowest rate of continuous cooling at which austenite can be cooled from above the **transformation range** to prevent its transformation above M_s. **A 941,** A01

critical crack length [L]—the crack length in a chevron-notch specimen at which the specimen's stress-intensity factor coefficient, *Y**, is a minimum, or equivalently, the crack length at which the maximum load would occur in a purely linear elastic fracture mechanics test. At the critical crack length, the width of the crack front is approximately one third the dimension B. **E 1823,** E08

critical current density—a current density above which a new and sometimes undesirable reaction occurs. **B 374,** B08

critical current density, J_c—the critical current divided by the total cross-sectional area of the conductor. **B 713,** B01

critical current, I_c—the maximum electrical current below which a superconductor exhibits superconductivity at some given temperature and magnetic field. **B 713,** B01

critical curve—in a binary, or higher order, phase diagram, a line along which the phases of a heterogeneous equilibrium become identical. **E 7,** E04

critical curve—in a binary, or higher order, phase diagram, a locus of points along which two or more phases exist in stable thermodynamic equilibrium. **E 1142,** E37

critical damping—the minimum viscous damping that will allow a displaced system to return to its initial position without oscillation. **D 653,** D18

critical damping (*in dynamic mechanical measurement*)—that damping required for the borderline condition between oscillatory and nonoscillatory behavior. **D 4092,** D20

critical defect, *n*—a serious defect that judgment and experience indicate is likely to prevent the usability or proper performance of a product from its intended purpose. **D 123,** D13

critical density—the unit weight of a saturated granular material below which it will lose strength and above which it will gain strength when subjected to rapid deformation. The critical density of a given material is dependent on many factors. **D 653,** D18

critical difference, *n*—the observed difference between two test results, which should be considered significant at the specified probability level. **D 123,** D13

critical flow—open channel flow in which the energy, expressed in terms of depth plus velocity head, is a minimum for a given flow rate and channel. The Froude number is unity at critical flow. **D 4410,** D19

critical flux at extinguishment, *n*—a flux level at the specimen surface corresponding to the distance of farthest advance and subsequent self-extinguishment of the flame on the centerline of a specimen. **E 176,** E05

critical frequency, f_c—frequency at which maximum or minimum amplitudes of excited waves occur. **D 653,** D18

critical half thickness, (a), *n*—an estimation of the half thickness of a sample in an *unstirred container,* in which the heat losses to the environment are less than the retained heat. This buildup of internal temperature leads to a thermal-runaway reaction. **E 1445,** E27

critical height *(ch), n*—the maximum exposed height of a cone or pyramid that will not cause a puncture failure of a geosynthetic at a specified hydrostatic pressure for a given period of time. **D 4439,** D35

critical height, H_c **(L)**—the maximum height at which a vertical or sloped bank of soil or rock will stand unsupported under a given set of conditions. **D 653,** D18

critical humidity—the relative humidity above which the atmospheric corrosion rate of some metals increases sharply. **G 15,** G01

critical hydraulic gradient—see **hydraulic gradient.** **D 653,** D18

critical illumination—see **illumination, critical.** **E 175,** E41

criticality—*in the nuclear industry*, the condition wherein a system or medium is capable of sustaining a nuclear chain reaction. **C 859,** C26

criticality controls—those mechanisms which provide a high level of assurance that the probability of occurrence of a critical excursion is acceptably low. **C 859,** C26

critical magnetic field strength, H_{c2}—the maximum magnetic field below which a superconductor exhibits superconductivity at zero current and temperature. **B 713,** B01

critical mass—*in the nuclear industry*, the minimum mass of fissile material that can be made critical with a specified geometrical arrangement and material composition. **C 859,** C26

critical path method—method of scheduling in a detailed manner the essential steps or actions that must be taken in sequence from the start to the completion of a construction project. **E 1605,** E06

critical pitting potential—the least noble potential where pitting corrosion will initiate and propagate. (See breakdown potential.) **G 15,** G01

critical point—in a phase diagram, that specific value of composition, temperature, pressure or combinations thereof at which the phases of a heterogeneous equilibrium become identical. (See also **transformation temperature.**) **E 7,** E04

critical point—in a binary phase diagram, that specific value of composition, temperature, pressure, or combinations thereof at which the phases of a heterogeneous equilibrium become identical. **E 1142,** E37

critical pressure—that pressure above which the liquid and vapor states are no longer distinguishable. **E 7,** E04

critical pressure—that pressure at the critical point. **E 1142,** E37

critical property, *n*—a quantitatively measurable characteristic which is absolutely necessary to be met if a material or product is to provide satisfactory performance for the intended use. **D 1711,** D09

critical radiant flux, *n*—the level of incident radiant heat energy on the floor covering system at the most distant flame-out point. It is reported as $W/cm^2(Btu/ft^2 \cdot s)$. **E 176,** E05

critical radiant flux, *n*—the level of incident radiant heat energy on the attic floor insulation system at the most distant flame-out point. It is reported as W/cm^2(or $Btu/ft^2 \cdot s$). **E 176,** E05

critical slip angle, (rad or degree), *n*—the value of the slip angle at the peak lateral force coefficient. **F 538,** F09

critical slope—the maximum angle with the horizontal at which a sloped bank of soil or rock of given height will stand unsupported. **D 653,** D18

critical slope ratio, r_c—the unloading slope ratio at the critical crack length. **E 1823,** E08

critical surface—see **critical circle.** **D 653,** D18

critical surface—in a ternary, or higher order, phase diagram, the area upon which the phases in equilibrium become identical. **E 7,** E04

critical surface—in a ternary or higher order phase diagram, the area upon which the phases in equilibrium become identical, (E 7, E-4). **E 1142,** E37

critical temperature—that temperature above which the vapor phase cannot be condensed to liquid by an increase in pressure. **E 7,** E04

critical temperature—that temperature at the critical point. **E 1142,** E37

critical temperature, T_c—the maximum temperature below which a superconductor exhibits superconductivity at zero magnetic field and current. **B 713,** B01

critical temperature, (T_c), *n*—an estimation of the lowest temperature of an *unstirred container* at which the heat losses to the environment are less than the retained heat leading to a buildup of internal temperature. This temperature buildup leads to a thermal-runaway reaction. **E 1445,** E27

critical transition temperature—see **transition temperature.** **B 713,** B01

critical void ratio—see **void ratio.** **D 653,** D18

critical wax strength number, *n*—*in paper surface strength,* the average highest numerical designation of wax that does not disturb the surface of the paper whose surface strength is tested by the wax pick method under specified conditions. See **pick.** **D 1968,** D06

crizzle—an imperfection in the form of a multitude of fine surface fractures. **C 162,** C14

CRL—abbreviation for constant-rate-of-loading. **D 123,** D13

crocidolite—common name for riebeckite asbestos. Also known as blue asbestos. **C 460,** C17

crocidolite, *n*—common name for **riebeckite asbestos**. Also known as blue asbestos. **D 2946,** C17

crocking, *n*—a transfer of color from the surface of a colored fabric to an adjacent area of the same fabric or to another surface principally by rubbing action. (Compare **color staining, color bleeding**.) **D 123,** D13

crocking—the transfer of finish or color when leather is rubbed with a wet or dry cloth. **D 1517,** D31

crook—a distortion of a piece of lumber in which there is a deviation in a direction perpendicular to the edge from a straight line from end to end of the piece. **D 9,** D07

Crookes glass—a glass having low transmission for ultraviolet light, and containing cerium and other rare earths. **C 162,** C14

crop—a side with the belly trimmed off. **D 1517,** D31

crop oil concentrate—an emulsifiable petroleum oil-based product containing 15 to 20 % w/w surfactant and a minimum of 80 % w/w phytobland oil. **E 1519,** E35

crop oil (emulsifiable)—an emulsifiable petroleum oil-based product containing up to 5 % w/w surfactant and the remainder of a phytobland oil. **E 1519,** E35

crop oil (non-emulsifiable)—See **phytobland oil.** **E 1519,** E35

cross band, *n*—inner layers of veneer whose grain direction is usually perpendicular to that of the face plies, applied particularly to plywood of five or more plies and lumber-core panels. **D 1038,** D07

cross band, *v*—to place the grain of the inner layer of veneer at right angles to that of the face and back. **D 1038,** D07

crossbend test—a test in which fired or bisque porcelain enamel panels are progressively distorted by bending to determine the resistance of the coating to cracking. **C 286,** B08

cross break—a separation of the wood cells across the grain. Such breaks may be due to internal stress resulting from unequal longitudinal shrinkage or external forces. **D 9,** D07

cross-chest width, *n—in body measurements*, the distance from front break-point to front break-point. **D 123,** D13

cross-country track, *n*—surface not subject to repeated traffic in addition, no roads, routes, well-worn trails or man-made improvements; can consist of tank trails with crushed rock or having large exposed obstacles (rocks, boulders, etc.). **F 538,** F09

crosscut—to cut a board, timber, or log generally transverse to the direction of the fibers. A synonym is "buck." **D 9,** D07

cross-cut chisel point—chisel point with beveled point faces parallel to staple-crown axis. (See **right-angle chisel point.**) **F 592,** F16

cross direction—the direction of the paper or paperboard at right angles to the machine direction. Sometimes referred to as CD, CMD (cross machine direction), and across machine direction. **D 1968,** D06

cross direction—one of three mutually perpendicular directions used to define a worked material, specifically that direction in the plane of working which is at right angles to the direction of maximum elongation. **E 7,** E04

cross-dye effect, *n*—variation in dye pick-up between yarns or fibers, resulting from their inherent dye affinities. **D 123,** D13

crossed head—cup head with raised lines in cup forming cross. Also known as X head. **F 547,** F16

cross-fiber—asbestos that is located in (or derives from) veins or seams in which the fibers are close-packed and are oriented at a large angle to the boundaries of the veins or seams. **D 2946,** C17

cross-fired furnace—See **side-fired furnace**. **C 162,** C14

crossflow membrane filtration—a separation of the components of a fluid by semipermeable membranes through the application of pressure and flow parallel to the membrane surface. Includes the processes of reverse osmosis, ultrafiltration, nanofiltration, and microfiltration. **D 6161,** D19

cross grain—See **grain.** **D 9,** D07

cross grains or reeves, *n—in mica*, tangled laminations causing imperfect cleavage. **D 1711,** D09

cross grating—a two-dimensional diffraction grating, patterns from which are similar in the effects produced by diffracted electrons from thin flakes and films, and whose theory may therefore be used to explain the latter. **E 7,** E04

cross laminate—a laminate in which some of the layers of material are oriented approximately at right angles to the remaining layers with respect to the grain or strongest direction in tension. (See also **parallel laminate**). **D 883,** D20

cross laminate, n— See *cross laminate* under **laminate,** *n*. **D 907,** D14

cross laminate—a laminate in which some of the layers of material are oriented approximately at right angles to the remaining layers with respect to the grain or strongest direction in tension. (See also **parallel laminate**.) (D-20) **F 412,** F17

crosslink, *v*—to form chemical bonds between molecules to produce a three-dimensional network. **D 907,** D14

crosslink, *n*—chemical bond bridging one polymer chain to another. **D 1566,** D11

crosslinking, *n*—the formation of a three dimensional polymer by means of interchain reactions resulting in changes in physical properties. **D 883,** D20

cross linking—the reaction of a difunctional molecule with each of two molecules of a polymer. This change of the polymer from linearity produces profound changes in the physical properties. **D 1695,** D01

crosslinking, *n*—the formation of a three dimensional polymer by means of interchain reactions resulting in changes in physical properties. (D-20) **F 412,** F17

crosslinking (the act of), *n*—formation of chemical bonds between polymer chains to give a network structure. **D 1566,** D11

cross-machine direction, *n*—the direction in the plane of the fabric perpendicular to the direction of manufacture. **D 4439,** D35

cross-machine direction, CD, *n*—the direction in the plane of the fabric perpendicular to the direction of manufacture. **D 123,** D13

crossply laminate—a laminate composed of only 0 and 90° plies. This is not necessarily symmetric. **D 3878,** D30

cross-product contamination, *n*—the unintentional mixing of powders with distinct differences in either physical characteristics or chemical composition, or both. **B 243,** B09

cross section—a section of a stem, board, or other piece of wood taken at right angles to its longitudinal axis. **D 9,** D07

cross section, σ—the quotient of P by Φ, where P is the probability of the interaction for one target entity when subjected to the particle fluence Φ (ICRU).

$$\sigma = P/\Phi$$

Unit: m^2

The special unit of cross section is the barn, *b*.

$$1\text{ b} = 10^{-28}\text{ m}^2$$

E 170, E10

cross section:

enhanced elastic—EIA, cross section for elastic scattering that is larger than that predicted by *Rutherford* due to partial penetration of a nucleus in the specimen by the incident particle.

nuclear reaction—EIA, the probability of a particular nuclear reaction as a function of energy and the emission direction of the detected product. Usually expressed as an area in units of barns = 10^{-28} m^2.

Rutherford—EIA, nuclear reaction cross section for the particular case of elastic scattering as calculated from classical mechanics. First evaluated by *Rutherford.*

stopping—EIA, the energy loss of a particle incident on the specimen per unit area density of specimen atoms. Usually expressed in units of eV·cm ^{2}/atom. **E 673,** E42

cross section—the apparent cross-sectional area of the nucleus as calculated on the basis of the probability of occurrence of a reaction by collision with a particle. It does not necessarily coincide with the geometrical cross-sectional area πr^2. It is given in units of area, 1 barn = 10^{-24} cm^2. **E 1316,** E07

cross section, *n*—shows the width and thickness of material, including the center section's shape, core diameter, and core wrap. **F 1379,** F14

cross-sectioning—a method in which a specimen is mechanically cleaved, cut, or polished in a plane perpendicular to the interface under study, so that associated compositional differences or gradients in the plane of the polished surface may be observed. **E 673,** E42

cross section of the cavity, *n*—in a mold for test specimens, the area of a planar section perpendicular to the flow pattern during filling of the mold that forms the critical portion of the test specimen. **D 883,** D20

crosstalk—the undesired effect of force readings appearing on an unloaded axis of a transducer while applying force to another. **E 867,** E17

cross talk—the signal leakage (acoustic or electric) across an intended acoustic or electric barrier. **E 1316,** E07

crotch, *n—in anatomy,* the body area adjacent to the vertex of the included angle between the legs. **D 123,** D13

crotch, *n—in anatomy,* the body area adjacent to the vertex of the included angle between the legs. **F 1494,** F23

croupon—untanned, whole cattlehide with the belly and shoulder cut off. Comparable to a butt bend in tanned leather. **D 1517,** D31

crowding—improper horizontal character spacing. **F 149,** F05

crowfoot weave, *n*—a broken-twill weave 1-up and 3-down or 3-up and 1-down with two ends to the right and two ends to the left, commonly referred to as four-harness satin or broken crow. **D 123,** D13

crown—the top or roof of a melter. **C 162,** C14

crown—the top or highest point of the internal surface of the transverse cross section of a pipe. **C 822,** C13

crown, *n—in anatomy,* the top of the head. **D 123,** D13

crown—also roof or back, that is, the highest point of the cross section. *In tunnel linings,* the term is used to designate either the arched roof above spring lines or all of the lining except the floor or invert. (ISRM) **D 653,** D18

crown—staple end opposite staple point, connecting both staple legs and providing bearing area. **F 592,** F16

crown optical glass—See **optical crown glass.** **C 162,** C14

crown process—{archaic} a method of making flat glass by blowing a large bulb, opening it, and then spinning it flat. **C 162,** C14

crown width—overall width of staple including both staple legs. **F 592,** F16

crowsfeet, *n—in fabrics,* fine wrinkles of varying degree of intensity, size, and shape. **D 123,** D13

crowsfeet, *n*—small flowmarks of V-shape on calendered sheeting. **D 1566,** D11

crow's-foot—*on semiconductor wafers,* intersecting cracks in a pattern resembling a "crow's foot" *Y* on {111} surfaces and a cross ''+" on {100} surfaces. **F 1241,** F01

crude asbestos—hand-cobbed (released from its ore by manual hammer impact) cross-vein asbestos in its natural or unfiberized form. **D 2946,** C17

crude mica—mica as mined; crude crystals with dirt and rock adhering. **D 1711,** D09

crude oil, *n—(Syn. crude petroleum, crude)* a naturally occurring hydrocarbon mixture, generally in a liquid state, which may also include compounds of sulfur, nitrogen, oxygen, metals, and other elements. **D 4175,** D02

crude stripper oil, *n*—a by-product of the manufacture of citrus juice, composed largely of d-limonene and containing up to 1.5 % of aldehydes. (See also **d-limonene.**) **D 804,** D01

crudiness (kroodiness)—the extent to which a sample of asbestos contains crudy elements. See **crudy.** **D 2946,** C17

crudy (kroodi)—the quality of processed asbestos with relatively low specific surface area and degree of fiberization, containing an appreciable portion of intact assemblages of fiber (derived from the term crude asbestos). **D 2946,** C17

crudy (kroodi) bundle—a heavy assemblage of asbestos in closed-packed parallel orientation, that may be partially crushed, and that has a transverse dimension exceeding 8 mm. **D 2946,** C17

crush—*on flat glass sheets,* a lightly pitted condition with a dull gray appearance. **C 162,** C14

crush, *n—in pile yarn floor coverings,* loss of tuft definition due to entanglement and compression of pile fibers. **D 5684,** D13

crushed feathers, *n*—feathers and feather fiber resulting from curling, crushing, or chopping feathers without removing the quill. **D 123,** D13

crushed gravel, *n*—the product resulting from the artificial crushing of gravel with substantially all fragments having at least one face resulting from fracture. **C 125,** C09

crushed leather—leather made from chrome-vegetable retanned kidskins with the grain accentuated by boarding or other process. **D 1517,** D31

crushed stone, *n*—the product resulting from the artificial crushing of rocks, boulders, or large cobblestones, substantially all faces of which have resulted from the crushing operation. **C 125,** C09

crushed stone—the product resulting from the artificial crushing of rocks, boulders, or large cobblestones, substantially all faces of which have resulted from the crushing operation. **D 1079,** D08

crusher-run, *n*—the total unscreened product of a stone crusher. **D 8,** D04

crushing strength—a property of solid material that indicates its ability to withstand collapse from external, compressive loads. **C 460,** C17

crushing strength, *n— for pipes,* a property of solid material that indicates its ability to withstand collapse from external, compressive loads. **C 1154,** C17

crushing strength—the property of a particle to resist physical breakdown when contained and subjected to a slowly increasing continuously applied force. **D 2652,** D28

crushing strength—property of solid material that indicates its ability to withstand collapse from external compressive loads. **D 2946,** C17

crush strength, *n*—in catalysis, a measure of the resistance of a single particle of a formed catalyst to loss of form when subjected to a compressive force. **D 3766,** D32

crust—used as an adjective or in the phrase "in the crust," refers to leather than has been tanned but not finished (see also **rough**). **D 1517,** D31

cryogenic temperatures—for purposes of this practice, cryogenic temperatures are temperatures at or below – 70°C. **E 1705,** E48

cryogenic temperatures—temperatures below or equal to – 100°C. **E 1705,** E48

cryology—the study of the properties of snow, ice, and frozen ground. **D 653,** D18

cryoprotectant—a chemical substance used to protect cells during freezing and rewarming. **E 1705,** E48

cryptoflorescence, *n*—crystalline deposit of water-soluble compounds in the pores of masonry. **C 1232,** C15

crystal—a solid composed of atoms, ions, or molecules arranged in a pattern which is periodic in three dimensions. **E 7,** E04

crystal—solid composed of atoms, ions, or molecules, arranged in a pattern which is periodic in three dimensions, (E 7, E-4). **E 1142,** E37

crystal—a solid composed of atoms, ions or molecules arranged in a pattern that is periodic in three dimensions. **F 1241,** F01

crystal analysis—a method for determining crystal structure, for example, the size and shape of the unit cell and the location of all atoms within the unit cell. **E 7,** E04

crystal, birefringent—a cr pertaining to the use of a microscope. **E 175,** E41

crystal defect—departure from the regular arrangement of atoms in the ideal crystal lattice. **F 1241,** F01

crystal glass—*(1)* colorless, highly transparent glass which is frequently used for art or tableware.

(2) colorless, highly transparent glass historically containing lead oxide. **C 162,** C14

crystal indices—See **Miller indices.** **F 1241,** F01

crystalline glaze—See *crystalline glaze* under **glaze.** **C 242,** C21

crystalline regions—see **crystallinity.** **D 1695,** D01

crystallinity—a regular arrangement of the atoms of a solid in space. In most polymers, including cellulose, this state is usually imperfectly achieved. The crystalline regions (ordered regions) are submicroscopic volumes in which there is more or less regularity of arrangement of the component molecules. In these regions there is sufficient geometric order to enable definite X-ray diffraction patterns to be obtained. (see also **degree of crystallinity; degree of order**) **D 1695,** D01

crystallinity—regular arrangement of the atoms of a solid in space. **E 1142,** E37

crystallinity index—a number used to represent the state of crystallinity of cellulose as a whole. Unfortunately, it has been differently defined by different investigators and should not be used unless it is clearly stated which crystallinity index is meant. (see also **crystallinity**) **D 1695,** D01

crystallite—a single crystalline region. (see also **crystallinity**) **D 1695,** D01

crystallite—a crystalline grain not bounded by habit planes. **E 7,** E04

crystallite—crystalline grain not bounded by habit planes, (E 7, E-4). **E 1142,** E37

crystallization—arrangement of previously disordered material segments of repeating patterns into geometric symmetry. **E 1142,** E37

crystallization, *n*—a condition where ink evaporates and forms crystals. **F 1857,** F05

crystallization, polymer, *n*—arrangement of previously disordered polymer segments of repeating patterns into geometric symmetry. **D 1566,** D11

crystallization temperature—that temperature at which a specimen undergoes crystallization upon cooling. **E 1142,** E37

crystallographic cleavage—the separation of a crystal along a plane of fixed orientation relative to the three-dimensional crystal structure within which the separation process occurs, with the separation process causing the newly formed surfaces to move away from one another in directions containing major components of motion perpendicular to the fixed plane. **E 1823,** E08

crystallographic discoloration, *n*—*in mica*, discoloration appearing as bands of lighter or darker shades of basic color of a block of mica. **D 1711,** D09

crystallographic notation—a symbolism based on Miller indices used to label planes and directions in a crystal as follows:

plane	(111)	direction	[111]
family of planes	{111}	family of directions	⟨111⟩

F 1241, F01

crystal membrane or solid-state electrode membrane—a thin single or mixed crystal that is an ionic conductor and that separates the internal reference element or internal reference solution from the sample solution. **D 4127,** D19

crystal originated particle (COP)—a surface depression that is formed during soft alkaline chemical treatment of silicon wafer surfaces that contain crystal defects at or close to the wafer surface and that scatters light similarly to a very small particle, also called **surface micro defect (SMD).** **F 1241,** F01

crystal (see transducer)—the piezoelectric element in an ultrasonic search unit. The term is used to describe single crystal piezoelectrics as well as polycrystalline piezoelectrics, such as ferroceramics. **E 1316,** E07

crystal system—one of seven groups into which all crystals may be divided; triclinic, monoclinic, orthorhombic, hexagonal, rhombohedral, tetragonal, and cubic. **E 7,** E04

C-scan—an ultrasonic data presentation which provides a plain view of the test object, and discontinuities therein. **E 1316,** E07

C-stage, *n*—the final stage in the reaction of certain thermosetting materials in which they have become practically insoluble and infusible. **D 883,** D20

C-stage, *n*—*in polymer chemistry*, the final stage in a thermosetting resin reaction in which the material does not soften when heated, and is essentially insoluble in most liquids. (Compare **A-stage** and **B-stage.**) **D 907,** D14

C-stage—the final stage in the reaction of certain thermosetting resins in which the material is relatively insoluble and infusible. Certain thermosetting resins in a fully cured adhesive layer are in this stage. Sometimes referred to as *resite.* **E 1749,** E06

C*(t)—Integral, C*(t) $[FL^{-1}\ T^{-1}]$—a mathematical expression, a line or surface integral that encloses the crack front from one crack surface to the other, used to characterize the local stress-strain rate fields at any instant around the crack front in a body subjected to extensive creep conditions. **E 1823,** E08

C test, *n*—a standard statistical test for homogeneity of variance. **D 121,** D05

Ct product, *n*—the concentration-time product in ppm · min obtained by integration of the area under a concentration-time curve. **E 176,** E05

CTS—abbreviation for **clear to send.** **F 1457,** F05

cuam viscosity—see **viscosity, cuprammonium.** **D 1695,** D01

cube, *n*—*in packaging,* the volume of space occupied by the unit under consideration, computed by multiplying overall exterior length, width, and height. For shipping purposes cube is expressed to the nearest 0.1 ft^3 (2830 cm^3). (Compare **displacement.**) **D 996,** D10

cube texture—a texture found in wrought metals in the cubic system, in which nearly all the crystal grains have a plane of the type (100) parallel or nearly parallel to the plane of working and a direction of the type [001] parallel or nearly parallel to the direction of elongation. **E 7,** E04

cubic—having three mutually perpendicular axes of equal length. **E 7,** E04

cuene viscosity—see **viscosity, cupriethylenediamine.** **D 1695,** D01

cuff, *n*—*in garment construction*, a finished edge at the end of either a garment sleeve or trouser leg created by turning back or rolling up and stitching the fabric. **F 1494,** F23

cull—in grading lumber or other wood products, pieces or parts thereof, that do not meet the lowest quality standards requirements. **D 9,** D07

cullet—glass product or portions of product usually suitable for addition to raw batch.

(a)—foreign cullet—cullet from an outside source.

(b)—domestic cullet (factory cullet)—cullet from within the plant.

(c)—a portion of a glass article that will later be cut off and discarded or remelted. **C 162,** C14

cullet cut—synonymous with **block reek.** **C 162,** C14

cultural object, *n*—any man-made or modified natural article that reflects the beliefs, social forms, and material traits of racial, social, religious, or other specific groups. **D 123,** D13

culture—controlled inoculation, growth, and harvesting of known microorganisms. **F 1600,** F20

culvert—a pipeline intended to convey water under a highway, railroad, canal, or similar facility. **C 822,** C13

cumene—(C_9H_{12}) mol weight 120.19; clear, flammable liquid; melting point, −96.0°C; boiling point, 152.4°C. **D 4790,** D16

cumulative (acoustic emission) amplitude distribution F(V)—see **distribution, amplitude, cumulative.** **E 1316,** E07

cumulative (acoustic emission) threshold crossing distribution F_t(V)—see **distribution, threshold crossing, cumulative.** **E 1316,** E07

cumulative droplet diameter ($D_{v.1}$ **and** $D_{v.9}$)—diameter of drop such that 10 % and 90 %, respectively, of the liquid volume is in drops of smaller diameter. **E 1102,** E35

cumulative erosion, *n— in cavitation and impingement erosion*, the total amount of material lost from a solid surface during all exposure periods since it was first exposed to cavitation or impingement as a newly-finished surface. (More specific terms that may be used are *cumulative mass loss, cumulative volume loss*, or *cumulative mean depth of erosion.* See also **cumulative erosion-time curve**.) **G 40,** G02

cumulative erosion rate, *n*—the cumulative erosion at a specified point in an erosion test divided by the corresponding cumulative exposure duration; that is, the slope of a line from the origin to the specified point on the cumulative erosion-time curve. (*Synonym:* **average erosion rate**) **G 40,** G02

cumulative erosion-time curve, *n—in cavitation and impingement erosion*, a plot of cumulative erosion versus cumulative exposure duration, usually determined by periodic interruption of the test and weighing of the specimen. This is the primary record of an erosion test. Most other characteristics, such as the incubation period, maximum erosion rate, terminal erosion rate, and erosion rate-time curve, are derived from it. **G 40,** G02

cumulative frequency spectrum—See **exceedances spectrum.** **E 1823,** E08

cumulative occurrences spectrum—See **exceedances spectrum.** **E 1823,** E08

cumulative oversize distribution graph, *n*—a graph obtained by plotting the total (cumulative) percentage by the mass retained on each of a set of sieves versus the corresponding aperture sizes. **E 1638,** E29

cumulative permeation, *n*—the total mass of chemical that permeates during a specified time from when the material is first contacted. **F 1494,** F23

cumulative sample—See **sample.** **D 1356,** D22

cumulative undersize distribution graph, *n*—a graph obtained by plotting the total (cumulative) percentage by the mass passing each of a set of sieves versus the corresponding aperture sizes. **E 1638,** E29

cup—a distortion of a board in which there is a deviation flatwise from a straight line across the width of the board. **D 9,** D07

cup, cupped head—head with concave or recessed, instead of flat, top surface to accommodate nail set and to prevent its slipping during countersinking. **F 547,** F16

cupel, *vt*—to refine precious metals in a cupel by exposure to high temperature in an oxidizing atmosphere. **E 135,** E01

cupel, *n*—a small, shallow, porous cup, used in assaying to separate precious metals from lead and other base metals. **E 135,** E01

cup gun—a spray gun with a fluid container as an integral part. **C 286,** B08

cupped oval head—head with oval or convex top surface and concave bearing surface. (See **umbrella head.**) **F 547,** F16

cupped point—incomplete needle point; resulting from breaking of tip of point during forming. **F 547,** F16

cupping—the pouring of slip over areas of a part during draining to produce uniform application. **C 286,** B08

cupping—the condition sometimes occurring in heavily cold worked rods and wires, in which the outside fibers are still intact and the central zone has failed in a series of cup-and-cone fractures. **E 7,** E04

cupping, *n*—a type of irregular wear characterized by a variation in wear rate that may be periodic (essentially cycloidally shaped) around the tread band circumference in one or more rows; the variation of loss is essentially independent of individual projections if the pattern contains these projections. **F 538,** F09

cuprammonium process—a process for making rayon by dissolving cellulose in an ammoniacal copper solution and spinning the resulting solution into an acid bath, thereby regenerating the cellulose as fibers. **D 1695,** D01

cuprammonium viscosity—see **viscosity, cuprammonium.** **D 1695,** D01

cupriethylenediamine viscosity (cuene or CED)—see **viscosity, cupriethylenediamine.** **D 1695,** D01

curb—a component for holding the tumblers in place. **F 471,** F12

cure—the reaction mechanism in which the physical, chemical, and mechanical properties of a hydraulic cement change through the phases of slurry-paste-solid with time, with or without external heat, in the presence of water. **C 460,** C17

cure, *v—in building construction*, to attain the intended performance properties of a compound by means of evaporation, chemical reaction, heat, radiation, or combinations thereof. **C 717,** C24

cure, *n—in building construction*, the process by which a compound attains its intended performance properties by means of evaporation, chemical reaction, heat, radiation, or combinations thereof. **C 717,** C24

cure, *n*—the reaction mechanism in which the physical, chemical and mechanical properties of a hydraulic cement change through the phases of slurry-paste-solid with time, with or without external heat, in the presence of water. **C 1154,** C17

cure—*in grouting*, the change in properties of a grout with time. **D 653,** D18

cure, *v*—to change the properties of a polymeric system into a more stable, usable condition by the use of heat, radiation, or reaction with chemical additives. **D 883,** D20

cure, *v*—to change the physical properties of an adhesive by chemical reaction, which may be condensation, polymerization, or vulcanization; usually accomplished by the action of heat and catalyst, alone or in combination, with or without pressure. (See also **dry** and **set**.) **D 907,** D14

cure, n— See **vulcanization**, the preferred term. **D 1566,** D11

cure—reaction mechanism in which the physical, chemical, and mechanical properties of a hydraulic cement change through the phases of slurry-paste-solid with time, with or without external heat, in the presence of water. **D 2946,** C17

cure—to change the properties of a polymeric system into a more stable, usable condition by the use of heat, radiation, or reaction with chemical additives. **E 1749,** E06

cure, *v*—to change the properties of a polymeric system into a more stable, usable condition by the use of heat, radiation, or reaction with chemical additives. (ISO, D-20) **F 412,** F17

cure, *v*—to change the properties of a polymeric system into a more stable, usable condition by the use of heat, radiation, or reaction with chemical additives. **F 1251,** F04

cure, air or ambient, *n*—the method of setting or hardening products wherein the portland cement is allowed to hydrate at atmospheric conditions of pressure, preferably under conditions to inhibit water and heat loss. **C 1154,** C17

cure, autoclave—a means for accelerating the cure reaction at elevated temperature and pressure in saturated steam, where reactive siliceous material has been incorporated into the cementitious matrix, such that a hydrothermal reaction takes place between the cement and silica yielding calcium silicate. **C 460,** C17

cure, autoclave, *n*—a means for accelerating the cure reaction at elevated temperature and pressure in saturated steam, where reactive siliceous material has been incorporated into the cementitious matrix, such that a hydrothermal reaction takes place between the cement and silica yielding calcium silicate. **C 1154,** C17

cure, autoclave—means for accelerating the cure reaction at elevated temperature and pressure in saturated steam, where reactive siliceous material has been incorporated into the cementitious matrix, such that a hydrothermal reaction takes place between the cement and silica yielding calcium silicate. **D 2946,** C17

cure cycle—the schedule of time periods, at specified conditions, to which a reacting thermosetting material is subjected to reach a specified property level. **D 883,** D20

cured, *adj—in building construction,* pertaining to the state of a compound that has attained its intended performance properties by means of evaporation, chemical reaction, heat, radiation, or combinations thereof. **C 717,** C24

cure, immersed—the method of setting or hardening asbestos-cement products by immersion in water or in an aqueous solution, resulting in modified chemical activity between the cementitious components with an increase or decrease in the rate of cure. **D 2946,** C17

cure meter, *n*—a testing device that measures the progress of vulcanization at a vulcanizing temperature. **D 1566,** D11

cure, normal cure—the method of setting or hardening asbestos-cement products wherein the portland cement is allowed to hydrate at atmospheric conditions of pressure, preferably under conditions to inhibit water loss. **C 460,** C17

cure, normal cure—method of setting or hardening asbestos-cement products wherein the hydraulic cement is allowed to hydrate at atmospheric conditions of pressure, preferably under conditions to inhibit water loss. **D 2946,** C17

cure (portland cement plaster or stucco), *v*—*(1)* to provide conditions conducive to the hydration process of portland cement plaster or stucco.
(*2*) to maintain proper temperature and a sufficient quantity of water within the plaster to ensure cement hydration. **C 11,** C11

cure time—*in grouting,* the interval between combining all grout ingredients or the formation of a gel and substantial development of its potential properties. **D 653,** D18

cure time—the period of time that a reacting thermosetting material is exposed to specific conditions to reach a specified property level. **D 883,** D20

Curie point—that temperature above which a substance becomes paramagnetic. **E 7,** E04

Curie point—see **Curie temperature**. **E 1142,** E37

Curie point—the temperature at which ferromagnetic materials can no longer be magnetized by outside forces, and at which they lose their residual magnetism (approximately 1200 to 1600°F (649 to 871°C) for many metals). **E 1316,** E07

Curie temperature—temperature above which a ferromagnetic or ferroelectric material becomes paramagnetic, or paraelectric, respectively. **E 1142,** E37

Curie temperature, T_c—the temperature above which a ferromagnetic material becomes paramagnetic. **A 340,** A06

curing, *n*—action taken to maintain moisture and temperature conditions in a freshly-placed cementitious mixture to allow hydraulic cement hydration and (if applicable) pozzolanic reactions to occur so that the potential properties of the mixture may develop (see ACI 308). **C 125,** C09

curing, *n*—see the preferred term **vulcanization**. **D 123,** D13

curing—treating raw hides or skins so as to minimize putrefaction and bacterial action, but to enable the skins to be wet back conveniently in preparation for tanning (see **brining, dry pickling, dry salting, green salting,** and **pickle**). **D 1517,** D31

curing—the development of hardness, strength, or other desired properties by evaporation, polymerization, or chemical and physical change in a material that is initially liquid or plastic in consistency. **D 3743,** D04

curing, *n*—chemical process of developing ultimate properties of a finish or other material over a specified period of time. Compare **drying.** **E 631,** E06

curing agent—an additive incorporated in the furnish of asbestos-cement products resulting in modified chemical activity between the cementitious components with an increase or decrease in the rate of cure. **C 460,** C17

curing agent, *n*—an additive incorporated in the constituents of products producing a change in chemical activity between the cementitious components with an increase or decrease in the rate of cure. **C 1154,** C17

curing agent, *n*— *relative to adhesives,* a substance or mixture of substances that is part of an adhesive and is used to promote curing by taking part in the reaction. (Compare **catalyst**.) (Compare **blocked curing agent**.) (Synonym **hardener.**) **D 907,** D14

curing agent, n—synonym for **vulcanizing agent**, the preferred term. **D 1566,** D11

curing agent—additive incorporated in the furnish of asbestos-cement products resulting in modified chemical activity between the cementitious components with an increase or decrease in the rate of cure. **D 2946,** C17

curing agent—anything such as heat, a catalyst, or a protective coating that assures, enhances, or accelerates curing. **D 3743,** D04

curing compound, *n*—a liquid that, when applied as a coating to the surface of newly-placed concrete, forms a membrane that retards the evaporation of water and, in the case of white pigmented compounds, reflects heat (see also **curing**). **C 125,** C09

curing retarder—a material added to cement to increase the time for curing. Sodium chloride in high concentrations is an example. **D 653,** D18

curing temperature, *n*—the temperature to which an adhesive or an assembly is subjected to cure the adhesive. (See also **drying temperature, setting temperature**.) **D 907,** D14

curing time—the period between application and the time when the material attains its intended physical properties. **C 717,** C24

curing time, *n*—the period of time during which an assembly is subjected to heat or pressure, or both, to cure the adhesive. (See also **joint-conditioning time, setting time**.) **D 907,** D14

curl—the degree of curvature of a sheet of paper or carbon paper. **F 221,** F05

curl, *n*—a phenomenon in which the edges of a paper substrate bend towards the center of the paper. **F 1857,** F05

curled selvage, *n*—self-descriptive. **D 123,** D13

curling—a defect similar to crawling. **C 286,** B08

curling temperature—the temperature at which noticeable curling occurs on gradually heating a leather specimen in water. **D 1517,** D31

curl resistance—the tendency of sheet carbon to be flat on exposure to varying conditions of temperature and humidity. **F 221,** F05

curls—long flat flakes manufactured by the cutting action of a knife in such a way that they tend to be in the form of a helix. **D 1554,** D07

current, ac core loss, I_c—the rms value of the in-phase component (with respect to the induced voltage) of the exciting current supplied to a coil which is linked with a ferromagnetic core. **A 340,** A06

current, ac exciting, I—the rms value of the total current supplied to a coil that is linked with a ferromagnetic core.

NOTE—Exciting current is measured under the condition that any other coil linking the same core carries no current. **A 340,** A06

current, ac, magnetizing, I_m—the rms value of the magnetizing component (lagging with respect to applied voltage) of the exciting current supplied to a coil that is linked with a ferromagnetic core. **A 340,** A06

current, constant, I_{dc}—the steady current which is located in a winding and which produces a magnetostatic condition. **B 713,** B01

current, dc, I_{dc}—a steady-state dc current. A dc current flowing in an inductor winding will produce a unidirectional magnetic field in the magnetic material. **A 340,** A06

current density—the electric current to or from a unit area of an electrode surface. **G 15,** G01

current density (*cd*)—current per unit area. **B 374,** B08

current density (cd)—current per unit area. **B 713,** B01

current dollars—dollars of purchasing power in which actual prices are stated, including inflation or deflation. **E 833,** E06

current efficiency—the proportion, usually expressed as a percentage, of the current that is effective in carrying out a specified process in accordance with Faraday's law. **B 374,** B08

current efficiency—the proportion, usually expressed as a percentage, of the current that is effective in carrying out a specific process in accordance with Faraday's law. **C 859,** C26

current efficiency—the ratio of the electrochemical equivalent current density for a specific reaction to the total applied current density. **G 15,** G01

current flow method—a method of magnetizing by passing a current through a component via prods or contact heads. The current may be alternating, rectified alternating, or direct. **E 1316,** E07

current good manufacturing practices (CGMP)—current regulations published by the United States Food and Drug Administration (FDA) regarding manufacturing, processing, packaging and storing of drug and biological products. **E 1705,** E48

current induction method—a method of magnetizing in which a circulating current is induced in a ring component by the influence of a fluctuation magnetic field that links the component. **E 1316,** E07

current integration—the measurement of total electric charge deposited into a specimen by the incident beam. **E 673,** E42

current response—change in freeboard or draft due to current forces acting to displace the boom from rest. **F 818,** F20

currying—a process of treating tanned hides with oils and greases to prepare them for belting, sole, harness leathers, etc. **D 1517,** D31

curtain grouting—injection of grout into a sub-surface formation in such a way as to create a barrier of grouted material transverse to the direction of the anticipated water flow. **D 653,** D18

curtaining—see **sag or sagging**. **D 16,** D01

curtains—a defect in sheet steel ground coatings characterized by a draped pattern of darkened areas that are sometimes blistered. May also appear in cover coats applied over the ground coat or direct-on. **C 286,** B08

"curtain type" boom—boom consisting of a flexible skirt supported by flotation. **F 818,** F20

curtain wall—See **wall.** **E 631,** E06

curvature, *n*—the shape of the needle viewed in profile. Some common shapes include, but are not limited to: straight, ½ curve or "ski", ⅛ circle, ¼ circle, ⅜ circle, ½ circle, ⅝ circle, and compound curvature. **F 1840,** F04

curvature of field—a property of a lens that causes the image of a plane to be focused into a curved surface instead of a plane. **E 7,** E04

curved head—head with hemispherical or nearly hemispherical bearing surface and flat top surface. Also, head of plain, flat-headed nail with two opposite edges cupped downward. **F 547,** F16

curve resolving—the construction of the individual peaks of a spectrum that consists of overlapping peaks, also called curve fitting or peak fitting. **E 673,** E42

curve, thermal, *n*—the plot of a parameter against temperature or time. (ICTAC) **E 473,** E37

cushion, *n*—*for inflatable restraints*, the inflatable fabric envelope portion of a module. **D 123,** D13

cushion, *v*—to use **cushioning material** to reduce shock and vibration transmitted to a packaged product from an externally applied force. **D 996,** D10

cushioned vinyl flooring, *n*—any vinyl sheet floor covering incorporating a foam layer as part of its construction. **F 141,** F06

cushioning material—a material used to isolate or reduce the effect of externally applied shock or vibration forces, or both. (See also **buffer, divider, molded shape.**)

creped cellulose wadding—a material consisting of cellulose fibers produced by the sulfite process, loosely matted into sheet form and then creped. It is available in single or multiple sheet form, either plain or embossed, and may be backed with various papers. It is also available impregnated with asphalt for water resistance.

filler pad—in packaging, a **pad** used to fill space; sometimes applied to a soft flexible pad made with various loose filling materials to provide cushioning effects. (Compare **buffer.**)

foam-in-place cushioning material, n—one formed by dispensing, usually into a box or mold, reactive chemical components that expand to envelop items packaged or occupy void areas.

macerated paper— generally, waste paper torn up mechanically for use as a **cushioning material.** **D 996,** D10

customary units—a set of industry-unique units from the cgs-emu system of units and U.S. inch-pound systems and units derived from the two systems.

NOTE—Examples of customary units used in ASTM A-6 standards include:

Quantity Name	Quantity Symbol	Unit Name	Unit Symbol
Magnetic field strength	H	oersted	Oe
Magnetic induction (magnetic flux density)	B	gauss	G
Specific core loss	$P_c(\beta;f)$	watt/pound	W/lb

A 340, A06

customer video, *n*—a printer control printed wiring board that receives video data and commands from a host to transfer them to the image output terminal control. (See **C-Video.**) **F 1457,** F05

cut—see **fraction.** **B 243,** B09

cut, *n*—*in asbestos and glass yarns*, the number of 100-yd lengths of yarn per pound; an indirect yarn numbering system. **D 123,** D13

cut, *n—in wool yarns*, the number of 300-yd lengths of yarn per pound; an indirect yarn numbering system. **D 123,** D13
cut, *n—as applied to woven fabric*, a length approximately 60 yard in the greige. **D 123,** D13
cutaneous sense, *n*—any of the senses whose receptors lie in the skin or immediately beneath it (or in the external mucous membranes): contact, pressure, warmth, cold, and pain. **E 253,** E18
cutback—solvent-thinned bitumen used in cold-process roofing adhesives, flashing cements, and roof coatings. **D 1079,** D08
cut-back asphalt, *n*—petroleum residuum (asphalt) which has been blended with petroleum distillates. **D 8,** D04
cut-back products, *n*—petroleum or tar residuums which have been blended with distillates. **D 8,** D04
cut glass—glassware decorated by grinding figures or patterns on its surface by abrasive means, followed by polishing. **C 162,** C14
cuticle, *n—in mammalian hair fibers*, the layers of flattened cells enclosing the cortex, which forms an envelope of overlapping scales surrounding the fiber. **D 123,** D13
cutinite—See *cutinite* under **maceral**. **D 121,** D05
cut-layers—as applied to laminated plastics, a condition of the surface of machined or ground rods and tubes and of sanded sheets in which cut edges of the surface layer or lower laminations are revealed. **D 883,** D20
cut nail—normally, wedge-shaped, ½ to 6-in. nails of various types sheared from stiff-stock-sheet steel, with sheared-square point end narrower than upset head end. **F 547,** F16
cut-off, *n—in zippers*, the measurement of a separate element from the head side to the pocket side of the legs. **D 123,** D13
cutoff—a detail designed to prevent lateral water movement into the insulation where the membrane terminates at the end of a day's work, or used to isolate sections of the roofing system. It is usually removed before the continuation of the work. **D 1079,** D08
cutoff frequency—*of an anechoic wedge or set of wedges*, the lowest frequency above which the normal incidence sound absorption coefficient is at least 0.990. **C 634,** E33
cutoff height—a specified distance of a high on the surface record from a chord representing 25 ft (7.5 m) on the longitudinal scale. The chord may represent less than 25 ft (7.5 m) if it is from the lows on each side of the high. **E 867,** E17
cut-off level—same as **rejection level. E 1316,** E07
cut-off scar—a mark on the base of a glass bottle caused by the cutting of the gob in the Owens process. **C 162,** C14
cut pile, *n—for pile yarn floor covering*, pile in which the legs of any one tuft element are not connected to the legs of any other tuft element. **D 123,** D13
cut pile floor covering, *n*—a pile floor covering in which the pile is composed of tufts in the form of cut loops. **D 123,** D13
cut pile yarn floor covering, *n*—a pile floor covering in which the pile is composed of adjacent tuft elements that are separated or cut. **D 5684,** D13
cut resistance, *n— in chainsaw testing*, the ability of a material, while in contact with the linked cutters, to resist cut through of the cutters of a moving saw chain, independent of either jamming or chain stop. **F 1494,** F23
cut selvage, *n*—cuts or breaks that occur in the selvage only. **D 123,** D13
cut sizes—flat glass sheets cut to specific dimensions. **C 162,** C14
cut stock—a term for softwood lumber indicating the product generally has been manufactured to dimensions suitable for a fabricating operation with little further processing required. **D 9,** D07
cut stock—a term generally applied to bottom stock for shoes, such as soles, taps, lifts, blocks, and strips cut from sole leather. Also applies to cut-to-pattern leather. **D 1517,** D31
cut stock—bottom stock for shoes such as soles, taps, lifts, blocks, and strips of sole leather. **F 869,** F08
cut stone—stone fabricated to specific dimensions. **C 119,** C18
cut strip test, *n—in fabric testing*, a strip test in which the specimen is cut to the specified testing width. **D 123,** D13
cutter—(*1*) a workman engaged in grinding designs on glass.
(*2*) one who cuts flat glass.
(*3*) the tool used in cutting glass. **C 162,** C14
cut thread—produced by removing material from the surface with a form cutting tool. **F 1789,** F16
cut-through time, *n—for chainsaw cut resistance*, the time required for a running chainsaw to effect complete breakthrough of a protective garment or protective device. **F 1494,** F23
cutting—(*1*) scoring flat glass with a diamond or a steel wheel, and breaking it along the scratch.
(*2*) producing cut glass. **C 162,** C14
cutting down—polishing or buffing for the purpose of removing roughness or irregularities. **B 374,** B08
cutting egde,, *n*—cutting edges are made of various goemetric shapes, that is, triangular, diamond, and hexagonal. The various edges may be sharpened by the manufacturer depending on the user performance. **F 1840,** F04
cuttings—small-sized rock fragments produced by a rock drill. (ISRM) **D 653,** D18
cutting wear, *n—in solid impingement erosion*, the erosive wear associated with the dissipation of kinetic energy of impact arising from the tangential component of the velocity of the impacting particles. **G 40,** G02
cut wire blasting—blasting with short, cut lengths of metal wire. See **abrasive blasting. B 374,** B08
C-Video—customer video. **F 1457,** F05
CVR—contrast variation ratio is the ratio between the maximum and minimum PCS within a graphic shape:

$$CVR = \frac{PCS,\max}{PCS,\min}$$

F 149, F05
CVT—abbreviation for **constant velocity transport. F 1457,** F05
cyan—one of the primary colors, a sky blue, in subtractive color theory. **F 1156,** F12
cyaniding, *n*—introducing carbon and nitrogen into a solid steel object by holding it above Ac_1 in contact with molten cyanide of suitable composition. **A 941,** A01
cycle—the firing period in a regenerative furnace. **C 162,** C14
cycle—*in fatigue*, one complete sequence of values of load (strain) that is repeated under constant amplitude loading (straining). The symbol *N* (see definition of **fatigue life**) is used to indicate the number of cycles. **E 1823,** E08
cycle annealing, *n*—**annealing** employing a predetermined and closely controlled time-temperature cycle to produce specific properties or a specific microstructure. **A 941,** A01
cycle length, *n—in braided rope*, the distance, parallel to the rope axis, of the strand to make one revolution around the rope. **D 123,** D13
cycle ratio, *D*— the ratio of cycles endured, *n*, to the estimated fatigue life, N_f, obtained from the stress versus fatigue life (*S-N*) or the strain versus fatigue life (ϵ-*N*) diagram for cycles of the same

character, that is, $D = n/N_f$. **E 1823,** E08

cycles endured, *n*—*in fatigue*, the number of cycles of specified character (that produce fluctuating load) which a specimen has endured at any time in its load history. **E 1823,** E08

cycle time, *n*—in molding, the total time used to carry out a complete sequence of operations making up the molding cycle. **D 883,** D20

cycle time—the time required by an alcohol plant to complete one cycle. **E 1705,** E48

cycle time, *n—in an intermittent contact test*, the interval of time from the start of one contact period to the start of the next contact period. **F 1494,** F23

cyclically magnetized condition, *CM*—a magnetic material is in a cyclically magnetized condition when, after having been subjected to a sufficient number of identical cycles of magnetizing field, it follows identical hysteresis or flux-current loops on successive cycles which are not symmetrical with respect to the origin of the axes. **A 340,** A06

cyclic loading—See **fatigue loading.** **E 1823,** E08

cyclic redundancy check, *n*—an error checking system used in data transmission. (See **CRC**.) **F 1457,** F05

cyclohexane-995—cyclohexane with a purity of 99.5 weight % minimum determined by analysis by gas chromatography. Refer to proposed specifications for complete requirements. **D 4790,** D16

cylinder—See **container.** **D 996,** D10

cylinder kraft—See **containerboard.** **D 996,** D10

cylinder molded watermark—watermark made in paper by using a wax mold of the desired image fastened to the wire mesh basket upon which the paper is being formed during rotation of the basket in and out of the pulp. Common in Europe, not so in the United States. **F 1156,** F12

cylinder process—{archaic} manufacture of window glass wherein molten glass is blown and drawn into the form of a cylinder, which is subsequently split longitudinally, reheated in a flattening kiln, and flattened. **C 162,** C14

cylinders—any of various rotating members in printing presses; especially, one of three on a rotary offset press (plate, blanket, or impression). **F 425,** F05

cylinder (test)—a cast cylindrical specimen of concrete. **C 822,** C13

cylindrical wet-scrub extraction, *n*—a carpet cleaning method in which (1) a cleaning agent is sprayed onto the carpet, and (2) soil and cleaning agents are removed by a machine which feeds water into two counter-rotating brushes. **D 123,** D13

D

d-10—the diameter of a soil particle (preferably in millimetres) at which 10 % by weight (dry) of the particles of a particular sample are finer. Synonymous with the effective size or effective grain size. **D 653,** D18

d-60—the diameter of a soil particle (preferably in millimetres) at which 60 % by weight (dry) of the particles of a particular sample are finer. **D 653,** D18

DAC (distance amplitude correction) (swept gain, time corrected gain, time variable gain, etc.)—electronic change of amplification to provide equal amplitude from equal reflectors at different depths. **E 1316,** E07

daily vehicle miles traveled, (DVMT), *n*—average daily traffic on a road segment, expressed as AADT, multiplied by the length of the road segment. **E 867,** E17

daisy wheel element—an element used in an impact printing device in which the fully formed characters are contained on the ends of finger-like projections radiating out from the center of a disk, the device resembling a daisy. **F 909,** F05

daisy wheel printer—an impact printing device utilizing a daisy wheel element. The paper on the carriage remains stationary while the element is positioned and caused to strike the paper to produce the characters in their proper positions. **F 909,** F05

Dalton—an arbitrary unit of molecular weight, 1½ the mass of the nuclide of carbon 12. Unit of measure for the smallest, size of the molecular retained by an ultrafilter. **D 6161,** D19

damage, *n—in structures and structural mechanics,* a structural anomaly in material or structure created by manufacturing or service usage. **D 3878,** D30

damage—a defect of the crystal lattice of a single crystal silicon specimen in the form of irreversible deformation that results from mechanical surface treatments such as sawing, lapping, grinding, sandblasting, and shot peening at room temperature without subsequent heat treatments. **F 1241,** F01

damage, *n—in cavitation or impingement,* any effect on a solid body resulting from its exposure to these phenomena. This may include loss of material, surface deformation, or any other changes in microstructure, properties, or appearance. **G 40,** G02

damage depth, T_z—the maximum thickness of the damage region as revealed by preferential etch that removes silicon in the region of the deformation. **F 1241,** F01

damaged feathers, *n*—feathers that have been broken, damaged by insects, by mildew or rot, or otherwise materially injured. **D 123,** D13

damage resistance, *n—in structures and structural mechanics,* a measure of the relationship between the force, energy, or other parameter(s) associated with an event or sequence of events and the resulting damage size and type.

discontinuous fiber-reinforced composite—any composite material consisting of a matrix reinforced by discontinuous fibers. The fibers may be whiskers or chopped fibers.

fabric-reinforced composite—any composite material consisting of a matrix reinforced by fabric (woven, knitted, or braided assemblages of fibers).

fiber-reinforced composite—any composite material consisting of a matrix reinforced by continuous or discontinuous fibers.

filament count—number of filaments in the cross section of a fiber bundle.

filamentary composite—a composite material reinforced with continuous fibers.

unidirectional fiber-reinforced composite—any fiber-reinforced composite with all fibers aligned in a single direction. **D 3878,** D30

damp—to cause a loss or dissipation of the oscillatory or vibrational energy of an electrical or mechanical system. **C 634,** E33

damp, *adj*—moist but not visibly wet. **D 4538,** D33

dampeners—rollers that carry the fountain solution from water fountain to press master. **F 425,** F05

dampening mechanism or system—any mechanism or system employed on an offset press to apply dampening solution, or its equivalent, to a press master. **F 425,** F05

dampening solution—the aqueous solution used to keep the non-image areas of the master moist so that they will not accept ink. Also called fountain solution. **F 425,** F05

damping—reduction in the amplitude of vibration of a body or system due to dissipation of energy internally or by radiation. (ISRM) **D 653,** D18

damping—the loss in energy, dissipated as heat, that results when a material or material system is subjected to an oscillatory load or displacement. **D 4092,** D20

damping—loss in energy, dissipated as heat, that results when a material or material system is subjected to an oscillatory load or displacement, (D 4092, D-20). **E 1142,** E37

damping ratio—for a system with viscous damping, the ratio of actual damping coefficient to the critical damping coefficient. **D 653,** D18

damping ratio, μ —the ratio of actual damping to critical damping. **D 4092,** D20

damping, search unit—limiting the duration of a signal from a search unit subject to a pulsed input by electrically or mechanically decreasing the amplitude of successive cycles. **E 1316,** E07

dampproofing—treatment of a surface or structure to resist the passage of water in the absence of hydrostatic pressure. **D 1079,** D08

D.A.N., *n*—Divers Alert Network. **F 1549,** F32

dangerously flammable textile, *n*—not defined. This term is implied in the Standard For The Flammability Of Clothing Textiles (16 CFR Part 1610) under the Flammable Fabrics Act (15 USC 1191 et seq.) from which a meaning can be inferred. (See also **flammable textile.)** **D 123,** D13

Danner process—a mechanical process for continuously drawing glass cane or tubing from a rotating mandrel. **C 162,** C14

DAPI staining—staining of DNA in particular by using DAPI fluorochrome stain (Serva 18860). **E 1705,** E48

dark adaptation—the adjustment of the eyes when one passes from a bright to a darkened place. **E 1316,** E07

dark adaptation—conditioning of photoelectrostatic copying papers in the absence of light to permit the recovery of photoconductive properties. **F 335,** F05

dark decay—loss of apparent surface voltage in the absence of light. **F 335,** F05

dark decay rate—the rate of loss of apparent surface voltage per unit time in the absence of light. **F 335,** F05

darkfield illumination—the illumination of an object such that it appears illuminated with the surrounding field dark. This results from illuminating the object with rays of sufficient obliquity so that none can enter the objective directly. As applied to electron microscopy, the image is formed using only electrons scattered by the object. **E 7,** E04

dark (or Townsend) discharge—a discharge which may or may not be self-sustaining. It is characterized by a uniform field, current in the microampere range, and a nonluminous inter-electrode space. **B 542,** B02

darts, *n*— See **comet.** **D 6488,** D01

dash-bond coat, *n*—a thick slurry of portland cement, sand, and water, machine-sprayed or hand-dashed with a paddle or stiff-bristled broom to any acceptable surface, to provide improved adhesion and a key for the subsequent portland cement plaster or stucco coat. **C 11,** C11

dashpot—a hydraulic cylinder device with a controlled leak designed to eliminate impact loading of mechanisms. Hardness testers may employ a dashpot to bring an indenter into contact with a specimen without impact or other disturbance. **E 7,** E04

data element, *n*—a single, atomic piece of data that cannot be subdivided and still retain any meaning. **F 1294,** F05

data identifier, DI, *n*—a specified character, or string of characters, that define the intended use of the data elements that follow. **F 1294,** F05

data monitor, *n*—a mode of printer operation in which the information transmitted to the printer is printed in hexadecimal numbers. **F 1457,** F05

data products interface, *n*—a parallel interface standard that is used to connect elements of a communications system. **F 1457,** F05

data set, *n*—data communications equipment for transmitting coded data over phone lines. **F 1457,** F05

data stream compatibility, *n*—the extent to which an electronic signal agrees with the signal requirements of a system. (See **DSC.**) **F 1457,** F05

data terminal equipment, *n*—any piece of equipment at which a communications path ends or begins, a terminal or a printer. (See **DTE.**) **F 1457,** F05

data terminal ready, *n*—an electronic signal from the printer to the host that the printer is ready to be used. (See **DTR.**) **F 1457,** F05

dating nail—galvanized, regular-stock-steel, copper or brass, 1½ by ¼ to 2½ by 3⁄16-in. nails with ⅜ to ½-in. flat numeral head and medium diamond point. **F 547,** F16

daughter—a nuclide, stable or radioactive, formed by radioactive decay. A synonym for decay product. **C 859,** C26

daylight, *n*—total (global) hemispherical solar radiation. See also **solar irradiance, global** **G 113,** G03

daylight illuminant, *n*—illuminant having the same, or nearly the same, relative spectral power distribution as a phase of daylight. **E 284,** E12

daylight 0.785 rad, 0 rad (45°, 0°) luminous directional reflectance , *n*—daylight 0.785 rad, 0 rad (45°, 0°) luminous directional reflectance (for brevity called reflectance) is the ratio of the luminous flux from a specimen illuminated at an angle of 0.785 rad (45°) by CIE Standard Source C (average daylight) and viewed perpendicularly by the CIE Standard Observer, to the luminous flux from the standard magnesium oxide layer, similarly illuminated and viewed (see Discussion). The combination of illumination at 0.785 rad (45°) and viewing at 0 rad (0°) (perpendicularly) has been selected as being representative of average conditions of illuminating and viewing. The property of reflectance determines which of two specimens will appear lighter when viewed in average daylight at an angle at which the observation of highlights is avoided. **D 2946,** C17

day tank—a periodic melting unit, which supplies glass for small volume applications. **C 162,** C14

dB_{AE}—a logarithmic measure of acoustic emission signal amplitude, referenced to 1 μV.

$$\text{Signal peak amplitude } (dB_{AE}) = 20 \log_{10}(A_1/A_0)$$

where:
A_0 = 1 μV at the sensor output (before amplification), and
A_1 = peak voltage of the measured acoustic emission signal.

Acoustic Emission Reference Scale:

dB_{AE} Value	Voltage at Sensor Output
0	1 μV
20	10 μV
40	100 μV
60	1 mV
80	10 mV
100	100 mV

E 1316, E07

dB control—a control that adjusts the amplitude of the display signal in dB units. **E 1316,** E07

DBP—disinfection by-products (a rule as part of the SDWA). **D 6161,** D19

dc—symbol used to designate an electric voltage or current whose amplitude does not vary periodically with respect to time, as for example the output of a chemical cell or that of a thermocouple. The term is also applied to the output of such devices as dynamos and rectifiers, whose amplitude is not strictly time-invariant. **D 2864,** D27

DC1/DC3, *n*—a control sequence used with asynchronous transmission that enables the printer to signal the host to start and to stop transmitting data. **F 1457,** F05

DCOF—dynamic coefficient of friction **F 1646,** F13

D-cracking, *n—in concrete,* a series of cracks near to and roughly parallel to features such as joints, edges, and structural cracks. **C 125,** C09

DCS, *n*—decompression sickness, a medical condition with a variety of symptoms that may result from gas or bubbles in the tissues of divers after pressure reduction. **F 1549,** F32

deactivation—the process of prior removal of the active corrosive constituents, usually oxygen, from a corrosive liquid by controlled corrosion of expendable metal or by other chemical means, thereby making the liquid less corrosive. **G 15,** G01

dead anneal—jargon for a state of negligible residual stress. **C 162,** C14

dead-burned, *adj*—the state of a basic refractory material resulting from a heat treatment that yields a product resistant to atmospheric hydration or recombination with carbon dioxide. **C 71,** C08

dead burned dolomite—dolomitic limestone that has been heated with or without additives to a temperature sufficiently high and for a long enough time to decompose the carbonate structure so as to form calcium oxide and periclase in a matrix that provides resistance to subsequent hydration and recombination with carbon dioxide. **C 51,** C07

dead-burned magnesite, n— see **magnesite, dead-burned.** **C 71,** C08

dead-burned refractory dolomite, n—see **dolomite, dead-burned refractory. C 71,** C08dead cotton, *n*—a small nep of cotton fibers which is gathered on the surface of the fabric and which is different in color from the surrounding fabric. **D 3990,** D13

dead end filtration—a process in which water is forced through a media which captures the retained particles on and within it, where the process involves one influent and one effluent stream. **D 6161,** D19

dead leg—any inactive, trapped or stagnant zone of a biological fluid that is to be sampled aseptically where this liquid zone would not be representative of the bulk fluid that is to be sampled. This "dead leg" zone could deviate from the bulk system in oxygen content, nutrients levels, material composition, temperature, bacterial contamination, and other process variables that would prevent any sample drawn through this system from representating the bulk fluid quality to be tested. **E 1705,** E48

dead level—absolutely horizontal, or zero slope (see also **slope**). **D 1079,** D08

dead level asphalt—a roofing asphalt conforming to the requirements of Specification D 312, Type I. **D 1079,** D08

dead level roofing—a roofing system applied on a surface with a 0 to 2 % incline. **D 1079,** D08

deadload—the load imposed on pipe, that is determined by depth and width of the trench at top of pipe, as well as unit weight and character of backfill material. **C 896,** C04

dead load—See **load.** **D 996,** D10

deadload—the static load imposed on the top of the pipe. **F 412,** F17

dead plate—*in automatic production of molded glass,* a stationary plate receiving a glass article awaiting transfer. **C 162,** C14

dead time—any interval during data acquisition when the instrument or system is unable to accept new data for any reason. (E 750) **E 1316,** E07

dead time correction—an adjustment to the elapsed counting time of a detector system to compensate for the time the system is processing signals and cannot accept new signals. **C 859,** C26

dead-weight loading—a method of loading in which a weight is supported solely by the specimen and has no other mechanical connection to the testing machine. In hardness testing, the weight is supported by the indenter. **E 7,** E04

dead zone—the distance in the material from the surface of the test object to the depth at which a reflector can first be resolved under specified conditions. It is determined by the characteristics of the search unit, the ultrasonic test instrumentation, and the test object. **E 1316,** E07

deaeration—*in environmentally affected fatigue testing*, the process of removal of air from the liquid environment before and during a test. **E 1823,** E08

deaerator—a device to remove air from water. **D 6161,** D19

deagglomeration—the process of breaking down, usually by physical means, the masses of particles that are held together by relatively weak cohesive forces resulting in a final system of aggregates or primary particles, or both. **C 242,** C21

deairing—the process of removing entrapped air, or absorbed air from a mass or slurry, usually by application of a vacuum. **C 242,** C21

deairing, *n*—the process of removing entrapped air or absorbed air from a mass or slurry, usually by application of a vacuum. **C 1145,** C28

dealloying—see **parting**. **G 15,** G01

debinding, n— see **binder removal.** **B 243,** B09

debiteuse—a slotted, floating clay block through which glass issues in the Fourcault process. **C 162,** C14

debond—a deliberate separation of a bonded joint or interface, usually for repair or rework purposes. **D 3878,** D30

debossment—the depth of a print impression into the surface of a document. **F 149,** F05

debris—as applied to geologic debris flows, a mixture of loose, poorly-sorted rock fragments or soil material, or both, potentially ranging from clay to boulder-size particles that may include fragmental organic matter and other exotic detritus. **D 4410,** D19

debris, *n*—*in tribology*, particles that have become detached in a wear or erosion process. **G 40,** G02

debulking—the application of a temporary vacuum bag, bleeder, vacuum, or pressure, with or without heat, to remove trapped air and possibly some resin, in order to compact a composite lay-up. (Syn. **pre-bleeding, compacting.**) **E 1749,** E06

deburring—the removal of burrs, sharp edges, or fins by mechanical, chemical, or electrochemical means. **B 374,** B08

debutanization, *n— of crude oil*, the removal of the light hydrocarbons up to and including *n*-butane, and retention of the heavier hydrocarbons. **D 4175,** D02

Debye ring—a continuous circle, concentric about the undeviated beam, produced by monochromatic X-ray diffraction from a randomly oriented crystalline powder. An analogous effect is obtained with electron diffraction. **E 7,** E04

Debye-Scherrer method—a method of X-ray diffraction employing monochromatic radiation and a polycrystalline specimen mounted on the axis of a cylindrical strip of film. (See **powder method.**) **E 7,** E04

decarbonator—a device to remove carbon dioxide from water. **D 6161,** D19

decarburization, *n*—the loss of carbon from the surface of a steel object as a result of its being heated in a medium that reacts with the carbon. **A 941,** A01

decarburization—loss of carbon from the surface layer of a carbon containing alloy due to reaction with one or more chemical substances in a medium that contacts the surface. **B 374,** B08

decarburization—loss of carbon from the surface of a carbon containing alloy due to a reaction with one or more chemical substances in a medium that contacts the surface. Decarburization may be either *(1) partial*, that is, where carbon content is less than the unaffected interior but greater than the room temperature solubility limit of carbon in ferrite or (2) *complete*, that is, where carbon content is less than the solubility limit of carbon in ferrite so that only ferrite is present. **E 7,** E04

decarburization—a loss of carbon from the surface layer of the fastener, normally associated with heat treatment. **F 1789,** F16

decarburized enameling steel—a special type of steel sheet of extremely low carbon content, suitable for porcelain enamel cover coat application direct to the metal (Type I of Specification A 424). **C 286,** B08

decarburized steel— see **decarburized enameling steel.** **C 286,** B08

decating mark, *n*—a crease mark or impression extending across the cloth near the beginning or end of a piece due to the thickness of the fabric leader seam. **D 123,** D13

decay—decomposition of wood substance caused by action of wood-destroying fungi, resulting in softening, loss of strength and weight, and often in change of texture and color.

advanced (or typical) decay—the older stage of decay in which the destruction is readily recognized because the wood has become punky, soft and spongy, stringy, ring-shaked, pitted, or crumbly. Decided discoloration or bleaching of the rotted wood is often apparent.

incipient decay—the early stage of decay that has not proceeded far enough to soften or otherwise perceptibly impair the hardness of the wood. It may be accompanied by a slight discoloration or bleaching of the wood.

pocket rot—advanced decay that appears in the form of a hole, pocket, or area of soft rot usually surrounded by apparently sound wood. **D 9,** D07

decay constant, λ—of a radioactive nuclide in a particular energy state, the quotient of dP by dt, where dP is the probability of a given nucleus undergoing a spontaneous nuclear transition from that energy state in the time interval d*t* (ICRU).

$$\lambda = dP/dt$$

Unit: s^{-1} **E 170,** E10

decay rate, *d;* **[T^{-1}]; dB/s**—*for airborne sound,* the rate of decrease of sound pressure level after the source of sound has stopped; *for vibration,* the rate of decrease of vibratory acceleration, velocity, or displacement level after the excitation has stopped. **C 634,** E33

decay time—the interval of time required for a pulse to decay from its maximum value to some specified fraction of that value. (ISRM) **D 653,** D18

decay/time curve—a plot of the relationship between time and intensity of emission, or similar interrelationships. **F 1156,** F12

deceleration period, *n— in cavitation or liquid impingement erosion,* the stage following the acceleration period or the maximum rate period (if any) during which the erosion rate has an overall decreasing trend although fluctuations may be superimposed on it. (See also **erosion rate-time pattern.**) **G 40,** G02

decibel, dB—the term used to identify ten times the common logarithm of the ratio of two like quantities proportional to power or energy. (See **level, sound transmission loss.**) Thus, one decibel corresponds to a power ratio of $10^{0.1}$ and *n* decibels corresponds to a power ratio of $(10^{0.1})^n$. **C 634,** E33

decibel (dB)—twenty times the base ten logarithm of the ratio of two ultrasonic signal amplitudes, dB = 20 $\log_{10}$(amplitude ratio). **E 1316,** E07

decision analysis—a technique for making economic decisions in an uncertain environment that allows a decision maker to include alternative outcomes, risk attitudes, or subjective impressions about uncertain events in an evaluation of investments. **E 833,** E06

deck—the horizontal structural substrate supporting the plaza deck system. See also **structural slab.** **C 717,** C24

deck—the structural surface to which the roofing or waterproofing system (including insulation) is applied. **D 1079,** D08

deck, *n*—the upholstered support under the seat cushion in a loose-seat construction. **E 176,** E05

deck, *n—in upholstered furniture,* the upholstered support under the seat cushion in a loose seat construction. **E 176,** E05

decking—the multiple layer loading of ware for firing. **C 286,** B08

deck oven—equipment where the cavity is generally wider than it is tall, and the bottom of the cavity is called the deck. See **oven.** **F 1827,** F26

decoder, *n—as part of the two-dimensional symbol and linear bar code reading systems,* the electronic package that receives the signals from the scanner, executes the algorithm to interpret the signals into meaningful data, and acts as the interface to other devices. **F 1294,** F05

decolorizing—the process of producing a colorless appearance in glass. **C 162,** C14

decommissioning (closure)—the engineered closure of a well, borehole, or other subsurface monitoring device sealed with plugging materials. Decommissioning also includes the planning and documenting of all associated activities. A synonym is abandonment. **D 653,** D18

decomposers—microorganisms, predominantly bacteria and fungi, that convert complex organic matter into simpler organic and inorganic molecules. **F 1600,** F20

decomposition—for peats and organic soils, see **humification.** **D 653,** D18

decomposition potential—the minimum potential, exclusive of *IR* drop, at which an electrochemical process can take place at an appreciable rate. **B 374,** B08

decompression, *n*—the technique used to allow the controlled removal of excess insert gas from the body during and after a dive to prevent decompression sickness (DCS). **F 1549,** F32

decontamination—the process of removing undesirable physical or chemical constituents, or both, from equipment to reduce the potential for cross-contamination.
The process of removing or reducing to a known level undesirable physical or chemical constituents, or both, from a sampling apparatus to maximize the representativeness of physical or chemical analysis proposed for a given sample. **D 653,** D18

decontamination, *n*—the removal of a contaminant or contaminants from the surface or matrix, or both, of CPC to the extent necessary for its next intended action (for example, reuse and disposal). **F 1494,** F23

deconvolution—*AES, XPS,* a mathematical procedure to (*1*) remove the contribution to a peak of one of the factors contributing to its line width, for example, X-ray linewidth, analyzer broadening; or (*2*) remove the energy loss background by deconvoluting the spectrum with an electron energy loss spectrum. **E 673,** E42

deconvolution—a procedure by which line-of-sight measurements of drop size distribution and optical extinction are converted into local representations of the distribution and number density. **E 1620,** E29

decorated—adorned, embellished, or made more attractive by means of color or surface detail. **C 242,** C21

decorating fire—See *decorating fire* under **firing.** **C 242,** C21

decoration:

inglaze decoration—a ceramic decoration applied on the surface of an unfired glaze and matured with the glaze.

overglaze decoration—a ceramic or metallic decoration applied and fired on the previously glazed surface of ceramic ware.

polychrome decoration—a multicolor decoration.

underglaze decoration—a ceramic decoration applied directly on the surface of ceramic ware and subsequently covered with a transparent glaze. **C 242,** C21

decoupling—the ratio of the radius of the blasthole to the radius of the charge. In general, a reducing of the strain wave amplitude by increasing the spacing between charge and blasthole wall. (ISRM) **D 653,** D18

de-enameling—the removal of porcelain enamel from the base metal. **C 286,** B08

deep buff—the first cut or split underneath the top grain or machine buff on which no traces of the grain remain. **D 1517,** D31

deep emission function decay length—the asymptotic emission function decay length for increasing depths from the surface. **E 673,** E42

deep etching—macroetching; etching preliminary to macroexamination, intended to develop gross features such as segregation, grain flow, cracks or porosity. **E 7,** E04

deep level impurity—a chemical element that when introduced into a semiconductor has an energy level (or levels) that lies in the mid-range of the forbidden energy gap, between those of the dopant impurity species. **F 1241,** F01

deep wood failure, n—failure that is invariably several to many cells away from the adhesive layer, in which the fracture path is strongly infuenced by the grain angle and the growth-ring structure. **D 907,** D14

deerskin—a deerskin tanned and finished with the grain surface intact. **D 1517,** D31

defect, *n*—an imperfection of sufficient magnitude to warrant rejection based on the specified requirements. **A 941,** A01

defect—*of a manufactured carbon or graphite product*, any irregularity in the chemistry, microstructure, or macrostructure. **C 709,** D02

defect—any irregularity or imperfection in a tree, log, piece, product, or lumber that reduces the volume of sound wood or lowers its durability, strength, or utility value. **D 9,** D07

defect, *n—specific for inflatable restraints*, an imperfection in a cut piece of fabric that judgment and experience indicate is likely to result in either the hazardous or improper deployment of the inflatable restraint module in which the imperfection is incorporated. **D 123,** D13

defect, *n*—a departure of a quality characteristic from its intended level, or state, that occurs with a severity sufficient to cause an associated product or service not to satisfy intended normal, or reasonably foreseeable, usage requirements. **D 1711,** D09

defect, *n*—one or more flaws whose aggregate size, shape, orientation, location, or properties do not meet specified acceptance criteria and are rejectable. **E 1316,** E07

defect, in inspection and grading, *n*—the departure or nonconformance of some characteristic from its intended level or state. **D 123,** D13

defective—having flaws or dimensional deviations greater than acceptable for the intended use. **C 709,** D02

defective paint surface—paint that is cracking, flaking, chipping, or peeling from a building component (for example, window sill, door, or door frame). **E 1605,** E06

defective unit—a unit that fails to conform to one or more of the visual, dimensional, or tactile requirements. **D 1517,** D31

defect resolution—a property of a test system that enables the separation of indications due to defects in a test specimen that are located in close proximity to each other. **E 1316,** E07

deficiency—departure from, or noncompliance with, specified accreditation criteria. **E 1187,** E36

define (X-rays)—limit a beam of X-rays by passage through apertures in order to obtain a parallel, divergent, or convergent beam. **E 7,** E04

defining fixed point, *n*—thermometric fixed point of an idealized system, to which a numerical value has been assigned, used in defining a temperature scale. **E 344,** E20

definition—the clarity or sharpness of a microscopical image. **E 7,** E04

definition, *n*—statement that describes a concept and permits its differentiation from other concepts within a system of concepts. **E 1992,** E02

definition, image definition—the sharpness of delineation of image details in a radiograph. Generally used qualitatively. **E 1316,** E07

definitive care—a level of therapeutic intervention capable of providing comprehensive health care services for a specific condition. **F 1177,** F30

deflagration index, (K_{St}), *n*—maximum *dP/dt* normalized to a 1.0 m^3 volume. It is measured at the optimum dust concentration. K_{St} is defined according to the following cubic relationship:

$$K_{St} = (dP/dt)_{max}V^{1/3}$$

where:

P = pressure, (bar)
t = time, (s)
V = volume, (m^3)
K_{St} = (bar m/s)

E 1445, E27

deflection—the linear distance that a test specimen bends at the center from no load to stated load when loaded as a beam. **C 460,** C17

deflection—the deviation of a structural element from its original shape or plane due to physical loading, temperature changes or rotation of its supports. **C 717,** C24

deflection—deformation or displacement from the original contour or shape. **C 904,** C03

deflection, *n*—the linear distance that a test specimen bends at the center from no load to stated load, when loaded as a beam with the load applied at the center of the span. **C 1154,** C17

deflection—linear distance that a test specimen bends at the center from no load to stated load when loaded as a beam. **D 2946,** C17

deflection temperature—the temperature at which a specimen will deflect a given distance at a given load under prescribed conditions of test. (See Test Method D 648.) Formerly called heat distortion. **F 412,** F17

deflector atomizer—an atomizer in which a liquid jet spreads out over a solid surface, forming a spray whose shape depends upon the solid surface. **E 1620,** E29

deflocculate—to separate agglomerates in a slurry by chemical and physical means to achieve and maintain particle-to-particle separation. **C 242,** C21

deflocculating—the thinning of the consistency of a slip by adding a suitable electrolyte. **C 286,** B08

deflocculating agent (deflocculant) (dispersing agent)—an agent that prevents fine soil particles in suspension from coalescing to form flocs. **D 653,** D18

defoaming agent—a material that eliminates or suppresses foam in the spray tank **E 1519,** E35

defoliant—a chemical that causes the foliage to drop from plants. **E 609,** E35

deformability—*in grouting*, a measure of the elasticity of the grout to distort in the interstitial spaces as the sediments move. **D 653,** D18

deformation, *n*—a change in shape of a material caused by forces of compression, shear, tension, or torsion. **D 123,** D13

deformation—change in shape or size. **D 653,** D18

deformation—a change in the shape or size of a solid body. (ISRM) **D 653,** D18

deformation bands—bands produced within individual grains during cold working which differ variably in orientation from the matrix. **E 7,** E04

deformation eutectic—See **eutectic, deformation.** **C 242,** C21

deformation lines—thin bands or lines produced in grains of some metals, particularly those of face-centered cubic structure, by cold working; they are not removed by repolishing and re-etching. **E 7,** E04

deformation, permanent, *n*—the net long-term change in a dimension of a specimen after deformation and relaxation under specified conditions. (Syn. **permanent set, nonrecoverable deformation, and nonrecoverable stretch.** **D 4848,** D13

deformation point—See **dilatometric softening point**. **C 162,** C14

deformation resolution (deformation sensitivity), R_d (L)—ratio of the smallest subdivision of the indicating scale of a deformation-measuring device to the sensitivity of the device. **D 653,** D18

deformation wear, *n— in solid impingement erosion*, the erosive wear of a material associated with the dissipation of kinetic energy of impact arising from the normal component of the velocity of the impacting particles. It is therefore the sole component of wear for particles impacting at a 90° angle of attack. **G 40,** G02

deformed—See **mechanically deformed.** **F 547,** F16

degasification—the process of removing dissolved gasses from water. **D 6161,** D19

degassing—removal of gases. **D 2652,** D28

degradable plastic, *n*—a plastic designed to undergo a significant change in its chemical structure under specific environmental conditions resulting in a loss of some properties that may vary as measured by standard test methods appropriate to the plastic and the application in a period of time that determines its classification. **D 883,** D20

degradation—a deleterious change in the physical or chemical properties, or both, of a material. **C 904,** C03

degradation, *n*—a deleterious change in the chemical structure, physical properties, or appearance of a plastic. **D 883,** D20

degradation—in general chemical use, the conversion of a complex compound to a simpler. Specifically for cellulose, the breakdown of the polymer chain, usually by hydrolysis or oxidation. Degradation is usually applied to changes in chemical structure. (see also **deterioration**) **D 1695,** D01

degradation, *n*—change of a chemical compound to a less complex compound (dictionary definition). **D 1968,** D06

degradation—the geologic process by which stream beds, flood plains, the bottoms of other water bodies, and other land surfaces are lowered in elevation by the removal of material by fluids. **D 4410,** D19

degradation—damage by weakening or loss of some property, quality, or capability. **E 1749,** E06

degradation, *n*—a deleterious change in chemical structure, physical properties, or appearance of a plastic. (D-20) **F 412,** F17

degradation, *n*—a deleterious change in the chemical structure, physical properties, or appearance of a plastic. **F 1251,** F04

degrade, *n*—a reduction in quality of lumber, logs, or other wood products due to processing. **D 9,** D07

degrained leather—leather from which the grain has been removed after tanning, by splitting, abrading, or other process. **D 1517,** D31

degras, moellon—the partially oxidized oil pressed out of sheepskin after tannage with cod or other marine oil. (See also **moellon.**) **D 1517,** D31

degreasing—the removal of grease and oils from a surface.

solvent degreasing—degreasing by immersion in liquid organic solvent.

vapor degreasing—degreasing by solvent vapors condensing on the parts being cleaned. **B 374,** B08

degree Celsius, °C, *n*—derived unit of temperature in the International System of Units (SI). (See **kelvin**). **E 344,** E20

degree centigrade, *n*—obsolete term. Use **degree Celsius.** **E 344,** E20

degree day—see **degree day, heating** and **degree day, cooling.** **E 772,** E44

degree-day, cooling, (DDC or DDF, Celsius or Fahrenheit respectively)—one cooling degree-day is counted for each degree that the daily mean temperature is higher than a base temperature; used to estimate energy requirements for air conditioning or refrigeration. **E 772,** E44

degree-day, heating, (DDC or DDF, Celsius or Fahrenheit respectively)—one heating degree-day is counted for each degree that the daily mean temperature is lower than a base temperature; used to estimate energy requirements for heating. **E 772,** E44

degree-days—the difference between the average temperature each day and 32°F (0°C). In common usage degree-days are positive for daily average temperatures above 32°F and negative for those below 32°F (see **freezing index**). **D 653,** D18

degree Fahrenheit, °F, *n*—non-SI unit of temperature commonly used in the United States of America. **E 344,** E20

degree of consolidation (percent consolidation), *U* **(D)**—the ratio, expressed as a percentage, of: (*1*) the amount of consolidation at a given time within a soil mass, to (*2*) the total amount of consolidation obtainable under a given stress condition. **D 653,** D18

degree of crystallinity—the fraction by mass of a cellulose sample occurring in crystalline regions. The method of determining crystallinity must be stated. (see also **crystallinity**) **D 1695,** D01

degree of freedom—in heterogeneous equilibrium, an external variable that may be adjusted independently without causing a change of state; the external variables usually considered are: temperature, pressure, and concentration parameters numbering one less than the order of the system. **E 7,** E04

degree of lateral order—the relative degree of molecular alignment. As for degree of crystallinity, quantitative values must be defined in terms of the experimental measurements. (see also **crystallinity**) **D 1695,** D01

degree of polymerization, DP—in general, the average number of base units, or of monomeric units per molecule in linear polymers. Specifically, the average number of anhydroglucose units (or derivative units) per molecule of cellulose (or cellulose derivative). The type of average obtained depends upon the method used for the determination. Hence, the method must always be specified. **D 1695,** D01

degree of saturation—see **percent saturation.** **D 653,** D18

degree of saturation—the extent or degree to which the voids in rock contain fluid (water, gas, or oil). Usually expressed in percent related to total void or pore space. (ISRM) **D 653,** D18

degree of sensitivity (*sensitivity ratio*)—see **remolding index.** **D 653,** D18

degree of substitution, DS—in a cellulose derivative, the average number of hydroxyl groups substituted per anhydroglucose unit. DS varies from zero to about 3. **D 1695,** D01

degrees of freedom, *n—for a set*, the number of values that can be assigned arbitrarily and still get the same value for each of one or more statistics calculated from the set of data. **D 123,** D13

degrees-of-freedom—the minimum number of independent coordinates required in a mechanical system to define completely the positions of all parts of the system at any instant of time. In general, it is equal to the number of independent displacements that are possible. **D 653,** D18

degrees of freedom—the number of observations minus the number of constraints imposed upon the system. In general, only one constraint (for example, the mean value) is imposed and the total degrees of freedom are one less than the number of observations. **D 4790,** D16

dehumidify—to reduce, by any process, the quantity of water vapor within a given space. **E 41,** G03

deionization—the removal of ions from a solution by ion exchange. **B 374,** B08

deionization (DI)—the removal of ions from a solution by ion exchange. **D 6161,** D19

delamination, *n— as related to metallic coated steel,* the separation of a coating (either full or partial thickness) from underlying layers; the separation can occur in small localized areas or large areas of surface. **A 902,** A05

delamination, *n*—the separation of the layers of material in a laminate. **D 883,** D20

delamination, *n*—the separation of layers in a laminate because of failure of the adhesive, either in the adhesive itself or at the interface between the adhesive and the adherend. **D 907,** D14

delamination, *n*—the separation of layers in a laminate because of failure of the adhesive, either in the adhesive itself or at the interface between the adhesive and the adherend. **D 1038,** D07

delamination—separation of plies in a laminate. This may be local or may cover a large area in the laminate. **D 3878,** D30

delamination—the separation of two or more layers or plies of reinforcing material within a pultrusion. **D 3918,** D20

delamination—*in protective coatings,* a separation of one coat from another coat within a coating system; or from the substrate. **D 4538,** D33

delamination—the separation of the layers (lamina) of material in a laminate. **E 1749,** E06

delamination—the separation of layers in a multilayered structure. **F 17,** F02

delamination—the separation of the layers of material in a laminate. (D 883, D-20) **F 869,** F08

delamination—See Terminology F 17. **F 1327,** F02

delamination strength, *n*—the tensile force required to separate the component layers under specified force. **D 123,** D13

delay—time interval (fraction of a second) between detonation of explosive charges. (ISRM) **D 653,** D18

delay distance *(D), n*—the distance the air flows past a wind vane during the time it takes the vane to return to 50 % of the initial displacement. **D 1356,** D22

delayed deformation, *n*—deformation which is time-dependent and exhibited by material subject to a continuing force. **D 123,** D13

delayed deformation, *n*—deformation which is time-dependload of a skein of yarn adjusted for the linear density of the yarn expressed in an indirect system. **D 4848,** D13

delayed fishscaling—a fishscaling defect that occurs after the final porcelain enamel processing (see also **fishscaling**). **C 286,** B08

delayed sweep—an A-scan or B-scan presentation in which an initial part of the time scale is not displayed. **E 1316,** E07

deleading—deprecated term. Use **lead-based paint hazard abatement.** **E 1605,** E06

delegated practice—the medical activities of providers performed under the authority and direction of a licensed physician. **F 1177,** F30

deleterious impurities—impurities that might be a health or safety concern, particularly with respect to toxicity, carcinogenicity, or immunogenicity. Deleterious impurities must be controlled and their levels determined using suitable analytical methods. **E 1705,** E48

delft ware—a calcareous earthenware having an opaque white glaze and monochrome overglaze decorations. (Originated in Delft, Holland.) **C 242,** C21

delicate or gentle cycle, *n*—a cycle in which agitation is slow and time is reduced. **D 123,** D13

delicate pretreatment procedure, *n*—*in textile conservation,* the washing, rinsing, drying, and pressing actions followed when fabrics are of certain yarn and fabric constructions or fiber contents, such as lightweight or sheer fabrics, fine yarns, silks, or wools. (Compare **sturdy pretreatment procedure.**) **D 123,** D13

delimited term, *n*—a term whose term entry includes a specified domain or area of application. **E 1992,** E02

delivery—(*1*) the final act of any glass-forming unit on a particular article; consisting of motion to remove the article from the mold.

(*2*) the process or equipment used for directing charges or gobs of glass to a forming machine. **C 162,** C14

delivery—the leather or fabricated leather articles presented at any one time for inspection or test. **D 1517,** D31

delivery rate—mass of mixture discharged from the dispenser per unit of time at a specified temperature, usually expressed in g/s at 80 °F (26 °C). **D 3064,** D10

delivery rate—use **sediment delivery ratio** or sediment yield, whichever is meant. **D 4410,** D19

delta—a sediment deposit formed where moving water is slowed by a slower moving body of water. **D 4410,** D19

delta code, *n*—*in encoding information in a one-dimensional medium,* intervals that are subdivided into modules that are assigned values of "1" or "0" where "1s" are bars and "0s" are spaces. Examples are UPC, Code 128, Code 93 and Code 49. **F 1294,** F05

Delta Distance bar code—See **bar code.** **F 1156,** F12

delta ferrite—designation commonly assigned to delta iron containing alloying elements in solid solution. Small amounts of carbon and large amounts of other alloying elements markedly affect the high-and-low-temperature limit of equilibrium. **E 7,** E04

delta iron (δ Fe)—solid phase of pure iron which is stable at temperatures between 1400 and 1539°C and possesses the body-centered cubic lattice. Strictly, there is no difference between delta and alpha iron. **E 7,** E04

deltas—points in a fingerprint formed between diverging type lines. **F 1156,** F12

demagnetization—the reduction of residual magnetism to an acceptable level. **E 1316,** E07

demagnetization curve—the portion of a flux versus dc current plot (dc hysteresis loop) that lies in the second or fourth quadrant, that is, between the residual induction point, B_r, and the coercive force point, H_c. Points on this curve are designated by the coordinates, B_d and H_d. **A 340,** A06

demagnetizing coefficient, D_B—is defined by the equation:

$$D_B = [\Gamma_m(H_a - H)]/B_i$$

where:

H_a = applied magnetic field strength,
H = magnetic field strength actually existing in the magnetic material,
B_i = intrinsic induction, and
Γ_m = 1 in the cgs system and $4\pi \times 10^{-7}$, henry/metre in the mksa system (rationalized).

NOTE—For a closed, uniform magnetic circuit, the demagnetizing coefficient is zero. **A 340,** A06

demagnetizing factor, N_D—defined as 4π times the demagnetizing coefficient, D_B. **A 340,** A06

demagnetizing field strength, H_d—a magnetic field strength applied in such a direction as to reduce the induction in a magnetized body. See **demagnetization curve.** **A 340,** A06

demineralization—the process of removing minerals from water. **D 6161,** D19

demulsibility, *n*— *in petroleum products,* the ability of a mixture of liquids (usually hydrocarbons and water) to separate into its components after the mixture has been vigorously agitated. **D 4175,** D02

denaturant—toxins or noxious materials added to ethanol to make it unfit for human consumption. **E 1705,** E48

denatured alcohol—ethyl alcohol containing an addition of a poisonous substance, making it unfit for human consumption. **E 7,** E04

denatured ethanol—ethanol that is mixed with other chemicals or denaturants to make it unsuitable for human consumption.

E 1705, E48

denatured fuel ethanol—fuel ethanol to which chemicals (denaturants) have been added to make the ethanol unfit for human consumption in accordance with regulations of the Bureau of Alcohol, Tobacco, and Firearms of the U. S. Treasury Department. **E 772,** E44

denatured fuel ethanol—fuel ethanol to which chemicals (denaturants) have been added to make the ethanol unfit for human consumption in accordance with the regulations of the Bureau of Alcohol, Tobacco, and Firearms of the U. S. Treasury Department. **E 1126,** E48

denatured fuel ethanol—fuel ethanol to which chemicals (denaturants) have been added to make the ethanol unfit for human consumption in accordance with the regulations of the Bureau of Alcohol, Tobacco, and Firearms of the U.S. Treasury Department. **E 1705,** E48

dendrites—crystals, usually formed during solidification or sublimation, which are characterized by a tree-like pattern composed of many branches; pine-tree or fir-tree crystals. **E 7,** E04

dendritic powder—particles, usually of electrolytic origin, having the typical pine tree structure. Synonymous with **arborescent powder.** **B 243,** B09

denier, *n*—the unit of linear density, equal to the mass in grams of 9000 m of fiber, yarn, or other textile strand that is used in a direct yarn numbering system. (See also **linear density**) **D 123,** D13

denim, *n*—a durable woven twill fabric, usually of all cotton or a blend of cotton and man-made fibers, made from a variety of yarn numbers, and in various fabric weights, colors, designs, and finishes. **D 123,** D13

dense—a term used for optical glass having a high index of refraction. **C 162,** C14

dense—*(1)* term used in stress grading of certain softwood species to signify a high specific gravity. *(2)* a visual estimate of high specific gravity. To be classified as dense, the softwood species shall average on one end or the other of each piece, not less than six annual rings per inch and one-third or more latewood. Pieces not less than four rings per inch shall be accepted as dense if they average one-half or more latewood. **D 9,** D07

dense-graded aggregate, *n*—an aggregate that has a particle size distribution such that when it is compacted, the resulting voids between the aggregate particles, expressed as a percentage of the total space occupied by the material, are relatively small. **D 8,** D04

dense rubber, *n—in building construction,* a solid rubber material substantially free of cells or porosity. **C 717,** C24

densification—see **finely sized biomass materials**. **E 1126,** E48

densified biomass fuels—fuel made by compressing biomass to increase the density and to form the fuel into a specific shape such as cubes, pressed logs, pellets, or briquettes. Also see **fuel pellet**. **E 1126,** E48

densified particulate biomass fuels—a fuel made by mechanical compression of biomass to increase the bulk density and to press the fuel into a specific shape, such as pellets and briquettes. The fuel can have a maximum volume of 16.39 cm^3(1 $in.^3$) such that the largest dimension is 7.62 cm (3 in.). **E 1126,** E48

densified particulate biomass fuels—a fuel made by mechanical compression of biomass to increase the bulk density and to press the fuel into a specific shape, such as pellets and briquettes. The fuel can have a maximum volume of 16.39 cm^3(1 $in.^3$) such that the largest dimension is 7.62 cm (3 in.). **E 1705,** E48

densitometer—an instrument which measures the relationship between incident light and transmitted or reflected light and, using a logarithmic scale, gives a numerical measurement that corresponds to a material's opacity or a film's photographic density. **E 7,** E04

densitometer—not recommended, see **microphotometer.** **E 135,** E01

densitometer, *n*—instrument designed for measuring optical density of a photographic negative or positive or a printed image. **E 284,** E12

densitometer—a device for measuring the optical density of radiograph film. **E 1316,** E07

densitometry, *n*—technique for measurement of optical density by use of a densitometer. **E 284,** E12

density, δ— the ratio of mass to volume of a material. In the cgs-emu system of units, g/cm^3. In SI units, kg/m^3. **A 340,** A06

density, *n*—the weight per unit volume of a material. **C 11,** C11

density, *n*—mass per unit volume (preferred over deprecated term **unit weight**). **C 125,** C09

density:

absolute or true density—the weight divided by the volume excluding open and closed pores.

apparent or pycnometric density—the weight divided by the volume excluding open pores, but including closed pores.

tap density—the apparent density of a powdered or granulated material resulting when the receptacle containing the material is vibrated or tapped under standard or specified conditions. **C 242,** C21

density—mass per unit volume, usually expressed in g/cm^3 or lb/ft^3. **C 460,** C17

density—the mass of a unit volume of the material at any given temperature. The unit of mass is expressed in grams, the unit of volume in cubic centimetres (unless otherwise specified), and the temperature scale in degrees Celsius. The SI unit of density is kilogram per cubic metre. **C 859,** C26

density—the weight per unit volume in air, expressed in pounds per cubic foot of a product. **C 904,** C03

density, *n*—mass per unit volume expressed in g/cm^3 or lb/ft^3. **C 1154,** C17

density, *n*—The mass per unit volume of a substrate at a specified temperature and pressure; usually expressed in g/mL, kg,L, g/cm^3, g/L, kg/m^3 or lb/gal. See **specific gravity**. **D 16,** D01

density, *n*—mass per unit volume. **D 123,** D13

density—the mass per unit, ρ (ML^{-3}) kg/m^3.

density of dry soil or rock, ρ_d (ML^{-3}) kg/m^3—the mass of solid particles per the total volume of soil or rock.

density of saturated soil or rock, ρ_{sat} (ML^{-3}) kg/m^3—the total mass per total volume of completely saturated soil or rock.

density of soil or rock (bulk density), ρ (ML^{-3}) kg/m^3—the total mass (solids plus water) per total volume.

density of solid particles, ρ_s(ML^{-3}) kg/m^3—the mass per volume of solid particles.

density of submerged soil or rock, ρ_{sub} (ML^{-3}) kg/m^3—the difference between the density of saturated soil or rock, and the density of water.

density of water, ρ_w(ML^{-3}) kg/m^3—the mass per volume of water. **D 653,** D18

density, *n*—mass per unit volume (D 3288, D-9). **D 996,** D10

density, *n*—the mass per unit volume of substance. **D 1356,** D22

density, *n*—the mass per unit volume of a material. **D 1566,** D11

density—the mass per unit volume at a specified temperature. **D 1695,** D01

density—mass per unit volume, usually expressed in g/cm^3 or lb/ft^3. **D 2946,** C17

density—mass of a given volume of material at a specified temperature. **D 3064,** D10

density—the mass of a substance per unit volume, ρ in kg/L or kg/m^3. Use ρ_s for density of solid particles, ρ_w for water, ρ_d for dry sediment with voids, ρ_{sat} for saturated sediment, ρ_{wet} for wet sediment, and ρ_b for submerged sediment (buoyant weight). **D 4410,** D19

density, n—see **reflectance density, reflection density, transmission density,** or **transmittance density**. **E 284,** E12

density—the mass of a unit volume of a material at a specified temperature. The units shall be stated, such as grams per millilitre, grams per cubic centimetre, pounds per cubic foot, or other. The form of the expression shall be the following:

Density at x . . .

where x = temperature of the material. See also **Baumé gravity** and **specific gravity**. **E 1547,** E15

density—weight per unit volume, usually expressed in pounds per cubic inch, pounds per cubic foot, or kilograms per cubic metre. **E 1749,** E06

density—see **image density.** **F 335,** F05

density, *n*—*in printed symbologies*, the number of data characters that can be contained in a given unit of measure. Linear bar code density is expressed in characters per inch (CPI) and two-dimensional symbol density is expressed in characters per square inch (CPSI). **F 1294,** F05

density, absolute or true—the mass under specified conditions of a unit volume of a solid sorbent excluding its pore volume and inter-particle voids. **D 2652,** D28

density, apparent—the weight in air of a unit volume of a material. **D 883,** D20

density, apparent—the mass per unit volume of a sheet of pulp or paper. It is commonly calculated by dividing the basis weight by the caliper, although it must be recognized that the numerical value thus obtained is dependent upon the definition of the ream. **D 1695,** D01

density, apparent—the weight in air of a unit volume of a material at a specified temperature. The units shall be stated. The form of expression shall be the following:

Apparent density at x

where x = temperature of the material. **E 1547,** E15

density, apparent, *n*—the weight in air of a unit of volume of a material. **F 1251,** F04

density, apparent (density, bulk)—the mass under specified conditions of a unit volume of a solid sorbent including its pore volume and inter-particle voids. **D 2652,** D28

density, apparent (of applied insulation), *n*—the mass per unit volume of in-place mass thermal insulation. **C 168,** C16

density, block—see **density, particle.** **D 2652,** D28

density, bulk—the weight per unit volume of a material including voids inherent in material as tested. **D 883,** D20

density, bulk, *n*—the mass per unit volume of a material, including any voids present. **D 1566,** D11

density, bulk—the mass of a cellulosic material that will fill a unit volume of a container under specified conditions. **D 1695,** D01

density, bulk—see **density, apparent.** **D 2652,** D28

density, bulk, *n*—the weight in air of a unit of volume of a material. **F 1251,** F04

density comparison strip—alternative term for **step-wedge comparison film.** **E 1316,** E07

density current—the movement of one fluid under, through, or over another fluid of differing density. **D 4410,** D19

density (dry)— the weight per unit volume of an unimpregnated powder metallurgy part. **B 243,** B09

density (film)—transmission density is the common logarithm of the ratio of the radiant flux incident on the sample to the radiant flux transmitted by the sample, assuming no reflection. **E 7,** E04

density (film)—the quantitative measure of film blackening when light is transmitted or reflected.

$$D = \log (I_0/I) \text{ or } D = \log (I_0/R)$$

where:
D = density,
I_0 = light intensity incident on the film,
I = light intensity transmitted, and
R = light intensity reflected.

E 1316, E07

density (film or line)—the quantitative measure of film blackening.

$$D = \log I_o/I$$

where:
D = density,
I_o = light intensity incident on the film, and
I = light intensity transmitted.

C 859, C26

density in air—the weight per unit volume in vacuum minus the weight of a volume of air equal to the difference between the volume of the sample and the volume of brass weights equivalent to weight in vacuum of the sample. **D 4790,** D16

density (ρ, (ML^{-3}), kg/m^3), *n*—mass per unit volume. **D 4439,** D35

density (of gases)—the mass of a unit volume of a gas at a stated temperature and pressure. The units shall be stated. The form of expression shall be the following:

Density at x, y

where:
x = temperature of the gas, and
y = pressure of the gas. **E 1547,** E15

density of plastics—the weight per unit volume of material at 23°C expressed as D23c, $g/cm^3(kg/m^3)$. **F 412,** F17

density, packing (bulk)—the ratio of the mass of a collection of discrete pieces of solid material to the sum of the volumes of: the solids in each piece, the voids within the pieces, and the voids among the pieces of the particular collection. **D 3766,** D32

density, particle (density, block)—the mass under specified conditions of a unit volume of a solid sorbent including its pore volume but excluding inter-particle voids. **D 2652,** D28

density, particle (envelope)—the ratio of the mass of a particle to the sum of the volumes of: the solid in each piece and the voids within each piece, that is, within close-fitting imaginary envelopes completely surrounding each piece. **D 3766,** D32

density ratio, *n*—the ratio, often expressed as a percentage, of the density of a porous material to the density of the same material completely free of porosity. **B 243,** B09

density, skeletal—the ratio of the mass of discrete pieces of solid material to the sum of the volumes of: the solid material in the pieces and closed (or blind) pores within the pieces. **D 3766,** D32

density, theoretical—the ratio of the mass of a collection of discrete pieces of solid material to the sum of the volumes of said pieces, the solid material having an ideal regular arrangement at the atomic level. **D 3766,** D32

density (wet)— the weight per unit volume of a powder metallurgy part impregnated with oil or other nonmetallic materials. **B 243,** B09

dents—depressions in the copper foil which do not significantly change the thickness of the copper foil. **B 846,** B05

dents per unit width, *n—for woven pile yarn floor covering,* the number of binding sites per unit width; dents being the reed spaces through which the warp yarns pass in the loom or the metal strips in the reed which form these spaces. **D 123,** D13

denuder, *n*—a device designed to collect or remove gases from an air stream by diffusion to a collecting surface or secondary air stream while permitting the passage of particles. **D 1356,** D22

deoxidation products—a term specifically applied to those nonmetallic inclusions formed as a result of the addition of deoxidizing agents to molten metal. **E 7,** E04

deoxidized copper, high-residual phosphorus—copper deoxidized with phosphorus residual in amounts 0.015 to 0.04 %. The copper is not susceptible to hydrogen embrittlement, as determined in Specification B 379. The copper is of relatively low-electrical conductivity due to the amount of phosphorus present.

NOTE—International Standards Organization specifications permit up to 0.050 % phosphorus. **B 846,** B05

deoxidized copper, low-residual phosphorus— copper deoxidized with phosphorus residual in amounts 0.004 to 0.012 %. The copper is not readily susceptible to hydrogen embrittlement, as determined in Specification B 379. The copper in the annealed condition has a minimum conductivity of 90 % IACS. **B 846,** B05

dependent variable, *n*—See **response variable**. **E 456,** E11

depleted uranium—less than the accepted value of the weight percent of the isotope ^{235}U in natural uranium. **C 859,** C26

depletion—selective removal of one component of an alloy, usually from the surface or preferentially from grain boundary regions. **E 7,** E04

depletion—decrease of toner concentration in a developer composition characterized by low-image density. **F 335,** F05

depletion layer—a region in which the mobile carrier charge density is not sufficient to neutralize the net fixed charge density of donors and acceptors. Also known as **barrier layer, blocking layer,** or **space-charge layer.** **F 1241,** F01

depletion width—the width of the region in a semiconductor near a metal-semiconductor contact in which the charge-carrier density is insufficient to neutralize the fixed-charge density of donors and acceptors. **F 1241,** F01

deployed length, *n*—the coils as extended for use. It is specified by the coverage where:

Helical coverage = number of coil loops per unit × helical loop spacing

Concertina coverage = number of coil loops divided by 2 × concertina loop spacing.

F 1379, F14

deployment, *n—for inflatable restraints,* the sequence of events related to the activation of a module. **D 123,** D13

deployment—placing a boom in the water and making it operational. **F 818,** F20

depolarization—a decrease in the polarization of an electrode at a specified current density. **B 374,** B08

depolarization—not a preferred term. (See **polarization**.) **G 15,** G01

depolarizer—a substance or a means that produces depolarization. **B 374,** B08

deposit—the amount of pesticide on a unit area of surface. **E 609,** E35

deposit corrosion—localized corrosion under or around a deposit or collection of material on a metal surface. (See also **crevice corrosion**.) **G 15,** G01

deposition, *n*—the transfer of an atmospheric constituent to a surface due to gravity or another mechanism, or the material which is transferred.

dry deposition, n—all forms of deposition derived from the net vertical transfer of chemical species to a surface that are not the result of precipitation.

wet deposition, n—the precipitation of water from the atmosphere in the form of hail, rain, sleet, and snow. **D 1356,** D22

deposition—the chemical, mechanical, or biological processes through which sediments accumulate in a resting place. **D 4410,** D19

deposition aid—a material that improves the ability of agrichemical sprays to deposit on targeted surfaces. **E 1519,** E35

depositional carbon, *n—as used in Test Method D 5061,* a group of carbon forms that are formed from cracking and nucleation of gas-phase hydrocarbon molecules during coal carbonization.

pyrolytic carbon, n—as used in Test Method D 5061, an anisotropic carbon form that is formed by the deposition of carbon parallel to an inert substrate causing the resulting texture to appear ribbon-like.

sooty carbon, n—as used in Test Method D 5061, an isotropic carbon form comprised of approximately spherical particles of less than 1-μm diameter sometimes referred to as combustion black.

spherulitic carbon, n—as used in Test Method D 5061, a spherical anisotropic carbon form sometimes referred to as thermal black that is formed by the deposition of carbon concentrically around a nucleus.

divided sample—See *divided sample* under **sample**. **D 121,** D05

deposit rate—the amount of any material deposited per unit area. **E 1102,** E35

depression, *n*—localized pavement surface areas at a lower elevation than the adjacent paved areas. **E 867,** E17

depression, *n*—localized pavement surface areas at a lower elevation than the adjacent paved areas. **E 1778,** E17

depth, *n*—in the case of a beam, the dimension parallel to the direction in which the load is applied. **D 883,** D20

depth, *n*—in the case of a beam, the dimension parallel to the direction in which the load is applied. (D-20) **F 412,** F17

depth-dose distribution—variation of absorbed dose with depth from the incident surface of a material exposed to a given radiation. **E 170,** E10

depth factor—the factor included in deriving the allowable bending stress for rectangular members which takes into account the somewhat lower unit strength developed in larger members as compared to smaller members. Archaic terminology, more correctly termed **size factor.** **D 9,** D07

depth-integrated sample—a discharge-weighted (velocity-weighted) sample of water-sediment mixture collected at one or more verticals in accordance with the technique of depth integration; the discharge of any property of the sample expressible as a concentration can be obtained as the product of the concentration and the water discharge represented by the sample. **D 4410,** D19

depth-integrating sediment sampler—a device that collects a representative water-sediment mixture at all points along the sampling vertical. **D 4410,** D19

depth integration—a method of sampling at every point throughout a given depth (the sampled depth) whereby the water-sediment mixture is collected isokinetically so that the contribution from each point is proportional to the stream velocity at the point. This process yields a sample with properties that are discharge weighted over the sampled depth. Ordinarily, depth integration is performed by traversing either a depth- or point-integrating sampler vertically at an acceptably slow and constant rate; however, depth integration can also be accomplished with vertical slot samplers. **D 4410,** D19

depth of field—the depth or thickness of the object space that is simultaneously in acceptable focus. **E 7,** E04

depth of field—the range of distance over which an imaging system gives satisfactory definition when it is in the best focus for a specific distance. **E 1316,** E07

depth of field, *n—in bar code reading,* the difference between the minimum and maximum distance from the aperture of the reader to the symbol throughout which the bar code can be accurately interpreted. **F 1294,** F05

depth of focus—the depth or thickness of the image space that is simultaneously in acceptable focus. **E 7,** E04

depth of gloss—the optical phenomenon of relative depth perceived when viewing reflective surfaces. **D 2825,** D21

depth of penetration, *n*—(1) the distance a penetrant has entered into a solid material as measured from the surface of the material; (2) the maximum depth at which a magnetic or ultrasonic indication can be measured in a test specimen. **C 1145,** C28

depth of penetration—in electromagnetic testing, the depth at which the magnetic field strength or intensity of induced eddy currents has decreased to 37 % of its surface value. The depth of penetration is an exponential function of the frequency of the signal and the conductivity and permeability of the material. Synonymous terms are standard depth of penetration and skin depth. (See also **skin effect.**) **E 1316,** E07

depth of penetration, d_p—in internal reflection spectroscopy, the distance into the less refractive medium at which the amplitude of the evanescent wave is e^{-1}(that is, 36.8 %) of its value at the surface:

$$d_p = \frac{\lambda_1}{2\pi(sin^2\theta - \eta_{21}{}^2)^{1/2}}$$

where: $n_{21} = n_2/n_1$ = refractive index of sample divided by that of the IRE; $\lambda_1 = \lambda/n_1$ = wavelength of radiant energy in the sample; and θ = angle of incidence. **E 131,** E13

depth profiling—*AES, XPS, SIMS,* monitoring the signal strength as a function of some variable (for example, sputtering time) that can be related to distance from the surface. **E 673,** E42

depth resolution—*AES, XPS, SIMS,* the depth range over which a signal increases (or decreases) by a specified amount when profiling an ideally sharp interface between two media. By convention, the depth resolution corresponds to the distance over which a 16 % to 84 % (or 84 % to 16 %) change in signal is measured. **E 673,** E42

depth scale—*EIA,* a relationship between energy loss and target depth that allows a direct correlation between multichannel analyzer channel number and depth in the specimen. **E 673,** E42

depuration—loss of a substance from an organism as a result of any active or passive process. **E 943,** E47

derivative, *adj*—pertaining to the first derivative (mathematical) of any curve with respect to temperature or time. **E 473,** E37

derivative absorption spectrum—a plot of rate of change of absorbance or of any function of absorbance with respect to wavelength or any function of wavelength, against wavelength or any function of wavelength. **E 131,** E13

derived differential curve—the curve derived from the data obtained by the use of the differential method of thermal analysis. The changes in the temperature difference $\Delta(\theta - \theta')$, between a specimen and a neutral body, for a constant interval of temperature $\Delta\theta$ are plotted against the temperature. An arithmetic treatment of the differential data resulting in a plot of $\Delta(\theta - \theta')/\Delta\theta$ *versus* θ. (See **differential curve.**) **E 7,** E04

dermal toxicity—the toxic effect to an organism resulting from contact of the pesticide with the skin. **E 609,** E35

desalination—see **demineralization.** **D 6161,** D19

descender, *n*—that portion of an alphabetic character that extends below the baseline. **F 1457,** F05

descending fork point—in a ternary phase diagram, the configuration at the convergence of the three divariant curves upon each of the three high-temperature phases associated in Class III univariant equilibrium; for example, the division of a descending liquidus surface valley into two descending liquidus-surface valleys. **E 7,** E04

descriptive analysis, *n*—any method to describe and quantify the sensory characteristics of stimuli by a panel of trained assessors. **E 253,** E18

desensitize—chemical treatment of non-image master areas to make the areas ink repellent. **F 425,** F05

desiccant, *n*—a hygroscopic substance used to absorb water vapor from the air to maintain a low relative humidity in a container. **D 996,** D10

desiccant, *n*—compounding material used to irreversibly absorb moisture present (in a rubber mix) particularly for the purpose of minimizing risk of porosity during vulcanization. **D 1566,** D11

designated person—an individual who is qualified by experience or training to perform an assigned task. **F 819,** F18

designated size—the dimensional name for a particular size that may or may not be equal to or related to the dimensions used for design purposes or of the manufactured product. **C 822,** C13

design contingency—in project design/construction, the amount of funds added to the estimated construction costs to cover unanticipated construction costs due to the incompleteness of the design, where the contingency is inversely proportional to the level of completeness of the design documentation. **E 833,** E06

design development—the phase of a project consisting of drawings and document preparation to fix and describe the size and character of the building systems, material, and elements. **E 833,** E06

design life—the period of time during which a system or component is expected to perform its intended function, without significant degradation of performance and without requiring major maintenance or replacement. **E 772,** E44

design load—the load at which the geosynthetic is required to operate in order to perform its intended function. **D 4439,** D35

design of experiments, *n*—the arrangement in which an experimental program is to be conducted, and the selection of the levels (versions) of one or more factors or factor combinations to be included in the experiment. Synonyms include experiment design and experimental design. **E 456,** E11

design of experiments, *n*—the arrangement in which an experimental program is to be conducted, and the selection of the levels (versions) of one or more factors or factor combinations to be included in the experiment. Synonyms include experiment design and experimental design. **E 1325,** E11

design program—the information detailing project function, purpose, and characteristics inclusive of floor area, functional spaces, equipment, and building systems. **E 833,** E06

design program—See **facility performance**. **E 1480,** E06

design strength—the minimum acceptable 0.01-in. (0.3-mm) crack D-load. **C 822,** C13

desktop forgery—counterfeits and forgeries using computerized systems to produce documents appearing genuine. **F 1156,** F12

desorption, *n*—a process in which a sorbed material is released from another material, as the desorption of moisture from fibers; the reverse of absorption, adsorption, or both. **D 123,** D13

desorption, *n*—the process of freeing from a sorbed state. **D 1356,** D22

desorption—the separation of an adsorbate as such from a sorbent. **D 2652,** D28

destructive test—See **test, destructive**. **E 1749,** E06

destructive test—a test to determine the properties of a material or the behavior of an item which results in the destruction of the sample or item. **F 1789,** F16

detached dwelling—See **dwelling.** **E 631,** E06

detected feature—*in image analysis*, an object or constituent of interest that is isolated for measurement by adjustment of the threshold setting to its particular range of gray level. **E 7,** E04

detection limit—a stated limiting value which designates the lowest concentration or mass that can be estimated or determined with confidence and which is specific to the analytical procedure used. **C 859,** C26

detection limit—a stated limiting value that designates the lowest concentration or mass that can be estimated or determined with confidence and that is specific to the analytical procedure used (see Practice E 876). **E 135,** E01

detection limit—the smallest concentration of an element or compound that can be measured for specific analysis conditions and data collection periods. **E 673,** E42

detection limit—stated limiting value that designates the lowest concentration or mass that is capable of being estimated or determined with confidence and that is specific to the analytical procedures used. **E 1605,** E06

detection monitoring—a program of monitoring for the express purpose of determining whether or not there has been a contaminant release to ground water. **D 653,** D18

detection process—a past-oriented strategy that attempts to identify unacceptable output after it has been produced and then separate it from the good output. **F 1789,** F16

detection system—a system which relies on final inspection as the primary means of controlling the quality of finished fasteners. **F 1789,** F16

detector, *n*—device to convert radiant energy into a neural signal (such as the eye) or an electrical signal (such as a phototube, photomultiplier tube, photocell, photodiode, or the like). **E 284,** E12

detector:

angle—EIA, see **angle** *of detector:*

efficiency—EIA, fraction of particles incident on the detector that actually generate a detectable signal.

foil—EIA, a thin sheet, usually metal or plastic, placed over a detector to absorb low energy or high mass products, or both, from nuclear reactions, while transmitting other reaction products.

solid angle—see **angle,** *solid, of detector.* **E 673,** E42

detector efficiency—in the measurement of radioactivity, the fraction of particles or photons striking the detector that gives rise to a detected response. (In many cases, it is defined to mean relative efficiency, the product of detector efficiency and detector geometry.) **C 859,** C26

detector geometry—in the measurement of radioactivity, the fraction of emissions from the source (particles or photons) that impinge on the detector. Also called solid angle. **C 859,** C26

detergency—the removal of soil, using a detergent. **D 459,** D12

detergent—a surface-active agent that possesses the ability to clean soiled surfaces.

anionic detergent—a detergent that produces aggregates of negatively-charged ions with colloidal properties.

cationic detergent—a detergent that produces aggregates of positively-charged ions with colloidal properties.

nonionic detergent—a detergent that produces aggregates of electrically-neutral molecules with colloidal properties. **B 374,** B08

detergent, *n*—*in textile product care*, a cleaning agent containing one or more surfactants as the active ingredient(s). **D 123,** D13

detergent—a composition that removes soil. **D 459,** D12

anionic detergent—a detergent that produces negatively charged colloidal ions in solution. **D 459,** D12

cationic detergent—a detergent that produces positively charged colloidal ions in solution. **D 459,** D12

dry-cleaning detergent **(charge-type)**—a dry-cleaning detergent used at a given percentage by volume that can pass through a diatomaceous earth-coated filter in the dry-cleaning system without change in composition. **D 459,** D12

dry-cleaning detergent **(dry-cleaning aid)**—a detergent that when added to a dry-cleaning solvent increases cleaning effectiveness. **D 459,** D12

dry-cleaning detergent **(non-charge type)**—any dry-cleaning detergent that is not of the charge type. **D 459,** D12

inorganic alkaline detergent—a water-soluble inorganic alkali or alkaline salt having detergent properties, but containing no soap or synthetics. **D 459,** D12

nonionic detergent—a detergent that produces electrically neutral-colloidal particles in solution. **D 459,** D12

synthetic detergent—a detergent produced by chemical synthesis and comprising an organic composition other than soap. **D 459,** D12

detergent—a formulated cleaning composition, generally containing one or more surfactant(s) as the essential component(s). However, under **detergent,** see *inorganic alkaline detergent.* Imprecisely, the terms *detergent* and *surfactant* have been used interchangeably. **D 459,** D12

dry-cleaning detergent—a formulated composition added to the solvent bath in dry-cleaning operations to improve cleaning. **D 459,** D12

inorganic alkaline detergent—a formulated cleaning composition containing water-soluble alkali or alkaline salts, but generally no surfactants. **D 459,** D12

detergent, *n*—*in textile product care*, any cleaning agent. **D 5253,** D13

detergent—a cleansing agent; any of numerous synthetic water soluble or liquid-organic preparations that are chemically different from soaps but resemble them in the ability to emulsify oils and hold dirt in suspension. **D 6161,** D19

detergent alkylate—a mixture of alkylated aromatic hydrocarbons which when sulfonated yields an alkyl aryl sulfonate detergent. The term usually refers to an alkyl benzene in which the alkyl radical is a mixture of straight-chain and isomeric branched-chain groups, averaging 10 or more carbon atoms. **D 459,** D12

detergent feeder—a device that automatically feeds detergents into wash tanks of spray-type commercial dishwashing and glasswashing machines. **F 1827,** F26

detergent remover—a penetrant remover that is a solution of a detergent in water. **E 1316,** E07

detergent resistance—the degree to which a polish film exhibits no apparent deterioration when spotted or cleaned with a solution of a nonabrasive, nonammoniacal detergent. **D 2825,** D21

deteriorated condition—condition of surfaces of such components as walls, windows, and baseboards that are in need of repair (or replacement) due to physical or mechanical breakdown of paint or other materials. **E 1605,** E06

deterioration—a permanent impairment of the physical properties. (see also **degradation**) **D 1695,** D01

determination, *n*—the application of the complete measurement procedure to one piece, specimen or object to produce *one* numerical measured value to be used to form an average or median. **F 538,** F09

determination value, *n*—the numerical quantity calculated by means of the test method equation from the measurement values obtained as directed in a test method. (See also **observation.**) **D 123,** D13

deterministic design, *n*—design based on the physical and mechanical properties of the materials, elements, and structures involved (compare **probabilistic design**). **E 631,** E06

deterrent—a security feature to discourage illicit duplication or counterfeiting by instilling anxiety, doubt, fear, etc., in the mind of the counterfeiter. **F 1156,** F12

detonation—an extremely rapid and violent chemical reaction causing the production of a large volume of gas. (ISRM) **D 653,** D18

developed footprint length, [*L*], *n*—the maximum footprint dimension in the circumferential direction of the tire, under stated conditions of measurement. **F 538,** F09

developed footprint width, [*L*], *n*—the maximum lateral dimension of a tire footprint under stated conditions of measurement. **F 538,** F09

developed length, *n*—the bottom contour length from the snowboard ski tip to the snowboard ski tail, sometimes called the material length. **F 1107,** F27

developed length, L_N—bottom contour length as measured from the ski tip to the ski tail, commonly called the material length. **F 472,** F27

developer—a material that is applied to the test surface to accelerate bleedout and to enhance the contrast of indications. **E 1316,** E07

developer—the material or combination of materials that renders visible a latent electrostatic image when brought into intimate contact with it. Electrostatic developers can be either liquid or dry and can consist of a toner and a carrier. **F 335,** F05

developer, *n*—acidic materials which react with leuco dyes to form color. **F 1623,** F05

developer, aqueous—a suspension of developer particles in water. **E 1316,** E07

developer, dry powder—a fine free-flowing powder used as supplied. **E 1316,** E07

developer, liquid film—a suspension of developer particles in a vehicle which leaves a resin/polymer film on the test surface after drying. **E 1316,** E07

developer, non-aqueous—developer particles suspended in a non-aqueous vehicle prior to application. **E 1316,** E07

developer, soluble—a developer completely soluble in its carrier, not a suspension of powder in a liquid, which dries to an absorptive coating. **E 1316,** E07

developing time—the elapsed time between the application of the developer and the examination of the part. **E 1316,** E07

development—the process of converting a latent electrostatic image into a visible image. **F 335,** F05

deviation—*in protective coatings,* a departure of a characteristic from established procedures or from specified requirements. **D 4538,** D33

deviation, *n*—the difference between a measurement or quasi-measurement and its stated value or intended level. **E 456,** E11

deviation (X-ray)—the angle between the diffracted beam and the transmitted incident beam. It is equal to twice the Bragg angle θ. **E 7,** E04

deviator of stress (strain)—the stress (strain) tensor obtained by subtracting the mean of the normal stress (strain) components of a stress (strain) tensor from each normal stress (strain) component. (ISRM) **D 653,** D18

deviator stress, Δ, σ (FL^{-2})—the difference between the major and minor principal stresses in a triaxial test. **D 653,** D18

devitrification—crystallization of glass. **C 162,** C14

devitrification—a surface defect manifested by loss of gloss as a result of crystallization. **C 286,** B08

devitrification—crystallization of an amorphous substance. **E 7,** E04

devitrification—crystallization of an amorphous substance, (E 7, E-4). **E 1142,** E37

dew, *n*—water vapor that has condensed onto a surface near the ground because of radiational cooling of that surface to a temperature that is below the dew point of the air surrounding the surface. **D 1356,** D22

dew cell, dew probe, *n*—an instrument that measures the temperature at which a saturated salt solution (usually of lithium chloride) is in equilibrium with the water vapor in moist air. **D 1356,** D22

dew cell, dew probe—an instrument that measures the temperature at which a saturated salt solution (usually of lithium chloride) is in equilibrium with the water vapor in moist air. **D 4023,** D22

dewetting—a condition that results when molten solder has coated a surface and then receded leaving irregular-shaped mounds of solder separated by areas covered with a thin film but base metal is not exposed. **B 846,** B05

dew-/frost-point hygrometer—See **hygrometer.** **D 1356,** D22

dew-/frost-point hygrometer—an instrument that measures the surface temperature at which ambient water vapor condenses. **D 4023,** D22

dew point, *n*—the temperature below which condensation of water vapor begins to take place when the atmosphere is cooled. **D 123,** D13

dew point—the temperature at any given pressure at which liquid initially condenses from a gas or vapor. It is specifically applied to the temperature at which water vapor starts to condense from a gas mixture (**water dew point**), or at which hydrocarbons start to condense (**hydrocarbon dew point**). **D 4150,** D03

dew point—the temperature to which water vapor must be reduced to obtain saturation vapor pressure, that is, 100 % relative humidity.

NOTE—As air is cooled, the amount of water vapor that it can hold decreases. If air is cooled sufficiently, the actual water vapor pressure becomes equal to the saturation water-vapor pressure, and any further cooling beyond this point will normally result in the condensation of moisture. **E 41,** G03

dewpoint temperature, *n*—the temperature at which condensation of water vapor in a space begins for a given state of humidity and pressure as the vapor temperature is reduced; the temperature corresponding to saturation (100 % relative humidity) for a given absolute humidity at constant pressure. **C 168,** C16

dew-point temperature—See **temperature.** **D 1356,** D22

dew-point temperature—the temperature (above 0°C) to which a gas or vapor must be cooled at constant pressure and constant water-vapor composition in order for saturation to occur. **D 2864,** D27

dew probe—See **dew cell.** **D 1356,** D22

dextrins—high molecular weight sugars, intermediates obtained in the conversion of starch to fermentable sugar. **E 1705,** E48

dezincification—a type of corrosion found with some copper-zinc alloys which occurs by solution of a small region and immediate redeposition of the copper in a spongy porous form, thus giving the impression of selective removal of zinc from the alloy. **E 7,** E04

dezincification—See **parting**; specific to copper-zinc alloys. **G 15,** G01

DGS (distance gain size-German AVG)—distance amplitude curves permitting prediction of reflector size compared to the response from a back surface reflection. **E 1316,** E07

D head—nail head with semi-circular rim and head segment omitted during heading, with omitted segment reaching from rim to shank projection, to allow tight collating of nails in strip form. **F 592,** F16

diagnostic verifier, *n*—a device that automatically evaluates the quality of a film master or a printed code by comparing the observed information to a standard for the printed symbol. **F 1294,** F05

diagonal—in hardness testing, a line joining two opposite corners of a diamond pyramid indentation. **E 7,** E04

diagonal bracing—See **loading.** **D 996,** D10

diagonal wear, *n*—a type of irregular wear characterized by an increased wear rate region or band oriented transversely (from shoulder to shoulder) at some non-90° angle with respect to the circumferential centerline of the tread band. **F 538,** F09

dial—a numerically calibrated part used to align the tumblers and operate the lock. **F 471,** F12

dial and ring (spy-resistant)—a dial and ring designed to restrict the unauthorized observation of combination dialing. **F 471,** F12

dial shield—a shroud to shield the dial from unauthorized observation of combination dialing. **F 471,** F12

dialysis—a separation process dependent on different diffusion rates of solutes across a permeable membrane without an applied hydraulic driving force. **D 6161,** D19

diamagnetic material—a material whose relative permeability is less than unity.

NOTE—The intrinsic induction, B_i, is oppositely directly to the applied magnetizing force *H*. **A 340,** A06

diamagnetic material—a material whose relative permeability is less than unity.

NOTE—The intrinsic induction, B_i, is oppositely directed to the applied magnetizing force *H*. **B 713,** B01

diamagnetic material—a material whose relative permeability is less than unity.

NOTE—The intrinsic induction Bi is oppositely directed to the applied magnetizing force *H*. **E 1316,** E07

diameter:

arithmetic mean diameter—that diameter located at the centroid of the distribution of size.

equivalent diameter (sphere)—the diameter of a theoretical sphere of a material which under identical physical conditions yields the same value of the particular fineness characteristic as the actual irregularly shaped dispersed particle of the same material.

median diameter—that diameter at which the area under the curve of size versus frequency is divided into two equal parts. **C 242,** C21

diameter—length of longest straight line through center of largest cross section of head. **F 547,** F16

diameter—length of longest straight line through center of cross section of wire from which nail is formed. Diameter of pointed nail is that immediately below the grip marks. Diameter of nail with roll-threaded shank is that of wire or nail prior to being threaded. In the case of fluted, roll-grooved, and twisted nails, formed diameter refers to crest diameter. Diameter does not include coating except cladding. (See **thread-crest diameter.**) **F 547,** F16

diameter—*of a semiconductor wafer*, the linear dimension across the surface of a circular wafer that contains the wafer center and excludes flats or other peripheral fiduciary geometries. **F 1241,** F01

diameter, *n*—specified size diameters shall be measured across the center line of the packaged coil with a tolerance of ±2 in.; installed diameters are always less than the packaged diameter of the coil. **F 1379,** F14

diameter, sedimentation—the diameter of a hypothetical sphere of the same specific gravity and the same settling velocity as the given particle in the same fluid. **D 4410,** D19

diamond, *n*—*in zippers*, the wedge-shaped portion of a slider between the throats. **D 123,** D13

diamond—the opening formed by the woven wires in chain link fence fabric. **F 552,** F14

diamond count—the number of diamond openings from one edge of fabric to the other. The count of a given fabric shall begin at the first completed diamond at one edge and continue to the unfinished (half) or full opening at the other edge. **F 552,** F14

diamond head—square head of trapezoidal longitudinal section, with flat upper and bearing surfaces and bearing surface larger than upper surface. **F 547,** F16

diamond paste, *n*—diamond dust dispersed in a paste or slurry for use as a grinding or polishing compound. **C 1145,** C28

diamond point—symmetrical point having four approximately equal bevelled planes forming a pyramid; its length measured along cut edge of point. Usually applied unless otherwise specified. **F 547,** F16

diamond tool, *n*—any tool in which the working area is inset with diamonds or diamond dust. **C 1145,** C28

diamond wheel, *n*—a bonded grinding wheel in which the abrasive grains are crushed and sized natural or synthetic diamonds. **C 1145,** C28

diaphragm—a porous or permeable membrane separating anode and cathode compartments of an electrolytic cell from each other or from an intermediate compartment. **B 374,** B08

diaphragm nail—stout, bright or galvanized, regular-stock-steel or hardened-steel, helically threaded, 1 ½ by 0.135 to 2½ by 0.148-in. nails with flat slightly countersunk 5⁄16-in. head and medium diamond point. **F 547,** F16

diaspore clay, *n*—a rock consisting essentially of diaspore bonded by fireclay. **C 71,** C08

diatom—single cell marine animal having a coating consisting principally of silica. **D 6161,** D19

diatomaceous earth—See **diatomite.** **C 242,** C21

diatomaceous earth (DE) filtration—filtration using an amphorous, lightweight siliceous earth medium occurring naturally as the fossil remains of diatoms. **D 6161,** D19

diatomaceous silica, *n*—insulation composed principally of diatomaceous earth with or without binders, and which usually contains reinforcing fibers. **C 168,** C16

diatomite (diatomaceous earth)—amorphous lightweight siliceous material having the theoretical formula $SiO_2{\cdot}nH_2O$, occurring naturally as the fossil remains of tiny plants termed diatoms; also known as *kiesel-guhr, tripolite,* and *infusorial earth.* **C 242,** C21

dibutyl phthalate (DBP) absorption number, *n*—the number of cubic centimetres of DBP absorbed by 100 g of carbon black under specified test conditions. **D 3053,** D24

dice—the more or less cubical fracture of tempered glass. **C 162,** C14

dicyclopentadiene resin, *n*—a resin produced by the thermal polymerization of a concentrated dicyclopentadiene stream. **D 6440,** D01

die—the part or parts making up the confining form in which a powder is pressed. The parts of the die may be some or all of the following: die body, punches, and core rods. Synonymous with **mold.** **B 243,** B09

die body—the stationary or fixed part of a die. **B 243,** B09

die casting, *n*—a casting process wherein molten metal is poured or injected under high pressure into the cavity of a metal mold and solidified; also, a product of such a process. **B 899,** B02

die cut—See **container.** **D 996,** D10

die insert—a removable liner or part of a die body. **B 243,** B09

dielectric, *n*—a medium in which it is possible to maintain an electric field with little supply of energy from outside sources. **D 1711,** D09

dielectric—a medium in which it is possible to maintain an electric field with little supply of energy from outside sources. The energy required to produce the electric field is recoverable, in whole or in part. A vacuum, as well as any insulating material is a dielectric. **D 2864,** D27

dielectric, *adj*—pertaining to the appearance of those materials for which the first surface reflectance is characteristic of the illuminant; compare **metal-like**. **E 284,** E12

dielectric breakdown—a threshold effect in a dielectric medium where, at some electric field strength across the medium, bound electrons become unbound and travel through the medium as a current. In solid media, the region of the current path is permanently damaged. The unit of measurement is usually volts per unit of thickness. **D 5077,** D10

dielectric breakdown voltage—the potential difference at which electrical failure occurs in an electrical insulating material or insulation structure, under prescribed test conditions. **D 2864,** D27

dielectric breakdown voltage (electric breakdown voltage), *n*—the potential difference at which dielectric failure occurs under prescribed conditions, in an electrical insulating material located between two electrodes. **D 1711,** D09

dielectric constant—see **relative permittivity.** **D 1711,** D09

dielectric constant— See **dielectric constant, relative** (See also **permittivity** .) **D 2864,** D27

dielectric constant— see **permittivity, relative**. **E 1142,** E37

dielectric constant, absolute—the same as **permittivity.** **D 2864,** D27

dielectric constant, relative—the same as **permittivity, relative.** **D 2864,** D27

dielectric dissipation factor, *D*—the ratio of the loss factor, ϵ'', to the absolute permittivity, ϵ', or

$$D = \epsilon''/\epsilon'$$

E 1142, E37

dielectric failure—the failure of an element in a dielectric circuit that exists when the insulating element becomes conducting. This event may take the form of a gradual increase in current exceeding a specified value, but it usually takes the form of an almost instantaneous charge transfer accompanied by collapse of the insulating properties and partial of complete localized destruction of the dielectric medium. In the case of liquids and gases the failure may be self-healing. **D 2864,** D27

dielectric failure (under test), *n*—an event that is evidenced by an increase in conductance in the dielectric under test limiting the electric field that can be sustained. **D 1711,** D09

dielectric loss angle—the angle whose tangent is the dissipation factor or arctan ϵ''/ϵ'. **E 1142,** E37

dielectric printing process—a nonimpact printing technique in which specially treated paper consisting of a conductive base layer coated with a nonconductive thermoplastic material is used to hold an electric charge usually applied directly by a set of electrode styli. The electric charge corresponds to the latent image of the original. Following the charging step, the paper is imaged by a toner system similar to that of electrostatic copying devices. This technique is sometimes called electrographic, and is currently employed on general purpose nonimpact printers, plotting and facsimile devices. **F 909,** F05

dielectric strength, *n*—the voltage gradient at which dielectric failure of the insulating material occurs under specific conditions of test. **D 1711,** D09

dielectric strength—a property of an insulating material described by the average voltage gradient at which electric breakdown occurs under specific conditions of test. **D 2864,** D27

dielectric thermal analysis, (*DETA or DEA*), *n*—a technique in which the dielectric constant (permittivity, or capacitance) and dielectric loss (conductance) of a substance under oscillating electric field are measured as a function of temperature or time while the substance is subjected to a controlled-temperature program in a specified atmosphere. (ICTAC) **E 473,** E37

die lubricant, *n*—a lubricant applied to the walls of the die and to the punches to facilitate the pressing and ejection of the compact. **B 243,** B09

die marks—See **grip marks.** **F 547,** F16

diene polymer, *n*—a polymer formed from one or more monomer species, at least one of which is a diolefin. **D 1566,** D11

die-parting line—a lengthwise flash or depression on the surface of a pultruded plastic part.

NOTE—The die-parting line is associated with the area where separate pieces of the die join together to form the cavity. **D 3918,** D20

diesel fuel oil, *n*—any petroleum liquid suitable for the generation of power by combustion in compression ignition (diesel) engines. **D 4175,** D02

diesel index, *n*—an approximation of the cetane number (the ignition performance) of diesel fuel, calculated from the density and the aniline point. (No longer widely used for distillate fuels but applicable to some blended distillate residual fuels (see also **cetane index**).) **D 4175,** D02

die set—the parts of a press that hold and locate the die in proper relation to the punches. **B 243,** B09

die swell, *n*—difference between the dimensions of the cross section of an extrudate and the corresponding dimensions of the die orifice by which the extrudate is formed. **D 1566,** D11

dietary accumulation—the net accumulation of a substance by an organism as a result of ingestion in the diet. **E 943,** E47

die-wall lubricant, *n*—synonymous with **die lubricant.**

B 243, B09

difference absorption spectrum—a plot of the difference between two absorbances or between any function of two absorbances, against wavelength or any function of wavelength. **E 131,** E13

difference limen, *n*—See **threshold, difference**. **E 253,** E18

differentia, *n*—a characteristic that distinguishes one species of a genus from all other species of that genus. **E 1992,** E02

differential, *adj*—pertaining to a difference in measured or measureable quantities usually between a substance and some reference or standard material. **E 473,** E37

differential (acoustic emission) amplitude distribution F(V)—see **distribution, differential (acoustic emission) amplitude f(V).** **E 1316,** E07

differential (acoustic emission) threshold crossing distribution f_t**(V)**—see **distribution, differential (acoustic emission) threshold crossing.** **E 1316,** E07

differential aeration cell (oxygen concentration cell)—a concentration cell caused by differences in oxygen concentration along the surface of a metal in an electrolyte. (See **concentration cell**.) **G 15,** G01

differential blackbody—an apparatus for establishing two parallel isothermal planar zones of different temperatures, and with effective emissivities of 1.0. (E 1213) **E 1316,** E07

differential coils—two or more coils electrically connected in series opposition such that any electric or magnetic condition, or both, that is not common to the areas of a specimen being electromagnetically tested will produce an unbalance in the system and thereby yield an indication. **E 1316,** E07

differential curve: (*1*) *in thermal analysis*——a curve resulting from the differential method of thermal analysis when the difference in temperature ($\theta - \theta'$) between a specimen and a neutral body is plotted against the temperature of the latter.

(*2*) *in dilatometry*—a curve produced by plotting against the temperature the difference in changes of length or volume between a body of known expansivity and a body (specimen) of unknown expansivity. **E 7,** E04

differential dyeing behavior, *n—of cotton*, the tendency of cotton fibers to absorb and retain selectively varying proportions of different dyes from a binary dye bath. **D 123,** D13

differential heating, *n*—heating that intentionally produces a temperature gradient within a steel object such that, after cooling, a desired stress distribution or variation in properties is present within the object. **A 941,** A01

differential heat of adsorption—the heat evolved during the adsorption of an incremental quantity of adsorbate at a given level of adsorption. **D 2652,** D28

differential interference contrast illumination—a microscopical technique employing a beam-splitting double-quartz prism; that is a modified Wollaston prism placed ahead of the objective with a polarizer and analyzer in the 90° crossed positions. The two light beams are made to coincide at the focal plane of the objective, thus rendering height differences visible as variations in color. The prism can be moved, shifting the interference image through the range of Newtonian colors. **E 7,** E04

differential leak detector—a leak detector employing two similar gage tubes in a bridge circuit with a trap which is selective for the tracer gas between the system and one of the tubes. **E 1316,** E07

differentially coated sheet, *n*—metallic coated sheet with specified difference in weight (mass) of metallic coating between the two surfaces. **A 902,** A05

differential measurements—in electromagnetic testing, measurements made in which the imbalance in the system is measured using differential coils in contrast to absolute and comparative measurements. (See also **differential coils**). **E 1316,** E07

differential Pirani gage—a leak detecting device employing two similar Pirani tubes as arms of a Wheatstone bridge. **E 1316,** E07

differential price escalation rate—the expected percent difference between the rate of increase assumed for a given item of cost (such as energy), and the general rate of inflation. **E 833,** E06

differential readout—in electromagnetic testing, the signal output of differential coils. (See also **differential coils**). **E 1316,** E07

differential scanning calorimetry *(DSC), n*—A technique in which the difference into a substance and a reference material is measured as a function of temperature while the substance and reference material are subjected to a controlled-temperature program. (ICTAC)

NOTE—The record is the differential scanning calorimetric or DSC curve. Two modes, power compensation differential scanning calorimetry, and heat flux differential scanning calorimetry can be distinguished, depending on the method of measurement used. **E 473,** E37

differential scanning calorimetry (DSC), *n*—a technique in which the difference in energy inputs into a substance and a reference material is measured as a function of temperature, while the substance and the reference material are subjected to a controlled temperature program. **E 1445,** E27

differential settlement—settlement that varies in rate or amount, or both, from place to place across a structure. **D 653,** D18

differential shrinkage, *n—in zippers*, the difference in longitudinal dimensional change between the zipper tape and the fabric to which the zipper is attached. **D 123,** D13

differential signal—in electromagnetic testing, an output signal that is proportional to the rate of change of the input signal. **E 1316,** E07

differential system—an electromagnetic testing system that uses coil assemblies and associated electronics to detect an electric or magnetic condition, or both, that is not common to the areas of the specimen being tested. (See also **differential coils**). **E 1316,** E07

differential thermal analysis *(DTA), n*—A technique in which the temperature difference between the substance and a reference material is measured as a function of temperature, while the substance and reference material are subjected to a controlled-temperature program. (ICTAC)

NOTE—The term *quantitative differential thermal analysis* covers those uses of DTA where the equipment is designed to produce quantitative results. **E 473,** E37

differential thermal analysis (DTA), *n*—a technique in which the temperature difference between a substance and reference material is measured as a function of temperature while the substance and the reference material are subjected to a controlled temperature program. **E 1445,** E27

differential thermocouple—two thermocouples placed in series opposition (bucking). **E 7,** E04

differential thermocouple— see **differential thermopile**. **E 1142,** E37

differential thermopile—a number of temperature sensors connected in series-opposing and arranged so that there is an increase in output signal for a given temperature difference between alternate junctions maintained at a reference temperature and the measured temperature. **E 1142,** E37

differential titration—determination of the concentration of two or more species in one sample by titration with a reagent containing a species that reacts with both sample species. In order to obtain distinct end-point breaks, the formation constants or solubility products of the two species formed must differ by several orders of magnitude. **D 4127,** D19

diffraction—a change in the direction of propagation of sound energy in the neighborhood of a boundary discontinuity, such as the edge of a reflective or absorptive surface. **C 634,** E33

diffraction:—(*1*) a modification which radiation undergoes, as in passing by the edge of opaque bodies or through narrow slits, in which the rays appear to be deflected.

(*2*) coherent scattering of X-radiation by the atoms of a crystal which necessarily results in beams in characteristic directions. Sometimes called reflection.

(*3*) the scattering of electrons, by any crystalline material, through discrete angles depending only on the lattice spacings of the material and the velocity of the electrons. **E 7,** E04

diffraction grating—an artificially produced periodic array of scattering centers capable of producing a pattern of diffracted energy, such as accurately ruled lines on a plane surface. **E 7,** E04

diffraction pattern (X-rays)—the spatial arrangement and relative intensities of diffracted beams. **E 7,** E04

diffraction ring—the diffraction pattern produced by a given set of planes from randomly oriented crystalline material. (See also **Debye ring.**) **E 7,** E04

diffuse, *adj*—*in optical propagation*, transmission or reflection of flux with diffusion. **E 284,** E12

diffuse, *adj*—referring to radiometric quantities, indicates that the flux propagates in many directions, as opposed to direct beam which refers to collimated flux. When referring to solar irradiance, it is the global irradiance less the direct beam irradiance. When referring to reflectance, it is the directional hemispherical reflectance less the specular reflectance.

NOTE—Diffuse has been used in the past to refer to hemispherical collection (including the specular component) or irradiation, with equal radiance for all directions over a hemisphere. This use is deprecated in favor of the more precise term hemispherical. **E 772,** E44

diffused layer—a region of opposite conductivity type formed near the surface of a semiconductor crystal as a result of the introduction of impurities into the silicon crystal by means of solid state diffusion. **F 1241,** F01

diffuse indications—indications that are not clearly defined as, for example, indications of subsurface defects. **E 1316,** E07

diffuse porous wood—certain hardwoods in which the pores exhibit little or no variation in size or distribution throughout the growth ring, only decreasing slightly in size, gradually toward the outer border of the ring. **D 9,** D07

diffuser, *n*—device used to alter the spatial distribution of flux by diffusion. **E 284,** E12

diffuse reflectance, ρ_d, *n*—the ratio of the reflected flux to the incident flux, where the reflection is at all angles within the hemisphere bounded by the plane of measurement except in the direction of the specular reflection angle. **E 284,** E12

diffuse reflectance factor, R_d, *n*—the ratio of the flux reflected at all angles within the hemisphere bounded by the plane of measurement except in the direction of the specular reflection angle, to the flux reflected from the perfect reflecting diffuser under the same geometric and spectral conditions of measurement. **E 284,** E12

diffuse reflection—reflection in which the flux is scattered in many directions by diffusion at or below the surface. **E 131,** E13

diffuse reflection, *n*—reflection in which flux is scattered in many directions by diffusion at or below the surface. See **diffusion**. **E 284,** E12

diffuse reflection, *n*—diffusion by reflection in which, on the macroscopic scale, there is no regular reflection. **E 349,** E21

diffuse reflection, *n*—the component of reflected light that emanates uniformly in all directions from the reflecting surface. **F 1294,** F05

diffuse sound field—the sound in a region where the intensity is the same in all directions and at every point. **C 634,** E33

diffuse spurce—emitting in more than one direction (from a relatively large area; sometimes referred to as a Lambertian source). **F 1156,** F12

diffuse transmission, *n*—transmission in which diffusion occurs, independently, on a macroscopic scale, of the laws of refraction. **E 284,** E12

diffuse transmission, *n*—transmission in which diffusion occurs independently, on the macroscopic scale, of the laws of refraction. **E 349,** E21

diffuse transmittance, T_d, *n*—the ratio of the flux transmitted by a specimen to the incident flux, the transmitted flux being measured at all forward angles except the regular transmission angle. **E 284,** E12

diffusion—(*1*) spreading of a constituent in a gas, liquid, or solid tending to make the composition of all parts uniform; (*2*) the spontaneous movement of atoms or molecules to new sites within a material. **B 374,** B08

diffusion, *n*—change of the angular distribution of a beam of radiant flux by a transmitting material or a reflecting surface such that flux incident in one direction is continuously distributed in many directions, the process not conforming (on a macroscopic scale) to the laws of Fresnel (regular) reflection and refraction and there being no change in frequency (wavelength) of the monochromatic components of the flux. **E 284,** E12

diffusion, *n*—change of the spatial distribution of a beam of radiation when it is deviated in many directions by a surface or a medium. **E 349,** E21

diffusion—in leak testing, the flow of the gas through a substance in which the gas actually migrates through the crystal lattice of the substance rather than through a geometrical leak (molecular diameters versus hole dimension). **E 1316,** E07

diffusion—the movement of a material such as a gas or liquid, in the body of a plastic. **F 412,** F17

diffusion-alloyed powder, *n*—a partially alloyed powder produced by means of a diffusion anneal. **B 243,** B09

diffusion coating, *n*—any process whereby a base metal is either coated with another metal and heated to a sufficient temperature in a suitable environment, or exposed to a gaseous or liquid medium containing the other metal, thereby causing diffusion of the coating or other metal into the base metal, with a resultant change in the composition and properties of its surface. **A 941,** A01

diffusion coating—an alloy coating produced by applying heat to one or more coatings deposited on a basis metal. **B 374,** B08

diffusion cycle—specific time and temperature to acquire a depth of diffusion and composition. **B 374,** B08

diffusion limited current density—the current density, often referred to as limiting current density, that corresponds to the maximum transfer rate that a particular species can sustain due to the limitation of diffusion. **G 15,** G01

diffusion, molecular—See **molecular diffusion.** **D 1356,** D22

diffusion-transfer process—a rapid photographic process in which a negative image is produced at one location, with unused imaging materials then diffusing across a thin fluid layer to produce a positive image on a receptor sheet. **E 7,** E04

diffusion treatment (or coating)—(*1*) process of producing a surface layer (diffusion layer) by diffusion of another metal or non-metal into the surface of the basis material.

(*2*) in electroplating, heat treatment applied to a work piece to achieve alloying or intermetallic compound formation between two or more coatings on a basis material. **B 374,** B08

diffusion zone—the zone of viable composition at the junction between two different materials, such as in welds or between the surface layer and the core of clad materials or bearings, in which interdiffusion between the various components has taken place. **E 7,** E04

diffusivity, thermal, *n*—the ratio of thermal conductivity of a substance to the product of its density and specific heat. (In SI units: $(W/(m\bullet K))/((kg/m^3)\bullet(J/(kg\bullet K))) = m^2/s$.) (In inch-pound units: $(Btu/(hr\bullet ft)\ F)/((lb/ft^3)(Btu/(lb\bullet F)) = ft^2/hr)$) **C 168,** C16

digester—a bioreactor in which anaerobic bacteria are used to decompose biomass or organic wastes into methane and carbon dioxide. **E 1126,** E48

digester—a bioreactor in which anaerobic bacteria are used to decompose biomass or organic wastes into methane and carbon dioxide. **E 1705,** E48

digital—the representation of data or physical quantities in the form of discrete codes, such as numerical characters, rather than a continuous stream. **E 1316,** E07

digital image—an image composed of discrete pixels each of which is characterized by a digitally represented luminance level. **E 1316,** E07

digital image acquisition system—a system of electronic components which, by either directly detecting radiation or converting analog radiation detection information, creates an image of the spatial radiation intensity map comprised of an array of discrete digital intensity values (see **pixel**). **E 1316,** E07

digital image enhancement—any operation used for the purpose of enhancing some aspect of the original image. **E 1316,** E07

digital image processing system—a system which uses algorithms to process digital image data. **E 1316,** E07

digital pH/mV meter—an instrument with digital display of millivolts or pH units. Less operator error is associated with digital instruments since there is no need for interpolation and no risk of confusing scales. **D 4127,** D19

digitization—the conversion of an analog signal to digital values using an analog-to-digital converter "sampling" or "digital sampling." **E 131,** E13

digitization—dividing into discrete elements, usually with the assignment of numerical values to each element. **F 1156,** F12

digitization noise—the noise generated in an interferogram through the use of an analog-to-digital converter whose least significant bit represents a value comparable to, or greater than, the peak-to-peak noise level in the analog data. **E 131,** E13

digitize (for radiology)—the act of converting an analog image or signal to a digital presentation. **E 1316,** E07

digs—deep, short scratches. **C 162,** C14

di-hydrated or double hydrated lime—dolomitic lime which has been hydrated under greater than atmospheric pressure and contains less than 8 % unhydrated oxides. **C 51,** C07

dilatancy—property of volume increase under loading. (ISRM) **D 653,** D18

dilatancy—the expansion of cohesionless soils when subject to shearing deformation. **D 653,** D18

dilatometer—the instrument used in dilatometry for measuring lengths or volume changes. **E 7,** E04

dilatometric softening point—the temperature at which the viscous sag of the glass specimen exactly counteracts the expansion as thermal expansion measurements proceed during heating. **C 162,** C14

dilatometry—the measurement of length or volume changes of a substance undergoing a change in temperature, pressure, or state. See Practice E 80. **E 7,** E04

dilatometry, *n*—see **Thermodilatometry.** **E 473,** E37

diluent, *n*—an ingredient added to an adhesive, usually to reduce the concentration of bonding materials. (See also **extender, thinner**.) **D 907,** D14

diluent—a gas, liquid, or solid used to reduce the concentration of an active ingredient in the formulation or application of a pesticide. **E 609,** E35

diluent—a gas, liquid, or solid used to reduce the concentration of an active ingredient in the formulation or application of a pesticide (see Definitions E 609). **E 1102,** E35

dilution factor—for a specified constituent, the ratio of the quantity of a diluted material to the quantity of the undiluted material. **C 859,** C26

dilution factor—the ratio of the volume of a diluted solution to the volume of original solution containing the same quantity of solute as the diluted solution. **E 131,** E13

dilution factor, f, *n*—the ratio of the volume of a diluted solution to the volume of the original solution containing the same quantity of solute as the diluted solution. **D 4175,** D02

dimensional change, *n*—a generic term for changes in length or width of a fabric specimen subjected to specified conditions. **D 123,** D13

dimensional change, *n—in pressing and finishing of garments*, the change in dimensions of a fabric caused by pressing and finishing during garment manufacture. **D 123,** D13

dimensional change in boiling water (felt), *n*—the change in length and width with any associated change in thickness produced by immersion in boiling water under specified conditions. **D 123,** D13

dimensional change of a compact, *n*— the difference, at room temperature, between the size of the sintered specimen and the die size. **B 243,** B09

dimensional stability, *n*—the ability of a resilient flooring to retain its original dimensions during the service life of the product. **F 141,** F06

dimensional stabilization—treatment of wood to reduce swelling and shrinking caused by changes in its moisture content. **D 9,** D07

dimension ratio—the average specified diameter of a pipe divided by the minimum specified wall thickness. **F 412,** F17

dimensions, *n—in packaging,* the measurement of length, width (or diameter), and depth of **containers,** expressed in that order; it should be stated as "inside" or "outside." **D 996,** D10

dimension stone—natural stone that has been selected and fabricated to specific sizes or shapes. **C 119,** C18

dimple—a shallow depression in the porcelain enamel, sometimes a defect. **C 286,** B08

dimple—a shallow depression in a wafer surface, with a concave, spheroidal shape and gently sloping sides. **F 1241,** F01

DIN color system, *n*—color order system developed for the Deutsche Industrie Normung (German Standardization Institute) to provide equality of visual spacing of colors in specified series, based on the attributes hue, saturation, and relative darkness degree. **E 284,** E12

dinnerware—ceramic whiteware made in a given pattern and in a full line of articles comprising a dinner service. **C 242,** C21

dip, *n*—a chemical composition that is applied to a textile cord or fabric to improve its adhesion to rubber or other elastomer. **D 123,** D13

dip application—application by direct immersion. **E 1102,** E35

dip encapsulation (a type of conformal coating), *n*—an embedding process in which the insulating material is applied by immersion and without the use of an outer container. **D 1711,** D09

dipentene, *n*—chemically defined as the optically inactive form of the monocyclic terpene hydrocarbon limonene.

chemically processed dipentene, n—recovered as a product or a by-product in connection with the chemical treatment and conversion of other terpenes.

destructively distilled dipentene, n—from the lighter portions of the oil recovered during the destructive distillation of pine wood.

steam-distilled dipentene, n—fractionated from the crude oleoresinous extract during the processing of related steam-distilled wood naval stores.

sulfate dipentene, n—from the crude condensate of the vapors generated in the digestion of wood in the sulfate paper pulp process. **D 804,** D01

diphase metal cleaner—a composition which produces two phases in the cleaning tank, namely, a solvent layer and an aqueous layer, which cleans by solvent action and emulsification. **D 459,** D12

diploid—in two parts; for example, an element with only two naturally occurring isotopes, such as lithium [Li^6 and Li^7], or pairs of chromosomes in human beings. **F 1156,** F12

dipole relaxation time, γ—the exponential decay time required for the electric polarization of any point of a suitably charged dielectric to fall from its original value to 1/e of that value, due to the loss of dipole orientation. **E 1142,** E37

dipped—dipped in bath of molten zinc for coating purposes, with excess zinc removed; resulting in coating essentially free from blisters, lumps, gritty areas, acid spots, dross warts, and flux. **F 547,** F16

dip pick-up, *n—in glass cords,* the amount of dip solids present as supplied. **D 123,** D13

dip pick-up, *n—in a textile cord or fabric,* the amount of dip or dip components present after processing, including drying, as determined by prescribed methods, and expressed as a percentage of the mass of the oven-dried dip-free material. **D 123,** D13

dipping—the process of coating a metal shape by immersion in slip, removal, and draining. In dry process enameling, the method of coating by immersing the heated metal shape for a short time in powdered frit. **C 286,** B08

dipping (latex), *n*—a process in which a layer of rubber is deposited on a mold or form as a result of immersion in a bath of compounded latex. **D 1566,** D11

dipping weight—see **pick-up.** **C 286,** B08

DIP switch—dual in-line package **switch.** **F 1457,** F05

dip tube—tubing connecting the lower portion of the container or dispenser with the valve. **D 3064,** D10

direct bonded basic brick, *n*—a fired refractory in which the grains are joined predominantly by a solid state diffusion mechanism. **C 71,** C08

direct contact magnetization—a technique of magnetizing in which the current is passed through a part via prods or contact heads. **E 1316,** E07

direct detection of mycoplasma—detection of mycoplasma by cultivation in culture media. **E 1705,** E48

direct development—see **positive development.** **F 335,** F05

directed application—an application to a restricted area such as a row, bed, or at the base of plants. **E 609,** E35

directed application—an application to a restricted area such as a row, bed, or at the base of plants (see Definitions). **E 1102,** E35

direct electrophotographic copying process—an electrophotographic copying process in which the photoconductor is an integral part of the final copy. **F 335,** F05

direct exposure imaging—in the direct exposure imaging method, the conversion screen and image recorder are simultaneously exposed to the neutron beam. **E 1316,** E07

direct fire—a method of maturing porcelain enamel wherein the products of combustion come in contact with the ware. **C 286,** B08

direct-fired furnace—a melting furnace having neither recuperator nor regenerator. **C 162,** C14

direct-image master—a master that is imaged by a mechanical marking device. **F 425,** F05

directional, *adj*—(1) so designed that performance depends on direction or is restricted in direction; more effective in some directions than others.

(2) referring to a beam, beam in which the flux measured is confined to directions that differ moderately from the centroid direction or axis of the beam. (E 179) **E 284,** E12

directional—over an infinitesimal element of solid angle in a given direction. For properties, a solid angle small enough that the property does not vary within the solid angle may be considered an element of solid angle. Indicated by the symbols θ, Φ, where θ is the angle between the given direction and the normal to the sample surface, and;uf is the azimuth angle of the direction measured counter-clockwise from a reference mark on the sample. See **radiometric properties** and **quantities.** **E 772,** E44

directional emissivity of a thermal radiator, *n*—ratio of the thermal radiance of the radiator in a given direction to that of a full radiator at the same temperature. Symbol: $\epsilon(\theta, \phi)$; $\epsilon(\theta, \phi) = L_{e,th}(\theta,\phi)/L_{e\ (\epsilon\ =\ 1)}$. **E 349,** E21

directionality, *n— (1) perceived,* the degree to which the appearance of a surface changes as the surface is rotated in its own plane, under fixed conditions of illumination and viewing.

(2) measured—(scattering indicatrix, azimuthally nonisotropic)—difference in pattern of near-specular and semidiffusely scattered light, dependent upon the azimuthal angles of the incident and viewing beams. **E 284,** E12

direction change—a change in the orientation of the principal dimension or of the support of adjoining units of the roofing system. **D 1079,** D08

direction of lay—the lateral direction, designated as left-hand or right-hand, in which the wires of a member or units of a conductor run over the top of the member or conductor as they recede from an observer looking along the axis of the member or conductor. **B 354,** B01

direction of lay, *n*—the helical disposition of the components of a strand or cord **D 123,** D13

direction of slippage, *n—in yarn slippage testing,* the line of movement parallel to either the filling or the warp on a woven fabric in which minimum force is required to produce yarn slippage. **D 123,** D13

direction of twist, *n*—the right or left direction of the helix formed in a twisted strand as indicated by superimposition of the capital letter "S" or "Z." **D 123,** D13

direct measurement—determination of sample concentration or activity by directly relating the electrode potential to the level of the species being measured. The level can be read from a calibration curve or the log scale of a selective ion meter. **D 4127,** D19

direct-measuring bed-load sampler—a device which physically collects and holds bed load. **D 4410,** D19

direct medical control—the process of providing immediate physician orders to EMS personnel through direct communication (a.k.a. on-line medical control, indirect medical direction). **F 1177,** F30

direct memory access, *n*—an input/output facility which allows transfers directly in or out of main storage without passing through the processors general registers. (See **DMA**.) **F 1457,** F05

direct-on—see **cover coat.** **C 286,** B08

director, *n*—optical components, such as mirrors, lenses, gratings, or other objects, such as ceilings or walls, that direct radiant flux from a source to a specimen to be observed or measured. **E 284,** E12

direct oxygen service—component surfaces in contact with oxygen-enriched environments during normal operations. Examples include oxygen compressor piston rings, control valve seats, pipes, regulators, vessels and fittings. **G 126,** G04

direct print—a photographic print of an original negative. **E 7,** E04

direct quenching, *n—in thermomechanical processing*, **quenching** immediately following the final hot deformation. **A 941,** A01

direct quenching, *n—in thermochemical processing*, **quenching** immediately following the thermochemical treatment. **A 941,** A01

direct-reading XRF—see **XRF direct-reading analyzer**. **E 1605,** E06

direct reduced iron, *n*—iron ores that have been reduced to essentially metallic iron by heat and reducing agents, but without melting, and processed into suitable shapes (typically pellets) for use as a charge material in a melting operation. **A 644,** A04

direct shear test—a shear test in which soil or rock under an applied normal load is stressed to failure by moving one section of the sample or sample container (shear box) relative to the other section. **D 653,** D18

direct sound field—the sound that arrives directly from a source without reflection. **C 634,** E33

direct thermal imaging product—paper, film, or other substrate upon which a coating is applied; the imaging components consist of a color former, a developer, a sensitizer and antioxidants which react to form an image when heated from a thermal printhead. **F 1623,** F05

direct thermal paper,, *n*—paper coated with a heat-reactive coating, which changes from a colorless form to an intense colored state upon contact with a thermal printhead. **F 1623,** F05

direct thermal product, *n*—substrate coated with a heat-sensitive formulation for the purpose of creating an image when heat is applied from a thermal printhead. **F 1623,** F05

direct-use facility, *n*—a facility which uses geothermal energy for purposes other than the generation of electricity (e.g., space heating, greenhouses, bathing, and industrial processes). **E 957,** E44

direct viewing—a view that is not redefined by optical or electronic means. **E 1316,** E07

direct weathering, n—a technique of weathering in which the test specimens are exposed to all prevailing elements of the atmosphere. **G 113,** G03

direct yarn numbering system, *n*—a system that expresses yarn linear density (number) in mass per unit length. (See also **denier, spyndle number**, and **tex**.) **D 123,** D13

dirt—a small particle of foreign material imbedded in the surface of flat glass sheets. **C 162,** C14

dirt, *n—general term*, any undesirable, extraneous, or contamination material visible in transmitted or reflected light in or on pulp, paper, or paperboard. **D 1968,** D06

dirt, *n—quantitative term*, any undesirable, extraneous, or contaminating material in or on pulp, paper or paperboard, that has marked contrasting color to the rest of the sheet when viewed at more than one angle by reflected light, and that has an equivalent black area of 0.04 mm^2 or more. See **dirt**, *equivalent black area of a dirt speck* (**EBA**). **D 1968,** D06

dirt, *n—equivalent black area of a dirt speck (***EBA***), the area of the black spot on the white background of the TAPPI Standard Dirt Chart that makes the same visual impression on its background as does the dirt speck on the particular background in which it is embedded.* **D 1968,** D06

dirt—*in paper,* refers to the presence of relatively nonreflective foreign particles embedded in the sheet. The size and lack of reflectance of the particles may be such that they will be mistaken for inked areas by an optical scanner. **F 149,** F05

dirt—*on semiconductor wafers*, surface contaminant that cannot be removed by pre-inspection solvent cleaning. **F 1241,** F01

dirt, *n—when referenced for scanning purposes*, the presence of non-reflective foreign particles embedded in a substrate. **F 1294,** F05

dirt (coatings)—see **soil.** **D 16,** D01

dirt pick-up, *n—in building construction*, soiling caused by a foreign material other than micro-organism growth that is deposited on, adhered to, or embedded into a sealant, coating, or membrane. **C 717,** C24

dirt receptacle first vacuum cleaner system, *n*—a vacuum cleaner construction in which the dirt laden air is passed through a dirt receptacle (bag type filter, bagless filter, or other type of dirt separator). The separated air is then pulled through the fan (by-pass) or fan and motor (flow through) and expelled from the cleaner. This type of construction is sometimes referred to as clean air or indirect system. **F 395,** F11

dirt resistance (*coatings*)—the ability of a coating to resist soiling by foreign material, other than microorganisms, deposited on or embedded in the dried coating. **D 16,** D01

disarticulation—separation at a joint; amputation through a joint. **F 869,** F08

disbond—an area within a bonded interface between two adherents in which an adhesive or cohesive failure has occurred. It may occur at any time during the life of the structure and may arise from a wide variety of causes. It is sometimes used to describe an area of separation between two laminae in the finished laminate (the term "delamination" is preferred). **D 3878,** D30

discharge—see **hazardous waste discharge**. **E 1605,** E06

discharge capacity, thermal—the amount of heat that can be removed from a storage device during a period of time and for a specific set of values for the initial and final temperatures of the storage device, the temperature of the entering fluid, and the mass flow rate of fluid through the storage system. **E 772,** E44

discharged area development (DAD)—a process in electrostatic copying where the photoconductive element is charged with a charge of the same sign as that of the toner. A light source is used to discharge only those areas that are to receive toner to form the image. In the development process, the charged background areas repel the like charged toner to the discharged areas on the photoconductor. **F 335,** F05

discharge pressure—in leak testing, same as **forepressure**. **E 1316,** E07

discharge test time—the duration of a single transient test in which energy is removed from the storage device. **E 772,** E44

discharge tube leak indicator—a glass tube attached to a system being leak tested, with the glass tube having electrodes attached to a source of high-frequency high voltage, such as a Tesla coil or induction coil, so that changes in the color of the electrical discharge can be observed when a suitable tracer gas (methane, carbon dioxide, alcohol) flows through the leak. **E 1316,** E07

discharge velocity, *v, q* **(LT^{-1})**—rate of discharge of water through a porous medium per unit of total area perpendicular to the direction of flow. **D 653,** D18

discharge (water)—the volume of water flowing through a cross-section in a unit of time, including sediment or other solids that may be dissolved in or mixed with the water; usually cubic feet per second (f^3/s), or cubic metres per second (m^3/s). **D 4410,** D19

discoloration—a streak or other pattern on the surface that causes a noticeable change of color from the rest of the pultruded surface. **D 3918,** D20

discontinuity, *n*—a lack of continuity or cohesion; an intentional or unintentional interruption in the physical structure or configuration of a material or component. **E 1316,** E07

discontinuity surface—any surface across which some property of a rock mass is discontinuous. This includes fracture surfaces, weakness planes, and bedding planes, but the term should not be restricted only to mechanical continuity. (ISRM) **D 653,** D18

discontinuous fiber—a polycrystalline or amorphous fiber that is discontinuous within the sample or component or that has one or both ends inside of the stress field under consideration. The minimum diameter of a discontinuous fiber is not limited, but the maximum diameter may not exceed 0.25 mm (0.010 in.). **D 3878,** D30

discontinuous fiber-reinforced composite, *n*—a ceramic matrix composite material reinforced by chopped fibers. **C 1145,** C28

discontinuous stringer—three or more Type B or C inclusions aligned in a plane parallel to the hot working axis and offset by no more than 15 μm with a separation of less than 40 μm (0.0016 in.) between the two nearest neighbor inclusions (see Practices E 45 and E 1122). **E 7,** E04

discontinuous yielding, *n*—in a uniaxial test, a hesitation or fluctuation of force, such as is sometimes observed at or near the onset of plastic deformation, due to localized yielding. (The stress-strain curve need not appear to be discontinuous.) **E 6,** E28

discounted payback (DPB) period—the time required for the cumulative benefits from an investment to pay back the investment cost and other accrued costs considering the time value of money. **E 833,** E06

discount factor—a multiplicative number (calculated from a discount formula for a given discount rate and interest period) that is used to convert costs and benefits occurring at different times to a common time. **E 833,** E06

discounting—a technique for converting cash flows that occur over time to equivalent amounts at a common time. **E 833,** E06

discount rate—the rate of interest reflecting the investor's time value of money, used to determine discount factors for converting benefits and costs occurring at different times to a base time. **E 833,** E06

discrete bar code, *n*—a bar code symbol in which the intercharacter space is not part of the code and is allowed to vary dimensionally within wider tolerances than those specified for bars and spaces. An example would be Code 39. **F 1294,** F05

discrete sample, *n*—one or more units taken from a material that consists of separately identifiable units. (Compare to **bulk sample.**) **D 123,** D13

discrete variate, *n*—a variate that is a measurement based on a scale that has a limited or finite number of steps; such as a count, a rating scale, or a ratio of successes to total observations. **D 123,** D13

discrimination, *n*—the process of qualitatively or quantitatively differentiating among stimuli. **E 253,** E18

discrimination test, *n*—any method to determine if differences among stimuli are perceptible. For example: triangle tests, duo-trio tests, paired comparison tests, etc. **E 253,** E18

dish (cross or transverse bow)—the departure from flatness across the full width of the strip in the form of a single arc, excluding burrs. **B 846,** B05

dished head— See **cup head.** **F 547,** F16

dish grinder, *n*—a grinding machine equipped with a dish-shaped abrasive wheel as a grinding mechanism **C 1145,** C28

dishload—a dishrack (see **dishwashing machine, commercial:** *dishrack*), loaded with ten 9 in. plates of a specified weight, used to put a thermal load in the dishwasher during the washing energy test. **F 1827,** F26

dish rack—a peg type, polypropylene, or other noncorrosive material rack designed to hold tableware in place during washing. **F 1827,** F26

dishwashing machine, commercial—machines that uniformly wash, rinse, and sanitize kitchen ware. The machines shall be capable of removing physical soil from properly racked and prescraped tableware, and sanitizing multiple-use tableware. **F 1827,** F26

dishwater inlet—incoming water to dishwater. **F 1827,** F26

dishwater inlet temperature—the temperature of the incoming water to dishwater, measured at the dishwater inlet connection. See **dishwashing machine, commercial**and *dishwasher inlet.* **F 1827,** F26

dish wheel, *n*—dish-shaped abrasive grinding wheel. **C 1145,** C28

disinfectants—physical and chemical agents used for inactivating or destroying microorganisms. **F 1600,** F20

disinfection—the process of killing organisms in a water supply or distribution system by means of heat, chemicals, or UV light. **D 6161,** D19

disinfestant—an agent that kills, inactivates, or repels organisms in or on plants, animals, or inanimate objects. **E 609,** E35

disintegration—the reduction of massive material to powder. **B 243,** B09

disk—a round, commercially flat solid blank made from a flat rolled product. **B 846,** B05

disk feeder, *n*—a rotating disk beneath the opening of a bin which delivers material from the bin at a specified rate by controlling the rate of rotation of the disk and the size of the gate opening of the bin. **C 1145,** C28

disk grinder, *n*—a grinding machine equipped with a large abrasive disk as the work mechanism. **C 1145,** C28

disk wheel, *n*—a bonded abrasive wheel mounted on a plate so that grinding may be done on the side of the wheel. **C 1145,** C28

dislocation—a line imperfection in a crystal that either forms the boundary between slipped and nonslipped areas of a crystals or that is characterized by a closure failure of Burger's circuit. **F 1241,** F01

disordered structure—the crystal structure of a solid solution in which the atoms of different elements are randomly distributed with respect to the available lattice sites. **E 7,** E04

dispatch life support—the knowledge, procedures, and skills used by trained emergency medical dispatchers in guiding care by means of post-dispatch (pre-arrival) instruction to callers. **F 1177,** F30

dispenser—commercial equipment designed to deliver a beverage or food product. **F 1827,** F26

dispersant, *n*—*in engine oil,* an additive that reduces deposits on oil-wetted engine surfaces primarily through suspension of particles. **D 4175,** D02

dispersant—the material, usually an organic solvent, in a liquid developer system that conveys toner particles, but does not itself become a part of the image in a finished print. (See also **carrier.**) **F 335,** F05

dispersed system—in laboratory analysis of grain sizes, an initial condition whereby the particles begin to settle from a stirred mixture; when stirring stops, each particle settles independently of other particles. **D 4410,** D19

dispersing agent—a material that increases the stability of a suspension of particles in a liquid medium. **B 374,** B08

dispersing agent—a material that increases the stability of a suspension of particles in a liquid medium. **D 459,** D12

dispersing agent—*in grouting,* an addition or admixture that promotes dispersion of particulate grout ingredients by reduction of interparticle attraction. **D 653,** D18

dispersing agent—see **deflocculating agent.** **D 653,** D18

dispersing agent (latex), *n*—a surface-active substance used to facilitate the suspension of solid compounding materials in a liquid medium and to stabilize the dispersion thereby produced. **D 1566,** D11

dispersion—the variation of refractive index with wavelength of light. See related term **Abbé value**. **C 162,** C14

dispersion—*in a fine particle suspension,* the condition which results when a stable suspension of particles is achieved by physical or chemical means in which no evidence of reflocculation or re-agglomeration of the particles is observed. **C 242,** C21

dispersion—the phenomenon of varying speed of transmission of waves, depending on their frequency. (ISRM) **D 653,** D18

dispersion, n— See **adhesive dispersion**. **D 907,** D14

dispersion, *n*—the most general term for a system consisting of particulate matter suspended in a fluid. **D 1356,** D22

dispersion, *n*—a system of particles distributed in a solid, liquid or gas. **E 1620,** E29

dispersion, *n*—the spread of values of a frequency distribution about an average (in statistics). (Quantitative measures of dispersion include range, variance, standard deviation, mean deviation and relative span.) **E 1620,** E29

dispersion coating—a coating consisting of particles of one material contained in a matrix of another metal or nonmetal. **B 374,** B08

dispersion coating—*in flexible barrier materials,* (*1*) a process of applying a material, suspended or dispersed in a vehicle, to a surface in such a way that a continuous, coalesced, adherent layer results when the vehicle liquid (usually water) is evaporated. (*2*) Also, the product resulting from such a process. **F 17,** F02

dispersion coating—See Terminology F 17. **F 1327,** F02

dispersion staining—the color effects produced when a transparent object, immersed in a liquid having a refractive index near that of the object is viewed under the microscope by a transmitted light and precise-aperture control. **D 1129,** D19

dispersion strengthened material—a material consisting of a metal and finely dispersed, substantially insoluble, metallic or nonmetallic phase. **B 243,** B09

dispersion (the act of), *n*—application of shearing forces to distribute one or more compounding materials uniformly throughout the mass of a continuum of material. **D 1566,** D11

dispersoid, *n*—the particles of a dispersion. **D 1356,** D22

dispersoid—in metallography, finely divided particles of relatively insoluble constituents which can be seen in the microstructure of certain alloys. **E 7,** E04

displacement—a change in position of a material point. (ISRM) **D 653,** D18

displacement, *n*— *in packaging,* the volume occupied by a **container,** calculated from its outside dimensions. (Compare **cube.**) **D 996,** D10

displacement dose (D$_d$)—the quotient of d $\bar{\epsilon}_d$ by dm, where d $\bar{\epsilon}_d$ is that part of the mean energy imparted by radiation to matter which produces atomic displacements (that is, excluding the part that produces ionization and excitation of electrons) in a volume element of mass dm.

$$D_d = d\,\bar{\epsilon}_d/dm$$

Unit: J · kg^{-1} **E 170,** E10

displacement grouting—injection of grout into a formation in such a manner as to move the formation; it may be controlled or uncontrolled. See also **penetration grouting**. **D 653,** D18

displacements per atom (dpa)—the mean number of times each atom of a solid is displaced from its lattice site during an exposure to displacing radiation, as calculated following standard procedures (see **displacement dose**). **E 170,** E10

disposal facility—facility or part of a facility at which hazardous waste is intentionally placed into or on any land or water, and at which waste will remain contained after closure. **E 1605,** E06

dispose, *v*—to discard, abandon, or manage as waste. **D 5681,** D34

dissipation factor— see **tan delta.** **D 4092,** D20

dissipation factor— see **tangent delta**. **E 1142,** E37

dissipation factor, *D*—the ratio of the loss index to its relative permittivity or

$$D = \kappa''/\kappa'$$

It is also the tangent of its loss angle, δ, or the cotangent of its phase angle, θ. The dissipation factor is related to the power factor, *PF*, by the following equation:

$$D = PF/\sqrt{1-(PF)^2}$$

D 2864, D27

dissipation factor, dielectric—same as **dissipation factor.** **D 2864,** D27

dissipation factor (loss tangent) (tan δ), *D, n*—the ratio of the loss index to its relative permittivity or

$$D = \kappa''/\kappa'$$

It is also the tangent of its loss angle, δ, or the cotangent of its phase angle, θ. **D 1711,** D09

dissipation factor, magnetic, D_m—the tangent of the hysteretic angle that is equal to the ratio of the core loss current, I_c, to the magnetizing current, I_m. Thus:

$$D_m = \tan\beta = \cot\gamma = I_c/I_m = \omega L_1/R_1 = 1/Q_m$$

NOTE—This dissipation factor is also given by the ratio of the energy dissipated in the core per cycle of a periodic *SCM* excitation (hysteresis and eddy current heat loss) to 2π times the maximum energy stored in the core. **A 340,** A06

dissociation—as applied to heterogeneous equilibria, the transformation of one phase into two, or more, new phases, all of different composition. **E 7,** E04

dissociation—as applied to heterogeneous equilibria, the transformation of one phase into two or more new phases, all of different composition, (E 7, E-4). **E 1142,** E37

dissociation—in leak testing, the breakdown of a substance into two or more constituents.

NOTE—Dissociation is sometimes referred to as cracking. **E 1316,** E07

dissociation constant—a number indicating the extent to which a substance dissociates in solution. [For a simple two-species complex *AB*, the constant is given by the product of the molar concentrations of *A* and of *B* divided by the molar concentrations of the undissociated species *AB*. For example, with hydrofluoric acid:

$$([H^+][F^-])/([HF]) = K = 6.7 \times 10^{-4} \text{ at } 25°C$$

The smaller the value of *K*, the less the complex is dissociated. *K* varies with temperature, ionic strength, and the nature of the solvent.] **D 4127,** D19

dissociation pressure—at a designated temperature, the pressure at which a phase will transform into two, or more new phases, of different composition. **E 7,** E04

dissolved gas—natural gas held in solution in reservoir liquids at the prevailing temperature and pressure of the reservoir. **D 4150,** D03

dissolved load—the part of the stream load that is carried as dissolved solids. **D 4410,** D19

dissolved matter—that matter, exclusive of gases, which is dispersed in water to give a single homogeneous liquid phase. **D 1129,** D19

dissolved solids—the mass of constituents in a filtered water sample. For operational purposes, the filter pore is usually 0.00045 mm. **D 4410,** D19

dissolved solids—the residual material remaining after filtering the suspended material from a solution and evaporating the solution to a dry state at a specified temperature. That matter, exclusive of gases, which is dissolved in water to give a single homogeneous liquid phase. **D 6161,** D19

dissolving pulps—see **pulps, dissolving. D 1695,** D01

distal end—the working end, comprised of two blades, that is furthest from the surgeon when in use. **F 1078,** F04

distance amplitude compensation (electronic)—The compensation or change in receiver amplification necessary to provide equal amplitude on the display of the ultrasonic flaw detector for reflectors of equal area which are located at different depths in the material. **E 1316,** E07

distance amplitude response curve—a curve showing the relationship between the different distances and the amplitudes of ultrasonic response from targets of equal size in an ultrasonic response from targets of equal size in an ultrasonic transmitting medium. (E 127) **E 1316,** E07

distance constant *(L, m), n*—the distance the air flows past a rotating anemometer during the time it takes the cup wheel or propeller to reach *(1 − 1/e)* or 63 % of the equilibrium speed after a step change in wind speed. **D 1356,** D22

distance linearity range—the range of horizontal deflection in which a constant relationship exists between the incremental horizontal displacement of vertical indications on the A-scan presentation and the incremental time required for reflected waves to pass through a known length in a uniform transmission medium. (E 494) **E 1316,** E07

distillate—the overhead product of distillation such as ethanol liquid from the top of a beer still. **E 1705,** E48

distillation—the process of condensing steam from boiling water on a cool surface. **D 6161,** D19

distillation—the act of vaporizing and condensing a liquid in sequential steps to effect separation from a liquid mixture. **E 1705,** E48

distillation pressure, *n*—the pressure measured as close as possible to the point where the vapor temperature is taken, normally at the top of the condenser. **D 4175,** D02

distillation temperature, *n*—(Syn. *head temperature, vapor temperature), in column distillation,* the temperature of the saturated vapor measured just above the top of the fractionating column. **D 4175,** D02

distillers grains—the insoluble solids that have been separated from the stillage bottoms or beer. Moisture content may range from 60 to 85 %, depending upon the level of dewatering during separation. **E 1705,** E48

distinctness of image—degree of clarity exhibited by images reflected from a surface. **D 2825,** D21

distinctness-of-image gloss—the sharpness with which image outlines are reflected by the surface of an object. **D 16,** D01

distinctness-of-image gloss, *n*—aspect of gloss characterized by the sharpness of images of objects produced by reflection at a surface. **E 284,** E12

distinctness of (reflected) image—see **distinctness-of-image gloss. E 284,** E12

distorted—physically changed from the natural and original shape, caused by stress of any type. **F 819,** F18

distorted image, *n*—a character, symbol, line drawing, or halftone that is deformed in shape or is vague and lacking a definite outline. **F 1457,** F05

distortion, *n*—*in fabrics,* a general term for a visible defect in the texture of a fabric. **D 123,** D13

distortion, *n*—*in textile battings,* defects such as holes, lumps, or thin areas caused by movement of fibers. **D 123,** D13

distortion—a change in shape of a solid body. (ISRM) **D 653,** D18

distortion—an aberration of lens systems where axial and marginal magnifications are unequal.

(*1)barrel distortion*—the distortion in the image which occurs when the magnification of the image in the center of the field is greater than in the edge of the field. This is also called negative distortion.

(2) *pincushion distortion*—the distortion in the image which results when the magnification in the center of the field is less than it is at the edge of the field. This is also called positive distortion. **E 7,** E04

distortion, *n*—*optical,* a defect in an image-forming system whereby the image is not the shape of an ideal image of the object. For example, a straight pole, viewed through a window having nonplanar surfaces, may appear to have bends in it. **E 284,** E12

distortion factor, *df*—a numerical measure of the distortion in any ac nonsinusoidal waveform. For example, if by Fourier analysis or direct measurement E_1, E_2, E_3, and so forth are the effective values of the pure sinusoidal harmonic components of a distorted voltage waveform, then the distortion factor is the ratio of the root mean square of the second and all higher harmonic components to the fundamental component.

$$df = [E_2^2 + E_3^2 + E_4^2 + \cdots]^{1/2} E_1$$

NOTE—There are no dc components (E_0) in the distortion factor. **A 340,** A06

distortion, harmonic—the departure of any periodically varying waveform from a pure sinusoidal waveform.

NOTE—The distorted waveform that is symmetrical about the zero amplitude axis and is most frequently encountered in magnetic testing contains only the odd harmonic components, that is fundamental, 3rd harmonic, 5th harmonic, and so forth. Nonsymmetrical distorted waveforms must contain some even harmonic components, in addition to the fundamental and, perhaps, some odd harmonic components. **A 340,** A06

distributed impact test, *n*—*in impingement erosion testing,* an apparatus or method that produces a spatial distribution of impacts by liquid or solid bodies over an exposed surface of a specimen. **G 40,** G02

distribution—the thicknesses of the walls of a glass article over its entire area. **C 162,** C14

distribution, amplitude, cumulative (acoustic emission) F(V)—the number of acoustic emission events with signals that exceed an arbitrary amplitude as a function of amplitude V. **E 1316,** E07

distribution, differential (acoustic emission) amplitude f(V)—the number of acoustic emission events with signal amplitudes between amplitudes of V and $V + \Delta V$ as a function of the amplitude V. $f(V)$ is the absolute value of the derivative of the cumulative amplitude distribution F(V). **E 1316,** E07

distribution, differential (acoustic emission) threshold crossing f_t(V)—the number of times the acoustic emission signal waveform has a peak between thresholds V and $V + \Delta V$ as a function of the threshold V. f_t(V) is the absolute value of the derivative of the cumulative threshold crossing distribution F_t (V). **E 1316,** E07

distribution, logarithmic (acoustic emission) amplitude g(V)—the number of acoustic emission events with signal amplitudes between V and αV (where α is a constant multiplier) as a function of the amplitude. This is a variant of the differential amplitude distribution, appropriate for logarithmically windowed data. **E 1316,** E07

distribution reinforcement—reinforcement, typically running 90° to the main or circumferential reinforcement, intended to disperse concentrated loads to larger areas of a structural member. **C 822,** C13

distribution subsystem—that portion of the solar system from the storage device to the point of ultimate use. **E 772,** E44

distribution temperature, *n*—*of a source*, temperature, usually expressed in kelvins, of the full radiator having a relative spectral power distribution in the visible region approximately the same as that of the source. **E 284,** E12

distribution, threshold crossing, cumulative (acoustic emission) F_t(V)—the number of times the acoustic emission signal exceeds an arbitrary threshold as a function of the threshold voltage (V). **E 1316,** E07

distribution tube (Type D)—a seamless or welded copper tube known as copper distribution tube (Type D). **B 846,** B05

distributor rollers—rollers that break down and distribute the ink. **F 425,** F05

diurnal, *adj*—recurring daily. **D 1356,** D22

dive computer, *n*—a microprocessor-based electronic instrument that provides data based on a specific decompression model. **F 1549,** F32

dive profile, *n*—the depth/time history of a dive that typically consists of bottom time, maximum depth, and surface interval. **F 1549,** F32

divergence loss—that part of transmitted energy lost due to spreading of wave rays in accordance with the geometry of the system. **D 653,** D18

divergent bevel point—points sheared obliquely to staple-leg axis, with beveled face in opposite direction on each leg, across thick leg side leading from lower to upper thick face; designed to lead staple legs into opposite directions perpendicular to staple plane during driving. **F 592,** F16

divergent chisel point—chisel point with beveled point faces at angle to staple crown in plane perpendicular to staple crown axis; designed to lead staple legs into opposite directions perpendicular to staple plane during driving. **F 592,** F16

divergent point—See **divergent bevel point.** **F 592,** F16

diversion—the distribution and sale of legitimate products through unauthorized dealers. **F 1156,** F12

diversion mode—placement of a boom to redirect the movement of a floating substance. **F 818,** F20

divided cell—a cell containing a diaphragm or other means for physically separating the anolyte from the catholyte. **B 374,** B08

divider, *n*—a device, made of various materials, that separates the space within a **container** into two or more spaces, cells, compartments, or layers. **D 996,** D10

division, *n*—*in sample preparation,* processes which divide a sample into two or more subsamples of equal mass and composition. **E 135,** E01

divorced eutectic—a structure in which the components of an eutectic appear to be entirely separate. **E 7,** E04

d-limonene, *n*—a purified optically active terpene hydrocarbon recovered from by-products of the citrus industry. **D 804,** D01

D-load—the supporting strength of a pipe loaded under three-edge-bearing test conditions expressed in pounds per linear foot per foot of inside diameter or horizontal span, or expressed in newtons per linear metre per millimetre of inside diameter or horizontal span. **C 822,** C13

D-load, 0.01-in. (0.3-mm) crack—the maximum three-edge-bearing test load supported by a concrete pipe before a crack having a width of 0.01 in. (0.3 mm) occurs, measured at close intervals, throughout a continuous length of 1 ft (300 mm) or more measured parallel to the longitudinal axis of pipe barrel expressed as D-load. **C 822,** C13

D-load ultimate *(Du)*—the maximum three-edge-bearing test load supported by a pipe, expressed as D-load. **C 822,** C13

DMA—abbreviation for **direct memory access.** **F 1457,** F05

2D matrix symbols—see **matrix symbols.** **F 1294,** F05

DNA fluorochrome stain—staining of DNA specifically by the use of bisbenzamide fluorochrome stain or other DNA fluorochromes of comparable quality and performance, such as DAPI (4′,6-diamidine-2-phenyl-indole-2HCl)-Serva 18860. **E 1705,** E48

D_{Nf}, D_{Lf}, D_{Af}, D_{Vf}—diameters such that the cumulative number of particles, (N), length of diameter, (L), surface area, (A), or volume, (V), from zero diameter to these respective diameters is the fraction, (f), of the corresponding sum for the total distribution.

Example:

$D_{V0.5}$ is the volume median diameter; that is, 50 % of the total volume of liquid is in drops of smaller diameter and 50 % is in drops of larger diameter. **E 1620,** E29

doctor-bar or blade, *n*—A scraper mechanism that regulates the amount of adhesive on the spreader rolls or on the surface being coated. **D 907,** D14

doctor blade streaks, *n*—lines in the image or non-image areas of a gravure print caused by a defect of the doctor blade. **D 6488,** D01

doctor-roll, *n*—a roller mechanism that is revolving at a different surface speed, or in an opposite direction, resulting in a wiping action for regulating the adhesive supplied to the spreader roll. **D 907,** D14

doctor streak, *n*—*in printed cloth,* a wavy white or colored streak in the warp direction of printed cloth caused by a defective doctor blade. **D 123,** D13

document—a form designed as input to a document reader. **F 149,** F05

documentary characteristic, *n*—any historic, stylistic, iconographic, technological, intellectual, aesthetic, or religious data pertaining to an object under consideration for conservation. **D 123,** D13

documentary standard, *n*—document, arrived at by open consensus procedures, specifying necessary details of a method of measurement, definitions of terms, or other practical matters to be standardized. (Compare **physical standard**.) **E 284,** E12

documentation, *n*—the record of information (historic, stylistic, iconographic, technological, intellectual, aesthetic, scientific, or spiritual) about any object of cultural or natural heritage. **D 123,** D13

document glass—{archaic} an ultraviolet absorbing glass used for protecting documents. **C 162,** C14

document reader—a scanning device that scans one to five lines of data in fixed locations on a document at a single pass. Generally, re-scanning of a portion of the document is not possible, one direction of the scan being provided by movement of the form past the reading head. The forms used generally don't exceed 8 to ¾ in. in width by 4 to ¼ in. in depth. Also see **page reader**. **F 149,** F05

DoD—Department of Defense. **F 1156,** F12

doeskin—commercial term for white leather from sheep or lambskin, tanned with alum or formaldehyde or both. **D 1517,** D31

doffing tube, *n*—a component of an open-end spinning machine, which is an extension to the navel and is used to guide the withdrawn yarn en route to the take-up rollers. (See also **navel**) **D 123,** D13

doghouse—a protrusion in or near the back wall through which batch is introduced into the melter. **C 162,** C14

dolly, *n*—a low platform or structure mounted on wheels or casters, designed primarily for moving bulky loads for short distances. (Compare **pallet.**) **D 996,** D10

dolly, *n*—an assembly equipped with a hitch and one or more axles, typically used to support the forward end of a semitrailer. **E 867,** E17

doloma-carbon refractory, *n*—a refractory brick manufactured predominantly from a mixture of refractory-grade doloma and 2 to 20 weight percent carbonaceous materials, with resin, tar, pitch or a combination of these materials as the bonding agent; the refractory-grade doloma may be either dead-burned dolomite, synthetic doloma, fused doloma or combinations of these materials, and the carbonaceous material may be either graphite, carbon black, or a combination of these materials. **C 71,** C08

doloma, fused, *n*—refractory-grade material consisting predominantly of lime and magnesia which has solidified from a fused or molten state. **C 71,** C08

doloma-magnesia refractory, *n*—a refractory, which may be burned or unburned, manufactured predominantly of a mixture of refractory-grade doloma and refractory-grade magnesia in which the refractory-grade doloma predominates by weight. **C 71,** C08

doloma refractory, *n*—a refractory manufactured predominantly of dead-burned dolomite, synthetic doloma, fused doloma or combinations of these materials. **C 71,** C08

doloma, refractory-grade, *n*—a dead-burned or fused refractory material consisting predominately of lime and magnesia; the three principal types are dolomite, dead-burned; doloma, synthetic; and doloma, fused. **C 71,** C08

doloma, synthetic, *n*—a refractory-grade doloma which has been derived from blending magnesia and lime or dolomite and dead-burning to form a dense, hydration-resistant material; and having a MgO content of 30 % to 80 % and maximum CaO content of 70 %. **C 71,** C08

dolomite, *n*—a carbonate rock consisting predominantly of magnesium carbonate ($MgCo_3$) and calcium carbonate ($CaCO_3$) in approximately an equal molecular ratio. **C 71,** C08

dolomite—a sedimentary carbonate rock (a variety of limestone) that consists largely or entirely of the mineral dolomite. **C 119,** C18

dolomite—a double carbonate of lime and magnesium having the general formula $CaCO_3 \cdot MgCO_3$. See also **limestone.** **C 162,** C14

dolomite—the double carbonate of lime and magnesia having the general formula $CaCO_3 \cdot MgCO_3$. **C 242,** C21

dolomite, dead-burned, *n*—a refractory grade doloma which is obtained by burning dolomite above 1450°C long enough to form a dense hydration-resistant material composed primarily of lime and magnesia. **C 71,** C08

dolomitic—indicates the presence of 35 to 46 % magnesium carbonate ($MgCO_3$) in the limestone from which the material was formed. **C 51,** C07

dolomitic limestone—see **limestone.** **C 51,** C07

domain, *n*—*as used in Test Method D 5061*, a region of anisotropy in a carbon form that is distinctively marked by its isochromatic boundary and cleavage. **D 121,** D05

domains, ferromagnetic—magnetized regions, either macroscopic or microscopic in size, within ferromagnetic materials. Each domain, in itself, is magnetized to intrinsic saturation at all times, and this saturation induction is unidirectional within the domain. **A 340,** A06

domain wall—a boundary region between two adjacent domains within which the orientation of the magnetic moment of one domain changes into a different orientation of the magnetic moment in the other domain. **A 340,** A06

dome, *n*—in reinforced plastics, an end of a filament-wound cylindrical container. **D 883,** D20

dominant wavelength, *n*—the wavelength of a spectrally pure light that, when added to a reference achromatic (white) light, will produce a combination that matches the color of a specimen light. **E 284,** E12

donor—relatively colorless material capable of generating color when transferred to a reactive receptor surface. **F 549,** F05

donor—an impurity or imperfection in a semiconductor that donates electrons to the conduction band, leading to electron conduction. **F 1241,** F01

donut fryer—see *specialty fryer*. **F 1827,** F26

door, *n*—usually swinging or sliding barrier by which an entry is closed and opened. **E 1605,** E06

door—combination nose and closure for nose-loading stapler. **F 592,** F16

door—the movable barrier that permits access to the cavity for placement or removal of food, and whose function is to prevent emission of energy from the passage or opening that provides access to the cavity. **F 1827,** F26

DOP, *n*—dioctyl phthalate (di-2-ethylhexyl phthalate). **D 1356,** D22

dopant density—*in an uncompensated extrinsic semiconductor*, the number of dopant impurity atoms per unit volume, usually given in atoms/cm^3 although the SI unit is atoms/m^3. Symbols: N_D for donor impurities and N_A for acceptor impurities. **F 1241,** F01

dope—slang for mold lubricant. **C 162,** C14

dope—a composition, usually a cellulosic lacquer, for application on textiles and leathers. **D 16,** D01

dope—a composition, usually a cellulose lacquer, for application on textiles and leathers. Also a very viscous crude reaction product, as acetylation dope. **D 1695,** D01

doping—the addition of foreign species into a photoconductive material to modify its semiconductor properties such as light sensitivity, response times, and dark conductivity. **F 335,** F05

doping—addition of impurities to a semiconductor to control the electrical resistivity. **F 1241,** F01

Doppler bubble detector, *n*—a device that uses the Doppler Effect to monitor specific areas of the circulatory system and detect the presence of bubbles as a measure of decompression stress. **F 1549,** F32

doré bead, *n*—a gold and silver alloy bead which results from cupellation. **E 135,** E01

dormant oil—a horticultural mineral oil applied during the dormant phase of the targeted plant. (See **horticultural spray oil.**) **E 1519,** E35

dosage—the quantity of substance applied per unit weight or volume of the fluid being treated. **D 2652,** D28

dose—number of beam particles per unit area that impinge on the specimen. Alternatively, the dose may be defined as the charge per unit area that impinges on the specimen.

rate—number of beam particles per unit area per second that impinge on the specimen. Alternatively, the dose rate may be defined as the current per unit area that impinges on the specimen. **E 673,** E42

dose, dosage—the quantity of substance applied per unit treated or applied to or entering an organism. **E 609,** E35

dosimeter—a device that, when irradiated, exhibits a quantifiable change in some property of the device that can be related to absorbed dose in a given material using appropriate analytical instrumentation and techniques. **E 170,** E10

dosimetry system—a system used for determining absorbed dose, consisting of dosimeters, measurement instruments and their associated reference standards, and procedures for the system's use. **E 170,** E10

dot distortion, *n*—an undesired change in shape of halftone dots during plate making or printing operations, or both. **D 6488,** D01

dot doubling, *n*—the unintentional printing of two images slightly out of register from a single form. **D 6488,** D01

dote—See **decay.** **D 9,** D07

dot gain, *n*—the increase in halftone dot size in relation to the dots on the lithographic film; it is the arithmetic difference between the dot area on the original film and dot area on the printed sheet. **D 6488,** D01

dot loss, *n*—the total or partial loss of the smallest dots on the sheets, usually in the areas of 25 % screen or less. **D 6488,** D01

dot matrix—a method of character generation in which each character is formed by a pattern of dots selected from a grid or matrix. **F 909,** F05

dot slur, *n*—dot gain that occurs in a pronounced direction and is not symmetrical. **D 6488,** D01

double, n—a half rope. **F 1773,** F08

double-action pressing—a method by which a powder is pressed in a die between opposing moving punches. **B 243,** B09

double aging, *n*—employment of two different aging treatments, in sequence, to control the type of precipitate formed from a supersaturated alloy matrix in order to obtain the desired properties. **A 941,** A01

double amplitude—total or peak to peak excursion. **D 653,** D18

double boiling system—a series of mixtures characterized by the vaporization of a liquid phase in one temperature (or pressure) range and the vaporization of another liquid phase within another temperature (or pressure) range; for example, a salt and water mixture which boils with the expulsion of water at moderately elevated temperature and then at higher temperature the molten salt itself boils to produce salt vapor. **E 7,** E04

double bow, *n*—two fabric bows, arcing in the same direction, as in a flattened *M* or *W* depending on the viewing angle. (Compare **double reverse bow)** **D 123,** D13

double braid, *n*—*in rope*, a braided construction consisting of two hollow braided ropes, one inside the other. **D 123,** D13

double-cavity mold—see **multiple cavity mold.** **C 162,** C14

double-cavity process—see **multiple cavity process.** **C 162,** C14

double coil, *n*—a configuration where a smaller diameter coil is placed inside a larger diameter coil. The assembled rolls may be either concertina style or helical style. **F 1379,** F14

double crest—crest having two ridges separated by slight depression. **F 547,** F16

double draining—a defect evidenced by flowing of the slip on the ware, which occurs after it appears that draining has been completed. **C 286,** B08

double-dressed—as applied to chamois skins, with the grain removed and buffed or sueded on both surfaces. **D 1517,** D31

double, duplex, dual head—thick flat head to be stuck by hammer, with a second head, flange, or collar spaced slightly below to act as bearing head to prevent nail head from sinking into material into which nail is driven and to facilitate withdrawal after temporary use of nail. On double and duplex heads, the first and second heads are flat; whereas the second head of dual head is a cone head. **F 547,** F16

double-face ware—ware that has a finish coat on both surfaces. **C 286,** B08

double fold, *n*— *of paper*, one complete oscillation of the paper test specimen, during which it is folded first forward, then backward about the same base. **D 1968,** D06

double glazing—insulated glazing that incorporates two panes separated by an air gap. **C 162,** C14

double glazing unit—two panes of glass separated by a permanently sealed cavity. **C 162,** C14

double-gob process—See **double-cavity process.** **C 162,** C14

double-headed, duplex-head, dual-head, nail—bright or coated, regular-stock-steel, 1 3/4 by 0.113 to 4 by 0.207-in. nails with double 3/16 to 7/16-in. head, medium diamond point, and 1/4 to 7/16-in. distance between head to be struck by hammer and bearing head. Length of nails measured from bearing surface of head. **F 547,** F16

double hooked bow, *n*—one hooked bow at each side of the fabric that are in opposite directions. (See also **hooked bow)** **D 123,** D13

double island canopy—used for all types of cooking equipment mounted back to back in an island configuration. See *canopy*. **F 1827,** F26

double-junction reference electrode—a reference electrode that utilizes two separate filling solutions: an inner solution to provide a constant level of the ion sensed by the reference element and an outer filling solution that provides a low-junction potential liquid junction with the sample solution. The liquid junction potential between the inner and outer solutions is constant but not necessarily zero. **D 4127,** D19

double layer flat—a coil in which the product is spirally wound into two connected disk-like layers such that one layer is on top of the other. (Sometimes called "double layer pancake coil" or "double layer spirally wound coil.") **B 846,** B05

double loop ties, *n*—twistable ties used to secure the barbed tape at the tie point (also known as *bag ties*). **F 1379,** F14

double melting system—a series of mixtures which, with rising temperature, first develops a liquid phase that is totally converted to vapor before a second liquid phase appears; for example, a salt and water mixture which, upon heating, first melts to an aqueous solution of the salt, the water then boils away and the salt residue itself melts. **E 7,** E04

double modulation, *n*—a technique in which a modulated signal is further varied by a second means. **E 131,** E13

double-pass internal reflection element— an internal reflection element in which the radiant power transverses the length of the optical element twice, entering and leaving via the same end. **E 131,** E13

double pick, *n*—*in woven fabrics,* two picks wrongly placed in the same shed. (See also **mispick.** Compare **jerk-in.**) **D 123,** D13

double pour—to apply two layers of aggregate and bitumen to a built-up roof. **D 1079,** D08

double press-double sinter, *v*—to repress and sinter a previously presintered or sintered compact. **B 243,** B09

doubler—an interlining between the toe lining and vamp of the shoe to aid in preserving the shape during wear. **F 869,** F08

double reverse bow, *n*—two fabric bows arcing in opposite directions. (See also **bow.** Compare **double bow.**) **D 123,** D13

doublers, n—an extra piece of facing attached to strength or stiffen the panel or to distribute the load more widely to the core. **C 274,** D30

double salt—a compound of two salts that crystallize together in a definite proportion. **B 374,** B08

double shoulder—the fore part of the hide cut off at right angles to the backbone line at the break of the fore flank, with the belly cut off and the head cut off behind the horn holes. **D 1517,** D31

double-sided griddle—equipment for cooking food by direct contact with two hot surfaces. See **griddle.** **F 1827,** F26

double spread, *n*—See *double spread* under **spread.** **D 907,** D14

double-stitched seam-finish, *n*—a finish for the raw edges of a plain seam, in which another row of machine stitching is made through both seam allowances placed together. **D 123,** D13

double tempering, *n*—a treatment in which a quench-hardened steel object is given two complete tempering cycles at substantially the same temperature for the purpose of ensuring completion of the tempering reaction and promoting stability of the resultant microstructure. **A 941,** A01

doublet (in characteristic X-ray spectra)— a separation of characteristic radiation into subspecies of slightly different wavelength. **E 7,** E04

doublet injector—an impinging jet atomizer in which there are two colliding liquid jets. **E 1620,** E29

double-welt seam, *n*—*in home sewing,* a complex seam formed on the inside of the product, in which one trimmed seam allowance is enclosed and two rows of stitching are visible on the face side. (Compare **welt seam.** Syn. *mock flat-felled seam.*) **D 4965,** D13

doughnuts, *hickeys, n*—round voids in a solid print with ink in the center. **D 6488,** D01

dough (rubber), *n*—a paste-like mass of smooth texture, consisting of a rubber mix and solvent; it is used for spreading. **D 1566,** D11

down, *n*—the fine, soft plumage of waterfowl, consisting of light, fluffy filaments having at least two barbs attached; that is, barbs growing from the quill point but without a quill shaft. **D 123,** D13

down discharge—makeup air method is used when spot cooling of the cooking staff is desired to help relieve the effects of severe radiant heat generated from equipment such as charbroilers. The makeup air must be heated or cooled, or both, depending on the climate. **F 1827,** F26

down draw—process of continuously drawing glass downward from an orifice. **C 162,** C14

down fibers, *n*—detached barbs from down; plumules and detached barbs from the basal end of waterfowl feather quill shafts that are indistinguishable from the barbs of down. **D 123,** D13

download, *v*—to transfer data from the host to a server, in this case a printer. **F 1457,** F05

down, nestling, *n*—a down not fully developed with a sheath and with soft barbs emanating from the sheath. **D 4523,** D13

down-tank—the direction in a melter from the batch feeding end toward the exit. **C 162,** C14

$(dP/dt)_{ex}$, *n*—the maximum rate of pressure rise during the course of a single deflagration. **E 1445,** E27

$(dP/dt)_{max}$, *n*—maximum value for the rate of pressure increase per unit time reached during the course of a deflagration for the optimum concentration of the dust tested. It is determined by a series of tests over a large range of concentrations. It is reported in bar/s. **E 1445,** E27

dpi—dots per inch. **F 1156,** F12

dpmm—dots per millimetre. **F 1156,** F12

draft—minimum vertical depth of the membrane below the water line. **F 818,** F20

drag—the resistance to shrinkage of the foot or base of a ceramic article during drying or firing as a result of friction with the setter, slab, or sagger on which it rests. **C 242,** C21

drag, *n*—the resistance of the foot or base of a ceramic article to shrinkage during firing time due to friction with the slab or sagger on which it rests. **C 1145,** C28

drag—physical resistance to spreading of a polish. **D 2825,** D21

dragade—{archaic} see **drag ladle.** **C 162,** C14

drag bit—a noncoring or full-hole boring bit, which scrapes its way through relatively soft strata. (ISRM) **D 653,** D18

dragging—a fringe effect attached to the trailing edge of the developed electrostatic image. (See also **tailing.**) **F 335,** F05

drag-in—the water or solution that adheres to the objects introduced into a bath. **B 374,** B08

drag ladle—{archaic} to produce cullet by ladling glass from the melt into water. **C 162,** C14

drag-out—the solution that adheres to the objects removed from a bath. **B 374,** B08

drag out, *n*—excessive ink around shadow areas of the image usually associated with excessively deep etches on less absorbent substrate. **D 6488,** D01

dragout—the carryout or loss of penetrant materials as a result of their adherence to the test pieces. **E 1316,** E07

drain—a means for intercepting, conveying, and removing water. **D 653,** D18

drainage, *n*—*of an insulating varnish,* a measure of the variation in thickness from top to bottom of a varnish film obtained on the surface of a vertically dipped coated panel after a specified time and temperature. **D 1711,** D09

drainage basin—See **watershed.** **D 4410,** D19

drainage course—See **percolation layer** . **C 717,** C24

drainage curtain—*in grouting,* a row of open holes drilled parallel to and downstream from the grout curtain of a dam for the purpose of reducing uplift pressures. **D 653,** D18

drainage gallery—*in grouting,* an opening or passageway from which grout holes or drainage curtain holes, or both, are drilled. See also **grout gallery.** **D 653,** D18

drainage hole—an opening in a construction provided for the escape of unwanted liquid, as in a retaining wall. Compare **vent hole, weep hole.** **E 631,** E06

drainage loss—oil accumulating and pooling against the boom skirt and escaping with the flow of water down and along the skirt. **F 818,** F20

drainback solar energy system—see **solar energy system, drainback.** **E 772,** E44

drain casting—See *drain casting* under **casting.** **C 242,** C21

draindown solar energy system—see **solar energy system, draindown.** **E 772,** E44

draining—*in ceramic manufacture*, the process of removing excess slip from dipped or cast items by gravity flow. **C 242,** C21

draining—the part of the dipping or flowcoating process in which the excess slip flows from suitably positioned ware. **C 286,** B08

drain line—a nonuniform thickness of coating appearing as a line or streak in dipped or flow-coated ware. **C 286,** B08

drains—a piping system used to collect and carry off surface and ground water. **C 896,** C04

drain tile—pipe for collecting and conveying surface and subsurface water from an area. **C 822,** C13

drain time—time required for porcelain enamel slip applied by dipping, slushing, or flow coating to complete movement across the surfaces of a coated part. **C 286,** B08

drain time—that portion of the dwell time during which the excess penetrant or emulsifier drains from the part. **E 1316,** E07

DRAM—**d**ynamic **r**andom **a**ccess **m**emory. **F 1457,** F05

draw—See **pull.** **C 162,** C14

draw-back, *n*—a weave distortion characterized by tight and slack places in the same warp yarn. **D 123,** D13

draw bar—a refractory member submerged in molten glass that defines the position of the sheet in a drawing process. **C 162,** C14

drawdown (L)—vertical distance the free water elevation is lowered or the pressure head is reduced due to the removal of free water. **D 653,** D18

draw gang—{archaic} people employed to cut and handle glass as it comes from the lehr. **C 162,** C14

drawing, *n*—*in textile processing*, the process of stretching or attenuating a material to increase the length per unit mass. **D 123,** D13

drawing, *n*—an architectural, structural, mechanical, or electrical plan, elevation, or section indicating in isometric perspective or in axonometric perspective the detailed location, dimension, quantity, or extent of material, product, or member to be furnished. Compare **shop drawing, working drawing.** **E 631,** E06

drawing, *n* (dessin):

record set drawing (as-built drawing), n (dessin de l'ouvrage fini (dessin d'après exécution))—construction drawing revised to show changes made during the construction process, usually based on marked-up prints, drawings, and other data furnished by the contractor.

working drawing, n (dessin d'exécution)—detail drawing, usually produced by a draftsperson under direction of an architect, engineer, or other designer showing the form, quantity, and relationship of construction elements and materials and indicating their location, identification, grades, dimensions, and connections. **E 1480,** E06

drawings, *n*—*in pipe laying*, drawings prepared by the purchaser to show the location and details for the construction of the pipeline and appurtenances. **C 460,** C17

drawings, *n*—*in pipe laying*, drawings prepared by the purchaser to show the location and details for the construction of the pipeline and appurtenances. **C 1154,** C17

drawings, *n*—*in pipe laying*, illustrations (such as drawings, sketches, graphs, or photographs) on paper, film, diskettes, magnetic tape or other media, blueprints, layouts, designs, figures, portrayals, pictures, charts, maps, images, patterns, diagrams, representations, and plans supplied by the purchaser to show the location and details for the construction of the pipeline and appurtenances. **D 2946,** C17

drawing stock—the as-rolled or extruded section, normally round, made from any of a variety of manufacturing processes, intended as feed stock for subsequent drawing operations. **B 354,** B01

drawn glass—glass made by a continuous drawing operation. **C 162,** C14

drawn grain—shrunken, shriveled, or wrinkled grain surface of leather. **D 1517,** D31

drawn stress relieved (DSR)—a thermal treatment of a cold-drawn tubular product to improve ductility without significantly affecting its tensile strength or microstructure. **B 846,** B05

draw ratio, *n*—the relation of the final (length per unit mass) to original length per unit mass of a material resulting from drawing. **D 123,** D13

draw ratio (DR), *n*—the relation of the final length per unit mass to original length per unit mass of a material resulting from drawing. **D 123,** D13

draw texturing, *n*—for processing thermoplastic fibers, the simultaneous or sequential process of drawing and imparting crimp, thus producing increased molecular orientation and increased bulk. **D 123,** D13

d-RDF—abbreviation for densified refuse-derived fuel. **E 856,** D34

dredge, dredging—*in dry process enameling,* (*1*) the application of dry, powdered frit to hot ware by sifting.

(*2*) The sieve used to apply powdered porcelain enamel frit to the ware. **C 286,** B08

dress career apparel, *n*—career apparel which is not generally subject to abusive wear and for which appearance is a more important attribute than durability. (See also **career apparel, vocational career apparel**) **D 123,** D13

dressed crude mica, *n*—crude mica from which the dirt and rock have been mainly removed. **D 1711,** D09

dressed size—See **lumber.** **D 9,** D07

dressed stone—See **cut stone, finished stone.** **C 119,** C18

dress glove, *n*—a covering for the hand, often extending part way up the arm, worn primarily for formal or dress occasions. **D 123,** D13

dressing, *n*—(1) the process of restoring the efficiency of an abrasive grinding wheel by removal of dulled grains; (2) reshaping the faces of grinding wheels to special contours. **C 1145,** C28

dress shirt, *n*—*for boys*, a shirt made with a specific collar size or numerical size and designed to be worn with a tie and jacket. **D 123,** D13

dress shirt, *n*—*for men*, a shirt made with a specific collar size and sleeve length where appropriate, and designed to be worn with a tie and jacket. **D 123,** D13

drier—a composition that accelerates the drying of oil, paint, printing ink, or varnish. Driers are usually metallic compositions and are available in both solid and liquid forms. **D 16,** D01

drift—this is the slow nonrandom change with time in the potential (emf) of an ion-selective electrode cell assembly maintained in a solution of constant composition and temperature. **D 4127,** D19

drift—in electron optics, motion of the electron beam or image due to current, voltage or specimen instabilities or charging of a projection such as dirt in or near the electron beam. **E 7,** E04

drift, *n—in instrumental methods of quantitative analysis,* a gradual change in instrument response from start to completion of a set of determinations. **E 135,** E01

drift—the movement of chemicals outside the intended target by air mass transport or diffusion. **E 1102,** E35

drift—in leak testing, the relatively slow change in the background output level of the leak detector due to the electronics rather than a change in the level of the tracer gas. **E 1316,** E07

drift control agent—a material used in liquid spray mixtures to reduce spray drift. **E 1519,** E35

drill—a machine or piece of equipment designed to penetrate earth or rock formations, or both. **D 653,** D18

drillability—index value of the resistance of a rock to drilling. (ISRM) **D 653,** D18

drill carriage; jumbo—a movable platform, stage, or frame that incorporates several rock drills and usually travels on the tunnel track; used for heavy drilling work in large tunnels. (ISRM) **D 653,** D18

drill cuttings—fragments or particles of soil or rock, with or without free water, created by the drilling process. **D 653,** D18

drilling fluid—a fluid (liquid or gas) that may be used in drilling operations in remove cuttings from the borehole, to clean and cool the drill bit, and to maintain the integrity of the borehole during drilling. **D 653,** D18

drilling pattern—the number, position, depth, and angle of the blastholes forming the complete round in the face of a tunnel or sinking pit. (ISRM) **D 653,** D18

drill mud—*in grouting,* a dense fluid or slurry used in rotary drilling; to prevent caving of the bore hole walls, as a circulation medium to carry cuttings away from the bit and out of the hole, and to seal fractures or permeable formations, or both, preventing loss of circulation fluid. **D 653,** D18

dripping—in testing thermal protective material, a response evidenced by flowing of the fiber polymer. **F 1494,** F23

drips—*in protective coatings,* the small drops of coating that collect on the edge of the coated work. **D 4538,** D33

drive cam—a cam that actuates the tumblers by a drive pin and is gated to retract or accept a locking mechanism. **F 471,** F12

drive nail, drive screw—terms applied to helically threaded nails, twisted nails, and annularly threaded shoe nail. (See **nail, thread.**) **F 547,** F16

drive pin—a protrusion on or through a drive cam or tumbler which drives or turns the next tumbler. **F 471,** F12

driver—tool component that pushes fastener from the driving chamber of the tool into the members being assembled. **F 592,** F16

driving coefficient, [*nd*], *n*—the ratio of the driving force to the normal force. **F 538,** F09

driving coefficient, peak[*nd*], *n*—the maximum value of the driving coefficient. **F 538,** F09

driving force, [*F*], *n—of a tire,* the positive longitudinal force resulting from the application of driving torque. **F 538,** F09

driving torque, [ML^2/T^2], *n—of a wheel,* the positive wheel torque. **F 538,** F09

drop, *n—in body measurements,* the difference between the chest girth and the waist girth. **D 123,** D13

drop, *n*—that part of a bedcovering that hangs perpendicular to the floor. **D 123,** D13

drop, *n*—a single liquid particle having a generally spheroidal shape. **E 1620,** E29

drop bar—component of a double gate latch assembly (sometimes called *drop rod*) (see Specification F 900). **F 552,** F14

drop cap—See **railing systems.** **E 631,** E06

drop cap—the cover of a railing post or newel that is exposed to view, usually below the stair stringer or floor. **E 1481,** E06

droplet, *n*—a small liquid particle of such size and density as to fall under still conditions but which may remain suspended under turbulent conditions. **D 1356,** D22

droplet, *n*—see **drop**; also a small drop. **E 1620,** E29

drop, liquid, *n*—see **liquid drop**. **G 40,** G02

drop out colors—See **reflective ink**. **F 149,** F05

drop out ink—See **reflective ink**. **F 149,** F05

dropped stitch, *n—in knitted fabrics,* an unknitted stitch. **D 123,** D13

dropping—{archaic} forming by heating in a mold without the use of pressure. **C 162,** C14

drop size, *n*—the diameter of a liquid drop if it is approximately spherical; otherwise, the approximate shape and appropriate dimensions must be described. **G 40,** G02

drop test—See **package testing.** **D 996,** D10

drop throat—See **submerged throat**. **C 162,** C14

drop weight, *n*—that weight which is raised to a selected height and released. This weight does not impact the sample directly; rather it strikes another stationary weight that is in contact with the sample. **E 1445,** E27

drop-weight tear test (DWTT)—a test of plain-carbon or low-alloy pipe steels over the temperature range where the fracture changes from a brittle to a ductile mode. The mode can be determined from the appearance of propagating fractures. **E 1823,** E08

dross—a mixture of metal oxide and metal on the surface of a float bath. **C 162,** C14

drum, *n*—(*1*) a cylindrical **shipping container** having straight sides, and flat, convex or embossed ends, designed for storage and shipment as an unsupported outer package that may be shipped without boxing or crating. It may be made of metal, or of plywood, or of fiber with wooden, metal or fiber ends. Drums are also made of rubber or plastics (Compare **barrel.**); (*2*) in set-up paper boxes, a shell or tube with paper or cellophane head, used for powder box. (See **divider.**) **D 996,** D10

drum dryer, *n*—a heated, rotating drum in which tumbling or cascading raw materials are dried. **C 1145,** C28

drum dyeing—the application of dyestuffs to leather by immersion of the leather in a revolving drum containing the dyestuff solution, as contrasted with table dyeing. **D 1517,** D31

drumhead leather—see **parchment.** **D 1517,** D31

drum printer—a type of printer that employs a rotating cylinder or drum, the outer surface of which contains fully formed characters. As the drum rotates in front of the paper form, hammers are caused to strike the back of the form synchronized in such a way as to produce the desired character in their proper position. **F 909,** F05

dry, *v*—to change the physical state of an adhesive on an adherend by the loss of solvent constituents by evaporation or absorption, or both. (See also **cure, set.**) **D 907,** D14

dry, ash-free—the test data calculated to a theoretical base of no moisture or ash associated with the sample. **E 856,** D34

dry, ash-free basis—See *dry, ash-free basis* under **reporting bases**. **D 121,** D05

dry ash-free basis, *n*—test data calculated to a theoretical base of no moisture or ash associated with the sample. **D 5681,** D34

dry basis—See *dry basis* under **reporting bases**. **D 121,** D05

dry basis—exclusive of any moisture which may be present. **D 2652,** D28

dry basis, *n*—test data calculated to a theoretical base of no moisture associated with the sample. **D 5681,** D34

dry basis—the test data calculated to a theoretical base of no moisture associated with the sample. **E 856,** D34

dry basis moisture content—of biomass fuels, the ratio of the weight of the water in a sample to the weight of the dry material. It is expressed as a percent. **E 1126,** E48

dry basis moisture content—of biomass fuels, the ratio of the weight of the water in a sample to the weight of the dry material. It is expressed as a percent. **E 1705,** E48

dry-blend, *n*—a dry compound prepared without fluxing or addition of solvent (also called powder blend). **D 883,** D20

dry-blend, *n*—a free-flowing dry compound prepared without fluxing or addition of solvent (also called powder blend). (D-20) **F 412,** F17

dry bond adhesive, n—Synonym for **contact adhesive**. **D 907,** D14

dry bright polish—a polish that dries to a gloss without buffing. **D 2825,** D21

dry-bulb temperature—temperature of the air as indicated by an accurate thermometer, corrected for radiation if significant. **D 9,** D07

dry-bulb temperature—See **temperature.** **D 1356,** D22

dry-bulb temperature, *n*—the temperature of the ambient air; for example, the temperature that is measured by the dry-bulb thermometer of a psychrometer. **G 113,** G03

dry-bulb temperature, *t*—the temperature of the ambient air, for example, the temperature that is measured by the drybulb thermometer of a psychrometer. **D 4023,** D22

dry carrier toner—dry powdered toner mixed with bead or granular particles as a carrier to form a developer. **F 335,** F05

dry chop—dried fiber glass strand which has been chopped. See **chopped fiber** and **wet chop**. **C 162,** C14

drycleanable button, *n*—a button that can be solvent-cleaned without damage such as dissolving or loss of finish. (See dry-cleaning, finish). **D 123,** D13

drycleaning, *n*—*in the care of textiles,* the cleaning of consumer textile products with organic solvents such as petroleum solvent or perchloroethylene. **D 123,** D13

dry cleaning—under **cleaning,** see *dry cleaning.* **D 459,** D12

dry-cleaning detergent—a formulated composition added to the solvent bath in dry-cleaning operations to improve cleaning. **D 459,** D12

dry-cleaning detergent—under **detergent,** see *dry-cleaning detergent.* **D 459,** D12

dry cockle, *n*—a wave pattern which is apparent after ink drying. **F 1857,** F05

dry coke—See *dry coke* under **coke.** **D 121,** D05

dry crock—same as wet crock without using water. **D 2825,** D21

dry deposition—See **deposition.** **D 1356,** D22

dry edging—rough edges and corners of glazed ceramic ware caused by insufficient glaze coating. **C 242,** C21

dryer—a kiln or chamber, or machine through which the green veneers are passed to remove excess moisture. **D 1038,** D07

dry extraction cleaning, *n*—a method in which an absorbent compound is dispersed over the surface of a textile product by hand or machine, thoroughly brushed through the pile, allowed to dry, and removed by suction. **D 123,** D13

dry felt—(*1*) see felt;
(*2*) a felt which has not been saturated with bitumen. **D 1079,** D08

dry fiber—a condition in which fibers are not fully encapsulated by resin during pultrusion. **D 3918,** D20

dry-film thickness, *n*—the thickness of a cured film, coating, or membrane. **C 717,** C24

dry foam extraction cleaning, *n*—a process by which a highly aerated, low moisture content shampoo is brushed through the textile floor covering pile or applied to the surface of upholstery. **D 123,** D13

dry gage (drigage) v.—to form cullet by running a stream of molten glass into a water bath. (See synonomous term **frit, v.**) **C 162,** C14

dry gas—natural gas containing little or no water vapor. **D 4150,** D03

dry impingement—See **impingement.** **D 1356,** D22

dry in, *n*—a situation where the cells in the anilox cylinder become partly or completely filled with dry ink. **D 6488,** D01

drying—removal by evaporation, of uncombined water or other volatile substance from a ceramic raw material or product, usually expedited by low-temperature heating. **C 242,** C21

drying—See **seasoning.** **D 9,** D07

drying, *n*—process of developing, solely by evaporation of volatile ingredients, ultimate properties of a finish or other material over a specified period of time. Compare **curing.** **E 631,** E06

drying crack—a defect characterized by a fissure in the bisque. **C 286,** B08

drying in, *n*—the accumulation of dried ink in the cells of the cylinder. **D 6488,** D01

drying oil—an oil that possesses to a marked degree the property of readily taking up oxygen from the air and changing to a relatively hard, tough, elastic substance when exposed in a thin film to the air. **D 16,** D01

drying oven, *n*—a closed unit in which specimens are dried by heating. **C 1145,** C28

drying oven—an oven used for increasing the evaporation rate of rinse water or an aqueous developer vehicle from test parts. **E 1316,** E07

drying shrinkage, *n*—the contraction of a moist body during the drying process, expressed as linear percent of the original length or volume percent of the original volume. **C 1145,** C28

drying shrinkage, *n*—in this test method, the change in linear dimension of the test specimen due to drying from a saturated condition to an equilibrium weight and length under specified accelerated drying conditions. **C 1209,** C15

drying temperature, *n*—the temperature to which an adhesive on an adherend or in an assembly or the assembly itself is subjected to dry the adhesive. (See also **curing temperature, setting temperature**.) **D 907,** D14

drying time, *n*—the period of time during which an adhesive on an adherend or an assembly is allowed to dry with or without the application of heat or pressure, or both. (See also **curing time, joint-conditioning time,** and **setting time**.) **D 907,** D14

drying time—the time required for a cleaned, rinsed or wet developed part to dry. **E 1316,** E07

drying, vacuum, *n*—the technique of expediting the removal of moisture from a material or body by the use of a vacuum in conjunction with a conventional drying system. **C 1145,** C28

dry method—magnetic particle inspection in which the ferromagnetic particles employed are in the dry powder form. **E 1316,** E07

dry milling—the grinding of porcelain enamel materials without a liquid vehicle. **C 286,** B08

dry milling, *n*—the process of reducing the particle size of a substance by milling without the use of a liquid medium. **C 1145,** C28

dry mix—See *dry process* under **process.** **C 242,** C21

dry objective—any microscopical objective designed for use without immersion liquids. **E 7,** E04

dry occupancy—an occupied space below the plaza deck system in which the computed or anticipated relative humidity is below 30 %. **C 717,** C24

dry pack—a cement-sand mix with minimal water content used to fill small openings or repair imperfections in concrete. **D 653,** D18

dry pickling—a method of curing skins from wool sheep with sodium sulfate and sodium chloride. **D 1517,** D31

dry point, *n*—*in batch distillation,* the temperature observed at the instant the last drop of liquid evaporates from the lowest point in the flask. **D 4175,** D02

dry point temperature—the temperature observed immediately after the liquid just disappears from the bottom of the flask during a distillation test. **D 4790,** D16

dry powder—finely divided ferromagnetic particles suitably selected and prepared for magnetic particle inspection. **E 1316,** E07

dry pressing—See *dry pressing* under **pressing.** **C 242,** C21

dry process—See *dry process* under **process.** **C 242,** C21

dry process enameling—a porcelain enameling process in which the metal article is heated to a temperature above the maturing temperature of the coating (usually 1600 to 1750°F, (approximately 870 to 955°C)), the coating materials applied to the hot metal as a dry powder, and fired. **C 286,** B08

dry salting—a method of curing hides in which the hides are first greensalted and then dried. **D 1517,** D31

dry scraping—*of coatings,* method of removing dried, often deteriorated, paint film using a blade or similar tool. **E 1605,** E06

dry screening, *n*—the process of separating small sizes of granular or powdered solids from coarser particles by passing them through a screen of desired mesh size while in the dry state. **C 1145,** C28

dry sieving—*for the purpose of Test Method D 4749,* the test method for the sieving of coal after the sample has been airdried under prescribed conditions; this is generally used when testing with coal particles larger than 600 μm (No. 30 U.S.A. Standard Sieve Series.) **D 121,** D05

dry solvent, *n*—any organic solvent used to dissolve another material. (See **solvent**). **D 123,** D13

dry-spot, *n*—an imperfection in reinforced plastics, an area of incomplete surface film where the reinforcement has not been wetted with resin. **D 883,** D20

dry spray—a defect confined to sprayed ware manifesting itself in the fired porcelain enamel as a rough, sandy texture. **C 286,** B08

dry spray—*in protective coatings,* a rough, powdery, non-coherent film produced when an atomized coating partially dries before reaching the surface. **D 4538,** D33

dry strength, n— See **bond strength**. **D 907,** D14

dry strength—See **strength, dry**. **E 1749,** E06

dry strength—the strength of materials and of adhesive joints determined immediately after drying under specified conditions or after a period of conditioning in a standard laboratory atmosphere. **F 869,** F08

dry tack, n—See *dry tack* under **tack**. **D 907,** D14

dry technique—in magnetic particle examination, the examination technique in which the ferromagnetic particles are applied in the dry powder form. **E 1316,** E07

dry time—the time that a copy must reside in the copy tray before it appears to be dry. **F 335,** F05

dry time, *n*—a measure of the rate of ink absorption into a substrate to the point at which no image transfer, smear, or surface damage occurs when in contact with another surface. **F 1857,** F05

dry toner, *n*—the material in a dry developer system which when deposited on a substrate by the field of an electrostatic charge pattern, becomes the visible record. **F 1457,** F05

dry unit weight (*dry density*)—see **unit weight**. **D 653,** D18

drywall nail—See **gypsum-wallboard nail.** **F 547,** F16

dry weight—the weight per unit area of the bisque. **C 286,** B08

dry weight—a term is usually applied to the mass of the ovendry material, but it is ambiguous unless the method of drying is specified. **D 1695,** D01

DSC—abbreviation for **data stream compatibility**. **F 1457,** F05

2D stacked symbols—see **multi-row symbols.** **F 1294,** F05

D2T2, *n*—abbreviation for **dye diffusion thermal transfer**. **F 1623,** F05

DTA (DSC) curve, *n*—a record of a thermal analysis where the temperature difference (ΔT) or the energy change (Δq) is plotted on the ordinate and temperature or time is plotted on the abscissa. **E 1445,** E27

DTE—abbreviation for **data terminal equipment**. **F 1457,** F05

DTR—abbreviation for **data terminal ready**. **F 1457,** F05

2D (two-dimensional) symbol, *n*—a machine-readable symbol that must be examined both vertically and horizontally to read the entire message. Two-dimensional symbols may be of two types: matrix symbols and multi-row symbols. **F 1294,** F05

dual-drum mixer, *n*—a mixer consisting of a long drum containing two compartments separated by a bulkhead with a swinging chute extending through the unit. **C 1145,** C28

dual-head nail—See **double-headed nail.** **F 547,** F16

dual in-line package switch, *n*—a device used to direct the signal line to or from one component or another. (See **DIP switch**.) **F 1457,** F05

dual metal, *n*—two metals of different composition that are fusion bonded at all interfacial surfaces by casting metal of one composition against metal of a second composition. **A 644,** A04

dual orifice nozzle—a swirl chamber atomizer containing a primary injector and a concentric annular secondary injector, each injector comprising a separate orifice and set of tangential slots. The nozzle is normally operated only with the primary injector at low flow rates, with secondary liquid introduced at a specified pressure. (This definition applies to devices used in the gas turbine industry.) **E 1620,** E29

dual ovenable—terms describing a food packaging container used to prepare food in either a conventional oven or a microwave oven. **F 1479,** F02

dual search unit—a search unit containing two elements, one a transmitter, the other a receiver. **E 1316,** E07

dubbing, (dubbin)—a mixture, primarily oils and fats, used for restoring fatty matter to military footwear in the field. **D 1517,** D31

duck, *n*—a compact, firm, heavy, plain-weave cotton fabric, mass per square yard 6 to 50 oz. (See also **flat duck**, and **plied yarn duck**) **D 123,** D13

duckbill point—end of wire flattened to thin elliptical cross section having sharp periphery, in appearance somewhat like bill of duck, to facilitate clinching at predetermined point, slight transverse depression may be formed across point. **F 547,** F16

ductile erosion behavior, *n*—erosion behavior having characteristic properties that can be associated with ductile fracture of the exposed solid surface; that is, considerable plastic deformation precedes or accompanies material loss from the surface which can occur by gouging or tearing or by eventual embrittlement through work hardening that leads to crack formation. (See also **brittle erosion behavior**.) **G 40,** G02

ductile failure—a pipe failure mode which exhibits material deformation (stretching, elongation, or necking down) in the area of the break. **F 412,** F17

ductile fracture, *n*—fracture that occurs with appreciable plastic deformation of the material. **A 644,** A04

ductile fracture area, *n*—The fraction or percent of the fracture surface that formed by ductile fracture. (When observed with no or low magnification, ductile fracture appears grayer and duller than brittle fracture.) **A 644,** A04

ductile iron, *n*—a cast iron that has been treated in the liquid state so as to cause substantially all of its graphitic carbon to occur as spheroids or nodules in the as-cast condition. **A 644,** A04

ductility—the ability of a material to deform plastically without fracturing. **B 374,** B08

ductility—condition in which material can sustain permanent deformation without losing its ability to resist load. (ISRM) **D 653,** D18

ductility, *n*—the ability of a material to deform plastically before fracturing. **E 6,** E28

ductility of externally threaded fasteners—the ability of a fastener to deform before it fractures. Machined test specimens made from a fastener allow the measurement of elongation and reduction of area which are criteria used to evaluate the specimen. However, since yielding and fracture normally occur in the screw threads, these are impractical for the actual fastener. Hardness and the wedge tensile test are ductility indicators for the actual fastener. The lower the ratio of its specified minimum yield strength to its specified minimum tensile strength, the greater the fastener ductility. **F 1789,** F16

ductor roller—an ink or water roller that alternately contacts the fountain roller and the distributing roller. **F 425,** F05

dullness—a lack of normal pultruded suface gloss or shine.

NOTE—This condition can be caused by insufficient cure locally or in large areas, resulting in the dull band created on a pultruded part within the die when the pultrusion process is interrupted briefly (see **stop mark**). **D 3918,** D20

dull point—end of point rounded in contrast to being sharp. **F 547,** F16

dumbbell specimen (rubber), *n*—a flat specimen having a narrow straight central portion of essentially uniform cross section with enlarged ends. **D 1566,** D11

dummy—a mechanical device, operated by the blower's feet, for wetting, raising, opening and closing the paste mold in mouth-blowing glassware. **C 162,** C14

dummy microphone—a microphone substitute which has electrical characteristics identical to a functional microphone, but which has essentially no sensitivity to incident sound pressure. **C 634,** E33

dummy (or dummy cathode)—a cathode in a plating solution that is not to be made use of after plating. Often used for removal or decomposition of impurities. **B 374,** B08

dummy specimen, *n*—a noncombustible (as defined by 46 CFR 164.009) specimen used for standardizing the operating condition of the equipment, roughly 20 mm in thickness with a density of 750 ± 100 kg/m^3. **E 176,** E05

dummy specimen, *n*—a noncombustible insulating board used for stabilizing the operating condition of the equipment, mounted in the apparatus in the position of the specimen and removed only when a test specimen is to be inserted. It shall be roughly 20 ± 5 mm in thickness with a density of 750 ± 100 kg/m^3. **E 176,** E05

dummy specimen—Use **blank sample**. **E 1605,** E06

dunes (stream)—bed forms of coarse sediment, generally transverse to the direction of flow, with a triangular profile having a gentle upstream slope. Dunes advance downstream by the movement of sediment along the upstream slope and by the deposition of sediment on the steep downstream slope. Dunes move downstream at low velocities compared to the stream flow velocity. **D 4410,** D19

dunnage—See **loading.** **D 996,** D10

dunting—the cracking that occurs in fired ceramic bodies as a result of thermally induced stresses. **C 242,** C21

duo-trio test, *n*—a method of discrimination testing comprised of two coded samples and one identified reference. One of the coded samples and the reference are identical. The assessor is asked to select which of the two coded samples is different from the reference or which of the two coded samples is the same as the reference. **E 253,** E18

duple nozzle—see **dual orifice nozzle**. **E 1620,** E29

duplex channel, *n*—a data transmission system capable of transmitting in both directions at once. **F 1457,** F05

duplex coating—See **composite plate.** **B 374,** B08

duplex copy—a sheet with copied images on both sides. **F 335,** F05

duplex dwelling—See **dwelling.** **E 631,** E06

duplex grain size—the simultaneous presence of two grain sizes, in substantial amounts, with one grain size appreciably larger than the other. (Synonymous with **mixed grain size.**) **E 7,** E04

duplex-head nail—See **double-headed nail.** **F 547,** F16

duplexing—*in a copy system*, the process by which images are placed on both sides of the copy sheet. **F 335,** F05

duplex microstructure—a two-phase structure. **E 7,** E04

duplex nozzle—a swirl chamber atomizer comprising a single discharge orifice and two sets of tangential slots, each with a separately controlled liquid supply. The smaller (primary) slots supply liquid at low flow rates, and both sets (primary and secondary) are utilized as flow increases. **E 1620,** E29

duplicate, *n*—*in experimenting or testing*, one of two or more runs with the same specified experimental or test conditions but with each experimental or test condition not being established independently of all previous runs. (Compare **replicate.**) **D 123,** D13

duplicate, *vt*—*in experimenting or testing*, to repeat a run so as to produce a duplicate. (Compare **replicate.**) **D 123,** D13

duplicate analysis, *n*—paired determinations on the same sample performed by one analyst at essentially the same time. **D 5681,** D34

duplicate analysis—paired determinations on the same sample performed by one analyst at essentially the same time. **E 856,** D34

duplicates—two independent determinations performed by one analyst at essentially the same time. **E 1547,** E15

duplicate sample—one of two (2) representative portions taken from the same sample or sample source. **D 1129,** D19

duplication—the reproduction of a document or part thereof by means of a photoreproductive device. **F 1156,** F12

durability—the measure of the ability of dimension stone to endure and to maintain its essential and distinctive characteristics of strength, resistance to decay, and appearance. Durability is based on the length of time that a stone can maintain its innate characteristics in use. This time will vary depending on the environment, the use, and the finish of the stone in question (for example, outdoor versus indoor use). **C 119,** C18 *durability*—See **chemical durability**. **C 162,** C14

durability, *n*—the property of an article of being resistant to physical or chemical damage, or both, under the usual conditions of service, and of being useful over extended periods of time and use. **C 1145,** C28

durability, *n*—the ability of a material to resist weathering action, chemical attack, abrasion, and other conditions of service. **C 1180,** C12

durability—a general term for permanence or resistance to deterioration. Frequently used to refer to the degree of resistance of a species of wood to attack by wood-destroying fungi under conditions that favor such attack. In this connection the term "decay resistance" is more specific. **D 9,** D07

durability—a relative term indicating degree of permanency. It may be applied to individual protective, decorative, or functional properties, for example, "the durability of gloss," but if used in a general way, for example, "the excellent durability of a paint," implies the ability of the described coating to retain, to the indicated degree, all the properties required for the continued service of the coating. **D 16,** D01

durability, *n—as related to adhesive joints*, the endurance of joint strength relative to the required service conditions. **D 907,** D14

durability—(*1*) *as applied to wood*, its lasting qualities or permanence in service with particular reference to decay.

(*2*) *as applied to the glue bond*, its resistance to deterioration related to exposure conditions. (See also **delamination.**) **D 1038,** D07

durability, *n—of paper*, the capacity of paper or paperboard to resist the effects of wear in performance situations. **D 1968,** D06

durability—See **building performance.** **E 631,** E06

durability—See **building performance**. **E 1480,** E06

durability—the quality of a component to perform as designed for its design life. **E 1705,** E48

durability—the measure of the ability of a material or structure to endure and maintain its essential and distinctive characteristics of strength, resistance to decay, and appearance, with relation to a specific environment of use. **E 1749,** E06

durability, *n*—in weathering, a measure of the retention of original condition and function of a material after exposure to a specified set of conditions. **G 113,** G03

durability "D" cracking, *n*—closely spaced crescent-shaped hairline cracking pattern that initiates adjacent to joints, cracks, or free edges, first manifesting itself at the intersection of joints, cracks, or free edges; dark coloring of the cracking pattern and surrounding area often exists with "D" cracking. **E 867,** E17

durability "D" cracking, *n*—closely spaced crescent-shaped hairline cracking pattern that initiates adjacent to joints, cracks, or free edges, first manifesting itself at the intersection of joints, cracks or free edges; dark coloring of the cracking pattern and surrounding area often exists with "D" cracking. **E 1778,** E17

durable-press, *adj*—having the ability to retain substantially the initial shape, flat seams, pressed-in creases, and unwrinkled appearance during use and after laundering or drycleaning. (See **wash and wear.**) **D 123,** D13

duration of load—the duration of stress or the time during which a load acts on a member. In wood, a design consideration for modifying allowable stresses, based on the accumulated loadings anticipated in the life of a structure. **D 9,** D07

durometer—(*1*) an instrument for measuring the hardness of rubber-like materials. (*2*) a term used to identify the relative hardness of rubber-like materials, for example "low durometer" (relatively soft) or "high durometer" (relatively hard). **C 717,** C24

durometer, *n*—an instrument for measuring indentation hardness. **D 883,** D20

durometer, *n*—an instrument for measuring the indentation hardness of rubber. **D 1566,** D11

durometer—an instrument for measuring hardness, that is, the resistance to the penetration (without puncturing) of the indentor into the surface of rubber or other shoe material. **F 869,** F08

durometer hardness—an arbitrary numerical value which measures the resistance to indentation of the blunt indentor point of the durometer. The value may be taken immediately or after a very short specified time. **F 869,** F08

dust, *n*—a general term, depending upon application, applied to solid particles predominantly larger than colloidal and capable of temporary suspension in air or other gases.

dust loading, n—an engineering term for *dust concentration,* usually applied to the contents of collection ducts and the emissions from stacks.

carpet-embedded dust, n—soil and other particulate matter, approximately 5-μm equivalent aerodynamic diameter and larger, embedded in carpet pile and normally removable by household vacuum cleaners.

surface dust, n—soil and other particulate matter, approximately 5-μm equivalent aerodynamic diameter and larger, adhering to floor surfaces and normally removable by household vacuum cleaners. **D 1356,** D22

dust—an imprecise term referring to particulates capable of temporary suspension in air or other gases; also, particles smaller than an arbitrarily selected size. **D 2652,** D28

dust binder, *n*—a light application of bituminous material for the express purpose of laying and bonding loose dust. **D 8,** D04

dust coat—a relatively thin, sprayed coating of slip. **C 286,** B08

dustfall—See *particle fall* under **particle.** **D 1356,** D22

dusting—(*1*) In dry-process enameling, see *dredging*.

(*2*) A spraying defect characterized by a piling up of almost dry slip in confined areas.

(*3*) The removal of extraneous material from the bisque before firing.

(*4*) See **dry spray.** **C 286,** B08

dusting, *n*—the application of a powder to a rubber surface, generally to prevent adhesion to another surface. **D 1566,** D11

dusting—the developer/toner particles deposited in and around the machine on other than the electrostatic copy. **F 335,** F05

dust loading—See **dust.** **D 1356,** D22

dust pressing, *n*—the process of forming ceramic bodies of 1.5 % or less water content by pressing in a mold. **C 1145,** C28

dust ruffle, *n*—a fabric which (*1*) lies flat over the box spring under the mattress on a bed, and (*2*) has a pleated, tucked, or gather drop to the floor. **D 123,** D13

dutchman—a thin wedge of leather or fiberboard inserted between the insole and outsole of a shoe, or between the lifts of a built-up heel, to throw the foot inward or outward and to correct foot posture. **F 869,** F08

Dutch or Scotch Method—a method of application for asbestos-cement roofing shingles which are rectangular in shape and lap at the top and one side to form either a square or rectangular pattern. **C 460,** C17

Dutch or Scotch method, *n— in shingles*, a method of application for roofing shingles which are rectangular in shape and lap at the top and one side to form either a square or rectangular pattern.

C 1154, C17

Dutch or Scotch method—method of application for asbestos-cement roofing shingles which are rectangular in shape and lap at the top and one side to form either a square or rectangular pattern. **D 2946,** C17

dwarf width—a condition in which the crosswise (of the direction of pultrusion) dimension of a flat surface of the part is less than that the die normally would yield for a particular composite.

NOTE—This condition is usually caused by a partial blockage of the pultrusion die cavity caused by "build-up" or particles of the composite adhering to the cavity surface. This condition is commonly called a "lost edge" when the flat surface has a free edge that is altered by the build-up. **D 3918,** D20

dwelling, *n*—a **building** designed or occupied as the living quarters for one or more families or households.

apartment—a separate part of a **building** intended as a **dwelling unit** for an individual, family, group, or small household (also used as a synonym for **apartment building**).

detached dwelling—a **dwelling unit** standing by itself.

duplex dwelling—one of a pair of **dwelling units,** generally joined by a common floor/ceiling.

modular dwelling—a manufactured **home** consisting completely or in part of **modules.**

semi-detached dwelling—one of a group of **dwelling units** joined by a common sidewall and occasionally by a garage, carport, or similar structure. **E 631,** E06

dwelling, *n*—building designed or occupied as the living quarters for one or more families or households. **E 1605,** E06

dwelling, *n* (logement)—**building** designed or occupied as the living quarters for one or more families or households. **E 1480,** E06

dwelling unit—a unit providing complete independent living facilities for one or more persons, including permanent provisions for living, sleeping, eating, cooking, and sanitation. (See also **house.**) **E 631,** E06

dwelling unit—unit providing complete independent living facilities for one or more persons, including permanent provisions for living, sleeping, eating, cooking, and sanitation. **E 1605,** E06

dwell mark—a fracture surface marking representing the site of a fracture discontinuity caused by a sudden shift in the stress distribution or by a fracture stopping for some length of time; also known as an arrest line. **C 162,** C14

dwell time—the total time that the penetrant or emulsifier is in contact with the test surface, including the time required for application and the drain time. **E 1316,** E07

dwell time—the duration of time beginning at the instant a tribometer's test foot is placed upon the surface to be tested, and ending at the instant a shear force is applied to the sensor. **F 1646,** F13

dye, *n*—a colorant added to an engine coolant to give it a distinctive color. **D 4725,** D15

dye sensitizing—the incorporation of dye-stuffs into a photoconductive coating to alter its spectral response. **F 335,** F05

dyestain, *n*—an area of discoloration due to uneven absorption of a colorant. **D 123,** D13

dye streak, *n*—an unintended stripe in a fabric due to uneven absorption of a colorant. **D 123,** D13

dye sublimination, *n*—an imaging process from thermal dye ribbons, which employ selected organic dyes, and have characteristics different from an imaging process using pigmented materials employed in thermal transfer wax ribbons. **F 1623,** F05

dye tracer—*in grouting*, an additive whose primary purpose is to change the color of the grout or water. **D 653,** D18

dynamic—exhibiting change or movement. **C 717,** C24

dynamic adsorptive capacity—the quantity of a given component adsorbed per unit of adsorbent from a fluid, or fluid mixture moving through a fixed bed at the breakpoint for that component. **D 2652,** D28

dynamical theory—the explanation of diffraction phenomena in terms of dynamical interaction between the incident beam, all scattered waves and the crystal lattice, where the latter is treated as a triply periodic field of potential. **E 7,** E04

dynamic coefficient of friction—the ratio of the parallel component of force applied to a moving body that maintains constant relative motion of two surfaces in physical contact one with another, but otherwise unconstrained, to the normal component of the force—usually the force caused by gravity—applied to the body under clean, dry conditions. **C 242,** C21

dynamic coefficient of friction (DCOF)— a coefficient of friction, μ_d, obtained during relative translation between two contacting solid bodies; used interchangeably with *kinetic coefficient of friction.* **F 1646,** F13

dynamic hold-up, *n*— *in column distillation*, the quantity of liquid held up in the column under normal operating conditions. **D 4175,** D02

dynamic hydroplaning, *n*—hydroplaning of pneumatic tires with separation caused by a thick fluid film due principally to the generation of fluid inertial forces. **E 867,** E17

dynamic intermediate precision—the precision of the measurement of a characteristic determined under dynamic intermediate precision conditions in which the same equipment moving at operating speed measures the characteristic at the same location repeatedly. **E 867,** E17

dynamic leakage measurement—leakage determined by measuring the tracer gas equilibrium partial pressure while the system is actively being pumped. **E 1316,** E07

dynamic leak test—a form of leak test in which some of the tracer gas entering through a leak is continually removed for sensing purposes. **E 1316,** E07

dynamic load—See **load.** **D 996,** D10

dynamic mechanical analysis (*DMA*), *n*—a technique in which the storage modulus (elastic response) and loss modulus (viscous response) of a substance under oscillatory load is measured as a function of temperature, time, or frequency of oscillation, while the substance is subjected to a controlled-temperature program in a specified atmosphere. (ICTAC) **E 473,** E37

dynamic mechanical measurement—a technique in which either the modulus or damping, or both, of a substance under oscillatory load or displacement is measured as a function of temperature, frequency, or time, or combination thereof. **D 4092,** D20

dynamic modulus—see **complex modulus.** **D 4092,** D20

dynamic modulus—see **complex modulus**. **E 1142,** E37

dynamic random access memory, *n*—RAM than cannot be retained without continuous or regular stimulation. (See **DRAM.**) **F 1457,** F05

dynamic range—the difference, in decibels, between the overload level and the minimum signal level (usually fixed by one or more of the noise levels, low-level distortion, interference, or resolution level) in a system or sensor. **E 1316,** E07

dynamic range—a measure of the capability of a test system to accept input signals of varying magnitudes, given by the ratio of the maximum to minimum input signals which at constant gain will produce distortion-free outputs having discernable changes with incremental variations in input.

NOTE—Dynamic range may be stated as the numerical value of the ratio; however, this is usually expressed in decibels.

NOTE—When the output indications can be related to the size of recognized targets, such as flat-bottomed holes, dynamic range is sometimes expressed in terms of the maximum and minimum hole sizes that can be displayed. **E 1316,** E07

dynamic range (for radiology)—the span of signal intensity which defines the system's range of performance. **E 1316,** E07

dynamic sensitivity of leak detector—the minimum leak rate that the detector is capable of detecting while the enclosure under test is actively being evacuated continuously under specified conditions. **E 1316,** E07

dynamic SIMS—SIMS analysis at a primary ion current density and dose such that more than one monolayer of material is removed during the analysis. **E 673,** E42

dynamic tear (DT) energy [J]—the total energy required to fracture standard DT specimens tested in accordance with the provisions of Test Method E 604. **E 1823,** E08

dynamic thermal response curve,, *n*—graphical representation of the response characteristic of a thermal printing system over a given range of print energies; the *y*-axis is optical density and the *x*-axis is print energy (watts/dot or millijoules) or energy density (millijoules/square millimetres). **F 1623,** F05

dynamic vulcanization, *n*—the process of intimate melt mixing a thermoplastic polymer and a suitable reactive rubbery polymer to generate a thermoplastic elastomer with a chemically cross-linked rubbery phase resulting in properties close to those of a thermoset rubber when compared to the same cross-linked composition. **D 1566,** D11

dynamography—the instrumentation method for recording forces. **F 869,** F08

dynamometer—an elastic calibration device used to verify the indicated loads applied by a fatigue testing system. It shall consist of an instrumented member having mass, stiffness, and end displacements such that the inertial effects of the specimen and its attachments to the testing machine for which the verification of loads is desired are duplicated within 5 %. The instrumentation shall permit an accurate determination of the magnitude of the average strain in a region of the uniform transverse cross section when the dynamometer is subjected to a tensile or compressive force along its longitudinal axis, within 1 % of the true strains. A strain gaged specimen is often used as a dynamometer. **E 1823,** E08

dynamometer dynamic loads [F]—the maximum and minimum loads (or the mean load and the load amplitude) that correspond to the readings obtained from the dynamometer output according to an existing static calibration. Such loads are considered true specimen dynamic loads for the purpose of this terminology. **E 1823,** E08

dynamometer range [F]—the range of loads for which the dynamometer may be used for verification purposes. A dynamometer for use in tension and in compression will have two dynamometer ranges, one in tension and one in compression. **E 1823,** E08

dystetic equilibrium—synonymous with **eutectoid equilibrium.** **E 7,** E04

E

EAN bar code—See **bar code**. **F 1156,** F12

EAN International, *n*—abbreviation for the (European) International Article Numbering Association. **F 1294,** F05

EAN symbology, *n*—symbologies specified by the International Article Numbering Association, EAN International. **F 1294,** F05

early stiffening, *n*—rapid loss of plasticity or rapid development of rigidity in freshly mixed hydraulic cement paste, mortar, or concrete.

false set, n—with little evolution of heat, which can be dispelled by further mixing without the addition of water.

flash set, n—with evolution of considerable heat, which cannot be dispelled by further mixing without the addition of water.

C 219, C01

earlywood—the less dense, large-celled, part of the growth layer formed first during the annual growth cycle. A synonym is *springwood.* **D 9,** D07

earth—see **soil**. **D 653,** D18

earthenware—a glazed or unglazed nonvitreous ceramic whiteware. **C 242,** C21

earth material—soil, bedrock, or fill. **D 653,** D18

earth pressure—the pressure or force exerted by soil on any boundary.

	Symbol	Unit
Pressure	p	FL^{-2}
Force	P	F or FL^{-1}

active earth pressure, P_A, p_A—the minimum value of earth pressure. This condition exists when a soil mass is permitted to yield sufficiently to cause its internal shearing resistance along a potential failure surface to be completely mobilized.

earth pressure at rest, P_o, p_o—the value of the earth pressure when the soil mass is in its natural state without having been permitted to yield or without having been compressed.

passive earth pressure, P_p, p_p—the maximum value of earth pressure. This condition exists when a soil mass is compressed sufficiently to cause its internal shearing resistance along a potential failure surface to be completely mobilized.

D 653, D18

ease, *n*—*in garment construction*, the difference between garment measurement and body measurement. **F 1494,** F23

easement—See **railing systems.** **E 631,** E06

easement, *n*—the curved portion of a rail and handrail forming a transition in the vertical plane between the horizontal and inclined sections of a handrail. **E 1481,** E06

ease of use—cumulative effect of drag encountered in application or removal, or both, and the amount of time required to achieve the desired finish. **D 2825,** D21

easily oxidized coals, *n*—low rank coals such as subbituminous or lignitic coals. **D 121,** D05

EBCDIC—**e**xtended **b**inary **c**ode **d**ecimal **i**nterchange **c**ode. **F 1457,** F05

Eberbach—See **micro penetration tester.** **E 7,** E04

EBL—elevated blood level. **E 1605,** E06

ebonite, *n*—a hard material made by sulfur vulcanization of rubber, in which the hardness is substantially obtained by the action of the sulfur. **D 1566,** D11

EC50—a statistically or graphically estimated concentration that is expected to cause one or more specified effects in 50 % of a group of organisms under specified conditions. **E 943,** E47

ECA—an abbreviation for *environmentally controlled area*; an area whose temperature and humidity is controlled within specified limits; the presence of grease, dirt, chemical contaminants, etc., are excluded. **E 1749,** E06

echo—indication of reflected energy. **E 1316,** E07

economic evaluation methods—a set of economic analysis techniques that consider all relevant costs associated with a project investment during its study period, comprising such techniques as life-cycle cost, benefit-to-cost ratio, savings-to-investment ratio, internal rate of return, and net savings. **E 833,** E06

economic life—that period of time over which an investment is considered to be the least-cost alternative for meeting a particular objective. **E 833** **E 631,** E06

economic life—that period of time over which an investment is considered to be the least-cost alternative for meeting a particular objective. **E 833,** E06

ecosystem—organisms and the surrounding environment combined in a community that is self-supporting. **F 1600,** F20

ecotoxicity, *n*—the propensity of a test material to produce adverse behavioral, biochemical, or physiological effects in non-human organisms or populations. **D 6384,** D02

ED50—a statistically or graphically estimated dose that is expected to cause one or more specified effects in 50 % of a group of organisms under specified conditions. **E 943,** E47

eddy current—an electric current developed in a material as a result of induced voltages developed in the material. **A 340,** A06

eddy current—an electrical current caused to flow in a conductor by the time or space variation, or both, of an applied magnetic field. **E 1316,** E07

eddy current testing—a nondestructive testing method in which eddy current flow is induced in the test object. Changes in the flow caused by variations in the specimen are reflected into a nearby coil, coils, or Hall effect device for subsequent analysis by suitable instrumentation and techniques. **E 1316,** E07

edge—a sharp, narrow, steel surface that is attached throughout the length of the sidecut on the bottom edge of the snowboard ski. **F 1107,** F27

edge angle—the included angle between two opposite edges of a hardness indenter. **E 7,** E04

edge clearance—the distance between the bottom of a channel of a lock-strip gasket and the edge of material installed in the channel. **C 717,** C24

edge close-outs, n—members placed around the panel sides to protect the sandwich from damage or to attach the panel to a support or another panel. **C 274,** D30

edge closures—structural members framing the periphery of a sandwich panel providing support and a means of attachment to the panel as well as an environmental seal. **E 1749,** E06

edge contrast, EC, *n*—the difference between the space reflectance (*Rs*) and the adjoining bar reflectance (*Rb*) as calculated by the equation: $EC = Rs - Rb$. **F 1294,** F05

edge cracking, *n*—crescent-shaped cracks or fairly continuous cracks that are located within 0.6 m (2 ft) of the pavement edge. **E 867,** E17

edge cracking, *n*—crescent-shaped cracks or fairly continuous cracks that are located within 0.6 m (2 ft) of the pavement edge. **E 1778,** E17

edge distance [L], *n*—the distance from the edge of a bearing specimen to the center of the hole in the direction of applied force. **E 6,** E28

edge distance ratio, *n*—the ratio of the edge distance to the pin diameter. **E 6,** E28

edge effect—in electromagnetic testing, the disturbance of the magnetic field and eddy currents due to the proximity of an abrupt change in specimen geometry (edge). This effect generally results in the masking of discontinuities within the affected region. (This effect is also termed the **end effect**.) **E 1316,** E07

edge effect—an image defect characterized by a low density or poor fill-in of solid areas coupled with a higher density outline. **F 335,** F05

edge effect—localized structure about the edge of a specimen. **F 1241,** F01

edge irregularity—a variation in the stroke width of a printed character. **F 149,** F05

edge joint—See **joint**. **D 907,** D14

edge (of gypsum board), *n*—the paper-bound edge as manufactured. **C 11,** C11

edge profile—*on edge contoured wafers* (*whose edges have been shaped chemically or mechanically*), a description of the contour of the boundary of the wafer that joins the front and back surfaces. **F 1241,** F01

edge protector—See **loading.** **D 996,** D10

edge spacer, *n*—*in building construction in glazing*, a spacer placed to prevent edge contact and to position laterally a pane of glass or a panel within the supporting frame. **C 717,** C24

edge-stitched seam finish, *n*—*in home sewing*, a seam finish in which machine stitching is placed close to the cut edge of each seam allowance. (Compare **zigzagged seam finish**.) **D 4965,** D13

edge stripping—application of felt strips cut to narrower widths than the normal felt-roll width to cover a joint between flashing and built-up roofing. **D 1079,** D08

edge-to-edge symbologies, *n*—a bar code scheme that can be decoded using edge-to-similar-edge measurements, such as from the start of one bar to the start of another or from the end of one bar to the end of another. Examples are Code 93, Code 128, Code 49 and Code 16K. **F 1294,** F05

edge-tracking (*coatings*)—a residual, discernible pattern in a roller-applied coating characterized by trails from either or both ends of the roller. **D 16,** D01

edge venting—the practice of providing regularly spaced protected openings at a roof perimeter to relieve water vapor pressure in the insulation. **D 1079,** D08

edgewise compressive strength—a term describing the load carrying capacity of flat sandwich constructions when a compressive load is applied uniformly to each facing, usually defined in terms of developed facing stresses as compared to the yield stress of the facings (see Test Method C 364). **E 1749,** E06

edging—grinding the edge of flat glass to a desired shape or size. See also **centering**. **C 162,** C14

edging—(*1*) the process of removing bisque from the edge of a piece of ware to expose the underlying porcelain enamel.

(*2*) The spraying of special slip onto the edge of the ware. **C 286,** B08

edging—the lateral current flow during exposure in the electrostatic copy process. **F 335,** F05

edging—beveling a narrow scarf from leather or fabric, particularly where upper parts are joined by other than lap seams or where upper edges are finished by burnishing. **F 869,** F08

edging brush—a stiff-bristled brush with metal guide, used to remove bisque from edges of ware before the firing operation. **C 286,** B08

edit-accepted, *adj*—having the unadjusted base data compliant with standards of practice. **E 867,** E17

edit-rejected, *adj*—having the unedited base data non-compliant with standards of practice. **E 867,** E17

EDTA—a term used to designate the compound ethylene diamine tetraacetic acid having the structural formula:

```
HOOC—CH2              CH2COOH
          \          /
           NC2H4N
          /          \
HOOC—CH2              CH2COOH
```

or any of its salts that may be specified, used as a sequestering agent. **D 459,** D12

effect diameter (effective size), D_{10}, D_e **(L)**—particle diameter corresponding to 10 % finer on the grain-size curve. **D 653,** D18

effective attenuation length—the average emission function decay length when the emission depth distribution function is sufficiently close to exponential for a given application. **E 673,** E42

effective cadmium cut-off energy (E_c)—the energy at which a specified cadmium container performs like a theoretically perfect filter and, therefore, has the following properties:

(*1*) for all energies below E_c, no neutron reactions occur, and

(*2*) for all energies above E_c, neutron reactions occur at the same rate as if the cadmium were not present. **E 170,** E10

effective carriage mass, *n*—*in CRL-type tensile testing machine*, the force actually applied to a specimen by the mass of the carriage, plus any added masses. **D 123,** D13

effective case depth—the perpendicular distance from the surface of a hardened case to the furthest point where a specified level of hardness is maintained. **F 1789,** F16

effective circuit permeability, μ_{eff}—when a magnetic circuit consists of two or more components, each individually homogeneous throughout but having different permeability values, the effective (overall) permeability of the circuit is that value computed in terms of the total magnetomotive force, the total resulting flux, and the geometry of the circuit. **A 340,** A06

effective crack size, a_e **[L]**—the physical crack size augmented to account for crack-tip plastic deformation. **E 1823,** E08

effective depth penetration (EDP)—in electromagnetic testing, for (a) thickness, the minimum depth beyond which a test system can no longer reliably detect a further increase in specimen thickness, or (b) defects, the limit for reliably detecting metallurgical or mechanical discontinuities by way of conventional continuous wave (CW) eddy current instrumentation and sensors. The EDP point is approximately three times the standard depth of penetration. **E 1316,** E07

effective drainage porosity—see **effective porosity**. **D 653,** D18

effective electrical contact radius, (cm)—*of a spreading resistance probe assembly*, an empirical quantity defined by

$$a = n\rho/4R_s$$

where:

n = number of current-carrying probes across which the potential drop is determined,

ρ = resistivity of a homogeneous semiconductor specimen, Ω · cm, and

R_s = measured spreading resistance, Ω.

F 1241, F01

effective fiber length, *n*—*in vibroscope test for linear density*, that portion of the fiber free to vibrate between fixed supports or holders. **D 123,** D13

effective force, $\bar{F}$ **(F)**—the force transmitted through a soil or rock mass by intergranular pressures. **D 653,** D18

effective gage length, *n—in tensile testing*, the estimated length of the specimen subjected to a strain equal to that observed for the true gage length. **D 123,** D13

effective heat of combustion, *n*—the measured heat release divided by the mass loss for a specified time period.) **E 176,** E05

effective heat of combustion, *n*—the measured heat release divided by the mass loss for a specific time period. **E 176,** E05

effective insulation ratio, *n—in thermal transmittance of textile only*, the increase in insulation afforded by the fabric in comparison to the uncovered test plate under specified conditions of test. **D 123,** D13

effective pathlength (or effective thickness), d_e*—in internal reflection spectroscopy,* the analog of the sample thickness in transmission spectroscopy that represents the distance of propagation of the evanescent wave within an absorbing sample in IRS. It is defined from the relationship: $R = 1 - \alpha d_e$, and is related to the absorption parameter by: $a = \alpha\ d_e$. **E 131,** E13

effective permeability—a hypothetical quantity that describes the magnetic permeability that is experienced under a given set of physical conditions such as a cylindrical test specimen in an encircling coil at a specific test frequency. This quantity may be different from the permeability of the particular metal being tested in that it takes into account such things as the geometry of the part, the relative position of the encircling coil, and characteristics of the magnetic field. **E 1316,** E07

effective porosity (effective drainage porosity), n_e **(D)**—the ratio of: (*1*) the volume of the voids of a soil or rock mass that can be drained by gravity, to (*2*) the total volume of the mass. **D 653,** D18

effective pressure—see **stress.** **D 653,** D18

effective rolling radius, *n*—the ratio of the linear velocity of the wheel center of the free rolling tire in the X′ direction to the spin velocity. **F 538,** F09

effective size—see **effective diameter.** **D 653,** D18

effective size—the particle size, in millimetres, which corresponds to 10 percent finer on the cumulative particle size distribution curve. **D 2652,** D28

effective stress—see **stress.** **D 653,** D18

effective thermal property, *n*—thermal properties derived from heat-conduction theory applied to ignition/flame-spread data treating the material as homogenous in structure. **E 176,** E05

effective thickness B_e **[L]**—for compliance-based extension measurements:

$$B_e = B - (B - B_N)^2/B$$

E 1823, E08

effective unit weight—see **unit weight.** **D 653,** D18

effective unloading slope ratio, *r*—the ratio of an effective unloading slope to that of the initial elastic loading slope on a test record of load versus specimen mouth opening displacement. **E 1823,** E08

effective velocity, *n*—velocity calculated on the basis of arrival times and propagation distances determined by artificial AE generation; used for computed location. **E 1316,** E07

effective yield strength, σ_Y **[FL^{-2}]**—an assumed value of uniaxial yield strength, that represents the influences of plastic yielding upon fracture test parameters. **E 1823,** E08

effect load XX (ELXX), *n*—a statistically or graphically estimated loading rate of test material that is expected to cause one or more specified effects in XX % of a representative subpopulation of organisms under specified conditions. **D 6384,** D02

effervescent atomizer—an internal mixing pneumatic atomizer in which gas bubbles are dispersed in the liquid stream. **E 1620,** E29

efficiency, *n*—a measure of the performance of a collector.

collection efficiency, n—the percentage of a specified substance retained by a gas cleaning or sampling device.

fractional efficiency, n—the mean collection efficiency for specific size fractions of a contaminant. **D 1356,** D22

efficiency—see **total efficiency** and **full-energy peak efficiency.** **E 170,** E10

efficiency, collector—*of a solar thermal collector*, the ratio of the amount of energy removed by the heat transfer fluid to the solar energy incident on the collector.

NOTE—For flat-plate collectors, the value of the incident solar energy used is usually based on gross collector area; for concentrating collectors the value is usually based on the aperture area. **E 772,** E44

efficiency, instantaneous collector—ratio of the amount of energy removed by the heat transfer fluid of a solar collector over a specified time period (usually 5 or 15 min) to the solar energy incident on the collector area in the same period, under steady-state or quasi-steady state.

NOTE—For flat plate collectors, the area used is usually the gross collector area; for concentrating collectors the area used is usually the gross aperture area. **E 772,** E44

efficiency, period system—ratio of the useful energy supplied by the solar energy system over a period of time to the solar energy incident on the collector area of the system in the same period.

NOTE—The period considered has to be of a suitable length for the type of system. For example, it would not be useful to define the efficiency of a solar space heating system over a month in the summer.

NOTE—For flat-plate collector systems, the value of incident solar energy used is usually based on the gross collector area; for concentrating collector systems, the value is usually based on the aperture area. **E 772,** E44

efficient vulcanizing (EV) systems, *n*— as applied to natural rubber and isoprene- and butadiene-based synthetic rubbers, a vulcanizing system making efficient use of sulfur and producing at optimum cure a network containing a preponderance of thermally stable monosulfidic crosslinks. **D 1566,** D11

efflorescence, *n*—crystalline deposit, usually white, of water-soluble compounds on the surface of masonry. **C 1232,** C15

efflorescence, *n*—a condition that occurs when soluble salts in a dry coating or the substrate migrate to the surface due to the movement of water through the film; characterized by a (commonly) white, nonuniform powder or crystalline incrustation, not removable with neutral water but usually removed with dilute mineral acid. **D 16,** D01

efflorescence, *n*—a powdery (usually white) exudation on the surface of a specimen caused by precipitation or crystallization of soluble material that has migrated to the surface. **E 284,** E12

efflorescence (bloom)—a white powdery substance occurring on the surface of asbestos-cement products and caused by the migration of soluble salts, followed by precipitation and carbonation of calcium hydroxide to the surface followed by an atmospheric carbonation. **C 460,** C17

efflorescence (bloom), *n*—a white powdery substance occurring on the surface of products and caused by the migration of soluble salts, followed by precipitation of calcium hydroxide at the surface followed by an atmospheric carbonation. **C 1154,** C17

efflorescence (bloom)—white powdery substance occurring on the surface of asbestos-cement products and caused by the migration of soluble salts, followed by precipitation and atmospheric carbonation of calcium hydroxide, to the surface followed by an atmospheric carbonation. **D 2946,** C17

effluent—the liquid discharge from a process. **F 1600,** F20

efflux time—time required for all grout to flow from a flow cone. **D 653,** D18

1/E fluence—that fluence in the portion of the neutron spectrum produced by moderating media for which the fluence per unit energy, ideally, is inversely proportional to the neutron energy. **E 170,** E10

egg-case nail—coated, regular-stock-steel, 1 ⅛ by 0.072-in. nail with flat 7⁄32-in. head and sharp medium diamond point. **F 547,** F16

eggshell, *adj*—semimatte, having a texture resembling that of the outer surface of the shell of a chicken egg. **E 284,** E12

eggshelling—the texture of a fired glaze similar in appearance to the surface of an eggshell. **C 242,** C21

eggshell or eggshelling—the texture of a fired ceramic coating similar in appearance to the surface of an eggshell. In porcelain enamel, usually a defect. **C 286,** B08

EIFS, n—See **exterior insulation and finish system.** **E 631,** E06

eight-harness satin, *n*—a warp-faced or filling-faced weave illustrating that the entire face of the fabric surface is covered with warp or filling yarn, respectively. **D 123,** D13

ejector, *n*—a device that uses a fluid under pressure, such as steam, air, or water, to move another fluid by developing suction through differential pressure. **D 1356,** D22

ejector air—air used to convey powder from pump to the part being coated. **C 286,** B08

elastic electron scatter—the scatter of electrons by an object without loss of energy, usually an interaction between electrons and atoms. **E 7,** E04

elastic fabric, *n*—a textile fabric made from an elastomer either alone or in combination with other textile materials. (See also **elastomeric yarn**.) **D 123,** D13

elasticity, *n*—that property of a material by virtue of which it tends to recover its original size and shape immediately after removal of the force causing deformation. **D 123,** D13

elasticity—property of material that returns to its original form or condition after the applied force is removed. (ISRM) **D 653,** D18

elasticity—that property of materials that causes them to return to their original form or condition after the applied force is removed. **D 4092,** D20

elasticity—that property of materials that causes them to return to their original form or condition after the applied force is removed, (D 4092, D-20). **E 1142,** E37

elastic limit, *n*—the greatest stress that can be applied to a material without permanent deformation (Compare **yield point**) **D 123,** D13

elastic limit—point on stress strain curve at which transition from elastic to inelastic behavior takes place. (ISRM) **D 653,** D18

elastic limit, *n*—the greatest stress that a material is capable of sustaining, without any permanent strain remaining, upon complete release of the stress. **D 1566,** D11

elastic limit, *n— of paper and paperboard*, the value of paper or paperboard tensile force above which the ratio of the rate of change in the tensile force to the rate of change in length is no longer constant. See **elongation** and **tensile strength**. **D 1968,** D06

elastic limit, *n—in mechanics*, the stress intensity at which stress and deformation of a material subjected to an increasing force cease to be proportional; the limit of stress within which a material will return to its original size and shape when the force is removed, and hence, not a permanent set. **D 4439,** D35

elastic limit, *n—in mechanics*, the maximum stress which can be obtained in a material without causing permanent deformation of the material. (Compare *yield point.*) **D 4848,** D13

elastic limit [FL^{-2}], *n*—the greatest stress which a material is capable of sustaining without any permanent strain remaining upon complete release of the stress. **E 6,** E28

elastic modulus—see **complex modulus and storage modulus.** **D 4092,** D20

elastic modulus—see **complex modulus** and **storage modulus**. **E 1142,** E37

elastic recovery—in hardness testing, the shortening of the original dimensions of the indentation upon release of the applied load. **E 7,** E04

elastic region, *n— of paper and paperboard*, the region of tensile force-elongation behavior of a specific paper or paperboard where the ratio of the rate of change in the tensile force to the rate of change in length is constant. See **elongation** and **tensile strength**. **D 1968,** D06

elastic state of equilibrium—state of stress within a soil mass when the internal resistance of the mass is not fully mobilized. **D 653,** D18

elastic strain energy—potential energy stored in a strained solid and equal to the work done in deforming the solid from its unstrained state less any energy dissipated by inelastic deformation. (ISRM) **D 653,** D18

elastic tape, *n*—a textile tape containing rubber or other elastomers to permit rubber-like stretch in at least one direction. **D 123,** D13

elastic true strain, ϵ_e, *n*—elastic component of the true strain. **E 6,** E28

elastic webbing, *n*—a textile webbing containing rubber or other elastomers to permit rubber-like stretch in at least one direction. **D 123,** D13

elastomer, *n*—a macromolecular material that returns rapidly to approximately its original dimensions and shape after substantial deformation by a weak force and release of the force. **C 717,** C24

elastomer, *n*—a macromolecular material that at room temperature returns rapidly to approximately its initial dimensions and shape after substantial deformation by a weak stress and release of the stress. **D 883,** D20

elastomer, *n*—a macromolecular material that returns rapidly to approximately the initial dimensions and shape after substantial deformation by a weak stress and release of the stress. (R) **D 907,** D14

elastomer—a macromolecular material that returns rapidly to its approximate initial dimensions and shape after substantial deformation by a weak stress and subsequent release of that stress. **D 1079,** D08

elastomer, *n*—a term often used for rubber and polymers that have properties similar to those of rubber. **D 1566,** D11

elastomer, *n*—a natural or synthetic polymer having the property of substantially recovering its size and shape after removal of a deforming force. **D 4175,** D02

elastomer, *n*—a macromolecular material that returns rapidly to approximately its initial dimensions and shape after substantial deformation by a weak stress and release of the stress. (D-11) **F 412,** F17

elastomer—a macromolecular material that at room temperature returns rapidly to approximately its initial dimensions and shape after substantial deformation by a weak stress and release of the stress. (D 883, D-20) **F 869,** F08

elastomer, *n*—a macromolecular material that at room temperature returns rapidly to approximately its initial dimensions and shape after substantial deformation by a weak stress and release of the stress. **F 1251,** F04

elastomer, *n*—a term often used for rubber and polymers that have properties similar to rubber. **F 1494,** F23

elastomeric, *adj*—having the characteristics of an elastomer. **C 717,** C24

elastomeric seal—a material or device that uses an elastomer to effect a seal between separable piping components. **F 412,** F17

elastomeric yarn, *n*—a nontextured yarn which can be stretched repeatedly at room temperature to at least twice its original length and which after removal of the tensile force will immediately and forcibly return to approximately its original length. **D 123,** D13

elbow, *n—in anatomy*, the joint which articulates between the upper arm and the lower arm. **D 123,** D13

elbow, *n—in anatomy*, the joint that articulates between the upper arm and the lower arm. **F 1494,** F23

elbow girth, *n—in body measurements*, with the arm bent at 90° and the clenched fist placed on the hip, the circumference of the elbow. **D 123,** D13

electrical calibration—the calibration of an instrument by use of known quantities of electrical energy. **C 859,** C26

electrical center—the center established by the electromagnetic field distribution within a test coil. A constant intensity signal, irrespective of the circumferential position of a discontinuity, is indicative of electrical centering. The electrical center may be different from the physical center of the test coil. **E 1316,** E07

electrical conductivity—the reciprocal of the resistance in ohms measured between opposite faces of a centimetre cube of an aqueous solution at a specified temperature. **D 1129,** D19

electrical conductivity, *n*—the property of a fluid or solid that permits the passage of an electrical current as a result of an impressed emf. **D 1356,** D22

electrical conductivity of a particulate substance—the current flowing through a unit cross section for an imposed unit potential gradient under specified conditions of packing. **D 2652,** D28

electrical contact, *n*—any physical contact between two or more parts which will permit the flow of electricity between the parts. **C 1145,** C28

electrical discharge—a discontinuous movement of electrical charges through an insulating medium, initiated by electron avalanches and supplemented by secondary processes. **D 2864,** D27

electrically continuous surface—a surface that is electrically conductive in that current can be passed at an applied voltage between any two points of its physical surface. **D 5077,** D10

electrical overstress (EOS)—overstress which may be due to ESD or the operation of items beyond their electrical specifications. **D 5077,** D10

electrical porcelain—vitrified whiteware having an electrical insulating function. **C 242,** C21

electrical steel—a term used commercially to designate strip or sheet used in electrical applications and historically has referred to flat-rolled, low-carbon steels or alloyed steels with silicon or aluminum, or both. Common types of electrical steels used in the industry are grain-oriented electrical steel, nonoriented electrical steel, and magnetic lamination steel. **A 340,** A06

electrical steel, grain oriented—a flat-rolled silicon-iron alloy usually containing approximately 3 % silicon, having enhanced magnetic properties in the direction of rolling and normally used in transformer cores. **A 340,** A06

electrical steel, nonoriented—a flat-rolled silicon-iron or silicon-aluminum-iron alloy containing 0.0 to 3.5 % silicon and 0.0 to 1.0 % aluminum and having similar core loss in all directions. **A 340,** A06

electrical system components, *n*—cable trays, conduits and other raceways, open run cables and conductors, cables, conductors, cabinets, and other components, as defined or used in the National Electrical Code, and air drops. **E 176,** E05

electrical testing facility—a location with qualified personnel, testing equipment, and procedures for the inspection and electrical testing of electrical insulating protective equipment. **F 819,** F18

electric boosting—an auxiliary method of adding heat to the glass in a gas- or oil-fired melter by passing electric current through the molten glass. **C 162,** C14

electric breakdown voltage—see **dielectric breakdown voltage. D 1711,** D09

electric-butt, cold-upset weld—a weld between two prepared ends of adjacent lengths of work-hardenable material that is pressed together in a resistance welding apparatus which melts the abutting ends to unite them and form the weld. In turn, the resulting weld zone is forced out in the form of flash by a cold-upsetting operation which, through cold-working, restores the material substantially to its original tensile strength. **B 354,** B01

electric constant— the same as **permittivity of free space. D 2864,** D27

electric creepage strength—the average voltage gradient under specific conditions of test and for a specific electrode configuration, at which dielectric failure occurs along the interface between a solid insulating material and the fluid in which it is immersed, or at the interface between two solids that are in close physical contact with each other but are not bonded chemically. Dielectric creepage failure may result in tracking. **D 2864,** D27

electric field strength, *n*—the magnitude of the vector force on a point charge of unit magnitude and positive polarity. **D 1711,** D09

electric furnace, *n*—a furnace or kiln in which the main source of heat is provided by electrical means. **C 1145,** C28

electric hygrometer—See **hygrometer. D 1356,** D22

electric hygrometer—an instrument that determines the water vapor content of an atmosphere by measuring the change in resistance or capacitance of hygroscopic material. **D 4023,** D22

electric log—a record or log of a borehole obtained by lowering electrodes into the hole and measuring any of the various electrical properties of the rock formations or materials traversed. **D 653,** D18

electric strength—see **dielectric strength. D 1711,** D09

electric typewriter—a typewriter in which an electromechanical device causes the type element to be activated when the keys are struck. **F 909,** F05

electrification time, *n*—the time during which a steady direct potential is applied to electrical insulating materials before the current is measured. **D 1711,** D09

electrification time—the time during which a steady direct potential is applied to electrical insulating materials before the current is measured. **D 2864,** D27

electrified shearling, (electricified lambskin)—shearling in which the wool has been straightened by a special process. **D 1517,** D31

electrochemical admittance—the reciprocal of the electrochemical impedance, $\Delta I / \Delta E$. **G 15,** G01

electrochemical cell—an electrochemical system consisting of an anode and a cathode in metallic contact and immersed in an electrolyte. (The anode and cathode may be different metals or dissimilar areas on the same metal surface.) **G 15,** G01

electrochemical equivalent—the weight of an element, compound, radical, or ion involved in a specified electrochemical reaction during the passage of unit quantity of electricity, such as a Faraday, ampere-hour, or coulomb. **B 374,** B08

electrochemical impedance—the frequency dependent, complex valued proportionality factor, $\Delta E/\Delta I$, between the applied potential (or current) and the response current (or potential) in an electrochemical cell. This factor becomes the impedance when the perturbation and response are related linearly (the factor value is independent of the perturbation magnitude) and the response is caused only by the perturbation. The value may be related to the corrosion rate when the measurement is made at the corrosion potential. **G 15,** G01

electrochemical noise—fluctuations of potential or current, or both, originating from uncontrolled variations in a corrosion process. **G 15,** G01

electrochemical potential (electrochemical tension)—the partial derivative of the total electrochemical free energy of the system with respect to the number of moles of the constituent in a solution when all other factors are constant. (Analogous to the chemical potential of the constituent except that it includes the electrical as well as the chemical contributions to the free energy.) **G 15,** G01

electrochemistry—the branch of science and technology which deals with transformations between chemical and electrical energy. **B 374,** B08

electrode—a conductor through which current enters or leaves an electrolytic cell, at which there is a change from conduction by electrons to conduction by charged particles of matter, or vice versa. **B 374,** B08

electrode, *n—in emission spectrometry,* either of two terminals between which an electrical discharge occurs. **E 135,** E01

electrode—the energized or grounded conductor portion of electrical test equipment which is placed near or in contact with the material or equipment being tested. **F 819,** F18

electrode clearance—the shortest path from the energized electrode to the ground electrode. **F 819,** F18

electrode gap—not recommended, see **analytical gap.** **E 135,** E01

electrode life—the length of time that an electrode functions usefully. Life-time of solid-state and glass electrodes is limited by mechanical failure of the electrode body or chemical attack on the sensing membrane, and can range from a few days, if the electrode is used continuously in hot or abrasive flowing streams, to several years under normal laboratory conditions. The life-time of liquid membrane electrodes is limited by loss of ion exchanger with use, and is generally 1 to 6 months. **D 4127,** D19

electrode pair—a sensing electrode and a reference electrode; the reference electrode may be separate or combined in one body with the sensing electrode. **D 4127,** D19

electrodeposition—the process of depositing a substance upon an electrode by electrolysis. See **electroforming, electroplating, electrorefining,** and **electrowinning.** **B 374,** B08

electrode potential—the difference in potential between an electrode and the immediately adjacent electrolyte referred to some standard electrode potential as zero.

dynamic E.P.—the electrode potential measured when current is passing between the electrode and the electrolyte.

equilibrium E.P.—a static electrode potential when the electrode and the electrolyte are in equilibrium with respect to a specified electrochemical reaction.

standard E.P.—an equilibrium electrode potential for an electrode in contact with an electrolyte in which all of the components of a specified chemical reaction are in their standard states. The standard state for an ionic constituent is unit ion activity.

static E.P.—the electrode potential measured when no net current is flowing between the electrode and the electrolyte. **B 374,** B08

electrode potential—the potential of an electrode in an electrolyte as measured against a reference electrode. (The electrode potential does not include any resistance losses in potential in either the solution or external circuit. It represents the reversible work to move a unit charge from the electrode surface through the solution to the reference electrode.) **G 15,** G01

electrodermal tracing, *n*—the display of physiological patterns of either skin resistance or skin conductance obtained through exosomatic recording with a galvanograph component. **E 2035,** E52

electrodialysis (ED)—a process in which ions are transferred through membranes from a less concentrated to a more concentrated solution using direct current electric power as the driving force. **D 6161,** D19

electrodialysis reversal (EDR)—same as ED with the addition of a polarity reversal step added to improve performance. **D 6161,** D19

electroforming—the production or reproduction of articles by electrodeposition upon a mandrel or mold that is subsequently separated from the deposit. **B 374,** B08

electro-galvanized, electro-zinc plated—See **zinc-plated.** **F 547,** F16

electrogalvanizing—electrodeposition of zinc coatings. **B 374,** B08

electrogoniometry—an instrumentation system for recording angular displacement at a joint continuously during movement. **F 869,** F08

electrographite, *n— in carbon and graphite technology,* a synonym for manufactured graphite. **C 709,** D02

electrokinetics—involves the application of an electric field to soil for the purpose of dewatering materials of very low permeability to enhance stability. The electric field produces negative pore pressures near a grout pipe that facilitates grout injection. **D 653,** D18

electroless plating—term in use but not recommended. See **autocatalytic plating**. **B 374,** B08

electroluminescence, *n*—luminescence resulting from electrical excitation. (See also **luminescence**.) **E 284,** E12

electrolysis—production of chemical changes by the passage of current through an electrolyte. **B 374,** B08

electrolysis—production of chemical changes of the electrolyte by the passage of current through an electrochemical cell. **G 15,** G01

electrolyte—(*1*) a conducting medium in which the flow of current is accompanied by movement of matter. Most often an aqueous solution of acids, bases, or salts, but includes many other media, such as fused salts, ionized gases, some solids, etc.

(*2*) a substance that is capable of forming a conducting liquid medium when dissolved or melted. **B 374,** B08

electrolyte—a substance that ionizes in aqueous solution; also, a solution containing ions. Weak electrolytes are only slightly dissociated into ions in solution (acetic acid), and strong electrolytes are highly dissociated (HCl, NaCl). **D 4127,** D19

electrolytic capacitor paper, *n*—very pure, porous paper, 17 to 100 μm thick, used to separate the metallic electrodes in electrolytic capacitors. **D 1711,** D09

electrolytic cell—a unit apparatus in which electrochemical reactions are produced by applying electrical energy, or which supplies electrical energy as a result of chemical reactions and which includes two or more electrodes and one or more electrolytes contained in a suitable vessel. **B 374,** B08

electrolytic cell—a unit apparatus in which electrochemical reactions are produced by applying electrical energy, or that supplies electrical energy as a result of chemical reactions and that includes two or more electrodes and one or more electrolytes contained in a suitable vessel. **C 859,** C26

electrolytic cleaning—a process of removing soil, scale, or corrosion products from a metal surface by subjecting it as an electrode to an electric current in an electrolytic bath. **G 15,** G01

electrolytic copper—copper of any origin, refined by electrolytic deposition including electrowinning. Usually when this term is used alone it refers to electrolytic tough pitch copper. This designation applies to the following:
—cathodes that are the direct product of the refining operation.
—electrodeposited copper cast in refinery shapes suitable for hot or cold working or both, and by extension, to fabricators' products made therefrom.
—electrodeposited copper cast into ingots or ingot bars suitable for remelting. **B 846,** B05

electrolytic extraction—See **extraction.** **E 7,** E04

electrolytic polishing—a metallographic preparation procedure where metal is preferentially dissolved from high points on an anodic surface by passage of an electric current through a conductive bath, to produce a specular reflecting surface. Used as an alternative to mechanical polishing. **E 7,** E04

electrolytic powder—powder produced by electrolytic deposition, or by the pulverization of an electrodeposit. **B 243,** B09

electrolytic printer—a facsimile printing device employing a moist, electrolyte-impregnated roll-fed paper. The current (signals) is applied by feeding the paper between a stationary contact and a revolving drum with a helical contractor. The revolving of the drum causes the point of electrical contact between the blade and helical wire to move laterally across the paper, one line per revolution. As the electrical current passes through the paper, it causes a change of coloration at all points of contact. **F 909,** F05

electrolytic process, *n*—the application of a metallic coating on a steel product by passing an electric current through a chemical solution in which the product is immersed; the coating may be applied in a continuous process or a batch process. **A 902,** A05

electromagnet—a soft iron core surrounded by a coil of wire that temporarily becomes a magnet when an electric current flows through the wire. **E 1316,** E07

electro-magnetic field—magnetic forces developed by the passage of an electric current through a conductor (wrapped around an iron (or similar metal or alloy) core). **F 1156,** F12

electromagnetic focusing device—See **focusing device.** **E 7,** E04

electromagnetic interference—See **EMI**. **E 1749,** E06

electromagnetic lens—an electromagnet designed to produce a suitably shaped magnetic field for the focusing and deflection of electrons or other charged particles in electron-optical instrumentation. **E 7,** E04

electromagnetic pulse—See **EMP**. **E 1749,** E06

electromagnetic shield—a screen or other housing placed around a device or circuit to reduce the effects on them from both electric and magnetic fields. **D 5077,** D10

electromagnetic testing—a nondestructive test method for materials, including magnetic materials, that uses electromagnetic energy having frequencies less than those of visible light to yield information regarding the quality of testing material. **E 1316,** E07

electromagnetic vibratory atomizer—a vibratory atomizer in which an electromagnetic transducer transmits high-frequency oscillations to the liquid. **E 1620,** E29

electromigration—(1) Current-induced atomic diffusion in a solid metal due to electron momentum and the potential gradient, (2) electrochemical process of growth of metallic path across an insulating surface under imposed electric field. **B 542,** B02

electromotive force (emf), *n*—the electrical potential difference which produces or tends to produce an electric current. **E 344,** E20

Electromotive Force Series (EMF Series)— a list of elements arranged according to their standard electrode potentials, with "noble" metals such as gold being positive and" active" metals such as zinc being negative. **G 15,** G01

electromotive series—a table that lists in order the standard electrode potentials of specified electrochemical reactions. **B 374,** B08

electromyography—the instrumentation method for recording action potentials from the muscles (MAPS). **F 869,** F08

electron—a subatomic particle having a negative charge of 4.8025×10^{-10} esu, and a charge-to-mass ratio or specific charge of $5.2737 \pm 0.0015 \times 10^{17}$ esu/g. **E 7,** E04

electron:

attenuation length—the average distance that an electron with a given energy travels between successive inelastic collisions as derived from a particular model in which elastic scattering is assumed to be insignificant. With this model, electrons are assumed to be scattered only inelastically and predominantly in the forward direction although, in reality, elastic scattering will modify the electron trajectories. For example, Auger electrons or photoelectrons reaching the surface in surface analysis experiments are thus assumed to have travelled along straight-line paths from their point of generation.

flooding—*in surface analysis,* irradiation of a specimen with low-energy electrons in order to change (generally to reduce) or stabilize the charging potential.

inelastic mean free path—the average of distances, measured along the trajectories, that particles with a given energy travel between inelastic collisions in a substance.

retardation—*AES, XPS,* a method of measuring the kinetic energy distribution by retarding the emitted electrons before or within the analyzer.

spectrometer—see **electron energy analyzer**. **E 673,** E42

electron beam—a stream of electrons in an electron optical system. **E 7,** E04

electron diffraction—the phenomenon, or the technique of producing diffraction patterns through the incidence of electrons upon matter. **E 7,** E04

electro-negative developer—a developer containing negatively charged toner particle. **F 335,** F05

electron energy analyzer—*AES, XPS,* a device for measuring the number of electrons as a function of kinetic energy. (See also **spectrometer**)

pass energy—*XPS, AES,* the mean kinetic energy of electrons in the energy dispersive portion of an electron energy analyzer that will allow them to traverse the analyzer and be counted.

fixed analyzer transmission—*AES, XPS,* a mode of analyzer operation that varies the electron retardation but keeps the pass energy constant in the final analyzer stage.

fixed retarding ratio—*AES, XPS,* a mode of operation in which the electron kinetic energy is analyzed by varying the retarding potential on lens elements preceding the analyzer and the analyzer pass energy so that the analyzer pass energy is a constant fraction of the kinetic energy. **E 673,** E42

electron energy loss spectrum—*XPS,* the energy spectrum of electrons from a monoenergetic electron source after interaction with the specimen, exhibiting peaks due to inelastic loss processes. The spectrum obtained using an incident electron beam of about the same energy as an XPS peak approximates the loss spectrum associated with that XPS peak. Also see **characteristic electron loss phenomena.** **E 673,** E42

electron equilibrium—charged-particle equilibrium for electrons. **E 170,** E10

electron flooding—see **electron.** **E 673,** E42

electron gun—a device for producing and accelerating a beam of electrons. **E 7,** E04

electronic data interchange, *n*—the computer to computer exchange of business information in a standardized format. **A 941,** A01

electronic data interchange, EDI, *n*—the communication of data between business trading partners accomplished in a standard format and syntax. **F 1294,** F05

Electronic Industries Association, EIA, *n*—an abbreviation for a standards organization in the United States specializing in the electrical and functional characteristics of interface equipment. **F 1294,** F05

electronic scanning system, *n*—a device used to look at readable pages and convert the text found there to an electronic signal. (See **ESS.**) **F 1457,** F05

electronic typewriter—an electric typewriter in which the keyboard input is received by an electronic processor built into the typewriter. This unit then controls the print head and other features to produce the typing action. **F 909,** F05

electron image—a representation of an object formed by a beam of electrons focused by an electron optical system (See **image.**) **E 7,** E04

electron lens—a device for focusing an electron beam to produce an image of an object. **E 7,** E04

electron micrograph—a reproduction of an image formed by the action of an electron beam on a photographic emulsion. **E 7,** E04

electron microscopy—the study of materials by means of the electron microscope. **E 7,** E04

electron microscopy impression—See **impression.** **E 7,** E04

electron optical axis—the path of an electron through an electron optical system along which it suffers no deflection due to lens fields. This axis does not necessarily coincide with the mechanical axis of the system. **E 7,** E04

electron optical system—a combination of parts capable of producing and controlling a beam of electrons to produce an image of an object. **E 7,** E04

electron optics—the science that deals with the propagation of electrons, as light optics deals with that of light and its phenomena. **E 7,** E04

electron probe—a narrow beam of electrons used to scan or illuminate an object or screen. **E 7,** E04

electron retardation—see **electron.** **E 673,** E42

electron spectrometer—see **electron energy analyzer.** **E 673,** E42

electron standard field—an electron field whose particle energy and direction, spatial uniformity, and particle fluence rate uniformity are well established and reproducible. **E 170,** E10

electron trajectory—the path of an electron. **E 7,** E04

electron velocity—the rate of motion of an electron. **E 7,** E04

electron volt—the kinetic energy gained by an electron after passing through a potential difference of 1 V. **E 1316,** E07

electron wavelength—the wavelength necessary to account for the deviation of electron rays in crystals by wave diffraction theory. It is numerically equal to the quotient of Planck's constant divided by the electron momentum. This is approximately $\lambda = (12.3/V)$ *A.*, where V = the accelerating potential in volts. **E 7,** E04

electro-optic (sensor)—device that converts optical energy to electrical signals. **F 1156,** F12

electrophoresis—the movement of colloidal particles produced by the application of an electric potential. **B 374,** B08

electrophoresis, *n*—the movement of colloidal particles or macromolecules through a solution under the action of an electromotive force applied through electrodes in contact with the solution. **C 1145,** C28

electrophoresis—migration of dispersed solid, liquid or gaseous material to one of two electrodes under the influence of an impressed direct-current voltage. **D 2652,** D28

electrophoretic deposition—the process of depositing material on a workpiece from a porcelain enamel slip suspension due to the movement of particles under the influence of an impressed direct current voltage. **C 286,** B08

electrophotographic copying process—a process in which a copy of an original document results from the combined action of light and externally applied electric forces. **F 335,** F05

electrophotographic material—photoconductor suitable for application in photographic processes. **F 335,** F05

electro-photographic (photography)—creation of an image onto a substrate by generation of a latent image on a photoreceptive surface. **F 1156,** F12

electrophotographic printer—a nonimpact printing technique that is similar to the technology employed in typical office copiers, which forms a copy by attracting toner particles to a static charge on the surface of a photoconductor, then transferring the toner image to the surface of a sheet of paper. In the normal office copier, the charged image (latent image) of the original document is formed on the photoconductor simply through exposure of the photoconductor to reflected light from the document. In an electrophotographic printer, the image is formed by a light-source (laser, CRT, LED, LCS, laser diode, or other controlled light source) which erases or discharges a static image charge on the photoconductor according to information being supplied through the input data stream. Each bit of data can be related to a character shape in the memory of the printing system, and in most cases characters are formed by a dot matrix method similar in concept to that of the matrix printer. Paper can be sheet- or roll-fed or continuous form. **F 909,** F05

electroplated—surface provided with usually thin electrochemical deposit of brass, cadmium, copper, nickel, tin, zinc, etc., as a result of immersion in electrolytic bath. **F 547,** F16

electroplated button, *n*—plastic buttons which have been made conductive by chemical treatment followed by the electroplating of metallic coatings. **D 123,** D13

electroplating—the electrodeposition of an adherent metallic coating upon an electrode for the purpose of securing a surface with properties or dimensions different from those of the basis metal. **B 374,** B08

electropolishing—the improvement in surface finish of a metal effected by making it anodic in an appropriate solution. **B 374,** B08

electro-positive developer—a developer containing positively charged toner particle. **F 335,** F05

electrorefining—the process of anodically dissolving a metal from an impure anode and depositing it cathodically in a purer form.

B 374, B08

electrosensitive printer—a nonimpact printing device employing a double layer coated paper. Imaging is accomplished by touching or nearly touching the paper surface with an electrically charged stylus which burns away the top coating, exposing an undercoating of a contrasting color to produce the desired images. This process is also known as electroerosion printing. **F 909,** F05

electrostatic atomizer—a device wherein an electric charge is the primary source of energy utilized to produce a spray.
emitting spray angle, n see initial spray angle. **E 1620,** E29

electro-static charged—charge, often negative, induced or transferred onto a surface. **F 1156,** F12

electrostatic copying process—a process in which a copy of an original document results from charged particle development of an electrostatically formed image. **F 335,** F05

electrostatic decay half-life, *n—in textiles,* the time in minutes for the maximum voltage induced on the textile to be reduced to one half of the maximum voltage by the various decay mechanisms: conduction and ionization of the air. **D 123,** D13

electrostatic discharge—the transfer of electrostatic charge between bodies at different electrostatic potentials. **D 5077,** D10

electrostatic discharge (ESD) protective—a property of materials capable of one or more of the following:
(1) preventing the generation of static electricity.
(2) dissipating electrostatic changes over its surface or volume.
(3) providing shielding from ESD or electrostatic fields.
D 5077, D10

electrostatic discharge sensitive (ESDS)—a property of items in which they are inherently sensitive (ESDS) susceptible to either catastrophic failure or latent damage when exposed to sources of ESD. Items are often categorized as to their levels of sensitivity but in all cases require some means of ESD protective packaging and handling. **D 5077,** D10

electrostatic focusing device—See **focusing device.**
E 7, E04

electrostatic immersion lens—See **immersion objective**
E 7, E04

electrostatic lens—a lens producing a potential field capable of deflecting electron rays to form an image of an object.
E 7, E04

electrostatic master—a master imaged electrostatically.
F 425, F05

electrostatic powder porcelain enamel— a mixture comprised of frit and additives ground and/or blended together to form a powder suitable for dry electrostatic application. **C 286,** B08

electrostatic precipitation—See **precipitation.** **D 1356,** D22

electrostatic precipitator—See **precipitator.** **D 1356,** D22

electrostatic propensity, *n*—the capacity of a nonconducting material to acquire and hold an electrical charge by induction (by means of corona discharge) or by triboelectric means (rubbing with another material). **D 123,** D13

electrostatic retention—the tenacity with which a charged, electrostatically deposited powder porcelain coating adheres to the work piece before it is fired. **C 286,** B08

electrostatics—the science of forces and fields of electric charges in a state of rest. **F 335,** F05

electrostatic shield—a barrier or enclosure that prevents the penetration of an electrostatic field. **D 5077,** D10

electrostatic spraying—a technique for attaining a uniform coating in which the material sprayed is given an electrical charge.
E 1316, E07

electrotyping—the production of printing plates by electroforming.
B 374, B08

electrowinning—the production of metals by electrolysis with insoluble anodes in solutions derived from ores or other materials.
B 374, B08

element, *n—in zippers,* a device designed for interlocking, capable of being affixed along the edge of a tape. (Compare **continuous element.**) **D 123,** D13

element—the component containing the membrane, generally replaceable, such as a spiral wound cartridge. **D 6161,** D19

element, *n*—an isolated (totally bounded by void) projection.
F 538, F09

element—the interchangeable type font of an impact printing device: see **thimble, ball, daisy wheel.** **F 909,** F05

element, *n—in coded symbologies,* a single bar or space in linear or stacked configurations or a single cell in matrix (2-D) configurations. **F 1294,** F05

element edge, *n*—the location where the scan reflectance profile intersects the midpoint between the space reflectance (Rs) and bar reflectance (Rb) of adjoining elements. **F 1294,** F05

element printer—an impact printer that generates copy via interchangeable "elements" that each contain a full set of characters. Characters are formed when the element strikes the paper itself through an ink ribbon. **F 909,** F05

element width, *n*—the thickness of an element measured from the leading edge of an element to the trailing edge of the same element. (See **X dimension.**) **F 1294,** F05

elevated blood level (EBL)—lead content in blood that exceeds the safe level established by regulation/local jurisdiction.
E 1605, E06

elevated temperature testing—tests on plastic pipe above 23°C (73°F). **F 412,** F17

elevator, *n*—a general term describing a mechanical device on the Motion Control, Inc. Fiber Information System which moves the specimen clamp while preparing the specimen or while taking a length/uniformity index or strength/elongation measurement.
D 123, D13

elevator—synonym for **bin**, commonly used in the grain industry.
D 653, D18

elimination—this, like the *definite conclusion of identity,* is the highest degree of confidence expressed by the document examiner in handwriting comparisons. By using this expression the examiner denotes no doubt in his opinion that the questioned and known writings were not written by the same individual.
Examples—It has been concluded that the John Doe of the known material did not write the questioned material, or it is my opinion (or conclusion or determination) that the John Doe of the known material did not write the questioned material. **E 1658,** E30

elk leather—trade term used to designate chrome-tanned cattlehide for uppers of work shoes, hunting boots, some children's shoes, and others requiring flexibility and durability. More properly **elk-finished cowhide.** Leather made from elkhide is known as "buckskin." **D 1517,** D31

elliptical reinforcement—a line of reinforcement in the approximate shape of an ellipse. **C 822,** C13

ellipticity—*in optics, of elliptically polarized light,* the angle *x* given by the inverse tangent of the ratio of the minor to the major axis of the ellipse described by the electric vector of the light.
F 1241, F01

Elmendorf test—See **package testing.** **D 996,** D10

elongated grain—a grain with one principal axis significantly longer than either of the other two. **E 7,** E04

elongated piece *(of aggregate), n*—a particle of aggregate for which the ratio of the length to width of its circumscribing rectangular prism is greater than a specified value (see also **flat piece** (*of aggregate*)). **C 125,** C09

elongation, *n*—extension produced by a tensile stress. **C 717,** C24

elongation, *n*—the ratio of the change in length of a rope during application of tension to the original length of the rope when new. **D 123,** D13

elongation, *n*—the ratio of the extension of a material to the length of the material prior to stretching. (Compare **extension.**) **D 123,** D13

elongation, *n*—extension produced by a tensile stress. **D 1566,** D11

elongation, *n*—*of paper and paperboard,* See **stretch**. **D 1968,** D06

elongation—extension produced by a tensile stress. (D 1566, D 11) **F 869,** F08

elongation—the increase in length of the gage length expressed as a percentage of the original gage length. **F 1789,** F16

elongation at break, *n*—the elongation corresponding to the breaking force. (Compare **elongation at rupture.** See also **elongation.**) **D 123,** D13

elongation at break, *n*—the elongation corresponding to the breaking load, that is, the maximum load. **D 4439,** D35

elongation at breaking load, *n*—*in fiber strength testing of cotton,* the elongation corresponding to the maximum load, and expressed as a percentage of the ⅛-in. (3.2-mm) gage length. **D 123,** D13

elongation at rupture, *n*—the elongation corresponding to the force-at-rupture. (Compare **elongation at break.**) **D 123,** D13

elongation at specified force, (EASF), *n*—the elongation associated with a specified force on the force-extension curve. **D 123,** D13

elongation *El*, *n*—the increase in gage length of a body subjected to a tension force, referenced to a gage length on the body. Usually elongation is expressed as a percentage of the original gage length. **E 6,** E28

elongation, percent, *n*—the extension of a uniform section of a specimen expressed as percent of the original length. **D 1566,** D11

elongation, ultimate, *n*—the elongation at the time of rupture. **D 1566,** D11

eluant—a liquid used to extract one material from another, as in chromatography. **E 1316,** E07

elute, *v*—to remove sorbed materials from a sorbent by means of a fluid. **D 1356,** D22

EM—abbreviation for **end of message**. **F 1457,** F05

embedding, *n*—a general term for all methods of surrounding or enclosing components and assemblies with a substantial thickness of electrically insulating solid or foam material with voids and interstices between the parts substantially filled. See **potting, encapsulation,** and **dip encapsulation.** **D 1711,** D09

embedding compound—see **taping compound.** **C 11,** C11

embedment—(*1*) the process of pressing a felt, aggregate, fabric, mat, or panel uniformly and completely into hot bitumen or adhesive to ensure intimate contact at all points;

(*2*) the process of pressing granules into coating in the manufacture of factory-prepared roofing, such as shingles. **D 1079,** D08

embedment—the placement of materials completely around the pipe to provide support. **F 412,** F17

embossed—decorated in relief on the surface of the ware. **C 242,** C21

embossed, *adj*—having a permanent multilevel surface produced by mechanical or chemical means. **F 141,** F06

embossed—any deformation of the back of the acceptor media due to imaging, usually by type action. **F 221,** F05

embossed head—head with raised or impressed figure or design. **F 547,** F16

embossed leather—leather that has been ornamented with a geometrical or fancy design by heavy pressure in a machine. **D 1517,** D31

embossing, *n*—the tendency of a paper to take a permanent pattern, either depressed or raised, as a result of contact with the blanket during printing. **[TAPPI]** **D 6488,** D01

embossing—imprinting under heavy pressure any design on the surface of the upper material. **F 869,** F08

embossment—a decoration in relief or excised on the ware surface. **C 242,** C21

embossment—the height of raised print or raised surface on a document. **F 149,** F05

embrittlement, *n*—the loss or partial loss of ductility in a steel, such that failure is characteristically by fracture without appreciable deformation. **A 902,** A05

embrittlement—the reduction of the normal ductility in a metal due to a physical or chemical change. As it relates to these test methods, embrittlement is the loss of ductility caused by the reaction of cuprous oxide in the copper product when exposed at elevated temperatures to a reducing atmosphere. **B 846,** B05

embrittlement, *n*—the formation of a brittle residue as the result of pyrolysis or incomplete combustion. **D 123,** D13

embrittlement—the formation of a brittle residue as a result of pyrolysis or incomplete combustion. **F 1494,** F23

embrittlement—the severe loss of ductility or toughness or both, of a material, usually a metal or alloy. **G 15,** G01

embrittlement, hydrogen—See **hydrogen embrittlement.** **B 374,** B08

emergency medical dispatcher (EMD)—a trained public safety telecommunicator with additional training and specific emergency medical knowledge essential for the efficient management of emergency communications. **F 1177,** F30

emergency medical facility—a physical structure, excluding mobile vehicles, which has been approved by the appropriate regulatory authority to receive emergency patients and which is equipped and staffed to evaluate and treat patients with life threatening conditions. **F 1177,** F30

emergency medical services—the provision of services to patients requiring immediate assistance due to illness or injury, including access, response, rescue, prehospital and hospital treatment, and transportation. **F 1177,** F30

emergency medical services (EMS) system—a coordinated arrangement of resources (including personnel, equipment, and facilities) organized to respond to medical emergencies, regardless of the cause. **F 1177,** F30

emf (electromotive force)—an electrical potential. **B 374,** B08

EMI—an abbreviation for *electromagnetic interference*; caused by electric and magnetic fields that emanate from a wide range of electrical circuitry. **E 1749,** E06

emission, *n*—release of radiant energy. **E 349,** E21

emission, burst—a qualitative description of the discrete signal related to an individual emission event occurring within the material.

NOTE—Use of the term burst emission is recommended only for describing the qualitative appearance of emission signals. **E 1316,** E07

emission, continuous—a qualitative description of the sustained signal level produced by rapidly occurring acoustic emission from structural sources, leaks, or both.

NOTE— Use of the term *continuous emission* is recommended only for describing the qualitative appearance of emission signals. **E 1316,** E07

emission depth distribution function (for a measured signal)—for particles or radiation emitted from a surface in a given direction, the probability that the particle or radiation leaving the surface in a specified state originated from a specified depth measured normally from the surface into the material. **E 673,** E42

emission function decay length—the negative reciprocal slope of the logarithm of the emission depth distribution function at a specified depth. **E 673,** E42

emission microscope—a type of electron microscope in which the specimen is the cathode source of the electrons. Sometimes used synonymously with shadow microscope. **E 7,** E04

emission mixture, *n*—the total mixture in the outside atmosphere of emissions from all sources. **D 1356,** D22

emission (or absorption) spectroscopy—measurement of energy spectrum emitted by or from, or absorbed by, an object under some form of energetic stimulation; for example, light, electrical discharge (heating), etc. **F 1156,** F12

emissions, *n*—substances discharged into the air from a stack, vent, or other discrete source.

emission rate, n—the mass emitted per unit of time from a source or, alternatively, per unit of material or energy produced or consumed by a process. **D 1356,** D22

emission spectrometry, *n*—the branch of spectroscopy pertaining to the emission of electromagnetic radiation by atoms, ions, radicals, and molecules. **E 135,** E01

emissive power—the use of the term emissive power is discouraged in favor of the preferred term **radiant exitance**. **E 772,** E44

emissivity—the ratio of the radiation given off by the surface of a body to the radiation given off by a perfect black body at the same temperature. **C 242,** C21

emissivity, *n*—the ratio of the radiation given off by the surface of a body to the radiation given off by a perfect black body at the same temperature. **C 1145,** C28

emissivity, *n*—the ratio of the power per unit area radiated from a material's surface to that radiated from a black body at the same temperature. **E 176,** E05

emissivity, ϵ—the ratio of the radiance of a body at a given temperature to the corresponding radiance of a blackbody at the same temperature. **E 1316,** E07

emissivity of a thermal radiator, *n*—ratio of the thermal radiant exitance of the radiator to that of a full radiator at the same temperature. Symbol: ϵ, $\epsilon = M_{e,th}/Me(\epsilon = 1)$.

NOTE—Formerly *"pouvoir émissif"* (fr.). **E 349,** E21

emittance, ϵ, *n*—the ratio of the radiant flux emitted by a specimen to that emitted by a blackbody at the same temperature and under the same conditions. **C 168,** C16

emittance, ϵ— for a sample at a given temperature, ratio of the radiant flux emitted by a sample to that emitted by a blackbody radiator at the same temperature, under the same spectral and geometric conditions of measurement. See **radiometric properties and quantities**. **E 772,** E44

emittance, directional $\epsilon(\theta; \phi)$, *n*—the ratio of the radiance from a surface in a particular direction to the radiance from a blackbody at the same temperature under the same conditions. **C 168,** C16

emittance, hemispherical ϵ_H or $\epsilon(2\pi)$, *n*—the average directional emittance over a hemispherical envelope covering a surface. **C 168,** C16

emittance, spectral ϵ_λ or $\epsilon(\lambda; \theta;\phi)$, *n*—an emittance based on the radiant energy emitted per unit wavelength interval (monochromatic radiant energy). **C 168,** C16

emittance, total ϵ_T or $\epsilon(t)$, *n*—an emittance that is an integrated average over all wavelengths of radiant energy emitted. **C 168,** C16

EMP—an abbreviation for *electromagnetic pulse*; a sudden intense discharge of electromagnetic energy that occurs naturally as a result of lightning discharge and can be induced by near-surface or high-altitude nuclear explosions. **E 1749,** E06

emplacement—the establishment of contaminant residence in the vadose zone in a particular phase. **D 653,** D18

EMS region—a defined geographic area used for EMS planning, development, and coordination. **F 1177,** F30

emu—the notation emu is an indicator of electromagnetic units. When used in conjunction with magnetic moment, m, it denotes units of ergs per oersted, erg/Oe. A moment of 1 erg/Oe is produced by a current of 10 amperes (1 abampere) flowing in a loop of area 1 cm^2. The work done to rotate a moment of 1 erg/Oe from parallel to perpendicular in a uniform field of 1 Oe is 1 erg. The conversion to the SI units of magnetic moment J/T (joule/tesla) or A m^2 is given by

$$\frac{erg/Oe\ (cgs-emu)}{J/T\ (SI)} \equiv \frac{10\ amperes\ cm^2\ (cgs-emu)}{A\ m^2\ (SI)} = 10^{-3}$$

Magnetization, M, the magnetic moment per unit volume, has units erg/(Oe-cm^3), often expresssed as emu/cm^3. **A 340,** A06

emulation, *n*—a technique where one device (in this case, a printer) is expected to behave exactly like another. **F 1457,** F05

emulsifiable concentrate—a single-phase liquid system having the property of forming an emulsion when mixed with water. **E 609,** E35

emulsification, *n*—the process of dispersing an immiscible liquid in another liquid. **C 1145,** C28

emulsification time—the time that an emulsifier is permitted to remain on the part to combine with the surface penetrant prior to removal. Also called emulsification dwell time. **E 1316,** E07

emulsifier—a substance that modifies the surface tension of colloidal droplets, keeping them from coalescing, and keeping them suspended. **D 653,** D18

emulsifier—a liquid that interacts with an oily substance to make it water-washable. **E 1316,** E07

emulsifier—a surfactant that promotes the suspension of one immiscible liquid in another. **E 1519,** E35

emulsifier, hydrophilic—a water-based liquid used in penetrant examination which interacts with the penetrant oil rendering it water-washable. **E 1316,** E07

emulsifier, lipophilic—an oil based liquid used in penetrant examination which interacts with the penetrant oil rendering it water-washable. **E 1316,** E07

emulsifying agent—a substance that increases the stability of an emulsion. **B 374,** B08

emulsifying agent—a material that increases the stability of a dispersion of one liquid in another. **D 459,** D12

emulsifying agent—a surfactant that promotes the suspension of one liquid in another. **E 609,** E35

emulsifying agent (latex), *n*—a surface-active substance used to facilitate the dispersion of an immiscible liquid compounding material in another liquid and to stabilize the emulsion thereby produced. **D 1566,** D11

emulsion—a suspension of fine particles or globules of one or more liquids in another liquid. **B 374,** B08

emulsion—a suspension of fine particles or globules of one or more liquids in another liquid. **D 459,** D12

emulsion—a system containing dispersed colloidal droplets. **D 653,** D18

emulsion, *n*—a two-phase liquid system in which small droplets of one liquid (the internal phase) are immiscible in, and are dispersed uniformly throughout, a second continuous liquid phase (the external phase). (See also **adhesive dispersion**.) **D 907,** D14

emulsion—an intimate mixture of bitumen and water, with uniform dispersion of the bitumen or water globules, usually stabilized by an emulsifying agent or system. **D 1079,** D08

emulsion—a two-phase liquid system in which small droplets of one liquid (the internal phase) are immiscible in, and are dispersed uniformly throughout, a second, continuous liquid phase (the external phase). **D 2507,** F07

emulsion—a suspension of fine particles or globules of one or more liquids in another liquid. **E 609,** E35

emulsion calibration curve—the plot of a function of the relative transmittance of the emulsion versus a function of the exposure. **C 859,** C26

emulsion calibration curve, *n—in photographic optical emission spectrometry,* the plot of the degree of blackening of the developed photographic emulsion as a function of the intensity of the spectral line to which it has been exposed. **E 135,** E01

emulsion cleaner—a composition which forms an emulsion capable of dissolving or suspending soil. **D 459,** D12

emulsion paint—under **paint,** see *emulsion paint.* **D 16,** D01

emulsion sealant—See **latex sealant.** **C 717,** C24

emulsion side, *n*—the image-bearing side of a film master. (See also the opposite **base side**.) **F 1294,** F05

enamel—see **porcelain enamel.** **C 286,** B08

enamel—a paint that is characterized by an ability to form an especially smooth film. **D 16,** D01

enamel, aluminum—see **aluminum enamel.** **C 286,** B08

enamel, beading—see **beading enamel.** **C 286,** B08

enamel, blackboard— see **chalkboard enamel.** **C 286,** B08

enamel, cast iron— see **cast iron enamel.** **C 286,** B08

enamel, chalkboard— see **chalkboard enamel.** **C 286,** B08

enamel, copper—see **copper enamel.** **C 286,** B08

enameled—coated with enamel of desired color and often baked. **F 547,** F16

enameling iron—a very low-carbon, low-metalloid, cold-rolled sheet steel, produced specifically for use as a base metal for porcelain enamel. **C 286,** B08

enamel, jewelers'— see **jewelers' enamel.** **C 286,** B08

enamel, reclaim—see **reclaim.** **C 286,** B08

enamel scrapings—see **scrapings.** **C 286,** B08

enantiotropic transformation—a reversible metastable phase change; for example, the freezing of sulfur directly to the rhombic phase, or the direct melting of the latter, without passing through the stable intermediate monoclinic phase. **E 7,** E04

encapsulation, *n*—an embedding process utilizing removable molds or other techniques in which the insulating material forms the outer surfaces of the finished unit. **D 1711,** D09

encapsulation—see **lead-based paint encapsulation.** **E 1605,** E06

encasement—special materials, their placement and configuration which are designed to fully surround the pipe, and develop a field supporting strength which exceeds that developed by other commonly used installation and bedding techniques. **C 896,** C04

encasement, *n*—see **incasement,** *n.* **F 412,** F17

encasement, *v*—see **incasement,** *v.* **F 412,** F17

encephalopathy, *n*—damage to the brain that is caused by severe lead toxicity that is capable of becoming permanent brain damage. **E 1605,** E06

encircling coils—in electromagnetic testing, coil(s) or coil assembly that surround(s) the part to be tested. Coils of this type are also referred to as annular, circumferential, or feed-through coils. **E 1316,** E07

enclosed carbon arc, *n*—a light source in which an arc is produced across a pair of carbon rods by a high energy electrical source, such that a high intensity light is emitted. The carbons are enclosed in an inverted glass dome which acts to prolong the life of the carbons, and to modify the spectral power distribution received by the specimens. **G 113,** G03

encryption—encoding. **F 1156,** F12

end, *n*—an individual sliver, roving, yarn, or cord. **D 123,** D13

end, *n—in fabric,* an individual warp yarn (single or ply) or cord. **D 123,** D13

end, *n—in fibrous composites,* a general term for a continuous, ordered assembly of essentially parallel, collimated filaments, with or without twist. **D 3878,** D30

end-centered—having an atom (or group of atoms) separated by a translation of the type ½, ½, 0 from a similar atom (or group of atoms). The number of atoms in an end-centered cell must be a multiple of two. **E 7,** E04

end connector—device permanently attached to the boom used for joining boom sections to one aother or to other accessory devices. **F 818,** F20

end-construction tile, *n*—tile designed to receive its compressive stress parallel to the axes of the cells. **C 43,** C15

end effect—see **edge effect.** **E 1316,** E07

end effect—the loss in sensitivity to discontinuities located near the extreme ends of the tube as the ends of the tube enter or leave the test coil. (E 215) **E 1316,** E07

end-fired furnace—a melter with fuel supplied from the back wall. **C 162,** C14

end-grain nailing—nailing in such a way that the point of the nail follows the grain of the wood so that the shank is parallel, or nearly parallel to the grain in that member holding the nail point. It is weaker than side-grain nailing in direct withdrawal. **D 996,** D10

end groups—the functional groups at either end of the cellulose molecule. It is frequently used for the terminal glucose units that contain these groups. It is usually clear from the context which is meant, and the differentiation is seldom important. One end group is reducing (probably a hemiacetal) and one nonreducing (a free hydroxyl in the 4-position). **D 1695,** D01

end match—matched lumber with a tongue on one end of the piece and a matching groove on the opposite end. **D 9,** D07

end (of gypsum board), *n*—the end perpendicular to the paper-bound edge. The gypsum core is always exposed. **C 11,** C11

end of message, *n*—an electronic signal from the sender to the receiver indicating the last byte has been transmitted. (See **EM**.) **F 1457,** F05

end of text, *n*—an electronic signal sent to indicate the end of the printable body of the message, it may be followed by control bytes. (See **ETX**.) **F 1457,** F05

endothermic—pertaining to a reaction that occurs with the adsorption of heat. **D 653,** D18

endothermic reaction, *n*—a chemical reaction in which heat is absorbed. **C 1145,** C28

end out, *n*—a void caused by a missing warp yarn. **D 123,** D13

end point—in chemical analysis, that stage during a titration at which a marked change is observed. This could be a change in indicator color, potential, or some other indicating parameter. **C 859,** C26

end point—the occurrence in the effluent of the maximum permissible concentration of an adsorbate of interest. **D 2652,** D28

end point inflection—the largest change in observed property per unit quantity of titrant. This is indicative of the end point. **C 859,** C26

end-port furnace—synonymous with **end-fired furnace**. **C 162,** C14

end post—the terminal post at the end of a line of fence. **F 552,** F14

end product—the chemical compound resulting from a particular metabolic process. **F 1600,** F20

end rail clamp—a two-piece clamp with carriage bolt and nut designed for horizontal rail connections to an end post. **F 552,** F14

ends—straight pieces, shorter than the nominal length, left over after cutting the product into mill lengths, stock lengths, or specific lengths. They are subject to minimum length and maximum weight requirements. **B 846,** B05

endurance limit—the maximum stress below which a bolt or screw can presumably endure an infinte number of strees cycles. **F 1789,** F16

endurance, thermal, *n*—the ability of a ceramic product to withstand thermal shock or to withstand deterioration during exposure to high temperatures. **C 1145,** C28

end-use, *n*—*in wear testing*, the use for which a textile is intended. **D 123,** D13

end user, *n*—*for the purpose of this guide*, this term is used to identify specifically the party requiring protective clothing (for example, the employer of the person wearing the garment). **F 1494,** F23

energizing cycle—the application of a magnetizing force to a conductor. **E 1316,** E07

energy:

per channel—EIA, energy differences between two successive channels.

edge—EIA, values of the backscattering energy in a BS spectrum for an element (or isotope) that is located at the surface of the specimen.

loss—EIA, energy dissipated by the particles of the incident beam as they penetrate through the specimen.

of incident beam—average energy of analyzing particle in the incident beam at the moment of impact.

pass—(See *pass energy* under **electron energy analyzer**.)

surface approximation—EIA, see **surface energy approximation.** **E 673,** E42

energy absorption—the dissipation or transfer of energy due to motion into heat energy. **F 869,** F08

energy, acoustic emission event—the total elastic energy released by an emission event. **E 1316,** E07

energy, acoustic emission signal—the energy contained in a detected acoustic emission burst signal, with units usually reported in joules and values which can be expressed in logarithmic form (dB, decibels). **E 1316,** E07

energy calibration—a process of establishing the relationship between photon or particle energy and channel number in the spectrometer. The energy calibration may be as simple as building a table of two or more energy-channel pairs or as complex as using a least squares algorithm to establish a function describing the energy versus channel relationship. **E 170,** E10

energy efficiency—the product of the current efficiency and the voltage efficiency for a specified electrochemical process. **B 374,** B08

energy equivalent, *n*—the energy required to raise the temperature of a calorimeter system 1°C (or 1°F) per gram of sample. **D 5681,** D34

energy equivalent, heat capacity, or water equivalent, *n*—the energy required to raise the temperature of the calorimeter an arbitrary unit. This is the quantity that, when multiplied by the corrected temperature rise, then adjusted for extraneous heat effects, and divided by the weight of the sample, gives the gross calorific value. **D 121,** D05

energy input rate—peak rate (Btu/h or kW) at which the equipment consumes energy; typically reflected during preheating, $E_{input\ rate}$.

$$E_{input\ rate} = \frac{E \times 60}{t}$$

where:

$E_{input\ rate}$	=	measured peak energy input rate. Btu/h, kW,
E	=	energy consumed during period of peak energy input, Btu, kWh or kJ and, and
t	=	period of peak energy input, min.

F 1827, F26

energy loss—the energy per unit volume that is lost in each deformation cycle. (ISO) **D 4092,** D20

energy product, B_dH_d—the product of the coordinate values of any point on a demagnetization curve. **A 340,** A06

energy-product curve, magnetic—the curve obtained by plotting the product of the corresponding coordinates, B_d and H_d, of points on the demagnetization curve as abscissa against the induction, B_d, as ordinates.

NOTE—The maximum value of the energy product, $(B_dH_d)_m$, corresponds to the maximum value of the external energy.

NOTE—The demagnetization curve is plotted to the left of the vertical axis and usually the energy-product curve to the right. **A 340,** A06

energy product, maximum $(B_dH_d)_m$—for a given demagnetization curve, the maximum value of the energy product. **A 340,** A06

energy-to-food—the energy (Btu or kWh) imparted to a food product during a cooking energy efficiency (see **cooking energy efficiency**) test, E_{food}. **F 1827,** F26

energy utilization—amount of energy used, E_{util}.

$$E_{util} = \frac{H_g}{P}$$

where:

E_{util}	=	energy utilization, Btu/lb
Hg	=	amount of energy burned, Btu, and
P	=	weight of product, lb.

F 1827, F26

engine coolant, *n*—a heat exchange fluid designed to transfer heat from the engine block and accessories to the air through the radiator, consisting of water plus a corrosion inhibitor package, or a blend of water and glycol engine coolant concentrate. **D 4725,** D15

engine coolant concentrate, *n*—a formulated liquid product intended to be diluted with water for use in engine cooling systems. **D 4725,** D15

engine dynamometer test, *n*—a laboratory full-scale engine test designed to evaluate corrosion protection and inhibitor stability of engine coolants under simulated operational conditions. **D 4725,** D15

engineer, *n*—*in pipe laying*, the person, firm, corporation or government agency acting for the owner as his duly authorized agent in the designing and engineering of the project. **C 460,** C17

engineer, *n—in pipe laying*, the person, firm, corporation, or government agency acting for the owner as his duly authorized agent in the designing and engineering of the project. **C 1154,** C17

engineer, *n*—in pipe laying, the person, firm, corporation, or government agency acting for the owner as his duly authorized agent in the designing and engineering of the project. **D 2946,** C17

engineered controls—measures implemented at the work site to contain, control, or otherwise reduce exposure to lead dust and debris. **E 1605,** E06

engineered plastic, *n*—a material that has been made by specific design and through use of particular monomers and monomer sequences to produce a plastic with desired properties, possibly for a specific application. **D 883,** D20

engineering economics—the application of economic techniques to the evaluation of design and engineering alternatives. **E 833,** E06

engineering economics, *n* (économie du génie)—application of engineering, mathematical, and economic techniques to the economic evaluation of engineering alternatives. **E 1480,** E06

engineering plastics, *n*—those plastics and polymeric compositions for which well-defined properties are available such that engineering rather than empirical methods can be used for the design and manufacture of products that require definite and predictable performance in structural applications over a substantial temperature range. **D 883,** D20

engineering plastics, *n*—those plastics and polymeric compositions for which well-defined properties are available, such that engineering rather than empirical methods can be used for design and manufacture of products requiring definite and predictable performance in structural applications over a substantial temperature range. (D-20) **F 412,** F17

engineering stress, S [FL^{-2}], *n*—the stress calculated on the basis of the original dimensions of the specimen. **E 6,** E28

English degree—number of grains of substance per one British imperial gallon of water. **D 6161,** D19

engobe—a slip coating applied to a ceramic body for imparting color, opacity, or other characteristics, and subsequently covered with a glaze. **C 242,** C21

engraving—the process of carving figures, letters, etc., upon glass by abrasive means. **C 162,** C14

enhancement factor, *n*—the correction for the departure of the mixture of air and water vapor from ideal gas laws. **D 1356,** D22

enhancement factor—the correction for the departure of the mixture of air and water vapor from ideal gas laws. **D 4023,** D22

enriched uranium—greater than the accepted value of the weight percent of the isotope ^{235}U in natural uranium. **C 859,** C26

enrichment culture—a liquid culture that enhances the growth of a given type of organism. **F 1600,** F20

entanglement, *n*—the extent or degree to which the filaments in a yarn are interlocked and cannot be readily separated. **D 123,** D13

enthalpy—a thermodynamic *function* defined by the equation $H = U + PV$ where H is the enthalpy, *U* is the *internal* energy, P is the pressure, and V the volume of the system. **E 1142,** E37

entity—that which can be individually described and considered. **E 1187,** E36

entrained air—see **air void. C 125,** C09

entrainment loss—oil droplets escaping with the flow of water diverted under the skirt. **F 818,** F20

entrance angle, β, *n—in retroreflection*, angle between the illumination axis and the retroreflector axis. **E 284,** E12

entrapped air—see **air void. C 125,** C09

envelope, *n—in packaging,* a container of flexible material having only two **faces** and joined at three edges to form a partial enclosure. The nonjoined edge provides a filling opening which later may be closed or sealed. **D 996,** D10

envelope—a continuous membrane edge seal formed at the perimeter and at penetrations by folding the base sheet or ply over the plies above and securing it to the top of the membrane. The envelope prevents bitumen seepage from the edge of the membrane. **D 1079,** D08

envelope—*of a particle in catalysis*, a theoretical surface that encloses the particle and that bridges openings to the interior of the particle. **D 3766,** D32

envelope—a layer of cell membrane-derived lipoprotein that surrounds the protein coat (capsid) of some viruses. **E 1705,** E48

envelope, drainage—the materials completely surrounding a pipe to provide support or protection or act as a filter. **F 412,** F17

envelope grouting—grouting of rock surrounding a hydraulic pressure tunnel for purpose of consolidation, and primarily, reduction of permeability. **D 653,** D18

environment, *n—as related to fire,* the conditions and surroundings that may influence the behavior of a material, product, or assembly when it is exposed to ignition sources or fire. **E 176,** E05

environment—*in fatigue testing*, the aggregate of chemical species and energy that surrounds a test specimen. **E 1823,** E08

environmental chamber—See **building space. E 631,** E06

environmental compartment, *n*—a subdivision of the environment based on physical or chemical properties, or both. **D 6384,** D02

environmental conditions, *n—in textile testing*, the atmosphere in which specified moisture levels, temperature ranges, and concentrations of gases are controlled. **D 123,** D13

environmentally-assisted cracking—the initiation or acceleration of a cracking process due to the conjoint action of a chemical environment and tensile stress. **G 15,** G01

environmentally controlled area—See **ECA. E 1749,** E06

environmental resistance, *n*—a measure of a direct thermal product's ability to resist the effects of exposure to environmental contaminants on either the imaged or unimaged areas. **F 1623,** F05

environmental stress cracking—the development of cracks in a material that is subjected to stress or strain in the presence of specific chemicals. **F 412,** F17

environment-assisted cracking, EAC—a cracking process in which the environment promotes crack growth or higher crack growth rates than would occur without the presence of the environment. **E 1823,** E08

environment chamber— *in fatigue testing*, the container of the bulk volume surrounding a test specimen. **E 1823,** E08

environment chamber volume [L^3]—*in fatigue testing*, that bulk volume surrounding a test specimen. **E 1823,** E08

environment composition [ML^{-3}]—*in corrosion fatigue testing*, the concentration of the chemical components in the fluid environment surrounding a test specimen. **E 1823,** E08

environment hydrogen content [ML^{-3}]—*in corrosion fatigue testing*, the hydrogen gas concentration of the fluid environment surrounding a test specimen. **E 1823,** E08

environment monitoring— *in fatigue testing*, the periodic or continuous measurement of fluid concentrations of the environment. **E 1823,** E08

environment oxygen content [ML^{-3}]—*in corrosion fatigue testing*, the oxygen concentration of the fluid environment surrounding a test specimen. **E 1823,** E08

environment pressure [FL^{-2}]—*in fatigue testing*, the pressure of the bulk volume surrounding a test specimen. **E 1823,** E08

environment temperature— *in fatigue testing*, the temperature of the bulk volume surrounding a test specimen. **E 1823,** E08

environment volume [L^3]—*in fatigue testing*, the total volume immediately surrounding a test specimen plus that contained in a circulating reservoir if applicable. **E 1823,** E08

enzyme—biological catalyst that is protein in nature. **E 1705,** E48

EPA—Environmental Protection Agency **E 1605,** E06

EPA–Environmental Protection Agency (USA)—an organization that has set the potable water standards. **D 6161,** D19

EPA identification—unique number assigned by EPA to each generator or transporter of hazardous waste, and to each treatment, storage, or disposal facility. **E 1605,** E06

ephemeral gully—a channel that is formed by gully erosion on cropland and that is routinely but temporarily obscured by mechanical operations such as tilling. **D 4410,** D19

epicondyles—the bony prominence above the condyles where the ligaments attach. **F 869,** F08

epidermis, *n*—*in mammalian hair fibers*, the outside or surface layer of the fiber consisting of flat, irregular, horny cells or scales. **D 123,** D13

epitaxial layer—*in semiconductor technology*, a layer of a single crystal semiconducting material grown on a host substrate which determines its orientation. **F 1241,** F01

epithermal neutrons—a general classification of neutrons with energies above those of thermal neutrons; or frequently, neutrons with energies in the resonance range, between the thermal limit and some upper limit, such as 0.1 MeV (see **thermal neutrons**). **E 170,** E10

epoxy—a multicomponent resin grout that usually provides very high, tensile, compressive, and bond strengths. **D 653,** D18

epoxy plastic, *n*—a thermoplastic or thermosetting plastic containing ether or hydroxyalkyl repeating units, or both, resulting from the ring-opening reactions of lower-molecular weight polyfunctional oxirane resins, or compounds, with catalysts or with various polyfunctional acidic or basic coreactants. **F 1251,** F04

epoxy plastics, *n*—thermoplastic or thermosetting plastics containing ether or hydroxyalkyl repeating units, or both, resulting from the ring-opening reactions of lower molecular weight polyfunctional oxirane resins, or compounds, with catalysts or with various polyfunctional acidic or basic coreactants. **D 883,** D20

epoxy resin—a viscous liquid or brittle solid containing epoxide groups that can be crosslinked into final form by means of a chemical reaction with a variety of setting agents used with or without heat. **C 904,** C03

EPROM—**e**rasable **p**rogrammable **r**ead **o**nly **m**emory; a type of memory chip used in computing equipment (in this case, a printer). **F 1457,** F05

EPS, *n*—expanded polystyrene. See **rigid cellular polystyrene thermal insulation board.** See also **cellular polystyrene.** **E 631,** E06

epsilon (ϵ)—designation generally assigned to intermetallic, metal-metalloid, and metal-nonmetallic compounds found in ferrous alloy systems (for example, Fe_3Mo_2, $FeSi$, Fe_3P). **E 7,** E04

epsilon carbide—carbide with hexagonal close-packed lattice which precipitates during the first stage of tempering of primary martensite. Its composition corresponds to the empirical formula $Fe_{2.4}C$. **E 7,** E04

epsilon structure—structurally analogous close-packed phases (similar to epsilon brass), or electron compounds, that have ratios of 7 valence electrons to 4 atoms. **E 7,** E04

equal-discharge-increment (EDI) method— a procedure for obtaining the discharge weighted suspended-sediment concentration of flow at a cross section whereby depth integration is performed at the centers of three or more equal flow segments of the cross section. If approximately equal volumes of water-sediment mixture are collected from each flow segment, the samples may be composited for analysis. If unequal volumes are collected, samples from each flow segment must be analyzed separately and the results combined mathematically. **D 4410,** D19

equal-energy illuminant, *n*—illuminant having the same spectral power at all wavelengths in the specified spectral region. **E 284,** E12

equal-energy source, *n*—a source having the same radiant exitance in each constant wavelength interval in the spectral region under consideration. **E 284,** E12

equal-width-increment (EWI) method—a procedure of obtaining the discharge weighted suspended-sediment concentration of flow at a cross section by performing depth integration at a series of verticals equally spaced across the cross section and using the same vertical transit rate at all sampling verticals. **D 4410,** D19

equatorial mount—a sun-tracking mount, usually clock-driven, whose axis of rotation is parallel to that of the earth. **E 772,** E44

equiaxed grain—a polygonal crystallite, in an aggregate, whose dimensions are approximately the same in all directions. **E 7,** E04

equilay conductor—see **concentric-lay conductor. B 354,** B01

equilibrium, *n*—condition reached in air drying in which change in weight of the sample, under conditions of ambient temperature and humidity, is no more than 0.1 %/h or 0.05 %/½ h. **D 121,** D05

equilibrium—a state of dynamic balance between the opposing actions, reactions, or velocities of a reversible process. **E 7,** E04

equilibrium adsorptive capacity—the quantity of a given component adsorbed per unit of adsorbent from a fluid or fluid mixture at equilibrium temperature and concentration, or pressure. **D 2652,** D28

equilibrium capacitance—that capacitance reached after an MOS specimen at a fixed bias is illuminated and then allowed to stabilize in darkness. **F 1241,** F01

equilibrium diagram— See **constitutional diagram. E 7,** E04

equilibrium diagram— see **constitutional diagram. E 1142,** E37

equilibrium eutectic—See **eutectic, equilibrium. C 242,** C21

equilibrium moisture basis—See *equilibrium moisture basis* under **reporting bases. D 121,** D05

equilibrium moisture content—a moisture content at which wood neither gains nor loses moisture to the surrounding air. **D 9,** D07

equilibrium moisture content—(*1*) the moisture content of a material stabilized at a given temperature and relative humidity, expressed as percent moisture by weight;

(*2*) the typical moisture content of a material in any given geographical area. **D 1079,** D08

equilibrium moisture content—the percentage mass of moisture in a material when it has attained equilibrium with water vapor at a specified pressure or partial pressure. It is determined by successive weighings of the sample, either on the adsorption or desorption curve, and plotting moisture content against the logarithm of time. It should be designated as adsorption equilibrium or desorption equilibrium. **D 1695,** D01

equilibrium (reversible) potential—the potential of an electrode in an electrolytic solution when the forward rate of a given reaction is exactly equal to the reverse rate. (The equilibrium potential can only be defined with respect to a specific electrochemical reaction.) **G 15,** G01

equilibrium surface composition—see **sputtering.** **E 673,** E42

equipment check, *n*—an indication of a problem in the printer, sent by the printer to the cluster controller. **F 1457,** F05

equipment test level accuracy—(*1*) For a single test equipment, using a large group of test specimens, the average percentage of test deviation from the correct average value.

(*2*) The average percentage deviation from the average value obtained from similar tests, on the same test specimen or specimens, when measured with a number of other test equipments that have previously been proven to have both suitable reproducibility of measurement and test level, and whose calibrations and quality have general acceptance for standardization purposes and where better equipment for establishing the absolute accuracy of test is not available. **A 340,** A06

equipotential line—a line connecting points of equal hydraulic head. A set of such lines provides a contour map of a potentiometric surface. **D 653,** D18

equitransference—equal diffusion rates of the positively and negatively charged ions of an electrolyte across a liquid junction. **D 4127,** D19

equitransferent filling solution—a reference electrode filling solution in which the diffusion rates of negatively and positively charged ions are equal. For low liquid junction potentials, the ionic strength of the filling solution must be high compared to the sample solution. **D 4127,** D19

equivalent alkalies, Na_2Oeq, *n*—*in hydraulic cement,* total combined sodium and potassium oxides, expressed as a percentage, calculated from the analysis, by means of the formula ($Na_2Oeq = \% Na_2O + 0.658 \times \% K_2O$). **C 219,** C01

equivalent boron content (EBC)—*in the nuclear industry,* the concentration of natural boron that would provide a thermal neutron absorption cross section equivalent to the concentration of a specific impurity element. **C 859,** C26

equivalent boron content (EBC) **factor**—the ratio of the product of the atomic weight of boron and a specified impurity thermal neutron absorption cross section to the product of the boron thermal neutron absorption cross section and the atomic weight of a specified impurity element.

$$\frac{(\text{at. wt. of boron})\ (\sigma a \text{ of impurity})}{(\sigma a \text{ of boron})\ (\text{at. wt. of impurity})}$$

This factor is used to convert an impurity element concentration to a neutron absorption equivalent to natural boron. **C 859,** C26

equivalent conductivity—in an electrolyte, the conductivity of the solution divided by the number of equivalents of conducting solute per unit volume, that is, the conductivity divided by the normality of the solution. **B 374,** B08

equivalent diameter (equivalent size), *D* **(L)**—the diameter of a hypothetical sphere composed of material having the same specific gravity as that of the actual soil particle and of such size that it will settle in a given liquid at the same terminal velocity as the actual soil particle. **D 653,** D18

equivalent diameter (sphere)—See *equivalent diameter (sphere)* under **diameter.** **C 242,** C21

equivalent-energy fracture toughness (*K-EE*) [$FL^{-3/2}$]—the crack extension resistance determined by the procedure specified in Practice E 992. **E 1823,** E08

equivalent fission fluence—the fluence of fission spectrum neutrons that would give a detector or material response for a particular reaction equal to that in a given neutron field. **E 170,** E10

equivalent fluid—a hypothetical fluid having a unit weight such that it will produce a pressure against a lateral support presumed to be equivalent to that produced by the actual soil. This simplified approach is valid only when deformation conditions are such that the pressure increases linearly with depth and the wall friction is neglected. **D 653,** D18

equivalent I.Q.I. sensitivity—that thickness of I.Q.I. expressed as a percentage of the section thickness radiologically examined in which a 2T hole or 2 % wire size equivalent would be visible under the same radiological conditions. **E 1316,** E07

equivalent monoenergetic neutron fluence, $\Phi_{eq}(E_o)$—characterizes an incident energy fluence spectrum, $\Phi(E)$, in terms of the fluence of monoenergetic neutrons at a specific energy, E_o, required to produce the same displacement kerma, K_o, in a specific material (for example, silicon) as Φ (E). **E 170,** E10

equivalent 2200 m/s fluence, Φ_o—a measure of the thermal neutron fluence made with a l/v detector and using the 2200 m/s cross section.

$$\Phi_o = nv_o t$$

where:

n = neutron density,
v_o = 2200 m/s, and
t = exposure time of the detector.

E 170, E10

equivalent nitrogen pressure—the calculated pressure that a gage or another device would indicate if the gas in the device were replaced by nitrogen at the same molecular density. **E 1316,** E07

equivalent orifice, *n*—the diameter of the sharp-edged circular opening in the plate mounted in an ASTM Plenum Chamber (see Specification F 431, for Air Performance Measurement Plenum Chamber for Vacuum Cleaners), the opening having a resistance to air flow equivalent to the resistance caused by a specific usage of the vacuum cleaner. The equivalent orifice diameter is expressed in inches or millimetres. **F 395,** F11

equivalent penetrameter sensitivity—that thickness of penetrameter, expressed as a percentage of the section thickness radiographed, in which a 2T hole would be visible under the same radiographic conditions. **E 1316,** E07

equivalent per million (epm)—a unit chemical equivalent weight of solute per million unit weights of solution. **D 1129,** D19

equivalent per million (EPM)—a unit chemical equivalent weight of solute per million unit weights of solution. Concentration in equivalents per million is calculated by dividing concentration in ppm by the equivalent weight of the substance or ion. Equivalent weight is the atomic weight of the substance divided by the valence of the substance. **D 6161,** D19

equivalent product, *n*—provides the same amount of product or number of recommended uses as contained in the package being replaced (as related to Subcommittee D10.46 package source reduction). **D 5488,** D10

equivalent single axle load, (ESAL), *n*—a numerical factor that expresses the relationship of a given axle load to another axle load in terms of their effect on the serviceability of a pavement structure.

E 867, E17

equivalent spherical diameter—See *equivalent diameter (sphere)* under **diameter.** **C 242,** C21

equivalent uniform annual value—See **annual value.** **E 833,** E06

equivalent uranium content (EUC)—a concentration of ^{238}U that would provide a fast neutron absorption cross section equivalent to the concentration of a specific impurity element. **C 859,** C26

equivalent volume sphere diameter—the diameter of a hypothetical sphere having the same volume as the actual particle. **E 1620,** E29

equivalent weight—the weight of an element that will combine with, react with, or replace one atomic weight of hydrogen or one half atomic weight of oxygen. **C 859,** C26

equivalent weight—the weight of an ion determined by dividing the sum of the atomic weights of its component atoms by its valence. **D 6161,** D19

equiviscous temperature (EVT), *n*—the temperature at which a bitumen attains the proper viscosity for built-up membrane application. EVT may or may not be applicable to polymer modified bituminous sheet materials. (See **EVT for asphalt** and **EVT for coal tar pitch**). **D 1079,** D08

equiviscous temperature for asphalt, *n*—the equiviscous temperature (EVT) for roofing asphalt (Specification D 312, Type I, II, III, or IV) is as follows: (*1*) *mop application*—the temperature at which the asphalt's apparent viscosity is 125 cP. (*2*) *mechanical spreader application*—the temperature at which the asphalt's apparent viscosity is 75 cP. See **equiviscous temperature** and **EVT application range**. **D 1079,** D08

equiviscous temperature for coal tar—the equiviscous temperature (EVT) for roofing coal tar (Specification D 450, Type I or III) is the temperature at which the coal tar's apparent viscosity is 25 cP. See **equiviscous temperature** and **EVT application range**. **D 1079,** D08

eradication—the complete elimination of a pest from an area. **E 609,** E35

erasability—ease with which the image may be removed without impairing the surface of the paper. **F 221,** F05

erasable optical medium—an erasable and rewritable storage medium where the digital data is represented by the degree of reflectivity of the medium recording layer; the data can be altered. **E 1316,** E07

erasable programmable read only memory, *n*—a type of memory chip used in printers where the content of the chip can be read only when in normal use, but with special treatment it can be erased and reprogrammed. **F 1457,** F05

erasure correction, *n*—a use of error correction characters to correct data errors that have unknown locations. **F 1294,** F05

ERD—energy recovery device. **D 6161,** D19

erosion—the progressive loss of material from a solid surface due to mechanical interaction between that surface and a fluid, a multicomponent fluid, or solid particles carried with the fluid. **G 15,** G01

erosion, *n*—*in tribology*, progressive loss of original material from a solid surface due to mechanical interaction between that surface and a fluid, a multicomponent fluid, or impinging liquid or solid particles. **G 40,** G02

erosion corrosion, *n*—nonuniform, accelerated corrosion characterized by a smooth appearance and caused by highvelocity coolant. **D 4725,** D15

erosion-corrosion—a conjoint action involving corrosion and erosion in the presence of a moving corrosive fluid, leading to the accelerated loss of material. **G 15,** G01

erosion-corrosion, *n*—a synergistic process involving both erosion and corrosion, in which each of these processes is affected by the simultaneous action of the other, and in many cases is thereby accelerated. **G 40,** G02

erosion, electrical, *n*—the progressive wearing away of electrical insulation by the action of electrical discharges. **D 1711,** D09

erosion of refractories, *n*—wearing away of refractory surfaces by the washing action of moving liquids. **C 71,** C08

erosion rate, *n*—any determination of the rate of loss of material (erosion) with exposure duration. (See also **rationalized erosion rate.**) **G 40,** G02

erosion rate-time curve, *n*—a plot of instantaneous erosion rate versus exposure duration, usually obtained by numerical or graphical differentiation of the cumulative erosion-time curve. (See also **erosion rate-time pattern.**) **G 40,** G02

erosion rate-time pattern, *n*—any qualitative description of the shape of the erosion rate-time curve in terms of the several stages of which it may be composed. **G 40,** G02

erosion resistance (*coatings*)—the ability of a coating to withstand being worn away by chalking or by the abrasive action of water or windborne particles of grit. The degree of resistance is measured by the amount of the coating retained. See **abrasion resistance.** **D 16,** D01

erosion resistance, electrical, *n*—the resistance of electrical insulating materials to erosion by the action of electrical discharges. **C 1145,** C28

erosion resistance, electrical, *n*—the quantitative expression of the amount of electrical erosion under specific conditions. **D 1711,** D09

ERR—external rOM receptacle. **F 1457,** F05

ERR, *n*—an error. **F 1457,** F05

error—difference of an observation from the best obtainable estimate of the true value. **D 121,** D05

error, *n*—*of a temperature measurement*, result of a temperature measurement minus a true value of temperature. **E 344,** E20

error—*in a statistical sense*, any deviation of an observed value from the true, but generally unknown, value. When expressed as a fraction or percentage of the value measured, it is called a relative error. All statements of precision or bias should indicate clearly whether they are expressed in absolute or relative sense. **E 1547,** E15

error—the substitution of one character for another. **F 149,** F05

error:

Type I—false rejection errors; those that do not recognize an authentic item as being authentic. (See **FR.**) **F 1156,** F12

error correction, *n*—use of error correction characters to locate and correct data errors that have unknown locations. **F 1294,** F05

error correction characters, *n*—special symbol characters that are calculated mathematically from other symbol characters for the purpose of error correction, error detection or erasure correction. **F 1294,** F05

error correction level, *n*—an indicator of the number of error correction characters in a symbol. **F 1294,** F05

error detection, *n*—use of error correction characters to determine that the number of errors in the symbol exceeds the error correction capacity. **F 1294,** F05

error of result, *n*—the test result minus the accepted reference value (of the characteristic).

NOTE—It is not possible to correct for random error. **E 456,** E11

error of the first kind, α, *n—in a statistical test,* the rejection of a statistical hypothesis when it is true. **D 123,** D13

error of the second kind, β, *n—in a statistical test,* the acceptance of a statistical hypothesis when it is false. **D 123,** D13

error, random—the chance variation encountered in all experimental work despite the closest possible control of variables. It is characterized by the random occurrence of both positive and negative deviations from the mean value for the method, the algebraic average of which will approach zero in a long series of measurements. **E 1547,** E15

error rate—the ratio of the number of character substitutions to the total number of characters read. **F 149,** F05

error rates:

Type II—false acceptance errors; those that accept a non-genuine item as if it were the real thing. (See **FA**.) **F 1156,** F12

ERT—energy recovery turbine. **D 6161,** D19

ESC, *n*—abbreviation for **escape**; a keyboard control character whose function varies with the software or type of terminal. **F 1457,** F05

ESCA—acronym for "electron spectroscopy for chemical analysis," a term historically used to describe a technique whereby one generates electron spectra by irradiating a specimen with narrow band characteristic X-rays. **E 673,** E42

escape or pair production peak—a peak in a gamma ray spectrum resulting from the pair production process within the detector, annihilation of the positron produced, and escape from the detector of one or more of the annihilation photons (see **pair production** and **annihilation radiation**).

single escape peak—the gamma ray spectrum peak corresponding to escape of one of the annihilation photons from the active volume of the detector. The energy of the single escape peak is equal to the original gamma ray energy minus 511 keV.

double escape peak—the gamma ray spectrum peak corresponding to escape of both of the annihilation photons from the active volume of the detector. The energy of the double escape peak is equal to the original gamma ray energy minus 1.022 MeV. **E 170,** E10

Esc key, *n*—the keyboard key which is used to generate the ESC. **F 1457,** F05

escutcheon—See **railing systems.** **E 631,** E06

escutcheon, *n*—a protective or ornamental cover located at the termination of a post, baluster, or rail against a tread, floor, or wall. (Syn. *collar, cover flange, cover plate,* or *cover ring*.) **E 1481,** E06

escutcheon pin—small, regular-stock-steel or nonferrous, ¼ by 0.035 to 2 by 0.092-in. nails with oval head and medium diamond point. **F 547,** F16

ESS—abbreviation for **electronic scanning system**. **F 1457,** F05

ester gum—under **resin, synthetic,** see *ester gum.* **D 16,** D01

ester gum, *n*—a resin made from rosin and a polyhydric alcohol, generally glycerol or pentaerythritol. **D 804,** D01

estimate—*in statistical analysis,* the particular value or values of a parameter computed by an estimation procedure for a given sample. **E 1823,** E08

estimated crack extension, Δa[L]—an increase in estimated crack size ($\Delta a = a - a_{oq}$). **E 1823,** E08

estimated crack size a[L]—the distance from a reference plane to the observed crack front developed from measurements of elastic compliance or other methods. The reference plane depends on the specimen form, and it is normally taken to be either the boundary, or a plane containing either the load line or the centerline of a specimen or plate. The reference plane is defined prior to specimen deformation. **E 1823,** E08

estimated texture depth, *n*—the estimate of mean texture depth, by means of a linear transformation of mean profile depth. **E 867,** E17

estimation—*in statistical analysis,* a procedure for making a statistical inference about the numerical values of one or more unknown population parameters from the observed values in a sample. **E 1823,** E08

ESWTR—enhanced surface water treatment rule. **D 6161,** D19

etch, *n*—a roughened surface produced by a chemical or electrochemical means. **B 374,** B08

etch, *v*—to dissolve unevenly a part of the surface of a metal. **B 374,** B08

etch—to attack the surface of glass with hydrofluoric acid or other agent, generally for marking or decoration. **C 162,** C14

etch—an aqueous solution used to define the non-image areas of a paper master or metal lithographic plate by making them water receptive and ink repellent. **F 425,** F05

etch, *n*—a solution, a mixture of solutions, or a mixture of gases that attacks the surfaces of a film or substrate, removing material either selectively or nonselectively. See also **anisotropic etch, preferential etch.** **F 1241,** F01

etched—(*1*) treated by etching.

(*2*) weathered so that surface is roughened. **C 162,** C14

etched—an altered surface texture resulting from chemical attack. **C 286,** B08

etched—cleaned of grease and oil, with slightly dulled, microscopically roughened surface, usually by a chemical process. **F 547,** F16

etch figures—markings formed on a crystal surface by etching or chemical solution and usually related geometrically to the crystal structure. **E 7,** E04

etching—controlled preferential attack on a metal surface for the purpose of revealing structural details. **E 7,** E04

etching—the removal of surface material by chemical or electrochemical methods. **E 1316,** E07

etch pit—a pit, resulting from preferential etching, localized on the surface of a wafer at a crystal defect or stressed region. **F 1241,** F01

ethanol—ethyl alcohol, the chemical compound C_2H_5OH, a two carbon alcohol. **E 1705,** E48

ethanol (ethyl alcohol, grain alcohol)— CH_3CH_2OH; can be produced chemically from ethylene or biologically from the fermentation of various sugars from carbohydrates found in agricultural crops and cellulosic residues from crops or wood. **E 1126,** E48

ethanol (ethyl alcohol, grain alcohol)—CH_3CH_2OH; can be produced chemically from ethylene or biologically from the fermentation of various sugars from carbohydrates found in agricultural crops and cellulosic residues from crops or wood. **E 1705,** E48

ethylbenzene (C_8H_{10})—mol weight 106.16; clear, colorless, flammable liquid; freezing point, −94.97°C; boiling point, 136°C. **D 4790,** D16

ethylcellulose—any of several ethyl ethers of cellulose. The one most generally used in industry has sufficient substitution to be soluble in organic solvents. **D 1695,** D01

ethylene plastic—a plastic based on polymers of ethylene or copolymers of ethylene with other monomers, the ethylene being in greatest amount by mass. **F 1251,** F04

ethylene plastics, *n*—plastics based on polymers of ethylene or copolymers of ethylene with other monomers, the ethylene being in greatest amount by mass. (ISO) **D 883,** D20

ethylene plastics, *n*—plastics based on polymers of ethylene or copolymers of ethylene with other monomers, the ethylene being in greatest amount by mass. (ISO, D-20) **F 412,** F17

ethylene-vinyl acetate (EVA)—copolymers from these two monomers form a class of plastic materials. **F 869,** F08

ETX—abbreviation for **end of text**. **F 1457,** F05

ETX/ACK, *n*—a standard for ''ready/not-ready" status communication between the host and a printer. **F 1457,** F05

eutectic:

deformation eutectic—the composition within a system of two or more components that, on heating under specified conditions, develops sufficient liquid to cause deformation at the minimum temperature.

equilibrium eutectic—the composition within any system of two or more crystalline phases that melts completely at the minimum temperature, or the temperature at which such a composition melts. **C 242,** C21

eutectic—mixture of two or more substances which solidifies as a whole when cooled from the liquid state, without change in composition. **E 1142,** E37

eutectic alloy—the alloy which has the composition of the eutectic point. **E 7,** E04

eutectic arrest—in a cooling curve (or heating curve) an approximately isothermal segment, corresponding to the time interval during which the heat of transformation from the liquid phase to two or more conjugate solid phases is being evolved, (or conversely). **E 7,** E04

eutectic carbides—in hypereutectic tool steels, the skeleton-like structure of the eutectic carbide. **E 7,** E04

eutectic colony, grain—a two-phase region which solidified progressively from a simple center and, therefore, has some uniformity of structural relationship. **E 7,** E04

eutectic equilibrium—a reversible univariant transformation in which a liquid, that is stable only at superior temperature, decomposes into two or more conjugate solid phases; for example, $L = \alpha + \beta$, $L = \alpha + \beta + \gamma$, etc. **E 7,** E04

eutectic point—the composition of a liquid phase that is in univariant equilibrium with two or more solid phases; the lowest melting alloy of a composition series. **E 7,** E04

eutectic point—see **eutectic**. **E 1142,** E37

eutectic structure—the structure resulting when an alloy has passed through a eutectic equilibrium upon freezing. **E 7,** E04

eutectic temperature—the temperature below which all liquid portions of an aqueous suspension have entered the solid phase. **E 1705,** E48

eutectoid equilibrium—a reversible univariant transformation in which one solid phase, that is stable only at superior temperature, decomposes into two or more conjugate solid phases; for example, $\alpha = \beta + \gamma$, $\alpha = \beta + \gamma + \delta$, etc. **E 7,** E04

eutectoid point—the composition of a solid phase which, upon cooling, undergoes univariant transformation into two, or more, other solid phases. **E 7,** E04

eutectoid reaction— See **eutectoid equilibrium.** **E 7,** E04

eutectoid structure—the microstructure resulting when an alloy has passed through an eutectoid equilibrium upon cooling. **E 7,** E04

evacuated tube collector— see **collector, evacuated tube.** **E 772,** E44

evacuation, *n*—the process used to move the patient/subject between the completion of extraction and the beginning of transportation. **F 1490,** F32

evaluate, *v* (évaluer (installations))—*(a facility)* to assess the capability of a facility to perform the function(s) for which it is designed, used, or required to be used. **E 1480,** E06

evaluation—a review, following interpretation of the indications noted, to determine whether they meet specified acceptance criteria. **E 1316,** E07

evaluation—See **facility evaluation**. **E 1480,** E06

evaluation matrix—two-dimensional listing of properties, results, or both, etc, used to compare and evaluate data. **F 1156,** F12

evaluation period, *n*—the perod of time an item is used before being evaluated on the specific performance properties. **D 123,** D13

evaluation threshold—a threshold value used for analysis of the examination data. Data may be recorded with a *system examination threshold* lower than the *evaluation threshold*. For analysis purposes, dependence of measured data on the *system examination threshold* must be taken into consideration. **E 1316,** E07

evaluator, *n*—a part of, or an attachment to, an unevenness testing instrument, which automatically gives an estimate of R_u. **D 123,** D13

evanescent wave—the standing wave that exists in the less refractive medium, normal to the reflecting surface of the IRE during internal reflection. **E 131,** E13

evaporation—process where a liquid (water) passes from a liquid to a gaseous state. **D 6161,** D19

evaporation—the vaporization of a material by heating it, usually in a vacuum. In electron microscopy this process is used for shadowing or to produce thin support films by condensation of the vapors of metals or salts. **E 7,** E04

evaporation reduction agent—a material that reduces the evaporation rate of a spray mix during or after application, or both. **E 1519,** E35

evaporation residue—the nonvolatile impurities remaining after vaporizing a substance. **D 4790,** D16

even flow atomizer—see **even spray atomizer**. **E 1620,** E29

even spray atomizer—a fan spray atomizer that produces a relatively uniform band of liquid, usually by means of injection, through an elliptical orifice. (This definition applies to devices utilized in the agricultural industry.) **E 1620,** E29

event, acoustic emission (emission event)— a local material change giving rise to acoustic emission. **E 1316,** E07

event count (Ne)—see **count, event.** **E 1316,** E07

event count rate ($\dot{N}e$)— see **rate, event count.** **E 1316,** E07

event sampling—See **sampling.** **D 1356,** D22

eversion—a turning outward, eversion of the foot; turning the sole away from the mid-line of the body. **F 869,** F08

evolutionary operation (EVOP), *n*—a sequential form of experimentation conducted in production facilities during regular production.

NOTE—The principal theses of EVOP are that knowledge to improve the process should be obtained along with a product, and that designed experiments using relatively small shifts in factor levels (within production tolerances) can yield this knowledge at minimum cost. The range of variation of the factors for any one EVOP experiment is usually quite small in order to avoid making out of tolerance products, which may require considerable replication, in order to be able to clearly detect the effect of small changes. **E 456,** E11

evolutionary operation (EVOP), *n*— a sequential form of experimentation conducted in production facilities during regular production.

NOTE—The principal theses of EVOP are that knowledge to improve the process should be obtained along with a product, and that designed experiments using relatively small shifts in factor levels (within production tolerances) can yield this knowledge at minimum cost. The range of variation of the factors for any one EVOP experiment is usually quite small in order to avoid making out-of-tolerance products, which may require considerable replication, in order to be able to clearly detect the effect of small changes. **E 1325,** E11

evolved gas analysis *(EGA)*, *n*—a technique in which the nature and amount, or both, of gas or vapor evolved by a substance is subjected to a controlled-temperature program. (ICTAC)

NOTE—Some specific forms of EGA have become established for investigating different aspects of catalysis, such as reduction, oxidation or desorption. In this context, EGA in a hydrogen atmosphere is known as temperature-programmed reduction (TPR); EGA in an oxygen atmosphere is temperature-programmed oxidation (TPO); and EGA in the absence of decomposition, in an inert atmosphere or vaccum, is temperature-programmed desorption (TPD). For each technique the method used for gas identification and quantification should always be clearly stated. **E 473,** E37

evolved gas detection, *(EGD)*, *n*—see **evolved gas analysis.** **E 473,** E37

EVT application range, *n*—the bitumen application temperature range. The range is from 25°F below the EVT to 25°F above the EVT (50°F span). The temperature is measured in the mop cart or mechanical spreader just prior to the application of bitumen to the substrate. **D 1079,** D08

Ewald sphere—a geometric construction, of radius equal to the reciprocal of the wavelength of the incident radiation, with its surface at the origin of the reciprocal lattice. Any crystal plane will reflect if the corresponding reciprocal lattice point lies on the surface of this sphere. **E 7,** E04

examination, *n*—analysis of (1) the materials, structure, embellishments, and condition of a cultural object, and (2) the impact of these factors along with documentary characteristics on the treatment, display, and storage of the object. **D 123,** D13

examination—an element of investigation, without the use of special laboratory appliances or procedures, of supplies and services to determine conformance to those specified requirements which can be determined by such investigations. Examination is generally nondestructive and includes, but is not limited to visual, auditory, olfactory, tactile, gustatory, and other investigations, simple physical manipulation, gaging, and measurement. **D 1517,** D31

examination, *n*—a procedure for determining a property (or properties) or other conditions or characteristics of a material or component by direct or indirect means. **E 1316,** E07

examination area—that portion of a structure being monitored with acoustic emission. **E 1316,** E07

examination medium—a powder or suspension of magnetic particles that is applied to a magnetized test surface to determine the presence or absence of surface or slightly subsurface discontinuities. **E 1316,** E07

examination region—that portion of the test article evaluated using acoustic emission technology. **E 1316,** E07

exceedances spectrum— *in fatigue loading*, representation of spectrum loading contents by the number of times specified values of a particular loading parameter (peak, range, and so forth) are equaled or exceeded (also known as *cumulative occurrences* or *cumulative frequency spectrum*). **E 1823,** E08

excess electrostatic charge, *n*—the algebraic sum of all positive and negative electric charges on the surface of, or in, a specific volume. **D 1711,** D09

excess hydrostatic pressure—see **hydrostatic pressure**. **D 653,** D18

excessive corrosion—corrosion that is not removed by cleaning as described in Practice E 864. **E 1749,** E06

excess moisture—synonym for *surface moisture*. **D 121,** D05

exchange capacity—the capacity to exchange ions as measured by the quantity of exchangeable ions in a soil or rock. **D 653,** D18

exchange capacity—see **ion-exchange capacity.** **D 1695,** D01

exchange current density—the rate of charge transfer per unit area when an electrode reaches dynamic equilibrium (at its reversible potential) in a solution; that is, the rate of anodic charge transfer (oxidation) balances the rate of cathodic charge transfer (reduction). **G 15,** G01

excitation depth distribution function—the probability that specified excitations are created at specified depths measured normally from a surface into the material by a beam of specified particles or radiation incident on the surface in a given direction. **E 673,** E42

excitation potential—the applied potential on an X-ray tube required to produce characteristic radiation from the target. **E 7,** E04

excitation potential (X-ray), *n*—the potential required to produce characteristic radiation from an element. **E 135,** E01

excitation purity, p_e, *n*—the ratio of the lengths of line segments NC to ND, obtained when a straight line is drawn on the CIE chromaticity diagram from the achromatic (neutral) point (N), through the point (C) representing the color stimulus considered, to the intersection (D) with the spectrum locus or the straight boundary of nonspectral colors, as the case may be. **E 284,** E12

excitation (stimulus)—an external force (or other input) applied to a system that causes the system to respond in some way. **D 653,** D18

exciting current, ac, I—See **current, ac exciting.** **A 340,** A06

exciting power, rms, P_z—the product of the ac rms exciting current and the rms voltage induced in the exciting (primary) winding on a magnetic core.

NOTE—This is the apparent volt-amperes required for the excitation of the magnetic core only. When the core has a secondary winding, the induced primary voltage is obtained from the measured open-circuit secondary voltage multiplied by the appropriate turns ratio. **A 340,** A06

exciting power, specific, $P_{z(B;f)}$—the value of the ac rms exciting power divided by the active mass of the specimen (volt-amperes/unit mass) taken at a specified maximum value of cyclically varying induction B and at a specified frequency f. **A 340,** A06

exciting voltage, E—the ac rms voltage across a winding linking the flux of a magnetic core. The voltage across the winding equals that across the assumed parallel combination of core inductance L_1, and core resistance, R_1. **A 340,** A06

excluded counters, *n*—permanent traffic counters for which monthly traffic adjustment factors may not be calculated for a given year. **E 867,** E17

exclusion booming—placement of a boom to protect an area from the entry of a floating substance. **F 818,** F20

exemplar, *n*—a specimen of physical evidence of known origin (source: Osterburg and Ward, *Criminal Investigation*, 1992, p. 837). **E 1732,** E30

exfiltration—the volume of pipeline flow leaving a sewer and its connections into the soil from pipe, joints, connections, and appurtenances. **C 822,** C13

exfiltration—the quality of water leaving the test section during a specified time period. **C 896,** C04

exfiltration, pipe—the passage of fluid from a pipe section through small holes or leaks. **F 412,** F17

exfoliation—corrosion that proceeds laterally from the sites of initiation along planes parallel to the surface, generally at grain boundaries, forming corrosion products that force metal away from the body of the material, giving rise to a layered appearance. **G 15,** G01

exhaust hood—a device that captures hot air, odors, and vapors produced in the cooking process and directs them to an exhaust fan. **F 1827,** F26

exhaustion—see **depletion.** **F 335,** F05

exhaust pressure—in leak testing, same as **fore-pressure**. **E 1316,** E07

exhaust tubulation—Same as **pump-out tubulation**. **E 1316,** E07

exinite—See *exinite* under **maceral**. **D 121,** D05

exitance, *M*, *n*—flux leaving a surface per unit area. **E 284,** E12

exitance, radiant— see **radiant exitance.** **E 772,** E44

exogenous inclusion—a nonmetallic constituent produced by entrapment of foreign material in the melt. (See **inclusions.**) **E 7,** E04

exotherm—the temperature rise resulting from the liberation of heat by any process of chemical reaction. **E 1749,** E06

exothermic—pertaining to a reaction that occurs with the evolution of heat. **D 653,** D18

exothermic reaction, *n*—a chemical reaction in which heat is evolved. **C 1145,** C28

expandable plastic, *n*—a plastic in a form capable of being made cellular by thermal, chemical, or mechanical means. **D 883,** D20

expandable plastic, *n*—a plastic in a form capable of being made cellular by thermal, chemical, or mechanical means. (D-20) **F 412,** F17

expanded bed—a bed of granular particles through which a fluid flows upward at a rate sufficient to slightly elevate and separate the particles without changing their relative positions. **D 2652,** D28

expanded blast-furnace slag, *n*—the lightweight cellular material obtained by controlled processing of molten blast-furnace slag with water or water and other agents, such as steam or compressed air or both. **C 125,** C09

expanded metal—See **screen**. **E 1481,** E06

expanded plastics—See **cellular plastic.** **D 883,** D20

expanded plastics—See **cellular plastics.** **F 412,** F17

expanded rubber, *n*—*in building construction*, an elastomeric closed-cell material made by incorporating a gas-producing ingredient into a rubber compound that decomposes during vulcanization, producing gas bubbles that expand the compound; the bubbles are predominately non-interconnecting. **C 717,** C24

expansion anchor, *n*—a component that forms a connection to bony element by means of a mechanism which enlarges once the component is inserted into the bony elements. **F 1582,** F04

expansion curve, thermal—the curve produced by plotting a dimension or the volume of a substance versus the temperature. **E 7,** E04

expansion joint, n—see **control (expansion-contraction) joint**. **C 11,** C11

expansion joint, *n*—*in building construction*, a formed or assembled joint at a predetermined location, which prevent the transfer of forces across the joint as a result of movement or dimensional change of different elements of a structure or building. **C 717,** C24

expansion joint—a structural separation between two building elements that allows free movement between the elements without damage to the roofing or waterproofing system. **D 1079,** D08

expansion joint—See **joints.** **E 631,** E06

expansion joint seal, *n*—*in building construction*, a seal or sealant system that provides an elastic, compressible, or flexible barrier to the passage of gases, liquids, or solids. **C 717,** C24

expansive cement—a cement that tends to increase in volume after it is mixed with water. **D 653,** D18

expansive hydraulic cement, *n*—a hydraulic cement that forms a paste when mixed with water, and increases in volume a controlled amount during the early hardening period occurring after setting. **C 219,** C01

expectation, error of, *n*—a bias due to preconceived ideas that influences an assessor's judgement. **E 253,** E18

expendable container— See **container.** **D 996,** D10

expendable pallet— See **pallet.** **D 996,** D10

experimental design, *n*—see **design of experiments.** **E 456,** E11

experimental design, n—see **design of experiments.** **E 1325,** E11

experimental error, *n*—variability attributable only to a test method itself. **D 123,** D13

experimental unit, *n*—a portion of the experiment space to which a treatment is applied or assigned in the experiment.

Note—The unit may be a patient in a hospital, a group of animals, a production batch, a section of a compartmented tray, etc. **E 456,** E11

experimental unit, *n*—a portion of the experiment space to which a treatment is applied or assigned in the experiment.

Note—The unit may be a patient in a hospital, a group of animals, a production batch, a section of a compartmented tray, etc. **E 1325,** E11

experiment space, *n*—the materials, equipment, environmental conditions and so forth that are available for conducting an experiment. **E 456,** E11

experiment space, *n*—the materials, equipment, environmental conditions and so forth that are available for conducting an experiment. **E 1325,** E11

expert, *n*—a common term for a person with extensive experience in a product category who performs perceptual evaluations to draw conclusions about the effects of variations in raw materials, processing, storage, aging, etc. Experts often operate alone. (See also **assessor and expert assessor.**) **E 253,** E18

expert assessor, *n*—an assessor with a high degree of sensory acuity who has experience in the test procedure and established ability to make consistent and repeatable sensory assessments. An expert assessor functions as a member of a sensory panel. (See also **assessor** and **expert.**) **E 253,** E18

exposed finish, *n*—units whose surfaces are intended to be left exposed or painted. **C 43,** C15

exposed tape width, *n*—*in zippers*, the part of the tape extending beyond the shoulders of the interlocking elements to the outer tape edge. **D 123,** D13

exposed underlayer—the underlying layer of mat or roving not covered by surface mat in a pultrusion.

Note—This condition can be caused by reinforcement shifting, too narrow surface mat, too wide underlying mat, uneven slitting of surface mat, or necking down of surface mat or excessive tension in pulling it off the spindle. **D 3918,** D20

exposure—(*1*) the transverse dimension of a roofing element not overlapped by an adjacent element in any roofing system. The exposure of any ply in a membrane may be computed by dividing the felt width minus 51 mm (2 in.), by the number of shingled plies; thus, the exposure of a 914-mm (36-in.) wide felt in a shingled, four-ply membrane should be 216 mm (8½ in.);

(*2*) the time during which a portion of a roofing element is exposed to the weather. **D 1079,** D08

exposure, *n*—contact with a chemical, biological, physical or other agent over a specified time period. **D 1356,** D22

exposure:—(*1*) The act of submitting material to radiation to which it is sensitive.

(*2*) the quantitative measure of exposure as a function of intensity and time of the radiation (often the product of $I \times t$) falling on a sensitive material.

(*3*) *X-ray*—the product of X-ray intensity and time. **E 7,** E04

exposure, *n*—the irradiance of a receiver integrated over the exposure time. **E 135,** E01

exposure—contact with a chemical or physical agent. **E 943,** E47

exposure—the product of irradiance and time required to produce a suitable pattern on the recording medium. **E 1316,** E07

exposure, *n*—the act of subjecting the test specimen to the test conditions. **G 113,** G03

exposure angle, *n*—the tilt from horizontal of the test specimen or any other exposed material, or both. **G 113,** G03

exposure duration, *n*— *in erosion or wear*, exposure time, or any other appropriate measure of the accumulation of exposure to an erosion or wear environment. **G 40,** G02

exposure index—the rating of a film for use in connection with exposure tables, exposure computers, and exposure meters. (See also **sensitivity.**) **E 7,** E04

exposure lattitude—range of light or time settings over which an acceptable copy can be produced. **F 335,** F05

exposure racks, at-latitude—in solar energy applications, racks that hold specimens at an inclination angle equal to the latitude of the rack location, facing south. **E 772,** E44

exposure, radiographic exposure—the subjection of a recording medium to radiation for the purpose of producing a latent image. Radiographic exposure is commonly expressed in terms of milliampere-seconds or millicurie-hours for a known source-to-film distance. **E 1316,** E07

exposure rate, X—the quotient of d*X* by d*t*, where d*X* is the increment of exposure in the time interval, d*t* (ICRU).

$$X = dX/dt$$

Unit: $C \cdot kg^{-1}\ s^{-1}$ **E 170,** E10

exposure scale—in a photographic process, the range of exposure over which substantially correct reproduction is obtained. This is measured by the ratio of the exposure corresponding to the minimum useful gradient at the high exposure end of the scale to that corresponding to the minimum useful gradient at the low exposure end. **E 7,** E04

exposure table—a summary of values of radiographic exposures suitable for the different thicknesses of a specified material. **E 1316,** E07

exposure time, *n*—the time during which a receiver is irradiated. **E 135,** E01

exposure, *X*—the quotient of d*Q* by d*m*, where the value of d*Q* is the absolute value of the total charge of the ions of one sign produced in air when all the electrons (negatrons and positrons) liberated by photons in air of mass d*m* are completely stopped in air (ICRU).

$$X = dQ/dm$$

Unit: $C \cdot kg^{-1}$ **E 170,** E10

extended binary code decimal interchange code, *n*—an eight-bit code set used on communications lines with IBM or IBM-compatible terminals and computers. (See **EBCDIC**.) **F 1457,** F05

extended life coolant, *n*—an engine coolant for light-duty service vehicles with recommended change-out of the coolant after 160 000 km (100 000 miles), 5 years, or 4000 operating hours. **D 4725,** D15

extended light scatterer (XLS)—a feature larger than the spatial resolution of the inspection equipment, on or in a wafer surface, resulting in increased light scattering intensity relative to that of the surrounding wafer surface. **F 1241,** F01

extended pigments—organic pigments diluted with an extender (for example, alumina trihydrate, blanc fixe, or calcium carbonate). **D 16,** D01

extended source—a source of infrared radiation whose image completely fills the field of view of a detector.

NOTE—The irradiance is independent of the distance from the source to the region of observation. In practice, sources that are not extended sources are considered to be point sources; see **point source**. **E 1316,** E07

extender—an additive whose primary purpose is to increase total grout volume. **D 653,** D18

extender, *n*—substance, generally having some adhesive action, added to an adhesive to reduce the amount of the primary binder required per unit area. (See also **binder, diluent, filler,** and **thinner**.) **D 907,** D14

extender, *n*—an organic material used to augment the polymer in a compound. **D 1566,** D11

extender—a material that increases the effective life of an agrichemical after application. **E 1519,** E35

extensibility, *n*—that property by virtue of which a material can undergo extension or elongation following the application of sufficient force. **D 123,** D13

extension, *n*—the change in length of a material due to stretching. (Compare **elongation.**) **D 123,** D13

extension—linear strain associated with an increase in length. (ISRM) **D 653,** D18

extension—movement by which the angle at the joint is increased in the sagittal plane; the opposite of flexion. **F 869,** F08

extensional definition, *n*—a definition of a term formed by listing all of the entities that comprise the concept. **E 1992,** E02

extension force, *n*—the force required to stretch a material to a given length. **D 123,** D13

extension lead wires—wires leading from a thermocouple to the voltage-measuring instrument. These wires should have the same temperature-emf relationship as the thermocouple wires. **E 7,** E04

extension-recovery chart, *n*—*in elastic materials testing*, a continuously plotted graph of tension versus extension resulting from an extension-recovery cycle. (Compare **tension-recovery chart.** See also **extension-recovery cycle** and **tension-recovery cycle**.) **D 123,** D13

extension-recovery cycle, *n*—*in tension testing*, the continuous extension of a specimen, with a momentary hold at a specified extension, followed by a controlled rate of return to zero extension. (Compare **tension-recovery cycle**. See also **extension-recovery chart** and **tension-recovery chart**.) **D 123,** D13

extension wires, *n*—those having temperature-emf characteristics that when connected to a thermocouple effectively transfer the reference junction to the other end of the wires (compare **compensating wires).** **E 344,** E20

extensometer, *n*—a device for determining elongation of a specimen as it is strained. **D 1566,** D11

extensometer, *n*—a device for measuring linear strain

E 6, E28

extensometer—a device for sensing the elongation of fastener material while it is subjected to tensile stress, for the purpose of measuring linear deformation under controlled test conditions. **F 1789,** F16

exterior glazed—glazed from the exterior of the building. **C 717,** C24

exterior insulation and finish system (EIFS), *n*—non-load-bearing outdoor wall finish system consisting of a thermal insulation board, an attachment system, a reinforced base coat, exterior joint sealant, and a compatible finish.

EIFS back-wrapping—process of enclosing exposed edges of thermal insulation board by applying the reinforced base coat around the edges and onto the backside of the insulation or onto the substrate.

EIFS base-coat adhesive—semi-liquid adhesive material providing a matrix within which reinforcing mesh is embedded as a base coat.

EIFS insulation board—rigid thermal insulation, interposed between lamina and substrate, providing thermal resistance to the wall assembly. See also **expanded polystyrene thermal insulation board.**

EIFS mechanical fastener, n—special mechanical device providing structural connection of the EIFS system, including the insulation board, to the wall structure.

EIFS reinforcement—mesh or other fibrous component of the EIFS base coat included to provide increased toughness and resistance to mechanical impact and cracking.

EIFS reinforcing mesh, n—woven or non-woven mesh fabric used as a reinforcement in the EIFS base coat.

EIFS substrate—surface or structure of a wall to which EIFS is adhesively or mechanically attached. **E 631,** E06

exterior pack—See **pack.** **D 996,** D10

exterior stop—a removable or integral stop on the outdoor side of glass or a panel. **C 717,** C24

exterior type plywood—a term applied to plywood that is capable of withstanding prolonged exposure to severe service conditions including prolonged and repeated wetting without failure in the glue bonds; the commercial classification is a function of veneer grade as well as glue durability. **D 1038,** D07

exterior work area—outdoor porch, stairway, or other element of trim or walls on the exterior of a building. **E 1605,** E06

external circuit—the wires, connectors, measuring devices, current sources, etc., that are used to bring about or measure the desired electrical conditions within the test cell. **G 15,** G01

external flotation (aka outboard flotation)—flotation element located external to the boom membrane. **F 818,** F20

external force—a force that acts across external surface elements of a material body. (ISRM) **D 653,** D18

external load-crushing strength test—a test of the pipe in which external crushing forces are exerted in specified directions and locations on a specified length of pipe. **C 822,** C13

external mixing pneumatic atomizer—a pneumatic atomizer in which pressurized gas is directed on a liquid film or jet outside the nozzle, so as to form a spray. **E 1620,** E29

external ROM receptacle, *n*—a printed wiring board that accepts read only memory cartridges containing fonts or emulations. (See **ERR**.) **F 1457,** F05

external sealing bands—flexible wrappings that are applied to the outside of a concrete pipe, box section, or manhole section joint intended to control the movement of fluids or solids through the joint. **C 822,** C13

external tension—external tension member separated from the boom membrane by bridles. **F 818,** F20

extinction—a decrease in the intensity of the diffracted beam caused by perfection or near perfection of crystal structure. (See also **primary extinction and secondary extinction.**) **E 7,** E04

extinction coefficient—the ratio of the diffracted beam intensity when extinction is present to the diffracted beam intensity when extinction is absent. It is applicable to either **primary** or **secondary extinction.** **E 7,** E04

extra-atomic relaxation energy—XPS, see **screening energy.** **E 673,** E42

extract—a liquid, powder, or solid concentrate of vegetable tannin obtained by extracting tannins from natural sources. **D 1517,** D31

extractable matter, *n*—nonfibrous material in or on a textile, not including water, which is removable by a specified solvent or solvents, as directed in a specified procedure. **D 123,** D13

extractable matter, *n*—nonfibrous material in or on a textile not including water that is removeable by a specified solvent or solvents, as directed in a specified procedure. **D 5684,** D13

extraction—a general term concerning chemical methods of isolating phases from the metal matrix.

(1) acid extraction—removal of phases by dissolution of the matrix metal in an acid.

(2) chlorine extraction/emph>—removal by formation of a volatile chloride.

(3) electrolytic extraction—removal by using an electrolytic cell containing an electrolyte which preferentially dissolves the metal matrix. **E 7,** E04

extraction, *n*—the process of removing the patient/subject from an immediately hazardous environment along with assessment, treatment, stabilization, and packaging. **F 1490,** F32

extraction cleaner, *n*—a portable, floor-supported cleaning system that can have the floor nozzle directly connected to the extractor or separated from the extractor housing by a hose, and is designed primarily for wet cleaning of carpet. The cleaning involves applying a solution on the carpet and its subsequent removal. The solution dispensing system may be totally self-contained or require hook up to a water supply when in use. The extractor may have some form of agitation to assist in the soil removal. The extractor may have attachments and provisions for cleaning other surfaces. **F 395,** F11

extraction cleaning, *n*—a general term for a number of refurbishing methods in which the cleaning agent is delivered onto the textile product, agitated, and simultaneously removed by suction. **D 123,** D13

extractives—compounds occurring in plant materials but not forming part of the structural elements, that are removed with neutral solvents such as ether, alcohol, and water. **D 9,** D07

extractives—compounds occurring in plant materials, but not forming part of the structural elements, that are removed with neutral solvents such as ether, alcohol, and water. **D 1695,** D01

extrados—the exterior curved surface of an arch, as opposed to intrados, which is the interior curved surface of an arch. (ISRM) **D 653,** D18

extra large head—usually more than 3 ½ times shank diameter, normally referring to head of regular-stock-steel nail. **F 547,** F16

extraneous ink—any spot appearing within the "read" area, but outside the **COL**, caused by smear, tracking, or splatter that can be caused either in the manufacturing or while entering data on the form and can result in less optimum readability. **F 149,** F05

extraneous moisture—synonym for *surface moisture.*

D 121, D05

extrapolated onset temperature, *n*—empirically, the temperature found by extrapolating the baseline (prior to the peak) and the leading side of the peak to their intersection. **E 1445,** E27

extra smooth sheet, *n*—product produced by cold rolling (skin passing) the metallic coated sheet with a small reduction in thickness to smooth the surface and impart resistance to stretcher strain. (Syn. *skin passed sheet*) **A 902,** A05

extra-steep asphalt— See **super-steep asphalt.** **D 1079,** D08

extreme pressure (EP) additive, *n—in a lubricant,* a substance that minimizes damage to metal surfaces in contact under high stress rubbing conditions. **D 4175,** D02

extreme weather conditions—environmental conditions that have occurred only once during the past 30 years. **E 1705,** E48

extrinsic fiber optic chemical sensor, *n*—a fiber optic chemical sensor in which modulation of the optical signal is not effected through a change in the properties of the fiber itself. **E 131,** E13

extrudability—force required to produce sealant flow through an orifice. **C 717,** C24

extrudate, *n*—the material that issues from an extruder. **D 1566,** D11

extrude, *v*—to shape a plastic body by forcing the body through a die. **C 1145,** C28

extruded—formed by being forced through a shaping orifice as a continuous body. **C 709,** D02

extruded—forced through the shaping die of an extruder. The extrusion may be of solid or hollow cross section. **F 869,** F08

extruded particleboard—a particleboard manufactured by forcing a mass of particles coated with an extraneous binding agent through a heated die with the applied pressure parallel to the faces and in the direction of extruding. **D 1554,** D07

extruder, *n*—a device, such as a pug mill, that forces plastic bodies through a die of appropriate shape and size in a continuous column. **C 1145,** C28

extruder, *n*—machine designed to force a rubber or rubber mix through an orifice, which is often shaped to the geometry of the desired product. **D 1566,** D11

extrusion—a uniform metal shape, long in relation to its cross-sectional dimensions, produced by forcing a suitably preheated billet or preform through an orifice (die) of the desired cross section. Extrusions generally are furnished in straight lengths. **B 846,** B05

extrusion, *n*—shaping of brick by pushing plastic clay or shale through a die opening that forms the peripheral dimensions of the brick. The column of extrudate is then cut into sections to provide the third dimension of the brick. Water is added to the clay or shale in sufficient quantities to permit laminar flow through the extrusion machine. The consistency of the extrudate may vary from stiff and capable of supporting several times its weight to soft and deformable under slight loads. **C 43,** C15

extrusion, *n*—a process in which heated or unheated plastic is forced through a shaping orifice (a die) in one continuously formed shape, as in film, sheet, rod, or tubing. **D 883,** D20

extrusion, *n*—the continuous shaping of a material during plastic passage through a die. **D 1566,** D11

extrusion, *n*—a process whereby heated or unheated plastic forced through a shaping orifice becomes one continuously formed piece. (D-20) **F 412,** F17

extrusion billet, *n*—solid or hollow form, commonly cylindrical, used as the final length of material charged into the extrusion press cylinder, and is usually a cast product, but may be a wrought product or sintered from powder compact. **B 881,** B07

extrusion coating—*in flexible barrier materials,* (*1*) a process of extrusion whereby a molten extrudate adheres to the surface of another (solid) material, forming a continuous layer upon cooling. (*2*) Also, the product resulting from such a process. **F 17,** F02

extrusion coating—See Terminology F 17. **F 1327,** F02

extrusion ingot, *n*—solid or hollow cast form, usually cylindrical, suitable for extruding. **B 881,** B07

extrusion log, *n*—starting stock for extrusion billet. Extrusion log is usually produced in lengths from which shorter extrusion billets are cut. **B 881,** B07

extrusion mark (score line), *n*—a mark formed on an extrudate during the extrusion process either by accident or intent. **D 1566,** D11

extrusion welding—a process in which heated plastic is forced through a shaping orifice (or die) and applied with pressure to suitably prepared, locally preheated plastic pieces of the same resin base, to join them. **C 904,** C03

exudation—the action by which all or a portion of the low melting constituent of a compact is forced to the surface during sintering. Sometimes referred to as "bleed out." Synonymous with **sweating.** **B 243,** B09

exudation—the migration of a substance to the surface of a coating or lining film. **D 4538,** D33

eye—the opening in the bottom of a pot furnace through which the flame enters. **C 162,** C14

eye bolt—a bolt having a head which is a closed or open ring which as a threaded shank and has a defined breaking strength, proof load, and tensile strength. **F 1789,** F16

eyebrow—used for direct mounting to ovens and some dishwashers. **F 1827,** F26

eye clearance—the distance from the back lens of an eyepiece to the proper location of the viewer's eye, typically about 8 mm (about 20 mm for high eyepoint eyepieces which permit the use of eyeglasses). **E 7,** E04

eye lens—the lens in an eyepiece nearest to the eye. **E 7,** E04

eyepiece—the lens system used in an optical instrument for magnification of the image formed by the objective. **E 7,** E04

eyepiece micrometer— See **ocular micrometer.** **E 7,** E04

F

F—an F factor that contains a portion of the *E. coli* genome. **E 1705,** E48

FA—false acceptance; incorrectly accepting an inaccurate or non-genuine item or fact as if it were real; also called **Type II error**. **F 1156,** F12

fabric, *n—in textiles,* a planar structure consisting of yarns or fibers. **D 123,** D13

fabric—a fibrous material containing natural or synthetic fibers, or both, in yarn form, interlaced in various configurations (woven, knitted, or nonwovens). **D 459,** D12

fabric—*for rock or soil,* the spatial configuration of all textural and structural features as manifested by every recognizable material unit from crystal lattices to large scale features requiring field studies. **D 653,** D18

fabric—the orientation in space of the elements composing the rock substance. (ISRM) **D 653,** D18

fabric, *n*—a woven or nonwoven cloth of organic or inorganic filaments, threads, or yarns. **D 1079,** D08

fabric—a planar textile (Synonym: cdth)

braided fabric, n—a cloth constructed by a braiding process.

knitted fabric, n—a cloth constructed by a knitting process.

nonwoven fabric, n—a cloth constructed by bonding or interlocking, or both (but not interlacing) fiber by any combination of mechanical, chemical, thermal, or solvent means.

plied yarn, n—a yarn formed by twisting together two or more single yarns in one operation.

woven fabric, n—a cloth constructed by a weaving process. **D 3878,** D30

fabricate, *v—in buttons,* the conversion of a blank into a complete button. **D 123,** D13

fabricate, *v*—to manufacture, form, construct, or assemble a product or **component.** **E 631,** E06

fabricating, *n*—the manufacture of plastic products from molded parts, rods, tubes, sheeting, extrusions, or other forms by appropriate operations such as punching, cutting, drilling, and tapping including fastening plastic parts together or to other parts by mechanical devices, adhesives, heat sealing, or other means. **D 883,** D20

fabricating, *n*—the manufacture of plastic products from molded parts, rods, tubes, sheeting, extrusions, or other forms by appropriate operations such as punching, cutting, drilling, and tapping including fastening plastic parts together or to other parts by mechanical devices, adhesives, heat sealing, or other means. (D-20) **F 412,** F17

fabricator, *n—as related to corrugated metal pipe,* (*1*) the organization that produces the finished pipe, or (*2*) for structural plate pipe, the organization that processes flat sheets and other items needed for the field assembly of the finished products. **A 902,** A05

fabric (chain link fence)—a fencing material made from wire helically wound and interwoven in such a manner as to provide a continuous mesh without knots or ties except in the form of knuckling or of twisting the ends of the wires to form the selvages of the fabric. **F 552,** F14

fabric count—the number of warp and woof ends per inch, and the number of filling picks per inch. **F 869,** F08

fabric dip, *n—for tire fabrics,* a chemical composition which is applied to a textile cord or fabric to improve its adhesion to rubber compounds. **D 123,** D13

fabric, *n* (édifice)—*of a building,* all the elements, components, parts, and materials of a building, at any scale and of any age.

historic fabric, n (édifice historique)—those portions of the building fabric that have historic significance. **E 1480,** E06

fabric growth, *n*—the difference between the original length of a specimen and its length after the application of a specified tension for a prescribed time and the subsequent removal of the tension **D 123,** D13

fabric ribbon—an inked ribbon wherein the substrate is a woven cloth material, such as nylon, cotton, silk, etc. **F 221,** F05

fabric roof-system, *n*—a system of coated fabric or laminated fabric along with support cables, edge ropes, clamps, neoprene, roof drains, arch wear strips, and anchor bolts that constitutes the outside top covering of a building. **D 123,** D13

fabric softener—a laundry auxiliary product or laundry detergent ingredient whose primary function is to give fabrics a soft feel, smooth surface, or reduce static electricity, or a combination thereof. **D 459,** D12

fabric stability, *n—in vinyl-coated glass screening and louver cloth,* the property denoting the ability to resist slippage of yarn segments in one direction over yarn segments in the opposite direction. **D 123,** D13

fabric stretch, *n*—the increase in length of a specimen of a fabric resulting from a load applied under specified conditions. **D 123,** D13

FAC—free available chlorine. **D 6161,** D19

facade paint—a decorative and protective coating for exterior masonry surfaces–usually for buildings and walls. **D 16,** D01

face, *n*—the surface designed to be left exposed to view or to receive decoration or additional finishes. **C 11,** C11

face—to cover with a new surface. **C 896,** C04

face—the wide surface of rectangular pieces of lumber. Often the surface that determines the grade of lumber destined for remanufacture. **D 9,** D07

face, *n—in buttons,* that portion which will be exposed after attaching to the substrate. **D 123,** D13

face—See **container.** **D 996,** D10

face—the better side of a panel in any grade of plywood calling for a face and back; also either side of a panel where the grading rules draw no distinction between faces. **D 1038,** D07

face, *adj*—pertaining to viewing a specimen at an angle close to its surface normal. **E 284,** E12

face angle—the included dihedral angle between two opposite faces of an indenter. **E 7,** E04

face angle, *n*—the aspecular angle when a specimen is viewed close to its surface normal and illuminated near 45°, or when those angles of illumination and viewing are interchanged. **E 284,** E12

face-centered—having atoms (or groups of atoms) separated by translations of ½, ½, 0; ½, 0, ½; and 0, ½, ½ from a similar atom (or group of atoms). The number of atoms in a face-centered cell must be a multiple of four. **E 7,** E04

face color, *n*—the color of a material when viewed at its face angle. **E 284,** E12

face (crystal)—an idiomorphic plane surface on a crystal. **E 7,** E04

face dimpling, n—buckling of the compressive facing into the individual cells of the honeycomb core due to compressive loading or the prepreg facing sagging into the individual honeycomb cells during cocure. **C 274,** D30

face glazing, *n*—the method of glazing in which a rabbeted glazing recess, with no removable stop, is used and a triangular bead of compound is applied to the face of the lite and extended onto the recess. **C 717,** C24

face (heading)—the advanced end of a tunnel, drift, or excavation at which work is progressing. (ISRM) **D 653,** D18

face nail—See **siding nail.** **F 547,** F16

face side, *n—in textile materials*, the side of the material that is outward in the completed object. (Ant. **back side**.) **D 123,** D13

face surface bend—procedure whereby load is applied perpendicular to the weld axis of the joint, at the weld, such that the weld face is subject to tension. **C 904,** C03

face wrinkling, n—buckling of the compressive facing into or away from the core. This progresses the width of the panel and causes failure. **C 274,** D30

facility, *n*—a physical setting used to serve a specific purpose.

facility durability, n—the capability of a **facility** to maintain serviceability for a specified time.

facility evaluation, n—comparison of the qualitative and quantitative results of observations, measurements, analyses, or other tests against criteria established for a specified purpose and to a specified precision and reliability.

facility function, n—the purpose or activity for which the **facility** is designed, used, or required to be used.

facility management—practice of planning and managing workplaces.

facility performance, n—the behavior in service of a **facility** for a specified use.

facility project brief (statement of work)—document describing services to be provided by the design consultant (architect, engineer, or interior designer) for a **facility,** in detail sufficient for the design to proceed.

facility serviceability, n—the capability of a **facility** to perform the function(s) for which it is designed, used, or required to be used.

facility use, n—the functions and activities that take place in a **facility.** **E 631,** E06

facility—see **hazardous-waste facility**. **E 1605,** E06

facility durability, *n* (durée de facilité)— *of a facility*, the capability of a facility to maintain serviceability for at least a specified period of time. Compare **durability**—*of a building*. **E 1480,** E06

facility evaluation, *n*—comparison of the qualitative and quantitative results of judgments, observations, measurements, analyses, or other tests against performance criteria established for a specified purpose, and to a specified precision and reliability. (See also **requirement statement**.) Compare **rating process**. **E 1480,** E06

facility-in-service, *n* (installation en service)—facility as completed and operational; for example, an occupied **building** or a road in service. **E 1480,** E06

facility, *n* (facilité)—a physical setting used to serve a specific purpose. **E 1480,** E06

facility operator, *n* (exploitant d'installations)—organization or agency having a contract with the owner or investor to operate a facility. **E 1480,** E06

facility performance, *n* (rendement d'installations)—behavior in service of a facility for a specified use. **E 1480,** E06

facility program, *n* (programme d'installations):

design program, n (programme de conception)—*(design brief)* document specifying what facilities will be provided to the **occupants**, and confirming to the owner the requirements for the facility.

functional program, n (programme fonctionnel—document that specifies functional facility servicability requirements of **occupants** and owner.

project brief, n (exposé de projet)—document describing the required **facility serviceability** in detail sufficient for the project to proceed. **E 1480,** E06

facility project brief (statement of work), *n* (exposé de projet d'installations (exposé de travaux))—document describing services to be provided by the design consultant (architect, engineer, or interior designer) for a facility, in detail sufficient for the design to proceed. **E 1480,** E06

facility scattered neutrons—neutrons scattered in the facility that contribute to the film exposure. **E 1316,** E07

facility servicability, *n* (fonctionnalité d'installations)—capability of a facility to perform the function(s) for which it is designed, used, or required to be used. **E 1480,** E06

facility serviceability profile, *n* (profil de fonctionnalité d'installations)—graphic display, usually as a bar chart, of the set of **rating scores** for aspects of the serviceability of a facility. **E 1480,** E06

facility use, *n* (utilisation d'installations)—functions and activities that take place in a facility. **E 1480,** E06

facing, *n*—a protective or decorative (or both) surface applied as the outermost layers of an insulation system. **C 168,** C16

facing—the outer layer of revetment. **D 653,** D18

facing, *n*—a form of *linerboard* used as a flat member of **corrugated fiberboard** (sometimes erroneously called a **liner**). **D 996,** D10

facing—the outermost layer or composite component of a sandwich construction, generally thin and of high density, that resists most of the edgewise loads and flatwise bending moments (Syn. *face; skin*). **E 1749,** E06

facing brick, *n*—brick made especially for facing purposes. Produced from selected clays and available in all typical face sizes, various colors, and in smooth, fine, medium, and coarse textures. See Specification C 216. **C 43,** C15

facing delamination, n—where the facing becomes disbonded from the core. **C 274,** D30

facing leathers—a lightweight leather generally used for facing seams, and binding the edges of shoe uppers. Also applied to lightweight, smooth calf and lamb, and to skivers, of which the inner surfaces of billfolds and wallets are frequently made. **D 1517,** D31

facing, n—the outermost layer or composite component of a sandwich construction, generally thin and of high density, which resists most of the edgewise loads and flatwise bending moments: synonymous with face, skin and facesheet. **C 274,** D30

facing unit—manufactured masonry unit designed for use where one or more faces will be exposed and for which the specification includes requirements on color, finish, and other properties affecting appearance. **C 1232,** C15

factor, *n—in experimenting*, a condition or circumstance that is being investigated to determine if it has an effect upon the result of testing the property of interest. **D 123,** D13

factorial experiment (general), *n*—in general, an experiment in which all possible treatments formed from two or more factors, each being studied at two or more levels (versions) are examined so that interactions (differential effects) as well as main effects can be estimated. **E 456,** E11

factorial experiment (general), *n*— in general, an experiment in which all possible treatments formed from two or more factors, each being studied at two or more levels (versions), are examined so that interactions (differential effects) as well as main effects can be estimated. **E 1325,** E11

factory-finished boards—boards with a factory-applied surface as, for example, powder or liquid coatings or overlays. These finished boards require no further field finishing. **D 1554,** D07

factory-primed boards—boards with a factory-applied primer that requires subsequent finishing in the field. **D 1554,** D07

factory sole leather—one of the two principal types of sole leather. It is tanned and finished to have more flexibility and compressibility than Finders' sole leather, and is more suitable for use in shoemaking machinery. (See also **Finders' sole leather.**) **D 1517,** D31

factory square—10 m^2(108 ft^2). **D 1079,** D08

fade—{archaic} attack of the surface of glass causing an oily or whitish surface. **C 162,** C14

fading, *n*—a change in color, usually to a lighter and less-saturated color. **E 284,** E12

fading unit (AATCC), *n*—a specific amount of exposure made under the conditions specified in various test methods where one Fading Unit is one-twentieth (1⁄20th) of the exposure required to produce a color change equal to Step No. 4 on the Gray Scale for Color Change or 1.7 ± 0.3 CIELAB units of color difference on Blue Wool Lightfastness Standard L 4 or 20 ± 1.7 CIELAB units of color difference on the Xenon Reference Fabric or combination thereof.(**AATCC Technical Manual)** **G 113,** G03

Fahrenheit, *adj*—pertaining to or denoting something related to the expression of temperature in degrees Fahrenheit. **E 344,** E20

Fahrenheit—designation of a degree on the Fahrenheit temperature scale that is related to the International Practical Temperature Scale by means of the equation: $T_F = 1.8\ T_C + 32$.

where:

T_F is the temperature in degree Fahrenheit and T_C is the temperature in degrees Celsius. **E 1142,** E37

Fahrenheit (°F)—designation of a degree on the Fahrenheit temperature scale that is related to the International Practical Temperature Scale. **D 6161,** D19

faience mosaics—faience tile that are less than 6 $in.^2$ (39 cm^2) in facial area, usually 5⁄16 to 3⁄8 in. (8 to 9.5 mm) thick, and usually mounted to facilitate installation. **C 242,** C21

faience tile—glazed or unglazed tile, generally made by the plastic process, showing characteristic variations in the face, edges, and glaze that give a handicrafted, nonmechanical, decorative effect. **C 242,** C21

faience ware—formerly a decorated earthenware with an opaque glaze, but currently designating a decorated earthenware having a transparent glaze. **C 242,** C21

failure, *n*—an arbitrary point beyond which a material ceases to be functionally capable of its intended use. (Compare **rupture.)** **D 123,** D13

failure, *n*—*in sewn fabric seams,* fabric or sewing thread rupture or seam (yarn) slippage. **D 123,** D13

failure, n—See **adhesive-joint failure, adhesion failure, cohesion failure** and **substrate failure.** **D 907,** D14

failure—see **dielectric failure.** **D 1711,** D09

failure, *n*—an arbitrary point beyond which a material ceases to be functionally capable of its intended use. **D 4439,** D35

failure, *n*—*in testing geosynthetics,* water or air pressure in the test vessel at failure of the geosynthetic. **D 4439,** D35

failure, adhesive—rupture of an adhesive bond, such that the plane of separation appears to be at the adhesive-adherend interface. **F 412,** F17

failure by rupture—see **shear failure.** **D 653,** D18

failure criterion—specification of the mechanical condition under which solid materials fail by fracturing or by deforming beyond some specified limit. This specification may be in terms of the stresses, strains, rate-of-change of stresses, rate-of-change of strains, or some combination of these quantities, in the materials. **D 653,** D18

failure criterion—theoretically or empirically derived stress or strain relationship characterizing the occurrence of failure in the rock. (ISRM) **D 653,** D18

failure (in rocks)—exceeding the maximum strength of the rock or exceeding the stress or strain requirement of a specific design. (ISRM) **D 653,** D18

failure (of a bulk solid)—plastic deformation of an overconsolidated bulk solid subject to shear, causing dilation and a decrease in strength. **D 653,** D18

fairing—a shape that produces a smooth transition from one direction to another. Also referred to as a *feathering*. **E 1749,** E06

fall—undesirable descent due to the force of gravity, usually from a standing posture or during ambulation, to a lower level, usually the ground or floor. **F 1646,** F13

fallback—shrinkage, settlement, or loss of plugging material placed in a borehole or well. **D 653,** D18

fallback—a reduction in bitumen softening point, sometimes caused by refluxing or overheating in a relatively closed container. (See also **softening point drift**). **D 1079,** D08

fall-off—tendency of an electrostatically deposited powder to fall off the work piece during normal processing. **C 286,** B08

fall velocity—the settling rate of a particle in a given medium. **D 4410,** D19

false brinelling, *n*—damage to a solid bearing surface characterized by indentations not caused by plastic deformation resulting from overload but thought to be due to other causes such as *fretting corrosion*. (See also **brinelling**.) **G 40,** G02

false indication, *n*—an NDT indication that is interpreted to be caused by a condition other than a discontinuity or imperfection. **E 1316,** E07

false set—see **early stiffening** **C 219,** C01

false set—*in grouting*, the rapid development of rigidity in a freshly mixed grout without the evolution of much heat. **D 653,** D18

family—a complete series of penetrant materials required for the performance of a liquid penetrant examination. **E 1316,** E07

family, *n*—a set of fonts sharing the same type style, but differing in height, weight, and posture. **F 1457,** F05

family (of crystal planes)—the planes in any one crystal that have common Miller indices, regardless of sign. **E 7,** E04

fan/control energy rate—the rate of energy consumption (Btu/h or kW) by the equipment's controls or fan motor, to both, E_{fan}.

$$E_{fan} = \frac{E_{fan} \times 60}{t_{preheat} \times 1000}$$

where:

E_{fan}	=	fan/control energy measured in Btu or Wh, and
$t_{preheat}$	=	preheat time, min.

F 1827, F26

fancy leather—leathers made from hides and skins of all kinds that have commercial importance and value primarily because of grain, or distinctive finish, whether natural or the result of processing. Such processing may be graining, printing, embossing, ornamenting (including in gold, silver, and aluminum finishes), or any other finishing operation enhancing the appeal of leather. **D 1517,** D31

fan first vacuum cleaner system, *n*—a vacuum cleaner construction in which the dirt laden air is passed through the fan system and then into the dirt receptacle (bag type filter, bagless filter or other type of dirt separator). The separated air is then expelled from the cleaner. This type of construction is sometimes referred to as dirty air or direct system. **F 395,** F11

fan pattern, *n*—a spray pattern in which the liquid flux is concentrated in a narrow oval or ellipse in a plane perpendicular to the spray axis. **E 1620,** E29

fan spray atomizer—a pressure atomizer that produces a flat sheet of liquid that collapses into particles. The angle or width of the sheet is controlled by the shape of a slot or oval discharge orifice, by an external deflector, or by impinging jets. **E 1620,** E29

faraday—the number of coulombs (96,490) required for an electrochemical reaction involving one chemical equivalent. **B 374,** B08

far field—the zone of the beam where equal reflectors give exponentially decreasing amplitudes with increasing distance. **E 1316,** E07

far-infrared—pertaining to the infrared region of the electromagnetic spectrum with wavelength range from approximately 25 to 1000 μm (wavenumber range 400 to 10 cm^{-1}). **E 131,** E13

fast atom bombardment mass spectrometry (FABMS)—the technique in which a mass spectrometer is used to measure the mass-to-charge ratio and abundance of secondary ions emitted from a target as a result of the bombardment of fast neutral atoms. **E 673,** E42

fast axis—*in optics*, of a doubly refracting crystal, that direction in which the velocity of light is a maximum. **F 1241,** F01

fastener, *n*—*in packaging,* a device that serves to secure one part to another; for example, nail, screw, **staple, strapping, stitch** or adhesive. **D 996,** D10

fastener, *n*—integral tooth of connector plate and/or separate nail used to fasten connector plate to wood member. **E 1807,** E06

fastener electroplating—the electro deposition by electrolysis of an adherent metallic coating upon a fastener serving as an electrode. This coating may function as protective, decorative, or in a defined engineering function such as wearability or abrasive resistance. **F 1789,** F16

fasteners:

self-sealing fastener—a fastener that provides a tight seal without the need for sealant material nor the use of a mechanical seal (for example, an interference fit fastener).

wet-installed fastener—a fastener that is coated on the shank and under the head with a curing-type sealant to provide a corrosion barrier and a secondary seal. **E 1749,** E06

fastener standard—a document which details the attributes of a finished fastener and includes such characteristics as geometry, material or chemistry, heat treatment, finish, lot size, and packaging. **F 1789,** F16

fast Fourier transform (FFT)—a method for speeding up the computation of a discrete FT by factoring the data into sparse matrices containing mostly zeroes. **E 131,** E13

fast neutrons—a term for designating neutrons of energy exceeding some threshold that must be specified (typically 0.1 or 1 MeV); often associated with those neutrons predominantly responsible for displacement damage of materials in neutron radiation fields. **E 170,** E10

fast pack container— See **container.** **D 996,** D10

Fast Walsh/Hadamard/Fourier transforms—See **transforms.** Complex mathematical procedures for data handling. **F 1156,** F12

fate, environmental—the form and location of a material resulting from transport and transformation. **E 943,** E47

fatigue—the process of progressive localized permanent structural change occurring in a material subjected to conditions that produce fluctuating stresses and strains at some point or points and that may culminate in cracks or complete fracture after a sufficient number of fluctuations. **D 653,** D18

fatigue—decrease of strength by repetitive loading. (ISRM) **D 653,** D18

fatigue, *n*—the decrease in response of a photoelectric radiant energy receiver caused by the accumulated exposure of the receiver to radiant energy. **E 135,** E01

fatigue—the process of progressive localized permanent structural change occurring in a material subjected to conditions that produce fluctuating stresses and strains at some point or points and that may culminate in cracks or complete fracture after a sufficient number of fluctuations. **E 1823,** E08

fatigue—any degradation in electrophotographic properties of a photoconductive material caused by repetitive charging, light exposure, or other operation in the imaging process. **F 335,** F05

fatigue, *n*—the process of progressive localized permanent structural change occurring in a material subjected to conditions that produce fluctuating stresses and strains at some point or points and that may culminate in cracks or complete fracture after a sufficient number of fluctuations. **F 1582,** F04

fatigue breakdown, *n*—the deterioration of an elastomeric product during repeated deformation. **D 1566,** D11

fatigue-crack-growth rate, *da/dN* , [L]—the rate of crack extension under fatigue loading, expressed in terms of crack extension per cycle of fatigue. **E 1823,** E08

fatigue-crack-growth threshold, ΔK_{th} [$FL^{-3/2}$]—that asymptotic value of ΔK at which da/dN is essentially zero. For most materials an operational, though arbitrary, definition ofΔ K_{th} is given as that ΔK which corresponds to a fatigue crack growth rate of 10^{-10} m/cycle. **E 1823,** E08

fatigue cycle—See **cycle.** **E 1823,** E08

fatigue ductility D_f—the ability of a material to deform plastically before fracturing, determined from a constant-strain amplitude, low-cycle fatigue test. **E 6,** E28

fatigue ductility exponent, *c*, *n*—the slope of a log-log plot of the plastic strain range and the fatigue life. **E 6,** E28

fatigue, dynamic, *n*—the deterioration of a material by repeated deformation. See fatigue breakdown. **D 1566,** D11

fatigue life, *n*—the number of deformations required to produce a specified state of fatigue breakdown in a test specimen or product that is deformed under a prescribed set of conditions. **D 1566,** D11

fatigue life, *n*— the number of loading cycles, *N*, of a specified character that a given specimen sustains before failure of a specified nature occurs. **F 1582,** F04

fatigue life for *p* %survival—an estimate of the fatigue life that *p* % of the population would attain or exceed under a given loading. The observed value of the median fatigue life estimates the fatigue life for 50 % survival. Fatigue life for *p* % survival values, where *p* is any number, such as, 95, 90, and so forth, also may be estimated from the individual fatigue life values. **E 1823,** E08

fatigue life, N_f, *n*—the numbers of cycles of stress or strain of a specified character that a given specimen sustains before failure of a specified nature occurs. **E 6,** E28

fatigue life, N_f —the number of cycles of a specified character that a given specimen sustains before failure of a specified nature occurs. Fatigue life, or the logarithm of fatigue life, is a dependent variable. **E 1823,** E08

fatigue limit—point on stress-strain curve below which no fatigue can be obtained regardless of number of loading cycles. (ISRM) **D 653,** D18

fatigue limit—(See *Endurance Limit*). **F 1789,** F16

fatigue limit for *p* %survival [FL^{-2}]—the limiting value of fatigue strength for *p* % survival as *N* becomes very large; *p* may be any number, such as 95, 90, and so forth. **E 1823,** E08

fatigue limit, S_f [FL^{-2}]—the limiting value of the median fatigue strength as the fatigue life, N_f, becomes very large. **E 1823,** E08

fatigue loading—periodic, or not periodic, fluctuating loading applied to a test specimen or experienced by a structure in service. (Also known as *cyclic loading*.) **E 1823,** E08

fatigue notch factor, k_f—the ratio of the fatigue strength of a specimen with no stress concentration to a specimen with a stress concentration for the same percent survival at *N* cycles and for the same conditions. **E 1823,** E08

fatigue notch sensitivity, *q*—a measure of the degree of agreement between fatigue notch factor, k_f, and theoretical stress concentration factor, k_t. **E 1823,** E08

fatigue strength—the maximum stress on an externally threaded fastener which can be tolerated for a specified number of repeated cycles prior to failure. **F 1789,** F16

fatigue strength at *N* cycles, S_N [FL^{-2}]—a value of stress for failure at exactly *N* cycles as determined from an $S - N$ diagram. The value of S_N thus determined is subject to the same conditions as those which apply to the $S - N$ diagram. **E 1823,** E08

fatigue strength at N Cycles, S_n[FL^{-2}], *n*—a value of stress for failure at exactly *N* cycles as determined from an *S-N* diagram. The value S_n thus determined is subject to the same conditions as those that apply to the *S-N* diagram. **F 1582,** F04

fatigue strength for *p* % survival at *N* cycles [FL^{-2}]—an estimate of the stress level at which *p* % of the population would survive *N* cycles; *p* may be any percent, such as 95, 90, and so forth. **E 1823,** E08

fatigue test, *n*— a test designed to evaluate the cyclic load properties of a material, component, interconnection, subconstruct, construct, subassembly, or assembly. **F 1582,** F04

fatigue testing system—a device for applying repeated load cycles to a specimen or component. **E 1823,** E08

fatigue wear, *n*—wear of a solid surface caused by fracture arising from material fatigue. **G 40,** G02

fatliquor—an emulsion of oils or greases in water, usually with an emulsifying agent, used to lubricate the fibers of leather. **D 1517,** D31

fatty alcohol sulfate—the product obtained by treating a one-chain fatty alcohol with a sulfonating agent, the major constituent being the half sulfuric acid ester of the fatty alcohol or a salt thereof. **D 459,** D12

fatty matter, free—the sum of the free rosin acids and free fatty acids plus the unsaponified and unsaponifiable fatty matter. **D 459,** D12

fatty matter, total—fatty and rosin acids plus unsaponified and unsaponifiable fatty matter. This fatty matter is usually isolated from an acidic 50/50 volume percent solution of ethanol and water by extraction with petroleum ether, as per Test Method D 460 and Test Methods D 820. **D 459,** D12

fatty matter, unsaponifiable—fatty matter (other than acids) that contain no saponifiable esters, such as fatty alcohols, and mineral oil. **D 459,** D12

fatty matter, unsaponified—fatty matter containing saponifiable esters, such as fatty oils, glycerides, and lanolin. **D 459,** D12

fat wrinkle—wrinkles on the grain of leather caused by fat deposits in the live animal. **D 1517,** D31

fault—a fracture or fracture zone along which there has been displacement of the two sides relative to one another parallel to the fracture (this displacement may be a few centimetres or many kilometres). (See also **joint fault set** and **joint fault system**. (ISRM) **D 653,** D18

fault breccia—the assemblage of broken rock fragments frequently found along faults. The fragments may vary in size from inches to feet. (ISRM) **D 653,** D18

fault gouge—a clay-like material occurring between the walls of a fault as a result of the movement along the fault surfaces. (ISRM) **D 653,** D18

faulting of joints and cracks, *n*—difference in elevation across a joint or crack. **E 867,** E17

faulting of joints and cracks, *n*— difference in elevation across a joint or crack. **E 1778,** E17

faying surface—the surface that makes contact with another surface. **E 1749,** E06

faying surface seal—a seal installed between two overlapping surfaces. **E 1749,** E06

FDA—Food and Drug Administration (USA). **D 6161,** D19

feasibility study, *n* **(étude de faisabilité)**—study of a planned scheme or development, the practicality of its achievement, and its projected financial outcome. **E 1480,** E06

feather—See **striation.** **C 162,** C14

feather fiber, *n*—detached barbs of feathers which are not joined or attached to each other. **D 123,** D13

feathering, *n*—a ragged edge on printed type. **D 6488,** D01

feathering, *n*—a ragged or finely trailing pattern at the sharp edge of type or solid areas. **D 6488,** D01

feathering—See **fairing.** **E 1749,** E06

feathering—an undesirable thread-like deposit extending radially from the edge of an image. **F 221,** F05

feathering—an undesirable thread-like deposit extending radially from the edge of an image (F 221). **F 335,** F05

feathering, *n*—a type of element irregular wear characterized by thin strips of rubber extending from the edge of the element. **F 538,** F09

feathering, *n*—instantaneous ink spread over substrate causing fuzzy edges, spidery lines and poor print quality. See also **wicking.** **F 1857,** F05

feathers—an imperfection consisting of clusters of fine seed caused by dirt or foreign material entering the glass at the time of casting or shaping. **C 162,** C14

feathers, *n*—the outgrowth forming the contour and external covering of fowl. **D 123,** D13

feathers, crushed, *n*—feathers and feather fiber resulting from curling, crushing, or chopping feathers without removing the quill. **D 4523,** D13

feathers, damaged, *n*—feathers that have been broken, damaged by insects, by mildew or rot, or otherwise materially injured. **D 4523,** D13

feathers, nestling, *n*—immature feathers in which the barbs are held together and covered by a sheath. **D 4523,** D13

feathers, nonwaterfowl, *n*—feathers derived from chickens, turkeys, or other landfowl. (Syn. *feathers, landfowl*) **D 4523,** D13

feathers, quill, *n*—feathers which are over 100 mm (4 in.) in length or which have a quill point exceeding 9.5 mm (⁹⁄₁₆ in.) in length. **D 4523,** D13

feathers, waterfowl, *n*—feathers from ducks or geese, or both. **D 4523,** D13

featured edge, *n*—an edge configuration of the paper bound edge of gypsum board that provides special design or performance. **C 11,** C11

feature, *n* (élément caractéristique)—of a facility, a building element, building component, building subsystem, unit of furnishing or

equipment, or aspects of design, arrangement, form of color, which helps or hinders the satisfaction of a requirement for serviceability.
combination of features, n (éléments caractéristiques combiné)—*of a facility*, features which, when present together in a facility, affect satisfying a requirement for serviceability. **E 1480,** E06

feature-specific measurement, *n*—an individual measurement of each detected feature in the field of view. **E 7,** E04

Federal Register (FR)—document published daily by the Federal Government that contains such entries as proposed or final regulations. See **CFR** and **Code of Federal Regulation**. **E 1605,** E06

feebly magnetic material—a material generally classified as" non-magnetic," whose maximum normal permeability is less than 4. **A 340,** A06

feed—the input solution to a treatment/purification system or device, including the raw water supply prior to any treatment. The liquid entering the module. **D 6161,** D19

feed—the mixture of particles and fluid that is introduced to the filter.

NOTE—This term is commonly used for liquids. **F 740,** F21

feed channel spacer—a plastic netting between membrane leaves which provides the flow channel for the fluid passing over the surface of the membrane and increases the turbulence of the feed-brine stream. **D 6161,** D19

feed chute—a device that can replace the carriage for the purpose of slicing multiple small cross-sectional food products. This device also includes an end weight or spring to permit slicing of product to within the last one-quarter inch. **F 1827,** F26

feed distributor—the plastic mesh cylinder at the core of the fiber bundle which distributes the feed evenly. **D 6161,** D19

feeder—a mechanical device for regularly producing and delivering gobs of glass to a forming unit. **C 162,** C14

feeder process—See **gob process**. **C 162,** C14

feeder shoe—See **pusher.** **F 592,** F16

feed, gravity, *n*—the movement of materials from one container to another container or location by force of gravity. **C 1145,** C28

feedstock, *n—in metal injection molding (MIM)*, a moldable mixture of metal powder and binder. **B 243,** B09

feedstock—the base raw material that is the source of carbohydrate, such as starch, for producing sugars that can be fermented into alcohol and carbon dioxide. **E 1705,** E48

feed-through coils— see **encircling coils.** **E 1316,** E07

feed unit, *n—in an open-end spinning machine*, the device which presents the feed stock to the opening roller by either a feed roller and feed plate combination or interacting feed rollers. (See also **opening device.**) **D 123,** D13

feed water—that water entering a device or process. **D 6161,** D19

feldspar—a mineral aggregate consisting chiefly of microcline, albite, or anorthite or combination thereof. **C 242,** C21

Felicity effect—the presence of acoustic emission, detectable at a fixed predetermined sensitivity level at stress levels below those previously applied. (E 1067) **E 1316,** E07

Felicity effect—the presence of detectable acoustic emission at a fixed predetermined sensitivity level at stress levels below those previously applied. **E 1316,** E07

Felicity ratio—the ratio of the stress at which the Felicity effect occurs to the previously applied maximum stress. (E 1067, E 1118)

NOTE—The fixed sensitivity level will usually be the same as was used for the previous loading or test. (E 1118) **E 1316,** E07

felt, *n*—a textile (fabric) characterized by the densely matted condition of most or all of the fibers of which it is composed. **D 123,** D13

felt, *n*—a textile structure characterized by interlocking and consolidation of its constituent fibers achieved by the interaction of a suitable combination of mechanical energy, chemical action, moisture, and heat but without the use of weaving, knitting, stitching, thermal bonding, or adhesives. **D 123,** D13

felt—a flexible sheet manufactured by the interlocking of fibers with a binder or through a combination of mechanical work, moisture, and heat. Felts are manufactured principally from vegetable fibers (organic felts), asbestos fibers (asbestos felts) or glass fibers (glass fiber felts); other fibers may be present in each type. **D 1079,** D08

felt layer—(1) a machine used for applying bitumen and built-up roofing felts;
(2) See **ply**. **D 1079,** D08

felt mill ream—the mass in pounds of 480 ft^2 of dry, unsaturated felt, also termed "point weight." **D 1079,** D08

felt side—the top side of the paper in the paper manufacturing process as opposed to **wire side**. Optical scanning forms should be printed on the felt side. **F 149,** F05

female end of pipe (bell, socket, groove, modified groove)—that portion of the end of the pipe, regardless of its shape or dimensions, which overlaps a portion of the end of the adjoining pipe. **C 822,** C13

FEMF—Fuel Ethanol Manufacturing Facility. **E 1705,** E48

fence—a part of the lever that fits into tumbler gates, permitting the lever to engage the drive cam. **F 471,** F12

fence—a type of barrier consisting of, but not limited to, posts, boards, wire, stakes, or rails that is used to inhibit patrons from coming into contact with the moving portion or restricted portion of an amusement ride or device. **F 747,** F24

fence, *n*—one who deals in the redistribution (resale) of goods (usually legitimate) obtained by illegitimate means, that is, theft, misappropriation, or diversion. **F 1156,** F12

fence, *v*—engaging in redistribution or resale of goods obtained illegally, usually to knowing buyers; most often applied to smaller scale operations. **F 1156,** F12

fence nail—stout, bright, regular-stock-steel, 1¾ by 0.135 to 4 by 0.225-in. nails with large flat 9⁄32 to 15⁄32-in. head and medium diamond point. **F 547,** F16

"fence type" boom—boom consisting of a self-supporting or stiffened membrane supported by flotation. **F 818,** F20

Feret's diameter—See **caliper diameter.** **E 7,** E04

fermentation—decomposition of organic compounds, by microorganisms, to fuels and chemicals such as alcohols, acids, and energy-rich gases. **E 1126,** E48

fermentation—decomposition of organic compounds, by microorganisms, to fuels and chemicals such as alcohols, acids, and energy-rich gases. **E 1705,** E48

fermentation—the biochemical reaction process where microorganisms in a nutrient medium convert a feedstock to a product. **E 1705,** E48

fermentation fuel—a fuel produced by fermentation of biomass.

NOTE—Ethyl alcohol is the most common form of fermentation fuel. **E 772,** E44

fermentation fuel—a fuel produced by fermentation of biomass. **E 1126,** E48

fermentation fuel—a fuel produced by fermentation of biomass. **E 1705,** E48

Fermi energy (level)—*for metals,* the energy of the top-most filled electron level at zero Kelvin. *For insulators and semiconductors,* the Fermi level is usually between the valence and conduction bands. **E 673,** E42

Fermi level referencing—*XPS*, a method of establishing the binding energy scale for a particular specimen by assigning the kinetic energy corresponding to the Fermi level, as determined by analysis of the specimen's XPS or UPS spectrum, as the point of zero binding energy. See also **Fermi energy (level)**. **E 673,** E42

ferric chloride—crystalline form of $FeCl_3$ $6H_2O$, a coagulant. **D 6161,** D19

ferric sulfate—$Fe_2(SO_4)_3$-$9H_2O$, a coagulant. **D 6161,** D19

ferrimagnetic material—a material whose atomic magnetic moments are both ordered and anti-parallel but being unequal in magnitude produce a net magnetization in one direction. **A 340,** A06

ferrite—a term referring to magnetic oxides in general, and especially to material having the formula M O Fe_2 O_3, where M is a divalent metal ion or a combination of such ions. Certain ferrites, magnetically "soft" in character, are useful for core applications at radio and higher frequencies because of their advantageous magnetic properties and high volume resistivity. Other ferrites, magnetically "hard" in character, have desirable permanent magnet properties. **A 340,** A06

ferrite—designation commonly assigned to alpha iron containing alloying elements in solid solution. Increasing carbon content markedly decreases the high-temperature limit of equilibrium. **E 7,** E04

ferrite grain size—the grain size of the ferrite in predominantly ferritic steels. See Test Methods E 112. **E 7,** E04

ferritize, *vt*—to increase the quantity of ferrite in the matrix of a ferrous casting through an appropriate heat treatment. **A 644,** A04

ferritizing anneal, *n*—the process of producing a predominantly ferritic matrix in cast iron through an appropriate heat treatment. **A 644,** A04

ferritizing anneal, *n*—a **heat treatment** that produces a predominantly ferritic matrix in a steel object. **A 941,** A01

ferritizing anneal—the process of producing a predominantly ferritic matrix in a ferrous alloy through an appropriate heat treatment. **E 7,** E04

ferromagnetic—a term applied to materials that can be magnetized or strongly attracted by a magnetic field. **E 1316,** E07

ferromagnetic material—a material whose magnetic moments are ordered and parallel producing magnetization in one direction. **A 340,** A06

ferromagnetic material—a material that, in general, exhibits the phenomena of magnetic hysteresis and saturation, and whose permeability is dependent on the magnetizing force. **E 1316,** E07

ferrous sulfate—$FeSO_4$-$7H_2O$, a coagulant. **D 6161,** D19

ferrous sulfate—cupric sulfate dosimeter—a liquid chemical radiation dosimetry system composed of water with ferrous sulfate or ferrous-ammonium sulfate and cupric sulfate in aqueous sulfuric acid solution and whose response is based quantitatively on the amount of oxidation of ferrous to ferric ions by ionizing radiation, as analyzed by spectrophotometry. It is considered to be a reference standard dosimeter. **E 170,** E10

festooning—loading a cartridge with a supply of imaging material without the use of carrier spools. **F 221,** F05

fetter ring nail—term applied to annularly threaded nail. (See **thread.**) **F 547,** F16

F factor—an episome of *E. coli*. Encoded on it are the functions necessary to produce an F pilus. **E 1705,** E48

fiber—attenuated glass generally with a diameter of less than a millimetre and a length/diameter ratio of more than 5. **C 162,** C14

fiber, *n*—*in fiber-cement*, any material in a form such that it has a minimum length to average maximum transverse dimension of 10 to 1, a maximum cross-sectional area of 5.06×10^{-2} mm^2 (corresponding to a circular cross section of 0.254 mm in diameter) and a maximum transverse dimension of 0.254 mm. **C 1154,** C17

fiber, *n*—*in fiber-reinforced cement*, any material in a form such that it has an aspect ration of at least 10 and a typical transverse dimension less than 2 mm. **C 1154,** C17

fiber, *n*—*in textiles*, a generic term for any one of the various types of matter that form the basic elements of a textile and that is characterized by having a length at least 100 times its diameter. **D 123,** D13

fiber—*for peats and organic soils*, a fragment or piece of plant tissue that retains a recognizable cellular structure and is large enough to be retained after wet sieving on a 100-mesh sieve (openings 0.15 mm). **D 653,** D18

fiber—the unit cell of vegetable growth, which is many times longer than its diameter and which consists largely of cellulose. It is the basic unit of pulps and papers. **D 1695,** D01

fiber, *n*—a thread-like body or filament many times longer than its diameter. For paper, fibers usually are of vegetable origin but may be derived from animal, mineral, or synthetic sources for special types of paper products. **D 1968,** D06

fiber, *n*—any material in a form such that it has a minimum length to average maximum transverse dimension of 10 to 1, a maximum cross-sectional area of 5.06×10^{-2} mm^2(corresponding to a circular cross section of 0.254 mm in diameter) and a maximum transverse dimension of 0.254 mm. **D 2946,** C17

fiber, *n*—one or more filaments in an ordered assemblage. **D 3878,** D30

fiber—a particle with a length to diameter ratio of ten or more. **F 740,** F21

fiber adhesion, *n*—*in asbestos,* force required to overcome the tendency of fibers to remain affixed to the host rock. **D 2946,** C17

fiber-axis—the preferred direction of fiber texture. **E 7,** E04

fiber beard, *n*—*in length testing of fibers*, fibers caught randomly on a comb which are subsequently straightened and parallelized without stretching or damaging. **D 123,** D13

fiber birefringence, *n*—the algebraic difference of the index of refraction of the fiber for plane polarized light vibrating parallel to the longitudinal axis of the fiber and the index of refraction for light vibrating perpendicular to the long axis. (See also **birefringence.**) **D 123,** D13

fiberboard—See **containerboard.** **D 996,** D10

fiberboard—a firm, somewhat flexible composition material in sheet form made from long vegetable fibers. Fiberboard is used for counters, insoles, midsoles, and heel lifts. Leatherboard is a type of fiberboard in which the fiber content is at least 75 % leather. **F 869,** F08

fiberboard container— See **container.** **D 996,** D10

fiberboard nail—bright or electroplated, regular-stock-steel or hardened-steel 1 by 0.054 to 2 by 0.062-in. nails with flat 3⁄32 or 7⁄64-in. head and medium needle point. **F 547,** F16

fiberboard partitions—a set of **corrugated** or **solid fiberboard** or **paperboard** pieces slotted so they will interlock when assembled to form a number of cells into which articles may be placed for shipment. (See also **divider.**) **D 996,** D10

fiberboard tube—See **tube.** **D 996,** D10

fiber bridging—reinforcing fiber material that is found bridging across on an inside radius of a pultruded product.

NOTE—This condition is caused by shrinkage stresses around such a radius during cure. **D 3918,** D20

fiber bundle—the heart of the permeator consisting of the hollow fiber polymer membrane, epoxy tube sheet, nub and feed distributor. **D 6161,** D19

fiber-cement products, *n*—manufactured thin section composites of hydraulic cementitious matrices and discrete non-asbestos fibers. **C 1154,** C17

fiber channel, *n*—a component of an open-end spinning machine through which the fibers are conveyed by an air current from the opening device to the rotor. (See also **opening device.**) **D 123,** D13

fiber chip, *n—in man-made textiles,* staple fibers that are massed together as a unit and that maintain a single geometry or alignment. **D 123,** D13

fiber cohesion, *n*—the resistance to separation of fibers in contact with one another. **D 123,** D13

fiber cohesion, *n—in asbestos,* force required to separate fibers from each other. **D 2946,** C17

fiber content—the amount of fiber present in a composite expressed either as percent by weight or percent by volume. This is sometimes stated as a fraction, that is, fiber volume fraction. **D 3878,** D30

fiber content—the amount of fiber present in a composite, usually expressed as volume percent of the composite. **E 1749,** E06

fiber density, *n*—mass per unit volume of the solid matter of which a fiber is composed, measured under specified conditions. **D 123,** D13

fiber (fibre)—a structural feature of wrought metal revealed by directional properties, manifested by the appearance of an etched longitudinal section, by the appearance of a fracture, or by an X-ray pattern of crystal orientations. **E 7,** E04

fiberfill, *n*—manufactured fibers especially engineered as to linear density, cut length, and crimp for use as a textile filling material. **D 123,** D13

fiberglass, reinforcement—continuously formed fibers added to enhance the overall performance of a matrix. Fiber glasses used for specific chemical or physical properties, or both, include:

A-glass—compositions typically with lower electrical resistivity, strength, and durability than E-glass.

AR-glass—compositions resistant to corrosion by alkali.

C-glass—compositions resistant to corrosion by most acids.

D-glass—compositions with a low dielectric constant.

E-glass—compositions with electrical properties, strength, and durability appropriate for most applications.

E-CR-glass—E-glass compositions modified for improved resistance to corrosion by most acids.

R-glass—compositions with high tensile strength, resistant to corrosion by most acids.

S-glass—compositions with high tensile strength, elastic modulus and service temperature. **C 162,** C14

fiber length—when applied to a pulp, this applies to the mean fiber length and both the experimental details and the calculations must be specified. **D 1695,** D01

fiber length distribution—the frequency distribution of the individual fiber lengths in a material about the mean fiber length. **D 1695,** D01

fiber-mat reinforced products, *n*— manufactured thin section composites of hydraulic cementitious matrices and non-asbestos fibers in two-dimensional scrim(s). **C 1154,** C17

fiber, optical—a fiber with high transmittance, low scattering, and minimal transverse loss of light. **C 162,** C14

fiber optic buffer, *n*—material placed on or around a cladded optical fiber to protect it from mechanical damage. **E 131,** E13

fiber optic chemical sensor, *n*—a fiber optic sensor that responds to a chemical stimulus. **E 131,** E13

fiber optics, *n*—the branch of science and technology devoted to the transmission of radiant energy through fibers made of transparent materials. **E 131,** E13

fiber optics—method by which light is transmitted through thin transparent fibers. **E 1316,** E07

fiber optic sensor, *n*—a device that responds to an external stimulus and transmits through an optical fiber a modulated optical signal, indicating one or more characteristics of the stimulus. **E 131,** E13

fiber or frit-type reference electrode—a reference electrode in which the filling solution contacts the sample solution by means of asbestos fibers or a small sintered piece of porous glass. **D 4127,** D19

fiber orientation—the direction or alignment of the longitudinal axis of the fiber with respect to a stated reference axis. **E 1749,** E06

fiber pick, *n*—the delaminating, splitting, or tearing of paper surface fibers that occurs when the force between the paper and blanket exceeds the surface strength of the paper. **D 6488,** D01

fiber prominence—a visible and measurable pattern of the reinforcing material on the surface of a pultruded plastic part. **D 3918,** D20

fiber-reinforced, *adj*—imparting improved properties to hydraulic cement matrices due to the inclusion of fibers. **C 1154,** C17

fiber-reinforced plastic (FRP), *n*— a plastic that contains fibers in various forms such as cloth, mat, strands, or chopped to enhance properties. **D 907,** D14

fiber rope, *n*—a rope produced primarily from textile fibers. **D 123,** D13

fibers—the slender threadlike elements or groups of wood fibers or similar cellulosic material resulting from chemical or mechanical defiberization, or both, and sometimes referred to as fiber bundles. **D 1554,** D07

fibers—particulate matter with a length of 100 μm or greater, and a length-to-width ratio of 10 to 1 or greater. **G 126,** G04

fiber saturation point—the moisture content at which the cell walls are saturated with water (bound water) and no water is held in the cell cavities by capillary forces. It usually is taken as 25 to 30 % moisture content, based on weight when ovendry. **D 9,** D07

fiber saturation point—the moisture content in the absorption of water by cellulose when the cell walls are saturated, but virtually no free water is present in the grosser capillary structure. The concept is based on the hypothesis that a discontinuity occurs in the sorption curve at this point. Some careful studies of the region have failed to show a discontinuity, but it is still possible to define "fiber saturation point" as an extrapolation of the adsorption curve. In any case, it should only be used with reference to the method of determination. **D 1695,** D01

fiber show, *n*—strands or bundles of fibers not covered by resin which are at or above the surface of a reinforced plastic. **D 883,** D20

fiber show—strands or bundles of fibers that are not covered by resin and that are at or above the surface of a reinforced plastic pultrusion. **D 3918,** D20

fiber spicule—rod-like piece composed of asbestos fibers that are in close-packed parallel orientation, undisturbed from their natural relative positions, and sufficiently numerous to render the piece inflexible. If the transverse dimension exceeds 1 mm the term "pencil" may be applied. **D 2946,** C17

fiber strength unit—see **strength unit.** **D 2946,** C17

fiber structure—the morphological structure of a fiber at the visual or microscopic level. (see also **fine structure**) **D 1695,** D01

fiber tear, *n—in an FRP adhesively bonded joint,* failure occurring exclusively within the fiber reinforced plastic matrix, charisterized by the appearance of reinforcing fibers on both ruptured surfaces. (Also called *fiber tear failure.)* **D 907,** D14

fiber-tear failure, n—in an FRP adhesively bonded joint—**see fiber tear.** **D 907,** D14

fiber texture—a texture characterized by having only one preferred crystallographic direction. **E 7,** E04

fiber volume fraction—see **fiber content.** **D 3878,** D30

Fiberweigh, *n*—an instrument having a capacity for weighing a 51-grain (3.30-g) specimen with a sensitivity of at least 0.2 % of the mass being weighed. **D 123,** D13

fiber, wood—a comparatively long (1/25 in. or less to 1/3 in.), narrow, tapering wood cell closed at both ends. **D 9,** D07

fibric peat—peat in which the original plant fibers are slightly decomposed (greater than 67 % fibers). **D 653,** D18

fibril—a fiberlike bundle of molecules, oriented in one direction either with the fiber axis or spirally around it. It makes up the fibrous structural unit of the single fiber. The cellulose fibrils are crystalline as shown by X-ray diffraction. **D 1695,** D01

fibril, *n—in asbestos,* fiber composed of a single crystal. The smallest structural component of an asbestiform mineral. **D 2946,** C17

fibrogram, *n—in testing cotton fibers for length,* the curve representing the second cumulation of the length distribution of the fibers sensed by the length measuring instrument in scanning the fiber board. **D 123,** D13

Fibronaire, *n*—an instrument which determines the micronaire reading of raw cotton fibers using the "porous-plug" air flow technique. **D 123,** D13

fibrous, *adj*—pertaining to, consisting of, or containing fibers. **C 1154,** C17

fibrous, *adj—in asbestos,* pertaining to fine acicular (needle-like), flexible crystalline morphology associated with the subdivision of macro-assemblages of an asbestiform mineral. **D 2946,** C17

fibrous-felted boards—a felted wood-base panel material manufactured of refined or partly refined lignocellulosic fibers characterized by an integral bond produced by an interfelting of fibers and in the case of certain densities and control of conditions of manufacture by ligneous bond, and to which other materials may have been added during manufacture to improve certain properties. **D 1554,** D07

fibrous particulate, *n— in asbestos,* fiber, fiber fragment, or fiber agglomerate. **D 2946,** C17

fibrous peat—see **fibric peat**. **D 653,** D18

Fick's laws of diffusion, *n*—mathematical descriptions of the movement of one type of molecule through another. **F 1494,** F23

fid, *n*—a wooden or hard plastic tool used as an aid in rope splicing. **D 123,** D13

fidutiary marks—marks placed on objects to assure proper optical orientation item-to-item. **F 1156,** F12

field—the portion of the object in view. **E 7,** E04

field, *n*—that portion of the surface of a specimen that is illuminated by the illuminator or viewed by the receiver. **E 284,** E12

field—any group of characters defined as a unit of information. **F 149,** F05

field adaption—the forced recovery of excited electrons (as in dark adaption) by means of an externally applied electrical field such as a charging corona. **F 335,** F05

field angle—the included angle between those points on opposite sides of a light beam at which the luminous intensity is 10 % of the maximum value. **E 1316,** E07

field, bipolar—longitudinal magnetic field within a part that has two poles. **E 1316,** E07

field blank—blank sample prepared at the field sampling location. **E 1605,** E06

field-box nail—coated, barbed, regular-stock-steel, 2 by 0.099 to 2¼ by 0.106-in. nails with ¼-in. oval head and medium diamond point. **F 547,** F16

field check—(*1*) a survey of existing conditions at a construction site (also called *field observation*). (*2*) verification of an existing structure and its dimensions compared with those shown on drawings (also called *field measure*). **E 631,** E06

field, circular magnetic—generally, the magnetic field surrounding any electrical conductor or part resulting from a current being passed through the part or conductor from one end to another. **E 1316,** E07

field delimiter—See **field separator**. **F 149,** F05

field joint—See **joints.** **E 631,** E06

field latex, *n*—See **latex, field** **D 1566,** D11

field lens—the lens nearest the field diaphragm in an eyepiece. **E 7,** E04

field, longitudinal magnetic—magnetic field wherein the flux lines traverse the component in a direction essentially parallel with its longitudinal axis. **E 1316,** E07

field, magnetic—the space, within and surrounding a magnetized part or a conductor carrying current, in which the magnetic force is exerted. **E 1316,** E07

field, magnetic leakage—the magnetic field that leaves or enters the surface of a part at a discontinuity or change in section configuration of a magnetic circuit. **E 1316,** E07

field mark—See **field separator**. **F 149,** F05

field measure—See **field check.** **E 631,** E06

field measurement, in image analysis—the aggregate measurement of the detected features in a field of view. **E 7,** E04

field moisture equivalent—see **moisture equivalent**. **D 653,** D18

field of testing—broad sphere of science, engineering, or technology used to describe a general area of testing for classification purposes only. (For accreditation purposes, fields of testing are subdivided into specific tests, groups of tests, or product areas.) **E 1187,** E36

field of view—angle subtended when viewing an image through an aperture, often expressed in terms of the chord of the angle at the image plane. **F 1156,** F12

field of view, *n*—the lineal dimension defining the length of a bar code that can be read in one scan, particularly significant in moving beam and array technologies. **F 1294,** F05

field of view (FOV)—the shape and angular dimensions of the cone or the pyramid which define the object space imaged by the system; for example, rectangular, 4° wide by 3° high. **E 1316,** E07

field, residual magnetic—the field that remains in a piece of magnetizable material after the magnetizing force has been removed. **E 1316,** E07

field, resultant magnetic—(sometimes called vector): a magnetic field that is the result of two magnetizing forces impressed upon the same area of a magnetizable object. **E 1316,** E07

field separator—a mark or symbol printed in scan ink that identifies fields to the scanner (Syn. *field mark*). **F 149,** F05

field sound transmission class, FSTC—sound transmission class calculated in accordance with Classification E 413 using values of field transmission loss. **C 634,** E33

fieldstone, *n*—natural building stone as found in the field. **E 631,** E06

field stop, *n*—the physical diameter that limits the angular field of view of an optical system. **E 284,** E12

field strength—see **magnetic field strength.** **E 1316,** E07

field testing, *n*—testing performed in the field under actual conditions of temperature and exposure to the fluids for which the immersion testing is being performed. **D 4439,** D35

field transmission loss, *FTL*—sound transmission loss measured in accordance with Annex A1 of Test Method E 336. **C 634,** E33

figure—the pattern produced in a wood surface by annual growth rings, rays, knots, deviations from regular grain such as interlocked and wavy grain, and irregular coloration. **D 9,** D07

figured glass—flat glass having a pattern on one or both surfaces. **C 162,** C14

figure of merit, magnetic, $\mu'/\cot\gamma$—the ratio of the real part of the complex relative permeability to the dissipation factor of a ferromagnetic material.

NOTE—The figure of merit index of the magnetic efficiency of the core in various ac electromagnetic devices. **A 340,** A06

filament—a single glass fiber of indefinite length. **C 162,** C14

filament, *n*—a long flexible thread of small cross section, usually extruded or drawn. **C 1145,** C28

filament, *n*—*in textiles*, a continuous fiber of extremely long length. **D 123,** D13

filament, *n*—*in steel cord*, the individual element in a steel strand or cord. **D 123,** D13

filament, *n*—a fibrous form of matter with an aspect ratio >10 and an effective diameter <1 mm. (See also **monofilament.**) **D 3878,** D30

filament—an electrically heated wire used as a source of a radiation, such as electrons, or as a source of heat, such as in the vaporization of a metal. **E 7,** E04

filament—a variety of fibers characterized by extreme length. Also known as *fibers* and used interchangeably. **E 1749,** E06

filamentary composites—See **composite, filamentary**. **E 1749,** E06

filamentary (multifilamentary) superconductor—a composite superconductor consisting of at least one superconductive wire embedded in a matrix. **B 713,** B01

filament yarn, *n*—a yarn composed of (continuous) filaments assembled with or without twist. **D 123,** D13

filar micrometer or filar eyepiece—an eyepiece equipped with a fiducial line in its focal plane, which is movable by means of a calibrated micrometer screw, in order to make accurate measurements of length. **E 7,** E04

file-grip nail, file-thread nail—terms applied to helically threaded nails provided with file threads. (See **thread.**) **F 547,** F16

file specimen, *n*—portion of the material to be tested which is stored under conditions in which it is stable, and is used for comparison between exposed and original state. **G 113,** G03

file thread, file-grip thread—large number of fine, closely spaced helical threads, with large or extra large lead angle, rolled onto shank. Usually applied to certain aluminum asbestos-shingle face nails. **F 547,** F16

filiform corrosion—corrosion that occurs under some coatings in the form of randomly distributed threadlike filaments. **G 15,** G01

filiform corrosion resistance (*coatings*)—the ability of a coating to resist that type of corrosion of metal substrates characterized by a definite thread-like structure and directional growth that occurs under coatings. **D 16,** D01

fill—the unit charge of batch into a melter or pot. **C 162,** C14

fill—man-made deposits of natural soils or rock products and waste materials. **D 653,** D18

fill—deprecated term, see **filling**. **D 4439,** D35

filled candle, *n*—a candle produced and used within the same vessel. **F 1972,** F15

filler—a material used to increase the bulk of a product without adding to its effectiveness in functional performance. **B 374,** B08

filler—an inert inorganic material used as an extender or diluent in the furnish of asbestos-cement which does not add to the cementitious value of the cement. **C 460,** C17

filler—*in manufactured carbon and graphite product technology*, carbonaceous particles comprising the base aggregate in an unbaked green-mix formulation. **C 709,** D02

filler—a general term for a material that is inert under the conditions of use and serves to occupy space and may improve physical properties. **C 859,** C26

filler, *n*—an inert inorganic material used as an extender or mineral diluent in the constituents of products which does not add to the cementitious value of the cement. **C 1154,** C17

filler—a pigmented composition for filling the pores or irregularities in a surface preparatory to application of other finishes. **D 16,** D01

filler, *n*—nonfibrous material, such as insoluble clays or gypsum, together with starches, gums, etc., added to a fabric to increase its weight or to modify the appearance or handle of the fabric. (Compare **sizing**.) **D 123,** D13

filler—a material added to soap or other detergent that does not improve its attractiveness or its effectiveness under the conditions of use. **D 459,** D12

filler, *n*—a relatively inert material added to a plastic to modify its strength, permanence, working properties, or other qualities, or to lower costs. (See also **reinforced plastic.**) **D 883,** D20

filler, *n*—a relatively nonadhesive substance added to an adhesive to improve its working properties, permanence, strength, or other qualities. (See also **binder** and **extender**.) **D 907,** D14

filler, *n*—a solid compounding material, usually in finely divided form, which may be added in relatively large proportions to a polymer for technical or economic reasons. **D 1566,** D11

filler, *n*—*for paper or paperboard*, a material, generally nonfibrous and inorganic, added to the fiber furnish. **D 1968,** D06

filler, *n*—*for paperboard*, the inner ply or plies of a multi-ply sheet. **D 1968,** D06

filler—inert inorganic material used as an extender or diluent in the furnish of asbestos-cement which does not add to the cementitious value of the cement. **D 2946,** C17

filler—in composite materials, a primarily inert solid constituent added to the matrix to modify the composite properties or to lower cost. **D 3878,** D30

filler, *n*—a relatively inert material added to a plastic to modify its strength, permanence, working properties, or other qualities or to lower costs. (See also **reinforced plastic**.) (D-20) **F 412,** F17

filler—a solid compounding material, usually in finely divided form, which may be added in relatively large proportions to a polymer for technical or economic reasons. (D 1566, D-11) **F 869,** F08

filler, *n*—a relatively inert material added to a plastic to modify its strength, performance, working properties, or other qualities, or to lower costs. (See also **reinforced plastic.**) **F 1251,** F04

filler character, *n*—a character inserted to extend an item of data to achieve a desired length. (See also **pad character**.) **F 1294,** F05

filler, inert, *n*—a filler having no reinforcing effect. **D 1566,** D11
filler pad—See **cushioning material.** **D 996,** D10
filler phase, *n—as used in Test Method D 5061,* a discontinuous solid formed from coal macerals and minerals that do not deform thermoplastically during carbonization. **D 121,** D05
filler sheet, *n*—a sheet of deformable or resilient material that, when placed between the assembly to be bonded and the pressure applicator, or when distributed within a stack of assemblies, aids in providing uniform application of pressure over the area to be bonded. **D 907,** D14
filler strip—See **lock-strip**, the preferred term. **C 717,** C24
fillet, *n*—that portion of an adhesive which fills the corner or angle formed where two adherends are joined. **D 907,** D14
fillet—curved intersection of head and shank, specified by its radius. **F 547,** F16
fillet bead joint, n—Use **fillet sealant joint** (preferred term). **C 717,** C24
fillet head—head with extra large curved intersection of head and shank. **F 547,** F16
fillet seal—a seal applied at the juncture of two adjoining parts or surfaces and along the edges of faying surfaces as a continuous bead of sealing material. **E 1749,** E06
fillet sealant joint, *n—in building construction,* a joint where sealant is applied over the joint to the face of substrates that are approximately perpendicular to each other. **C 717,** C24
fill factor—the ratio of the weight of a given length of a stranded conductor to the weight of the solid conductor of the same diameter, length, and material. It may also be expressed by the ratio of the nominal area of the conductor times (1 plus the stranding factor) and the area of the circle that encloses the strands of the conductor. **B 354,** B01
fill factor—for internal probe electromagnetic testing, the ratio of the effective cross-sectional area of the primary internal probe coil to the cross-sectional area of the tube interior. **E 1316,** E07
fill factor—for encircling coil electromagnetic testing, the ratio of the cross-sectional area of the test specimen to the effective cross-sectional core area of the primary encircling coil (outside diameter of coil form, not inside diameter which is adjacent to specimen.) **E 1316,** E07
fill factor—in magnetic particle examination, the ratio of the cross sectional area of the part being tested to the cross sectional area of the encircling coil. **E 1316,** E07
fill in, *n*—a condition where the halftone dots and fine type lose definition because the non-image areas of the plate between the dots or characters become plugged with dry ink or a mixture of dry ink and paper fibers. **D 6488,** D01
fill-in—image spread within nonimage areas of characters, digits, or designs. **F 549,** F05
filling, *n*—yarn running from selvage to selvage at right angles to the warp in a woven fabric. **D 123,** D13
filling—generally, the material occupying the space between joint surfaces, faults, and other rock discontinuities. The filling material may be clay, gouge, various natural cementing agents, or alteration products of the adjacent rock. (ISRM) **D 653,** D18
filling, *n*—yarn running from selvage to selvage at right angles to the warp in a woven fabric. **D 4439,** D35
filling band, *n—in woven fabrics,* a visual defect across the width due to a change occurring in the yarn for a large number of picks. (Compare **filling bar**) **D 123,** D13
filling bar, *n—in woven fabrics,* a visual defect across the width which contains a limited number of picks of different appearance than normal. (Compare **filling band**) **D 123,** D13
filling elongation and tension, *n*—stretch or tension measured at right angles to the warp direction of the fabric. **D 123,** D13
filling in—undesirable printing in small nonimage areas, such as halftones and small letters due for example, too much ink being run on the press. **F 425,** F05
filling material, *n—for feathers or down,* the contents of an industry product containing feathers or down of any kind or type, with or without natural or synthetic materials. **D 123,** D13
filling point—the level at which a glass bottle has the nominal capacity. **C 162,** C14
filling solution—the solution inside a sensing or reference electrode that is replenished periodically. Solutions that are permanently sealed within the electrode (like the buffer inside a pH electrode) are usually called internal reference solutions to differentiate them from filling solutions. **D 4127,** D19
filling tests, *n—in cloth testing,* tests in which the filling yarns are torn. **D 123,** D13
filling-to-filling seam, *n*—a sewn seam in which the yarns in the filling direction on both sides of the sewn seam are perpendicular to the seam. **D 123,** D13
fill leakage, *n—in comforters,* either partial or total penetration of the stuffing material through the outer or face fabric. **D 123,** D13
fill material, n—material used in the construction of a structural fill. **D 5681,** D34
fill ratio—see **compression ratio.** **B 243,** B09
film, *n*—a thin coating or layer of a substance over the surface of another material. **C 1145,** C28
film, *n*—in plastics, an optional term for sheeting having a nominal thickness not greater than 0.25 mm (0.01 in.). **D 883,** D20
film—sheeting, including that composed of cellulose or its derivatives, having nominal thickness not greater than 10 mils. (see also **films**) **D 1695,** D01
film, *n*—in plastics, term for sheeting having a nominal thickness not greater than 0.25 mm (0.01 in.). (See also **sheeting.**) **F 1251,** F04
film cassette—See **cassette.** **E 7,** E04
film clarity—characteristic of a deposited film which permits an unobstructed view of the substrate. **D 2825,** D21
film contrast—a qualitative expression of the slope or steepness of the characteristic curve of a film; that property of a photographic material which is related to the magnitude of the density difference resulting from a given exposure difference. **E 1316,** E07
film master, *n*—a master negative or positive film transparency of a specific bar code symbol from which a printing plate is made. **F 1294,** F05
film resistance—that portion of the contact resistance that is due to the presence of contaminants on one or both of the contact members. **B 542,** B02
films—transparent sheeting. If this is used as a generic term, regardless of sheet thickness, the plural should be used and the singular restricted to a specified size. (see also **film**) **D 1695,** D01
films, *n*—trimmed mica split to specific ranges of thickness under 0.15 mm processed from block and thins. **D 1711,** D09
film speed—a numerical value expressing the response of animage receptor to the energy of penetrating radiation under specified conditions. **E 1316,** E07
film strength—the relative resistance of the bisque to mechanical damage. **C 286,** B08
film weight—*in the classification of film adhesives,* weight per unit area of film adhesive usually expressed in pounds per square foot, kilograms per square metre, etc. **E 1749,** E06
filter—see **plugging compound.** **C 286,** B08

filter, *n*—a porous medium for collecting particulate matter.

absolute filter, n—a filter or filter medium of ultra-high collection efficiency for very small particles (submicrometre size) so that essentially all particles of interest or of concern are collected.

controlled-pore filter, n—a filter of various plastics or metals having a structure of controlled uniform pore size. **D 1356,** D22

filter—a device which modifies the light coming from the light source either chromatically or with regard to intensity.

(*1*) *color*—a device which transmits principally a predetermined range of wavelengths.

(*2*) *contrast*—a color filter, usually with strong absorption, whose function is to utilize the spectral absorption bands of the subject to control the contrast of the image by exaggerating or diminishing the brightness difference between areas of different color. Maximum contrast is obtained when the transmission of the filter is entirely within the absorption band of an area but not of its surroundings.

(*3*) *interference*—a combination of several thin optical films to form a layered coating for transmitting or reflecting a narrow band of wavelengths by virtue of interference effects.

(*4*) *neutral*—(*a*) a color filter that reduces the intensity of the transmitted illumination without affecting its hue.

(*b*) a color filter having identical transmission at all wavelengths throughout the spectrum. Such an ideal filter does not exist in practice.

(*5*) *orthochromatic*—a color filter whose function is to modify the illumination quality reaching the film so that the brightness of colored objects will be relatively the same in the resultant black-and-white positive.

(*6*) *photometric*—a color filter whose function is to convert the quality of illumination from that of one source to that of another. Most frequently the term is used for a filter altering the illumination quality from that of one color temperature to that of another.

(*7*) *X-Ray*—a material that preferentially absorbs certain wavelengths. **E 7,** E04

filter—a substance that attenuates the radiant power reaching the detector in a definite manner with respect to spectral distribution. **E 131,** E13

filter, *n*—a substance that attenuates the radiant power reaching the detector in a definite manner with respect to spectral distribution. **E 135,** E01

filter—a network that passes electromagnetic wave energy over a described range of frequencies and attenuates energy at all other frequencies. **E 1316,** E07

filter—uniform layer of material, usually of higher atomic number than the specimen, placed between the radiation source and the film for the purpose of preferentially absorbing the softer radiations. **E 1316,** E07

filter, *v*—to pass a fluid containing particles through a filter medium whereby particles are separated from the fluid. **F 740,** F21

filter, *n*—a device for carrying out filtration which consists of the combination of the filter medium and suitable hardware for constraining and supporting it in the path of the fluid. **F 740,** F21

filterability—the rate at which particles can be separated from a slurry by means of a permeable medium under specified conditions. **D 2652,** D28

filter aid—an inert, insoluble material, more or less finely divided, used as a filter medium or to assist in filtration by preventing excessive packing of the filter cake. **B 374,** B08

filter bedding stone—(generally 6-in. minus material) stone placed under graded riprap stone or armor stone in a layer or combination of layers designed and installed in such a manner as to prevent the loss of underlying soil or finer bedding materials due to moving water. **D 653,** D18

filter block—a cellular vitrified clay block unit, of proprietary configuration, designed to underbed the media in trickling filters. **C 896,** C04

filter cake—the accumulated particles on a filter surface, usually from a slurry mixture. **D 6161,** D19

filter coefficients—values which defind a mask filter in image processing. **E 1316,** E07

filter medium—the permeable material used for a filter that separates particles from a fluid passing through it. **F 740,** F21

filter, neutral—a filter that attenuates the radiant power reaching the detector by the same factor at all wavelengths within a prescribed wavelength region. **E 131,** E13

filter, neutral, *n*—a filter that attenuates the radiant power reaching the detector by the same factor at all wavelengths within a prescribed wavelength region. **E 135,** E01

filter, nonselective—not recommended, see **filter, neutral.** **E 135,** E01

filter (protective filter)—a layer or combination of layers of pervious materials designed and installed in such a manner as to provide drainage, yet prevent the movement of soil particles due to flowing water. **D 653,** D18

filter system—the combination of one or more filters with all the associated process hardware required for filtration. **F 740,** F21

filtrate—the fluid that has passed through a filter. **D 4410,** D19

filtrate—the portion of the feed stream which has passed through a filter. **D 6161,** D19

filtrate—the fluid that has passed through the filter.

NOTE—This term is commonly used for liquids. **F 740,** F21

filtration—the process of passing a liquid through a porous medium for the removal of suspended matter. **D 4410,** D19

filtration—the process by which particles are separated from a fluid by passing the fluid through a permeable material. **F 740,** F21

filtration efficiency—that fraction of suspended particles retained by the filter. **F 740,** F21

filtration ratio—a particle concentration in the feed stream divided by that in the filtrate. **F 740,** F21

fin—(*1*) an imperfection caused by entrance of glass into a seam between mold parts during forming.

(*2*) the feather edge obtained when cutting flat glass. **C 162,** C14

fin—fine feather-edge protrusion from the surface (Syn. *flash*).

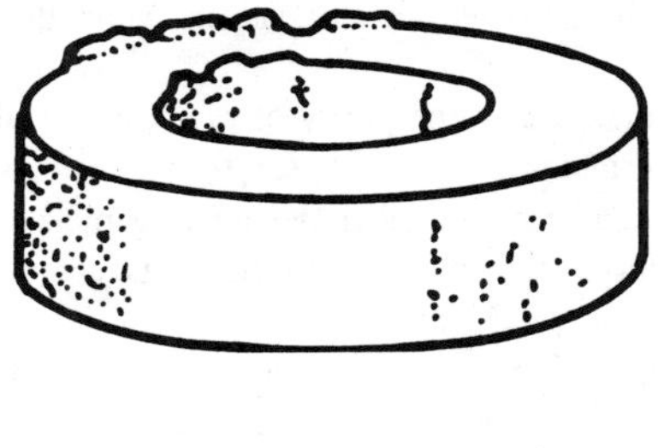

F 109, C21

fin—thin projection from cut edge of point, a defect. Occasionally found on rim of head. **F 547,** F16

fin—See **burr.** **F 547,** F16

fin—portion of the boom membrane above the float. **F 818,** F20

final inspection—inspection by a qualified inspector, industrial hygienist, or local public health official to determine whether abatement and cleanup meet applicable standards. **E 1605,** E06

final set—*in grouting*, a degree of stiffening of a grout mixture greater than initial set, generally stated as an empirical value indicating the time in hours and minutes that is required for cement paste to stiffen sufficiently to resist the penetration of a weighted test needle. **D 653,** D18

final twist, *n*—the number of turns per unit length in a single yarn component of a plied yarn or the plied yarn component of a cabled yarn as the component lies in the more complex structure. **D 123,** D13

financial management rate-of-return (FMRR)—See **adjusted internal rate-of-return (AIRR).** **E 833,** E06

finder patterns, *n*—a unique pattern, containing no data, that is specific to each symbology, whose purpose is to locate the symbol within the reader's field of view. **F 1294,** F05

Finders' sole leather—one of the two principal types of sole leather. It has less flexibility and compressibility than factory sole leather and is more suitable for use in shoe repair. (See also **factory sole leather.**) **D 1517,** D31

fine—very small, as in fine powder; narrow, as in fine lines; (high quality, as in fine gold or fine wine). **F 1156,** F12

fine aggregate, *n*—*(1)* aggregate passing the ⅜-in. (9.5-mm) sieve and almost entirely passing the 4.75-mm (No. 4) sieve and predominantly retained on the 75-μm (No. 200) sieve: or *(2)* that portion of an aggregate passing the 4.75-mm (No. 4) sieve and retained on the 75-μm (No. 200) sieve. **D 8,** D04

fine annealing—to attain more uniform structure than ordinarily required, maximizing homogeneity of physical properties by minimizing prior thermally induced variations. **C 162,** C14

fine coal—See *fine coal* under **sample.** **D 121,** D05

fine end, *n*—a smaller than normal diameter warp end. (*Syn.* light end, thin end. (Compare **coarse end.**) **D 123,** D13

fine-grained—containing predominately small grains that are less than 0.1 mm in size. **C 709,** D02

fine hackle—See **mist.** **C 162,** C14

fine-material load—that part of the total sediment load that is composed of particles of a finer size than the particles present in appreciable quantities in the bed material; normally, the fine-material load consists of material finer than 0.062 mm. **D 4410,** D19

fine mineral surfacing—water-insoluble inorganic material, more than 50 % of which passes the 500-μm (No. 35) sieve, used on the surface of roofing. **D 1079,** D08

fine nail—slender, bright, regular-stock-steel, ½ by 0.035 to 1 ⅛ by 0.072-in. nails with flat ⅛ to ¹¹⁄₆₄-in. head and medium diamond point. Also, slender, electroplated, hardened-steel, 1¼ by 0.054 to 2½ by 0.083-in. nails with brad head and short diamond point. **F 547,** F16

fineness, *n*—a measure of the purity of precious metals expressed in parts per thousand. **B 899,** B02

fineness—a measurement number designating the particle size of a material, usually reported as percent passing a screen of a particular standard size. **C 242,** C21

fineness, *n*—a measurement number designating the particle size of a material, usually reported as passing a screen of a particular standard size. **C 1145,** C28

fineness, *n*—*of fibers,* a relative measure of size, diameter, linear density, or mass per unit length expressed in a variety of units. **D 123,** D13

fineness—a measure of particle-size. **D 653,** D18

fineness modulus, *n*—an empirical factor obtained by adding total percentages of a sample of aggregate retained on each of a specified series of sieves and dividing by 100. The sieve sizes used are: No. 100 (150 μm), No. 50 (300 μm), No. 30 (600 μm), No. 16 (1.18 mm), No. 8 (2.36 mm), No. 4 (4.75 mm),⅜ in. (9.5 mm), ¾ in. (19.0 mm), 1½ in. (38.1 mm) and larger, increasing in the ratio of 2 to 1. **C 11,** C11

fineness modulus, *n*—a factor obtained by adding the percentages of material in the sample that is coarser than each of the following sieves (cumulative percentages retained), and dividing the sum by 100: 150-μm (No. 100), 300-μm (No. 50), 600-μm (No. 30), 1.18-mm (No. 16), 2.36-mm (No. 8), 4.75-mm (No. 4), 9.5-mm (⅜-in.), 19.0-mm (¾-in.), 37.5-mm (1½-in.), 75-mm (3-in.), 150-mm (6-in.). **C 125,** C09

fineness modulus—an empirical factor obtained by adding the total percentages of an aggregate sample retained on each of a specified series of sieves, and dividing the sum by 100; in the United States, the U.S. Standard sieve sizes are: No. 100 (149 μm), No. 50 (297 μm), No. 30 (590 μm), No. 16 (1,190 μm), No. 8 (2,380 μm), and No. 4 (4,760 μm) and ⅜ in. (9.5 mm), ¾ in. (19 mm), 1½ in. (38 mm), 3 in. (76 mm), and 6 in. (150 mm). **D 653,** D18

fineness of enamel—a measurement of the degree to which a frit has been milled in wet or dry form, usually expressed in grams residue retained on a certain mesh screen from a 50-cm^3 or a 100-g sample. **C 286,** B08

fines—the portion of a powder composed of particles which are smaller than a specified size, currently less than 44 μm. See also **superfines.** **B 243,** B09

fines—the portions of a powder composed of particles smaller than a specified size. **C 242,** C21

fines, *n*—the portions of a powder composed of particles smaller than a specified size. **C 1145,** C28

fines—portion of a soil finer than a No. 200 (75-μm) U.S. standard sieve. **D 653,** D18

fines—particles smaller than the smallest nominal specification particle size. **D 2652,** D28

fines, *n*—*in asbestos,* the class of material having the smallest range of particle size, that is segregated by a particle size classification of asbestos by any relevant test method. Customarily, the fraction that passes through the finest aperture screen used in Test Method D 2589. **D 2946,** C17

fines, *n*—that portion of pelleted carbon black that passes through a specified sieve under standard conditions. **D 3053,** D24

fine sprays—distribution of droplets with 100 μm $< D_{V.5} \leq$ 400 μm. **E 1102,** E35

fine structure—a generic term denoting the submicroscopic structure that depends on molecular arrangement. **D 1695,** D01

fine transfer—see **material transfer, bridge.** **B 542,** B02

finger blisters—finger shaped blisters or wrinkles in the plies of a built-up roofing or waterproofing membrane. **D 1079,** D08

finger joint, *n*—see *finger joint* under **joint.** **D 907,** D14

finger mark, *n*—an irregular spot showing variation in picks per inch for a limited width. **D 123,** D13

finger rings—the feature of the scissors that forms the gripping surface for the surgeon (commonly classified as the ring-handled feature). **F 1078,** F04

fingertip towel, *n*—a textile product with fringes and side hems or selvages which is smaller than a hand towel, generally used as decoration and to dry hands. **D 123,** D13

finger wrinkles—See **finger blisters.** **D 1079,** D08

finial—See **railing systems.** **E 631,** E06

finial, *n*—an ornamental piece on the top of a post, newel, or railing; frequently in the form of an urn or pineapple, and so named. **E 1481,** E06

fining, also refining—the process by which molten glass becomes relatively free of undissolved gases. **C 162,** C14

finish—*(1)* the part of a bottle for holding the cap or closure.
(2) stage in melting process after glass appears free of seeds.
(3) the quality of a glass surface as influenced by very fine pits and cracks remaining from a grinding and polishing process. **C 162,** C14

finish—the exposed top surface of the plaza deck system, or traffic, or wearing surface. **C 717,** C24

finish—*(1)* final coat in a paint system; at the termination of cure or drying *(2)* sometimes refers to the entire coating system: the texture, color, and smoothness of a surface, and other properties affecting appearance. **D 16,** D01

finish, *n—in buttons*, the surface condition or texture. **D 123,** D13

finish, n—See **paperboard.** **D 996,** D10

finish—materials applied to the grain and sometimes split surface of the leather to cover blemishes, create smoothness and give uniformity of color and appearance which may vary from dull to glossy. **D 1517,** D31

finish, *n*—*(1)* the final treatment or **coating** of a surface, *(2)* the fine or decorative work required to make a **building** or its parts complete. **E 631,** E06

finish—the final surface visual appearance of the instrument classified as follows:
(1) **bright or mirror finish**—highly reflective surfaces.
(2) **satin, matte, or black finish**—reduced reflected surfaces (as compared to bright or mirror finish). **F 1078,** F04

finish, *n*—a chemical or mechanical modification, or both, of the fabric for a specific performance result. **F 1494,** F23

finish—the final surface visual appearance of a needle, that may be classified as follows: **F 1840,** F04

finish coat, *n*—the final layer of plaster applied over a basecoat or other substrate. **C 11,** C11

finish coat floating, *n*—the finishing act of spreading, compacting, and smoothing the finish coat plaster or stucco to a specified surface texture. **C 11,** C11

finished, *adj—for glass laminates*, a descriptive term for woven fabrics that have passed through a treating procedure which is compatible with a resin matrix or facilitates manufacturing, or both. **D 123,** D13

finished, *adj—in textile floor covering materials*, the completion of all manufacturing operations. **D 123,** D13

finished fabric weight, *n*—mass per unit area expressed in grams per square metre (ounces per square yard), grams per linear metre (ounces per linear yard), or inversely as metres per kilogram (linear yards per pound), or square metres per kilogram (square yards per pound). **D 123,** D13

finished grade—See **grade.** **E 631,** E06

finished metal connector plate—galvanized or chemically surfaced steel connector plate, with or without prepunched holes. **E 1807,** E06

finished pile yarn floor covering, *adj—in textile floor covering materials*, the pile yarn floor covering that has undergone all of the steps of the manufacturing process. **D 5684,** D13

finished stone—dimension stone with one or more mechanically exposed surfaces.
bush-hammered—a rough uniformly patterned surface produced by an impact tool.
chat sawn—a rough finish produced by gangsawing with course chat sand.
flamed—see *thermal*
honed—a superfine, smooth, satin-like, nonreflective finish.
polished—a smooth, reflective finish.
rock face—a rough nonuniform convex surface that projects from the plane of the wall (stone).
split face—a rough nonuniform cleft surface which may be slightly concave or convex.
thermal—a rough nonreflective finish (with a few reflections from cleavage planes) produced by applying a flame. **C 119,** C18

finished yield, *n—in knitted fabrics*, the number of finished square yards per pound (square metres per kilogram) of finished fabric. **D 123,** D13

finisher—*(1)* person in charge of a day-tank during the melting and fining of the glass.
(2) the workman who does the final work, such as polishing or putting the handle or foot on a piece of ware. **C 162,** C14

finish grinding, *n*—the completion of a grinding operation to obtain a desired surface appearance or accurate dimensions. **C 1145,** C28

finishing bar, *n*—uneven appearance across the entire fabric width. **D 123,** D13

finishing compound, *n—(sometimes called topping compound)* a compound specifically formulated and manufactured for use over taping or all purpose compounds to provide a smooth and level surface for the application of decoration. **C 11,** C11

finishing head—brad head, as found on finishing nail. **F 547,** F16

finishing hydrated lime—hydrated lime suitable for use in the finish coat of plaster. **C 51,** C07

finishing nail—slender, bright, regular-stock-steel, 1 by 0.058 to 4 by 0.135-in. nails with flat or cupped 0.086 to 0.177-in. brad head and medium diamond point for countersinking where concealment of head is important. (See also, **fine nail, moulding and trim nails, wallboard nails.**) **F 547,** F16

finishing quicklime—quicklime suitable (after slaking to a lime putty) for use in the finish coat of plaster. **C 51,** C07

finishing technique, *n— as applies to laundry and dry cleaning procedures*, the mechanical means by which the garment is put in its final state (for example, pressing, drying, wrinkle removal, and so forth). **F 1494,** F23

finish wearing surface—See *traffic surface.* **C 717,** C24

finite element—one of the regular geometrical shapes into which a figure is subdivided for the purpose of numerical stress analysis. (ISRM) **D 653,** D18

fire—See *bisque fire; decorating fire; glost fire; single fire* under **firing.** **C 242,** C21

fire, *n—as related to textile flammability*, an uncontrolled conflagration in which materials are destroyed by burning as evidenced by flames of varying size and shape, and a high-intensity heat source of 5 kw or greater, such as burning contents of a room, a burning basket, burning building, or forest fire. **D 123,** D13

fire, *n*—destructive burning as manifested by any or all of the following: light, flame, heat, smoke. **E 176,** E05

firebrick, *n*—any type of refractory brick specifically fireclay brick. **C 71,** C08

firebrick, insulating, *n*—a refractory brick characterized by low thermal conductivity and low heat capacity. **C 71,** C08

fire-characteristic profile, *n*—an array of fire-test-response characteristics, all measured using tests relevant to the same fire scenario, for a material, product, or assembly to address, collectively, the corresponding fire hazard. (See also **fire hazard, fire risk,** and **fire-test-response characteristic.**) **E 176,** E05

fire check—a check resulting from localized thermal stress during forming. **C 162,** C14

fire clay, *n*—a sedimentary clay of low flux content. **C 43,** C15

fireclay, *n*—an earthy or stony mineral aggregate that has as the essential constituent hydrous silicates of aluminum with or without free silica, plastic when sufficiently pulverized and wetted, rigid when subsequently dried, and of suitable refractoriness for use in commercial refractory products. **C 71,** C08

fire clay—a sedimentary clay of low-flux content. **C 896,** C04

fire clay—a silicious clay rich in hydrous aluminum silicates. **D 653,** D18

fireclay, nodular, *n*—a rock containing aluminous or ferruginous nodules, or both, bonded by fireclay. **C 71,** C08

fireclay, plastic or bond, *n*—a fireclay of sufficient natural plasticity to bond nonplastic materials. **C 71,** C08

fireclay plastic refractory, *n*—a fireclay material tempered with water and suitable for ramming into place to form a monolithic furnace lining that will attain satisfactory physical properties when subjected to the heat of furnace operation. **C 71,** C08

fire crack—cracking, frequently intergranular in nature, that occurs in some metallic materials when too rapidly heated or when stressed and heated rapidly. Not to be confused with "quench crack." **E 7,** E04

fire cracks—see synonomous term **fire check**. **C 162,** C14

fired bond, *n*—bond developed between particulate constituents of brick solely as the result of the firing process. The bond may result from fusion or melting of one or more constituents of the composition or the surface of particles. Other thermal mechanisms such as sintering and interparticle reaction may be responsible for the bond.

The higher the heat treatment, the greater the extent of bonding and consequently the greater the developed strength and the lower the resulting porosity. The bond development should be sufficient to provide the specified strength, porosity, and durability for any particular product. **C 43,** C15

fire-door nail—bright, regular-stock-steel, barbed, 1 ¼ by 0.092-in. nail (3d shingle nail) with flat ¼-in. head and medium diamond point and 2 by 0.099-in. nail (6d box nail) with flat 17/64-in. head and medium diamond point. **F 547,** F16

fire endurance, *n*—a measure of the elapsed time during which a material or assemblage continues to exhibit fire resistance. **E 176,** E05

fire exposure, *n*—process by which or extent to which humans, animals, materials, products, or assemblies are subjected to the conditions created by fire. **E 176,** E05

fire gases, *n*—the airborne products emitted by a material, product, or assembly undergoing pyrolysis or combustion, that exist in the gas phase at the relevant temperature. **E 176,** E05

fire hazard, *n*—the potential for harm associated with fire. **E 176,** E05

fire marks—a defect characterized by tiny indentations similar in appearance to shallow pinholes. **C 286,** B08

fire model, *n*—a physical representation or set of mathematical equations that approximately simulate the dynamics of burning and associated processes. **E 176,** E05

fire over—idling a melting unit at operating temperature. **C 162,** C14

fire performance, *n*—response of a material, product, or assembly in a particular fire, other than in a fire test involving controlled conditions (different from **fire-test-response characteristic**). **E 176,** E05

fire performance characteristic, *n*—this term is deprecated. (See **fire-test-response characteristic** and **fire performance** (q.v.).) **E 176,** E05

fire performance test, *n*—this term is deprecated. (See **fire-test-response characteristic** and **fire performance** (q.v.).)

fireproof, adj—an inappropriate and misleading term. Do not use. (See commentary in X1.2.) **E 176,** E05

fire point—the lowest temperature at which a specimen will sustain burning for 5 s under specified conditions of test. **D 2864,** D27

fire point, *n*—the lowest temperature at which a liquid or solid specimen will sustain burning for 5 s. **D 4175,** D02

fire-polish—to make glass smooth, rounded, or glossy by heating the glass surface. **C 162,** C14

fireproofing tile, *n*—tile for use as a protection for structural members against fire. **C 43,** C15

fire-refined copper—copper of any origin or type finished by furnace refining without having been processed at any stage by electrolytic or chemical refining. Usually when the term fire-refined copper is used alone it refers to fire-refined tough pitch copper. This designation applies to the following:
—copper cast in refinery shapes suitable for hot or cold working or both, and by extension, to fabricators' products made therefrom.
—ingots or ingot bars suitable for remelting. **B 846,** B05

fire resistance, *n*—the property of a material or assemblage to withstand fire or give protection from it. **E 176,** E05

fire resistance—as applied to buildings, the property of a material or assembly to withstand fire or to give protection from it, characterized by the ability to confine a fire or to continue to perform a structural function, or both. **E 631,** E06

fire-resistance classification, *n*—a standard rating of fire-resistance and protective characteristics of a building construction or assembly. **C 11,** C11

fire resistant, adj—See **fire resistive,**the preferred term. **E 176,** E05

fire resistant boom (aka fire containment boom)—boom intended for containment of burning oil slicks. **F 818,** F20

fire resistive, *adj*—having fire resistance (TCG-01). **E 176,** E05

fire-resistive barrier system, *n*—a specific construction of devices, materials, or coatings installed around, or applied to, the electrical system components. **E 176,** E05

fire retardance, *n*—the resistance to combustion of a material when tested under specified conditions. **D 123,** D13

fire retardant—having or providing comparatively low flammability or flame spread properties. **D 9,** D07

fire-retardant—a descriptive term which implies that the described product, under accepted methods of test, will significantly: (*a*) reduce the rate of flame spread on the surface of a material to which it has been applied, or (*b*) resist ignition when exposed to high temperatures, or (*c*) insulate a substrate to which it has been applied and prolong the time required to reach its ignition, melting, or structural-weakening temperature. **D 16,** D01

fire-retardant, *adj*—having or providing comparatively low flammability or flame spread properties (E 176, E-5). **D 996,** D10

fire retardant, adj—not a defined term. Use as a modifier only with defined compound terms: **fire-retardant barrier, fire-retardant chemical, fire-retardant coating,** and **fire-retardant treatment.** **E 176,** E05

fire-retardant barrier, *n*—a layer of material which, when secured to a combustible material or otherwise interposed between the material and a potential fire source, delays ignition and combustion of the material when the barrier is exposed to fire. **E 176,** E05

fire-retardant chemical, *n*—a chemical, which when added to a combustible material, delays ignition and combustion of the resulting material when exposed to fire. **E 176,** E05

fire-retardant coating—a coating that will do one or more of the following: (*1*) reduce the flame spread on the substrate over which the coating is applied, sometimes at the sacrifice of the coating (see intumescent coating);

(*2*) resist ignition of the substrate when exposed to high temperature; or

(*3*) insulate the substrate to which the coating is applied and thereby prolong the time required to reach its ignition, melting or structural-weakening temperature. **D 16,** D01

fire-retardant coating, *n*—a fluid-applied surface covering on a combustible material which delays ignition and combustion of the material when the coating is exposed to fire. (See also **flame-retardant coating.** Compare **fire-retardant barrier.**) **E 176,** E05

fire retardant, n—a deprecated term. Do not use. **E 176,** E05

fire-retardant treatment, *n*—the use of a fire-retardant chemical or a fire-retardant coating. (See also **flame-retardant treatment.**) **E 176,** E05

fire risk, *n*—an estimation of expected fire loss that combines the potential for harm in various fire scenarios that can occur with the probabilities of occurrence of those scenarios. **E 176,** E05

fire scale—copper oxide subscale formed just under the surface of silver-copper alloys when they are annealed in air. **E 7,** E04

fire scenario, *n*—a detailed description of conditions, including environmental, of one or more of the stages from before ignition to the completion of combustion in an actual fire, or in a full scale simulation. **E 176,** E05

fire stop—a through-penetration fire stop is a specific construction consisting of the materials that fill the opening around penetrating items such as cables, cable trays, conduits, ducts, and pipes and their means of support through the wall or floor opening to prevent spread of fire. **E 176,** E05

fire test, *n*—a procedure, not necessarily a standard test method, in which the response of materials to heat or flame (or both) under controlled conditions is measured or otherwise described. **E 176,** E05

fire test exposure severity, *n*—a measure of the degree of fire exposure; specifically in connection with Test Methods E 119, E 152, and E 163, the ratio of the area under the curve of average furnace temperature to the area under the standard time/temperature curve, each from the start of the test to the end or time of failure, and above the base temperatures 68°F (20°C). **E 176,** E05

fire-test-response characteristic, *n*—a response characteristic of a material, product, or assembly, to a prescribed source of heat or flame, under controlled fire conditions; such response characteristics may include but are not limited to ease of ignition, flame spread, heat release, mass loss, smoke generation, fire endurance, and toxic potency of smoke. **E 176,** E05

fire-test-response-characteristic index, *n*—a single quantitative measure that combines two or more fire-test-response characteristics for a material, product, or assembly, all developed under test conditions compatible with a common fire scenario, addressing collectively the corresponding threat. See also *fire-test-response-characteristic profile, fire hazard, fire risk, fire-test-response characteristic.* **E 176,** E05

fire-test-response-characteristic profile, *n*—array of fire-test-response characteristics for a material, product, or assembly, all developed under test conditions compatible with a common fire scenario, addressing collectively the corresponding threat. See also *fire hazard, fire risk, fire-test-response characteristic.* **E 176,** E05

firing, *v*—process of heating the material to elevated tempera-tures. The temperatures are usually in excess of 930°C. The extent of firing is a function of both time and temperature. The firing develops the inter-particulate bond, the strengths, the pore structure, and the color of the product. The extent of firing should be sufficient to produce the levels of these properties required by the specifications for the particular product. **C 43,** C15

firing—the controlled heat treatment of ceramic ware in a kiln or furnace, during the process of manufacture, to develop the desired properties.

bisque fire—the process of kiln-firing ceramic ware before glazing.

decorating fire—the process of firing ceramic or metallic decorations on the surface of glazed ceramic ware.

firing curve—a diagram or table showing the time and temperature planned or experienced by ware going through a firing operation.

firing cycle—the time required for one complete firing operation (cold-to-cold).

firing range—the range of firing temperature within which a ceramic composition develops properties which render it commercially useful.

glost fire—the process of kiln-firing bisque ware to which glaze has been applied.

single fire—the process of maturing an unfired ceramic body and its glaze in one firing operation. **C 242,** C21

firing—the controlled heat treatment of ceramic ware in a kiln or furnace to develop the desired final properties. **C 286,** B08

firing expansion, *n*—the increase in the dimensions of a substance or product during thermal treatment. **C 1145,** C28

firing range—the time-temperature interval in which a porcelain enamel or ceramic coating is satisfactorily matured. **C 286,** B08

firing temperature—the degree of sensible heat attained by the ware during the maturing of the coating. **C 286,** B08

firing time—the period during which the ware remains in the firing zone of the furnace to mature the coating. **C 286,** B08

firing zone—that portion of the furnace, usually a continuous furnace, through which the ware passes and that remains at or near the firing temperature of the coating. **C 286,** B08

first cost—costs incurred in placing a building or building subsystem into service, including, but not limited to, costs of planning, design, engineering, site acquisition and preparation, construction, purchase, installation, property taxes and interest during the construction period, and construction related fees (Syn. *initial investment cost, initial cost*). **E 833,** E06

first floor—See **floor.** **E 631,** E06

first hit location—a zone location method defined by which channel among a group of channels first detects the signal. **E 1316,** E07

first-loss tow/current velocity—minimum tow/current velocity normal to the membrane at which oil escapes past a boom. **F 818,** F20

first point of no break—the amount (weight-mass) of porcelain enamel slip retained when it stops sliding off an enameled pick-up panel and is observed to drain smoothly from the panel without showing a wavey pattern on the wet surface (known also as "yield point"). **C 286,** B08

first side—{archaic} the surface of plate which is ground and polished first. **C 162,** C14

first story—See **building space.** **E 631,** E06

fishbone—a striation that does not reach entirely across the fracture surface. **C 162,** C14

fish-eye, *n*—small globular mass that has not blended completely into the surrounding material. See **gel.** **D 883,** D20

fish-eye—a small globular mass that has not blended completely into the surrounding pultruded material.

NOTE—This condition is particularly evident in a transparent or translucent material. **D 3918,** D20

fish-eye, *n*—small globular mass that has not blended completely into the surrounding material. (D-20) **F 412,** F17

"fish-hook-like" thread— See **buttess-type thread.** **F 547,** F16

fishing tool—*in grouting*, a device used to retrieve drilling equipment lost or dropped in the hole. **D 653,** D18

fishmouth—(*1*) a half-cylindrical or half-conical opening formed by an edge wrinkle or failure to embed a roofing felt;

(*2*) *in shingles*, a half-conical opening formed at a cut edge. **D 1079,** D08

"fish-mouth" point—incomplete chisel wedge point with two narrow ridges at end, a defect resulting from premature breaking of wire during forming. **F 547,** F16

fishscaling—a defect appearing as small half-moon shaped fractures somewhat resembling the scales of a fish. **C 286,** B08

fishscaling, delayed— see **delayed fishscaling.** **C 286,** B08

fishtails—structures, attributed to substrate defects, either above or below the surface plane after epitaxial deposition, which have their ''tails" aligned in a particular crystallographic direction (see also **crescents**). **F 1241,** F01

fission chamber—an ionization chamber containing one or more surfaces coated with fissionable material. **E 170,** E10

fissure—a surface irregularity in the form of a narrow opening or crack. **C 859,** C26

fissure—a gapped fracture. (ISRM) **D 653,** D18

fissures, *n*—surface defects consisting of narrow openings or cracks. **C 1145,** C28

fit, *n*—the quality, state or manner in which the length and closeness of clothing, when worn, relates to the human body. **F 1494,** F23

fit—the degree to which an article of hosiery or footwear conforms to the shape of the foot upon which it is worn. **F 1646,** F13

fitted sheet, *n*—*in textiles*, a product usually made with boxed corners, sometimes elasticized with shape and size to conform to the contours of the mattress and used for covering the mattress on a bed. **D 123,** D13

fitting, *n*—*for pipes*, component such as wyes, tees, and adaptors for use in laying asbestos-cement pipe, such that, when properly installed yields as an assembly equivalent in serviceability and strength to the pipe sections. **C 460,** C17

fitting, *n*—*for conduit*, component such as adapters, reducers, increasers, bends and bell ends, for use in laying asbestos-cement conduit and made to such dimensions as will provide equivalent strength and silt-tight joints when assembled with the conduit. **C 460,** C17

fitting—products such as wyes, tees, elbows, adapters, etc. used in the installation of vitrified clay pipelines. **C 896,** C04

fitting, *n*—*for pipes*, component such as wyes, tees, and adaptors for use in laying pipe, such that, when properly installed yields an assembly equivalent in serviceability and strength to the pipe sections. **C 1154,** C17

fitting, *n*—*for conduit*, component such as adapters, reducers, increasers, bends and bell ends, for use in laying conduit and made to such dimensions as will provide equivalent strength and silt-tight joints when assembled with the conduit. **C 1154,** C17

fitting, *n*—*for conduit*, component such as adaptors, reducers, increasers, bends, and bell ends, for use in laying asbestos-cement conduit and made of equivalent strength and to such dimensions as will provide silt-tight joints when assembled with the conduit. **D 2946,** C17

fitting, *n*—*for pipes*, components such as wyes, tees, elbows, and adaptors for use in laying asbestos-cement pipe, such that, when properly installed, yield as an assembly equivalent inserviceability and strength to the pipe sections. **D 2946,** C17

fitting, *n*—a piping component used to join or terminate sections of pipe or to provide changes of direction or branching in a pipe system. **F 412,** F17

fitting stool—a low stool, generally having a forward projection, the top of which recedes at an approximate angle of 30° and is utilized for the fitting of shoes. **F 869,** F08

fit-up, *n* (aménagement)—alterations and improvements to the **base building** and to the **building systems** including demolition, where required, to prepare the accommodation for occupancy. **E 1480,** E06

fixed—See **windows and doors.** **E 631,** E06

fixed analyzer transmission—see **electron energy analyzer.** **E 673,** E42

fixed-angle internal reflection element— an internal reflection element which is designed to be operated at a fixed angle of incidence. **E 131,** E13

fixed beam scanner, *n*—either a visible light or laser scanner reading in a set plane. **F 1294,** F05

fixed bed—a bed of granular particles through which a fluid flows without causing substantial movement of the bed. **D 2652,** D28

fixed carbon, *n*—*in the case of coal, coke, and bituminous materials*, the solid residue other than ash, obtained by destructive distillation, determined by definite prescribed methods. **D 121,** D05

fixed carbon, *n*—the ash-free carbonous material that remains after volatile matter is driven off during the proximate analysis of a dry sample. **D 5681,** D34

fixed carbon—carbon remaining after heating in a prescribed manner to decompose thermally unstable components and to distill volatiles. **E 1126,** E48

fixed carbon—carbon remaining after heating in a prescribed manner to decompose thermally unstable components and to distill volatiles. **E 1705,** E48

fixed-feed grinding, *n*—the process of feeding a material to be ground to a grinding wheel at a given rate or in specific increments. **C 1145,** C28

fixed-load or fixed-displacement applied *K* curves—curves obtained from a fracture mechanics analysis of a specific specimen configuration. **E 1823,** E08

fixed matter—residues from the ignition of particulate or dissolved matter, or both. **D 1129,** D19

fixed matter—residues from the ignition of particulate or dissolved matter, or both. **D 6161,** D19

fixed mount reader, *n*—a bar code reader that is mounted in a stationary fashion to intersect the plane of a bar code symbol passing before the reader. Often employed in conveyor and transport systems. **F 1294,** F05

fixed pitch, *adj*—describes a character set in which all character cells are of equal width. (See **proportional spacing.**) **F 1457,** F05

fixed point, *n*—*in thermometry*, reproducible temperature of equilibrium of a system of two or more phases under specified conditions. **E 344,** E20

fixed quality area (FQA)—central area of a wafer surface, defined by a nominal edge exclusion, *X*, over which the specified values of a perimeter apply. **F 1241,** F01

fixed retainer, *n*—*in zippers*, a device permanently attached to the retainer pin at the bottom of one stringer. **D 123,** D13

fixed retarding ratio—see **electron energy analyzer.** **E 673,** E42

fixing—the process of rendering a developed image permanent. **F 335,** F05

flagging, *n*—*in sewn seams*, a mode of failure evidenced by slippage of one or more yarns entirely out of the original seam. **D 123,** D13

flake—a small wood particle of predetermined thickness specifically produced as a primary function of specialized equipment of various types, with the cutting action across the direction of the grain (either radially, tangentially, or at an angle between), the action being such as to produce a particle of uniform thickness, essentially plane of the flakes, in over-all character resembling a small piece of veneer. **D 1554,** D07

flake chip—See **peripheral chip**. **F 1241,** F01

flake-galvanized—See **hot-galvanized.** **F 547,** F16

flake-glass—hot glass squeezed thin between two metal rolls or blown into a thin film, followed by fracturing into small particles. **C 162,** C14

flake powder—flat or scale-like particles whose thickness is small compared with the other dimensions. **B 243,** B09

flakes—in wrought ferrous products, flakes appear as short discontinuous internal cracks attributed to stresses produced by localized transformation and hydrogen solubility effects during cooling after hot working. They appear in a fracture surface as bright silvery areas with a coarse texture. In deep acid-etched transverse section they appear as discontinuities which are usually located in the midway to center location of the section. Known also as shatter cracks and hairline cracks. **E 7,** E04

flaking—See **delamination**. **A 902,** A05

flaking—*in protective coatings*, the detachment of small pieces of the coating film. **D 4538,** D33

flaking—that phenomenon manifested in film carbon ribbons and carbon paper by the actual detachment of pieces of the ink itself from the substrate. Flaking is the result of loss of adhesion, usually due to strain or extension of the film ribbon substrate due to tension or to contact with ribbon guides or parts other than the type font, or both. Flaking of carbon paper is usually due to stress or cracking of the ink film by folding, handling, etc. **F 221,** F05

flaking resistance (*coatings*)—the ability of a coating to resist the actual detachment of film fragments either from the previously applied coating or the substrate. Flaking is generally preceded by cracking, checking, or blistering and is the result of loss of adhesion. Also known as scaling resistance. **D 16,** D01

flaking (scaling), *n*—phenomenon manifested in paint films by the actual detachment of pieces of the film itself either from its substrate or from paint previously applied. **E 1605,** E06

flaky finish—appearance of crazing, checking or flaking with or without separation of finish film. **D 1517,** D31

flame, *n*—*as related to ignition of textiles*, a controlled hot, luminous zone of gas or matter in gaseous suspension, or both, of constant size and shape that is undergoing combustion as evidenced by a low-intensity heat flux of less than 1 kw such as candle flame or match flame. (Compare **fire**.) **D 123,** D13

flame, *n*—*as related to textile flammability*, a hot luminous zone of gas or matter in gaseous suspension, or both, that is undergoing combustion, that is relatively constant in size and shape, and that produces a relatively low heat flux. (Compare **fire.**) **D 4391,** D13

flame, *n*—a hot, usually luminous zone of gas that is undergoing combustion. **E 176,** E05

flame, *n*—*as related to ignition of textiles*, a controlled hot luminous zone of gas or matter in gaseous suspension, or both, of constant size and shape that is undergoing combustion as evidenced by a low-intensity heat source of less than 5 kW, such as a burner flame on a gas stove. **F 1494,** F23

flame annealing, *n*—**annealing** in which the heat is applied directly by a flame. **A 941,** A01

flame application time, *n*—the time interval for which the ignition flame is applied to a material. **D 123,** D13

flame front, *n*—the leading edge of a flame propagating through a gaseous mixture or across the surface of a liquid or solid.

flameproof, adj—an inappropriate and misleading term. Do not use. **E 176,** E05

flame hardening, *n*—a process in which only the surface layer of a suitable steel object is heated by flame to above Ac_3 or Ac_{cm}, and then the object is **quenched**. **A 941,** A01

flame impingement, *n*—direct contact between a flame and a material. **F 1494,** F23

flame-out, *n*—the time at which the last vestige of flame or glow disappears from the surface of the test specimen, frequently accompanied by a final puff of smoke; Time 0 is the time at which the specimen is moved into the chamber and the door closed. **E 176,** E05

flameproof, *adj*—an inappropriate and misleading term. Do not use. **E 176,** E05

flame resistance, *n*—the property of a material whereby flaming combustion is prevented, terminated, or inhibited following application of a flaming or nonflaming source of ignition, with or without subsequent removal of the ignition source. **D 123,** D13

flame resistance, *n*—the ability to withstand flame impingement or give protection from it. **E 176,** E05

flame resistance—the property of a material whereby flaming combustion is prevented, terminated, or inhibited following application of a flaming or non-flaming source of ignition, with or without subsequent removal of the ignition source. **E 1749,** E06

flame resistant, *adj*—having flame resistance. **D 123,** D13

flame resistant, *adj*—having flame resistance. **E 176,** E05

flame resistive, n—See **flame resistant,** the preferred term. **E 176,** E05

flame retardant, *adj*—not defined. This term should not be used as an adjective except in the terms "flame-retardant-treated" and "flame-retardant treatment." **D 123,** D13

flame retardant, *n*—a chemical used to impart flame resistance. **D 123,** D13

flame retardant, *adj*—not a defined term. Use only as a modifier with defined compound terms: **flame-retardant chemical, flame-retardant coating,** and **flame-retardant treatment.** **E 176,** E05

flame-retardant chemical, *n*—a chemical, which when added to a combustible material, delays ignition and reduces flame spread of the resulting material when exposed to flame impingement. (See also **fire-retardant chemical.**) **E 176,** E05

flame-retardant coating, *n*—a fluid-applied surface covering on a combustible material which delays ignition and reduces flame spread when the covering is exposed to flame impingement. (See also **fire-retardant coating.**) **E 176,** E05

flame retardant, n—a deprecated term. Do not use. **E 176,** E05

flame-retardant-treated, *adj*—having received a flame-retardant treatment. **D 123,** D13

flame-retardant treatment, *n*—a process for incorporating or adding flame retardant(s) to a material or product. **D 123,** D13

flame-retardant treatment, *n*—the use of a flame-retardant chemical or a flame-retardant coating. (See also **fire-retardant treatment.**) **E 176,** E05

flame speed, *n*—the velocity of propagation of a flame front through a gaseous mixture (fuel and oxidizer) relative to a reference point. **E 176,** E05

flame spread, *n*—the propagation of a flame away from the source of ignition. **D 123,** D13

flame spread classification, *n*—a standard rating of relative surface burning characteristics of a building material as compared to a standard material. **C 11,** C11

flame spread index, *n*—a number or classification indicating a comparative measure derived from observations made during the progress of the boundary of a zone of flame under defined test conditions. **E 176,** E05

flame spread, n—See **surface flame spread, volumetric flame spread.** **E 176,** E05

flame-spread time, *n*—the time taken by a flame on a burning material to travel a specified distance under specified conditions. **D 123,** D13

flaming debris, *n*—material which continues to flame as it separates and moves away from the flaming source. **D 123,** D13

flammability, *n*—those characteristics of a material that pertain to its relative ease of ignition and relative ability to sustain combustion. **D 123,** D13

flammability, *n*—those characteristics of a material that pertain to its ignition and support of combustion. **F 1494,** F23

flammable, *adj*—(*1*) capable of burning with a flame under specified conditions, or (*2*) when used to designate high hazard, subject to easy ignition and rapid flaming combustion. **E 176,** E05

flammable liquid—a liquid having a flash point below 37.8°C (100°F) and having a vapor pressure not exceeding 40 psi (absolute) at 37.8°C and shall be known as a Class I liquid. **E 772,** E44

flammable textile, *n*—any combustible textile that burns with a flame. (See also **flammability.** Compare **combustible textile, noncombustible textile.**) **D 123,** D13

flange—that part of a lock-strip gasket which extends to form one side of a channel. **C 717,** C24

flange, n—See **box.** **D 996,** D10

flange, *n*—a flat plate or formed piece at the end of a railing or rail element for attachment to the adjoining construction or supporting member. **E 1481,** E06

flange, *n*—that part of the rim which gives lateral support to the tire. **F 538,** F09

flange—in chain link fencing, a plate or casting for securing a post to a floor or a horizontal rail to a wall. **F 552,** F14

flange, *n*—the extension of the wrap beyond the core wire or stiffening groove. **F 1379,** F14

flanged joint—see **joint, flanged.** **F 412,** F17

flange lock slider, *n*—*in zippers*, a slider with notches in the flanges of the slider that block the shoulders of the elements when the stringers are pulled apart, thus preventing further separation of the chain. **D 123,** D13

flange plate—See **railing systems.** **E 631,** E06

flanges, *n*—*in zippers*, the edges of the slider formed to contain the chain. **D 123,** D13

flanking transmission—transmission of sound from the source to a receiving location by a path other than that under consideration. **C 634,** E33

flank, side—surface between root and crest; "leading" flank being on point side of crest and "following" flank being on head side of crest. **F 547,** F16

flanky—a characteristic of loose grain leather that forms coarse wrinkles on bending with the grain inward. **D 1517,** D31

flannel, *n*—*as applied to bed sheeting*, a napped fabric used in the fabrication of sheeting products. **D 123,** D13

flare, *n*— the spreading of the filament ends or the strand ends at the cut end of a steel tire cord, expressed as the unravelled length. **D 123,** D13

flare—extraneous light in the dark area. **F 335,** F05

flared—staple legs spread into outward opposite directions 90° with crown plane. **F 592,** F16

flare decay—image contrast loss due to flare. **F 335,** F05

flare joint—see **joint, flare.** **F 412,** F17

flash—as in welding, the metal that protrudes at the weld of the tube, internally, externally, or both, as a result of the pressure applied when a forge-type seam is produced; the two types of flash are internal flash and external flash. **B 846,** B05

flash, *n*—the excess material protruding from the surface of a molded article at the mold junctions. **D 1566,** D11

flash, *n*—appearance of a material when viewed close to the angle at which it is lightest. **E 284,** E12

flash—excess material that forms at the parting line of a mold or die, or the overflow of excess adhesive outside the area of attachment in a bonded assembly. **E 1749,** E06

flashed finish, *n*—units whose surface faces have a range of color produced by the control of the atmospheric conditions in the kiln during firing. **C 43,** C15

flashing—applying a thin layer of opaque or colored glass to the surface of clear glass or vice versa. See also **striking**. **C 162,** C14

flashing—a generic term describing the transitional area between the waterproofing membrane and surfaces above the wearing surface of the building deck; a terminal closure or barrier to prevent ingress of water into the system. **C 717,** C24

flashing—the system used to seal membrane edges at walls, expansion joints, drains, gravel stops, and other places where the membrane is interrupted or terminated. Base flashing covers the edges of the membrane. Cap or counterflashing shields the upper edges of the base flashing. **D 1079,** D08

flashing cement—a trowelable mixture of cutback bitumen and mineral stabilizers including asbestos or other inorganic fibers. **D 1079,** D08

flash magnetization—magnetization by a current flow of very brief duration. **E 1316,** E07

flash (or flash plate)—a very thin electrodeposit used for a final coating; for intermediate coatings of the same nature use **strike.** **B 374,** B08

flashover, *n*—a disruptive electrical discharge at the surface of electrical insulation or in the surrounding medium, which may or may not cause permanent damage to the insulation. **D 1711,** D09

flashover, *n*—the rapid transition to a state of total surface involvement in a fire of combustible materials within an enclosure. **E 176,** E05

flashover—the electrical discharge or arc occurring between electrodes and over or around, but not through, the equipment being tested. **F 819,** F18

flash point, *n*—the lowest temperature of a specimen, corrected to a pressure of 760 mm Hg (101.3 kPa), at which application of an ignition source causes any vapor from the specimen to ignite under specified conditions of test. **D 1711,** D09

flash point—the lowest temperature corrected to a barometric pressure of 101.3 kPa (760 mm Hg), at which application of a test flame causes the vapor of a specimen to ignite under specified conditions of test. **D 2864,** D27

flash point—of a liquid, the minimum temperature at which it gives off vapor in sufficient concentration to form an ignitable mixture with air near the surface of the liquid within the vessel as specified by appropriate test procedure and apparatus. **E 772,** E44

flash point—the lowest temperature at which vapors above a volatile combustible substance ignite in air when exposed to a flame. **E 1316,** E07

flashpoint, *n*—the lowest temperature, corrected to a pressure of 101.3 kPa (760 mm Hg, 1013 mbar), at which application of an ignition source causes the vapors of the specimen to ignite under specified conditions of test. **E 1445,** E27

flash point—the temperature at which a combustible liquid ignites. **E 1705,** E48

flash set—see **early stiffening** **C 219,** C01

flash set—*in grouting*, the rapid development of rigidity in a freshly mixed grout, usually with the evolution of considerable heat; this rigidity cannot be dispelled nor can the plasticity be regained by further mixing without addition of water; also referred to as quick set or grab set. **D 653,** D18

flat—See **container.** **D 996,** D10

flat, *adj*—*(1) of a coating material*, a material that is capable of imparting a finish free of gloss.

(2) of a surface finish, free of gloss. **E 284,** E12

flat, *n*—a rectangular metal bar of width greater than thickness. **E 631,** E06

flat—*on a semiconductor wafer*, a portion of the periphery of a circular wafer that has been removed to a chord. See also **primary orientation flat, secondary flat.** **F 1241,** F01

flat asphalt—a roofing asphalt conforming to the requirements of Specification D 312, Type II. **D 1079,** D08

flatband capacitance, C_{fb}—the capacitance of an MOS structure at the flatband voltage. **F 1241,** F01

flatband condition—*in microelectronics*, the point at which an external applied voltage causes there to be no internal potential difference across an MOS structure. **F 1241,** F01

flatband voltage, V_{fb}—the applied voltage necessary to produce the flatband condition. **F 1241,** F01

flat cable, *n*—any cable with two smooth or corrugated, but essentially flat, surfaces. **D 1711,** D09

flat clinch—clinch formed by folding staple legs parallel to crown with movable clincher. **F 592,** F16

flat conductor, *n*—a conductor with a width-to-thickness ratio arbitrarily chosen as 5 to 1 or greater. **D 1711,** D09

flat conductor cable, *n*—a cable of flat conductors. **D 1711,** D09

flat crown—straight staple crown in contrast to rounded, formed, or offset staple crown. **F 592,** F16

flat cure, n— a synonym for **plateau vulcanization** **D 1566,** D11

flat duck, *n*—duck having the warp of two single yarns woven as one and either single or plied filling yarn. (See also **duck.**) **D 123,** D13

flat-felled seam, *n*—*in home sewing*, a complex seam formed on the outside of a product with cut edges enclosed and two rows of machine stitching visible from the face side. **D 4965,** D13

flat glass—a general term covering sheet glass, plate glass, float glass, and various forms of rolled glass. See related term **bent glass.** **C 162,** C14

flat grain—See **grain.** **D 9,** D07

flat head—most common head, having flat, circular, parallel top and bearing surfaces and slightly rounded edge or rim. **F 547,** F16

flat jet atomizer—see **flat spray atomizer**. **E 1620,** E29

flatness—*for wafer surfaces*, the deviation of the front surface, expressed in TIR or maximum FPD, relative to a specified reference plane when the back surface of the wafer is ideally flat, as when pulled down by a vacuum onto an ideally clean flat chuck. **F 1241,** F01

flatness of field—a qualitative term describing how well the image of a planar specimen is reproduced as a plane in the image field. (See **curvature of field.**) **E 7,** E04

flat piece *(of aggregate), n*—a particle of aggregate for which the ratio of the width to thickness of its circumscribing rectangular prism is greater than a specified value (see also **elongated piece** (*of aggregate*)). **C 125,** C09

flat-platen pressed—a method of consolidating and hot pressing a panel product in which the applied pressure is perpendicular to the faces. **D 1554,** D07

flat product—a rectangular or square solid section of relatively great length in proportion to thickness. Included in the designation "flat product" depending on the width and thickness, are plate, sheet, strip, and bar. Also included is the product known as "flat wire." **B 846,** B05

flat-ride—an amusement ride that operates on a single level whether over a controlled, fixed course or track, or confined to a limited area of operation. **F 747,** F24

flats—a longitudinal flat area on a normally convex surface of a pultrusion.

NOTE—This condition may be caused by shifting in the reinforcement, lack of sufficient reinforcement, or local fouling of the die surface. **D 3918,** D20

flat sheet, *n*—sheet with sheared, silt, or sawed edges that has been flattened or leveled. **B 899,** B02

flat sheet, *n*—*in textiles*, a flat, hemmed product, usually rectangular, used for covering the mattress on a bed and used for sleeping on or under. **D 123,** D13

flat sheet membrane—a sheet type membrane may be coated onto a fabric substrate. **D 6161,** D19

flat sheets, Type A, *n*—sheet intended for exterior applications, where it may be subjected to the direct action of sun, rain, or snow, and when tested in accordance with Test Methods C 1185, does demonstrate compliance with Specification C 1186. **C 1154,** C17

flat sheets, Type B, *n*—sheet intended for exterior applications where it will not be subjected to the direct action of sun, rain, or snow, and when tested in accordance with Test Methods C 1185, does demonstrate compliance with Specification C 1186. **C 1154,** C17

flat sheets, Type F (flexible)—an asbestos-cement flat sheet suitable for exterior or interior use, where a board having higher strength and density, smoother surface, greater flexibility, and lower moisture absorption is desired. **C 460,** C17

flat sheets, Type F (flexible)—asbestos-cement flat sheet suitable for exterior or interior use, where a board having higher strength and density, smoother surface, greater flexibility, and lower moisture absorption is desired. **D 2946,** C17

flat sheets, Type U (utility)—an asbestos-cement flat sheet suitable for exterior or interior use, having sufficient strength for general utility and construction purposes, and where a board having maximum flexibility, highest density, smoother surface, and lower moisture absorption is not essential. **C 460,** C17

flat sheets, Type U (utility)—asbestos-cement flat sheet suitable for exterior or interior use, having sufficient strength for general utility and construction purposes, and where a board having maximum flexibility , highest density, smoother surface, and lower moisture absorption is not essential. **D 2946,** C17

flat spray atomizer—a fan spray atomizer that produces a planar spray pattern.

NOTE—In agricultural applications, the spray patterns have tapered edges. **E 1620,** E29

flattened-shank nail—round wire nail with portion of shank flattened for a certain distance between point and head to facilitate driving of nail between steel members and wrapping of flattened portion of shank around steel rod during driving. **F 547,** F16

flattened thread, flat-bottom thread—thread having flat root. Flattened thread occasionally referring to flattened thread crest. **F 547,** F16

flatting agent—a material added to paints, varnishes, and other coating materials to reduce the gloss of the dried film. **D 16,** D01

flat wire—a product up to and including 0.188 in. (4.78 mm) in thickness and up to and including 1¼ in. (31.8 mm) in width. **B 846,** B05

flavor, *n*—the total of the sensations perceived by means of the taste buds, olfactory organ, and the buccal cavity which may include pain, temperature, and tactile sensations. **E 253,** E18

flaw—in dry process enameling, a defect of the ware that is cause for rejection. **C 286,** B08

flaw—a defect sufficiently greater than those typical of the morphology of a carbon or graphite body to influence a property. **C 709,** D02

flaw, *n*—an imperfection or discontinuity that may be detectable by nondestructive testing and is not necessarily rejectable. **E 1316,** E07

flaw characterization, *n*—the process of quantifying the size, shape, orientation, location, growth, or other properties, of a flaw based on NDT response. **E 1316,** E07

fleece, *n*—the wool of one sheep as obtained by shearing. **D 123,** D13

flesh—the inner side of hide or skin. Also, adjective referring to that side. **D 1517,** D31

flesher—the flesh split or under-cut of a sheepskin, split before tanning. (See also **chamois leather**.) **D 1517,** D31

flex cracking, *n*—cracking primarily caused by application of mechanical stress-strain cycling. **F 538,** F09

flexibility, *n*—that property of a material by virtue of which it may be flexed or bowed repeatedly without undergoing rupture. **D 123,** D13

flexibility—the ability of applied shoe polish to flex and bend in creases of the shoe without powdering, peeling, or cracking, and without damaging the shoe finish. **D 2825,** D21

flexibility, *n*—the ability to be bent, turned, or twisted without cracking, breaking or showing other permanent damage and with or without returning of itself to its former shape. **F 141,** F06

flexible—easily hand-folded, flexed, twisted, and bent. **F 17,** F02

flexible—See Terminology F 17. **F 1327,** F02

flexible foam—a preformed, cured, flexible, cellular material made from foamed rubber or plastic. **C 717,** C24

flexible hinges or joints—nonrigid or nonfixed center hinges used in prostheses. **F 869,** F08

flexible polypropylene, *n*—a material having a 2 % secant modulus of less than 300 MPa (40,000 psi) as determined by Test Method D 5323 produced by polymerization of propylene with or without other alpha olefin monomers. **D 4439,** D35

flexion—the condition of decreasing the angle between two body segments; opposite of extension. **F 869,** F08

flex life, *n*—the number of cycles required to produce a specified state of failure in a specimen that is flexed in a prescribed method. **D 1566,** D11

flex life—the number of cycles required to produce a specified state of failure in a specimen that is flexed in a prescribed method. (D 1566, D-11) **F 869,** F08

flex life test—a laboratory method used to determine the life of a product when subjected to dynamic bending stresses. **F 869,** F08

flexometer, *n*—a machine that subjects a test specimen to repeated deformation by compression, tension, shear, bending, torsion or any combination thereof. **D 1566,** D11

flexural rigidity, *n*—
general—resistance to bending.
specific—the couple on either end of a strip of unit width bent into unit curvature in the absence of any tension. **D 123,** D13

flexural strength, *n*—the maximum load sustained by a standard specimen of a sheet material when subjected to a bending force. **C 11,** C11

flexural strength, *n—in asbestos-cement bulkheading*, the transverse breaking load in newtons per meter of width (lbf/ft of width) of saturated sheets when loaded on a span of 1.524 m (5 ft) with the load applied equally and simultaneously at both one-third points of the span. **C 460,** C17

flexural strength, *n—in asbestos-cement plastic-foam core insulating panels*, the average breaking load in newtons (or lbf) of specimens conditioned seven days at 23± 2°C and 50 ± 10 % relative humidity, loaded as simple beams with the load applied equally and simultaneously at both one-third points of the span. **C 460,** C17

flexural strength, *n—in flat asbestos-cement sheet*, the average breaking load in newtons (or lbf) of dried specimens loaded as simple beams on a span of 254 mm (10 in.) with the load applied equally and simultaneously at both one-third points of the span. **C 460,** C17

flexural strength, *n—in corrugated asbestos-cement sheets*, the average breaking load in newtons per meter of width (or lbf/ft of width) of dried specimens loaded as simple beams on a 762 mm (30 in.) span with the load applied equally and simultaneously at both one-third points of the span. **C 460,** C17

flexural strength, *n—in pipe*, the ability of a standard pipe longitudinal section to withstand external loads that bear on the pipe transversely to its longitudinal axis and induce bending. **C 460,** C17

flexural strength, *n—in plastic foam core asbestos-cement insulating panels*, see **flexural strength,** *n—in asbestos-cement plastic-foam core insulating panels.* **C 460,** C17

flexural strength—a property of solid material that indicates its ability to withstand a flexural or transverse load. **C 709,** D02

flexural strength, *n*—a measure of the ultimate strength of a specified beam in bending. **C 1145,** C28

flexural strength, *n— in bulkheading*, the transverse breaking load in newtons per meter of width (lbf/ft of width) of saturated sheets when loaded on a span of 1.524 m (5 ft) with the load applied equally and simultaneously at both one-third points of the span. **C 1154,** C17

flexural strength, *n— in plastic-foam core insulating panels*, the average breaking load in newtons (or lbf) of specimens conditioned at 50 % relative humidity, loaded as simple beams with the load applied equally and simultaneously at both one-third points of the span. **C 1154,** C17

flexural strength, *n— in flat sheet*, the average breaking load in newtons (or lbf) per metre (foot) of width of dried specimens loaded as simple beams on a span of 254 mm (10 in.) with the load applied at the center of the span. **C 1154,** C17

flexural strength, *n— in corrugated sheets*, the average breaking load in newtons per meter of width (or lbf·ft of width) of dried specimens loaded as simple beams on a 762-mm (30-in.) span with the load applied equally and simultaneously at both one-third points of the span. **C 1154,** C17

flexural strength, *n— in pipe,* the ability of a standard pipe longitudinal section to withstand external loads that bear on the pipe transversely to its longitudinal axis and induce bending. **C 1154,** C17

flexural strength, *n—in asbestos-cement bulkheading,* the transverse breaking load in newtons per metre of width (lbf/ft of width) of saturated sheets when loaded on a span of 1.524 m (5 ft) with the load applied equally and simultaneously at both one-third points of the span. **D 2946,** C17

flexural strength, *n—in asbestos-cement plastic-foam core insulating panels,* the average breaking load in newtons (or lbf) of specimens conditioned seven days at 23 ± 2°C and 50 ± 10 % relative humidity, loaded as simple beams with the load applied equally and simultaneously at both one-third points of the span. **D 2946,** C17

flexural strength, *n—in corrugated asbestos-cement sheets,* the average breaking load in newtons per metre of width (or lbf/ft of width) of dried specimens loaded as simple beams on a 762 mm (30 in.) span with the load applied equally and simultaneously at both one-third points of the span. **D 2946,** C17

flexural strength, *n—in flat asbestos-cement sheet,* the average breaking load in newtons (or lbf) of dried specimens loaded as simple beams on a span of 254 mm (10 in.) with the load applied equally and simultaneously at both one-third points of the span. **D 2946,** C17

flexural strength, *n—in pipe,* the ability of a standard pipe longitudinal section to withstand external loads that bear on the pipe transversely to its longitudinal axis and induce bending. **D 2946,** C17

flexural strength, *n—in plastic foam core asbestos-cement insulating panels,* see **flexural strength,** *in asbestos-cement plastic-foam core insulating panels.* **D 2946,** C17

flexural strength, *n*—property of solid material that indicates its ability to withstand a flexural or transverse bending load. **D 2946,** C17

flexure plate pivot—a type of pivot or hinge in which the motion occurs through the bending of a thin elastic plate. **E 7,** E04

flexure stress—the tensile component of the bending stress produced on the surface of a glass section opposite to that experiencing a locally impinging force. **C 162,** C14

flicker method, *in image analysis* —the procedure of alternating between the live video image and the detected image while altering the gray-level threshold range to establish the optimum discrimination and detection. **E 7,** E04

flint—usually in phrases "flint-dried" or "flint hides." Air or sun-dried without other curing. **D 1517,** D31

flint fireclay, *n*—a hard or flint-like fireclay occurring as an unstratified massive rock, practically devoid of natural plasticity and showing a conchoidal fracture. **C 71,** C08

flint glass—(*1*) a lead-containing glass.

(*2*) term used by container industry for colorless glass. **C 162,** C14

flint glass cullet, *n*—a particulate glass material that contains no more than 0.1 mass percent Fe_2O_3, or 0.0015 mass percent Cr_2O_3, as determined by chemical analysis. **D 5681,** D34

flint optical glass—See **optical flint glass.** **C 162,** C14

flip—deprecated term, do not use. This term may have various meanings that are not clearly defined. **E 284,** E12

flitch—a portion of a log sawed on two or more sides and intended for remanufacture into lumber or sliced or sawed veneer. The term is also applied to the resulting sheets of veneer laid together in sequence of cutting. **D 9,** D07

flitch—a portion of a log sawed on two or more sides and intended for remanufacture into sliced or sawn veneer. The term is also applied to the resulting sheets of veneer stacked together in sequence of cutting. **D 1038,** D07

float, *n—in woven fabric,* the portion of a warp or filling yarn that extends unbound over two or more filling or warp yarns. **D 123,** D13

float, *n*—a defect in which a warp or filling yarn extends unbound over the ends with which it should be interlaced. **D 123,** D13

float—separable component of a boom that provides buoyancy. **F 818,** F20

float bath—a pool of molten metal, commonly tin, contained within a refractory receptacle and protected from oxidation by an inert atmosphere, upon which molten glass is drawn into a flat sheet. See **float glass.** **C 162,** C14

floated finish—a concrete finish provided by consolidating and leveling the concrete with only a power driver or hand float, or both. A floated finish is coarser than a troweled finish. For specifications, See ACI 301-72. **C 717,** C24

floater—(*1*) a floating clay shape to skim foreign materials or control their passage in a melter.

(*2*) an object, generally a porous silica brick, introduced into a melting furnace, which will float on the surface of the molten glass for tens of minutes to several hours, thus revealing the surface flow. **C 162,** C14

floater hole—{archaic} an opening in a melter through which floaters are placed. **C 162,** C14

float glass—flat glass that has been formed on molten metal, commonly tin. **C 162,** C14

floating controlled load— See **load.** **D 996,** D10

floating roller peel test—See **test, floating roller peel.** **E 1749,** E06

floating threshold—any threshold with amplitude established by a time average measure of the input signal. (E 750) **E 1316,** E07

floats (as in asbestos floats)—air conveyed asbestos fractions such as accumulate in the air filtration system of an asbestos mill. **D 2946,** C17

float/sink, *n*—a reference to the physical action that particles undergo when immersed in a liquid of a predetermined specific gravity. **D 121,** D05

floc—loose, open-structured mass formed in a suspension by the aggregation of minute particles. **D 653,** D18

floc—a loose, open-structured mass produced by the aggregation of minute particles. **D 6161,** D19

flocculant—an agent that produces flocs or aggregates from small suspended particles. **D 4410,** D19

flocculate—to aggregate into larger particles, to increase in size to the point where precipitation occurs. **B 374,** B08

flocculate—a grouping of primary particles, aggregates, or agglomerates having weaker bonding than either the aggregate or agglomerate structures. **C 242,** C21

flocculating—the thickening of the consistency of a slip by adding a suitable electrolyte. **C 286,** B08

flocculating agent—a coagulating substance such as alum, ferrous sulfate, or lime which, when added to water, forms a precipitate that expedites the settling of suspended matter. **D 4410,** D19

flocculation—the process of forming flocs. **D 653,** D18

flocculation, *n*—synonymous with agglomeration. **D 1356,** D22

flocculation, *n*—formation (sometimes reversible) of loosely coherent, partially agglomerated rubber, distributed in the liquid phase of a latex. **D 1566,** D11

flocculation—the process of agglomerating fine particles into larger groupings called flocs. **D 6161,** D19

flocculent—chemical(s) which, when added to water, form bridges between suspended particles causing them to agglomerate into larger groupings (flocs) which then settle or float by specific gravity differences. **D 6161,** D19

flocculent structure—see **soil structure**. **D 653,** D18

flock, *n*—a material obtained by reducing textile fibers to fragments as by cutting, tearing, or grinding, to give various degrees of comminution. **D 123,** D13

flocked blanket, *n*—a blanket made with a fishnet-type scrim sandwiched between two thin layers of foam with flock adhered to the outside of the foam. **D 123,** D13

flocs or floccules—masses of solids formed in a liquid by addition of coagulants (flocculants), or through biochemical processes, or by agglomeration of individual particles. **D 4410,** D19

flood coat—the top layer of bitumen used to hold the aggregate on an aggregate-surfaced, built-up roofing membrane. **D 1079,** D08

flooded system—a system which, while being tested, becomes so filled with tracer gas as to make impracticable further leak testing. **E 1316,** E07

flooding—a means of compacting trench backfill by the introduction of water by gravity. **C 896,** C04

flooding nozzle—see **deflector atomizer**. **E 1620,** E29

flood nozzle—see **deflector atomizer**. **E 1620,** E29

flood point, *n*— *(in column distillation)* the point at which the upflowing vapor flow obstructs the down-coming reflux and the column suddenly loads with liquid. **D 4175,** D02

floor, *n*—the rock material immediately underlying a coal bed. **D 121,** D05

floor—bottom of near horizontal surface of an excavation, approximately parallel and opposite to the roof. (ISRM) **D 653,** D18

floor, *n*—in a building, a supporting structure (generally horizontal) and constituting the bottom level of each **story.**

first floor—in a building, (*1*) (in the United States) the floor of a **building** that is at, or closest to, **finished grade** (also used as a synonym for **ground floor**). (*2*) (except in the United States) the **floor** of a **building** that is next above the **floor** at, or closest to, **finished grade.**

flooring, n—a material used to construct the uppermost layer of a **floor.**

sub-floor, n—a part of a **floor** over which one or more components may be added to complete the **floor.**

sub-flooring, n—the material used in constructing a **sub-floor.**

underlayment, n—in flooring, a layer of material usually placed upon the **sub-floor** that provides a smooth, even base for **flooring.** **E 631,** E06

floor area:

gross floor area—the entire area within the inside perimeter of the exterior **walls.**

net floor area—that part of the **gross floor area** located within **occupiable space.** **E 631,** E06

floor area, *n* (superficie de plancher):

gross floor area (superficie brute de plancher)—entire area within the inside perimeter of the exterior walls.

net floor area (superficie nette de plancher)—that part of the **gross floor area** located within occupiable space. **E 1480,** E06

floor brick, *n*—smooth, dense brick, highly resistant to abrasion, used as finished floor surfaces. See Specification C 410. **C 43,** C15

floor covering, *n*—an essentially planar material, having a relatively small thickness in comparison to its length or width, which is laid on a floor to enhance the beauty, comfort, and utility of the floor. **D 123,** D13

flooring—stone used as in interior pedestrian wearing surface. **C 119,** C18

flooring brad——bright, regular-stock-steel, 2 by 0.120 to 4 by 0.192–in. nails with deep (32 deg) countersunk flat or cupped 0.162 to 0.244–in. head and medium diamond point. Also, slender, bright, regular-stock-steel, 1 ¼ by 0.076 to 2 ½ by 0.113–in. nails with deep (32 deg) countersunk flat or cupped or brad 0.128 to 0.155–in. head and blunt diamond point.

machine flooring brad ——bright, regular stock-steel, 1 by 0.072-in. nail with special 0.113-in. brad head with cylindrical rim and medium diamond point. **F 547,** F16

flooring material, *n*—any pliable planar structure used as a base surface in camping tentage, but excluding such things as rugs or carpets placed in the tent that are not integral parts of the item. **D 123,** D13

flooring nail—bright, stiff-stock or hardened-steel, helically threaded, 1 by 0.072 to 3 ½ by 0.148-in. nails with flat or checkered 9/64 to 9/32-in. countersunk or casing head and blunt diamond point. **F 547,** F16

floor, *n* (plancher)—(in a building) supporting structure (generally horizontal) and constituting the bottom level of each story. **E 1480,** E06

floor surface—See *walkway surface.* **F 1646,** F13

flop, *n*—a difference in appearance of a material viewed over two widely different aspecular angles. **E 284,** E12

flop, *adj*—pertaining to the appearance of a material when viewed from a direction far from the specular angle, typically 70° or more. **E 284,** E12

flop angle, *n*—the aspecular angle when a material is viewed from a direction far from the specular, typically 70° or more. **E 284,** E12

flop color, *n*—color of a material when viewed from a specified direction far from the specular angle, typically 70° or more. **E 284,** E12

flop contrast, *n*—a measure of the degree of dissimilarity in appearance of a specimen when it is viewed at two widely different aspecular angles. **E 284,** E12

flop index, *n*—a numerical scale of flop obtained by visual experiments. **E 284,** E12

flotation—portion of a boom that provides buoyancy. **F 818,** F20

flotation cell, *n*—the vessel or compartment in which the flotation test is performed. **D 121,** D05

flounce, *n*—a ruffled drop on a bedcovering. **D 123,** D13

flow, *v*—*in building construction*, the deformation of a tape sealant under stress. **C 717,** C24

flow, *n*—a laboratory measured mortar property that indicates the percent increase in diameter of the base of the truncated cone of mortar when it is placed on a flow table, and mechanically raised and dropped specified times under specified conditions. **C 1180,** C12

flow, *n*—movement of an adhesive during the bonding process, before the adhesive is set. **D 907,** D14

flow—Same as **flow rate**. **E 1316,** E07

flowback nozzle—see **by-pass nozzle**. **E 1620,** E29

flow balancing—the use of an imposed pressure drop (flow balancing tube), to minimize conversion differences of modules operating in parallel. **D 6161,** D19

flow balancing tube—see **flow balancing**. **D 6161,** D19

flow brightening—the melting of an electrodeposit, followed by solidification, especially of tin plate. **B 374,** B08

flow-button—the pellet of frit used in the Fusion Flow Test. **C 286,** B08

flow cavitation, *n*—cavitation caused by a decrease in static pressure induced by changes in velocity of a flowing liquid. Typically, this may be caused by flow around an obstacle or through a constriction, or relative to a blade or foil. A cavitation cloud or "cavitating wake" generally trails from some point adjacent to the obstacle or constriction to some distance downstream, the bubbles being formed at one place and collapsing at another. **G 40,** G02

flow channel—the portion of a flow net bounded by two adjacent flow lines. **D 653,** D18

flow coating—the process of coating a metal shape by causing the slip to flow over its surface and allowing it to drain. **C 286,** B08

flow cone—*in grouting*, a device for measurement of grout consistency in which a predetermined volume of grout is permitted to escape through a precisely sized orifice, the time of efflux (flow factor) being used as the indication of consistency. **D 653,** D18

flow curve—the locus of points obtained from a standard liquid limit test and plotted on a graph representing water content as ordinate on an arithmetic scale and the number of blows as abscissa on a logarithmic scale. **D 653,** D18

flow failure—failure in which a soil mass moves over relatively long distances in a fluid-like manner. **D 653,** D18

flow function, FF—the plot of unconfined yield strength versus major consolidation stress for one specified bulk solid. **D 653,** D18

flow hole—{archaic} See **throat**. **C 162,** C14

flow index, F_w, I_f **(D)**—the slope of the flow curve obtained from a liquid limit test, expressed as the difference in water contents at 10 blows and at 100 blows. **D 653,** D18

flow line—a defect induced by discontinuous flow velocities during forming of molded or extruded bodies. **C 709,** D02

flow line—a line formed by the inverts of pipe. **C 822,** C13

flow line—the path that a particle of water follows in its course of seepage under laminar flow conditions. **D 653,** D18

flow line—one or more streaks distinguished by a difference in light reflectance from the surrounding area, charateristic of injection-molded parts. (See also **weld mark**.)

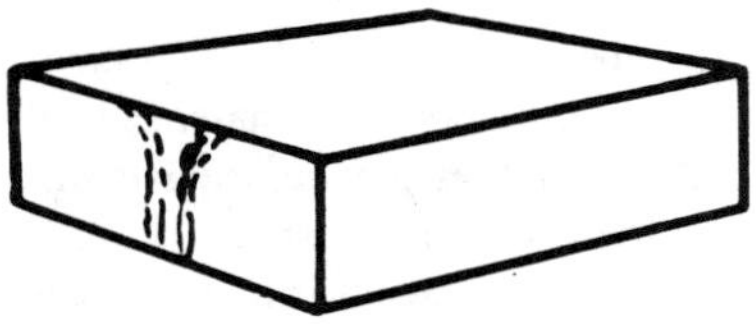

F 109, C21

flow lines—a fiber pattern, frequently observed in wrought metal, which indicates the manner in which the metal flowed during deformation. **E 7,** E04

flow log—a borehole geophysical log used to record vertical movement of ground water and movement of water into or out of a well or borehole and between formations within a well. **D 653,** D18

flow marks, *n*—marks or line on a molded product, caused by imperfect fusion of flowing fronts. **D 1566,** D11

flowmeter, *n*—an instrument for measuring the rate of flow of a fluid (that is, liquid or gas) moving through a system. **D 1356,** D22

flow net—a graphical representation of flow lines and equipotential (piezometric) lines used in the study of seepage phenomena. **D 653,** D18

flow path—represents the area between two flow lines along which ground water can flow. **D 653,** D18

flow process—See **gob process**. **C 162,** C14

flow-proportioned sample—a sample obtained by collecting an aliquot volume in proportion to the rate of flow of the stream sampled. **D 1129,** D19

flow rate—the time required for a powder sample of standard weight to flow through an orifice in a standard instrument according to a specified procedure. **B 243,** B09

flow rate—in leak testing, *(1)* the rate at which gas passes a given cross section of a system, determined by the product of the volume passing per unit time and its (partial) pressure at the cross section; *(2)* a product of the (partial) pressure difference of a gas at the ends of a conduit or across the face of an orifice, and the conductance of the gas for the conduit or orifice. Expressed in pressure-volume per unit time, such as pascal cubic metres per second. **E 1316,** E07

flow rate—rate of extrusion, weight per unit of time, g/10 min (kg/s), of molten resins through a die of specified length and diameter, under prescribed conditions of temperature, load, and piston position in the barrel as the timed measurement is being made. **F 412,** F17

flow slide—the failure of a sloped bank of soil in which the movement of the soil mass does not take place along a well-defined surface of sliding. **D 653,** D18

flow steady, *n*—a characteristic of a flow system where the specific discharge is constant in time at any point. **D 653,** D18

flowthrough electrodes—ion-selective and reference electrodes designed for anaerobic measurements. The two electrodes are connected by plastic tubing to a syringe or peristaltic pump, and the sample is pumped through the electrodes at a constant rate. Ion-selective electrodes can be made in a flow through configuration for the measurement of very small samples (0.2 to 0.3 mL) or samples that must be measured anaerobically. **D 4127,** D19

flow value, N_ϕ **(degrees)**—a quantity equal to tan [45 deg + (ϕ/2)]. **D 653,** D18

flow velocity—see **specific discharge**. **D 653,** D18

flue lining—a manufactured tubular non-load bearing fired clay unit, normally used for conveying hot gases in chimneys. **C 896,** C04

fluence—the number of beam particles per unit area, where that area is defined perpendicular to the direction of the beam. **E 673,** E42

fluid-applied elastomer—an elastomeric material, fluid at ambient temperature, that dries or cures after application to form a continuous membrane. Such systems normally do not incorporate reinforcement. **D 1079,** D08

fluid carrier, *n*—a fluid in which particles are suspended to facilitate their movement or application. **C 1145,** C28

fluid coke, *n*—petroleum coke with a granular, microscopic layered structure resulting from injection of petroleum feedstock into a flowing, loose bed of coke particles. **D 4175,** D02

fluid-energy mill, *n*—a size-reduction apparatus in which grinding is achieved by the collision of the particles being ground in a high-velocity steam of air, steam, or other fluid. **C 1145,** C28

fluidifier—*in grouting*, an admixture employed in grout to increase flowability without changing water content. **D 653,** D18

fluidity—the reciprocal of viscosity. **D 1695,** D01

fluidized bed—a bed of granular particles in which the fluid flows upward at a rate sufficient to suspend the particles completely and randomly in the fluid phase. **D 2652,** D28

fluid migration, *n*—*in building construction*, accumulation of a fluid from a sealant, coating, or membrane on or in an adjacent material. **C 717,** C24

fluid temperature (FT), *n*—*in reference to the fusibility of coal and coke ash according to Test Method D 1857*, the temperature at which the fused mass has spread out in a nearly flat layer with a maximum height of 1.6 mm (1⁄16 in.). **D 121,** D05

fluid temperature, FT, *n*—in ash fusion determinations, the temperature at which a fused mass has spread out in a nearly flat layer with maximum height of 1.6 mm (1/16 in.). **D 5681,** D34

fluorescence—photoluminescence in which the emitted optical radiation results from direct transitions from a photo-excited singlet energy level to a lower singlet level, these transitions taking place generally within 10 nanoseconds after excitation. **D 2864,** D27

fluorescence—the emission of radiant energy from an atom, molecule, or ion resulting from absorption of a photon and a subsequent transition to the ground state without a change in total spin quantum number. **E 131,** E13

fluorescence, *n*—photoluminescence that ceases when excitation ceases. **E 284,** E12

fluorescence—the emission of light by a substance as a result of the absorption of some other radiation of shorter wavelengths only as long as the stimulus producing it is maintained. **E 1316,** E07

fluorescence—the emission of visible radiation by a substance as a result of, and only during, the absorption of black light radiation. **E 1316,** E07

fluorescence—the emission of visible radiation by a substance as the result of, and only during, the absorption of black light radiation. **E 1316,** E07

fluorescence—the property of emitting radiation in the visible range as a result of absorption of radiation in the ultraviolet range from some other source. Optical brighteners that have this property are sometimes added to paper to enhance its whiteness or brightness to the eye in normal lighting. The emitted radiation can cause erratic reflectance values. **F 149,** F05

fluorescent, *adj*— *(1)* exhibiting fluorescence.
(2) of a source, having a primary radiant flux emitter that excites fluorescence in a phosphor to produce light. **E 284,** E12

fluorescent brightener, n—discouraged term for **fluorescent whitening agent**. **E 284,** E12

fluorescent examination method—the magnetic particle examination method employing a finely divided fluorescent ferromagnetic inspection medium. **E 1316,** E07

fluorescent illuminant, *n*—illuminant representing the spectral distribution of the radiation from a specified type of fluorescent lamp. **E 284,** E12

fluorescent magnetic particle inspection—the magnetic particle inspection process employing a finely divided fluorescent ferromagnetic inspection medium that fluoresces when activated by black light (3200 to 4000 Å (320 to 400 nm)). **E 1316,** E07

fluorescent penetrant, *n*—an inspection penetrant which fluoresces or glows in ultraviolet light. **C 1145,** C28

fluorescent screen—a sheet of material which emits visible light when exposed to invisible radiation. **E 7,** E04

fluorescent screen—alternative term for **intensifying screen** *(b)*. **E 1316,** E07

fluorescent ultraviolet lamp, *n*—a lamp in which the irradiance from a low pressure mercury arc is transformed to a higher wavelength UV by a phosphor. The spectral power distribution of a fluorescent lamp is determined by the emission spectrum of the phosphor and the UV transmittance of the glass tube. **G 113,** G03

fluorescent whitening agent, *FWA, n*—fluorescent dye or pigment that absorbs near-ultraviolet radiant flux and re-emits the power as visible light (violet-blue), thereby causing a whiter appearance when added to a yellowish-white material. **E 284,** E12

fluorescent whitening agents (FWA)—(optical bleach, fluorescent brightener) complex, organic molecules that adhere to fabrics as though they were dyes. Ultraviolet (UV) energy is absorbed, converted, and emitted as visible blue light to enhance fabric appearance and maintain whiteness or brightness. **D 459,** D12

fluorescent X-rays (fluorescent analysis)— characteristic X-rays excited by radiation of wavelength shorter than the corresponding absorption edge. **E 7,** E04

fluorination—the process of chemically reacting a material with fluorine or a fluorine-containing compound to produce a desired product, for example, the reaction of UO_3 with fluorine to produce UF_6. **C 859,** C26

fluorite (CaF_2) (fluorspar)— an inorganic mineral of the isometric form, used as a source of fluorine for fluxing of glasses, and glazes. **C 242,** C21

fluorocarbon plastic, *n*—a plastic based on polymers made with perfluoromonomers. **D 883,** D20

fluorocarbon plastic, *n*—a plastic based on polymers made with perfluoromonomers. (ISO, D-20) **F 412,** F17

fluorocarbon plastic, *n*—a plastic based on polymers made with monomers composed of fluorine and carbon only. **F 1251,** F04

fluorohydrocarbon plastic, *n*—a plastic based on polymers made with monomers composed of fluorine, hydrogen, and carbon only. **F 1251,** F04

fluorohydrocarbon plastics, *n*—plastics based on polymers made with monomers composed of fluorine, hydrogen, and carbon only. (ISO) **D 883,** D20

fluorohydrocarbon plastics, *n*—plastics based on polymers made with monomers composed of fluorine, hydrogen, and carbon only. (ISO, D-20) **F 412,** F17

fluoroplastic, *n*—a plastic based on polymers made from monomers containing one or more atoms of fluorine, or copolymers of such monomers with other monomers, the fluorine-containing monomer(s) being in greatest amount by mass. **D 883,** D20

fluoroplastic, *n*—a plastic based on polymers made from monomers containing one or more atoms of fluorine, or copolymers of such monomers with other monomers, the fluorine-containing monomer(s) being in the greatest amount by mass. (D-20) **F 412,** F17

fluoroplastic, *n*—a plastic based on polymers with monomers containing one or more atoms of fluorine or copolymers of such monomers with other monomers, the fluorine-containing monomer(s) being in greatest amount by mass. (See also **fluorocarbon plastic, chlorofluorocarbon plastic, fluorohydrocarbon plastic,** and **chlorofluorohydrocarbon plastic.**) **F 1251,** F04

fluoroptic temperature measurement—temperature measurement based on the variation in total luminescence of a fluoroptic phosphor which has been previously calibrated versus a known temperature standard. **F 1479,** F02

fluoroscopy—the visual observation on a fluorescent screen of the image of an object exposed to penetrating, ionizing radiation. **E 1316,** E07

flush joint or flush coupled—casing or riser with ends threaded such that a consistent inside and outside diameter is maintained across the threaded joints or couplings. **D 653,** D18

flute—See **corrugated fiberboard.** **D 996,** D10

flute—helical and, sometimes, vertical deformation rolled onto shank. **F 547,** F16

fluted—deformed with continuous, symmetrical, helical, or longitudinal depressions. Four or five of these flutes are usually formed onto wire from which fluted nail is made, in contrast to threading where shapes are rolled onto nail shank in a thread-rolling process. Flutes resulting from this manufacturing process extend the full length of the nail shank. (See **roll-grooved** and **mechanically**

deformed.) **F 547,** F16

flutes—continuous symmetrical depressions along nail shank. **F 547,** F16

fluvial sediment—particles derived from rocks, biological materials, or chemical precipitants, that are transported by, suspended in, or deposited by flowing water. **D 4410,** D19

flux—an ingredient that reduces batch melting temperature. **C 162,** C14

flux—a substance that promotes fusion in a given ceramic mixture. **C 242,** C21

flux—a substance that promotes fusion in a given ceramic mixture. **C 286,** B08

flux, *n*—a bituminous material, generally liquid, used for softening other bituminous materials. **D 8,** D04

flux—a bituminous material used as a feed stock for further processing and as a material to soften other bituminous materials. **D 1079,** D08

flux—the membrane throughput, usually expressed in volume of permeate per unit time per unit area, such as gallons per day per ft^2 or litres per hour per m^2. **D 6161,** D19

flux—see **radiant flux, luminous flux.** **E 284,** E12

flux—the number of beam particles per unit area per second, where the area is defined to be perpendicular to the direction of the beam. **E 673,** E42

flux—the number of particles flowing through a given area per unit time. **E 1620,** E29

flux, *n*—the combining of photons of light energy to create radiant power. **F 1294,** F05

flux block—a refractory furnace block used in contact with glass in melting. **C 162,** C14

flux-current loop, incremental (biased)—the curve developed by plotting magnetic induction, *B*, versus magnetic field strength, *H*, when the magnetic material is cyclically magnetized while under dc bias condition. This loop will not be symmetrical about the *B* and *H* axes. **A 340,** A06

flux-current loop, normal—the curve developed by plotting magnetic induction, *B*, versus magnetic field strength, *H*, when the magnetic material is symmetrically cyclically magnetized.

NOTE—The area of the loop is proportional to the sum of the static hysteresis loss and all dynamic losses. **A 340,** A06

flux density—see **magnetic induction.** **B 713,** B01

flux density, magnetic—the strength of a magnetic field, expressed in flux lines per unit area. **E 1316,** E07

fluxing agent, *n*—any substance which will promote fusion of ceramic materials. **C 1145,** C28

fluxing lime—a term referring to quicklime used as an agent in the manufacture of steel or glass. **C 51,** C07

flux leakage field—the magnetic field that leaves or enters the surface of a part as the result of a discontinuity or a change in section. **E 1316,** E07

flux line—See **metal line.** **C 162,** C14

flux lines—see **lines of force.** **E 1316,** E07

flux linkage, $\mathcal{L}$—the sum of all flux lines in a coil.

$$\mathcal{L} = \phi_1 + \phi_2 + \phi_3 + \cdots\phi_N$$

where:
ϕ_1 = flux linking turn 1;
ϕ_2 = flux linking turn 2, and so forth; and
ϕ_N = flux linking the *N*th turn.

NOTE—When the coupling coefficient, *k'*, is less than unity, the flux linkage equals the product of the average flux linking the turns and the total number of turns. When the coupling coefficient is equal to unity, the flux linkage equals the product of the total flux linking the coil and the total number of turns. **A 340,** A06

flux linkage, mutual, $\mathcal{L}_m$—the flux linkage existing between two windings on a magnetic circuit. Mutual linkage is maximum when the coupling coefficient is unity. **A 340,** A06

flux path length, effective, ℓ_1—the calculated length of the flux paths in a magnetic core, which is used in the calculations of certain magnetic parameters. **A 340,** A06

flux path length, ℓ—the distance along a flux loop. **A 340,** A06

flux penetration—the depth to which a magnetic flux exists in a part. **E 1316,** E07

flux profile, *n*—the curve relating incident radiant heat energy on the specimen plane to distance from the point of initiation of flaming ignition, that is, 0 cm. **E 176,** E05

flux, radiant—see **radiant flux.** **E 772,** E44

flux-sensitive—a sampling process where the magnitude measured responds to the flux of particles through the sampling region. **E 1620,** E29

flux size distribution—the size distribution of particles passing through a sampling zone during a given interval of time, wherein individual particles are counted and sized. **E 1620,** E29

fluxstone—a term referring to limestone (high-calcium, magnesian, or dolomitic) used as an agent in the manufacture of iron and steel or glass. **C 51,** C07

flux volts, E_f—the voltage induced in a winding of a magnetic component when the magnetic material is subjected to repeated magnetization under *SCM* or *CM* conditions.

$$E_f = 4.443\ B_{max}A'Nf \times 10^{-8}\ \text{V}\ (SCM\ \text{excitation})$$
$$E_f = 2.221\ \Delta\ BA'\ Nf \times\ 10^{8}\ \text{V}\ (CM\ \text{excitation})$$
$$E_f = 1.1107\ E_{avg}$$

which
A' = solid cross-sectional area of the core in cm^2,
N = number of winding turns, and
f = the frequency in hertz.

A 340, A06

fly—a part that makes contact with the drive pin for rotation of the tumblers. **F 471,** F12

fly ash, *n*—the finely divided residue that results from the combustion of ground or powdered coal and that is transported by flue gases from the combustion zone to the particle removal system. **C 125,** C09

fly ash—the finely divided residue resulting from the combustion of ground or powdered coal and which is transported from the firebox through the boiler by flue gases. **D 653,** D18

fly ash, *n*—the finely divided particles of ash entrained in flue gases arising from the combustion of fuel. **D 1356,** D22

flying spot scanning—*in optical character recognition,* a device employing a moving spot of light to scan a sample space, the intensity of the transmitted or reflected light being sensed by the photoelectric transducer. **F 149,** F05

FMC-2 color difference, *n*—color difference calculated by use of the Friele-MacAdam-Chickering, Version 2, equations based on the MacAdam chromaticity-difference-perceptibility ellipses and the Munsell value function. **E 284,** E12

foam—a layer of bubbles on the surface of molten glass.

C 162, C14

foam, *n—in cleaning textiles,* a frothy mass of fine bubbles generated by whipping or agitating a shampoo. **D 123,** D13

foam—a mass of bubbles formed on liquids by agitation. **D 459,** D12

foam cleaning, *n*—a process in which a prepared foam is applied to a textile product, scrubbed in, allowed to dry, and the encapsulated soil is removed by suction. **D 123,** D13

foam core—a lightweight cellular structure (rigid foam) material used in sandwich panel construction; innermost portion of a multilayer adherend assembly. **E 1749,** E06

foamed adhesive, *n*—an adhesive, the apparent density of which has been decreased substantially by the presence of numerous gaseous cells dispersed throughout its mass. (Synonym **cellular adhesive**.) **D 907,** D14

foamed adhesive—See **adhesive, foamed**. **E 1749,** E06

foamed plastics, n—See **cellular plastics** (the preferred terminology). (D-20) **F 412,** F17

foamed plastics, n—See **cellular plastics** (the preferred terminology). **D 883,** D20

foam glass—a thermally insulating material consisting of a high volume fraction of gaseous inclusions distributed throughout a glass matrix. **C 162,** C14

foaming agent—a material that increases the stability of a suspension of gas bubbles in a liquid medium. **D 459,** D12

foaming agent—a material that increases the volume or stability of the foam formed in a spray mixture. **E 1519,** E35

foaming agent (latex), *n*—a substance used to facilitate the formation of suspended gas in latex during the production of latex foam. **D 1566,** D11

foaming tendencies, *n*—a laboratory test conducted in glassware to evaluate the tendency of an engine coolant to foam under standard conditions of aeration and temperature. **D 4725,** D15

foam-in place cushioning material—See **cushioning material.** **D 996,** D10

foam line—a boundary in a melting furnace beyond which foam no longer appears on the glass surface. **C 162,** C14

foam stabilizer (latex), *n*—a substance used in the preparation of latex foam, before gelation, drying, and vulcanization to help maintain the foam cell structure. **D 1566,** D11

foam suppressant—See **defoamer**. **E 1519,** E35

foam tear, *n*—a condition wherein the foam portion of a laminated fabric ruptures prior to the failure of the bond. **D 123,** D13

focal length—the distance from the second principal point to the point on the axis where parallel rays entering the lens will converge or focus. **E 7,** E04

focal plane deviation (FPD)—*on a wafer surface,* the distance parallel to the optical axis from a point on the wafer surface to the focal plane of the optical system. **F 1241,** F01

focal spot—that area on the target of an X-ray tube which is bombarded by electrons. **E 7,** E04

focal spot—for x-ray generators, that area of the anode (target) of an x-ray tube which emits x-ray when bombarded with electrons. **E 1316,** E07

focus—a point at which rays originating from a point in the object converge or from which they diverge or appear to diverge under the influence of a lens or diffracting system. **E 7,** E04

focused beam—converging energy of the sound beam at a specified distance. **E 1316,** E07

focusing camera (X-rays)—a diffraction camera in which the X-ray source of a divergent X-ray beam, specimen (polycrystalline) and film all lie on one circle, which results in the diffracted beams all being focused on the film. **E 7,** E04

focusing device (electrons)—a device which effectively increases the angular aperture of the electron beam illuminating the object, rendering the focusing more critical. **E 7,** E04

focusing magnifier: *(1)*—a low-power microscope, telescope or simple lens used to observe the electron image formed on a fluorescent screen.

(*2*) a magnifying lens mounted so that its focal plane is coincident with its base, used to obtain a sharp focus in the plane of the sensitive material in a camera. **E 7,** E04

focusing (X-rays)—the operation of producing a convergent beam in which all rays meet in a point or line. **E 7,** E04

fog, *n*—a visible aggregate of condensed water vapor or ice crystals suspended in the atmosphere near the earth's surface. **D 1356,** D22

fog—a general term used to denote any increase in optical density of a processed photographic emulsion caused by anything other than direct action of the image forming radiation and due to one or more of the following:

(*a*) *aging*—deterioration, before or after exposure, or both, resulting from a recording medium that has been stored for too long a period of time, or other improper conditions.

(*b*) *base*—the minimum uniform density inherent in a processed emulsion without prior exposure.

(*c*) *chemical*—resulting from unwanted reactions during chemical processing.

(*d*) *dichroic*—characterized by the production of colloidal silver within the developed sensitive layer.

(*e*) *oxidation*—caused by exposure to air during developing.

(*f*) *exposure*—arising from any unwanted exposure of an emulsion to ionizing radiation or light at any time between manufacture and final fixing.

(*g*) *photographic*—arising solely from the properties of an emulsion and the processing conditions, for example, the total effect of inherent fog and chemical fog.

(*h*) *threshold*—the minimum uniform density inherent in a processed emulsion without prior exposure. **E 1316,** E07

fog, *n*—a quiescent cloud near the earth's surface. **E 1620,** E29

fog—see **fatigue** and **pre-exposure.** **F 335,** F05

fog density—a general term used to denote any increase in the optical density of a processed film caused by anything other than the direct action of the image-forming radiation. **E 1316,** E07

fog nozzle—a high capacity pressure atomizer that produces a dispersion of fine drops. **E 1620,** E29

fog quenching, *n*—**quenching** in a mist. **A 941,** A01

fog room, *n*—a moist room in which the humidity is controlled by atomization of water. **C 219,** C01

fog seal, *n*—a light application of bituminous material to an existing pavement as a seal to inhibit raveling, or to seal the surface, or both. Medium and slow-setting bituminous emulsions are usually used and may be diluted with water. **D 8,** D04

foil—a term often applied to thin sheet or strip usually 0.005 in. (0.13 mm) or less in thickness. **B 846,** B05

foil, *n*—rolled wrought product, rectangular in cross section, and of thickness less than 0.006 in. [up through 0.15 mm].

bright two-side foil, n—foil having a uniform bright specular finish on both sides.

matte one-side foil, n—foil having a diffuse reflecting finish on one side and a bright specular finish on the other. **B 881,** B07

foil—a thin sheet of a material, usually a metal, not exceeding 0.005 in. in thickness. **E 7,** E04

fold—See **lap**. **C 162,** C14

fold—a bend in the strata or other planar structure within the rock mass. (ISRM) **D 653,** D18

fold—a doubling over of metal which occurs during forging at or near the intersection of diameter changes which are found on the shoulders, heads, or shanks of bolts and screws, or on nuts at the intersection of diameter changes on the top face or on the bottom face. **F 1789,** F16

folded reinforcement—an unintentional or unspecified misalignment of mat or fabric reinforcing material in relation to the contour of a pultruded section.

NOTE—Such folds may or may not affect the surface appearance of the pultrusion and are chiefly visible in a cut cross section of the product. Such reinforcement irregularities are usually due to shifting and crowding of the reinforcing material during pultrusion. **D 3918,** D20

folder stock, *n*—a paperboard used for the manufacture of folders for filing purposes. **D 1968,** D06

folding—see **aliasing.** **E 131,** E13

folding—turning over the edge of a shoe part to form a finished edge, performed by hand or machine. **F 869,** F08

folding endurance, *n— of paper*, the average of the logarithms to the base 10 of the individual folding numbers. See **folding number**. **D 1968,** D06

folding number, *n*—the number of double folds required to cause failure of a paper test specimen when it is subjected to a prescribed folding procedure. **D 1968,** D06

fold number, *n*—See **folding number**. **D 1968,** D06

folds—defects in metal, usually on or near the surface, caused by continued fabrication of overlapping surfaces. **E 7,** E04

foliar application—application of a chemical to the stems, fruit, leaves, needles, or blades of a plant. **E 1102,** E35

foliation—the somewhat laminated structure resulting from segregation of different minerals into layers parallel to the schistosity. (ISRM) **D 653,** D18

follower—See **pusher.** **F 592,** F16

follow-up program, *n*—the sampling, inspection, tests, or other measures conducted by the certification organization on a periodic basis to determine the continued compliance of products that are being made by the manufacturer to the requirements of the standard specification. **F 1494,** F23

font—a set of graphic shapes that may be alphabetic, numeric, or both and may include other symbols. **F 149,** F05

font—a style of typeface, such as Helvetica or Gothic. **F 1156,** F12

font, *n*—a set of characters that share the same type style and size. **F 1457,** F05

food cutters—machine that uniformly reduces food products to a small particle size for salads, spreads, bread crumbs, and other food service recipes. Reduction of the food product is accomplished by combining the rotation of the product bowl with the perpendicular high speed rotation of a set of stainless steel cutlery knives. **F 1827,** F26

food mixing machines—vertical electric machine that shall be adaptable for mixing, whipping, and beating food products. The size range (as expressed by bowl capacity) from 12 qt to 140 qt and does not include special purpose machines that are intended solely for mixing dough. **F 1827,** F26

food service equipment manuals—manuals for food service equipment shall contain as a minimum, complete installation, operating, preventive maintenance, safety warnings, cleaning, and service instructions, including applicable charts and diagrams. **F 1827,** F26

food simulant—a well-characterized substance used in place of food for investigative studies. **F 1479,** F02

food slicing machines—a machine with electrically-driven rotating slicing blades. **F 1827,** F26

food waste disposal, commercial—a device intended for grinding food waste into small particles that are then flushed by water into a sanitary sewer system. Food wastes can be cooked or noncooked soils from the preparation or serving of foods. Use not intended for grinding glass, china, metals, clam or oyster shells, large bones, wood, paper, cardboard, or plastic. **F 1827,** F26

food waste pulper and waterpress assembly, commercial—a device intended for grinding food waste, food service paper, cardboard products, food service plastic products, documents including computer printouts, general office and retail store paper, and cardboard waste. Materials are ground in a water-filled tank (*pulper*) to produce a slurry which is then passed to the waterpress to be dewatered. Use not intended for grinding glass, china, metal, wood, clam, or oyster shells. Any small pieces of metal inadvertently placed in the pulper, such as cardboard box staples, aluminum refreshment cans, or tin food cans, shall be removable from the outside of the pulper tank. **F 1827,** F26

foot, *n*—the terminal part of the vertebrate leg, including the ankle, upon which an individual stands (see **foot protective device**). **F 1494,** F23

footcandle, *n*—unit of illuminance equal to one lumen per square foot. **E 284,** E12

footcandle (fc)—the illumination on a surface, 1 ft^2 in area, on which is uniformly distributed a flux of 1 lm (lumen). It equals 10.8 lm/m^2. **E 1316,** E07

footing—portion of the foundation of a structure that transmits loads directly to the soil. **D 653,** D18

foot length, *n—in body measurements*, with the subject standing barefoot, the distance from the most prominent part of the heel where it touched the floor to the end of the most prominent toe. **D 5219,** D13

footprint area, [L^2], *n*—the gross contact area of a tire that is loaded (under stated conditions) against a smooth flat surface. **F 538,** F09

foot protective device, *n—for chain saw cut resistance*, an article of personal equipment which is worn over the foot and ankle for the purpose of providing limited protection from injury due to contact with a moving saw chain. **F 1494,** F23

foot traffic units, *n—for pile floor covering*, the number of passes by human walkers over a specific group of carpet samples. **D 123,** D13

footwall—the mass of rock beneath a discontinuity surface. (ISRM) **D 653,** D18

footwear, *n*—a boot or shoe of any construction. **F 1494,** F23

footwear, *n*—wearing apparel for the feet (such as shoes, boots, slippers, or overshoes), usually excluding hosiery. **F 1646,** F13

force, *n*—a physical influence exerted by one body on another which produces acceleration of bodies that are free to move and deformation of bodies that are not free to move. (Compare **strength.**) **D 123,** D13

force account—term describing PHA self-performance of modernization work by the use of employees, as distinct from performance by a contractor. **E 1605,** E06

force-at-rupture, *n*—the force applied to a material immediately preceding rupture. (Compare **breaking force.** See also **rupture.**) **D 123,** D13

force at specific elongation, FASE, *n*—the force associated with a specific elongation on the force-elongation curve. **D 4439,** D35

force-at-specified-elongation (FASE), *n*—the force associated with a specific elongation on the force-extension or force-elongation curve. **D 123,** D13

forced convection— see **convection, forced.** **E 772,** E44

forced drying temperature—a temperature between room temperature and 150°F (65°C). **D 16,** D01

force-deformation curve, *n*—a graphical representation of the force and deformation relationship of a material under conditions of compression, shear, tension, or torsion. (Compare **force-elongation curve, force-extension curve,** and **stress-strain curve.**) **D 123,** D13

forced vibration (forced oscillation)—vibration that occurs if the response is imposed by the excitation. If the excitation is periodic and continuing, the oscillation is steady-state. **D 653,** D18

force-elongation curve, *n*—a graphical representation of the force and elongation relationship of a material under tension. (Compare **force-deformation curve, force-extension curve,** and **stress-strain curve.**) **D 123,** D13

force-elongation curve, *n—in a tensile test,* a graphical representation of the relationship between the magnitude of an externally applied force and the change in length of the specimen in the direction of the applied force. (*Synonym* for stress-strain curve.) **D 4439,** D35

force-extension curve, *n*—a graphical representation of the force and extension relationship of a material under tension. (Compare **force-deformation curve, force-elongation curve,** and **stress-strain curve.**) **D 123,** D13

forehearth—a section of a melting furnace in one of several forms from which glass is taken for forming. **C 162,** C14

foreign matter, *n—in cotton,* non-lint material commonly referred to as waste or trash such as dust, sand, seed-coat fragments, leaves, and stems normally present in raw and partially processed cotton. **D 123,** D13

foreign matter—*in protective coatings,* insoluble foreign particles such as sand, lint, dust, and dirt that get mixed with the coating material before, during, or after application; causing the formation of raised specks in the dried film. **D 4538,** D33

fore-line—in leak testing, the line between a fore pump and the pump it backs. **E 1316,** E07

fore-line valve—in leak testing, a vacuum valve placed in the fore-line to permit isolation of the diffusion pump from its backing pump. **E 1316,** E07

forensic psychophysiology, *n*—the scientific discipline dealing with the relationship and applications of PDD tests within the legal system. It encompasses the academic discipline that provides the student, the practitioner, and the researcher with the theoretical and applied psychological, physiological, and psychophysiological fundamentals for a thorough understanding of PDD tests, and the skills and qualifications for conducting PDD examinations. The modifier A forensic≅ delineates and delimits this discipline from the broader discipline of psychophysiology. **E 2035,** E52

forepoling—driving forepoles (pointed boards or steel rods) ahead of the excavation, usually over the last set erected, to furnish temporary overhead protection while installing the next set. (ISRM) **D 653,** D18

forepressure—in leak testing, the total pressure on the outlet side of a pump measured near the outlet port. Sometimes called the back pressure, backing pressure, outlet pressure, exhaust pressure, or discharge pressure. In discussing the action of a vapor jet, the term forepressure may be used to designate the total pressure of the gas against which the jet impinges. **E 1316,** E07

fore pump—in leak testing, the pump that produces the necessary fore vacuum for a pump which is incapable of discharging gases at atmospheric pressure. Sometimes called the backing pump. **E 1316,** E07

foreset bed—the advancing and relatively steep frontal slope of a delta, that progressively covers the bottomset bed and in turn is covered by the topset bed. Foreset beds represent the greater part of the volume of a delta. **D 4410,** D19

forest products laboratory etch—See **FPL etch and sulfochromate etch.** **E 1749,** E06

forged structure—the macrostructure through a suitable section of a forging which reveals direction of working. **E 7,** E04

forgery—the replication or alteration of a document's data with the intent to defraud, such as check amount, check signature, and date. **F 1156,** F12

forging, *n*—metal part worked to a predetermined shape by one or more processes such as hammering, upsetting, pressing, rolling, etc.

blocker-type forging, n—forging made in a single set of impressions to the general contour of a finished part.

die forging, n—forging formed to the required shape and size by working in impression dies.

hammer forging, n—forging produced by repeated blows of a forging hammer.

hand forging, n—forging, manually-manipulated, which is worked, through repeated strokes or blows, between flat or simply shaped dies.

rolled ring forging, n—cylindrical product of relatively short height, circumferentially rolled from a hollow section. **B 881,** B07

forging cracks—occur during fastener manufacturing at the cutoff or forging operations and are located on the top of the head or on the raised periphery of indented head bolts and screws. **F 1789,** F16

forging preform—an unsintered, presintered, or fully sintered compact intended to be subjected to hot or cold forming and densification. **B 243,** B09

forging stock, *n*—wrought or cast rod, bar or other section suitable for forging. **B 881,** B07

fork—a piece of metal equipment used during the firing operation for placing ware in, and removing it from a box furnace. **C 286,** B08

forked blade clincher—special clincher (anvil) for hooking plastic bags to facilitate placement of encircling staple. **F 592,** F16

forking—a mechanism whereby a propagating fracture branches to form two new fractures separated by an acute angle. **C 162,** C14

form—a set of equivalent planes in a crystal. In general, they will have the same spacing but different Miller indices. For example, in the cubic system, the planes (101), (110), (011), etc. are planes of the form (110). In the tetragonal system, however, the planes (101) and (110) belong to different forms. Equivalent directions are also spoken of as directions of a form. **E 7,** E04

formaldehyde tannage—tannage used especially for white leathers and washable glove leathers. **D 1517,** D31

format—preprogrammed identification of fields to be read by an optical scanner. **F 149,** F05

formed—See **mechanically deformed.** **F 547,** F16

formed crown—staple crown formed during driving, for example, for carding or fastening wire. **F 592,** F16

former (latex), *n*—a shaped object on which a rubber article is produced by dipping into a latex, from which the article is subsequently removed. **D 1566,** D11

form factor—a factor used in the calculation of the bending strength of wood members to compensate for nonrectangular shape. Defined as the ratio of the modulus of rupture of a member of desired shape to that of a "standard" member. **D 9,** D07

form factor, *ff*—the ratio of the rms value of a periodically alternating quantity to its average absolute value.

NOTE—For a sinusoidal variation, the form factor is:

$$\pi/2\sqrt{2} = 1.1107$$

A 340, A06

forming—the shaping of hot glass. **C 162,** C14

forming—the shaping or molding of ceramic ware. **C 242,** C21

forming, *n*—a process in which the shape of plastic pieces such as sheets, rods, or tubes is changed to a desired configuration. **D 883,** D20

forming, *n*—a process in which the shape of plastic pieces such as sheets, rods, or tubes is changed to a desired configuration. **F 412,** F17

forming hood—the partially enclosed volume in which individual glass fibers and groups of fibers are collected into a wool pack. **C 162,** C14

forming rolls—rolls used in forming flat glass. **C 162,** C14

form rollers—ink or water rollers that contact the master on an offset press. **F 425,** F05

formula, *n*—a list of the materials and their amounts used in the preparation of a compound. **D 1566,** D11

formulation—the form in which a chemical is offered to the user. **E 1102,** E35

formulation deposit rate—the amount of formulation deposited per unit area. **E 1102,** E35

formulation rate—the amount of chemical formulation applied per unit treated, expressed in terms of mass or volume per relevant unit treated. (For area treatment, kg/ha, lb/A, or oz/1000 ft of row; for space application, mg/m^3, or oz/1000 ft^3; for individual units, kg/plant or animal). **E 1102,** E35

formula weight—the weight, in grams, pounds, or other units, obtained by adding the atomic weights of all elemental constituents in a chemical formula. **B 374,** B08

Formvar—a plastic material used for the preparation of replicas, or specimen supporting membranes. Trade name for poly(vinyl formal) 15/95. **E 7,** E04

Formvar replica—See **replica.** **E 7,** E04

forsterite ($2MgO{\cdot}SiO_2$)—a magnesium silicate mineral, usually produced synthetically as a ceramic raw material; may be a reaction-produced phase in fired ceramics. **C 242,** C21

forsterite porcelain—See *fosterite porcelain* under **porcelain.** **C 242,** C21

forsterite whiteware—See *fosterite whiteware* under **ceramic whiteware.** **C 242,** C21

forward error correction—data handling process that anticipates the possible occurrence of an error in a data stream before it happens, and makes suitable provision for accommodating that error; can handle more than one error (in a given stream). **F 1156,** F12

forward scattered radiation—radiation which is scattered less than 90° with respect to the incident beam, that is, forward in the general direction of the radiation source. **E 1316,** E07

fossil resin—under **resin, natural,** see *fossil resin.* **D 16,** D01

fouling—the reduction of flux due to a build-up of solids on the surface or within the pores of the membrane, resulting in changed element performance. **D 6161,** D19

fouling index (FI)—see **SDI.** **D 6161,** D19

foundation—lower part of a structure that transmits the load to the soil or rock. **D 653,** D18

foundation soil—upper part of the earth mass carrying the load of the structure. **D 653,** D18

foundry casting, *n*—a casting process wherein a molten metal is poured by gravity into the cavity of a mold and solidified; also, a product of such a process. **B 899,** B02

foundry nail, smooth foundry nail—bright, regular-stock-steel, ¾ by 0.120 to 9 by 0.162-in. nails with large thin flat 7⁄16 to ½-in. head and medium diamond point. **F 547,** F16

fountain—a reservoir for ink or water. **F 425,** F05

fountain roller—the roller that revolves in the fountain. **F 425,** F05

fountain solution—the chemical solution carried in the water fountain of the offset press and used to keep blank areas of the master free of ink. **F 425,** F05

Fourcault process—the method of making sheet glass by drawing vertically upward from a slotted debiteuse block. **C 162,** C14

Fourdrinier kraft— See **containerboard.** **D 996,** D10

Fourier transform (FT)—the mathematical process used to convert an amplitude-time spectrum to an amplitude-frequency spectrum, or *vice versa.* **E 131,** E13

Fourier transform infrared (FT-IR) spectrometry—a form of infrared spectrometry in which an interferogram is obtained; this interferogram is then subjected to a Fourier transform to obtain an amplitude-wavenumber (or wavelength) spectrum. **E 131,** E13

four-point-¼point flexure, *n*—configuration of flexural strength testing where a specimen is symmetrically loaded at two locations that are situated one quarter of the overall span, away from the outer two support bearings. **C 1145,** C28

four-point probe—an electrical probe arrangement for determining the resistivity of a material in which separate pairs of contacts are used (*1*) for passing current through the specimen and (*2*) measuring the potential drop caused by the current. **F 1241,** F01

four-side structural sealant glazing, *n*—*in building construction,* a glazing system wherin all sides of a rectangular-shaped lite or panel are structural sealant glazed. **C 717,** C24

foxing—a shoe component that reinforces or covers the shoe at a point of particular wear or stress, such as the heel area of the joint between sole and upper. **F 869,** F08

foxing, *n*—a piece of material applied to the upper or extending around the outsole of a boot or shoe. **F 1646,** F13

F pilus—a protrusion on *E. coli* that is necessary for mating. The F pilus also contains the receptor for phage M13. **E 1705,** E48

FPL etch—an abbreviation for *forest products laboratory etch*; an etchant used for preparing the surface of aluminum alloys for adhesive bonding. (Syn. **sulfochromate etch.**) **E 1749,** E06

FR, *n*—a designation noting that an electrical insulating material has been subjected to a standard test for flammability and has a rating in accordance with that standard. **D 1711,** D09

FR—False Rejection, of a genuine (real, authorized) individual person or item on the fallacious assumption that it (he/she) was represented improperly or falsely. **F 1156,** F12

fraction—that portion of a powder sample which lies between two stated particle sizes. Synonymous with **cut.** **B 243,** B09

fractional efficiency—See **efficiency.** **D 1356,** D22

fractional exposure dose (FED), *n*—the ratio of the Ct product for a gaseous toxicant produced in a given test to that Ct product of the toxicant which has been determined statistically from independent experimental data to produce lethality in 50 % of test animals within a specified exposure and postexposure time. Since the time values in this ratio numerically cancel, the FED also is simply the ratio of the average concentration of a gaseous toxicant to its LC_{50} value for the same exposure time. When not used with reference to a specific toxicant, the term FED represents the summation of FEDs for individual toxicants in a combustion atmosphere. **E 176,** E05

fractional factorial design, *n*—a factorial experiment in which only an adequately chosen fraction of the treatments required for the complete factorial experiment is selected to be run.

NOTE—This procedure is sometimes called fractional replication. **E 456,** E11

fractional factorial design, *n*—a factorial experiment in which only an adequately chosen fraction of the treatments required for the complete factorial experiment is selected to be run.

NOTE—This procedure is sometimes called fractional replication. **E 1325,** E11

fractional ion yield—*SIMS,* the ratio of the number of secondary ions of a particular species to the total number of secondary ions emitted by a specimen. **E 673,** E42

fractional reflectance, R_f, *n*—the ratio of the flux reflected from a specimen in a specified solid angle to that incident on the specimen. **E 284,** E12

fractional transmittance, T_f, **n**—the ratio of the flux transmitted through a specimen in a specified solid angle to the incident flux. **E 284,** E12

fractionation, *n*—the process of separating a mixture into components having different properties (as by distillation, precipitation, or screening). **D 1356,** D22

fractionation, elastic, *n*—a process in which soft aggregate is separated from harder aggregate by hurling the composite aggregate against a steel plate, the hard particles rebounding farther from the plate than the softer, more friable particles. **C 1145,** C28

fractography, *n*—means and methods for characterizing a fractured specimen or component. **C 1145,** C28

fracture—a complete break in the stone (see crack, microcrack, seam). **C 119,** C18

fracture—that portion of a vitrified clay pipe from which a fragment has been broken. It is distinguished by well-defined fracture faces and sharp edges where the fracture faces meet the surface of the pipe. **C 896,** C04

fracture—the general term for any mechanical discontinuity in the rock; it therefore is the collective term for joints, faults, cracks, etc. (ISRM) **D 653,** D18

fracture—a break in the mechanical continuity of a body of rock caused by stress exceeding the strength of the rock. Includes joints and faults. **D 653,** D18

fracture—a rupture or break of the polish film (usually multiple). **D 2825,** D21

fracture—cracks, crazing, or delamination, or a combination thereof, resulting from physical damage to the pultrusion. **D 3918,** D20

fracture (as in fractured asbestos)—a clearly defined transverse break in the asbestos in situ. **D 2946,** C17

fractured face, *n*—an angular, rough, or broken surface of an aggregate particle created by crushing, by other artificial means, or by nature. **D 8,** D04

fracture ductility, ϵ_f, *n*—the true plastic strain at fracture. **E 6,** E28

fracture frequency—the number of natural discontinuities in a rock or soil mass per unit length, measured along a core or as exposed in a planar section such as the wall of a tunnel. **D 653,** D18

fracture grain size—grain size determined by comparing a fracture of a specimen with a set of standard fractures. For steel, a fully martensitic specimen is generally used and the depth of hardening as well as the prior austenitic grain size is determined. **E 7,** E04

fracture origin, *n*—the source from which brittle fracture commences. **C 1145,** C28

fracture pattern—spatial arrangement of a group of fracture surfaces. (ISRM) **D 653,** D18

fracture, spontaneous, *n*—cracking or chipping which occurs without immediately apparent external causes. **C 1145,** C28

fracture strength, S_f [FL^{-2}], *n*—the normal stress at the beginning of fracture. Fracture strength is calculated from the load at the beginning of fracture during a tension test and the original cross-sectional area of the specimen. **E 6,** E28

fracture stress [FL^{-2}], *n*—the true normal stress on the minimum cross-sectional area at the beginning of fracture. **E 6,** E28

fracture surface markings—fracture surface features that may be used to determine the fracture origin location and the nature of the stress that produced the fracture. **C 162,** C14

fracture system—that family of related fracture surfaces lying within an object, having a common cause and origin. **C 162,** C14

fracture test: (*1*) —*general*the production of a fracture in a metal sample to determine such things as discontinuities, grain size, and composition.

(*2*) *steel*—a test which utilizes a hardened steel disk section prepared from billet or bar stock which is fractured parallel to the grain flow so that, among other things, discontinuities due to inclusion segregates can be detected visually. **E 7,** E04

fracture toughness—a conventional fracture mechanics strength parameter indicating the resistance of a material to crack extension. **D 9,** D07

fracture toughness—a generic term for measures of resistance to extension of a crack. **E 1823,** E08

fracturing—*in grouting*, intrusion of grout fingers, sheets, and lenses along joints, planes of weakness, or between the strata of a formation at sufficient pressure to cause the strata to move away from the grout. **D 653,** D18

fragmentation—the breaking of rock in such a way that the bulk of the material is of a convenient size for handling. (ISRM) **D 653,** D18

frame, *n*—a list, compiled for sampling purposes, which designates the items (units) of a population or universe to be considered in a study. **E 456,** E11

frame—See **windows and doors.** **E 631,** E06

frame hinge—a fitting that attaches to the gate frame and functions with the post hinge. **F 552,** F14

framework—the basic structure supporting installed fabric, namely the posts, rails, and tension wire. **F 552,** F14

frame yarn, *n*—pile yarn in a Wilton-type carpet. **D 123,** D13

framing member, *n*—stud, plate, track, joist, furring, and other support to which a gypsum panel product, or metal plaster base is attached. **C 11,** C11

framing nail—See **common nail, threaded common nail, threaded nail.** **F 547,** F16

frangible—easily broken. (See **friable**.) **F 1156,** F12

FRC—free residual chlorine. **D 6161,** D19

free (available) chlorine—chlorine existing as hypochlorous acid or its dissociated ions. Chlorine remaining after the demand has been satisfied. **D 6161,** D19

free bend, *n*—the bend obtained by applying forces to the ends of a specimen without the application of force at the point of maximum bending. **E 6,** E28

free-blown—See **offhand glass.** **C 162,** C14

freeboard—the space above a filter bed in a filtration vessel to allow for expansion of the bed during back washing. **D 6161,** D19

freeboard—minimum vertical height of the boom above the waterline. **F 818,** F20

freeboard—the distance between the level of liquid in the kettle and the top of the lip of the kettle. **F 1827,** F26

free bottom camber, h_F—the maximum height of the running surface measured from a plane vertical surface with the ski on an edge, free from the deflection caused by its weight. X_{hF} is the location of h_F from the tail of the ski. **F 472,** F27

free bottom camber, H_f—the height of the running surface from a vertical plane surface measured at the highest point, with the snowboard ski held vertically on edge, free from the effect of the snowboard ski weight. **F 1107,** F27

"free-carbon" in tars, *n*—the hydrocarbon fraction that is precipitated from a tar by dilution with carbon disulfide or benzene. **D 8,** D04

"free carbon" in tars—the hydrocarbon fraction that is precipitated from a tar by dilution with carbon disulfide. **D 1079,** D08

free choice sensory profiling, *n*—a form of sensory profiling in which each assessor chooses his or her own attributes to describe a group of samples. The assessors' sensory profiles are combined statistically (for example, by generalized Procrustes analysis) to produce a map of the samples. **E 253,** E18

free convection—see **convection, natural.** **E 772,** E44

free cyanide: (*1*) *true*—the actual concentration of cyanide radical, or equivalent alkali cyanide, not combined in complex ions with metals in solution.

(*2*) *calculated*—the concentration of cyanide, or alkali cyanide, present in solution in excess of that calculated as necessary to form a specified complex ion with a metal or metals present in solution.

(*3*) *analytical*—the free cyanide content of a solution, as determined by a specified analytical method.

NOTE—The true value of free cyanide is rarely known with certainty and is therefore usually only dealt with in discussions of theory. The calculated or analytical value is usually used in practice. **B 374,** B08

free edge, *n*—an unrestrained pavement boundary. **E 867,** E17

free edge, *n*—an unrestrained pavement boundary. **E 1778,** E17

free electrons—an electron, not directly associated with the structure of an atom or molecule, free to move under the influence of an applied electric or magnetic field. **D 2864,** D27

free-energy diagram—a graphical representation of the variation with concentration of the Gibbs Free Energy, at constant pressure and temperature. **E 7,** E04

free-energy surface—in a ternary, or higher order, free energy diagram, the locus of points representing the Gibbs Free Energy as a function of concentration, with pressure and temperature constant. **E 7,** E04

free form (unformatted form)—a form on which the data appears in variable length fields. Preprinted symbols and guides are absent or minimal. Field delimiters are entered with the data. **F 149,** F05

free impurity, *n*—the impurities in a coal that exist as individual discrete particles that are not a structural part of the coal and that can be separated from it by coal preparation methods. **D 121,** D05

free lime, *n*—*in asbestos-cement,* uncombined calcium hydroxide. **C 460,** C17

free lime, *n*—uncombined calcium hydroxide. **C 1154,** C17

free lime, *n*—*in asbestos-cement,* uncombined calcium hydroxide. **D 2946,** C17

free moisture—that water, which is not chemically bound, and that is loosely bound to a material, but which can be removed by drying at 105°C, for a time to achieve constant weight, expressed as a percent of the initial weight of the material. **C 242,** C21

free moisture—synonym for *surface moisture.* **D 121,** D05

free product—liquid phase contaminants released into the environment. **D 653,** D18

free rolling tire, *n*—a loaded tire rolling without applied driving or braking torque. **F 538,** F09

freestanding candle, *n*—a rigid candle (that is, pillar-shaped, column-shaped, or figurine) recommended to be used on a heat–resistant, nonflammable surface or on a candle accessory. **F 1972,** F15

free vapor phase—a condition of contaminant residence in which volatilized contaminants occur in porosity that is effective to free and open gaseous flow and exchange, such porosity generally being macroporosity. **D 653,** D18

free vibration—vibration that occurs in the absence of forced vibration. **D 653,** D18

free vibration (*in dynamic mechanical measurement*)—a technique for performing dynamic mechanical measurements in which the sample is deformed, released, and allowed to oscillate freely at the system's natural resonant frequency. **D 4092,** D20

free water, *n*—all water contained by gypsum board or plaster in excess of that chemically held as water of crystallization. **C 11,** C11

free water elevation (water table) (ground water surface) (free water surface) (ground water elevation)—elevations at which the pressure in the water is zero with respect to the atmospheric pressure. **D 653,** D18

free water (gravitational water) (ground water) (phreatic water)—water that is free to move through a soil or rock mass under the influence of gravity. **D 653,** D18

freeze-drying—sublimation of water from a frozen aqueous suspension. **E 1705,** E48

freeze-thaw cycle—the freezing and subsequent thawing of a material. **C 717,** C24

freeze thaw resistance, *n*—the ability of masonry to maintain integrity under the forces caused by cyclic action of freezing and thawing in the presence of moisture. **C 1232,** C15

freezing—lowering the temperature of an aqueous suspension to a point at or below the temperature of ice crystal formation. **E 1705,** E48

freezing index, *F* **(degree-days)**—the number of degree-days between the highest and lowest points on the cumulative degree-days—time curve for one freezing season. It is used as a measure of the combined duration and magnitude of below-freezing temperature occurring during any given freezing season. The index determined for air temperatures at 4.5 ft (1.4 m) above the ground is commonly designated as the air freezing index, while that determined for temperatures immediately below a surface is known as the surface freezing index. **D 653,** D18

freezing point, *n*—the temperature at which ice crystals begin to form in an engine coolant when tested in accordance with Test Method D 1177 for Freezing Point of Aqueous Engine Coolant Solution. **D 4725,** D15

freezing point—the temperature at which the liquid and solid states of a substance are in equilibrium at a given pressure (usually atmospheric). For pure substances it is identical with the melting point of the solid form.**Lange, 10th Ed.** **D 4790,** D16

freezing point—See **melting point**. **E 7,** E04

freezing point, *n*—fixed point of a single component system in which liquid and solid phases are in equilibrium at a specified pressure, usually 101 325 Pa, and the system is losing heat slowly. (Compare

melting point.) **E 344,** E20

freezing temperature— see **crystallization temperature**. **E 1142,** E37

French antelope lambskin—when tanned in France, a lustrous suede finish applied to French, Spanish, Algerian, and Balkan skins; in the United States "French antelope finish" is applied to suede leathers made from South American and New Zealand pickled lambskins. **D 1517,** D31

french cording—finished top edge of shoe upper, usually made by folding and stitching in place a narrow strip of fabric over top edge of upper after lining is stitched in. Full face of binding or cording is visible inside and outside of upper edge. **F 869,** F08

French degree—calcium carbonate equivalents expressed in parts per hundred thousand. Concentration in French degree is calculated by dividing concentration in calcium carbonate equivalents by ten. **D 6161,** D19

French kid—leather tanned from kidskin by an alum or vegetable process. **D 1517,** D31

French or hexagonal method—a method of application for asbestos-cement roofing, whereby the shingles have at least three corners clipped so that when they are laid with their diagonals perpendicular to the eave of the roof, they lap at the top and sides to form a hexagonal pattern. **C 460,** C17

French or hexagonal method, *n—in shingles*, a method of application for asbestos-cement roofing, whereby the shingles have at least three corners clipped so that when they are laid with their diagonals perpendicular to the eave of the roof, they lap at the top and sides to form a hexagonal pattern. **C 1154,** C17

French or hexagonal method—a method of application for asbestos-cement roofing, whereby the shingles have at least three corners clipped so that when they are laid with their diagonals perpendicular to the eaves of the roof, they lap at the top and sides to form a hexagonal pattern. **D 2946,** C17

French seam, *n—in home sewing*, a complex seam formed on the inside of a product with both cut edges enclosed and no stitching rows visible on the face side. (Compare **mock French seam**.) **D 4965,** D13

frequency, *ν*— the number of cycles per unit time. **E 131,** E13

frequency, *n*—reciprocal of the period. Symbol; *f*, *ν*.

NOTE—When the independent variable is time, the unit of frequency is the hertz. Symbol: Hz (1 Hz = 1 s^{-1}). (This unit is also called "cycle per second," c/s.) **E 349,** E21

frequency—the number of cycles per second of alternating electric current induced into the tubular product. For eddy-current testing described herein, the frequency is normally 1 to 125 kHz, inclusive. (E 215) **E 1316,** E07

frequency, angular, **ω**—the number of radians per second traversed by a rotating vector that represents any periodically varying quantity.

NOTE—Angular frequency, ω, is equal to 2π times the cyclic frequency, *f*. **A 340,** A06

frequency, cyclic, *f*—the number of hertz (cycles/second) of a periodic quantity. **A 340,** A06

frequency distribution, *n—of a sample*, a table giving for each value of a discrete variate, or for each group of values of a continuous variate, the corresponding number of observations. **D 123,** D13

frequency distribution, *n—of a population*, a function that, for a specific type of distribution, gives for each value of a random discrete variate, or each group of values of a random continuous variate, the corresponding probability of occurrence. **D 123,** D13

frequency distribution—the way in which the frequencies of occurrence of members of a population, or a sample, are distributed in accordance with the values of the variable under consideration. **E 1823,** E08

frequency domain filtering, *n*—a filtering operation performed by first calculating the spectrum of the profile record and then multiplying the spectral components by the frequency response transfer function of the filter. **E 867,** E17

frequency, f—the number of cycles per unit time of periodic process. **E 1142,** E37

frequency factor—See **multiplicity factor**. **E 7,** E04

frequency, ***f*** **(T^{-1})**—number of cycles occurring in unit time. **D 653,** D18

frequency (fundamental)—in resonance testing, the frequency at which the wave length is twice the thickness of the examined material. **E 1316,** E07

frequency (inspection)—effective ultrasonic wave frequency of the system used to inspect the material. **E 1316,** E07

frequency profile , *n*—a plot of the dynamic properties of a material, at a constant temperature, as a function of test frequency. **D 4092,** D20

frequency profile—*in dynamic mechanical measurement*, plot of the dynamic properties of a material, at a constant temperature, as a function of test frequency, (D 4092, D-20). **E 1142,** E37

frequency (pulse repetition)—the number of times per second an electro-acoustic search unit is excited by the pulse generator to produce a pulse of ultrasonic energy. This is also called pulse repetition rate. **E 1316,** E07

frequency (X-ray)—the number of alternations per second of the electric vector of the X-ray beam. It is equal to the velocity divided by the wavelength. **E 7,** E04

fresh concrete, *n*—concrete which possesses enough of its original workability so that it can be placed and consolidated by the intended methods. **C 125,** C09

fresh water environment, *n*—the aerobic, aqueous compartment, characteristically with a salinity of less than five parts per thousand. **D 6384,** D02

Fresnel fringes—a class of diffraction fringes formed when the source of illumination and the viewing screen are at a finite distance from a diffracting edge. In the electron microscope these fringes are best seen when the object is slightly out of focus. **E 7,** E04

Fresnel (lens)—See **lens**. **F 1156,** F12

Fresnel lens, circular—a sheet of transparent material into which concentric grooves have been formed in such a pattern that light will be focused as with a lens. (Focusing mirrors of similar design are also available.) **E 772,** E44

Fresnel lens, linear—a sheet of transparent material into which parallel grooves have been formed in such a pattern that light will be focused as by a cylindrical lens. (Focusing mirrors of similar design are also available.) **E 772,** E44

Fresnel reflection, *n*—the process by which radiant flux is reflected from an optically smooth interface between two dielectric media. **E 284,** E12

Fresnel-reflector system—flat mirrors arranged in an array such that they reflect onto a target, the illuminated area of which simulates the shape and size of the flat mirror. (Such an array simulates the ray-tracing of a parabolic trough of the same aperture angle.) **E 772,** E44

fresnel-reflector system, ***n***—flat mirrors arranged in an array such that they reflect onto a target, the illuminated area of which simulates the size and shape of the flat mirror. Such an array simulates the ray-tracing of a parabolic trough of the same aperture angle. **G 113,** G03

fretting, *n*—small amplitude oscillatory motion, usually tangential, between two solid surfaces in contact. **B 542,** B02

fretting, *n*—*in tribology*, small amplitude oscillatory motion, usually tangential, between two solid surfaces in contact. **G 40,** G02

fretting corrosion—a form of fretting wear in which corrosion plays a significant role. **B 542,** B02

fretting corrosion—the deterioration at the interface between contacting surfaces as the result of corrosion and slight oscillatory slip between the two surfaces. **G 15,** G01

fretting corrosion, *n*—a form of fretting wear in which corrosion plays a significant role. **G 40,** G02

fretting wear—wear arising as a result of fretting. See **fretting**. **B 542,** B02

fretting wear, *n*—wear arising as a result of fretting (see **fretting**). **G 40,** G02

Freundlich adsorption isotherm—a logarithmic plot of quantity of component adsorbed per unit of adsorbent *versus* concentration of that component at equilibrium and at constant temperature, which approximates the straight line postulated by the Freundlich adsorption equation

$$X/M = kC^n$$

where:

X = quantity adsorbed,
M = quantity of adsorbent,
C = concentration,
k = constants.
and
n

D 2652, D28

friable—easily crumbled; brittle. (See **frangible** for comparison.) **F 1156,** F12

Fricke dosimeter—a liquid chemical radiation dosimetry system composed of water and ferrous sulfate or ferrous ammonium sulfate in aqueous sulfuric acid solution and whose response is based quantitatively on the amount of oxidation of ferrous to ferric ions by ionizing radiation, as analyzed by spectrophotometry. It is considered to be a reference-standard dosimetry system. **E 170,** E10

friction—the resistance developed between the physical contacting, but otherwise unconstrained, surfaces of two bodies when there is movement or tendancy for movement of one body relative to the other parallel to the plane of contact. **C 242,** C21

friction, *n*—the resistance to the relative motion of one body sliding, rolling, or flowing over another body with which it is in contact. **D 123,** D13

friction, *n*—resistance to relative motion between two bodies in contact (*Websters*).

coefficient of friction—the ratio of the force required to move one surface over another, to the total force applied normal to those surfaces.

kinetic coefficient of friction—the ratio of the force required to move one surface over another, to the total force applied normal to those surfaces, once that motion is in progress.

static coefficient of friction—the ratio of the force required to move one surface over another, to the total force applied normal to those surfaces, at the instant motion starts. **D 996,** D10

friction, *n*—resistance to the relative motion of one body sliding, rolling, or flowing over another with which it is in contact. **F 1646,** F13

friction coat, calender, *n*—a layer of rubber compound applied to a fabric by shearing action so that it impregnates the fabric. **D 1566,** D11

friction coating, *n*—applying a rubber coat (friction coat) on a textile by shearing action so that the coat impregnates the textile. **D 1566,** D11

friction force, *n*—the resisting force tangential to the interface between two bodies when, under the action of an external force, one body moves or tends to move relative to the other. (See also **coefficient of friction**.) **G 40,** G02

friction polymerization—the process by which organic compounds (such as adsorbed air pollutants) on mating surfaces that move relative to each other polymerize to yield compounds of high molecular weight.

NOTE—When this occurs on electrical contacts and the materials formed remain on the surface, contact resistance may increase substantially. The solid materials that are produced by the polymerization process are called "friction polymers" or "frictional polymers." **B 542,** B02

friction ratio, *n*—ratio of surface speeds of two adjacent rolls (mill, calender, or refiner). **D 1566,** D11

frictive track—a series of crescent cracks lying along a common axis, paralleling the direction of frictive contact; also known as a chatter sleek. **C 162,** C14

frigorifico hides—cattlehides from South American slaughtering and freezing plants, cured in brine and salted. **D 1517,** D31

fringe, *n*—one of the light or dark bands produced by the interference of the light scattered by the real object and the virtual image of the object. **E 1316,** E07

fringe effect—toner deposition in non-image areas adjacent to the image. Heading and tailing are fringe effects. **F 335,** F05

frit, *n*—glass in particulate form, generally of controlled size distribution. **C 162,** C14

frit, *n*—(*1*) the process of pouring molten glass into water or between cooled rollers. See **shrend, dry gage**.

(*2*) to decorate or seal with glass in particulate form. **C 162,** C14

frit—a product made by quenching and breaking up a glass of a specific composition, used customarily used as a component of a glaze, body, or porcelain enamel. **C 242,** C21

frit, *n*—a porous material permeable to gas flow usually made by sintering microbeads of an appropriate material. **D 1356,** D22

frit, clear—see **clear frit.** **C 286,** B08

frit, colored—see **colored frit.** **C 286,** B08

frit, porcelain enamel—the small friable particles produced by quenching a molten glassy material (see also **clear frit** and **colored frit**). **C 286,** B08

fritted bubbler—See **bubbler.** **D 1356,** D22

fritted glaze—See *fritted glaze* under **glaze.** **C 242,** C21

fritting—the rapid chilling of the molten glassy material to produce frit. **C 286,** B08

fritting (A-fritting)—an electric breakdown between mating metallic contacts, separated by an insulating film, which occurs when the field strength exceeds approximately 1 000 000 V/cm. Metallic bridges are produced through the film if the fritting voltage is above the contact melting voltage. **B 542,** B02

fritting voltage—the voltage at which fritting occurs. **B 542,** B02

frizing (friezing)—in tanning Mocha glove leather, a process of removing the grain surface involving severe liming for not less than a month, during which the elastin structure of the grain layer is destroyed. **D 1517,** D31

frog, *n*—an indentation in one bed surface of a brick manufactured by molding or pressing. The identation may be shallow (not exceeding ⅜ in. (9.5 mm)) and is termed a panel. A deeper indentation is called a deep frog. Limits on the extent of frogging are given in Specifications C 62, C 216, and C 652. **C 43,** C15

frog, *n*—an indentation in a bed surface of a masonry unit. Indentations not exceeding ⅜ in. are termed a frog, sometimes called a panel or panel frog. Indentations exceeding ⅜ in. are termed a deep frog. **C 1232,** C15

front—the forepart of a hide or skin. Particularly in horsehide leathers, the front is used for garments, baseballs, etc. It is the part left when the butt is cut off about 22 in. (559 mm) from the root of the tail. **D 1517,** D31

front break-point, *n—in anatomy*, the location on the front of the body where the arm separates from the body. **D 123,** D13

front end breakaway, *n—in cornering vehicle*, the point in the curved trajectory of a vehicle when it can no longer be maintained on its intended path because of front wheel departure toward the outside of the curve. **F 538,** F09

front face discharge—makeup air method that is very flexible, but should be directed away from the hood, but the closer the air outlet's lower edge is to the bottom of the hood, the lower the velocity must be to avoid drawing effluent out of the hood. **F 1827,** F26

front high-hip, *n—in body measurements,* the distance from one imaginary side seam to the other imaginary side seam at the high-hip level. (Syn. **abdominal extension.)** **D 5219,** D13

front side—Not preferred, use **front surface.** **F 1241,** F01

front surface—*of a semiconductor wafer*, the exposed surface upon which active semiconductor devices have been or will be fabricated. **F 1241,** F01

frost, *n*—ice crystals resulting from the direct sublimation of water vapor onto a surface that is below freezing. **D 1356,** D22

frost action—freezing and thawing of moisture in materials and the resultant effects on these materials and on structures of which they are a part or with which they are in contact. **D 653,** D18

frost boil—(*a*) softening of soil occurring during a thawing period due to the liberation of water form ice lenses or layers.

(*b*) the hole formed in flexible pavements by the extrusion of soft soil and melt waters under the action of wheel loads.

(*c*) breaking of a highway or airfield pavement under traffic and the ejection of subgrade soil in a soft and soupy condition caused by the melting of ice lenses formed by frost action. **D 653,** D18

frosted—surface treated to scatter light or to simulate frost. **C 162,** C14

frosted area—{archaic} See **hackle.** **C 162,** C14

frost heave—the raising of a surface due to the accumulation of ice in the underlying soil or rock. **D 653,** D18

frosting, *n*—a change in color in a limited area of a fabric caused by abrasive wear. **D 123,** D13

frosting, *n*—formation of a matte, whitish appearance on a rubber surface exposed to air, resulting from the action of ozone; often confused with bloom. **D 1566,** D11

frosting, *n*—a light-scattering surface resembling fine crystals. See also **chalking, haze, bloom**. (D-20) **F 412,** F17

frost-point hygrometer—See **dew-/frost-point hygrometer** under **hygrometer.** **D 1356,** D22

frost-point temperature—See **temperature.** **D 1356,** D22

froth, *n*—a collection of bubbles and particles on the surface of a pulp in a froth flotation cell. **D 121,** D05

frother, *n*—a reagent used in froth flotation to control the size and stability of the air bubbles, principally by reducing the surface tension of water. **D 121,** D05

froth flotation, *n*—a process for cleaning fine coal in which hydrophobic particles, generally coal, attach to air bubbles in a water medium and rise to the surface to form a froth. The hydrophilic particles, generally the ash-forming matter, remain in the water phase. **D 121,** D05

Froude number—dimensionless number expressing the ratio of inertial to gravity forces in free-surface flow. It is equal to the average velocity divided by the square root of the product of the average depth and the acceleration due to gravity. **D 4410,** D19

FRP—fiberglass reinforced plastic. **D 6161,** D19

FRP insulating tubes and rods—fiberglass reinforced plastic (FRP) products manufactured so that the tubes and rods produced will meet the electrical and mechanical requirements in the standard. **F 819,** F18

fruit-box nail—coated, regular-stock-steel, 1 ⅜ by 0.072-in. nail with flat 7⁄32-in. head and medium diamond point. **F 547,** F16

frustrated total reflection (FTR)—the reflection which occurs when a nonabsorbing coupling mechanism acts in the process of total internal reflection to make the reflectance less than unity. **E 131,** E13

fryer—equipment, including a cooking vessel, in which oils are placed to such a depth that the cooking food is essentially supported by displacement of the cooking fluid within a perforated container rather than by the bottom of the vessel. Heat delivery to the cooking oils varies with fryer models. **F 1827,** F26

F-test, *n*—a test of statistical significance based on the use of George W. Snedecor's F-distribution and used to compare two sample variances or a sample variance and a hypothetical value. **D 123,** D13

fuel alcohol—ethyl, methyl, or higher alcohols with impurities (including water but excluding denaturants) produced for use as a fuel alone or as an addition to other fuels, such as gasoline. **E 772,** E44

fuel alcohol—ethyl, methyl, or higher alcohols with impurities (including water but excluding denatureants) produced for use as a fuel alone or as an addition to other fuels, such as gasoline. **E 1126,** E48

fuel alcohol—ethyl, methyl, or higher alcohols with impurities (including water but excluding denaturants) produced for use as a fuel alone or as an addition to other fuels, such as gasoline. **E 1705,** E48

fuel assembly—*in the nuclear industry*, a grouping of fuel elements that is not taken apart during the charging and discharging of a reactor core. **C 859,** C26

fuel element—*in the nuclear industry*, the smallest structurally discrete part of a reactor that has fuel as its principal constituent. **C 859,** C26

fuel ethanol—ethanol with impurities (including water but excluding denaturants). **E 1126,** E48

fuel ethanol—ethanol with impurities (including water but excluding denaturants). **E 1705,** E48

fuel ethanol manufacturing facility—a manufacturing facility of any size designed to produce fuel ethanol by a fermentation process. **E 1126,** E48

fuel ethanol manufacturing facility—a manufacturing facility of any size designed to produce fuel ethanol by a fermentation process. **E 1705,** E48

fuel pellet—a densified fuel form, usually cylindrical, die-formed, usually by extrusion, with random lengths and open broken ends. Generally of high density. **E 1126,** E48

full annealing, *n*—**annealing** a steel object by **austenitizing** it and then cooling it slowly through the **transformation range**. **A 941,** A01

full annealing—heating steel above the upper critical transformation temperature, holding it there long enough to fully transform the steel to austenite, and then cooling it at a controlled rate, in a furnace, to below a specified temperature. A full anneal refines grain structure and provides a relatively soft, ductile material that is free of internal stresses. **F 1789,** F16

full cone nozzle—see **solid cone atomizer**. **E 1620,** E29

full cone pattern, *n*—a cone pattern wherein the liquid is distributed throughout the pattern. **E 1620,** E29

full-energy peak—the peak in an energy spectrum recorded by a photon detector that occurs when the full energy of an incident photon is absorbed by the detector. This is sometimes referred to as the photopeak. **E 170,** E10

full-energy peak efficiency—the ratio of the net count rate in the full-energy peak to the emission rate of the photons from a sample giving rise to the peak. The value is dependent on the source-detector-shield geometry and the photon energy. This is sometimes referred to as the photopeak efficiency. **E 170,** E10

full grain—having the original grain surface of the skin. **D 1517,** D31

full-impulse-voltage wave, *n*—an aperiodic transient voltage that rises rapidly to a maximum value, then falls less rapidly to zero. **D 1711,** D09

full radiator, *n*—a thermal radiator that completely absorbs all incident radiant flux, whatever the wavelength, direction of incidence, or polarization. **E 284,** E12

full radiator: blackbody (USA), Planckian radiator, *n*—thermal radiator which absorbs completely all incident radiation, whatever the wavelength, the direction of incidence or the polarization. This radiator has, for any wavelength, the maximum spectral concentration of radiant exitance at a given temperature. **E 349,** E21

full radiator—see **blackbody.** **E 772,** E44

full scale—See **analyzer.** **D 1356,** D22

full size specimen—tension test specimen consisting of a completed fastener for testing in the ready to use condition without altering the configuration. **F 1789,** F16

full trailer, *n*—a freight trailer supported at both ends by attached axles. **E 867,** E17

full-wave direct current (FWDC)—a rectified three-phase alternating current. **E 1316,** E07

fully dense material—a material completely free of porosity and voids. **B 243,** B09

fully nested experiment, *n*—a nested experiment in which the second factor is nested within levels (versions) of the first factor and each succeeding factor is nested within versions of the previous factor. **E 456,** E11

fully nested experiment—a nested experiment in which the second factor is nested within levels (versions) of the first factor and each succeeding factor is nested within versions of the previous factor. **E 1325,** E11

fully tempered glass—flat glass that has been tempered to a high surface or edge compression to meet the requirements of Specification C 1048. See **heat-strengthened glass**. **C 162,** C14

fully transposed conductor—a conductor in which every strand occupies every relative position in the conductor at regularly specified intervals along its length. **B 713,** B01

fumarole, *n*—a vent at the earth's surface that emits steam or gaseous vapor. **E 957,** E44

fume, *n*—properly, the solid particles generated by condensation from the gaseous state, generally after volatilization from melted substances, and often accompanied by a chemical reaction such as oxidation. **D 1356,** D22

fume stack, *n*—a box-like duct with thermocouples and baffles through which flames and hot fumes from a burning specimen pass whose purpose is to permit measurement of the heat release from the burning specimen. **E 176,** E05

fumigant—a chemical used in gaseous form to kill or inhibit pests. **E 609,** E35

function—a purpose of the entire project or some portion thereof determined by the needs or desires of the user/owner and expressed in two words, an active verb and a measurable noun. **E 833,** E06

functional barrier—*in food packaging*, a material that effectively restricts passage of solids, liquids, semi-solids, vapors, or forms of energy such as ultraviolet light through itself, across its borders, or interface with another material or substance. **F 1479,** F02

functional classification, *n*—the grouping of streets and highways into classes, or systems, according to the character of service they are intended to provide. **E 867,** E17

functional, *adj* (fonctionnel)—performing or able to perform a regular function; designed or developed chiefly from the point of view of use. **E 1480,** E06

functionality, *adj* (fonctionnalité)—being suitable for a particular use or function. **E 1480,** E06

functional program— See **facility program**. **E 1480,** E06

functional requirement, n—See **serviceability requirement**. **E 1480,** E06

function analysis—an examination of the project consisting of (*1*) the determination of the project functions; (*2*) the examination and sorting of these functions into categories; (*3*) the selection of the critical functions and arrangement into a logical order; and (*4*) the determination of the project cost allocated to performing each critical function. **E 833,** E06

function, basic—a function that is necessary to achieve the primary purpose of a building system or element. **E 833,** E06

function, *n* (fonction)—the action for which a person or thing is specially fitted or used or for which a thing exists. **E 1480,** E06

function performance, n—See **facility serviceability**. **E 1480,** E06

fundamental frequency—lowest frequency of periodic variation. **D 653,** D18

fundamental resonant frequency, *n*—*in linear density testing*, the lowest frequency at which free oscillations can exist in a fiber tensioned between two fixed points. **D 123,** D13

fungi resistance—the ability of plastic pipe to withstand fungi growth or their metabolic products, or both, under normal conditions of service or laboratory tests simulating such conditions. **F 412,** F17

fungistat—an agent that inhibits the germination of fungus spores or the development of mycelium. **E 609,** E35

fungus—primitive plants distinguished from algae by the absence of chlorophyll. **D 6161,** D19

fungus resistance—the ability of a sandwich construction to withstand fungi growth or their metabolic products, or both, under normal conditions of service or laboratory test simulating such conditions. **E 1749,** E06

furanic compounds—a class of chemical compounds characterized by the presence of heterocyclic structures consisting of a five-membered ring containing four carbon atoms and one oxygen atom.

These compounds may be found dissolved in electrical insulating fluids, either as residual contaminants of refinery extraction processes in which furfural is used, or from the degradation of cellulose insulation. **D 2864,** D27

furan plastics—plastics based on furan resins. (ISO) **D 883,** D20

furan resin—a thermosetting catalyzed condensation reaction product from furfuryl alcohol, furfural or combination thereof. **C 904,** C03

furan resin, *n*—a resin in which the furan ring is an integral part of the polymer chain and represents the greatest amount by mass. (ISO) **D 883,** D20

fur brush development—development of electrostatic images by transporting the toner to the image area through the use of fur-like fibers. **F 335,** F05

furfural—an aldehyde derivative of certain biomass conversion processes, used as a solvent. **E 1126,** E48

furfural—an aldehyde derivative of certain biomass conversion processes, used as a solvent. **E 1705,** E48

furnace, arc-image, *n*—a furnace in which high temperatures are produced by focusing radiation from high-temperature arcs into the furnace chamber. **C 1145,** C28

furnace, box—see **box furnace.** **C 286,** B08

furnace carbon black, *n*—type of carbon black produced by the decomposition reaction of hydrocarbons when injected into a high velocity stream of combustion gases under controlled conditions. **D 1566,** D11

furnace, continuous— see **continuous furnace.** **C 286,** B08

furnace, image, *n*—a furnace in which high temperatures are generated by focusing radiation from a high-temperature source, such as the sun or an electric arc. **C 1145,** C28

furnace, recuperative, *n*—a furnace equipped with a heat exchanger in which heat is conducted from the combustion products through a system of ducts or through flue walls in a manner so as to preheat the air as it enters the burner to unite with the fuel. **C 1145,** C28

furnace, regenerative, *n*—a furnace having a cyclic heat exchanger which alternately receives heat from gaseous combustion products and transfers heat to the air or gas of the fuel mixture before combustion takes place. **C 1145,** C28

furnace, solar, *n*—an image-type furnace in which solar radiation is focused into a relatively small area as a source of heat producing extremely high temperatures. **C 1145,** C28

furnace, thermal gradient, *n*—a tubular furnace in which a controlled temperature gradient is maintained along its length. **C 1145,** C28

furnish, *n*—in any papermaking process, all of the materials added prior to sheet formation **D 1968,** D06

furniture-carton nail—bright, regular-stock-steel, ¾ to 2-in. nails with circular 1-in. cap head and medium diamond point. (See **cap nail.**) **F 547,** F16

furniture covering, *n*—a general term for attached upholstery fabric, slipcovers and throws. **D 123,** D13

furniture mock-up or assembly, *n*—a representation of production furniture that uses the same upholstery cover material and upholstery material, constructed in the same manner as in production furniture, but with straight, vertical sides. **E 176,** E05

furniture nail—plated, regular-stock-steel or brass, ⅜ to ¾-in. nails with extra large, decorative head and long diamond or needle point. **F 547,** F16

furniture unit, *n*—*in upholstered furniture*, a complete single piece of upholstered seating, such as a sofa, love seat, lounge chair, rocker, or recliner. **D 123,** D13

furring—buildup or bristling of magnetic particles due to excessive magnetization of the component under examination resulting in a furry appearance. **E 1316,** E07

furring nail—See **self-furring nail.** **F 547,** F16

furring tile, *n*—tile for lining the inside of walls and carrying no superimposed loads. **C 43,** C15

fusain—See *fusain* under **coal**. **D 121,** D05

fuse, *v*—to melt or join by the use of heat. **C 1145,** C28

fuse, *v*—(*1*) to convert plastic powder or pellets into a homogeneous mass through heat and pressure; (*2*) to make a plastic piping joint by heat and pressure. **F 412,** F17

fuse—to melt; to join together by melting. **F 1156,** F12

fused fabric, *n*—a type of bonded fabric made by adhering a fusible fabric to another fabric, such as for use in an interlining. **D 123,** D13

fused grain refractory, *n*—a refractory made predominantly from grain that has solidified from a fused or molten condition. **C 71,** C08

fused or fusion cast refractory, *n*—a solidified material made by melting refractory ingredients and pouring it into molds (see also **molten cast refractory**). **C 71,** C08

fused quartz—vitreous silicon dioxide produced by melting silica, generally in the form of granular quartz. See related terms **fused silica** and **vitreous silica**. **C 162,** C14

fused silica—vitreous silicon dioxide produced by flame hydrolysis of silicon tetrachloride (or similar compounds) or by thermal consolidation of a silica gel. See related terms **vitreous silica** and **fused quartz**. **C 162,** C14

fused silica refractory, *n*—a product composed predominantly of fused, noncrystalline silica. **C 71,** C08

fusel oil—a clear, colorless, poisonous, liquid mixture of alcohols obtained as a by-product of grain fermentation; generally amyl, isoamyl, propyl, isopropyl, butyl, and isobutyl alcohols and acetic and lactic acids. **E 1126,** E48

fusel oil—a clear, colorless, poisonous, liquid mixture of alcohols obtained as a by-product of grain fermentation; generally amyl, isoamyl, propyl, isopropyl, butyl, and isobutyl alcohols and acetic and lactic acids. **E 1705,** E48

fuse oils—complex group of higher molecular weight materials including ketones and aldehydes produced as a byproduct by the yeast fermentation during ethanol production. **E 1705,** E48

fusible fabric, *n*—a utilitarian fabric which has a thermoplastic adhesive applied to one side, sometimes in a pattern of dots, so that the surface can be bonded to another fabric surface by the use of heat and pressure. **D 123,** D13

fusing—see **heat fixing.** **F 335,** F05

fusinite—see *fusinite* under **maceral**. **D 121,** D05

fusion—the process of melting; usually the result of interaction of two or more materials. **C 242,** C21

fusion bonded, *n*—a method of creating a carpet pile construction by adhering loops or lengths of yarn to the face of a primary backing. **D 123,** D13

fusion casting, *n*—the process of forming items by casting molten materials in mold. **C 1145,** C28

fusion flow—the relative flow of various glasses or frits in the molten state. **C 286,** B08

fusion point, *n*—the temperature or range of temperatures at which melting or softening, as a result of partial melting, of a composition, will occur. **C 1145,** C28

fusion seal—a bond formed by combining two or more materials through melting or other means so that the joining layers become indistinguishable at the interface. **F 17,** F02 *fusion seal*—See Terminology F 17. **F 1327,** F02

fusion test, *n*—any test to determine the temperature or range of temperatures at which fusion takes place, or to determine the flow or other properties of a material at fusion temperatures. **C 1145,** C28

fusion test, button— see **button test.** **C 286,** B08

fusion welding—a joining process in which the mating surfaces of two thermoplastic parts are melted by induced heat and rapidly pressed together while still molten to form a homogeneous bond. **C 904,** C03

future value—the value of a benefit or a cost at some point in the future, considering the time value of money (Syn. *future worth*). **E 833,** E06

future worth—See **future value**. **E 833,** E06

fuzz, *n*—untangled fiber ends that protrude from the surface of a yarn or fabric. **D 123,** D13

fuzz ball, *n*—loose and frayed fibers that have formed into a ball and have then been woven or knitted into the fabric. (Compare **pills.**) **D 123,** D13

fuzz balls—broken or abraded filaments which have collected as loose bundles or balls during the manufacture of impregnated material, occasionally incorporated into the impregnated material. **E 1749,** E06

fuzzy, *adj*—characterized by a hairy appearance due to broken fibers or filaments. **D 123,** D13

fuzzy texture—a defect characterized by a myriad of minute bubbles, broken bubbles, and dimples in the porcelain enamel surface. **C 286,** B08

FWA buildup—the course of change in fluorescence emission intensity or fluorescence shade or both, using specified exhaust procedure:

(*1*) for a specified number of successive applications of FWA, or

(*2*) by varying the FWA concentration in a series of single applications. **D 459,** D12

FWA exhaust efficiency—a measure of FWA substantivity as expressed by:

(*1*) exhaust coefficient (E.C.)—the ratio of FWA concentration taken up by unprewhitened substrate, (wt of FWA (s)/wt of substrate) to that concentration of FWA remaining in the bath, (wt of FWA (b)/wt of bath) under specified application conditions.

$$\text{E.C.} = \frac{\text{wt FWA (s)/wt substrate}}{\text{wt FWA (b)/wt bath}}$$

(2) percent exhaust (%E)—the ratio of FWA on the substrate (wt FWA (s)) obtained under specified conditions to the total FWA introduced in the original bath (wt FWA (o)).

FWA fastness (on substrate)—degree of change in fluorescence emission intensity or fluorescence shade or both, when a substrate containing FWA is exposed for a specific length of time to any specified natural or artificial environment. **D 459,** D12

FWA fluorescence emission intensity—the difference between the Z (CIE standard observer) tri-stimulus value of a sample treated with FWA and that of the untreated sample under standardized illumination conditions (D_{65}) and viewing conditions (CIE approved geometry) for any specified substrate and specimen presentation techniques. **D 459,** D12

FWA fluorescence shade—(*1*) the perceived direction of the shift in hue caused by the addition of an FWA to any specified near-white substrate (psychological definition), or (*2*) the wave length at which an extension of the line connecting the points on a CIE diagram corresponding to the chromaticity coordinates (measured under standardized illumination conditions (D_{65}) (CIE approved geometry)) of the untreated substrate to those of the treated substrate intersects the spectrum locus (psychophysical definition). **D 459,** D12

FWA formulation-dependent fluorescence emission intensity ratio—the fluorescence emission intensity obtained with a given FWA on a specified substrate under specified conditions in a designated formulation system relative to that obtained with the same FWA under identical conditions in a different formulation. **D 459,** D12

FWA levelness—the uniformity of distribution of FWA on substrate when applied by a specified method. **D 459,** D12

FWA rate of exhaust index—the time required for an FWA bath of specified composition to be half-depleted by exhaustion onto a particular substrate under specified conditions. **D 459,** D12

FWA stability (in solution)—degree of resistance of FWA in solution under specified exposure condition to specific bath additives. **D 459,** D12

FWA substrate selectivity ratio—the fluorescence emission intensity exhibited by a substrate, relative to that obtained on a reference substrate, after treating these in a specified mixed load, using a given FWA, a designated formulation system, and specified conditions. **D 459,** D12

G

G—symbol for the dimensionless ratio between an acceleration in length per time-squared units, and the acceleration of gravity in the same units (D 1596). **D 996,** D10

g—symbol for the acceleration of gravity at the earth's surface. **D 996,** D10

µg—microgram. **E 1605,** E06

g, *n*—a unit of acceleration where 1 *g* is equal to the acceleration of gravity, 9.8 m/s^2 (32.2 ft/s^2). **F 538,** F09

GaAs—gallium arsenide, a semiconductor material; used in making integrated circuits and light-emitting diodes (**LEDs**). **F 1156,** F12

gable wall—{archaic} the charging end wall of a glass-melting furnace. **C 162,** C14

GAC—granular activated carbon. **D 6161,** D19

gaffer—head workman, foreman, or blower of a glass hand shop. **C 162,** C14

gage, *n—of a tufting machine*, the average centerline distance between the needles. **D 123,** D13

gage, *n—of tufted pile yarn floor covering*, the average distance between adjacent binding sites in the widthwise direction. **D 123,** D13

gage, *n—in knitted fabrics*, a measure of fineness expressing the number of needles per unit of width (across the wales). **D 123,** D13

gage, *n—in full-fashioned hosiery*, a measure of fineness expressing the number of needles per 1.5 in. *on the needle bar.* **D 123,** D13

gage, *n—in warp knitting*, for simplex, tricot, milanese, number of needles per English inch; for raschel, kayloom, twice the number of needles per English inch. **D 123,** D13

gage (also *gauge*), *n—(1) in metal products*, a number designating a specific thickness of metal sheet, or diameter of wire, cable, or fastener shank tabulated in a standardized series, each of which represents a decimal fraction of an inch (or millimetre). (*2*) distance in inches (or millimetres) between adjacent lines of holes or fasteners. **E 631,** E06

gage height—the height of a water surface above an established or arbitrary datum at a particular gaging station; also termed stage. **D 4410,** D19

gage length, *n—in cement testing*, the nominal length between the innermost ends of metal studs that are molded into a test specimen with the axis of the stud in each end of the test specimen coincident with the longitudinal axis of the test specimen and with each other. **C 219,** C01

gage length, *n—in tensile testing*, the length of a specimen measured between the points of attachment to clamps while under uniform tension. **D 123,** D13

gage length, *n*—the known distance between bench marks. **D 1566,** D11

gage length [L], *n*—the original length of that portion of the specimen over which strain or change of length is determined. **E 6,** E28

gage length, *L* **(L)**—distance over which the deformation measurement is made. **D 653,** D18

gage plate—a device that determines the thickness of an individual slice of food product. **F 1827,** F26

gage plate adjuster—a manually operated dial or a lever that is used by the operator to set the *gage plate*. Usual practice is to indicate the thickness by a graduated scale with a 0 (zero) setting to indicate that the *gage plate* is completely closed. **F 1827,** F26

gage pressure—See **pressure.** **D 1356,** D22

gage pressure—pressure measured from atmospheric pressure as the base. **E 631,** E06

gage pressure—difference between the absolute pressure and atmospheric pressure. **E 1316,** E07

gage pressure—the difference in pressure existing within a system and that of the atmosphere. Zero gage pressure is equal to atmospheric pressure. **E 1749,** E06

gage protector—*in grouting*, a device used to transfer grout pressure to a gage without the grout coming in actual contact with the gage. **D 653,** D18

gage saver—see **gage protector**. **D 653,** D18

gaging station—a particular site on a stream, canal, lake, or reservoir at which systematic observations of hydrologic data are obtained. **D 4410,** D19

gaiter, *n—for chain saw cut resistance*, a foot protective device worn outside the footwear. **F 1494,** F23

gaiting, *n—in warp knitting*, the setting of a guide bar one or more needle spaces to the right or left in order to increase the pattern possibilities. **D 123,** D13

galactan—a polysaccharide composed essentially of galactose units. (see also **arabinogalactan**) **D 1695,** D01

galactoglucomannan—one of the hemicelluloses of softwoods, containing three types of sugar units—galactose, glucose, and mannose. **D 1695,** D01

galactomannan—a polysaccharide containing galactose and mannose units. Galactomannans usually have a long chain of mannose units with galactose side chains and are found in seed gums (guar, locust bean). **D 1695,** D01

gall—layer of molten sulfates floating upon glass. **C 162,** C14

galling, *n*—a form of surface damage arising between sliding solids, distinguished by macroscopic, usually localized, roughening and creation of protrusions above the original surface; it often includes plastic flow or material transfer or both. **G 40,** G02

gallon, U. S.—a volume equal to 231 in.^{3} For paint, varnish, lacquer, and related products this is measured at 77°F (25°C). **D 16,** D01

galvanic anode, *n*—a metal electrode that sacrificially corrodes when coupled to a more noble metal in a conducting medium, thereby supplying a protective electric current to the more noble electrode. **B 899,** B02

galvanic cell—an electrolytic cell capable of producing electrical energy by electrochemical action. **B 374,** B08

galvanic corrosion—accelerated corrosion of a metal because of an electrical contact with a more noble metal or non metallic conductor in an electrolyte. **D 6161,** D19

galvanic corrosion—accelerated corrosion of a metal because of an electrical contact with a more noble metal or nonmetallic conductor in a corrosive electrolyte. **E 1749,** E06

galvanic corrosion—accelerated corrosion of a metal because of an electrical contact with a more noble metal or nonmetallic conductor in a corrosive electrolyte. **G 15,** G01

galvanic couple—a pair of dissimilar conductors, commonly metals, in electrical contact. (See **galvanic corrosion.**) **G 15,** G01

galvanic current—the electric current between metals or conductive nonmetals in a galvanic couple. **G 15,** G01

galvanic series—a list of metals and alloys arranged according to their relative potentials in a given environment. See **electromotive series.** **B 374,** B08

galvanic series—a list of metals and alloys arranged according to their relative corrosion potentials in a given environment. **G 15,** G01

galvanized—See **zinc-coated.** **F 547,** F16

galvanized—pertaining to zinc-coated articles. **F 552,** F14

galvanized after weaving—*in chain link fabric*, describes steel fabric that is hot-dip coated with zinc after weaving (see Specification A 392). **F 552,** F14

galvanized before weaving—*in chain link fabric*, describes fabric that is woven from steel wire galvanized before weaving. **F 552,** F14

galvanized coating, *n*—a coating of virtually pure zinc on steel, applied by various methods or processes including hot-dip process and electrodeposition (electrolytic process). **A 902,** A05

galvanized steel wire—steel wire coated with zinc. **B 354,** B01

galvanizing—application of a coating of zinc. **B 374,** B08

galvannealed coating, *n*—a coating on steel of zinc-based alloy, containing about 6 to 15 % iron, produced by hot-dip immersion in a high-zinc content coating bath, followed by heating the steel to induce diffusion alloying between the molten zinc coating and the steel. **A 902,** A05

galvanodynamic—refers to a technique wherein current, continuously varied at a selected rate, is applied to an electrode in an electrolyte. **G 15,** G01

galvanostaircase—refers to a galvanostep technique for polarizing an electrode in a series of constant current steps wherein the time duration and current increments or decrements are equal for each step. **G 15,** G01

galvanostatic—an experimental technique whereby an electrode is maintained at a constant current in an electrolyte. **G 15,** G01

galvanostep—refers to a technique in which an electrode is polarized in a series of current increments or decrements. **G 15,** G01

gamma activity—the spontaneous emission from the nucleus of high energy, short wavelength, electromagnetic radiation. **C 859,** C26

gamma-cellulose—(*1*) Historically, a term used to indicate impurities of short chain lengths found in pulps, predominately hemicelluloses. (*2*) Gama-cellulose content, as measured by TAPPI Method T 203 om-93, is the pulp fraction soluble in caustic, which remains in solution upon acidification. **D 1695,** D01

gamma iron (γ Fe)—solid nonmagnetic phase of pure iron which is stable at temperatures between 910 and 1400°C and possesses the face-centered cubic lattice. **E 7,** E04

gamma (γ) loss peak , *n*—the third peak in the damping curve below the melt, in the order of decreasing temperature or increasing frequency. **D 4092,** D20

gamma (γ) loss peak—*in dynamic mechanical measurement*, third peak in the damping curve below the melt, in the order of decreasing temperature or increasing frequency, (D 4092, D-20). **E 1142,** E37

gamma number—degree of substitution (DS) × 100. This term is of widespread use in the cellulose xanthate field. **D 1695,** D01

gamma (photography)—the tangent of the angle which the straight-line part of the characteristic curve makes with the log exposure axis and in a photographic film or plate is a measure of the extent of development. **E 7,** E04

gamma-radiography—a technique of producing radiographs using gamma-rays. **E 1316,** E07

gamma ray—electromagnetic penetrating radiation having its origin in the decay of a radioactive nucleus. **E 1316,** E07

gamma ray—electromagnetic radiation having its origin in an atomic nucleus. **E 1316,** E07

gamma-ray standard field—a gamma ray field produced by a particular radioactive nuclide (such as ^{60}Co) that is well established and reproducible as to the absorbed dose rate produced in a specific material at a designated location within the field at any given time. **E 170,** E10

gap—an open joint or split in the inner plies which results when crossband or center veneers are broken or not tightly butted. **D 1038,** D07

gap-filling adhesive, *n*—an adhesive capable of forming and maintaining a bond between surfaces that are not close-fitting. **D 907,** D14

gap length, ℓ_g—the distance that the flux transverses in the central region of a gap in a core having an "air" (nonmagnetic) gap in the flux path may be considered unity in the gap. **A 340,** A06

gap ratio—sweep width divided by boom length. **F 818,** F20

gap scanning—short fluid column coupling technique. **E 1316,** E07

Gardner color scale, *n*—a color scale for clear, light-yellow fluids, defined by the chromaticities of glass standards numbered from 1 for the lightest to 18 for the darkest. **E 284,** E12

garment, *n*—a single item of clothing (for example, shirt). **F 1494,** F23

garment leather—a name covering material for leather coats, jackets, hats, and breeches made most commonly from sheep, cabretta, pig, or goat skins, cattle, or horse hides. **D 1517,** D31

gas, *n*—one of the states of matter, having neither independent shape nor volume and tending to expand indefinitely. **D 1356,** D22

gas—the state of matter in which the molecules are practically unrestricted by intermolecular forces so that the molecules are free to occupy all space within an enclosure. In vacuum technology, the word gas has been loosely applied to the uncondensed gas and vapor within a vacuum system. **E 1316,** E07

gas atomizer—see **pneumatic atomizer**. **E 1620,** E29

gas chromatography—method of chemical analysis involving the separation of volatile constituents of a mixture by means of gas flow entrainment, vapor pressure differences, and affinity of specific compounds for various liquids or solid materials. **F 1156,** F12

gas chromatography, GC—all chromatographic methods in which the moving phase is gaseous. The stationary phase may be either a dry granular solid or a liquid supported by the granules or by the wall of the column, or both. Separation is achieved by differences in the partition-distribution of the components of a sample between the mobile and stationary phases, causing them to move through the column at different rates and from it at different times. (E 355) **D 2864,** D27

gas classification—the separation of powder into particle size fractions by means of a gas stream of controlled velocity. **B 243,** B09

gas electrode membrane—a plastic film, permeable to gases but impermeable to water, separating the electrode from the simple solution. **D 4127,** D19

gaseous inclusion—a round or elongated bubble in the glass. See **blister** and **seed**. **C 162,** C14

gas holes—blow holes, channels, or porosity produced by gas evolution, usually during solidification. **E 7,** E04

gas-house coal tar, *n*—coal tar produced in gas-house retorts in the manufacture of illuminating gas from bituminous coal. **D 8,** D04

gasification—any chemical or heat process used to convert a feedstock to a gaseous fuel. **E 1126,** E48

gasification—any chemical or heat process used to convert a feedstock to a gaseous fuel. **E 1705,** E48

gasifier—a device that converts solid fuel to gas. Generally refers to thermochemical processes. Major types are moving bed (fixed bed), entrained bed, and fluidized bed. **E 1126,** E48

gasifier—a device that converts solid fuel to gas. Generally refers to thermochemical processes. Major types are moving bed (fixed bed), entrained bed, and fluidized bed. **E 1705,** E48

gasket, *n*—any preformed, deformable device designed to be placed between two adjoining parts to provide a seal. **C 717,** C24

gasket glazing—a method of setting glass or panels in prepared openings, using a preformed gasket to obtain a weathertight seal. **C 717,** C24

gasket joint—see **joint, compression gasket** and **joint, bell-and-spigot gasket.** **F 412,** F17

gasket leather—a collective term sometimes used for the cattlehide leather, either vegetable, chrome or combination tannages, with special stuffing added, and employed in pump valves, as piston packing, and so forth. **D 1517,** D31

gasket, lock-strip, H-type—two channel recesses, of equal or unequal sizes, one on either side of a central web. **C 717,** C24

gasket, lock-strip, reglet-type—a channel recess on the inner edge and a tongue, or spline, on the outer edge, the latter being designed for insertion in a reglet. **C 717,** C24

gasket (mechanical), *n*—a deformable material clamped between essentially stationary faces to prevent the passage of matter through an opening or joint. **D 1566,** D11

gasket, structural—See **lock-strip gasket**, the preferred term. **C 717,** C24

gasket, zipper—See **lock-strip gasket**, the preferred term. **C 717,** C24

gas–liquid chromatography, GLC—gas chromatographic method utilizing a liquid as the stationary phase, which acts as a solvent for the sample components. **D 2864,** D27

gas-liquid nozzle—see **pneumatic atomizer**. **E 1620,** E29

gas meter, *n*—an instrument for measuring the quantity of a gas passing through the meter. **D 1356,** D22

gasohol—a blend of unleaded gasoline and up to 10 % anhydrous ethanol. **E 1126,** E48

gasoline, *n*—a volatile mixture of liquid hydrocarbons, normally containing small amounts of additives, suitable for use as a fuel in spark-ignition internal combustion engines. **D 4175,** D02

gasoline-alcohol blend, *n*—a spark-ignition engine fuel consisting primarily of gasoline along with a substantial amount of one or more alcohols. **D 4175,** D02

gasoline-ether blend, *n*—a spark-ignition engine fuel consisting primarily of gasoline along with a substantial amount of one or more ethers. **D 4175,** D02

gasoline-oxygenate blend, *n*—a spark-ignition engine fuel consisting primarily of gasoline along with a substantial amount of one or more oxygenates. **D 4175,** D02

gasometer, *n*—an apparatus employing a calibrated volume which is used to calibrate gas-measuring devices. **D 1356,** D22

gas phase ignition, *n*—ignition of pyrolysis products leaving a heated surface by a pilot flame or other ignition source that does not impinge on nor significantly affect (by re-radiation) the heated surface. **E 176,** E05

gassing—the evolution of gases from one or more of the electrodes during electrolysis. **B 374,** B08

gassing—(*1*) the formation of gas bubbles due to bacterial contamination in the milled porcelain enamel slip.

(*2*) See **boiling.** **C 286,** B08

gassing tendency—the capability of an insulating liquid either to absorb or generate gases when exposed to voltage stress. The measure of the gassing tendency is the volume of gas evolved or absorbed per unit time by an insulating liquid subjected to electrical stress under prescribed conditions (by Method D 2300, Test Method for Gassing of Insulating Oils Under Electrical Stress and Ionization (Modified Pirelli Method)). It is commonly expressed in units of microlitres per minute (μL/min) with a positive value indicating gas is evolved and a negative value indicating gas is absorbed. The SI unit is cubic millimetres per minute (mm^3/min). **D 2864,** D27

gas–solid chromatography, GSC—gas chromatographic method utilizing an active (absorbant) solid as the stationary phase. **D 2864,** D27

gassy surface—a defect characterized by poor gloss and fuzzy surface texture. **C 286,** B08

gate—a shut-off device for flow of glass and combustion gases in a forehearth, more commonly a water-cooled member rather than a refractory body. **C 162,** C14

gate, *n*—in an injection mold, a constriction in the flow channel between the runner and the mold cavity. **D 883,** D20

gate—a valve placed in a vacuum system to facilitate the isolation of a selected section of the system. **E 7,** E04

gate—same as **rejection level**. **E 1316,** E07

gate—an electronic means of selecting a segment of the time range for monitoring or further processing. **E 1316,** E07

gate, *n*—in an injection mold, a constriction in the flow channel between the runner and the mold cavity. (D-20) **F 412,** F17

gate clip—a galvanized pressed steel or pressed aluminum fitting designed to secure chain link fabric to the vertical sides of a gate frame. **F 552,** F14

gate cover—a component to prevent the lever from entering the tumbler gateway while changing combination. **F 471,** F12

gate frame—the structural member(s) of the gate. **F 552,** F14

gate holdback or keeper—a device to receive and hold the gate leaf in an open position. **F 552,** F14

gate (injection or transfer mold), *n*— the orifice through which a shaped cavity in a mold is filled with material. **D 1566,** D11

gate latch—a fitting to hold the gate in place when closed. **F 552,** F14

gate opening—the clear distance between the gate posts. **F 552,** F14

gate post—the post to which a gate is attached via hinges. It may also be used as a terminal post. **F 552,** F14

gate transom—a panel of chain link fabric over the top of a gate opening, framed with top and bottom rail. Typically used in tennis court fence (see Practice F 969). **F 552,** F14

gather, *n*—the mass of glass picked up by the hand shop working on the punty or blowing iron. **C 162,** C14

gather, *v*—to get glass from a pot or day-tank on the pipe or punty. **C 162,** C14

gauge, *n*—*in knitted fabrics*, a measure of fineness expressing the number of needles per unit of width (across the wales). **D 4850,** D13

gauge, gage—instrument used to measure wire diameter. (See **wire gage.**) **F 547,** F16

gauging plaster, *n*—a calcined gypsum plaster used with lime for a finish coat. **C 11,** C11

gauntlet—the area of a glove between the wrist and the reinforced edge of the opening. **F 819,** F18

gauss (plural gausses), G—the unit of magnetic induction in the cgs-emu system of units. The gauss is equal to 1 maxwell per square centimetre of 10^{-4} tesla. See **magnetic induction (flux density).** **A 340,** A06

GD—gallons per day. See **GPD**. **D 6161,** D19

GE—graphic escape. **F 1457,** F05

Geiger-Müller counter—See **counter**. **E 7,** E04

gel, *n*—a semisolid system consisting of a network of solid aggregates in which liquid is held. **C 1145,** C28

gel—*in grouting*, the condition where a liquid grout begins to exhibit measurable shear strength. **D 653,** D18

gel, *n*—(*1*) a semisolid system consisting of a network of solid aggregates in which liquid is held.

(*2*) the initial jelly-like solid phase that develops during the formation of a resin from a liquid.

(*3*) with respect to vinyl plastisols, gel is a state between liquid and solid that occurs in the initial states of heating, or upon prolonged storage. **D 883,** D20

gel, *n*—a semisolid system consisting of a network of solid aggregates in which liquid is held. **D 907,** D14

gel—a liquid containing a colloidal structural network that forms a continuous matrix and completely pervades the liquid phase. **D 2507,** F07

gel, *n*—(*1*) a semisolid system consisting of a network of solid aggregates in which liquid is held; (*2*) the initial jelly-like solid phase that develops during the formation of a resin from a liquid; (*3*) with respect to vinyl plastisols, gel is a state between liquid and solid that occurs in the initial stages of heating, or upon prolonged storage. (D-20) **F 412,** F17

gel, *n*—*in polymer*, a semisolid system consisting of a network of solid aggregates in which liquid is held. **F 1251,** F04

gel, *n*—*in polymerization*, the initial jelly-like solid phase that develops during the formation of a resin from a liquid. **F 1251,** F04

gel, *n*—*with vinyl plastisols*, a state between liquid and solid that occurs in the initial stages of heating, or upon prolonged storage. **F 1251,** F04

gelatinization—treatment of starch grains with heat and water to cause the swelling and expansion of the starting material. **E 1705,** E48

gelatin replica—See **replica**. **E 7,** E04

gelation, *n*—formation of a gel. **D 907,** D14

gel, dry rubber, *n*—the portion of unvulcanized rubber insoluble in a chosen solvent. **D 1566,** D11

gel, latex, *n*—a semi-solid system consisting of a network of aggregates in which liquid is held. **D 1566,** D11

gelling (latex), *n*—the formation of a uniform coagulum from which the aqueous phase has not been separated. **D 1566,** D11

gel point, *n*—the stage at which a liquid begins to exhibit pseudo-elastic properties. **D 883,** D20

gel point—the stage at which a liquid begins to exhibit pseudo-elastic properties. **F 412,** F17

gel time—*in grouting*, the measured time interval between the mixing of a grout system and the formation of a gel. **D 653,** D18

gel time, *n*—the period of time from the initial mixing of the reactants of a liquid material composition to the time when gelation occurs, as defined by a specific test method. **D 883,** D20

gel time, *n*—the period of time from the initial mixing of the reactants of a plastic or rubber composition to the time when gelation occurs as defined by a specified test. **D 1566,** D11

gel time, *n*—*of solventless varnish*, the time required, at a specified temperature, for a solventless varnish to be transformed from a liquid state to a gel, as measured with a suitable gel time apparatus. **D 1711,** D09

gel time—the time from the initial mixing of the reactants of a plastic or rubber composition to the time when gelation occurs, as measured by a specific test. **F 412,** F17

general indices of metamerism, n—see **indices of metamerism potential**. **E 284,** E12

general precipitate—a precipitate which is dispersed throughout the matrix. **E 7,** E04

general-purpose nail— See **common nail.** **F 547,** F16

general rate law—$dC/dt = k\,(1 - C)^n$ where C is fractional conversion, t is the time in minutes, and n is the reaction order. **E 1445,** E27

general shear failure—see **shear failure**. **D 653,** D18

generic class, *n*—*as used with textile fibers*, a grouping having similar chemical compositions or specific chemical characteristics. **D 123,** D13

generic criteria, *n*—common characteristics pertaining to organizations' human resources, material resources, and quality systems which provide a basis for assessing the qualifications of testing or inspection agencies. **D 1356,** D22

generic criteria, *for laboratory accreditation, n*—accreditation criteria expressed in general terms which address organization, human and material resources, operating procedures, calibration and quality assurance practices of a laboratory. **E 1187,** E36

genome (of a virus)—the genetic material consisting of nucleic acid (RNA or DNA). **E 1705,** E48

geochemistry, *n*—the study of the chemistry of the rocks and fluids of the earth for the purpose of understanding their composition, their temperature, and their origin. **E 957,** E44

geocomposite, *n*—a product composed of two or more materials, at least one of which is a geosynthetic. **D 4439,** D35

geofoam, *n*—block or planar rigid cellular foamed polymeric material used in geotechnical engineering applications. **D 4439,** D35

geogrid, *n*—a geosynthetic formed by a regular network of integrally connected elements with apertures greater than 6.35 mm (¼ in.) to allow interlocking with surrounding soil, rock, earth, and other surrounding materials to function primarily as reinforcement. **D 4439,** D35

geologic erosion—the erosion process on a given land form that is not associated with the activities of man. **D 4410,** D19

geologic or natural erosion—the erosion process on or in a given land form undisturbed by activities of man and his agents. **D 4410,** D19

geomembrane, *n*—an essentially impermeable geosynthetic composed of one or more synthetic sheets. **D 4439,** D35

geometric lathe—very complicated device for the generation of complex geometric patterns in surfaces, often under computer control of the cutting engine. **F 1156,** F12

geometric metamerism, *n*—deprecated term; do not use. The phenomenon it is used to describe does not conform to the internationally accepted definitions of metamerism. Use **gonioappearance**. **E 284,** E12

geometric unsharpness— the penumbral shadow in a radiological image which is dependent upon 1) the radiation source dimensions, 2) the source to object distance, and 3) object to detector distance. **E 1316,** E07

geonet, *n*—a geosynthetic consisting of integrally connected parallel sets of ribs overlying similar sets at various angles for planar drainage of liquids or gases. **D 4439,** D35

geophysical borehole log—a log obtained by lowering an instrument into a borehole and continuously recording a physical property of native or backfill material and contained fluids. Examples include resistivity, induction, caliper, sonic, and natural gamma logs. **D 653,** D18

geophysical log, *n*—a graphic record of the measured or computed physical characteristics of the rock section encountered in a borehole, plotted as a continuous function of depth. Measurements are made by a sonde which contains the detectors, as it is withdrawn from the borehole by a wire line. Several measurements are usually made simultaneously, and the resulting curves are displayed side by side on the common depth scale. A common suite of logs used in coal exploration include caliper, density (gamma-gamma), natural gamma, and resistivity.

caliper log, n—a continuous mechanical measurement of the diameter and thus the rugosity of the borehole. The tool identifies zones where swelling or cavings (washouts) have occurred during drilling. The tool's value is in allowing qualitative or quantitative corrections to be made to other geophysical logs which are affected by borehole size (especially density).

density log (gamma-gamma log), n—measures electron density within lithologic units which is related to their bulk density. The wireline tool records the intensity of gamma radiation (in counts per second) from a nuclear source within the tool after it has been attenuated and backscattered by lithologies within the borehole. Due to the distinctly low density of coals, the density log is essential in coal exploration for identifying coal seams and coal-seam partings. The bias/resolution of density logs can be affected by source-detector spacing (closer spacing increases resolution), borehole size, and irregularities (see *caves* or *washouts*), and the presence of casing and logging speed.

natural gamma-ray log, n—a record of the natural radioactivity of the lithologies encountered in the borehole environment. During recording of geophysical logs, the amount of natural radiation is recorded and presented in either counts per second (CPS) or American Petroleum Institute (API) units. Unlike many other log types, a representative natural gamma log can be obtained where borehole or fluid conditions, or both, are not optimal or where casing is present. The natural gamma log is most often used in the coal environment for identifying clastic lithologies and differentiating coal seams and coal-seam partings.

resistivity log, n—a measure of the voltage differential of strata along the walls of a borehole when electrical current is passed through the strata. The resistivity log requires a fluid-filled hole to provide a conductive medium constantly between electrodes on the tool. The spacing between the electrodes determines the precision of the bed boundary relationships in much the same manner as with the density log. The resistivity log is useful primarily in conjunction with other log types. The logs are affected by casing, logging speed, electrode spacing, formation porosity, and resistivity changes in the borehole fluid. **D 121,** D05

geosynthetic, *n*—a planar product manufactured from polymeric material used with soil, rock, earth, or other geotechnical engineering related material as an integral part of a man-made project, structure, or system. **D 4439,** D35

geosynthetic clay liner, *n*—a manufactured hydraulic barrier consisting of clay bonded to a layer or layers of geosynthetic materials. **D 4439,** D35

geotechnical engineering, *n*—the engineering application of geotechnics. **D 4439,** D35

geotechnics, *n*—the application of scientific methods and engineering principles to the acquisition, interpretation, and use of knowledge of materials of the earth's crust to the solution of engineering problems. **D 4439,** D35

geotextile, *n*—a permeable geosynthetic comprised solely of textiles. **D 4439,** D35

geothermal, *adj*—relating to or derived from the natural heat of the earth. **E 957,** E44

geothermal anomaly, *n*—a conspicuous deviation of the earth's temperature, geothermal gradient, or heat flow from average values; an area where such a deviation exists. **E 957,** E44

geothermal energy, *n*—the thermal energy contained in the rocks and fluids of the earth. **E 957,** E44

geothermal facility, *n*—the physical components necessary for the utilization of geothermal energy, including the reservoir, production and injection wells, pipelines, and the power plant or direct-use facility. **E 957,** E44

geothermal fluid, *n*—water in a vapor or liquid phase or in a mixture of these phases that exists within or has been emitted from a geothermal reservoir, together with any entrained or dissolved substances. **E 957,** E44

geothermal gradient, *n*—the change in temperature of the earth with depth, expressed either in degrees of temperature per unit depth, or units of depth per degree. **E 957,** E44

geothermal heat pump, *n*—a heat pump that transfers energy to or from the earth. **E 957,** E44

geothermal power plant, *n*—a facility for the production of electricity using geothermal energy, typically including a turbine, a generator, and associated surface equipment. **E 957,** E44

geothermal reserves, *n*—the amount of energy anticipated to be economically recoverable from a geothermal facility over a specified time period (e.g., the project life) using existing technology. Geothermal reserves are expressed in units of energy (e.g., terajoules in SI units), which are dimensionally equivalent to units of power multiplied by units of time (e.g., Megawatt-years or kilowatt-hours). Geothermal reserves may also be expressed as an equivalent amount of another energy source (e.g., barrels of oil equivalent).

Example of Usage:
This facility has geothermal reserves of 4,000 Megawatt-years, recoverable over a project life of 30 years. **E 957,** E44

geothermal reservoir, *n*—an aquifer of sufficient temperature and permeability to support the economic use of geothermal energy. **E 957,** E44

geothermal steam, *n*—a geothermal fluid in the vapor phase. **E 957,** E44

geothermometer, *n*—a method of estimating the temperature of a geothermal reservoir based on the minerals in the reservoir rock or the concentration of chemical species in geothermal fluids that have come from the reservoir. **E 957,** E44

German degree—Calcium oxide equivalents expressed in parts per hundred thousand. Concentration in German degree is calculated by dividing concentration in calcium carbonate equivalents by 17.86 . **D 6161,** D19

germanium detector, lithium drifted—a semi-conductor detector that can be used for high-resolution gamma-ray spectrometry. **C 859,** C26

gettering—the process that immobilizes impurities at locations away from the region of the specimen to be investigated. **F 1241,** F01

geyser, *n*—a spring that intermittently blows forth hot water and steam. **E 957,** E44

GFD (GPDSF)—unit of permeate rate or flux; gallons per day per square foot of effective membrane area. **D 6161,** D19

ghosting—a gas-chromatographic interference, showing as a peak, which appears at the same elution time as a component from previous injection. **D 1129,** D19

ghosting, *n*—the presence of a faint mirror image of a design appearing in areas not intended to receive that portion of the image. **D 6488,** D01

ghosting—in carbonless copy products, the transferring of a secondary image from the CB sheet to the CF sheet. **F 549,** F05

ghosting, *chemical, n*—the appearance of gloss or dull mirror images that are printed on the reverse side of the sheet. **D 6488,** D01

ghosting, *mechanical, n*—the appearance of a phantom image on the printed side of the sheet. **D 6488,** D01

ghosting, *starvation, n*—mechanical ghosting that appears in solids and shadows as either a lighter image or a darker image than the background. **D 6488,** D01

Gibbs free energy—the maximum useful work that can be obtained from a chemical system without net change in temperature or pressure; $\Delta F = \Delta H - T\Delta S$. **E 7,** E04

Gibbs Phase Rule—maximum number of phases (*P*) that may coexist at equilibrium is equal to two, plus the number of components (*C*) in the mixture, minus the number of degrees of freedom (*F*): $P + F = C + 2$ (E 7, E-4). **E 1142,** E37

Gibbs triangle—an equilateral triangle, used for plotting composition in a ternary system. **E 7,** E04

gilbert, Gb—the unit of magnetomotive force in the cgs-emu system of units. The gilbert is a magnetomotive force of $4\pi/10$ ampere-turns. See **magnetomotive force.** **A 340,** A06

gill box leather—a leather used in textile machinery, similar to comber leather. **D 1517,** D31

ginned lint, *n*—cotton fibers that have been separated from their seeds by ginning but not subjected to any further processing after ginning. (See **lint cotton**.) **D 123,** D13

girder—a horizontal member used to support heavy loads such as other beams along its length. **D 9,** D07

girth—distance around; circumference.

(a) joint—around metatarso—phalangeal joint,

(b) waist—smallest girth behind joint,

(c) instep—smallest girth passing over prominence on middle cuneiform,

(d) long heel—seat to instep to give "pass line" in riding boot,

(e) short heel—seat to lowest crease in front of ankle,

(f) ankle—around and above ankle bones,

(g) calf, thigh as necessary. (manual of shoemaking: c. & k. clark ltd. 1976, p. 44) **F 869,** F08

girth, *n*—in body measurements, a circumferential measurement around some part of the body, such as neck, chest, waist, and so forth. **F 1494,** F23

glacial till (till)—material deposited by glaciation, usually composed of a wide range of particle sizes, which has not been subjected to the sorting action of water. **D 653,** D18

glancing angle—the angle (usually small) between an incident X-ray beam and the surface of the specimen. **E 7,** E04

glancing exit—*AES, EIA, XPS,* geometrical arrangement in which the scattered (or emitted) particles are near 90° from the normal to the specimen surface. This results in improved depth resolution. **E 673,** E42

glancing incidence—*AES, EIA,* geometrical arrangement in which the incident particles are near 90° from the normal to the specimen surface. This results in improved depth resolution. **E 673,** E42

glare, *n*—condition of vision in which there is discomfort or a reduction in ability to see details, objects, or both, caused by an unsuitable distribution or range of luminance, or by extreme contrasts in space. **E 284,** E12

glare—excessive brightness which interferes with clear vision, critical observation, and judgment. **E 1316,** E07

glare—See **lighting**. **E 1480,** E06

glass—an inorganic product of fusion that has cooled to a rigid condition without crystallizing. **C 162,** C14

glass—a term sometimes used for porcelain enamel or frit. **C 286,** B08

glass, *n*—an inorganic product of fusion which has cooled to a rigid condition without crystallizing. **D 883,** D20

glass blowing—the shaping of hot glass by air pressure. **C 162,** C14

glass ceramic—solid material, partly crystalline and partly glassy, formed by the controlled crystallization of a glass. **C 162,** C14

glass-coated steel, glass-lined steel, glassed steel—designations generally applied to a class of porcelain enamels that have high resistance to chemical attack at elevated temperatures and pressures. **C 286,** B08

glass container—general term applied to glass bottles and jars. **C 162,** C14

glass container—See **container.** **D 996,** D10

glass count, *n*—an indirect yarn numbering system, equal to the number of 100 yd. lengths per pound. **D 123,** D13

glass electrode—a half cell in which the potential measurements are made through a glass membrane. **B 374,** B08

glass eye—a defect consisting of a large unbroken blister. **C 286,** B08

glass felt—glass fibers bonded into a sheet with resin and suitable for impregnation in the manufacture of bituminous waterproofing, roofing membranes, and shingles. **D 1079,** D08

glass finish—a material applied to the surface of glass fibers used to reinforce plastics and intended to improve the physical properties of such reinforced plastics over that obtained using glass reinforcement without finish. **D 883,** D20

glassiness—a glassy, marbleized, streaked appearance at the pultruded surface.

Note—This condition is visually evident, but reinforcement is in fact fully encapsulated with resin. **D 3918,** D20

glass mat—a thin mat of glass fibers with or without a binder. **D 1079,** D08

glass paper—a glass fiber product made by dispersing chopped glass fibers in a liquid medium followed by settling and drying to produce a thin sheet. **C 162,** C14

glass redox, redox ratio—a measure of the relative oxidation/reduction state of a glass as indicated typically by any one of the ratios: FeO/total iron as Fe_2O_3, FeO/Fe_2O_3, Fe^{2+}/Fe^{3+}, or $Fe^{2+}/(Fe^{2+} + Fe^{3+})$. **C 162,** C14

glass transition—the reversible change in an amorphous polymer or in amorphous regions of a partially crystalline polymer from (or to) a viscous or rubbery condition to (or from) a hard and relatively brittle one. **D 883,** D20

glass transition—the reversible change in amorphous polymer, or in amorphous regions of a partially crystalline polymer, from (or to) a viscous or rubbery condition to (or from) a hard and relatively brittle one. **D 4092,** D20

glass transition—reversible change in an amorphous material or in amorphous regions of a partially crystalline material, from (or to) a viscous or rubbery condition to (or from) a hard and relatively brittle one. **E 1142,** E37

glass transition—the reversible change in an amorphous polymer or in amorphous regions of a partially crystalline polymer from (or to) a viscous or rubbery condition to (or from) a hard and relatively brittle one. **F 412,** F17

glass transition temperature—on heating, the temperature at which a glass transforms from an elastic to a viscoelastic material, characterized by the onset of a rapid change in thermal expansivity. **C 162,** C14

glass transition temperature—a temperature chosen to represent the temperature range over which the glass transition takes place. **E 1142,** E37

glass transition temperature (Tg)—the approximate midpoint of the temperature range over which the glass transition takes place. **D 883,** D20

glass transition temperature, T_g —the approximate midpoint of the temperature range over which the glass transition takes place. **D 4092,** D20

glass transition temperature (Tg)—the approximate midpoint of the temperature range over which the glass transition takes place. **F 412,** F17

glassware corrosion test, *n*—a laboratory screening test for evaluating the corrosion protection properties of engine coolants on metal test specimens under controlled conditions of aeration and temperature. **D 4725,** D15

glaze, *n*—a hard, glassy, fused ceramic coating which may have a matte or glossy surface. **C 43,** C15

glaze—a ceramic coating matured to the glassy state on a formed ceramic article, or the material or mixture from which the coating is made.

bright glaze—a colorless or colored ceramic glaze having high gloss.

clear glaze—a colorless or colored transparent ceramic glaze.

crystalline glaze—a glaze containing macroscopic crystals.

fritted glaze—a glaze in which a part or all of the fluxing constituents are prefused.

leadless glaze—a ceramic coating matured to a glassy state on a formed article, or the material or the mixture from which the coating is made, to which no lead has been deliberately added.

mat glaze—a colorless or colored ceramic glaze having low gloss.

opaque glaze—a nontransparent colored or colorless glaze.

raw glaze—a glaze compounded primarily from raw constituents, that is, containing no prefused materials.

semi-mat glaze—a colorless or colored glaze having moderate gloss.

slip glaze—a glaze consisting primarily of a readily fusible clay or silt.

vellum glaze—a semi-mat glaze having a satin-like appearance. **C 242,** C21

glaze—a hard glassy fused coating. **C 896,** C04

glaze—a very thin coating of a paint product usually a semi-transparent coating tinted with Van Dyke brown, burnt sienna, or a similar pigment, applied on a previously painted surface to produce a decorative effect. **D 16,** D01

glaze—See **windows and doors.** **E 631,** E06

glaze coat—(*1*) the top layer of asphalt in a smooth-surfaced built-up roof assembly;

(*2*) a thin protective coating of bitumen applied to the lower plies or top ply of a built-up membrane, when application of additional felts, or the flood coat and aggregate surfacing are delayed. **D 1079,** D08

glazed ceramic mosaic tile—ceramic mosaic tile with glazed faces. **C 242,** C21

glazed finish—produced by polishing grain surface under heavy pressure of a roller of agate, glass or steel. Infrequently made by a varnish or shellac coating. **D 1517,** D31

glazed (glace) kid—chrome tanned goatskin and kidskin leather, in either black or colors, which has a glazed finish. **D 1517,** D31

glazed manifold, *n*—a manifold paper having a high gloss, or polish, formed on the surface of the paper by methods such as friction glazing, calendering, plating, etc. **D 1968,** D06

glazed pot—generally, a new pot coated inside with a thin layer of glass to protect it from the raw batch. **C 162,** C14

glazed tile—tile with a fused impervious facial finish composed of ceramic materials, fused with the body of the tile which may be a nonvitreous, semivitreous, vitreous, or impervious body resulting in a surface that may be clear, white, or colored.

glazed interior tile—a glazed tile with a body that is suitable for interior use and which is usually nonvitreous, and is not required or expected to withstand excessive impact or be subject to freezing and thawing conditions.

glazed tile, extra duty glaze—tile with a durable glaze that is suitable for light-duty floors and all other surfaces on interiors where there is no excessive abrasion or impact. **C 242,** C21

glaze fit—the stress relationship between the glaze and body of a fired ceramic product. **C 242,** C21

glazing, *v*—*in building construction,* the installation of glass or other materials in prepared openings. **C 717,** C24

glazing—See **windows and doors.** **E 631,** E06

glazing, *n*—material installed in a window sash, ventilator, or panel such as glass, plastic. **E 1605,** E06

glazing bead—See **windows and doors.** **E 631,** E06

glazing compound—a dough-like material consisting of pigment and vehicle, used for sealing window glass in frames. It differs from putty in that it retains its plasticity for an extended period. **D 16,** D01

glazing material—See **windows and doors.** **E 631,** E06

glitch, *n*—a print defect that displaces the laser scan line so that it appears to start and stop late. **F 1457,** F05

glitter, *n*—the appearance attributable to brilliant reflection from many small, discrete reflecting elements. **E 284,** E12

global—indicates measurements or observations of a total dispersion of particles (such as, a sample representative of an entire liquid spray). **E 1620,** E29

global testing, *n*—testing conducted at two or more laboratories or test sites for the purpose of comparing candidate tire performance at each location for selected characteristic properties. **F 538,** F09

global threshold, GT, *n*—the reflectance level that discriminates bars from spaces in a scan reflectance profile. **F 1294,** F05

glory hole—an opening exposing the hot interior of a furnace used to reheat the ware in hand-working. **C 162,** C14

gloss—the shine or luster of a porcelain enamel. **C 286,** B08

gloss, *n*—the luminous fractional reflectance of a material in the specular direction. (See **specular gloss**) **D 123,** D13

gloss—the geometrically selective reflectance of a surface responsible for its shiny or lustrous appearance. Surface reflectance is commonly at a maximum in or near the geometric directions in which a mirror would reflect light. **D 1695,** D01

gloss, *n*—angular selectivity of reflectance of surface-reflected light responsible for the degree to which reflected highlights or images of objects may be seen as superimposed on a surface. **D 2946,** C17

gloss, *n*—a shiny or lustrous appearance resulting from the tendency of a surface to reflect light energy at one angle more than at others. See **reflectance, directional.** **E 253,** E18

gloss, *n*—angular selectivity of reflectance, involving surface-reflected light, responsible for the degree to which reflected highlights or images of objects may be seen as superimposed on a surface. (See also **distinctness-of-image gloss, haze** (*in reflection*), **luster, sheen, specular gloss.**) **E 284,** E12

glossary, *n*—a collection of specialized terms and their meanings. **E 1992,** E02

glossmeter—an instrument for measuring the ratio of the light regularly or specularly reflected from a surface to the total light reflected. **E 1316,** E07

gloss oil, *n*—a solution of limed rosin or limed rosin acids in a volatile solvent, used chiefly in surface coatings. (When made from tall oil, the source is usually indicated.) **D 804,** D01

gloss reflectance factor, R_S, *n*—ratio of the specularly reflected part of the (whole) flux reflected from the specimen to the flux reflected from a specified gloss standard under the same geometric and spectral conditions of measurement. **E 284,** E12

gloss retention—maintenance of gloss of a film under normal use conditions. **D 2825,** D21

gloss retention, *n*—portion of original gloss retained by a specimen after treatment such as aging or abrasion under specified conditions, calculated by a specified equation. **E 284,** E12

gloss, specular, *n*—ratio of flux reflected in a specular direction to incident flux for specific source and receptor apertures (usually measured relative to a standard of specified index of refraction). **D 2946,** C17

glost fire—See *glost fire* under **firing.** **C 242,** C21

glove leather—term covering two distinct classes:

(*1*) the leather used for dress gloves, including those for street, riding, and sports wear. Tanned predominately from hair sheep, wool sheep, and lamb skins and to a lesser degree from deer, pig, goat, and kid skins, and

(*2*) the leather used for utilitarian or work gloves and made of a variety of hides and skins, of which the most important are horsehides, cattlehide splits and bellies, pigskins, and cowhides. **D 1517,** D31

glove splits—split chrome-tanned cattlehide leather used for work gloves. **D 1517,** D31

glow, *n*—visible, flameless combustion of the solid phase of a material. (See also **afterglow** and **smoldering.**) **D 123,** D13

glow, *n*—(*1*) the visible light emitted by a substance because of its high temperature. (*2*) visible light, other than from flaming, emitted by a solid undergoing combustion. **E 176,** E05

glow, *n*—visible, flameless combustion of the solid phase of a material. **F 1494,** F23

glow discharge—a self-sustaining discharge characterized by essentially symmetrical electrodes, low current density, and a high cathode fall of about 200 V. **B 542,** B02

glucan—a macromolecular substance that can be hydrolyzed to give almost exclusively glucose. **D 1695,** D01

glucoamylase—enzyme that acts specifically to convert dextrins to glucose by hydrolysis. **E 1705,** E48

glucomannan—a hemicellulose consisting essentially of glucose and mannose. **D 1695,** D01

glucose—the most prominent simple sugar (6-membered $C_6H_{12}O_6$) produced from starches and cellulose material by hydrolysis. **E 1705,** E48

glucuronoxylan—a common designation for the xylose-containing hardwood hemicelluloses. (see also **methylglucuronoxylan**) **D 1695,** D01

glue, *n*—originally, a hard gelatin obtained from hides, tendons, cartilage, bones, etc., of animals, and also an adhesive prepared from this substance by heating with water. (See also **adhesive, gum, mucilage, paste, resin,** and **sizing.**) **D 907,** D14

glue, v—See **bond,** *v.* **D 907,** D14

glue, n—See **adhesive.** **D 1038,** D07

glued seam, *n*—*in home sewing,* a seam formed by an adhesive. (Compare **sewn seam, stapled seam, thermally bonded seam.**) **D 123,** D13

glue-laminated wood, *n*—See *glue-laminated wood* under **wood laminate.** **D 907,** D14

glue line, n— Synonym for **bondline.** **D 907,** D14

GLULAM, *n*—Synonym for **structural-glued-laminated timber.** **D 907,** D14

glulam rivet nail—plain or galvanized, flat, hardened-steel, 2⅜ by ¼-in. nail with sheared V-shaped point and flat upset wedge-shaped head; designed to be driven through undersize truncated apertures in regular-stock-steel connector plates from which nails cantilever into wood. **F 547,** F16

glycol engine coolant concentrate, *n*— an engine coolant concentrate in which the freeze point depressant is ethylene or propylene glycol, with inhibitors to minimize foaming and corrosion. **D 4725,** D15

gmax—the maximum value of acceleration experienced during impact expressed in units of g's. **F 869,** F08

gnomonic projection—a projection in which the orientation of a crystal plane at the center of the unit sphere is represented by the point where the plane normal intersects the plane of projection which is tangent to the unit sphere at the zenith. **E 7,** E04

gob—(*1*) a portion of hot glass delivered by a feeder.

(*2*) a portion of hot glass gathered on a punty or pipe. **C 162,** C14

gob process—a process whereby glass is delivered to a forming unit in "gob" form. **C 162,** C14

gold decoration—*XPS,* a method whereby a very thin coat of evaporated gold on an insulator is used as a charge reference; the gold should be deposited as unconnected islands covering the area analyzed. **E 673,** E42

gonioapparent, *adj*—pertaining to change in apperance with change in illumination angle or viewing angle. **E 284,** E12

gonioappearance, *n*—the phenomenon in which the appearance of a specimen changes with change in illumination or viewing angle. **E 284,** E12

goniochromatic match, *n*—match between a pair of specimens that holds under all sets of angular illuminating-viewing conditions. **E 284,** E12

goniochromatism, *n*—change in any or all attributes of color of a specimen on change in angular illuminating-viewing conditions but without change in light source or observer. **E 284,** E12

goniometer—an instrument devised for measuring the angle through which a specimen is rotated. **E 7,** E04

goniometer, *n*—*in X-ray spectrometry,* a device used to adjust the angular relationships among a sample, crystal, and detector in an X-ray spectrometer. **E 135,** E01

goniometer, *n*—an instrument for measuring or setting angles. **E 284,** E12

goniometer—an instrument to measure angles. **F 1241,** F01

goniophotometer, *n*—instrument that measures flux as a function of angles of illumination or observation. **E 284,** E12

goniophotometer, *n*—photometer for measuring the directional light distribution characteristics of sources, lighting fittings, media, and surfaces.

NOTE—A goniophotometer for measuring the spatial distribution of luminous intensity is also called a distribution photometer. **E 349,** E21

goniospectrophotometer, *n*—spectrophotometer having the capability of measuring with a variety of illuminating and viewing angles using bidirectional geometry; also known as multi-angle spectrophotometer. **E 284,** E12

good engineering practices—include design practices and criteria accepted in professional societies (ASTM, AIChE, ASME, ACS, etc.), proved by experience, verified by actual data, etc., that will meet the process, safety, and environmental requirements of the system. **E 1705,** E48

good laboratory practices (GLP), *n*—guidelines for the management of laboratory experiments which are published by regulatory agencies or other recognized groups, and are concerned with the organizational process and the conditions under which laboratory studies are planned, performed, monitored, recorded, and reported. **D 6384,** D02

gouge—a surface defect in which material has been removed (scooped out by a sharp instrument) that causes a decrease in strength in a highly stressed area. **E 1749,** E06

gouge, *n*—a groove or cavity in the flooring surface accompanied by material removal and penetration below the immediate flooring surface. **F 141,** F06

gouges—see **chips**. **D 3918,** D20

gout, *n*—foreign matter trapped in a fabric by accident, usually lint or waste. (See also **slug.**) **D 123,** D13

GPD—unit of flow rate; gallons per day. See **GD**. **D 6161,** D19

grab bar—See **railing systems.** **E 631,** E06

grab bar—Synonym for **grab rail**. **E 1481,** E06

grab rail—See **railing systems.** **E 631,** E06

grab rail—a short length of rail located for safety or convenience to assist a person in movement at a specific location. (Syn. *grab bar.*) **E 1481,** E06

grab test, *n—in fabric testing,* a tensile test in which the central part of the width of the specimen is gripped in the clamps. **D 123,** D13

grab test, *n—in fabric testing,* a tension test in which only a part of the width of the specimen is gripped in the clamps. **D 4439,** D35

gradation—the distribution of particles of granular material among standard sizes usually expressed in terms of cumulative percentages larger or smaller than each of a series of sieve openings. **C 822,** C13

gradation, *n*—the particle size distribution of aggregate as determined by separation with standard screens. Gradation of aggregate is expressed in terms of the individual percentages passing standard screens. Sieve analysis and screen analysis are synonyms when referring to gradation of aggregate. **C 1180,** C12

gradation (grain-size distribution) (texture)—the proportions by mass of a soil or fragmented rock distributed in specified particle-size ranges. **D 653,** D18

grade, *n—in asbestos,* asbestos fiber that has the same physical, chemical, and mechanical properties and which is designated by a particular code corresponding to any given specifications. **C 460,** C17

grade—the designation given a material by a manufacturer such that it is always reproduced to the same specifications established by the manufacturer. **C 709,** D02

grade, *n*—subdivision of flat sheets based on minimum wet and equilibrium flexural strengths. **C 1154,** C17

grade—the designation of the quality of logs or of a manufactured piece of wood. **D 9,** D07

grade, *n—in warp knitting,* a term used to indicate the defect index evaluation of fabric determined by the number of defects per unit, for example per pound, per linear yard, or per square yard. **D 123,** D13

grade, *n—in wool and mohair,* a numerical designation used in classifying wool and mohair in their raw, semi-processed, and processed forms based on average fiber diameter and variation of fiber diameter. **D 123,** D13

grade, *v*—to assign a numerical value based on number, size, and severity of defects seen during a visual inspection. **D 123,** D13

grade, *n—in asbestos,* asbestos fiber that has the same chemical, physical, and mechanical properties and which is designated by a particular code corresponding to any given specifications. **D 2946,** C17

grade, *n*—a level or elevation of a land or water surface.

average grade—the arithmetic mean of the elevations of various ground surfaces within a stated area of **building construction.**

finished grade—the surface elevation of lawns, walks, drives, or other improved surfaces after completion of construction or grading operations.

natural grade—the elevation of the original or undisturbed surface of the ground.

sub-grade—the ground elevation established to receive an additional surfacing. **E 631,** E06

graded sediment—in geology, a sediment consisting chiefly of grains of the same size range. In engineering, a sediment having a uniform or equable distribution of particles from coarse to fine. **D 4410,** D19

graded stream—a stream in which a steady state has been reached such that over a period of time the discharge and sediment load entering the system are balanced by the discharge and sediment load leaving the system. **D 4410,** D19

grade/recovery, *n*—the relationship between quality and quantity of the clean coal product. The quality can be defined in terms of ash, sulfur, or Btu content. The quantity can be designated as yield or heating value recovery (Btu or combustibles). **D 121,** D05

grade rings—precast concrete rings used for vertical adjustment at the top of a manhole to set manhole casting to proper grade. **C 822,** C13

gradient furnace—a furnace within which a known temperature gradient is maintained between the two ends. Sometimes known as a Rosenhain Furnace. **E 7,** E04

gradient ratio, *n—in geotextiles,* the ratio of the hydraulic gradient through a soil-geotextile system to the hydraulic gradient through the soil alone. **D 4439,** D35

grading, *n*—the procedure used to identify and quantify the number of imperfections in a roll of fabric detected during visual inspection. **D 123,** D13

grading—the degree of mixing of size classes in sedimentary material. **D 4410,** D19

graduated glassware—glassware that is marked with one or more graduations for volumetric measuring purposes. **C 162,** C14

graft copolymer—a copolymer in which polymeric side chains have been attached to the main chain of a polymer of different structure. **D 883,** D20

graft copolymer—a copolymer in which polymeric side chains have been attached to the main chain of a polymer of different structure. (D-20) **F 412,** F17

grain—(*1*) a distinguishable rock constituent which itself has a distinct identity, for example, a mineral crystal, an oolith, a rock fragment (in sedimentary rocks), or clast.

(*2*) a direction in a rock body along which it is more easily broken, split, or cut. See rift. **C 119,** C18

grain, *n—in manufactured carbon and graphite product technology,* a region in a carbon or graphite body that is identifiable as being derived from a particle of filler. **C 709,** D02

grain—the direction, size, arrangement, appearance, or quality of the fibers in lumber or other wood products. To have a specific meaning the term must be qualified.

bastard sawn grain—grain pattern in hardwood lumber in which the annual rings make angles of 30 to 60 deg with the surface of the piece.

chipped grain—a machine defect of surfaced lumber, where the grain of the wood has been torn out in small particles by the action of the planer knives.

close grain—(*1*) narrow, inconspicuous annual rings. The term is sometimes used to designate wood having small and closely spaced pores, but in this sense the term "fine textured" is more often used.

(*2*) in stress grading, wood averaging on one end or the other of each piece not less than six nor more than 30 annual rings/in. Pieces averaging at least five or more than 30 rings/in. are accepted as close grain if containing one third or more summerwood.

coarse grain—wide conspicuous annual rings in which there is considerable difference between earlywood and latewood. The term is sometimes used to designate wood with large pores, such as oak, ash, chestnut, and walnut, but in this sense the term "coarse textured" is more often used.

cross grain—fiber deviation from a line parallel to the sides of the piece. Cross grain may be either diagonal or spiral grain or a combination of the two.

curly grain—grain distortion with an irregular curled appearance; "birdseye" is an extreme case of curly grain.

diagonal grain—grain pattern in which the annual rings are at an angle with the axis of a piece as a result of sawing at an angle with the bark of the tree or log. A form of *cross grain.*

edge grain—grain pattern in which the wide surfaces of the sawn piece extend approximately at right angles to the annual growth rings. Lumber is considered edge grained when the rings form an angle of 45 to 90 deg with the wide surface of the piece.

end grain—the grain pattern exposed when ends of logs or timbers, dimension, boards, and other wood products are cut perpendicular to the fiber direction.

fine grain—a synonym for *close grain.*

flat grain—the grain pattern resulting when lumber has been sawed in a plane approximately perpendicular to the radius of the log. Lumber is considered flat grained when the annual growth rings make an angle of less than 45 deg with the surface of the piece.

interlocked grain—wood in which the fibers are inclined in one direction in a number of rings of annual growth, then gradually reverse and are inclined in an opposite direction in succeeding growth rings, then reverse again.

loosened grain—a separation or loosening of the earlywood from the latewood due to defects in the wood or processing such as planing.

medium grain—used in stress grading to denote wood averaging on one end or the other of each piece not less than four annual rings/in.

mixed grain—lumber and other wood products unrestricted or unsegregated as to the grain angle.

open grain—(*1*) common classification related to finishing of woods with large pores, such as oak, ash, and chestnut. Also known as "coarse textured."

(*2*) used in stress grading to denote no limitations on rate of growth as measured by annual rings per inch.

plainsawn—a synonym for *flat grain.*

quartersawn—a synonym for *edge grain.*

raised grain—a condition of the surface of dressed lumber in which the hard latewood is raised above the softer earlywood but not torn loose from it.

rift sawn—a synonym for *edge grain.*

slash grain—a synonym for *flat grain.*

spiral grain—wood in which the fibers take a spiral course about the trunk of a tree instead of the normal vertical course. The spiral may extend in a right-handed or left-handed direction around the tree trunk. Spiral grain is a form of cross grain.

straight-grained wood—wood in which the fibers run parallel to the axis of a piece.

torn grain—a machine defect of surfaced lumber, where the fibers of the wood have been torn out around knots and curly places by the action of the planer knives.

vertical grain—a synonym for *edge grain.*

wavy-grained wood—wood in which the fibers form a pattern of fairly uniform waves or undulations. **D 9,** D07

grain, *n*—an inch-pound unit of weight, equal to 0.002285 avoirdupois oz (0.0648 g). **D 16,** D01

grain, *n*—*in yarn spinning,* a direct yarn numbering system for sliver, top or roving, equal to the mass in grains of 120 yds. (See **American grain count**) **D 123,** D13

grain, *n*—*in measuring mass,* 1/7000 lb avoirdupois. **D 123,** D13

grain—the outer or hair side of a hide or skin. Also used as an adjective referring to that side. **D 1517,** D31

grain, *n*—anisotropy introduced into rubber during processing operations. **D 1566,** D11

grain, *n*—the machine direction of paper. **D 1968,** D06

Grain—unit of weight, 0.648 g, 0.000143 lb. **D 6161,** D19

grain—an individual crystallite in metals. **E 7,** E04

grain boundary—an interface separating two grains, where the orientation of the lattice changes from that of one grain to that of the other. When the orientation change is very small the boundary is sometimes referred to as a subboundary. **E 7,** E04

grain boundary—the surface that may be closed over which a crystal is in contact with other crystals in a solid body. **F 1241,** F01

grain dropping—the dislodgement and loss of a grain or grains (crystals) from a metal surface as a result of intergranular corrosion. **G 15,** G01

grained leather—any leather on which the original natural grain, through any method, process, or manipulation, has been changed or altered. **D 1517,** D31

grain growth, *n*—an increase in the grain size of a steel object, usually as a result of exposure to elevated temperatures. **A 941,** A01

grain growth—an increase in the grain size of a metal usually as a result of heating at an elevated temperature. **E 7,** E04

graininess—the visual impression of irregularity of silver deposit in a processed film. **E 1316,** E07

graining—a process for producing a decorative finish by transferring a pattern to the porcelain enamel surface by means of rolls. **C 286,** B08

graining paste—a mixture of color oxides, fluxes, and oils. **C 286,** B08

graining roll—a specialized type of roll used for transferring the grain pattern to the porcelain enamel. **C 286,** B08

grain long—paper grain direction in sheets of paper is parallel to the long dimension of the sheet. **F 149,** F05

grain long paper, *n*—paper in which the machine direction parallels the longest sheet dimension. **D 1968,** D06

grain magnesite, n—see **magnesite, grain.** **C 71,** C08

grain pattern, *n*—Synonyn for *spangle.* **A 902,** A05

grain short—paper grain direction in sheets of paper is parallel to the short dimension of the sheet. **F 149,** F05

grain short paper, *n*—paper in which the machine direction parallels the shortest sheet dimension. **D 1968,** D06

grain size, *n*—the dimensions of the grains or crystals in a polycrystalline metal, exclusive of twinned regions and subgrains when present. **A 941,** A01

grain size—the dimensions of the grains or crystals in a polycrystalline metal exclusive of twinned regions and subgrains when present. Grain size is usually estimated or measured on the cross section of an aggregate of grains. Common units are: (*1*) average diameter, (*2*) average area, (*3*) number of grains per linear unit, (*4*) number of grains per unit area, and (*5*) number of grains per unit volume. See Test Methods E 112.

(*1*) *ASTM grain size number*—a grain size designation bearing a relationship to average intercept distance at 100 diameters magnification according to the equation: G = ASTM grain size number = $10.0 - 2 \log_2 \bar{L}$, where $\bar{L}$ is the average intercept distance in millimetres at 100 magnification.

(*2*) *average grain diameter*—the mean diameter of an equiaxed grain section whose size is representative of all the grain sections in the aggregate being measured. **E 7,** E04

grain-size analysis (mechanical analysis) (particle-size analysis)—the process of determining grain-size distribution. **D 653,** D18

grain size comparison eyepiece—an eyepiece provided with calibrated patterns representing a series of standard sizes of grains. The eyepiece must be used at a total magnification for which the patterns have been calibrated. **E 7,** E04

grains per U.S. gallon (GPG)—number of grains of substance per one U.S. gallon of water. Concentration in GPG is calculated by dividing concentration in ppm of the ion by 17.1. One grain weighs 1/7000 lb and one U.S. gallon weighs 8.3 lb. **D 6161,** D19

granite (commercial definition)—a visibly granular, igneous rock generally ranging in color from pink to light or dark gray and consisting mostly of quartz and feldspars, accompanied by one or more dark minerals. The texture is typically homogeneous but may be gneissic or porphyritic. Some dark granular igneous rocks, though not geologically granite, are included in the definition. **C 119,** C18

graniteware—a one-coat porcelain enameled article with a mottled pattern produced by controlled corrosion of the metal base prior to firing. **C 286,** B08

Gran's plot paper—a type of graph paper designed to eliminate calculations when measurements are made by the Gran's plot technique. Electrode potentials in millivolts are plotted on the vertical antilogarithmic axis and either volume of reagent added or sample concentration on the horizontal linear axis. Special Gran's plot papers can be constructed that automatically correct for volume changes occurring during titrations. **D 4127,** D19

Gran's plots—a method of plotting apparent concentration (as derived from the electrode potential) versus the volume of reagent added to the sample. Gran's plots are especially useful for plotting titrations that would give poor end-point breaks if plotted conventionally. They can also be used to determine concentration by known addition with greater precision than can be obtained by a single addition measurement. **D 4127,** D19

granular—composed of particles visible to the unaided eye. For sedimentary stone, the predominant particle distribution is less than 4 mm. in size. **C 119,** C18

granular activated carbon—activated carbon in particle sizes predominantly greater than 80 mesh. **D 2652,** D28

granular applicator—an apparatus consisting of a hopper, a metering device, and a device for spreading or placing the granules in the target area. **E 1102,** E35

granularity—presence of voids or discontinuities within the image giving the image a grainy appearance. **F 549,** F05

granular material—synonym for **bulk solid.** **D 653,** D18

granular powder—particles having approximately equidimensional nonspherical shapes. **B 243,** B09

granulated blast-furnace slag, *n*—the glassy, granular material formed when molten blast-furnace slag is rapidly chilled, as by immersion in water. **C 125,** C09

granulated blast-furnace slag, *n*—the glassy granular material formed when molten blast-furnace slag is rapidly chilled, as by immersion in water. **C 219,** C01

granulated rubber, *n*—particulate rubber composed of mainly nonspherical particles that span a broad range of maximum "particle dimension" from below 425 μm (40 mesh) to 12 mm (0.47 in.); the key feature of this type of particulate rubber is the fraction of the material in the greater than 1 mm (0.08 in.) up to 12 mm (0.47 in.) maximum "particle dimension" range. See **particulate rubber**. **D 1566,** D11

granulation—the production of coarse metal particles by pouring the molten metal through a screen into water (shotting), or by violent agitation of the molten metal while solidifying. **B 243,** B09

granule—See **mineral granules.** **D 1079,** D08

granules—small ceramic or natural colored mineral pellets or grains applied to asbestos-cement to lend color to the surface. **C 460,** C17

granules, *n*—small ceramic or natural colored mineral pellets or grains applied to products to lend color to the surface. **C 1154,** C17

granules—small ceramic or natural colored mineral pellets or grains applied to asbestos-cement to lend color to the surface. **D 2946,** C17

graphic escape, *n*—an electronic signal sent to the printer to take it out of the graphic mode. (See **GE**.) **F 1457,** F05

graphic protection—procedure for the identification of counterfeit graphic images; based upon the inclusion of deliberate errors (for example, broken fonts, variable line weight, line imperfections, etc.) or special graphic devices (for example, rainbow printing) into the original design of the object. **F 1156,** F12

graphite—an allotropic crystalline form of the element carbon, occurring as a mineral, commonly consisting of a hexagonal array of carbon atoms (space group P 6_3/mmc) but also known in a rhombohedral form (space group R 3m). **C 709,** D02

graphite—*in carbon and graphite technology*, an artifact consisting predominantly of the element carbon and possessing extensive long range order. **C 709,** D02

graphite-base carbon refractory, n—see **carbon refractory, graphite-base.** **C 71,** C08

graphite, compacted, *n*—a graphite shape that is intermediate between flake graphite and nodular graphite that typically appears in a polished section as thick flakes with blunt (compacted) ends. **A 644,** A04

graphite flake, *n*—an irregularly shaped particle of graphite, usually appearing in a polished section as curved plates, such as found in gray cast irons. **A 644,** A04

graphite, flake—an irregularly shaped body, usually appearing as long curved plates of graphitic carbon such as found in gray cast irons. **E 7,** E04

graphite, nodular, *n*—spheroidal shaped graphite typically found in ductile irons and compact clusters of graphite typically found in malleable irons. (See **graphite, spheroidal**, and **temper carbon**.) **A 644,** A04

graphite permanent mold casting, *n*—a metal object produced by introducing molten metal by gravity or low pressure into a graphite mold and allowing it to solidify. **B 899,** B02

graphite, primary, *n*—graphite precipitated in cast iron during solidification. **A 644,** A04

graphite rosette, *n*—arrangement of graphite flakes in which the flakes extend radially from centers of crystallization in gray cast iron. **A 644,** A04

graphite, spheroidal, *n*—spheroidal shaped graphite having a polycrystalline radial structure, usually found in ductile iron and to a controlled, limited extent in compacted graphite iron. **A 644,** A04

graphitic corrosion—the deterioration of metallic constituents in gray cast iron, which leaves the graphitic particles intact. (The term "graphitization" is commonly used to identify this form of corrosion but is not recommended because of its use in metallurgy for the decomposition of carbide to graphite.) **G 15,** G01

graphitization annealing, *n*—**annealing** a steel object in such a way that some or all of the carbon is precipitated as graphite. **A 941,** A01

graphitize, *vt*—to precipitate graphite in an iron-carbon alloy. **A 644,** A04

GRAS—materials "generally regarded as safe," as listed by the FDA. **D 6161,** D19

grassers—calf skins or kips that have a coarser grain due to poor feeding of the animals. **D 1517,** D31

graticule—a scale on glass or other transparent material placed in the eyepiece or at an intermediate plane on the optic axis of a light microscope for the location and measurement of objects (a graticule is different than a reticle, see **reticle**). **E 7,** E04

grating, concave, *n*—a diffraction grating on a concave mirror surface. **E 135,** E01

grating, diffraction, *n*—a series of a large number of narrow, close, equally spaced, diffracting slits or grooves capable of dispersing light into its spectrum. **E 135,** E01

grating frequency—the number of lines per unit length engraved into that grating; often expressed as lines per millimetre. **F 1156,** F12

grating, plane, *n*—a transmission or reflecting grating whose surface is flat. **E 135,** E01

grating, reflection, *n*—a diffraction grating from which the incident light is reflected to form a spectrum. **E 135,** E01

grating, transmission, *n*—a transparent diffraction grating through which light is transmitted. **E 135,** E01

gravel, *n*—coarse aggregate resulting from natural disintegration and abrasion of rock or processing of weakly bound conglomerate. **C 125,** C09

gravel—rounded or semirounded particles of rock that will pass a 3-in. (76.2-mm) and be retained on a No. 4 (4.75-μm) U.S. standard sieve. **D 653,** D18

gravel—coarse, granular aggregate, with pieces larger than sand grains, resulting from the natural erosion of rock. **D 1079,** D08

gravel in—to spread aggregate into hot bitumen on the surface of the roofing membrane. **D 1079,** D08

gravel pack—common nomenclature for the terminology, primary filter of a well (see *primary filter pack*). **D 653,** D18

gravel road, *n*—two lane, all-weather, occasionally maintained, hard or loose surface (for example, large rock, paved, crushed rock, gravel) intended for medium-weight, low-density traffic, in accordance with the U.S. Federal Highway Administration. **F 538,** F09

gravel size (fluvial sediment)—between 2.0 and 64 mm in diameter. **D 4410,** D19

gravel stop—a flanged device, frequently metallic, designed to prevent loose aggregate from washing off the roof and to provide a continuous finished edge for the roofing. **D 1079,** D08

gravitational water—see **free water**. **D 653,** D18

gravity filter—a filter through which water flows through it by gravity. **D 6161,** D19

gravity flow, *n*—flow in a direction parallel to the plane of a geotextile or related product driven predominately by a difference in elevation between the inlet and outflow points of a specimen. **D 4439,** D35

gravity grouting—grouting under no applied pressure other than the height of fluid in the hole. **D 653,** D18

gray area—{archaic} See **mist**. **C 162,** C14

graybody, *n*—a body having the same spectral emittance at all wavelengths. **C 168,** C16

gray body, *n*—nonselective radiator whose spectral emissivity is less than 1. **E 349,** E21

gray felt, *n*—a blend of white fibers with naturally colored or dyed fibers, *or both*, that has an overall gray appearance. **D 123,** D13

gray iron, *n*—cast iron that has a relatively large proportion of the graphitic carbon present in the form of flake graphite. The metal has a gray fracture. **A 644,** A04

gray level, in image analysis—a specific neutral color value existing within the range from black to white. **E 7,** E04

gray market—goods available through diversion, usually more cheaply and without service guarantee. **F 1156,** F12

gray scale, n—the scale consists of nine pairs of standard gray chips each pair representing a difference in color or contrast (shade and strength) corresponding to a numerical fastness rating. The results of colorfastness tests are rated by visually comparing the difference in color represented by the scale. **G 113,** G03

gray water—the waste water of a system that may be a combination of the liquid and water-carried wastes except human wastes. **F 412,** F17

grazing exit (incidence)—EIA, same as **glancing exit (incidence).** **E 673,** E42

grazing incidence—immersion inspection with the beam directed at a glancing angle to the test surface. **E 1316,** E07

grease extractors—a system of components designed for integration within the exhaust hood for the removal of the airborne grease particles and the condensate of grease vapors, or both, for immediate or future disposal. See **exhaust hood**. **F 1827,** F26

grease removal devices: **F 1827,** F26

grease-resistant barrier— See **barrier material.** **D 996,** D10

grease wool, *n*—wool taken from the living sheep and which has not been commercially scoured. **D 123,** D13

green—unsintered (not sintered); for example: green compact, green density, green strength. **B 243,** B09

green—(*1*) freshly sawed wood, or wood that has received no drying; unseasoned. Lumber that may have become wet to above the fiber saturation point may be referred to as being in the "green condition."

(*2*) wood above a stipulated moisture content, as lumber above 19 % moisture content in accordance with the American Softwood Lumber Standard. **D 9,** D07

green, *adj*—hue description applied to light of wavelengths from 495 to 550 nm and to visually similar stimuli. **D 2946,** C17

green carbon—a formed, but unfired carbon body. **C 709,** D02

green coke, *n*—*as used in Test Method D 5061*, carbonaceous binder or filler phase material that has exceeded the temperature of thermoplasticity, but has not obtained the temperature of metallurgical coke. **D 121,** D05

green density, *n*—the mass per unit volume of an unsintered compact.

B 243, B09

green expansion, *n*—the increase in dimensions of an ejected compact relative to the die dimensions, measured at right angles to the direction of pressing. **B 243,** B09

green salting—a process of curing hides in which they are treated with salt on the flesh side and stacked in piles to cure for a period of ten days or more. **D 1517,** D31

greensand—a mineral (glauconite), used as a filtration medium. See **manganese greensand.** **D 6161,** D19

greenstone—a metamorphic rock of basic or ultrabasic composition, of very fine grain size, ranging in color from medium green to yellowish green to almost black. **C 119,** C18

greige cord, *n*—*in tire cords,* a cord that has not been adhesive treated, or otherwise treated before use. (See **cord** **D 123,** D13

greige goods, *n*—textile fabrics that have received no bleaching, dyeing, or finishing treatment after being produced by any textile process. **D 123,** D13

greige thread, *n*—undyed or unfinished sewing thread in the state following final plying or equivalent step in processing sequence, such as extruding, or braiding. **D 123,** D13

greige yield, *n*—*in knitted fabrics,* the number of finished square yards per pound (square metres per kilogram) of greige fabric. **D 123,** D13

grex, *n*—an obsolete direct numbering system for fiber yarn or other textile strand equal to the mass in grams per 10 000 m. **D 123,** D13

grey, *adj*—color description applied to achromatic stimuli of moderate relative luminance. **D 2946,** C17

griddle—equipment for cooking food in its own juices or oil by direct contact with a hot surface. **F 1827,** F26

grin, *v*—*in sewn seams,* to stress a seam so that the individual stitches can be seen. **D 123,** D13

grindability:

absolute grindability index—a characteristic number expressed as the change in specific surface area of a material per unit of time in a specific comminution system.

relative grindability index—a characteristic number expressed as the change in specific surface area or other criteria such as particle size or fineness of a material per unit of time, mill revolutions or other standard with respect to a known standard sample in a specific comminution system. **C 242,** C21

grinding—the removal of metal by means of rotating rigid wheels containing abrasive. **B 374,** B08

grinding—the removal of material from the surface of a specimen by abrasion through the use of randomly oriented hard-abrasive particles bonded to a suitable substrate, such as paper or cloth, where the abrasive particle size is generally in the range of 60 to 600 grit (approximately 150 to 15 μm) but may be finer. **E 7,** E04

grinding japan—see **japan, grinding.** **D 16,** D01

grinding mark—a pattern of fine striations or scoring, usually directional, resulting from machining, as distinct from **surface marks**

F 109, C21

grip, *v*—*in tensile testing,* to hold, grasp, or secure, for example, to grip the specimen by the jaws of the clamps. **D 123,** D13

grip mark, gripe mark, gripper mark—indentations and ridges along shank, usually near head, made by gripping devices that hold wire during heading; often used with other markings to classify nail manufacturer. **F 547,** F16

grit—the coarse foreign particles in paint materials and coatings, often of irregular shape, that are hard, abrasive, and resistant to disintegration. **D 16,** D01

grit blasting—abrasive blasting with small irregular pieces of steel or malleable cast iron. **B 374,** B08

grog, *n*—a granular material produced from calcined or burned refractories, usually alumina-silica. **C 71,** C08

grog fireclay mortar, *n*—raw fireclay mixed with calcined fireclay, or with broken fireclay brick, or both, all ground to suitable fineness. **C 71,** C08

groin—bank or shore-protection structure in the form of a barrier placed oblique to the primary motion of water, designed to control movement of bed load. **D 653,** D18

grooming, *v*—*in tire testing,* mechanically reworking a snow test surface in order to obtain a surface with more consistent properties. **F 538,** F09

groove—see **female end of pipe.** **C 822,** C13

groove—a decorative face treatment, consisting of narrow parallel channels formed into the surface of the panel; such as machined.

V-groove—narrow and shallow V- or U-shaped channels machined on the plywood face to achieve a decorative effect. **D 1038,** D07

groove, *n*—a void that is relatively narrow compared to its length. **F 538,** F09

groove—*in a semiconductor wafer,* a shallow scratch with rounded edges, that is usually the remnant of a scratch not completely removed by polishing. **F 1241,** F01

groove angle—the angle of support for a pipe when a formed groove is made in bedding or foundation. **F 412,** F17

groove, average depth, [*L*], *n*—the average of all tire groove depth measurements in a single groove. **F 538,** F09

grooved—general term sometimes used to denote threaded, fluted, twisted, knurled, barbed, etc. **F 547,** F16

grooved griddle—equipment with a grooved cooking surface that imitates characteristic striped sear mark of a charbroiler. See **griddle.** **F 1827,** F26

grooved guide body—formed guiding device used for placing staple over work as in carding or wiring. **F 592,** F16

groove (void) area fraction, [*nd*], *n*—the ratio of the groove (void) area to the footprint area of a tire. **F 538,** F09

groove (void) area, [L^2], *n*—that portion of tire footprint area which is not contacted by ribs or elements. **F 538,** F09

groove (void) depth, [*L*], *n*—a measurement of the perpendicular distance from a real or calculated reference plane defined by edges of two adjacent ribs (lugs) to the lowest point of contact in the groove (void). **F 538,** F09

grooving—long narrow grooves or depressions in a surface of a pultrusion parallel to its length.

NOTE—This condition is usually caused by die fouling or by a spot of resin build-up on the die surface, effectively changing the shape of the cross section. **D 3918,** D20

gross buoyancy—weight of fresh water displaced by a boom totally submerged. **F 818,** F20

gross buoyancy to weight ratio—gross buoyancy divided by boom weight. **F 818,** F20

gross calorific value—the energy released by combustion of a unit quantity of refuse-derived fuel at constant volume or constant pressure in a suitable calorimeter under specified conditions such that all water in the products is in the liquid form. This is the measure of calorific value predominantly used in the United States. **Synonym:** higher heating value. **E 856,** D34

gross calorific value (gross heat of combustion at constant volume), $Q_v(gross)$ **,** *n*—the heat produced by combustion of unit quantity of a solid or liquid fuel when burned at constant volume in an oxygen bomb calorimeter under specified conditions, with the resulting water condensed to a liquid. **D 121,** D05

gross calorific value, (gross heat of combustion), Q_v **(gross),** *n*—the heat produced by combustion of unit quantity of a solid or liquid specimen when burned at constant volume in an oxygen bomb calorimeter under specified conditions with the resulting water condensed to a liquid. **D 5681,** D34

gross collector area— see **area, gross collector.** **E 772,** E44

gross cross-sectional connector plate area—cross-sectional area of metal connector plate determined by multiplying gross thickness of plate by gross dimension of plate perpendicular to direction of load application. **E 1807,** E06

gross erosion—the total of all sheet, gully, and channel erosion in a watershed. **D 4410,** D19

gross floor area—See **floor area.** **E 1480,** E06

gross heat of combustion, *n*—the maximum amount of heat per unit mass that theoretically can be released by the combustion of a material, product, or assembly; it can be determined experimentally only under conditions of high pressure and in pure oxygen (See **effective heat of combustion**). **E 176,** E05

gross sample—See *gross sample* under **sample.** **D 121,** D05

gross sample—a sample representing one lot, normally composed of a number of increments, on which neither reduction nor division has been performed. **E 856,** D34

gross sample *n*—a sample representing one lot, normally composed of a number of increments, on which neither reduction nor division has been preformed. **D 5681,** D34

gross-vehicle weight—the total mass of the vehicle or the vehicle combination including all connected components. **E 867,** E17

ground—a metallic connection with the earth to establish zero potential. **D 5077,** D10

ground arch—the theoretical stable rock arch that develops some distance back from the surface of the opening and supports the opening. (ISRM) **D 653,** D18

ground coat—*(1)* a porcelain enamel applied directly to the base metal to function as an intermediate layer between the metal and the cover coat.

(*2*) *on sheet steel,* a porcelain enamel coating containing adherence-promoting agents which may be used either as an intermediate layer between the metal and the cover coat or as a single coat over the base metal. **C 286,** B08

ground-coat boiling— see **boiling.** **C 286,** B08

ground fireclay, *n*—fireclay or a mixture of fireclays that have been subjected to no treatment other than grinding or weathering, or both. **C 71,** C08

ground fireclay mortar, *n*—a refractory mortar consisting of finely ground raw fireclay. **C 71,** C08

ground floor—See **floor.** (Synonym for **first floor, first story.**) **E 631,** E06

ground-glass focusing screen—a glass screen, one side of which is ground or made diffusing and mounted for use in a camera, in place of photosensitive material, for the purpose of intercepting, viewing, and focusing a real image formed on it. **E 7,** E04

grounding—connecting to ground or to a conductor that is grounded. **D 5077,** D10

ground reaction forces—the forces, both shearing and normal, acting on the foot during contact with the ground. **F 869,** F08

ground refractory material, double-screened, *n*—a refractory material that contains its original gradation of particle sizes resulting from crushing, grinding, or both, and from which particles coarser and finer than two specified sizes have been removed by screening. **C 71,** C08

ground refractory material, single-screened, *n*—a refractory material that contains its original gradation of particle sizes resulting from crushing, grinding, or both, and from which particles coarser than a specified size have been removed by screening. **C 71,** C08

ground rubber, *n*—particulate rubber composed of mainly non-spherical particles that span a range of maximum "particle dimension" from well below 425 μm (40 mesh) to 2 mm (0.08 in.) as a maximum "particle dimension." See **particulate rubber.** **D 1566,** D11

ground vulcanized rubber, *n*—vulcanized rubber in particulate form; used as an extender or filler. **D 1566,** D11

ground water—that part of the subsurface water that is in the saturated zone. **D 653,** D18

groundwater—water confined in permeable sand layers between rock or clay; that part of the subsurface water that is in the saturated zone. **D 6161,** D19

ground-water barrier—soil, rock, or artificial material which has a relatively low permeability and which occurs below the land surface where it impedes the movement of ground water and consequently causes a pronounced difference in the potentiometric level on opposite sides of the barrier. **D 653,** D18

ground-water basin—a ground-water system that has defined boundaries and may include more than one aquifer of permeable materials, which are capable of furnishing a significant water supply. **D 653,** D18

ground-water discharge—the water released from the zone of saturation; also the volume of water released. **D 653,** D18

ground-water divide—a ridge in the water table or other potentiometric surface from which ground water moves away in both directions normal to the ridge line. **D 653,** D18

ground-water elevation—see **free water elevation.** **D 653,** D18

ground-water flow—the movement of water in the zone of saturation. **D 653,** D18

ground-water level—the level or the water table surrounding a borehole or well. The ground-water level can be represented as an elevation or as a depth below the ground surface. **D 653,** D18

ground-water, perched—see **perched ground-water.** **D 653,** D18

ground-water recharge—the process of water addition to the saturated zone; also the volume of water added by this process. **D 653,** D18

ground-water surface—see **free water elevation.** **D 653,** D18

groundwood pulp, *n*—a type of mechanical pulp produced by grinding wood logs against a rotating stone. **D 1968,** D06

group, *n*—*in upholstered furniture,* a number of individual upholstered furniture units that are related by one or more physical characteristics such as styling, color, shape or covering. **D 123,** D13

group—*in fatigue,* specimens of the same type tested at a specific time, or consecutively, at one stress level. A group may comprise one or more specimens. **E 1823,** E08

group AQL, *n*—the AQL assigned to a group of material properties. **D 1711,** D09

group delta, *n*—the maximum allowable difference between axle spacings within an axle group and the average axle spacing within the group. **E 867,** E17

group erase—an **OCR** graphic shape that will delete a group or string of three or more characters. **F 149,** F05

group limit, *n*—the maximum axle spacing permitted for axles within the axle group. **E 867,** E17

group spacing, *n*—for each axle group, the horizontal distance between the midpoint of that axle group and the preceding axle group; the group spacing of the vehicle's front axle group is assumed to be zero. **E 867,** E17

grout, *n*—gypsum or portland cement plaster used to fill crevices or to fill hollow metal frames. **C 11,** C11

grout, *n*—a cementitious mixture, with or without admixtures, that is used primarily to fill voids. **C 125,** C09

grout—concrete containing no coarse aggregates; a thin mortar. **C 717,** C24

grout, *n*—a mixture of cementitious materials, aggregates, water, with or without admixtures, initially produced to pouring consistency without segregation. Requirements for grout are contained in Specification C 476. **C 1180,** C12

grout—*in soil and rock grouting*, a material injected into a soil or rock formation to change the physical characteristics of the formation. **D 653,** D18

groutability—the ability of a formation to accept grout. **D 653,** D18

groutability ratio of granular formations—the ratio of the 15 % size of the formation particles to be grouted to the 85 % size of grout particles (suspension-type grout). This ratio should be greater than 24 if the grout is to successfully penetrate the formation. **D 653,** D18

groutable rock bolts—rock bolts with hollow cores or with tubes adapted to the periphery of the bolts and extending to the bottom of the bolts to facilitate filling the holes surrounding the bolts with grout. **D 653,** D18

grout cap—a "cap" that is formed by placing concrete along the top of a grout curtain. A grout cap is often used in weak foundation rock to secure grout nipples, control leakage, and to form an impermeable barrier at the top of a grout curtain. **D 653,** D18

grouted-aggregate concrete—concrete that is formed by injecting grout into previously placed coarse aggregate. See also **preplaced aggregate concrete**. **D 653,** D18

grout gallery—an opening or passageway within a dam utilized for grouting or drainage operations, or both. **D 653,** D18

grout header—a pipe assembly attached to a ground hole, and to which the grout lines are attached for injecting grout. Grout injector is monitored and controlled by means of valves and a pressure gate mounted on the header; sometimes called grout manifold. **D 653,** D18

grout, hydraulic-cement, *n*—a grout made with hydraulic cement. **C 125,** C09

grout mix—the proportions or amounts of the various materials used in the grout, expressed by weight or volume. (The words "by volume" or "by weight" should be used to specify the mix.) **D 653,** D18

grout (monitoring wells)—a low permeability material placed in the annulus between the well casing or riser pipe and the borehole wall (that is, in a single-cased monitoring well), or between the riser and casing (that is, in a multicased monitoring well), to maintain the alignment of the casing and riser and to prevent movement of ground water or surface water within the annular space. **D 653,** D18

grout nipple—*in grouting*, a short length of pipe, installed at the collar of the grout hole, through which drilling is done and to which the grout header is attached for the purpose of injecting grout. **D 653,** D18

grout (nonshrink), hydraulic-cement, *n*—a hydraulic-cement grout that produces a volume that, when hardened under stipulated test conditions, is greater than or equal to the original installed volume, often used as a transfer medium between load-bearing members. **C 125,** C09

grout pipe—a pipe or tube that is used to transport cement, bentonite, or other plugging materials from the ground surface to a specified depth in a well or borehole. The material may be allowed to flow freely or it may be injected under pressure. The term tremie pipe is frequently used interchangeably. **D 653,** D18

grout shoe—a plug fabricated of relatively inert materials that is positioned with the lowermost section of a permanent casing and fitted with a passageway, often with a flow check device, through which grout is injected under pressure to fill the annular space. After the grout has set, the grout shoe is usually drilled out. **D 653,** D18

grout slope—the natural slope of grout injected into preplaced-aggregate or other porous mass. **D 653,** D18

grout system—formulation of different materials used to form a grout. **D 653,** D18

grout take—the measured quantity of grout injected into a unit volume of formation, or a unit length of grout hole. **D 653,** D18

growth—an increase in dimensions of a compact which may occur during sintering. (Converse of shrinkage.) **B 243,** B09

growth, *n*—an increase in one or more dimensions of an object or a material. **D 123,** D13

growth, *n*—the increase in length of a specimen caused by the application of a continuing load or force under specified conditions. (See also **delayed deformation**.) **D 123,** D13

growth, *n*—*of a fabric*, the difference between the original length of a specimen and its length after the application of a specified load for a prescribed time, and the subsequent removal of the load. (See also **permanent deformation.**) **D 123,** D13

grub hole—a hole through the hide caused by the penetration of the warble fly. **D 1517,** D31

grunerite asbestos—a member of the mineral group of monoclinic amphiboles derived from grunerite or cummingtonite-grunerite. Its representative chemical formula is approximately $(Mg,Fe)_6(Si_8O_{22})(OH)_2$. Its Chemical Abstracts number is 12172-73-5 (see Practice E 849). **C 460,** C17

grunerite asbestos, *n*—a member of the mineral group of monoclinic amphiboles derived from grunerite or cummingtonite-grunerite. Its representative chemical formula is approximately $(Mg,Fe)_6(Si_8{\cdot}O_{22})(OH)_2$. Its Chemical Abstracts number is 12172-73-5. **D 2946,** C17

guard electrode, *n*—one or more electrically conducting elements, arranged and connected in an electric instrument or measuring circuit so as to divert unwanted conduction or displacement currents from, or confine wanted currents to, the measurement device. **D 1711,** D09

guard electrode—one or more electrically conducting elements, arranged and connected in an electrical instrument or measuring circuit so as to divert unwanted conduction or displacement currents from, or confine wanted currents to, the measuring device. **D 2864,** D27

guard frame—in video-based automatic image analysis, an internal border, smaller than the monitor image frame, used to restrict the measurement area and thus eliminate errors in sizing features that interesect the measurement area border when used in conjunction with specific feature selection rules (also called the active frame). **E 7,** E04

guardrail—See **railing systems.** **E 631,** E06

guardrail system—a railing system, providing protection for building users against accidental fall and injury, located at or near the outer edge of a stair, ramp, landing, platform, deck, balcony, hatchway, manhole, floor opening, porch, or accessible roof; at the perimeter of an opening or accessible surface, such as the opening of a stair; or at a location at which an operating condition requires access limitation to a designated area. (Compare **railing system**.) **E 1481,** E06

guard region—in video based automatic image analysis, that portion of the imaged area between the guard (active) frame and the image frame which is employed in a variety of ways to eliminate sizing errors of features that intersect the guard frame. **E 7,** E04

guayule rubber, *n*—naturally occurring *cis* polyisoprene obtained from the shrub, *Parthenium Argentatum.* **D 1566,** D11

gudgeon—bright, regular-stock-steel nails with annular flange or collar located along shank. (See **tile nail, acoustical.**) **F 547,** F16

guide body—tool component that aligns and supports staple during driving. **F 592,** F16

guide bushing, *n*—the steel bushing that surrounds, aligns, and holds the stationary intermediate weight in place. **E 1445,** E27

guided bend, *n*—the bend obtained by using a mandrel to guide and force the portion of the specimen being bent between two faces of a die. **E 6,** E28

guide for rating, *n* (guide d'évaluation)—a document which explains how to rate the serviceability of an existing or planned facility for a specific purpose. A guide identifies typical requirements and provides a rating scale for comparison with the relevant combinations of features present in the facility. **E 1480,** E06

guideline, *n*—a written statement or outline of policy, practice, or conduct. **E 631,** E06

guide pin—a pin affixed to the inside of one of the forceps halves that aligns with a hole on the other tweezer half without protruding through when closed. **F 1638,** F44

guide pin hole—the hole in one forceps half into which the guide pin fits without passing through when closed. **F 1638,** F44

guide rollers—those rollers whose function is to control and transport the paper or substrate through the press. **F 425,** F05

guide system, *n*—the rails, wires, and shaft that guide the drop weight during its fall. **E 1445,** E27

guilloche pattern—pattern formed of two or more bands interlaced in such a way as to repeat a (curved, flowing) design; the design can be non-symmetrical. Often used with documents of value to increase the difficulty of engraving counterfeit printing plates; modern photoreproduction methods have lessened the effectiveness of the technique. **F 1156,** F12

Guinier-Preston streak—an elongated spot on the Laue pattern from a single crystal which first appears during age hardening. **E 7,** E04

Guinier-Preston zones—those regions of a crystal which give rise to Guinier-Preston streaks. **E 7,** E04

gum, *n*—any of a class of colloidal substances, exuded by or prepared from plants, sticky when moist, composed of complex carbohydrates and organic acids, which are soluble or swell in water. (See also **adhesive, glue,** *n*, and **resin**.) **D 907,** D14

gum compound, *n*—a rubber compound containing only those ingredients necessary for vulcanization and small amounts of other ingredients for processing, coloring, and for improving the resistance to ageing. **D 1566,** D11

gummed—See **coated.** **F 547,** F16

gummed paper tape— See **tape.** **D 996,** D10

gumming up—applying a solution of gum arabic, or its equivalent, to the metal master to prevent oxidation and to protect it from damage during washout, make-ready operations, or plate storage. **F 425,** F05

gum thus, *n*—botanically, the oleoresin from trees of Boswellia species native to Arabia and Somaliland. **D 804,** D01

gun-grade, *n*—*in building construction*, a material of a suitable viscosity to extrude through the nozzle of a hand- or power-operated device. **C 717,** C24

gunning, *v*—an application technique that uses a pneumatic means to transport a refractory material and place it onto a cold or hot surface. **C 71,** C08

gunning materials, refractory, *n*—mixtures of refractory aggregate and bond(s) specially prepared for gunning. **C 71,** C08

gusset, *n*—*(1)* a piece used to give additional size or strength in a particular location of an object.

(*2*) the folded-in portion of flattened tubular film. **D 883,** D20

gusset, *n*—a plate used to connect two or more members or to reinforce a joint. **E 631,** E06

gusset, *n*—*(1)* a piece used to give additional size or strength in a particular location of an object.

(*2*) the folded-in portion of flattened tubular film. (D-20) **F 412,** F17

gusset leather—a soft, flexible leather used for gussets in shoes, bags, and cases. **D 1517,** D31

gustation, *n*—the sense of taste. **E 253,** E18

gustiness, *adj*—now referred to as intensity of turbulence which is defined as the ratio of the root mean square of wind velocity fluctuations to the mean wind velocity. **D 1356,** D22

gutta-percha, *n*—hard thermoplastic substance, mainly *trans* polyisoprene obtained from trees of the *Sapotaceae* family. **D 1566,** D11

G value—see **radiation chemical yield.** **E 170,** E10

gypsum, *n*—the mineral consisting primarily of fully hydrated calcium sulfate, $CaSO_4 \cdot 2H_2O$ or calcium sulfate dihydrate. **C 11,** C11

gypsum, *n*—see **calcium sulfate**. **C 219,** C01

gypsum backing board, *n*—a ¼ in. to ⅝ in. gypsum board for use as a backing for gypsum wallboard, acoustical tile, or other dry cladding.

water resistant gypsum backing board—a gypsum board designed for use on walls primarily as a base for the application of ceramic, or plastic tile. **C 11,** C11

gypsum base for veneer plasters, *n*—a gypsum board used as the base for application of a gypsum veneer plaster. **C 11,** C11

gypsum board, *n*—the generic name for a family of sheet products consisting of a noncombustible core primarily of gypsum with paper surfacing. **C 11,** C11

gypsum concrete, *n*—a calcined gypsum mixed with wood chips, or aggregate, or both, used primarily for poured roof decks . **C 11,** C11

gypsum core board, *n*—a ¾ in. (19.0 mm) to 1 in. (25.4 mm) gypsum board consisting of a single board or factory laminated multiple boards, used as a gypsum stud or core in semisolid or solid gypsum board partitions. **C 11,** C11

gypsum-deck nail—helically threaded cap nails with special type of thread with extra large lead angle. **F 547**, F16

gypsum formboard, *n*—a gypsum panel product used as the permanent form for poured gypsum roof decks. **C 11**, C11

gypsum lath, *n*—a gypsum board used as the base for application of gypsum plaster.

perforated gypsum lath—a gypsum lath having perforations to provide mechanical keying of the basecoat plaster.

foil-backed gypsum lath—the same as plain gypsum lath except that in addition, the back surface shall be covered with a continuous sheet of pure bright finished aluminum foil.

type X lath—a gypsum lath specially manufactured to provide specific fire-resistant characteristics. **C 11**, C11

gypsum-lath nail—bright or blued, regular-stock-steel, 1 by 0.092 to 1¼ by 0.166 and 1 ¾ by 0.092-in. nails with large flat 19/64 to ⅜-in. head and long diamond point. Also, regular-stock-steel, 1 by 0.120 to 1 ½ by 0.148-in. nails with flat ½-in. head and medium diamond point. Also, aluminum-alloy, 1⅛ by 0.099 to 1 ½ by 0.105-in. nails with flat 19/64 or 5/14-in. head and medium diamond point. **F 547**, F16

gypsum molding plaster, *n*—a calcined gypsum plaster used primarily for plaster casts or molds, sometimes used as a gauging plaster. **C 11**, C11

gypsum neat plaster, *n*—a calcined gypsum plaster without aggregate; common usage is for gypsum plaster used for base coats **C 11**, C11

gypsum panel products, *n*—the general name for a family of sheet products consisting essentially of gypsum. **C 11**, C11

gypsum plaster, *n*—the generic name for a family of powdered cementitious products consisting primarily of calcined gypsum with additives to modify physical characteristics, and having the ability, when mixed with water, to produce a plastic mortar or slurry which can be formed to the desired shape by various methods and will subsequently set to a hard, rigid mass. **C 11**, C11

gypsum sheathing, *n*—a gypsum board used as a backing for exterior surface materials, manufactured with water-repellant paper and may be manufactured with a water-resistant core. **C 11**, C11

gypsum-sheathing nail—galvanized, barbed, regular-stock-steel, 1¾ by 0.120-in. nail with flat 7/16-in. head and medium diamond point. **F 547**, F16

gypsum tile or block, *n*—a cast gypsum building unit **C 11**, C11

gypsum wallboard, *n*—a gypsum board used primarily as an interior surfacing for building structures.

foil-backed gypsum wallboard—a gypsum wallboard with the back surface covered with a continuous sheet of pure bright finished aluminum foil.

type X gypsum wallboard—a gypsum wallboard specially manufactured to provide specific fire-resistant characteristics. **C 11**, C11

gypsum-wallboard nail, gypsumboard nail, drywall nail—bright or blued, regular-stock-steel, annularly threaded, 1 ⅛ by 0.098 to 2 by 0.105-in. nails with flat, nub, or crossed slightly countersunk ¼ to 19/64-in. head and long diamond point. Also, slender, colored (baked-lacquer finished), regular-stock-steel, smooth or annularly threaded, 1⅛ by 0.062 to 2 by 0.083-in. nails with slightly countersunk 0.181-in. head and medium diamond or long needle point. **F 547**, F16

gyromagnetic ratio, proton, γ_p—the ratio of the magnetic moment of a hydrogen nucleus to its angular momentum.

NOTE—The gyromagnetic ratio is used to calculate the magnetic field from a measured resonance frequency when using the nuclear magnetic resonance technique.
The relationship is:

$$B = (2\pi f/\gamma_p)\text{ gausses} = (2\pi f/\gamma_p) \times 10^{-4}\text{ teslas}$$

where:

f = resonance frequency in cycles per second (hertz) and
γ_p = gyromagnetic ratio (the accepted value at present for water is $2.675\,12 \times 10^4$ gauss^{-1} s^{-1}).

A 340, A06

gyro-stabilized accelerometer, *n*—a precision vertical gyroscope fitted with one to three accelerometers to provide orthogonal measurements referenced to the earth-fixed axis system. **F 538**, F09

H

HAA—A group of six halo acetic acids regulated in drinking water (mono, di and tri-chloroacetic acid, mono and di bromoacetic acid and chlorobromoacetic acid). **D 6161,** D19

habitable space—See **building space.** **E 631,** E06

habit plane—crystallographic plane in a parent phase along which a new phase (or phases) is (are) generated. **E 7,** E04

hackle—a finely structured fracture surface marking giving a matte or roughened appearance to the surface, having varying degrees of coarseness. Finely structured hackle is variously known as fine hackle, frosted area, gray area, matte, mist, and stippled area. Coarsely structured hackle is also known as striation. **C 162,** C14

hackle marks—fine ridges on the fracture surface of the glass, parallel to the direction of propagation of the fracture. **C 162,** C14

hair, *n*—natural animal fiber other than sheep's wool or silk. **D 123,** D13

hairline crack—*in protective coatings*, a very fine crack (having a hairlike appearance) that is visible on the surface of a dried coating film. **D 4538,** D33

hairline or hairlining—a defect manifested in finished ware as a line or system of lines in a strain pattern, having the appearance of cracks healed by fusion. **C 286,** B08

hair-on leather—leather tanned without removing the hair from the skins or hide. **D 1517,** D31

halation, *n*—*(1)* the apparent increase in size of a primary or secondary light source due to scattering of light toward the observer, the surround being significantly darker than the light source.

(2) in retroreflection, the apparent increase in size of a retroreflector, viewed by directional illumination (for example, automobile headlights) in a dark surround. **E 284,** E12

half, n—a climbing rope that must be used with another half rope with one or both ropes being attached to protection anchors. **F 1773,** F08

half bath—See **building space.** **E 631,** E06

half-car roughness index (HRI), *n*—an index resulting from a mathematical simulation of vehicular response to the longitudinal profile of a pavement using the half-car simulation model described in Practice E 1170 and a measured speed of 50 mph (80 km/h). **E 867,** E17

half cell—an electrode immersed in a suitable electrolyte. It may be designed to yield a known constant potential, in which case unknown potentials may be measured against it; for example, the calomel half cell. **B 374,** B08

half cell—a standard reference electrode used to measure the electrical potential between it and the reinforcing steel. Most commonly used on bridge decks is the saturated copper-copper sulfate half cell (CSE). **D 3743,** D04

half-duplex channel, *n*—a data transmission channel capable of transmitting in both directions, but in one direction at a time. **F 1457,** F05

half-hard wire—as applied to aluminum, wire that has been processed so as to produce a strength approximately midway between that of soft wire and that of hard-drawn wire. **B 354,** B01

half-life—see **decay constant.** **E 170,** E10

half-life—the time required for one half of a given number of radioactive atoms to undergo decay. **E 1316,** E07

half-life—the time required for one half a given number of radioactive atoms to undergo decay. **E 1316,** E07

half-tone—process in which a series of dots of variable diameter and (regular) interdot spacing is used to print photographs and full drawings. **F 1156,** F12

half-tone image—image produced using a **half-tone process**. **F 1156,** F12

half-value layer—the thickness of an absorbing material required to reduce the intensity of a beam of incident radiation to one-half of its original intensity. **E 1316,** E07

half-value layer (HVL)—the thickness of an absorbing material required to reduce the intensity of a beam of incident radiation to one half of its original intensity. **E 1316,** E07

half-value thickness—the thickness of a specified substance which, when introduced into the path of a given beam of radiation, reduces its intensity to one half. **E 1316,** E07

half-wave current (HW)—a rectified single-phase alternating current that produces a pulsating unidirectional field. **E 1316,** E07

halo, *n*—a shadow image created by ink interaction. See also **corona**. **F 1857,** F05

halocarbon plastics—plastics based on resins made by the polymerization of monomers composed only of carbon and a halogen or halogens. **D 883,** D20

halo effect—a condition characterized by an unusually clean background region at the boundary of the image areas. **F 335,** F05

halogen—any element of the family of the elements fluorine, chlorine, bromine and iodine (definition for purpose of this standard). **D 6161,** D19

halogen—any element of the family of the elements fluorine, chlorine, bromine, and iodine. Compounds do not fall under the strict definition of halogen. However, for the purpose of this standard, this word provides a convenient descriptive term for halogen-containing compounds. Of significance in halogen leak detection are those which have enough vapor pressure to be useful as tracer gases. **E 1316,** E07

halogen leak detector—a leak detector that responds to halogen tracer gases. Also called halogen-sensitive leak detector or halide leak detector. *(1)* The copper-flame detector or halide torch consists of a bunsen burner with flame impinging on a copper plate or screen, and a hose with sampling probe to carry tracer gas to the air intake of the burner. *(2)* The alkali-ion diode halogen detector depends on the variation of positive ion emission from a heated platinum anode when halogen molecules enter the sensing element. **E 1316,** E07

haloing—the formation of a contrasting discoloration around the edges of the workpiece when compared to interior areas. **C 286,** B08

haloing, *n*—the peripheral lines around print detail caused by excessive ink, excessive roller impression or faulty plate shoulders. **D 6488,** D01

haloing, *n*—a shadow effect around the entire printed segment or around the leading edge of a printed segment caused by excessive pressure between the printing plate and the printed surface. **F 1294,** F05

Hamming code—See **code.** **F 1156,** F12

hamper, *n*—a **container** (commonly used for shipping fruits and vegetables) circular, elliptical or polygonal in horizontal cross section, the tube dimensions being usually greater than the bottom. It has slotted sides and a bottom that may be loose, stapled, or nailed in place. The top may or may not be open. (Compare **basket**.) **D 996,** D10

H and D curve—See **characteristic curve**. **E 7**, E04

handful, *n—in sampling,* as much fiber as the hand can contain. **D 2946,** C17

handgrip, *n*—the part of a handrail designed to provide a secure grasp. **E 1481,** E06

hand-held scanner, *n*—a scanner held and operated by a human, thus enabling the scanner to be brought to the symbol. **F 1294,** F05

hand-held vacuum cleaner, *n*—a small, portable, hand-held, hand-supported vacuum cleaner usually with a nozzle as an integral part of the cleaner. The cleaner is primarily designed for light-duty, all-purpose, small-area cleaning and may be designed for liquid pickup. In use, the usually integrally designed nozzle is guided over the surface area to be cleaned. The cleaner may contain a driven agitator to assist in dirt removal and may have attachments or provisions, or both, for both floor and above-the-floor cleaning. **F 395,** F11

handhold—any strap, handle, depression, or other provision for grasping the boom by hand. **F 818,** F20

handling damage, *n— as used in fractography,* scratches, chips, cracks, etc., due to the handling of the specimen/component. **C 1145,** C28

handling point, *n*—an indication of the minimum temperature to which an oil should be heated in any part of the suction or delivery line of an oil-handling installation when the installation is operating. **D 4175,** D02

handling reinforcement—reinforcement intended to reduce the risk of collapse of the pipe or section during handling or storage prior to and during final placement. **C 822,** C13

hand of combination, lock installation—a type of lock installation that is determined by the position of the lock bolt viewed from the cover side of the door, that is, vertical up, vertical down, right hand, or left hand. **F 471,** F12

hand-overcast seam finish, *n—in home sewing*, a seam finish in which hand stitches are sewn such that the thread wraps around the cut edges of each seam allowance. **D 4965,** D13

hand print boxes—restraints for controlling entry of scannable information by hand. Controlling of size, shape, and configuration of hand printed entries on an optical scanning form. **F 149,** F05

handrail—See **railing systems.** **E 631,** E06

handrail, *n*—a horizontal, sloping, or vertical member normally grasped by hand for guidance or support. **E 1481,** E06

handrail bracket—See **railing systems.** **E 631,** E06

handrail bracket—a device attached to a wall, post, or other surface to support the handrail. **E 1481,** E06

handrail height—See **railing systems.** **E 631,** E06

handrail height—the vertical distance from the top surface of the top rail to the surface of the finished floor, top of a ramp, or the nosing line of stair treads. **E 1481,** E06

handshaking, *v*—an exchange of signals between two devices in a computer network, as prelude to data exchange, it is to determine the readiness of each device to a data exchange. **F 1457,** F05

handsheet, *n*—a sheet of fibrous material produced by a specified procedure, generally in a laboratory. **D 1968,** D06

hand washing, *n*—the most gentle form of home laundering using hand manipulation without the use of a machine or device such as a scrubbing board. **D 123,** D13

hanging rack—see **burning bars, points, or tools.** **C 286,** B08

hanging wall—the mass of rock above a discontinuity surface. (ISRM) **D 653,** D18

hang pick, *n*—a pick, caught on a warp yarn knot for a short distance, producing a triangular-shaped hole in the fabric. **D 123,** D13

hank—specifically, a coiled or looped bundle (as of yarn, rope or wire) ususally containing a definite aggregate measure of the material. **D 2864,** D27

hardboard—a generic term for a panel manufactured primarily from inter-felted lignocellulosic fibers (usually wood), consolidated under heat and pressure in a hot-press to a density of 31 lb/ft^3 (specific gravity 0.50) or greater, and to which other materials may have been added during manufacture to improve certain properties.

medium-density hardboard—a hardboard as previously defined with a density between 31 and 50 lb/ft^3 (specific gravity between 0.50 and 0.80).

high-density hardboard—a hardboard as previously defined with a density greater than 50 lb/ft^3 (specific gravity 0.80). **D 1554,** D07

hardboard nail—slender, usually colored (baked-lacquer finished), stiff-stock or usually hardened-steel, usually annularly threaded, 1 to 1 ⅝ by 0.058-in. nails with small flat head and long needle point for fastening plain or prefinished ⅛ and ¼-in. hardboard for interior applications. Also, slender bright or colored (baked-lacquer finished), galvanized, stiff-stock, or usually hardened-steel, usually helically threaded, 2 to 3 by 0.105 and 0.120-in. nails with countersunk 3⁄16 or 13⁄64-in. head and pilot needle point for fastening hardboard for exterior applications. **F 547,** F16

hardboard underlayment—a service-grade hardboard made or machined to close thickness tolerances for use as a leveling course and to provide a smooth surface under floor covering materials. **D 1554,** D07

hard chromium—chromium plate for engineering rather than decorative applications. Not necessarily harder than the latter. **B 374,** B08

hard-coating—*in anodizing aluminum*, an anodic oxide coating on aluminum with a higher apparent density and thickness and a greater resistance to wear than conventional coatings. **B 374,** B08

hard-coat system—type of finish system designed to withstand increased impact loads by increasing the strength of the base coat. Also called high-impact system. **E 631,** E06

hard-drawn wire—as applied to aluminum and copper, wire that has been cold drawn to final size so as to approach the maximum tensile strength obtainable. **B 354,** B01

hard edge—an edge reinforcement used to either maintain edge integrity under load or at attachment points. **E 1749,** E06

hardenability, *n*—the property that determines the depth and distribution of hardness induced by **quenching** a steel object. **A 941,** A01

hardened—heat-treated medium-carbon or medium-high-carbon-steel, with treating process resulting in toughened nail with greater stiffness at high flexural loads. (See **heat-treated.**) **F 547,** F16

hardened nail—heat-treated medium-low or medium-high carbon-steel nail. **F 547,** F16

hardener—*in grouting*, in a two component epoxy or resin, the chemical component that causes the base component to cure. **D 653,** D18

hardener, n— synonym for **curing agent**. **D 907,** D14

hardening, *n*—increasing the hardness by suitable treatment, usually involving heating and cooling. **A 941,** A01

hard glass—*(1)* a glass of relatively high viscosity at elevated temperatures.

(2) a glass with a high softening point.

(3) commonly refers to a glass difficult to melt. **C 162,** C14

hardmetal, cemented carbide, n—sintered material characterized by high strength and wear resistance and comprising one or more carbides of refractory metals as the main component bonded by metallic binder phase. **B 243,** B09

hard mica, *n*—mica which when slightly bent shows no tendency to delaminate. **D 1711,** D09

hardness—the relative refractoriness of a porcelain enamel or frit. **C 286,** B08

hardness—the resistance of a material to deformation, particularly permanent deformation, indentation, or scratching. **C 709,** D02

hardness, *n*—the resistance to indentation as measured under specified conditions. **C 717,** C24

hardness—a term relating to the capacity of wood to withstand denting and abrasion; for purposes of comparison, hardness in wood is often measured as the load in pounds required to embed a 0.444-in. ball to one-half its diameter (Methods D 143). **D 9,** D07

hardness, *n*—*in water*, dissolved salts of calcium, magnesium, and other cations that destroy the action of soap; expressed as parts per million (ppm) or grains per gallon (gr/gal) of calcium carbonate. (Compare **softness**.) **D 123,** D13

hardness—resistance of a material to indentation or scratching. (ISRM) **D 653,** D18

hardness—the polyvalent-cation concentration of water (generally calcium and magnesium). **D 1129,** D19

hardness, *n*—a physical property of a rubber vulcanizate, characterized by resistance to indentation. **D 1566,** D11

hardness—a term referring to the degree of cooking of a pulp. A hard pulp is one in which the residual lignin content is relatively high. **D 1695,** D01

hardness—a generic term referring to the resistance of a particle to breakdown as measured by specific tests. **D 2652,** D28

hardness, *n*—resistance to permanent deformation or indentation. **D 4175,** D02

hardness—the polyvalent-cation concentration of water (generally calcium and magnesium). Usually expressed as mg/L as $CaCO_3$. **D 6161,** D19

hardness, *n*—the resistance of a material to deformation, particularly permanent deformation, indentation, or scratching. **E 6,** E28

hardness—a measurement of the resistance to indentation. **F 1078,** F04

hardness—a measure of a material's ability to resist abrasion and/or indentation. **F 1789,** F16

hardness impression—See **impression**. **E 7,** E04

hardpan—a hard impervious layer, composed chiefly of clay, cemented by relatively insoluble materials, that does not become plastic when mixed with water and definitely limits the downward movement of water and roots. **D 653,** D18

hard particle erosion, *n*—deprecated term; use the preferred synonyms *solid impingement erosion* or *solid particle erosion.* **G 40,** G02

hard points—reinforced points within a sandwich construction to distribute stresses, resist concentrated compression loads, and maintain integrity of an attachment. **E 1749,** E06

hard rubber, n—deprecated term, see **ebonite**. **D 1566,** D11

hard size, *n*—sections of cloth containing an excessive quantity of sizing. **D 123,** D13

hard water, *n*—*in textile conservation*, water having a concentration of more than 60 ppm (3.5 gr/gal) hardness calculated as calcium carbonate. **D 123,** D13

hardwood dimension—hardwood stock processed to the specified thickness, width, and length, or in multiples thereof. **D 9,** D07

hardwoods—generally one of the botanical groups of trees that have broad leaves in contrast to the conifers or softwoods. The term has no reference to the actual hardness of the wood. **D 9,** D07

hard (X-rays)—of short wavelength. **E 7,** E04

Haring cell—A rectangular box of non-conducting material, with principal and auxiliary electrodes so arranged as to permit estimation of throwing power or electrode polarizations and potentials between them. **B 374,** B08

harmonic analysis—an analytical technique whereby the amplitude or phase, or both, of the frequency components of a complex periodic signal is determined. **E 1316,** E07

harmonic distortion—nonlinear distortion characterized by the appearance in the output of harmonics other than the fundamental component when the input wave is sinusoidal. **E 1316,** E07

harmonics—those vibrations which are integral multiples of the fundamental frequency. **E 1316,** E07

harness, *n*—one or more hookup bundles tied, clamped, or otherwise fitted together for final installation; used for interconnecting electrical circuits. **D 1711,** D09

harness leather—a self-explanatory term sometimes so defined as to include collar and saddlery leathers. Harness leather, including the related items mentioned, is practically all made of cattlehides, vegetable-tanned, except for a considerable quantity of pigskins used for making saddle seats. **D 1517,** D31

harsh (as in harsh asbestos)—description for asbestos with relatively high harshness. **D 2946,** C17

harshness—apparent property of asbestos that exhibits a significant degree of stiffness or rigidity. **D 2946,** C17

hat leather—usually sheepskin or calfskin for sweatbands of hats. The grain splits of sheepskin are vegetable-tanned for this purpose. **D 1517,** D31

haunch—that portion of the pipe barrel extending from bottom to springline. **C 896,** C04

haunch—that portion of the pipe barrel extending from bottom to springline. **F 412,** F17

haunching—the act of placing bedding material around the haunch of the pipe. **C 896,** C04

haunching—the act of placing bedding material around the haunch of the pipe. **F 412,** F17

hawser twist, *n*—the construction of cabled yarn, cord, or rope in which the single and first-ply twist are in the same direction and the second-ply twist is in the opposite direction, and S/S/Z or Z/Z/S construction. **D 123,** D13

hazard—the adverse effect(s) that may result from exposure(s). **E 943,** E47

hazard assessment, *n*—the determination of the lack of safety or degree of risk based on all integral parts of an exposure situation, including the characteristics of the chemical(s) to which one is exposed and the conditions that determine degree of exposure. **F 1494,** F23

hazardous biological materials—living biological materials, and products derived therefrom, that pose a potential threat to human health. **E 1705,** E48

hazardous chemical—any solid, liquid, gas, or mixture thereof that can potentially cause harm to the human body through inhalation, ingestion, or skin absorption. **F 1494,** F23

hazardous waste—liquid or solid waste, or combination of solid wastes, which because of its quantity, concentration, or physical, chemical, or infectious characteristics is capable of: (*a*) causing, or significantly contributing to an increase in mortality or to an increase in serious irreversible, or incapacitating reversible, illness; or (*b*) posing a substantial present or potential hazard to human health or to the environment when improperly treated, stored, transported, or disposed of, or otherwise managed. **E 1605,** E06

hazardous-waste discharge—accidental or intentional spilling, hazardous-waste leaking, pumping, pouring, emitting, discharge emptying, or dumping of hazardous wastes onto any land or water. **E 1605,** E06

hazardous-waste facility—all contiguous land, structures, or other appurtenances, and improvements on the land, used for treating, storing, or disposing of hazardous waste. **E 1605,** E06

hazardous-waste landfill—disposal facility or part of a facility where hazardous waste is placed in or on land and that is not a land surface treatment facility, a surface impoundment, or an injection well. **E 1605,** E06

hazardous-waste manifest—shipping document (EPA Form 8700-22) used for identifying the quantity, composition, origin, routing, and destination of hazardous waste during its transportation from the point of generation to the point of treatment, storage, or disposal. **E 1605,** E06

hazardous-waste small quantity generator—generator who produces less than 100 kg of hazardous waste per month (or accumulates less than 100 kg at any one time) or one who produces less than 1 kg of acutely hazardous waste per month (or accumulates less than 1 kg of acutely hazardous waste at any one time). **E 1605,** E06

hazardous-waste storage—holding on-site of hazardous waste for a temporary period, at the end of which the hazardous waste is treated, disposed of, or stored elsewhere. **E 1605,** E06

hazardous-waste treatment—any method, technique, or process, including neutralization, designed to change the physical, chemical, or biological character or composition of any hazardous waste so as to neutralize it, or render it nonhazardous or less hazardous, or to recover it, make it safer to transport, store, or dispose of, or amenable for recovery, storage, or volume reduction. **E 1605,** E06

haze—the cloudy or turbid aspect or appearance of an otherwise transparent specimen caused by light scattered from within the specimen or from its surfaces. **D 883,** D20

haze—the turbidity produced by the material in some cellulose derivatives that is not completely soluble in the medium. This turbidity may refer to solutions or to the solid plastic. **D 1695,** D01

haze—film whose clarity is impaired with varying degrees of opacity. **D 2825,** D21

haze, *n—in reflection,* (*1*) scattering of light at the glossy surface of a specimen responsible for the apparent reduction in contrast of objects viewed by reflection at the surface.

(*2*) percent of reflected light scattered by a specimen having a glossy surface so that its direction deviates more than a specified angle from the direction of specular reflection. (*3*) cloudy appearance attributable to light scattering. **E 284,** E12

haze, *n—in transmission,* (*1*) the scattering of light by a specimen responsible for the apparent reduction in contrast of objects viewed through it.

(*2*) the percent of transmitted light that is scattered so that its direction deviates more than a specified angle from the direction of the incident beam. **E 284,** E12

haze—the cloudy or turbid aspect or appearance of an otherwise transparent specimen caused by light scattered from within the specimen or from its surfaces. (D-20) **F 412,** F17

haze—see **smudge**. **F 549,** F05

haze—*on a semiconductor wafer,* non-localized light scattering resulting from surface topography (microroughness) or from dense concentrations of surface or near-surface imperfections. See also **laser light scattering event**. **F 1241,** F01

haze, *n—in plastics,* the cloudy or turbid aspect or appearance of an otherwise transparent material caused by light scattered from within the specimen or from its surfaces. **F 1251,** F04

Hazen color, n— see **platinum-cobalt color scale**. **E 284,** E12

hazing, *n*—the fine deposit of ink on the non-image area of the gravure cylinder. It is also referred to as *scumming* or dirty wipe. **D 6488,** D01

HDLC—abbreviation for **high-level data link control**. **F 1457,** F05

head, *n—in zippers,* the portion of the element that engages the pocket. **D 123,** D13

head—pressure at a point in a liquid, expressed in terms of the vertical distance of the point below the surface of the liquid. (ISRM) **D 653,** D18

head—that portion of the hide from the snout to the flare into the shoulder. **D 1517,** D31

head, *n*—pressure at a point in a liquid, expressed in terms of the vertical distance of the point below the surface of the liquid. **D 4439,** D35

head—See **windows and doors.** **E 631,** E06

head—upset or deformation of shank, usually at or near end of shank opposite point end; formed during manufacture of nail to provide area to be struck by hammer during driving and to offer bearing resistance. **F 547,** F16

header—see **manifold**. **D 6161,** D19

header tile, *n*—tile designed to provide recesses for brick header units in masonry faced walls. **C 43,** C15

head girth, *n—in body measurements,* the maximum circumference of the head above the ears. **D 123,** D13

head height—total height of head; the sum of all head elements measured parallel to nail axis; for practical purposes, measured from top of fillet to top of head. **F 547,** F16

heading—a fringe effect appearing on the leading edge (relative to copy machine feed direction) of the developed electrostatic image. **F 335,** F05

head joint—the mortar joint perpendicular to the substrate and perpendicular to the direction of the course being laid. On a floor it may be called a cross joint. **C 904,** C03

head lap, *n*—(*1*) the distance between the lower (nose) edge of an overlapping tile and the upper edge of the lapped unit in the course immediately below; (*2*) *for shingle tile only*—the distance between the lower (nose) edge of an overlapping shingle tile and the upper edge of the lapped unit in the second course below. **C 43,** C15

headlap—the shortest distance between the lower edge of an overlapping shingle or sheet and the upper edge of the lapped unit in the second course below. **C 460,** C17

headlap, *n—in shingles,* the shortest distance between the lower edge of an overlapping shingle or sheet and the upper edge of the lapped unit in the second course below. **C 1154,** C17

headlap—the minimum distance, measured at 90 deg to the eave along the face of a shingle or felt as applied to a roof, from the upper edge of the shingle or felt, to the nearest exposed surface. **D 1079,** D08

headlap—the shortest distance between the lower edge of an overlapping shingle or sheet and the upper edge of the lapped unit in the second course below. **D 2946,** C17

head length—on a hook-head or similarly headed nail, projected distance between shank and extremity of head. **F 547,** F16

headless—nail without upset or bend at head end. **F 547,** F16

head loss—the reduction in liquid pressure usually associated with the passage of a solution through a filter media bed.

D 6161, D19

head margin—the space between the edge of the image and the gripper edge of the press sheet. **F 425,** F05

head rim—peripheral part of head. **F 547,** F16

head signature, *n*—a condition where one pass of jets overlaps the next pass causing positive stitch lines. See also **stitch lines**. **F 1857,** F05

head space—the unfilled space in a glass container fitted with a closure device. **C 162,** C14

head space—volume in the upper portion of the dispenser not filled with liquid contents usually expressed as percent of total volume of dispenser at a specified temperature. **D 3064,** D10

head, static—the height above a standard datus of the surface of a column of water (or other liquid) that can be supported by the static pressure at a given point. The static head is the sum of the elevation head and the pressure head. **D 653,** D18

head (total)—the sum of three components at a point: (1) elevation head, h which is equal to the elevation of the point above a datum; (2) pressure head, hp, which is the height of a column of static water than can be supported by the static pressure at the point; and (3) velocity head, hv, which is the height the kinetic energy of the liquid is capable of lifting the liquid. **D 653,** D18

healing—the ability of a polish film to return to original state after being disturbed by fingerprints, marks, etc. **D 2825,** D21

healing power—the ability of a glaze to heal surface blemishes during firing. **C 242,** C21

health care provider—an organization, institution, or individual authorized to provide direct patient care. **F 1177,** F30

heartwood—the inner layer of a woody stem wholly composed of nonliving cells and usually differentiated from the outer enveloping layer (sapwood) by its darker color. It is usually more decay resistant than sapwood. **D 9,** D07

heat, *n*—the total molten metal output from a single heating in a batch melting process or the total molten metal output from essentially a single heating in a continuous melting operation using basically constant charge and processing conditions and targeted at a fixed metal chemistry at the furnace spout. A heat can also be defined as a fixed time period for a continuous melting operation provided that it is shorter than the time period covered by the above definition. **A 644,** A04

heat, *n*—a specific lot of material representing a single melt of steel produced to a specified chemical analysis. **A 902,** A05

heat, *n*—a generic term denoting a specific **lot** of **steel**, based upon steelmaking and casting considerations. **A 941,** A01

heat, *n*—refer to melt. **B 899,** B02

heat-absorbing glass—glass having the property of absorbing a substantial percentage of radiant energy in the near infrared of the spectrum. **C 162,** C14

heat-activated adhesive, *n*—a dry adhesive film that is rendered tacky or fluid by application of heat or heat and pressure to the assembly. (Compare **hot-melt adhesive**.) **D 907,** D14

heat-actuated cooling—the use of thermal energy to initiate a thermodynamic cycle which results in a local decrease in temperature. **E 772,** E44

heat analysis, *n*—a chemical analysis representative of a specific heat of steel. **A 902,** A05

heat analysis, *n*—the chemical analysis determined by the steel producer as being representative of a specific **heat** of **steel**. **A 941,** A01

heat analysis—a chemical analysis of a given heat by the producer which determines the percentages of its elements. **F 1789,** F16

heat buildup, *n*—the accumulation of thermal energy generated within a material as a result of hysteresis, evidenced by an increase in temperature. **D 1566,** D11

heat capacity, *n*—Synonym for **energy equivalent**. **D 121,** D05

heat capacity, *n*—the quantity of heat required to raise a system one degree in temperature either at constant volume or constant pressure. **D 5681,** D34

heat capacity—see **thermal capacity.** **E 772,** E44

heat capacity—quantity of heat necessary to change the temperature of an entity, substance or system by one Kelvin of temperature. **E 1142,** E37

heat capacity (energy equivalent, or water equivalent), *n*—the energy required to raise the temperature of a calorimeter one arbitrary unit; the quantity that when multiplied by the corrected temperature rise, then adjusted for extraneous heat effects and divided by the mass of the sample, gives the gross calorific value. **D 5681,** D34

heat, chemical, *n*—sensation of increased temperature resulting from exposure to substances such as capsaicin or hot peppers. The sensation tends to persist after the stimulus is removed. **E 253,** E18

heat durability, *n*—the extent to which a material retains its useful properties at ambient air conditions, following its exposure to a specified temperature and environment for a specified time and its return to the ambient air conditions. (Compare **heat resistance**.) **D 123,** D13

heat durable, *adj*—having heat durability. **D 123,** D13

heat exposure, *v*— the process of subjecting a specimen to an elevated temperature at atmospheric pressure for a specified period of time with the intent of determining properties by subsequent testing. **D 1079,** D08

heat fixing—the process of making a developed image permanent by heating. **F 335,** F05

heat flow; heat flow rate, Q, *n*—the quantity of heat transferred to or from a system in unit time. (Q in SI units: W.) (Q in inch-pound units: Btu/h.) **C 168,** C16

heat flow, *n*—dissipation or transfer of heat coming from within the earth by conduction, convection or radiation at the surface; usually reported in units of energy per unit time per unit area, for example, joules per second per square metre or watts per square metre. **E 957,** E44

heat flux, *n*—the thermal intensity indicated by the amount of power per unit area. **D 123,** D13

heat flux, *n*—the thermal intensity indicated by the amount of energy transmitted per unit area and per unit time (cal/cm^2-s) (watts/cm^2). **F 1494,** F23

heat flux, q, *n*—the heat flow rate through a surface of unit area perpendicular to the direction of heat flow.
(q in SI units: W/m^2)
(q in inch-pound units: Btu/h/ft^2 = Btu/h ft^2) **C 168,** C16

heat flux transducer, HFT, *n*—a device containing a thermopile (or equivalent) that produces an output which is a function of the heat flux. **C 168,** C16

heat for ignition, *n*—the product of time from initial specimen exposure until the flame front reaches the 150-mm position and the flux level at this position, the latter obtained in prior calibration of the apparatus. **E 176,** E05

heat for sustained burning, *n*—the product of time from initial specimen exposure until the arrival of the flame front, and the incident flux level at that same location as measured with a dummy specimen during calibration. **E 176,** E05

heat-fused joint—see **joint, heat-fused.** **F 412,** F17

heat gun—blower-equipped apparatus that emits heat with sufficient intensity to soften dried paint to permit scraping from the surface. **E 1605,** E06

heating curve—graphical representation of the course of temperature rise of a sample or body as a function of time. **E 7,** E04

heating flux, *n*—the prescribed incident flux imposed externally from the heater onto the specimen at the initiation of the test. **E 176,** E05

heating flux, *n*—the incident flux imposed externally from the heater on the specimen at the initiation of the test. **E 176,** E05

heating loss, *n*—mass loss, in percent, when carbon black is heated at 125°C for 1 h; the heating loss is primarily attributed to moisture content. **D 3053,** D24

heat joining—making a joint by heating the mating surfaces of the pipe components to be joined and pressing them together so that they fuse and become essentially one piece. **F 412,** F17

heat loss rate—the rate at which heat is lost from a system or component of a system, per degree temperature difference between its average temperature and the average ambient air temperature. **E 772,** E44

heat mark—extremely shallow depression or groove in the surface of a plastic visible because of a sharply defined rim or a roughened surface. (See also **shrink mark.**) **D 883,** D20

heat mark—extremely shallow depression or groove in the surface of a plastic visible because of a sharply defined rim or a roughened surface. (See also **sink-mark.**) (D-20) **F 412,** F17

heat number, *n*—the alpha, numeric, or alphanumeric designator used to identify a specific **heat** of **steel**. **A 941,** A01

heat of adsorption—the heat evolved during adsorption. **D 2652,** D28

heat of hydration—heat evolved by chemical reactions with water, such as that evolved during the setting and hardening of Portland cement. **D 653,** D18

heat, physical, *n*—sensation experienced as a result of exposure to thermally hot substances such as water above 120°F. The duration of the sensation is usually limited to the time of direct contact with the stimulus. **E 253,** E18

heat release rate, *n*—the heat evolved from the specimen, per unit of time. **E 176,** E05

heat release rate, *n*—the heat evolved from the specimen, expressed per unit area of exposed specimen area per unit of time. **E 176,** E05

heat release rate, *n*—the heat evolved from the specimen per unit of time. **E 176,** E05

heat release rate, *n*—the heat evolved from the specimen per unit of time and area. **E 176,** E05

heat resistance, *n*—the extent to which a material retains useful properties as measured during exposure of the material to a specified temperature and environment for a specified time. (Compare **heat durability**.) **D 123,** D13

heat resistant, *adj*—having heat resistance. **D 123,** D13

heat-resisting glass—glass able to withstand a relatively high thermal shock, because of low expansion coefficient or a high mechanical strength, or both. **C 162,** C14

heat sanitize—fresh water, heater to a minimum specified temperature (usually a minimum of 180°F), to destroy or kill any residual bacteria from tableware. **F 1827,** F26

heat sanitizing, continuous oval-conveyor type—dishwashing machine and conveyor-table when assembled, shall form an oval-shaped dish handling system. Machines shall automatically convey racks of soiled tableware through the treatment stages, including a final heat sanitizing rinse of the machine, conveying them out to the clean tableware removal area of the conveyor: additional parts: recirculating pre-wash chamber, rinse chamber, conveying mechanisms, heating equipment and horizontal conveyor tables. See **dishwashing machine, commercial**. **F 1827,** F26

heat sanitizing, conveyor rack type—machines that automatically convey racks of soiled tableware through treatment stages and final heat sanitizing rinse, conveying them out at the clean end of the machine; additional parts: rinse chamber, heating equipment, and conveying mechanism. See **dishwashing machine, commercial.** **F 1827,** F26

heat sanitizing, rackless conveyor type—machines shall automatically convey unracked soiled tableware through the treatment stages, includes a final heat sanitizing rinse, conveying them out at the clean end of the machine; additional parts: rinse chamber, conveying mechanisms, and heating equipment. See **dishwashing machine, commercial**. **F 1827,** F26

heat sanitizing, stationary rack type—manually fed machines, includes a final heat sanitizing rinse; additional parts: heating equipment. See **dishwashing machine, commercial**. **F 1827,** F26

heat seal—See **seal.** **D 996,** D10

heat seal—the result of bonding surfaces by controlled application of heat, pressure, and dwell time. **F 1327,** F02

heat sealing adhesive tape—a strip of material (usually fabric, metal foil, paper, or plastic film) coated with an adhesive activated with the application of heat. **E 1749,** E06

heat sensitive—a material composition that will produce an image from application of localized heat. **F 221,** F05

heat-sensitive dipping (latex), *n*— dipping process in which a heated form is immersed in latex containing a substance that causes the latex to gel at an elevated temperature. **D 1566,** D11

heat sensitizer, *n*—gelling agent effective only at elevated temperature. **D 1566,** D11

heat shrinkage, *n*—a decrease in one or more dimensions of an object or material exposed to heat. **D 123,** D13

heat-strengthened glass—flat glass that has been tempered to a moderate surface or edge compression to meet the requirements of Specification C 1048. (See **fully tempered glass**.) **C 162,** C14

heat stress, *n*—(physiological) adverse condition caused by exposure to elevated temperature, radiant heat flux, or combinations of these factors. **E 176,** E05

heat transfer fluid—(*1*) *in solar energy systems*, a liquid or gas that passes through the solar collector and carries the absorbed thermal energy away from the collector. (*2*) any fluid that is used to transfer thermal energy between subsystems in solar energy systems. **E 772,** E44

heat-treated—a term sometimes used for tempered glass. (See **tempered glass**.) **C 162,** C14

heat-treated—heated above critical temperature and subsequently quenched, which may be followed by tempering for the purposes of obtaining certain desirable conditions or properties, such as hardness, toughness, and stiffness at high flexural loads. **F 547,** F16

heat-treating—the process of subjecting a wood-base panel material (usually hardboard) to a special heat treatment after hot pressing to increase some strength properties and water resistance. **D 1554,** D07

heat treatment, *n*—heating and cooling a steel object in such a way as to obtain desired conditions or properties. **A 941,** A01

heat-up efficiency—the energy imparted to a food product, expressed as a percentage of energy consumed by the equipment during a cooking event, in which a food product is heated from a predetermined initial temperature to a predetermined final temperature, η *heat-up*. **F 1827,** F26

heat-up temperature response—temperature rise on the surface of a steel plate during the test period in accordance with the heat-up temperature-response test. **F 1827,** F26

heat welded seam, *n*—a seam produced by grooving abutting edges of resilient flooring and filling said groove with heated, fused or melted material to provide a bond and seal. **F 141,** F06

heave—upward movement of soil caused by expansion or displacement resulting from phenomena such as: moisture absorption, removal of overburden, driving of piles, frost action, and loading of an adjacent area. **D 653,** D18

heave response—ability of the boom to react to the vertical motion of the water surface. **F 818,** F20

heavy-centered spray pattern—*in protective coatings*, an uneven spray pattern having more coating in the center, and less at the edges. **D 4538,** D33

heavy-duty, *adj— in internal combustion engine operation*, characterized by average speeds, power output, and internal temperatures that are generally close to the potential maximums. **D 4725,** D15

heavy-duty engine, *n— in internal combustion engine types*, one that is designed to allow operation continuously at or close to its peak output. **D 4725,** D15

heavy goods, *n*—name applied in Calcutta quotations to material called "sacking" in the United States. **D 123,** D13

heavy leather—a somewhat indefinite term, generally understood to include vegetable-tanned sole, belting, strap, and mechanical leathers made from unsplit cattlehides. More recently it also refers to thick side leathers. **D 1517,** D31

heavy metals—elements having a high density or specific gravity of approximately 5.0 or higher. A generic term used to describe contaminants such as cadmium, lead, mercury, etc. Most are toxic to humans in low concentration. **D 6161,** D19

heavy nail, heavy-gage nail—See **stout nail.** **F 547,** F16

HeCd—helium-cadmium; most often used to refer to a laser system using helium gas and cadmium electrodes to support the internal gas discharge. **F 1156,** F12

hedonic scale, *n*—a scale on which liking or disliking of a stimulus is expressed. **E 253,** E18

heel, *n*—a solid part of a shoe or boot projecting downward and attached to or forming the back part of the sole under the heel of the foot. **F 1646,** F13

heel-ankle circumference, *n—in body measurements*, with the subject standing barefoot, the distance around the foot from the point where the back of the heel contacts the floor and over the juncture of the foot and leg at the front of the ankle and back to the starting point. **D 5219,** D13

heel bead—glazing compound applied at the base of a glazing channel, on the removable stop side, *after* setting glass or panel and before the removable stop is installed. **C 717,** C24

heel, b_H—the widest part of the ski in the tail section of the ski. **F 472,** F27

heel contact—the region on a glass container surface, located between the bearing surface and the sidewall sections, which normally contacts other similarly shaped and upright containers. **C 162,** C14

heel (of the snowboard ski)—the widest part of the tail section of the snowboard ski. **F 1107,** F27

heel seat, *n*—the part of a shoe to which the heel is attached. **F 1646,** F13

heel tap—an imperfection in which the base or bottom of a bottle is very thick in one area and very thin in another. **C 162,** C14

heel-toe wear, *n*—a type of irregular wear characterized by different wear rates at the leading and trailing edges of a projection (element). **F 538,** F09

height, *n*—the vertical dimension of the unit in the face of a wall. **C 43,** C15

height, *n*—vertical dimension of the face of a unit when the unit is positioned as a stretcher. **C 1232,** C15

height, *n—in body measurements,* the vertical distance from the crown of a standing subject to the soles of the feet. (Syn. **stature**) **D 5219,** D13

height, *n*—of a building, the vertical distance measured from the **finished grade** to average level of the roof above the level of the highest **wall.** **E 631,** E06

height—distance between root and crest, measured perpendicular to nail axis. **F 547,** F16

height—the distance before stretching from one outer edge of a knuckle or twist to the outer edge of the knuckle or twist on the opposite edge of fabric. **F 552,** F14

height—sum of draft and freeboard. **F 818,** F20

height of capillary rise—see **capillary rise.** **D 653,** D18

helical, *n*—the most simple pattern in a barbed tape coil where there are no concertina attachments and each coil loop is left free in its natural spiral (also called *spiral*). **F 1379,** F14

helical loop spacing, *n*—average distance between each coil loop in a helical coil configuration. **F 1379,** F14

helical thread—continuous multiple helical depressions rolled onto nail shank with resulting expansion approximately equal to depression. Unless otherwise specified, medium lead angle is implied. **F 547,** F16

heliostat—a reflector that is mechanically moved so that solar flux is reflected in a constant direction to a stationary receiver or target. **E 772,** E44

helium accumulation fluence monitor (HAFM)—a passive neutron dosimeter whose measured reaction product is helium. The neutron fluence is obtained by dividing the helium concentration by the spectrum-averaged cross section (see **spectrum-averaged cross section**). (See also ASTM Test Method E 910, for Application and Analysis of Helium Accumulation Fluence Monitors for Reactor Vessel Surveillance, and ASTM Master Matrix E 706, for Light-Water Reactor Pressure Vessel Surveillance Standards.) **E 170,** E10

helium bombing—a pressure-evacuation test in which helium is used as the test gas. **E 1316,** E07

helium drift—*(1)* in leak testing with a probe, the drift from a leak or permeable gasket located at some distance from the end of the probe but which is detected by the probe and can mislead the operator into suspecting the area near the probe; *(2)* a gradual wandering of the output meter needle of the leak detector due to slowly changing helium concentrations (either due to a leak or outgassing) in the detector tube. Expressed in scale divisions per unit time. **E 1316,** E07

helium leak detector—a leak detector using helium as the tracer gas. **E 1316,** E07

helix angle—See **lead angle.** **F 547,** F16

hem, *n—in garment construction,* a simple finish in which the raw fabric edge is turned under and stitched to a garment. (See **cuff.**) **F 1494,** F23

hemiacetal groups—functional groups derived from carbonyl groups by addition of one molecule of an alcohol, of the general structure:

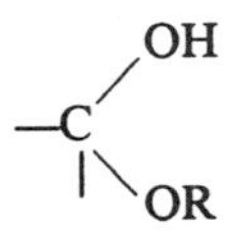

D 1695, D01

hemicellulose—any of a number of cell-wall polysaccharides that are removable by extraction with aqueous alkali and that may be hydrolyzed by boiling with dilute acids to give constituent monosaccharide units; any of the noncellulosic cell-wall polysaccharides. **D 1695,** D01

hemic peat—peat in which the original plant fibers are moderately decomposed (between 33 and 67 % fibers). **D 653,** D18

hemihydrate, *n*—the dry powder, calcium sulfate hemihydrate, resulting from calcination of $CaSO_4{\cdot}2H_2O$, calcium sulfate dihydrate. See **calcined gypsum**. **C 11,** C11

hemispherical, adj— see **hemispherical optical measuring system.** **E 284,** E12

hemispherical, 2 π—over an entire hemisphere. It is assumed that the incident radiance is uniform in density over the hemisphere. **E 772,** E44

hemispherical optical measuring system, *n*—an optical measuring system for measuring the reflecting or transmitting properties of specimens, wherein either the specimen is uniformly irradiated from all directions within the hemisphere on one side of the specimen or the flux emanating from the specimen is evaluated uniformly for all directions within a hemisphere. **E 284,** E12

hemispherical temperature (HT), *n—in reference to the fusibility of coal and coke ash according to Test Method D 1857*, the temperature at which the cone has fused down to a hemispherical lump at which point the height is one half the width of the base. **D 121,** D05

hemispherical temperature, HT, *n*—the temperature at which a pyrometric cone has fused down to a hemispherical lump where the height is one half the width of the base. **D 5681,** D34

henry (plural henries), H—the unit of self- or mutual inductance. The henry is the inductance of a circuit in which a voltage of 1 V is induced by a uniform rate of change 1 A/s in the circuit. Alternatively, it is the inductance of a circuit in which an electric current of 1 A/s produces a flux linkage of one weber turn (Wb turn) or 10^8 maxwell-turns. See **inductance, mutual, and inductance, self.** **A 340,** A06

Henry's Law, *n*—the principle that the mass of a gas dissolved in a liquid is proportional to the pressure of the gas above the liquid. **D 4175,** D02

HEPA filter—see **high-efficiency particulate air filter**. **E 1605,** E06

herbaceous plants—nonwoody species of vegetation, usually of low lignin content such as grasses. **E 1126,** E48

herbaceous plants—nonwoody species of vegetation, usually of low lignin content such as grasses. **E 1705,** E48

hermetically sealed aseptic container—a container that is designed and intended to be secure against the entry of microorganisms and thereby to maintain the commercial sterility of its contents. **F 17,** F02

hermetically sealed aseptic container—See Terminology F 17. **F 1327,** F02

hermetically tight seal—a seal which does not exhibit leakage when dynamically tested with commercially built leak detectors that are sensitive to a gas on the pressure side opposite to the side on which the leak detector is located, or which does not exhibit leakage with any form of liquid test. **E 1316,** E07

hermetic seal—See **seal.** **D 996,** D10

herringbone—a series of long continuous waves running at various angles to the rolling direction. **B 846,** B05

herringbones, *n—in mica,* numerous rulings that intersect to form a series of "V's" with included angles of about 120°. **D 1711,** D09

hertz, Hz—the unit of cyclic frequency, *f.* **A 340,** A06

Hertzian cone crack—See **percussion cone**. **C 162,** C14

Hertzian contact area, *n*—the apparent area of contact between two nonconforming solid bodies pressed against each other, as calculated from Hertz' equations of elastic deformation. **G 40,** G02

Hertzian contact pressure, *n*—the magnitude of the pressure at any specified location in a Hertzian contact area, as calculated from Hertz' equations of elastic deformation. **G 40,** G02

Hertzian stress—See **contact stress**. **C 162,** C14

hessians, *n*—name applied in United Kingdom, India, and on the Continent to material called "burlap" in the United States. (See also **burlap.**) **D 123,** D13

heterogeneity—having different properties at different points. (ISRM) **D 653,** D18

heterogeneous—non-uniform in microstructure or composition. **E 7,** E04

heterogeneous equilibrium—in a chemical system, a state of dynamic balance amongst two or more homogeneous phases which are capable of stable coexistence in mutual or sequential contact. **E 7,** E04

hex, *n*—abbreviation for **hexadecimal**. **F 1457,** F05

hexadecimal, *n*—a base-16 numbering system consisting of 16 different digits, 0 through 9 and A through F. (See **hex**.) **F 1457,** F05

hexagonal close packed:—(*1*) a structure containing two atoms per unit cell located at (0, 0, 0) and (⅓, ⅔, ½) (or ⅔, ⅓, ½).

(*2*) one of the two ways in which spherical objects can be most closely packed together, such that the close-packed planes are alternately staggered in the order A-B-A-B-A-B. **E 7,** E04

hexagonal (concerning lattices for crystals)—Having two equal coplanar axes, a_1 and a_2, at 120 deg to each other and a third axis, *c*, at right angles to the other two; *c* may or may not be equal to a_1 and a_2. **E 7,** E04

hexosan—frequently used in contradistinction to pentosan, for a polysaccharide consisting mainly of hexose units. **D 1695,** D01

Heyn method—an intercept method for determining grain size. See Test Methods E 112. **E 7,** E04

H-gasket—See **gasket, lock-strip, and H-type**. **C 717,** C24

hickey, *n*—an imperfection caused by dirt, hardened ink, paper debris or other unwanted particles that cling to the press or plate during printing. **D 6488,** D01

hickey, *halo, n*—a doughnut shaped hickey with an inked center, often caused by foreign particles such as ink skin. **[TAPPI]** **D 6488,** D01

hickey, *void, n*—a totally un-inked spot on the printed sheet, often caused by dust or pick-outs that adhere to the offset blanket or plate, thereby interfering with transfer of ink. **D 6488,** D01

hide—the pelt of a large animal, such as cow, horse, etc. Also used interchangeably with Skin. **D 1517,** D31

hide grades—standard hide grades, take-up, and delivery practice are given in the booklet, "Approved Standard Practice Governing the Take-Up and Delivery of Domestic Packer Hides," published by The Leather Industries of America, Inc., 1000 Thomas Jefferson Street NW, Suite 515, Washington, DC 20007. **D 1517,** D31

hide powder—purified, shredded rawhide as a reagent in the determination of tannins. **D 1517,** D31

Hide Powder, Standard—any lot of hide powder officially approved by the American Leather Chemists Association. **D 1517,** D31

hide substance—nitrogen content of hide or leather multiplied by 5.62. **D 1517,** D31

hiding—the ability of the applied polish to cover scuff marks made by scratches in the shoe finish. **D 2825,** D21

hiding power—the ability of a paint, or paint material as used, to hide or obscure a surface to which it has been uniformly applied. When expressed numerically, it is generally in terms of the number of square feet over which a gallon of paint, or pound of pigment, as used, can be uniformly spread to produce a specified contrast ratio (see **contrast ratio**). The term **covering power** has no specific relationship to hiding power, and actually has no precise meaning. **D 16,** D01

hiding power, *n*—*(1)* the ability of a coating material to hide the surface coated by producing a specified opacity.

(2) the area over which a specified volume of paint can be spread to produce a specified contrast, C_c, between areas where the substrate is black and where it is white. **E 284,** E12

hierarchical experiment, *n*—see **nested experiment.** **E 456,** E11

high brackish water—water with an approximate concentration of total dissolved solids ranging from 10 000 to 30 000 mg/L. See **brackish water** and **sea water.** **D 6161,** D19

high-bust girth—see **chest girth.** **D 5219,** D13

high-calcium—indicates the presence of 0 to 5 % magnesium carbonate ($MgCO_3$) in the limestone from which the material was formed. **C 51,** C07

high-calcium limestone—see **limestone.** **C 51,** C07

high-carbon steel— See **steel grades.** **F 547,** F16

high-clearance sprayer—an apparatus consisting of the components of a boom sprayer mounted on a self-propelled vehicle whose frame is constructed to permit the vehicle to pass over plants with minimum damage. **E 1102,** E35

high-conductivity copper—copper that in the annealed condition has a minimum electrical conductivity of 100 % IACS. **B 846,** B05

high crown—staple crown with inverted "V" wire cross-section prior to staple driving; designed to provide rigidity during driving and flattened when fully driven. (see **"V" Crown.**) **F 592,** F16

high-density fiberboard nail—See **shake nail, shingle nail, wood shingle** and **asbestos shingle nail.** **F 547,** F16

high-density particleboard—a particleboard as previously defined with a density greater than 50 lb/ft^3 (specific gravity 0.80).

NOTE—It is the industry practice to measure density of particleboards on the basis of moisture content and volume at time of test. **D 1554,** D07

high-density polyethylene plastics (HDPE), *n*—those linear polyethylene plastics, having a standard density of 0.941 g/cm^{3}or greater. (D-20) **F 412,** F17

high density polyethylene plastics, (HDPE) *n*—those linear polyethylene plastics, g. v., having a standard density of 0.941 g/cm^3 or greater. **D 883,** D20

high-efficiency particulate air (HEPA) filter—filter capable of separating out particles of 0.3 μm or greater from a body of air at 99.97 % efficiency or greater. See also **ULPA filter.** **E 1605,** E06

high elongation, *adj*—*in steel tire cord*, a cord with an average elongation at break greater thatn 3.0% **D 123,** D13

higher heating value—the energy content of gas (Btu/ft^3), measured at standard conditions, *HV*. **F 1827,** F26

higher heating value, HHV, *n*—a synonym for gross calorific value. **D 5681,** D34

higher heating value (HHV)—a synonym for gross calorific value. **E 856,** D34

high-hip girth, *n*—*in body measurements*, the circumference of the body at a point approximately 7.5 cm (3 in.) below the waist and parallel to the floor. (Compare **hip girth.**) **D 123,** D13

high-level data link control, *n*—a CCITT standard for communication line protocol. (See **HDLC.**) **F 1457,** F05

high life expectancy, LE-100, *n*—*of paper*, a paper is expected to be usable for 100 years. **D 1968,** D06

high life expectancy paper, LE-100, *n*— a paper expected to be usable for 100 years under prescribed conditions. **D 1968,** D06

highlighting lines, *n*—*in label printing*, horizontal, thick, divider lines placed above and below the package identification building block. **F 1294,** F05

high lights—those portions of a metal article most exposed to buffing or polishing operations, and, hence, having the highest luster. **B 374,** B08

highloft nonwoven fabric, *n*—a low-density fiber network structure characterized by a high ratio of thickness to mass per unit area. **D 123,** D13

high-modulus aramid, *n*—*for the purpose of these test methods*, those aramid yarns with an initial modulus of at least 400 gf/den (35 N/tex). **D 123,** D13

high-phosphate detergent—detergent that contains the phosphate equivalent of at least 5 % tri-sodium phosphate (TSP). **E 1605,** E06

high point, High—the point on a load-displacement plot, at the start of an unloading-reloading cycle, at which the displacement reverses direction, that is, the point at which the specimen mouth begins closing due to unloading. **E 1823,** E08

high-pressure, (*HP...*), *adj*—a prefix for different thermoanalytical techniques in which the pressure in the apparatus is above ambient. (ICTAC)

NOTE—As an example, high-pressure thermogravimetric analysis is designated HPTGA. **E 473,** E37

high-pressure molding, *n*—a method of molding or laminating in which the pressure used is greater than 1400 kPa (200 psi). **D 883,** D20

high-pressure steam cooker—equipment wherein the cooking compartment operates between 10 psig and 15 psig. See **steam cooker.** **F 1827,** F26

high profile tile, *n*—tile having a rise to width ratio greater than 1:4. **C 43,** C15

high-purity water—highly treated water with attention to microbiological, particle, organics and mineral reduction or elimination. **D 6161,** D19

high referral, *adj*— *in paper*, descriptive of any grade of paper designed for use in situations involving frequent handling. **D 1968,** D06

high resistance—contact resistance exceeding an arbitrary, specified limit. **B 542,** B02

high-resolution NMR spectrometer—an NMR apparatus that is capable of producing, for a given isotope, line widths that are less than the majority of the chemical shifts and coupling constants for that isotope. **E 131,** E13

high ride—an amusement ride whose motion is in a fixed or variable plane from horizontal to vertical. **F 747,** F24

high strength bolts—a term which is used commercially to denote ASTM A325 or A490 bolts which are primarily used in construction applications. **F 1789,** F16

high-strength low-alloy steel, *n*—a **steel**, other than a **carbon steel** or an **interstitial-free steel**, that conforms to a specification that requires the minimum content for each specified alloying element to be lower than the applicable limit in the definition for **alloy steel**, and the yield point or yield strength of the product to be at least 36 ksi or 250 MPa. **A 941,** A01

high-tenacity fiber, *n*—a manufactured fiber either (*1*) belonging to a generic class of fibers having exceptional breaking strength; or (*2*) having a breaking strength significantly greater than the average strength of other (regular tenacity) fibers in the same generic class and of equivalent linear densities. **D 123,** D13

high-transmission glass—glass that transmits an exceptionally high percentage of the visible light. **C 162,** C14

high usage, *adj— in paper folders*, descriptive of any grade of folder designed for use in situations where folders are handled frequently. **D 1968,** D06

high volatile A bituminous coal—See *high volatile A bituminous coal* under **rank.** **D 121,** D05

high volatile B bituminous coal—See *high volatile B bituminous coal* under **rank.** **D 121,** D05

high volatile C bituminous coal—See *high volatile C bituminous coal* under **rank.** **D 121,** D05

high-voltage power supply, *n*—a source of electrical power which provides voltage above logic levels, usually internal to the component. (See **HVPS**.) **F 1457,** F05

hillock—See **pyramid.** **F 1241,** F01

hinge—the minimum thickness of gasket material between the channel recess and the lock-strip cavity; the plane at which bonding occurs when the flange is bent open to receive or release installed material. **C 717,** C24

hinge—location between boom segments at which the boom can be folded back 180° upon itself. **F 818,** F20

hinge nail—light or heavy, bright, regular-stock-steel, 1¼ by 3⁄16 to 4 by ⅜-in. nails with flat or oval countersunk (95 deg) or oval ¼ to ½-in. head and long diamond or chisel point. **F 547,** F16

hinge stress—the tensile component of the bending stress generated on the same surface of a glass section as, but displaced from, the site of a locally impinging force. **C 162,** C14

hip, *n—in anatomy*, the laterally projecting region formed by the lateral parts of the pelvis and the upper part of the femur together with the flesh covering them. **D 123,** D13

hip, *n—in anatomy*, the laterally projecting region formed by the lateral parts of the pelvis and the upper part of the femur together with the flesh covering them. **F 1494,** F23

hip and ridge, finishing pieces—rectangular pieces of roofing shingles cut to a flare or taper and applied with a side lap to conceal the joint of roofing shingles along the hips and ridge of a roof. **C 460,** C17

hip and ridge, finishing pieces, *n—in shingles*, rectangular pieces of roofing shingles cut to a flare or taper and applied with a side lap to conceal the joint of roofing shingles along the hips and ridge of a roof. **C 1154,** C17

hip and ridge, finishing pieces—rectangular pieces of roofing shingles cut to a flare or taper and applied with a side lap to conceal the joint of roofing shingles along the hips and ridge of a roof. **D 2946,** C17

hip and ridge tile, *n*—tile designed for application to the hip or ridge of a roof. **C 43,** C15

hip girth, *n—in body measurements*, the maximum circumference of the body at the level of maximum prominence of the buttocks. **D 123,** D13

hip girth, *n—in body measurements*, the maximum circumference of the body at the level of maximum prominence of the buttocks. **F 1494,** F23

historical sample—See preferred term, **archive sample.** **C 859,** C26

historic fabric—See **fabric.** **E 1480,** E06

hit—the detection and measurement of an AE signal on a channel. **E 1316,** E07

hit-and-miss—a manufacturing imperfection on planed lumber, defined by grade rules as a series of planer skips with surfaced areas between. Permitted depth of misses is usually specified. **D 9,** D07

hit-or-miss—a manufacturing imperfection on planed, rough, or partially planed lumber, providing a maximum permitted scantness. **D 9,** D07

Hi-Vol (high-volume air sampler)—See **sampler.** **D 1356,** D22

HOBE—an abbreviation for *honeycomb before expansion*; honeycomb made by layering sheets, usually of thin aluminum or paper, containing adhesive at the cell nodes. **E 1749,** E06

hob nail—stout, regular-stock-steel, ⅜ to ⅝-in. nails with large decorative (high square, fancy, round bevel, checkered, grooved, etc.) head and sheared-bevel point. **F 547,** F16

hockle, *n—in rope*, a strand kink in a rope causing yarn displacement in the strand resulting in rope deformation and damage. **D 123,** D13

hogged fuel—ground wood fuel that is usually a by-product of a wood products manufacturing process. **E 772,** E44

hogged fuel—ground wood fuel that is usually a by-product of a wood products manufacturing process. **E 1126,** E48

hogged fuel—ground wood fuel that is usually a by-product of a wood products manufacturing process. **E 1705,** E48

hog ring—a preformed open wire clip designed to close up into a ring to secure chain link fabric to horizontal tension wire (see Specification F 626). **F 552,** F14

hog ring, hog-ring staple—open-ended, rounded, ring, or rectangular "U"-formed staple; used for encircling applications, that is, for attaching materials to rounded or rectangular base material by closing or wrapping hog ring around base material. **F 592,** F16

hold down—see **lockdown.** **F 335,** F05

holding pump—a fore pump used to hold a vapor pump at operating conditions while a roughing pump reduces the system pressure to a point at which the valve between the vapor pump and the system can be opened without stopping the flow of vapor from the nozzles. **E 1316,** E07

hold pressure, *n*—in molding, the melt pressure during the hold time interval in injection molding. **D 883,** D20

hold time [T]—*in fatigue testing*, the amount of time in the cycle where the controlled test variable (for example, load, strain, displacement) remains constant with time. **E 1823,** E08

hole, *n—in fabric*, an imperfection where one or more yarns are sufficiently damaged to create an aperture. **D 123,** D13

hole—a deep depression or void, the bottom of which is not visible by normal (20/20) vision under 200 fc illumination.

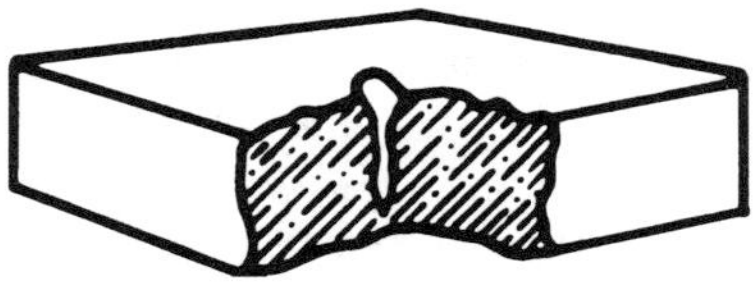

F 109, C21

hole—a mobile vacancy in the electronic valence structure of a semiconductor that acts like a positive electron charge with positive mass; the majority carrier in *p*-type material. **F 1241,** F01

hole-burning, *n*—in luminescence, the photo-induced disappearance of a narrow segment within a broader absorption or emission band. **E 131,** E13

holes—openings in or through lumber that may extend partially or entirely through a piece and may be from any cause. **D 9,** D07

hole spacing, *n*—*on a button*, the distance from the center of one hole to another. **D 123,** D13

holiday—an area where a liquid-applied material is missing. **D 1079,** D08

holland cloth, *n*—a completely filled woven fabric having a smooth gloss finish on both sides used as a separating medium for sheeted rubber compounds. **D 123,** D13

holland cloth, *n*—a completely filled woven fabric (usually starch-filled linen) with a smooth glass-like finish on both sides, used as a separating medium for raw rubber and rubber compounds. **D 1566,** D11

hollow braid, *n*—*in rope*, a braided construction of either plain or twill braid, having an empty center. **D 123,** D13

hollow brick, *n*—a cellular masonry unit made from clay or shale and kiln-fired to produce designated properties. See Specification C 652. The hollow brick has larger percentages of void areas than permissible with solid brick. See Specification C 62 and Specification C 216. **C 43,** C15

hollow casting—See *drain casting* under **casting.** **C 242,** C21

hollow characters—characters or lines that print as outlines. **F 425,** F05

hollow characters, *n*—print defects in which the center of the dark area is lighter than the edge, such that the character appears to have a heavy outline. **F 1457,** F05

hollow cone atomizer—a cone atomizer wherein most of the liquid is directed toward the outside of the spray pattern. **E 1620,** E29

hollow cone pattern, *n*—a cone pattern wherein most of the liquid is concentrated near the outside of the pattern. **E 1620,** E29

hollow fiber (HF) membrane—self-supporting membrane fibers which have a hollow bore like a cylinder. In reverse osmosis, the membrane is usually on the outside with the bore conveying the permeate. In ultra and micro filtrations the membrane is generally on the inside. **D 6161,** D19

hollow masonry unit—unit whose net cross-sectional area in any plane parallel to the bearing surface is 75 % or less of its gross cross-sectional area measured in the same plane. **C 1209,** C15

hollow masonry unit—unit whose net cross-sectional area in any plane parallel to the surface containing cores, cells, or deep frogs is less than 75 % of its gross cross-sectional area measured in the same plane. **C 1232,** C15

hollow ware—a class of utensils such as pots, pans, and kettles. **C 286,** B08

holocellulose—the total polysaccharide fraction of extractive-free wood. The method of isolation or of determination should always be given. **D 1695,** D01

holography (acoustic)—an inspection system using the phase interface between the ultrasonic wave from an object and a reference signal to obtain an image of reflectors in the material under test. **E 1316,** E07

holography (optical)—a technique for recording, and reconstructing, the amplitude and phase distributions of a wave disturbance; widely used as a method of three-dimensional optical image formation. The technique is accomplished by recording the pattern of interference between coherent light reflected from the object of interest (object beam), and light that comes directly from the same source (reference beam). **E 1316,** E07

Homal eyepiece—See **amplifier**. **E 7,** E04

home, *n*—a place of residence. (See also **dwelling.**) **E 631,** E06

home laundering, *n*—a process by which textile products or parts thereof may be washed, bleached, dried, and pressed by any customary method designed for use in a residence, or nonprofessional use. (See also **professional care**.) **D 123,** D13

home laundering—the process of cleaning and restoring textile materials to a serviceable condition using the washing and drying equipment commonly found in the home. **D 459,** D12

homogeneity—having different properties at different points. (ISRM) **D 653,** D18

homogeneity—having the same properties at all points. (ISRM) **D 653,** D18

homogeneous carburizing, *n*—a process that converts a low-carbon steel to one of substantially uniform and higher carbon content throughout the section, so that a specific response to **hardening** may be obtained. **A 941,** A01

homogeneous mass—a mass that exhibits essentially the same physical properties at every point throughout the mass. **D 653,** D18

homogeneous material, *n*—a material in which relevant properties are not a function of the position within the material. **C 168,** C16

homogeneous (radiation) (monochromatic)—of the same wavelength. **E 7,** E04

homogeneous surface—surface that exhibits essentially the same appearance and properties at every point throughout the area examined. **E 1605,** E06

homogeneous vinyl flooring, *n*—a floor surfacing unit in sheet or tile form that is of uniform structure and composition throughout, usually consisting of vinyl plastic resins, plasticizers, fillers, pigments and stabilizers. **F 141,** F06

homogenization (raw rubber), *n*—repeated passage of raw rubber through a mill or other mixing device, under specified conditions, to ensure uniformity. **D 1566,** D11

homogenizing, *n*—holding a steel object at high temperature to eliminate or decrease chemical segregation by diffusion. **A 941,** A01

homologous lines, *n*—*in optical emission spectrometry*, spectral lines that exhibit minimal change in their intensity ratios with variations in excitation conditions. **E 135,** E01

homologues of phenol—compounds of the phenol series whose structure differs regularly by some radical (for example, $-CH_3$) from that of its adjacent neighbor in the series. Also *cresols* and *xylenols*. **D 4790,** D16

homopolymer, *n*—a polymer resulting from polymerization involving a single monomer. **D 883,** D20

homopolymer, *n*—a polymer formed from a single monomer species. **D 1566,** D11

homopolymer, *n*—a polymer resulting from polymerization involving a single monomer. (D-20) **F 412,** F17

honeycomb—advanced stage of decay caused by *Fomes pini.* **D 9,** D07

honeycomb before expansion—See **HOBE**. **E 1749,** E06

honeycomb core—a sheet material, formed into cell structure (usually hexagonal) similar to honeycomb and used as core material in the construction of sandwich panel assemblies. **E 1749,** E06

honeycombing—in lumber and other wood products, separation of the fibers in the interior of the piece, usually along the wood rays. The failures often are not visible on the surfaces, although they can be the extensions of surface and end checks. **D 9,** D07

honeycomb sandwich panel—sandwich constructions consisting of honeycomb core adhesively clad with face sheets. **E 1749,** E06

honeycomb structure—see **soil structure**. **D 653,** D18

Hong Kong seam finish, *n—in home sewing*, a seam finish in which a binding fabric is used to encase the cut edge of each seam allowance separately and in which the binding fabric has one cut edge enclosed and the other cut edge exposed. (Compare **bound seam finish**.) **D 4965,** D13

hood test—an overall test in which an object under vacuum test is enclosed by a hood which is filled with tracer gas so as to subject all parts of the test object to examination at one time. A form of dynamic leak test in which the entire enclosure or a large portion of its external surface is exposed to the tracer gas while the interior is connected to a leak detector with the objective of determining the existence of leakage. **E 1316,** E07

hood with integrated makeup air—an exhaust hood that introduces makeup air through a plenum, as an integral part of the hood. Integral makeup air options include *down discharge, front face discharge*, or *internal discharge* (*short-circuit*), or both. **F 1827,** F26

hook, *n*—an anchoring component that fastens to the spine by means of a curved blade passed under or over lamina, transverse or spinous processes or into an anatomic or surgically created notch or opening. **F 1582,** F04

hook and loop fastener—a touch fastener, comprised of two flexible mating strips, the surface of one mating strip being covered with tiny, stiff protrusions shaped like hooks which engage the other mating strip which is covered with pliable loops. **D 123,** D13

hook blade, *n*— that portion of a spinal hook that is placed under, over, or into a bony structure to provide attachment. **F 1582,** F04

hook body—that portion of a spinal hook that connects the hook blade to the longitudinal element. **F 1582,** F04

hooked bow, *n*—a fabric condition in which the filling or course yarns are in the proper position for most of the fabric width but are pulled out of alignment at one side of the fabric. (See also **double hooked bow**) **D 123,** D13

hook head—head, usually flattened; formed by bending wire at right angle to shank; resulting in fastener shaped like "L". **F 547,** F16

hook-head metal-lath nail— See **metal-lath nail.** **F 547,** F16

hooks, *n—in fiber testing*, curved or bent fiber ends caused by the carding or specimen preparation processes. **D 123,** D13

hookup bundle, *n*—a group of insulated conductors or hookup cables grouped into an assembly prior to installation, usually with multiple breakouts. **D 1711,** D09

hookup cable, *n*—two or more insulating conductors in a common covering, or two or more insulated conductors twisted or molded together without a common covering, or one or more insulated conductors with a conductive shield with or without an outer covering. **D 1711,** D09

hookup wire, *n*—an insulated conductor that is used to make point-to-point connections in an electrical or electronic system. **D 1711,** D09

hoop fastener nail—bright, blued or galvanized, regular-stock-steel, 3⁄8 by 0.092 to 1 by 0.162-in. nails with thin flat 3⁄8 to 11⁄16-in. long, 0.207 to 3⁄8-in. wide hook head and medium diamond point. **F 547,** F16

hoop stress—the tensile stress in the wall of the pipe in the circumferential orientation due to internal hydrostatic pressure. **F 412,** F17

hopper—the converging portion of a bin. **D 653,** D18

horizon (soil horizon)—one of the layers of the soil profile, distinguished principally by its texture, color, structure, and chemical content.

"A" horizon—the uppermost layer of a soil profile from which inorganic colloids and other soluble materials have been leached. Usually contains remnants of organic life.

"B" horizon—the layer of a soil profile in which material leached from the overlying "A" horizon is accumulated.

"C" horizon—undisturbed parent material from which the overlying soil profile has been developed. **D 653,** D18

horizontal bar code, *n*—a bar code symbol presented in such a manner that its overall length dimension is parallel to the horizon; also called, **picket fence.** **F 1294,** F05

horizontal ladder gasket—a ladder gasket installed in a vertical plane in such a way that the intermediate cross members (muntins) are vertical members. **C 717,** C24

horizontal shear—shear that occurs in planes parallel to the longitudinal axis of the member; sometimes referred to as longitudinal shear. **D 9,** D07

horizontal sliding window—See **windows and doors.** **E 631,** E06

horizontal traction (traction)—a force acting in a horizontal axis through the wheel transducer; that is, locked wheel drag force. **E 867,** E17

hornification—the behavior of certain forms of cellulose and related materials on drying, which results in a lowered reactivity. This is apparently caused by the formation of new hydrogen bonds with a corresponding decrease in accessibility. **D 1695,** D01

horsehide leather—leather made from the hide of a horse or a colt. (See also **cordovan** and **front.**) **D 1517,** D31

horseshoe, *n*—a length of sliver folded in a manner such that the two ends can be fed simultaneously into the needle field of the Fibroliner FL-101. **D 123,** D13

horticultural mineral oil—An EPA registered pesticidal paraffinic spray oil for application on growing crops having a minimum unsulfonated residue of 92.0 % v/v by Test Method D 483. The distillation range should be narrow (80° F or less) for the 90 % minus 10 % distillation points by Test Method D 1160 at 10 mm Hg and complying with FDA 21CFR178.3620(c). **E 1519,** E35

hose, *n*—a flexible conduit consisting of a tube, reinforcement, and usually an outer cover. **D 1566,** D11

hose—an electrical insulating tube with a longitudinal slit designed to be installed temporarily on energized electrical wires. **F 819,** F18

hose assembly, *n*—a length of hose with a coupling attached to one or each end. **D 1566,** D11

hose end sprayer—an apparatus designed to be attached to standard garden hose, consisting of a hand-held container for spray mixture with an integral metering head through which water from the garden hose flows. The metering head uses water pressure, siphon effect, or some other water powered means to meter the spray mixture into the water stream which is then atomized. **E 1102,** E35

hose reinforcing wire, *n*—a single filament of steel wire with a metallic coating (usually brass) used in the reinforcement of a rubber or other elastomer hose **D 123,** D13

hosiery—any covering of the foot not intended for ground contact during ambulation, such as socks, stockings, etc. **F 1646,** F13

hospital—a special department in the porcelain enamel shop where damage to fired ware may be repaired. **C 286,** B08

host, *n*—a device or computer delivering the information to a server, in this case a printer. **F 1457,** F05

hot-applied sealant, *n*—a compound that is applied in a molten state and cures primarily by cooling to ambient temperatures. **C 717,** C24

hot-applied sealant, *n*—a compound that is applied in a molten state and cures primarily by cooling to ambient temperatures. **D 5535,** D04

hot cathode—a heated element of a vacuum enclosed electrical system emitting electrons thermionically. It is maintained at a potential negative with respect to a second element to accelerate the emitted electrons. **E 7,** E04

hot-cathode ionization gage—see **ionization vacuum gage. E 1316,** E07

hot chocolate dispenser—commercial equipment designed to deliver a predetermined amount of hot chocolate flavored beverage. **F 1827,** F26

hot coining—see **powder forging. B 243,** B09

hot-cold working, *n*—the mechanical deformation of austenitic and precipitation hardening steels at a temperature just below the **recrystallization temperature** to increase the yield strength and hardness by plastic deformation or precipitation hardening effects induced by plastic deformation, or both. **A 941,** A01

hot corrosion—an accelerated corrosion of metal surfaces that results from the combined effect of oxidation and reactions with sulfur compounds and other contaminants, such as chlorides, to form a molten salt on a metal surface which fluxes, destroys, or disrupts the normal protective oxide. **G 15,** G01

hot densification, *n*—the consolidation, at elevated temperature and pressure, of an unsintered, presintered, or sintered powder preform, or encapsulated or loose powder, in order to reduce porosity. **B 243,** B09

hot dip coating—a metallic coating obtained by dipping the basis metal into a molten metal. **B 374,** B08

hot dip galvanizing—immersion of fasteners in a bath of molten zinc for a controlled time period to obtain specified coating weight or thickness. **F 1789,** F16

hot-dipped, hot-dip galvanized, hot-dipped zinc-coated—See **dipped. F 547,** F16

hot-dip process, *n*—the application of a metallic coating on a steel product by immersion of the product in a bath of the molten metal which forms the coating; the coating may be applied in a continuous process or a batch process. **A 902,** A05

hot, dry climate—desert climate distinguished as sunny, hot, and dry year round, with rare scattered showers. **G 113,** G03

hot end—those manufacturing operations concerned with hot glass, that is, melting, forming, annealing. **C 162,** C14

hot-filament ionization gage—see **ionization vacuum gage. E 1316,** E07

hot-flame reaction, *n*—a rapid, self-sustaining, luminous, sometimes audible, reaction of the sample or its decomposition products with the atmosphere. **D 4175,** D02

hot forging—see **powder forging. B 243,** B09

hot forming—heat is applied to wire or rod to enhance metal flow into dies using machine applied pressures as opposed to metal removal by cutting for forming purposes. **F 1789,** F16

hot-galvanized, barrel-galvanized, flake-galvanized, tumbler-galvanized, wean-galvanized—zinc coating applied in heated tumbling barrel containing zinc-flakes. **F 547,** F16

hot-gas welding—a technique for joining thermoplastics (usually sheets) in which the materials are first softened by a jet of hot gas from a welding gun. A rod of the same plastic is used to fill the heated gap and join the sheets at the same time pressure is applied by either the rod or the tip of the gun. Sometimes referred to as string bead welding. **C 904,** C03

hot hold—maintaining the temperature of an idled melter below its operating temperature but above the liquidus temperature of the glass. **C 162,** C14

hot isostatic pressing, *n*—the pressing of a powder, compact or sintered object by subjecting it, at elevated temperature, to nominally equal pressure from every direction. **B 243,** B09

hot junction—See **measuring junction. E 7,** E04

hot-melt adhesive, *n*—an adhesive which is rendered fluid by heat and forms a bond upon cooling. **D 907,** D14

hot mold—the process of forming glassware in hot uncoated molds. **C 162,** C14

hot-plate welding—see *butt welding (machine).* **C 904,** C03

hot pressing, *n*—the simultaneous heating and pressing of a powder preform, compact, or encapsulated or loose powder. **B 243,** B09

hot pressing—See *hot pressing* under **pressing. C 242,** C21

hot pressing, *n*—a fabrication of a ceramic component utilizing temperature and uniaxial pressure to achieve the desired density and shape formation. Densification is achieved by particle rearrangement, viscous/plastic flow, or diffusional transport, or both. **C 1145,** C28

hot-pressing—process for increasing the density of a wet-felted or air-felted mat of fibers or particles by pressing the dried, damp, or wet mat between platens of hot-press to compact and set the structure by simultaneous application of heat and pressure. **D 1554,** D07

hot pressing—the curing of thermosets by heat and pressure application. **E 1749,** E06

hot quenching, *n*—an imprecise term used to cover a variety of quenching procedures in which the quenching medium is maintained at a prescribed temperature above 160°F or 70°C. **A 941,** A01

hot repress powder forging—hot densification of a P/M preform by forging where material flow is mainly in the direction of pressing. Also referred to as *hot restriking* or *hot coining.* **B 243,** B09

hot restriking—see **powder forging. B 243,** B09

hot-rolled drawing stock—the as-rolled section, normally round, produced by hot rolling in a rod mill from a billet or wire bar. **B 354,** B01

hot-setting adhesive, *n*—an adhesive that requires a temperature at or above 100°C (212°F) to set it. (Compare **cold-setting adhesive, intermediate-setting adhesive**, and **room-temperature-setting adhesive**.) **D 907,** D14

hot spot—the zone of highest temperature within a melter. **C 162,** C14

hot spring, *n*—a thermal spring whose temperature is above that of the human body. **E 957,** E44

hot stamped—mark made on an item with the use of heat to effect formation or transfer of (an) image. **F 1156,** F12

"hot stuff" or "hot"—a roofer's term for hot bitumen. **D 1079,** D08

hot tack, *n*—the cohesive strength during the cooling stage before solidification of a heat seal bond formed by a wax-polymer blend. **D 4175,** D02

hot top range—equipment with a heavy cast-iron top on which pots and pans are placed, intended for heavy duty continuous cooking. See **range. F 1827,** F26

hot upset powder forging—hot densification of a P/M preform by forging where there is a significant amount of lateral material flow. **B 243,** B09

hot water extraction cleaning, *n*—a process in which a heated solution of detergent is sprayed into the textile material and immediately removed by a wet suction nozzle behind the spray-head. **D 123,** D13

hot worked structure—the structure of a material worked at a temperature higher than the recrystallization temperature. **E 7,** E04

hot working, *n*—mechanical deformation of a metal at temperatures above its **recrystallization temperature**. **A 941,** A01

hot working—controlled mechanical operations for shaping a product at temperatures above the recrystallization temperature. **B 846,** B05

hot zone—the area surrounding the heating element(s) or heat exchanger surface. **F 1827,** F26

hours of operation, *n* (heures d'exploitation):

active hours, n (heures d'activité)—times when a facility is normally fully occupied and operational.

silent hours, n (heures d'inactivité)—period when a facility is essentially unoccupied and only security and building operations staff are present.

transitional hours, n (heures de transition)—times in the morning after the first workers normally arrive, until a facility is fully operational, and in the evening from the end of normal work until the occupants have left. **E 1480,** E06

house—See **building.** **E 631,** E06

house depressurization, *n*—the situation, pertaining to a specific location in a house, whereby the static pressure at that location is lower than the static pressure in the immediate vicinity outside the house. **D 1356,** D22

household dirt, *n*—dirt commonly found in residential households, the specific compositions of which will vary from one environment to another and is removable with a household vacuum cleaner. **F 395,** F11

household vacuum cleaner, *n*—a vacuum cleaner suitable for the normal cleaning tasks encountered in residences, in contrast to a commercial vacuum cleaner. **F 395,** F11

house, *n* (maison)—**building** intended in its entirety as a **dwelling**. **E 1480,** E06

HPC—heterotrophic plate count. Formerly called SPC. **D 6161,** D19

huck towel, *n*—a plain weave nonterry foundation product constructed with small warp and filling floats, having hems or selvages, which is used to dry a person's hands and utensils such as glasses, plates, bowls, and flatware. **D 123,** D13

HUD—U.S. Department of Housing and Urban Development. **E 1605,** E06

hue—under **color of an object,** see *hue*. **D 16,** D01

hue, *n*—attribute of color perception by means of which objects are judged to be red, yellow, blue, or intermediate between some adjacent pair of these. **D 2946,** C17

hue, *n*—that attribute of a colored object by which it is identified as red, green, blue, etc. It is the psychological correlate of the physical dimension of dominant or complementary wavelength. **E 253,** E18

hue, *n*—the attribute of color perception by means of which a color is judged to be red, orange, yellow, green, blue, purple, or intermediate between adjacent pairs of these, considered in a closed ring (red and purple being an adjacent pair.) (See also **Munsell hue**.) **E 284,** E12

hue—(a particular) color (as distinct from other colors; a quality that distinguishes colors in the visible portion of the spectrum; the first of the three dimensions of color: hue, lightness, and saturation.) **F 1156,** F12

Hull cell—a trapezoidal box of nonconducting material with electrodes arranged to permit observation of cathodic or anodic effects over a wide range of current densities. **B 374,** B08

Hull-Davey charts—charts for indexing the lines of powder patterns on which a function of the interplanar spacing of the Bragg angle is plotted against the axial ratio for a number of different lattice planes. **E 7,** E04

Hull method (for X-ray crystal analysis)—See **Debye-Scherrer method**. **E 7,** E04

human-readable interpretation, *n*—the letters, digits or other characters associated with specific symbol characters and printed along with the linear bar code or two-dimensional symbol. **F 1294,** F05

human resources, *n*—those elements of support or capability that are provided by humans using their mental and physical capabilities. **D 1356,** D22

human tissue burn tolerance—in the testing of thermal protective materials, the amount of thermal energy predicted to cause a second-degree burn in human tissue. **F 1494,** F23

humectant—a material which increases the equilibrium water content and increases the drying time of an aqueous spray deposit. **E 1519,** E35

humic acid—a variety of water-soluble organic compounds, formed by the decayed vegetable matter, which is leached into a water source by runoff or percolation. Present in most surface and some ground waters. Higher concentrations cause a brownish tint. Difficult to remove except by adsorption, ultrafiltration, nanofiltration or reverse osmosis. **D 6161,** D19

humic peat—see **sapric peat**. **D 653,** D18

humidify—to increase, by any process, the quantity of water vapor within a given space. **E 41,** G03

humidistat—a regulatory device, activated by changes in humidity, used for the automatic control of relative humidity. **E 41,** G03

humidity, *n*—the condition of the atmosphere in respect to water vapor. (Compare **absolute humidity** and **relative humidity**.) **D 123,** D13

humidity, *n*—a measure of the amount of water vapor in a gas. Also see **absolute humidity** and **relative humidity.** **D 1356,** D22

humidity—the condition of the atmosphere in respect to water vapor. (See also **humidity, absolute; humidity, relative.**) **E 41,** G03

humidity, absolute, *n*—the mass of water vapor per unit volume. **C 168,** C16

humidity, absolute—the mass of water vapor per unit volume of the atmosphere usually measured as grams per m^3. **D 6161,** D19

humidity, absolute—the weight of water vapor present in a unit volume of air, for example, grains per cubic foot, or grams per cubic metre.

NOTE—The amount of water vapor is also reported in terms of weight per unit weight of dry air, for example, grains per pound of dry air. This value differs from values calculated on a volume basis and should not be referred to as absolute humidity. It is designated as **humidity ratio, specific humidity,** or **moisture content,** which also see. **E 41,** G03

humidity indicator—an instrument or device that displays the approximate humidity condition within a **package.** **D 996,** D10

humidity range—See **operating humidity range of analyzer** under **analyzer.** See also **operating humidity range of sample** under **sample.** **D 1356,** D22

humidity ratio—in a mixture of water vapor and air, the mass of water vapor per unit mass of dry air. **E 41,** G03

humidity, relative, *n*—the ratio of the mol fraction of water vapor present in the air to the mol fraction of water vapor present in saturated air at the same temperature and barometric pressure. Approximately, it equals the ratio of the partial pressure or density of the water vapor in the air to the saturation pressure or density, respectively, at the same temperature. **C 168,** C16

humidity, relative—the ratio of the actual pressure of existing water vapor to the maximum possible (saturation) pressure of water vapor in the atmosphere at the same temperature, expressed as a percentage. **D 6161,** D19

humidity, relative—the ratio of the actual pressure of existing water vapor to the maximum possible (saturation) pressure of water vapor in the atmosphere at the same temperature, expressed as a percentage. **E 41,** G03

humidity, specific—in a mixture of water vapor and air, the mass of water vapor per unit mass of moist air. **E 41,** G03

humification—a process by which organic matter decomposes. **D 653,** D18

humus—a brown or black material formed by the partial decomposition of vegetable or animal matter; the organic portion of soil. **D 653,** D18

hung window—See **windows and doors. E 631,** E06

Hunter color difference, *n*—color difference calculated by the use of the Hunter equations, based on the opponent-color coordinates, *L, a, b,* applied to CIE 1931 tristimulus values for CIE standard illuminant *C,* and by extension to the CIE 1964 standard observer and other CIE standard illuminants. **E 284,** E12

Huygens eyepiece—an achromatic eyepiece invented by Huygens and consisting of a plano-convex eyelens and a plano-convex collective, between which is a field diaphragm. **E 7,** E04

H_{50} value, *n*—a drop height with a 50 % probability of reaction, as determined experimentally by the Bruceton up-and-down method. **E 1445,** E27

HVPS—abbreviation for **high-voltage power supply. F 1457,** F05

hybar wire—flat wire normally used in box stitching. Specifically:

Gage No.	Cross-Sectional Dimensions, in. (mm)
000	0.060 by 0.017 (1.52 by 0.43)
00	0.060 by 0.018 (1.52 by 0.46)
0	0.060 by 0.019 (1.52 by 0.48)
1	0.060 by 0.020 (1.52 by 0.51)
2	0.060 by 0.024 (1.52 by 0.61)
3	0.060 by 0.028 (1.52 by 0.71)

F 592, F16

hybrid—(for composite materials) containing at least two distinct types of matrix or reinforcement. Each matrix or reinforcement type can be distinct because of its a) physical or mechanical properties, or both, b) material form, or c) chemical composition. **D 3878,** D30

hybrid asymmetrical—this refers to a snowboard ski shape that is asymmetrical from side to side but symmetrical from tip to tail, allowing the same board to be used by both regular-foot and goofy-foot riders by reversing the direction of travel, but retaining the non-mirror image sidecut shapes of a full asymmetrical. **F 1107,** F27

hybrid longitudinal element, *n*—a longitudinal element consisting of two or more types of longitudinal elements of different size or cross-section manufactured into a single element. **F 1582,** F04

hydrate—a solid, crystalline material composed of water and components of natural gas formed under pressure at temperatures above the freezing point of water. **D 4150,** D03

hydrated lime—a dry powder obtained by treating quicklime with water enough to satisfy its chemical affinity for water under the conditions of its hydration. It consists essentially of calcium hydroxide or a mixture of calcium hydroxide and magnesium oxide or magnesium hydroxide, or both. **C 51,** C07

hydrated lime—dry calcium hydroxide. **D 6161,** D19

hydration, *n*—the chemical reaction between hydraulic cement and water forming new compounds most of which have strength-producing properties. **C 219,** C01

hydration—formation of a compound by the combining of water with some other substance. **D 653,** D18

hydration resistance, *n*—the degree to which a refractory material resists chemical combination with water. This resistance is measured by either Test Method C 456 or Test Method C 492, whichever is applicable. **C 71,** C08

hydraulic atomizer—see **pressure atomizer. E 1620,** E29

hydraulic cement, *n*—a cement that sets and hardens by chemical reaction with water and is capable of doing so under water. **C 125,** C09

hydraulic cement, *n*—a cement that sets and hardens by chemical interaction with water and that is capable of doing so under water. **C 219,** C01

hydraulic communication—the migration of fluids from one zone to another, with reference to this guide, especially along a casing grout plug, or through back fill materials. **D 653,** D18

hydraulic conductivity (field aquifer tests)—the volume of water at the existing kinematic viscosity that will move in a unit time under a unit hydraulic gradient through a unit area measured at right angles to the direction of flow. **D 653,** D18

hydraulic conductivity *(k),* *n*—the rate of discharge of water under laminar flow conditions through a unit cross-sectional area of a porous medium under a unit hydraulic gradient and standard temperature conditions (20°C). **D 4439,** D35

hydraulic conductivity ratio (HCR), *n*—the ratio of the hydraulic conductivty of the soil/geotextile system, k_{sg}, at any time during the test, to the initial hydraulic conductivity, k_{sgo}, measured at the beginning of the test (new). **D 4439,** D35

hydraulic fluid, *n*—liquid used in hydraulic systems for transmitting power. **D 4175,** D02

hydraulic fracturing—the fracturing of an underground strata by pumping water or grout under a pressure in excess of the tensile strength and confining pressure; also called hydrofracturing. **D 653,** D18

hydraulic gradient—the change in total hydraulic head of water per unit distance of flow.

critical hydraulic gradient, i_c (D)—hydraulic gradient at which the intergranular pressure in a mass of cohesionless soil is reduced to zero by the upward flow of water. **D 653,** D18

hydraulic gradient, *i, s (D)*—the loss of hydraulic head per unit distance of flow, dH/dL. **D 4439,** D35

hydraulic hydrated lime—the hydrated dry cementitious product obtained by calcining a limestone containing silica and alumina to a temperature short of incipient fusion so as to form sufficient free lime (CaO) to permit hydration, and at the same time, leaving unhydrated sufficient calcium silicates to give a dry powder meeting hydraulic property requirements. **C 51,** C07

hydraulic jump—an abrupt transition from supercritical flow to subcritical or tranquil flow, accompanied by considerable turbulence or gravity waves, or both. **D 4410,** D19

hydraulic leathers—see **gasket leather. D 1517,** D31

hydraulic mortar, *n*—deprecated term; see mortar. **C 219,** C01

hydraulic mortar—a mortar that is capable of setting and hardening due to the interaction of water and the constituents of the mortar. **C 904,** C03

hydraulic pressure test—Same as **hydrostatic test. E 1316,** E07

hydraulic transmissivity, θ (L^2 T^{-1}), *n—for a geotextile or related product,* the volumetric flow rate of water per unit width of specimen per unit gradient in a direction parallel to the plane of the specimen. **D 4439,** D35

hydrocarbon dew point—(see **dew point**) **D 4150,** D03

hydrocarbon plastics—plastics based on resins made by the polymerization of monomers composed of carbon and hydrogen only. **D 883,** D20

hydrocelluloses—water-insoluble products of the hydrolysis of cellulose with acids. They are molecularly heterogeneous in the sense that they are composed of molecules varying in degree of polymerization. The average degree of polymerization (DP) and the DP distribution depend on the nature of the acid treatment and of the original cellulose. The term may also be applied to any insoluble polysaccharide so formed and separated as a more or less homogeneous fraction from the mixture of products, but the singular form" hydrocellulose" should not be used without an article, to avoid the implication of a molecularly homogeneous species. **D 1695,** D01

hydrodynamic specific surface—the specific surface of a fibrous material as measured by the filtration resistance of a compacted pad formed from a fiber suspension under specified conditions. **D 1695,** D01

hydrogenated resin, *n*—a resin that has been subjected to some degree of reaction with hydrogen to decrease the level of residual unsaturation. **D 6440,** D01

hydrogen blistering—the formation of blisters on or below a metal surface from excessive internal hydrogen pressure. (Hydrogen may be formed during cleaning, plating, corrosion, etc.) **G 15,** G01

hydrogen cycle—the operation of a cation-exchange cycle wherein the removal of specified cations from the influent water is accomplished by exchange with an equivalent amount of hydrogen ion from the exchange material. **D 1129,** D19

hydrogen embrittlement—embrittlement of a metal or alloy caused by absorption of hydrogen which may occur, for example, during pickling, cathodic cleaning, electroplating, and autocatalytic plating processes. **B 374,** B08

hydrogen embrittlement—hydrogen-induced cracking or severe loss of ductility caused by the presence of hydrogen in the metal. **G 15,** G01

hydrogen loss—the loss in weight of metal powder or of a compact caused by heating a representative sample for a specified time and temperature in a purified hydrogen atmosphere—broadly a measure of the oxygen content of the sample, when applied to materials containing only such oxides as are reducible with hydrogen and no hydride forming element. **B 243,** B09

hydrogen overvoltage—overvoltage associated with the liberation of hydrogen. **B 374,** B08

hydrogen sulfide (H_2S)—mol weight 34.08; flammable, poisonous gas with characteristic odor of rotten eggs. **D 4790,** D16

hydrogen treating—a refining process in which an unfinished petroleum insulating oil is contacted with hydrogen gas at elevated temperatures and pressures in the presence of a catalyst, to improve its color, odor, stability, and other properties. **D 2864,** D27

hydrokinetic roof system, *n*—a roof system that depends on quick drainage via water shedding to prevent water entry into or through the system. **D 1079,** D08

hydrologic unit—geologic strata that can be distinguished on the basis of capacity to yield and transmit fluids. Aquifers and confining units are types of hydrologic units. Boundaries of a hydrologic unit may not necessarily correspond either laterally or vertically to lithostratigraphic formations. **D 653,** D18

hydrolysis—the act of cleaving or splitting of complex molecules by the chemical addition of a water molecule. Acid hydrolysis is defined as the chemical addition of water to a compound such as starch in the presence of an acid as a catalyst that will form another compound such as glucose. **E 1705,** E48

hydrolytic adsorption—the adsorption of a weakly ionized acid or base formed by the hydrolysis of some types of salts in aqueous solution. **D 2652,** D28

hydrolytically degradable plastic, *n*—a degradable plastic in which the degradation results from hydrolysis. **D 883,** D20

hydrolytically degradable plastics, n—See **degradable plastic.** **D 883,** D20

hydrolytic stability, *n*—the ability to withstand the environmental effects of high humidity. **D 123,** D13

hydrolytic stability, *n*—the resistance of a material to permanent changes in properties due to reaction with water. **D 4175,** D02

hydrometer-thermometer field tester, *n*—a device designed to indicate the freezing point of an engine coolant by measurement of the relative density of the coolant at a specific temperature. **D 4725,** D15

Hydrophelicity—the property that defines a material as attracting water. Water exhibits an acute contact angle with hydrophilic materials. **D 653,** D18

hydrophilic—(*1*) tending to absorb water.

(*2*) tending to concentrate in the aqueous phase. **B 374,** B08

hydrophilic—having an affinity for water. **D 1695,** D01

hydrophilic—having an affinity for water. **D 6161,** D19

hydrophilic—wettable by water; as opposed to oleophilic (see **oleophilic**); important materials' property in relation to their use in ink-making. **F 1156,** F12

hydrophilic—having a strong affinity for water, wettable. **F 1241,** F01

hydrophilic emulsifier—see **emulsifier.** **E 1316,** E07

hydrophilic (literally, water-loving)—a descriptive term applied to the group or radical of a surfactant molecule that makes or tends to make it soluble in water. **D 459,** D12

hydrophobic—(*1*) tending to repel water.
(*2*)lacking affinity for water. **B 374,** B08

hydrophobic—having an antagonism to water. **D 1695,** D01

hydrophobic—lacking an affinity to water. **D 6161,** D19

hydrophobic—having little affinity for water, nonwettable. **F 1241,** F01

hydrophobicity—the property that defines a material as being water repellent. Water exhibits an obtuse contact angle with hydrophobic materials. **D 653,** D18

hydrophobic (literally, water-averting)—a descriptive term applied to the moiety of a surfactant molecule that makes it, or tends to make it, insoluble in water. **D 459,** D12

hydroplaning (aquaplaning) of pneumatic tires, *n*—a phenomenon that occurs when the load-bearing surface of a pneumatic tire is separated from a solid surface by a substance (usually a fluid and usually water). **E 867,** E17

hydroplaning, dynamic—see **dynamic hydroplaning.** **E 867,** E17

hydroplaning, rubber reversion—see **rubber reversion hydroplaning.** **E 867,** E17

hydroplaning speed, *n*—the initial speed at which a pneumatic tire begins to full dynamic hydroplaning under a given set of conditions. **E 867,** E17

hydroplaning, viscous—see **viscous hydroplaning.** **E 867,** E17

hydrostatic design stress—the recommended maximum hoop stress that can be applied continuously with a high degree of certainty that failure of the pipe will not occur. **F 412,** F17

hydrostatic head—the fluid pressure of formation water produced by the height of water above a given point. **D 653,** D18

hydrostatic pressure—see **hydrostatic strength.** **C 460,** C17

hydrostatic pressure, *n*—a state of stress in which all the principal stresses are equal (and there is no shear stress), as in a liquid at rest; the product of the unit weight of the liquid and the difference in elevation between the given point and the free liquid elevation. **C 717,** C24

hydrostatic pressure—the pressure exerted by water at rest. **C 822,** C13

hydrostatic pressure—a state of stress in which all the principal stresses are equal (and there is no shear stress). (ISRM) **D 653,** D18

hydrostatic pressure—see **hydrostatic strength**. **D 2946,** C17

hydrostatic pressure, *n*—a state of stress in which all the principal stresses are equal (and there is no shear stress), as in a liquid at rest; induced artificially by means of a gaged pressure system; the product of the unit weight of the liquid and the difference in elevation between the given point and the free water elevation. **D 4439,** D35

hydrostatic pressure, u_o **(FL^{-2})**—a state of stress in which all the principal stresses are equal (and there is no shear stress), as in a liquid at rest; the product of the unit weight of the liquid and the different in elevation between the given point and the free water elevation.

excess hydrostatic pressure (hydrostatic excess pressure), $\bar{u}$, u (FL^{-2})—the pressure that exists in pore water in excess of the hydrostatic pressure. **D 653,** D18

hydrostatic roof system, *n*—a roof system that is capable of holding water without allowing water to penetrate the system. **D 1079,** D08

hydrostatic strength—the ability of a pipe, under specified conditions to withstand internal pressure of specified magnitude. **C 460,** C17

hydrostatic strength, *n*— *in pipes*, the ability of the pipe and coupling sleeve to withstand the forces resulting from internal pressure. **C 1154,** C17

hydrostatic strength—the ability of a pipe, under specified conditions, to withstand internal pressure of specified magnitude. **D 2946,** C17

hydrostatic test—a test of the ability of a pipe or its joint to withstand internal hydrostatic pressure. **C 822,** C13

hydrostatic test—in leak testing, a pressure test in which the component being tested is filled completely with water or another liquid. Pressure, if required, is then applied to the liquid for the required time and the outside of the component is examined visually for leaks. **E 1316,** E07

hydrotropy—the increase in solubility of a substance which is only slightly soluble in an aqueous system by the addition of a third substance. This third substance is called a "hydrotrope" or "hydrotropic agent." **D 459,** D12

(2-hydroxyethyl)cellulose—any of several cellulose ethers in which some of the hydroxyl groups have been substituted with hydroxyethyl groups. Hydroxyethyl cellulose, except at very low degrees of substitution, is water-soluble. **D 1695,** D01

(2-hydroxyethyl)ethylcellulose—a mixed ether of cellulose containing both hydroxyethyl and ethyl groups. **D 1695,** D01

(2-hydroxyethyl)methylcellulose—a mixed ether of cellulose containing both hydroxyethyl and methyl groups. **D 1695,** D01

hydroxyl alkalinity—see **alkalinity**. **D 6161,** D19

hydroxyl groups—the monovalent group —OH, characteristic of hydroxides and alcohols. **D 1695,** D01

hydroxyl number—the number of milligrams of potassium hydroxide (KOH) equivalent to the hydroxyl content of 1 g of sample. **D 16,** D01

hydroxyl number—the milligrams of potassium hydroxide equivalent to the hydroxyl content of 1 g of sample. In the case of a pure compound, the hydroxyl number is inversely proportional to the hydroxyl equivalent weight:

$$\text{equivalent weight (g/equivalent)} = 56100/\text{hydroxyl number}$$

E 1547, E15

(2-hydroxypropyl)methylcellulose—a mixed ether of cellulose containing both hydroxypropyl and methyl groups. **D 1695,** D01

hygrochromic, *adj*—known to change color when the moisture content or relative humidity varies. **E 284,** E12

hygrometer, *n*—any instrument for measuring the humidity of the atmosphere. **D 123,** D13

hygrometer, *n*—an instrument for measuring the humidity of a gas.

dew-/frost-point hygrometer, n—an instrument that measures the surface temperature at which ambient water vapor condenses.

electric hygrometer, n—an instrument that determines the water vapor content of an atmosphere by measuring the change in resistance or capacitance of hygroscopic material.

mechanical hygrometer, n—an instrument for determining the water vapor content of an atmosphere by measuring the dimensional change produced in an hygroscopic material. **D 1356,** D22

hygroscopic—attracting, absorbing, and retaining atmospheric moisture. **D 1079,** D08

hygroscopic—attracting, absorbing, and retaining atmospheric moisture. **E 1749,** E06

hygroscopic—to absorb moisture from the atmosphere; having a tendency to absorb moisture from the atmosphere. **F 1156,** F12

hygroscopic capacity (hygroscopic coefficient), w_e **(D)**—ratio of: (*1*) the weight of water absorbed by a dry soil or rock in a saturated atmosphere at a given temperature, to (*2*) the weight of the oven-dried soil or rock. **D 653,** D18

hygroscopic water content, w_H **(D)**—the water content of an air-dried soil or rock. **D 653,** D18

hyperbaric chamber, *n*—a chamber in which the internal pressure can be elevated by the introduction of compressed air (or other gasses) that is used primarily for medical treatment (including decompression sickness and arterial gas embolism) and physiological studies. **F 1549,** F32

hypereutectic alloy—any composition between the eutectic point and the composition of that solid phase, of the pair into which the eutectic liquid decomposes, which is considered the less important; thus, the identification of the hypereutectic range, as distinguished from the hypoeutectic range, is a matter either of personal preference, or a common usage for each alloy system. **E 7,** E04

hypereutectoid alloy—any composition between the eutectoid point and the composition of that solid phase, of the pair into which the eutectoid solid phase decomposes, which is considered the less important; for the meaning of "important," see **hypereutectic alloy**. **E 7,** E04

hypereutectoid structure—the microstructure of a hypereutectoid alloy. For example, microstructural aggregate found in iron-carbon alloys which consist of primary crystals of cementite together with nodules of pearlite. **E 7,** E04

hyperfiltration—separation of dissolved ions from a feed stream as in nanofiltration and reverse osmosis. **D 6161,** D19

hypochlorite bleach—solution containing the hypochlorite ion (ClO^-) in a reactive solution. **F 1156,** F12

hypoeutectic alloy—any composition between the eutectic point and the composition of that solid phase, of the pair into which the eutectic liquid decomposes, which is considered the more important; for the meaning of "important," see **hypereutectic alloy**. **E 7, E04**

hypoeutectoid alloy—any composition between the eutectoid point and the composition of that solid phase, of the pair into which the eutectoid solid phase decomposes, which is considered the more important; for the meaning of "important," see **hypereutectic alloy**. **E 7, E04**

hypoeutectoid structure—the microstructure of a hypoeutectoid alloy. For example, the microstructural aggregate found in iron-carbon alloys which consist of primary crystals of ferrite together with nodules of pearlite. **E 7, E04**

hysteresis—the residual strain in stone after the stress causing such strain is changed. **C 119, C18**

hysteresis—incomplete recovery of strain during unloading cycle due to energy consumption. (ISRM) **D 653, D18**

hysteresis, *n*—the conversion of mechanical energy to heat in rubber undergoing strain. **D 1566, D11**

hysteresis—the maximum difference between corresponding transducer outputs (of the wheel force transducer system) at increasing and decreasing applied calibration force, expressed as a percentage of maximum output. Proven outliers are excluded. **E 867, E17**

hysteresis—(1) the lagging of the magnetic effect when the magnetic force acting upon a ferromagnetic body is changed.

(2) the phenomenon exhibited by a magnetic system wherein its state is influenced by its previous history. **E 1316, E07**

hysteresis—failure of a property that has been changed by an external agent to return to its original value when that external agent has been removed; especially if the original change is not reversed exactly, even though the end point is the same as the starting point. **F 1156, F12**

hysteresis diagram—*in fatigue*, the stress-strain path during a cycle. **E 1823, E08**

hysteresis (electrode memory)—hysteresis is said to have occurred if, after the concentration has been changed and restored to its original value, there is a different potential observed. The reproducibility of the electrode will consequently be poor. The systematic error is generally in the direction of the concentration of the solution in which the electrode was previously immersed. **D 4127, D19**

hysteresis loop—the divergence between the paths of the adsorption and desorption isotherms. **D 2652, D28**

hysteresis loop, biased—an incremental hysteresis loop that lies entirely in any one quadrant.

NOTE—In this case, both of the limiting values of H and B are in the same direction. **A 340, A06**

hysteresis loop, incremental—the hysteresis loop, nonsymmetrical with respect to the B and H axes, exhibited by a ferromagnetic material in a *CM* condition.

NOTE—In this case, both of the limiting values H may have opposite polarity, but definitely have different absolute values of H_m. An incremental loop may be initiated at either some point on a normal hysteresis loop or at some point on the normal induction curve of the specimen. **A 340, A06**

hysteresis loop (*in dynamic mechanical measurement*)—the closed curve representing successive stress-strain status of the material during a cyclic deformation. (ISO) **D 4092, D20**

hysteresis loop, intrinsic—a hysteresis loop obtained with a ferromagnetic material by plotting (usually to rectangular coordinates) corresponding dc values of intrinsic induction, B_i, for ordinates and magnetic field strength H for abscissae. **A 340, A06**

hysteresis loop loss, W_h—the power expended in a single slow excursion around a normal hysteresis loop. The energy is the integrated area enclosed by the loop measured in gauss-oersteds. Using the cgs-emu system of units:

$$W_h = (\int HdB/4\pi) \text{ ergs}$$

where the integrated area enclosed by the loop is measured in gauss-oersteds. **A 340, A06**

hysteresis loop, normal—a closed curve obtained with a ferromagnetic material by plotting (usually to rectangular coordinates) corresponding dc values of magnetic induction (B) for ordinates and magnetic field strength (H) for abscissa when the material is passing through a complete cycle between equal definite limits of either magnetic field strength, $\pm H_m$, or magnetic induction, $\pm B_m$. In general, the normal hysteresis loop has mirror symmetry with respect to the origin of the B and H axes, but this may not be true for special materials. **A 340, A06**

hysteresis loss, *n*—the loss of mechanical energy due to hysteresis. **D 1566, D11**

hysteresis loss, incremental, $P_{\Delta h}$—the power (watts) as a result of hysteresis expended in a ferromagnetic material while being driven through an incremental flux-current loop by a *CM*-type of excitation. **A 340, A06**

hysteresis loss, normal, P_h—(*1*) the power expended in a ferromagnetic material, as a result of hysteresis, when the material is subjected to a *SCM* excitation.

(2) The energy loss/cycle in a magnetic material as a result of magnetic hysteresis when the induction is cyclic (but not necessarily periodic). **A 340, A06**

hysteresis loss, rotational—the hysteresis loss that occurs in a body when subjected to a constant magnetizing force, the direction of which rotates with respect to the body, either in a continuously cyclic or in a repeated oscillatory manner. **A 340, A06**

hysteresis, magnetic—the property of a ferromagnetic material exhibited by the lack of correspondence between the changes in induction resulting from increasing magnetic field strength and from decreasing magnetic field strength. **A 340, A06**

hysteretic angle, magnetic, β—the mean angle by which the fundamental component of exciting current leads the fundamental component of magnetizing current, I_m, in an inductor having a ferromagnetic core.

NOTE—Because of hysteresis, the instantaneous value of the hysteretic angle will vary during the cycle of *SCM* excitation. However, β is taken to be the mean effective value of this angle. **A 340, A06**

I

IACS—the International Annealed Copper Standard; an international standard of electrical conductivity. **E 1316,** E07

IC50—a statistically or graphically estimated concentration of test material that, under specified conditions, is expected to cause a 50 % inhibition of a biological process (such as growth or reproduction) for which the data are not dichotomous. **E 943,** E47

IC—integrated circuit. **F 1156,** F12

ice—a glass enamel frit which, when applied to glassware and fired, produces a textured surface. **C 162,** C14

ice boom—boom intended for use in ice-infested waters, designed to withstand effects of ice contact. **F 818,** F20

ice-bulb temperature—See **temperature.** **D 1356,** D22

ice-bulb temperature, t_i—the temperature that a thermometer indicates when its bulb is surrounded by a thin film of ice (or a frozen moistened covering), and allowed to cool by sublimation of the ice into the surrounding air. **D 4023,** D22

ice dam—a mass of ice formed at the transition from a warm to a cold roof surface. Frequently formed by refreezing meltwater at the overhang of a steep roof, an ice dam may cause ice and water to back up under shingles or other roofing materials. **D 1079,** D08

ice, dry, *n*—smooth ice without loose surface materials. **F 538,** F09

ice point, *n*—thermometric fixed point of ice and water saturated with air at a pressure of 101 325 Pa. **E 344,** E20

ICI, *n*—trademarked abbreviation for Imperial Chemical Industries. Do not use as the abbreviation for International Commission on Illumination; see **CIE**. **E 284,** E12

IC/TC, *n*—abbreviation for Intelligent Color/Trash Coordinator. **D 123,** D13

ID—inside dimensions or inside diameter. **D 996,** D10

ID coil—a coil or coil assembly used for electromagnetic testing by insertion into the test piece as in the case of an inside probe for tubing. Coils of this type are also referred to as inside coils, inserted coils, or bobbin coils. **E 1316,** E07

ideal black, *n*—object or material that absorbs all light impinging on it. **D 2946,** C17

ideal crack—a simplified model of a crack. In a stress-free body, the crack has two smooth surfaces that are coincident and join within the body along a smooth curve called the crack front; in two-dimensional representations the crack front is called the crack tip. **E 1823,** E08

ideal-crack-tip stress field—the singular stress field, infinitesimally close to the crack front, that results from loading an ideal crack. In a linear-elastic homogeneous body, the significant stress components vary inversely as the square root of the distance from the crack tip. **E 1823,** E08

ideal gas—a gas that obeys Boyle's law and has zero heat of free expansion (or also obeys Charles' law). Also known as a perfect gas. **E 1316,** E07

identification—See **grip mark.** **F 547,** F16

identification (definite conclusion of identity)—this is the highest degree of confidence expressed by document examiners in handwriting comparisons. The examiner has no reservations whatever, and although prohibited from using the word "fact," the examiner is certain, based on evidence contained in the handwriting, that the writer of the known material actually wrote the writing in question.

Examples—It has been concluded that John Doe wrote the questioned material, or it is my opinion [or conclusion] that John Doe of the known material wrote the questioned material. **E 1658,** E30

identifyee—person or thing being identified. **F 1156,** F12

identity period—the distance between equivalent points in adjacent unit cells, usually measured in a direction parallel to a crystal axis. **E 7,** E04

ideographic—(an) image conveying an idea, such as Chinese writing. **F 1156,** F12

idiomorphic crystal—single crystals that have grown without restraint so that the habit planes are clearly developed. **E 7,** E04

idle energy rate—the average rate of energy consumed (Btu/h or kW) by the equipment while it is "holding" the cooking medium, cavity, or surface (see **cooking medium, cooking cavity or cooking surface**) at the equipment's thermostat set point, $E_{idle\ rate}$.

$$E_{idle\ rate} = \frac{E_{idle} \times 60}{t_{idle}}$$

where:

$E_{idle\ rate}$ = idle energy rate,
E_{idle} = idle energy consumption, Btu, kWh or kJ, and,
t_{idle} = length of idle test period, min.

F 1827, F26

idler rollers—rollers in the inking device of a lithographic duplicator that are rotated by frictional contact with gear-driven rollers. **F 425,** F05

IDT, *n*—abbreviation for Intelligent Data Terminal. **D 123,** D13

IEC—International Electrotechnical Commission. **F 1156,** F12

IF/F—Identification Friend/(or)Foe. **F 1156,** F12

ignitability, *n*—the propensity to ignition, as measured by the time to sustained flaming, in seconds, at a specified heating flux. **E 176,** E05

ignitability, *n*—the propensity for ignition, as measured by the time to sustained flaming at a specified heating flux. **E 176,** E05

ignition, *n*—the initiation of combustion. **D 123,** D13

ignition, *n*—the initiation of combustion. **E 176,** E05

ignition, *n*—the initiation of combustion. **E 1445,** E27

ignition, *n*—the initiation of combustion. **F 1494,** F23

ignition delay time, *n*—the time lapse between application of heat to a material and its ignition. It is the time in seconds between insertion of the sample into the flask and ignition. It is maximum at the minimum autoignition temperature and also referred to as ignition lag. **E 1445,** E27

ignition delay time, t_d, *n*—the time between the onset of dispersion of the dust sample into a cloud and the activation of the ignition source. **E 1445,** E27

ignition loss, *n*—*in glass textiles*, the amount of organic material consumed by ignition. **D 123,** D13

ignition temperature, *n*—the lowest temperature at which sustained combustion of a material can be initiated under specified test conditions. **E 176,** E05

ignition temperature (kindling point)— the lowest temperature at which combustion will occur spontaneously under specified conditions. **D 2652,** D28

illegal upgrade—substitution of a component of lesser grade or quality for one of higher grade or quality during manufacture, rework, or repair in contravention of **OEM** specifications. **F 1156,** F12

ill-fitting—a term descriptive of footwear that is either too loose or tight on the foot for its intended purpose or protection thereof. **F 869,** F08

illicit duplication—making copies without the permission to do so; simple counterfeiting. **F 1156,** F12

illuminance—luminous irradiance. **E 772,** E44

illuminance—the density of luminous flux on a surface per unit area. Measured in the SI system by lux. **E 1316,** E07

illuminance, *E, E* $_v$, n—luminous flux incident per unit of area. **E 284,** E12

illuminant, *n—in color determination,* incident luminous flux specified only by its spectral distribution. **D 2946,** C17

illuminant, *n*—radiant flux that may be specified by its spectral power distribution, and that can, in illuminating objects, affect their perceived colors. **E 284,** E12

illuminant metamerism, *n*—the property of specimens having different spectral characteristics and having the same color when viewed by a normal observer under a given illuminant, but different colors when viewed under a different illuminant, other conditions remaining the same. **E 284,** E12

illuminant mode, *n*—color seen as ascribed to a source of illumination. **E 284,** E12

illumination, *n—in lighting*, the density or flux of light on a unit area of surface. **D 123,** D13

illumination—See **brightfield, conical, darkfield, polarized light**. **E 7,** E04

illuminator, *n*—the portion of a radiometric or photometric instrument that provides the illuminating beam on the specimen, including the source, occasionally the monochromator or spectral filters, a diffuser such as an integrating sphere, if used, and associated optics. **E 284,** E12

illuminator plane, *n*—the plane containing the specimen normal and the axis of the illuminator. **E 284,** E12

ilmenite—a mineral having the theoretical composition $FeO{\cdot}TiO_2$ used principally in the production of titanium oxide. **C 242,** C21

image—a representation of an object produced by means of radiation, usually with a lens or mirror system. **E 7,** E04

image—the optical counterpart of an object produced by means of an image-producing device. **F 221,** F05

image—the optical counterpart of an object produced by a lens or mirror system or the graphic representation of an object. **F 335,** F05

image area, *n*—that portion of the page which is printed, including the space between letters and lines. (See **percent coverage** and **maximum image area**.) **F 1457,** F05

image color—appearance of the image dependent upon the spectral reflectance of the image, the spectral composition of the incident light, and the spectral response of the observer. **F 221,** F05

image data file—a digital file containing radiological image and text information. **E 1316,** E07

image definition—see **definition.** **E 1316,** E07

image density—the contrast between image and background as measured by densitometer. **F 221,** F05

image density—the optical density of the developed image as detected by the eye, or measured by a reflection densitometer. **F 335,** F05

image density, *n*—a perception of the image darkness as detected by the eye or measured by a reflection densitometer, where the image density is the log to the base ten of the reciprocal of the reflectance. (See **reflectance**.) **F 1457,** F05

image density, *n*—a measure of the reflectance or density of the imaged portion of a thermal imaging product after exposure to heat energy; typically measured with a densitometer, reflectometer, or opacimeter. **F 1623,** F05

image depth profile—*AES, XPS, SIMS*, a three-dimensional representation of the spatial distribution of a particular elemental or molecular species (as indicated by emitted secondary ions or electrons) as a function of depth or material removed by sputtering. **E 673,** E42

image processing—a method whereby digital image data is transformed through a mathematical function. **E 1316,** E07

image processing, *n*—techniques for filtering, storing and retrieving images, and for processing pictorial information by computer. **F 1294,** F05

image processing, in image analysis—the computer modification of a digitized image on a pixel-by-pixel basis to emphasize or de-emphasize certain aspects of the image. **E 7,** E04

image quality indicator—a device or combination of devices whose image or images on a neutron radiograph provide visual or quantitative data, or both, concerning the radiographic sensitivity of the particular neutron radiograph. **E 1316,** E07

image quality indicator (IQI)—in industrial radiology, a device or combination of devices whose demonstrated image or images provide visual or quantitative data, or both, to determine radiologic quality and sensitivity. Also known as a penetrameter (disparaged).

Note—It is not intended for use in judging size nor establishing acceptance limits of discontinuities. **E 1316,** E07

image rotation—in electron optics, the angular shift of the electron image of an object about the optic axis induced by the tangential component of force exerted on the electrons perpendicular to the direction of motion in the field of a magnetic lens. **E 7,** E04

image well—an imaginary well located opposite a control well such that a boundary is the perpendicular bisector of a straight line connecting the control and image wells; used to simulate the effect of a boundary on water-level changes. **D 653,** D18

imaging line scanner—an apparatus that scans in a single dimension and is moved perpendicular to the scan direction to produce a two-dimensional thermogram of a scene. **E 1316,** E07

imbedded abrasive grains—*on a semiconductor wafer*, abrasive particles mechanically forced into the surface. **F 1241,** F01

IMFD diameter—the diameter of the circumscribed circle of the IMFD's uniform cross section, intended to fit in the reamed portion of the diaphyseal bone (see Specification F 339, Dimension A). **F 1611,** F04

IMFD for reamed application—any IMFD that is recommended for use in the medullary canal of a diaphyseal bone that has been prepared by reaming. **F 1611,** F04

IMFD reamer diameter—the diameter of the circumscribed circle of the cutting portion of the reamer's cross section. **F 1611,** F04

imitation—see **associative counterfeit**. **F 1156,** F12

immature fibers, *n*—(*1*) (cotton fibers treated with sodium hydroxide solution)—fibers that either (*a*) have swollen and assumed a spiral form, or (*b*) remained flat, thinly outlined, and almost transparent. Total wall width is less than the lumen width.

(*2*) (cotton fibers observed under polarized light)—fibers that appear purple, indigo, or blue, turn orange or yellow-orange upon rotation to the subtractive position, and upon removal of the selenite plate show parallel extinction. (Compare **mature fibers.**) **D 123,** D13

immediate elastic recovery, *n*—recoverable deformation which is essentially independent of time, that is, occurring in (a time approaching) zero time and recoverable in (a time approaching) zero time after removal of the applied force. (See **delayed deformation**.) **D 123,** D13

immersion lens—See **immersion objective**. **E 7,** E04

immersion objective—an objective in which a medium of high refractive index is used in the object space to increase the numerical aperture and hence the resolving power of the lens. **E 7,** E04

immersion objective (electron optics)—a lens system in which the object space is at a potential (or in a medium of index of refraction) different from that of the image space. **E 7,** E04

immersion plate—a metallic deposit produced by a displacement reaction in which one metal displaces another from solution, for example:

$$Fe + Cu^{++} \rightarrow Cu + Fe^{++}$$

B 374, B08

immersion rinse—a means of removing surface penetrant, in which the test part is immersed in a tank of either water or remover. **E 1316,** E07

immersion rinse—a means of removing excess penetrant in which the test parts are dipped into an agitated tank of water or remover. **E 1316,** E07

immersion testing—an ultrasonic examination method in which the search unit and the test part are submerged (at least locally) in a fluid, usually water. **E 1316,** E07

immunity—a state of resistance to corrosion or anodic dissolution of a metal caused by thermodynamic stability of the metal. **G 15,** G01

impact angle, *n— in impingement erosion,* an angle that could be either the **angle of attack** or the **angle of incidence**, which see. Because of this ambiguity, this term should be specially defined when used or, preferably, used only in contexts where the ambiguity does not matter. **G 40,** G02

impact apparatus or machine, *n*—the total apparatus including the foundation parts, guide rails, electromagnet lift, winch, and tools. **E 1445,** E27

impact atomizer—a pressure atomizer in which a spray pattern is achieved by the impaction of one or more liquid jets against a solid surface. **E 1620,** E29

impact bending—application of an impact load in bending to obtain a measure of the ability of wood to absorb shock loads. **D 9,** D07

impact bruise—See **percussion cone.** **C 162,** C14

impact ignition resistance—the resistance of a material to ignition when struck by an object in an oxygen atmosphere under a specific test procedure. **G 126,** G04

impact insulation class, IIC—a single-number rating derived from measured values of normalized impact sound pressure levels in accordance with Annex A1 of Test Method E 492. It provides an estimate of the impact sound insulating performance of a floor-ceiling assembly. **C 634,** E33

impaction, *n*—a forcible contact of particles of matter, a term often used synonymously with impingement. **D 1356,** D22

impact, Izod—a specific type of impact test made with a pendulum-type machine on a cantilever beam specimen and also the values obtained by this method. **F 412,** F17

impact loading—in hardness testing, a phenomenon in which a momentary overload is inadvertently applied to the indenter by the inertia of parts of the tester subjected to large accelerations. **E 7,** E04

impactor, *n*—a device for collecting airborne or emission particulate matter in which the air or gas being sampled is impacted or impinged against a surface.

cascade impactor, n—a type of impactor which employs several stages of impaction in series to collect successively smaller sizes of particles. **D 1356,** D22

impact pressure range—the variation in an imaging device due to type font width, on feet, impact, and other adjustments inherent in the device. **F 221,** F05

impact printer—a printer in which printing is the result of mechanical impacts. See **chain printer, drum printer, element printer, dot matrix.** **F 909,** F05

impact resistance, *n*—resistance to fracture under the sudden application of an external force. **D 123,** D13

impact resistance, *n*—resistance to fracture under shock force. **D 1566,** D11

impact sensitive—the property of a carbon paper, ribbon, or paper to produce an image from impact (as with a typewriter). **F 221,** F05

impact strength, *n— as related to adhesives,* the kinetic energy per area absorbed by an adhesively-bonded joint when fractured by an impacting body. **D 907,** D14

impact strength—the property of a particle to resist physical breakdown when subjected to a rapidly increasing applied force. **D 2652,** D28

impact test—a test to determine the resistance of a porcelain enamel to fracture caused by a sudden blow. **C 286,** B08

impact tester, *n*—a device for dropping a cylindrical weight from a variable height onto a coated metal test panel; the greater the height required to produce cracks in the coating, the greater its impact resistance. **D 16,** D01

impact tools, *n*—the drop weight, intermediate weight, and anvil. **E 1445,** E27

impact, tup—a falling weight (tup) impact test developed specifically for pipe and fittings. **F 412,** F17

impact velocity, *n— in impingement erosion,* the relative velocity between the surface of a solid body and an impinging liquid or solid particle. **G 40,** G02

impact wear, *n*—wear due to collisions between two solid bodies where some component of the motion is perpendicular to the tangential plane of contact. **G 40,** G02

impedance—the total opposition that a circuit presents to the flow of an alternating current, specifically the complex quotient of voltage divided by current. **E 1316,** E07

impedance, acoustic—the product of the density and sonic velocity of a material. The extent of wave energy transmission and reflection at the boundary of two media is determined by their acoustic impedances. (ISRM) **D 653,** D18

impedance (acoustic)—a mathematical quantity used in computation of reflection characteristics at boundaries; product of wave velocity and material density. **E 1316,** E07

impedance analysis—in electromagnetic testing, an analytical method that consists of correlating changes in the amplitude, phase, or quadrature components, or all of these, of a complex test signal voltage to the electromagnetic conditions within the test specimen. **E 1316,** E07

impedance plane diagram—a graphical representation of the locus of points, indicating the variations in the impedance of a test coil as a function of basic test parameters. **E 1316,** E07

impedance ratio, $z/\rho c \equiv r/\rho c + jx/\rho c$; [dimensionless]—the ratio of the specific normal acoustic impedance at a surface to the characteristic impedance of the medium. The real and imaginary components are called, respectively, **resistance ratio** and **reactance ratio.** **C 634,** E33

impedance, Z—the ratio of the time dependent voltage, $v(t)$, across a circuit, a circuit element, or material to the time dependent current, $i(t)$, through it; that is:

$$Z = v(t)/i(t)$$

E 1142, E37

imperfection, *n*—a material discontinuity or irregularity that is detectable by **inspection**. **A 941,** A01

imperfection—any defect or blemish in or on wood that detracts from its appearance or lowers its utility. In grading rules, a class of limitation on manufacturing practice. **D 9,** D07

imperfection, *n*—a departure of a quality characteristic from its intended level or state. **D 123,** D13

imperfection, *n*—a departure of a quality characteristic from its intended condition. **E 1316,** E07

imperial gallon (IG)—1.2 times U.S. gallon. **D 6161,** D19

impermable boundary—the conceptual representation of a natural feature such as a fault or depositional contact that places a boundary of significantly less-permeable material laterally adjacent to an aquifer. **D 653,** D18

impervious—that degree of vitrification evidenced visually by complete resistance to dye penetration. **C 242,** C21

impervious carbon—the same as impervious graphite with the exception that the base stock has not been graphitized. **C 709,** D02

impervious graphite—manufactured graphite that has been impregnated with a resinous material to make the final article impervious to liquids in the recommended operating range. **C 709,** D02

impingement, *n*—the act of bringing matter forcibly in contact.

dry impingement, n—the process of impingement carried out so that particulate matter carried in the gas stream is retained upon the surface against which the stream is directed.

wet impingement, n—the process of impingement carried out within a body of liquid, the latter serving to retain the particulate matter. **D 1356,** D22

impingement, *n*— *in tribology*, a process resulting in a continuing succession of impacts between (liquid or solid) particles and a solid surface. **G 40,** G02

impingement atomizer—in rocketry, an impinging jet atomizer; in industrial processing, an impact atomizer.

Note—Since this term is ambiguous, it is not recommended for describing either device. **E 1620,** E29

impingement attack, n—deprecated term for **impingement corrosion**. (The latter term is preferred so as to avoid confusion with **liquid impingement erosion**.) **G 40,** G02

impingement corrosion—a form of erosion-corrosion generally associated with the local impingement of a high-velocity, flowing fluid against a solid surface. **G 15,** G01

impingement corrosion, *n*—a form of erosion-corrosion generally associated with the impingement of a high-velocity, flowing liquid containing air bubbles against a solid surface. **G 40,** G02

impinger, *n*—broadly, a sampling instrument employing impingement of the collection of particulate matter.

midget impinger, n—a specific instrument employing wet impingement, using a liquid volume of 10 mL and a gas flow of 2.8 L/min.

standard impinger, n—a specific instrument employing wet impingement, using a liquid volume of 75 mL and a gas flow of 28 L/min. **D 1356,** D22

impinging jet atomizer—a pressure atomizer in which atomization is achieved by the external collision of two or more liquid jets. **E 1620,** E29

importance factor, *n* (coefficient d'importance)— *of a facility requirement*, a numerical indication of the relative importance of a requirement for serviceability, expressed on a scale of 0 to 9, with 0 being not related, 1 being relevant but least important, and 9 being most important. **E 1480,** E06

impreg—wood impregnated with synthetic resin that is cured in place so as to reduce materially swelling and shrinking of the wood on exposure to varying environmental conditions. **D 1038,** D07

impregnate, *v*— in roofing materials manufacture, to completely surround the fibers in a felt or mat with bitumen, with the spaces between the fibers partially or completely filled without a continuous coating of bitumen on the surface. **D 1079,** D08

impregnated fabric, *n*—a fabric in which the interstices between the yarns are completely filled with the impregnating compound throughout the thickness of the material, as distinguished from sized or coated materials, where these interstices are not completely filled. **D 123,** D13

impregnation—a process of filling the pores of a sintered compact, with a nonmetallic material such as oil, wax, or resin. **B 243,** B09

impregnation—partial filling of the open pore structure with another material. **C 709,** D02

impregnation—filling the void structure of hardened concrete by saturating it with a liquid system. **D 3743,** D04

impregnation time of paper, *n*—the time in seconds required for a liquid of specified composition and viscosity to penetrate completely from one face of a sheet of paper to the other under certain prescribed conditions. **D 1711,** D09

impressed current—an electric current supplied by a device employing a power source that is external to the electrode system. (An example is d-c current for cathodic protection.) **G 15,** G01

impressed current protection—application of cathodic protection by means of an external dc power source. **D 3743,** D04

impression: (*1*) *electron microscopy*—the reproduction of the surface contours of a specimen formed in a plastic material after the application of pressure and heat, or both.

(*2*) *hardness*—the imprint or dent made in the specimen by the indenter of a hardness-measuring device. **E 7,** E04

impression replica—See **replica**. **E 7,** E04

impressions per ribbon—the average number of characters which can be printed from a given ribbon on a given typewriter or printer of specified print quality under specific usage conditions. **F 221,** F05

imprinter, pressure—any device used to produce or impress marks or patterns on a surface, for example, a device such as that used with credit cards and address plates. **F 909,** F05

imprinting—(*1*) The act of using an imprinter;

(*2*) The output of any imprinter. **F 909,** F05

improve—See **building modification.** **E 631,** E06

impulsive sound, *n*—a brief, intrusive sound, such as that associated with a tire blowout, operation of a punch press, the discharge of a firearm, a door slam, or a shout, usually characterized by a rapid rise time in the initial pressure pulse of less than a few milliseconds, and by a decay time of less than a few seconds. **C 634,** E33

impure coal—See *impure coal* under **coal**. **D 121,** D05

impurities, of a biological drug product—all process-related (non-adventitious) substances present in the raw materials, bulk drug, or final drug product that are not considered to be the active material, additives, or excipients. **E 1705,** E48

inadequate stimulus, *n*—a stimulus which is not regarded as normally affecting a particular sense, but which may actually do so (for example, pressure on the eye ball producing light, or electrical stimulation generating a taste). **E 253,** E18

incandescence, *n*—the self-emission of radiant energy in the visible spectrum due to the thermal excitation of atoms or molecules. **E 284,** E12

incandescene, *n*—emission of light produced by a material when intensely heated. It can be produced with or without combustion. **E 176,** E05

incandescent illuminant, *n*—illuminant representing the spectral distribution of radiation from an incandescent lamp of specified color temperature. **E 284,** E12

incasement, *n*—a rigid structure or pipe surrounding a buried pipe to provide additional support or protection. **F 412,** F17

incasement, *v*—placement of a rigid structure or pipe surrounding a buried pipe to provide additional support or protection. **F 412,** F17

inches per rack (IPR), *n*—*in warp knitting*, the length of fabric in one rack measured on the machine under operating take-up tension. **D 123,** D13

inch-pound units—the units of length, area, volume, weight, and temperature in common use in the United States at the present time. These include, but are not limited to: (*1*) length—feet, inches, and fractional inches, (*2*) area—square feet and square inches, (*3*) volume—cubic feet, cubic inches, gallons, and ounces, (*4*) weight—pounds and ounces, and (*5*) temperature—degrees Fahrenheit. **C 896,** C04

inch threaded Class 2A coating thickness—a coating thickness which does not exceed ⅙ of the allowance for Class 2A threads to avoid interference. **F 1789,** F16

incident angle—see **angle of incidence.** **E 772,** E44

incident irradiation, *n*—the amount of flux per unit area that is normal (perpendicular) to a surface or detector. **F 1294,** F05

incident particle energy—the effective energy of the primary particles incident on the specimen surface, usually expressed in kiloelectronvolts (keV) per atomic particle. **E 673,** E42

incineration, *n*—controlled burning of waste products or other combustible material. **D 5681,** D34

incinerator, *n*—a device constructed for the purpose of containing a material for thermal oxidation. **D 5681,** D34

incinerator—enclosed device using controlled-flame combustion in which waste materials are pyrolyzed or burned. **E 1605,** E06

incipient anisotropic phase, *n*—*as used in Test Method D 5061*, a binder-phase carbon texture having a domain size (less than 0.5 μm) that is near the measuring resolution of the light microscope. **D 121,** D05

incipient fusion, *n*—beginning of the development of fired bond. **C 43,** C15

incise—to make slitlike lacerations generally parallel to the grain in the lateral surface of timbers that are resistant to treatment, so that deeper and more uniform penetration of preservative may be obtained. **D 9,** D07

incised—decorated by cutting or indenting the ware surface. **C 242,** C21

inclination angle, *n*—*of a tire*, the angle between the Z'-axis and the wheel plane. **F 538,** F09

incline—the slope of a roof expressed in percent or in the number of vertical units of rise per horizontal unit of run. **D 1079,** D08

incline impact (Conbur) test—See **package testing.** **D 996,** D10

included angle—angle between the sides of the point in the longitudinal section through the nail in the direction perpendicular to the point sides. **F 547,** F16

included counters, *n*—permanent traffic counters for which monthly traffic adjustment factors may be calculated for a given year, based on the quantity of edit-accepted data. **E 867,** E17

inclusion—a foreign solid or gas within or bounded by the glass matrix. **C 162,** C14

inclusion, *n*—*as used in fractography,* a foreign body from other than the normal composition of the bulk advanced ceramic. **C 1145,** C28

inclusion—any foreign matter or particles that are either encapsulated or imbedded in the pultrusion. **D 3918,** D20

inclusion—embedded foreign material or a stain other than from normal composition or forming, or both (see **blemish**). **F 109,** C21

inclusion count—determination of the number, kind, size, and distribution of non-metallic inclusions. See Practice E 45. **E 7,** E04

inclusions—foreign particles which may or may not be enclosed in the copper foil. **B 846,** B05

inclusions, *n*—foreign matter in the mica.

air inclusions appear by transmitted light as grayish areas and as silvery areas by reflected light. These are gaseous inclusions.

clay inclusions appear by any light as areas of blue, gray, brown, etc., and are intrusions of earthy materials.

mineral inclusions appear by transmitted light as areas of deep distinct and highly saturated colors such as black, brown, green, red, and so forth. These are concentrated metallic oxides.

vegetable and smokey inclusions appear by transmitted light as areas of pastel colors of low to medium saturation such as pale yellow, pale brown, pale green, and so forth. These are dispersed metallic oxides. The term "vegetable" is a misnomer. **D 1711,** D09

inclusions—foreign material held mechanically, usually referring to non-metallic particles, such as oxides, sulfides, silicates, etc. **E 7,** E04

incoherent electron scatter—the deflection of electrons by either electrons or atoms which results in a loss of kinetic energy by the incident electron. **E 7,** E04

incomplete block design, *n*—a design in which the experiment space is subdivided into blocks in which there are insufficient experimental units available to run a complete set of treatments or replicate of the experiment. **E 456,** E11

incomplete block design, *n*—a design in which the experiment space is subdivided into blocks in which there are insufficient experimental units available to run a complete set of treatments or replicate of the experiment. **E 1325,** E11

incomplete head—nail head with semi-circular rim and portion of head omitted during heading, to allow tight collating of nails in strip form. (See **D head, notched head.**) **F 592,** F16

incomplete thread—thread not fully formed. **F 547,** F16

inconclusive—this is commonly used synonymously with no conclusion when the examiner is at the zero point on the scale of confidence. A potential problem is that some people understand this term to mean something short of definite (or conclusive), that is, any degree of probability, and the examiner should be aware of this ambiguity. **E 1658,** E30

incongruent transformation—a nonisothermal, or nonisobaric, phase change in which one, or both, of the phases involved undergo composition change during the process. **E 7,** E04

increment, *n*—a small portion of a lot collected by one operation of a sampling device and normally combined with other increments from the lot to make a gross sample. **D 121,** D05

increment, *n*—a portion of a lot as collected by one individual manual or mechanical sampling operation and normally combined with other increments from the lot to make a gross sample. **D 5681,** D34

increment, *n—in sampling,* a portion of material removed from a lot by a single operation. **E 135,** E01

increment—a portion of the lot collected by one individual manual or mechanical sampling operation and normally combined with other increments from the lot to make a gross sample. **E 856,** D34

incremental cost (benefit)—the additional cost (benefit) resulting from an increase in the investment in a building project (Syn. *marginal cost (benefit)*). **E 833,** E06

incremental permeability—the ratio of the change in magnetic induction to the corresponding change in magnetizing force when the mean induction differs from zero. **E 1316,** E07

increment borer—an augerlike instrument with a hollow bit, used to extract cores from trees for study of growth and age and for other research purposes. **D 9,** D07

increments (solid sample)—portions of material selected from various parts of a lot, which may be tested individually or composited and tested as a unit. **E 1547,** E15

incubation period, *n— in cavitation and impingement erosion,* the initial stage of the erosion rate-time pattern during which the erosion rate is zero or negligible compared to later stages. Also, the exposure duration associated with this stage. (Quantitatively it is sometimes defined as the intercept on the time or exposure axis, of a straight line extension of the maximum-slope portion of the cumulative erosion-time curve.) **G 40,** G02

indent—See **impression.** **E 7,** E04

indent—*on a semiconductor wafer,* an edge defect that extends from the front surface to the back surface. **F 1241,** F01

indentation—See **impression.** **E 7,** E04

indentation—*1*) the extent of deformation by the indentor point of any one of a number of standard hardness testing instruments; *2*) a recess in any surface of a rubber article. **F 869,** F08

indentation hardness, *n*—the hardness as evaluated from measurements of area or depth of the indentation made by pressing a specified indenter into the surface of a material under specified static loading conditions. **E 6,** E28

indentation hardness—the resistance of a material to indentation. This is the usual type of hardness test in which a pointed or rounded indenter is pressed into a surface under a substantially static load. **F 1789,** F16

indenter—in hardness testing, a tool, usually of diamond and having a definite geometrical shape, which is forced into the surface of the specimen. **E 7,** E04

indenter constant—in hardness testing, a numerical constant relating the area of the indentation to the square of the measured diagonal. **E 7,** E04

independent variable—the selected and controlled variable (namely, stress or strain). It is denoted *X* when plotted on appropriate coordinates. **E 1823,** E08

index, *n—(synonymous with "number in Committee E-17 usage, for example; PSI, RN),* a number or formula expressing some property, form, ratio, etc. of the relation or proportion of one amount or dimension to another. **E 867,** E17

index, changing—the mark on the dial ring used when changing the combination. **F 471,** F12

index, L.O.B.C. (locked on by combination)—the mark on the dial ring used to set the combination so the back cover can be removed. **F 471,** F12

index of refraction, *n*—See *refractive index.* **D 4175,** D02

index of refraction, *n*—the numerical expression of the ratio of the velocity of light in a vacuum to the velocity of light in a substance, at a specified wavelength. **E 284,** E12

index of refraction—the relative index of refraction defined by Snell's law as the ratio of the sine of the angle of incidence to the sine of the angle of refraction. **F 1241,** F01

index, opening—the mark on the dial ring used when dialing the combination to open the lock. **F 471,** F12

index test, *n*—a test procedure which may contain a known bias but which may be used to establish an order for a set of specimens with respect to the property of interest. **D 4439,** D35

indian tanned—combination tanned with alum and vegetable tannins. **D 1517,** D31

India-tanned—Term applied to hides and skins from India, considered as a semitanned raw material and generally retanned in the United States before finishing. **D 1517,** D31

indicated dynamic loads [F]—the maximum and minimum loads (or the mean load and the load amplitude) that correspond to the readings obtained from the load cell associated with the fatigue testing system, according to an existing static calibration. The load cell calibration may have been furnished by the machine manufacturer or may have been developed by the user. **E 1823,** E08

indication—the response or evidence from a nondestructive examination. **E 1316,** E07

indication, *n*—the response or evidence from a nondestructive examination that requires interpretation to determine relevance. **E 1316,** E07

indication—that which marks or denotes the presence of a reflector. **E 1316,** E07

indications—eddy-current signals caused by any change from uniformity of a tube. These changes from uniformity affect the electrical characteristic of the tube but may not be detrimental to the end use of the product. (E 215) **E 1316,** E07

indications did not—this carries the same weight as the indications term that is, it is a very weak opinion.

Examples—There is very little significant evidence present in the comparable portions of the questioned and known writings, but that evidence *suggests* that the John Doe of the known material did not write the questioned material, or I found *indications* that the John Doe of the known material did *not* write the questioned material but the evidence is far from conclusive.

See Discussion after **indications.** **E 1658,** E30

indications (evidence to suggest)—a body of writing has few features which are of significance for handwriting comparison purposes, but those features are in agreement with another body of writing.

Examples—There is evidence which *indicates* (or *suggests*) that the John Doe of the known material may have written the questioned material but the evidence falls far short of that necessary to support a definite conclusion. **E 1658,** E30

indicator (pH)—a substance that changes color when the pH of the medium is changed. In the case of most useful indicators, the pH range within which the color changes is narrow. **B 374,** B08

indicator titration—in ion-selective electrode work, a type of differential complexometric titration in which a metal ion complex is added to the sample as an electrode indicator, and a different metal ion in the sample is determined by titration with the complexing agent contained in the indicator. An electrode that responds to the indicator metal ion is used to detect the point at which the titrant has complexed all of the ion being determined. An abrupt change in the indicator ion concentration occurs when excess complexing agent is added. **D 4127,** D19

indices—See **Miller indices.** **E 7,** E04

indices of metamerism potential, *n*— indices indicating the degree to which two metameric specimens may develop metamerism, derived solely from their different spectral characteristics. **E 284,** E12

indifferent point—in a phase diagram, a maximum point, or minimum point, that is, a composition and temperature, or pressure, at which congruent transformation occurs. **E 7,** E04

indigenous (endogenous) inclusion—a nonmetallic material that precipitates from the melt. (See **inclusions.**) **E 7,** E04

indigenous inclusions—See **deoxidation products**. **E 7,** E04

indigneous species—a species that is likely, due to historical presence, to occur at a specified site for some portion of its life span. **E 943,** E47

indirect detection of mycoplasma—detection of mycoplasma by DNA staining or any method other than cultivation. **E 1705,** E48

indirect electrophotographic copying process—an electrophotographic copying process in which the photoconductor is not an integral part of the final copy. **F 335,** F05

indirect exposure—a method in which only a gamma-insensitive conversion screen is exposed to the neutron beam. After exposure, the conversion screen is placed in contact with the image recorder. **E 1316,** E07

indirect magnetization—magnetization induced in a part when no direct electrical contact is made. **E 1316,** E07

indirect medical direction—the physician management of all clinical aspects of an EMS system, including but not limited to planning, training, implementation, and evaluation (a.k.a. off-line medical control direct medical direction). **F 1177,** F30

indirect oxygen service—service in which oxygen is not normally contacted but in which it might be as a result of a reasonably foreseeable malfunction, operator error, or process disturbance. Examples: liquid oxygen tank insulation, liquid oxygen pump motor bearings. **G 126,** G04

indirect yarn numbering system, *n*—a system that expresses yarn linear density (number) in length per unit mass. (See also **cotton count, metric count, worsted count, cut, run.**) **D 123,** D13

individual pellet hardness, *n*—the force required to fracture or crush a carbon black pellet. **D 3053,** D24

indoor furniture, *n*—furniture manufactured for use in the interior of a building. **D 123,** D13

induced current method—see **current induction method.** **E 1316,** E07

induced field—see **indirect magnetization.** **E 1316,** E07

inductance, core, L_1—the effective parallel circuit inductance of a ferromagnetic core based upon a hypothetical nonresistive path that is exclusively considered to carry the magnetizing current, I_m.

NOTE—The product $I_m^2 \omega L_1$ equals the quadrature power delivered to the core. **A 340,** A06

inductance, incremental, L_Δ—the self-inductance of an electrical circuit when the ferromagnetic core has an ac cyclic magnetization produced by specified values of both ac and dc components of the exciting current. **A 340,** A06

inductance, initial, L_0—the limiting value of the core inductance, L_1 reached in a ferromagnetic core when, under ac symmetrical cyclic excitation, the magnetizing current has been progressively and gradually reduced from a comparatively high value to a zero value.

NOTE—Initial inductance may be obtained by highly sensitive ASTM bridge methods working in the range in which μ_L is a linear function of H. A series of decreasing values of μ_L is measured and plotted versus corresponding values of magnetizing current, I_m (or other suitable excitation parameter), and the data extrapolated to zero excitation. See **permeability, initial dynamic.** **A 340,** A06

inductance, intrinsic (ferric), L_i—that portion of the self-inductance which is due to the intrinsic induction in a ferromagnetic core.

NOTE—It is determined at a specified value of the magnetizing current. **A 340,** A06

inductance, mutual, L_m—the common property of two electrical circuits that determines the flux linkage in one circuit (the secondary) produced by a given current in the other circuit (the primary). The mutual inductance, L_m, is defined by the equation:

$$L_m = \mathcal{L}_2 I_1$$

where:
$\mathcal{L}_2$ = flux linkage in the secondary and
I_1 = current in the primary, assuming no current in the secondary.

NOTE—If $\mathcal{L}_2$ is in maxwell-turns and I_1 is in amperes, then the mutual inductance in henries is defined by the equation:

$$L_m = (\mathcal{L}_2/I_1) \times 10^{-8}$$

NOTE—If the linkage is proportional to the current (no ferromagnetic material present), the inductance is constant and may be obtained from the equation:

$$e_2 = L_m(di_1/dt)$$

where:
e_2 = instantaneous induced emf in the secondary and
di_1/dt = time rate of change of the current in the primary.

NOTE—If ferromagnetic materials or eddy currents are present, the mutual inductance must be regarded as a function of the primary current, its rate of change, and the magnetic history of the material. Thus:

$$e_2 = -(d(L_m i_1)/dt) = -[L_m(di_1/dt) + i_1(dL_m/dt)]$$

A 340, A06

inductance, self, L—that property of an electric circuit that determines the flux linkage produced by a given current in the circuit. The self-inductance, L, is defined by the equation:

$$L = \mathcal{L}/I$$

where:
$\mathcal{L}$ = flux linkage and
I = current.

NOTE—If $\mathcal{L}$ is in maxwell-turns and I in amperes, then the self-inductance in henries is defined by the equation:

$$L = (\mathcal{L}/I) \times 10^{-8}$$

NOTE—If the linkage is proportional to the current (no ferromagnetic material present), the inductance is constant and may be obtained from the equation:

$$e_2 = -L_m(d\,i/dt)$$

where:
e = instantaneous induced emf and
di/dt = time rate of change of the current.

NOTE—If ferromagnetic material or eddy currents are present, the self-inductance must be regarded as a function of the circuit current, its rate of change, and the magnetic history of the material. Thus:

$$e = -(d(Li)/dt) = -[L(di/dt) + i(dL/dt)]$$

A 340, A06

inductance, series, L_s—the effective series ac self-inductance exhibited by an inductor having a ferromagnetic core and subjected to an *SCM* excitation after the core has been demagnetized.

NOTE—The value of series inductance is a function of the level of excitation. **A 340,** A06

inductance, winding, L_w—the linear inductance of the magnetizing winding as a result of the flux caused by the ac symmetrical cyclic magnetization exciting current, *I*. The flux linking the winding is that flux outside of the ferromagnetic core material. **A 340,** A06

induction—the relief of repression of transcription of lysogenic phage genes encoding the functions for lytic growth, so that the phage will grow lytically. **E 1705,** E48

induction, B—See **magnetic induction (flux density).** **A 340,** A06

induction, biased, B_b—the value of the apparent dc magnetic induction around which the ac cyclic changes are occurring in a magnetic material resulting from the biasing magnetizing field. This value is a function of the incremental magnetizing field and is not determined by the normal induction curve. **A 340,** A06

induction curve, intrinsic (ferric)—a curve of a previously demagnetized specimen depicting the relation between intrinsic induction and corresponding ascending values of magnetic field strength. This curve starts at the origin of the B_i and H axes. **A 340,** A06

induction curve, normal—a curve of a previously demagnetized specimen depicting the relation between normal induction and corresponding ascending values of magnetic field strength. This curve starts at the origin of the B and H axes. **A 340,** A06

induction fryer—a thermostatically controlled piece of equipment, including a cooking vessel powered by electromagnetic coils inside stainless steel immersion tubes. The electromagnetic field created by these coils causes eddy currents to form in the metal surrounding them, in which the amount of heat in the metal is controlled by changing the magnetic field surrounding the induction coils in response to the fryer thermostat. See **fryer**. **F 1827,** F26

induction hardening, *n—in surface hardening*, a process in which only the surface layer of a suitable steel object is heated by electrical induction to above Ac_3 or Ac_{cm}, and then the object is **quenched**. **A 941,** A01

induction hardening, *n—in through hardening*, a process in which a suitable steel object is heated by electrical induction to above Ac_3 or Ac_{cm} throughout its section, and then the object is **quenched**. **A 941,** A01

induction heating, *n*—heating by electrical induction. **A 941,** A01

induction, incremental, B_Δ—one half the algebraic difference of the extreme values of the magnetic induction during a cycle in a magnetic material that is subjected simultaneously to a biasing magnetizing field and a symmetrically cyclically varying magnetizing field. Twice the incremental induction is indicated by the symbol ΔB, thus:

$$B_\Delta = \Delta B/2$$

A 340, A06

induction, intrinsic, B_i—the vector difference between the dc magnetic induction in a magnetic material and the magnetic induction that would exist in a vacuum under the influence of the same magnetic field strength. This is expressed by the equation:

$$B_i = B - \Gamma_m H$$

NOTE—In the cgs-emu system of units, $B_i/4\pi$ is often called magnetic polarization. **A 340,** A06

induction, maximum:

(*1*) B_m—the maximum value of induction, B, in a dc hysteresis loop. The tip of this loop has the magnetostatic coordinates H_m, B_m, which exist simultaneously.

(*2*) B_{max}—the maximum value of induction, B, in an ac flux-current loop.

NOTE—In a flux-current loop, the magneto-dynamic values B_{max} and H_{max} do not exist simultaneously; B_{max} occurs later than H_{max}. **A 340,** A06

induction, normal, B—the maximum induction, in a magnetic material that is in a symmetrically cyclically magnetized condition. **A 340,** A06

induction range—an electric range that causes heat to be generated directly in the magnetic material comprising the cooking container by means of an induced electromagnetic field. The amount of heat generated in the cooking container is controlled by varying the strength of the magnetic field. See **range**. **F 1827,** F26

induction, remanent, B_d—the magnetic induction that remains in a magnetic circuit after the removal of an applied magnetic field.

NOTE—If there are no air gaps or other inhomogeneities in the magnetic circuit, the remanent induction, B_d, will equal the residual induction, B_r; if air gaps or other inhomogeneities are present, B_d will be less than B_r. **A 340,** A06

induction, residual, B_r—the value of magnetic induction corresponding to zero magnetizing field when the magnetic material is subjected to symmetrically cyclically magnetized conditions. **A 340,** A06

induction, saturation, B_s—the maximum intrinsic induction possible in a material. **A 340,** A06

inductive coil—a (partial) loop (or one with one or more turns) of wire, or other conductive material, in which an electrical signal is induced when the coil is placed into an electromagnetic field. **F 1156,** F12

industrial grade—a quality of aromatic hydrocarbons suitable for many industrial applications that have a tolerance for nonreactive impurities. **D 4790,** D16

industrial hygienist—person certified by the American Board of Industrial Hygiene, or an industrial hygienist in training, or a person with equivalent education or experience or both. **E 1605,** E06

industrial hygienist, *n*—a person who, by experience and academic training, is qualified to recognize, evaluate, and control chemical, physical, and biological agents in the workplace, or a person certified by the American Board of Industrial Hygiene. **F 1494,** F23

industrialized building—See **building.** **E 631,** E06

industrialized building system—See **building system.** **E 631,** E06

industrial talc—a mineral product varying in composition from that approaching the theoretical formula of talc, $Mg_3Si_4O_{10}(OH)_2$, to mixtures of talc and other naturally associated minerals, some of which may be fibrous as defined in ASTM Definitions D 2946, Terms Relating to Asbestos. These fibrous minerals may or may not be asbestos. **D 16,** D01

industrial talc, nonasbestos type—industrial talc of which less than 2 particles per 100 particles (by light microscopy) are asbestos fibers, where "asbestos fiber" is defined as being both a fiber by Definitions D 2946 and one of the asbestiform varieties of serpentine, riebeckite, cummingtonite (which are chrysotile, crocidolite and amosite, respectively), anthophyllite, tremolite, or actinolite. The nonasbestiform varieties of these same minerals are not asbestos. **D 16,** D01

industrial waste—the water-conveyed residues resulting from manufacturing or processing operations. **C 896,** C04

industrial yarn, *n*—a yarn composed of continuous filaments, usually of high breaking tenacity, produced with or without twist, and intended for applications in which functional properties are of primary importance; for example, in reinforcing material in elastomeric products (tires, hose, belting), in protective coverings, and in cordage and webbing, etc. **D 123,** D13

industry products, *n—for feathers or down,* products such as furniture, pillows, comforters, sleeping bags, and wearing apparel which are wholly or partially filled with feathers or down; bulk stocks of processed feathers or down intended for use in the manufacture of such products. **D 123,** D13

inelastic:

inelastic mean free path—see **electron.**

scattering correction to background—XPS, a method of correcting background for contributions of inelastic scattering processes, most often approximated by simulating the background through a peak by assuming that the rise in background is proportional to the peak area at higher kinetic energy. A more accurate correction is done by deconvolving the energy loss spectrum itself.

scattering cross-section—AES, XPS, a measure of the probability that an electron traversing a material will undergo an inelastic scattering process, expressed as an area per unit event.

scattering event—ISS, a collision process in which a fraction of the kinetic energy imparted by the probe ion contributes to an increase in the internal energy of the target material, and is not recovered as kinetic energy of the scattered probe ion or target atom recoil. **E 673,** E42

inelastic deformation—the portion of deformation under stress that is not annulled by removal of stress. (ISRM) **D 653,** D18

inelastic electron scatter—See **incoherent electron scatter**. **E 7,** E04

inelastic strain, ϵ_{in} — the strain that is not elastic. **E 1823,** E08

inert—not participating in any fashion in chemical reactions. **D 653,** D18

inert anode—an anode that is insoluble in the electrolyte under the conditions prevailing in the electrolysis. **B 374,** B08

inert components—those elements or components of natural gas (fuel gas) that do not contribute to the heating value. **D 4150,** D03

inertial loading—See **impact loading**. **E 7,** E04

inert (inactive) ingredient—component of an aerosol formulation that does not contribute to the specific effect of the formulation. In some cases, it may be quite arbitrarily defined (for example with insecticides, only the propellants are considered as inert ingredients). **D 3064,** D10

inertinite—See *inertinite* under **maceral**. **D 121,** D05

inertodetrinite—See *inertodetrinite* under **maceral**. **D 121,** D05

inferior—lower, also toward the foot. **F 869,** F08

infill, *n*—a series of balusters of a baluster railing system and the structural as well as the decorative elements, including the panels, mesh, or similar elements, of a panel railing system; located between top and bottom rails and posts; to serve the twofold purpose of (*a*) protecting bodies from penetrating and falling through the baluster and panel infill areas and (*b*) providing a specified resistance to horizontal thrusts as are potentially encountered within the infill area. **E 1481,** E06

infill area—the field of baluster and panel railing systems, bordered by top and bottom rails and posts. **E 1481,** E06

infiltrant efficiency, *n*—the ratio of the mass of infiltrant absorbed by the part to the mass of infiltrant originally used, expressed as a percentage. **B 243,** B09

infiltration—a process of filling the pores of a sintered, or unsintered, compact with a metal or alloy of lower melting point. **B 243,** B09

infiltration—the volume of groundwater entering a sewer and its connections from the soil through pipe, joints, connections, or appurtenances. **C 822,** C13

infiltration—the quality of ground water entering the test section during a specified time period. **C 896,** C04

infiltration erosion—the pitting, channeling, and coarsening of the surface porosity that results from the dissolution of the base metal by the liquid infiltrant, as the infiltrant flows into the matrix. **B 243,** B09

infiltration loading density—infiltrant weight per unit area of contact between infiltrant and part. **B 243,** B09

infiltration, pipe—the passage of fluid into a pipe section through small holes or leaks. **F 412,** F17

infiltration residue—material that remains on the surface of the part after infiltration. **B 243,** B09

infinite pad method—*in optical character recognition,* a method of measuring reflectance of a paper stock such that doubling the number of backing sheets of the same stock will not change measured reflectance. **F 149,** F05

infinite thickness, *n*—term applied to a layer of material so thick that increasing its thickness does not change its reflectance or other optical properties. **E 284,** E12

inflatable boom—boom that uses inflated gas-filled chambers as the flotation. **F 818,** F20

inflatable restraint, *n*—a vehicular safety device designed to cushion an occupant or equipment during collision; airbag. **D 123,** D13

inflation—a rise in the general price level, usually expressed as a percentage rate. **E 833,** E06

inflation pressure loss rate, *n*—rate of change of normalized inflation pressure, determined from the slope of the linear portion of the log pressure versus time curve. **F 538,** F09

inflator, *n—for inflatable restraints,* a device for generating and directing expansion gases into a cushion. **D 123,** D13

inflection point, *n*—the first point of the force-elongation curve at which the second derivative equals zero. **D 4439,** D35

inflection point—position on a curved line, such as a phase boundary, where the direction of curvature is reversed. **E 7,** E04

inflow—the volume of any kind of water entering a sewer and its connections from outside sources not including those sources described under "infiltration." **C 822,** C13

influence value, *I* **(D)**—the value of the portion of a mathematical expression that contains combinations of the independent variables arranged in dimensionless form. **D 653,** D18

influent stream, *n*—see preferred term **losing stream**. **D 653,** D18

information data plane—a surface on which data is recorded. **F 1156,** F12

information depth—maximum depth, normal to the specimen surface, from which useful signal information is obtained.

E 673, E42

infrared, *adj*—pertaining to the region of the electromagnetic spectrum from approximately 0.78 to 300 μm. **D 1711,** D09

infrared—invisible light and heat radiation, adjacent to the red end of the visible spectrum, with wavelengths from 700 to about 3000 nm (nanometres). **E 7,** E04

infrared—pertaining to the region of the electromagnetic spectrum with wavelength range from approximately 0.78 to 1000 μm (wavenumber range 12 800 to 10 cm^{-1}). **E 131,** E13

infrared, *adj*—referring to radiant flux having wavelengths longer than the wavelengths of light, usually wavelengths from about 780 nm to about 1 mm. **E 284,** E12

infrared imaging system—an apparatus that converts the two-dimensional spatial variations in infrared radiance from any object surface into a two-dimensional thermogram of the same scene, in which variations in radiance are displayed in gradations of gray tone or in color. **E 1316,** E07

infrared radiation, *n*—radiation for which the wavelengths of the monochromatic components are greater than those for vissible radiation, and less than about 1 mm.

NOTE—The limits of the spectral range of infrared radiation are not well defined and may vary according to the user. Committee E-2.1.2 of the CIE distinguishes in the spectral range between 780 nm and 1 mm:

IR-A	780 to 1400 nm
IR-B	1.4 to 3 μm
IR-C	3μ to 1mm

E 349, E21

infrared-radiation opaque (ir-opaque)—a hiding quality of ink coating or other substances that causes obstruction of an image from the human eye, but allows detection or sensing by a photodiode operating with a light source that is usually specified at a peak response of about 900 nm. **F 1156,** F12

infrared reflector—a material with a reflectance in the infrared region as close as possible to 1.00. **E 1316,** E07

infrared response—a particular type of optical system used in some scanners. As a general rule, nonscan inks for this purpose are in the red portion of the color spectrum. **F 149,** F05

infrared sensing device—one of a wide class of instruments used to display or record, or both information related to the thermal radiation received from any object surfaces viewed by the instrument. The instrument varies in complexity from spot radiometers to two-dimensional real-time imaging systems. **E 1316,** E07

infrared spectroscopy—pertaining to spectroscopy in the infrared region of the electromagnetic spectrum. **E 131,** E13

infrared thermographer—the person qualified or trained to use infrared imaging radiometer. **E 1316,** E07

infrared thermography—see **thermography, infrared.**

E 1316, E07

inglaze decoration—See *inglaze decoration* under **decoration.**

C 242, C21

ingot, *n*—cast form suitable for fabricating (rolling, forging, extruding, etc.) or remelting. **B 881,** B07

ingot, *n*—a casting of simple shape suitable for hot-working or remelting. **B 899,** B02

ingot—cast metal in a form intended for subsequent fabrication.

E 7, E04

ingot—*in silicon technology,* a solid section of polycrystalline or single crystal silicon, generally of slightly irregular cylindrical or rectangular shape. **F 1241,** F01

ingot and ingot bar—refinery shapes used for remelting (not fabrication). Ingots normally range in weight from 20 to 35 lb (9 to 16 kg) and ingot bars from 50 to 70 lb (23 to 32 kg). Both are usually notched to facilitate breaking into smaller pieces. **B 846,** B05

ingot scum—slag, dross, or oxidation appearing on the top surface of ingots during pouring, which, when entrapped, is a source of inclusions. **E 7,** E04

inherent ash, *n*—the residue remaining from the inherent impurities after ignition under conditions specified for the ash determination.

D 121, D05

inherent flame-resistance, *n*—*as applied to textiles,* flame resistance that derives from an essential characteristic of the fiber from which the textile is made. (Compare **flame resistance.**) **D 123,** D13

inherent fluorescence—fluorescence that is an intrinsic characteristic of a material. **E 1316,** E07

inherent impurity, *n*—the inorganic material in coal that is structurally part of the coal and cannot be separated from it by coal preparation methods. **D 121,** D05

inherently flame-resistant, *adj*—having inherent flame-resistance.

D 123, D13

inherent moisture—See *inherent moisture* under **moisture.**

D 121, D05

inherent viscosity—see **viscosity, inherent.** **D 1695,** D01

inhibition load XX (ILXX), *n*—a statistically or graphically estimated loading rate of test material that is expected to cause a XX % inhibition of a biological process (such as growth or reproduction) of a representative subpopulation of organisms under specified conditions and is expressed as an analog as opposed to digital measure. **D 6384,** D02

inhibitor—a substance used to reduce the rate of a chemical or electrochemical reaction, commonly corrosion or pickling.

B 374, B08

inhibitor—a material that stops or slows a chemical reaction from occurring. **D 653,** D18

inhibitor, *n*—a substance used in low concentration which suppresses a chemical reaction. **D 883,** D20

inhibitor, *n*—a substance that slows down chemical reaction. Inhibitors are sometimes used in certain types of adhesives to prolong storage or working life. (Also *retarder.*) (Compare **catalyst** and **hardener.**) **D 907,** D14

inhibitor, *n*—material used to suppress a chemical reaction.

D 1566, D11

inhibitor—any substance which when added to an electrical insulating fluid retards or prevents undesirable reactions. **D 2864,** D27

inhibitor, *n*—a chemical compound added to engine coolant to mitigate cooling system degradation. **D 4725,** D15

inhibitor—a substance added to a material to retard or prevent deterioration. **D 4790,** D16

inhibitor, *n*—a substance used in low concentration which suppresses chemical reaction. (D-20) **F 412,** F17

initial backfill—location for placement of selected material, native or import, extending from the top of the bedding material to an elevation 1 ft above top of pipe. **C 896,** C04

initial boiling point—the temperature observed immediately after the first drop of distillate falls into the receiving cylinder during a distillation test. **D 4790,** D16

initial consolidation (*initial compression*)—see **consolidation.**

D 653, D18

initial cost—See **first cost.** **E 833,** E06

initial deformation temperature (IT), *n*—*in reference to the fusibility of coal and coke ash according to Test Method D 1857,* the temperature at which the first rounding of the apex of the cone occurs. Shrinkage or warping of the cone is ignored if the tip remains sharp. **D 121,** D05

initial deformation temperature, IT, *n*—the temperature at which the first rounding of the apex of a pyrometric cone occurs; shrinking or warping of the cone is ignored if the tip remains sharp. **D 5681,** D34

initial investment cost—See **first cost.** **E 833,** E06

initial modulus, *n*—the slope of the initial straight portion of a stress-strain or force-elongation curve. **D 123,** D13

initial modulus, *n*—*in a stress-strain curve*, the slope of the initial straight-line portion of the curve. **D 4848,** D13

initial permeability—the slope of the induction curve at zero magnetizing force as the test specimen is being removed from a demagnetizing condition (slope at origin of BH curve before hysteresis is observed). **E 1316,** E07

initial potential—see **charge acceptance.** **F 335,** F05

initial pulse—the response of the ultrasonic system display to the transmitter pulse (sometimes called main bang). **E 1316,** E07

initial rate of absorption, *n*—a distinct property that offers different information from absorption. It is a measure of the suction of water upward into a dry brick from a bed face during one minute of exposure. It is expressed as grams of water picked up in one minute by a net area of 30 in.^{2}(194 cm^2).

Initial rate of absorption is one factor influencing the quality of bond between brick and mortar. It is used in brick standards to recommend construction practices for enhancing mortar to brick bonding. **C 43,** C15

initial recovery, *n*—the decrease in strain in a specimen resulting from the removal of force, before creep recovery takes place. **E 6,** E28

initial set—a degree of stiffening of a grout mixture generally stated as an empirical value indicating the time in hours and minutes that is required for a mixture to stiffen sufficiently to resist the penetration of a weighted test needle. **D 653,** D18

initial setting time—the time interval from the start of mixing the component parts at a specified temperature, *(a)* to that time when a Gillmore needle weighing 1 lb (454 g) and having a tip 1/24 in. (1 mm) in diameter by 3/16 in. (5 mm) long will penetrate mortar 5/8 in. (16 mm) thick to a depth of 3/16 in. (5 mm) in 1 min, or, *(b)* to that time at which a 1/4 in. (6.4 mm) wide joint of the mortar between bricks is indented less than 1/24 in. (1 mm) by a Vicat needle during a 10-min period. **C 904,** C03

initial spray angle, *n*—the plane angle of the spray emerging from the nozzle discharge orifice. **E 1620,** E29

initial strain, *n*—the strain introduced into a specimen by the given loading conditions, before creep takes place. **E 6,** E28

initial stress, *n*—the stress introduced into a specimen by imposing the given constraint conditions before stress relaxation takes place. **E 6,** E28

initial survey—systematic inspection of a dwelling unit by a qualified inspector, using a portable XRF analyzer, atomic absorption spectroscopy, or other approved testing techniques, to determine whether a lead-based paint hazard is present. **E 1605,** E06

initial tangent modulus, *n*—the slope of the stress-strain curve at the origin. **D 907,** D14

initial temperature—See **activation temperature**. **F 1623,** F05

initial tensile modulus, J_p (FL^{-1}), Nm^{-1}, *n*—*for geosynthetics*, the ratio of the change in force per unit width to the change in elongation of the initial portion of a force-elongation curve. **D 4439,** D35

injectability—see **groutability.** **D 653,** D18

injection molded button, *n*—a method of forming which requires the filling of a cavity under pressure with polymer that will take the form of the mold when cooled. **D 123,** D13

injection molding, *n*—the process of forming a material by forcing it, in a fluid slate under pressure, through a runner system (sprue, runner, gate(s) into the cavity of a closed mold. **D 123,** D13

injection molding, *n*—the process of forming a material by forcing it, in a fluid state and under pressure, through a runner system (sprue, runner, gate(s)) into the cavity of a closed mold. **D 883,** D20

injection molding, *n*—the process of forming a material by forcing it, under pressure, from a heated cylinder through a sprue (runner, gate) into the cavity of a closed mold. (D-20) **F 412,** F17

injection molding—a molding procedure whereby a heat-softened plastic material is forced from a cylinder into a cavity which gives the article a desired shape. **F 869,** F08

injection seal—a seal accomplished by injecting sealant into holes, joggles, channels, grooves, and other voids caused by buildup of structure boundaries. **E 1749,** E06

injection time, *n*—the time interval from the beginning of screw forward movement until switching over to hold pressure. **D 883,** D20

ink deposit—a coating containing a coloring material which transfers in part or entirely to a copy sheet at the point of pressure contact. (F 129) **F 221,** F05

inked ribbon—a ribbon composed of a supporting substrate of film, fabric, or paper and a coating or impregnation of a coloring material. The coloring material is of such nature that it will transfer in part or entirely to a copy sheet at the point of pressure contact. (F 129) **F 221,** F05

ink eradicator—chemical used to remove ink from a writing surface, often used in the illicit alteration of documents. **F 1156,** F12

ink fill-in, *n*—expansion of a mark beyond specified tolerances. **F 1294,** F05

ink-form roller—the roll in an offset press that delivers ink to the master in the lithographic copy process. **F 425,** F05

ink jet printer—a nonimpact printer in which the characters are formed by projecting droplets of ink onto a substrate. **F 909,** F05

ink kick-out, *n*—the settling of pigment or resin in the bulk ink. **D 6488,** D01

ink layer—*(1)* the components of the image coating used in thermal transfer products which contain waxes, resins, pigments and other materials deposited or coated onto a substrate; *(2)* the layer of thermally sensitive material which is transferred during the thermal transfer printing process. **F 1623,** F05

ink misting, *n*—the ejection of fine particles of ink into the air and onto the press, occurring when the ink film splits in more than one place. The centrifugal forces generated by high speed rollers can cause this problem. **D 6488,** D01

ink mottle, *n*—a condition in which large solid images show very localized variations in color strength or gloss, or both. **D 6488,** D01

ink over-emulsification, *n*—a condition that occurs when a lithographic ink picks up too much dampening solution resulting in a weak print or snowflake pattern. **D 6488,** D01

ink refusal, *n*—the failure of a roller or plate to accept ink. Alternative term: *stripping*—see *Gravure Printing*. **D 6488,** D01

ink sedimentation, *n*—the settling out of undissolved resin or improperly ground pigments. **D 6488,** D01

ink setoff, *n*—the undesirable transfer of an ink in any printing process from fresh prints to any other surface. **D 6488,** D01

ink setoff, *n*—the transfer of ink to the back side of the sheet or area of the web that is the result of slow drying. **D 6488,** D01

inlaid sheet flooring, *n*—a floor surfacing material in which the decorative pattern or design is formed by colored areas set in to the surface. The design so formed may or may not extend through to a backing. **F 141,** F06

in-leakage rate—the combined leakage rate from all existing leaks in a specified evacuated vessel in pressure volume units per unit of time. **E 1316,** E07

inlet—the opening, flange, connection, or coupling on a leak detector or leak testing system through which the tracer gas may enter due to a leak in an object under test. **E 1316,** E07

inlet flange—see **inlet.** **E 1316,** E07

inlet port—see **inlet.** **E 1316,** E07

in-line clinch—flat clinch with both staple legs in straight alignment. **F 592,** F16

in-line coagulation—a filtration process performed by continually adding a coagulant to the raw feedwater and then passing the water through a filter(s) to remove the microfloc which has been formed. **D 6161,** D19

inner liner, *n—of a tire,* the innermost layer(s) of a tubeless tire that limit(s) diffusion of the inflation medium into the carcass. **F 538,** F09

inner packing—See **packing.** **D 996,** D10

inner plies—plies other than face or back plies in a panel construction. Subface, subback, crossband, and center are classed as inner plies. **D 1038,** D07

inner spindle assembly—a component in some manipulation-resistant locks that connects the knob with the cam slide. **F 471,** F12

innocuous impurities—impurities that are not a health or safety concern in the product. The route of administration of the drug may be a significant criterion in the determination of whether an impurity is innocuous. **E 1705,** E48

inoculated iron, *n*—cast iron, either liquid or solid, to which one or more inoculating alloys have been added while the iron was in the molten state. **A 644,** A04

inoculated iron, fully, *n*—cast iron, either liquid or solid, to which all molten metal additions, including all inoculating alloys, have been added. **A 644,** A04

inoculating alloy, *n*—an alloy added to molten iron for the principle purpose of nucleating a primary phase such as graphite. Inoculating alloys are frequently used to avoid the formation of primary carbide by enhancing the nucleation of graphite. **A 644,** A04

inoculum, *n*—living spores, bacteria, single celled organisms, or other live materials that are introduced into a test medium. **D 6384,** D02

inorganic, *adj*— being or composed of matter other than hydrocarbons and their derivatives, or matter that is not of plant or animal origin. **D 1079,** D08

inorganic alkaline detergent—a formulated cleaning composition containing water-soluble alkali or alkaline salts, but generally no surfactants. **D 459,** D12

inorganic alkaline detergent—under **detergent,** see *inorganic alkaline detergent.* **D 459,** D12

inorganics—being or composed of matter other than hydrocarbons and their derivatives, or matter that is not of plant or animal origin. **G 126,** G04

inorganic silt—see **silt.** **D 653,** D18

in-plane flow, *n*—fluid flow confined to a direction parallel to the plane of a geotextile or related product. **D 4439,** D35

inplane loads, n—loads which are parallel to the facings. **C 274,** D30

in-plant cleaning, *n—for textile floor coverings,* a cleaning process performed in a facility away from the location where the product is used. **D 123,** D13

in-process sampling inspection—the performance of specific inspections and/or tests, on samples of product drawn from a prescribed point in the processing stream (usually characteristic sensitive) for the purpose of determining conformance of the product at that point with the requirements. **F 1789,** F16

inquartation, *vt*—the addition of silver to an assay sample to facilitate parting. **E 135,** E01

inrush current—a transient current that exists at the instant of contact closure and persists for a relatively short time. **B 542,** B02

inseam—the hidden seam of a welted shoe holding together the welt, upper, lining, and insole. **F 869,** F08

inseam, *n—in garment construction,* with the trousers folded by mating the inside leg seams, measure from center of crotch to bottom edge of trouser's leg or cuff. **F 1494,** F23

inseam length, *n— in body measurements,* from center of crotch to 25.4 mm (1 in.) below top of the shoe. **F 1494,** F23

insect screening, *n—in coated glass yarn fabrics,* a woven netting having an approximately even spaced mesh of 12 by 12 yarns or more per 25.4 mm (1 in.) **D 123,** D13

insert, *n*—a part consisting of metal or other material which may be molded into position or may be pressed into the molding after the completion of the molding operation. (ISO) **D 883,** D20

insert, *n*—usually a thin filler or frame of wood, **fiberboard,** plastic, or other suitable material used to take up space, or separate articles within a **package.** (Compare **buffer, divider, separator.**) **D 996,** D10

insert, *n*—a part consisting of metal or other material that may be molded into position or may be pressed into the molding after the completion of the molding operation. (D-20, ISO) **F 412,** F17

insert—a reusable, threaded attachment point fixed permanently in the snowboard ski at the time of manufacture, used to mount the bindings to the board. It is typically arranged in a pattern corresponding to a particular binding manufacturer's pattern. **F 1107,** F27

inserted coil—see **ID coil.** **E 1316,** E07

insert-fitting joint—see **joint, clamped insert-fitting.** **F 412,** F17

insertion force—the force required to mate two connector halves. **B 542,** B02

insertion loss, *IL—of a silencer or other sound-reducing element, in a specified frequency band,* the decrease in sound power level, measured at the location of the receiver, when a sound insulator or a sound attenuator is inserted in the transmission path between the source and the receiver. **C 634,** E33

inserts, n—apparatus placed into the sandwich for attaching items: synonymous with hard points. **C 274,** D30

in service coefficient of friction—a coefficient of friction measured under a specified condition of use, which may not be clean and dry, and hence, not a property of the ceramic surface. **C 242,** C21

in-service conditions—the normal conditions to which a system and its components will be exposed during their operational lifetimes. This does not include stagnation conditions; see **stagnation conditions.** **E 772,** E44

inside bevel point—bevel point with its beveled face on staple inside; used to produce an outward clinch or to provide additional penetration in the base material, or both. **F 592,** F16

inside coil—see **ID coil.** **E 1316,** E07

inside knuckle—the juncture between the inside sidewall and the inside bottom surface of a glass article. **C 162,** C14

inside-leg length, *n—in body measurements,* the vertical distance from the crotch to the soles of the feet. **D 123,** D13

inside-out testing—see **bell jar testing.** **E 1316,** E07

inside seam, *n—in home sewing,* a seam having seam allowances located within the interior, or the underside, of the product. (Compare **outside seam.**) **D 4965,** D13

insipid, *adj*—lacking in distinctive sensory characteristics. **E 253,** E18

in situ—applied to a rock or soil when occurring in the situation in which it is naturally formed or deposited. **D 653,** D18

insolation—the use of the term *insolation* is discouraged in favor of the preferred term, **solar irradiance.** **E 772,** E44

insole—a sole of leather or other material cut to the size and shape of the bottom of the last. In some shoe constructions, the insole surface forms the inside of the bottom of the shoe; in others it is covered with a sock lining of thin material which conceals stitching, nails, etc. (Also known as innersole). **F 869,** F08

inspection, *n*—the process of measuring, examining, testing, gaging, or otherwise comparing the unit of product with the applicable requirements. **A 941,** A01

inspection, *n*—the process of measuring, examining, testing, gaging, or otherwise comparing a characteristic or property of a material with applicable requirements. In this case only by visual examination. **D 123,** D13

inspection, *n*—*in fabric grading*, the process of viewing, measuring, examining, or otherwise comparing the visual characteristics of a fabric with applicable requirements. **D 123,** D13

inspection, *n*—the process of measuring, examining, testing, gaging, or otherwise evaluating materials, products, services, systems, or environments. **D 1356,** D22

inspection—the process of measuring, examining, testing or otherwise comparing the unit of product. **D 1517,** D31

inspection—process of measuring, examining, testing, gaging, or using other procedures to ascertain the quality or state, detect errors or defects, or otherwise appraise materials, products, services, systems, or environments to a pre-established standard. **E 1187,** E36

inspection, *n*—a procedure for viewing or observing visual characteristics of a material or component in a careful, critical manner. **E 1316,** E07

inspection—visual examination of the test part after completion of the liquid penetrant processing steps. **E 1316,** E07

inspection—process of measuring, examining, testing, gaging, or using other procedures to ascertain the quality or state of, detect errors or defects in, or otherwise appraise materials, products, services, systems, or environments to a pre-established standard. **F 1789,** F16

inspection agency—see *inspection body*, (ISO Guide 2). **E 1187,** E36

inspection band, *n*—an area of the bar code symbol where measurements are taken spanning from 10 to 90 % of the average bar height. **F 1294,** F05

inspection body—body that performs inspection services on behalf of a certification body, (ISO Guide 2). **E 1187,** E36

inspection medium—see **examination medium.** **E 1316,** E07

inspection torque—the torque necessary to maintain tightening motion in a fastener at its fully preloaded installed tension. **F 1789,** F16

inspector, *n*—*in pipe laying*, an authorized representative of the engineer, or owner, assigned to make any and all necessary inspections of the work performed, including materials and equipment furnished. **C 460,** C17

inspector, *n*—*in pipe laying*, an authorized representative of the engineer, or owner, assigned to make any and all necessary inspections of the work performed, including materials and equipment furnished. **C 1154,** C17

inspector—*in pipe laying*, an authorized representative of the engineer, or owner, assigned to make any and all necessary inspections of the work performed, including materials and equipment furnished. **D 2946,** C17

inspector—person appointed by an authority to examine and evaluate designated domains for conformance to established rules and regulations. **E 1605,** E06

installation—the process of assembling dimension stone into a structure. **C 119,** C18

instantaneous coefficient of thermal expansion—See **coefficient of thermal expansion.** **E 7,** E04

instantaneous collector efficiency—see **efficiency, instantaneous collector.** **E 772,** E44

instantaneous erosion rate, *n*—the slope of a tangent to the cumulative erosion-time curve at a specified point on that curve. **G 40,** G02

instantaneous field of view (IFOV)—for a scanning system, the angular dimensions in object space within which objects are imaged by an individual detector (unit = deg or rad). **E 1316,** E07

instantaneous sampler—a suspended-sediment sampler that takes a representative specimen of the water-sediment mixture in a stream at a desired depth and moment of time. **D 4410,** D19

instantaneous sampling—See **sampling.** **D 1356,** D22

instep—top part of the arch of a foot over the metatarsal bones from back of the toes to the ankle, also the corresponding part of a shoe last. **F 869,** F08

instrumental detection efficiency—*SIMS,* the ratio of ions for a particular species detected to ions produced. **E 673,** E42

instrumentation dead time— see **dead time, instrumentation.** **E 1316,** E07

instrument line shape (ILS) function—the FT of the function by which an interferogram is weighted. **E 131,** E13

instrument response time, *n*—the time required for an indicating or detecting device to undergo a defined displacement following an abrupt change in the property being measured. **D 4175,** D02

instrument response time—the time required for an indicating or detecting device to undergo a defined displacement following an abrupt change in the quantity being measured. **E 131,** E13

instrument traceability—the ability to demonstrate that a particular measuring instrument or artifact standard has been calibrated at acceptable time intervals against a national or international standard or against a secondary standard which has been in turn calibrated against the national standard or transfer standard. **E 170,** E10

insufficient cure—a pultrusion abnormality created by lack of, or incomplete, cross-linking of the resin.

NOTE—This condition can usually be detected by dull surface appearance, low Barcol hardness, and low physical properties. Thick sections, cured from the outside in, can reveal insufficient cure in the center of the section even though completely cured on the surface. This condition can be caused by insufficient die temperature, improper catalyst, or pulling too fast for the die temperature. **D 3918,** D20

insulated—separated from other conducting surfaces by a dielectric substance (including air space) offering a high resistance to the passage of current.

NOTE—When any object is said to be insulated, it is understood to be insulated in a suitable manner for the conditions to which it is subjected. Otherwise, it is, within the purpose of this definition, uninsulated. Insulating covering of conductors is one means of making the conductor insulated. **F 819,** F18

insulated conductor, *n*—a conductor covered by a layer or layers of insulating material and whose prime function is to carry current in an electric circuit. **D 1711,** D09

insulating concrete—a lightweight concrete made with lightweight coarse aggregate and having relatively low insulating characteristics. **C 717,** C24

insulating firebrick, n— see **firebrick, insulating.** **C 71,** C08

insulating formboard—a specially fabricated cellulosic fiberboard designed for use as a permanent form for certain poured-in-place roof constructions. **D 1554,** D07

insulating liquid, fluid or gas—a fluid (liquid or gaseous) which does not readily conduct electricity. Electrical insulating fluids typically provide both electrical insulation and heat transfer in electrical equipment. **D 2864,** D27

insulating material—a material of relatively low electrical conductivity and high dielectric strength, usually used to support or provide electrical separation for conductors, in which a voltage applied between two points on or within the material produces a small and sometimes negligible current. **D 2864,** D27

insulating material (insulator), *n*—a material in which a voltage applied between two points on or within the material produces a small and sometimes negligible current. **D 1711,** D09

insulating roof deck—a cellulosic fiberboard product designed for use in open-beam ceiling roof construction. The product is composed of multiple layers of structural insulating board laminated together with water-resistant adhesive. **D 1554,** D07

insulating-siding nail— See **brick-siding nail.** **F 547,** F16

insulation—See **thermal insulation**. **D 1079,** D08

insulation building-board nail, tileboard nail—galvanized, electrogalvanized or cadmium or nickel-plated, regular-stock-steel, 1¼ and 1¾ by 0.054-in. nails with flat 3/32-in. head and medium needle point. Also, plain or colored (baked-lacquer finished) hardened-steel, smooth or annularly threaded, 1¼ by 0.054 to 1¾ by 0.062-in. nails with slightly countersunk 0.109-in. head and medium diamond or long needle point. **F 547,** F16

insulation-lath nail—blued, regular-stock-steel, 1⅛ and 1¾ by 0.092-in. nails with flat ⅜-in. head and long diamond point. (See **gypsum-lath nail.**) **F 547,** F16

insulation resistance—the apparent resistance between adjacent contacting laminations, calculated as a ratio of the applied voltage to conduction current. This parameter is normally a function of the applied force and voltage. **A 340,** A06

insulation resistance—see **resistance, insulation.** **D 1711,** D09

insulation-sheathing nail—galvanized, barbed, regular-stock-steel, 1¾ and 2 by 0.115 or 0.120-in. nails with flat 7/16 or ½-in. head and medium diamond point. **F 547,** F16

insulator (electric)—a device having high electrical resistance and used for supporting or separating conductors to prevent undesired flow of current from them to other objects. (Also known as electrical insulator.) **D 5077,** D10

integral, *adj—in geosynthetics,* forming a necessary part of the whole; constituent. **D 4439,** D35

integral heat of adsorption—the sum of the differential heats of adsorption from zero to a given level of adsorption. **D 2652,** D28

integrally mold—to join and cure two or more uncured (B-staged) composite details to create an assembly during a single autoclave cure cycle. **E 1749,** E06

integral neutron fluence—the fluence of neutrons integrated over all energies.

$$\Phi = \int_0^\infty \Phi(E)\, dE$$

E 170, E10

integral tooth of metal connector plate—plate projection punched from metal connector plate at right angle to its surface which remains attached to plate and serves as a fastening element. **E 1807,** E06

integral tripack—See **color film: monopack**. **E 7,** E04

integrated-fountain system—a system that delivers both the ink and fountain solution to the offset master by the same form roller. **F 425,** F05

integrating camera—A diffraction camera in which the specimen is moved relative to the incident beam in order to cause diffraction to occur from an extended area of the specimen surface. **E 7,** E04

integrating sphere, *n*—an optical device used either to collect flux reflected or transmitted from a specimen into a hemisphere or to provide isotropic irradiation of a specimen from a complete hemisphere, consisting of an approximately spherical cavity with apertures (ports) for admitting and detecting flux, and usually having additional apertures over which sample and reference specimens are placed and for including or excluding the specularly reflected components. **E 284,** E12

integrating (Ulbrecht) sphere, *n*—part of an integrating photometer. A sphere which is coated internally with a white diffusing paint as nonselective as possible, and which is provided with an associated equipment for making a photometric measurement at a point of the inner surface of the sphere. A screen placed inside the sphere prevents the point under observation from receiving any radiation directly from the source. **E 349,** E21

integration period, τ—the time, in seconds, required for the pen or other indicator to move 98.6 % of its maximum travel in response to a step function. **E 131,** E13

integrator, *n—in the tensile testing of textiles,* a device for obtaining the time integral of the load. **D 123,** D13

integrator, *n—in textile unevenness testing,* a device that calculates the coefficient of variation unevenness or the mean deviation unevenness. **D 123,** D13

integrator (count rate)—a device for obtaining the time integral of the counts as from a radioactive material. **C 859,** C26

intended trajectory, *n*—the intended or ideal path (rectilinear or curvilinear) to bring a vehicle to a stop, that is, under controlled angular orientation. **F 538,** F09

intensifying screen—a sheet of a substance which emits visible light or X-rays or photoelectrons, or combinations of these, under the action of X-rays, thus enhancing the darkening of a film placed in contact with it. **E 7,** E04

intensifying screen—a material that converts a part of the radiographic energy into light or electrons and that, when in contact with a recording medium during exposure, improves the quality of the radiograph, or reduces the exposure time required to produce a radiograph, or both. Three kinds of screens in common use are:

(a) metal screen—a screen consisting of dense metal (usually lead) or of a dense metal compound (for example, lead oxide) that emits primary electrons when exposed to X- or gamma-rays.

(b) fluorescent screen—a screen consisting of a coating of phosphors which fluoresces when exposed to X or gamma radiation.

(c) fluorescent-metallic screen—a screen consisting of a metallic foil (usually lead) coated with a material that fluoresces when exposed to X or gamma radiation. The coated surface is placed next to the film to provide fluorescence; the metal functions as a normal metal screen. **E 1316,** E07

intensional definition, *n*—a two-part definition of a term consisting of a class name for the concept, the genus, and one or two more characteristics that differentiate the concept from others in the same class. **E 1992,** E02

intensiostatic—See **galvanostatic.** **G 15,** G01

intensity—ideally, the radiant power per unit area. In spectrochemical analysis, the radiant power as measured by a photographic or photoelectric receiver, or as estimated visually. **C 859,** C26

intensity, *n*—the perceived magnitude of a stimulus. **E 253,** E18

intensity—apparent depth of a black or colored image produced by a carbon or ribbon. **F 221,** F05

intensity, *n*—the amount of radiant or luminous flux per unit solid angle that diverges from a light source. **F 1294,** F05

intensity, I_v, I_e, *n*—flux per unit solid angle. (See also **luminous intensity, radiant intensity**.) **E 284,** E12

intensity (line)—the light intensity of an image of a slit formed in the focal plane of a spectrograph or spectrometer and having a width approximately equal to that formed by monochromatic radiation. Also see **spectral line.** **C 859,** C26

intensity of scattering—the energy per unit time per unit area of the general radiation which is diffracted by matter. Its value depends upon the scattering power of the individual atoms of the material, upon the scattering angle, and upon the wavelength of the radiation. **E 7,** E04

intensity ratio (relative intensity ratio), *n*—the ratio of two (relative) intensities. **E 135,** E01

intensity (*X-rays*)—the energy per unit of time of a beam per unit area perpendicular to the direction of propagation. **E 7,** E04

interaction, *n*—the condition that exists among factors when a test result obtained at one level of a factor is dependent on the level of one or more additional factors. **D 123,** D13

interatomic Auger process—*AES, XPS,* an Auger transition in which final electron vacancies are in valence levels or molecular orbitals, some of which may be predominantly orbitals of a neighboring bonded atom. **E 673,** E42

interbody spacer, *n*—a structure (biologic or synthetic) to replace (partially or totally) the vertebral body or intervertebral disk(s), or both. **F 1582,** F04

interception count—the number of particles (or clusters of particles) of a phase or constituent of interest that are crossed by the lines of a test grid. **E 7,** E04

intercept method—See **Heyn method.** **E 7,** E04

intercharacter space, *n*—the space between the last bar of one character and the first bar of the next that separates two adjacent characters. Also called intercharacter gap. **F 1294,** F05

intercoat contamination—*in protective coatings,* the presence of foreign matter between successive coats. **D 4538,** D33

intercolor feathering, *n*—instantaneous ink spread into adjacent inks. **F 1857,** F05

interconnected porosity—a network of pores in and extending to the surface of a sintered compact. Usually applied to powder metallurgy materials where the interconnected porosity is determined by impregnating the specimens with oil. **B 243,** B09

interconnection, *n*—the mechanical interface or connection mechanism between at least two components or between components and bony elements of the spine, pelvis, or ribs. **F 1582,** F04

interconnector—a device to connect adjacent membrane elements in series and to seal the product channel from the feed-brine channel. **D 6161,** D19

intercorrelation coefficient, (r_{XX})—a measure of the linear association between values of the same type of variable expressed as a correlation coefficient, (*r*). **E 131,** E13

intercritical annealing/isothermal annealing—heating a steel above the lower critical transformation temperature, but below the upper-critical transformation temperature, to dissolve all the iron carbides, but not transform all the ferrite to austenite. Cooling slowly from this temperature, through the lower critical temperature, produces a structure of ferrite and pearlite that is free of internal stresses. In *intercritical annealing,* the steel continues to cool slowly in the furnace, similarly to full annealing. In *isothermal annealing,* cooling is stopped just below the lower critical, assuring complete transformation to ferrite and coarse pearlite, and eliminating the potential for bainite formation. The coarse pearlite structure greatly improves machinability of medium carbon steels. **F 1789,** F16

intercrystalline corrosion— See **intergranular corrosion.** **G 15,** G01

intercrystalline cracks—cracks or fractures that occur between the grains or crystals in a polycrystalline aggregate. **E 7,** E04

interdendritic—located within the branches of a dendrite or between the boundaries of two or more dendrites. **E 7,** E04

interdendritic porosity—voids occurring between the dendrites in cast metal. **E 7,** E04

interface—a boundary between two phases with different chemical or physical properties. **E 673,** E42

interface—the boundary between two materials. **E 1316,** E07

interface—*of an epitaxial wafer,* the boundary between the substrate and the epitaxial layer as determined by a specific measurement method such as infrared reflectance or spreading resistance profiling. **F 1241,** F01

interface, *n*—the area in a device where different subsystems connect electrically or a device used to establish a connection between two different devices. **F 1457,** F05

interface, *n*— one of the two mating surfaces, lines or points of contact within an interconnection between two components, between any component and bone, or between two bony elements. **F 1582,** F04

interface width, observed—*AES, XPS, SIMS,* the distance over which a 16 % to 84 % (or 84 % to 16 %) change in signal is measured at the junction of two dissimilar matrices. **E 673,** E42

interfacial chemical coacervation—process for creating many very small droplets of microencapsulated materials by a chemical reaction at the interface between two or more (immiscible) compounds. **F 1156,** F12

interfacial region—that volume adjacent to an interface having physical or chemical properties different from either bulk phase as a result of its proximity to the interface. **E 673,** E42

interfacial tension—the contractile force of an interface between two phases. See **surface tension.** **B 374,** B08

interfacial tension—the force existing in a liquid-liquid phase interface that tends to diminish the area of the interface. This force, which is analogous to the surface tension of liquid-vapor interfaces, acts at each point on the interface in the plane tangent at that point. **D 459,** D12

interfacial tension, *n*—the force existing in a liquid-liquid phase interface that tends to diminish the area of the interface. This force, which is analogous to the surface tension of liquid-vapor interfaces, acts at each point on the interface in the plane tangent at that point. (*Compilation of ASTM Standard Definitions,* 7th Edition.) **D 2864,** D27

interference, *n*—*in testing,* an effect due to the presence of a constituent or characteristic that influences the measurement of another constituent or characteristic. **D 123,** D13

interference, *n*—an undesired output caused by a substance or substances other than the one being measured. **D 1356,** D22

interference—the effect of a combination of wave trains of various phases and amplitudes. **E 7,** E04

interference—the variation with distance or time of the amplitude of a wave which results from the superposition of two or more waves having the same, or nearly the same frequency. **E 1316,** E07

interference contrast microscope—microscope in which surface details of an object in which there is no appreciable absorption are revealed by the interference between two beams of light. **F 1241,** F01

interference filter, *n*—filter constructed of extremely thin alternate layers of high and low refractive-index material and capable of transmitting narrow spectral bands formed by constructive interference within the desired waveband and destructive interference at other wavelengths. **E 284,** E12

interference, *n*—any activity or event that could produce anomalous measurements. **C 634,** E33

interference seal—a seal produced between a fastener and its hole when a fastener of a given diameter is driven into a hole of a smaller diameter. An interference seal is also produced when a fastener shank is expanded by the installation process. **E 1749,** E06

interference signal—*SIMS,* signal measured at the mass position of interest due to another, undesired species. **E 673,** E42

interfering substance—any species, other than the ion being measured, whose presence in the sample solution affects the measured potential of a cell. Interfering substances fall into two classes: "electrode" interferences and" method" interferences. Examples of the first class would be those substances which give a similar response to the ion being measured and whose presence generally results in an apparent increase in the activity (or concentration) of the ion to be determined (for example, Na^+ for the Ca^{++} electrode), those species which interact with the membrane so as to change its chemical composition (that is, organic solvents for the liquid or poly(vinyl chloride) (PVC) membrane electrodes) or electrolytes present at a high concentration giving rise to appreciable liquid-junction potentials. The second class of interfering substance is that which interacts with the ion being measured so as to decrease its activity or apparent concentration, but where the electrode continues to report the true activity (that is, CN^- present in the measurement of Ag^+). **D 4127,** D19

interferogram, double-sided—interferogram measured with approximately equal retardation on either side of the centerburst. **E 131,** E13

interferogram, I(δ)—record of the modulated component of the interference signal measured as a function of retardation by the detector. **E 131,** E13

interferogram, laser reference—sinusoidal interferogram of a laser source measured at the same time as the signal interferogram. **E 131,** E13

interferogram, signal—interferogram of the beam of radiant energy whose spectrum is desired. **E 131,** E13

interferogram, single-sided—interferogram in which sampling is initiated close to the centerburst and continues through that point to the maximum retardation desired. **E 131,** E13

interferogram, white light—reference interferogram of a broadband light source measured at the same time as the signal interferogram and used to initiate data acquisition of consecutive scans for signal-averaging. **E 131,** E13

interferometer—device used to divide a beam of radiant energy into two or more paths, generate an optical path difference between the beams, and recombine them in order to produce repetitive interference maxima and minima as the optical retardation is varied. **E 131,** E13

interferometer, Genzel—interferometer in which the beam is focused in the plane of the beamsplitter and collimated before the moving mirror(s). **E 131,** E13

interferometer, lamellar grating—interferometer in which the beam is reflected from two interleaved mirrors, one of which is stationary while the other is movable. **E 131,** E13

interferometer, Michelson—interferometer in which an approximately collimated beam of radiant energy is divided into two paths by a beamsplitter; one beam is reflected from a movable mirror and the other from a stationary mirror, and they are then recombined at the beamsplitter. **E 131,** E13

interferometer, rapid-scanning—interferometer in which the retardation is varied rapidly enough that the modulation frequencies in the interferogram are sufficiently high that the interferogram signal can be amplified directly without additional modulation by an external chopper. **E 131,** E13

interferometer, refractively scanned—interferometer in which the retardation between two beams is generated by the movement of a wedged optical element. **E 131,** E13

interferometer, slow-scanning—interferometer in which the retardation is continuously varied, but so slowly that an external chopper is needed to modulate the beam at a frequency which is high enough for a-c signal amplification. **E 131,** E13

interferometer, stepped-scanning—interferometer in which the movable element is held stationary for the length of time required for signal integration and digitization of each sample point, and then translated to the next sample point. **E 131,** E13

intergranular corrosion—a preferential attack at the grain boundaries. **E 7,** E04

intergranular corrosion—preferential corrosion at or adjacent to the grain boundaries of a metal or alloy. **G 15,** G01

intergranular pressure—see **stress.** **D 653,** D18

interior foam-filled tube—a tube filled with homogeneous unicellular thermosetting foam having closed cells blown with noncombustible gases, with the filling bonded to the interior wall.

NOTE—The foam filling should be free of voids, separations, holes, cracks, or the like. **F 819,** F18

interior glazed—glazed from the interior of the building. **C 717,** C24

interior packing—See **packing.** **D 996,** D10

interior pipe size (IPS)—See **iron pipe size.** **E 631,** E06

interior stop—a removable or integral stop on the indoor side of the glass or panel. **C 717,** C24

interior type plywood—a term frequently applied to plywood bonded with adhesives that maintain adequate bonds under conditions usually existing in the interior of buildings; the commercial classification is a function of veneer grade as well as glue durability. **D 1038,** D07

interior work-area—hallway, room, part of a room, or group of rooms in which a lead-hazard abatement activity takes place on the inside of a building. **E 1605,** E06

interlaboratory test comparison—organization, performance and evaluation of tests on the same or similar items or materials by two or more laboratories in accordance with predetermined conditions, (ISO Guide 2). **E 1187,** E36

interlaboratory testing—the evaluating of a test method in more than one laboratory by analyzing data obtained from one or more materials that are as homogeneous as practical. **D 123,** D13

interlaminar—describing objects (for example, voids), events (for example, fracture), or fields (for example, stress) between the laminae of a laminate. **D 3878,** D30

interlaminar—descriptive term pertaining to some object (voids), event (fracture), or potential field (shear stress) referenced as existing or occurring between two or more adjacent laminae. **E 1749,** E06

interlayer—a material used for bonding glass lites together. See related term **laminated glass**. **C 162,** C14

interlayer paper—see **layer insulation.** **D 1711,** D09

Interleaved 2 of 5 bar code—See **bar code**. **F 1156,** F12

interleaving—process in bar code symbology of assigning information content to the spaces in the code as well as to the bars. **F 1156,** F12

interlining, *n*—any textile which is intended for incorporation into an article of wearing apparel as a layer between an outer shell and inner lining. **D 123,** D13

interlock—See **windows and doors.** **E 631,** E06

interlock—a device or system, either electrical, mechanical, or electromechanical, that serves to prevent exposure to an electric shock, physical injury, or excessive radiation emission when a door, cover, or access panel is opened or removed. **F 1827,** F26

interlocking tile, *n*—tile with a system of ribs or grooves enabling the lateral joining of adjacent tiles in the same course with the overlocking edge of one tile covering the underlocking edge of another forming the side lap. In some tile designs, provision is also made for interlocking in the overlapping area of the head lap. **C 43,** C15

intermediary plane—any plane in a microscope where a real image of a specimen is formed. Reticles can be inserted at intermediary planes for superposition on the image. **E 7,** E04

intermediate annealing, *n*—**annealing** wrought steel objects at one or more stages during manufacture prior to final thermal treatment. **A 941,** A01

intermediate fiberboard sheathing—a cellulosic fiberboard sheathing product, approximately 22 lb/ft^3, used in frame construction under masonry veneer, siding, shingles, and stucco. **D 1554,** D07

intermediate image—any image of an object formed by a lens other than the final projector lens. **E 7,** E04

intermediate pack— See **pack.** **D 996,** D10

intermediate package— See **package.** **D 996,** D10

intermediate phase—in a chemical system, a distinguishable homogeneous substance whose composition range of existence does not extend to any of the pure components of the system. **E 7,** E04

intermediate precision conditions, *n*—conditions under which test results are obtained with the same test method using test units or test specimens (see Practice E 691, 10.3) taken at random from a single quantity of material that is as nearly homogeneous as possible, and with changing conditions such as operator, measuring equipment, location within the laboratory, and time. **E 456,** E11

intermediate precisions, *n*—the closeness of agreement between test results obtained under specified intermediate precision conditions.

Note—The specific measure and the specific conditions must be specified for each intermediate measure of precision; thus, "standard deviation of test results among operators in a laboratory," or "day-to-day standard deviation within a laboratory for the same operator."

Note—Because the training of operators, the agreement of different pieces of equipment in the same laboratory and the variation of environmental conditions with longer time intervals all depend on the degree of within-laboratory control, the intermediate measures of precision are likely to vary appreciably from laboratory to laboratory. Thus, intermediate precisions may be more characteristic of individual laboratories than of the test method. **E 456,** E11

intermediate principal plane—see **principal plane**. **D 653,** D18

intermediate principal stress—see **stress**. **D 653,** D18

intermediate rail—one of two or more rails between the top rail and floor. **E 1481,** E06

intermediate rail—horizontal member of the framework running continuously at any point between the top and bottom of the fence (sometimes called *middle rail*). **F 552,** F14

intermediate rib, *n*—one or more rib(s) located between the centerline and the shoulder ribs of the tread band. **F 538,** F09

intermediate row, *n*—a row located between the circumferential centerline and the shoulder ribs/rows of the tread band. **F 538,** F09

intermediate temper—as applied to aluminum, any temper between soft and hard drawn. **B 354,** B01

intermediate-temperature-setting adhesive, *n*—an adhesive that sets in the temperature range from 31 to 99°C (87 to 211°F). (Synonym *warm-setting adhesive.*) (Compare **cold-setting adhesive, hot-setting adhesive**, and **room-temperature-setting adhesive**.) **D 907,** D14

intermetallic phases—compounds, or intermediate solid solutions, containing two or more metals, which usually have characteristic properties and crystal structures different from those of the pure metals or the terminal solid solutions. **E 7,** E04

intermittency effect, *n*—*in optical emission spectrometry*, the departure from the reciprocity law that may occur when the exposure of a photographic emulsion is made in a series of discrete increments, rather than in a continuous exposure of the same total energy. **E 135,** E01

intermittent disfigurement—disfigurement of the shape of the pultrusion for a limited length.

Note—This condition usually results from die fouling or from folded or crowded reinforcing material and is corrected after a short distance. **D 3918,** D20

intermittent moving bed (pulse, slug)— an adsorption process characterized by upward flow of a fluid through a fixed bed of granular adsorbent with periodic withdrawal of spent adsorbent from the bottom of the bed and additions of reprocessed or virgin adsorbent to the top of the bed. **D 2652,** D28

intermittent sampling—See **sampling.** **D 1356,** D22

intermodal container— See **container.** **D 996,** D10

internal absorptance, α_i, *n*—ratio of the absorbed radiant or luminous flux to the flux that has entered the first surface of the layer of material. **E 284,** E12

internal absorptance of a homogeneous nondiffusing plate, *n*—ratio of the radiant or luminous flux absorbed between the entry and exit surfaces of the plate to the flux which leaves the entry surface. Symbol: a_i, $a_i + \tau_i = 1$.

Note—For a given plate the internal absorptance is a function of the path length of the radiation in the plate, and thus of the angle of incidence. The fundamental concept is spectral internal absorptance. a_i (λ). **E 349,** E21

internal carbon referencing—*XPS*, a method of determining the charging potential of a specimen by comparing the experimentally determined binding energy of the C1s peak maximum from a specific carbon group within the specimen to a standard binding energy value for that carbon group. A hydrocarbon group within the specimen is often used for this purpose. **E 673,** E42

internal conversion, *n*—a transition between electronic states of the same total spin quantum number (multiplicity). **E 131,** E13

internal discharge—method in which makeup air is introduced inside the hood. This design has limited application, and the amount of supply air able to be introduced varies considerable with the type of cooking equipment and the exhaust flow rate. Makeup air may be untempered depending on climatic conditions. **F 1827,** F26

internal filling solution of sensing electrode—in liquid membrane electrodes, an aqueous internal filling solution contacts the internal reference element and the membrane, which is saturated with ion exchanger. The filling solution normally contains a fixed level of chloride and of the ion for which the electrode was designed; the concentration of this ion determines the zero potential point of the electrode. In addition, the filling solution is saturated with silver chloride to prevent the silver chloride of the internal reference element from dissolving. **D 4127,** D19

internal flotation—flotation element located within the boom membrane. **F 818,** F20

internal friction (shear resistance), (FL^{-2})—the portion of the shearing strength of a soil or rock indicated by the terms $p \tan \phi$ in Coulomb's equation $s = c + p \tan \phi$. It is usually considered to be due to the interlocking of the soil or rock grains and the resistance to sliding between the grains. **D 653,** D18

internal mixing pneumatic atomizer—a pneumatic atomizer in which gas and liquid are introduced and mixed within the nozzle, and are then discharged as a two-phase dispersion through a common orifice. **E 1620,** E29

internal oxidation—preferential oxidation of certain components or phases within the bulk of a solid metal. **E 7,** E04

internal oxidation—the formation of isolated particles of corrosion products beneath the metal surface. (This occurs as the result of preferential oxidation of certain alloy constituents by inward diffusion of oxygen, nitrogen, sulfur, etc.) **G 15,** G01

internal rate of return (IRR)—the compound rate of interest that, when used to discount study period costs and benefits of a project, will make the two equal. **E 833,** E06

internal reference electrode—a reference electrode that is contained inside an ion-selective electrode assembly. Comment: The system frequently consists of a silver-silver chloride electrode in contact with an appropriate solution containing chloride and a fixed concentration of the ion for which the membrane is selective. **D 4127,** D19

internal, reflection attachment, IRA—the transfer optical system which supports the IRE, directs the energy of the radiant beam into the IRE, and then redirects the energy into the spectrometer or onto the detector. The IRA may be part of an internal reflection spectrometer or it may be placed into the sampling space of a spectrometer. **E 131,** E13

internal reflection element (IRE)—the transparent optical element used in internal reflection spectroscopy for establishing the conditions necessary to obtain the internal reflection spectra of materials. **E 131,** E13

internal reflection spectroscopy (IRS)— the technique of recording optical spectra by placing a sample material in contact with a transparent medium of greater refractive index and measuring the reflectance (single or multiple) from the interface, generally at angles of incidence greater than the critical angle. **E 131,** E13

internal reflux, *n*—(Syn. *reflux*) the liquid normally running down inside a distillation column. **D 4175,** D02

internal seam—See **stepdown test.** **E 7,** E04

internal shrinkage cracks—longitudinal cracks in the pultrusion that are found within sections of roving reinforcement.

NOTE—This condition is caused by shrinkage strains during cure that show up in the roving portion of the pultrusion where transverse strength is low. **D 3918,** D20

internal standard—a material present in or added to samples that serves as an intensity reference for spectral measurements. **C 859,** C26

internal standard—a material present in or added to samples in known amount to serve as a reference measurement. **D 1129,** D19

internal standard—a compound of known behavior added to a sample to facilitate the analysis. **D 4790,** D16

internal standard, *n*—*in spectrometry*, a material present in specimens or added to test samples that serves as an intensity reference for spectral measurements. **E 135,** E01

internal standard line—a spectral line of an internal standard, with which the radiant energy of an analytical line is compared. **C 859,** C26

internal standard line, *n*—a spectral line of an internal standard, with which the radiant energy of an analytical line is compared. **E 135,** E01

internal transmission density—logarithm to the base 10 of the reciprocal of the internal transmittance. Symbol: D_i, $D_i = -\log_{10} \tau_i$.

NOTE—See Note to **internal transmittance.**

NOTE—In German, the symbol E is still in use and the natural logarithm is also used sometimes instead of the common logarithm; the corresponding quantity is then called "natürliches Absorptionsmass." (= ln $1/\tau i$). **E 349,** E21

internal transmittance, τ_i, *n*—the ratio of flux reaching the exit surface of a specimen to the flux that penetrates the entry surface. **E 284,** E12

internal transmittance of a homogeneous nondiffusing plate, *n*—ratio of the radiant or luminous flux reaching the exit surface of the plate to the flux which leaves the entry surface.

NOTE—For a given plate the internal transmittance is a function of the path length of the radiation in the plate, and thus of the angle of incidence. The fundamental concept is "spectral internal transmittance" $\tau(\lambda)$. **E 349,** E21

International Organization for Standards, *n*—an organization that develops and publishes international standards for a variety of technical applications, including data processing and communications. (See **ISO.**) **F 1457,** F05

international phyrheliometric scale—pyrheliometric scale decreed as being in operation from July 1, 1957, in order to meet an urgent need for a single international scale. **E 772,** E44

International Practical Temperature Scale (IPTS-48), *n*—the temperature scale adopted by the 11th General Conference on Weights and Measures in 1960 and replaced in 1968 by the International Practical Temperature Scale of 1968. **E 344,** E20

International Practical Temperature Scale of 1968 (IPTS-68), *n*—the temperature scale adopted by the 13th General Conference on Weights and Measures in 1968. **E 344,** E20

international roughness index (IRI), *n*—an index computed from a longitudinal profile measurement using a quarter-car simulation at a simulation speed of 50 mph (80 km/h). **E 867,** E17

international roughness index, true—see **true international roughness index.** **E 867,** E17

International System of Units, SI—a complete coherent system of units whose base units are the metre, kilogram, second, ampere, kelvin, mole, and candela. Other units are derived as combinations of the base units or supplementary units. **A 340,** A06

International Temperature Scale of 1990 (ITS-90), *n*—the temperature scale prepared in accordance with instructions of the 18th General Conference on Weights and Measures, and adopted on

January 1, 1990. **E 344,** E20

interphase, *n—in an adhesive joint,* a region of finite dimension extending from a point in the adherend where the local properties (chemical, physical, mechanical, and morphological) begin to change from the bulk properties of the adherend to a point in the adhesive where the local properties equal the bulk properties of the adhesive. **D 907,** D14

interphase—the region between two distinct phases over which there is a variation of a property. **E 673,** E42

interplanar distance—the perpendicular distance between adjacent parallel lattice planes. **E 7,** E04

interpretation—the determination of whether indications are relevant or nonrelevant. **E 1316,** E07

interpretation, *n*—the determination of whether indications are relevant, nonrelevant, or false. **E 1316,** E07

interpreter—that part of the OCR system which analyzes the input data and determines what the individual characters are and what their relation is to each other. **F 149,** F05

inter-projection wear, *n*—a type of irregular treadwear characterized by different wear rates on one or more adjacent projections (either transverse or circumferential orientation); this results in a step-off in tread depth between the adjacent projections. **F 538,** F09

interpupillary distance—spacing between the pupils of the eyes; eyepieces on binocular microscopes should be set at this distance for comfortable and accurate viewing. **E 7,** E04

interrupted aging, *n*—**aging** at two or more temperatures, by steps, and cooling to room temperature after each step. **A 941,** A01

interrupted quenching, *n*—**quenching** in which the object being quenched is removed from the quenching medium while the object is at a temperature substantially higher than that of the quenching medium. **A 941,** A01

intersection count—the number of boundaries between the matrix phase and the phase or constituent of interest that are crossed by the lines of a test grid. For isolated particles in a matrix, the number of feature intersections will equal twice the number of feature interceptions. **E 7,** E04

interstice—see preferred term **void.** **D 653,** D18

interstitial—occurring between the grains or in the pores in rock or soil. **D 653,** D18

interstitial-free steel, *n*—a **steel** that has essentially all of its carbon and nitrogen chemically combined with stabilization elements rather than being present interstitially. **A 941,** A01

interstitial water—water occupying space between sediment or soil particles (Syn. **pore water**). **E 943,** E47

intersystem crossing—a transition between electronic states that differ in total spin quantum number (multiplicity). **E 131,** E13

interval, arrival time (Δt_{ij})—the time interval between the detected arrivals of an acoustic emission wave at the *ith* and *jth* sensors of a sensor array. **E 1316,** E07

interval erosion rate, *n*—the slope of a line joining two specified points on the cumulative erosion-time curve. **G 40,** G02

interval estimate—the estimate of a parameter given by two statistics, defining the end points of an interval. **E 1823,** E08

intervenor physician—a licensed M.D. or D.O., having not previously established a doctor/patient relationship with the emergency patient, who is willing to accept responsibility for patient care, and who can provide proof of a current medical license. **F 1177,** F30

intervention required, *n*—a message from the host computer indicating to the operator that some action is required before printing can continue. **F 1457,** F05

intervertebral body fusion cage, *n*—a hollow device which contains graft material. **F 1582,** F04

intervertebral body fusion devices, *n*—a structure which is placed in the disc space between two adjacent vertebral bodies to provide support for eventual arthrodeses of the two adjacent vertebral bodies. **F 1582,** F04

intracrystalline cracking—See **transcrystalline cracking**. **E 7,** E04

intralaminar—describing objects (for example, voids), events (for example, fracture), or fields (for example, stress) within the laminae of a laminate. **D 3878,** D30

intra-projection wear, *n*—a type of irregular wear characterized by a different wear rate at two or more locations within a given projection. **F 538,** F09

intrinsic fiber optic chemical sensor, *n*—a fiber optic chemical sensor in which the modulation of the optical signal is effected through a change in the properties of the optical fiber itself, and such modulation occurs while the radiant energy is guided by the optical fiber. **E 131,** E13

intrinsic linewidth, of specimen—*AES, XPS,* the linewidth contribution arising from the specimen. The measured linewidth is a convolution of this function and broadening contributions of the instrument (for example, X-ray source radiation linewidth, spectrometer energy resolution). **E 673,** E42

intrinsic shear strength, S_o (FL^{-2})—the shear strength of a rock indicated by Coulomb's equation when $p \tan \phi$ (shear resistance or internal friction) vanishes. Corresponds to cohesion, *c,* in soil mechanics. **D 653,** D18

intrinsic viscosity—see **viscosity, intrinsic.** **D 1695,** D01

intumescent coating—a fire-retardant coating (which see) that when heated forms a foam produced by nonflammable gases, such as carbon dioxide and ammonia. This results in a thick, highly insulating layer of carbon (about fifty times as thick as the original coating) that serves to protect the coated substrate from fire. **D 16,** D01

invariant equilibrium—a stable state amongst a number of phases exceeding by two the number of components in the system and in which more of the external variables (pressure, temperature, or concentrations) may be varied without causing a decrease in the number of phases present. **E 7,** E04

invariant equilibrium—stable state among a number of phases exceeding by two the number of components in the system and in which more of the external variables (pressure, temperatures, or concentrations) may be varied without causing a decrease in the number of phases present, (E 7, E-4), . **E 1142,** E37

invariant point—a point defined by the unique values of temperature, pressure, and concentrations in a system with the maximum number phases which can coexist in equilibrium. **E 7,** E04

invariant point—point defined by the unique values of temperature, pressure, and concentrations in a system with the maximum number of phases that can coexist in equilibrium, (E 7, E-4), . **E 1142,** E37

inverse rate curve—in thermal analysis, the curve that is obtained when the length of time required by the specimen to pass through successive and constant intervals of temperature is plotted against the temperature. **E 7,** E04

inverse segregation—a concentration of low melting constituents in those regions in which solidification first occurs. **E 7,** E04

inverse wear rate, [*nd*], *n*—the distance traveled by a tire, after break-in, per unit loss in tread depth. **F 538,** F09

inversion, *n*—a reversal of the normal atmospheric temperature gradient, thus an increase of temperature of the air with increasing altitude. **D 1356,** D22

inversion—the act of turning inward, turning the sole of the foot toward the midline of the body. **F 869,** F08

invert—the bottom or lowest point of the internal surface of the transverse cross section of a pipe. **C 822,** C13

invert—on the cross section, the lowest point of the underground excavation or the lowest section of the lining. (ISRM) **D 653,** D18

inverted microscope—a microscope so arranged that the line of sight is directed upwards through the objective to the object. **E 7,** E04

investment cost—first cost and later expenditures which have substantial and enduring value (generally more than one year) for upgrading, expanding, or changing the functional use of a building or building subsystem. **E 833,** E06

invisible waste, *n*—*in cotton testing*, weight loss due to dust, moisture, loose fibers, etc., carried away by the air stream during the test. **D 123,** D13

iodine adsorption number, *n*—the number of grams of iodine adsorbed per kilogram of carbon black under specified conditions. **D 3053,** D24

iodometry—an application of iodine chemistry to oxidation-reduction titrations for the quantitative analysis of chemical compounds. **C 859,** C26

ion—an electrified portion of matter of atomic or molecular dimensions. **B 374,** B08

ion—a charged portion of matter of atomic or molecular dimensions. **D 6161,** D19

ion beam—a directed flux of charged atoms or molecules.
current—the measured rate of flow of charged atoms or molecules incident upon the specimen per unit time, usually expressed in amperes (A).
current density—the ion beam current incident on the specimen per unit cross-sectional area, usually expressed in amperes per square centimetre (A/cm^2).
energy—in surface analysis, the mean kinetic energy of the ions in the beam (see **beam,** *energy*). **E 673,** E42

ion content, *n*—a measure of the free sodium, potassium, and chloride content in a direct thermal or thermal transfer product; trace amounts may cause chemical attack to the thermal printhead elements. **F 1623,** F05

ion deposition printer—a printer where ion charges are gated onto a dielectric drum. Toner is picked up by the charge, then transferred to the paper. Once the toner is deposited on the paper, it can be affixed by either pressure or heat fusion. **F 149,** F05

ion exchange—a reversible process by which ions are interchanged between a solid and a liquid with no substantial strucdural changes of the solid. **B 374,** B08

ion exchange—a reversible process by which ions are interchanged between an insoluble material and a liquid with no substantial structural changes of the material. **D 1129,** D19

ion exchange—a reversible process by which ions are interchanged between a solid and a liquid with no substantial structural changes in the solid. **D 1695,** D01

ion exchange—a reversible process by which ions are interchanged between a solid and a liquid with no substantial structural changes in the solid; ions removed from a liquid by chemical bonding to the media. **D 6161,** D19

ion-exchange capacity—for cellulosic ion-exchange materials, the number of milliequivalents of ions that can be exchanged by 1 g of the ion-exchange material. **D 1695,** D01

ion-exchange capacity (volume basis)—the number of milliequivalents of exchangeable ions per millilitre of backwashed and settled bed of ion-exchange material in its standard form. **D 1129,** D19

ion-exchange capacity (volume basis)—the number of milliequivalents of exchangeable ions per millilitre of backwashed and settled bed of ion-exchange material in its standard form. **D 6161,** D19

ion-exchange capacity (weight basis)—the number of milliequivalents of exchangeable ions per dry gram of ion-exchange material in its standard form. **D 1129,** D19

ion-exchange capacity (weight basis)—the number of milliequivalents of exchangeable ions per dry gram of ion-exchange material in its standard form. **D 6161,** D19

ion-exchange material—a water insoluble material that has the ability to exchange reversibly certain ions in its structure, or attached to its surface as functional groups, with ions in a surrounding medium. **D 1129,** D19

ion-exchange material—a water insoluble material that has the ability to reversibly exchange ions in its structure, or attached to its surface as functional groups, with ions in a surrounding medium. **D 6161,** D19

ion-exchange materials—insoluble solid materials with the ability to exchange reversibly certain ions, in the structure or attached to the surface as functional groups, with ions in a surrounding medium. **D 1695,** D01

ion-exchange membrane—an ion-exchange material in a form suitable for use as a barrier between two fluids. **D 1129,** D19

ion-exchange membrane—an ion-exchange material in a form suitable for use as a barrier between two fluids. **D 6161,** D19

ion-exchange particle—an ion-exchange material in the form of spheroids or granules with an average diameter less than 10 mm. **D 1129,** D19

ion-exchange particle—an ion-exchange material in the form of spheroids or granules. **D 6161,** D19

ion-exchange resin—a synthetic, organic-ion-exchange material. **D 1129,** D19

ion-exchange resin—an organic ion-exchange material substrate, usually synthetic. **D 6161,** D19

ionic cellulose ethers—those water-soluble cellulose ethers which contain ionizable groups, in more than trace amounts. Sodium carboxymethyl cellulose is an example. **D 1695,** D01

ionic strength—the weighted concentration of ions in solution, computed by multiplying the concentration (*c*) of each and every ion in solution by the corresponding square of the charge (*Z*) on the ion, summing and dividing by 2: ionic strength = $(1/2)\Sigma Z^2 C$. Conductivity measurements give a rough estimate of ionic strength. The ionic strength (and to a lesser extent, the concentration of nonionic dissolved species) largely determines the activity coefficient of each ion in the solution. **D 4127,** D19

ionic strength—measure of the overall electrolytic potential of a solution, the strength of a solution based on both the concentrations and valencies of the ions present. **D 6161,** D19

ionic-strength adjustment buffer—a pH buffered solution of high ionic strength added to samples and calibration solutions before measurement in order to achieve identical ionic strength and hydrogen ion activity. In addition, complexing agents and other components are often added to minimize the effects of certain interferences. **D 4127,** D19

ion image—*SIMS,* a two-dimensional representation of the spatial distribution of a particular secondary ion emitted from a specific area of the specimen. **E 673,** E42

ion implantation—the injection of ions into a specimen. **E 673,** E42

ionization, *n*—the process by which electrons are lost from or transferred to neutral molecules or atoms to form positively or negatively charged particles. **D 1711,** D09

ionization—the process by which neutral atoms or molecules, such as air, acquire a positive or negative charge. **D 5077,** D10

ionization—the disassociation of molecules into charged particles (ions). **D 6161,** D19

ionization—a process in which a charged particle is created from a parent atom or molecule or other bound state. **E 170,** E10

ionization chamber—a device for the measurement of radiation intensity by means of determining the degree to which the radiation ionizes a gas. **E 7,** E04

ionization cross-section—the probability that an incident particle traversing a gas or solid will produce an ionizing collision. The total ionization cross-section includes all electron vacancies produced by a primary collision and subsequent Coster-Kronig or Auger decay process. The partial ionization cross-section results from one particular process such as a primary collision to produce an initial innershell vacancy in a particular shell, a Coster-Kronig process, or an Auger ejection process to produce particular distributions of electron vacancies. **E 673,** E42

ionization method—a method of X-ray diffraction in which the intensity of the diffracted beam is measured by means of an ionization chamber. **E 7,** E04

ionization potential—the minimum energy, expressed in (electron) volts, required to remove an electron from an atom or molecule to form a positive ion. **E 1316,** E07

ionization vacuum gage—an indicating device used in vacuum systems to determine the pressure of the residual gas in a system by measuring the positive ion current produced in the process of ionization in an electrical field. **E 7,** E04

ionization vacuum gage—a vacuum gage comprising a means of ionizing the gas molecules, electrodes to facilitate the collection of the positive ions formed, and means of indicating the magnitude of the collected ion current. Various types of ionization gages are distinguished according to the method of producing the ionization. The common types are as follows:

(*a*) *hot-cathode ionization gage*—the ions are produced by collisions with electrons emitted from a hot filament (or cathode) and accelerated by an electric field. Also called hot-filament ionization gage, or simply ion gage. The Bayard-Alpert ionization gage employs a tube with an electrode structure designed to minimize X-ray-induced electron emission from the ion collector.

(*b*) *cold-cathode ionization gage*—the ions are produced by a cold-cathode discharge, usually in the presence of a magnetic field which lengthens the path of the electrons between cathode and anode. The discharge tube is a transparent tube in which the color and form of a cold-cathode discharge (without the presence of a magnetic field) gives an indication of the pressure and the nature of the gas. The Phillips ionization gage is a cold-cathode ionization gage in which a magnetic field is directly parallel to the axis of an annular electrode (normally the anode) located between two plate electrodes perpendicular to the axis. Various modifications of the Penning gage are named after the inventors, and certain types are referred to as magnetron vacuum gages.

(*c*) *radioactive ionization gage*—the ions are produced by radiations (usually alpha particles) emitted from a radioactive source. **E 1316,** E07

ionizing radiation—any type of radiation consisting of *charged* particles or *uncharged* particles, or both, that as a result of physical interaction, creates ions by either primary or secondary processes. (For example, *charged* particles could be positrons or electrons, protons, or other heavy ions, and *uncharged* particles could be X rays, gamma rays, or neutrons.) **E 170,** E10

ion lifetime—the average time that an ion exists in a particular electronic configuration, for example, a vacancy in a particular shell in an atom. **E 673,** E42

ion line, *n*—a spectral line resulting from radiation emitted during electron transition as an ionized atom decays to a lower, but still ionized, energy level. **E 135,** E01

ion neutralization—*ISS, SIMS,* the charge exchange processes in which a probe is neutralized by the material surface or gas phase species with which it interacts. **E 673,** E42

ion pump—an electrical device for pumping gas comprising a means for ionizing the gas and a system of electrodes at suitable potentials, and in some cases also a magnetic field, which causes the ions formed to move towards a surface on which they are absorbed or buried. **E 1316,** E07

ion-scattering:

spectrometer—*ISS,* an instrument capable of generating a beam of principally monoenergetic, singly charged, low-energy ions and determining the energy distribution of the probe ions that have been scattered from the solid surface through a known angle.

spectrometry—*ISS,* a technique to elucidate composition and structure of the outermost atomic layers of a solid material, in which principally monoenergetic, singly charged, low-energy (less than 10 keV) probe ions are scattered from the surface and are subsequently detected and recorded as a function of the energy or scattering angle, or both.

spectrum—*ISS,* a representation in which the scattered ion intensity is presented as a function of the ratio of the scattered ion energy to the incident ion energy. **E 673,** E42

ion-selective electrode—electrochemical sensors, the potentials of which are linearly dependent on the logarithm of the activity of a given ion in solution. Such devices are distinct from systems that involve redox reactions.

(*A*) Primary Electrodes

(*1*) *crystalline electrodes*—may be homogeneous or heterogeneous.

(*a*) *homogeneous membrane electrodes*—are ion-selective electrodes in which the membrane is a crystalline material prepared from either a single compound or a homogeneous mixture of compounds (that is, Ag_2S, AgI/Ag_2S).

(*b*) *heterogeneous membrane electrodes*—are formed when an active substance, or mixture of active substances, is mixed with an inert matrix, such as silicone rubber or PVC, or placed on hydrophobized graphite, to form the sensing membrane that is heterogeneous in nature.

(*2*) *noncrystalline electrodes*—In these electrodes a support, containing an ionic (either cationic or anionic) species or an uncharged species, forms the ion-selective membrane which is usually interposed between two aqueous solutions. The support used can be either porous (for example, filter, glass frit, etc.) or nonporous (for example, glass or inert polymeric material such as PVC, yielding with the ion-exchanger and the solvent a *solidified* homogeneous mixture). These electrodes exhibit a response due to the presence of the ion-exchange material in the membrane.

(*a*) *rigid matrix electrodes*—Glass electrodes are ion-selective electrodes in which the sensing membrane is a thin piece of glass whose chemical composition determines the selectivity of the electrode. In this group are: hydrogen ion-selective electrodes and monovalent cation-selective electrodes.

(*b*) *electrodes with a mobile carrier:*

(*1*) *positively charged*—bulky cations (for example, those of quaternary ammonium salts or salts of transition metal complexes such as derivatives of 1,10-phenanthroline) which, when dissolved in a suitable organic solvent and held on an inert support (for example, filter or PVC), provide membranes that are sensitive to changes in the activities of anions.

(2) *negatively charged*—complexing agents (for example, of type $(RO)_2PO_2^-$) or bulky anions (for example, tetra-*p*-chlorophenylborate anions) that when dissolved in a suitable organic solvent and held in an inert support (for example, filter or PVC), provide membranes that are sensitive to changes in the activities of cations.

(3) *uncharged (neutral) carrier*—electrodes based on solutions of molecular carriers of cations (for example, antibiotics, macrocyclic compounds or other sequestering agents) that can be used in membrane preparations which show sensitivity and selectivity to certain cations.

(*B*) Sensitized Ion-selective Electrodes

(*1*) *gas-sensing electrodes:*

(*a*) *secondary reaction type*—sensors composed of an indicating and a reference electrode that uses a gas-permeable membrane or an air gap to separate the sample solution from a thin film of an intermediate solution, which is held between the gas membrane and the ion-sensing membrane of the electrode, or placed on the surface of the electrode using a wetting agent (for example, air-gap electrode). This intermediate solution interacts with the gaseous species in such a way to produce a change in a measure value (for example, pH) of the intermediate solution. This change is sensed by the ion-selective electrode and is proportional to the partial pressure of the gaseous species in the sample.

(*b*) *redox reaction type*—a gas sensing electrode using a platinum redox electrode, typified by the hydrogen gas electrode. This responds to the redox equilibrium between hydrogen gas and hydrogen ion in solution.

(*c*) *amperometric detector*—this is a type that uses an amperometric detector, generally behind a gas permeable membrane. Such devices have been used for oxygen and for free chlorine.

(*2*) *enzyme substrate electrodes*—sensors in which an ion-selective electrode is covered with a coating containing an enzyme that causes the reaction of an organic or inorganic substance (substrate) to produce a species to which the electrode responds. Alternatively, the sensor could be covered with a layer of substrate that reacts with the enzyme to be assayed.

NOTE—The term *ion-specific electrode* is not recommended. The term *specific* implies that the electrode does not respond to additional ions. Since no electrode is truly specific for one ion, the term *ion-selective* is recommended as more appropriate. *Selective ion-sensitive electrode* is a little-used term to describe an ion-selective electrode. **D 4127,** D19

ion source—in leak testing, that part of a leak detector tube in which tracer gas is ionized preliminary to being detected. **E 1316,** E07

ion species—type and charge of ion such as Ar^+, O^-, and H_2^+. If an isotope is used, it should be specified. **E 673,** E42

IOT controller, *n*—a printed wiring board in the image output terminal that controls the operation of the image output terminal and communicates with the electronic scanning system. **F 1457,** F05

ipsative scaling, *n*—a method of assigning scale values that takes the individual's own characteristic behavior as the standard of comparison. **E 253,** E18

ipsilateral, *adj*—located, occurring, or acting on the same side of the body. **F 1646,** F13

IQI sensitivity—in radiography, the minimum discernible image and the designated hole in the plaque-type, or the designated wire image in the wire type image quality indicator. **E 1316,** E07

IR drop—the voltage across a resistance in accordance with Ohm's law:

$$E = IR$$

where:
E = potential (voltage),
I = current, and
R = resistance.

B 374, B08

iron—a term used for measuring thickness of sole leather. One iron equals 1⁄48 in. (0.53 mm). **D 1517,** D31

iron—a term used for measuring thickness of sole leather. One iron equals 1⁄48 in. (0.53 mm). (D 1517, D-31) **F 869,** F08

iron, enameling—see **enameling iron.** **C 286,** B08

ironing, *n*—a method of pressing using a heated hand iron, sometimes together with mositure or steam, and a gliding motion. **D 123,** D13

iron pipe size, IPS, *n*—the nominal inside dimension of **pipe** in inches (or millimetres). (Also called *interior pipe size, standard pipe size.*) **E 631,** E06

iron-silicon alloys—a material composition containing up to 5 % silicon with balance iron. **A 340,** A06

iron-sodium tartrate, EWNN or FeTNa—a complex tartaric acid salt of sodium and ferric iron. Solutions in aqueous sodium hydroxide will dissolve cellulose. Both the German abbreviation, EWNN, and the English, FeTNa, are to be found in English articles. The details of preparation of the solution should always be given. **D 1695,** D01

iron-sodium tartrate viscosity—the viscosity of a solution or dispersion of cellulose or pulp under standardized condition when dissolved in the iron-sodium tartrate solvent. (see also **iron -sodium tartrate**) **D 1695,** D01

ironstone ware—(stone china, white granite ware)—historic terms for a durable English earthenware. **C 242,** C21

Iron tannage—tannage with salts of iron. **D 1517,** D31

irradiance, *n*—the radiant power per unit area incident on a receiver, typically reported in watts per square metre, $W\ m^{-2}$ **G 113,** G03

irradiance at a point on a surface, *n.*—quotient of the radiant flux incident on an element of the surface containing the point, by the area of that element. Symbol: E_e, E; $E_e = d\Phi_e/dA$; Unit: watt per square metre, $W{\cdot}m^{-2}$.

NOTE—In ultraviolet radiation therapy and photobiology, this quantity is called dose rate (International Photobiology Committee, 1954). **E 349,** E21

irradiance, E—the radiant flux (power) per unit area incident on a given surface (unit = W/m^2). **E 1316,** E07

irradiance, *E, E_e, n*—the radiant flux incident per unit area. **E 284,** E12

irradiance, *H*, **(of a receiver),** *n*—the radiant power per unit area incident on a receiver. See **exposure.** **E 135,** E01

irradiation, *n*—application of radiation to an object. **E 349,** E21

irradiation—(*1*) process of exposing a surface or material to radiant flux; (*2*) at a point on surface, see **radiant exposure.** **E 772,** E44

irregularity factor— *in fatigue loading*, the ratio of the number of zero crossings with positive slope (or mean crossings) to the number of peaks or valleys in a given, load-time history. **E 1823,** E08

irregular loading— See **spectrum loading.** **E 1823,** E08

irregular powder—particles lacking symmetry. **B 243,** B09

irregular wear, *n*—a type of treadwear characterized by substantial variations of tread loss both from projection to projection and frequently from point to point on a given projection. **F 538,** F09

irreversible adsorption—adsorption in which the desorption isotherm is displaced toward higher equilibrium adsorption capacities from the adsorption isotherm. **D 2652,** D28

irreversible fiber optic chemical sensor, *n*—a fiber optic chemical sensor that undergoes a permanent depletion or degradation of the transduction element as a result of the transduction process. **E 131,** E13

irrigation pipe—pipe intended for the distribution of irrigation water by pumping or gravity. **C 822,** C13

ISCC-NBS color name, *n*—two- to four-word descriptive phrase for a surface color, such as "vivid orange" or "dark grayish reddish brown," defined by sections of the Munsell color solid. Developed cooperatively by the Inter-Society Color Council and the National Bureau of Standards. **E 284,** E12

islanding—a phenomenon that occurs when applied film is not adequately anchored, resulting in migration of film-forming materials to myriads of small pools scattered over the surface and surrounded by completely uncoated areas. **D 2825,** D21

ISO—International Organization for Standardization. **F 1156,** F12

ISO—abbreviation for **International Organization for Standards.** **F 1457,** F05

isoabsorptive point—a wavelength at which the absorptivities of two or more substances are equal. **E 131,** E13

isobar—a plot of quantity adsorbed per unit of adsorbent against equilibrium temperature when concentration or pressure is held constant. **D 2652,** D28

isobar—section, at constant pressure through a phase diagram. **E 7,** E04

isochor—in a phase diagram, a section, or contour, at constant volume. **E 7,** E04

isochrome—a curve showing the distribution of the excess hydrostatic pressure at a given time during a process of consolidation. **D 653,** D18

ISO equation—an equation showing the interrelations between stress, pressure and dimensions in pipe, namely:

$$S = \frac{P\,(ID + t)}{2t}$$

or

$$S = \frac{P\,(OD - t)}{2t}$$

where:

S = hoop stress,
P = pressure,
ID = average inside diameter,
OD = average outside diameter, and
t = minimum wall thickness.

(See ISO R 161.) **F 412,** F17

isohel—curve on a map showing areas of equal sunshine duration during a given interval of time, normally a year. **E 772,** E44

isokinetic sampling—See **sampling.** **D 1356,** D22

isokinetic sampling—to sample in such a way that the water-sediment mixture moves with no change in velocity as it leaves the ambient flow and enters the sampler intake. **D 4410,** D19

isolated—an object that is not readily accessible to persons unless special means of access are used. **F 819,** F18

isolated ramps, *n*—an interstate interchange ramp at which the traffic flow is not influenced by upstream or downstream ramps. **E 867,** E17

isolation joint, *n*—*in building construction*, a formed or assembled joint specifically intended to separate and prevent the bonding of one element of a structure to another and having little or no transference of movement or vibration across the joint. **C 717,** C24

isolation test—in leak testing, a method of determining whether a leak is present in a system, or of obtaining an estimate of its magnitude, by observing the rate of rise of pressure in the evacuated system when the system is isolated from the pump. See also **rate of rise.** **E 1316,** E07

isometric—a crystal form in which the unit dimension on all three axes is the same. **E 7,** E04

isomorphous—having the same crystal structure; usually referring to intermediate phases which form a continuous series of solid solutions. **E 7,** E04

isomorphous system—a complete series of mixtures in all proportions of two, or more, components, wherein unlimited mutual solubility exists in both the liquid and solid states. **E 7,** E04

isoperibol, *adj*—a term used in combustion calorimetry meaning constant temperature jacket (environment). **D 121,** D05

isoperibol calorimeter, *n*—a calorimeter that has a jacket of uniform and constant temperature. **D 5681,** D34

isopleth—in a ternary, or higher order, temperature-concentration, or pressure-concentration, phase diagram, a (vertical) two-dimensional section, having a linear composition series along one axis and temperature, or pressure, along the other axis. **E 7,** E04

isopleth—a line on a chart or graph connecting points having a specified constant value of a single variable as a function of two other specified variables. **E 772,** E44

isopotential point—for a cell containing an ion-selective electrode and a reference electrode there is often a particular activity of the ion concerned for which the potential of the cell is independent of temperature. That activity, and the corresponding potential, define the isopotential point. The identity of the reference electrode, and the composition of the filling solution of the measuring electrode, must be specified. **D 4127,** D19

isopropylbenzene—see **cumene**. **D 4790,** D16

isosbestic point—the wavelength at which the absorptivities of two substances, one of which can be converted into the other, are equal. **E 131,** E13

isostatic pressing, *n*—the pressing of a powder, compact, or sintered object by subjecting it to a nominally equal pressure from every direction. **B 243,** B09

isostilbic point, *n*—in luminescence, the wavelength at which the intensity of emission of a sample does not change during a physical interaction or chemical reaction. **E 131,** E13

isotactic, *adj*—pertaining to a type of polymeric molecular structure containing a sequence of regularly spaced asymmetric atoms arranged in like configuration in a polymer chain. **D 883,** D20

isotactic, *n*—a polymeric molecular structure containing a sequence of regularly spaced asymmetric atoms arranged in like configuration in the polymer chain. **D 1566,** D11

isotactic, *adj*—pertaining to a type of polymeric molecular structure containing a sequence of regularly spaced asymmetric atoms arranged in like configuration in a polymer chain. **F 412,** F17

isotactic, *adj*—pertaining to a type of polymeric molecular structure containing a sequence of regularly spaced asymmetric atoms arranged in like configuration in a polymer chain. **F 1251,** F04

isotere—a plot of equilibrium concentration or pressure against temperature when the quantity adsorbed per unit of adsorbent is held constant. **D 2652,** D28

isotherm—a plot of quantity adsorbed per unit of adsorbent against equilibrium concentration, or pressure, when temperature is held constant. **D 2652,** D28

isotherm—section, at constant temperature, through a phase diagram. **E 7,** E04

isothermal annealing, *n*—**austenitizing** a steel object and then cooling it to, and holding it at, a temperature at which austenite transforms to a ferrite-carbide aggregate. **A 941,** A01

isothermal transformation, *n*—a change in phase at any constant temperature. **A 941,** A01

isotope—nuclides having the same atomic number but different mass numbers. **C 859,** C26

isotope ratios—the relative amounts of two isotopes of an element, expressed as a ratio, in a sample or material. **C 859,** C26

isotopes—atoms of an element that have the same number of protons (atomic number) in the nucleus and undergo the same chemical reactions, but have differing numbers of neutrons in their nuclei and different atomic weights. **F 1156,** F12

isotopic assay—the determination of the percentage (by weight or by atoms) of isotopic constituents of an element. **C 859,** C26

isotopic composition—the relative amounts, expressed in atom or weight fraction of the various isotopes of an element in a sample or material. **C 859,** C26

isotropic—having the same value for a property in all directions. **E 7,** E04

isotropic—having uniform properties in all directions. The measured properties of an isotropic material are independent of the axis of testing. **E 1749,** E06

isotropic diffuser— see **Lambertian diffuser**. **E 284,** E12

isotropic mass—a mass having the same property (or properties) in all directions. **D 653,** D18

isotropic material—material in which the magnetic properties are the same for all directions. **A 340,** A06

isotropic material—a material whose properties do not vary with direction. **D 653,** D18

isotropic phase, *n*—*as used in Test Method D 5061*, a binder-phase carbon texture that exhibits optical properties that are the same in all directions when viewed with an optical microscope having mutually exclusive polarized light, for example, crossed nicols. **D 121,** D05

isotropic radiant energy— see **radiant energy, isotropic**. **E 772,** E44

isotropy—having the same properties in all directions. (ISRM) **D 653,** D18

isotropy—the condition of having the same values of properties in all directions. **E 7,** E04

I.S. process—a bottle-making process using removable, individual forming sections, in which the gob enters the blank or parison mold by gravity. **C 162,** C14

item, *n*—*(1)* an object or quantity of material on which a set of observations can be made: *(2)* an observed value or test result obtained from an object or quantity of material. **E 456,** E11

J

jacket, *n*—a form of facing applied over insulation. **C 168,** C16

jacket, *n*—a textile product, woven or felted into tubular or sleeve form, ready for covering and shrinking on a machine roll. **D 123,** D13

jacket, *n*—an integral covering (sometimes fabric, reinforced), which is applied over the insulation, core, shield, or armor of a cable and whose prime function is to provide mechanical or environmental protection for the component(s) that it covers. **D 1711,** D09

jackhammer—an air driven percussion drill that imparts a rotary hammering motion to the bit and has a passageway to the bit for the injection of compressed air for cleaning the hole of cuttings. **D 653,** D18

jacking—a method of installing pipe by the trenchless method using equipment and pipe designed for this purpose. **C 896,** C04

jacking force—the force applied to the pipe along the longitudinal axis of the pipeline by the pipe jacking equipment. **C 896,** C04

jack-leg—a portable percussion drill of the jack-hammer type, used in underground work; has a single pneumatically adjustable leg for support. **D 653,** D18

jackson turbidity unit, JTU—unit of measure used with the jackson candle turbidimeter. **D 6161,** D19

jamb—the vertical structural member forming the sidewall of an opening or port in a furnace superstructure, supporting the crown load over that opening. **C 162,** C14

jamb—See **windows and doors.** **E 631,** E06

jamming, *n—for chainsaw cut resistance*, the clogging action manifested by a protective garment which can produce a chain stop. **F 1494,** F23

japan—a varnish yielding a hard, glossy, dark-colored film. Japans are usually dried by baking at relatively high temperatures. **D 16,** D01

japan, a vehicle for japan colors; frequently contains shellac. **D 16,** D01

japan color—a paste containing pigment and a grinding japan vehicle used for lettering and decoration. **D 16,** D01

japan drier—a resinate-base liquid drier. **D 16,** D01

japanned—tumble-pointed. **F 547,** F16

japanned leather—see **patent leather.** **D 1517,** D31

jar mill—a small ball mill (see also **ball mill**). **C 286,** B08

jar test—a laboratory procedure for the evaluation of a treatment to reduce dissolved, suspended colloidal and non settleable matter from water (see Practice D 2035). **D 6161,** D19

jasper ware—a vitreous, opaque, colored, unglazed ceramic ware having white or contrasting relief decorations and containing a substantial amount of barite. **C 242,** C21

jaw face, *n—in tensile testing machines*, the surface of a jaw which in the absence of a liner contacts the specimen. **D 123,** D13

jaw liner, *n—in tensile testing machines*, any material placed between the jaw face and the specimen to improve the holding power of the jaws. **D 123,** D13

jaws, *n—in tensile testing machines*, the elements of a clamp which grip the specimen. **D 123,** D13

J-clinched point—pointed end of nail having been driven through member against curved anvil plate, bent sideways and driven back into member. **F 547,** F16

"J" configuration—boom positioned in a "J" shape. **F 818,** F20

Jeffries' method—a method for determining grain size based on counting grains in a prescribed area. See Test Methods E 112. **E 7,** E04

Jeffries' multiplier—a factor used in the Jeffries' method for grain size determinations. See Test Methods E 112. **E 7,** E04

jerk-in, *n—in woven fabric*, an extra filling thread dragged into the shed with the regular pick and extending only part of the way across the cloth. (Compare **double pick, mispick.**) **D 123,** D13

jet-blast erosion, *n—(airfields only)* darkened areas on the pavement surface where bituminous binder has been burned or carbonized; localized burned areas may vary in depth up to approximately 15 mm (½ in.) **E 867,** E17

jet-blast erosion, *n*—(*airfields only*) darkened areas on the pavement surface where bituminous binder has been burned or carbonized; localized burned areas may vary in depth up to approximately 15 mm (½ in.). **E 1778,** E17

jet fuel, *n*—any liquid suitable for the generation of power by combustion in aircraft gas turbine engines. **D 4175,** D02

jet grouting—technique utilizing a special drill bit with horizontal and vertical high speed water jets to excavate alluvial soils and produce hard impervious columns by pumping grout through the horizontal nozzles that jets and mixes with foundation material as the drill bit is withdrawn. **D 653,** D18

jet, liquid, n— See **liquid jet.** **G 40,** G02

jet segment, n— See **liquid jet.** **G 40,** G02

jetting—a means of compacting trench backfill by the introduction of water under pressure through a nozzle. **C 896,** C04

jetting—when applied as a drilling method, water is forced down through the drill rods or casings and out through the end aperture. The jetting water then transports the generated cuttings to the ground surface in the annulus of the drill rods or casing and the borehole. The term jetting may also refer to a development technique (see **well screen jetting**). **D 653,** D18

jetty—an elongated artificial obstruction projecting into a body of water from a bank or shore to control shoaling and scour by deflection of the force of water currents and waves. **D 653,** D18

jewelers' enamel—a special type of porcelain enamel used in the manufacture of jewelry, insignia, and art objects. **C 286,** B08

jiggering—forming ceramic ware from a plastic body by differential rotation of a profile tool and mold, the mold having the contour of one surface of the ware and the profile tool that of the other surface. **C 242,** C21

***J*-integral, *J* [FL^{-1}]**—a mathematical expression, a line or surface integral that encloses the crack front from one crack surface to the other, used to characterize the local stress-strain field around the crack front. **E 1823,** E08

joggle—a displacement machined or formed in a structural member to accommodate the base of an adjacent member. **E 1749,** E06

joining, *n*—the juncture of two separate plaster applications of the same coat, usually within a single surface plane. **C 11,** C11

joint—that portion of the conductor where the ends of two wires, rods, or groups of wires are joined by brazing, soldering, welding or by mechanical means. **B 354,** B01

joint, *n—in building construction*, the space or opening between two or more adjoining surfaces. **C 717,** C24

joint—a connection between two pipe sections, made either with or without the use of additional parts. **C 822,** C13

joint—an individual length of pipe, or the means of closure to form a pipeline. **C 896,** C04

joint—the junction of two or more pieces of wood. Usually implies a structural junction in which stress is transmitted from one piece to another. **D 9,** D07

joint—a break of geological origin in the continuity of a body of rock occurring either singly, or more frequently in a set or system, but not attended by a visible movement parallel to the surface of discontinuity. (ISRM) **D 653,** D18

joint , *n*

adhesive joint, n—location at which two adherends are held together with a layer of adhesive. (See also **bond,** *n.*)

edge joint, n—in wood bonding, a joint made by bonding adherends edge to edge (with grain directions parallel) to form wider stock.

finger joint, n—a joint formed by bonding two precut members shaped like fingers.

laminate joint, n—in wood bonding, a joint made by bonding layers of adherends face to face to form thicker stock.

lap joint, n— a joint made by placing one adherend partly over another and bonding together the overlapped portions.

scarf joint, n— a joint made by cutting away similar angular segments of two adherends and bonding the adherends with the cut areas fitted together.

starved joint, n— a joint that has an insufficient amount of adhesive to produce a satisfactory bond.

joint aging time, n—synonym for **joint conditioning time**. **D 907,** D14

joint—the junction of two adjacent pieces of wood or veneer.

butt joint—the place where two pieces of wood are joined together end to end.

edge joint—the place where two pieces of wood are joined together edge to edge.

glue joint—the place where two pieces of wood are joined together by means of glue.

open joint—failure of bond or separation of two adjacent pieces of veneer so as to leave veneers.

starved joint—a glue joint that is poorly bonded because of an insufficient quantity of glue.

sunken joint—in the case of plywood, a depression in the surface of the face ply directly above an edge joint in a lumber core or crossband. Usually the result of localized shrinkage in the edge-jointed layer. **D 1038,** D07

joint, *n*—a designed and constructed or sawn space or opening between adjoining pavement surfaces. **D 5535,** D04

joint, *n*—general term. See particular joint of interest. Compare **connection.**

butt joint—a joint having the edge or end of one member matching the edge, end, or face of another member without overlap.

cold joint, n—boundary between later-applied and previously-applied coatings, plaster, mortar, or concrete.

construction joint—in the construction of members intended to be continuous, a predetermined, intentionally created discontinuity between or within constructions and having the ends of the discontinuous members fastened to each other to provide structural continuity.

control joint—in concrete, concrete masonry, stucco, or coating systems; a formed, sawed, or assembled joint acting to regulate the location of cracking, separation, and distress resulting from dimensional or positional change.

expansion joint—a discontinuity between two constructed elements, or **components,** allowing for differential movement (such as expansion) between them without damage.

field joint—a connection between adjoining members or parts, made at the time of installation. Compare **construction joint.**

slip joint—a joint allowing axial sliding movement of joined parts. **E 631,** E06

joint, *n*—a discontinuity made necessary by design or by interruption of a paving operation. **E 867,** E17

joint, *n*—a discontinuity made necessary by design or by interruption of a paving operation. **E 1778,** E17

joint—the location at which two pieces of pipe or a pipe and a fitting are connected together. **F 412,** F17

joint—the junction where the scissor blades are secured by a screw allowing the instrument to pivot. **F 1078,** F04

joint acceptance test—test procedure, utilizing compressed air or pressurized water, intended to determine the acceptability of an individual installed pipeline joint. **C 822,** C13

joint, adhesive-bonded—a joint made using an adhesive to bond the piping components. **F 412,** F17

joint backing—See **sealant backing.** **C 717,** C24

joint, bell and spigot gasket—a connection between piping components consisting of a bell end on one component, an elastomeric gasket between the components, and a spigot end on the other component. See *joint, push on.* **F 412,** F17

joint, butt-fused—a joint in which the prepared ends of the joint components are heated and then placed in contact to form the joint. **F 412,** F17

joint, clamped insert-fitting—a mechanical joint using external metal clamps or other mechanical devices to form a pressure seal between an inset fitting and the outside surface of the pipe. **F 412,** F17

joint compound, *n*—a compound used for taping or finishing gypsum board, or both. **C 11,** C11

joint, compression fitting—a mechanical joint made by compressing the components either externally, internally, or axially to form a pressure seal between the fitting and the surface of the pipe. **F 412,** F17

joint, compression gasket—a mechanical joint that utilizes a compression nut or a gland nut against a gasket to develop a pressure seal. **F 412,** F17

joint-conditioning time—the time interval between the removal of the joint from the conditions of heat or pressure, or both, used to accomplish bonding and the attainment of approximately maximum bond strength. (Synonym *joint-aging time.*) (See also **curing time, drying time,** and **setting time.**) **D 907,** D14

joint diagram—a diagram constructed by accurately plotting the strike and dip of joints to illustrate the geometrical relationship of the joints within a specified area of geologic investigation. (ISRM) **D 653,** D18

jointed—veneer or other ply components that have machined edges for tightest possible layup. **D 1038,** D07

jointed concrete pavement (JCP), *n*—Portland cement concrete pavement which has transverse joints placed at planned intervals. **E 867,** E17

jointed concrete pavement (JCP), *n*— Portland cement concrete pavement that has transverse joints placed at planned intervals. **E 1778,** E17

joint (fault) set—a group of more or less parallel joints. (ISRM) **D 653,** D18

joint (fault) system—a system consisting of two or more joint sets or any group of joints with a characteristic pattern, that is, radiating, concentric, etc. (ISRM) **D 653,** D18

joint filler, *n*—a compressible material used in a partial or totally filled expansion, control, or isolation joint by its permanent placement in or between building materials such as concrete or masonry during construction; sometimes used as a sealant backing in a partially filled joint. **C 717,** C24

joint, flanged—a mechanical joint using pipe flanges, a gasket, and bolts. **F 412,** F17

joint, flare—a mechanical compression connection between flared-end plastic pipe and a fitting specifically designed to accept flared-end plastic pipe. **F 412,** F17

joint, heat-fused—a joint made using heat and pressure only. **F 412,** F17

jointing yard—{archaic} place between the grinding and polishing operations in the continuous manufacture of plate glass, where the plaster joints are remade, sometimes called "middle yard." **C 162,** C14

joint line—See **parting line.** **C 162,** C14

joint, mechanical—a connection between piping components employing physical force to develop a seal or produce alignment. **F 412,** F17

joint pattern—a group of joints that form a characteristic geometrical relationship, and which can vary considerably from one location to another within the same geologic formation. (ISRM) **D 653,** D18

joint, push on—a joint in which an elastomeric ring gasket is compressed in the annular space between a bell end or socket and a spigot end of pipe. **F 412,** F17

joint reinforcement—reinforcement, in or near the joint, intended to enhance the structural characteristics of the joint area of a concrete pipe or box section. **C 822,** C13

joint reinforcing metal, *n*—strips of expanded metal, woven or welded wire mesh used to reinforce corners and other areas of plaster and lath. **C 11,** C11

joint, saddle-fused—a joint in which the curved base of the saddle fitting and a corresponding area of the pipe surface are heated and then placed together to form the joint. **F 412,** F17

joint seal deterioration, *n*—any condition which enables incompressible materials or water to infiltrate into the joint from the surface. **E 867,** E17

joint seal deterioration, *n*—any condition which enables incompressible materials or water to infiltrate into a previously sealed joint from the surface. **E 1778,** E17

joint, socket-fused or insert-fused—a joint in which the joining surfaces of the components are heated, and the joint is made by inserting one component into the other. **F 412,** F17

joint, solvent-cemented—a joint made using a solvent cement to unite the components. **F 412,** F17

joint spalling, *n*—cracking, breaking, or chipping of concrete pavement edges within 0.6 m (2 ft) of a joint. **E 867,** E17

joint spalling, *n*—cracking, breaking, or chipping of concrete pavement edges within 0.6 m (2 ft) of a joint. **E 1778,** E17

joint tape, *n*—a type of paper, metal, fabric, glass mesh, or other material, commonly used with a cementitious compound, to reinforce the joints between adjacent gypsum boards. **C 11,** C11

joint test apparatus—sealing device used on the inside of the pipe to isolate a single pipe joint so that it can be tested for leakage without filling the entire pipe with compressed air or water. See joint acceptance test. **C 822,** C13

joint, threaded—a mechanical joint that utilizes threaded pipe and fittings. **F 412,** F17

Jordan diagram—a graph showing the variation of some magnetic parameter versus frequency when the excitation is within the Rayleigh range. **A 340,** A06

joule, J—the unit of energy in the SI system of units. One joule is one watt-second. **A 340,** A06

***J-R* curve**—a plot of resistance to stable crack extension, Δa_p. **E 1823,** E08

judge, *n*—See **assessor.** **E 253,** E18

jumbo—a specially built mobile carrier used to provide a work platform for one or more tunneling operations, such as drilling and loading blast holes, setting tunnel supports, installing rock bolts, grouting, etc. **D 653,** D18

jumpers or jumping— see **poppers.** **C 286,** B08

junction potential—the portion of the total observed potential developed between the sensing and reference electrodes that is formed at the liquid/liquid junction between the reference electrode filling solution and the sample solution. For accuracy, the junction potential should be as low and as constant as possible despite variations in the composition of the sample solution. Reference electrode filling solutions should be judiciously chosen to minimize liquid junction potential. **D 4127,** D19

juncture, *n*—the interface between two different tire components or different compounds within the same component. **F 538,** F09

juncture cracking, *n*—a crack with opening originating at a juncture between two components. **F 538,** F09

juncture opening, *n*—a separation developing in a juncture. **F 538,** F09

just noticeable difference, *n*—See **threshold, difference.** **E 253,** E18

just noticeable difference, n—see **just perceptible difference.** **E 284,** E12

just-perceptible difference, *n*—color difference that is just large enough to be perceived in almost every trial. **E 284,** E12

jute, *n*—soft fibers from the inner bark of the round pod jute (*Corchorus capsularis*), the long pod jute (*Corchorus olitorius*), and from the inner bark of other closely related plants, such as kenaf, sometimes referred to as Meshta (*Hibiscus cannabinus*). **D 123,** D13

juvenile wood—the wood formed adjacent to the pith, characterized by progressive change in cell dimension, different microstructure than mature wood, and greater shrinkage parallel to the grain.

NOTE—In softwoods, juvenile wood is characterized by lower specific gravity and strength. It may vary in quantity from tree to tree and extend from the pith 5 to 20 growth rings. **D 9,** D07

J-valve, *n*—a manually operated, spring-loaded, SCUBA cylinder check valve that incorporates a low air warning/reserve air mechanism. **F 1549,** F32

K

k—abbreviation for *kilo*; 1000 units. **F 1156,** F12

Kaiser effect—the absence of detectable acoustic emission at a fixed sensitivity level, until previously applied stress levels are exceeded. **E 1316,** E07

kangaroo—leather made from the hide of the kangaroo. **D 1517,** D31

kaolin—a variety of clay containing a high percentage of kaolinite. **D 653,** D18

kaolin (china clay)—a refractory clay consisting essentially of minerals of the kaolin group and which fires to a white or nearly white color. **C 242,** C21

kaolinite—a common clay mineral having the general formula $Al_2(Si_2O_5)$ (OH_4); the primary constituent of kaolin. **D 653,** D18

kappa number—the number of millilitres of 0.1 *N* potassium permanganate solution consumed by 1 g of ovendry pulp under specified conditions. It is an indication of the hardness or bleachability of a pulp. An indirect estimate of lignin content. **D 1695,** D01

karat—a twenty-fourth part by weight; thus 18-karat gold is 18/24 pure. **B 374,** B08

karst—a geologic setting where cavities are developed in massive limestone beds by solution of flowing water. Caves and even underground river channels are produced into which surface runoff drains and often results in the land above being dry and relatively barren. (ISRM) **D 653,** D18

***K*-decreasing test**—a test in which the normalized *K*-gradient is nominally negative. In Test Method E 647, *K*-decreasing tests involve load shedding as the crack grows either continuously or by a series of decremental steps. **E 1823,** E08

Keene's cement, *n*—an anhydrous gypsum plaster characterized by a low mixing water requirement and special setting properties, primarily used with lime to produce hard, dense finish coats **C 11,** C11

keg—See **barrel.** **D 996,** D10

keg, nail keg—synonymous with 100 lb (45.36 kg) of nails. Round, bulging container for bulk nails, made of wooden staves, steel hoops, and flat heads, usually holding 100 lb of nails. Terms no longer in common usage. **F 547,** F16

Kellner eyepiece—a positive eyepiece consisting of an achromatic eyelens and a single collective, in which the image plane and field diaphragm is external and near the collective. **E 7,** E04

kelly—a heavy-wall tube or pipe, usually square or hexagonal in cross section, which works inside the matching center hole in the rotary table of a drill rig to impart rotary motion to the drill string. **D 653,** D18

kelvin, *n*—the unit of thermodynamic temperature; the SI unit of temperature for which an interval of one Kelvin (K) equals exactly an interval of one degree Celsius (1°C) and for which a level of 273.15 K equals exactly 0°C. **D 123,** D13

Kelvin—designation of the thermodynamic temperature scale and the degree on this scale, (E 344, E-20). **E 1142,** E37

kelvin, K, *n*—base unit of temperature in the International System of Units (SI). **E 344,** E20

kemp fiber, *n*—a medullated animal fiber in which the diameter of the medulla is 60 %, or more, of the diameter of the fiber. **D 123,** D13

kerf, *n*—synonym for *sipe*. **F 538,** F09

kerma, *K*—the quotient of dE_{tr} by dm, where dE_{tr} is the sum of the initial kinetic energies of all the charged ionizing particles liberated by uncharged ionizing particles in a material of mass dm (ICRU).

$$K = dE_{tr}/dm$$

The special name of the unit of kerma is the gray (Gy).

$$1\ Gy = 1\ J \cdot kg^{-1}$$

E 170, E10

kerosine number—the millilitres of kerosine held per 100 g of felt as determined by Test Method D 727. To obtain the percentage saturating capacity of the felt for any bituminous saturant, obtain the specific gravity of the saturant and multiply that figure by the kerosine number of the felt. **D 1079,** D08

ketones—a class of organic compounds possessing a carbonyl group attached to two hydrocarbon groups. Acetone is the first member of this series. **D 4790,** D16

keV (kilo electron volt)—a unit of energy equal to one thousand electron volts, used to express the energy of X rays, gamma rays, electrons, and neutrons. **E 1316,** E07

key, *n*—the grip or mechanical bond of one coat of plaster to another coat, or to a plaster base. It may be accomplished physically by the penetration of wet mortar or crystals into paper fibers, perforations, scoring irregularities, or by the embedment of the lath. **C 11,** C11

keyboard input seed—seed provided by operator intervention; can be item serial number, the data point, per se, or any other character sequence desired; see computer generated seed. **F 1156,** F12

key locking dial—a dial that can be locked with a key to prevent rotation. **F 471,** F12

key locking dial ring—a dial ring containing a key lock to lock dial and prevent rotation. **F 471,** F12

key species—a species of special concern for ecological reasons. **E 943,** E47

***K*-factor**—the strength constant in the formula for **radial crushing strength** of a plain sleeve specimen of sintered metal. See also **radial crushing strength.** **B 243,** B09

kHz—kilohertz; 1000 Hertz. **F 1156,** F12

kickplate—See **railing systems.** **E 631,** E06

kick plate—Synonym for **toe board**. **E 1481,** E06

kid—originally referring to leathers made from the skins of immature goats, the term is now rather loosely applied to glove and shoe leathers made from goatskins. **D 1517,** D31

kid—originally referred to leathers made from the skins of immature goats, term is now used rather loosely to apply to glove and shoe leathers made from goat skins. **F 869,** F08

kiddie ride—an amusement ride designed primarily for use by children up to 12 years of age. **F 747,** F24

Kikuchi lines—light and dark lines superimposed on the background of a single crystal electron diffraction pattern caused by diffraction of diffusely scattered electrons within the crystal. **E 7,** E04

killed steel, *n*—a **steel** deoxidized to such a level that essentially no reaction occurred between carbon and oxygen during solidification. **A 941,** A01

kiln—a chamber used for drying and conditioning lumber, veneer, and other wood products in which the temperature and relative humidity of the circulated air can be varied and controlled, often steam heated and vented. **D 9,** D07

kiln schedule—in kiln drying, the time schedule of predetermined or actual dry-bulb and wet-bulb temperatures used in drying a kiln charge of lumber or other wood products. **D 9,** D07

***K*-increasing test**—a test in which the normalized *K*-gradient is nominally positive. For the standard specimens in Test Method E 647, the constant-load-amplitude test will result in a *K*-increasing test where the normalized *K*-gradient increases but is always positive. **E 1823,** E08

kinesthesis, *n*—perception of pressure, position, or motion in muscles, tendons, or joints. **E 253,** E18

kinetic coefficient of friction, *n*— the coefficient of friction under conditions of macroscopic relative motion between two bodies. **G 40,** G02

kinetic friction, *n*—friction developed between two bodies in motion. (Compare **static friction**.) **D 123,** D13

kink, *n—in fabric,* a short length of yarn that has spontaneously doubled back on itself to form a loop. **D 123,** D13

kink, *n—in rope*, an abrupt bend or loop in the rope which is the result of an unbalanced twist relationship in the rope structure. **D 123,** D13

kink—a type of waviness occurring interior to the edges, not to be confused with the more abrupt departures as ridges or surface marks. (See also **waviness**.)

F 109, C21

kink (in asbestos)— definite, well defined change in direction of the asbestos in situ that could lead to points of weakness in the fibers when separated. **D 2946,** C17

kip—skin from a bovine animal in size between a calf and a cow, weighing in green-salted condition approximately from 15 to 35 lb (6.8 to 15.9 kg). **D 1517,** D31

kitchen—See **building space.** **E 631,** E06

knapsack sprayer—a sprayer apparatus, carried on the operator's back, consisting of a spray solution tank, pressure source, and an atomizing device that forms and distributes the spray. Spray pressure is supplied by a lever operated manual or engine powered pumps or a compressed air tank. Some knapsack sprayers have air carrier blowers to distribute the spray. **E 1102,** E35

knee, *n—in anatomy*, the joint between the lower and upper leg. **D 123,** D13

knee, *n—in anatomy*, the joint between the lower and upper leg. **F 1494,** F23

knee girth, *n—in body measurements*, with the leg straight, the circumference of the knee over the knee cap and parallel to the floor. **D 123,** D13

knife edge pivot—a type of pivot or hinge in which the motion occurs through the rotation of a knife edge resting on a plane or a vee notch. **E 7,** E04

knife-grade, *n—in building construction*, a material of a soft, plastic consistency that is applied with a knife or spatula hand tool. **C 717,** C24

knife-line attack—intergranular corrosion of an alloy, usually stabilized stainless steel, along a line adjoining or in contact with a weld after heating into the sensitization temperature range. **G 15,** G01

knife marks—a series of surface imprints or markings made by the machine knives in dressed lumber. **D 9,** D07

knit—a textile process that interlocks, in a specific pattern loop of yarn by means of needles or wires. **D 3878,** D30

knit-line—see **weld-line** (preferred terminology). (D-20) **F 412,** F17

knit-line, n—See **weld-line** (the preferred terminology). **D 883,** D20

knitted fabric, *n*—a structure produced by interlooping one or more ends of yarn or comparable material. **D 123,** D13

knitted fabric—See *knitted fabric* under **fabric.** **D 3878,** D30

knob—a component of some manipulation-resistant-locks that is rotated on the dial to operate the cam slide. **F 471,** F12

knock, *n—in a spark ignition engine, abnormal combustion*, often producing audible sound, caused by autoignition of the air/fuel mixture. **D 4175,** D02

knockings—the oversize residue obtained in screening a ceramic slip. **C 242,** C21

knock-off—a product that copies another design closely but carries its own name; slang for counterfeit. **F 1156,** F12

knock-on—the movement of a constituent of the specimen deeper into the specimen matrix as a result of collisions with the primary particle. **E 673,** E42

knockout—(*1*) a portion of a piece of pressed ware which has been so designed that it can be knocked out to make a hole.

(*2*) in glass containers, a protrusion of glass caused by missing metal at a junction of mold parts. **C 162,** C14

Knoop—See **micro penetration tester**. In a more restricted sense, a type of diamond hardness indenter having edge angles of 172° 30 min, and 130°. **E 7,** E04

Knoop Hardness Number (HK), *n*—an expression of hardness obtained by dividing the force applied to the Knoop indenter by the projected area of the permanent impression made by the indenter. **C 1145,** C28

Knoop hardness number, HK $[FL^{-2}]$, *n*—A number related to the applied load and to the projected area of the permanent impression made by a rhombic-based pyramidal diamond indenter having included edge angles of 172° 30 min and 130° 0 min computed from the equation:

$$HK = P/0.07028d^2$$

where:

P = applied load, kgf, and

d = long diagonal of the impression, mm.

In reporting Knoop hardness numbers, the test load is stated. **E 6,** E28

Knoop hardness test, *n*—an indentation hardness test using calibrated machines to force a rhombic-based pyramidal diamond indenter having specified edge angles, under specified conditions, into the surface of the material under test and to measure the long diagonal after removal of the load. **E 6,** E28

Knoop indenter, *n*—a rhombic-based pyramidal-shaped diamond indenter with edge angles of 172° 30′ and 130° 00′. **C 1145,** C28

knot—an imperfection; an inhomogeneity in the form of a vitreous lump. **C 162,** C14

knot—that portion of a branch or limb which has been surrounded by subsequent growth of the wood of the tree. As a knot appears on the cut surface it is merely a section of the entire knot, its shape depending upon the direction of the cut.

branch knots—two or more knots diverging from a common point at or near the pith.

centerline knot—a knot whose centroid lies at the center of the width of any lumber face; sometimes referred to as a center knot.
corner knot—an edge knot containing the intersection of adjacent faces.
decayed knot—a knot that, due to advanced decay, is softer than the surrounding wood.
edge knot—*(1)* a knot located at the edge of the face in a piece of lumber.
(2) in stress grading under Practice D 245, a knot whose perimeter falls within one-sixth of the knot width from the edge of the piece. The knot width is measured along a line transverse to the piece.
elsewhere knot—a knot that is not a centerline, edge, or corner knot.
encased knot—a knot whose rings of annual growth are not intergrown with those of the surrounding wood.
firm knot—a knot that is solid across its face, but which contains incipient decay.
fixed knot—a knot that will hold its place in dry lumber under ordinary conditions, but can be moved under pressure, although not easily pushed out.
group knots—two or more single knots grouped together.
hollow knot—an apparently sound knot containing a hole more than 1/4-in. in diameter.
intergrown knot—a knot whose rings of annual growth are completely intergrown with those of the surrounding wood.
knot cluster—two or more knots grouped together as a unit, the fibers of the wood being deflected around the entire unit. Distinct from a group of single knots in which each is a unit.
loose knot—a knot that is not held firmly in place by growth or position and that cannot be relied upon to remain in place.
oval knot—a knot cut at from 45 to 90° to the long axis of the limb.
pith knot—a sound knot having a pith hole not over 1/4 in. in diameter.
round knot—a knot that is cut at approximately right angles to its long axis of the limb.
single knot—a knot having adjoining wood fibers deflected around it alone and not around another knot.
sound knot—a knot that is solid across its face, at least as hard as the surrounding wood, and shows no indication of decay.
spike knot—a knot cut at from 0 to 45° to the long axis of the limb.
star-checked knot—a knot having radial checks.
tight knot—a knot so fixed by growth or position that it will firmly retain its place in the piece.
unsound knot—a knot which is not solid across the face as a result of decay and is not as hard as the surrounding wood.
watertight knot—having sound and watertight wood completely intergrown with the surrounding wood on one surface or on the entire projection of one end of the knot. **D 9,** D07

knot—an imperfection or nonhomogeneity in materials used in fabric construction, the presence of which causes surface irregularities. **D 1079,** D08

knot breaking force, *n*—*in tensile testing*, the breaking force of a strand having a specified knot configuration tied in the test method portion of the strand mounted between the clamps of a tensile testing machine. (Compare **knot breaking strength.** See also **breaking force.**) **D 123,** D13

knot-breaking strength, *n*—strength expressed in terms of knot breaking force. (See **knot breaking force.**) **D 123,** D13

knotholes—voids remaining after removal of knots. **D 1038,** D07

known, *n*—of established origin associated with the matter under investigation. **E 1732,** E30

knuckle—the selvage obtained by interlocking adjacent pairs of wire ends and bending the wire back into a loop. **F 552,** F14

knuckle area—in reinforced plastics, the area of transition between sections of different geometry in a filament-wound part. **D 883,** D20

knuckles, *n*—small tough rubber pieces scattered throughout a bale of raw rubber that do not disperse easily or accept carbon black and other compounding materials during mixing. **D 1566,** D11

Knudsen vacuum gage—an absolute manometer based on the principle of the transfer of momentum from a hot to a cold surface by gas molecules as in a radiometer. **E 7,** E04

knurled—loose term used to denote threaded, fluted, or grooved parallel or nearly parallel to nail axis with deformations not passing around body (see **thread**); also, barbed or deformed in repetitive pattern along surface. **F 547,** F16

knurling—in glass containers, a raised pattern of beads, ridges, crescents, or other shapes, molded on a glass surface for the purpose of decoration and/or improvement in the mechanical strength of the glass in service. Also known as **stippling**. **C 162,** C14

Koehler illumination—a specular illumination system. In reflected-light microscopy, used directly for the brightfield mode, and as a preliminary setup for all other modes except darkfield. The image of the field diaphragm is focused on the specimen surface and the image of an undiffused lamp source is focused in the plane of the aperture diaphragm. **E 7,** E04

Konowalow's law—the vapor of a binary mixture contains the larger proportion of that component, which, upon addition to the liquid, will raise its vapor pressure. **E 7,** E04

Koopmans energy—a calculated energy of an electron in an orbital, on the assumption that its removal to infinity is unaccompanied by electronic relaxation. **E 673,** E42

kosher hide—hide of an animal that has been slaughtered according to Jewish religious custom by having its throat cut crosswise, resulting in a different pattern of the hide sometimes referred to as a "cutthroat" or "stuckthroat". **D 1517,** D31

K radiation—characteristic X-rays produced by an atom when a vacancy in the *K* shell is filled by one of the outer electrons. **E 7,** E04

K radiation, *n*—characteristic X rays produced by an atom or ion when a vacancy in the K shell is filled by one of the outer electrons. **E 135,** E01

K-radiation, *n*—characteristic X-rays produced by an atom or ion when a vacancy in the K-shell is filled by one of the outer electrons. **E 1605,** E06

kraft, n—See **paperboard.** **D 996,** D10

kraft pulp—pulp cooked by the alkaline liquor consisting essentially of a mixture of caustic soda and sodium sulfide. The make-up chemical is traditionally sodium sulfate, which is reduced to the sulfide in the chemical recovery process; hence the alternative designation, sulfate pulp. **D 1695,** D01

Krypton 85—a tracer gas used to test for leakage when the radioisotope leak test method is used. **E 1316,** E07

K series—the set of X-ray wavelengths making up *K* radiation. **E 7,** E04

K-series, *n*—the set of X-ray wavelengths making up K radiation. **E 135,** E01

Kubelka-Munk absorption coefficient, *K*, *n*—for a thin layer within an isotropic absorbing and scattering material over a black backing, the limit as the layer thickness approaches zero of the fraction of the incident radiation absorbed by the layer, divided by its thickness. **E 284,** E12

Kubelka-Munk scattering coefficient, *S, n*—for a thin layer within an isotropic scattering and absorbing material over a black backing, the limit as the layer thickness approaches zero of the fraction of the incident radiation scattered (reflected) by the layer, divided by its thickness. **E 284,** E12

Kubelka-Munk theory, *n*—phenomenological turbid-medium theory relating the reflectance and transmittance of scattering and absorbing materials to optical constants (*Kubelka-Munk absorption coefficient, Kubelka-Munk scattering coefficient*) and the concentrations of their colorants. **E 284,** E12

K-value—*EIA,* a kinematic factor (between 0 and 1) that relates the backscattered energy to the incident energy. **E 673,** E42

K-valve, *n*—a simple, manually operated on-off SCUBA cylinder valve. **F 1549,** F32

kV (kilo volt)—a unit of electrical potential difference equal to one thousand volts, used to described the accelerating potential of an X-ray tube. **E 1316,** E07

K_{xy}—hydraulic conductivity in the horizontal plane, radially from the control well. **D 653,** D18

kyanite ($Al_2O_3{\cdot}SiO_2$)—the most abundant of the mineral polymorphs that include andalusite and sillimanite, commonly used as a source of mullite in ceramics. **C 242,** C21

L

label, *n*—a piece of paper or other material to be affixed to a container or article, on which is printed a legend, information concerning the product, or addresses. It may also be printed directly on the container. (Compare **tag.**) **D 996,** D10

label, *n*—a piece of paper, cloth, polymer, metal, or other material affixed to something and indicating its contents, destination, or other information. **F 1294,** F05

label, *v*—*for protective clothing,* to attach a symbol or other identifying mark, the use of which has been authorized by a certification organization. **F 1494,** F23

laboratory accreditation—formal recognition that a testing laboratory is competent to carry out specific tests or specific types of tests, (ISO Guide 2, *B*). **E 1187,** E36

laboratory accreditation system—system having its own rules of procedure and management, for carrying out laboratory accreditation, (ISO Guide 2, *A*). **E 1187,** E36

laboratory assessor—person who carries out some or all functions related to laboratory assessment, (ISO Guide 2). **E 1187,** E36

laboratory bias—See **bias.** **D 1356,** D22

laboratory sample—See *laboratory sample* under **sample.** **D 121,** D05

laboratory sample, *n*—a portion of material taken to represent the lot sample, or the original material, and used in the laboratory as a source of test specimens. **D 123,** D13

laboratory sample, *n*—*in wool top,* portions drawn from the lot in accordance with the described procedure. **D 123,** D13

laboratory sample, *n*—a portion of material taken to represent the lot sample, or the original material, and used in the laboratory as a source of test specimens. **D 4439,** D35

laboratory sample, *n*—a representative portion of a gross sample received by a laboratory for analysis. **D 5681,** D34

laboratory sample—a representative portion of the gross sample received by the laboratory for analysis. **E 856,** D34

laboratory sampling unit, *n*—a portion of material taken to represent one of the lot sampling units or the original material and used in the laboratory as a source of test specimens. **D 123,** D13

lace leather—a form of rawhide leather (from cattlehides) for lacing together sections of power-transmission belts, sometimes prepared also with an alum and oil, chrome, or combination tannage. **D 1517,** D31

lack of resin fillout—a pultrusion condition where an area contains reinforcement not wetted with sufficient quantity of resin.

Note—This condition usually occurs at the edge of a pultrusion. **D 3918,** D20

lacquer—a coating composition that is based on synthetic thermoplastic film-forming material dissolved in organic solvent that dries primarily by solvent evaporation. Typical lacquers include those based on nitrocellulose, other cellulose derivatives, vinyl resins, acrylic resins, etc. **D 16,** D01

lacquer—see **cellulose lacquer.** **D 1695,** D01

lacquer-finished—coated with lacquer and often baked, usually in such colors as to match or blend with color of item to be fastened. **F 547,** F16

ladder code—See **vertical bar code.** **F 1294,** F05

ladder gasket—a lock-strip gasket in the form of a subdivided frame having one or more integrally formed intermediate cross members. **C 717,** C24

ladle—a long-handled, cup-shaped tool for transferring batch or molten glass. **C 162,** C14

ladle analysis, *n*—Deprecated term. Use preferred term *heat analysis.* **A 902,** A05

ladle analysis—(See *Heat Analysis*). **F 1789,** F16

ladle brick, *n*—brick suitable for lining ladles used to contain molten metal. **C 71,** C08

lag deposits—the larger and heavier particles that are sorted out and left behind in stream channels. **D 4410,** D19

lagging, *n*—*in mining or tunneling,* short lengths of timber, sheet steel, or concrete slabs used to secure the roof and sides of an opening behind the main timber or steel supports. The process of installation is also called lagging or lacing. **D 653,** D18

lag time—See **analyzer.** **D 1356,** D22

laid fabric, *n*—a fabric made without filling yarn, the parallel warp yarns being held together by means of rubber latex or other binding material. **D 123,** D13

laitance, *n*—a layer of weak material derived from cementitious material and aggregate fines either: 1) carried by bleeding to the surface or to internal cavities of freshly placed concrete, or 2) separated from the concrete and deposited on the concrete surface or in internal cavities during placement of concrete under water. **C 125,** C09

laitance, *n*—*in building construction,* a weak layer of cement and aggregate fines on a concrete surface that is usually caused by an overwet mixture, overworking the mixture, improper or excessive finishing, or combinations thereof. **C 717,** C24

lake—a special type of pigment consisting essentially of an organic soluble coloring matter combined more or less definitely with an inorganic base or carrier. It is characterized generally by a bright color and a more or less pronounced translucency when made into an oil paint.

Under this term are included two (and perhaps three) types of pigment: (*a*) the older original type composed of hydrate of alumina dyed with a solution of the natural organic color, (*b*) the more modern and far more extensive type made by precipitating from solution various coal-tar colors by means of a metallic salt, tannin, or other suitable reagent, upon a base or carrier either previously prepared or coincidently formed, and (*c*) a number combining both types in varying degree might be regarded as a third class. **D 16,** D01

Lambertian diffuser, *n*—ideal surface that reflects or transmits radiation completely in accordance with Lambert's cosine law. When illuminated from any direction, its radiance is the same for every direction of view. **E 284,** E12

Lambert's law, *n*—the intensity (flux per unit solid angle) emitted in any direction from a surface varies as the cosine of the angle between the normal to the surface and the direction of the emitted flux (also called Lambert's cosine law). See **Lambertian diffuser.** **E 284,** E12

Lambert's (thickness) law, *n*—see **Bouguer's law.** **E 284,** E12

lambskin leather—term applied to leather from either lambskins or sheepskins, which are practically indistinguishable after tanning. **D 1517,** D31

lamb's tongue—See **railing systems.** **E 631,** E06

lamb's tongue—an ornamental curved or tapered fitting terminating a handrail, usually tapered to the tip. **E 1481,** E06

Lamb wave—a specific mode of propagation in which the two parallel boundary surfaces of the material under examination (such as a plate or the wall of a tube) establish the mode of propagation. The Lamb wave can be generated only at particular values of frequency, angle of incidence and material thickness. The velocity of the wave is dependent on the mode of propagation and the product of the material thickness and the examination frequency. **E 1316,** E07

lamina—a subunit of a laminate consisting of one or more adjacent plies of the same material with identical orientation. **D 3878,** D30

lamina orientation—same as **ply orientation.** **D 3878,** D30

laminar flow, *n*—flow in which the head loss is proportional to the first power of the velocity. **D 4439,** D35

laminar flow (streamline flow) (viscous flow)—flow in which the head loss is proportional to the first power of the velocity. **D 653,** D18

laminate, *n*—a product made by bonding together two or more layers of material or materials. (See also **wood laminates**.)

cross laminate, n—a laminate in which some of the layers of material are oriented at right angles to the remaining layers with respect to the grain or strongest direction in tension. (Compare **parallel laminate**.)

parallel laminate, n—a laminate in which all the layers of material are oriented approximately parallel with respect to the grain or strongest direction in tension. (Compare **cross laminate**.) **D 907,** D14

laminate, *v*—to unite layers of material with adhesive. (Compare **bond,** v.) **D 907,** D14

±45° laminate—a balanced symmetric laminate composed of only +45° plies and −45° plies. **D 3878,** D30

laminate—any fiber- or fabric-reinforced composite consisting of laminae (plies) with one or more orientations with respect to some reference direction. **D 3878,** D30

laminate—a product made by bonding together two or more layers of material or materials. **E 1749,** E06

laminate—a product made by bonding together two or more layers of material or materials. (See also **multilayered structure**) **F 17,** F02

laminate, *n*—a product made by bonding together two or more layers of material or materials. (See also **cross laminate** and **parallel laminate**.) **F 412,** F17

laminate, *n*—object or material structure formed by the process of lamination. **F 1156,** F12

laminate, *v*—join layers of materials together using fusion; the process of joining layers of materials together using adhesion. **F 1156,** F12

laminate, *n*—a product made by bonding together two or more layers of material or materials. **F 1251,** F04

laminate—See Terminology F 17. **F 1327,** F02

laminate coordinate axes—a set of coordinate axes, usually right-handed Cartesian, used as reference in describing the directional properties and geometrical structure of the laminate. Usually the x-axis and the y-axis lie in the plane of the laminate and the x-axis is the reference axis from which ply angle is measured. **D 3878,** D30

laminate, cross-plied, *n*—a nonparallel laminate. **F 1251,** F04

laminated fabric, *n*—a layered fabric structure wherein a face or outer fabric is joined to a continuous sheet material, such as polyurethane foam, in such a way that the identity of the continuous sheet material is retained, either by the flame method or by an adhesive, and this in turn normally, but not always, is joined on the back with a backing fabric such as tricot. (See also **bonded fabric, coated fabric.**) **D 123,** D13

laminated fabric, *n*—*in fabric roof systems*, a flexible material composed of superimposed layers of textile fabric firmly united by bonding or impregnating with an adherent polymeric material to one or more surfaces. **D 123,** D13

laminated glass—(*1*) in flat glass, an assembly consisting of two or more lites of glass bonded together by an interlayer (see also Specification C 1172) and (*2*) in tableware, two or more distinct layers of glass fused together. **C 162,** C14

laminated veneer lumber (LVL), *n*— lumber made by laminating veneers in which the grain of all the veneers is essentially parallel to the lengthwise dimension. **D 907,** D14

laminated wood product, n—See under **wood laminates.** **D 907,** D14

laminate joint—See **joint.** **D 907,** D14

laminate midplane—the plane that is equidistant from both surfaces of the laminate. **D 3878,** D30

laminate, *n*—a product made by bonding together two or more layers of material or materials. (See also **cross laminate** and **parallel laminate.**) (ISO) **D 883,** D20

laminate, parallel, *n*—a laminate in which all layers or plies are oriented with their principal direction (grain or strongest direction in tension) parallel with the principal direction of the laminate. **F 1251,** F04

laminate principal axis—the laminate coordinate axis that coincides with the direction of maximum inplane Young's modulus. **D 3878,** D30

laminate, symmetrical—a composite laminate in which the ply orientation is symmetrical about the laminate midplane. **E 1749,** E06

lamination—see **pressing crack.** **B 243,** B09

lamination—line of demarcation or elongated void generally parallel to the principal grain direction of a carbon or graphite body. **C 709,** D02

lamination—a stratification of the material in the plane of the wall of a unit. **C 896,** C04

lamination, *n*—*1*) the process of preparing a laminate. *2*) any layer in a laminate. (Compare **laminate,** *n* and **wood laminate.**) **D 907,** D14

lamination—imperfection in flat products resulting from the presence of voids or inclusions aligned approximately parallel to the worked surface. **E 7,** E04

lamination—*in flexible barrier materials*, the process of preparing a laminate which consists of two or more flexible barriers bonded together (see also **laminate**). **F 17,** F02

lamination—See Terminology F 17. **F 1327,** F02

lamination factor, (space factor, stacking factor), *S*—a numeric, less than unity and usually expressed as a percentage, which is defined as the ratio of the uniform solid height h of the magnetic material in a laminated core to the actual height h' (core buildup) when measured under a specified pressure. S is thus equal to the ratio of the volume of magnetic material in a uniform laminated core to the overall geometric volume of the core. **A 340,** A06

lamination stack resistance—the electrical resistance measured in the direction perpendicular to the plane of lamination in a stack of laminations. **A 340,** A06

lamination surface insulation—the insulation between core laminations produced by a surface condition or layer either formed or applied for this purpose.

NOTE—In commercial practice, this insulating layer is frequently designated as core plate. **A 340,** A06

lamination thickness, *d*—the active thickness of a single lamination cut from sheet stock, including any core plate material. **A 340,** A06

lamination width, *w*—the width of a core lamination perpendicular to the direction of the induction therein. **A 340,** A06

laminelle—any one of the layers being joined together to form a laminate by the process of lamination. **F 1156,** F12

lampworking—forming glass articles from tubing and cane by heating in a gas flame. **C 162,** C14

lance—(*1*) see **hackle**.

(*2*) a piece of hardware used for introducing a controlled atmosphere gas flow into molten glass or into parts of the furnace structure. **C 162,** C14

landfill—see **hazardous waste landfill**. **E 1605,** E06

landfill—a land disposal technique that uses excavated pits to contain the oil spill waste material. The waste is placed in the excavation, covered over, and left to degrade. **F 1600,** F20

landfill gas—biogas produced from the natural degradation of the organic material in landfills. **E 1126,** E48

landfill gas—biogas produced from the natural degradation of the organic material in landfills. **E 1705,** E48

landfill liner—continuous layer of natural or man-made materials, beneath or on the sides of a surface impoundment, landfill, or landfill cell, which restricts the downward or lateral escape of hazardous waste, hazardous waste constituents, or leachate. **E 1605,** E06

landscape mode, *adj*—a printer output orientation in which the printed lines run parallel to the direction of movement of the paper. (See **portrait mode.**) **F 1457,** F05

landslide—the perceptible downward sliding or movement of a mass of earth or rock, or a mixture of both. (ISRM) **D 653,** D18

landslide (slide)—the failure of a sloped bank of soil or rock in which the movement of the mass takes place along a surface of sliding. **D 653,** D18

lane-to-shoulder dropoff, *n—(highways, roads and streets only)* difference in elevation between the traveled surface and the shoulder surface. **E 867,** E17

lane-to-shoulder dropoff, *n—(highways, roads and streets only)* difference in elevation between the traveled surface and the shoulder surface. **E 1778,** E17

lane-to-shoulder separation, *n—(highways, roads and streets only)* widening of the joint between the edge of the slab and the shoulder. **E 867,** E17

lane-to-shoulder separation, *n—(highways, roads and streets only)* widening of the joint between the edge of the slab and the shoulder. **E 1778,** E17

langelier saturation index, LSI—an index calculated from total dissolved solids, calcium concentration, total alkalinity, pH, and solution temperature that shows the tendency of a water solution to precipitate or dissolve calcium carbonate. **D 6161,** D19

Langmuir adsorption theory—the surface of an adsorbent has only uniform energy sites and adsorption is limited to a monomolecular layer. **D 2652,** D28

Langmuir isotherm—a plot of isothermal adsorption data which to a reasonable degree fit the Langmuir adsorption equation. **D 2652,** D28

lap—(*1*) an imperfection; a fold in the surface of a glass article caused by incorrect flow during forming.

(*2*) a tool used for polishing glass. **C 162,** C14

lap—a condition in which adjacent veneers overlap one another instead of making a smooth edge joint. **D 1038,** D07

lap—(*1*) a surface imperfection on worked metal caused by folding over a fin overfill, or similar surface condition and then impressing this into the surface by subsequent working without welding it.

(*2*) A flat surface which holds an abrasive for polishing operations. **E 7,** E04

lap cement—the cementitious material used to seal the side and end laps of corrugated roofing. **C 460,** C17

lap cement, *n*—the cementitious material used to seal the side and end laps of corrugated roofing. **C 1154,** C17

lap cement—the cementitious material used to seal the side and end laps of corrugated roofing. **D 2946,** C17

lap (*coatings*)— the region where one area of a coated surface merges into an adjacent freshly-coated area during application of a single coat to the entire surface. **D 16,** D01

lap joint, *n*—See *lap joint* under **joint**. **D 907,** D14

lapped seam,— except the topstitching is farther from the fold than that of a lapped seam 6 to 10 mm (¼ to ⅜ in.). **D 4965,** D13

lapped seam for leather or nonwoven material, *n—in home sewing*, a complex seam formed on the outside of the product, with neither cut edge enclosed and having one or two rows of stitching and a cut edge visible from the outside. (Compare **lapped seam for woven or knitted fabric**.) **D 4965,** D13

lapped seam for woven or knitted fabric, *n—in home sewing*, a complex seam formed on the inside of the product with neither raw cut edge enclosed, and having one visible line of topstitching on the face side and a small fold formed by the topstitching. (Compare **tucked seam**.) **D 4965,** D13

lapping—rubbing two surfaces together, with or without abrasives, for the purpose of obtaining extreme dimensional accuracy or superior surface finish. **B 374,** B08

lapping—the abrasive removal of material using graded abrasive particles in a loose form as in a liquid slurry on a platen. **E 7,** E04

lap sealant joint, *n—in building construction*, a joint where sealant is applied within the joint between approximately parallel substrates that are face-to-face. **C 717,** C24

lap seam—the simplest form of seam used in sewing, made by lapping one part slightly over the other and sewing through both. **F 869,** F08

lapse rate, *n*—the rate of change of the absolute value of any meteorological element with increase of height. **D 1356,** D22

large grain(s), *n— as used in fractography,* a single (or cluster of) grain(s) having a size significantly greater than that encompassed by the normal grain size distribution. **C 1145,** C28

large head—larger than standard head, usually more than 3 to 3½ times shank diameter. **F 547,** F16

large volume testing—using a large volume of the material to be tested as an inoculum in direct detection of mycoplasma. **E 1705,** E48

larrigan leather—oil-tanned light cattlehides, used for moccasins. **D 1517,** D31

LAS—an abbreviation for alkyl benzene sulfonate in which the alkyl radical is a straight chain. **D 459,** D12

LASER—**l**ight **a**mplification by **s**timulated **e**mission of **r**adiation. **F 1457,** F05

laser beam welding, *n*—a welding process that uses a laser beam as the heat source. **A 941,** A01

laser copier—a copier that uses a laser scanner to generate the image. **F 1156,** F12

laser diffraction, *n*—in particle measurement, the creation of a spatial pattern of light produced by the impingement of a laser beam on one or more dispersed particles, wherein the particle size distribution may be inferred from the nature of the pattern. **E 1620,** E29

laser etching—etching by ablation with a controlled high energy laser beam to create a pattern, mark, or alphanumeric phrase for use in subsequent identification of the item so marked. **F 1156,** F12

laser glass—a glass of special composition that emits amplified electromagnetic radiation upon proper electromagnetic stimulation. **C 162,** C14

laser light scattering—a phenomenon suitable for the measurement of particle size in that particles illuminated by a collimated laser beam cause the light to be scattered through angles which are inversely proportional to the size (generally expressed as a diameter) of the particles. **C 242,** C21

laser-light scattering event—a signal pulse that exceeds a preset threshold, generated by the interaction of a laser beam with a localized light scatterer (LLS) at a wafer surface as sensed by a detector. See also **haze**. **F 1241,** F01

laser marking, *n*—etching of data by lasers that directly mark a surface such as metal, wood, and fiberboard. **F 1294,** F05

laser printer—a nonimpact printer that employs a laser light source drive by digital signals to create images on a photoconductor. See **electrophotographic printer.** **F 909,** F05

laser printers—xerographic printers using a laser scanning device (**laser ROS**) as the writing device. **F 1156,** F12

laser ROS—raster output scanner using a laser beam for the output device. **F 1156,** F12

laser scanner—an optical scanning device that uses the intense monochromatic light beam given off by a laser as its source of illumination. **F 149,** F05

laser scanner, *n*—an optical bar code reading device using a low energy laser light beam as its source of illumination. **F 1294,** F05

Laskin nozzle—a type of portable air-operated aerosol generator capable of producing droplets in the size range from 1 to 2 μm. **E 1620,** E29

last *n*—a piece of wood, metal, or synthetic material roughly following the shape of the foot and acting as a form on which a shoe is made.

v.—to shape a fitted upper to the last using the stretch of the leather (or other material) and then fix it temporarily or permanently to the insole. **F 869,** F08

latent image—a condition produced and persisting in the image receptor by exposure to radiation and able to be converted into a visible image by processing. **E 1316,** E07

latent images—images created on or in a substance (substrate) that have not yet been visualized (for example, undeveloped film). **F 1156,** F12

lateral, *adj*—a descriptive term for a textile fiber composed of two or more polymers at least two of which have a continuous longitudinal external surface. **D 123,** D13

lateral—away from the midline. (Opposite: medial) **F 869,** F08

lateral accretion deposits—See **point bar.** **D 4410,** D19

lateral crack—a crack produced beneath and generally paralleling a glass surface during the unloading phase of mechanical contact with a hard, sharp object. (See **cleavage crack**.) **C 162,** C14

lateral force coefficient, *n*—*of a tire*, the ratio of lateral force to normal force. **F 538,** F09

lateral force, [*F*], *n*—*of a tire*, the component of the tire force vector in the Y' direction. **F 538,** F09

lateral groove, *n*—a groove that has its long dimension oriented at a direction non-parallel to the tire circumferential centerline; it most frequently opens into a void at both ends. **F 538,** F09

lateral holding strength, *n*—the force required to disengage a snap fastener resulting from a pull in the plane parallel to the material to which the snap fastener is attached. **D 123,** D13

lateral order—the degree of regularity of arrangement of atoms and atomic groups in the direction normal to the molecular chain axes in linear polymers. Quantitative evaluation is impossible without further specification and without description of the particular experimental technique. **D 1695,** D01

lateral resistance of metal connector plate—resistance to slip or pulling, or both, from wood, in direction of applied external shear force, of integral teeth or separate nails, or both, fastening connector plate to wood members. Also called *tooth holding resistance, grip resistance,* and *peel resistance* to cover specific events; yet, preferably called *lateral resistance*. See **shear strength of metal connector plate**. **E 1807,** E06

lateral scroll—See **railing systems.** **E 631,** E06

lateral scroll—a fitting that curves in a horizontal plane, used to terminate a handrail; often ending as a round plate covering the top of a post. **E 1481,** E06

latewood—the denser, smaller-celled, later-formed part of a growth layer. A synonym is *summerwood.* **D 9,** D07

latex, *n*—a stable dispersion of polymeric substance in an essentially aqueous medium. **D 907,** D14

latex, creamed, *n*—a latex, the rubber concentration of which has been increased by creaming and removal of the separated serum. **D 1566,** D11

latex, creaming, *n*—a reversible process consisting of gravitational accumulation of rubber particles surrounded by serum, near the top or bottom of the latex. **D 1566,** D11

latex, creaming agent, *n*—a material added to latex to promote creaming. See **latex, creaming**. **D 1566,** D11

latex, field, *n*—natural rubber latex with or without a preservative and prior to concentration or any other processing. **D 1566,** D11

latex foam rubber, *n*—*in building construction*, an elastomeric open-cell material made by whipping or stirring air or some other gas into rubber latex compound as it is gelled and before it is subsequently cured. **C 717,** C24

latex, mechanical stability, *n*—the ability of latex to resist coagulation under the influence of mechanical agitation **D 1566,** D11

latex-modified portland cement concrete—concrete that includes polymer emulsion, resulting in a dense hardened concrete resisting the movement of moisture and chloride ions. Usually used as a relatively thin overlay (1 to 1½ in. (25 to 38 mm)) on conventional concrete bridge decks. **D 3743,** D04

latex paint—under **paint**, see *latex paint.* **D 16,** D01

latex rubber, *n*—colloidal aqueous dispersion of rubber. **D 1566,** D11

latex sealant, *n*—a compound that cures primarily through water evaporation. **C 717,** C24

lath—see **gypsum lath**. **C 11,** C11

lath—thin, narrow strips of rough wood. **D 9,** D07

lathe—the machine on which rotary, half-round, and rift veneer is cut. **D 1038,** D07

lather—a foam or froth when a detergent is agitated in water or other liquid. **D 459,** D12

lath nail—See **metal-lath nail, wood-lath nail.** **F 547,** F16

latigo leather—a type of lace leather, alum and vegetable tanned, used in saddlery. **D 1517,** D31

latitude—when the photographic process is represented by an H and D curve, the latitude is the projection on the exposure axis of that part of the curve which approximates a straight line within the tolerance permitted for the purpose at hand. **E 7,** E04

lattice—(*1*) a space lattice is a set of equal and adjoining parallelopipeds formed by dividing up space by three sets of parallel planes, the planes in any one set being equally spaced. There are seven ways of so dividing space, corresponding to the seven crystal systems. The unit parallelopiped is usually chosen as the unit cell of the system.

(2) a point lattice is a set of points in space so located that each point has identical surroundings. There are fourteen ways of so arranging points in space, corresponding to the 14 Bravais lattices. **E 7,** E04

lattice parameter—the term is used for the fractional coordinates *x, y, z* of lattice points when these are variable. Also used to indicate the lengths of the axes *a, b, c,* and their included angles α, β, γ. **E 7,** E04

lattice pattern—in reinforced plastics, a pattern of filament winding with a fixed arrangement of open voids. **D 883,** D20

Laue equations—the three simultaneous equations which state the conditions which must be met for diffraction from a three-dimensional network of diffraction centers. **E 7,** E04

Laue method (for crystal analysis)—a method of X-ray diffraction employing a beam of white radiation, a fixed single crystal specimen and a flat photographic film usually normal to the incident beam. If the film is located on the same side of the specimen as the X-ray source, the method is known as the back reflection Laue method, if on the other side as the transmission Laue method. **E 7,** E04

launderability, *n—in buttons,* the ability of a button to undergo multiple cycles of laundering without damage such as cracks or loss of finish. **D 123,** D13

laundering, *n—in textile product care,* a process intended to remove soil or stains by treatment (washing) with an aqueous detergent solution (and possibly bleach) and normally including subsequent rinsing, extracting, and drying.. **D 123,** D13

laundering—a process intended to remove soils or stains, or both, by washing in an aqueous detergent solution, that normally includes subsequent rinsing, extracting, and drying. (See Guide D 5548.) **D 459,** D12

laundering, *v*—a process used to refurbish a textile product or parts thereof by (1) cleaning it in water containing a cleaning agent and possibly bleach, (2) drying it, and (3) if required, ironing or pressing it. **D 5253,** D13

laundering, *n*—a process used to refurbish a textile product or parts thereof by (1) cleaning it in water containing a cleaning agent, and possibly bleach, (2) drying it, and (3) usually ironing or pressing it. **D 5497,** D13

laundry additive—a separately added product that contributes to the effectiveness of laundering or provides a specialized function, or both. (See Guide D 5548.) **D 459,** D12

laundry formula, *n*—a list of chemicals, amounts, and procedures used in a laundry operation. **F 1494,** F23

lawn, *n—as in microbiology,* a cloudy, uniform growth of bacteria in a thin layer of top agar in a petri dish. **F 1494,** F23

lay, *n*—one turn of the helix of a wire or member of a standard conductor. See also **direction of lay** and **length of lay.** **B 354,** B01

lay, *v*—to arrange the wires or members of a conductor either by twisting them or by forming them into one or more layers helically applied. **B 354,** B01

lay, *n*—(*1*) the length of twist produced by stranding filaments, such as fibers, wires, or roving; (*2*) the angle that such filaments make with the axis of the strand during a stranding operation. **D 883,** D20

lay—the predominant direction of the surface texture. **F 1241,** F01

layer—{archaic} person who has charge of laying the glass in plaster on the grinding and polishing tables. **C 162,** C14

layer, *n—as related to veneer and plywood,* a single veneer ply or two or more plies laminated with grain direction parallel. **D 907,** D14

layer—a single veneer ply or two or more plies laminated with grain direction parallel. Two or more plies laminated with grain direction parallel is a parallel laminated layer. **D 1038,** D07

layer insulation, *n*—paper, 5 to 1200 μm thick, used to insulate between layers of conductors in transformers or other inductive apparatus. **D 1711,** D09

layer lattice—a type of structure found in crystals which tend to form in thin sheets. **E 7,** E04

layer of reinforcement—circumferential reinforcement that is one bar or wire in thickness. **C 822,** C13

lay factor—the ratio of the length of lay to the external diameter of the corresponding layer of wires or members in the stranded conductor. **B 354,** B01

laying length—the centerline length of an installed pipeline system, section, or fitting. **F 412,** F17

laying yard—{archaic} place where the rough glass is laid on grinding and polishing tables with plaster. **C 162,** C14

lay up, *n*—in reinforced plastics, an assembly of layers of resin-impregnated material ready for processing. **D 883,** D20

lay up, *v*—in reinforced plastics, to assemble layers of resin-impregnated material for processing. **D 883,** D20

lay-up—a process of fabrication involving the placement of successive layers of materials. **E 1749,** E06

lay-up code—a designation system for abbreviating the stacking sequence of laminated composites. **D 3878,** D30

LBP—lead-based paint. **E 1605,** E06

LC_{50}, *n*—a measure of lethal toxic potency; the concentration of gas or smoke calculated statistically from concentration-response data to produce lethality in 50 % of test animals within a specified exposure and postexposure time. **E 176,** E05

LC50—a statistically or graphically estimated concentration that is expected to be lethal to 50 % of a group of organisms under specified conditions. **E 943,** E47

LC—a type of electronic circuit element combining an inductor (L) and a capacitor (C), usually to provide sensitivity or output at some specific frequency related to the product of the values of L and C. **F 1156,** F12

LD-50—concentration required for 50 % mortality (lethal dose). **D 6161,** D19

LD50—a statistically or graphically estimated dose that is expected to be lethal to 50 % of a group of organisms under specified conditions. **E 943,** E47

L/D ratio—one measure of the resolution capability of a neutron radiographic system. It is the ratio of the distance between the entrance aperture and the image plane (*L*) to the diameter of the entrance aperture (*D*). **E 1316,** E07

lea, *n—in cotton yarns,* the number of 120-yd lengths of yarn per pound; an indirect yarn numbering system. **D 123,** D13

lea, *n—in linen yarns,* the number of 300-yd lengths of yarn per pound; an indirect yarn numbering system. **D 123,** D13

leachate—the liquid residue from the migration of environmental pollutants mediated by surface or ground water. **F 1600,** F20

leaching—the removal in solution of the more soluble materials by percolating or moving waters. (ISRM) **D 653,** D18

leaching—the removal of soluble soil material and colloids by percolating water. **D 653,** D18

leaching (latex), *n*—a process in which latex articles are washed in water to remove water-soluble substances. **D 1566,** D11

lead abatement contractor—any business entity, public unit, or person performing the actual abatement for a lead abatement project. **E 1605,** E06

lead angle, helix angle—angle made by helix of thread with plane perpendicular to nail axis; measured at pitch line; small lead angle usually refers to less than 20 deg; medium lead angle usually refers to 55 to 65 deg; large lead angle refers to 65 to 75 deg; extra large lead angle refers to 80 to 85 deg. **F 547,** F16

lead-based paint removal—action or process of abatement, that entails stripping lead-based paint from surfaces of components. **E 1605,** E06

lead-containing paint—lead-based paint or other similar surface-coating materials containing lead or a lead compound, and in which the lead content is in excess of 0.06 % by weight of the total nonvolatile content of the paint, or the weight in the dried paint film. **E 1605,** E06

leaded—characteristic of a metallic body containing metallic lead in dispersed form. **E 7,** E04

leaded dust—dust containing lead or lead compounds at potentially hazardous concentrations. **E 1605,** E06

leaded glass—pieces of glass fixed together at their edges with soldered strips of lead or other ductile metal. **C 162,** C14

leaded paint—paint or other coatings containing lead compounds at potentially hazardous concentrations. **E 1605,** E06

lead exposure—subjection to the presence of a lead hazard that could result in ingestion or inhalation of lead-containing substances. **E 1605,** E06

lead head, cast lead head, compressed lead head—lead encasement pressed or cast onto or around small 9⁄32-in. steel head, or both. **F 547,** F16

lead-head nail—See **roofing nail.** **F 547,** F16

leading—a typographical term indicating the amount of white space between lines of printed characters. **F 1457,** F05

leading edge—the edge of a form that is used as a base for locating the first line of data to be scanned. **F 149,** F05

leading [pronounced *leeding*]—front or top edge as leading edge of master or blanket. **F 425,** F05

leadless glaze—See *leadless glaze* under **glaze**. **C 242,** C21

lead paint hazard—presence of lead-based paint in places and site conditions posing a danger of ingestion or inhalation. **E 1605,** E06

lead poisoning—toxic condition caused by excessive exposure to lead. **E 1605,** E06

lead positive/negative—qualitative (not quantitative) judgment of lead content that determines abatement action measures. **E 1605,** E06

lead screen—see **intensifying screen (a).** **E 1316,** E07

leaf—the sandwich layer of flat-sheet membrane/product channel spacer/flat-sheet membrane, glued together on the sides and across the outer end in a spiral wound element. **D 6161,** D19

leak—a hole, or void in the wall of an enclosure, capable of passing liquid or gas from one side of the wall to the other under action of pressure or concentration differential existing across the wall, independent of the quantity of fluid flowing. **E 1316,** E07

leak—any opening in a flexible package that is contrary to intention and either lets contents escape or permits substances to enter. **F 1327,** F02

leakage, *n*—the flow of liquid from one hydrogeolic unit to another. **D 653,** D18

leakage field—see **field, magnetic leakage.** **E 1316,** E07

leakage flux—the flux outside the boundary of the practical magnetic circuit. **A 340,** A06

leakage rate—the flow rate of a liquid or gas through a leak at a given temperature as a result of a specified pressure difference across the leak. Standard conditions for gases are 25°C and 100 kPa. Leakage rates are expressed in various units such as pascal cubic metres per second or pascal litres per second. **E 1316,** E07

leakance, *n* (T^{-1})—the ratio K′/b′, in which K′ and b′ are the vertical hydraulic conductivity and the thickness, respectively, of the confining beds. **D 653,** D18

leak artifact—a device used to introduce gas into a system at a controlled rate, usually 10^{-7} mol/s or less. **E 1316,** E07

leak detector—a device for detecting, locating, or measuring, or combination thereof, leakage. **E 1316,** E07

leak exit—the point where a leak appears. **E 1749,** E06

leak path—the path a leak follows from the leak source to the leak exit. **E 1749,** E06

leak source—the point where a leak starts. **E 1749,** E06

leak testing—comprises procedures for detecting or locating or measuring leakage, or combinations thereof. **E 1316,** E07

leaky aquifier, *n*—aquifiers, whether artesian or unconfined, that lose or gain water through adjacent less permeable beds. **D 653,** D18

lean gas—natural gas containing little or no hydrocarbons commercially recoverable as liquid products. **D 4150,** D03

lease (bail)—contract between the owner of real property (lessor) and another party (lessee) for the possession and use of the property for a specified term in return for rent or other income. **E 1480,** E06

leash—a cord-like device wherein one end is attached to the top surface of the snowboard ski, or the binding, and the other end provides an apparatus to attach to one of the rider's legs. **F 1107,** F27

least count, *n*—*in tensile testing machines*, the smallest change in the indicated property that can customarily be determined. (See **sensitivity**) **D 123,** D13

least count, *n*—the smallest change in indication that can customarily be determined and reported. **E 6,** E28

least difference of practical importance, δ, *n*—the smallest difference based on engineering judgment deemed to be of practical importance when considering whether a significant difference exists between two statistics or between a statistic and a hypothetical value. **D 123,** D13

leather—a general term for hide or skin that still retains its original fibrous structure more or less intact, and that has been treated so as to be imputrescible even after treatment with water. The hair or wool may or may not have been removed. Certain skins, similarly treated or dressed, and without the hair removed, are termed "fur." No product may be described as leather if its manufacture involves breaking down the original skin structure into fibers, powder or other fragments by chemical or mechanical methods, or both, and reconstituting these fragments into sheets or other forms. **D 1517,** D31

leatherboard—a type of fiberboard in which the fiber content is at least 75 % leather, usually with asphaltic or resinous binder. **D 1517,** D31

Le Chatelier's theorem—if a system in equilibrium is subjected to a constraint by which the equilibrium is altered, a reaction takes place which opposes the constraint, that is, one by which its effect is partially annulled. **E 7,** E04

LED—Light Emitting Diode. **F 1156,** F12

LED—**l**ight **e**mitting **d**iode.

LF, see line feed. **F 1457,** F05

ledeburite—intimate mixture of austenite and cementite in metastable equilibrium, formed on rapid cooling during the eutectic reaction in alloys of iron and carbon containing greater than 2 percent but less than 6.67 percent carbon. Further slow cooling causes decomposition of the austenite into ferrite and cementite (pearlite) as a result of the eutectoid reaction. **E 7,** E04

ledge—see **bedrock**. **D 653,** D18

ledger paper, *n*—a paper characterized by strength, high tearing resistance, erasability, water resistance, ink receptivity, uniformity of surface, and smoothness. **D 1968,** D06

leeches—permanent magnets or electromagnets that are attached to the electrodes carrying magnetizing current and that are strong enough to hold electrode contact firmly. **E 1316,** E07

leg—staple part connecting staple crown with staple point; driven through and into or through materials being fastened. **F 592,** F16

legging, *n*—the drawing of filaments or strings when adhesive-bonded substrates are separated. (See also **stringiness** and **webbing**.) (Compare **teeth**.) **D 907,** D14

legs, *n—in zippers*, the two portions of a separate element that affix the element to the bead. **D 123,** D13

leg thickness—maximum dimension of staple-leg cross section measured parallel to staple-crown axis. **F 592,** F16

leg width—maximum dimension of staple-leg cross section measured perpendicular to staple-crown axis. **F 592,** F16

lehr, leer—a long, tunnel-shaped oven for heat treating glass by continuous passage. **C 162,** C14

lehr loader—a device for properly placing and spacing glass articles on a continuous lehr belt. **C 162,** C14

length, *n*—the horizontal dimension of the unit in the face of the wall. **C 43,** C15

length, *n*—the maximum dimension of the tile measured parallel to the water channels or perpendicular to the eave of the roof. **C 43,** C15

length, *n*—horizontal dimension of the face of a unit when the unit is positioned as a stretcher. **C 1232,** C15

length, *n—of a fabric*, the distance from one end of a fabric to the other measured parallel to the side edge of the fabric while it is under zero tension and is free of folds or wrinkles. **D 123,** D13

length—distance between beginning and end of point, measured along edge of point. **F 547,** F16

length—distance measured parallel to shank axis from maximum diameter of bearing surface of head to extreme end of point; except in case of cement-coated, brad-headed, and oval-countersunk-headed nails where measurement includes complete head.

Discussion—In evaluating nail performance, length is measured to and including one third of length of point. **F 547,** F16

length analyzer, *n*—an instrument which determines the upper-half-mean length and length uniformity index of a test beard of cotton. **D 123,** D13

length between, L_b, *n*—in textile unevenness testing, the length between which unevenness is measured; the equivalent of the length of strand segments weighed in a direct method of measuring unevenness. **D 123,** D13

length change, *n—in cement testing*, an increase or decrease in linear dimension due to causes other than applied load, usually measured along the longitudinal axis of a test specimen and expressed as a percentage of a gage length. **C 219,** C01

length/depth—the distance between the two edges of a form, reached by moving at right angle to a nominal data line. **F 149,** F05

length distribution, *n—of staple fibers*, a graphic or tabular presentation of the proportion or percentage (by number or by weight) of fibers having different lengths. **D 123,** D13

length group, *n*—all fibers, or pulls, whose lengths fall within a given length interval. **D 123,** D13

length interval, *n*—a class interval of ⅛ in. (3 mm), usually designated by its midpoint length in odd-numbered sixteenths of an inch. **D 123,** D13

length of lay—the axial length of one turn of the helix of a wire or member. **B 354,** B01

length of metal connector plate—dimension of metal connector plate parallel to longitudinal axis of coiled metal strip from which plate was sheared during its fabrication, not necessarily along the long plate dimension. **E 1807,** E06

length of tear, *n—in fabric*, the measured distance propagated by a tearing force from the initiation of the force until its termination. **D 123,** D13

lengths—straight pieces of the product. **B 846,** B05

lengths, standard—uniform lengths recommended in a Simplified Practice Recommendation or established as a Commercial Standard. **B 846,** B05

length within, L_w, **n**—in textile unevenness testing, the length over which unevenness is measured. **D 123,** D13

leno weave, *n*—a weave in which two adjacent warp yarns cross each other between the picks. **D 123,** D13

lens:

Fresnel—optical device which is modified in configuration to lessen its mass or bulk by division into segments comprised primarily of the active surfaces. **F 1156,** F12

lenticular anisotropic phase, *n—as used in Test Method D 5061*, a group of binder-phase anisotropic carbon textures distinguished by their lens-shaped domains (that is, length (L) to width (W) ratio of $2W < L < 4W$) and subdivided based on domain widths as fine lenticular (1.0- to 3.0-μm), medium lenticular (3.0- to 8.0-μm), and coarse lenticular (8.0- to 12.0-μm) size categories. **D 121,** D05

lenticular lens—a biconvex lens; in this instance, a plate containing a series of parallel cylindrical lens engraved into its surface; the lenses might, or might not, all be of the same optical character and activity. **F 1156,** F12

let-go, *n*—an area in laminated glass over which an initial adhesion between interlayer and glass has been lost. **D 883,** D20

lethal load XX (LLXX), *n*—a statistically or graphically estimated loading rate of test material that is expected to be lethal to XX % of a subpopulation of organisms under specified conditions. **D 6384,** D02

let-off mark, *n—in woven fabrics*, a corrugated defect pattern distributed across the fabric width. (See **crack mark** and **shier**.) **D 123,** D13

lettered head—head with raised or depressed, identifying letter or letters on top surface of head. **F 547,** F16

letter quality printer—term implies that printer output quality matches that of a typical office electric typewriter. **F 909,** F05

leuco—colorless. **F 1156,** F12

leuco dye, *n*—color precursors, examples of which are triphenyl methane and fluoran classes which react with a developer to form a colored image. **F 1623,** F05

Levant—term applied to goatskin on which the grain pattern is accentuated in tannage. Goatskin embossed to give a Levant pattern is properly described as "Levant-grained goatskin." Sheep, seal, and other skins bearing this pattern should not be described as "Levant leather" but as "Levant-grained sheepskin," etc. **D 1517,** D31

level crossings—*in fatigue loading*, the number of times that the load-time (strain-time) history crosses a given load (strain) level with a positive slope or a negative slope, or both, as specified during a given period. **E 1823,** E08

leveling, *n*—(*1*) the process whereby a film of liquid coating flows out after application so as to minimize any surface irregularities such as brush marks, orange peel, peaks, or craters, that have been produced by the mechanical process of application. (*2*) a measure or rating of the leveling ability of a coating. **D 16,** D01

leveling—the property of a freshly spread polish to dry to a uniform and streak-free appearance. **D 2825,** D21

leveling action—the ability of a plating solution to produce a surface smoother than that of the substrate. **B 374,** B08

leveling-off degree of polymerization, LODP—the nearly constant degree of polymerization of cellulose reached after very prolonged mild hydrolysis or short drastic hydrolysis. **D 1695,** D01

level, *L*—ten times the common logarithm of the ratio of a quantity proportional to power or energy to a reference quantity of the same kind. (See **sound power level, sound pressure level.**) The quantity so obtained is expressed in decibels. **C 634,** E33

level (of a factor), *n*—a given value, a specification of procedure or a specific setting of a factor.

NOTE—"Version" is a general term applied both to quantitative and qualitative factors. The more restrictive term "level" is frequently used to express more precisely the quantitative characteristic. For example, two versions of a catalyst may be presence and absence. Four levels of a heat treatment may be 100°C, 120°C, 140°C, and 160°C. **E 456,** E11

level (of a factor), *n*—a given value, a specification of procedure or a specific setting of a factor.

NOTE—*Version* is a general term applied both to quantitative and qualitative factors. The more restrictive term *level* is frequently used to express more precisely the quantitative characteristic. For example, two versions of a catalyst may be presence and absence. Four levels of a heat treatment may be 110°C, 120°C, 140°C, and 160°C. **E 1325,** E11

level reduction, *LR*—*in a specified frequency band,* the decrease in sound pressure level, measured at the location of the receiver, when a barrier or other sound-reducing element is placed between the source and the receiver. **C 634,** E33

lever—a part or component that retracts the bolt after alignment of proper combination. **F 471,** F12

lever principle—in a phase diagram, the relative proportions of the conjugate phases, at a stated value of temperature and pressure, or both, is such that a state of mechanical balance would obtain, if the corresponding weight of each phase were placed upon its composition point upon the tie-element (tie-line, tie-triangle, etc.) and the fulcrum were located at the gross composition point of the mixture. **E 7,** E04

lever screw—attached lever to the bolt. **F 471,** F12

lever spring—applies directional force to the lever. **F 471,** F12

Libbey-Owens sheet process—{archaic} See **Colburn sheet process**. **C 162,** C14

lid—abbreviation for **liquid development.** **F 335,** F05

life—the length of time a filter medium can remain in service under specified conditions. **F 740,** F21

life cycle—See **study period**. **E 833,** E06

life-cycle cost (LCC) method—a technique of economic evaluation that sums over a given study period the costs of initial investment (less resale value), replacements, operations (including energy use), and maintenance and repair of an investment decision (expressed in present or annual value terms). **E 631,** E06

life-cycle test—a comparative study in which organisms, that are subjected to different treatments, are observed at least from a life stage in one generation to the same life stage in the next generation. **E 943,** E47

life expectancy designation, *n*—*for paper records,* a rating in years for the life expectancy of paper, under prescribed conditions, primarily for records. **D 1968,** D06

life expectancy (LE), *n*—*for paper,* length of time a product can be expected to maintain its functional (that is, physical, chemical, appearance, and so forth) characteristics when stored under prescribed conditions. **D 1968,** D06

lift—a defect characterized by the spontaneous separation of large pieces of porcelain enamel from the base metal. **C 286,** B08

lift, *n*—the concrete placed between two consecutive horizontal construction joints. **C 717,** C24

lift hole—a small hole cast or drilled in the wall of the pipe or section for inserting a bolt, loop of cable or other device used in handling the pipe or section. **C 822,** C13

lifting point—structural point on the end connector or along the length of a boom section designed for the attachment of a lifting device, such as a crane. **F 818,** F20

lift off—the removal of one or more images of copy from the substrate by transferring to an intermediate member. **F 221,** F05

lift-off effect—the effect observed in an electromagnetic test system output due to a change in magnetic coupling between a test specimen and a probe coil whenever the distance between them is varied. **E 1316,** E07

light—(*1*) a term used for optical glass having a low index of refraction. (*2*) an architectural term for a panel or sheet of glass. See also **lite**. **C 162,** C14

light, n—Use **lite** (preferred term). **C 717,** C24

light, *adj*—*highly reflecting,* as in the term light green. **D 2946,** C17

light, *n*—electromagnetic radiation in the spectral range detectable by the normal human eye (approximately 380 to 760 nm). **D 2946,** C17

light, *n*—*in color determination,* radiant energy evaluated according to the CIE photopic spectral luminous efficiency function. **D 2946,** C17

light, *n*—(*1*) electromagnetic radiant power that is visually detectable by the normal human observer, radiant power having wavelengths from about 380 nm to about 780 nm.

(*2*) radiant power evaluated with respect to wavelength according to the CIE spectral luminous efficiency function. **E 284,** E12

light, *adj*—*referring to the color of a non-self-luminous body,* having a high luminous reflectance factor, as "light green" or "light gray." **E 284,** E12

light—See **windows and doors.** **E 631,** E06

light—electromagnetic radiation in the spectral range detectable by the normal human eye (wavelengths of approximately 380 to 780 nm). **E 1316,** E07

light amplification by stimulated emission of radiation, *n*—a device that generates a narrow beam of coherent light. (See **LASER**.) **F 1457,** F05

light decay—reduction of apparent surface voltage caused by exposure to light. **F 335,** F05

light decay rate—the reduction of apparent surface voltage at specific exposure conditions. **F 335,** F05

light-duty, *adj*— *in internal combustion engine operation*, characterized by average speeds, power output, and internal temperatures that are generally much lower than the potential maximums. **D 4725,** D15

light duty engine, *n*— *in internal combustion engine types*, one that is designed to be normally operated at substantially less than its peak output. **D 4725,** D15

light emitting diode, *n*—a solid-state electronic device or transistor which emits light. (See **LED**.) **F 1457,** F05

lightfastness, *n*—the ability of a material to withstand color change on exposure to light. **E 284,** E12

lightfastness, *n*—ability of an image to resist fading on exposure to different lighting conditions. **F 1857,** F05

lightfastness standards, n—See **AATCC blue wool lightfastness standards**. **E 284,** E12

lightfield illumination—See **brightfield illumination**. **E 7,** E04

light filter—See **color filter**. **E 7,** E04

light (glazing) n—use **lite**. **E 1605,** E06

lighting (éclairage):

ambient light (lumière ambiante)—surrounding light, such as that reaching an object in a room from all light sources in the room.

glare (éblouissement)—effect of brightness or brightness differences within the visual field sufficiently high to cause annoyance, discomfort, or loss of visual performance.

task lighting (éclairage (de travail) localisé)—localized lighting system consisting of a functional arrangement of luminaires to accommodate the specific visual task or work area needs. **E 1480,** E06

light intensity—the light energy reaching a unit area of surface per unit time. **E 1316,** E07

lightness—under **color of an object,** see *lightness.* **D 16,** D01

lightness, *n*—(*1*) the attribute of color perception by which a non-self-luminous body is judged to reflect more or less light.

(*2*) the attribute by which a perceived color is judged to be equivalent to one of a series of grays ranging from black to white. **E 284,** E12

lightness—the quality of a color that expresses various shades of tones within the same color range; for example, light pink to intense red. **F 1156,** F12

light pen, *n—in a bar code system,* a handheld scanning wand that is used as a contact bar code reader where the operator traverses the reader across the bar code symbol. **F 1294,** F05

light point defect (LPD)—*not preferred,* see **localized light scatterer (LLS)**. **F 1241,** F01

light-reducing glass—a term applied to flat glass having reduced light transmittance. **C 162,** C14

light resistance—the ability to resist the deleterious action of light. **F 869,** F08

light source, *n—for symbology readers,* light energy can be emitted in straight lines from a point source or from several points as an extended source. **F 1294,** F05

light stability—*in optical character recognition,* the resistance to change of the color of the image when exposed to radiant energy. **F 149,** F05

light stability—resistance to change of color of the image with exposure to radiant energy. **F 221,** F05

lightweight concrete, *n*—concrete with a density of less than 115 lb/ft^3(1840 kg/m^3). **F 141,** F06

lightweight concrete masonry unit—unit whose oven-dry density is less than 105 lb/ft^3(1680 kg/m^3). **C 1209,** C15

ligne, *n*—a unit of measure for buttons, one ligne equals 0.635 mm (0.025 in.) **D 123,** D13

ligne size, *n*—a unit of measure for button diameter; one ligne equals 0.635 mm (0.025 in.). **D 123,** D13

lignin—that part of plant material which is not saccharified by the action of 72 % sulfuric acid or 42 % hydrochloric acid, after the resins, waxes, and tannins have been removed. **D 1695,** D01

lignin, *n*—an amorphous, noncarbohydrate polymer of high molecular weight, primarily composed of aromatic organic molecules, which is found within and between the cell walls of most plants. **D 1968,** D06

lignin—the noncarbohydrate, polyphenolic, structural constituent of wood and some other plant tissues that encrusts the cell walls and cements the cells together. **E 1126,** E48

lignin—the noncarbohydrate, polyphenolic, structural constituent of wood and some other plant tissues that encrusts the cell walls and cements the cells together. **E 1705,** E48

lignin plastics—plastics based on lignin resins. (ISO) **D 883,** D20

lignin resin—a resin made by heating lignin or by reaction of lignin with chemicals or resins, the lignin being in greatest amount by mass. (ISO) **D 883,** D20

lignite A—See *lignite A* under **rank**. **D 121,** D05

lignite B—See *lignite B* under **rank**. **D 121,** D05

lignitic class—See *lignitic class* under **rank**. **D 121,** D05

lime—a general term which includes the various chemical and physical forms of quicklime, hydrated lime, and hydraulic lime. It may be high-calcium, magnesian, or dolomitic. **C 51,** C07

lime—calcium oxide (CaO), or a mixture of calcium oxide (CaO) and magnesium oxide (MgO). **C 162,** C14

lime—specifically, calcium oxide (CaO_2); also loosely, a general term for the various chemical and physical forms of quicklime, hydrated lime, and hydraulic hydrated lime. **D 653,** D18

lime—$Ca(OH)_2$, calcium hydroxide, a common water treatment chemical. **D 6161,** D19

lime glass—jargon for soda-lime-silicate glass. **C 162,** C14

lime mortar—a lime putty mixed with an aggregate, suitable for masonry purposes. **C 51,** C07

lime putty—the product obtained by slaking quicklime with water according to the directions of the manufacturer or by mixing hydrated lime and water to a desired consistency. **C 51,** C07

lime soda softening—use of lime and Na_2CO_3 for softening water. **D 6161,** D19

limestone—an initially sedimentary rock consisting chiefly of calcium carbonate or of the carbonates of calcium and magnesium. Limestone may be of high calcium, magnesian, or dolomitic.

(*1*) *dolomitic limestone*—limestone containing from 35 to 46 % magnesium carbonate ($MgCO_3$).

(*2*) *magnesium limestone*—a limestone containing from 5 to 35 % $MgCO_3$.

(*3*) *high-calcium limestone*—a limestone containing from 0 to 5 % $MgCO_3$. **C 51,** C07

limestone—a rock of sedimentary origin composed principally of calcium carbonate (the mineral calcite), or the double carbonate of calcium and magnesium (the mineral dolomite), or some combination of these two minerals. **C 119,** C18

limestone—either calcitic limestone ($CaCO_3$) or dolomitic limestone ($CaCO_3 \cdot MgCO_3$). **C 162,** C14

limestone—a sedimentary carbonate rock, composed chiefly of calcite ($CaCO_3$), but sometimes containing appreciable dolomite. **C 242,** C21

limestone—either calcite limestone ($CaCO_3$) or dolomitic limestone ($CaCO_3$-$MgCO_3$). **D 6161,** D19

limestone marble—compact, dense limestone that will take a polish is classified as marble in trade practice. Limestone marble may be sold as limestone or as marble. **C 119,** C18

liming material—a general term which includes the various chemical and physical forms of lime, limestone, mollusk shells, marl, and slag whose calcium and magnesium compounds are capable of neutralizing acidity. **C 51,** C07

95 % limit (difference between two results)—the maximum absolute difference expected for approximately 95 % of all pairs of results from laboratories similar to those in the interlaboratory study. **E 1547,** E15

limited solid solution—a crystalline miscibility series whose composition range does not extend all the way between the components of the system, that is, the system is not isomorphous. **E 7,** E04

limiting current density: (*1*) *cathodic*—the maximum current density at which satisfactory deposits can be obtained.

(*2*) *anodic*—the maximum current density at which the anode behaves normally, without excessive polarization. **B 374,** B08

limiting quality level *(LQL* **or** , *n—in acceptance sampling*, the fraction of nonconforming items at which the process average can be considered barely tolerable; the process average at which the risk of acceptance is called the **consumer's risk.** **D 123,** D13

limiting resolution—the highest spatial frequency of a target that an imaging sensor is able to resolve. **E 1316,** E07

limiting viscosity number—see **viscosity, intrinsic.** **D 1695,** D01

limit of detection—a calibration curve usually has the shape shown in Fig.1 of D 4127.

By analogy with definitions adopted in other fields, the limit of detection should be defined as the concentration for which, under the specified conditions, the potential E deviates from the average potential in region I by some arbitrary multiple of the standard error of a single measurement of the potential in region I.

In the present state of the art, and for the sake of practical convenience, a simple (and more convenient) definition is recommended at this time. The practical limit of detection may be taken as the activity (or concentration) of A at the point of intersection of the extrapolated linear segments of the calibration curve, as shown in Fig. 2 of D 4127. Since many factors affect the detection limit, the experimental conditions used should be reported, that is composition of the solution, the history and preconditioning of the electrode, stirring rate, etc. **D 4127,** D19

Lindemann glass—a high X-ray transmittance glass composed of the oxides of lithium, beryllium, and boron. (No elements with an atomic number above 8.) **C 162,** C14

line, *n—of a tire*, a group of similar tires of different sizes but common construction type (bias, belted bias or radial) all with a common tire name. **F 538,** F09

lineage—a low-angle grain boundary resulting from an array of dislocations. **F 1241,** F01

lineage structure—orientation deviations, of the order of minutes or a few degrees at most, from perfect alignment of the crystal axes of parallel arms of a dendrite. **E 7,** E04

linear absorption coefficient of an absorbing medium, *n*—quotient of the internal absorptance of a path element traversed by the radiation, by the length d of this element. Symbol: a; $-\,d\Phi = a\Phi dl$; Unit: m^{-1}; $al = \ln 10D_i$.

NOTE—The linear absorption coefficient is also the part of the linear attenuation coefficient that is due to absorption.

NOTE—In German practice, a linear absorption coefficient is also defined for a homogeneous medium of finite thickness d, as the quotient of the "Absorptions-mass" (logarithm of the reciprocal of the internal transmittance), by the thickness d of the layer. According to whether the natural logarithm or the logarithm to the base 10 is used, one may distinguish the "natürliche Absorptionskoeffizient:" (m_n): quotient of the "natürliche Absorptionsmass" (See **internal transmission density**.) by the thickness d of the layer traversed by the radiation, and the "dekadische Absorptionskoeffizient" (m): quotient of the internal transmission density by the thickness d of the layer.

NOTE—a/ρ, where ρ is the density of the medium, is called "mass absorption coefficient." **E 349,** E21

linear accelerator—an electron generator in which the acceleration of the particles is connected with the propagation of a high-frequency field inside a linear or corrugated waveguide. **E 1316,** E07

linear attenuation coefficient—a measure of the fractional decrease in radiation beam intensity per unit of distance traveled in the material (cm^{-1}). **E 1316,** E07

linear attenuation (extinction) coefficient of an absorbing and diffusing medium, for a collimated beam of radiation—quotient of the relative decrease in spectral concentration of radiant or luminous flux of a collimated beam of radiation during traversal with normal incidence of an infinitesimal layer of the medium, by the thickness of that layer. Symbol: μ; $-\,d\Phi = \mu\Phi dl$; Unit: m^{-1}.

NOTE—This concept only applies strictly to slightly diffusing edia.

NOTE—μ/ρ, where ρ is the density of the medium, is called the "mass attenuation coefficient." **E 349,** E21

linear CCD arrays—a line (linear array) of charge-coupled devices (CCD), semiconductor devices that may be used as circuit elements or as a sequence of photosensitive cells. **F 1156,** F12

linear density—mass per unit length. **B 354,** B01

linear density, *n*—mass per unit length. **D 123,** D13

linear density, *n*—mass per unit length; the quotient obtained by dividing the mass of a fiber or yarn by its length. **D 4439,** D35

linear dispersion—*in spectrochemical analysis*, the derivative $dx/d\lambda$ where x is the distance along the spectrum and λ is the wavelength. Linear dispersion is usually expressed as millimetres per angstrom. **C 859,** C26

linear dispersion—the derivative, $dx/d\lambda$, where x is the distance along the spectrum, in the plane of the exit slit, and λ is the wavelength. **E 131,** E13

linear dispersion, *n*—the derivative $dx/d\lambda$ where x is the distance along the spectrum and λ is the wavelength. **E 135,** E01

linear expansion, coefficient of, $\bar{\alpha}$ —the change in length per unit length per degree change in temperature or

$$\bar{\alpha} = \frac{L_1 - L_0}{L_0 \cdot \Delta T}$$

where:
L_1 = length of specimen at the higher temperature,
L_0 = length at lower temperature, and
ΔT = difference between temperatures.

A 340, A06

linear expansion, L_e **(D)**—the increase in one dimension of a soil mass, expressed as a percentage of that dimension at the shrinkage limit, when the water content is increased from the shrinkage limit to any given water content. **D 653,** D18

linear integrator, *n—in textile unevenness testing*, an integrator that operates continuously and reports unevenness for a certain, and unchanging, time past. (*Syn.* fading memory integrator) **D 123,** D13

linearity—See **analyzer.** **D 1356,** D22

linearity (amplitude)—a measure of the proportionality of the amplitude of the signal input to the receiver, and the amplitude of the signal appearing on the display of the ultrasonic instrument or on an auxiliary display. **E 1316,** E07

linearity (time or distance)—a measure of the proportionality of the signals appearing on the time or distance axis of the display and the input signals to the receiver from a calibrated time generator or from multiple echos from a plate of material of known thickness. **E 1316,** E07

linear lea, *n*—an indirect yarn numbering system in the linen spinning system equal to the number of 300-yd lengths per pound.

D 123, D13

linear low density polyethylene plastics, (LLDPE) *n*—those linear polyethylene plastics, q.v., having a standard density of 0.919 to 0.925 g/cm^3. **D 883,** D20

linear magnification—See **magnification.** **E 7,** E04

linear medium density polyethylene plastics, (LMDPE) *n*—those linear polyethylene plastics, q. v., having a standard density of 0.926 to 0.940 g/cm^3. **D 883,** D20

linear (normal) strain—the change in length per unit of length in a given direction. (ISRM) **D 653,** D18

linear polyethylene plastics, *n*—those containing insignificant amounts of long-chain branching but which may contain significant amounts, by design, of short-chain branching. **D 883,** D20

linear shrinkage, L_s (D)—decrease in one dimension of a soil mass, expressed as a percentage of the original dimension, when the water content is reduced from a given value to the shrinkage limit. **D 653,** D18

linear symbol, *n*—a one-dimensional, linear sequence of rectangular bars and spaces that are arranged in a predetermined pattern following specific rules to represent elements of data that are defined as characters. **F 1294,** F05

linear (tensile or compressive) strain, n—the change per unit length due to force in an original linear dimension. **E 6,** E28

line-at-a-time printer—syn. **line printer.** **F 909,** F05

lineation—the parallel orientation of structural features that are lines rather than planes; some examples are parallel orientation of the long dimensions of minerals; long axes of pebbles; striae on slickensides; and cleavage-bedding plane intersections. (ISRM) **D 653,** D18

line ending, *n*—an electronic code sequence denoting the end of a printed line or of a command. **F 1457,** F05

line feed, *n*—an ASCII command, which when sent to a printer will advance the printing to the next line. (See **LF.**) **F 1457,** F05

line-focus collection— see **collector, line-focus.** **E 772,** E44

line focus (in X-ray tubes)—a long-narrow focal spot. **E 7,** E04

line indices—the Miller indices of the set of planes producing a diffraction line. **E 7,** E04

line (in X-ray diffraction patterns)—an array of small diffraction spots so arranged that they appear to form a continuous line on the film. **E 7,** E04

line, neutral, *n—of a tire,* a line of tires to which a brand name may be added after their manufacture. **F 538,** F09

line of creep (path of percolation)—the path that water follows along the surface of contact between the foundation soil and the base of a dam or other structure. **D 653,** D18

line of reinforcement—circumferential reinforcement comprised of one or more layers. **C 822,** C13

line of seepage (seepage line) (phreatic line)—the upper free water surface of the zone of seepage. **D 653,** D18

line pair, *n*—an analytical line and the internal standard line with which it is compared. **E 135,** E01

line pairs per millimetre—a measure of the spatial resolution of an image conversion device. A line pair test pattern consisting of one or more pairs of equal width, high contrast lines and spaces is utilized to determine the maximum density of lines and spaces that can be success-fully imaged. The value is expressed in line pairs per millimetre. **E 1316,** E07

line pair test pattern—a pattern of one or more pairs of objects with high contrast lines of equal width and equal spacing. The pattern is used with an imaging device to measure spatial resolution. **E 1316,** E07

line post—the intermediate post in a fence. line supporting the top rail or tension wire and barb arm where applicable **F 552,** F14

line post cap—a cap or top with a loop used to position the top rail or tension wire on top of the line posts. **F 552,** F14

line printer—(*1*) a device that prints a line of characters as a unit. *Syn.* **line-at-a-time printer;**
(*2*) contrast with character printer, page printer. **F 909,** F05

liner, *n—in packaging,* (*1*) generally, any linear material that separates a product within a **container** from the basic walls of the **container,** (*2*) in **fiberboard containers,** a creased fiberboard sheet inserted in a container and usually fitting against the side and end panels (*liner* is sometimes erroneously used for *linerboard* or **facing**). (Compare **divider.**)
case liner (bag liner)—a lining, usually paper, or treated materials placed inside a shipping container for the purpose of preventing sifting, or entrance of moisture, dust, or dirt. **D 996,** D10

liner, *n*—a foot protective device worn inside the footwear. **F 1494,** F23

line rail clamp—See **boulevard clamp.** **F 552,** F14

linerboard—See **paperboard.** **D 996,** D10

line reflector, *n*—reflector in the form of a line, the width of which subtends a very small solid angle at the observer's eye, so that the observer cannot readily distinguish its width. **E 284,** E12

liner pitting, *n*—cavities that develop on the coolant side of cast iron cylinder liners as a result of cavitation corrosion. **D 4725,** D15

lines—fine cords or strings, usually on the surface of sheet glass. **C 162,** C14

line scanner—an apparatus that scans along a single line of a scene to provide a one-dimensional thermal profile of the scene. **E 1316,** E07

line skew—the angular displacement of a line in relation to its intended position. **F 149,** F05

lines of force—a conceptual representation of magnetic flux based upon the line pattern produced when iron filings are sprinkled on paper laid over a permanent magnet. **E 1316,** E07

line spacing—the distance between the average base line of one line to the average base line of the next line. **F 149,** F05

line tender, *n*—the individual who controls the diver's search pattern. **F 1549,** F32

lining—coating or layer adhered to or in intimate contact with the interior surface and ends of asbestos-cement pipe and related fittings, with said coating or layer being more chemically resistant than the pipe and related fittings. **D 2946,** C17

lining fabric, *n—for apparel,* a textile used to cover the inner sides of garments. **D 123,** D13

lining leather—any leather used for making shoe linings which includes sheep, lamb, kid, goat, cattle, calf, kip and splits. **D 1517,** D31

linoleum, *n*—a surfacing material composed of a solidified mixture of linseed oil, pine rosin, fossil or other resins or rosins, or an equivalent oxidized oleoresinous binder, ground cork, wood flour, mineral fillers, and pigments, bonded to a fibrous or other suitable backing. **F 141,** F06

linoleum cement—the binder in linoleum consisting of a mixture of linseed oil, pine rosin, fossil or other resins or rosins, or an equivalent oxidized oleoresinous binder. **F 141,** F06

lino-nail—bright, regular-stock-steel, ⅝ by 0.062-in. nail with oval head and medium diamond point. **F 547,** F16

lint, *n—in loose cotton,* fibers mostly of spinnable length. (See also **linters.**) **D 123,** D13

lint content, *n*—that portion of a mass of cotton fiber consisting of fiber, including normal moisture content, but excluding foreign matter. **D 123,** D13

lint cotton, *n*—loose cotton fibers in any form, either raw or processed, free of seeds and not bound together in yarn or fabric. (See also **ginned lint.**) **D 123,** D13

linters, *n*—the short fibrous material adhering to the cotton seed after the spinnable lint has been removed by ginning and which is subsequently recovered from the seed by a process called "delinting." **D 123,** D13

linters—the short fibrous material adhering to cotton seed after the ginning operation. After removal from the seed it is used to a limited extent as a fibrous raw material for special papers. The principal use, however, is for chemical cellulose, that is, as the raw material for the manufacture of cellulose derivatives. **D 1695,** D01

lip—the inner face of the tip of a flange on a lock-strip gasket. **C 717,** C24

lip—overhang of tool nose when magazine rests flush with work surface. **F 592,** F16

lipophilic emulsifier—see **emulsifier lipophilic.** **E 1316,** E07

lip pressure—the pressure exerted by the lip of a lock-strip gasket on material installed in the channel, when the lock-strip is in place. **C 717,** C24

lip seal pressure—the lip pressure required to effect a seal against the passage of water and air. **C 717,** C24

liptinite—See *liptinite* under **macerals**. **D 121,** D05

liquefaction—the process of transforming any soil from a solid state to a liquid state, usually as a result of increased pore pressure and reduced shearing resistance. **D 653,** D18

liquefaction potential—the capability of a soil to liquefy or develop cyclic mobility. **D 653,** D18

liquefaction (spontaneous liquefaction)—the sudden large decrease of the shearing resistance of a cohesionless soil. It is caused by a collapse of the structure by shock or other type of strain and is associated with a sudden but temporary increase of the prefluid pressure. It involves a temporary transformation of the material into a fluid mass. **D 653,** D18

liquefied natural gas (LNG)—natural gas that has been liquefied, after processing, for storage or transportation purposes. (This definition is from ISO NP 14532.) **D 4150,** D03

liquefied petroleum gas (LPG), *n*— a mixture of normally gaseous hydrocarbons, predominantly propane or butane or both, that has been liquefied by compression or cooling, or both, to facilitate storage, transport, and handling. **D 4175,** D02

liquid—(flammability regulations) a substance that has a definite volume but no definite form, except such given by its container. It has a viscosity of 1×10^{-3} to 1×10^{3} St (1×10^{-7} to 1×10^{-1} $m^2 s^{-1}$) at 104°F (40°C) or an equivalent viscosity at agreed upon temperature. (This does not include powders and granular materials.) Liquids are divided into two classes:

Class A, low viscosity—a liquid having a viscosity of 1×10^{-3} to 25.00 St (1×10^{-7} to 25.00×10^{-4} m^2s^{-1}) at 104°F (40°C) or an equivalent viscosity at an agreed upon temperature.

Class B, high viscosity—a liquid having a viscosity of 25.01 to 1×10^{3} St (25.01×10^{-4} to $1 \times 10^{-1} m^2 s^{-1}$) at 104°F (40°C) or an equivalent viscosity at an agreed upon temperature. **D 16,** D01

liquid bituminous material—one having a definite volume but no definite form, except as provided by its container. It has a viscosity of 0.1 to 1×10^5 cSt (mm^2/s) at 40°C. This does not include powders or granular materials. **D 1079,** D08

liquid curing medium (LCM), *n*—a molten phase, generally a molten mixture of sodium nitrate, that is used as a heating medium for the continuous vulcanization of a rubber mix, usually following extrusion. **D 1566,** D11

liquid development—development by means of a toner dispersed in an organic liquid carrier. **F 335,** F05

liquid drop, *n*—a small body of liquid held together primarily by surface tension. **G 40,** G02

liquid impingement, *n*—impingement by liquid particles. **G 40,** G02

liquid impingement damage, n—See **damage**. **G 40,** G02

liquid impingement erosion, *n*—progressive loss of original material from a solid surface due to continued exposure to impacts by liquid drops or jets. **G 40,** G02

liquid infiltration, *v*—densification of a composite by infiltration with a liquid. **C 1145,** C28

liquid-in-glass thermometer, *n*—a temperature-measuring instrument whose indications are based on the temperature coefficient of expansion of a liquid relative to that of its containing glass bulb. **E 344,** E20

liquid ion-exchange electrode membrane—a porous plastic disk, permeable to the ion exchanger, and impermeable to water, which allows the ion exchanger to contact the sample solution and separates the internal filling solution from the sample. **D 4127,** D19

liquidity index (water-plasticity ratio) (relative water content), *B*, R_w, I_L **(D)**—the ratio, expressed as a percentage, of: (*1*) the natural water content of a soil minus its plastic limit, to (*2*) its plasticity index. **D 653,** D18

liquid jet, *n*—a body of liquid projected into motion, usually of approximately cylindrical shape, such as could be produced by discharging the liquid through an orifice. In liquid impingement testing two kinds of liquid jet are used:

(*1*) **continuous jet**—a continuous flow of liquid in the form of a jet.

(*2*) **slug, or jet segment**—a body of liquid projected into motion, in the form approximately of a finite cylinder whose length is usually no more than several times its diameter and which moves in a direction approximately parallel to its length. **G 40,** G02

liquid level—the level of liquid in a borehole or well at a particular time. The liquid level can be reported as an elevation or as a depth below the top of the land surface. If the liquid in ground water is known as water level. **D 653,** D18

liquid, limit, *LL*, L_w, w_L **(D)**—(*a*) the water content corresponding to the arbitrary limit between the liquid and plastic states of consistency of a soil.

(*b*) the water content at which a pat of soil, cut by a groove of standard dimensions, will flow together for a distance of ½ in. (12.7 mm) under the impact of 25 blows in a standard liquid limit apparatus. **D 653,** D18

liquid nitrogen freezers—freezers that operate by a refrigeration system in which cooling is provided by a refrigerant such as liquid nitrogen. **E 1705,** E48

liquid nitrogen storage—storage directly in liquid nitrogen or in the vapor phase above liquid nitrogen. **E 1705,** E48

liquid penetrant examination—a nondestructive test that uses suitable liquids that penetrate discontinuities open to the surface of solid materials and, after appropriate treatment, indicate the presence of discontinuitites. **E 1316,** E07

liquid phase—contaminant residing as a liquid in vadose zone pore space, often referred to as "free product." **D 653,** D18

liquid phase sintering—sintering of a compact, or loose powder aggregate under conditions where a liquid phase is present during part of the sintering cycle. **B 243,** B09

liquid suspension—the system resulting from the intimate, physical mixing of a liquid and particulate solids or dispersable materials, to form a time-stable, uniform, fluid mixture. **C 242,** C21

liquid-tight—*for the purpose of this practice*, referring to the absence of detectable liquid penetration inside the chemical protective suit when sprayed with the liquid at specified pressure, direction and duration. **F 1494,** F23

liquidus, *n*—the lowest temperature at which an alloy under equilibrium conditions begins to freeze on cooling or is completely melted on heating. **B 899,** B02

liquidus—the locus of points in a phase diagram, representing the temperature, under equilibrium conditions, at which each composition in the system begins to freeze during cooling, or completes melting during heating. **E 7,** E04

liquidus—locus of points in a phase diagram, representing the temperature, under equilibrium conditions, at which each composition in the system begins to freeze during cooling, or completes melting during heating, (E 7, E-4), . **E 1142,** E37

liquidus temperature—the maximum temperature at which equilibrium exists between the molten glass and its primary crystalline phase. **C 162,** C14

liquid-volume measurement—*in grouting*, measurement of grout on the basis of the total volume of solid and liquid constituents. **D 653,** D18

liquor finish—very thin wire coating produced by wire immersion in metallic salts, usually copper, offering very limited corrosion resistance. **F 547,** F16

liquor-finished wire—wire, usually of ferrous material, drawn through a wet solution of metallic salts, that is, tin or copper, to facilitate drawing and to enhance the appearance and surface characteristics of the finished wire, and not to provide the properties of a coated wire. **B 354,** B01

lisle, *n*—for a definition of cotton lisle see Rule 3 of the "Amended Trade Practice Rules for the Hosiery Industry," as promulgated by the Federal Trade Commission, Aug. 30, 1960 and amended June 10, 1964. **D 123,** D13

list, *v*—*for protective clothing*, to publish a register of equipment or materials which has been verified by a certification organization as being acceptable and meeting the requirements of standard specifications. **F 1494,** F23

lite, *n*—one piece of glazing (also spelled light), synonym: pane. **C 717,** C24

lite—See **windows and doors.** **E 631,** E06

lite, *n*—one piece of glazing (preferred term); (also spelled light) (*synonym*—pane). **E 1605,** E06

lite, light—a panel or sheet of glass. **C 162,** C14

litho crayon—a greasy black crayon for drawing an image directly onto an offset master. **F 425,** F05

lithography—the process of printing from a plane surface on which the image to be printed is ink receptive and water repellent and the non-image area is ink repellent and water receptive. **F 425,** F05

lithology—the description of rocks, especially sedimentary clastics and especially in hand specimens and in outcrops, on the basis of such characteristics as color, structures, mineralogy, and particle size. **D 653,** D18

live field—in video-based automatic image analysis, the visible, real-time gray level image from the television camera. **E 7,** E04

live load—the portion of the load transmitted to pipe from wheel or tread impacts. **C 896,** C04

live load—See **load.** **D 996,** D10

live load—portion of load transmitted to the pipe from wheel or traveling loads or other surcharged load. **F 412,** F17

liver—in dry process enameling, a defect characterized by a wave-like form of abnormally thick porcelain enamel. **C 286,** B08

load—See **pull**. **C 162,** C14

load, *vt*—to apply a force. **D 123,** D13

load, *n*—(*1*) the force in weight units applied to a body; (*2*) the weight of the contents of a **container** or transportation device; (*3*) a qualitative term denoting the contents of a container.

constant load, n—a load that is invariable or unchanging.

dead load, n— a constant load that, in structures (as a bridge, building, or machines) is due to the weight of the members, the supporting structure, and permanent attachments or accessories (*Webster*).

dynamic load, n—an imposed force in motion; that is, one that may vary in magnitude, sense, and direction.

floating load— (*1*) a shipment, usually a *unitized load* (or loads), or a large individual article, so prepared that it may move in the carrying vehicle, the movement being restricted or retarded by friction between the load and the vehicle; (*2*) a method of packing in which the contents of the **container** are supported within the container by **cushioning** devices or **materials**; (*3*) *floating controlled load*—a **floating load** within a vehicle in which the movement is retarded, snubbed, or restrained by suitable devices.

live load, n— a moving load on a structure.

palletized load— a load made up of articles, loose or in **containers** placed on **pallets** or skids.

palletized unit load—a unitized **load** fixed to a pallet.

static load, n—an imposed stationary force, constant in magnitude, sense and direction.

unitized load—*in packaging,* a type of unit load consisting of articles or containers secured together so as to be handled as an entity.

unit load—*in distribution,* an item or assembly of items assembled or restrained for handling and transportation as a single entity. **D 996,** D10

load—(*1*) the amount of nonprotein material in vegetable-tanned leather.

(*2*) the amount of tannin in vegetable-tanned leather. **D 1517,** D31

load—See **static load.** **E 631,** E06

load—the amount of food to be cooked during a cooking energy efficiency (see **cooking energy efficiency**) test. **F 1827,** F26

load-bearing partition, *n*—a partition designed to support a portion of the building structure. **C 11,** C11

load-bearing tile, *n*—tile for use in masonry constructions designed to carry superimposed loads. See Specification C 34. **C 43,** C15

load cell—a device which indicates the applied force by means of an electrical voltage. Usually the electrical voltage increases linearly with applied load. **E 1823,** E08

load cell—device for determining the magnitude of a force (load) applied to it; usually either a semiconductive material or a resistive wire strain gauge. **F 1156,** F12

load cycle—See **cycle.** **E 1823,** E08

loaded radius, [*L*], *n*—*of a tire*, the wheel plane distance from the center of tire contact in the footprint to the wheel center, specified as a static or dynamic (rolling) measurement. **F 538,** F09

load [F]—used in Practices E 1049 to denote force, stress, strain, torque, acceleration, deflection, or other parameters of interest. **E 1823,** E08

load index, *n*—a numerical code associated with the maximum load a tire can carry at the speed indicated by its speed symbol under specified conditions. **F 538,** F09

loading, *n*—the act of placing a load on or in; to load a car, a vessel, or a test specimen. (See also **unitization.**)

anti-skid plate— a device, generally metal, about 4 by 6 in. (100 by 150 mm), with sharp projections on each face, placed under, against, or between containers and car floor, to retard shifting of load in transit.

blocking, n— material used to prevent or control movement of the unit or *load* or to facilitate handling (A 700, A-1).

bracing, n— material or devices used to hold articles or sections of loads in position and prevent shifting within a transportation vehicle or within a **container.** (See **molded shape, strapping.**)

diagonal bracing— (*1*) a member reaching at an angle from a gate or other structure to the wall or floor of a freight car or truck to strengthen and reinforce the gate or structure; (*2*) a member, single or multiple, of a **crate** or **box** attached diagonally or at an angle to add reinforcement to the container.

dunnage, n— (*1*) in a carrying vehicle, the temporary **blocking,** flooring or lining, racks, standards, strips, stakes, or similar **bracing,** or supports not constituting a part of the carrying vehicle, used to protect and make freight secure in, or on a carrying vehicle (Classification, Rule 30);in a **container,** materials not constituting a part of the container, frequently by-product or scrap, used for filling space, for **blocking** or **bracing,** or otherwise to protect and secure the contents.

dynamic compression loading—in packaging, the application of a force in motion that usually occurs in 10 s or less.

edge protector— a right-angle piece placed over the edge of **boxes, crates, bundles** and gates, or **bracing** to distribute the pressure from metal bands or ties in order to prevent cutting.

spreader, n— (*1*) a horizontal wooden member placed between two gates in a car to hold the gates in position against the load; (*2*) a rigid device placed between sling legs, lifting cables, or ropes to prevent them from damaging cargo being loaded or unloaded.

static compression loading—in packaging, the application of a force for an extended period, often days, weeks, or months. **D 996,** D10

loading—the addition of glucose, magnesium sulfate, or other materials necessary to give leather the physical properties needed for working in modern shoe machinery. (Also known as filling or stuffing.) **D 1517,** D31

loading rate, *n*—the ratio of test material to aqueous medium used in the preparation of WAF, WSF, or mechanical dispersion and in the interpretation of the results of a toxicity study with a poorly water-soluble lubricant or lubricant component. **D 6384,** D02

loading (unloading) rate [F T^{-1}]—the time rate of change in the monotonic increasing (decreasing) portion of the load-time function. **E 1823,** E08

load limit—(*1*) the maximum force, in weight units, a body can withstand without damage; (*2*) the greatest allowable weight that may be placed in a **container** or vehicle. For containers, load limit is often expressed in terms of the gross weight of the container and its contents. **D 996,** D10

load-line displacement resulting from creep, V_c [L]— displacement at the loading pins from cracking associated with creep strain accumulation. **E 1823,** E08

load, *P* [F]—the force applied to a test specimen or to a component. **E 1823,** E08

load range, *n*—a letter designation (A, B, C, D) or, for P-metric tires, standard load (SL) or extra load (XL), used to identify a given size tire with its load and inflation limits when used in a specific type of service. **F 538,** F09

load range, ΔP [FL^{-2}]—*in fatigue loading*, the algebraic difference between successive valley and peak loads (positive range or increasing load range) or between successive peak and valley loads (negative range or decreasing load range). In constant amplitude loading, the range is given as follows:

$$\Delta P = P_{max} - P_{min}$$

E 1823, E08

load rating, [*M*], *n*—the maximum load a tire is rated to carry for a given usage at a specified cold inflation pressure. **F 538,** F09

load ratio (also stress ratio), *R, A*—*in fatigue*, the algebraic ratio of the two loading parameters of a cycle. The most widely used ratios are as follows

$$R = \frac{\text{minimum load}}{\text{maximum load}} = \frac{P_{min}}{P_{max}} = \frac{S_{min}}{S_{max}}, \text{ and}$$

$$A = \frac{\text{loading amplitude}}{\text{mean load}} = \frac{P_a}{P_m} = \frac{S_a}{S_m}$$

E 1823, E08

load ratio, R, A, *n*—in fatigue loading, the algebraic ratio of the two loading parameters of a cycle. **F 1582,** F04

load-recovery cycle, *n*—*in elastic fabric testing*, a continuous curve or plot of load versus elongation (with movement stopped momentarily at point of reversal) describing the elongation and recovery of an elastic fabric; also known as the loading and unloading cycle. **D 123,** D13

load (strain) amplitude, P_a (S_a or ϵ_a) [F or FL^{-2}]—*in fatigue loading*, one half of the range of a cycle (also known as *alternating load*). **E 1823,** E08

load symbol, *n*—a code associated with the maximum load a tire can carry at the speed indicated by its speed symbol under specified conditions. **F 538,** F09

load type, *n*—a qualitative description of the contents of a **container** with respect to **density,** fragility, and degree of **blocking, bracing,** and **cushioning** necessary. Load types are further divided into the following categories:

Type I load (easy)—a qualitative term denoting contents of low or moderate **density** conforming to the shape of the **container** and lending support to all faces of the container.

Type II load (average)—a qualitative term denoting contents of low or moderate density providing, when packed directly into a **shipping container,** nonshifting support at several points on the face of the container.

Type III load (difficult)—a qualitative term denoting contents characterized by irregular shape not lending support to the container, or by great density or extreme fragility. **D 996,** D10

load (water load)—an amount of water used to moderate the microwave radiation absorbed by a susceptor during simulated microwave heating tests. **F 1479,** F02

load-wear index, *n*—(Syn. *load-carrying property of a lubricant*) an index of the ability of a lubricant to prevent wear at applied loads. **D 4175,** D02

loam—a mixture of sand, silt, or clay, or a combination of any of these, with organic matter (see **humus**). **D 653,** D18

local—indicates measurements or observations of a small part of a larger region of interest. **E 1620,** E29

local action corrosion—corrosion caused by local corrosion cells on a metal surface. **G 15,** G01

local corrosion cell—an electrochemical cell created on a metal surface because of a difference in potential between adjacent areas on that surface. **G 15,** G01

localized corrosion—corrosion at discrete sites, for example, pitting, crevice corrosion, and stress corrosion cracking. **G 15,** G01

localized light scatterer (LLS)—an isolated feature, such as a particle or a pit, on or in a wafer surface, resulting in increased light scattering intensity relative to that of the surrounding wafer surface; sometimes called **light point defect**. **F 1241,** F01

local magnetization—magnetization of a prescribed volume or surface of a part. **E 1316,** E07

local shear failure—see **shear failure**. **D 653,** D18

local testing, *n*—testing conducted at one laboratory or test site for the purpose of comparing a number of candidate tires for selected characteristic properties. **F 538,** F09

local thickness—mean of the thickness measurements of which a specified number is made within a reference area. **F 1789,** F16

locating boss—an integral part of the case used to align the lock in mounting (not used in all locks). **F 471,** F12

location accuracy, *n*—a value determined by comparison of the actual position of an AE source (or simulated AE source) to the computed location. **E 1316,** E07

location, cluster, *n*—a location technique based upon a specified amount of AE activity located within a specified length or area, for example: 5 events within 12 linear inches or 12 square inches. **E 1316,** E07

location, computed, *n*—a source location method based on algorithmic analysis of the difference in arrival times among sensors. **E 1316,** E07

location, continuous AE signal, *n*— a method of location based on continuous AE signals, as opposed to hit or difference in arrival time location methods. **E 1316,** E07

location marker—a number or letter made of lead (Pb) or other highly radiation attenuative material that is placed on an object to provide traceability between a specific area on the image and the part. **E 1316,** E07

location, source, *n*—any of several methods of evaluating AE data to determine the position on the structure from which the AE originated. Several approaches to source location are used, including zone location, computed location, and continuous location. **E 1316,** E07

location, zone, *n*—any of several techniques for determining the general region of an acoustic emission source (for example, total AE counts, energy, hits, and so forth). **E 1316,** E07

lock case—the housing into which all components parts are installed or attached. **F 471,** F12

lockdown—adhesion of original or electrostatic copy to bearing surface as due to static charges. **F 335,** F05

locking ability—a characteristic intentionally manufactured or added to a fastener to resist loosening. **F 1789,** F16

lock signal (NMR)—the NMR signal used to control the field-frequency ratio of the spectrometer. It may or may not be the same as the reference signal. **E 131,** E13

lockstitch, *n*—*in home sewing*, a type of stitch formed with two threads: one needle thread and one bobbin thread interlaced with each other. **D 5646,** D13

lock stitch—a double stitch used at beginning and end of manufacturer's seam on corrugated boxes. **F 592,** F16

lockstitch—a double thread stitch that locks the threads together within the material. Breaking one stitch does not permit the seam to be ravelled out. **F 869,** F08

lock-strip cavity—the groove in the face of a lock-strip gasket designed to receive and retain the lock-strip. **C 717,** C24

lock-strip gasket, *n*—*in building construction*, a gasket in which sealing pressure is attained by inserting a keyed locking strip into a mating keyed groove in one face of the gasket. **C 717,** C24

lock strip or locking strip—the strip that is designed to be inserted in the lock-strip cavity to force the lips against material placed in the channel. **C 717,** C24

lockup, *n*—*of a wheel*, the condition of a wheel in which its rotational velocity about the wheel spin axis is zero and it is prevented from rotating in the presence of applied wheel torque. **F 538,** F09

loess—a uniform aeolian deposit of silty material having an open structure and relatively high cohesion due to cementation of clay or calcareous material at grain contacts. **D 653,** D18

loftiness—the measure of the loose specific volume of an asbestos fiber. This is inversely related to the dry bulk density. **D 2946,** C17

log—a section of the trunk of a tree usually referring to a length suitable for conversion to commercial products. **D 9,** D07

logarithmic (acoustic emission) amplitude distribution g(V)—see **distribution, logarithmic (acoustic emission) amplitude.** **E 1316,** E07

logarithmic decrement—the natural logarithm of the ratio of any two successive amplitudes of like sign, in the decay of a single-frequency oscillation. **D 653,** D18

logarithmic decrement—*in dynamic mechanical measurement*, natural logarithm of the ratio of any two (or more) successive amplitudes of like sign, in the decay of a single oscillation: $\Delta = \frac{1}{k} \ln \frac{A_n}{A_{(n+k)}}$.

where:

A_n and $A_{(n+k)}$ are amplitudes (in radians of rotation) of two oscillations and K is the number of oscillations separating the two amplitude measurements. **E 1142,** E37

logarithmic decrement, Δ (*in dynamic mechanical measurement*) —the natural logarithm of the ratio of any two (or more) successive amplitudes of like sign, in the decay of single frequency oscillation:

$$\Delta = \frac{1}{k} \ln \frac{A_n}{A_{n+k}}$$

where A_n and A_{n+k} are amplitudes (in radians of rotation) of two oscillations, and k is the number of oscillations separating the two amplitude measurements. **D 4092,** D20

logbook, *n*—notebook that accompanies each XRF analyzer instrument, for recording such information as daily performance, maintenance problems, and average reading time. **E 1605,** E06

log normal distribution—drop size distribution described by the following equation:

$$f_n(D) = (1/\sqrt{\pi}) \int_{-\infty}^{K \times Ln(D/D_{gm})} exp(-z^2)dz$$

where:

$f_n(D)$ = the number fraction of drops of diameter less than D,
D_{gm} = the geometric mean diameter,
K = $1/[(\sqrt{2})\sigma_{gm}]$, and
σ_{gm} = the standard deviation of the geometric mean diameter.

E 1620, E29

log-normal distribution—the distribution of N when log (N) is normally distributed. (Accordingly, it is convenient to analyze log (N) using methods based on the normal distribution.) **E 1823,** E08

logo,—see **logotype**. **F 1457,** F05

logotype, *n*—a single image which generally contains a symbol, trademark, or identifying name of a business, association, or product. (See **logo**.) **F 1457,** F05

log run—in relation to lumber manufacture in softwoods, the total yield of a log or group of logs in lumber without any grading; in hardwood, certain lower grades are excluded. **D 9,** D07

log scale—the measure of the potential lumber content of a log or logs based on particular cuttings and sizes, with the measure varying according to the exact nature of the log scale or log rule involved. **D 9,** D07

long—a comparative term signifying a slow-setting glass. **C 162,** C14

long barb, *n*—barbs having an average length of 1.2 ± 5 %. **F 1379,** F14

longitudinal—generally, the direction parallel to the grain of wood. Sometimes the long axis of a wood specimen. **D 9,** D07

longitudinal corrugation—a condition similar to dish except that the sense of curvature changes sign at least once across the width of the strip. **B 846,** B05

longitudinal cracking, *n*—cracks in the pavement predominantly parallel to the direction of traffic. **E 867,** E17

longitudinal cracking, *n*—cracks in the pavement predominantly parallel to the direction of traffic. **E 1778,** E17

longitudinal direction—that direction which is parallel to the direction of maximum elongation in a worked material. (See also **cross direction**). **E 7,** E04

longitudinal element, *n*—a component whose long axis is parallel, or nearly so, to the long axis of the spine. **F 1582,** F04

longitudinal force, [*F*], *n*—*of a tire*, the component of the tire force vector in the *X′* direction. **F 538,** F09

longitudinal force, tire (F_x)—the component of a tire force vector in the *X′* direction. **E 867,** E17

longitudinal groove, *n*—an endless groove that has its major (long) dimension substantially parallel to the tire circumferential centerline; the walls of the groove may not be perfectly parallel planes, but may have short alternating sections of the wall at angles to the tire circumferential centerline. **F 538,** F09

longitudinal magnetization—a magnetic field wherein the lines of force traverse the part in a direction essentially parallel with its longitudinal axis. **E 1316,** E07

longitudinal profile, *n*—the perpendicular deviations of the pavement surface from an established reference parallel to the lane direction, usually measured in the wheel tracks. **E 867,** E17

longitudinal profile—the perpendicular deviations of the pavement surface from an established reference parallel to the lane direction, usually measured in the wheel tracks. **E 867,** E17

longitudinal profile measurement, *n*—a series of elevation values taken at a constant interval along a wheel track. **E 867,** E17

longitudinal reinforcement—reinforcement, in a concrete pipe or box section, running parallel to the intended flow. **C 822,** C13

longitudinal rod wave—see **compression wave**. **D 653,** D18

longitudinal slip velocity, [*L/T*], *n*—the effective rolling radius multiplied by the difference between the spin velocity (in rad/unit time) of a driven or braked tire and that of a free rolling tire when each is traveling in a straight line. **F 538,** F09

longitudinal sonic pulse—a sonic pulse in which the displacements are in the direction of propagation of the pulse. **C 709,** D02

longitudinal thread—multiple flutes parallel or nearly parallel to nail axis, rolled onto nail shank, having a lead angle of or closely approaching 90 deg. **F 547,** F16

longitudinal wave—those waves in which the particle motion of the material is essentially in the same direction as the wave propagation. (E 494) **E 1316,** E07

longitudinal wave, v_l **(LT^{-1})**—wave in which direction of displacement at each point of medium is normal to wave front, with propagation velocity, calculated as follows:

$$y_l = \sqrt{(E/\rho)[(1-\nu)/(1+\nu)(1-2\nu)]} = \sqrt{(\lambda + 2\mu)/\rho}$$

where:

E	=	Young's modulus,
ρ	=	mass density,
λ and μ	=	Lamé's constants, and
ν	=	Poisson's ratio.

D 653, D18

long knot, *n*—*in raw silk*, knots which have loose ends from 3 to 25 mm in length. **D 123,** D13

long-line current—electric current through the earth from an anodic to a cathodic area of a continuous metallic structure. (Usually used only where the areas are separated by considerable distance and where the current results from concentration-cell action.) **G 15,** G01

long magazine—magazine providing space for more than regular capacity of a particular tool style. **F 592,** F16

long point—point with 20-deg or smaller included angle for 0.040 to 0.065-in. wire diameter; 25-deg or smaller included angle for 0.072 to 0.225-in. wire diameter; 30-deg or smaller included angle for 0.250 to 0.325-in. wire diameter. **F 547,** F16

long slug, *n*—*in raw silk*, a slug which exceeds 10 mm (½ in.) in length or which is very much larger in diameter than the yarn. **D 123,** D13

long-term hydrostatic strength (LTHS)—the hoop stress that when applied continuously will cause failure of the pipe at 100 000 h (11.43 years). **F 412,** F17

long wave (quer wave), *W* **(LT^{-1})**—dispersive surface wave with one horizontal component, generally normal to the direction of propagation, which decreases in propagation velocity with increase in frequency. **D 653,** D18

loom fly, *n*—waste fibers created during weaving, that are woven into a fabric. **D 123,** D13

loopback test, *n*—a test in which a known signal is sent from a source to a remote device or interface, and the received signal is then returned and checked against the transmitted signal to verify that no change occurred. **F 1457,** F05

loop breaking force, *n*—*in tensile testing*, the breaking force of a specimen consisting of two lengths of strand from the same supply looped together in a specified configuration and mounted between the clamps of a tensile testing machine. (Compare **loop breaking strength.** See also **breaking force.**) **D 123,** D13

loop-breaking strength, *n*—strength expressed in terms of loop breaking force. (See **loop breaking force, strength.**) **D 123,** D13

loop cap—See **line post cap.** **F 552,** F14

looped pile floor covering, *n*—a pile floor covering in which the pile is composed of uncut loops only. **D 123,** D13

loop pile, *n*—*for pile yarn floor covering*, pile in which, for each loop, a tuft leg of one tuft element is connected to a tuft leg of another tuft element at another binding site so as to form a loop which projects above the backing fabric between the binding sites of the connected tuft elements. **D 123,** D13

loop spacer wire, *n*—wire that is attached loop-to-loop (or on consecutive attachment points) on the barbed tape to preset the loop spacing. **F 1379,** F14

loop tension, *n*—*in elastic materials testing*, the total tension at any specified extension that is exerted on a specimen in a loop formation. **D 123,** D13

loopy selvage, *n*—an improperly woven selvage of uneven width or a selvage containing irregular filling loops extending beyond the outside selvages. **D 123,** D13

loose course, *n—in knitted fabrics*, a row of loops in the widthwise direction that is larger, looser, or longer than the stitches in the main body of the fabric. **D 123,** D13

loose fill insulation, *n*—insulation in granular, nodular, fibrous, powdery, or similar form designed to be installed by pouring, blowing, or hand placement. **C 168,** C16

loose-laid membrane—a ballasted roofing membrane that is attached to the substrate only at the edges and penetrations through the roof. **D 1079,** D08

loose paint—see **deteriorating paint**. **E 1605,** E06

loose RO—see **nanofiltration**. **D 6161,** D19

loose side—*in knife-cut veneer,* the side of the sheet that was in contact with the knife as the sheet was being cut, and that contains cutting checks. (See **tight side**.) **D 1038,** D07

losing stream, *n*—a stream or reach of a stream in which water flows from the stream bed into the ground. **D 653,** D18

loss angle, δ— the angle whose tangent is the dissipation factor or arctan κ''/κ'. It is also the difference between 90 deg and the phase angle. **D 2864,** D27

loss angle, δ— see **phase angle.** **D 4092,** D20

loss angle, δ— see **phase angle**. **E 1142,** E37

loss angle, dielectric— same as **loss angle.** **D 2864,** D27

loss angle, magnetic, γ—the mean angle by which the fundamental component of core loss current leads the fundamental component of exciting current, *I*, in an inductor having a ferromagnetic core.

NOTE—The loss angle, γ, is the complement of the hysteretic angle, β.

NOTE—Because of hysteresis, the instantaneous value of the loss angle will vary during the cycle of *SCM* excitation; however, γ is taken to be the mean effective value of this angle. **A 340,** A06

loss angle (phase defect angle), δ, *n*—the angle whose tangent is the dissipation factor or arctan κ''/κ'. It is also the difference between 90° and the phase angle. **D 1711,** D09

loss factor—obsolete term; see **loss index.** **D 1711,** D09

loss factor—see **tan δ.** **D 4092,** D20

loss factor—see **tangent delta**. **E 1142,** E37

loss factor, ϵ''—the magnitude of the imaginary part of the complex permittivity. (See **permittivity, complex**.) **E 1142,** E37

loss index—the same as **loss index, dielectric.** **D 2864,** D27

loss index, dielectric, $\kappa''(\epsilon_r'')$—the product of the "relative permittivity" and the dissipation factor and is a measure of the ac dielectric loss. It is also the magnitude of the imaginary part of the "relative complex permittivity." **D 2864,** D27

loss index, κ'' **(**ϵ_r''**),** *n*—the magnitude of the imaginary part of the relative complex permittivity. It is the product of the relative permittivity and dissipation factor.

D 1711, D09

loss modulus—*M*″**(Pa)**—[loss compliance O″ (Pa^{-1})]: the imaginary part of the complex modulus (compliance). **D 4092,** D20

loss modulus—quantitative measure of energy dissipation, defined as the ratio of stress 90° out of phase with oscillating strain to the magnitude of strain, (D 4092, D-20). **E 1142,** E37

loss of back reflection—an absence or significant reduction in the amplitude of the indication from the back surface of the part under examination. **E 1316,** E07

loss of circulation—the loss of drilling fluid into strata to the extent that circulation does not return to the surface. **D 653,** D18

loss on ignition—the percent loss in weight of a material on being calcined at a temperature sufficiently high, and for a time long enough, to achieve constant weight without melting, expressed as a percent of the initial weight of the dry material (without free moisture). **C 242,** C21

loss rate—rate at which oil is lost past a boom (m^3/h). **F 818,** F20

loss tangent—same as **dissipation factor.** **D 2864,** D27

loss tangent—see **tan delta.** **D 4092,** D20

loss tangent—in dielectric measurements, see **tangent delta**. **E 1142,** E37

lot, *n*—a finite quantity of a given product manufactured under production conditions that are considered uniform. **A 644,** A04

lot, *n*—a finite quantity of a given product, produced under conditions that are considered uniform for sampling purposes. **A 902,** A05

lot, *n*—a definite quantity of product manufactured under conditions that are considered uniform. **A 941,** A01

lot, *n*—a quantity of metal made under conditions that, for sampling purposes, are considered uniform. **B 899,** B02

lot, *n*—a quantity of a single material that is considered as a unit, such as that manufactured during a single production run, offered at one time for sale, or offered at one time for inspection, or the contents of one or more transport containers drawn from one or more bins of material from a single production run, sequentially packaged from one or more bins, or a group of samples secured from one of the above. **C 219,** C01

lot, *n—in asbestos-cement conduit*, each 1000 lengths of conduit or less, of a given class, type and size manufactured on each machine during a 24-h period. **C 460,** C17

lot, *n—in asbestos-cement nonpressure sewer pipe*, for pipe sizes 150 mm (6 in.) and smaller, those lengths of pipe of that size, class and type manufactured during the same work shift. For pipe sizes 200 through 525 mm (8 through 21 in.), each 300 lengths of pipe or less, of identical size, class and type manufactured on each machine during a 24-h period. For pipe sizes larger than 525 mm (21 in.) each 300 lengths of pipe or less, of identical size, class and type manufactured on each machine during a period of consecutive working days not exceeding seven days. **C 460,** C17

lot, *n—in asbestos-cement pressure pipe, storm drain and transmission pipe*, for pipe sizes 525 mm (21 in.) in diameter and smaller, each 300 lengths of pipe or less, of identical class and size manufactured on each machine during a 24-h period. For pipe larger than 525 mm (21 in.), each 300 lengths of pipe or less, of identical class and size manufactured on each machine during a period of consecutive working days not exceeding seven days. **C 460,** C17

lot, *n—in asbestos*, not less than 2 and not more than 2000 bags of asbestos fiber of the same type and grade offered at any one time for delivery or testing. **C 460,** C17

lot—an assemblage of concrete pipe, all being of like size, material, and strength designation, manufactured by the same process. The lot size may differ from the quantity designated in the contract or order. **C 822,** C13

lot—a quantity of material that is thought to be uniform in one or more stated properties such as isotopic, chemical, or physical characteristics. **C 859,** C26

lot—specific group of clay products having characteristics of sufficient similarity that individual specimens selected from that group may be considered representative of the whole group. **C 896,** C04

lot, *n—in conduit*, each 1000 lengths of conduit or less, of a given class, type and size manufacture on each machine during a 24-h period. **C 1154,** C17

lot, *n—in nonpressure sewer pipe*, for pipe sizes 150 mm (6 in.) and smaller, those lengths of pipe of that size, class and type manufactured during the same work shift. For pipe sizes 200 mm (8 in.)

through 525 mm (21 in.), each 300 lengths of pipe or less, of identical size, class, and type manufactured on each machine during a 24-h period. For pipe sizes larger than 525 mm (21 in.) each 300 lengths of pipe or less, of identical size, class, and type manufactured on each machine during a period of consecutive working days not exceeding seven days. **C 1154,** C17

lot, *n—in underdrain pipe*, each 400 m (1300 ft) of pipe or less of a given type and size manufactured on each machine during a 24-h period. **C 1154,** C17

lot—any number of concrete masonry units or related units designated by the producer of any configuration or dimension manufactured by the producer using the same materials, concrete mix design, manufacturing process, and curing method. **C 1209,** C15

lot, *n*—a discrete quantity of coal for which the overall quality to a particular precision needs to be determined. **D 121,** D05

lot, *n—in acceptance sampling*, that part of a consignment or shipment consisting of material from one production lot. **D 123,** D13

lot, *n—in acceptance sampling of cotton*, the main stock, supply, or source of fibers to be sampled. **D 123,** D13

lot, *n—in bonded, fused, or laminated fabric*, a single run on the bonding or laminating machine in which the processing is carried out without stopping or changing processing conditions, and consisting of either a single dye lot or a single gray goods lot. **D 123,** D13

lot, *n—in wool top*, the entire quantity, not exceeding 20 000 lb (9100 kg) of a single combing, that comprises a single unit for which a test for neps, vegetable matter, or colored fiber, or all three combined is desired. **D 123,** D13

lot—in roofing, *(1)* production lot—all material produced in one eight-hour shift of the same type (and color when applicable);

(2) delivery lot—all material of the same type delivered at one time by one truck or railroad car. **D 1079,** D08

lot—("inspection lot")—a collection of units of product from which a sample is to be drawn and inspected to determine conformance with the acceptability criteria, and is to be accepted or rejected as a whole. It may differ from a collection of units designated as a lot for other purposes for example, production, shipment, etc. **D 1517,** D31

lot, *n*—a mass or collection of articles of similar composition and characteristics. **D 1566,** D11

lot, *n*—an entity of electrical insulating material or product which, insofar as is practicable, consists of a single type, grade, class, size, or composition that was manufactured under essentially the same conditions and is available to the user for sampling at one time. **D 1711,** D09

lot, *n—of paper or paperboard*, a quantity of paper or paperboard of a single type, grade, grammage, thickness, and composition about which it is desired to make a judgment (usually as to conformance to specification) by examining or testing a small fraction called the sample. **D 1968,** D06

lot, *n—in asbestos,* not fewer than 2 and not more than 2000 bags of asbestos fiber of the same type and grade offered at any one time for delivery or testing. **D 2946,** C17

lot, *n—in asbestos-cement conduit*, each 1000 lengths of conduit or less, of a given class, type and size manufactured on each machine during a 24-h period. **D 2946,** C17

lot, *n—in asbestos-cement nonpressure sewer pipe*, for pipe sizes 150 mm (6 in.) and smaller, those lengths of pipe of that size, class and type manufactured during the same work shift. For pipe sizes 200 through 525 mm (8 through 21 in.), each 300 lengths of pipe or less, of identical size, class and type manufactured on each machine during a 24-h period. For pipe sizes larger than 525 mm (21 in.) each 300 lengths of pipe or less, of identical size, class and type manufactured on each machine during a period of consecutive working days not exceeding seven days. **D 2946,** C17

lot, *n—in asbestos-cement pressure pipe, storm drain, and transmission pipe*, for pipe sizes 525 mm (21 in.) in diameter and smaller, each 300 lengths of pipe or less, of identical class and size manufactured on each machine during a 24-h period. For pipe larger than 535 mm (21 in.), each 300 lengths of pipe or less, of identical class and size manufactured on each machine during a period of consecutive working days not exceeding seven days. **D 2946,** C17

lot, *n*—a quantity of carbon black that is essentially uniform in composition and characteristics. **D 3053,** D24

lot, *n*—a unit of production, or a group of other units or packages, taken for sampling or statistical examination, having one or more common properties and being readily separable from other similar units. **D 4439,** D35

lot, *n*—a large designated quantity of a material which can be represented by a properly selected gross sample. **D 5681,** D34

lot, *n—in sampling*, a collection of material regarded as a unit. **E 135,** E01

lot—a definite quantity of a product or material accumulated under conditions that are considered uniform for sampling purposes. **E 456,** E11

lot—a designated quantity of refuse-derived fuel that can be represented by a properly selected gross sample. **E 856,** D34

lot—a batch or fraction thereof, in which each unit is identical in chemical composition, physical properties, and dimensions. **E 1749,** E06

lot, *n*—a collection of units of product manufactured under conditions of production that are considered uniform. **F 412,** F17

lot—a defined quantity of fasteners of the same size, configuration, and material that has been processed at one time from the same heat of material and submitted for inspection at one time. **F 1789,** F16

lot number, *n*—a unique alphanumeric designation for a lot that is traceable to manufacturing records. **B 899,** B02

lot number, *n*—the number used by a producer to identify an entity of electrical insulating material or product. **D 1711,** D09

lot sample, *n*—one or more shipping units taken at random to represent an acceptance sampling lot and used as a source of laboratory samples. (See **bulk sample.**) **D 123,** D13

lot sample, *n—in cotton*, a relatively large sample taken in the field to represent a consignment, shipment, or lot, for use in the preparation of the laboratory samples. **D 123,** D13

lot sample, *n*—a quantity of carbon black selected to represent a lot for testing purposes and taken in accordance with Practice D 1799 or D 1900. **D 3053,** D24

lot sample, *n*—one or more shipping units taken at random to represent an acceptance sampling lot and used as a source of laboratory samples. **D 4439,** D35

lot sampling inspection—a random sample drawn from a lot and performance of specified inspections and tests to determine the acceptability of the lot. **F 1789,** F16

lot sampling unit, *n*—a portion of material taken to represent a lot and used as a source of laboratory sampling units or test specimens, or both. (See **primary sampling unit.**) **D 123,** D13

lot size—the number of units of product in a lot. **D 1517,** D31

lot (solid sample)—a discrete quantity of material. It may contain a single batch or several batches or be the product of continuous process broken into units on the basis of time or shipment. It is very desirable that individual batches in a lot be specifically identified so that they may become individual or stratified units for inspection. **E 1547,** E15

lot tolerance fraction defective, *n*—the process average at which quality is considered barely tolerable; the process average at which the risk of acceptance is called the consumer's risk. **D 123,** D13

louis heel—high heels with a gracefully curved back, with a lip extending part way under the shank resulting in a curved breast. Louis heels are usually 16/8 to 24/8 in. in height and are sometimes called french heels. **F 869,** F08

louver blade—a shaped asbestos-cement product used to guard ventilation ports. **C 460,** C17

louver blade, *n*—a shaped asbestos-cement product used to guard ventilation ports. **C 1154,** C17

louver blade—a shaped asbestos-cement product used to guard ventilation ports. **D 2946,** C17

louver cloth, *n*—*in coated glass yarn fabrics,* a woven netting having an approximately even-spaced mesh of fewer than 12 by 12 yarns per 25.4 mm (1 in.) **D 123,** D13

Lovibond color system, *n*—a system of color specification based on the optical densities of yellow, red, and blue filter glasses required to modify light from a standard source to obtain light matching some given light. **E 284,** E12

Lovibond tintometer, *n*—instrument for evaluating the colors of materials by visual comparison with the colors of glasses of the Lovibond color system. **E 284,** E12

low-acid food—any food, other than alcoholic beverages, with a finished equilibrium pH greater than 4.6 and a water activity (a_w) greater than 0.85. Tomatoes and tomato products having a finished equilibrium pH less than 4.7 are not classed as low-acid foods. **F 17,** F02

low-alloy steel, *n*—a steel, other than a **carbon steel** or an **interstitial-free steel**, that conforms to a specification that requires the minimum content for each specified alloying element to be lower than the applicable limit in the definition for **alloy steel**. **A 941,** A01

low-carbon-steel—See **steel grades.** **F 547,** F16

low-density particleboard—a particleboard as previously defined with a density of less than 37 lb/ft^3 (specific gravity 0.59). **D 1554,** D07

low-density polyethylene plastics (LDPE), *n*—those branched polyethylene plastics, having a standard density of 0.910 to 0.925 g/cm^3. (D-20) **F 412,** F17

low density polyethylene plastics, (LDPE) *n*—those branched polyethylene plastics, q. v., having a standard density of 0.910 to 0.925 g/cm^3. **D 883,** D20

low-energy gamma radiation—gamma radiation having energy less than 200 keV. **E 1316,** E07

low-energy photon radiation—gamma- and X-ray photon radiation having energy less than 200 keV (excluding visible and ultraviolet light). **E 1316,** E07

lower body, *n*—that part of the human body which includes all portions between the waist and feet, not to include the feet. **F 1494,** F23

lower capability index (CPL), *n*—the difference between the sample mean ($\bar{x}$) and the lower specification limit divided by three times the standard deviation.

$$CPL = \frac{(\bar{x} - LSL)}{3 \times s}$$

A 644, A04

lower control limit (LCL), *n*—control limit for points below the central line. **E 456,** E11

lower critical point—in a phase diagram, a specified value of composition, temperature and pressure or combinations thereof occurring as a minimum in temperature, or pressure, for the coexistence of two, or more, conjugate phases and at which the conjugate phases become identical. **E 7,** E04

lower heating value, LHV, *n*—a synonym for net calorific value. **D 5681,** D34

lower heating value (LHV)—a synonym for net calorific value. **E 856,** D34

lower limit of flammability or lower flammable limit (LFL), *n*—the minimum concentration of a combustible substance that is capable of propagating a flame through a homogeneous mixture of the combustible and a gaseous oxidizer under the specified conditions of test. **E 1445,** E27

lower punch—the member of the compacting tool set or die that forms the bottom of the part being produced, and the position of which determines the volume of powder fill. Secondary or subdivided lower punches may be necessary to facilitate filling, forming and ejecting of multiple level parts. **B 243,** B09

lower range value, *n*—the lowest quantity that an instrument is adjusted to measure. **E 344,** E20

lower specification limit (LSL), *n*—the lowest specified value. **A 644,** A04

lower temperature limit of flammability, (LTL), *n*—the lowest temperature, corrected to a pressure of 101.3 kPa (760 mm Hg, 1013 mbar), at which application of an ignition source causes a homogeneous mixture of a gaseous oxidizer and vapors in equilibrium with a liquid (or solid) specimen to ignite and propagate a flame away from the ignition source under the specified conditions of test. **E 1445,** E27

lower tolerance limit (LTL) (lower specification limit), *n*—a tolerance limit that defines the lower conformance boundary for an individual unit of a manufacturing or service operation. **E 456,** E11

lower yield strength, *LYS* [FL^{-2}], *n*—in a uniaxial test, the minimum stress recorded during discontinuous yielding at or near the onset of plastic deformation, ignoring transient effects. **E 6,** E28

lowest-observed-effect concentration (LOEC)—in a toxicity test, the tested concentration of one or more chemicals immediately above the highest tested concentration that did not result in a statistically significant change in the particular toxicological variable compared to that value in the control (s). **E 943,** E47

low point, Low—the point on the reloading portion of an unloading-reloading cycle where the load is one half the high point load. **E 1823,** E08

low power stretch, *n*—that property of a fabric whereby it exhibits high fabric stretch and good recovery from low tension. **D 123,** D13

low-pressure air test—testing procedure, utilizing compressed air, intended to determine the acceptability of a pipe section, joint, or pipeline. **C 822,** C13

low-pressure molding, *n*—a method of molding or laminating in which the pressure is 1400 kPa (200 psi) or less. **D 883,** D20

low-pressure steam cooker—equipment wherein the cooking compartment operates between 3 and 9.9 psig. See **steam cooker**. **F 1827,** F26

low profile tile, *n*—tile having a rise to width ratio equal to, or less than 1:4. **C 43,** C15

low reflecting coating—a dielectric coating applied to an air-glass surface to reduce light reflection to a minimum. **E 7,** E04

low-silicate coolant concentrate, *n*— an engine coolant concentrate containing not more than 250 ppm silicon. **D 4725,** D15

low temperature preservation—stabilizing viable or biologically active material by freezing or freeze-drying. **E 1705,** E48

low volatile bituminous coal—See *low volatile bituminous coal* under **rank**. **D 121,** D05

low-voltage power supply, *n*—a source of electrical power which provides voltage at logic levels, usually internal to the component. **F 1457,** F05

lpi—lines per inch or line pairs per inch, depending upon usage. **F 1156,** F12

lpmm—line pairs per millimetre (mm). **F 1156,** F12

L-radiation—characteristic X-rays produced by an atom when a vacancy in the L-shell is filled by one of the outer electrons. **E 7,** E04

L radiation, *n*—characteristic X rays produced by an atom or ion when a vacancy in the L shell is filled by one of the outer electrons. **E 135,** E01

L-radiation, *n*—characteristic X-rays produced by an atom or ion when a vacancy in the L-shell is filled by one of the outer electrons. **E 1605,** E06

L-series—the set of X-ray wavelengths making up L-radiation. **E 7,** E04

L-series, *n*—the set of X-ray wavelengths making up L radiation. **E 135,** E01

LSI—langelier saturation index, measure of $CaCO_3$ solubility in brackish waters. See **S&DSI**. **D 6161,** D19

Lubbock nozzle—a variable-area nozzle in which a moveable plunger exposes additional inlet area as pressure increases. **E 1620,** E29

lubricant—a solution that, when applied to glass fibers, facilitates their handling by reducing mutual abrasion. (See also **mold lubricant**.) **C 162,** C14

lubricant, *n*—any material interposed between two surfaces that reduces the friction or wear between them. **D 4175,** D02

lubricant, *n*—(*1*) a material used to reduce the friction between two mating surfaces that are being joined by sliding contact. (*2*) an additive that is added to a plastic compound to lower the viscosity or otherwise improve the processing or product characteristics. **F 412,** F17

lubricant, *n*—any substance interposed between two surfaces for the purpose of reducing the friction or wear between them. **G 40,** G02

lubricant base stock, *n*—a liquid that may be used alone as a lubricant, but normally is used as a major ingredient in formulated lubricants. **D 4175,** D02

lubricant bloom—See **bloom.** **D 883,** D20

lubricant (mold), *n*—See **release agent (mold)**, the preferred term. **D 1566,** D11

lubricating—mixing with, or incorporating in, a powder, some agent to facilitate pressing and ejecting the compact from the die body; applying a lubricant to the die walls and punch surfaces. **B 243,** B09

lubricity—*in grouting*, the physico-chemical characteristic of a grout material flow through a soil or rock that is the inverse of the inherent friction of that material to the soil or rock; comparable to "wetness." **D 653,** D18

Lüder's lines—markings which appear on the surface of metals stretched past the yield point. The markings are approximately parallel to the direction of maximum shear stress and essentially independent of crystal orientation; also called stretcher strains, flow figures, Hartmann lines. **E 7,** E04

lugeon—a measure of permeability defined by a pump-in test or pressure test, where one Lugeon unit is a water take of 1 L/min per metre of hole at a pressure of 10 bars. **D 653,** D18

lugging, *adj*—in *internal combustion engine operation*, characterized by a combined mode of relatively low speed and high power output. **D 4175,** D02

Luggin probe or Luggin-Haber capillary— a device used in measuring the potential of an electrode with a significant current density imposed on its surface. (The probe minimizes the IR drop that would otherwise be included in the measurement and without significantly disturbing the current distribution on the specimen.) **G 15,** G01

lumber—the product of the sawmill and planing mill usually not further manufactured other than by sawing, resawing, passing lengthwise through a standard planing machine, crosscutting to length, and matching.

blanked lumber—lumber planed to a size in excess of the corresponding standard-dressed size to permit remanufacture or special use.

boards—lumber less than 2 in. in nominal thickness. Boards less than 6 in. in width may be classified as strips.

common lumber—in softwoods, a general term for nonstress graded lumber that has appearance quality less than select grade but is suitable for general construction and utility purposes.

dimension—lumber from nominal 2 in. through 4 in. thick and 2 or more inches wide.

dressed lumber—lumber that is surfaced by a planing machine on one side (S1S), two sides (S2S), one edge (S1E), two edges (S2E), or any combination of sides and edges (S1S1E, S2S1E, S1S2E, or S4S). Dressed lumber may also be referred to as planed or surfaced.

dressed size—the dimensions of lumber after surfacing with a planing machine. Usually 1/4 or 3/4 in. less than nominal size. The American Softwood Lumber Standard lists standard dressed sizes.

factory and shop lumber—lumber that is produced or selected primarily for manufacture into nonstructural products. It is graded on the basis of the percentage of the area that will produce a limited number of cuttings of a specified minimum size and quality.

finish—lumber suitable for millwork or for the completion of the interior of a building. Chosen particularly because of appearance or ability to accept a high quality finish.

framing lumber—lumber used for the structural members of a building, such as studs and joists.

matched lumber—lumber that is edge or end dressed and shaped to make a close tongued-and-grooved joint at the edges or ends when laid edge-to-edge or end-to-end.

patterned lumber—lumber that is shaped to a pattern in addition to being dressed, matched, or shiplapped, or any combination of these workings.

plain end lumber—worked lumber without end matching or with plain trimming and square ends.

planed lumber—See *dressed lumber.*

remanufactured lumber—lumber that has been further processed to change its size or shape after grading.

resawn lumber—the product of sawing any thickness of lumber to develop thinner lumber. The term as used in commercial transactions is mostly to denote the product of resawing dressed and graded lumber.

ripped lumber—the product of sawing any width of lumber to develop narrower lumber. The term as used in commercial

transactions is mostly to denote the product of ripping dressed and graded lumber.
rough lumber—lumber as it comes from the saw prior to any dressing operation.
saw-sized lumber—sawn rough lumber of near-exact, uniform, net size, with a plus or minus sawing tolerance permitted.
select lumber—in softwoods a general term for lumber of good appearance and finishing qualities.
shiplapped lumber—lumber that is edge dressed to make a lapped joint.
shop lumber—see *factory and shop lumber.*
stress-graded lumber—lumber of any thickness and width that is graded for its mechanical properties.
surfaced lumber—see *dressed lumber.*
timbers—lumber 5 or more inches in least dimension.
worked lumber—lumber which, in addition to being dressed, has been matched, shiplapped, or patterned.
yard lumber—lumber generally intended for ordinary construction and general building purposes, without further manufacture, but where design properties are not required. Terminology becoming obsolete. **D 9,** D07

lumber, *n*—the product of the saw and planing wood mill, not further manufactured than by sawing, resawing, planing, crosscutting to length, and matching.
matched lumber—**lumber** that is shaped to make a close-tongued and grooved joint at its edges or ends
plank, n— a broad board usually more than 1 in. thick, laid with its wide dimension horizontal, and used as a bearing surface.
resawn lumber— rough or surfaced **lumber** that has been divided into two or more thicknesses by sawing.
rough lumber— undressed **lumber** as it comes from the saw.
shiplapped lumber—**lumber** that is shaped at its edges to make a close-rabbeted, or lapped joint.
surfaced lumber—**lumber** that is dressed on one or more faces. **D 996,** D10

lumber, *n*—a sawn piece of wood smaller than 100 mm (4 in.) in its least dimension. **E 1807,** E06

lumen, *n*—*in vegetable fibers,* the central canal of the fiber. **D 123,** D13

lumen—lumininous flux emitted within one steradian by a point source having a spatially uniform luminous intensity of one candela. SI unit of luminous flux. **E 1316,** E07

lumen, *lm, n*— luminous flux emitted within one steradian by a point source having a spatially uniform luminous intensity of 1 candela. SI unit of luminous flux. **E 284,** E12

luminance—the ratio of the luminous intensity of a surface in a given direction to a unit of projected area. Measured in candela per square meter. **E 1316,** E07

luminance factor, *Y, n*—ratio of the luminance of a specimen to that of a perfect diffuser, when illuminated and viewed under specified geometric conditions. **E 284,** E12

luminance, *L, L_v, n*—luminous flux in a beam, emanating from a surface, or falling on a surface, in a given direction, per unit of projected area of the surface as viewed from that direction, per unit solid angle. **E 284,** E12

luminescence—the emission of radiant energy during a transition from an excited electronic state of an atom, molecule, or ion to a lower electronic state. **E 131,** E13

luminescence, *n*—emission of light ascribable to nonthermal excitation. (See also **electroluminescence**, **chemical luminescence**, and **photoluminescence**.) **E 284,** E12

luminometer number, *n*—a measure of the flame temperature in a wick lamp burning the candidate material as fuel at a specified flame radiation level in the green-yellow band of the visible spectrum. **D 4175,** D02

luminophores—materials that emit light energy of one wavelength when activated by light of another wave length; includes both fluors and phosphors. **F 1156,** F12

luminosity—a measure of emitted light intensity. **E 1316,** E07

luminosity function, n—see **spectral luminous efficiency function**. **E 284,** E12

luminous, *adj*—*in color determination,* indicates that the radiant flux is evaluated by weighting according to the luminous efficiency function of the CIE 1931 standard observer. **D 2946,** C17

luminous, *adj*—weighted according to the spectral luminous efficiency function $V(\lambda)$ of the CIE. **E 284,** E12

luminous efficacy, *K_M, n*—quotient of total luminous flux divided by total radiant flux; lumens per watt. K_M = 683 lm/W. **E 284,** E12

luminous flux, ϕ, ϕ_v—the time rate of flow of light energy; luminous power. **E 284,** E12

luminous intensity, *I, I_v, n*— the light flux per unit solid angle. **E 284,** E12

luminous (photometric)—referring to a radiometric quantity, indicates the weighted average of the spectral radiometric quantity, with the photopic spectral luminous efficiency function given in Annex A1 of Practice E 971, for Calculation of Photometric Transmittance and Reflectance of Materials to Solar Radiation, being the weighting function. **E 772,** E44

luminous reflectance, *n*—see **reflectance, luminous**. **D 2946,** C17

luminous transmittance, *n*—the ratio of the luminous flux transmitted by a body to the flux incident upon it. **D 883,** D20

lump—*in porcelain enamels,* a rounded projection in the enamel surface, usually a defect. **C 286,** B08

lump—a raised area on the surface having the appearance of being solid. **F 109,** C21

lusec—a unit of flow rate equal to 0.133 mPa·m^3/s. **E 1316,** E07

luster—an iridescent decorative surface appearance. **C 286,** B08

luster, *n*—*in buttons*, the degree of brilliance exhibited in pearlized or pearl buttons. **D 123,** D13

luster—in the pulp and paper industry, synonymous with "gloss" or "sheen." (see also **gloss**) **D 1695,** D01

luster, *n*—the appearance characteristic of a surface that reflects more in some directions than it does in other directions, but not of such high gloss as to form clear mirror images. **E 284,** E12

lusters—(*1*) dilute solutions of metallic salts that, when applied and fired to glass, produce an iridescent effect on the surface.
(*2*) solutions of metallic resinates that, when applied and fired to glass, produce an opaque, mother-of-pearl iridescent effect. **C 162,** C14

lux, lx, *n*— illuminance corresponding to a luminous flux density of one lumen per square metre. SI unit of illuminance. **E 284,** E12

LVPS—see **low-voltage power supply**. **F 1457,** F05

lysimeter—a device to measure the quantity or rate of water movement through a block of soil, usually undisturbed or in-situ; or to collect such percolated water for analyses. **D 653,** D18

lysis, *n*—the disintegration or destruction of whole bacterial cells. **F 1494,** F23

lysogen—a bacterial strain that has a phage stably maintained. In the case of lambda, the phage is integrated into the host genome. The integrated phage is called a prophage. **E 1705,** E48

M

M, *n*—the number of standard deviations, mutually concurred by the supplier and purchaser, to be used for calculations of statistical conformance to such items as minimums, maximums, specification ranges, and process capability indices. **A 644,** A04

MA—Commonwealth of Massachusetts. **E 1605,** E06

MacAdam color difference, *n*—nonspecific term for color difference calculated by using one of several equations and methods based on MacAdam's color-difference-perceptibility ellipses or ellipsoids. **E 284,** E12

macadam, dry-bound and water bound, *n*—a pavement layer containing essentially one-size coarse aggregate choked in place with an application of screenings or sand; water is applied to the choke material for water-bound macadam. Multiple layers must be used. **D 8,** D04

MacAdam limits, *n*— see **optimal colors**. **E 284,** E12

maceral, *n*—a microscopically distinguishable organic component of coal, but including any mineral matter not discernible under the optical microscope.

exinite—Synonym for *liptinite*.

liptinite, n—a group of macerals composed of alginite, cutinite, resinite, and sporinite.

alginite, n—a maceral derived from waxy walls (thimble cups) of *Botryococcus* or allied types of algae.

cutinite, n—a maceral derived from the waxy coatings (cuticles) of leaves and other plant parts.

resinite, n—a maceral derived from the resinous secretions and exudates of plant cells, occurring as discrete homogeneous bodies or clusters, individuals of which are usually round, oval, or rod-like in cross section.

sporinite, n—a maceral derived from the waxy coatings (exines) of spores and pollen.

inertinite, n—a group of macerals composed of fusinite, inertodetrinite, macrinite, micrinite, sclerotinite, and semifusinite.

fusinite, n—the maceral distinguished by the well-preserved original form of plant cell wall structure, intact or broken, with open or mineral-filled cell lumens (cavities), and by having a reflectance (except in meta-anthracite), well above that of associated vitrinite. The particle size is generally greater than 50 μm except when isolated from other macerals.

inertodetrinite, n—the maceral occurring as individual, angular, clastic fragments of other inertinite macerals, surrounded by other macerals, commonly vitrinite, or minerals, and also distinguished by a reflectance higher than that of associated vitrinite. The size of the particles is generally greater than 2 μm and less than 50 μm.

macrinite, n—the maceral that is distinguished by a reflectance higher than that of associated vitrinite, absence of recognizable plant cell structure, and a nonangular shape. The particle size is generally greater than 10 μm.

micrinite, n—the maceral that is distinguished by reflectance higher than that of associated vitrinite, absence of recognizable plant cell structure, and occurring as individual particles of size less than 10 μm and commonly about 1 μm.

sclerotinite, n—maceral having reflectance between that of fusinite and associated vitrinite and occurring as round or oval cellular bodies of varying size (20 to 300 μm) or as interlaced tissues derived from fungal remains.

semifusinite, n—the maceral that is intermediate in reflectance between fusinite and associated vitrinite, that shows plant cell wall structure with cavities generally oval or elongated in cross section, but in some specimens less well defined than in fusinite. The particle size is generally greater than 50 μm except when isolated from other macerals. Often, it occurs as a transitional material between vitrinite and fusinite.

vitrinite, n—the maceral and maceral group composing all or almost all of the vitrain, and like material occurring in attrital coal as the component of reflectance intermediate between those of exinite and inertinite.

macrinite—See *macrinite* under **maceral**. **D 121,** D05

macerated paper—See **cushioning material. D 996,** D10

machinability—a measure of the ease with which a material can be shaped with the aid of cutting or abrasive tools. **C 709,** D02

machine bite—a depressed cut at the end of a piece of lumber made by the machine knives in dressing; sometimes called snipe. **D 9,** D07

machine buffs—that cut of the hide from which a buffing of approximately 1⁄64 in. (0.4 mm) (1 oz) in thickness has been removed from the grain. This should leave a portion of the grain on approximately the entire hide. **D 1517,** D31

machine burn—a darkening or charring of the wood caused by frictional heat generated by machine knives or rolls. **D 9,** D07

machine damage, *n*—as used in fractography, chips and surface-subsurface microcracks, striations, and scratches created during the machining process. **C 1145,** C28

machine direction, *n*—the direction parallel to the paper-bound edge of the gypsum board. **C 11,** C11

machine direction, *n*—the direction of a paper or paperboard corresponding or parallel to the direction of flow of the stock along the paper machine; sometimes referred to as MD or along machine direction. (See **cross direction**.) **D 1968,** D06

machine direction, *n*—the direction in the plane of the fabric parallel to the direction of manufacture. **D 4439,** D35

machine direction, *n*—the direction in which a product moves through the manufacturing process. **F 141,** F06

machine direction, MD, *n*—the direction in the plane of the fabric parallel to the direction of manufacture. **D 123,** D13

machined specimen—a test specimen machined from a full-size fastener to specific dimensions to standardize test results; often specified when a full-size fastener cannot be reasonably or practically tested. **F 1789,** F16

machine gouge—a groove below the regular surface of cut made by machine knives in dressing lumber. **D 9,** D07

machine language—a language designed for use by a machine, without translation. **F 149,** F05

machine offset—an abrupt dressing variation in the edge surface of lumber, occurring usually near the end of the piece without reducing the width or without changing the plane of the wide surface. **D 9,** D07

machine quality—term applied to nails manufactured with closer tolerances than usual and selected for close adherence to specification and freedom from foreign matter. **F 547,** F16

machine-readable, *n*—that characteristic of printed information that permits direct transfer of information from a printed surface to a data-processing system without operator intervention. **F 1294,** F05

machine run—the product of sawing, dressing, or patterning lumber without further grading. **D 9,** D07

machine stitch, *n*—*in home sewing,* the resulting arrangement of sewing threads in a repeating unit formed by a sewing machine.

D 123, D13

macrocell corrosion—corrosion of a metal embedded in porous media (for example, concrete or soil) caused by concentration or galvanic cells which exist on a scale at least as large as the smallest major dimension of the corroding item (for example, the diameter of a bar or pipe). **G 15,** G01

macroetch—controlled etching of the surface of a metallic specimen, intended to reveal a structure which is visible at low magnifications (not usually greater than 10 times). **E 7,** E04

macrograph—a graphic reproduction of an object, slightly reduced in size, unmagnified, or magnified ten diameters or less (photomacrograph). **E 7,** E04

macrograph—a photographic reproduction of any object that has not been magnified more than ten times. **F 1789,** F16

macromolecular material, n—See **polymer** the preferred term **D 1566,** D11

macropore—interaggregate cavities that service as the principal avenues for the infiltration and drainage of water and for aeration. **D 653,** D18

macropore—pores with widths exceeding 50 nanometres (500 angstrom units). **D 2652,** D28

macroporosity—large intergranular porosity with large pore throats, including soil cracks, moldic porosity, animal burrows and other significant void space. **D 653,** D18

macroscopic—visible either with the naked eye or under low magnification (as great as about ten diameters). **F 1789,** F16

macroscratch—a scratch that is visible to the unaided eye under either incandescent (high intensity) or fluorescent (diffuse) illumination. **F 1241,** F01

macrostrain, n—the mean strain over any finite gage length of measurement large in comparison with interatomic distances. **E 6,** E28

macrostructure—the structure of metal as revealed by macroscopic examination. **F 1789,** F16

MADT—see **monthly average daily traffic.** **E 867,** E17

magazine—mechanism for storing and feeding fasteners. **F 592,** F16

magma, *n*—molten rock within the earth or within other planets. **E 957,** E44

magnesia—the chemical compound magnesium oxide (MgO). **C 51,** C07

magnesia—magnesium oxide (MgO), calcined or hard burned as periclase loosely applied also to the hydrate $Mg(OH)_2$, and made synthetically from seawater or brine, or (impure) from magnesite. **C 242,** C21

magnesia brick, *n*—a refractory brick, which may be burned or unburned, manufactured predominantly or entirely of either seawater magnesia, dead-burned magnesite, fused magnesia, or combinations of these materials. **C 71,** C08

magnesia-carbon brick, *n*—a refractory brick manufactured predominantly of a mixture of refractory-grade magnesia and 5 to 30 weight percent carbonaceous material, with resin, tar, pitch, or a combination of these materials as the bonding agent; the refractory-grade magnesia may be either sea-water magnesia, dead-burned magnesite, fused magnesia, or a combination of these materials; and the carbonaceous material may be either flake or vein graphite, carbon black, or petroleum or metallurgical coke, or a combination of these materials. **C 71,** C08

magnesia-chrome brick, *n*—a refractory brick, which may be burned or unburned, manufactured predominantly of a mixture of refractory-grade magnesia and refractory-grade chrome ore in which the refractory-grade magnesia predominates by weight; the refractory-grade magnesia may be either seawater magnesia, dead-burned magnesite, fused magnesia, or a combination of these materials. **C 71,** C08

magnesia-doloma refractory, *n*—a refractory, which may be burned or unburned, manufactured predominantly of a mixture of refractory-grade magnesia and refractory-grade doloma in which the refractory-grade magnesia predominates by weight. **C 71,** C08

magnesia, fused, *n*—a refractory material consisting predominantly of crystalline magnesium oxide which has solidified from a fused or molten state. **C 71,** C08

magnesian—indicates the presence of 5 to 35 % magnesium carbonate ($MgCO_3$) in the limestone from which the material was formed. **C 51,** C07

magnesian limestone—see **limestone.** **C 51,** C07

magnesia, refractory, *n*—a dead-burned refractory material consisting predominantly of crystalline magnesium oxide. **C 71,** C08

magnesia, refractory-grade, *n*—a dead-burned or fused refractory material consisting predominantly of crystalline magnesium oxide; the three principal types are magnesia, sea water; magnesite, dead-burned; and magnesia, fused. **C 71,** C08

magnesia, sea water, *n*—a deadburned refractory-grade magnesia which has been derived from the treatment of sea water or other brines. **C 71,** C08

magnesia-spinel brick, *n*—a refractory brick whose predominant component is sintered or fused magnesia and whose lesser component is sintered, fused, or in situ formed magnesium aluminate spinel ($MgO \cdot Al_2O_3$). **C 71,** C08

magnesite, dead-burned, *n*—the granular product obtained by burning (firing) magnesite or other substances convertible to magnesia, upon heating above 1450°C long enough to form dense, weather-stable granules suitable for use as a refractory or in refractory products. **C 71,** C08

magnesite, grain, *n*—dead-burned magnesia in granular form of size suitable for refractory purposes. **C 71,** C08

magnesite, seawater, *n*—dead-burned magnesia made by a chemical process using seawater or other saline solutions. **C 71,** C08

magnet—a body that produces a magnetic field external to itself.

NOTE—By convention, the north-seeking pole of a magnet is marked with an *N*, +, or is colored red.

NOTE—Natural magnets consist of certain ores such as magnetite (loadstone); artificial (permanent) magnets are made of magnetically hard materials; electromagnetics have cores made of magnetically soft materials which are energized by a current carrying winding. **A 340,** A06

magnetic alignment—See **alignment**. **E 7,** E04

magnetic anomaly device—device for the detection of variations in a magnetic field, especially those caused by the presence of a foreign body in the field; for example, a submarine in the ocean, or a gun in a passenger's coat pocket, or a special piece of metal attached to an object that is not to be removed from an area without prior permission. **F 1156,** F12

magnetic brush development—development in which the toner is carried by means of a brush-like array of magnetic particles held erect under the influence of a magnetic field. **F 335,** F05

magnetic circuit—a region at whose surface the magnetic induction is tangential.

NOTE—A practical magnetic circuit is the region containing the flux of practical interest, such as the core of a transformer. It may consist of ferromagnetic material with or without air gaps or other feebly magnetic materials such as porcelain, brass, and so forth. **A 340,** A06

magnetic constant (permeability of space), Γ_m—the dimensional scalar factor that relates the mechanical force between two currents to their intensities and geometrical configurations. That is:

$$dF = \Gamma_m I_1 I_2 dl_1 \times (dl_2 \times r_1)/nr^2$$

where:

Γ_m = magnetic constant when the element of force, dF, of a current element $I_1 dl_1$ on another current element $I_2 dl_2$ is at a distance r,

r_1 = unit vector in the direction from dl_1 to dl_2, and

n = dimensionless factor. The symbol n is unity in unrationalized systems and 4π in rationalized systems.

NOTE—The numerical values of Γ_m depend upon the system of units used. In the cgs-emu system of units, $\Gamma_m = 1$, in the SI system, $\Gamma_m = 4\pi \times 10^{-7}$ H/m.

NOTE—The magnetic constant expresses the ratio of magnetic induction to the corresponding magnetizing force at any point in a vacuum and therefore is sometimes called the permeability of space, μ_v.

NOTE—The magnetic constant times the relative permeability is equal to the absolute permeability.

$$\mu_{abs} = \Gamma_m \mu_r$$

A 340, A06

magnetic domains—small areas on the surface or in the body of thin films of a magnetic medium each of which maintains a discrete magnetic field orientation relative to the others around it. Used to store the information in a binary format. **F 1156,** F12

magnetic excursion range, ΔB, ΔH—the excursion ranges equaling the algebraic differences between the upper and lower values of B, and between the upper and lower values of H, in a hysteresis or flux-current loop. **A 340,** A06

magnetic field—the volume within and surrounding either a magnetized part or a current-carrying conductor wherein a magnetic force is exerted. **E 1316,** E07

magnetic field indicator—a pocket meter that is used to locate or determine the relative intensity of leakage field emanating from a part. **E 1316,** E07

magnetic field meter—an instrument designed to measure the flux density of magnetic fields. **E 1316,** E07

magnetic field of induction—the magnetic flux field induced in a region such that a conductor carrying a current in the region would be subjected to a mechanical force, and an electromotive force would be induced in an elementary loop rotated with respect to the field in such a manner as to change the flux linkage. **A 340,** A06

magnetic field strength—the measured intensity of a magnetic field at a point, usually expressed in oersteds or amperes per metre. **B 713,** B01

magnetic field strength—the measured intensity of a magnetic field at a point, expressed in oersteds or amperes per metre. **E 1316,** E07

magnetic field strength, ac—the value of one of three dynamic magnetic field strength parameters in common use. They are:

(a) H_L—an assumed peak value computed in terms of peak magnetizing current (considered to be sinusoidal).

(b) H_z—an assumed peak value computed in terms of measured rms exciting current (considered to be sinusoidal).

(c) H_p—computed in terms of a measured peak value of exciting current, and thus equal to the value H'_{max}. **A 340,** A06

magnetic field strength, biasing, H_b—the algebraic mean value of the magnetic field strength in a magnetic material that is subjected simultaneously to a constant magnetizing field and a periodically varying magnetizing field.

NOTE—The biasing magnetizing field and the biased magnetic induction are corresponding coordinates of a single point on the B-H plane but not necessarily on the normal induction curve.

NOTE—The biasing magnetic field strength, H_b, is equal to the applied constant magnetizing field only when the applied periodically varying magnetizing field is symmetrical. **A 340,** A06

magnetic field strength, H—the magnetic vector quantity at a point in a magnetic field which measures the ability of electric currents or magnetized bodies to produce magnetic induction at the given point.

NOTE—The magnetic field strength, H, may be calculated from the current and the geometry of certain magnetizing circuits. For example, in the center of a uniformly wound long solenoid.

$$H = C\,(NI/l)$$

where:

H = magnetic field strength,

H = constant whose value depends on the system of units,

N = number of turns,

I = current, and

l = axial length of the coil.

If I is expressed in amperes and l is expressed in centimetres, then $C = 4\pi/10$ to obtain H in the cgs-emu system of units, the oersted.

If I is expressed in amperes and l is expressed in metres, then $C = 1$ to obtain H in the SI units, ampere-turn per metre.

NOTE—The magnetic field strength, H, at a point in air may be calculated from the measured value of induction at the point by dividing this value by the magnetic constant Γ_m.

A 340, A06

magnetic field strength, incremental, H_Δ—a value equal to one half the algebraic difference of the maximum and minimum values of the magnetic field strength during a cycle in a magnetic material that is subjected simultaneously to a biasing magnetic field strength and a symmetrical periodically varying magnetic field strength.

Twice the incremental magnetic field strength is indicated by the symbol ΔH.

Thus:

$$H_\Delta = \Delta H/2$$

A 340, A06

magnetic field strength, maximum—*(a)* H_m—the maximum value of H in a dc hysteresis loop.

(b) H_{max}—the maximum value of H in an ac flux-current loop. **A 340,** A06

magnetic flux, ϕ—the product of the magnetic induction, B, and the area of a surface (or cross section), A, when the magnetic induction B is uniformly distributed and normal to the plane of the surface.

$$\phi = BA$$

where:

ϕ = magnetic flux,

B = magnetic induction, and

A = area of the surface.

NOTE—If the magnetic induction is not uniformly distributed over the surface, the flux, ϕ, is the surface integral of the normal component of B over the area.

$$\phi = \int\int_s B \cdot dA$$

NOTE—Magnetic flux in a scalar and has no direction. **A 340,** A06

magnetic flux, Φ—the product of the magnetic induction, B, and the area of a surface (or cross section), A, when the magnetic induction B is uniformly distributed and normal to the plane of the surface.

$$\Phi = BA$$

where:
Φ = magnetic flux,
B = magnetic induction, and
A = area of the surface.

NOTE—If the magnetic induction is not uniformly distributed over the surface, the flux, Φ, is the surface integral of the normal component of B over the area.

$$\Phi = \int \int_s B \cdot dA$$

NOTE—Magnetic flux is a scalar and has no direction. **B 713,** B01

magnetic flux density, B—an alternate term for **magnetic induction.** **A 340,** A06

magnetic flux jump—the collective, discontinuous motion of magnetic flux lines in a superconductor, produced by mechanical, thermal, magnetic, or electrical disturbances. **B 713,** B01

magnetic flux pinning—the trapping of magnetic flux lines at defects in the superconducting material. **B 713,** B01

magnetic history—magnetic condition of a ferromagnetic part under test based on previous exposures to magnetic fields. (E 566) **E 1316,** E07

magnetic hysteresis—in a magnetic material, as iron, a lagging in the values of resulting magnetization due to a changing magnetic force. (See also **hysteresis**.) **E 1316,** E07

magnetician—one skilled in the theory and practice of magnetics. **A 340,** A06

magnetic induction (also flux density), *B*—that magnetic vector quantity which at any point in a magnetic field is measured either by the mechanical force experienced by an element of electric current at the point, or by the electromotive force induced in an elementary loop during any change in flux linkages with the loop at the point.

NOTE—If the magnetic induction, B, is uniformly distributed and normal to a surface or cross section, then the magnetic induction is:

$$B = \frac{\Phi}{A}$$

where:
B = magnetic induction,
Φ = total flux, and
A = area.

NOTE—B_{in} is the instantaneous value of the magnetic induction and B_m is the maximum value of the magnetic induction. **B 713,** B01

magnetic induction (flux density), *B*—that magnetic vector quantity which at any point in a magnetic field is measured either by the mechanical force experienced by an element of electric current at the point, or by the electromotive force induced in an elementary loop during any change in flux linkages with the loop at the point.

NOTE—If the total flux, ϕ is uniformly distributed and normal to a surface or cross section, then the magnetic induction is:

$$B = \phi / A$$

where:
B = magnetic induction,
ϕ = total flux, and
A = area.

NOTE—B_{in} is the instantaneous value of the magnetic induction and B_m is the maximum value of the magnetic induction. **A 340,** A06

magnetic ink character recognition, MICR—a recognition technology that utilizes ink capable of being magnetized and sensed. A practical application is E-13B, which is used primarily within the North American financial industry. E-13B consists of 14 characters printed to high specifications using ink with iron oxide pigments, or other inks utilizing ingredients capable of being magnetized. **F 149,** F05

magnetic lamination steel—a flat-rolled, low-carbon (usually below 0.06 %) steel containing 0.0 to 1.0 % silicon and up to 0.4 % aluminum and having similar core loss in all directions. **A 340,** A06

magnetic leakage flux—the excursion of magnetic lines of force from the surface of a test specimen. **E 1316,** E07

magnetic lens—a device for focussing an electron beam by means of a magnetic field. **E 7,** E04

magnetic line of force—an imaginary line in a magnetic field which at every point has the direction of magnetic induction at that point.

NOTE—Extended lines of force must always form nonintersecting closed loops. **A 340,** A06

magnetic medium—material coated with a magnetically susceptible material upon which a series of individual magnetic domains can be created, usually at will, for the storage of information; for example, magnetic tape, computer disks, etc. **F 1156,** F12

magnetic moment, m—a measure of the magnetic field strength, H, produced at points in space by a plane current loop or a magnetized body.

NOTE—The magnetic moment of a plane current loop is a vector, the magnitude of which is the product of the area of the loop and the current; the direction of the vector is normal to the plane of the loop in that direction around which the current has a clockwise rotation when viewed along the vector.

NOTE—The magnetic moment of a magnetized body is the volume integral of the magnetization, M.

NOTE—In the cgs-emu system of units, magnetic moment is usually defined as the pole strength multiplied by the distance between poles. This is sometimes called the magnetic dipole moment. **A 340,** A06

magnetic ohm—the unit of reluctance sometimes used in the cgs-emu system of units. One magnetic ohm equals one gilbert/maxwell or $4\pi/10^9$ ampere-turns/weber. **A 340,** A06

magnetic particle examination—a nondestructive test method utilizing magnetic leakage fields and suitable indicating materials to disclose surface and near-surface discontinuity indications. **E 1316,** E07

magnetic particle examination flaw indications—the accumulation of ferromagnetic particles along the areas of flaws or discontinuities due to the distortion of the magnetic lines of force in those areas. **E 1316,** E07

magnetic particle field indicator—an instrument, typically a bi-metal (for example, carbon steel and copper) octagonal disk, containing artificial flaws used to verify the adequacy or direction, or both, of the magnetizing field. **E 1316,** E07

magnetic particle inspection method—a method for detecting magnetic discontinuities or inhomogeneities on or near the surface in suitably magnetized materials that uses finely divided magnetic particles that tend to congregate in regions of magnetic nonuniformity associated with the magnetic discontinuities or inhomogeneities.

NOTE—Magnetic particle inspection is an accepted method for the detection of defects. **A 340,** A06

magnetic particles—finely divided ferromagnetic material capable of being individually magnetized and attracted to distortion in a magnetic field. **E 1316,** E07

magnetic polarization, *J*—in the cgs-emu system of units, the intrinsic induction divided by 4π is sometimes called magnetic polarization or magnetic dipole moment per unit volume. **A 340,** A06

magnetic pole—the magnetic poles of a magnet are those portions of the magnet toward which or from which the external magnetic induction appears to converge or diverge, respectively.

NOTE—In the hypothetical case of a uniformly magnetized body of constant cross-sectional area, the poles would be located at its ends.

NOTE—By convention, the north-seeking pole is marked with an *N*, or +, or is colored red. **A 340,** A06

magnetic pole—one of two or more areas of flux leakage on a part. **E 1316,** E07

magnetic pole strength, *p*—the magnetic moment divided by the distance between the poles.

$$p = {}_m/l$$

where:
p = pole strength,
m = magnetic moment, and
l = distance between the poles.

A 340, A06

magnetic printer—a printer in which magnetic signals are recorded onto a magnetic belt or drum. A magnetic toner is attracted to the drum and transferred to paper where it is fused to the sheet. **F 149,** F05

magnetic rating (MR), *n*—an empirical value reflecting the effect of the magnetic particles, such as magnetic iron compounds, in asbestos material as measured by a magnetic effect analyzer. It is not a quantitative measure of the magnetic particles in the material. **D 2946,** C17

magnetic saturation—that degree of magnetization where a further increase in magnetizing force produces no significant increase in magnetic flux density (permeability) in a specimen. **E 1316,** E07

magnetic shielding—in electron microscopy, shielding for the purpose of preventing extraneous magnetic fields from affecting the electron beam in the microscope. **E 7,** E04

magnetics (magnetism)—that branch of science which deals with the laws of magnetic phenomena and their application to practice. **A 340,** A06

magnetic storage medium—a storage medium that uses magnetic properties (magnetic dipoles) to store digital data (for example, a moving drum, disk, or tape or a static core or film). **E 1316,** E07

magnetic transformation—an intensive property change from a ferromagnetic to a paramagnetic state, or the reverse, which occurs in certain solid materials under applied pressure and temperature, or both. (See also **Curie point**.) **E 7,** E04

magnetic transformation—intensive property change from a ferromagnetic to a paramagnetic state, or the reverse, which occurs in certain solid materials under applied pressure and temperature, or both. (See also **Curie temperature**), (E 7, E-4), . **E 1142,** E37

magnetic writing—a form of nonrelevant indication sometimes caused when the surface of a magnetized part comes in contact with another piece of ferromagnetic material. **E 1316,** E07

magnetization, circular—see **field, circular.** **E 1316,** E07

magnetization, longitudinal—see **field, longitudinal.** **E 1316,** E07

magnetization, *M*—the component of the total magnetizing force that produces the intrinsic induction in a magnetic material.

$$M = (B - \Gamma_m H)/\Gamma_m \mu_r = B_i/\mu_{abs}$$

where:
M = magnetization,
H = applied magnetizing force,
Γ_m = magnetic constant,
B = total magnetic induction,
μ_r = relative permeability,
μ_{abs} = absolute permeability, and
B_i = intrinsic induction.

NOTE—The magnetization can be interpreted as the volume density of magnetic moment. **A 340,** A06

magnetizing current—the flow of either alternating or direct current used to induce magnetism into the part being inspected. **E 1316,** E07

magnetizing current, ac, I_m—See **current, ac magnetizing.** **A 340,** A06

magnetizing force—the magnetizing field applied to a ferromagnetic material to induce magnetization. **E 1316,** E07

magnetizing force, **H**—an alternate term for **magnetic field strength.** **A 340,** A06

magnetodynamic—the magnetic condition when the values of magnetic field strength and induction vary, usually periodically and repetitively, between two extreme limits. **A 340,** A06

magnetometer—device for measuring and comparing the direction and magnitude of magnetic fields. **F 1156,** F12

magnetomotive force, $\mathcal{F}$—the line integral of the magnetizing field around any flux loop in space.

$$\mathcal{F} = \int H \cdot dl$$

where:
$\mathcal{F}$ = magnetomotive force,
H = magnetic field strength, and
dl = unit length along the loop.

NOTE—The magnetomotive force is proportional to the net current linked with any closed loop of flux or closed path.

$$\mathcal{F} = CNI$$

where:
$\mathcal{F}$ = magnetomotive force,
N = number of turns linked with the loop,
I = current in amperes, and
C = constant whose value depends on the system of units. In the cgs-emu system of units, $C = 4\pi/10$. In the SI system, $C = 1$.

A 340, A06

magneto-optic—optical properties of certain materials that can be altered by exposure to magnetic fields; most common form of this is the alteration of the angle of polarization of light passing through a material, or reflecting from it, as a function of the magnetic field strength in which it is immersed—this is used in one version of the **CD** music player. **F 1156,** F12

magneto-optic polarization—polarization of light using the magneto-optic effect; see **magneto-optic**. **F 1156,** F12

magnetostatic—the magnetic condition when the values of magnetic field strength and induction are considered to remain invariant with time during the period of measurement. This is often referred to as a dc (direct current) condition. **A 340,** A06

magnetostriction—the change in dimensions of a body resulting from magnetization. **A 340,** A06

magnetostrictive cavitation test device, *n*—a vibratory cavitation test device driven by a magnetostrictive transducer. **G 40,** G02

magnet, permanent—see **permanent magnet.** **E 1316,** E07

magnetron—a type of microwave energy generator usually used in microwave cooking equipment. **F 1827,** F26

magnet wire—a metal electrical conductor, covered with electrical insulation, for use in the assembly of electrical inductive apparatus such as coils for motors, transformers, generators, relays, magnets, and so forth. **D 1711,** D09

magnification—a ratio of the size of an image to its corresponding object. This is usually determined by linear measurement. **E 7,** E04

magnification factor, *n*—the constant multiplier applied to the nominal dimensions of a bar code symbol to obtain the actual dimensions at which it must be produced. **F 1294,** F05

magnitude estimation, *n*—a ratio scaling technique in which numerical values are assigned to the intensity or hedonic tone of attributes of products. The values assigned are proportional to the assessor's perception of the attributes. See **ratio scaling.** **E 253,** E18

mailing tube—See **tube.** **D 996,** D10

main components, *n*—*in textiles,* those sections of a consumer product which constitute the largest continuous homogeneous areas of the product that are typical of the product. **D 123,** D13

main effect, average effect, *n*—a term describing a measure for the comparison of the responses at each level (version) of a factor averaged over all levels (versions) of other factors in the experiment.

NOTE—The term "main effect" may describe the parameter in an assumed model or the estimate of this parameter. **E 456,** E11

main effect, average effect, *n*—a term describing a measure for the comparison of the responses at each level (version) of a factor averaged over all levels (versions) of other factors in the experiment.

NOTE—The term *main effect* may describe the parameter in an assumed model or the estimate of this parameter. **E 1325,** E11

maintain, *vt*—to monitor the usage of test methods and to take remedial steps when necessary. **D 123,** D13

maintain—See **building modification.** **E 631,** E06

maintainability (entretenabilité)—capability of a system or facility to be maintained to a specified level of **reliability**, at a specified measure of cost or economy. **E 1480,** E06

maintenance and repair cost—the total of labor, material, and other related costs incurred in conducting corrective and preventative maintenance and repair on a building, or on its systems and components, or on both. **E 833,** E06

maintenance mix, *n*—a mixture of bituminous material and mineral aggregate applied at ambient temperature for use in patching holes, depressions, and distress areas in existing pavements using appropriate hand or mechanical methods in placing and compacting the mix. These mixes may be designed for immediate use or for use out of a stockpile at a later time without further processing. **D 8,** D04

majolica—formerly an earthenware with an opaque luster glaze and overglaze colored decorations, but currently designating any decorated earthenware having an opaque glaze. **C 242,** C21

major defect, *n*—a defect other than critical, that judgment and experience indicate is likely to materially reduce the usability of a product for its intended purpose. **D 123,** D13

major imperfection, *n*—*in fabric grading,* a deviation in a roll of fabric that judgment and experience indicate is likely to have an adverse affect upon subsequent processing of the fabric. **D 123,** D13

majority carrier—type of charge carrier constituting more than one half the total charge-carrier concentration (for example, holes in *p*-type material). **F 1241,** F01

major modification—any change in either the structural or operational characteristics of the ride or device which will alter its performance from that specified in the manufacturer's design criteria. **F 747,** F24

major package defect—a defect that is likely to result in failure or reduce significantly the usability of the package for its intended use. **F 17,** F02

major package defect—See Terminology F 17. **F 1327,** F02

major principal plane—see **principal plane**. **D 653,** D18

major principal stress—see **stress**. **D 653,** D18

major tile facial dimensions—the overall length or width of the tile, including the lugs on opposite sides. **C 242,** C21

major tile thickness—the thickness of tile, including any maximum protuberances or ridges on the back. **C 242,** C21

makeup air—the supply of outside air, tempered or untempered, to a building in a controlled manner for the replacement of air exhausted through the hood, or the supply of outside air, untempered, into the hood to reduce the amount of tempered air exhausted from the building. **F 1827,** F26

male end of pipe (spigot, tongue, modified tongue)—that portion of the end of the pipe, regardless of its shape or dimensions, which is overlapped by a portion of the end of the adjoining pipe. **C 822,** C13

maleic resin—under **resin, synthetic,** see *maleic resin.* **D 16,** D01

malleable, ferritic, *n*—a ferrous alloy that is cast as white iron but which is converted by an appropriate heat treatment to a microstructure of temper carbon embedded in a ferritic matrix essentially free of pearlite and carbide. **A 644,** A04

malleable iron, *n*—a cast iron of such composition that it solidifies as white iron, which upon proper heat treatment is converted to a metallic matrix with nodules of temper carbon. **A 644,** A04

malleableize, *vt*—to convert white iron into malleable iron through an appropriate graphitizing heat treatment. **A 644,** A04

malleable, pearlitic, *n*—a ferrous alloy that is cast as white iron but which is converted by an appropriate heat treatment to a microstructure of temper carbon embedded in a matrix containing a controlled quantity, form, and distribution of pearlite or tempered martensite. **A 644,** A04

mA (milli ampere)—a unit of current equal to 0.001 amperes, used to express the tube current of an X-ray tube. **E 1316,** E07

mandrel—a form used as a cathode in electroforming; a mold or matrix. **B 374,** B08

mandrel—a form, fixture, or male mold used as the base for production of a part in processes such as lay-up or filament winding. **D 3878,** D30

mandrel—a form, fixture, or male mold used in the production of a part by lay-up or filament winding. **E 1749,** E06

mandrel (in bend testing), *n*—the tool used to control the strain on the concave side of a bend in a wrap-around bend test and also to apply the bending force in a semi-guided or guided bend test.

E 6, E28

manganese greensand—a manganese dioxide coated greensand used as a filter medium for removal of manganese and iron. See **greensand**. **D 6161,** D19

manhole—a precast concrete structure for vertical access to a pipeline or other closed structure. **C 822,** C13

manhole base—a concrete slab foundation for the bottom manhole riser section or a bottom manhole riser section with an integrally cast concrete floor. **C 822,** C13

manhole reducer section—a concrete pipe section used as a transition between different-diameter manhole riser sections. **C 822,** C13

manhole riser section—a concrete pipe section used to construct a manhole exclusive of the base, reducers, and top section. **C 822,** C13

manhole top—the concrete slab top or conical top used to reduce the manhole riser diameter to that of the desired access hole. **C 822,** C13

manifest—see **hazardous-waste manifest**. **E 1605,** E06

manifold—see **grout header**. **D 653,** D18

manifold—an enlarged pipe with connections available to the individual feed, brine and product ports of a desalination device. **D 6161,** D19

manifold—to make many or several copies. **F 221,** F05

manifolding performance—the intensity and sharpness of the image as a function of the number of carbon copies produced in one typing or writing. **F 221,** F05

manifold paper, *n*—a lightweight paper used primarily for copies by interleaving with carbon paper. **D 1968,** D06

manipulation resistant lock—a combination lock designed so as to resist manipulation. **F 471,** F12

man-made fiber, *n*—a class name for various genera of filament, tow or staple produced from fiber forming substances which are chemically synthesized or modified. (Syn. **manufactured fiber**) **D 123,** D13

manmades—materials made by chemical processes such as synthetic textile fibers and most plastics, or regenerated natural polymers such as cellulose and collagen to form paper, rayon, or reconstituted leather. **F 869,** F08

man-made staple fiber, *n*—fiber of spinnable length manufactured directly or by cutting filaments. **D 123,** D13

mannan—strictly, a polysaccharide composed entirely of mannose units, but used conventionally to distinguish the hexosan wood hemicelluloses from the pentosans (xylan). (see also **galactoglucomannan** and **glucomannan**) **D 1695,** D01

mannogalactan—see **galactomannan. D 1695,** D01

manual count, *n*—measurement of traffic characteristics based on human observation, which may or may not be electronically recorded. **E 867,** E17

manufactured building—See **building. E 631,** E06

manufactured carbon—a bonded granular carbon body whose matrix has been subjected to a temperature typically between 900 and 2400°C. **C 709,** D02

manufactured fiber, *n*—a class name for various genera of filament, tow, or staple produced from fiber forming substance which may be (1) polymers synthesized from chemical compound, (2) modified or transformed natural polymers, or (3) glass. (See **man-made fiber**) **D 123,** D13

manufactured graphite—a bonded granular carbon body whose matrix has been subjected to a temperature typically in excess of 2400°C and whose matrix is thermally stable below that temperature. **C 709,** D02

manufactured home—See **building. E 631,** E06

manufactured sand, *n*—fine aggregate produced by crushing rock, gravel, iron blast-furnace slag, or hydraulic-cement concrete. **C 125,** C09

manufactured unit, *n*—a quantity of finished adhesive or finished adhesive component, processed at one time. (Compare **batch**.) **D 907,** D14

manufacturer, *n— as related to corrugated metal pipe*, the organization that produces the metal sheet from which pipe is made. **A 902,** A05

manufacturer, *n*—the organization responsible for the conversion of materials into products meeting the requirements of a product specification. **A 941,** A01

manufacturer, *n—in pipe laying*, the party that manufactures, fabricates, or produces materials or products. **C 460,** C17

manufacturer—the group, corporation, partnership, or individual that manufactures or supplies a product. **C 822,** C13

manufacturer, *n*—the party that manufactures, fabricates, or produces materials or products. **C 1154,** C17

manufacturer, *n—in pipe laying*, the party that manufactures, fabricates, or produces materials or products. **D 2946,** C17

manufacturer, *n—of a tire*, the name of a company or wholly owned subsidiary making the tire. **F 538,** F09

manufacturer's joint— See **container. D 996,** D10

manufacturer's leather—see **factory sole leather. D 1517,** D31

manufacturer's seam—joint produced by corrugated box manufacturer during container fabrication. **F 592,** F16

manufacturing imperfections—includes all defects or blemishes that are produced in manufacturing, such as chipped grain, loosened grain, raised grain, torn grain, skips in dressing, hit and miss, variation in sawing, miscut lumber, machine burn, machine gouge, mismatching, and insufficient tongue or groove. **D 9,** D07

map cracking, *n*—a series of interconnected cracks that extend only into the upper portion of the slab. **E 867,** E17

map cracking, *n*—a series of interconnected cracks that extend only into the upper portion of the slab. **E 1778,** E17

mar—mutilation of polish film reparable only by recoating. **D 2825,** D21

mar, *n*—a mark made on the flooring surface by the deposition of material from friction or rubbing of traffic bodies against the surface. **F 141,** F06

maraging, *n*—a precipitation hardening treatment applied to a special group of **alloy steels** to precipitate one or more intermetallic compounds in a matrix of essentially carbon-free martensite. **A 941,** A01

marbleized finish—a surface appearance, obtained by coloring and graining, that resembles variegated marble. **C 286,** B08

marching modulus curve, *n*—a type of vulcanization during which the increasing modulus does not reach a maximum value. **D 1566,** D11

margin—the distance between any boundary of the printing area and the nearest parallel paper edge. **F 149,** F05

marginal cost (benefit)—See **incremental cost (benefit)**. **E 833,** E06

marine—relating to the ocean. **F 1600,** F20

marine board, *n*—an insulation board of 750 ± 100 kg/m^3 density that meets the noncombustibility criteria of 46 CFR 164.009. **E 176,** E05

markers—the electronically generated time pulses or other indicators that are used on the instrument display to measure distance or time. **E 1316,** E07

marking, *n—in packaging,* the application of numbers, letters, **labels, tags,** symbols or colors to provide identification and to expedite handling during shipment and storage. (Compare **code.**) **D 996,** D10

marking, *n*—the removal of printed ink from the web to idler rolls, turn bars, or nose cone on a printing press. **D 6488,** D01

marking—marking or stamping the size, case number, stock number, and similar information on the lining of a shoe. The stamping of trade names and numbers on sock linings, heel pads, and soles is called branding or embossing because it is done with a heated brand or die. **F 869,** F08

marking, marker head— See **lettered head, numeral head.** **F 547,** F16

marking nail—See **dating nail, marking head, numeral head.** **F 547,** F16

marking position—the area designated to mark information on a mark read form. Also called a **response position**. **F 149,** F05

mark of conformity, (*for certification*)—protected mark, applied or issued under the rules of a certification system, indicating that adequate confidence is provided that the relevant product, process or service is in conformity with a specific standard or other normative document, (ISO Guide 2). **E 1187,** E36

markoff—an indentation or imprinting of the skin surface due to any cause (such as foreign matter between mating parts, improper tooling, mismatch of detail parts, etc.). **E 1749,** E06

mark reading—machine reading, by optical means, of marks (usually vertical or horizontal bars) that have been manually entered. **F 149,** F05

mark scanning—the automatic optical sensing of marks usually recorded manually on a data medium. **F 149,** F05

mark sensing—machine reading of marks (usually pencil strokes) on a punched card, by using the conductive properties of the mark itself. **F 149,** F05

marl—calcareous clay, usually containing from 35 to 65 % calcium carbonate ($CaCO_3$). **D 653,** D18

mar resistance, *n*—(1) ability of a coating to resist visual damage caused by light abrasion, impact, or pressure. (2) resistance of the surface of the coating to permanent deformation resulting from the application of a dynamic mechanical force. **D 16,** D01

marsh—a wetland characterized by grassy surface mats which are frequently interspersed with open water or by a closed canopy of grasses, sedges, or other herbacious plants. **D 653,** D18

martempering, *n*—**quenching** an austenitized steel object in a medium at a temperature in the upper part of, or slightly above, the **martensite range**, holding it in the medium until its temperature is substantially uniform throughout, and then cooling it in air through the **martensite range**. **A 941,** A01

martensite—metastable phase resulting from the diffusionless athermal decomposition of austenite below a certain temperature known as the M_s temperature (martensite start temperature). It is produced during quenching when the cooling rate of a steel, in the austenitic condition, is such that the pearlite and bainite, or both, transformation is suppressed. The composition of the martensite is identical with that of the austenite from which it transformed. Hence, martensite is a super-saturated solid solution of carbon in alpha iron (ferrite) having a body-centered tetragonal lattice. It is a magnetic plate-like constituent formed by a diffusionless shear type of transformation. These plates may appear needle-like or veriform in cross-section. **E 7,** E04

martensite range, *n*—the temperature interval between M_s and M_f. **A 941,** A01

martensitic—a plate-like constituent having a similar appearance and mechanisms of formation to that of martensite. **E 7,** E04

marver—(*1*) a flat plate on which a hand gather of glass is rolled, shaped, and cooled.

(*2*) also the process of doing same. **C 162,** C14

mash—the mixture of sugars, nutrients, and water that is capable of being fermented by microorganisms such as yeast in ethanol fermentation. **E 1705,** E48

masked area, *n*—a portion of the exposed specimen which is protected from light exposure by masking. (See also **control**.) **G 113,** G03

masking—various materials applied to specific areas of parts to prevent coating from being deposited. **B 374,** B08

masking, *n*—the phenomenon where one quality within a mixture obscures one or several other qualities present. **E 253,** E18

masking—in leak testing, the covering of a section of a test object so as to prevent tracer gas from entering leaks that may exist in the covered section. **E 1316,** E07

masking power—the ability of a fired glaze to mask visually the body on which it is applied. **C 242,** C21

masonry, *n*—construction, usually set in mortar, of natural building stone or manufactured units such as brick, concrete block, adobe, glass block, tile, manufactured stone, or gypsum block. **E 631,** E06

masonry cement, *n*—a hydraulic cement for use in mortars for masonry construction, containing one or more of the following materials: portland cement, portland blast furnace slag cement, portland-pozzolan cement, natural cement, slag cement, or hydraulic lime; and in addition usually containing one or more materials such as hydrated lime, limestone, chalk, calcareous shell, talc, slag, or clay as prepared for this purpose. **C 11,** C11

masonry cement, *n*—a hydraulic cement manufactured for use in mortars for masonry construction or in plasters, or both, which contains a plasticizing material and, possibly, other performance-enhancing addition(s). **C 219,** C01

masonry nail—plain, electro-zinc-plated or galvanized, hardened-steel, knurled (longitudinally or nearly longitudinally threaded or fluted), ½ to 4 by 0.148 or up to 0.250-in. nails with flat or checkered 5⁄16 to -in. head and medium diamond point. **F 547,** F16

masons hydrated lime—hydrated lime suitable for use for masonary purposes. **C 51,** C07

masons quicklime—quicklime suitable (after slaking to a lime putty) for use for masonry purposes. **C 51,** C07

mass, *n*—the quantity of matter in a body. (See also **weight**.) **D 123,** D13

mass:

analyzer—a device for dispersing ions as a function of their mass-to-charge ratio.

resolution—the ratio $M/\Delta M$ where ΔM is the full width at half-maximum peak height for the ion peak of mass M.

resolving power—the peak-to-valley ratio between adjacent, equal-sized peaks, separated by one mass unit.

spectrum—a plot of the measured ion signal as a function of mass-to-charge ratio. **E 673,** E42

mass-absorption coefficient—See **absorption coefficient**. **E 7,** E04

mass, active, m_1—the effective value of mass, which may be used with values of ℓ_1, and A' to evaluate a magnetic core as though it has an equivalent uniform flux path having the same induction at all points. **A 340,** A06

mass attenuation coefficient—a measure of the fractional decrease in radiation beam intensity per unit of surface density $cm^2 \cdot gm^{-1}$. **E 1316,** E07

mass burning rate, *n*—mass loss per unit time by materials burning under specified conditions. **E 176,** E05

mass color—the color, when viewed by reflected light, of a pigment-vehicle mixture of such thickness as to obscure completely the background. Sometimes called over-tone or mass-tone. **D 16,** D01

mass concentration—See **concentration.** **D 1356,** D22

mass concentration, *n*— *in multi-component or multi-phase mixtures,* the mass of a specified component or phase per unit mass or unit volume of the total. (See also **particle concentration.**) **G 40,** G02

mass energy-absorption coefficient, μ_{en}/ρ—of a material for uncharged ionizing particles, the product of the mass energy transfer coefficient, μ_{tr}/ρ, and $(1\text{-}g)$, where g is the fraction of the energy of secondary charged particles that is lost to bremsstrahlung in the material (ICRU).

$$(\mu_{en}/\rho) = (\mu_{tr}/\rho)(1 - g)$$

Unit: $m^2 \cdot kg^{-1}$ **E 170,** E10

mass energy-transfer coefficient (μ_{tr}/ρ)—of a material for uncharged ionizing particles, the quotient of dE_{tr}/EN by ρdl, where E is the energy of each particle (excluding rest energy), N is the number of particles, and dE_{tr}/EN is the fraction of incident particle energy that is transferred to kinetic energy of charged particles by interactions in transversing a distance dl in the material of density ρ (ICRU).

$$(\mu_{tr}/\rho) = 1/\rho EN \cdot dE_{tr}/dl$$

Unit: $m^2 \cdot kg^{-1}$ **E 170,** E10

mass-fiber—asbestos that is randomly oriented in situ. **D 2946,** C17

mass loss concentration, *n*—the mass loss of a test specimen per unit exposure chamber volume in $g \cdot m^{-3}$. **E 176,** E05

mass median size, *n*—a measurement of particle size for samples of particulate matter, consisting of that diameter such that the mass of all larger particles is equal to the mass of all smaller particles. **D 1356,** D22

mass number—the whole number nearest to the atomic mass expressed in either atomic mass units or as (chemical) atomic weight. **E 1316,** E07

mass scattering coefficient—that part of the mass absorption coefficient due to scattering. **E 7,** E04

mass spectrometer leak detector—a mass spectrometer adjusted to respond only to the tracer gas. **E 1316,** E07

mass spectrometer (M.S.)—an instrument that is capable of separating ionized molecules of different mass to charge ratio and measuring the respective ion currents. The mass spectrometer may be used as a vacuum gage that relates an output which is proportioned to the partial pressure of a specified gas, as a leak detector sensitive to a particular tracer gas, or as an analytical instrument to determine the percentage composition of a gas mixture. Various types are distinguished by the method of separating the ions. The principal types are as follows:

(a) Dempster (M.S.)—The ions are first accelerated by an electric field through a slit, and are then deflected by a magnetic field through 180° so as to pass through a second slit.

(b) Bainbridge-Jordan (M.S.)—The ions are separated by means of a radial electrostatic field and a magnetic field deflecting the ions through 60° so arranged that the dispersion of ions in the electric field is exactly compensated by the dispersion in the magnetic field for a given velocity difference.

(c) Bleakney (M.S.)—The ions are separated by crossed electric and magnetic fields. Also called cross fields (M.S.).

(d) Nier (M.S.)—A modification of the Dempster (M.S.) in which the magnetic field deflects the ions.

(e) Time of Flight (M.S.)—The gas is ionized by a pulse-modulated electron beam and each group of ions is accelerated toward the ion collector. Ions of different mass to charge ratios traverse their paths in different times.

(f) Radio-Frequency (M.S.)—The ions are accelerated into a radio-frequency analyzer in which ions of a selected mass to charge are accelerated through openings in a series of spaced plates alternately attached across a radio-frequency oscillator. The ions emerge into an electrostatic field which permits only the ions accelerated in the analyzer to reach the collector.

(g) Omegatron (M.S.)—The ions are accelerated by the cyclotron principle. **E 1316,** E07

mass spectrum—a record, graph, table, etc., that shows the relative number of ions of various mass that are produced when a given substance is processed in a mass spectrometer. **E 1316,** E07

mass strength, *n*—a measure of the tendency for carbon black pellets to pack together and to influence flow in a bulk handling system. **D 3053,** D24

mass-tone—see **mass color**. **D 16,** D01

masstone, *n*—in paint technology, a pigment-vehicle mixture containing a single colorant only. **E 284,** E12

mass, total, *m*—the actual mass of a magnetic core. **A 340,** A06

mass transfer coefficient (MTC)—mass (or volume) transfer through a membrane based on driving force. **D 6161,** D19

mass transfer zone (adsorption wave) (adsorption zone)—the region in which the concentration of the adsorbate of interest in the fluid decreases from influent concentration to the lowest detectable concentration. **D 2652,** D28

mass unit weight—see **unit weight**. **D 653,** D18

master—an intermediate, temporary, and usually expendable image vehicle for use on duplicators. **F 335,** F05

master—a coated sheet of imaged material (paper, plastic, or metal) which receives the ink onto the image and which is attached to the plate cylinder; the ink in turn is transferred to create the final document. **F 425,** F05

master alloy—an alloy rich in one or more desired addition elements that is added to a melt to raise the percentage of a desired constituent. **B 374,** B08

master-alloy powder, *n*—a powder with high alloy concentration, designed to be diluted when mixed with a base powder to produce the desired composition. **B 243,** B09

masterbatch, *n*—a homogeneous mixture of rubber and one or more materials in known proportions for use as a raw material in the preparation of the final compounds. **D 1566,** D11

master composite sample, *n*— *in asbestos,* not less than two and not more than ten composite samples so combined and reduced as to form a test sample that represents a lot that consists of more than 200 bags but not more than 2000 bags. **D 2946,** C17

MasterFormat—a standard sequence of numbers and titles for organizing information about construction requirements, products, and activities (source 1995 edition of MasterFormat, published by CSI). **E 833,** E06

master mold—a form or template utilized for multiple fabrication. **F 869,** F08

master set, *n*—a selected group of witness tires, each with different test response characteristics to provide a range of values for the measured property or properties. **F 538,** F09

mastic, *n*—a material of relatively viscous consistency that dries or cures to form a protective finish, suitable for application to thermal insulation in thickness greater than 30 mils (0.76 mm) per coat. **C 168,** C16

mastic, n—See **mastic adhesive**. **D 907,** D14

mastic—See **flashing cement** and **asphalt mastic**. **D 1079,** D08

mastic—a thick adhesive material used as a cementing agent for holding waterproofing membrane in place or used as a caulking for sealing. **D 3743,** D04

mastic adhesive, *n*—a gap-filling adhesive applied as a paste or putty-like material. (Also called *mastic*.) **D 907,** D14

mastication, *n*—a breakdown or softening of raw rubber, or a mix by the combined action of mechanical work (shear) and atmospheric oxygen, sometimes accelerated by the use of a peptizer and frequently at elevated temperatures. **D 1566,** D11

mat—a glass-fiber product of felt-like nature. **C 162,** C14

mat, *n*—a fibrous material consisting of randomly oriented chopped or swirled filaments loosely held together with a binder. **D 883,** D20

mat—a random arrangement of fine fibers uniformly distributed to form a thin, highly porous, felt-like material. **E 1749,** E06

Matadero hides—Argentinian cattlehides corresponding roughly to city butcher or small packer hides in the United States. **D 1517,** D31

match, n—see color match. **E 284,** E12

match, *v*—to provide, by selection, formulation, adjustment, or other means, a trial color that is indistinguishable from, or within specified tolerances of, a specified standard color under specified conditions. **E 284,** E12

matched lumber—See **lumber.** **D 996,** D10

matched test sieve, *n*—a test sieve that reproduces the results of a master test sieve within defined limits for a designated material. **E 1638,** E29

matching—Under **lumber** see *matched lumber.* **D 9,** D07

matching—*in plywood*, the arrangement of strips of veneer to obtain a particular repetitive pattern.

book matching—turning alternative adjacent strips of veneer of a flitch over.

mismatching—making sheets of face veneer from specially selected dissimilar (in color or grain or both) strips of veneer.

reversed matching—turning alternate adjacent strips of veneer of a flitch end for end; also called "swing matching."

slip matching—laying adjacent strips of veneer tight side up without turning; also called "slide matching." **D 1038,** D07

matching, *n*—the experimental process of equating or relating stimuli, usually to determine the similarity between standard and unknown, or between unknowns. **E 253,** E18

mated mechanical transfer—a system requiring a CB donor sheet, usually a colored or black pigmented layer which may be bound or covered by a protective material and a CF receptor sheet on which an image is produced on application of pressure. **F 549,** F05

mated transfer paper—a transfer paper requiring two different coatings to come into contact to develop an image. **F 549,** F05

material, *n*—a quantity of carbon black with unique composition and characteristics. **D 3053,** D24

material, *n*—in home sewing, a planar structure such as textile fabric, plastic film, or leather. **D 4965,** D13

material form—the contour, arrangement, and structure of an unconsolidated composite material, especially with regard to the geometry and nature of the reinforcement. Factors considered part of the material form include, but are not limited to: reinforcement diameter, reinforcement length (for discontinuous reinforcements), tow size or count, fabric areal weight, fabric style, reinforcement content, and ply thickness. **D 3878,** D30

materials engineer—person who works with the properties of materials, usually to alter them in some manner so as to make them more useful for some application. **F 1156,** F12

material test report—a written document or electronic record, signed by an authorized party, attesting that the raw material is in accordance with specified requirements, including the actual results of all required chemical analyses, tests, and examinations. **F 1789,** F16

material tests—the quality and property tests of component materials. **C 822,** C13

material transfer—a general term to describe the carry-over of material from one electrical contact to another.

NOTE—When the discussion becomes specific, the term "gain" or "loss" is used with respect to a particular contact (for d-c application, anode or cathode; for a-c application, stationary contact or movable contact). For example, anode gain, anode loss, stationary contact gain. **B 542,** B02

material transfer, bridge—material transfer that occurs without the presence of a gaseous electric discharge. The filament of molten contact material that connects the two separating contacts does not rupture in the middle; thus there is a gain of material on one contact and a loss of material from the other. **B 542,** B02

material transfer, cathode—movement of contact metal from the cathode by means of a cathode arc. **B 542,** B02

material transfer, needle—material transfer that results in a buildup with a smaller diameter and a relatively great length. **B 542,** B02

material transfer, negative—see **material transfer.** **B 542,** B02

material transfer, positive—see **material transfer.** **B 542,** B02

maternity wear, *n*—clothing worn during pregnancy. **F 1494,** F23

mat finish—smooth dull finish applied to leather for shoe uppers, handbags, belts, etc. **D 1517,** D31

mat finish (matte finish)—a dull finish. **B 374,** B08

mat-formed particleboard—a particleboard in which the coated particles are formed first into a mat having substantially the same length and width as the finished board before being flat-platen pressed. **D 1554,** D07

mat glaze—See *mat glaze* under **glaze.** **C 242,** C21

mathematical/analytical (M/A) technique—a technique of obtaining probability functions for economic measures of project worth without the repeated trials of simulation. **E 833,** E06

mathematical model—the representation of a physical system by mathematical expressions from which the behavior of the system can be deduced with known accuracy. (ISRM) **D 653,** D18

mat reinforcement— see **quadrant mat.** **C 822,** C13

matric potential—the energy required to extract water from a soil against the capillary and adsorptive forces of the soil matrix. **D 653,** D18

matric suction—for isothermal soil systems, matric suction is the pressure difference across a membrane separating soil solution, in-place, from the same bulk (see **soil-water pressure**). **D 653,** D18

matrix—See **mandrel.** **B 374,** B08

matrix, *adj*—a descriptive term for a textile fiber in which one or more polymeric fibrous material(s) is dispersed in another. **D 123,** D13

matrix—*in grouting*, a material in which particles are embedded, that is, the cement paste in which the fine aggregate particles of a grout are embedded. **D 653,** D18

matrix, *n*—the part of an adhesive which surrounds or engulfs embedded filler or reinforcing particles and filaments. **D 907,** D14

matrix—substance in which the analyte or property exists. **D 1129,** D19

matrix—the continuous phase. **E 7,** E04

matrix, *n*—*in methods of chemical analysis,* all components of a material except the analyte. **E 135,** E01

matrix—the essentially homogeneous phase in a composite material in which reinforcements such as fibers, filaments, particles, etc., are embedded. **E 1749,** E06

matrix array, *n*—a set of rows and columns of transductive elements able to convert the quantity of light received to a relatively-sized modulated electrical signal. Distinguished from a linear array. (See **photodiode array (matrix)**.) **F 1294,** F05

matrix content—the amount of matrix present in a composite expressed either as percent by weight or percent by volume. For polymer matrix composites this is resin content. **D 3878,** D30

matrixed binary codes—two-dimensional binary notation symbology; similar to an encrypted or scrambled bar code except that it is two-dimensional, not limited to only one dimension. **F 1156,** F12

matrix effects:

Auger—any change of an Auger spectrum (for example, shape or signal strength) due to the physical environment (for example, amorphous/crystalline, thin layer/thick layer, or rough/smooth surface) of the emitting element and not due to chemical bonding or changes in concentration.

SIMS—any change in the secondary ion yields which are caused by changes in the chemical composition or structure of a particular specimen.

AES—see **Auger**. **E 673,** E42

matrix material (preparation of reference materials)—the principal materials in which one or more constituents may be dispersed. **C 859,** C26

matrix metal—the continuous phase of a polyphase alloy or mechanical mixture; the physically continuous metallic constituent in which separate particles of another constituent are embedded. **B 243,** B09

matrix of composite superconductor—the continuous longitudinal phase of a pure metal, a polyphase alloy, or mechanical mixture that is not in the superconducting state at the normal operating conditions of the embedded superconductor. **B 713,** B01

matrix symbols, *n*—an arrangement of regular polygon shaped cells where the center-to-center distance of adjacent elements is uniform and the arrangement of cells represents data or symbology functions, or both. **F 1294,** F05

matte, *n*—lacking luster or gloss. Synonymous with "flat" in paint terminology. **E 284,** E12

matting, *n*—*for pile yarn floor coverings,* loss of tuft definition due to entangle. **D 5684,** D13

mattress, *n*—a mattress is a ticking (outermost layer of fabric or related material) filled with a resilient material, used alone or in combination with other products, intended or promoted for sleeping upon. **E 176,** E05

mattress blade—special clinching blade for fastening sisal pads in bedding plants, permitting stapling inside edging wire. **F 592,** F16

maturation (latex), *n*—controlled storage of compounded latex before processing, which allows stabilization and escape of air bubbles, optimizes dispersion of ingredients, and allows partial pre-vulcanization of rubber particles. **D 1566,** D11

mature fibers, *n*—(cotton fibers treated with sodium hydroxide solution)—fibers that have swollen into unconvoluted and almost rod-like shapes, where total wall width is equal to or greater than the lumen width. **D 123,** D13

mature fibers, *n*—(cotton fibers observed under polarized light)—fibers that appear yellow, yellow green, or green and are yellow or light yellow upon rotation to the subtractive position (through 90°) and show little or no parallel extinction on removal of the selenite plate. (Compare **immature fibers.**) **D 123,** D13

maturing range—the time-temperature range within which a ceramic body, glaze, or other composition may be fired to yield specified properties. **C 242,** C21

maturing temperature—the temperature at which porcelain enamel must be held for a selected time to achieve the desired properties. **C 286,** B08

maturing temperature, *n*—the temperature, as a function of time and bonding condition, that produces desired characteristics in bonded components. **D 907,** D14

maturity, *n*—the extent of the development of a property of a cementitious mixture. **C 125,** C09

maturity function, *n*—a mathematical expression which uses the measured temperature history of a cementitious mixture during the curing period to calculate an index that is indicative of the maturity at the end of that period. **C 125,** C09

maturity index, *n*—an indicator of maturity which is calculated from the temperature history of the cementitious mixture by using a maturity function. **C 125,** C09

maturity index, *n*—a relative indication of cotton fiber maturity. **D 123,** D13

maximum amplitude (L, LT^{-1}, LT^{-2})—deviation from mean or zero point. **D 653,** D18

maximum bend moment—the greatest moment applied to a needle during a bend test. **F 1840,** F04

maximum curve—in a phase diagram, a univariant line, tracing the meetings of a pair of bivariant surfaces at intermediate composition and coinciding with their highest temperature at each pressure level, or their highest pressure at each temperature level; congruent transformation occurs everywhere along a maximum curve. **E 7,** E04

maximum density (*maximum unit weight*)—see **unit weight**. **D 653,** D18

maximum draft—maximum vertical dimension of the boom below the water line. **F 818,** F20

maximum dynamic load—sum of all instantaneous dynamic loads including those due to acceleration, wave forces, and so forth. **F 818,** F20

maximum elongation, El_{max}, *n*—the elongation at the time of fracture, including both elastic and plastic deformation of the tensile specimen. **E 6,** E28

maximum erosion rate, *n*— *in cavitation and liquid impingement erosion,* the maximum instantaneous erosion rate in a test that exhibits such a maximum followed by decreasing erosion rates. (See also **erosion rate-time pattern**.) **G 40,** G02

maximum hardness—a hardness specified in fastener standards above which the fastener is considered nonconforming to the standard. **F 1789,** F16

maximum heating temperature, *n*—the maximum temperature, as recommended by the manufacturer, to which a hot-applied sealant or filler for pavement cracks or joints can be heated while conforming to all specification requirements and result in appropriate application characteristics. **D 5535,** D04

maximum image area, *n*—the portion on a page which can be printed. (See **percent coverage** and **image area.**) **F 1457,** F05

maximum inflation pressure, *n—for inflatable restraints*, the maximum internal cushion pressure occurring after breakout pressure. **D 123,** D13

maximum life expectancy, LE-1000, *n—for paper*, a paper is expected to be usable for 1000 years. **D 1968,** D06

maximum life expectancy paper, LE-1000, *n—for paper*, a paper expected to be usable for 1000 years under prescribed conditions. **D 1968,** D06

maximum load, P_{max} [F]—*in fatigue*, the highest algebraic value of applied load in a cycle. By convention, tensile loads are positive and compressive loads are negative. **E 1823,** E08

maximum load rating, [*M*], *n—of a passenger tire,* the load rating at the maximum permissible cold inflation pressure for that tire. **F 538,** F09

maximum (non-statistical), *n*—the highest acceptable actual test result; any valid individual test result above the maximum is cause for rejection of the component or material lot being tested.

$$\text{All } x_i \leq \text{Maximum} = USL$$

A 644, A04

maximum operating distance from outlet, *n*—the maximum operating distance of the vacuum cleaner from the supply outlet is the straight line measured distance from the front of the nozzle to the face of the cord input plug or to the face of the wall outlet (see Discussion) in the case of a central vacuum cleaning system, with the maximum straight line extension from the front of the cleaning system nozzle in the operating position parallel to and touching the floor. **F 395,** F11

maximum page length, *n*—the maximum number of lines of text or graphic equivalent which a printer can output onto a sheet of paper; this value can be set by either the printer or the computer, or both. **F 1457,** F05

maximum permissible errors, *n—of a thermometer or thermometric system*, extreme values permitted by regulation or specification of the difference between the indication of a thermometer or thermometeric system and the true value of temperature. **E 344,** E20

maximum plunger travel, [*L*], *n—in tire testing*, the relative displacement of tread surface by a plunger, measured from the point of initial contact of the plunger with the tread surface to the point of maximum force at rupture or at the bottom-out point. **F 538,** F09

maximum point—that composition and temperature, or pressure, at which a heterogeneous equilibrium occurs at its highest temperature, or pressure, when this does not coincide with one of the composition limits of the equilibrium, that is, when it occurs at an intermediate composition; the equilibrium becomes congruent (univariant) at the maximum point. **E 7,** E04

maximum print position, *n*—the rightmost point at which the printer can mark the paper. **F 1457,** F05

maximum rate period, *n— in cavitation and liquid impingement erosion*, a stage following the acceleration period, during which the erosion rate remains constant (or nearly so) at its maximum value. (See also **erosion rate-time pattern.**) **G 40,** G02

maximum self-registering clinical thermometer, *n*—clinical thermometer designed to retain the indication of its maximum measured temperature until reset. **E 344,** E20

maximum size (of aggregate), *n—in specifications for, or descriptions of aggregate*, the smallest sieve opening through which the entire amount of aggregate is required to pass. **D 8,** D04

maximum sound level, L_{AFmax}*[nd], (dB) n*—Ten times the common logarithm of the square of the ratio of the largest frequency-weighted and exponential-time-weighted (or other wise time-averaged) sound pressure during the measurement period to the square of the reference-sound-pressure of 20 micro pascals. The subscripts designate the frequency weighting (A or C), and time the weighting or averaging (F for fast, S for slow, I for impulse, or a number with proper units to indicate time interval). **C 634,** E33

maximum (statistical), *n*—the highest acceptable statistical test result; for compliance, the sample mean ($\bar{x}$) plus *M* standard deviations(s), where *M* is a matter of agreement between the supplier and purchaser, must be less than, or equal to, the upper specification limit.

$$\bar{x} + M \times s \leq \text{Maximum} = USL$$

A 644, A04

maximum stress-intensity factor, K_{max} [$FL^{-3/2}$]—*in fatigue*, the maximum value of the stress-intensity factor in a cycle. This value corresponds to P_{max}. **E 1823,** E08

maximum sublimation point—in a PT phase diagram, the highest pressure and temperature at which a solid species of intermediate composition may exist in equilibrium with vapor of identical composition. **E 7,** E04

maximum use pressure—the maximum pressure to which a material can be subjected due to a reasonably foreseeable malfunction, operator error, or process upset. **G 126,** G04

maximum use temperature—the maximum temperature to which a material can be subjected due to a reasonably forseeable malfunction, operator error, or process upset. **G 126,** G04

maxwell, φ—the unit of magnetic flux in the cgs-emu system of units. One maxwell equals 10^{-8} weber. See **magnetic flux.**

NOTE—

$$e = -N d\phi/dt \times 10^{-8}$$

where:

e	=	induced instantaneous emf in volts,
$d\phi/dt$	=	time rate of change of flux in maxwells per second, and
N	=	number of turns surrounding the flux, assuming each turn is linked with all the flux.

A 340, A06

MCL—maximum contaminant level. **D 6161,** D19

McQuaid-Ehn grain size—the austenitic grain size developed in steels by carburizing at 1700°F (927°C) followed by slow cooling. See Test Methods E 112. **E 7,** E04

MD—State of Maryland. **E 1605,** E06

MDSA—see **message data storage area.** **F 1457,** F05

mealiness, *n*—an uneven or grainy solid or halftone dot caused by improperly ground ink or the presence of foreign matter. **D 6488,** D01

mean—arithmetical average of a set of numbers. **D 1517,** D31

mean coefficient of thermal expansion, α (mm/mm·°C or in./in.·°C)—from temperature T_1 to temperature $TT_1 < T_2$):

$$\alpha = \frac{0.01P}{T_2 - T_1}$$

where P = percent linear thermal expansion, as defined below. **C 242,** C21

mean crossings—*in fatigue loading*, the number of times that the load-time history crosses the mean load level with a positive slope or a negative slope, or both, as specified during a given period. **E 1823,** E08

mean depth of erosion, *n— in cavitation and impingement erosion*, the average thickness of material eroded from a specified surface area, usually calculated by dividing the measured mass loss by the density of the material to obtain the volume loss and dividing that by the area of the specified surface. (Also known as *mean depth of penetration* or *MDP*. Since that might be taken to denote the average value of the depths of individual pits, it is a less preferred term.) **G 40,** G02

mean depth of penetration, n—See **mean depth of erosion.** **G 40,** G02

mean deviation unevenness, *U %, n—in textiles*, the average of the absolute values of the deviations of the linear densities of the integrated lengths between which unevenness is measured and expressed as a percentage of the average linear density for the total length within which unevenness is measured. (See also **unevenness, coefficient of variation unevenness.**) **D 123,** D13

mean diameter, arithmetic—See *arithmetic mean diameter* under **diameter.** **C 242,** C21

mean diameters—the family of diameters, $\bar{D}_{pq}$, such that:

$$\bar{D}_{pq}{}^{(p-q)} = (\Sigma_i D_i{}^p)/(\Sigma_i D_i{}^q)$$

where:

p and q	=	dissimilar positive integers. (The value of q may also be "zero.")

E 1620, E29

mean droplet diameter ($\bar{D}_{pq}$)—is represented by the following:

$$\bar{D}_{pq}^{(p-q)} = \frac{\Sigma D_i^p}{\Sigma D_i^q}$$

where:

D_i	=	the diameter of the ith particle,
ΣD_i^o	=	the total number of drops in the sample. Thus:
$\bar{D}_{10}$	=	length mean diameter,
$\bar{D}_{20}$	=	area mean diameter,
$\bar{D}_{30}$	=	volume mean diameter,
$\bar{D}_{31}$	=	volume/length mean diameter (mean evaporative diameter),
$\bar{D}_{32}$	=	SAUTER mean diameter, and
$\bar{D}_{43}$	=	DeBROUKERE or HERDAN mean diameter.

E 1102, E35

mean escape depth—the average depth normal to the surface from which the specified particles or radiations escape as defined by:

$$\int_0^\infty z\phi(z,\theta)dz \,/ \int_0^\infty \phi(z,\theta)dz$$

where $\phi(z,\theta)$	=	the emission depth distribution function for depth *z* from the surface into the material and for direction of emission θ with respect to the surface normal.

E 673, E42

mean free path—See **electron**, *inelastic mean free path.* **E 673,** E42

mean free path—the average distance that a molecule travels between successive collisions with other molecules. **E 1316,** E07

mean ionic activity coefficient—*for a salt that is composed of two monovalent ions,* the geometric mean of the individual ionic activity coefficients. (The geometric mean is obtained, in this case, by multiplying the two individual ionic activity coefficients and then taking the square root.) It is important because, unlike individual ionic activity coefficients, it can be measured by a variety of techniques, such as freezing point depression and vapor pressure, as well as by paired sensing electrodes. **D 4127,** D19

mean length, *n—in testing of cotton fibers*, the average length of all the fibers in the test specimen based on mass-length data. **D 123,** D13

mean load, P_m (or S_m or ϵ_m) [F or FL^2]— *in fatigue loading*, the algebraic average of the maximum and minimum loads in constant amplitude loading, or of individual cycles in spectrum loading,

$$P_m = \frac{P_{max} + P_{min}}{2}$$

or the integral average of the instantaneous load values of a spectrum loading history. **E 1823,** E08

mean modulus—the ratio of mean stress to mean strain. (ISO) **D 4092,** D20

mean panel rating (MPR), *n*—the average value, for each section of highway pavement, of ride quality ratings assigned by a ride rating panel. **E 867,** E17

mean particle diameter—the weighted average particle size, in millimetres, of a granular adsorbent computed by multiplying the percent retained in a size fraction by the respective mean sieve openings, summing these values and dividing by 100. **D 2652,** D28

mean profile depth (MPD), *n*—the average of all mean segment depths of all segments of the profile. **E 867,** E17

mean roughness index (MRI), *n*—the average of the international roughness index (IRI) values for the right and left wheel tracks. **E 867,** E17

mean segment depth, *n*—the average value of the profile depth of the two halves of a segment having a given baselength. **E 867,** E17

mean specific heat, *n*—the quantity of heat required to change the temperature of a unit mass of a substance one degree, measured as the average quantity over the temperature range specified. (It is distinguished from true specific heat by being an average rather than a point value.) (In SI units: J/kg•K) (In inch-pound units: Btu/lb•F) **C 168,** C16

mean square, *n—in analysis of variance*, a contraction of the expression "mean of the squared deviations from the appropriate average(s)" where the divisor of each sum of squares is the appropriate degrees of freedom. **D 123,** D13

mean strain—the average value of strain during a single complete hysteresis loop of cyclic deformation. (ISO) **D 4092,** D20

mean stress—the average value of the stress during a single complete hysteresis loop of cyclic deformation. (ISO) **D 4092,** D20

mean temperature, *n—in thermal transmittance of textiles only*, the average of the hot plate temperature and the temperature of the calm, cool air that prevailed during the test. **D 123,** D13

mean texture depth (MTD), *n*—the mean depth of the pavement macrotexture profile determined by the volumetric technique of Test Method E 965. **E 867,** E17

mean value—arithmetic average of data values; calculated as the algebraic sum of the data values divided by the number of data values. **E 1605,** E06

mean-variance criterion—a technique for evaluating the relative risk and return when choosing among competing projects that dictates that the project value with the higher mean (that is, expected value of project worth) and lower standard deviation be chosen. **E 833,** E06

mean ($\bar{x}$), *n*—the sum of the individual data points (*x*'s) divided by the number of data points (*n*).

$$\bar{x} = \frac{\Sigma x_i}{n}$$

A 644, A04

measured energy input rate—peak rate at which the equipment consumes energy, measured during a period when the equipment is operating at full input. **F 1827,** F26

measured heat release of specimen, *n*—the observed heat release under the variable flux field imposed on the specimen and measured. **E 176,** E05

measured inflation pressure, *n*—gage pressure of a tire measured at a given time under ambient temperature and barometric pressure. **F 538,** F09

measured sediment load—that part of the total sediment discharge that can be measured with available suspended-sediment samplers; does not include bed-load discharge and suspended sediment discharge very near the bed. **D 4410,** D19

measurement, *n*—act of quantifying a property or dimension. **E 1605,** E06

measurement accuracy—the numerical or percentage deviation of a measured value (or a value computed from one or more measurements) from its true value or from some absolute or standardized value. This deviation may depend upon the procedures used and is caused chiefly by systematic errors in the calibrations of the equipment used, which errors, if known, may be removed from the measured data to enhance the accuracy of the measured or computed value. **A 340,** A06

measurement area—the surface area that is examined for conformance to one or more specified requirements. **B 374,** B08

measurement control standard—a material or item used for monitoring the performance of a measurement system. **C 859,** C26

measurement interval, *n*—the distance travelled, in kilometres (miles), between two successive groove (void) depth measurements. **F 538,** F09

measurement plan, *n*—a document formally describing the specific steps to be taken during a measurement, including any unique requirements. **C 634,** E33

measurement quality assurance plan—a documented program for the measurement process that ensures, on a continuing basis, that the overall uncertainty meets the requirements of the specific application. This plan requires traceability to, and consistency with, nationally or internationally recognized standards. **E 170,** E10

measurement set, *n*—the set of acoustical measurements and related data obtained at a single measurement location during a specified time interval. **C 634,** E33

measurement system—the specific combinations of instrumentation, operator, and procedure used to make a particular measurement. **E 170,** E10

measurement traceability—the ability to demonstrate by means of an unbroken chain of comparisons that a measurement is in agreement within acceptable limits of uncertainty with comparable nationally or internationally recognized standards. **E 170,** E10

measurement uncertainty—an estimate of potential inaccuracies in a measured or derived quantity based on explicit evaluation and combination of all sources of errors. Note - Quantitative uncertainty estimates are typically given in the form of variances (or standard deviations) and covariances (or correlations) derived from statistical procedures combining random, systematic, and calculational (modeling) uncertainties. **E 7,** E04

measurement value, *n*—the numerical result of quantifying a particular property or dimension. (See **observation**). **D 123,** D13

measuring junction—the junction of a thermocouple placed at a location where an unknown temperature is to be measured or where an established temperature is to be maintained. When the measuring junction is at a higher temperature than the reference junction it is commonly called the hot junction. **E 7,** E04

measuring junction, *n*—that junction of a thermocouple which is subjected to the temperature to be measured. **E 344,** E20

mechanical adhesion, *n*—See *mechanical adhesion* under **adhesion.** **D 907,** D14

mechanical alignment— See **alignment.** **E 7,** E04

mechanical analysis—see **grain-size analysis.** **D 653,** D18

mechanical analysis—a determination of the particle-size distribution of a sample by mechanical separation. **D 4410,** D19

mechanical bonds, *n*—the attachment created when plaster penetrates, into or through, the substrate, or envelops irregularities in the surface of the substrate. **C 11,** C11

mechanical cell, *n*—a type of flotation cell that uses mechanical agitation of a pulp by means of an immersed impeller (rotor) and stator stirring mechanism. Aeration to the cell can be from an external pressurized air source or self-induced air. **D 121,** D05

mechanical cleaner—see **cleaner, mechanical.** **F 412,** F17

mechanical cleaning—process for removing dirt, scale, or other deposits from surfaces through the use of mechanical means such as abrasive blasting. **B 374,** B08

mechanical connection—a joining of two or more elements by means of mechanical fasteners, such as screws, bolts, or rivets but not by welding or adhesive bonding. **E 631,** E06

mechanical count, *n*—measurement of traffic characteristics by sensors and electronic recording of the measurements, independent of human observation. **E 867,** E17

mechanical deposition—a coating process in which particles of the plating metal are impacted against the fastener surface such that cold welding of the plating metal to the fastener surface is accomplished. **F 1789,** F16

mechanical disk scanner—a rotating scanning disk that breaks light reflection during the optical reading operation into a series of light points that are directed through the slit of a fixed aperture and onto the surface of a photomultiplier tube. **F 149,** F05

mechanical dispersion, *n*—a low energy aqueous medium produced by continuous stirring of the test solution and containing both dissolved and undissolved components of the test material. **D 6384,** D02

mechanical durability—resistance of the carbon paper or ribbon to cutting after repeated use. **F 221,** F05

mechanical freezers—freezers that operate by a refrigeration system in which cooling is provided by mechanical means such as a compressor. **E 1705,** E48

mechanical hygrometer—See **hygrometer.** **D 1356,** D22

mechanical hygrometer—an instrument for determining the water vapor content of an atmosphere by measuring the dimensional change produced in an hygroscopic material. **D 4023,** D22

mechanical hysteresis, *n*—the energy absorbed in a complete cycle of loading and unloading. **E 6,** E28

mechanical joint—see **joint, mechanical.** **F 412,** F17

mechanical leather—a collective term for many types of leather used in connection with textile and other machinery. **D 1517,** D31

mechanically alloyed powder, *n*—a composite powder produced by mechanically incorporating other constituents which are generally insoluble within the deformable particles of the matrix metal. **B 243,** B09

mechanically deformed—nails with rolthreaded or formed shanks to improve holding power.

roll-threaded— with helical, annular, or longitudinal deformations rolled on fastener shank.

formed—with barbed, angular, serrated, longitudinally or helically fluted, longitudinally or helically roll-grooved or twisted deformations formed on wire from which nail is made. **F 547,** F16

mechanically foamed plastic, *n*—a cellular plastic in which the cells are formed by the physical incorporation of gases. **D 883,** D20

mechanically plated, peen-coated, peen-galvanized—covered with coat of zinc through peen-coating, that is, by tumbling in a container holding powdered zinc and numerous glass beads. **F 547,** F16

mechanical plating—the application of a metallic layer by impacting spherical objects (for example glass beads) against the work surface in the presence of powdered coating metal and appropriate chemicals. **B 374,** B08

mechanical polishing, *n*— *of metallic coatings*, the loosening and detachment of superficial, small particles of coating metal during processing or testing, due to mechanical abrasion. **A 902,** A05

mechanical properties—those properties of a material which are associated with elastic and inelastic reaction when force is applied, or which involve the relationship between stress and strain. **C 859,** C26

mechanical properties, *n*—those properties of a material that are associated with elastic and inelastic reaction when force is applied, or that involve the relationship between stress and strain. **E 6,** E28

mechanical properties—fastener characteristics which relate to its reaction to applied loads; these properties may be those of the basic raw material or result from the manufacturing process. **F 1789,** F16

mechanical property—any property of wood that relates to its ability to support load or resist deflection. **D 9,** D07

mechanical pulp, *n*—fibrous material obtained by treating wood or other plant material with a predominantly mechanical process, such as stone groundwood (SGW), pressurized groundwood (PGW), refiner mechanical pulp (RMP), and thermomechanical pulp (TMP). (See **chemical pulp** and **semichemical pulp**.) **D 1968,** D06

mechanical refrigeration—a refrigeration system in which cooling is provided by mechanical means such as a compressor. **E 1705,** E48

mechanical stage—a device provided for adjusting the position of a specimen, usually by translation in two directions at right angles to each other. **E 7,** E04

mechanical testing—determination of mechanical properties. **B 374,** B08

mechanical testing, *n*—the determination of mechanical properties. **E 6,** E28

mechanical transfer papers—papers that produce a visible image by the transfer of a coating to another sheet which may or may not be specially treated. **F 221,** F05

mechanical transfer papers—papers that produce a visible image by the transfer of a coating to another sheet which may or may not be specially treated. (F 221) **F 549,** F05

mechanical twins—See **twin bands.** **E 7,** E04

med fiber, *n*—a medullated animal fiber in which the diameter of the medulla is less than 60 % of the diameter of the fiber. **D 123,** D13

medial—toward the inside or center. (Opposite: lateral) **F 869,** F08

median, *n*—*for a series of observations*, after arranging them in order of magnitude, the value that falls in the middle when the number of observations is odd or the arithmetic mean of the two middle observations when the number of observations is even. **D 123,** D13

median crack—damage produced in glass by the static or translational contact of a hard, sharp object on the surface. (See also **cleavage crack**.) **C 162,** C14

median diameter—See *median diameter* under **diameter.** **C 242,** C21

median diameter—the grain diameter such that half of the sediment by mass is composed of particles of larger size and half by mass is composed of particles of smaller size; commonly denoted by the symbol ″D_{50}”. **D 4410,** D19

median droplet diameter (D_{xf})—For cumulative distributions, D_{xf}, where x is V, A, L, or N, are diameters such that the fraction (f) of the total of volume, surface area, length of diameter, or number of drops respectively, is in drops of smaller diameter.

Thus:
$D_{V.5}$ = volume median diameter,
$D_{A.5}$ = area median diameter,
$D_{L.5}$ = length median diameter, and
$D_{N.5}$ = number median diameter.

Also, $D_{V.1}$ and $D_{V.9}$ = diameter of drop such that 10 % and 90 % respectively, of the liquid volume is in drops of smaller diameter. **E 1102,** E35

median fatigue life—the middle value of the observed fatigue lives, arranged in order of magnitude, of the individual specimens in a group tested under essentially identical conditions. If the sample size is even, it is the average of the two middlemost values. **E 1823,** E08

median fatigue strength at *N* cycles [FL^{-2}]—an estimate of the stress level at which 50 % of the population would survive N cycles. **E 1823,** E08

median load, *n*—*in tensile testing*, that load level that is exceeded by half the recorded peaks and which in turn exceeds the other half of the recorded peaks, in a specified distance of cross-head travel. **D 123,** D13

median particle size, *n*—the particle size at which half the distribution (by mass, volume, number, etc.) is larger than and half smaller than the stated size. **E 1638,** E29

median surface—*of a semiconductor wafer*, the locus of points in the wafer equidistant between the front and back surfaces. **F 1241,** F01

medical direction—physician responsibility for the development, implementation, and evaluation of the clinical aspects of an EMS system (a.k.a. medical direction). **F 1177,** F30

medical director—off-line—a physician responsible for all aspects of an EMS system dealing with the provision of medical care (also known as System Medical Director). **F 1177,** F30

medical protocol—pre-established physician authorized procedures or guidelines for medical care of a specified clinical situation, based on patient presentation (a.k.a. standing orders). **F 1177,** F30

medical removal (of workers)—temporary removal of workers from a job site due to elevated blood-lead levels as defined in HUD interim guidelines. Compare **administrative removal**. **E 1605,** E06

medical transportation services—the moving of patients from one location to another. Specific services include any or all of the following: emergency and non-emergency medical response and transportation; scheduled and non-scheduled interfacility transfers, medical standbys, long-distance medical transfers, air medical response and transport (helicopter and fixed wing aircraft); and stretcher and wheelchair transport services. **F 1177,** F30

medical transportation system—a sub-system of the emergency medical services system consisting of an organization or collection of medical transport services which provide transportation, treatment, and observation of patients for a specific geographic area. **F 1177,** F30

medium—the material that supports the growth and reproduction of microorganisms. **F 1600,** F20

medium barb, *n*—barbs having an average length of 0.4 ± 5 %. **F 1379,** F14

medium buffs—see **upholstery leather.** Formerly often referred to as "special machine buffs." **D 1517,** D31

medium crown—staple crown usually larger than 5⁄16 in. (8 mm) and up to and including 11⁄16 in. (17 mm) in width. **F 592,** F16

medium-density particleboard—a particleboard as previously defined with a density between 37 and 50 lb/ft^3 (specific gravity between 0.59 and 0.80). **D 1554,** D07

medium density polyethylene plastics (MDPE) , *n*—those branched polyethylene plastics, having a standard density of 0.926to 0.940 g/cm^3. (D-20) **F 412,** F17

medium density polyethylene plastics, (MDPE) *n*—those branched polyethylene plastics, q. v., having a standard density of 0.926 to 0.940 g/cm^3. **D 883,** D20

medium grain—See **grain.** **D 9,** D07

medium-hard drawn wire—as applied to copper wire, having tensile strength less than the minimum for hard-drawn wire, but greater than the maximum for soft wire. **B 354,** B01

medium-high carbon— See **steel grades.** **F 547,** F16

medium life expectancy, LE-50, *n—for paper*, a paper is expected to be usable for 50 years. **D 1968,** D06

medium life expectancy paper, LE-50, *n—for paper*, a paper expected to be usable for 50 years under prescribed conditions. **D 1968,** D06

medium-low carbon— See **steel grades.** **F 547,** F16

medium (plural, media), *n*—a nutrient system for the cultivation of cells or organisms, and especially bacteria. **F 1494,** F23

medium point—point with 28 to 35-deg included angle for 0.040 to 0.065-in. wire diameter; 32 to 38-deg included angle for 0.072 to 0.225-in. wire diameter; 37 to 44-deg included angle for 0.250 to 0.325-in. wire diameter. **F 547,** F16

medium sprays—distribution of droplets with 400 μm < $D_{V.5} \leq$ 1200 μm. **E 1102,** E35

medium volatile bituminous coal—See *medium volatile bituminous coal* under **rank.** **D 121,** D05

medium weight concrete masonry unit—unit whose oven-dry density is at least 105 lb/ft^3(1680 kg/m^3) and less than 125 lb/ft^3(2000 kg/m^3). **C 1209,** C15

medulla, *n—in mammalian hair fibers*, the more or less continuous cellular marrow inside the cortical layer in most medium and coarse fibers. **D 123,** D13

medullated fiber, *n*—an animal fiber that in its original state includes a medulla. **D 123,** D13

meeting rail—See **windows and doors.** **E 631,** E06

megohm—unit of measurement of water purity by electrical resistance; One million ohms; reciprocal of conductivity. See **microsiemens, ohm.** **D 6161,** D19

melamine plastic, *n*—plastic based on resins made by the condensation or melamine and aldehydes. **F 1251,** F04

melamine plastics, *n*—plastics based on resins made by the condensation of melamine and aldehydes. **D 883,** D20

melamine resin—under **resin, synthetic** see *melamine resin.* **D 16,** D01

melt, *n*—the total molten metal produced in a single heat. **A 644,** A04

melt, *n*—all the metal that, while molten, was held at the same time in the same holding vessel. **B 899,** B02

melt—a specific quantity of glass made at one time. **C 162,** C14

melt—to change a solid into a liquid by the application of heat; or the liquid resulting from such action. **C 242,** C21

melter—(*1*) a melting unit constructed of refractory materials, with a glass level maintained essentially constant by feeding batch to replace glass continuously withdrawn. Also known as a tank, continuous furnace, and glass melting furnace.

(*2*) {archaic} person in charge of the melting furnace during the early stages of filling and melting of batch. **C 162,** C14

melter, *n*—a piece of equipment designed specifically to melt and heat hot-applied sealant and filler accurately and controllably. **D 5535,** D04

melter-applicator, *n*—a piece of equipment designed specifically to melt, heat accurately and controllably, and apply hot-applied sealants and fillers to pavement cracks or joints uniformly. **D 5535,** D04

melter tender—{archaic} person who regulates the temperature of a melting furnace in order that the glass will have the proper viscosity for casting. **C 162,** C14

melt index—the flow rate of PE material when measured in accordance with Test Method D 1238. **F 412,** F17

melting—the thermal process by which the charge is completely converted into molten glass free from undissolved batch. **C 162,** C14

melting, n—the liquefaction of material under the influence of heat. **D 123,** D13

melting—thermal process by which a material changes from a crystalline to a liquid form. **E 1142,** E37

melting—a response evidenced by softening of the material, resulting in a nonreversible change. **F 1494,** F23

melting area—the area of a melter under fire. Usually the area up to the bridgewall or floater. **C 162,** C14

melting furnace—the complete unit of a glass melting assembly, including the melter, regenerators or recuperators, flues, refiners, forehearths, channels, throats, etc. **C 162,** C14

melting point—in a phase diagram, the temperature at which the liquidus and solidus coincide at an invariant point. **E 7,** E04

melting point, *n*—fixed point of a single component system in which liquid and solid phases are in equilibrium at a specified pressure, usually 101 325 Pa, and the system is gaining heat slowly. (Compare **freezing point.**) **E 344,** E20

melting point—in a phase diagram, the temperature at which the liquidus and solidus coincide at an invariant point, (E 7, E-4). **E 1142,** E37

melting point, final—the temperature at which the last crystal disappears into the melt. **E 1547,** E15

melting point, initial—the temperature at which positive evidence of liquefaction is observed. **E 1547,** E15

melting pressure—at a stated temperature, the pressure at which the solid phases of an element, or congruently melting compound, may coexist at equilibrium with liquid of the same composition. **E 7,** E04

melting temperature—an arbitrarily defined reference point used for comparison purposes at which the glass viscosity is 10 Pa-s. **C 162,** C14

melting temperature—at a stated pressure, the temperature at which a solid phase of an element, or congruently melting compound, may coexist at equilibrium with liquid of the same composition. **E 7,** E04

melting temperature—at a given pressure, the temperature at which a crystalline solid undergoes a phase transition to the isotopic state when heat is added. **E 1142,** E37

melt pressure, *n*—the pressure applied to the plastic material in front of the screw driving the injection molding process on a reciprocating screw machine that is calculated based on the hydraulic force acting axially on the screw. **D 883,** D20

melt temperature, *n*—the temperature of the molten plastic. **D 883,** D20

member—a group of wires stranded together for combination with other stranded groups into a multiple-membered conductor. **B 354,** B01

membrane—a flexible or semiflexible roof covering or waterproofing, whose primary function is the exclusion of water. **D 1079,** D08

membrane—a thin space of material covering a structure of separating solutions and permitting selection transport of a chemical species between the two solutions. **D 4127,** D19

membrane—engineered thin semipermeable film which serves as a barrier permitting the passage of materials only up to a certain size, shape, or electro-chemical character. Membranes are used as the separation agent in reverse osmosis, electrodialysis, ultrafiltration, nanofiltration, and microfiltration, as disc filters in laboratories, and as pleated filter cartridges, particularly for microfiltration. **D 6161,** D19

membrane—any thin sheet or layer. **E 7,** E04

membrane—continuous portion of a boom that serves as a barrier to the movement of a substance. **F 818,** F20

membrane area—the active area available for micro, nano and ultrafiltration and reverse osmosis. **D 6161,** D19

membrane color, *n*—a visual rating of particulates on a filter membrane against ASTM Color Standards. **D 4175,** D02

membrane compaction—see **compaction**. **D 6161,** D19

membrane configuration—the design and shape of a given membrane element (cartridge) such as tubular, spiral wound or hollow fiber. **D 6161,** D19

membrane element—a bundle of spiral membrane envelopes or hollow fiber membranes bound together as a discrete entity. **D 6161,** D19

membrane filter—a thin, nonfibrous filtration medium for fluids, with mean pore size larger than 0.01 μm in diameter, with which particles larger than the rated pore size are retained at or near the delivery surface. **D 1129,** D19

membrane filter—geometrically regular porous matrix; removes particles above pore size rating by physical size exclusion. **D 6161,** D19

membrane, liquid—a liquid material that, when applied to a surface, cures to form a continuous waterproof film by cooling, evaporation of solvent, or by polymerization. **D 3743,** D04

membrane, preformed elastomeric sheet—a sheet membrane composed of cured or vulcanized rubber. **D 3743,** D04

membrane, preformed modified bitumen—a membrane composed of asphalt or coal tar modified with rubber or other modifiers and reinforced with woven or nonwoven mesh or scrim. **D 3743,** D04

membrane salt passage—SP_m is the concentration of a compound in the permeate related to its average concentration on the feed/concentrate side. **D 6161,** D19

membrane, sheet—any functionally continuous flexible structure of felt, fabric, or mat, or combinations thereof, and plying cement. **D 3743,** D04

membrane softening—use of crossflow membrane to substantially reduce hardness ions in water. See **nanofiltration**. **D 6161,** D19

memory color, *n*—color of an object that, according to the judgment of the observer, would match the color of another object previously seen by that observer. **E 284,** E12

meniscus—See **onion**. **C 162,** C14

MEPS—See **rigid cellular polystyrene thermal insulation board.** **E 631,** E06

mercerization—the process of subjecting a vegetable fiber to the action of a fairly concentrated aqueous solution of a strong base so as to produce great swelling with resultant changes in fine structure, dimensions, morphology, and mechanical properties. **D 1695,** D01

merchant pig iron, *n*—pig iron produced for commercial sale to foundries. **A 644,** A04

merino, *adj*—from pure-bred merino sheep. **D 123,** D13

mesh, *n*—the square opening of a sieve. **D 8,** D04

mesh, *n*—*in coated glass yarn fabrics,* the number of warp yarns or ends per linear 25.4 mm (1 in.) followed by the number of filling yarns or picks per linear 25.4 mm (1 in.). **D 123,** D13

mesh—the square opening of a sieve. **D 1079,** D08

mesh, *n*—*in a piece of woven wire cloth,* the nominal number of apertures or fraction thereof per in. **E 1638,** E29

mesh—*in chain link fabric*, the clear distance between parallel wires forming a diamond. **F 552,** F14

mesh—an alignment of opposing teeth. The teeth may be in-line or angled. **F 1638,** F44

mesh-lapping—process of overlapping one piece of mesh onto another that has been applied to a surface previously. **E 631,** E06

mesic peat—see **hemic peat**. **D 653,** D18

mesomorphous cellulose—those portions of cellulose in which the segments of the individual molecules have some regularity of arrangement, but not enough to permit strict lattice order to give a distinct X-ray diffraction pattern. **D 1695,** D01

mesophile—an organism with an optimum growth temperature range of 20 to 45°C. **F 1600,** F20

mesopic, *adj*—pertaining to vision at intermediate levels of illumination, at which both retinal cones and retinal rods are stimulated. **E 284,** E12

mesopore—pores of widths between 2 and 50 nanometres (20 and 500 angstrom units). **D 2652,** D28

message data storage area, *n*—a RAM in the receiving device where the incoming signal is stored until it is ready to be taken into the device's working memory. **F 1457,** F05

messenger wire—a metallic supporting member either solid or stranded which may also perform the function of a conductor. **B 354,** B01

meta-anthracite—See *meta-anthracite* under **rank**. **D 121,** D05

metabolic pathway—the sequence of biochemical reactions that allows conversion for carbon or energy assimilation. **F 1600,** F20

metal—slang for molten glass in a melter. **C 162,** C14

metal blister—bloating of the metal sheet. **C 286,** B08

metal cast button, *n*—a button produced by the casting of molten metals and metal alloys into single cavity or multiple cavity molds. **D 123,** D13

metal connector plate—finished (coated or galvanized) steel or bare stainless-steel plate of specified thickness with or without integral multiple plate projections (teeth) or nail holes, or a combination of both, with projections partially sheared from solid sheet during its fabrication and projecting from the plate in a single direction or both directions perpendicular to the plate surface area; plate of specified thickness to which appropriate tolerances apply. Metal connector plates are manufactured from coiled strips of structural quality sheet metal, produced in various lengths and widths, and designed to connect wood members so as to transmit forces from one wood member (or section) to another one or more wood members (or section). Other common terms include *plate, metal plate, metal-plate connector, nail plate, truss plate*, but the preferable term is **metal connector plate**. **E 1807,** E06

metal connector plate with integral teeth—metal connector plate with integral multiple projections (teeth) partially sheared from solid sheet during its fabrication and projecting from the plate in a single direction or in both directions perpendicular to the plate surface area. See **metal connector plate**. **E 1807,** E06

metal distribution ratio—the ratio of the thickness of metal upon two specified areas of a cathode. See **throwing power.** **B 374,** B08

metal dusting—accelerated deterioration of metals in carbonaceous gases at elevated temperatures to form a dust-like corrosion product. **G 15,** G01

metal filter—a metal structure having controlled interconnected porosity produced to meet filtration or permeability requirements. **B 243,** B09

metal flashing—See **flashing**; frequently used as through-wall, cap-, or counter-flashing. **D 1079,** D08

metal injection molding (MIM), *n*— a process in which a mixture of metal powders and a binder system is forced under pressure into a die. See also **powder injection molding**. **B 243,** B09

metal-lath nail, hook-head metal-lath nail—bright, blued or galvanized, regular-stock-steel, 1⅛ by 0.106-in. nail with thin flat 7⁄16 or -in. hook head and medium diamond point. (See **self-furring nail.**) **F 547,** F16

metallic, *adj*—pertaining to the appearance of a gonioapparent material containing metal flakes. **E 284,** E12

metallic driers—metallic or organometallic compounds used to accelerate the drying of inks. **F 1156,** F12

metallic resinates, *n*—rosin in which part or all of the rosin acids have been chemically reacted with those metals that give soaps or salts which are water insoluble. **D 804,** D01

metallic yield, *n*—the mass percentage of a ferrous waste stream that is generally recoverable as metal or alloy. **D 5681,** D34

metal-like, *adj*—pertaining to the appearance of a bare metal. **E 284,** E12

metal line—the line on the refractory of a melting furnace which marks the boundary between the glass and the atmosphere above it. **C 162,** C14

metallizing—(*1*) the application of a metal layer to the surface of non-conducting or non-metallic materials.

(*2*) the application of metallic coatings by nonelectrolytic procedures such as spraying of molten or heat-softened metal or deposition from the vapor phase. **B 374,** B08

metallizing—See **thermal spraying.** **G 15,** G01

metallography—that branch of science which relates to the constitution and structure, and their relation to the properties, of metals and alloys. **E 7,** E04

metallography—the study of the structure of fastener metals using optical or electronic microscopes that produce a magnified image of the material structure of the fastener. **F 1789,** F16

metallurgical-coke-base carbon refractory, n—see **carbon refractory, metallurgical-coke-base.** **C 71,** C08

metal marking resistance—the ability of a coating to withstand streaking or marking when a metal object is rubbed against or dragged across the surface of the coating. **D 16,** D01

metal (or metallic) coating—generally denotes all metal coatings obtained by any method applied to a basis material. **B 374,** B08

metal powder, *n*—particles of elemental metals or alloys, normally less than 1000 μm (1 mm) in size. **B 243,** B09

metal shadowing—See **shadowing**. **E 7,** E04

metal spraying—application of a metallic coating by projecting molten or heat-softened metal from a source (gun) onto the workpiece. **B 374,** B08

metal substrate—see **base metal.** **C 286,** B08

metal test specimens, *n*—standard samples of metals commonly used in engine cooling systems. **D 4725,** D15

metameric, *adj*—(*1*) pertaining to spectrally different objects or color stimuli that have the same tristimulus values.

(*2*) pertaining to objects, having different spectrophotometric curves, that match when illuminated by at least one specific spectral composition and observed by a specific observer. (See also **parameric.**) **E 284,** E12

metameric color pairs—colors that appear the same to the human eye, but differently to a machine. **F 1156,** F12

metamerism, *n*—property of two specimens that match under a specified illuminator and to a specified observer and whose spectral reflectances or transmittances differ in the visible wavelengths. See also **illuminant metamerism**, **observer metamerism**, **paramerism.** **E 284,** E12

metamerism indices, n— see **special indices of metamerism**, **indices of metamerism potential**. **E 284,** E12

metamers, *n—(1)* spectrally different objects or color stimuli that have the same tristimulus values.

(2) specimens differing in spectral reflectance but having colors that match in light of one spectral composition, when viewed by one observer, but may not match in light of other spectral compositions, or when viewed by another observer. See also **paramers**. **E 284,** E12

metastable—a state of apparent equilibrium which has a higher free energy than has the true equilibrium state; usually applied to a phase existing outside its temperature and pressure span of equilibrium existence, by reason of a greatly delayed transformation. **E 7,** E04

metatectic equilibrium— See **peritectoid equilibrium**. **E 7,** E04

meta-xylene—1,3-dimethylbenzene (C_8H_{10}) mol weight 106.16; clear, colorless, flammable liquid; freezing point, −47.87°C; boiling point, 139.3°C. **D 4790,** D16

meteorological precipitation—See **precipitation.** **D 1356,** D22

metering pump—a mechanical arrangement that permits pumping of the various components of a grout system in any desired proportions or in fixed proportions. (*Syn.* proportioning pump, variable proportion pump.) **D 653,** D18

metering valve—a valve that delivers a definite limited amount of aerosol formulation each time the valve mechanism is operated. **D 3064,** D10

meter leather—a specialty leather made from sheepskins, treated to make it impermeable and used for the measuring bags of gas meters. **D 1517,** D31

methanogenic bacteria—microorganisms capable of producing methane. **E 1126,** E48

methanogenic bacteria—microorganisms capable of producing methane. **E 1705,** E48

methanol (methyl alcohol, wood alcohol)— an alcohol, CH_3OH, formed by catalytically combining carbon monoxide (CO) with hydrogen (H_2) in a 1:2 ratio, under high temperature and pressure. Commercially, it is often manufactured by steam reforming natural gas. It is also formed in the destructive distillation of wood. **E 1126,** E48

methanol (methyl alcohol, wood alcohol)—an alcohol, CH_3OH, formed by catalytically combining carbon monoxide (CO) with hydrogen (H_2) in a 1:2 ratio, under high temperature and pressure. Commercially, it is often manufactured by steam reforming natural gas. It is also formed in the destructive distillation of wood. **E 1705,** E48

method bias—See **bias.** **D 1356,** D22

method detection limit, *n*—the minimum concentration of an analyte that can be reported with a 99 % confidence that the valve is above zero, based on a standard deviation of greater than seven replicate measurements of the analyte in the matrix of concern at a concentration near the low standard. **D 1356,** D22

method of least squares, *n*—a technique of estimation of a parameter which minimizes Σe^2, where *e* is the difference between the observed value and the predicted value derived from the assumed model. **E 456,** E11

method of least squares, *n*—a technique of estimation of a parameter which minimizes Σe^{2}, where *e* is the difference between the observed value and the predicted value derived from the assumed model. **E 1325,** E11

methoxyl group—the monovalent group —OCH_3, characteristic of methyl alcohol and its esters or ethers. **D 1695,** D01

methoxy propanol, *n*—a glycol ether used as an engine coolant concentrate. **D 4725,** D15

methylcellulose—any of several methyl ethers of cellulose. Commercially, the water-soluble ether (degree of substitution approximately 1.5 to 2.0). **D 1695,** D01

methylene blue active substances (MBAS)—compounds that react with methylene blue, a cationic dye, causing it to transfer from an aqueous solution to an immiscible organic liquid upon equilibration under conditions such as those described in Test Method D 2330. The reactive compounds are principally surfactants of the sulfonate type $(RSO_3)^-Na^+$, the sulfate ester type $(ROSO_3)^-Na^+$, and sulfated nonionics $(RE_nOSO_3)^-Na^+$. Soaps are not included among the surfactants reactive to MBAS. **D 459,** D12

methylethylcellulose—a cellulose ether containing both methyl and ethyl groups. **D 1695,** D01

methylglucuronoxylan—the main hemicellulose of hardwood pulps; a polysaccharide containing xylose and 4-O-methylglucuronic acid groups. In the wood it is partially acetylated. **D 1695,** D01

metric count, *n*—an indirect yarn numbering system for sliver roving, and yarn, equal to the number of 1000 m per kilogram. **D 123,** D13

metric sabin, $[L^2]$—the unit of measure of sound absorption in the metre-kilogram-second system of units. **C 634,** E33

MeV (mega or million electron volts)—a unit of energy equal to one million electron volts, used to express the energy of X rays, gamma rays, electrons, and neutrons. **E 1316,** E07

MFFT—abbreviation of minimum film forming temperature. **D 16,** D01

M_f, M_s—See **transformation temperature**. **A 941,** A01

MGD (MGPD)—millions of gallons per day. **D 6161,** D19

Mho—a measure of water purity by conductance, reciprocal of ohms. See **ohm**. **D 6161,** D19

MHz—MegaHertz; millions of Hertz. **F 1156,** F12

mica splittings, *n*—trimmed or untrimmed mica split to thickness under 0.003 mm produced from block, thins, and splitting block. **D 1711,** D09

MICR—(*1*) An abbreviation commonly applied to the character set (E-13B) contained in ANSI X3.2-76 and X9.13-83.
(*2*) See **magnetic ink character recognition**. **F 149,** F05

MICR—Magnetic Ink Character Recognition. **F 1156,** F12

micrinite—See *micrinite* under **maceral**. **D 121,** D05

microaerophilic bacteria—aerobic bacteria that require 2-10 % oxygen in order to grow. See **bacteria (microaerophiles)** . **D 6161,** D19

microbe—bacteria and other organisms that require the aid of a microscope to be seen. **D 6161,** D19

microbes—microscopic organisms, including algae, bacteria, fungi, protozoa, and viruses. **F 1600,** F20

microbial contamination—See Terminology F 17. **F 1327,** F02

microbial corrosion—corrosion that is affected by the action of microorganisms in the environment. **G 15,** G01

microbial degradation, *n*—synonym for biodegradation. **D 6384,** D02

microbiological aerosol, *n*—an airborne particle partially or exclusively composed of microorganisms including bacteria and fungi. **D 5681,** D34

microbiological contamination (of packaged products)—the entry of viable microorganisms into a finished package due to lack of or loss of package integrity. **F 17,** F02

microbiologically influenced corrosion (MIC)—corrosion inhibited or accelerated by the presence or activity, or both, of microorganisms. **G 15,** G01

microbiological package integrity—the physical condition of a finished package, including, but not limited to, the security of package seals, which ensures the maintenance of the package contents in a commercially sterile condition. **F 17,** F02

micro-capsules—very small beads or capsules (typically a few micrometers, μm, in size) containing specific materials, used as a method of delivery of those materials. **F 1156,** F12

microcharacter—See **micro penetration tester**. **E 7,** E04

microclimatology, *n*—the science that deals with the climate of restricted areas and investigates their phenomena and causes. **D 1356,** D22

microcrack—a crack too small to be seen with the unaided eye (see crack, fracture, seam). **C 119,** C18

microcrystalline cellulose—a commercial name for cellulose that has been hydrolyzed to the limiting DP and that consists essentially of microcrystals. **D 1695,** D01

microcrystalline limestone—a limestone that consists largely or wholly of crystals that are so small as to be recognizable only under magnification. If it is capable of taking a polish, it is classified commercially as a marble. **C 119,** C18

microcut—the scratch made by a microcharacter test. **E 7,** E04

microfiber batting, *n*—a textile filling material containing fibers, such as polyester or olefin, which have a diameter of less than 10 μm. **D 123,** D13

microfiltration (MF)—filtration designed to remove particles in the approximate range of 0.05 to 2 μm. **D 6161,** D19

micro focus X-ray tube—an X-ray tube having an effective focal spot size not greater than 100 μm. **E 1316,** E07

microgram, μg, *n*—one millionth of a gram. **E 1605,** E06

micrograph—a graphic reproduction of an object as seen through the microscope or equivalent optical instrument, at magnifications greater than ten diameters. (**photomicrograph**). **E 7,** E04

microhardness, *n*—hardness measured with an instrument having a smaller indentor and applying a lower force than the standard instrument, permitting measurements on smaller specimens or thinner sheets that are not amenable to measurement by normal instruments. **D 1566,** D11

microinch—one millionth of an inch, 0.000001 in. = 0.001 mil. **B 374,** B08

microindent—See **indentation**. **E 7,** E04

micrometeorology, *n*—the study of the meteorological characteristics of a local site that is usually small and often is confined to a shallow layer of air next to the ground. **D 1356,** D22

micrometer eyepiece—an eyepiece that has a scale permanently positioned in its focal plane, thus, in effect, superimposing the scale on the image of the field being observed. **E 7,** E04

micrometer (μm)—one millionth of a meter, 0.001 mm. **B 374,** B08

micrometre—a unit of length equal to one millionth of a metre. **E 1316,** E07

micron—obsolete (see **micrometer**). **B 374,** B08

micron—a term for micrometre. **E 1316,** E07

micronaire reading, *n*—a relative measurement of fiber fineness derived from the porous plug air-flow method. **D 123,** D13

micron (micrometre)—a metric unit of measurement equivalent to 10^{-6} metres, 10^{-4} centimetres. Symbol is μm. **D 6161,** D19

micron of mercury—a unit of pressure equal to that exerted by a column of mercury standing one micrometre high. **E 1316,** E07

microorganism—see **microbe**. **D 6161,** D19

micro-particles—tiny particles used to mark or identify an item. **F 1156,** F12

micro penetration hardness—the hardness number obtained by use of a low load tester whose indentation is usually measured with a high power microscope. **E 7,** E04

micro penetration tester—a testing machine capable of applying low loads, usually in the range from 1 g to 5 kg to form an indentation or a scratch or both, as a basis for measuring hardness. Typical names associated with commercial testers are Bergsman, Eberbach, Knoop, Microcharacter, Tukon, Vickers, etc. **E 7,** E04

microphotograph—a microscopically small photograph. **E 7,** E04

microphotometer, *n*—an instrument for measuring the relative transmittance of small areas on a photographic emulsion such as spectral lines and X-ray diffraction patterns. **E 135,** E01

micropore—intraaggregate capillaries responsible for the retention of water and solutes. **D 653,** D18

micropore—pores with widths not exceeding 2 nanometres (20 angstrom units). **D 2652,** D28

microporosity—intragranular porosity and microscopic intergranular porosity with submicroscopic pore throats. **D 653,** D18

microporosity—extremely fine porosity in castings. **E 7,** E04

micro-printing—the use of very small characters, often in the range of 0.004 to 0.008 in. in height. **F 1156,** F12

microprocessor, *n*—a central processing chip within the electronic controller section of the apparatus. **D 121,** D05

microprocessor unit/image generator, *n*—a set of components on the printer PWB which convert coded data (ASCII) into video data. **F 1457,** F05

microroughness—surface roughness components with spacing between irregularities (spatial wavelength) less than about 100 μm. **F 1241,** F01

microscope—an instrument capable of producing a magnified image of a small object. **E 7,** E04

microscope field-of-view—that area observed with a particular set of optics. **F 1241,** F01

microscopy—the science of the interpretive use, and applications of microscopes. **E 175,** E41

microscratch—a scratch that is not visible to the unaided eye under fluorescent (diffuse) illumination but is visible to the unaided eye under incandescent (high intensity) illumination. **F 1241,** F01

microseism—seismic pulses of short duration and low amplitude, often occurring previous to failure of a material or structure. (ISRM) **D 653,** D18

micro separometer rating (MSEP), *n*— a numerical value indicating the ease of separating emulsified water from a sample by coalescence. **D 4175,** D02

microsiemens—unit of measurement of water purity by electrical conductivity; one micromho; reciprocal of resistivity. See **megohm, ohm**. **D 6161,** D19

microstrain, n—the strain over a gage length comparable to interatomic distances. **E 6,** E28

microstructure—the structure of a suitably prepared specimen as revealed by a microscope. **E 7,** E04

microthrowing power—the ability of a plating solution or a specified set of plating conditions to deposit metal in pores or scratches. **B 374,** B08

microtunneling—trenchless installation of pipe by jacking the pipe behind a remotely controlled, steerable, laser guided, microtunnel boring machine that provides continuous support to the excavated face under various geotechnical conditions including the presence of groundwater. **C 896,** C04

microwave extraction cell—a polytetrafluoroethylene cell used for evaluating microwave active materials. Refer to Test Method F 1349 for schematics of cell construction. **F 1479,** F02

microwave only food package—a container used to heat foods only in a microwave oven. **F 1479,** F02

microwave oven—equipment that uses ultrahigh frequency electromagnetic radiation in the bands of 915 ± 25 and 2450 ± 50 MHz to prepare food. **F 1827,** F26

microwave radiation emission—the microwave energy to which persons might be exposed during operation or servicing of microwave cooking equipment. **F 1827,** F26

microwave susceptor—packaging material that, when placed in a microwave field, is designed to interact with the field and provide substantial heat to the package contents. **F 1479,** F02

micrurgy—the use of a micromanipulator in combination with a microscope. **E 175,** E41

middle rail—See **intermediate rail**. **F 552,** F14

midget impinger—See **impinger.** **D 1356,** D22

mid-infrared—pertaining to the infrared region of the electromagnetic spectrum with wavelength range from approximately 2.5 to 25 μm (wavenumber range 4000 to 400 cm^{-1}). **E 131,** E13

mid-neck girth, *n*—*in body measurements*, the circumference of the neck approximately 25 mm (1 in.) above the neck base (See neck base girth.). **D 123,** D13

mid rail—See **railing systems.** **E 631,** E06

mid rail, *n*—a rail located between top rail and bottom rail or between top rail and floor if there is no bottom rail. **E 1481,** E06

midsole—a sole of leather or other material placed between the outsole and the insole. **F 869,** F08

mid-thigh girth, *n*—*in body measurements*, the circumference of the upper leg between the hip and the knee. **D 123,** D13

migration, *n*—ink spreading over a long period of time. See also **feathering**. **F 1857,** F05

migration development—premature color development on the CF side of a CFB sheet caused by free CB oil-dye penetrating through the sheet until it reacts with the CF receptor coating. **F 549,** F05

mil—a unit of length equal to 1⁄1000 of an inch [0.0254 mm].

B 354, B01

mil—one thousandth of an inch, 0.001 in. = 25.4 μm. **B 374,** B08

mil—1 mil = 0.001 in. **F 1156,** F12

mildew (fungus) resistance (*coatings*)—the ability of a coating to resist fungus growth that can cause discoloration and ultimate decomposition of a coating's binding medium. **D 16,** D01

mildewstat—a chemical agent that inhibits the growth of mildew. **D 16,** D01

milk glass—a translucent or opaque white glass. **C 162,** C14

milkiness—a condition of pronounced cloudiness in glass. **C 162,** C14

milking, *n*—a white colored buildup on the non-image areas of the offset blanket that usually occurs when the paper coating or paper filler (in the case of uncoated paper) softens due to inadequate water resistance. **D 6488,** D01

milk of lime—a suspension of hydrated lime (or slaked quicklime) in water in such proportions as to resemble milk in appearance. **C 51,** C07

mill, *n*—a machine used for rubber mastication, mixing, or sheeting, having two counter-rotating rolls with adjustable longitudinal axis separation that usually rotate at different speeds. **D 1566,** D11

mill addition—any of the materials added to the ball mill charge of a frit. **C 286,** B08

mill broke, *n*—any paper generated in a paper mill prior to the completion of a specific paper manufacturing process which is unsuitable for subsequent applications but can be reused in the paper manufacturer process. **D 1968,** D06

mill certification—producing mill certificate or proof of conformance with specified minimum allowable stresses for heat identification number of metal coil(s) from which metal connector plates were fabricated. **E 1807,** E06

milled asbestos—all grades of asbestos that are recovered as a result of mechanical comminution and screen classification or air classification of asbestos ore. **D 2946,** C17

Miller-Bravais indices—indices used for the hexagonal system which involve the use of a fourth axis a_3, coplanar with and at 120° to a_1 and a_2. **E 7,** E04

Miller indices—*of a crystallographic plane*, the smallest integers proportional to the reciprocals of the intercepts of the plane on the three crystal axes. **F 1241,** F01

Miller indices (for lattice planes)—the reciprocals of the fractional intercepts which a plane makes on the three axes. The symbols are (*hkl*). **E 7,** E04

mill finish, *adj*—having a nonuniform finish which may vary from piece to piece and within a piece, and which may not be entirely free of stains or oil. See also *mill finish sheet*. **B 881,** B07

mill grain, *n*—*in rubber*, grain which is imparted to rubber sheeting while being mixed or conditioned in a rubber mill and which is parallel to the direction the rubber moves in the mill. **D 123,** D13

milliamperes (mA)—the technical term is **tube current** and is defined as the current passing between the cathode and anode during the operation of an x-ray tube, measured in milliamperes (mA) and usually taken as a measure of x-ray intensity. **E 1316,** E07

milliequivalent per litre (meq/L)—a weight-volume measurement obtained by dividing the concentration expressed in milligrams per litre by the equivalent weight of the substance or ion. If specific gravity is unity meq/L is the same as epm. **D 6161,** D19

milligram per litre (mg/L)—a weight-volume measurement which expresses the concentration of a solute in milligrams per litre of solution. When specific gravity is unity mg/L = ppm. When specific gravity is not unity, mg/L divided by specific gravity of solution equals ppm. **D 6161,** D19

millimetre of mercury—a unit of pressure corresponding to a column of mercury exactly 1 mm high under standard acceleration of gravity. Sometimes call torr. **E 1316,** E07

millimicron (Mu)—a unit of length used in measuring light waves. The peak spectral response of a scanner is expressed in Mu. **F 149,** F05

milling—the mechanical treatment of metal powder, or metal powder mixtures, as in a ball mill, to alter the size or shape of the individual particles, or to coat one component of the mixture with another. **B 243,** B09

milling—a process by which asbestos ore is treated mechanically to produce grades of asbestos with specified properties. **D 2946,** C17

milling, *n*—in waste derived fuels, reduction in particle size by shearing, cutting, grinding to a suitable particle size for analysis and characterization. **D 5681,** D34

milling—a reduction in particle size of refuse-derived fuel by shearing, cutting, or grinding to a suitable particle size for analysis and characterization (see **sample reduction**). **E 856,** D34

millrun—the total yield of lumber from a mill or a group of logs without regard to grades. **D 9,** D07

mill scale—the heavy oxide layer that forms during hot fabrication or heat treatment of metals. **B 374,** B08

millwork—generally building materials made of finished wood and manufactured in millwork plants and planing mills often assembled for installation upon delivery to the job site. Includes such items as inside and outside doors, window and door frames, molding, and interior trim. **D 9,** D07

MIM—see **metal injection molding**. **B 243,** B09

mineral fiber, *n*—insulation composed principally of fibers manufactured from rock, slag, or glass, with or without binders. **C 168,** C16

mineral fiber felt—a felt with rock wool as the principal component. **D 1079,** D08

mineral granules—opaque, natural, or synthetically colored aggregate commonly used to surface cap sheets, granule-surfaced sheets, and roofing shingles. **D 1079,** D08

mineral insulating oil, *n*—an oil of mineral origin, refined from petroleum crude oil, possessing electrical insulating properties. **D 2864,** D27

mineralization—the microbial-mediated breakdown of organic materials into inorganic materials. **F 1600,** F20

mineralized coal—See *mineralized coal* under **coal**. **D 121,** D05

mineralizer—a processing additive that promotes either the recrystallization or the partial fusion or sintering of certain mineral or ceramic materials, often facilitating the desired conversion at a lower temperature. **C 242,** C21

mineral parting, *n*—discrete layer of mineral or mineral-rich sediment interbedded with coal along which, in mining, separation commonly occurs. Layers of bone coal having indefinite boundaries usually are not considered to be partings because they do not form planes of physical weakness. They may merge vertically or horizontally with layers that are bony or coaly shale and that do form planes of physical weakness. **D 121,** D05

mineral rubber, *n*—a compounding material (not a rubber) prepared from petroleum asphalt and used as a tackifier, softener, or extender. **D 1566,** D11

mineral stabilizer—a fine, water-insoluble inorganic material, used in admixture with solid or semisolid bituminous materials.

D 1079, D08

mineral streak—an olive to greenish-black or brown discoloration of undetermined cause; commonly associated with bird pecks and other injuries. Occurs in streaks usually containing accumulations of mineral matter. **D 9,** D07

mineral-surfaced roofing—built-up roofing whose top ply consists of a granule-surfaced sheet. **D 1079,** D08

mineral-surfaced sheet—a felt that is coated on one or both sides with asphalt and surfaced with mineral granules. **D 1079,** D08

mineral tanned—leathers which have been tanned with chemical compounds of mineral origin without the use of vegetable tanning materials, notably the salts of chromium, aluminum, and zirconium. **D 1517,** D31

minifocus X-ray tube—an X-ray tube having an effective focal spot size between 100 and 400 μm. **E 1316,** E07

minimum acceptable rate of return—the minimum percentage return required for an investment to be economically acceptable. **E 833,** E06

minimum application temperature, *n*—the minimum temperature, as recommended by the manufacturer, to which a hot-applied sealant or filler for pavement cracks or joints must be heated while conforming to all specification requirements and result in appropriate application characteristics. **D 5535,** D04

minimum curve—in a phase diagram, a univariant line tracing the meeting of a pair of bivariant surfaces at intermediate composition and coinciding with their lowest temperature at each pressure level, or their lowest pressure at each temperature level; congruent transformation occurs everywhere along a minimum curve. **E 7,** E04

minimum detectable leakage rate—the magnitude of the smallest leakage rate that can be unambiguously detected by a given leak detector in the presence of conditions existing at time of test. **E 1316,** E07

minimum detectable mass permeated, *n*— the smallest mass of permeant that is detectable with the complete permeation test system. **F 1494,** F23

minimum detectable permeation rate, *n*— the lowest rate of permeation that is measurable with the complete permeation test system. **F 1494,** F23

minimum detectable temperature difference (MDTD)—a measure of the compound ability of an infrared imaging system and an observer to detect a target of unknown location at one temperature against a large uniform background at another temperature when displayed on a monitor for a limited time. **E 1316,** E07

minimum detection limit—See **analyzer.** **D 1356,** D22

minimum determinability—the lowest value that can be determined within the stated precision of a method expressed quantitatively in the same dimension that is used for reporting results of the test. **D 1129,** D19

minimum edge contrast, ECmin, *n*—the smallest edge contrast in a scan reflectance profile. **F 1294,** F05

minimum hardness—the hardness value of a fastener below which it is not in conformance with the specification. **F 1789,** F16

minimum load, P_{min} **[F]**—*in fatigue*, the lowest algebraic value of applied load in a cycle. By convention, tensile loads are positive and compressive loads are negative. **E 1823,** E08

minimum local thickness—the lowest local thickness value on the significant surface of a single article. **F 1789,** F16

minimum (non-statistical), *n*—the lowest acceptable actual test result; any valid individual test result below the minimum is cause for rejection of the component or material lot being tested.

$$\text{All } x_i \geq \text{Minimum} = LSL$$

A 644, A04

minimum point—that composition and temperature, or pressure, at which a heterogeneous equilibrium occurs at its lowest temperature, or pressure, when this does not coincide with one of the composition limits of the equilibrium, that is, when it occurs at an intermediate composition; the equilibrium becomes congruent (univariant) at the minimum point. **E 7,** E04

minimum reflectance, Rmin, *n*—the smallest reflectance value in a scan reflectance profile. **F 1294,** F05

minimum reflectivity difference, *n*—the difference between the smallest minimum space reflectance value and the largest maximum bar reflectance value as measured across the entire symbol. According to the equation:

$$MAD = \text{Min}\ (As \min) - \text{Max}\ \ (Ab \max)$$

where:
As = Minimum Space Reflectance, and
Ab = Maximum Bar Reflectance.

F 1294, F05

minimum resolvable temperature difference (MRTD)—a measure of the ability of an infrared imaging system and the human observer to recognize periodic bar targets on a display. The MRTD is the minimum temperature difference between a standard periodic test pattern (7:1 aspect ratio, 4 bars) and its blackbody background at which an observer can resolve the pattern as a four-bar pattern. **E 1316,** E07

minimum size *(of aggregate)*, *n*—in specifications for, or description of aggregate, the smallest sieve opening through which the entire amount of aggregate is required to pass. **C 125,** C09

minimum (statistical), *n*—the lowest acceptable statistical test result; for compliance, the sample mean ($\bar{x}$) minus *M* standard deviation(s), where *M* is a matter of agreement between the supplier and purchaser, must be greater than, or equal to, the lower specification limit (LSL).

$$\bar{x} - M \times s \geq \text{Minimum} = LSL$$

A 644, A04

minimum stress-intensity factor coefficient, Y^*_m — the minimum value of the stress intensity factor coefficient, *Y**. **E 1823,** E08

minimum stress-intensity factor, K_{min} **[$FL^{-3/2}$]**—*in fatigue*, the minimum value of the stress-intensity factor in a cycle. This value corresponds to P_{min} when $R > 0$ and is taken to be zero when $R \leq 0$. **E 1823,** E08

minimum test value, *n—for geosynthetics*, the lowest sample value from documented manufacturing quality control test results for a defined population from one test method associated with one specific property. **D 4439,** D35

minimum thickness, *n— of steel sheet*, an ordering designation which indicates that the applicable tolerance for thickness is all plus from the ordered thickness. **A 902,** A05

minor defect, *n*—a defect that is not likely to materially reduce the usability of the product from its intended purpose, or is a departure from established standards having little bearing on the effective use of operation of a product. **D 123,** D13

minor imperfection, *n—in fabric grading*, a deviation in a roll of fabric that judgment and experience indicate is likely to have no bearing on subsequent processing of the fabric. **D 123,** D13

minority carrier—type of charge carrier constituting less than one half the total charge-carrier concentration (for example, electrons in *p*-type material). **F 1241,** F01

minority carrier lifetime—*of a homogeneous semiconductor*, the average time interval between the generation and recombination of minority carriers. **F 1241,** F01

minor modification—any change that does not alter the structural or operational characteristics of the ride or device nor change its performance from that specified in the manufacturer's design criteria. **F 747,** F24

minor package defect—a defect that does not significantly reduce the usability of the package for its intended purpose, or that is a departure from established standards having little or no bearing on the effective use of the package. **F 17,** F02

minor package defect—See Terminology F 17. **F 1327,** F02

minor principal plane—see **principal plane**. **D 653,** D18

minor principal stress—see **stress**. **D 653,** D18

minor tile facial dimension—the length or width of the tile exclusive of the lugs. **C 242,** C21

minor tile thickness—the thickness of tile that does not include maximum protuberances or ridges. **C 242,** C21

minus sieve—the portion of a powder sample which passes through a standard sieve of specified number. (See **plus sieve**.) **B 243,** B09

mirror assembly, *n*—a mirror, marked and aligned with the viewing rakes, used as an aid in quickly identifying and tracking the flame front progress. **E 176,** E05

mirror, first or front surface—an optical mirror on which the reflecting surface is applied to the front surface of the mirror instead of to the back, that is, to the first surface of incidence. **E 175,** E41

mirror illuminator—a thin, half-round opaque mirror interposed in a microscope for the purpose of directing an intense oblique beam of light to the object. The light incident on the object passes through one half the aperture of the objective and the light reflected from the object passes through the other half aperture of the objective. **E 7,** E04

miscibility gap—a region of multi-phase equilibrium; commonly applied to the specific case in which an otherwise continuous series of liquid, or solid, solutions is broken, over a limited temperature range, by the intrusion of a two phase field that terminates at a critical point. See **binodal curve**. **E 7,** E04

mismatch—an uneven fit in worked lumber when adjoining pieces do not meet tightly at all points of contact or when the surfaces of adjoining pieces are not in the same plane. **D 9,** D07

mispick, *n—in woven fabrics,* a pick not properly interlaced which causes a break in the weave pattern. (See **double pick**.) **D 123,** D13

misprint, *n—in printed fabric,* colors or patterns, or both, either missed, or partially missed, or incorrectly positioned relative to each other. **D 123,** D13

misread, *n*—a disparity between the data encoded in a symbol and the data output from a code reader. Also known as bad read or mis-scan. Such errors will not be detected by test routines in the decode algorithm. (See also, **non-read**.) **F 1294,** F05

misregister, *n—in printed fabric,* colors or patterns not correctly positioned. (Compare **misprint.**) **D 123,** D13

missing dots, *n*—a noticeable area void of halftones or an area of the cylinder that has cells that have not transferred ink to the paper. It can be caused by a worn cylinder or rough paper surface, or both. (Also referred to as *skipped dots*). **D 6488,** D01

miss-knit, *n—in knitted fabrics,* a deviation from the designated knitting pattern. **D 123,** D13

mist—a finely structured fracture marking giving a matte or roughened appearance to a surface; having continuous degrees of coarseness. Mist is variously known as fine hackle, frosted area, matte, and stippled area. **C 162,** C14

mist, *n*—liquid, usually water in the form of particles suspended in the atmosphere at or near the surface of the earth; small water droplets floating or falling, approaching the form of rain, and sometimes distinguished from fog as being more transparent or as having particles perceptibly moving downward. **D 1356,** D22

mist, *n*—liquid, usually water, in the form of particles suspended in the atmosphere at or near the surface of the earth; small water droplets floating or falling, approaching the form of rain, and sometimes distinguished from fog as being more transparent or as having particles perceptively moving downward. **E 1620,** E29

misting—see **dusting.** **F 335,** F05

mists—distribution of droplets with $50\ \mu m < D_{V.5} \leq 100\ \mu m$. **E 1102,** E35

mitered fittings—fittings manufactured by using beveled pipe segments. **C 896,** C04

mitered pipe—See **beveled pipe.** **C 896,** C04

miter ending—See **railing systems.** **E 631,** E06

miter ending—an angular or dovetailed member end, designed to fit an adjacent matching member, thereby providing continuity of profile at the connection. **E 1481,** E06

mix—See **batch (3)**. **C 162,** C14

mix, *n*—an adequate mixture of rubber in any form with other material(s). **D 1566,** D11

mixed aniline point, *n*—the minimum equilibrium solution temperature of a mixture of two volumes of aniline (aminobenzene), one volume of sample, and one volume of *n*-heptane of specified purity. **D 4175,** D02

mixed bed—a physical mixture of anion-exchange and cation-exchange materials. **D 1129,** D19

mixed-bed—a physical mixture of anion-exchange and cation-exchange materials. **D 6161,** D19

mixed end, *n—in woven fabrics,* a warp yarn differing from that normally being used in the fabric. **D 123,** D13

mixed filling, *n—in woven fabrics*, a filling yarn differing from that normally being used in the fabric. (See also **filling band**. Compare **barré**.) **D 123,** D13

mixed grain size—See **duplex grain size**. **E 7,** E04

mixed-in-place pile—a soil-cement pile, formed in place by forcing a grout mixture through a hollow shaft into the ground where it is mixed with the in-place soil with an auger-like head attached to the hollow shaft. **D 653,** D18

mixed-in-place (road mix), *n*—a bituminous surface or base course produced by mixing mineral aggregate and cut-back asphalt, bituminous emulsion, or tar at the job-site by means of travel plants, motor graders, drags, or special road-mixing equipment. Open or dense-graded aggregates, sand, and sandy soil may be used. **D 8,** D04

mixed liquor, *n—in sewage treatment,* the contents of an aeration tank including the activated sludge mixed with primary effluent or the raw wastewater and return sludge. **D 6384,** D02

mixed matrix of composite superconductor—matrix composed of more than one component. **B 713,** B01

mixed potential—the potential of a specimen (or specimens in a galvanic couple) when two or more electrochemical reactions are occurring simultaneously. **G 15,** G01

mixed powder—a powder made by mixing two or more powders as uniformly as possible. The constituent powders will differ in chemical composition or in particle size or shape, or a combination thereof. **B 243,** B09

mixed reflection, *n*—partly specular and partly diffuse reflection. **E 284,** E12

mixed reflection, *n*—partly regular and partly diffuse reflection.

NOTE—The irradiance or illuminance received from a point source after regular (diffuse) reflection, varies inversely as the square of the distance to the source (diffuser). **E 349,** E21

mixed transmission, *n*—a combination of diffuse and regular transmission. **E 284,** E12

mixed transmission, *n*—partly regular and partly diffuse transmission.

NOTE—The irradiance or illuminance received from a point source, after regular (diffuse) transmission, varies inversely as the square of the distance to the source (diffuser). **E 349,** E21

mixer—a machine employed for blending the constituents of grout, mortar, or other mixtures. **D 653,** D18

mixer, *n*—a machine that incorporates and disperses compounding ingredients into rubber to form a mix or a compound, through the action of mechanical work (shear). **D 1566,** D11

mixer, internal, *n*—a machine with a closed cavity in which a specially shaped rotor(s) masticates the rubber or incorporates and disperses compounding materials, or both, into the rubber. **D 1566,** D11

mixing—the thorough intermingling of powders of two or more materials. **B 243,** B09

mixing cycle—the time taken for the loading, mixing, and unloading cycle. **D 653,** D18

mixing ratio ***(r)***, *n*—the ratio of the mass of water vapor m_v to the mass of dry air m_a, present in the moist air;

$$r=\frac{m_v}{m_a}$$

D 1356, D22

mixing ratio, *r*—the ratio of the mass of water vapor m_v to the mass of dry air m_a present in the moist air;

$$r=\frac{m_v}{m_a}$$

D 4023, D22

mixing speed—the rotation rate of a mixer drum or of the paddles in an open-top, pan, or trough mixer, when mixing a batch; expressed in revolutions per minute. **D 653,** D18

mixture design, *n*—a design in which two or more ingredients or components shall be mixed and the response is a property of the resulting mixture that does not depend upon the amount of the mixture.

NOTE—The proportions of each of the *q* components (X_i) in the mixture shall satisfy the conditions $O \leq X_i \leq 1$ and $\sum_{i=1}^{c} X_i = 1$; and each experimental point is defined in terms of these proportions.

NOTE—In some fields of application the experimental mixtures are described by the terms ''formulation'' or ''blend.'' The use of mixture designs is appropriate for experimenting with the formulations of manufactured products, such as paints, gasoline, foods, rubber, and textiles.

NOTE—In some applications, the proportions of the components of the mixture may vary between 0 and 100 % of the mixture (''complete domain''). In others, there may be operative restraints, so that at least one component cannot attain 0 or 100 % (''reduced domain''). **E 456,** E11

mixture design, *n*—a design in which two or more ingredients or components shall be mixed and the response is a property of the resulting mixture that does not depend upon the amount of the mixture.

NOTE—The proportions of each of the *q* components (X_i) in the mixture shall satisfy the conditions $O \leq X_i \leq 1$ and $\sum_{i=1}^{c} X_i = 1$; and each experimental point is defined in terms of these proportions.

NOTE—In some fields of application the experiment mixtures are described by the terms *formulation* or *blend*. The use of mixture designs is appropriate for experimenting with the formulations of manufactured products, such as paints, gasoline, foods, rubber, and textiles.

NOTE—In some applications, the proportions of the components of the mixture may vary between 0 and 100 % of the mixture (*complete domain*). In others, there may be operative restraints, so that at least one component cannot attain 0 or 100 % (*reduced domain*). **E 1325,** E11

mm—millimetre. **F 1156,** F12

mnemonic—an aid to memory. **F 1156,** F12

mobile home—Obsolete term. Use **manufactured home.** (See **building.**) **E 631,** E06

mobility, *μ—of a carrier in a semiconductor*, the ensemble average of drift velocity of a charge carrier per unit electric field. The unit that is customarily used is $cm^2/V·s$, but the SI unit is $m^2/V·s$. **F 1241,** F01

mobius loop—a one-sided surface formed by holding one end of a rectangle fixed, rotating the opposite end through 180° and then applying it to the first end. **F 221,** F05

mocha leather—leather from any variety of hair sheep; after the grain has been removed by a liming process known as frizing, the fine fibers below the grain are sueded. **D 1517,** D31

mocha suede—a washable and durable suede produced from chrome-tanned Arabian blackhead hair sheepskins (commonly called blackhead mochas) mechanically abraded on the flesh side. Tightness of fiber structure found in these skins imparts the characteristic fineness of finish associated with mocha suede. **D 1517,** D31

mock French seam, *n—in home sewing*, a complex seam formed on the inside of the product with cut edges enclosed and no stitching visible on the face side; similar in appearance to the French seam but constructed differently. (Compare **French seam.**) **D 4965,** D13

mock leno weave, *n*—a weave in which the warp yarns remain parallel but form open warp stripes by programmed interlacing of warp and filling yarns simulating a leno appearance. **D 123,** D13

mockup, *n*—a section or a structure or assembly, built full-size or to scale, for the purpose of studying construction details, testing performance, judging appearance, or any combination thereof. **E 631,** E06

modality, *n*—any of the sensory systems (for example, auditory, taste, or visual modality). **E 253,** E18

mode, *n*—the value of the variate for which the relative frequency in a series of observations reaches a local maximum. **D 123,** D13

mode—the type of ultrasonic wave propagating in the materials as characterized by the particle motion (for example, longitudinal, transverse, etc.). **E 1316,** E07

mode—one of the three classes of crack (surface) displacements adjacent to the crack tip. These displacement modes are associated with the stress-strain fields around the crack tip and are designated one, two, and three. Arabic numerals 1, 2, and 3 are used for the general case, and they represent opening, sliding, and tearing displacements, respectively. Roman numerals are used to specialize the mode to plane strain (I and II) or to antiplane-strain (III). **E 1823,** E08

mode conversion—phenomenon by which an ultrasonic wave that is propagating in one mode can reflect or refract at an interface to form ultrasonic wave(s) of other modes. **E 1316,** E07

model—frequently used to describe the following: *(1)* an individual style in a line of footwear; *(2)* the paper patterns from which shoe cutting patterns are made; or *(3)* the last from which a complete set of lasts is graded. **F 869,** F08 *model code*—See **code.** **E 631,** E06

MODEM—**mod**ulator/**dem**odulator. **F 1457,** F05

moderately conductive, *adj*—describes a solid material having a volume resistivity between 1 and 10 000 000 Ω–cm. **D 1711,** D09

moderator—a material used to reduce neutron energy by scattering without appreciable capture. **C 859,** C26

moderator—a material used to slow fast neutrons. Neutrons are slowed down when they collide with atoms of light elements such as hydrogen, deuterium, beryllium, and carbon. **E 1316,** E07

modernize—See **building modification.** **E 631,** E06

modes, *n*—*of appearance*, various manners in which colors can be perceived (see also **aperture mode, illuminant mode, object mode**.) **E 284,** E12

modified Auger parameter—the Auger parameter plus the photon energy, which equals the kinetic energy of the sharpest Auger peak plus the binding energy of the most intense photoelectron peak. **E 673,** E42

modified design—a concrete pipe design changed from a standard design. **C 822,** C13

modified grab test, *n*—*in fabric testing*, a tensile test in which the control part of the width of the specimen is gripped in the clamps and in which lateral slits are made midlength of the specimen severing all yarns bordering that portion of the specimen held between the two clamps. **D 123,** D13

modified groove—see **female end of pipe.** **C 822,** C13

modified internal rate of return (MIRR)—See **adjusted internal rate of return (AIRR)**. **E 833,** E06

modified rosin, *n*—rosin that has been treated with heat or catalysts, or both with or without added chemical substances, so as to cause substantial change in the structure of the rosin acids, as isomerization, hydrogenation, dehydrogenation, or polymerization, without substantial effect on the carboxyl group. **D 804,** D01

modified tongue—see **male end of pipe.** **C 822,** C13

modified vegetable oil—an oil, extracted from seeds, that has been chemically modified (for example, methylated). **E 1519,** E35

modified vegetable oil concentrate—an emulsifiable, chemically modified vegetable oil product containing 5 to 20 % w/w surfactant and the remainder chemically modified vegetable oil. **E 1519,** E35

modified worsted system, *n*—a worsted system for spinning man-made fibers which relies on pin control of fibers during sliver weight reduction, but which bypasses the system of combing required with wool to remove noil. **D 123,** D13

modifier—*in grouting*, an additive used to change the normal chemical reaction or final physical properties of a grout system. **D 653,** D18

modifier, *n*—any chemically inert ingredient added to an adhesive formulation to change its properties. (Compare **filler, plasticizer, extender**.) **D 907,** D14

modifier code, *n*—a character code appended to the axle configuration code to define the approximate load carrying capacity of the vehicle. **E 867,** E17

modular container— See **container.** **D 996,** D10

modular dwelling—See **dwelling unit.** **E 631,** E06

modulate, *v*—to vary a characteristic or parameter of an entity in accordance with a characteristic or parameter of another entity. **E 131,** E13

modulated current plating—a method of electroplating in which the cathode current density is changed periodically. See **pulse plating, ripple plating, periodic reverse plating.** **B 374,** B08

modulation—*AES,* the periodic waveform added to the spectrometer pass energy to obtain the desired Auger spectrum display. The modulation should be given as eV peak-to-peak, thereby including the geometrical factor of the spectrometer, rather than volts peak-to-peak. The frequency and waveform should also be given. **E 673,** E42

modulation analysis—an analytical method used in electromagnetic testing that separates responses due to various factors influencing the total magnetic field by separating and interpreting, individually, frequencies or frequency bands in the modulation envelope of the (carrier frequency) signal. **E 1316,** E07

modulation frequency, f_v—the frequency, in Hz, at which radiant energy of a given wavenumber is modulated by a rapid-scanning interferometer. **E 131,** E13

modulation, MOD, *n*—the ratio of minimum edge contrast (ECmin) to symbol contrast (SC) expressed as:

$$MOD = ECmin/SC$$

F 1294, F05

modulation transfer function (MTF)—in infrared imaging systems, the modulus of a Fourier transform that describes the spatial distribution of the overall attenuation in amplitude of a thermal imaging system.

NOTE—MTF is a sensitive function of spatial frequency. **E 1316,** E07

modulator/demodulator, *n*—a device that converts electronic signals from the form used in data processing (digital) to the form used in communication (analog), and vice versa. (See **MODEM**.) **F 1457,** F05

module, *n*—*for inflatable restraints*, an assembly composed of an inflator, a cushion, a mounting device, a trigger, and a cover. **D 123,** D13

module—a membrane element combined with the element's housing. Pressure vessel containing membrane element(s). **D 6161,** D19

module, *n*—a unit of structure based on a standard pattern of standard dimensions. (See also **modular dwelling.**)

core module—a **module** containing electrical, plumbing, heating, and related **subsystems.** **E 631,** E06

module, *n*—the narrowest nominal width of bar or space in a symbol. **F 1294,** F05

modulo, *n*—an arithmetic operation in which the result is the remainder after division. The type of algorithm used to calculate the check digit for certain bar code symbols. **F 1294,** F05

modulus, *n*—as related to sealants used in building construction, the stress (force/unit area) at a corresponding strain (elongation) expressed as a percent of the original dimension. **C 717,** C24

modulus, *n*—the property of a material representative of its resistance to deformation. (See also **chord modulus, initial modulus, tangent modulus, Young's modulus**). **D 123,** D13

modulus, complex—see **complex modulus.** **D 4092,** D20

modulus, complex—see **complex modulus**. **E 1142,** E37

modulus, elastic—see **complex modulus and storage modulus.** **D 4092,** D20

modulus, elastic—see **complex modulus** and **storage modulus**. **E 1142,** E37

modulus, loss—see **complex modulus and loss modulus.** **D 4092,** D20

modulus, loss—see **complex modulus** and **loss modulus**. **E 1142,** E37

modulus of deformation—see **modulus of elasticity**. **D 653,** D18

modulus of elasticity— see **complex modulus and storage modulus.** **D 4092,** D20 *modulus of elasticity*— see **complex modulus** and **storage modulus**. **E 1142,** E37

modulus of elasticity [FL^{-2}], *n*—the ratio of stress to corresponding strain below the proportional limit. **E 6,** E28

modulus of elasticity (modulus of deformation), *E, M* **(FL^{-2})**—the ratio of stress to strain for a material under given loading conditions; numerically equal to the slope of the tangent or the secant of a stress-strain curve. The use of the term **modulus of elasticity** is recommended for materials that deform in accordance with Hooke's law; the term **modulus of deformation** for materials that deform otherwise. **D 653,** D18

modulus of elasticity, *MPa (FL^{-2}), n*—the ratio of stress (nominal) to corresponding strain below the proportional limit of a material, expressed in force per unit area, such as megapascals (pounds-force per square inch). **D 4439,** D35

modulus of rupture in bending [FL^{-2}], *n*—the value of maximum tensile or compressive stress (whichever causes failure) in the extreme fiber of a beam loaded to failure in bending computed from the flexure equation:

$$S_b = Mc/I$$

where:
M = maximum bending moment, computed from the maximum load and the original moment arm,
c = initial distance from the neutral axis to the extreme fiber where failure occurs, and
I = initial moment of inertia of the cross section about the neutral axis.

E 6, E28

modulus of rupture in torsion [FL^{-2}], *n*—the value of maximum shear stress in the extreme fiber of a member of circular cross section loaded to failure in torsion computed from the equation:

$$S_s = Tr/J$$

where:
T = maximum twisting moment,
r = original outer radius, and
J = polar moment of inertia of the original cross section.

E 6, E28

modulus of subgrade reaction—see **coefficient of subgrade reaction.** **D 653,** D18

modulus of volume change—see **coefficient of volume compressibility.** **D 653,** D18

modulus, secant—slope of the line connecting the origin and a given point of the stress-strain curve. **C 717,** C24

modulus, storage—see **complex modulus and storage modulus.** **D 4092,** D20

modulus, storage—see **complex modulus** and **storage modulus**. **E 1142,** E37

modulus, tangent, *n*—the slope of the line touching (tangent to) the stress-strain curve at a given point on the curve. **C 717,** C24

modulus, tensile, n—See **tensile stress at given elongation**, the preferred term. **D 1566,** D11

modulus, Young's, *n*—the ratio of normal stress to corresponding strain for tensile or compressive stresses below the proportional limit of the material. **D 1566,** D11

moellon—synthetic moellon is made by direct oxidation of cod or other fish oils (see also **degras**). **D 1517,** D31

mohair, *n*—the hair of the Angora goat, *Capra* species. **D 123,** D13

mohair paint roller cover—a cover in which the paint applicating material is woven of short pile velour that contains wool or angora goat hair. **D 16,** D01

Mohr circle—a graphical representation of the stresses acting on the various planes at a given point. **D 653,** D18

Mohr circle of stress (strain)—a graphical representation of the components of stress (strain) acting across the various planes at a given point, drawn with reference to axes of normal stress (strain) and shear stress (strain). (ISRM) **D 653,** D18

Mohr envelope—the envelope of a sequence of Mohr circles representing stress conditions at failure for a given material. (ISRM) **D 653,** D18

Mohr envelope (rupture envelope) (rupture line)—the envelope of a series of Mohr circles representing stress conditions at failure for a given material. **D 653,** D18

moil—(*1*) the glass remaining on a punty or blowpipe after a gob has been cut off or after a piece of ware has been blown and severed.
(*2*) the glass originally in contact with the blowing mechanism or head, which becomes cullet after the desired article is severed from it. **C 162,** C14

moiré, *n*—an undesirable, unintended interference pattern caused by the out-of-register overlap of two or more regular patterns such as dots or lines. **D 6488,** D01

moire patter—a pattern developed from interference or light blocking, when gratings, screens, or regularly spaced patterns are superimposed on one another. **E 175,** E41

Moiré (pattern)—an effect caused by the superimposition of a repetitive design(s) to produce a pattern distinct from that of its components; often an unwanted side-effect of optical manipulation, as in the creation of screening patterns during color separation procedures; from the watery, wavy pattern created in Moiré silks with engraved rollers. **F 1156,** F12

moist room, *n*—an enclosed room for storage and curing of paste, mortar, and concrete specimens in which temperature and high relative humidity can be controlled within specified limits. **C 219,** C01

moisture:

inherent moisture, n—in coal, the moisture that exists as an integral part of the coal in its natural state, including water in pores, but not that present in macroscopically visible fractures.

residual moisture, n—that moisture remaining in the sample after air drying.

total moisture, n—all of the moisture in and on a consignment or sample of coal. **D 121,** D05

moisture, *n—as used with textiles*, water absorbed, adsorbed or resorbed by a material. (See also **water**.) **D 123,** D13

moisture—the volatile substances evolved during volatilization and purging of the sample-residue flask and absorbed on the desiccant contained in the absorption tubes under the conditions of the test. **E 1547,** E15

moisture, atmospheric—ambient humidity that may be absorbed by hygroscopic material during sampling and testing and may lead to erroneous results. **D 4790,** D16

moisture content—amount of water contained within a concrete masonry unit or related unit at a given time expressed as a percentage of the total amount of water in the unit under saturated conditions. **C 1209,** C15

moisture content—the amount of water contained in the wood, usually expressed as a percentage of the mass of the ovendry wood. **D 9,** D07

moisture content, *n*—that part of the total mass of a material that is absorbed or adsorbed water, compared to the total mass. (Compare **moisture pick-up** and **moisture regain.**) **D 123,** D13

moisture content—see **water content.** **D 653,** D18

moisture content—the moisture present in a cellulosic material, as determined by prescribed methods, conventionally expressed as a percentage of the total mass of the "wet" material. **D 1695,** D01

moisture content—the water content of a substance as measured under specified conditions. **D 2652,** D28

moisture content, *n*—the percentage, by mass, of water absorbed and adsorbed by carbon black. **D 3053,** D24

moisture content, *n—at moisture-equilibrium,* the moisture content of a material in equilibrium with air of known, or specified, temperature and relative humidity. **D 4920,** D13

moisture content—the moisture present in a material, as determined by definite prescribed methods, expressed as a percentage of the weight of the sample on either of the following bases:

(*1*) Original weight.

(*2*) Moisture-free weight.

NOTE—This is variously referred to as moisture content or moisture "as is" or "as received."

NOTE—This is also referred to as moisture regain (frequently contracted to "regain") or moisture content on the "oven-dry," "moisture-free," or "dry" basis. **E 41,** G03

moisture content—the amount of water contained in the biomass, expressed as either a percentage of the mass of the oven-dry biomass or of the wet biomass.

$$\text{moisture content, dry basis} = \frac{(\text{weight, wet sample} - \text{weight, dry sample})}{\text{weight, dry sample}} \times 100$$

$$\text{moisture content, wet basis} = \frac{(\text{weight, wet sample} - \text{weight, dry sample})}{\text{weight, wet sample}} \times 100$$

E 1126, E48

moisture content—the amount of water contained in the biomass, expressed as either a percentage of the mass of the oven-dry biomass or of the wet biomass, moisture content, dry basis. **E 1705,** E48

moisture content, n—the amount of moisture in a material determined under prescribed conditions and expressed as a percentage of the mass of the moist material, that is, the original mass comprising the oven-dried substance plus any moisture present. **D 123,** D13

moisture-controlled concrete masonry unit— concrete masonry unit whose moisture content conforms to the requirements for Type I classification of Specifications C 55, C 90, or C 129. **C 1209,** C15

moisture-density curve—see **compaction curve.** **D 653,** D18

moisture-density test—see **compaction test.** **D 653,** D18

moisture equilibrium, *n*—the condition reached by a material when it no longer takes up moisture from, or gives up moisture to, the surrounding atmosphere. (Compare **moisture-free.**) **D 123,** D13

moisture equilibrium—the condition reached by a sample when it no longer takes up moisture from, or gives up moisture to the surrounding air. **D 1695,** D01

moisture equilibrium—the condition reached by a sample when the net difference between the amount of moisture absorbed and the amount desorbed, as shown by a change in weight, shows no trend and becomes insignificant. **E 41,** G03

moisture equilibrium for preconditioning, *n*—the moisture condition reached by a material during free exposure to moving air in the standard atmosphere for preconditioning. **D 123,** D13

moisture equilibrium for preconditioning—the moisture equilibrium condition reached by a sample after free exposure to air controlled at the standard condition for preconditioning. **D 1695,** D01

moisture equilibrium for testing, *n*—the condition reached by a sample or specimen during free exposure to moving air controlled at specified conditions. **D 123,** D13

moisture equilibrium for testing, *n—for industrial yarns and tire cords,* the condition reached when, after free exposure to a test atmosphere which is in motion, two successive weighings not less than 4 h apart, show not more than 0.1 % progressive change in mass of the specimen or sample. **D 123,** D13

moisture equilibrium for testing—the equilibrium moisture condition reached by a sample after free exposure to moving air controlled at standard conditions. **D 1695,** D01

moisture equivalent:

centrifuge moisture equivalent, W_e, CME (D)—the water content of a soil after it has been saturated with water and then subjected for 1 h to a force equal to 1000 times that of gravity.

field moisture equivalent, FME—the minimum water content expressed as a percentage of the weight of the oven-dried soil, at which a drop of water placed on a smoothed surface of the soil will not immediately be absorbed by the soil but will spread out over the surface and give it a shiny appearance. **D 653,** D18

moisture expansion—an increase in dimension or bulk volume of a ceramic article caused by reaction with water or water vapor. **C 242,** C21

moisture-free, *adj*—the condition of a material that has been exposed in an atmosphere of desiccated air until there is no further significant change in its mass. **D 123,** D13

moisture-free, *adj—in textiles,* a descriptive term for a material that (*1*) has been exposed to a flow of desiccated air at a specified temperature until there is no further significant change in mass, or (*2*) has been treated by a distillation process using a suitable solvent. (Compare **moisture equilibrium.**) **D 123,** D13

moisture gradient—a condition of graduated moisture content between successive thickness zones of wood that may be losing or absorbing moisture. During seasoning the gradations are between the relatively dry surface zones and the wet zones at the center of the piece. **D 9,** D07

moisture pick-up, *n*—the mass of absorbed and adsorbed water that is held by material, compared to the mass of the dried material. (Compare **moisture content,** and **moisture regain.**) **D 123,** D13

moisture pick-up, *n—at moisture-equilibrium,* the moisture pick-up of a material in equilibrium with air of known, or specified, temperature and relative humidity. **D 123,** D13

moisture regain, *n*—the amount of moisture in a material determined under prescribed conditions and expressed as a percentage of the mass of the moisture-free material. (See **moisture content.** **D 123,** D13

moisture regain, *n*—the amount of water resorbed by a dried material at specified equilibrium conditions of temperature and humidity, compared to the mass of the dried material. (See **standard moisture regain.** Compare **commercial moisture regain, moisture content,** and **moisture pick-up.**) **D 123,** D13

moisture regain—the moisture present in a cellulosic material, as determined by prescribed methods, expressed as a percentage of the ovendry mass. **D 1695,** D01

moisture regain—the moisture in a material determined under prescribed conditions and expressed as a percentage of the weight of the moisture-free specimen.

NOTE—Moisture regain may result from either absorption or desorption, and differs from moisture content only in the basis used for calculation.

E 41, G03

moisture regain, commercial—an arbitrary figure formally adopted as the regain to be used in calculating the commercial or legal mass of shipments or deliveries of any specific material. **D 1695,** D01

moisture regain, standard—the moisture regain of sample brought from a lower moisture regain into equilibrium with the standard atmosphere. **D 1695,** D01

moisture resistant paper—a category of optical scanning paper developed to meet unusual ambient or climatic conditions, for example, census forms or meter reading forms. **F 149,** F05

molality ($^{-}m_1$)— Moles (gram molecular weight) of solute per 1000 g of solvent. **D 6161,** D19

molar absorptivity—*in spectrochemical analysis,* the product of the absorptivity, *a*, and the molecular weight of the substance. **C 859,** C26

molar absorptivity, ϵ— see **absorptivity, molar.** **E 131,** E13

molarity (m_1)—moles (gram molecular weight) of solute per litre of total solution. **D 6161,** D19

molar linear absorption coefficient (ϵ_m)—a constant relating the spectrophotometric absorbance, A_λ, of an optically absorbing molecular species at a given wavelength, λ, per unit pathlength, *d*, to the molar concentration, *c*, of that species in solution:

$$\epsilon_m = A_\lambda/(d \times c)$$

SI Unit: $m^2\ mol^{-1}$ **E 170,** E10

mold—see **die.** **B 243,** B09

mold—a form (usually metal) in which glass is shaped. **C 162,** C14

molded—formed in a closed die by the application of external pressure. **C 709,** D02

molded glass—glass that is formed in a mold as distinct from cast, rolled, drawn, or offhand ware. **C 162,** C14

molded pulp—See **packaging.** **D 996,** D10

molded shape—*in packaging,* a material formed to specific contours for **cushioning,** or **blocking** or **bracing** purposes. **D 996,** D10

mold form—the cavity or shape that uncured composite laminae are placed into or onto and from which they derive their form. **E 1749,** E06

molding—the pressing of powder to form a compact. **B 243,** B09

molding, *v*—shaping of brick by dropping, throwing, or vibrating wet clay or shale in a mold cavity shaped to provide the peripheral dimensions of the brick. Sufficient water is mixed with the clay or shale to produce a soft consistency.

When insides of molds are sanded to prevent sticking of clay, the product is sand-struck brick. When the molds are wetted to prevent sticking, the product is water-struck brick. **C 43,** C15

molding, bag—See **bag molding.** **D 883,** D20

molding, blow—See **blow molding.** **D 883,** D20

molding, compression—See **compression molding.** **D 883,** D20

molding, compression, *n*—the process of forming a material to a desired shape by flow induced by a force applied after a material is placed in the mold cavity. **D 1566,** D11

molding, contact pressure—a method of molding or laminating in which the pressure is only slightly more than necessary to hold the materials together during the molding operation. This pressure is usually less than 69 kPa (10 psi). (D-20) **F 412,** F17

molding, contact pressure, n—See **contact pressure molding.** **D 883,** D20

molding, high-pressure—molding or laminating in which the pressure used is greater than 1380 kPa (200 psi). (D-20) **F 412,** F17

molding, high-pressure, n—See **high-pressure molding.** **D 883,** D20

molding, injection—See **injection molding.** **D 883,** D20

molding, injection, *n*—the process of forming a material by forcing it from an external heated chamber through a sprue (runner, gate) into the cavity of a closed mold by means of a pressure gradient that is independent of the mold clamping force. **D 1566,** D11

molding, low-pressure—molding or laminating in which the pressure used is 1380 kPa (200 psi) or less. (D-20) **F 412,** F17

molding, low-pressure, n—See **low-pressure molding.** **D 883,** D20

molding pressure, compression—the calculated fluid pressure applied to the material in the mold. **D 883,** D20

molding pressure, compression—the calculated fluid pressure applied to the material in the mold. (D-20) **F 412,** F17

molding pressure, injection—the pressure applied to the cross-sectional area of the material cylinder. **D 883,** D20

molding pressure, injection—the pressure applied to the cross-sectional area of the material cylinder. (D-20) **F 412,** F17

molding pressure, transfer—the pressure applied to the cross-sectional area of the material pot or cylinder. **D 883,** D20

molding pressure, transfer—the pressure applied to the cross-sectional area of the material pot or cylinder. (D-20) **F 412,** F17

molding shrinkage, *n*—the difference in dimensions between a molded product and the mold cavity in which it was molded, both the mold and product being at normal room temperature when measured. **D 1566,** D11

molding, transfer—See **transfer molding.** **D 883,** D20

molding, transfer, *n*—the process of forming a material by forcing it from an auxiliary heated chamber through a sprue hole (runner, gate) into the cavity of a closed mold by means of a pressure gradient that is dependent on the mold clamping force. **D 1566,** D11

mold lubricant—a substance applied on or into molds to reduce friction or prevent adhesion. **C 162,** C14

mold lubricant, n—synonym for **release agent (mold).** **D 1566,** D11

mold mark—mark or seam on glassware resulting from a mold joint. **C 162,** C14

mold marks, *n*—surface imperfection transferred to a molded product from corresponding marks on a mold. **D 1566,** D11

mold open time, *n*—the time interval from the instant the mold begins to open until it is closed again. **D 883,** D20

mold temperature, *n*—the mean temperature of the mold cavity surface measured after the system has obtained thermal equilibrium and immediately after opening the mold. **D 883,** D20

molecular contaminant (non-particulate contamination), *n*—the molecular contaminant may be in a gaseous, liquid, or solid state. **G 126,** G04

molecular diffusion, *n*—a process of spontaneous intermixing of different substances, attributable to molecular motion and tending to produce uniformity of concentration. **D 1356,** D22

molecular flow—the flow of gas through a passage under conditions such that the mean-free path is greater than the largest dimension of a transverse section of the passage. **E 1316,** E07

molecular leak—a leak of such geometric configuration that gas flow through it obeys the laws of molecular flow (Knudsen's law). The flow is proportional to the difference of the end pressures and inversely proportional to the square root of the molecular weight of the gas. **E 1316,** E07

molecular replica— See **replica.** **E 7,** E04

molecular SIMS—the SIMS technique when applied to molecular or polyatomic secondary ions. **E 673,** E42

molecular weight cut off (MWCO)—the rating of a membrane for the size of uncharged solutes it will reject. Also referred to as nominal molecular weight cut off (NMWCO). **D 6161,** D19

mole fraction, *n*—the ratio of the number of molecules (or moles) of a compound or element to the total number of molecules (or moles) present.

mole fraction of water vapor (x_v), n—the ratio of the number of moles of water vapor, n_v, to the total number of moles of water and dry air:

$$x_v=\frac{n_v}{n_v+n_a}$$

where:

$$n_v=\frac{m_v}{M_v}$$

$$n_a=\frac{m_a}{M_a}$$

and where M_v and M_a, = molecular weights of water vapor and air, respectively. **D 1356,** D22

mole fraction of water vapor, x_v—the ratio of the number of moles of water vapor n_v to the total number of moles of water and dry air;

$$x_v = \frac{n_v}{n_v + n_a}$$

where:

$$n_v = \frac{m_v}{M_v}$$

$$n_a = \frac{m_a}{M_a}$$

and where M_v and M_a are the molecular weights of watervapor and air, respectively. **D 4023,** D22

mole run—a meandering ridge in a membrane not associated with insulation or deck joints. **D 1079,** D08

moles of substituent combined, MS—in a cellulose derivative, the average number of substituent molecules per glucose unit. For most derivatives, MS = DS, but for a few, such as hydroxyethylcellulose, the substituted group may also be reactive and MS may be greater than DS. **D 1695,** D01

molten cast refractory, *n*—a solidified material made by melting refractory ingredients and pouring into molds (see also **fused or fusion cast refractory**). **C 71,** C08

molten state—the liquid phase of a solid substance existing above its melting point temperature. **D 4790,** D16

molten substances—metals in their liquified, elevated temperature state, as well as related non-metallic substances also handled at elevated temperatures such as slag, dross, and salt. Excluded are liquid hot substances that may be associated with metal processing such as water, oil, and caustic solutions. **F 1494,** F23

monitor, *n*—a device that continually measures or intermittently samples and analyzes atmospheres or emissions for the concentration of a specific constituent or constituents, or for the level of a physical property (such as temperature) to provide either a real-time read-out or an electrical signal.

continuous monitor, n—a device for the uninterrupted measurement of atmospheric or emission concentrations or properties in real or near-real time. **D 1356,** D22

monitoring—the continual sampling, measuring, recording, and/or signaling, of the characteristics of water or waterborne material. **D 1129,** D19

monitoring path—See **point analyzer.** **D 1356,** D22

monitoring path length—See **point analyzer.** **D 1356,** D22

monitoring well (observation well)—a special well drilled in a selected location for observing parameters such as liquid level or pressure changes or for collecting liquid samples. The well may be cased or uncased, but if cased the casing should have openings to allow flow of borehole liquid into or out of the casing. **D 653,** D18

monochromatic, *adj*—characterized by a single wavelength or, by extension, by a small range of wavelengths that can be described by stating a single wavelength. **E 284,** E12

monochromatic—a property of a beam of electromagnetic radiation in which all waves in the beam have the same wavelength. **E 1316,** E07

monochromatic—a single color, although saturation may vary from point to point within the image. **F 1156,** F12

monochromatic (homogeneous)—of the same wavelength. **E 7,** E04

monochromatic objective—an objective, usually made of fused quartz, which has been corrected for use only with monochromatic light. **E 7,** E04

monochromatic radiation—radiation characterized by a single frequency. By extension, radiation of a very small range of frequency or wavelength which can be described by stating a single frequency or wavelength. **E 349,** E21

monochromator—a device or instrument that, with an appropriate energy source, may be used to provide a continuous calibrated series of electromagnetic energy bands of determinable wavelength or frequency range. **E 131,** E13

monochromator, *n*—a device for isolating monochromatic radiation from a beam of polychromatic radiation. **E 135,** E01

monochromator, *n*—a device for isolating monochromatic radiation from a beam of radiation including a broad range of wavelengths. **E 284,** E12

monochromator—a device for isolating monochromatic radiation from a beam of radiation which includes a broad range of wavelengths. **E 1316,** E07

monochromator (X-rays)—a device for producing monochromatic radiation from heterochromatic radiation. It usually consists of a crystal so arranged as to diffract one wavelength of particularly high intensity, such as the characteristic, out of a beam of mixed white and characteristic radiation. **E 7,** E04

monochrome decoration—a single color decoration. **C 242,** C21

monoclinic—having three axes of any length with two included angles equal to 90° and one included angle not equal to 90°. **E 7,** E04

monocrystal—See **single crystal.** **F 1241,** F01

monocyclic terpenes, *n*—a designation sometimes used in the trade to describe a heterogeneous mixture of monocyclic, bicyclic, and other related terpene $C_{10}H_{16}$ hydrocarbons recovered or removed in the fractionation of certain terpenes or other essential oils, or as a by-product in the chemical conversion of pinenes generally sold under trade names. **D 804,** D01

monodisperse, *adj*—describing a population of drops of substantially equal diameter. **E 1620,** E29

monofilament, *n*—a single filament which can function as a yarn in commercial textile operations, that is, it must be strong and flexible enough to be knitted, woven, or braided, etc. (See **yarn.**) **D 123,** D13

monofilament—a continuous filament strong enough to function as a yarn in commercial textile operations or as an entity in other operations. **D 3878,** D30

mono-hydrated lime—dolomitic lime which has been hydrated at atmospheric pressure and contains more than 8 % unhydrated oxides. **C 51,** C07

monolithic, *n*—a material of uniform composition applied as a continuous surface or structure. **C 904,** C03

monolithic refractory, *n*—a refractory which may be installed *in situ,* without joints to form an integral structure. **C 71,** C08

monolithic refractory construction, *n*—a refractory installation utilizing monolithic refractories. **C 71,** C08

monomer, *n*—a low-molecular-weight substance consisting of molecules capable of reacting with like or unlike molecules to form a polymer. **D 883,** D20

monomer, *n*—a relatively simple compound which can react to form a polymer. (Compare **polymer**.) **D 907,** D14

monomer, *n*—a low molecular weight substance consisting of molecules capable of reacting with like or unlike molecules to form a polymer. **D 1566,** D11

monomer, *n*—a relatively simple compound which can react to form a polymer. (See also **polymer**.) (D-20) **F 412,** F17

monomer, *n*—a relatively simple compound which can react to form a polymer. (See also **polymer.**) **F 1251,** F04

monomolecular layer—an adsorbed film, one molecule thick. **D 2652,** D28

monopack—See **color film**. **E 7,** E04

monotectic equilibrium—a reversible binary univariant transformation in which a liquid phase, that is stable only at higher temperatures, decomposes, with lowering temperature, into a new liquid phase and a solid phase, for example: $L_1 = L_2 + \alpha$. **E 7,** E04

monotectic point—the composition and temperature in a binary system at which exists a liquid that is capable of univariant decomposition, with lowering temperature, into another liquid phase and a solid phase. **E 7,** E04

monotropic transformation—a nonreversible metastable phase change. **E 7,** E04

month, *n*—for reporting analyses of outdoor air on a monthly rate, results are calculated to a base of thirty days. **D 1356,** D22

monthly average daily traffic, (MADT), *n*—the mean traffic volume for a month, calculated by the sum of Monthly Average Days of the Week (MADWs) divided by seven; or, in the absence of a MADW for each day of the week, divided by the number of MADWs during the month. **E 867,** E17

monthly average days of the week, (MADWs), *n*—a traffic volume mean statistic for each day of the week, over the period of one month, calculated from included permanent counter data as the sum of all traffic for each day of the week (Sunday, Monday, and so forth through the week), during a month, divided by the occurrences of that day during the month. **E 867,** E17

monthly average weekday traffic, (MAWDT), *n*—the four day average of traffic for the period Monday through Thursday in each month, calculated as the sum of MADWs for Monday through Thursday, divided by four. **E 867,** E17

monthly average weekend traffic, (MAWET), *n*—the two-day average of traffic for the period Saturday through Sunday in each month, calculated as the sum of MADWs for Saturday through Sunday, divided by the number of MADWs for Saturday through Sunday during the month. **E 867,** E17

monthly classification ratio, (MCR), *n*—the monthly average volume by vehicle classification divided by the annual average volume by vehicle classification, calculated for vehicle classification annual summary statistics. **E 867,** E17

monthly traffic ratio, (MTR), *n*—the monthly average traffic volume divided by the annual average traffic volume, calculated for traffic volume annual summary statistics. **E 867,** E17

monthly weight ratio, (MWR), *n*—the monthly average ESALs by vehicle classification divided by the annual average ESALs by vehicle classification, calculated for the vehicle weigh annual summary statistics. **E 867,** E17

montmorillonite—a group of clay minerals characterized by a weakly bonded sheet-like internal molecular structure; consisting of extremely finely divided hydrous aluminum or magnesium silicates that swell on wetting, shrink on drying, and are subject to ion exchange. **D 653,** D18

monumental stone—rock of adequate quality to be quarried and cut as dimension stone as it exists in nature, as used in the monument and memorial industry. **C 119,** C18

moon crater—*on a semiconductor wafer,* surface texture that results when a wafer floats during the initial stages of chemical polishing in a rotating cup etcher. **F 1241,** F01

moonstone glass—a type of opal glass resembling the mineral moonstone. **C 162,** C14

mop-and-flop—a procedure in which roofing elements (insulation boards, felt plies, cap sheets, etc.) are initially placed upside down adjacent to their ultimate locations, are coated with adhesive, and are then turned over and adhered to the substrate. **D 1079,** D08

mopping—the application of hot bitumen with a mop or mechanical applicator to the substrate or to the plies of a built-up roof. There are four types of mopping: *(1) solid*—a continuous coating; *(2) spot*—bitumen is applied in roughly circular areas, generally about 460 mm (18 in.) in diameter, leaving a grid of unmopped, perpendicular areas; *(3) strip*—bitumen is applied in parallel bands, generally 200 mm (8 in.) wide and 300 mm (12 in.) apart; *(4) sprinkle*—bitumen is shaken onto the substrate from a broom or mop in a random pattern. **D 1079,** D08

Morocco grain—embossed imitation of the natural goat grain on other kinds of leather. **D 1517,** D31

Morocco leather—vegetable-tanned fancy goatskin leather having a distinctive pebbled grain. **D 1517,** D31

morphology, *n*—the form and structure of a particular organism. **F 1494,** F23

mortar, *n*—a mixture of gypsum plaster with aggregate or hydrate lime, or both, and water to produce a trowelable fluidity. **C 11,** C11

mortar, *n*—a mixture of finely divided hydraulic cementitious material, fine aggregate, and water in either the unhardened or hardened state; hydraulic mortar. **C 219,** C01

mortar, *n*—a mixture consisting of cementitious materials, fine aggregate, water, and with or without admixtures, that is used to construct unit masonry assemblies. **C 1180,** C12

mortar, air-setting, *n*—a composition of finely ground materials, marketed in either a wet or dry condition, which may require tempering with water to attain the desired consistency and which is suitable for laying refractory brick and bonding them strongly upon drying and upon subsequent heating at furnace temperatures. **C 71,** C08

mortar bond or grout bond, *n*—adhesion between mortar or grout and masonry units, reinforcement, or connectors. **C 1180,** C12

mortar cement, *n*—a hydraulic cement manufactured for use in masonry mortar designed for specific bond and air content criteria. **C 219,** C01

mortar, heat setting, *n*—a refractory mortar of finely ground materials whose potential strength is dependent on use at furnace or process temperatures. **C 71,** C08

mortar, refractory, *n*—a finely ground preparation which becomes plastic and trowelable when tempered with water, and is suitable for laying and bonding refractory brick. **C 71,** C08

mosaic crystal—an imperfect single crystal composed of regions each very slightly disoriented with respect to its neighbor. **E 7,** E04

mosaic structure—the structure of a material containing mosaic crystals. **E 7,** E04

mote, *n*—a whole, immature cotton seed. **D 123,** D13

motion control—the restriction of extraneous motion of the foot by the shoe. **F 869,** F08

motion segment, *n*—two adjacent vertebrae, the intervening disc, and the associated ligamentous structures. **F 1582,** F04

motor-generator (MG set)—a machine that consists of one or more motors mechanically coupled to one or more generators. In plating, such a machine in which the generator delivers dc of appropriate amperage and voltage. **B 374,** B08

motorized nozzle, *n*—an attachment for a vacuum cleaner containing an electric motor-driven agitator that assists in dirt removal from a floor surface. **F 395,** F11

mottle, *n*—the nonuniform transfer or absorption of the ink into the substrate creating an uneven ink film. **D 6488,** D01

mottle, *n*—a spotty nonuniformity of color appearance on a scale that is larger than the colorant particles, typically 1 to 10 mm. **E 284,** E12

mottle—a gross random nonuniformity in the visual density of a printed area. **F 335,** F05

mottle, *back trap, n*—a nonuniform density variation of a printed ink film due to nonuniform ink absorption into the paper. **D 6488,** D01

mottled iron, *n*—a cast iron containing a mixed structure of gray iron and white iron of variable proportions. The fracture has a mottled appearance. **A 644,** A04

mottle, *halftone, n*—a nonuniform transfer of halftone dots. **D 6488,** D01

mottle, *print, n*—an uneven appearance within the continuous ink film solid areas of a print, with respect to density, gloss or color. Also referred to as "solid area mottle". **D 6488,** D01

mottle, *trapping, n*—nonuniform print due to improper tack sequence of the inks. **D 6488,** D01

mottle, *water interference, n*—nonuniform print caused by poor ink transfer due to the printing substrates' inability to absorb fountain solution. **D 6488,** D01

mottling, *vt*—the presence in the surface of a film, of irregularly shaped, randomly distributed areas that vary in color, gloss, or sheen, causing the film to be non-uniform in appearance, also known as **blotching**. **D 16,** D01

mottling—physical migration of polish film from smooth, even distribution at time of application to small discrete pools of material, resulting in a blotchy or spotty appearance. **D 2825,** D21

mottling, *n*—non-uniformity of image density which follows patterns in the substrate or by non-uniform ink-substrate interaction. **F 1857,** F05

moulding—a specially worked wood member used mostly for decoration but often serves a useful purpose in other ways; generally worked from lumber of strip size; may be a plane surface but often curved or patterned. **D 9,** D07

moulding and trim nail—bright zinc-plated, slim, hardened-steel, 1¼ by 0.054 to 2½ by 0.083-in. nails with blunt point and button head. (See **finishing nail, fine nail, hardboard nail, insulation building-board nail, tileboard nail, wallboard nail.**) **F 547,** F16

mound—irregularly shaped projection on a semiconductor wafer surface, with one or more irregularly developed facets (see also **pyramid**). **F 1241,** F01

mouse teeth—distal tip teeth that interdigitate. **F 1638,** F44

mouth, *n*—*in zippers*, the opening in a slider that receives the chain. **D 123,** D13

mouthfeel, *n*—a mixed experience deriving from sensations in the oral cavity that relate to physical (for example, density, viscosity, particulate) or chemical (for example, astringency, cooling) properties of a stimulus material. Trained panelists differentiate the physical sensations as texture properties and the chemical sensations as flavor properties. **E 253,** E18

mouth width, *n*—*of zippers*, the measurement between the slider flanges at the point where they bear against the shoulders of the interlocked elements or at the outermost edges of the bead if the bead extends beyond the elements. **D 123,** D13

mouton—a sheepskin shearling tanned and finished for use as a fur, usually with wool straightened. **D 1517,** D31

movable bed—a stream bed made up of materials readily transportable by the stream flow. **D 4410,** D19

movable by-pass clincher—clinching mechanism for forming a by-pass clinch. **F 592,** F16

movable in-line clincher—clinching mechanism for forming an in-line clinch. **F 592,** F16

movable retainer, *n*—*in zippers*, a movable or sliding device performing a similar function to that of the fixed retainer, the purpose being to permit separation of the two stringers from the bottom, without the necessity of opening the zipper from the top. **D 123,** D13

moving beam scanner, *n*—a scanning device where scanning motion is achieved by mechanically moving the light beam through the bars and spaces of a bar code symbol. **F 1294,** F05

moving range, *n*—the difference without regard to sign between two successive observations. **D 123,** D13

MPP—see **maximum print position.** **F 1457,** F05

MPU/IG—see **microprocessor unit/image generator.** **F 1457,** F05

MR-8—the original optical scanning test device that measures the amount of reflected light in millivolts. **F 149,** F05

2200 m/s cross section, $\sigma(v_o)$—the neutron cross section at $v_o = 2200$ m/s (*E* is about 0.0253 eV). **E 170,** E10

MTBF—Mean Time Between Failures; a quality assurance term; the larger the number, the better. **F 1156,** F12

MTTR—Mean Time To Repair; a quality assurance term; the shorter, the better. **F 1156,** F12

mucilage, *n*—an adhesive prepared from a gum and water, and also in a more general sense, a liquid adhesive which has a low order of bonding strength. (See also **adhesive, glue, paste,** and **sizing.**) **D 907,** D14

muck—stone, dirt, debris, or useless material; or an organic soil of very soft consistency. **D 653,** D18

mud—a mixture of soil and water in a fluid or weakly solid state. **D 653,** D18

mud-cracking, , *n*—an irregular broken network of cracks in the film, which occurs due to volatile loss while drying or curing. **D 16,** D01

mud cracking—surface cracking resembling a dried mud flat. **D 1079,** D08

mudcracking—*in protective coatings*, a particular pattern of cracking in a coating with the appearance of a dried mud puddle (see cracking and checking). **D 4538,** D33

mudflow—a mass of water-sediment mixture which, because of its high viscosity, moves more slowly than water. **D 4410,** D19

mudguard, *n*—a strip of material applied to a shoe upper just above the sole intended as a protection against dampness or as an ornament. **F 1646,** F13

mudjacking—see **slab jacking**. **D 653,** D18

mud mat, *n*—in construction design: (*1*) a surfacing layer suitable for the application of membrane waterproofing, installed on a prepared subgrade; (*2*) an area of fill compacted to no less than 95 % of the maximum dry density of the fill based upon the appropriate laboratory compaction test, such as defined in Test Methods D 698 or D 1557. **D 1079,** D08

mud pit—usually a shallow, rectangular, open, portable container with baffles into which drilling fluid and cuttings are discharged from a borehole and that serves as a reservoir and settling tank during recirculation of the drilling fluids. Under some circumstances, an excavated pit with a lining material may be used. **D 653,** D18

mud slab, *n*—a 2 in. (50 mm) minimum thickness non-reinforced concrete slab suitable as the substrate for membrane waterproofing. **D 1079,** D08

mud-up—to seal a furnace structure with wet clay or castable refractory material. **C 162,** C14

muffle—an enclosure in a furnace to protect the ware from the flame and products of combustion. **C 162,** C14

mukluk leather—leather usually made from deer, elk, or similar skins. It is tanned white with aldehydes, alum, or syntans. It is highly permeable to moisture vapor and retains its flexibility at very low temperatures. **D 1517,** D31

mulch treatment, *n*—a spray application of bituminous material used to temporarily stabilize a recently seeded area. The bituminous material can be applied to the soil or to straw or hay mulch as a tie-down, also. **D 8,** D04

mull—See **windows and doors.** **E 631,** E06

Mullen test—See **package testing.** **D 996,** D10

Mullins Effect, *n*—the softening of a vulcanized rubber (reduction in the stress at a given strain) as a result of previous deformation. **D 1566,** D11

mullion—See **windows and doors.** **E 631,** E06

mullite—a rare mineral of theoretical composition $3Al_2O_3{\cdot}2SiO_2$, a relatively stable phase in ceramics produced by the high temperature reaction of alumina and silica or by the thermal decomposition of alumina-silica minerals such as kyanite, sillimanite, andalusite, and various clay minerals. **C 242,** C21

mullite porcelain—a vitreous ceramic whiteware for technical application in which mullite ($3Al_2O_3{\cdot}2SiO_2$) is the essential crystalline phase. **C 242,** C21

mullite refractories, *n*—refractory products consisting predominantly of mullite ($3Al_2O_3{\cdot}2SiO_2$) crystals formed either by conversion of one or more of the sillimanite group of minerals, or by synthesis from appropriate materials employing either melting or sintering processes. **C 71,** C08

mullite whiteware—any ceramic whiteware in which mullite ($3Al_2O_3{\cdot}2SiO_2$) is the essential crystalline phase. **C 242,** C21

multi-axial tension, *n*—stress in more than one direction. **D 4439,** D35

multi-axis accelerometer—device for determining the rate(s) of motion (velocity) and acceleration, and the direction of motion in dimensions or multi-dimensional space. **F 1156,** F12

multibench blasting—the blasting of several benches (steps) in quarries and open pits, either simultaneously or with small delays. (ISRM) **D 653,** D18

multi-cased well—a well constructed by using successively smaller diameter casings with depth. **D 653,** D18

multi-colored band structures—alternating bands of many colors in a nonregular arrangement that conveys encoded information; used on particles that are tags for commercial dynamites. **F 1156,** F12

multicomponent system—the complete series of compositions produced by mixing, in all proportions, two or more components. **E 7,** E04

multidirectional magnetization—the alternative application of magnetic fields in different directions during the same time frame. **E 1316,** E07

multifont reader—a reading device that can read forms containing intermixed characters printed in a number of fonts. Multifont reading eliminates the need to prebatch the input data by font prior to submission to the scanner. **F 149,** F05

multigroup cross section—average particle fluence-weighted cross section in an energy interval of a multigroup model. (See also **multigroup model**).

$$\sigma_i = \int_{E_i}^{E_{i+1}} \Phi(E)\,\sigma(E)\ dE \,/ \int_{E_i}^{E_i+1} \Phi(e)\,dE$$

E 170, E10

multigroup model—subdivision of an energy spectrum into a number of subintervals: (E_i, E_{i+1}), where: i = 1, 2, ... n. **E 170,** E10

multigroup particle fluence—particle fluence in the energy intervals of a multigroup model (see also **multigroup model**).

$$\Phi = \int_{E_i}^{E_{i+1}} \Phi(E)\,dE$$

E 170, E10

multilateral arrangement—recognition arrangement that covers the acceptance of each other's results by more than two parties, (ISO Guide 2) **E 1187,** E36

multilayer deposit—a deposit consisting of two or more layers of metal deposited successively, consisting of either different metals or layers of the same metal with different characteristics. **B 374,** B08

multilayered structure—*in flexible barrier materials*, a structure that consists of two or more continuous layers or plies of material. **F 17,** F02

multilayered structure—See Terminology F 17. **F 1327,** F02

multilevel pile, *n—for pile yarn floor covering*, pile in which some tuft legs are substantially longer than others, resulting in a sculptured appearance or pattern. **D 5684,** D13

multimedia filter—filter with a bed consisting of three or more separate filter media. The coarsest, lowest density at the top and the finest, highest density at the bottom. **D 6161,** D19

multimolecular layer—an adsorbed film more than one molecule thick. **D 2652,** D28

multiphase coolant, *n*—an engine coolant composed of immiscible liquids or undissolved solids, or both. **D 4725,** D15

multiple—straight lengths of integral multiples of a base length, with suitable allowance for cutting, if and when specified. **B 846,** B05

multiple back reflections—successive reflections from the back surface of the material under examination. **E 1316,** E07

multiple-cavity mold—a mold possessing multiple cavities for simultaneous fabrication of multiple articles of glass. **C 162,** C14

multiple-cavity process—a glass-molding process that uses multiple charges of glass and forms them simultaneously. **C 162,** C14

multiple cloning site—DNA that contains several contiguous restriction enzyme recognition sites; also called a polylinker. **E 1705,** E48

multiple correlation coefficient, (R)—the correlation, $r_{y\hat{y}}$, between the accepted reference values, Y_i, and the values determined using the calibration equation, $\hat{Y}_i$, equal to the square root of the coefficient of multiple determination, R^2. **E 131,** E13

multiple font reader—a reading device that can read more than one type font, but only one font may be read at a time. **F 149,** F05

multiple-gob process—See **multiple-cavity process.** **C 162,** C14

multiple heat, *n*—two or more molten **primary heats**, in whole or in part, combined in a common ladle or in a common non-oscillating mold. **A 941,** A01

multiple laboratories operational precision—the standard deviation of the results of a series of determinations by several laboratories employing the method with its associated sample container preparation, collection, splitting, preservation, transmission, and storage on a homogeneous sample. **D 1129,** D19

multiple-layer adhesive, *n*—a dry-film adhesive, usually supported, with a different adhesive composition on each side; designed to bond dissimilar materials such as the core to face bond of a sandwich composite. **D 907,** D14

multiple length staple fibers, *n*—man-made staple fibers that are two or more times the nominal cut fiber length. **D 123,** D13

multiple pressing—a method of pressing whereby two or more compacts are produced simultaneously in separate die cavities. **B 243,** B09

multiple reflections—successive echoes of ultrasonic energy between two surfaces. **E 1316,** E07

multiple-row blasting—the drilling, charging, and firing of several rows of vertical holes along a quarry or opencast face. (ISRM) **D 653,** D18

multiple scattering event—*ISS,* a collision process that may be described as a sequence of binary scattering events which may or may not be elastic. **E 673,** E42

multiple-screened wells—two or more monitoring wells situated in the same borehole. These devices can be either individual casing strings and screen set at a specific depth, a well with screens in more than one zone, or can consist of devices with screens with tubing or other collecting devices attached that can collect a discrete sample. **D 653,** D18

multiple stitch zigzag, *n—in home sewing,* a simple machine stitch pattern of alternating diagonal segments with each segment of two or more stitches having equal length and width. **D 5646,** D13

multiplet or exchange splitting—*XPS,* splitting of a photoelectron line caused by the interaction of the unpaired electron created by photoemission with other unpaired electrons in the atom. **E 673,** E42

multiplicity factor—a factor used in calculating the intensity of diffraction from a polycrystalline specimen. It is equal to the number of sets of planes of the same family. **E 7,** E04

multiplicity of infection—the ratio of infecting phage to host bacteria. **E 1705,** E48

multi-ribbed—See **thread, vertical.** **F 547,** F16

multi-row symbology, *n*—symbologies where a long symbol is broken into sections and "stacked" one upon another similar to sentences in a paragraph. Examples are: Code 16K, Code 49 and PDF417. **F 1294,** F05

multi-strike film ribbon—is a ribbon wherein the substrate film such as polyester is coated or impregnated with an ink which allows several different imprints be made from multiple overstrikes on the same location on the ribbon, and still result in full characters being printed. **F 221,** F05

multivariant—having two or more degrees of freedom. **E 7,** E04

municipal ferrous scrap, *n*—ferrous waste that is collected from industrial, commercial, or household sources and destined for disposal facilities. **D 5681,** D34

municipal solid wastes (MSW)—the refuse materials collected from urban areas in the form of organic matter, glass, plastics, waste paper, etc., not including human wastes. **E 1126,** E48

municipal solid wastes (MSW)—the refuse materials collected from urban areas in the form of organic matter, glass, plastics, waste paper, etc., not including human wastes. **E 1705,** E48

Munsell Book of Color, *n*—current Munsell Color Company physical exemplification of the Munsell color order system, consisting of about 1600 color chips arranged in a cylindrical coordinate system of planes of constant Munsell hue on which Munsell value is displayed vertically and Munsell chroma horizontally. **E 284,** E12

Munsell chroma, *n*—an attribute of color used in the Munsell color system to indicate the degree of departure of a color from a gray of the same Munsell value, in steps that are visually approximately equal in magnitude. **E 284,** E12

Munsell color system, *n*—a system of specifying colors of surfaces illuminated by daylight and viewed by an observer adapted to daylight, in terms of three attributes: hue, value, and chroma, using scales that are perceptually approximately uniform. **E 284,** E12

Munsell hue, *n*—an attribute of color used in the Munsell color system to indicate the hue of a specimen viewed in daylight. **E 284,** E12

Munsell notation, *n*—(*1*) the Munsell hue, value, and chroma assigned to the color of a specimen by visually comparing the specimen to the chips in the Munsell Book of Color.

(*2*) a notation in the Munsell color system, derived from luminous reflectance Y and Chromaticity Coordinates x and y in the 1931 CIE system for Standard Illuminant C, by the use of scales defined by the Optical Society of America Subcommittee on the Spacing of the Munsell Colors. **E 284,** E12

Munsell value, *n*—an attribute of color used in the Munsell color system to indicate the lightness of a specimen viewed in daylight, on a scale extending from 0 for ideal black to 10 for ideal white, in steps that are visually approximately equal in magnitude. **E 284,** E12

muntin, *n*—a secondary intermediate member subdividing a glazed area. **C 717,** C24

muntin—See **windows and doors.** **E 631,** E06

Murgatroyd belt—{archaic} that portion of the sidewall of a bottle near the bottom. **C 162,** C14

muskeg—level, practically treeless areas supporting dense growth consisting primarily of grasses. The surface of the soil is covered with a layer of partially decayed grass and grass roots which is usually wet and soft when not frozen. **D 653,** D18

muslin, *n—as applied to bed sheeting,* a plain weave fabric with not fewer than 128 yarns/in.^{2}(645 mm^2). **D 123,** D13

mussiness, *n*—surface distortion in a fabric characterized by undesirable unevenness due to many minor deformations. **D 123,** D13

mutagen—a substance that increases the normal mutation rate. **F 1600,** F20

mutual aid—the furnishing of resources, from one individual or agency to another individual or agency, including but not limited to facilities, personnel, equipment, and services, pursuant to an agreement with the individual or agency, for use within the jurisdiction of the individual or agency requesting assistance. **F 1177,** F30

m-value—the negative slope of a curve plotting log flow versus log time. A measurement of the degrees of membrane compaction as a result of temperature, pressure and time. **D 6161,** D19

MV (mega or million volt)—a unit of electrical potential difference equal to one million volts, used to describe the accelerating potential of an X-ray tube. **E 1316,** E07

mycoplasma—the smallest prokaryotes capable of living freely, lacking a cell wall, having a circular double-stranded DNA relatively rich in adenine and thymine, and containing 16s and 23s ribosomal RNAs. They can be found as contaminants in cell cultures. **E 1705,** E48

mylonite—a microscopic breccia with flow structure formed in fault zones. (ISRM) **D 653,** D18

mythical image—the assumed intersections of the extended sides of a microindentation which is established in order that a true diagonal can be approximated. **E 7,** E04

N

NaHMP—sodium hexametaphosphate, an antiscalant. **D 6161,** D19

nail, *n*—straight, slender fastener, usually pointed and headed; designed to be driven through connector plate or plates with or without nail holes; serving as separate supplementary or primary fastener. **E 1807,** E06

nail—straight, slender fastener, usually pointed and headed; 6 in. or less in length; designed to be driven; to hold two or more pieces together or to act as support. (See **screw nail; drive screw.**) **F 547,** F16

nail-base fiberboard sheathing—a specially manufactured cellulosic fiberboard product, approximately 25 lb/ft^3, designed for use in frame construction to permit the direct application of certain exterior siding materials such as wood or cement-asbestos shingles. **D 1554,** D07

nail hole, *n*—an appropriately sized opening such that tile shall not be fractured by the fastener and fixing process used to attach the tile to the roof deck. **C 43,** C15

nail hole—round perforation in metal connector plate through which a nail can be driven to fasten plate to wood members (or section) and to transmit shear loads; providing predetermined location for appropriately locating nails to be driven. See **plate hole**. **E 1807,** E06

nailing—(*1*) exposed-nailing of roofing wherein nail heads are bare to the weather;

(*2*) concealed-nailing of roofing wherein nail heads are concealed from the weather. (See also **blind nailing**). **D 1079,** D08

nail-on plate—solid or prepunched (or predrilled) metal connector plate of specified thickness (gage); manufactured to various sizes, that is, lengths and widths; designed to be fastened with nails (or staples) to wood members and to transmit forces from one wood member (or section) to another one or more wood members (or section). **E 1807,** E06

NAK, *n*—a signal from the printer which indicates that the previous data block was in error and that retransmission can begin; also used as a not ready signal. **F 1457,** F05

Nanofiltration (NF)—a crossflow process with pore sizes designed to remove selected salts and most organics above about 300 molecular weight range, sometimes referred to as loose RO. **D 6161,** D19

nanometre (nm)—a unit, 10^{-9} metre, used in the measurement of the wavelength of light. It is identical to the older term millimicron, mμ. **C 859,** C26

nap—the wooly or fuzzy surface finish of some fabrics and some leathers such as suede and antelope, reversed calf and side leather. **F 869,** F08

napa leather—chrome, alum, or combination tanned sheepskin glove leather, drum colored. **D 1517,** D31

nap finish—a process in which the natural grain layer of the leather is removed and the outer surface of the leather is then given a napped finish. (See also **suede finish**.) **D 1517,** D31

naphtha, aromatic solvent—a concentrate of aromatic hydrocarbons including C_8, C_9 and C_{10} homologs. **D 4790,** D16

naphtha-based oil—a petroleum oil containing a majority of the naphtha fraction. **E 1519,** E35

naphthalene ($C_{10}H_6$)—mol weight 128.16; monoclinic prismatic plates; commercially available as white scales, powder, balls, or cakes; odor of moth balls; solidification point, 80.1°C; sublimes above melting point. **D 4790,** D16

naphthene-aromatics, *n*—a mixture of naphthenic and aromatic hydrocarbons which are adsorbed from a paraffinic solvent on an adsorbent during percolation and then desorbed with an aromatic solvent such as toluene. **D 8,** D04

naphthenic oil, *n*—an hydrocarbon process oil containing more than 30 %, by mass, of naphthenic hydrocarbons. **D 1566,** D11

naphthenic oil—a term applied to mineral insulating oil derived from special crudes having very low, naturally occurring *n*-paraffin (wax) contents. Such an oil has a low natural pour point and does not need to be dewaxed nor does it usually require the use of a pour depressant. **D 2864,** D27

narrow band-pass filter—filter that passes only a narrow range of wavelengths or frequencies. (Narrow is a relative term, the value of which depends upon the wavelength region of interest.) **F 1156,** F12

narrow crown—staple crown usually 5⁄16 in. (8 mm) in width or smaller. **F 592,** F16

narrow elastic fabric, *n*—an elastic fabric that is less than 150 mm, or 6 in., in width. (Compare **wide elastic fabric.**) **D 123,** D13

narrow fabric, *n*—a textile fabric not exceeding 12 in. (300 mm) in width. **D 123,** D13

national standard—an artifact, such as a well-characterized instrument or radiation source, that embodies the international definition of primary physical measurement standard for national use. **E 170,** E10

national standards laboratory—the laboratory which maintains a nation's measurement standards, such as the National Institute of Standards and Technology in the United States. **E 170,** E10

native asphalt, *n*—asphalt occurring as such in nature. **D 8,** D04

native hide—a cattlehide without a brand. **D 1517,** D31

native material—in place geologic (or soil) materials encountered at a site. **D 653,** D18

native water—water from a sampled medium; this water has been unaffected by sampling, handling, and preservation. **D 4410,** D19

natural aging, *n*—spontaneous aging of a super-saturated solid solution at room temperature. **A 941,** A01

natural aging—aging in which the material is exposed to conditions replicating those that are present in actual service in a component for oxygen service. **G 126,** G04

natural cement, *n*—a hydraulic cement produced by calcining a naturally occurring argillaceous limestone at a temperature below the sintering point and then grinding to a fine powder. **C 219,** C01

natural clay tile—a tile made by either the dust-pressed method or the plastic method, from clays that produce a dense body having a distinctive, slightly textured appearance. **C 242,** C21

Natural Color System, *n*—color order system based on resemblances of colors to up to four of six "elementary" colors red, yellow, green, blue, black, and white, in which the attributes of the colors are hue, chromaticness, and blackness. **E 284,** E12

natural convection— see **convection, natural.** **E 772,** E44

natural fiber, *n*—a class name for various genera of fibers (including filaments) of (*1*) animal, (*2*) mineral, or (*3*) vegetable origin. (Compare **man-made fiber.**) **D 123,** D13

natural finish, *n*—units having unglazed or uncoated surfaces burned to the natural color of the material used in forming the body. **C 43,** C15

natural frequency—the frequency at which a body or system vibrates when unconstrained by external forces. (ISRM) **D 653,** D18

natural frequency (displacement resonance) f_n—frequency for which phase angle is 90° between the direction of the excited force (or torque) vector and the direction of the excited excursion vector. **D 653,** D18

natural gas—a naturally occurring mixture of hydrocarbon and nonhydrocarbon gases found in porous geological formations (reservoirs) beneath the earth's surface, often in association with petroleum. The principal constituent of natural gas is methane. **D 4150,** D03

natural gas, processed—a methane-rich commercial gaseous product derived from naturally occurring gas mixtures by processing (also referred to as merchantable natural gas). **D 4150,** D03

natural grade—See **grade.** **E 631,** E06

natural grain—refers to a surface that is not filled with pigments and the haircells are easily viewed with the naked eye. **D 1517,** D31

natural levee—raised berms or crests above the flood-plain surface adjacent to the channel usually containing coarser materials deposited by flood flows. **D 4410,** D19

natural linewidth—See **intrinsic linewidth.** **E 673,** E42

naturally dispersed sample—a sample having sediment that will not settle in about 4 h due to the character of fineness of particles or due to the nature of the dissolved constituents, or both. **D 4410,** D19

natural pH, *n*—the measured pH of the pulp prior to the addition of collector, frother, or any conditioning agents. **D 121,** D05

natural process limits (NPL), *n*—limits which include a stated fraction of the individuals in a population.

NOTE—*Natural process limits* will not ordinarily be the dimensional limits shown on an engineering drawing. They are mostly used to compare the natural capability of the process to tolerance limits. **E 456,** E11

natural resin—see **resin, natural.** **D 16,** D01

natural resin, *n*—a resin of vegetable or animal origin. **D 6440,** D01

natural spreading rate—the spreading rate that occurs when a coating is applied in a manner natural to the operator's technique, perceptions, and expectations, as they relate to coating tools, substrate, and characteristics of the coating itself. **D 16,** D01

natural state—condition of woodwork before priming and painting. **E 1605,** E06

natural-type environment—in solar energy applications, the natural aspects of the outdoor exposure environment elements (or simulation), including changes with time, that may affect the thermal performance of a collector through degradation of collector materials or physical damage to the collector configuration. (Typical aspects include radiant exposure, ambient temperature, and rain impingement.) **E 772,** E44

natural uranium—uranium whose isotopic composition, as it occurs in nature, has not been altered. **C 859,** C26

natural weathering—long-term exposure of materials to unconcentrated sunlight on fixed-angle (or seasonally adjusted) racks, the purpose of which is to assess the effects of environmental factors in various functional and decorative parameters of interest. **E 772,** E44

natural weathering, *n*—outdoor exposure of materials to unconcentrated sunlight, the purpose of which is to assess the effects of environmental factors on various functional and decorative parameters of interest. **G 113,** G03

naval stores, *n*—chemically reactive oils, resins, tars, and pitches derived from the oleoresin contained in, exuded by, or extracted from wood chiefly of the pine species (*Genus Pinus*). **D 804,** D01

navel, *n*—a component of an open-end spinning machine located on the axis of the rotor through which the yarn is withdrawn from the rotor and which modifies the twist of the yarn inside the rotor. (See **doffing tube.**) **D 123,** D13

NBS color difference, *n*—color difference calculated by use of the Judd-Hunter National Bureau of Standards equations, which are unique in including terms taking account of (*1*) the masking effect of gloss on the detection of color differences and (*2*) the relative importance of chromaticness and lightness in a particular viewing arrangement, such as variation in the separation between the two specimens compared. **E 284,** E12

NC—effective thermal neutron content or neutron radiographic contrast. NC is the percent background film exposure due to unscattered thermal neutrons. **E 1316,** E07

NC—Numerical Control. **F 1156,** F12

Nd-YAG laser—a laser operating with a neodymium-yttrium-aluminum-garnet crystal; primary emission wavelength = 1.06 μm. **F 1156,** F12

near field—the region of the ultrasonic beam adjacent to the transducer and having complex beam profiles. Also known as the Fresnel zone. **E 1316,** E07

near-infrared—pertaining to the infrared region of the electromagnetic spectrum with wavelength range from approximately 0.78 to 2.5 μm (wavenumber range 12 800 to 4000 cm^{-1}). **E 131,** E13

near size particle, *n*—a particle of a size approximately equal to the aperture size of the sieve. **(ISO/DIS 2395)** **E 1638,** E29

near-specular, *adj*—pertaining to the appearance of a material when viewed from a direction close to the specular angle, typically within 25°. **E 284,** E12

near-specular angle, *n*—the aspecular angle when a material is viewed close to the specular direction, typically within 25°. **E 284,** E12

near-specular color, *n*—color of a material when viewed at a specified near-specular angle. **E 284,** E12

near surface discontinuity—a discontinuity not open to, but lying near, the surface of a part undergoing examination which produces broad, fuzzy, lightly held powder patterns. **E 1316,** E07

neat cement grout—a mixture of hydraulic cement and water without any added aggregate or filler materials. **D 653,** D18

neat gypsum plaster—see **gypsum neat plaster.** **C 11,** C11

neat petroleum—oil visibly free of contaminants. **D 1129,** D19

nebulize, *v*—to produce droplets. **E 1620,** E29

nebulizer, *n*—a device for converting a sample solution into a gas-liquid aerosol for atomic absorption, emission, and fluorescence analysis. This may be combined with a burner to form a nebulizer burner. **E 135,** E01

neck—(*1*) the part of a bottle between finish and shoulder.

(*2*) {archaic} the structure that connects the melting and working chambers of a melting furnace.

(*3*) the structure connecting the uptake and port in a melting furnace. **C 162,** C14

neck base girth, *n*—*in body measurements,* the circumference of the neck over the cervicale at the back and at the top of the collar bone at the front. **D 123,** D13

neck base girth, *n*— *in body measurements,* the circumference of the neck over the cervical at the back and at the top of the collar bone at the front. **F 1494,** F23

neck formation, *n*—during sintering, the development of a necklike bond between particles. **B 243,** B09

neck girth, *n*—*in garment construction,* with shirt open in front and collar fully extended and laid out flat, measure from center of collar button to the far end of the button hole. **F 1494,** F23

necking, *n*—*in building construction,* the localized reduction in cross section that may occur in a material under stress. **C 717, C24**

necking, *n*—the localized reduction in cross-section which may occur in a material under tensile stress. **D 883,** D20

necking, *n*—the localized reduction in cross section that may occur in a material under tensile stress. **D 1566,** D11

necking, *n*—the localized reduction of the cross-sectional area of a specimen which may occur during stretching. **E 6,** E28

necking, *n*—the localized reduction in cross section that may occur in a material under tensile stress. (D-20) **F 412,** F17

neck ring—a metal mold part used to form the finish of a hollow glass article. **C 162, C14**

necktie, *n*—a decorative band of fabric worn around the neck and tied in a knot or a bow. **D 123,** D13

needle—the vertical, reciprocating, refractory part of a feeder which alternatively forces the glass through the orifice and pulls it up after shearing. **C 162, C14**

needle damage, *n*—*in sewn fabrics*, the partial or complete yarn severance or fiber fusing caused by a needle passing through a fabric during sewing. **D 123,** D13

needled felt, *n*—a textile structure composed entirely of fibers principally interlocked and reoriented through the action of felting needles. **D 123,** D13

needle ducility—a measure of the amount of plastic bending a needle can withstand. **F 1840,** F04

needle glazing—*in building construction*, the application of a small bead of a sealant at the sight line adhering to the sash and glass or panel by means of a nozzle with an orifice not exceeding ⅛ in. (3.18 mm) in diameter. **C 717, C24**

needle length, *n*—the distance measured along the needle curvature from end to end. **F 1840,** F04

needle point—point forming circular cone. **F 547,** F16

needle-punched batting, *n*—a textile filling material which is stabilized by mechanically entangling the fibers. **D 123,** D13

needle radius, *n*—the radius of the uniformly curved portion or portions of the needle measured from the centerline of the needle body. **F 1840,** F04

needles—elongated rod-like particles. **B 243,** B09

needles per unit width, *n*—*for tufted pile yarn floor covering*, the number of binding sites per unit of floor covering width; needles being the means of inserting the pile yarn into the backing fabric. **D 123,** D13

needle wire diameter, *n*—the gage or thickness of the needle wire, measured at a location between the needle body and the attachment area, where either no or minimal work has taken place. **F 1840,** F04

Neel temperature—the temperature above which an antiferromagnetic material becomes paramagnetic. **E 1142,** E37

Néel wall—in a thin magnetic film (less than about 10^{-6} cm thick for iron), a domain wall in which the magnetic moment at any point is substantially parallel to the film surface. See also **domain wall**. **A 340,** A06

negative—an image in which the dark tones of the original appear light and the light tones appear dark. **F 335,** F05

negative air pressure (vacuum) test—test procedure utilizing air at a pressure less than atmospheric pressure, intended to determine the acceptability of a pipe section or multiple pipe sections, or an installed pipeline or manhole before or after backfill. **C 822,** C13

negative developer—see **reversal developer.** **F 335,** F05

negative development—see **reversal development.** **F 335,** F05

negative distortion— See **distortion**. **E 7,** E04

negative eyepiece—an eyepiece in which the real image of the object is formed between the lens elements of the eyepiece. **E 7,** E04

negative image—a developed image in which dark areas correspond to light area of the original subject. **F 335,** F05

negative, photographic—a sensitized plate or film which has been exposed in a camera and which upon development has the lights and shades inverse to those of the original subject. The plate or film does not become a negative until it is exposed, after which it may be an undeveloped or a developed negative. **E 7,** E04

negative print—a photograph having approximately the opposite rendition of light and shade as the original subject. **E 7,** E04

negative replica—See **replica**. **E 7,** E04

negative side waterproofing, *n*—an application wherein the waterproofing system and the source of hydrostatic pressure are on opposite sides of the structural element. **D 1079,** D08

neoprene—a synthetic rubber (polychloroprene) used in liquid- or sheet-applied elastomeric roofing membranes or flashing. **D 1079,** D08

neoprene—an elastomer, polychloroprene, formed by adding hydrogen chloride to monovinylacetylene. **F 869,** F08

neoprene treated, *adj*—*in glass fiber*, a descriptive term for the application of polychloroprene rubber compound to improve the stability, knot holding properties, and abrasion resistance of the cord. **D 123,** D13

nep, *n*—a tightly tangled knot-like mass of unorganized fibers. **D 123,** D13

neper, *n*—a division of the logarithmic scale wherein the number of nepers is equal to the natural logarithm of the scalar ratio of either two voltages or two currents. **D 1711,** D09

nepheline syenite—a mineral aggregate consisting chiefly of albite, microcline, and nephelite, each in significant amount. **C 242,** C21

nephelometer—an instrument that measures the amount of light scattered in a suspension. **D 4410,** D19

nephelometer—a device used to measure mainly the turbidity of water with results expressed in nephelometric turbidity units (NTU). Measures light at 90°. **D 6161,** D19

nephelometry—measurement of the intensity of light scattered at 90 deg to the incident beam of light. **C 859,** C26

Nernst equation—a mathematical description of electrode behavior:

$$E = E_x + 2.3\, RT/zF \log A$$

where:

E	=	total potential developed between the sensing and reference electrodes, mL,
E_x	=	potential dependent on the choice of reference electrodes,
RT/zF	=	Nernst factor,
R and F	=	constants,
z	=	charge on the ion, including sign,
T	=	temperature, and,
A	=	activity of the ion to which the electrode is responding.

D 4127, D19

Nernst factor—the term $2.3RT/nF$ in the Nernst equation, which is equal (at $T = 25°C$) to 59.16 mV when $n = 1$ and 29.58 mV when $n = 2$, and which includes the sign of the charge on the ion in the term n. The Nernst factor varies with temperature. This factor is often referred to as the *electrode slope*. **D 4127,** D19

Nernstian response—the response of an ion-selective electrode is over a given range of activity (or concentration) in which a plot of the potential of such an electrode in conjunction with a reference electrode versus the logarithm of the ionic activity of a given species (a_A) is linear with a slope of $2.303 \times 10^3 RT/z_A{}^F$ mV/decade ($59.16/z_A$ mV per unit of pa_A at 25°C). **D 4127,** D19

nerve, *n*—the elastic resistance of unvulcanized rubber or rubber mixes to permanent deformation. **D 1566,** D11

nest, *n*—a group of articles, **cans, baskets,** or **boxes** of various sizes, that fit one within the other. **D 996,** D10

nested experiment, *n*—an experiment to examine the effect of two or more factors in which the same level (version) of a factor cannot be used with all levels (versions) of other factors. Synonym: hierarchical experiment. **E 456,** E11

nested experiment, *n*—an experiment to examine the effect of two or more factors in which the same level (version) of a factor cannot be used with all levels (versions) of other factors. Synonym: hierarchical experiment. **E 1325,** E11

nestling down, *n*—a down not fully developed with a sheath and with soft barbs emanating from the sheath. **D 123,** D13

nestling feathers, *n*—immature feathers in which the barbs are held together and covered by a sheath. **D 123,** D13

net assignable area— See **space categories**. **E 1480,** E06

net benefits (savings)—the difference between the benefits and the costs—where both are discounted to present or annual value dollars. **E 833,** E06

net boom—special purpose boom in which all or part of the membrane material is netting. **F 818,** F20

net calorific value—the energy released by combustion of a unit quantity of refuse-derived fuel at constant volume or constant pressure in a suitable calorimeter under specified conditions such that all water in the products remains in the gaseous form. The net calorific value is the lower heating value that can be calculated from the gross calorific value by making a correction for the heat of vaporization of the water. This is the measure of calorific value in Europe but is rarely used in the United States. **Synonym:** lower heating value. **E 856,** D34

net calorific value (net heat of combustion at constant pressure), *n*—the heat produced by combustion of unit quantity of a solid or liquid fuel when burned, at a constant pressure of 1 atm (0.1 MPa), under conditions such that all the water in the products remains in the forms of vapor. **D 121,** D05

net calorific value (net heat of combustion at constant pressure), Q_p, *n*—the heat produced by combustion of unit quantity of a solid or liquid specimen when burned at a constant pressure of 0.1 MPa (1 atm), under conditions such that all the water in the products remains in the form of vapor. **D 5681,** D34

net count rate—the count rate recorded in a spectrometer after correcting for background and coincidence events. **E 170,** E10

net density—total density less fog and support (film base) density. **E 1316,** E07

net floor area—See **floor area**. **E 1480,** E06

net heat of combustion, *n*—the oxygen bomb (see Test Method D 3286) value for the heat of combustion, corrected for gaseous state of product water. **E 176,** E05

net molded edge—an edge, not physically altered after molding, in final form ready for use. **E 1749,** E06

net positive suction head, NPSH, *n*— the difference between total pressure and vapor pressure in a fluid flow, expressed in terms of equivalent height of fluid, or "head," by the equation:

$$\text{NPSH} = (P_o/w) + (V^2/2g) - (P_v/w)$$

where:

P_o = static pressure,
P_v = vapor pressure,
V = flow velocity,
w = specific weight of fluid, and
g = gravitational acceleration.

This quantity is used in pump design as a measure of the tendency for cavitation to occur at the pump inlet. It can be related to the cavitation number. **G 40,** G02

net programmable area— See **space categories**. **E 1480,** E06

net pyrgeometer—see **pyranometer, net.** **E 772,** E44

net pyrradiometer— see **pyranometer, net.** **E 772,** E44

net-section area, A_N **[L^2]**—area of the net remaining ligament. **E 1823,** E08

net thickness, B_N **[L]**—distance between the roots of the side grooves in side-grooved specimens. **E 1823,** E08

network, *n*—a three dimensional reticulate structure formed by interchain or intrachain bonding of polymer molecules in combination with chain entanglements. **D 1566,** D11

network level analysis, *n*—evaluation of a network of pavement to enable selection of candidate projects, project scheduling, and budget estimates. **E 867,** E17

network structure—a structure in which one constituent occurs primarily at the grain boundaries, thus partially or completely enveloping the grains of the other constituents; on a two-dimensional section cut through such a structure, the grain boundary constituent will appear as a network. **E 7,** E04

Neumann bands—See **twin bands**. **E 7,** E04

neutral, *adj*—achromatic or without hue. **E 284,** E12

neutral body—a comparison piece used in the differential method of thermal analysis, which has nearly the same thermal properties as the test specimen, and which produces no heat effects within the temperature range through which the specimen is being tested. See Practice E 14. **E 7,** E04

neutral content, *n*—the total amount of material contained in naval stores, such as rosin, tall oil, and their derivatives that do not contain any acidic functionality. **D 804,** D01

neutral fat—the unsaponified and unsaponifiable matter that becomes included along with fatty acids in a diethyl ether extract of samples under analysis for soap content, as described in 25.2 of Test Method D 460. **D 459,** D12

neutral filter—see **filter, neutral.** **E 131,** E13

neutral filter—see **filter, neutral.** **E 135,** E01

neutralization value—a number used as a measure of the acidic or basic constituents present in an insulating liquid, usually expressed in terms of equivalent milligrams of potassium hydroxide per gram of sample. **D 2864,** D27

neutralizer—a dilute alkaline solution with which sheet metal ware is treated as a part of the pickling process subsequent to the acid treatment. A chemical or mixture of chemicals which, when added to water, produces the dilute alkaline solution. **C 286,** B08

neutral refractories, n—see **refractories, neutral.** **C 71,** C08

neutral sized paper, *n*—paper that has been manufactured using a procedure or process at a pH value of 7 (with a normal range of 6.5 to 7.5) that results in paper that has resistance to water penetration. See **sizing**. **D 1968,** D06

neutral soap—under **soap,** see *neutral soap*. **D 459,** D12

neutral solution—a fluid environment containing an equal amount of hydrogen and hydroxyl ions, that is, pH = 7. **E 1823,** E08

neutral stress—see **stress**. **D 653,** D18

neutral white surface—a highly reflecting surface whose diffuse reflectance makes it a colorimetric match of a nonselective reflecting white surface under specified viewing conditions. (psychophysical definition). **D 459,** D12

neutron—an elementary particle having no electric charge, a rest mass of 1.67495×10^{-27} kg and a mean life of 1000 s. **C 859,** C26

neutron—a neutral elementary particle having an atomic mass close to 1. In the free state outside of the nucleus, the neutron is unstable having a half-life of approximately 10 min. **E 1316,** E07

neutron activation detector—a neutron fluence sensing device, well-characterized with respect to geometry, mass, composition and cross section which produces activity with a sufficiently long half life to permit its measurement after withdrawal from the neutron field. **E 170,** E10

neutron cross section—a measure of the probability that a neutron will interact with a nucleus. The cross section is a function of the neutron energy and the structure of the interacting nucleus. Also see **barn.** **C 859,** C26

neutron fluence—see **particle fluence.** **E 170,** E10

neutron fluence rate—see **particle fluence rate.** *neutron flux*—see **particle fluence rate.** **E 170,** E10

neutron flux—the number of neutrons passing per second through an area of one square centimetre. **C 859,** C26

neutron radiography—the process of producing a radiograph using neutrons as the penetrating radiation. **E 1316,** E07

neutron radiography (NRT)—a process of making an image of the internal details of an object by the selective attenuation of a neutron beam by the object. **E 1316,** E07

newel—See **railing systems.** **E 631,** E06

newel, *n*—a decorative or structural post member at the start or end of a stair run, often extending above a handrail; usually square or rectangular in cross section, supporting the end of a stair railing or serving as the common support for two stair railings, often supporting a stair stringer and platform; also, the center post of a spiral stair. **E 1481,** E06

Newtonian behavior, *n*—the property of a liquid in which its viscosity is constant over a stated range of strain rates. (Compare **non-Newtonian behavior**.) **D 907,** D14

newtonian fluid—a true fluid that tends to exhibit constant viscosity at all rates of shear. **D 653,** D18

Newtonian fluid, *n*—a fluid that at a given temperature exhibits a constant viscosity at all shear rates or shear stresses. **D 4175,** D02

newton (N)—the SI unit of force (kg·m/s^2). **E 1316,** E07

2^n factorial experiment, *n*—a factorial experiment in which *n* factors are studied, each of them in two levels (versions). **E 456,** E11

2^n factorial experiment, *n*—a factorial experiment in which *n* factors are studied, each of them in two levels (versions). **E 1325,** E11

nick—an indentation on the surface of a bolt, nut, screw, or stud. Also referred to as a gouge. **F 1789,** F16

nickel dipping, nickel flashing, or nickel pickling—a process for depositing metallic nickel on steel by galvanic action, reduction, or both. **C 286,** B08

Nicol prism—a prism, used for polarizing or analyzing light, made by cementing together, with Canada balsam, two pieces of calcite in such a way that the extraordinary ray from the first piece passes through the second piece while the ordinary ray is reflected to the side into an absorbing layer of black print. When two Nicol prisms are crossed, therefore, no light passes through. **E 7,** E04

Nicol prism—a prism, used for polarizing or analyzing light, made by cementing together, with Canada balsam, two pieces of calcite in such a way that the extraordinary ray from the first piece passes through the second piece while the ordinary ray is reflected to the side into an absorbing layer of black paint. When two Nicol prisms are crossed, therefore, no light passes through. **E 175,** E41

nineteen-inch selvage—a prepared roofing sheet with a 432-mm (17-in.) granule-surfaced exposure and a 483-mm (19-in.) selvage. **D 1079,** D08

ninety-pound—a prepared roll roofing with a granule-surfaced exposure that has a mass of approximately 4400 g/m^2(90 lb/108 ft^2). **D 1079,** D08

nip, *n*—the radial clearance between rolls of a mill or calendar on a line of centers. **D 1566,** D11

nitrating pulps—pulps used for the manufacture of cellulose nitrate and subject to various specifications by the manufacturers, including those of alpha-cellulose content and viscosity. **D 1695,** D01

nitre—an industrial term for $NaNO_3$. **C 162,** C14

nitride-carbide inclusion types—a compound with the general formula $M_x(C, N)_y$ observed generally as colored idiomorphic cubic crystals, where *M* includes Ti, Cb, Ta, Zr. **E 7,** E04

nitriding, *n*—introducing nitrogen into a solid steel object by holding it at a suitable temperature in contact with a nitrogenous environment. **A 941,** A01

nitrocellulose—see **cellulose nitrate.** **D 1695,** D01

nitrogen alloying, *n*—the transfer of nitrogen from a furnace atmosphere to powder or a P/M part, in such a way as to increase the nitrogen content of the material within controlled limits. **B 243,** B09

nitrogen narcosis, *n*—a distinct anesthetic effect, characterized by loss of judgment and disorientation, caused from breathing nitrogen at increased partial pressures. **F 1549,** F32

nitrogen surface area (NSA), *n*—the total surface area of carbon black that is calculated from nitrogen adsorption data using the B.E.T. theory. **D 3053,** D24

nits or nitty enamel—a porcelain enamel blemish in dry process enameling characterized by minute surface pits visible only on close examination. **C 286,** B08

Nixdorf bar code—See **bar code.** **F 1156,** F12

(n, k) symbology, *n*—a class of bar code symbologies in which each symbol character is *n* modules in width and is composed of *k* bar and space pairs. A subset of these is the *n, k, m,* class of symbologies, where *m* represents the maximum width of an element in modules. (See **module**.) **F 1294,** F05

NMR absorption band; NMR band—a region of the spectrum in which a detectable signal exists and passes through one or more maxima. **E 131,** E13

NMR absorption line—a single transition or a set of degenerate transitions is referred to as a line. **E 131,** E13

NMR apparatus; NMR equipment—an instrument comprising a magnet, radio-frequency oscillator, sample holder, and a detector that is capable of producing an electrical signal suitable for display on a recorder or an oscilloscope, or which is suitable for input to a computer. **E 131,** E13

noble—the positive (increasingly oxidizing) direction of electrode potential. **G 15,** G01

noble metal—a metal that does not readily tend to furnish ions, and therefore does not dissolve readily, nor easily enter into such reactions as oxidations, etc. The opposite of base metal.

NOTE—Since there is no agreement over the sign of electrode potentials, the words noble and base are often preferred because they are unambiguous. **B 374,** B08

noble metal—a metal with a standard electrode potential that is more noble (positive) than that of hydrogen. **G 15,** G01

no conclusion (totally inconclusive, indeterminable)—This is the zero point of the confidence scale. It is used when there are significantly limiting factors, such as disguise in the questioned and/or known writing or a lack of comparable writing, and the examiner does not have even a leaning one way or another.

Examples—No conclusion could be reached as to whether or not the John Doe of the known material wrote the questioned material, or I could not determine whether or not the John Doe of the known material wrote the questioned material. **E 1658,** E30

node—point, line, or surface of standing wave system at which the amplitude is zero. **D 653,** D18

node—the bonded portion of the honeycomb flat sheet material; the honeycomb cell's double wall. **E 1749,** E06

nodular fireclay, n—see **fireclay, nodular.** **C 71,** C08

nodular graphite, *n*—graphite in the form of nodules or spheroids in iron castings. **A 644,** A04

nodularity, *n*—the volumetric proportion of spheroidal or nodular graphite to total graphite in a ductile iron or a compacted graphite iron matrix (see Test Method A 247, for Evaluating the Microstructure of Graphite in Iron Castings, Types I and II). **A 644,** A04

nodularity, degree of, *n*—the volumetric proportion of spheroidal or nodular graphite to total graphite in a ductile iron matrix (see Test Method A 247, Types I and II). **A 644,** A04

nodular powder—irregular particles having knotted, rounded, or similar shapes. **B 243,** B09

nodule—a rounded projection formed on a cathode during electrodeposition. **B 374,** B08

nodules—small irregularly shaped imperfections in the electrolytic copper foil which do not cause the foil to fail the thickness tolerance. **B 846,** B05

nodulizing alloy, *n*—an alloy added to molten iron for the primary purpose of causing the formation of spheroidal graphite during solidification. **A 644,** A04

no-flow point, *n*—the temperature corresponding to a specified degree of blockage of a screen by separated solids. **D 4175,** D02

no identification—this expression could be understood to mean anything from a strong probability that the suspect wrote the questioned writing; to a complete elimination. It is not only confusing but also grammatically incorrect when used informally in sentences such as "I no identified the writer" or "I made a no ident in this case." **E 1658,** E30

noil, *n*—the short fibers removed in combing; applied particularly to wool, but also to other fibers as cotton, silk, and rayon. **D 123,** D13

noise—See **contact noise.** **B 542,** B02

noise—an extraneous electronic signal that effects baseline stability. **D 1129,** D19

noise—See **analyzer.** **D 1356,** D22

noise—*in surface analysis,* the random fluctuation of the measured intensity at a particular location in a spectrum, usually expressed as an RMS (root-mean-square), standard deviation, or a peak-to-peak value. **E 673,** E42

noise—in electromagnetic inspection, any nonrelevant signal that tends to interfere with the normal reception or processing of a desired flaw signal. It should be noted that such noise signals may be generated by inhomogeneities in the inspected part that are not detrimental to the end use of the part. **E 1316,** E07

noise—the data present in a radiological measurement which is not directly correlated with the degree of radiation attenuation by the object being examined. **E 1316,** E07

noise—any undesired signal (electrical or acoustic) that tends to interfere with the reception, interpretation, or processing of the desired signal. **E 1316,** E07

noise—(*1*) random variations of one or more characteristics of any entity such as voltage, current, or data.

(*2*) a random signal of known statistical properties of amplitude, distribution, and spectral density.

(*3*) loosely, any disturbance tending to interfere with the normal operation of a device or system. **F 149,** F05

noise equivalent temperature difference (NETD)—the target-to-background temperature difference between a blackbody target and its blackbody background at which the signal-to-noise ratio of a thermal imaging system or scanner is unity. **E 1316,** E07

noise isolation class, NIC—a single-number rating calculated in accordance with Classification E 413 using measured values of noise reduction. It provides an estimate of the sound isolation between two enclosed spaces that are acoustically connected by one or more paths. **C 634,** E33

noise reduction coefficient, NRC—a single-number rating derived from measured values of sound absorption coefficients in accordance with 11.7 of Test Method C 423. It provides an estimate of the sound absorptive property of an acoustical material. **C 634,** E33

noise reduction, *NR—in a specified frequency band,* the difference between the average sound pressure levels measured in two enclosed spaces or rooms due to one or more sound sources in one of them. **C 634,** E33

NOM—natural organic matter. **D 6161,** D19

nominal, *adj*—name or identifying value of a measurable property by which a conductor or component or property of a conductor is identified, and to which tolerances are applied. **B 354,** B01

nominal , *n*—representative value of a measurable property determined under a set of conditions, by which a product may be described. **D 4439,** D35

nominal contact pressure, [FL^{-2}] , *n—in tribology,* an average contact pressure between two conforming bodies, calculated by dividing the contact force by the **apparent area of contact.** **G 40,** G02

nominal diameter—references the internal diameter in name only to the nearest unit dimension. **C 896,** C04

nominal diameter—the diameter of a sphere that has the same volume as the sediment particle. Sometimes called *equivalent spherical diameter.* **D 4410,** D19

nominal dimension—dimension that is greater than the specified dimension by the thickness of a mortar joint. It is usually expressed as a whole number. **C 1232,** C15

nominal discount rate—the rate of interest reflecting the time value of money stemming both from inflation and the real earning power of money over time. **E 833,** E06

nominal filtration rating, *n*—(Deprecated term) an arbitrary micrometre value for a filter medium indicated by a filter manufacturer. **D 4175,** D02

nominal gage length, *n—General — in tensile testing,* the length of a specimen under specified pretension measured from nip-to-nip of the jaws of the holding clamps in their starting position at the beginning of the test, and including any portion of the specimen in contact with bollard or snubbing surfaces.

Specific 1, the length of a specimen under specific pre-tension between frets, in instruments where the specimen is not held by clamps, for example, in a vibroscope.

Specific 2, the length of a specimen measured between the points of attachment to the tabs while under specified pretension. **D 123,** D13

nominal maximum size *(of aggregate),* *n*—in specifications for, or description of aggregate, the smallest sieve opening through which the entire amount of the aggregate is permitted to pass. **C 125,** C09

nominal maximum size (of aggregate), *n—in specifications for, or descriptions of aggregate,* the smallest sieve opening through which the entire amount of the aggregate is permitted to pass. **D 8,** D04

nominal mesh size, *n*—the designated dimension of the openings of a sieve. **E 1638,** E29

nominal (net-section) stress, σ_N [FL^{-2}]—*in fracture testing,* a measure of the stress on the net cross section calculated in a simplified manner and without taking into account stress gradients produced by geometric discontinuities such as holes, grooves, fillets, and so forth. **E 1823,** E08

nominal plunger energy, *½ [ML^2/T^2],* *n—in tire testing,* one half of the product of a peak force (required to rupture the tire structure in tread area) and maximum plunger travel into a tire at the time of rupture. **F 538,** F09

nominal pressure—the intended operating pressure. **E 1749,** E06

nominal size—as applied to products such as lumber, traditionally the approximate rough-sawn commercial size by which it is known and sold in the market. Actual rough-sawn sizes may vary from the nominal. Reference to standards or grade rules is required to determine nominal/actual finished size relationships. **D 9,** D07

nominal stress, *n*—the stress at a point calculated on the net cross section by simple elastic theory without taking into account the effect on the stress produced by discontinuities such as holes, grooves, fillets, or any combination of them. Taken in part from **D 907,** D14

nominal stress [FL^{-2}], *n*—the stress at a point calculated on the net cross section by simple elastic theory without taking into account the effect on the stress produced by geometric discontinuities such as holes, grooves, fillets, and so forth. **E 6,** E28

nominal temperature—the intended operating temperature. **E 1749,** E06

nominal thickness, *n— of steel sheet,* an ordering designation which indicates that the applicable tolerance for thickness is both plus and minus from the ordered thickness. **A 902,** A05

nominal value, *n*—representative value of a measurable property by which a product may be identified **D 4439,** D35

nominal wall, *n*—specified wall thickness with a published plus and minus tolerance from the specified thickness at any point. **B 899,** B02

nominal width, *n—in reference to bar codes,* the intended width of an element. **F 1294,** F05

nonaromatic hydrocarbons—one or more types of hydrocarbons identified as paraffins, cycloparaffins (naphthenes), and olefins. Generally, the saturated types, paraffins and cycloparaffins, constitute the impurities in the commercial grades of aromatic hydrocarbons. **D 4790,** D16

nonassociated gas—natural gas not in contact with, nor dissolved in, reservoir liquids. **D 4150,** D03

nonbanded coal—See *nonbanded coal* under **coal.** **D 121,** D05

nonbearing wall—See **wall.** **E 631,** E06

noncanopy—a fixed device located in close proximity over cooking equipment and in some cases having a shelf-like appearance. The purpose of which is to channel air movement to contain and capture the unwanted by-products resulting from cooking activities. **F 1827,** F26

noncapacitive, intermittent dc arc, *n*—a series of electrical discharges in which dc pulses are initiated by separate means, either mechanically or electrically. **E 135,** E01

noncarbonated mechanically refrigerated visible product, beverage dispenser—counter-top equipment, mechanically refrigerated, with a transparent, impact-resistant container designed to afford a visual display of the beverage. **F 1827,** F26

non-carbonate hardness—hardness caused by chlorides, sulfates, and nitrates of calcium and magnesium. **D 6161,** D19

non-chlorine bleach, *n*—a bleach that does not release the hypochlorite ion in solution, for example, sodium perborate, sodium percarbonate. **D 123,** D13

nonchromatic, adj— see **achromatic.** **E 284,** E12

noncohesive sediments—discrete particles, the movement of which for given erosive forces depends only upon the properties of shape, size, and density and upon the relative position of the particle with respect to surrounding particles. **D 4410,** D19

non-collocated—not in the same location. **F 1156,** F12

noncombustible, *adj*—not combustible. (See **combustible.)** **E 176,** E05

noncombustible—that fraction of a macrosample remaining after moisture and combustibles are driven off by heat and combustion. It is composed of metallic and glass particles in addition to the residue from the combustion of organic substances. **E 856,** D34

noncombustible textile, *n*—a textile that will neither ignite nor give off vapors that will ignite when subjected to external sources of ignition. (Compare **combustible textile**) **D 123,** D13

noncombustive reaction, *n*—a reaction other than combustion or thermal degradation undergone by certain materials upon exposure to elevated temperatures. **D 4175,** D02

non-compression glazing, n—see **non-compression glazing system**. **C 717,** C24

non-compression glazing, *v—in building construction,* the act of installing components into a non-compression glazing system. **C 717,** C24

non-compression glazing system, *n—in building construction,* a sealant or combination of sealant(s) used to provide weather-tightness by way of the adhesive/cohesive characteristics of the components in the joint. **C 717,** C24

noncondensable gas—a gas whose temperature is above its critical temperature, so that it cannot be liquefied by increase of pressure alone. **E 1316,** E07

noncondensible gases, *n*—in geothermal, chemical species (such as carbon dioxide or hydrogen sulfide) that are constituents of geothermal fluids, that partition primarily into the vapor phase when geothermal fluids boil, and that do not condense along with geothermal steam when put through a condenser in a geothermal power plant. **E 957,** E44

nonconforming, *adj*—a description of a unit or a group of units that does not meet the unit or group tolerance. **D 123,** D13

nonconforming item, *n*—an item that does not satisfy the requirements of the applicable specification. **D 123,** D13

nonconforming unit, *n*—a unit of product containing at least one nonconformity. **D 1711,** D09

nonconformities per hundred units, *n*—a calculated ratio of nonconforming units to the number of units inspected, the quotient being multiplied by 100 (see **percent nonconforming**). **D 1711,** D09

nonconformity, *n*—an occurrence of failing to satisfy the requirements of the applicable specification; a condition that results in a nonconforming item. **D 123,** D13

nonconformity, *n*—a departure of a quality characteristic from its intended level or state that occurs with a severity sufficient to cause a test result not to meet a specification requirement. **D 1711,** D09

nonconformity—the nonfulfillment of specified requirements, (ISO 8402). **E 1187,** E36

non-contact, *n*—a method of bar code reading typified by fixed or moving beam scanners having a greater optical throw and depth of field than contact, or wand scanners. **F 1294,** F05

nondestructive assay (NDA)—an elemental or isotopical analysis of a sample in which the chemical and physical properties of that sample remain essentially unaltered. **C 859,** C26

Nondestructive Evaluation—see *Nondestructive Testing.* **E 1316,** E07

Nondestructive Examination—see *Nondestructive Testing.* **E 1316,** E07

Nondestructive Inspection—see *Nondestructive Testing.* **E 1316,** E07

nondestructive test—See **test, nondestructive**. **E 1749,** E06

Nondestructive Testing (NDT), *n*—the development and application of technical methods to examine materials or components in ways that do not impair future usefulness and serviceability in order to detect, locate, measure and evaluate flaws; to assess integrity, properties and composition; and to measure geometrical characteristics. **E 1316,** E07

nondrying oil—an oil that does not of itself possess to a perceptible degree the power to take up oxygen from the air and lose its liquid characteristics. **D 16,** D01

non-elastic elongation, (*NE*), *n*—*of rope*, elongation after cyclic tensioning the rope to a specified force for a specified number of cycles. **D 123,** D13

non-erasable optical data—a non-erasable, non-rewriteable storage medium where the digital data is represented by the degree of reflectivity of the mediums recording layer. The data cannot be altered. **E 1316,** E07

non-ferromagnetic, *adj.*—magnetic character that does not derive from the presence of iron or other ferritic materials, for example, barium titanate. Also, the absence of magnetic character deriving from the presence of ferritic materials such as iron, chromium, nickel, cobalt, etc. **F 1156,** F12

nonferromagnetic material—a material that is not magnetizable and hence, essentially not affected by magnetic fields. This would include paramagnetic materials and diamagnetic materials. **E 1316,** E07

nonfibrous spicules—acicular particles resembling assemblages of asbestos fibers composed of nonfibrous, or semifibrous minerals such as picrolite. **D 2946,** C17

non-fill, *n*—defect resulting from the failure of the rubber to fill out all the mold pattern detail. **D 1566,** D11

nonflammable textile, *n*—any combustible textile that burns without a flame. (See also **glow, smoldering.** Compare **flammable textile, combustible textile,** and**noncombustible textile.**) **D 123,** D13

non-hygroscopic material, *n*—material which neither absorbs nor retains water vapor. **D 1356,** D22

nonimpact printer—a printer in which image formation is not the result of mechanical impacts. Examples are thermal printers, electrostatic printers, electrophotographic printers, and ink jet printers. **F 909,** F05

non-interlocking tile, *n*—tile without restrictive ribs, grooves, or channels at the side lap or head lap. **C 43,** C15

nonionic cellulose ethers—that class of cellulose ethers which does not contain any ionizable groups. **D 1695,** D01

nonionic detergent—under **detergent,** see *nonionic detergent.* **D 459,** D12

nonionic polyelectrolyte—neutral charged polymers, usually polyacrylamides, used for coagulation/flocculation. See **polyelectrolytes**. **D 6161,** D19

nonionic surfactant—a surface-active agent having no ionizable polar end groups but comprised of hydrophilic and lipophilic segments. **E 1519,** E35

nonlinearity—the maximum deviation of the transducer output(s) (of the wheel force transducer system) from the best-fit linear relation to the applied calibration force, expressed as a percentage of full scale. Proven outliers are excluded. **E 867,** E17

non-lint content, *n*—that portion of a mass of cotton fiber which is essentially foreign matter. **D 123,** D13

non-load-bearing tile, *n*—tile for use in masonry constructions carrying no superimposed loads. See Specification C 56. **C 43,** C15

nonmagnetic—a relative term describing a material which, for practical purposes, may be considered to have a relative permeability close to unity.

Note—Certain materials may be nonmagnetic only under limited conditions. **A 340,** A06

nonmetallic—any material, other than a metal, or any composite in which the metal is not the most easily ignited component and for which the individual constituents cannot be evaluated independently. **G 126,** G04

nonmetallic inclusions—particles of impurities (usually oxides, sulfides, silicates and such) that are held mechanically or are formed during solidification or by subsequent reaction within the solid metal (See **exogenous inclusions**). **E 7,** E04

nonmetallics—organic and inorganic materials such as ceramics, glasses, polymers, rubber, cloth, wood, and paper products. **G 126,** G04

non-Newtonian behavior, *n*—the property of a liquid in which its viscosity is not constant over a stated range of strain rates. **D 907,** D14

non-Newtonian oil or fluid, *n*—an oil or fluid that at a given temperature exhibits a viscosity that varies with changing shear stress or shear rate. **D 4175,** D02

nonoperational mode—the condition that exists when a solar collector has been filled, purged of heat transfer fluid (if a liquid), and capped (but not sealed) to prevent contamination by foreign substances prior to exposure. **E 772,** E44

nonoriented electrical steel—a flat-rolled electrical steel which has approximately the same magnetic properties in all directions. **A 340,** A06

nonparametric, *adj*—a term referring to a statistical technique that does not assume the nature of the underlying frequency distribution is known. **D 123,** D13

nonplastic ceramics—nonclay ceramic materials that when mixed with water do not exhibit the rheological property-plasticity. **C 242,** C21

non-pressure pipe—pipe designed for gravity-conveyed medium which must resist only intermittent static pressures and does not have a pressure rating. **F 412,** F17

non-read, *n*—the failure of a bar code scanner to recognize or decode a bar code symbol after a scanning attempt. Such a non-read may be caused by bar code symbol defects, scanner defects, or operator error. **F 1294,** F05

nonread ink—See **reflective ink**. **F 149,** F05

nonreducing end groups—those terminal glucose units of cellulose or its derivatives which contain a free hydroxyl in the 4-position and do not reduce Fehling's solution or similar reagents. **D 1695,** D01 *nonreflective ink*—See **scan ink**. **F 149,** F05

nonreinforced pipe—concrete pipe designed without reinforcement. **C 822,** C13

nonrelevant indication, *n*—an NDT indication that is caused by a condition or type of discontinuity that is not rejectable. False indications are non-relevant. **E 1316,** E07

nonresonant forced and vibration technique , *n*—a technique for performing dynamic mechanical measurements, in which the sample is oscillated mechanically at a fixed frequency. **D 4092,** D20

nonrigid plastic, *n*—for purposes of general classification, a plastic that has a modulus of elasticity either in flexure or in tension of not over 70 MPa (10 000 psi) at 23°C and 50 % relative humidity when tested in accordance with Test Methods D 790, Test Method D 747, Test Method D 638, or Test Methods D 882. **D 883,** D20

non-sag sealant, *n*—a compound that exhibits little or no flow when applied in vertical or inverted joints. **C 717,** C24

non-sample contacting equipment—related equipment associated with the sampling effort, but that does not directly contact the sample (for example, augers, drilling rods, excavations machinery). **D 653,** D18

nonscan ink—See **reflective ink**. **F 149,** F05

nonscreen-type film (direct-type film)— X-ray film designed for use with or without metal screens, but not intended for use with salt screens. **E 1316,** E07

nonselective radiator, *n*—thermal radiator whose spectral emissivity is independent of wavelength over the range considered. **E 349,** E21

nonselective surface—a surface for which the spectral optical properties reflectance, absorptance, transmittance, and emittance are essentially independent of wavelength over a particular wavelength range.

NOTE—For solar absorbers, the absorption of solar energy is largely confined to the wavelength range from 0.3 to 3.0 μm, but there is significant flux emitted at wavelengths out to about 30 μm. **E 772,** E44

nonselectivity reflecting white surface—a surface that exhibits equal, diffuse and high reflectance at all wavelengths in the visible region [380 to 700 nm.] (physical definition). **D 459,** D12

nonseparable zipper, *n*—a zipper having two stringers that are permanently attached to each other at either or both ends. (Compare **separable zipper**.) **D 123,** D13

non-standard virgin material—a plastic resin or compound in the form of powder or pellets which does not meet the specification requirements for which it was manufactured, and has not been subjected to use or processing other than that required for its initial manufacture. **F 412,** F17

non-sulfur vulcanizing system, *n*— a vulcanizing system not requiring free or donated sulfur. **D 1566,** D11

nonvitreous (nonvitrified)—that degree of vitrification evidenced by relatively high water absorption. **C 242,** C21

non-volatile—term used to denote the calculated chemical basis of a material in which the volatile fraction of that material is removed, relative to a specific temperature. For example, in lime and limestone, the loss on ignition is considered to be the volatile fraction. **C 51,** C07

nonvolatile content—the portion of a coating that does not evaporate during drying or curing under specified conditions, comprising the binder and, if present, the pigment. (The percent volatile content is obtained by subtracting the nonvolatile content from 100.) **D 16,** D01

nonvolatile content, *n*—the portion of a material that remains after volatile matter has been evaporated under specified ambient or accelerated conditions. (See **solids content.**) **D 907,** D14

nonvolatile extractables—*for purposes of the ASTM tests for volatile extractables and nonvolatile extractables,* nonvolatiles shall be defined as those compounds that give more than 50 % recovery in spike and recovery studies using the applicable nonvolatile extractables method. **F 17,** F02

nonvolatile extractables—those chemical species which are released from microwave food packaging under simulated use conditions and are detected using an applicable nonvolatile extractables method. **F 1479,** F02

nonvolatile ingredient—component of an aerosol formulation with a vapor pressure less than atmospheric pressure (14.7 psia (101 kPa)) at 105 °F (40.6 °C). **D 3064,** D10

nonvolatile matter, *n*—*in insulating varnish,* that portion of a varnish which is not volatilized when exposed to specified conditions. **D 1711,** D09

nonvolatile matter—the oily, gummy, or resinous residue remaining after evaporating volatile hydrocarbon materials. **D 4790,** D16

nonvolatile organic chemical—See **organic chemical.** **D 1356,** D22

nonvolatile residue (NVR), *n*—residual molecular and particulate matter remaining following the filtration of a solvent containing contaminants and complete evaporation of the solvent at a specified temperature. **G 126,** G04

nonvolatiles—materials remaining after the loss of volatile components. **D 2825,** D21

nonvolatile vehicle—the liquid portion of a paint excepting its volatile thinner and water. **D 16,** D01

nonwaterfowl feathers, *n*—feathers derived from chickens, turkeys, or other landfowl. **D 123,** D13

nonwoven blanket, *n*—a blanket produced by bonding or interlocking of fibers, or both, accomplished by mechanical, chemical, thermal, or solvent means, or combination therefore. **D 123,** D13

nonwoven fabric, *n*—a textile structure produced by bonding or interlocking of fibers, or both, accomplished by mechanical, chemical, thermal, or solvent means and combinations thereof. **D 123,** D13

nonwoven fabric—See *nonwoven fabric* under **fabric.** **D 3878,** D30

no-observed-effect concentration (NOEC)—in a toxicity test, the tested concentration of one or more chemicals immediately below the lowest tested concentration that resulted in a statistically significant change in a particular toxicological variable compared to that value in the control (s). **E 943,** E47

normal—an imaginary line forming a right angle with the tangent to a curved surface at a particular point. It is used as a basis for determining angles of incidence, reflection, and refraction. **E 175,** E41

normal, $\perp$—directional in a direction normal (perpendicular) to a surface. **E 772,** E44

normal consistency, *n*—a degree of plasticity of a hydraulic-cement paste that is appropriate for testing as measured by a stipulated method. **C 219,** C01

normal direction, *n*—*for geotextiles,* the direction perpendicular to the plane of a geotextile. **D 4439,** D35

normal direction—that direction which is perpendicular to the plane of working in a worked material. (See **cross direction**). **E 7,** E04

normal distribution, *n*—the distribution that has the probability function:

$$f(x) = (1/\sigma)(2\pi)^{|n\$}\exp[-(x-\mu)^2/2\sigma^2]$$

where:
x = random variate,
μ = mean of the distribution, and
σ = standard deviation of the distribution.

(*Syn.* Gaussian distribution, law of error) **D 123,** D13

normal distribution—drop size distribution described by the following equation:

$$f_n(D) = \{1/[\sigma\sqrt{(2\pi)}]\} \int_{-\infty}^{D} exp\{-[x - \bar{D}_1 o)^2]/(2\sigma^2)\}\, dx$$

where:
$f_n(D)$ = the (D) fraction of drops of diameter less than D. **E 1620,** E29

normal force—that component of the force between contacting bodies perpendicular to their interface. **B 542,** B02

normal force—a force directed normal to the surface element across which it acts. (ISRM) **D 653,** D18

normal force, [*F*], *n—of a tire,* the component of a tire force vector in the Z' direction. **F 538,** F09

normal impact velocity, n—See **impact velocity**. **G 40,** G02

normal incidence (also see straight beam)—a condition in which the axis of the ultrasonic beam is perpendicular to the entry surface of the part under examination. **E 1316,** E07

normal incidence sound absorption coefficient, α_n; [dimensionless] — *of a surface, at a specified frequency,* the fraction of the perpendicularly incident sound power absorbed or otherwise not reflected. **C 634,** E33

normal inspection—inspection that is used when there is no statistically significant evidence that the quality of the product being submitted is better or poorer than the specified quality level. **D 1517,** D31

normalization—converting actual data to a set of reference conditions in order to "standardize" operation to common base. **D 6161,** D19

normalization, *n—in spectrometric analysis, (1)* the process of adjusting instrument output to conform to an established condition using one or more homogeneous specimens or reference materials; *(2)* the adjustment of the analysis total to 100 %, or some other total. **E 135,** E01

normalize—by calculation, to revert a given thickness (actual) of cured composite to a standard thickness (that is, a specific per ply thickness standard) to yield equivalent fiber stress (based upon the standard). **E 1749,** E06

normalized crack size, *a/W*—the ratio of crack size, *a*, to specimen width, *W*. **E 1823,** E08

normalized erosion resistance, N_e, *n*—the volume loss rate of a test material, divided into the volume loss rate of a specified reference material similarly tested and similarly analyzed. "Similarly analyzed" means that the two erosion rates must be determined for corresponding portions of the erosion rate-time pattern; for instance, the maximum erosion rate or the terminal erosion rate. **G 40,** G02

normalized incubation resistance, N_o, *n—in cavitation and liquid impingement erosion,* the incubation period of a test material, divided by the incubation period of a specified reference material similarly tested and similarly analyzed. (See also **normalized erosion resistance.**) **G 40,** G02

normalized inflation pressure, *n*—measured pressure of a tire adjusted, according to the ideal gas law, to the nominal test temperature and one atmosphere barometric pressure. **F 538,** F09

normalized *K*-gradient, $C = (1/K)\cdot dK/da[L^{-1}]$—the fractional rate of change of K with increasing crack length. **E 1823,** E08

normalized noise isolation class, NNIC—a single-number rating calculated in accordance with Classification E 413 using measured values of normalized noise reduction. (See **normalized noise reduction.**) **C 634,** E33

normalized noise reduction, *NNR—between two rooms, in a specified frequency band,* the value that the noise reduction in a given field test would have if the reverberation time in the receiving room were 0.5 s. NNR is calculated as follows:

$$NNR = NR + 10 \log (T/0.5)$$

where:
NR = noise reduction, dB and
T = reverberation time in receiving room, s. **C 634,** E33

normalizing, *n*—heating a steel object to a suitable temperature above the **transformation range** and then cooling it in air to a temperature substantially below the **transformation range**. **A 941,** A01

normalizing—a variation of full annealing in which steel is heated above the upper critical temperature and is then air cooled in air, rather than in a furnace. Normalizing relieves the internal stresses caused by previous working, and while it produces sufficient softness and ductility for many purposes, it leaves the steel harder and with a higher tensile strength than full annealing. To remove cooling stresses, normalizing if often followed by tempering. **F 1789,** F16

normally consolidated soil deposit—a soil deposit that has never been subjected to an effective pressure greater than the existing overburden pressure. **D 653,** D18

normal mode—*of a room,* one of the possible ways in which the air in a room, considered as an elastic body, will vibrate naturally when subjected to an acoustical disturbance. With each normal mode is associated a resonance frequency and, in general, a group of wave propagation directions comprising a closed path. **C 634,** E33

normal operating conditions—the usual range of physical operating conditions (flow, pressure, temperature, etc.) for component or system. **E 1705,** E48

normal operating conditions—the usual range of physical conditions for which a facility was designed to operate. **E 1705,** E48

normal permeability—the ratio of the induction (when cyclically made to change symmetrically about zero) to the corresponding change in magnetizing force. **E 1316,** E07

normal segregation—a concentration of alloying components or constituents that have lower melting points in those regions which are the last to solidify. **E 7,** E04

normal state—the thermodynamic state in which a superconductive material no longer exhibits any of the characteristics of the superconducting state. **B 713,** B01

normal stress—see **stress.** **D 653,** D18

normal stress, (FL^{-2}), *n*—the component of applied stress that is perpendicular to the surface on which the force acts. **D 4439,** D35

normal stress [FL^{-2}], *n*—the stress component perpendicular to a plane on which the forces act. Normal stress may be either. **E 6,** E28

normal stress, s—the stress component perpendicular to a plane on which the forces act. (E-28) **D 4092,** D20

normal temperature, *n*—as applied to laboratory observations of the physical characteristics of bituminous materials, 25°C (77°F).

D 8, D04

normal uranium—uranium containing the same weight percentage $_{235}U$ as occurs in nature. It may be attained by blending uranium of different isotopic compositions or by processing in a diffusion plant. See **natural uranium.** **C 859,** C26

normal weather conditions—the range of environmental conditions in a local climatic region that occurred during the past 30 years. This excludes extreme conditions that have occurred only once during that period. **E 1705,** E48

normal weight concrete masonry unit—unit whose oven-dry density is 125 lb/ft^3(2000 kg/m^3) or greater. **C 1209,** C15

normative, *n*—information that is part of the standard. **F 1294,** F05

normative terminology, *n*—a terminology that is standardized or authoritatively prescribed. **E 1992,** E02

nose, *n*—the lower visible edge of tile as applied on the roof. **C 43,** C15

nose—{archaic} the working end or refining chamber of a tank. **C 162,** C14

nose—guide-body area where fastener is driven from tool. **F 592,** F16

nose extension—See **lip.** **F 592,** F16

nose lugs, *n*—projections on the underside of the nose of each tile contoured to fit into the main water courses of the tile immediately below, inhibiting the entry of wind-driven rain. **C 43,** C15

no-slump grout—grout with a slump of 1 in. (25 mm) or less according to the standard slump test (Test Method C 143). See also **slump** and slump test. **D 653,** D18

notch, *n*—a groove smaller in both width and length than a lateral groove that contains one closed end. **F 538,** F09

notch—*on a semiconductor wafer*, an intentionally fabricated indent of specified shape and dimensions oriented such that the diameter passing through the center of the notch is parallel with a specified low index crystal direction. **F 1241,** F01

notched head—nail head with semi-circular rim and "V"-notch, having rounded "V" corners, wide notch part at rim, and narrow notch part at nail-shank projection, formed during heading, to allow tight collating of nails in strip form. **F 592,** F16

notch length, a_n (L)—the distance from a reference plane to the front of the machined notch. The reference plane depends on the specimen form, and normally is taken to be either the boundary, or a plane containing either the loadline or the centerline of a specimen or plate. The reference plane is defined prior to specimen deformation. **E 1823,** E08

notch tensile strength (*NTS*) [FL^{-2}]—the maximum nominal (net-section) stress that a notched tensile specimen is capable of sustaining. **E 1823,** E08

not ready signal, *n*—a signal from a device in a system indicating that it will not be able to receive or send the next data segment. **F 1457,** F05

novolac (or novolak), *n*—a phenolic-aldehyde resin which, unless a source of methylene groups is added, remains permanently thermoplastic. (See also **resinoid** and **thermoplastic.**) **D 883,** D20

novolak, *n*—a phenolic-aldehydic resin that, unless a source of methylene groups is added, remains permanently thermoplastic. (See also **thermoplastic**, *n.*) (Compare **resinoid.**) **D 907,** D14

nozzle, *n*—a part of a vacuum cleaner system that is applied to a surface to be cleaned and may incorporate an agitation device to assist dirt removal. **F 395,** F11

nozzle refractory, *n*—a refractory shape containing an orifice for the purpose of transmitting molten metal from a refractory-lined container. **C 71,** C08

n-propylbenzene—1-phenyl propane (C_9H_{12}) mol weight 120.19; used as a reference standard for identifying and determining C_9 aromatics; melting point, −99.2°C; boiling point, 159.2°C. **D 4790,** D16

NTU—see **nephelometer**. **D 6161,** D19

***n*-type**—a variety of semiconductive material in which the majority current carriers are electrons, formed when donor impurities are incorporated into the crystal structure in small concentrations. **F 1241,** F01

nuance, *n*—a two-dimensional attribute that distinguishes among colors having the same hue. **E 284,** E12

nub head—head with protruding knob. (See **washer head.**) **F 547,** F16

nubuck—type of leather where the grain layer is very lightly buffed (snuffed), to create a very fine nap and still called top grain leather. **D 1517,** D31

nuclear activity—the number of disintegrations occurring in a given quantity of material per unit of time. "Curie" is the unit of measurement. One curie is equivalent to 3.7×10^{10} disintegrations per second. **E 1316,** E07

nuclear grade—material of a quality adequate for use in nuclear application. **C 859,** C26

nuclear magnetic resonance (NMR) spectroscopy—that form of spectroscopy concerned with radio-frequency-induced transitions between magnetic energy levels of atomic nuclei. **E 131,** E13

nuclear material license—a permit issued by the Nuclear Regulatory Commission or a state to receive, process, store, or ship source or special nuclear material. **C 859,** C26

nuclear poison—a substance that, because of its high neutron absorption cross section, can reduce reactivity. The term is commonly called poison. **C 859,** C26

nucleation—the preplating step in which a catalytic material, often a palladium or gold compound, is absorbed on a surface to act as sites for initial stages of deposition. **B 374,** B08

nucleocapsid—the outer protein coat or shell (capsid) of a virus plus its inner core of nucleic acid and proteins. **E 1705,** E48

nuclide—an atomic species characterized by the constitution of its nucleus, specifically by the number of protons and neutrons. **C 859,** C26

nuclide—an atomic species characterized by the constitution of its nucleus, specifically by the number of protons and neutrons. **D 1129,** D19

nuclide—a species of an atom characterized by its mass number, atomic number, and its nuclear energy state. **E 170,** E10

Nukiyama-Tanasawa distribution—drop size distribution described by the following equation:

$$dN/dD = a \times D^m \exp(-b \times D^n)$$

where:
N = the number of drops of diameter smaller than D, and a, m, b, and n are parameters (usually m = 2 and n$\neq$ 1). **E 1620,** E29

Nukiyama-Tanasawa nozzle—an internal mixing pneumatic atomizer consisting of two concentric tubes. Liquid emerging from the inner tube is atomized by air flowing through the annulus between the tubes. **E 1620,** E29

NUL—see **null signal**. **F 1457,** F05

null signal, *n*—an electronic signal with no data. **F 1457,** F05

number density—the number of particles contained in a specified volume of space at a given instant. **E 1620,** E29

number of pieces of trash, *n*—*in testing cotton with the Trash Meter*, a number correlated with the total number of pieces of trash on the surface of the sample of cotton over the viewing window. **D 123,** D13

number system character, *n*—the first, or left-hand, digit in a UPC number that identifies a particular numbering system. Similar to AI or DI numbers, but used specifically for retail applications according to UCC rules **F 1294,** F05

numeral head—head with raised or depressed, identifying number or numbers on top surface of head. **F 547,** F16

numeric—a machine **vocabulary** that includes only the primary numbers as contrasted to **alphanumeric**, which includes both letters and numerals. **F 149,** F05

numerical aperture—the product of the lowest index of refraction in the object space multiplied by the sine of half the angular aperture of the objective. **E 175,** E41

numerical aperture (NA)—the sine of half the angular aperture of an objective lens multiplied by the refractive index of the medium between the lens and the sample. **E 7,** E04

numerical aperture (NA), *n*—the sine of one half of the vertex angle of the largest cone of meridional rays that can enter or leave an optical system or element, multiplied by the refractive index of the medium in which the cone is located. **E 131,** E13

nut—an internally threaded product intended for use on external or male screw threads such as a bolt or a stud for the purpose of tightening or assembling two or more components. **F 1789,** F16

nutating—nodding; reciprocating mirror assembly used for scanning a beam. **F 1156,** F12

nutrient—a substance that supports organismal growth. **F 1600,** F20

nutrient broth, *n*—a liquid medium. **F 1494,** F23

nu-value—see synonymous term **Abbé value**. **C 162,** C14

N (Wide to Narrow Ratio), *n*—*in symbologies with two element widths*, the wide to narrow ratio of elements is calculated by summing the average wide bar width and average wide space width and dividing the sum by 2 times *Z*. Intercharacter gaps, if applicable, are not included.

$$N = (\text{average wide bar} + \text{average wide space})\ (2 * Z)$$

F 1294, F05

nylon plastic, *n*—a plastic based on resins composed principally of a long-chain synthetic polymeric amide which has recurring amide groups as an integral part of the main polymer chain. **F 1251,** F04

nylon plastics, *n*—plastics based on resins composed principally of a long-chain synthetic polymeric amide which has recurring amide groups as an integral part of the main polymer chain. **D 883,** D20

nylon plastics—plastics based on resins composed principally of a long-chain synthetic polymer amide which has recurring amide groups as an integral part of the main polymer chain. (D-20) **F 412,** F17

Nyquist frequency—modulation frequency or wavenumber above which aliasing occurs. **E 131,** E13

O

"O"—See **windows and doors.** **E 631,** E06

object beam—the portion of laser radiation which illuminates the test object surface, is scattered, and carries object information to the recording medium. **E 1316,** E07

object beam angle—the angle between a line from the center of the object to the center of the recording medium and the normal to the center of the recording medium. **E 1316,** E07

object evaporation—See **shadowing**. **E 7,** E04

object-film distance—the distance between the surface of the source side object and the plane of the recording medium.

NOTE—In the case where the recording medium is placed directly in contact with the object being examined, the distance is equal to the thickness of the object. **E 1316,** E07

objective—the primary magnifying system of a microscope. A system, generally of lenses, less frequently of mirrors, forming a real, inverted, and magnified image of the object. **E 175,** E41

objective aperture—See **aperture**. **E 7,** E04

objective, fluorite—an objective using the mineral fluorite in its construction. It is usually intermediate between achromatic and apochromatic in correction, but may be more highly corrected. **E 175,** E41

object mode, *n*—color seen as ascribed to an object. **E 284,** E12

object plane resolution—the dimension in the object plane that corresponds to the product of a system's instantaneous field-of-view and a specified distance from the system to the object. **E 1316,** E07

object scattered neutrons—neutrons scattered by the test objects that contribute to the film exposure. **E 1316,** E07

objects of cultural heritage, *n*—any items taken from nature, or modified or created by humankind that are significant for cultural interpretation and scientific research. **D 123,** D13

objects of natural heritage, *n*—any items taken from nature, modified or unmodified, that are significant for the interpretation of the natural environment, or that are significant for cultural and scientific research. **D 123,** D13

oblique illumination—nonspecular illumination under which the light impinges at an oblique angle to the optical axis. (See **conical illumination; diffuse illumination; darkfield illumination**.) **E 7,** E04

oblique illumination—see **illumination, oblique.** **E 175,** E41

observation, *n*—the process of determining the presence or absence of attributes or making measurements of a variable. **D 123,** D13

observation, *n*—a result of the process of determining the presence or absence of an attribute or making a measurement of a variable. (Compare **measurement value, determination value,** and **test result**) **D 123,** D13

observation, *n*—(*1*) the process of obtaining information regarding the presence or absence of an attribute of a test specimen, or of making a reading on a characteristic or dimension of a test specimen, or (*2*) the attribute or measurement information obtained from the process. (The term ″observed value" is preferred for this second usage.) **E 456,** E11

observation angle, *n*—angle between the axes of the incident beam and the observed (reflected) beam, (in *retroreflection,* α, angle between the illumination axis and the observation axis). **E 284,** E12

observation well—typically, a small diameter well used to measure changes in hydraulic head, usually in response to a nearby pumping well. **D 653,** D18

observed fluorescence lifetime, τ—the time required for the fluorescence intensity to decay to 1/e of its initial value after the termination of excitation. **E 131,** E13

observed value, *n*—the value obtained by carrying out the complete protocol of the test method once, being either a single test determination or an average or other specified combination of a specified number of test determinations. **E 456,** E11

observer, *n*—an assessor in a visual sensory test. (See also **assessor.**) **E 253,** E18

observer metamerism, *n*—the property of specimens having different spectral characteristics and having the same color when viewed by one observer, but different colors when viewed by a different observer under the same conditions. **E 284,** E12

observer, standard, CIE 1931, *n—in color determination,* hypothetical observer based on color mixture data obtained for a 2° field of view for 17 real observers, adopted by the CIE in 1931. **D 2946,** C17

observer, supplementary, CIE 1964, *n—in color determination,* hypothetical observer based on color mixture data obtained for a 10° field of view for 76 real observers, adopted by the CIE in 1964. **D 2946,** C17

obsidian—a highly siliceous natural glass. **C 162,** C14

obvious ignition, *n*—pronounced continuous and self-sustaining combustion of the test system accompanied by rapid generation of heat and smoke. It is a matter of operator judgment based upon experience in this type of operation. **E 176,** E05

occluded cell—an electrochemical cell created at a localized site on a metal surface which has been partially obstructed from the bulk environment. **G 15,** G01

occluded vapor phase—condition of contaminant residence in which volatilized contaminants occur in porosity that is ineffective to free and open gaseous flow and exchange, such porosity generally being microporosity; frequently termed dead-end pore space. **D 653,** D18

occlusion—the trapping of undissolved gas in a solid during solidification. **E 1316,** E07

occlusion, *n*—the physical process of covering a chemical that has been applied to or spilled on the skin, thereby disallowing its evaporation and generally increasing its absorption through the skin. **F 1494,** F23

occupancy instrument (OI) (accord d'occupation (AO))—agreement between a prospective or current occupant and the manager or owner of a facility concerning occupancy in that facility. **E 1480,** E06

occupancy, *n* (occupation)—discouraged term, to be used only in connection with building codes, where the term refers to the number of occupants in a space, or other specific classification of use. **E 1480,** E06

occupant (occupant)—(*1*) Department, agency, corporation, or other organization, or a part thereof, that is or will be occupying space in a particular facility. (*2*) Individual or family living in a housing **dwelling**.

building occupant (occupant d'un immeuble)—one who has certain rights to, possession of, or control over the premises occupied, such as **tenant** or owner. **E 1480,** E06

occupational exposure limit to asbestos—the maximum time weighted average (TWA) number of asbestiform fibers to which nearly all workers may be repeatedly subjected over a normal day or a 40-h week and the peak number of asbestiform fibers to which an individual may be subjected during 15 min or less, without known adverse health effects. Lifetime exposure is based on 99 % probability of protecting all workers from acquiring clinically significant disease for a working life of 50 years. **E 1542,** E34

occupational exposure limit to carbon disulfide—the maximum time weighted average (TWA) of carbon disulfide to which nearly all workers may be repeatedly subjected over a normal 8-h day or a 40-h week, the ceiling concentration of carbon disulfide or the maximum peak above ceiling concentration of carbon disulfide that cannot be exceeded, without known adverse health effects. Lifetime exposure is undefined. **E 1542,** E34

occupational exposure limit to ozone—the maximum time weighted average (TWA) of ozone to which nearly all workers may be repeatedly subjected over a normal 8-h day or a 40-h week or the ceiling limit of ozone that cannot be exceeded, without known adverse health effects. Lifetime exposure is undefined. **E 1542,** E34

occupational exposure limit to quartz dust—the maximum time weighted average (TWA) concentration of respirable quartz particles to which nearly all workers may be repeatedly subjected over a normal 10-h day or a 40-h week without known adverse health effects. Ceiling and lifetime exposures are undefined. **E 1542,** E34

occupational exposure limit to synthetic amorphous silica—the maximum time weighted average (TWA) concentration of respirable synthetic amorphous silica to which nearly all workers may be repeatedly subjected over a normal 8-h day or a 40-h week without known adverse health effects. Ceiling and lifetime exposures are undefined. **E 1542,** E34

occupational exposure limit to water-insoluble chromates—the maximum time weighted average (TWA) concentration of water-insoluble chromates as chromium to which nearly all workers may be repeatedly subjected over a normal 8-h day or a 40-h week without known adverse health effects. Ceiling and lifetime exposures are undefined. **E 1542,** E34

occupational exposure limit XXXXX, "generic"—the maximum time weighted average (TWA) concentration to which nearly all workers may be repeatedly subjected for a normal 8- to 10-h workday, 40-h workweek without known adverse health effects. **E 1542,** E34

occupiable space—See **building space.** **E 631,** E06

occurrences spectrum— *in fatigue loading*, representation of spectrum loading contents by the number of times a particular loading parameter (peak, range, and so forth) occurs within each specified loading interval between lower and upper bound values. **E 1823,** E08

OCR—See **optical character recognition**. **F 149,** F05

OCR—Optical Character Recognition. **F 1156,** F12

OCR-A—an abbreviation commonly applied to the character set contained in ANSI X3.17-81. **F 149,** F05

OCR-B—an abbreviation commonly applied to the character set contained in ANSI X3.49-75 (R-1982). **F 149,** F05

octane number, *n*—*for spark ignition engine fuel*, any one of several numerical indicators of resistance to knock obtained by comparison with reference fuels in standardized engine or vehicle tests. **D 4175,** D02

octave band, *n*—a band of sound frequencies for which the highest frequency in the range is (within 2%) twice the lowest frequency. The position of the band is identified by the rounded geometric mean of the highest frequency and the lowest frequency of the band. The nominal mid-band frequencies of "preferred" octave bands as defined in ANSI S1.6 fall in the series 16, 31.5, 63, 125, 250, 500, 1000 Hz etc. **C 634,** E33

octave band sound pressure level, OBSPL—Lpl/lf where f indicates the nominal center frequency of a specific band if applicable, [nd], (dB), *n*—sound pressure level for sound filtered using an octave-band filter meeting the requirements of ANSI S1.11. **C 634,** E33

ocular—see **eyepiece.** **E 175,** E41

ocular micrometer—a glass disk, of a diameter which permits introducing it into standard oculars, upon one surface of which a fine scale is engraved accurately. **E 7,** E04

OD—the outside dimension of a **package** or part, or outside diameter of a cylindrical container or wire or rod. **D 996,** D10

odor, *n*—that property of a substance which affects the sense of smell; any smell; scent; perfume.

odor threshold, n—the concentration of an odorous compound at which the physiological effect elicits a response 50 % of the time. **D 1356,** D22

odor, *n*—See **aroma.** **E 253,** E18

odorant, *n*—odorous substance. **D 1356,** D22

odorant, *n*—a substance that stimulates the olfactory receptors. **E 253,** E18

odorimetry, *n*—measurement of olfactory sensations. **E 253,** E18

odor-intensity index—the number of times the concentration of the original sample is halved by addition of odor-free water to obtain the least definitely perceptible odor. **D 1129,** D19

odor threshold—See **odor.** **D 1356,** D22

odor threshold number—the greatest dilution of the sample with odor-free water to yield the least definitely perceptible odor. **D 1129,** D19

OEM—original equipment manufacturer. **D 6161,** D19

OEM coatings—original equipment manufacturers coatings, which include automotive, marine, furniture, appliance, as well as many other miscellaneous consumer and industrial applications. **D 16,** D01

OE/OEM—original equipment/original equipment manufacturer. Also: the actual manufacturer of end items, major parts, or subassemblies which are subject to repair or remanufacture using parts or materials that may have been obtained from vendors, and not manufactured or fabricated in house by the **OEM**. **F 1156,** F12

Oersted—unit of measure for magnetic coercivity; amperes per metre. **F 1156,** F12

oersted, Oe—the unit of magnetic field strength in the cgs-emu system of units. One oersted equals a magnetic field strength of 1 Gb/cm of flux path. One oersted equals $1000/4\pi$ or 79.58 ampere-turns per metre. See **magnetic field strength.** **A 340,** A06

offal—parts of hides not used for standard grades of leathers, the heads, shoulders, and bellies of heavy leather. **D 1517,** D31

off-axis character—type piece slightly mis-oriented, usually by rotation. **F 1156,** F12

off-axis laminate—a laminate whose principal axis is oriented at an angle theta other than 0 or 90° with respect to a reference direction, usually related to principal load or stress direction. **D 3878,** D30

off-axis response ($U/(U_f \cos \theta)$), *n*—the ratio of the indicated wind speed (U) at various angles of attack θ to the indicated wind speed at zero angle of attack (U_f) multiplied by the cosine of the angle of attack. **D 1356,** D22

off grade, *adj*—not meeting predetermined specifications. **D 1968,** D06

offhand glass—glass produced by the offhand process.

C 162, C14

offhand process—the process of forming glassware by a craftsman working without the aid of molds. **C 162,** C14

office—See **building space.** **E 631,** E06

office copies, *n— (also quick copies, imaged copies),* reproductions made by direct or indirect electrostatic printing, ink jet printing, thermographic processing or similar processes, as contrasted with conventional printing, such as letterpress or offset. **D 1968,** D06

official test aerosol, OTA—a standard insecticide dispenser and formulation prepared by the Chemical Specialties Manufacturers Association (CSMA) for use in aerosol test methods for flying insects. **D 3064,** D10

off-line—pertaining to the operation of a functional unit not under the direct control of the computer. **F 149,** F05

off-line testing—eddy current tests conducted on equipment that includes the test coil and means to propel individual tubes under test through the coil at appropriate speeds and conditions. **E 1316,** E07

offset—an imperfection resulting from mold parts not properly matched, that is, a finish or base offset from the body or neck. **C 162,** C14

offset—see **set-off.** **F 335,** F05

offset—a printing process in which the inked impression from a printing member is transferred to a rubber blanket and then transferred to the paper being printed. **F 425,** F05

offset, *n*—unintentional transfer of ink (as from a freshly printed substrate). **F 1857,** F05

off set barbs, *n*—angling of the barb cluster from the center line of the tape in opposite directions. **F 1379,** F14

offset hinge—a swing gate hinge that permits the gate to swing 180° from the closed to the open position (see Specification F 900). **F 552,** F14

offset modulus, J_o, *(FL^{-1})*, Nm^{-1}, *n—for geosynthetics,* the ratio of the change in force per unit width to the change in elongation below an arbitrary offset point at which there is a proportional relationship between force and elongation, and above the inflection point on the force-elongation curve. **D 4439,** D35

offset paper, *n*—an uncoated or coated paper designed for use in offset lithography. **D 1968,** D06

offset point—See **side point.** **F 547,** F16

offset press—an indirect rotary press having a plate cylinder, a blanket cylinder, and an impression cylinder. **F 425,** F05

offset printing—printing produced on an offset press from a planographic master or plate. **F 425,** F05

ogee—See **railing systems.** **E 631,** E06

ogee, *n*—a molding with a profile having a double curve formed by a convex line turning into a concave line, resulting in an S-shape. **E 1481,** E06

ohm—unit of electric resistance equal to the resistance of a circuit in which a potential difference of one volt produces a current of one ampere. **D 6161,** D19

oil—See **petroleum, vegetable, paraffinic,** and so forth. **E 1519,** E35

oil air filter—a filter or series of filters placed in the air flow line from an air compressor to reduce the oil content of the air. **D 653,** D18

oil canning—a form of buckling; *in flat sandwich constructions,* a defect occasioned by excessive compressive loads and represented by waviness of the product. **E 1749,** E06

oil color—an oil paint containing a high concentration of colored pigment, commonly used for tinting paint. **D 16,** D01

oil content—the measured amount of oil contained in an oil-impregnated object, for example, a self-lubricating bearing. **B 243,** B09

oil furnace process, *n*—a process for producing furnace carbon blacks that uses oil as the source of hydrocarbons for decomposition by injection into a high-velocity stream of combustion gases. **D 3053,** D24

oil-gas tars, *n*—tars produced by cracking oil vapors at high temperatures in the manufacture of oil gas. **D 8,** D04

oiling, *n*—a coating applied to metallic coated steel sheet alone or in addition to chemical treatment for further protection against the onset of storage corrosion; the oil coating is intended as a corrosion inhibitor only and not as a rolling or drawing lubricant. **A 902,** A05

oiling off—coating the surface of leather with oil. **D 1517,** D31

oil of (pine) tar, *n*—certain heavier fractions of the volatile oil recovered by distilling pine-tar oil to convert it into pine tar. **D 804,** D01

oil of turpentine, *n*—the pharmaceutical name for spirits of turpentine that conforms to the requirements of the National Formulary. **D 804,** D01

oil paint—under **paint,** see *oil paint.* **D 16,** D01

oil resistance, *n—of insulating varnish,* a measure of the retention of properties after exposure to a specified oil under specified conditions of test. **D 1711,** D09

oil resistance, *n*—the ability of a substrate to resist oil present on human fingers. **F 1857,** F05

oil spillage, *n*—a localized deterioration or softening of a bituminous pavement surface caused by the spilling of oil, fuel, or other solvents. **E 867,** E17

oil spillage, *n*—a localized deterioration or softening of the pavement surface caused by the spilling of oil, fuel, or other solvents. **E 1778,** E17

oil tannage—tannage with cod oil or other oxidizable oil, usually of marine origin. **D 1517,** D31

oil-tempered—heated above the critical temperature, quenched in oil, and tempered. **F 547,** F16

oil trap—a device used to remove oil from the compressed air discharged from an air compressor. **D 653,** D18

oil varnish—under **varnish,** see *oil varnish.* **D 16,** D01

OL—abbreviation for **open loop**. **F 1457,** F05

old growth—timber in or from a mature, naturally established forest. If the trees have grown during most of their lives in active competition for sunlight, the bole is usually straight and relatively free of limbs. **D 9,** D07

olefin fiber—a manufactured fiber in which the fiber-forming substance is any long-chain synthetic polymer composed of at least 85 % by weight of ethylene, propylene, or other olefin units, except amorphous (noncrystalline) polyolefins qualify under rubber. **D 123,** D13

olefin plastic, *n*—a plastic based on polymers made by the polymerization of olefins or copolymerization of olefins with other monomers, the olefins being at least 50 mass %. **F 1251,** F04

olefin plastics—plastics based on polymers made by the polymerization of olefins or copolymerization of olefins with other monomers, the olefins being at least 50 mass %. **D 883,** D20

olefin plastics—plastics based on polymers made by the polymerization of olefins or copolymerization of olefins with other monomers, the olefins being at least 50 mass %. (D-20) **F 412,** F17

oleophilic—likes oil; wettable with oils. **F 1156,** F12

oleoresin, *n*—pine gum, the nonaqueous secretion of resin acids dissolved in a terpene hydrocarbon oil that is produced or exuded from the intercellular resin ducts of a living tree, and is present, together with oxidation products, in the dead wood of weathered limbs and stumps. **D 804,** D01

oleoresin, *n*—a plant product containing chiefly essential oil and resin. **F 141,** F06

oleoresins—nonaqueous secretions of resin acids dissolved in terpenic hydrocarbons that are produced or exuded from the intercellular resin ducts of living trees, especially the conifers, and accumulated, together with oxidation products, in the wood of weathered limbs or stumps. **D 1695,** D01

olfactory, *adj*—of, relating to, or connected with the sense of smell. **D 1356,** D22

olfactory, *adj*—pertaining to the sense of smell. **E 253,** E18

oligomer, *n*—a substance composed of only a few nonomeric units repetitively linked to each other, such as a dimer, trimer, tetramer, etc., or their mixtures. **D 883,** D20

oligomer, *n*—a polymer consisting of only a few monomer units such as a dimer, trimer, tetramer, etc., or their mixtures. **D 1566,** D11

oligomer, *n*—a polymer consisting of only a few monomer units such as a dimer, trimer, tetramer, etc., or their mixtures. (D-20) **F 412,** F17

oligomer, *n*—a polymer consisting of only a few monomer units such as a dimer, trimer, tetramer, etc., or their mixtures. **F 1251,** F04

oligomerization, *n*—the process of converting a monomer or mixture of monomers into a oligomer. **D 883,** D20

oligosaccharides—polymeric carbohydrates containing relatively few (compared to the polysaccharides) sugar units connected by glycosidal linkages. Two to nine units has been suggested as a suitable range. For longer chains the polymers cannot be readily separated into individual molecular species. **D 1695,** D01

O&M—operation and maintenance. **D 6161,** D19

omnidirectional, *adj*—refers to a code format that can be read, regardless of orientation, from a multiplicity of angles. **F 1294,** F05

OMR (optical mark reading)—the process of identification of marks by an optical scanner. **F 149,** F05

one-coat ware, one-coat work—(*1*) articles finished in a single coat of porcelain enamel.

(*2*) Sometimes a contraction of one-cover-coat ware, in which the finish consists of a single cover coat applied over ground coat. **C 286,** B08

one-fire finish—a porcelain enamel on the finished product processed in a single firing. **C 286,** B08

one-on-one—the application of a single ply of roofing over the substrate, followed by the application of a second single ply over the first (phased application). **D 1079,** D08

one-percent length *(L1 %N)*, *n*—*in fiber testing*, the length exceeded by 1 % of the number of fibers in a test specimen. **D 123,** D13

on-feet—the desired and optimum uniformity of contact between the type font line and the platen. **F 221,** F05

on-grade—in contact with the ground, *as related to floor location*, in contact with the ground or with less than 18 in. of well-ventilated space between the bottom of the lowest horizontal structural member and any point of the ground. **F 141,** F06

onion—(*1*) a term sometimes used for the bulklike mass of glass at the origin of the drawn sheet in the Fourcault process.

(*2*) the spread of molten glass as it flows from a canal onto a tin bath. **C 162,** C14

on-line—pertaining to the operation of a functional unit when under the direct control of the computer. **F 149,** F05

on-line analysis—analytical procedure performed in a process environment using automatic or continuous sampling. **D 4790,** D16

on-line medical physician—a physician immediately available, when medically appropriate for communication of medical direction to non-physician prehospital care providers in remote location. **F 1177,** F30

on-line testing—eddy current tests conducted on equipment that includes the test coil and means to propel tubes under test through the coil at appropriate speeds and conditions as an integral part of a continuous tube manufacturing sequence. **E 1316,** E07

on-location cleaning, *n*—*for textile floorcoverings and upholstered furniture*, a cleaning process performed in the location where a product is used. **D 123,** D13

onset temperature, *n*—the temperature at which a deflection from the established baseline is first observed. **E 1445,** E27

onset temperature, *n*—in a thermal transfer ribbon, the temperature at which the imaging material begins to change physical form (that is, start to melt). **F 1623,** F05

onyx marble—translucent, generally layered, cryptocrystalline calcite with colors in pastel shades, particularly yellow, brown, and green. **C 119,** C18

oolitic limestone—a limestone composed largely of the spherical or subspherical particles called oolites or ooliths. **C 119,** C18

ooze—traditionally, vegetable-tanned suede leather. Now also refers to other tannages sueded or napped on the grain side. **D 1517,** D31

opacifier—a material that imparts or increases the diffuse reflectance of porcelain enamel. **C 286,** B08

opacity—the property of reflecting light diffusely and nonselectively; properly defined in Test Method C 347 under the term *contrast ratio*. **C 286,** B08

opacity—the degree of obstruction to the transmission of visible light.

In this sense "opacity" is a relative term, it being considered that given a film sufficiently thin, in paint technology at least, there is no absolutely opaque substance. **D 16,** D01

opacity, *n*—a measure of the degree to which the intensity of light is reduced as it passes through a gas, due to absorption and scattering. **D 1356,** D22

opacity, *n*—*of paper*, one hundred times the ratio of the light reflected by a paper specimen when the specimen is backed by a black body of 0.5 % reflectance or less to that when the specimen is backed by a thick stack of the same kind of paper. **D 1968,** D06

opacity, *n*—(*1*) *optical*, the ability of a specimen to prevent the transmission of light; the reciprocal of the transmittance factor.

(*2*) *paper backing*, the ability of a sheet of paper to hide a surface behind and in contact with it, expressed as the ratio of the reflectance factor R_b when the sheet is backed by a black surface to the reflectance factor R_∞ when it is backed by a pile of sheets of the same kind, and of such number that further addition of sheets does not affect the measured opacity.

(*3*) *white backing*, the ability of a thin film or sheet of material, such as paint or paper, to hide a surface behind and in contact with it, expressed as the ratio of the reflectance factor R_b when the material is backed by a black surface to the reflectance factor R_w when it is backed by a white surface (usually having a reflectance factor of 0.89). **E 284,** E12

opacity—the property of paper that minimizes the show-through of printing from the back side or the next sheet. The ratio of the paper reflectance with a black backing to the paper reflectance with a white backing. **F 149,** F05

opacity (printing), *n*— See **opacity** (*2*) *paper backing*. **E 284,** E12

opacity value, *n*—the calculated value of a material by means of two reflectance measurements, *R*1 and *R*2. The first measurement, *R*1, is taken on a sample of a blank material using an infinite pad backing. The second measurement, *R*2, is taken of the same blank material with black backing. The reflectance value of the black backing should not exceed 5 %. The opacity value is calculated according to the following equation:

$$\text{Opacity value} = 1.00 - (R1 - R2/R1) = R2/R1$$

F 1294, F05

opal glass—glass with fiery translucence; typically a nearly opaque white glass. **C 162,** C14

opaque, *adj*—transmitting no optical radiation. **E 284,** E12

opaque body, *n*—body which transmits practically no light. **E 349,** E21

opaque glaze—See *opaque glaze* under **glaze.** **C 242,** C21

open—the quality of asbestos with a relatively high specific surface area and degree of fiberization, free from an appreciable portion of crudy material. **D 2946,** C17

open area, *n*—the ratio of the total area of the apertures to the total functional area of the screening media, expressed as a percentage. **E 1638,** E29

open assembly time, *n*—See *open assembly time* under **assembly time**. **D 907,** D14

open bubble—a surface cavity formed by a gaseous inclusion which was vented to ambient atmosphere during processing. **C 162,** C14

open cell, *n*—a cell not totally enclosed by its walls and hence interconnecting with other cells. **C 717,** C24

open cell, *n*—a cell not totally enclosed by its walls, and hence interconnecting with other cells. (See **closed cell.**) **F 1251,** F04

open-cell cellular plastic, *n*—a cellular plastic in which there is a predominance of interconnected cells. **D 883,** D20

open-cell cellular plastic, *n*—a cellular plastic in which there is a predominance of interconnected cells. (D-20) **F 412,** F17

open cell material, *n*—a cellular material in which substantially all cells in the mass are open cells. **C 717,** C24

open cell sealant backing, *n*—*in building construction*, a flexible and very compressible fabricated shape (usually cylindrical), without a surface skin, that is composed of open cell material. **C 717,** C24

open-circuit potential—the potential of an electrode measured with respect to a reference electrode or another electrode when no current flows to or from it. **G 15,** G01

open construction—See **building constructions.** **E 631,** E06

open crate—See **crate.** **D 996,** D10

open cut—an excavation through rock or soil made through a hill or other topographic feature to facilitate the passage of a highway, railroad, or waterway along an alignment that varies in topographic relief. An open cut can be comprised of single slopes or multiple slopes, or multiple slopes and horizontal benches, or both. (ISRM) **D 653,** D18

open deep fat fryer—equipment with a deep kettle containing oil or fat for cooking food within a perforated container. See **fryer**. **F 1827,** F26

open-end spinning machine, *n*—a textile machine for converting staple fiber into spun yarn by a continuous process in which the individual fibers or groups of fibers are caused to assemble at the open end of the forming yarn. (See **rotor-type open-end spinning machine**.) **D 123,** D13

open-end yarn, *n*—a continuous strand of fibers, produced directly from sliver or roving, in a single continuous operation, by opening and reassembling them in a spinning element to form the yarn. **D 3888,** D13

open-face fabric, *n*—a face or shell fabric constructed with specifically designed open areas to show the substrate when joined to another material. **D 123,** D13

open flame sunshine carbon arc, *n*—a light source in which an arc is produced across a pair of copper coated carbon rods filled with rare earth elements intended to produce a specific spectral power distribution. The carbons are open to the atmosphere and may be surrounded by a glass lantern arrangement which acts to modify the spectral power distribution received by the specimens. **G 113,** G03

open-graded aggregate, *n*—an aggregate that has a particle size distribution such that when it is compacted, the voids between the aggregate particles, expressed as a percentage of the total space occupied by the material, remain relatively large. **D 8,** D04

open grain—See **grain.** **D 9,** D07

opening—*for the purpose of Test Method D 4749*, openings and apertures shall be regarded as synonomous terms. Dimensions for round and square openings shall be determined as follows: for round holes, dimensions shall refer to the opening diameter; for square holes, dimensions shall refer to the distance between parallel wires. **D 121,** D05

opening device, *n*—*in open-end spinning*, either a drafting system or an opening roller which separates the feed stock into individual fibers or very small tufts prior to their reassembly into yarn. (See **feed unit**.) **D 123,** D13

opening force—the force available to open the contacts. **B 542,** B02

opening roller, *n*—*a component of the opening device in open-end spinning machines*, a roller covered with pins or teeth or similar device used to separate the feed stock into individual fibers or very small tufts by a continuous combing action. **D 123,** D13

open loop, *n*—a way of electronically interconnecting the components of a system. **F 1457,** F05

open loop, *adj*—refers to a testing mode in which fresh collection medium flows continuously through the collection chamber of the test cell. **F 1494,** F23

open-plan workstation—See **building space.** **E 631,** E06

open pore, *n*—a pore communicating with an exterior surface. **B 243,** B09

open porosity—See *open porosity* under **porosity.** **C 242,** C21

open pot—a pot open to the flames and gases of combustion. **C 162,** C14

open pot—a pot wherein the glass surface is not protected from the furnace atmosphere. **C 162,** C14

open seams—unfilled fissures or naturally occuring cracks in stone. **C 119,** C18

open system—See **building system.** **E 631,** E06

open time—length of time a coating remains wet enough to allow for brushing-in at the laps; also called wet edge time. **D 16,** D01

open water boom—boom intended for use in open waters (see Practice F 625 for environmental descripters). **F 818,** F20

operable—See **windows and doors.** **E 631,** E06

operating characteristic curve (OC-curve), *n*—*in acceptance sampling*, the curve which has as its abscissa a hypothesized lot average, and which has as its ordinate the probability of accepting the lot, when the plan is used. (See also **Type A operating characteristic curve** and **Type B operating characteristic curve.**) **D 123,** D13

operating conditions, extreme—unusual physical conditions to which a component or system may be exposed and for which it is not designed or intended to withstand, nor is it required to withstand by a local regulatory agency. **E 772,** E44

operating conditions, normal—the usual range of physical conditions (for example, temperature, pressure, wear and tear, weather) for which the component or system was designed. **E 772,** E44

operating cost—the expenses incurred during the normal operation of a building or a building system or component, including labor, materials, utilities, and other related costs. **E 833,** E06

operating cycle—an ion-exchange process consisting of a backwash, regeneration, rinse, and service run. **D 1129,** D19

operating humidity range of analyzer—See **analyzer.** **D 1356,** D22

operating humidity range of sample—See **sample.** **D 1356,** D22

operating pressure—the gage hydraulic pressure at which feedwater enters a device. **D 6161,** D19

operating pressure—the pressure expected under normal operating conditions. **G 126,** G04

operating site, ***n***—in waste management, a location or facility where waste is treated, stored, of disposed as part of an on-going operation. **D 5681,** D34

operating temperature—the temperature expected under normal operating conditions. **G 126,** G04

operating temperature range of analyzer—See **analyzer.** **D 1356,** D22

operating temperature range of sample—See **sample.** **D 1356,** D22

operational classification, *n*—the grouping of permanent-counter sites, and roadways, on the basis of observed variation in traffic measurements, rather than designed purpose of the roadway. **E 867,** E17

operational definition, *n*—a definition that describes the procedure or process by which an entity is made or calculated. **E 1992,** E02

operational period—See **analyzer.** **D 1356,** D22

operator—person having direct control of equipment or operation. **E 1605,** E06

operator—the person having direct control of the starting, stopping, or speed of an amusement ride. **F 747,** F24

ophthalmic glass—glass used in eyeglasses. **C 162,** C14

opponent-color scales, *n*—scales that denote one color by positive scale values, the neutral axis by zero value, and an approximately complementary color by negative scale values. Common examples include scales that are positive in the red direction and negative in the green direction (CIE *a**, Hunter *a*) and scales that are positive in the yellow direction and negative in the blue direction (CIE *b**, Hunter *b*). **E 284,** E12

opportunity cost of capital—the rate of return available on the next best available investment of comparable risk. **E 833,** E06

optic, *adj*—having variations in wall thickness, producing refractive effects. **C 162,** C14

optic, *n*—a lens or prism in an optical instrument. **C 162,** C14

optical axis, *n*—an imaginary line joining the centers of curvature of the surfaces of lenses or mirrors in an optical system. **E 135,** E01

optical bleach—see **fluorescent whitening agent.** **D 459,** D12

optical brightener, n—discouraged term for **fluorescent whitening agent.** **E 284,** E12

optical brightener—a material often added to paper during its manufacture to improve its brightness or whiteness. These materials can cause erratic reflectance values when used with optical scanners that are sensitive to the short wavelength portions of the spectrum. See **fluorescence**. **F 149,** F05

optical character reader—an information processing device that accepts prepared forms and converts data from them to computer output media via **optical character recognition**. **F 149,** F05

optical character recognition (OCR)—character recognition that uses optical means to identify graphic characters. **F 149,** F05

optical crown glass—optical glass with a low dispersion relative to its index of refraction, usually forming the converging element of an optical system. Generally, optical glasses possessing an Abbé value greater than 50. **C 162,** C14

optical density, n— see **reflectance density, reflection density, transmission density, transmittance density.** **E 284,** E12

optical density—the degree of opacity of a translucent medium (darkening of film) expressed as follows:

$$OD = \log(I_o/I)$$

where:
OD = optical density,
I_o = light intensity incident on the film, and
I = light intensity transmitted through the film.
E 1316, E07

optical density—the image intensity or density in terms of or measured by a reflectance densitometer. **F 221,** F05

optical density—a measure of image intensity by reflectance densitometer. **F 335,** F05

optical density of smoke, *D, n*—a measure of the attenuation of a light beam passing through smoke, expressed as the common logarithm of the ratio of the incident flux, I_o, to the transmitted flux, I. ($D = \log_{10}(I_o/I)$). **E 176,** E05

optical disk—data storage device; can be alterable or permanent. **F 1156,** F12

optical emission spectrograph—a spectrograph for recording electromagnetic radiation in one or more of the following regions: ultraviolet, visible, or infrared. **C 859,** C26

optical emission spectroscopy, *n*—pertaining to emission spectrometry in the ultraviolet, visible, or infrared wavelength regions of the electromagnetic spectrum. **E 135,** E01

optical fiber—See **fiber, optical**. **C 162,** C14

optical fiber, *n*—a filament-shaped dielectric material that guides radiant energy. **E 131,** E13

optical flint glass—optical glass with a high dispersion relative to its index of refraction, usually forming the diverging element of an optical system. Generally, optical glasses possessing an Abbé value less than 50. **C 162,** C14

optical glass—glass of high quality having closely specified optical properties, used in the manufacture of optical systems. **C 162,** C14

optical glass numerical designation—the numerical designation in common usage is based on the index of refraction for the helium line (n_d) and the Abbé value. The unity factor for the index is dropped (that is, 1.496 becomes 496) and the decimal point for the Abbé value is also dropped (Abbé value 64.4 becomes 644). Thus a glass with an index of refraction of 1.496 and an Abbé value of 64.4 is specified 496-644 without reference to chemical composition. **C 162,** C14

optical interference filter—optical filter that passes only a very narrow range of wavelengths. **F 1156,** F12

optical line pair test pattern—see **line pair test pattern.** **E 1316,** E07

optical measuring path length, *n*—the length of the optical beam over which the atmosphere or emission concentration is measured and averaged. **D 1356,** D22

optical opacity—an expression for the amount of light absorbed and scattered by a suspension reported as: extinction coefficient, or percent of incident light scattered in 90°, or percent of incident light transmitted at 180° over a standard distance, or all three. **D 4410,** D19

optical path difference— see **retardation.** **E 131,** E13

optical pyrometer—an instrument with the temperature of an object is determined by comparing its brightness at some fixed wavelength with that of a standardized source. **E 7,** E04

optical resolution—measure of the minimum separation between two adjacent characters that still permits their optical recognition as discrete objects; often expressed in terms of lpi, lpmm, dpi, dpmm, etc. **F 1156,** F12

optical retardation— see **retardation.** **E 131,** E13

optical-rotatory power—related to the magnitude of the optical rotation of a polarized beam of light. **F 1156,** F12

optical scanner—(*1*) a scanner that uses light for examining patterns. (*2*) a device that scans optically and usually generates an analog or digital signal. **F 149,** F05

optical throw, *n*—the distance from the aperture of a bar code reader to the leading vertical plane of the depth of field. Also, the minimum distance a bar code symbol can be away from a scanner and still be successfully read. **F 1294,** F05

optical whitening agent—see **fluorescent whitening agent.** **D 459,** D12

optic axis—*of a doubly refracting crystal,* that direction through the crystal along which no double refraction occurs. **F 1241,** F01

optimal colors, *n*—object colors having the maximum possible luminance factor for each chromaticity. **E 284,** E12

optimum concentration range, *n*—in analysis of trace metals, a range, defined by limits expressed in concentration, below which scale expansion must be used and above which curve correction should be considered. **D 5681,** D34

optimum cure, *n*—the state of vulcanization at which a desired property value or combination of property values is obtained. **D 1566,** D11

optimum energy, *n*—energy (millijoules/square millimetres) which achieves the best optical print density and consistent, well-defined image; dependent on ribbon thickness, coating formulation, ribbon speed in the printer, and print head temperature. **F 1623,** F05

optimum exposure—the time-light intensity relationship that produces the most satisfactory print. **F 335,** F05

optimum frequency—in electromagnetic testing, that frequency which provides the largest signal-to-noise ratio obtainable for the detection of an individual material property. Each property of a given material may have its own optimum frequency. **E 1316,** E07

optimum moisture content (optimum water content), *OMC,* w_o **(D)** —the water content at which a soil can be compacted to a maximum dry unit weight by a given compactive effort. **D 653,** D18

orange-box nail—coated, regular-stock-steel, 1¼ by 0.072-in. nail with flat 7⁄32-in. head and medium diamond point. **F 547,** F16

orange peel—a finish resembling the dimpled appearance of an orange peel. **B 374,** B08

orange peel—a pattern of roughness or waviness on a vitreous or glassy surface which resembles the skin of an orange in texture. **C 162,** C14

orangepeel—a pitted texture of a fired glaze resembling the surface of rough orange peel. **C 242,** C21

orange peel—a surface condition characterized by an irregular waviness of the porcelain enamel resembling an orange skin in texture; sometimes considered a defect. **C 286,** B08

orange peel—*in protective coatings,* the dimpled appearance of a dried coating film resembling the surface of an orange. **D 4538,** D33

orange peel, *n*—a pattern of roughness somewhat like that of the outer surface or an orange, appearing when the film split pattern is not uniform over the printed surface. **D 6488,** D01

orange peel, *n*—the appearance of irregularity of a surface resembling the skin of an orange. **E 284,** E12

orange-peel—uneven surface somewhat resembling an orange peel. (D-20) **F 412,** F17

orange peel—*on a semiconductor wafer,* large-featured, roughened type of surface visible to the unaided eye (see also **spiral**). **F 1241,** F01

orbital energy—*XPS,* Koopmans energy corrected for intra-atomic relaxation. **E 673,** E42

order-disorder transformation—a phase change among two solid solutions having the same crystal structure but in which the atoms of one phase, the disordered one, are randomly distributed and in the other, the different kinds of atoms occur in a regular sequence upon the crystal lattice, that is, in an ordered arrangement. **E 7,** E04

order distribution—see **lateral order.** **D 1695,** D01

ordered structure—that crystal structure of a solid solution in which the atoms of different elements seek preferred lattice positions. **E 7,** E04

order (in X-ray reflection)—the factor *n* in the Bragg equation (see **Bragg equation**). In X-ray reflection from a crystal, the order is an integral number which is the path difference measured in wavelengths between reflections from adjacent planes. **E 7,** E04

order of a system—the number of components; for example, a system of binary order is made up of two components. **E 7,** E04

order strengthening—a thermal treatment of a cold-worked product at a temperature below its recrystallization temperature causing ordering to occur to obtain an increase in yield strength. **B 846,** B05

ordinance, *n*—a rule or law adopted by local governmental authority. **E 631,** E06

organic, *adj*— being or composed of hydrocarbons or their derivatives, or matter of plant or animal origin. **D 1079,** D08

organic chemical, *n*—a carbon-based compound in which the element carbon is attached to other carbon atom(s), hydrogen, oxygen, or other elements in a chain, ring, or three-dimensional structure.

nonvolatile organic chemical, n—an organic compound with a saturation vapor pressure less than 10^{-8} kPa at 25°C.

polar organic chemical, n—an organic compound that may exhibit a relatively high electric dipole moment or may be readily ionized, typically containing heteroatoms, such as oxygen, sulfur, nitrogen, phosphorus, chlorine, and bromine.

semivolatile organic chemical, n—an organic compound with a saturation vapor pressure between 10^{-2} and 10^{-8} kPa at 25°C.

volatile organic chemical, n—an organic compound with a saturation vapor pressure greater than 10^{-2} kPa at 25°C. **D 1356,** D22

organic clay—a clay with a high organic content. **D 653,** D18

organic fiber—natural or synthetic fiber, having a length-to-diameter (or to maximum transverse dimension) ratio of at least 100 to 1. **D 2946,** C17

organic fiber content, *n—in asbestos-cement,* the percentage of organic matter expressed as the equivalent percentage of wood cellulose in a dried sample. **C 460,** C17

organic fiber content, *n—in asbestos-cement,* the percentage of organic matter expressed as the equivalent percentage by mass of wood cellulose in a dried sample. **D 2946,** C17

organic (inorganic) SIMS—the SIMS technique when applied to organic (inorganic) specimens or organic (inorganic) molecules placed on a solid. **E 673,** E42

organics—being or composed of hydrocarbons or their derivatives, or matter of plant or animal origin. **G 126,** G04

organic silt—a silt with a high organic content. **D 653,** D18

organic soil—soil with a high organic content. **D 653,** D18

organic terrain—see **peatland**. **D 653,** D18

organizational component, *n*—a portion of an organization with specific tasks and activities that constitutes a part of the total effort and accomplishment of the organization. **D 1356,** D22

organoleptic, *adj*—relating to a property of a sample perceived by the sense organs (obsolete, see **sensory**). **E 253,** E18

organosol, *n*—a suspension of a finely divided polymer in a plasticizer, together with a volatile organic liquid. **D 883,** D20

organosol, *n*—a suspension of a finely divided polymer in a plasticizer together with a volatile organic liquid. **D 1566,** D11

organosol, *n*—a suspension of a finely divided plastic in a plasticizer, together with a volatile organic liquid. **F 1251,** F04

orientation, *n*—*in buttons*, the degree of order and spatial alignment of pearlescent pigment crystals internally or in a coating. **D 123,** D13

orientation—the angle made by the crystallites of the cellulose with the fiber axis. This is approximately the same angle as that made by the molecules or the fibrils. **D 1695,** D01

orientation—the angular position of a crystal described by the angles which certain crystallographic axes make with the frame of reference. In hardness measurements, the relationship between the direction of the axes of the indenter of a hardness tester and the direction of non-homogeneous properties of the specimen. **E 7,** E04

orientation, *n*—the plane in which the exposed face of the specimen is located during testing. **E 176,** E05

orientation, *n*—the plane in which the exposed face of the specimen is located during testing, either vertical or horizontal facing up. **E 176,** E05

orientation—*of a single crystal surface*, the crystallographic plane, described in terms of its Miller indices, with which the surface is ideally coincident. **F 1241,** F01

orientation of a crystal—the angular position of a crystal described by the angles which certain crystallographic axes make with the frame of reference. **C 709,** D02

orientation of a grain—the angular position of a grain described by the angles which a defined set of axes of the grain make with the frame of reference. **C 709,** D02

orientation of an object—the angular position of an object described by the angles which a defined set of axes or surfaces of the object make with the frame of reference. **C 709,** D02

orifice—an opening through which glass flows. In a feeder, an opening in the bottom of the spout formed by the orifice ring. **C 162,** C14

orifice atomizer—see **plain jet atomizer**. **E 1620,** E29

orifice meter, *n*—a flowmeter, employing as the measure of flow rate the difference between the pressures measured on the upstream and downstream sides of the orifice (that is, the pressure differential across the orifice) in the conveying pipe or duct. **D 1356,** D22

orifice ring—that ring that forms the hole through which glass flows in the feeder process. (See also **bushing**.) **C 162,** C14

original—the object to be copied. **F 335,** F05

original crack size, a_o, **[L]**—the physical crack size at the start of testing. **E 1823,** E08

original twist, *n*—the twist in a single or plied yarn component of a plied or cabled yarn as the component was before incorporation into the more complex structure. **D 123,** D13

original uncracked ligament, b_o **[L]**—distance from the original crack front to the back edge of the specimen, that is:

$$b_o = W - a_o$$

E 1823, E08

O-ring, n— See **seal, O-ring.** **D 1566,** D11

O-ring gasket—a solid gasket of circular cross section. **C 822,** C13

orthochromatic—(*1*) of, pertaining to, or producing tone values (of light or shade) in a photograph, corresponding to the tones of nature.

(*2*) photographic use; designating a film made sensitive to green and blue, but not red, light. **E 7,** E04

orthogonal array, *n*—a table of coefficients identifying the levels, or some weight associated with the levels, for each factor to be used in the analysis of specified effects, which are arranged in such a manner that each effect will be independent of the other effects. **E 456,** E11

orthogonal array, *n*—a table of coefficients identifying the levels, or some weight associated with the levels, for each factor to be used in the analysis of specified effects, which are arranged in such a manner that each effect will be independent of the other effects. **E 1325,** E11

orthogonal contrasts, *n*—two contrasts are orthogonal if the contrast coefficients of the two sets satisfy the condition that, when multiplied in corresponding pairs, the sum of the products is equal to zero. See **contrast** and **contrast analysis.** **E 456,** E11

orthogonal contrasts, *n*—two contrasts are orthogonal if the contrast coefficients of the two sets satisfy the condition that, when multiplied in corresponding pairs, the sum of the products is equal to zero. See **contrast** and **contrast analysis.** **E 1325,** E11

orthogonal trajectory deviation, *n*—the perpendicular deviation or distance from the center of the vehicle to the TGL at the end of a stopping test. **F 538,** F09

orthopedic and surgical felt, *n*—a white, soft, low-density, highly resilient felt. **D 123,** D13

orthopedic devices—any device of leather, metal or other material included in the construction of a shoe or inserted in a shoe to prevent or correct foot defects and deformities. **F 869,** F08

orthopedic leathers—a general term for the types of leather used in the manufacture of artificial limbs, braces, etc., for orthopedic purpose. The leathers may range from chamois and horsehide glove to case and strap leathers. **D 1517,** D31

orthorhombic—having three mutually perpendicular axes of unequal lengths. **E 7,** E04

orthotic—an accessory inserted in a shoe to prevent or correct foot defects and deformities. **F 869,** F08

orthotropic—having three mutually perpendicular planes of elastic symmetry. **E 1749,** E06

orthotropic material—a material in which a property of interest, at a given point, possesses three mutually perpendicular planes of symmetry, which taken together define the principal material coordinate system. **D 3878,** D30

ortho-xylene—1,2-dimethylbenzene (C_8H_{10}) mol weight 106.16; clear, colorless, flammable liquid; freezing point, −25.18°C; boiling point, 144°C. **D 4790,** D16

OSA-UCS color system, *n*—Optical Society of America Uniform Color Scales color order system based on equality of visual spacing, which uses the opponent-color scales $\pm L$ (lightness), $\pm j$ (yellowness-blueness), and $\pm g$ (greenness-redness).

E 284, E12

OSA-UCS samples, *n*—current Optical Society of America physical exemplification of the OSA-UCS color system, consisting of about 550 samples displayed on a face-centered-cubic lattice such that each interior sample has 12 nearest neighbors at equal intervals from it. **E 284,** E12

oscillating mirror scanner, *n*—a single beam scanner with a mirror that oscillates in a plane at right angles to the scanner beam which causes the field of view to be swept by the beam. **F 1294,** F05

oscillation—the variation, usually with time, of the magnitude of a quantity with respect to a specified reference when the magnitude is alternately greater and smaller than the reference. **D 653,** D18

osmosis—the spontaneous flow of water from a less concentrated solution to a more concentrated solution through a semipermeable membrane until chemical potential equilibrium is achieved. **D 6161,** D19

osmotic pressure—a measurement of the potential energy difference between solutions on either side of a semipermeable membrane. A factor in designing the operating pressure of reverse osmosis equipment. The applied pressure must first overcome the osmotic pressure inherent in the chemical solution in order to produce any flux. **D 6161,** D19

Ostwald coefficient, *n*—the solubility of a gas expressed as the volume of gas dissolved per volume of liquid when the gas and liquid are in equilibrium at the specified partial pressure of gas and at the specified temperature. **D 4175,** D02

Ostwald color system, *n*—color order system in which colors are specified in terms of the attributes hue, blackness, and whiteness, and are spaced according to the results of spinning-disk mixing of specified amounts of ideal black, white, and maximally chromatic samples. **E 284,** E12

other alkali-insoluble impurities, *n*—*in scoured wool*, the oven-dried, ash-free, alcohol-extractives-free, alkali-insoluble substances other than vegetable matter base, such as skin, cotton or other fibers, paper, string, tag (dung) pieces, and paint pieces, etc. **D 123,** D13

ounce—an ounce is equivalent in thickness to 1⁄64 in. (approximately 0.4 mm). **D 1517,** D31

outcrop—the exposure of the bedrock at the surface of the ground. (ISRM) **D 653,** D18

outdoor exposure—normal weather conditions, that is, the sun's rays, rain, air, temperature changes, and wind. **F 412,** F17

outdoor-indoor transmission loss, *OITL—of a building facade, in a specified frequency band*, ten times the common logarithm of the ratio of the airborne sound power incident on the exterior of the facade to the sound power transmitted by the facade and radiated to the interior. The quantity so obtained is expressed in decibels. **C 634,** E33

outgassing, *n*—the emission of occluded gasses from a material by vacuum, heat, or pressure. **C 717,** C24

outgassing—the upward and outward emission of air or moisture vapor from concrete. **D 3743,** D04

outgassing—the process by which materials expel gases. **E 772,** E44

outgassing—the evolution of gas from a material in a vacuum. **E 1316,** E07

outlet pressure—see **forepressure.** **E 1316,** E07

out-of-roundness—the allowed difference between the maximum measured diameter and the minimum measured diameter (stated as an absolute deviation). **F 412,** F17

output—See **analyzer.** **D 1356,** D22

outside bevel divergent point—staple point with two-plane beveled face on staple outside and along thick leg side; designed to lead staple legs into opposite directions and to result at the same time in their crossing during driving. **F 592,** F16

outside bevel point—bevel point with its beveled face on staple outside. **F 592,** F16

outside diameter, *n*—the maximum diameter of a tire when it is mounted and inflated. **F 538,** F09

outside drive—a lock that is gear-driven by a spindle located outside the lock case on either the end, top, or bottom of the lock case. **F 471,** F12

outside-leg length, *n*—*in body measurements*, the distance from the side waist to the soles of the feet, following the curve of the body. **D 123,** D13

outside seam, *n*—*in home sewing*, a seam in which the seam allowance of the completed seam is visible from the face side of the sewn product. (Compare **inside seam.**) **D 4965,** D13

outsole—the bottom sole thickness. The surface of which is exposed to wear. **F 869,** F08

outward clinch—clinch with both staple legs flared outwardly during driving. **F 592,** F16

oval head, oval rivet head—circular head having convex top surface, with its height smaller than its radius. **F 547,** F16

ovality—(%),

$$\frac{(\text{maximum measured diameter} - \text{minimum measured diameter})}{\text{average measured diameter}}$$

$\times$ 100

F 412, F17

oven—general purpose equipment, operating in dry heat mode, used primarily for baking and roasting food products. **F 1827,** F26

oven-dried, *adj*—the condition of a material that has been heated under prescribed conditions of temperature and humidity until there is no further significant change in its mass. **D 123,** D13

oven-dried wool, *n*—wool dried to moisture equilibrium under specified conditions. **D 123,** D13

oven drying loss—the reduction in weight resulting when a substance is heated in an oven under specified conditions. **D 2652,** D28

oven dry wood—wood that has been dried to constant weight at about 217°F ± 2°F (103°C ± 1°C) and low humidity. By definition, oven dry wood has zero moisture content. **E 1126,** E48

oven glass—(*1*) glass suitable for manufacture of articles to be used in baking and roasting foods.

(*2*) glassware made from oven glass. **C 162,** C14

oven ware—ceramic whiteware for culinary oven use. **C 242,** C21

overaging, *n*—**aging** under conditions of time and temperature greater than those required to obtain maximum change in a certain property, so that the property is altered away from the maximum. **A 941,** A01

overall cleaning, *n*—*for upholstered furniture*, the application of an appropriate cleaning agent to the entire fabric covering, accomplished while the upholstery fabric remains attached to the furniture unit. **D 123,** D13

overall coefficient of heat transfer—See **transmittance, thermal.** **C 168,** C16

overall gas penetration resistance, *n*— the integrity of a totally encapsulating chemical protective suit to resist the inward leakage of gases when exposed to a hazardous chemical environment. **F 1494,** F23

overall height—maximum vertical dimension of boom. **F 818,** F20

overall length—the total length of the individual pipeline system, section, or fitting prior to installation. **F 412,** F17

overall liquid penetration resistance, *n*—the integrity of a chemical protective suit to resist the inward leakage of liquids when exposed to a hazardous chemical environment. **F 1494,** F23

overall magnetization—magnetization of an entire part with a single energizing cycle. **E 1316,** E07

overall precision—See **precision.** **D 1356,** D22

overall profile grade, *n—as defined in ANSI X3.182*, the lowest grade received by measurement of the following parameters: edge determination, overall profile reflectance grade, decode and decodability; grades may be denoted by letters (*A* to *F*) or numbers (4.0 to 0). **F 1294,** F05

overall rate of return (ORR)—See **adjusted internal rate of return (AIRR).** **E 833,** E06

overall serviceability score—See **serviceability score.** **E 1480,** E06

overall width, *n*—the maximum cross-sectional width of a tire, including protective or decorative ribs. **F 538,** F09

overbreak—the quantity of rock that is excavated or breaks out beyond the perimeter specified as the finished excavated tunnel outline. (ISRM) **D 653,** D18

overburden—the loose soil, sand, silt, or clay that overlies bedrock. In some usages it refers to all material overlying the point of interest (tunnel crown), that is, the total cover of soil and rock overlying an underground excavation. (ISRM) **D 653,** D18

overburden load—the load on a horizontal surface underground due to the column of material located vertically above it. (ISRM) **D 653,** D18

overcoating—the act of making a developed electrostatic image permanent by spraying or laminating a protective film or similar layer over the surface. **F 335,** F05

overconsolidated soil deposit—a soil deposit that has been subjected to an effective pressure greater than the present overburden pressure. **D 653,** D18

overconsolidation ratio, OCR—the ratio of preconsolidation vertical stress to the current effective overburden stress. **D 653,** D18

overcure, *n*—a state of vulcanization cure beyond the state of optimum cure. **D 1566,** D11

overdamped-well response—characterized by the water level returning to the static level in an approximately exponential manner following a sudden change in water level (see for comparison **underdamped well response**). **D 653,** D18

overdrawn structure—a condition, sometimes found to exist in very heavily cold-worked rods or wires. (**cupping**). Cupping is overdrawn material, but the converse is not true. **E 7,** E04

overdrilling—the process of drilling out a well casing and any material placed in the annular space. **D 653,** D18

overdry—the state of a cellulosic material that has been dried to constant mass at a temperature of 100 to 105°C. **D 1695,** D01

overemulsification—excessive emulsifier dwell time which results in the removal of penetrants from some discontinuities. **E 1316,** E07

over-flow drain—a drain for eliminating the excess foam and starch created during the cooking process. **F 1827,** F26

overglaze decoration—See *overglaze decoration* under **decoration.** **C 242,** C21

overhang—See **lip.** **F 592,** F16

overhead, *n—in a bar code system*, the fixed number of symbol characters required for start, stop, and checking in addition to data carrying characters. As an example, a symbol requiring a start, stop and single check character contains three characters of overhead. **F 1294,** F05

overhead slide gate—any horizontal slide gate supported only from above (see Specification F 1184). **F 552,** F14

overheating, *n*—heating a steel object to such a high temperature that excessive grain growth occurs. **A 941,** A01

overheating—(*1*) *in ferrous alloys*, heating to an excessively high temperature such that the properties/structure undergo modification. The resulting structure is very coarse-grained. Unlike burning, it may be possible to restore the original properties/structure by further heat treatment or mechanical working, or a combination thereof.

(*2*) *in aluminum alloys*, overheating produces structures that show areas of resolidified eutectic or other evidence that indicates the metal has been heated within the melting range. **E 7,** E04

over-laminate, *n—in reference to code printing*, a coating or material added to protect a printed symbol. **F 1294,** F05

overland flow—rainfall runoff from a surface containing concentrated flow no larger than rill flow. **D 4410,** D19

overlength staple fibers, *n*—man-made staple fibers that are at least 10 % longer than the nominal or average cut length. **D 123,** D13

overload recovery time—an interval of nonlinear operation of an instrument caused by a signal with amplitude in excess of the instrument's linear operating range. **E 1316,** E07

overpacking, *n*—(*1*) **packaging** that exceeds minimum requirements, (*2*) additional packaging used to increase protection. **D 996,** D10

overpress—an imperfection; projecting excess glass resulting from imperfect closing of mold joints. **C 162,** C14

overpressed metal connector plate—metal connector plate with teeth, fully penetrating wood member, with tooth side of plate pressed more than one half of plate thickness below surface of wood member; in contrast to underpressed metal connector plate, the surface of which is not in contact with the surface of the wood members. **E 1807,** E06

overrun—the relation between the liquid volume of the cream mix, and that of the dispensed aerated product. **D 3064,** D10

overshoot, *n* (θ_n)—the amplitude of a deflection of a wind vane as it oscillates about θ_B after release from an initial displacement.

overshoot ratio, n (Ω)—the ratio of two successive overshoots of a wind vane, as expressed by the following equation:

$$\Omega = \frac{\theta_{(n+1)}}{\theta_n}$$

where θ_n and $\theta_{(n+1)}$= n and n + 1 overshoots, respectively. **D 1356,** D22

oversize powder—particles coarser than the maximum permitted by a given particle size specification. **B 243,** B09

overspray—the slip from the spray gun not deposited on the ware. Also, spray application of a light coat of slip to an unfired porcelain enamel. **C 286,** B08

overspray—*in protective coatings*, any material not deposited on the surface being coated. **D 4538,** D33

overstriking, *v*—printing characters over each other, in some cases this is desired, in others it is a defect. **F 1457,** F05

overt—obvious. **F 1156,** F12

over-tone—see **mass color.** **D 16,** D01

overtoning, *n*—any of the conditions occurring in the developing unit when the toner concentration is too high. **F 1457,** F05

overturning moment, [*FL*], *n—of a tire*, the component of a tire moment vector tending to rotate a tire about the *X*′-axis, positive clockwise when looking in the positive direction of the *X*′-axis.

F 538, F09

overvoltage—the irreversible excess of potential required for an electrochemical reaction to proceed actively at a specified electrode, over and above the reversible potential characteristics of that reaction. **B 374,** B08

overvoltage—the change in potential of an electrode from its equilibrium or steady state value when current is applied. **G 15,** G01

overwashing—too long or too vigorous washing, or both, which results in removal of penetrants from some discontinuities. **E 1316,** E07

Owens process—{archaic} a bottle-making process in which the blank or parison mold is filled by suction and subsequently blown. **C 162,** C14

owner, *n—in pipe laying,* the person, firm corporation, or government agency entering into contract with the contractor for the installation of asbestos-cement pipe and accessories. **C 460,** C17

owner—the public agency or authority, group, corporation, partnership, or individual that specifies products or services for use on a project that it presently or eventually will own or administrate. **C 822,** C13

owner, *n—in pipe laying,* the person, firm, corporation, or government agency entering into contract with the contractor for the installation of pipe and accessories. **C 1154,** C17

owner, *n—in pipe laying,* the person, firm corporation, or government agency entering into contract with the contractor for the installation of asbestos-cement pipe and accessories. **D 2946,** C17

owner—see **property owner**. **E 1605,** E06

oxbow lake—a lake formed when a meander bend is cut off and its ends filled in, thus isolating the lake from the main channel of the stream. **D 4410,** D19

oxidation—a reaction in which electrons are removed from a reactant. Sometimes, more specifically the combination of a reactant with oxygen. **B 374,** B08

oxidation—loss of electrons by a constituent of a chemical reaction. (Also refers to the corrosion of a metal that is exposed to an oxidizing gas at elevated temperatures.) **G 15,** G01

oxidation grain size—*(1)* grain size determined by holding a specimen at a suitably elevated temperature in a mildly oxidizing atmosphere. The specimen is polished before oxidation and etched afterwards.

(*2*) refers to the method involving heating of polished steel specimen to a specified temperature, followed by quenching and repolishing. The grain boundaries are sharply defined by the presence of iron oxide. Grain size is expressed as an ASTM Number. **E 7,** E04

oxidation inhibitor—any substance added to an insulating fluid to improve its resistance to deleterious attack in an oxidizing environment. For example, 2,6-ditertiary-butyl paracresol is sometimes added to petroleum insulating oil to improve its oxidation stability. **D 2864,** D27

oxidation life—a measure of the ability of an insulating liquid to resist oxidation under a prescribed set of conditions. Often the changes in color, neutralization number, interfacial tension, initial appearance of sludge, or rate of sludge formation are the criteria used to measure this quality. **D 2864,** D27

oxidation of carbon—the chemical combination of carbon with oxygen or oxygen-containing compounds. **C 709,** D02

oxidation-reduction potential—the electromotive force developed by a noble metal electrode immersed in the water, referred to the standard hydrogen electrode. **D 1129,** D19

oxidation-reduction potential—the electromotive force developed by a noble metal electrode immersed in the water, referred to the standard hydrogen electrode. **D 6161,** D19

oxidation stability— see **oxidation life.** **D 2864,** D27

oxidative induction time, OIT—an accelerated aging index for relative resistance to oxidative decomposition. **E 1142,** E37

oxidatively degradable plastic, *n*—a degradable plastic in which the degradation results from oxidation. **D 883,** D20

oxidatively degradable plastic, *n*—See **degradable plastic.** **D 883,** D20

oxide defect—an area of missing oxide on the back surface of back-sealed wafers discernible to the unaided eye. **F 1241,** F01

oxide film replica—See **replica**. **E 7,** E04

oxide network—continuous or discontinuous oxides that follow prior particle boundaries. **B 243,** B09

oxide type inclusions—oxide compounds occurring as non-metallic inclusions in metals usually as a result of deoxidizing additions. In wrought products, that is, steel, they may occur as a "stinger" formation composed of distinct granular or crystalline appearing particles. **E 7,** E04

oxidized—darkened or dulled by surface treatment or by the natural oxidizing of metal. **F 547,** F16

oxidizing agent—a compound that causes oxidation, thereby itself becoming reduced. **B 374,** B08

oxidizing agent—a batch ingredient that raises the state of oxidation of the melt. **C 162,** C14

oxycelluloses—water-insoluble substances formed by the action of oxidizing agents on cellulose. The chemical nature of oxycelluloses varies with the oxidant used, and the type is indicated by attaching the name of the oxidant adjectivally to "oxycelluloses" as in "hypochlorite oxycelluloses." Any such mixture is "an oxycellulose" and the word should not be used in the singular without either the definite or indefinite article. In many respects, the phrases "oxidized cellulose" or "partially oxidized cellulose" are preferable. **D 1695,** D01

oxygenate, *n*—an oxygen-containing ashless organic compound, such as an alcohol or ether, which may be used as a fuel or fuel supplement. **D 4175,** D02

oxygen consumption principle, *n*—the expression of the relationship between the mass of oxygen consumed during combustion and the heat released. **E 176,** E05

oxygen consumption principle, *n*—the expression of the relationship between the mass of oxygen consumed during combustion and the heat released. **E 176,** E05

oxygen demand—the amount of oxygen required, under the specified test conditions for the oxidation of waterborne organic and inorganic matter. **D 1129,** D19

oxygen demand—the amount of oxygen required for the oxidation of waterborne organic and inorganic matter under the specified test conditions. **D 6161,** D19

oxygen depletion, *n—in a fire,* reduction of oxygen (O_2) content of an atmosphere as a result of combustion. **E 176,** E05

oxygen-enriched—a fluid mixture (gas or liquid) that contains more than 25 mol % oxygen. **G 126,** G04

oxygen-enriched atmosphere—a fluid mixture (gas or liquid) that contains more than 25 mol % oxygen. **G 126,** G04

oxygen-free copper, extra-low phosphorus— oxygen-free copper containing 0.001 to 0.005 % phosphorus. The copper is not readily susceptible to hydrogen embrittlement, as determined in Specification B 379. The copper in the annealed condition has a minimum conductivity of 98.16 % IACS. **B 846,** B05

oxygen-free copper, low phosphorus—oxygen-free copper containing 0.005 to 0.12 % phosphorus. The copper is not susceptible to hydrogen embrittlement, as determined in Specification B 379. The copper in the annealed condition has a minimum conductivity of 90 % IACS.

—deoxidized, phosphorus-arsenical copper
—arsenical, tough-pitch copper
—silver-bearing copper
—sulfur-bearing copper
—deoxidized phosphorus-tellurium copper
—zirconium-bearing copper
—tellurium-bearing copper

NOTE—Coppers listed contain the designated element or elements in amounts as agreed upon between the manufacturer or supplier and the purchaser. **B 846,** B05

oxygen-free electronic copper—high-purity, high-conductivity oxygen-free copper normally intended for electronic applications. The copper has high resistance to hydrogen embrittlement, as determined in Specification B 170. The copper in the annealed condition has a minimum electrical conductivity of 101 % IACS. **B 846,** B05

oxygen number, *n—for plumage,* a measure of the degree of cleanliness; the amount of oxidizable water soluble and fine suspended matter present in a water extract. **D 123,** D13

oxygen/uranium ratio—the mole ratio of oxygen to uranium in a sample of uranium oxide. **C 859,** C26

ozone cutting and checking—cracks produced by ozone in a material under mechanical stress. **F 819,** F18

P

"P"—professionally clean only, an alternate to the *"S"* dry cleaning term although not in customary usage. **D 5253,** D13

P—effective pair production content. *P* is the percent background exposure caused by pair production in 2 mm of lead. **E 1316,** E07

PA—abbreviation for **program attention.** **F 1457,** F05

PAC—powdered activated carbon or poly-aluminum chloride. **D 6161,** D19

pack—*(1)* the quantity of ware packed.

(2) the ratio of packed ware to theoretical. **C 162,** C14

pack, *n*—the final configuration of material with necessary protection afforded for the distribution system.

blister pack, n—encapsulation of an item inside a preformed semi-rigid transparent, or opaque plastic bubble that is affixed to a semi-rigid backing.

exterior pack— a **container, bundle,** or assembly that is sufficient by reason of material, design, and construction to protect its contents during shipment or storage.

intermediate pack—in packaging, a **wrap, box,** *or* **bundle** *that contains two or more unit packs of identical items, and that is in turn enclosed by an outer* **shipping container.** **D 996,** D10

pack, *v*—to place material into a container for handling, storage, and transportation. **D 996,** D10

package—*(1) n,* a container providing means of protection and handling to a product; (2) *v,* to design, manufacture, or provide protection to a product; (3) *when referring to a fiber container,* a container not necessarily complying with all the requirements for a "box" in accordance with the *Uniform Freight Classification* and *National Motor Freight Classification*; (*4*) when referring to a fiber container, one of the special authorized containers described in detail in the Classifications in the section titled" Authorized Packages or Shipping Containers," of the above Freight Classifications.

aerosol package— a package that contains in addition to its contents, a gas under pressure. It is fitted with a valve which enables the contents to be dispensed.

consumer package—a primary container (which may be enclosed in a secondary container) designed to contain, store, and protect from the point of manufacture to the point of use, a product intended for household or individual use. (See *unit package* under **package.**)

industrial package, n—a package used for the transportation or storage of commodities, the contents of which are not meant for retail sale without being repackaged.

intermediate package—a wrap, box, or bundle (that is, a container) that contains two or more unit packages of identical items (also called a *secondary package*).

primary package—a container in direct contact with and enclosing the product along with any required protective material(s).

secondary package, n—a container enclosing one or more primary packages along with any required protective material(s).

unit package— the first tie, wrap, or container applied to a single item, a quantity of the same item, a set, or an item with all its component parts, that constitutes a complete and identifiable package containing the unit of issue of a product for ultimate use (also called a *primary package*). **D 996,** D10

package, *v*—to prepare goods for distribution by enclosing in a container or covering. (See Terminology D 996.) **D 1968,** D06

package—a container providing protection to a product during distribution, storage, retailing, and use. **F 1479,** F02

packaged building—See **building.** **E 631,** E06

package integrity—the physical capability of a given package to protect its contents with the desired level of protection over a defined period of service; for example, as a barrier to physical, microbiological, or chemical challenges. **F 1327,** F02

package material, *n*—each separate and distinct material which by itself or in combination with other materials, forms a package component. **D 996,** D10

package source reduction, *n*—the reduction of the weight or volume of the packaging materials used in a package containing an equivalent product. **D 5488,** D10

package testing:

bursting strength— the strength of a material in pounds-force per square inch (or kilopascals), measured by the Cady or Mullen tester. (See **Cady test,** and **Mullen test.**)

bursting strength test—a test for measuring the resistance of a material to bursting measured in pounds per square inch (or kilopascals). (See **Mullen tester.**)

Cady test—a test for **bursting strength** made on a specific type of machine.

caliper—*(1) n,* thickness (as related to **paperboard**) of a sheet measured under specified procedures expressed in thousandths of an inch (or millimetres). Thousandths of an inch are sometimes termed "points"; (2) *n,* the precision instrument used in the paperboard industry to measure thickness; *(3) v,* to measure with a caliper.

compression test— a test for measuring resistance to external compressive forces. (See **loading.**)

drop test—a test for measuring the durability of an article, or the protection, or the retention properties of a container, or all three by subjecting the packaged product to a free fall from predetermined heights onto a surface with prescribed characteristics.

Elmendorf test— a test for measuring the tearing resistance of tape, paperboard, or other materials.

incline impact (Conbur) test—a test for determining principally the resistance of a packaged product to damage from impacts.

Mullen test—a test for bursting strength made on a specific type of machine.

puncture test— *(1)* a test for measuring the resistance of board to puncture; (2) the strength of a material, primarily involving tear and stiffness, expressed in units as measured by the puncture tester.

revolving drum test—a test for measuring the protection to contents, or the retention properties of a container, or both, by subjecting the packaged products to rough handling in a standard revolving drum.

testing machine compression—in package testing, an imposed force in motion applied at a predetermined rate and condition until a predetermined end point is achieved. **D 996,** D10

package unit, *n*—*in pipe laying*, several units of asbestos-cement pipe bound together for the purpose of being transported and, when desired, unloaded at the job site. **C 460,** C17

package unit, *n*— *in pipe laying*, several units of pipe bound together for the purpose of being transported and, when desired, unloaded at the job site. **C 1154,** C17

package unit, *n*—*in pipe laying*, several units of asbestos-cement pipe bound together for the purpose of being transported and, when desired, unloaded at the job site. **D 2946,** C17

packaging, *n*—*(1)* the technique of preparing goods for distribution;

(*2*) the design criteria, processes, and procedures used to protect material from deterioration and damage from the time manufacturing is completed until ultimate use or disposal;

(*3*) the processes and procedures used to protect an item in a unit package.

bulk packaging— (*1*) a method of containing loose or granular materials for shipping or storage, (*2*) a method of assembling many items into a **container** for shipment or storage.

child-resistant packaging—packaging designed or constructed to be significantly difficult for most children under 5 years of age to open, or to obtain a toxic or harmful amount of the contents within a reasonable time; and not difficult for normal adults to use properly.

commercial packaging—the methods and materials used by a supplier to satisfy the requirements of the distribution system.

foam-in-place packaging— See **cushioning material.**

industrial packaging—packaging of partially manufactured or finished goods for distribution from manufacturer to manufacturer, and to users other than retail customers.

molded pulp—a type of **packaging** which has been formed from wet paper pulp to predetermined shape to generally fit one or more individual items. Used as apple trays, egg flats, egg carton. (See **molded shape.**)

skin packaging— a process whereby a product placed on a backing material is covered by a closely fitting thermoplastic film bonded to the backing material.

vacuum packaging— the technique of **packaging** in **containers** from which substantially all air has been removed prior to sealing. **D 996,** D10

packaging, *n*—(See Terminology D 996.) **D 1968,** D06

packaging—See Terminology D 996. **D 5077,** D10

packaging—See Practices A 700for Packaging, Marking, and Loading Methods for Steel Products for Domestic Shipment. **F 547,** F16

pack cementation—*in diffusion coatings*, a coating process similar to pack carburizing that involves packing the parts to be coated into a retort with a carefully blended mixture of powders and then exposing the pack under hermetically sealed conditions or in an inert atmosphere to elevated temperatures. **B 374,** B08

packed distillation column—a column or tube constructed with internals of ceramic, steel, or fiberglass-type materials to separate one or more volatile liquids by distillation. **E 1705,** E48

packer—*in grouting*, a device inserted into a hole in which grout or water is to be injected which acts to prevent return of the grout or water around the injection pipe; usually an expandable device actuated mechanically, hydraulically, or pneumatically. **D 653,** D18

packer hides—hides from meatpacking houses. **D 1517,** D31

packer (monitoring wells)—a transient or dedicated device placed in a well that isolates or seals a portion of the well, well annulus, or borehole at a specific level. **D 653,** D18

pack-hardened, pack-carburized—See **case-hardened.** **F 547,** F16

packing, *n*—the selection or construction of the **shipping container** and assembling of items or **packages** therein, including any necessary **blocking, bracing,** or **cushioning,** weatherproofing, exterior **strapping,** and **marking** of shipping container for identification of contents. (Compare **packaging**).

absorbent packing, n—a material included within a **package** to soak up liquids resulting from leakage or liquefaction of contents.

interior (inner) packing—material or parts used in supporting, positioning, or cushioning an item in its outer shipping container. (See also **buffer, cushioning material, divider, liner, separator.**)

saddle, n— a form of **interior packing** shaped to fit the article in order to increase the area of contact between the **blocking** and the article.

shell, (when referring to a *fiberboard packaging), n*—a specific form of inner **packing** consisting of a sheet of **corrugated** or **solid fiberboard** scored and folded to form a joined or unjoined **tube** open at both ends. **D 996,** D10

packing leather—see **hydraulic leathers.** **D 1517,** D31

packing material—any material in which compacts are embedded during the presintering or sintering operation. **B 243,** B09

packing (mechanical), *n*—a deformable material used to prevent or control the passage of matter between surfaces that move in relation to each other. **D 1566,** D11

pac leather—highly water-resistant leather used by lumbermen, hunters, and others for outdoor wear. **D 1517,** D31

pad, *n*—(when referring to a **fiberboard container**) a **corrugated** or **solid fiberboard** sheet or other authorized material used for extra protection or for separating tiers or layers of articles when packed for shipment. (Compare **buffer, divider, separator**). See **cushioning material.** **D 996,** D10

pad character, *n*—See **filler character.** **F 1294,** F05

paddle mixer—a mixer consisting essentially of a trough within which mixing paddles revolve about the horizontal axis, or a pan within which mixing blades revolve about the vertical axis. **D 653,** D18

paddling, puddling—{archaic} the rough shaping of a piece of glass in a furnace by means of paddles or tools preparatory to the pressing operation for making optical glass blanks. **C 162,** C14

page—a form on which many lines of data may be entered for reading by a page reader. Pages are larger in size than **documents**. **F 149,** F05

page reader—an electronic machine capable of reading full pages of printed data. **F 149,** F05

pail, *n*—a **container,** usually cylindrical, with a handle, available in sizes of about 1 to 12 gal (3 to 50 L). **D 996,** D10

paint, *n*—*in general*, a pigmented coating. **E 1605,** E06

paint brush—a paint application tool consisting of a flexible brushing part composed of long filamentary material (brushing material) bound to a handle. **D 16,** D01

paint brush bristle—hair of the swine (for example: pig, hog, boar), used in brushing material. **D 16,** D01

paint brush ferrule—outer band that joins the brushing material to the handle. **D 16,** D01

paint brush filament—a synthetic polymer extrusion used in brushing material. **D 16,** D01

paint brush head—brush without the handle. **D 16,** D01

paint brush length clear—also called "length out", the exposed length of the brushing material from the ferrule to the tip end. **D 16,** D01

paint brush thickness—measurement of the brushing material across the narrow opening of the ferrule. **D 16,** D01

painted—coated with paint or plastic by dipping or barrel-tumbling. **F 547,** F16

paint *n, general*— a pigmented coating. See **coating.** **D 16,** D01

paint *n, specific*— a classification sometimes employed to distinguish pigmented drying oil coatings ("paints") from synthetic enamels and lacquers.

emulsion paint— a paint, the vehicle of which is an emulsion of binder in water. The binder may be oil, oleoresinous varnish, resin, or other emulsifiable binder.

latex paint—a paint containing a stable aqueous dispersion of synthetic resin, produced by emulsion polymerization, as the principal constituent of the binder. Modifying resins may also be present.

oil paint—a paint that contains drying oil or oil varnish as the basic vehicle ingredient.

paste paint—a paint in which the pigment is sufficiently concentrated to permit a substantial reduction with vehicle before use.

water paint—a paint, the vehicle of which is a water emulsion, water dispersion, or ingredients that react chemically with water. **D 16,** D01

paint pad—a paint application tool consisting of short filamentary material usually bonded to a flat, resilient backing connected to a handle designed to apply paint by a wiping action. **D 16,** D01

paint residue—dry or chemically softened paint that remains on a surface after initial paint removal. **E 1605,** E06

paint roller—a complete paint application tool consisting of a roller frame and a roller cover designed to apply paint by a rolling action. **D 16,** D01

paint roller core—a structural tube that forms the base of the roller cover to which paint applicating material is attached. **D 16,** D01

paint roller cover—a tubular sleeve consisting of a paint applicating material secured to a core. **D 16,** D01

paint roller cover pile height—also called" nap length"; the length of the roller cover paint applicating material from pile backing to pile face. **D 16,** D01

paint roller frame—a frame and handle assembly designed to hold a roller cover. **D 16,** D01

paint scrapings—waste material consisting of paint removed by wet or dry scraping during an abatement process. **E 1605,** E06

paint *vb*—to apply a thin layer of a coating to a substrate by brush, spray, roller, immersion, or any other suitable means. **D 16,** D01

paired comparison, *n*—a method in which stimuli are presented in pairs for comparison on the basis of some defined criterion. **E 253,** E18

pair production—a process that corresponds to the creation of an electron-positron pair by the interaction of a gamma ray in the field of a nucleus. The energy of the gamma ray must exceed 1.022 MeV (twice the rest mass of an electron). **E 170,** E10

pair production—the process whereby a gamma photon with energy greater than 1.02 MeV is converted directly into matter in the form of an electron-positron pair. Subsequent annihilation of the positron results in the production of two 0.511 MeV gamma photons. **E 1316,** E07

pair production—the process whereby a gamma photon with energy greater than 1.02 MeV is converted directly into matter in the form of an electron-positron pair. Subsequent annihilation of the positron results in the production of two 0.511 MeV gamma photons. **E 1316,** E07

pajamas, *n*—a two piece ensemble consisting of a top and a bottom generally used for, but not limited to, nightwear. **D 123,** D13

palatable, *adj*—sufficiently pleasant to be consumed. **E 253,** E18

palladium barrier leak detector—a leak detector using hydrogen as the tracer gas and using the principle of hydrogen diffusing through a hot palladium barrier into an evacuated vacuum gage. **E 1316,** E07

pallet, *n*—horizontal platform device used as a base for assembling, storing, handling, and transporting materials and products in a unit load. (Compare **skid, dolly.**)

expendable pallet— a **pallet** intended to be discarded after a single use.

reusable pallet— a **pallet** intended for more than one shipment (reusable for economic life).

stringer, n— a wooden member to which the face or faces of a **pallet** or the deck of a platform are fastened and running at right angles to the members making up the face or the deck. **D 996,** D10

palletized load—See **load.** **D 996,** D10

palletized unit load— See **load.** **D 996,** D10

pallet nail—bright, stiff-stock or hardened-steel, helically threaded (with medium lead angle) or annularly threaded, 1½ by 0.105 to 4 by 0.177-in. nails with smooth or checkered flat 9⁄32 to 7⁄16-in. head and medium or blunt diamond or blunt chisel point. **F 547,** F16

panchromatic—sensitive, as a film or plate, to light of all hues. **E 7,** E04

pane, n—See **lite.** **C 717,** C24

pane—See **windows and doors.** **E 631,** E06

pane, n—use **lite.** **E 1605,** E06

panel—cut stone with face dimensions large in relation to its thickness, for placement in a building structure or frame assembly. **C 119,** C18

panel—a sheet of plywood of any construction. **D 1038,** D07

panel, *n*—a group of assessors chosen to participate in a sensory test. **E 253,** E18

panel, *n*—in a building, (*1*) a portion of a surface flush with or recessed from, or sunk below the surrounding area, sometimes set off by distinct molding or other decorative measure. (*2*) a usually flat and rectangular piece of construction material made to form part of a surface (as of a **wall,** ceiling, or **floor**). (See also **railing systems** and **windows and doors.**) **E 631,** E06

panel, *n*—a flat infill element between a top rail, bottom rail, and posts. **E 1481,** E06

panel clamp—a two-piece clamp with carriage bolt and nut, designed to secure prefabricated panels on free standing enclosures. **F 552,** F14

panelist, *n*—See **assessor.** **E 253,** E18

panelized construction—See **building constructions.** **E 631,** E06

panel member, *n*—See **assessor.** **E 253,** E18

panel, n—(*1*) a **face** or *side* of a **box** or **crate,** (*2*) any flat area between folds, or edges of a **box, carton,** or interior part. **D 996,** D10

pan head—flat head with rim having rounded upper edge and squared lower edge, usually slotted. **F 547,** F16

pan mixer—a mixer comprised of a horizontal pan or drum in which mixing is accomplished by means of the rotating pan of fixed or rotating paddles, or both; rotation is about a vertical axis. **D 653,** D18

pans—containers used to hold the food product in the oven capacity: (*1*) a full-size bake pan is nominally 18 by 26 by 1 in. (457 by 660 by 25 mm), (*2*) a half size bake pan is nominally 18 by 13 by 1 in. (457 by 330 by 25 mm), and (*3*) a steam pan is nominally 12 by 20 by 2.5 in. (305 by 508 by 64 mm). **F 1827,** F26

paper:

creped duplex paper—a double sheet composed of two layers of **creped paper** united with asphalt, latex, or other adhesives, generally with 20, 25, or 30-lb paper and 20, 25, or 30 lb of laminating material.

creped paper— paper, usually **kraft** (bleached sulfate) in various basis weights, that has been machined to resemble crepe and thus has stretch. It is used for wrapping purposes and for bag and barrel **liners.**

water-resistant paper—paper that is treated by the addition of materials to provide a degree of resistance to damage or deterioration by water in liquid form.

wet-strength paper—paper that has been treated with chemical additives to aid in the retention of resistance to bursting, tearing, or rupturing when wet. **D 996,** D10

paper acidity, *n*—the extent to which water-soluble materials in paper alter the hydrogen-hydroxyl ion equilibrium of reagent water causing an excess of hydrogen ions, as measured by titrating with a standardized solution of alkali under specified conditions. **D 1968,** D06

paper alkalinity, *n*—the extent to which water-soluble materials in paper alter the hydrogen-hydroxyl ion equilibrium of reagent water causing an excess of hydroxyl ions, as measured by titrating with a standardized solution of acid under specified conditions. **D 1968,** D06

paper bleed, *n*—an optical phenomenon that causes bars to appear larger and spaces to appear narrower than actually printed because of the scattering of incident light rays within the media. **F 1294,** F05

paperboard, *n*—one of the two broad subdivisions of paper (general term, the other being paper as specific term). Paperboard is, in general, heavier and thicker than paper and is constructed primarily for strength properties such as stiffness, tearing resistance and the like. Sheets with thickness greater than 0.010 in. (0.25 mm) (with some exceptions) fall in this class. (See also **containerboard.**)

boxboard, n— a general term designating the grades of **paperboard** used for fabrication of folding and set-up **boxes (cartons).** Customarily shipped in sheets.

chipboard, n— a **paperboard** generally made from reclaimed paper stock.

finish, n— in packaging, a term descriptive of the surface of **paperboard** depending on the final machining in manufacture:

(*1*) *dry finish*—not dampened before going through the calender rolls. The surface is not as dense nor as smooth as water-finished board;

(*2*) *water finish*—dampened on one (or both) sides in the calendering opening to provide a relatively hard and glossy finish.

kraft, n— a term (derived from a German word meaning strength) applied to pulp, paper, or **paperboard** produced from virgin wood fibers by the sulphate process.

linerboard, n—(*1*) **paperboard** used for the flat facings in **corrugated fiberboard;** also paperboard used as the outer plies of solid fiberboard, (*2*) **containerboard** made of blends of re-worked paper fibers usually made on cylinder machines but can be made on Fourdrinier or Inverform machines. It is sometimes called jute—a misnomer, since jute fibers are no longer used (See **facing.**) **D 996,** D10

paperboard—one of the two broad subdivisions of paper (general term), the other being paper (specific term). The distinction between paperboard and paper is not sharp but, broadly speaking, paperboard is heavier in basis weight, thicker and more rigid than paper. In general, all sheets 12 points (0.012 in., 3.0 mm) or more in thickness are classified as paperboard. There are a number of exceptions based upon traditional nomenclature. For example, blotting paper, felts, and drawing paper in excess of 12 points are classified as paper while corrugating medium, chipboard, and linerboard less than 12 points are classified as paperboard. Paperboard is made from a wide variety of furnishes on a number of types of machines, principally cylinder and fourdrinier. The board classes are: (*1*) containerboard, which is used for corrugated boxes; (*2*) boxboard, which is principally used to make cartons; and (*3*) all other paperboard. **D 1968,** D06

paperboard blank—a flat piece of **paperboard** that has been cut to size, die-cut, corner cut, scored, printed, or otherwise prepared and ready to be formed or folded and joined, or the flat pieces of parts thereof. **D 996,** D10

paperboard step—a term applied to trays, shells, or platforms positioned inside a **container** to raise one portion of contents higher than others. It may be attached to a lid for the purpose of holding certain articles positioned when the lid is in closed position. **D 996,** D10

paperboard stop—pieces of **boxboard** attached to the inside surface of the ends of a set-up paper box lid to hold the base in proper position when the height of the base is less than the height of the lid. **D 996,** D10

paperboard tube—See **tube.** **D 996,** D10

paper grain—the paper machine direction of paper. **F 149,** F05

paper grain direction—see **paper machine direction**. **F 149,** F05

paper linting, *n*—the picking off of loosely bonded paper surface fibers, or dust, or both, that can accumulate on an offset plate or blanket interfering with print quality. **D 6488,** D01

paper machine direction—the direction of paper grain parallel with the direction of movement on the paper machine. It is also called *grain direction*. The direction at right angles to the paper machine direction is called the *cross-machine direction*, or simply, *cross direction*. **F 149,** F05

papermaker's felt, *n*—a fabric, made from wool or man-made fibers or mixtures of both, fabricated as an endless belt for use on a paper making machine. (See also *felt* and *needled felt.*) **D 123,** D13

paper manifold set—a business form prepared by interleaving one-time carbon paper and a receiving paper to give the number of parts required for comparison. A standard test set is 8 ½ by 11 in. (21.59 by 27.94 cm) and most frequently contains six plies six bond sheets, usually 12 lb (17 by 22-500) (45 g/m^2) and five carbons, usually 8 lb (20 by 30-500) (18.75 g/m^2). (F 497) **F 221,** F05

paper manufacturing process, *n*—an operation that begins with the pulping of fibrous and nonfibrous raw materials and ends after the first slitter/winder, with the cutting and trimming of the reel into smaller rolls. In an operation in which the finished product is sheeted directly off the machine, the production of rough sheets constitutes the end of the process. In an operation that involves supercalendaring, the end of the process is at the slitter/winder following the supercalendar. In an operation that involves off-machine coating, the process ends at the slitter/winder following the coater or the supercalendar associated with the coater. **D 1968,** D06

paper mill sludge, *n*—a slurry of solid residues from a paper mill's waste water treatment system containing fibers and other material. **D 1968,** D06

paper multiwall-sack— See **bag.** **D 996,** D10

paper, OCR—paper used in **OCR** systems; refer to ANSI X3.62-79. **F 149,** F05

paper product, *n*—any item manufactured from paper or paperboard. **D 1968,** D06

paper smoothness—the degree of irregularity of the surface of paper determined by the measurement of the flow of air between the paper surface and a plain surface under specified clamping and air pressure. The resistance to air flow increases as the paper goes from rough to smooth. With an instrument measuring the rate of flow, as the paper goes from rough to smooth, the number rating goes down. With an instrument measuring the time for a given volume to flow, as the paper goes from rough to smooth, the number rating increases. **F 149,** F05

paper weight—see **basis weight.** **F 149,** F05

paraffinic oil, *n*—an hydrocarbon process oil, most or all of which is composed of alkanes **D 1566,** D11

paraffinic oil—a term applied to mineral insulating oil derived from crudes having substantial contents of naturally occurring *n*-paraffins (wax). Such an oil must be dewaxed and may need the addition of a pour depressant in order to exhibit a low pour point. **D 2864,** D27

paraffinic oil—a petroleum oil (derived from paraffin crude oil) whose paraffinic carbon type content is typically greater than 60 %. **E 1519,** E35

parallax factor (used in electron stereomicroscopy)—parallax factor (f) is the rate of change of elevation with respect to parallax:

f = $dx/dy = \csc \sigma / 2M \times 10^{3} \mu m/mm$
x = elevation
y = parallax, the apparent lateral displacement of an image point.

where σ is the stereo angle and M is the final magnification of the image. **E 7,** E04

parallel communication, *n*—the transmission system where code for a character is sent over a set of channels, all bits at once, one bit per channel. (See **serial communication**.) **F 1457,** F05

parallel-core conductor—see **concentric-lay conductor.** **B 354,** B01

parallel laminate—a laminate in which all the layers of material are oriented approximately parallel with respect to the grain or strongest direction in tension. (See also **cross laminate**) **D 883,** D20

parallel laminate, n— See *parallel laminate* under **laminate,** *n.* **D 907,** D14

parallel laminate—a laminate in which all the layers of material are oriented approximately parallel with respect to the grain or strongest direction in tension. (See also **cross laminate**.) (D-20) **F 412,** F17

paramagnetic material—a material having a relative permeability which is slightly greater than unity, and which is practically independent of the magnetizing force. **A 340,** A06

paramagnetic material—a material that has a relative permeability slightly greater than unity and that is practically independent of the magnetizing force. **E 1316,** E07

parameric, *adj*—pertaining to specimens having different spectrophotometric curves that produce approximately the same color sensation under the same illuminating and viewing conditions. (See also **metameric**.) **E 284,** E12

paramerism, *n*—phenomenon in which specimens having different spectrophotometric curves produce approximately the same color sensation under the same illuminating and viewing conditions. (See also **metamerism**.) **E 284,** E12

paramers, *n*—specimens having different spectrophotometric curves that produce approximately the same color sensation under the same illuminating and viewing conditions. (See also **metamers**.) **E 284,** E12

parameter, *n—in statistics*, a variable that describes a characteristic of a population or mathematical model. **D 123,** D13

parameter—*in statistics*, a constant (usually to be estimated) defining some property of the population frequency distribution, such as the population median or the population standard deviation. **E 1823,** E08

parameter (in crystals)—See **lattice parameter**. **E 7,** E04

parametric, *adj*—a term referring to a statistical technique that assumes the nature of the underlying frequency distribution is known. **D 123,** D13

parametric system—a system that uses the response of a sensor to produce an output that is correlatable with the selected parameter. **D 1129,** D19

parasite—an organism living in or on another living organism, obtaining from it part or all of its nutriment. **E 609,** E35

para-xylene—1,4-dimethylbenzene (C_8H_{10}) mol weight 106.16; clear, colorless, flammable liquid; freezing point, 13.26°C; boiling point, 137 to 138°C. **D 4790,** D16

parchment—tanned sheepskins. Vellum is practically the same as parchment except it is made from calfskins. In addition to its use as "parchment" for diplomas and records, it is also utilized for banjo and drum heads, lampshades, etc. **D 1517,** D31

parent coil or plate, *n*—coil of sheet or a plate that has been processed to final temper as a single unit and may subsequently be cut into two or more smaller coils or into individual sheets or smaller plates to provide the required width and length. **B 881,** B07

parent material—material from which a soil has been derived. **D 653,** D18

parfocal eyepiece—see **eyepiece, parfocal.** **E 175,** E41

parison—a preliminary shape or blank from which a glass article is to be formed. **C 162,** C14

parison, *n*—the shaped plastic mass, generally in the form of a tube, used in blow molding. (ISO) **D 883,** D20

parison, *n*—the shaped plastic mass, generally in the form of a tube, used in blow molding. (D-20, ISO) **F 412,** F17

parison mold—See **blank mold**. **C 162,** C14

parity, *n*—a system for encoding characters as 'odd' (having an odd number of binary ones in their structure) or 'even' (having an even number of binary ones in their structure), used as a self-checking mechanism in bar codes. **F 1294,** F05

parity, *n*—an error checking system in which an extra bit may be added to the character data bits to show whether the number of bits in the character transmitted is even or odd. **F 1457,** F05

parity error, *n*—a parity error occurs when the receiving device counts the number of bits in the received character, including the parity bit, and finds that the number does not agree with the predetermined odd/even requirement. **F 1457,** F05

parkerized—chemically treated to provide iron and steel with dark corrosion-resistant protective coating by boiling in solution of manganese dihydrogen phosphate and subsequently applying coating of paraffin oil. **F 547,** F16

parody—the intentional mockery of a product through imitation (often resulting in an infringement of the original product trademark.) **F 1156,** F12

parquet flooring nail—hardened-steel, annularly threaded, 1⅛ by 0.062 to 1¼ by 0.072-in. nails with deep countersunk 0.080 to 0.113-in. casing head and diamond or needle point. **F 547,** F16

part, *n—in upholstered furniture*, one component of a furniture unit, such as a seat, back, pillow, or arm cushion. **D 123,** D13

partial cleavage, *n—in textiles*, a transverse gouge, cut or other cross-wise rent in the fiber. Clearly penetrating at least the cuticle of the fiber. **D 123,** D13

partial discharge—an electrical discharge that only partially bridges the insulation between conductors. It may or may not occur adjacent to a conductor. **D 2864,** D27

partial discharge apparent power loss (P_a), *n*—the summation over a period of time of all corona pulse amplitudes multiplied by the rms test voltage.

$$P_a = I_t V_s$$

where:
P_a = apparent power loss in time interval $(t_1 - t_0)$, W,
I_t = average corona current, A, and
V_s = applied rms test voltage, V.

D 1711, D09

partial discharge (corona), *n*—an electrical discharge that only partially bridges the insulation between conductors. **D 1711,** D09

partial discharge (corona) energy (*W*), *n*—the energy drawn from the test voltage source as the result of an individual discharge. **D 1711,** D09

partial discharge (corona) extinction voltage (CEV), *n*—the highest voltage at which partial discharges above some stated magnitude no longer occur as the applied voltage is gradually decreased from above the inception voltage. **D 1711,** D09

partial discharge (corona) inception voltage (CIV), *n*—the lowest voltage at which continuous partial discharges above some stated magnitude (which may define the limit of permissible background noise) occur as the applied voltage is gradually increased. **D 1711,** D09

partial discharge (corona) level, *n*—the magnitude of the greatest recurrent discharge during an observation of continuous discharges. **D 1711,** D09

partial discharge (corona) power loss (*P*), *n*—the summation of the energies drawn from the test voltage source by individual discharges occurring over a period of time, divided by that time period.

$$P = \frac{1}{T}\sum_{i=1}^{i=m} Q_i V_i$$

where:
P = discharge power, W,
T = time period, s,
m = number of the final pulse during T, and
$Q_i V_i$ = discharge energy of the *i*th pulse.

D 1711, D09

partial discharge (corona) pulse rate (*n*), *n*—the average number of discharge pulses that occur per second or in some other specified time interval. **D 1711,** D09

partial discharge pulse, *n*—a voltage or current pulse that occurs at some designated location in a circuit as a result of a partial discharge. **D 1711,** D09

partial discharge pulse voltage (V_t), *n*—the terminal pulse voltage resulting from a partial discharge represented as a voltage source suddenly applied in series with the capacitance of the insulation system under test, and that would be detected at the terminals of the system under open-circuit conditions. **D 1711,** D09

partial discharge quantity (terminal corona charge) (Q_t), *n*—the magnitude of an individual discharge in an insulation system expressed in terms of the charge transfer measured at the system terminals. **D 1711,** D09

partial immersion thermometer, *n*—a liquid-in-glass thermometer designed to indicate temperatures correctly when the bulb and a specified part of the stem are exposed to the temperatures being measured. (Compare **complete immersion thermometer** and **total immersion thermometer.**) **E 344,** E20

partially alloyed powder, *n*—a powder in which the alloy addition or additions are metallurgically bonded to an elemental or pre-alloyed powder. **B 243,** B09

partially balanced incomplete block design (PBIB), *n*—an incomplete block design in which each block contains the same number *k*, of different versions from the *t* versions of the principal factor.

NOTE—The arrangement is such that not all pairs of versions occur together in the same number of the blocks; some versions can therefore be compared with greater precision than others. **E 456,** E11

partially balanced incomplete block design (PBIB), *n*—an incomplete block design in which each block contains the same number *k*, of different versions from the *t* versions of the principal factor.

NOTE—The arrangement is such that not all pairs of versions occur together in the same number of the blocks; some versions can therefore be compared with greater precision than others. **E 1325,** E11

partially nested experiment, *n*—a nested experiment in which several factors may be crossed as in factorial experiments and other factors nested within the crossed combinations.

NOTE—It is not unusual to find that experiments consist of both factorial and nested segments. See **nested experiment.** **E 456,** E11

partially nested experiment—a nested experiment in which several factors may be crossed as in factorial experiments and other factors nested within the crossed combinations.

NOTE—It is not unusual to find that experiments consist of both factorial and nested segments. See **nested experiment.** **E 1325,** E11

partially oriented yarn, *n*—filament yarn in which polymer molecules are only partially aligned. **D 123,** D13

partially preformed—not fully formed prior to driving. **F 592,** F16

partial pressure—the contribution of one component of a system to the total pressure of its vapor at a specified temperature and gross composition. **E 7,** E04

partial pressure—the pressure caused by a gas, either by itself, or in the presence of other gases. When a second gas is not present, the partial pressure is the same as the total pressure. **E 1316,** E07

partial replacement disc—a structure intended to restore a portion of the support and motion or a portion thereof, between adjacent vertebral bodies. **F 1582,** F04

partial sensory profile, *n*—a profile comprising certain selected attributes, with their intensity values. Examples are flavor profile, odor profile, and texture profile. **E 253,** E18

participant, *n*—*in wear testing,* any individual that uses a test or control textile during a wear test. **D 123,** D13

particle—a minute quantity or fragment of matter whose size and shape depend on the forces of cohesion. It is usually only a single crystal or a unit of matter with a specific gravity approximating that of a single crystal. **C 242,** C21

particle, *n*—a small discrete mass of solid or liquid matter.

particle concentration, n—concentration expressed in terms of number of particles per unit volume of air or other gas.

particle fall, n—a measurement of air contamination consisting of the mass rate at which solid particles deposit from the atmosphere.

particle size, n—an expression for the size of liquid or solid particles expressed as the average or equivalent diameter.

particle size distribution, n—the relative percentage by mass, surface area, volume, number, or other property of each of the different size fractions of particulate matter. **D 1356,** D22

particle—the aggregate component of a particle board manufactured by mechanical means from wood or other lignocellulosic material (comparable to the aggregate in concrete) including all small subdivisions of wood such as chips, curls, flakes, sawdust, shavings, slivers, strands, wood flour, and wood wool. Particle size may be measured by the screen mesh that permits passage of the particles and another screen upon which they are retained, or by the measured dimensions as for flakes and strands. **D 1554,** D07

particle, *n*—a small discrete mass of solid or liquid matter. **E 1620,** E29

particle—a small discrete mass of solid or liquid matter.

NOTE—Gas bubbles in a liquid may be considered to be particles. **F 740,** F21

particle—*on a silicon wafer surface*, a small, discrete piece of foreign material or silicon not connected crystallographically to the wafer. **F 1241,** F01

particleboard corestock—common name given to particle board manufactured for use as a core for overlaying. **D 1554,** D07

particleboard panel stock—common name given to particle board manufactured primarily for use as panel material, and in which the surfaces may be treated to obtain decorative effects. **D 1554,** D07

particleboards—a generic term for a panel manufactured from lignocellulosic materials (usually wood) primarily in the form of discrete pieces or particles, as distinguished from fibers, combined with a synthetic resin or other suitable binder and bonded together under heat and pressure in a hot-press by a process in which entire interparticle bond is created by the added binder, and to which other materials may have been added during manufacture to improve certain properties. Particleboards are further defined by the method of pressing. When the pressure is applied in the direction perpendicular to the faces as in a conventional multi-platen hot press, they are defined as flat-platen pressed and when the applied pressure is parallel to the faces, they are defined as extruded. **D 1554,** D07

particleboard underlayment—an underlayment grade particleboard made or machined to close thickness tolerances for use as a leveling course and to provide a smooth surface under floor covering materials. **D 1554,** D07

particle concentration—See **particle.** **D 1356,** D22

particle concentration, *n*—a measure of the liquid or solid particle content in a mixture of particles and fluid. The following more specific terms are in use:

(*1*) **rain density**—the mass of liquid per unit volume of mixture in an actual or simulated rainfield.

(*2*) **solids loading ratio**—the mass of solid particles per unit volume of mixture in a solid impingement environment.

(*3*) **volume concentration**—the volume of the liquid or solid particles per unit volume of mixture.

(*4*) **quality**—the mass of vapor phase per unit mass of a liquid-vapor two-phase single-component fluid.

(*5*) **mass concentration**—which see. **G 40,** G02

particle drift deposits—the deposition of chemical particles outside the intended target. **E 1102,** E35

particle fall—See **particle.** **D 1356,** D22

particle fluence, Φ—the quotient of dN by da, where dN is the number of particles incident on a sphere of cross-sectional area da (ICRU).

$$\Phi = \frac{dN}{da}$$

Unit: m^{-2} **E 170,** E10

particle fluence rate, ϕ—the quotient of dϕ by dt, where dΦ is the increment of particle fluence in the time interval dt (ICRU).

$$\phi = \frac{d\Phi}{dt} = \frac{d^2N}{da\,dt}$$

Unit: $m^{-2} \cdot s^{-1}$ **E 170,** E10

particle number—the number of particles emitted, transferred or received (ICRU). **E 170,** E10

particle (particulate contaminant), , *n*—a piece of matter in a solid state with observable length, width, and thickness. **G 126,** G04

particle shape—a characterization of the shape or configuration of a particle fitting it into any one of ten basic classes, as follows:

Class	
A	acicular—needle shaped
B	angular—sharp-edged or having roughly polyhedral shape
C	crystalline—of geometric shape freely developed
D	dendritic—having a branched crystalline shape
E	fibrous—regularly or irregularly threadlike
F	flakey—lamellar, plate-like
G	granular—having an approximately equidimensional but irregular shape
H	irregular—lacking any symmetry
I	nodular—having a rounded irregular shape
J	spherical—globule shaped

C 242, C21

particle sieve diameter—a measure of the size of a sediment particle; the smallest standard sieve opening through which the particle will pass. For an elongated particle, it is a measure of its intermediate axis. **D 4410,** D19

particle size—the controlling lineal dimension of an individual particle as determined by analysis with sieves or other suitable means. **B 243,** B09

particle size—See **average particle size.** **C 242,** C21

particle size—the controlling lineal dimension of individual particles. **C 859,** C26

particle size—See **particle.** **D 1356,** D22

particle size—the diameter of solid or liquid particles, expressed in micrometres. **D 3064,** D10

particle size—a linear dimension, usually designated as *diameter*, used to characterize the size of a particle; the dimension may be determined by any of several different techniques, including sedimentation, sieving, micrometric measurement, or direct measurement. **D 4410,** D19

particle-size analysis—see **grain-size analysis.** **D 653,** D18

particle size distribution—the percentage by weight, or by number, of each fraction into which a powder sample has been classified with respect to sieve number or microns. (Preferred usage: "particle size distribution by frequency.") **B 243,** B09

particle-size distribution—a profile of the sizes of particles contained in a material in which the quantities must be expressed on some basis which may be total number, total surface, or total weight or volume of the particles in the material. **C 242,** C21

particle size distribution—the percentage by mass or by number of each fraction into which a powder sample has been classified with respect to sieve number or microns. **C 286,** B08

particle-size distribution—see **gradation, grain-size distribution.** **D 653,** D18

particle size distribution—See **particle.** **D 1356,** D22

particle-size distribution—the relative amount of a sediment sample of a range in specific sizes in terms of percentages by mass finer than a given size, D, often shown on a semilog plot. **D 4410,** D19

particle size distribution, *n*—*in sieve analysis*, the percentages, by mass or number, of all fractions into which various sizes of particles are classified. **E 1638,** E29

particle-size, intermediate axis—the size of a sediment particle determined by direct measurement of the axis normal to a plane representing the longest and shortest axes. **D 4410,** D19

particle sizing—segregation of granular material into specified particle size ranges. **C 709,** D02

particle velocity, *u;* [LT^{-1}]; m/s—a fluctuating velocity superimposed by the presence of sound on the other velocities the particles of the medium may have. In analogy with alternating voltage its magnitude can be expressed in several ways, such as instantaneous particle velocity or peak particle velocity, but the unqualified term means root-mean-square particle velocity. In air, the other velocities are those due to thermal agitation and wind currents. **C 634,** E33

particulate, *adj*—pertaining to or composed of particles.

personal sample—See **sample.**

*p***H,** *n*—the negative logarithm to the base ten of the conventional hydrogen ion activity. **D 1356,** D22

particulate, *adj*—of or relating to minute separate particles. **D 4175,** D02

particulate, *adj*—of or relating to particles. **F 740,** F21

particulate—discrete particle of dirt or other material (see **dirt**). **F 1241,** F01

particulate—a general term used to describe a finely divided solid of organic or inorganic matter, including metals. These solids are usually reported as the amount of contaminant by the population of a specific micrometer size. **G 126,** G04

particulate contamination—*on a semiconductor wafer*, a particle or particless on the surface of the wafer. **F 1241,** F01

particulate grout—any grouting material characterized by undissolved (insoluble) particles in the mix. See also **chemical grout**. **D 653,** D18

particulate matter, *n*—see **powder**. **B 243,** B09

particulate matter—that nonliquid matter, exclusive of gases, which is heterogeneously dispersed in water. **D 1129,** D19

particulate reinforced ceramic matrix composite, *n*—a ceramic matrix composite in which the reinforcing components are particles of equiaxed or platelet geometry (in contrast to whiskers or short fibers). **C 1145,** C28

particulate rubber, *n*—rubber, vulcanized or unvulcanized, that has been transformed into a collection of particles, with or without a coating of a partitoning agent to prevent agglomeration during production, transportation, or storage (see **buffing, rubber, granulated rubber, ground rubber**, and **powdered rubber**). **D 1566,** D11

particulate solid—synonym for **bulk solid**. **D 653,** D18

particulate wood fuel—any wood fuel with a maximum particle volume of 16.39 cm^3(1 $in.^3$) such that the largest dimension is 7.62 cm (3 in.). **E 1126,** E48

particulate wood fuel—any wood fuel with a maximum particle volume of 16.39 cm^3(1 $in.^3$) such that the largest dimension is 7.62 cm (3 in.). **E 1705,** E48

parting, *vt*—separating silver from gold by selectively dissolving the silver in acid. **E 135,** E01

parting—the selective corrosion of one or more components of a solid solution alloy. **G 15,** G01

parting agent—a material applied to one or both surfaces of a sheet to prevent blocking. **D 1079,** D08

parting limit—the minimum concentration of a more noble component in an alloy, above which parting does not occur in a specific environment. **G 15,** G01

parting line—line or seam on glassware resulting from the joint of two mold parts. **C 162,** C14

parting line, n—another name for spew line. **D 1566,** D11

partition—See **fiberboard partition.** **D 996,** D10

partition coefficient—when, to a polyphase mixture there is added a definite quantity of a new component that is insufficient to bring about a phase change, the proportioning of the new component among the several phases at equilibrium may be expressed by means of a ratio known as a partition coefficient; since the ratio often changes very little over small ranges of variation in the amount of the added component, it is customary simply to refer to the partition coefficient of the added component without reference to the quantity of the addition, provided the latter is small. **E 7,** E04

partitioning—the act of movement of contaminants from one soil residence phase to another. **D 653,** D18

partition tile, *n*—tile for use in building interior partitions, subdividing areas into rooms, or similar constructions, and carrying no superimposed loads. **C 43,** C15

parts per billion (ppb)—a measure of proportion by weight, equivalent to a unit weight of solute per billion unit weights of solution (approximate μg/L or mg/m^3 in dilute solutions). **D 6161,** D19

parts per million—mass unit of any substance in a million mass-units of the water-substance mixture. **D 4410,** D19

parts per million (ppm)—a measure of proportion by weight, equivalent to a unit weight of solute per million unit weights of solution (approximate mg/L or g/m^3 in dilute solutions). **D 6161,** D19

parts per trillion (ppt)—a measure of proportion by weight, equivalent to a unit weight of solute per trillion unit weights of solution (approximate g/L or μg/m^3 in dilute solutions). **D 6161,** D19

part wool felt, *n*—a felt composed of any one of or a combination of new and recycled wool fibers mixed with one or more man-made fibers, vegetable fibers, or animal fibers other than wool. **D 123,** D13

pascal cubic metres per second (Pa·m³/s)—the preferred unit of gas flow in the SI system. One Pa·m^3/s is approximately equal to 10 atm cm^3/s or, more precisely, 1 Pa·m^3/s = 9.8692 atm·cm^3/s. **E 1316,** E07

pascal (Pa)—One pascal is approximately equal to 1×10^{-5} atm or, more precisely, 1 Pa = 0.98692×10^{-5} atm. **E 1316,** E07

pass—a treatment step or one of multiple treatment steps producing in a membrane system a product stream. **D 6161,** D19

passing off—the simulation of a trademark to closely resemble the original. **F 1156,** F12

passivating—a process for creating passivity (q.v.). **B 374,** B08

passivation—a process to render the surface condition of stainless steel chemically inactive. **F 1078,** F04

passivation—the process in metal corrosion by which metals become passive. (See **passive**.) **G 15,** G01

passivator—a type of inhibitor which appreciably changes the potential of a metal to a more noble (positive) value. **G 15,** G01

passive—the state of the metal surface characterized by low corrosion rates in a potential region that is strongly oxidizing for the metal. **G 15,** G01

passive-active cell—a corrosion cell in which the anode is a metal in the active state and the cathode is the same metal in the passive state. **G 15,** G01

passive earth pressure—see **earth pressure**. **D 653,** D18

passive fiber optic chemical sensor, *n*—a fiber optic sensor that utilizes the intrinsic spectroscopic properties of the analyte to modulate the optical signal. **E 131,** E13

passive refrigeration—a refrigeration system in which cooling is provided by a refrigerant such as liquid nitrogen. **E 1705,** E48

passive solar energy system—See **building subsystem.** **E 631,** E06

passive state of plastic equilibrium—see **plastic equilibrium**. **D 653,** D18

passivity—the condition of a metal that retards its normal reaction in a specified environment and associated with the assumption of a potential more noble than its normal potential. **B 374,** B08

pass-over style—used over counter-height equipment when pass-over configuration (from the cooking side to the serving side) is required. **F 1827,** F26

pasta cooker—equipment, including a cooking vessel, in which the cooking medium, water or a water-oil mixture, (see **cooking medium**) is placed to such a depth that the cooking of food is essentially supported by displacement of the cooking medium rather than by the bottom of the vessel. Heat is delivered to the cooking medium by means of an immersed electric element or band wrapped vessel (electric pasta cooker), or by heat transfer from gas burners through either the walls of the pasta cooker or through tubes passing through the cooking fluid (gas pasta cooker). **F 1827,** F26

paste—a prepared mixture consisting of a suspension of undissolved solid(s) in a liquid medium sufficiently viscous that it cannot achieve a level surface without application of external force; not a slurry. **C 242,** C21

paste, *n*—an adhesive composition having a characteristic plastic-type consistency, that is, a high order of yield value, such as that prepared by heating a mixture of starch and water and subsequently cooling the hydrolyzed product. (Compare **adhesive, glue, mucilage,** and **sizing.**) **D 907,** D14

paste mold—a mold lined with adherent carbon, used wet for blown ware. **C 162,** C14

paste paint—under **paint,** see *paste paint.* **D 16,** D01

patch, *n*—a portion of pavement surface which has been replaced or additional material has been applied to the pavement after original construction. **E 867,** E17

patch, *n*—a portion of pavement surface which has been replaced or where additional material has been applied to the pavement after original construction. **E 1778,** E17

patches—insertions of sound wood or synthetic material in veneer or panels from which defective portions have been removed. **D 1038,** D07

patching compound, *n*—compound used to fill or smooth minor depressions or irregularities in a flooring surface. **F 141,** F06

pate dure (hard paste)—a French term designating ceramic whitewares fired at relatively high temperatures. **C 242,** C21

patenting, *n*—*in wire making,* heating a medium-carbon or high-carbon steel before wire drawing, or between drafts, to a temperature above the **transformation range**, and then cooling it in air, or a bath of molten lead or salt, to a temperature below Ae_1. **A 941,** A01

patent leather—Any leather with a finish system that produces a high gloss with great depth and clarity. Usually obtained by one or more applications of a high solids coating, but is also made by film lamination. **D 1517,** D31

pate tendre (soft paste)—a French term designating ceramic whitewares fired at relatively low temperatures. **C 242,** C21

path length—the distance traveled by the laser radiation from the beam splitter to the recording medium. **E 1316,** E07

path length difference—the difference in path length between the object beam and the reference beam. **E 1316,** E07

pathogenic—disease causing. **E 1705,** E48

pathogens—organisms that are capable of causing disease. **F 1600,** F20

path percolation (line of creep)—the path that water follows along the surface of contact between the foundation soil or rock and the base of a dam or other structure. **D 653,** D18

pattern—See **diffraction pattern**. **E 7,** E04

pattern—see **components pattern**. **E 1605,** E06

pattern—metal, fiberboard, wood or paper forms from the outlines of which are cut the various pieces of the shoe. **F 869,** F08

patternate, *v*—to measure the pattern of a spray. **E 1620,** E29

patternation, *n*—the measurement and characterization of spray patterns. **E 1620,** E29

patternator, *n*—a device designed to patternate sprays. **E 1620,** E29

paved road, *n*—two or more lanes, all-weather, maintained, hard surface roads with good driving visibility used for heavy and high-density traffic, in accordance with the U.S. Federal Highway Administration. **F 538,** F09

pavement, bituminous—see **bituminous pavement. E 867,** E17

pavement characteristic, *n*—a physical feature or property of a pavement surface such as type, roughness, texture, and skid resistance **E 867,** E17

pavement condition, *n*—a quantitative representation of distress in pavement at a given point in time. **E 867,** E17

pavement, continuously reinforced concrete (CRCP)—see **continuously reinforced concrete pavement. E 867,** E17

pavement distress, *n*—external indications of pavement defects or deterioration. **E 867,** E17

pavement distress, *n*—external indications of pavement defects or deterioration. **E 1778,** E17

pavement, jointed concrete (JCP)—see **jointed concrete pavement. E 867,** E17

pavement macrotexture, *n*—the deviations of a pavement surface from a true planar surface with the characteristic dimensions of wavelength and amplitude from 0.5 mm up to those that no longer affect tire-pavement interaction. **E 867,** E17

pavement management section/segment, *n*—a contiguous pavement area considered to have uniform construction, maintenance, usage history, and condition. **E 867,** E17

pavement-micro texture (micro-rugosity), *n*—the deviations of a pavement surface from a true planar surface with characteristic dimensions of wavelength and amplitude less than 0.5 mm. It is suggested that the symbol *mitx* be used. **E 867,** E17

pavement performance, *n*—ability of a pavement to fulfill its purpose over time. **E 867,** E17

pavement, Portland cement concrete—see **Portland cement concrete pavement. E 867,** E17

pavement pumping—ejection of soil and water mixtures from joints, cracks, and edges of rigid pavements, under the action of traffic. **D 653,** D18

pavers—the glass-contact refractory blocks used in the bottom of a melter. See **tank block**. **C 162,** C14

pavers—unglazed porcelain or natural clay tile formed by the dust-pressed method and similar to ceramic mosaics in composition and physical properties but relatively thicker with 6 $in.^2$ (39 cm^2) or more of facial area. **C 242,** C21

paving—stone used in an interior pedestrian wearing surface as in patios, walkways, driveways, and the like. (See *flooring*) **C 119,** C18

paving brick, *n*—brick made to provide the wearing surface of highways, streets, driveways, walkways, patios, and similar applications. See Specifications C 902 and C 1272. **C 43,** C15

payback method—a technique of economic evaluation that determines the time required for the cumulative benefits from an investment to recover the investment cost and other accrued costs (see **discounted payback period; simple payback period**). **E 833,** E06

Pb—chemical symbol for the element, lead. **E 1605,** E06

Pb-positive/negative—see **lead positive/negative**. **E 1605,** E06

PCB,—see **printed circuit board**. **F 1457,** F05

PCS—See **print contrast signal.** **F 149,** F05

PDD examination, *n*—a process that encompasses all activities that take place between a PDD examiner and an examinee during a specific series of interactions. These interactions may include the pretest interview, the use of the polygraph to collect physiological data from the examinee while presenting a series of tests, the test data analysis phase, which may be supplemented in a post-test interview phase. **E 2035,** E52

PDD examiner, *n*—someone who has successfully completed formal education and accredited training in conducting PDD examinations and, if appropriate, is certified or licensed by their agency or state to conduct such examinations. **E 2035,** E52

PDPA, *n*—phase Doppler particle analyzer. **E 1620,** E29

pea gravel, *n*—in roofing, small, natural aggregate, 6 to 10 mm (¼ to ⅜ in.) in diameter, screened to comply with Specification C 33, size 8. **D 1079,** D08

peak, *n*—in constant rate of elongation load-strain testing, the point of maximum load. **D 1079,** D08

peak, *n*—that part of a DTA or DSC curve which is attributable to the occurrence of a single process. It is normally characterized by a deviation from the established base line, a maximum deflection, and a reestablishment of a base line not necessarily identical to that before the peak. **E 1445,** E27

peak, *n*—that portion of a heating curve which is attributable to the occurrence of a single process. It is normally characterized by a deviation from the established baseline, a maximum deflection, and a reestablishment of a baseline not necessarily identical to that before the peak. **E 1445,** E27

peak—*in fatigue loading,* the point where the first derivative of the load-time history changes from positive to negative sign; the point of maximum load in constant amplitude loading. **E 1823,** E08

peak braking coefficient, [*nd*], *n—of a tire,* the maximum value of the braking coefficient that occurs prior to wheel lockup as the braking torque is progressively increased; this measure is influenced by operating conditions. **F 538,** F09

peak force, *n—in tear testing,* the force required to break one or more yarn components in a woven fabric specimen, or break the fiber bonds or interlocks in other manufactured forms of a fabric specimen. **D 123,** D13

peak load, *n*—maximum force recorded during a strength test. **D 907,** D14

peak melting point, *n*—in a thermal transfer ribbon, the maximum temperature to which the imaging material should be subjected. **F 1623,** F05

peaks—extraneous marks extending from the character outward past the COL. **F 149,** F05

peak shear strength—maximum shear strength along a failure surface. (ISRM) **D 653,** D18

peak sound pressure level, L_{PK} [*nd*], (dB), *n*—ten times the common logarithm of the square of the ratio of the largest absolute value of the instantaneous sound pressure in a stated frequency band during a specified time interval to the reference sound pressure of 20 micropascals. **C 634,** E33

peak temperature, *n*—the temperature corresponding to the maximum deflection of the DTA or DSC curve. **E 1445,** E27

peak value (of an impulse voltage wave), *n*—the maximum value of voltage. **D 1711,** D09

peak width, FWHM—full width at half-maximum peak height above background. **E 673,** E42

peaned point—point formed like that usually found on rivets. **F 547,** F16

pearl ash—{archaic} commercial potassium carbonate. **C 162,** C14

pearlescent, *adj*—exhibiting various colors depending on the angles of illumination and viewing, as observed in mother-of-pearl. **E 284,** E12

pearlite—a metastable microstructure formed, when local austenite areas attain the eutectoid composition, in alloys of iron and carbon containing greater than 0.025 percent but less than 6.67 percent carbon. The structure is an aggregate consisting of alternate lamellae of ferrite and cementite formed on slow cooling during the eutectoid reaction. In an alloy of given composition, pearlite may be formed isothermally at temperatures below the eutectoid temperature by quenching austenite to a desired temperature (generally above 550°C) and holding for a period of time necessary for transformation to occur. The interlamellar spacing varies directly with the transformation temperature; that is, the higher the temperature the greater the spacing. **E 7,** E04

pearlite colony—a circumscribed aggregate within which lamellae of corresponding phases have the same orientation. **E 7,** E04

pearlite nodule—cluster of wedge-shaped pearlite colonies. **E 7,** E04

pearlitic structure—a microstructure resembling that of the pearlite constituent in steel, therefore, a lamellar type of structure of varying degrees of coarseness. **E 7,** E04

peat—a naturally occurring highly organic substance derived primarily from plant materials. **D 653,** D18

peatland—areas having peat-forming vegetation on which peak has accumulated or is accumulating. **D 653,** D18

pebbled grain—an embossed or naturally shrunken leather grain finish resembling a pebbled surface, ranging from fine pebbled Morocco goat to heavy Scotch grain upper leather. **D 1517,** D31

pebble mill—see **ball mill.** **C 286,** B08

peccary—a wild boar found in Central and South America. The skin is usually chrome-tanned and shaved to light weight for glove leathers. It is distinguishable from pigskin and carpincho leather by the fact that bristle holes occur in straight-line groups of three. **D 1517,** D31

pectic substrates—complex polysaccharides containing a large proportion of galacturonic acid units. **D 1695,** D01

pectins—colorless, amorphous, water-soluble polysaccharides occurring in plant tissues that yield pectic acid and methanol on hydrolysis. **D 1695,** D01

pedestrian—a person using legs or leg surrogates (for example, prosthetic limbs, crutches, etc.) as the principal mechanism for locomotion. **F 1646,** F13

peelable seal—the opening characteristic of forcibly separating two package substrates, which have been joined together by a sealing process, without tearing the substrates. **F 1327,** F02

peeler log—a log selected as suitable for cutting into rotary veneer. **D 1038,** D07

peeling—See **delamination**. **A 902,** A05

peeling—the detachment or partial detachment of an electrodeposited coating from a basis metal or undercoat. **B 374,** B08

peeling—See **orangepeel; shivering.** **C 242,** C21

peeling—a defect characterized by the spontaneous detachment of pieces of porcelain enamel from cast iron. **C 286,** B08

peeling resistance of metal connector plate with integral teeth —resistance to consecutive withdrawal of adjacent teeth of metal connector plate from wood member during shear load application starting at the last row of teeth and progressing toward the center of the connection. See **lateral resistance of metal connector plate**. **E 1807,** E06

peel ply—a removable ply molded onto the surface of a laminate to provide a chemically clean surface for bonding or painting after removal. **E 1749,** E06

peel strength, *n*—the average load per unit width of bondline required to separate progressively a flexible member from a rigid member or another flexible member. **D 907,** D14

peen-coated, peen-galvanized—See **mechanically plated.** **F 547,** F16

peening—See **shot peening.** **B 374,** B08

peen plating—See **mechanical plating.** **B 374,** B08

peephole—a small opening in a furnace wall for observation purposes. **C 162,** C14

peerless cut nail—name for small, regular-stock-steel, cut nails with broad flat circular head and sheared long-tapered square point to facilitate clinching. **F 547,** F16

Pel—picture **el**ement. **F 1457,** F05

pellet—a briquet or compact, usually cylindrical, formed by pressing a powder in a die. The pellet may or may not have been sintered following compaction. **C 859,** C26

pellet—see **fuel pellet**. **E 1126,** E48

pelleted activated carbon—a form of granular activated carbon that is cylindrical in shape. **D 2652,** D28

pelletized biomass fuel— see **densified particulate biomass fuels**. **E 1126,** E48

pelletized biomass fuel—see **densified particulate biomass fuels.** **E 1705,** E48

pellet size distribution, *n*—the percentage, by mass, of carbon black retained on each of a specified series of sieve screens arranged with progressively smaller openings. **D 3053,** D24

pellicle mirror—a thin transparent membrane used in place of a transparent flat glass disk of a "plane glass illuminator." **E 7,** E04

pelt—a raw skin with hair. Usually refers to fur animals. **D 1517,** D31

Peltier coefficient, *n*—the reversible heat which is absorbed or evolved at a thermocouple junction when unit current passes in unit time. **E 344,** E20

PEM—abbreviation for **printer emulation module**. **F 1457,** F05

pencil—rod-like assemblage of asbestos fibers in close-packed parallel orientation, of generally uniform diameter that can be fiberized readily. **D 2946,** C17

pencil beam—a radiation beam which has little divergence, usually created by collimating an intense source of radiation. **E 1316,** E07

pencil point—long needle point. **F 547,** F16

pendulum hardness tester, *n*—a device for measuring the hardness of a dry film, based on the damping time required for a specified decrease in oscillation (swing) amplitude; the shorter the damping time the lower the hardness. **D 16,** D01

penetrability—a grout property descriptive of its ability to fill a porous mass; primarily a function of lubricity and viscosity. **D 653,** D18

penetrameter— alternative term for **image quality indicator**. **E 1316,** E07

penetrameter sensitivity—alternative term for **IQI sensitivity**. **E 1316,** E07

penetrant—a solution or suspension of dye. **E 1316,** E07

penetrant—a material that enhances the ability of an agrichemical to enter a substrate or penetrate a surface. **E 1519,** E35

penetrant comparator—an intentionally flawed specimen having separate but adjacent areas for the application of different liquid penetrant materials so that a direct comparison of their relative effectiveness can be obtained.

Note—It can also be used to evaluate liquid penetrant techniques, liquid penetrant systems, or test conditions. **E 1316,** E07

penetrant, fluorescent—a penetrant that emits visible radiation when excited by black light. **E 1316,** E07

penetrant, post emulsifiable—a liquid penetrant that requires the application of a separate emulsifier to render the excess surface penetrant water-washable. **E 1316,** E07

penetrant, solvent-removable—a liquid penetrant so formulated that most of the excess surface penetrant can be removed by wiping with a lint-free material, with the remaining surface penetrant traces removable by further wiping with a lint-free material lightly moistened with solvent remover. **E 1316,** E07

penetrant, visible—a liquid penetrant that is characterized by an intense color, usually red. **E 1316,** E07

penetrant, water-washable—a liquid penetrant with a built-in emulsifier. **E 1316,** E07

penetrating agent—a material that increases the penetration of a liquid medium into a porous material. **D 459,** D12

penetration—the depths to which one material extends into or penetrates another. **C 709,** D02

penetration, *n*—the consistency of a bituminous material expressed as the distance in tenths of a millimetre (0.1 mm) that a standard needle penetrates vertically a sample of the material under specified conditions of loading, time, and temperature. **D 8,** D04

penetration—depth of hole cut in rock by a drill bit. (ISRM) **D 653,** D18

penetration, *n*—the entering of an adhesive into an adherend. **D 907,** D14

penetration—the consistency of a bituminous material expressed as the distance in tenths of a millimetre (0.1 mm) that a standard needle penetrates vertically a sample of material under specified conditions of loading, time, and temperature. A cone is sometimes used for special purposes instead of a needle. **D 1079,** D08

penetration, *n*— *of a propellant or of petrolatum*, the depth, in tenths of a millimetre, that a standard cone penetrates the sample under prescribed conditions of weight, time, and temperature. **D 4175,** D02

penetration, *n*— *of petroleum wax*, the depth, in tenths of a millimetre, to which a standard needle penetrates into the wax under defined conditions. **D 4175,** D02

penetration depth—the maximum depth in a material from which usable ultrasonic information can be obtained and measured. **E 1316,** E07

penetration grouting—filling joints or fractures in rock or pore spaces in soil with a grout without disturbing the formation; this grouting method does not modify the solid formation structure. See also **displacement grouting**. **D 653,** D18

penetration macadam, *n*—a pavement layer containing essentially one-size coarse aggregate, penetrated in place by a heavy application of bituminous material, followed by an application of a smaller size coarse aggregate, and compacted. Multiple layers containing still smaller coarse aggregate may be used. **D 8,** D04

penetration resistance curve (Proctor penetration curve)—the curve showing the relationship between: (*1*) the penetration resistance, and (*2*) the water content. **D 653,** D18

penetration resistance (standard penetration resistance) (Proctor penetration resistance), p_R, *N* **(FL^{-2} or Blows L^{-1})**—(*a*) number of blows of a hammer of specified weight falling a given distance required to produce a given penetration into soil of a pile, casing, or sampling tube.

(*b*) unit load required to maintain constant rate of penetration into soil of a probe or instrument.

(*c*) unit load required to produce a specified penetration into soil at a specified rate of a probe or instrument. For a Proctor needle, the specified penetration is 2½ in. (63.5 mm) and the rate is ½ in. (12.7 mm)/s. **D 653,** D18

penetration time—same as **dwell time**. **E 1316,** E07

penetrator—See **indenter**. **E 7,** E04

Penning gage—see **ionization vacuum gage.** **E 1316,** E07

Penn-Vernon process—see synonymous term **Pittsburgh sheet process**. **C 162,** C14

penny size, penny weight—denoting length of nail, indicated by "d" (not indication of diameter). Because of standardization of sizes of common nails, sinkers, and cooler nails, penny weight denotes, in addition to length, the head and shank diameter of these nails. For other nail types, sizes are referred to by length and diameter. **F 547,** F16

penta resin—under **resin, synthetic,** see *penta resin*. **D 16,** D01

pentosans—one of the groups of amorphous carbohydrates included under the general term "hemicellulose." Pentosans yield principally pentoses on acid hydrolysis. The principal pentosan in wood is xylan. **D 1695,** D01

peptizer (dry rubber), *n*—a compounding material used in small proportions to accelerate, by chemical action, the softening of rubber under the influence of mechanical action, or heat, or both. **D 1566,** D11

percale, *n*—*as applied to bed sheeting*, a plain weave fabric with not fewer than 180 yarns/in.^{2}(645 mm^2). **D 123,** D13

percent active FWA—the absolute concentration of specified **FWA** in (*1*) raw materials, or (*2*) finished product, expressed in terms of specific structure or generic name, when known. **D 459,** D12

percentage elongation, *n*— *of paper*, a mathematical quantity used to express elongation (stretch) as a percentage increase in the length of a paper test specimen at rupture, in comparison to its length at the beginning of a tensile test carried to rupture under specified conditions. See **elongation**. **D 1968,** D06

percentage point, *n*—a difference of 1 % of a base quantity. **D 123,** D13

percent area, *n*—*in testing cotton with the Trash Meter*, the ratio of total area of trash on the surface of a sample of cotton to that of the area of the viewing window, expressed in a percentage of the area of the viewing window. **D 123,** D13

percent compaction—the ratio, expressed as a percentage, of: (*1*) dry unit weight of a soil, to (*2*) maximum unit weight obtained in a laboratory compaction test. **D 653,** D18

percent conductivity—the ratio of the resistivity of the International Annealed Copper Standard (IACS) at 20°C to the resistivity of a material at 20°C, expressed in percent. Results are calculated on a weight basis or volume basis and so specified. **B 354,** B01

percent consolidation—see **degree of consolidation**. **D 653,** D18

percent coverage, *n*—the ratio of the area actually covered by the ink (or print material) to the area of the page times one hundred. (See **image area** and **maximum image area**.) **F 1457,** F05

percent elongation, *n*—the increase in length of a specimen expressed as a percentage of the original length.. **D 123,** D13

percent evaporated, *n*— *in batch distillation*, the sum of the percent recovered and the percent loss. **D 4175,** D02

(*2*) *percent exhaust* (*%E*)—the ratio of FWA on the substrate (wt FWA (s)) obtained under specified conditions to the total FWA introduced in the original bath (wt FWA (o)). **D 459,** D12

percent fines—amount, expressed as a percentage by weight, of a material in aggregate finer than a given sieve, usually the No. 200 (74 μm) sieve. **D 653,** D18

percent hydrolysis-resistant cellulose—the residue after treatment of cellulose with acid under specified conditions, expressed as a percentage of the original. **D 1695,** D01

percentile—one of the values in a series dividing the distribution of the variable in the series into 100 groups of equal frequency or size. **C 242,** C21

percent linear thermal expansion—the change in length per unit length as temperature is changed from temperature T_1 to temperature $T_2 (T_1 < T_2)$, expressed as a percent:

$$P = [(L_2 - L_1)/L_o] \times 100 + A$$

where:
L_o = sample length at T_o (between 20 and 30°C),
L_1 = sample length at T_1,
L_2 = sample length at T_2, and
A = instrument correction.

C 242, C21

percent loss, *n*— *in batch distillation*, 100 minus the percent total recovery. **D 4175,** D02

percent nonconforming, *n*—a calculated ratio of nonconforming units to the number of units inspected, the quotient being multiplied by 100. **D 1711,** D09

percent nozzle pattern overlap—calculated by subtracting the nozzle spacing from the single nozzle pattern width and then dividing by the nozzle spacing and multiplying the result by one hundred. **E 1102,** E35

percent saturation (degree of saturation), S, S_r**(D)**—the ratio, expressed as a percentage, of: (*1*) the volume of water in a given soil or rock mass, to (*2*) the total volume of intergranular space (voids). **D 653,** D18

percept, *n*—the result of the process of perception. **E 284,** E12

perception, *n*—the awareness of the effects of stimuli. **E 253,** E18

perception, *n*—See **visual perception.** **E 284,** E12

perched ground water—unconfined ground water separated from an underlying body of ground water by an unsaturated zone. **D 653,** D18

perched water table—a water table usually of limited area maintained above the normal free water elevation by the presence of an intervening relatively impervious confining stratum. **D 653,** D18

perched water table—groundwater separated from an underlying body of groundwater by unsaturated soil or rock. Usually located at a higher elevation than the groundwater table. (ISRM) **D 653,** D18

percolation—the movement of gravitational water through soil (see **seepage**). **D 653,** D18

percolation—the movement of water through the vadose zone, in contrast to infiltration at the land surface and recharge across a water table. **D 653,** D18

percolation layer (drainage course)—a layer of washed gravel or of a manufactured drainage media that allows water to filter through to the drain. **C 717,** C24

percussion cone—damage produced by contact stresses generated by mechanical contact of a hard, blunt object with a glass surface. **C 162,** C14

percussion drilling—a drilling technique that uses solid or hollow rods for cutting and crushing the rock by repeated blows. (ISRM) **D 653,** D18

percussion drilling—a drilling process in which a hole is advanced by using a series of impacts to the drill steel and attached bit; the bit is normally rotated during drilling. See **rotary drilling**. **D 653,** D18

perfect gas—see **ideal gas.** **E 1316,** E07

perfect reflecting diffuser—ideal reflecting surface that neither absorbs nor transmits light, but reflects diffusely, with the radiance of the reflecting surface being the same for all reflecting angles, regardless of the angular distribution of the incident light. **E 284,** E12

perfect transmitting diffuser, *n*— ideal transmitting specimen that neither absorbs nor reflects light, but transmits diffusely, with the radiance of the specimen being the same for all transmitting angles, regardless of the angular distribution of the incident light. **E 284,** E12

perforated hardboard—hardboard with closely spaced factory punched or drilled holes. **D 1554,** D07

perforated metal connector plate—metal connector plate with prepunched or predrilled holes. Also called *punched metal connector plate.* **E 1807,** E06

perforation—a slot or hole made in well casing to allow for communication of fluids between the well and the annular space. **D 653,** D18

perforations, *n*— *in paper*, a line of slits scored in a sheet of paper so that it may be easily folded, refolded, and afterwards torn off with ease. **D 1968,** D06

performance—See **building performance.** **E 631,** E06

performance—See **building performance**. (See also **facility performance**) **E 1480,** E06

performance—ability of a boom to contain or deflect oil under a given set of environmental conditions. **F 818,** F20

performance criterion, of a facility—See **requirement statement**. **E 1480,** E06

performance properties—design feature(s) manufactured into the fastener to achieve a specific characteristic relative to the fastener application, such as torque-tension. **F 1789,** F16

performance property, *n*—*in wear testing*, any chemical or physical property of a fiber, yarn, or fabric that is evaluated during the wear-refurbishing cycles. **D 123,** D13

performance property, *n*—a result obtained by conducting a performance test. **D 4439,** D35

performance standard, *n*—in building constructions, a standard that defines the required **performance** of a building material, element, **subsystem,** or **system.** **E 631,** E06

performance test—a test of capacity to achieve a desired result. **C 859,** C26

performance test, *n*—a test which simulates in the laboratory as closely as practicable selected conditions experienced in the field and which can be used in design. (Synonym for **design test.**) **D 4439,** D35

performance test, *n*—*in geosynthetics,* a laboratory procedure which simulates selected field conditions which can be used in design. **D 4439,** D35

performance test method of a facility—See **requirement statement**. **E 1480,** E06

periclase grain, *n*—a crystalline magnesium oxide in granular form; this contains at least 85 % magnesia. **C 71,** C08

period—time interval occupied by one cycle. **D 653,** D18

period, *n*—size of the minimum interval of the independent variable after which the same characteristics of a periodic phenomenon recur.

Note—In radiation the independent variable is the time, and the corresponding quantity is the periodic time: Symbol: T; Unit: second (s). **E 349,** E21

periodic reverse plating—a method of plating in which the current is reversed periodically. The cycles are usually no longer than a few minutes and may be much less. **B 374,** B08

period system efficiency— see **efficiency, period system.** **E 772,** E44

peripheral chip—shallow crater formed in the periphery of the specimen through conchoidal fracture and resultant spalling. Also known as **flake chip** or **surface chip.** **F 1241,** F01

peripheral indent—localized deviation from an otherwise smooth peripheral profile that shows no sign of conchoidal fracture. **F 1241,** F01

peritectic equilibrium—a reversible univariant transformation in which a solid phase, that is stable only at lower temperature, decomposes into a liquid and a solid phase that are conjugate at higher temperature, or the reverse; for example: $\alpha + L = \beta$, $\alpha + \beta + L = \gamma$, etc. **E 7,** E04

peritectoid equilibrium—a reversible univariant transformation in which a solid phase, that is stable only at low temperature, decomposes with rising temperature into two or more conjugate solid phases; for example: $\alpha + \beta = \gamma$, $\alpha + \beta + \gamma = \delta$, etc. **E 7,** E04

perlite, *n*—insulation composed of natural perlite ore expanded to form a cellular structure. **C 168,** C16

perlite—an aggregate used in lightweight insulating concrete and in preformed perlite insulating board; formed by heating and expanding siliceous volcanic glass. **D 1079,** D08

perm, *n*—a unit of measurement of water vapor permeance; a metric perm, or 1 g/24 h·m^2·mm Hg. U.S. unit, 1 grain/h·ft^2·in. Hg. **C 11,** C11

perm, *n*—the mass rate of water vapor flow through one square foot of a material or construction of one grain per hour induced by a vapor pressure gradient between two surfaces of one inch of mercury or in units that equal that flow rate. **C 168,** C16

perm, *n*—empirical unit of **water-vapor permeance** (mass flow rate), equal to one grain (avoirdupois) of water vapor per hour flowing through one square foot of a material or construction induced by a vapor-pressure difference of one inch of mercury between the two surfaces. **E 631,** E06

permafrost—perennially frozen soil. **D 653,** D18

permanence, *n*—the resistance of an adhesive bond to deteriorating influences. **D 907,** D14

permanence, *n*—*of paper*, the tendency to resist changes in any or all of its properties with the passage of time. **D 1968,** D06

permanent boom—boom intended for long-term or permanent deployment. **F 818,** F20

permanent care label, *n—as applied to textile products,* a care label that remains legible and attached to a textile product throughout its useful life. **D 123,** D13

permanent counter, *n*—a device intended to continuously measure and periodically record traffic volume, vehicle classification, or weigh, or combination thereof. **E 867,** E17

permanent deformation, *n*—the net long-term change in a dimension of a specimen after deformation and relaxation under specified conditions. **D 123,** D13

permanent hardness—the total milliequivalents of hardness minus the milliequivalents of bicarbonate alkalinity in a water. See **hardness, alkalinity**. **D 6161,** D19

permanent linear change, *n*—the percent dimensional change in length (based on original length) of a refractory specimen free of externally applied stresses, after being subjected to a prescribed heat treatment. **C 71,** C08

permanent magnet—a magnet that retains a high degree of magnetization virtually unchanged for a long period of time (characteristic of materials with high retentivity). **E 1316,** E07

permanent magnet lens—an electron lens consisting of permanent magnets. **E 7,** E04

permanent mold casting, *n*—a metal object produced by introducing molten metal by gravity or low pressure into a mold constructed of durable material, usually iron or steel, and allowing it to solidify. See also graphite permanent mold casting. **B 899,** B02

permanent plugging—a seal that has a hydraulic conductivity that is equivalent or less than the hydraulic conductivity of the geologic formation. This term is often used with uncased boreholes. **D 653,** D18

permanent set—See **residual deflection.** **E 631,** E06

permanent set—the amount by which an elastic material fails to return to its original form after deformation. **F 869,** F08

permanent strain—the strain remaining in a solid with respect to its initial condition after the application and removal of stress greater than the yield stress (commonly also called "residual" strain). (ISRM) **D 653,** D18

permanganate number—the number of millilitres of 0.1 *N* potassium permanganate solution that is consumed by 1 g of ovendry pulp under certain specified and carefully controlled conditions. It is an indication of the hardness or bleachability of a pulp. It also is an indirect estimate of lignin content. **D 1695,** D01

permeability—a property measured as the rate of passage of a liquid or gas through a compact; measured under specified condition. **B 243,** B09

permeability, *n*—the property of a porous material that permits a fluid (or gas) to pass through it; in construction, commonly refers to water vapor permeability of a sheet material or assembly and is defined as water vapor permeance per unit thickness. Metric unit of measurement, metric perms per centimetre of thickness. See **water vapor transmission, perm, permeance.** **C 11,** C11

permeability—the measure through a material of fluid flow, gas, or liquid. **C 242,** C21

permeability—a property measured by the rate of passage of a fluid under a pressure gradient through a material. **C 709,** D02

permeability—that property which permits movement of a liquid through the pores and interstices of the concrete. **C 822,** C13

permeability, *n*—the rate of flow of fluid under a differential pressure through a material. (See also **air permeability, porosity**.) **D 123,** D13

permeability—see **coefficient of permeability**. **D 653,** D18

permeability—the capacity of a rock to conduct liquid or gas. It is measured as the proportionality constant, k, between flow velocity, v, and hydraulic gradient, I; $v = k{\cdot}I$. (ISRM) **D 653,** D18

permeability, *n*—the permeation rate divided by the pressure gradient of the gas or vapor. For a homogeneous material that obeys Fick's law, the permeability is equal to the product of the diffusion coefficient and the solubility coefficient of the gas or vapor. **D 1566,** D11

permeability, *n*—the rate of flow of a liquid under a differential pressure through a material. **D 4439,** D35

permeability, *n—of geotextiles,* hydraulic conductivity. **D 4439,** D35

permeability, *n*—the ability of a rock to transmit fluid. **E 957,** E44

permeability—the ratio of flux density produced to magnetizing force (the ease with which a material can become magnetized). **E 1316,** E07

permeability—the property of a filter medium that permits a fluid to pass through under a pressure differential. **F 740,** F21

permeability, a-c—a generic term used to express various dynamic relationships between magnetic induction, B, and magnetizing force, H, for magnetic material subjected to a cyclic excitation by alternating or pulsating current. The values of a-c permeability obtained for a given material depend fundamentally upon the excursion limits of dynamic excitation and induction, the method and conditions of measurement, and also upon such factors as resistivity, thickness of laminations, frequency of excitation, etc.

NOTE—The numerical value for any permeability is meaningless unless the corresponding B or H excitation level is specified. For incremental permeabilities not only must the corresponding d-c B or H excitation level be specified, but also the dynamic range (ΔB or ΔH). **E 1316,** E07

permeability, ac, impedance, incremental, $\mu_{\Delta z}$—the value of impedance permeability obtained when ac excitation is superimposed on a dc excitation. **A 340,** A06

permeability, ac, inductance, incremental, $\mu_{\Delta L}$—the value of inductance permeability, μ_L, obtained when the ac excitation is superimposed on a dc excitation. **A 340,** A06

permeability, ac, inductance, μ_L—the value developed from the measured inductive component of the electrical circuit for a material in an *SCM* condition, the permeability is evaluated from the measured inductive component of the electrical circuit representing the magnetic specimen. This circuit is assumed to be composed of paralleled linear inductive and resistive elements, ωL_1 and R_1. **A 340,** A06

permeability, ac, magnetic—a generic term used to represent a dynamic material property. It is expressed as the ratio of the magnetic induction, B, to the magnetic field strength, H, that produced the induction. The value of H may be calculated from several different component values of the exciting current. (See **magnetic field strength, ac,** and various permeabilities.)

NOTE—The numerical value for any permeability is meaningless unless the corresponding B or H excitation level is specified. For incremental permeabilities not only the corresponding dc B or H excitation level must be specified, but also the dynamic excursion limits of dynamic excitation range (ΔB or ΔH). **A 340,** A06

permeability, ac, peak, μ_p—the ratio of the measured peak value of magnetic induction to the peak value of the magnetic field strength, H_p, calculated from the measured peak value of the exciting current. **A 340,** A06

permeability, ac, rms, impedance, μ_z—the ratio of the measured peak value of magnetic induction, B, to the apparent magnetic field strength, H_z, calculated from the rms value of the total exciting current.

NOTE—The value of the current used to compute H_z is obtained by multiplying the measured value of rms exciting current by 1.414. This assumes that the total exciting current is magnetizing current and is sinusoidal.
A 340, A06

permeability coefficient—the steady-state rate of flow of gas through unit area and thickness of a solid barrier per unit pressure differential at a given temperature. **E 1316,** E07

permeability coefficient—a dimensional factor, K, used in the equation:

$$K = q\eta L/\Delta PA,$$

where:
q = flow rate of a given fluid,
η = viscosity of the fluid,
A = area of the medium,
L = thickness of the medium, and
ΔP = pressure differential.

F 740, F21

permeability, dc, μ—a generic term used to represent a number of magnetostatic material properties. The value represented is the ratio of the induction, B, to the dc magnetic field strength, H, producing magnetic flux under the specific magnetizing conditions.

NOTE—The magnetic constant Γ_m is a scalar quantity differing in value and uniquely determined by each electromagnetic system of units. In the cgs–emu system of units, Γ_m is 1 gauss/oersted, and in the SI system, $\Gamma_m = 4\pi \times 10^{-7}$ H/m.

NOTE—Relative permeability is a pure number which is the same in all unit systems. The value and dimension of absolute permeability depends on the system of units used.

NOTE—For any ferromagnetic material permeability is a function of the degree of magnetization. However, initial permeability, μ_0, and maximum permeability, μ_m, are unique values for a given specimen under specified conditions.

NOTE—Except for initial permeability, μ_0, a numerical value for any of the dc permeabilities is meaningless unless the corresponding B or H excitation level is specified.

NOTE—For the incremental permeabilities, μ_Δ and $\mu_{\Delta i}$, a numerical value is meaningless unless both the corresponding values of mean excitation level (B or H) and the excursion range (ΔB or ΔH) are specified.
A 340, A06

permeability, d-c—permeability is a general term used to express relationships between magnetic induction, B, and magnetizing force, H, under various conditions of magnetic excitation. These relationships are either *(1)* absolute permeability, which in general is the quotient of a change in magnetic induction divided by the corresponding change in magnetizing force, or *(2)* relative permeability, which is the ratio of the absolute permeability to the magnetic constant (γ_m).

NOTE—The magnetic constant γ_m is a scalar quantity differing in value and uniquely determined by each electromagnetic system of units. In the unrationalized cgs system γ_m is 1 gauss/oersted and the mksa rationalized system $\gamma_m = 4\pi \times 10^{-7}$ H/m.

NOTE—Relative permeability is a pure number which is the same in all unit systems. The value and dimension of absolute permeability depends on the system of units employed.

NOTE—For any ferromagnetic material, permeability is a function of the degree of magnetization. However, initial permeability, μ_o, and maximum permeability, μ_m, are unique values for a given specimen under specified conditions.

NOTE—Except for initial permeability, μ_o, a numerical value for any of the d-c permeabilities is meaningless unless the corresponding B or H excitation level is specified.

NOTE—For the incremental permeabilities $\mu\Delta$ and $\mu\Delta\ i$, a numerical value is meaningless unless both the corresponding values of mean excitation level (B or H) and the excursion range (ΔB or ΔH) are specified.
E 1316, E07

permeability, dc, absolute, μ_{abs}—the ratio of the total induction, ΔB, to the dc magnetic field strength, ΔH, which produced it. Also described as:

$$\mu_{abs} = \Gamma_m + \mu_i = \Gamma_m \cdot \mu_r$$

A 340, A06

permeability, dc, reversible, μ_{rev}—the ratio of magnetic induction, ΔB, to the dc magnetic field strength increase, ΔH, when the magnetic field strength is first established at a value, H, then reduced by a small increment H, and then reestablished to the value, H. **A 340,** A06

permeability, differential, μ_d—the ratio of an increment of induction, ΔB, to an increment of magnetic field strength, ΔH, for any point on a dc hysteresis loop. It is also the absolute slope ($\Delta B/\Delta H$) of the curve at any point on the normal magnetizing curve.

NOTE—For a symmetrical series circuit in which each component has the same cross-sectional area, reluctance values add directly giving:

$$\mu_{eff} = \frac{\ell_1 + \ell_2 + \ell_3 + \cdots}{\frac{\ell_1}{\mu_1} + \frac{\ell_2}{\mu_2} + \frac{\ell_3}{\mu_3} + \cdots}$$

For a symmetrical parallel circuit in which each component has the same flux path length, permeance values add directly giving:

$$\mu_{eff} = \frac{\mu_1 A_1 + \mu_2 A_2 + \mu_3 A_3 + \cdots}{A_1 + A_2 + A_3 + \cdots}$$

A 340, A06

permeability, ideal, μ_a—the ratio of the magnetic induction to the corresponding magnetic field strength after the material has been simultaneously subjected to a value of ac magnetizing field approaching saturation superimposed on a given dc magnetizing field, and the ac magnetizing field has thereafter been gradually reduced to zero. The resulting ideal permeability is thus a function of the incremental field and residual strongly polarized domains that remain after the ac field is reduced to zero.

NOTE—Ideal permeability, sometimes called anhysteretic permeability, is principally significant to feebly magnetic material and to the Rayleigh range of soft magnetic material. **A 340,** A06

permeability, incremental, μ_Δ—the ratio of the change of magnetic induction, B, to the corresponding change in magnetic field strength, H, under dc biasing conditions and when B is not equal to zero. This value is also the slope of a straight line joining the excursion limits of an incremental hysteresis loop.

NOTE—When the change in H is reduced to zero, the incremental permeability, μ_Δ, becomes the reversible permeability, μ_{rev}.
A 340, A06

permeability, incremental intrinsic, $\mu_{\Delta i}$—the ratio of the change in the intrinsic induction B_i to the corresponding change in magnetic field strength when the mean induction differs from zero.
A 340, A06

permeability, initial, μ_0—the limiting value approached by the normal permeability as the applied magnetic field strength, H, is reduced to zero. The permeability is equal to the slope of the normal induction curve at the origin of linear B and H axes.

A 340, A06

permeability, initial dynamic, μ_{0d}—the limiting value of each of the various ac permeabilities reached in a magnetic material as the magnetizing current is first raised to a moderate value then is progressively and gradually reduced to a zero value. See **initial inductance**.

NOTE—This same value, μ_{0d}, is also equal to the initial values of both impedance permeability, μ_z, and peak permeability, μ_p. **A 340,** A06

permeability, instantaneous—(Coincident with B_{max}), μ_t—with *SCM* excitation, the ratio of the maximum induction B_{max} to the instantaneous magnetic field strength, H_t, which is the value of apparent magnetic field strength, H', determined at the instant when B reaches a maximum. **A 340,** A06

permeability, intrinsic, μ_i—the ratio of the calculated value of intrinsic induction B_i, to the corresponding magnetic field strength, H.

NOTE—See definition of susceptibility. **A 340,** A06

permeability intrinsic, *n* **(L^{-3})**—a measure of the ease with which a porous medium can transmit a fluid under a potential gradient. **D 653,** D18

permeability, maximum, μ_m—the highest value of permeability achieved when the magnetic material is subjected to a symmetrically cyclically magnetized condition.

NOTE—Under dc test conditions the maximum permeability, μ_m, is the highest value of normal permeability μ, developed by the magnetic material.

NOTE—Under ac test conditions, the maximum permeability is the highest value of ac permeability achieved under symmetrically cyclically magnetized conditions and with no biasing magnetic field in the magnetic material. **A 340,** A06

permeability, normal, dc, μ—the ratio of any magnetic induction, B, to the corresponding dc magnetic field strength, H, when the magnetic material has been subjected to *SCM* conditions. **A 340,** A06

permeability of refractories, *n*—the capacity of a refractory for transmitting a fluid (gas or liquid). **C 71,** C08

permeability, relative, μ_r—the ratio of the absolute permeability of a material to the magnetic constant Γ_m, giving a pure numeric parameter.

NOTE—In the cgs-em system of units, the relative permeability is numerically the same as the absolute permeability. **A 340,** A06

permeability test—a test to determine the movement of a liquid through concrete under a hydraulic or pressure gradient. **C 822,** C13

permeability, unoccupied space, μ_v—the permeability of space (vacuum), identical with the magnetic constant, Γ_m. **A 340,** A06

permeability, water vapor— See **water vapor permeability**. **C 168,** C16

permeable—allowing material to pass through. **D 6161,** D19

permeance—the rate of water vapor transmission per unit area at a steady state through a membrane or assembly, expressed in ng/Pa·s·m^2(grain/ft^2·h·in. Hg). **D 1079,** D08

permeance, *n*—the permeation rate divided by the pressure differential of a gas or vapor between opposite faces of a solid body. **D 1566,** D11

permeance, $\mathcal{P}$—the reciprocal of the reluctance of a magnetic circuit. **A 340,** A06

permeance (water vapor), *n*—the ratio of the rate of water vapor transmission (WVT) through a material or assembly between its two parallel surfaces to the vapor pressure differential between the surfaces. Metric unit of measurement is the metric perm, 1 g/24 h·m^2·mm Hg; U.S. unit, 1 grain/h·ft^2·in. Hg. See **water vapor transmission**, **permeability, perm** **C 11,** C11

permeance, water vapor— See **water vapor permeance**. **C 168,** C16

permeate—that portion of the feed stream which passes through a membrane. **D 6161,** D19

permeate collector fabric—see **product (permeate) channel spacer**. **D 6161,** D19

permeation, *n*—the transmissioin of a fluid through a porous medium (NEW). **D 4439,** D35

permeation grouting—filling joints or fractures in rock or pore spaces in soil with a grout, without disturbing the formation. **D 653,** D18

permeation rate, *n*—the flow rate of a gas or vapor, under specified conditions, through a prescribed area of a solid body, divided by that area. **D 1566,** D11

permeator—a reverse osmosis module of the hollow fiber configuration consisting of membrane(s) and pressure vessel. **D 6161,** D19

permittivity—see **relative permittivity** **D 1711,** D09

permittivity, $\kappa(\epsilon)$—a factor giving the influence of an extensive, isotropic dielectric medium on the forces of attraction or repulsion between two electrified bodies (see Discussion 1 of this definition). It is the product of the "relative permittivity," $\kappa'(\epsilon_r)$, and the "permittivity of free space (vacuum)," κ_o (ϵ_o).

$$\kappa = \kappa' \cdot \kappa_o$$

The SI unit is: farad per metre (F/m);

where:

farad (F)	=	s^4A^2/m^2kg,
s	=	seconds,
A	=	amperes,
m	=	metres, and
kg	=	kilograms.

D 2864, D27

permittivity, absolute— the same as **permittivity.** **D 2864,** D27

permittivity, absolute, ϵ'—the magnitude of the real part of the complex permittivity expressed in F/m. **E 1142,** E37

permittivity, complex, ϵ^*—a material property deduced from the ratio of the admittance, Y_p, of a given electrode configuration separated by that material, to the admittance of the identical electrode configuration separated by a vacuum (or air for most practical purposes), Y_v:

$$\epsilon^* = Y_p / Y_v = Y_p / iWC_v = \epsilon' - i\epsilon''$$

where:

ϵ'	=	absolute permittivity
ϵ''	=	loss factor,
C_v	=	capacitance with vacuum separating the electrodes, and
i	=	$\sqrt{-1}$.

E 1142, E37

permittivity, high frequency, $\epsilon\infty'$—the permittivity of a material that is appproached asymptotically as the applied frequency is increased to the point at which dipoles can no longer reorient. **E 1142,** E37

permittivity, low frequency, $\epsilon\infty'$— the permittivity of a material that is approached asymptotically as the applied frequency is decreased to the point at which all dipoles can reorient. **E 1142,** E37

permittivity of free space (vacuum), $\kappa_0(\epsilon_0)$—is defined by the following equation derived from wave theory:

$$\kappa_0 = 1/\mu_0 c^2 = 8.854 \times 10^{-12} \text{ farad per metre}$$

where: the magnetic permeability of free space,

$$\mu_0 = 4\pi \times 10^{-7} \text{ henry per metre; 1 henry} = \text{m}^2\text{kg/s}^2\text{A}^2$$

and the speed of electromagnetic waves in free space,

$$c = 2.998 \times 10^{8} \text{ me tres per second.}$$

D 2864, D27

permittivity of free space (vacuum), ϵ_o—is defined by the following equation derived from wave theory:

$$\epsilon_o = 1/\mu_o \cdot c^2 = 8.854 \times 10^{-12} \text{ F/m}$$

where the magnetic permeability of free space, μ_o is;

$$\mu_o = 4\pi \times 10^{-7} \text{ H/m}$$

and the speed of electromagnetic waves in free space, c, is given by;

$$c = 2.998 \times 10^{8} \text{ m/s.}$$

E 1142, E37

permittivity, relative, κ'—a material property deduced from the ratio of capacitance of a given configuration of electrodes separated by the material, C_p, to the capacitance of the same electrode configuration separated by a vacuum (or air for most practical purposes), C_v:

$$\kappa' = C_p/C_v = \epsilon'/\epsilon_o$$

$$\kappa'_{vac} = 1$$

Experimentally, the material separating the electrodes must replace the vacuum (or air) at all points where it makes a significant change in capacitance. Relative permittivity is commonly referred to as the dielectric constant. The use of κ' (kappa prime) for *relative* permittivity, and ϵ' (epsilon prime) for *absolute* permittivity is recommended and is consistent with the strict definition of terms; however, the scientific literature is not consistent.

E 1142, E37

permittivity, relative complex, $\kappa^*(\epsilon_r^*)$—the ratio of the admittance, Y, of a given set of electrodes with a material as dielectric to the admittance, Y_v, of the same configuration with vacuum as dielectric:

$$\kappa^* = Y/Y_v = Y/j\omega C_v = \kappa' - j\kappa''$$

where:
κ' = relative permittivity,
κ'' = dielectric loss index, and
C_v = capacitance with vacuum as the dielectric.

D 2864, D27

permittivity, relative, $\kappa'(\epsilon_r)$—a dimensionless number, dependent upon the nature and condition of the dielectric medium and upon the frequency of the applied electromagnetic field. For a vacuum $\kappa' \equiv 1.0$. For other dielectrics it is the quotient obtained by dividing the equivalent parallel capacitance, C_p, of a given configuration of electrodes with a material as a dielectric by the capacitance, C_v, of the same configuration of electrodes with vacuum (or air for most practical purposes) as the dielectric:

$$\kappa' = C_p/C_v$$

D 2864, D27

permittivity, (Ψ), (T^{-1}), *n—of geotextiles,* the volumetric flow rate of water per unit cross sectional area per unit head under laminar flow conditions, in the normal direction through a geotextile. **D 4439,** D35

peroxide number, *n*—the milliequivalents of constituents in 1000 g of wax that will oxidize potassium iodide. **D 4175,** D02

peroxides—a class of oxygen-containing compounds possessing a peroxyl-functional group. Hydrogen peroxide is the lowest member of this series. **D 4790,** D16

Persian Morocco—a name given to leather produced with Morocco grain, natural or embossed from hair sheepskin. An ambiguous term, applied in the bag trade to sheepskins and in the book trade mostly to goatskin (as distinguished from "Turkey Morocco"). It should be correctly confined to hair sheepskins. **D 1517,** D31

persians—India-tanned hair sheepskins. **D 1517,** D31

personal air samples—*airborne particulates,* samples of air collected from within the breathing zone of a person, but outside a respirator, if worn. **E 1605,** E06

personal computer, *n*—an electronic device programmable to carry out a variety of functions and intended primarily for use by one operator at a time. **F 1457,** F05

pertinent patient information—information obtained from all available resources that relates to the patient's condition and problems. This information must be continuously updated. All information must be recorded and reported. **F 1177,** F30

pervaporation—a separation process involving vaporization of one liquid from a mixture of two or more liquids, with the aid of a membrane which functions as a barrier to the liquid phase. **D 6161,** D19

pest—an organism existing under circumstances that makes it undesirable. **E 609,** E35

petalite—a lithium mineral of theoretical composition $Li_2O \cdot Al_2O_3 \cdot 8SiO_2$ which transforms on heating to a beta spodumene-silica solid solution product of very low or nil thermal expansion. **C 242,** C21

P2 etch—an etchant used for preparing the surface of aluminum alloys for adhesive bonding. (Syn. **sulfoferric etch.**) **E 1749,** E06

petri plate—a dish designed to contain solid growth medias. **F 1600,** F20

petrographic examination—methods of examining nonmetallic matter under suitable microscopes to determine structural relationships and to identify the phases or minerals present. With transparent materials, the determination of the optical properties, such as the indices of refraction and the behavior in transmitted polarized light, serve as means of identification. With opaque materials, the color, hardness, reflectivity, shape and etching behavior in polished sections serve as means of identification. Metallographic applications include examination of particles mechanically or chemically separated from the metal by these methods. **E 7,** E04

petroleum coke, *n*—a solid carbonaceous residue produced by thermal decomposition of heavy petroleum fractions or cracked stocks, or both. **D 4175,** D02

petroleum-coke-base carbon refractory, n—see **carbon refractory, petroleum-coke-base.** **C 71,** C08

petroleum color scale, *n*—a color scale for petroleum products, defined by 16 glass standards of specified luminous transmittance and chromaticity, graduated in steps of 0.5 from 0.5 for the lightest color to 8.0 for the darkest. **E 284,** E12

petroleum oil—oil derived from petroleum; contains a mixture of hydrocarbons that are broadly classified as paraffins, napthenes, aromatics, or other unsaturates, or combination thereof. **E 1519,** E35

petroleum pitch—a dark brown to black, predominantly aromatic, solid cementitious material obtained by the processing of petroleum, petroleum fractions, or petroleum residuals. **D 1079,** D08

P_{ex}, *n*—the maximum explosion pressure (above the pressure in the vessel at the time of ignition) reached during the course of a single deflagration test. **E 1445,** E27

P/F—the acronym for powder forging. See **powder forging.** **B 243,** B09

pH—the cologarithm (negative logarithm) of the hydrogen ion activity, less precisely, concentration, of a medium, as determined by inhibitors or electrometric means. **B 374,** B08

pH—the negative logarithm of the hydrogen ion concentration, which can be greatly affected by temperature, particularly under alkaline conditions. It is therefore important to measure alkaline earth solutions (AES) at a specific recommended temperature of 25°C. **C 51,** C07

pH—the cologarithm (negative logarithm) of the hydrogen ion activity; less precisely, concentration. **C 859,** C26

pH, *n*—*in common usage*, a measure of the acidity or alkalinity of a solution, on a logarithmic scale, with neutrality represented by a value of 7, with increasing acidity represented by decreasingly smaller values, and with increasing alkalinity represented by increasingly larger values. **D 123,** D13

pH—the negative logarithm of the hydrogen-ion activity in an aqueous solution, or, the logarithm of the reciprocal of the hydrogen-ion activity. **D 1129,** D19

pH—negative logarithm of the effective hydrogen-ion activity, approximately $-\log_{10}[H^+]$ where $[H^+]$ is equivalents per litre concentration. **D 6161,** D19

pH—defined formally as the negative logarithm to the base 10 of the conventional hydrogen ion activity. The pH of an aqueous solution is derived from *E*, the electromotive force (emf) of the cell:

reference | | solution | glass electrode

(where the double vertical line represents a liquid junction) when the electrodes are immersed in the solution, and E_s, the electromotive force obtained when the electrodes are immersed in a standard solution (whose assigned pH is designated pH(S)), by the following equation:

$$pH = pH(S) + \frac{(E - E_s)F}{(RT \ln 10)}$$

where:
F = faraday, 96 487 $C{\cdot}mol^{-1}$,
R = gas constant, 8.314 33 $J{\cdot}K^{-1}{\cdot}mol^{-1}$, and
T = absolute temperature, (t °C + 273.15). **E 1547,** E15

pH—chemical symbol meaning the (p)otential) of (H)ydrogen ion, indicating a measure of acidity or alkalinity of an aqueous solution. **E 1605,** E06

pH—the measurement of the acid concentration of a solution. Range is 0 to 14 (acid to basic), with pH 7 being neutral. **E 1705,** E48

PHA—see **public housing agency**. **E 1605,** E06

phase—a state of matter, either solid, liquid, or gaseous. **D 6161,** D19

phase—a physically homogeneous, mechanically separable portion of a material system. **E 7,** E04

phase—homogeneous, distinguishable portion of a material system. **E 1142,** E37

phase analysis—an analytical technique that discriminates between variables in a part undergoing electromagnetic testing part by the different phase angle changes that these conditions produce in the test signal. See also **phase detection.** **E 1316,** E07

phase angle, θ, *n*—the angle whose cotangent is the dissipation factor, arccot κ″/κ′ and is also the angular difference in the phase between the sinusoidal alternating voltage applied to a dielectric and the component of the resulting current having the same frequency as the voltage. **D 1711,** D09

phase angle, θ—(*1*) the angle whose cotangent is the dissipation factor, arcot κ″/κ′. (*2*) The angular difference in the phase between the sinusoidal alternating voltage applied to a dielectric and the component of the resulting current have the same frequency as the voltage. **D 2864,** D27

phase angle, δ—the angle between a sinusoidally applied strain and the resultant sinusoidal stress. **D 4092,** D20

phase angle, δ—angle between a sinusoidally applied strain and the resultant sinusoidal stress, (D 4092, D-20), . **E 1142,** E37

phase angle, Θ—the angular difference in phase between the sinusoidal alternating voltage applied to a dielectric and the component of the resulting current (to ground) having the same frequency as the voltage. **E 1142,** E37

phase angle—the angular equivalent of the time displacement between corresponding points on two sine waves of the same frequency. **E 1316,** E07

phase angle, dielectric— same as **phase angle.** **D 2864,** D27

phase contrast microscopy—a special method of controlled illumination, ideally suited for observing thin, transparent objects whose structural details vary only slightly in thickness or refractive index. This can also be applied to the examination of opaque materials to determine surface elevation changes. **E 7,** E04

phase correction—the operation in which the effects of an asymmetrical or chirped interferogram are corrected to eliminate instrumental phase contributions. **E 131,** E13

phased application—the installation of a roofing or waterproofing system during two or more separate time intervals; a roofing system not installed in a continuous operation. **D 1079,** D08

phase defect angle— same as **loss angle.** **D 2864,** D27

phase detection—the derivation of a signal whose amplitude is a function of the phase angle between two alternating currents, one of which is used as a reference. **E 1316,** E07

phase diagram—See **constitutional diagram**. **E 7,** E04

phase diagram—see **constitutional diagram**. **E 1142,** E37

phase difference—difference between phase angles of two waves of same frequency. **D 653,** D18

phase distribution, *n*—the partitioning of a semivolatile chemical compound between the gaseous (vapor) and the particle-associated phases in the atmosphere. **D 1356,** D22

phase hologram, *n*—a recording of the variations in light intensity caused by the interference of the reference beam with the object beam as variations in the thickness or index of refraction of the recording medium. The variations in thickness or index refract coherent light to produce the reconstruction. **E 1316,** E07

phase modulation—modulation produced by rapid oscillation of one mirror of a scanning interferometer through an amplitude which is smaller than the shortest wavelength in the spectrum to produce an interferogram which is, to a good approximation, the first derivative of the conventional interferogram. **E 131,** E13

phase of periodic quantity—fractional part of period through which independent variable has advanced, measured from an arbitrary origin. **D 653,** D18

phase rule—the maximum number of phases (*P*) that may coexist at equilibrium is equal to 2, plus the number of components (*C*) in the mixture, minus the number of degrees of freedom (*F*); $P + F = C + 2$. **E 7,** E04

phase rule—see **Gibbs Phase Rule**. **E 1142,** E37

phase-sensitive system—a system whose output signal is dependent on the phase relationship between the voltage returned from a pickup or sensing coil and a reference voltage. **E 1316,** E07

phase separation—the formation of coexisting glassy phases as a result of thermal history within an originally homogeneous glass. **C 162,** C14

phase shift—a change in the phase relationship between two alternating quantities of the same frequency. **E 1316,** E07

pH electrode—an ion-selective electrode, made of glass, that responds to hydrogen ion activity. Wide-range pH electrodes function over the activity range 1 *M* H^+(pH 0) to $10^{-14}M$ H^+(pH 14). pH electrodes may be subject to "acid error" in strongly acidic solutions and are also subject to "alkaline error" caused by response to sodium or other cations in basic solutions. **D 4127,** D19

phenol—hydroxy benzene, carbolic acid (C_6H_5OH) mol weight 94.11; colorless acicular crystals or white crystalline mass; characteristic odor; solidification point, 40.85°C; boiling point, 182°C. **D 4790,** D16

phenolic compounds—hydroxy derivatives of benzene and its condensed nuclei. **D 1129,** D19

phenolic plastic, *n*—a plastic based on resins made by the condensation of phenols, such as phenol and cresol, with aldehydes. **F 1251,** F04

phenolic plastics, *n*—plastics based on resins made by the condensation of phenols, such as phenol or cresol, with aldehydes. **D 883,** D20

phenolic resin—a thermosetting condensation product obtained by reacting phenol with an aldehyde. **C 904,** C03

phenolic resin—under **resin, synthetic,** see *phenolic resin.* **D 16,** D01

phenolic resin compound, single-stage—a phenolic material in which the resin, because of its reactive groups, is capable of further polymerization by application of heat. (See also **phenolic resin compound, two-stage**). **D 883,** D20

phenolic resin compound, two-stage—a phenolic material in which the resin is essentially not reactive at normal storage temperatures, but contains a reactive additive which causes further polymerization upon the application of heat. **D 883,** D20

Philips (Penning) vacuum gage—a sensitive cold cathode ionization gage in which the electron path length, and therefore the number of ionizing collisions, is increased by means of a magnetic field. Also known as a cold-cathode discharge gage. **E 7,** E04

Phillips ionization gage—see **ionization vacuum gage.** **E 1316,** E07

phloem—inner bark; the principal tissue concerned with the translocation of elaborate foodstuffs. A synonym is "bast." **D 9,** D07

pH/mV meter—an instrument that measures and displays the voltage developed between a sensing electrode and a reference electrode placed in a solution, and can convert the voltage developed by a pH electrode to a corresponding pH value. **D 4127,** D19

phosphate-coated, phosphatized—chemically treated to provide iron and steel with gray protective ferric-phosphate coating; to provide increased nail holding power by surface roughening. **F 547,** F16

phosphatized, *adj*—pertains to chemical treatment, in a phosphate solution, of uncoated and metallic coated sheet to prepare the surface for painting without further treatment except normal cleaning. (Syn. *phosphated, phosphate coated*) **A 902,** A05

phosphor—any substance that can be stimulated to emit light by incident radiation. **E 1316,** E07

phosphorescence—the emission of radiant energy from an atom, molecule, or ion resulting from absorption of a photon and a subsequent transition to the ground state with a change in total spin quantum number (see also **intersystem crossing**). **E 131,** E13

phosphorescence, *n*—photoluminescence that continues after excitation ceases. **E 284,** E12

phosphorescent—emitting light of some specific wavelength after being excited by light of a different, usually more energetic wavelength. **F 1156,** F12

photocell, photodetector, photodiode, phototube, n—See **detector.** **E 284,** E12

photocell/photoelectric cell/photo-sensor, *n*—a solid-state, photosensitive, electronic device in which use is made of the variation of current-voltage characteristics as a function of incident radiation (light). **F 1294,** F05

photochemical reaction, *n*—any chemical reaction that is initiated as a result of absorption of light. **D 1356,** D22

photochemical smog, *n*—a type of air pollution resulting from photochemical reactions. **D 1356,** D22

photochromic glass—glass that changes color or optical density under exposure to light; usually a reversible process. **F 1156,** F12

photochromic ink—ink that changes color or optical density under exposure to light; usually a reversible process. **F 1156,** F12

photochromism, *n*—a reversible change in color of a specimen due to exposure to light (or other radiant flux) without appreciable heating of the specimen. **E 284,** E12

photoconductor—a material that exhibits increased electrical conductivity when exposed to light. **F 335,** F05

photodegradable, *adj*—capable of undergoing a significant loss of properties that can be measured by standardized tests after exposure to representative amounts of sunlight. **D 5488,** D10

photodegradable plastic, *n*—a degradable plastic in which the degradation results from the action of natural daylight. **D 883,** D20

photodegradable plastic, *n*—See **degradable plastic.** **D 883,** D20

photodegradation, *n*—photochemically induced changes in the condition of the material. **G 113,** G03

photodiode array (linear) (PDA), *n*—a group of very small photodiodes, arranged in a line, that detects photon energy (light) from the radiation that strikes a surface and measures the amount of reflected light as electrical pulses. **F 1294,** F05

photodiode array (matrix) (PDA), *n*—an area photodiode array, having a two-dimensional configuration. **F 1294,** F05

photoelectric colorimeter, *n*—color-stimulus-measuring instrument using photoelectric detectors in which source-filter-detector response characteristics are adjusted so that the instruments read directly the tristimulus values or related quantities. **D 2946,** C17

photoelectric cross-section—the probability that an incident photon traversing a material will produce a photoelectron from a given subshell, expressed as an area unit per event. **E 673,** E42

photoelectric effect—a dipole interaction involving the interaction of photons with bound electrons in atoms, molecules, and solids, resulting in production of photoelectrons and excited ions. **E 673,** E42

photoelectron satellite peaks—See **photoelectron X-ray satellite peaks,** and **shakeup lines or shakeup satellites.** **E 673,** E42

photoelectron X-ray satellite peaks—photoelectron peaks in a spectrum resulting from photoemission induced by characteristic minor X-ray lines associated with the X-ray spectrum of the anode material. **E 673,** E42

photoelectron X-ray satellite subtraction—the removal of photoelectron X-ray satellite peaks from a spectrum. **E 673,** E42

photoemission—the emission of electrons from atoms or molecules caused by photoelectric effects. **E 673,** E42

photo fluorography—a photograph of the image formed on a fluorescent screen. **E 1316,** E07

photographic density—See **density**. **E 7,** E04

photoluminescence, *n*—luminescence produced by the absorption of radiant flux; distinguished from ordinary reflection by a time delay and, usually, a shift toward longer wavelengths. See **fluorescence** and **phosphorescence**. **E 284,** E12

photomacrograph—a macrograph made by photographic means. **E 7,** E04

photometer—an instrument for measuring photometric quantities, such as luminous intensity or luminous flux. **C 859,** C26

photometer—a device so designed that it furnishes the ratio, or a function of the ratio, of the radiant power of two electromagnetic beams. These two beams may be separated in time, space, or both. **E 131,** E13

photometer, *n*—an instrument for measuring light. **E 284,** E12

photometer, *n*—instrument used for measuring photometric quantities. **E 349,** E21

photometer—an instrument for measuring photometric quantitities, such as luminous intensity or luminous flux. See Terminology C 859, Relating to Nuclear Materials. **E 772,** E44

photometer, *n*—a device that measures the intensity and brightness of a light beam. **F 1294,** F05

photometric, *adj*—pertaining to measurement of quantities in which radiation is evaluated according to the spectral luminous efficiency function $V(\lambda)$. **E 284,** E12

photometric linearity—the ability of a photometric system to yield a linear relationship between the radiant power incident on its detector and some measurable quantity provided by the system. **E 131,** E13

photometry—See the preferred term, **spectrophotometry.** **C 859,** C26

photometry, *n*—the measurement of quantities associated with light, that is, radiation evaluated according to its visual effect, by weighting with the CIE spectral luminous efficiency function $V(\lambda)$. **E 284,** E12

photometry, *n*—measurement of quantities referring to radiation, evaluated according to the visual effect which it produces, as based on certain conventions. **E 349,** E21

photomicrograph—a micrograph made by photographic means. **E 7,** E04

photopic, *adj*—*(1)* pertaining to vision at sufficiently high levels of illumination that only the retinal cones are stimulated.

(2) pertaining to detectors with spectral responsivity weighted according to the CIE spectral luminous efficiency function $V(\lambda)$. **E 284,** E12

photoreceptor—device sensitive to light energy, often producing an electrical signal proportional in magnitude to the intensity of the incident light. **F 1156,** F12

photosedimentation—a technique of fine particle measurement wherein the size and number (or volume) of particles in a sedimenting suspension are determined by the effect of the presence of sedimenting particles on the intensity of a beam of light or X-ray or laser beam transmitted through the suspension as a function of settling time of the particles. **C 242,** C21

photo-stability—durability of the perceived properties of a color [dyestuff] or a material whilst under exposure to light, usually intense. **F 1156,** F12

photostimulable luminescence—the physical phenomenon of phosphors absorbing incident ionizing radiation, storing the energy in quasi-stable states and emitting luminescent radiation proportional to the absorbed energy when stimulated by radiation of a different wavelength. **E 1316,** E07

photostimulable luminescent phosphor—a phosphor capable of storing a latent radiological image which upon laser stimulation will generate luminescence proportional to the radiation intensity. **E 1316,** E07

photozone counter—a stream counter in which the interrogation zone is monitored for changes in a light signal because of the presence of a particle. **C 242,** C21

pH, *pH* **(D)**—an index of the acidity or alkalinity of a soil in terms of the logarithm of the reciprocal of the hydrogen ion concentration. **D 653,** D18

Phragmen method—See **focusing camera**. **E 7,** E04

phreatic eruption, *n*—an explosion of the surface of the earth that results from a sudden increase in the volume of groundwater when it flashes to steam due to contact with hot rocks. **E 957,** E44

phreatic line—the trace of the phreatic surface in any selected plane of reference. **D 653,** D18

phreatic line—see **line of seepage**. **D 653,** D18

phreatic surface—see **free water elevation**. **D 653,** D18

phreatic water—see **free water**. **D 653,** D18

phthalic anhydride—anhydride of phthalic acid ($C_8H_4O_3$) mol weight 148.11; white lustrous needles; odorless; solidification point, 130.8°C; boiling point, 295°C; sublimes. **D 4790,** D16

phthalic anhydride-1308, refined—phthalic anhydride with impurities limited to ultra trace amounts as indicated by a solidification point of 130.8°C minimum refined to improve color characteristics. Refer to Specification D 2403 for complete specifications. **D 4790,** D16

pH unit—a tenfold change in hydrogen ion activity, expressed as the negative logarithm of the hydrogen ion activity. Thus a solution of pH 7 is $10^{-7}M$ in hydrogen ion activity, and a pH 8 solution is $10^{-8}M$. The lower the solution pH, the more acid the solution. Occasionally, ion levels are expressed in p*X* units. Thus pNa = 3.5 is equivalent to $10^{-3.5}M$ Na^+, or $3.0 \times 10^{-4}M$. **D 4127,** D19

physical adsorption (van der Waals adsorption)—the binding of an adsorbate to the surface of a solid by forces whose energy levels approximate those of condensation. **D 2652,** D28

physical-chemical parameters, *n*—values for physical or chemical properties of a permeant or polymer, or both, such as solubility parameters, molecular weight, vapor pressure, and so forth. **F 1494,** F23

physical crack extension, Δa_p [L]—an increase in physical crack size.

$$\Delta ap = ap - a_o$$

E 1823, E08

physical crack size, a_p [L]—the distance from a reference plane to the observed crack front. This distance may represent an average of several measurements along the crack front. The reference plane depends on the specimen form, and normally is taken to be either the boundary, or a plane containing either the load line or the centerline of a specimen or plate. The reference plane is defined prior to specimen deformation. **E 1823,** E08

physical objective aperture—See **aperture, electron**. **E 7,** E04

physical properties—those properties inherent in the raw material which remain unchanged or exhibit only slight alteration in the fastener following manufacture. **F 1789,** F16

physical protection (protection physique)—barriers that will delay or deter someone attempting unauthorized physical access to assets at a specific location. **E 1480,** E06

physical stability—*in asbestos-cement products*, the ability of the product to maintain its physical dimensions and properties when in contact with liquid, gaseous, or solid media normally encountered in its service environment. **D 2946,** C17

physical standard, *n*—stable specimen having a value of a physical quantity assigned by accurate measurements under specified conditions, usually in a standards laboratory. **E 284,** E12

physiology, *n*—the branch of biology dealing with the functions and vital processes of living organisms or their parts and organs. **E 2035,** E52

phytobland oil—a highly refined paraffinic material with a minimum unsulfonated residue of 92 % v/v. **E 1519,** E35

piano leather—a name covering highly specialized leathers used for various purposes in the manufacture of pianos, both manual and mechanical. The leathers are made of selected skivers, full-grain sheepskins, or deerskins. **D 1517,** D31

pick, *n*—an individual filling yarn. **D 123,** D13

pick, *n*—an individual filling yarn of a fabric. **D 1566,** D11

pick, *n*—*in paper*, a defect that occurs in paper when its surface or the coating on its surface (or both) lifts from the paper's surface. **D 1968,** D06

pick count, *n*—*in woven fabrics*, the number of filling yarns per inch of fabric. **D 123,** D13

pick count, *n*—*in braided rope*, the number of strands rotating in one direction in one cycle length. **D 123,** D13

picker leather—leathers used for pickers in textile machinery, and having a wide range of properties. Some are hard rawhide buffalo leathers, others glycerin-treated rawhide and still others belting leather. **D 1517,** D31

picket—See **railing systems.** **E 631,** E06

picket, *n*—synonym for **baluster, baluster bar**. **E 1481,** E06

picket—a helically formed steel wire interwoven in the manufacture of chain link fabric or used for field splicing rolls of fabric into continuous chain link mesh from terminal post to terminal post. **F 552,** F14

picket casting—See **railing systems.** **E 631,** E06

picket fence code, *n*—a bar code printed so that individual bars are vertical (perpendicular to the horizon); the bars look like pickets in a fence. See the reverse, **vertical bar code** or **ladder code.** **F 1294,** F05

picket railing system—See **railing systems.** **E 631,** E06

picking—a rolling up in the lap (see *lap*) when the previously applied paint film is in a semisolid (gel-like) state of drying, causing a tacky resistance to the brush or roller and resulting in an unsightly nonuniform appearance in the final dried film. **D 16,** D01

picking, *n*—the removal of material (fiber, filler or coating) from the paper surface or underlying ink film during printing. **D 6488,** D01

pickle—an acid solution used to remove oxides or other compounds from the surface of a metal by chemical or electrochemical action. **B 374,** B08

pickle—to treat unhaired hides with a solution of salt and acid in order to prepare them for mineral tannage or for temporary preservation until they reach the tannery. **D 1517,** D31

pickle basket—a basket fabricated from corrosion-resistant material to hold ware during pickling. **C 286,** B08

pickle pills—small gelatin capsules containing chemicals used for testing the strength of pickling solutions. **C 286,** B08

pickling—the removal of oxides or other compounds from a metal surface by means of a pickle. **B 374,** B08

pickling—the chemical process of preparing the metal surface for porcelain enameling. **C 286,** B08

pick off, *n*—the undesirable removal of a previously printed ink film in subsequent units of the press due to insufficient drying of the ink. **D 6488,** D01

pick out mark, *n*—a fillingwise band or bar characterized by a chafed or fuzzy appearance. **D 123,** D13

pick-up—the amount of slip retained per unit area on dipped ware. **C 286,** B08

pick-up roll, *n*—a spreading device where the roll for picking up the adhesive runs in a reservoir of adhesive. **D 907,** D14

pictograph—a picture conveying information; for example, traffic control signs, such as those warning of children or pedestrians ahead. **F 1156,** F12

picture element, *n*—in pictures or displays which are made up of tiny dots, each dot is a picture element. **F 1457,** F05

picture framing—a perimeter thickness or color difference (usually darker) relative to the rest of the painted surface. **D 16,** D01

picture framing—a rectangular pattern of ridges in a membrane over insulation or deck joints. **D 1079,** D08

piecing, *n*—a thick place in a spun yarn caused by poor splicing. **D 123,** D13

pie disk, *n*—*in tire X-ray testing*, a circular disk of a specified diameter having six pie-shaped sections, each containing cords of different materials; it is used for demonstrating the discernment capability of an X-ray imaging system. **F 538,** F09

piezoelectric vibratory atomizer—a vibratory atomizer in which a piezoelectric transducer transmits high-frequency oscillations to a liquid, creating relatively uniform drops whose size is a function of the frequency of oscillation. **E 1620,** E29

piezometer—a device used to measure head at a point in the subsurface. **D 653,** D18

piezometric line (equipotential line)—line along which water will rise to the same elevation in piezometric tubes. **D 653,** D18

piezometric surface—the surface at which water will stand in a series of piezometers. **D 653,** D18

piezometric surface—an imaginary surface that everywhere coincides with the static level of the water in the aquifer. (ISRM) **D 653,** D18

pig, *n*—an oblong or square mass of metal that has been cast while still molten into a mold that gives the metal its particular shape; most commonly used for lead and tin in weights that can be handled manually. **B 899,** B02

pig—a shaped object on which to rest a blowpipe or punty during the gathering operation. **C 162,** C14

pig iron, *n*—the high carbon iron product obtained by the reduction of iron ores, typically in a blast furnace or an electric furnace, and cast into uniform shapes (pigs) having physical and chemical characteristics suitable for end use as foundry melting stock. **A 644,** A04

pigment—the fine solid particles used in the preparation of paint or printing ink and substantially insoluble in the vehicle. Asphaltic materials are not pigments except when they contain substances substantially insoluble in the vehicle in which they are used. **D 16,** D01

pigment, *n*—an insoluble compounding material used to impart color. **D 1566,** D11

pigment—water-insoluble solid coloring matter pulverized to a fine particle size and added to asbestos-cement products. **D 2946,** C17

pigment—a substance, generally in the form of fine particles, that is substantially insoluble in the matrix and is used because of its color or decorative properties. **D 3878,** D30

pigment bleed—a condition in which some of the ink pigment is dispersed in the dampening solution and deposits on the non-image area of the plate and then prints as a light background tint. **F 425,** F05

pigment-finished leather—leather finished with compounds containing opaque pigments which more or less conceal the grain pattern. Split leathers are often finished with pigments and embossed to simulate a grain. **D 1517,** D31

pigment volume—the percent by volume of pigment in the nonvolatile portion of a paint or printing ink, as calculated from bulking value and composition data. The letters PV are commonly used as an abbreviation. **D 16,** D01

pigskin—a surface defect characterized by a texture similar to that of pigskin. **C 286,** B08

pigskin—leather made from the skin of pigs or hogs. In the glove leather trade, the term pigskin includes peccary and carpincho. **D 1517,** D31

pile—relatively slender structural element that is driven, or otherwise introduced, into the soil, usually for the purpose of providing vertical or lateral support. **D 9,** D07

pile, *n—for pile yarn floor covering*, the texture surface composed of many tuft legs bound to a backing fabric in an orderly and repetitive array. **D 123,** D13

pile, *n—in pile fabric*, the raised loops or tufts (cut loops) that form all or part of the surface. (See also **cut pile floor covering** and **looped pile floor covering**.) **D 123,** D13

pile—relatively slender structural element which is driven, or otherwise introduced, into the soil, usually for the purpose of providing vertical or lateral support. **D 653,** D18

pile floor covering, *n*—a pile fabric intended for use as a floor covering. The pile may be in the form of cut loops or loops, or both. Both the cut loops and the loops may vary in height. **D 123,** D13

pile height, *n—in determination of tuft height*, a measurement that uses a small graduated ruler insetted until it touches the backing to measure the pile from the top surface of the primary backing to the top of the tuft. **D 123,** D13

pile lay, *n—in floor covering*, the direction in which most of the pile fibers lean in the original, uncrushed carpet. **D 123,** D13

pile lifting, *n*—the process of raising the pile on a textile. **D 123,** D13

pile retention, *n—in corduroy*, the degree to which cut-pile yarns are held secure and intact during wear. **D 123,** D13

pile thickness, *n—in pile yarn floor covering*, the difference in the unextended height of the tuft elements above the backing measured as the difference between two parallel plates exerting a specified compression on the pile and backing and on the backing with the pile removed. **D 123,** D13

pileup—*EIA*, counts in a backscattering spectrum arising from two separate events that occur so closely in time that the signals are not resolved by the detection system and cause counts to be recorded in erroneous channels. **E 673,** E42

pile yarn floor covering, *n*—a textile product in which yarn or yarn segments are attached intermittently to a backing fabric to project above the backing fabric and form a pile, in the form of cut loops or loops, or both, with the yarn entering the backing fabric substantially perpendicular to the plane of the backing fabric. See *cut pile, cut pile floor covering, cut-loop pile, level pile, loop pile, loop pile floor covering, multilevel pile.* **D 5684,** D13

pile yarn mass, *n—for backcoated pile yarn floor covering*, the mass per unit area of pile yarn. **D 123,** D13

piling, *n*—a buildup of paper, ink or coating on the offset blanket, plate or rollers in such a quantity that it interferes with print quality. [GATF] **D 6488,** D01

piling, *n*—a build up of paper or ink on the plate. **D 6488,** D01

piling, *n*—the build up of ink or other materials on the gravure cylinder or impression cylinder. **D 6488,** D01

piling—an accumulation of material on the blanket, rollers, and plate from various sources such as paper, ink, dampeners, and air contaminants. **F 425,** F05

Pilkington process—a process for making flat glass in which the glass continuously pours from a tank onto a spout and thence between forming rolls and is subsequently annealed as one continuous sheet. **C 162,** C14

pillar—in-situ rock between two or more underground openings: crown pillars; barrier pillars; rib pillars; sill pillars; chain pillars; etc. (ISRM) **D 653,** D18

pilling resistance, *n*—resistance to the formation of pills on a textile fabric. **D 123,** D13

pills, *n*—bunches or balls of tangled fibers which are held to the surface of a fabric by one or more fibers. (Compare **fuzz ball.**) **D 123,** D13

pilot—plain-shank section between point and threaded portion of shank. **F 547,** F16

pilot drift (pioneer tunnel)—a drift or tunnel first excavated as a smaller section than the dimensions of the main tunnel. A pilot drift or tunnel is usually used to investigate rock conditions in advance of the main tunnel, to permit installation of bracing before the principal mass of rock is removed, or to serve as a drainage tunnel. (ISRM) **D 653,** D18

piloted airblast nozzle—an airblast nozzle combined with a lower capacity pressure nozzle. **E 1620,** E29

piloted ignition, *n*—ignition of combustible gases or vapors by a pilot source of ignition (compare **spontaneous ignition, unpiloted ignition**). **E 176,** E05

pilot energy rate—average rate of energy consumption (Btu/h or kW) by the equipment's continuous, $E_{pilot\ rate}$.

$$E_{pilot\ rate} = \frac{E_{pilot} \times 60}{t_{pilot}}$$

where:

$E_{pilot\ rate}$	=	pilot energy rate,
E_{pilot}	=	pilot energy consumption during the test period, Btu, kWu or kJ, and
t_{pilot}	=	length of pilot of test period, min.

F 1827, F26

pilot point—point of threaded nail with plain portion of shank between top of point and threaded portion of shank. **F 547,** F16

pilot source of ignition, *n*—a discrete source of energy, such as, for example, a flame, spark, electrical arc, or glowing wire (compare **piloted ignition, unpiloted ignition**). **E 176,** E05

PIM—Powder Injection Molding. **B 243,** B09

pimple—a small solid bump or protrusion on the pipe surface. **C 896,** C04

pimple, *n*—an imperfection, a small, protuberance of varied shape on the surface of a plastic product. **D 883,** D20

pimple, *n*—small, sharp, or conical elevation on the surface of a plastic. **F 412,** F17

pin—small headless nail. **F 547,** F16

PIN—Personal Identification Number. **F 1156,** F12

pin-and-collar specimen, *n*—an assembly of a metal pin bonded inside a metal collar, used to measure the shear strength of an adhesive. **D 907,** D14

pinch, *n*—*in sampling,* as much fiber as can be taken up between the tips of the index finger and thumb. **D 2946,** C17

pincushion distortion—See **distortion.** **E 7,** E04

pineapple—See **railing systems.** **E 631,** E06

pineapple, *n*—a finial in the form of a pineapple. **E 1481,** E06

pine needle oil, *n*—an essential oil of typical fragrance obtained by steam distillation of the leaves (needles) of certain species of pine. **D 804,** D01

pinenes, *n*—bicyclic terpene hydrocarbons, the principal constituent of all turpentines and existing therein in two isomeric forms, alpha-pinene and beta-pinene. **D 804,** D01

pine oil, *n*—a colorless to amber colored volatile oil with characteristic pinaceous odor, consisting principally of isomeric tertiary and secondary cyclic terpene alcohols, with variable quantities of terpene hydrocarbons, ethers, ketones, phenols, and phenolic ethers, the amount and character of which depend on the source and method of manufacture. **D 804,** D01

pine tar, *n*—A product of the destructive distillation of pine wood.

kiln burned, pine tar, n—the heavy, oily liquid resulting from controlled carbonization (slow burning) of pine knots and stumpwood to charcoal in earth-covered piles or ''kilns,'' with introduction of insufficient air to permit complete combustion; contains undecomposed resin acids along with the decomposition products. This product is sometimes called "country tar."

pine tar oil, n—the oil obtained by condensing the vapors from the retorts in which resinous pine wood is destructively distilled (carbonized).

retort, pine tar, n—the tar produced by removal of volatile oils from pine tar oil by steam distillation. Several grades are marketed, namely: Thin, Medium, Heavy, and Extra Heavy, so classified on the basis of viscosity, and depending upon the quantity of volatile oils removed.

Stockholm, pine tar, n—kiln-burned pine tar produced in Scandinavian countries from wood of the Northern European pine, *Pinus sylvestris.* **D 804,** D01

pinhole, *n*—*in fabrics,* a very small hole, approximately the size of the cross section of a pin. **D 123,** D13

pinhole—a tiny hole in a film, foil, or laminate comparable in size to one made by a pin. **D 1079,** D08

pinhole—a hole through a liquid-applied membrane caused by the bursting of blisters or a pin point protrusion on the deck surface. **D 3743,** D04

pinhole—*in protective coatings,* minute holes through a coat or coats that expose an underlying coat or the substrate. **D 4538,** D33

pinhole—an opening in the carbon tissue that permits the coating to seep through to the uncoated side. **F 221,** F05

pinhole eyepiece—an eyepiece, or a cap to place over an eyepiece, which has a small central aperture instead of an eye lens. Used in adjusting or aligning microscopes. **E 7,** E04

pinhole ocular—See **pinhole eyepiece.** **E 7,** E04

pinhole, pinholing—a porcelain enamel surface defect caused by gas evolution and characterized by a smll hole resembling a pin prick that may extend to the base metal. **C 286,** B08

pinholes—small holes occurring as imperfections which penetrate entirely through the copper foil. **B 846,** B05

pinholes—imperfections in the surface of a ceramic body or glaze resembling pin pricks. **C 242,** C21

pinholes—small pore-like flaws in a coating that extend entirely through the applied film and have the general appearance of pin pricks when viewed by reflected light. **D 16,** D01

pinholes—small pore-like flaws in a coating that extend entirely through the applied film and have the general appearance of pin pricks when viewed by reflected light. **D 4538,** D33

pinholes, *n*—small voids in the printed area caused by the failure of ink to form a continuous ink film. **D 6488,** D01

pinholes—very small holes; sometimes found as a type of porosity in a casting because of microshrinkage or of gas evolution during solidification; or, in wrought products, due to removal of inclusions or microconstituents during macroetching of transverse sections. In photography, a very small circular aperture. **E 7,** E04

pinhole system—a group of two or more pinholes arranged to define a beam. **E 7,** E04

pinholing, *n*—the presence of a series of fine holes or voids in a film. **D 16,** D01

pinholing, *n*—the appearance of fine voids in the print area caused by nonuniform wetting of the substrate or some other mechanical problem. **D 6488,** D01

pinked seam finish, *n*—in home sewing, a seam finish in which a zigzagged or scalloped cut edge is produced. (Compare **stitched and pinked seam finish**). **D 4965,** D13

pink noise—noise with a continuous frequency spectrum and with equal power per constant percentage bandwidth. For example, equal power in any one-third octave band. **C 634,** E33

pin lock slider, *n*—a slider that incorporates a projection on the pull that fits between adjacent interlocking elements of a zipper when a pin lock slider is in the locked position. **D 123,** D13

pin mark, *n*—a series of holes near the edge parallel with the lengthwise direction of a fabric caused by the holding device on a pin tenter frame. (See also **clip mark.**) **D 123,** D13

pin mark or point mark—a visible imprint on the back of ware left by processing tools; sometimes synonymous with burning tool mark. **C 286,** B08

pin seal—natural grain sealskin tanned for fancy leather. Imitations on other skins should be described as "pin-grain sheepskin," "pin-grain goatskin," etc. **D 1517,** D31

pintle atomizer—an impact atomizer in which a liquid jet is directed against a fixed axisymmetric surface. (See also **variable-orifice poppet nozzle.**) **E 1620,** E29

pipe, *n*—tube in standardized combination of outside diameter and wall thickness, commonly designated by "Nominal Pipe Sizes" and "ANSI Schedule Numbers."

drawn pipe, n—pipe brought to final dimensions by drawing through a die.

extruded pipe, n—pipe formed by hot extruding.

seamless pipe, n—extruded or drawn pipe which does not contain any line junctures resulting from the method of manufacture.

structural pipe, n—pipe commonly used for structural purposes. **B 881,** B07

pipe, *n*—a tubular metal product, cast or wrought, of dimensions that conform to those referred to commercially as standard pipe sizes. **B 899,** B02

pipe—a tube or hollow concrete structure. **C 822,** C13

pipe—See **vitrified clay pipe.** **C 896,** C04

pipe, *n*—a tubular conduit for transport of fluids or finely divided solids; also, a hollow structural member or safety barrier; a hollow product of round cross section. **E 631,** E06

pipe , *n*—hollow round section of metal or other material, the size of which is usually designated by nominal size, in inches (millimeters), as influenced by inside diameter and wall thickness. **E 1481,** E06

pipe body—the clay material or mixture of clay materials from which vitrified clay pipe is made. **C 896,** C04

pipe bursting—process by which existing pipelines are broken by mechanical fracturing from either inside or outside with the remains being pushed into the surrounding soil while simultaneously inserting a new pipeline of equal or larger diameter. **C 896,** C04

pipe diameter—the inside diameter of a concrete pipe. **C 822,** C13

pipe insulation, *n*—insulation in a form suitable for application to cylindrical surfaces. **C 168,** C16

pipeline—pipe sections joined together. **C 822,** C13

pipeline—pipes joined to provide a conduit through which fluids flow. **C 896,** C04

pipe nipple—a short length of pipe with male threads at both ends. **B 846,** B05

pipe railing system—See **railing system.** **E 631,** E06

pipe railing system—a railing system fabricated of pipe or round tubing. **E 1481,** E06

pipe, seamless—seamless tube conforming to the particular dimensions commercially known as "standard pipe sizes." **B 846,** B05

pipe section—a single pipe. **C 822,** C13

pipe spigot—portion of a pipe or fitting which fits into a bell or socket of a preceeding pipe or fitting. **F 412,** F17

pipe, threadless—a seamless copper tube of standard pipe outside diameters conforming to particular dimensions commercially known as threadless pipe (TP). **B 846,** B05

pipe zone backfill—backfill in the area of the pipe, may be specified for depth and compaction. **F 412,** F17

piping—the progressive removal of soil particles from a mass by percolating water, leading to the development of channels. **D 653,** D18

piping, *n*—a system of **pipes.** **E 631,** E06

pipy—see **flanky.** **D 1517,** D31

piracy—the unauthorized publication, reproduction, use or imitation of an invention, creation, or product with the objective of having the result pass as genuine; similar to **counterfeiting**, but broader. **F 1156,** F12

Pirani gage—see **thermal conductivity vacuum gage.** **E 1316,** E07

pirani vacuum gage—a thermal conductivity or hot-wire gage in which the temperature of an electrically heated fine wire varies as the thermal conductivity of the residual gas. The wire has a high temperature coefficient of electrical resistance and the change in resistance is usually measured in a bridge circuit. (See **thermocouple vacuum gage**.) **E 7,** E04

piscicide—a chemical used to kill, control, or cause other adverse effects on fish. **E 609,** E35

pit—a small depression or cavity produced in a metal surface during electrodeposition or by corrosion. **B 374,** B08

pit—a defect similar to a dimple but slightly smaller. **C 286,** B08

pit—an excavation in the surface of the earth from which ore is obtained as in large open pit mining or as an excavation made for test purposes, that is, a testpit. (ISRM) **D 653,** D18

pit, *n*—an imperfection, a small crater in the surface of the plastic, with its width of approximately the same order of magnitude as its depth. **D 883,** D20

pit—tiny depression or hole on the grain surface of leather, due to natural causes or manufacturing. **D 1517,** D31

pit—a shallow depression or crater in which all surfaces are visible by normal (20/20) vision under 200 fc of illumination.

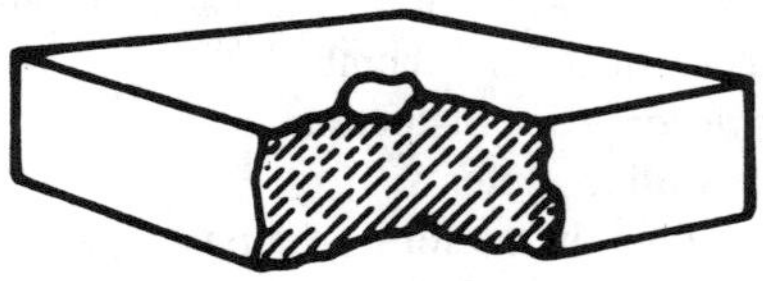

F 109, C21

pit, *n*—small crater in the surface of the plastic, with its width approximately the same order of magnitude as its depth. (D-20) **F 412,** F17

pit—a depression in a wafer surface that has steeply sloped sides that meet the surface in a distinguishable manner in contrast to the rounded sides of a dimple. **F 1241,** F01

pit, *n*—*in plastics*, an imperfection, a small crater in the surface, the depth and width of which are approximately the same order of magnitude. **F 1251,** F04

pitch—the distance from center point to center point of adjacent crests of an asbestos-cement product of corrugated or grooved cross-section. **C 460,** C17

pitch, *n*—*in cooling tower fill*, the wave length of corrugated products. **C 460,** C17

pitch, *n*—*in cooling tower fill*, the wave length of corrugated products. **C 1154,** C17

pitch—a term applied to the resin occurring in the wood of certain conifers. **D 9,** D07

pitch—a term to describe a variety of solid or semi solid products isolated from trees as follows:

Archangel pitch, n—originally a genuine pine pitch made from pine tar in the Archangel district of Russia; in this country a similar product is made from residues of pine origin blended with various oils to make a pitch for caulking boats.

Brewer's pitch, n—a term used to designate a type of pitch made by blending certain oils, waxes or other ingredients with rosin for the coating of beer barrels.

Burgundy pitch, n—originally the solidified resin obtained by heating and straining the air-dried solid oleoresin exuded by the Norway spruce (*Picea excelsa*) and European silver fir (*Abies pectinata*); now denotes an artificial mixture made by heating rosin with certain fixed oils, the combination being used for adhesive plasters.

Navy pitch, n—a pitch obtained by melting rosin with pine tar, with or without rosin distillation residues. The terms Archangel pitch, Brewer's pitch, Burgundy pitch, and Navy pitch have been in use in the naval stores industry for many years. They cover a wide variety of compounded products in which the one constituent common to all is a large proportion of rosin. **D 804,** D01

pitch—See **incline**; **coal-tar pitch**; or **petroleum pitch**. **D 1079,** D08

pitch—the distance from center point to center point of adjacent crests of an asbestos-cement product of corrugated or grooved cross-section. **D 2946,** C17

pitch, *n*—*in cooling tower fill*, the wave length of corrugated products. **D 2946,** C17

pitch, *n*—an inclination or slope measured in degrees, or percent, or by the ratio of rise and run. **E 631,** E06

pitch—the distance between one character reference point and the corresponding point on the next adjacent character. **F 149,** F05

pitch angle, (rad or degree), *n*—*in a vehicle*, the angle between its *X*-axis and the ground plane. **F 538,** F09

pitch-bonded basic brick, *n*—unburned basic refractory shapes bonded with pitch; if subsequently heat-treated sufficiently to minimize softening of the bond on reheating, they are referred to as *tempered.* **C 71,** C08

pitch diameter—for all practical purposes, the diameter of the smooth shank prior to threading. **F 547,** F16

pitches, *n*—black or dark-brown solid cementitious materials which gradually liquefy when heated and which are obtained as residua in the partial evaporation or fractional distillation of tar. **D 8,** D04

pitch-impregnated basic brick, *n*—burned basic refractory shapes impregnated with pitch after firing. **C 71,** C08

pitch line—line parallel to nail axis located at distance equal to one-half of pitch diameter from nail axis. **F 547,** F16

pitch pocket—an opening extending parallel to the annual growth rings containing, or that has contained, pitch, either solid or liquid. **D 9,** D07

pitch pocket—a flanged, open-bottomed metal container placed around a column or other roof-penetration, and filled with hot bitumen or flashing cement to seal the joint. **D 1079,** D08

pitch polishing—polishing operation in which pitch rather than felt is the resilient carrier for the polishing agent. **C 162,** C14

pitch seam—a shake or check filled with pitch. **D 9,** D07

pitch streak—a well-defined accumulation of pitch in a more or less regular streak in the wood of certain conifers. **D 9,** D07

pith—the small, soft tissue occurring in the structural center of a tree trunk, branch, twig, or log. **D 9,** D07

pith fleck—a narrow streak, resembling pith on the surface of a piece; usually brownish, up to several inches in length; resulting from burrowing of larvae in the growing tissues of the tree. **D 9,** D07

pit, ***n*****, (PT^{s})**—*as used in fractography*, a surface-distributed cavity created on the specimen/component due to an interaction/reaction between the material and the testing/service environment, for example, corrosion, oxidation. **C 1145,** C28

pits—small holes having jagged edges, occurring as imperfections which do not penetrate entirely through the copper foil. **B 846,** B05

pits—small depressions, voids or pinholes in stone, especially on a finished surface. **C 119,** C18

pitting, *n*—small defects in the surface of the photoreceptor that produce spots or voids on the printout. **F 1457,** F05

pitting—corrosion of a metal surface, confined to a point or small area, that takes the form of cavities. **G 15,** G01

pitting, *n—in tribology*, a form of wear characterized by the presence of surface cavities the formation of which is attributed to processes such as fatigue, local adhesion, or cavitation. **G 40,** G02

pitting factor—ratio of the depth of the deepest pit resulting from corrosion divided by the average penetration as calculated from weight loss. **G 15,** G01

Pittsburgh sheet process—the method of making sheet glass by drawing vertically upward from a free bath surface wherein definition of draw is established by a submerged refractory member. **C 162,** C14

pixel—The smallest addressable element in an electronic image. **E 1316,** E07

pixel—**picture element.** **F 1457,** F05

pixel, display size—the dimensions of the smallest picture element comprising the displayed image, given in terms of the imaged object's dimensions being represented by the element. **E 1316,** E07

pixel (picture element)—smallest spatial unit of an image. **E 7,** E04

pixels—picture elements, or cells. **F 1156,** F12

pixel size—the length and width of a pixel. **E 1316,** E07

pizza oven—See *conveyor oven* or *deck oven.* **F 1827,** F26

placement, *n*—the process of placing and consolidating concrete; a quantity of concrete placed and finished during a continuous operation; also inappropriately called pouring. **C 717,** C24

placket, *n—in garment construction*, a finished garment opening that is usually dependent on mechanical devices called closures to secure the opening. **F 1494,** F23

Plackett-Burman designs, *n*—a set of screening designs using orthogonal arrays that permit evaluation of the linear effects of up to $n = t - 1$ factors in a study of t, treatment combinations. **E 456,** E11

Plackett-Burman designs, *n*—a set of screening designs using orthogonal arrays that permit evaluation of the linear effects of up to $n = t - 1$ factors in a study of t, treatment combinations. **E 1325,** E11

plain—{archaic} relatively free of gaseous inclusions. **C 162,** C14

plain braid, *n—in rope*, a braided construction in which one strand of one direction of rotation about the axis of the rope passes over one strand of the opposite direction. (See **braided rope**.) **D 123,** D13

plain conductor—a conductor made up of one or more plain wires of the same metal. **B 354,** B01

plain jet atomizer—a pressure atomizer comprising a simple orifice designed to produce an unstable liquid jet. **E 1620,** E29

plain orifice atomizer—see **plain jet atomizer.** **E 1620,** E29

plain paper copying—a form of indirect electrophotographic copying in which the electrostatic image of the original is first formed on a photoconductor element, then developed with a toner, and finally transferred onto a plain sheet of paper where it is fixed (see also Terminology F 1156). **F 335,** F05

plainsawn—A synonym for *flat grain.* **D 9,** D07

plain seam, *n—in home sewing*, a seam formed by a single line of stitching. (Ant. **complex seam.**) **D 4965,** D13

plain-shank—term applied to nail without shank deformations, disregarding grip-marks. **F 547,** F16

plain steel—bright steel. **F 547,** F16

plain washer—a fastener accessory that accepts a bolt or screw through its center hole and provides a surface to distribute bearing stress. It also serves to provide a surface for head or nut rotation during tightening. **F 1789,** F16

plain weave, *n*—a fabric pattern in which each yarn of the filling passes alternately over and under a yarn of warp and each yarn of the warp passes alternately over and under a yarn of the filling. **D 123,** D13

plain wire—wire consisting of one metal only. **B 354,** B01

plaited rope, *n*—rope made from eight strands arranged in four pairs in which one strand is placed adjacent to a second in each pair and in which each strand of each pair has been twisted in one direction while each strand in each alternate pair has been twisted in the opposite direction and the four pairs of strands are intertwined maypole fashion in a manner such that each pair of strands passes over and under adjacent pair of strands. **D 123,** D13

planar grinding—the first step in a preparation procedure used to bring all specimens into the same plane of polish. It is unique to semi or fully automatic preparation equipment that utilize specimen holders. **E 7,** E04

planchets—tiny, thin, flat particles used to mark something by their presence in or on an item. **F 1156,** F12

Planckian locus, *n*—locus in a chromaticity diagram that represents the chromaticities of the radiation of Planckian radiators at different temperatures. **E 284,** E12

Planckian radiator, n— See **full radiator**. **E 284,** E12

plane, *adj*—having a flat, smooth surface with no significant variations such as elevations or depressions. **E 284,** E12

plane bed—a sedimentary stream bed without elevations or depressions larger than the maximum size of the bed material. **D 4410,** D19

plane (crystal)—an idiomorphic face of a crystal. Any atom-containing plane in a crystal. **E 7,** E04

planed-to-caliper hardboard—hardboard that is machined to a close thickness tolerance. **D 1554,** D07

plane, focal—a plane through the focal point perpendicular to the principal axis of a lens or mirror. **E 175,** E41

plane glass illuminator—a thin transparent flat glass disk interposed in a microscope or a lens imaging system for the purpose of directing light to the object without causing a reduction of the useful aperture of the lens system. **E 7,** E04

plane of weakness—surface or narrow zone with a (shear or tensile) strength lower than that of the surrounding material. (ISRM) **D 653,** D18

plane of working—the plane of maximum area extension. **E 7,** E04

plane-strain (chevron-notch) fracture toughness, K_{IvM} [$FL^{-3/2}$]—determined similarly to K_{Iv} or K_{Ivj} using the same specimen, or specimen geometries, but using a simpler analysis based on the maximum test load. Unloading-reloading cycles are not required in a test to determine K_{IvM} . **E 1823,** E08

plane-strain (chevron-notch) fracture toughness, K_{Iv} or K_{Ivj} [$FL^{-3/2}$]—under conditions of crack-tip plane strain in a chevron-notched specimen: K_{Iv} relates to extension resistance with respect to a slowly advancing steady-state crack. K_{Ivj} relates to crack extension resistance with respect to a sporadically advancing crack. **E 1823,** E08

plane-strain crack-arrest fracture toughness, K_{Ia} [$FL^{-3/2}$]—the value of crack-arrest fracture toughness, K_a , for a crack that arrests under conditions of crack-front plane-strain. **E 1823,** E08

plane-strain fracture toughness, K_{Ic} [$FL^{-3/2}$], J_{Ic} [FL^{-1}]—the crack-extension resistance under conditions of crack-tip plane strain. **E 1823,** E08

plane-stress fracture toughness, K_c [$FL^{-3/2}$]—crack-extension resistance under conditions that do not approach crack-tip plane strain to the degree required by an empirical criterion. **E 1823,** E08

plane stress (strain)—a state of stress (strain) in a solid body in which all stress (strain) components normal to a certain plane are zero. (ISRM) **D 653,** D18

plane wave—wave in which fronts are parallel to plane normal to direction of propagation. **D 653,** D18

planimetric method— See **Jeffries' method**. **E 7,** E04

plank—See **lumber. D 996,** D10

planographic printing—a method of printing in which the image and non-image areas of the printing plate are in the same plane. **F 425,** F05

plans and specifications, *n—in pipe laying*, documents prepared by the engineer or owner, or both, stipulating work to be done and materials to be used which, combined with other contract documents and Guide C 966, form the basis for a comprehensive contract between the owner and the contractor. **C 460,** C17

plans and specifications, *n—in pipe laying*, documents prepared by the engineer or owner, or both, stipulating work to be done and materials to be used which, combined with other contract documents and Guide C 966, form the basis for a comprehensive contract between the owner and the contractor. **C 1154,** C17

plans and specifications, *n—in pipe laying*, documents prepared by the engineer or owner, or both, stipulating work to be done and materials to be used which, combined with other contract documents and Guide C 966, form the basis for a comprehensive contract between the owner and the contractor. **D 2946,** C17

plantar—pertaining to the sole of the foot. **F 869,** F08

plant capacity—manufacture of product per unit time, expressed as m^3/day, m^3/h, GPD, MGD. **D 6161,** D19

plant mix, cold-laid, *n*—a mixture of cut-back asphalt, bituminous emulsion, or tar and mineral aggregate prepared in a central bituminous mixing plant and spread and compacted at the job-site when the mixture is at or near ambient temperature. **D 8,** D04

plant mix, hot-laid bituminous emulsion mixtures, *n*—a mixture of emulsion and heated mineral aggregate usually prepared in a conventional asphalt plant or drum mixer and spread and compacted at the job site at a temperature above ambient. **D 8,** D04

plant tests—the quality assurance tests performed prior to delivery as a basis of acceptance. **C 822,** C13

plaque—a round, clear area in a layer of host cells caused by virus growth and resultant killing or lysis of the cells. **E 1705,** E48

plaque, *n—as in virology*, a visible, clear area, which is theoretically the result of the infection and lysis of host cells by a single viable virus. **F 1494,** F23

plaque forming unit (PFU), *n*—a virus particle capable of producing plaques by infecting and lysing bacteria in a lawn in top agar. **F 1494,** F23

plasma—a partially or totally ionized gas or vapor. **B 542,** B02

plasmids—independently-replicating extrachromosomal genetic material. **F 1600,** F20

plasmon loss lines—peaks in an electron spectrum that are due to certain characteristic energy losses of electrons emitted from the specimen. These losses occur as a result of the excitation of collective oscillations among the valence band electrons. **E 673,** E42

plaster—see **gypsum plaster, gypsum neat plaster. C 11,** C11

plaster—{archaic} partially dehydrated gypsum used with water for bedding plates of glass to tables during the grinding and polishing of plate glass. **C 162,** C14

plaster-base finish, *n*—units whose surfaces are intended for the direct application of plaster. **C 43,** C15

plaster-base nail, plasterboard nail, plaster-board nail—See **gypsum-lath nail, wood-lath nail. F 547,** F16

plaster bond, *n*—the state of adherence between plaster coats or between plaster and a plaster base, produced by adhesive or mechanical interlock of plaster with base or special supplementary materials. **C 11,** C11

plastic—a descriptive term applied to a material that exhibits the property of plasticity or stickiness, where plasticity is the ability of a material to undergo substantial deformation without fracturing. **C 242,** C21

plastic—(*1*) capable of being deformed continuously and permanently in any direction without rupture, under a stress exceeding the yield value. (*2*) made of, consisting of or pertaining to plastics. (*3*) a material containing as an essential ingredient an organic substance of large molecular mass, which is solid in its finished state and, at some stage in its manufacture or in its processing into finished articles, can be shaped by flow. **D 1695,** D01

plastic, *n*—any of numerous polymeric materials that are usually thermoplastic or thermosetting, of high molecular weight and that can be molded, cast extruded, drawn, laminated, or otherwise fabricated into objects, powders, beads, films, filaments, fibers, or other shapes (*Webster Modified*). **F 1251,** F04

plastic, *n*—a material that contains, as an essential ingredient, one or more organic polymeric substances of large molecular weight, is solid in its finished state, and, at some stage in its manufacture of processing into finished articles, can be shaped by flow. **F 1494,** F23

plastic cement, *n*—a hydraulic cement to which one or more plasticizing agents (but not more than 12 % by volume) have been added during intergrinding or blending to increase the workability and molding qualities of the resultant cement paste, mortar, or plaster. **C 11,** C11

plastic cement—See **flashing cement**. **D 1079,** D08

plastic coated—covered fully or partially with polymer to provide ease of driving, increased holding power, or corrosion resistance, or a combination of these. **F 547,** F16

plastic conduit—plastic pipe or tubing used as an enclosure for electrical wiring. **F 412,** F17

plastic deformation—see **plastic flow**. **D 653,** D18

plastic deformation—permanent distortion of a material under the action of applied stresses. **F 1789,** F16

plastic equilibrium—state of stress within a soil or rock mass or a portion thereof, which has been deformed to such an extent that its ultimate shearing resistance is mobilized.

active state of plastic equilibrium—plastic equilibrium obtained by an expansion of a mass.

passive state of plastic equilibrium—plastic equilibrium obtained by a compression of a mass. **D 653,** D18

plastic flow (plastic deformation)—the deformation of a plastic material beyond the point of recovery, accompanied by continuing deformation with no further increase in stress. **D 653,** D18

plastic foam, n—See **cellular plastic** (the preferred terminology). **D 883,** D20

plasticity, *n*—that property of freshly mixed cement paste, mortar, or plaster which determines its workability and molding qualities. **C 11,** C11

plasticity—the property of a soil or rock which allows it to be deformed beyond the point of recovery without cracking or appreciable volume change. **D 653,** D18

plasticity—property of a material to continue to deform indefinitely while sustaining a constant stress. (ISRM) **D 653,** D18

plasticity, *n*—a property of adhesives that allows the material to be deformed continuously and permanently without rupture upon the application of a force that exceeds the yield value of the material. **D 907,** D14

plasticity, *n*—a characteristic of unvulcanized rubber indicated by the degree of retention of deformation after removal of the deforming force. **D 1566,** D11

plasticity—a property of a material that allows it to be deformed continuously and permanently without rupture upon the application of a force that exceeds the yield value of the material. (D 907, D-14) **F 869,** F08

plasticity—the ability of the metal to undergo permanent deformation without rupture. **F 1789,** F16

plasticity index, I_p, *PI*, I_w **(D)**—numerical difference between the liquid limit and the plastic limit. **D 653,** D18

Plasticity Retention Index (PRI), *n*— for natural rubber, the ratio of a "plasticity number" measured after air oven ageing, under specified conditions, to a "plasticity number" measured before ageing. **D 1566,** D11

plasticizer—a substance added to paint, varnish, or lacquer to impart flexibility. **D 16,** D01

plasticizer—in grouting, a material that increases the plasticity of a grout, cement paste, or mortar. **D 653,** D18

plasticizer, *n*—a substance incorporated in a material to increase its workability, flexibility, or distensibility. **D 883,** D20

plasticizer, *n— in an adhesive*, a substance added to increase softness, flexibility, and extensibility. **D 907,** D14

plasticizer, *n*—a compounding material used to enhance the deformability of a polymeric compound. **D 1566,** D11

plasticizer, *n*—a substance incorporated in a material to increase its workability, flexibility, or distensibility. (D-20) **F 412,** F17

plasticizer—a compounding material used to enhance the deformability of a polymeric compound. (D 1566, D-11) **F 869,** F08

plasticizer, *n*—a substance incorporated into a material to increase its workability, flexibility, or distensibility of the material. **F 1251,** F04

plastic limit, w_p, *PL*, P_w **(D)**—(*a*) the water content corresponding to an arbitrary limit between the plastic and the semisolid states of consistency of a soil. (*b*) water content at which a soil will just begin to crumble when rolled into a thread approximately ⅛ in. (3.2 mm) in diameter. **D 653,** D18

plastic or bond fireclay, n— see **fireclay, plastic or bond.** **C 71,** C08

plastic pipe—a hollow cylinder of a plastic material in which the wall thicknesses are usually small when compared to the diameter and in which the inside and outside walls are essentially concentric. **D 883,** D20

plastic pipe—a hollow cylinder of a plastic material in which the wall thicknesses are usually small when compared to the diameter and in which the inside and outside walls are essentially concentric. (D-20) **F 412,** F17

plastic pressing—See wet pressing under **pressing.** **C 242,** C21

plastic refractory, *n*—a refractory material, tempered with water, that can be extruded and that has suitable workability to be pounded into place to form a monolithic structure. **C 71,** C08

plastic replica—See **replica**. **E 7,** E04

plastic(s), *n*—a material that contains as an essential ingredient one or more organic polymeric substances of large molecular weight, is solid in its finished state, and, at some stage in its manufacture or processing into finished articles, can be shaped by flow. **D 883,** D20

plastic(s), *n*—a material that contains as an essential ingredient one or more organic polymeric substances of large molecular weight, is solid in its finished state, and, at some stage in its manufacture or processing into finished articles, can be shaped by flow.

NOTE—Rubber, textiles, adhesives, and paint, which may in some cases meet this definition, are not considered plastics. See ASTM definitions of these terms.

NOTE—The above definition may be used as a separate meaning to the definitions contained in the dictionary for the adjective "plastic."

NOTE—The plural form may be used as an adjective to refer to two or more plastic materials, for example, plastics industry. However, when the intent is to distinguish "plastic products" from "wood products" or "glass products," the singular form should be used. As a general rule, if the adjective is to restrict the noun modified with respect to the type of material, "plastic" should be used; if the adjective is to indicate that more than one type of plastic material is or may be involved, "plastics" is permissible. (D-20) **F 412,** F17

plastic(s)—a material that contains as an essential ingredient one or more organic polymeric substances of large molecular weight, is solid in its finished state, and, at some stage in its manufacture or processing into finished articles, can be shaped by flow. (F 412, F-17) **F 869,** F08

plastic soil—a soil that exhibits plasticity. **D 653,** D18

plastic state (plastic range)—the range of consistency within which a soil or rock exhibits plastic properties. **D 653,** D18

plastic true strain, ϵ_p, *n*—the inelastic component of true strain. **E 6,** E28

plastic tubing, *n*—(*1*) a particular size of plastic pipe in which the outside diameter is essentially the same as the corresponding size of copper tubing.
(*2*) small diameter flexible pipe. (See **plastic pipe.**) **D 883,** D20

plastic tubing, *n*—a particular size of smooth wall plastic pipe in which the outside diameter is essentially the same as the corresponding size of copper tubing. **F 412,** F17

plastic-zone adjustment, r_Y **[L]**—an addition to the physical crack size to account for plastic crack-tip deformation enclosed by a linear-elastic stress field. **E 1823,** E08

plastisol—a suspension of a finely divided resin in a plasticizer, that can be converted to a continuous film by the application of heat. Distinct from baking enamels etc., in that substantially all the original mixture becomes a part of the film; there is no significant evaporation of solvent. The films are usually much thicker than obtainable from coatings which depend on the evaporation of a volatile solvent. **B 374,** B08

plastisol, *n*—a liquid suspension of a finely divided PVC polymer or copolymer in a plasticizer. **D 883,** D20

plastisol, *n*—a suspension of a finely divided polymer in a plasticizer. **D 1566,** D11

plastisol—a suspension of a finely divided polymer in a plastizer. (D 1566, D-11) **F 869,** F08

plastisol, *n*—a liquid suspension of a finely divided PVC polymer or copolymer in a plasticizer. **F 1251,** F04

plastometer, *n*—an instrument for measuring the plasticity of a material. **D 1566,** D11

plastometer—an instrument for measuring the plasticity of a material. (D 1566, D-11) **F 869,** F08

plate—*as in inch-pound specifications,* a wrought flat product over 0.188-in. thick and over 12-in. wide, in straight lengths or coils (rolls). **B 846,** B05

plate—*as in SI specifications,* a wrought flat product over 5-mm thick and over 300-mm wide, in straight lengths or coils (rolls). **B 846,** B05

plate:
Alclad plate, n—composite plate product comprised of an aluminum-alloy core having on both surfaces a metallurgically bonded aluminum or aluminum-alloy cladding that is anodic to the core, thus electrolytically protecting the core against corrosion.
mill finish plate, n—plate having a non-uniform finish which may vary from piece to piece and within a piece, and which may not be entirely free of stains or oil.
tread plate, n—plate or sheet having a raised, figured pattern on one surface to provide improved traction. **B 881,** B07

plate, *n*—a flat-rolled metal product of same minimum thickness and width arbitrarily dependent on the type of metal. **B 899,** B02

plate, *n*—a flat, rolled sheet having a width and length much greater than thickness. (If of steel, having dimensions of 0.180 in. (4.6 mm) or greater in thickness, and greater than 8.0 in. (203 mm) in width.) **E 631,** E06

plate, *n*—abbreviated term for **metal connector plate**. **E 1807,** E06

plate—an image vehicle, usually permanent and reusable for use on printing presses. **F 335,** F05

plate, *n*—*as in microbiology*, a Petri dish containing culture medium. **F 1494,** F23

plate, *n*—a longitudinal element asymmetrical in the transverse plane and designed to resist tension, compression, bending, and torsion. **F 1582,** F04

plate-as-rolled, *n*—the quantity of plate product rolled at one time, either from an individual slab or directly from an ingot. **A 941,** A01

plateau, vulcanization, *n*—vulcanization during which the value of a desired property proceeds to a maximum or minimum and then remains essentially constant for a substantial period after the initial change. **D 1566,** D11

plate binding, *n*—the refusal of part of a lithographic plate to produce an image due to loss of ink receptivity. **D 6488,** D01

plate cassette—See **cassette**. **E 7,** E04

plate chamber—the chamber within the vacuum system of an electron microscope in which is placed the plate cassette. **E 7,** E04

plate-clinched point—pointed end of nail, having been driven through member against anvil plate, bent sideways. **F 547,** F16

plate cylinder—the cylinder to which the plate or master is attached. **F 425,** F05

plated—See **electro-plated.** **F 547,** F16

plate distillation column—column constructed with perforated plates to separate one or more volatile liquids by distillation. **E 1705,** E48

plate glass—flat glass formed by a rolling process, ground and polished on both sides, with surfaces essentially plane and parallel. **C 162,** C14

plate hole—opening in metal connector plate resulting from punching integral plate projection(s) from, or nail hole in, connector plate during its fabrication. See **nail hole**. **E 1807,** E06

plate-mark—any imperfection in a pressed plastic sheet resulting from the surface of the pressing plate. **D 883,** D20

platen—a plate of metal, especially one that exerts or receives pressure, as in a press used for gluing plywood. **D 1038,** D07

platen—the mounting plate of a press to which the mold assembly is bolted. **F 869,** F08

platen—flat plate or roller used as a support for printing or copying a document. **F 1156,** F12

plates—flat particles of metal powder having considerable thickness. **B 243,** B09

plate scumming, *n*—the deposit of ink on the non-image area of the plate that can be in the form of minute particles or larger continuous solids. **D 6488,** D01

plate shear—a term describing the stresses associated with shear distortion of planes parallel to the edge plane of a sandwich construction or core material when loaded in shear parallel to the plane of the facings (see Test Method C 273). **E 1749,** E06

plate swelling, *n*—an enlargement of rubber or polymer plates due to contact with an incompatible ink, coating, or wash up solvent. **D 6488,** D01

plate tinting, *n*—the existence of a weak coloration from the fountain solution onto the nonimage area of the plate. **D 6488,** D01

plate wave—see **Lamb wave.** **E 1316,** E07

plate wear, *n*—a physical abrading away of the non-image or image area, or both, of the plate. **D 6488,** D01

plate wear blinding, *n*—the failure of the litho plate to print due to erosion in the image area. **D 6488,** D01

plating—pressing leather with a heated metal plate, usually smooth, under high pressure. **D 1517,** D31

plating—deposition of an adherent metal onto the surface of the base metal of the fastener. A specific process should be specified; that is, electroplating, hot dip galvanizing, mechanical deposition, etc. **F 1789,** F16

plating range—the current density range over which a satisfactory electroplate can be deposited. **B 374,** B08

platinum-cobalt color—color measured in reference to color standards prepared with solutions of cobalt chloride and potassium chloroplatinate. Color standards are identical with the description given in APHA publications and is referred to as "APHA Color". **D 4790,** D16

platinum-cobalt color scale, *n*—a color scale for clear, light-yellow liquids, defined by specified dilutions of a platinum-cobalt stock solution, ranging from 5 for the lightest color to 500 for the darkest. **E 284,** E12

platinum 27 (Pt-27), *n*—the platinum standard to which the National Bureau of Standards referred thermoelectric measurements prior to 1973. **E 344,** E20

platinum 67 (Pt-67), *n*—the platinum standard used by the National Bureau of Standards after 1972 as the reference to which thermoelectric measurements are referred. **E 344,** E20

plattner mortar (diamond mortar)—a mortar consisting of a block, cylinder, and pestle made of surface-hardened tooled steel having dimensions such that the cylinder fits snugly into a depression in the block and the pestle fits in the cylinder. The crushing is done with the pestle, generally with the aid of a hammer. **C 859,** C26

pleat, *n*—three layers of fabric involving two folds or reversals of direction; the back fold may be replaced by a seam. **D 123,** D13

Plessey bar code—See **bar code**. **F 1156,** F12

plied yarn—See *plied yarn* under **yarn.** **D 3878,** D30

plied yarn duck, *n*—duck with plied yarns in both warp and filling (army, numbered, and special use ducks). (See **flat duck**.) **D 123,** D13

plier—portable stapler with attached clincher for placing staples away from edges of work. **F 592,** F16

plowing, *n*—*in tire testing,* a type of uncontrollability defined by a loss of steering control with no substantial vehicle yaw; the vehicle moves on a trajectory that is dictated by vehicle dynamics as determined by velocity, mass, and the available traction at each tire. **F 538,** F09

plowing, *n*—*in tribology,* the formation of grooves by plastic deformation of the softer of two surfaces in relative motion. **G 40,** G02

plow layer—the depth typically reached by a plow or other commonly used earth turning device used in agriculture. This depth is commonly one to two feet (0.3 to 0.61 m) below land surface. **D 653,** D18

plug—See **closure.** **D 996,** D10

plugging compound—a putty-like mixture of inorganic materials used to fill holes in iron castings to ensure an even surface for porcelain enameling. **C 286,** B08

plugging factor—see **fouling factor** and **SDI**. **D 6161,** D19

plugging material—a material that has a hydraulic conductivity equal to or less than that of the geologic formation(s) to be sealed. Typical materials include portland cement and bentonite. **D 653,** D18

plugging operation—cutting out a piece of data from a document and replacing it with another piece of substrate that matches the cutout in size and shape, containing altered (or different) information. **F 1156,** F12

plugs—straight-sided insertions of sound wood or synthetic material replacing defective portions of veneers. They usually are held in place by friction until veneers are bonded into plywood. **D 1038,** D07

plumage, *n*—the outgrowth of fowl, consisting of feathers and down (waterfowl) or feathers only (nonwaterfowl). **D 123,** D13

plumules, *n*—downy waterfowl plumage with underdeveloped soft and flaccid quill with barbs indistinguishable from those of down. **D 4523,** D13

plunger—the reciprocating metal part that forces glass into the contours of a mold, or which, in a blank mold, forms the initial cavity for subsequent blowing. (See also **needle**.) **C 162,** C14

plunger, *n*—*in tire testing,* a cylindrical rod with a hemispherical end. **F 538,** F09

plunging water jet barrier—special purpose barrier created by a series of coherent streams of water directed vertically downward into a body of water. **F 818,** F20

plus sieve—the portion of a powder sample retained on a standard sieve of specified number. (See **minus sieve.**) **B 243,** B09

plutonium-rich particle—any area of a sample that has a plutonium content identifiably greater than the surrounding matrix. **C 859,** C26

ply—a single layer of membrane reinforcement in the bituminous membrane waterproofing system. **C 717,** C24

ply, *n*—*(1)* any of the several layers of solid **fiberboard;** *(2)* any of the layers in plywood or laminated **paperboard;** *(3)* any of the walls of a **multiwall bag.** (See also **veneer.**) **D 996,** D10

ply—a single sheet of veneer, or several strips laid with adjoining edges, that may or may not be glued, which forms one veneer lamina in a glued plywood panel. (See also **layer**.) **D 1038,** D07

ply—a layer of felt in a built-up roofing membrane; a four-ply membrane has at least four plies of felt at any vertical cross section cut through the membrane. **D 1079,** D08

ply—(in laminar composites) the constituent single layer as used in fabricating, or occurring within a composite structure. **D 3878,** D30

ply, *n*—*in a tire,* a layer of rubber-coated parallel cords. **F 538,** F09

ply coordinate axes—a set of Cartesian coordinates two of which lie within the plane of the ply, one axis of which is parallel to the principal fiber direction and the other axis perpendicular to the principal fiber direction (the third axis is through the ply's thickness). **D 3878,** D30

plying cement—any bituminous material used for adhering layers of felts, fabrics, or mats to structural surfaces and to each other. **D 1079,** D08

ply orientation—the acute angle (theta) including 90° between a reference direction and the ply principal axis. The ply orientation is positive if measured counterclockwise from the reference direction and negative if measured clockwise. **D 3878,** D30

plyout—a single wrap of ribbon extending out of its normal position in relation to the core or the next wrap. **F 221,** F05

ply principal axis—the ply coordinate axis that coincides with the direction of maximum inplane Young's modulus. For balanced weave fabric either warp or fill direction may be chosen. **D 3878,** D30

ply rating, *n*—the term is used to identify a given tire with its maximum recommended load when used in a specific type of service, giving an index of tire strength, not necessarily representing the number of cord plies in the tire. **F 538,** F09

plywood, *n*—See *plywood* under **wood laminate**. **D 907,** D14

plywood—usually a crossbanded assembly made of layers of veneer or veneer in combination with a lumber core or other wood-based panel material joined with an adhesive. Plywood generally is constructed of an odd number of layers with grain of adjacent layers perpendicular to one another. Outer layers and all odd-numbered layers generally have the grain direction oriented parallel to the long dimension of the panel. **D 1038,** D07

plywood box—See **box.** **D 996,** D10

P/M—the acronym representing powder metallurgy. Used as P/M Part, P/M Product, P/M Process, etc. **B 243,** B09

P_{max}, *n*—the maximum pressure (above pressure in the vessel at the time of ignition) reached during the course of a deflagration for the optimum concentration of the dust tested. P_{max} is determined by a series of tests over a large range of concentrations. It is reported in bar. **E 1445,** E27

PM cycle—see **preventative maintenance cycle**. **F 1457,** F05

P/M forging—see **powder forging.** **B 243,** B09

P/M hot forming—see **powder forging.** **B 243,** B09

P/M part—see **powder metallurgy part.** **B 243,** B09

pneumatic atomizer—a device wherein the movement of gas or vapor is the primary source of energy utilized to produce a spray. **E 1620,** E29

pneumatic tire, *n*—a hollow tire that becomes load-bearing upon inflation with air, or other gas, to a pressure above atmosphere. **D 123,** D13

POA—see **printer output area**. **F 1457,** F05

pock—a partially closed surface cavity.

where:
w = width.

F 109, C21

pocket, *n*—*in zippers*, the cavity of an element designed to receive the head. **D 123,** D13

pocket, *n*—an opening in a structure to accept a construction member. **E 631,** E06

pocket-shaped—as applied to chamois skins, a skin trimmed in the form of a rectangle with the two corners at one end rounded. **D 1517,** D31

point—a thin, flat, triangular or diamond-shaped piece of zinc used to hold glass in wood sash by driving it into the wood. **C 717,** C24

point, *n*—a term used to describe the thickness of **paperboard,** a point being one thousandth of an inch. (See **caliper.**) **D 996,** D10

point—end of shank opposite head, usually tapered; formed during manufacture of nail to facilitate driving. **F 547,** F16

point, *n*—portion of the needle intended to initiate tissue penetration. **F 1840,** F04

point analyzer, *n*—See also **analyzer.**

monitoring path, n—the actual path over which an atmospheric or an emission compound concentration is measured and averaged.

monitoring path length, n—the length of the monitoring path over which the average atmosphere or emission compound concentration is measured and averaged using an open path analyzer. **D 1356,** D22

point bar—one or a series of low ridges, usually of coarse sediment, deposited on the inner (convex) side of a river bend. **D 4410,** D19

point bars—see **burning bars, points, or tools.** **C 286,** B08

point configuration, *n*—the shape of the point. Some common point configurations include, but are not limited to: taper, trocar, blunt, spatulated, conventional cutting edge, reverse cutting edge, cutting taper, and side cutting needle. **F 1840,** F04

point defect—a localized crystal defect such as a lattice vacancy, interstitial atom, or substitutional impurity. Contrast with **light point defect.** **F 1241,** F01

pointed—provided with sharp point. **F 547,** F16

pointed blade—See **pointed clincher.** **F 592,** F16

pointed clincher—blade clincher with sharp point to facilitate piercing of corrugated board. **F 592,** F16

point estimate—the estimate of a parameter given by a single statistic, for example, sample average (see **sample average**) (**arithmetic average**). **E 1823,** E08

point focus collector— see **collector, point focus**. **E 772,** E44

point-integrated sample—a sample of water-sediment mixture collected at a relatively fixed point in accordance with the technique of point integration. A point-integrated sample is discharge weighted. However, because the sample is obtained from a single point, the concentration of any component of the mixture that is transported exactly at stream velocity can be considered as either a spatial or a discharge-weighted concentration. Samples collected with instruments that instantaneously capture a quantity of water-sediment mixture are not true point-integrated samples. (See **point sample**.) **D 4410,** D19

point-integrating sediment sampler—an instrument capable of collecting a water-sediment mixture isokinetically for a specified period of time by opening and closing under water; an instrument suitable for performing point integration. **D 4410,** D19

point integration—a method of sampling at a relatively fixed point whereby the water-sediment mixture is withdrawn isokinetically for a specified period of time. **D 4410,** D19

pointless—See **sheared-square point.** **F 547,** F16

point projection X-ray microscopy—a method of producing enlarged images by means of X rays. The specimen is placed close to a point source of X rays and the magnification achieved is the ratio of source-image to source-object distance. Resolution depends primarily on the diameter of the source. **E 175,** E41

point reflector, *n*—reflector subtending a very small solid angle at the observer's eye, so that the observer cannot readily distinguish its size or shape. **E 284,** E12

point sample—sample of water-sediment mixture taken at a single point, either with an instantaneous or a point-integrating sampler. **D 4410,** D19

points, conjugate—the pair of points on the principal axis of a mirror or lens so located that light emitted from either point will be focused at the other. Related points in the object and image are located optically so that one is the image of the other. **E 175,** E41

point size, *n*—a measurement of the height of a type font, where a point is approximately 1⁄72 in. or 0.353 mm. **F 1457,** F05

point source—a source whose linear dimensions are very small compared with the distance from the source to the region of observation.

NOTE—The irradiance varies inversely with the square of the distance; a unique property of point sources. **E 1316,** E07

point source of light—a light source small in size relative to the object that it is illuminating, or is a great distance from the object so that the light rays incident upon it are essentially collimated. **F 1156,** F12

point value, *n*—*in asbestos*, an index of commercial value of asbestos fiber used in asbestos-cement products. Point value = (SU – 10)/1.39 where SU stands for strength unit. **C 460,** C17

point value, *n— in asbestos,* an index of commercial value of asbestos fiber used in asbestos-cement products. Point value = (SU – 10)/1.39 where SU stands for strength unit. **D 2946,** C17

point weight—See **felt mill ream.** **D 1079,** D08

Poiseuille flow—the particular case of laminar viscous flow through a long pipe of circular cross section. **E 1316,** E07

poison, *v*—interfere with the action of, such as poisoning of a phosphor. **F 1156,** F12

Poisson distribution, *n*—the distribution which has as its probability function:

$$P(r) = e^{-\mu}\mu^r/r!$$

where:

$P(r)$ = probability of obtaining exactly r occurrences of an event in one unit, such as a unit of time or area,
μ = both mean and variance of distribution, and
e = base of natural logarithms.

D 123, D13

Poisson's ratio, ν—the absolute value of the ratio of transverse strain to the corresponding axial strain resulting from uniformly distributed axial stress below the proportional limit of the material. **D 4092,** D20

Poisson's ratio, μ, *n*—the absolute value of the ratio of transverse strain to the corresponding axial strain resulting from uniformly distributed axial stress below the proportional limit of the material. **E 6,** E28

Poisson's ratio, (ν)—ratio between linear strain changes perpendicular to and in the direction of a given uniaxial stress change. **D 653,** D18

polar—see **polarizing element.** **E 175,** E41

polar-aromatics, *n*—a polar aromatic hydrocarbon fraction that is adsorbed on an adsorbing medium from a paraffinic solvent during percolation and then desorbed with a chlorinated hydrocarbon solvent such as trichloroethylene. **D 8,** D04

polarimeter, Senarmount (Friedel)—See **Senarmount (Friedel)** polarimeter. **C 162,** C14

polarimetry—measuring the polarization angles of light waves. **F 1156,** F12

polariscope—an optical device consisting of a light source, mutually perpendicular polarizing elements, and generally equipped with one or more retardation plates for qualitative observations of relative optical retardation by color differentiation. **C 162,** C14

polarizability—the inability of an electrode to reproduce a reading after a minute electrical current has been passed through the membrane. Glass pH electrodes are especially prone to polarization errors caused by small currents flowing from the pH meter input circuit and from static electrical charges built up as the electrodes are removed from the sample solution, or when the electrodes are wiped. **D 4127,** D19

polarization—the change in the potential of an electrode during electrolysis, such that the potential of an anode always becomes more noble and that of a cathode less noble than their respective static potentials. Equal to the difference between the static potential and the dynamic potential. **B 374,** B08

polarization—see **concentration polarization.** **D 6161,** D19

polarization—for plane polarized beams the plane of polarization is taken as the electric vector. For elliptically or partially polarized beams, polarization must be handled by use of the Stokes parameters, which is rather complex and must be handled in the text. The direction of polarization of a plane polarized beam is taken as the angle between the plane of polarization and the plane of incidence or of reflection or transmission. For other than parallel or perpendicular polarization, the subject is most easily handled in the text. **E 772,** E44

polarization—*in optics*, the term used to describe the orientation of the time-varying electric field vector in an electromagnetic wave. **F 1241,** F01

polarization—the change from the open-circuit electrode potential as the result of the passage of current. **G 15,** G01

polarization admittance—the reciprocal of polarization resistance (di/dE). **G 15,** G01

polarization energy—see **screening energy.** **E 673,** E42

polarization, parallel, ||—with the plane of polarization parallel to the plane of incidence, reflectance, or transmittance. **E 772,** E44

polarization, perpendicular, $\perp$— with the plane of polarization perpendicular to the plane of incidence, reflectance, or transmittance. **E 772,** E44

polarization resistance—the slope (dE/di) at the corrosion potential of a potential (E)–current density (i) curve. (It is inversely proportional to the corrosion current density when the polarization resistance technique is applicable.) **G 15,** G01

polarized light—light with waves that are all oscillating in the same plane, and moving in the same direction. **F 1156,** F12

polarized light—*in optics*, light exhibiting different properties in different directions at right angles to the line of propagation. **F 1241,** F01

polarized light illumination—a method of illumination in which the incident light is plane polarized before it impinges on the specimen. **E 7,** E04

polarizer—a substance or a means that produces or increases polarization. **B 374,** B08

polarizer—a Nicol prism, polarizing film, or similar device into which normal light passes and from which polarized light emerges. **E 7,** E04

polarizing element—a general term for a device for producing or analyzing plane polarized light. It may be a Nicol prism, some other form of calcite prism, a reflecting surface, or a polarizing **E 175,** E41

pole—a timber in the round, often to support power or telephone lines. **D 9,** D07

pole—the area on a magnetized part from which the magnetic field is leaving or returning into the part. **E 1316,** E07

pole-barn nail—bright or galvanized, stiff-stock or hardened-steel, helically threaded, 2 ½ to 9-in. nails with flat head and medium diamond point. (See **threaded nail.**) **F 547,** F16

pole figure (for crystalline aggregates)— a graphical representation of the crystal orientations present in an aggregate. **E 7,** E04

polepiece, *n*—in reinforced plastics, the supporting part of the mandrel used in filament winding, usually on one of the axes of rotation. **D 883,** D20

pole piece—a part of a magnetic electron lens made of a magnetically permeable material for the purpose of concentrating and shaping the magnetic field within the lens. **E 7,** E04

pole piece spacer—the central part of a magnetic electron lens pole piece assembly made of nonmagnetic material for the purpose of rigidly defining the separation of the pole pieces of a lens. The length of the spacer affects the strength and shape of the magnetic field in the lens. **E 7,** E04

polish—a temporary coating that enhances the appearance and may protect the substrate to which it is applied. **D 2825,** D21
polished—See **tumbled.** **F 547,** F16
polished aggregate, *n*—exposed aggregate worn smooth sufficiently to affect frictional characteristics. **E 867,** E17
polished aggregate, *n*—exposed aggregate worn sufficiently smooth to affect frictional characteristics. **E 1778,** E17
polished finish—a surface that has high luster and strong reflection of incident light. **C 119,** C18
polished plate glass—See **plate glass.** **C 162,** C14
polished wire glass—wire glass, ground and polished on both sides. **C 162,** C14
polishing—the smoothing of a metal surface by means of the action of abrasive particles attached by adhesive to the surface of wheels or endless belts usually driven at a high speed. **B 374,** B08
polishing—a mechanical, chemical, or electrolytic process or combination thereof used to prepare a smooth reflective surface suitable for microstructure examination, free of artifacts or damage introduced during prior sectioning or grinding. **E 7,** E04
poll, *n*—an electronic method the cluster controller uses to sequentially select each connected device. **F 1457,** F05
polling, *n*—automated transfer of traffic measurements from permanent counters to a computer for editing and summarization. **E 867,** E17
pollution—the condition caused by the presence of substances of such character and in such quantities that the quality of the environment is impaired. **D 4410,** D19
polyaddition—See **addition polymerization.** **D 883,** D20
polyamide plastics—See **nylon plastics.** **D 883,** D20
polyarylate, n—See **aromatic polyester.** **D 883,** D20
polyaryletherketone, *n*—a polymer in which aryl groups are connected by one or more ether as well as one or more Ketone linkages. **D 883,** D20
polyatomic ion—a charged multi-atom species. **E 673,** E42
polybutylene, *n*—a polymer prepared by the polymerization of butene as the sole monomer. (See **polybutylene plastics** and **butylene plastics.**) **D 883,** D20
polybutylene, *n*—a polymer prepared by the polymerization of butene as the sole monomer. (D-20) **F 412,** F17
polybutylene, *n*—a polymer prepared by the polymerization of butene as the sole monomer. (See also **polybutylene plastic** and **butylene plastic.**)
polybutylene plastic, plastic based on polymers made with butene as essentially the sole monomer. **F 1251,** F04
polybutylene plastics, *n*—plastics based on polymers made with butene as essentially the sole monomer. **D 883,** D20
polybutylene plastics, *n*—plastics based on polymers with butene as essentially the sole monomer. (D-20) **F 412,** F17
polycarbonate, *n*—a polyester polymer in which the repeating structural unit in the chain is of the carbonate type. **D 883,** D20
polycarbonate, *n*—a polymer in which the repeating structural unit in the chain is a carbonic acid ester of Bisphenol A. **F 1251,** F04
polycarbonate plastics—polyester plastics based on polymers in which the repeating structural units in the chains are essentially all of the carbonate type. **D 883,** D20
polychlorinated biphenyl (PCB),, *n*—PCB's belong to a broad family of organic chemicals known as chlorinated hydrocarbons and are produced by attaching one or more chlorine atoms to a biphenyl molecule. **D 2864,** D27
polychromator, *n*—a device for simultaneously isolating several rays of monochromatic radiation from a beam of polychromatic radiation. **E 135,** E01
polychrome decoration—See *polychrome decoration* under **decoration.** **C 242,** C21
polycondensation—See **condensation polymerization.** **D 883,** D20
polycondensation, n— See **condensation.** **D 907,** D14
polycrystalline—characteristic of an aggregate composed of more than one, and usually of a large number, of crystals. **E 7,** E04
polycrystalline—a form of semiconductor material made up of randomly oriented crystallites and containing large-angle grain boundaries, twin boundaries, or both. **F 1241,** F01
polydicyclocpentadiene plastic, n—a crosslinked thermoset polymer formed by the ring-opening metathesis polymerization of dicyclopentadiene. **D 883,** D20
polyelectrolyte—synthetic (or natural) molecules, containing multiple ionic groups, used as coagulants and flocculants; available as anionic, cationic and nonionic. **D 6161,** D19
polyester, *n*—a polymer in which the repeated structural unit in the chain is of the ester type. (ISO) **D 883,** D20
polyester, *n*—a polymer in which the repeated structural unit in the chain is of the ester type. **F 1251,** F04
polyester plastics—synonymous with **alkyd plastics.** **D 883,** D20
polyester resin—a condensation product resulting from a chemical reaction between a dicarboxylic acid and a dihydroxy alcohol or by the polymerization of a hydroxy carboxylic acid. **C 904,** C03
polyether, *n*—a polymer in which the repeated structural unit in the chain is of the ether type. **D 883,** D20
polyether, *n*—a polymer in which the repeated structural unit in the chain is of the ether type. **F 1251,** F04
polyethylene, *n*—a polymer prepared by the polymerization of ethylene as the sole monomer. (See **polyethylene plastics** and **ethylene plastics.**) **D 883,** D20
polyethylene, *n*—a polymer prepared by the polymerization of ethylene as the sole monomer. (D-20) **F 412,** F17
polyethylene—a polymer prepared by the polymerization of ethylene as the sole monomer. (D 883, D-20) **F 869,** F08
polyethylene, *n*—a polymer prepared by the polymerization of ethylene as the sole monomer. (See also **polyethylene plastic** and **ethylene plastic.**) **F 1251,** F04
polyethylene plastic—a plastic based on polymers made with ethylene as essentially the sole monomer. **F 1251,** F04
polyethylene plastics—plastics based on polymers made with ethylene as essentially the sole monomer. **D 883,** D20
polyethylene plastics—plastics based on polymers made with ethylene as essentially the sole monomer. (D-20) **F 412,** F17
polyethylene terephthalate, *n*—a polymer derived from terephthalic acid and ethylene glycol by condensation polymerization. **F 1251,** F04
polygraph examiner, *n*—a term used synonymously with PDD examiner. In the evolution of terminology within this discipline, the term APolygraph Examiners≅ replaced the antiquated appellation ALie Detector Operator ≅ (see definition of **PDD examiner**). **E 2035,** E52
polygraph instrument, *n*—a diagnostic instrument used during a PDD examination, which is capable of simultaneously monitoring, recording, and measuring at a minimum, respiratory, electrodermal, and cardiovascular activity as a response to auditory or visual stimuli. **E 2035,** E52
polygraphy, *n*—see definition of **PDD** **E 2035,** E52

polymer, *n*—a macromolecular material formed by the chemical combination of monomers having either the same or different chemical composition. **D 123,** D13

polymer, *n*—a substance consisting of molecules characterized by the repetition (neglecting ends, branch junctions and other minor irregularities) of one or more types of monomeric units. (IUPAC) **D 883,** D20

polymer, *n*—a compound formed by the reaction of simple molecules having functional groups which permit their combination to proceed to high molecular weights under suitable conditions. Polymers may be formed by polymerization (addition polymer) or polycondensation (condensation polymer). When two or more monomers are involved, the product is called a copolymer. (Compare **monomer**.) **D 907,** D14

polymer, *n*—a substance consisting of molecules characterized by the repetition (neglecting ends, branch junctions, and other minor irregularities) of one or more types of monomeric units. **D 1566,** D11

polymer, *n*—a substance consisting of molecules characterized by the repetition (neglecting ends, branch junctions, and other minor irregularities) of one or more types of monomeric units. (IUPAC, D-20) **F 412,** F17

polymer, *n*—a substance consisting of molecules characterized by the repetition (neglecting ends, branch junctions, and other minor irregularities) of one or more types of monomeric units. (See **copolymer.**) **F 1251,** F04

polymer, *n*—a substance consisting of molecules characterized by repetition (neglecting ends, branches, junctions, and other minor irregularities) of one or more chemically bonded types of monomeric units. **F 1494,** F23

polymer coating—a term used to describe color coatings such as PVC (poly(vinyl chloride)), polyester, and clear coatings such as polyurethane and acrylic urethane (see Specifications F 668, F 934, and F 1234). **F 552,** F14

polymeric poured (seamless) floors, *n*—a floor surfacing material composed of polymeric materials applied to the substrate in liquid form alone or in combination with mineral or plastic aggregates, desiccants, or fillers. **F 141,** F06

polymerization—a chemical reaction in which the molecules of monomers are linked together to form polymers. (See also **polycondensation** and **polyaddition.**) **D 883,** D20

polymerization, *n*—a chemical reaction in which the molecules of a monomer(s) are linked together in repeating units to form larger molecules. (R) **D 907,** D14

polymerization—the uniting of two or more structural chemical moieties to create a larger molecule, of greater mass. **F 1156,** F12

polymerization—a chemical reaction in which monomers are linked together to form polymers. (See also **polycondensation** and **polyaddition.**) **F 1251,** F04

polymerized rosin, n—rosin that has been treated by chemical or physical means, or both, in a manner so as to cause a union of a part of the rosin acids to form dimers to such an extent that the average molecular weight of such rosin will be measurably greater than that of the original rosin. Also known as "dimerged resin." **D 804,** D01

polymer network, *n*—a three-dimensional reticulate structure formed by chemical or physical linking of polymer chains. **D 1566,** D11

polymer(s)—a large molecule formed by the chemical union of reactive units called monomers. **D 4790,** D16

polymers—a substance consisting of molecules characterized by the repetition of one or more types of monomeric units. **D 6161,** D19

polymer sheet, *n*—a continuous polymeric planar structure. **F 1494,** F23

polymer technique—the examination technique in which a polymer is used as the particle suspension vehicle. **E 1316,** E07

polymethylmethacrylimide (PMMI), *n*—a thermoplastic polymer formed from a reaction of poly(methyl methacrylate) and monomethyl amine. **D 883,** D20

polymorphic substance—an element, or compound, capable of stable existence in different temperature and pressure ranges in two, or more, different crystalline states. **E 7,** E04

polymorphic substance—element, or compound, capable of stable existence in different temperature and pressure ranges in two, or more, different crystalline forms. **E 1142,** E37

polynary separator, *n*—a device that separates a single input feed stream into three or more output product streams. **D 5681,** D34

polynary system—See **multicomponent system.** **E 7,** E04

polyol, *n*—an alcohol having many hydroxyl groups, also known as a polyhydric alcohol or polyalcohol. **D 883,** D20

polyolefin, *n*—any long-chain synthetic polymer composed of at least 85 weight % of ethylene, propylene, or other olefin units (monomers), except amorphous (non-crystalline) polyolefin qualifying under Rubber 1, as defined by the Federal Trade Commission. **D 123,** D13

polyolefin, *n*—a polymer prepared by the polymerization of an olefin(s) as the sole monomer(s). (See **polyolefin plastics** and **olefin plastics.**) **D 883,** D20

polyolefin, *n*—a polymer prepared by the polymerization of an olefin(s) as essentially the sole monomer(s). (D-20) **F 412,** F17

polyolefin, *n*—a polymer prepared by the polymerization of an olefin(s) as the sole monomer(s). (See also **polyolefin plastic, olefin plastic.**) **F 1251,** F04

polyolefin-material cleanliness, *n*—the degree to which a polymer melt is free of filterable particles which remain insoluble in the melt under the specified test condition. **D 123,** D13

polyolefin monofilament, *n*—*as used in this specification*, a flat single filament of the slit-film type, which can function as a yarn in commercial textile operations. **D 123,** D13

polyolefin plastic—a plastic based on polymers made with an olefin(s) as essentially the sole monomer(s). **F 1251,** F04

polyolefin plastics, *n*—plastics based on polymers made with an olefin(s) as essentially the sole monomer(s). **D 883,** D20

polyolefin plastics, *n*—plastics based on polymers made with an olefin(s) as essentially the sole monomer(s). (D-20) **F 412,** F17

polyoxymethylene, *n*—a polymer in which the repeated structural unit in the chain is oxymethylene. **D 883,** D20

polyoxymethylene, *n*—a polymer in which the repeated structural unit in the chain is oxymethylene. **F 1251,** F04

polyoxymethylene plastics, *n*—acetal plastics based on polymers in which oxymethylene is essentially the sole repeated structural unit in the chains. (ISO) (See also **acetal plastics.**) **D 883,** D20

polyphthalamide (PPA), *n*—a polyamide in which residues of terephthalic acid or isophthalic acid or a combination of the two comprise at least 60 molar percentage of the dicarboxylic acid part of the repeating structural units in the polymer chain. **D 883,** D20

polypropylene, *n*—a polymer prepared by the polymerization of propylene as the sole monomer. (See **polypropylene plastics** and **propylene plastics.**) **D 883,** D20

polypropylene, *n*—a polymer prepared by the polymerization of propylene as the sole monomer. (D-20) **F 412,** F17

polypropylene, *n*—a polymer prepared by the polymerization of propylene as the sole monomer. (See also **polypropylene plastic, propylene plastic.**) **F 1251,** F04

polypropylene plastics—plastics based on polymers made with propylene as essentially the sole monomer. **D 883,** D20

polypropylene plastics—plastics based on polymers made with polypropylene as essentially the sole monomer. (D-20) **F 412,** F17

polysilicon (poly)—polycrystalline silicon. **F 1241,** F01

polystyrene, *n*—a polymer prepared by the polymerization of styrene as the sole monomer. (See **styrene plastics.**) **D 883,** D20

polystyrene—a plastic based on a resin made by polymerization of styrene as the sole monomer. For use as a standard in colorimetry, the polystyrene shall contain no internal or external additives. **D 4790,** D16

polystyrene, *n*—a polymer prepared by the polymerization of styrene as the sole monomer. (D-20) **F 412,** F17

polystyrene, *n*—a polymer prepared by the polymerization of styrene as the sole monomer. (See also **styrene plastic.**) **F 1251,** F04

polyterephthalate, *n*—a thermoplastic polyester in which the terephthalate group is one of the repeating structural unit in the polymer chain. **D 883,** D20

polyterephthalate, *n*—a thermoplastic polyester in which the terephthalate group is a repeated structural unit in the polymer chain. **F 1251,** F04

polyterephthalate plastics—a thermoplastic polyester in which the terephthalate group is a repeated structural unit in the chain, the terephalate being in greater amount than other dicarboxylates which may be present. **D 883,** D20

polyterpene resin, *n*—a resin produced by the polymerization of terpene fractions or mixtures of terpenes obtained from naval stores, or paper pulp production, or citrus juice production, or combinations thereof. **D 6440,** D01

polyurethane, *n*—a polymer prepared by the reaction of an organic diisocyanate with compounds containing hydroxyl groups. **D 883,** D20

polyuronides—polysaccharides containing uronic acid groups. **D 1695,** D01

poly(vinyl acetate), *n*—a polymer prepared by the polymerization of vinyl acetate as the sole monomer. **D 883,** D20

poly(vinyl acetate)—a polymer prepared by the polymerization of vinyl acetate as the sole monomer. (D-20) **F 412,** F17

poly(vinyl acetate), *n*—a polymer prepared by the polymerization of vinyl acetate as the sole monomer. **F 1251,** F04

polyvinyl acetate emulsion adhesive, *n*—a latex adhesive in which the polymeric portion comprises polyvinyl acetate, copolymers based mainly on polyvinyl acetate, or a mixture of these, and which may contain modifiers and secondary binders to provide specific properties. **D 907,** D14

poly(vinyl alcohol)—polymers prepared by the essentially complete hydrolysis of polyvinyl esters. **D 883,** D20

poly(vinyl alcohol), *n*—a polymer prepared by the essentially complete hydrolysis of polyvinyl ester. **F 1251,** F04

poly(vinyl chloride)—a polymer prepared by the polymerization of vinyl chloride as the sole monomer. **D 883,** D20

poly(vinyl chloride)—a polymer prepared by the polymerization of vinyl chloride as the sole monomer. (D-20) **F 412,** F17

poly(vinyl chloride), *n*—a polymer prepared by the polymerization of vinyl chloride as the sole monomer. **F 1251,** F04

poly(vinyl chloride) (pvc)—a polymer prepared by the polymerization of vinyl chloride as the sole monomer. (D 883, D-20) **F 869,** F08

pond—a surface which is incompletely drained. **D 1079,** D08

pontil—See **punty.** **C 162,** C14

pooling—the existence of excessive amounts of penetrant, emulsifier or developer in an incompletely drained area. **E 1316,** E07

pop-in—a discontinuity in the load against clip gage displacement record. This discontinuity is characterized by a sudden increase in displacement and, generally, a decrease in load. Subsequently, the displacement and load increase to above their respective values at pop-in. **E 1823,** E08

pop-off—in dry process enameling, a defect appearing as a small conical piece of porcelain enamel, either partially or entirely separated from the ware. **C 286,** B08

popouts, *n*—small holes in the pavement surface, normally ranging in diameter from 25 mm (1 in.) to 100 mm (4 in.) and depth from 13 mm (0.5 in.) to 50 mm (2 in.). **E 867,** E17

popouts, *n*—small holes in the pavement surface, normally ranging in diameter from 25 mm (1 in.) to 100 mm (4 in.) and depth from 13 mm (0.5 in.) to 50 mm (2 in.). **E 1778,** E17

poppers—a defect characterized by randomly occurring, relatively small, circular shaped areas of ground coat appearing in the first cover coat sheet steel porcelain enamel. **C 286,** B08

population, *n*—the totality of items or units of material under consideration. **E 456,** E11

population (or universe)— *in fatigue testing*, the totality of the set of test specimens, real or conceptual, that could be prepared in the specified way from the material under consideration. **E 1823,** E08

porcelain—a glazed or unglazed vitreous ceramic whiteware made by the porcelain process, and used for technical purposes, designating such products as electrical, chemical, mechanical, structural, and thermal wares when they are vitreous.

alumina porcelain—a vitreous ceramic whiteware for technical application in which alumina (Al_2O_3) is an essential crystalline phase.

chemical porcelain—vitreous ceramic whitewares used for containing, transporting, or reacting of chemicals.

cordierite porcelain—a vitreous ceramic whiteware for technical application in which cordierite ($2MgO{\cdot}2Al_2O_3{\cdot}5SiO_2$) is the essential crystalline phase.

forsterite porcelain—a vitreous ceramic whiteware for technical application in which forsterite ($2MgO{\cdot}SiO_2$) is the essential crystalline phase.

steatite porcelain—a vitreous ceramic whiteware for technical application in which magneisum metasilicate ($MgO{\cdot}SiO_2$) is the essential crystalline phase.

titania porcelain—a vitreous ceramic whiteware for technical application in which titania (TiO_2) is the essential crystalline phase.

zircon porcelain—a vitreous ceramic whiteware for technical application in which zircon ($ZrO_2{\cdot}SiO_2$) is the essential crystalline phase. **C 242,** C21

porcelain enamel—a substantially vitreous or glassy, inorganic coating bonded to metal by fusion at a temperature above 800°F (425°C approximate). **C 286,** B08

porcelain process—the method of producing glazed ware by which a ceramic body and glaze are matured together in the same firing operation. **C 242,** C21

porcelain tile—a ceramic mosaic tile or paver that is generally made by the dust-pressed method of a composition resulting in a tile that is dense, fine-grained, and smooth with sharply formed face, usually impervious and having colors of the porcelain type which are usually of a clear, luminous type or granular blend thereof.

C 242, C21

pore—an inherent or induced cavity within a particle or within an object. **B 243,** B09

pore—a discontinuity, essentially circular in cross section, in a coating extending through to the underlying coating or basis material. **B 374,** B08

pore—*in catalysis,* a small opening in a material permitting admission, adsorption, or passage of a fluid. **D 3766,** D32

pore—an opening or void in a membrane or filter matrix. **D 6161,** D19

pore—an internal cavity which may be exposed by cutting, grinding, or polishing to become a pit, pock, or hole.

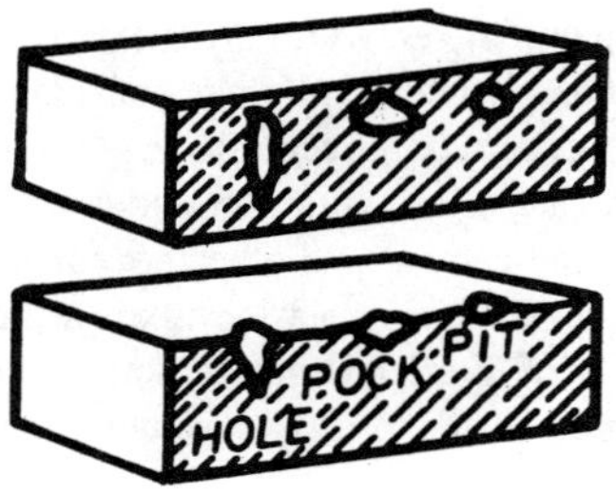

F 109, C21

pore diameter—the diameter of a pore in a model in which the pores in a sorbent are assumed to be cylindrical in shape and which is calculated from data obtained by a specified procedure. **D 2652,** D28

pore-forming material—a substance included in a powder mixture which volatilizes during sintering and thereby produces a desired kind and degree of porosity in the finished compact. **B 243,** B09

pore-liquid—liquid that occupies an open space between solid soil particles. Within this guide, pore-liquid is limited to aqueous pore-liquid; that includes water and its solutes. **D 653,** D18

pore-liquid tension—see **matric-suction** or **soil water pressure**. **D 653,** D18

pore, *n,* **(P^v)**—*as used in fractography,* a volume-distributed cavity created in a solid material during fabrication. **C 1145,** C28

pore pressure (*pore water pressure*)—see **neutral stress** under stress. **D 653,** D18

pores—in wood anatomy, a term applied to the cross section of a vessel or of a vascular tracheid. **D 9,** D07

pores—the complex network of channels in the interior of a particle of a sorbent. **D 2652,** D28

pore size distribution—*in catalysis,* the volume of pores of a material expressed or displayed as a function of pore diameters. **D 3766,** D32

pore volume—volume of the pores in a unit weight of a sorbent. **D 2652,** D28

pore volume distribution—the distribution of pore volume among pores of different size or diameter. **D 2652,** D28

pore volume distribution—*in catalysis,* an alternative term for pore size distribution. **D 3766,** D32

pore volume of flow (V_{pq}), *n*—the cumulative volume of flow through a test specimen divided by the volume of voids within the specimen. **D 4439,** D35

pore water—water contained in the voids of the soil or rock. **D 653,** D18

pore water—water occupying space between sediment or soil particles (Syn. **interstitial water**). **E 943,** E47

poromer, n— See **poromeric material,** the preferred term. **D 1566,** D11

poromeric material, *n*—a flexible synthetic leather-like material that is permeable to air and water vapor and usually resistant to water penetration and abrasion. **D 1566,** D11

porosity—the amount of pores (voids) expressed as a percentage of the total volume of the powder metallurgy part. **B 243,** B09

porosity—in metallic coatings, the presence of any disontinuity, crack, or hole in the coating that exposes a different underlying metal. **B 542,** B02

porosity, *n*—the percentage of the total volume of a material occupied by both open and closed pores. **C 71,** C08

porosity—the volume fraction of voids contained in a solid, often expressed as a percent.

closed porosity—the volume fraction of all pores within a solid mass that are closed off by surrounding solid and, hence, are inaccessible to each other and to the external surface: they thus are not detectable by gas or liquid penetration.

connected porosity—the volume fraction of all pores, voids, and channels within a solid mass that are interconnected with each other.

open porosity—the volume fraction of all pores, voids, and channels within a solid mass that are interconnected with each other and communicate with the external surface, and thus are measurable by gas or liquid penetration (Syn. *apparent porosity*). **C 242,** C21

porosity—the percentage of the total volume of a material occupied by both open and closed pores. **C 709,** D02

porosity—the percentage of the total volume of a material occupied by both open and closed pores. **C 859,** C26

porosity, *n*—the ratio of the volume of air or void contained within the boundaries of a material to the total volume (solid matter plus air or void) expressed as a percentage. (See **air permeability, permeability.**) **D 123,** D13

porosity—the ratio of the aggregate volume of voids or interstices in a rock or soil to its total volume. (ISRM) **D 653,** D18

porosity, *n*—the presence of numerous small cavities. **D 1566,** D11

porosity—the existence in a material of connected air voids. It is frequently expressed as the ratio of void volume to total volume. **D 1695,** D01

porosity—*of particles in catalysis,* the ratio of the volumes of the pores in the particles to the volumes enclosed by their envelopes. **D 3766,** D32

porosity: porosity, internal—the presence of numerous pits or pin holes beneath the pultruded surface; usually observable only in a cut cross section. **D 3918,** D20

porosity—that portion of a membrane filter volume which is open to fluid flow, also known as void volume. **D 6161,** D19

porosity—holes in a solid, not necessarily connected. **E 7,** E04

porosity, *n*—the ratio of the aggregate volume of interstices in a rock or soil to its total volume, usually stated as a percent. **E 957,** E44

porosity—*in paper,* the property that allows the passage of air through the sheet—an important factor in ink penetration, and also a quality that may affect paper feeding in some readers that have vacuum feeding mechanisms. **F 149,** F05

porosity—the ratio of void volume to total volume. **F 740,** F21

porosity, apparent—See *open porosity* under **porosity.** **C 242,** C21

porosity, *n* **(D)**—the ratio, usually expressed as a percentage, of: (*1*) the volume of voids of a given soil or rock mass, to (*2*) the total volume of the soil or rock mass. **D 653,** D18

porosity, surface—the presence of numerous visible pits or pin holes at or near the pultruded surface. **D 3918,** D20

porous—substances containing pores for fluids to pass due to an open physical structure. **D 6161,** D19

porous area—an area that will retain dye when tested in accordance with Practice E 165 and, if broken through at the porous area, will show evidence of dye penetration into the body.

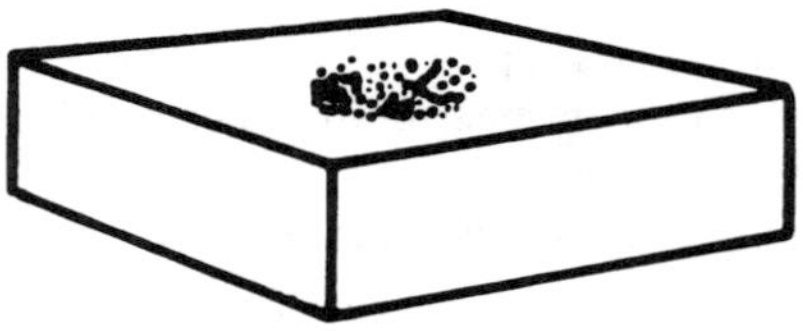

F 109, C21

porous packaging material—a material used in medical packaging which is intended to provide an environmental and biological barrier, while allowing sufficient air flow to be used in gaseous sterilization methods (for example, EtO, steam, gas plasma). **F 1327,** F02

porous region, *n*— *as used in fractography,* a three-dimensional zone of porosity or microporosity. **C 1145,** C28

porous seam, *n*— *as used in fractography,* a two-dimensional area of porosity or microporosity. **C 1145,** C28

porous woods—another name for hardwoods, which frequently have vessels or pores large enough to be seen readily without magnification. **D 9,** D07

port—an opening in a melting furnace designed for the passage of fuel, air/oxygen, flame, or exhaust gases or combination thereof. **C 162,** C14

port, *n*—an opening or aperture in an integrating sphere. **E 284,** E12

portable counter, *n*—a device which may be moved to various locations and periodically measures and records traffic volume, vehicle classification, or weigh, or combination thereof. **E 867,** E17

portable vacuum cleaner operational weight, *n*—the unit weight of the cleaner, including cord, as usually used to vacuum carpets plus the accessories such as hose, motorized nozzle, tools, and tool caddy if they are normally attached to or stored on the cleaner. Weight is with clean filters and the electric cord. **F 395,** F11

portable vacuum cleaner total weight, *n*—the weight of the cleaner, electric cord, all hoses, and attachments included with the model, but not including any packaging. Weight is with clean filters. **F 395,** F11

portal—the surface entrance to a tunnel. (ISRM) **D 653,** D18

portal—doorway; entrance. **F 1156,** F12

portfolio analysis—a technique used to seek the combination of assets with the maximum return for any given degree of risk (that is, variance of the return), or the minimum risk for any given rate of return. **E 833,** E06

portfolio (portefeuille)—group of securities, buildings, or other properties held by an individual or institutional investor. **E 1480,** E06

portion size—the number of lengths of pipe to be used for a specific test. **B 846,** B05

portland blast-furnace slag cement, *n*—a hydraulic cement consisting of an intimately interground mixture of portland-cement clinker and granulated blast-furnace slag or an intimate and uniform blend of portland cement and fine granulated blast-furnace slag in which the amount of the slag constituent is within specified limits. **C 219,** C01

portland cement, *n*—a hydraulic cement produced by pulverizing clinker consisting essentially of hydraulic calcium silicates, and usually containing one or more forms of calcium sulfate as an interground addition. **C 11,** C11

portland cement, *n*—a hydraulic cement produced by pulverizing portland-cement clinker, and usually containing calcium sulfate. **C 219,** C01

portland-cement clinker, *n*—a clinker, partially fused by pyroprocessing, consisting predominantly of crystalline hydraulic calcium silicates. **C 219,** C01

Portland cement concrete pavement, *n*—a pavement having a surface of aggregate mixed with Portland cement paste binder or a mixture of Portland cement and other pozzolans. **E 867,** E17

portland cement concrete pavement, *n*— a pavement having a surface of aggregate mixed with portland cement paste binder or a mixture of portland cement and other pozzolans. **E 1778,** E17

portland cement plaster, *n*—a plaster mix in which portland cement or combinations of portland and masonry cements or portland cement and lime are the principal cementitious materials mixed with aggregate. **C 11,** C11

portland-pozzolan cement, *n*—a hydraulic cement consisting of an intimate and uniform blend of portland cement or portland blast-furnace slag cement and fine pozzolan produced by intergrinding portland cement clinker and pozzolan, by blending portland cement or portland blast-furnace slag cement and finely divided pozzolan, or a combination of intergrinding and blending, in which the amount of the pozzolan constituent is within specified limits. **C 219,** C01

portrait mode, *adj*—a printer output orientation in which the printed lines run perpendicular to the direction of movement of the paper. (See **landscape mode.**) **F 1457,** F05

POS—an abbreviation for *p*oint *o*f *s*ale data entry systems where actual transactions are recorded by terminals operating on-line to a central computer. These systems frequently employ optical scanning as a means of capturing data. **F 149,** F05

positive column—that region of an electric discharge between the cathode and anode falls. **B 542,** B02

positive development, direct development—charged area development by means of toner particles having appropriate charge polarity so that dark areas of the print correspond to dark areas of the original. **F 335,** F05

positive displacement pump—a pump that will continue to build pressure until the power source is stalled if the pump outlet is blocked. **D 653,** D18

positive distortion— See **distortion**. **E 7,** E04

positive eyepiece—an eyepiece in which the focal plane is external to its lenses. **E 7,** E04

positive identification—This phrase is inappropriate because it seems to suggest that some identifications are more positive than others. **E 1658,** E30

positive image—a developed image in which dark areas correspond to dark areas of the original subject. **F 335,** F05

positive (negative) ion yield—the total number of positive (negative) secondary ions sputtered from the specimen per incident primary particle. **E 673,** E42

positive, photographic—a photograph having approximately the same rendition of light and shade as the original subject. **E 7,** E04

positive replica—See **replica**. **E 7,** E04

positive side waterproofing, *n*—an application wherein the waterproofing system and the source of hydrostatic pressure are on the same side of the structural element. **D 1079,** D08

positive transparency—a photographic print made on a transparent base from a negative. In electron microscopy this is used as an intermediate step to prepare a negative print. **E 7,** E04

possible/could have—these terms have no place in expert opinions on handwriting because the examiner's task is to decide to what degree of certainty it can be said that a handwriting sample is by a specific person. If the evidence is so limited or unclear that no definite or qualified opinion can be expressed, then the proper answer is *no conclusion.* To say that the suspect "could have written the material in question" says nothing about probability and is therefore meaningless to the reader or to the court. The examiner should be clear on the different meanings of "possible" and "probable," although they are often used interchangeably in everyday speech. **E 1658,** E30

post—short timber used in upright position for supporting structures of fencing. It may be round, split, or sawn. **D 9,** D07

post—See **railing systems.** **E 631,** E06

post, *n*—a vertical supporting member. **E 1481,** E06

post—post holding clincher to facilitate assembly of container bottoms. **F 592,** F16

post, *n*—a non-threaded anchor component that connects to the bony elements of the spine, pelvis, or ribs by means of a non-threaded hole in the bony element. **F 1582,** F04

post cap—a fitting atop a post usually to exclude water from tubular posts. **F 552,** F14

post-cleaning—the removal of residual liquid penetrant examination materials from the test part after the penetrant examination has been completed. **E 1316,** E07

post-cure, *n*—a treatment (normally involving heat) applied to an adhesive assembly following the initial cure, to modify specific properties. **D 907,** D14

post-cure, *v*—to expose an adhesive assembly to an additional cure, following the initial cure, for the purpose of modifying specific properties. **D 907,** D14

post cure, *n*—heat or radiation treatment, or both, to which a cured or partially cured thermosetting plastic or rubber composition is subjected, to enhance the level of one or more properties. **D 1566,** D11

post cure—heat or radiation treatment, or both, to which a cured or partially cured thermosetting plastic or rubber composition is subjected to enhance the level of one or more properties. **E 1749,** E06

postemergence herbicide—a chemical applied after emergence of the specified weed or crop. **E 609,** E35

post emulsification—a penetrant removal technique employing a separate emulsifier. **E 1316,** E07

post-fab, n—fabrication process where close-outs and inserts are attached or put into the panel after the facings are bonded to the core. **C 274,** D30

post-flashover, *adj*—the stage of a fire at which the average air temperature in the upper half of the room exceeds 600°C. **E 176,** E05

postforming, *n*—the forming of cured or partially cured thermosetting plastics. **D 883,** D20

post hinge—a fitting that attaches to the gate post, the pintle of which fits the frame hinge permitting the gate to swing. **F 552,** F14

post-nucleation—the step where, if necessary, the catalyst is converted to its final form. This is the final step prior to electroless plating. **B 374,** B08

post treatment—a process of applying chemical(s) to a membrane after formation to improve its performance. **D 6161,** D19

post treatment—the addition of chemicals to the product or concentrate stream to make it suitable for the desired end use application. **D 6161,** D19

post treatment—utilization of equipment such as degasifiers to make the product or concentrate stream, or both, suitable for the desired end use application. **D 6161,** D19

post-weld heat treatment, *n*—heating weldments immediately after welding, to provide **tempering**, **stress relieving**, or a controlled rate of cooling to prevent formation of a hard or brittle microstructure. **A 941,** A01

pot—a one-piece refractory container for molten glass. Types of pots include: **C 162,** C14

potable water—water that is satisfactory for drinking and culinary purposes, meeting the requirements of the health department having jurisdiction. **E 772,** E44

pot arch—a furnace for firing or preheating a pot. **C 162,** C14

potash—potassium oxide (K_2O). Loosely, a carbonate of potassium. **C 162,** C14

potential drop, Δh (L)—the difference in total head between two equipotential lines. **D 653,** D18

potentiodynamic—refers to a technique wherein the potential of an electrode with respect to a reference electrode is varied at a selected rate by application of a current through the electrolyte. **G 15,** G01

potentiometer—an instrument for the measurement of electromotive force by balancing against it an equal and opposite electromotive force across a calibrated resistance carrying a definite current. Potentiometers can be made manual or with automatic self-balancing features. **E 7,** E04

potentiometric surface—an imaginary surface representing the static head of ground water. The water table is a particular potentiometric surface. **D 653,** D18

potentiostaircase—refers to a potentiostep technique for polarizing an electrode in a series of constant potential steps wherein the time duration and potential increments or decrements are equal for each step. **G 15,** G01

potentiostat—an instrument for automatically maintaining an electrode in an electrolyte at a constant potential or controlled potentials with respect to a suitable reference electrode. **G 15,** G01

potentiostatic—the technique for maintaining a constant electrode potential. **G 15,** G01

potentiostep—refers to a technique in which an electrode is polarized in a series of potential increments or decrements. **G 15,** G01

pot furnace—a furnace for melting glass in pots. **C 162,** C14

pot furnace—a furnace used to smelt porcelain enamel raw batch contained in a crucible. **C 286,** B08

potholes, *n*—bowl-shaped holes in the pavement surface, grater than 100 mm (4 in.) in diameter, and more than 25 mm (1 in.) in depth. **E 867,** E17

potholes, *n*—bowl-shaped holes in the pavement surface, greater than 0.1 m (4 in.) in diameter, and more than 25 mm (1 in.) in depth. **E 1778,** E17

pot life—See **working life.** **C 717,** C24

pot life—the period of time during which a reacting thermosetting composition remains suitable for its intended processing after mixing with reaction-initiating agents. **D 883,** D20

pot life, n— Synonym for **working life**. **D 907,** D14

pot life, *n*—the period of time during which a reacting thermosetting plastic or rubber composition remains suitable for its intended use, after mixing with a reaction-initiating agent. **D 1566,** D11

pot-life—See **working life**. **E 1749,** E06

pot life—the period of time during which a reacting thermosetting plastic or rubber composition remains suitable for its intended processing after mixing with reaction-initiating agents. (D-20) **F 412,** F17

pot, pan, and utensils washing machine, commercial—machines that uniformly wash, rinse, and heat-sanitize food preparation utensils. The machines are capable of removing physical soil and sanitizing multiple pots, pans, and utensils from properly racked and pre-scraped multiple items. **F 1827,** F26

pot, pan, and utensil washing machine, commercial—see **pot, pan, and utensil washing machines, commercial**. **F 1827,** F26

pottery—all fired ceramic wares that contain clay when formed, except technical, structural, and refractory products. **C 242,** C21

potting, *n*—an embedding process for parts that are assembled in a container (or "can") into which the insulating material is poured, and the container remains an integral part of the finished unit as the outer surface. **D 1711,** D09

pot wagon—a vehicle used for transferring a pot from a pot arch to a pot furnace. **C 162,** C14

pouch, *n*—a small or moderate-sized bag, sack, or receptacle for carrying or containing parts. **D 996,** D10

pouching—enclosing in an envelope; usually, a transparent one. **F 1156,** F12

poultice corrosion— see **deposit corrosion.** **G 15,** G01

Pourbaix diagram (electrode potential-pH diagram)—a graphical representation showing regions of thermodynamic stability of species in metal-water electrolyte systems. **G 15,** G01

pour density—the mass of an amount of powder divided by its volume and measured by pouring the powder into a graduated container. **C 859,** C26

pour density, *n*—the mass per unit volume of pelleted carbon black. **D 3053,** D24

pouring pit refractory, *n*—a refractory associated with the transfer or flow control of molten steel between furnace and mold. **C 71,** C08

pour-on application—application by pouring a chemical onto the target. **E 1102,** E35

pour point—the lowest temperature at which a liquid can be observed to flow under specified conditions. **D 2864,** D27

powder, *n*—particles that are usually less than 1000 μm (1 mm) in size. **B 243,** B09

powder, *n*—particles of a solid characterized by small size, nominally within the range of from 0.1 to 1000 u.m. **B 899,** B02

powder—synonym for **bulk solid**, particularly when the particles of the bulk solid are fine. **D 653,** D18

powder—see **dry powder.** **E 1316,** E07

powder adhesion—the ability of an electrostatically charged powder to remain attached by static attraction to a grounded substrate. **C 286,** B08

powder blend—See **dry-blend.** **D 883,** D20

powder blower—a compressed air device used to apply magnetic powder over the surface of a part undergoing inspection. **E 1316,** E07

powder cleaner, *n*—a cleaning agent in which an absorbent compound is the principal ingredient. See **absorbent compound**. **D 123,** D13

powder coating, *n*—a coating produced by the use of a heat-fusible coating powder. **D 1711,** D09

powdered activated carbon—activated carbon in particle sizes predominantly smaller than 80 mesh. **D 2652,** D28

powdered iced tea dispenser—commercial equipment designed to deliver a portion of instant tea, usually mixed with tap water and dispensed into a container with ice. **F 1827,** F26

powdered rubber, *n*—particulate rubber composed of mainly non-spherical particles that have a maximum particle dimension equal to or below 425 μm (40 mesh). See **particulate rubber.** **D 1566,** D11

powder flow meter—an instrument for measuring the rate of flow of a powder according to a specified procedure. **B 243,** B09

powder forging—hot densification by forging of an unsintered, presintered, or sintered preform made from powder. Also referred to by the acronym P/F, P/M forging, or P/M hot forming. There are two basic forms of powder forging: **hot upset powder forging** and **hot repress powder forging.** **B 243,** B09

powdering, *n*—*as related to metallic coatings*, microcracking and fine particle separation of generally brittle coatings when the coating is severely stressed, resulting in a fine powder residue. **A 902,** A05

powdering—partial or total disintegration of the polish film resulting in a fine, light-colored material. **D 2825,** D21

powdering—a phenomenon that occurs in the creases of shoes during wear when polish film does not have adequate flexibility and breaks up into small powdery particles. **D 2825,** D21

powdering, *n*—a condition where ink easily rubs off a print after the normal drying period. Also referred to as" chalking". **D 6488,** D01

powder injection molding (PIM), *n*— a process in which a mixture of powders and a binder system is forced under pressure into a die. See also **metal injection molding**. **B 243,** B09

powder lubricant—an agent mixed with or incorporated in a powder to facilitate the pressing and ejecting of the compact. **B 243,** B09

powder metallurgy, *n*—the production and utilization of metal powders. **B 243,** B09

powder metallurgy part—a shaped object that has been formed from metal powders and bonded by heating below the melting point of the major constituent. A structural or mechanical component, bearing, or bushing, made by the powder metallurgy process. Synonymous with **P/M part** and **sintering.** **B 243,** B09

powder method—any method of X-ray diffraction involving a polycrystalline and preferably randomly oriented powder specimen and a narrow beam of monochromatic radiation. **E 7,** E04

powder porcelain enameling—process by which the application of porcelain enamel is achieved by dry electrostatic spraying. **C 286,** B08

powder porcelain resistivity—the opposition that a porcelain powder offers to the flow of direct current, equal to the voltage drop across the powder, divided by the current through the powder. (Also known as **electrical resistance**) **C 286,** B08

powder retention—Same as **electrostatic retention**. **C 286,** B08

powder rolling—see **roll compacting.** **B 243,** B09

powder to air ratio—ratio of the mass of powder delivered to the spray gun to the total volume of air used to convey and aspirate it. **C 286,** B08

power, active (real), *P*—the product of the rms current in a circuit, the rms voltage across the circuit and the cosine of the angular phase difference between the current and voltage. **A 340,** A06

power cable insulating paper, *n*—paper used, in conjunction with an insulating liquid, as the primary electrical insulation on conductors for transmission of electric energy. **D 1711,** D09

power factor, dielectric— same as **power factor.** **D 2864,** D27

power factor, magnetic, cos γ—(*a*) the cosine of the angle between vectors representing the rms values of the applied voltage of a circuit and the current in circuit.

(*b*) the ratio of the active (real) power to the apparent power in an ac circuit. **A 340,** A06

power factor, ***PF***, *n*—the ratio of the power in watts, *W*, dissipated in a material to the product of the effective sinusoidal voltage, *V*, and current, *I*, in volt-amperes. **D 1711,** D09

power factor, *PF*— the ratio of the power in watts, *W*, dissipated in a material, to the product of the effective values of voltage, *V*, and current, *I*, in volt-amperes:

$$PF = W/VI$$

The power factor is related to the **dissipation factor,** *D*, as follows:

$$PF = D/\sqrt{1 + D^2}$$

D 2864, D27

power loss—the power, per unit volume, that is transformed into heat through hysteresis. (ISO) **D 4092,** D20

power, reactive (quadrature), P_q—the product of the rms current in an electrical circuit, the rms voltage across the circuit, and the sine of the angular phase difference between the current and the voltage.

$$P_q = EI \sin \theta$$

where:

P_q = reactive power in vars,
E = voltage in volts,
I = current in amperes, and
θ = angular phase by which E leads I.

NOTE—The reactive power supplied to a magnetic core having an *SCM* excitation is the product of the magnetizing current and the voltage induced in the exciting winding. **A 340,** A06

power spectral density—the limiting mean-square value (for example, of acceleration, velocity, displacement, stress, or other random variable) per unit bandwidth, that is the limit of the mean-square value in a given rectangular bandwidth divided by the bandwidth, as the bandwidth approaches zero. **D 653,** D18

power spectral density—the limiting mean-square value (for example, of acceleration, velocity, displacement, stress, or other random variable) per unit bandwidth of frequency, that is the limit of the mean-square value of a given rectangular bandwidth divided by the bandwidth, as the bandwidth approaches zero. **E 1823,** E08

power unit, *n*—the part of a central vacuum system that contains the suction source and dirt receptacle. **F 395,** F11

pozzolan, *n*—a siliceous or siliceous and aluminous material, which in itself possesses little or no cementitious value but will, in finely divided form and in the presence of moisture, chemically react with calcium hydroxide at ordinary temperatures to form compounds possessing cementitious properties. **C 125,** C09

pozzolan, *n*—a siliceous or siliceous and aluminous material, which in itself possesses little or no cementitious value but which will, in finely divided form and in the presence of moisture, chemically react with calcium hydroxide at ordinary temperatures to form cementitious hydrates. **C 219,** C01

pozzolan—a siliceous or siliceous and aluminous material, which in itself possesses little or no cementitious value but will, in finely divided form and in the presence of moisture, chemically react with calcium hydroxide at ordinary temperatures to form compounds possessing cementitious properties. **D 653,** D18

ppb(v)—See **concentration.** **D 1356,** D22

ppm—parts per million. **F 1156,** F12

ppm(v)—See **concentration.** **D 1356,** D22

ppt—parts per thousand. **F 1156,** F12

pptt—parts per ten thousand. **F 1156,** F12

practical response time—the length of time that elapses between the instant at which an ion-selective electrode and a reference electrode are brought into contact with a sample solution (or at which the concentration of the ion of interest in a solution in contact with an ion-selective electrode and a reference electrode is changed) and the first instant at which the potential of the cell becomes equal to its steady-state value within 1 mV. The experimental conditions used should be stated, that is, the stirring rate, the composition of solution of which the response time is measured, the composition of the solution to which the electrode was exposed prior to this measurement, the history and preconditioning of the electrode, and the temperature. **D 4127,** D19

practical skills instructor—an individual who assists with practical skills instruction under the direction of the course instructor/coordinator. **F 1177,** F30

practice, *n*—a definitive procedure for performing one or more specific operations or functions that does not produce a test result. **D 123,** D13

pre-alloyed powder, *n*—powder composed of two or more elements that are alloyed in the powder manufacturing process in which the particles are of the same nominal composition throughout. **B 243,** B09

prebake resistance, n—deprecated term, see **precure heat tolerance.** **D 1566,** D11

prebleeding—See **debulking.** **E 1749,** E06

pre-bond treatment, n—Synonym for **surface preparation**. **D 907,** D14

preburn period, *n*—*in optical emission spectrometry*, the time interval after the initiation of a discharge during which the emitted radiation energy is not recorded for analytical purposes. **E 135,** E01

preceramic polymer, *n*—inorganic or organometallic polymers that can be converted (after polymer curing) to a ceramic by a thermal treatment. **C 1145,** C28

precious metals, *n*—the eight noble metals: gold, silver, palladium, platinum, rhodium, iridium, osmium, and ruthenium. **B 899,** B02

precipitate—an insoluble product of a chemical reaction of soluble compounds in water. **D 6161,** D19

precipitates—regions of silicon oxide contained within the silicon lattice, frequently manifested as etch pits. **F 1241,** F01

precipitation, *n*—separation of a new phase from solid, liquid, or gaseous solutions, usually with changing conditions of temperature or pressure, or both.

electrostatic precipitation, n—a process consisting of the separation of particulate matter from air or other gases under the influence of an electrostatic field.

meteorological precipitation, n—the precipitation of water from the atmosphere in the form of hail, mist, rain, sleet, and snow.

thermal precipitation, n—a process consisting of the separation of particulate matter from air and other gases under the influence of a relatively large temperature gradient extending over a short distance.

ultrasonic precipitation, n—a process consisting of the separation of particulate matter from air and other gases following agglomeration induced by an ultrasonic field. **D 1356,** D22

precipitation—separation of a new phase from solid, liquid, orgaseous solutions, usually with changing conditions of temperature or pressure, or both. **E 7,** E04

precipitation hardening, *n*—**hardening** caused by the precipitation of a constituent from a supersaturated solid solution. **A 941,** A01

precipitation heat treatment, *n*—**artificial aging** in which a constituent precipitates from a supersaturated solid solution. **A 941,** A01

precipitation heat treatment—a thermal treatment of a solution heat-treated product to produce property changes such as hardening, strengthening, and conductivity increase by precipitation of constituents from the supersaturated solid solution. This treatment has also been called "age hardened" and "precipitation hardened." **B 846,** B05

precipitation titration—a titration in which the titrant and the sample species form a precipitate. An electrode can be used to follow the titration if it senses either of the ions involved in the reaction. The success of the titration depends on the lower limit of detection of the electrode, the solubility of the precipitate, the reaction rate, and the possible presence of other species which are also precipitated by the titrant. Examples of precipitation titrations are the determination of halides with silver nitrate and sulfate with lead ion. **D 4127,** D19

precipitator, *n*—a device or apparatus for the separation of particulate matter from air or other gases.

electrostatic precipitator, n—apparatus employing electrostatic precipitation for the separation of particles from a gas stream.

thermal precipitator—See **precipitation.** **D 1356,** D22

precision—the agreement of repeated measurements of the same parameter expressed quantitatively as the standard deviation computed from the results of a series of controlled determinations. **C 242,** C21

precision—a generic term used to describe the dispersion of a set of measured values. **C 859,** C26

precision, *n*—the degree of agreement within a set of observations or test results obtained as directed in a test method. **D 123,** D13

precision, *n—under conditions of single-operator precision,* the single-operator-laboratory-sample-apparatus-day precision of a method; the precision of a set of statistically independent observations all obtained as directed in the method and obtained over the shortest practical time interval in one laboratory by a single operator using one apparatus and randomly drawn specimens from one sample of the material being tested. **D 123,** D13

precision, *n—under conditions of within-laboratory precision with multiple operators,* the multi-operator, single-laboratory-sample, single-apparatus-day (within-operator) precision of a method; the precision of a set of statistically independent test results all obtained in one laboratory using a single sample of material and with each test result obtained by a different operator, with each operator using one apparatus to obtain the same number of observations by testing randomly drawn specimens over the shortest practical time interval. **D 123,** D13

precision, *n—under conditions of between laboratory precision,* the multi-laboratory, single-sample, single-operator-apparatus-day (within-laboratory) precision of a method; the precision of a set of statistically independent test results all of which are obtained by testing the same sample of material and each of which is obtained in a different laboratory by one operator using one apparatus to obtain the same number of observations by testing randomly drawn specimens over the shortest practical time interval. **D 123,** D13

precision—the degree of agreement of repeated measurements of the same property, expressed in terms of dispersion of test results about the arithmetical mean result obtained by repetitive testing of a homogeneous sample under specified conditions. The precision of a method is expressed quantitatively as the standard deviation computed from the results of a series of controlled determinations. **D 459,** D12

precision—the degree of agreement of repeated measurements of the same parameter expressed quantitatively as the standard deviation computed from the results of a series of controlled determinations. **D 1129,** D19

precision, *n*—the degree of agreement of repeated measurements of the same property, expressed in terms of dispersion of test results about the mean result obtained by repetitive testing of a homogenous sample under specified conditions.

overall precision, n—a value including components of within-laboratory and between-user variability.

single-operator precision, n—a measure of the replication of repeated measurements obtained by a single operator on a given sample. **D 1356,** D22

precision—the degree of agreement of repeated measurement of the same property, expressed in terms of dispersion of test results about the arithmetical mean result obtained by repetitive testing of a homogeneous sample under specified conditions. The precision of a test method is expressed quantitatively as a standard deviation computed from the results of a series of controlled determinations. **D 4790,** D16

precision—the closeness of agreement between randomly selected individual measurements or test results . **E 131,** E13

precision—*of methods of chemical analysis,* a characteristic manifested by agreement among individual results at a given analyte content. **E 135,** E01

precision—the closeness of agreement between measurements obtained under prescribed conditions (see Terminology E 456). **E 170,** E10

precision, *n*—the closeness of agreement between test results obtained under prescribed conditions. See also **repeatability, reproducibility.** **E 284,** E12

precision, *n*—the scatter between individual values of test data within the subset, normally computed with respect to the mean of the subset. See **bias.** **E 344,** E20

precision, *n*—the closeness of agreement between independent test results obtained under stipulated conditions.

NOTE—Precision depends on random errors and does not relate to the true value or the specified value.

NOTE—The measure of precision usually is expressed in terms of imprecision and computed as a standard deviation of the test results. Less precision is reflected by a larger standard deviation.

NOTE—"Independent test results" means results obtained in a manner not influenced by any previous result on the same or similar test object. Quantitative measures of precision depend critically on the stipulated conditions. Repeatability and reproducibility conditions are particular sets of extreme stipulated conditions. **E 456,** E11

precision—a term used to indicate the capability of a person, an instrument, or a method to obtain reproducible results; specifically, a measure of the random error as expressed by the variance, the standard error, or a multiple of the standard error. **E 856,** D34

precision—the degree of agreement of repeated measurements of the same property. Precision statements in ASTM methods for industrial chemicals will be derived from the estimated standard deviation or coefficient of variation of a series of measurements and will be expressed in terms of the repeatability, the within-laboratory, between days variability, and the reproducibility of the method. **E 1547,** E15

precision—degree of mutual agreement between individual measurements, namely repeatability and reproducibility. Compare **accuracy**. **E 1605,** E06

precision—the closeness of agreement between randomly selected individual measurements or test results for a given set of experimental variables. **E 1823,** E08

precision, *n*—a measurement concept that expresses the ability to generate test results that *agree with each other* in absolute magnitude. **F 538,** F09

precision (as distinguished from accuracy)—the degree of mutual agreement between individual measurements, namely repeatability and reproducibility. (E 380) **F 221,** F05

precleaning—the removal of surface contaminants from the test part so that they will not interfere with the examination process. **E 1316,** E07

precoagulum, *n*—coagulum resulting from the partial inadvertent coagulation of a latex. **D 1566,** D11

precoat—the initial coating of the septum in a diatomaceous earth filter to provide initial straining medium. **D 6161,** D19

precoat—see **basecoat**. **F 1623,** F05

precondition, *v*—to bring a sample or specimen of a textile material to a relatively low moisture content (approximate equilibrium in an atmosphere between 3 and 25 % relative humidity) prior to conditioning in a controlled atmosphere for testing. **D 123,** D13

pre-conditioning—an activity conducted prior to placing plugging material into a borehole in order to stabilize the hole. **D 653,** D18

preconditioning—any preliminary exposure of a material to the influence of specified atmospheric conditions for the purpose of favorably approaching equilibrium with a prescribed atmosphere. **E 41,** G03

preconsolidation pressure (prestress), p_e **(FL^{-2})**—the greatest effective pressure to which a soil has been subjected. **D 653,** D18

precrack load, P_M **[F]**—the allowable precrack load. **E 1823,** E08

precure heat tolerance (adhesives), *n*— the amount of exposure (at a certain time and temperature) prior to bonding that a cure-activated adhesive system can tolerate without significant impairment of its adhesion potential. **D 1566,** D11

precut building—See **building.** **E 631,** E06

prediluted aqueous ethylene glycol coolant, *n*—a commercially prepared uniform solution (50 % by volume minimum) of ethylene-glycol-based engine coolant concentrate and deionized water (described in Specification D 1193, Type IV). **D 4725,** D15

predrilled hole—hole drilled through metal connector plate during its fabrication. **E 1807,** E06

preemergence herbicide—a chemical applied before emergence of a specified weed or planted crop. **E 609,** E35

pre-etching—the chemical treatment of the surface of a metal plate by the plate maker, prior to application of the photosensitive coating, to permit the more perfect removal of coating from non-image areas. **F 425,** F05

pre-exposure—exposure of a photoconductive medium to a light source prior to imaging. **F 335,** F05

prefabricated building—See **building.** **E 631,** E06

prefabricated panel system—See **building system.** **E 631,** E06

preference, *n*—choice of one product, treatment, or item over others in a given set based upon hedonic criteria. **E 253,** E18

preferential adsorption—adsorption in which a certain component or certain components are adsorbed to a much greater extent than others. **D 2652,** D28

preferential etch—a selective etch that etches regions of different lattice strain or conductivity at different rates, used to delineate crystal defects or regions of differeing conductivity on wafer surfaces. **F 1241,** F01

preferential sputtering—See **sputtering.** **E 673,** E42

preferred orientation— *in manufactured carbon and graphite product technology*, an alignment in the crystal or defect structure of a body leading to variations in physical properties as a function of direction; normally referenced to an orthogonal system where one of the axes is the working direction. **C 709,** D02

preferred orientation—a condition of polycrystalline aggregate in which the crystal orientations are not distributed at random. **E 7,** E04

preferred white, *n*— *in color determination,* the white color, usually bluish, that is judged by a given group of observers looking at a given series of specimens to be the whitest color attainable. **D 2946,** C17

preferred white surface(s)—the particular surface which under specified viewing conditions is considered whitest by a statistically significant majority of representative observers in a series of paired comparisons performed at a specified time and geographical locale (psychological definition). **D 459,** D12

prefilming airblast nozzle—an airblast nozzle in which solid surfaces are configured to produce a liquid film that subsequently is broken up by adjacent airstreams. **E 1620,** E29

prefit—a process to check the fit of mating detail parts in an assembly prior to adhesive bonding in order to ensure proper bondlines. **E 1749,** E06

pre-flame reaction, *n*— *in fuel testing*, a slow, nonluminous reaction of the sample or its decomposition products with the atmosphere in the combustion chamber. **D 4175,** D02

preform, *n*—a preshaped mat or woven structure formed from fibers of whiskers to the desired configuration and reinforcement architecture. **C 1145,** C28

preform, *n*—a coherent, shaped mass of powdered, granular or fibrous plastic molding compound, or of fibrous filler material with or without resin. (ISO) **D 883,** D20

preform, *n*—a preshaped fibrous reinforcement, normally without matrix, but often containing a binder to facilitate manufacture, formed by distribution of fibers to the approximate contour and thickness of the finished part. **D 3878,** D30

preform, *adj*—a descriptive word applied to a commercially produced spectroscopic electrode which is purified after cutting or machining. **E 135,** E01

preformed—formed prior to driving. **F 592,** F16

preformed flexible joint sealant—pliable material, formed into a defined cross section, that is applied to the surface of a pipe, box section, or manhole section joint, intended to control the movement of fluids or solids through the joint. **C 822,** C13

preformed gasket—an elastomeric compound molded in the form of a continuous strip, channel, or other shape, for use in filling joints and providing weathertight seals in glazing or between building components. **C 717,** C24

preforming—the initial pressing of a metal powder to form a compact which is subjected to a subsequent pressing operation other than coining or sizing. Also, the preliminary shaping of a refractory metal compact after presintering and before the final sintering. **B 243,** B09

preheat—to heat the equipment's cooking medium, cavity, or surface (see **cooking medium, cooking cavity,** or **cooking surface**) from a predetermined ambient temperature to the equipment's thermostatic set point. **F 1827,** F26

preheat duration—total time required for preheat (see **preheat**), from preheat initiation at controls to when the equipment is ready to cook. **F 1827,** F26

preheat energy—amount of energy consumed (Btu, kJ or kWh) by the equipment while heating the equipment's cooking medium, cavity, or surface (see **cooking medium, cooking cavity**, or **cooking surface**) from an ambient temperature to a thermostat set point, E_p. **F 1827,** F26

preheating, *n—for tool steels*, heating to an intermediate temperature immediately before final **austenitizing**. **A 941,** A01

preheating, *n*—heating before welding, a mechanical treatment, or some further thermal treatment. **A 941,** A01

preheating, solar—the use of solar energy to partially heat a substance, such as domestic potable water, prior to heating it to a higher desired temperature with auxiliary fuel. **E 772,** E44

preheat rate—average rate at which the temperature of the equipment's cooking medium, cavity, or surface (see **cooking medium, cooking cavity,** or **cooking surface**) is heated from a predetermined ambient temperature to the equipment's thermostat set point. **F 1827,** F26

preheat time—time required for the equipment to heat from the ambient room to a specified and calibrated operating temperature, or thermostat set point, T_p. **F 1827,** F26

preheat zone—that portion of a continuous furnace through which the ware passes before entering the firing zone. **C 286,** B08

prehospital emergency medical services—a sub-system of the emergency medical services system which provides medical services to patients requiring immediate assistance due to illness or injury, prior to the patients' arrival at an emergency medical facility. **F 1177,** F30

prehospital provider—all personnel providing emergency medical care in a location which is remote from facilities which are capable of providing definitive medical care. **F 1177,** F30

pre-launch production plan—a written description of the dimensional, mechanical, chemical, and performance testing that will be carried out during initial production, prior to full production. **F 1789,** F16

premix—a uniform mixture of ingredients to a prescribed analysis, prepared by the powder producer, for direct use in compacting powder metallurgy products. **B 243,** B09

premix, *n*—in reinforced thermosetting plastics, the admixture of resin, reinforcements, fillers, etc., not in web or filamentous form, ready for molding. **D 883,** D20

premix burner, *n—in flame atomic absorption and emission spectrometry*, a burner in which the fuel gas is mixed with the oxidizing gas before reaching the combustion zone. **E 135,** E01

prepack seal—a preassembly seal installed to fill voids or provide a support seal for subsequent fillet sealing. **E 1749,** E06

preparation time—the time from the removal of a cooked load (see **load**) from the equipment until the next load is entered in the equipment. **F 1827,** F26

preparatory coat—an initial coat of the liquid-applied membrane which is applied at cracks, joints, or terminal points to provide reinforcement to the membrane at these critical areas. **C 717,** C24

prepared coal—*for the purpose of Test Method D 4749*, any coal, regardless of its topsize, that has been manually or mechanically cleaned. This includes coal that has been processed over a picking table or air tables through a breaker, jig, or other device which segregates according to size or density (specific gravity). **D 121,** D05

preplaced aggregate concrete—concrete produced by placing coarse aggregate in a form and later injecting a portland cement-sand or resin grout to fill the interstices. **D 653,** D18

prepolymer, *n*—a polymer of degree of polymerization between that of the monomer or monomers and the final polymer. (ISO) **D 883,** D20

prepolymer, *n*—a polymer of degree of polymerization between that of the monomer or monomers and the final polymer. **F 1251,** F04

prepreg, *n*—in reinforced thermosetting plastics, the admixture of resin, reinforcements, fillers, etc., in web or filamentous form, ready for molding. **D 883,** D20

prepreg—the admixture of fibrous reinforcement and polymeric matrix used to fabricate composite materials. Its form may be sheet, tape, or tow. For thermosetting matrices, it has been partially cured to a controlled viscosity called "B stage." **D 3878,** D30

prepreg—a combination of mat, fabric, nonwoven material, or roving with resin usually advanced to the B-stage, ready for curing. **E 1749,** E06

preprinted symbol, *n*—a symbol printed in advance of application either on a label or on the article to be identified. **F 1294,** F05

preproduction test, *n*—a test or series of tests conducted by (*1*) an adhesive manufacturer to determine conformity of an adhesive batch to established production standards, (*2*) a fabricator to determine the quality of an adhesive before parts are produced, or (*3*) an adhesive specification custodian to determine conformance of an adhesive to the requirements of a specification not requiring qualification tests. (Compare **acceptance test** and **qualification test**.) **D 907,** D14

prepunched hole—hole punched through metal connector plate during its fabrication. **E 1807,** E06

presensitized master—offset master coated by the master supplier and ready for exposure when received by the offset printers. These masters are imaged photographically. **F 425,** F05

present serviceability, *n*—the current condition of a pavement (traveled surface) as perceived by the traveling public. **E 867,** E17

present serviceability rating (PSR), *n*—a mean rating of the serviceability of a pavement (traveled surface) established by a rating panel under controlled conditions. The accepted scale for highways is 0 to 5, with 5 being excellent. **E 867,** E17

present value—the value of a benefit or cost found by discounting future cash flows to the base time (Syn. *present worth*). **E 833,** E06

present value factor—The discount factor used to convert future values (benefits and costs) to present values (Syn. *present worth factor*). **E 833,** E06

present worth—See **present value.** **E 833,** E06

present worth factor—See **present value factor**. **E 833,** E06

preservation, *n*—action taken to retard or prevent deterioration of or damage to cultural objects. **D 123,** D13

preservation, *n— in packaging,* the application of protective measures such as cleaning, drying, and the use of preservatives, **barrier materials, cushioning,** and **containers.** **D 996,** D10

preservation, *n*—the act or process of applying measures to sustain the existing form, integrity, or materials of a **building,** structure, or **artifact** and the existing form or vegetative cover of a cite.

conservation, n—management of a natural resource, structure, or **artifact** to prevent misuse, destruction, or neglect. It may include detailed characterization and recording (technical or inventory) or provenance and history and application of measures.

protection, n—the act or process of applying measures designed to affect the physical condition of a **building,** structure, or **artifact** by guarding it from deterioration, loss, or attack; or, to cover or shield it from damage.

rehabilitation, n—of a structure, the act or process of returning a structure to a state of utility through **repair** or **alteration** which makes possible an efficient contemporary use.

restoration, n—the act or process of reestablishing accurately the form and details of a structure, site, or **artifact** as it appeared at a particular period in time, by means of removal of later work or by the **reconstruction** of missing earlier work. **E 631,** E06

preservative—a chemical substance which, when suitably applied to wood, makes the wood resistant to attack by fungi, insects, marine borers, or weather conditions. **D 9,** D07

preservative (latex), *n*—substance added to uncompounded latex before or after concentration to inhibit putrefaction and accompanying coagulation. **D 1566,** D11

preshadowed replica— See **replica**. **E 7,** E04

preshimmed sealant, *n—in building construction,* a sealant having encapsulated therein solids or discrete particles that limit its deformation within a joint under compression. **C 717,** C24

preshimmed tape sealant, *n—in building construction,* a preshimmed sealant in tape sealant form. **C 717,** C24

presintering—the heating of a compact at a temperature below the normal final sintering temperature, usually to increase the ease of handling or shaping the compact, or to remove a lubricant or binder prior to sintering. **B 243,** B09

press, *v*—to apply force to a mass of powder, generally while confined in a die or container, to form a compact. **B 243,** B09

press—an apparatus for applying and maintaining pressure on an assembly of veneers and adhesive in the manufacture of plywood. It may be operated mechanically or hydraulically and the platens may be cold or heated depending on the type of adhesive used. **D 1038,** D07

press—mechanical device that removes liquids from solids by mechanically pressing the solids against a porous surface. **E 1705,** E48

press-and-blow process—a process of glass manufacture in which the finish and parison are pressed and the parison is subsequently blown to form the final shape. **C 162,** C14

pressed bar—a compact in the form of a bar; a green compact. **B 243,** B09

pressed density, *n*—synonymous with **green density.** **B 243,** B09

pressed glass—glassware formed by pressure between a mold and a plunger. **C 162,** C14

pressed-in crease, *n*—a sharp crease inserted intentionally in a fabric usually by application of pressure, heat, and moisture. **D 123,** D13

pressed product, *n—in cooling tower fill,* that which is pressed singly or in stacks interlayered with templates, at a minimum pressure of 12 MPa (1740 psi). **C 460,** C17

pressed product, *n—in cooling tower fill,* that which is pressed singly or in stacks interlayered with templates, at a minimum pressure of 12 MPa (1740 psi). **D 2946,** C17

pressed products, *n— in cooling tower fill,* includes those that are pressed singly or in stacks interlayered with templates, at a minimum pressure of 12 MPa (1740 psi). **C 1154,** C17

pressing, *v*—shaping of brick by pressing clay or shale into a mold cavity which forms the peripheral dimensions of the brick. Different sub-classifications of pressing are defined by the quantity of water mixed with the clay or shale.

Dry pressing uses high forming pressures and low water contents usually between 0 and 5 %.

Plastic pressing uses low pressures and sufficient water to produce a plastic mixture.

Semi-dry pressing uses intermediate pressures and water quantities nominally between 5 and 14 %. **C 43,** C15

pressing:

dry pressing—forming ceramic ware in dies from powdered or granular material by direct pressure.

hot pressing—a jiggering process wherein a heated profile tool or plunger is used.

wet pressing (plastic pressing)—forming ceramic ware in dies from a plastic body by direct pressure. **C 242,** C21

pressing, *n—in the care of textiles,* a process of smoothing and shaping by heat and pressure, with or without the presence of steam. **D 123,** D13

pressing and finishing, *n*—this term takes into account all of the industrial pressing and finishing treatments used in garment production. **D 123,** D13

pressing crack—a rupture in the pressed compact caused by the mass slippage of a part of the compact. Synonymous with **lamination.** **B 243,** B09

pressing (n)—an as-pressed, oversized piece of glass used as the precursor for finishing to an optical element blank. **C 162,** C14

pressing (v)—the operation of forming pressed glass. **C 162,** C14

press-off, *n—in knitted fabric,* a condition in which the yarn fails to knit and either the fabric falls off the needles or the design is distorted or incomplete. **D 123,** D13

pressure, *n*—the force or load per unit area. **D 123,** D13

pressure, *n*—the force or load per unit area.

gage pressure, n—the difference in pressure between that existing within a system and that of the surrounding atmosphere.

static pressure, n—the pressure of a fluid at rest, or in motion, exerted perpendicularly to the direction of flow.

total pressure, n—the pressure representing the sum of static pressure and velocity pressure at the point of measurement.

velocity pressure, n—that pressure caused by and related to the velocity of the flow of fluid; a measure of the kinetic energy of the fluid. **D 1356,** D22

pressure—the internal force per unit area exerted by any material. Since the pressure is directly dependent on the temperature, the latter must be specified. The pressure may be reported in either of two ways:

(*1*) **absolute pressure**—the total pressure with zero as a reference point, usually expressed as pounds-force per square inch absolute (psia).

(*2*) **gage pressure**—the pressure in excess of atmospheric pressure. Under standard conditions at sea level, the numerical value of the absolute pressure is 14.7 higher than that of the gage pressure, which is usually expressed in pounds-force per square inch gage (psi). **D 3064,** D10

pressure—when expressed with reference to pipe, the force per unit area exerted by the medium in the pipe. **F 412,** F17

pressure atomizer—a device wherein pressurized liquid is the primary source of energy utilized to produce a spray. **E 1620,** E29

pressure atomizing nozzle—see **pressure atomizer**. **E 1620,** E29

pressure-break, *n*—as applied to a defect in a laminated plastic a break apparent in one or more outer sheets of the paper, fabric, or other base visible through the surface layer of resin which covers it. **D 883,** D20

pressure bulb—the zone in a loaded soil or rock mass bounded by an arbitrarily selected isobar of stress. **D 653,** D18

pressure check—an imperfection; a check or crack in a glass article resulting from too much pressure in forming. **C 162,** C14

pressure correction factor—a multiplier that adjusts the measured gas volume to the actual volume at standard gas pressure, P_{cf}

$$= \frac{\textit{absolute actual gas pressure psia}}{\textit{absolute standard pressure psia}}$$

$$= \frac{\textit{gas gage pressure psig + barometric pressure psia}}{\textit{absolute standard pressure psia}}$$

NOTE—Absolute standard gas temperature and pressure used in this calculation should be the same values used for determining the higher heating value. PG&E standard conditions are 519.67°R and 14.73 psia.
F 1827, F26

pressure, design—pressure used in design to determine the required minimum thickness and minimum mechanical properties.
E 1316, E07

pressure die-casting, *n*—Same as die casting. **B 899,** B02

pressure difference—in leak testing, the difference between the pressure on the inlet side of the leak and the pressure on the exit side of the leak. **E 1316,** E07

pressure dye test—*(1)* a form of leak test in which the item or items to be tested are filled with a liquid dye or fluorescent oil which is then pressurized for the purpose of driving the liquid through possible leakage paths with the presence of the leaks being visible when viewed from the exterior; *(2)* a form of leak test in which the item or items to be tested are immersed in a liquid dye or fluorescent oil which is then pressurized for the purpose of driving liquid into possible leakage paths with their presence being visible when the excess liquid has been removed from the exterior.
E 1316, E07

pressure-evacuation test—a leak test in which one or more devices are placed under gas pressure for a period of time, the objective being to accumulate enough gas in those devices that may leak to permit an indication on a leak detector sensitive to the gas when the devices are placed in an evacuated system joined to the leak detector. **E 1316,** E07

pressure filling—pressurizing a container by injecting a propellant through the valve under high pressure. The operation is usually used where the propellant content is small, and it is carried out under ambient conditions (that is, refrigeration equipment is not needed). **D 3064,** D10

pressure filtration—filtration performed in an enclosed pressurized filter vessel. **D 6161,** D19

pressure fixing—the process of making the image permanent by means of pressure. **F 335,** F05

pressure flow, *n*—flow in a direction parallel to the plane of a geotextile or related product driven predominately by a differential fluid pressure. **D 4439,** D35

pressure fryer—equipment with a deep kettle containing oil or fat covered by a heavy, gasketed lid with a pressure valve. The equipment kettle operates between 10 and 12 psig. See **fryer.**
F 1827, F26

pressure head—the head of water at a point in a porous system; negative for unsaturated systems, positive for saturated systems. Quantitatively, it is the water pressure divided by the specific weight of water. **D 653,** D18

pressure inflated—inflatable boom that requires pressurized gas for its flotation. **F 818,** F20

pressure-less steam cooker—equipment where in the cooking compartment operates between 0 and 2.9 psig. See **steam cooker.**
F 1827, F26

pressure nozzle—see **pressure atomizer.** **E 1620,** E29

pressure, *p* **(FL^{-2})**—the load divided by the area over which it acts.
D 653, D18

pressure pipe—pipe designed to resist continuous pressure exerted by the conveyed medium. **F 412,** F17

pressure probe—see **probe.** **E 1316,** E07

pressure rating—the estimated maximum pressure that the medium in the pipe can exert continuously with a high degree of certainty that failure of the pipe will not occur. **F 412,** F17

pressure relief device—a pressure-activated valve designed to automatically relieve excessive pressure. **E 772,** E44

pressure, saturation—the pressure, for a pure substance at any given temperature, at which vapor and liquid, or vapor and solid, coexist in stable equilibrium. **E 41,** G03

pressure sensitive—a carbon paper, ribbon, or paper capable of producing an image from pressure as with pen or pencil.
F 221, F05

pressure-sensitive adhesive, *n*—a viscoelastic material which in solvent-free form remains permanently tacky and will adhere instantaneously to most solid surfaces with the application of very slight pressure. **D 907,** D14

pressure sensitive adhesive—See **adhesive, contact pressure.**
E 1749, E06

pressure-sensitive-adhesive tape—See **tape.** **D 996,** D10

pressure-sensitive article, *n*—a combination of a pressure-sensitive adhesive with a backing or with a backing and release liner.
D 907, D14

pressure-sensitive copy paper—paper systems that develop a visible image on the application of localized pressure. **F 221,** F05

pressure tank—a container from which slip is removed by air pressure. **C 286,** B08

pressure testing—a method of permeability testing with water or grout pumped downhole under pressure. **D 653,** D18

pressure testing—a method of leak testing in which the component being tested is filled completely with a gas or liquid which is then pressurized. The outside of the component is examined for the detection of any leaks. **E 1316,** E07

pressure-treated wood—wood treated by applying pressure to force the preservative into it. **D 9,** D07

pressure, vapor—the pressure exerted by a vapor.

NOTE—If a vapor is kept in confinement over its source so that the vapor can accumulate, the temperature being held constant, the vapor pressure approaches a fixed limit called the maximum, or saturated, vapor pressure, dependent only on the temperature and the liquid.
E 41, G03

pressure vessel—the vessel containing one or more individual membrane elements and designed to withstand safely the hydraulic pressure driving the separation mechanism. **D 6161,** D19

pressure-void ratio curve (compression curve)—a curve representing the relationship between effective pressure and void ratio of a soil as obtained from a consolidation test. The curve has a characteristic shape when plotted on semilog paper with pressure on the log scale. The various parts of the curve and extensions to the parts of the curve and extensions to the parts have been designated as recompression, compression, virgin compression, expansion, rebound, and other descriptive names by various authorities.
D 653, D18

pressure washing—the cleaning of soil or rock surfaces accomplished by injection of water, air, or other liquids, under pressure.
D 653, D18

pressure, water vapor—the component of atmospheric pressure caused by the presence of water vapor, usually expressed in millimetres, inches of mercury, or pascals. **E 41,** G03

pretension, *n*—the specified tension applied to a specimen preparatory to making a test. **D 123,** D13

pretreat, *vt*—preapplication of cleaning agent to spots, stains, and areas of high soil concentration to maximize activation time and therefore facilitate soil removal during overall cleaning. **D 123,** D13

pretreatment—processes such as chlorination, filtration, coagulation, clarification, acidification which may be used on feedwater ahead at membrane devices to improve quality, minimize scaling and corrosion potential, control biological activity. **D 6161,** D19

prevailing torque—torque necessary to rotate a fastener relative to its mating component with the torque being measured with the fastener in motion and zero axial load in the assembly. **F 1789,** F16

preventative maintenance cycle—the predetermined interval through which specified servicing events take place. (See **PM cycle**.) **F 1457,** F05

prevention process—a future-oriented strategy that improves quality through continuous improvement activities by directing analysis and action toward correcting the process itself. Statistical process control and other statistical techniques are used. **F 1789,** F16

prevention system—a system which outlines advance quality planning, in–process inspection, process controls, and statistical methods to control the processes and seek continuous improvement. **F 1789,** F16

prevulcanization inhibitor (PVI), *n*— a compounding material that increases the time to incipient vulcanization of a rubber mix. **D 1566,** D11

primary accelerator, *n*—the principal highest concentration accelerator used in a vulcanizing system. **D 1566,** D11

primary atomization, *n*—the initial disintegration of a liquid jet, sheet or film. **E 1620,** E29

primary backing, *n*—*for tufted pile yarn floor covering*, the fabric through which the pile yarn is carried by needles to form tufts; the backing fabric. **D 123,** D13

primary biodegradation, *n*—degradation of the test material by microorganisms resulting in a change in the test material's physical or chemical properties, or both. **D 6384,** D02

primary biodegradation test, *n*—a test that monitors the disappearance of a test material by measuring some physical attribute of the material. **D 6384,** D02

primary boiling—the evolution of gas during the initial firing of porcelain enamel; sometimes a defect. **C 286,** B08

primary breakup, *n*—see **primary atomization**. **E 1620,** E29

primary circulation area—See **building space.** **E 631,** E06

primary circulation area— See **space categories**. **E 1480,** E06

primary clay (residual clay)—a clay that remains geologically at its site of formation. **C 242,** C21

primary colorants, *n*—a small number of colorants (dyes or pigments) that may be mixed subtractively to produce a large gamut of colors. **E 284,** E12

primary color stimuli, *n*—three selected colored lights used to specify the color of any light presented by the amounts of the three that must be mixed additively to produce light matching the light presented. **E 284,** E12

primary consolidation (primary compression) (primary time effect)—see **consolidation.** **D 653,** D18

primary container—a container which is in contact with the contents. **D 996,** D10

primary crystals—the first type of crystals that separates from a melt on cooling. **E 7,** E04

primary current distribution—the distribution of the current over the surface of an electrode in the absence of polarization. **B 374,** B08

primary extinction—a decrease in intensity of a diffracted X-ray beam caused by perfection of crystal structure extending over such a distance (about 10^{-4} cm or greater) that interference between multiply reflected beams inside the crystal causes a decrease in the intensity of the externally diffracted beam. **E 7,** E04

primary filter pack—a clean silica sand or sand and gravel mixture of selected grain size and gradation that is installed in the annular space between the borehole wall and the well screen, extending an appropriate distance above the screen, for the purpose of retaining and stabilizing the particles from the adjacent strata. The term is used in place of gravel pack. **D 653,** D18

primary flow-rate standard—See **standard.** **D 1356,** D22

primary heat, *n*—the product of a single cycle of a batch melting process. **A 941,** A01

primary hole—*in grouting*, the first series of holes to be drilled and grouted, usually at the maximum allowable spacing. **D 653,** D18

primary insulation, *n*—the first layer of two or more layers of insulating materials over a conductor. **D 1711,** D09

primary jacket, *n*—a layer of insulating material applied over the primary insulation for the purpose of providing mechanical protection for the primary insulation. **D 1711,** D09

primary light source, *n*—surface of an object emitting light produced by a transformation of energy. **E 284,** E12

primary lining—the lining first placed inside a tunnel or shaft, usually used to support the excavation. The primary lining may be of wood or steel sets with steel or wood lagging or rock bolts and shot-crete. (ISRM) **D 653,** D18

primary orientation flat—the flat of longest length on the wafer, oriented such that the chord is parallel with a specified low index crystal plane; major flat. **F 1241,** F01

primary passive potential (passivation potential)—the potential corresponding to the maximum active current density (critical anodic current density) of an electrode that exhibits active-passive corrosion behavior. **G 15,** G01

primary permeability—internal permeability of intack rock; intergranular permeability (not permeability due to fracturing). **D 653,** D18

primary porosity—the porosity that developed during the final stages of sedimentation or that was present within sedimentary particles at the time of deposition. **D 653,** D18

primary prevention—abatement of accessible lead-based paint and paint residue to prevent the possibility of lead poisoning. **E 1605,** E06

primary protective clothing—protective clothing designed to be worn for work activities during which significant exposure to molten substance splash, radiant heat, and flame is likely to occur. **F 1494,** F23

primary radiation—radiation coming directly from the source. **E 1316,** E07

primary sampling unit, *n*—the sampling unit containing all of the sources of variability which should be considered in acceptance testing; the sampling unit taken in the first stage of selection in any procedure for sampling a lot or shipment. (See **lot sampling unit**.) **D 123,** D13

primary sampling unit, *n*—the sampling unit containing all the sources of variability which should be considered in acceptance testing; the sampling unit taken in first stage of selection in any procedure for sampling a lot or shipment. **D 4439,** D35

primary seal—a seal that in combination with the structure and optional brush coat or secondary seal forms a continuous, durable, and absolute seal in the sealing plane and requires no additional seals. **E 1749,** E06

primary standard—See **standard.** **D 1356,** D22

primary standard, *n*—a physical standard calibrated by an absolute method. **E 284,** E12

primary-standard dosimeter—a dosimeter of the highest metrological quality, established and maintained as an absorbed-dose standard by a national or international standards organization. **E 170,** E10

primary state of stress—the stress in a geological formation before it is disturbed by man-made works. (ISRM) **D 653,** D18

primary (X-ray)—the beam incident on the sample. **E 7,** E04

primary X rays, *n*—*in spectrometry*, the emergent beam from the X-ray source. **E 135,** E01

prime coat, *n*—an application of a low-viscosity bituminous material to an absorptive surface, designed to penetrate, bond, and stabilize this existing surface and to promote adhesion between it and the construction course that follows. **D 8,** D04

primer—*in building construction*, a compatible coating designed to enhance adhesion. **C 717,** C24

primer—the first of two or more coats of a paint, varnish, or lacquer system. **D 16,** D01

primer, *n*—a coating applied to a surface, prior to the application of an adhesive, to improve the performance of the bond. **D 907,** D14

primer—a thin liquid bitumen applied to a surface to improve the adhesion of heavier applications of bitumen and to absorb dust. **D 1079,** D08

primer—a coating applied to a surface prior to the application of an adhesive, sealant, or paint to improve the adhesive bonding characteristics or corrosion resistance, or both, of the surface. **E 1749,** E06

primer—an organic solvent, which enhances adhesion, applied to plastic pipe and fittings prior to application of a solvent cement. **F 412,** F17

primer, coal tar—a liquid coal-tar base material, dispersed in a suitable solvent, for application to concrete to improve the bonding of coal-tar membrane sheets to the bridge deck. **D 3743,** D04

primer, neoprene base—a liquid polychloroprene rubber, dispersed in a suitable solvent, for application to concrete to improve the bonding of rubberized asphalt membrane sheets to the bridge deck. **D 3743,** D04

primer, resin-based—a liquid synthetic resin, dispersed in a suitable solvent, for application to concrete to improve the bonding of modified bitumen membrane sheets to the bridge deck. **D 3743,** D04

principal component analysis—a mathematical procedure for resolving sets of data into orthogonal components whose linear combinations approximate the original data to any desired degree of accuracy. **E 131,** E13

principal plane—each of three mutually perpendicular planes through a point in a soil mass on which the shearing stress is zero.

intermediate principal plane—the plane normal to the direction of the intermediate principal stress.

major principal plane—the plane normal to the direction of the major principal stress.

minor principal plane—the plane normal to the direction of the minor principal stress. **D 653,** D18

principal point—a point on the axis of symmetry of a lens or lens system from which any ray entering the lens or lens system will emerge in a parallel direction. **E 7,** E04

principal stress—see **stress.** **D 653,** D18

principal stress (normal) [FL^{-2}], *n*—the maximum or minimum value of the normal stress at a point in a plane considered with respect to all possible orientations of the considered plane. On such principal planes the shear stress is zero. **E 6,** E28

principal stress (strain)—the stress (strain) normal to one of three mutually perpendicular planes on which the shear stresses (strains) at a point in a body are zero. (ISRM) **D 653,** D18

printability gage, *n*—a printer's tool used to determine the amount of print gain under given printing conditions. **F 1294,** F05

printability range, *n*—the range of print gain found under actual working conditions, based on press sheets selected at random during a press run. **F 1294,** F05

print contrast, *n*—a differential measurement of light reflectance between image area and background area. **F 1457,** F05

print contrast ratio—*in optical character recognition,* the ratio obtained by subtracting the reflectance at an inspection area from the maximum reflectance found within a specified distance from that area, and dividing the result by that maximum reflectance. See **print contrast signal.** **F 149,** F05

print contrast signal—the relative value of the contrast of printing in relation to the paper background on which it is printed as defined by the following equation:

$$PCS_p = \frac{R_w - R_p}{R_w}$$

where:

R_w = maximum reflectance found within the area of interest to which the PCS of point p is referenced. (In measuring printed images, this area of interest should be a rectangle approximately twice the nominal character height by twice the nominal character width and centered on the character being measured), and

R_p = reflectance from a small measurement area centered on point p.

The reflectance R_w and R_p are measured within a circular area of 0.008 in. (0.2 mm) in diameter. **F 149,** F05

print contrast signal (PCS), *n*—a comparison between the reflectance of printed bars and intervening spaces. PCS, under a given set of illumination conditions, is defined by the following equation:

$$PCS = \frac{Rs - Rb}{Rs}$$

where:

Rs = Space relectance, and

Rb = Bar reflectance.

F 1294, F05

printed circuit board—see **printed wiring board and PWB.** **F 1457,** F05

printed sheet vinyl flooring, *n*—a floor surfacing material which has a pattern printed on a backing and is protected with a wear layer of transparent or translucent vinyl plastic. **F 141,** F06

printed wiring board, *n*—a circuit board on which the conductors have been set by a print-like process. (See **PWB.**) **F 1457,** F05

printer—an output unit that produces durable hard-copy record of data in the form of a sequence of discrete graphic characters belonging to a predetermined character set. **F 909,** F05

printer emulation module, *n*—a device used to simulate the presence of a printer in a system. (See **PEM.**) **F 1457,** F05

printer output area, *n*—the maximum area on the page to which the printer will print. **F 1457,** F05

printer ready—see **data terminal ready**. **F 1457,** F05

print gain, *n*—gain in bar width of the printed bar code symbol, compared to the original precision film master; caused in the plate-making process by ink spread during printing. **F 1294,** F05

printing—the act of producing an image on a document or other receptor media from a machine or printer designed for that purpose, such as a printing press, thermal printer, or a computer printer, such as a laser or ink jet. **F 335,** F05

printing—the act of producing an image on a document or other receptor media from a machine or printer designed for that purpose, such as a printing press, thermal printer, or a computer printer such as a laser or ink jet. **F 909,** F05

printing ink—a colored or pigmented liquid or paste composition that dries to a solid film after application as a thin layer by printing machinery. **D 16,** D01

printing module, *n*—those components in the laser printer that together drive the laser scanner, create the image on the page, and deliver the page to the stacker. **F 1457,** F05

printout—information printed on paper. **F 1457,** F05

printout queuing, *n*—a feature which allows a number of documents to be lined up or queued in memory for subsequent printout while other documents are being processed or printed, or both. **F 1457,** F05

print quality—the interrelationship of printed material and imprinted material that affects the optimum performance of the scanner. Refer to OCR Print Quality Guideline, ANSI X3.99-83> **F 149,** F05

print quality, *n*—as described in ANSI X3.182, the grade of a printed bar code symbol as rated by measurements and calculations of various parameters. **F 1294,** F05

print resistance (*coatings*)—the ability of a coating to resist taking on the imprint due to the pressure of another surface placed against it. **D 16,** D01

print speed rating, *n*—highest speed of image development under fixed operating conditions to obtain optimum print quality. **F 1623,** F05

print speed response, *n*—the response curve of optical density versus print speed with a fixed energy level input. **F 1623,** F05

print void, *n*—a defective spot of unprinted matter in a print solid or halftone that can be caused by a buildup of foreign matter on the plate or blanket. **D 6488,** D01

prism illuminator—a 45 to 90° prism interposed in a microscope for the purpose of directing an intense oblique beam of light to the object. The prism illuminator utilizes only one half the aperture of the microscope as does the mirror illuminator with an attending loss of resolution over that obtainable with either a plane glass illuminator or a pellicle mirror. **E 7,** E04

probabilistic design—design that accounts for the uncertainties due to statistical variabilities in physical and mechanical properties of the materials, elements, or structures, and in the applied loads (compare **deterministic design**). **E 631,** E06

probability function, *n—of a discrete variate*, the mathematical expression which gives the probability that a variate will take a particular value. **D 123,** D13

probability function, *n—of a continuous variate*, the mathematical expression whose definite integral gives the probability that a variate will take a value within the two limits of integration. **D 123,** D13

probability level, *n*—a general term that reflects the stated proportion of times an event is likely to occur. (Compare to *confidence level* and *significance level.*) **D 123,** D13

probability sample, *n*—a sample of which the sampling units have been selected by a chance process such that, at each step of selection, a specified probability of selection can be attached to each sampling unit available for selection.

NOTE—These probabilities of selection need not be equal. If equal, see simple random sample. See the general term—sample. Also, see Practice E 105. **E 456,** E11

probable—the evidence contained in the handwriting points rather strongly toward the questioned and known writings having been written by the same individual; however, it falls short of the" virtually certain" degree of confidence.

Examples—It has been concluded that the John Doe of the known material probably wrote the questioned material, or it is my opinion (or conclusion or determination) that the John Doe of the known material *probably* wrote the questioned material. **E 1658,** E30

probably did not—the evidence points rather strongly against the questioned and known writings having been written by the same individual, but, as in the probable range above, the evidence is not quite up to the "virtually certain" range.

Examples—It has been concluded that the John Doe of the known material probably did not write the questioned material, or it is my opinion (or conclusion or determination) that the John Doe of the known material probably did not write the questioned material. **E 1658,** E30

probe, *n*—a tube used for sampling or for measuring pressures at a distance from the actual collection or measuring apparatus. **D 1356,** D22

probe—in leak testing, a tube having an opening at one end, used for directing or collecting a stream of tracer gas. **E 1316,** E07

probe—see **search unit.** **E 1316,** E07

probe coil—in electromagnetic testing, a small coil or coil assembly that is placed on or near the surface of test objects. **E 1316,** E07

probe coil clearance—the perpendicular distance between adjacent surfaces of the probe and test part; also lift-off. **E 1316,** E07

probe cover and sheath, *n*—a device provided for the purpose of preventing biological contact between the patient and the probe or thermometer. **E 344,** E20

probe gas—in leak testing, a tracer gas which issues from an orifice so as to impinge on a restricted test area. **E 1316,** E07

probe ion—an ionic species intentionally produced by an ion source and directed onto the specimen surface at a known incident angle with a known energy. **E 673,** E42

probe test—a leak test in which the tracer gas is applied by means of a probe so that the area covered by the tracer gas is localized. This enables the individual leaks to be located. **E 1316,** E07

process:

dry process (dry mix)—the method of preparation of a ceramic body wherein the constituents are blended dry, following which liquid may be added as required for subsequent processing.

wet process (slip process)—the method of preparation of a ceramic body wherein the constituents are blended in sufficient liquid to produce a fluid suspension for use as such or for subsequent processing. **C 242,** C21

processability, *n*—the relative ease with which raw or compounded rubber can be handled in rubber machinery. **D 1566,** D11

process anneal—a heat treatment used to soften metal for further cold working; in ferrous sheet and wire industries, heating to a temperature close to but below the lower limit of the transformation range and subsequently cooling for working; in the nonferrous industries, heating above the recrystallization temperatures at a time and temperature sufficient to permit the desired subsequent cold working. **E 7,** E04

process annealing, *n—in the sheet and wire industries,* heating a steel object to a temperature close to, but below, Ac_1 and then cooling it, in order to soften it for further cold working. **A 941,** A01

process annealing—sometimes called subcritical annealing or stress relieving, performed at temperatures just below the lower critical temperature. Process annealing neither refines grains nor redissolves cementite, but does improve the ductility and decreases residual stress in work-hardened steel. **F 1789,** F16

process average, *n*—for the items produced, the true and unknown level of (*1*) the fraction of nonconforming items or (*2*) a characteristic of the items as determined by a specific test method. **D 123,** D13

process blacks—black dyestuffs or pigments made by the mixing of the three subtractive color theory primary colors: yellow, magenta, and cyan; so-called because the combining of the three primaries results in the subtraction of color, leaving only black. **F 1156,** F12

process control radiograph—a radiograph which images a beam purity indicator and sensitivity indicator under identical exposure and processing procedures as the test object radiograph. A process control radiograph may be used to determine image quality parameters in circumstances of large or unusual test object geometry. **E 1316,** E07

process fishscaling—fishscaling that appears during the drying or firing cycle of cover coat application. **C 286,** B08

process flow—the sequential process steps required to produce a fastener. **F 1789,** F16

processing—the work involved in transforming quarry blocks into dimension stone, including sawing, drilling, grinding, honing, polishing, carving, and all other operations necessary for installation. **C 119,** C18

processing, aid, *n*—a compounding material that improves processability of a polymeric compound by reducing nerve, providing better dispersion of dry material, increasing the extension rate, reducing power consumption during mixing, producing smoother surfaces on calendered and extruded products and improving knitting to name a few examples. **D 1566,** D11

processing capacity—the number of hits that can be processed at the processing speed before the system must interrupt data collection to clear buffers or otherwise prepare for accepting additional data. **E 1316,** E07

processing speed—the sustained rate (hits/s), as a function of the parameter set and number of active channels, at which AE signals can be continuously processed by a system without interruption for data transport. **E 1316,** E07

process oil, *n*—an hydrocarbon oil derived from petroleum or other sources and used as an extender or process aid. **D 1566,** D11

processor, *n—as applies to garment maintenance,* the party performing the care and maintenance operation. **F 1494,** F23

Proctor compaction curve—see **compaction curve**. **D 653,** D18

Proctor penetration curve—see **penetration resistance curve**. **D 653,** D18

Proctor penetration resistance—see **penetration resistance**. **D 653,** D18

prods—hand-held electrodes. **E 1316,** E07

producer, *n—as related to corrugated metal pipe,* deprecated term. Use preferred term **manufacturer**. **A 902,** A05

producer, *n*—primary manufacturer of the material. **B 881,** B07

producer, *n*—the primary manufacturer of the material. **B 899,** B02

producer's risk, *n*—the probability of rejecting a lot when the process average is at the acceptable quality level or AQL. **D 123,** D13

product, *n*—the upholstered furniture for which information is required. **E 176,** E05

product, *n*—mattress, or mattress with foundation, for which fire-test-response characteristics are to be measured. **E 176,** E05

product analysis, *n*—a chemical analysis of the semifinished or finished steel. **A 902,** A05

product channel spacer (permeate carrier)— the fabric or other material through which permeate water flows after it passes through the flat sheet membrane. **D 6161,** D19

product deterioration—chemical reaction or physical change within or between components considered compatible in the original formulation, that may be due to time or temperature of storage, or other factors. **D 3064,** D10

product formulation—the specific formulation of the completed product, including propellant(s), usually expressed as mass per mass percent. **D 3064,** D10

production capacity—maximum rate (lb/h) at which the equipment can bring the specified food product to a specified" cooked" condition in accordance with the cooking energy efficiency test (see **cooking energy efficiency**), *PC.*

$$PC = \frac{W \times 60}{t}$$

where:

PC = production capacity of the equipment, lb/h (kg/h),
W = total weight of food cooked during cooking test, lb (kg), and
t = total time of cooking test, min.

F 1827, F26

production cycle—the series of operations required to process through the facility a quantity of feedstock mixed with water having a volume equal to the typical volume of the fermentation system and return the facility to the configuration at the start of the cycle. The quantity of water mixed with the feedstock shall be as per specification for normal operation. This volume is equal to the sum of the working volumes of all fermenters in a batch fermentation process. This volume is equal to the sum of the working volumes of each stage of fermentation in a continuous fermentation process. **E 1705,** E48

production lot, *n*—that part of one manufacturer's production made from the same nominal raw material under essentially the same conditions and designed to meet the same specifications. **D 123,** D13

production plan—a complete written plan of fastener and process characteristics, process controls, tests, and acceptance procedures that will occur during full production. **F 1789,** F16

production rate—rate (lb/h) at which the equipment brings the specified food product to a specified "cooked" condition, *PR.* **F 1827,** F26

production unit—*as referred to in this practice,* is a quantity of geotextile agreed upon by the purchaser and seller for the purpose of sampling. **D 4439,** D35

productivity—flow rate of product water. **D 6161,** D19

product staging—a process in which the permeate from one membrane plant is used as the feed to another membrane plant in order to further improve product quality. **D 6161,** D19

product standard, *n*—material having a color designated as standard for a specified product. **E 284,** E12

product substitution—the willful misrepresentation of products including those items referred to as bogus parts, counterfeit parts or assemblies, or all three, and products with unauthorized remarking or rebranding, or both, in the fabrication or repair or both, of articles. **F 1156,** F12

product tube—the tube at the center of the spiral wound cartridge which collects permeate water. **D 6161,** D19

product water—purified water produced by a process. See **permeate**. **D 6161,** D19

proeutectoid carbide—primary crystals of cementite formed directly from the decomposition of austenite exclusive of that cementite which results from the eutectoid reaction. **E 7,** E04

proeutectoid ferrite—primary crystals of ferrite formed directly from the decomposition of austenite exclusive of that ferrite which results from the eutectoid reaction. **E 7,** E04

professional care, *n*—*for textile products*, cleaning and maintenance procedures requiring the services of a person specially trained and skilled in their use. **D 123,** D13

professional care, *n*—*for consumer textile products*, overall cleaning and maintenance procedures requiring the services of a person specially trained or skilled in their use and usually earning a livelihood by providing such services. **D 5253,** D13

proficiency testing—determination of laboratory testing performance by means of interlaboratory test comparisons, (ISO Guide 2). **E 1187,** E36

profile, *n*—wrought product that is long in relation to its cross sectional dimensions which is of a form other than that of sheet, plate, foil, rod, bar, tube, or wire.

extruded profile, n—profile produced by hot extruding.

structural profile, n—profile, rolled or extruded, commonly used for structural purposes such as angles, channels, H-beams, I-beams, tees, and zees. **B 881,** B07

profile, *n*—the contour of the top surface of the tile when viewed from the nose end. **C 43,** C15

profile—see **soil profile**. **D 653,** D18

profile, *vt*—*in optical emission spectrometry*, to scan and set the deflection of the grating, or actual or apparent position of the entrance slit, or actual or apparent location of the exit slits, to produce optimum measurement of intensity. **E 135,** E01

profile depth, *n*—the difference between the pavement macrotexture profile and a horizontal line through the top of the highest peak within a given baselength. **E 867,** E17

profile, longitudinal—see **longitudinal profile.** **E 867,** E17

profile record, *n*—a data record of the surface elevation, slope or acceleration, of arbitrary length. **E 867,** E17

profile segment, *n*—that part of a profile record for which the profile index will be calculated. **E 867,** E17

profile, transverse—see **transverse profile.** **E 867,** E17

profilometer, *n*—equipment used to measure profile of traveled surface roughness. **E 867,** E17

profilometer—an instrument for measuring the topographical profile of a surface. **F 1241,** F01

program—See facility program. **E 833,** E06

program, *n*—a documented policy with procedures for selection and use of CPC. **F 1494,** F23

program administrator, *n*—a person responsible for the formulation and implementation of a CPC program. **F 1494,** F23

program attention, *n*—an application-dependent signal sent by the operator to the cluster controller from the printer to signify that a task was performed. **F 1457,** F05

program authority, *n*—a person responsible for enforcing the requirements of a CPC program. **F 1494,** F23

program controller—a device which can automatically execute a pre-determined schedule of control. **E 7,** E04

programmable read only memory, *n*—an information storage area that can be recorded on and read by an operator, but modified only through special physical processes. (See **PROM**.) **F 1457,** F05

programmed temperature gas chromatography, PTGC—utilized changes in column temperature with time. **D 2864,** D27

progressive aging, *n*—**aging** by increasing the temperature in steps, or continuously, during the aging cycle. **A 941,** A01

progressive failure—failure in which the ultimate shearing resistance is progressively mobilized along the failure surface. **D 653,** D18

progressive failure—formation and development of localized fractures which, after additional stress increase, eventually form a continuous rupture surface and thus lead to failure after steady deterioration of the rock. (ISRM) **D 653,** D18

project brief—See **facility program**. **E 1480,** E06

projected length, L_P—length of the projection of the ski, measured between the ski tip and the ski tail parallel to the ski body pressed against a plane surface. **F 472,** F27

projected length, L_p—the length of the projection of the snowboard ski, measured between the snowboard ski tip and the snowboard ski tail with the snowboard ski unweighted on a plane surface (unweighted meaning solely under the influence of its own weight). **F 1107,** F27

projected treadlife, [*L*], *n*—the test distance that gives *h* as the average tread depth; where *h* is the height of treadwear indicator above groove (or void) base. **F 538,** F09

projection—a calculation, usually performed by a software package, which predicts the performance of parts or all of a water plant. **D 6161,** D19

projection, *n*—a pavement contacting area of the tread band, bounded by void. **F 538,** F09

projection distance—distance from the eyepiece to the image screen. **E 7,** E04

projection head—flange or collar spaced at short distance from end of wire, with wire projecting from upper side of flange or collar forming part of head and wire projecting from other side of flange or collar forming the shank. (See **gudgeon.**) **F 547,** F16

projection lens—the final lens in the electron microscope corresponding to an ocular or projector in a compound light microscope. This lens forms a real image on the viewing screen or photographic film. **E 7,** E04

project level pavement analysis, *n*—evaluation of a pavement section to select the type and timing of corrective actions. **E 867,** E17

project life, *n*—the time period over which the economic viability of a geothermal facility is evaluated. **E 957,** E44

project (projet)— resources and activities used to achieve a specific set of objectives within a specified time schedule. **E 1480,** E06

project record—use **as-built,** the preferred term. **E 631,** E06

PROM—**p**rogrammable **r**ead **o**nly **m**emory. **F 1457,** F05

proof—See **rod proof**. **C 162,** C14

proof, *n*—*in fire assay*, a synthetic verifier having a precious metal content similar to that expected in the test sample. **E 135,** E01

proof—measurement term of concentration of ethanol in water solutions. **E 1705,** E48

proof correction, *n*—*in fire assay*, the adjustment to the final assay obtained by analyzing the proof concurrently with the test sample. **E 135,** E01

proofing, *n*—process of coating a fabric with raw or compounded rubber to impede penetration by a liquid, usually water. **D 1566,** D11

proofing (the act of), *n*—process of coating a fabric with rubber. **D 1566,** D11

proof load—a tension-applied load that the fastener must support without evidence of permanent deformation. (For most carbon or alloy steel fastener strength grades or property classes, proof loads are established at approximately 90% to 93% of the expected minimum yield strength.) **F 1789,** F16

proof-test current—the current measured during ac proof tests.

NOTE—This current is an indication of the validity of the dielectric constant of the type of material used and the thickness of the total contact area under test. **F 819,** F18

propagance, *n*—ratio of propagated flux to incident flux; a general term that includes the results of reflection, transmission, and similar effects that modify the flux in a beam. **E 284,** E12

propagation of flame, *n*—the upward and outward movement of the flame front from the ignition source to the vessel walls, that is determined by visual observation. **E 1445,** E27

propagation of flames, *n*—a combustion reaction that produces at least a 7 % rise of the initial absolute pressure,

$$\frac{P_2}{P_1} \geqq 1.07.$$

E 1445, E27

propellant—liquefied gas with a vapor pressure greater than atmospheric pressure (14.7 psia (101 kPa)) at 105°F (40.6°C). **D 3064,** D10

propene concentrate, *n*— *in liquefied natural gas technology*, concentrate containing more than 50 % propene (Syn. *propylene*). **D 4175,** D02

property class—a system of strength classifications used for bolts, nuts, and screws manufactured to metric standards. **F 1789,** F16

property owner—person, firm, corporation, guardian, conservator, receiver, trustee, executor, or other judicial officer who, alone or jointly or severally with others, owns, holds, or controls the whole or any part of the freehold or leasehold title to any property, with or without accompanying actual possession of it, and shall include in addition to the holder of legal title, any vendee in possession of it, but may not include a mortgagee or an owner of a reversionary interest under a ground rent lease. **E 1605,** E06

proportional counter—a gas-filled tube operated in that range of applied voltage in which the electrical charge collected per isolated count is proportional to the electrical charge liberated by the initial ionizing event. The range of applied voltage depends upon the type and energy of the incident radiation. **C 859,** C26

proportional limit, *n*—the maximum stress that a material is capable of sustaining without significant deviation from proportionality of stress to strain. **D 907,** D14

proportional limit, *n*—the greatest stress which a material is capable of sustaining without any deviation from proportionality of stress to strain (Hooke's law). **D 4439,** D35

proportional limit [FL^{-2}], *n*—the greatest stress which a material is capable of sustaining without any deviation from proportionality of stress to strain (Hooke's law). **E 6,** E28

proportional spacing—a system of printing where the character spacing is set in accordance with the character width; this is in contrast to a fixed number of characters per inch such as 10 pitch (10 characters/inch), 12 pitch (12 characters/inch), 15 pitch (15 characters/inch), etc. **F 1457,** F05

proportional spacing, *n*—a system of printing wherein the character spacing is set in accordance with the character width. (See **fixed pitch**.) **F 1457,** F05

proportioning probe—in leak testing, a probe that can vary sample to pure air ratios between 100 % sample and 100 % pure air without substantially changing the total flow from the probe. **E 1316,** E07

proportioning pump—see **metering pump**. **D 653,** D18

proprietary—made and marketed by one having the exclusive right to manufacture and sell; privately owned and managed. **D 653,** D18

propylene glycol ether of methylcellulose—see **(2-hydroxypropyl)methylcellulose.** **D 1695,** D01

propylene plastic, *n*—a plastic based on polymers of propylene or copolymers of propylene with other monomers, the propylene being in the greatest amount by mass. **F 1251,** F04

propylene plastics, *n*—plastics based on polymers of propylene or copolymers of propylene with other monomers, the propylene being in the greatest amount by mass. (ISO) **D 883,** D20

propylene plastics, *n*—plastics based on polymers of propylene or copolymers of propylene with other monomers, the propylene being in the greatest amount by mass. (ISO, D-20) **F 412,** F17

protected water boom—boom intended for use in protected waters with moderate environmental conditions (see Practice F 625 for environmental descripters). **F 818,** F20

protecting tube, *n*—a tube designed to enclose a temperature-sensing device and protect it from the deleterious effects of the environment. **E 344,** E20

protection—See **preservation.** **E 631,** E06

protection course—semi-rigid sheet material placed on top of the waterproofing membrane to protect it against damage during subsequent construction and to provide a protective barrier against compressive and shearing forces induced by materials placed above it. **C 717,** C24

protection potential—the most noble potential where pitting and crevice corrosion will not propagate. **G 15,** G01

protection tube—a tube made of a specially selected material which can be used to protect a thermocouple from adverse effects of the environment. **E 7,** E04

protective atmosphere—*(1)* a gas envelope surrounding the part to be brazed, welded, or thermal sprayed, with the gas composition controlled with respect to chemical composition, dew point, pressure, flow rate; *(2)* the atmosphere in a heat treating or sintering furnace formulated to protect the parts or compacts from oxidation, nitridation, or other contamination from the environment. **B 374,** B08

protective clothing material, *n*—any element, constituent, or substance from which protective clothing is composed or can be made. **F 1494,** F23

protective filter—see **filter**. **D 653,** D18

protein—general term used to cover single cell microorganisms, extract of the microorganisms, (bacteria or fungi or algae) that is used for food or feed to animals and humans. **E 1705,** E48

protocol, *n*—a formal set of conventions governing a communication process. **F 1457,** F05

prototype plan—a written description of the dimensional, mechanical, chemical, and performance tests that will be used to facilitate the building of a prototype. **F 1789,** F16

protrusion, *n*—*in fabrics*, a general term for a visible group of fibers, a yarn, or a yarn segment that extends above the fabric surface. **D 123,** D13

proximal—nearest to the body, or center, or some other point considered as the center of a system. **F 869,** F08

proximal end—that portion of the instrument that is closest to the surgeon when in use. **F 1078,** F04

proximate analysis, *n*—*in the case of coal and coke*, the determination, by prescribed methods, of moisture, volatile matter, fixed carbon (by difference), and ash. **D 121,** D05

proximate analysis, *n*—the determination, by prescribed methods, of moisture, volatile matter, fixed carbon (by difference), and ash. **D 5681,** D34

proximate analysis—the determination, by prescribed methods, of moisture, volatile matter, fixed carbon (by difference), and ash. Unless otherwise specified, the term proximate analysis does not include determinations of chemical elements or any determinations other than those named. **E 856,** D34

proximate analysis—the determination, by prescribed methods, of moisture, volatile matter, fixed carbon (by difference), and ash. The term **proximate analysis** does not include determinations of chemical elements or determinations other than those named. **E 1126,** E48

proximate analysis—the determination, by prescribed methods, of moisture, volatile matter, fixed carbon (by difference), and ash. The term **proximate analysis** does not include determinations of chemical elements or determinations other than those named. **E 1705,** E48

proximate analysis—an assay of the moisture, ash, volatile matter, and fixed carbon as determined by prescribed test methods. Other constituents such as sulfur and phosphorus are not included. **E 1705,** E48

pseudobinary—a term of indefinite meaning, sometimes used synonymously with "quasibinary," sometimes used to designate an "isopleth." **E 7,** E04

pseudoglow discharge, *n*—a type of partial discharge characterized by pulses of relatively small amplitude, and generally, a long rise time. **D 1711,** D09

pseudomonotropy—an irreversible solid state transformation which occurs below the melting points of both the stable and metastable solid states. **E 7,** E04

pseudoreplica—See **replica**. **E 7,** E04

psychology, *n*—the science dealing with the mind and with the mental and emotional processes. **E 2035,** E52

psychometrics, *n*—application of measurement principles to psychological phenomena. **E 253,** E18

psychophysical methods, *n*—procedures for establishing relationships between measurable physical stimuli and sensory responses. **E 253,** E18

psychophysical power law, *n*—a power function describing the relationship between physical amounts of stimuli and their respective perceived intensities. The mathematical relationship is:

$$R = kS^n$$

where:
R = perceived intensity of a given stimulus,
S = physical amount of that stimulus,
k = constant reflecting the unit of measurement, and
n = exponent which is an empirical constant characteristic of the given sensory system.

E 253, E18

psychophysics, *n*—the study of the functions relating the physical measurements of stimuli and the sensations and perceptions the stimuli evoke. **E 284,** E12

psychophysiological detection of deception (PDD), *n*—the academic discipline that provides the student, the practitioner, and the researcher with the theoretical and applied psychological, physiological, and psychophysiological fundamentals for a thorough understanding of PDD tests and the skills and qualifications for conducting PDD examinations (see **polygraphy**). **E 2035,** E52

psychophysiological veracity (PV) examination, *n*—see definition of **PDD** **E 2035,** E52

psychophysiology, *n*—the study of interactions between mental and physiological processes. **E 2035,** E52

psychrometer, *n*—a variety of hygrometer comprising a dry bulb temperature indicator and a wet bulb temperature indicator which is cooled to the wet bulb temperature by the spontaneous evaporation of moisture. **D 123,** D13

psychrometer, *n*—a variety of hygrometer comprising a dry bulb thermometer and a wet bulb thermometer, which, when suitably aspirated, indicates the thermodynamic wet- and dry bulb temperature of the gas.

aspirated psychrometer, n—an hygrometer comprising wet- and dry-bulb thermometers that are mounted in a housing to which is attached a motor-driven fan or blower that draws air over the thermometer bulbs at a rate which produces the minimum wet-bulb reading.

sling (whirling) psychrometer, n—an hygrometer comprising wet- and dry-bulb thermometers that are mounted on a frame that can be rotated or whirled. **D 1356,** D22

psychrometer, aspirated—an hygrometer comprising wet- and dry-bulb thermometers that are mounted in a housing to which is attached a motordriven fan or blower that draws air over the thermometer bulbs at a rate which produces the mimimum wet-bulb reading. **D 4023,** D22

psychrometer, sling (whirling)—an hygrometer comprising wet- and dry-bulb thermometers that are mounted on a frame that can be rotated or whirled. **D 4023,** D22

P-T diagram—a two-dimensional, graphical representation of phase relationships in a system of any order, by means of the pressure and temperature variables. **E 7,** E04

PTFE tape—joint sealing tape composed of polytetrafluoroethylene. **D 653,** D18

P-T-X diagram—a three-dimensional, graphical representation of the phase relationships in a binary system, by means of the pressure, temperature and concentration variables. **E 7,** E04

***p*-type**—a variety of semiconductive material in which the majority current carriers are holes, formed when acceptor impurities are incorporated into the crystal structure in small concentrations. **F 1241,** F01

public and private key encryption—a method of encoding information in such a manner that anyone can create a message to a specific person, use the public key to encrypt the message content, and then transmit it to that person; only the receiving individual, who holds the private key, can decode the message back into plain language—knowledge of only the public key is not enough to permit decryption of the message. **F 1156,** F12

public housing agency (PHA)—any state, county, municipality, or other government entity or public body (or agency or instrumentality thereof) that is authorized to engage or assist in the development or operation of housing for low-income families. **E 1605,** E06

puckering, *n*—*in bonded, fused, or laminated fabrics,* a wavy, three-dimensional effect typified by closely spaced wrinkles, on either the face fabric or the backing fabric, or both. **D 123,** D13

puddling—soil consolidation by agitating by means of poles, a mixture of soil and sufficient water to leave a puddle on the surface. **C 896,** C04

puffed compact—a compact expanded by internal gas pressure. **B 243,** B09

pull—the quantity of glass delivered by a melting furnace in a given time, usually 24 h. **C 162,** C14

pull, *n*—a group of fibers grasped by the forceps at one time and drawn from the specimen in the combs. **D 123,** D13

pull, *n*—*in zippers*, a part connected to a zipper slider by which the slider is operated. **D 123,** D13

pulled-surface, *n*—as applied to a defect in a laminated plastic imperfections in the surface ranging from a slight breaking or lifting in spots to pronounced separation of its surface from its body. **D 883,** D20

pulled wool, *n*—wool taken from the pelt of a slaughtered sheep and which has not been commercially scoured. **D 123,** D13

pulling over—operation in which upper of a shoe is pulled over the last. **F 869,** F08

pull-out, *n*—void existing on the plane of polish of a metallographic specimen caused by the dislodging of a particle or constituent during the grinding or polishing operation. **E 7,** E04

pull post—a terminal post used in a line of fence to brace a long stretch or to effect a change in elevation along the fence line. **F 552,** F14

pulp, *n*—a fluid mixture of solids and water, also known as slurry. **D 121,** D05

pulper—a water filled tank with a motor driven grinding disk to grind and cut waste material, and mixes this material with water to produce a slurry that is pumped to the waterpress through a sizing screen. **F 1827,** F26

pulping, *n*—the operation of altering a cellulosic raw material such as pulpwood, rags, straw, reclaimed paper, and so forth, into a form suitable for further processing into paper or paperboard; may vary from simple mechanical action to rather complex, chemical digestive sequences and combinations of the two. (See **chemical pulp, semichemical pulp,** and **mechanical pulp**.) **D 1968,** D06

pulps, dissolving—chemical cellulose from wood pulp. **D 1695,** D01

pulps for chemical conversion—chemical cellulose from wood pulp. **D 1695,** D01

pulps for manufacture of cellulose derivatives—chemical cellulose from wood pulp. **D 1695,** D01

pulpwood—any wood cut or prepared primarily for the production of wood pulp. **D 9,** D07

pulse—a short wave train of mechanical vibrations. **E 1316,** E07

pulsed cavitation test, *n*—a test using a vibratory cavitation device in which the cavitation is generated intermittently, with alternating vibratory periods and quiescent periods of controlled relative duration. **G 40,** G02

pulse discharge, *n*—a type of partial-discharge phenomenon characterized by a spark-type breakdown. **D 1711,** D09

pulsed simulator—simulator whose irradiance output at the test plane area is in a single short duration pulse of 100 ms or less. **E 772,** E44

pulse echo method—an inspection method in which the presence and position of a reflector are indicated by the echo amplitude and time. **E 1316,** E07

pulse Fourier transform NMR—a form of NMR in which the sample is irradiated with one or more pulse sequences of radiofrequency power spaced at uniform time intervals, and the averaged free induction decay following the pulse sequences is converted to a frequency domain spectrum by a Fourier transformation. **E 131,** E13

pulse-height analyzer—an instrument for measuring the number of voltage pulses falling within each of one or more voltage ranges. **C 859,** C26

pulse length—a measure of the duration of a signal as expressed in time or number of cycles. **E 1316,** E07

pulseless-glow discharge, *n*—a type of partial-discharge phenomenon characterized by a diffused glow. **D 1711,** D09

pulse plating—a method of electroplating in which the current is frequently interrupted or periodically decreased. **B 374,** B08

pulse repetition rate—see **frequency (pulse repetition).** **E 1316,** E07

pulse travel time (T_t)—the total time, measured in seconds, required for the sonic pulse to traverse the specimen being tested, and for the associated electronic signals to reverse the circuits of the pulse-propogation circuitry. **C 709,** D02

pulse tuning—a control used on some ultrasonic examination equipment to optimize the response of the search unit and cable to the transmitter by adjusting the frequency spectrum of the transmitted pulse. **E 1316,** E07

pultruded, adj—made by pultrusion. **D 3918,** D20

pultrude, v—to draw resin-impregnated reinforcement through a die. **D 3918,** D20

pultrusion—(*1*) a process described as the reversed "extrusion" of resin-impregnated reinforcements in the manufacture of rods, tubes, sheets, and shapes of uniform cross section. The reinforcement, after being wet-out by the resin-application system, is drawn through a die to form the desired cross section. (*2*) a term that is applied to the product of the above process. (*3*) a term used to show association with the above process. **D 3918,** D20

pultrusion—a process to continuously fabricate composite structural shapes or flat sheet by drawing prepreg materials through forming dies to produce the desired constant cross-sectional shape and simultaneously curing the resin. **E 1749,** E06

pulverization—the reduction in particle size of metal powder by mechanical means, a specific type of disintegration. **B 243,** B09

pumpability—*in grouting*, a measure of the properties of a particular grout mix to be pumped as controlled by the equipment being used, the formation being injected, and the engineering objective limitations. **D 653,** D18

pump-down time—time of evacuation. **E 1316,** E07

pumping, *n*—ejection of water, material or both from beneath the pavement through a crack or joint. **E 867,** E17

pumping, *n*— ejection of liquid or solid material or both from beneath the pavement through a crack or joint. **E 1778,** E17

pumping, *n*—ejection of water, material, or both from beneath the pavement through a crack or joint. **E 1778,** E17

pumping of pavement (pumping)—see **pavement pumping**. **D 653,** D18

pumping sampler—a device that draws the water-sediment mixture through a pipe or hose, the intake of which is placed at the desired sampling point in a stream. **D 4410,** D19

pumping test—a field procedure used to determine in situ permeability or the ability of a formation to accept grout. **D 653,** D18

pump-out tubulation—a tube extending from an evacuated device through which gas is pumped and which is usually permanently sealed off after the device has been evacuated. Sometimes called exhaust tubulation. **E 1316,** E07

punch—part of a die or compacting tool set which is used to transmit pressure to the powder in the die cavity. See **upper punch** and **lower punch.** **B 243,** B09

punchouts, *n*—(*CRCP only*) a broken area of a concrete slab bounded by closely spaced cracks (usually less than 1 m (3 ft)). **E 867,** E17

punchouts, *n*—a broken area of a concrete slab bounded by closely spaced cracks (usually less than 1 m (3 ft)). **E 1778,** E17

punch ware—handmade, thin, blown glassware, especially tumblers. **C 162,** C14

puncture resistance, (F), *n*—the inherent resisting mechanism of the test specimen to the failure by a penetrating or puncturing object. **D 4439,** D35

puncture test—See **package testing.** **D 996,** D10

pungency, *n*—irritating, piercing, or sharp trigeminal sensation, experienced primarily in the nasal cavity, as a result of exposure to ammonia or to the volatiles of such substances as freshly prepared mustard or horseradish. **E 253,** E18

punty—(*1*) a gathering iron of solid cross-section.

(*2*) a device to which ware is attached for holding during fire polishing or finishing. **C 162,** C14

purchaser, *n—as related to corrugated metal pipe*, the person or agency that purchases the finished pipe. **A 902,** A05

purchaser, *n—in pipe laying*, the person, company, or organization that purchases any materials or work to be performed. **C 460,** C17

purchaser, *n—in pipe laying*, the person, company, or organization that purchases any materials or work to be performed. **C 1154,** C17

purchaser, *n—in pipe laying*, the person, company, or organization that purchases any materials or work to be performed. **D 2946,** C17

pure monomer resin, *n*—a resin produced by the polymerization of essentially pure compounds such as styrene, one of the methylstyrenes, or controlled mixtures of pure monomers. **D 6440,** D01

pure shear—a state of strain resulting from that stress condition most easily described by a Mohr circle centered at the origin. (ISRM) **D 653,** D18

pure stimulus, *n*—a form of excitation (for example, light waves, chemical substances, sound waves) that affects a receptor without having any concomitant or complicating forms of stimulation. **E 253,** E18

purge time, *n—in an intermittent contact test*, the time immediately following the termination of the contact time when the test chemical is removed from the challenge side chamber and air or nitrogen is blown over the outside surface of the protective clothing material. **F 1494,** F23

purified cotton cellulose—see **cotton cellulose, purified.** **D 1695,** D01

purified wood cellulose—see **wood cellulose, purified.** **D 1695,** D01

purity, *n*—the percentage of $CaSO_4 \cdot \frac{1}{2}H_2O$ in the calcined gypsum portion of a gypsum plaster or gypsum concrete, as defined by Specification C 28, for Gypsum Plasters. The percentage of $CaSO_4 \cdot 2H_2O$ in the gypsum or the gypsum portion of fully hydrated, dry, set gypsum plaster **C 11,** C11

purity, n—See **colorimetric purity, excitation purity**. **E 284,** E12

purity, of a biological drug product—the measure of the biologically active drug in relation to the total substances (not including additives) present in the drug product, usually expressed on a percentage basis. **E 1705,** E48

purlin—a wood member, commonly rectangular in cross section and comparatively long and slender, used most often in a horizontal position in heavy roof construction to support rafters and decking. **D 9,** D07

purlin nail, "straw"—galvanized, regular-stock-steel, aluminum-alloy or copper, 4 to 16 by ⅛-in., 0.135 or 0.148-in. nails of desired length with flat 11⁄32-in., curved or 15⁄32-in. head, 9⁄16-in. cast lead head or plastic washer and sheared-square or diamond point; for securing corrugated roofing to I-beams. **F 547,** F16

pusher—tool mechanism for forcing fastener into driving channel. **F 592,** F16

push-on joint—see **joint, push-on.** **F 412,** F17

putty—{archaic}a white polishing compound. **C 162,** C14

putty, *n*—a knife-grade sealant. **C 717,** C24

putty—a dough-like material consisting of pigment and vehicle, used for sealing glass in frames, and for filling imperfections in wood or metal surfaces. See **glazing compound**. **D 16,** D01

PVC—poly(vinyl chloride) coating on fence materials. **F 552,** F14

PVC—poly(vinyl chloride). **F 1156,** F12

P-V diagram—a graphical representation of the variation of the specific volume of a substance, with change in pressure. **E 7,** E04

PV limit, *n—in tribology*, the upper value of a load-bearing material's PV product above which a material fails to function satisfactorily. (See also **PV product**.) **G 40,** G02

PV product, *n*—**in tribology**, the product of the nominal contact pressure on a load-bearing surface and the relative surface velocity between the load-bearing material and its conterface.(See also **PV limit.**) **G 40,** G02

P-V-T diagram—a three-dimensional, graphical representation of a surface, describing the variation of the specific volume of a substance, with independent change of pressure and temperature. **E 7,** E04

PWB—see **printed wiring board**. **F 1457,** F05

P-X diagram—a two-dimensional, graphical representation of the isothermal phase relationships in a binary system; the coordinates of the graph are pressure and concentration. **E 7,** E04

P-X projection—a two-dimensional, graphical representation of the phase relationships in a binary system produced by making an orthographic projection of the phase boundaries of a P-T-X diagram upon a pressure-concentration plane. **E 7,** E04

pyramid—a structure displaying {111} facets that appears on surfaces after epitaxial growth (see also **mound**). Also known as "hillock." **F 1241,** F01

pyranometer—(*1*) a radiometer to measure the global solar irradiance in the plane of its aperture; (*2*) A radiometer used to measure the total solar radiant energy incident upon a surface per unit time per unit area. This energy includes the direct radiant energy, diffuse radiant energy, and reflected radiant energy from the background. **E 772,** E44

pyranometer, *n*—a radiometer used to measure the total solar radiant energy incident upon a surface per unit time per area. This energy includes the direct radiant energy, diffuse radiant energy, and reflected radiant energy from the background. **G 113,** G03

pyranometer, field—a pyranometer essentially meeting the World Meteorological Organization Class II specifications, appropriate to field use and typically exposed continuously. **E 772,** E44

pyranometer, net—an instrument for measuring the difference between the irradiance falling on the top and bottom of a horizontal surface. **E 772,** E44

pyranometer, reference—a pyranometer essentially meeting the World Meteorological Organization Class I specifications and used principally to calibrate other instruments. **E 772,** E44

pyranometer, spherical—instrument for measuring the solar flux falling from a solid angle 4π on a spherical surface. **E 772,** E44

pyrgeometer—an instrument for measuring net atmospheric irradiance on a horizontal upward facing black surface at the ambient air temperature. **E 772,** E44

pyrheliometer—a radiometer used to measure the direct or beam solar irradiance incident on a surface normal to the sun's rays. **E 772,** E44

pyrheliometer, ***n***—a radiometer used to measure the direct or beam solar irradiance incident on a surface normal to the sun's rays. **G 113,** G03

pyrheliometer, compensated—pyrheliometer based on the comparison of the heating of two identical metal strips, one exposed to a solar radiant energy, the other to a joule effect. **E 772,** E44

pyrheliometer, secondary reference—a pyrheliometer essentially meeting the World Meteorological Organization Class I specifications but not having self-calibrating capability. **E 772,** E44

pyridine bases—a mixture of pyridine and substituted pyridines. The pyridine bases in tar acids refer to those that react with 0.02 *N* perchloric acid. **D 4790,** D16

pyridine, refined—pyridine (including its impurities) having a total distillation range of 1.0°C. Refined to improve color characteristics. Refer to Specification D 2323 for complete specifications. **D 4790,** D16

pyro—a common expression for the compound tetrasodium pyrophosphate ($Na_4P_2O_7$), either hydrous or anhydrous. **C 286,** B08

pyrogenic steam—water vapor generated by combining hydrogen and oxygen in the furnace working chamber to produce high purity steam. **F 1241,** F01

pyrogens—any substance capable of producing a fever in mammals. Often a bacterial endotoxin such as lipo polysaccaride generated by gram negative bacteria at destruction. Chemically and physically stable, pyrogens are not necessarily destroyed by conditions that kill bacteria. **D 6161,** D19

pyrohydrolysis—decomposition by the action of heat and water vapor. **C 859,** C26

pyrolysis, *n*—process of simultaneous phase and chemical species change caused by heat (compare **smoldering**). **E 176,** E05

pyrolysis—the breaking apart of complex molecules by heating (over the range from 392° to 932° (200° to 500°C)) in the absence of oxygen, producing solid, liquid, and gaseous fuels. **E 1126,** E48

pyrolysis—the breaking apart of complex molecules by heating (over the range from 392° to 932° (200° to 500°C)) in the absence of oxygen, producing solid, liquid, and gaseous fuels. **E 1705,** E48

pyrolytic coating—a thin coating produced by the decomposition of a volatile compound on a hot surface. **C 162,** C14

pyrometer—an instrument for measuring temperatures (see **pyrometry**). An autographic or recording pyrometer automatically measures and records temperatures. **E 7,** E04

pyrometric cone equivalent (PCE), *n*—the number of that Standard Pyrometric Cone whose tip would touch the supporting plaque simultaneously with a cone of the refractory material being investigated when tested in accordance with Test Method C 24. **C 71,** C08

pyrometric cones, *n*—elongated trigonal pyramids of standard size made from specified mixtures of ceramic materials which when heated under stated conditions, may be used as an index of heat treatment. **C 71,** C08

pyrometry—the measurement of temperatures: for example, by measuring the electrical resistance of wire, the thermoelectric force of a couple, the expansion of solids, liquids or gases, the specific heat of solids, or the intensity of radiant energy per unit area. **E 7,** E04

pyrophyllite—a hydrated aluminum silicate mineral of the theoretical composition $Al_2O_3{\cdot}4SiO_2{\cdot}H_2O$, having physical properties in the raw state resembling mineral talc. **C 242,** C21

pyroxylin finish—a plastic finish based on nitrocellulose used on leather. **D 1517,** D31

pyrradiometer, spherical—instrument for measuring total flux falling from a solid angle 4π on a spherical surface. **E 772,** E44

Q

QC—quality control. **F 1156,** F12

QD—abbreviation for quick disassembly zipper. (See **releasing stop**.) **D 123,** D13

QR—abbreviation for quick release zipper. (See **releasing stop**.) **D 123,** D13

quad—one quadrillion (10^{15}) Btu. **E 1126,** E48

quad—one quadrillion (10^{15}) Btu. **E 1705,** E48

quadrant mat—additional tension zone circumferential reinforcement secured to a layer of reinforcement in the concrete pipe wall. **C 822,** C13

quadratic integrator, *n*—*in textile unevenness testing*, an integrator that operates continuously and reports unevenness for the time during which it has been active, giving equal weight to all portions of the input. **D 123,** D13

quadratic rate, *n*—the sum of the squares of the individual discharge magnitudes during a certain time interval divided by that time interval and expressed as $(\text{coulombs})^2$ per second. **D 1711,** D09

quadrivariant equilibrium—a stable state among a number of conjugate phases equal to two less than the number of components, that is, having four degrees of freedom. **E 7,** E04

quadruple curve—in a P-T diagram, a line representing the sequence of pressure and temperature values along which three conjugate phases occur in univariant equilibrium. **E 7,** E04

quadruple point—in a P-T diagram the pressure and temperature at which four conjugate phases occur in invariant equilibrium. **E 7,** E04

qualification—*in protective coatings:* The characteristics or abilities gained through training or experience, or both, that enable an individual to perform a required function. **D 4538,** D33

qualification (personal)—the characteristics or abilities gained through training or experience, or both, that enable an individual to perform a required function. **F 1789,** F16

qualification test, *n*—a series of tests conducted by the procuring activity, or an agent thereof, to determine conformance of materials, or materials system, to the requirements of a specification which normally results in a qualified products list under the specification. (Compare **acceptance test** and **preproduction test**.) **D 907,** D14

qualification test—an evaluation, generally nonrepetitive, conducted on an existing, altered, or new product to determine acceptability. **F 412,** F17

qualification testing—testing performed on a product to determine whether or not the product conforms to requirements of an applicable specification. **F 412,** F17

qualified identification—An *identification* is not qualified. However, opinions may be qualified when the evidence falls short of an *identification* or *elimination*. **E 1658,** E30

qualified technical personnel—persons such as engineers and chemists who, by virtue of education, training, or experience, know how to apply physical and chemical principles involved in the reactions between oxygen and other materials. **G 126,** G04

qualitative analysis—an analysis in which some or all of the components of a sample are identified. **C 859,** C26

qualitative analysis, *n*—the identification of some or all of the constituents of a sample. **E 135,** E01

qualitative sensory profile, *n*—a description of a sample consisting of sensory attributes, but without intensity values. **E 253,** E18

quality, *n*—the totality of features and characteristics of a product or service that bear on its ability to satisfy a given need.

quality assurance, n—a system of activities whose purpose is to provide assurance that the overall quality control job is in fact being done effectively.

quality control, n—the overall system of activities whose purpose is to provide a quality of product or service that meets the needs of users; also, the use of such a system. **D 1356,** D22

quality, *n*—collection of features and characteristics of a product, process, or service that confers its ability to satisfy stated or implied needs. **E 253,** E18

quality—totality of characteristics of an entity that bear on its ability to satisfy stated or implied needs. **E 1187,** E36

quality, n—See **particle concentration**. **G 40,** G02

quality assurance—all planned and systematic actions necessary to provide adequate confidence that an item or a facility will perform satisfactorily in service. **C 859,** C26

quality assurance—See **quality.** **D 1356,** D22

quality assurance, *n*—all those planned or systematic actions necessary to provide adequate confidence that a material, product, system, or service will satisfy given needs. **D 4439,** D35

quality assurance—all systematic actions necessary to provide adequate confidence that a calibration, measurement, or process is performed to a predefined level of quality. **E 170,** E10

quality assurance—all the planned and systematic activities implemented within the quality system, and demonstrated as needed, to provide adequate confidence that an entity will fulfill requirements for quality. **E 1187,** E36

quality assurance/quality control (QA/QC)—the efforts completed to evaluate the accuracy and precision of a sampling or testing procedure, or both. **D 653,** D18

quality assurance system—a manufacturing system for assuring quality that incorporates either a written control plan or employs other acceptable methods for controlling quality. This may include provisions for prototype development, initial production, and full production including advanced quality planning, continuous improvement, defect prevention, and in-process controls of dimensional, mechanical, and performance characteristics of the fastener. **F 1789,** F16

quality assurance test—a test in a program which is conducted to determine the quality level. **F 412,** F17

quality audit—systematic and independent examination to determine whether quality activities and related results comply with planned arrangements and whether these arrangements are implemented effectively and are suitable to achieve objectives. **E 1187,** E36

quality control—those quality assurance actions which provide a means to control and measure the characteristics of an item, process, or facility to established requirements. **C 859,** C26

quality control—See **quality.** **D 1356,** D22

quality control, *n*—the operational techniques and the activities which sustain a quality of material, product, system, or service that will satisfy given needs; also the use of such techniques and activities. **D 4439,** D35

quality control—operational techniques and activities that are used to fulfill requirements of quality. **E 1187,** E36

quality control test—an in-plant test that is conducted on a given test frequency to determine whether product is in accordance with the appropriate specification(s). **F 412,** F17

quality factor, *Q*, *n*—the reciprocal of the dissipation factor. **D 1711,** D09

quality factor, *Q*— the reciprocal of the dissipation factor (when applied to insulating materials). **D 2864,** D27

quality factor, Q—the ratio of elastic modulus to loss modulus, measured in tension, compression, flexure, or shear. **D 4092,** D20

quality management—all activities of the overall management function that determine the quality policy objectives and responsibilities, and implement them by means such as quality planning, quality control, quality assurance and quality improvement within the quality system. **E 1187,** E36

quality manual—a document stating the quality policy, quality system and quality practices of an organization, (ISO Guide 25). **E 1187,** E36

quality plan—a document setting out the specific quality practices, resource and sequence of activities relevant to a particular product, service, contract or project, (ISO 8402). **E 1187,** E36

quality policy—overall intentions and direction of an organization with regards quality, as formally expressed by top management. **E 1187,** E36

quality surveillance—continual monitoring and verification of the status of an entity and analysis of records to ensure that specified requirements are being fulfilled. **E 1187,** E36

quality system—organizational structure, procedures, processes and resources needed to implement quality management. **E 1187,** E36

quality verification tests—tests performed to check the degree of conformance to a standard or specification. **C 859,** C26

quantitative analysis—a measurement in which the amount of one or more components of a sample is determined. **C 859,** C26

quantitative analysis—the determination of the concentration and distribution of elements within the Auger analysis volume. **E 673,** E42

quantitative sensory profile, *n*—description of a sample consisting of both attributes and their intensity values. **E 253,** E18

quantum efficiency—measure of the ratio of energy input to converted energy output from a system, or a thermodynamic universe. **F 1156,** F12

quarry—an excavation in the surface of the earth from which stone is obtained for crushed rock or building stone. (ISRM) **D 653,** D18

quarry tile—glazed or unglazed tile, made by the extrusion process from natural clay or shale usually having 6 $in.^2$ (39 cm^2) or more of facial area. **C 242,** C21

quarter—the complete upper part of a shoe upper above the vamp line. **F 869,** F08

quarter, *n*—one side of the upper of a shoe or boot from heel to vamp. **F 1646,** F13

quartersawn—a synonym for *edge grain.* **D 9,** D07

quarter wave plate—a device used with a polarizer and analyzer designed to produce circularly polarized light. **E 7,** E04

quartz glass—See **fused silica.** **C 162,** C14

quartzite, ganister, *n*—for refractories, a rock consisting predominantly of the mineral quartz suitable for the manufacture of silica brick and characterized by a high silica (SiO_2) content and a low percentage of impurities (see Classification C 416). **C 71,** C08

quartzite (III)—highly indurated, typically metamorphosed sandstone containing at least 95 % free silica, which fractures conchoidally *through* the grains.

conglomerate—a sedimentary rock consisting of rounded pebbles and cobbles in a sandstone matrix, typically strongly cemented.

siltstone—a fine-grained, noncarbonated clastic rock composed mostly of detrital quartz and clay minerals in which the particles have an approximate size range from 0.06 to 0.005 mm. Siltstone may be designated fine-grained sandstone, and is texturally transitional between sandstone and shale. **C 119,** C18

quartzitic sandstone (II)—sandstone containing at least 90 % free silica (quartz grains plus siliceous cement), which may fracture around *or* through the constituent grains. **C 119,** C18

quasi-binary system—in a ternary or higher order system, a linear composition series between two congruent substances, wherein all equilibria, at all temperatures or pressures, involve only phases having compositions occurring in the linear series, so that the series may be represented as binary on a phase diagram. **E 7,** E04

quasi-isotropic—See **isotropic.** **E 7,** E04

quasi-isotropic laminate—a balanced and symmetric laminate for which a constitutive property of interest, at a given point, displays isotropic behavior in the plane of the laminate. Common quasi-isotropic laminates are $(0/\pm60)s$ and $(0/\pm45/90)s$. **D 3878,** D30

quasi-steady state—state of the solar collector test when the flow rate and temperature of the fluid entering the collector are constant. The exit fluid temperature changes are small and due only to the normal change in irradiance that occurs with time for clear sky conditions. **E 772,** E44

quasi-ternary system—in a quaternary or higher order system, a planar composition series among three congruent substances, wherein all equilibria, at all temperatures and pressures, involve only phases having compositions occurring upon the plane, so that they may be represented completely upon a ternary phase diagram. **E 7,** E04

quaternary system—the complete series of compositions produced by mixing four components in all proportions. **E 7,** E04

quebracho—a tanning material extracted from the wood of a South American tree. **D 1517,** D31

quench—the abrupt and uncontrolled loss of superconductivity produced by a disturbance./para> **B 713,** B01

quench aging, *n*—**aging** associated with **quenching** after **solution heat treatment**. **A 941,** A01

quench annealing, *n*—**annealing** an austenitic steel object by **solution heat treatment**. **A 941,** A01

quench cracks—surface discontinuities which usually transverse an irregular or erratic course on the surface of the fastener which may occur because of excessive high thermal and/or transformation stresses during fastener heat treatment. **F 1789,** F16

quench-hardened—See **hardened.** **F 547,** F16

quench hardening, *n*—**hardening** a steel object by **austenitizing** it, and then cooling it rapidly enough that some or all of the austenite transforms to martensite. **A 941,** A01

quench hardening—a treatment for copper-aluminum alloy products consisting of heating above the betatizing temperature followed by quenching to produce a hard martensitic structure. **B 846,** B05

quenching, *n*—rapid cooling. **A 941,** A01

quenching—see **fritting.** **C 286,** B08

quenching, *n*—the reduction of fluorescence by a competing deactivating process resulting from specific interaction between a fluorophor and another substance present in the system. **E 131,** E13

quenching crack—a crack formed as a result of thermal stresses produced by rapid cooling from a high temperature—not to be confused with fire crack. **E 7,** E04

quenching oil, *n*—oil used for cooling metals during a heat-treating operation. **D 4175,** D02

Quer-wave (love wave), *W*—dispersive surface wave with one horizontal component, generally normal to the direction of propagation, which decreases in propagation velocity with increase in frequency. **D 653,** D18

questioned, *n*—associated with the matter under investigation about which there is some question, including, but not limited to, whether the questioned and known items have a common origin. **E 1732,** E30

queue, *n*—a set of tasks stored in memory for use by a device in a specified order. (See **printout queuing**.) **F 1457,** F05

queue-rail system— Synonym for **traffic-rail system**. **E 1481,** E06

quick break—a sudden interruption of the magnetizing current. **E 1316,** E07

quick burst—Not a preferred term (see **quick burst test, quick burst pressure,** and **quick burst strength**). **F 412,** F17

quick burst pressure—the internal pressure required to bring a piping component to failure when subjected to a quick burst test. **F 412,** F17

quick burst strength—the hoop stress resulting from the quick burst pressure. **F 412,** F17

quick burst test—an internal pressure test designed to produce failure of a piping component over a relatively short period of time, usually measured in seconds. **F 412,** F17

quick condition (quicksand)—condition in which water is flowing upwards with sufficient velocity to reduce significantly the bearing capacity of the soil through a decrease in intergranular pressure. **D 653,** D18

quicklime—a calcined limestone, the major part of which is calcium oxide or calcium oxide in association with magnesium oxide, capable of slaking with water. **C 51,** C07

quicklime—CaO, calcium oxide. **D 6161,** D19

quicklime sizes—the different sizes depending upon the type of limestone, kind of kiln used, or treatment subsequent to calcining. The sizes commonly recognized are as follows:

(*1*) *large lump*—8 in. (203 mm) and smaller,

(*2*) *pebble or crushed*—2½ in. (64 mm) and smaller,

(*3*) *ground, screened or granular*—¼ in. (6.4 mm) and smaller, and

(*4*) *pulverized*—substantially all passing a No. 20 (850-μm) sieve. **C 51,** C07

quick test—see **unconsolidated undrained test**. **D 653,** D18

quiet zones, *n*—areas of high reflectance (spaces) immediately preceding the start character and following the stop character of a machine-readable symbol as defined for a particular symbology. (See also **clear area**.) **F 1294,** F05

quill, *n—in feathers*, the stem or central shaft. **D 123,** D13

quill feathers, *n*—feathers which are over 100 mm (4 in.) in length or which have a quill point exceeding 9.5 mm (⁹⁄₁₆ in.) in length. **D 123,** D13

quill point, *n—in feathers*, the section of quill extending beyond the section of barb attachment. **D 123,** D13

quill shaft, *n—in feathers*, the section of quill from which the barbs emanate. **D 123,** D13

quilt, *n*—a bedcovering assembly used primarily for warmth, consisting of an insulating filler secured between two layers of fabric, but generally lighter in weight and thinner than a comforter. **D 123,** D13

quilted, *n*—fused or stitched with thread through the upholstery cover material and one or more layers of upholstery material. **E 176,** E05

quinoline—benzo(b)pyridine (C_9H_7N) mol weight 129.15; colorless, refractive oil which darkens on storage; hygroscopic; penetrating odor not as offensive as pyridine; freezing point, −15°C; boiling point, 237.7°C. **D 4790,** D16

R

R_{10}—the portion of a cellulose pulp that is insoluble in 10 % sodium hydroxide using Test Method D 1696 or its equivalent. **D 1695,** D01

rabbet, *n*—a two-sided, L-shaped, recess in a sash or frame designed to receive glazing. **C 717,** C24

rack, *n*—*in warp knitting*, a unit of length measure consisting of 480 courses. **D 123,** D13

rack—longitudinal offset between adjacent fasteners within strip of collated fasteners. **F 592,** F16

racked fastener—fastener offset in strip. **F 592,** F16

rack, plating—a frame for suspending and carrying current to articles during plating and related operations. **B 374,** B08

radial—coincident with the radius from the pith to the circumference of the tree or log. A radial section is a longitudinal section in a plane that passes through the pith of the tree. **D 9,** D07

radial crushing strength—the relative capacity of a plain sleeve specimen of sintered metal to resist fracture induced by a load applied between flat parallel plates in a direction perpendicular to the axis of the specimen. **B 243,** B09

radial sectioning—a method of specimen preparation in which a specimen is polished by a cylinder in order to expose compositional changes below the original surface of a specimen with the intent that the depth of these layers can be related to the position on the surface created by the cylinder. **E 673,** E42

radial surface—a longitudinal surface or plane extending wholly or in part from the pith to the bark. **D 9,** D07

radian, *n*—the plane angle between two radii of a circle which intersects the circumference of the circle making an arc equal in length to the radius. **D 123,** D13

radiance, *n*—the rate of radiant emission per unit solid angle and per unit projected area of a source in a stated angular direction from the surface (usually the normal). **C 168,** C16

radiance factor at a point on the surface of a non-self-radiating body, in a given direction under specified conditions of irradiation, *n*—ratio of the radiance of the body to that of a perfect reflecting or transmitting diffuser, identically irradiated. Symbol: β. **E 349,** E21

radiance factor, β, β_e, *n*—ratio of the radiance from a point on a specimen, in a given direction, to that from the perfect reflecting or transmitting diffuser, similarly irradiated and viewed. **E 284,** E12

radiance (in a given direction, at a point on the surface of a source or receptor or at a point in the path of a beam), *n*—quotient of the radiant flux leaving, arriving at, or passing through an element of surface at this point, and propagated in directions defined by an elementary cone containing the given direction by the product of the solid angle of the cone and the area of the orthogonal projection of the element of surface on a plane perpendicular to the given direction. Symbol: L_e, L; $L_e = d^2\Phi$ (dω dA cos Θ); Unit: watt per steradian and per square metre, $W{\cdot}sr^{-1}\ m^{-2}$.

NOTE—Three special cases may be noted:

Case 1—At a point on the surface of a source, in a given direction, radiance is also the quotient of the radiant intensity in the given direction of an element of the surface at this point, by the area of the orthogonal projection of this element on a plane perpendicular to this direction (radiant intensity per unit projected area). $L_e = dI_e/(dA \cos \Theta)$.

Case 2—At a point on the surface of a receptor, in a given direction, radiance is also the quotient of the irradiance which is received at this point on a surface perpendicular to the given direction, by the solid angle of the elementary cone containing this direction and surrounding the beam which produces this irradiance (perpendicular irradiance per unit solid angle). $L_e = dE_e/d\omega$.

Case 3—On the path and in the direction of an element of a beam, in a nondiffusing, nonabsorbing medium, the radiance is also the quotient of the radiant flux $d\Phi_e$ which transports the beam, by the geometric extent dG of the beam. The geometric extent, which may be defined by two sections of the beam of areas dA and dA' of separation l, and having angles Θ and Θ′ between their normals and the direction of the beam is $dG = dA \cos \Theta\ d\omega$ where the numerical value in steradians of dω is $dA' \cos \Theta' l^{-2}$. $L_0 = d\Phi_0/dG = d^2\Phi_e/(d\omega\ dA \cos \Theta)$. In the absence of diffusion, it can be demonstrated in geometrical optics that the optical extent, product of the geometric extent of an element of a beam and the square of the refractive index of the medium of propagation, is an invariant along the length of the beam whatever the deviations which it undergoes by reflection or refraction ($dG{\cdot}n^2$ = constant). In consequence, the basic radiance, quotient of the radiance by the square of the refractive index, is invariant along the length of an element of a beam if losses by absorption or by reflection are taken as zero ($L_e{\cdot}n^{-2}$ = constant). **E 349,** E21

radiance, L—the flux per unit projected area per unit solid angle leaving a source or, in general, any reference surface. If $d^2\Phi$ is the flux emitted into a solid angle $d\omega$ by a source element of projected area dA cos θ, the radiance is defined as:

$$L = \frac{d^2\Phi}{d\omega{\cdot}dA \cos \theta}$$

where, θ is the angle between the outward surface normal of the area element dA and the direction of observation (unit = $W/sr{\cdot}m^2$). **E 1316,** E07

radiance, *L*, L_e, *n*—radiant flux in a beam, emanating from a surface, or falling on a surface, in a given direction, per unit of projected area of the surface as viewed from that direction, per unit of solid angle. **E 284,** E12

radiant, *adj*—pertaining to electromagnetic radiation, with the contributions at all wavelengths weighted equally. **E 284,** E12

radiant efficiency of a source of radiation, *n*—ratio of the radiant flux emitted to the power consumed. Symbol: η_e, η.

NOTE—The radiant efficiency of a source in a limited region of the spectrum may also be considered, that is, the ratio of the radiant flux emitted in this spectral region to the power consumed. **E 349,** E21

radiant emissive power— see **radiant exitance, emitted.** **E 772,** E44

radiant energy—energy transmitted as electromagnetic waves. **E 131,** E13

radiant energy, *n*—energy transmitted as electromagnetic radiation. **E 135,** E01

radiant energy, *n*—energy transmitted as electromagnetic radiation. **E 284,** E12

radiant energy, *n*—energy emitted, transferred, or received as radiation. Symbol: Q_e, Q; Unit: joule J (1 J = W·s).

NOTE—In ultraviolet radiation therapy and photobiology, this quantity is called "integral dose" (International Photobiology Committee, 1954). **E 349,** E21

radiant energy—energy in the form of photons or electromagnetic waves. **E 772,** E44

radiant energy, atmospheric—the part of terrestrial radiation that is emitted by the atmosphere. **E 772,** E44

radiant energy, blackbody—radiant energy emitted by a (laboratory) blackbody, or radiant energy having that spectral distribution. (Refer to Planck's law as defined in standard tests). **E 772,** E44

radiant energy, circumsolar—radiation scattered by the atmosphere so that it appears to originate from an area of the sky immediately adjacent to the sun. It causes the solar aureole, and its angular extent is directly related to the atmospheric turbidity, being greater with higher turbidity. **E 772,** E44

radiant energy, effective nocturnal—energy transfer required to maintain a horizontal upward-facing blackbody surface at the ambient air temperature, in the absence of solar irradiance. **E 772,** E44

radiant energy, infrared—radiant energy with wavelengths longer than 770 nm and less than 1 mm (approximately). **E 772,** E44

radiant energy, isotropic—diffuse radiant energy that has the same radiance in all directions. See **radiometric properties and quantities**. **E 772,** E44

radiant energy, terrestrial—radiant energy emitted by the earth, including its atmosphere. **E 772,** E44

radiant exitance at a point on a surface, *n*—quotient of the radiant flux leaving an element of the surface containing the point, by the area of that element. Symbol: M_e, M; $M_e = d\Phi_e/dA = \int_2 L_e \cos \theta d\omega$. Unit: watt per square metre, $W{\cdot}m^{-2}$.

NOTE—The name radiant emittance previously given to this quantity is abandoned because it has given rise to confusion. Thus, the term "emittance" has been used to designate either the flux per unit area leaving a surface (whatever the origin of the flux), the flux per unit area emitted by a surface (flux originating in the surface), or, principally, in certain circles in the United States of America, a quantity without dimensions similar to "emissivity," but applicable only to a specimen.

NOTE—The expression "self-radiant exitance" ($M_{e,s}$) indicates that the flux considered does not include reflected or transmitted flux.

The expression "thermal-radiant exitance" ($M_{e,th}$) indicates that the flux considered is produced by thermal radiation. These same adjectives (self, thermal) are equally applicable to other quantities, such as radiance, etc.

NOTE—In the case of a full radiator (blackbody), the radiance L_e is uniform in all directions. In consequence, when the solid angle is measured in steradians, the radiant exitance has the numerical value $M_e = \pi l_e$.

E 349, E21

radiant exitance at a point on a surface, $M = d\Phi_e/dA$— the quotient of the radiant flux leaving an element of the surface containing the point, by the area of that element. The exitance may be emitted, transmitted and/or reflected flux, and is measured in watts per square metre ($W \cdot m\Sigma^{-2}$). **E 772,** E44

radiant exitance, emitted—radiant flux emitted per unit surface area. **E 772,** E44

radiant exitance, M—the radiant flux per unit area leaving a surface That is,

$$M = \frac{d\Phi}{dA}$$

where:

$d\Phi$ = flux leaving a surface element dA (unit = W/m^2).

E 1316, E07

radiant exposure, *n*—time integral of the irradiance at a given point over a specified time interval. **E 284,** E12

radiant exposure, *n*—time integral of irradiance, typically reported in joules per square metre, J m^{-2} **G 113,** G03

radiant exposure at a point on a surface, *n*—surface density of the energy received. Symbol: H_e, H; $H_e = dQ_e/dA = \int E_e\, dt$; Unit: joule per square metre, $J{\cdot}m^{-2}$.

NOTE—Formerly "irradiation."

NOTE—Equivalent definition: Product of an irradiance and its duration.

NOTE—In ultraviolet radiation therapy and photobiology, this quantity is called dose (International Photobiology Committee, 1954).

E 349, E21

radiant exposure, H—time integral of irradiance, measured in joules per square metre. **E 772,** E44

radiant flux, Φ, *n*—the time rate of flow of radiant energy; radiant power. **E 284,** E12

radiant flux: radiant power, *n*—power emitted, transferred, or received as radiation: Symbol: Φ_e,Φ , P; $\Phi_e = dQ_e/dt$; Unit: watt (W). **E 349,** E21

radiant flux, $\Phi = dQ/dt$**[Watt (W)]**—power emitted, transferred, or received in the form of electromagnetic waves or photons. See **radiometric properties and quantities.** **E 772,** E44

radiant flux; radiant power, ϕ_e—radiant energy per unit time (unit = *W*). **E 1316,** E07

radiant flux density, *n*—the rate of radiant energy emitted from unit area of a surface in all radial directions of the overspreading hemisphere. **C 168,** C16

radiant flux, net—difference between downward and upward (total solar and terrestrial) radiant flux; net flux of all radiant energy across an imaginary horizontal surface. **E 772,** E44

radiant flux, net terrestrial—difference between downward and upward terrestrial radiant fluxes; net flux of terrestrial radiant energy. **E 772,** E44

radiant flux profile, *n*—the graph relating incident radiant heat energy on the specimen plane to distance from the point of initiation of flaming ignition, that is, 0 mm. **E 176,** E05

radiant flux (surface) density at a point of a surface, *n*—quotient of the radiant flux at an element of the surface containing the point, by the area of that element. (See also **irradiance** and **radiant exitance**.) Unit: watt per square metre, $W{\cdot}m^{-2}$. **E 349,** E21

radiant heat—heat communicated by energy propagated through space and transmitted by electromagnetic waves. **F 1494,** F23

radiant intensity, *I*, I_e, *n*—the radiant flux per unit solid angle. **E 284,** E12

radiant intensity, *J*, *n*—the radiant power emitted per unit solid angle in a specified direction. **E 135,** E01

radiant intensity of a point source in a given direction, $I = d\Phi/d\Omega$ — quotient of the radiant flux emitted by a source, or by an element of source, in an infinitesimal cone containing the given direction, by the solid angle of the cone, measured in watts per steradian. (Refer to the inverse square law.) **E 772,** E44

radiant intensity of a source, in a given direction, *n*—quotient of the radiant flux leaving the source propagated in an element of solid angle containing the given direction, by the element of solid angle. Symbol: I_e, I; $I_e = d\Phi_e/d\omega$; Unit: watt per stearadian, $W{\cdot}sr^{-1}$.

NOTE—For a source which is not a point source: The quotient of the radiant flux received at an elementary surface by the solid angle which this surface subtends at any point of the source, when this quotient is taken to the limit as the distance between the surface and the source is increased.

E 349, E21

radiant power—see **radiant flux.** **E 772,** E44

radiant power, P, *n*—the rate at which energy is transported in a beam of radiant energy. **D 4175,** D02

radiant power, *P*— the rate at which energy is transported in a beam of radiant energy. **E 131,** E13

radiant power, *P*, *n*—the rate at which energy is transported in a beam of radiant energy, preferably expressed in ergs per second or watts. **E 135,** E01

radiation, *n*—(*1*) emission or transfer of energy in the form of electromagnetic waves or particles.

(*2*) the electromagnetic waves or particles.

NOTE—In general, nuclear radiations and radio waves are not considered in this vocabulary, only optical radiations, that is, electromagnetic radiations (photons) of wavelengths lying between the region of transition to X rays (1 nm) and the region of transition to radio waves (1 mm).
E 349, E21

radiation—the process by which energy is emitted or transferred in the form of photons or electromagnetic waves. (See also **radiant energy**.) **E 772,** E44

radiation chemical yield, $G(x)$—the quotient of n(x) by $\bar{\epsilon}$ where n(x) is the mean amount of a specified entity, x, produced, destroyed, or changed by the mean energy, $\bar{\epsilon}$, imparted to the matter (see ICRU Report 33).

$$G(x) = (\mathrm{n}\ (x)/\ \bar{\epsilon})$$

SI Unit: mol J^{-1} **E 170,** E10

radiation coefficient—the quotient of the net radiant exitance of a blackbody (full radiator), by the temperature difference between the blackbody and the surroundings with which it is exchanging radiation. **E 772,** E44

radiation induced (or enhanced) diffusion—atom movement in the solid, well beyond the typical penetration depth of an incident particle, due to particle beam damage or bombardment induced defects. **E 673,** E42

radiation pattern, *n*—the optical pattern of light that leaves the media surface, as described by the radiated light intensity at various angles. **F 1294,** F05

radiation pyrometer—an instrument for determining temperatures by measuring the radiance (radiant energy per unit area) from an object. **E 7,** E04

radiation thermometer, *n*—a radiometer calibrated to indicate the temperature of a blackbody. **E 344,** E20

radiative field loader—device that absorbs energy when placed in a radiated field, for example, an **LC** circuit placed in a field of the properly tuned frequency will absorb radiated field energy at that frequency. **F 1156,** F12

radioactivation—an interaction process in which the product nucleus becomes radioactive. **E 170,** E10

radioactive daughter—the direct radioactive decay product of a radionuclide. **D 1129,** D19

radioactive decay—the spontaneous transformation of a nuclide into one or more nuclides accompanied by the release of radiation. **C 859,** C26

radioactivity—spontaneous nuclear disintegration with emission of corpuscular or electromagnetic radiation, or both. **D 1129,** D19

radioactivity half-life—the unvarying characteristic period of time in which one half of the radioactive atoms of a given radionuclide decay. **D 1129,** D19

radio frequency (r-f) display—the display of an unrectified signal on the CRT or recorder. **E 1316,** E07

radiograph—a permanent, visible image on a recording medium produced by penetrating radiation passing through the material being tested. **E 1316,** E07

radiograph—a permanent, visible image on a recording medium produced by penetrating radiation passing through the material being tested. **E 1316,** E07

radiographic contrast—the difference in density between an image and its immediate surroundings on a radiograph. **E 1316,** E07

radiographic equivalence factor—that factor by which the thickness of a material must be multiplied in order to determine what thickness of a standard material (often steel) will have the same absorption. **E 1316,** E07

radiographic exposure— see **exposure.** **E 1316,** E07

radiographic inspection—the use of X rays or nuclear radiation, or both, to detect discontinuities in material, and to present their images on a recording medium. **E 1316,** E07

radiographic inspection—the use of X rays or nuclear radiation, or both, to detect discontinuities in material, and to present their images on a recording medium. **E 1316,** E07

radiographic quality—a qualitative term used to describe the capability of a radiograph to show flaws in the area under examination. **E 1316,** E07

radiographic sensitivity—a general or qualitative term referring to the size of the smallest detail that can be seen on a radiograph, or the ease with which details can be seen. **E 1316,** E07

radiography—the art, act, or process of making radiographs. **E 1316,** E07

radiography—the process of producing a radiograph using penetrating radiation. **E 1316,** E07

radioisotope leak test system—a leak test system which uses a radioactive tracer gas and a detector for measuring the emission from the tracer. **E 1316,** E07

radioisotopes—radionuclides having the same atomic number. **D 1129,** D19

radiological examination—the use of penetrating ionizing radiation to display images for the detection of discontinuities or to help ensure integrity of the part. **E 1316,** E07

radiological examination—the use of penetrating ionizing radiation to display images for the detection of discontinuities or to help ensure integrity of the part. **E 1316,** E07

radiology—the science and application of X rays, gamma rays, neutrons, and other penetrating radiations. **E 1316,** E07

radiology—the science and application of X rays, gamma rays, neutrons, and other penetrating radiations. **E 1316,** E07

radiometer, *n*—a device for measuring radiant power that has an output proportional to the intensity of the input power. **E 344,** E20

radiometer, *n*—instrument for measuring radiation in energy or power units. **E 349,** E21

radiometer—instrument for measuring irradiance in energy or power units. **E 772,** E44

radiometer—an instrument for measuring the intensity of radiant energy. In infrared thermography, an apparatus that measures the average apparent temperature of the surface subtended by its field of view. **E 1316,** E07

radiometer, *n*—a general class of instruments designed to detect and measure radiant energy. **G 113,** G03

radiometric, *adj*—pertaining to measurement of quantities associated with radiant energy. **E 284,** E12

radiometric properties and quantities modifiers—modifiers that can be used to indicate the geometric, spectral, and polarization conditions under which radiometric properties and quantities are evaluated. Radiometric properties and quantities vary with the direction and geometric extent (solid angle) over which the incident or exitant flux, or both, is evaluated, and with the relative spectral distribution of the incident flux and the spectral response of the detector for exitant flux. For reflectance and transmittance, the direction and geometric extent of both the incident beam and exitant beam must be specified. For emittance, only the exitant beam need be specified, and for absorptance, only the incident beam need be specified. The properties also vary with the polarization of the incident flux and the sensitivity to polarization of the collector-detector system for flux incident or exitant at angles greater than about 15° from normal. See the following modifiers:

conical; directional; hemispherical; normal; and **spectral.** E 772, E44

radiometry, *n*—measurement of the quantities associated with radiation. **E 349,** E21

radio-opaque—not transparent to x-rays. **F 1156,** F12

radioscopic inspection—the use of penetrating radiation and radioscopy to detect discontinuities in material. **E 1316,** E07

radioscopy—the electronic production of a radiological image that follows very closely the changes with time of the object being imaged. **E 1316,** E07

radioscopy—the electronic production of a radiological image that follows very closely the changes with time of the object being imaged. **E 1316,** E07

radiosonde, *n*—a miniature radio transmitter with instruments that is carried aloft (as by an unmanned balloon) for broadcasting by means of precise tone signals or other suitable method the humidity, temperature, pressure, or other parameter every few seconds. **D 1356,** D22

radius of bend, *n*—the radius of the cylindrical surface of the pin or mandrel that comes in contact with the inside surface of the bend during bending. In the case of free or semi-guided bends to 180° in which a shim or block is used, the radius of bend is one half the thickness of the shim or block. **E 6,** E28

radius of influence of a well—distance from the center of the well to the closest point at which the piezometric surface is not lowered when pumping has produced the maximum steady rate of flow. **D 653,** D18

raggle—same as reglet. **C 717,** C24

raid end band—See **brace band.** **F 552,** F14

rail—See **railing systems, windows and doors.** **E 631,** E06

rail, *n*—a horizontal, inclined, or vertical member of a railing system, such as top, intermediate, or bottom member connecting balusters or posts, or both, at specified intervals. (Compare **handrail.**) **E 1481,** E06

rail—magazine component required for alignment of fastener and directing it into driving channel. **F 592,** F16

rail cap—See **railing systems.** **E 631,** E06

rail cap—Synonym for **cap rail**. **E 1481,** E06

rail end—a cup-shaped fitting used with a brace band to connect the top rail or brace to a post. **F 552,** F14

railing—See **banding.** **D 1038,** D07

railing—See **railing systems.** **E 631,** E06

railing, n— Use **railing system**. **E 1481,** E06

railing return—a bent at the end of a handrail, turning toward the wall or post to which the handrail is attached. **E 1481,** E06

railing system—See **railing systems.** **E 631,** E06

railing system—a framework of horizontal, inclined, vertical, and infill members, including panels and grillwork, for protection of building occupants against fall or injury and for offering safety and convenience in their movement. (Compare **guardrail**.) **E 1481,** E06

railing-system penetration limitation—See **railing systems.** **E 631,** E06

railing-system penetration limitation— an arrangement of railing elements designed to prevent passage of a sphere of specified diameter through a railing system. **E 1481,** E06

railing systems:

baluster, n—(baluster bar) same as **picket.**

balustrade, n—a **railing system** consisting of a row of **pickets** capped by a **handrail.**

bottom rail—the lowest member of a **railing system,** supporting **pickets** or **panels.**

cap, n—a fitting used to close the end of a **pipe,** tubular **post, newel,** or **rail.**

cap rail—a **handrail** fastened to the **top rail** of a **railing system.**

collar, n—Same as **escutcheon.**

cover flange—Same as **escutcheon.**

drop cap—the cover of a railing **post** or **newel** that is exposed to view, usually below stringer or **floor.**

easement, n—the curved portion of a **handrail** forming a transition in the vertical plane between horizontal and inclined sections of **handrail.**

escutcheon, n—a protective or ornamental cover located at the termination of a **post, picket,** or **rail** against a tread, **floor,** or wall. Same as **collar, cover flange.**

finial, n—an ornamental piece on the top of a **post, newel,** or **railing;** frequently in the form of an **urn** or **pineapple,** and so named.

flange plate—a flat (piece) between the end of a **railing** or rail element and the adjoining construction or supporting member.

grab bar—Same as **grab rail.**

grab rail—a short length of **rail** or similar member located for safety or convenience to assist a person in movement.

guardrail system—a **railing system** located for protection of building users at or near the outer edge of a **stair,** ramp, landing, platform, balcony, or accessible roof; at the perimeter of an opening or accessible surface, such as the opening of a **stair;** or at a location where an operating condition requires access limitation to guard against accidental fall and injury. Compare **railing system.**

handgrip, n—the part of a **handrail** designed to provide a secure grasp.

handrail, n—a horizontal or sloping member normally grasped by hand for guidance or support.

handrail bracket—a device attached to a wall, **post,** or other surface to support the **handrail.**

handrail height—the vertical distance from the top surface of the **top rail** to the surface of the finished **floor,** top of a ramp, or the nosing line of stair treads.

kick plate—Same as **toe board.**

lamb's tongue—an ornamental curved or tapered fitting terminating a **handrail,** usually tapered to the tip.

lateral scroll—a fitting that curves in a horizontal plane, used to terminate a **handrail;** often ending as a round plate covering the top of a **post.**

mid rail—a rail located between **top** and **bottom rails.**

miter ending—a square, angular, or dovetailed member end, designed to fit an adjacent matching member, thereby providing continuity of profile at the joint.

newel, n—*(1)* a decorative or structural **post** member, often extending above a **handrail** at the start or end of a **stair** run; usually square or rectangular in cross section, supporting the end of a stair railing or serving as the common support for two stair railings; often supporting a stair stringer and platform. *(2)* the center **post** of a spiral **stair.**

ogee, n—a molding with a profile having a double curve formed by a convex line turning into a concave line, and resulting in an S-shape.

panel n—in a **railing system,** a flat element between a **post, top rail** and **bottom rail.**

picket, n—one of a series of closely spaced upright members that support the **handrail** in a **railing system.** Same as **baluster, baluster bar.**

picket casting—an ornamental element attached to a **picket.**

picket railing system—a system consisting of **posts, pickets, top rail** and **bottom rail.**
pineapple, n—a **finial** in the form of a pineapple.
pipe railing system—a **railing system** fabricated from **pipe.**
post, n—in **railing systems,** a vertical supporting member. Compare **column.**
rail, n—a horizontal, inclined, or vertical member of a **railing system** connecting **pickets** or **posts,** or both, at specified intervals. Compare **handrail.**
rail cap—Same as **cap rail.**
railing, n—Use **railing system.**
railing system—a framework of horizontal, inclined, or vertical members, **panels,** grillwork, or their combinations that support a **handrail;** located at the edge of a **stair** flight, platform, or **floor** as a safety barrier. Compare **guardrail system, stair rail system.**
railing-system penetration limitation—an arrangement of railing elements designed to prevent passage of a sphere of specified diameter through a **railing system.**
ramp-rail system—a **railing system** located alongside a ramp.
screen, n—in a **railing system,** a perforated sheet or **wire mesh** in sheet form serving as a **panel.**
scroll, n—in a **railing system,** a cast or forged ornamental spiral of convoluted shape serving, for example, as a decorative **panel** or insert.
side mount—a railing support that anchors the **post** of a **railing system** to a vertical surface, such as fascia or stair stringer.
spindle, n—in a **railing system,** a tapered, round **picket** having a center portion larger in diameter than its ends.
stair-rail system—a **rail** or **railing system** located along a **stair** or landing.
toe board—in a **railing system,** a vertical plate forming a low curb at the open edge of a stairwell, platform, or floor; to provide a barrier preventing objects from falling beyond it. Same as **kick plate, toe plate.**
toe plate—Same as **toe board.**
top rail—the uppermost member of a **railing system.**
traffic-rail system—a **railing system** designed to control the movement of people and requiring special consideration for given conditions.
transfer-rail system—a **railing system** designed to support and to permit transfer of body weight in such locations as toilets, showers, and tub enclosures.
urn, n—a **finial** in the form of an urn.
volute, n—in a **railing system,** an ornamental spiral or *scroll*-shaped form serving, for example, as **handrail** termination.
wall clip—a bracket used for anchoring to a wall. Same as **wall flange.**
wall flange—Same as **wall clip.**
wall handrail—a **handrail** attached to a wall adjacent to a stair, and paralleling the **pitch** of the stair flight; also used along landings, walkways, ramps, and corridors. Same as **wall rail.**
wall rail—Same as **wall handrail.**
wall railing return—in a **railing system,** a bend at the end of a **wall handrail,** turning toward the wall to which the **handrail** is attached.
wire mesh—Same as **screen.** **E 631,** E06

railroading, *n*—the printing of a continuous mark or line in the plain area of the design that often results in a definite marking or scratching of the engraved cylinder. **D 6488,** D01

railroad tracks, *n*—a pattern on the printed sheet extending in the machine direction that is comprised of one or more linear regions. **[TAPPI]** **D 6488,** D01

rainbow printing—multi-colored printing in which discrete colors are merged into one another during the printing to give the appearance of a continuous spectrum of color, changing smoothly from one tone to the next with no sense of interruption or abruptness. **F 1156,** F12

rain density, n— See **particle concentration.** **G 40,** G02

raise—upwardly constructed shaft; that is, an opening, like a shaft, made in the roof of one level to reach a level above. (ISRM) **D 653,** D18

raise, *v*—to augment the value, usually of a document, by altering the stated or indicated value; for example, inserting a "0" after the "1" in each corner of the $1 to make it appear to be a $10 bill, or inserting additional digits into the amount line of a money order to the left of the decimal point so as to increase the apparent value of the money order. **F 1156,** F12

raised fiber surface, *n*—*in textile fabrics*, intentionally lifted fibers or yarns such as pile, napped, tufted, flocked, or similar surfaces. **D 123,** D13

rake—the sloped edge of a roof at the first or last rafter. **D 1079,** D08

RAM—**r**andom **a**cess **m**emory. **F 1457,** F05

Raman line (band)—a line (band) that is part of a Raman spectrum. **E 131,** E13

Raman shift—the displacement in wavenumber of a Raman line (band) from the wavenumber of the incident monochromatic beam. **E 131,** E13

Raman spectrum—the spectrum of the modified frequencies resulting from inelastic scattering when matter is irradiated by a monochromatic beam of radiant energy. **E 131,** E13

ramming mix, *n*—a refractory material, usually tempered with water, that cannot be extruded but that has suitable properties to permit ramming into place to form a monolithic structure. **C 71,** C08

ramp-rail system—See **railing systems.** **E 631,** E06

ramp-rail system—a railing system located along the open sides of a ramp. **E 1481,** E06

ranch type—a type of asbestos-cement roofing, rectangular in shape, which is lapped at the top and one side. **C 460,** C17

ranch type, *adj— in shingles*, pertaining to a type of roofing, rectangular in shape, that is lapped at the top and one side. **C 1154,** C17

ranch type—a type of asbestos-cement roofing shingle, rectangular in shape, which is lapped at the top and one side. **D 2946,** C17

random—run of mill lengths without any indicated preferred length. **B 846,** B05

random assess memory, *n*—an electronic data storage system in which the data can be stored and retrieved directly from any memory location. (See **ROM**.) **F 1457,** F05

random cause, *n*—one of many factors which contribute to variation but which are not feasible to detect and identify since they are random in origin and usually small in effect. **D 123,** D13

random composite sample, *n— in asbestos,* a composite sample in which each individual bag in the lot that is being sampled has an equal chance of being included in the composite sample. **D 2946,** C17

random error of result, *n*—a component of the error which, in the course of a number of test results for the same characteristic, varies in an unpredictable way. **E 456,** E11

random errors, *n*—errors introduced by the printer, reader, operator, or other system element; errors that are encountered during one scan that may not be encountered during the next scan. **F 1294,** F05

randomization, *n*—the procedure used to allot treatments at random to the experimental units so as to provide a high degree of independence in the contributions of experimental error to estimates of treatment effects.

NOTE—An essential element in the design of experiments is to provide estimates of effects free from biases due to undetected assignable causes within the experimental space. Randomization is a process to minimize this risk. The operational procedure for assignment "at random" involves the use of random numbers or some similar method for assuring that each unit has an equal chance of being selected for each treatment. **E 456,** E11

randomization, *n*—the procedure used to allot treatments at random to the experimental units so as to provide a higher degree of independence in the contributions of experimental error to estimates of treatment effects.

NOTE—An essential element in the design of experiments is to provide estimates of effects free from biases due to undetected assignable causes within the experimental space. Randomization is a process to minimize this risk. The operational procedure for assignment *at random* involves the use of random numbers or some similar method for assuring that each unit has an equal chance of being selected for each treatment. **E 1325,** E11

randomized block design, *n*—a design in which the experiment space is subdivided into blocks of experimental units, the units within each block being more homogeneous than units in different blocks.

NOTE—In each block the treatments are allocated randomly to the experimental units within each block. Replication is obtained by the use of two or more blocks, depending on the precision desired, and a separate randomization is made in each block. **E 456,** E11

randomized block design, *n*—a design in which the experiment space is subdivided into blocks of experimental units, the units within each block being more homogeneous than units in different blocks.

NOTE—In each block the treatments are allocated randomly to the experimental units within each block. Replication is obtained by the use of two or more blocks, depending on the precision desired, and a separate randomization is made in each block. **E 1325,** E11

randomized block experiment, *n*—a kind of experiment which compares the averages of *k* different treatments that appear in random order in each of *b* blocks. **D 123,** D13

randomized block factorial design, *n*—a factorial experiment run in a randomized block design in which each block includes a complete set of factorial combinations. **E 456,** E11

randomized block factorial design, *n*— a factorial experiment run in a randomized block design in which each block includes a complete set of factorial combinations. **E 1325,** E11

random loading—*in fatigue loading*, a spectrum loading (straining) where the peak and valley loads (strains) and their sequence result from a random process; the loading (straining) is usually described in terms of its statistical properties, such as the probability density function, the mean, the root mean square, the irregularity factor, and others as appropriate. **E 1823,** E08

random-ordered loading— *in fatigue loading*, a spectrum loading that is generated from a distinct set of peak and valley loads into a loading sequence by using a specific random sequencing process; a sequence of finite length is usually repeated identically. **E 1823,** E08

random orientation—a condition of a polycrystalline aggregate in which the constituent crystals have orientations completely random with respect to one another. **E 7,** E04

random sample, *n*— *in asbestos,* a sample in which each individual bag in the lot examined has an equal chance of being included in the sample. **D 2946,** C17

random sample—see **representative sample**. **E 1605,** E06

random sampling, *n*—the process of selecting units for a sample of size *n* in such a manner that all combinations of *n* units under consideration have an equal or ascertainable chance of being selected as the sample. **D 123,** D13

random sampling—technique for selecting a sample of *n* elements from a population of elements in such a way that each combination of *n* elements has the same probability of being selected. **E 1605,** E06

random sampling—*in statistical sampling*, the process of selecting sample units in such a way that all units under consideration have the same probability of being selected. **E 1605,** E06

random sampling (simple random sampling)—a technique for selecting a sample of *n* elements from a population of elements in such a way that each combination of *n* elements has the same probability of being selected. **C 859,** C26

random variance of increment collection (unit variance)—See *random variance of increment collection (unit variance)* under **variance**. **D 121,** D05

range—See **analyzer.** **D 1356,** D22

range, *n*—the region between the limits within which a quantity is measured, and is expressed by stating the lower and upper range-values. **E 344,** E20

range—the maximum sound path length that is displayed. **E 1316,** E07

range—the absolute value of the algebraic difference between the highest and the lowest values in a set of data. **E 1547,** E15

range, *n*—region between the limits within which a quantity is measured, and is expressed by stating the lower and upper range-values. **E 1605,** E06

range—equipment for cooking food by direct or indirect heat transfer from one or more cooking units, to one or more cooking containers. **F 1827,** F26

range, data, *n*—the absolute value of the difference between the highest and lowest values in a set of data. **A 644,** A04

range, exposure—See **exposure scale.** **E 7,** E04

range (of a deformation-measuring instrument)—the amount between the maximum and minimum quantity an instrument can measure without resetting. In some instances provision can be made for incremental extension of the range. **D 653,** D18

range, specification (non-statistical), *n*—the absolute value of the difference between the highest (USL) and lowest (LSL) specified values; for compliance with a non-statistical range, each valid individual test result must lie at, or within, the specification limits.

$$LSL \leq \text{All } x_i \leq USL$$

A 644, A04

range, specification (statistical), *n*—the absolute value of the difference between the highest (USL) and lowest (LSL) specified values; for compliance with a statistical range, all calculated values from the mean ($\bar{x}$) minus M times the standard deviation(s) to the mean plus M times the standard deviation, where M is a matter of agreement between the supplier and purchaser, must not lie outside of the specification limits.

$$LSL \leq \bar{x} - M \times s \quad \text{and} \quad \bar{x} + M \times s \leq USL$$

A 644, A04

rank, *n—of coal*, a classification designation that indicates the degree of metamorphism, or progressive alteration, from lignite to anthracite.

anthracite, n—the rank of coal, within the anthracitic class of Classification D 388, such that on the dry and mineral-matter-free basis, the volatile matter content of the coal is greater than 2 % but equal to or less than 8 % (or the fixed carbon content is equal to or greater than 92 % but less than 98 %), and the coal is nonagglomerating.

anthracitic class, n—a class of rank consisting of semianthracite, anthracite, and meta-anthracite.

bituminous class, n—a class of rank consisting of high volatile C bituminous coal, high volatile B bituminous coal, high volatile A bituminous coal, medium volatile bituminous coal, and low volatile bituminous coal.

high volatile A bituminous coal, n—the rank of coal, within the bituminous class of the Classification D 388, such that on the dry and mineral-matter-free basis, the volatile matter content of the coal is greater than 31 % (or the fixed carbon content is less than 69 %), and its gross calorific value is equal to or greater than 14 000 Btu/lb (32.65 Mj/kg) of coal on the moist, mineral-matter-free basis, and the coal is commonly agglomerating.

high volatile B bituminous coal, n—the rank of coal, within the bituminous class of Classification D 388, such that, on the moist, mineral-matter-free basis, the gross calorific value of the coal in British thermal units per pound is equal to greater than 13 000 (30.24 Mj/kg) but less than 14 000 (32.54 Mj/kg) and the coal commonly agglomerates.

high volatile C bituminous coal, n—the rank of coal, within the bituminous class of Classification D 388, such that, on the moist, mineral-matter-free basis, the gross calorific value of the coal in British thermal units per pound is equal to or greater than 11 500 (26.75 Mj/kg) but less than 13 000 (30.24 Mj/kg) and the coal commonly agglomerates, or equal to or greater than 10 500 (24.42 Mj/kg) but less than 11 500 (26.75 Mj/kg) and the coal agglomerates.

lignite A, n—the rank of coal, within the lignitic class of Classification D 388, such that, on the moist, mineral-matter-free basis, the gross calorific value of the coal in British thermal units per pound is equal to greater than 6300 (14.65 Mj/kg) but less than 8300 (19.31 Mj/kg), and the coal is nonagglomerating.

lignite B, n—the rank of coal, within the lignitic class of Classification D 388, such that, on the moist, mineral-matter-free basis, the gross calorific value of the coal in British thermal units per pound is less than 6300 (14.65 Mj/kg), and the coal is nonagglomerating.

lignitic class, n—a class of rank consisting of lignite A and lignite B.

low volatile bituminous coal, n—the rank of coal, within the bituminous class of Classification D 388, such that, on the dry and mineral-matter-free basis, the volatile matter content of the coal is greater than 14 % but equal to or less than 22 % (or the fixed carbon content is equal to or greater than 78 % but less than 86 %), and the coal commonly agglomerates.

medium volatile bituminous coal, n—the rank of coal, within the bituminous class of Classification D 388, such that, on the dry and mineral-matter-free basis, the volatile matter content of the coal is greater than 22 % but equal to or less than 31 % (or the fixed carbon content is equal to or greater than 69 % but less than 78 %), and the coal commonly agglomerates.

meta-anthracite, n—the rank of coal, within the anthracite class of Classification D 388, such that, on the dry and mineral-matter-free basis, the volatile matter content of the coal is equal to or less than 2 % (or the fixed carbon is equal to or greater than 98 %), and the coal is nonagglomerating.

semianthracite, n—the rank of coal, within the anthracitic class of Classification D 388, such that, on the dry and mineral-matter-free basis, the volatile matter content of the coal is greater than 8 % but equal to or less than 14 % (or the fixed carbon content is equal to or greater than 86 % but less than 92 %), and the coal is nonagglomerating.

subbituminous class, n—a class of rank consisting of subbituminous C coal, subbituminous B coal, and subbituminous A coal.

subbituminous C coal, n—the rank of coal, within the subbituminous class of Classification D 388, such that, on the moist, mineral-matter-free basis, the gross calorific value of the coal in British thermal units per pound is equal to greater than 8300 (19.31 Mj/kg) but less than 9500 (22.10 Mj/kg), and the coal is nonagglomerating.

subbituminous B coal, n—the rank of coal, within the subbituminous class of Classification D 388, such that, on the moist, mineral-matter-free basis, the gross calorific value of the coal in British thermal units per pound is equal to greater than 9500 (22.10 Mj/kg) but less than 10 500 (24.42 Mj/kg) and the coal is nonagglomerating.

subbituminous A coal, n—the rank of coal, within the subbituminous class of Classification D 388, such that, on the moist, mineral-matter-free basis, the gross calorific value of the coal in British thermal units per pound is equal to greater than 10 500 (24.42 Mj/kg) but less than 11 500 (26.75 Mj/kg) and the coal is nonagglomerating. **D 121, D05**

ranking, *n*—a method in which a series of three or more samples is presented at the same time and arranged in order of intensity or degree of some designated attribute. **E 253, E18**

ranney collector—an underground water collection system sometimes called ranney wells. **D 6161, D19**

rare earths—a group of chemical elements comprising those with atomic numbers ranging from 58 through 71, and referred to as the lanthanides; so-called because of their rarity of occurrence. **F 1156, F12**

rare earth screens— see **intensifying screen**. **E 1316, E07**

raspberry—See **strawberry**. **D 1079, D08**

raster—*SIMS,* the two-dimensional pattern swept out by the deflection of a primary ion beam. **E 673, E42**

raster output scanner—a two-dimensional scanning system used for data presentation. **F 1156, F12**

raster output scanner, *n*—an output peripheral, that converts computer data into a bit mapped image, which is sent to the host for storage or a printer for output. (See **ROS**.) **F 1457, F05**

ratchet lock slider, *n*—a slider with a locking mechanism that permits the slider to slip along the chain upon application of a predetermined force so as to prevent damage that would impair either the service or use of either the slider or chain. **D 123, D13**

ratchet nail—bright, regular-stock-steel, ¾ to 2 by 0.120-in. nails with single-crest annular ratchet thread, flat ⅜-in. head and medium diamond point. **F 547, F16**

ratchet thread—type of buttress thread, usually annular for engagement with nailing channels, retaining clips, etc. **F 547, F16**

ratch-setting by number *(L1 % N), n*—the basis for setting roll spacing in the drafting zone, namely, the length exceeded by 1 % of the number of fibers in a test specimen. **D 123, D13**

rate—the quantity of pesticide applied per unit treated. **E 609, E35**

rated strength—the strength in tension of a stranded conductor calculated in accordance with specification requirements. **B 354, B01**

rate, event count (Ṅe)—the time rate of the event count. **E 1316, E07**

rate of creep, *n*—the slope of the creep-time curve at a given time. **D 907,** D14

rate of creep, *n*—the slope of the creep-time curve at a given time. **D 4439,** D35

rate of creep, *n*—the slope of the creep-time curve at a given time. **E 6,** E28

rate of growth—the rate at which a tree has grown. The unit of measure is the number of annual growth rings per inch, measured radially in the trunk or in lumber cut from the trunk. **D 9,** D07

rate-of-oil-flow—the rate at which a specified oil will pass through a sintered porous compact under specified test conditions. **B 243,** B09

rate of return—the percentage yield on an investment per unit time. **E 833,** E06

rate of rise—in leak testing, the time rate of pressure increase at a given time in a vacuum system which is suddenly isolated from the pump by a valve. The volume and temperature of the system are held constant during the rate of rise measurement. See **isolation test.** **E 1316,** E07

rate of roughness—sum of the roughness divided by longitudinal distance covered by the blanking band. **E 867,** E17

rating, *n*—a quantitative or qualitative scale for evaluation of a specific property. **D 123,** D13

rating curve, sediment—a graph of the relationship between stream discharge and sediment discharge at a stream cross section. The graph is sometimes called a *sediment transport curve*. **D 4410,** D19

rating process, *n* (processus d'évaluation)—the process of determining the serviceability of a facility for a specified purpose. **E 1480,** E06

rating scale (échelle d'évaluation)—(*for an* ***aspect of facility serviceability***) set of descriptions of combinations of features, in which each combination has been selected to indicate a specific level of serviceability on a scale from the lowest to the highest level likely to be encountered. **E 1480,** E06

rating score (résultat d'évaluation)—result (expressed as a number) of finding the combination of features described in a specified rating scale (for one aspect of serviceability) that matches most closely the attributes present in a facility. **E 1480,** E06

ratioed spectrum, *n*—the calculated ratio of two single-beam spectra, one of which is a background spectrum. **E 131,** E13

rationalized erosion rate, *n*—*in liquid impingement erosion*, an erosion rate for impingement tests expressed in dimensionless form as follows: the volume of material lost per unit volume of (liquid or solid) particles impinging, both determined for the same area. **G 40,** G02

ratio scaling, *n*—scaling method in which values are assigned to the intensity of attributes in proportion to the assessor's perception of the intensity of the attributes with reference to a selected standard. **E 253,** E18

raveled strip test, *n*—*in fabric testing*, a strip test in which the specimen is cut wider than the specified testing width and an approximately even number of yarns are removed from each side to obtain the required testing width. **D 123,** D13

raveling, *n*—loss of pavement surface material involving the dislodging of aggregate particles and degradation of the bituminous binder. **E 867,** E17

raveling, *n*—loss of pavement surface material involving the dislodging of aggregate particles and degradation of the bituminous binder. **E 1778,** E17

raw batch—a glass charge without cullet. **C 162,** C14

raw coal—*for the purpose of Test Method D 4749*, any coal, regardless of its topsize, that has not been manually or mechanically cleaned. Crushed coal that has not been mechanically cleaned (including coal that has not been through a breaker which normally rejects oversize) is considered to be raw coal. Coal delivered to the surface from an underground mine is considered to be raw coal even when crushing and grinding is done underground. Coal removed from the pit of a surface mine is considered to be raw coal even when breaking and crushing facilities are provided *in the pit*. **D 121,** D05

raw cotton, *n*—ginned lint that has not been subjected to any textile manufacturing process. (See also **ginned lint**.) **D 123,** D13

raw cullet—a glass charge made totally of cullet. **C 162,** C14

raw glaze—See *raw glaze* under **glaze.** **C 242,** C21

rawhide—cattlehide that has been dehaired, limed, often stuffed with oil or grease, and has sometimes undergone other preparation, but has not been tanned. It is used principally for mechanical purposes, such as belt lacings, loom pickers, gaskets, pinions, gears, and for hand luggage, shoe laces, snowshoes, etc. **D 1517,** D31

raw petroleum coke, *n*—petroleum coke that has not been calcined. **D 4175,** D02

raw refractory dolomite, *n*— see **dolomite, raw refractory.** **C 71,** C08

raw streak—an undertanned center layer of leather, visible in cross section as a light-colored streak, especially as applied to heavy leather. **D 1517,** D31

raw water—water which has not been treated. Untreated water from wells, surface sources, the sea or public water supplies. **D 6161,** D19

raw wool, *n*—wool or hair of the sheep in the grease, pulled, or scoured state. **D 123,** D13

Rayleigh wave—an ultrasonic surface wave in which the particle motion is elliptical and the effective penetration is approximately one wavelength. **E 1316,** E07

Rayleigh wave, v_R **(LT^{-1})**—dispersive surface wave in which element has retrograding elliptic orbit with one major vertical and one minor horizontal component both in plane of propagation velocity:

$$v_R = \alpha v_t \text{ with } 0.910 < \alpha < 0.995 \text{ for } 0.25 < \nu < 0.5$$

D 653, D18

rayon pulps—pulps used in the manufacture of rayon, and subject to various specifications by the manufacturers, including those of purity, moisture content, sheet properties, and viscosity. **D 1695,** D01

RCRA—Resource Conservation and Recovery Act of 1976. **E 1605,** E06

***R*-curve**—a plot of crack-extension resistance as a function of stable crack extension, Δa_p or Δa_e. **E 1823,** E08

R_d and b, *n*—*for the purpose of this test method for color*, the daylight color of opaque cotton specimens represented by points in a space as described by Hunter in terms of three color scales: reflectance, R_d, and the chromaticity coordinates for redness or greenness, + *a* and yellowness or blueness, + *b*. **D 123,** D13

reach—the section of a sewer between structures. **C 896,** C04

reactant—*in grouting*, a material that reacts chemically with the base component of grout system. **D 653,** D18

reaction—any change in chemical composition accompanied by a change of enthalpy. **E 1142,** E37

reaction, *n*—any transformation of material accompanied by a change of enthalpy which may be endothermic or exothermic. **E 1445,** E27

reaction effect—the personnel injury, facility damage, product loss, downtime, or mission loss that could occur as the result of an ignition. **G 126,** G04

reaction isotherm—in a temperature-concentration phase diagram, a tie-element at constant temperature, representing univariant equilibrium among three or more phases. **E 7,** E04

reaction isotherm—a temperature-concentration phase diagram, a tie-element at constant temperature representing univariant equilibrium among three or more phases, (E 7, E-4). **E 1142,** E37

reaction rate—the number of a given reaction that occur per target nucleus per unit time. When appropriate this term may be interpreted as a probability of reaction per unit time. **E 170,** E10

reactivation (revivification)—oxidation processes for restoring the adsorptive properties of a spent sorbent. **D 2652,** D28

reactive aggregate—an aggregate containing siliceous material (usually in amorphous or crypto-crystalline state) which can react chemically with free alkali in the cement. **D 653,** D18

reactive monomer, *n—in solventless electrical varnish,* a substance that, when added to a resin, will combine chemically with that resin under specified conditions. **D 1711,** D09

reactive particulates, *n*—a particle or particles present in a clay body, which when near the surface may flake off or cause an eruption (pop-outs) of the surface when exposed to the weather. **C 43,** C15

reactivity—rate at which another material will form compounds with carbon or graphite. **C 709,** D02

reactivity—*in the nuclear industry,* a parameter giving the deviation from criticality of a nuclear chain-reacting medium such that positive values correspond to increasing reaction intensity and negative values correspond to decreasing intensity. **C 859,** C26

reactivity—the ability to react. For proper use of the term, the reaction in question and the conditions should be stated and the parameter used in measuring reactivity indicated, such as rate, uniformity, or the like. **D 1695,** D01

read area—one of several terms used to refer to the scan path or scan area. **F 149,** F05

read area, *n*—area covered by a scanner. Bar codes must reliably pass through the read area with the length of the symbol parallel to the scan plane. **F 1294,** F05

reader, *n*—a device used for machine reading of bar codes that usually consists of a scanner, a decoder, and a data communications interface. **F 1294,** F05

reading, *n*—a numerical value obtained from a digital display or indicated on a scale or dial of an apparatus or instrument. **E 135,** E01

reading, *n*—numerical value obtained from a digital display or indicated on a scale or dial of an apparatus or instrument. **E 1605,** E06

read ink—See **scan ink**. **F 149,** F05

read only memory, *n*—a form of electronic memory from which the system can extract information, but which cannot be altered or added to. **F 1457,** F05

readout instrumentation—See **analyzer.** **D 1356,** D22

ready-mixed plaster, *n*—a calcined gypsum plaster with aggregate added during manufacture **C 11,** C11

ready temperature—a predetermined starting temperature point for the washing energy test, which may vary with each dishwasher. **F 1827,** F26

reagent—a chemical used to react with another chemical, often to confirm or deny the presence of the second chemical. **F 1156,** F12

reagent blank—the contribution to the analytical result caused by substances in the reagents (or apparatus) as determined by carrying through all of the operations of the analysis with the same quantities of reagents that are used in the actual analysis of a material but without addition of sample. **C 859,** C26

reagent blank—a reference standard or correction factor obtained by subjecting one or more reagents to test conditions. **D 4790,** D16

reagent grade—See **analytical reagent grade.** **C 859,** C26

real area of contact, *n— in tribology,* the sum of the local areas of contact between two solid surfaces, formed by contacting asperities, that transmit the interfacial force between the two surfaces. (Contrast with **apparent area of contact**.) **G 40,** G02

real discount rate—the rate of interest reflecting that portion of the time value of money related to the real earning power of money over time. **E 833,** E06

real dollars—See **constant dollars**. **E 833,** E06

real image—a reproduction of an object by an optical system which gathers light from an object point and transforms it into a beam that converges toward another point. **E 1316,** E07

real-time radioscopy—radioscopy that is capable of following the motion of the object without limitation of time. **E 1316,** E07

real-time radioscopy—radioscopy that is capable of following the motion of the object without limitation of time. **E 1316,** E07

ream—in flat glass, layers of glass that are not homogeneous with the main body of the glass. **C 162,** C14

reamed diameter—the diameter of the hole cut by the reamer. **F 1611,** F04

rearfoot stability—the stability of the heel seat. (See **stability**) **F 869,** F08

rearm delay time—see **time, rearm delay.** **E 1316,** E07

reboil—reappearance of bubbles in molten glass after it previously appeared plain. **C 162,** C14

reboiling—gas evolution occurring and recurring during repeated firing of the ground coat; sometimes a defect. **C 286,** B08

rebonded fused grain refractory, *n*—a fired refractory brick or shape made predominantly or entirely from fused grain. **C 71,** C08

rebuild—See **building modification.** **E 631,** E06

rebuilt (rebuild)—to make extensive structural repairs or substantial component exchange, or both, so as to bring a device back into conformance with the description and performance specifications developed and described by the **OEM**. **F 1156,** F12

recalcitrant—a substance that is resistant to microbial degradation. **F 1600,** F20

recalescence—the increase in temperature which occurs after undercooling because the rate of liberation of heat during transformation of a material exceeds the rate of dissipation of heat. **E 7,** E04

receiver, *n*—the portion of a photometric instrument that receives the viewing beam from the specimen, including a collector such as an integrating sphere, if used, often the monochromator or spectral filters, the detector, and associated optics and electronics. **E 284,** E12

receiver—in solar energy systems, that part of the solar collector to which the solar irradiance is finally directed or redirected, and includes the absorber and any associated glazings through which the redirected energy must pass. **E 772,** E44

receiver plane, *n*—the plane containing the specimen normal and the axis of the receiver. **E 284,** E12

receiving room—*in architectural acoustical measurements,* the room in which the sound transmitted from the source room is measured. **C 634,** E33

receptor, *n*—a cellular structure mediating the physiological response to the presence of physical or chemical agents. **E 253,** E18

receptor, n—see **receiver**. **E 284,** E12

receptor—the component of the carbonless papers which is chemically coated or treated to develop or receive the colorless dyes or pigments respectively. **F 549,** F05

receptor, *n*—the material which receives on its surface an image from a thermal printing process. **F 1623,** F05

recipe, *n*—a formula, mixing procedure, and any other instructions needed for the preparation of a product. **D 1566,** D11

reciprocal lattice—a lattice of points each of which represents a set of planes in the crystal lattice, such that a vector from the origin of the reciprocal lattice to any point is normal to the crystal planes represented by that point and has a length which is the reciprocal of the plane spacing. **E 7,** E04

reciprocal linear dispersion—*in spectrochemical analysis*, the derivative $d\lambda/dx$ where λ is the wavelength and x is the distance along the spectrum. Also see **linear dispersion.** **C 859,** C26

reciprocal linear dispersion, *n*—the derivative $d\lambda/dx$ where λ is the wavelength and x is the distance along the spectrum. **E 135,** E01

reciprocity—bilateral relationship where both parties have the same rights and obligations towards each other, (ISO Guide 2). **E 1187,** E36

reciprocity law, *n*—the statement that in a photochemical reaction a constant effect is produced if the product of time and radiant power is a constant. **E 135,** E01

recirculating dip tank—a dip tank provided with a means for keeping the slip in constant circulation. **C 286,** B08

recirculating nozzle—see **by-pass nozzle**. **E 1620,** E29

reclaim—overspray that is removed from the spray booth and reconditioned for use. **C 286,** B08

reclaim, n— See **reclaimed rubber**, the preferred term. **D 1566,** D11

reclaimed asphalt pavement (RAP), *n*—asphalt pavement or paving mixture removed from its original location for use in recycled asphalt paving mixture. **D 8,** D04

reclaimed, rosin, *n*—rosin that has been recovered or reclaimed by any means from waste or deteriorated material, provided that the concentration of rosin acids is not below that normal for rosin, and any residual or contaminating component from the waste material itself or from any article used in the recovery process is not in sufficient quantity to cause the physical or chemical properties of the reclaimed product to differ materially from those of rosin. **D 804,** D01

reclaimed rubber, *n*—vulcanized rubber that has been thermally, mechanically, and/or chemically plasticized for use as a rubber diluent, extender, or processing aid. **D 1566,** D11

reclaiming—the removal of contaminants and products of degradation such as polar, acidic, or colloidal materials from used electrical insulating liquids by chemical or adsorbent means. **D 2864,** D27

reclaiming, *n*—the use of cleaning methods during recycling primarily to remove insoluble contaminants, thus making the oil suitable for further use. The methods may include settling, heating, dehydration, filtration, and centrifuging. **D 4175,** D02

recoatability—the application characteristics of a polish and the appearance of the film after successive coatings to a surface. **D 2825,** D21

recognition agreement—agreement that is based on the acceptance by one party of results, presented by another party, from implementation of one or more designated functional elements of a certification system, (ISO Guide 2). **E 1187,** E36

recognition algorithms—computer programs or instruction sets for the recognition of specific phenomena from a processing of data acquired for the system from some external source. **F 1156,** F12

recognition threshold, *n*—the lowest physical intensity at which a stimulus is correctly identified a specified percent of the time. **E 253,** E18

recoil implantation or knock-on—the injection, due to collisions caused by incident particles, of surface or near surface atoms into the bulk along the path of the incident beam. **E 673,** E42

recommendation, *n*—in building constructions, a written suggestion for policy, practice, conduct, design, or material, implying endorsement but not requiring compliance (see **guideline**). **E 631,** E06

reconciliation—a comparison of two or more estimate values by independent cost professionals for the purpose of reaching consistency in the estimate assumptions, parameters, and scope of work. **E 833,** E06

reconditioning—the removal of insoluble contaminants, moisture, and dissolved gases from used electrical insulating liquids by mechanical means. **D 2864,** D27

reconstruct—See **building modification.** **E 631,** E06

reconstruction—See **building modification.** **E 631,** E06

recording media—material capable of capturing or storing, or both, a radiological image in digital or analog form. **E 1316,** E07

recording medium—a film or detector that converts radiation into a visible image. **E 1316,** E07

recording medium—a light-sensitive material which detects the interference between the object beam and the reference beam. Typical recording media used in holography are silver halide film, thermoplastic film and electronic detectors, such as video tubes and CCD arrays. **E 1316,** E07

record set drawing— See **drawing**. **E 1480,** E06

recoverable elongation (*CE*), *n*—*of rope*, elongation which may be reclaimed after a period of relaxation after the rope was cyclic tensioned. **D 123,** D13

recovered material, *n*—materials and byproducts that have been separated, diverted, or removed from the solid waste stream, but not including those materials and byproducts generated from and reused within an original manufacturing process. **D 1968,** D06

recovered paper material, *n*—paper materials that have been separated, diverted, or removed from the solid waste stream excluding the virgin content of mill broke, for the pupose of use, reuse, or recycling, whether or not such materials require subsequent separation and processing. **D 1968,** D06

recovering—the process of covering an existing roofing system with a new roofing system. **D 1079,** D08

recovery, *n*—the percent of the valuable component (that is, Btu or combustible) from the feed that reports to the froth concentrate product. **D 121,** D05

recovery, *n*—the degree to which a rubber product returns to its normal dimensions after being distorted. **D 1566,** D11

recovery, percent, *n*—the amount of a material actually recovered by an assay using a prescribed procedure, or obtained from a process, as a percentage of the as-received material. **D 5681,** D34

recovery time—the time required for a test system to return to its original state after it has received a signal. **E 1316,** E07

recovery time—the time from the removal of the food product from the cooking equipment until the cooking medium, cavity, or surface (see **cooking medium, cooking cavity,** or **cooking surface**) is back up to temperature as indicated by thermostat(s) controlling the cooking medium, cavity, or surface cycling at their set point(s), $t_{recovery}$. **F 1827,** F26

recovery—Y (conversion)—the ratio of product quantity (permeate stream flow rate) over the feed quantity (feed stream flow rate), given as fraction or in percent. **D 6161,** D19

recreational vehicle, *n*—a vehicle with light or medium duty chassis and suspension designed for recreational living or hauling. **E 867,** E17

recrystallization, *n*—the formation of a new grain structure through a nucleation and growth process. **A 941,** A01

recrystallization—the formation of a new grain structure through nucleation and growth commonly produced by subjecting a metal, that may be strained, to suitable conditions of time and temperature. **E 7,** E04

recrystallization annealing, *n*—**annealing** a cold-worked steel object to produce a new grain structure without a change in phase. **A 941,** A01

recrystallization temperature, *n*—the approximate minimum temperature at which recrystallization of a cold-worked steel object occurs within a specified time. **A 941,** A01

recrystallized grain size—(*1*) the grain by heating following cold work where the time and temperature are so chosen that, while recrystallization is complete, essentially no grain growth has occurred.

(*2*) in aluminum and magnesium alloys, the grain size after recrystallization, without regard to grain growth or the recrystallization conditions. **E 7,** E04

recrystallized limestone—a limestone in which a new pattern of crystallinity has pervasively replaced the crystal orientation in the original clastic particles, fossils or fossil fragments, and interstitial cement. The new generation of crystals, encompassing both fragmental and matrix materials, extends across boundaries between former crystals. The new crystals generally are larger than those of the original rock. Evidence of original textures may or may not be retained. See also **marble** (next section). **C 119,** C18

rectification—the conversion of alternating into direct current. **B 374,** B08

rectified tar oil, *n*—the volatile oil that is rectified by steam distillation of pine tar, obtained by destructive distillation of various species of *Pinus*. **D 804,** D01

rectifier—a device that converts alternating into direct current by virtue of a characteristic permitting appreciable flow of current in only one direction. **B 374,** B08

recuperative furnace—a melting furnace having a recuperator. **C 162,** C14

recuperator—a continuous heat exchanger in which heat from exhaust gases is conducted through flue walls to incoming air. **C 162,** C14

recyclable, *adj*—capable of being removed, separated, or diverted from the solid waste stream in an available program, established by, but not limited to, manufacturers, retailers, or municipalities; processed and returned to use in the form of raw materials or products. **D 5488,** D10

recycle, *v*—a multiphased processing that includes removal, separation, and/or diversion, of materials from the solid waste stream; use of such materials as raw materials for the manufacture of new products; and the use of the new product. **D 1968,** D06

recycled, *adj*—diverted, separated, or removed from the solid waste stream, processed and returned to use in the form of raw materials or products. **D 5488,** D10

recycled asphalt paving mixture, *n*—a mixture of reclaimed asphalt pavement with the inclusion, if required, of asphalt cement, emulsified asphalt, cut-back asphalt, recycling agent, mineral aggregate, and mineral filler. **D 8,** D04

recycled content, *n*—the percent of a packaging material that is composed of recycled material. **D 5488,** D10

recycled content paper, *n*—a paper product containing recycled fiber expressed as a percentage of total fiber weight, or a paper product containing recovered material (excluding wood residues and sawdust) expressed as a percentage of total product weight. **D 1968,** D06

recycled engine coolant, *n*—engine coolant formulated using other than virgin materials. **D 4725,** D15

recycled fiber, *n*— *in paper*, fiber derived from recovered material, excluding wood residues and sawmilling residues, which has been repulped or reintroduced into the paper manufacturing process and made into a product or form usable in the manufacture of a product. **D 1968,** D06

recycled fiber content, *n*—the percentage of recycled fiber, by total fiber weight, of a paper product. **D 1968,** D06

recycled paper, *n*—a paper product whose fiber content consists totally of recycled fiber. **D 1968,** D06

recycled plastic, *n*—those plastics composed of post-consumer material or recovered material only, or both, that may or may not have been subject to additional processing steps of the types used to make products such as recycled-regrind or reprocessed or reconstituted plastics. **D 883,** D20

recycled wool, *n*—*as defined in the Wool Products Labeling Act as amended in 1980*, "the resulting fiber when wool has been woven or felted into a wool product which, without ever having been utilized in any way by the ultimate consumer, subsequently has been made into a fibrous state, or the resulting fiber when wool or reprocessed wool has been spun, woven, knitted, or felted into a wool product which, after having been used in any way by the ultimate consumer, subsequently has been made into a fibrous state." **D 123,** D13

recycling, *n*—the processing of oil that has become unsuitable for its intended use, in order to regain useful material. **D 4175,** D02

recycling agent (RA), *n*—a blend of hydrocarbons with or without minor amounts of other materials that is used to alter or improve the properties of the aged asphalt in a recycled asphalt paving mixture. **D 8,** D04

red edge—{archaic} numerous rouge pits located around the edges of a large sheet of polished plate glass. **C 162,** C14

redox (ORP) electrode—a metallic electrode, usually platinum, used to follow reversible oxidation-reduction reactions. The potential developed between the redox electrode and reference electrode is a Nernstian function of the ratio of a species in two different oxidation states. Redox (ORP) electrodes are primarily used to determine a species in one oxidation state by titration with a reagent that oxidizes or reduces the species being determined. **D 4127,** D19

redox (ORP) potential—the potential developed by a metallic electrode when placed in a solution containing a species in two different oxidation states. Some sensing electrodes, notably the early silver billet electrodes, act as redox (ORP) electrodes as well as sensing electrodes in certain types of solutions, leading to measurement errors. **D 4127,** D19

redox (ORP) titration—a titration in which the ratio of two different oxidation states of a species is changed by the incremental addition of an oxidizing or reducing agent. A redox (ORP) electrode, which responds to the logarithm of the ratio of the activities of the two oxidation states, is used to follow the titration. Among the many substances that can be determined in this manner are H_2O_2, I_2, Fe^{+3}, MnO_4^-, and Sn^{++}. Differential titrations are possible in some cases. **D 4127,** D19

redox potential—the potential of a reversible oxidation-reduction electrode measured with respect to a reference electrode, corrected to the hydrogen electrode, in a given electrolyte. **G 15,** G01

reduced metal powder—metal powder produced, without melting, by the chemical reduction of metal oxides or other compounds. **B 243,** B09

reduced viscosity—see **viscosity, reduced.** **D 1695,** D01

reducible sulfur—any form of sulfur or sulfur compounds in paper or paperboard that can be converted to hydrogen sulfide on treatment with a metal such as aluminum and an acid under the conditions of a specified test; it is a measure of the quantity of sulfur compounds in the paper or paperboard that may react with metals to cause tarnishing. **D 1968,** D06

reducing agent—a compound that causes reduction, thereby itself becoming oxidized. **B 374,** B08

reducing agent—a batch ingredient that lowers the state of oxidation of the melt. **C 162,** C14

reducing end groups—those terminal glucose units in cellulose or its derivatives in which the 1-position is not substituted or involved in a glycosidic linkage. These end groups will reduce Fehling's solution or similar reagents. **D 1695,** D01

reducing substances—impurities in pyridine that decolorize a solution of potassium permanganate of specified composition in Method D 2031. **D 4790,** D16

reduction—a reaction in which electrons are added to a reactant. More specifically, the addition of hydrogen or the abstraction of oxygen. Such a reaction takes place, for example, at the cathode in electrolysis. **B 374,** B08

reduction—the gain of electrons by a constituent of a chemical reaction. **G 15,** G01

reduction of area, *n*—the difference between the original cross-sectional area of a tension test specimen and the area of its smallest cross section. The reduction of area is usually expressed as a percentage of the original cross-sectional area of the specimen. **E 6,** E28

reed mark, *n—in woven fabrics*, a crack between groups of warp ends, either continuous or at intervals. **D 123,** D13

reel or spool—a cylindrical device that has a rim at each end and an axial hole for a shaft or spindle, and on which the product is wound to facilitate handling and shipping. **B 846,** B05

reentrant corner—an inside corner of a surface, producing stress concentrations in the roofing or waterproofing membrane. **D 1079,** D08

reentry—purchase of goods from a domestic manufacturer, allegedly for overseas destinations followed by return to this country for (re)sale. **F 1156,** F12

referee sample, *n— in asbestos,* a sample on which acceptability is based. Acceptability shall be based on a composite sample if the lot is 200 bags or less, or on a master composite sample if over 200 bags, but not in excess of 2000 bags. **D 2946,** C17

referee test—a test made to settle a disagreement as to conformance to specified requirements. **F 412,** F17

referee tests—tests performed by an umpire or third party appointed to investigate and report the facts and to frequently make recommendations. See **third party.** **C 859,** C26

reference beam—laser radiation impinging directly upon the recording medium through optical components and which typically does not contain information about the test object. In some tests, the reference beam may be reflected or scattered from a portion of the object surface. In this case, any object information contained in the reference beam is cancelled in the object beam by the interference between the object beam and the reference beam. **E 1316,** E07

reference beam angle—the angle formed between the center line of the reference beam and the normal to the recording medium. **E 1316,** E07

reference block—a block that is used both as a measurement scale and as a means of providing an ultrasonic reflection of known characteristics. **E 1316,** E07

reference coil—a coil or probe, which may be used in conjunction with the appropriate material, to electrically balance a comparative system. **E 1316,** E07

reference compound (NMR)—a selected material to whose signal the spectrum of a sample may be referred for the measurement of chemical shift (see also **chemical shift**). **E 131,** E13

referenced standards—those which contain guidelines or nondated requirements germane to one or more elements of the host standard. **F 1789,** F16

reference edge—the edge of the form used to align the form so that the nominal reading line will be parallel to the direction of scanning. Depending on the equipment used, this may be any edge of the form. **F 149,** F05

reference electrode—that half of the electrode pair which provides a constant potential regardless of solution composition. The potential developed by the sensing electrode is measured against this reference potential to give an overall system potential which can be converted to the level of the species sensed. **D 4127,** D19

reference electrode—the electrode (for example, hydrogen electrode, normal calomel electrode, or saturated calomel electrode) against which the electrical potential of a specimen is measured. **E 1823,** E08

reference electrode—electrode having a stable and reproducible potential, which is used in the measurement of other electrode potentials. **G 15,** G01

reference electrode filling solution—a concentrated salt solution contacting the internal reference element and the sample solution. The composition of the filling solution is chosen to maximize stability of the potentials developed at the internal reference element/filling solution interface and the filling solution/sample junction. In general, filling solutions for AgCl internal construction reference electrodes should: (*1*) contain Cl^- and be saturated with AgCl to prevent the reference element from dissolving; (*2*) be at least ten times higher in total ionic strength than the sample; (*3*) be equitransferent; (*4*) not contain the ion being measured or an ion that interferes with the measurement. **D 4127,** D19

reference junction—that junction of a thermocouple which is held at a known temperature. **E 7,** E04

reference junction, *n*—that junction of a thermocouple which is at a known temperature. **E 344,** E20

reference junction—that junction of a thermocouple which is maintained at a known temperature. **E 1142,** E37

reference junction correction—a correction in terms of electromotive force (millivolts) to be applied to the electromotive force generated by a thermocouple to compensate for the difference between the actual temperature of the reference junction and that used as the basic reference junction temperature in standard conversion tables. **E 7,** E04

reference junction correction—correction in the terms of electromotive force to be applied to the electromotive force generated by a thermocouple to compensate for the difference between the actual temperature of the reference junction and that used as the basic reference junction temperature in the standard conversion tables, (E 7, E-4). **E 1142,** E37

reference load [F]—*for spectrum loading*, used in Practices E 1049 to denote the loading level that represents a steady-state condition upon which load variations are superimposed. The reference load may be identical to the mean load of the history, but this is not required. **E 1823,** E08

reference material—a material or substance one or more properties of which are sufficiently well established to be used for the calibration of an apparatus, the assessment of a measurement method, or assigning values to materials. (ISO definition: ISO Guide 30, *Terms and Definitions Used in Connection with Reference Materials.*)[2] **C 859,** C26

reference material—a material or substance, one or more properties of which are sufficiently well established to be used for the assessment of a measurement method or for assigning values to materials. **E 7,** E04

reference material—a material or substance one or more properties of which are sufficiently well established to be used for the calibration of an apparatus, the assessment of a measurement method, or for assigning values to materials (ISO Guide 30–1981 (E)). **E 131,** E13

reference material—material or substance, one or more properties of which are sufficiently well established to be used for the calibration of an apparatus, the assessment of a measurement method, or for assigning values to materials, (ISO Guide 2, *A*). **E 1187,** E36

reference material—material of definite composition that closely resembles in chemical and physical nature the material with which the analyst expects to deal, and that is employed for calibration or standardization. **E 1605,** E06

reference material, *n*—a material with known performance. **G 113,** G03

reference material (RM), *n*—material or substance one or more of whose property values are sufficiently homogeneous and well established to be used for the calibration of an apparatus, the assessment of a measurement method, or for assigning values to materials. **E 135,** E01

reference neutron field—see **benchmark neutron field.** **E 170,** E10

reference sample—a matrix whose analytes of interest are of known or accepted concentration or property. **D 1129,** D19

reference sample, *n*—a sample designated as the one to which all others are to be compared. **E 253,** E18

reference sediment—a whole sediment near an area of concern used to assess sediment conditions exclusive of material(s) of interest. **E 943,** E47

reference specimen, *n*—a portion of the reference material that is to be exposed. **G 113,** G03

reference spectrum, *n*—an established sample spectrum. **E 131,** E13

reference standard, *n*—*in cotton testing,* a homogeneous lot of cotton having a known or accepted value for one or more physical properties. **D 123,** D13

reference standard—primary or secondary standards used to calibrate testing apparatus and methods. **D 4790,** D16

reference standard—a material or device whose properties are determined by comparison to another standard, such as a certified reference material. **E 7,** E04

reference standard, *n*—a physical standard used to calibrate a group of laboratory standards. **E 284,** E12

reference standard—a tube, plate or part with artificial discontinuities used for establishing the test sensitivity setting and for periodically checking and adjusting sensitivity setting as required. (See also **standard (1)**). (E 215) **E 1316,** E07

reference-standard dosimeter—a dosimeter of high metrological quality, used as a standard to provide measurements traceable to and consistent with measurements made using primary-standard dosimeters. **E 170,** E10

reference temperature, *n*—a fixed, reproducible temperature, to which a value is assigned, that can be used for the calibration of thermometers or other purposes. **E 344,** E20

reference tension, *n*—a low tensile force, generally about 1 % of the rope breaking strength, and used for initial rope tension determination. **D 123,** D13

reference test fluid, *n*—a standard glycol-base test fluid, described in Specification D 3585 Reference Fluid for Coolant Tests, used by laboratories to evaluate test methods and procedures. **D 4725,** D15

reference tire, *n*—a special tire included in a test program; the test results for this tire have significance as a base value or internal benchmark. **F 538,** F09

reference viscosity, *n*—the viscosity of Newtonian standard reference fluids certified at each of several temperatures by the supplier. **D 4175,** D02

refillable, *adj*—intended to be refilled for its original purpose one or more times in an available program established by, but not limited to, manufacturers, distributors, or retailers. **D 5488,** D10

refined—treated to reduce impurities. **D 4790,** D16

refined tar, *n*—tar freed from water by evaporation or distillation which is continued until the residue is of desired consistency; or a product produced by fluxing tar residuum with tar distillate. **D 8,** D04

refiner—a component of a melting furnace, for the purpose of conditioning the glass. See **nose.** **C 162,** C14

refiner, *n*—a two-roll mill with a high friction ratio, used for such operations as processing of reclaimed rubber or slightly scorched mixes and the crushing of impurities. **D 1566,** D11

refining—See **fining.** **C 162,** C14

reflectance, *n*—the fraction of the incident radiation upon a surface that is reflected from the surface. **C 168,** C16

reflectance—the fraction of incident light that is diffusely reflected, measured relative to magnesium oxide under standard conditions. **C 286,** B08

reflectance, *n*— *in optical measurement of paper,* the ratio of the reflected radiant of luminous flux to the incident flux, measured under specified conditions, expressed as a percentage. **D 1968,** D06

reflectance, ρ, *n*—ratio of the reflected radiant or luminous flux to the incident flux in the given conditions. **E 284,** E12

reflectance, *n*—ratio of the reflected radiant or luminous flux to the incident flux. Symbol: ρ_e, ρ_v, ρ; $\pi = \rho_r + \rho_d$.

NOTE—When mixed reflection occurs, the (total) reflectance may be divided into two parts, regular (ρ_r) and diffuse reflectance (ρ_d), corresponding, respectively, to the two modes of reflection referred to above.

In general, the values of the various reflectances depend upon the mode of irradiation, the spectral composition and state of polarization of the incident radiation. **E 349,** E21

reflectance—the ratio of the radiant flux reflected from a surface to that incident upon it. **E 1316,** E07

reflectance—the ratio of the response of a light sensor illuminated by diffuse reflection from the paper compared to that when the paper is replaced by a perfect diffuse reflector. A specially prepared surface of *barium sulfate* is considered to be a perfect diffuse reflector. **F 149,** F05

reflectance, *n*—the percentage of incident light reflected from an image area, where zero percent reflectance is black. (See **image density.**) **F 1457,** F05

reflectance, absolute—the ratio of the total reflectance by a document to the total light incident on the document. **F 149,** F05

reflectance, absolute, *n*—the ratio of the total reflectance from a document to the total light incident on that document. **F 1294,** F05

reflectance density, D_ρ, *n*—the negative logarithm to base ten of the reflectance. **E 284,** E12

reflectance, diffuse—reflected light whose angle of reflection varies from the angle of incidence of the illuminating light, such as reflection from a rough surface. **F 149,** F05

reflectance, diffuse, *n*—reflected light whose angle of reflection varies from the angle of incidence of the illuminating light, such as in reflection from a rough surface. **F 1294,** F05

reflectance, diffuse, or Rd, *n*—light energy diffusely reflected by a substance. As opposed to specular or directional reflectance, diffuse reflectance emanates in all directions, thus causing no gloss. **E 253,** E18

reflectance, directional, *n*—light energy reflected from a surface at a specified, characteristic angle at greater intensity than at other angles. **E 253,** E18

reflectance factor, *n*—ratio of the flux reflected from the specimen to the flux reflected from the perfect reflecting diffuser under the same geometric and spectral conditions of measurement. **E 284,** E12

reflectance factor, *n*—*at a point on a surface, for the part of the reflected radiation contained in a given cone with apex at the point of the surface, and for incident radiation of given spectral composition and geometric distribution,* ratio of the radiant flux reflected in the directions delimited by the cone to that reflected in the same directions by a perfect reflecting diffuser identically irradiated.

NOTE—For specularly reflecting surfaces that are irradiated by a source of small solid angle, the reflectance factor may be much larger than unity if the cone includes the mirror image of the source.

NOTE—If the solid angle of the cone approaches zero, or 2π sr, the reflectance factor approaches radiance factor or reflectance, respectively. In instruments called "reflectance spectrophotometers" the geometrical distribution is, in general, intermediate between these two extreme cases. The readings of these instruments, corrected for photometric-scale errors, wavelength-scale errors, and for deviations of the reflecting standard used from a perfect reflecting diffuser, are spectral reflectance factors. For a given sample, these values depend on the geometrical characteristics of the instrument.

NOTE—The term "directional reflectance" is used currently in the United States in this sense. **E 349,** E21

reflectance, luminous, *n*—ratio of luminous flux reflected by a specimen to that incident on it. **D 2946,** C17

reflectance, ρ— the ratio of the reflected flux to the incident flux. See **radiometric properties and quantities.** **E 772,** E44

reflectance, ***R***— the ratio of the radiant power reflected by the sample to the radiant power incident on the sample. **E 131,** E13

reflectance, specular—reflected light whose angle of reflection is equal, or nearly equal, to the angle of incidence of the illuminating light, such as in reflectance from a mirror. **F 149,** F05

reflectance, specular, *n*—reflected light whose angle of reflection is equal, or nearly equal, to the angle of incidence of the illuminating light, such as in reflection from a mirror. **F 1294,** F05

reflected (or refracted) wave—components of wave incident upon second medium and reflected into first medium (or refracted) into second medium. **D 653,** D18

reflected temperature—the temperature of the energy incident upon and reflected from the measurement surface of a specimen. **E 1316,** E07

reflection, *n*—processes by which incident flux leaves a surface from the incident side. **D 2946,** C17

reflection, *n*—*of radiant energy,* the process by which radiant energy is returned from a material or object. See also **diffuse reflection, mixed reflection, specular reflection.** **E 284,** E12

reflection, *n*—return of radiation by a surface without change of frequency of the monochromatic components of which the radiation is composed. **E 349,** E21

reflection—the process by which incident flux is returned from the surface of incidence of a stationary sample, without change in frequency. **E 772,** E44

reflection—see **echo.** **E 1316,** E07

reflection—the process by which the incident flux leaves a surface or medium from incident side, without a change in frequency, but may change polarity. Reflection is usually a combination of specular and diffuse reflection. **E 1316,** E07

reflection and refraction loss—that part of transmitted energy lost due to nonuniformity of mediums. **D 653,** D18

reflection angle—**angle of reflection.** **E 772,** E44

reflection cracking at joints, *n*—cracks in bituminous overlay surfaces that occur over concrete pavements at joints. **E 867,** E17

reflection cracking at joints, *n*— cracks in bituminous overlay surfaces that occur over concrete pavements at joints. **E 1778,** E17

reflection density, D_R, **n**—the negative logarithm to base ten of the reflectance factor. **E 284,** E12

reflection, diffuse, *n*—process by which incident flux is distributed by reflection over a wide range of angles. **D 2946,** C17

reflection method—the technique of producing a diffraction pattern by X-rays or electrons which have been reflected from a specimen surface. **E 7,** E04

reflection (optical) density, *n*—logarithm to the base 10 of the reciprocal of the reflectance. Symbol: D, $D = -\log_{10}\rho$. **E 349,** E21

reflection (X-ray)—See **diffraction.** **E 7,** E04

reflective ink—ink not sensed by the optical scanner, but visible to the human eye. Syn. *blind ink, drop-out colors, drop-out ink, nonread ink, nonscan ink.* **F 149,** F05

reflective insulation, *n*—insulation depending for its performance upon reduction of radiant heat transfer across air spaces by use of one or more surfaces of high reflectance and low emittance. **C 168,** C16

reflectivity—the reflectance of a coating so thick that additional thickness does not change the reflectance. **C 286,** B08

reflectivity, *n*—reflectance of a layer of material of such a thickness that there is no change of reflectance with increased thickness. Symbol: $\rho\ \infty$. **E 349,** E21

reflectivity, ρ_∞, R_∞—the reflectance of a material represented by a specimen so thick that further increase in thickness does not significantly change the reflectance. **E 284,** E12

reflectivity, $R\ \infty$—the reflectance of a microscopically homogeneous sample with a clean optically smooth surface and of thickness sufficient to be a completely opaque.

NOTE—Reflectivity is a property of a material and reflectance is a property of a sample of the material, with no restriction on thickness or surface topography. **E 772,** E44

reflectometer, *n*—instrument for the measurement of quantities pertaining to reflection. **E 349,** E21

reflectometer—an instrument for measuring reflectance. **E 772,** E44

reflectometry, *n*—technique for measurement of reflectance or reflectance factor. **E 284,** E12

reflector, n— See **area reflector, line reflector, point reflector.** **E 284,** E12

reflector—an interface at which an ultrasonic beam encounters a change in acoustic impedance and at which at least part of the energy is reflected. **E 1316,** E07

reflowing—See **flow brightening.** **B 374,** B08

reflux, in distillation processes—reflux is the liquid condensate recycle to the top of a distillation column to aid in purification of the overhead product (ethanol). **E 1705,** E48

reflux ratio, R, *n—in a distillation column,* the ratio of the condensate at the head of the column that is returned to the column (*reflux*) to that withdrawn as product. **D 4175,** D02

refraction, *n*—the deflection from a straight path undergone by a light ray in passing obliquely from one medium (as air) into another (as glass) in which its velocity is different. **D 123,** D13

refraction, *n*—change in the direction of propagation of radiation determined by change in the velocity of propagation in passing from one medium to another. **E 349,** E21

refraction angle—see **angle of refraction.** **E 772,** E44

refractive dispersion, *n*—the difference between the refractive indexes of a substance for light of two different wavelengths, both indexes being measured at the same temperature. For convenience in calculations, the value of the difference thus obtained is usually multiplied by 10 000. **D 4175,** D02

refractive index, *n*—the ratio of the velocity of light (of specified wavelength) in air, to its velocity in the substance under examination. This is the relative index of refraction. If absolute refractive index (that is, referred to vacuum) is desired, this value should be multiplied by the factor 1.00027, the absolute refractive index of air. **D 4175,** D02

refractive index, *n*—the phase velocity of radiant power in a vacuum divided by the phase velocity of the same radiant power in a specified medium. When one medium is a vacuum, *n* is the ratio of the sine of the angle of incidence to the sine of the angle of refraction. **E 131,** E13

refractive index (electrons)—the ratio of electron wavelength in free space to its wavelength in a material medium. **E 7,** E04

refractive index (index of refraction), *n*—the ratio of the velocity of radiation (as light) in the first of two media to its velocity in the second as it passes from one into the other. **D 123,** D13

refractometer, *n*—an instrument used to indicate the freezing point of engine coolants by refractive index. **D 4725,** D15

refractories, acid, *n*—refractories containing a substantial amount of silica that may react chemically with basic refractories, basic slags, or basic fluxes at high temperatures. **C 71,** C08

refractories, basic, *n*—refractories whose major constituent is lime, magnesia, or both, and which may react chemically with acid refractories, acid slags, or acid fluxes at high temperatures. **C 71,** C08

refractories, neutral, *n*—refractories that are resistant to chemical attack by both acid and basic slags, refractories, or fluxes at high temperatures. **C 71,** C08

refractoriness, *n—in refractories,* the capability of maintaining a desired degree of chemical and physical identity at high temperatures and in the environment and conditions of use. **C 71,** C08

refractory, *adj*—resistant to high temperature. **C 71,** C08

refractory block, *n*—a generally rectangular refractory shape of a size larger than usually can be laid with one hand. All sides may not be plane and parallel. **C 71,** C08

refractory brick, *n*—a generally rectangular refractory shape with one dimension greater than the other two and of a size that usually can be laid with one hand. All sides may not be plane and parallel such as wedges, arches, keys, and circle brick, as shown in Specification C 909. **C 71,** C08

refractory composite coating—a combination of heat-resistant ceramic materials applied to a metallic substrate which may or may not require heat treatment prior to service. This term may also be used for coatings applied to nonmetallic substrates, for example, graphite. **C 286,** B08

refractory fibers, *n*—nonmetallic, inorganic, continuous, or noncontinous filaments having those chemical and physical properties that make them applicable for structures, or as components of systems, that are exposed to environments above 1000°F (538°C). **C 71,** C08

refractory lime—lime (usually of a dolomitic type) that has been extremely hard burned so that it will possess little or no tendency for conversion of the oxides to hydroxides. **C 51,** C07

refractory metal thermocouple, *n*—one whose thermoelements have melting points above 1935 °C (3515 °F). **E 344,** E20

refractory (pl. -ies), *n*—nonmetallic materials having those chemical and physical properties that make them applicable for structures, or as components of systems, that are exposed to environments above 1000°F (538°C). **C 71,** C08

refractory shape, *n*—any refractory piece forming a stable mass with specific dimensions. **C 71,** C08

refurbish, *v—as applied to textile products,* to brighten or freshen up and restore to wearability or use by cleaning such as drycleaning, laundering, or steam cleaning. **D 123,** D13

refusal—*in grouting,* when the rate of grout take is low, or zero, at a given pressure. **D 653,** D18

refuse-derived fuel (RDF-3)—refuse-derived fuel-3 (RDF-3) is defined as a shredded fuel derived from municipal solid waste (MSW) which has been processed to remove metal, glass, and other inorganic materials. This material has a particle size such that 95 weight % passes through a 2-in. square mesh screen.

NOTE—Other refuse-derived fuel may be classified as follows:
RDF-1—Wastes used in as-discarded form.
RDF-2—Wastes processed to coarse particle size with or without ferrous metal separation.
RDF-4—Combustible waste processed into powder form, 95 weight % passing 10-mesh screening.
RDF-5—Combustible waste densified (compressed) into the form of pellets, slugs, cubettes, or briquetes.
RDF-6—Combustible waste processed into liquid fuels.
RDF-7—Combustible waste processed into gaseous fuel. **E 856,** D34

refuse-derived fuel (RDF)—fuel processed from industrial waste, municipal waste, garbage, or sewage sludge. **E 1126,** E48

refuse derived fuel 3 (RDF-3)—as defined by Committee E-38 on Resource Recovery, RDF-3 is a shredded fuel derived from municipal solid waste (MSW) that has been processed to remove metal, glass, and other inorganics. The material has a particle size such that 95 % weight passes through a 2 in. square mesh screen. **E 1126,** E48

refuse-derived fuel (RDF)—fuel processed from industrial waste, municipal waste, garbage, or sewage sludge. **E 1705,** E48

refuse derived fuel 3 (RDF-3)—as defined by Committee E-38 on Resource Recovery, RDF-3 is a shredded fuel derived from municipal solid waste (MSW) that has been processed to remove metal, glass, and other inorganics. The material has a particle size such that 95 % weight passes through a 2 in. square mesh screen. **E 1705,** E48

regain—see **moisture regain.** **D 1695,** D01

regenerable fiber optic chemical sensor, *n*—an active fiber optic chemical sensor that can be used for repetitive measurements by reviving an otherwise permanently depleted or degraded transduction element by chemical or physical means. **E 131,** E13

regenerated cellulose—cellulose regenerated from a solution of cellulose or from a cellulose derivative. **D 1695,** D01

regeneration—that part of the operating cycle of an ion-exchange process in which a specific chemical solution is passed through the ion-exchange bed to prepare it for a service run. **D 1129,** D19

regeneration—distillation or elution-type processes for restoring the adsorptive properties of a spent sorbent. **D 2652,** D28

regeneration—in ion exchange systems, the process of using either an acid, alkali, or salt solution to remove the accumulated cations or anions. The cation exchange resins take on hydrogen or sodium ions and the anion exchange resins take on hydroxide ions to restore themselves to the original hydrogen or hydroxide form when using strong acid and strong alkali solutions for the process. **D 6161,** D19

regeneration level—the total weight of regenerant used per unit quantity of ion-exchange material in a single regeneration. **D 1129,** D19

regenerative furnace—a melting furnace having regenerators. **C 162,** C14

regenerator—a cyclic heat interchanger that alternately receives heat from gaseous combustion products and transfers heat to air or gas before combustion. **C 162,** C14

regimen of a stream—characteristics of a stream with respect to flow duration, form of and changes in channel, capacity to transport sediment, and amount of material supplied for transportation. **D 4410,** D19

registered quality assurance system—a system that a registration body has found to be in compliance with a designated quality system standard. **F 1789,** F16

registration—procedure by which a body indicates relevant characteristics of a product, process or service, or particulars of a body or person, in an appropriate, publicly available list, (ISO Guide 2). **E 1187,** E36

registration—an evaluation of a fastener manufacturing facility's quality assurance system by an accredited registration body resulting in a certification of full compliance with a designated quality system standard; the registration body shall be accredited by a third party registration accreditation body, for example, ANSI/RAB. **F 1789,** F16

reglet, *n*—a continuous groove, slot or recess within a building component surface which receives other components such as flashing, gaskets or anchors; a continuous prefabricated metal or plastic device containing a groove, slot or recess which can be cast into (as a form) or mounted onto a building component surface. **C 717,** C24

reglet—a groove in a wall or other surface adjoining a roof surface for the attachment of counterflashing. **D 1079,** D08

reglet gasket—See **gasket, lock-strip, reglet type**. **C 717,** C24

regression line slope—the slope of a linear regression fit of acceptable J and $\Delta\ a$ values. **E 1823,** E08

regular, *adj*—*in color determination,* used to indicate flux transmitted or reflected in the image-forming state. **D 2946,** C17

regular, *adj*—denoting flux reflected or transmitted without diffusion in accordance with the laws of optics. **E 284,** E12

regular (direct) transmission, *n*—transmission without diffusion. **E 349,** E21

regular point—denoting medium diamond point. **F 547,** F16

regular reflection, n— see the preferred term, **specular reflection.** **E 284,** E12

regular (specular) reflection, *n*—reflection without diffusion in accordance with the laws of optical reflection. **E 349,** E21

regular transmission, *n*—transmission without diffusion. **E 284,** E12

regular transmittance—ratio of the light flux transmitted without diffusion to the flux incident. **D 883,** D20

regular transmittance factor, T_r, *n*—the ratio of the flux transmitted by a specimen and evaluated by a receiver to the flux passing through the same optical system and evaluated by the receiver when the specimen is removed from the system. **E 284,** E12

regular transmittance, τ_r, *n*—ratio of undiffused transmitted flux to incident flux. **E 284,** E12

regular wear, *n*—synonym for *uniform wear.* **F 538,** F09

regulation, *n*—a rule prescribing a set of conditions and requirements that have been made mandatory for those under its control, by an executive (administrative) authority. **E 631,** E06

regulation, *n* (règlement)—rule prescribing a set of conditions and requirements that have been made mandatory for those under its control, by an executive (administrative) authority. **E 1480,** E06

rehabilitation—See **preservation.** **E 631,** E06

reheat behavior, *n*—the changes in length or volume taking place in a fired refractory when subjected to a reheat test. **C 71,** C08

reheat test, *n*—the prescribed heat treatment of a fired refractory free of externally applied stresses to determine its linear or volume stability by measurements before and after the heating. **C 71,** C08

rehydrated mashed potato dispenser—commercial equipment designed to deliver whipped or mashed potatoes. **F 1827,** F26

reinforced concrete pipe—concrete pipe designed with reinforcement as a composite structure. **C 822,** C13

reinforced gummed tape— See **tape.** **D 996,** D10

reinforced joint, *n*—a concrete joint bridged by reinforcing steel embedded in both joining parts. **C 717,** C24

reinforced membrane—a roofing or waterproofing membrane reinforced with felts, mats, fabrics, or chopped fibers. **D 1079,** D08

reinforced plastic—a plastic with high strength fillers imbedded in the composition, resulting in some mechanical properties superior to those of the base resin. (See also **filler**.) **D 883,** D20

reinforced plastic—a plastic with high-strength fillers imbedded in the composition, resulting in some mechanical properties superior to those of the base resin. (See also **filler**.) **F 412,** F17

reinforced plastic, *n*—a plastic with high strength fillers imbedded in the composition, resulting in some mechanical properties superior to those of the base resin. (See also **filler.**) **F 1251,** F04

reinforced reaction injection molding (RRIM), *n*—the process of using solid reinforcements, such as, glass fiber, mica, or talc in the reaction injection molding process. **D 883,** D20

reinforced seam, *n*—*in sewn seams,* a seam that includes an additional layer of material on the face or back side of the seam allowance. **D 123,** D13

reinforcement—steel embedded in concrete in such a manner that the two materials act together to resist forces. **C 822,** C13

reinforcement, *n*—the act of increasing the mechanical performance capability of a rubber by the incorporation of materials that do not participate significantly in the vulcanization process. **D 1566,** D11

reinforcement distortion—knotted, tangled, widely spaced, or otherwise abnormal but local irregularities in reinforcement distribution throughout the pultruded cross section.

NOTE—This condition usually causes noticeable changes in the local reinforcement content with crushing of the reinforcement or resin-richness in isolated areas. **D 3918,** D20

reinforcement-rich area—an over-concentration of reinforcement in the pultruded cross section.

NOTE—This usually occurs where mat or other reinforcement is folded, creased, or bunched in a portion of the cross section. **D 3918,** D20

reinforcing agent, *n*—a material, not basically involved in the vulcanization process, used in rubber to increase the resistance of the vulcanizate to mechanical forces. **D 1566,** D11

reinforcing cord, *n*— a cord made from industrial yarns and used to provide added support to other materials, such as tires, hose, belting, protective coverings, webbing, etc. **D 123,** D13

reinforcing resin, , *n*—a resin that raises the upper service temperature of an adhesive blend. **D 6440,** D01

reinjection/injection, *n*—the process of conveying geothermal fluids to sub-surface formations through wells. **E 957,** E44

reins, *n*—a solid or pseudosolid organic material often of high molecular weight, which exhibits a tendency to flow when subjected to stress, usually has a softening or melting range, and usually fractures conchoidally. **F 1251,** F04

reject—brine, (concentrate) stream from a desalination device. Portion of the feed stream which does not pass through the membrane. **D 6161,** D19

reject—a character located but not identified by an optical scanner. **F 149,** F05

rejected material—material that fails to meet specifications. **C 859,** C26

rejection level—the value established for a test signal above or below which test specimens are rejectable, or otherwise distinguished from the remaining specimens. **E 1316,** E07

rejection number, *n*—*in acceptance sampling*, the minimum number of nonconforming items in a sample that requires the conclusion that the lot does not conform to specification. **D 123,** D13

rejection number—a number *R,* such that if the number of defective units in the sample is equal to or greater than *R,* the lot shall be rejected. **D 1517,** D31

rejection number, *n*—the minimum number of nonconformities for a given AQL and sample size (lot sample size) which will subject a lot to rejection. **D 1711,** D09

rejection of a lot—the disapproval of a lot as not conforming to contract or specification or both. **D 1517,** D31

reject rate—the number of rejects stated as a percentage of total items. Scanner items can be characters, marks, fields, documents, pages, etc. **F 149,** F05

reject (suppression)—a control for minimizing or eliminating low amplitude signals (electrical or material noise) so that larger signals are emphasized. **E 1316,** E07

related color, *n*—color perceived to belong to an area seen in relation to other colors. **E 284,** E12

related standards—those standards which possess certain relevance to the host standard in terms of understanding its concepts, but do not of necessity specify any mandated requirements. **F 1789,** F16

relative complex permittivity (relative complex dielectric constant) (relative complex capacitivity), κ^*, ϵ_r^*, *n*—the ratio of the admittance of a given configuration of the material to the admittance of the same configuration with vacuum as dielectric:

$$\kappa^* = Y/Y_v = Y/j\omega C_v = \kappa' - j\kappa''$$

where Y is the admittance with the material and $j\omega C_v$ is the admittance with vacuum. **D 1711,** D09

relative consistency, I_c, C_r **(D)**—ratio of: (*1*) the liquid limit minus the natural water content, to (*2*) the plasticity index. **D 653,** D18

relative density, *n*—see **density ratio.** **B 243,** B09

relative density, n—see **specific gravity** **C 125,** C09

relative density—see **specific gravity.** **D 2864,** D27

relative density, *n*—the ratio of the mass of a given volume of liquid to the mass of an equal volume of pure water at the same temperature. **D 4725,** D15

relative density, D_d, I_D **(D)**—the ratio of (*1*) the difference between the void ratio of a cohesionless soil in the loosest state and any given void ratio, to (*2*) the difference between the void ratios in the loosest and in the densest states. **D 653,** D18

relative density (specific gravity)—ratio of the density of the gaseous fuel, under specified conditions of temperature and pressure, to the density of normal dry air, at the same temperature and pressure. **D 4150,** D03

relative density (specific gravity)—the ratio of the mass of a given volume of liquid at 15.56°C (60°F) to the mass of an equal volume of pure water at the same temperature. **D 4790,** D16

relative efficiency—the rating of the adsorptive capacity of an adsorbent based on a comparison of its performance with that of a reference adsorbent in a defined test. **D 2652,** D28

relative humidity, *n*—the ratio of actual water vapor pressure to the saturation water vapor pressure at the same temperature, expressed as a percentage. **C 11,** C11

relative humidity, *n*—*of air*, the ratio of the pressure of water vapor present to the pressure of saturated water vapor at the same temperature. (See **absolute humidity, humidity**.) **D 123,** D13

relative humidity—the ratio of the mass per unit volume (or partial pressure) of water vapor in an air-vapor mixture to the saturated mass per unit volume (or partial pressure) of the water vapor at the same temperature, expressed as a percentage. **D 1079,** D08

relative humidity, *n*—the ratio of the actual pressure of existing water vapor to the maximum possible (saturation) pressure of water vapor in the atmosphere at the same temperature, expressed as a percentage. **G 113,** G03

relative humidity with respect to ice—See **relative humidity.** **D 1356,** D22

relative humidity with respect to ice—See **humidity.** **D 1356,** D22

relative humidity with respect to ice, U_i—the ratio in percent of the mole fraction of water vapor x_v in moist air to the mole fraction of water vapor x_{vi} that moist air would have if it were saturated with respect to ice at the same pressure p and temperature t;

$$U_i = \frac{x_v}{x_{vi}} \times 100$$

If the water vapor and air assumed to behave as ideal gases, then

$$U_i = \frac{e}{e_i} \times 100$$

where e is the partial pressure of the water vapor in the moist air and e_i is the saturation vapor pressure with respect to ice at the same temperature t. **D 4023,** D22

relative humidity with respect to water—See **relative humidity.** **D 1356,** D22

relative humidity with respect to water—See **humidity.** **D 1356,** D22

relative humidity with respect to water, U_w—the ratio in percent of mole fraction of water vapor x_v in moist air to the mole fraction of water vapor x_{vw} that the moist air would have if it were saturated with respect to water at the same pressure p and temperature t:

$$U_w = \frac{x_v}{x_{vw}} \times 100$$

If water vapor and air are assumed to behave as ideal gases, then

$$U_w = \frac{e}{e_w} \times 100$$

where e is the partial pressure of the water vapor in the moist air and e_w is the saturation vapor pressure with respect to water at the same temperature t. **D 4023,** D22

relative minimum—*in optics*, a minimum in the amount of light transmitted through a polarizer and analyzer combination that results from varying either the polarizing angle or the analyzing angle (with the other angle fixed). **F 1241,** F01

relative permittivity (relative dielectric constant) (SIC) $\kappa'(\epsilon_r)$, *n*—the real part of the relative complex permittivity. It is also the ratio of the equivalent parallel capacitance, C_p, of a given configuration of electrodes with a material as a dielectric to the capacitance, C_v, of the same configuration of electrodes with vacuum (or air for most practical purposes) as the dielectric:

$$\kappa' = C_p/C_v$$

D 1711, D09

relative retention ratio—in chromatography, the column retention time of a component divided by the column retention time of a standard. **D 1129,** D19

relative retention time (RRT)—See **retention time.** **D 1356,** D22

relative rigidity—*in dynamic mechanical measurement*, ratio of modulus at any temperature, frequency, or time to the modulus at a reference temperature, frequency, or time. **E 1142,** E37

relative rigidity (*in dynamic mechanical measurement*)—the ratio of modulus at any temperature, frequency, or time to the modulus at a reference temperature, frequency, or time. **D 4092,** D20

relative span—dimensionless parameter indicative of the uniformity of the distribution, and defined as follows:

$$\text{relative span} = (D_{V0.9} - D_{V0.1})/ D_{V0.5}$$

E 1620, E29

relative spectral energy (power) distribution, *n*—description of the spectral character of a radiation (description of an illuminant) by the way in which the relative spectral concentration of radiant energy varies throughout the spectrum. Symbol: $S(\lambda)$. **E 349,** E21

relative standard deviation—the standard deviation expressed as a percentage of the mean value. **C 859,** C26

relative standard deviation, RSD—the standard deviation expressed as a percentage of the mean value.

$$RSD = (100/\bar{X})\sqrt{\Sigma d^2/(n-1)}$$

where:
d = difference between individual results and the average,
n = number of individual results, and
$\bar{X}$ = average of the individual results.

E 135, E01

relative viscosity—see **viscosity, relative.** **D 1695,** D01

relative water content—see **liquidity index**. **D 653,** D18

relaxation energy—*XPS*, the energy associated with intra-atomic or extra-atomic electronic readjustment to the removal of an atomic electron, so as to minimize the energy of the final state of the system. **E 673,** E42

relaxation rate, *n*—the absolute value of the slope of the relaxation curve at a given time. **E 6,** E28

relaxation time—the time interval between cessation of application of a stimulus and the end of the stimulated activity. **F 1156,** F12

relaxed stress, *n*—the initial stress minus the remaining stress at a given time during a stress-relaxation test. **E 6,** E28

relay steel—soft magnetic iron-based alloy used in the construction of electromechanical relays and solenoid switches. High flux densities, low coercive fields, suitable mechanical hardness, and ease of fabrication are primary concerns. **A 340,** A06

release agent, *n*—a material added to a compound or applied to the mold cavity, or both, to reduce parts sticking to the mold. **D 883,** D20

release agent—See **parting agent**. **D 1079,** D08

release agent, *n*—a material used to keep a molding material from adhering to a mold. (D-20) **F 412,** F17

release agent (mold), *n*—a substance applied to the inside surface of a mold or added to a material to be molded, to facilitate removal of the product from the mold. **D 1566,** D11

release characteristic—see **sensitivity range.** **F 221,** F05

release film—a sheet of paper or plastic applied to membrane sheeting during manufacture that protects the membrane sheeting from adhering to itself or becoming contaminated by foreign materials but which can be easily removed from the membrane. **D 3743,** D04

release paper, *n*—a sheet, serving as a protectant or carrier, or both, for an adhesive film or mass, which is easily removed from the film or mass prior to use. **D 907,** D14

releasing slider, *n*—*in zippers*, a slider with a mechanical means for loosening the slider on the chain. **D 123,** D13

releasing stop, *n*—*in zippers*, a device attached at or near the top of the stringer on the separable pin side which limits the travel of the slider at the open end of the chain under normal closing operations. **D 123,** D13

relevant indication, *n*—an NDT indication that is caused by a condition or type of discontinuity that requires evaluation. **E 1316,** E07

reliability, *n* (fiabilité)—the probability of performing without failure a specified function under normal conditions for a specified period of time. **E 1480,** E06

relieving—the removal of material from selected portions of a colored metal surface by mechanical means, to achieve a multicolored effect. **B 374,** B08

relocking device—a mechanism separate from the combination lock designed to block the safe bolt work if the lock is punched. It may also incorporate a thermal or shock-releasing device. **F 471,** F12

relocking trigger—a component of the combination lock designed to block the bolt if the spindle is punched. **F 471,** F12

reluctance, $\mathcal{R}$—that quantity which determined the magnetic flux, ϕ, resulting from a given magnetomotive force, $\mathcal{F}$, around a magnetic circuit.

$$\mathcal{R} = \mathcal{F}/\phi$$

where:
$\mathcal{R}$ = magnetic reluctance,
$\mathcal{F}$ = magnetomotive force, and
ϕ = flux.

The reluctance is measured in gilberts per maxwell (magnetic ohms) in the SI system and in ampere-turns per weber in the mksa system. **A 340,** A06

reluctivity, ν—the reciprocal of the permeability of a medium. **A 340,** A06

remaining ligament, b [L]—distance from the physical crack front to the back edge of the specimen, that is

$$b = W - a_p$$

E 1823, E08

remaining stress, *n*—the stress remaining at a given time during a stress-relaxation test. **E 6,** E28

remanence, B_{dm}—the maximum value of the remanent induction for a given geometry of the magnetic circuit.

NOTE—If there are no gaps or other inhomogeneities in the magnetic circuit the remanence, B_{dm}, is equal to the retentivity, B_{rs}; if air gaps or other inhomogeneities are present, B_{dm} will be less than B_{rs}. **A 340,** A06

remanent induction, B_d—See **induction, remanent.** **A 340,** A06

re-mass, *v*—to form, during storage, large blocks from particulate forms, such as flakes, pastilles, or powder. **D 6440,** D01

remelted heat, *n*—the product of the remelting of a **primary heat,** in whole or in part. **A 941,** A01

remodel—See **building modification.** **E 631,** E06

remodeled soil—soil that has had its natural structure modified by manipulation. **D 653,** D18

remodeling sensitivity (sensitivity ratio), S_t (D)—the ratio of: (*1*) the unconfined compressive strength of an undisturbed specimen of soil, to (*2*) the unconfined compressive strength of a specimen of the same soil after remolding at unaltered water content. **D 653,** D18

remolding index, I_R (D)—the ratio of: (*1*) the modulus of deformation of a soil in the undisturbed state, to (*2*) the modulus of deformation of the soil in the remolded state. **D 653,** D18

remote fire tool—remotely controlled tool. **F 592,** F16

removable dial—a dial that is removable from the spindle. **F 471,** F12

removal extractor—a series of horizontal baffles, usually constructed of stainless steel, designed to remove grease and drain it away to a container. They are cleaned by running them through a dishwasher or by soaking and rinsing. *See grease extractors.* **F 1827,** F26

renewable energy resources—sources of energy that are regenerative or virtually inexhaustible, such as solar, wind, ocean, biomass, municipal wastes, and hydropower energy. Geothermal energy is sometimes also included in the term. **E 1126,** E48

renewable energy resources—sources of energy that are regenerative or virtually inexhaustible, such as solar, wind, ocean, biomass, municipal wastes, and hydropower energy. Geothermal energy is sometimes also included in the term. **E 1705,** E48

repair—See **building modification.** **E 631,** E06

repeatability—the standard deviation of results obtained by the same operator using the same instrument in successive measurements on the same sample. **C 242,** C21

repeatability, *n*—a measure of the precision of the analyzer to repeat its results on independent introductions of the same sample at different time intervals. **D 1356,** D22

repeatability—the precision of a test method expressed in terms of the agreement attainable between measurements made by a single operator using the same apparatus and techniques. **D 4790,** D16

repeatability, *n*—the precision of a method expressed as the agreement attainable between independent determinations performed by a single analyst using the same apparatus and techniques. **E 135,** E01

repeatability, *n*—the closeness of agreement between the results of successive measurements of the same test specimen, or of test specimens taken at random from a homogeneous supply, carried out on a single laboratory, by the same method of measurement, operator, and measuring instrument, with repetition over a specified period of time. See also **reproducibility.** **E 284,** E12

repeatability, *n*—*of results of temperature measurements,* closeness of agreement between the results of successive measurements of the same temperature carried out under the same conditions of measurement. **E 344,** E20

repeatability, *n*—precision under repeatability conditions.

NOTE—Repeatability is one of the concepts or categories of the precision of a test method.

NOTE—Measures of repeatability defined in this compilation are **repeatability standard deviation** and **repeatability limit.** **E 456,** E11

repeatability—the precision of a method expressed as the agreement attainable between independent determinations performed at essentially the same time (duplicates) by one analyst using the same apparatus and techniques. **E 1547,** E15

repeatability conditions, *n*—conditions where independent test results are obtained with the same method on identical test items in the same laboratory by the same operator using the same equipment within short intervals of time.

NOTE—See **precision** . **E 456,** E11

repeatability limit (r), *n*—the value below which the absolute difference between two individual test results obtained under repeatability conditions may be expected to occur with a probability of approximately 0.95 (95 %).

NOTE—The repeatability limit is 2.8 ($\approx 1.96 \sqrt{2}$) times the repeatability standard deviation. This multiplier is independent of the size of the interlaboratory study, as explained in Practice E 177.

NOTE—The approximation to 0.95 is reasonably good (say 0.90 to 0.98) when many laboratories (30 or more) are involved, but is likely to be poor when fewer than eight laboratories are studied. **E 456,** E11

repeatability, *r*, *n*—an established value, below which the absolute difference between two "within-laboratory" or "within test-site" test results may be expected to lie, with a specified probability. **F 538,** F09

repeatability, relative (*r*), *n*—a repeatability estimate expressed as percentage of the average of the property for which the estimate was obtained. **F 538,** F09

repeatability standard deviation, *n*—the standard deviation of test results obtained under repeatability conditions.

NOTE—It is a measure of the dispersion of the distribution of test results under repeatability conditions.

NOTE—Similarly, "repeatability variance" and "repeatability coefficient of variation" could be defined and used as measures of the dispersion of test results under repeatability conditions. **E 456,** E11

repertoire—includes all of the characters and graphic shapes in an OCR imaging device system. **F 149,** F05

repetitive diving, *n*—any dive conducted while the decompression model being followed for the dives indicates that inert gases are still present in the tissues from a previous dive. This is within the context of a fixed time period, usually 12 to 24 h, depending on the model being followed. **F 1549,** F32

repetitive group designation, *n*—an assigned letter on a decompression table that relates directly to the amount of residual inert gas in the diver's tissues following a dive. This group changes with time as the diver's body off-gases. **F 1549,** F32

repetitive impact erosion test, *n*—*in impingement erosion testing,* an apparatus or method that produces a controlled or countable number of impacts by liquid or solid particles of uniform size, shape, and impact velocity, all on the same location of the test specimen. One example of such a test is the "wheel-and-jet" type of liquid impact apparatus. **G 40,** G02

replacement cost—building component replacement and related costs, included in the capital budget, that are expected to be incurred during the study period. **E 833,** E06

replacement disc, *n*—a structure intended to restore support and motion between adjacent vertebral bodies. **F 1582,** F04

replenisher—a toning material added to the developer mix to replace that which is consumed during the copying process. **F 335,** F05

replica—a reproduction of a surface in a material, for example, a plastic.

(*1*) *atomic*—a thin replica devoid of structure on the molecular level, prepared by the vacuum or hydrolytic deposition of metals or simple compounds of low molecular weight.

(*2*) *cast*—a reproduction of a surface in plastic made by the evaporation of the solvent from a solution of the plastic or by polymerization of a monomer on the surface.

(*3*) *collodion*—a replica of a surface cast in nitro-cellulose.

(*4*) *Formvar*—a reproduction of a surface in a plastic Formvar film.

(*5*) *gelatin*—a reproduction of a surface prepared in a film composed of gelatin.

(*6*) *impression*—a surface replica which is made by impression. The results of making an impression.

(*7*) *molecular*—the reproduction of a surface in a high polymer such as collodion and other plastics.

(*8*) *negative*—that replica which is obtained by the direct contact of the replicating material with the specimen. In it, the contour of the replica surface is reversed with respect to that of the original.

(*9*) *oxide film*—a thin film of an oxide of the specimen to be examined. The replica is prepared by air, oxygen, chemical, or electrochemical oxidation of the parent metal and is subsequently freed either mechanically or chemically for purposes of examination.

(*10*) *plastic*—a reproduction in plastic of the surface to be studied, prepared by evaporation of the solvent from a solution of plastic, by polymerization of a monomer, or solidification of a plastic on the surface.

(*11*) *positive*—a replica, the contours of which correspond directly to the surface being replicated; that is, elevations on the surface are elevations on the replica.

(*12*) *preshadowed*—a replica formed by the application of the shadowing material to a surface to be replicated, before the thin replica film is cast or otherwise deposited on the surface.

(*13*) *pseudo*—a replica which has portions of the material being replicated embedded in it.

(*14*) *tape replica method* (faxfilm)—a method of producing a replica by pressing the softened surface of a tape or sheet of a plastic material on the surface to be replicated.

(*15*) *vapor deposited*—a replica formed of a metal or a salt by the condensation of the vapors of the material onto the surface to be replicated. **E 7,** E04

replicate, *n*—*in experimenting or testing*, one of two or more runs with the same specified experimental or test conditions and with each experimental or test condition being established independently of all previous runs. (Compare **duplicate.**) **D 123,** D13

replicate, *v*—*in experimenting or testing*, to repeat a run so as to produce a replicate. (Compare **duplicate.**) **D 123,** D13

replicate—in electron microscopy, to reproduce by means of a replica. **E 7,** E04

replicate, *n*—either (1) an individual test object from a sample of *n* objects or (2) one of *m* individual test values for a test object. **F 538,** F09

replicate (repeat) tests—nominally identical tests on different randomly selected test specimens conducted at the same nominal value of the independent variable *X*. **E 1823,** E08

replicates—two or more repetitions of a test determination. **E 1547,** E15

replication, *v*—the act of selecting and testing a number of replicates. **F 538,** F09

reporting bases:

as-determined basis, n—analytical data obtained from the analysis sample of coal or coke after conditioning and preparation to No. 60 (250-μm) sieve in accordance with Method D 2013. As-determined data represents the numerical values obtained at the particular moisture level in the sample at the time of analysis. These are normally converted, according to formulae contained in Practice D 3180, to conventional reporting bases.

as-received basis, n—analytical data calculated to the moisture condition of the sample as it arrived at the laboratory and before any processing or conditioning. If the sample has been maintained in a sealed state so that there has been no gain or loss, the as-received basis is equivalent to the moisture basis as sampled.

dry, ash-free basis, n—data calculated to a theoretical base of no moisture or ash associated with the sample. Numerical values as established by Test Methods D 3173 and D 3174 are used for converting the as-determined data to a moisture- and ash-free basis.

dry basis, n—data calculated to a theoretical base of no moisture associated with the sample. The numerical value as established by Test Method D 3173 is used for converting the as-determined data to a dry basis.

equilibrium moisture basis, n—data calculated to the moisture level established as the equilibrium moisture. Numerical values as established by Test Method D 1412 are used for the calculation. **D 121,** D05

representative quality indicator—(RQI)—an actual part or similar part of comparable geometry and attenuation characteristics to that of the test part(s), that has known or measurable features, or both, representing the facets of nonconformance for which the test part is to be examined. **E 1316,** E07

representative sample—a generic term implying that the population has been properly sampled. **C 859,** C26

representative sample—See *representative sample* under **sample**. **D 121,** D05

representative sample—a sample collected in such a manner that it has characteristics equivalent to the lot. **E 856,** D34

representative sample—sample of a universe or whole (for example, waste sample pile, lagoon, ground water, or waste stream), selected in a random sampling process, which represents the total properties in the universe or whole. **E 1605,** E06

representative sample—a sample containing enough measured elements that the effect of random fluctuations is acceptably small. **E 1620,** E29

representative sample, *n*—a sample taken from a quantity of material that exhibits the characteristics of the original material from which it was taken. **E 1638,** E29

re-press, *v*—to apply pressure to a previously pressed and either sintered or presintered compact. It includes **re-strike, coin,** and **size**. **B 243,** B09

repressing—the application of pressure to a previously pressed and sintered compact, usually for the purpose of improving some physical property. **B 243,** B09

reprocessed plastic—a thermoplastic prepared from usually melt processed scrap or reject parts by a plastics processor, or from non-standard or non-uniform virgin material. **D 883,** D20

reprocessed plastic—a thermoplastic prepared from usually melt processed scrap or reject parts by a plastics processor, or from non-standard virgin material or non-uniform virgin material. **F 412,** F17

reproducibility—the standard deviation of results obtained by different operators using the same or different instruments in different laboratories on the same sample. **C 242,** C21

reproducibility, *n*—a measure of the precision of different analyzers to repeat results on the same sample. **D 1356,** D22

reproducibility—the precision of a test method expressed in terms of agreement expected between measurements made in different laboratories using similar apparatus and the same procedure. **D 4790,** D16

reproducibility, *n*—the precision of a method expressed as the agreement attainable between determinations performed in different laboratories. **E 135,** E01

reproducibility, *n*—the closeness of agreement between the results of successive measurements of the same test specimen, or of test specimens taken at random from a homogeneous supply, but changing conditions such as operator, measuring instrument, laboratory, or time. The changes in conditions must be specified. See also **repeatability.** **E 284,** E12

reproducibility, *n—of results of temperature measurements*, closeness of agreement between the results of measurements of the same temperature carried out under changed conditions of measurement. **E 344,** E20

reproducibility, *n*—precision under reproducibility conditions. **E 456,** E11

reproducibility—the precision of a method expressed as the agreement attainable between determinations performed in different laboratories. **E 1547,** E15

reproducibility conditions, *n*—conditions where test results are obtained with the same method on identical test items in different laboratories with different operators using different equipment. **E 456,** E11

reproducibility limit, *n*—(*R*) the value below which the absolute difference between two test results obtained under reproducibility conditions may be expected to occur with a probability of approximately 0.95 (95 %).

NOTE—The reproducibility limit is 2.8 ($\approx 1.96 \sqrt{2}$) times the reproducibility standard deviation. The multiplier is independent of the size of the interlaboratory study (that is, of the number of laboratories participating), as explained in Practice E 177.

NOTE—The approximation to 0.95 is reasonably good (say 0.90 to 0.98) when many laboratories (30 or more) are involved but is likely to be poor when fewer than eight laboratories are studied. **E 456,** E11

reproducibility R, *n*—quantitative expression of the random error associate with operators working in different laboratories, each obtaining single results on identical test material when applying the same method. **D 4175,** D02

reproducibility, R, *n*—an established value, below which the absolute difference between two "between-laboratory" or "between test-site" test results may be expected to lie, with a specified probability. **F 538,** F09

reproducibility, relative (*R*), *n*—a reproducibility estimate expressed as percentage of the average of the property for which the estimate was obtained. **F 538,** F09

reproducibility standard deviation (S_R), *n*—the standard deviation of test results obtained under reproducibility conditions.

NOTE—Other measures of the dispersion of test results obtained under reproducibility conditions are the ''reproducibility variance" and the ''reproducibility coefficient of variation."

NOTE—The reproducibility standard deviation includes, in addition to between-laboratory variability, the repeatability standard deviation and a contribution from the interaction of laboratory factors (that is, differences between operators, equipment and environments) with material factors (that is, the differences between properties of the materials other than that property of interest). **E 456,** E11

request to send, *n*—an electronic signal sent by one component to another to determine if the second component is ready to receive a transmission from the first. **F 1457,** F05

requirement—a translation of the needs into a set of individual quantified or descriptive specifications for the characteristics of an entity in order to enable its realization and examination, (ISO Guide 25). **E 1187,** E36

requirement statement (enoncé de boisons)—*for a facility*, the serviceability requirements for a facility, together with applicable performance criteria, performance test methods, and optional explanatory comments.

criterion, n (critéré)—an established precedent, rule, measure, norm, or code upon which a decision may be based.

performance criterion, n (critére de rendement)—a quantitative statement of the level of performance needed to satisfy a serviceability requirement.

performance test method, n (méthode d'essai de rendement)—a method of determining whether the performance of a facility is in accordance with a specified performance criterion.

serviceability requirement, n (besoin de fontionalité)—*for a facility*, a qualitative statement of the serviceability required from a facility.

score, n—See **serviceability score.** **E 1480,** E06

reradiation—loss of energy by radiation from a surface previously heated by absorption. **E 772,** E44

rerefining—the use of primary refining processes on used electrical insulating liquids to produce liquids that are suitable for further use as electrical insulating liquids. **D 2864,** D27

re-refining, *n*—the use of refining processes during recycling to produce base stock for lubricants or other petroleum products from used oil. **D 4175,** D02

reroofing—the process of recovering or replacing an existing roofing system. See **recovering.** **D 1079,** D08

resale value—the monetary sum expected from the disposal of an asset at the end of its economic life, its useful life, or at the end of the study period. **E 833,** E06

resawn lumber—See **lumber.** **D 996,** D10

reserve alkalinity, *adj*—a term applied to engine coolant concentrates and antirusts to indicate the amount of alkaline inhibitors present in the product. **D 4725,** D15

reserve buoyancy—gross buoyancy minus boom weight. **F 818,** F20

reserve buoyancy to weight ratio—reserve buoyancy divided by boom weight. **F 818,** F20

reservoir—a man-made impounded body of water or controlled lake where water is collected and stored. **D 4410,** D19

resident species—a species that is regularly present at a specified site for some portion of its life span. **E 943,** E47

residual contamination R $_c$, *n*—the absolute mass of contaminant remaining after the cleaning process and expressed in milligrams per square centimetre of area or optionally as milligrams per square foot. **G 126,** G04

residual deflection—permanent deformation of a building element, component, or structure after complete or partial removal of applied force. Also called **permanent set** or **residual deformation.** **E 631,** E06

residual deformation—See **residual deflection.** **E 631,** E06

residual density, *n*—optical density of image after exposure to light or water. **F 1857,** F05

residual drawdown—the difference between the projected prepumping water-level trend and the water level in a well or piezometer after pumping or injection has stopped. **D 653,** D18

residual elongation (*RE*), *n*—*of rope*, elongation after cyclic tensioning the rope to a specified force for a specified period of time. **D 123,** D13

residual error, *n*—the difference between the observed result and the predicted value (estimated treatment response); Observed Result minus Predicted Value. **E 456,** E11

residual error, *n*—the difference between the observed result and the predicted value (estimated treatment response); Observed Result minus Predicted Value. **E 1325,** E11

residual induction, B_r—See **induction, residual.** **A 340,** A06

residual magnetic field—the field that remains in ferromagnetic material after the magnetizing force has been removed. **E 1316,** E07

residual moisture—See *residual moisture* under **moisture**. **D 121,** D05

residual moisture—the moisture content remaining in an RDF sample after it has been air-dried and milled down to an analysis sample. **E 856,** D34

residual nitrogen, *n*—nitrogen gas that is still dissolved in a diver's tissues after he has surfaced from a dive using compressed air as a breathing gas. **F 1549,** F32

residual potential—apparent surface voltage remaining in light exposed areas of the photoconductive surface. **F 335,** F05

residual soil—soil derived in place by weathering of the underlying material. **D 653,** D18

residual strain—the strain in a solid associated with a state of residual stress. (ISRM) **D 653,** D18

residual strain, n—strain associated with residual stress. **E 6,** E28

residual strength, σ_r **[FL^{-2}]**—the maximum value of the gross stress, neglecting the area of the crack, that a cracked specimen is capable of sustaining. **E 1823,** E08

residual stress—stresses that remain within a body as the result of plastic deformation or casting. **B 846,** B05

residual stress—stress remaining in a solid under zero external stress after some process that causes the dimensions of the various parts of the solid to be incompatible under zero stress, for example, (*1*) deformation under the action of external stress when some parts of the body suffer permanent strain; or (*2*) heating or cooling of a body in which the thermal expansion coefficient is not uniform throughout the body. (ISRM) **D 653,** D18

residual stress [FL^{-2}], *n*—stress in a body which is at rest and in equilibrium and at uniform temperature in the absence of external and mass forces. **E 6,** E28

residual technique—the application of the magnetic particles after the magnetizing force has been discontinued. **E 1316,** E07

residual torsion, *n*—revolutions made by a specified length of steel tire cord when one end is held in a fixed position and the other is allowed to turn freely. **D 123,** D13

residue, *n*—*for plumage*, quill pith, quill fragments, trash, or foreign matter. **D 123,** D13

residue—those substances that remain in the sample flask after sample volatilization under the conditions of the test. **E 1547,** E15

residue bottle—a storage vessel for small amounts of special nuclear materials that accumulate in the course of chemical analysis. **C 859,** C26

resilience, *n*—that property of a material by virtue of which it is able to do work against restraining forces during return from a deformed state. **D 123,** D13

resilience, *n*—the ratio of energy output to energy input in a rapid (or instantaneous) full recovery of a deformed specimen. **D 1566,** D11

resilience, impact, *n*—the ratio of output to input mechanical energy in a rapid deformation and recovery cycle of a rubber specimen. **D 1566,** D11

resilient connector—a flexible connection for joining pipe to structures capable of being deformed and deflected without rupture or leakage. **C 822,** C13

resilient flooring, *n*—an organic floor surfacing material made in sheet or tile form or formed in place as a seamless material of which the wearing surface is non-textile. The resilient floor covering classification by common usage includes, but is not limited to asphalt, cork, linoleum, rubber, vinyl, vinyl composition, and polymeric poured seamless floors. Resilient in this sense is used as a commonly accepted term, but does not necessarily define a physical property. **F 141,** F06

resin—*in grouting*, a material that usually constitutes the base of an organic grout system. **D 653,** D18

resin, *n*—a solid or pseudosolid organic material often of high molecular weight, which exhibits a tendency to flow when subjected to stress, usually has a softening or melting range, and usually fractures conchoidally. **D 883,** D20

resin, *n*—(*1*) a solid, semisolid, or pseudosolid organic material that has an indefinite and often high molecular weight, exhibits a tendency to flow when subjected to stress, usually has a softening or melting range, and usually fractures conchoidally. (*2*) liquid resin—an organic polymeric liquid which, when converted to its final state for use, becomes a resin. (See also **adhesive, glue, gum,** and **rosin.**) **D 907,** D14

resin, *n*—an organic material of indefinite and relatively high molecular mass that may be used as a softener, processing aid, vulcanizing agent, or reinforcing agent. **D 1566,** D11

resin—a solid or pseudosolid organic material often of high molecular weight, which exhibits a tendency to flow when subjected to stress, usually has a softening or melting range, and usually fractures conchoidally. **D 3878,** D30

resin, *n*—a solid or pseudosolid organic material, often of high molecular weight, which exhibits a tendency to flow when subjected to stress, usually has a softening or melting range, and usually fractures conchoidally. **F 412,** F17

resinates, metallic—See **metallic resinates**. **D 804,** D01

resin batch—the quantity of resin that has been formulated in a single continuous operation and subjected to chemical processing or physical mixing to produce a homogeneous material. **E 1749,** E06

resin binder—see **binder.** **F 335,** F05

resin bonded batting, *n*—a textile filling material which is stabilized by spraying it with an acrylic, polyvinyl acetate, or other suitable resin emulsion after which the batting is dried and cured. **D 123,** D13

resin-coated—covered with natural resin to provide ease of driving, increased holding power, or corrosion resistance, or a combination of these. **F 547,** F16

resin content—see **matrix content.** **D 3878,** D30

resin content—the amount of matrix present in a composite usually expressed in units of weight percent. **E 1749,** E06

resin grout—a grout system composed of essentially resinous materials such as epoxys, polyesters, and urethanes. **D 653,** D18

resinite—See *resinite* under **maceral**. **D 121,** D05

resin, natural—a solid organic substance, originating in the secretion of certain plants or insects, which is thermoplastic, flammable, nonconductive of electricity; breaks with a conchoidal fracture (when hard); and dissolves in certain specific organic solvents but not water.

fossil resin— a natural resin of ancient origin usually found in the earth. **D 16,** D01

resinoid, *n*—any of the class of thermosetting synthetic resins, either in their initial temporarily fusible state or in their final infusible state. (See also **novolak** and **thermosetting.**) **D 907,** D14

resinous—containing a polymer as a binder that is either hardened by chemical action (thermosetting), by the evaporation of a solvent, or by melting for application (thermoplastic). **C 904,** C03

resin particle—specially manufactured polymer beads used in the ion exchange process to remove dissolved salts from water. **D 6161,** D19

resin-rich area—an area of the pultrusion that lacks sufficient reinforcement.

NOTE—The fiber pattern may not be visible. **D 3918,** D20

resin streak, *n*—a streak of excess resin on the surface of a laminated plastic. **D 883,** D20

resin, synthetic—a synthetic substance physically similar to natural resin.

acrylic resin— a synthetic resin made from derivatives of acrylic acid.

alkyd resin—a synthetic resin made from polyhydric alcohols and polybasic acids; generally modified with resins, fatty oils or fatty acids.

ester gum—a resin made from rosin or rosin acids and a polyhydric alcohol, such as glycerine or pentaerythritol.

maleic resin— a resin made from a natural resin and maleic anhydride or maleic acid.

melamine resin— a synthetic resin made from melamine and aldehyde.

penta resin—ester gum made from rosin and pentaerythritol.

phenolic resin— a synthetic resin made from phenols and aldehydes.

styrene resin— a synthetic resin made from vinyl benzene.

urea resin—a synthetic resin made from urea and an aldehyde.

vinyl resin—a synthetic resin made from vinyl compounds. **D 16,** D01

resist, *n*—(*1*) a material applied to a part of a cathode or plating rack to render the surface nonconductive.

(*2*) A material applied to a part of the surface of an article to prevent reaction of metal from that area during chemical or electrochemical processes. **B 374,** B08

resistance—the scaler property of an electric circuit or of any body that may be used as part of an electric circuit which determines for a given current the rate at which electric energy is converted into heat or radiant energy and which has a value such that the product of the resistance and the square of the current gives the rate of conversion of energy.

In a d-c circuit:

$$P = I^2R$$

where:
P = power, W,
I = current, A, and
R = resistance, Ω.

B 354, B01

resistance—the ratio of the potential difference applied to a specimen to the current passed through by the applied potential. It is the reciprocal of **conductance**. The unit is ohm. **D 2864,** D27

resistance—the difficulty required to counterfeit, alter, duplicate, or simulate a secure document or an article. **F 1156,** F12

resistance, abrasion, *n*—the ability to withstand scuffing, scratching, rubbing, or wind-scouring. **C 168,** C16

resistance, apparent dc—the dc resistance measured at the end of a specified electrification time. The "apparent dc resistance" is the reciprocal of the "apparent dc conductance." The unit is ohm. **D 2864,** D27

resistance, core, R_1—the effective ac resistance of a hypothetical parallel resister that is considered to carry exclusively the core loss current, I_c, when a voltage is applied to the terminals of a coil encircling a magnetic core.

NOTE—The product, $I_c^2R_1$, equals the total core loss, P_c. **A 340,** A06

resistance, dc—the ratio of the dc voltage (in volts) to the total current (in amperes) carried through the material between two electrodes that are in contact with, or immersed in a specimen. The "dc resistance" is the reciprocal of the "dc conductance." The unit is ohm. **D 2864,** D27

resistance, freeze-thaw, *n*—resistance to cycles of freezing and thawing that could affect application, appearance, or performance. **C 168,** C16

resistance, impact (toughness), *n*— ability to withstand mechanical blows or shock without damage seriously affecting the effectiveness of the material or system. **C 168,** C16

resistance, insulation, *n*—the ratio of the dc voltage applied to two electrodes (on or in a specimen) to the total volume and surface current between them. **D 1711,** D09

resistance, surface, *n*—the ratio of the dc voltage applied to two electrodes (on the surface of a specimen) to the current between them. **D 1711,** D09

resistance, thermal—the average temperature difference between two defined surfaces of a particular body or assembly when unit thermal transmission in unit time through unit area is established between the surfaces. $R = K{\cdot}m^2/W$ ($R = °F{\cdot}h{\cdot}ft^2/Btu$). **D 1079,** D08

resistance, thermal, *R*, *n*—the quantity determined by the temperature difference, at steady state, between two defined surfaces of a material or construction that induces a unit heat flow rate through a unit area.

$$R = \Delta T/q$$

A resistance (*R*) associated with a material shall be specified as a material *R*. A resistance (*R*) associated with a system or construction of materials shall be specified as a system *R*. (*R* in SI units: $K/(W/m^2) = Km^2/W$.)(*R* in inch-pound units: $F/(Btu/h/ft^2) = F\ ft^2\ h/Btu$.) **C 168,** C16

resistance thermometer—an instrument for determining temperature by measuring the electrical resistance of a standardized material exposed to that temperature. **E 7,** E04

resistance thermometer, *n*—a temperature-measuring device comprised of a resistance thermometer element, internal connecting wires, a protective shell with or without means for mounting, a connection head, or connecting wire or other fittings, or both. **E 344,** E20

resistance thermometer element, *n*—the temperature-sensitive portion of the thermometer composed of resistance wire, film or semiconductor material, its supporting structure, and means for attaching connecting wires. **E 344,** E20

resistance to bending, *n— in paper*, the force required to deflect a rectangular test piece of paper, clamped at one end, through a specified angle when the force is applied near the free end of the test piece, normal to the plane which includes the near edge of the test piece, the clamp, and the point or line of application of the force. **D 1968,** D06

resistance to breakage—the resistance of the inked ribbon or carbon ribbon to rupture resulting from tension before or after use. **F 221,** F05

resistance to delamination, *n—for pile yarn floor covering*, the force/unit width exceeded when separating component layers. **D 5684,** D13

resistance (to flow)—the reciprocal of conductance. **E 1316,** E07

resistance to slippage, *n*—the force required to separate the parts of a standard seam by a specified amount. **D 123,** D13

resistance to ultraviolet radiation, *n—in polyolefin tape yarn*, the time-to-failure of yarns exposed to xenon-arc weathering. **D 123,** D13

resistance to yarn slippage, *n*—the force required to separate the parts of a standard seam by a specified amount. **D 123,** D13

resistance to yarn slippage, *n—at the seam*, the force required to displace one or more yarns in a fabric from the original position, causing differences in alignment, or spacing, or both. **D 123,** D13

resistance, volume, *n*—the ratio of the dc voltage applied to two electrodes (on or in a specimen) to the current in the volume of the specimen between the electrodes. **D 1711,** D09

resistance, winding, R_w—the effective ac series resistance of an inductor when no ferromagnetic materials are present.

NOTE—At low frequencies, R_w is only slightly greater than the dc resistance of the winding.

NOTE—The product I^2R_w equals the sum of the copper, eddy current, and dielectric losses in the winding.

NOTE—The total active power, *P*, delivered to an inductor having a ferromagnetic core is:

$$P = P_c + I^2R_w$$

A 340, A06

resistazone counter—the generic name used to describe stream counters in which the interrogation zone is monitored for changes in electrical resistance as a result of the presence of a particle. **C 242,** C21

resistivity, ρ—that property of a material which determines its resistance to the flow of an electric current, expressed by:

$$\rho = R{\cdot}A/\ell$$

where:
R = resistanceof the specimen, Ω,
A = cross sectional area, cm^2, and
ℓ = length of specimen, cm.

Units of electrical resistivity are ohm-centimetre (cgs) and ohm-metre (SI).

NOTE—This value is equivalent to the resistance between opposite faces of a cube of unit dimensions, and is designated "specific resistivity" or, by usage, "volume resistivity." **A 340,** A06

resistivity—the electrical resistance of a body of unit length, and unit cross-sectional area or unit weight. **B 354,** B01

resistivity—the ratio of the potential gradient paralleling the current passing through the specimen, to the current density. This is numerically equal to the resistance between opposite faces of a unit cube. It is the reciprocal of **conductivity**. The unit commonly used is: ohm-centimetre. The SI unit is ohm-metre. **D 2864,** D27

resistivity—the property of a substance (in this case, water) to resist the flow of electricity; the measurement of that resistance. The inverse of conductivity. Measured by a resistivity monitor, and described in ohms·cm. **D 6161,** D19

resistivity, ρ—the reciprocal of conductivity. **E 1142,** E37

resistivity—see **apparent surface resistivity** and **volume resistivity.** **F 335,** F05

resistivity, apparent dc volume—the "dc volume resistivity" measured at the end of a specified electrification time. It is the reciprocal of the "apparent dc volume conductivity." The unit most commonly used is ohm-centimetre. The SI unit is ohm-metre. **D 2864,** D27

resistivity, dc—the ratio of the dc potential gradient paralleling the current to the current density at a given instant in time and under prescribed conditions. It is the reciprocal of the" dc conductivity." In common practice the "dc resistivity" is numerically equal to the dc resistance between opposite faces of a centimeter cube of liquid. The unit is ohm-centimetre. The SI unit is ohm-metre. **D 2864,** D27

resistivity, dc volume—the property of a material that impedes the flow of electricity through its volume. It is numerically equal to the ratio of: the steady direct voltage gradient parallel to the current; to the steady-state current density within the material. It is the reciprocal of the *dc volume conductivity*. In common usage where the voltage gradient is in volts per centimetre and the current density is in amperes per square centimetre, the unit of dc volume resistivity is (ohm-centimetre). The SI unit is ohm-metre. **D 2864,** D27

resistivity (electrical)—the measure of difficulty with which charge carriers flow through a material; the reciprocal of conductivity. **F 1241,** F01

resistivity, surface, *n*—the surface resistance multiplied by that ratio of specimen surface dimensions (width of electrodes defining the current path divided by the distance between electrodes) which transforms the measured resistance to that obtained if the electrodes had formed the opposite sides of a square. **D 1711,** D09

resistivity, surface insulation (of a single-strip specimen)—the effective resistivity of a single insulative layer tested between applied bare metal contacts and the base metal of the insulated test specimen. **A 340,** A06

resistivity, surface insulation (of multi-strip specimens)—the resistance of a unit area per test strip calculated from a measurement of the electrical resistance of a stack of strips with test current perpendicular to the strip surface. **A 340,** A06

resistivity, thermal, *r*, *n*—the quantity determined by the temperature difference, at steady state, between two defined parallel surfaces of a homogeneous material of unit thickness, that induces a unit heat flow rate through a unit area. (*r* in SI units: m K/W.) (*r* in inch-pound units: h ft F/Btu or, h ft^2 F/Btu in.) **C 168,** C16

resistivity, volume ρ—See **resistivity.** **A 340,** A06

resistivity, volume, *n*—the ratio of the electric potential gradient to the current density when the gradient is parallel to the current in the material. **D 1566,** D11

resistivity, volume, *n*—the volume resistance multiplied by that ratio of specimen volume dimensions (cross-sectional area of the electrodes divided by the distance between electrodes) which transforms the measured resistance to that resistance obtained if the electrodes had formed the opposite sides of a unit cube. **D 1711,** D09

resistivity, water vapor— See **water vapor resistivity.** **C 168,** C16

resite—See **C-stage.** **E 1749,** E06

resitol—See **B-stage.** **E 1749,** E06

resol—See **A-stage.** **E 1749,** E06

resolution—the fineness of detail in an object which is revealed by an optical device. Resolution is usually specified as the minimum distance by which two lines or points in the object must be separated before they can be revealed as separate lines or points in the image (see **resolving power** and **shape resolution**). The theoretical limit of resolution is determined from the equation:

$$d = 0.61\ \lambda/(n \sin A.A./2)$$

where:

d	=	minimum distance between object points observed as distinct points in the image.
λ	=	wavelength of the radiation employed.
n	=	the minimum refractive index of the media between the object and the objective lens.
$A.A.$	=	the angular aperture.

E 7, E04

resolution, *n—in atomic spectrometry*, the minimum distance by which two spectral lines must be separated before they can be distinguished as being separate. **E 135,** E01

resolution:

depth—EIA, energy resolution translated into an equivalent resolution of depth in the specimen.

energy—EIA, the full width at half-maximum (FWHM) of the measured energy distribution when the energy distribution of the backscattered particles is monoenergetic.

lateral—EIA, the distance measured on the surface of a specimen over which changes in composition can be established with confidence by BS. This resolution is generally determined by the size of the beam spot.

system—EIA, the energy or depth resolution measured in a BS spectrum for a monoenergetic incident beam. **E 673,** E42

resolution—the smallest increment that a characteristic measuring process must distinguish and display. **E 867,** E17

resolution—the ability of ultrasonic equipment to give simultaneous, separate indications from discontinuities having nearly the same range and lateral position with respect to the beam axis. **E 1316,** E07

resolution, *n—in a bar code system*, the narrowest element dimension that can be distinguished by a particular reading device or method. **F 1294,** F05

resolution (of a deformation-measuring instrument)—the ratio of the smallest divisional increment of the indicating scale to the sensitivity of the instrument. Interpolation within the increment may be possible, but is not recommended in specifying resolution. **D 653,** D18

resolving power—the power of a lens or optical system to form separate images of close object points. The quantity $\lambda/d\lambda$ is commonly used as the measure of the resolving power of a spectrograph. **C 859,** C26

resolving power—the ability of a given lens system to reveal fine detail in an object. (See also Resolution.) **E 7,** E04

resonance—the reinforced vibration of a body exposed to the vibration, at about the frequency, of another body. **D 653,** D18

resonance method—a technique in which continuous ultrasonic waves are varied in frequency to identify resonant characteristics in order to discriminate some property of a part such as thickness, stiffness, or bond integrity. **E 1316,** E07

resonance reaction—*EIA,* a nuclear reaction that has a narrow peak in the nuclear reaction cross section, which is so much larger than the nuclear reaction cross sections at adjacent energies both above and below the peak that essentially all the particles detected from the reaction are due to the peak. **E 673,** E42

resonant forced vibration technique (*in dynamic mechanical measurement*)—a technique for performing dynamic mechanical measurements, in which the sample is oscillated mechanically at the system's natural resonant frequency. **D 4092,** D20

resonant frequency—a frequency at which resonance exists. **D 653,** D18

resorption, *n*—the process by which a material that has given material by desorption takes up some more of the material given up. **D 123,** D13

Resource Conservation and Recovery Act of 1976—an amendment to the Solid Waste Disposal Act of 1965. **E 1605,** E06

respiration rate, *n*—in an microbial aqueous system, the quantitative consumption of oxygen, generally expressed as mg O_2/L/h. **D 5681,** D34

respiratory tracing, *n*—a display of physiological patterns of the subject=s breathing activity as recorded by the pneumograph component. **E 2035,** E52

respondent, *n*—See **assessor.** **E 253,** E18

response—the motion (or other output) in a device or system resulting from an excitation (stimulus) under specified conditions. **D 653,** D18

response curve for ***N*** **cycles**—a curve fitted to observed values of percentage survival at *N* cycles for several stress levels, where *N* is the preassigned number such as 10^6, 10^7, and so forth. It is an estimate of the relationship between applied stress and the percentage of the population that would survive *N* cycles. **E 1823,** E08

response factor—in leak testing, the response of the halogen leak detector 0.3 MPa·m³/s of refrigerant-12 (dichlorodifluoromethane, CCl2F2) or less, divided by the response to the same quantity of another halogen test gas. Thus, the actual leak rate of a detected leak will be the indication of the detector multiplied by the response factor. The response of mixture of a tracer and nonhalogen gases will be the response factor of the tracer divided by the fraction of tracer gas in the test gas. **E 1316,** E07

response position—the area designated to mark information on a mark read form. **F 149,** F05

response surface, *n*—the pattern of predicted responses based on the empirical model derived from the experiment observations. **E 456,** E11

response surface, *n*—the pattern of predicted responses based on the empirical model derived from the experiment observations. **E 1325,** E11

response time, *n—in tensile testing machines*, the time required by the indicating or recording device to reflect an instantaneous change in force, usually 0 to 90 % of full scale. **D 123,** D13

response time—See **analyzer.** **D 1356,** D22

response time—the length of time necessary to obtain a stable electrode potential when the electrode is removed from one solution and placed in another of different concentration. Response time depends on the electrode type, the magnitude and direction of the concentration change, temperature, and the presence of electrode interferences, if any. After exposure to a change in concentration, values of apparent concentration asymptotically approach the true concentration. Under most conditions, an electrode will exhibit a value within 1 mV of the final value within 1 min. **D 4127,** D19

response time—the time required for a leak detector or leak testing system to yield a signal output equal to 63 % of the maximum signal attained when tracer gas is applied continuously to the system under test. Also called response. **E 1316,** E07

response to molten substance pour—*in testing thermal protective material*, the observed effect of molten substance contact on textile properties or deterioration of the material. **F 1494,** F23

response-type system number (RTSN), *n*—the raw measured output from a response-type system. **E 867,** E17

response type system number (RTSN)—the raw measured output from a response type system being calibrated. Units are arbitrary, being whatever the road meter in the response type system measures. **E 867,** E17

response variable, *n*—the variable that shows the observed results of an experimental treatment. Synonym dependent variable. **E 456,** E11

response variable, *n*—the variable that shows the observed results of an experimental treatment. Synonym: dependent variable. **E 1325,** E11

responsivity, *s*, *n*—quotient of the output of a detector *Y* by its input *X*: $s = Y/X$. (See also **spectral responsivity**.) **E 284,** E12

restoration, *n*—action directed towards returning the condition of an object to its original state. **D 123,** D13

restoration—See **preservation.** **E 631,** E06

rest potential—See **open-circuit potential**. **G 15,** G01

restriction endonuclease—a bacterial enzyme that cuts double-stranded DNA at positions consisting of specific short sequences of nucleotides. **E 1705,** E48

re-strike, *v*—to re-press for the purpose of increasing the density of a sintered compact (not to be confused with **coin** or **size**). **B 243,** B09

result, *n*—the value obtained by following the complete set of instructions of a test method. **D 4175,** D02

result—a value, that is, a single determination, an average of duplicates, or other specified grouping of replicates, obtained by carrying out the test method. **E 1547,** E15

resultant field—see **field, resultant.** **E 1316,** E07

resultant traction force, [ML/T^2], *n*—the vector sum of lateral and longitudinal traction forces. **F 538,** F09

resultant yarn number, *n*—the yarn number based on the observed mass per unit length of a plied yarn, a cabled yarn, or a yarn whose number has been changed by processing such as twisting or bulking. **D 123,** D13

retainer basket, *n*—in sampling, a one-way gate on a sampling device that minimizes loss of sample when retrieving a sampler; also called a core catcher. **D 5681,** D34

retainer pin, *n*—*in zippers*, a tube-like element, similar to the separable pin, attached over the bead at the bottom end of the stringer opposite to the separable pin and that is designed to hold the fixed retainer in position. **D 123,** D13

retaining ring—a component that holds the tumbler pack on the tumbler post. **F 471,** F12

retaining wall—See **wall.** **E 631,** E06

retannage—a modifying second tannage, applied after intermediate operations following the primary tannage. **D 1517,** D31

retard—bank-protection structure designed to reduce the riparian velocity and induce silting or accretion. **D 653,** D18

retardation—delay in deformation. (ISRM) **D 653,** D18

retardation, δ—optical path difference between two beams in an interferometer; also known as "optical path difference" or "optical retardation." **E 131,** E13

retardation, maximum, Δ—the greatest retardation generated by an interferometer in a given scan. **E 131,** E13

retardation plate—a plate placed in the path of a beam of polarized light for the purpose of introducing a difference in phase. Usually quarter-wave or half-wave plates are used, but if the light passes through them twice the phase difference is doubled. **E 7,** E04

retarder, *n*—a material that will lengthen the setting time of gypsum plaster. **C 11,** C11

retarder—a material that slows the rate at which chemical reactions would otherwise occur. **D 653,** D18

retarder, n— Synonym for **inhibitor**. **D 907,** D14

retarder, *n*—a material used to reduce the tendency of a rubber compound to vulcanize prematurely. **D 1566,** D11

re-temper, *v*—to add more water to a hydraulic-setting compound after the initial mixing, but before partial set has occurred. **E 631,** E06

retentate—see **concentrate**. **D 6161,** D19

retention time—the time that elapses from the introduction of the sample until the component peak maximum is reached. **D 1129,** D19

retention time *(RT), n*—time to elute a specific chemical from a gas chromatographic column, for a specific carrier gas flow rate, measured from the time the chemical is injected into the gas stream to when it appears in the detector.

relative retention time (RRT), n—a ratio of RTs for two chemicals for the same chromatographic column and carrier gas flow rate, where the denominator represents a reference chemical. **D 1356,** D22

retentivity—the ability of an adsorbent to resist desorption of an adsorbate. **D 2652,** D28

retentivity—the ability of a material to retain a portion of the applied magnetic field after the magnetizing force has been removed. **E 1316,** E07

retentivity, B_{rs}— the property of a magnetic material which is measured by its maximum value of the residual induction.

NOTE—Retentivity is usually associated with saturation induction. **A 340,** A06

retest—the tests given after the initial acceptance test, usually performed at regular periodic intervals or as required because of physical inspection. **F 819,** F18

reticle—a system of lines, circles, dots, cross hairs or wires, or some other pattern, placed in the eyepiece or at an intermediate plane on the optic axis which is used as a measuring reference, focusing target, or to define a camera field of view (a reticle is different than a graticule, see **graticule**). **E 7,** E04

reticulation, *n*—the cracking of a print (mud cracking) due to poorly plasticized ink, over-thinning of an ink or too little print pressure. **D 6488,** D01

retortable—capable of withstanding specified thermal processing in a closed retort at temperatures above 100°C. **F 17,** F02

retortable—See Terminology F 17. **F 1327,** F02

retorts—containers fabricated from various metals in which parts are loaded to be diffusion coated or diffusion heat treated. **B 374,** B08

retractable anvil—curved clincher used in stapling from outside of container. **F 592,** F16

retracted spew, n—deprecated term, see **backrinding.**

D 1566, D11

retraction, *n—in yarns and cords*, the reduction in length when previous restraint is removed and relaxation is allowed, thus causing a directionally proportional increase in linear density. **D 123,** D13

retrieval—removing a deployed boom from the water. **F 818,** F20

retrofit—See **building modification.** **E 631,** E06

retrofit—the modification of an existing building or facility to include new systems or components. **E 833,** E06

retrogradation, *n*—a change of starch pastes from low to high consistency on aging. **D 907,** D14

retrograde condensation—where the critical point of multicomponent liquid-vapor equilibrium occurs below the maximum in temperature and pressure of the two-phase region, a sequence of increasing pressure change, at a temperature between the temperature maximum and the critical point, will cause a partial condensation of the vapor to liquid and then a full return to the vapor state, called retrograde condensation. **E 7,** E04

retrograde vaporization—the inverse of retrograde condensation. **E 7,** E04

retroreflection, *n*—reflection in which the reflected rays are preferentially returned in directions close to the opposite of the direction of the incident rays, this property being maintained over wide variations of the direction of the incident rays. **E 284,** E12

retroreflection; reflex reflection, *n*—reflection in which light is returned in directions close to the direction from which it came, this property being maintained over wide variations in the direction of incident light. **E 349,** E21

retro-reflective, *adj*—a characteristic of material that reflects light back to its source regardless of the angle of incidence. **F 1294,** F05

retroreflectivity, *n*—property of a material or device in which, when directionally irradiated, the reflected rays are preferentially returned in directions close to the opposite of the direction of the incident rays, this property being maintained over wide variations in the direction of the incident rays. **E 284,** E12

retroreflector, *n*—a reflecting surface or device from which, when directionally irradiated, the reflected rays are preferentially returned in directions close to the opposite of the direction of the incident rays, this property being maintained over wide variations of the direction of the incident rays. **E 284,** E12

retroreflector axis, *n*—a designated line segment from the retroreflector center that is used to describe the angular position of the retroreflector. **E 284,** E12

retroreflector center, *n*—a point on or near a retroreflector that is designated to be the center of the device for the purpose of specifying its performance. **E 284,** E12

returnable, *adj*—intended to be returned one or more times in an available program established by, but not limited to manufacturers, distributors or retailers. **D 5488,** D10

returnable container— See **container.** **D 996,** D10

return flow nozzle—see **by-pass nozzle**. **E 1620,** E29

reusable, *adj*—capable of being used again or repeatedly. **D 1968,** D06

reusable, *adj*—intended to be used for its original purpose one or more times in an available program established by, but not limited to, manufacturers, distributors, or retailers (see also **refillable** and **available program** for further clarification). **D 5488,** D10

reusable container— See **container.** **D 996,** D10

reusable pallet—See **pallet.** **D 996,** D10

reverberant sound field—the sound in an enclosed or partially enclosed space that has been reflected repeatedly or continuously from the boundaries. **C 634,** E33

reverberation—the persistence of sound in an enclosed or partially enclosed space after the source of sound has stopped; *by extension,* in some contexts, the sound that so persists. **C 634,** E33

reverberation room—a room so designed that the reverberant sound field closely approximates a diffuse sound field, both in the steady state when the sound source is on, and during decay after the source of sound has stopped. **C 634,** E33

reverberatory wires, *n*—a wire mesh located in front of, but close to, the radiating surface of the panel heat source which serves to enhance the combustion efficiency and increase the radiance of the panel. **E 176,** E05

reversal developer (negative developer)—uncharged area development by means of toner particles having appropriate charge polarity so that dark areas of the print correspond to light areas of the original. **F 335,** F05

reversal development (negative development)—uncharged area development by means of toner particles having appropriate charge polarity so that dark areas of the print correspond to light areas of the original. **F 335,** F05

reversal plate, film, or paper—a photographic material which, after exposure to a subject, is processed to give a positive reproduction without transfer of the image to another sheet material. **E 7,** E04

reversal (slope reversal)— *in fatigue loading*, the occurrence where the first derivative of the load-time (strain-time) history changes sign. **E 1823,** E08

reverse—to reverse the direction of flow of gas and air in a regenerative furnace. **C 162,** C14

reverse circulation—a drilling system in which the circulating medium flows down through the annulus and up through the drill rod, that is, in the reverse of the normal direction of flow. **D 653,** D18

reversed calf—a term applied to calf leather of heavier weight, suede finished on the flesh surface, containing oils to increase water-resistance, used for shoes where a nappy leather is required. Originally called "trench calf" in England, the term "hunting calf" is also used in that country. The term "service leather" may also be used though this term is usually reserved for sueded split and side leather. **D 1517,** D31

reversed print—a print prepared from a positive transparency. (See also **negative print**.) **E 7,** E04

reversed transparency—See **positive transparency.** **E 7,** E04

reverse engineering—dismantling an item to determine why and how it functions and how it is made; often done by competitors to ascertain how their rivals' products are made, or by governments to determine how their opponents' weapons systems operate, and how they were designed and fabricated. **F 1156,** F12

reverse image, *n*—a symbol in which the dark areas are represented by the material substrate (background) and the light areas are represented by the printed portion of the symbol. **F 1294,** F05

reverse osmosis (RO)—the separation process where one component of a solution is removed from another component by flowing the feed stream under pressure across a semipermeable membrane. RO removes ions based on electro chemical forces, colloids, and organics down to 150 molecular weight. May also be called hyperfiltration. **D 6161,** D19

reverse retan—a term applied to leather tanned first with vegetable tannin and then with chromium compounds. **D 1517,** D31

reversible adsorption—adsorption in which the desorption isotherm approximates the adsorption isotherm. **D 2652,** D28

reversible fiber optic chemical sensor, *n*—a fiber optic chemical sensor in which the transduction element does not undergo a permanent depletion or degradation as a result of the transduction process. **E 131,** E13

reversion, *n*—*in building construction in joint sealing,* a loss of elastomeric properties and a decrease in durometer hardness of a seal or cured sealant following environmental exposure. **C 717,** C24

reversion—in general, the tendency of pulp or paper properties to return to those of the material at some other stage. It is frequently specifically applied to the loss of brightness with time, after bleaching has produced a high brightness. **D 1695,** D01

reversion (vulcanization), *n*—deterioration of vulcanizate properties that may occur when vulcanization time is extended beyond the optimum. **D 1566,** D11

revetment—bank protection by armor, that is, by facing of a bank or embankment with erosion-resistant material. **D 653,** D18

review—a deliberately critical examination, including observation of plant operation, evaluation of audit results, procedures, certain contemplated actions, and after-the-fact investigations of abnormal conditions. **F 1789,** F16

revivification—see **reactivation.** **D 2652,** D28

revolving drum test— See **package testing.** **D 996,** D10

revolving pot—{archaic} a rotating circular container from which the Owens machine gathers glass. **C 162,** C14

revolving tube—a hollow cylinder, concentric with the needle of a feeder, revolving in the glass. **C 162,** C14

reworked plastic, *n*—a plastic from a processor's own production that has been reground, pelletized, or solvated after having been previously processed by molding, extrusion, etc. See **recycled plastic** and **reprocessed plastic.** **D 883,** D20

rework plastic (thermoplastic)—a plastic from a manufacturer's own production that has been reground or pelletized for reuse by that same manufacturer. **F 412,** F17

REX hardness—*in sealants,* the hardness of a sealant as measured by a REX hardness gage. **E 1749,** E06

reynolds number—a dimensionless number expressing the ratio of inertia forces to viscous forces in a moving fluid. The number is given by VLr/m where "V", is the fluid's velocity," L" is a characteristic length or distance such as pipe diameter, "r" is the fluid's mass density, and "m" is the fluid's dynamic viscosity. **D 4410,** D19

RF—Radio Frequency. **F 1156,** F12

rheology—the science of measuring the flow and deformation properties of matter. For porcelain enamel slips, the most important parameter is their yield point. **C 286,** B08

rheology—the flow characteristics of a material, including viscosity. **F 1156,** F12

rhombohedral—having three equal axes, with the included angles equal to each other but not equal to 90°. **E 7,** E04

rib, *n*—a continuous circumferential projection. **F 538,** F09

ribbon—in some slate, narrow bands of contrasting color or appearance differing in some degree in chemical composition from the main body. **C 119,** C18

ribbon—a continuous strip of glass in process. **C 162,** C14

ribbon, *n*—*in textiles,* a fine-textured, narrow fabric which weighs less than 2.6 lb/100 yd per inch of width (approximately 15 oz/yd^2 or 510 g/m^2) and which is used primarily for trimming or decorative purposes. (See also **narrow fabric.**) **D 123,** D13

ribbon anisotropic phase, *n*—*as used in Test Method D 5061,* a group of binder-phase anisotropic carbon textures distinguished by their ribbon-like domains (that is, length (L) to width (W) ratio of $L > 4W$), and subdivided based on domain width as fine ribbon (2.0 to 12.0-μm), medium ribbon (12.0 to 25.0-μm), and coarse ribbon (>25.0-μm) size categories. **D 121,** D05

ribbon machine—a forming machine for the manufacture of blown glass products from a ribbon of molten glass. **C 162,** C14

ribbon process—a process whereby molten glass is delivered to a forming unit in a ribbon form. **C 162,** C14

ribbon wire—box stay wire of nominal 0.103 in. (2.62 mm) in width. **F 592,** F16

rib or element area, [L^2], *n*—that area within the outer periphery of a tire footprint that is contacted by ribs or elements. **F 538,** F09

Richard's outflow principle—the principle that states that pore-liquid will not generally flow into an air-filled cavity (at atmospheric pressure) in unsaturated soil. **D 653,** D18

rich gas—natural gas containing commercially recoverable amounts of condensable hydrocarbons. **D 4150,** D03

ride—the edge which acts as a cam. **F 1078,** F04

rideability, *n*—a subjective judgement of the comparative discomfort induced by traveling over a specific section of highway pavement in a vehicle. **E 867,** E17

rideability index (RI), *n*—an index derived from controlled measurements of the longitudinal profile in the wheel tracks and correlated with panel ratings of rideability. **E 867,** E17

ride number (RN), *n*—rideability index of a pavement using a scale of 0 to 5, with 5 being perfect and 0 being impassable. **E 867,** E17

ride quality rating, *n*—a numerical value subjectively assigned to a section of highway pavement by an individual quantifying his judgement of the level of ride quality for that section based on a psychophysical scale. **E 867,** E17

rider arch—one of a series of arches that support the checkerwork in a regenerator. **C 162,** C14

ride rating panel, *n*—a group of highway users, statistically representative of the total expected highway user population, in rating the ride quality of pavements. **E 867,** E17

ride relief—the contoured area between the shank and ride **F 1078,** F04

ridge—long, narrow protrusion on any surface.

F 109, C21

ridge roll—a half-round section of asbestos-cement applied along the hips and ridge of a roof to conceal and waterproof the apex joint of the roofing material. **C 460,** C17

ridge roll, *n*—a half-round section applied along the hips and ridge of a roof to conceal and waterproof the apex joint of the roofing material. **C 1154,** C17

ridging—an upward, tenting displacement of a membrane, frequently over an insulation joint. **D 1079,** D08

ridging indentation—a hardness indentation around which metal has been piled up above the plane of the specimen. **E 7,** E04

riebeckite asbestos—belongs to the mineral group of monoclinic amphiboles derived from riebeckite (glaucophane). Its empirical formula is $Na_{24}Fe_3^{2+}Fe_2^{3+}(Si_8O_{22})\text{-}(OH)_2$. Its Chemical Abstracts number is 12001-28-4. **C 460,** C17

riebeckite asbestos, *n*—a member of the mineral group of monoclinic amphiboles derived from riebeckite (glaucophane) of the alkali amphibole series. Usually referred to by the varietal name *crocidolite.* Its empirical formula is $Na_2Fe_3^{+2}Fe_2^{+3}(Si_8O_{22})(OH,F)_2$. Its Chemical Abstracts number is 12001-28-4. **D 2946,** C17

riffle, *n*—a hand-fed sample divider device that divides the sample into two parts of approximately the same weight. **D 121,** D05

rift—(*1*) a consistent direction or trend in a rock body along which the rock is most easily split or broken.

(*2*) The grain orientation in stone, particularly in sedimentary stones, showing more or less clearly how the stone was originally bedded, and with or without color or grain-size changes, or voids and hollow. **C 119,** C18

rift sawn—a synonym for *edge grain.* **D 9,** D07

rigging leather—a strong, flexible, vegetable-tanned leather. **D 1517,** D31

right-angle chisel point—chisel point with beveled point faces parallel to staple-crown axis. (See **cross-cut chisel point.**) **F 592,** F16

right side, n—See **face side**. **D 4965,** D13

rigid cellular polystyrene thermal insulation (RCPS)—rigid thermal insulation board formed by expansion of polystyrene resin beads or granules in a closed mold (EPS), or by the expansion of polystyrene resin in an extrusion process (XPS). **E 631,** E06

rigid hinge or joint—a steel hinge consisting of body limb segments which have a fixed or constant pivot location. **F 869,** F08

rigid plastic, *n*—for purposes of general classification, a plastic that has a modulus of elasticity, either in flexure or in tension, greater than 700 MPa (100 000 psi) at 23°C and 50 % relative humidity when tested in accordance with Test Method D 747, Test Methods D 790, Test Method D 638M, or Test Methods D 882. **D 883,** D20

rigid plastic—for purposes of general classification, a plastic that has a modulus of elasticity either in flexure or in tension greater than 700 MPa (100 000 psi) at 23°C and 50 % relative humidity when tested in accordance with Test Methods D 747, D 790, D 638, or D 882. **F 1251,** F04

rim—a protrusion, usually of base material, bordering either partially or completely a hole, pit, or pock.

F 109, C21

rim, *n*—the specially shaped circular periphery to which a tire may be mounted with appropriate bead filament. **F 538,** F09

rimmed steel, *n*—a **steel** that contained sufficient oxygen to generate carbon monoxide at the boundary between the solid metal and the remaining molten metal during solidification, resulting in an outer layer low in carbon. **A 941,** A01

rim strip, *n*—a layer of compound, with or without fabric reinforcement, that is applied at the bead to protect the carcass plies against damage from mounting tools and from rim chafing during service. **F 538,** F09

rim, test, *n*—a rim having the configuration and dimensions suitable for use with a test tire in accordance with publications of a tire and rim standards organization current at the time of tire manufacture. **F 538,** F09

rim thickness—thickness of peripheral part of head, measured parallel to nail axis. **F 547,** F16

ring—(*1*) a floating refractory body surrounding a glass surface from which gathers are made.

(*2*) a mold part that forms the rim of a pressed article. **C 162,** C14

ring, *n*—*in buttons,* a split ring used to fasten a staple attached button to the substrate. **D 123,** D13

ring, *n*—*in hosiery,* a narrow, visually different horizontal band. **D 123,** D13

ring-down count—see **count, acoustic emission, the preferred term.** **E 1316,** E07

ringhole—an opening in a melter through which glass is gathered. **C 162,** C14

ring nail, ring-barbed nail, ring-grooved nail, ring-grip nail, ring-shank nail—terms applied to annularly threaded nail. (See **thread, annular.**) **F 547,** F16

ring or toggle attached staple button, *n*—a button attached to one part of a flexible substrate by means of a ring or toggle rather than a needle or thread. The staple passes through an eyelet in the flexible substate and is secured by the ring or toggle that passes through the staple eye. **D 5497,** D13

ring-porous woods—hardwoods in which the pores of the earlywood are large compared to the latewood, thus forming a distinct zone or ring of pores. **D 9,** D07

ring section—narrow, peripheral section cut from a glass article for optical examination. **C 162,** C14

ring thread, ring-barb thread, ring-groove thread, ring-grip thread, ring-shank thread—See **annular thread.** **F 547,** F16

rinse—that part of the operating cycle of an ion exchange process in which a specified water is passed through a bed of the ion exchange material to remove the residual regenerant solution. **D 1129,** D19

rinse—the process of removing liquid penetrant examination materials from the surface of a test part by means of washing or flooding with another liquid, usually water. The process is also termed **wash.** **E 1316,** E07

rinse additive feeder—a device that automatically feeds rinse additives into recirculated or non-recirculated rinse water of spray-type commercial dishwashing and glasswashing machines. **F 1827,** F26

rinse, *v* **or** *n*—a process or treatment in an aqueous solution for the purpose of removing extraneous matter. (See Guide D 5548.) **D 459,** D12

ripening—in the manufacture of viscose rayon, that stage where the cellulose xanthate solution is stored several days under controlled conditions to reach a state from which coagulation is easier. **D 1695,** D01

ripple—in dry process enameling, a surface defect characterized by pronounced waviness, uniform over a considerable area. **C 286,** B08

ripple—small, triangular-shaped bed forms that are similar to dunes but smaller. **D 4410,** D19

ripple (dc)—regular modulations in the dc output wave of a rectifier unit, or a motor-generator set, originating from the harmonics of the ac input system in the case of a rectifier, or from the harmonics of the induced voltage of a motor generator set. **B 374,** B08

ripple mark—See **Wallner line.** **C 162,** C14

ripple plating—a method of electroplating in which the current is changed periodically by superimposing surges, ripples, pulses, or alternating current to the d-c electroplating current. **B 374,** B08

riprap stone—material generally less than 2 tons (1814 kg) in mass, specially selected and graded, when properly placed prevent erosion through minor wave action, or strong currents and thereby preserves the shape of a surface, slope, or underlying structure. **D 653,** D18

rise, *n*—the maximum dimension of the cross-sectional profile of the tile measured perpendicular to the roof surface as installed. **C 43,** C15

riser—the pipe extending from the well screen to or above the ground surface. **D 653,** D18

rise time, *n*—the time required for a free-rise cellular plastic to achieve its ultimate expansion under controlled conditions. **D 883,** D20

rise time—See **analyzer.** **D 1356,** D22

rise time (pulse rise time)—the interval of time required for the leading edge of a pulse to rise from some specified small fraction to some specified larger fraction of the maximum value. **D 653,** D18

rising damp, *n*—upward-moving moisture in a **wall** or other structure standing in water or in wet soil. (Compare: **capillary migration,** wicking). **E 631,** E06

risk—the probability or likelihood an adverse effect will occur. **E 943,** E47

risk-adjusted discount rate (RADR)—a discount rate that has been adjusted to account for risk. **E 833,** E06

risk analysis—the body of theory and practice that has evolved to help decision makers assess their risk exposures and risk attitudes so that the investment that is *best for them* is selected. **E 833,** E06

risk attitude—the willingness of decision makers to take chances or gamble on investments of uncertain outcome. **E 833,** E06

risk averse (RA)—See **risk attitude**. **E 833,** E06

risk exposure—the probability of investing in a project whose economic outcome is different from what is desired (the target) or what is expected. **E 833,** E06

risk neutral (RN)—See **risk attitude**. **E 833,** E06

risk taking (RT)—See **risk attitude**. **E 833,** E06

river boom (aka fast water boom)—boom intended for use in currents greater than 1 knot. **F 818,** F20

rivet head—See **oval head.** **F 547,** F16

RNT, *n*—residual nitrogen time. **F 1549,** F32

road meter, *n*—equipment that measures the vehicle axle vertical motion relative to the vehicle frame during travel to yield a measure of roughness, for example, Mays, PCA, Soiltest, Cox. **E 867,** E17

roan—a sheepskin, not split. **D 1517,** D31

robber—See **thief.** **B 374,** B08

robustness, *n*—insensitivity of a statistical test to departures from underlying assumptions. **E 456,** E11

rock—a naturally occurring, consolidated aggregation of one or more minerals constituting the crust of the Earth. **C 119,** C18

rock—natural solid mineral matter occurring in large masses or fragments. **D 653,** D18

rock—any naturally formed aggregate of mineral matter occurring in large masses or fragments. (ISRM) **D 653,** D18

rock anchor—a steel rod or cable installed in a hole in rock; in principle the same as rock bolt, but generally used for rods longer than about four metres. (ISRM) **D 653,** D18

rock asphalt—see **asphalt rock.** **D 8,** D04

rock bolt—a steel rod placed in a hole drilled in rock used to tie the rock together. One end of the rod is firmly anchored in the hole by means of a mechanical device or grout, or both, and the threaded projecting end is equipped with a nut and plate that bears against the rock surface. The rod can be pretensioned. (ISRM) **D 653,** D18

rock burst—a sudden and violent expulsion of rock from its surroundings that occurs when a volume of rock is strained beyond the elastic limit and the accompanying failure is of such a nature that accumulated energy is released instantaneously. **D 653,** D18

rock burst—sudden explosive-like release of energy due to the failure of a brittle rock of high strength. (ISRM) **D 653,** D18

rock crystal—(*1*) transparent quartz.

(*2*) highly polished blown glassware, hand-cut or engraved. **C 162,** C14

rocker—an imperfection; a bottle with bottom deformed so it wobbles (rocks). **C 162,** C14

rocker bottom—See **rocker**. **C 162,** C14

rock flour—see **silt**. **D 653,** D18

Rockingham ware—a semivitreous ware or earthenware having a brown or mottled brown bright glaze. **C 242,** C21

rock mass—rock as it occurs in situ, including its structural discontinuities. (ISRM) **D 653,** D18

rock mechanics—the application of the knowledge of the mechanical behavior of rock to engineering problems dealing with rock. Rock mechanics overlaps with structural geology, geophysics, and soil mechanics. **D 653,** D18

rock mechanics—theoretical and applied science of the mechanical behaviour of rock. (ISRM) **D 653,** D18

Rockwell hardness number, HR, *n*—a number derived from the net increase in the depth of indentation as the force on an indenter is increased from a specified preliminary test force to a specified total test force and then returned to the preliminary test force. **E 6,** E28

Rockwell hardness test, *n*—an indentation hardness test using a verified machine to force a diamond sphero-conical indenter (diamond indenter) or a hard steel ball indenter, under specified conditions into the surface of the material under test in two operations, and to measure the difference in depth of the indentation under the specified conditions of preliminary and total test forces (minor and major loads, respectively). **E 6,** E28

Rockwell superficial hardness number, n—See **Rockwell hardness number**. **E 6,** E28

Rockwell superficial hardness test, *n*—same as the Rockwell hardness test except that smaller preliminary and total test forces are used. **E 6,** E28

rod, *n*—solid wrought product that is long in relation to its circular cross section, which is 0.375 in. or greater [over 10.00 mm] in diameter.

cold-finished rod, n—rod brought to final dimensions by cold working to obtain improved surface finish and dimensional tolerances.

cold-heading rod, n—rod of a quality suitable for use in the manufacture of cold-headed products such as bolts and rivets.

extruded rod, n—rod produced by hot extruding.

rivet rod, n—See *cold-heading rod.* **B 881,** B07

rod, *n*—wrought material of round, solid straight lengths. **B 899,** B02

rod, *n*—a longitudinal element symmetrical in the transverse plane designed to resist tension, compression, bending, and torsion. **F 1582,** F04

rod cast button, *n*—a button fabricated from a disc sliced or sawed from a cast rod of formulated styrene modified polyester resin. **D 123,** D13

rod, for staybolts—a round solid section furnished in straight lengths. **B 846,** B05

rod, piston finish—a round rod having a special surface produced by turning or grinding to close tolerances for diameter and straightness. **B 846,** B05

rod proof—a test specimen taken from the melt on an iron rod. **C 162,** C14

rod, shafting—a round rod specially manufactured to the close straightness tolerances required for use in shafting. **B 846,** B05

roll—a length of the product 25 ft (7.62 m) to 50 ft (15.24 m) inclusive, spirally wound into a series of connected turns, with successive turns on top of one another. **B 846,** B05

roll, *v*—*in a vehicle*, the angular motion of a vehicle about its longitudinal axis through the center of gravity. **F 538,** F09

roll angle, (rad or degree), *n*—*in a vehicle*, the angle between the vehicle y-axis and the ground plane. **F 538,** F09

roll compacting—the progressive compacting of metal powders by the use of a rolling mill. Synonymous with **powder rolling. B 243,** B09

rolled clinch—clinch formed by solid clincher against which staple point is driven; obtained normally with desk stapler. **F 592,** F16

rolled compact—a compact made by passing metal powder continuously through a rolling mill so as to form relatively long sheets of pressed material. **B 243,** B09

rolled glass—(*1*) optical glass formed by rolling into plates at time of manufacture, as distinguished from transfer glass.
(*2*) flat glass formed by rolling. **C 162,** C14

rolled seam finish, *n*—*in home sewing*, a seam finish in which both cut edges are enclosed by rolling them to one side and hand stitching the rolled edge close to the seam line within the seam allowance. **D 4965,** D13

roller—{archaic} a blown cylinder for making window glass by the hand process. **C 162,** C14

roller-compacted concrete (RCC), *n*—concrete compacted while fresh by a roller, often a vibratory roller. **C 125,** C09

roller leather—vegetable-tanned sheep or calfskins used for cots or covers on the upper rolls of cotton-spinning machinery. **D 1517,** D31

rollers—any of various rotating members, free or gear driven in an offset press whose functioin is to transfer, apply, or spread fluids onto the cylinders or to guide the paper through the printer. (See also **ductor roller, form rollers, fountain roller, guide rollers,** etc.) **F 425,** F05

roller spatter resistance—the ability of a paint to resist the formation of various size droplets generated during paint roller application, which results in undesirable spots splashing on areas that are not being painted including the person applying the paint. **D 16,** D01

roll-grooved nail—bright or plated, helically grooved, round-wire, stiff-stock, 1 by 0.086 to 4 by 0.164-in. drive-screw nails with no clearance between flutes and head, with flat or slightly countersunk head and medium or long diamond point, with crest diameter being referred to as diameter. **F 547,** F16

roll-grooved, rolled-grooved—provided with four or more continuous symmetrical, longitudinal or helical flutes; resulting from roll-grooving of round wire prior to heading and pointing of nail. Flutes resulting from this manufacturing process extend all the way from head to point (See **fluted.**) **F 547,** F16

rolling—a tannery operation in which the grain surface is compressed and smoothed under pressure by a metal roller. **D 1517,** D31

rolling contact fatigue, *n*—a damage process in a triboelement subjected to repeated rolling contact loads, involving the initiation and propagation of fatigue cracks in or under the contact surface, eventually culminating in surface pits or spalls. **G 40,** G02

rolling direction (*in rolled metals*)—See **longitudinal direction. E 7,** E04

rolling resistance moment, [*FL*], *n*—*of a tire*, the component of a tire moment vector tending to rotate a tire about the Y'-axis, positive clockwise when looking in the positive direction of the Y'-axis. **F 538,** F09

rolling shear—in wood, shear in a longitudinal-transverse plane, with stresses perpendicular to the orientation of the fibers lying in the plane. This shearing force tends to roll the fibers. **D 9,** D07

rolling wear, *n*—wear due to the relative motion between two non-conforming solid bodies whose surface velocities in the nominal contact location are identical in magnitude, direction, and sense. **G 40,** G02

roll response—rotation of the boom from rest due to wave, wind, or current forces. **F 818,** F20

roll roofing—coated felts, either smooth or mineral-surfaced. **D 1079,** D08

roll straightening—*in diffusion coatings*, straightening of metal stock of various shapes by passing it through a series of staggered rolls, the rolls usually being in horizontal and vertical planes. **B 374,** B08

roll thread—a thread produced by action of a form tool which, when pressed into the surface of a blank, displaces material radially. **F 1789,** F16

ROM—**r**ead **o**nly **m**emory. **F 1457,** F05

roof, *n*—the rock material immediately overlying a coal bed. **D 121,** D05

roof—top of excavation or underground opening, particularly applicable in bedded rocks where the top surface of the opening is flat rather than arched. (ISRM) **D 653,** D18

roof and rock bolts—headed hot-rolled bars with cold-rolled or machine-cut threads at the end to be used with anchorage devices to hold up mine roofs, hold back walls, or hold down equipment or foundations. **F 1789,** F16

roof cement—See **flashing cement. D 1079,** D08

roof-deck nail—galvanized, regular-stock-steel and bright, hardened-steel, plain or annularly threaded, 3 by 0.135 to 4 ½ by 0.177-in. nails with flat or slightly countersunk 9⁄32 to 25⁄64-in. head and medium diamond point. **F 547,** F16

roofing fastener—See **purlin nail. F 547,** F16

roofing nail, asphalt-shingle nail, corrugated-roofing nail, sheet-roofing nail—bright or galvanized, plain-shank, barbed or threaded, regular-stock-steel, ¾ by 0.092 to 3 by 0.148-in. nails with flat, checkered, large, reinforced, umbrella lead or cast-lead ¼ to 9⁄16-in. head and medium or long diamond or needle point, often provided with lead conical neoprene or flat plastic washer. Also, aluminum-alloy, plain-shank or helically threaded, ¾ by 0.120 to 2 ¾ by 0.150-in. nails with flat or checkered ⅜ to 9⁄16-in. head and medium diamond or needle point, often provided with conical neoprene washer. (See **cap nail.**) **F 547,** F16

roofing system—an assembly of interacting components designed to weatherproof, and normally to insulate, a building's top surface. **D 1079,** D08

roofing system—assembly of interacting components designed to weatherproof, and sometimes to insulate, the roof surface of a building. **E 631,** E06

roof insulation board—structural insulating board fabricated for use as above-deck roof insulation. **D 1554,** D07

room temperature—a temperature in the range from 20 to 30°C

(68°F). **D 2864,** D27

room temperature—a temperature in the range of 20 to 30°C (68 to 85°F).

NOTE—The term "room temperature" is usually applied to an atmosphere of unspecified relative humidity. **E 41,** G03

room-temperature-setting adhesive, *n*— an adhesive that sets in the temperature range from 20 to 30°C (68 to 86°F), in accordance with the limits for Standard Room Temperature specified in Practice D 618. (Compare **cold-setting adhesive, hot-setting adhesive**, and **intermediate-setting adhesive**.) **D 907,** D14

root—innermost part of thread, rounded or flattened; joining flanks of adjacent threads. **F 547,** F16

root bend—procedure whereby a loading nose is applied along the weld face such that the weld root is subject to tension. **C 904,** C03

root diameter—twice the distance between nail axis and root, measured perpendicular to nail axis. **F 547,** F16

root mean square difference, (*RMSD*)—a measure of accuracy determined by the following equation:

$$RMSD = \left(\frac{1}{n}\sum_{i=1}^{n} e_i^2\right)^{1/2}$$

where:

n = the number of observations for which the accuracy is determined, and

e_i = the difference between a measured value of a property and its accepted value.

E 131, E13

root-mean-square strain—the square root of the mean value of the square of the strain, averaged over one cycle of deformation. (ISO). **D 4092,** D20

root-mean-square stress—the square root of the mean value of the square of the stress, averaged over one cycle of deformation. (ISO) **D 4092,** D20

rope, *n*—a compact and flexible, generally torsionally balanced continuous structure, greater than 4-mm (5⁄32-in.) diameter capable of applying or transmitting tension between two points. **D 123,** D13

rope-lay conductor—see **concentric-lay conductor.** **B 354,** B01

rope-lay conductor, *n*—a conductor composed of a central core surrounded by one or more layers of helically laid groups of strands. **D 1711,** D09

rope mark, *n—in dyed or finished fabrics,* a long irregularly shifting longitudinal mechanically induced streak. **D 123,** D13

roping, *n*—a term used for roving in the woolen system of spinning. **D 123,** D13

ROS—abbreviation for **raster output scanner**. **F 1457,** F05

rose head—pyramidal head with four triangular faces meeting at common vertex, having a square, rectangular, or circularized flat bearing surface. **F 547,** F16

rosin, *n*—a specific kind of natural resin obtained as a vitreous water-insoluble material from pine oleoresin by removal of the volatile oils, or from tall oil by the removal of the fatty acid components thereof. **D 804,** D01

rosin, *n*—a resin obtained as a residue in the distillation of crude turpentine from the sap of the pine tree (gum resin) or from an extract of the stumps and other parts of the tree (wood rosin). (Compare **resin.**) **D 907,** D14

rosin, *n*—a translucent amber to almost black brittle friable resin that is obtained by chemical means from the oleoresin or dead wood of pine trees or from tall oil. **F 141,** F06

rosin acids—acids derived from rosin (colophony) consisting mainly of isomers of abietic acid ($C_{20}H_{30}O_2$) with small amounts of hydroabietic acids ($C_{20}H_{32}O_2$ and $C_{20}H_{28}O_2$). **D 459,** D12

rosin acids or resin acids, *n*—principally monocarboxylic acids with the empirical formula $C_{19}H_{20}$—COOH. **D 804,** D01

rosin adducts, *n*—the addition product between rosin and an α, β unsaturated carboxylic acid such as fumaric acid, acrylic acid or maleic anhydride. **D 804,** D01

rosin based resins, *n*—resinous products derived from rosin or rosin adducts through chemical reaction with raw materials such as alcohols (especially polyhydric alcohols), formaldehyde, α,β unsaturated carboxylic acids, phenols etc., or combinations of these materials. **D 804,** D01

rosin oil, *n*—the relatively viscous, oily portion of the condensate obtained when rosin is subjected to dry destructive distillation; also used to describe specially compounded oils having a rosin oil base. **D 804,** D01

Rosin-Rammler distribution—drop size distribution described by the following equation:

$$f_v(D) = 1 - exp[-(D/D_{RR})]^n$$

where:

$f_v(D)$ = the volume fraction of liquid in drops of diameter less than D_{max},

D_{RR} = the "Rosin-Rammler" diameter ($D_{V0.632}$ from "Rosin-Rammler" distribution, f_v), and

n = parameter indicating the size span of the distribution.

E 1620, E29

rosin reclaimed—See **reclaimed rosin**. **D 804,** D01

rosin spirits, *n*—the relatively light, volatile portion of the condensate obtained in the first stages when rosin is subjected to dry destructive distillation. **D 804,** D01

rosin standards, *n*—the combinations of assembled colored glasses having the colors designated as representative of the established U.S. grades used in classifying rosin. **D 804,** D01

rosin type (sample), *n*—a sample of rosin, or a mold of thermosetting plastic material, used as an unofficial standard in grading rosin. **D 804,** D01

rot—See **decay.** **D 9,** D07

rotameter, *n*—a device, based on the principle of Stoke's law, for measuring rate of fluid flow, consisting of a tapered vertical tube having a circular cross section, and containing a float that is free to move in a vertical path to a height dependent upon the rate of fluid flow upward through the tube. **D 1356,** D22

rotary atomizer—see **centrifugal atomizer**. **E 1620,** E29

rotary conveyor type—See **pot, pan, and utensils washing machine**. **F 1827,** F26

rotary cup atomizer—a centrifugal atomizer in which liquid is fed to the interior surface of a spinning cup, from which it is flung out to form a spray. **E 1620,** E29

rotary-cut veneer—See under **veneer.** **D 1038,** D07

rotary disk atomizer—a centrifugal atomizer in which liquid is introduced near the center of one or more spinning disks, spreads out toward the rim, and is flung out to form a spray. **E 1620,** E29

rotary drilling—a drilling process in which a hole is advanced by rotation of a drill bit under constant pressure without impact. See **percussion drilling**. **D 653,** D18

rotary extraction cleaning, *n*—a cleaning procedure which uses a rotating brush machine or a series of rotating jets through which shampoo is fed or sprayed into the carpet pile and simultaneously removed by suction. **D 123,** D13

rotary press—a machine fitted with a rotating table carrying multiple dies in which a material is pressed. **B 243,** B09

rotary shampoo, *n*—a cleaning procedure which uses a rotating brush machine through which shampoo is fed into the pile of a textile floor covering. See also **rotary extraction cleaning**. **D 123,** D13

rotary smelter—any of the cylindrical smelters that depend on slow rotation about a horizontal axis for agitation of the molten mass. **C 286,** B08

rotary wheel atomizer—a centrifugal atomizer that consists of a rotating, hollow cylinder in which liquid is introduced to the interior and flows through passages to openings, where it is broken up into drops. **E 1620,** E29

rotating synchronized polygons—many faceted polygons, often 8 to 12 facets each, with reflecting coatings deposited on the facets; the devices are rotated at some set of synchronized rates that cause a light beam impinging onto the first polygon to be scanned over the entire surface of an object. **F 1156,** F12

rotation—movement about an axis; such as a body segment about a joint, also used to denote movement about the longitudinal axis of the bone (medial or lateral rotation). **F 869,** F08

rotation angle, ϵ, *n*—angle indicating the orientation of the specimen when it is rotated about a selected axis fixed in it (for plane specimens, usually the specimen normal); *in retroreflection,* angle indicating orientation after rotation about the retroreflector axis. **E 284,** E12

rotation cast button, *n*—a button fabricated from a disc blanked from a partially polymerized sheet formed in a rotating cylinder. **D 123,** D13

rotation plan, *n*—*in tire testing,* the scheduled change of wheel positions for all tires on a vehicle, and between vehicles in a convoy, and scheduled change of vehicle position in a convoy during a road test as appropriate. **F 538,** F09

rotisserie—a device fitted with one more mechanically rotated spits that hold the meat or poultry in position near a fixed heat source while the food is slowly being cooked on all sides. **F 1827,** F26

rotisserie broiler—equipment designed for continuous loading and cooking with vertically stacked spits. See **rotisserie**. **F 1827,** F26

rotisserie oven—equipment designed for batch cooking, with the individual spits arranged on a rotating wheel within a closed cavity. See **rotisserie**. **F 1827,** F26

rotor, *n*—*in open-end spinning machines,* a device resembling a centrifuge, in which the fibers are assembled and in which, by virtue of its rotation, real twist is inserted in the forming yarn. **D 123,** D13

rotor-type open-end spinning machine, *n*—an open-end spinning machine wherein the assembly of individual fibers and the insertion of real twist are effected by a rotor. (See also **open-end spinning machine**.) **D 123,** D13

rotovinyl, *n*—a vinyl flooring with a pattern produced by rotogravure printing. **F 141,** F06

RO train—one of two or more complete RO installations, including membranes and high pressure pump operating in parallel. **D 6161,** D19

rouge pits—{archaic} an imperfection; traces of rouge remaining in an incompletely polished glass surface. **C 162,** C14

rough, *adj*—a descriptive term for a fabric surface which has the feel of sandpaper. **D 123,** D13

rough cut—irregular-shaped areas of uneven corrugations on the veneer surface, occurring as the veneer is cut by the lathe or slicer. **D 1038,** D07

roughened finish, *n*—units whose plane die surfaces are entirely broken by mechanical means, such as wire cutting or wire brushing. **C 43,** C15

rough glass—(*1*) glass pieces obtained by cutting the original sheet of flat glass into workable sizes.

(*2*) rolled or patterned glass. **C 162,** C14

roughing—in leak testing, the initial evacuation of a vacuum system. **E 1316,** E07

roughing—the exposure and erection of the fibers of the material on the lasted over margin of the upper and the corresponding edges of the outsole in order to provide a more grippable surface to be cemented for the purpose of sole attaching. **F 869,** F08

roughing line—in leak testing, a line running from a mechanical pump to a vacuum chamber through which preliminary pumping is conducted in the rough vacuum range. **E 1316,** E07

roughing pump—in leak testing, a vacuum pump used for the initial evacuation of a vacuum system. **E 1316,** E07

rough lumber—See **lumber.** **D 996,** D10

roughness—height of each continuous scallop rounded to the nearest 0.05 in. (1 mm), except those less than 0.03 in. (0.8 mm) vertically and 2 ft (0.6 m) longitudinally. **E 867,** E17

roughness—the more narrowly spaced components of surface texture. **F 1241,** F01

roughness index, half-car (HRI)—see **half-car roughness index.** **E 867,** E17

roughness index, international (IRI)—see **international roughness index.** **E 867,** E17

roughness index, mean (MRI)—see **mean roughness index**. **E 867,** E17

roughometer, *n*—a road meter that measures the unidirectional vertical movement of damped, leaf-sprung wheel relative to the road meter's trailer frame during travel to yield a measure of roughness. **E 867,** E17

rough or burred edge, *n*—*in mica,* a frayed or serrated edge usually 0.8 mm deep or greater, or an edge turned up or down as caused by trimming with scissors, and so forth, or by rubbing the edge against sandpaper, stone, and so forth. **D 1711,** D09

rough, rough-tanned and in-the-rough—terms applied to cattlehide leathers tanned but not finished. (See also **crust.**) **D 1517,** D31

rough sawn—*in plywood,* a decorative treatment produced by scoring across the grain of the panel surface to provide a saw-roughened texture. **D 1038,** D07

round—a set of holes drilled and charged in a tunnel or quarry that are fired instantaneously or with short-delay detonators. (ISRM) **D 653,** D18

round, *n*—a solid member, circular in cross section. **E 631,** E06

round conductor flat cable, *n*—a flat cable made with parallel, round conductors in the same plane. **D 1711,** D09

rounded blade—a blade having a radius on its outer surface which forms a transition between the outer edge and the cutting edges. **F 1078,** F04

round edge, *n*—a rounded, paper-bound edge formation on gypsum board, commonly used for gypsum lath. **C 11,** C11

rounded thread, round-bottom thread—thread with rounded root extending to pitch line. Rounded thread also referring to rounded thread crest. **F 547,** F16

round head—circular head having convex top surface with height equal to its radius. **F 547,** F16

roundness—the ratio of the average radius of curvature of the individual edges of a particle to the radius of the maximum circle that can be inscribed within the particle. **D 4410,** D19

round or rounded crown—curved staple crown used in wiring, carding, and encircling operations; also, in spring-up applications in furniture manufacturing. **F 592,** F16

round-robin test, *n*—the practice of planning, conducting, analyzing, and interpreting the results of interlaboratory tests on the chemical and physical properties of engine coolants. **D 4725,** D15

round shank—term applied to nail with shank of circular cross section, made from round wire. **F 547,** F16

round table—{archaic} a type of laying table used for supporting plate glass for grinding and polishing. **C 162,** C14

routine dosimeter—dosimeter calibrated against a primary-, reference-, or transfer-standard dosimeter and used for routine absorbed-dose measurement. **E 170,** E10

routine maintenance, *n*—*in textile cleaning*, superficial daily or weekly cleaning to remove particulate soil and dust. **D 123,** D13

rove, *n*—*in jute processing*, a continuous strand of spun-jute fiber having a spyndle number of 50 or greater. **D 123,** D13

roving, *n*—in glass textiles, a multiplicity of filaments or yarns gathered together into an approximately parallel arrangement without twist. **C 162,** C14

roving, *n*—a loose assemblage of fibers drawn or rubbed into a single strand, with very little twist. In spun yarn systems, the product of the stage, or stages, just prior to spinning. **D 123,** D13

roving, *n*—*in glass textiles*, a multiplicity of filaments or yarns gathered together into an approximately parallel arrangement without twist. **D 123,** D13

roving, *n*—*in fibrous composites,* large filament count tow. (See also **tow** and **filament count.**) **D 3878,** D30

roving knot—a knotted or entangled section of roving found in a pultrusion.

NOTE—Such a knot may cause high fiber concentration locally and may or may not be visible as a white or light spot on the surface of the section. **D 3918,** D20

row, *n*—a rib or a continuous collection of elements that lie on a circumferential line parallel to the circumferential centerline of the tread band. **F 538,** F09

row/rib wear, *n*—a type of irregular wear characterized by a greater wear rate in one or more rows/ribs; the increased wear rate may occur at one or more circumferential locations in (on) a given row/rib and is independent of (ie. occurs across) individual projections resulting in a step-off in tread depth between adjacent rows/ribs. **F 538,** F09

RS—abbreviation for raw stock. (see **BP**) **F 549,** F05

RS232, *n*—a serial interface standard to connect elements of a communication system. **F 1457,** F05

RSM—Rapidly Solidified Materials. **B 243,** B09

RSP—Rapid Solidification Processing or Rapidly Solidified Powders. **B 243,** B09

RST—Rapid Solidification Technology. **B 243,** B09

R titration—a titration in which the sensing electrode responds to an ion introduced by a reagent prior to titration. A synonym for *indicator titration.* **D 4127,** D19

RTS—abbreviation for **request to send**. **F 1457,** F05

RTV, *n*—room temperature vulcanization (A) **D 907,** D14

rub—abrasion of a glass surface producing a frosted appearance. Also known as a scuff. **C 162,** C14

rubber, *n*—a material that is capable of recovering from large deformations quickly and forcibly, and can be, or already is, modified to a state in which it is essentially insoluble (but can swell) in boiling solvent, such as benzene, methylethyl ketone, and ethanol-toluene azeotrope. **D 123,** D13

rubber, *n*—term not defined by Committee D-20. Definition approved by Committee D-11 on Rubber and Rubber-Like Materials is as follows:

"**rubber**—a material that is capable of recovering from large deformations quickly and forcibly, and can be, or already is, modified to a state in which it is essentially insoluble (but can swell) in boiling solvent, such as benzene, methylethylketone, and ethanol-toluene azeotrope."

"A rubber in its modified state, free of diluents, retracts within 1 min to less than 1.5 times its original length after being stretched at room temperature (18 to 29°C) to twice its length and held for 1 min before release." **D 883,** D20

rubber—a material that is capable of recovering from large deformations quickly and forcibly, and can be, or already is, modified to a state in which it is essentially insoluble (but can swell) in boiling solvent such as benzene, methyl ethyl ketone, and ethanol-toluene azeotrope. **D 1079,** D08

rubber, *n*—a material that is capable of recovering from large deformations quickly and forcibly, and can be, or already is, modified to a state in which it is essentially insoluble (but can swell) in boiling solvent, such as benzene, methyl ethyl ketone, or ethanol-toluene azeotrope. **D 1566,** D11

rubber—a material that is capable of recovering from large deformations quickly and forcibly, and can be, or already is, modified to a state in which it is essentially insoluble (but can swell) in boiling solvent, such as benzene, methylethylketone, and ethanoltoluene azeotrope.

A rubber in its modified state, free of diluents, retracts within 1min to less than 1.5 times its original length after being stretched at room temperature (18 to 29°C) to twice its length and held for 1 min before release. (D-11) **F 412,** F17

rubber—a generic term that includes elastomers and elastomer compounds, regardless of origin. **F 819,** F18

rubber—a material that is capable of recovering from large deformations quickly and forcibly, and can be, or already is, modified to a state in which it is essentially insoluble (but can swell) in boiling solvent, such as benzene, methyl ethyl ketone, and ethanoltoluene azeotrope. (D 1566, D-11) **F 869,** F08

rubber, *n*—an elastic substance derived from various tropical plants, such as the general Hevea and Ficus, essentially a polymer of isoprene; the term is frequently applied to both natural and synthetic elastic substances. (*Webster Modified*)

saran plastic—see **vinylidene chloride plastic.** **F 1251,** F04

rubber, bound, *n*—the portion of the rubber in a mix that is so closely associated with the filler as to be unextractable by the usual rubber solvents. **D 1566,** D11

rubber, cellular, closed cell, *n*— a cellular material in which practically all the individual cells are nonconnecting. **D 1566,** D11

rubber compound, *n*—*as used in the manufacture of rubber articles*, an intimate mixture of elastomer(s) with all the materials necessary for the finished article. **D 123,** D13

rubber, crude, n—See **rubber raw**, the preferred term. **D 1566,** D11

rubber, expanded, *n*—cellular rubber having closed cells made from a solid rubber compound. **D 1566,** D11

rubber flooring, *n*—a floor surfacing material in sheet or tile form consisting of compounded natural or synthetic rubbers or both in combination with mineral fillers, pigments, and other additives. **F 141,** F06

rubber gasket—rubber formed and used as a seal in concrete pipe joints. **C 822,** C13

rubber, gel, *n*—the portion of rubber insoluble in a chosen solvent. (See also **rubber, sol.**) **D 1566,** D11

rubber hardness degree, international, *n*—a measure of hardness, the magnitude of which is derived from the depth of penetration of a specified indentor into a specimen as described in D 1415. **D 1566,** D11

rubber heel marking—the mechanical transfer of coloring matter from rubber heels to surfaces. **D 2825,** D21

rubberize, *v*—to impregnate or coat with rubber compound or both. **D 123,** D13

rubberize, *n*—to impregnate or coat, or both, a substrate with rubber. **D 1566,** D11

rubber latex, centrifuged, *n*—latex, the rubber concentration of which has been increased by the removal of serum by centrifugal force. **D 1566,** D11

rubber latex, evaporated, *n*—latex, the rubber concentration of which has been increased by evaporation of some of the water. **D 1566,** D11

rubber latex, preserved, *n*—rubber latex treated to inhibit putrefaction and accompanying coagulation. **D 1566,** D11

rubber latex, prevulcanized, *n*—a rubber latex in which the particles have been sufficiently vulcanized to produce films and useful articles by drying only. **D 1566,** D11

rubber latex, stabilized, *n*—rubber latex treated to inhibit premature coagulation. **D 1566,** D11

rubber, natural, *n*—a *cis*-1,4-polyisoprene that is obtained from a botanical source, usually *Hevea brasiliensis*. **D 1566,** D11

rubber, oil-extended, *n*—a grade of raw rubber containing a relatively high proportion of processing oil. **D 1566,** D11

rubber products, *n*—items of commerce in which the major portion of the filler-bearing material is a rubber. **D 1566,** D11

rubber, raw, *n*—natural or synthetic rubber, usually in bales or packages that is the starting material for the manufacture of rubber articles. **D 1566,** D11

rubber reversion hydroplaning, *n*—hydroplaning of pneumatic tires with separation caused by devulcanized rubber. **E 867,** E17

rubber, skim, *n*—rubber obtained by coagulating the rubber in the serum that separates out during the concentration of natural rubber latex. **D 1566,** D11

rubber, sol, *n*—the portion of rubber soluble in a chosen solvent. (See also **rubber, gel.**) **D 1566,** D11

rubber, sponge, *n*—cellular rubber consisting predominantly of open cells and made from a dry rubber compound. **D 1566,** D11

rubber, synthetic, *n*—rubber produced by polymerizing one or more monomers with or without post-polymerization chemical modification. **D 1566,** D11

rubber, virgin(s), *n*—obsolete term. See **wet spots**. **D 1566,** D11

rubbing stone—a shaped abrasive used in stoning porcelain enamel. **C 286,** B08

rubbing strips—boards nailed to the underside of **skids,** or floor boards of **crates** or **boxes** to provide a surface upon which to slide the containers. (See also **skid.**) **D 996,** D10

rubble, *n*—rough stones of irregular shapes and sizes, broken from larger masses either naturally or artificially, as by geological action, in quarrying, or in stone cutting or blasting. **D 8,** D04

rug, *n*—a textile floor covering of limited area which is complete in itself and is intended for use as a partial covering of a floor or another floor covering. **D 123,** D13

ruggedness, *n*—insensitivity of a test method to departures from specified test or environmental conditions. **E 456,** E11

ruggedness test, *n*—an experiment in which environmental or test conditions are deliberately varied to evaluate the effect of such variations. **D 123,** D13

ruggedness test, *n*—a factorial test designed to explore the sensitivity of the method to variations in the procedure (See Youden and Steiner, 1975). **D 1356,** D22

ruggedness test, *n*—a planned experiment in which environmental factors or test conditions are deliberately varied in order to evaluate the effects of such variation. **E 456,** E11

run, *n*—*in experimenting or testing*, a single performance or determination using one of a combination of experimental or test conditions. **D 123,** D13

run, *n*—*in knitted fabrics*, a series of dropped stitches. (See **dropped stitch.**) **D 123,** D13

run, *n*—*in the American woolen system*, the number of 1600-yd lengths of yarn per pound; an indirect yarn numbering system generally used for yarns spun on the woolen system. **D 123,** D13

run, *n*—*(1)* an uninterrupted sequence of occurrences of the same attribute or event in a series of observations, and *(2)* a consecutive set of successively increasing run-up or successively decreasing run-down values in a series of variable measurements. **E 456,** E11

run—the printing of the specific number of copies required on a particular job. **F 425,** F05

rundown, *n*—*in building construction*, bloom or fluid migration from a sealant, coating, or membrane that due to natural forces is carried down and accumulates on vertical and sloping surfaces. **C 717,** C24

run-in, *n*—*in tribology*, an initial transition process occurring in newly established wearing contacts, often accompanied by transients in coefficient of friction, or wear rate, or both, which are uncharacteristic of the given tribiological system's long term behavior. (*Synonym:* **break-in, wear-in.**) **G 40,** G02

run in, *v*—*in tribology*, to apply a specified set of initial operating conditions to a tribological system to improve its long term frictional or wear behavior, or both. (*Synonym:* **break in,** *v.* and **wear in,** *v.* See also **run-in,** *n:*) **G 40,** G02

runner, *n*—*(1)* the secondary feed channel in an injection or transfer mold that runs from the inner end of the sprue or pot to the cavity gate.

(2) the piece formed in a secondary feed channel or runner. **D 883,** D20

runner, *n*—*(1)* the secondary feed channel in an injection or transfer mold that runs from the inner end of the sprue to the cavity gate.

(2) the piece formed in a secondary feed channel or runner. (D-20) **F 412,** F17

runner bar—{archaic} iron casting attached to a circular grinding runner or head for abrasive grinding of plate glass. **C 162,** C14

runner cut—{archaic} an imperfection in plate glass resulting from fracture of the surface by the runner bar. **C 162,** C14

runner (injection or transfer mold), *n*—the secondary feed channel for transferring material under pressure from the inner end of the sprue to the cavity gate. **D 1566,** D11

running batch—the regular batch formulated to produce the desired composition when used with its own cullet. **C 162,** C14

running ground—*in tunneling*, a granular material that tends to flow or "run" into the excavation. **D 653,** D18

running sample—See **sample.** **D 1356,** D22

running surface—the entire bottom surface of the ski bordered by the side geometry. **F 472,** F27

running surface—the entire bottom surface of the snowboard ski bordered by the side geometry. **F 1107,** F27

runoff—that part of precipitation appearing in surface streams. **D 4410,** D19

run-off application—application of a liquid material using sufficient volume such that it begins to drip from the target. **E 1102,** E35

run-of-kiln quicklime—quicklime as drawn or discharged from a kiln. **C 51,** C07

run-of-mine (ROM) coal—*for the purpose of Test Method D 4749,* (*1*) *in the case of an underground mine*, it is that coal delivered to the surface by a slope belt, hoist, and so forth

(*2*) *in the case of a surface mine*, it is that coal as it exists after it has been removed from the pit and placed into the initial means of transportation whether it be an on-the-road or off-the-road haul truck, dump hopper which feeds a pit-to-plant conveyor, and so forth. For both underground and surface mines, ROM coal is as-mined and has not been exposed to any treatment such as breaking, crushing, or cleaning except for that done by the normal operations used to extract the coal from the ground, that is, blasting, ripping, loading, cutting, and so forth. **D 121,** D05

run-out—no fatigue failure at a specified number of load cycles. **E 1823,** E08

rupture, *n*—the breaking or tearing apart of a material. (Compare **failure.**) **D 123,** D13

rupture—that stage in the development of a fracture where instability occurs. It is not recommended that the term rupture be used in rock mechanics as a synonym for fracture. (ISRM) **D 653,** D18

rupture envelope (*rupture line*)—see **Mohr envelope**. **D 653,** D18

rural areas, *n*—individual agencies may have a political boundary definition of rural, which would be the definition for traffic-monitoring purposes; in the absence of such an individual agency definition, a rural area is defined as an area outside of a community with a population of 5000 persons or more. **E 867,** E17

russet—a term of varied meaning in the leather trade, since it connotes both color and tunnage:

(*1*) *russet calf*—The natural color of unfinished calf leather resulting from tannage by vegetable extracts.

(*2*) *russet harness*—A completely finished leather of bright, clean, uniform color, and finish.

(*3*) *russet sheepskin*—Leather tanned in cold-leached hemlock bark, used for shoe linings, with color resulting from the hemlock.

(*4*) *russet upholstery*—leather tanned but not finished. **D 1517,** D31

Russian leather—originally a Russian calfskin shoe leather, distinguished by its odor of birch oil. Now, in the United States a fancy calfskin stock. **D 1517,** D31

rust—a corrosion product consisting primarily of hydrated iron oxide. (A term properly applied only to ferrous alloys.) **G 15,** G01

rust (*coatings*)— the reddish material, primarily hydrated iron oxide, formed on iron or its alloys resulting from exposure to humid atmosphere or chemical attack. See **white rust.** **D 16,** D01

rustproof—made of nonferrous material or protectively coated, plated, galvanized or aluminized. (See **corrosion-resistant, rust-resistant.**) **F 547,** F16

rust resistance (*coatings*)—the ability of a coating to protect the substrate of iron or its alloys from rusting. **D 16,** D01

rust-resistant—term used to describe a material or treatment designed to inhibit rusting; not synonymous with rustproof. **F 547,** F16

rut, *n*—a contiguous longitudinal depression deviating from a surface plane defined by transverse cross slope and longitudinal profile. **E 867,** E17

rut, *n*—a contiguous longitudinal depression deviating from a surface plane defined by transverse cross slope and longitudinal profile. **E 1778,** E17

rut-depth—the maximum measured perpendicular distance between the bottom surface of the straightedge and the contact area of the gage with the pavement surface at a specific location. **E 867,** E17

rutile—a mineral form of titanium oxide (TiO_2) (tetragonal crystallization), but usually produced chemically for use in ceramics and other products. **C 242,** C21

R-value—a measure of the resistance to heat flow through a material calculated as $R = t/k$, where t is the thickness of the material and k is the thermal conductivity. **C 162,** C14

ryznar stability index (RSI)—an index indicating if a water has a tendency to corrode or precipitate $CaCO_3$; equals 2 pH ($CaCO_3$ saturation) - pH (actual), RSI < 6.0 scale formation, > 7.0 corrosive. **D 6161,** D19

S

S_{10}—the portion of a cellulose pulp that is soluble in 10 % sodium hydroxide using Test Method D 1696, or its equivalent. **D 1695,** D01

S_{18}—the portion of a cellulose pulp that is soluble in 18 % sodium hydroxide using Test Method D 1696, or its equivalent. **D 1695,** D01

"S"—clean with dry solvent cleaner only. **D 5253,** D13

S—effective scattered neutron content. *S* is the percent background film darkening caused by scattered neutrons. **E 1316,** E07

sabin, [L^2]—the unit of measure of sound absorption in the inch-pound system. **C 634,** E33

saccharification—the breaking of dextrins (starch) into simple sugars (hydrolysis). **E 1705,** E48

sack—See **bag.** **D 996,** D10

sacrificial protection—the form of corrosion protection wherein one metal corrodes in preference to another, thereby protecting the latter from corrosion. **B 374,** B08

saddle—See **packing.** **D 996,** D10

saddle, *n*—a construction to channel or direct surface water to drains or off the roof. A saddle may be located between drains or in a valley, and is often like a small hip roof or a pyramid with a diamond-shaped base. (See **cricket**). **D 1079,** D08

saddle curve—in a phase diagram, the locus of a series of maximum points which itself passes through a minimum value. **E 7,** E04

saddle-fused joint—see **joint, saddle-fused.** **F 412,** F17

saddle leather—vegetable-tanned cattlehide leather for harness and saddlery, usually of a natural tan shade and rather flexible. **D 1517,** D31

saddle point—the minimum point on a saddle curve, which is a univariant point, a composition of congruent melting. **E 7,** E04

saddle stapling—pamphlet or book stapling on a "V"-shaped table, which permits placement of staple in center fold. **F 592,** F16

saddle stitching—Similar to saddle stapling; however, accomplished with wire stitcher. **F 592,** F16

saddle-tie, *n, for furring*—a single or double strand of wire used to attach furring members to framing members of wall or ceiling assemblies. **C 11,** C11

saddle-tie, *v, for furring*—to attach furring members to framing members of wall or ceiling assemblies using a single or double strand of wire. **C 11,** C11

saddle-tie, *v, for wire hangers*—to attach wire hangers to main runners. **C 11,** C11

safety diver, *n*—diver who is dressed completely in dive gear and who remains on the surface to go to the immediate aid of a distressed diver underwater. **F 1549,** F32

safety glass—flat (including bent) glass so constructed, treated, or combined with other materials that, if broken by human contact, the likelihood and/or severity of cutting and piercing injuries that might result from such contact is reduced. See related terms "laminated glass," "fully tempered glass," "tempered glass," and "wired glass." **C 162,** C14

safety mechanism—a device intended to prevent accidental actuation of tool. **F 592,** F16

safety requirements—those regulations designed to protect against personal injury or property damage resulting from system failure, breakage, or accident. **C 859,** C26

safety stop, *n*—additional time spent at shallow depths even though the dive computer or dive table indicates that a direct ascent to the surface is possible. **F 1549,** F32

sag flow test—See **test, sag flow.** **E 1749,** E06

sagging—process of forming glass by reheating until it conforms to the shape of the mold or form on which it rests. **C 162,** C14

sagging—(*1*) a defect characterized by a wavy line or lines appearing on those surfaces of porcelain enamel that have been fired in a vertical position.

(*2*) A defect characterized by irreversible downward bending in an article insufficiently supported during the firing cycle.

NOTE—See the Proposed Method of Test for Sag Resistance of Steel Sheets for Porcelain Enameling, *Proceedings*, Am. Soc. Testing Mats., Vol 55, 1955, p. 431. **C 286,** B08

sagging—usually occurs in sedimentary rock formations as a separation and downward bending of sedimentary beds in the roof of an underground opening. (ISRM) **D 653,** D18

sagittal—pertaining to the anteroposterior median plane of the body (sagittal plane), the median vertical plane of the body dividing it into right and left halves that is the major plane for walking and running. **F 869,** F08

sag or sagging—nonuniform downward flow of a wet paint film that occurs between the times of application and setting, resulting in an uneven coating having a thick lower edge. **D 16,** D01

sags—downward movement of a paint film between the time of application and setting, resulting in an uneven coating having a thick lower edge. **D 4538,** D33

saladero hides—Argentinian hides corresponding to small-packer hides in the United States. **D 1517,** D31

salamander—a medium duty broiler, with approximately half the depth of an upright, and generally mounted above a commercial range. See **broiler, overfired**. **F 1827,** F26

sales code diameter—the overall diameter of a tin-plate fabricated aerosol can, taken from the double seam at the bottom of the container. **D 3064,** D10

sales code height—the height of a tin-plate fabricated aerosol can, taken from between the double seams at the top and bottom of the cylinder. **D 3064,** D10

sales square—the quantity of prepared roofing required to cover 9.3 m^2 (100 ft^2) of deck. **D 1079,** D08

saline electrolyte, *n*—a solution customarily consisting of the chlorides of the alkali metals. **B 899,** B02

salinity—the concentration of dissolved matter found in water after bromide and iodide have been replaced by an equivalent quantity of chloride, all carbonate converted to oxide, and all organic matter destroyed. **D 1129,** D19

salinity—the concentration of inorganic salts in water. **D 6161,** D19

salt bath, *n*—a heat transfer apparatus, utilizing molten salts as the heating medium, generally used for vulcanization. See **liquid curing medium (LCM).** **D 1566,** D11

salt cake—an industrial term for sodium sulfate (Na_2SO_4). **C 162,** C14

salt flux—amount of dissolved salt passing through the membrane, moles per day per square unit of membrane area. **D 6161,** D19

salt glaze, *n*—units whose surface faces have a lustrous glazed finish from the thermo-chemical reaction of the silicates of the clay body with vapors of salt or chemicals. **C 43,** C15

salt glaze—a glaze produced by the reaction, at elevated temperature, between the ceramic body surface and salt fumes produced in the kiln atmosphere. **C 242,** C21

salt passage, SP—

$$SP = \frac{CP \times 100,}{CF}$$

the ratio of product (*CP*) and feed (*CF*) salt concentrations expressed as percent. **D 6161,** D19

salt rejection, SR—(100 - salt passage) expressed as percent:

$$SR = 100\left(1 - \frac{CP}{CF}\right)$$

D 6161, D19

salt stain—discoloration on the surface of hides and skins, developed during the curing process. **D 1517,** D31

salt water—See **gall.** **C 162,** C14

salt water, *n—*the aerobic, aqueous compartment, characteristically with a salinity equal to or greater than five parts per thousand. **D 6384,** D02

salty, (taste), *adj—*taste produced by substances such as sodium chloride when in solution. **E 253,** E18

salvage—waste material or used material—often having reached or exceeded the originally intended service life but that still has some residual utility; often either after rebuilding or reworking by someone other than the **OEM**, or one designated by the **OEM**, or an item resold as first quality but with less than its original utility or one that has been obtained by means of illegitimate means. **F 1156,** F12

salvage value—the value of an asset, assigned for tax computation purposes, that is expected to remain at the end of the depreciation period. **E 833,** E06

sample, *n—*one or more portions of a liquid or solid material taken in an unbiased manner from a batch, heat, lot, or process stream to be representative of the whole, for subsequent testing to determine the chemical, physical, mechanical, or other quality characteristics of the material, or combination thereof. **A 644,** A04

sample, *n—*a portion of the material in a lot, selected according to a specific sampling plan, intended to represent the lot. **A 902,** A05

sample—the final form of a material submitted for chemical analysis (drillings, millings, and so forth) or a prepared test specimen to be used for mechanical testing. **B 846,** B05

sample—the number of specimens drawn from a lot. **C 822,** C13

sample—each piece or group of pieces selected from a lot and used to determine whether the product complies with the specification criteria. **C 896,** C04

sample, *n—*a quantity of material taken from a larger quantity for the purpose of estimating properties or composition of the larger quantity.

*analysis sample, n—*final subsample prepared from the original gross sample but reduced to 100 % through No. 60 (250-μm) sieve and divided to not less than 50 g.

*coarse coal, n—*that portion of a coal sample being subjected to a washability study that is larger than a specific predetermined particle size, generally between 2.36 mm (No. 8 USA Standard Sieve Series) and 9.5 mm (3⁄8 in.) round in diameter.

*divided sample, n—*a sample that has been reduced in quantity.

*fine coal, n—*that portion of a coal sample being subject to a washability study that is smaller than the predetermined particle size, generally between 2.36 mm (No. 8 USA Standard Sieve Series) and 9.5 mm (3⁄8 in.) round in diameter, which is specified in Test Method D 4371.

*gross moisture sample, n—*a sample representing one lot of coal and composed of a number of increments on which neither reduction nor division has been performed or a subsample for moisture testing taken.

*gross sample, n—*a sample representing one lot of coal and composed of a number of increments on which neither reduction nor division has been performed.

*laboratory sample, n—*the sample, not less than the permissible weight delivered to the laboratory for further preparation and analysis.

*representative sample, n—*a sample collected in such a manner that every particle in the lot to be sampled is equally represented in the gross or divided sample.

*subsample, n—*a sample taken from another sample.

*unbiased sample (representative sample)—*a sample free of bias. **D 121,** D05

sample, *n—*(*1*) a portion of a lot of material which is taken for testing or for record purposes.

(*2*) a group of specimens used, or observations made, which provide information that can be used for making statistical inferences about the population(s) from which the specimens were drawn. (See also **lot sample, laboratory sample,** and **specimen**) **D 123,** D13

sample—piece or quantity of bulk material that has been selected by some sampling process. **D 653,** D18

sample, *n—*a small part or portion of a material or product intended to be representative of the whole. **D 883,** D20

sample—a discreet portion of matrix intended to represent its source. **D 1129,** D19

sample, *n—*a portion of a population intended to be representative of the whole.

*cumulative sample, n—*a sample obtained over a period of time with (*1*) the collected atmosphere being retained in a single vessel, or (*2*) with a separated component accumulating into a single whole.

*operating humidity range of sample, n—*the range of ambient relative humidity of air which passes through the analyzer's sensing system, over which the monitor will meet all performance specifications.

*operating temperature range of sample, n—*the range of ambient temperatures of air, which passes through the analyzer's sensing system, over which the analyzer will meet all performance specifications.

*personal sample, n—*a sample representative of air-borne dust that is likely to enter the test subject's respiratory system and which is therefore taken by a collection apparatus (membrane filter) positioned near the nose and mouth of the subject or in the breathing zone.

*running sample, n—*withdrawal of a portion of the atmosphere over a period of time with continuous analysis or with separation of the desired material continuously and in a "linear" form. **D 1356,** D22

sample—a sample consists of one or more units of product drawn from a lot, the units of the sample being selected at random without regard to their quality. **D 1517,** D31

sample, *n—*the portion or unit(s) selected to represent the lot. **D 1566,** D11

sample, *n—*one or more units of product taken from a lot without regard to the quality of the unit. (Also often termed lot sample). **D 1711,** D09

sample, *n—of paper or paperboard,* a specified number of test units selected in accordance with a prescribed procedure to represent the lot. See **lot**. **D 1968,** D06

sample, *n*—a portion of carbon black selected for use in obtaining a test result. **D 3053,** D24

sample, *n*—(*1*) a portion of material which is taken for testing or for record purposes. (*2*) a group of specimens used, or of observations made, which provide information that can be used for making statistical inferences about the population(s) from which the specimens are drawn. (See also **laboratory sample, lot sample,** and **specimen.**) **D 4439,** D35

sample, *n—in methods of chemical analysis,* a portion of a material selected and processed to render its composition representative of the composition of the whole. (Contrast **specimen**). **E 135,** E01

sample, *n*—a small part or portion of a material or product intended to be representative of the whole. **E 284,** E12

sample, *n*—a group of items, observations, test results, or portions of material, taken from a large collection of items, observations, test results, or quantities of material, which serves to provide information that may be used as a basis for making a decision concerning the larger collection. **E 456,** E11

sample—a portion of material taken from a larger quantity for the purpose of estimating properties or composition of the larger quantity. See **analysis sample, gross sample, laboratory sample, and representative sample.** **E 856,** D34

sample, *n*—a portion of material taken from a larger quantity for the purpose of estimating properties or composition of the larger quantity. **E 1605,** E06

sample, *n*—a portion, or part, of a quantity of material. **E 1638,** E29

sample—the specimens from the population selected for test purposes. **E 1823,** E08

sample—one or more units of product randomly selected from a lot to represent that lot. **F 412,** F17

sample, *n*—a selected number of *n* test objects that accurately represent the lot or population of interest. **F 538,** F09

sample, *n*—a group of units or portion of material, taken from a larger collection of units or quantity of material, which serves to provide information that can be used as a basis for action on the larger quantity. **G 113,** G03

sample, all-levels (liquid sample)—one obtained by submerging a closed sampler to a point as near as possible to the draw-off level, then opening the sampler and raising it at a rate such that it is about three-fourths full as it emerges from the liquid. An all-levels sample is not necessarily an average sample because the tank volume may not be proportional to the depth and because the operator may not be able to raise the sampler at the variable rate required for proportionate filling. The rate of filling is proportional to the square root of the depth of immersion. **E 1547,** E15

sample average (arithmetic average)—the sum of all the observed values in a sample divided by the sample size. It is a point estimate of the population mean. **E 1823,** E08

sample, average (liquid sample)—one that consists of proportionate parts from all sections of the container. **E 1547,** E15

sample, bottom (liquid sample)—one obtained from the material on the bottom surface of the tank, container, or line at its lowest point. (Bottom samples are usually taken to check for water, sludge, scale, etc.) **E 1547,** E15

sample, composite, compartment-tank (liquid sample) (ship, barge, etc.)—a blend of individual all-levels samples from each compartment that contains the product being sampled in proportion to the volume of material in each compartment. **E 1547,** E15

sample, composite, single-tank (liquid sample)—a blend of the upper, middle, and lower samples. For a tank of uniform cross section, such as an upright cylindrical tank, the blend consists of equal parts of the three samples. For a horizontal cylindrical tank, the blend consists of the three samples in the proportions shown in E 1547. **E 1547,** E15

sample contacting equipment—equipment that comes in direct contact with the sample or portion of sample that will undergo chemical analyses or physical testing (for example, ground water well bailer, split-spoon sampler, soil gas sampling probe). **D 653,** D18

sample, continuous (liquid sample)—one obtained from a pipeline conveying the product in such a manner as to give a representative average of the stream throughout the period of transit. **E 1547,** E15

sample division, *n*—the process whereby a sample is reduced in weight without change in particle size. **D 121,** D05

sample division—the process of obtaining a smaller sample from a larger sample so that the representative properties of the larger sample are retained. During this process it is assumed that no change in particle size or other characteristics occurs. **E 856,** D34

sample, drain (liquid sample)—one obtained from the draw-off or discharge valve. Occasionally, a drain sample may be the same as a bottom sample, as in the case of a tank car. **E 1547,** E15

sampled zone—that part of a transect presumed to be wholly represented by sediment samples. **D 4410,** D19

sample for examination—a specified number of units taken from a lot for the purpose of visual, dimensional or tactile inspection. **D 1517,** D31

sample for examination—a specified number of units taken from a lot for the purpose of visual, dimensional or tactile inspection. **F 869,** F08

sample for test—a specified number of sample units taken from a lot for the purpose of testing the lot for all physical and chemical properties for which requirements are specified. **D 1517,** D31

sample for test—a specified number of sample units taken from a lot for the purpose of testing the lot for all physical and chemical properties for which requirements are specified. **F 869,** F08

sample, gross, *n*—a single large sample obtained by placing in a single container two or more increments taken from a lot. **E 135,** E01

sample, gross (solid sample)—a composite prepared by mixing the increments. **E 1547,** E15

sample integrity—the unimpaired chemical composition of a test sample upon the extraction of said test sample for analysis. **E 176,** E05

sample, jar (liquid sample)—one obtained by placing a jar into the path of a free-flowing stream so as to collect a definite volume from the full cross section of the stream. **E 1547,** E15

sample, laboratory—See **laboratory sample**. **D 4439,** D35

sample, laboratory, *n*—a subsample selected from the properly prepared gross sample of a lot for submission to one or more laboratories for chemical analysis. See **select**. **E 135,** E01

sample, laboratory (solid sample)—that portion of the subsample that is sent to the laboratory for testing. **E 1547,** E15

sample, lot—See **lot sample**. **D 4439,** D35

sample median—the (*1*) middle value when all observed values in a sample are arranged in order of magnitude if an odd number of items (units) are tested or (*2*) the average of the two middle-most values if an even number of items (units) are tested. It is a point estimate of the population median, or 50 % value. **E 1823,** E08

sample, middle (liquid sample)—one obtained from the middle of the tank contents. **E 1547,** E15

sample, mixed (liquid sample)—one obtained after mixing or vigorously stirring the contents of the original container, and then pouring out or drawing off the quantity desired. **E 1547,** E15

sample, outlet (liquid sample)—one normally obtained at the level of the tank outlet (either fixed or a swing line outlet). **E 1547,** E15

sample pathlength, *b*—*in a spectrophotometer*, the distance, measured in the direction of propagation of the beam of radiant energy, between the surface of the specimen on which the radiant energy is incident and the surface of the specimen from which it is emergent. **E 131,** E13

sample percentage—the percentage of observed values between two stated values of the variable under consideration. It is a point estimate of the percentage of the population between the same two stated values. (One stated value may be ''minus infinity" or ''plus infinity.") **E 1823,** E08

sample preparation, *n*—the process that may include air drying, crushing, division, and mixing of a gross sample for the purpose of obtaining an unbiased analysis sample. **D 121,** D05

sample preparation—the process that includes drying, size reduction, division, and mixing of a laboratory sample for the purpose of obtaining an unbiased analysis sample. **E 856,** D34

sample, prepared, *n*—a sample on which comminution, division, blending, or other procedures have been performed to make the sample ready for analysis. **E 135,** E01

sampler, *n*—a device in which or through which atmospheric or emission samples are collected for analysis.

constant flow high-volume sampler, n—a sampler that is equipped with a constant flow device.

Hi-Vol (high-volume air sampler), n—a device for sampling large volumes of an atmosphere; collecting the contained particulate matter by filtration; and consisting of a high-capacity vacuum pump, a filter to collect suspended particles, and means for measuring or controlling the flow rate, or both. **D 1356,** D22

sample reduction, *n*—the process whereby a sample is reduced in particle size by crushing or grinding without significant change. **D 121,** D05

sample reduction—the process whereby sample particle size is reduced without change in sample weight (see milling). **E 856,** D34

sample size—the number of units of product in the sample. **D 1517,** D31

sample size, *n*—the number of units of product taken to make up the sample. **D 1711,** D09

sample size, *n*—the number of units in a sample or the number of observations in a sample. **E 456,** E11

sample size—the number of units of product in the sample. **F 869,** F08

sample skein, *n*—skein reeled from the package or beam of the laboratory sample, and used in the laboratory as a source of specimens. **D 123,** D13

sample spectrum, *n*—a spectrum, either single-beam or ratioed, that contains spectral features due to an analyte of interest. **E 131,** E13

sample standard deviation, *s*—the square root of the sample variance. It is a point estimate of the population standard deviation, a measure of the "spread" of the frequency distribution of a population. **E 1823,** E08

sample, sub (solid sample)—a smaller sample produced in a specified manner by the reduction in volume or quantity of the gross sample. **E 1547,** E15

sample system—See **analyzer.** **D 1356,** D22

sample, test, *n*—a subsample selected from the properly prepared laboratory sample which has a suitable sample weight or volume for one or more determinations by chemical analysis. **E 135,** E01

sample thief—a device used to obtain samples from containers of material. **C 859,** C26

sample, top (liquid sample)—one normally obtained 6 in. (152 mm) below the top surface of the tank contents. **E 1547,** E15

sample, tube or thief (liquid sample)—one obtained with a sampling tube or special thief, either as a core sample or spot sample from the specified point in the container. **E 1547,** E15

sample, umpire, *n*—a laboratory sample submitted to a laboratory of recognized capability chosen to resolve a difference. **E 135,** E01

sample unit (for test purposes)—the total quantity of material necessary to obtain one test result for each of the properties and characteristics specified in the material specification or procurement document. In testing of small package units, the Sample Unit may be a package unit randomly selected from the material representing the lot. In testing commodities in which the units are individually too small to provide sufficient material for evaluating all the properties specified in the material specification, the Sample Unit may be a sufficient amount of the material, taken as an aggregate to provide the quantity of material required. **D 1517,** D31

sample unit (for test purposes)—the total quantity of material necessary to obtain one test result for each of the properties and characteristics specified in the material specification or procurement document. **F 869,** F08

sample, upper (liquid sample)—one obtained from the middle of the upper third of the tank contents. **E 1547,** E15

sample variance, s^2—the sum of the squares of the differences between each observed value and the sample average divided by the sample size minus one. It is a point estimate of the population variance. **E 1823,** E08

sample weight, *n*—the amount of test material determined by weighing with a balance. **E 135,** E01

sampling—process of selecting samples from a lot for use in testing. **C 896,** C04

sampling—obtaining a representative portion of the material concerned. **D 1129,** D19

sampling, *n*—a process consisting of the withdrawal or isolation of a fractional part of the whole.

active sampling, n—a means of collecting an airborne or emission substance that employs a mechanical device such as a pump or vacuum-assisted critical orifice to draw air or emissions into or through the sampling device.

collocated sampling, n—the simultaneous collection of two or more air or emission samples by samplers placed side-by-side (often mistakenly termed *co-located* sampling), placed close enough to each other to ensure that comparable samples are collected, but separated sufficiently to prevent cross-interference.

condensation sampling, n—a process consisting of the collection of one or several components of a gaseous mixture by simple cooling of the gas stream in a device which retains the condensate.

continuous sampling, n—sampling without interruptions throughout an operation or for a predetermined time.

cryogenic sampling, n—the collection of an air or emission substance by condensation in a trap cooled to a temperature sufficient to condense or freeze the substance being collected, usually used in near real-time or sequential monitoring. Also called *freeze-trapping*.

event sampling, n—a special form of intermittent sampling where the duration of a sampling period is defined as a single, discrete occurrence of precipitation, dew, fog, or frost.

grab sampling, n—the collection of an atmospheric or emission substance without regard to variations, whether temporal or spatial.

instantaneous sampling, n—obtaining a sample of an atmosphere in a very short period of time such that this sampling time is insignificant in comparison with the duration of the operation or the period being studied.

intermittent sampling, n—sampling successively for limited periods of time throughout an operation or for a predetermined period of time.

isokinetic sampling, n—sampling in which the linear velocity and direction of the fluid entering the sampling nozzle is equal to the undisturbed fluid stream at the sample point.

reactive sampling, n—the collection of an air or emission substance by reacting it with a chemical reagent (for example, derivatization).

sampling period(s), n—the record length or interval over which data collection occurs.

sampling rate (Hz), n—the rate at which data collection occurs, usually presented in samples per second (Hertz).

sampling train, n—the assemblage of equipment necessary to sample atmospheres.

sequential sampling—See **sample, running.**

source sampling, n—withdrawal, with or without simultaneous isolation of specific components, of a portion of the offgases from a source of pollutants. **D 1356,** D22

sampling—see **digitization.** **E 131,** E13

sampling—a process whereby a test sample is extracted from a fire test environment. **E 176,** E05

sampling, *v*—the act of selecting samples. **F 538,** F09

sampling fraction, f, *n*—the ratio f of the number of sampling units selected for the sample to the number of sampling units available.

NOTE—For the simple random sample case, f = n/N where n is the sample size and N is the number of sampling units available. When f > 0.10 estimation of the precision of an estimator should take account of this magnitude of f. See finite population correction. **E 456,** E11

sampling frequency—number of interferogram data points digitized per second in a single scan. **E 131,** E13

sampling interval—difference in retardation between successive sample points in an interferogram. **E 131,** E13

sampling period—See **sampling.** **D 1356,** D22

sampling plan, *n*—a procedure for obtaining a sample. **D 123,** D13

sampling plan result, *n*—the number obtained for use in judging the acceptability of a lot when applying a sampling plan. **D 123,** D13

sampling probe—in leak testing, a device used to collect tracer gas from an area of the test object and feed it to the leak detector at the reduced pressure required. Also called a sniffing probe. **E 1316,** E07

sampling rate—See **sampling.** **D 1356,** D22

sampling (solid sample)—the process of extracting a small fraction of material from a larger bulk, so that it will be sufficiently representative of the bulk for the intended purpose. **E 1547,** E15

sampling train—See **sampling.** **D 1356,** D22

sampling unit, *n*—an identifiable, discrete unit or subunit of material that could be taken as part of a sample. **D 123,** D13

sampling unit, *n*—*in wool*, a portion of material that is taken at one time from one physical location and that is combined with similar portions to make up the laboratory sample. **D 123,** D13

sampling unit, *n*—an identifiable, discrete unit or subunit of material that could be taken as part of a sample. (See also **primary sampling unit, laboratory sample,** and **specimen.**) **D 4439,** D35

sampling unit, primary—See **primary sampling unit.** **D 4439,** D35

sampling vertical—an approximately vertical path from the water surface to the bottom along which one or more samples are collected to define various properties of the flow, such as sediment concentration. **D 4410,** D19

sampling with replacement, *n*—a procedure used with some probability sampling plans in which a selected unit is replaced after any step in selection so that this sampling unit is available for selection again at the *next* step of selection, or at any other succeeding step of the sample selection procedure. **E 456,** E11

sand, *n*—fine aggregate resulting from natural disintegration and abrasion of rock or processing of completely friable sandstone. **C 125,** C09

sand—particles of rock that will pass the No. 4 (4.75-mm) sieve and be retained on the No. 200 (75-μm) U.S. standard sieve. **D 653,** D18

sand blasting—abrasive blasting with sand. **B 374,** B08

sand boil—the ejection of sand and water resulting from piping. **D 653,** D18

sand casting, *n*—a casting process wherein molten metal is poured by gravity into the cavity of a sand mold and solidified; also, a product of such a process. **B 899,** B02

sanded grout—grout in which sand is incorporated into the mixture. **D 653,** D18

sand equivalent—a measure of the amount of silt or clay contamination in fine aggregate as determined by test (Test Method D 2419). **D 653,** D18

sand finish, *n*—units whose surface faces are covered with sand, applied either to the clay column in the stiff mud process or as the lubricant to the molds in the soft mud process. **C 43,** C15

sand holes—{archaic} small fractures in the surface of glass, produced by the rough grinding operation, that have not been removed by subsequent fine grinding. **C 162,** C14

sand-lime brick, *n*—See **calcium-silicate brick.**

C 1209, C15

sand size (fluvial sediment)—0.062 to 2 mm in diameter. **D 4410,** D19

sandstone (I)—sedimentary rock composed mostly of mineral and rock fragments within the sand size range (from 2 to 0.06 mm) and having a minimum of 60 % free silica, cemented or bonded to a greater or lesser degree by various materials including silica, iron oxides, carbonates, or clay, and which fractures around (not through) the constituent grains. **C 119,** C18

sandwich panel—a structure consisting of relatively dense high-strength facing(s) bonded to a less dense low-strength intermediate material or core. **E 1749,** E06

sanforizing mark, *n*—a crimped, rippled, wavy, pebbled, or cockled place in the cloth showing distortion of the texture. **D 123,** D13

sanitary ware—See **china sanitary ware.** **C 242,** C21

sanitary ware—porcelain enameled ware such as sinks, lavatories, and bathtubs. **C 286,** B08

sanitization—reduction in the number of bacterial contaminants to safe levels. See **disinfection**. **D 6161,** D19

sanitizing agent feeder—a device that automatically feeds sanitizer into recirculated or non-recirculated rinse water of spray-type commercial dishwashing and glasswashing machines. **F 1827,** F26

sap—the moisture in unseasoned wood, containing nutrients and other chemicals in solution. **D 9,** D07

saponification—the alkaline hydrolysis of fats whereby a soap is formed; more generally, the hydrolysis of an ester by an alkali with the formation of an alcohol and a salt of the acid portion. **B 374,** B08

saponification—alkaline hydrolysis of esters to produce the component soaps and alcohols. **D 459,** D12

saponification number, *n*— *in petroleum technology*, the number of milligrams of potassium hydroxide that is consumed by 1 g of oil under the conditions of the test. **D 4175,** D02

sapric peat—peat in which the original plant fibers are highly decomposed (less than 33 % fibers). **D 653,** D18

sapwood—the wood containing some living cells and forming the initial wood layer beneath the bark of the log. The thickness of the sapwood layer varies by species and may be lighter in color than heartwood. Under most conditions the sapwood is more susceptible to decay than heartwood. **D 9,** D07

saran plastics—See **vinylidene chloride plastics.** **D 883,** D20

sash—See **windows and doors.** **E 631,** E06

sash, *n*—single frame in a door or window that holds one or more lites (panes) of glass. See **window sash**. **E 1605,** E06

satellite, *n*—extraneous or undesirable ink droplets. See also **spatter, spray**. **F 1857,** F05

satellite drops, *n*—drops formed during atomization that are associated with a much larger drop. **E 1620,** E29

satellite peaks—See **photoelectron X-ray satellite peaks** and **shake-up lines or shake-up satellites.** **E 673,** E42

satin etch—decorative glass surface that scatters light upon reflection, yielding a dull sheen finish. **C 162,** C14

satin finish—a surface finish that behaves as a diffuse reflector and which is lustrous but not mirrorlike. **B 374,** B08

satin, matte, or black finish—a reduced reflective surface varying from a dull appearance to a blackened surface. **F 1840,** F04

saturated felt—a felt that has been immersed in hot bitumen; the felt adsorbs and absorbs as much bitumen as it can retain under the processing conditions, but remains porous and contains voids. **D 1079,** D08

saturated gun—a self-biased electron gun in which electron emission is limited by space charge rather than filament temperature. **E 7,** E04

saturated unit weight—see **unit weight**. **D 653,** D18

saturates, *n*—a mixture of paraffinic and naphthenic hydrocarbons that on percolation in a paraffinic solvent are not adsorbed on the adsorbing medium. Other compounds such as naphthenic and polar aromatics are adsorbed thus permitting the separation of the saturate fraction. **D 8,** D04

saturation—under **color of an object,** see saturation. **D 16,** D01

saturation, *n*—the condition existing when a vapor is in equilibrium with the plane surface of a condensed phase of the same substance (liquid or solid).

saturation mixing ratio, n—the ratio of the mass of water vapor, m_v, to the mass of dry air, m_a, present in saturated air.

saturation vapor pressure, n—the vapor pressure of a system at a given temperature, wherein the vapor of the substance is in equilibrium with a plane surface of that substance's pure liquid or solid phase.

saturation vapor pressure of water, n—the pressure of water vapor in equilibrium with plane surface of a condensed phase.

saturation mixing ratio—See **saturation.**

saturation vapor pressure of water—See **saturation.** **D 1356,** D22

saturation—the condition existing when a vapor is in equilibrium with the plane surface of a condensed phase of the same substance (liquid or solid). **D 4023,** D22

saturation—the point at which a solution contains enough of a dissolved solid, liquid, or gas so that no more will dissolve into the solution at a given temperature and pressure. **D 6161,** D19

saturation—the condition of coexistence in stable equilibrium of a vapor and a liquid or a vapor and solid phase of the same substance at the same temperature. **E 41,** G03

saturation, *n*—*in color,* the dimension of color that describes its purity; if highly saturated it appears to be pure hue and free of gray, but if of low saturation it appears to have a great deal of gray mixed with it. **E 253,** E18

saturation—a condition in which an increase in input signal produces no increase in amplitude on the display. **E 1316,** E07

saturation—relative or comparative color characteristic resulting from a hue's dilution with white light. **E 1316,** E07

saturation—the characteristic of colors that expresses the amount of a given hue in a color, that is, the intensity of the hue, for example, a grayish red versus an intense red. **F 1156,** F12

saturation activity—the number of disintegrations per unit time for the steady-state condition in which the rate of production of a nuclide is equal to the rate of loss by radioactive decay or nuclear transmutation. **E 170,** E10

saturation charge—the maximum electrostatic charge that can be held on a photoconductive layer without the occurrence of dielectric breakdown. **F 335,** F05

saturation curve—see **zero air voids curve**. **D 653,** D18

saturation, degree of—the ratio of the weight of water vapor associated with a pound of dry air to the weight of water vapor associated with a pound of dry air saturated at the same temperature. **E 41,** G03

saturation density, *n*—the optical density value at which no additional density increase is seen on the thermal image as the energy delivered to thermal paper is increased. **F 1623,** F05

saturation level—see **vertical limit.** **E 1316,** E07

saturation, magnetic—the total magnetization produced in a ferromagnetic material, at which point the incremental permeability has progressively decreased to approach unity. **E 1316,** E07

saturation mixing ratio—the ratio of the mass of water vapor m_v to the mass of dry air m_a present in saturated air. The saturation mixing ratio is designated by r_w when saturation is with respect to the plane surface of liquid water, and by r_i when saturation is with respect to the plane surface of ice. **D 4023,** D22

saturation, *s, n—(1) in the CIE 1976 L*, u*, v** system, the quotient of the CIE 1976 *u, v,* chroma C^*_{uv} by the CIE 1976 lightness L^*, $s = C^*_{uv}/L^*$.

(*2*) attribute of a visual sensation that permits a judgment to be made of the proportion of pure chromatic color in the total sensation. (See also **chroma** (*3*).) **E 284,** E12

saturation temperature, *n*—the temperature at which a thermal image reaches its maximum optical density. **F 1623,** F05

saturation vapor pressure of water—the pressure of water vapor in equilibrium with plane surface of a condensed phase. When the condensed phase is liquid, the saturation vapor pressure is designated by e_w, and when the condensed phase is solid, the saturation vapor pressure is designated by e_i. The saturation vapor pressure is a function of temperature. **D 4023,** D22

saucer head—circular, cupped head with concave top surface and convex bearing surface. **F 547,** F16

saucer pits—See **shallow etch pits.** **F 1241,** F01

Saunderson correction, *n*—mathematical expression relating, in turbid medium theory, the fluxes on opposite sides of a boundary at which there is a change in refractive index. **E 284,** E12

sauter mean diameter, SMD [L], *n*—the diameter of a drop that has the same ratio of volume to surface area as the ratio of total volume to total surface area in a distribution of drops, as computed from the equation:

$$SMD = \Sigma_i n_i d_i^{\,3} / \Sigma_i n_i d_i^{\,2}$$

where:

i = a sampling size interval,
d_i = drop diameter, and
n_i = number of drops in that interval.

G 40, G02

savings-to-investment ratio (SIR)—either the ratio of present value savings to present value investment costs, or the ratio of annual value savings to annual value investment costs. **E 833,** E06

saw blade defect—*on semiconductor wafers*, a roughened area visible after polishing with a pattern characteristic of the saw blade travel. **F 1241,** F01

saw burn—blackening or carbonization of a cut surface of a pultruded section.

NOTE—This condition is usually caused by cutting with a dull saw blade, cutting too slowly, or cutting a highly reinforced material with a diamond blade without water. **D 3918,** D20

saw chain, *n*—a closed loop of cutters linked together for use in a portable power-operated tool. **F 1494,** F23

sawdust—wood particles resulting from the cutting and breaking action of saw teeth. **D 1554,** D07

saw exit mark—a ragged edge at the periphery of the wafer consisting of numerous small adjoining edge chips resulting from saw blade exit. **F 1241,** F01

saw kerf—(*1*) grooves or notches made in cutting with a saw; (*2*) that portion of a log, timber, or other piece of wood removed by the saw in parting the material into two pieces. **D 9,** D07

saw-log size—a log large enough to produce lumber or other products that can be sawed. Its size and quality vary with the utilization practices of the region. **D 9,** D07

saw marks—surface irregularities in the form of a series of alternating ridges and depressions in arcs whose radii are the same as those of the saw blade used for slicing. **F 1241,** F01

saw milling residue, *n*—sawdust, chips, slabs, bark, and other debris generated during the processing of logs into finished lumber or lumber products. **D 1968,** D06

sawn veneer—See under **veneer.** **D 1038,** D07

Saybolt color, *n*—an empirical definition of the color of a clear petroleum liquid. **D 4175,** D02

Saybolt color, *n*—an empirical definition of the color of a clear petroleum liquid based on a scale of −16 (darkest) to +30 (lightest). **E 284,** E12

SBS—sodium bisulfite, $NaHSO_3$. **D 6161,** D19

SC—an abbreviation for self-contained. **F 549,** F05

scab—in porcelain enameling metal sheets or castings, a defect having the appearance of a loose piece of metal, tongue, or flap on the surface. **C 286,** B08

scaffold nail—See **double-headed nail.** **F 547,** F16

scale—an adherent oxide coating that is thicker than the superficial film referred to as tarnish. **B 374,** B08

scale—a small particle of foreign material embedded in the surface of molded glass articles. **C 162,** C14

scale—the oxide formed on the surface of the metal during heating. **C 286,** B08

scale—a deposit formed from solution directly upon a surface. **D 1129,** D19

scale—a condition wherein resin plates or particles are on the surface of a pultrusion.

NOTE—Scales can often be readily removed, sometimes leaving surface voids or depressions. **D 3918,** D20

scale, *n*—a defined arrangement of the elements of a set of stimuli or responses. **E 284,** E12

scale, *v*—to assess the content of one or more appearance attributes in the members of a set of stimuli. **E 284,** E12

scale factor—percentage difference between a full-sized original and its reduced or enlarged image; often expressed as the ratio between the two. **F 1156,** F12

scale inhibitor—a chemical which inhibits the growth of micro-crystals (inhibits precipitation of sparingly soluble salts). See **antiscalant**. **D 6161,** D19

scale of particle sizes—based on AGU (American Geophysical Union) scale. **D 4410,** D19

scale suppressant, *n*—a substance added to engine coolant concentrates, corrosion inhibitor packages, or supplemental coolant additives that helps prevent the formation of hard water mineral scale. **D 4725,** D15

scaling—the process of forming scale with or without acid fumes; sometimes refers to spontaneous detachment of scale. **C 286,** B08

scaling—same as spalling. **C 717,** C24

scaling—the build-up of precipitated salts on a surface, such as membranes, pipes, tanks, or boiler condensate tubes. **D 6161,** D19

scaling, *n*—the deterioration of the upper concrete slab surface, normally 3 mm (0.125 in.) to 13 mm (0.5 in.) in depth, resulting in the loss of surface mortar. **E 867,** E17

scaling—see **flaking**. **E 1605,** E06

scaling, *n*—the deterioration of the upper concrete slab surface, normally 3 mm (0.125 in.) to 13 mm (0.5 in.) in depth, resulting in the loss of surface mortar. **E 1778,** E17

scaling resistance (coatings)—See **flaking resistance.**

D 16, D01

scalloped selvage, *n*—an abrupt, narrow indentation in the selvage. **D 123,** D13

scallops—excursions of the surface record above and below the blanking band. **E 867,** E17

scan—a search for information to be recognized by the recognition unit of the optical scanner, and the conversion of the optical signal to an electrical signal. **F 149,** F05

scan, *n*—the search for a symbol which is to be optically recognized; also, a search for marks to be recognized by the interpretation unit of an optical scanner. **F 1294,** F05

scan area—the area of a form that contains information to be scanned. **F 149,** F05

scan band—a strip across a document that passes directly beneath a scanning head of a reader. **F 149,** F05

scan ink—ink that is sensed by the optical scanner. Synonym for *nonreflective ink, read ink.* **F 149,** F05

scanner—(*1*) a device that examines a spatial pattern one part after another, and generates analog or digital signals corresponding to the pattern. Scanners are often used to mark reading, pattern recognition, or character recognition.

(*2*) See **flying spot scanner, optical scanner, mark scanning, optical scanner**. **F 149,** F05

scanner—device for observing or writing information; sometimes used to describe a device that covers a range of frequencies or other parameters for a set of signals. **F 1156,** F12

scanner, *n*—a device that examines a spatial pattern, part by part, and generates analog or digital signals corresponding to the pattern. **F 1294,** F05

scanning—the movement of a search unit relative to the test piece in order to examine a volume of the material. **E 1316,** E07

scanning index—the distance the search unit is moved between scan paths after each traverse of the part. **E 1316,** E07

scanning microscope—an electron microscope in which the image is formed by a beam operating in synchronism with an electron probe scanning the object. The intensity of the image forming beam is proportional to the scattering or secondary emission of the specimen where the probe strikes it. **E 7,** E04

scanning/reading range (also called Operating Range (OR), *n*—the combined distance of optical ability of a lens, a photographic material, or a photographic system to distinguish detail under specific conditions. Also, the narrowest element dimension that can be distinguished by a particular reading device. **F 1294,** F05

scan reflectance profile, *n*—*according to ANSI X3.182,* a record of the reflectance measured using the reference reflectivity method across the entire bar code symbol. **F 1294,** F05

scarfing—the removing of flash or bead by a cutting operation. **B 846,** B05

scarf joint, *n*—See *scarf joint* under **joint**. **D 907,** D14

SCAS—see **semicontinuous activated sludge.** **D 459,** D12

scattered energy—energy that is reflected in a random fashion by small reflectors in the path of a beam of ultrasonic waves. **E 1316,** E07

scattered ion:

energy—ISS, for a binary elastic collision, the kinetic energy of the probe ion following a binary elastic collision, E_s, is given by:

$$E_s = E_0[M_0/(M_0 + M_1)]^2(\cos\theta + [(M_1/M_0)^2 - \sin^2\theta]^{1/2})^2$$

where:

E_s = kinetic energy for the scattered probe ion,
E_0 = energy of the incident probe ion prior to collision, determined from the product of ionic charge and accelerating potential,
M_0 = mass of the probe ion,
M_1 = mass of the target atom, and
θ = angle between the initial and final velocity vectors for the probe ion, as determined from a common origin in the laboratory coordinate system, expressed as a value between 0 and 180°.

energy ratio—ISS, the value E_s/E_0 which may be used as the abscissa of an ion-scattering spectrum. For definition of E_s and E_0, see **scattered ion** *energy.*

intensity, experimental—ISS, the measured response of the energy filtering and detection system as a consequence of bombarding the specimen material with a beam of probe ions, usually presented as the ordinate of an ion-scattering spectrum.

intensity, theoretical—ISS, defined by an equation of the form:

$$I_i(\theta) = I_0 N_i P_i \alpha_i (d\sigma_i/d\Omega)\theta \Delta\Omega T$$

where:

$I_i(\theta)$ = scattered ion intensity from atoms of species, *i*, at a given scattering angle, θ, ions s^{-1},
I_0 = intensity of incident probe ions, ions s^{-1},
N_i = number of scattering centers of species *i* per unit area of surface, or per unit volume accessible to the incident beam, atoms $metre^{-2}$,
P_i = probability that the probe ion remains ionized after interacting with an atom of species *i*,
α_i = geometric or shadowing factor for species *i* in the given environment and geometry,
$(d\sigma_i/d\Omega)\theta$ = differential scattering cross section per unit solid angle, for species *i*, taken at the angle for which scattering is measured; that is, the angular distribution of scattered ion intensity per unit flux of incident ions, per atom of species *i*, $metre^2$ $atom^{-1}$ $steradian^{-1}$,
$\Delta\Omega$ = solid angle of acceptance determined by the entrance aperture of the filtering and detection system, steradians, and
T = fractional transmission of the analyzing and detection system.

E 673, E42

scattered neutrons—neutrons that have undergone a scattering collision but still contribute to film exposure. **E 1316,** E07

scattering, *n*—the process by which light or other electromagnetic radiant flux passing through matter is redirected over a range of angles. **E 284,** E12

scattering—in leak testing, dispersion or diffusion in various directions due to intermolecular or ionic collisions as applied to the effect of the residual gas in a mass spectrometer tube or an ion beam traversing the tube. **E 1316,** E07

scattering—the dispersion, deflection, or redirection of the energy in an ultrasonic beam caused by small reflectors in the material being examined. **E 1316,** E07

scattering loss—that part of transmitted energy lost due to roughness of reflecting surface. **D 653,** D18

scattering (of radiant energy), *n*—*in optical emission spectrometry,* reflection of radiant energy in random directions by matter located between the source and the detector. **E 135,** E01

scattering, 90° (or 180°)—scattering which is observed at an angle of 90° (or 180°) to the direction of the incident beam. **E 131,** E13

scattering tinting strength, *n*—relative change in the scattering properties of a standard black material (with no scattering colorant present) when a specified amount of a white or chromatic scattering colorant is added to it. **E 284,** E12

scattering (X-ray)—a general term which includes both Compton and coherent scattering. **E 7,** E04

scatter rug, *n*—a small rug which is designed to be flexible and is usually cleaned by laundering. **D 123,** D13

scavenger—a chemically active substance added to a system or mixture to consume or inactivate an undesirable substance. **D 2864,** D27

schedule—a pipe size system (outside diameters and wall thicknesses) originated by the iron pipe industry. **F 412,** F17

schematic design—the phase of a project during which the initial solutions to the program are prepared by the design professional through the use of plans and sections and system's descriptions. **E 833,** E06

schist—a foliated metamorphic quartz-feldspar-containing rock characterized by thin foliae of platy or prismatic minerals such as mica or chlorite. Schists split readily along these planes of foliation. This rock exists in many graduations, some of them progressing into a gneiss. **C 119,** C18

schistosity—the variety of foliation that occurs in the coarser-grained metamorphic rocks and is generally the result of the parallel arrangement of platy and ellipsoidal mineral grains within the rock substance. (ISRM) **D 653,** D18

Schlieren system—an optical system used for visual display of an ultrasonic beam passing through a transparent medium. **E 1316,** E07

scintillation, *n*—the multiple discharges or small arcs that originate in the more conductive areas of the insulation surface, and span less conductive areas. **D 1711,** D09

scintillator—*in the nuclear industry*, a transparent substance that emits visible or near ultraviolet light when traversed by an ionizing particle. **C 859,** C26

scintillator, *n*—a transparent substance that emits visible or near ultraviolet light when traversed by an ionizing particle. **C 1145,** C28

scintillators and scintillating crystals— a detector that converts ionizing radiation to light. **E 1316,** E07

scissoring—lateral misalignment. **F 1638,** F44

sclerotinite—See *sclerotinite* under **maceral**. **D 121,** D05

SCOF—static coefficient of friction **F 1646,** F13

scorch, *n*—premature vulcanization of a rubber compound. **D 1566,** D11

scorch, Mooney, *n*—the time to incipient cure of a compound when tested in the Mooney shearing disk viscometer under specific conditions (see Test Method D 1646). **D 1566,** D11

score, *n*—an impression or crease in **paperboard, corrugated** or **solid fiberboard** to locate and facilitate folding. **D 996,** D10

score, *v*—to form a score. **D 996,** D10

score—a cut made by a flaying knife on the flesh side. **D 1517,** D31

scored finish, *n*—units whose face surfaces are grooved as they come from the die. **C 43,** C15

scoring, *n*—*in tribology*, a severe form of wear characterized by the formation of extensive grooves and scratches in the direction of sliding. **G 40,** G02

scorotron, *n*—a corotron with a screen. **F 1457,** F05

scotch grain—a pebbled pattern embossed on cattlehide or calf leather. **D 1517,** D31

scotopic, *adj*—pertaining to vision at sufficiently low levels of illumination that only the retinal rods are stimulated. **E 284,** E12

scour—the enlargement of a flow section by the removal of the boundary material by the motion of a fluid. **D 4410,** D19

scoured wool, *n*—wool from which the bulk of impurities has been removed by an aqueous or solvent washing process. **D 123,** D13

scouring—a wet process of cleaning by chemical or mechanical means, or both. **D 459,** D12

scrambling—mixing up; encrypting. **F 1156,** F12

scrap—*in the nuclear industry*, residues that contain sufficient quantities of source or special nuclear material to be worthy of recovery. **C 859,** C26

scrape, *n*—the crystallized pine oleoresin collected from the scarified faces of trees being worked for turpentine. **D 804,** D01

scrapings—the overspray that has been recovered from a spray booth. **C 286,** B08

scratch—damage on a glass surface in the form of a line caused by the relative movement of an object across and in contact with the glass surface. **C 162,** C14

scratch—damage resulting from the movement of a hard pointed object. **D 2825,** D21

scratch—in micro-indentation hardness testing, a mark or groove cut in the specimen by moving a loaded indenter across the surface. **E 7,** E04

scratch, *n*—a shallow cut or narrow groove in the flooring surface. A line or furrow made in the flooring surface by rasping or rubbing with a pointed or jagged object. **F 141,** F06

scratch—*on semiconductor wafers*, a shallow groove or cut below the established plane of the surface, with a length to width ratio greater than 5:1. **F 1241,** F01

scratching, *v*—a form of wear, in which a minute groove-like break in a flooring surface is made by a rubbing contact with a tool or particle, the total deformation being confined to the most immediate surface level. **F 141,** F06

scratching, *n*—*in tribology*, the mechanical removal or displacement, or both, of material from a surface by the action of abrasive particles or protuberances sliding across the surfaces. (See also **plowing**.) **G 40,** G02

scratch-resistant coatings—coatings applied to glass surfaces to reduce the effects of frictive contact. **C 162,** C14

screen, *n*—in laboratory work an apparatus, in which the apertures are circular, for separating sizes of material. **D 8,** D04

screen—an apparatus with circular apertures for separating sizes of material. **D 1079,** D08

screen—See **railing systems. E 631,** E06

screen—alternative term for intensifying screen. **E 1316,** E07

screen, *n*—a perforated sheet consisting of wire mesh, woven or welded wire fabric, or standard or flattened expanded metal; serving as an infill panel and being an integral part of a panel railing or guardrail system. **E 1481,** E06

screen, *v*—a cursory examination of a document or article to determine whether it appears to be genuine. **F 1156,** F12

screen analysis—see **sieve analysis. B 243,** B09

screen-back hardboard (S1S)—hardboard with a reverse impression of a screen on the back produced when a damp or wet mat is hot-pressed into a board and dried in the press. **D 1554,** D07

screen dump, *n*—a command that causes information showing on a terminal display to be printed just as it appears on the display. **F 1457,** F05

screening, *n*—the appearance of a cross-hatch pattern on the print that can be caused by ink drying too rapidly. **D 6488,** D01

screening, *n*—a pattern of vertical and diagonal lines on the printed substrate where the ink flowout between cells in the cylinder is such that an uneven ink film is formed. **D 6488,** D01
screening, *n*—a preliminary selection procedure. **E 253,** E18
screening design, *n*—a balanced design, requiring relatively minimal amount of experimentation, to evaluate the lower order effects of a relatively large number of factors in terms of contributions to variability or in terms of estimates of parameters for a model.

NOTE—In screening designs, the term lower order effects is sometimes limited to first order terms such as linear components of main effects, but often includes both first order terms and second order terms such as two factor interactions and quadratic curvature components of main effects.
E 456, E11

screening design, *n*—a balanced design, requiring relatively minimal amount of experimentation, to evaluate the lower order effects of a relatively large number of factors in terms of contributions to variability or in terms of estimates of parameters for a model.

NOTE—In screening designs, the term lower order effects is sometimes limited to first order terms such as linear components of main effects, but often includes both first order terms and second order terms such as two factor interactions and quadratic curvature components of main effects.
E 1325, E11

screening energy—the diminished energy of an ion due to coulombic attraction of electrons in the immediate environment. **E 673,** E42
screenings, *n*—a residual product resulting from the artificial crushing of rock, boulders, cobble, gravel, blast-furnace slag or hydraulic cement concrete, all of which passed the smallest screen used with the crushing operation and most of which passed the 2.36-mm (No. 8) sieve. **D 8,** D04
screening test, *n*—*as related to fire,* a fire-response test performed to determine whether a material, product, or assembly (*a*) exhibits any unusual fire-related characteristics, (*b*) has certain expected fire-related characteristics, or (*c*) is capable of being preliminarily categorized according to the fire characteristic in question. **E 176,** E05
screen lens—an electrostatic electron lens consisting of a combination of screens or foils at different potentials. **E 7,** E04
screen marks—a mesh pattern left by a silk screen process in the applied ceramic glass enamel before or after firing. **C 162,** C14
screen mesh, *n*—the average number of openings per linear inch of material. **C 1145,** C28
screen plate or film— See **color film.** **E 7,** E04
screen test—a standard test for fineness of porcelain enamel slip or powder. **C 286,** B08
screw—the fastener which joins the scissor halves **F 1078,** F04
screw, *n*—an anchor component that connects to the bony elements of the spine, pelvis, or ribs by means of threads. **F 1582,** F04
screw—a mechanical fastener having basic design characteristics which facilitate its assembly into a tapped hole or to form its own threads during installation. **F 1789,** F16
screw contact—see **contact, screw.** **B 542,** B02
screw head—See **slotted head.** **F 547,** F16
screw interconnection, *n*—an interconnection having an implant component sandwiched between the screw head (or screw thread) and bony element or other implant components. **F 1582,** F04
screw nail, screwnail, screw-grip nail, screw-shank nail, scru-shank nail, screw-thread nail—terms applied to helically threaded nail and screw-threaded nail. (See **thread.**) **F 547,** F16
screw-on-bead or stop—stop, molding, or bead secured in place by screws or nails. **C 717,** C24
screw, screw-grip thread, screw-shank thread, scru-shank thread, screw thread—See **helical thread, screw thread.** **F 547,** F16
screw stock—metal in the form of wire or rod, usually a free-machining alloy used for automatic screw machine work. **E 7,** E04
screw thread—helical thread with small lead angle rolled onto nail shank. **F 547,** F16
scrim, *n*—lightweight woven or non-woven fabrics that are manufactured from yarns. **D 1079,** D08
scrim—a reinforcing fabric woven into an open mesh construction, used in the processing of tape or other B-stage material to facilitate handling and control bondline thicknesses. Also referred to as a *carrier.* **E 1749,** E06
scroll—See **railing systems.** **E 631,** E06
scroll, *n*—a cast or forged ornamental spiral of convoluted shape, serving, for example, as a decorative panel or insert. **E 1481,** E06
scroll—a decorative ornament on top of a gate. **F 552,** F14
scrubber, *n*—a type of apparatus used in sampling and in gas cleaning in which the gas is passed through a space containing wetted packing or spray. **D 1356,** D22
SCS—abbreviation for **system network architecture character string**. **F 1457,** F05
scud—remnants of epithelial tissue, hair, dirt, etc., left in the hair follicles after unhairing. **D 1517,** D31
scudding—removal of scud from unhaired hides by scraping with a blade, either by hand or machine. **D 1517,** D31
scuff—See **rub**. **C 162,** C14
scuff—disfigurement of polish film resulting from the abrading or scraping action repairable without recoating. **D 2825,** D21
scuff, *v*—a form of wear, in which a mark, gall, roughness or other damage is caused by the rubbing of traffic bodies against a flooring surface and may involve deposition of a foreign material onto the flooring surface. **F 141,** F06
scuff—the color of the CF or CFB sheet caused by capsule damage on the CB sheet when it is pulled under pressure across the mate sheet. **F 549,** F05
scuffing—long white scrape marks on the surface of the pultrusion.

NOTE—This condition usually results from mechanical scraping or scratching of the pultrusion in the machine or in handling it afterwards.
D 3918, D20

scuffing, *n*—*in lubrication*, surface damage resulting from localized welding at the interface of rubbing surfaces with subsequent fracture in the proximity of the weld area. **D 4175,** D02
scuffing—marks on coated surface due to movement between sheets under pressure. **F 335,** F05
scuffing—*n*, a form of wear occurring in inadequately lubricated tribosystems that is characterized by macroscopically-observable changes in surface texture, with features related to the direction of relative motion. **G 40,** G02
scuff marks—scratches or other damage to the shoe finish or polish coating. **D 2825,** D21
scum—a floating layer of unmelted material on the molten glass surface. **C 162,** C14
scum, *n*—a condition that occurs when a lithographic plate has become sensitized in the nonimage areas and these areas begin to take ink. **[GATF]** **D 6488,** D01
scum—*in lithographic images,* undesirable printing of ink in the non-image areas as a result of those areas on the plate becoming sensitized and thus ink receptive. **F 425,** F05
scumming—a defect characterized by areas of poor gloss on the

surface of porcelain enamel. **C 286,** B08

scumming, *n*—the deposit of ink in the non-imaging area. **D 6488,** D01

scuttle—a hatch that provides access to the roof from the interior of the building. **D 1079,** D08

scye depth, *n—in body measurements*, the distance from the cervicale to a point level with the armpit. **D 123,** D13

SDI—silt density index—an index calculated from the rate of plugging of 0.45 μm membrane filter. It is an indication of the amount of particulate matter in water, sometimes called fouling index. **D 6161,** D19

SDLC—abbreviation for **synchronous data link control**. **F 1457,** F05

S&DSI—Stiff and Davis saturation index, measure of $CaCO_3$ solubility in seawater or highly saline water. See **LSI**. **D 6161,** D19

SDWA—Safe Drinking Water Act of the United States, specifying required purity levels of municipal potable water. **D 6161,** D19

seal, *n—in building construction*, a barrier against the passage of liquids, solids, or gases. **C 717,** C24

seal:

heat seal—in packaging, a method of bonding mating surfaces under controlled application of heat, pressure, and dwell time.

hermetic seal—in packaging, a seal that prevents passage of air and other gases. **D 996,** D10

seal—the closure of a structure to make it leakproof by the application of sealant to fasteners, seams, and any other possible leak path. **E 1749,** E06

seal—a continuous joint of two or more surfaces of sheet material such as made by fusion or adhesion. **F 17,** F02

seal—See Terminology F 17. **F 1327,** F02

sealant—*in building construction*, a material that has the adhesive and cohesive properties to form a seal. **C 717,** C24

sealant—a mixture of polymers, fillers, and pigments used to fill and seal joints where moderate movement is expected; unlike caulking, it cures to a resilient solid. **D 1079,** D08

sealant, *n*—a material that has adhesive and cohesive properties to seal joints, cracks, or other narrow openings (generally less than 3-in. (76-mm) wide) in pavements against the entrance or passage of water or other debris. **D 5535,** D04

sealant backing—*in building construction*, a compressible material placed in a joint before applying a sealant. **C 717,** C24

sealant-sealed seam finish, *n—in home sewing*, a seam finish in which a clear liquid seam-sealant is used as the treatment to prevent fraying. **D 4965,** D13

seal contamination—foreign matter in the seal area such as, but not limited to, water, grease, or food. **F 17,** F02

seal contamination—See Terminology F 17. **F 1327,** F02

seal creep—the reduction in width of a seal due to a force being exerted on it, such as a bulky product, pouch distortion, or internal air pressure. **F 1327,** F02

seal creep resistance—a measure of the ability of a sealed package or seal to remain intact when subjected to a constant force. **F 1327,** F02

sealed (anodic) coating—*in anodizing aluminum*, an anodic oxide coating on aluminum that has been treated in an aqueous or steam medium resulting in reduced porosity of the coating. **B 374,** B08

sealed insulating glass—See **windows and doors.** **E 631,** E06

sealer—a liquid composition to prevent excessive absorption of finish coats into porous surfaces; also a composition to prevent bleeding (see **size**). **D 16,** D01

sealing—permanent joining by heating an interface to reduce the viscosity of glass(es) for geometrical accommodation and bonding. **C 162,** C14

sealing, absolute—See **absolute sealing**. **E 1749,** E06

sealing blade—See **blade clincher.** **F 592,** F16

sealing element—a separate or bonded material between the sleeve and the pipe that forms a seal. **C 896,** C04

sealing glass—a glass with special thermal expansion characteristics intended for sealing to another glass or solid. **C 162,** C14

sealing material, n—See **seal**. **C 717,** C24

sealing of anodic coating—a process which, by absorption, chemical reaction, or other mechanism, increases the resistance of an anodic coating to staining and corrosion, improves the durability of colors produced in the coating, or imparts other desirable properties. **B 374,** B08

sealing surface—that portion of the finish of a glass container which makes contact with the sealing gasket or liner. See **finish**. **C 162,** C14

sealing tape—See **tape sealant**. **C 717,** C24

seal (mechanical), *n*—any material or device that prevents or controls the passage of matter across the separable members of a mechanical assembly. **D 1566,** D11

seal, O-ring, *n*—a product of precise dimensions molded in one piece to the configuration of a torus with circular cross section, suitable for use in a machined groove for static or dynamic service. **D 1566,** D11

seal plane—all surfaces of a shelter that establish seal continuity and are in immediate contact with the environment. **E 1749,** E06

seal strength—a measure of the mechanical strength of the bond between sealed materials of a package. **F 1327,** F02

seam, *n—in wire*, a longitudinal discontinuity that extends radially into the wire from its surface. **A 902,** A05

seam—a naturally filled or bonded crack which does not adversely affect the strength of a stone (see crack, fracture, microcrack). **C 119,** C18

seam, *n*—a mark on a glass surface resulting from the joint between matching mold parts. **C 162,** C14

seam, *v*—to slightly grind the sharp edges of a piece of glass. **C 162,** C14

seam—See **container.** **D 996,** D10

seam, *n*—a permanent joining of two or more materials. **D 4439,** D35

seam, *n*—the connection of two or more pieces of material by mechanical, chemical, or fusion methods to provide the integrity of a single piece of the material. **D 4439,** D35

seam, *n—in home sewing*, a line along which two or more sections of a product have been joined. (Syn. *joining line*. See also **glued seam, sewn seam, stapled seam, thermally bonded seam**.) **D 4965,** D13

seam—an unwelded fold or lap on the surface of a metal which appears as a crack, usually the result of defects in casting or working which have not welded shut. **E 7,** E04

seam—a noncontinuous joint of two or more surfaces of sheet material such as made by stitching, spot adhesions, or intermittent fusion. **F 17,** F02

seam—See Terminology F 17. **F 1327,** F02

seam—a straight or smooth curved line surface discontinuity running longitudinally on the fastener thread, shank, and head. **F 1789,** F16

seam allowance, *n*—the width of fabric used in making a seam assembly, bounded by the edge of the fabric and the furthest stitch line. **D 4439,** D35

seam allowance, *n—in sewn materials*, the width of material between the cut edge of a fabric section and the line parallel to the cut edge along which that section will be joined to another section during assembly of the finished product. **D 4965,** D13

seam assembly, *n*—the composite structure obtained when fabric(s) are joined by means of a seam. **D 123,** D13

seam assembly, *n*—the unit obtained by joining fabrics with a seam, including details such as fabric direction(s), seam allowance, sewing threads used, and number of stitches per unit length; and sometimes additional details of fabrication such as sewing-machine type and speed, needle type and size, etc. **D 4439,** D35

seam damage, *n—in sewn fabrics*, an adverse change in the physical condition of one or more of the components in a seam which would reduce the seam acceptability such as yarn slippage, needle damage, or fabric rupture. **D 123,** D13

seam damage, *n—in sewn fabrics*, any change in the physical condition of one or more of the components in a seam which reduces seam efficiency. **D 123,** D13

seam design engineering, *n*—the procedures used to select a specific thread, a specific stitch type, and a specific seam type to achieve the required seam strength. **D 4439,** D35

seam efficiency, *n—in sewn fabrics*, the ratio of seam strength to fabric strength. **D 123,** D13

seam efficiency, sewn, *n—in sewn fabrics*, the ratio expressed as a percentage of seam strength to fabric strength. **D 4439,** D35

seam engineering, *n—in sewn fabrics*, the procedures used to select a specific combination of sewing thread, stitch type, seam type, and stitch density to achieve the maximum sewn seam strength for a particular fabric type. **D 123,** D13

seam failure, *n—in sewn fabrics*, that point at which an external force (1) ruptures the sewing thread, (2) ruptures the fabric, (3) causes excessive yarn slippage adjacent to the stitches, or (4) causes any combination of these unacceptable conditions. **D 123,** D13

seam finish, *n—in home sewing*, a treatment of the fabric cut edges of the seam allowance in a plain seam. **D 4965,** D13

seam interaction, *n—in sewn fabrics*, the net effect of the relationship between the combination of fabric, seam type, stitch type, and stitch density on seam efficiency. **D 123,** D13

seam interaction, *n*—the result of combining a specific textile, a specific stitch type, and a specific seam type. **D 4439,** D35

seamless pipe, *n*—a round, hollow product made with a continuous periphery in all stages of manufacture and produced to the particular dimensions commercially known as standard pipe sizes. **B 899,** B02

seam mark, *n—in finished cloth*, a pressure mark caused by the thickness of the seam being pressed against the cloth. **D 123,** D13

seam moisture, *n*—synonym for *inherent moisture*. **D 121,** D05

seam slippage, *n—in sewn fabrics*, the displacement of the fabric yarn parallel and adjacent to the stitch line. **D 123,** D13

seam type, *n—in sewn fabrics*, an alphanumeric designation relating to the essential characteristics of fabric positioning and rows of stitching in a specified sewn fabric seam. **D 123,** D13

seam type, *n—in sewn fabrics*, an alphanumeric designation relating to the essential characteristics of fabric positioning and rows of stitching in a specific sewn fabric seam (see Federal Standard 751). **D 4439,** D35

search-gas—Same as **tracer gas**. **E 1316,** E07

search unit—an electro-acoustic device used to transmit or receive ultrasonic energy, or both. The device generally consists of a nameplate, connector, case, backing, piezoelectric element, wearface, or lens, or wedge. **E 1316,** E07

seasonal factor, *n*—a summarized short term count adjustment for variability by period of year, which can be based on monthly, weekly or daily factors when estimating typical annual traffic characteristics. **E 867,** E17

season cracking—See **stress-corrosion cracking**. **G 15,** G01

seasoning—drying; the term often applied to the process of removing moisture from wood to achieve a moisture content appropriate for the performance expected of the final product.

air-dried—dried by exposure to air, usually in a yard, without artificial heat.

dry—seasoned; in softwood lumber, the abbreviation S-Dry means not in excess of 19 % moisture content at time of surfacing, in accordance with recognized standards.

kiln-dried—dried in a closed chamber in which temperature and relative humidity of the circulated air can be controlled. The drying temperature is usually greater than outside air temperature. In softwood lumber grading rules under the jurisdiction of the Southern Pine Inspection Bureau, the abbreviation KD indicates seasoned in a kiln to a moisture content not in excess of 15 % moisture content.

ovendry—dried in an oven to remove all moisture. The temperature employed usually is 101 to 105°C or 214 to 221°F in accordance with ASTM Methods D 2016 Test for Moisture Content of Wood.

shipping-dry—dried to a moisture content judged low enough to prevent stain, mold, and decay in transit. **D 9,** D07

seat—prepared position on the siege where the pot rests. **C 162,** C14

sea water—water with an approximate concentration of total dissolved solids ranging from 30 000 to 60 000 mg/L. See **brackish water, high brackish water**. **D 6161,** D19

secant modulus—See **modulus, secant**. **C 717,** C24

secant modulus—the slope of a line drawn from the origin to any specified point on the stress-strain curve. **C 904,** C03

secant modulus—slope of the line connecting the origin and a given point on the stress-strain curve. (ISRM) **D 653,** D18

secant modulus, *n*—the slope of the secant drawn from the origin to any specified point on the stress-strain curve. **D 907,** D14

secant modulus, *n*—the ratio of stress (nominal) to corresponding strain at any specified point on the stress-strain curve. **D 4439,** D35

secant modulus, J_{sec}, *(*FL^{-1}*)*, Nm^{-1}, *n—for geosynthetics*, the ratio of change in force per unit width to the change in elongation between two points on a force-elongation curve. **D 4439,** D35

secondary accelerator, *n*—accelerator used in smaller concentrations compared to the primary accelerator, to achieve a faster rate of vulcanization. **D 1566,** D11

secondary atomization, *n*—the disintegration of unstable drops or liquid fragments resulting from primary atomization to produce stable drops. **E 1620,** E29

secondary bonding—the joining together, by the process of adhesive bonding, of two or more cured composite parts, during which the only chemical (or thermal) reaction taking place is the curing of the adhesive itself. **E 1749,** E06

secondary circulation area—See **building space.** **E 631,** E06

secondary circulation area—See **space categories**. **E 1480,** E06

secondary clay (sedimentary clay)—a clay that has been geologically transported from its place of formation. **C 242,** C21

secondary (concerning X-rays)—the X-rays emitted by a specimen irradiated by a primary beam. **E 7,** E04

secondary consolidation (*secondary compression*) (*secondary time effect*)—see **consolidation**. **D 653,** D18

secondary container—a container which encloses one or more primary containers. **D 996,** D10

secondary electrons—*AES,* electrons leaving a surface, produced through various mechanisms of energy transfer from the incident beam. By convention, electrons with energies ≤50 eV are considered as secondary electrons. **E 673,** E42

secondary extinction—a decrease in the intensity of a diffracted X-ray beam caused by parallelism or near-parallelism of mosaic blocks in a mosaic crystal; the lower blocks are partially screened from the incident radiation by the upper blocks, which have reflected some of it. **E 7,** E04

secondary filter pack—a clean, uniformly graded sand that is placed in the annulus between the primary filter pack and the over-lying seal, or between the seal and overlying grout backfill, or both, to prevent movement of seal or grout, of both, into the primary filter pack. **D 653,** D18

secondary flat—a flat of length shorter than the primary orientation flat, whose position with respect to the primary orientation flat identifies the type and orientation of the wafer; minor flat. **F 1241,** F01

secondary flow-rate standard—See **standard.** **D 1356,** D22

secondary hardening, *n*—the hardening phenomenon that occurs during high-temperature **tempering** of certain **steels** containing one or more carbide-forming alloying elements. **A 941,** A01

secondary hole—*in grouting,* the second series of holes to be drilled and grouted usually spaced midway between primary holes. **D 653,** D18

secondary ion—ions ejected from a specimen surface as a result of energy transfer from a primary beam.

angular distribution—*SIMS,* the secondary ion yield as a function of emission angle.

energy distribution—the number of secondary ions as a function of the energy at a specified collection angle.

mass spectrometry—the technique in which a mass spectrometer is used to measure the mass-to-charge ratio of secondary ions emitted from a target as a result of particle (ion or neutral) bombardment.

signal gating—the process of accepting secondary ion signal from only a portion of the sputtered area of the specimen to minimize crater edge effects.

yield—the total number of ions sputtered from the specimen per incident ion of given mass, energy, charge, and angle of incidence. **E 673,** E42

secondary light source, *n*—surface or object that is not self-emitting but receives light and re-directs it, at least in part, by reflection or transmission. **E 284,** E12

secondary lining—the second-placed, or permanent, structural lining of a tunnel, which may be of concrete, steel, or masonry. (ISRM) **D 653,** D18

secondary protective clothing—protective clothing designed for continuous wear for work activities in designated locations in which intermittent exposure to molten substance splash, radiant heat, and flame sources is possible. **F 1494,** F23

secondary radiation—radiation emitted by any substance as the result of irradiation by the primary source. **E 1316,** E07

secondary road, *n*—two lane, occasionally maintained, hard or loose surface (for example, large rock, paved crush rock, gravel, soil aggregate) intended for medium-weight, low-density traffic, in accordance with the U.S. Federal Highway Administration. **F 538,** F09

secondary seal—a seal that alone cannot provide a dependable absolute seal. **E 1749,** E06

secondary sources (of lead)—sources of lead exposure other than lead-based paint. **E 1605,** E06

secondary standard, *n*—a standard calibrated by reference to another standard, such as a primary standard. Often, a secondary standard is prepared from typical production-type material. **C 1145,** C28

secondary standard—See **standard.** **D 1356,** D22

secondary standard thermocouple, *n*—a thermocouple that has had its temperature-emf relationship determined by reference to a primary standard of temperature. **E 344,** E20

secondary state of stress—the resulting state of stress in the rock around man-made excavations or structures. (ISRM) **D 653,** D18

secondary X rays, *n*—the X rays emitted by a specimen irradiated by primary X rays. **E 135,** E01

second generation—second major evolution of a design, etc. Also, the second level of repetition or copying made by copying a first level (generation) reproduction or copy; usually of poorer quality than the first generation copy from which it was made. **F 1156,** F12

second growth—timber that has grown after the removal, whether by cutting, fire, wind, or other agency, of all or a large part of the previous stand. Often limited to that growth following removal of old-growth timber. **D 9,** D07

second-hand filling material, *n*—an industry product which contains any filling material which has previously been used should not be offered for sale unless a clear and conspicuous disclosure of that fact is made on the label thereof and in all advertising and invoices relating to such product. **D 123,** D13

second side—{archaic} the final side of plate glass to be ground and polished. **C 162,** C14

section—in a phase diagram, a planar cut through a space diagram; sections are commonly isotherms, isobars or isopleths. **E 7,** E04

section height, [*L*], *n*—the radial height of a tire section, expressed as one half the difference between the outside diameter of the unloaded tire and the nominal rim diameter; the outside diameter is measured on a tire-wheel assembly with the tire inflated to rated inflation pressure. **F 538,** F09

section mark, *n*—*in woven fabrics,* warp bands of different color, texture, or luster. **D 123,** D13

section width, *n*—the width of a new tire, including 24-h inflation growth and including normal sidewalls, but not including protective side ribs, bars, or decorations. **F 538,** F09

sedigraph, *n*—an instrument for determining the particle size distribution of a particulate solid, making use of a physical relation between rate of settling (sedimentation) in a liquid and the particle size. **C 1145,** C28

sediment—See **fluvial sediment.** **D 4410,** D19

sediment—(*1*) particulate material that usually lies below water, and (*2*) formulated particulate material that is intended to lie below water in a test. **E 943,** E47

sedimentation—(a) consists of five fundamental processes: (*1*) weathering, (*2*) erosion, (*3*) transportation, (*4*) deposition, and (*5*) diagenesis, or consolidation into rock; (b) deposition of particles, especially in engineering. **D 4410,** D19

sedimentation—the precipitation or settling of insoluble materials from a suspension, either by gravity or artificially. For example, centrifuge, pressure. **D 6161,** D19

sediment basin—a structure created by construction of a barrier or small dam-like structure across a waterway or by excavating a basin or a combination of both to trap or restrain sediment. **D 653,** D18

sediment delivery—See **sediment yield.** **D 4410,** D19

sediment delivery ratio—the ratio of sediment yield to gross erosion expressed in percent. **D 4410,** D19

sediment discharge—the mass or volume of sediment passing a stream cross section in a unit of time. (The term may be qualified as suspended-sediment discharge, bedload discharge or total-sediment discharge.) **D 4410,** D19

sediment load—a general term that refers to material in suspension or in transport, or both; it is not synonymous with either discharge or concentration. (See **bed-load** and **suspended-sediment load**.) **D 4410,** D19

sedimentology—the scientific study of sediment, sedimentary rocks, and the processes by which they were formed. **D 4410,** D19

sediment particle—fragment of mineral or organic material in either a singular or aggregate state. **D 4410,** D19

sediment sample—a quantity of water-sediment mixture or deposited sediment that is collected to represent some property or properties of the sampled medium. **D 4410,** D19

sediment sump—a blank extension beneath the well screen used to collect fine-grained material from the filter pack and adjacent strata. The term is synonymous with rat trap or tail pipe. **D 653,** D18

sediment transport rate—See **sediment discharge**. **D 4410,** D19

Seebeck coefficient, *n*—the rate of change of thermal emf with temperature at a given temperature, normally expressed as emf per unit of temperature (same as **thermoelectric power**). **E 344,** E20

Seebeck emf, *n*—the net emf set up in a thermocouple under condition of zero current. It represents the algebraic sum of the Peltier and Thomson emf (same as **thermal electromotive** force). **E 344,** E20

Seebeck emf—See **thermal emf.** **F 1241,** F01

seed—a relatively small gaseous inclusion in glass. **C 162,** C14

seed coat fragment, *n*—*in cotton*, a portion of a cotton seed, usually black or dark brown in color, broken from a mature or immature seed, and to which fibers and linters may or may not be attached. **D 123,** D13

seed cotton, *n*—cotton, as harvested and before ginning, consisting of seeds with the fibers attached and usually including measurable amounts of foreign matter. **D 123,** D13

seep—a small area where water oozes from the soil or rock. **D 653,** D18

seepage—the infiltration or percolation of water through rock or soil to or from the surface. The term seepage is usually restricted to the very slow movement of ground water. (ISRM) **D 653,** D18

seepage face, *n*—a boundary between the saturated flow field and the atmosphere along which a subsurface liquid discharges, either by evaporation or movement "downhill" along the land surface or in a well as a thin film in response to the force of gravity. **D 653,** D18

seepage force—the frictional drag of water flowing through voids or interstices in rock, causing an increase in the intergranular pressure, that is, the hydraulic force per unit volume of rock or soil which results from the flow of water and which acts in the direction of flow. (ISRM) **D 653,** D18

seepage force, J **(F)**—the force transmitted to the soil or rock grains by seepage. **D 653,** D18

seepage line—see **line of seepage**. **D 653,** D18

seepage line, *n*—the uppermost level at which a flowing liquid emerges along a seepage face. **D 653,** D18

seepage (percolation)—the slow movement of gravitational water through the soil or rock. **D 653,** D18

seepage velocity, V_a, $V_1(LT^{-1})$—the rate of discharge of seepage water through a porous medium per unit area of void space perpendicular to the direction of flow. **D 653,** D18

segmental testing—a method of isolating and testing portions of an installed pipeline to determine the location of an air loss in excess of the standard. **C 896,** C04

segment die, *n*—a die fabricated by the assembly of several die sections within a retaining bolster or shrinkage ring. **B 243,** B09

segmented testing—a method of testing vitrified clay pipe using segmented bearings as detailed in Test Method C 301. **C 896,** C04

segregation, *n*—the unintentional separation of the constituents of concrete or particles of an aggregate, causing a lack of uniformity in their distribution. **C 125,** C09

segregation—*in grouting*, the differential concentration of the components of mixed grout, resulting in nonuniform proportions in the mass. **D 653,** D18

segregation—concentration of alloying elements in specific regions in a metallic object. **E 7,** E04

seismic support—mass (heavy) supported on springs (weak) so that mass remains almost at rest when free end of springs is subjected to sinusoidal motion at operating frequency. **D 653,** D18

seismic velocity—the velocity of seismic waves in geological formations. (ISRM) **D 653,** D18

seismometer—instrument to pick up linear (vertical, horizontal) or rotational displacement, velocity, or acceleration. **D 653,** D18

seizure, *n*—*in lubrication*, welding between surfaces in relative motion that results in immobilization of the parts. **D 4175,** D02

SEL—see **substrate equivalent lead concentration**. **E 1605,** E06

select, *v*—*in sample preparation*, to mix and divide a sample (with or without comminution) to ensure that subsamples equally represent the original lot. **E 135,** E01

selected area aperture—*SIMS*, the mechanical equivalent of electronic signal gating, commonly used in stigmatic mass spectrometers. **E 673,** E42

selected material—finely divided material free of debris, organic material, and large stones. **C 896,** C04

selective heating, *n*—intentionally heating only certain portions of a steel object. **A 941,** A01

selective ion meter—an instrument that measures the potential of pH and ion-selective electrodes and displays it directly in concentration or activity units, such as pH, moles per litre, or milligrams per litre, on a logarithmic scale. Millivolt scales are provided for titrations and to verify proper electrode operation. **D 4127,** D19

selective quenching, *n*—**quenching** only certain portions of a steel object. **A 941,** A01

selective radiator, *n*—radiator whose spectral emissivity depends on the wavelength over the range considered. **E 349,** E21

selective surface—a surface for which the spectral optical properties reflectance, absorptance, emittance, or transmittance vary significantly with wavelength, which enhances the collection (or rejection) of radiant energy in a restricted portion of the spectrum.

NOTE—An example of a selective surface would be a collector cover glazing that has a high transmittance over the solar spectrum (300 to 2500 nm) and high reflectance over the spectral region of principal thermal infrared emission from the absorber. **E 772,** E44

selectivity—the characteristic of a test system that is a measure of the extent to which an instrument is capable of differentiating between the desired signal and disturbances of other frequencies or phases. **E 1316,** E07

selectivity constant—a measure of a liquid ion exchange or glass electrode's response to an interfering ion compared to its response to the ion being measured. The smaller the selectivity constant, the greater the electrode's preference for the ion being measured. The principal use of selectivity constants is in determining whether an electrode can be used in a particular application. Since selectivity constants vary somewhat with the ratio of the two ions, they should not be used to correct for errors caused by interferences. The term *selectivity constant* is sometimes mistakenly applied to solid-state electrodes (that are based on solubility product) to indicate the maximum allowable ratio of interference to ion being measured. Unlike liquid and glass electrodes, solid-state electrodes exhibit an *all-or-nothing* response to an interference, with no interference as long as the critical ratio is not exceeded. **D 4127,** D19

selector valve, *n*—device that permits the switching of the route taken by a fluid. **F 2002,** F29

selector valve, *n*—*in anesthesia breathing systems,* device that is used to direct gases to an anesthesia reservoir bag or a ventilator intended for use during anesthesia. **F 2002,** F29

selenide-type inclusions— See **sulfide-type inclusions.** **E 7,** E04

self-absorption, *n*—*in optical emission spectrometry*, the reduction in relative intensity in the central portion of spectral lines resulting from selective absorption by the cooler outer vapor of the source envelope of radiation emitted by the hot central core. **E 135,** E01

self-balancing potentiometer—See **potentiometer.** **E 7,** E04

self-biased gun—an electron gun in which the cathode cap is biased with respect to the filament by means of a bias resistor through which the emission current flows between the filament and cap. This type of gun provides high intensity at low angular apertures. (See **biased gun, unbiased gun, saturated gun.**) **E 7,** E04

self-bound seam finish, *n*—*in home sewing*, a seam finish in which one seam allowance encloses both cut edges. **D 4965,** D13

self-checking bar code, *n*—a bar code that will not permit a single printing defect to be interpreted as a valid character within the symbology algorithm. **F 1294,** F05

self-clinching—term applied to nail with point or shank designed in such a way that nail clinches automatically while fully driven. **F 547,** F16

self-clocking, *n*—a bar code symbol designed to be read with a single aperture reader over a range of velocities, with allowances included for change in velocity after reading has commenced. **F 1294,** F05

self-contained chemical paper—a sheet that has two materials added during the paper manufacturing process or coated on the formed sheet, that react under pressure to form a visible image. **F 549,** F05

self-contained mechanical paper—a sheet that has a pigmented substrate that becomes apparent when a surface coating is rendered transparent by application of pressure. **F 549,** F05

self-correcting—algorithm that modifies itself during repetitive operations to take advantage of newly acquired information; akin to curve-fitting. **F 1156,** F12

self-curing, adj— See **self-vulcanizing**. **D 907,** D14

self-damping conductor—a conductor with internal vibration damping properties which will protect itself from damaging strains caused by aeolian vibration. **B 354,** B01

self-electrodes, *n*—*in optical emission spectrometry*, a pair of electrodes composed of the material being analyzed. **E 135,** E01

self-extinguishing, *adj*—not defined; the Board of Directors of ASTM has ruled that the term "self-extinguishing" shall not be used in ASTM standards. It has no meaning except in association with a specific test method or specific conditions of burning. **D 123,** D13

self-flow, *n*—the characteristic of a refractory castable to flow readily under its own weight without showing signs of segregation or separation. **C 71,** C08

self-furring nail—galvanized, regular-stock-steel, 1¼ to 2 ½ by 0.106-in. nails with ⅜-in. flat head, medium diamond point, and washer or spacer on shank; for fastening reinforcing wire mesh and spacing it from nailing member. **F 547,** F16

self heating, *n*—a rise in the temperature of a material, assemblage, or product caused by internal, exothermic chemical reaction. **E 176,** E05

self ignition, n—See **spontaneous ignition**, the preferred term. **E 176,** E05

self-inflating—boom that automatically inflates as it is deployed. **F 818,** F20

self-leveling sealant, *n*—a compound that exhibits flow sufficient to seek gravitational leveling. **C 717,** C24

self-limiting powder porcelain— the maximum thickness of electrostatically-charged powder that can be deposited as a surface film. **C 286,** B08

self-noise, *n*—extraneous non-acoustical signals, generated or induced in a measurement system. **C 634,** E33

self-polishing-type floor polish—a floor polish that dries to a shine. **D 2825,** D21

self-quenching, *n*—in luminescence, the reduction of luminescence through the depletion of an excited atomic or molecular entity by interaction with another entity of the same species in the ground state. **E 131,** E13

self-reversal, *n*—*in optical emission spectrometry*, the extreme case of self-absorption in which intensity decreases with increasing concentration. **E 135,** E01

self-spreading—term applied to nail with split shank designed in such a way that two or more legs penetrate material in divergent directions. **F 547,** F16

self-stressing grout—expansive-cement grout in which the expansion induces compressive stress in grout if the expansion movement is restrained. **D 653,** D18

self-sustaining discharge—a discharge in which all carriers necessary for the transport of current in the discharge are produced by this discharge itself. **B 542,** B02

self-timing indicator—assorted versions of a **Hamming code**, which permits the system to keep track of where it is at any moment in time. **F 1156,** F12

self-vulcanizing, *adj*—pertaining to an adhesive that undergoes vulcanization without the application of heat. (See also **vulcanization**, *n.*) **D 907,** D14

self-weighted bottom camber, H_b—the height of the running surface from a plane surface, measured at the highest point, with only the influence of the snowboard ski weight. **F 1107,** F27

selvage, *n*—the woven edge portion of a fabric parallel to the warp. **D 123,** D13

selvage—an edge or edging which differs from the main part of: (*1*) a fabric; or (*2*) granule-surfaced roll roofing. **D 1079,** D08

selvage, *n*—the woven edge portion of a fabric parallel to the warp. **D 4439,** D35

selvage—the edge finish on woven chain link fabric joining pairs of pickets. The selvage may be knuckled or twisted. **F 552,** F14

selvage mark, *n*—*in finished cloth,* a lengthwise crease mark along the selvage caused by an edge being folded or doubled. **D 123,** D13 *semianthracite*—See *semianthracite* under **rank**. **D 121,** D05

semiapochromatic objective—a compromise, in the correction for chromatic and spherical aberration, between achromatic and apochromatic objectives; frequently called fluorite objectives. **E 7,** E04

semiapochromatic objective lens—an objective lens with both longitudinal and spherical chromatic correction for green and blue. Should be used with green or blue filters. **E 7,** E04

semiautomatic image analyzer—a device which can detect and measure features of interest in an image but requires the operator to perform feature discrimination. **E 7,** E04

semichemical pulp, *n*—fibrous material obtained by treating wood or other plant material with a combination of chemical and mechanical processes, such as chemimechanical pulp (CMP), chemithermomechanical pulp (CTMP), bleached chemithermomechanical pulp (BCTMP), and neutral sulfite semichemical pulp (NSSC). (See **chemical pulp** and **mechanical pulp**.) **D 1968,** D06

semicontinuous activated sludge, SCAS—a term used to designate a test procedure described in Test Method D 2667 that is used to confirm determinations of the degree of biodegradability of alkylbenzene sulfonates. **D 459,** D12

semi-detached dwelling—See **dwelling.** **E 631,** E06

semidrying oil—an oil that possesses the characteristics of a drying oil but to a lesser degree. There is no definite line of demarcation between drying and semidrying oils. **D 16,** D01

semi-efficient vulcanizing (semi-EV) system, *n*—as applied to natural rubber and isoprene- and butadiene-based synthetic rubbers, a vulcanizing system having sulfur and accelerator concentrations between those of a conventional sulfur vulcanizing system and an EV system. **D 1566,** D11

semifusinite—See *semifusinite* under **maceral**. **D 121,** D05

semi-guided bend, *n*—the bend obtained by applying a force directly to the specimen in the portion that is to be bent. **E 6,** E28

semikilled steel, *n*—an incompletely deoxidized **steel** that contained sufficient oxygen to form enough entrapped carbon monoxide during solidification to offset solidification shrinkage. **A 941,** A01

semi-mat glaze—See *semi-mat glaze* under **glaze.** **C 242,** C21

semi-muffle furnace—a furnace with a partial muffle, in which the products of a combustion come in contact with the ware. **C 286,** B08

semi-permanent mold casting, *n*—mold casting that is made with an expendable core such as sand. **B 899,** B02

semipermeable membrane—a membrane which preferentially allows the passage of specific compounds while rejecting others. **D 6161,** D19

semi-porcelain—a trade term designating semivitreous dinnerware. **C 242,** C21

semi-restraint, *adj*—of or relating to a method of mounting that allows an object a limited degree of movement (for example, contraction or expansion of a fabric). **D 123,** D13

semirigid plastic, *n*—for purposes of general classification, a plastic that has a modulus of elasticity either in flexure or in tension of between 70 and 700 MPa (10 000 and 100 000 psi) at 23°C and 50 % relative humidity when tested in accordance with Test Method D 747, Test Methods D 790, Test Method D 638M, or Test Methods D 882. **D 883,** D20

semirigid plastic, *n*—for the purposes of general classification, a plastic that has a modulus of elasticity either in flexure or in tension of between 70 and 700 MPa (10 000 and 100 000 psi) at 23°C and 50% relative humidity when tested in accordance with Test Methods D 638, D 747, D 790, or D 882. (D-20) **F 412,** F17

semirigid plastic—for purposes of general classification, a plastic that has a modulus of elasticity either in tension of between 70 and 700 MPa (10 000 and 100 000 psi) at 23°C and 50 % relative humidity when tested in accordance with Test Methods D 747, D 790, or D 882. **F 1251,** F04

semitrailer, *n*—a freight trailer supported at its forward end by a truck tractor or another trailer and at its rearward end by attached axles. **E 867,** E17

semivitreous (semivitrified)—that degree of vitrification evidenced by a moderate or intermediate water absorption. **C 242,** C21

semivolatile organic chemical—See **organic chemical.** **D 1356,** D22

Senarmont (Friedel) polarimeter—an optical device for measuring relative optical retardation consisting of a light source, a polarizing filter, a retardation plate exhibiting one-quarter wave retardation with its optical axis perpendicular to the polarizing axis, and a rotatable polarizing element (the analyzer) whose angular position can be read on a scale. **C 162,** C14

sensitive ink—an ink that changes its appearance or characteristics, or both, under stress; a sympathetic ink. **F 1156,** F12

sensitive tint plate—a gypsum plate (better known as a red I or full wave plate), used with cross polarized light producing circularly polarized light. For anisotropic materials, colors are enhanced, for isotropic materials, a magenta hue is produced. **E 7,** E04

sensitivity, *n*—*in electronic systems*, the minimum change in the input signal that produces a change in the output signal that can be reliably measured. **D 123,** D13

sensitivity, *n*—*in mechanical systems*, the smallest change that can be induced on a material by the system and be reliably measured. (See also **least count.**) **D 123,** D13

sensitivity—the effect of remolding on the consistency of a cohesive soil. **D 653,** D18

sensitivity, *n*—the change of instrument response with change in analyte concentration. **E 135,** E01

sensitivity, *n*—ability to perceive, identify or differentiate, qualitatively or quantitatively, or both, one or more stimuli by means of the sense organs. **E 253,** E18

sensitivity—see **contrast sensitivity, equivalent IQI sensitivity, equivalent penetrameter sensitivity, IQI sensitivity, radiographic sensitivity.** **E 1316,** E07

sensitivity—in the case of a leak detector, the response of the detector to tracer gas leakage (that is, scale divisions per unit of leakage rate). **E 1316,** E07

sensitivity—the degree of capability of a magnetic particle examination technique for indicating surface or near surface discontinuities in ferromagnetic materials. **E 1316,** E07

sensitivity—a measure of the smallest ultrasonic signal which will produce a discernible indication on the display of an ultrasonic system. **E 1316,** E07

sensitivity analysis—a test of the outcome of an analysis by altering one or more parameters from an initially assumed value(s). **E 833,** E06

sensitivity control—the control in the instrument that adjusts the amplifier gain, and is one of the factors that determines the capacity to detect discontinuities. **E 1316,** E07

sensitivity criterion, *n*—a quantitative measure of the relative merit of two test methods, that combines the estimated variance of each method with the ability of the method to measure differences in the property of interest. **D 123,** D13

sensitivity factor:

elemental—*XPS,* intensities of peaks relative to those of a standard, for example, F1s, for atoms in typical homogeneous environments. Division of peak height or peak area intensities by the appropriate sensitivity factors gives the relative number of atoms detected, on the assumption of sample homogeneity.

relative Auger elemental—AES, the ratio of the Auger signal strength of a specified Auger transition from a single element to that from a selected standard element (for example, silver), as measured under identical conditions.

SIMS—the factor used to convert the net counts per unit time, for a particular species, matrix and experimental conditions, to concentration. **E 673,** E42

sensitivity (of an instrument)—the differential quotient dQ_0/dQ_1, where Q_0 is the scale reading and Q_1 is the quantity to be measured. **D 653,** D18

sensitivity (of a transducer)—the differential quotient dQ_0/dQ_1, where Q_0 is the output and Q_1 is the input. **D 653,** D18

sensitivity of leak test—the smallest leakage rate that an instrument, method, or system is capable of detecting under specified conditions. See **minimum detectable leakage rate.** **E 1316,** E07

sensitivity, photographic—the degree to which a photographic material responds by a change in its chemical or physical state to the action of both light and chemical development. The sensitivity of a given photographic material varies with wavelength of the incident radiation. The average degree of response is commonly referred to as speed. **E 7,** E04

sensitivity range—the impact pressure range over which a ribbon or carbon will produce an image of specified quality. **F 221,** F05

sensitivity ratio, SR, *n—in comparing two test methods,* the ratio of the sensitivities of the test methods with the large sensitivity in the numerator. **D 123,** D13

sensitivity value—the value determined by the smallest standard discontinuity in any given sensitivity indicator observable in the radiographic image. Values are defined by identification of type of indicator, size of defect, and the absorber thickness on which the discontinuity is observed. **E 1316,** E07

sensitization—the absorption of a reducing agent, often a stannous compound, on the surface. **B 374,** B08

sensitization—a process resulting in a metallurgical condition which causes susceptibility of an alloy to intergranular corrosion or intergranular environmentally assisted cracking in a specific environment. **G 15,** G01

sensitizer, *n*—a material which forms an eutectic with the developer to lower the melt point; acting as a solvent with the developer, the leuco dye is solubilized and reacts to form color. **F 1623,** F05

sensitizing—see **dye sensitizing.** **F 335,** F05

sensitometric curve— See **characteristic curve.** **E 7,** E04

sensor, *n*—a device designed to respond to a physical stimulus (as temperature, illumination, and motion) and transmit a resulting signal for interpretation or measurement, or for operating a control. **D 1356,** D22

sensor, *n—of a thermometer or thermometric system,* element of the thermometer or thermometric system that is directly affected by the temperature to be measured. **E 344,** E20

sensor, *n*—part of the monitor that is sensitive to the variable being measured. **F 2002,** F29

sensor, acoustic emission—a detection device, generally piezoelectric, that transforms the particle motion produced by an elastic wave into an electrical signal. **E 1316,** E07

sensor area, *n*—part of the sensor at which the variable being measured is sensed. **F 2002,** F29

sensory, *adj*—pertaining to the sense organs. **E 253,** E18

sensory panel, *n*—a group of individuals that may be selected on the basis of sensitivity to stimuli, reliability, or whose perceptions are judged to be representative of some larger population. It is used to obtain information concerning the sensory attributes of physical stimuli. **E 253,** E18

sensory profile, *n*—a description of the sensory properties of a sample, consisting of the sensory attributes in the order of perception, and with assignment of an intensity value for each attribute. **E 253,** E18

separable pin, *n—in zippers,* a tube-like element attached over the bead at the bottom end of one stringer. **D 123,** D13

separable zipper, *n*—a zipper fitted with special components at the bottom of the chain, so as to permit complete disengagement and then reengagement of the two stringers. (Compare **nonseparable zipper**.) **D 123,** D13

separate-application adhesive, *n*— a term used to describe an adhesive consisting of two parts, one part being applied to one adherend and the other part to the other adherend and the two brought together to form a joint. **D 907,** D14

separate-element zipper, *n*—a zipper consisting of two series of separately formed elements, each attached to one of the opposing edges of two tapes which are engaged and disengaged by the movement of a slider. (Compare **continuous element zipper**.) **D 123,** D13

separator, *n*—a component of some open-end spinning machines located inside the rotor to direct the incoming fibers to the slide surface. **D 123,** D13

separator, *n*—any material interposed between articles or packing components. (Compare **buffer, divider, insert, liner.**) **D 996,** D10

separator cloth—a fabric, coated with TFE-fluorocarbon or similar release agent, placed between the lay-up assembly and the bleeder system to facilitate subsequent bleeder-system removal from the laminate after it has been cured. **E 1749,** E06

separator paper, *n*—thin paper applied to a stranded conductor to prevent migration of extruded insulation into the strands and, subsequently, to facilitate stripping of the insulating material from the conductor. **D 1711,** D09

SE probe—see **dual search unit (twin probe).** **E 1316,** E07

sequential excitation NMR; continuous wave (CW) NMR—a form of high-resolution NMR in which nuclei of different field-frequency ratio at resonance are successively excited by sweeping the magnetic field or the radio frequency. **E 131,** E13

sequential response—The assignment, according to local protocols, of emergency medical resources with varying levels of care capability to the scene of an illness or injury based on information received from previously arrived, medically trained, on-scene responders. A sequential response differs from a simultaneous response. **F 1177,** F30

sequential sampling—See *running sample* under **sample.** **D 1356,** D22

sequestering agent—an agent that forms soluble complex compounds with, or sequesters, a simple ion, thereby suppressing the activity of that ion. Thus, in water treatment the effects of hardness can be suppressed by adding agents to sequester calcium and magnesium. See **chelating agent.** **B 374,** B08

sequestering agent—any compound that, in aqueous solution, combines with a metallic ion to form a water-soluble combination in which the ion is substantially inactive. **D 459,** D12

serial communication, *n*—in data communications, refers to sending the code for a character over a single line one bit at a time. (See **parallel communication**.) **F 1457,** F05

serial printer—a printer where characters are printed one at a time. **F 149,** F05

series collection, *n*—an operation involving the use of two or more collectors joined in series. **D 1356,** D22

series grouting—similar to stage grouting, except each successively deeper zone is grouted by means of a newly drilled hole, eliminating the need for washing grout out before drilling the hole deeper. **D 653,** D18

series (in x-ray spectra)—the group of characteristic X-ray lines which results when a vacancy in one particular electron level is filled from outside levels. Thus, the K series lines are emitted when a K level vacancy is filled. **E 7,** E04

serpentine marble (III)—a rock consisting mostly or entirely of serpentine (hydrated magnesium silicate), green to greenish-black in color, commonly veined with calcite, and dolomite or magnesite, or both (magnesium carbonate).

NOTE—The term *verd* (or *verde*) *antique* is popularly used for serpentine marble, but it may also refer to green marble or limestone marble devoid of serpentine. Because of its composition, "serpentine marble" is not a true marble in the geologic sense. **C 119,** C18

serrated—See **barbed.** **F 547,** F16

serrations—corrugations in the cutting edge of the blades. **F 1078,** F04

serum, (latex), *n*—the dispersion medium of a rubber latex or the solution remaining, after either coagulation or centrifugation. **D 1566,** D11

serviceability—See **building performance.** **E 631,** E06

serviceability, n— See **facility serviceability**. (See also **building performance**.) **E 1480,** E06

serviceability requirement— See **requirement statement**. **E 1480,** E06

serviceability requirements profile (SRP) (profit de besoins génériques de fonctionnalite (PBGF))—listing of the levels of serviceability (capability to perform, including quality) required in a facility. **E 1480,** E06

service damage, *n— as used in fractography,* scratches, chips, cracks, etc., created during use of the component. **C 1145,** C28

service factor—a factor which is used to reduce a strength value to obtain an engineering design stress. The factor may vary depending on the service conditions, the hazard, the length of service desired, and the properties of the pipe. **F 412,** F17

service hardboard—a hardboard of about 55 lb/ft^3 (specific gravity 0.88) density intended for use where standard strength board is not required and better dimensional stability is desired. **D 1554,** D07

service life—the period of time required under use conditions to change the appearance of a surface treated with a floor polish sufficiently to require retreatment. **D 2825,** D21

service life (service time)—the elapsed time until the end point is reached in an adsorption process. **D 2652,** D28

service run—that part of the operating cycle of an ion-exchange process in which a water is passed through a bed of the ion-exchange material in order to remove specific ions from the water or to exchange them for an equivalent amount of a specific ion from the bed material. **D 1129,** D19

service strength setting time—the time interval, after the initial mixing of the composition, at a specific temperature, at which time the product has acquired 90 % of its specified strength. **C 904,** C03

service test—a test in which the product is used under actual service conditions. **F 869,** F08

service time—see **service life.** **D 2652,** D28

serving, *n—on magnet wire or conductor*, a uniform wrapping of insulation around a conductor. **D 1711,** D09

servo accelerometer, *n*—an accelerometer containing servo mechanisms, electronics, and a seismic element to sense inertial reaction. **F 538,** F09

set, *n*—the hardening and hydration of a gypsum plaster. See **setting time.** **C 11,** C11

set—a flow property of porcelain enamel slip affecting the rate of draining, residual thickness, and uniformity of coating. **C 286,** B08

set—*in grouting*, the condition reached by a cement paste, or grout, when it has lost plasticity to an arbitrary degree, usually measured in terms of resistance to penetration or deformation; initial set refers to first stiffening and final set refers to an attainment of significant rigidity. **D 653,** D18

set, *n*—strain remaining after complete release of the force producing the deformation. **D 883,** D20

set, *v*—to convert an adhesive into a fixed or hardened state by chemical or physical action, such as condensation, polymerization, oxidation, vulcanization, gelation, hydration, or evaporation of volatile constituents. (See also **cure** and **dry**.) **D 907,** D14

set, *n*—strain remaining after complete release of the force producing the deformation. **D 1566,** D11

set—strain remaining after complete release of the load producing the deformation. **E 6,** E28

set, *n*—strain remaining after complete release of the force producing the deformation. (D-11) **F 412,** F17

set, *v*—to convert an adhesive into a fixed or hardened state by chemical or physical action, such as condensation, polymerization, oxidation, vulcanization, gelation, hydration, or evaporation of volatile constituents. (D-14) **F 412,** F17

set, *n—in tire testing*, a selected number of test or control tires having nominally identical properties. **F 538,** F09

set—the positioning of the blade for proper cutting action. **F 1078,** F04

set—the at rest position of the instrument halves that will provide the intended closing relationship of fit and force. **F 1638,** F44

set after break, *n*—the set of a test piece after stretching it to rupture. **D 1566,** D11

set mark, *n*—a stop mark resulting from a prolonged loom shutdown. **D 123,** D13

set-off—the unintentional transfer of part of an image or ink from its intended location to another surface. **F 221,** F05

set-off—the unintentional transfer of part of an image or ink from its intended location to another surface. **F 335,** F05

set-off—the unintentional transfer of part of an image or ink from its intended location to another surface. **F 425,** F05

set time—(*1*) the hardening time of portland cement; or (*2*) the gel time for a chemical grout. **D 653,** D18

setting, *n*—the process, due to chemical reactions, occurring after the addition of mixing water, that results in a gradual development of rigidity of a cementitious mixture. **C 125,** C09

setting, *n*—process by which, after application, a liquid (wet-state) material changes to a serviceable condition by **curing** or **drying.** **E 631,** E06

setting agent—the component of a mixture that catalyzes or reacts with the resin component to cause hardening by polymerization. **C 904,** C03

setting block, *n—in building construction in glazing*, a piece of resilient material placed to support, to distribute the load, to prevent edge contact, and to align a pane of glass or a panel within its supporting frame. **C 717,** C24

setting rate—a comparative term referring to the time required for the glass surface to cool between the limits of the working range. **C 162,** C14

setting shrinkage—*in grouting*, a reduction in volume of grout prior to the final set of cement caused by bleeding, by the decrease in volume due to the chemical combination of water with cement, and by syneresis. **D 653,** D18

setting temperature, *n*—the temperature to which an adhesive or an assembly is subjected to set the adhesive. (See also **curing temperature** and **drying temperature**.) **D 907,** D14

setting time, *n*—the elapsed time required for a gypsum plaster to attain a specified hardness and strength after mixing with water **C 11,** C11

setting time, *n*—the elapsed time from the first addition of liquid during the mixing of the refractory castable until the time that the refractory castable has developed enough strength so that forms can be removed without any slumping or deformation. **C 71,** C08

setting time, *n*—the period of time during which an assembly is subjected to heat or pressure, or both, to set the adhesive. (See also **curing time, joint-conditioning time,** and **drying time**.) **D 907,** D14

setting-up agent or set-up agent—an electrolyte used to increase the measured pick-up of a slip. **C 286,** B08

settle mark—See **chill mark**. **C 162,** C14

settling—the process of depositing, by gravity, matter suspended in water. **D 4410,** D19

settling velocity, *n*—the terminal rate of fall of a particle through a fluid as induced by gravity or other external force; the rate at which frictional drag balances the accelerating force (or the external force). **D 1356,** D22

sew, *v*—to unite or fasten with stitches. **D 123,** D13

sewage—waste matter carried off by sewers. **C 896,** C04

sewage (wastewater)—the spent water of a community which is a combination of liquid and water-carried wastes. **C 822,** C13

sewer—a pipeline intended to convey sewage. **C 822,** C13

sewer—generally, an underground conduit usually carrying waste matter in a liquid medium. **C 896,** C04

sewerage—system for collection, treatment, and disposal of sewage. **C 896,** C04

sewer brick, *n*—low absorption, abrasive-resistant brick intended for use in drainage structures. See Specification C 32. **C 43,** C15

sewer line—See **sewer.** **C 896,** C04

sewer pipe—vitrified clay pipe as described in Specification C 700. **C 896,** C04

sewing force, *n*—the force applied to a sewing thread at the needle eye during penetration of a material by the needle. **D 123,** D13

sewing hole, *n*—*in buttons*, a hole in either the flange or shank used to attach the button to the substrate by means of a needle and thread. **D 123,** D13

sewing machine, *n*—*in home sewing*, a machine that uses needles and bobbin threads in the stitch formation and is primarily for home sewing use rather that industrial use. **D 123,** D13

sewing thread, *n*—a flexible, small-diameter yarn or strand, usually treated with a surface coating lubricant, or both, intended to be used to stitch one or more pieces of material or an object to a material. **D 123,** D13

sewing thread, *n*—a flexible, small diameter yarn or strand, usually treated with a surface coating, lubricant, or both, intended to be used to stitch one or more pieces of material or an object to a material. **D 4439,** D35

sewn seam, *n*—*in sewn fabrics*, a series of stitches joining two or more separate plies of a material or materials of planar structure such as textile fabric. **D 4439,** D35

sewn seam, *n*—*in sewn materials*, a juncture along which two or more planar structures, such as textile material, are joined by sewing. (Compare **glued seam, stapled seam, thermally bonded seam**. Syn. **stitched seam**.) **D 4965,** D13

sewn seam strength, *n*—*in sewn fabrics*, the maximum resistance to rupture of the junction formed by stitching together two or more planar structures. **D 123,** D13

sewn seam strength, *n*—*for geotextiles*, the maximum resistance, measured in kilonewtons per metre, of the junction formed by stitching together two or more planar structures. **D 4439,** D35

sew-through flange button, *n*—a button that has two or more holes in its flange for passage of a needle and thread so that the button can be attached to a flexible substrate. **D 123,** D13

sew through flange button, *n*—a button attached to one part of a flexible substrate by means of a needle and thread passed through two or more holes in its flange and through the substrate. (Compare **sew-through shank button**). **D 123,** D13

sew-through flange button, *n*—a button attached to one part of a flexible substrate by means of needle and thread passed through two or more holes in its flange and through the substrate. (Compare *sew-through shank button*.) **D 5497,** D13

sew through shank button, *n*—a button attached to one part of a flexible substrate by means of a needle and thread passed through a hole or loop in the intergral shank and through the substrate. (Compare **sew-through flange button).** **D 123,** D13

sew-through shank button, *n*—a button attached to one part of a flexible substrate by means of needle and thread passed through a hole or loop in the integral shank and through the substrate. (Compare *sew-through flange button*.) **D 5497,** D13

shade—a term descriptive of a lightness difference between surface colors, the other attributes of color being essentially constant. A lighter shade of a color is one that has higher lightness but approximately the same hue and saturation; and a darker shade is one that has a lower lightness.

Primarily, the term "shade" is derived from shadow and designates a change in appearance analogous to that produced by a local reduction in illumination. It should, therefore, when strictly used, express only the change toward a darker color. Shade of a color has been defined by several authorities as the mixture of black with that color, thus establishing its opposite character to "tint," but by extension of its relative sense it has been frequently and widely used to include lighter shades by use of the adjective "lighter" or" paler." Although such expressions apparently involve a contradiction, it is clear that while we may have a shade of a color or darker color of the same sort, it is easy to conceive of another shade not quite so dark and therefore lighter. **D 16,** D01

shade, *n*—(*1*) a color produced by a dye or pigment mixture including black dye or pigment. See also **shade,** *v*; **tint,** *n*; **tint,** *v*.

(*2*) an expression of color difference from a reference dyeing such that another dye must be added to produce a match.

(*3*) a color slightly different from a reference color. **E 284,** E12

shade, *v*—to adjust the color of a test specimen to be a closer color match to the standard. See also **shade,** *n*; **tint,** *v*; **tint,** *n*. **E 284,** E12

shade bracket nail—bright, regular-stock-steel, ¾ to 1 by 0.080 or 0.092-in. nails with slightly countersunk ½ or 9/32-in. head and needle point. **F 547,** F16

shade sorting, *n*—process of grouping together, often by instrumental measurement, similarly colored materials so that the materials within each group may be used together in a finished product without perceived color variation. **E 284,** E12

shading coefficient, *n*—the ratio of the solar heat gain through a glazing system under a specific set of conditions to the solar gain through a single light of double-strength sheet glass under the same conditions. **D 123,** D13

shading or shadowing—the act of casting a shadow across any surface. **E 772,** E44

shadow—a region in a body that cannot be reached by ultrasonic energy traveling in a given direction because of the geometry of the body or a discontinuity in it. **E 1316,** E07

shadow angle—the angle between the line of motion of the evaporated atoms and the surface being shadowed. The angle is analogous to the angle of incidence in optics. May be specified as an arc tangent "a" so that "a" is in the ratio between the height of the object casting the shadow over the length of the shadow (see **shadowing**). **E 7,** E04

shadow cast replica—a replica which has been shadowed. (See also **shadowing.)** **E 7,** E04

shadowing—a process by which a metal or salt is deposited on a specimen at an angle (see **shadow angle**) from a heated filament in a vacuum to enhance image contrast by inhibiting the deposition of the shadowing material behind projections.

(*1*) *metal shadowing*—the enhancement of contrast in a microscope specimen by the vacuum deposition of a dense metal at an angle generally not perpendicular to the surface of the object.

(*2*) *oblique evaporation*—the condensation of evaporated material onto a substrate which is inclined to the direct line of the vapor stream in order to produce shadows. **E 7,** E04

shadow microscope—an electron microscope which forms a shadow image of an object using electrons emanating from a point source located close to the object. **E 7,** E04

shadow wall—a structure built on top of the bridgewall or suspended from the crown to limit the flow of heat from the melter to the refiner. **C 162,** C14

shaft—generally a vertical or near vertical excavation driven downward from the surface as access to tunnels, chambers, or other underground workings. (ISRM) **D 653,** D18

shake—(*1*) a longitudinal separation of the wood. Generally two forms of shake are recognized, although variations and combinations may be used in industrial definitions.

heart shake—a shake that starts out at or near the pith and extends radially. Synonyms are *heart crack, rift crack.* A heart shake in which several radial cracks are present is termed a star shake.

ring shake—a shake occurring in the plane of the growth rings in the outer portion of the latewood for partial or entire encirclement of the pith, occasionally moving radially to an adjacent latewood ring. A synonym is "cup shake."

(*2*) A rectangular, board-like element for roof cover construction, similar to shingles, generally one of three types.

handsplit and resawn shakes—a shake having a split face and a sawn back.

tapersplit shake—a shake having two split faces and a natural, shingle-like taper.

straightsplit shake—a shake having two split faces and with no pronounced taper. **D 9,** D07

shake nail, cedar-shake or shingle nail, wood-shake face nail—enameled (baked-lacquer finished), galvanized, regular-stock-steel, plain-shank or annularly threaded, aluminum-alloy, plain-shank or helically threaded, or stainless-steel, annularly threaded, 1 ½ by 0.080 to 2½ by 0.092-in. nails with flat ⅛ to 9⁄64-in. head and medium diamond point. Also, aluminum-alloy 1 ¼ by 0.086 to 1¾ by 0.092-in. nails with flat 5⁄32-in. head and blunt diamond point. Also, aluminum-alloy, annularly threaded, 1 ⅜ and 2 by 0.099-in. nails with special type of thread and with flat slightly countersunk 5⁄32-in. head and long needle point; for fastening cedar shakes to fiberboard nail-base sheathing. **F 547,** F16

shakeoff process—*XPS,* a multi-electron photoelectric process in which two or more electrons are emitted, partitioning between them the excess kinetic energy. **E 673,** E42

shakeup lines or shakeup satellites—*XPS,* photoelectrons originating from photoelectric processes in which the final ion is left in an excited state, so that the photoelectron has a characteristic energy slightly less than that of the normal photoelectron. **E 673,** E42

shaking test—a test used to indicate the presence of significant amounts of rock flour, silt, or very fine sand in a fine-grained soil. It consists of shaking a pat of wet soil, having a consistency of thick paste, in the palm of the hand; observing the surface for a glossy or livery appearance; then squeezing the pat; and observing if a rapid apparent drying and subsequent cracking of the soil occurs. **D 653,** D18

shale, *n*—a thinly stratified, consolidated, sedimentary clay with well-marked cleavage parallel to the bedding. **C 43,** C15

shale—a laminated sedimentary rock composed of clay minerals.

alabaster—a soft, easily carved massive form of gypsum (calcium sulfate), often pleasingly blotched and stained. A banded stalagmitic calcite is also called alabaster. **C 119,** C18

shale—a thinly stratified, consolidated, sedimentary clay with well-marked cleavage parallel to the bedding. **C 896,** C04

shallow etch pits—etch pits that are small and shallow in depth under high magnification >200× (see also **haze**). **F 1241,** F01

shallow wood failure, n—failure that is invariably within the first one or two layers of cells beyond the adhesive layer in which the fracture path is not influenced by the wood-grain angle or growth-ring structure. **D 907,** D14

sham, *n*—a fabric covering for a bed pillow, usually of decorative fabric that matches the bedcovering. **D 123,** D13

shampoo, *n*—a solution of detergent in water formulated for specialized cleaning tasks. **D 123,** D13

shank—refers to leg positions of hide pattern. **D 1517,** D31

shank, *n—in buttons,* that part positioned perpendicular to and at the center back of the flange, and having a hole or loop for use in attaching the button to one part of a flexible substrate by means of a needle or thread, a ring, or a toggle. **D 5497,** D13

shank—main body of nail extending from head to point.

shank, formed or deformed—See **barbed, fluted, grooved, knurled, threaded or twisted.**

shank diameter— See **diameter.**

shear-resistant— See **threaded shear-resistant nail.**

smooth shank— See **plain shank.** **F 547,** F16

shank—that part of the sole of the shoe between the heel and the ball or tread; also, the steel, plastic, or wood piece inserted in the arch of a shoe for reinforcement; also, the leg segment between the thigh and the foot. **F 869,** F08

shank—(*1*) the part of either scissor half that yields configuration, length, and leverage (*2*) the part of the scissor between the finger ring and joint. **F 1078,** F04

shank eye, *n—in buttons,* the hole or loop in the shank of a sew-through shank button or the hole in the loop of the staple of a staple button. **D 5497,** D13

shape, *n*—this term is no longer recommended; the term **profile** is preferred. **B 881,** B07

shape calibration—a process of establishing a relationship between the expected peak shape and energy. The shape calibration may be as simple as the user supplying a few peak full-width at half maximum (FWHM)/energy (or channel) pairs or as complex as using a least squares algorithm to establish several parameters characterizing the peak width and its deviation from a pure

Gaussian. Shape calibration may be an explicitly defined function of an analysis program or implicitly done in connection with some other operation (typically the energy calibration). **E 170,** E10

shaped stone—dimension stone processed by carving, grinding, sawing, or other means into specific nonplanar configurations. **C 119,** C18

shape factor—a dimensionless ratio of lengths, surface areas, or volumes of the particles, useful for characterizing or comparing particles that otherwise have similar physical properties. **C 242,** C21

shape factor—generally defined as the ratio of the major dimension of a particle to the minor dimension. **C 859,** C26

shape resolution—an electron image exhibits shape resolution when a polygon can be recognized as such in the image. Roughly, the particle diameter (defined as the diameter of a circle of the same area as the particle) must exceed the resolution by a factor equal to the number of sides on the polygon. **E 7,** E04

shapes, *n*—materials of solid section in such forms as angles, channels, tees, I-beams, and four-fluted bars. **B 899,** B02

shark fin—an upward-curled felt sidelap or endlap. **D 1079,** D08

sharkskin—leather made from the top grain of the skins of sharks. It has various natural markings. The term should not be applied to leather made from other skins and embossed. **D 1517,** D31

sharks teeth—a term relative to the appearance of the cut edge of a glass sheet having dagger-like striations starting at the scored edge and extending toward the opposite surface. **C 162,** C14

sharpener—the device for holding the sharpening and truing stones. The sharpener can be either attached to the slicer or removable from the slicer's sharpener attachment point. If it is removable, a storage location within the machine structure shall be provided. **F 1827,** F26

sharp fire—combustion with excess air and short flame. **C 162,** C14

sharpness—*in carbon paper and inked ribbon images*, similarity of the geometry of the image to the type face and the rate of change of image density at the edge. **F 221,** F05

sharp-notch strength, σ_s [FL^{-2}]—the maximum nominal (net-section) stress that a sharply notched specimen is capable of sustaining. **E 1823,** E08

sharp point—end of point being keen, instead of being slightly rounded or dull. **F 547,** F16

shattering—a phenomenon observed in hardness testing in which fissures or subsurface cracks originate in a hardness indentation and spread to adjacent parts of the specimen. **E 7,** E04

shaving—a small wood particle of indefinite dimensions developed incidental to certain woodworking operations involving rotary cutterheads usually turning in the direction of the grain; and because of this cutting action, producing a thin chip of varying thickness, usually feathered along at least one edge and thick at another and usually curled. **D 1554,** D07

shavings—the thin slices of wood removed in dressing. **D 9,** D07

shear—relative displacement of adjacent planes in a member. **D 9,** D07

shear, *n*—*in an adhesively-bonded joint*, stress, strain or failure resulting from applied forces that tends to cause adjacent planes of a body to slide parallel in opposite directions. **D 907,** D14

shear—an action or stress, resulting from applied forces, which causes (or tends to cause) two contiguous parts of a body to slide, relative to each other, in a direction parallel to their plane of contact. **D 4092,** D20

shear-cake—a counterweighted refractory slab used as a gate or door to a small furnace or oven. **C 162,** C14

shear coaxial injector—a pneumatic atomizer consisting of a central liquid stream surrounded by a high-velocity co-annular gas stream. **E 1620,** E29

shear crimping, n—buckling of the compressive facing due to low core shear modulus. Usually causes the core to fail in shear at the crimp. **C 274,** D30

sheared-bevel point—one directional point having one flat surface; formed by shearing nail wire off at angle to shank axis, inducing nail to diverge from straight penetration unless countermeasures are introduced. **F 547,** F16

sheared point—See **sheared-bevel point, sheared-square point.** **F 547,** F16

sheared-square point—extremity of nail shank opposite head end of nail sheared at right angle to shank axis. **F 547,** F16

shear failure (failure by rupture)—failure in which movement caused by shearing stresses in a soil or rock mass is of sufficient magnitude to destroy or seriously endanger a structure.

general shear failure—failure in which the ultimate strength of the soil or rock is mobilized along the entire potential surface of sliding before the structure supported by the soil or rock is impaired by excessive movement.

local shear failure—failure in which the ultimate shearing strength of the soil or rock is mobilized only locally along the potential surface of sliding at the time the structure supported by the soil or rock is impaired by excessive movement. **D 653,** D18

shear force—a force directed parallel to the surface element across which it acts. (ISRM) **D 653,** D18

shear fracture, *n*—a mode of fracture in crystalline materials resulting from translation along slip planes which are preferentially oriented in the direction of the shearing stress. **E 6,** E28

shearling—leather made from sheepskin that has been sheared shortly before slaughter, the short wool being left on the skin when tanned. **D 1517,** D31

shear mark—a mark appearing in glass, caused by the action of the cutting shear. **C 162,** C14

shear modulus, *n*—the ratio of shear stress to corresponding shear strain below the proportional limit. (Compare **secant modulus**.) (See **proportional limit**.) **D 907,** D14

shear modulus, *G* (Pa)—the quotient of shear stress and shear strain. (See ISO 537.)

$$G = \frac{\sigma}{y}$$

D 4092, D20

shearogram/hologram, *n*—the common term for an interferometric image provided by S/H systems. **F 538,** F09

shearographic or holographic (S/H) systems, *n*—a shearographic or holographic system using interferometric laser imaging to nondestructively inspect tires. **F 538,** F09

shear or torsion, G [FL^{-2}], *n*—commonly designated as modulus of rigidity, shear modulus, or torsional modulus. **E 6,** E28

shear plane—a plane along which failure of material occurs by shearing. (ISRM) **D 653,** D18

shear rate, γ—the time rate of change of shear strain. For a one-dimensional shear flow, it is the velocity gradient. **D 4092,** D20

shear rate, *n*—*in fluid flow*, the velocity gradient across the fluid. **D 4175,** D02

shear resistance—see **internal friction**. **D 653,** D18

shear rigidity, n—the sandwich property which resists shear distortions: synonymous with shear stiffness. U=hG; the core thickness (approximate) times the core shear modulus. **C 274,** D30

shears strength, n—See **bond strength.**. **D 907,** D14

shear strain—the change in shape, expressed by the relative change of the right angles at the corner of what was in the undeformed state an infinitesimally small rectangle or cube. (ISRM) **D 653,** D18

shear strain, *n*—the tangent of the angular change, due to force between two lines originally perpendicular to each other through a point in the body. **D 907,** D14

shear strain, γ—the tangent of the angular change, due to force, between two lines originally perpendicular to each other through a point in a body. (E-28) **D 4092,** D20

shear strain, n—the tangent of the angular change, due to force, between two lines originally perpendicular to each other through a point in a body. **E 6,** E28

shear strength, *n*—the resistance to forces that cause, or tend to cause, two contiguous parts of a body to slide relatively to each other in a direction parallel to their contact. **D 123,** D13

shear strength—(monitoring wells) a measure of the shear or gel properties of a drilling fluid or grout. **D 653,** D18

shear strength, *n—in an adhesive joint,* the maximum average stress when a force is applied parallel to the joint. (See **bond strength.**) **D 907,** D14

shear strength—the maximum load applied normally to a fastener's axis that can be supported prior to fracture. Single shear is load occurring in one transverse plane, thus cutting the fastener into two pieces; double shear is load applied in two planes so that, at fracture, the fastener would be cut into three pieces. **F 1789,** F16

shear strength [FL^{-2}], *n*—the maximum shear stress which a material is capable of sustaining. Shear strength is calculated from the maximum load during a shear or torsion test and is based on the original dimensions of the cross section of the specimen. **E 6,** E28

shear strength of metal connector plate—resistance to shear force by net plate cross section, expressed as force per unit of length of full cross section of connector plate, when used in connections composed of pairs of plates. **E 1807,** E06

shear strength, *s*, T_f **(FL^{-2})**—the maximum resistance of a soil or rock to shearing stresses. See **peak shear strength.** **D 653,** D18

shear stress—a state of stress where adjacent planes in a member tend to slip on one another. **D 9,** D07

shear stress, τ—a stress acting parallel to the surface of the plane being considered. **D 653,** D18

shear stress—stress directed parallel to the surface element across which it acts. (ISRM) **D 653,** D18

shear stress, τ—the stress component tangential to the plane on which forces act. (E-28) **D 4092,** D20

shear stress [FL^{-2}], *n*—the stress component tangential to the plane on which the forces act. **E 6,** E28

shear stress (*shearing stress*) (*tangential stress*)—see **stress.** **D 653,** D18

shear test—an experiment to determine the flow properties of a bulk solid by applying different states of stress and strain to it. **D 653,** D18

shear tester—an apparatus for performing shear tests. **D 653,** D18

shear transfer plate—metal connector plate with integral teeth projecting from plate in both directions perpendicular to plate surfaces; designed to be placed between adjacent wood members and to connect and transmit forces from one wood member to another. **E 1807,** E06

shear wave—wave motion in which the particle motion is perpendicular to the direction of propagation. **E 1316,** E07

shear wave (rotational, equivoluminal)—wave in which medium changes shape without change of volume (shear-plane wave in isotropic medium is transverse wave). **D 653,** D18

shear wave search unit (Y cut quartz search unit)—a straight beam search unit used for generating and detecting shear waves. **E 1316,** E07

sheath, *n—in feathers,* a covering at the quill point end of nestling feathers or nestling down which holds the barbs together. **D 123,** D13

sheath-core, *adj*—a descriptive term for a multicomponent textile fiber consisting of a continuous envelope which encases a continuous, central, internal region. **D 123,** D13

sheathed crate—See **crate.** **D 996,** D10

sheathed thermocouple, *n*—a thermocouple having its thermoelements, and sometimes its measuring junction, embedded in ceramic insulation compacted within a metal protecting tube. **E 344,** E20

sheathed thermocouple wire, *n*—one or more pairs of thermoelements (without measuring junction(s)) embedded in ceramic insulation compacted within a metal protecting tube. **E 344,** E20

sheathed thermoelement, *n*—a thermoelement embedded in ceramic insulation compacted within a metal protecting tube. **E 344,** E20

sheath-enclosed-scale thermometer, *n*—the cylindrical glass envelope which encloses the scale and capillary tube. **E 344,** E20

sheathing, *n*—**lumber** or sheet materials, such as plywood, **fiberboard** or other suitable **panel** or sheet material fastened to the frame across the openings of a crate to deter pilferage, or entry of dirt or water or of other articles. **D 996,** D10

sheathing—cellulosic fiberboard for use in housing and other building construction, which may be integrally treated, impregnated or coated to give it additional water resistance. **D 1554,** D07

sheathing nail—See **common nail, threaded common nail, diaphragm nail, fiberboard nail, gypsum-sheathing nail, hardboard nail.** **F 547,** F16

sheathing/siding nail—stainless-steel, annularly threaded, 1½ by 0.120-in. nail with slightly countersunk 7⁄64-in. head and sharp medium diamond point. **F 547,** F16

sheen, *n*—the specular gloss at a large angle of incidence for an otherwise matte specimen. **E 284,** E12

sheepskin paint roller cover—a cover in which the paint applicating material is wool fleece still attached to its tanned natural skin. **D 16,** D01

sheer, *n*—a fabric that is transparently thin or diaphanous. **D 123,** D13

sheet—*as in inch-pound specifications,* a rolled flat product up to and including 0.188-in. thick and over 24-in. wide, in straight lengths or coils (rolls). **B 846,** B05

sheet—*as in SI specifications,* a rolled flat product up to and including 5-mm thick and over 600-mm wide, in straight lengths or coils (rolls). **B 846,** B05

sheet, *n*—rolled wrought product that is rectangular in cross section, with thickness 0.006 in. and greater [over 0.15 mm] but less than 0.250 in. [up through 6.30 mm], and with sheared, slit, or sawed edges.

Alclad sheet, n—composite sheet product comprised of an aluminum-alloy core having on both surfaces (if on one side only, it is *Alclad one-side sheet*) a metallurgically bonded aluminum or aluminum-alloy coating that is anodic to the core, thus electrolytically protecting the core alloy against corrosion.

Alclad one-side sheet, n—alclad sheet with only one side coated.

coiled sheet, n—sheet in coils with slit edges.
flat sheet, n—sheet with sheared, slit, or sawed edges, which has been flattened or leveled.
mill finish sheet, n—sheet having a nonuniform finish which may vary from sheet to sheet and within a piece, and which may not be entirely free of stains or oil.
one-side bright mill finish sheet, n—sheet having a moderate degree of brightness on one side, and a mill finish on the other.
standard one-side bright finish sheet, n—sheet having a uniform bright finish on one side, and a mill finish on the other.
standard two-side bright finish sheet, n—sheet having a uniform bright finish on both sides. **B 881,** B07

sheet, *n*—a flat-rolled metal product of some maximum thickness and minimum width arbitrarily dependent on the type of metal; it is thinner than plate. **B 899,** B02

sheet, *n—in textiles,* a large rectangular usually plain woven fabricated product which is used over a mattress on a bed. The product may be carded or combed yarn in a wide range of constructions. **D 123,** D13

sheet, *n*—an individual piece of sheeting. (See also **film, sheeting.**) **D 883,** D20

sheet, *n*—a thin, flat, rolled metal product having mill or cut edges. (If of steel, having dimensions of less than 0.229 in. (5.8 mm) thickness, and greater than 12.0 in. (305 mm) width and length.) **E 631,** E06

sheet—*in honeycomb core material,* a slice of honeycomb cut from a production block. **E 1749,** E06

sheet blanket, *n*—a thin blanket of cotton or cotton and synthetic blend having a nap on both sides. **D 123,** D13

sheet cast button, *n*—a button fabricated from a disc blanked from a cast sheet of formulated styrene-modified polyester resin. **D 123,** D13

sheet disintegration, *n*—the breakup of an unstable liquid film into particles. **E 1620,** E29

sheet glass—flat glass made by continuous drawing. **C 162,** C14

sheeting, *n*—a form of plastic in which the thickness is very small in proportion to length and width and in which the plastic is present as a continuous phase throughout, with or without filler. (See also **film.**) **D 883,** D20

sheeting, *n*—the process of converting a rubber, rubber mix, rubber dough, or latex into a sheet. **D 1566,** D11

sheeting, *n*—a form of plastic in which the thickness is very small in proportion to length and width and in which the plastic is present as a continuous phase throughout, with or without filler. (See also **film.**) **F 1251,** F04

sheet-metal nail—nail stamped out of sheet metal and formed to desired shape. (See **roofing nail.**) **F 547,** F16

sheet molding compound (SMC), *n*—a fiber-reinforced thermosetting compound, in sheet form. **D 883,** D20

sheet, resilient flooring, *n*—flexible resilient flooring, packaged in roll form, in which the length substantially exceeds the width. **F 141,** F06

sheet resistance, R_s **[Ω or Ω per square]**—*of a semiconductor or thin metal film,* the ratio of the potential gradient (electric field) parallel with the current to the product of the current density and thickness. **F 1241,** F01

sheet-roofing fastener nail—See **purlin nail.** **F 547,** F16

sheet-roofing nail— See **roofing nail.** **F 547,** F16

shelf-life—*in building construction,* the maximum time pack aged materials can be stored under specified conditions and still meet the performance requirements specified. **C 717,** C24

shelf life—maximum time interval during which a material may be stored and remain in a usable condition; usually related to storage conditions. **D 653,** D18

shelf life, n—Synonym for **storage life.** **D 907,** D14

shelf life—the length of time packaged materials can be stored under specified conditions and still remain effective in use (C 717, C-24). **D 996,** D10

shelf life, *n*—See **storage life, shelf.** **D 1566,** D11

shelf life—See **storage life.** **E 1749,** E06

shelf roughness—roughness on upward facing surfaces where undissolved solids have settled on parts during a plating operation. **B 374,** B08

shell, *n*—the outer walls of a hollow masonry unit. Shell can either be an end shell or a face shell. **C 1232,** C15

shell—See **packing.** **D 996,** D10

shell—a portion from the butt end of a horsehide, from which leather of tight, firm, fiber structure and fine grain is made. (See also **cordovan.**) **D 1517,** D31

shell hardening, *n*—a surface hardening process in which a suitable steel object, when heated through and quench hardened, develops a martensitic layer or shell that closely follows the contour of the piece and surrounds a **core** of essentially pearlitic transformation product. **A 941,** A01

shells, *n*—the outer walls of tile or hollow brick. **C 43,** C15

shelters:
expandable shelters—those shelters that are expanded from the transport size to a larger size, at expansion ratios of three-to-one or less and perhaps to a different shape. (See Specification PS 27.)
highly expandable shelters—as classified by the U.S. Department of Defense (MIL-STD-907B), those shelters that have expansion ratios greater than three-to-one from their transport size.
knockdown shelters—as classified by the U.S. Department of Defense (MIL-STD-907B), those shelters that are reduced in height and nested with identical items for transportation.
large area shelters—as classified by the U.S. Department of Defense (MIL-STD-907B), those shelters that are disassembled and packed in dedicated or general-purpose containers for shipment.
nonexpandable shelters—as classified by the U.S. Department of Defense (MIL-STD-907B), those shelters that are used in the same size and shape in which they are transported. **E 1749,** E06

shield, *n*—a nonconducting medium for altering the current distribution on an anode or cathode. **B 374,** B08

shield, *v*—to alter the normal current distribution on an anode or cathode by the interposition of a nonconductor. **B 374,** B08

shield, *n*—a conducting layer placed around an insulated conductor or cable to limit the penetration of electric or electromagnetic fields. **D 1711,** D09

shield—a conductive protective member, partly or wholly enclosing one or more elements of electric equipment or test specimen, for the purpose of reducing or eliminating the electric or magnetic flux, or both, within or beyond that element or elements. **D 2864,** D27

shield, electric or magnetic, *n*—a conductive protective component that encloses, in whole or in part, one or more elements of electric equipment or test specimen in order to reduce or eliminate the electric or magnetic flux, or both, within or beyond that element or elements. **D 1711,** D09

shielding—in an electron-optical instrument, the protection of the electron beam from distortion due to extraneous electric and magnetic fields. Since the metallic column of the microscope is at ground potential, it provides electrostatic shielding. Magnetic shields may be made of a high permeability material. **E 7,** E04

shielding effectiveness—the ability of a sandwich panel of suitable thickness and physical characteristics to exclude (protect) sensitive components or units from electromagnetic radiation (interference). **E 1749,** E06

shielding glass—a transparent glass containing oxides of heavy elements, such as lead, to absorb ionizing radiation. **C 162,** C14

shield (radiation)—a device that, when placed in a radiation field, reduces (attenuates) the radiation flux. **C 859,** C26

shier, *n—in woven fabrics*, fine fillingwise cracks randomly distributed across the fabric width. (Compare **let-off mark** and **crack mark.**) **D 123,** D13

shim—See **shim spacer**; **spacer.** **C 717,** C24

shim—*in plywood*, a long narrow repair in the panel surface not more than 3⁄16 in. (4.8 mm) wide made of wood or suitable synthetic compound. **D 1038,** D07

shim—a material, typically placed under the IQI which is radiogically similiar to the object being imaged. **E 1316,** E07

shim spacer, *n—in building construction in glazing*, a spacer designed and placed to position the face surfaces of a pane of glass or a panel between the stops and to prevent glass or panel contact with the stops. **C 717,** C24

shiner, *n*—a streak, usually short, caused by a lustrous section of a filament yarn. **D 123,** D13

shiner or shiner-scale—a defect characterized by minute fishscaling occurring in overfired ground coat. **C 286,** B08

shingle—a thin, oblong piece of wood with one end thinner than the other, lapped lengthwise in covering roofs and outer walls of buildings produced by sawing. **D 9,** D07

shingle—(*1*) a small unit of prepared roofing designed for installation with similar units in overlapping rows on inclines normally exceeding 25 %;

(*2*) to cover with shingles; and

(*3*) to apply any sheet material in overlapping rows like shingles. **D 1079,** D08

shingle backer—a specially fabricated sheathing-grade cellulosic fiberboard used as a backer strip in coursed shingle construction. **D 1554,** D07

shingle-backer nail— See **shake nail.** **F 547,** F16

shingle nail

wood–shingle nail—bright or galvanized, regular-stock-steel, plain-shank or annularly threaded, 1 1⁄4 by 0.076 to 2 by 0.106-in. nails with flat 7⁄32 to 9⁄32-in. head and medium or blunt diamond point. Also, bright or colored (baked-lacquer finished), aluminum-alloy, plain-shank or helically threaded, 7⁄8 by 0.080 to13⁄4 by 0.099-in. nails with flat 5⁄32 to 9⁄32-in. head and medium or blunt diamond point. Also, aluminum-alloy, annularly threaded, 1 3⁄8 and 2 by 0.101-in. nails with special type of thread and with flat slightly countersunk 5⁄32-in. head and long needle point for fastening cedar shingles to fiberboard nail-base sheathing.

asbestos-shingle nail—bright or galvanized, regular-stock-steel, annularly threaded, 1 to 2 by 0.113-in. nails with flat 13⁄32-in. head and long diamond or needle point. Also, aluminum-alloy, helically threaded, 11⁄8 to 2 1⁄2 by 0.099 to 0.135-in. nails with smooth or striated flat 3⁄16 to 7⁄16-in. head or 9⁄64 to 5⁄32-in. casing head and medium diamond or needle point. Also, aluminum-alloy, annularly threaded, 13⁄8 and 2 by 0.101-in. nails with special type of thread and with flat slightly countersunk 0.190-in. striated head and long needle point; for fastening asbestos shingles to fiberboard nail-base sheathing.

asphalt-shingle nail—See **roofing nail.**

siding nail— See **brick-siding nail, hardboard nail, roofing nail, sheathing/siding nail.**

aluminum-siding nail—plain-shank or helically threaded, aluminum-alloy 1 by 0.099 to 2 1⁄2 by 0.135-in. nails with flat 1⁄4 to 5⁄16-in. head and medium diamond point.

asbestos-siding face nail—bright, colored (baked-lacquer finished) or galvanized, regular-stock-steel, hardened-steel, bronze, aluminum or stainless-steel, annularly threaded, file-grip or screw-thread, 1 1⁄4 by 0.080 to 2 by 0.105-in. nails with smooth or striated flat 3⁄16-in. or button head and medium diamond or needle point.

common siding nail—bright or colored (baked-lacquer finished), galvanized, regular-stock-steel or hardened-steel, plain-shank or threaded, 1 3⁄4 by 0.080 to 3 by 0.128-in. nails with flat 5⁄32 to 19⁄64-in. head and medium diamond point.

corrugated-siding nail—See **roofing nail.**

hardboard-siding nail—See **hardboard nail.**

insulated-siding nail—bright or colored (baked-lacquer finished) aluminum-alloy, 11⁄2 by 0.113 to 2 1⁄2 by 0.135-in. nails with flat 7⁄32 to 9⁄32-in. head and medium diamond point.

insulating-siding nail—See **brick siding nail.**

wood-siding nail— bright and colored (baked-lacquer finished), plain-shank or helically threaded, aluminum-alloy, 1 7⁄8 by 0.106 to 27⁄8 by 0.148-in. nails with 9⁄64 to 11⁄32-in. casing or 17⁄64 to 5⁄16-in. sinker head and medium or blunt diamond point. Also, bright or colored (baked-lacquer finished), stainless-steel, annularly threaded, 21⁄8and 2 3⁄8 by 0.083 and 0.095-in. nails with slightly countersunk 3⁄16-in. head and medium diamond point. (See **common siding nail.**) **F 547,** F16

shingling—(*1*) the procedure of laying parallel felts so that one longitudinal edge of each felt overlaps, and the other longitudinal edge underlaps, an adjacent felt. (See also **ply**). Normally, felts are shingled on a slope so that the water flows over rather than against each lap;

(*2*) the application of shingles to a sloped roof. **D 1079,** D08

ship and galley tile—a special quarry tile having an indented pattern on the face of the tile to produce an antislip effect. **C 242,** C21

shiplapped lumber— See **lumber.** **D 996,** D10

shipment, *n—in asbestos*, any consignment formed of one or several lots of asbestos fiber. **C 460,** C17

shipment, *n*—goods or commodities which are transported together as a unit; a quantity of product for which a bill of lading has been signed by the carrier. **D 123,** D13

shipment, *n—in asbestos,* any consignment formed of one or several lots of asbestos fiber. **D 2946,** C17

shipper–receiver difference—*in the nuclear industry*, the difference between the quantity stated by the shipper as having been shipped and the quantity stated by the receiver as having been received. **C 859,** C26

shipping container— See **container.** **D 996,** D10

shirt, *n*—a cloth garment for the upper part of the body made of either woven or knitted fabric usually having sleeves, a neck opening, a front opening and a tail long enough to be tucked inside trousers or skirt. **F 1494,** F23

shirt back length, *n— in garment construction*, a measurement centered below the collar band to bottom edge of shirttail. **F 1494,** F23

shirt front length, *n— in garment construction*, a measurement from the placket top, below the collar band to the bottom edge of the shirt's hem in front. **F 1494,** F23

shirt sides, *n— in garment construction*, a measurement of length from below armhole to the bottom hem. **F 1494,** F23

shivering (peeling)—the splintering that occurs in fired glazes or other ceramic coatings as a result of critical compressive stresses. **C 242,** C21

SHMP—sodium hexametaphosphate. (NaHMP) **D 6161,** D19

shock—the short duration high force intial part of an impact. **F 869,** F08

shock absorption—the reduction of peak force by increasing the time over which the force is applied. The reduction of ground reaction forces is of primary importance in athletic footwear. (*Syn.* shock attenuation) **F 869,** F08

shock attenuation—see **shock absorption.** **F 869,** F08

shock load, *n*—the sudden application of an external force. **D 1566,** D11

shock pulse—a substantial disturbance characterized by a rise of acceleration from a constant value and decay of acceleration to the constant value in a short period of time. **D 653,** D18

shock wave—a wave of finite amplitude characterized by a shock front, a surface across which pressure, density, and internal energy rise almost discontinuously, and which travels with a speed greater than the normal speed of sound. (ISRM) **D 653,** D18

shoe finish—coatings applied to leather or other shoe materials at tanneries and shoe manufacturers before retailing to the consumer. **D 2825,** D21

shoe fold, *n*—a fabric folded from both ends into twelve or sixteen folds to the piece, the length of the fold depending upon the length of the piece. (Compare **book fold**.) **D 123,** D13

shoe leather—this term is self-explanatory and embraces a variety of leathers including:

(*1*) *sole leather*—made from cattle hides and to a small extent from horsehides and buffalo hides, comprising both the heavier grade, used for outer soles of shoes and the lighter grades and offal (heads, shoulders and bellies) used to a greater or less extent for heels, insoles, toecaps, counters, etc.;

(*2*) *upper leather*—made principally from calfskins, goatskins, cattlehides, horsehides, and other classes of animal skins, going into shoe uppers, and

(*3*) *miscellaneous shoe leathers,* including welting, lining stock, tongue stock, facing stock, etc. **D 1517,** D31

shoot—see **XRF-shoot**. **E 1605,** E06

shop drawing—a drawing prepared by the fabricator based on a **working drawing** and used in a shop or on a site for assembly. **E 631,** E06

shop drawing—See **drawing**. **E 1480,** E06

Shore A hardness—a measurement of hardness for rubbers and plastics using a Shore A hardness gage (durometer). **E 1749,** E06

shorelines—a defect characterized by a series of rings or lines in the surface of porcelain enamel similar in appearance to the lines on the shore produced by receding water. **C 286,** B08

shorn pile, *n*—pile that is removed when a specimen is sheared. **D 123,** D13

shorn pile, *n*—pile that is removed when the face pile is sheared. **D 5684,** D13

short—a comparative term signifying a fast-setting glass. **C 162,** C14

short, *n*—an imperfection in a molded plastic part due to an incompletely filled out condition. **D 883,** D20

short, *n*—in a molded material, an incompletely filled out condition. **F 412,** F17

short barb, *n*—barbs having an average length of 0.1875 ± 5 %. **F 1379,** F14

short-circuit—see *internal discharge.* **F 1827,** F26

short-fiber content (SFC), *n*—that percentage of fibers (by number or by weight) in a test specimen, that is shorter than 12.5 mm (0.5 in.) in length. **D 123,** D13

short-finish—an imperfection resulting from incomplete polishing. **C 162,** C14

shortness, *n*—a qualitative term that describes an adhesive that does not string cotton, or otherwise form filaments or threads during application. **D 907,** D14

short plumbing (hospital) threads—a special modified form of the tapered pipe thread basically achieved by elimination of the imperfect threads. This thread is normally used only on chromium-plated nipples to ensure that a minimum of thread shall remain exposed after the nipple is threaded into a fitting. **B 846,** B05

short point—point with 40-deg or greater included angle for 0.040 to 0.065-in. wire diameter; 45-deg or greater included angle for 0.072 to 0.225-in. wire diameter; 55-deg or greater included angle for 0.250 to 0.325-in. wire diameter. **F 547,** F16

short seal boom—boom that, when grounded, seals against the shoreline. **F 818,** F20

short-term exposure limit (STEL), *n*—the airborne concentration of a substance in a continuous 15-minute time period which should not be exceeded at any time during a workday. **D 1356,** D22

short-term exposure limit (STEL)—the 15-min time-weighted average exposure which shall not be exceeded at any time during a work day, even if the Occupational Exposure Limit is not exceeded. Exposures above the Occupational Exposure Limit up to the STEL should not be longer than 15 min and should not occur more than 4 times per day. The minimum interval between these exposures should be 60 min.

The following specified periods of time must be addressed either by determining a value or by stating the "time" is not addressed:

a) workday (number of hours specified)

b) workweek (number of hours specified)

c) lifetime (that is, cumulative)

d) very short term maximum (for example, short-term exposure limit, ceiling, etc.)

Other specified periods of time *may* be addressed. **E 1542,** E34

short term weld factor—a dimensionless number that provides a relative measure of the tensile strength of a welded thermoplastic test specimen to the tensile strength of the manufactured sheet. **C 904,** C03

shot, *n*—small spherically shaped particles of metal. **B 899,** B02

shot—a short energizing cycle in a magnetic particle examination. **E 1316,** E07

shot blasting—blasting with small spherical objects, such as metallic shot, propelled against a metallic surface. See **abrasive blasting.** **B 374,** B08

shotcrete, *n*—a mortar or concrete that is projected pneumatically at high velocity onto a surface.

dry–mixture shotcrete, n—shotcrete in which most of the mixing water is added at the nozzle.

wet–mix shotcrete, n—shotcrete in which most of the ingredients, including water, are mixed prior to introduction into the delivery hose. **C 125,** C09

shotcrete—mortar or concrete conveyed through a hose and pneumatically projected at high velocity onto a surface. Can be applied by a "wet" or "dry" mix method. (ISRM) **D 653,** D18

shotcreting, *n*—an installation technique using a pump to convey tempered refractory castable to a nozzle where an admixture and air pressure are injected, spraying the castable stream onto a substrate where the castable becomes sufficiently stiff to withstand the force of gravity. **C 71,** C08

shot peening—a process whereby hard, small spherical objects (such as metallic shot) are propelled against a metallic surface for the purpose of introducing compressive stresses into that surface, hardening it or obtaining decorative effects. **B 374,** B08

shot volume, *n*—the total hollow space of a mold including cavity or cavities, runner(s), and sprue. **D 883,** D20

shoulder, *n—in zippers*, the bearing surface of an interlocking element by which the chain is contained inside the flanges of the slider. **[D13.54] D 2050**

shoulder circumference, *n, in body measurements*, with arms down at sides, the maximum distance around the shoulders at the top of the arm. **D 123,** D13

shoulder—half of a double shoulder, **D 1517,** D31

shoulder, *n—of a tire*, that region of a tire formed by the conjunction of the tread and sidewall. **F 538,** F09

shoulder—edge, when present, at junction of thread crest and flank. If double crested, primary shoulder is nearer head and should be equal to or larger in diameter than secondary shoulder, which is closer to point. **F 547,** F16

shoulder, b_V—the widest point, of the ski in the shovel section of the ski. **F 472,** F27

shoulder contact—the region on a glass container surface located between the neck and sidewall sections, which normally contacts other similarly shaped and upright containers. **C 162,** C14

shoulder joint, *n—in anatomy*, the junction of the collarbone and the shoulder blade. (See also **acromion**.) **D 123,** D13

shoulder length, *n—in body measurements*, the distance from the side neck base to the armscye line at the shoulder joint. **D 123,** D13

shoulder rib, *n*—a rib at or near the outer edge or shoulder of the tread band. **F 538,** F09

shoulder row, *n*—a row located at or near the shoulder of the tread band. **F 538,** F09

shoulder slope, *n—in body measurements*, the angle formed when the slant of the shoulder line deviates from the horizontal line that originates at the side neck base. **D 123,** D13

shoulder wear, *n*—a type of irregular wear characterized by an increased wear rate in the outer edge of the shoulder rib or row compared to the inner shoulder edge. **F 538,** F09

shovel length, l_S—the projected length of the forward turn-up, measured from the tip to the contact point where a 0.5-mm feeler gage intersects the running surface with the ski body pressed against a plane surface. **F 472,** F27

shovel length, L_s—the projected length of the forward turn-up, measured from the tip to the contact point where an 0.1-mm feeler gage intersects the running surface with the snowboard ski unweighted on a plane surface. **F 1107,** F27

shovel slicing—mechanical action of causing bedding material to uniformly contact the pipe haunches. **C 896,** C04

shovel surface area—that surface forward of the shovel contact point. The shovel contact point is located at l_S from the tip. **F 472,** F27

shovel surface area—that surface forward of the shovel contact point. The shovel contact point is located at L_s from the tip. **F 1107,** F27

shoving, *n*—the displacement of a localized area of the pavement surface generally associated with turning, braking or accelerating vehicles. **E 867,** E17

shoving, *n*— the horizontal displacement of a localized area of the pavement surface which may also include some vertical displacement. **E 1778,** E17

shower curtain, *n*—a hanging fabric used to prevent water spillage from a shower area. **D 123,** D13

showering—a particular form of corona discharge characterized by strongly ionized streamers or streams of luminous plasma. It generally occurs at a field value just below that which is required for a complete breakdown. **B 542,** B02

show-through, *n*—the printed image can be observed on the opposite side of the sheet under normal lighting conditions. **D 6488,** D01

show through, *n*—optical density of an image on the backside of a printed area. **F 1857,** F05

shrend—{archaic} **term for dry gage, v.** **C 162,** C14

shrink, *vt*—to cause to contract, to compact cloth by causing to contract when subjected to washing, boiling, steaming or other processes. **F 1494,** F23

shrinkage—a decrease in dimensions of a compact which may occur during sintering. (Converse of **growth**.) **B 243,** B09

shrinkage, *n*—the decrease in dimension of a refractory material during manufacture or service. **C 71,** C08

shrinkage—a decrease in length, area, or volume. **C 717,** C24

shrinkage—a reduction in size of a composition which occurs during its hardening process, curing process, or both. **C 904,** C03

shrinkage, *n*—a decrease in volume due to chemical reaction or drying. **C 1180,** C12

shrinkage—reduction in dimensions due to lowering the moisture content below the fiber saturation point. **D 9,** D07

shrinkage, *n*—a decrease in one or more dimensions of an object or material. **D 123,** D13

shrinkage, *n*—a decrease in one or more dimensions of an object or material. **F 1494,** F23

shrinkage-compensating—*in grouting*, a characteristic of grout made using an expansive cement in which volume increase, if restrained, induces compressive stresses that are intended to offset the tendency of drying shrinkage to induce tensile stresses. See also **self-stressing grout**. **D 653,** D18

shrinkage index, *SI* **(D)**—the numerical difference between the plastic and shrinkage limits. **D 653,** D18

shrinkage limit, *SL*, w_s **(D)**—the maximum water content at which a reduction in water content will not cause a decrease in volume of the soil mass. **D 653,** D18

shrinkage, linear—a reduction in the length of a cast bar of a composition during its hardening process, curing process, or both. **C 904,** C03

shrinkage ratio, *R* **(D)**—the ratio of: (*1*) a given volume change, expressed as a percentage of the dry volume, to (*2*) the corresponding change in water content above the shrinkage limit, expressed as a percentage of the weight of the oven-dried soil. **D 653,** D18

shrinkage temperature—the temperature at which measurable shrinkage occurs when leather is gradually heated in an aqueous medium. **D 1517,** D31

shrinkage, unrestrained—a reduction in size of a composition which occurs during its hardening process, curing process, or both, with no external forces applied that can inhibit such reduction. **C 904,** C03

shrinkage, volume—a reduction in occupied space of a composition during its hardening process, curing process, or both. **C 904,** C03

shrink mark—an imperfection, a depression in the surface of a molded material where it has retracted from the mold. **D 883,** D20

shrink mark—depression in the surface of a molded material where it has retracted from the mold. (D-20) **F 412,** F17

shrink wrap—plastic material that shrinks in size when heated; used to wrap delicate materials, or those subject to attack from the environment. **F 1156,** F12

shroud, *n—in packaging,* a protective **cover** placed over a load unit, or **package** to cover the top and four sides. **D 996,** D10

shutter—a mechanical device which permits regulation of the time during which light is allowed to act on a light sensitive medium. **E 7,** E04

shuttle mark, *n—in woven fabrics,* a fine fillingwise line caused by damage to a group of warp yarns by shuttle abrasion. **D 123,** D13

SI—an abbreviation for the International System of Units. **A 340,** A06

side—a side is half of a hide along the back bone line and with the tail not more than 6 in. (152 mm) long. **D 1517,** D31

side camber, *W*—the maximum distance from a line drawn between the widest points of the ski and the sidecut of the ski. **F 472,** F27

side centered—equivalent to end centered with a different choice of axes. (See **end centered.**) **E 7,** E04

side-construction tile, *n*—tile designed to receive its compressive stress at right angles to the axes of the cells. **C 43,** C15

side cut—a board, plank, cant, or timber which does not contain the pith. **D 9,** D07

side cut—that line describing the curved portion of the ski contour limited by the lines at the b_H and b_V dimensions, and defined by the bottom edge. **F 472,** F27

sidecut—that line describing the curved portion of the snowboard ski contour limited by the lines at the b_H and b_V dimensions and defined by the bottom edge. **F 1107,** F27

sidecut geometry—the configuration of the curve bordering the running surface and defined by the bottom edge. **F 1107,** F27

sidecut magnitude—the maximum distance from a line drawn between the widest points of the snowboard ski and the edge of the snowboard ski. **F 1107,** F27

side-fired furnace—a furnace with fuel supplied from the side. **C 162,** C14

side geometry—the configuration of the curve bordering the running surface and defined by the bottom edge. **F 472,** F27

side lap, *n*—the distance by which the side edge of one tile overlaps the side edge of an adjacent tile in the same course; for interlocking tile this corresponds to the width of that part of the tile that contains the ribs, grooves, or channels which provide for interlocking. (See *Discussion* under **head lap**.) **C 43,** C15

side lap—the shortest horizontal distance between the exposed side edge of a course of asbestos-cement roofing or siding material and the most proximate underlying area of roof deck or side wall not covered by the preceding adjacent course. **C 460,** C17

side lap, *n*—the shortest horizontal distance between the exposed side edge of a course of roofing or siding material and the nearest underlying area of roof deck or side wall not covered by the preceding adjacent course. **C 1154,** C17

side lap—the shortest horizontal distance between the exposed side edge of a course of asbestos-cement roofing or siding material and the most proximate underlying area of roof deck or side wall not covered by the preceding adjacent course. **D 2946,** C17

side leather—shoe upper leather made from the grain side of cattlehides. The name comes from the practice of splitting the hide along the backbone into two sides before tanning. The skins are usually shaved on the flesh side to uniform thickness and the grain is corrected. **D 1517,** D31

side mount—See **railing systems.** **E 631,** E06

side mount—a railing-system support that anchors the post or baluster of a railing system to a vertical or inclined surface, such as a fascia or stair stringer. **E 1481,** E06

side point—an eccentric nonsymmetrical point. **F 547,** F16

side-port furnace—a furnace with ports on the sides. **C 162,** C14

sideslip angle, (rad or degree), *n—in a vehicle,* the angle between the projection of a test vehicle's longitudinal axis on the road plane and the velocity vector at some specified point in the test vehicle. **F 538,** F09

sidestrike—activation of stapling head resulting from forcing container resting on table against head. **F 592,** F16

side surface bend—procedure whereby a load is applied on the weld in line with the weld axis. **C 904,** C03

side-tone, *n*—appearance at the flop angle. **E 284,** E12

sidewall, *n—of a tire,* that portion of a tire between the tread and the bead. **F 538,** F09

sidewall component, *n*—an individual part of the sidewall construction, either a separate compound or a separately assembled piece. **F 538,** F09

sidewall, glass container—the midsection of a glass container located between the heel and shoulder contact regions; also referred to as barrel. **C 162,** C14

sidewall rubber, *n*—the exterior rubber layer of a tire that extends over the sidewall part of the carcass. **F 538,** F09

siege—{archaic} floor of a pot furnace, often called "bench." **C 162,** C14

Siemens—a measure of electrical conductance in water, equivalent to a mho. See **Mho, Ohm.** **D 6161,** D19

sieve—a standard wire mesh or screen, especially when used in graded sets to determine the mesh size or particulate size distribution of particulate or granular solids. **C 242,** C21

sieve, *n*—a standard wire mesh or screen, especially when used in graded sets to determine the mesh size or particular size distribution of particulate and granular solids. **C 1145,** C28

sieve, *n*—in laboratory work an apparatus, in which the apertures are square, for separating sizes of material. **D 8,** D04

sieve—an apparatus with square apertures for separating sizes of material. **D 1079,** D08

sieve, *n*—a plate, sheet, or woven wire cloth, or other device, with regularly spaced apertures of uniform size, mounted in a suitable frame or holder, for use in separating material according to size. The terms sieve and screen can be used interchangeably. **E 1638,** E29

sieve analysis—particle size distribution; usually expressed as the weight percentage retained upon each of a series of standard sieves of decreasing size and the percentage passed by the sieve of finest size. Synonymous with **screen analysis.** **B 243,** B09

sieve analysis—the particle size distribution of a particulate or granular solid or sample thereof, when determined by weight percent passage through, or retention on, a graded set of sieves. **C 242,** C21

sieve analysis, *n*—the particle size distribution of a particulate or granular solid or sample thereof, when determined by passage through and retention on a graded set of sieves. **C 1145,** C28

sieve analysis—determination of the proportions of particles lying within certain size ranges in a granular material by separation on sieves of different size openings. **D 653,** D18

sieve analysis, *n*—the act of obtaining a particle size distribution using sieves. **E 1638,** E29

sieve classification—the separation of powder into particle size ranges by the use of a series of graded sieves. **B 243,** B09

sieve fraction—that portion of a powder sample which passes through a standard sieve of specified number and is retained by some finer sieve of specified number. **B 243,** B09

sight glass—the glass window for a viewing port, generally for a pressurized system. **C 162,** C14

sight line—*in building construction in glazing*, the line of intersection of an opaque material with a transparent material. **C 717,** C24

sigma (σ)—solid phase found originally in binary iron-chromium alloys which is in stable equilibrium at temperatures below 820°C. Now used to identify any structure which possesses the same complex body-centered crystal structure. **E 7,** E04

signal—the data present in a radiological measurement which is directly correlated with the degree of radiation attenuation by the object being examined. **E 1316,** E07

signal, acoustic emission (emission signal)—an electrical signal obtained by detection of one or more acoustic emission events. **E 1316,** E07

signal amplitude, acoustic emission—the peak voltage of the largest excursion attained by the signal waveform from an emission event. **E 1316,** E07

signal conditioner—the component that receives the output from a sensor and processes it for subsequent use. **D 1129,** D19

signal gradient—same as **differential readout**. **E 1316,** E07

signal height—*EIA,* the number of counts in the channels of a backscattering spectrum due to a specific element in the target. **E 673,** E42

signal overload level—that level above which operation ceases to be satisfactory as a result of signal distortion, overheating, or damage. **E 1316,** E07

signal overload point—the maximum input signal amplitude at which the ratio of output to input is observed to remain within a prescribed linear operating range. **E 1316,** E07

signal-to-background ratio—*AES,* the ratio of signal (above background) to that of the nearby background on the high kinetic energy side of the elastically scattered Auger electrons. **E 673,** E42

signal to background ratio—*SIMS, XPS, for a spectral peak,* the ratio of the maximum counts in the peak above the background to the magnitude of the background. **E 673,** E42

signal-to-noise ratio—the ratio of the signal intensity (above background) to that of noise in determining that signal. **E 673,** E42

signal-to-noise ratio—the ratio of values to signal (response containing relevant information) to that of noise (response containing nonrelevant information). **E 1316,** E07

signal-to-noise ratio—the ratio of the amplitude of an ultrasonic indication to the amplitude of the maximum background noise. **E 1316,** E07

signal-to-noise ratio, *S/N*—the ratio of the signal, *S*, to the noise, *N*, as indicated by the instrumental read-out indicator. **E 131,** E13

signature, acoustic emission (signature)— a characteristic set of reproducible attributes of acoustic emission signals associated with a specific test article as observed with a particular instrumentation system under specified test conditions. **E 1316,** E07

signature plate—a transparent plate bearing that specific information which lends a cloak of authority to a photographic document such as an identification badge or an access control passport; the plate is placed just in front of the film plane in the camera being used to obtain the photograph of the individual being authorized access, or whose identity is being verified. **F 1156,** F12

significance level, α, *n*—the stated upper limit for the probability of a decision being made that a hypothesis about the value of a parameter is false when in fact it is true. (See **confidence level, probability level**.) **D 123,** D13

significance level—the decimal probability that a result will exceed the critical value. **E 1547,** E15

significance level—the stated probability (risk) that a given test of significance will reject the null hypothesis (that a specified effect is absent) when the hypothesis is true. **E 1823,** E08

significant—statistically significant. An effect or difference between populations is said to be present if the value of a test statistic is significant, that is, lies outside of selected limits. **E 1823,** E08

significant loss, *n*—any loss that introduces a bias in final results that is of appreciable economic importance to the concerned parties. **D 121,** D05

significant loss, *n*—any loss that introduces a bias in final results and that is of appreciable importance to concerned parties. **D 5681,** D34

significantly alter—any action which would change the mechanical or performance capabilities of the fastener following its original manufacture. **F 1789,** F16

significant surface—a surface area where the minimum thickness to be met shall be designated on the applicable drawing or by the provision of a suitably marked sample. However, if not designated, significant surfaces shall be defined as those normally visible, directly or by reflection, which are essential to the appearance or serviceability of the fastener when assembled in normal position, or which can be the source of corrosion products that deface visible surfaces on the assembled fastener. **F 1789,** F16

silent hours—See **hours of operation**. **E 1480,** E06

silica, *n*—*in asbestos cement*, pulverized silicon dioxide (SiO_2) used as a filler or a part of the cementitious material. **C 460,** C17

silica, *n*—pulverized silicon dioxide (SiO_2) used as a filler or a part of the cementitious material. **C 1154,** C17

silica, *n*—*in asbestos cement*, pulverized silicon dioxide (SiO_2) used as a filler or a part of the cementitious material. In the latter case, the material shall meet the requirements of Specification C 618. **D 2946,** C17

silica glass—See **fused silica**. **C 162,** C14

silica (SiO_2)—the common oxide of silicon usually found naturally as quartz or in complex combination with other elements as silicates. **C 242,** C21

silicate gel, *n*—a jelly-like substance consisting of polymerized silicate and entrapped coolant. **D 4725,** D15

silicate type inclusions—inclusions composed essentially of silicate glass, normally plastic at foraging and hot-rolling temperatures, which appear in steel in the wrought condition as small elongated inclusions usually dark in color under reflected light as normally observed. **E 7,** E04

silicon carbide refractories, *n*—refractory products consisting predominantly of silicon carbide. **C 71,** C08

silicone elastomer, *n*—an elastomer containing cross-linked silicone polymer and fillers, usually silica. **F 1251,** F04

silicone oil—a generic term for a family of relatively inert liquid organosiloxane polymers used as electrical insulation. They have the generic formula:

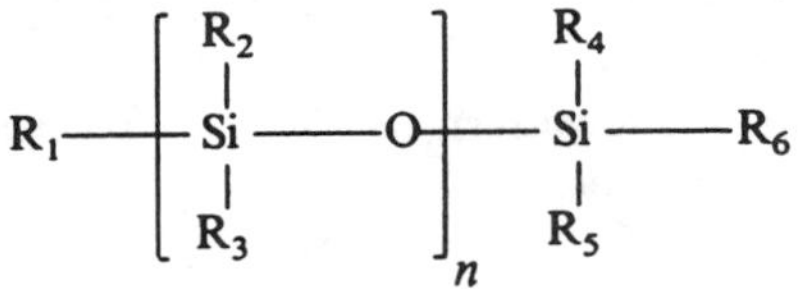

where the attached R groups may be H, methyl, vinyl, phenyl, alkyl, or substituted phenyl or alkyl radicals. **D 2864,** D27

silicone plastics—plastics based on polymers in which the main polymer chain consists of alternating silicone and oxygen atoms.

D 883, D20

silicone polymer, *n*—a polymer of alternating silicon-oxygen atoms consisting of repeating of diorganosiloxy groups. **F 1251,** F04

silicone sealant, *n*—a liquid-applied curing compound based on polymer(s) of polysiloxane structures. **C 717,** C24

siliconizing—*in diffusion coatings*, diffusion of silicon into solid metal at elevated temperatures. **B 374,** B08

silk screen process—a decorating process in which a design is printed on glass through a silk mesh, woven wire, or similar screen. **C 162,** C14

sill—generally a horizontal wood member forming the lowest part of the framework of a construction, from relatively small size as in a window frame to a much larger size as in a railroad bridge; a threshold or door sill. **D 9,** D07

sill—See **windows and doors.** **E 631,** E06

sillimanite—See**andalusite.** **C 242,** C21

silo—synonym for **bin**. **D 653,** D18

siltation—See **deposition.** **D 4410,** D19

silt barrier—boom with very deep skirt used to control the movement of suspended sediments. **F 818,** F20

silt (inorganic silt) (rock flour)—material passing the No. 200 (75-μm) U.S. standard sieve that is nonplastic or very slightly plastic and that exhibits little or no strength when air-dried. **D 653,** D18

silt size—that portion of the soil finer than 0.02 mm and coarser than 0.002 mm (0.05 mm and 0.005 mm in some cases). **D 653,** D18

silt size (fluvial sediment)—0.004 to 0.062 mm in diameter. **D 4410,** D19

silver/silver chloride reference electrode—a reference electrode with an internal reference element of silver and silver chloride. The filling solution must contain chloride and be saturated with respect to silver chloride. Saturated KCl, 1 *M* KCl, and equitransferent filling solutions are all commonly used. Ag/AgCl electrodes have better thermal characteristics than calomel electrodes and are therefore preferable for specific ion measurements and precision pH measurements. **D 4127,** D19

similarities were noted/differences as well as similarities—these expressions are meaningless without an explanation as to the extent and significance of the similarities or differences between the known and questioned material. These terms should never be substituted for gradations of opinions. **E 1658,** E30

simple (concerning lattices)—having similar atoms or groups of atoms separated by integral translations only. **E 7,** E04

simple linear regression—a statistical method of estimating the linear relationship between a dependent variable *y* and an independent variable *x* using the linear model

$$y = b_o + b_z X + \epsilon$$

E 131, E13

simple liquid—a single-phase liquid having a vapor pressure of less than 16 psi Reid vapor pressure at 100°F (830 mm Hg at 37.8°C) and a Saybolt viscosity of less than 10 000 s (2160 cSt) at 25°C. **E 1547,** E15

simple machine stitch pattern, *n*—*in home sewing*, a repeating segment of machine stitches, with each repeat consisting of one or more stitches long. (Compare **complex machine stitch pattern**.) **D 123,** D13

simple machine stitch pattern, *n*—*in home sewing*, a repeating segment of machine stitches, with each repeat consisting of one or more stitches long. (Compare complex machine stitch pattern.) **D 5646,** D13

simple payback (SPB) period—the time required for the cumulative benefits from an investment to pay back the investment cost and other accrued costs, not considering the time value of money. **E 833,** E06

simple seam, n—See **plain seam**. **D 4965,** D13

simple shear—shear strain in which displacements all lie in one direction and are proportional to the normal distances of the displaced points from a given reference plane. The dilatation is zero. (ISRM) **D 653,** D18

simplex airblast nozzle—see **piloted airblast nozzle**. **E 1620,** E29

simplex channel, *n*—a data transmission system capable of transmitting in only one direction. (See **duplex channel**.) **F 1457,** F05

simplex nozzle—a swirl chamber atomizer comprising a single set of tangential liquid inlets (or slots) combined with a single circular discharge orifice. **E 1620,** E29

simulated service corrosion test, *n*— evaluation of the effects of a circulating engine coolant on metal test specimens and automotive cooling system components under controlled, essentially isothermal laboratory conditions. **D 4725,** D15

simulation—the imitation of a document or article, including similar security features, in a form that is intended to pass as genuine in circumstances of ordinary use. **F 1156,** F12

simultaneous, *adj*—the application of two or more techniques to the same sample at the same time. (ICTAC)

NOTE—A hyphen is used to separate the abbreviations of the techniques; for example, simultaneous thermogravimetric analysis and differential scanning calorimetry would be TGA-DSC. **E 473,** E37

simultaneous response—The assignment of multiple emergency medical resources to the scene of an illness or injury based on initially available information and local opera-tional policies. These may have varying levels of care capability (for example ALS and BLS, ground and air). Subsequent care and/or transportation of the patient is provided by the unit which most closely meets the patient's needs. A simultaneous response differs from a sequential response. **F 1177,** F30

single, n—a climbing rope that may be used alone. **F 1773,** F08

single-action pressing, *n*—a method by which a powder is pressed in a stationary die between one moving and one fixed punch. **B 243,** B09

single-beam spectrum, *n*—a spectrum determined through one physical path. **E 131,** E13

single-cased well—a monitoring well constructed with a riser but without an exterior casing. **D 653,** D18

single coil, *n*—a configuration where there is one coil with one diameter, the coil being either helical or concertina style. **F 1379,** F14

single-component developer—see **single-component toner.** **F 335,** F05

single-component toner—a dry powder in an electrostatic copying process that does not contain a carrier and is used to form a visible image. **F 335,** F05

single crystal—a body of crystalline material that contains no large-angle boundaries or twin boundaries. Also known as **monocrystal.** **F 1241,** F01

single fire—See *single fire* under **firing.** **C 242,** C21

single-fluid atomizer—see **pressure atomizer**. **E 1620,** E29

single-grained structure—see **soil structure**. **D 653,** D18

single-impact damage, n—See **damage**. **G 40,** G02

single input wire construction—a stranded conductor design which varies the number of wires within a range of conductor sizes in order to permit that range of conductor sizes to be constructed from wires of a single diameter. **B 354,** B01

single island canopy—used for all types of cooking equipment in a single line island configuration. See *canopy*. **F 1827,** F26

single jet atomizer—see **plain jet atomizer**. **E 1620,** E29

single-junction reference electrode—a reference electrode containing a single electrolyte. The electrolyte (*1*) provides a constant level of the ion sensed by the reversible reference element and (*2*) forms a low-junction potential liquid junction with the sample solution. **D 4127,** D19

single layer flat—a coil in which the product is spirally wound into a single disk-like layer. (Sometimes called "pancake coil" or "single layer spirally wound coil.") **B 846,** B05

single-level pile, *adj—in floor coverings*, having all pile tufts at the same level. **D 123,** D13

single-operator operational precision, SOOP—the standard deviation of the results of a series of determinations by a single operator employing the method with its associated sample container preparation, collection, splitting, preservation, transmission, and storage on a homogeneous sample. **D 1129,** D19

single-operator precision—See **precision.** **D 1356,** D22

single-pass internal reflection element— in internal reflection spectroscopy, an internal reflection element in which the radiant power transverses the length of the element only once; that is, the radiant power enters at one end of the optical element and leaves via the other end. **E 131,** E13

single sampling, *n—in acceptance sampling*, a sampling plan for which the decision to accept or reject a lot is based on a single sample. **D 123,** D13

single-sided griddle—See **griddle**. **F 1827,** F26

single spread, *n*—See *single spread* under **spread**. **D 907,** D14

single stimulus method, *n*—any psychophysical method in which a judgment follows the presentation of only one stimulus at a time. **E 253,** E18

single stitch zigzag, *n—in home sewing*, a simple machine stitch pattern made by the needle moving up and down and alternately from one side to the other while the fabric moves through the feed mechanism in either the forward or reverse direction with all segments having equal length. Compare **multiple stitch zigzag.** **D 123,** D13

single-strand breaking force, *n—in tensile testing*, the breaking force of one strand that follows a specified path, usually a straight line, between the clamps of a tensile testing machine. **D 123,** D13

single-strike film ribbon—an inked ribbon wherein the substrate is a plastic film material such as polyethylene, where each area of the ribbon is capable of producing only one image. **F 221,** F05

single-strike paper ribbon—an inked ribbon wherein the substrate is paper, where each area of the ribbon is capable of producing only one image. **F 221,** F05

singlet state—an electronic state with a total spin quantum number of zero. **E 131,** E13

single twist, *n*—the amount of twist in each individual single yarn element in a tire cord structure based on the length of the element after twist has been removed from the cord. **D 123,** D13

single yarn, *n*—the simplest strand of textile material suitable for operations such as weaving, knitting, etc. **D 123,** D13

single yarn—See *single yarn* under **yarn.** **D 3878,** D30

sinker—bright or coated, slender, regular-stock-steel, 1 ⅛ by 0.067 to 5 ¾ by 0.244-in. nails with 11⁄64 to ½-in. sinker head and medium diamond point, with diameter of head smaller than that of cooler and common nail of same designation. **F 547,** F16

sinker head—flat, slightly countersunk (115 to 130 deg) head, as found on sinkers and corker nails; smaller in diameter than head on cooler and common nails. **F 547,** F16

sinking indentation—a hardness indentation around which the metal has been depressed below the plane of the specimen. **E 7,** E04

sinter, *v.*—the bonding of adjacent surfaces of particles in a mass of powder, or a compact, by heating. **B 243,** B09

sinter—a ceramic material or mixture fired to less than complete fusion, resulting in a coherent mass, or the process involved. **C 242,** C21

sinterable powder—a powder in which the bonding of adjacent surfaces of particles can be accomplished by heating. **C 859,** C26

sintered glass—glassware made by fusing glass particles. **C 162,** C14

sinter forging—see **powder forging.** **B 243,** B09

sintering—the bonding of adjacent surfaces of particles in a mass of powder or a compact by heating. **C 859,** C26

sintering, *n*—in catalysis, a reduction of the surface area of a catalyst carrier or other solid catalytic material caused by heat. **D 3766,** D32

sipe, *n*—a molded or cut rectangular void which is substantially narrower than the major grooves or voids. **F 538,** F09

siphon nozzle—a pneumatic atomizer in which an air stream aspirates and atomizes liquid from a reservoir located beneath the nozzle. **E 1620,** E29

sisal plier—See **plier; mattress blade.** **F 592,** F16

sitting spread, *n—in body measurements*, the circumference of the widest part of the hips with the subject seated. **D 5219,** D13

sixel encoding, *n*—a method of grouping data, so that each byte resembles a code for an ASCII character. **F 1457,** F05

size, *v*—to re-press a sintered compact to decrease the dimensional variation (not to be confused with **coin** or **re-strike**). **B 243,** B09

size—the representative dimension that best describes the extent in space of a particle, agglomerate, or aggregate. **C 242,** C21

size—usually a liquid composition to prevent excessive absorption of all paints into plaster, old wall paint, and similar porous surfaces; also a liquid composition used as a first coat on metal to improve adhesion of succeeding coats (latter usage is limited to the metal decorating industry).

The terms **sealer** and **size** are almost synonomous, but usage has established certain differences. A **sealer** is ordinarily a thin varnish or clear lacquer and is usually applied on wood and metal surfaces. Ordinary painter's **size** is a thin solution of glue, starch or other water-soluble substance and is usually applied on plaster surfaces, but **size** used in metal decorating is a thin varnish. **D 16,** D01

size, n— Synonym for **sizing**. **D 907,** D14

size—asphalt, rosin, wax, or other additive introduced to the stock for a fibrous-felted board, prior to forming, or added to the blend of particles and resin for a particle board, to increase water resistance. **D 1554,** D07

size, *n*—one of a series of graduated measurements in manufactured articles of clothing conventionally identified by numbers, letters, or words. **F 1494,** F23

size consist, *n*—the particle size distribution of a coal. **D 121,** D05

size effect—influence of specimen size on its strength or other mechanical parameters. (ISRM) **D 653,** D18

size factor—the factor included in deriving the allowable bending stress for rectangular members which takes into account the somewhat lower unit strength developed in larger members as compared to smaller members. **D 9,** D07

sizing—final pressing of sintered compacts to decrease dimensional variation. **B 243,** B09

sizing, *n*—a generic term for compounds which, when applied to yarn or fabric, form a more or less continuous solid film around the yarn and individual fibers. (See also **filler**.) **D 123,** D13

sizing, *n*—the process of applying a material on a surface in order to fill pores and thus reduce the absorption of the subsequently applied adhesive or coating or to otherwise modify the surface properties of the substrate to improve the adhesion, and also, the material used for this purpose. (Synonym *size.*) (See also **primer.**) **D 907,** D14

sizing, *n*—the addition of materials to a papermaking furnish, or the application of materials to a surface or board, which results in a paper or board that exhibits some resistance to liquid penetration. **D 1968,** D06

sizing, size—an organic coating applied to glass fibers directly below a bushing to provide some or all of the following: abrasion resistance, lubricity, antistatic characteristics, and chemical bonding to a subsequent matrix. **C 162,** C14

sizing system, *n—in garment construction*, a method of designating garment sizes. **D 123,** D13

skater's cracks—curvilinear cracks in a roofing membrane that appear to relate neither to the direction of application of the membrane components nor the substrate components. **D 1079,** D08

skein, *n*—a continuous strand of yarn, wound on a hand or motorized reel. **D 123,** D13

skein break factor, *n*—the comparative breaking strength of a skein of yarn adjusted for the linear density of the yarn expressed in an indirect system; the product of the breaking load of the skein and the yarn number expressed in an indirect system. **D 123,** D13

skein breaking tenacity, *n*—the skein breaking strength divided by the product of the yarn number in a direct numbering system and the number of strands placed under tension. **D 123,** D13

skein loop-length, *n*—the inside length of a coil of yarn mounted vertically as measured under a specified load. **D 123,** D13

skein shrinkage, *n*—a measure of true or intrinsic yarn shrinkage not including crimp contraction. **D 123,** D13

skein strength, *n*—the force required to rupture a skein of yarn, expressed in units of force, as breaking load. **D 123,** D13

skew, *n*—a fabric condition resulting when filling yarns or knitted courses are angularly displaced from a line perpendicular to the edge or side of the fabric. **D 123,** D13

skew—rotational deviation from correct horizontal and vertical orientation; may apply to a single character, line, or entire document. **F 149,** F05

skew, *n*—rotation of a bar code symbol about an axis parallel to the symbol's length. **F 1294,** F05

ski afterbody—that portion of the ski aft of point *O*, a distance of $l_C/2$. **F 472,** F27

ski body—that portion of the ski within the dimensions of l_C. **F 472,** F27

ski body center—point *O*, which is located at a distance of $l_C/2 + l_T$ from the tail of the ski. **F 472,** F27

skid, *n*—one of a pair or series of parallel runners usually wood, affixed to the underside of **boxes, crates,** or an item to allow entry of truck forks, or to facilitate sliding. (Compare **rubbing strip.**) **D 996,** D10

skid box—See **box.** **D 996,** D10

skid number (friction number), *n*—the number that is used to report the results of a pavement skid test conducted in accordance with Test Method E 274. **E 867,** E17

skid number-percent normalized gradient, *n*—the speed gradient divided by the skid number, both at the same speed and multiplied by 100. The percent normalized gradient is usually designated by the symbol PNG_v where v is the speed at which the percent normalized gradient is determined.

$$PNG_v = 100(G/SN)_v$$

E 867, E17

skid number (*SN*), *n*—slide braking coefficient multiplied by 100. **F 538,** F09

skid number-speed gradient, *n*—the slope of skid number versus speed multiplied by −1. The gradient is normally designated by the symbol G_v, where v is the speed at which the slope is determined, *SN* is the skid number, and *V* is the speed:

$$G_v = -(dSN / dV), \text{ that may be approximated by:}$$

$$-(SN_1 - SN_2)/(V_1 - V_2)$$

E 867, E17

skid platform—a single platform of wood, metal, plastic, or combination of these materials, elevated a short distance above the floor by runners or legs to facilitate mechanical handling. (Compare **pallet.**) **D 996,** D10

skid resistance, *n*—the ability of the traveled surface to prevent the loss of tire traction. **E 867,** E17

ski forebody—that portion forward of point *O*, a distance of $l_C/2$. **F 472,** F27

skim coat, calender, *n*—a layer of rubber compound applied to a fabric by pressure normal to the surface. **D 1566,** D11

skim coating (the act of), *n*—the process of applying a thin layer of rubber or rubber mix to a sheet material by means of a calender, without shear forces, between the rubber and the sheet. **D 1566,** D11

skimmer block—a partially submerged object near the exit end of a melting furnace, designed to (*1*) keep foam, unmelted batch, etc. from exiting the tank; (*2*) isolate atmospheres on either side; or (*3*) interrupt surface flow of glass from the melter, or combinations of these. **C 162,** C14

skimming—abstraction of information by electronic means from a magnetic stripe or other such device for use in counterfeiting other magnetic stripes or devices. Removal of finished product from the normal production flow after testing, but before shipment of the goods. (Also, the surreptitious removal of a portion of the money received from some cash-generating or handling operation or activity, such as gambling, prior to the formal accounting of the income for purposes of concealing the true value of funds from the authorities or others, or both, with legitimate interest in knowing the correct amount of the income.) **F 1156,** F12

skimmings (tall oil), *n*—the curd, not acidified or otherwise processed, skimmed from the black liquor of the alkaline paper pulp industry, from which tall oil is obtained. **D 804,** D01

skim rubber, *n*—rubber obtained from the dilute latex that is separated during the concentration of natural rubber latex. **D 1566,** D11

skin, *n—in building construction*, a thin layer at the surface of a sealant, cellular rubber gasket, sealant backing, or joint filler that differs in physical properties from the material beneath it. **C 717,** C24

skin, *n*—a relatively dense layer at the surface of a cellular polymeric material. **D 883,** D20

skin—the pelt of a small animal, such as calf, pig, sheep, etc. Also used interchangeably with **hide.** **D 1517,** D31

skin, *n*—a relatively dense layer at the surface of a cellular polymeric material. **D 1566,** D11

skin—See **facing.** **E 1749,** E06

skin, *n*—a relatively dense layer at the surface of a cellular polymeric material. (D-20) **F 412,** F17

skin, applied, *n*—a thin surface layer of elastomeric material applied to a cellular product. **D 1566,** D11

skin depth—see **depth of penetration.** **E 1316,** E07

skin effect—the phenomenon wherein the depth of penetration of electric currents into a conductor decreases as the frequency of the current is increased. At very high frequencies, the current flow is restricted to an extremely thin outer layer of the conductor. (See also **depth of penetration.**) **E 1316,** E07

skin effect—the phenomenon that causes the magnetization produced by alternating current to be contained near the surface of a ferromagnetic part. **E 1316,** E07

skin effect, magnetic—the nonuniform magnetodynamic term applies to the nonuniform distribution of induction existing at various points in the cross section of a magnetic core. Skin effect is produced primarily by eddy current phenomena and it increases with the frequency of ac excitation. It can ordinarily be neglected in testing at commercial power frequencies. **A 340,** A06

skin friction, *f* **(FL^{-2})**—the frictional resistance developed between soil and an element of structure. **D 653,** D18

skin packaging—See **packaging.** **D 996,** D10

skin passed sheet, *n*—Synonym for *extra smooth sheet.* **A 902,** A05

skip—an unsurfaced area on dressed lumber; sometimes designated as skip dressing. **D 9,** D07

skip distance—in angle beam examination, the distance along the test surface, from sound entry point to the point at which the sound returns to the same surface. It can be considered the top surface distance of a complete vee path of sound in the test material. **E 1316,** E07

skirt—continuous portion of the boom below the floats. **F 818,** F20

skirting leather—a specialized vegetable-tanned cattlehide leather used for skirts or hanging portions of saddles that come between the legs of a rider and the horse. **D 1517,** D31

ski size — **F 472,** F27

ski tail, T—the extreme rear edge of the ski. **F 472,** F27

ski tip, S—the extreme forward point or edge of the ski. **F 472,** F27

skiver—the grain split of a sheepskin used for hat sweatbands and small leather goods. **D 1517,** D31

skiving—cutting off a thin layer of leather to bring it to uniform thickness. **D 1517,** D31

skylight—flat glass installed at an angle greater than 15° from the vertical in a building exterior. **C 162,** C14

slab—a piece of stone produced by shaving or splitting in the first milling or quarrying operation. A slab has two parallel surfaces. **C 119,** C18

slab—in belting leather, the parts of a bend left after the centers are cut out. (See also **split.**) **D 1517,** D31

slabbing—the loosening and breaking away of relatively large flat pieces of rock from the excavated surface, either immediately after or some time after excavation. Often occurring as tensile breaks which can be recognized by the subconchoidal surfaces left on remaining rock surface. (ISRM) **D 653,** D18

slab glass—optical glass obtained by cutting or forming the chunk glass into plates or slabs. **C 162,** C14

slabjacking—*in grouting,* injection of grout under a concrete slab in order to raise it to a specified grade. **D 653,** D18

slack end, *n*—a warp yarn woven under insufficient tension. **D 123,** D13

slack pick, *n*—a single filling yarn woven under insufficient tension. **D 123,** D13

slack quenching, *n*—the incomplete **hardening** of a steel object due to **quenching** from the austenitizing temperature at a rate slower than the **critical cooling rate** for the particular steel composition, resulting in the formation of one or more transformation products in addition to martensite. **A 941,** A01

slack selvage, *n*—slack ends in the fabric edge. **D 123,** D13

slack tannage—(*1*) incompletely tanned leather, evidenced by a raw or undertanned streak in the central layer of a piece.

(*2*) A light tannage, that is, deliberately less than usual. **D 1517,** D31

slag—partially fused mixture of spilled batch, overflowed glass, breeze coal, and clay from the siege. **C 162,** C14

slag—the fused agglomerate which separates in metal smelting and floats on the surface of the molten metal. (See also **blast-furnace slag.**) **D 1079,** D08

slag cement, *n*—a hydraulic cement consisting mostly of an intimate and uniform blend of ground, granulated blast-furnace slag and, one or both of portland cement or hydrated lime, in which the amount of the slag constituent is within specified limits. **C 219,** C01

slagging of refractories, *n*—destructive chemical reaction between refractories and external agencies at high temperatures resulting in the formation of a liquid. **C 71,** C08

slaking—deterioration of rock on exposure to air or water. **D 653,** D18

slaking—the process of breaking up or sloughing when an indurated soil is immersed in water. **D 653,** D18

slam-off, *n*—*in woven fabrics,* a distortion due to the entrapment of the filling carrier in the shed. **D 123,** D13

slant—a piece of vitrified clay pipe made so that one end has a plane of approximately 45° or 60° to its longitudinal axis. The end may be made with a contoured surface to fit another pipe. **C 896,** C04

slanted magazine—a magazine attached at an angle to tool; used for storing clips of slanted nails or staples. **F 592,** F16

slat, *n*—a thin flat strip of material used as a member of a **box, crate, pallet, basket,** or other similar structure. A basket **stave.** **D 996,** D10

slate—microcrystalline metamorphic rock most commonly derived from shale and composed mostly of micas, chlorite, and quartz. The micaceous minerals have a subparallel orientation and thus impart strong cleavage to the rock which allows the latter to be split into thin but tough sheets. **C 119,** C18

slaters' cement—a type of caulking compound, usually gray in color, and used to cover exposed bolt heads or at the side and end laps of corrugated roofing and in other places where water-resistant putty-like material is desired. **C 460,** C17

slater's cement, *n*—a type of caulking compound, usually gray in color, and used to cover exposed bolt heads or at the side and end laps of corrugated roofing and in other places where waterresistant putty-like material is desired. **C 1154,** C17

slaters' cement—a type of caulking compound, usually gray in color, and used to cover exposed bolt heads or at the side and end laps of corrugated roofing and in other places where water-resistant putty-like material is desired. **D 2946,** C17

slating nail—galvanized, regular-stock-steel, 1 by 0.106 to 2 by

0.148-in. nails with slightly countersunk 5⁄16 to 7⁄16-in. flat head and medium diamond point. Also, aluminum-alloy, 7⁄8 by 0.106 to 1 1⁄2 by 0.135-in. nails with large flat 5⁄16 to 3⁄8-in. head and medium diamond point. Also, solid copper, 7⁄8 by 0.109 to 3 by 0.148-in. nails with large flat head and medium diamond point. **F 547,** F16

slats—dried, untanned sheepskins, with little or no wool. **D 1517,** D31

sleek—an imperfection; a fine scratchlike mark having smooth boundaries, usually produced by a foreign particle in the polishing operation. **C 162,** C14

sleeve—a coupling which contains or compresses the sealing element and meets the requirements of the standard. The sleeve may be affixed to one end of the pipe at the factory. **C 896,** C04

sleeve, *n*—*in packaging,* a form, made of various materials and in various shapes, usually open at both ends, that is slipped over an item. **D 996,** D10

sleeve, *n*—*(1) in concrete, masonry, or other construction,* a tubular section of sheet metal or other material placed to provide a pocket or opening for the insertion of a railing or other member. *(2)* an internal or external tubular splice between abutting sections of **pipe, tubing,** or similar members. **E 631,** E06

sleeved grout pipe—see **tube A manchette**. **D 653,** D18

sleeve interconnection, *n*—an interconnection in which an implant component passes through any opening that limits motion in one or more planes. **F 1582,** F04

sleeve, post—a specified length of tube or pipe set into a concrete wall, grade beam, or slab, into which fence posts are later placed. **F 552,** F14

sleeve roll-up—a sleeve carrier formed of flat canvas-like material in which a pair of sleeves is rolled lengthwise. **F 819,** F18

sleeves, *n*—the tubular fireclay shapes that encase an immersed metal rod in the valve assembly of a bottom-pouring ladle. **C 71,** C08

sleeve shoulder roll—the rolled or reinforced edge of an insulating sleeve nearest to the shoulder. **F 819,** F18

sleeve, toprail—a fitting used to join two pieces of toprail when a swedged toprail is not used. **F 552,** F14

sleeve-type reference electrode—a reference electrode in which the filling solution contacts the sample solution by means of a narrow ring-shaped opening between a removable outer sleeve and the inner body of the electrode. The space between the body and sleeve widens above the tip to form a reservoir for the filling solution. Sleeve-type electrodes provide exceptionally stable junction potentials, making them especially suitable for specific ion and precision pH measurements. The junction area, due to its size and the high-leak rate of the internal filling solution, does not easily become clogged. **D 4127,** D19

sleeving, *n*—braided, knitted, or woven fabric of cylindrical form having a width less than 100 mm (4 in.) (circumference less than 200 mm (8 in.). (See also **tubing**.) **D 123,** D13

slenderness ratio, *n*—the effective unsupported length of a uniform column divided by the least radius of gyration of the cross-sectional area. **E 6,** E28

slender, slim nail—nails with shank diameter usually at least one gage smaller than common nails of same length. **F 547,** F16

sley, *n*—the number of warp ends per inch of fabric width, exclusive of selvage. **D 123,** D13

sliced veneer—See under **veneer.** **D 1038,** D07

slicer—machine for producing veneer by slicing. **D 1038,** D07

slide pump sprayer—a sprayer consisting of a telescoping pump operated by both hands. On the outlet end of the pump is mounted a spray nozzle. On the inlet of the pump is attached a line leading to a container containing the spray solution. **E 1102,** E35

slider—that member of a sliding contact pair, normally the smaller, which may be moved through a range of locations on the opposing member. **B 542,** B02

slider, *n*—*in zippers,* the part that opens a zipper when it is moved in one direction and closes the zipper when it is moved in the opposite direction. **D 123,** D13

slide surface, *n*—*in the rotor of an open-end spinning machine,* that part of the internal surface of the rotor on which the fibers are deposited and are caused to slide to the collecting surface. **D 123,** D13

sliding—a defect in the draining characteristics of slip wherein patches of the coating slide, producing an uneven coating. **C 286,** B08

sliding—relative displacement of two bodies along a surface, without loss of contact between the bodies. (ISRM) **D 653,** D18

sliding braking coefficient, *n*—*of a tire,* the braking coefficient for a non-rotating tire that occurs after wheel lock-up. **F 538,** F09

sliding electrical contacts—contacting members that perform their function while undergoing relative tangential motion. **B 542,** B02

sliding glass door—See **windows and doors.** **E 631,** E06

sliding wear, *n*—wear due to the relative motion in the tangential plane of contact between two solid bodies. **G 40,** G02

slightly cloudy pyridine—quantitative index of turbidity observed with aqueous solution of pyridine, the latter containing trace amounts of oil. **D 4790,** D16

slime—biological deposits of gelatinous or filamentous matter. **D 6161,** D19

slinger—see **centrifugal atomizer**. **E 1620,** E29

sling psychrometer—See **psychrometer.** **D 1356,** D22

slip—a slurry containing chemical additives to control rheology. **C 242,** C21

slip—translation of a portion of a crystal relative to the adjacent portion. **E 7,** E04

slip—a process of plastic deformation in which one part of a crystal undergoes a shear displacement relative to another in a fashion that preserves the crystallinity of the material. **F 1241,** F01

slip angle, *n*—*of a tire,* the angle between the X'-axis and direction of travel of the center of tire contact. **F 538,** F09

slip angle, critical, *n*—the value of the slip angle at the maximum of lateral force coefficient. **F 538,** F09

slip bands—See **slip lines.** **E 7,** E04

slip casting—See *drain casting* and *solid casting* under **casting.** **C 242,** C21

slip coating—a ceramic material or mixture other than a glaze, applied to a ceramic body and fired to the maturity required to develop specified characteristics. **C 242,** C21

slipcover, *n*—a removable, fitted protective textile cover, often decorative and specifically made for upholstered furniture. (See **furniture covering.**) **D 123,** D13

slip crack—a rupture in the pressed compact caused by the mass slippage of a part of the compact. Synonymous with **pressing crack.** **B 243,** B09

slip-fiber, *n*—*in asbestos,* asbestos that is located in, or derives from, veins or seams in which the fibers are oriented approximately parallel to each other and inclined at a small angle to the boundaries of the vein or seam and usually not as closely packed as are cross-fibers. **C 460,** C17

slip-fiber—asbestos that is located in, or derives from, veins or seams in which the fibers are oriented approximately parallel to each other and inclined at a small angle to the boundaries of the vein or seam and usually not as closely packed as are cross-fibers. **D 2946,** C17

slip glaze—See *slip glaze* under **glaze.** **C 242,** C21

slip joint—See **joint.** **E 631,** E06

slip line—a step occurring at the intersection of a slip plane with the surface. **F 1241,** F01

slip lines (slip bands)—traces of slip planes observed at low magnifications on the polished surface of a crystal which has been deformed after polishing; since no differences in orientation exist, repolishing will remove the traces. With increasing resolving power and magnification, an individual line may be revealed as a series of parallel lines. The "line" which is visible at low magnifications is then described as a slip band. **E 7,** E04

sliplining—a method of inserting new pipe into an existing pipeline. **C 896,** C04

slippage, *n*—the movement of adherends with respect to each other during the bonding process. **D 907,** D14

slippage—relative lateral movement of adjacent components of a built-up membrane. It occurs mainly in roofing membranes on a slope, sometimes exposing the lower plies or even the base sheet to the weather. **D 1079,** D08

slippage cracking, *n*—crescent or half-moon-shaped cracks having two ends pointed opposite to the direction of traffic. **E 867,** E17

slippage cracking, *n*— cracking associated with the horizontal displacement of a localized area of the pavement surface. **E 1778,** E17

slipping layer, *n*—a layer applied to a dye diffusion thermal transfer ribbon which promotes smooth flow of the ribbon through the printer. **F 1623,** F05

slip plane—the crystallographic plane on which the dislocations forming the slip move. **F 1241,** F01

slip planes—in a given metal, slip occurs most easily along certain crystallographic planes. Hence, these planes are termed slip planes. **E 7,** E04

slip process—See *wet process* under **process.** **C 242,** C21

slip resistance—the frictional force opposing movement of an object across a surface. **C 242,** C21

slip resistance—frictional force opposing movement of an object across a surface, usually with reference to the sole or heel of a shoe on a floor. **D 2825,** D21

slip resistance—resistance of sheet carbon to slippage when placed between multiple sheets of copy paper. **F 221,** F05

slip ring—a continuous metal ring by means of which electrical current can be conducted to or from brush contacts. One electrical contact member is designed to rotate with respect to the other. **B 542,** B02

slip ring assembly—two or more slip rings with connecting leads or terminals that have been mounted to a common structure. **B 542,** B02

slip ring capsule—an assembly that includes a slip ring assembly, brushes, and bearings for conducting current on multiple circuits from a stationary body to one that may rotate. **B 542,** B02

slip roll ink buildup, *n*—an undesirable accumulation of ink on the dampening rolls of a continuous type dampening system. **D 6488,** D01

slip, slurry—a suspension of finely divided ceramic material in liquid. **C 286,** B08

slipsole—a half-sole extending from the toe of the shoe to the shank on the bottom surface. **F 869,** F08

slit—a narrow aperture, usually rectangular in shape. **E 7,** E04

slit failure—a form of brittle failure which exhibits only a very small crack through the wall of the pipe with no visible (to the naked eye) material deformation in the area of the break. **F 412,** F17

slit system—a group of two or more slits arranged to define a beam. **E 7,** E04

sliver—(slī-ver) a bundle of untwisted, substantially parallel glass fibers. **C 162,** C14

sliver, *n*—a continuous strand of loosely assembled fibers that is approximately uniform in cross-sectional area and without twist. **D 123,** D13

sliver—a continuous strand of loosely assembled fibers that is approximately uniform in cross-sectional area and without twist. **D 1695,** D01

sliver knitted fabric, *n*—a single-jersey fabric in which untwisted staple fibers are knitted in at each loop to form a pile surface on the technical back of the jersey structure. **D 123,** D13

slivers—particles of nearly square or rectangular cross-section with a length parallel to the grain of the wood of at least four times the thickness. **D 1554,** D07

slope—the incline of a roof expressed as a ratio of the height in mm (in.) of vertical rise per horizontal distance in m (ft). **C 460,** C17

slope, *n*—*in roofing*, the incline of a roof expressed as a ratio of the height in millimetres (inches) of vertical rise per horizontal distance in metres (feet). **C 1154,** C17

slope—the excavated rock surface that is inclined to the vertical or horizontal, or both, as in an open-cut. (ISRM) **D 653,** D18

slope—the tangent of the angle between the roof surface and the horizontal plane, expressed as a percentage, or in inches of rise per foot of horizontal distance. (See also **incline**). **D 1079,** D08

slope—the incline of a roof expressed as a ratio of the height in millimetres (inches) of vertical rise per horizontal distance in metres (feet). **D 2946,** C17

slope—the angle between the horizontal axis and the line formed by plotting electrode potentials against ion level on semilogarithmic graph paper. By analogy, slope has also come to be a measure of electrode response to the ion being detected. Theoretical Nernstian slope, which is temperature-dependent, is 59.16 mV at 25°C for a ten-fold change in the activity of a monovalent ion and 29.58 mV for a divalent ion. Less than theoretical slopes (that is, smaller millivolt changes per decade) are observed if potentials are plotted against concentration, if interferences are present, or if liquid electrodes need renewal. Greater than theoretical slopes are relatively rare and usually indicate that more than one electrode process is occurring. **D 4127,** D19

slope, n—See **pitch.** **E 631,** E06

slot scanner, *n*—the scanning portion of a point-of-sale system embedded within a retail checkout counter. **F 1294,** F05

slot seam, *n*—*in home sewing*,a complex seam formed on the inside of the product, having a decorative seam underlay slightly visible from the face side held in place by two visible rows of stitching. **D 123,** D13

slotted head—head provided with cut or struck slot for insertion of screwdriver or to simulate head of wood screw. **F 547,** F16

slough-off, *n*—*in woven fabrics*, a defect caused by several coils of yarn slipping off the filling bobbin simultaneously and being woven into the fabric in a group. **D 123,** D13

sloughs—a stagnant or sluggish channel of water in a flood plain. **D 4410,** D19

slow stable crack extension [L]—a displacement controlled crack extension beyond the stretch zone width. The extension stops when the applied displacement is held constant. **E 1823,** E08

slow test—see **consolidated-drain test**. **D 653,** D18

slub, *n*—an abruptly thickened place in a yarn. (See **piecing, slug**. Compare **cockles**) **D 123,** D13

sludge—a water charged sedimentary deposit. **D 653,** D18

sludge—a water-formed sedimentary deposit. **D 1129,** D19

sludge, *n*—an undesirable residue in rubber latex. **D 1566,** D11

sludge—a water-formed sedimentary deposit. **D 6161,** D19

sludge blanket—suspended bed of solids in a solids contact or sludge blanket clarifier. **D 6161,** D19

sluffing—a condition wherein scales peel off or become loose, either partially or entirely, from the pultrusion.

NOTE—This term is applied to an occurrence during the pultrusion process and is not to be confused with scraping, prying, or physically removing the scale from the pultrusion. "Sluffing" is sometimes spelled "sloughing". **D 3918,** D20

slug—any nonfibrous glass in an insulating glass fiber product. **C 162,** C14

slug, *n*—*in raw silk*, a thickened place several times the diameter of the yarn, 3 mm (⅛ in.) or over in length. **D 123,** D13

slug—a volume of water or solid object used to induce a sudden change of head in a well. **D 653,** D18

slug, *n*—*in glass,* unattenuated particles of glass of substantially larger diameter than the average filament diameter. (*Syn*. lump, piecing, slough-off, slub)

absolute humidity, vapor concentration, vapor density, d_v— the ratio of the mass of water vapor m_v to the total volume of the moist air v:

$$d_v = \frac{m_v}{v}$$

D 4023, D22

slug, n—See **liquid jet**. **G 40,** G02

slugged bottom—See **heel tap**. **C 162,** C14

slump—a measure of consistency of freshly mixed concrete or grout. See also **slump test**. **D 653,** D18

slump—See **block flow**. **E 1749,** E06

slumping, *n*—the lack of shape retention of a molded part, during subsequent processing, due to the effect of gravity. **B 243,** B09

slumping—See **sagging**. **C 162,** C14

slump test—a test to determine consistency of slip whereby measurement is made of the spreading of a specified volume of slip over a flat plate. **C 286,** B08

slump test—the procedure for measuring slump (Test Method C 143).[2] **D 653,** D18

slunk—the skin of an unborn or prematurely born animal, especially calf. **D 1517,** D31

slurry—a suspension of solids in water. **B 374,** B08

slurry—a prepared mixture consisting of a free-flowing suspension of undissolved solid(s) in a liquid medium; not a paste. **C 242,** C21

slurry, *n*—any pourable suspension of a high content of insoluble particulate solids in a liquid medium, most often water. **C 1145,** C28

slurry—a suspension of solid particles in a liquid that can be separated by filtration or sedimentation (does not include emulsions). **E 1547,** E15

slurry, *n*—a suspension of solids in liquid. **E 1620,** E29

slurry, *n*—a suspension of solid material in liquid **(C 242)** **G 40,** G02

slurry-broadcast resin monolithic surfacing—a flooring system whereby a dry inert filler is added to a catalyzed resin binder to yield a semi-flowable consistency. This mixture is then poured onto the substrate and spread to desired thickness, followed immediately by seeding the fresh surface to excess with a dry inert filler. Upon hardening, any unbonded filler is removed. The application may be repeated once or several times to yield a greater thickness of surfacing. A top coat may be applied. **C 904,** C03

slurry cutoff wall—a vertical barrier constructed by excavating a vertical slot under a bentonite slurry and backfilling it with materials of low permeability for the purpose of the containment of the lateral flow of water and other fluids. **D 653,** D18

slurry grout—a fluid mixture of solids such as cement, sand, or clays in water. **D 653,** D18

slurry seal, *n*—an application of a fluid mixture of bituminous emulsion, fine aggregate, mineral filler, and water to an existing pavement. Single or multiple applications may be used. **D 8,** D04

slurry trench—a trench that is kept filled with a bentonite slurry during the excavation process to stabilize the walls of the trench. **D 653,** D18

slush grouting—application of cement slurry to surface rock as a means of filling cracks and surface irregularities or to prevent slaking; it is also applied to riprap to form grouted riprap. **D 653,** D18

slushing—the manipulation of dipped ware to distribute the slip uniformly and remove excess material. **C 286,** B08

slush molding—a method for casting rubber and thermoplastics in which the resin in liquid form is poured into a hot mold where a viscous skin forms. The excess slush is drained off, the mold is cooled, and the molding stripped out. **F 869,** F08

small quantity—see **hazardous-waste small quantity generator**. **E 1605,** E06

small timbers—a term used mostly to designate square or near-square dimension and timber sizes over 2 and under 9 in. in nominal thickness. **D 9,** D07

smash, *n*—*in woven fabrics*, a relatively large hole characterized by broken warp ends and floating picks. (Compare **slam-off**) **D 123,** D13

SMBS—sodium metabisulfite, $Na_2S\ _2O_5$. **D 6161,** D19

smear fastness, *n*—the ability of an ink and substrate combination to resist ink transfer from a printed area to an unprinted area when swiped by objects or instruments. **F 1857,** F05

smear resistance—the ability of a polished surface to remain unscuffed when touched. **D 2825,** D21

smell, *n*—See **aroma.** **E 253,** E18

smelt—a specific batch or lot of frit. **C 286,** B08

smelter—a furnace in which the raw materials of a frit batch are melted. **C 242,** C21

smelter—a furnace in which the raw materials of the frit batch are melted. **C 286,** B08

smelter drippings—drippings of molten glassy material formed on the crown of the smelter. **C 286,** B08

smelt (*n*)—a specific batch or lot of frit.

(*v*)—the act of melting a batch of frit. **C 242,** C21

smog, *n*—a term derived from smoke and fog, applied to extensive atmospheric contamination by aerosols, these aerosols arising partly through natural processes and partly from the activities of human subjects. **D 1356,** D22

smoke—*in flat glass,* streaked areas appearing as slight discoloration. **C 162,** C14

smoke, *n*—small gas-borne particles resulting from incomplete combustion, consisting predominantly of carbon and other combustible material, and present in sufficient quantity to be observable independently of the presence of other solids. **D 1356,** D22

smoke, *n*—the airborne solid and liquid particulates and gases evolved when a material undergoes pyrolysis or combustion. **E 176,** E05

smoked—(*1*) a term applied to the discoloring of glass in a reducing flame.

(*2*) glass covered with smoky film from open-fired lehrs. **C 162,** C14

smoke developed index, *n*—a number or classification indicating a comparative measure derived from smoke obscuration data collected during the test for surface burring characteristics. **E 176,** E05

smoke obscuration, *n*—reduction of light transmission by smoke, as measured by light attenuation. **E 176,** E05

smoke obscuration, *n*—reduction of light transmission by smoke, as measured by light attenuation. **E 176,** E05

smoke point, *n*—the maximum height of a smokeless flame of fuel burned in a wick-fed lamp. **D 4175,** D02

smoke toxicity, *n*—the propensity of smoke to produce adverse biochemical or physiological effects. (See **smoke.**) **E 176,** E05

SMOKE unit, *n*—the concentration of smoke particulates in a cubic metre of air that reduces the percent transmission of light through a 1-m path to 10 %. SMOKE = standard metric optical kinetic emission. **E 176,** E05

smoldering, *n*—the combustion of a solid material without accompaniment of flame but generally with the production of smoke. (See also **afterglow** and **glow**.) **D 123,** D13

smoldering, *n*—combustion of a solid without flame, often evidenced by visible smoke. **E 176,** E05

smooth—{archaic} finely ground surface prior to polishing. **C 162,** C14

smooth crack growth behavior—generally, crack extension in chevron-notch specimens which is characterized by slow, continuously advancing crack growth, and a relatively smooth load displacement record. However, any test behavior is automatically characterized as smooth crack-growth behavior unless it satisfies the conditions for crack-jump behavior. **E 1823,** E08

smooth-edge carpet plywood strip nail— hardened-steel, 11⁄16 by 0.105-in. nail with countersunk flat 7⁄32-in. head and long diamond point. **F 547,** F16

smooth finish, *n*—units whose surfaces are not altered or marked in manufacture, but left as a plane surface as formed by the die. **C 43,** C15

smoothing—*XPS,* a mathematical treatment of the data to reduce the noise. **E 673,** E42

smoothness, *n*—a measure of the surface roughness or topography of a thermal substrate, critical for optimum printhead contact. **F 1623,** F05

smooth-surfaced roof—a built-up roof without mineral aggregate surfacing. **D 1079,** D08

smooth - two - side hardboard (S2S)—hardboard produced from a dry mat pressed between two smooth hot platens. **D 1554,** D07

smooth (-wall) blasting—a method of accurate perimeter blasting that leaves the remaining rock practically undamaged. Narrowly spaced and lightly charged blastholes, sometimes alternating with empty dummy holes, located along the breakline and fired simultaneously as the last round of the excavation. (ISRM) **D 653,** D18

smudge—the tendency of an image to smear or streak onto an adjacent area when rubbed; involves the redeposition of abraded material. **F 221,** F05

smudge—a coloration on the CF surface caused by contact with ruptured capsules on the CB surface. **F 549,** F05

smudge—dense local area of contamination usually caused by handling or fingerprints. **F 1241,** F01

smudge resistance—the ability of an image to withstand smudging. **F 221,** F05

smudging—the tendency of a developed electrostatic image to smear upon rubbing. **F 335,** F05

SNA—abbreviation for **system network architecture**. **F 1457,** F05

snag, *n*—*in fabrics,* a yarn or part of a yarn pulled or plucked from the surface. **D 123,** D13

snagging resistance, *n*—*in textile fabrics,* the resistance to the formation of snags. **D 123,** D13

snake, snaking—(*1*) progressive longitudinal cracking in continuous flat glass operation.

(*2*) *archaic,* variation in the width of the sheet during the sheet glass-drawing process.

(*3*) movement from side to side of the flat glass ribbon. **C 162,** C14

snap—{archaic} a device for gripping a piece of formed glass for fire polishing and finishing. **C 162,** C14

snap, *n*—vivid color imagery, high chroma color. **F 1857,** F05

snap action, *n*—the force required to disengage a snap fastener resulting from a pull exerted perpendicular to the plane of material to which the snap fastener is attached. **D 123,** D13

snap fastener, *n*—a device for attaching one material to another consisting of matching male and female parts, each of which is attached to a separate material so that the parts can be joined by a low compressive force and separated by a low perpendicular tensile force. **D 123,** D13

snap temper, *n*—a precautionary interim stress-relieving treatment applied to a high-hardenability steel immediately after **quenching** to prevent cracking because of delay in **tempering** it at the prescribed higher temperature. **A 941,** A01

***S-N* curve**—a plot of stress against the number of cycles to failure. The stress can be maximum stress, S_{max} ; minimum stress, S_{min} ; stress range, ΔS or S_r ; or alternating stress, S_a. The curve indicates the *S-N* relationship for a specified value of S_m, *A*, or *R* and a specified probability of survival. For *N*, a log scale is commonly used. For *S*, either a logarithmic or a linear scale is used. **E 1823,** E08

***S-N* curve for *p* %survival**—a curve fitted to the fatigue life for *p* % survival values at each of several stress levels. It is an estimate of the relationship between applied stress and the number of cycles-to-failure that *p* % of the population would survive where *p* may be any percent, such as 95, 90, and so forth. **E 1823,** E08

***S-N* curve for 50 % survival**—a curve fitted to the median values of fatigue life at each of several stress levels. It is an estimate of the relationship between applied stress and the number of cycles-to-failure that 50 % of the population would survive. **E 1823,** E08

S-N diagram, *n*— a plot of stress against the number of cycles to failure. The stress can be maximum stress S_{max}, minimum stress S_{min}, stress range *S* or S_r, or alternating stress S_a. The diagram indicates the *S-N* relationship for a specified value of S_m (*mean stress*) *A*, or *R* (*load or stress ratio*), and a specified probability of survival. For *N*, a log scale is almost always used. For *S*, a linear scale is used most often, but a log scale is sometimes used. **F 1582,** F04

Snell's law, *n*—the product of the sine of the angle of refraction by the refractive index of the refracting medium is equal to the product of the sine of the angle of incidence by the index of refraction of the medium containing the incident beam. **E 284,** E12

sniffing probe—Same as **sampling probe**. **E 1316,** E07

snip—the concave surface from which a chip has been dislodged. **C 119,** C18

snipe—See **machine bite.** **D 9,** D07

snow, *n*—a solid form of wet deposition composed of white or translucent ice crystals chiefly in complex hexagonal form and often agglomerated into snowflakes. **D 1356,** D22

snowball—*on semiconductor wafer*, a track with the appearance of a snowball rolled through snow under magnification. **F 1241,** F01

snowboard bindings (feet retainers)—apparatuses that attach the rider's feet to the snowboard ski. The two types of snowboard bindings have been identified as follows:

hard boot snowboard bindings—these are designed to be used with hard boots, retaining the boots by means of attaching the boot sole to the binding.

soft boot snowboard bindings—these are designed to be used with soft boots, retaining the boots by means of binding straps across the toe and ankle areas. **F 1107,** F27

snowboard boots—footwear appropriate for use with snowboard skis and snowboard bindings. The two types of snowboard boots have been identified as follows:

hard snowboard boots—plastic shell alpine ski-type outer boots with the inner boot consisting of an alpine ski boot-type bladder for comfort and warmth. They are designed to be used with hard boot snowboard bindings.

soft snowboard boots—outer boots typically consisting of a rubber or plastic lower portion and a leather or nylon upper portion; the inner boot consists of an alpine ski boot-type bladder for warmth and support. They are designed to be used with soft boot snowboard bindings. **F 1107,** F27

snowboard ski—a single plane device ridden with a sideways stance with the feet somewhat perpendicular to the longitudinal axis of the device. **F 1107,** F27

snowboard ski afterbody—that portion of the snowboard ski aft of Point *O*, a distance of $L_c/2$. **F 1107,** F27

snowboard ski body—the portion of the snowboard ski within the dimension of L_c . **F 1107,** F27

snowboard ski body center—Point *O*, located a distance of $L_c/2 + L_t$ from the tail of the snowboard ski . **F 1107,** F27

snowboard ski forebody—that portion forward of Point *O*, a distance of $L_c/2$. **F 1107,** F27

snowboard ski longitudinal centerline—*(1) symmetrical snowboard skis*: the longitudinal axis of symmetry; *(2) assymetrical snowboard skis*: the longitudinal line that passes through the center of the insert pattern or through the manufacturer-indicated centers of binding mount areas on snowboard skis without inserts. **F 1107,** F27

snowboard ski shoulder—the widest point in the shovel section of the snowboard ski. **F 1107,** F27

snowboard ski tail—the extreme rear edge of the snowboard ski. **F 1107,** F27

snowboard ski tip—the extreme forward point of the snowboard ski. **F 1107,** F27

snowboard ski width, *W* —the horizontal perpendicular distance between two vertical parallel planes, placed on either edge of the snowboard ski, parallel to the longitudinal centerline of the snowboard ski. **F 1107,** F27

snow flaking, *n*—the appearance of tiny, white unprinted specks (voids) in type or solids, or both, in offset printing that can be caused by excessive emulsification. **D 6488,** D01

snow flaking, *n*—small white spots in a printed area caused by the absence of ink. **D 6488,** D01

snow, hard pack, *n*—*in tire testing*, packed base without loose snow. **F 538,** F09

snow, medium hard pack, *n*—in tire testing, packed base with some loose snow **F 538,** F09

snow, medium pack, *n*—*in tire testing*, groomed packed base with 2.5 to 5.0 cm (1 to 2 in.) loose snow. **F 538,** F09

snow, soft pack, *n*—*in tire testing*, freshly fallen or deeply groomed base snow with 5.0 to 7.5 cm (2 to 3 in.) loose snow. **F 538,** F09

S/N ratio—signal-to-noise ratio; often used as a measure of signal quality or strength. **F 1156,** F12

snuffed top grain, top grain snuffed—see **corrected grain.** **D 1517,** D31

soaking, *n*—prolonged holding at a selected temperature. **A 941,** A01

soaking heat, *n*—a test condition in which the specimen is completely immersed in an atmosphere maintained at a controlled temperature. **C 168,** C16

soaking pit—a conditioning furnace used to bring the glass in open pots to a uniform temperature for casting. **C 162,** C14

soap, *n*—a cleaning agent usually consisting of sodium or potassium salts of fatty acids. **D 123,** D13

soap—the product formed by the saponification or neutralization of fats, oils, waxes, rosins, or their acids with organic or inorganic bases. **D 459,** D12

anhydrous soap—soap, free of water and all other concomitants. **D 459,** D12

blended soap **(for example, blended palm oil soap)**—a soap in which more than half but not all of the fatty acid stock is from the source stated. **D 459,** D12

built soap—a mixture of soap and one or more builders containing not less than 50 % of anhydrous soap. **D 459,** D12

neutral soap—an essentially unbuilt soap substantially free from uncombined alkali or fatty matter. **D 459,** D12

soap powder—a mixture in powdered form of soap and one or more alkaline detergents, but composed principally of the latter. **D 459,** D12

straight soap **(for example, straight palm oil soap)**—a soap in which the fatty acid stock is solely from the source stated. **D 459,** D12

superfatted soap—a toilet soap containing any one or more of the following: unsaponified oils, fats (triglycerides), fatty acids, lanolin and its derivatives, fatty esters or alcohols, and similar agents. See **fatty matter, free.** **D 459,** D12

soapstone (steatite)—a talc-rich rock with a characteristic slippery feel. Soapstone is quarried for special purposes, such as fireplaces and laboratory counter tops, because of its refractory nature and resistance to acids. **C 119,** C18

socket—the portion of a jointing system that is designed to accept a plain-end pipe or a spigot-end pipe. **C 896,** C04

socket—the portion of a jointing system that is designed to accept a plain-end pipe or spigot-end pipe. **F 412,** F17

socket end—the end portion of a piping component which is designed to accept a plain-end piping component or spigot-end piping component. **F 412,** F17

socket-fused joint—see **joint, socket-fused.** **F 412,** F17

socket welding (machine)—a joining technique for thermoplastic pipe whereby the joining surfaces of the pipe and the fitting are inserted into a heating mandrel that is equipped with appropriate sized male and female heater bushings. The surfaces are heated to the melt point, removed, and held together under pressure until fused. **C 904,** C03

sock lining—a piece of material placed over the entire insole on the inside of a shoe to protect the foot from stitches on the inside in certain types of constructions. **F 869,** F08

soda—an industrial term for sodium oxide (Na_2O). Loosely, a carbonate of sodium. **C 162,** C14

soda lime glass—an abbreviated term for soda-lime-silicate glass. **C 162,** C14

soda-lime-silicate glass—glass compositions containing soda (Na_2O), lime (CaO), and silica (SiO_2) as the main ingredients. **C 162,** C14

sodium carboxymethylcellulose—see **carboxymethylcellulose.** **D 1695,** D01

sodium cycle—the operation of a cation-exchange unit wherein the removal of specific cations from the influent water is accomplished by exchange with an equivalent amount of sodium ion from the exchange material. **D 1129,** D19

sodium iodide detector, thallium activated, NaI(Tl)—a scintillation detector crystal used for detection of gamma rays. **C 859,** C26

sodium rhodizonate method—*for lead detection*, use of a dilute solution of sodium rhodizonate to spot test a painted surface qualitatively for the presence of lead. **E 1605,** E06

sodium sulfide method—*for lead detection*, use of a dilute solution of sodium sulfide to spot test a painted surface qualitatively for the presence of lead. **E 1605,** E06

soft (as in soft asbestos)—description of asbestos with relatively high softness. **D 2946,** C17

softener, *n*—a compounding material used to produce a mix of reduced viscosity, which facilitates incorporation of rubber additives. **D 1566,** D11

softener—water treatment equipment that uses a sodium based ion-exchange resin principally to remove cations as calcium and magnesium. **D 6161,** D19

softening—see **membrane softening**. **D 6161,** D19

softening point—the temperature at which a bitumen becomes soft enough to flow as determined by an arbitrary, closely defined method. **D 1079,** D08

softening point drift—a change in the softening point during storage or application. (See also **fallback**). **D 1079,** D08

softening point (S.P.)—that temperature at which a glass fiber of uniform diameter elongates at a specific rate under its own weight when measured by Test Method C 338. The viscosity at the softening point depends on the density and surface tension. For example, for a glass of density 2.5 g/cm^3 and surface tension 300 dynes/cm, the softening point temperature corresponds to a viscosity of $10^{6.6}$ Pa · s. **C 162,** C14

softening temperature—the temperature, under specified conditions, at which porcelain enamel or frit begins to flow. **C 286,** B08

softening temperature (ST), *n*—*in reference to the fusibility of coal and coke ash according to Test Method D 1857*, the temperature at which the cone has fused down to a spherical lump in which the height is equal to the width at the base. **D 121,** D05

softening temperature, ST, *n*—the temperature at which a pyrometric cone has fused down to a spherical lump in which the height is equal to the width at the base. **D 5681,** D34

soft fire—a flame with a deficiency of air. **C 162,** C14

soft glass—(*1*) a glass of relatively low viscosity at elevated temperatures.

(*2*) a glass with a low softening point.

(*3*) commonly refers to a glass easy to melt. **C 162,** C14

soft mica, *n*—mica which when slightly bent shows a tendency to delaminate. **D 1711,** D09

softness, *n*—*in water*, the relative absence of dissolved calcium, magnesium, and other salts that react with soluble soaps to form insoluble precipitates. (Compare **hardness**.) **D 123,** D13

softness—inherent property of asbestos implying a high degree of flexibility and low cohesion. **D 2946,** C17

soft window coverings, *n*—curtains, draperies, or other accessories on wall or window openings that are either lined or unlined and primarily constructed of textile fabrics. **D 123,** D13

soft wire—wire that has been drawn or rolled to final size and then heated to remove the effects of cold working. **B 354,** B01

softwoods—generally, one of the botanical groups of trees that in most cases have needlelike or scalelike leaves; the conifers; also the wood produced by such trees. The term has no reference to the actual hardness of the wood. **D 9,** D07

soft (X-rays)—of long wavelength. **E 7,** E04

soil—*of coatings*, disfiguring foreign materials such as dirt, soot, or stain, other than microorganisms, deposited on or embedded in a dried film of applied coating material; also called *dirt.* **D 16,** D01

soil—solid foreign matter embedded in or adhered on the surface. **D 2825,** D21

soilability—the relative ease with which extraneous matter attachs to or builds up on the surface of a material. **C 286,** B08

soil aggregate, *n*—natural or prepared mixtures consisting predominantly of stone, gravel, or sand which contain a significant amount of minus 75-μm (No. 200) silt-clay material. **D 8,** D04

soilant, *n*—a discoloring substance with a dispersed color component that is not in solution, and therefore can cling to the surface of a coating without penetrating into the film. **D 16,** D01

soil application—application of chemical made primarily to the soil rather than to vegetation. **E 609,** E35

soil binder—see **binder**. **D 653,** D18

soil compaction—see **compaction, soil.** **F 412,** F17

soil (earth)—sediments or other unconsolidated accumulations of solid particles produced by the physical and chemical disintegration of rocks, and which may or may not contain organic matter. **D 653,** D18

soiled cloth—see **artificially soiled cloth.** **D 459,** D12

soil-forming factors—factors, such as parent material, climate, vegetation, topography, organisms, and time involved in the transformation of an original geologic deposit into a soil profile. **D 653,** D18

soil gas—vadose zone atmosphere. **D 653,** D18

soil horizon—see **horizon**. **D 653,** D18

soil injection—the mechanical placement of a pesticide beneath the soil surface with a minimum disturbance of the soil. **E 609,** E35

soil injection—the mechanical placement of a pesticide beneath the soil surface with a minimum disturbance of the soil (see Definitions E 609). **E 1102,** E35

soil (in reference to detergency)—matter out of place. **D 459,** D12

soil mechanics—the application of the laws and principles of mechanics and hydraulics to engineering problems dealing with soil as an engineering material. **D 653,** D18

soil physics—the organized body of knowledge concerned with the physical characteristics of soil and with the methods employed in their determinations. **D 653,** D18

soil profile (profile)—vertical section of a soil, showing the nature and sequence of the various layers, as developed by deposition or weathering, or both. **D 653,** D18

soil-redeposition—deposition of removed soil on a surface during a cleaning process. **D 459,** D12

soil retention—the property of holding foreign matter in or on the surface after a cleaning process. **D 2825,** D21

soil stabilization—chemical or mechanical treatment designed to increase or maintain the stability of a mass of soil or otherwise to improve its engineering properties. **D 653,** D18

soil structure—the arrangement and state of aggregation of soil particles in a soil mass.

flocculent structure—an arrangement composed of flocs of soil particles instead of individual soil particles.

honeycomb structure—an arrangement of soil particles having a comparatively loose, stable structure resembling a honeycomb.

single-grained structure—an arrangement composed of individual soil particles; characteristic structure of coarse-grained soils. **D 653,** D18

soil suspension—highly diffused mixture of soil and water. **D 653,** D18

soil texture—see **gradation.** **D 653,** D18

soil-water pressure—the pressure on the water in a soil-water system, as measured by a piezometer for a saturated soil, or by a tensiometer for an unsaturated soil. **D 653,** D18

sol, *n*—a liquid dispersion of colloidal solid particles, commonly between 5 and 100 nm in size. **C 1145,** C28

solar, *adj*—(*1*) *referring to radiometric quantities*, indicates that the radiant flux involved has the sun as its source, or has the relative spectral distribution of solar flux. (*2*) *referring to an optical property*, indicates a weighted average of the spectral property, with a standard solar spectral irradiance distribution as the weighting function. **E 772,** E44

solar constant—the total solar irradiance at normal incidence on a surface in free space at the earth's mean distance from the sun (1 AU = 1.496×10^{11} m).

NOTE—The current accepted value of the solar constant at 1 AU is 1353 ± 21 $W \cdot m^{-2}$ and is subject to change. See Definitions E 491.

E 772, E44

solar cooling systems—the complete assembly of subsystems and components necessary to convert solar energy into other forms of energy for space cooling purposes. **E 772,** E44

solar degradation—(*1*) the process by which exposure to solar energy deteriorates the properties of materials and components; (*2*) The deterioration produced by exposure to solar energy. **E 772,** E44

solar energy—the radiant energy originating from the sun. Approximately 99 % of solar energy lies between the wavelengths of 300 to 3500 nm. **E 772,** E44

solar energy system—See **building subsystem.** **E 631,** E06

solar energy system, active—a solar energy system that uses mechanical equipment (pumps, fans) that is not an integral part of a structure to collect and transfer thermal energy, either to the point of use or to be stored for later use. **E 772,** E44

solar energy system, drainback—a solar energy system in which the heat transfer fluid is drained out of the collector and exposed piping, and into a storage tank, a holding tank, or expansion tank in order to protect the collector and piping from damage due to freezing. **E 772,** E44

solar energy system, draindown—a solar energy system in which the heat transfer fluid is drained out of the collector and exposed piping to an external drain in order to protect the collector and piping from damage due to freezing. **E 772,** E44

solar energy system, hybrid—any solar energy system that combines the characteristics of two separate systems. Particularly, a solar energy system supplemented by a conventional energy system may be termed a hybrid system. **E 772,** E44

solar energy system, open—a solar energy system that has its storage tank exposed (open) to atmospheric pressure. **E 772,** E44

solar energy system, passive—a solar energy system that uses natural convection, conduction, or radiation to distribute thermal energy through a structure, or a portion of that structure within the limits of the indoor design temperature conditions. It can include movable components such as dampers, insulation, or blinds, which may be moved periodically either manually or automatically. **E 772,** E44

solar energy system, thermosiphon—a solar energy system in which the heat transfer fluid circulates by convection as the less dense, warm fluid rises and is displaced by the denser, cooler fluid. **E 772,** E44

solar fraction—ratio of the amount of input energy contributed by the solar energy system to the total input energy required for the application. **E 772,** E44

solar heating and cooling systems—the complete assembly of subsystems and components necessary to convert solar energy into thermal energy and use this energy in combination with auxiliary energy, where required, for combined heating and cooling purposes. **E 772,** E44

solar heating system—the complete assembly of subsystems and components necessary to convert solar energy into thermal energy and use this energy in combination with auxiliary energy, where required, for heating purposes. **E 772,** E44

solar irradiance at a point of a surface, $E_s = d\Phi_s/dA$—the quotient of the solar flux incident on an element of a surface containing the point, by the area of that element, measured in watts per square metre. **E 772,** E44

solar irradiance, average, E_s—the time integral of solar irradiance over a specified time period divided by the duration of that time period. **E 772,** E44

solar irradiance, diffuse, $E_s(d)$—the downward scattered solar flux as received on a horizontal surface from a solid angle of 2π steradian (hemisphere) with the exception of a conical solid angle with a 100 milliradions (approximately 6°) included plane angle centered upon the sun's disk, measured in watts per square metre. **E 772,** E44

solar irradiance, direct, E_s—solar flux coming from the solid angle of the sun's disk incident on a surface perpendicular to the axis of that solid angle. In conventional instruments the acceptance cone includes a plane angle of about 6°. **E 772,** E44

solar irradiance duration—(*1*) a bright sunshine duration: time interval during which direct solar energy casts distinct shadows; (*2*) Geographically or topographically possible sunshine duration: maximum interval during which solar energy can reach a given surface. **E 772,** E44

solar irradiance, global $E_S(2\pi)$, *n*—solar irradiance received on an upward facing horizontal surface directly from the solid angle of the sun's disk and scattered or diffusely reflected in traversing the atmosphere, measured in watts per square metre. **G 113,** G03

solar irradiance, global, $E_s(2\pi)$—solar irradiance received on an upward-facing horizontal surface directly from the solid angle of the sun's disk and scattered or diffusely reflected in traversing the atmosphere, measured in watts per square metre. **E 772,** E44

solar irradiance, instantaneous—solar irradiance at a point in time measured in watts per square metre. **E 772,** E44

solar irradiation— see **radiant exposure,** the preferred term. **E 772,** E44

solarization—a change in appearance of glass as a result of exposure to sunlight. **C 162,** C14

solarization, *v*—change in transmittance, reflectance, or absorptance property of a material, such as glass, as a result of exposure to sunlight or other light sources. **G 113,** G03

solar noon—that instant of any day at which the sun reaches its zenith, or crosses the meridian. **E 772,** E44

solar radiation—see **solar energy.** **E 772,** E44

solar radiation, infrared—radiation for which the wavelengths of the components are longer than those for visible radiation, and shorter than about 1 mm. **G 113,** G03

solar radiation, ultraviolet—radiation for which the wavelengths of the components are shorter than those for visible radiation and longer than about 1 nm. **G 113,** G03

solar radiation, visible—any radiation capable of causing a visual sensation. **G 113,** G03

solar rights—the legal right of a person who uses a solar energy device not to have his or her sunlight blocked by another person's new structure or foliage. **E 772,** E44

solar screening, *n—of coated fiber glass yarn solar screening*, a woven fabric that imparts a shielding or protection from light, heat, wind, and insects without excessive alteration or impairment of visual viewing, and that has a mesh in excess of 12 by 12 in. with a rib pattern in the warp direction formed by the weaving of two or more contiguous yarns with a minimum of space between such yarns followed by space equivalent to the width of one or more of the yarns in the rib. **D 123,** D13

solar spectrum—spectral distribution of typical terrestrial sunlight at air mass 1.5 as defined in Standard E 891, for Terrestrial Direct Normal Solar Spectral Irradiance Tables for Air Mass 1.5. **E 772,** E44

solar thermal collector— see **collector, solar thermal.** **E 772,** E44

solar water heating system—the complete assembly of subsystems and components necessary to convert energy into thermal energy and use this energy in combination with auxiliary energy, where required, to provide hot water. **E 772,** E44

solar water heating system, direct—a solar water heating system in which the potable water passes directly from the water supply, through the collectors and storage, to the residential hot water supply. **E 772,** E44

solar water heating system, indirect—a solar water heating system in which a closed circulation loop isolates one fluid from contact with others in the system. This closed loop may contain a nonpotable fluid. **E 772,** E44

solderability—the capability of a metal to be wetted by solder. **B 846,** B05

solder embrittlement—reduction in ductility of a metal or alloy associated with local penetration by molten solder long grain boundaries. **E 7,** E04

solder sealing glass—a sealing glass characterized by a low softening point for use as an intermediate bonding material. **C 162,** C14

sole—the bottom of the foot or shoe. **F 869,** F08

sole leather butt bend—a double bend. **D 1517,** D31

solenoid—an electrical conductor formed into a coil. **E 1316,** E07

sol-gel processing, *v*—the chemical synthesis of oxides based on the hydrolysis of metal alkoxides to form sols and gels; as liquids, the sols are suitable for casting and infiltration. **C 1145,** C28

solid—(flammability regulations) a substance that has a viscosity greater than 1×10^3 St (1×10^{-1} m^2s^{-1}) at 104°F (40°C) or an equivalent viscosity at an agreed upon temperature. (This includes powders and granular materials.) **D 16,** D01

solid—a state of matter in which the relative motion of molecules is restricted and in which molecules tend to retain a definite fixed position relative to each other. A solid may be said to have a definite shape and volume. **E 1547,** E15

solid area, *A'*—the effective solid portion of the cross section of a core (perpendicular to the induction) which is composed of magnetic material. **A 340,** A06

solid bituminous material—one having a viscosity of over 1×10^5 cSt (mm^2/s) at 40°C or an equivalent viscosity at an agreed-upon temperature. This includes powders and granular materials. **D 1079,** D08

solid braid, *n—in rope*, a braided construction in which each strand alternately passes under and over one or more of the other strands while all strands are rotating around the axis with the same direction of rotation. **D 123,** D13

solid casting—See *solid casting* under **casting**. **C 242,** C21

solid clincher—non-moving clincher. **F 592,** F16

solid-color, *adj*—not containing flake or gonioapparent pigments. **E 284,** E12

solid conductor—a conductor consisting of one wire. **B 354,** B01

solid conductor, *n*—a conductor consisting of one strand. **D 1711,** D09

solid cone atomizer—a cone atomizer in which a significant quantity of liquid is directed into the center region of the spray pattern. **E 1620,** E29

solid cone pattern, *n—Syn.* for **full cone pattern**. **E 1620,** E29

solid fiberboard—See **containerboard.** **D 996,** D10

solid flotation—boom that uses solid buoyant material for the flotation element. **F 818,** F20

solidification point, *n— of petroleum wax*, that temperature in the cooling curve of the wax where the slope of the curve first changes significantly as the wax sample changes from a liquid to a solid state. **D 4175,** D02

solidification point—an empirical constant defined as the temperature at which the liquid phase of a substance is in approximate equilibrium with a relatively small portion of the solid phase. **D 4790,** D16

solidification range—the temperature range between the liquidus and the solidus. **E 7,** E04

solidification shrinkage crack—a crack that forms, usually at elevated temperature because of the shrinkage stresses built up during solidification of a metal casting; a hot crack. **E 7,** E04

solid impingement, *n*—impingement by solid particles. **G 40,** G02

solid impingement erosion, *n*—progressive loss of original material from a solid surface due to continued exposure to impacts by solid particles. (*Synonym:* **solid particle erosion, hard particle erosion.**) **G 40,** G02

solid masonry unit—unit whose net cross-sectional area in any plane parallel to the surface containing cores or cells is 75 % or more of its gross cross-sectional area measured in the same plane. **C 1209,** C15

solid masonry unit—unit whose net cross-sectional area in any plane parallel to the surface containing cores, cells, or deep frogs is 75 % or more of its gross cross-sectional area measured in the same plane. **C 1232,** C15

solid metal connector plate—metal connector plate without any prepunched or predrilled plate holes. **E 1807,** E06

solid metal-coupon control specimen—solid metal connector plate sample of same material as metal connector plate under scrutiny of dimensions meeting the requirements of Test Methods E 8 (or other applicable standard or specification) without plate holes or integral plate projections. **E 1807,** E06

solid mopping—See **mopping**. **D 1079,** D08

solid particle erosion, *n*—Synonym for *solid impingement erosion.* **G 40,** G02

solid-phase chemical dosimeter—plastic, dyed plastic, or glass whose optical density, usually in the visible range, changes when exposed to ionizing radiation. Examples that are in use include dyed polymethylmethacrylate (red perspex), undyed poly(vinyl) chloride (PVC), dyed polyamide (blue dye in Nylon matrix), and dyed polychlorostyrene (green dye in chlorostyrene matrix). It is generally considered to be a secondary standard dosimetry system. **E 170,** E10

solids—two types of solids are present in mash. First, insoluble solids are present as solid matter present in the liquid portion of the mash. Secondly, soluble solids are dissolved in the liquid portion of the mash. **E 1705,** E48

solids concentration, *n*—the ratio, expressed as a percent, of the weight (mass) of solids to the sum of the weight of solids plus water. **D 121,** D05

solids contact clarifier—water treating device used in lime softening, waste water treatment and coagulation processes. **D 6161,** D19

solids content, *n*—the percentage by weight of the nonvolatile matter in an adhesive. (See **nonvolatile content.**) **D 907,** D14

solids loading ratio, n—See **particle concentration**. **G 40,** G02

solid solution—a solid phase in which the composition and properties including lattice parameter can vary continuously without changing the crystal structure; a primary or terminal solid solution is limited by and has the crystal structures of a pure metal, a secondary or intermediate solid solution has the basic crystal structure of an intermetallic compound but does not necessarily include its stoichiometric composition. **E 7,** E04

solid-state matrix array TV camera—two-dimensional CCD array using solid-state technology, usually a silicon-based semiconductor photoreceptor; the array structure facilitates digitization of the image. **F 1156,** F12

solid-state sintering—sintering of a powder or compact without formation of a liquid phase. **B 243,** B09

solid state track recorder (SSTR)—a plastic, mineral, or emulsion material that records the passage of energetic ionizing particles as latent tracks that may be chemically enlarged or developed for microscopy or other techniques. **E 170,** E10

solidus, *n*—the highest temperature at which under equilibrium conditions an alloy begins to melt on heating or is completely solid on cooling. **B 899,** B02

solidus—the locus of points in a phase diagram, representing the temperature, under equilibrium conditions, at which each composition in the system begins to melt during heating, or completes freezing during cooling. **E 7,** E04

solidus—locus of points in a phase diagram representing the temperature, under equilibrium conditions, at which each composition in a system begins to melt during heating or completes freezing during cooling, (E 7, E-4). **E 1142,** E37

solid vinyl flooring, *n*—See **homogeneous vinyl flooring**. **F 141,** F06

solid vinyl tile, *n*—a resilient tile flooring composed of binder, fillers and pigments compounded with suitable stabilizers and processing aids. The tile meets requirements of Specification F 1700. The binder consists of polymers and/or copolymers of vinyl chloride, other modifying resins, and plasticizers which comprise at least 34 % by weight of the finished tile. The polymers and copolymers of vinyl chloride comprise at least 60 % of the weight of the binder. **F 141,** F06

solid waste—discarded solid materials, excluding recovered materials. **D 1968,** D06

solid waste stream, *n*—discarded material moving from the point of discard to ultimate disposition. **D 1968,** D06

soller slit—a slit containing a set of thin, closely spaced, parallel metal plates used for the purpose of largely eliminating convergent and divergent rays. **E 7,** E04

soller slit—a slit containing a set of thin, closely placed parallel metal plates (or tubes) used for the purpose of largely eliminating convergent and divergent rays. **E 135,** E01

solubility—the extent that one material will dissolve in another, generally expressed as mass percent, or as volume percent or parts per 100 parts of solvent by mass or volume. The temperature should be specified. **D 3064,** D10

solubility, excessive—the tendency of a porcelain enamel frit to dissolve, as a function of time and temperature, in the medium in which it is present in amounts sufficient to adversely affect the rheology of the porcelain enamel slip. **C 286,** B08

solubility parameter, *n*— *of liquids*, the square root of the heat of vaporization minus work of vaporization (cohesive energy density), per unit volume of liquid, at 298 K. **D 4175,** D02

solubility product—$[M^+]\ ^a[X^-]^b/[MX]$ where the brackets indicate the concentrations of the components of the ionization equilibrium $M\ _aX_b \rightleftarrows aM^+ + bX$. For sparingly soluble salts [MX] is essentially unity. **D 6161,** D19

soluble boron in boron carbide—the empirically determined boron that dissolves from boron carbide under specified test conditions. **C 859,** C26

soluble iron—the determination of "soluble iron" used in this test method corresponds operationally to the "complexed and dissolved" iron determination described by Vuorinen et al. in their study of the species of iron released from pyrite oxidation by *T. ferrooxidans*. They found that values of complexed and dissolved iron corresponded closely with "total iron" as determined after hot sulfuric acid digestion of samples, particularly at 1 to 2 % pulp density. **E 1705,** E48

soluble oil, *n*—an oil-rich concentrate that will mix with water to form an emulsion imparting such properties as lubrication, cooling, and corrosion inhibition. **D 4175,** D02

soluble oil, *n*—a combination of mineral oil base and one or more polar organic materials, such as petroleum sulfonates, sulfated vegetable oils, and fatty acids or their salts. **D 4725,** D15

solute phase—a condition of contaminant residence in which contaminants are dissolved in ground water in either the saturated or the vadose zone. **D 653,** D18

solutes—matter dissolved in a solvent. **D 6161,** D19

solution, *n*—a homogeneous or single-phase, variable-composition mixture of one substance (solute) in another (solvent), in which the former is dispersed as separated molecules, ions, or atoms. The solvent or the solution may be solid, liquid, or gas. **C 1145,** C28

solution—in a chemical system, a phase existing over a range of composition. **E 7,** E04

solution annealing—heating an austenitic stainless steel to a temperature that puts the carbides into solution. The steel is held at this temperature long enough to achieve grain growth. It is then quenched in a medium for fast cooling, which prevents most of the carbides from reprecipitating. The process achieves optimum creep strength. **F 1789,** F16

solution cavern—openings in rock masses formed by moving water carrying away soluble materials. **D 653,** D18

solution coating—*in flexible barrier materials*, (*1*) a process in which a substrate is covered with a homogeneous solution containing the coating material, followed by removal of the (usually organic) solvent(s). (*2*) Also, the product resulting from such a process. **F 17,** F02

solution coating—See Terminology F 17. **F 1327,** F02

solution heat treatment, *n*—heating a steel object to a suitable temperature, holding it at that temperature long enough to cause one or more constituents to enter into solid solution, and then cooling it rapidly enough to hold such constituents in solution. **A 941,** A01

solution heat treatment—a thermal treatment of a product to put alloying elements into solution in the base metal by heating into the temperature range of solid solubility, followed by cooling at a sufficient rate to retain them in a supersaturated solid solution. **B 846,** B05

solvent, *n*—any liquid used to dissolve another material. **D 123,** D13

solvent—the liquid part of an aerosol formulation used to dissolve solid or other liquid parts. **D 3064,** D10

solvent—here defined as water. **D 6161,** D19

solvent-activated adhesive, *n*—a dry-film adhesive that is rendered tacky just prior to use by application of a solvent. **D 907,** D14

solvent adhesive, *n*—an adhesive having a volatile organic liquid as a vehicle. **D 907,** D14

solvent cement—an adhesive made by dissolving a plastic resin or compound in a suitable solvent or mixture of solvents. The solvent cement dissolves the surfaces of the pipe and fittings to form a bond between the mating surfaces provided the proper cement is used for the particular materials and proper techniques are followed. **F 412,** F17

solvent-cemented joint—see **joint, solvent-cemented.** **F 412,** F17

solvent cementing—making a pipe joint with a solvent cement. (See **solvent cement**.) **F 412,** F17

solvent cement joint—a joint made by using a solvent cement to unite the components. **F 412,** F17

solvent extraction—the selective separation of one or more constituents contained in a given solvent by intimately contacting the solvent with a second immiscible solvent in which the selected constituents(s) is more soluble. **C 859,** C26

solvent extraction—a process used in refining some mineral insulating oils in which an unrefined or a partially refined petroleum distillate is contacted countercurrently with an immiscible solvent so as to selectively remove undesirable materials from the mineral oil. **D 2864,** D27

solvent holdout—the degree to which a paper base or other substrate can resist penetration of organic liquids either during coating or in the process of liquid development. **F 335,** F05

solvent joint—a joint made by using a solvent to unite the components. **F 412,** F17

solvent relative humidity, *n*—the humidity of air over a drycleaning bath and in equilibrium with the solvent and its small amount of water. **D 123,** D13

solvent-release sealant, *n*—a compound that cures primarily through solvent evaporation. **C 717,** C24

solvent remover—a volatile liquid penetrant used to remove excess penetrant from the surface being examined. **E 1316,** E07

solvent residue—type of dirt found on wafer surfaces after solvent evaporation from the surface that arises either from the solvent itself or material that the solvent has removed from the surface and redeposited. **F 1241,** F01

solvent resistance—a measure of the inertness of base papers and coatings toward the solvent used in the electrostatic coatings and liquid toner systems. **F 335,** F05

solvent-soluble material in paper, *n*—the mass of material that can be extracted from a dry specimen by a specified solvent under prescribed conditions, expressed as a percentage of the original dry mass. **D 1711,** D09

solvent tannage—a tanning system utilizing organic solvents, such as acetone, in place of water to carry the tanning agents. **D 1517,** D31

solvus—the locus of points in a phase diagram, representing the temperature, under equilibrium conditions, at which each composition of a solid phase becomes capable of coexistence with another solid phase, that is, a solid-solubility limit. Usually applied to the terminal solid solution. **E 7,** E04

solvus—locus of points in a phase diagram representing the temperature under equilibrium conditions at which each composition of a solid phase becomes capable of coexistence with another solid phase, that is, a solid-solubility limit usually applied to the terminal solid solution, (E 7, E-4). **E 1142,** E37

sonde, *n*—an elongate cylindrical tool assembly used in a borehole to acquire a geophysical log. **D 121,** D05

sonic anemometer/thermometer, *n*—an instrument consisting of a transducer array containing paired sets of acoustic transmitters and receivers, a system clock, and microprocessor circuitry to measure intervals of time between transmission and reception of sound pulses. **D 1356,** D22

sonication, *n*—the act of subjecting a material to the shearing forces of high-frequency sound waves. **D 6384,** D02

sonic nozzle—a pneumatic or vibratory atomizer in which energy is imparted, at frequencies below 20 kHz, to the liquid.

Note—For frequencies greater than 20 kHz, see **ultrasonic nozzle**.

Note—Alternatively, a pneumatic atomizer in which gas velocities reach or exceed the local speed of sound. **E 1620,** E29

sonic temperature—See **temperature.** **D 1356,** D22

sonic-whistle atomizer—a pneumatic type of sonic or ultrasonic nozzle. **E 1620,** E29

soot, *n*—agglomerations of particles of carbon impregnated with tar, formed in the incomplete combustion of carbonaceous material. **D 1356,** D22

soot fall—See **particle fall.** **D 1356,** D22

sorbed phase—a condition of contaminant residence in which contaminants are adsorbed into the surface of soil particles or absorbed by soil organic matter. **D 653,** D18

sorbent, *n*—a solid or liquid medium in or upon which materials are collected by absorption, adsorption, or chemisorption. **D 1356,** D22

sorbent boom—sorbent material contained or arranged in the form of a long cylinder. **F 818,** F20

sorbent sampling, n—the collection of chemicals from an air or emission sample by allowing the air or emissions to contact a sorbent. **D 1356,** D22

sorbite—an aggregate of carbide and ferrite produced by tempering martensite at temperatures in the vicinity of 600°C and which may be resolved readily at relatively low magnification (for example, 500×).

(Contemporary) With reference to tool steels, an aggregate of carbide and ferrite produced by cooling at a rate too slow for martensite formation and too fast for pearlite formation. **E 7,** E04

sorption, *n*—in general, the taking up of some substance (sorbate) into or on the surface of another (sorbent), without specification of the type of process. **C 1145,** C28

sorption, *n*—the process of taking up or holding a material by adsorption or absorption, or both. **D 123,** D13

sorption, *n*—a process by which one material (the sorbent) takes up and retains another material (the sorbate) by the processes of absorption, adsorption, or chemisorption. **D 1356,** D22

sorption—a process in which fluid molecules are taken up by absorption and adsorption. **D 2652,** D28

sorption—the taking up of gas by absorption, adsorption, chemisorption, or any combination of these processes. **E 1316,** E07

sorting—the process by which sedimentary particles are selectively separated from associated but dissimilar particles by flowing water. **D 4410,** D19

sorting sample, *n*—in waste management, a 100 to 150 kg (200 to 300 lb) portion of a vehicle load of municipal solid waste that is deemed to represent the characteristics of that load. **D 5681,** D34

SOS—abbreviation for **start of scan.** **F 1457,** F05

sound absorption—(*1*) the process of dissipating sound energy. (*2*) the property possessed by materials, objects and structures such as rooms of absorbing sound energy. (*3*) *A;* [L^2]; metric sabin—*in a specified frequency band,* the measure of the magnitude of the absorptive property of a material, an object, or a structure such as a room. **C 634,** E33

sound absorption coefficient, α; [dimensionless]; metric sabin/m^2—*of a surface, in a specified frequency band,* the measure of the absorptive property of a material as approximated by the method of Test Method C 423. *Ideally,* the fraction of the randomly incident sound power absorbed or otherwise not reflected. **C 634,** E33

sound attenuation—the reduction of the intensity of sound as it travels from the source to a receiving location. Sound absorption is often involved as, for instance, in a lined duct. Spherical spreading and scattering are other attenuation mechanisms. **C 634,** E33

sound-deadening board—a specially manufactured cellulosic fiberboard product for use in building construction in wall and floor assemblies to reduce sound transmission. **D 1554,** D07

sound energy density, *D;* [$ML^{-1}T^{-2}$]; J/m^3—the quotient obtained when the sound energy in a region is divided by the volume of the region. The sound energy density at a point is the limit of that quotient as the volume that contains the point approaches zero. **C 634,** E33

sound energy, *E;* [ML^2T^{-2}]; J—energy added to an elastic medium by the presence of sound, consisting of potential energy in the form of deviations from static pressure and of kinetic energy in the form of particle velocity. **C 634,** E33

sounding well—*in grouting,* a vertical conduit in a mass of coarse aggregate for preplaced aggregate concrete which contains closely spaced openings to permit entrance of grout. **D 653,** D18

sound insulation—the capacity of a structure to prevent sound from reaching a receiving location. Sound energy is not necessarily absorbed; impedance mismatch, or reflection back toward the source, is often the principal mechanism. **C 634,** E33

sound intensity, *I;* [MT^{-3}]; W/m^2—the quotient obtained when the average rate of energy flow in a specified direction and sense is divided by the area, perpendicular to that direction, through or toward which it flows. The intensity at a point is the limit of that quotient as the area that includes the point approaches zero. **C 634,** E33

sound isolation—the degree of acoustical separation between two locations, especially adjacent rooms. **C 634,** E33

sound level—*of airborne sound,* a sound pressure level obtained using a signal to which a standard frequency-weighting has been applied. **C 634,** E33

sound level, L_{AF}—where the A designates the frequency weighting and the F designates fast exponential time weighting (the A is replaced by C to designate C-weighting, and the F by either S or I to designate slow or impulse time weighting), [nd], (dB), *n- of airborne sound,* a sound pressure level obtained using a signal to which a standard frequency-weighting and exponential time weighting has been applied.

NOTE—Standard frequency-weightings designated A and C, and exponential time weightings designated fast, slow, and impulses, are defined in ANSI S1.4, Specification for Sound Level Meters.

NOTE—The frequency-weighting and exponential time weighting must be specified unless made clear from the context.

NOTE—The frequency-weighting modifies the amplitude of the signal as a function of frequency to adjust for differences in perception of sound at different frequencies.

NOTE—In symbols, A-weighted sound level L_{AF}, at running time t is:

$$L_{AF}(t) = 10\ log\ \{[1/0.125)\ \int^{t}_{-\infty} P_A^2(v)^{-(t-v)/0.125} dv]/Po^2$$

where 0.125 is the time constant in seconds for fast time weighting (1.0 for slow time weighting), *v* is a dummy variable of integration, $P_A^2(v)$ is the squared, instantaneous, time varying, A-weighted sound pressure in Pascals, and P_o is the reference sound pressure of 20 μPa. **C 634,** E33

sound power level, L_W—*of airborne sound,* ten times the common logarithm of the ratio of the sound power under consideration to the standard reference power of 1 pW. The quantity so obtained is expressed in decibels. **C 634,** E33

sound power, *W;* [ML^2T^{-3}]; W— *in a specified frequency band,* the rate at which acoustic energy is radiated from a source. *In general,* the rate of flow of sound energy, whether from a source, through an area, or into an absorber. **C 634,** E33

sound pressure level, L_p—*of airborne sound,* ten times the common logarithm of the ratio of the square of the sound pressure under consideration to the square of the standard reference pressure of 20 μPa. The quantity so obtained is expressed in decibels. **C 634,** E33

sound pressure, *p;* [$ML^{-1}T^{-2}$]; Pa—a fluctuating pressure superimposed on the static pressure by the presence of sound. In analogy with alternating voltage its magnitude can be expressed in several ways, such as instantaneous sound pressure or peak sound pressure, but the unqualified term means root-mean-square sound pressure. In air, the static pressure is barometric pressure. **C 634,** E33

sound stone—stone which is free of cracks, fissures, or other physical defects. **C 119,** C18

sound transmission class, STC—a single-number rating calculated in accordance with Classification E 413 using values of sound transmission loss. It provides an estimate of the performance of a partition in certain common sound insulation problems. **C 634,** E33

sound transmission coefficient, τ; [dimensionless]—*of a partition, in a specified frequency band,* the fraction of the airborne sound power incident on the partition that is transmitted by the partition and radiated on the other side. **C 634,** E33

sound transmission loss, *TL*—*of a partition, in a specified frequency band,* ten times the common logarithm of the ratio of the airborne sound power incident on the partition to the sound power transmitted by the partition and radiated on the other side. The quantity so obtained is expressed in decibels. **C 634,** E33

sound wood—wood free of any form of decay, incipient or advanced, and from insect holes. **D 9,** D07

sour, *v—in refurbishing textiles,* to neutralize the alkalinity of a material after cleaning by using a weak acid such as acetic acid or an acid-forming salt. **D 123,** D13

source, *n—in color determination,* that which furnishes light or other radiation; real device by which radiant flux is produced (see **illuminant**). **D 2946,** C17

source, *n*—an object that produces light or other radiant flux, or the spectral power distribution of that light. (See also **standard source, CIE standard source *A*, CIE standard source *B*, CIE standard source *C*.**) **E 284,** E12

source—a machine or radioactive material that emits penetrating radiation. **E 1316,** E07

source, CIE standard, *n*—see **standard source**. **D 2946,** C17

source documents—the original materials or facsimiles from which input for a data processing system is derived. **F 149,** F05

source-film distance—the distance between the radiation-producing area of the source and the film. **E 1316,** E07

source marking, *n*—the bar code marking of a specific item at the point of initial production of the item. **F 1294,** F05

source material—any material, except special nuclear material, that contains 0.05 or more of uranium, thorium, or any combination of the two. **C 859,** C26

source room—*in architectural acoustical measurements,* the room that contains the noise source or sources. **C 634,** E33

source sampling—See **sampling.** **D 1356,** D22

source-separated steel cans, *n*—post-consumer products that are generated as separated can fractions by commercial or household sources.

(a) all other steel cans, n—steel containers for food products or liquids, with a maximum capacity of 5 gal, that are not included in one of the other definitions.

(b) bi-metal beverage cans, n—steel cans with nonferrous metal convenience ends (normally made of aluminum), originally containing beer or carbonated beverages, but not including other contaminants.

(c) bi-metal food cans, n—steel cans with nonferrous metal (usually aluminum) convenience ends, originally containing snack foods, but not including other contaminants. **D 5681,** D34

source (X-rays)—the area emitting primary X-rays in a diffraction experiment. The actual source is always the focal spot of the X-ray tube but the virtual source may be a slit or pinhole, depending on the conditions of the experiment. **E 7,** E04

sour gas—natural gas containing concentrations of sulfur compounds which make it impractical to use without purification because of toxicity or corrosive effects, or both, on piping and equipment. **D 4150,** D03

sour, (taste), *adj*—taste produced by substances such as citric acid when in solution. **E 253,** E18

Soxhlet apparatus, *n*—an apparatus for use in extracting organic or inorganic material with a suitable solvent in which the solvent is recirculated by evaporation and subsequent condensation. **D 1356,** D22

space, *n*—the element of a bar code symbol whose reflectance is greater than the global threshold (equivalent to a binary 0) as opposed to the bar or dark element. **F 1294,** F05

space application—dispersion of liquid, gas, or dry particles in an air space in such a manner that target pests are exposed to the chemical. **E 1102,** E35

space categories, *n* **(catégories de locaux) (See also floor area):**

assignable area, n (zone assignable)—**floor areas** of a facility assigned to or available for assignment to occupant groups of functions, including interior walls, building columns, and building projections, and excluding circulation.

basement (sous-sol)—space partly below average grade having less than one half of its clear height (measured from floor level to ceiling level) below average grade.

building core and service area, n (noyau d'un immeuble et aire de service)—**floor area** of a facility necessary for the general operation of a building that is not available for general occupancy, including: **primary circulation areas**, mechanical, electrical, telephone, and custodial rooms serving individual floors; toilet rooms, building lobbies and atria, stairways, elevators, vertical shafts and chases, loading docks; and also central, mechanical, electrical, telephone, and custodial spaces and penthouses, but excluding interstitial area.

building gross area, n (superficie brute d'un immeuble)—sum of the floor areas of a building included within the outside face of outer building walls for all stories.

building service area (aire de service d'un immeuble)—Use preferred term **building core and service area**.

cellar (cave)—space wholly or partly below average grade having more than one half of its clear height (measured from floor level to ceiling level) below average grade.

circulation space— See **primary circulation; secondary circulation**.

primary circulation area (aire de circulation principale)—portion of a **building** that is a public corridor, lobby, or atrium; or is required for access by all occupants on a floor to stairs, elevators, toilet rooms, or building entrances.

secondary circulation area (aire de circulation secondaire) —portion of a **building** required for access to some subdivision of space, whether bounded by walls or not, that does not serve all **occupants** on a **floor**, and that is not defined as **primary circulation area**.

support space (locaux de soutien)—(*in offices*) part of **usable area** not assigned or dedicated to a specific task or function. Support space includes meeting rooms, waiting areas, storage, lounges, operational equipment (for example, computer rooms), copy areas, libraries, and similar areas.

usable area, n (superficie utilisable)—**floor area** of a facility assigned to, or available for assignment to, occupant groups or functions, including interior walls, building columns and projections, and **secondary circulation**.

workplace (poste de travail)—part of a **usable area**, intended for an individual or group to work in.

workspace (aire de travail)—part of **usable area**, intended for a specific function or type of work.

workstation (poste de travail) (aménagement ouvert)—all or part of a **workplace**, suitable for carrying out one function or type of work. **E 1480,** E06

space charge aberration—an aberration resulting from the mutual repulsion of the electrons in a beam. This aberration is most noticeable in low-voltage, high-current beams. This repulsion acts as a negative lens causing rays which were originally parallel to diverge. **E 7,** E04

space compression, *n*—a method of packing a series of space characters into a shorter sequence; to save time for transmission, the receiving device then expands the shorter sequence to its original length. **F 1457,** F05

space lattice—See **lattice**. **E 7,** E04

spacer, *n—in building construction in glazing,* a piece of resilient material placed to maintain space between a pane of glass or a panel and its supporting frame. **C 717,** C24

spacer, *n*—a device made of any material that serves to maintain a predetermined distance between any two points in a **shipping container,** or between any interior part and articles contained therein. (Compare **divider.**) **D 996,** D10

space reflectance (Rs), *n*—in reference to bar codes, the highest reflectance value in a space or quiet zone. **F 1294,** F05

spacing—the distance between adjacent blastholes in a direction parallel to the face. (ISRM) **D 653,** D18

spacing—distance between centers of adjacent crests, measured perpendicular to crests. **F 547,** F16

spacing (between lattice planes)—See **interplanar distance.** **E 7,** E04

spacing of increments, *n*—pertains to the kind of intervals between increments. Two spacing methods are recognized: systematic and random. Systematic spacing is usually preferable.

systematic spacing 1, n—in which the movements of individual increment collection are spaced evenly in time or in position over the lot.

random spacing 2, n—in which the increments are spaced at random in time or in position over the lot. **D 121,** D05

spading—see **shovel slicing.** **C 896,** C04

spall, *n*—a fragment or chip as from concrete, brick, stone or other similar materials. **C 717,** C24

spall, *v*—to break off fragments or chips, as from concrete, brick, stone or other similar materials, by water freezing within the material, corrosion expansion of embedded metal, movement pressures, or other physical or chemical processes. **C 717,** C24

spalling—See **delamination.** **A 902,** A05

spalling, *adj*—the development of spalls. **C 717,** C24

spalling—(*1*) longitudinal splitting in uniaxial compression, or (*2*) breaking-off of plate-like pieces from a free rock surface. (ISRM) **D 653,** D18

spalling, *n*—*in tribology*, the separation of macroscopic particles from a surface in the form of flakes or chips, usually associated with rolling element bearings and gear teeth, but also resulting from impact events. **G 40,** G02

spalling of refractories, *v*—the cracking or rupturing of a refractory unit, which usually results in the detachment of a portion of the unit. **C 71,** C08

spalling of refractories, mechanical, *n*—the spalling of a refractory unit caused by stresses resulting from impact or pressure. **C 71,** C08

spalling of refractories, structural, *n*—the spalling of a refractory unit caused by stresses resulting from differential changes in the structure of the unit. **C 71,** C08

spalling of refractories, thermal, *n*—the spalling of a refractory unit caused by stresses resulting from nonuniform changes of the unit produced by a difference in temperature. **C 71,** C08

spalls—(*1*) fragments or chips from a piece of dimension stone. (*2*) waste stone usually of small size from the quarrying and milling of dimension limestone. **C 119,** C18

spall, spalling, or spontaneous spalling— a defect characterized by chipping that occurs without apparent external causes. **C 286,** B08

span, *n*—the algebraic difference between the upper and lower range-values of an instrument. **E 344,** E20

spandrel glass—architectural glass that is used in a nonvision area or charting of a building. **C 162,** C14

span drift—See **analyzer.** **D 1356,** D22

spangle, *n*—*in hot-dip coatings*, the crystalline structure that develops on a metallic-coated surface when the molten coating metal solidifies, especially on steel sheet and articles coated after fabrication. (Syn. *grain pattern*) **A 902,** A05

span length, *n*—*in length testing of cotton with the Fibrograph*, the distance spanned by a specified percentage of the fibers in the test beard, taking the amount reading at the starting point of the scanning as 100 %. **D 123,** D13

spark, *n*—*in optical emission spectrometry*, a high voltage capacitor discharge. **E 135,** E01

spark coil leak detector—a high-frequency discharge coil of the Tesla type which indicates pin holes in glass vacuum systems by a spark jumping between the core of the coil and the pin hole. **E 1316,** E07

spark, discharge—a non-self-sustaining discharge characterized by high luminosity and a ratio of lateral dimension to length that is substantially smaller than unity. **B 542,** B02

sparking fuels, *n*—*within the context of Test Method D 3175*, fuels that do not yield a coherent cake as residue in the volatile matter determination but do evolve gaseous products at a rate sufficient to mechanically carry solid particles out of the crucible when heated at the standard rate. Such coals normally include all low-rank noncaking coals and lignites but may also include those anthracites, semianthracites, bituminous, chars and cokes that lose solid particles as described above. These are defined as *sparking fuels* because particles escaping at the higher temperatures may become incandescent and spark as they are emitted. **D 121,** D05

sparkle, *n*—the visual contrast between the appearance of highlights on the particles of a gonioapparent pigment and their immediate surround. **E 284,** E12

spark line—not recommended, see **ion line.** **E 135,** E01

spark-source mass spectrometry—measurement of the charge-to-mass ratio of particles ionized by RF spark source heating in which a target is bombarded by sparks at a high frequency; used to determine the elemental composition of a sample in cases where the constituents are mostly of high atomic weight or relatively nonvolatile, or both (for example, heavy metals). **F 1156,** F12

spark test—an electrical test in which a spark is used to detect discontinuity of coating. **C 286,** B08

spar varnish—under **varnish,** see *spar varnish.* **D 16,** D01

spatial averaging—the combination of particle size distributions for regions or locations within a liquid dispersion into a distribution representative of a larger sampling region. **E 1620,** E29

spatial domain filtering, *n*—a filtering operation performed directly on the profile record. **E 867,** E17

spatial frequency—a measure of detail in terms of equivalent, uniformly spaced, cyclical patterns. In an object or image plane, it may be expressed in units of cycles per millimeter (cy/mm) or line pairs per millimeter (lp/mm). In an imaging system, it may be expressed in units of cycles per milliradian (cy/mrad) or line pairs per milliradian (lp/mrad). **E 1316,** E07

spatial grain size—the average size of the three-dimensional grains, as opposed to the more conventional grain size determined by a simple average of observations made on a cross section of the material. **E 7,** E04

spatial resolution—because of ambiguities due to alternative definitions, it is recommended that "spatial resolution" not be used unless a specific definition is provided by the user. **E 1620,** E29

spatial size distribution—the size distribution of particles in a given volume of space, wherein there is no significant variation in the distribution during the sampling interval. **E 1620,** E29

spatter, *n*—a type of extraneous or undesirable ink droplet originating when a portion of an ink droplet strikes the intended area and is deflected to an unintended area. **F 1857,** F05

SPC—standard (heterotrophic) plate count—measurement method for enumerating bacteria. **D 6161,** D19

spear point—symmetrical point with four bevel faces meeting at point center. **F 592,** F16

special calibration board, *n*—a specially assembled noncombustible insulating board used for standardizing the operating condition of the equipment which is used only to measure the flux distribution at specified intervals along the specimen surface. It shall be roughly 20 ± 5 mm in thickness with a density of 750 ± 100 kg/m^3. **E 176,** E05

special calibration dummy specimen, *n*—a dummy specimen made of the same material as the dummy specimen, intended only for use in calibration of flux gradient along the specimen. **E 176,** E05

special cause variation—an intermittent, unpredictable, and unstable variation signaled by a point beyond the control limits, a run, or some other nonrandom pattern of points within the control limits. **F 1789,** F16

special design—a concrete pipe design for sizes, loads, or service conditions not covered by a standard design. **C 822,** C13

special indices of metamerism, *n*— indices of degree of metamerism associated with specific changes in illuminating or viewing conditions, such as change of illuminant or change of observer. **E 284,** E12

specially-shaped brick, *n*—a brick manufactured to a basic shape of other than a rectangular prism. **C 43,** C15

special nuclear material—plutonium, ^{233}U, uranium enriched in ^{233}U or ^{235}U to greater than its natural abundance, and any other materials defined as SNM under the Atomic Energy Act of 1954, as amended. This term does not include source materials. **C 859,** C26

special purpose boom—boom that departs from the general characteristics of "fence type" and "curtain type" booms, either in design or intended use. **F 818,** F20

special-purpose tile—a tile, either glazed or unglazed, made to meet or to have specific physical design or appearance characteristics such as size, thickness, shape, color, or decoration; keys or lugs on backs or sides; special resistance to staining, frost, alkalies, acids, thermal shock, physical impact, high coefficient of friction, or electrical properties. **C 242,** C21

special symbol/character—*in a character set,* a character that is neither a numeral, letter, or a blank, for example, virgule, asterisk, dollar sign, comma, period, etc. **F 149,** F05

specialty felt, *n*—one of a number of special purpose felt structures available for, but not limited to, a specific end-use application. **D 123,** D13

specialty fryer—equipment with a wide and shallow cooking area that allows a layer of food to be lowered into oil on a screen or shallow basket (that is, donut and fish/chicken fryers). See **fryer**. **F 1827,** F26

specific—straight lengths that are uniform in length, as specified, and subject to established length tolerances. **B 846,** B05

specific activity—the number of nuclear disintegrations of a radionuclide per unit time per unit mass. **C 859,** C26

specific adhesion, *n*—See *specific adhesion* under **adhesion**. **D 907,** D14

specific airflow resistance, *r;* [$ML^{-2}T^{-1}$]; mks rayl (Pa·s/m)—the product of the airflow resistance of a specimen and its area. This is equivalent to the quotient of the air pressure difference across the specimen divided by the linear velocity, measured outside the specimen, of airflow through the specimen. **C 634,** E33

specific area, *n—of wool,* the ratio of the fiber surface to fiber volume. **D 123,** D13

specification—a precise statement of a set of requirements to be satisfied by a material, product, system, or service indicating, whenever appropriate, the procedure by means of which it may be determined whether the requirements given are satisfied. As far as practicable, it is desirable that the requirements be expressed numerically in terms of appropriate units together with their limits. **C 859,** C26

specification, *n*—a precise statement of a set of requirements to be satisfied by a material, product, system, or service that indicates the procedures for determining whether each of the requirements is satisfied. **D 123,** D13

specification, *n*—a precise statement of a set of requirements to be satisfied by a material, product, system or service that indicates the procedures for determining whether each of the requirements is satisfied. **D 4439,** D35

specification, *n*—a precise statement of a set of requirements to be satisfied by a material, product, system, or service. **E 631,** E06

specification—a document setting forth pertinent details of a product, such as performance, chemical composition, physical properties and dimensions, prepared for use in, or to form the basis for, an agreement between negotiating parties. **F 869,** F08

specification, *n* (caractérisation (caractérisation technique))—precise statement of a set of requirements to be satisfied by a material, product, system, or service. **E 1480,** E06

specification limits, *n*—see **tolerance limits.** **E 456,** E11

specific capacity—the rate of discharge from a well divided by the drawdown of the water level within the well at a specific time since pumping started. **D 653,** D18

specific criteria , *n*—accreditation criteria related to the ability of a laboratory to conduct specific tests or specific types of tests. **E 1187,** E36

specific discharge, *n* **(LT^{1})**—the rate of flow of water through a porous medium per unit area measured at a right angle to the direction of flow. **D 653,** D18

specific energy for cutting wear [$M\ L^{-1}\ T^{-2}$], *n—in solid impingement erosion,* the kinetic energy of impinging particles associated with removal of unit volume of target material by cutting wear. **G 40,** G02

specific energy for deformation wear [$M\ L^{-1}\ T^{-2}$], *n—in solid impingement erosion,* the kinetic energy of impinging particles associated with removal of unit volume of target material by deformation wear. **G 40,** G02

specific flux—flux divided by net pressure driving force. **D 6161,** D19

specific gravity, *n*—the ratio of mass of a volume of a material at a stated temperature to the mass of the same volume of distilled water at a stated temperature. **C 125,** C09

specific gravity—as applied to wood, the ratio of the ovendry weight of a sample to the weight of a volume of water equal to the volume of the sample at some specific moisture content, as green, air-dry, or ovendry. **D 9,** D07

specific gravity, *n*—(*1*) ratio of the weight of a given volume of substance to the weight of an equal volume of water. (*2*) ratio of the density of a substance at a specified temperature to the density of water. See **density**. **D 16,** D01

specific gravity, *n—of felt,* the relative mass per unit volume of felt expressed as a percentage of the mass per unit volume of water. **D 123,** D13

specific gravity:

specific gravity of solids, G, G_s, S_s (D)—ratio of: (*1*) the weight in air of a given volume of solids at a stated temperature to (*2*) the weight in air of an equal volume of distilled water at a stated temperature.

apparent specific gravity, G_a, S_a (D)—ratio of: (*1*) the weight in air of a given volume of the impermeable portion of a permeable material (that is, the solid matter including its impermeable pores or voids) at a stated temperature to (*2*) the weight in air of an equal volume of distilled water at a stated temperature.

bulk specific gravity (specific mass gravity), G_m, S_m (D)—ratio of: (*1*) the weight in air of a given volume of a permeable material (including both permeable and impermeable voids normal to the material) at a stated temperature to (*2*) the weight in air of an equal volume of distilled water at a stated temperature. **D 653,** D18

specific gravity, *n*—the ratio of the density of the substance in question to the density of a reference substance at specified conditions of temperature and pressure. **D 1356,** D22

specific gravity—the ratio of the mass of a specimen to the mass of an equal volume of water, both at the same specified temperature. Generally, density is more useful. (see also **density**) **D 1695,** D01

specific gravity,, *n*—the ratio of a mass of a given volume of liquid at a given temperature to the mass of an equal volume of pure water at the same temperature. When reporting results, explicitly state the standard reference temperature, for example, specific gravity 15/15°C. (D 1298) **D 2864,** D27

specific gravity, *n— deprecated term,* the ratio of the density of a substance to that of a reference substance such as water (for solids and liquids) or hydrogen (for gases) under specified conditions. See *relative density.* **D 4175,** D02

specific gravity—ratio of the mass of any volume of a substance to the mass of an equal volume of water at 4°C. **D 4410,** D19

specific gravity, *n*—the ratio of the density of the substance in question to the density of a reference substance at specified conditions of temperature and pressure. **D 4439,** D35

specific gravity—the ratio of the mass (density) of a sample material to the mass (density) of an equal volume of water at the same specified temperature. **D 6161,** D19

specific gravity—the ratio of the mass of a unit volume of a material at a stated temperature to the mass of the same volume of gas-free distilled water at a stated temperature. The form of expression shall be the following:

Specific gravity *x*/*y* °C . . .

where:
x = temperature of the material, and
y = temperature of the water.
See also **Baumé gravity** and **density**. **E 1547,** E15

specific gravity, apparent (of solids and liquids)—the ratio of the weight in air of a unit volume of a material at a stated temperature to the weight in air of equal density of an equal volume of gas-free distilled water at a stated temperature. If the material is a solid, the volume shall be that of the impermeable portion. The form of expression shall be the following:

Apparent specific gravity *x*/*y* °*C*

where:
x = temperature of the material, and
y = temperature of the water. **E 1547,** E15

specific gravity in air—the ratio of the weight in air of a unit volume of a material at a stated temperature to the weight in air of equal density of an equal volume of gas free distilled water at a stated temperature. **D 4790,** D16

specific gravity (of solids and liquids)—the ratio of the mass of a unit volume of a material at a stated temperature to the mass of the same volume of gas-free distilled water at a stated temperature. If the material is a solid, the volume shall be that of the impermeable portion. The form of expression shall be the following:

Specific gravity *x*/*y* °C

where:
x = temperature of the material, and
y = temperature of the water. **E 1547,** E15

specific heat, *n*—the ratio of the specific heat capacity of a material to the specific heat capacity of a reference material at the same temperature. **E 1142,** E37

specific heat capacity, C—quantity of heat required to provide a unit temperature change to a unit mass of material. **E 1142,** E37

specific humidity, *q*—the ratio of the mass of water vapor m_v to the total mass $m_v + m_a$ of the moist air:

$$q = \frac{m_v}{m_v + m_a}$$

D 4023, D22

specific humidity *(q), n*—the ratio of the mass of water vapor, m_v, to the total mass, m_v+ m_a, of the moist air:

$$q = \frac{m_v}{m_v + m_a}$$

D 1356, D22

specific inductive capacity—the same as **permittivity, relative**. **D 2864,** D27

specific luminance, n— see **coefficient of retroreflected luminance**. **E 284,** E12

specific normal acoustic admittance, $y \equiv g - jb$; $[M^{-1}L^{2}T]$; reciprocal mks rayl (m/Pa·s)—the reciprocal of the specific normal acoustic impedance. The real and imaginary components are called, respectively, **specific normal acoustic conductance** and **specific normal acoustic susceptance.** **C 634,** E33

specific normal acoustic impedance, $z \equiv r + jx$; $[ML^{-2}T^{-1}]$; mks rayl (Pa·s/m)—*at a surface,* the complex quotient obtained when the sound pressure averaged over the surface is divided by the component of the particle velocity normal to the surface. The real and imaginary components of the specific normal acoustic impedance are called, respectively, **specific normal acoustic resistance** and **specific normal acoustic reactance.** **C 634,** E33

specific storage—the volume of water released from or taken into storage per unit volume of the porous medium per unit change in head. **D 653,** D18

specific storage, *n* **(L^{-1})**—the volume of water released from or taken into storage per unit volume of the porous medium per unit change in head. **D 653,** D18

specific surface—the surface area of one gram of powder, usually expressed in square centimetres. **B 243,** B09

specific surface—the surface per unit mass (or less frequently per unit volume) of a moisture-free sample. The specific external surface is used as a measure of the degree of fineness of fibrillation. **D 1695,** D01

specific surface (L^{-1})—the surface area per unit of volume of soil particles. **D 653,** D18

specific viscosity—see **viscosity, specific.** **D 1695,** D01

specific wear rate, n— see **wear factor**. **G 40,** G02

specific weight (of sediment deposits)— the dry weight of sediment solids per unit volume of deposit in place. Synonymous with **volume-weight.** **D 4410,** D19

specific with ends—specific lengths, including ends. **B 846,** B05

specific yield—the ratio of the volume of water that the saturated rock or soil will yield by gravity to the volume of the rock or soil. In the field, specific yield is generally determined by tests of unconfined aquifers and represents the change that occurs in the volume of water in storage per unit area of unconfined aquifer as the result of a unit change in head. Such a change in storage is produced by the draining or filling of pore space and is, therefore, mainly dependent on particle size, rate of change of the water table, and time of drainage. **D 653,** D18

specified dimensions—dimensions to which masonry units or constructions are required to conform. Actual (measured) dimensions may differ from the specified dimensions by permissible variations. **C 1232,** C15

specified temperature—application temperature specified by the sealant manufacturer. **C 717,** C24

specifying agency—the individual engineer, firm, or political subdivision charged with and having the responsibility for the design of a facility, product, equipment, or material requirements. **C 896,** C04

specifying agency—the individual engineer, firm, or political subdivision charged with and having responsibility for the design of a facility, product, equipment, or material requirements. **F 412,** F17

specimen, *n*—a portion of a sample on which a specific test is performed. (Syn. *test specimen*) **A 902,** A05

specimen—an individual unit on which a test can be made. **C 822,** C13

specimen—sample, or portion thereof, which is to be tested and the test results to be reported. **C 896,** C04

specimen, *n*—a specific portion of a material or laboratory sample upon which a test is performed or which is selected for that purpose. (Syn. *test specimen*). **C 1145,** C28

specimen, *n*—a specific portion of a material or a laboratory sample upon which a test is performed or which is selected for that purpose. **D 123,** D13

specimen—pieces or quantity taken or prepared from a sample for testing. **D 653,** D18

specimen, *n*—a piece or portion of a sample used to make a test. (ISO) **D 883,** D20

specimen—that portion of a sample unit required for a single measurement of a given property or characteristic. **D 1517,** D31

specimen , *n*—a piece of material appropriately shaped and prepared so that it is ready to use for a test. **D 1566,** D11

specimen, *n*—a specific portion of a material or laboratory sample upon which a test is performed or which is taken for that purpose. (*Syn.* test specimen) **D 4439,** D35

specimen, *n*—*in methods of chemical analysis*, a piece of material selected to be typical of the whole under the assumption that the whole is composed of pieces of similar composition. (Contrast **sample**). **E 135,** E01

specimen, *n*—manufactured item of the product, representative prototype of the product, or mock-up of the product. **E 176,** E05

specimen, *n*—the manufactured item of the product, or representative prototype of the product. **E 176,** E05

specimen, *n*—a construction consisting of electrical system components and a fire-resistive barrier system. **E 176,** E05

specimen, *n*—a piece or portion of a sample used to make a test. **E 284,** E12

specimen—See **test specimen**. **E 1605,** E06

specimen, *n*—a piece or portion of a sample used to make a test. (D-20, ISO) **F 412,** F17

specimen center of pin hole distance, H* [L], *n*—the distance between the center of the pin holes on a pin-loaded specimen. **E 1823,** E08

specimen chamber (electron optics)—the compartment located in the column of the electron microscope in which the specimen is placed for observation. **E 7,** E04

specimen charge (electron optics)—the electrical charge resulting from the impingement of electrons on a nonconducting specimen. **E 7,** E04

specimen charging—the accumulation of electrical charge on the specimen caused by particle bombardment. **E 673,** E42

specimen clamp, *n*—the device which is used to transport the fiber test beard of cotton through the gathering, combing, brushing, length, length uniformity, strength, and elongation measurement operations. **D 123,** D13

specimen contamination (electron optics)— the contamination of the specimen caused by the condensation upon it of residual vapors in the microscope under the influence of electron bombardment. **E 7,** E04

specimen distortion (electron optics)— a physical change in the specimen caused by desiccation or heating by the electron beam. **E 7,** E04

specimen gage length, δd [L], *n*— the distince between the points of displacment measure (for example, clip gage, gage length). **E 1823,** E08

specimen grid—See **specimen screen**. **E 7,** E04

specimen holder (electron optics)—a device which supports the specimen and specimen screen in the correct position in the specimen chamber of the microscope. **E 7,** E04

specimen loader, *n*—an instrument which obtains the test specimen by lowering an open specimen clamp onto the surface of the raw cotton samples, closing the jaws of the specimen clamp and ascending with a portion of the fibers retained in the jaws awaiting the combing and brushing procedure. **D 123,** D13

specimen normal, n— see **surface normal**. **E 284,** E12

specimen pathlength— see **sample pathlength**. **E 131,** E13

specimen screen (electron optics)—a disk of fine screen, usually 200-mesh stainless steel, copper, or nickel, which supports the replica or specimen support film for observation in the microscope. **E 7,** E04

specimen span, *S* [L]—distance between specimen supports. **E 1823,** E08

specimen stage—the part of the microscope which supports the specimen holder and specimen in the microscope, and can be moved in a plane perpendicular to the optic axis from outside the column. **E 7,** E04

specimen stage controls—the external controls by means of which the stage can be moved. **E 7,** E04

specimen strain—a distortion of the specimen resulting from stresses occurring during preparation or observation. In electron metallography, strain may be caused by stretching during removal of a replica or during subsequent washing or drying. Also, electrical and thermal stresses caused by the electron beam may arise during observation. **E 7,** E04

specimen temperature, *T* [θ]—*in fatigue testing*, the average temperature in the specimen test section during isothermal testing, or the temperature in the specimen test section at any instant of time during cyclic-temperature testing. **E 1823,** E08

specimen thickness, *B* [L]—the distance between the parallel sides of a test specimen. **E 1823,** E08

specimen width, *W* [L]—the distance from a reference position (for example, the front edge of a bend specimen or the load line of a compact specimen) to the rear surface of the specimen. **E 1823,** E08

specking—the discoloration of an enamel surface due to foreign particles in the fired glass. **C 286,** B08

speckle—the random interference pattern which results from the illumination of an optically rough surface with coherent radiation. In laser systems, it results in the granular effect which can be seen in a scattered beam. **E 1316,** E07

speckled ware—a decorative finish with particles of one color appearing in a uniform background of another color or shade. **C 286,** B08

specks, *n—in woolen fabrics,* small pieces of undyed vegetable matter which can be removed by carbonizing or can be covered by dyeing or inking. **D 123,** D13

spectral, *adj—in color determination,* indicates either a function of wavelength as in spectral transmittance, or spectral concentration, as in spectral flux. **D 2946,** C17

spectral, *adj—(1) modifying a quantity,* descriptor that the quantity is a function of wavelength; (*2*) *for radiometric quantities*, pertaining to monochromatic radiant energy at a specified wavelength or, by extension, to radiant energy within a narrow wavelength band about a specified wavelength. **E 284,** E12

spectral—(*1*) for a radiometric quantity, concentration per unit wavelength (or frequency), indicated by the subscript λ following the symbol for the quantity, as $L\lambda = dL/d\lambda$; at a specific wavelength, indicated by the subscript λ with the wavelength in parentheses, as $L\lambda$ (500 nm). (*2*) For a radiometric property, at a specific wavelength (or frequency), indicated by the wavelength in parentheses, as ρ (750 nm), or as a function of wavelength, indicated by the symbol (λ) following the symbol for the property, as ε(λ). See **radiometric properties** and **quantities.** **E 772,** E44

spectral, *adj*—of or relating to, or made by an array of components of an emission or wave separated and arranged in the order of some varying characteristic, such as wavelength, mass, or energy. **F 1294,** F05

spectral analysis—determination of the distribution of frequencies or wavelengths of transmission or absorption, or both, within the electromagnetic spectrum; analytical method for identification of materials, or of electromagnetic waves (radio or microwaves). **F 1156,** F12

spectral background, *n*—non-specific radiation within the spectrum that is not directly related to the observed line or overlapping lines. **E 135,** E01

spectral band, *n*—an arrangement of a specific set of adjacent wavelengths. **F 1294,** F05

spectral bandwidth, $\Delta\ \bar{\nu}$—the wavelength interval, $\Delta\ \bar{\nu}$, or wavenumber interval, $\bar{\nu}$, of radiant energy leaving the exit slit of a monochromator measured at half the peak detected radiant power. **E 131,** E13

spectral bandwidth, $\Delta\lambda$, *n*—the wavelength interval, $\Delta\lambda$, of radiant energy leaving the exit slit of a monochromator measured at half the peak detected power. **E 284,** E12

spectral characteristic, *n*—the reflectance, reflectance factor, transmittance, or transmittance factor as a function of wavelength, used to characterize a specimen. **E 284,** E12

spectral comparator—an instrument for the inspection and wavelength measurement of spectrograms. **C 859,** C26

spectral comparator, *n*—an instrument for the inspection and wavelength measurement of spectrograms. **E 135,** E01

spectral concentration of a radiometric quantity, *n*—quotient of the quantity, taken over an infinitesimal range on either side of a given wavelength, by the range. $X_e, \lambda = dX_e/d\lambda$.

NOTE—Frequencies, wavenumbers, or their logarithms may also be used; if there is a risk of ambiguity, this should be avoided by means of the wording: "spectral concentration in terms of frequency," etc. (See preliminary remarks at the beginning of "quantities" regarding use of the adjective spectral.) **E 349,** E21

spectral distribution curve, *n*—the curve showing the absolute or relative radiant power emitted or absorbed by a substance as a function of wavelength, frequency, or any other directly related variable. **E 135,** E01

spectral distribution curve of a radiometric quantity (radiant flux, radiant intensity, etc.), *n*—curve representing the spectral concentration of the quantity as a function of wavelength.

NOTE—Commonly the relative spectral distribution curve is used; that is, the curve representing the ratio of the spectral concentration of the quantity to a certain value of the same quantity. **E 349,** E21

spectral line—an image of a slit formed in the focal plane of a spectrograph or spectrometer and having a width approximately equal to that formed by monochromatic radiation. **C 859,** C26

spectral line, *n—in optical emission spectrometry*, an image of an entrance slit formed at the focal plane of a spectrograph or spectrometer by the passage of a discrete wavelength of radiant energy. **E 135,** E01

spectral luminous efficiency function, *V*(**λ**), *n*—the relative effectiveness of radiant power to stimulate the perception of light by the normal human observer, as a function of wavelength: the function adopted as standard by the CIE. **E 284,** E12

spectral position—the effective wavelength or wavenumber of an essentially monochromatic beam of radiant energy. **E 131,** E13

spectral power distribution, *n*—the absolute or relative radiant power emitted by a source, or incident upon a receiver as a function of wavelength. **G 113,** G03

spectral power distribution, SPD, *S*(λ), *n*—specification of an illuminant by the spectral composition of a radiometric quantity, such as radiance or radiant flux, as a function of wavelength. **E 284,** E12

spectral (referring to radiometric quantities) *adj.* —for monochromatic radiation at a specified wavelength (or frequency), or, by extension, for radiation within a narrow wavelength band about a specified wavelength.

NOTE—When certain quantities, such as absorptance for transmittance, etc., are considered for monochromatic radiation, they are functions of wavelength (or frequency, or wave number, etc.). They then may be designated by the same term preceded by the adjective "spectral," and by the same symbol followed by λ (or ν, or σ, etc.) in parentheses; example: spectral transmittance, $\tau(\lambda)$.

Spectral quantities are frequently plotted as a function of wavelength (or frequency) to produce a spectral curve.

If the spectral concentration of a quantity *X* is considered, it also may be designated by way of abbreviation by the name of the quantity preceded by the adjective "spectral" as before, and by the symbol for the quantity with the subscript λ (or ν, or σ, etc.), but it must be remembered that *X* and $X\lambda$are quantities of a different kind because $X\lambda = dX/d\lambda$. **E 349,** E21

spectral resolution, *R*—the ratio $\lambda/\Delta\lambda$ where λ is the wavelength of radiant energy being examined and $\Delta\lambda$ is the spectral bandwidth expressed in wavelength units; or, alternatively, the ratio $\bar{\nu}/\Delta\ \bar{\nu}$ where $\bar{\nu}$ is the wavenumber of radiant energy being examined and $\Delta\ \bar{\nu}$ is the spectral bandwidth expressed in wavenumber units. **E 131,** E13

spectral response—the variation in sensitivity of a device to light of different wavelengths. **F 149,** F05

spectral response—the relative light sensitivity of an electrophotographic layer to light of differing wavelengths. **F 335,** F05

spectral response, *n*—the variation in sensitivity of a reading device to light of different wavelengths. **F 1294,** F05

spectral responsivity, *s(λ), n—of a detector,* quotient of the detector output by the monochromatic input as a function of wavelength. **E 284,** E12

spectral slit width—the mechanical width of the exit slit, divided by the linear dispersion in the exit slit plane. **E 131,** E13

spectrochemical buffer—in spectrochemical analysis, a substance which by its addition or presence tends to minimize the effects of one or more of the elements on the emission of other elements. **C 859,** C26

spectrochemical carrier—in emission spectrochemical analysis, a material added to a sample to facilitate its vaporization into the analytical gap. **C 859,** C26

spectrochemical carrier, *n—in dc-arc spectrometry,* a material added to a specimen to facilitate selective vaporization of analytes into the analytical gap. **E 135,** E01

spectrocolorimeter, *n*—spectrophotometer, one component of which is a dispersive element (such as a prism, grating, or interference filter or wedge) that is normally capable of producing as output only colorimetric data (such as tristimulus values and derived color coordinates) but not the underlying spectral data from which colorimetric data are derived. **E 284,** E12

spectrogoniophotometer, *n*—goniophotometer having the capability of measuring as a function of wavelength; see the preferred term, **goniospectrophotometer**. **E 284,** E12

spectrograph—an instrument with one slit that uses photography to obtain a record of a spectral range simultaneously. The radiant power passing through the optical system is integrated over time, and the quantity recorded is a function of radiant energy. **E 131,** E13

spectrograph, *n*—a spectrometer for optical radiation measurements in which the receiver system detects radiant power simultaneously at many points across the spectral region of interest, for example by use of a photographic medium or an array-type photometric detector. **E 284,** E12

spectrographic—pertains to an analytical determination based upon an indirect measurement of properties of a spectrum (light emission, mass, energy) characteristic of the substance being analyzed. Measurements are usually made from a photographic display of the spectrum. **C 859,** C26

spectrograph (X-ray)—an instrument for recording photographically at predetermined angles the results of diffraction experiments. **E 7,** E04

spectrometer—an instrument for measuring some function of power, or other physical quantity, with respect to spectral position within a spectral range. **E 131,** E13

spectrometer, *n*—an instrument for measuring a specified property as a function of a spectral variable. *In optical radiation measurements,* the spectral variable is wavelength or wavenumber and the measured property is (or is related to) absorbed, emitted, reflected, or transmitted radiant power. See **spectrograph, spectrophotometer, spectroradiometer.** **E 284,** E12

spectrometer:electron energy analyzer

dispersion—AES, the change in electron image position at the exit of a spectrometer per unit change in electron energy.

energy resolution—AES, XPS, the ratio of the full width at half-maximum intensity of the response curve for monoenergetic electrons at a given energy to the energy of the electrons.

transmission—AES, XPS, the ratio of the number of electrons at a given energy transmitted through the spectrometer to the number entering the entrance aperture of the spectrometer at that energy. **E 673,** E42

spectrometer tube—the sensing element of a mass spectrometer leak detector. **E 1316,** E07

spectrometer (X-ray)—(*1*) an instrument similar to a spectrograph but employing a movable X-ray measuring device, such as a Geiger-Müller counter or ionization chamber, instead of a photographic film. The measuring device moves on a circle centered on the spectrometer axis.

(*2*) An instrument for recording, similar to the spectrograph, except that a Geiger-Müller counter, scintillation counter, proportional counter, or ionization chamber substitutes for the photographic recording. **E 7,** E04

spectrometric—pertains to an analytical determination based upon the direct measurement of properties of a spectrum (usually light emission, mass, energy) characteristic of the substance being analyzed. Measurements are made of electrical signals produced that are proportional to the properties being measured. **C 859,** C26

spectrometry, *n*—an analytical technique for the quantitative or qualitative characterization of a sample, based on a comparison of the sample's spectrum with the spectrum of a known compound or with a standard(s) of known composition. **D 1356,** D22

spectrometry, *n*—The branch of physical science treating the theory and practice of the measurement of spectra. **E 131,** E13

spectrophotometer—a spectrometer with associated equipment, so designed that it furnishes the ratio, or a function of the ratio, of the radiant power of two beams as a function of spectral position. The two beams may be separated in time, space, or both. **E 131,** E13

spectrophotometer, *n*—a spectrometer for optical radiation measurements in which the receiver system is a photometer. See **photometric.** **E 284,** E12

spectrophotometer, *n*—instrument for measuring the ratio of two spectral radiometric quantities. **E 349,** E21

spectrophotometry—a method for identification of substances and determination of their concentration by measuring light transmittance in different parts of the spectrum. **C 859,** C26

spectrophotometry, *n*—a method for identification of substances and determination of their concentration by measuring light transmittance in different parts of the spectrum. **D 1356,** D22

spectrophotometry, *n*—quantitative measurement of reflection or transmission properties as a function of wavelength. (See also **abridged spectrophotometry**.) **E 284,** E12

spectroradiometer, *n*—a spectrometer for measuring emitted optical radiant power. **E 284,** E12

spectroradiometer, *n*—instrument for measuring the spectral concentration of radiant energy or radiant power. **E 349,** E21

spectroscopy, *n*—the branch of physical science treating the theory and interpretation of spectra (see Terminology E 135). **E 131,** E13

spectrum, *n*—an actual or notational arrangement of the component parts of any phenomenon, as electromagnetic waves or particles, ordered in accordance with the magnitude of a common physical property, as wavelength, frequency, or mass. **E 131,** E13

spectrum, *n*—the spatial arrangement of components of radiant power in order of wavelength. **E 284,** E12

spectrum:

aligned incidence—EIA, a backscattering spectrum recorded with the analyzing beam aligned with crystallographic axes or planes of the specimen that produce channeling.

random (incidence)—EIA, a backscattering spectrum recorded with the analyzing beam incident on the specimen in a direction such as to produce no channeling. **E 673,** E42 *spectrum analyzer XRF*—see **XRF spectrum analyzer**. **E 1605,** E06

spectrum averaged cross section, $\bar{\sigma}$—the cross section averaged over the energy distribution of the neutron fluence, where the energy limits of integration are chosen according to the neutron spectrum and reaction cross section considered. It is defined as:

$$\bar{\sigma} = \int \sigma(E)\,\Phi(E)\mathrm{d}E \,/ \int \Phi(E)\mathrm{d}E$$

E 170, E10

spectrum, internal reflection—the spectrum obtained by the technique of internal reflection spectroscopy. **E 131,** E13

spectrum loading—*in fatigue loading*, a load-time program consisting of some (or all) unequal peak and valley loads. (Also known as *variable amplitude loading* or *irregular loading*.) **E 1823,** E08

spectrum locus, *n*—the locus of points on a chromaticity diagram representing chromaticities of monochromatic lights of various wavelengths. **E 284,** E12

spectrum of radiation, *n*—(*1*) spatial display of a complex radiation produced by separation of its monochromatic components.

(*2*) composition of a complex radiation. **E 349,** E21

specular, *adj*—same as regular when applied to reflection. **D 2946,** C17

specular, *adj*—pertaining to flux reflected from the surface of an object, without diffusion, at the specular angle. **E 284,** E12

specular—refers to reflection without distortions, for example, from a high quality first surface mirror. **F 1156,** F12

specular angle, *n*—the angle of reflection equal and opposite to the angle of incidence. **E 284,** E12

specular gloss, *n*—the relative luminous fractional reflectance of a specimen in the specular direction. (See **gloss**.) **D 123,** D13

specular gloss, *n*—see **gloss, specular**. **D 2946,** C17

specular gloss, *n*— *in waxed paper and paperboard technology*, the degree to which a surface simulates a mirror in its capacity to reflect incident light. **D 4175,** D02

specular gloss, *n*—(*1*) ratio of flux reflected in specular direction to incident flux for a specified angle of incidence and source and receptor angular apertures.

(*2*) perceived surface brightness associated with the luminous specular (regular) reflection of a surface. **E 284,** E12

specular reflection—reflection without diffusion, in accordance with the laws of optical reflection, as in a mirror. **E 131,** E13

specular reflection, *n*—reflection without diffusion, in accordance with the laws of optical reflection, as in a mirror. **E 284,** E12

specular reflection, *n*—the mirror-like reflection of light from a surface. Also, reflection of light from a surface at an angle equal, but opposite to, the angle of incidence. **F 1294,** F05

specular transmittance, n—See **regular transmittance** (the preferred terminology). **D 883,** D20

speed—commonly used term to describe the relative light sensitivity of photoconductive papers. **F 335,** F05

speed effect—the phenomenon in electromagnetic testing of which the evidence is a change in the signal voltage resulting from a change in the relative motion between the specimen and a test coil assembly. **E 1316,** E07

speed (photographic or film) (see sensitivity)—a measure of the response of sensitivity of the material to light, often expressed numerically according to one of several systems, for example, H. and D., D.I.N., Scheiner, and American Standard speed. **E 7,** E04

speed relating to rotary disc floor machines—low speed: up to 800 R.P.M., high speed: more then 800 R.P.M. but less than 1500 R.P.M., and ultra high speed: 1500 R.P.M. or more.

NOTE—Effectiveness of the floor machine depends upon machine weight and pad diameter as well as R.P.M.'s. **D 2825,** D21

spelk—rod-like close packed assemblage, of asbestiform structure, and of generally uniform diameter, that can be fiberized readily. **D 2946,** C17

spew line, *n*—line on the surface of a molded product at the junction of the mold parts. **D 1566,** D11

spew, spue—any constituent of leather that comes to the surface in the form of a white crystallized or dark gummy deposit. **D 1517,** D31

spherical aberration—a lens defect in which image-forming rays passing through the outer zones of the lens focus at a distance from the principal plane different from that of the rays passing through the center of the lens. **E 7,** E04

spherical powder—globular shaped particles. **B 243,** B09

spherical projection—a projection in which the orientation of a crystal plane is represented by the point where the plane normal intersects a sphere drawn with the crystal as the center. **E 7,** E04

spherical wave—wave in which wave fronts are concentric spheres. **D 653,** D18

sphericity—the ratio of the surface area of a hypothetical sphere of the same volume as the particle to the actual surface area of the particle. (A more convenient expression is the ratio of the diameter of a circle with an area equal to that of the projection of a grain when it rests on its larger face to the diameter of the smallest circle circumscribing this projection). (Shape factor). **D 4410,** D19

spheroidite—a coarse aggregate of carbide and ferrite usually produced by tempering martensite at temperatures slightly below the eutectoid temperature. Generally, any aggregate of ferrite and large spheroidal carbide particles no matter how produced. **E 7,** E04

spheroidize annealing—a type of subcritical annealing used to soften steel and improve machinability. Heat treating fine pearlite for a long time just below the lower critical temperature of the steel, followed by a very slow cooling, produces a spheroidal or gobular form of the pearlite. **F 1789,** F16

spheroidizing, *n*—heating and cooling a steel object to produce a spheroidal or globular form of carbide in its microstructure. **A 941,** A01

sphincter seal—*in the nuclear industry*, a ring-shaped closure through which cylindrical objects are introduced into a glovebox without violating the contamination barrier. **C 859,** C26

spicule (in asbestos)—acicular particle of nonfibrous minerals, superficially resembling a spelk, but which shatters rather than fiberize when subjected to mechanical force. **D 2946,** C17

spider—(*1*) a wheel-like casting consisting of a rim and radial spokes on which are mounted felt polishing pads.

(*2*) assembly of radiating tie rods on the top of a furnace. **C 162,** C14

spider—a defect characterized by a starshaped fracture in the porcelain enamel. **C 286,** B08

spigot—see **male end of pipe.** **C 822,** C13

spigot—that portion of a vitrified clay pipe that fits into the bell or socket of the preceding pipe. **C 896,** C04

spike—*in an epitaxial wafer surface*, a tall, thin dendrite or crystalline filament that often occurs at the center of a recess. **F 1241,** F01

spiked sample—a sample portion into which a known amount of analyte has been added. **D 1129,** D19

spiked sediment—a sediment to which a material has been added for experimental purposes. **E 943,** E47

spill—see **flare.** **F 335,** F05

spillback nozzle—see **by-pass nozzle.** **E 1620,** E29

spill decay—see **flare decay.** **F 335,** F05

spill nozzle—see **by-pass nozzle.** **E 1620,** E29

spill return nozzle—see **by-pass nozzle.** **E 1620,** E29

spin axis, *n*—*of a wheel*, the axis of rotation of a wheel. **F 538,** F09

spindle—See **railing systems.** **E 631,** E06

spindle, *n*—a tapered baluster of circular cross-section(s), having a center diameter larger or smaller than the diameter at its ends. **E 1481,** E06

spindle—a component or part that connects the dial and drive cam. **F 471,** F12

spinning cup atomizer—see **rotary cup atomizer**. **E 1620,** E29

spinning disk atomizer—see **rotary disk atomizer**. **E 1620,** E29

spinodal curve—a graphical representation of the realizable limit of the super-saturation of a solution. **E 7,** E04

spinodal heat treatment—a thermal treatment of a solution-heat-treated product to produce property changes such as hardening, strengthening, and conductivity increase by spinodal decomposition of a solid solution. This treatment has also been called "age hardened," "spinodal hardened," or "spinodally decomposed." **B 846,** B05

spin orbit splitting—the splitting of p-, d-, or f-levels arising from coupling of the spin and orbital angular momentum. **E 673,** E42

spinout, *n*—*in tire testing*, a type of uncontrollability defined by a loss of steering control due to rapid or substantial yaw, or both. **F 538,** F09

spin-spin coupling constant (NMR) *J*—a measure, expressed in hertz (Hz), of the indirect spin-spin interaction of different magnetic nuclei in a given molecule. **E 131,** E13

spin velocity, *n*—the angular velocity of the wheel about its spin axis. **F 538,** F09

spin welding (machine)—a joining technique used for bonding thermoplastic parts by frictional heat caused by differential rotation of one or both pieces. Rotation is then stopped and pressure held until heat is dissipated. Also known as frictional welding. **C 904,** C03

spiral—"orange peel" type surface texture with a spiral pattern. **F 1241,** F01

spiral grain—See **grain.** **D 9,** D07

spiral thread—misnomer for helical thread. **F 547,** F16

spiral wound cartridge—a crossflow membrane element design consisting of a product tube, flat membrane leaves, feed channel spacers, anti-telescoping devices, and brine (concentrate) seal. **D 6161,** D19

spiral wound membrane—a flat sheet membrane with one or more feed channel spacers and barrier layers, all of which are rolled into a spiral configuration. **D 6161,** D19

spirits of turpentine, *n*—the volatile oil consisting primarily of a number of terpene hydrocarbons of the general formula $C_{10}H_{16}$.

Canada turpentine—same as Canada balsam, the oleoresin obtained from trees of the Canadian silver fir (*Abies balsamea*).

sulfite turpentine—this term is not in good usage, because the volatile oil recovered in the conversion of wood to pulp by the sulfite process consists chiefly of cymene ($C_{10}H_{14}$) rather than pinene and other terpenes.

Venice turpentine—same as larch turpentine, the oleoresin of the European larch tree, *Larix europea* or *Larix decidua*—the term is also now used widely to describe the clear yellow liquid portion of pine oleoresin, or a synthetic product of similar composition made by dissolving rosin in a terpene solvent. **D 804,** D01

spirit varnish—under **varnish,** see *spirit varnish.* **D 16,** D01

spirometer, *n*—a displacement gasometer consisting of an inverted bell resting upon or sealed by liquid (or other means) and capable of showing the amount of gas added to or withdrawn from the bell by the displacement (rise or fall) of the bell. **D 1356,** D22

spitout—a glaze defect of the pinhole type developed in the decorating kiln, as a result of the evolution of minute gas bubbles from body or glaze. **C 242,** C21

splash cup atomizer—an impact atomizer in which one or more liquid jets are directed against the internal surface of a cup. **E 1620,** E29

splashover—oil splashing over a boom's freeboard. **F 818,** F20

splash plate atomizer—an impact atomizer in which one or more liquid jets are directed against a flat plate. **E 1620,** E29

splay—deposits of flood debris, usually of sand, scattered on the flood plain. **D 4410,** D19

splice—see **joint.** **B 354,** B01

splice—the joint between two lengths of base film, fabric, or paper. **F 221,** F05

splice, *n*—the joint formed either by overlapping or butting the ends of a given tire component in the course of assembling the tire. **F 538,** F09

splice connections, *n*—joining of consecutive coils. **F 1379,** F14

splice crack, *n*—a crack originating at a splice. **F 538,** F09

splice opening, *n*—a parting of a splice along the interface of the assembled ends of a given component. **F 538,** F09

splice plate—a plate used for fastening and joining members. See also **gusset.** **E 631,** E06

splice (reinforcement)—junction of two sections of parallel reinforcement intended to provide continuity and to transfer forces between the two sections. **C 822,** C13

splicing tail, *n*—the barb clusters extending beyond the last attachment at both ends of the coil. **F 1379,** F14

spline key—a part that secures the position of drive cam on a spindle in relation to correct dial calibrations. **F 471,** F12

spline or tongue—that part of a reglet-type lock-strip gasket which is designed to be installed in a reglet in supporting material. **C 717,** C24

split—an imperfection; a crack or check going from surface to surface of a glass article. **C 162,** C14

split—a separation of the wood parallel to the fiber direction, due to the tearing apart of the wood cells. **D 9,** D07

split—a membrane tear resulting from tensile stress. **D 1079,** D08

split—a term used to describe the portion of hide or skin, split into two or more thicknesses, other than the grain or hair side. Splits are usually named according to their sequence of production, such as "main," "second," or "slab" split (in case of upholstery leather), or for the use to which they are to be put, such as "flexible" (for innersoles); "glove," "waxed" (for cheap shoe-uppers); "bag and case" (finished with pyroxylin or pigment finish), sole, etc. **D 1517,** D31

split die, *n*—a die made of parts which can be separated for ready removal of the compact. **B 243,** B09

split feed—a liquid-phase adsorption process in which a powdered adsorbent is added to the solution to be treated in two or more steps, with or without intermediate separation of the adsorbent. **D 2652,** D28

split-level house—See **building.** **E 631,** E06

split sample—a single sample separated into two or more individual parts in a manner that each part is representative of the original sample. **D 4410,** D19

split screen presentation—two or more discrete images displayed on the same screen at the same time. **F 1156,** F12

split sheet—See **nineteen-inch selvage**. **D 1079,** D08

split spacing grouting—a grouting sequence in which initial (primary) grout holes are relatively widely spaced and subsequent grout holes are placed midway between previous grout holes to ''split the spacing.''; this process is continued until a specified hole spacing is achieved or a reduction in grout take to a specified value occurs, or both. **D 653,** D18

split-stitch, *n—in knitted fabrics*, a stitch in which one part of the yarn is knit and the other part is dropped. **D 123,** D13

split-μ test, *n*—a wet traction or stopping distance test conducted on a test course with substantially different wet friction levels for the left and right tire test lanes. **F 538,** F09

splitting—(*1*) cutting leather into two or more layers.

(*2*) cutting a hide into two sides preparatory to tanning. **D 1517,** D31

splitting resistance, *n—of felt*, the force required to overcome the interfacial strength of a material and specifically to separate a felt into two layers (of approximately equal thickness). **D 123,** D13

spodumene (alpha spodumene)—a lithium mineral of the theoretical composition $Li_2O{\cdot}Al_2O_3{\cdot}4SiO_2$ (monoclinic crystallization) which on heating inverts to beta spodumene, a form having very low or nil thermal expansion. **C 242,** C21

sponge, *n*—a form of metal characterized by a porous condition that is the result of decomposition or reduction of a compound without fusion. **B 899,** B02

sponge iron—a coherent, porous mass of substantially pure iron produced by solid state reduction of iron oxide (for example, iron ore or mill scale). **B 243,** B09

sponge iron powder—ground and sized sponge iron, which may have been purified or annealed or both. **B 243,** B09

sponge rubber, *n—in building construction*, an elatomeric open-cell material made by incorporating a gas-producing ingredient into a rubber compound, which expands the rubber compound during vulcanization into a material containing cells that are predominately interconnecting. **C 717,** C24

spongy—a porous condition in metal powder particles usually observed in reduced oxides. **B 243,** B09

spongy enamel—a defect characterized by masses of bubbles occurring in local areas giving rise to a spongy appearance. **C 286,** B08

spontaneous chipping— see **spall, spalling, or spontaneous spalling.** **C 286,** B08

spontaneous heating or self heating, *n*—an exothermic reaction of a material due to slow or incomplete reaction that results in a temperature rise above that of its surroundings. **E 1445,** E27

spontaneous heating temperature or self-heating temperature, *n*—the lowest temperature at which spontaneous or self heating occurs under the specified test conditions. This temperature refers to a much earlier stage of reaction than that associated with the autoignition temperature of the material. **E 1445,** E27

spontaneous ignition, *n*—unpiloted ignition caused by an internal exothermic reaction (compare **piloted ignition**). **E 176,** E05

spoon proof—test ladle specimen taken during various stages of melting and fining. **C 162,** C14

sporinite—See *sporinite* under **maceral**. **D 121,** D05

sport shirt, *n—for boys*, a shirt made in numerical sizes, designed for informal wear and may be worn with or without a jacket. **D 123,** D13

sport shirt, *n—for men*, a shirt designed for informal wear and made with body sizes such as small, medium, or large; they may be worn with or without a jacket. **D 123,** D13

spot, *n*—a small discolored area on, or in, a fabric. **D 123,** D13

spot and stain removal, *n*—a cleaning procedure for localized areas with cleaning agents and mechanical action specific to the removal of the foreign substances present. **D 123,** D13

spot clean, *n*—to remove spots and stains by treating them with cleaning agens and mechanical actions specific to the fiber, fabric, product type, and the foreign material present **D 123,** D13

spot mopping—See **mopping**. **D 1079,** D08

spots—areas outside the maximum **COL**, which are contrasting with the background. **F 149,** F05

spots, *n—in reference to bar code*, unwanted dark areas in the spaces, quiet zones and intercharacter gaps (if applicable) of a symbol that may be caused by presence of extraneous ink, printing errors, or dirt. **F 1294,** F05

spot size, *n*—the diameter of the focused image of the emitter in scanners that use apertured optical systems. **F 1294,** F05

spotting out—the delayed appearance of spots and blemishes on plated or finished surfaces. **B 374,** B08

spot treatment—application of a chemical to a small restricted area, usually to control the spread of a pest. **E 1102,** E35

spot welds, *n*—means of permanently joining two adjacent coil loops by fusing them through a heat or electrical process. **F 1379,** F14

spout—the part of a feeder that carries the orifice, revolving tube, needle, etc. **C 162,** C14

spray—the dispersed discharge from an aerosol-type dispenser in the form of small droplets or particles. (This does not include foam-type dispensers.) **D 3064,** D10

spray, *n*—a dynamic collection of particles, usually generated by the process of atomization. **E 1620,** E29

spray, *v*—to disperse or apply as a spray. **E 1620,** E29

spray, *n*—a type of extraneous or undesirable ink dot near the printed zones which originate from the printhead **F 1857,** F05

spray angle, *n*—the plane angle formed by the profile of a spray pattern. **E 1620,** E29

spray buffing—the restorative maintenance of a previously polished floor, by the action of a suitable floor polishing machine immediately following the mist-spraying of an appropiate product onto the surface whereby the wet application is buffed to dryness. **D 2825,** D21

spray characterization, *n*—the process of describing a spray, based on theory or measurement, in terms of parameters such as liquid flow or flux, patternation, particle size and velocity. **E 1620,** E29

spray classification—classification of sprays by droplet size. **E 1102,** E35

spray coating—an aerosol spray product for surface application, that leaves a residual clear or pigmented finish for protective or decorative purposes. **D 3064,** D10

spray deposit rate—the amount of spray liquid deposited per unit area. Mean deposit rate is the average amount of deposit over the entire spray swath. Effective spray deposit rate is the mean deposit from center to center of adjoining swaths. **E 1102,** E35

spray drift—the movement of airborne spray particles from the intended application (target) area. **E 609,** E35

sprayed width per nozzle—the effective width sprayed by a single nozzle. (For broadcast spraying it is the nozzle spacing; for band spraying it is the band width, for row crop spraying it is the number of nozzles per row divided by the row width). **E 1102,** E35

sprayer attachment—an apparatus consisting of the individual components of a sprayer (tank, pressure source, pressure controls, spray liquid lines, pressure nozzles, etc.) in an arrangement to permit its mounting on another implement to permit application of spray at the same time another field operation is performed. **E 1102,** E35

spray lime—a hydrated lime of such fineness that at least 95 % of the particles will pass a No. 325 (45-μm) sieve. **C 51,** C07

spray pattern, *n*—the flux distribution of liquid discharge by an atomizer. **E 1620,** E29

spray probe—in leak testing, a device for directing a small jet of tracer gas on an object under vacuum testing. **E 1316,** E07

spray quenching, *n*—**quenching** in a spray of liquid. **A 941,** A01

spray rate—the amount of spray liquid emitted by an application unit during treatment, expressed in volume per unit treated. (For area treatment, L/ha, or gal/A; for space treatment, mL/m^3, or oz/1000 ft^3; for individual units, L/plant, mL/animal, or gal/tree). **E 1102,** E35

spray sagging—a process defect characterized by a wavy line or lines appearing on vertical surfaces of sprayed ware prior to drying. **C 286,** B08

spread, *n*—the quantity of adhesive per unit joint area applied to an adherend, usually expressed in pounds of adhesive per thousand square feet of joint area.

double spread, n—application of adhesive to both adherends of a joint.

single spread, n—application of adhesive to only one adherend of a joint. **D 907,** D14

spread—extent of image expansion in comparison to dimension of imaging device. **F 549,** F05

spreader—See **loading.** **D 996,** D10

spreader—a material which increases the area that a droplet of a given volume of spray mixture will cover on a target. **E 1519,** E35

spreader/sticker—a material that has the properties of both a spreader and a sticker. **E 1519,** E35

spreading—the action of flowing out over a surface during application. **D 2825,** D21

spreading rate—the area covered by a unit volume of coating material frequently expressed as square feet per gallon. **D 16,** D01

spreading resistance, R_s (Ω)—*of a semiconductor*, the ratio of (*1*) the potential drop between a small-area conductive metal probe and a reference point on the semiconductor, to (*2*) the current through the probe. **F 1241,** F01

spready hide—a hide of large area in proportion to the mass. **D 1517,** D31

springback—see **green expansion.** **B 243,** B09

spring characteristics, c (FL^{-1})—ratio of increase in load to increase in deflection:

$$c = l/C$$

where:

C = compliance.

D 653, D18

spring-clip staple—staple for fastening flat metal clip holding undulated upholstery spring. Also, staple used to attach coiled springs to frame in bedding manufacture where staple serves as clip. **F 592,** F16

spring crown—hip crown; designed to flatten during driving in order to provide optimum lateral guidance in driving channel of stapler. (See **high crown; "V" crown.**) **F 592,** F16

spring head—See **cupped oval head, umbrella head.** **F 547,** F16

springline—the points on the internal surface of the transverse cross section of a pipe intersected by the line of maximum horizontal dimension; or in box sections, the mid-height of the internal vertical wall. **C 822,** C13

spring line—the line of maximum horizontal dimension of the transverse cross section. **C 896,** C04

spring line—a line along the length of the pipe at its maximum width along a horizontal plane. **F 412,** F17

spring wire, *n*—round wire intended especially for the manufacture of springs. **B 899,** B02

springwood—See **earlywood.** **D 9,** D07

spring zone—that location within a melting furnace having the highest temperatures and an up-welling convective flow of molten glass due to density differences. See **hot spot**. **C 162,** C14

sprinkle mopping—See **mopping**. **D 1079,** D08

sprue, *n*—(*1*) the primary feed channel that runs from the outer face of an injection or transfer mold to the mold gate in a single cavity mold.

(*2*) the piece formed in a primary feed channel or sprue. **D 883,** D20

sprue, *n*—material remaining in the sprue hole after molding. **D 1566,** D11

sprue, *n*—(*1*) the primary feed channel that runs from the outer face of an injection or transfer mold, to the mold gate in a single cavity mold or a runner in multiple-cavity mold; (*2*) the piece of material formed in the primary feed channel opening. **F 412,** F17

sprue hole, *n*—the passageway through which a rubber is forced into a mold. **D 1566,** D11

sprue mark—a mark, usually elevated, left on the surface of an injection or transfer molded part after removal of the sprue or material in filling channel. **F 869,** F08

sprung mass—the total mass minus the unsprung mass. **E 867,** E17

spud—to remove the roofing aggregate and most of the bituminous top coating by scraping and chipping. **D 1079,** D08

spun glass—See **fiber**. **C 162,** C14

spun yarn, *n*—*in a staple system*, a continuous strand of fibers held together by some binding mechanism. **D 123,** D13

sputter depth profile—*AES*, the compositional depth profile obtained when material is removed by sputtering as a result of ion bombardment. **E 673,** E42

sputter depth profile—*SIMS*, the compositional depth profile obtained when material is removed by sputtering as a result of primary bombardment. **E 673,** E42

sputtered neutral mass spectrometry (SNMS)—a technique in which a mass spectrometer is used to measure the mass-to-charge ratio and abundance of secondary neutral species emitted from a target as the result of particle bombardment. These neutral species are analyzed by using plasma, electron, or photon ionization methods. **E 673,** E42

sputtering—the production of specimens in the form of thin films by deposition from a cathode subjected to positive ion bombardment. **E 7,** E04

sputtering—the phenomenon which occurs when atoms and ions are ejected from the specimen as a result of particle bombardment.

equilibrium surface composition—the steady-state surface composition produced by sputter-etching a homogeneous specimen under nonvarying conditions for the ambient vacuum and the primary beam.

fractional yield—*SIMS*, the ratio of the number of atoms and ions of a particular element to the total number of atoms and ions ejected from the specimen.

preferential—the phenomenon which may occur when the sputtering of multicomponent specimens causes a change in the equilibrium surface composition of the specimen.

rate—the amount of specimen material removed per unit time as a result of particle bombardment.

yield—the number of atoms and ions ejected from the specimen per incident ion. **E 673,** E42

spyndle number, *n*—*in jute*, a direct yarn-numbering system for jute rove and jute yarns in which the number of pounds per spyndle, or 14 400-yd (13 167-m) length, is expressed as pounds per spyndle. (See also **direct yarn numbering system.**) **D 123,** D13

square—a roof area of 9.29 m^2 (100 ft^2), or enough material to cover 9.29 m^2 of deck. **D 1079,** D08

square, *n*—an equal-sided rectangular **bar** or **tube** having sharp or slightly rounded edges. **E 631,** E06

square-cut glass—optical glass cut in small squares, separated and designated by weight. **C 162,** C14

squareness, *n*—*of paper dimensions*, the variation of the angle of the corner of a sheet of paper from 90°. **D 1968,** D06

square point—sheared square point; also known as no point or square-cut point. **F 547,** F16

square point—point sheared perpendicular to staple-leg axis to form a pointless staple-leg end; known as blunt point. **F 592,** F16

square root normal distribution—drop size distribution described by the following equation in derivative form:

$$df_n/dD = \{1/[2\sqrt{(2 \times \pi \times \sigma_s \times D)}]\} \times exp\{-[(\sqrt{D}) - (D)_{med}]^2/[2 \times \sigma^2]\}$$

where:

f_n = the number fraction of drops of diameter less than D, and
σ_s = the "standard deviation" from $(\sqrt{D})_{med}$.

E 1620, E29

square shank—term applied to nails with shank of essentially square cross section with or without longitudinal flutes or diagonal barbs; usually found on nails made from square wire. Shank-diameter of nails measured across diagonal (see **diameter**). **F 547,** F16

square spray nozzle—a pressure atomizer with a special orifice outlet configuration designed to create a spray whose cross section approaches a square. **E 1620,** E29

square-wire nail—bright, diagonally barbed, square-wire, regular-stock-steel, 2 by 0.113 to 4 by 0.192-in. common nails with ¼ to ⅜-in. flat head and medium diamond point; also, 2 by 0.099 and 2½ by 0.113-in. box nails with ¼ and 17⁄64-in. flat head and medium diamond point; also, 2 by 0.091 and 2 ½ by 0.099-in. finish nails with 0.124 and 0.131-in. flat head and medium diamond point; also, 1 ½ by 0.131-in. truss nail with 9⁄32-in. flat head and medium diamond point. **F 547,** F16

squealer—Same as audible leak indicator. **E 1316,** E07

squeegee oil—a liquid mixture of organic materials used as the vehicle in squeegee paste. **C 286,** B08

squeegee paste, screening ink, screening paste—a mixture of squeegee oil and finely divided inorganic materials such as color oxides and fluxes. **C 286,** B08

squeeze-out, *n*—adhesive pressed out of the joint during the bonding process. **D 907,** D14

S_{10}–S_{18}—an estimate of the portion of degraded cellulose in a pulp obtained by subtracting S_{18} from S_{10}. **D 1695,** D01

stability—(*1*) resistance to devitrification.
(*2*) chemical durability, resistance to weathering. **C 162,** C14

stability—the condition of a structure or a mass of material when it is able to support the applied stress for a long time without suffering any significant deformation or movement that is not reversed by the release of stress. (ISRM) **D 653,** D18

stability—resistance to overturning moment. **F 818,** F20

stability factor (stability number), N_s **(D)**—a pure number used in the analysis of the stability of a soil embankment, defined by the following equation:

$$N_s = H_c\gamma_e/c$$

where:

H_c = critical height of the sloped bank,
γ_e = effective unit of weight of the soil, and
c = cohesion of the soil

NOTE—Taylor's "stability number" is the reciprocal of Terzaghi's" stability factor." **D 653,** D18

stability to thermal oxidation, *n*—*for polyolefin monofilaments*, the time-to-failure, when polyolefin monofilaments exposed to circulating air, at 125°C. **D 123,** D13

stabilization—a treatment of magnetic material designed to increase the permanency of its magnetic properties or conditions. **A 340,** A06

stabilization—see **soil stabilization**. **D 653,** D18

stabilization annealing—heating an austenitic stainless steel used in severe aqueous corrosion environments by first solution annealing and then reheating to about 1600°F, and holding at that temperature. The treatment causes grain boundary precipitation of carbides, but also the hold time permits the chromium remaining in the austenite solution to redistribute within the grains, restoring corrosion resistance, even adjacent to the grain boundaries. **F 1789,** F16

stabilized borehole liquid level—the borehole liquid level which remains essentially constant with time, that is, liquid does not flow into or out of the borehole. **D 653,** D18

stabilizer—a metal, but not necessarily the matrix, in electrical contact with a superconductor, to act as an electric shunt in the event that the superconductor reverts to the normal state. **B 713,** B01

stabilizer—an ingredient added to a plastic to retard possible degradation. **F 412,** F17

stabilizer, *n*—materials added to the coating mixture or used in a topcoat to reduce the reversibility and degradation of the color reaction and unreacted background. **F 1623,** F05

stabilizer, dry rubber, *n*—a substance present in or added to raw rubber to maintain properties at or near their initial values during its production, processing, and storage. **D 1566,** D11

stabilizer, (latex), *n*—a substance occurring naturally in or added to latex that prevents agglomeration or coagulation of the rubber particles during processing and/or compounding. **D 1566,** D11

stabilizer (rubber), *n*—a substance present in or added to raw rubber to maintain the properties at or near their initial values during drying, processing, and storage. **D 1566,** D11

stabilizing treatment, *n*—any treatment intended to stabilize the microstructure or dimensions of a steel object. **A 941,** A01

stable crack extension [L], *n*—a displacement-controlled crack extension beyond the stretch-zone width. The extension stops when the applied displacement is held constant. **E 1823,** E08

stable fabric, *n*—a textile fabric, the dimensions of which do not change significantly with multiple passes through measuring devices. **D 123,** D13

stacker—See **lehr loader**. **C 162,** C14

stacker—a device for accumulating processed documents in optical scanners and card readers. **F 149,** F05

stacking chair, *n*—chair that is intended to be stacked when not in use. **E 176,** E05

stacking fault—a two dimensional defect that results from a deviation from the normal stacking sequence of atoms in a crystal. **F 1241,** F01

stacking sequence—the arrangement of ply orientations and material components in a laminate specified with respect to some reference direction. **D 3878,** D30

stacking sequence—the order in which each individual ply is layed up, or stacked, on the tool. **E 1749,** E06

stack vent—a vertical outlet in a built-up roofing system to relieve the pressure exerted by water vapor between the roofing membrane and the vapor retarder or deck. **D 1079,** D08

stage—*in grouting*, the length of hole grouted at one time. See also **stage grouting**. **D 653,** D18

stage—a sequestial arrangement of pressure vessels, usually reject staged such as 2:1 array, sometimes permeate staged as in double pass RO. **D 6161,** D19

stage—a device for holding a sample in the desired position in the optical path. **E 7,** E04

stage grouting—sequential grouting of a hole in separate steps or stages in lieu of grouting the entire length at once; holes may be grouted in ascending stages by using packers or in descending stages downward from the collar of the hole. **D 653,** D18

stage micrometer—a graduated scale used on the stage of a microscope for calibration. **E 7,** E04

stage micrometer—calibrated scale for length measurement in conjunction with a microscope. **F 1241,** F01

staggered nested experiment, *n*—a nested experiment in which the nested factors are run within only a subset of the versions of the first or succeeding factors. **E 456,** E11

staggered nested experiment, *n*—a nested experiment in which the nested factors are run within only a subset of the versions of the first or succeeding factors. **E 1325,** E11

stagger wound—a coil in which the turns are positioned into layers approximately parallel to the axis of the coil, but not necessarily with the fixed regularity of a level or traverse wound coil. **B 846,** B05

staging—see **brine staging** and **product staging**. **D 6161,** D19

stagnation conditions—in solar energy systems, the conditions (that is, temperature and pressure) existing when energy system has attained a quasi-steady state after the flow of heat-transfer fluid has stopped, but the absorber continues to receive significant solar irradiance. **E 772,** E44

stain—(*1*) an imperfection; chemical corrosion on the surface of glass.

(*2*) transparent coloration achieved by application of metallic ions on glass at an elevated temperature. **C 162,** C14

stain—a discoloration in wood that may be caused by such diverse agencies as microorganisms, metal, or chemicals. The term also applies to materials used to impart color to wood.

blue stain—a bluish or grayish discoloration of the sapwood caused by the growth of certain dark-colored fungi on the surface and in the interior of the wood which is not accompanied by deterioration of the wood. While not caused by a wood-destroying fungi, the presence of blue stain indicates conditions possibly conducive to growth of destructive fungi.

brown stain—a rich brown to deep chocolate-brown discoloration of the sapwood of some pines caused by a fungus that acts much like the blue-stain fungus. See comments on conditions conducive to staining.

chemical brown stain—a brown discoloration of chemical origin that sometimes develops on wood in the course of air seasoning or kiln drying, probably from the oxidation of extractives in the wood.

sapstain—See *blue stain*. **D 9,** D07

stain—a discoloration, arising from foreign materials, that penetrates into the coating. **D 16,** D01

stain—a penetrating composition that changes the color of a surface, usually transparent and leaving practically no surface film. **D 16,** D01

stain, *n*—an area of discoloration that penetrates the fabric surface. **D 123,** D13

stain—a local deposit of soil or visible discoloration on a substrate. **D 459,** D12

stain—discoloration by foreign matter. **D 2825,** D21

stain—area contamination that is chemical in nature and cannot be removed except through further lapping or polishing. **F 1241,** F01

stainability—the relative ease with which a material is penetrated and discolored by a foreign material. **C 286,** B08

stain, contact, *n*—discoloration of a product by another material or product in the area of direct contact. **D 1566,** D11

stain, diffusion, *n*—discoloration of a light colored veneer or coating, caused by the diffusion of staining materials from the rubber through the veneer or coating. **D 1566,** D11

staining—the discoloration of image receptor sheet during contact with a carbon sheet. **F 221,** F05

staining power—the ability of the polish to impart color to leather during normal polishing without hiding the leather grain. **D 2825,** D21

stainless steel, *n*—a **steel** that conforms to a specification that requires, by mass percent, a minimum chromium content of 10.5 or more, and a maximum carbon content of less than 1.20. **A 941,** A01

stainless steel—the raw material of the instrument that is in accordance with Specification F 899. **F 1078,** F04

stain, migration, *n*—discoloration, caused by volatile constituents of a rubber, that occurs on any portion of the surface of any object in the proximity of or adjacent to but not in direct contact with the rubber causing the stain. **D 1566,** D11

stains—see **inclusions.** **D 1711,** D09

stair, *n*—an uninterrupted series of level steps, or connecting flights of steps, extending between two or more **floors** or landings. **E 631,** E06

stair-rail system—See **railing systems.** **E 631,** E06

stair-rail system—a railing system located along the open sides of a stair or landing. (Compare **guardrail/railing/stair-rail system**.) **E 1481,** E06

stairway, n—See **stair.** **E 631,** E06

stalagmometer—an apparatus for determining surface tension. The mass of a drop of a liquid is measured by weighing a known number of drops or by counting the number of drops obtained from a given volume of the liquid. **B 374,** B08

stand-alone—capable of functioning alone, without supervision or interconnection to the main system. **F 1156,** F12

standard—uniform lengths recommended in a simplified practice recommendation or established as a commercial standard.

B 846, B05

standard, *n*—an accepted reference sample or device used for establishing measurement of a physical quantity.

primary flow-rate standard, n—a device or means of measuring flow rate based on direct primary observations, such as time and physical dimensions.

primary standard, n—a standard directly defined and established by some authority, against which all secondary standards are compared.

secondary flow-rate standard, n—a flow-rate-measuring device, such as an orifice meter, that has been calibrated against a primary standard.

secondary standard, n—a standard used as a means of comparison, but checked against a primary standard.

transfer standard, n—a type of secondary standard.

working flow-rate standard, n—a flow rate measuring device, such as an orifice meter, that has been calibrated against a secondary flow-rate standard.

working standard, n—a standard used in the laboratory or field for periodic standardization of a measurement instrument. **D 1356,** D22

standard—1) a physical reference used as a basis for comparison or calibration;
2) a concept that has been established by authority, custom, or agreement to serve as a model or rule in the measurement of quality or establishment of a practice or procedure. **E 7,** E04

standard, n—see **documentry standard, physical standard, primary standard, product standard, reference standard, transfer standard, verification standard, working standard.** **E 284,** E12

standard—(1) a physical reference used as a basis for comparison or calibration; (2) a concept that has been established by authority, custom, or agreement to serve as a model or rule in the measurement of quality or the establishment of a practice or procedure. **E 1316,** E07

standard, *n*—material of established origin with certified properties. **E 1732,** E30

standard addition or known addition—a procedure for the determination of the concentration of a particular species in a sample by adding known amounts of that species to the sample solution and recording the change in potential of an ion-selective electrode versus a suitable reference electrode. **D 4127,** D19

standard air—See **air at normal conditions.** **D 1356,** D22

standard air density, *n*—atmospheric air density of 0.075 lb/ft^3(1.201 kg/m^3). **F 395,** F11

standardant, *n*—a material used for standardization. **E 135,** E01

standard atmosphere for preconditioning, *n—in textiles*, an atmosphere having a relative humidity of 5 to 25 %, ±2 % tolerance for the selected relative humidity, and a temperature of not over 50°C (122°F), with ±1°C (±2°F) tolerance for the selected temperature and used to partially dry the material before further treatment or conditioning. **D 123,** D13

standard atmosphere for preconditioning, *n*—a set of controlled conditions having a temperature not over 50°C (122°F), with respective tolerances of ±1°C (±2°F), and a relative humidity of 5–25 % ± 2 % for the selected humidity that drying can be achieved prior to conditioning in the standard atmosphere for testing textiles. **D 4920,** D13

standard atmosphere for testing, *n*—an atmosphere for testing in which the conditions for relative humidity and temperature are specified and controlled. (Compare atmosphere for testing). **D 123,** D13

standard atmosphere for testing, *n—in textiles*, an atmosphere for testing in which the air is maintained at a relative humidity of 65 ± 2 % and at a temperature of 21 ± 1°C (70 ± 2°F). (See **atmosphere for testing.**) **D 123,** D13

standard atmosphere for testing, *n—in glass textiles*, an atmosphere for testing in which the air is maintained at a relative humidity of at least 45 % and no greater than 67 %, tolerance of ±2 % for the selected relative humidity, and a temperature of at least 20°C (68°F) and no greater than 25°C (77°F), with a tolerance of ±1°C (±2°F) at the selected temperature. (Compare **atmosphere for testing.**) **D 4920,** D13

standard atmosphere for testing, *n—in tire cord and industrial yarn textiles*, an atmosphere for testing in which the air is maintained at a relative humidity of 55 ± 2 % and at a temperature of 24 ± 1°C (75 ± 2°F). (Compare **atmosphere for testing.**) (See **standard atmosphere for testing.**) **D 4920,** D13

standard atmosphere for testing textiles,, *n*—laboratory conditions for testing fibers, yarns, and fabrics in which air and relative humidity are maintained at specific levels within established tolerances. **D 123,** D13

standard atmospheric conditions, *n*—29.92 in. Hg (101 325 Pa), 68°F (20°C), and 30 % relative humidity. **F 395,** F11

standard compaction—see **compaction test.** **D 653,** D18

standard depth of penetration—see **depth of penetration.** **E 1316,** E07

standard depth of penetration (SDP)—the depth at which the eddy current density is reduced to approximately 37 % of the density at the surface, Eddy-current testing is most effective when the wall thickness does not exceed the SDP or in heavier tube walls when discontinuities of interest are within one SDP. (E 215) **E 1316,** E07

standard design—a published and proven concrete pipe design. **C 822,** C13

standard deviation—the positive square root of the variance. **C 859,** C26

standard deviation, *n*—the most usual measure of the dispersion of observed values or results expressed as the positive square root of the variance. **D 121,** D05

standard deviation—a measure of the dispersion of a series of results around their mean, computed as the positive square root of the variance. The standard deviation is the basis for most statements of precision and may be obtained from an analysis of variance of results of an interlaboratory test program. **D 4790,** D16

standard deviation, *n*—the most usual measure of the dispersion of observed values or results, expressed as the positive square root of the sum of the squared deviations from the sample average, divided by the degrees of freedom, usually $n - 1$, where n is the number of observed values. **E 284,** E12

standard deviation, *n*—the most usual measure of the dispersion of observed values or results expressed as the positive square root of the variance. **E 456,** E11

standard deviation—the square root of the variance. **E 856,** D34

standard deviation—a measure of the dispersion of a series of results around their average, expressed as the square root of the quantity obtained by summing the squares of the deviations from the average of the results and dividing by the number of observations minus one. It is also the square root of the variance and can be calculated as follows:

$$s = \sqrt{\frac{\Sigma(X_i - \bar{X})^2}{n-1}}$$

where:
s = estimated standard deviation of the series of results,
X_i = each individual value,
$\bar{X}$ = average (arithmetic mean) of all values, and
n = number of values. **E 1547,** E15

standard deviation (in statistics)—a measure of the precision of the readings, the average deviation of the deviations from the mean value. **E 1605,** E06

standard deviation of the mean (sample)—the sample standard deviation divided by the square root of the number of measurements used in the calculation of the mean. **C 859,** C26

standard deviation (*s*), *n*—a measure of the dispersion of a series of results around their average, expressed as the square root of the quantity obtained by summing the squares of the deviations from the average of the results and dividing by the number of observations minus one; it is also the square root of the variance and is calculated as follows:

$$s = \sqrt{\frac{\Sigma(x_i - \bar{x})^2}{(n-1)}}$$

where:
s = estimated standard deviation of the series of results,
x_i = each individual value,
$\bar{x}$ = mean (average) of all values, and
n = number of values.

A 644, A04

standard deviation, *s*, *n—of a sample*, a measure of the dispersion of variates observed in a sample expressed as the positive square root of the sample variance. **D 123,** D13

standard dimension ratios (SDR)—a specific ratio of the average specified outside diameter to the minimum specified wall thickness (D_o/t) for outside diameter-controlled plastic pipe, the value of which is derived by adding one to the pertinent number selected from the ANSI Preferred Number Series 10. Some of the values are as follows:

ANSI Preferred Number Series 10	SDR
5.0	6.0
6.3	7.3
8.0	9.0
10.0	11.0
12.5	13.5
16.0	17.0
20.0	21.0
25.0	26.0
31.5	32.5
40.0	41.0
50.0	51.0
63.0	64.0

(See reference: ANSI Preferred Numbers, Z17.1 (Designated as R 10 in ISO 3 and ISO 497).) **F 412,** F17

standard error of calibration, (SEC)—a measure of calibration accuracy determined by the following equation:

$$SEC = \left(\frac{1}{n-p-1}\sum_{i=1}^{n} e_i^2\right)^{1/2}$$

where:
n = the number of observations in the calibration data set,
p = the number of independent variables in the calibration, and
e_i = the difference between a measured value of a property and its accepted value.

E 131, E13

standard error of performance, (SEP)—a measure of accuracy determined by the following equation:

$$SEP = \left[\frac{1}{n-1}\sum_{i=1}^{n}(e_i - \bar{e})^2\right]^{1/2}$$

where:
n = the number of observations for which the accuracy is determined,
e_i = the difference between a measured value of a property and its accepted reference value, and
$\bar{e}$ = is the mean of all the e_i.

E 131, E13

standard-fall diameter—the diameter of a sphere with a specific gravity of 2.65 and the same standard-fall velocity as the particle. **D 4410,** D19

standard-fall velocity—the rate of fall that a particle would finally attain if falling alone in quiescent distilled water of infinite extent and a temperature of 24°C. **D 4410,** D19

standard grain-size micrograph—a micrograph taken of a known grain size at a known magnification, which is used to determine grain size by direct comparison with another micrograph or with the image of a specimen. **E 7,** E04

standard gravity, *n*—as adopted by the International Committee on Weights and Measures, an acceleration of 9.80665 m/s^2. **D 1356,** D22

standard hardboard—hardboard substantially as manufactured at the end of hot pressing, except for humidification to adjust moisture content, trimming to size, and other subsequent machining, and having the properties associated with hardboard meeting specifications for that quality product. **D 1554,** D07

standard (hardness) block—a carefully prepared metal block used to calibrate hardness test machines. **E 7,** E04

standard hide powder—any lot of hide powder officially approved by the American Leather Chemists Association. **D 1517,** D31

standard illuminant, *n*—a luminous flux, specified by its spectral distribution, meeting specifications adopted by a standardizing organization. (See, for example, **CIE standard illuminant *A*, CIE standard illuminant *B*, CIE standard illuminant *C*, CIE standard illuminant D_{65}.**) **E 284,** E12

standard impinger—See **impinger.** **D 1356,** D22

standard inside diameter dimension ratio (SIDR)—a specific ratio of the average specified inside diameter to the minimum specified wall thickness (D_i/t) for inside diameter-controlled plastic pipe, the value of which is derived by subtracting one from the pertinent number selected from the ANSI Preferred Number Series 10. Some of the values are as follows:

ANSI Preferred Number Series 10	SIDR
5.0	4.0
6.3	5.3
8.0	7.0
10.0	9.0
12.5	11.5
16.0	15.0
20.0	19.0
25.0	24.0
31.5	30.5
40.0	39.0
50.0	49.0
63.0	62.0

(See reference: ANSI Preferred Numbers, Z17.1 (Designated as R 10 in ISO 3 and ISO 497).) **F 412,** F17

standard instrument—a measuring instrument intended to define, represent physically, maintain, or reproduce the unit of measurement of a quantity (or a multiple or submultiple of that unit) in order that other measuring instruments may be calibrated against it by comparison (see **calibrated instrument**). **E 170,** E10

standardization, *n*—(*1*) the process of adjusting instrument output to a previously established calibration; (*2*) the experimental establishment of the concentration of a reagent solution. **E 135,** E01

standardization—correlation of an instrument response to a standard of known accuracy. **E 1187,** E36

standardization, instrument, *n*—the adjustment of an instrument, prior to use, to an arbitrary reference value. (See also **calibration, instrument**.) **E 1316,** E07

standardize, *v*—to adjust instrument output to correspond to a previously established calibration using one or more homogeneous specimens or reference materials. (See **calibrate, verify**). **E 284,** E12

standard kaolin—kaolin from any lot officially approved by the American Leather Chemists Association. **D 1517,** D31

standard laboratory atmosphere—an atmosphere, the temperature and relative humidity of which is specified, with tolerances on each. **E 41,** G03

standard leak—a device that permits a tracer gas to be introduced into a leak detector or leak testing system at a known rate to facilitate calibration of the leak detector. **E 1316,** E07

standard leakage rate—the rate of flow of atmospheric air under conditions in which: inlet pressure is 0.1 MPa ± 5 %; outlet pressure is less than 1 kPa; temperature is 25°C± 5°C; and dew point is less than −25°C. **E 1316,** E07

standard match—with a tongue and groove on opposite edges but not centered. **D 9,** D07

standard moisture regain, *n*—the moisture regain of a material at equilibrium with the standard atmosphere for testing textiles. (See **moisture regain.**) **D 123,** D13

standard moisture regain—see **moisture regain, standard.** **D 1695,** D01

standard neutron field—see **benchmark neutron field.** **E 170,** E10

standard observer, *n— in color determination*—see **observer, standard**. **D 2946,** C17

standard observer, *n*—an ideal observer having visual response described by the CIE color-matching functions. (See also **CIE 1931 standard observer, CIE 1964 supplementary standard observer**.) **E 284,** E12

standard penetration resistance—see **penetration resistance**. **D 653,** D18

standard pipe size—See **iron pipe size.** **E 631,** E06

standard polyolefin monofilament, *n—as used in this specification*, a flat polyolefin strand, approximately 0.05 mm (2 mil) thick by 2.5 mm (100 mil) wide and oriented with a draw ratio between 5:1 and 7:1. **D 123,** D13

standard, primary, *n*—one whose calibration is determined by measurement according to specified parameters. **C 1154,** C17

standard, primary, *n*—one whose calibration is determined by measurement according to specified parameters. **D 2946,** C17

standard reference material—a reference material, the composition or properties of which are certified by a recognized standardizing agency or group. **E 131,** E13

standard reference material, (SRM), *n*—see **certified reference material**. **E 135,** E01

standard reference test tire, (SRTT), *n*—a tire that meets the requirements of Specification E 1136, commonly used as a control tire or surface monitoring tire. **F 538,** F09

standard, secondary, *n*—standard calibrated by reference to another standard such as a primary, reference, laboratory, or working standard. **C 1154,** C17

standard, secondary, *n*—standard calibrated by reference to another standard such as a primary, reference, laboratory, or working standard. **D 2946,** C17

standard-sedimentation diameter—the diameter of a sphere with the same specific gravity and fall velocity as the given particle. **D 4410,** D19

standard source, *n— in color determination,* light source whose spectral energy distribution is known or defined. **D 2946,** C17

standard source, *n*—a source of radiant flux meeting specifications adopted by a standardizing organization. (See also **CIE standard source *A*, CIE standard source *B*, CIE standard source *C*.**) **E 284,** E12

standard staple—staple with nominal ½ in. (13 mm)-wide crown, ¼ in. (6 mm)-long legs, made of 0.019 in. (0.48 mm) wire, commonly used in desk-type staplers. **F 592,** F16

standard subtraction or known subtraction—a variation of the standard addition method. In this procedure, changes in the potential resulting from the addition of a known amount of a species (for example, a complexing agent) that reacts stoichiometrically with the ion of interest are employed to determine the original activity or concentration of the ion. **D 4127,** D19

standard temperature/time curve (standard time/temperature curve), *n—in fire testing,* a graphical representation derived from prescribed time-temperature relationships and used to control furnace temperature with progressing time. **E 176,** E05

standard test conditions—the parameters under which a membrane manufacturer tests devices for flow and salt rejection. **D 6161,** D19

standard thermoelement, *n*—a thermoelement that has been calibrated with reference to platinum 67 (Pt-67). **E 344,** E20

standard volumetric solution—a solution of accurately determined concentration used in the quantitative analysis of chemicals and other products. The concentration of such solutions is usually expressed in terms of normality or molarity. **E 1547,** E15

standard weathering reference material (SWRM), *n*—a Weathering Reference Material (WRM) whose well documented weathering degradation properties have been certified by a recognized standards agency or group and which are identical when exposed to identical test conditions. **G 113,** G03

standby diver, *n*—a diver at the dive location available to assist a diver in the water. **F 1549,** F32

standing orders—a type of medical protocol which provides specific-written orders for actions, techniques, or drug administration when communication has not been established for direct medical direction. **F 1177,** F30

standing wave—a wave produced by simultaneous transmission in opposite directions of two similar waves resulting in fixed points of zero amplitudes called nodes. **D 653,** D18

standoff—*in abrasive blasting*, distance from blast nozzle to part when abrasive blasting. **B 374,** B08

staple, *n*—natural fibers or cut lengths from filaments. **D 123,** D13

staple, *n—in grease wool*, a tuft or lock of fibers which naturally cling together, as found in a fleece. **D 123,** D13

staple, *n*—a U-shaped round or flat wire fastener that is: (*1*) used to secure parts together. (It may or may not be clinched); (*2*) driven over wires and bands to hold parts in position; (*3*) used to fasten together the flaps of **fiberboard** boxes. (Compare **stitch.**) **D 996,** D10

staple, *v*—to fasten with staples. **D 996,** D10

staple, *n*—a collection of short filaments of spinnable length that are natural, manufactured directly in a short length, or cut from continuous filaments. **D 3878,** D30

staple, *n*—*in buttons,* a looped metal shank securely positioned perpendicular to and at center back of the button flange for use in attaching the button to one part of a flexible substrate by means of a needle and thread, a ring, or a toggle. **D 5497,** D13

staple—"U"-shaped wire fastener usually with two same-size pointed or pointless legs connected by crown located opposite staple-point ends; designed to be driven by strike, pneumatic, electric, manual or spring tools through or through and into layers of penetrable material and to hold two or more pieces together. **F 592,** F16

staple, *n*—an anchor component that connects the bony elements of the spine, pelvis, or ribs to each other or to other implant components by using at least two interconnected posts. **F 1582,** F04

stapled seam, *n*—a seam formed by shaped metal devices such as U-shaped staples. (Compare **glued seam, sewn seam.**) **D 123,** D13

staple fiber—fibers produced in short, random lengths as distinguished from chopped fiber. **C 162,** C14

staple glass yarn, *n*—yarn made from filaments that are nominally 200 to 380 mm (8 to 15 in.) in length. **D 123,** D13

staple length, *n*—*in grease wool,* the length of a staple obtained by measuring the natural staple without stretching or disturbing the crimp of the fibers. **D 123,** D13

staple length—distance from top of staple crown to tip of staple point. Also referred to as leg length. In contrast, length of bulk staple, driven by hand-hammer, is measured from bottom of staple crown to tip of staple point. **F 592,** F16

staple point—See **bevel point, chisel point, clinch point, cross-cut chisel point, divergent point, inside bevel point, outside bevel point, outside bevel divergent point, right-angle chisel point, spear point, step point.** **F 592,** F16

staple spacing—dimension used to describe the relative location of staple or staples in workpiece; often, the dimension to center or center-to-center of staple. **F 592,** F16

stardusting—an extremely fine form of roughness on the surface of a metal deposit. **B 374,** B08

star head—slightly raised pyramidal head with multiple triangular faces meeting at common vertex, having a polygonal or circularized flat bearing surface. **F 547,** F16

star marks—a defect sometimes occurring in sheet steel iron cover coats where the dried ware is set down too hard on the firing fixture points and the enamel coating is fractured. **C 286,** B08

starring—see **back emission** and **self-limiting**. **C 286,** B08

stars—a defect similar to **star marks** appearing in the surface as a series of small hairlines radiating from a common center. They are typical of porcelain enamel powder systems. **C 286,** B08

start bi, *n*—the first element in a character in asynchronous serial transmission. **F 1457,** F05

starters—lateral sections of roofing shingles, usually 75 mm (3 in.) wide, and applied beneath the first course of shingles with slight overhang at the eaves. **C 460,** C17

starters, *n*—*in shingles,* lateral sections of roofing shingles applied beneath the first course of shingles with slight overhang at the eaves. **C 1154,** C17

starters—lateral sections of roofing shingles, usually 75 mm (3 in.) wide, and applied beneath the first course of shingles with slight overhang at the eaves. **D 2946,** C17

starting solution—see **etch.** **F 425,** F05

starting threshold (U_0, m/s), *n*—the lowest wind speed at which a rotating anemometer starts and continues to turn and produce a measurable signal when mounted in its normal position. **D 1356,** D22

start of scan, *n*—an electronic signal in a transmission which indicates to the receiver that the next series of signals are from an image and not text code. **F 1457,** F05

start of text, *n*—an electronic signal in a transmission which indicates to the receiver that the next series of signals are from text and not image code. **F 1457,** F05

start/stop characters or patterns, *n*—distinct characters or patterns used at the beginning and end of bar code symbols that provide timing references and direction-of-read information to the decoding logic. **F 1294,** F05

starved joint, *n*—See *starved joint* under **joint**. **D 907,** D14

starwheel, *n* (also called pick off) —when the feed mechanism of a printer digs into the inked surface. **F 1857,** F05

state of statistical control, *n*—a condition in which a process, including a measurement process, is subject only to random variation. **D 123,** D13

static—exhibiting little or no change or movement. **C 717,** C24

static calorimeter, *n*—a calorimeter without a thermostated jacket. **D 5681,** D34

static coefficient of friction—the ratio of the parallel component of force applied to a stationary body that just overcomes the friction or resistance to relative motion of two surfaces in physical contact one with another, but otherwise unconstrained, to the normal component of the force—usually the force caused by gravity—applied to the body under clean, dry conditions. **C 242,** C21

static coefficient of friction, *n*— the coefficient of friction corresponding to the maximum friction force that must be overcome to initiate macroscopic motion between two bodies. **G 40,** G02

static friction, *n*—friction developed between two touching bodies at the time one body starts to move relative to another. (Compare **kinetic friction.**) **D 123,** D13

static hold-up, *n*—(Syn. *wellage*) the quantity of liquid retained in a distillation column after draining at the end of a distillation. **D 4175,** D02

static load, *n*—*in textile testing,* a mass which exerts a force by means of the mass alone without motion. **D 123,** D13

static load—See **load.** **D 996,** D10

static load, *n*—an imposed stationary force that is constant in magnitude, direction, and sense. **E 631,** E06

static pressure—See **pressure.** **D 1356,** D22

static SIMS—SIMS analysis at a primary ion current density and dose such that during the analysis each sputtering event occurs from a previously unsputtered region. **E 673,** E42

static test, *n*— single cycle loading tests designed to evaluate the mechanical properties of materials, components, interconnections, subconstructs, constructs, subassemblies, or assemblies. **F 1582,** F04

static thermal response curve, *n*—the relationship of the response characteristic of a thermal printing system over a given range of print energies; optical (reflectance) density (*y*-axis) versus temperature (*x*-axis). **F 1623,** F05

static water level—the elevation of the top of a column of water in a monitoring well or piezometer that is not influenced by pumping or conditions related to well installation, hydrologic testing, or nearby pumpage. **D 653,** D18 *stationary extractor*—a series of horizontal baffles that run the full length of the exhaust hood and are not removable for cleaning. It includes one or more water manifolds with spray nozzles that, upon activation, wash the grease extractor with hot, detergent-injected water, removing the accumulation of grease from the extractor. See *grease extractors*. **F 1827,** F26

stationary repeatability precision—the precision of the measurement of a characteristic determined under repeatability condition with the instrumented equipment stationary. **E 867,** E17

station rack type, water-driven rotary spray type—See **pot, pan, and utensils washing machine**. **F 1827,** F26

station/work uniform, *n*—a nonprimary protective clothing ensemble consisting of a shirt and pants that is intended to be worn by members of the fire and rescue services while on duty. **F 1494,** F23

statistic, *n*—a quantity that is calculated from observations on a sample and that estimates a parameter of a population. **D 123,** D13

statistic, *n*—a quantity calculated from a sample of observations, most often to form an estimate of some population parameter. **E 456,** E11

statistic—a summary value calculated from the observed values in a sample. **E 1823,** E08

statistical bias—See **bias.** **C 859,** C26

statistical control—a condition when all special causes of variation have been eliminated from a process and only common causes remain. **F 1789,** F16

statistical measure, *n*—statistic or mathematical function of a statistic. **E 456,** E11

statistical noise—*XPS,* the noise in the spectrum due solely to the statistics of randomly detected single events; the root mean square of the deviations in neighboring channels is equal to the square root of the average counts per channel. **E 673,** E42

statistical process control—the use of statistical techniques, such as control charts, to analyze a process or its outputs so as to take appropriate actions to achieve and maintain a state of statistical control and to improve the process capability. **F 1789,** F16

statistical thickness surface area (STSA), *n*—the external surface area of carbon black that is calculated from nitrogen adsorption data using the de Boer theory and a carbon black model. **D 3053,** D24

status, *n*—a message sent by a device to the host describing its condition, the condition can be any of several things, such as ready, able to execute the command, etc. **F 1457,** F05

status switches, *n*—switches located on the mainframe electronic chassis in the console used to set the number of specimens tested on each laboratory sample and turn off instruments not being used in operating the system. **D 123,** D13

stave, *n*—(*1*) one of the shaped vertical units of a coopered wooden **container** which, when placed edge-to-edge and bound with hoops around a circular bottom (sometimes top and bottom), forms a **barrel** or other staved container; (*2*) one of the thin, wood or veneer **slats** forming the **body,** sides, or **cover** of a **basket** or **hamper.** **D 996,** D10

stay—paper or metal material used to join parts of a **box** together at a corner, or to reinforce a cut-score on **solid fiberboard.** It is most commonly used for set-up boxes. **D 996,** D10

stay log—a device used on a veneer lathe to which is fastened a flitch or segment of a bolt to secure desired grain effects in the veneer. **D 1038,** D07

steadite—(*1*) ternary eutectic found in alloys of iron, carbon and phosphorus. In cast irons it consists of austenite, cementite and iron phosphide or austenite and iron-phosphide. If conditions are such that carbon is deposited as cementite, the three constituents of the eutectic are present. If, however, the carbon is deposited as graphite, that precipitated from the ternary eutectic crystallizes on existing flakes and the eutectic consists of two constituents only, namely, austenite and iron-phosphide.

(*2*) Binary eutectic found in alloys of iron and phosphorus consisting of ferrite and iron-phosphide. **E 7,** E04

steady-state crack—a crack that has advanced slowly until the crack-tip plastic zone size and crack-tip sharpness remain constant with further crack extension. Although crack-tip conditions can be a function of crack velocity, the steady-state crack-tip conditions for metals have appeared to be independent of the crack velocity within the range attained by the loading rates specified in Test Method E 1304. **E 1823,** E08

steady-state permeation, *n*—the constant rate of permeation that occurs after breakthrough when the chemical contact is continuous and all forces affecting permeation have reached equilibrium. **F 1494,** F23

steady state simulator—simulator whose irradiance output at the test plane area is continuous for periods of a second or greater. **E 772,** E44

steady state (thermal), *n*—a condition for which all relevant parameters in a region do not vary over two consecutive steady-state time periods by more than the steady-state tolerance, and no long-term monotonic drifts are present.

Where, the steady-state time period is the time constant of the apparatus-specimen system with additional time necessary if physical phenomena are present, such as moisture transport, which could cause a long-term monotonic drift.

Steady-state tolerance consists of (possibilities in order of increasing magnitude):

(*1*) The imprecision of the mean of a set of data points. This can be defined as twice the standard deviation of a set of N independent data points divided by the square root of N, $2\sigma / \sqrt{N}$,

(*2*) The scatter of the data. This would be 2σ, or,

(*3*) Some larger value may be chosen resulting in less precision. **C 168,** C16

steady-state vibration—vibration in a system where the velocity of each particle is a continuing periodic quantity. **D 653,** D18

steam cooker—cooking equipment wherein heat is imparted to food in a closed compartment. The compartment can be at or above atmospheric pressure and the steam can be static or circulated. The pressure occurring in the food compartment of these steamers during cooking ranges from zero gage psi pressure to 15 psig. **F 1827,** F26

steam jacketed kettle—cooking equipment comprised of two hemispherical metal shells of different sizes, the smaller shell inserted into the larger shell and fastened in a way that provides a space around its bottom and up its sides to form a pressure vessel with one-half, two-thirds, or full jacketing, wherein heat is imparted to food by steam circulating through the jacket. **F 1827,** F26

steatite porcelain—See *steatite porcelain* under **porcelain.** **C 242,** C21

steatite talc—massive talc or the pulverized product thereof having the general formula $3MgO{\cdot}4SiO_2{\cdot}H_2O$. **C 242,** C21

steatite whiteware—See *steatite whiteware* under **ceramic whiteware.** **C 242,** C21

steel, *n*—a material that conforms to a specification that requires, by mass percent, more iron than any other element and a maximum carbon content of generally less than 2. **A 941,** A01

steel, cold-rolled— see **cold-rolled steel.** **C 286,** B08

steel cord, *n*—a formed structure made by twisting together two or more steel filaments when used as an end product or a combination of strands or filaments and strands. **D 123,** D13

steel cord wrap, *n*—a filament wound helically around a steel cord. **D 123,** D13

steel filament, *n*—the individual element in a steel strand or cord. **D 123,** D13

steel grades—steel classified by carbon content. The following grades are generally used for wire or raw material purchase; but do not restrict nails to specific analysis. When the steel grade is referred to in the definition of a nail type, it is intended only as a general indication of the likely material used. Any steel of suitable analysis may be used in nail manufacture unless specified otherwise in the product standard.

low-carbon steel— a grade of steel (see Specification A 510 for General Requirements for Wire Rods and Coarse Round Wire, Carbon Steel) wherein the maximum of the carbon range is up to and including 0.15 %.

medium low-carbon steel—a grade of steel (Specification A 510) wherein the maximum of the carbon range exceeds 0.15 % up to and including 0.23 %.

medium high-carbon steel—a grade of steel (Specification A 510) wherein the maximum of the carbon range exceeds 0.23 % up to and including 0.44 %.

high-carbon steel— a grade of steel (Specification A 510) wherein the maximum of the carbon range exceeds 0.44 %. **F 547,** F16

steel scrap, *n*—discarded steel or steel products, generally segregated by composition and size or "grade," suitable for melting. **A 644,** A04

steel sheet designation, *n*—a title given to a steel sheet product, associated with unique requirements for chemical composition and with mandatory or nonmandatory (typical) mechanical properties; the specific titles include *commercial steel, drawing steel, deep drawing steel, extra deep drawing steel, forming steel, high strength-low alloy steel, high temperature steel,* and *structural steel.* **A 902,** A05

steel slag, *n*—the nonmetallic product consisting essentially of calcium silicates and ferrites combined with fused oxides of iron, aluminum, manganese, calcium and magnesium, that is developed simultaneously with steel in basic oxygen, electric, or open hearth furnaces. **D 8,** D04

steel strand, *n*—a group of steel filaments combined together to form a unit product. **D 123,** D13

steep asphalt—a roofing asphalt conforming to the requirements of Specification D 312, Type III. **D 1079,** D08

steeple head—See **cone head.** **F 547,** F16

steer hide—see **hide grades.** **D 1517,** D31

steering wheel angle, (rad or degree), *n—in a vehicle*, the angular displacement of the steering wheel, measured from its orientation in the vehicle's straight-ahead position. **F 538,** F09

stem—the bole or trunk of a tree. **D 9,** D07

stem, *n—of a liquid-in-glass thermometer*, capillary tube through which the meniscus of the thermometric liquid moves with change of temperature. **E 344,** E20

stemming—(*1*) the material (chippings, or sand and clay) used to fill a blasthole after the explosive charge has been inserted. Its purpose is to prevent the rapid escape of the explosion gases. (2) the act of pushing and tamping the material in the hole. (ISRM) **D 653,** D18

stepdown test—a test involving the preparation of a series of machined steps progressing inward from the surface of a bar for the purpose of detecting by visual inspection internal laminations caused by inclusion segregates. See Practice E 45. **E 7,** E04

step-drawdown test—a test in which a control well is pumped at constant rates in "steps" of increasing discharge. Each step is approximately equal in duration, although the last step may be prolonged. **D 653,** D18

step exposure (emission spectrographic)—an emission spectrographic exposure using a step filter or step-sector disk that interrupts the radiation passing through the spectrograph slit and falling on the photographic emulsion. The length of time of the interruption varies in steps along the vertical length of the spectral line as determined by the "steps" in the filter or disk. **C 859,** C26

step fracture—See **striation.** **C 162,** C14

stepladder code, *n*—see **vertical bar code.** **F 1294,** F05

stepped gray scale—means of differentiating between discrete, discernible levels of gray in a black and white image; a measure of photodetection sensitivity; the greater the number of levels or steps, the better the sensitivity of the system. **F 1156,** F12

step point—notched point with step faces perpendicular to staple crown; designed to facilitate self-clinching of staple legs when driven against clinching plate. **F 592,** F16

step wear, *n*—a synonym for *row/rib wear.* **F 538,** F09

step wedge—a device with discrete step thickness increments used to obtain an image with discrete density step values. **E 1316,** E07

step-wedge calibration film—a step-wedge comparison film the densities of which are traceable to a nationally recognized standardizing body. **E 1316,** E07

step-wedge comparison film—a strip of processed film carrying a stepwise array of increasing photographic density. **E 1316,** E07

step wedge comparison film—a radiograph with discrete density steps that have been verified by comparison with a calibrated step wedge film. **E 1316,** E07

stereo angle—one half of the angle through which the specimen is tilted when taking a pair of stereomicrographs. The axis of rotation lies in the plane of the specimen. **E 7,** E04

stereographic projection—the projection to a plane from a spherical projection, customarily using the South Pole as the eye point. **E 7,** E04

stereology—the study of mathematical procedures used to derive three-dimensional parameters describing a structure from two-dimensional measurements. **E 7,** E04

stereomicroscope—a light optical microscope that permits each eye to examine the specimen at a slightly different angle, thereby retaining its three-dimensional relationship. **E 7,** E04

stereoscopic micrographs—a pair of micrographs of the same area but taken from different angles so that the two micrographs when properly mounted and viewed reveal the structures of the objects in their three-dimensional relationships. **E 7,** E04

stereoscopic specimen holder—a specimen holder designed for the purpose of making stereomicrographs. It makes possible the tilting of the specimen through the stereo angle. **E 7,** E04

sterilant—an agent used to achieve commercial sterility. **F 17,** F02

sterilant—See Terminology F 17. **F 1327,** F02

sterile—free from any viable organism, either active or dormant. **D 1129,** D19

sterile—free of any living organism. **E 1705,** E48

sterile—free of any viable microorganisms, either active or dormant (D 1129, D-19). **F 17,** F02

sterile—See Terminology F 17. **F 1327,** F02

sterile, *adj*—free from viable microorganisms. **F 1494,** F23

sterilization—destruction or removal of all viable organisms.

D 6161, D19

stick, *n*—extruded form substantially uniform in cross section cut to desired length or weight. **B 881,** B07

stick—See **strip.** **F 592,** F16

sticker—a material that assists the spray deposit to adhere or stick to the target and may be measured in terms of resistance to time, wind, water, mechanical action, or chemical action. **E 1519,** E35

stick-fed tool—See **strip-fed tool.** **F 592,** F16

stickies, *n*—particulate, generally non-fibrous, contaminants suspended in pulps of the type used in papermaking which adhere to themselves, other components of the paper structure, components of the papermaking machinery, and in the case where the contaminants are incorporated into the finished paper or paper products, cause the unexpected and generally undesirable adherence one to another of units of the finished paper or paper product. **D 1968,** D06

sticking—a response evidenced by softening and adherence of the material to the hot surface or other material. **F 1494,** F23

stick-slip, *n*—a phenomenon occurring when boundary lubrication is deficient, manifested by alternate periods of sticking and slipping of the surfaces in contact. **D 123,** D13

stick-slip—rapid fluctuations in shear force as one rock mass slides past another, characterized by a sudden slip between the rock masses, a period of no relative displacement between the two masses, a sudden slip, etc. The oscillations may be regular as in a direct shear test, or irregular as in a triaxial test. **D 653,** D18

stick-slip, *n—in tribology*, a cyclic fluctuation in the magnitudes of friction force and relative velocity between two elements in sliding contact, usually associated with a relaxation oscillation dependent on elasticity in the tribosystem and on a decrease of the coefficient of friction with onset of sliding or with increase of sliding velocity. **G 40,** G02

stick vacuum cleaner, *n*—a non-self-standing, hand-held, floor-supported, portable vacuum cleaner which has the floor nozzle directly connected to the cleaner. The cleaner is generally designed for light-duty cleaning of household dirt. In use, the cleaner is guided over the floor by means of a handle attached to the cleaner housing. The cleaner may contain a driven agitator to assist in dirt removal on floor surfaces. The cleaner may have attachments or provisions, or both, for both floor and above-the-floor cleaning. **F 395,** F11

sticky limit, T_w **(D)**—the lowest water content at which a soil will stick to a metal blade drawn across the surface of the soil mass. **D 653,** D18

stiction, *n—in tribology*, a force between two solid bodies in nominal contact, acting without the need for an external normal force pressing them together, which can manifest itself by resistance to tangential motion as well as resistance to being pulled apart. **G 40,** G02

Stiff & Davis stability index, S&DSI—an index calculated from total dissolved solids, calcium concentration, total alkalinity, pH and solution temperature that shows the tendency of a water solution to precipitate or dissolve calcium carbonate. S&DSI is used primarily for seawater RO applications. **D 6161,** D19

stiffener, *n*—a compounding material used to increase the viscosity of an unvulcanized rubber mix. **D 1566,** D11

stiffener, *n*—a reinforcing member designed to limit or prevent the deformation of an attaching member. **E 631,** E06

stiffener—component that provides support to the membrane. **F 818,** F20

stiffness, *n*—resistance to bending. **D 123,** D13

stiffness, *n—with regard to circular bending of textiles*, resistance to multidirectional bending. **D 123,** D13

stiffness—the ratio of change of force (or torque) to the corresponding change in translational (or rotational) deflection of an elastic element. **D 653,** D18

stiffness, *n*—resistance to bending. **D 4439,** D35

stiffness (as applied to copper)—the property of a conductor that causes it to resist permanent deformation by bending. **B 354,** B01

stiffness, bending, *n*—the force required to produce a bent configuration under specified conditions. **D 1566,** D11

stiffness-force—displacement ratio. (ISRM) **D 653,** D18

stiffness (or bending stiffness), *n—of paper or paperboard*, the degree to which paper or board resists bending when subjected to a bending force in its intended use, or when using a defined testing procedure. **D 1968,** D06

stile—See **windows and doors.** **E 631,** E06

stillage—the liquid products or waste remaining after distillation of a beer. The soluble residue are water, proteins, etc. **E 1705,** E48

stimulation—the application of a stimulus such as force, pressure, heat, etc., to a test article to cause activation of acoustic emission sources. **E 1316,** E07

stimulus, *n*—anything that has the potential for activating a receptor. **E 253,** E18

stimulus, *n*—any action or condition that has the potential for evoking a response. **E 284,** E12

sting-out—the stream of hot air or flame, or both, driven by positive internal pressure, exhausted through openings in melting furnaces. **C 162,** C14

stippled area—{archaic} See **hackle.** **C 162,** C14

stippled finish—a pebbly textured porcelain enamel, often multicolored. **C 286,** B08

stippling—see synonomous term **knurling**. **C 162,** C14

stitch, *n—in sewing*, the configuration of the interlacing of sewing thread in a specific repeated unit. (See **stitching, stitch type**) **D 123,** D13

stitch, *v—in making rubberized articles*, to press uncured rubber compound into or around yarns or cords to form a composite of the materials and to remove entrapped air. **D 123,** D13

stitch, *n—in packaging*, a metal fastener. Stitches are machine-formed using wire drawn from a spool. (Compare **staple. Staples** are preformed.) **D 996,** D10

stitch, *v*—(*1*) to unite, by means of **staples** or stitching wire, the joints or parts of containers, or the flaps for **closure;** (*2*) to unite by sewing with thread, as a **sack.** **D 996,** D10

stitch, *n*—the repeated unit formed by the sewing thread in the production of seams in a sewn fabric (see Federal Standard 751a). **D 4439,** D35

stitch, *n*—the repeated unit formed by a sewing thread(s) through a material or materials such as a textile fabric. **D 5646,** D13

stitch—staple cut and formed from wire immediately prior to driving by same machine. **F 592,** F16

stitch, *n*—the image laid down in a single traverse of a printhead across the printer. **F 1857,** F05

stitch density, *n—in sewn seams*, the number of stitches per unit length in one row of stitching in the seam. **D 123,** D13

stitched and pinked seam finished, *n—in home sewing*, a seam finish in which a zigzagged or scalloped cut edge is reinforced by a line of stitching. (Compare **pinked seam finish**). **D 4965,** D13

stitched seam, *n*—See **seam seam.** **D 4965,** D13

stitcher—machine that cuts, forms, and drives wire stitches. **F 592,** F16

stitches, *n—in tufted pile floor covering*, the number of pile tufts per inch in the lengthwise direction. **D 123,** D13

stitches, *n—in tufted pile yarn floor covering*, the number of pile tufts per unit length in the lengthwise direction. **D 5684,** D13

stitch gage, *n—in sewn seams*, the perpendicular distance between adjacent parallel rows of stitching. **D 123,** D13

stitching, *n*—a series of stitches embodied in a material or materials of planar structure such as woven textile fabrics, usually for ornamental purposes or finishing an edge, or both. **D 123,** D13

stitching, *n*—a series of stitches embodied in a material or materials such as a woven, knitted or textile fabric for ornamental purposes, forming a seam, attaching functional items, or finishing an edge. **D 5646,** D13

stitching wire—See **box stay wire.** **F 592,** F16

stitch lines, *n*—the relationship between consecutive passes of printhead when printing a solid image area. **F 1857,** F05

stitch type, *n*—a numerical designation relating to the essential characteristics of the interlacing of sewing thread(s) in a specified seam in sewn fabric. **D 123,** D13

stitch type, *n—in sewn seams*, a numerical designation relating to the essential characteristics of the interlacing of sewing thread(s) in a specified stitch. **D 123,** D13

S titration—a titration in which the sensing electrode responds to the ion being removed from the sample by the titrant. **D 4127,** D19

stock, n— See **compound**, the preferred term. **D 1566,** D11

stock in process, *n—in textiles*, staple fibers at any stage of manufacture between the opening of the bale and the completion of the spinning process. **D 123,** D13

stock solution—a homogeneous liquid mixture from which aliquants, aliquots, or unmeasured portions can be taken as needed. **C 859,** C26

stock steel—standard steel of regular or stiff stock.

regular—bright, non-hardened, usually low or medium low-carbon steel.

stiff—bright, non-hardened, usually medium low or medium high-carbon steel, having higher hardness, toughness, and stiffness than regular steel. **F 547,** F16

stock with ends—stock lengths, including ends. **B 846,** B05

stoichiometric—having the precise weight relationship of the elements in a chemical compound; or (quantities of reacting elements or compounds) being in the same weight relationship as the theoretical combining weight of the elements involved. **C 859,** C26

stoichiometric, *adj*—characterized by or being a proportion of substances or energy in a specific chemical reaction in which there is no excess of any reactant or product. **D 1356,** D22

Stokes' diameter—the diameter of a hypothetical sphere having the same density as the actual particle, as computed from the following Stokes' drag force formula:

$$\text{Drag force} = 3\,\pi\,\mu\,v_p\,(\text{Stokes' diameter})$$

where:

μ = viscosity of fluid medium creating drag on the particle, and

v_p = velocity of the particle relative to the medium. **E 1620,** E29

Stokes line (band)—a Raman line (band) that has a frequency lower than that of the incident monochromatic beam. **E 131,** E13

stoking—see **continuous sintering.** **B 243,** B09

stone—an imperfection; a crystalline inclusion in glass. **C 162,** C14

stone—crushed or naturally angular particles of rock. **D 653,** D18

stone chips, *n*—small angular fragments of stone containing no dust. **D 8,** D04

stoneware—a vitreous or semivitreous ceramic ware of fine texture, made primarily from nonrefractory fire clay. **C 242,** C21

stoning—the operation of removing by abrasion the undesirable portions of porcelain enamel. **C 286,** B08

stop, *n*—in building construction, in sealants and glazing, a strip of metal, plastic, or wood used around the periphery of a lite of glass or a panel to secure it in place (also referred to improperly as a bead). **C 717,** C24

stop, *n—in a zipper*, the device at the top and bottom of chain or stringer that prevents the slider from leaving the chain. **D 123,** D13

stop—*in grouting*, a packer setting at depth. **D 653,** D18

stop, *n*—any window or diaphragm that restricts the passage of light rays in an optical device. (See also **aperture stop, field stop**.) **E 284,** E12

stop bit, *n*—in asynchronous serial transmission, the last element(s) in a character. **F 1457,** F05

stop grouting—the grouting of a hole beginning at the lowest packer setting (stop) after the hole is drilled to total depth. **D 653,** D18

stop-leak additives, *n*—a compound containing particulates that is added to the cooling system for the purpose of stopping or minimizing leaks. **D 4725,** D15

stopless glazing—See **structural sealant glazing**, the preferred term. **C 717,** C24

stop mark, *n*—a visible change in the density of the weave across the width of the fabric caused by the tension on the warp not being adjusted properly after the loom has been stopped. (See **set mark**) **D 123,** D13

stop mark—a band, either dull or glossy, on the surface, approximately ½ to 3 in. (12 to 76 mm) wide and extending around the periphery of a pultruded shape.

NOTE—This condition is the result of an interruption in the normal continuous pulling operation. **D 3918,** D20

stop-off—See **resist.** **B 374,** B08

stopper head, *n*—a rounded refractory shape, usually made from clay and graphite, providing a valve head seating into a nozzle brick, this assembly forming a metal flow control for bottom-pouring ladles. **C 71,** C08

stop pin—a pin of preset length affixed to the inside of one of the tweezer halves designed to limit teeth contact upon closure and prevent their damage. **F 1638,** F44

stopping:

cross section—EIA, see **cross section,** *stopping.*

cross section factor—EIA, the stopping factor expressed per unit volume density of the constituent molecular entity of the specimen at the appropriate depth.

factor—EIA, the ratio between a (differential) energy interval in a backscattering spectrum and the corresponding (differential) depth interval in the specimen.

power—EIA, same as **stopping,** *cross section.* **E 673,** E42

stopping distance, *n*—the path distance (rectilinear or curved) needed to bring a vehicle to a stop from some selected initial brake application speed. **F 538,** F09

stopping off—the application of a resist to any part of an electrode—cathode, anode, or rack. **B 374,** B08

stop, removable—a removable molding or bead used around the periphery of a light of glass or a panel to secure it in place.

C 717, C24

stop, stationary or integral—the face of a rabbet or fin formed as an integral part of a frame or sash and against which glass or a panel is set. **C 717,** C24

storage—see **hazardous-waste storage.** **E 1605,** E06

storage coefficient—the volume of water an aquifer releases from or takes into storage per unit surface area of the aquifer per unit change in head. For a confined aquifer, the storage coefficient is equal to the product of the specific storage and aquifer thickness. For an unconfined aquifer, the storage coefficient is approximately equal to the specific yield. **D 653,** D18

storage component, thermal—a component of a building used for storing thermal energy. Includes all identifiable elements that serve an architectural as well as thermal function. **E 772,** E44

storage device, thermal—the container(s) plus all contents of the container(s) used for storing thermal energy. The transfer fluid and accessories such as heat exchangers, flow switching devices, valves, and baffles which are integral with the thermal storage container(s) are considered a part of the storage device. **E 772,** E44

storage factor—see **quality factor,** ***Q.*** **D 1711,** D09

storage factor—a name formerly used at times for the *quality factor.* **D 2864,** D27

storage factor, magnetic, Q_m—the cotangent of the hysteretic angle that is equal to the ratio of the magnetizing current, I_m, to the core loss current I_c.

$$Q_m = \cot \beta = \tan \gamma = 1/D_m = I_m/I_c = R_1/\omega L_1$$

NOTE—The storage factor is also given by the ratio of 2π times the maximum energy stored in the core to the energy dissipated in the core (hysteresis and eddy current heat loss) per cycle of a periodic *SCM* excitation. **A 340,** A06

storage hardening, *n*—the increase in the viscosity of raw or unvulcanized rubber during storage. **D 1566,** D11

storage life, *n*—the period of time during which a packaged adhesive can be stored under specified temperature conditions and remain suitable for use. (Synonym **shelf life.**) (Compare **working life.**) **D 907,** D14

storage life—the length of time that a packaged adhesive, sealant, or other product can be stored under specified temperature conditions and remain suitable for use (Syn. **shelf-life.**) **E 1749,** E06

storage life, shelf, *n*—the period of time after production during which a material or product, that is stored under specified conditions, retains its intended performance capabilities. **D 1566,** D11

storage medium, thermal—the material in the storage device, independent of the containing structure, in which the major portion of the thermal energy is stored. **E 772,** E44

storage modulus—***M′*** **(Pa),** [storage compliance *C* (Pa $^{-1}$)]—the real part of the complex modulus (complex compliance). **D 4092,** D20

storage modulus—quantitative measure of elastic properties defined as the ratio of the stress, in-phase with strain, to the magnitude of the strain. **E 1142,** E37

storage phosphor imaging plate—a flexible or rigid reusable detector that stores a radiological image as a result of exposure to penetrating radiation. **E 1316,** E07

storage point, *n*—an indication of the minimum temperature to which an oil should be heated in any part of an oil-handling installation when starting up after a shutdown. It is also an indication of the minimum temperature at which the oil should be stored in a tank fitted with an outflow heater. **D 4175,** D02

storage stability—the ability of a product to maintain its original characteristics over extended storage periods, under normal variations in temperature conditions. **D 3064,** D10

storm anchor (clip), *n—in shingles,* a corrosion-resistant metal fastener with a flat base and a shank that fastens the concealed lower corner of each shingle to the exposed edge of the adjacent shingle. **C 460,** C17

storm anchor (clip), *n— in shingles,* a corrosion-resistant metal fastener with a flat base and a shank that fastens the concealed lower corner of each shingle to the exposed edge of the adjacent shingle. **C 1154,** C17

storm anchor (clip), *n—in shingles,* a corrosion-resistant metal fastener with a flat base and a shank that fastens the concealed lower corner of each shingle to the exposed edge of the adjacent shingle. **D 2946,** C17

storm drain—a pipeline intended to convey storm water. **C 822,** C13

storm sewer—a pipeline intended to convey storm water. **C 822,** C13

storm water—precipitation run-off. **C 822,** C13

story—See **building space.** **E 631,** E06

stout nail—nails with shank diameter usually at least one gage larger than common nails of same length. **F 547,** F16

stover—the dried stalks and leaves of a crop remaining after the grain has been harvested. **E 1126,** E48

stover—the dried stalks and leaves of a crop remaining after the grain has been harvested. **E 1705,** E48

STP—sodium triphosphate - $Na_5P_3O_{10}$, a cleaning agent. **D 6161,** D19

STPP—sodium tripolyphosphate. See **STP.** **D 6161,** D19

straight air, *adj*—the description of a vacuum cleaning system that does not have a mechanically driven agitator (sometimes referred to in the industry as *straight suction*). **F 395,** F11

straight beam—a vibrating pulse wave train traveling normal to the test surface. **E 1316,** E07

straight line drag forces—load on a boom that results from towing it from one end. **F 818,** F20

straight nail—nail not bent or bowed. **F 547,** F16

straightness, *n—in steel cord,* the property of a cord characterized by a lack of deviation from its central axis over short lengths of a cord. **D 123,** D13

straight-run pitch, *n*—a pitch run to the consistency desired in the initial process of distillation and without subsequent fluxing. **D 8,** D04

straight soap—under **soap,** see *straight soap.* **D 459,** D12

straight stitch, *n—in home sewing,* a simple machine stitch pattern of straight, single stitch segments of equal length. The simple machine stitch pattern is formed by the down and up movement of the needle while the fabric moves through the feed mechanisms in either the forward or reverse direction. **D 123,** D13

straight stream nozzle—see **plain jet atomizer.** **E 1620,** E29

straight throat—a throat whose floor is at the same level as the floor of the melter. **C 162,** C14

strain—jargon for optical retardation. **C 162,** C14

strain, *n*—deformation of a material caused by the application of an external force. (See also **tensile strain.**) **D 123,** D13

strain, *n*—the unit change due to stress, in the size or shape of a body referred to its original size or shape. **D 907,** D14

strain, *n*—the unit change, due to force, in the size or shape of a body referred to its original size or shape. **D 1566,** D11

strain, ϵ—the unit change, due to force, in the size or shape of a body referred to its original size or shape. **D 4092,** D20

strain, *n*—the change in length per unit of length in a given direction. **D 4439,** D35

strain—the change per unit of length in a linear dimension of a body, that accompanies a stress. **F 412,** F17

strain aging, *n*—**aging** induced by cold working. **A 941,** A01

strain amplitude—the ratio of the maximum deformation, measured from the mean deformation, to the free length of the unstrained test specimen. (ISO) **D 4092,** D20

strain, ϵ (D)—the change in length per unit of length in a given direction. **D 653,** D18

strain disk—a disk of glass having a calibrated amount of birefringence at a specified location, and used as a comparative measure of the degree of retardation. **C 162,** C14

strain, *e, n*—the per unit change, due to force, in the size or shape of a body referred to its original size or shape. Strain is a nondimensional quantity, but it is frequently expressed in inches per inch, metres per metre, or percent. **E 6,** E28

strain ellipsoid—the representation of the strain in the form of an ellipsoid into which a sphere of unit radius deforms and whose axes are the principal axes of strain. (ISRM) **D 653,** D18

strainer, *n*—a machine designed to force a rubber or rubber mix through a sieve or sieves to remove extraneous material. **D 1566,** D11

strain gage accelerometer, *n*—an accelerometer using strain gages to sense the motion of the seismic element. **F 538,** F09

strain hardening, *n*—an increase in hardness and strength of a metal caused by plastic deformation at temperatures below its **recrystallization temperature**. (Syn. *work hardening*) **A 941,** A01

strain hardening—an increase in hardness and strength caused by plastic deformation at temperatures lower than the recrystallization range. **B 354,** B01

strain hardening—the increase in strength and hardness and decrease in ductility due to permanent deformation of the structure by cold working. **B 846,** B05

straining block, *n*—a rigid aluminum test block with a rubber surface on one side capable of simulating an anomaly in a tire. **F 538,** F09

straining block holding fixture, *n*—a device for holding one or more straining blocks in the S/H system during the calibration process. **F 538,** F09

strain, linear—the change in length of a specimen due to an applied stress relative to the original length of the specimen before the stress was applied. Linear strain is often shown as:

$$\frac{L_2 - L_1}{L_1}$$

where:
L_1 = original length of specimen, and
L_2 = length of the specimen when subjected to an applied stress.

C 904, C03

strain (linear or normal), ϵ (D)—the change in length per unit of length in a given direction. **D 653,** D18

strainline or strainlining— see **hairline or hairlining.** **C 286,** B08

strain lines—visualized as schlieren, which are a manifestation of the refraction of light waves occurring at the interface between phases of differing indices of refraction; most readily observed using polarized light. **F 1156,** F12

strain point (St.P.)—that temperature corresponding to a specific rate of elongation of a glass fiber when measured by Test Method C 336 or a specific rate of midpoint deflection of a glass beam when measured by Test Method C 598. **C 162,** C14

strain, *r*—the change in length, due to an applied force, per unit length of a specimen. **E 1142,** E37

strain rate—the rate of relative length deformation with time due to an applied stress. Strain rate is often shown as:

$$\frac{L_2 - L_1}{L_1(T_2 - T_1)}$$

where:
L_1 = original length of specimen,
L_2 = length of the specimen when subjected to an applied stress,
T_1 = time when stress is applied, and
T_2 = time when L_2 is measured.

C 904, C03

strain resolution (strain sensitivity), R_s **(D)**—the smallest subdivision of the indicating scale of a deformation-measuring device divided by the product of the sensitivity of the device and the gage length. The deformation resolution, R_d, divided by the gage length. **D 653,** D18

strain (stress) rate—rate of change of strain (stress) with time. (ISRM) **D 653,** D18

strain (stress) tensor—the second order tensor whose diagonal elements consist of the normal strain (stress) components with respect to a given set of coordinate axes and whose off-diagonal elements consist of the corresponding shear strain (stress) components. (ISRM) **D 653,** D18

strand—See **wire strand.** **A 902,** A05

strand, *n*—one of the wires of any stranded conductor. **B 354,** B01

strand, *n*—one of the wires of any stranded conductor. **B 713,** B01

strand—a multiplicity of continuous glass filaments combined in a single compact unit, without twist. **C 162,** C14

strand, *n*—a single fiber, filament, or monofilament. **D 123,** D13

strand, *n*—an ordered assemblage of textile fibers having a high ratio of length to diameter and normally used as a unit, including slivers, rovings, single yarns, plied yarns, cords, braids, ropes, etc. **D 123,** D13

strand, *n*—*in fiber rope*, an ordered assemblage of textile yarns used to make fiber rope. **D 123,** D13

strand, *n*—*in braided rope*, one of a number of similar units which are intertwined to produce a braided rope, each unit consisting of one or more yarns which are not twisted together and which follow a prescribed path through the braided rope. **D 123,** D13

strand—a wood flake having a minimum predetermined length-to-width ratio of 2:1. **D 1554,** D07

strand—in fibrous composites, Synonym for tow. **D 3878,** D30

stranded conductor—a conductor composed of a group of wires, usually twisted, or of any combination of such groups of wires. **B 354,** B01

stranded conductor—a conductor composed of a group of wires, usually twisted, or of any combination of such groups of wires. **B 713,** B01

stranding factor—the ratio of the mass or electrical resistance of a unit length of stranded conductor to that of a solid conductor of the same cross-sectional area. The stranding factor *m* for the completed stranded conductor is the numerical average of the stranding factors for each of the individual wires in the conductor, including the straight core wire, if any (for which the stranding factor is unity). The stranding factor (m_{ind}) for any given wire in a concentric-lay-stranded conductor is:

$$m_{ind} = \sqrt{1 + (9.8696/n^2)}$$

where:

n = length of lay/diameter of helical path of the wire

B 354, B01

strand irregularity, *n—in textiles*, variation in a property along a strand. **D 123,** D13

strap bellies—thin, lightweight, vegetable-tanned cattlehide bellies, rather flexible and with a low load, processed for the strap trade. **D 1517,** D31

strap nail—bright, regular-stock-steel, 1 ¼ by 0.092 to 2 by 0.113-in. nails with oval 15⁄64 to 17⁄64-in. head and short diamond point. **F 547,** F16

strapping, *n*—a flexible strip material, generally of flat or round cross section, used to: (*1*) fasten merchandise within a **container,** (*2*) hold together a **bundle** or **bale,** (*3*) reinforce a **shipping container,** (*4*) secure a shipping container or an article to a pallet. (Compare **bracing.**) **D 996,** D10

straw—See **purlin nail.** **F 547,** F16

strawberry—a small bubble or blister in the flood coating of a gravel-surfaced membrane. **D 1079,** D08

stray current—current through paths other than the intended circuit, such as through heating coils or the tank. **B 374,** B08

stray current corrosion—the corrosion caused by electric current from a source external to the intended electrical circuit, for example, extraneous current in the earth. **G 15,** G01

stray radiant energy—all radiant energy that reaches the detector at wavelengths that do not correspond to the spectral position under consideration. **E 131,** E13

stray radiant power, P_s—the total detected radiant power outside a specified wavelength (wave number) interval each side of the center of the spectral band passed by the monochromator under stated conditions for wavelength (wave number), slit dimensions, light source, and detector. **E 131,** E13

stray radiant power ratio, P_s/P_t—the ratio of stray radiant power to the total detected radiant power. **E 131,** E13

streak, *n*—an extended unintentional stripe of narrow width, often a single yarn. **D 123,** D13

streaking—nonuniform deposition of a polish film. **D 2825,** D21

streaking, *n*—undesirable lines in a print caused by damaged rollers, cylinder bounce or bottoming gears. **D 6488,** D01

streaking, *n*—linear deposits of ink that run across both image and non-image areas. **D 6488,** D01

streaks, *n*—a print defect which appears as light bands in the printed area in the direction of the printhead travel. **F 1857,** F05

streambank erosion—the removal of bank material by flowing water. **D 4410,** D19

stream counter—an instrument in which the particles to be characterized are presented to an interrogation zone as a one-dimensional stream, the size of the particle being deduced from the physical changes it causes in the interrogation zone. (See also **resistazone** and **photozone.**) **C 242,** C21

stream discharge—the quantity of flow passing through a cross section in a unit of time. **D 4410,** D19

streaming potential—a potential that is generated in dilute solutions moving at high velocity through small channels. [Liquid junction potential problems are often wrongly confused with streaming potentials.] **D 4127,** D19

streamline flow—see **laminar flow.** **D 653,** D18

strength—a term to indicate relative thickness in sheet glass. **C 162,** C14

strength—(*1*) the ability of a member to sustain stress without failure. (*2*) in a specific mode of test, the maximum stress sustained by a member loaded to failure. **D 9,** D07

strength, *n*—the property of a material that resists deformation induced by external forces. (Compare **force.**) **D 123,** D13

strength—maximum stress which a material can resist without failing for any given type of loading. (ISRM) **D 653,** D18

strength, n—See **bond strength.** **D 907,** D14

strength, *n—of paper*, the force at which paper ruptures. **D 1968,** D06

strength, *n—dyer's*, the color quality that increases with an increase in the amount of dye present, other conditions remaining constant. (See also **tinting strength.**) **E 284,** E12

strength, *n*—resistance to external force or load or generation of internal strain, expressed in terms of units of force, lbf, pounds force (N, newtons).

ultimate strength, n—maximum resistance to applied force, load, or stress that a material, member, component, or assembly of a structure can withstand without failure. **E 631,** E06

strength:

compressive strength—the maximum compressive strength that a material is capable of sustaining. Compressive strength is calculated from the maximum load during a compressive test and the original cross-sectional area of the specimen.

dry strength—the strength of an adhesive joint or composite structure determined immediately after drying under specified conditions or after a period of conditioning in a standard laboratory atmosphere.

wet strength—the strength of an adhesive bond or composite measured after exposing the test specimen to moisture/water vapor until saturated. **E 1749,** E06

strength, *n*—resistance to external force or load or generation of internal strain, expressed in terms of units of force, newtons (pound-force). **E 1807,** E06

strength—the stress required to break, rupture, or cause a failure. **F 412,** F17

strength analyzer, *n*—an instrument which determines tensile strength and elongation at breaking load for a test beard of cotton. **D 123,** D13

strength at rupture, *n*—strength expressed in terms of the force at rupture. (Compare **breaking strength.**) **D 123,** D13

strength, double—sheet glass between 0.115 and 0.134 in. (2.92 and 3.40 mm) thick. **C 162,** C14

strength grade—a system of strength classifications used for bolts, nuts, and screws made to inch standards. **F 1789,** F16

strength ratio—the hypothetical ratio of the strength of a member to the strength it would have if no weakening defects were present. **D 9,** D07

strength-reducing defects—imperfections affecting strength, such as checks, compression wood, cross grain, decay, knots, shakes, splits, etc. **D 9,** D07

strength, single—sheet glass between 0.085 and 0.101 in. (2.16 and 2.57 mm) thick. **C 162,** C14

strength, transverse (or flexural), *n*— the breaking load applied normal to the neutral axis of a beam. **C 168,** C16

strength unit, *n—in asbestos*, unit of reinforcing potential of asbestos fiber in asbestos-cement products. An asbestos fiber that yields a flexural modulus of rupture of 27 MPa at a product density of 1.6 g/cm^3 when used as 10 % of the furnish (dry ingredients) is arbitrarily defined as having 100 strength units. Therefore, the number of strength units of a given asbestos is equal to 1000/(% fiber required in the dry mix to yield 27 MPa at 1.6 g/cm^3). **C 460,** C17

strength unit, *n— in asbestos,* unit of reinforcing potential of asbestos fiber in asbestos-cement products. An asbestos fiber that yields a flexural modulus of rupture of 27 MPa at a product density of 1.6 g/cm^3 when used as 10 % of the furnish (dry ingredients) is defined as having 100 strength units. Therefore, the number of strength units of a given asbestos is equal to 1000/(% fiber required in the dry mix to yield 27 MPa at 1.6 g/cm^3). **D 2946,** C17

stress—force per unit of area. **D 9,** D07

stress, *n*—the resistance to deformation developed within a material subjected to an external force. **D 123,** D13

stress, *n*—force exerted per unit area at a point within a plane. **D 907,** D14

stress, *n*—the intensity, at a point in a body, of the internal forces (or components of force) that act on a given plane through the point. **D 1566,** D11

stress, σ—the intensity at a point in a body of the internal forces or components of force that act on a given plane through the point. (E-28) **D 4092,** D20

stress, *n*—internal force developed by application of external force or load or generation of internal strain, expressed in terms of units of force per unit area, psi, pounds force per square inch (Mpa, megapascals). **E 631,** E06

stress, *n*—internal force developed by application of external force or load or generation of internal strain expressed in terms of unit of force per unit of area, megapascals (pound-force per square inch). When the forces are parallel to the plane on which it acts, the stress is called shear stress; when the forces are normal to the plane on which it acts, the stress is called normal stress; when the normal stress is directed toward the plane on which it acts, it is called compressive stress; when the normal stress is directed away from the plane on which it acts, it is called tensile stress. Sometimes referred to as unit stress. **E 1807,** E06

stress amplitude—the ratio of the maximum applied force, measured from the mean force, to the cross-sectional area of the unstressed test specimen. (ISO) **D 4092,** D20

stress concentration factor—See **theoretical stress concentration factor (or stress concentration factor)** k_t. **E 1823,** E08

stress corrosion crack—spontaneous failure of metals by cracking under combined action of corrosion and stress, residual or applied. **B 846,** B05

stress-corrosion crack—a crack which may be intergranular or transgranular depending on the material, resulting from the combined action of corrosion and stress, either external (applied) or internal (residual). **E 7,** E04

stress-corrosion cracking—a cracking process that requires the simultaneous action of a corrodent and sustained tensile stress. (This excludes corrosion-reduced sections which fail by fast fracture. It also excludes intercrystalline or transcrystalline corrosion which can disintegrate an alloy without either applied or residual stress.) **G 15,** G01

stress-corrosion cracking, SCC—a cracking process that requires the simultaneous action of a corrodent and sustained tensile stress. **E 1823,** E08

stress-crack, *n*—an external or internal crack in a plastic caused by tensile stresses less than its short-time mechanical strength. **D 883,** D20

stress crack, *n*—an external or internal crack in a plastic caused by tensile stresses less than its short-time mechanical strength. **D 4439,** D35

stress-crack, *n*—an external or internal crack in a plastic caused by tensile stresses less than its short-time mechanical strength. **F 1251,** F04

stress crack, environmental, *n*—a stress crack, the development of which has been accelereated by the environment to which the plastic is exposed. (See **stress-crack.**) **F 412,** F17

stress cycle—See **cycle.** **E 1823,** E08

stress decay, *n—in mechanics,* the reduction in force to hold a material at a fixed deformation over a period of time. **D 4848,** D13

stress ellipsoid—the representation of the state of stress in the form of an ellipsoid whose semi-axes are proportional to the magnitudes of the principal stresses and lie in the principal directions. The coordinates of a point *P* on this ellipse are proportional to the magnitudes of the respective components of the stress across the plane normal to the direction *OP*, where *O* is the center of the ellipsoid. (ISRM) **D 653,** D18

stress [FL^{-2}], *n*—the intensity at a point in a body of the forces or components of force that act on a given plane through the point. Stress is expressed in force per unit of area (pounds-force per square inch, megapascals, and so forth). **E 6,** E28

stress, [FL^{-2}]—force acting over a unit area. Traditionally, the symbol for stress is either S or σ , as a matter of choice. **E 1823,** E08

stress-intensity factor at crack initiation, K_o[$FL^{-3/2}$]—the value of K at the onset of rapid fracturing. **E 1823,** E08

stress-intensity factor coefficient, Y^*—a dimensionless parameter that relates the applied load and specimen geometry to the resulting crack-tip stress-intensity factor in a chevron-notch specimen test. **E 1823,** E08

stress-intensity factor, K, K_1, K_2, K_3, K_I, K_{II}, K_{III} [$FL^{-3/2}$]—the magnitude of the mathematically ideal, crack-tip stress field (stress-field singularity) for a particular mode in a homogeneous, linear-elastic body. **E 1823,** E08

stress-intensity factor range, ΔK [$FL^{-3/2}$]—*in fatigue*, the variation in the stress-intensity factor in a cycle, that is

$$\Delta K = K_{max} - K_{min}$$

E 1823, E08

stress intensity factor threshold for environment assisted cracking, K_{EAC} [$FL^{-3/2}$]—the highest value of the stress intensity factor (K) at which crack growth is not observed for a specified combination of material and environment and where the measured value may depend on specimen thickness. **E 1823,** E08

stress intensity factor threshold for plane strain environment-assisted cracking, K_{IEAC} [$FL^{-3/2}$]— the highest value of the stress intensity factor (K) at which crack growth is not observed for a specified combination of material and environment and where the specimen size is sufficient to meet requirements for plane strain as described in Test Method E 399. **E 1823,** E08

stress, σ, p, f (FL^{-2})—the force per unit area acting within the soil mass.

effective stress (effective pressure) (intergranular pressure), $\bar{\sigma}$, f (FL^{-2})—the average normal force per unit area transmitted from grain to grain of a soil mass. It is the stress that is effective in mobilizing internal friction.

neutral stress (pore pressure) (pore water pressure), u, u_w (FL^{-2}) —stress transmitted through the pore water (water filling the voids of the soil).

normal stress, σ, p (FL^{-2})—the stress component normal to a given plane.

principal stress, σ_1, σ_2, σ_3(FL^{-2})—stresses acting normal to three mutually perpendicular planes intersecting at a point in a body, on which the shearing stress is zero.

major principal stress, σ_1(FL^{-2})—the largest (with regard to sign) principal stress.

minor principal stress, σ_3(FL^{-2})—the smallest (with regard to sign) principal stress.

intermediate principal stress, σ_2(FL^{-2})—the principal stress whose value is neither the largest nor the smallest (with regard to sign) of the three.

shear stress (shearing stress) (tangential stress), τ, *s* FL^{-2})—the stress component tangential to a given plane.

total stress, σ, *f* (FL^{-2})—the total force per unit area acting within a mass of soil. It is the sum of the neutral and effective stresses. **D 653,** D18

stress ratio—ratio of ultimate stress of metal connector plate to ultimate stress of matched solid metal-coupon control specimen. Also called *effectiveness ratio* and *efficiency ratio.* **E 1807,** E06

stress relaxation—stress release due to creep. (ISRM) **D 653,** D18

stress relaxation, *n*—the decrease in stress after a given time at constant strain. **D 1566,** D11

stress relaxation, *n*—the time-dependent decrease in stress in a solid under given constraint conditions. **E 6,** E28

stress relaxation—the decrease in stress, at constant strain, with time. **F 412,** F17

stress relief—a treatment of a product to reduce residual stresses. **B 846,** B05

stress relieving, *n*—heating a steel object to a suitable temperature, holding it long enough to reduce residual stresses, and then cooling it slowly enough to minimize the development of new residual stresses. **A 941,** A01

stress retention—the residual load expressed as a percentage of the original load on a material after the test specimen has been maintained at a constant elongation for a specified time. **D 996,** D10

stress, *s*—force per unit area. **E 1142,** E37

stress, S, *n*— the intensity at a point in a body of the forces or components of force that act on a given plane through the point. **F 1582,** F04

stress-strain curve, *n*—a graphical representation of the stress and strain relationship of a material under conditions of compression, shear, tension, or torsion. (Compare **force-deformation curve, force-elongation curve,** and **force-extension curve.**) **D 123,** D13

stress-strain diagram, *n*—a diagram in which corresponding values of stress and strain are plotted against each other. **D 907,** D14

stress-strain diagram, *n*—a diagram in which corresponding values of stress and strain are plotted against each other. Values of stress are usually plotted as ordinates (vertically) and values of strain as abscissas (horizontally). **E 6,** E28

stress (strain) field—the ensemble of stress (strain) states defined at all points of an elastic solid. (ISRM) **D 653,** D18

stretch, *n—of paper*, the maximum tensile strain (elongation) developed in a paper test specimen prior to rupture in a tensile test that has been carried to rupture under the conditions specified. **D 1968,** D06

stretch—*1*) an increase or elongation in dimension; *2*) mechanically enforcing an elongation of belting during the vulcanization to render it less extensible in service. **F 869,** F08

stretching—the process of putting tension on the fabric to make it hang uniformly along the line of fence between terminal posts. **F 552,** F14

stretch stitch, *n—in home sewing*, a complex machine stitch pattern or of various combinations of straight stitch, single stitch zigzag, or multiple stitch zigzag. It is produced by coordinated motions of needle and feed as the fabric moves through the feed mechanism in the forward or reverse direction. **D 123,** D13

stretch woven fabric, *n*—a woven fabric which is capable of at least 20 % stretch in either warp or filling direction, or both, under loads and conditions encountered in use, and of almost complete recovery after removal of the force. **D 123,** D13

stretch zone width (*SZW*) [L]—the length of crack extension that occurs during crack-tip blunting, for example, prior to the onset of unstable brittle crack extension, pop-in, or slow stable crack extension. The *SZW* is co-planar with the original (unloaded) fatigue precrack and refers to an extension of the original crack. **E 1823,** E08

stria—a cord of low intensity generally of interest only in optical glass. (See also **cord**.) **C 162,** C14

striated—a term used to describe plywood with a face veneer that has been grooved or scored parallel to the grain. **D 1038,** D07

striated head—head with parallel ridges and grooves to identify high-density fiberboard nail. **F 547,** F16

striations—ridges or steps on the fracture surface parallel to the local direction of the crack front propagation, generally having the appearance of feathers. **C 162,** C14

striations—*in semiconductor technology*, helical features on the surface of a silicon wafer associated with local variations in impurity concentration. **F 1241,** F01

strike: (*1*) *n*—a thin film of metal to be followed by other coatings.
(*2*) *n*—a solution used to deposit a strike.
(*3*) *v*—to plate for a short time, usually at a high initial current density. **B 374,** B08

strike—the direction or azimuth of a horizontal line in the plane of an inclined stratum, joint, fault, cleavage plane, or other planar feature within a rock mass. (ISRM) **D 653,** D18

strike tacker—stapler activated by striking it with hand or mallet. **F 592,** F16

strike-through, *n*—the penetration of the vehicle of a printing ink through the sheet, so that the ink is apparent on the reverse side under normal lighting conditions. **D 6488,** D01

strike through, *n*—ink vehicle penetrates the substrate and is visible on the opposite side. **F 1857,** F05

striking—development of color or opacity during cooling or reheating. **C 162,** C14

striking surface, *n*—the hardened, smooth, circular bottom surface of the intermediate tool that is in contact with the test sample. **E 1445,** E27

string—an imperfection; a straight or curled line, usually resulting from slow solution of a large grain of sand or foreign material. **C 162,** C14

stringer, *n—in zippers*, the tape, bead, and element assembly that constitutes one side of a chain. **D 123,** D13

stringer—See **pallet.** **D 996,** D10

stringer—a single, high-aspect ratio, elongated inclusion, two or more elongated inclusions, or a number of small non-deformable inclusions aligned in a linear pattern due to deformation. **E 7,** E04

stringiness, *n*—the property of an adhesive that results in the formation of filaments or threads when adhesive transfer surfaces are separated. (See also **legging** and **webbing**.) (Compare **teeth.**) **D 907,** D14

strip, *n*—a process or solution used for the removal of a coating from a basis metal or an undercoat. **B 374,** B08

strip, *v*—to remove a coating from the basis metal or undercoat. **B 374,** B08

strip—*as in inch-pound specifications,* a rolled flat product, other than flat wire, up to and including 0.188-in. thick, in straight lengths, coils (rolls), or traverse wound on reels or spools having either slit, sheared or slit and rolled edges in widths up to 24-in. inclusive, or having finished drawn or rolled edges, in widths over 1¼-in. to 12-in. inclusive. **B 846,** B05

strip—*as in SI specifications,* a rolled flat product other than flat wire up to and including 5-mm thick in straight lengths, coils (rolls), or traverse wound on reels or spools having either slit or sheared edges in widths up to 600-mm inclusive, or having finished drawn or rolled edges, in widths over 30-mm to 300-mm inclusive. **B 846,** B05

strip, *n*—a flat-rolled metal product of some maximum thickness and width arbitrarily dependent on the type of metal, it is narrower than sheet. **B 899,** B02

strip—a thin batten or narrow, thin board. Specific descriptions may vary between softwood and hardwood. **D 9,** D07

strip, *n*—a flat, thin member, much longer than wide, having width greater than thickness (if of steel, having dimensions of 0.229 in. (5.8 mm) or less thickness and 12.0 (305 mm) or less width). **E 631,** E06

strip—staples, nails, or pins collated and cohered to facilitate automatic driving with appropriate tool. **F 592,** F16

strip-fed tool—tool using staples or nails that are collated and cohered in strip form. **F 592,** F16

striping, *n*—a printing imperfection observed when the print becomes alternatively more and less intense across the web. **D 6488,** D01

strip mopping—See **mopping**. **D 1079,** D08

strip nail—steel, 1 ¼ to 2 ½-in. nails spot welded to disposable metal strip that feed nails into nailing machine provided with staple-type magazine. During punching of nail from strip, small washer is formed under head. Also, nail stored in special strip to serve as magazine for feeding nailing machine. **F 547,** F16

stripper, *n*—*in textiles*, a product, usually the reducing agent, that changes the coloring material dye, or soil stain to reduced color. (See also **bleach**). **D 123,** D13

stripper punch—a punch which, in addition to forming the top or bottom of the die cavity, later moves further into the die to eject the compact. **B 243,** B09

stripping—strip flashing:

(*1*) the technique of sealing a joint between metal and built-up membrane with one or two plies of felt or fabric and hot- or cold-applied bitumen;

(*2*) the technique of taping joints between insulation boards or deck panels. **D 1079,** D08

stripping yard—{archaic} the place where glass plates are removed from tables after grinding and polishing. **C 162,** C14

strip test, *n*—*in fabric testing*, a tensile test in which the full width of the specimen is gripped in the clamps. **D 123,** D13

stroke—*in character recognition,* a straight line or arc used as a segment of a graphic character. Each character is made up of a variable number of the strokes. **F 149,** F05

stroke, *n*—a movement in a single direction, either forward or backwards. **F 395,** F11

stroke—*in character recognition*, a straight line or arc used as a segment of a graphic character. **F 909,** F05

stroke average width—the average of actual stroke widths taken at points along the length of a stroke. **F 149,** F05

stroke centerline—*in character recognition*, a line midway between the two stroke edges. **F 149,** F05

stroke centerline—*in character recognition*, a line midway between the two edges of a stroke. **F 909,** F05

stroke device—an input device providing a set of coordinates that record the path of the device. **F 909,** F05

stroke edge—*in character recognition,* the line of discontinuity between a side of a stroke and the background, obtained by averaging, over the length of the stroke, the irregularities resulting from the printing and detecting processes. **F 149,** F05

stroke width—*in character recognition,* the distance measured perpendicularly to the stroke centerline between the two stroke edges. **F 149,** F05

strong probability did not—this carries the same weight as strong probability on the identification side of the scale; that is, the examiner is virtually certain that the questioned and known writings were not written by the same individual.

Examples—There is strong probability that the John Doe of the known material did not write the questioned material, or in my opinion (or conclusion or determination) it is highly probable that the John Doe of the known material did not write the questioned material. **E 1658,** E30

strong probability (highly probable, very probable)—the evidence is very persuasive, yet some critical feature or quality is missing so that an *identification* is not in order; however, the examiner is virtually certain that the questioned and known writings were written by the same individual.

Examples—There is *strong probability* that the John Doe of the known material wrote the questioned material, or it is my opinion (or conclusion or determination) that the John Doe of the known material *very probably* wrote the questioned material. **E 1658,** E30

struck surface, *n*—the surface of a molded brick that is not in contact with the mold and from which the excess clay/shale mixture is removed. **C 43,** C15

structural adhesive, *n*—a bonding agent used for transferring required loads between adherends exposed to service environments typical for the structure involved. **D 907,** D14

structural boards—boards graded for structural applications requiring stress grading for assignment of allowable properties. **D 9,** D07

structural bolt—a heavy hex head bolt having a controlled thread length intended for use in structural connections and assembly of such structures as buildings and bridges. **F 1789,** F16

structural clay facing tile, *n*—tile designed for use in interior and exterior unplastered walls, partitions or columns. See Specification C 212. **C 43,** C15

structural clay tile, *n*—hollow burned-clay masonry building units with parallel cells or cores or both. **C 43,** C15

structural gasket—See *lock-strip gasket* **C 717,** C24

structural-glued-laminated timber, *n*— an engineered, stress-rated product of a timber laminating plant comprising assemblies of specially selected and prepared wood laminations securely bonded together with adhesives, with the following characteristics: (*1*) The grain of all laminations is approximately parallel longitudinally, and (*2*) The laminations may be comprised of pieces end-joined to form any length, of pieces placed or glued edge-to-edge to make wider ones, or of pieces bent to curved form during gluing. (Synonym **GLULAM**) ANSI/AITC A190.1—1992, American National Standard for Wood Products—Structural Glued Laminated Timber (Edited to conform with ASTM format.) **D 907,** D14

structural quality sheet coil—coiled sheet metal used for production of metal connector plates meeting minimum specified grade properties for yield and ultimate stresses and elongation. If Specification A 653/A 653M Grade 40 is specified, the steel properties shall exceed 16 pct for elongation for a 50 mm (2.0 in.) gage length of the 275 MPa (40 ksi) yield point and 380 MPa (55 ksi) ultimate tensile stress, according to Specification A 653/A 653M. **E 1807,** E06

structural sandwich construction, n—a laminar construction comprising a combination or alternating dissimilar simple or composite materials assembled and intimately fixed in relation to each other so as to use the properties of each to attain specific structural advantages for the whole assembly. **C 274,** D30

structural sealant, *n—in building construction*, a sealant capable of transferring dynamic or static ("live" or "dead," or both) loads, or both, across joint members exposed to service environments typical for the structure involved. **C 717,** C24

structural sealant glazing, *n—in building construction*, a glazing system wherein a structural sealant is used to transfer loads between a lite or panel and a supporting framework, without mechanical fasteners or other methods of attachment. **C 717,** C24

structural slab—a horizontal, supporting, cast-in-place, concrete building deck. **C 717,** C24

structural system—a combination of load supporting and transmitting construction elements or **components** of an assembly or **building** including connections. **E 631,** E06

structural test, *n*—determination of one or more values for load, stress, and deflection characteristics of a material or assembly. **E 631,** E06

structure—one of the larger features of a rock mass, like bedding, foliation, jointing, cleavage, or brecciation; also the sum total of such features as contrasted with texture. Also, in a broader sense, it refers to the structural features of an area such as anti-clines or synclines. (ISRM) **D 653,** D18

structure—see **soil structure**. **D 653,** D18

structure—as applied to a crystal, the shape and size of the unit cell and the location of all atoms within the unit cell. As applied to microstructure, the size, shape and arrangement of phases. **E 7,** E04

structure, *n*—the order of data elements in a message. **F 1294,** F05

structureborne sound—sound that arrives at the point of interest, such as the edge of a partition, by propagation through a solid structure. **C 634,** E33

structure factor—the ratio of the amplitude of the wave scattered by all the atoms of a unit cell to the amplitude of the wave scattered by a single electron. Symbol = *F*. **E 7,** E04

strut, *n*—a short member of a **crate** or **bracing** used as reinforcement and designed to resist compressive forces in the direction of its length. **D 996,** D10

stubble, *n—in shorn floor covering*, the portion of the pile that remains after shearing. **D 123,** D13

stubble, *n—in pile yarn floor covering*, the number of pile tufts per unit length in the lengthwise direction. **D 5684,** D13

stubble height, *n*—the distance the stubble extends above the backing fabric. **D 123,** D13

stucco, *n*—a portland cement-aggregate plaster mix designed for use on exterior surfaces. See **portland cement plaster**. **C 11,** C11

stud—one of a series of slender wood structural members used as supporting elements in walls and partitions. In softwood grading, a stress grade to describe lumber suitable for stud use. **D 9,** D07

study period—the length of time over which an investment is analyzed (Syn. *life cycle, time horizon*). **E 833,** E06

stuffer yarn, *n*—an extra backing yarn running in the warp direction through a woven pile floor covering. **D 123,** D13

stuffing—the process of incorporating grease in leather by drumming the wet leather with warm, molten grease and oils. **D 1517,** D31

stump—the part of a tree remaining above and below ground after the main stem is cut off. **D 9,** D07

sturdy pretreatment procedure, *n—in textile conservation*, the washing, rinsing, drying, and ironing actions followed when backing fabrics are medium to heavy weight. (Compare **delicate pre-treatment procedure.**) **D 123,** D13

STX—abbreviation for **start of text**. **F 1457,** F05

stylus—pen-like device for writing, sometimes by inscription. **F 1156,** F12

styrene monomer—ethenylbenzene, vinylbenzene, phenylethylene, ($C_6H_5CH = CH_2$) mol weight 104.15; clear, colorless, flammable liquid; characteristic odor; solidification point −30.6°C; boiling point 145.2°C. Refer to Specification D 2827 for complete specifications. **D 4790,** D16

styrene plastic, *n*—a plastic based on polymers of styrene or copolymers of styrene with other monomers, the styrene being greatest amount by mass. **F 1251,** F04

styrene plastics, *n*—plastics based on polymers of styrene or copolymers of styrene with other monomers, the styrene being in greatest amount by mass. (ISO) **D 883,** D20

styrene plastics, *n*—plastics based on polymers of styrene or copolymers of styrene with other monomers, the styrene being the greatest amount by mass. (ISO, D-20) **F 412,** F17

styrene resin—under **resin, synthetic,** see *styrene resin*. **D 16,** D01

styrene-rubber plastics, *n*—plastics based on styrene polymers and rubbers, the styrene polymers being in the greatest amount by mass. (ISO) **D 883,** D20

styrene-rubber plastics, *n*—plastics based on styrene polymers and rubbers, the styrene polymers being in the greatest amount by mass. (ISO, D-20) **F 412,** F17

styrene-rubber (SR) pipe and fitting plastics—plastics containing at least 50% styrene plastics combined with rubbers and other compounding materials, but not more than 15 % acrylonitrile. **F 412,** F17

styrenic thermoplastic elastomer (TES), *n*—a copolymer material comprising styrenic end-block segments bound to a rubbery segment that may be saturated or unsaturated. **D 883,** D20

subassembly, *n*— any portion of an implant assembly that is composed of two or more components. **F 1582,** F04

subbase—a layer used in a pavement system between the subgrade and base coarse, or between the subgrade and portland cement concrete pavement. **D 653,** D18

subbituminous A coal—See *subbituminous A coal* under **rank**. **D 121,** D05

subbituminous B coal—See *subbituminous B coal* under **rank**. **D 121,** D05

subbituminous C coal—See *subbituminous C coal* under **rank**. **D 121,** D05

subbituminous class—See *subbituminous class* under **rank**. **D 121,** D05

subconstruct, *n*—any portion of an implant construct that is composed of two or more components including the spine, pelvis, ribs, or substitute structure. **F 1582,** F04

subcritical annealing, *n*—**annealing** at a temperature slightly below Ac_1. **A 941,** A01

subcritical flow—open channel flow that is deeper and at lower velocity than critical flow for the same flow rate; sometimes called tranquil flow. A Froude number less than one exists. **D 4410,** D19

sub-face (sub-back)—the ply adjacent to the exposed face (or back) of a parallel laminated outer layer. **D 1038,** D07

sub-floor—See **floor.** **E 631,** E06

subfloor, *n*—that structural layer intended to provide support for design loadings which may receive resilient floor coverings directly if the surface is appropriate or indirectly via an underlayment if its surface is not suitable. **F 141,** F06

sub-flooring—See **floor.** **E 631,** E06

subflooring nail—See **threaded sinker (annularly threaded), diaphragm nail.** **F 547,** F16

subgrade—the soil prepared and compacted to support a structure or a pavement system. **D 653,** D18

sub-grade—See **grade.** **E 631,** E06

subgrade surface—the surface of the earth or rock prepared to support a structure or a pavement system. **D 653,** D18

subgroup, *n*—(*1*) **object sense,** *n*—a set of units or quantity of material obtained by subdividing a larger group of units or quantity of material, and (*2*) **measurement sense,** *n*—a set of groups of observations obtained by subdividing a larger group of observations. See **rational subgroup.** **E 456,** E11

subgroup—one or more events or measurements used to analyze the performance of a process. **F 1789,** F16

subject, *n*—the individual to whom a stimulus is applied. **E 253,** E18

subject contrast—the ratio (or the logarithm of the ratio) of the radiation intensities transmitted by selected portions of the specimen. **E 1316,** E07

sublimation, *n*—the volatilization of a solid directly to the vapor state, without passing through the liquid state. **C 1145,** C28

sublimation pressure—at a stated temperature, that pressure at which congruent equilibrium between a solid substance and its vapor occurs. **E 7,** E04

sublimation pressure—in a binary system, at a stated temperature, that pressure at which congruent equilibrium between a solid substance and its vapor occurs. **E 1142,** E37

sublimation temperature—at a stated pressure, that temperature at which congruent equilibrium between a solid substance and its vapor occurs. **E 7,** E04

sublimation temperature—in a binary system, at a stated pressure, that temperature at which congruent equilibrium between a solid substance and its vapor occurs. **E 1142,** E37

submerged throat—a throat with its floor below the floor of the melter and its top below the metal line. **C 162,** C14

submerged unit weight—see **unit weight.** **D 653,** D18

submersible boom—boom that normally resides on the seabed and is positioned by inflating with air, causing it to rise to the water surface. **F 818,** F20

submicroscopic—See **ultramicroscopic.** **E 7,** E04

subsample—See *subsample* under **sample.** **D 121,** D05

subsample, *n*—a portion of a material that is collected by subdividing or trimming of a sample. **D 5681,** D34

subsealing—*in grouting*, grouting under concrete slabs for the purpose of filling voids without raising the slabs. **D 653,** D18

subsidence—the downward displacement of the overburden (rock or soil, or both) lying above an underground excavation or adjoining a surface excavation. Also the sinking of a part of the earth's crust. (ISRM) **D 653,** D18

subsieve fraction—particles all of which will pass through a 44-μm (No. 325) standard sieve. **B 243,** B09

subsoil—(*a*) soil below a subgrade of fill. (*b*) that part of a soil profile occurring below the "A" horizon. **D 653,** D18

substandard dwelling—dwelling or residence that does not meet minimum local housing codes, or is unsafe because of violation of fire, safety, or health codes. **E 1605,** E06

substitution, *v*—to put or use one document or item in place of another; within this context, the substituted document or item often is of lesser quality or value. **F 1156,** F12

substitution error, *n*—the replacement of a bar code marked character(s), by an erroneous character(s) usually traceable to poor quality printing, decoding logic error, human input error, or any combination thereof. **F 1294,** F05

substitution error rate, SER, *n*—the ratio of the number of invalid or incorrect characters entered into the data base to the number of valid characters entered. **F 1294,** F05

substrate—*in diffusion coatings*, material on which coating is deposited. **B 374,** B08

substrate—a body, board, or layer of material on which some other active or useful material or component may be deposited or laid, as for example, an electronic circuitry laid on an alumina ceramic board. **C 242,** C21

substrate, *n*—*in building construction*, (*1*) a material upon which films, treatments, adhesives, sealants, membranes, and coatings are applied; (*2*) materials that are bonded or sealed together by adhesives or sealants. **C 717,** C24

substrate, *n*—a body, board, or layer of material on which some other active or useful material or component may be deposited or laid, as for example, an electronic circuitry laid on an alumina ceramic board. In catalysts, the formed, porous, high-surface-area carrier on which the catalytic agent is widely and thinly distributed for reasons of performance and economy. **C 1145,** C28

substrate—the soiled surface that is being cleaned. **D 459,** D12

substrate, *n*—*as related to adhesives*, a material upon which an adhesive is applied. (Compare **adherend.**) **D 907,** D14

substrate—the surface upon which the roofing or waterproofing membrane is placed (structural deck or insulation). **D 1079,** D08

substrate—(substratum) that which lies under; foundation. **E 7,** E04

substrate—material or object upon which something resides; for example, paper is a substrate for writing; fiberglass-epoxy composites are substrates for printed circuit boards. **F 1156,** F12

substrate—*in semiconductor technology*, a wafer that is the basis for subsequent processing operations in the fabrication of semiconductor devices or circuits. **F 1241,** F01

substrate, *n*—the material (usually paper, plastic or metal) upon which a bar code symbol is printed or reproduced. **F 1294,** F05

substrate base—unfinished building-material surface to which finishes are applied. Examples include wood, plaster, drywall, masonry, and metal. **E 1605,** E06

substrate effect—see **XRF substrate effect.** **E 1605,** E06

substrate equivalent lead concentration (SEL)—average of at least three XRF analyzer readings on a surface from which the coating has been removed. **E 1605,** E06

substrate failure, n—in characterizing the locus of an adhesvie-joint failure, the fracture occurring within the substrate. (See *cohesion failure.*) **D 907,** D14

subsurface corrosion— See **internal oxidation.** **G 15,** G01

subsurface discontinuity—any defect that does not open onto the surface of the part in which it exists. **E 1316,** E07

subsystem—See **building subsystem.** **E 631,** E06

sub-threshold, *adj*—pertaining to a stimulus below the specified threshold. **E 253,** E18

subtractive color mixture, *n*—mixture of absorbing media or superposition of filters so that the spectral composition of light passing through the combination is determined by simultaneous or successive absorption. **E 284,** E12

subtractive color theory—the three primary colors of this theory are yellow, magenta, and cyan; when mixed together, they create black by the subtraction of the color elements, not white as happens when the three primary colors of additive color theory: red, yellow, and green, are mixed together additively. **F 1156,** F12

subtractive primaries, n—see **primary colorants.** **E 284,** E12

suction, *n*—the absolute difference between ambient and subatmospheric pressure expressed in inches or millimetres of water. **F 395,** F11

suction process—any process in which glass is gathered by vacuum into the mold. **C 162,** C14

suds—a foam or lather generated on or in a detergent solution. **D 459,** D12

suede finish—a finish produced by running the surface of leather on a carborundum or emery wheel to separate the fibres in order to give the leather a nap. The grain side of the leather may be suede-finished, but the process is most often applied to the flesh surface. The term "suede" is applied to chrome or alum-tanned leather, while "ooze" is applied to vegetable-tanned suede. The term "suede" when used alone refers to leather only. *The term denotes a finish, not a type of leather.* **D 1517,** D31

sugars—molecules of carbohydrate, namely monosaccharides and disaccharides such as glucose, galactose, mannase, sucrose or fructose, etc. **E 1705,** E48

sulfate attack—*in grouting,* harmful or deleterious reactions between sulfates in soil or groundwater and the grout. **D 653,** D18

sulfated oil—a newer term signifying the same type of material as sulfonated oil. **D 459,** D12

sulfate pulp—see **kraft pulp.** **D 1695,** D01

sulfation—the introduction into an organic molecule of the sulfuric ester group (or its salts) —O—SO_3H, where the sulfur is linked through an oxygen atom to the parent molecule. **D 459,** D12

sulfidation—the reaction of a metal or alloy with a sulfur-containing species to produce a sulfur compound that forms on or beneath the surface of the metal or alloy. **G 15,** G01

sulfide creep—in electrical contacts, spontaneous migration of a sulfide based corrosion product that occurs at elevated humidity across a gold rich surface. **B 542,** B02

sulfide-type inclusions—in steels, nonmetallic inclusions composed essentially of manganese iron sulfide solid solutions (Fe, Mn) S. They are characterized by plasticity at hot-rolling and forging temperatures and, in the hot worked product, appear as dove gray elongated inclusions varying from a threadlike to oval outline. Selenide type inclusions may behave similarly. **E 7,** E04

sulfite cellulose—a by-product of paper mills, produced in sulfiting wood pulp, used as a tanning material; more correctly named lignosulfonate since it does not contain cellulose. **D 1517,** D31

sulfite pulp—wood pulp produced by cooking with a sulfite liquor made by dissolving sulfur dioxide in an aqueous base. **D 1695,** D01

sulfochromate etch—etchant used for preparing the surface of aluminum alloys for adhesive bonding (sulfuric acid/sodium dichromate). Also known as the *Forest Products Laboratory (FPL) etch.* **E 1749,** E06

sulfoferric etch—etchant used for preparing the surface of aluminum alloys for adhesive bonding (sulfuric acid/ferric sulfate). Also known as the *P2 etch.* **E 1749,** E06

sulfonated oil—a water dispersible or soluble surface active material obtained by treating an unsaturated or hydroxylated fatty oil, acid, or ester with an agent capable of sulfating or sulfonating it at least partially. **D 459,** D12

sulfonation—the introduction into an organic molecule of the sulfonic acid group (or its salts) —SO_3H where the sulfur atom is joined to a carbon atom of the parent molecule. **D 459,** D12

sulfur, combined, *n*—the sulfur remaining in a vulcanizate after extraction by a prescribed method. **D 1566,** D11

sulfur dioxide (SO_2)—mol weight 64.07; colorless, nonflammable gas with strong suffocating odor soluble in water and organic solvents. **D 4790,** D16

sulfur donor vulcanizing system, *n*— a vulcanizing system in which there is no elemental sulfur present and all of the sulfur available for the crosslinking is provided by the partial decomposition of sulfur-containing materials. **D 1566,** D11

sulfur, extractable, *n*—all the sulfur removed by a solvent from a rubber mix or vulcanizate. **D 1566,** D11

sulfur, free, *n*—ideally, the uncombined sulfur in a rubber mix or vulcanizate. **D 1566,** D11

sulfuric anhydride (organically combined)—that portion of sulfur, calculated as sulfur trioxide, bound through oxygen to carbon in an organic sulfate, or bound directly to carbon in an organic sulfonate. **D 459,** D12

sulfur modifier—the concentrated product produced by reacting sulfur with a hydrocarbon material, used with additional elemental sulfur to produce a binder for sulfur polymer cement concrete. **C 904,** C03

sulfur mortar—a product consisting of fillers, such as carbon or silica flour, dispersed in sulfur. Small amounts of modifying additives may be included. **C 904,** C03

sulfur polymer cement—the product obtained by reacting sulfur with chemical modifiers to produce a binder for sulfur polymer cement concrete. **C 904,** C03

sulfur polymer cement concrete—a thermoplastic chemical-resistant construction material composed of sulfur polymer cement binder or a binder composed of elemental sulfur and sulfur modifier, hot-mixed with appropriate types and amounts of aggregate and mineral filler. Commonly referred to as sulfur concrete or sulfur polymer concrete. **C 904,** C03

sulfur, total, *n*—all the sulfur present in a material irrespective of its chemical form or origin. **D 1566,** D11

summerwood—See **latewood.** **D 9,** D07

sum of squares, *n*—*in analysis of variance,* a contraction of the expression "sum of the squared deviations from the appropriate average(s)" where the average(s) of interest may be the average(s) of a specific subset(s) of data or of the entire set of data. **D 123,** D13

sump—a depression around a drain. **D 1079,** D08

sump throat—{archaic} See **submerged throat.** **C 162,** C14

sunk cost—a cost that has already been incurred and which should not be considered in making a new investment decision. **E 833,** E06

superconducting—adjective describing a material exhibiting superconductivity. **B 713,** B01

superconducting state—the thermodynamic state in which the material exhibits superconductivity. **B 713,** B01

superconducting transition—that combination of values of temperature, *T*, electrical current density, *J*, and magnetic field, *H*, at which a transition from the superconducting to the normal state takes place. **B 713,** B01

superconductive—adjective describing a material exhibiting the characteristics of normal conductivity, but which shows superconductivity under appropriate conditions. **B 713,** B01

superconductivity—a property of many elements, alloys, and compounds by virtue of which their electrical resistivity vanishes and they become strongly diamagnetic under appropriate conditions. **B 713,** B01

superconductor—a material that exhibits superconductivity under appropriate conditions. **B 713,** B01

supercooling, *n*—an unstable state in which an engine coolant exists as a liquid below its normal freezing point. **D 4725,** D15

supercooling—to cool a substance below a transition temperature without the transition occurring, especially to cool below the freezing point without solidification. **E 1142,** E37

supercritical flow—open channel flow that is shallower and at higher velocity than critical flow for the same flow rate. A Froude number greater than one exists. **D 4410,** D19

superfatted soaps—under **soap,** see *superfatted soap.* **D 459,** D12

superfines—the portion of a powder composed of particles which are smaller than a specified size, currently less than 10 μm. **B 243,** B09

superimposed ac—a form of current in which an alternating current component is superimposed on the direct plating current. **B 374,** B08

superimposed load—load imposed by travel over, or by material brought and placed over the trench area, after pipe installation. **C 896,** C04

superimposed load, *n*—force applied to a specimen or structure other than that associated with its own mass. **E 176,** E05

superior—higher, denoting the upper of two parts, also means towards the head. **F 869,** F08

superlattice—See **ordered structure**. **E 7,** E04

supernatant—floating on the surface; often used to describe the liquid above a precipitate or solid. **C 859,** C26

supernatant, *n*—the liquid above settled solids. **D 6384,** D02

supernatant—that liquid remaining after separation of a liquid/solid mixture. **E 1705,** E48

supernate or supernatant—the liquid above the surface of settled sediment. **D 4410,** D19

supersaturation—a state in which the inorganic salt (s) are in solution at a level higher than the respective solubility product. **D 6161,** D19

super-steep asphalt—a roofing asphalt conforming to the requirements of Specification D 312, Type IV. **D 1079,** D08

superstructure—those parts of a melter above the sidewall tank blocks. **C 162,** C14

supplemental coolant additive (SCA) maintenance dose, *n*—smaller periodic additions of SCA, subsequent to the precharge dose of SCA, required to maintain protection against general corrosion, cylinder liner pitting, and scaling in heavy-duty engines. **D 4725,** D15

supplemental coolant additive (SCA) precharge dose, *n*—initial concentration of SCA in engine coolant required to enable the coolant to provide adequate protection against cylinder liner pitting and scaling in heavy-duty engines plus general corrosion protection when water only is used as a coolant. **D 4725,** D15

supplier, *n*—person, company, or organization that furnishes products as a jobber or distributor, as distinct from a producer. **B 881,** B07

supplier, *n—in pipe laying,* the party who supplies material or services. A supplier may or may not be the manufacturer. **C 460,** C17

supplier, *n—in pipe laying,* the party who supplies material or services. A supplier may or may not be the manufacturer. **C 1154,** C17

supplier, *n—in pipe laying,* the party who supplies material or services. A supplier may or may not be the manufacturer. **D 2946,** C17

supplier—the party that is responsible for the product, process or service and is able to ensure that quality assurance is exercised. The definition may apply to manufacturers, distributors, importers, assembles, service organizations, etc., (EN 45020). **E 1187,** E36

supplier declaration—procedure by which a supplier gives written assurance that a product, process or service conforms to specified requirements, (ISO Guide 2). **E 1187,** E36

support—structure or structural feature built into an underground opening for maintaining its stability. (ISRM) **D 653,** D18

supported film adhesive—See **adhesive, supported film.** **E 1749,** E06

supported gasket member—a gasket member held in place by a supporting frame member. **C 717,** C24

supported needled felt, *n*—a needled felt that is composed entirely of fibers physically interlocked and reoriented in combination with interlay, scrim, or foundation of knitted, stitched, bonded, or extruded structure. **D 123,** D13

supporter—cam mechanism, especially spring-loaded; designed to provide crown and lateral guidance for staple in driving channel of stapler. **F 592,** F16

support (foot)—the degree to which the shoe upper provides sufficient containment of the motions of the foot. **F 869,** F08

support space—See **space categories**. **E 1480,** E06

suppression—see **reject (suppression).** **E 1316,** E07

suppressor device—an ion-exchange-based device placed between the analytical column and the conductivity detector of an ion chromatography system, that exchanges counter-ions with hydronium or hydroxyl ions. This process enchances the analyte signal by lowering eluant background conductivity, and, for strong acid and strong base ions, forms a more conductive species. **D 1129,** D19

supra-threshold, *adj*—pertaining to a stimulus above the specified threshold. **E 253,** E18

surface—an interface between a condensed phase and a non-condensed phase (or free space). **E 673,** E42

surface active agent—a substance that affects markedly the interfacial or surface tension of solutions even when present in very low concentrations. **B 374,** B08

surface-active agent—(surfactant) an organic compound that reduces the surface tension of a liquid, or the interfacial tension between two liquids or between a liquid and a solid. It can also modify the properties of the liquid to which it is added, generally water. **D 459,** D12

ampholytic or amphoteric (zwitterionic) surfactant—a chemical compound capable of producing either (or both) positively or negatively charged surface active ions in an aqueous solution. The charge carried on the *ampholytic*surface active ions depends upon the pH of the solution but that on *amphoteric (zwitterionic)* surfactants does not. **D 459,** D12

anionic surfactant—a chemical compound that produces negatively charged surface active ions in solution. **D 459,** D12

cationic surfactant—a chemical compound that produces positively charged surface active ions in solution. **D 459,** D12

nonionic surfactant—a chemical compound that produces uncharged surface active particles in solution. **D 459,** D12

surface-active agent—a material that when added to a liquid medium modifies the properties of the medium at a surface or interface.

NOTE—**surface-active agent** is the general term which includes soluble detergents in liquid medium, dispersing agents, emulsifying agents, foaming agents, penetrating agents, and wetting agents. **E 1519,** E35

surface area—the total area of the surface of a powder or solid including both external and accessible internal surfaces (from voids, cracks, open porosity, and fissures). **C 242,** C21

surface area—the total area of the surface. As usually used for reactive pulps, this applies not to external surface, but to the internal surface as well, the total surface available to gases or penetrating liquids. It is useful to express this as specific surface, area per gram. **D 1695,** D01

surface area—*in catalysis*, the measure of the total surface of a material per unit mass or volume. **D 3766,** D32

surface area (B.E.T.)—the total surface area of a solid calculated by the B.E.T. (Brunauer, Emmett, Teller) equation, from nitrogen adsorption or desorption data obtained under specified conditions. **D 2652,** D28

surface area distribution—the distribution of surface area according to some parameter such as pores of different size or diameter. **D 2652,** D28

surface area, specific, *n*—the area, per unit mass of a granular or powdered or formed porous solid, of all external plus internal surfaces that are accessible to a penetrating gas or liquid. **C 1145,** C28

surface characteristics—a set of terms considered to be the minimum needed to precisely describe the condition of a surface especially regarding slip resistance (material, contamination, slope, texture, hardness, coating, temperature). **F 1646,** F13

surface chip—See **peripheral chip.** **F 1241,** F01

surface clay, *n*—an unconsolidated, unstratified clay, occurring on the surface. **C 43,** C15

surface clay—an unconsolidated, unstratified clay, occurring on the surface. **C 896,** C04

surface coefficient, *n*—the ratio of the steady-state heat exchange rate (time rate of heat flow per unit area of a particular surface by the combined effects of radiation, conduction, and convection) between a surface and its external surroundings (air or other fluid and other visible surfaces) to the temperature difference between the surface and its surroundings. (See **conductance, film.**) **C 168,** C16

surface color, *n*—color perceived as belonging to the surface of a specimen, without the specimen appearing to be self-luminous. **E 284,** E12

surface contour, *n*—divergence of a surface from planeness rough (high) to smooth (low). **D 123,** D13

surface-crack length, 2*c* [L]—in part-through surface crack (PS) specimens, a distance measured on the specimen surface between the two points at which the crack front intersects the specimen surface. Crack length is less than the specimen width. **E 1823,** E08

surfaced lumber—See **lumber.** **D 996,** D10

surface dust—See **dust.** **D 1356,** D22

surface energy approximation—*EIA,* a simplification of calculations involving the energy of an ion passing through a solid specimen. The energy of the ion at the surface is used in place of a properly averaged energy. This approximation is used to determine the energy at which scattering or stopping cross sections, or both, are evaluated. **E 673,** E42

surface finger oxide—the oxide that follows prior particle boundaries into a part from the surface and cannot be removed by physical means, such as rotary tumbling. **B 243,** B09

surface finish—the geometric irregularities in the surface of a solid material. Measurement of surface finish shall not include inherent structural irregularities unless these are the characteristics being measured. **C 709,** D02

surface flame spread, *n*—the propagation of a flame away from the source of ignition across the surface of a liquid or a solid. Compare: **volumetric flame spread** and **burning velocity.** **E 176,** E05

surface flame spread, *n*—the propagation of a flame away from the source of ignition across the surface of the specimen. **E 176,** E05

surface force—any force that acts across an internal or external surface element in a material body, not necessarily in a direction lying in the surface. (ISRM) **D 653,** D18

surface friction, *n*—resistance to slipping offered by surface harsh (high) to slippery (low). **D 123,** D13

surface hardening, *n*—a generic term covering any of several processes that, by **quench hardening** only, produce in a steel object a surface layer that is harder or more wear resistant than the **core.** **A 941,** A01

surface marks—relatively long, narrow, shallow grooves or cuts in the surface, such as scratches, score marks, and machining marks. (See also **grinding mark**

F 109, C21

surface monitoring tire, *n*—a reference tire used to evaluate changes in a test surface over a selected time period. **F 538,** F09

surface normal, *n*—the direction normal or perpendicular to the surface of a plane specimen. **E 284,** E12

surface oxides—oxygen containing compounds and complexes formed at the surface of an adsorbent. **D 2652,** D28

surface preparation, *n*—a mechanical or chemical method used to make a substrate more receptive to forming an adhesive bond. (Synonym *pre-bond treatment.*) **D 907,** D14

surface preparation—a physical or chemical preparation, or both, of an adherend surface to render it suitable for adhesive joining. **E 1749,** E06

surfacer—a pigmented composition for filling minor irregularities to obtain a smooth uniform surface preparatory to applying finish coats; usually applied over a primer and sandpapered for smoothness. **D 16,** D01

surface reflection—reflection in which radiant energy is returned exclusively at the surface of the specimen. **E 131,** E13

surface resistance—see **resistance, surface.** **D 1711,** D09

surface resistivity, *n*—*of paper*, the resistance to electric current between electrodes placed on the same paper surface along the current path between the electrodes when a defined dc voltage is applied and the gap between the electrodes is specified. **D 1968,** D06

surface resistivity, ρ_s—the surface resistivity of a material is the ratio of the potential gradient parallel to the current along its surface to the current per unit width of the surface. **D 5077,** D10

surface roughness—*AES,* the deviation of the topography of an actual surface from an ideal atomically smooth and planar surface. The rms deviation from the center line average is a measure of surface roughness. **E 673,** E42

surface roughness, R_a, *n*—the arithmetic average deviation profile from the centerline, normally reported in micrometres. **G 126,** G04

surface segregation—a diffusion controlled process (as opposed to evaporation, preferential sputtering, or other processes) that causes the surface composition of a homogeneous solid to differ from the bulk composition. **E 673,** E42

surface tension—that property, due to molecular forces, that exists in the surface film of all liquids and tends to prevent the liquid from spreading. **B 374,** B08

surface tension, *n*—the property, due to molecular forces, by which the surface film of all liquids tends to bring the contained volume into a form having the least area. **C 1145,** C28

surface tension—the force existing in a liquid-vapor phase interface that tends to diminish the area of the interface. This force acts at each point on the interface in the plane tangent at that point. **D 459,** D12

surface tension—a property arising from the molecular forces of the surface film of all liquids which tend to alter the contained volume of liquid into a form of minimum superficial area, expressed as work in newtons per millimetre. **D 1129,** D19

surface texture—the topographic deviations of a real surface from a reference surface. **F 1241,** F01

surface topography, *n*—the geometrical detail of a solid surface, relating particularly to microscopic variations in height. **G 40,** G02

surface treatment, *n*—an application of bituminous material followed by a layer of mineral aggregate. Multiple applications of bituminous material and mineral aggregate may be used. **D 8,** D04

surface void, *n,* **(SV^s)**—*as used in fractography,* a surface-distributed cavity created on a specimen/component during fabrication as a consequence of a reaction/interaction between the material and the processing environment, that is, a surface reaction layer. **C 1145,** C28

surface water absorption, *n*—*by a fabric,* the process of removing liquid water from a surface such as human skin, dishes, or furniture. **D 123,** D13

surface wave—a wave confined to a thin layer at the surface of a body. (ISRM) **D 653,** D18

surface wave—see **Rayleigh wave.** **E 1316,** E07

surface wetting and adhesion—See **wetting and adhesion, surface.** **C 168,** C16

surfacing mat, *n*—a thin mat of fine fibers used primarily to produce a smooth surface on a reinforced plastic. **D 883,** D20

surfactant—a contraction of the term surface-active agent. **D 459,** D12

surfactant—a material that improves the emulsifying, dispersing, spreading, wetting, or other surface-modifying properties of liquids. **E 609,** E35

surfactant—See **surface-active agent.** **E 1519,** E35

surfactants, *n*—surface active molecular species that exhibit both water soluble and oil soluble properties, and affect the physical behavior at the interface between water and oil phases by forming emulsions or changing the wetting characteristics of solid surfaces exposed to water and oil. **D 4175,** D02

surge magnetization—use of a high initial current for a short period (less than a second), then a continuous reduced current while the inspection medium is applied. **E 1316,** E07

surgical scissors with inserts—a stainless steel instrument, available in various sizes and configurations, used in surgical procedures for cutting body tissue, gauze, and suture. An instrument of this type has tungsten carbide, stellite, or other inserts. **F 1078,** F04

surrogate microbe, *n*—a microorganism which is used to act as a simulant for other microorganisms which are pathogenic to humans. **F 1494,** F23

surrogate species—a species that is tested to estimate responses of other species, for which direct testing is impractical. **E 943,** E47

surround, *n*—portion of the visual field immediately around the object or light source of interest. See **ambient field.** **E 284,** E12

surveillance—the act of maintaining supervision or vigilance over a well-specified portion of water so that detailed information is provided concerning the state of that portion. **D 1129,** D19

susceptibility, κ—a ratio of the intrinsic induction, B_i, as a result of the magnetization of a material to the induction in space because of the influence of the corresponding magnetic field strength, H.

$$\kappa = B_i/\Gamma_m H = \mu_r - 1$$

where:

Γ_m = magnetic constant and
μ_r = relative permeability.

NOTE—The preceding equations apply to an isotropic material if the SI, an abbreviation for the international system of units, are used.

NOTE—In the classical cgs-emu system of units:

$$\kappa = B_i/4\pi\Gamma_m H = (\mu_r - 1)/4\pi$$

A 340, A06

susceptibility, *n*—the ratio of the intrinsic induction due to the magnetization of a material to the induction in space due to the influence of the corresponding magnetizing force. **C 1145,** C28

susceptibility—when not otherwise qualified, the degree of change in viscosity with temperature. **D 1079,** D08

susceptibility, initial, κ_0—the limiting value of susceptibility when the intrinsic induction approaches zero. **A 340,** A06

susceptibility, mass, χ—the susceptibility divided by the density of a body is called the susceptibility per unit mass, χ, or simply the mass susceptibility.

$$\chi = \kappa/\delta$$

where δ = density. **A 340,** A06

suspended sediment—sediment that is carried in suspension by the turbulent components of the fluid or by Brownian movement. **D 4410,** D19

suspended-sediment concentration—See **concentration of sediment (by mass).** **D 4410,** D19

suspended-sediment discharge—the quantity of suspended-sediment passing through a stream cross section per unit of time. **D 4410,** D19

suspended-sediment load—that part of the sediment load which is suspended sediment. **D 4410,** D19

suspended-sediment sampler—a device that collects a representative portion of the water with its suspended-sediment load. **D 4410,** D19

suspended solids (SS)—solid organic and inorganic particles that are held in suspension in a liquid. **D 6161,** D19

suspension—a mixture of liquid and solid materials. **D 653,** D18

suspension—a two-phase system consisting of a finely divided solid dispersed in a solid, liquid, or gas. **E 609,** E35

suspension—a two-phase system consisting of a finely divided solid dispersed in a liquid. **E 1316,** E07

suspension agent—an additive that decreased the settlement rate of particles in liquid. **D 653,** D18

suspension deflection—the change in the vertical distance between the axle at its centerline and a hypothetical reference line directly above the axle centerline on the rigid frame structure. **E 867,** E17

suspension effect—the change in junction potential when a reference electrode is immersed to different depths in a solution containing resins or charged colloids, such as certain clays and soils. The suspension effect, which can be as large as 100 mV, does not occur with sensing electrodes. **D 4127,** D19

suspension, liquid—See **liquid suspension.** **C 242,** C21

sustainable capacity, *n*—the power which a component of a geothermal facility (such as a reservoir or a power plant) is capable of sustaining for a specified period of time.

Examples of Usage:

This reservoir has a sustainable capacity of 100 MW for at least 30 years.

This reservoir has a capacity of 50 MW, sustainable for at least the first 20 years of a 30-year project life.

With proper maintenance, this power plant has a sustainable capacity of 30 MW for 30 years. **E 957,** E44

sustained flaming, *n*—existence of flame on or over most of the specimen surface for periods of at least 4 s. **E 176,** E05

sustained flaming, *n*—the existence of flame on or over the surface of the specimen for a period of 4 s or more. **E 176,** E05

sustained flaming, *n*—existence of flame on or over the surface of the specimen for periods of 5 s. **E 176,** E05

sustained pressure test—a constant internal pressure test for an extended period of time. **F 412,** F17

swab test—a low-voltage electrical test used to evaluate continuity of porcelain enamel. **C 286,** B08

swage, *n*—the term used to describe any attachment method that uses mechanical force to crimp the end of the needle and firmly hold the suture in place **F 1840,** F04

swamp—a forested or shrub covered wetland where standing or gently flowing water persists for long periods on the surface. **D 653,** D18

swath, effective width—the center to center distance between overlapping broadcast applications. **E 1102,** E35

sweating—see **exudation.** **B 243,** B09

swedge—the formed end of a piece of toprail so that it will fit into and join another piece of toprail. **F 552,** F14

sweep—the uniform and repeated movement of an electron beam across the CRT. **E 1316,** E07

sweep—see **bend**. **F 412,** F17

sweeping mode—movement of a boom relative to the water for the purpose of controlling or collecting a floating substance. **F 818,** F20

sweep width (aka swath)—width intercepted by a boom in collection mode, the projected distance between the ends of a boom deployed in a "U," "V," or "J" configuration. **F 818,** F20

sweet—a term applied to easily workable glass. **C 162,** C14

sweet gas—natural gas with sulfur compounds low enough that it can be used without further purification. **D 4150,** D03

sweet, (taste), *adj*—taste produced by substances such as sucrose when in solution. **E 253,** E18

swell, *n*—a hump in the pavement surface that may occur over a small area or as a longer, gradual wave; either type of swell can be accompanied by surface cracking. **E 867,** E17

swell, *n*—a hump in the pavement surface that may occur over a small area or as a longer, gradual wave; either type of swell can be accompanied by surface cracking. **E 1778,** E17

swelling, *n*—the increase in volume of a specimen immersed in a liquid or exposed to a vapor. **D 1566,** D11

swelling—increase in volume and dimensions caused by penetration of a liquid. **D 1695,** D01

swept gain—see **DAC.** **E 1316,** E07

swimwear, *n*—textile garments intended for wear in fresh, chlorinated, or salt water. **D 123,** D13

swinging field—see **multidirectional magnetization.** **E 1316,** E07

swirl—helical or concentric features that are visible to the unaided eye after preferential etch, and appear to be discontinuous under 100× magnification. **F 1241,** F01

swirl atomizer—a pressure atomizer in which the liquid is swirled, typically by means of tangential inlets, slotted distributors, vanes, or cores. **E 1620,** E29

swirl chamber atomizer—a swirl atomizer in which a chamber is located between the swirl generating devices and a discharge orifice. **E 1620,** E29

swirl coaxial injector—a pneumatic atomizer consisting of a central liquid stream surrounded by a high-velocity co-annular gas stream, wherein swirl is unparted to at least one of the streams. **E 1620,** E29

sword-point anvil— See **pointed clincher.** **F 592,** F16

SWRO—seawater reverse osmosis. **D 6161,** D19

symbol, *n*—a combination of linear bar code or two-dimensional symbol characters including start/stop characters, quiet zones, data characters, special function characters and error detection or correction characters, or both, required by a particular symbology to form a complete, scannable entity. **F 1294,** F05

symbol character, *n*—a unique bar or space pattern, or both, which is defined for a particular symbology. **F 1294,** F05

symbol check character, *n*—a symbol character calculated from other characters in a bar code symbol determined by an algorithm defined in the symbology specification and used to verify that the bar code has been correctly composed and read. The symbol check character does not include data encoded in the symbol. **F 1294,** F05

symbol contrast, SC, *n*—in reference to bar codes, the difference between the largest and smallest reflectances in a scan reflectance profile.

$$SC = R\text{max} - R\text{min}$$

F 1294, F05

symbol density, *n*—*in reference to bar codes*, the number of characters per lineal inch, limited by the width of the narrowest bar or space. **F 1294,** F05

symbol grade, *n*—*in reference to ANSI X3.182*, the simple average of all profile grades using the standard weighting 4.0 = *A*, 3.0 = *B*, 2.0 = *C*, 1.0 = *D*, and 0.0 = *F*. **F 1294,** F05

symbol length, *n*—the physical length of a bar code symbol, including quiet zones. **F 1294,** F05

symbology, *n*—*in reference to readable codes*, a discrete set of characters used to represent and transmit information by use of a set of rules for encoding information in the symbol by an arrangement of parallel, rectangular, bars and spaces arranged in a predetermined pattern or by two dimensional cells arranged according to unambiguous rules. **F 1294,** F05

symbology identifier (SI), *n*—an optional three character code that may prefix transmitted data from a bar code reader identifying the symbology read and any options enabled in the reader or special features of a symbology encountered. **F 1294,** F05

symbology reference decode algorithm, *n*—a decoding algorithm that may be found in a particular application or symbology specification, or both. **F 1294,** F05

symmetrical—this refers to a snowboard ski shape that has a longitudinal line of symmetry, thereby allowing a single board to be used for both regular-foot and goofy-foot binding mounting positions. **F 1107,** F27

symmetrical laminate—See **laminate, symmetrical**. **E 1749,** E06

symmetrically cyclically magnetized condition, *SCM*—a magnetic material is in a *SCM* condition when, under the influence of a magnetic field strength that varies cyclically between two equal positive and negative limits, its successive hysteresis loops or flux-current loops are both identical and symmetrical with respect to the origin of the axes. **A 340,** A06

symmetrical thread—thread where following flank is a mirror-image of leading flank about line through crest and perpendicular to nail axis. **F 547,** F16

symmetrical weld—a fused joint joining two sheets in which the joint and weld configuration on either side of the sheets is identical. **C 904,** C03

symmetric laminate—a laminate in which the stacking sequence for the plies located on one side of the geometric midplane are the mirror image of the stacking sequence on the other side of the midplane. **D 3878,** D30

symmetric membrane—membrane and bulk polymer have equivalent characteristic (isotropic). **D 6161,** D19

symmetry—a property of a crystal in virtue of which equivalent lattice points can be brought into coincidence by operations such as rotation, inversion, or reflection. Such operations, are called symmetry operations or elements. **E 7,** E04

sympathetic ink—See **sensitive ink**. **F 1156,** F12

synchronous, *adj*—data sent together with a time base that allows the sender and receiver to operate locked together. **F 1457,** F05

synchronous data link control, *n*—the set of electronic signals used in the system network architecture. **F 1457,** F05

synchrotron radiation—*XPS,* a continuous radiation created by the acceleration of high energy electrons, as in a synchrotron or storage ring. Monochromatized, it is a practical variable energy source of photons for photoelectron spectroscopy. **E 673,** E42

syndet—a contraction of the term synthetic detergent. Now used to characterize personal washing bars containing surfactants in combination with soap. **D 459,** D12

syneresis—*in grouting*, the exudation of liquid (generally water) from a set gel which is not stressed, due to the tightening of the grout material structure. **D 653,** D18

syneresis, *n*—the contraction of a gel accompanied by the separation of a liquid. **D 883,** D20

syneresis, *n*—the exudation of small amounts of liquid by gels on standing. **D 907,** D14

syneresis, *n*—the contraction of a gel accompanied by the separation of a liquid. (D-20) **F 412,** F17

synergism, *n*—the joint action of two or more stimuli whose combination elicits a level of sensation greater than the result of combining the effects of each stimulus taken separately. **E 253,** E18

synergist—an auxiliary material that has the property of increasing the effect of the active ingredient, even though it may have little specific activity itself.

NOTE—In the case of insecticides, synergists are considered as active ingredients. **D 3064,** D10

syngas—the synthetic gas resulting from incomplete combustion or pyrolysis of organic material to primarily carbon monoxide and hydrogen. (See also **synthesis gas**.) **E 1126,** E48

syngas—the synthetic gas resulting from incomplete combustion or pyrolysis of organic material to primarily carbon monoxide and hydrogen. (See also **synthesis gas**.) **E 1705,** E48

syntactic cellular plastics—materials consisting of hollow sphere fillers in a resin matrix. **D 883,** D20

syntan—a synthetic organic tanning material. **D 1517,** D31

syntax, *n*—in reference to codes, the rules governing the use of appropriate identifiers, delimiters, separator character(s), and other non-data characters as well as data positioning within the message. **F 1294,** F05

syntectic equilibrium—a reversible univariant transformation in which a solid phase, that is stable only at lower temperature, decomposes into two conjugate liquid phases that remain stable at higher temperature; for example: $L_1 + L_2 = \alpha$. **E 7,** E04

synthesis gas—mixtures of gas in suitable proportions for the production of synthetic products without adding further reactants, such as carbon monoxide and hydrogen, for synthesis of methanol. **E 1126,** E48

synthesis gas—mixtures of gas in suitable proportions for the production of synthetic products without adding further reactants, such as carbon monoxide and hydrogen, for synthesis of methanol. **E 1705,** E48

synthesis grade—a quality of aromatic hydrocarbons and related chemicals representing the highest purity available on a commercial scale. It encompasses those materials identified with a quantitative index of purity. **D 4790,** D16

synthetic, *adj— in lubricants*, originating from the chemical synthesis of relatively pure organic compounds from one or more of a wide variety of raw materials. **D 4175,** D02

synthetic atmosphere—See **atmosphere.** **D 1356,** D22

synthetic blood, *n*—a mixture of a red dye/surfactant, thickening agent, and distilled water having a surface tension and viscosity, and a red color representative of blood and other body fluids, and making it more usable for visible detection. **F 1494,** F23

synthetic detergent—a detergent containing surfactants other than soap. Historically, the term differentiated the first detergents from the earlier soap-based detergents. **D 459,** D12

synthetic detergent—under **detergent,** see *synthetic detergent.* **D 459,** D12

synthetic fillers—generally of the "wood dough" type, limited to the repair of minor defects in panels. **D 1038,** D07

synthetic gypsum, *n*—a chemical product, consisting primarily of calcium sulfate dihydrate ($CaSO_4{\cdot}2H_2O$) resulting primarily from an industrial process. **C 11,** C11

synthetic hydrocarbon resin, *n*—a low molecular-weight thermoplastic polymer derived from steam-cracker fractions, coal tar fractions, from a variety of pure hydrocarbon monomers, or mixtures thereof. **D 6440,** D01

synthetic paint roller cover—a cover in which the paint applicating material is man-made material. **D 16,** D01

synthetic patches, plugs, and shims—generally composed of two-component formulations of materials which will provide high level performance in terms of bond characteristics, weatherability and durability, used to repair defects in veneer or plywood panels. **D 1038,** D07

synthetic resin—see **resin, synthetic.** **D 16,** D01

synthetic test solution—a solution of two or more components prepared under specified conditions for use in evaluation of adsorbents. **D 2652,** D28

sysgen, *n*—the definition of all elements in a computing system.

F 1457, F05

system—See **roofing system.** **D 1079,** D08

system—See **building system.** **E 631,** E06

system— of a building—See **building system**. **E 1480,** E06

systematic composite sample, *n—in asbestos,* a composite sample that is obtained by taking every *n*th bag in the lot being sampled. **D 2946,** C17

systematic error—an error caused by undetected fundamental flaws in the test equipment by inadequate understanding of the theory underlying the measurement or by repetitious errors on the part of the analyst. **E 856,** D34

systematic error, *n*—errors introduced by the printer, reader, or other system elements that are consistent for all bars or all spaces. **F 1294,** F05

systematic error of result, *n*—a component of the error, which in the course of a number of test results for the same characteristic, remains constant or varies in a predictable way.

NOTE—Systematic errors and their causes may be known or unknown. **E 456,** E11

systematic sample, *n— in asbestos,* the sample that is formed from every *n*th bag of the lot to be examined. **D 2946,** C17

systematic sampling, *n*—the process of selecting units in a sample in accordance with a specific order or location in time or space, or both. **D 123,** D13

systematic sampling, *n*—sample selection procedure in which every kth element is selected from the universe or population; for example, u, u + k, u + 2k, u + 3k, etc., where u is in the interval 1 to k. **E 456,** E11

system (*crystal*)—See **crystal system**. **E 7,** E04

system examination threshold—the electronic instrument threshold (see **evaluation threshold**) which data will be detected. **E 1316,** E07

systemic disease—a disease in which the pathogen or its products spread throughout the organism. **E 609,** E35

system induced artifacts—anomalies that are created by a system during the acquisition, display processing, or storage of a digital image. **E 1316,** E07

system network architecture, *n*—a proprietary communication system developed by IBM. (See **SNA**.) **F 1457,** F05

system noise—the noise present in a radiological measurement resulting from the individual elements of the radiological system. **E 1316,** E07

T

tabby sample, *n*—the section of tire cord fabric between two tabbies that have been woven separately with a distance of 0.5 to 1.0 m (18 to 36 in.) between them. **D 123,** D13

Taber stiffness unit, *n— of paper,* the common unit measure used with Taber instruments. **D 1968,** D06

table dyeing—see **brush coloring.** **D 1517,** D31

table nut—a hexagon nut used to attach the dial ring to the tube. **F 471,** F12

table run—used to describe leather that has not been sorted and graded before selling by the tanner. (Also known as tannery run, or TR). **D 1517,** D31

tableware—all utensils and decorative articles used on the table for meal service. **C 242,** C21

tack—sticky or adhesive quality of the surface of applied sealant. **C 717,** C24

tack, *n—for rubber or rubber compounds,* a property that causes two layers of these materials, when pressed together, to adhere at the area of contact. **D 123,** D13

tack, *n*—the property of an adhesive that enables it to form a bond of measurable strength immediately after adhesive and adherend are brought into contact under low pressure. (See also **tack range** and **tacky-dry**.)

aggressive tack, n—Synonym for **dry tack**.

dry tack, n—the property of certain adhesives, particularly nonvulcanizing rubber adhesives, to adhere on contact to themselves at a stage in the evaporation of volatile constituents, even though they seem dry to the touch. (Synonym *aggressive tack.*) **D 907,** D14

tack—sticky, gummy character of a polish film, rendering the surface conducive to dust accumulation, fingerprinting, etc. **D 2825,** D21

tack—sticky, gummy character of a polish film, rendering polish surfaces conducive to dust accumulation. **D 2825,** D21

tack—the ability of a material to adhere to itself. **F 869,** F08

tack coat (bond coat), *n*—an application of bituminous material to an existing relatively nonabsorptive surface to provide a thorough bond between old and new surfacing. **D 8,** D04

tacker—electric, manual, or pneumatic tool for driving light-wire staples without clinching. **F 592,** F16

tackifier, *n*—a compounding material that enhances the ability of vulcanized rubber to adhere to itself or another material. **D 1566,** D11

tackifying resin, *n*—a resin, which when mixed with certain compatible polymers, imparts tack, under use conditions, to the finished formulation. **D 6440,** D01

tacking cut, *n*—small holes or cuts along the selvage. **D 123,** D13

tack range, *n*—the period of time in which an adhesive will remain in the tacky-dry condition after application to an adherend, under specified conditions of temperature and humidity. **D 907,** D14

tack, rubber, *n*—the property that causes contacting surfaces of unvulcanized rubber to adhere to each other. **D 1566,** D11

tack tear, *n*—the measurement of the resistance of a coated fabric to tearing under conditions simulating an installation that has been tacked in place. **D 123,** D13

tacky-dry, *adj*—pertaining to the condition of an adhesive when the volatile constituents have evaporated or been absorbed sufficiently to leave it in a desired tacky state. **D 907,** D14

Tafel slope—the slope of the straight line portion of a polarization curve, usually occurring at more than 50 mV from the open-circuit potential, when presented in a semi-logarithmic plot in terms of volts per logarithmic cycle of current density (commonly referred to as volts per decade). **G 15,** G01

tag, *n*—a paper item, such as ticket or slip, which is marked to provide information to the consumer at the time of purchase, which is permanently affixed to the product. **D 123,** D13

tag, *n*—a **paperboard** card, metal plate, or plastic card on which shipping or identification data are given. (Compare **label.**) **D 996,** D10

tag board, *n*—a paperboard used for shipping tags, file folders, printed forms, envelopes, etc. **D 1968,** D06

taggant, n—See **tags.** **F 1156,** F12

tags—labels or marks—overt or covert, to identify or to recognize someone or something. **F 1156,** F12

tail height, h_T—the height of the underside of the tail from a plane surface with the center of the ski body pressed against that surface. **F 472,** F27

tail height, H_t—the height of the underside of the tail from a plane surface with the snowboard ski unweighted on a plane surface. **F 1107,** F27

tailing—a fringe effect appearing on the trailing edge of the developed electrostatic image, but not directly attached to it. (See also **dragging**.) **F 335,** F05

tailings—the (size) residue, coarse or fine, removed from a separation process. **C 242,** C21

tailings, *n*—the underflow product from coal froth flotation. **D 121,** D05

tail length, L_t—the projected length of the tail turn-up, measured from the snowboard ski tail to the contact point where an 0.1-mm feeler gage intersects the running surface of the snowboard ski resting unweighted on a plane surface. **F 1107,** F27

tail surface area—that surface from the tail contact point aft. The tail contact point is located l_T from the tail. **F 472,** F27

tail surface area—that surface from the tail contact point aft. The tail contact point is located L_t from the tail. **F 1107,** F27

tail turn-up length, l_T—the projected length of the tail turn-up, measured from the ski tail to the contact point where a 0.5-mm feeler gage intersects the running surface with the ski body pressed against a plane surface. **F 472,** F27

taint, *n*—a taste or odor foreign to a product. **E 253,** E18

take—see **grout take.** **D 653,** D18

takeoff rate, *n— in column distillation,* the volume of product withdrawn from the reflux divider over a specified period. **D 4175,** D02

take-out—a mechanical device for removing a finished article from any glass-forming unit. **C 162,** C14

take-up, *n—in textile fabrics,* the difference in distance between two points in a yarn as it lies in a fabric and the same two points after the yarn has been removed from the fabric and straightened under a specified tension, expressed as a percentage of the straightened length. **D 123,** D13

take-up rollers, *n—in open-end spinning machines,* a pair of closely set, continuously rotating cylinders which withdraw the spun yarn from the rotor. **D 123,** D13

talc—a phyllosilicate mineral having the general formula $3MgO$-$4SiO_2$-H_2O and noted for its extreme softness, low thermal and electrical conductivity, and fire resistance. (See also **steatite talc.**) **C 242,** C21

talc—see **industrial talc** and **industrial talc, nonasbestos type.** **D 16,** D01

tall oil, *n*—a generic name for a number of products obtained from the manufacture of wood pulp by the alkali (sulfate) process or more popularly known as the kraft process.

acid refined tall oil, n—the product obtained by treating crude tall oil in solvent solution with sulfuric acid under controlled conditions to remove dark color bodies and odoriferous materials. Removal of the solvent yields a product with lighter color and higher viscosity than crude tall oil with approximately the same fatty acids-to-rosin ratio.

crude tall oil, n—a dark brown mixture of fatty acids, rosin, and neutral materials liberated by the acidification of soap skimmings. The fatty acids are a mixture of oleic acid and linoleic acid with lesser amounts of saturated and other unsaturated fatty acids. The rosin is composed of resin acids similar to those found in gum and wood rosin. The neutral materials are composed mostly of polycyclic hydrocarbons, sterols, and other high-molecular weight alcohols.

distilled tall oil, n—the class of products obtained by distilling crude tall oil in fractionating equipment under reduced pressure under such conditions that the ratio of rosin acids to fatty acids is varied over a wide range. The products that generally contain less than 90 % of fatty acids, are known as distilled tall oils. **D 804,** D01

tall oil, distilled— See *distilled tall oil* under **tall oil**. **D 804,** D01

tall oil fatty acids, *n*—the class of products containing 90 % or more fatty acids obtained by fractionation of crude tall oil. **D 804,** D01

tall oil heads (light ends), *n*—the low-boiling fractions obtained by the fractional distillation of crude tall oil under reduced pressure. **D 804,** D01

tall oil pitch, *n*—the residue from the distillation of crude tall oil. It is generally recognized that tall oil pitches contain some high-boiling esters and neutral materials with lesser amounts of rosin and fatty acids. **D 804,** D01

tall oil rosin—See **rosin**. **D 804,** D01

tall oil soap, *n*—the product formed by the saponification or neutralization of tall oil with organic or inorganic bases. **D 804,** D01

talus—rock fragments mixed with soil at the foot of a natural slope from which they have been separated. **D 653,** D18

tamper—a heavy cylindrical metal section of tubing that is operated on a wire rope or cable. It slips over the riser and fits inside the casing or borehole annulus. It is generally used to tamp annular sealants or filter pack materials into place and prevent bridging. **D 653,** D18

tamper resistant device, *n*—an indication or barrier to entry to a package or container which if breached or missing can reasonably be expected to provide visible evidence to a concerned person that tampering has occurred. **F 1156,** F12

tan,δ, *n*—is the dimensionless ratio of energy lost to energy returned during one cycle of a periodic process. Tan δ is normally calculated by dividing the loss component of the property measured by a periodic method by the storage component (for example, tan δ= E"/E' as used in DMA). **E 473,** E37

tan delta, tan δ—the ratio of the loss modulus to the storage modulus, measured in compression, tension, flexure, or shear.

$$Tan\,\delta = \frac{K''}{K'} = \frac{E''}{E'} = \frac{G''}{G'}$$

D 4092, D20

tangent delta, tan δ— *in dielectric measurements*, the ratio of the loss factor, ϵ'', to the absolute permittivity, ϵ'; that is:

$$\tan\delta = \epsilon''/\epsilon'$$

E 1142, E37

tangent erosion rate, *n— in cavitation or liquid impingement erosion*, the slope of a straight line drawn through the origin and tangent to the knee of the cumulative erosion-time curve, when that curve has the characterisitic S-shaped pattern that permits this. In such cases, the tangent erosion rate also represents the maximum cumulative erosion rate exhibited during the test. **G 40,** G02

tangential—strictly, coincident with a tangent at the circumference of a tree or log, or parallel to such a tangent. In practice, however, it often means roughly coincident with a growth ring. A tangential section is a longitudinal section through a tree or limb perpendicular to a radius. Flat-grained lumber is sawed tangentially. **D 9,** D07

tangential stress—see **stress**. **D 653,** D18

tangent modulus—See **modulus, tangent**. **C 717,** C24

tangent modulus—the slope of a tangent line drawn at the steepest initial portion of the stress-strain curve. **C 904,** C03

tangent modulus, *n*—the ratio of change in stress to change in strain derived from the tangent to any point on a stress-strain curve. **D 123,** D13

tangent modulus—slope of the tangent to the stress-strain curve at a given stress value (generally taken at a stress equal to half the compressive strength). (ISRM) **D 653,** D18

tangent point, *n—for geotextiles*, the first point of the force-elongation curve at which a major decrease in slope occurs. **D 4439,** D35

tangle sheet, *n*—a piece of mica that splits well in places but tears in others, producing a large percentage of partial films. **D 1711,** D09

tank—synonymous with **melter**. **C 162,** C14

tank block—a generic term for large refractory blocks used in the structure of melters. See **pavers**. **C 162,** C14

tank heater—heating system that maintains the dishwasher's wash tank temperature. **F 1827,** F26

tank vacuum cleaner—See **canister vacuum cleaner**. **F 395,** F11

tank voltage—the total voltage between the anode and cathode of a plating bath or electrolytic cell during electrolysis. It is equal to the sum of: (*1*) the equilibrium reaction potential, (*2*) the *IR* drop, and (*3*) the electrode potentials. **B 374,** B08

tannery run—see **table run.** **D 1517,** D31

tan δ (tangent delta)— same as **dissipation factor**. **D 2864,** D27

tap—(*1*) to drain a furnace.

(*2*) to remove excess slag from the floor of a pot furnace. **C 162,** C14

tap density, *n*—the mass per unit volume of a powder that has been placed in a specified container and has been settled under specified conditions by mechanical action. **B 243,** B09

tap density—See *tap density* under **density.** **C 242,** C21

tap density—the apparent density of a powder obtained when the receptacle is tapped or vibrated during loading under specified conditions. **C 859,** C26

tape—a composite superconductor in the form of a flat ribbon or strip. **B 713,** B01

tape, *n—in textiles*, a narrow fabric with a mass per unit area of less than 0.5 kg/m^2(0.1 lb/ft^2) for each 25.4 mm (1 in.) of width and which is used primarily for utilitarian purposes. (See also **narrow fabric** and **textile ribbon**.) **D 123,** D13

tape, *n—in zippers*, a strip of material along one edge of which the bead and elements are attached. **D 123,** D13

tape:

gummed paper tape— a **kraft** paper in a wide range of basic weights, usually 35, 60 and 90 lb (24 by 35–500) that is gummed on one side and slit into rolls of various widths.

pressure-sensitive-adhesive tape—in packaging, a backing, as tape, coated so as to permit it to adhere instantaneously to most surfaces with the application of slight pressure.

reinforced gummed tape—an adhesive-coated tape that contains glass, rayon, or sisal fibers embedded in latex, resin or asphaltic laminants between two sheets of **kraft.** **D 996,** D10

tape—ribbons, usually of paper or cloth, coated with adhesive that are used to fasten veneers together for convenience in handling during the gluing operation. **D 1038,** D07

tape—materials in which the reinforcing filaments or fibers are laid in a single direction within a resin matrix in the B-stage. **E 1749,** E06

tape ends, *n—in zippers*, the tape extending beyond the stops at either or both ends of the stringers. **D 123,** D13

tapeless measure—a device which measures length using the number of rotations of a wheel in contact with the surface of the material to be measured. **D 996,** D10

taper—included angle of countersink. **F 547,** F16

taper—half of the difference between b_V and b_H or $(b_V - b_H)/2$. **F 1107,** F27

taper—*of a semiconductor wafer*, the linear component of the variation in thickness across a wafer, indicated by the angle between the best fit plane to the front surface and the ideally flat back surface of the wafer. Also known as **wedge.** **F 1241,** F01

taper candle, *n*—a slender candle produced to be used with a candle accessory for support. **F 1972,** F15

tapered edge, *n*—an edge formation of gypsum board which provides a shallow depression at the paper-bound edge to receive joint reinforcement **C 11,** C11

tapered edge strip—a tapered insulation strip used to elevate the roofing at the perimeter and at penetrations of the roof. **D 1079,** D08

tapered wedge point—point with two major planes forming "V" and with single minor plane on each flank; forming rectangular cross section. **F 547,** F16

tape replica—See **replica**. **E 7,** E04

taper, *V*—half of the difference between b_V and b_H or $(b_V - b_H)/2$. **F 472,** F27

tape sealant—*in building construction*, a sealant having a preformed shape, and intended to be used in a joint initially under compression. **C 717,** C24

tape yarn, *n*—a yarn of a flat, tape-like character produced by slitting an extruded film. **D 123,** D13

taping compound, *n—(sometimes called embedding compound)* a compound specifically formulated and manufactured for use in embedding of joint reinforcing tape at gypsum board joints. **C 11,** C11

tap test—See **test, tap**. **E 1749,** E06

tar, *n*—brown or black bituminous material, liquid or semisolid in consistency, in which the predominating constituents are bitumens obtained as condensates in the destructive distillation of coal, petroleum, oil-shale, wood, or other organic materials, and which yields substantial quantities of pitch when distilled. **D 8,** D04

tar—a brown or black bituminous material, liquid or semisolid in consistency, in which the predominating constituents are bitumens obtained as condensates in the processing of coal, petroleum, oil-shale, wood, or other organic materials. **D 1079,** D08

tar concrete, cold-laid, *n*—a plant mix containing a medium-viscosity grade of tar and a graded mineral aggregate, designed to be laid either shortly after mixing or when the mixture is at or near ambient temperature. **D 8,** D04

tar concrete, hot laid, *n*—a plant mix containing a high-viscosity grade of tar and a densely graded mineral aggregate designed to be laid at or near the elevated temperature of mixing. **D 8,** D04

tare, *n*—the mass of all external and internal packing materials (including bobbins, tubes, etc.) of a case, bale, or other type of container. **D 123,** D13

tare weight, *n*—the weight of a **container** or of packaging materials excluding the contents. **D 996,** D10

target, *n*—that part of an X-ray tube which the electrons strike and from which X rays are emitted. **E 135,** E01

target—specimen under investigation.

thick—EIA, specimen whose thickness produces backscattered particles whose energies, for each constitutive element, vary greatly with respect to the system resolution.

thin—EIA, specimen whose thickness is sufficiently small that the variations in energies of particles backscattered from atoms of each constitutive element is small with respect to the system resolution. **E 673,** E42

target—that part of the anode of an X-ray emitting tube hit by the electron beam. **E 1316,** E07

target cost—the planning expenditure, determined in cooperation with the cost professionals and the design team, for project elements. **E 833,** E06

target (in X-ray tubes)—that part of an X-ray tube which the electrons strike and from which X-rays are emitted. **E 7,** E04

target monitoring zone—the ground water flow path from a particular area or facility in which monitoring wells will be screened. The target monitoring zone should be a stratus (strata) in which there is a reasonable expectation that a vertically placed well will intercept migrating contaminants. **D 653,** D18

target octane number, *n*—the research or motor octane number quality desired for a specific product. **D 4175,** D02

target plane, *n*—the plane, perpendicular to the line of sight of a radiation thermometer, that is in focus for that instrument. **E 344,** E20

tarnish—corrosion products in the form of thin films or spots that do not protrude significantly from the surface of the metallic coating; for example, reaction products of copper from oxygen or reduced sulfur. **B 374,** B08

tarnish—the chemical compound on the surface of a contact resulting from the reaction of the contact material and the inorganic constituents of the surrounding atmosphere. **B 542,** B02

tarnish—a thin film of stain on the surface of glass. **C 162,** C14

tar or pitch bearing basic ramming mix, *n*— a tar or pitch bearing refractory that has suitable properties to permit ramming into place to form a monolithic structure. **C 71,** C08

tar or pitch bearing basic refractories, *n*— refractories consisting principally of basic grains to which either tar or pitch has been added during manufacture. **C 71,** C08

task lighting—See **lighting**. **E 1480,** E06

task outline—a general plan of the estimating work to be performed. **E 833,** E06

taste, *n*—perception resulting from stimulating the gustatory receptors in the taste buds. **E 253,** E18

taster, *n*—an assessor in a taste test. (See also **assessor.**) **E 253,** E18

tawing—the old English term applied to the process of making leather with alum as distinguished from tanning which was originally confined to vegetable tanning. **D 1517,** D31

TBC—total bacteria count, the total number of viable microorganisms present in the sample, excluding anaerobic organisms. **D 6161,** D19

TCC—total colony count. **D 6161,** D19

TDS—total dissolved solids, usually expressed as mg/L or ppm (parts per million). **D 6161,** D19

tealight candle, *n*—a cylindrical filled candle produced with a diameter and height of approximately 1.5 in. (38 mm) and 0.75 in. (19 mm) respectively **F 1972,** F15

tear—an imperfection; a small surface section of glass torn out by adherence to another hot solid. **C 162,** C14

tear drop, *n—in woven fabrics,* short elliptical deviations of one or more adjoining picks. **D 123,** D13

tearing—a defect in the surface of porcelain enamel, characterized by short breaks or cracks which have been healed. **C 286,** B08

tearing energy, *n*—the work done in tearing a material. **D 123,** D13

tearing force, *n*—the average force required to continue a tear previously started in a fabric. **D 123,** D13

tearing force, *n—in fabric,* the force applied to propagate a tear initiated under specified conditions. **D 123,** D13

tearing strength, *n—in fabrics,* the capacity of a material to withstand the ultimate tearing force required to propagate a tear after its initiation. **D 123,** D13

tearing strength, *n*—the force required either *(1)* to start or *(2)* to continue or propagate a tear in a fabric under specified conditions. **D 123,** D13

tearing strength, (F, (F), kN) **,** *n*—the force required either (*1*) to start or (*2*) to continue or propagate a tear in a fabric under specified conditions. **D 4439,** D35

tear off—to remove an existing roofing system down to the structural deck. **D 1079,** D08

tear offs—small pieces of leather, less than half a skin, that are torn from a skin during the staking or other tanning operations. **D 1517,** D31

tear resistance, *n—in textiles,* the resistance to the tearing force. **D 123,** D13

tear (rubber), *n*—mechanical rupture initiated and propogated at a site of high stress concentration caused by a cut, defect, or localized deformation. **D 1566,** D11

tear strength, *n*—the maximum force required to tear a specified specimen, the force acting substantially parallel to the major axis of the test specimen. **D 1566,** D11

teaser—the worker in direct charge of furnace operations who regulates the charging of batch and adjusts fires. **C 162,** C14

technical performance—discouraged term for **performance**—*of a building* or **performance**—*of a facility.* **E 1480,** E06

technician—an employee of the testing laboratory assigned to perform the actual operations of testing. **C 717,** C24

teemer—{archaic} the person who teems or casts the pot of glass. **C 162,** C14

teeming—See **casting**. **C 162,** C14

teeth, *n*—the resultant surface irregularities or projections formed by the breaking of filaments or strings which may form when adhesive-bonded substrates are separated. (Compare **legging, stringiness,** and **webbing**.) **D 907,** D14

teeth—serrations formed on the inside faces of the distal end of the tweezer halves. **F 1638,** F44

telco lines—**teleccommunication lines**, such as telephone and other communication pathways that are used to transmit information from one location to another. **F 1457,** F05

telecommunication lines, *n*—any communication pathways such as telephone and others that are used to transmit information from one location to another. **F 1457,** F05

telegraphing, *n*—in a laminate or other type of composite construction, a condition in which irregularities, imperfections, or patterns of an inner layer are visibly transmitted to the surface. **D 907,** D14

telescoping—the movement of the outer layers of a spiral wound cartridge in the direction of the feed flow caused by excessive pressure drop through the feed channel spacer. **D 6161,** D19

telomer, *n*—a polymer composed of molecules having terminal groups incapable of reacting with additional monomers, under the conditions of the synthesis, to form larger polymer molecules of the same chemical type. (IUPAC, ISO) **D 883,** D20

telomer, *n*—a polymer composed of molecules having terminal groups incapable of reacting with additional monomers, under the conditions of the synthesis, to form larger polymer molecules of the same chemical type. (D-20, IUPAC, ISO) **F 412,** F17

telomer, *n*—a polymer composed of molecules having terminal groups incapable of reacting with additional monomers, under the conditions of the synthesis, to form larger polymer molecules of the same chemical type. **F 1251,** F04

temper, *n—as related to metallic-coated steel wire,* stiffness or resistance to bending, typically described by reference to tensile strength. **A 902,** A05

temper—the metallurgical structure and properties of a product resulting from thermal or mechanical processing treatments. **B 846,** B05

temper, *v*—to bring a refractory mixture to a usable state by mixing with a liquid. **C 71,** C08

temper—(*1*) the degree of residual stress in annealed glass measured polarimetrically or by polariscopic comparison with a standard such as one or more strain disks. (See also **strain disk.**)

(*2*) term sometimes used in referring to tempered glass. (See also **tempered glass.**) **C 162,** C14

temper, *v*—in hydraulic-setting compounds, to bring to a usable state by mixing in or adding water. **E 631,** E06

temper annealing—a thermal treatment above the eutectoid temperature for copper-aluminum alloy products to minimize the presence of the stable eutectoid structure. **B 846,** B05

temperate bacteriophage—a bacteriophage that can grow lytically, killing the host, or can exist stably in the host. **E 1705,** E48

temperature—the thermal state of matter as measured on a definite scale. **B 713,** B01

temperature—

absolute temperature, n—(*1*) temperature measured on the thermodynamic scale, designated as Kelvin (K). (2) temperature measured from absolute zero (−273.15°C or 459.67°F).

dry-bulb temperature (t), n—the temperature of the ambient air, for example, the temperature that is measured by the dry-bulb thermometer of a psychrometer.

ice-bulb temperature (t_i)*, n*—the temperature that a thermometer indicates when its bulb is surrounded by a thin film of ice (or a frozen moistened covering), and allowed to cool by sublimation of the ice into the surrounding air.

sonic temperature (T_s) *(K), n*—an equivalent temperature that accounts for the effects of temperature and moisture on acoustic wavefront propagation through the atmosphere.

thermodynamic dew-point temperature (T_d)*, n*—the temperature at which moist air with mixing ratio, r_w, and total pressure, p, when saturated with respect to water at the same pressure, p, will have a saturation mixing ratio, r_w, equal to the given mixing ratio, r.

thermodynamic frost-point temperature (T_f), n—the temperature at which moist air with mixing ratio, r, and total pressure, p, when saturated with respect to ice at the same pressure, p, will have a saturation mixing ratio, r_i, equal to the given mixing ratio, r.

thermodynamic ice-bulb temperature (T_i), n—the temperature that moist air at pressure, p, temperature, t, and mixing ratio, r, will attain when brought adiabatically to saturation at pressure, p, by sublimation of pure ice into the moist air.

thermodynamic wet-bulb temperature (T_w), n—the temperature that moist air at pressure, p, temperature, t, and mixing ratio, r, will attain when brought adiabatically to saturation at pressure, p, by the evaporation of pure liquid water into the moist air.

virtual temperature, n—the temperature, T_v, which dry air must have at the given barometric pressure, p, in order to have the same density as moist air at the same pressure, P, given temperature, T, and mixing ratio, r, provided that the dry and moist air behave in accordance with the perfect gas equation of state.

$$T_v = T\frac{(1+r/\epsilon)}{(1+r)}$$

where:

r = mixing ratio (mass of water vapor per mass of dry air, and

ϵ = ratio of the molecular weight of water vapor to that of dry air.

wet-bulb temperature (t_w), n—the temperature indicated by the wet-bulb thermometer of a psychrometer. **D 1356,** D22

temperature—the thermal state of matter as measured on a definite scale. **E 41,** G03

temperature, absolute—(*a*) temperature measured on the thermodynamic scale, designated as kelvins (K).

(*b*) Temperature measured from absolute zero (−273.15°C or − 459.67°F). The numerical values are the same for both the Kelvin scale and the ideal gas scale. **B 713,** B01

temperature correction factor—a multiplier that adjusts the measured gas volume to the actual volume at standard gas temperature, T_{cf}

$$= \frac{\textit{absolute standard gas temperature}\ ^\circ R}{\textit{absolute actual gas temperature}\ ^\circ R}$$

$$= \frac{\textit{absolute standard gas temperature}\ ^\circ R}{[\textit{gas temperature}\ ^\circ F + 459.67]\ ^\circ R}$$

F 1827, F26

temperature correction factor (TCF)—defines the effect of temperature on permeate flow relative to a base temperature (25°C), is mainly a function of fluid characteristics but also membrane polymer. **D 6161,** D19

temperature, dew point—see dew point. **E 41,** G03

temperature, dry-bulb—the temperature of the air as indicated by an accurate thermometer, corrected for radiation if significant. **E 41,** G03

temperature effects—changes in potential developed between a sensing and reference electrode due to changes in temperature. The slope of a sensing electrode varies with temperature, as does the potential developed by the internal reference element in the reference electrode. In addition, the solubility of salts such as AgCl or Hg_2Cl_2 changes with temperature. It is therefore advisable to measure all samples and standardizing solutions at the same temperature. **D 4127,** D19

temperature envelope—the temperature range over which a particular penetrant inspection test will operate. **E 1316,** E07

temperature index, *n*—a number which permits comparison of the temperature/time characteristics of an electrical insulating material, or a simple combination of materials, based on the temperature in degrees Celsius which is obtained by extrapolating the Arrhenius plot of life versus temperature to a specified time, usually 20 000 h. **D 1711,** D09

temperature range—those temperatures at which electrode measurements can be made. The lower temperature limit is set by the freezing point of the solution that has the highest freezing point (including the sample as well as all filling solutions). The upper temperature limit is set by the boiling point of the solutions, except for liquid ion exchange electrodes, whose upper temperature limit is determined by the solubility of the ion exchange solution. Electrode life decreases with exposure to high temperatures. In many cases, the limit of detection also rises with increasing temperature. For these reasons, it is preferable to cool hot solutions before measurement. **D 4127,** D19

temperature stability, *n—for a geotextile*, the percent change in tensile strength or in percent elongation as measured at a specified temperature and compared to values obtained at the standard conditions for testing geotextiles. **D 4439,** D35

temperature uniformity—the comparison of individual temperatures measured on the surface of a steel plate at the end of the test period in accordance with the heat-up temperature-response (see **heat-up temperature-response**) test. **F 1827,** F26

temperature, wet-bulb—wet bulb temperature (without qualification) is the temperature indicated by a wet-bulb psychrometer constructed and used according to specifications. **E 41,** G03

temper brittleness, *n*—brittleness that results when certain **steels** are held within, or are cooled slowly through, a certain range of temperature below the **transformation range**. **A 941,** A01

temper carbon, *n*—compact aggregates or nodules of graphite found in malleable iron as a result of heat treatment. **A 644,** A04

temper carbon—clusters of finely divided in malleable iron, that are formed as a result of decomposition of cementite, for example, by heating white cast iron to temperatures above the ferrite-austenite transformation temperature and holding at these temperatures for a considerable period of time (**graphite, nodular**). **E 7,** E04

tempered—reheated after hardening to some temperature below the critical range and subsequently cooled to increase toughness and ductility. **F 547,** F16

tempered glass—a general term for glass that has been subjected to a thermal treatment characterized by rapid cooling to produce a compressively stressed surface layer. See **fully tempered glass** and **heat-strengthened glass**. **C 162,** C14

tempered hardboard—a hardboard subjected to tempering as previously defined or specially manufactured with other variation in usual process so that the resulting product has special properties of stiffness, strength, and water-resistance associated with boards meeting specifications for that quality product. **D 1554,** D07

tempered martensite—the decomposition products which result from heating martensite to temperatures below the ferrite austenite (Ae_1) transformation temper. Under the light microscope, darkening of the martensite needles is observed in the initial stages of tempering. Prolonged tempering at high temperatures produces spheroidized carbides in a matrix of ferrite. At the higher resolution of the electron microscope, the initial stage of tempering is observed to result in a structure containing a precipitate of fine epsilon iron carbide particles. At about 500°F (260°C), there is a transition to a structure of larger and elongated cementite particles in a ferrite matrix. With further tempering at higher temperature, the cementite particles become spheroidal, decreased in number, and increased in size. **E 7,** E04

tempered refractory castable, *n*—a refractory castable to which liquid has been added and sufficiently mixed to produce the desired consistency for placement. **C 71,** C08

tempered service hardboard—service hardboard, as previously defined, which has been given a tempering treatment to improve such properties as stiffness, strength, and water resistance. **D 1554,** D07

tempering, *n*—reheating a quench hardened or normalized steel object to a temperature below Ac_1, and then cooling it at any desired rate. **A 941,** A01

tempering—a thermal treatment of a quench-hardened product to improve ductility. **B 846,** B05

tempering—the manufacturing process of adding to a fiber or particle panel material a siccative material such as drying oil blends of oxidizing resin which are stabilized by baking or other heating after introduction. **D 1554,** D07

temper number—a relative evaluation of the annealing process of glassware as determined with strain disks (see Test Methods C 148). **C 162,** C14

tempest—term used to describe techniques used to reduce emanation of electronic data or intelligence from a tactical shelter. **E 1749,** E06

template, *n*—(*1*) a pattern used as a guide in fabricating elements. (*2*) a precise, detailed pattern or layout to provide essential fabrication details. **E 631,** E06

temple mark, *n—in woven fabrics,* small holes or distortions adjacent to the selvage. **D 123,** D13

temporal averaging—the combination of particle size distributions obtained at different points in time into a distribution representative of a longer time interval. **E 1620,** E29

temporal resolution—because of ambiguities due to alternative definitions, it is recommended that "temporal resolution" not be used unless a specific definition is provided by the user. **E 1620,** E29

temporal size distribution—see flux size distribution. In context with particle size distributions, the use of *temporal* is not recommended because of possible confusion with other meanings for this term. **E 1620,** E29

temporary coating—a coating designed to protect or decorate a substrate for a limited time that can be readily removed either by mechanical or chemical means. **D 16,** D01

temporary decommissioning—the engineered closure of a well intended to be returned to service at some later date (generally no more than six months). Temporary plugging should not damage the structural integrity of the well. Plugging materials consist of sand, bentonite, or other easily removed materials. **D 653,** D18

temporary hardness—usually the bicarbonate salts of calcium and magnesium. **D 6161,** D19

tenacity, *n*—the tensile stress expressed as force per unit linear density of the unstrained specimen. **D 123,** D13

tenacity, *n*—the tensile test of a material based on the linear density of the unstrained material. **D 123,** D13

tenacity, *n—in a tensile test,* the force exerted on the specimen based on the linear density of the unstrained material. **D 4848,** D13

tenacity at rupture, *n*—the tenacity at the force-at-rupture. (See also **force-at-rupture, rupture, tenacity.**) **D 123,** D13

tenacity-at-specified-elongation, TASE, *n*—the tenacity of a material at its force-at-specified-elongation. (Compare **breaking tenacity.**) **D 123,** D13

tenant (locataire)— organization that has rights and obligations of occupancy in a facility, as specified in a lease or occupancy agreement. **E 1480,** E06

tensile, *adj*—relating to tension in, or on, a material. **D 123,** D13

tensile creep rupture strength, [FL^{-1}], *n— for geosynthetics,* the force per unit width that will produce failure by rupture in a creep test in a given time, at a specified constant environment **D 4439,** D35

tensile creep strain, *n*—the total strain at any given time. **D 4439,** D35

tensile energy absorption (TEA), *n—of paper,* a mathematical quantity used to express the energy absorbed by a paper specimen prior to rupture in a tensile test carried to rupture under conditions specified. **D 1968,** D06

tensile hysteresis, *n—in mechanics,* hysteresis resulting from the extension of a material. (See *hysteresis.*) **D 4848,** D13

tensile hysteresis curve, *n*—a complex load-elongation or stress-strain curve obtained under either of two conditions:

(*1*) When a specimen is successively subjected to the application of a load or force less than that causing rupture, and the removal of the load or force according to a predetermined procedure.

(*2*) When a specimen is stretched less than the breaking elongation and allowed to relax by removal of the strain according to a predetermined procedure. **D 123,** D13

tensile index, *n*—a mathematical quantity calculated by dividing the tensile strength of a sample by its mass per unit area, both terms expressed in SI units. See **tensile strength**. **D 1968,** D06

tensile modulus, *J*, *(FL^{-1})*, *Nm^{-1}*, *n—for geotextiles,* the ratio of the change in tensile force per unit width to a corresponding change in strain (slope). **D 4439,** D35

tensile set, *n*—the extension remaining after a specimen has been stretched and allowed to retract in a specified manner expressed as a percentage of the original length. **D 1566,** D11

tensile stiffness, *n— of paper,* a mathematical quantity expressing the ratio of the tensile force on paper to its tensile strain in the elastic region of the tensile force-elongation behavior of the material **D 1968,** D06

tensile strain, *n*—the strain on a material subjected to tension. **D 123,** D13

tensile strain recovery, *n*—the percent of recoverable extension to the total extension impressed on a fiber under specified conditions. **D 123,** D13

tensile strength—the maximum load per unit of original cross-sectional area that a conductor attains when tested in tension to rupture. **B 354,** B01

tensile strength—a property of solid material that indicates its ability to withstand a uniaxial tensile load. **C 709,** D02

tensile strength, *n*—the strength of a material under tension as distinct from compression, torsion, or shear. **D 123,** D13

tensile strength, *n*—the strength shown by a specimen subjected to tension, as distinct from torsion, compression, or shear. **D 123,** D13

tensile strength, *n*—the breaking load (or force) per unit cross-sectional area of the unstrained specimen. **D 123,** D13

tensile strength, *n*—the strength of a material under tension as distinct from compression, torsion or shear. **D 123,** D13

tensile strength, *n— in an adhesive joint,* the maximum tensile stress that a material is capable of sustaining calculated from the maximum load applied perpendicular to the joint divided by the original cross-sectional area of the joint. **D 907,** D14

tensile strength, *n*—the maximum tensile stress applied during stretching a specimen to rupture. **D 1566,** D11

tensile strength, *n— of paper,* the maximum tensile force developed per unit width of a paper test specimen prior to rupture in a tensile test which has been carried to rupture under conditions specified. **D 1968,** D06

tensile strength, *n—for geotextiles,* the maximum resistance to deformation developed for a specific material when subjected to tension by an external force. **D 4439,** D35

tensile strength of metal connector plate—resistance to tensile force by net plate cross section normal to the direction of load application, expressed as force per unit of width of full cross section of connector plate, when used in pairs of plates. **E 1807,** E06

tensile strength, S_u [FL^{-2}], *n*—the maximum tensile stress which a material is capable of sustaining. Tensile strength is calculated from the maximum load during a tension test carried to rupture and the original cross-sectional area of the specimen. **E 6,** E28

tensile strength, σ_{TS} **[FL^{-2}]**—the maximum tensile stress that a material is capable of sustaining. Tensile strength is calculated from the maximum load during a tension test carried to rupture and the original cross-sectional area of the specimen. **E 1823,** E08

tensile strength (unconfined or uniaxial tensile strength), T_o **(FL^{-2})** —the load per unit area at which an unconfined cylindrical specimen will fail in a simple tension (pull) test. **D 653,** D18

tensile stress, *n*—the stress within a material subjected to tension. **D 123,** D13

tensile stress—normal stress tending to lengthen the body in the direction in which it acts. (ISRM) **D 653,** D18

tensile stress, *n*—a stress applied to stretch a test piece (specimen). **D 1566,** D11

tensile stress at given elongation, *n*— the stress required to stretch the uniform cross section of a test specimen to a given elongation. **D 1566,** D11

tensile stress [FL^{-2}], *n*—normal stress due to forces directed away from the plane on which they act, or **E 6,** E28

tensile test, *n—in textiles,* a test in which a material is extended in one direction to determine one or more of its force-extension, or stress-strain, characteristics; for example, breaking force, elongation at break. **D 123,** D13

tensile test, *n—in textiles,* a test in which a textile material is stretched in one direction to determine the force-elongation characteristics, the breaking force, or the breaking elongation. **D 4439,** D35

tensile testing machine, *n*—an apparatus designed to impart, or transmit, force/extension, or stress/strain, to a material and to measure the effect of the action. (See also **constant-rate-of-extension tensile testing machine, constant-rate-of-load tensile testing machine, andconstant-rate-oftraverse tensile testing machine.**) **D 123,** D13

tensile work, *n— of paper,* a term having the same meaning as **tensile energy absorption** when used for paper. **D 1968,** D06

tensiometer—a device for measuring soil-water matric potential (or tension or suction) of water in soil in-situ; a porous, permeable ceramic cup connected through a water filled tube to a pressure measuring device. **D 653,** D18

tension, *n*—a uniaxial force tending to cause the extension of a body or the balancing force within that body resisting the extension. **D 123,** D13

tension, *n*—a uniaxial force tending to cause the stretching of a material. **D 4848,** D13

weight, *n—in textile testing,* an object, having a known mass, used in such procedures as weighing, calibrating, and applying a force. **D 4848,** D13

tension band—an offset strip of metal shaped to fit around the terminal post and used with a carriage bolt and nut to attach the tension bar to the post. **F 552,** F14

tension bar—the bar used with tension bands or other post connectors to secure the fabric to a terminal post. **F 552,** F14

tension fatigue, *n*—fracture, through crack growth, of a component or test specimen subjected to a repeated tensile deformation. **D 1566,** D11

tension member—any component that carries horizontal (axial) tension loads imposed upon the boom. **F 818,** F20

tension or compression, E [FL^{-2}], *n*—Young's modulus (modulus in tension or modulus in compression). **E 6,** E28

tension parallel to grain—the imposition of a tensile stress which acts in a direction parallel to the fiber direction of the wood. **D 9,** D07

tension-recovery chart, *n—in tension testing,* a continuously plotted graph of tension versus extension resulting from a tension-recovery cycle. (Compare **extension-recovery chart**. See also **extension-recovery cycle** and **tension-recovery cycle**.) **D 123,** D13

tension-recovery cycle, *n—in tension testing,* the continuous application of tension on a specimen with a momentary hold at the maximum tension, followed by return to zero tension at a controlled rate. Compare **extension-recovery cycle**.) (See also **extension-recovery chart load-recovery cycle** and **tension-recovery chart**.) **D 123,** D13

tension set, n—see **tensile set**, the preferred term. **D 1566,** D11

tension-supported roof, *n*—a fabric roof-system, that is properly secured and primarily held in place by tensile forces applied across the system.**[D13.59] D 4851** **D 123,** D13

tension test, *n—in textiles,* a test designed to measure the tautness in a textile strand or fabric. (See also **tensile test**.) **D 123,** D13

tension wire—a wire, with or without a marcelled pattern, typically used along the top of a chain link fence in place of a top rail and along the bottom of a chain link fence for additional security (see Specification A 824 for *marcelled tension wire*). (Sometimes incorrectly called *coil spring wire*.) **F 552,** F14

tension wood—an abnormal form of wood found in the upper side of the bole and branches of leaning trees of some hardwood species and characterized by the presence of gelatinous fibers and excessive longitudinal shrinkage. Tension wood fibers hold together tenaciously so that sawed surfaces usually have projecting fibers, and planed surfaces often are torn or have raised grain. Tension wood may cause warping. **D 9,** D07

tenter mark, *n*—a visible deformation on the side edge or body of a fabric due to pressure from clips or pins. (See **clip mark, pin mark.**) **D 123,** D13

tenth-value-layer (TVL)—the thickness of the layer of a specified substance which, when introduced into the path of a given narrow beam of radiation reduces the intensity of this radiation by a factor of ten. **E 1316,** E07

teratogen—a chemical agent that causes congenital malformations in a developing embryo or fetus. **E 609,** E35

term, compound, n—see **compound term** **E 1992,** E02

term, delimited, n—see **delimited term** **E 1992,** E02

terminal erosion rate, *n— in cavitation or liquid impingement erosion,* the final steady-state erosion rate that is reached (or appears to be approached asymptotically) after the erosion rate has declined from its maximum value. (See also **terminal period** and **erosion rate-time pattern**.) **G 40,** G02

terminal period, *n— in cavitation or liquid impingement erosion,* a stage following the deceleration period, during which the erosion rate has levelled off and remains approximately constant (sometimes with superimposed fluctuations) at a value substantially lower than the maximum rate attained earlier. This occurs in some, but not all, cavitation and liquid impingement tests. (See also **erosion rate-time pattern**.) **G 40,** G02

terminal post—the basic load-bearing component for a line of fence, to or from which the fabric is stretched; it may be an end, gate, corner or pull post. **F 552,** F14

terminal solid solution—in a multicomponent system, any solid phase, of limited composition range, which includes the composition of one of the components of the system. **E 7,** E04

terminal velocity—the limiting velocity reached by a particle falling under the action of gravity in a still liquid at a specified temperature. **D 4410,** D19

terminator, *n*—a character or sequence of characters, such as a line ending, used to mark the end of a command whose length is variable. **F 1457,** F05

terminology, *n*—a set of terms representing the concepts of a field. **E 1992,** E02

ternary system—the complete series of compositions produced by mixing three components in all proportions. **E 7,** E04

terne coating, *n*—a lead-based coating, most commonly applied to steel sheet; the coating typically contains from 3 to 20 % tin, and other minor (<1 %) alloying elements may be present. **A 902,** A05

terpene alcohol, *n*—an alcohol directly related to or derived from a terpene hydrocarbon. **D 804,** D01

terpene resins, *n*—the products formed by polymerization of β -pinene, α-pinene, limonene and other terpene hydrocarbons. **D 804,** D01

terpenes, *n*—a class of unsaturated organic compounds having the empirical formula $C_{10}H_{16}$ occurring in most essential oils and oleoresinous plants. **D 804,** D01

terpolymer, *n*—a polymer formed from three monomer species. **D 1566,** D11

terracing—a network of contours that are associated with pyramid-like defects on epitaxially deposited surfaces and are related to the orientation of the surface. **F 1241,** F01

terra sigillata—a porous, red clay ware characterized by embossed decorations of the same color and a satin-like unglazed surface. **C 242,** C21

terrazzo, *n*—a mosaic flooring made by embedding marble, onyx, granite, or glass in portland cement which is poured in place, then polished. **F 141,** F06

terrestrial—consisting of land, as distinguished from water. **F 1600,** F20

terrestrial (or soil) environment, *n*—the aerobic environmental compartment which is found in and on natural soils. **D 6384,** D02

terry fabric, *n*—a material with a woven warp pile or a knitted pile, with uncut loops on a single side or uncut loops on both sides, and which is used for such products as toweling, beachwear, and bathrobes. **D 123,** D13

terry towel, *n*—a textile product with end hems or fringes and side hems or selvages which is made with loop pile on one or both sides generally covering the entire surface or forming strips, checks, or other patterns. **D 123,** D13

4-tertiary butyl catechol—*p-tert*-butylcatechol ($C_{10}H_{14}O_2$) mol weight 166.2; colorless crystals; polymerization inhibitor for styrene, butadiene, and other olefins; boiling point, 285°C; melting point, 52°C. **D 4790,** D16

tertiary hole—*in grouting*, the third series of holes to be drilled and grouted usually spaced midway between previously grouted primary and secondary holes. **D 653,** D18

tervariant equilibrium—a stable state among a number of conjugate phases equal to one less than the number of components, that is, having three degrees of freedom. **E 7,** E04

tesla, T—the SI unit of magnetic induction. One tesla is equal to 1.0 Wb/m^2 or 10^4 gausses. **A 340,** A06

Tessar—trade name for a photographic objective made by combining a positive lens, a negative lens, and a doublet, which is used for making macrographs. **E 7,** E04

test—technical operation that consists of the determination of one or more characteristics of a given product, process or service according to a specified procedure, (ISO Guide 2). **E 1187,** E36

test, *n*—a procedure for determining a property or characteristic of a material or a component by direct measurement. **E 1316,** E07

test, *n*—critical examination of the quality, composition, or properties of a material, usually involving standardized test procedures. **E 1605,** E06

test:

accelerated test—the testing of materials by exposure to intensified simulation of service conditions, for example, weathering, radiation, etc.

climbing drum peel test—a method of determining the relative peel resistance of adhesive bonds between a relatively flexible adherend and a rigid adherend, and the relatively flexible facing of a sandwich structure and its core (see Test Method D 1781-).

destructive test—a test involving the destruction of assemblies or parts in order to evaluate the maximum performance of the assembly or part.

floating roller peel test—a method of determining the relative peel resistance of adhesive bonds between one rigid and one flexible adherend (see Test Method D 3167).

nondestructive test—an inspection test for the evaluation of structural quality without damaging the assembly, for example, ultrasonics, visual inspection, etc.

sag flow test—a method of determining the maximum thickness to which a material can be applied without sagging and that provides a means of measuring the amount of sag flow at a given thickness (for highly viscous resins) (see Test Method D 2730). **E 1749,** E06

test, acceptance—one made at the option of the purchaser to verify that a product meets design criteria. **F 819,** F18

test assembly—the wall or floor into which the test sample(s) is (are) mounted or installed. **E 176,** E05

test assembly, *n*—horizontal or vertical construction on which test specimens are to be mounted together with associated instrumentation. **E 176,** E05

test bar, *n*—a bar-shaped coupon that is tested with or without subsequent preparation for the determination of physical or mechanical properties. **A 644,** A04

test beard, *n*—*in length testing of cotton*, the portion of the test specimen that has been combed and brushed into a" beard" which protrudes from the outside of the comb(s) or the clamp(s). **D 123,** D13

test chemical, *n*—the liquid or gas that is used to challenge the protective clothing material specimen. **F 1494,** F23

test coil—the section of the coil assembly that examines the material under test in a comparative system; the coil used to examine the material in an absolute or differential comparative system (E 566). **E 1316,** E07

test coupon, *n*—specially designed casting, or portion thereof, that is used to provide a representative sample of the iron from which it was cast. **A 644,** A04

test cycle, *n*—*in tire testing*, one complete circuit of the specified test course. **F 538,** F09

test, design—one made on a sample treated as representative of an industrial product. These tests will not generally be repeated in quantity production. **F 819,** F18

test determination, *n*—(*a*) the process of carrying out the series of operations specified in a test method whereby one or more readings (observations) are made on a test specimen and the observations combined to obtain the value of a property of the test specimen, or (*b*) the value obtained by the process. **D 1968,** D06

test determination, *n*—(*1*) the process of deriving from one or more test observations (observed values) the presence or absence of an attribute or the value of a characteristic or dimension of a single test specimen, or (*2*) the attribute (presence or absence) or value derived from the process (see *test specimen*). **E 456,** E11

test distance, *n*—distance traveled by a vehicle after tire break-in. **F 538,** F09

test frequency—in electromagnetic testing, the number of complete cycles per unit time of the alternating current applied to the primary test coil. **E 1316,** E07

testing, *n*—the determination by technical means of properties; performance; or elements of materials, products, services, systems, or environments which involve application of established scientific principles and procedures. **D 1356,** D22

testing—an element of inspection which generally denotes the determination by technical means of the properties or elements of supplies, or components thereof, and involves the application of established scientific principles and procedures. **D 1517,** D31

testing—action of carrying out one or more tests, (ISO Guide 2). **E 1187,** E36

testing—an element of inspection which generally denotes the determination by technical means of the properties or elements of supplies, or components thereof, and involves the application of established scientific principles and procedures. **F 869,** F08

testing agency—see *testing laboratory*. **E 1187,** E36

testing laboratory—a qualified organization authorized to test building sealants for compliance with specified standards. **C 717,** C24

testing laboratory—laboratory that measures, examines, tests, calibrates, or otherwise determines the characteristics or performance of materials or products, (ISO Guide 2, *A*). **E 1187,** E36

testing programs—programs involving test procedures to determine chemical/protective clothing material interactions or compatibilities. Testing programs include, but are not limited to, testing protective clothing materials for resistance to degradation, penetration, and permeation. **F 1494,** F23

testing, ultrasonic—a nondestructive method of examining materials by introducing ultrasonic waves into, through or onto the surface of the article being examined and determining various attributes of the material from effects on the ultrasonic waves. **E 1316,** E07

test lug, *n*—a sample produced as an appendage on a casting, that may be removed and tested to qualify the casting or the iron from which it was produced. **A 644,** A04

test matrix, *n*—a group of candidate tires usually with specified reference tires; all tests are normally conducted in one test program. **F 538,** F09

test measurement, *n*—a quantitative expression of one value determined for a property of interest by a single application of a specified test procedure. **D 1711,** D09

test method, *n*—a definitive procedure for the identification, measurement, and evaluation of one or more qualities, characteristics, or properties of a material, product, system, or service that produces a test result. (Compare **practice**.) **D 123,** D13

test method—defined technical procedure to determine one or more specified characteristics of a material or product, (ISO Guide 2, *A*). **E 1187,** E36

test observation, *n*—see **observation.** **E 456,** E11

test of significance—a statistical test that, by use of a specified test statistic, purports to provide a test of a null hypothesis (under certain assumptions); for example, that an imposed treatment in the experiment is without effect. **E 1823,** E08

test (or testing), *n*—a prodedure performed on an object (or set of nominally identical objects) using specified equipment that produces data unique to the object (or set). **F 538,** F09

test (performance test of a facility)—See **requirement statement**. **E 1480,** E06

test piece—a specimen containing known artificial or natural defects used for checking the efficiency of magnetic particle flaw detection processes. **E 1316,** E07

test piece—member of test specimen. **E 1807,** E06

test pit—a shallow excavation made to characterize the subsurface. **D 653,** D18

test portion, *n*—a quantity of material that is representative of the analysis sample and sufficient to obtain a single test result for the property or properties to be measured. **D 121,** D05

test program, *n*—an ordered series of tests grouped together using a predefined plane. **F 538,** F09

test quality level— see **rejection level.** **E 1316,** E07

test record, *n*—a document or electronic record that contains the observations and derived data obtained by applying a given test method. **A 941,** A01

test report, *n*—a document that presents the applicable qualitative or quantitative results obtained by applying one or more given test methods. **A 941,** A01

test report—document that presents test results and other information relevant to a test, (ISO Guide 2, *A*). **E 1187,** E36

test report—a written document or electronic record, signed by an authorized party, which contains sufficient data and information to verify that the tested fastener properties conform to the particular specification requirements. **F 1789,** F16

test result, *n*—a value obtained by applying a given test method, expressed as a single determination or a specified combination of a number of determinations. (See **observation**.) **D 123,** D13

test result, *n*—the value that expresses the level of a property of the test unit. **D 1711,** D09

test result, *n*—the value obtained for a given property from one test unit, which may be a single observation or the combination of multiple observations, as required by a specific test method. See **test unit**. **D 1968,** D06

test result, *n*—a value obtained by applying a given test method, expressed either as a single observation or a specified combination of a number of observations. **D 4439,** D35

test result, *n*—a value obtained by applying a method one time to a test material. **E 135,** E01

test result, *n*—the value of a characteristic obtained by carrying out a specified test method.

NOTE—The test method should specify that one or a number of individual observations be made and their average or another appropriate function, such as the medium or the standard deviation, be reported as the test result. It also may require standard corrections to be applied, such as correction of gas volumes to standard temperature and pressure. A test result, therefore, can be a result calculated from several observed values. In the simple case, the test result is the observed value itself. **E 456,** E11

test result, *n*—the average or median of a specified number of determinations; it is the reported value for a test. **F 538,** F09

test ring—a ring specimen containing artificial subsurface discontinuities which is used to evaluate and compare the overall performance and sensitivity of magnetic particle examination techniques. **E 1316,** E07

test, routine—a type of test made regularly on production material. **F 819,** F18

test run, *n*—a single pass of a loaded tire over a given test surface. **F 538,** F09

test sample—a representative part of the experimental environment (gases, liquids, or solids) for purposes of analysis. **E 176,** E05

test sample, *n*—a quantity suitable for use directly in a selected testing device. **E 1638,** E29

test section—the portion of pipeline under test. **C 896,** C04

test section, *n*—a distinct area of construction. **D 4439,** D35

test section—portion(s) of a pipe, fitting, or pipeline under test. **F 412,** F17

test sieve, *n*—a sieve that meets all of the minimum requirements of ASTM Committee E-11. **E 1638,** E29

test skein, *n*—a small skein which has a prescribed length of yarn and is used for the determination of linear density or breaking force, or both. **D 123,** D13

test specimen, *n*—a test object, suitably prepared from a sample, for evaluation of the chemical, physical, mechanical, or metallurgical quality of the sample. **A 644,** A04

test specimen, *n*—the specific portion of a test sample upon which a test is to be performed, and that is obtained by systematically reducing the size of the sample until a representative specimen of the required mass is obtained. **C 460,** C17

test specimen—specimen, or portion thereof, which is to be tested and the test results reported, or which is to be prepared for further testing, and the test results reported. **C 896,** C04

test specimen, *n*—the specific portion of a test sample upon which a test is to be performed, and that is obtained by systematically reducing the size of the sample until a representative specimen of the required mass is obtained. **C 1154,** C17

test specimen, n—synonym for **test portion. D 121,** D05

test specimen, *n—(Fibrograph),* the cotton fibers placed randomly on a pair of Fibrograph combs for fiber length measurements. **D 123,** D13

test specimen, *n—in cotton length tests with the Length Analyzer,* the cotton fibers protruding randomly from the base of a Motion Control Inc. specimen clamp for fiber length measurement before being combed or brushed. **D 123,** D13

test specimen, *n—in cotton maturity tests,* the series of slides observed by one technician as one half of the test. **D 123,** D13

test specimen, *n—for wool top,* a length of specified mass taken at random from a length of wool top selected as a laboratory sample. **D 123,** D13

test specimen, *n*—a portion of a test unit upon which one or more test measurements are made. **D 1711,** D09

test specimen, *n— of paper,* test unit, or portion of the test unit, upon which single or multiple test determinations are to be made. See **test unit. D 1968,** D06

test specimen, *n*—the specific portion of a test sample upon which a test is to be performed, and that is obtained by systematically reducing the size of the sample until a representative specimen of the required mass is obtained. **D 2946,** C17

test specimen, *n*—the fire stop being tested. **E 176,** E05

test specimen, *n*—stack of five identical stacking chairs. **E 176,** E05

test specimen, *n*—the portion of a test unit needed to obtain a single test determination.

NOTE—When used for a physical test, this is sometimes called "test piece." For a chemical test, it is sometimes called *test portion* or *test sample*. For optical and other tests, it is also sometimes called *test sample*. In interlaboratory evaluation of test methods and other statistical procedures, it is best to reserve the word **sample** for the whole amount of material involved and not the individual test specimens, pieces or portions being tested. **E 456,** E11

test specimen—portion of a test unit needed to obtain a single test determination. **E 1605,** E06

test specimen—sample connections to be tested to determine a particular plate strength characteristic; fabricated by connecting two or more butted wood members with two parallel metal connector plates at each connection, placed symmetrically on opposite sides along the butted ends. **E 1807,** E06

test specimen, *n*—a specific portion of the samples upon which the testing is to be performed. **G 113,** G03

test statistic—a function of the observed values in a sample that is used in a test of significance. **E 1823,** E08

test surface—that surface of a part through which the ultrasonic energy enters or leaves the part. **E 1316,** E07

test, tap—a nondestructive evaluation procedure for detecting areas of panel delamination in sandwich or other composite constructions; outer surface of the panel is tapped with a hammer or coin. Changes in acoustic emissions (sound) resulting from tapping are used to distinguish between delaminated and nondelaminated sections of the panel (see Test Method E 492). **E 1749,** E06

test thermoelement, *n*—a thermoelement that is to be calibrated with reference to platinum 67 (Pt-67) by comparing its thermal emf with that of a standard thermoelement. **E 344,** E20

test tire, *n*—a tire used in a test. **F 538,** F09

test tire set, *n*—one or more test tires as required by the test equipment or procedure, to perform a test thereby producing a single test result. **F 538,** F09

test unit, n—synonym for **analysis sample. D 121,** D05

test unit, *n*—a fraction of a unit of product from which one or more test specimens are taken for each property. **D 1711,** D09

test unit, *n—of paper,* a unit or portion of the sample sufficient to obtain the test result(s) for the property or properties to be measured. **D 1968,** D06

test unit, *n*—the total quantity of material (containing one or more test specimens) needed to obtain a test result as specified in the test method. See *test result.* **E 456,** E11

test wheel—a wheel and test tire assembly mounted to a test vehicle by means of a force or torque transducer. **E 867,** E17

tetragonal—having three mutually perpendicular axes, two equal in length and unequal to the third. **E 7,** E04

tex, *n*—the unit of linear density, equal to the mass in grams of 1000 meters of fiber, yarn, or other textile strand, that is used in a direct yarn numbering system. (See also **linear density** and **direct yarn numbering system.**) **D 123,** D13

Texas steers—usually side-branded steer hides of a narrow close compact pattern, plump, and not necessarily from Texas. **D 1517,** D31

tex ticket number, *n*—the designator assigned to a serving thread to indicate the approximate linear density. (Compare **tex**) **D 123,** D13

textile, *n*—originally a woven fabric, now generally applied to: (*1*) staple fibers and filaments suitable for conversion to or use as yarns, or for the preparation of nonwoven fabrics, (*2*) yarns made from natural or manufactured fibers, (*3*) fabrics and other manufactured products made from fibers as defined above, and from yarns, and (*4*) garments and other articles fabricated wholly from one or more of the above elements, and articles made principally from the above when the products retain the characteristic flexibility and drape of the original fabrics. **D 123,** D13

textile, *adj*—of or pertaining to textiles. **D 123,** D13

textile fiber, *n—general,* a generic term for the various types of matter that form the basic elements of textile fabrics and other textile structures. **D 123,** D13

textile fiber, *n—specific,* a unit of matter that is characterized by having a length at least 100 times its diameter or width and which can be spun into a yarn or made into a fabric by interlacing in a variety of methods, including weaving, knitting, braiding, felting, and twisting. **D 123,** D13

textile floor covering, *n*—a system having a use-surface composed of textile material and generally used for floor covering. **D 123,** D13

textural component, *n—as used in Test Method D 5061,* the collective term used to describe carbon forms and recognizable coal- and process-derived components (binder-phase, filler-phase, and miscellaneous material), in coke. **D 121,** D05

texture—*(1)* a modified appearance of dimension stone resulting from one or several mechanical surface treatments. Untreated stone surfaces have textural characteristics described under *(2)*.

(2) that aspect of the physical appearance of a rock that is determined by size, shape, and mutual relations of the component grains or crystals. Textures related to dimension stone include *equigranular* (grains of approximately the same size); *inequigranular* (grains of markedly unequal sizes); *porphyritic* ;*interlocking* (in which grains with irregular boundaries interlock by mutual penetration); interlocking and porphyritic textures are characteristic of granites and marbles; *clastic* (naturally cemented fragmental grains but without mosaic or interlocking relations; this texture is typical of sandstones and some limestones); *mosaic* (closely packed grains with smooth to moderately irregular, noninterlocking mutual boundaries); *granoblastic* (a megascopically granular mosaic texture in which the grains are tightly compacted and the minerals are dominantly equidimensional and present irregular mutual boundaries; mosaic and granoblastic textures are characteristic of metamorphic rocks). **C 119,** C18

texture—a surface pattern as compared with a smooth finish. **C 460,** C17

texture, *n*—a surface pattern as compared with a smooth finish. **C 1154,** C17

texture—in wood anatomy, the sizes, distribution, and proportional volumes of the cellular elements of which wood is composed; often used interchangeably with grain. Depending on the relative size and distribution of the cellular elements, texture may be coarse (open grain) or fine, even, or uneven. **D 9,** D07

texture, *n*—the surface appearance and hand of a textile material. **D 123,** D13

texture, *n—in carpets or pile floor coverings,* the detailed configuration of loops, cut pile ends, and individual fibers in the pile layer. **D 123,** D13

texture—*of soil and rock,* geometrical aspects consisting of size, shape, arrangement, and crystallinity of the component particles and of the related characteristics of voids. **D 653,** D18

texture—the arrangement in space of the components of a rock body and of the boundaries between these components. (ISRM) **D 653,** D18

texture, *n*—a surface pattern as compared with a smooth finish. **D 2946,** C17

texture—the geometric aspects of the component particles of a sediment deposit or rock including size, shape, and arrangement. **D 4410,** D19

texture, *n—in pile yarn floor coverings,* in the case of pile yarn floor coverings, the detailed configuration of loops, cut pile ends, and individual fibers in the pile. **D 5684,** D13

texture, *n*—the rheological and structural (geometric and surface) attributes of a product perceptible by the kinesthetic, tactile, visual, and auditory senses. **E 253,** E18

texture, *n*—the visible surface structure depending on the size and organization of small constituent parts of a material; typically, the surface structure of a woven fabric. **E 284,** E12

texture (cube)—See **preferred orientation.** **E 7,** E04

textured, *adj—in evaluating the color of a surface,* structural quality of a surface determined by the topography of its constituents. **D 2946,** C17

textured boards—boards that are factory produced with a nonplaner surface by, for example, molding, embossing, machining, or post-forming or any combination of these processes. **D 1554,** D07

texture (deformation)—See **preferred orientation.** **E 7,** E04

textured glass yarn, *n*—a yarn processed from continuous filament yarn in such a manner to induce bulk to the yarn by disorientation of the filaments. **D 123,** D13

textured product, *n—in cooling tower fill,* product with bas-relief motifs that extend the external surface of a sheet sufficiently to increase its heat transfer capacity by at least 10 % under conditions of turbulent flow. **C 460,** C17

textured products, *n— in cooling tower fill,* those with bas-relief motifs that extend the external surface of a sheet sufficiently to increase its heat transfer capacity by at least 10 % under conditions of turbulent flow. **C 1154,** C17

texture product, *n—in cooling tower fill,* a product with bas-relief motifs that extend the external surface of a sheet sufficiently to increase its heat transfer capacity by at least 10 % under conditions of turbulent flow. **D 2946,** C17

texture shape factor, *n*—average of weighted sum of the ratios of amplitude to wavelength, as determined from an amplitude versus wavenumber (reciprocal of wavelength) spectrum. **E 867,** E17

TFE fluorocarbon—tetrafluoroethylene or its polymer, polytetrafluoroethylene $(C_2F_4)_n$. **C 859,** C26

thalweg—the line connecting the lowest or deepest point along a stream bed, valley, or reservoir, whether underwater or not. **D 4410,** D19

the atmosphere—See **atmosphere.** **D 1356,** D22

theoretical CO_2 (carbon dioxide), *n*—the amount of carbon dioxide which could hypothetically be produced from the complete biological oxidation of all the carbon in a material. **D 6384,** D02

theoretical density—the density of a material calculated from the number of atoms per unit cell and measurement of the lattice parameters. **C 859,** C26

theoretical elastic stress concentration factor (or stress concentration factor) k_t—the ratio of the greatest stress in the region of a notch or other stress concentrator as determined by the theory of elasticity (or by experimental procedures that give equivalent values) to the corresponding nominal stress. **E 1823,** E08

theoretical O2 (oxygen), *n*—the amount of oxygen that is theoretically required to oxidize a material. **D 6384,** D02

theoretical time curve—see **consolidation time curve.** **D 653,** D18

thermal analysis *(TA)* **,** *n*—a group of techniques in which a physical property of a substance is measured as a function of temperature or time while the substance is subjected to a controlled-temperature program. (ICTAC) **E 473,** E37

thermal arrest—See **arrest.** **E 7,** E04

thermal blanket, *n*—a blanket woven so that cells or openings are created in the fabric so that air warmed by the body is trapped between the yarns, such as textured or leno weaves; this product can be napped or unnapped. **D 123,** D13

thermal bonded batting, *n*—a textile filling material which contains low-melting point fibers or polymers which, when heated, fuse the batting materials together. **D 123,** D13

thermal break, *n*—a solid or cellular material or combination of materials of low thermal transmission placed between components of high thermal transmission in order to reduce the heat flow across the assembly. **D 883,** D20

thermal capacity, *n*—the quantity of heat required to change the temperature of the body one degree. For a homogeneous body, it is the product of mass and specific heat. For a nonhomogeneous body, it is the sum of the products of mass and specific heat of the individual constituents. (May also be seen as heat capacity.) (In SI units: J/K) (In inch-pound units: Btu/F) **C 168,** C16

thermal capacity—the amount of thermal energy that can be stored in a storage device during a period of time and for a specific set of values (that is, initial temperature of the storage device, the temperature of the entering fluid, and the mass flow rate of fluid through the storage system). **E 772,** E44

thermal capacity, theoretical—the amount of energy that can be stored in the storage device if all its components undergo an increase in temperature from the original value to a final value. **E 772,** E44

thermal character, *n*—apparent difference in temperature of the fabric and the skin of the observer touching it. **D 123,** D13

thermal conditioning—adjusting the average temperature and reducing the thermal gradients in glass. **C 162,** C14

thermal conductance, Γ—time rate of heat flow through a unit area of a body induced by a unit temperature difference between the body surfaces. **E 1142,** E37

thermal conductivity, *n*—time rate of unidirectional heat transfer per unit area, in the steady-state, between parallel planes separated by unit distance, per unit difference of temperature of the planes. **D 123,** D13

thermal conductivity, λ—time rate of heat flow, under steady conditions, through unit area, per unit temperature gradient in the direction perpendicular to the area. **E 1142,** E37

thermal conductivity vacuum gage—a vacuum gage containing two surfaces at different temperatures between which heat can be transported by the gas molecules so that changes in the temperature (or in the heating power required to maintain constant temperature) of one of the surfaces can be correlated with the gas pressure. Various types of thermal conductivity gages are distinguished according to the method of indicating the temperature change. The common types are listed below:

(1) Pirani Gage—An increase of pressure from the zero point causes a decrease in the temperature of a heated filament of material having a large temperature coefficient of resistance thus unbalancing a Wheatstone bridge circuit (or the circuit is adjusted to maintain the filament temperature constant).

(2) Thermocouple Gage—The decrease in temperature of a heated filament as the pressure rises is indicated by decreased emf in a thermocouple circuit having the junction in thermal contact with the center of the heated filament.

(3) Thermistor Gage—A form of Pirani gage employing a thermistor as the heated element.

(4) Bimetallic Strip Gage—Deflection of a bimetallic strip with changing temperature indicates the changes in pressure. **E 1316,** E07

thermal curve—*in thermal analysis,* plot of a dependent variable (such as, heat flow or weight loss) against an independent variable, time or temperature, under defined time-temperature conditions. **E 1142,** E37

thermal decomposition, *n*—a process of extensive chemical species change caused by heat (different from thermal degradation, q.v.; compare **pyrolysis**). **E 176,** E05

thermal degradation, *n*—irreversible and undesirable change in the properties of a material due to exposure to heat. **D 1566,** D11

thermal degradation, *n*—a process whereby the action of heat or elevated temperature on a material, product, or assembly causes a loss of physical, mechanical, or electrical properties (different from **thermal decomposition,** q.v.). **E 176,** E05

thermal diffusivity—ratio of thermal conductivity of a substance to the product of its density and specific heat capacity. **E 1142,** E37

thermal electromotive force—the voltage generated when one junction of two dissimilar metal wires is at a different temperature than the other junction. **E 7,** E04

thermal electromotive force—voltage generated when one junction of two dissimilar electrical conductors is at a different temperature than the other junction. **E 1142,** E37

thermal electromotive force (thermal emf), *n*—the net emf set up in a thermocouple under conditions of zero current. Same as **Seebeck emf.** **E 344,** E20

thermal emf—the net emf set up in a thermocouple under conditions of zero current. Also known as **Seebeck emf.** **F 1241,** F01

thermal end point—in testing of thermal protective materials, the point of where the sensor response on the recorder chart intersects the human tissue burn tolerance criteria overlay. **F 1494,** F23

thermal endurance—the relative ability of glassware to withstand thermal shock. **C 162,** C14

thermal endurance, *n*—an expression for the stability of an electrical insulating material, or a simple combination of materials, when maintained at elevated temperatures for extended periods of time. **D 1711,** D09

thermal endurance—property of a material to resist changes in chemical, physical, mechanical, or electrical properties upon exposure to temperatures for extended periods of time. **E 1142,** E37

thermal expansion, *n*—in refractories, the reversible change in size of materials due to temperature changes. **C 71,** C08

thermal expansion—See **mean coefficient of thermal expansion; percent linear thermal expansion.** **C 242,** C21

thermal expansion—the increase in the dimensions or the volume of a body due to a change in temperature. (See **coefficient of thermal expansion.**) **E 7,** E04

thermal expansion— see **coefficient of linear thermal expansion.** **E 1142,** E37

thermal hazard—relates to the laboratory test methods employed to measure thermal characteristics and to predict burn injury potential. **F 1494,** F23

thermal imaging transfer ribbon, *n*—plastic film or other material, upon which a dye or pigmented coating is applied; imaging results when a thermal printhead transfers the coating onto a suitable substrate or receptor media. **F 1623,** F05

thermal insulation, *n*—a material or assembly of materials used to provide resistance to heat flow. **C 168,** C16

thermal insulation—a material or assembly of materials used primarily to provide resistance to heat flow. **C 634,** E33

thermal insulation—a material applied to reduce the flow of heat.

D 1079, D08

thermal insulation, *n*—the resistance to dry heat transfer via conduction, convection, and radiation. **F 1494,** F23

thermal insulation system, *n*—applied or installed thermal insulation complete with any accessories, vapor retarder, and facing required. **C 168,** C16

thermal ionization—a process in which ions are formed when an atom or molecule interacts with a heated surface. Ionization only occurs when the work function of the surface and the temperature of the surface provide the necessary ionization energy to the atom or molecule. **C 859,** C26

thermalization—the process of slowing neutron velocities by permitting the neutrons to come to thermal equilibrium with a moderating medium. **E 1316,** E07

thermalization factor—the inverse ratio of the thermal neutron flux obtained in a moderator, per source neutron. **E 1316,** E07

thermal life, *n*—the time necessary for a specific property of a material, or a simple combination of materials, to degrade to a defined end point when aged at a specified temperature. **D 1711,** D09

thermal life—aging time, at a particular temperature, at which the material fails. **E 1142,** E37

thermal life curve, *n*—a graphical representation of thermal life at a specified aging temperature in which the value of a property of a material, or a simple combination of materials, is measured at room temperature and the values plotted as a function of time. **D 1711,** D09

thermally bonded seam, *n—in home sewing*, a seam formed by the use of a heat-sensitive material that fuses two material sections together along a seam line when heat and pressure are applied. (Compare **glued seam, sewn seam, stapled seam**.) **D 4965,** D13

thermally foamed plastic—a cellular plastic produced by applying heat to effect gaseous decomposition or volatilization of a constituent. **D 883,** D20

thermally foamed plastic—a cellular plastic produced by applying heat to effect gaseous decomposition or volatilization of a constituent. (D-20) **F 412,** F17

thermally stimulated current analysis, *n*—a technique in which the current generated when dipoles change their alignment in a substance is measured as a function of temperature or time while the substance is subjected to a controlled-temperature program in a specified atmosphere. (ICTAC)

NOTE—The technique can be applied in several ways: for example; the substance can be pre-conditioned by heating and cooling in a nonoscillating electric field to create aligned, frozen dipoles. The substance may then generate a thermally stimulated current during subsequent heating with no field applied. **E 473,** E37

thermally thick, *n*—the thickness of a medium that is large enough to have the predominate thermal (temperature) effects experienced within that distance, that is, negligible heat is lost from its unexposed side. **E 176,** E05

thermal neutrons—neutrons in thermal equilibrium with the medium through which they are traveling or diffusing. **E 170,** E10

thermal neutrons—neutrons having energies ranging between 0.005 eV and 0.5 eV; neutrons of these energies are produced by slowing down fast neutrons until they are in equilibrium with the moderating medium at a temperature near 20°C. **E 1316,** E07

thermal operating level, *n*—the operating condition at which the radiance of the heat source produces a specified constant heat flux to some specified position at the specimen surface. **E 176,** E05

thermal overshoot—a measure of the extent to which the control band of a temperature controller is exceeded. **F 1156,** F12

thermal precipitation—See **precipitation.** **D 1356,** D22

thermal printing head—mechanism for printing on, or with, a temperature-sensitive or labile substance; heat transfer is effected by touching heated pins tips to the material to cause a reaction of some sort. **F 1156,** F12

thermal processing—the application of heat to achieve a reduced level of microbiological activity. **F 17,** F02

thermal processing—See Terminology F 17. **F 1327,** F02

thermal radiation, *n*—process of emission in which the radiant energy originates in the thermal agitation of the particles of matter (atoms, molecules, ions).

NOTE—The terms "thermal radiation" and, in German "Temperaturstrahlung" apply not only to the process of emission, but also to the radiation itself. **E 349,** E21

thermal radiator, *n*—source emitting by thermal radiation. in parentheses indicates the year of last reapproval. A superscript epsilon (ϵ) indicates an editorial change since the last revision or reapproval. **E 349,** E21

thermal resistance, *n*—the reciprocal of thermal transmittance. **D 123,** D13

thermal resistance, R,—*n—in asbestos-cement plastic-foam core insulating panels*, the thermal gradient that will develop under a given steady state heat flow between two parallel unit area surfaces for a given material thickness. **C 460,** C17

thermal resistance, R, *n*—the difference in temperature which will develop under a given steady state heat flow between two parallel unit area surfaces for a given material thickness. **C 1154,** C17

thermal resistance, *R, n—in asbestos-cement plastic-foam core insulating panels*, the thermal gradient that will develop under a given steady state heat flow between two parallel unit area surfaces for a given material thickness. **D 2946,** C17

thermal resistance, R—Under steady-state conditions, the temperature difference required to produce a unit of heat flux through a specimen. **E 1142,** E37

thermal resistivity, *n*—the reciprocal of thermal conductivity. **D 123,** D13

thermal resistivity, r—under steady-state conditions, the temperature gradient, in the direction perpendicular to the isothermal surface, per unit heat flux. **E 1142,** E37

thermal resolution—the smallest apparent temperature difference between two blackbodies that can be measured by an infrared sensing device. **E 1316,** E07

thermal response, dynamic, *n*—the relationship between the thermal image optical density (*y*-axis) versus printhead energy (*x*-axis); printhead energy density (millijoules/square millimetres) is a function of printhead power times the pulse width (milliseconds) divided by the area of the printhead dot (square millimetres). **F 1623,** F05

thermal shock—a rapid change in temperature imposed on a glass body. **C 162,** C14

thermal shock—a condition of stress brought about by a large temperature difference across a body or glaze. (See also **thermal shock failure; thermal shock resistance testing.**) **C 242,** C21

thermal shock, *n*—a large and rapid temperature change, resulting in large temperature differences within or across a body. **C 1145,** C28

thermal shock—the stress-producing phenomenon resulting from sudden temperature drops in a roof membrane when, for example, a rain shower follows brilliant sunshine. **D 1079,** D08

thermal shock failure—mechanical failure of a glaze or body, as a result of the stress caused by a large temperature difference across the ware. **C 242,** C21

thermal shock resistance testing—the act of exposing ware to a rapid temperature change to determine the temperature difference a glaze or body can withstand without mechanical failure.

$$\Delta T_{max} = \sigma/T\alpha$$

where:
σ = stress,
E = Young's Modulus, and
α = mean coefficient of thermal expansion.

C 242, C21

thermal shrinkage, *n—of textile yarns and cords,*contraction in length caused by heat **D 123,** D13

thermal spalling—the breaking of rock under stresses induced by extremely high temperature gradients. High-velocity jet flames are used for drilling blast holes with this effect. (ISRM) **D 653,** D18

thermal spraying—a group of processes wherein finely divided metallic or nonmetallic materials are deposited in a molten or semimolten condition to form a coating. (The coating material may be in the form of powder, ceramic rod, wire, or molten materials.) **G 15,** G01

thermal stability, *n*—the resistance to permanent changes in properties caused solely by heat. **D 4175,** D02

thermal stability—resistance to permanent changes in property caused solely by heat, (D 2160, D-2). **E 1142,** E37

thermal storage medium— see **storage medium, thermal.** **E 772,** E44

thermal stress—the stress produced by a temperature differential within a glass body. **C 162,** C14

thermal transmittance, *n*—time rate of unidirectional heat transfer per unit area, in the steady-state, between parallel planes, per unit difference of temperature of the planes. **D 123,** D13

thermal transmittance, n—time rate of unidirectional heat transfer per unit area in the steady state, between parallel planes, per unit difference of temperature of the planes. **E 1142,** E37

thermal treatment—a controlled heating; time at maximum temperature-cooling cycle as needed to satisfy the property and grain structure requirements of the temper. **B 846,** B05

thermal vaporizer—an apparatus consisting of a container for chemical and a heater to maintain the vessel at a temperature sufficiently high to accelerate evaporation of the pesticide. The apparatus may contain a blower to disperse the pesticide vapor into a treated area or may rely upon natural turbulent diffusion for dispersion. **E 1102,** E35

thermionic cathode gun—(hot cathode gun). An electron gun which derives its electrons from a heated filament which may also serve as the cathode. **E 7,** E04

thermionic emission—the ejection of a stream of electrons from a hot cathode, usually under the influence of an electrostatic field. **E 7,** E04

thermistor, *n*—a semiconductor, the primary function of which is to exhibit a monotonic decrease in electrical resistance with an increase in sensor temperature, that is, a semiconductor for which the temperature coefficient of resistance is negative and exhibits neither discontinuities nor changes in sign. **E 344,** E20

thermoanalytical, *adj*—of, or pertaining to, thermal analysis. (ICTAC) **E 473,** E37

thermochemical treatment, *n*—a **heat treatment** carried out in a medium suitably chosen to produce a change in the chemical composition of the steel object by exchange with the medium. **A 941,** A01

thermochromic—material, ink, that changes color as a function of temperature. **F 1156,** F12

thermochromism, *n*—a change in color with temperature change. **E 284,** E12

thermocouple—two dissimilar electrical conductors so joined as to produce a thermal electromotive force when the junctions are at different temperatures. **E 7,** E04

thermocouple, *n—in thermometry,* the sensor of a thermoelectric thermometer, consisting of electrically conducting circuit elements of two different thermoelectric characteristics joined at a junction. **E 344,** E20

thermocouple—Two dissimilar electrical conductors joined as to produce a thermal electromotive force. **E 1142,** E37

thermocouple assembly, *n*—an assembly consisting of two thermocouple elements and one or more associated parts such as terminal block, connection head, and protecting tube. **E 344,** E20

thermocouple calibration, *n*—the process of determining the emf developed by a thermocouple with respect to temperature established by a standard. **E 344,** E20

thermocouple gage—see **thermal conductivity vacuum gage.** **E 1316,** E07

thermocouple vacuum gage—a thermal conductivity or hot-wire gage in which the temperature of an electrically heated fine wire varies as the thermal conductivity of the residual gas. The thermocouple measures the temperature change. **E 7,** E04

thermodilatometry, *n*—a technique in which a dimension of a substance under negligible load is measured as a function of temperature while the substance is subjected to a controlled-temperature program in a specified atmosphere. (ICTAC)

NOTE—Linear thermodilatometry and volume thermodilatometry are distinguished on the basis of the dimension measured. **E 473,** E37

thermodynamic dew-point temperature—See **temperature.** **D 1356,** D22

thermodynamic dew-point temperature, T_d—the temperature at which moist air with mixing ratio *r* and total pressure *p*, when saturated with respect to water at the same pressure *p*, will have a saturation mixing ratio r_w equal to the given mixing ratio *r*. **D 4023,** D22

thermodynamic frost-point temperature—See **temperature.** **D 1356,** D22

thermodynamic frost-point temperature—See **temperature.** **D 1356,** D22

thermodynamic frost-point temperature, T_f—the temperature at which moist air with mixing ratio *r* and total pressure *p*, when saturated with respect to ice at the same pressure *p*, will have a saturation mixing ratio r_i equal to the given mixing ratio *r*. **D 4023,** D22

thermodynamic ice-bulb temperature, T_i—the temperature that moist air at pressure *p*, temperature *t*, and mixing ratio *r* will attain when brought adiabatically to saturation at pressure *p* by sublimation of pure ice into the moist air. The ice-bulb temperature t_i is approximately, but *not* identically, equal to the thermodynamic ice-bulb temperature T_i. **D 4023,** D22

thermodynamic wet-bulb temperature—See **temperature.** **D 1356,** D22

thermodynamic wet-bulb temperature, T_w—the temperature that moist air at pressure *p*, temperature *t*, and mixing ratio *r* will attain when brought adiabatically to saturation at pressure *p* by the evaporation of pure liquid water into the moist air. The wet-bulb temperature t_w is approximately, but *not* identically, equal to the thermodynamic wet-bulb temperature T_w. **D 4023,** D22

thermoelectric power, *n*—the rate of change of thermal emf with temperature at a given temperature. Same as **Seebeck coefficient.** **E 344,** E20

thermoelectric pyrometer—a device with which temperatures are measured by utilizing the thermoelectric effects. In its simplest form, it consists of a thermocouple of two dissimilar metals which develop an emf when the junctions are at different temperatures and an instrument for measuring the emf developed by the thermocouple. **E 7,** E04

thermoelectric thermometer, *n*—thermometer for which the thermometric quantity is an emf produced by the Seebeck effect. **E 344,** E20

thermoelement, *n*—one of the circuit elements comprising a thermocouple. **E 344,** E20

thermoelement, *n*—one of the two dissimilar electrical conductors comprising a thermocouple. **E 344,** E20

thermogalvanic corrosion—the corrosive effect resulting from the galvanic cell caused by a thermal gradient across the metal surface. **G 15,** G01

thermogram—see **thermal curve**. **E 1142,** E37

thermogram—a visual image which maps the apparent temperature pattern of an object or scene into a corresponding contrast or color pattern. **E 1316,** E07

thermography, infrared—the process of displaying variations of apparent temperature (variations of temperature or emissivity, or both) over the surface of an object or a scene by measuring variations in infrared radiance.

NOTE—In general, *passive thermography* refers to examination of an object or system during its normal operational mode, without the application of any additional energy source for the express purpose of generating a thermal gradient in the object or system; *active thermography* refers to the examination of an object upon intentional application of an external energy source. The energy source (active or passive) may be a source of heat, mechanical energy (vibration or fatigue testing), electrical current, or any other form of energy. **E 1316,** E07

thermogravimetric analysis, *(TGA)*, *n*—a technique in which the mass of a substance is measured as a function of temperature or time while the substance is subjected to a controlled-temperature program in a specified atmosphere. (ICTAC)

NOTE—The record is the thermogravimetric or TG curve. This definition is similar to the ICTAC definition of the term *thermogravimetry*. **E 473,** E37

thermogravimetry (TG), *n*—see **thermogravimetric analysis.** **E 473,** E37

thermoluminescence dosimeter (TLD)—a dosimeter made of a material that stores energy when irradiated by ionizing radiation and then releases that energy in the form of visible light when heated. The light output of heated TLD is measured photometrically. TLDs are secondary-standard dosimetry systems. Examples of commonly available TLDs include lithium fluoride (LiF), calcium fluoride with various trace activators (such as $CaF_2 : Mn$ and $CaF_2 : Dy$), calcium sulfate ($CaSO_4 : Mn$ and $CaSO_4 : Dy$), and lithium borate ($Li_2B_4O_7 : Mn$). **E 170,** E10

thermomagnetometry, *n*—a family of thermoanalytical techniques in which a magnetic characteristic of a substance is measureed as a function of temperature or time while the substance is subjected to a controlled-temperature program in a specified atmosphere. (ICTAC)

NOTE—Thermogravimetric analysis with a magnetic field acting on the specimen is the most common example. **E 473,** E37

thermomechanical analysis, (*TMA*), *n*—a technique in which the deformation of a substance under nonoscillatory load is measured as a function of temperature or time while the substance is subjected to a controlled-temperature program in a specified atmosphere. (ICTAC)

NOTE—The load on the substance may be compressive, tensile, flexural or torsional. When the applied load is too low to cause deformation, TMA measures a dimension of the substance and in this mode is called **thermodilatometry.** **E 473,** E37

thermomechanical process, *n*—controlled thermal and deformation treatment performed at an elevated temperature. **E 7,** E04

thermometric fixed point, *n*—fixed point useful in the practice of thermometry. **E 344,** E20

thermomicroscopy,, *n*—see **thermoptometry**. **E 473,** E37

thermo-osmosis—the process by which water is caused to flow in small openings of a soil mass due to differences in temperature within the mass. **D 653,** D18

thermopile, *n*—a number of similar thermocouples connected in series, arranged so that alternate junctions are at the reference temperature and at the measured temperature, to increase the output for a given temperature difference between reference and measuring junctions. **E 344,** E20

thermopile—a number of temperature sensors connected in series and arranged such that there is an increase in output signal for a given temperature difference between alternate junctions maintained at a reference temperature and the measured temperature. **E 1142,** E37

thermoplastic—any one of a group of materials capable of being repeatedly softened or melted by increases in temperature followed by subsequent solidification on cooling. **C 904,** C03

thermoplastic, *n*—a plastic that repeatedly can be softened by heating and hardened by cooling through a temperature range characteristic of the plastic, and that in the softened state can be shaped by flow into articles by molding or extrusion. **D 883,** D20

thermoplastic, *adj*—capable of being repeatedly softened by heating and hardened by cooling through a temperature range characteristic of the plastic, and that in the softened state can be shaped by flow into articles by molding or extrusion for example. **D 883,** D20

thermoplastic, *adj*—capable of being repeatedly softened by heat and hardened by cooling. **D 907,** D14

thermoplastic, *n*—a material that will repeatedly soften when heated and harden when cooled. (See also **novolak**.) **D 907,** D14

thermoplastic, *n*—a polymer that repeatedly can be softened by heating and hardened by cooling through a temperature range characteristic of the polymer, and in the softened state can be shaped into articles. **D 1566,** D11

thermoplastic, *adj*—a property of a polymer that permits it to be repeatedly softened by heating and hardened by cooling through a characteristic temperature range, and in the softened state can be shaped into articles. **D 1566,** D11

thermoplastic—a plastic that repeatedly can be softened by heating and hardened by cooling through a temperature range characteristic of the plastic, and that in the softened state can be shaped by flow into articles by molding or extrusion. **D 3878,** D30

thermoplastic—a polymer material that will repeatedly soften when heated and harden when cooled. **E 1749,** E06

thermoplastic, *n*—a plastic that can be repeatedly softened by heating and hardened by cooling through a temperature range characteristic of the plastic, and that in the softened state can be shaped by flow into articles by molding or extrusion. (D-20) **F 412,** F17

thermoplastic, *adj*—capable of being repeatedly softened by heating and hardened by cooling through a temperature range characteristic of the plastic, and that in the softened state can be shaped by flow into articles by molding or extrusion for example. (D-20) **F 412,** F17

thermoplastic—capable of being repeatedly softened by heating and hardened by cooling through a temperature range characteristic of the plastic, and that in the softened state can be shaped by flow into articles by molding or extrusion. **F 869,** F08

thermoplastic, *n*—a plastic that repeatedly can be softened by heating and hardened by cooling through a temperature range characteristic of the plastic, and that in the softened state can be shaped by flow into articles by molding or extrusion. **F 1251,** F04

thermoplastic elastomer, *n*—a material having the general properties of an elastomer and capable of being repeatedly softened by heat and hardened to shape by cooling without significant degradation of the polymer system. **C 717,** C24

thermoplastic elastomers (TPE)—a diverse family of rubber-like materials that, unlike conventional vulcanized rubbers, can be processed and recycled like thermoplastic materials. **D 883,** D20

thermoplastic elastomer (TPE), *n*— a diverse family of rubber-like materials, that, unlike conventional vulcanized rubbers, can be processed and recycled like thermoplastic materials. **D 1566,** D11

thermoplastic piping compound—a mixture of a thermoplastic polymer with other ingredients such as fillers, stabilizers, catalysts, processing aids, lubricants, modifiers, pigments, or curing agents, but not plasticizers except in the case of CAB piping compound. **F 412,** F17

thermoplastic vulcanizate, *n*—a thermoplastic elastomer with a chemically cross-linked rubbery phase, produced by dynamic vulcanization. **D 1566,** D11

thermoplastic weld—the joining together of two plastic surfaces by a combination of heat and pressure, with or without filler material. **C 904,** C03

thermoptometry, *n*—a family of techniques in which an optical characteristic of a substance is measured as a function of temperature or time while the substance is subjected to a controlled-temperature program in a specified atmosphere. (ICTAC)

NOTE—Measurement of total light, light of specific wavelength(s), refractive index, and luminescence leads, respectively, to *thermophotometry, thermospectrometry, thermorefractometry,* and *thermoluminescence.* Observations under the microscope lead to *thermomicroscopy.*

E 473, E37

thermoset, *n*—a plastic that, after having been cured by heat or other means, is substantially infusible and insoluble. **D 883,** D20

thermoset, *adj*—pertaining to the state of a plastic in which it is substantially infusible. **D 883,** D20

thermoset, *n*—a crosslinked polymeric material. **D 907,** D14

thermoset—a class of polymers that, when cured using heat, chemical, or other means, changes into a substantially infusible and insoluble material. **D 3878,** D30

thermoset—a polymer material that will undergo or has undergone a chemical reaction by the action of heat, catalysts, ultraviolet light, etc., leading to a relatively infusible state. **E 1749,** E06

thermoset, *n*—a plastic that, when cured by application of heat or chemical means, changes into a substantially infusible and insoluble product. (D-20) **F 412,** F17

thermoset elastomer, *n*—an elastomeric material that attains its intended properties by an irrevisible cross-linking reaction caused by heat alone, heat with a cross-linking agent, chemical interaction, or by radiation; whereby after cross-linking, any reprocessing with heat and pressure will severely degrade the material. **C 717,** C24

thermoset plastic, *n*—a plastic that, after having been cured by heat or other means, is substantially infusible and insoluble. **F 1251,** F04

thermosetting, *adj*—capable of being changed into a substantially infusible or insoluble product when cured by heat or other means. **D 883,** D20

thermosetting, *adj*—having the property of undergoing a chemical reaction by the action of heat, catalysts, ultraviolet light, etc., leading to a relatively infusible state. **D 907,** D14

thermosetting, *adj*—capable of being changed into a substantially infusible or insoluble product when cured by heat or other means. (D-20) **F 412,** F17

thermosetting resin, *n*—a polymeric material capable of crosslinking under the influence of heat, pressure, radiation, ultraviolet light, or chemical agents to form a thermoset. (See also **A-stage, B-stage,** and **C-stage.**) **D 907,** D14

thermosiphon solar energy system— see **solar energy system, thermosiphon.** **E 772,** E44

thermowell, *n*—a closed-end reentrant tube designed for the insertion of a temperature-sensing element, and provided with means for pressure-tight attachment to a vessel. (See also **protecting tube.**) **E 344,** E20

thick edge, *n*—a mica splitting with an edge or end thicker than 1½ times the maximum thickness measured at any other point on the splitting or if the thickness of the edge or end exceeds the maximum average thickness allowed for the grade of splittings. **D 1711,** D09

thickener—a vessel designed to concentrate treatment sludges; similar to a clarifier. **D 6161,** D19

thickener (latex), *n*—a substance used to raise or control the viscosity of latex without making major changes in the total solids content. **D 1566,** D11

thickness, *n*—a measurement of the cross section of the tile made perpendicular to its surface. **C 43,** C15

thickness—that dimension designed to lie at right angles to the face of the wall, floor, or other assembly. **C 1232,** C15

thickness—the distance between one surface and its opposite. **D 123,** D13

thickness—the distance between one surface and its opposite; the distance between the upper and lower surfaces of the material, measured under a specified pressure. **D 123,** D13

thickness—the perpendicular distance between bounding surfaces such as bedding or foliation planes of a rock. (ISRM) **D 653,** D18

thickness, *n*—*of an electrical insulating material,* the perpendicular distance between the two surfaces of interest, determined in accordance with a standard method. **D 1711,** D09

thickness, *n*—*of paper or paperboard,* the perpendicular distance between the two principal surfaces of the paper or paperboard as measured under the specified conditions. See **caliper**. **D 1968,** D06

thickness—*of an epitaxial layer,* the distance from the surface of the wafer to the layer-substrate interface **F 1241,** F01

thickness—*of a semiconductor wafer,* the distance through the wafer between corresponding points on the front and back surfaces. **F 1241,** F01

thickness, compressed—See **compressed thickness**. **D 4439,** D35

thickness, *t*—thickness, measured perpendicular to the running surface. X_A indicates the location of thickness measurement from the tail of the ski. **F 472,** F27

thick place, *n*—a yarn defect characterized by a diameter greater than that of the adjoining segments and extending for 6 mm (¼ in.). (See also thin place.) **D 123,** D13

thick place, *n—in fabric,* an unintentional change in fabric appearance characterized by a small area of more closely spaced yarns, or by a congregation of thick yarns as compared to the adjacent construction. **D 123,** D13

thick splitting—a mica splitting whose thickness in the major portion of its area (or over the entire area) exceeds the following:

(*1*) A bookform splitting that exceeds the maximum average thickness allowed for the grade.

(*2*) A loose splitting that exceeds 0.03 mm in thickness.

(*3*) A loose with powder splitting that exceeds 0.025 mm in thickness. **D 1711,** D09

thief—an auxiliary cathode so placed as to divert to itself some current from portions of the work which would otherwise receive too high a current density. **B 374,** B08

thief—See the preferred term, **sample thief.** **C 859,** C26

thigh, *n—in garment construction,* with trouser leg laid out flat, measure from crotch seam straight across to leg side seam. **F 1494,** F23

thigh girth, *n—in body measurements,* the maximum circumference of the upper leg close to the crotch. (Compare **mid-thigh girth**.) **D 123,** D13

thigh girth, *n— in body measurements,* the maximum circumference of the upper leg close to the crotch. **F 1494,** F23

thimble—a refractory shape used for stirring a pot-made optical glass. **C 162,** C14

thimble element—an element used in an impact printing device in which the fully formed characters are located on the ends of finger-like devices that are similar to a daisy wheel except that the device is formed to produce a cup-like or thimble structure. **F 909,** F05

thimble printer—an impact printing device utilizing a thimble element. The paper on the carriage remains stationary while the element is positioned and caused to strike the paper to produce the characters in their proper positions. **F 909,** F05

thin filling, *n—in woven fabrics,* a filling yarn smaller in diameter than normal. **D 123,** D13

thin film—a layer of material, typically less than 100 nm in thickness deposited or grown on a substrate. **E 673,** E42

thin film composite (TFC)—see **composite membrane**. **D 6161,** D19

thinner—the portion of a paint, varnish, lacquer, or printing ink, or related product that volatilizes during the drying process. **D 16,** D01

thinner, *n*—a volatile liquid added to an adhesive to modify the consistency or other properties. (See also **diluent** and **extender**.) **D 907,** D14

thin place, *n*—a yarn defect characterized by a segment that is substantially (at least 25 %) smaller in diameter than the average diameter of the yarn. (See also **thick place**.) **D 123,** D13

thin place, *n—in fabric,* an unintentional change in fabric appearance characterized by a small area of loosely spaced yarns or by a congregation of thin yarns as compared to the adjacent construction. **D 123,** D13

thin splitting, *n*—a mica splitting whose thickness in the major section of its area, or over the entire area, is less than the minimum average for the grade. **D 1711,** D09

thin stone/thin veneer—a cladding under 2-in. thick. **C 119,** C18

thin-wall tube, *n*—tube with specified wall thickness 3 % or less of the specified outside diameter. **B 899,** B02

thiophene—thiofuran (C_4H_4S) mol weight 84.13; a cyclic organosulfur; colorless, highly reactive liquid; freezing point, –38.5°C; boiling point, 84.12°C. **D 4790,** D16

third party—a person or organization other than the principals. See **referee tests.** **C 859,** C26

third party—person or body that is recognized as being independent of the parties involved, as concerns the issue in question, (ISO Guide 2). **E 1187,** E36

thirtieth highest hour, *n*—for all edit-accepted hours of data during a one-year period, the thirtieth highest hourly traffic volume; this volume is commonly used as a representative hour of traffic volume in roadway design. **E 867,** E17

thixotropy—the property of a material to thin upon isothermal agitation and to thicken upon subsequent rest. **C 904,** C03

thixotropy—the property of a material that enables it to stiffen in a relatively short time on standing, but upon agitation or manipulation to change to a very soft consistency or to a fluid of high viscosity, the process being completely reversible. **D 653,** D18

thixotropy, *n—in a liquid,* the property of thinning when subjected to strains greater than the yield strain and of rethickening with time upon subsequent rest. (See **yield strain** and **viscosity**.) **D 907,** D14

thixotropy—a property of nonsag materials that display a reduction in viscosity when a shearing action is applied but resist seeking their own level when left undisturbed. **E 1749,** E06

THM—trihalomethanes; a group of low molecular weight molecules which can result from chlorination of organics typically found in surface water. **D 6161,** D19

THMP—trihalomethane precursors; organic molecules found in water which have the potential of reacting with chlorine to form THMs. **D 6161,** D19

thong hole, *n—in zippers,* the opening at the end of a pull. **D 123,** D13

thread—annular and helical and, sometimes, longitudinal deformations rolled onto shank; in general, with deformations passing entirely around body; usually resulting in expanded ridges and depressions, larger and smaller, respectively, than wire diameter. (See **knurled.**) **F 547,** F16

thread break, *n—in sewn seams,* a mode of failure evidenced by rupture of the sewing thread. **D 123,** D13

thread count—the total number of warp and filling threads in one square inch of fabric. **F 221,** F05

threaded—annular, helical or longitudinal, symmetrical or nonsymmetrical, flat-bottom or round-bottom deformations with single or double crest shoulders and with rounded or flat flanks; formed onto nail shank after heading by passing through roll-threading dies. (See **mechanically deformed, knurled, thread.**) **F 547,** F16

threaded nail

threaded common nail, threaded nail, threaded sinker, n—bright or galvanized, regular-stock-steel, stiff-stock or hardened-steel, annularly or helically threaded, 1 by 0.072 to 6 by 0.262-in. nails with flat 3⁄16 to 17⁄32-in. head and 1 ⅛ by 0.072 to 5¾ by 0.238-in. sinkers with11⁄64 to ½-in. sinker head and medium diamond point.

threaded shear-resistant nail—stout, short, helically threaded, 1 ½ to 2 ½ by 0.135 and 0.148-in. nails with 5⁄16-in. sinker head and medium diamond point. **F 547,** F16

thread holder, *n*—the support package on which a sewing yarn is wound. **D 123,** D13

thread lap—a doubling over of metal on the thread which is created during roll threading operations. **F 1789,** F16

threadless pipe (TP)—seamless tube conforming to the particular dimensions commercially known as "threadless pipe (TP)."

B 846, B05

three-body abrasive wear, *n*—a form of abrasive wear in which wear is produced by loose particles introduced or generated between the contacting surfaces. **G 40,** G02

three-edge-bearing method—a method for applying the load to a pipe in an external load-crushing strength test. **C 822,** C13

three-phase system—a vapor phase and two liquid phases, one of which is usually the propellant inside the container. **D 3064,** D10

three-point flexure, *n*—configuration of flexural strength testing where a specimen is loaded at a location midway between two support bearings. **C 1145,** C28

three-quarter-hard wire—as applied to aluminum, wire that has been processed to produce a strength approximately midway between that of half-hard wire and that of hard-drawn wire. **B 354,** B01

three way sort—an electromagnetic sort based on a signal response from the material under test above or below two levels established by three or more calibration standards. **E 1316,** E07

threshold, absolute, *n*—See **threshold, detection.** **E 253,** E18

threshold concentration—the minimum concentration at which a substance can be detected by the taste or odor test method employed. **D 2652,** D28

threshold, detection, *n*—the minimum physical intensity of a stimulus which has a probability of 50 % above chance of eliciting a response. **E 253,** E18

threshold, difference, *n*—the minimum difference in physical intensity between a stimulus and a comparison stimulus such that there is a specified probability (most frequently 0.50) that an assessor will perceive the two stimuli as different. The difference threshold is often called the difference limen (DL) or the just noticeable difference (JND). **E 253,** E18

threshold level—the setting of an instrument that causes it to register only those changes in response greater or less than a specified magnitude. **E 1316,** E07

threshold limit value (TLV), *n*—threshold limit value-time weighted average (TLV-TWA) — the time-weighted average concentration for a conventional 8–h workday and 40–h workweek, to which it is believed that nearly all workers may be repeatedly exposed, day after day, without adverse effect. **D 1356,** D22

threshold neutron activation detector—a device containing a particular nuclide that is not significantly activated by neutrons below a certain threshold energy. **E 170,** E10

threshold odor test—a method of evaluating the odor level in a fluid by dilution under specified conditions with an odor-free fluid. **D 2652,** D28

threshold, population, *n*—the median or other measure of central tendency of the distribution of detection or recognition thresholds for a specified population. **E 253,** E18

threshold potential—the minimum charge level on the surface of an electrostatic copy that is necessary to attract a particular toner by overcoming the attracting force between the toner and the carrier. **F 335,** F05

threshold, recognition, *n*—the minimum physical intensity of a stimulus for which there is a specified probability (most frequently 0.50) that an assessor will assign the same descriptor each time it is presented. **E 253,** E18

threshold setting—the setting of the instrument that causes it to register only those changes in eddy-current response greater than a specified magnitude. (E 215)

NOTE—Sensitivity and threshold settings usually are indicated by arbitrary numbers on the control panel of the testing instrument. These numerical settings differ among instruments of different types. It is, therefore, not proper to translate a numerical setting on one instrument to that of another type. Even among instruments of the same design and from the same manufacturer, sensitivity and threshold settings may vary slightly when detecting the same discontinuity. Therefore, undue emphasis on the numerical value of sensitivity and threshold settings is not justified. (E 215) **E 1316,** E07

threshold setting, in image analysis—the selected range of gray levels corresponding to a constituent in the field of the view. **E 7,** E04

threshold stopping speed (TSS), *n*— the maximum saw chain speed measured in metres per second (m/s) (feet per minute (f/m)) that does not produce a cut through when the chain saw is dropped onto the test specimen. **F 1494,** F23

threshold strain (ozone testing), *n*— the highest tensile strain at which a rubber specimen can be exposed continually to an ozone-containing environment without development of ozone cracks. **D 1566,** D11

threshold temperature—temperature on the ΔT versus T curve where the slope changes in the direction indicating an exothermic reaction, that is, the sample is beginning to self-heat. **E 1445,** E27

threshold, terminal, *n*—(*1*) the maximum intensity of a stimulus that will produce a given type of sensory experience without change in modality. (*2*) the intensity of stimulation above which increase in intensity cannot be detected. **E 253,** E18

threshold treatment—the process of stopping precipitation at the start of occurrence; usually does not stop the formation of nuclei but does inhibit growth. See **antiscalant.** **D 6161,** D19

throat—a fully or partially submerged, narrow passageway between the melter and refiner of a tank. **C 162,** C14

throat—the central part of a shoe vamp opening where it is seamed to the front of the quarter. **F 869,** F08

throat depth—distance at which a tool may place fastener from edge of the work. **F 592,** F16

throats, *n*—*in zippers,* the two openings in a slider that receive the stringers. **D 123,** D13

throttling—in leak testing, reducing the net pumping speed of a pumping system by partially closing a valve or installing a section of pipeline with low conductance. **E 1316,** E07

through-coil technique—see **coil technique.** **E 1316,** E07

throughput—the vector product of the area and solid angle of a beam at its focus and the square of the refractive index of the medium in which the beam is focused. **E 131,** E13

throughput—same as **flow rate** (1). **E 1316,** E07

throughput—the rate at which documents can be processed through an optical scanner. Usually expressed as "documents per minute." **F 149,** F05

through transmission technique—a test procedure in which the ultrasonic vibrations are emitted by one search unit and received by another at the opposite surface of the material examined. **E 1316,** E07

through-wall flashing—a water-resistant membrane or material assembly extending totally through a wall and its cavities, positioned to direct any water within the wall to the exterior. **D 1079,** D08

throw, *n*—a removable, unfitted protective textile cover, used over upholstered furniture. (See **furniture covering.**) **D 123,** D13

throw—the projection of broken rock during blasting. (ISRM) **D 653,** D18

throw—a quantity of liquid or powder ingredient that is augured, pumped, or dispensed into a liquid of larger mass and makes up the basic flavoring and/or solid of a finished product. **F 1827,** F26

throwing power—the improvement of the coating (usually metal) distribution over the primary current distribution on an electrode (usually cathode) in a given solution, under specified conditions. The term may also be used for anodic processes for which the definition is analogous. **B 374,** B08

thrust—force applied to a drill in the direction of penetration. (ISRM) **D 653,** D18

thrust collar—a plastic cylinder placed between the last spiral wound cartridge and vessel end plate to support the last cartridge in a pressure vessel against telescoping. **D 6161,** D19

ticket number, *n*—the tex ticket number assigned to a sewing thread to designate its approximate linear density of the sewing thread. **D 123,** D13

ticking—marks caused by the bottom of the upper case character while printing in the lower case, or opposite. **F 149,** F05

tie—wire, clip, or band used to attach the fabric to the top rail and line posts (see Specification F 626). **F 552,** F14

tie line—in a binary or higher order phase diagram, an isothermal, isobaric straight line connecting the compositions of a pair of conjugate phases. **E 7,** E04

tie line—in a binary or higher order phase diagram, an isothermal, isobaric straight line connecting the compositions of a pair of conjugate phases, (E 7, E-4). **E 1142,** E37

tie-off, *n*—in waterproofing, the transitional seal used to terminate a waterproofing application at the top or bottom of flashing or by forming a watertight seal with the substrate, membrane, or waterproofing system(s). **D 1079,** D08

tie points, *n*—the location where ties are placed securing the coils to brackets, wires, or cables including fence fabric, at the specified spacing. **F 1379,** F14

tiered response—a predetermined, protocol driven, level of medical care and vehicle operation mode based on multiple levels of resource response. The two types of tiered response are sequential responsee (q.v.) and simultaneous response (q.v.). **F 1177,** F30

ties and slits, *n*— *in paper*, ties are areas of uncut paper between slits of cut paper within a perforation. **D 1968,** D06

tie stitch—See **lock stitch. F 592,** F16

tie tetrahedron—in a quaternary or higher order phase diagram, an isothermal, isobaric four-cornered space figure connecting the compositions of four conjugate phases. **E 7,** E04

tie triangle—in a ternary or higher order phase diagram, an isothermal, isobaric plane three-cornered, straight-sided figure connecting the compositions of three conjugate phases. **E 7,** E04

tight—rock remaining within the minimum excavation lines after completion of a blasting record. (ISRM) **D 653,** D18

tight—in leak testing, free from leaks according to a given specification. **E 1316,** E07

tight end, *n—in woven fabrics*, a yarn which was woven under excessive tension or has shrunk more than a normal amount. **D 123,** D13

tightened inspection—inspection under a sampling plan using the same quality level as for normal inspection, but requiring more stringent acceptance criteria. **D 1517,** D31

tight pick, *n—in woven fabrics*, a filling yarn which was woven under excessive tension or has shrunk more than a normal amount, which may cause puckering at the junction with normal picks. (Compare **wavy cloth.**) **D 123,** D13

tight selvage, *n—in woven fabrics*, selvage yarns shorter than warp yarn in the body of the fabric. **D 123,** D13

tight side—*in knife-cut veneer,* the side of the sheet that was farthest from the knife as the sheet was being cut and that contains no cutting checks. (See **loose side**.) **D 1038,** D07

tight twist end, *n*—a single end with higher than normal twist. **D 123,** D13

tile—a thin modular stone unit. **C 119,** C18

tile—a ceramic surfacing unit, usually relatively thin in relation to facial area, made from clay or a mixture of clay and other ceramic materials, called the body of the tile, having either a "glazed" or "unglazed" face and fired above red heat in the course of manufacture to a temperature sufficiently high to produce specific physical properties and characteristics. (See also **ceramic mosaic tile; conductive ceramic tile; faience tile; glazed ceramic mosaic tile; glazed interior tile; glazed tile; glazed tile, extra duty glaze; major tile facial dimensions; major tile thickness; minor tile facial dimension; minor tile thickness; porcelain tile; unglazed tile; wedging of tile.**) **C 242,** C21

tile nail

acoustical-tile nail—slender, electroplated, regular-stock-steel or stiff-stock, 1 to 1 ¾ by 0.062-in. nails with ¼-in. projection head with 0.135-in. collar and sharp, blunt, or medium diamond point.

asbestos-tileboard nail—stainless steel or other nonferrous metal nails with casing or oval head. (See also **asbestosboard nail.**)

roofing-tile nail— galvanized, regular-stock-steel, 5 to 7 by 0.148-in. nails with flat 5⁄16-in. head and medium diamond point.

tileboard nail— See **asbestosboard nail, fiberboard nail, insulation building-board nail. F 547,** F16

tile, resilient flooring, *n*—resilient flooring which is packaged in flat pieces which can be installed as individual units. **F 141,** F06

till—see **glacial till. D 653,** D18

tilt, *n*—rotation of a bar code symbol about an axis perpendicular to the substrate. Compare **pitch, skew. F 1294,** F05

tilt angle—*in solar energy applications*, the angle between the horizontal and the plane of the detector (collector, photovoltaic array, instrument) surface. **E 772,** E44

tilt angle, *n*—the angle between the horizontal and the plane of the test fixture, glass frame, or sensory instrumentation. **G 113,** G03

tilting skillet—see **braising pan. F 1827,** F26

tilt, of target—same as **angle** *of incidence.* **E 673,** E42

timbers—See **lumber. D 9,** D07

time angle of internal friction, ϕ_t—inclination of the time yield locus of the tangency - point with the Mohr stress circle passing through the origin. **D 653,** D18

time-average sound level, TAV—L_{AT} where T is the time of the interval of measurement, [nd], (dB), *n*—ten times the logarithm to the base ten of the ratio of mean-square instantaneous A-weighted sound pressure, during a stated time interval T, to the square of the standard reference sound pressure. **C 634,** E33

time constant, *n*—the time required for the magnitude of change in a signal to reach a value of 63.2 % of its final value in response to a step function input. **D 1711,** D09

time constant—*AES,* the time required for a signal to change by $1 - (1/e)$ (63.2 %) of its final value in response to a step function input. **E 673,** E42

time constant—of a solar collector, the time required for the fluid leaving a solar collector to attain 63.2 % of the resulting change in equilibrium outlet temperature following a step change in solar irradiance or inlet fluid temperature.

NOTE—The step change involved should be spelled out in the procedure. **E 772,** E44

time curve—see **consolidation time curve. D 653,** D18

time factor, T_v, T **(D)**—dimensionless factor, utilized in the theory of consolidation, containing the physical constants of a soil stratum influencing its time-rate of consolidation, expressed as follows:

$$T = k\,(1 + e)t/(a_v \gamma_w \cdot H^2) = (c_v \cdot t)/H^2$$

where:
k = coefficient of permeability (LT^{-1}),
e = void ratio (dimensionless),
t = elapsed time that the stratum has been consolidated (T),
a_v = coefficient of compressibility (L^2F^{-1}),
γ_w = unit weight of water (FL^{-3}),
H = thickness of stratum drained on one side only. If stratum is drained on both sides, its thickness equals 2H (L), and
c_v = coefficient of consolidation (L^2T^{-1}).
D 653, D18

time horizon—See **study period.** **E 833,** E06

time interval, *n*—the time between weighings of the permeation cup. **F 1494,** F23

time of drying, *n*—*of insulating varnish,* the time required for a film of varnish to dry to a tackfree state under specified conditions. **D 1711,** D09

time of integration, *n*—*in yarn evenness testing,* the time during which a point-to-point integrator stays switched on. **D 123,** D13

time of setting, *n*—the elapsed time from the addition of mixing water to a cementitious mixture until the mixture reaches a specified degree of rigidity as measured by a specific procedure.

time of final setting (of concrete), n—the elapsed time, after initial contact of cement and water, required for the mortar sieved from the concrete to reach a penetration resistance of 4000 psi (27.6 MPa). (C 403/C 403M).

time of initial setting (of concrete), n—the elapsed time, after initial contact of cement and water, required for the mortar sieved from the concrete to reach a penetration resistance of 500 psi (3.5 MPa). (C 403/C 403M) **C 125,** C09

time-of-wetness, *n*—the total amount of time that a surface is wet. This is typically reported in hours. **G 113,** G03

time profile—a plot of the modulus or damping, or both of a material versus time. **D 4092,** D20

time-proportioned sample—a sample collected at preselected time intervals. **D 1129,** D19

time quenching, *n*—interrupted **quenching** in which the duration of holding in the quenching medium is controlled. **A 941,** A01

time response paper—a special graph paper on which electrode potentials are plotted on the vertical antilogarithmic axis and time is plotted on the horizontal axis as the reciprocal of time. The plot is extrapolated to time-equals-infinity to determine the final electrode potential that would be observed at complete system equilibrium. Time response paper is useful for evaluating electrode performance and for rapidly estimating solution concentration in cases where electrode response is slow, such as close to the lower limit of detection. **D 4127,** D19

time-temperature curve—in thermal analysis, a curve produced by plotting time against the temperature. **E 7,** E04

time-temperature curve—in thermal analysis, a curve produced by plotting time against the temperature, (E 7, E-4), . **E 1142,** E37

time-to-break, *n*—the time interval during which a specimen is under prescribed conditions of tension and is absorbing the energy required to reach maximum load. **D 123,** D13

time-to-thermal-runaway, (t_c), *n*—an estimation of the time required for an exothermic reaction, in an adiabatic container, (that is, no heat gain or loss to the environment), to reach the point of thermal runaway. **E 1445,** E27

time value of money—the time-dependent value of money stemming both from changes in the purchasing power of money (that is, inflation or deflation), and from the real earning potential of alternative investments over time. **E 833,** E06

time yield locus—the yield locus of a bulk solid which has remained at rest under a given normal stress for a certain time. **D 653,** D18

timing mark—a printed mark that controls the reading of a mark read field. The mark printed in scan ink tells the reader the location of information to be scanned. **F 149,** F05

tinned—See **electroplated, wash-tinned.** **F 547,** F16

tinned wire—see **coated wire.** **B 354,** B01

tin oxide (SnO_2)—in finely ground form used in glazes as an opacifier. **C 242,** C21

tint—a color produced by the mixture of white pigment or paint in predominating amount with a colored pigment or paint, not white. The tint of a color is, therefore, much lighter and much less saturated than the color itself. **D 16,** D01

tint, *n*—a color produced by the mixture of white pigment or paint with a chromatic pigment or paint. (See also **tint,** *v*, **shade,** *n*, **shade,** *v*.) **E 284,** E12

tint, *v*—to adjust the color of a test specimen to be a closer color match to the standard. (See also **tint,** *n*; **shade,** *vt*; **shade,** *n*.) **E 284,** E12

tinting, *n*—a background wash of color seen uniformly across the non-image area of a print. See **plate tinting**. **D 6488,** D01

tinting—the occurrence in lithographic copy of a more or less uniform tint of the color being printed on the unprinted areas of the sheet. **F 425,** F05

tinting strength—the power of coloring a standard paint or pigment. **D 16,** D01

tinting strength, *n*—measure of the effectiveness with which unit quantity of a colorant alters the color of a material. **E 284,** E12

tint plate—a retardation plate introduced into the polariscope, which adds one wave (565 nm) of retardation to the field which it covers, and which shifts the observed retardation color scale to more easily distinguishable tints. **C 162,** C14

tint strength, *n*—the ratio, expressed as tint units, of the reflectance of a standard paste to a sample paste, both prepared and tested under specified conditions. **D 3053,** D24

tip height, h_S—the height of the underside of the tip from a plane surface with the center of the ski body pressed against that surface. **F 472,** F27

tip height, H_s—the height of the underside of the tip from a plane surface with the snowboard ski unweighted. **F 1107,** F27

tire, *n*—a load-bearing, ground-contacting circumferential attachment to a vehicle wheel. **D 123,** D13

tire, *n*—a load-bearing ground-contacting circumferential attachment to a vehicle wheel. **F 538,** F09

tire-axis system—the origin of the tire-axis system is the center of the tire contact. The X′ axis is the intersection of the wheel plane and the road plane with a positive direction forward. The Z′ axis is perpendicular to the road plane with a positive direction downward. The Y′ axis is in the road plane, its direction being chosen to make the axis system orthogonal and right-hand. **E 867,** E17

tire axis system, *n*—the origin of the tire axis system is the center of the tire contact where the X'-axis is the intersection of the wheel plane and the road plane with a positive direction forward, the Z'-axis is perpendicular to the road plane with a positive direction downward, and the Y'-axis is in the road plane, its direction being chosen to make the axis system orthogonal and right-hand. **F 538,** F09

tire bead, *n*—that part of a tire that comes in contact with the rim and that is shaped to secure the tire to the rim. **D 123,** D13

tire bead wire, *n*—a monofilament steel wire with a metallic coating, usually bronze, used in forming a tire bead. **D 123,** D13

tire, belted bias, *n*—a bias tire containing a belt. **F 538,** F09

tire, bias, *n*—a pneumatic tire in which the ply cords that extend to the beads are laid at angles substantially less than 90° to the center line of the tread. **F 538,** F09

tire cord, *n*—a twisted or formed structure composed of two or more single or plied industrial yarn elements having the same nominal twist, direction of twist, length, and tension. **D 123,** D13

tire cord fabric, *n*—a fabric consisting of tire cord warp with widely spaced (usually 1 to 5 picks/in.) single yarn filling. **D 123,** D13

tire fabric, *n*—a textile fabric, other than tire cord fabric, which is used as a reinforcement in tires. **D 123,** D13

tire forces—the external forces acting on the tire by the road. **E 867,** E17

tire forces, [*F*], *n*—the external forces acting on a tire by the road. **F 538,** F09

tire-load—the portion of the gross-vehicle weight imposed upon the static tire at the time of weighing, expressed in units of mass, pounds (kilograms), due only to the vertically-downward force of gravity acting on the mass of the static vehicle. **E 867,** E17

tire moments, [*FL*], *n*—the external moments acting on the tire by the road. **F 538,** F09

tire, pneumatic, *n*—a hollow tire that becomes load-bearing upon inflation with air, or other gas, to a pressure above atmospheric. **F 538,** F09

tire, radial, *n*—a pneumatic tire in which the ply cords that extend to the beads are laid substantially at 90° to the center line of the tread, the tire being stabilized by a belt. **F 538,** F09

tire, snow (also mud and snow tire), *n*—a pneumatic tire designed for, or shown to have, good traction on roads covered with mud or snow. **F 538,** F09

tire weight, *n*—the weight of an unmounted tire without tube or flap. **F 538,** F09

tire-wet pavement interaction, zone concept, *n*—a division of the load-bearing surface of a moving pneumatic tire into three basic zones; noncontact, partial contact, and contact. **E 867,** E17

tissue forceps—a device formed in two generally symmetrical halves with their proximal ends secured together and set so their distal ends will stay separated unless pressed together. **F 1638,** F44

tit—an imperfection; a small protrusion on a glass article. **C 162,** C14

titania porcelain—See *titania porcelain* under **porcelain.** **C 242,** C21

titania whiteware—See *titania whiteware* under **ceramic whiteware.** **C 242,** C21

titer, *n*—the quantity of a substance required to react with, or to correspond to, a given amount of another substance. **F 1494,** F23

titer (pronounced tē'ter) (of fatty acids)—the maximum temperature achieved during the solidification of fatty acids, which have been cooled below the melting point by a standardized procedure. **D 459,** D12

titration—*in chemical analysis,* the determination of the reactive capacity, usually of a solution, especially, the analytical process of successively adding measured amounts of a reagent (as a standard solution) to a known volume or weight of a sample or sample solution until a desired end point is reached. **C 859,** C26

titration—a quantitative analytical technique for measuring the concentration of a species by incremental addition of a reagent (titrant) containing a species that reacts with the sample species. Sensing electrodes can be used to follow titrations if they respond to either the species being determined or the titrant ion. Electrode potentials are plotted against millilitres of titrant added, either on standard coordinate graph paper or on Gran's plot paper, and the equivalence point (when equivalent amounts of the two species are present) is determined from the curve. **D 4127,** D19

t-joint, *n*—the condition created by the overlapping intersection of three or four sheets in the membrane. **D 1079,** D08

T nail—bright, etched, coated, galvanized, aluminum-coated, plastic-coated, knurled or annularly threaded, stiff-stock or aluminum-alloy, round-wire, 1 by 0.080 to 2 ½ by 0.131-in. nails of T shape with 5⁄32 or 17⁄64-in. round, square or oval-finish head of sinker, with or without heavy fillet, and with diamond or chisel point; driven with special nailing machine provided with staple-type magazine. **F 547,** F16

TOC—total organic carbon, a measure of the level of organic constituents in water. **D 6161,** D19

TOCl—total organic chlorine. **D 6161,** D19

toe area cut zone, *n*— *in the testing of foot protective devices,* that area excluding the sole which extends from the front most part of the footwear to a vertical plane 15 + 0.25 mm behind the toe box; or in the absence of a toe box, the area which extends to a vertical plane 65 + 6.25 mm from the front of the footwear. **F 1494,** F23

toe bead—a bead of glazing or sealing compound applied at the base of a glazing channel, on the integral stop side of the glass, *before* setting the glass or panel, to prevent leakage past the stop (see also *heel bead*). **C 717,** C24

toe board—See **railing systems.** **E 631,** E06

toe board—a vertical plate at the bottom of a railing system located at the open edge of a stairwell, platform, ramp, or floor; forming a low curb to provide a barrier preventing objects from falling beyond it. (Syns. *kick plate* and *toe plate.*) **E 1481,** E06

toe box, *n*—*in testing of foot protective devices,* a component inserted into the toe area of footwear. **F 1494,** F23

toe-in—a small reduction of the outside diameter at the cut end of a length of thermoplastic pipe. **F 412,** F17

toe plate—See **railing systems.** **E 631,** E06

toe plate—Synonym for **toe board.** **E 1481,** E06

toggle, *n*—*in buttons,* a clip used to fasten a staple button to the flexible substrate. **D 123,** D13

toggle, *n*—*in buttons,* a clip used to fasten a staple button to the flexible substrate. **D 5497,** D13

tolerance, *n*—the allowable deviation from a value or standard; especially the total range of variation permitted in maintaining a specified dimension in machining, fabricating, or constructing a member or assembly. **E 631,** E06

tolerance—the defined limit of allowable departure from the true value of a quantity measured or estimated by a WIM system. **E 867,** E17

tolerance—the defined limits of allowable (acceptable) departure from the true value of a measured quantity. **E 867,** E17

tolerance—the total range of variation (usually bilateral) permitted for a size, position, or other required quantity; the upper and lower limits between which a dimension must be held. (E 380) **F 221,** F05

tolerance—the dimensional standard met by the reamer at the time of manufacture. **F 1611,** F04

tolerance interval, *n*—a range constructed from an experimental data sample so as to statistically enclose P % or more of the population from which the sample was drawn with a confidence level of 100 $(1-\alpha)$ %. **A 644,** A04

tolerance interval—an interval computed so that it will include at least a stated percentage of the population with a stated probability. **C 859,** C26

tolerance interval, *n*—an interval computed so that it will include at least a stated percentage of the population with a stated probability. **C 1145,** C28

tolerance interval—an interval computed so that it will include at least a stated percentage of the population with a stated probability. **E 1823,** E08

tolerance level—the stated probability that the tolerance interval includes at least the stated percentage of the population. It is not the same as a confidence level but the term confidence level is frequently associated with tolerance intervals. **E 1823,** E08

tolerance limits—bounds of a tolerance interval. **C 859,** C26

tolerance limits, *n*—bounds of a tolerance interval. **C 1145,** C28

tolerance limits—the two statistics that define a tolerance interval. (One value may be "minus infinity" or "plus infinity.") **E 1823,** E08

tolerance limits (specification limits), *n*—limits that define the conformance boundaries for an individual unit of a manufacturing or service operation. **E 456,** E11

tolerance limits, specification or calibration—

(1) The permitted degree of departure of the value of some parameter, X, from its nominal value X_n.
(2) The guaranteed maximum error in the reading of some instrument scale, or in the calibration of some circuit component, or in the value of any parameter, from its correct value (which may be assumed to be a true value within the resolution of the calibration). Symmetrical tolerance limits, which do not involve the measurement of any parameter, may be quoted in two ways:

(a) **incremental tolerance,** $\Delta\chi$—this is satisfied by the following limits:

$$X_n + \Delta\chi \geqq X \geqq X_n - \Delta\chi$$

(b) **fractional tolerance,** T_x—this is defined by the following absolute ratio:

$$T_x = |\Delta\chi / X_n|$$

which is usually as a percentage, so that the allowed limits of X becomes $X_n| 1 \pm T_x|$. For a sum or difference function, the incremental tolerance, ΔF, equals the absolute sum of the component Δ values. For a product or ratio function, the fractional tolerance, T_F, equals the absolute sum of the component T values. For any other function, using calculus as follows:

$$\Delta F\,(x, y, x \cdots) = \left(\frac{\delta F}{\delta x}\right)\Delta x + \left(\frac{\delta F}{\delta y}\right)\Delta y + \left(\frac{\delta F}{\delta z}\right)\Delta\, z \cdots$$

NOTE—The preceding tolerance limits are symmetrical limits, such as ±5 %. Occasionally, unsymmetrical limits may be specified such as + 5 %, − 2 % or 0 %, − 5 % or 10 %, − 0 %, and so forth.

A 340, A06

tolerances, *n*—*in mathematics,* prescribed limits of variation for specified properties of a particular material based on observed values obtained by specified test methods and on samples that are representative of the material. **D 123,** D13

tolerances—dimensional limitations established by manufacturers, customers, associations, and government agencies. **F 547,** F16

tolerance specification, *n*—the total allowable variation around a level or state (upper limit minus lower limit), or the maximum acceptable excursion of a characteristic. **E 456,** E11

toluene—methyl benzene, toluol (obsolete) (C_7H_8) mol weight 92.13; clear, colorless, highly flammable liquid; odor somewhat like benzene; freezing point, −94.99°C; boiling point, 110.6°C. **D 4790,** D16

toluene discoloration, *n*—the transmittance, at 425 nm, of the filtrate obtained from the toluene extract of carbon black, compared to that of pure toluene. **D 3053,** D24

toluene insolubles, *n*—that portion of the pentane insolubles not soluble in toluene (methylbenzene). **D 4175,** D02

toluene, nitration grade—toluene with maximum paraffin impurities of 1.5 volume % having a total distillation range of 1°C maximum. Refer to Specification D 841 for complete specifications. **D 4790,** D16

tomography—any radiologic technique that provides an image of a selected plane in an object to the relative exclusion of structures that lie outside the plane of interest (see **tomogram** and **(CT) computed tomography**). **E 1316,** E07

tonal, *adj—in reference to audible sound,* capable of exciting an auditory sensation having pitch. **C 634,** E33

toner—an organic pigment that does not contain inorganic pigment or inorganic carrying base. **D 16,** D01

toner—a material used to increase the intensity and to control the color of the ink image transferred from ribbon or carbon paper. **F 221,** F05

toner—the material in a developer system which when deposited by the field of an electrostatic charge pattern, becomes the visible record. **F 335,** F05

toner—material used to create the actual image in image generation. **F 1156,** F12

toner, *n*—a dispersion of concentrated pigment or dye used to manufacture, strengthen or modify the color of an ink. **F 1294,** F05

toner throwout—see **dusting.** **F 335,** F05

tongue—see **male end of pipe.** **C 822,** C13

tongue-and-grooved, *adj*—type of **lumber** joint consisting of a tongue and a groove on opposite edges to provide close fitting into an adjacent grooved or tongued piece. **D 996,** D10

tongue tile—*in a port,* the projecting partition between gas and air stream. **C 162,** C14

toning, *n*—the deposit of ink in the non-image area caused by poor wiping of doctor blade. **D 6488,** D01

toning—the build-up of ink in non-image areas of the master from the imbalance of ink and fountain solution and its subsequent transfer to the copy. **F 425,** F05

tool—machine for driving fasteners. **F 592,** F16

tooling—*in building construction,* the act of compacting and contouring a sealant in a joint. **C 717,** C24

tooling—a term used to describe the shaping of a fillet bead of applied sealant to a feathered edge where it meets the substrate(s). **E 1749,** E06

tooling time, *n*—*of a sealant,* the time interval after application of a one-component sealant or after mixing and application of multi-component sealant during which tooling is possible. **C 717,** C24

tool marks—longitudinal or circumferential grooves of shallow depth produced by the movement of manufacturing tools over the bolt, nut, or screw surface. **F 1789,** F16

tooth, *n*—integral projection of metal connector plate formed in direction perpendicular to plate surface(s) during punching process. Also called prong, barb, plug, and nail, yet, preferably called tooth. **E 1807,** E06

toothed nail—flat, L-shaped, ½ to 1¹⁵⁄₁₆-in. cleats, sheared from

16-gage steel sheet; provided with toothed serrations along narrow sides of long shank and with slightly tapered, dull point; driven with special nailing machine provided with staple-type magazine. **F 547,** F16

top, *n—in man-made fibers or tow to top process,* a sliver obtained by drafting, along with breaking or cutting a multifilament tow. **D 123,** D13

top, *n—in wool,* a continuous untwisted strand of wool fibers from which the shorter fibers or noils have been removed by combing. **D 123,** D13

top, *n—in worsted processes,* a sliver in which the fibers have been parallelized, and usually combed. (See also **top,** *in wool.*) **D 123,** D13

top grain—the first cut taken from the grain side of a split hide from which nothing except the hair and associated epidermis have been removed. **D 1517,** D31

toplap—the shortest distance between the lower edge of an overlapping shingle or sheet, and the upper edge of the lapped unit in the first course below. **C 460,** C17

toplap, *n*—the shortest distance between the lower edge of an overlapping shingle or sheet, and the upper edge of the lapped unit in the first course below . **C 1154,** C17

toplap—the shortest distance between the lower edge of an overlapping shingle or sheet, and the upper edge of the lapped unit in the first course below. **D 2946,** C17

top lift—the top or outer layer or wearing surface of leather, rubber, plastic or composition on the heel of the shoe. **F 869,** F08

top loader—a vertically oriented agitation machine which is used for home laundry. **D 459,** D12

top of borehole—the surface of the ground surrounding the borehole. **D 653,** D18

topology, *n—in a computer network,* the physical layout including the method of connection between nodes. **F 1294,** F05

topping compound—see **finishing compound.** **C 11,** C11

topping (the act of), *n*—See **skim coating**, the preferred term. **D 1566,** D11

top rail—See **railing systems.** **E 631,** E06

top rail—the uppermost member of a railing system. **E 1481,** E06

top rail—horizontal member of the framework running from terminal post to terminal post on top of the line posts. **F 552,** F14

top scissor half—the component which contains the screw head at assembly. **F 1078,** F04

topset bed—a layer of sediment deposited on the top surface of an advancing delta that is continuous with the landward alluvial plain. **D 4410,** D19

topsize, *n*—the opening of the smallest screen in the series upon which is retained less than 5 % of the sample (see Test Method D 4749). **D 121,** D05

topsize, nominal—*for the purpose of Test Method D 4749,* the sieve designating the upper limit or topsize shall be that sieve of the series given in the Standard Series of Sieves section with the smallest openings upon which is cumulatively retained a total of less than 5 % of the sample. This defined topsize is not to be confused with the size of the largest particle in the lot. **D 121,** D05

topsoil—surface soil, usually containing organic matter. **D 653,** D18

topstitching, *n—in home sewing,* a line of stitching that is visible on the face side of the finished product, usually stitched while having the face side of the material uppermost. **D 4965,** D13

top stop, *n—in zippers,* a part affixed between or immediately above the interlocking elements, on either or both stringers, to prevent the slider from leaving the chain. **D 123,** D13

top story—See **building space.** **E 631,** E06

torque, *n*—a moment (of forces) which produces or tends to produce rotation or torsion. **D 123,** D13

torque, *[FL], n—of a wheel,* the external torque applied to a tire from a vehicle about the wheel spin axis. **F 538,** F09

torque (T), wheel—the external torque applied to a tire from a vehicle about the wheel spin axis. Driving torque is positive wheel torque; braking torque is negative wheel torque. **E 867,** E17

torr—a unit of pressure equal to 1/760th of an atmosphere. **E 1316,** E07

torsional braid analysis, (*TBA*), *n*—a particular case of dynamic mechanical analysis in which the material is supported on a braid and the specimen is examined in torsion. (ICTAC) **E 473,** E37

torsional pendulum—a device for performing dynamic mechanical analysis, in which the sample is deformed torsionally and allowed to oscillate in free vibration. **D 4092,** D20

torsional shear test—a shear test in which a relatively thin test specimen of solid circular or annular cross-section, usually confined between rings, is subjected to an axial load and to shear in torsion. In-place torsion shear tests may be performed by pressing a dentated solid circular or annular plate against the soil and measuring its resistance to rotation under a given axial load. **D 653,** D18

torsional strength—a load, usually expressed in terms of applied torque, at which the fastener fails by being twisted off about its axis. **F 1789,** F16

torsional stress—the shear stress on a transverse cross section, resulting from a twisting action. (E-28) **D 4092,** D20

torsional stress [FL^{-2}], *n*—the shear stress on a transverse cross section resulting from a twisting action. **E 6,** E28

torsion resistance, *n—in tire beadwire,* the number of turns of twist in a short length of wire that causes rupture **D 123,** D13

total—see **radiometric properties and quantities.** **E 772,** E44

total burn technique—in emission spectrographic analyses, continued excitation of the sample until the electrode charge is consumed (totally burned). **C 859,** C26

total carbon, *n*—carbon content remaining in the solid products derived from the combustion or reaction of coal, coal by-products, or coke, inclusive of carbonate in any form. **D 121,** D05

total case depth—the distance measured perpendicularly from the surface of a hardened case to a point where differences in chemical or physical properties of the case and core no longer can be distinguished. **F 1789,** F16

total cold wall heat flux, *n*—the heat flux that would be transferred to an object whose temperature is 70°F (21°C). **E 176,** E05

total combustibles, n—combustible materials that include paints, lacquers, coatings, plastics, and so forth, associated with an original metal product, as well as combustible materials which become associated with the product after it is manufactured. **D 5681,** D34

total cross section—the sum of the absorption and scattering cross sections. **E 1316,** E07

total crotch length, *n—in body measurements,* the distance from the waist level at center front through the crotch to the waist level at center back. **D 123,** D13

total crotch length, *n— in body measurements,* the distance from the waist level at center front through the crotch to the waist level at center back. **F 1494,** F23

total cyanide—the total content of cyanide expressed as the radical

CN^-, or alkali cyanide whether present as simple or complex ions. The sum of both the combined and free cyanide content of a solution. **B 374,** B08

total cycle period, τ_t[T]— the time for the completion of one cycle. The parameter τ_t can be separated into hold and non-hold (that is, steady and dynamic) components:

$$\tau_t = \Sigma\tau_h + \Sigma\tau_{nh}$$

where:
$\Sigma\tau_h$ = sum of all the hold portions of the cycle and,
$\Sigma\tau_{nh}$ = sum of all the non-hold portions of the cycle.

τ_t also is equal to the reciprocal of the overall frequency when the frequency is held constant. **E 1823,** E08

total distillation range—the temperature range, expressed in degrees Celsius, observed in vaporizing a material under specified conditions. **D 4790,** D16

total efficiency—the ratio of the net count rate for all energies in a gamma ray spectrometer system to the gamma ray emission rate of monoenergetic photons from a sample. The value is dependent on the source-detector-shield geometry and the photon energy. If gamma rays of more than one energy are emitted by the source, it may be necessary to determine a separate total efficiency for each such energy. **E 170,** E10

total elongation, El_t, *n*—the elongation determined after fracture by realigning and fitting together of the broken ends of the specimen. **E 6,** E28

total elongation (TE), *n*—**of rope**, the entire elongation at any given applied force. **D 123,** D13

total equivalent boron content (TEBC)—the sum of the individual equivalent boron content (EBC) values. **C 859,** C26

total fixed charge density, N_{tf}—the sum of the nonmobile charge densities: oxide fixed charge density, oxide trapped charge density, and interface trapped charge density. **F 1241,** F01

total flux meter, *n*—the instrument used to measure the level of radiant heat energy incident on the specimen plane at any point. **E 176,** E05

total image unsharpness—the blurring of test object features, in a radiological image resulting from any cause(s). **E 1316,** E07

total immersion thermometer, *n*—a liquid-in-glass thermometer designed to indicate temperatures correctly when just that portion of the thermometer containing the liquid is exposed to the temperature being measured. (Compare **complete immersion thermometer** and **partial immersion thermometer.**) **E 344,** E20

total indicator runout (TIR)—*on a wafer surface*, the smallest perpendicular distance between two planes, both parallel with the reference plane, which enclose all points on the front surface of a wafer within the flatness quality area or the site, depending on which is specified. **F 1241,** F01

total insulation (I_T), *n*—the total resistance to dry heat loss from the manikin, that includes the resistance provided by the clothing and the air layer around the clothed manikin. **F 1494,** F23

total internal surface—the surface of a cellulosic material available for absorption of a given substance. A useful substance for this purpose is water; the value obtained with water is close to the potential maximum surface that can be developed without disruption of the crystalline structure. **D 1695,** D01

total Kjeldahl nitrogen—the sum of the nitrogen contained in the free ammonia and other nitrogen compounds which are converted to ammonium sulfate under specified digestion conditions. **D 1129,** D19

total mass, *n*—*in pile yarn floor covering*, the mass of all matter in the mass per unit area expressed in grams per square metre (ounces per square yard) or in grams per linear metre (ounces per linear yard). **D 5684,** D13

total mass stopping power, S/ρ—of a material for charged particles, the quotient of dE by ρ dl, where dE is the energy lost by a charged particle in traversing a distance, dl, in the material of density ρ (ICRU).

$$S/\rho = (1/\rho)\, dE/dl$$

Unit : $J{\cdot}m^2{\cdot}kg^{-1}$
($eV{\cdot}m^2{\cdot}kg^{-1}$ is also used). **E 170,** E10

total matter—the sum of the particulate and dissolved matter. **D 1129,** D19

total metal basis—analytical results calculated by reference to the sum of the masses of uranium and plutonium rather than the sum of their oxides. **C 859,** C26

total moisture—See *total moisture* under **moisture**. **D 121,** D05

total moisture—the water contained in a sample. The determination of the total moisture is made by drying a sample under controlled conditions of temperature, time, and air flow. The determination may consist of a single-stage or a two-stage drying process. **E 856,** D34

total or gross-contact area, [L^2], *n*—that area encompassed by the outer periphery of a tire footprint. **F 538,** F09

total pressure—See **pressure.** **D 1356,** D22

total reflectance, ρ, *n*—the ratio to the incident flux of the radiant or luminous flux reflected at all angles within the hemisphere bounded by the plane of measurement. **E 284,** E12

total reflectance factor, *n*—the ratio of the radiant or luminous flux, reflected at all angles within the hemisphere bounded by the plane of the specimen, to the flux reflected from the perfect reflecting diffuser under the same geometric and spectral conditions of measurement. **E 284,** E12

total-sediment discharge—the total quantity of sediment passing a section per unit of time. **D 4410,** D19

total-sediment load (total load)—all of the sediment in transport; that part moving as suspended load plus that moving as bedload. **D 4410,** D19

total soil-water potential—the sum of the energy-related components of a soil-water system; for example, the sum of the gravitational, matrix and osmotic potentials. **D 653,** D18

total solar ultraviolet, *adj*—solar energy above the solar cut on and below the visible, when received after transmittance through the atmosphere. **G 113,** G03

total stress—see **stress.** **D 653,** D18

total thickness variation, (TTV)—*of a semiconductor wafer*, the difference between the maximum and minimum values of the thickness of the wafer. **F 1241,** F01

total transmittance, τ_t, *n*—the ratio of the flux transmitted at all forward angles to the incidence flux. **E 284,** E12

total variance—See *total variance* under **variance**. **D 121,** D05

total vertical trunk length, *n*—*in body measurements*, the distance from the right shoulder line midway between the neck base and the shoulder joint, down the back through the crotch and over the projection of the right breast to the starting point. **D 123,** D13

total weight basis moisture content—of biomass fuels, the ratio of the weight of the water in a sample to the weight of the wet material. It is expressed as a percent (also called wet basis moisture content). **E 1126,** E48

total weight basis moisture content—of biomass fuels, the ratio of the weight of the water in a sample to the weight of the wet material. It is expressed as a percent (also called wet basis moisture content). **E 1705,** E48

touch sanding—*in plywood,* a light surface sanding to control thickness; not intended as a full surfacing operation. **D 1038,** D07

touch trip—See **bottom trip.** **F 592,** F16

toughened glass—See **tempered glass.** **C 162,** C14

toughness—See **resistance, impact (toughness).** **C 168,** C16

toughness, *n*—the capacity of a material to absorb energy. (Compare **work-to-break, work-to-rupture.**) **D 123,** D13

toughness at rupture, *n*—toughness of a material to breaking or tearing apart. **D 123,** D13

toughness index, I_T, T_w—the ratio of: (*1*) the plasticity index, to (*2*) the flow index. **D 653,** D18

tough pitch copper—copper of any origin cast in the form of refinery shapes, containing a controlled amount of oxygen in the form of cuprous oxide. By extension the term is also applicable to fabricators' products made therefrom. **B 846,** B05

tow, *n*—**in manufactured fibers**, a twistless multifilament strand suitable for conversion into staple fibers or sliver, or for direct spinning into yarn. **D 123,** D13

tow—*in fibrous composites,* a continuous, ordered assembly of essentially parallel, collimated filaments, normally without twist and of continuous filaments (Synonyms: strand and silver). **D 3878,** D30

towel, *n*—an absorbent textile product used for drying or wiping. **D 123,** D13

towing—transporting a boom from one place to another by pulling from one end. **F 818,** F20

TOX—total organic halides. **D 6161,** D19

TOXFP—total organic halide formation potential. **D 6161,** D19

toxic hazard, *n*—*as related to fire,* the potential for physiological harm from toxic products of combustion. **E 176,** E05

toxicity, *n*—the propensity of a test material to produce adverse behavioral, biochemical, or physiological effects in a living organism. **D 6384,** D02

toxicity, *n*—the propensity of a substance to produce adverse biochemical or physiological effects. **E 176,** E05

toxicity—the property of a chemical, or combination of chemicals, to adversely affect organisms, tissues or cell. **E 943,** E47

toxicity, *n*—the propensity of a substance to produce adverse biochemical or physiological effects. **F 1494,** F23

toxicity—the property of a material, or combination of materials, to adversely affect organisms. **F 1600,** F20

toxicity characteristic leaching procedure (TCLP)—a mild laboratory extraction procedure designed to determine the mobility of organic and inorganic constituents present in liquid, solid, and multiphasic wastes. **E 1605,** E06

toxicity test—an experiment used to study the adverse effect (s) of one or more chemicals on whole organisms, tissues, or cells. **E 943,** E47

toxic potency, *n*—*as applied to inhalation of smoke or its component gases,* a quantitative expression relating concentration and exposure time to a particular degree of adverse physiological response, for example, death, on exposure of humans or animals. **E 176,** E05

trace—a constituent or impurity making up only a small portion of the sample, the upper limit of the trace or microconstituent being about 100 μg/g. This upper boundary is not rigidly fixed. **C 859,** C26

trace, *n*—a constituent or impurity making up only a small portion of the sample, the upper limit of the trace or microconstituent being about 100 μg/g; this upper boundary is not rigidly fixed. **C 1145,** C28

traceability—the ability to demonstrate by means of an unbroken chain of comparisons that a measurement is in agreement within acceptable limits of uncertainty with comparable nationally or internationally recognized standards. **E 7,** E04

traceability—the property of a result of a measurement whereby it can be related to appropriate standards, generally international or national standards, through an unbroken chain of comparison, (ISO Guide 2). **E 1187,** E36

traceability, *n*—the ability to trace the history, application, or location of an item or activity and like items or activities by means of recorded identification. (ANSI/ASQC-1987, *A*) **E 1187,** E36

traceability—ability to track an item back through the procurement cycle to the true manufacturer. Also: ability to track an item through its entire manufacturing process to verify the use of appropriate procedures at each step in the process, and that the item meets all applicable quality requirements. **F 1156,** F12

traceability—the ability to verify the manufacturing history, raw material, heat number, location, or application of an item by means of recorded identification. **F 1789,** F16

tracer gas—a gas which, passing through a leak, can then be detected by a specific leak detector and thus disclose the presence of a leak. Also called search gas. **E 1316,** E07

tracer probe leak location—Same as **probe test.** **E 1316,** E07

track, *n*—a partially conducting path of localized deterioration on the surface of an insulating material. **D 1711,** D09

track—the undesired appearance of ink on the copy in a repeat pattern. **F 425,** F05

track—See **rail.** **F 592,** F16

track etch technique—a measurement method consisting of a dielectric material that records the passage of energetic heavy ionizing particles as latent tracks that can be developed or enlarged by various chemical methods and observed by microscopy methods or other techniques. **E 170,** E10

tracking, *n*—the process that produces tracks as a result of the action of electric discharges on or close to the insulation surface. **D 1711,** D09

tracking collector— see **collector, tracking.** **E 772,** E44

tracking, contamination, *n*—tracking caused by scintillations that result from the increased surface conduction due to contamination. **D 1711,** D09

tracking error—(*1*) for a two-axis tracking collector, the angular deviation between the collector-sun line and a line that is normal to the aperture plane. (*2*) for a single-axis tracking collector, the angular deviation between two planes that intersect along the axis of rotation. One plane contains the optical axis of the collector and the other contains the center of the sun. **E 772,** E44

tracking resistance, *n*—the quantitative expression of the voltage and the time required to develop a track under specified conditions. **D 1711,** D09

traction—transport of debris by running water in which the particles are swept along close to the bed of the stream by rolling, sliding, or saltation. **D 4410,** D19

traction—the coefficient of friction generated between the outsole and typical jogging or running surface. **F 869,** F08

traction, S_1, S_2, S_3(FL^{-2})—applied stress. **D 653,** D18

traction test, *n*—in tire testing, a series of *n* test runs at a selected operational condition; a traction test is characterized by an average value for the measured performance parameter. **F 538,** F09

traction test—method of evaluating the traction characteristics of footwear. **F 869,** F08

traction vector angle, (rad or degree), *n*—the angle between the resultant traction force vector and the X'-axis. **F 538,** F09

tractive force coefficient, *n—of a tire,* the ratio of tractive force to normal force on a tire footprint. **F 538,** F09

tractor, *n*—a powered unit capable of propelling itself and towing other (unpowered) units on a highway. **E 867,** E17

traffic marking—marring or discoloration, or both, of a floor surface by traffic. **D 2825,** D21

traffic-rail system—See **railing systems.** **E 631,** E06

traffic-rail system—a railing system designed to control the movement of people and requiring special consideration for given use conditions. Also, *queue-rail system.* **E 1481,** E06

traffic surface—a surface exposed to traffic, either pedestrian or vehicular, also described as finish wearing surface. **C 717,** C24

traffic volume, *n*—the number of motorized highway vehicles with two or more axles passing a point on a roadway in a unit of time. **E 867,** E17

trailing—see **dragging.** **F 335,** F05

train—a grouping of devices. See **array, bank, block.** **D 6161,** D19

trained assessor, *n*—an assessor with a high degree of sensory acuity who has experience with the test procedure and an established ability to make consistent and repeatable sensory assessments. (See also **assessor** and **expert.**) **E 253,** E18

trajectory, *n*—the rectilinear or curvilinear path of a vehicle during a stopping maneuver; it is defined by the center of gravity and the transient angular orientation of the vehicle. **F 538,** F09

trajectory guide line (TGL), *n*—the centerline marked on the test course pavement that constitutes the intended trajectory; it is used by the driver to guide or steer the vehicle on its intended path. **F 538,** F09

trammage, *n—in woven crepes,* a puckered area in which a filling yarn has twist running in the same direction for several picks instead of alternating *S* and *Z* twist. **D 123,** D13

tranquil flow—see **subcritical flow**. **D 4410,** D19

transcrystalline cracking—cracking or fracturing which occurs through or across a crystal; intracrystalline cracking. **E 7,** E04

transducer—an electromagnetic device for converting electrical energy into magnetic or mechanical energy and vice versa. (E 1033) **E 1316,** E07

transducer—an electroacoustical device for converting electrical energy into acoustical energy and vice versa. See also **crystal.** **E 1316,** E07

transducer—device for the conversion of some form of energy into another. **F 1156,** F12

transducers, acoustic emission—see **sensor, acoustic emission.** **E 1316,** E07

transducer shadow correction, *n*—the ratio of the *true* along-axis velocity, as measured in a wind tunnel or by another accepted method, to the instrument along-axis wind measurement. **D 1356,** D22

transfer—the act of moving a developed image, or a portion thereof, from one surface to another as by electrostatic or adhesive forces. **F 335,** F05

transfer efficiency—the amount (weight/mass) deposited on a specified target divided by the spray gun output (weight/mass) per unit of time. **C 286,** B08

transference, *n*—process by which blocking results. **F 1857,** F05

transference number (transport number)—the proportion of the total current carried by the ions of a given kind. **B 374,** B08

transference (or transport, or migration)—the movement of ions through the electrolyte associated with the passage of the electric current. **B 374,** B08

transference, thermal, *n*—the steady-state heat flow from (or to) a body through applied thermal insulation and to (or from) the external surroundings by conduction, convection, and radiation. It is expressed as the time rate of heat flow per unit area of the body surface per unit temperature difference between the body surface and the external surroundings. **C 168,** C16

transfer failure, n—in characterizing the locus of failure in a pressure-sensitive article, separation at the interface of adhesive and backing. **D 907,** D14

transfer function (U_t= a + bR, m/s)—the linear relationship between wind tunnel speed and the rate of rotation of the anemometer throughout the specified working range. **D 1356,** D22

transfer glass—optical glass cooled in the pot in which it was melted. **C 162,** C14

transfer molding—a method of forming articles by fusing a plastic material in a chamber and then forcing essentially the whole mass into a hot mold where it solidifies. **D 883,** D20

transfer molding—a method of forming articles by fusing a plastic material in a chamber and then forcing essentially the whole mass into a hot mold where it solidifies. (D-20) **F 412,** F17

transfer-rail system—See **railing systems.** **E 631,** E06

transfer-rail system—a railing system designed to support and to permit the transfer of body weight in such locations as toilets, showers, and tub enclosures. **E 1481,** E06

transfer standard—See **standard.** **D 1356,** D22

transfer standard, *n*—a physical standard used to transfer a calibration from one instrument to another, usually from a reference instrument in a standards laboratory to an instrument in the field. **E 284,** E12

transfer-standard dosimeter—a dosimeter, often a reference-standard dosimeter suitable for transport between different locations, used to compare absorbed-dose measurements. **E 170,** E10

transfinite algebra—a branch of higher mathematics dealing with the algebra of infinity. **F 1156,** F12

transflection—an experimental method whereby radiant energy that is transmitted through the specimen is returned through the specimen by means of an external reflector. **E 131,** E13

transformation, *n*—the change from one set of variables, *x*, to another set, x', by the use of a function, $x' = f(x)$. **D 123,** D13

transformation ranges, *n*—those ranges of temperature within which austenite forms during heating and transforms during cooling. **A 941,** A01

transformation temperature, *n*—the temperature at which a change in phase occurs, with the limiting temperatures of the **transformation ranges** designated using the following symbols:

Ac_{cm}—the temperature at which the solution of cementite in austenite is completed during heating.

Ac_1—the temperature at which austenite begins to form during heating.

Ac_3—the temperature at which transformation of ferrite to austenite is completed during heating.

Ac_4—the temperature at which austenite transforms to delta ferrite during heating.

Ae_1, Ae_3, Ae_{cm}, Ae_4—the temperatures of phase change at equilibrium.

Ar_{cm}—the temperature at which precipitation of cementite starts during cooling.

Ar_1—the temperature at which transformation of austenite to ferrite or to ferrite plus cementite is completed during cooling.

Ar_3—the temperature at which austenite begins to transform to ferrite during cooling.

Ar_4—the temperature at which delta ferrite transforms to austenite during cooling.

M_f—the temperature at which transformation of austenite to martensite is substantially completed during cooling.

M_s—the temperature at which transformation of austenite to martensite starts during cooling. **A 941,** A01

transformation temperature—the temperature at which a change in phase occurs. The term is sometimes used to denote the limiting temperature of a transformation range. Sometimes incorrectly and loosely referred to as Critical Point. **E 7,** E04

transformation temperature—temperature at which a change in phase occurs, (E 7, E-4). **E 1142,** E37

transformed flow net—a flow net whose boundaries have been properly modified (transformed) so that a net consisting of curvilinear squares can be constructed to represent flow conditions in an anisotropic porous medium. **D 653,** D18

transformer oil, oxidation inhibited—a suitably refined mineral insulating oil to which an oxidation inhibitor has been added. **D 2864,** D27

transformer oil, uninhibited—a suitable refined mineral insulating oil containing no additives and only such as remain in the oil. **D 2864,** D27

transforms:

Fast Walsh/Hadamard/Fourier—complex mathematical procedures for data handling. **F 1156,** F12

transhipped goods—diverted or hijacked goods; goods that are being shipped by a transfer agent. **F 1156,** F12

transitional hours— See **hours of operation**. **E 1480,** E06

transition curve—in a P-T diagram, the locus of the temperature and pressure values at which a congruent equilibrium between two solid phases exists. **E 7,** E04

transition, first order—a change of state, associated with crystallization or melting in a polymer. **D 883,** D20

transition, first order, *n*—a reversible change in phase of a material; in the case of polymers, usually crystallization or melting. **D 1566,** D11

transition, first order—a change of state, associated with crystallization or melting in a polymer. (D-20) **F 412,** F17

transition flow—in leak testing, the flow of gases under conditions intermediate between laminar viscous flow and molecular flow. **E 1316,** E07

transition, glass, *n*—the reversible physical change in a material from a viscous or rubbery state to a brittle, glassy state. **D 1566,** D11

transition metals—referring to the periodic table, those elements shown as being members of groups IB, IIB, IIIB, IVB, VB, VIB, VIIB, and VIII of the periodic table. **F 1156,** F12

transition phase—a non-equilibrium state that appears in a chemical system in the course of transformation between two equilibrium states. **E 7,** E04

transition point—at a stated pressure, the temperature, or at a stated temperature, the pressure, at which two solid phases exist in congruent equilibrium, that is, an allotropic transformation temperature, or pressure. **E 7,** E04

transition point, *n*—*in bar code reading*, the edge of a space or bar where continued movement to an adjacent and complementary module causes a photodetector to reverse its bias from dark current to saturation and vice versa. **F 1294,** F05

transition, second order, *n*—see **transition temperature, glass** **D 1566,** D11

transition structure—in precipitation from solid solution, a metastable precipitate which is coherent with the matrix. **E 7,** E04

transition temperature, *n*—the test temperature for which the fracture surface of the test specimen shows 50 % ductile and 50 % brittle fracture. **A 644,** A04

transition temperature (also critical transition temperature)—the maximum temperature below which a superconductor exhibits superconductivity at a given magnetic field and current. **B 713,** B01

transition time, t_T**[T]**—time required for extensive creep conditions to develop in a cracked body. For specimens, this is the time required for the creep deformation zone to spread through a substantial portion of the uncracked ligament. **E 1823,** E08

transit time (*t*, (s)), *n*—the time required for an acoustic wavefront to travel from the transducer of origin to the receiving transducer. **D 1356,** D22

translucency, *n*—the property of a specimen by which it transmits light diffusely without permitting a clear view of objects beyond the specimen and not in contact with it. **E 284,** E12

translucent, *adj*—transmitting light diffusely, but not permitting a clear view of objects beyond the specimen and not in contact with it. **E 284,** E12

translucent base media—materials with properties that allow radiological interpretation by transmitted or reflected light. **E 1316,** E07

translucent body, *n*—body which transmits light principally by diffuse transmission. Objects are not seen distinctly through such a body. **E 349,** E21

transmembrane pressure—the net driving force across the membrane. The hydraulic pressure differential from the feed side to permeate side less the osmotic pressure differential on each side. **D 6161,** D19

transmission, *n*— *of radiant energy*, the process whereby radiant energy passes through a material or object. (See also **diffuse transmission, mixed transmission, regular transmission**.) **E 284,** E12

transmission, *n*—passage of radiation through a medium without change of frequency of the monochromatic components of which the radiation is composed. **E 349,** E21

transmission—passage of radiant energy through a material so that it emerges from a surface other than the surface of incidence, without change in frequency. **E 772,** E44

transmission coefficient—fraction of monochromatic flux internally transmitted by unit path length of a medium. One minus the absorption coefficient.

NOTE—The transmission coefficient changes with wavelength.

E 772, E44

transmission densitometer—an instrument that measures the intensity of the transmitted light through a radiographic film and provides a readout of the transmitted film density. **E 1316,** E07

transmission density, D_T, *n*—the negative logarithm to base ten of the transmittance factor. **E 284,** E12

transmission, heat, *n*—the quantity of heat flowing through unit area due to all modes of heat transfer induced by the prevailing conditions. **C 168,** C16

transmission method—a method of X-ray or electron diffraction in which the recorded diffracted beams emerge on the same side of the specimen as the transmitted primary beam. **E 7,** E04

transmission microscope—a microscope in which the image forming rays pass through (are transmitted by) the specimen being observed. **E 7,** E04

transmission (optical) density, *n*—logarithm to the base 10 of the reciprocal of the transmittance. Symbol: D, $D = -\log_{10} \tau$. **E 349,** E21

transmission rate, water vapor—See **water vapor transmission rate.** **C 168,** C16

transmission/reflection interaction—see **transflection.** **E 131,** E13

transmission window—the (electromagnetic) wavelength region in which a material is transparent, and the extent to which this is so. **F 1156,** F12

transmissivity—the volume of water at the existing kinematic viscosity that will move in a unit time under a unit hydraulic gradient through a unit width of the aquifer. **D 653,** D18

transmissivity—the rate at which water of the prevailing kinematic viscosity is transmitted through a unit width of the aquifer under a unit hydraulic gradient. **D 653,** D18

transmissivity—the volume of water of the prevailing kinematic viscosity transmitted in a unit time through a unit width of the aquifer under a unit hydraulic gradient. **D 653,** D18

transmissivity of an absorbing material, *n*—internal transmittance of a layer of the material such that the path of the radiation is of unit length. **E 349,** E21

transmittance, *n*—the ratio of the radiant power transmitted by a material to the radiant power incident upon it. **E 135,** E01

transmittance, *n*—*of light,* that fraction of the incident light of a given wavelength which is not reflected or absorbed, but passes through a substance. **E 253,** E18

transmittance, τ, *n*—the ratio of transmitted flux to incident flux, under specified geometric and spectral conditions. (See also **diffuse transmittance, internal transmittance, regular transmittance, total transmittance.**) **E 284,** E12

transmittance, *n*—ratio of the transmitted radiant or luminous flux to the incident flux. Symbol: τ_e, τ_v, τ; $\tau = \tau_r + \tau_d$.

NOTE—Where mixed transmission occurs, the (total) transmittance may be divided into two parts, regular transmittance (τ) and diffuse transmittance (τ_d), corresponding, respectively, to the two modes of transmission referred to above.

In general, the values of the various transmittances depend upon the mode of irradiation, the spectral composition, and the state of polarization of the incident radiation. **E 349,** E21

transmittance, τ—the ratio of the transmitted radiant flux to the incident flux (Practice E 491). See **radiometric properties and quantities.** **E 772,** E44

transmittance, τ—the ratio of the radiant flux transmitted through a body to that incident upon it. **E 1316,** E07

transmittance, *n*—the ratio of transmitted flux to incident flux under specified geometric and spectral conditions, expressed either as a percentage or a decimal fraction. **G 113,** G03

transmittance density, D_t, *n*—the negative logarithm to base ten of the transmittance. **E 284,** E12

transmittance factor, T, *n*—ratio of the flux transmitted by the specimen to the flux transmitted by the perfect transmitting diffuser under the same geometric and spectral conditions of measurement. **E 284,** E12

transmittance (of light)—that fraction of the emitted light of a given wavelength which is not reflected or absorbed, but passes through a substance. **C 859,** C26

transmittance, spectral, *n*—the transmittance as a function of wavelength. **G 113,** G03

transmittance, T, *n*—the fraction of radiant energy entering a substance that reaches its further boundary. **D 4175,** D02

transmittance, *T*— the ratio of radiant power transmitted by the sample to the radiant power incident on the sample. **E 131,** E13

transmittance, thermal, *n*—the heat transmission in unit time through unit area of a material or construction and the boundary air films, induced by unit temperature difference between the environments on each side. **C 168,** C16

transmitted film density—the density of radiographic film determined by measuring the transmitted light. **E 1316,** E07

transmitted light scanner—an optical scanner that operates by sensing light transmitted through paper instead of reflected from its surface. **F 149,** F05

transparency, *n*—the degree of regular transmission, thus the property of a material by which objects may be seen clearly through a sheet of it. **E 284,** E12

transparent, *adj*—transmitting radiant energy without diffusion. **E 284,** E12

transparent body, *n*—body in which the light transmission is mainly regular and which has a high regular transmittance. Objects are seen distinctly through such a body if its geometrical form is suitable. **E 349,** E21

transpassive region—the region of an anodic polarization curve, noble to and above the passive potential range, in which there is a significant increase in current density (increased metal dissolution) as the potential becomes more positive (noble). **G 15,** G01

transponders—transducers with wireless links for data transmission, usually low power radio or microwave, when triggered by a broadcast signal of inquiry. **F 1156,** F12

transportation—the complex process of moving sediment particles by water. The principal factors affecting transportation are turbulence; ratio of settling velocity to water velocity; shape, size, density, and quantity of particles; and saltation. **D 4410,** D19

transported soil—soil transported from the place of its origin by wind, water, or ice. **D 653,** D18

transposed conductor—a composite conductor in which filaments or strands occupy different relative positions about the conductor axis in a regular manner. **B 713,** B01

transposition length—the length in which a filament or strand returns to its original relative position in a transposed conductor. **B 713,** B01

transuranium element—an element above uranium in the periodic table, that is, with an atomic number greater than 92. **C 859,** C26

transverse—directions in wood at right angles to the wood fibers. Includes radial and tangential directions. A transverse section is a section through a tree or timber at right angles to the pitch. **D 9,** D07

transverse—crosswise; at right angles to the longitudinal axis of the body. **F 869,** F08

transverse construction joint deterioration, *n*—(*CRCP only*) series of closely spaced transverse cracks or a large number of interconnecting cracks occurring near a construction joint. **E 867,** E17

transverse construction joint deterioration, *n*—(*CRCP only*) series of closely spaced transverse cracks or a large number of interconnecting cracks occurring near a construction joint. **E 1778,** E17

transverse cracking, *n*—cracks in the pavement that are predominantly perpendicular to the direction of traffic. **E 867,** E17

transverse cracking, *n*—cracks in the pavement that are predominantly perpendicular to the direction of traffic. **E 1778,** E17

transverse direction— See **cross direction.** **E 7,** E04

transverse element, *n*—a component or subassembly that links longitudinal members together. **F 1582,** F04

transverse loads, n—loads which are perpendicular to the facings: synonymous with flatwise load. **C 274,** D30

transverse profile, *n*—the vertical deviations of the pavement surface from a horizontal reference perpendicular to the lane direction. **E 867,** E17

transverse rupture strength—the stress calculated from the flexure formula, required to break a specimen supported near the ends as a simple beam; the load is applied midway between the center lines of the supports. **B 243,** B09

transverse strain, n—linear strain in a plane perpendicular to the axis of the specimen. **E 6,** E28

transverse wave—see **shear wave**. **E 1316,** E07

transverse wave—wave motion in which the particle displacement at each point in a material is perpendicular to the direction of propagation. (E 494) **E 1316,** E07

transverse wave (shear wave)—a wave in which the displacement at each point of the medium is parallel to the wave front. (ISRM) **D 653,** D18

transverse wave, v_t **(LT^{-1})**—wave in which direction of displacement of element of medium is parallel to wave front. The propagation velocity, v_t, is calculated as follows:

$$v_t = \sqrt{G/\rho} = \sqrt{\mu/\rho} = \sqrt{(E/\rho)[1/2(1 + v)]}$$

where:
G = shear modulus,
ρ = mass density,
v = Poisson's ratio, and
E = Young's modulus.

D 653, D18

trap (dust, sulfur, water, gas, etc.)—any of various devices for preventing passage of dust, sulfur, water, gas, etc., often while allowing other matter to pass through. **C 859,** C26

trap efficiency—the percent of the incoming sediment load that is deposited. **D 4410,** D19

trapezoidal wire—a wire with a keystone-shaped (wedge) cross section with two circular parallel sides. **B 354,** B01

trapped magnetic flux—the magnetic flux retained in a superconductor when the applied magnetic field is reduced or reversed in direction. **B 713,** B01

traps—items, marks, potential tell-tales, that can be used to indicate when some thing has been tampered with in some manner that causes the trap to visualize, for example, the EZ-Rub or fugitive inks on checks that are destroyed when the check is altered. **F 1156,** F12

trash, *n—in cotton,* undeveloped seed, motes, small bits of seed coat, or particles of leaf appearing as specks. **D 123,** D13

trash, *n—in testing cotton with the Trash Meter,* foreign matter having a distinct difference, as seen by a video camera, between light and dark color from that of cotton. **D 123,** D13

Trash Meter, *n*—an instrument which optically measures the amount of trash on the surface of a raw cotton sample as presented to the viewing window. **D 123,** D13

trash removal device, *n—in open-end spinning machines,* a system for removing impurities from the opened feed stock before the fibers are conveyed to the rotor. **D 123,** D13

trauma care system—a subsystem within the EMS system designed to manage the treatment of the trauma patient. **F 1177,** F30

travel, *n*—a change in appearance of a material as it is viewed over a wide range of aspecular angles. **E 284,** E12

traveled surface, *n*—any man-made, solid surface for vehicular travel, for example, highways, runways, rails, guideways. **E 867,** E17

traveled surface roughness, *n*—the deviations of a surface from a true planar surface with characteristic dimensions that affect vehicle dynamics, ride quality, dynamic loads, and drainage, for example, longitudinal profile, transverse profile, and cross slope. **E 867,** E17

travertine—a variety of crystalline or microcrystalline limestone distinguished by layered structure. Pores and cavities commonly are concentrated in some of the layers, giving rise to an open texture.

marble (I calcite, II dolomite)—carbonate rock that has acquired a distinctive crystalline texture by recrystallization, most commonly by heat and pressure during metamorphism, and is composed principally of the carbonate minerals calcite and dolomite, singly or in combination. **C 119,** C18

travertine marble (IV)—a porous or cellularly layered, partly crystalline calcite of chemical origin.

NOTE—Travertine is formed by precipitation from generally hot solutions of carbonated spring water, usually at the bottom of shallow pools.

C 119, C18

TRC—total residual chlorine. **D 6161,** D19

tread, *n—of a tire,* the part of a tire that comes in contact with the ground. **F 538,** F09

tread arc width, *n*—the length of the arc measured from one extreme of tread design proper to the opposite extreme; that is, from shoulder to shoulder perpendicular to the circumferential center line. **F 538,** F09

tread band, *n*—an annular volume of rubber that encompasses the outer pavement contacting periphery of a tire; the width is normally much greater than the thickness, and both of these dimensions vary with tire size. **F 538,** F09

tread depth average, *n— in a single tire,* the average of all tire groove (void) depth measurements. **F 538,** F09

tread depth, [*L*], *n*—synonym for *groove (void) depth.* **F 538,** F09

tread hardness, *n*—the hardness of an element in the tread design as measured by a designated standard gage. **F 538,** F09

treadlife, *n*—the distance required to produce wear-out. **F 538,** F09

tread radius, *n*—the radius of a circle whose arc best fits the tread surface when radius template used is held perpendicular to the circumferential center line of an inflated tire. **F 538,** F09

treadwear indicator, *n*—a raised portion of a groove bottom or a void bottom that is molded in a tire at fairly regular intervals around the circumference to provide a visual indication that most of a tread has been worn away. **F 538,** F09

treated carbon—carbon that is coated or impregnated on the side not carbonized. The treatment is ordinarily applied to improve curl resistance or slip resistance of the sheet or both. **F 221,** F05

treated iron, *n*—molten cast iron to which all basic alloys and nodulizing alloys have been added but not necessarily all inoculating alloy additions. **A 644,** A04

treatment—a process that is applied to one or both sides of the copper foil to enhance the adhesion of the foil to the base substrate that shall not degrade the foil or the base substrate. **B 846,** B05

treatment—the act or manner of treating wood; the quantity of preservative or other substance specified or used to treat wood. **D 9,** D07

treatment, *n*—a combination of the levels (versions) of each of the factors assigned to an experimental unit, synonym **treatment combination.** **E 456,** E11

treatment—see **hazardous-waste treatment**. **E 1605,** E06

treatment combination, *n—in experimenting,* one set of experimental conditions. **D 123,** D13

treatment combination, *n*—see **treatment.** **E 456,** E11

treatment table, *n*—a recompression schedule used to treat decompression sickness or embolisms. **F 1549,** F32

tree—a woody plant having one well-defined stem and a more or less definitely formed crown, usually attaining a height of at least 8 ft. **D 9,** D07

trees—branched or irregular projections formed on a cathode during electrodeposition especially at edges and other high current density areas. **B 374,** B08

tremie—material placed under water through a tremie pipe in such a manner that it rests on the bottom without mixing with the water. **D 653,** D18

tremie method—the method whereby materials are emplaced in the bottom of a borehole with a small diameter pipe. **D 653,** D18

tremolite asbestos, *n*—asbestiform variety of the monoclinic amphibole silicate minerals of the tremolite-actinolite (ferroactinolite) series. Its empirical formula is $Ca_2Mg_5Si_8O_{22}(OH)_2$. Its Chemical Abstracts number is 77536-66-4. **D 2946,** C17

trench—usually a long, narrow, near vertical sided cut in rock or soil such as is made for utility lines. (ISRM) **D 653,** D18

triangle bars—burning bars of triangular cross section (see also **burning bars, points, or tools**). **C 286,** B08

triangle test, *n*—a method of discrimination testing comprised of three coded samples, two of which are identical. The assessor is asked to select the odd sample. **E 253,** E18

triaxial compression—compression caused by the application of normal stresses in three perpendicular directions. (ISRM) **D 653,** D18

triaxial shear test (triaxial compression test)—a test in which a cylindrical specimen of soil or rock encased in an impervious membrane is subjected to a confining pressure and then loaded axially to failure. **D 653,** D18

triaxial state of stress—state of stress in which none of the three principal stresses is zero. (ISRM) **D 653,** D18

triboelectric effect—the phenomenon of producing an electrostatic voltage differential between dissimilar materials by contact or friction as between the toner and the carrier. **F 335,** F05

triboelement, *n*—one of two or more solid bodies that comprise a sliding, rolling, or abrasive contact, or a body subjected to impingement or cavitation. (Each triboelement contains one or more **tribosurfaces**.) **G 40,** G02

tribology, *n*—the science and technology concerned with interacting surfaces in relative motion, including friction, lubrication, wear, and erosion. **G 40,** G02

tribosurface, *n*—any surface (of a solid body) that is in moving contact with another surface or is subjected to impingement or cavitation. **G 40,** G02

tribosystem, *n*—any system that contains one or more triboelements, including all mechanical, chemical, and environmental factors relevant to tribological behavior. (See also **triboelement**.) **G 40,** G02

trichromatic system, *n*—system for specifying color stimuli in terms of tristimulus values based on matching colors by additive mixture of three suitably chosen reference color stimuli. **E 284,** E12

trickle collector— see **collector, trickle.** **E 772,** E44

trickling filter—a facility for the waste waters which utilizes a flow of liquid over fixed-media. **C 896,** C04

triclinic—having three axes of any length, none of the included angles being equal to one another or equal to 90°. **E 7,** E04

tricomponent fiber, *n*—a fiber consisting of three polymers which are chemically different, physically different, or any combination of such differences. **D 123,** D13

triggered capacitor discharge—a series of electrical discharges from capacitors, the energy of which is obtained from either an ac or dc electrical supply. **E 135,** E01

trigger trip—tool activation by trigger operation in conjunction with bottom trip activation. **F 592,** F16

trim, *v*—*in textiles*, to cut off a portion of a material. **D 123,** D13

trim—the removal of parts of a raw hide not suitable for making leather, such as portions from the outer edges of heads, shanks and bellies. **D 1517,** D31

trim, *v*—*in home sewing*, to cut off a narrow strip, width, or portion of a material, usually along a cut edge of a seam allowance. **D 4965,** D13

trimmed block, *n*—dressed crude mica that has been split into thickness of 0.18 mm and over, and has side trimming done to remove irregularities, imperfections, and residues of dirt and rock. **D 1711,** D09

trimmers—units of various shapes consisting of such items as bases, caps, corners, mouldings, angles, and so forth, necessary or desirable to make a complete installation and to achieve sanitary purposes as well as architectural design for all types of tile work. **C 242,** C21

trim nail—See **moulding nail.** **F 547,** F16

triple-cavity mold—See **multiple-cavity mold**. **C 162,** C14

triple-cavity process—See **multiple-cavity process**. **C 162,** C14

triple curve—in a P-T diagram, a line representing the sequence of pressure and temperature values among which two conjugate phases occur in univariant equilibrium. **E 7,** E04

triple-gob process—See **multiple-cavity process**. **C 162,** C14

triple point—in a P-T diagram, the temperature and pressure at which three phases occur in invariant equilibrium. **E 7,** E04

triple point, *n*—fixed point of a system in which three phases are in equilibrium. **E 344,** E20

triple point of water, *n*—triple point of the liquid, solid, and vapor phases of water. **E 344,** E20

triplet injector—an impinging jet atomizer in which there are three colliding liquid jets. **E 1620,** E29

triplet state—an electronic state with a total spin quantum number of one. **E 131,** E13

tripoli—friable and dustlike silica used as an abrasive. **B 374,** B08

trisodium phosphate (TSP)—$Na_3PO_4 12H_2O$, a cleaning agent. **D 6161,** D19

tristimulus colorimeter, *n*—instrument that measures psychophysical color, in terms of tristimulus values, by the use of filters to convert the relative spectral power distribution of the illuminator to that of a standard illuminant, and to convert the relative spectral responsivity of the receiver to the responsivities prescribed for a standard observer. **E 284,** E12

tristimulus filters, *n*—*in cotton fiber color testing with the Color Meter*, optical filters used in conjunction with specific color lamps to obtain a response function approximating the tristimulus functions of the CIE Standard Observer for Source C. **D 123,** D13

tristimulus values—the amounts of three specified stimuli required to match a color. (See also **CIE spectral tristimulus values**.) **E 284,** E12

tristimulus values—the CIE tristimulus values, X, Y, and Z, are a mathematical transformation that represents the relative amounts of the (non-real) primary color stimuli required to match some color stimulus the specification of which they meet. **F 1156,** F12

tristimulus weighting factors, $S\bar{x}$, $S\bar{y}$, $S\bar{z}$, *n*—factors obtained from products of the spectral power *S* of an illuminant and the spectral color matching functions $\bar{x}$, $\bar{y}$, $\bar{z}$ of an observer, usually tabulated at wavelength intervals of 10 or 20 nm, used to compute tristimulus values by multiplication by the spectral reflectance, transmittance, or radiance (or the corresponding factors) and summation. **E 284,** E12

trivial name—a name that is not produced by any systematic procedure of naming. **D 1129,** D19

trochanter—bony prominence near the proximal end of the femur at the base of the femoral neck. **F 869,** F08

trolley wire—a round or shaped, solid bare, hard conductor ordinarily used to supply current to motors through traveling current collectors. **B 354,** B01

troostite—a previously unresolvable, fine aggregate of carbide and ferrite produced by tempering martensite at temperatures in the vicinity of 400°C. Term variously and erroneously applied to bainite and nodular fine pearlite. Confusion arose because of similarity in appearance among the three structures before the advent of high-power microscopy.

(Contemporary). With reference to tool steels, synonymous with upper bainite. **E 7,** E04

tropical—the region lying between the tropics of Cancer and Capricorn. **F 1600,** F20

trough—See **valley**. **E 1823,** E08

trouser back rise, *n— in garment construction*, a measurement from the crotch seam to bottom edge of waistband at center of the back. **F 1494,** F23

trouser front rise, *n— in garment construction*, a measurement from the crotch seam to the bottom edge of waistband at center front. **F 1494,** F23

trouser hips, *n— in garment construction*, the garment's circumference measured at the bottom of pockets or bartack on fly. **F 1494,** F23

trousers,—*n, pl*—an outer garment extending from the waist to the ankle covering each leg separately. (Syn. pants). **F 1494,** F23

trouser waist, *n— in garment construction*, with trousers folded in half by the crease or mating the leg inseams, measure across waist-band's width and double the measurement. **F 1494,** F23

troweled finish—a concrete finish provided by smoothing the surface with power driven or hand trowels or both, after the float finishing operation. A troweled finish is smoother than the floated finish. For specifications, see ACI 301-72. **C 717,** C24

true attenuation—that portion of the observed ultrasound energy loss which is intrinsic to the medium through which the ultrasound propagates. True attenuation losses may be attributed to the basic mechanisms of absorption and scattering. (E 664) **E 1316,** E07

true continuous technique—magnetic particle examination in which the magnetizing current is applied prior to the application of the magnetic particles and is maintained without interruption throughout the examination. **E 1316,** E07

true density—a term used when considering the density of a porous solid. It is defined as the ratio of the mass of the material to its true volume. **C 859,** C26

true gage length, *n—in tensile testing,* a precise length between well-defined bench marks located on the specimen while under known tension in the unsupported portion between the holding clamps and free from contact with any snubbing surfaces or other sources which could result in nonuniform strain. **D 123,** D13

true international roughness index, *n*—the value of international roughness index that would be computed for a longitudinal profile measurement with the constant interval approaching zero. **E 867,** E17

trueness, *n*—the closeness of agreement between the population mean of the measurements or test results and the accepted reference value.

NOTE—The measure of trueness usually is expressed in terms of bias. Greater bias means less favorable trueness.

NOTE—"Population mean" is, conceptually, the average value of an indefinitely large number of test results.

NOTE—Trueness is the systematic component of accuracy. **E 456,** E11

trueness—the lack of significant curvature, inclination, note-worthy elevations, or depressions. **E 867,** E17

true rise, *n—in body measurements*, the vertical distance (plumb line) from the waist level at the side to the crotch. **D 123,** D13

true solution—one in which the components are 100 % dissolved in the base solvent. **D 653,** D18

true strain, ϵ, *n*—in a body subjected to axial force, the natural logarithm of the ratio of the gage length at the moment of observation to the original gage length. **E 6,** E28

true stress, σ $[FL^{-2}]$, *n*—in a tension or compression test the axial stress calculated on the basis of the cross-sectional area at the moment of observation instead of the original cross-sectional area. **E 6,** E28

true value, μ, *n—in statistics*, the value towards which the average of single results obtained by N labortories tends, when N becomes very large. **D 4175,** D02

true value—value of measurand that would be obtained by a perfect measurement. **E 170,** E10

true value, *n—of a temperature*, value attributed to a particular temperature and accepted, sometimes by convention, as having an uncertainty appropriate for a given purpose. **E 344,** E20

true volume—the volume of the solid material only, excluding the volume of any pores or voids. **C 859,** C26

truncated point—needle point with its tip sheared square. **F 547,** F16

truncation—*in fatigue loading*, the exclusion of cycles with values above or below a specified level (referred to as truncation level) of a loading parameter (peak, valley, range, and so forth). **E 1823,** E08

truncation, *n—in bar code printing*, decreasing the length of bars to reduce the height of the symbol below normal specifications. **F 1294,** F05

trunk—the main stem or body of a tree. **D 9,** D07

trunk nail—bright, regular-stock-steel, ½ to 1 ¾-in. nails with rivet or oval head and extra long "V" point. **F 547,** F16

trunnions, *n—in zippers*, the two pivots at the end of the pull that fit into the bail. **D 123,** D13

trussed rafter nail— See **threaded nail.** **F 547,** F16

truss plate—See metal connector plate. **E 1807,** E06

truss rod—a tensioning rod, used for bracing in gates and at terminal posts. May be threaded at one or both ends or contain a tightener or turnbuckle for adjusting tension. **F 552,** F14

truss rod tightener—a tightening device used with an adjusting nut and a truss rod (sometimes incorrectly called *turnbuckle*). **F 552,** F14

TSD facility—facility for treatment, storage, or disposal of hazardous waste. **E 1605,** E06

TSP—tri-sodium phosphate. **E 1605,** E06

TSS—total suspended solids. Concentration of undissolved solids in a liquid, usually expressed in mg/L or ppm. **D 6161,** D19

t-test, *n*—a test of statistical significance based on the use of Student's *t*-distribution and used to compare two sample averages or a sample average and a hypothetical value. **D 123,** D13

TTHM—total trihalomethane. **D 6161,** D19

T titration—a titration in which the sensing electrode responds to the titrant ion, not to the sample species being determined. **D 4127,** D19

tube, *n*—hollow wrought product that is long in relation to its cross section, which is symmetrical and is round, elliptical, a regular hexagon or octagon, or square or rectangular with sharp or rounded corners, and has uniform wall thickness except as affected by corner radii.

Alclad tube, n—a composite tube product comprised of an aluminum-alloy core having on either the inside surface (only) or outside surface (only) a metallurgically bonded aluminum or aluminum-alloy coating that is anodic to the core, thus electrolytically protecting the core against corrosion.

arc-welded tube, n—tube made from sheet or plate formed by positioning two opposite edges of the metal together and butt welded by either the gas-tungsten or gas-metal arc-welding method, with or without the use of filler metal; individually fabricated tube may be welded together to produce the ordered length.

drawn tube, n—tube brought to final dimensions by drawing through a die.

extruded tube, n—tube formed by hot extruding.

finned tube, n—tube which has integral fins or projections protruding from its outside surface.

heat exchanger tube, n—tube used in apparatus in which fluid inside the tube will be heated or cooled by fluid outside the tube, but the term is usually not applied to coiled tube or to tube for use in refrigerators or radiators.

seamless tube, n—a tube that does not contain any junctures (metallurgical welds) resulting from the method of manufacture and may be produced by die and manrel, or by hot piercer processes.

sized tube, n—tube that, after extrusion, has been cold drawn a slight amount to minimize ovality.

structural tube, n—tube commonly used for structural purposes.

welded tube, n—tube produced by forming and seam-welding sheet longitudinally. **B 881,** B07

tube, *n*—a hollow product of round or any other cross section having a continuous periphery of uniform shape. **B 899,** B02

tube, *n*—a cylindrical paperboard **container,** convolutely or spirally wound, having paper, wood, metal, or combination ends.

collapsible tube—in packaging, a flexible **container** having a dispensing tip and cap at one end, and closed at the opposite end, serving as both package and dispenser.

composite tube—in packaging, a **tube** consisting usually of multiple layers of **fiberboard,** film, or foil providing barrier properties, and an outer layer of paper.

fiberboard tube—in packaging, a cylindrical or multisided form.

paperboard (mailing) tube—in packaging, a cylindrical **container** wherein the length does not exceed ten times the girth with a minimum ⅛-in. (3-mm) wall thickness. **D 996,** D10

tube, *n*—a tubular conduit for transport of fluids or finely divided solids; also, a hollow structural member; a hollow product of round or other cross section. **E 631,** E06

tube—the component that encloses the spindle. It is attached to the lock horn and dial ring (not used in all lock designs). **F 471,** F12

tube, air conditioning—a seamless copper tube conforming to a standard series of sizes, and to specified internal cleanliness requirements normally furnished in straight lengths, with the ends capped or sealed. **B 846,** B05

tube A manchette—*in grouting*, a grout pipe perforated with rings of small holes at intervals of about 12 in. (305 mm). **D 653,** D18

tube, automotive and general service—a seamless copper tube of small diameter conforming to a standard series of sizes commercially known as automotive and general service tube. **B 846,** B05

tube, capillary—a tube of small inside diameter with an inside surface of highest quality and conforming to close-diameter tolerances. It is subject to special tests to ensure precision and uniformity of bore and is specially cleaned and packed. **B 846,** B05

tube, condenser—see **tube, heat exchanger.** **B 846,** B05

tube, copper service—a bendable copper water tube for underground water service. **B 846,** B05

tube, copper water—a seamless copper tube conforming to the particular dimensions commercially known as copper water tube and designated as Types K, L, and M. **B 846,** B05

tube current—the current, measured in milliamperes, passing between the cathode and anode during the operation of an X-ray tube. **E 1316,** E07

tube current—the transfer of electricity, created by the flow of electrons, from the filament to the anode target in an X-ray tube; usually expressed in unit of milliamperes. **E 1316,** E07

tube, ferrule—a tube from which metal rings or collars (ferrules) are made for use in installing condenser tubes. **B 846,** B05

tube furnace—a muffle furnace in which combustion occurs within alloy tubes. **C 286,** B08

tube, heat exchanger—a tube manufactured to special requirements as to dimensional tolerances, finish, and temper for use in condensers and other heat exchangers. **B 846,** B05

tube, heat exchanger with integral enhanced surface—an external or internal surface, or both, modified by a cold forming operation, to produce an enhanced surface for improved heat transfer. The enhancement may take the form of longitudinal or helical fins or ridges, or both, or modifications thereto. **B 846,** B05

tube, heat exchanger with integral fins— a tube having a series of metallic ribs on the outside or inside surface either parallel to the longitudinal axis or circumferentially extended from the tube to increase the effective surface area for heat transfer applications. The fins may be mechanically applied, drawn, or integrally extended from the tube wall. **B 846,** B05

tube horn—an integral part of the case to which the tube is attached (not used in all locks). **F 471,** F12

tuberculation—the formation of localized corrosion products that appear on a surface as knoblike prominences (tubercules). **G 15,** G01

tube, refrigeration service—a seamless copper tube conforming to a standard series of sizes, and to special internal cleanliness and dehydration requirements, normally furnished to soft temper coils and with ends capped or sealed. **B 846,** B05

tube, seamless—a tube produced with a continuous periphery in all stages of the operations. **B 846,** B05

tube, seamless water—a tube conforming to the particular dimensions of tube commercially known as copper water tube and designated as Types K, L, and M in inch-pound units and Types A, B, and C in SI units. **B 846,** B05

tube terminal staple—staple designed for use as a wiring terminal in electronic assembly. **F 592,** F16

tube/tubing—hollow section of metal or other material having a round, square, rectangular, or other cross-sectional form, its size being designated by outside dimension(s) and wall thickness, in inches (millimeters). **E 1481,** E06

tube, waveguide—a tube used as a transmission line to electronic equipment. **B 846,** B05

tube, welded—tube made by processing strip into a tubular shape and welding the edges to make a longitudinal seam. **B 846,** B05

tube, welded water—a tube conforming to the particular dimensions of tube commercially known as copper water tube and designated as Types WK, WL, and WM in inch-pound units, and Types WA, WB, and WC in SI units. **B 846,** B05

tubing, *n*—braided, knitted, or woven fabric of cylindrical form having a width of 4 in. or more (circumference of 8 in. or more). (See also **sleeving**.) **D 123,** D13

tubing, *n*—a system of **tubes.** **E 631,** E06

tubing nail—See **conduit nail.** **F 547,** F16

tuck, *n*—*in rope*, a free strand of the rope placed between the rope strands during splicing. **D 123,** D13

tucked seam, *n*—*in home sewing*, a complex seam formed on the inside of the product with neither cut edge enclosed and having one visible line of topstitching on the face side and a visible free folded edge (tuck). (Compare **lapped seam.**) **D 4965,** D13

tucking defect, *n*—*in knitted fabrics*, one or more unwanted tuck loops. **D 123,** D13

tuck point, *v*—*(1) (historical)* to point masonry with a flush mortar joint that approximates the color of the masonry units and then add a mortar strip of contrasting color such that a narrow mortar joint width is simulated. *(2)* regional term for repoint. **C 1180,** C12

tuckstone—blocks placed on top of flux blocks to protect them from flame and gases and as a seal between them and the supported side and end walls. **C 162,** C14

tuck wall—the series of tuckstones or a wall serving the same purpose made of courses of brick. **C 162,** C14

tuft, *n*—*in pile fabrics*, those cut or uncut loops which are attached to the backing fabric at one binding site and which form part of the fabric face. **D 123,** D13

tuft, *n*—*in pile yarn floor coverings*, those cut or uncut loops which form part of the fabric face that are attached or bound to the backing fabric at binding sites. **D 5684,** D13

tuft bind, *n*—the force required to pull a tuft from a cut pile floor covering or to pull free one leg of a loop from a looped pile floor covering. **D 123,** D13

tuft bind, *n*—*in pile fabrics*, the force required to pull a tuft element from a pile yarn floor covering. **D 5684,** D13

tufted, *n*—buttoned or laced through the upholstery cover material and upholstery material. **E 176,** E05

tufted fabric, *n*—a fabric with a pile consisting of tufts or loops formed by inserting yarn into a previously prepared backing fabric. **D 123,** D13

tuft element, *n*—*for pile yarn floor covering*, a segment of yarn bound to a backing fabric at a binding site so that two portions (legs) of the yarn project above the backing fabric, one portion on each side of the binding site. **D 123,** D13

tuft height, *n*—for cut pile yarn floor covering, the length of a tuft leg. **D 5684,** D13

tufting, *n*—*in upholstered furniture*, localized indentation of upholstered furniture surfaces and cushions, by the use of or appearance of buttons, providing an aesthetic treatment. **D 123,** D13

tuft leg, *n*—*for pile yarn floor covering*, one of the two portions of a tuft element that project above the backing fabric on the pile side of the floor covering. **D 123,** D13

tuft length, *n*—*for pile yarn floor covering*, the length of a tuft element measured while extended in a straight line under zero tension. **D 123,** D13

tumbled—cleaned and polished by agitation in rotating drum containing polishing compound, such as sawdust or other burnishing media. **F 547,** F16

tumbler-galvanized— See **hot-galvanized.** **F 547,** F16

tumbler, hand change—an assembly, usually in the form of a circular disk, with a gate designed to accept a fence. When released by hand, the position of the gate may be set in order to change the combination. **F 471,** F12

tumbler, hole change—a component, usually in the form of a circular disk, into which a drive pin may be inserted at a selected position, to change the combination. **F 471,** F12

tumbler, key change—an assembly, usually in the form of a circular disk, with a gate designed to accept a fence. When released by a key, the position of the gate may be set to change the combination. **F 471,** F12

tumbler pack—an assembly of tumblers and components for assembly on a tumbler post. **F 471,** F12

tumbler post—a component upon which the tumbler pack is assembled and about which the tumblers rotate. **F 471,** F12

tumbler, screw change—an assembly, usually in the form of a circular disk, with threaded perforations into which a screw serving as a drive pin may be inserted at selected positions for change of combination. **F 471,** F12

tumbler, wheel—an assembly, usually in the form of a circular disk, with a gate and drive pin, the locations of which may be set to selected positions for changing the combination. **F 471,** F12

tumbling—See **barrel finishing.** **B 374,** B08

tundra—the vast, treeless, nearly level plains of the arctic regions. **F 1600,** F20

tuned circuit—electronic circuit designed to be responsive to or at some preselected frequency. **F 1156,** F12

tunnel—a man-made underground passage constructed without removing the overlying rock or soil. Generally nearly horizontal as opposed to a shaft, which is nearly vertical. (ISRM) **D 653,** D18

turbidity—reduction of transparency of a sample due to the presence of particulate matter. **D 1129,** D19

turbidity—an expression of the optical properties of a sample which causes light rays to be scattered and absorbed rather than transmitted in straight lines through the sample. Turbidity of water is caused by the presence of suspended and dissolved matter such as clay, silt, finely divided organic matter, plankton, other microscopic organisms, organic acids, and dyes. **D 4410,** D19

turbidity—a suspension of fine particles that scatters or absorbs light rays. **D 6161,** D19

turbidity, *n*—reduction of transparency of a specimen due to the presence of particulate matter. **E 284,** E12

turbidity, absolute—the fractional decrease of incident monochromatic light through the sample, integrating both scattering and transmitted light. **D 1129,** D19

turbidity current—See **density current.** **D 4410,** D19

turbidity, Jackson candle—an empirical measure of turbidity in special apparatus, based on the measurement of the depth of a column of water sample that is just sufficient to extinguish the image of a burning standard candle observed vertically through the sample. **D 1129,** D19

turbidity, jackson candle (JTU)—an empirical measure of turbidity in special apparatus, based on the measurement of the depth of a column of water sample that is just sufficient to extinguish the image of a burning standard candle observed vertically through the sample. **D 6161,** D19

turbidity, nephelometric—an empirical measure of turbidity based on a measurement of the light-scattering characteristics (Tyndall effect) of the particulate matter in the sample. **D 1129,** D19

turbidity, nephelometric (NTU)—an empirical measure of turbidity based on a measurement of the light-scattering characteristics (tyndall effect) of the particulate matter in the sample. **D 6161,** D19

turbine power nozzle, *n*—an attachment for a vacuum cleaner containing a turbine driven agitator that assists in dirt removal from a floor surface. **F 395,** F11

turbulence—the irregular motion of a flowing fluid. **D 4410,** D19

turbulent flow—that type of flow in which any water particle may move in any direction with respect to any other particle, and in which the head loss is approximately proportional to the second power of the velocity. **D 653,** D18

turbulent flow, *n*—that type of flow in which any water particle may move in any direction with respect to any other particle, and in which the head loss is approximately proportional to the second power of the velocity. **D 4439,** D35

turnaround document—a form produced by an electronic data processing system intended for future re-entry, possibly with added data, via an optical scanner. **F 149,** F05

turnbuckle—a tightening device with left hand and right hand threads at opposite ends, used with a truss rod. **F 552,** F14

turn insulating paper, *n*—paper used to insulate conductors that will become coils in a transformer or other inductive apparatus. The conductors are commonly rectangular in cross section. **D 1711,** D09

turpentine, oil of— See **oil of turpentine**. **D 804,** D01

turpentine, spirits of— See **spirits of turpentine**. **D 804,** D01

turpentine substitute, *n*—this term is not recommended, because its use to describe a paint thinner of mineral oil origin is illegal under the U.S. Federal Naval Stores Act. **D 804,** D01

tweak—to make a mechanical adjustment of a spring arm of an electrical contact assembly to adjust the force or position of the contact. **B 542,** B02

tweel, tuille—a counterweighted furnace door, opening vertically. **C 162,** C14

tweezer mark—*on semiconductor wafers*, marks caused by scraping tweezers across the wafer surface. **F 1241,** F01

tweezer residue—contamination placed upon the wafer surface while holding the wafer with unclean tweezers. **F 1241,** F01

twelve-harness satin, *n*—a weave similar to eight-harness satin except in warp-faced fabrics warp yarns show on the face of the fabric eleven out of twelve adjacent yarns and in filling-faced fabrics filling yarns show on the face eleven out of twelve adjacent yarns. **D 123,** D13

twill braid, *n—in rope*, a braided construction in which one strand of one direction of rotation about the axis passes over two strands of the opposite direction and it in turn passes under the next two strands of opposite direction. **D 123,** D13

twill weave, *n*—a weave characterized by diagonal lines produced by a series of floats staggered in the warp direction, which are normally formed by the filling (a filling-faced twill). (See also **warp-faced twill**.) **D 123,** D13

twin, n—a special type of half rope in which both rope strands are used as one with both running through the same protection anchors. **F 1773,** F08

twin band—a volume within a crystal bounded by twinning planes. **F 1241,** F01

twin bands—bands across a crystal grain, observed on a polished and etched section, the crystallographic orientations of which have a mirror image relationship to the orientation of the matrix grain across a composition plane which usually is parallel to the sides of the band.

(*1*) *annealing twin*—twin bands which are produced during annealing following cold work.

(*2*) *mechanical twins*—twin bands which are produced by cold work.

(*3*) *Neumann bands*—mechanical twins in ferrite. **E 7,** E04

twine, *n*—(*1*) a term applied loosely to a variety of textile strands used for tying such articles as parcels, bundles, or bales.

(*2*) an aggregate of fibers or yarns compacted into a partially or completely balanced twisted structure of indefinite length, generally used for tying or binding. **D 123,** D13

twin-fluid atomizer—see **pneumatic atomizer**. **E 1620,** E29

twinned crystal—a crystal in which the lattice is of two parts related to each other in orientation as mirror images across a coherent planar interface known as the twinning plane or twin boundary. **F 1241,** F01

twist, *n*—the number of turns per unit length made by a filament or strand about a conductor axis. **B 713,** B01

twist—a distortion caused by the turning or winding of the edges of a board so that the four corners or any face are no longer in the same plane. **D 9,** D07

twist, *n—in textile strands*, the helical or spiral configurations induced by turning a strand about its longitudinal axis. **D 123,** D13

twist, *n*—the number of turns about its axis per unit of length observed in a yarn or other textile strand. **D 123,** D13

twist, *n—in fibrous composites,* a measure of the number of turns per unit length a fiber bundle makes around its axis. **D 3878,** D30

twist—a condition of longitudinal progressive rotation found in pultruded parts.

NOTE—This term describes a condition that can be easily detected for a noncircular cross section by placing the pultrusion on a plane surface, holding one end flat with the surface, and observing if one edge or side of the other end does not lie parallel with that surface. **D 3918,** D20

twist—the type of selvage obtained by twisting adjacent pairs of wire ends together in a close helix of 1½ machine turns, which is equivalent to three full twists. **F 552,** F14

twist (amount of), *n*—the number of turns per unit length of yarn or textile strand (or of a product such as cable or hose) around its axis. **D 1566,** D11

twist balance, *n—in glass fiber cord and sewing thread*, the relationship of primary and final twist to each other and to the cord size such that residual torsional effects are nullified. **D 123,** D13

twisted—provided with four or more continuous, symmetrical, helical deformations; resulting from twisting of square nail wire about its own longitudinal axis prior to heading and pointing of nail. Helical deformations resulting from this manufacturing process extend all the way from head to point, thus preventing inclusion of clearance between head and thread. **F 547,** F16

twisted conductor—a composite conductor in which the filaments or strands are displaced about the conductor axis. **B 713,** B01

twisted length—same as **twist pitch.** **B 713,** B01

twisted nail—helically twisted, squarewire, ½ by 0.072 to 6 by 0.250-in. drivescrew nails, usually of tempered stiff-stock, with flat or countersunk head and medium diamond point, with crest diameter being referred to as diameter. **F 547,** F16

twisted or laid rope, *n*—rope made from three or more strands which are laid or twisted together in a twist direction opposite to the twist direction in the strands. **D 123,** D13

twist factor, *n*—the product obtained when the twist expressed in turns per centimetre is multiplied by the square root of the yarn number expressed in tex.

$$\text{Twist factor (TF)} = \text{tpcm} \times \sqrt{T}$$

where:
T = yarn number expressed in tex.

D 123, D13

twist multiplier, *TM, n*—the quotient of the twist expressed in turns per inch and the square root of the yarn number in an indirect system.

$$\text{Twist multiplier (TM)} = \text{tpi}/\sqrt{N}$$

where:
N = yarn number in an indirect system, the cotton system unless otherwise specified.

D 123, D13

twist nail—slender, copper or aluminum nails with flat head and medium needle point for twist clinching, that is, for having part of nail shank twisted to form a clinched point. **F 547,** F16

twist pitch (also twisted length)—the length in which a filament or strand returns to its original relative position in a twisted conductor. **B 713,** B01

twist take-up, *n*—the change in length of a yarn or other textile strand caused by twisting, expressed as a percent of the original untwisted length. **D 123,** D13

twist, yarn, *n*—the number of turns, about its axis, per unit of length, observed in a yarn or other textile strand. **D 1566,** D11

two-body abrasive wear, *n*—a form of abrasive wear in which the hard particles or protuberances which produce the wear of one body are fixed on the surface of the opposing body. **G 40,** G02

two coat-one fire—the application of two different coats of enamel followed by a single firing step. **C 286,** B08

two-dimensional symbology—see **2D symbol**. **F 1294,** F05

two-fluid atomizer—see **pneumatic atomizer**. **E 1620,** E29

two-front button, *n*—a button in which the face and back shape are identical. **D 5497,** D13

two-of-five test, *n*—a method of discrimination testing comprised of five samples: two are identical samples of one product and three are identical samples of a second product. The assessor's task is to determine which two samples are different from the other three. **E 253,** E18

two-phase system—a vapor phase and a single liquid phase, usually containing dissolved propellant, inside the container. **D 3064,** D10

two-point threshold (touch), *n*—the minimum normal distance between two objects such that they are perceived as two a specified percent of the time. **E 253,** E18

two-side structural sealant glazing, *n*—*in building construction,* a glazing system wherein two sides of a rectangular-shaped lite or panel, usually opposing, are structural sealant glazed. **C 717,** C24

two-way sort—an electromagnetic sort based on a signal response from the material under test above or below a level established by two or more calibration standards. **E 1316,** E07

T-X diagram—a two-dimensional, graphical representation of the isobaric phase relationships in a binary system; the coordinates of the graph are temperature and concentration. **E 7,** E04

tyndall effect—the path of light through a heterogeneous medium made visible by the solid particles. **D 6161,** D19

type—the designation of plywood as a function of veneer grade and adhesive durability. **D 1038,** D07

Type A operating characteristic curve, *n*—an operating characteristic curve which describes the operation of a sampling plan where the size of the lot being sampled is taken into consideration. **D 123,** D13

type bar—a bar, mounted on an impact printer, that holds type slugs. **F 909,** F05

Type B operating characteristic curve, *n*—an operating characteristic curve which describes the operation of a sampling plan where items are drawn at random from a theoretically infinite process. **D 123,** D13

type font—type of a given size and style, for example, 10-point Bodoni Modern. **F 909,** F05

Type I—exhaust hood designed for removal of grease and smoke. It includes both listed grease filter, baffles, or extractors for removal of the grease and fire suppression equipment. See **exhaust hood**. **F 1827,** F26

Type I apparel—apparel designed for general heavy work; it may be subjected to breaking and tearing stresses indoors or out. **D 123,** D13

Type I error—false rejection error(s); the improper rejection of an authentic and genuine article by individual(s) attempting to validate authenticity; see also FR (False Rejection). **F 1156,** F12

Type II—exhaust hood designed for collection and removal of steam, vapor, heat, and odors where grease is not present. It may or may not have filters or baffles and typically does not have a fire suppression system. **F 1827,** F26

Type II apparel—apparel designed for light work or leisure activities; it will not be expected to undergo severe physical stresses. **D 123,** D13

Type II error—See **error rates**. False acceptance errors; the incorrect acceptance of a non-authentic and non-genuine article, or both, by individuals attempting to validate authenticity; see also FA (False Acceptance). **F 1156,** F12

Type II mineral oils, *n*— *for steam and gas turbine lubricating systems,* oils for machinery that do require lubricants with enhanced load carrying capacity. **D 4175,** D02

Type I mineral oils, *n*— *for steam and gas turbine lubricating systems,* oils for machinery that do *not* require lubricants with enhanced load carrying capacity. **D 4175,** D02

typewriter—a machine for writing in characters similar to those produced by printers' type by means of keyboard-operated types striking through an inked ribbon. **F 909,** F05

typical metal connector plate—metal connector plate representative of single shipment of plate to be tested; with plate manufacturing procedure simulating actual production conditions anticipated during plate fabrication as well as during member and component assembly. **E 1807,** E06

typical serviceability score—See **serviceability score**. **E 1480,** E06

typical value, *n*—*for geosynthetics,* the mean value calculated from documented manufacturing quality control test results for a defined population obtained from one test method associated with on specific property. **D 4439,** D35

typp, *n*—an obsolete indirect yarn numbering system equal to the number of 1000-yd lengths per pound. **D 123,** D13

U

UCC/EAN, *n*—a consortium of UCC and EAN that promulgates international standards. **F 1294,** F05

"U" configuration (aka catenary configuration)—boom positioned in a "U" shape. **F 818,** F20

ullage, *n*—that volume of a closed system or container which is filled with vapor. **D 4175,** D02

ULPA filter—ultra-low-penetration air filter. **E 1605,** E06

ultimate analysis, *n*—*in the case of coal and coke*, the determination of carbon and hydrogen in the material, as found in the gaseous products of its complete combustion, the determinations of sulfur, nitrogen, and ash in the material as a whole, and the calculation of oxygen by difference. **D 121,** D05

ultimate analysis, *n*—in analysis of combustible materials, the determination of the percentages of carbon, hydrogen, sulfur, nitrogen, chlorine, ash, and oxygen in a moisture-free sample; the percentage of oxygen is usually determined by difference. **D 5681,** D34

ultimate analysis—the determination of the percentages of carbon, hydrogen, sulfur, nitrogen, chlorine, ash, and oxygen in a dry refuse-derived fuel sample. The percentage of oxygen may be obtained by difference between 100 % and the sum of the other determined elemental analyses. **E 856,** D34

ultimate analysis—the determination of the elemental composition of the organic portion of carbonaceous materials as well as the total ash and moisture. Determined by prescribed methods. **E 1126,** E48

ultimate analysis—the determination of carbon and hydrogen in the material, as found in the gaseous products of its complete combustion, the determination of sulfur, nitrogen, and ash in the material as a whole, and the calculation of oxygen by difference. **E 1705,** E48

ultimate analysis—the determination of the elemental composition of the organic portion of carbonaceous materials as well as the total ash and moisture. Determined by prescribed methods. **E 1705,** E48

ultimate bearing capacity, q_e, q_{ult} **(FL^{-2})**—the average load per unit of area required to produce failure by rupture of a supporting soil or rock mass. **D 653,** D18

ultimate biodegradation, *n*—degradation achieved when the test material is totally utilized by microorganisms, resulting in the production of carbon dioxide (and possibly methane, in the case of anerobic biodegradation), water, inorganic compounds, and new microbial cellular constituents (biomass or secretions, or both). **D 6384,** D02

ultimate biodegradation test, *n*—a test which estimates the extent to which the carbon in a material is converted to carbon dioxide or methane, either directly by measuring the production of carbon dioxide O_2 or methane, or indirectly by measuring the consumption of oxygen. **D 6384,** D02

ultimate elongation—the elongation at the time of rupture. (D 3196, D-11) **F 869,** F08

ultimate load—maximum load the pole will carry in the specified direction, applied to produce bending and torsional stresses. **C 1213,** C27

ultimate resilience, R_u [FL^{-2}], *n*—a material parameter defined by the equation:

$$R_u = S_u^2/2E$$

where:

S_u = ultimate tensile strength, and
E = elastic modulus, as determined from a conventional tension test.

G 40, G02

ultimate strength—the maximum three edge bearing test load supported by a pipe. **C 822,** C13

ultimate strength—See **strength.** **E 631,** E06

ultimate strength, *n*—maximum resistance to external force, load, or generation of internal strain of a material (of a given size), member, connection, component, or assembly at which failure occurs; expressed in terms of units of force, lbf, pounds of force (N, newtons).

window, n—an assembly consisting of a surrounding **frame** and one or more **sashes, ventilators,** or fixed **lites** of glass, or a combination of these, designed to be installed in a **wall** opening for the purpose of admitting light or air, or both.

"X"—in window and door design, a designation used to indicate an **operable sash, ventilator,** or **panel.** **E 631,** E06

ultimate strength—maximum resistance to external force, load, or generation of internal strain of a material, member, connection, component, or assembly at which failure occurs; expressed in terms of units of force, newtons (pound-force); as compared to ultimate stress which is expressed in units of force per unit of area. Often referred to as *maximum load, ultimate load, maximum strength*, or *nominal strength*, and incorrectly referred to as ultimate stress. **E 1807,** E06

ultimate strength—tensile strength. **F 869,** F08

ultimate tensile strength—the highest load attained during a tensile test, converted to unit stress based on the original cross-section area of the tensile test specimen. **C 709,** D02

ultra accelerator, *n*—a highly active accelerator for fast vulcanization and often used for low-temperature vulcanization. **D 1566,** D11

ultrafiltration, , *n*—a process employing semipermeable membrane under a hydraulic pressure gradient for the separation of components in a solution. The pores of the membrane are of a size which allow passage of the solvent(s) but will retain non-ionic solutes based primarily on physical size, not chemical potential. **D 1129,** D19

ultrafiltration (UF)—a process employing semipermeable membrane under a hydraulic pressure gradient for the separation of components in a solution. The pores of the membrane are of a size which allow passage of the solvent(s) but will retain non-ionic solutes based primarily on physical size, not chemical potential. **D 6161,** D19

ultra-low-penetration air (ULPA) filter—filter capable of separating out particles of 0.13 μm or greater from a body of air at 99.9995 % efficiency or greater. See also **HEPA filter**. **E 1605,** E06

ultramicroscopic—below the resolution of the microscope. **E 7,** E04

ultra pure water—water generally used in semiconductor industry having specifications (chemical, physical and biological) for extremely low contaminant levels. **D 6161,** D19

ultrasonic—pertaining to mechanical vibrations having a frequency greater than approximately 20 000 Hz. **E 1316,** E07

ultrasonic cavitation test device, *n*— a vibratory cavitation test device whose driving frequency is in the ultrasonic range, about 20 kHz or greater. (For lower frequencies, the term *vibratory cavitation test device* is preferred.) **G 40,** G02

ultrasonic leak detector—an instrument that detects ultrasonic energy produced by molecular turbulence that occurs in the transition from laminar to turbulent flow of a gas through an orifice and that converts this energy to a usable signal. **E 1316,** E07

ultrasonic noise level—the large number of unresolved indications resulting from structure or possibly from numerous small discontinuities, or both. (E 127) **E 1316,** E07

ultrasonic nozzle—a pneumatic or vibratory atomizer in which energy is imparted, at high frequency, to the liquid.

NOTE—Common usage refers to frequencies within or above the human auditory range, but recommended practice should be restricted to frequencies above 20 kHz.

NOTE—For frequencies below 20 kHz, see **sonic atomizer**.

E 1620, E29

ultrasonic precipitation—See **precipitation.** **D 1356,** D22

ultrasonic spectroscopy—analysis of the frequency spectrum of an ultrasonic wave. **E 1316,** E07

ultratrace—conveniently used to designate the region below 1. **C 859,** C26

ultraviolet—invisible light radiation, adjacent to the violet end of the visible spectrum, with wavelengths from about 200 to 400 nm (nanometres). **E 7,** E04

ultraviolet—pertaining to the region of the electromagnetic spectrum from approximately 10 to 380 nm. The term ultraviolet without further qualification usually refers to the region from 200 to 380 nm. **E 131,** E13

ultraviolet, *adj*—referring to radiant flux having wavelengths shorter than the visible wavelengths about 10 to 380 nm. **E 284,** E12

ultraviolet-activated fluorescence, *n*— fluorescence resulting from the absorption of ultraviolet radiant flux, that is, flux with wavelengths shorter than 380 nm. (See also **visible-activated fluorescence.**) **E 284,** E12

ultraviolet radiation, *n*—radiation for which the wavelengths of the monochromatic components are smaller than those for visible radiation, and more than about 1 nm.

NOTE—The limits of the spectral range of ultraviolet radiation are not well defined and may vary according to the user. Committee E-2.1.2 of the CIE distinguishes in the spectral range between 100 and 400 nm:

UV-A	315 to 400 nm
UV-B	280 to 315 nm
UV-C	100 to 280 nm

E 349, E21

ultraviolet response—a particular type of optics system used in some optical scanners. As a general rule, nonscan inks for this response will be in the violet portion of the color spectrum. **F 149,** F05

ultraviolet (UV) radiation—wave lengths between 200 to 300 nm. These wave lengths have a strong germicidal effect. The maximum effect is at 253.7 nm. **D 6161,** D19

umbilical, *n*—in surface-supplied diving, diver's gas and a communication and safety line all in one bundle. **F 1549,** F32

umbrella head—extra-large cupped oval head, provided with slight projection of nail shank above head to provide striking surface during driving. **F 547,** F16

umpire—a laboratory of recognized capability chosen to resolve a difference, for instance, a disputed shipper-receiver difference. **C 859,** C26

unaided eye—visual inspection, without the use of special equipment or enhancement excepting the use of corrective lenses. **B 846,** B05

unaided eye—visual inspection, without the use of special equipment or enhancement excepting the use of corrective lenses. **C 896,** C04

unary system—composed of one component. **E 7,** E04

unassigned blankets—blankets that are in storage prior to being issued for use. **F 819,** F18

unbacked exposure, *n*—a technique of weathering in which the test specimens are exposed such that the portion of the specimen being evaluated is open to the effects of the weather on all sides. **G 113,** G03

unbiased gun—(zero biased gun) an electron gun in which the cathode cap is at the same potential as the filament (see **self-biased gun**). This type of gun provides a low-intensity illumination with large angular aperture. **E 7,** E04

unbiased sample—See *unbiased sample* under **sample**. **D 121,** D05

unbonded batting, *n*—a textile filling material which is neither needle-punched, resin bonded, or thermal bonded. (See also **needle-punched batting, resin bonded batting,** and **thermal bonded batting.**) **D 123,** D13

unburned brick, *n*—brick manufactured by processes that do not involve firing of the finished product. **C 71,** C08

uncertainty—a parameter associated with the result of a measurement, that characterizes the dispersion of the values that could reasonably be attributed to the measurand or derived quantity. **E 170,** E10

uncertainty, *n*—*of a temperature measurement*, parameter, derived from an analysis of a measurement and its result, that characterizes the range in which the true value of temperature is estimated to lie, generally with a given confidence. **E 344,** E20

uncertainty, *n*—an indication of the variability associated with a measured value that takes into account two major components of error: (*1*) bias, and (*2*) the random error attributed to the imprecision of the measurement process. **E 456,** E11

uncertainty—lack of certain, deterministic, values for the variable inputs used in an economic analysis of a building or building system. **E 833,** E06

uncertainty—measure of the combination of the bias and precision error in specified instrumentation, or the measure of the repeatability of a reported test result. **F 1827,** F26

uncombined calcium hydroxide, *n*—*in asbestos-cement*, the content of unreacted calcium hydroxide that remains in the sample of cured product when tested. **C 460,** C17

uncombined calcium hydroxide, *n*— the content of unreacted calcium hydroxide that remains in the sample of cured product when tested. **C 1154,** C17

uncombined calcium hydroxide, *n*—*in asbestos-cement*, the content of unreacted calcium hydroxide that remains in the sample of cured product when tested. **D 2946,** C17

unconfined aquifer, *n*—an aquifer that has a water table. **D 653,** D18

unconfined compressive strength—the load per unit area at which an unconfined prismatic or cylindrical specimen of material will fail in a simple compression test without lateral support. **D 653,** D18

unconfined compressive strength—see **compressive strength**. **D 653,** D18

unconfined test, *n*—a test in which the test sample is placed directly upon the anvil with no lateral confinement. **E 1445,** E27

unconfined yield strength, f_c— the major principal stress of the Mohr stress circle being tangential to the yield locus with the minor principal stress being zero. A synonym for **compressive strength**. **D 653,** D18

unconsolidated-undrained test (quick test)—a soil test in which the water content of the test specimen remains practically unchanged during the application of the confining pressure and the additional axial (or shearing) force. **D 653,** D18

uncontrollability, *n*—any deviation of the vehicle from the intended trajectory (TGL) during or at the end of a test, or both. **F 538,** F09

undamped natural frequency—*of a mechanical system,* the frequency of free vibration resulting from only elastic and inertial forces of the system. **D 653,** D18

underarm length, *n*—*in body measurements,* with the arm down, the distance from the armpit to the inner wrist bone. **D 123,** D13

underarm length, *n*—*in body measurements,* with arm hanging down, the distance from the armpit to the inner wrist bone. **D 5219,** D13

underconsolidated soil deposit—a deposit that is not fully consolidated under the existing overburden pressure. **D 653,** D18

under-cooling—a decrease in temperature below that at which an equilibrium phase change exists, without the occurrence of the transformation. **E 7,** E04

under cooling—see **supercooling.** **E 1142,** E37

undercure, *n*—a state of vulcanization between the onset of vulcanization and the state of optimum cure. **D 1566,** D11

underdamped-well response—response characterized by the water level oscillating about the static water level following a sudden change in water level. (See for comparison **overdamped well response.**) **D 653,** D18

underdrain—a type of asbestos-cement pipe having a multiplicity of perforations along its length, intended for use in surface or below-surface drainage. **C 460,** C17

underdrain, *n*—a type of pipe having a multiplicity of perforations along its length, intended for use in surface or below-surface drainage. **C 1154,** C17

underdrain—a type of asbestos-cement pipe having a multiplicity of perforations along its length, intended for use in surface or below-surface drainage. **D 2946,** C17

underglaze decoration—See *underglaze decoration* under **decoration.** **C 242,** C21

underlay, *n*—a resilient layer of material placed under rugs or carpets to increase comfort and improve service life of the carpet. **D 5684,** D13

underlayment—See **floor.** **E 631,** E06

underlayment, *n*—a material placed under resilient flooring to provide a suitable installation surface. **F 141,** F06

underlay nail, underlayment nail—bright, regular-stock-steel, stiff-stock or hardened-steel, annularly threaded, 1 by 0.080 to 3 by 0.148-in. nails with flat or slightly countersunk 3⁄16 to 5⁄16-in. head and medium diamond point. **F 547,** F16

under-the-cap filling—a process for adding the propellant to the container by sealing off the head of the container and injecting the propellant at ambient temperature and high pressure, under the valve cap before crimping. This process is usually used where the propellant content is large and cold filling is not desired. **D 3064,** D10

undertone—the color of a thin layer of pigment-vehicle mixture applied on a white background. **D 16,** D01

underwear, *n*—clothing worn next to the skin under outer clothes. **D 123,** D13

undisturbed sample—a soil sample that has been obtained by methods in which every precaution has been taken to minimize disturbance to the sample. **D 653,** D18

undulated staple—staple with curves in crown for better driving stability. **F 592,** F16

UN/EDIFACT—abbreviation for United Nations, Electronic Data Interchange for Administration, Commerce and Transport, an international group dedicated to standards for electronic interchange. **F 1294,** F05

uneven dyeing, *n*—cloth which shows variations in shade due either to incorrect dyeing methods or faulty materials. **D 123,** D13

unevenness, *n*—*in textiles,* variation in the linear density of a continuous strand or of a portion of a strand. (See also **coefficient of variation unevenness, mean deviation unevenness.**) **D 123,** D13

unglazed tile—a hard, dense tile of homogeneous composition throughout, deriving color and texture from the materials of which the body is made. The colors and characteristics of the tile are determined by the materials used in the body, the method of manufacture, and the thermal treatment. **C 242,** C21

UNH—uranyl nitrate hexahydrate, $UO_2\ (NO_3)_2{\cdot}6H_2O$. **C 859,** C26

uniaxial state of stress—state of stress in which two of the three principal stresses are zero. (ISRM) **D 653,** D18

uniaxial (unconfined) compression—compression caused by the application of normal stress in a single direction. (ISRM) **D 653,** D18

unicellular rubber, n—See **expanded rubber.** **C 717,** C24

unidirectional conductor—see concentric-lay conductor. **B 354,** B01

uniform-chromaticity-scale diagram, *n*— chromaticity diagram on which all pairs of just-perceptibly different colors of equal luminance are represented by pairs of points separated by nearly equal distances. **E 284,** E12

Uniform Code Council, Inc., U.C.C., *n*—a standard setting organization originally chartered by the National Grocer's Association; it supervises the standards for the Universal Product Code, UPC. **F 1294,** F05

uniform color space, *n*—schematic arrangement of colors in space in which spatial intervals between points correspond to visual differences between colors represented by those points. **E 284,** E12

uniform corrosion—corrosion that proceeds at about the same rate over a metal surface. **G 15,** G01

uniform diffuser, n— see **Lambertian diffuser.** **E 284,** E12

uniform diffuse reflection, *n*—diffuse reflection in which the spatial distribution of the reflected radiation is such that the radiance or luminance is the same in all directions in which the radiation is reflected. **E 349,** E21

uniform diffuse transmission, *n*—diffuse transmission in which the spatial distribution of the transmitted radiation is such that the radiance or luminance is the same in all directions in which the radiation is transmitted. **E 349,** E21

uniform elongation, El_u, *n*—the elongation determined when the maximum load is reached. **E 6,** E28

uniformity coefficient—the ratio of the particle diameter corresponding to 60 % finer on the cumulative particle size distribution curve to the particle diameter corresponding to 10 % finer on the same distribution curve. **D 2652,** D28

uniformity index, *n*—*in fiber length testing of cotton,* the ratio between the mean length and the upper-half-mean length expressed as a percentage of the upper-half-mean length. **D 123,** D13

uniformity ratio, *n*—*in cotton length testing with the Fibrograph,* the ratio between two span lengths expressed as a percentage of the longer length. **D 123,** D13

uniformly graded—a quantitative definition of the particle size distribution of a soil which consists of a majority of particles being of the same approximate diameter. A granular material is considered uniformly graded when the uniformity coefficient is less than about five (see Test Method D 2487). Comparable to the geologic term well sorted. **D 653,** D18

uniform wear, *n*—a type of treadwear characterized by equal tread loss both from projection to projection and from point to point on a given projection, resulting in a smooth appearance of all parts of the tread pattern. **F 538,** F09

unilateral arrangement—recognition arrangement that covers the acceptance of one party's results by another party, (ISO Guide 2). **E 1187,** E36

unilaterally punched metal connector plate—metal connector plate with integral teeth projecting from plate in single direction perpendicular to plate surface area. **E 1807,** E06

uniplanar, *adj*—descriptor for illuminating and viewing geometry in which the axes of the illuminator and the receiver and the normal to the specimen surface are in the same plane; thus directional illumination or viewing is provided by one beam, or by two beams spaced 180° apart in azimuth. The number and angular distribution of the beams should be specified. **E 284,** E12

unique hue, *n*—perceived hue that cannot be described by a hue name other than its own. (E 1360) **E 284,** E12

unit—*measurement,* a precisely specified quantity in terms of which the magnitudes of other quantities of the same kind can be stated. **C 634,** E33

unit, *n*—an object on which a measurement or observation may be made. **E 456,** E11

unit, *n*—an individual, detachable assembly of chassis, body, and axles comprising part of a complete vehicle; units include tractors, full trailers, and semitrailers. **E 867,** E17

unit—see **dwelling unit**. **E 1605,** E06

unit—the smallest single portion of material received in any one lot (for example, a single roll of material). **E 1749,** E06

unit, *n*—a single object (for example, a vacuum cleaner) from the total population of like objects on which a measurement or observation may be made. **F 395,** F11

unitary hue, n—see **unique hue**. **E 284,** E12

unit bed-load discharge—bed-load discharge per unit width of river bed. Units are reported in mass of discharge per unit time per unit width, such as tons per day per foot. **D 4410,** D19

unit cell—a parallelepiped element of crystal structure, containing a certain number of atoms, the repetition of which through space will build up the complete crystal. **D 1695,** D01

unit cell—a parallelepiped element of crystal structure, containing a certain number of atoms, the repetition of which through space will build up the complete crystal. (See **lattice**.) **E 7,** E04

united millimetres (or united inches)—the sum of the dimensions in millimetres (or inches) of the length and width of a light of glass or panel. **C 717,** C24

unitization, *n*—the assembled group of containers or items in a single load that can be handled as a unit throughout the distribution system. **D 996,** D10

unitized load—See **load.** **D 996,** D10

unit length of instrument, L_c, *n*—in textile unevenness testing, the length of strand being measured between the sensing elements at any moment. **D 123,** D13

unit load—See **load.** **D 996,** D10

unit magnetic rating (1 MR), *n*—a calibrating standard with an MR of one is defined as containing 0.18 g of U.S. Institute of Standards and Technology standard sample No. 29 (a) or iron ore (magnetite) uniformly distributed over the space specified for a 10-g test specimen [14 mm (0.75 in.) diameter by 73 mm (2.875 in.) long], by dispersion in a magnetically inert material. A 10-g specimen has one unit magnetic rating when it produces a magnetic effect equivalent to that of 0.18 g of standard magnetite as described above. Conversely, a 10-g specimen producing a magnetic effect such that the resulting induced current in the magnetic analyzer galvanometer is *k* times that of the 0.18 g of standard magnetite as described above, would have a magnetic rating (MR) of *k*. **D 2946,** C17

unit of product—a piece of leather in the form in which it is purchased, such as a single hide, side, skin or part thereof; or a single fabricated-leather article in the form in which it is purchased, such as a counter, a pair of shoes, gasket, etc. **D 1517,** D31

unit of product, *n*—an entity of electrical insulating material or product for inspection to determine its classification as conforming or non-conforming. **D 1711,** D09

unit of structure (in crystals)—the unit cell, or the group of atoms associated with a unit cell. **E 7,** E04

unit package—See **package.** **D 996,** D10

unit sample, *n*— *in asbestos,* a sample drawn from one bag. **D 2946,** C17

units placed in usage—manufactured masonry units that have been installed in masonry. **C 1232,** C15

unit weight, *n*—*of aggregate,* mass per unit volume. (Deprecated term—use preferred term **bulk density**). **C 125,** C09

unit weight, γ (FL^{-3})—weight per unit volume (with this, and all subsequent unit-weight definitions, the use of the term weight means force).

dry unit weight (unit dry weight), γ_d, γ_e (FL^{-3})—the weight of soil or rock solids per unit of total volume of soil or rock mass.

effective unit weight, γ_e (FL^{-3})—that unit weight of a soil or rock which, when multiplied by the height of the overlying column of soil or rock, yields the effective pressure due to the weight of the overburden.

maximum unit weight, γ_{max}(FL^{-3})—the dry unit weight defined by the peak of a compaction curve.

saturated unit weight, γ_G, γ_{sat}(FL^{-3})—the wet unit weight of a soil mass when saturated.

submerged unit weight (buoyant unit weight), γ_m, γ', γ_{sub} (FL^{-3})—the weight of the solids in air minus the weight of water displaced by the solids per unit of volume of soil or rock mass; the saturated unit weight minus the unit weight of water.

unit weight of water, γ_w (FL^{-3})—the weight per unit volume of water; nominally equal to 62.4 lb/ft^3 or 1 g/cmt^3.

wet unit weight (mass unit weight), γ_m, γ_{wet} (FL^{-3})—the weight (solids plus water) per unit of total volume of soil or rock mass, irrespective of the degree of saturation.

zero air voids unit weight, γ_z, γ_s (FL^{-3})—the weight of solids per unit volume of a saturated soil or rock mass. **D 653,** D18

univariant equilibrium—a stable state among a number of phases equal to one more than the number of components, that is, having one degree of freedom. **E 7,** E04

Universal Product Code, UPC, *n*—a bar code symbology specified by the Uniform Code Council. **F 1294,** F05

universe (population), *n*—the totality of the set of items, units, or measurements, etc., real or conceptual, that is under consideration.

NOTE—This definition of universe is being revised to incorporate the concept of including one or more populations. *Use with caution.*

E 456, E11

unloading modulus—slope of the tangent to the unloading stress-strain curve at a given stress value. (ISRM) **D 653,** D18

unmated mechanical transfer paper—a sheet that has a clean pigmented coating that is physically transferred to a second sheet upon the application of pressure.

F 549, F05

unmeasured-sediment discharge—the difference between the total sediment discharge and the measured suspended-sediment discharge. **D 4410,** D19

unpiloted ignition, *n*—ignition caused by one or more sources of energy without the presence of a pilot source of ignition (compare **piloted ignition, spontaneous ignition**). **E 176,** E05

unprocessed municipal solid waste, *n*—municipal solid waste in its as-discarded form and that has not been size-reduced, separated, or otherwise processed. **D 5681,** D34

unrelated color, *n*—color perceived to belong to an area seen in isolation from other colors. **E 284,** E12

unsampled depth—the unsampled part of the sampling vertical; usually within 0.8 to 0.15 m of the stream bed, depending on the kind of suspended-sediment sampler used. **D 4410,** D19

unsampled zone—the unsampled part of the sampling vertical, usually assumed to be 90 to 150 mm above the stream bed, depending on the kind of sampler used. **D 4410,** D19

unsaponifiable matter—see **fatty matter, unsaponifiable.** **D 459,** D12

unsaponifiable matter, *n*—the total amount of non-acidic organic material, both free and combined, present in naval stores products such as rosin, tall oil, and their derivatives after saponification. Unsaponifiable matter is composed primarily of alcohols, sterols, aldehydes, and hydrocarbons. For example, it is material that will not form a soluble soap when refluxed with alcoholic potassium hydroxide. **D 804,** D01

unsaponified matter—see **fatty matter, unsaponified.** **D 459,** D12

unscheduled cessation—a rare, unusual, or irregular stopping of an amusement ride operation, intentional or otherwise, that is considered abnormal or potentially hazardous in nature, or both, due to its cause, method, or result. **F 747,** F24

unstable brittle crack extension [L]—an abrupt crack extension occurring with or without prior stable crack extension in a standard fracture test specimen under crosshead or clip gage displacement control. **E 1823,** E08

unsupported film adhesive—See **adhesive, unsupported film.** **E 1749,** E06

unsupported gasket member—a gasket member joining infill materials without being held in place by a supporting frame member. **C 717,** C24

unsupported needled felt, *n*—a needled felt that is composed entirely of fibers physically interlocked and reoriented with, and of themselves without an interlay, scrim, or foundation of knitted, stitched, bonded, or extruded structure. **D 123,** D13

untreated, *n*—a descriptive term for glass fiber yarns having no applied chemicals or coatings, other than the minimal lubricant or binder used to control intra-fiber abrasion. **D 123,** D13

UPC bar code—bar code symbology, Universal Product Code, similar to **EAN** bar code; (often used for marking products to be sold in commercial operations such as supermarkets). **F 1156,** F12

updraw—the process of continuously drawing glass of various cross sections, such as cane or tubing, by a method similar to the Fourcault process. **C 162,** C14

upholstered, *adj*—covered with material (as fabric or padding) to provide a soft surface. **E 176,** E05

upholstered, *n*—covered with material (as fabric or padding) to provide a soft surface. **E 176,** E05

upholstered, *adj*—covered with material (as fabric or padding) to provide a soft surface. **E 176,** E05

upholstered furniture, *n*—furniture covered with such materials as textiles or leather, and generally with padding or cushions, or both. **D 123,** D13

upholstered furniture, *n*—a unit of interior furnishing that (*1*) contains any surface that is covered, in whole or in part, with a fabric or related upholstery cover material, (*2*) contains upholstery material, and (*3*) is intended or promoted for sitting or reclining upon. **E 176,** E05

upholstery cleaning instructions, *n*—any of the various letter codes supplied by the fabric manufacturer to provide acceptable cleaning methods. **D 5253,** D13

upholstery cover material, *n*—the outermost layer of fabric or related material used to enclose the main support system or upholstery materials, or both, used in the furniture item. **E 176,** E05

upholstery cover material, *n*—the outmost layer of fabric or related material used to enclose the main support system or upholstery materials, or both, used in the furniture item. **E 176,** E05

upholstery fabric, *n*—the exterior fabric covering applied to a furniture unit. **D 123,** D13

upholstery leather—a general term for leathers processed for use in furniture, airplanes, buses, and automobiles. The staple raw material in this country consists of spready cattlehides, split at least once and in many cases two or three times. The top or grain cuts go into the higher grades and the splits into the lower grades. **D 1517,** D31

upholstery material, *n*—the padding, stuffing, or filling material used in a furniture item, which may be either loose or attached, enclosed by an upholstery cover material, or located between the upholstery cover material and support system, if present. (This includes, but is not limited to, material, such as foams, cotton batting, polyester fiberfill, bonded cellulose, or down). **E 176,** E05

upholstery nail—bright, regular-stock-steel, two-piece nails with extra-large specially formed head and medium diamond or needle point. **F 547,** F16

uplift—the upward water pressure on a structure.

	Symbol	Unit
unit symbol	u	FL^{-2}
total symbol	U	F or FL^{-1}

D 653, D18

uplift—the hydrostatic force of water exerted on or underneath a structure, tending to cause a displacement of the structure. (ISRM) **D 653,** D18

uplift—*in grouting*, vertical displacement of a formation due to grout injection. **D 653,** D18

upper—all of the upper parts of a shoe stitched together and ready for lasting and bottoming. It includes both the outside and lining of the shoe. **F 869,** F08

upper, *n*—that area of the footwear above the sole. **F 1494,** F23

upper, *n*—the parts of a shoe or boot that are above the sole. **F 1646,** F13

upper-arm girth, *n*—*in body measurements*, the maximum circumference of the arm usually midway between the elbow and the shoulder joint. **D 123,** D13

upper-arm length, *n*—*in body measurements*, with the arm bent, the distance from the shoulder joint along the outside of the arm to the prominence of the elbow. **D 123,** D13

upper capability index (CPU), *n*—the difference between the upper specification limit and the sample mean divided by three times the sample standard deviation.

$$CPU = \frac{(USL - \bar{x})}{3 \times s}$$

A 644, A04

upper control limit (UCL), *n*—control limit for points above the central line. **E 456,** E11

upper critical point—in a phase diagram, a specific value of composition, temperature and pressure, or combinations thereof, occurring as a maximum in temperature, or pressure, for the coexistence of two or more conjugate phases and at which the conjugate phases become identical. **E 7,** E04

upper cut zone, *n*— *in the testing of foot protective devices,* the area starting at the top of the area of protection on the footwear and extending downward to include the entire upper, excluding the toe area cut zone. **F 1494,** F23

upper-half-mean length, *n*—*in fiber length testing of cotton,* the mean length by number, of the longer one half of the fibers by weight. **D 123,** D13

upper leather—the shoe leather that covers the upper part of the foot. It is made principally from calfskins, goatskins, cattlehides, horsehides, and other classes of animal skins. **D 1517,** D31

upper limit log normal distribution, (ULLN)—drop size distribution described by the following equation:

$$f_v(D) = \{1/\sqrt{\pi}\} \int_{-\infty}^{del \times log[A \times D/(D_{max} - D)]} exp(-t^2)\, dt$$

where:

$f_v(D)$	=	the volume fraction of liquid in drops of diameter less than D,
D_{max}	=	the upper bound of drop diameters, and
de-l and *A*	=	ULLN parameters determining dispersion and skewness.

E 1620, E29

upper limit of flammability or upper flammable limit (UFL)—the maximum concentration of a combustible substance that is capable of propagating a flame through a homogeneous mixture of the combustible and a gaseous oxidizer under the specified conditions of test. **E 1445,** E27

upper punch—the member of the compacting tool set or die that closes the die and forms the top of the part being produced. **B 243,** B09

upper quartile length, *n*—*in testing of cotton fibers,* that length which is exceeded by 25 % of the fibers, by weight, in the test specimen. **D 123,** D13

upper range-value, *n*—the highest quantity that an instrument is adjusted to measure. **E 344,** E20

upper specification limit (USL), *n*—the highest specified value. **A 644,** A04

upper tolerance limit (UTL) (upper specification limit), *n*—a tolerance limit applicable to the upper conformance boundary for an individual unit of a manufacturing or service operation. **E 456,** E11

upper yield strength, *UYS* [FL^{-2}], *n*—in a uniaxial test, the first stress maximum (stress at first zero slope) associated with discontinuous yielding at or near the onset of plastic deformation. **E 6,** E28

upright broiler—a heavy duty freestanding piece of equipment with a high input rate and production capacity among overfired broilers. See **broiler, overfired**. **F 1827,** F26

upright vacuum cleaner, *n*—a self-standing, floor-supported, portable vacuum cleaner which has the floor nozzle directly connected to the cleaner and is designed for normal-duty cleaning of household dirt. In use, the cleaner is guided over the floor by means of a handle attached to the cleaner housing. The cleaner contains a driven agitator to assist in dirt removal on floor surfaces. The cleaner may have attachments or provisions, or both, for both floor and above-the-floor cleaning. **F 395,** F11

UPS—acronym for "ultra-violet photoelectron spectroscopy," similar to XPS but employing characteristic ultra-violet radiation. **E 673,** E42

upset head—metal slightly upset, with deformed portion serving as head. **F 547,** F16

uptake—acquisition of a substance from the environment by an organism as a result of any active or passive process. **E 943,** E47

urania—oxides of uranium, generally UO_2. **C 859,** C26

urban area, *n*—individual agencies may have a political boundary definition of urban, which would be the definition for traffic monitoring purposes; in the absence of such an individual agency definition, an urban area is defined as a community with a population of 5000 persons or more. **E 867,** E17

urea-formaldehyde adhesive, *n*—(1) an aqueous colloidal dispersion of urea-formaldehyde polymer which may contain modifiers and secondary binders to provide specific adhesive properties, (2) a type of adhesive, based on a dry urea-formaldehyde polymer and water. **D 907,** D14

urea plastics—plastics based on resins made by the condensation of urea and aldehydes. **D 883,** D20

urea resin—under **resin, synthetic,** see *urea resin.* **D 16,** D01

urethane coatings—coatings based upon vehicles containing a minimum of 10 percent by weight (nonvolatile vehicle basis) of a polyisocyanate monomer reacted in such a manner as to yield polymers containing any ratio, proportion or combination of urethane linkages, active isocyanate groups, or polyisocyanate monomer. The reaction products may contain excess isocyanate groups available for further reaction at time of application or may contain essentially no free isocyanate as supplied.

Type I, one-package prereacted—urethane coatings characterized by the absence of any significant quantity of free isocyanate groups. They are usually the reaction product of a polyisocyanate and a polyhydric alcohol ester of vegetable oil acids and are hardened with the aid of metallic soap driers.

Type II, one-package moisture cured—urethane coatings characterized by the presence of free isocyanate groups and capable of conversion to useful films by the reaction of these isocyanate groups with ambient moisture.

Type III, one-package heat cured—urethane coatings that dry on cure by thermal release of blocking agents and regeneration of active isocyanate groups that subsequently react with substances containing active hydrogen groups.

Type IV, two-package catalyst—urethane coatings that comprise systems wherein one package contains a prepolymer or adduct having free isocyanate groups capable of forming useful films by combining with a relatively small quantity of catalyst, accelerator, or crosslinking agent such as a monomeric polyol or polyamine contained in a second package. This type has limited pot-life after the two components are mixed.

Type V, two-package polyol—urethane coatings that comprise systems wherein one package contains a prepolymer or adduct or other polyisocyanate capable of forming useful films by combining with a substantial quantity of a second package containing a resin having active hydrogen groups with or without the benefit of catalyst. This type has limited pot-life after the two components are mixed.

Type VI, one-package nonreactive lacquer—urethane solution coatings characterized by the absence of any significant quantity of free isocyanate or other functional groups. Such coatings convert to solid films primarily by solvent evaporation. **D 16,** D01

urethane foam, (flexible), *n*—an open cell, elastomeric material made by the addition reaction of a polyol with a polyisocyanate. **D 1566,** D11

urethane plastic—a plastic based on polymers in which the repeated structural units in the chains are of the urethane type, or on copolymers in which urethane and other types of repeated structural units are present in the chains. **F 1251,** F04

urethane plastics—plastics based on polymers in which the repeated structural units in the chains are of the urethane type, or on copolymers in which urethane and other types of repeated structural units are present in the chains. (ISO) **D 883,** D20

urn—See **railing systems.** **E 631,** E06

urn, *n*—an ornamental vase used as a finial. **E 1481,** E06

uronic acids—oxidation products of the sugars in which a primary alcohol group has been oxidized to a carboxyl without changing the reducing functional group. **D 1695,** D01

usable area—See **space categories**. **E 1480,** E06

use— of a facility—See **facility use**. **E 1480,** E06

used oil, *n—in petroleum product recycling*, oil whose characteristics have changed since being originally manufactured, and which is suitable for recycling (see also **waste oil**). **D 4175,** D02

used oil, *n*—any oil that has been in a piece of equipment (for example, an engine, gearbox, transformer, or turbine), whether operated or not (see also **used oil,** *in petroleum product recycling*). **D 4175,** D02

useful ion yield—*SIMS,* the ratio of the number of ions of a particular isotope detected to the total number of atoms and ions of the same element sputtered. **E 673,** E42

useful life—the period of time over which an investment is considered to meet its original objective. **E 833** **E 631,** E06

useful life—the period of time over which an investment is considered to meet its original objective. **E 833,** E06

USEPA—U.S. Environmental Protection Agency. **D 6161,** D19

use-surface, *n—for pile yarn floor covering*, that part of a textile floor covering directly exposed to traffic. **D 123,** D13

utility core—Use **core module.** **E 631,** E06

utility function—a function that shows how utility (that is, satisfaction) varies with money or income. **E 833,** E06

utility trailer, *n*—a full trailer, typically used for light duty hauling and towed by passenger vehicles or light trucks. **E 867,** E17

utility vacuum cleaner, *n*—a portable, floor-supported vacuum cleaner having a nozzle separated from the cleaner housing by a hose. The cleaner is designed to be used in places other than normal living areas such as utility rooms, basements, workshops, and garages. Besides household dirt, it is designed to pick up various types of larger debris and may be designed for liquid pickup. In use, only the nozzle is guided over the surface area to be cleaned. The cleaner may have detachable nozzles, attachments, and wands for both floor and above-the-floor cleaning. **F 395,** F11

U-type furnace—a continuous furnace wherein the ware travels in a U-shaped path. **C 286,** B08

UV absorber, *n*—a compounding material that, through its ability to absorb ultraviolet radiation and render it harmless, retards the deterioration caused by sunlight and other UV light sources. **D 1566,** D11

U-value—the total heat transfer coefficient of a window at steady state and standardized conditions. **C 162,** C14

V

vacuum, *v*—to clean using an electrically powered machine to create suction in order to remove loose, particulate soil and lint. **D 123,** D13

vacuum—a degree of rarefaction below atmospheric pressure: negative pressure. **D 653,** D18

vacuum, *n*—pressure below atmospheric pressure. **D 3053,** D24

vacuum—in vacuum technology a given space filled with gas at pressures below atmospheric pressure. **E 1316,** E07

vacuum-and-blow process—a bottle manufacturing process whereby glass is gathered by vacuum and subsequently blown. **C 162,** C14

vacuum bag molding—a process in which an adhesive or composite assembly is cured under pressure generated by drawing a vacuum in the space between the lay-up and a flexible sheet placed over it and sealed at the edge. (Compare with **bag molding**.) **E 1749,** E06

vacuum cassette—a flexible light-tight container that, when operated under a vacuum, holds film and screen in intimate contact during a radiographic exposure. **E 1316,** E07

vacuum cassette—a light-tight device having a flexible entrance window, which when operated under a vacuum, holds the film and conversion screen in intimate contact during exposure. **E 1316,** E07

vacuum chamber, *n*—a device that allows a vacuum to be applied to a surface. **D 4439,** D35

vacuum cleaner, *n*—a system or device that removes foreign material from surfaces by means of the air flow caused by subatmospheric pressure, having an intake intended to be moved in proximity to the surface, a means of separating the material from the air, and a receptacle for collecting the separated material. The inlet may be fixed or attached to other equipment and provision is made for removing collected material. **F 395,** F11

vacuum distillation—the separation of two or more liquids under reduced vapor pressure; reduces the boiling points of the liquids being separated. **E 1126,** E48

vacuum distillation—to affect separation of two or more liquids under reduced pressure operation of a distillation column. Vacuum reduces the boiling points of the liquids being separated. **E 1705,** E48

vacuum distillation—the separation of two or more liquids under reduced vapor pressure; reduces the boiling points of liquids being separate. **E 1705,** E48

vacuum forming—a forming process in which a heated plastic sheet is drawn against the mold surface by evacuating the air between it and the mold. **D 883,** D20

vacuum level—the potential of the vacuum space at a sufficiently large distance outside the specimen such that electric fields caused by different work functions of different parts of the surface are zero or extremely small. **E 673,** E42

vacuum level referencing—*XPS*, a method of establishing the binding energy scale for a particular specimen by assigning the kinetic energy corresponding to the vacuum level as the point of zero binding energy. See also **vacuum level**. **E 673,** E42

vacuum lock—See **air lock**. **E 7,** E04

vacuum packaging—See **packaging.** **D 996,** D10

vacuum plated button, *n*—a button that is flash metal coated in vacuum chambers and subsequently colored to simulate other metal finishes. **D 123,** D13

vacuum testing—*(1)* a method of testing for leaks in which the object under test is evacuated and the tracer gas applied to the outside surface of the object; *(2)* a leak-testing procedure in which the enclosure under examination is evacuated, the tracer gas applied to the outside surface of the enclosure, and the gas detected after entering the enclosure. **E 1316,** E07

vadose zone—the hydrogeological region extending from the soil surface to the top of the principle water table; commonly referred to as the "unsaturated zone" or "zone of aeration". These alternate names are inadequate as they do not take into account locally saturated regions above the principle water table (for example, perched water zones). **D 653,** D18

valence band spectrum—*XPS,* photoelectron energy distribution arising from the less tightly bound electrons involved in the chemical bonds of the specimen material. **E 673,** E42

validation, *n*—proof, with reference materials or materials traceable to them, that a method is acceptable for all elements in its scope. **E 135,** E01

validation—the quality assurance evaluation of an item of equipment or overall process wherein the equipment or process, or both, is challenged to perform under the "worst case" conditions of process variables and applicable micro-organism contamination to meet preestablished acceptance criteria. **E 1705,** E48

validation—confirmation of the results of the verification process. **F 1156,** F12

validation count, *n*—measurement of traffic characteristics for acceptance testing of a separate traffic measurement on the same road segment. **E 867,** E17

valley—*in fatigue loading*, the occurrence where the first derivative of the load-time history changes from negative to positive sign; (also known as *trough*); the point of minimum load in constant amplitude loading. **E 1823,** E08

valleys—edge irregularities which result in an indentation in the side of a strike. **F 149,** F05

valley trenching—gully erosion occurring in flood plains. **D 4410,** D19

value, n—see **Munsell value.** **E 284,** E12

value analysis team leader (VATL)—the individual who facilitates the **value analysis**. **E 833,** E06

value analysis (VA)—the procedure for developing and evaluating alternatives to a proposed economical design that best fulfills the needs and requirements of the user/owner of the building. (Also referred to as *value engineering*.) **E 833,** E06

valve—a mechanism for discharging products from aerosol-type dispensers. **D 3064,** D10

valve leather—see **hydraulic leathers.** **D 1517,** D31

vamp—the lower part of a shoe upper which is attached to the sole or welting. **F 869,** F08

van der Waals adsorption— see **physical adsorption.** **D 2652,** D28

vane, *n*—*in feathers*, the section that consists of a solid, stiff collection of barbs, as distinguished from the section near the quill point that has soft, fluffy barbs. **D 123,** D13

vane shear test—an in-place shear test in which a rod with thin radial vanes at the end is forced into the soil and the resistance to rotation of the rod is determined. **D 653,** D18

Vant Hoff's law—equilibrium shifts with increasing temperature so as to absorb heat, or with decreasing temperature so as to liberate heat. **E 7,** E04

vapor, *n*—the gaseous phase of matter that normally exists in a liquid or solid state.

vapor concentration—See **concentration.**

vapor pressure of moist air (e), n—the product of the mole fraction of water vapor x_v and the total pressure p at temperature t:

$$e=p\,x_v$$

D 1356, D22

vapor—the gaseous form of substances that are normally in the solid or liquid state, and that can be changed to these states either by increasing the pressure or decreasing the temperature. **E 41,** G03

vapor barrier—See **water vapor retarder (barrier)**. **C 168,** C16

vapor barrier—See **vapor retarder**. **D 1079,** D08

vapor-deposited replica— See **replica**. **E 7,** E04

vapor deposition:

chemical—ion, induced by heat or gaseous reduction of a vapor condensing on the substrate.

physical—a process for depositing a coating by evaporating and subsequently condensing an element or compound, usually in a high vacuum. **B 374,** B08

vapor drift—the dispersion of vaporized chemical to the atmosphere and areas surrounding the target area during and following application. **E 1102,** E35

vapor fixing—the act of making a developed electrostatic image permanent by submitting the image-forming toner to the vapor of a solvent for the toner. **F 335,** F05

vaporization curve—in a P-T diagram the locus of pressure and temperature values at which a congruent liquid is in equilibrium with its vapor. **E 7,** E04

vaporization point—at a stated pressure, the temperature at which a congruent liquid is in equilibrium with its vapor, or, at a stated temperature, the pressure at which the same event occurs. **E 7,** E04

vaporization point—at a given pressure, the temperature at which the vapor pressure of the liquid is equal to the external pressure or at a stated temperature the external pressure on the liquid that is equal to its vapor pressure. **E 1142,** E37

vaporization temperature—at a given pressure, the temperature at which the vapor pressure of the liquid is equal to the external pressure. **E 1142,** E37

vapor-liquid ratio, *n— of a liquid,* the ratio, at a specified temperature and pressure, of the volume of vapor in equilibrium with liquid to the volume of liquid sample charged, at 32°F (0°C). **D 4175,** D02

vapor migration—the movement of water vapor from a region of high vapor pressure to a region of lower vapor pressure. **D 1079,** D08

vapor pressure—the pressure exerted by the vapor of a solid or liquid when in equilibrium with the solid or liquid. **C 859,** C26

vapor pressure, *n*—the pressure exerted by the vapor of a liquid when in equilibrium with the liquid. **D 4175,** D02

vapor pressure—in a binary system, that pressure at which a liquid and its vapor are in equilibrium at a definite temperature and total pressure. **E 1142,** E37

vapor pressure—the pressure exerted by the vapor of a solid or liquid when in equilibrium with the solid or liquid. **E 1316,** E07

vapor pressure of moist air, *e*—the product of the mole fraction of water vapor x_v and the total pressure p at temperature t:

$$e = px_v$$

If the moist air is assumed to behave in accordance with the ideal gas laws, then the vapor pressure e is identically equal to the partial pressure of the water vapor in the moist air. **D 4023,** D22

vapor retarder—a layer of material or a laminate used to appreciably reduce the flow of water vapor into the roofing system. **D 1079,** D08

var—the unit of reactive (quadrature) power. One var is the product of one volt and one ampere in phase quadrature. **A 340,** A06

variability—also called "variance"; the number of degrees of freedom of a heterogeneous phase equilibrium. **E 7,** E04

variable, *n*—a quantity to which any of the values in a given set may be assigned. (See **parameter** and **variate**) **D 123,** D13

variable amplitude loading— See **spectrum loading**. **E 1823,** E08

variable-angle internal reflection element— an internal reflection element which can be operated over a range of angles of incidence. **E 131,** E13

variable-area nozzle—a pressure atomizer in which the open area of one or more liquid flow passages may be varied so as to control the discharge characteristics. **E 1620,** E29

variable-orifice pintle nozzle—see **variable-orifice poppet nozzle**. **E 1620,** E29

variable-orifice poppet nozzle—a variable-area nozzle in which a moveable poppet determines the annular orifice area. **E 1620,** E29

variables data, *n*—measurements which vary and may take any of a specified set of numerical values. **D 123,** D13

variables data, *n*—measurements which vary and may take any of a specified set of numerical values. **E 456,** E11

variables, method of, *n*—measurement of quality by the method of variables consists of measuring and recording the numerical magnitude of a quality characteristic for each of the units in the group under consideration.

NOTE—This involves reference to a continuous scale of some kind. **E 456,** E11

variable weight line—a line drawn such that it is of different line widths (thicknesses) at various points along its length. **F 1156,** F12

variance, *n*—the mean square of deviations (or errors) of a set of observations; the sum of squared deviations (or errors) of individual observations with respect to their arithmetic mean divided by the number of observations less one (degrees of freedom); the square of the standard deviation (or standard error).

random variance of increment collection (unit variance), S_r^2, *n* —the theoretical variance calculated for a uniformly mixed lot and extrapolated to 0.5-kg (1-lb) increment size.

segregation variance of increment collection, S_s^2, *n*—the variance caused by nonrandom distribution of ash content or other constituent in the lot. For a method of estimating this variance, see Annex A1 of Test Methods D 2234.

total variance, S_o^2, *n*—the overall variance resulting from collecting single increments and including division and analysis of the single increments.

variance of analysis, S_a^2, *n*—the variance caused by chance errors (deviations) of analysis.

variance of division, S_d^2, *n*—the variance caused by chance errors (deviations) of sample division.

variance of division and analysis, S_{da}^2, *n*—the variance caused by the combined chance errors of division and analysis. **D 121,** D05

variance, σ^2, *n—of population,* a measure of the dispersion of members of the population expressed as a function of the sum of the squared deviations from the population mean. **D 123,** D13

variance—See **variability**. **E 7,** E04

variance, *n*—a measure of the squared dispersion of observed values or measurements expressed as a function of the sum of the squared deviations from the population mean or sample average.

NOTE—The sample variance, or variance of a sample of n observed values, is computed as $s^2 = [1/(n - 1)][\Sigma(y_i - \bar{y})^2]$. The sample standard deviation s is the positive square root of the sample variance. The population variance $\sigma^2 = \int_R (y - \mu)^2 f(y)dy$, where R is the region over which the random variable y is defined, and where $f(y)$ is the probability density function and μ is the population mean of y. The population standard deviation (σ) is the positive square root of the population variance. **E 456,** E11

variance—the mean square of deviations (or errors) of a set of observations; the sum of square deviations (or errors) of individual observations with respect to their arithmetic mean divided by the number of observations less one (degrees of freedom); the square of the standard deviation (or standard error). **E 856,** D34

variance—a measure of the dispersion of a series of results around their average. It is the sum of the squares of the individual deviations from the average of the results, divided by the number of results minus one. **E 1547,** E15

variance, s **,** *n—of a sample,* a measure of the dispersion of variates observed in a sample expressed as a function of the squared deviations from the sample average. **D 123,** D13

variance (sample)—a measure of the dispersion of a set of results. Variance is the sum of the squares of the individual deviations from the sample mean divided by one less than the number of results involved. It may be expressed as

$$S^2 = \frac{\Sigma (x_i - \bar{x})^2}{n - 1}$$

C 859, C26

variate, *n*—a measured value that includes a random error of measurement; a variable with which a probability distribution is associated. (See also **variable** and **attribute data.**) **D 123,** D13

variation in sawing—a deviation from the anticipated line of cut in sawing rough lumber or other rough wood surfaces. **D 9,** D07

varistor, *n*—a material having an electrical resistance that is sensitive to changes in applied voltage. **C 1145,** C28

varnish—a liquid composition that is converted to a transparent or translucent solid film after application as a thin layer.

bituminous varnish—a dark-colored varnish containing bituminous ingredients. The varnish may be either of the oil or spirit type.

oil varnish—a varnish that contains resin and drying oil as the basic film-forming ingredients and is converted to a solid film primarily by chemical reaction.

spar varnish— a varnish for exterior surfaces. The name originated from its use on spars of ships.

spirit varnish— a varnish that is converted to a solid film primarily by solvent evaporation. **D 16,** D01

varnish, electrical insulating, *n*—a liquid resin system that is applied to and cured on electrical components providing electrical, mechanical, and environmental protection. **D 1711,** D09

varved clay—alternating thin layers of silt (or fine sand) and clay formed by variations in sedimentation during the various seasons of the year, often exhibiting contrasting colors when partially dried. **D 653,** D18

vat dyeing—see **drum dyeing.** **D 1517,** D31

V-board—see **corrugated fiberboard.** **D 996,** D10

"V" configuration—boom positioned in a "V" shape. **F 818,** F20

"V"-crown—staple crown with inverted "V" cross section on staple; used in spring-up work in furniture and bedding. **F 592,** F16

V-cuts, *n—in mica,* edge cuts in the form of a "V" with an included angle of 120° or less. **D 1711,** D09

1/v detector—a detector made of a nuclide whose cross section for a particular reaction of interest (for example, (n, γ), (n, p), (n,α)) varies as the reciprocal of the velocity of the impinging neutrons. The 1/v property of the detector is valid only over a restricted energy range and usually below 10 keV (see **2200 m/s cross section**). **E 170,** E10

veal—a large calfskin, almost as large as a kip. **D 1517,** D31

vector—a fragment of DNA usually containing an origin of replication that is engineered to accept a foreign piece of DNA. **E 1705,** E48

vector—mathematical term completely specified by a magnitude and a direction; a one-dimensional array. **F 1156,** F12

vee path—the angle-beam path in materials starting at the search-unit examination surface, through the material to the reflecting surface, continuing to the examination surface in front of the search unit, and reflection back along the same path to the search unit. The path is usually shaped like the letter V. **E 1316,** E07

vegetable matter, *n—in wool top,* the pieces of burrs, seeds, shive, leaves, twigs, and grasses which have escaped removal in processing, also foreign vegetable fibers such as hemp, sisal, etc., if present. **D 123,** D13

vegetable matter base, *n—in raw wool,* oven-dried scoured burrs, seeds, twigs, leaves, and grasses, free of mineral matter and alcohol-extractable matter. **D 123,** D13

vegetable matter present, *n—in raw wool,* the weight of vegetable matter base present in the raw wool, adjusted to a moisture content of 12 %, an alcohol-extractives content of 1.5 %, and a mineral matter content of 0.5 %. **D 123,** D13

vegetable oil—oil extracted from seeds; typically those of corn, cotton, peanut, rapeseed, sunflower, canola, or soybean. **E 1519,** E35

vegetable oil concentrate—an emulsifiable vegetable oil product containing 5 to 20 % w/w surfactant and a minimum of 80 % w/w vegetable oil. **E 1519,** E35

vegetable peeling machine, electric—machine consisting of the following: a cylinder having an abrasive or ribbed wall; an abrasive disk with lobes; a peel trap (when waste disposer is not specified); a waste outlet, a water inlet and air-gap type sprayer; and a cylinder cover. **F 1827,** F26

vegetable tanning—the conversion of rawhides into leather by treating with water solutions of tannin extracted from materials of vegetable origin. **D 1517,** D31

vehicle—the liquid portion of a paint or printing ink. Anything that is dissolved in the liquid portion of a paint or printing ink is a part of the vehicle. **D 16,** D01

vehicle, *n*—an assembly of one or more units coupled together for travel on a highway; vehicles include one powered unit and may include one or more unpowered full trailer or semitrailer units. **E 867,** E17

vehicle—a liquid medium for the suspension of magnetic particles. **E 1316,** E07

vehicle—a solvent, or a suspension medium, for use in mixing inks or paints. **F 1156,** F12

vehicle classification, *n*—the identification, summarization and reporting of traffic volume by vehicle configuration. **E 867,** E17

vehicle classification, *n*—the process of characterizing vehicles by axle or unit configuration. **E 867,** E17

vehicle miles traveled, (VMT), *n*—average Sunday through Saturday vehicle movement on a specific road segment multiplied by the length of the road segment, reported in the form of daily and annual VMT; this statistic is also referred to as "vehicle miles of travel."

E 867, E17

vehicle operation mode—the manner of operation of an emergency medical vehicle, involving the use of warning devices and the exercise of driving privileges legally allowed for emergency vehicles. **F 1177,** F30

veiny—appearance of leather characterized by many clearly visible blood vessels, either closed or cut open by buffing or shaving operations. **D 1517,** D31

Vello process—a process for continuously drawing glass tubing (or cane) in which glass is fed downward to the draw through an annular orifice. **C 162,** C14

vellum—see **parchment.** **D 1517,** D31

vellum glaze—See *vellum glaze* under **glaze.** **C 242,** C21

velocity head—the square of the average velocity divided by twice the acceleration due to gravity. **D 4410,** D19

velocity pressure—See **pressure.** **D 1356,** D22

veneer—a nonload-bearing facing of stone attached to a backing for the purpose of ornamentation, protection, or insulation.
C 119, C18

veneer—the decorative surface of an asbestos-cement shingle or sheet, usually pigmented or granuled for color. **C 460,** C17

veneer, *n*—the decorative surface of a shingle or sheet, usually pigmented or granuled for color. **C 1154,** C17

veneer, *n*—thin sheets of wood, produced on a rotary lathe, or by slicing, or sawing. **D 996,** D10

veneer—thin sheets of wood from which plywood is made; also referred to as plies in the glued panel.

figured veneer—veneer containing irregular grain formations that add to its value for furniture panel faces and other decorative uses. Various figures are referred to as rift-cut, comb-grained, stripe, rope, mottle, fiddle-back, cross-fire, quartered, crossbar, curley, blister, birds' eye, feather, crotch, ribbon, stump, burl, etc.

half-round veneer—a manner of cutting veneer to bring out a certain beauty of figure. The flitch is mounted on a stay log and is cut on a lathe. It differs from rotary-cut veneer in that the flitch is cut with a wider sweep than when mounted at the lathe center, and the center of the tree is not near the center of rotation.

plain sliced veneer—veneer sliced approximately tangent to the growth rings; also termed "flat cut."

quartered veneer—veneer produced by slicing or sawing a log on a plane approximately perpendicular to the growth rings. This may bring out a certain figure resulting from the presence of rays which are especially conspicuous in oak.

rift cut veneer—veneer cut at approximately 45 deg to the rays.

rotary-cut veneer—veneer cut in a continuous ribbon by centering the entire log or bolt in a lathe and rotating it against a knife.

sawn veneer—veneer produced by sawing.

sliced veneer—veneer that is sliced off by moving a log, bolt, or flitch laterally against a knife or vice versa. **D 1038,** D07

veneer—the decorative surface of an asbestos-cement shingle or sheet, usually pigmented or granuled for enhanced appearance.
D 2946, C17

veneer, *n*—a thin layer of rubber covering the surface of the tire sidewall. **F 538,** F09

veneer-box nail—coated, regular-stock-steel, 1 ½ by 0.080-in. nail with flat $^{15}/_{64}$ or ¼-in. head and medium needle point.
F 547, F16

veneer plaster, *n*—a calcined gypsum plaster specially manufactured to provide high strength, hardness, and abrasion resistance when applied in thin coats over a gypsum base for veneer plasters
C 11, C11

vent—an opening designed to convey water vapor or other gas from inside a building or a building component to the atmosphere.
D 1079, D08

vented cap—a cap with a small hole that is installed on top of the riser. **D 653,** D18

vent hole—*in grouting*, a hole drilled to allow the escape of air and water and also used to monitor the flow of grout. **D 653,** D18

vent hole—an opening for the escape of gases of relief of pressure, often required in fabricated, immersion-coated, or hot-dip galvanized steel members. Compare **drainage hole, weephole.**
E 631, E06

ventilator—See **windows and doors.** **E 631,** E06

ventilator—see **exhaust hood**. **F 1827,** F26

vent pipe—*in grouting*, a small-diameter pipe used to permit the escape of air, water, or diluted grout from a formation.
D 653, D18

verification, *n*—the act or process of verifying. **D 123,** D13

verification—confirmation by examination and provision of evidence that an instrument, material, reference, or standard is in conformance with a specification. **E 7,** E04

verification, *n*—confirmation that the calibration or standardization of an instrument is acceptable. **E 135,** E01

verification—checking or testing to assure conformance with the specification, (Test Method E 384, Committee E-4; Test Method E 10, Committee E-28; Test method E 18, Committee E-28; Test Method E 92, Committee E-28). **E 1187,** E36

verification—see **calibrate**. **E 1605,** E06

verification—the determination of whether a document or article is genuine, or if it has been counterfeited, altered, duplicated, or simulated. **F 1156,** F12

verification, *n*—a technical process used to evaluate whether or not a specific symbol meets a standard specification. **F 1294,** F05

verification—an act of confirming, substantiating, and assuring that an activity or condition has been implemented in conformance with the specified requirements. **F 1789,** F16

verification of conformity—confirmation by examination of evidence, that a product, process or service fulfills specified requirements, (ISO Guide 2).
E 1187, E36

verification standard, *n*—calibrated physical standard used to verify the accuracy of calibration of measurement scales, operating characteristics, or systems responses of color-measuring instruments. **E 284,** E12

verifier, *n*—*in emission spectrometry*, a material used to determine whether standardization is required. **E 135,** E01

verifier, *n*—a device that measures bars, images, spaces, quiet zones, locator points and optical characters of a symbol to determine whether or not a standard specification is met. **F 1294,** F05

verify, *v*—*specific-in textile testing,* to determine whether a previously calibrated instrument, standard solution, or other standard is still properly calibrated. **D 123,** D13

verify, *v*—*general,* to establish that an operation has been completed correctly. **D 123,** D13

verify, *v*—to assess the overall reliability and accuracy of an instrument or method of measurement by use of material standards for which the measurable quantities have accepted values. See **verification standard**. **E 284,** E12

vermiculite, *n*—insulation composed of natural vermiculite ore expanded to form an exfoliated structure. **C 168,** C16

vermiculite—an aggregate used in lightweight insulating concrete, formed by heating and expanding a micaceous mineral.
D 1079, D08

vernonia oil—a low-viscosity epoxidized drying oil from the seed of an African plant, *Vernonia galamensis*, containing three reactive epoxy groups and three carbon—carbon double bonds per triglyceride molecule and is characterized by its very low viscosity and melting point. **D 16,** D01

version A, *n—when applied to UPC codes*, the standard 12-digit symbol. **F 1294,** F05

version E, *n—when applied to UPC codes*, a 6-digit symbol formed by zero suppression. **F 1294,** F05

vertebral body replacement device, *n*—a structure which is designed to restore anatomic position and support to a section of spine lacking one or more vertebral bodies and intervening disc(s). **F 1582,** F04

vertebral span, *n*—the number of vertebra that are spanned by the longitudinal element, including the vertebrae containing anchor components. **F 1582,** F04

vertical—an approximately vertical path from water surface to stream bed along which one or more samples are taken to define sediment concentration or distribution. May also be referred to as *sampling vertical.* **D 4410,** D19

vertical accretion deposits—flood-plain deposits formed by deposition of suspended sediment from overbank flood waters. **D 4410,** D19

vertical bar code, *n*—a symbol oriented so that the bars are parallel to the horizon; also termed ladder or step ladder bar code. **F 1294,** F05

vertical field separators—*(1)* A vertical line separating data fields. (*2*) See **field separator**. **F 149,** F05

vertical grain—See **grain.** **D 9,** D07

vertical illumination—light incident on an object from the objective side such that smooth planes perpendicular to the optical axis of the objective appear bright. **E 7,** E04

vertical ladder gasket—a ladder gasket installed in a vertical plane in such a way that the intermediate cross members (muntins) are horizontal members. **C 717,** C24

vertical limit—the maximum readable level of vertical indications determined either by an electrical or a physical limit of an A-scan presentation. **E 1316,** E07

vertical load, *n*—the normal reaction of the tire on the road which is equal to the negative of normal force. **F 538,** F09

vertical load (F_z)—the downward vertical component of force between the tire and the road. **E 867,** E17

vertical load (load)—force acting in a vertical axis through the wheel transducer; that is, weight. **E 867,** E17

vertically oscillating arm type—See **pot, pan, and utensils washing machine**. **F 1827,** F26

vertical redundancy, *n*—the availability of more than a single scan path through the elements of a bar code symbol. **F 1294,** F05

vertical strip drain, *n*—a geocomposite consisting of a geotextile cover and drainage core installed vertically into soil to provide drainage for accelerating consolidation of soils. **D 4439,** D35

vertical thread—misnomer for longitudinal thread. **F 547,** F16

vessels—wood cells of comparatively large diameter that have open ends and are set one above the other so as to form continuous tubes. The openings of the vessels on the surface of a piece of wood are usually referred to as pores. **D 9,** D07

viability—the ability to grow and reproduce. **F 1600,** F20

viable—ability to live or grow. For example, bacteria, plants. **D 6161,** D19

vial, *n*—a small, usually cylindrical, vessel capable of closure, such as for medications, perfumes, essences, and samples. **D 996,** D10

vibrated beam wall (injection beam wall)—barrier formed by driving an H-beam in an overlapping pattern of prints and filling the print of the beam with cement-bentonite slurry or other materials as it is withdrawn. **D 653,** D18

vibrating needle atomizer—a vibratory atomizer in which liquid under pressure is supplied through a hollow vibrating needle. **E 1620,** E29

vibrating reed atomizer—a vibratory atomizer in which individual drops are produced from a liquid reservoir by an oscillating reed. **E 1620,** E29

vibration—an oscillation wherein the quantity is a parameter that defines the motion of a mechanical system (see **oscillation**). **D 653,** D18

vibration isolation—a reduction, attained by the use of a resilient coupling, in the capacity of a system to vibrate in response to mechanical excitation. **C 634,** E33

vibrative atomizer—see **vibratory atomizer**. **E 1620,** E29

vibratory atomizer—a device wherein an oscillating solid surface is the primary source of energy. **E 1620,** E29

vibratory cavitation, *n*—cavitation caused by the pressure fluctuations within a liquid, induced by the vibration of a solid surface immersed in the liquid. **G 40,** G02

vibratory cavitation test device, *n*— a device used to generate cavitation in a liquid through the vibrations of a solid surface in contact with the liquid. Usually such devices are driven at a frequency roughly in the range from 10 to 30 kHz by a magnetostrictive or a piezo-electric transducer. **G 40,** G02

vibratory finishing—a process for deburring and surface finishing in which the product and an abrasive mixture are placed in a container and vibrated. **B 374,** B08

vibrothermography—a thermographic technique for examining an object in which temperature differences are produced by excitation. **E 1316,** E07

Vickers—See **micro penetration tester**. In a more restricted sense, the 136° diamond pyramid indenter used in microindentation hardness tests.type="med"> **E 7,** E04

Vickers Hardness Number (HV), *n*—an expression of hardness obtained by dividing the force applied to a Vickers indenter by the surface area of the permanent impression made by the indenter. **C 1145,** C28

Vickers hardness number, HV [FL^{-2}], *n*—a number related to the applied load and the surface area of the permanent impression made by a square-based pyramidal diamond indenter having included face angles of 136°, computed from the equation:

$$HV = 2P \sin\ (\alpha/2)/d^2 = 1.8544P/d^2$$

where:
P = applied load, kgf,
d = mean diagonal of the impression, mm, and
α = face angle, of diamond = 136°.

E 6, E28

Vickers hardness test, *n*—an indentation hardness test using calibrated machines to force a square-based pyramidal diamond indenter having specified face angles, under a predetermined load, into the surface of the material under test and to measure the diagonals of the resulting impression after removal of the load. **E 6,** E28

Vickers hardness test—standard method for measuring the hardness of metals, particularly those with extremely hard surfaces; the surface is subjected to a standard pressure for a standard length of time by means of a pyramid-shaped diamond. The diagonal of the resulting indention is measured under a microscope and the Vickers hardness value is then read from a conversion table. **F 1789,** F16

Vickers indenter, *n*—a square-based pyramidal-shaped diamond indenter with face angles of 136° 00′. **C 1145,** C28

video data, *n*—a signal sent to the image output terminal by the electronic subsystem that turns the laser on and off to create an image on the photoreceptor. **F 1457,** F05

video presentation—display of the rectified, and usually filtered, r-f signal. **E 1316,** E07

viewing angle, *v*, *n*—*in retroreflection,* the angle between the retroreflector axis and the observation axis. **E 284,** E12

viewing conditions, *n*—the conditions under which a visual observation is made, including the angular substance of the specimen at the eye; the geometric relationship of source, specimen and eye; the photometric and spectral character of the field of view surrounding the specimen; and the state of adaptation of the eye. **C 1154,** C17

viewing conditions, *n*—the conditions under which a visual observation is made, including the angular substance of the specimen at the eye; the geometric relationship of source, specimen and eye; the photometric and spectral character of the field of view surrounding the specimen; and the state of adaptation of the eye. **D 2946,** C17

viewing conditions, *n*—the conditions under which a visual observation is made, including the angular subtense of the specimen at the eye, the geometric relationship of source, specimen, and eye, the photometric and spectral character of the source, the photometric and spectral character of the field of view surrounding the specimen, and the state of adaptation of the eye. **E 284,** E12

viewing rakes, *n*—a set of bars with wires spaced at 50-mm intervals for the purpose of increasing the precision of timing the flame front progress along the specimen. **E 176,** E05

viewing screen—feature of microwave equipment, usually part of the door assembly, that is opaque to microwave energy, but visually transparent to provide viewing of the oven contents. **F 1827,** F26

vignetting, *n*—loss of light rays at stops of an optical instrument, other than the aperture stop. **E 284,** E12

vinyl acetate plastics—plastics based on polymers of vinyl acetate or copolymers of vinyl acetate with other monomers, the vinyl acetate being in greatest amount by mass. (ISO) **D 883,** D20

vinyl asbestos tile, *n*—an obsolete form of resilient tile composed of vinyl plastic binders, crysotile asbestos fibers, mineral fillers and pigments. **F 141,** F06

vinyl chloride plastics—plastics based on polymers of vinyl chloride or copolymers of vinyl chloride with other monomers, the vinyl chloride being in greatest amount by mass. (ISO) **D 883,** D20

vinyl chloride plastics—plastics based on polymers of vinyl chloride or copolymers of vinyl chloride with other monomers, the vinyl chloride being in the greatest amount by mass. (ISO, D-20) **F 412,** F17

vinyl-coated glass yarn, *n*—glass continuous filament yarn, coated with a pigment and plasticized vinyl chloride resin. **D 123,** D13

vinyl composition tile, *n*—a resilient floor covering composed of binder, fillers, and pigments. The binder shall consist of one or more resins of poly (vinyl chloride), or vinyl chloride copolymers, or both, compounded with suitable plasticizers and stabilizers. Other polymeric resins may be incorporated as part of the binder. **F 141,** F06

vinyl ester resin—a thermosetting reaction product of epoxy resin with a polymerizable unsaturated acid usually methacrylic acid, which is then diluted with a reactive monomer—usually styrene. **C 904,** C03

vinylidene chloride plastics—plastics based on polymer resins made by the polymerization of vinylidene chloride or copolymerization of vinylidene chloride with other unsaturated compounds, the vinylidene chloride being in the greatest amount by weight. **D 883,** D20

vinylidene chloride plastics—plastics based on polymer resins made by the polymerization of vinylidene chloride or copolymerization of vinylidine chloride with other unsaturated compounds, the vinylidene chloride being in the greatest amount by weight. (D-20) **F 412,** F17

vinyl resin—under **resin, synthetic,** see *vinyl resin.* **D 16,** D01

viral penetration, *n*—the penetration of a material by a virus. **F 1494,** F23

viral resistant, *adj*—referring to materials which impede viral penetration under specified laboratory test conditions and detection methods. **F 1494,** F23

virgin compression curve—see **compression curve.** **D 653,** D18

virgin fiber, *n*— *in paper,* fiber that has been derived from sources other than paper or textile materials that have been recovered. **D 1968,** D06

virgin glycol, *n*—glycol that has not been used previously. **D 4725,** D15

virgin plastic—a plastic material in the form of pellets, granules, powder, floc, or liquid that has not been subjected to use or processing other than that required for its initial manufacture. **D 883,** D20

virgin plastic—a plastic material in the form of pellets, granules, powder, floc, or liquid that has not been subjected to use or processing other than that required for its initial manufacture. (D-20) **F 412,** F17

virgin wool, *n*—*as defined in the Wool Products Labeling Act,* "the terms 'virgin' or 'new' as descriptive of a wool product, or any fiber or part thereof, shall not be used when the product or part so described is not composed wholly of new or virgin fiber which has never been reclaimed from any spun, woven, knitted, felted, braided, bonded, or otherwise manufactured or used product." **D 123,** D13

virtual-front time (of an impulse-voltage wave), *n*—a time equal to 1.67 times the interval t_f between the instants when the voltage is 0.3 and 0.9 times the peak value. **D 1711,** D09

virtual image—a reproduction of an object by an optical system which gathers light from an object point and transforms it into a beam that appears to diverge from another point. **E 1316,** E07

virtual leak—(1) the semblance of a leak in a vacuum system caused by slow release of trapped gas.

(2) during a rate-of-rise test, the semblance of a leak in a vacuum system caused by slow release of sorbed or occluded gas or gases on or in the surfaces and pores of all materials in a system which has been exposed to atmospheric pressure prior to evacuation. **E 1316,** E07

virtual origin (of an impulse voltage wave), *n*—the point of intersection (o_1) with the line of zero voltage of a line drawn through the points of 0.3 and 0.9 times the peak voltage on the front of an impulse voltage wave. **D 1711,** D09

virtual-peak value (of an impulse-voltage wave), *n*—a value derived from a recording of an impulse wave on which high-frequency oscillations or overshoot of limited magnitude may be present. **D 1711,** D09

virtual temperature—See **temperature.** **D 1356,** D22

virtual time to half-value (of an impulse-voltage wave), *n*—the time interval (t_2) between the virtual origin (o_1) and the instant on the tail when the voltage has decreased to half the peak value. **D 1711,** D09

virus, *n—a minute infectious agent,* which lacks independent metabolism and is only able to replicate within a living host cell. **F 1494,** F23

viscoelasticity—property of materials that strain under stress partly elastically and partly viscously, that is, whose strain is partly dependent on time and magnitude of stress. (ISRM) **D 653,** D18

viscoelasticity—a combination of viscous and elastic properties in a material, with the relative contribution of each dependent upon time, temperature, stress, and strain rate. **D 1079,** D08

viscoelasticity, *n*—a combination of viscous and elastic properties in a material with the relative contribution of each being dependent on time, temperature, stress, and strain rate. **D 1566,** D11

viscose—a solution of sodium cellulose xanthate prepared by dissolving the reaction product formed by the interaction of carbon disulfide and alkali cellulose in an aqueous solution of sodium hydroxide. Viscose is used mainly in the manufacture of rayon or cellophane, where it is extruded through fine openings of the proper shape into a coagulating bath. **D 1695,** D01

viscosity—the property of fluids that opposes the relative motion of adjacent portions of a given fluid producing a type of internal friction and exhibiting a resistance to flow. **C 242,** C21

viscosity—the internal fluid resistance of a substance which makes it resist a tendency to flow. **D 653,** D18

viscosity—the property of resistance of flow exhibited within the body of a material. **D 883,** D20

viscosity, *n—of a liquid,* the resistance to flow expressed as the ratio of the applied shear stress to the resulting rate of shearing strain. (See **Newtonian behavior, non-Newtonian behavior**, and **thixotropy**.) (Compare **consistency**.) **D 907,** D14

viscosity, *n*—the resistance of a material to flow under stress. **D 1566,** D11

viscosity—in general, the resistance of a fluid to flow or motion within itself. As applied to cellulose or its derivatives, the viscosity of a solution of the substance. **D 1695,** D01

viscosity—the ratio of shear stress to shear rate. For non-Newtonian fluids, it is preferable to report shear stress and shear rate. **D 2507,** F07

viscosity—same as **viscosity, absolute**. **D 2864,** D27

viscosity—the internal resistance to flow of a solid (powder), liquid, or gas at a specified temperature. Viscosity is a definite measurement for the consistency of a material. **D 3064,** D10

viscosity, η—the property of resistance to steady flow exhibited within the body of the material. **D 4092,** D20

viscosity, η—the property of resistance to steady flow exhibited within the body of the material, (D 4092, D-20). **E 1142,** E37

viscosity—the property of a fluid that presents a resistance to shearing flow. **E 1316,** E07

viscosity—the property of resistance to flow exhibited within the body of a material. **F 412,** F17

viscosity, *n*—the property of resistance of flow exhibited with the body of a material. **F 1251,** F04

viscosity, absolute, η—the ratio of shear stress to shear rate. It is the property of internal resistance of a fluid that opposes the relative motion of adjacent layers. The unit most commonly used for insulating fluids is the centipoise. **D 2864,** D27

viscosity coefficient—the shearing stress necessary to induce a unit velocity flow gradient in a material. **D 883,** D20

viscosity coefficient, *n*—the shearing stress tangentially applied that will induce a velocity gradient. A material has a viscosity of one poise when a shearing stress of one dyne per square centimetre produces a velocity gradient of (1 cm/s)/cm. (See also **viscosity**.) **D 907,** D14

viscosity coefficient—the shearing stress necessary to induce a unit velocity gradient in a material. **D 4092,** D20

viscosity coefficient—the shearing stress necessary to induce a unit velocity flow gradient in a material. **F 412,** F17

viscosity, coefficient of— same as **viscosity, absolute**. **D 2864,** D27

viscosity, cuprammonium—the viscosity of a solution or dispersion of cellulose or pulp in cuprammonium hydroxide under standardized conditions. **D 1695,** D01

viscosity, cupriethylenediamine—the viscosity of a solution or dispersion of cellulose or pulp under standard conditions when dissolved in a solution of copper hydroxide in ethylenediamine. **D 1695,** D01

viscosity, dynamic— same as **viscosity, absolute**. **D 2864,** D27

viscosity in absolute units, CGS poises—the force in dynes required to move, at a velocity of 1 cm/s, one surface having an area of 1 cm^2 past another parallel-like surface 1 cm away, overcoming the resistance to shear of the material filling the space between. **D 1695,** D01

viscosity index—the ratio of the viscosity of a highly concentrated solution to that of a dilute solution. It is a measure of solvent power and, in derivatives, of uniformity of substitution. **D 1695,** D01

viscosity, inherent—the quotient of the natural logarithm of relative viscosity by the concentration, that is, ln η_{rel}/c. The concentration should be specified. **D 1695,** D01

viscosity, intrinsic—the limiting value of reduced viscosity, η_{sp}/c, as c (concentration) approaches zero. In the cellulosic field the concentration is generally expressed as grams per decilitre. The IUPAC Committee of Nomenclature has recommended the expression "Limiting viscosity number" for this and the concentration is generally expressed as grams per millilitre. **D 1695,** D01

viscosity, kinematic—the quotient of the absolute (dynamic) viscosity divided by the density, η/ρ, both at the same temperature. For insulating liquids the unit most commonly used is the centistokes (100 cSt = 1 St). **D 2864,** D27

viscosity, Mooney, *n*—a measure of the viscosity of a rubber or rubber compound determined in a Mooney shearing disk viscometer (see Test Methods D 1646). **D 1566,** D11

viscosity number—see **viscosity, reduced.** **D 1695,** D01

viscosity ratio—see **viscosity, relative.** **D 1695,** D01

viscosity, reduced—the specific viscosity divided by the concentration. In the cellulosic field the concentration is generally expressed as grams per decilitre. The IUPAC Committee on Nomenclature has recommended the expression "viscosity number" for this, in which case the preferred expression of concentration is grams per millilitre. **D 1695,** D01

viscosity, relative—the ratio of the viscosity of a solution to that of the pure solvent. The IUPAC Committee on Nomenclature has recommended the expression "Viscosity ratio" for this. **D 1695,** D01

viscosity, Saybolt Universal—the efflux time in seconds of 60 mL of sample flowing through a calibrated Saybolt Universal orifice under specified conditions. **D 2864,** D27

viscosity, specific—the difference between the viscosity of a solution and that of a solvent, divided by the latter. **D 1695,** D01

viscosity, xanthate—for a cellulosic pulp, the viscosity of the viscose that will be obtained as a result of a specified series of processes. **D 1695,** D01

viscous damping—the dissipation of energy that occurs when a particle in a vibrating system is resisted by a force that has a magnitude proportional to the magnitude of the velocity of the particle and direction opposite to the direction of the particle. **D 653,** D18

viscous flow—see **laminar flow**. **D 653,** D18

viscous flow—flow, usually at low velocity, in which fluid elements flow in a straight line parallel to the direction of flow of the liquid in the absence of turbulence. **D 1695,** D01

viscous flow—the flow of gas through a duct under conditions such that the mean free path is very small in comparison with the smallest dimension of a transverse section of the duct. This flow may be either laminar or turbulent. **E 1316,** E07

viscous hydroplaning, *n*—hydroplaning of pneumatic tires with separation caused by a thin fluid film due principally to the generation of fluid viscous forces. **E 867,** E17

viscous leak—a leak of such geometric configuration that gas flow through it is viscous in nature; that is, the flow obeys Poiseuille's Law. The flow rate is proportional to the difference of the squares of the end pressures, and inversely proportional to the gaseous viscosity. **E 1316,** E07

visibility, *n*—the properties and behavior of light waves and objects interacting in the environment to produce light signals capable of evoking visual sensation. **E 284,** E12

visibility—the quality or state of being perceived by the eye. In many outdoor applications, visibility is defined in terms of the distance at which an object can be reliably resolved from its surroundings. In outdoor applications it usually is defined in terms of contrast or size of a standard test object, observed under standardized viewing conditions, having the same threshold as the given object. **E 1316,** E07

visibility distance, *n*—the greatest distance at which an object can be seen. **E 284,** E12

visible—pertaining to radiant energy in the electromagnetic spectral range visible to the normal human eye (approximately 380 to 780 nm). **E 131,** E13

visible, *adj*—pertaining to that portion of the electromagnetic spectrum to which the eye is sensitive, approximately 380 to 780 nm in wavelength. **E 284,** E12

visible-activated fluorescence, *n*— fluorescence resulting from the absorption of visible radiant flux, that is, flux with wavelengths 380 to 780 nm. (See also **ultra-violet-activated fluorescence.**) **E 284,** E12

visible laser diode, VLD, *n*—a light source operating in the human visible light spectrum. **F 1294,** F05

visible light—electromagnetic radiation in the 400–700 (4000–7000 Å) wavelength range. **E 1316,** E07

visible light—radiant energy generated in 400 to 700 nm (4000 to 7000 Å) wavelength range. **E 1316,** E07

visible radiation, *n*—any radiation capable of causing a visual sensation.

NOTE—The limits of the spectral range of visible radiation are not well defined and may vary according to the user. The lower limit is generally taken between 380 and 400 nm and the upper limit between 760 and 790 nm (1 nanometer, nm = 10^{-9} m). **E 349,** E21

visible response—a particular type of optical response system used in some scanners. There are very few reflective inks for this type of system. **F 149,** F05

visible waste, *n*—*in cotton testing*, foreign matter deposited in the waste boxes of the machine during the test. **D 123,** D13

vision, *n*—the sense of sight. **E 253,** E18

visitor (visitor)— (*in a facility*) person present who is not an **occupant** of that facility. **E 1480,** E06

visual adaptation—the adjustment of the eyes when one passes from a bright to a darkened place. **E 1316,** E07

visual colorimeter, *n*—an instrument, using the eye as detector, that measures color stimuli produced by mixing one or more of at least three primary colors. **E 284,** E12

visual field—point or points in space that can be perceived when the head and eyes are kept fixed. The field may be monocular or binocular. **E 1316,** E07

visual inspection—a visual check made to detect constructional defects. **F 819,** F18

visual perception, *n*—the visual experience resulting from stimulation of the retina and the resulting activity of associated neural systems. **E 284,** E12

vitrain—See *vitrain* under **coal**. **D 121,** D05

vitreous enamel—see **porcelain enamel.** **C 286,** B08

vitreous silica—a generic term for vitreous silicon dioxide. See related terms **fused quartz** and **fused silica**. **C 162,** C14

vitreous slip—a slip coating matured on a ceramic body producing a vitrified surface. **C 242,** C21

vitreous (vitrified)—that degree of vitrification evidenced by low water absorption. (See also **impervious; nonvitreous; semivitreous.)** **C 242,** C21

vitrification—the progressive reduction and elimination of porosity of a ceramic composition, with the formation of a glass phase, as a result of heat treatment. **C 242,** C21

vitrification—solidification of an aqueous suspension at low temperatures without the formation of ice crystals. **E 1705,** E48

vitrification range—the maturing range of a vitreous body. **C 242,** C21

vitrified bond, *n*—a bond created by the fusion of ceramic materials, principally clays and feldspar. **C 1145,** C28

vitrified clay pipe—a pipe made from various clays or combinations thereof which are shaped, dried, and fired to a point where the glass-forming components fuse to form a bond between the crystalline grains. **C 896,** C04

vitrinite—See *vitrinite* under **maceral**. **D 121,** D05

vitrinite type, *n*—*as used in Test Method D 5061*, reflectance classes of vitrinite which span 0.1 % reflectance intervals. **D 121,** D05

VLSI circuit—Very Large Scale Integrated circuit device. **F 1156,** F12

"V" nail—headless nails with central V-shaped slot at head end. **F 547,** F16

VOC—an abbreviation for *volatile organic compound*; an organic compound with the tendency to become vapor at specified conditions of temperature and pressure. **E 1749,** E06

vocabulary—See **repertoire**. **F 149,** F05

vocational career apparel, *n*—career apparel which is generally subject to abusive wear and for which durability is a more important attribute than appearance. (See also **career apparel** and **dress career apparel**) **D 123,** D13

VOC (viable organism count)—a measure of biological activity (living or growing) in water. **D 6161,** D19

VOC (volatile organic compound)—an organic compound with a vapor pressure higher than water. **D 6161,** D19

void—a defective area in which a part of the basis material or underlayer is visible after final coating. **B 374,** B08

void—an unfilled space enclosed within an apparently solid carbon or graphite body. **C 709,** D02

void—space in a soil or rock mass not occupied by solid mineral matter. This space may be occupied by air, water, or other gaseous or liquid material. **D 653,** D18

void, *n*—(*1*) in a solid plastic, an unfilled space of such size that it scatters radiant energy such as light.
(*2*) a cavity unintentionally formed in a cellular material and substantially larger than the characteristic individual cells. (ISO) **D 883,** D20

void—any pocket of enclosed gas or air within a composite. **D 3878,** D30

void—any pocket of gas entrapment within or between the plies of reinforcement (see **blister** and **porosity**). **D 3918,** D20

void—*in structural members*, any opening, small crack, or crevice occurring at the juncture of structural members (such as chambers, reliefs, joggles, butt joints, or fasteners). **E 1749,** E06

void—the absence of ink, or an area of significantly lower density of ink, within the confines of a character. **F 149,** F05

void, *n*—(*1*) in a solid plastic, an unfilled space of such size that it scatters radiant energy such as light.

(*2*) a cavity unintentionally formed in a cellular material and substantially larger than the characteristic individual cells. (D-20, ISO) **F 412,** F17

void, *n*—a volume (in the tread band) defined by the lack of rubber; the depth dimension of this volume may vary from point to point in (on) the tread band. **F 538,** F09

void, *n*—(*1*) in a solid plastic, an unfilled space of such size that it scatters radiant energy such as light, (*2*) a cavity unintentionally formed in a cellular material and substantially larger than the characteristic individual cells. **F 1251,** F04

void, cellular material, *n*—a cavity unintentionally formed in a cellular material and substantially larger than the characteristic individual cells. **D 1566,** D11

void ratio, *e* **(D)**—the ratio of: (*1*) the volume of void space, to (*2*) the volume of solid particles in a given soil mass.
critical void ratio, e_c (D)—the void ratio corresponding to the critical density. **D 653,** D18

void ratio (e, (D)) , *n*—the ratio of the volume of void space to the volume of solids. **D 4439,** D35

void(s), *n*—in bar code standards, unwanted light areas contained within the dark elements of a positive bar. **F 1294,** F05

void space—ratio of the volume of voids in a powder bed to that of the overall volume of the powder bed. **C 242,** C21

volatile, *adj*—a relative term expressing the tendency to form vapor; that is, at room temperature or in some other temperature domain. **C 1145,** C28

volatile corrosion inhibitor (*VCI*)—See **corrosion.** **D 996,** D10

volatile extractables—*for purposes of the ASTM tests for volatile extractables and nonvolatile extractables*, volatiles shall be defined as those compounds that give more than 50 % recovery in spike and recovery studies using the applicable volatile extractables method. **F 17,** F02

volatile extractables—those chemical species which are released in the vapor state from microwave food packaging under simulated use conditions and are detected using an applicable volatile extractables method. **F 1479,** F02

volatile fatty acid (VFA) number (of latex), *n*—the number of grams of potassium hydroxide equivalent to the volatile fatty acids in a latex sample containing 100 g of total solids. **D 1566,** D11

volatile ingredients—the components of an aerosol formulation with a vapor pressure greater than atmospheric pressure (14.7 psia (101 kPa)) at 105°F (40.6°C). **D 3064,** D10

volatile materials—materials, particularly chemicals such as organic solvents that are readily vaporized at room temperature. **E 1605,** E06

volatile matter, *n*—those products, exclusive of moisture, given off by a material such as gas or vapor, determined by definite prescribed methods which may vary according to the nature of the material. **D 121,** D05

volatile matter—that matter that is changed under conditions of the test to the gaseous state. **D 1129,** D19

volatile matter, *n*—those products, exclusive of moisture, given off by a material as gas or vapor as determined by definite prescribed methods. **D 5681,** D34

volatile matter—those products, exclusive of moisture, given off by a material as gas or vapor, determined by definite prescribed methods. **E 856,** D34

volatile matter—those products, exclusive of moisture, given off by a material as gas or vapor, determined by definite prescribed methods that may vary according to the nature of the material. **E 1126,** E48

volatile matter—those products, exclusive of moisture, given off by a material as gas or vapor, determined by definite prescribed methods that may vary according to the nature of the material. **E 1705,** E48

volatile organic chemical—See **organic chemical.** **D 1356,** D22

volatile organic compound—See **VOC**. **E 1749,** E06

volatiles, *n*—materials readily vaporizable at relatively low temperatures. **D 123,** D13

volatile solvent—a nonaqueous liquid that evaporates readily at room temperature and atmospheric pressure. **D 2825,** D21

volatile thinner—see **thinner.** **D 16,** D01

volatility—the tendency of a solid or liquid material to pass into the vapor state at ordinary temperature. **E 1142,** E37

volatility—the capability of evaporating into a gas. **E 1749,** E06

volcanoes, *n*—printing imperfection resembling the crater of a volcano with a ring of heavy ink surrounding an area of no or very little ink. **D 6488,** D01

voltage alignment— See **alignment**. **E 7,** E04

voltage efficiency—the ratio, usually expressed as a percentage, of the equilibrium reaction potential in a given electrochemical process to the bath voltage. **B 374,** B08

voltage, induced secondary, E_2—the rms value of the open circuit voltage induced in the secondary winding N_2 of an inductor as a result of cyclic variations of the flux linkages with N_2. **A 340,** A06

voltage, maximum retest—voltage, either ac rms or dc average, that is equal to the proof test voltage for new protective equipment. **F 819,** F18

voltage, maximum use—the ac voltage (rms) classification of the protective equipment that designates the maximum nominal design voltage of the energized system that may be safely worked. The nominal design voltage is equal to phase-to-phase voltage on multiphase circuits.

NOTE—If there is no multiphase exposure in a system area, and the voltage exposure is limited to phase (polarity on dc systems) to ground potential, the phase (polarity on dc systems) to ground potential shall be considered to be the nominal design voltage.

NOTE—If electrical equipment and devices are insulated, or isolated, or both, such that the multiphase exposure on a grounded wye circuit is removed, then the nominal design voltage may be considered as the phase-to-ground voltage on that circuit. **F 819,** F18

voltage, nominal design—a nominal value consistent with the latest revision of ANSI C84.1, assigned to the circuit or system for the purpose of conveniently designating its voltage class. **F 819,** F18

voltage, retest—voltage, either ac rms or dc average, that used protective equipment must be capable of withstanding for a specific test period without breakdown. **F 819,** F18

voltage spike—rapid voltage excursion. **F 1156,** F12

voltage threshold—a voltage level on an electronic comparator such that signals with amplitudes larger than this level will be recognized. The voltage threshold may be user adjustable, fixed, or automatic floating. (E 750) **E 1316,** E07

volt-ampere, P_a—the unit of apparent power. **A 340,** A06

volume concentration—See **concentration.** **D 1356,** D22

volume concentration— See **particle concentration.** **G 40,** G02

volume density—See **concentration.** **D 1356,** D22

volume percent solids—the portion of a coating that remains as part of the dry film expressed as percent by volume. **D 16,** D01

volume percent superconductor—that percentage by volume of a composite superconductor which is superconducting under normal operating conditions. **B 713,** B01

volume resistivity—the resistance in ohms of a body of unit length and unit cross-sectional area. **B 354,** B01

volume resistivity, *n*—*of paper,* the resistance to electric current along a current path between electrodes placed on opposite sides of a sheet of paper when a defined dc voltage is applied and the thickness of the paper between the electodes is specified. **D 1968,** D06

volume resistivity—the volume resistance (in ohm-centimetres) between opposite faces of a centimetre cube of the material. A practical relative value for paper is obtained by the measure of resistance to electrical current electrodes placed in contact with the opposing surfaces of the sample. **F 335,** F05

volume resistivity, ρ_v—the volume resistivity of a material is the ratio of the potential gradient parallel to the current in the material to the current density. **D 5077,** D10

volumeter (powder)—an apparatus for the measurement of the volume of a known mass of power for the purpose of determining its density. **C 859,** C26

volumetric—chemical analysis based upon the reaction of a volume of standard solution with the material being analyzed. **C 859,** C26

volumetric flame spread, *n*—flame propagation through the volume of a gaseous mixture. **E 176,** E05

volumetric glassware—See **graduated glassware.** **C 162,** C14

volumetric shrinkage (volumetric change), V_s **(D)**—the decrease in volume, expressed as a percentage of the soil mass when dried, of a soil mass when the water content is reduced from a given percentage to the shrinkage limit. **D 653,** D18

volume-weight—See **specific weight.** **D 4410,** D19

volute—See **railing systems.** **E 631,** E06

volute, *n*—an ornamental spiral or scroll-shaped form serving, for example, as the handrail termination. **E 1481,** E06

Von Mises yield theory—stated ratio between shear and tension yield stresses for an isotropic, solid material. Theoretical yielding in shear is assumed to occur at a stress equal to 0.577 of the yield stress in tension. **E 1807,** E06

von Post humification scale—a scale describing various stages of decomposition of peat ranging from H1, which is completely undecomposed, to H10, which is completely decomposed. **D 653,** D18

vortex loss—oil escaping past a boom due to drainage vortices produced at the boom. **F 818,** F20

votive candle, *n*—a candle produced for use fully within a candle accessory, specifically, a votive holder. **F 1972,** F15

votive holder, *n*—a small open vessel designed to hold a votive candle during use; the votive holder shall be capable of containing the melted candle. **F 1972,** F15

"V" thread—thread with leading flank of one thread intersecting with following flank of adjacent thread at thread root. **F 547,** F16

vulcanizate, *n*—the product of vulcanization, a crosslinked rubber. **D 1566,** D11

vulcanization, *n*—an irreversible chemical reaction during which a rubber compound's chemcial structure is changed to make it less plastic, more resistant to solvents, and have improved physical and mechanical properties. **C 717,** C24

vulcanization, *n*—an irreversible process, usually accomplished through the application of heat, during which a rubber compound through a change in its chemical structure (for example, crosslinking) becomes less plastic and more resistant to swelling by organic liquids while elastic properties are conferred, improved, or extended over a greater range of temperatures. **D 123,** D13

vulcanization, *n*—an irreversible process during which a rubber compound, through a change in its chemical structure (for example, cross-linking), becomes less plastic and more resistant to swelling by organic liquids and elastic properties are conferred, improved, or extended over a greater range of temperature. **D 883,** D20

vulcanization, *n*—an irreversible chemical change (for example, cross-linking) in which a rubber compound becomes less plastic with greater elasticity, strength, stability, and chemical resistance over a greater range of temperatures. (R) **D 907,** D14

vulcanization, *n*—an irreversible process during which a rubber compound, through a change in its chemical structure (for example, crosslinking), becomes less plastic and more resistant to swelling by organic liquids while elastic properties are conferred, improved, or extended over a greater range of temperature. **D 1566,** D11

vulcanization, *n*—an irreversible process during which a rubber compound, through a change in its chemical structure (for example, cross-linking), becomes less plastic and more resistant to swelling by organic liquids and elastic properties are conferred, improved, or extended over a greater range of temperature. (D-20) **F 412,** F17

vulcanization—an irreversible process during which a rubber compound, through a change in its chemical structure (for example, crosslinking) becomes less plastic and more resistant to swelling by organic liquids. Elastic properties are conferred, improved or extended over a greater range of temperature. **F 869,** F08

vulcanize, *v*—to subject to vulcanization. **D 907,** D14

vulcanized vegetable oil, *n*—a compounding material made by heating unsaturated vegetable oil with cross-linking agents; it is used as a processing aid or extender. **D 1566,** D11

vulcanizing agent, *n*—compounding material that produces crosslinking in rubber. **D 1566,** D11

vulcanizing system, *n*—the combination of vulcanizing agent and, as required, accelerators, activators, and retarders used to produce the desired vulcanization characteristics or vulcanizate characteristics. **D 1566,** D11

V-X diagram—a graphical representation of the isothermal or isobaric phase relationships in a binary system, the coordinates of the graph being specific volume and concentration. **E 7,** E04

W

"W"—clean with water-based shampoo only. **D 5253,** D13
WAD—See **worst area difference.** **F 149,** F05
wadding—See **cushioning material.** **D 996,** D10
wafer—a wood flake having a predetermined length of at least 1$^3/_{16}$ in. (30 mm). **D 1554,** D07
wafer—*in semiconductor technology*, a thin slice with parallel faces cut from a semiconductor crystal. **F 1241,** F01
wagon nail—annealed, barbed, regular-stock-steel nails with round, oval, cone, flat countersunk, or steeple head and medium diamond point. **F 547,** F16
waist, *n—in anatomy*, the part of the body at the location between the lowest rib and hip identified by bending the body to the side. **D 123,** D13
waist—the smallest part of the foot between the instep and the ball, also the corresponding part of a shoe last. **F 869,** F08
waist, *n—in anatomy*, the part of the body at the location between the lowest rib and hip identified by bending the body to the side. **F 1494,** F23
waist, b_M—the narrowest point of the ski body between the heel and shoulder. **F 472,** F27
waist girth, *n—in body measurements*, the circumference of the waist immediately below the lowest rib. **D 123,** D13
waist girth, *n— in body measurements*, the circumference of the waist immediately below the lowest rib. **F 1494,** F23
waist (of the snowboard ski)—the narrowest point of the snowboard ski body between the heel and the shoulder. **F 1107,** F27
wale, *n—in knitted fabrics*, a column of loops in successive courses that is parallel with the loop axes. **D 123,** D13
wale, *n—in woven fabrics*, one of a series of raised portions or ribs lying warpwise in the fabric. **D 123,** D13
walk in—to embed insulation panels in hot bitumen or adhesive by walking on them immediately after application. **D 1079,** D08
walking off—the disappearance or weakening of image parts on the press master during the press run. Usually a blind image is slightly visible but ink repellent, whereas an image that has walked off is not there. **F 425,** F05
walking stick—long arm attached to tool allowing operator to use it without bending over. **F 592,** F16
walkway surface—a structure intended to be used by a person attempting to walk. **F 1646,** F13
wall, *n*—a part of a **building** that divides spaces vertically.
bearing wall—a **wall** supporting a vertical load in addition to its own weight.
curtain wall—a **nonbearing** exterior **wall,** secured to and supported by the structural members of the **building.**
nonbearing wall—a **wall** that does not support a vertical load other than its own weight.
retaining wall—a **wall** not enclosing portions of a **building,** designed to resist the lateral displacement of soil or other material. **E 631,** E06
wall, *n*—part of a building that divides spaces vertically. **E 1605,** E06
wallaby—leather from skins of the wallaby, small and medium-sized kangaroo. **D 1517,** D31
wall and top material, *n*—any pliable planar structure used as a nonbase surface in camping tentage including roofs, sides, windows, screens, doors, awnings, flies, and canopies. **D 123,** D13
wallboard.—see **gypsum wallboard** **C 11,** C11
wallboard nail—slender, colored (baked-lacquer finished), regular-stock-steel or hardened-steel, smooth or annularly threaded, 1 ⅛ by 0.062 to 2 by 0.083-in. nails with slightly countersunk 0.109 to 0.181-in. head and medium diamond or long needle point. **F 547,** F16
wall bracket—a bracket used for anchoring the central portion of a handrail to a wall. **E 1481,** E06
wall clip—See **railing systems.** **E 631,** E06
wall clip—a bracket used for anchoring the end of a handrail to a wall. (Syn. *wall flange.*) **E 1481,** E06
wall flange—See **railing systems.** **E 631,** E06
wall flange—Synonym for **wall clip**. **E 1481,** E06
wall friction, f' **(FL^{-2})**—frictional resistance mobilized between a wall and the soil or rock in contact with the wall. **D 653,** D18
wall handrail—See **railing systems.** **E 631,** E06
wall handrail—a handrail attached to a wall or other building element adjacent to a stair and along landings, walkways, ramps, and corridors. (Syn. *wall rail.*) **E 1481,** E06
wall-mounted canopy—used for all types of cooking equipment located against a wall. See *canopy*. **F 1827,** F26
Wallner line—a fracture surface marking, having a wavelike profile in the fracture surface. Such marks frequently appear as a series of curved lines, indicating the direction of propagation of the fracture from the concave to the convex side of a given wallner line. Also known as ripple mark. **C 162,** C14
wall (pipe)—the structural element composed of concrete or concrete and steel between the inside and outside surfaces of a concrete pipe. **C 822,** C13
wall rail—See **railing systems.** **E 631,** E06
wall rail—Synonym for **wall handrail**. **E 1481,** E06
wall railing return—See **railing systems.** **E 631,** E06
wall railing return—a bend at the end of a wall handrail, turning toward the wall to which the handrail is attached. **E 1481,** E06
walls—exterior vertical sides of vitrified clay filter block. **C 896,** C04
walls, veneered—See *veneer.* **C 119,** C18
wall yield locus—a plot of the wall shear stress versus wall normal stress. The angle of wall friction is obtained from the wall yield locus as the arctan of the ratio of the wall shear stress to wall normal stress. **D 653,** D18
walrus—leather from the hides of walrus. Walrus hide is very thick, and is used for buffing wheels. When split, it is used for bag leather. Split walrus and seal leather are practically indistinguishable, and "walrus leather" in the traveling-goods industry is used to refer to sealskin leather on which a simulated walrus grain is embossed. **D 1517,** D31
wand scanner—a hand-held optical scanner used in applications where it's impractical to transport data past a fixed read head. **F 149,** F05
wand scanner, *n*—a hand-held device used as a contact bar code or OCR reader. **F 1294,** F05
wane—bark or lack of wood from any cause on edge or corner of a piece. **D 9,** D07
warming-in—reheating glass to permit further working, also for striking. **C 162,** C14
warm, moist climate—subtropical climate distinguished as warm and humid year round, with frequent rain showers. **G 113,** G03
warm-setting adhesive, *n*—Synonym for **intermediate-temperature-setting adhesive.** **D 907,** D14
warm spring, *n*—thermal spring whose temperature is appreciably above the local mean annual temperature but below that of the

human body. **E 957,** E44

warm-up (rubber processing), *n*—the reduction in viscosity of a rubber or rubber mix, by mechanical work and heat to render it suitable for further processing. **D 1566,** D11

warp—any variation from a true or plane surface. Warp includes **bow, crook, cup,** and **twist,** or any combination thereof. **D 9,** D07

warp, *n*—(1) the yarn running lengthwise in a woven fabric (2) a group of yarns in long lengths and approximately parallel, put on beams or warp reels for further textile processing including weaving, knitting, twisting, dyeing, etc. **D 123,** D13

warp, *n*—a significant variation from the original true, or plane surface. **D 907,** D14

warp, *n*—the yarn running lengthwise in a woven fabric. **D 4439,** D35

warp—*of a semiconductor wafer*, the difference between the maximum and minimum distances of the median surface of a free, unclamped wafer from a reference plane. **F 1241,** F01

warpage—distortion which may occur in a compact during sintering. **B 243,** B09

warpage—curvature of a flat specimen measured as deviation of the specimen surface from a true plane along the edges or the diagonals and at the mid-length of an edge or diagonal, expressed as a percent of the length of the edge or diagonal, and called convex or concave with respect to the face of the specimen. **C 242,** C21

warpage, *n*—distortion caused by nonuniform change of internal stresses. (See also **dome.**) **D 883,** D20

warp elongation and tension, *n*—stretch or tension measured in the warp direction of the fabric. **D 123,** D13

warp-faced twill, *n*—a twill weave in which the warp yarns produce the diagonal effect. (See also **twill weave**.) **D 123,** D13

warping—a change or deformation in the original flatness of an asbestos-cement product. **C 460,** C17

warping, *n*—change or deformation in the original flatness of an asbestos-cement product. **C 1154,** C17

warping—deformation in the original flatness of an asbestos-cement flat product. **D 2946,** C17

warp streak, *n*—*in woven fabric*, a narrow band running lengthwise and characterized by apparent differences in color from adjoining ends. **D 123,** D13

warp test—see Test Method C 314. **C 286,** B08

warp tests, *n*—*in cloth testing*, tests in which the warp yarns are torn. **D 123,** D13

warp-to-filling seam, *n*—a sewn seam in which the warp yarns are perpendicular to the sewn seam on one side of the seam and parallel to the seam on the opposite side of that seam. **D 123,** D13

warp-to-warp seam, *n*—a sewn seam in which the yarns in the warp direction on both sides of the seam are perpendicular to the seam. **D 123,** D13

wash—same as **rinse**. **E 1316,** E07

washability analysis of coal, *n*—the analysis of the specific gravity distribution of chemical and physical characteristics of coal. **D 121,** D05

washable (as applied to a garment)—a term applied to a garment that will be restored to wearability by laundering according to an acceptable procedure in the absence of irreparable damage. **D 459,** D12

wash-and-wear, *adj*—a generic term applied to garments which satisfactorily retain their original neat appearance after repeated wear and suitable home laundering with little or no pressing or ironing. (Compare **durable-press.**) **D 123,** D13

washboard—an imperfection; ripples, waves, etc., on the surface of glassware. **C 162,** C14

washboard, *n*—*in hosiery*, a ridgy effect caused by uneven tension between feeds on the knitting machine. **D 123,** D13

washed out solid, *n*—a decrease in the density of solids caused by ink taking on an excessive amount of water or by the insufficient transfer of ink. **D 6488,** D01

washer head—head with washer-like flange to serve as bearing surface for head. (See **nub head.**) **F 547,** F16

washer-spacer—a component located between the tumblers and intended to prevent the transfer of movement by friction between adjacent tumblers. **F 471,** F12

washer-tension—a component used to apply tension to the tumbler pack. **F 471,** F12

washing—*in grouting*, the physical act of cleaning the sides of a hole by circulating water, water and air, acid washes, or chemical substances through drill rods or tremie pipe in an open hole. **D 653,** D18

washing period—the period of time from the beginning of the dishwasher's wash cycle until the end of the final rinse. **F 1827,** F26

wash load—the portion of the stream sediment load composed of particles, usually finer than 0.062 mm in diameter which are found only in relatively small quantities in the bed. **D 4410,** D19

wash, *v* **or** *n*—a cleaning process carried out in an aqueous medium. **D 459,** D12

washout nozzle—a tubular extension with a check valve utilized at the end of a string of casing through which water can be injected to displace drilling fluids and cuttings from the annular space of a borehole. **D 653,** D18

waste—*in the nuclear industry*, any residue containing radioactive materials not currently considered useful or economically recoverable. **C 859,** C26

waste, *n*—a material that is unwanted at its present location; that is no longer useful for its original purpose; that has been disposed, or any combination thereof. **D 5681,** D34

waste composition, *n*—of a solid waste, characterization of multi-constituent waste by a breakdown into specified waste components on the basis of mass or volume fraction or percentage. (*Syn.***solid waste composition**.) **D 5681,** D34

waste oil, *n*—*in petroleum technology*, oil having characteristics making it unsuitable either for further use or for economic recycling. **D 4175,** D02

waste water—water discharged from a process as a result of its formation or use in that process. **D 1129,** D19

water, *n*—the chemical compound H_2O. **D 123,** D13

water absorption, *n*—the amount of water absorbed by a material under specified test conditions commonly expressed as weight percent of the test specimen. **C 11,** C11

water absorption, *n*—the amount of water absorbed by a material under specified test conditions. **D 1566,** D11

water absorption—increase in mass of a product during immersion in water under specified conditions of duration and temperature, expressed as a percent of its dry mass. **D 2946,** C17

water accommodated fraction (WAF), *n*—the predominantly aqueous portion of a mixture of water and a material poorly soluble in water which separates in a specified period of time after the mixture has undergone a specified degree of mixing and which includes water, dissolved components, and dispersed droplets of the poorly water soluble material. **D 6384,** D02

water-based coating, *n*—deprecated term. Use the preferred term, waterborne coating. **D 16,** D01

water beading—surface property that causes the formation of discrete water droplets on the polished surface. **D 2825,** D21

water boil efficiency—the energy (latent heat of vaporization) required to boil water from the cooking medium (see **cooking medium**), expressed as a percentage of the quantity of energy input to the equipment during the boil-off period, $\eta water\ boil$.

$$\eta waterboil(\%) = \frac{W \times E_{vap} \times 100}{E_{input}}$$

where:

W = weight loss of water, lb (g),
E_{vap} = heat of vaporization of water, Btu/lb (J/kg),
= 970 Btu/lb (2256 kJ/kg) at standard pressure (29.921 in. Hg), and
E_{input} = energy consumed by the equipment, Btu.

F 1827, F26

waterborne coating, *n*—a coating in which the principal volatile constituent is water. See also **water-reducible coating**. **D 16,** D01

water break—the appearance of a discontinuous film of water on a surface signifying nonuniform wetting and usually associated with a surface contamination. **B 374,** B08

water-break—failure of water to maintain a continuous film on metallic, vitreous, or similar surfaces on withdrawing from clean water. **D 459,** D12

water break test—a quality control test of conditioned water. **E 1316,** E07

water-cementitious material ratio, *n*—the ratio of the mass of water, exclusive only of that absorbed by the aggregates, to the mass of cementitious material (*hydraulic*) in concrete, mortar, or grout, stated as a decimal (see also **water-cement ratio**). **C 125,** C09

water-cement ratio, *n*—the ratio of the mass of water, exclusive only of that absorbed by the aggregates, to the mass of portland cement in concrete, mortar, or grout, stated as a decimal. **C 125,** C09

water-cement ratio—the ratio of the weight of water to the weights of Portland cement in a cement grout or concrete mix. See also **grout mix**. **D 653,** D18

water consumption—the total amount of water consumed by the equipment during an event such as cooling water for condensing, cooking, or preheating. **F 1827,** F26

water content, *n*—the water content of a substance, as measured under specified conditions. (*Compilation of ASTM Standard Definitions*, 7th Edition.) **D 2864,** D27

water content, *w* **(D)**—the ratio of the mass of water contained in the pore spaces of soil or rock material, to the solid mass of particles in that material, expressed as a percentage. **D 653,** D18

water dam—synonymous with **water gate**. **C 162,** C14

water depth-nominal, *n*—the nominal thickness of the water layer, that is, the volume of water divided by the area of the wetted pavement surface. **E 867,** E17

water depth-positive, *n*—the distance from the water surface to the reference plane which is the top of the pavement asperities. **E 867,** E17

water dew point—(see **dew point**) **D 4150,** D03

water discharge—the quantity of water passing a stream cross section per unit of time. **D 4410,** D19

water-emulsion floor polish—an emulsion-based floor polish in which water is the continuous or external phase and falling into two categories:

(*1*) **polymer-emulsion floor polish**—a water-emulsion floor polish containing a predominance of synthetic emulsion polymers.

(*2*) **wax-emulsion floor polish**—a water-emulsion floor polish containing a predominance of natural or syn thetic waxes, or both. **D 2825,** D21

water equivalent, *n*—Synonym for **energy equivalent**. **D 121,** D05

water-extractable material—substances dissolved from other substances by water under specified conditions. **D 2652,** D28

water filtration vacuum cleaner, *n*—a system or device that removes foreign material from surfaces by means of the air flow caused by subatmospheric pressure, having an intake intended to be moved in proximity to the surface, and a water reservoir which separates and collects the material from the air. The inlet may be fixed or attached to other equipment. It also may be designed for liquid pick up. **F 395,** F11

water finish, *n*—compact glossy finish produced on one or both surfaces of a paper or paper-board by wetting with water as the sheet is being calendered. **D 1711,** D09

water-formed deposits—any accumulation of insoluble material derived from water or formed by the reaction of water upon the surfaces in contact with the water. **D 1129,** D19

waterfowl feathers, *n*—feathers from ducks or geese, or both. **D 123,** D13

water gain—see **bleeding**. **D 653,** D18

water-gate—See gate. Also known as a water-dam. **C 162,** C14

water glass—sodium silicate glass that is readily soluble in water. **C 162,** C14

water-hardened—quenched in water after heating to critical temperature. **F 547,** F16

water hardness—describes a quality of water arising from the presence of dissolved mineral salts (usually calcium and magnesium and sometimes iron and manganese). The degree of hardness is measured in grains of calcium carbonate ($CaCO_3$) per gallon (gpg) or parts per million (ppm), the latter being more accurately expressed as milligrams per litre (mgL). One grain per gpg of $CaCO_3$ equals 17.1 ppm or mgL. Water essentially free of calcium and magnesium is described as soft; if appreciable amounts of both minerals are present, it is called hard. The U.S. Geological Survey categories of hardness are:

	Soft	Moderately Hard	Hard	Very Hard
Grains per gallon	0.0–3.5	3.6–7.0	7.1–10.5	10.6+
Parts per million or milligrams per litre	0.0–60	61–120	121–180	over 180

D 459, D12

water-holding capacity (D)—the smallest value to which the water content of a soil or rock can be reduced by gravity drainage. **D 653,** D18

water jet barrier—barrier created by stream of pressurized water spray directed across the water surface. **F 818,** F20

water mark, water spot—an appearance defect characterized by a depressed spot. **C 286,** B08

water migration resistance—the ability of either facing or core materials to prevent migration of water in sandwich panels. **E 1749,** E06

water of hydration or combined water—that water in a material that cannot be removed by drying at 110°C, as it is chemically bound, expressed as a percent of the weight of the material. **C 242,** C21

water paint—under **paint,** see *water paint*. **D 16,** D01

water path—the distance from the transducer to the test surface in immersion or water column testing. **E 1316,** E07

water-plasticity ratio (*relative water content*) (*liquidity index*)—see **liquidity index**. **D 653,** D18

water pollution—the harmful or objectionable material introduced into water in sufficient quantities to adversely affect its usefulness. **D 4410,** D19

waterpress—a device that de-waters the slurry generated in the pulper by use of a tapered compression cone and perforated screen, then discharges the pulp down a chute to a waste container. Water removed during this process is recirculated to the *pulper* tank in order to conserve fresh water use. **F 1827,** F26

waterproof, *adj*—deprecated term. **D 996,** D10

waterproof—*as applied to plywood,* the term is synonymous with **exterior type plywood**. **D 1038,** D07

waterproofing, *n*—*in building construction*, treatment of a surface or structure to prevent the passage of liquid water under hydrostatic, dynamic, or static pressure. **C 717,** C24

waterproofing—treatment of a surface or structure to prevent the passage of water under hydrostatic pressure. **D 1079,** D08

waterproofness (coated fabric), *n*— the property of impenetrability by liquid water. **D 1566,** D11

water-reducible coating, *n*—a coating that can be reduced in viscosity by the addition of water. **D 16,** D01

water-repellant, *n*—a material or treatment for surfaces to provide resistance to penetration by water. **E 631,** E06

water repellant substances—materials such as waxes, soaps, or silicones that render the surface of shingles more water repellant. **D 2946,** C17

water repellant system, *n*—an exterior coating system for above grade concrete or masonry which temporarily repels water but which is not intended to prevent the passage of moisture under hydrostatic pressure.

(1) film forming type—latex or oil based paints or similar materials that may require repair or renewal because of climatic action;

(2) pore penetrating type—types that fill or coat the interior of the surface pores of concrete or masonry and are temporarily retained in these surface pores thus providing water repelling action; periodic repair or renewal may be required because of climatic action. **D 1079,** D08

water repellency (coated fabric), *n*— the property of being resistant to wetting by liquid water. **D 1566,** D11

water-repellent leather—a leather that has been treated with any of a variety of chemicals to reduce the ability of water to spread on or wet the surface. Treatments can be for grain or suede leathers. **D 1517,** D31

water-repellent paper, *n*—gypsum board paper surfacing which has been formulated or treated to resist water penetration. **C 11,** C11

water repellent substances—materials such as waxes, soaps, or silicones which render the surface of shingles more water repellent. **C 460,** C17

water-repellent substances, *n*—*in shingles*, materials such as waxes, soaps, or silicones which render the surface of shingles more water repellent. **C 1154,** C17

water-resistance, *n*—measured ability to retard both penetration and wetting by water in liquid form. **D 996,** D10

water resistance—ability of the image to resist deformation or change in color with immersion in water. **F 221,** F05

water resistance, *n* (also called water fastness) —ability of an image to resist water damage. **F 1857,** F05

water resistance (coated fabric), *n*— the property of retarding both penetration and wetting by liquid water. **D 1566,** D11

water resistant—a term frequently applied to plywood that is capable of withstanding limited exposure to water or to severe conditions without failure in the glue bonds. **D 1038,** D07

water-resistant barrier— See **barrier material.** **D 996,** D10

water-resistant core, *n*—a gypsum board specially formulated to resist water penetration. **C 11,** C11

water-resistant leather—a term formerly applied to leather heavily stuffed with oils, greases, and waxes, used primarily in work shoes and boots. Currently, the term may also, and more commonly, include upper leathers treated with any of a variety of chemicals to reduce the penetration or absorption of liquid water as measured in D 2098, Test for Dynamic Water Resistance of Shoe Upper Leather by the Dow Corning Leather Tester, or D 2099, Test for Dynamic Water Resistance of Shoe Upper Leather by the Maeser Water Penetration Tester. **D 1517,** D31

water retained, *n*—*in textiles*, the amount of water absorbed by the fibers, adsorbed on the surface of the fibers, and held within the voids of the fabric after immersion, measured under specified conditions. **D 123,** D13

water retention, *n*—the moisture remaining in and on a material after a specified mechanical treatment. **D 123,** D13

water service—the pipe from the water main or other source of water supply to the building or other point of use or distribution. **F 412,** F17

watershed—all lands enclosed by a continuous hydrologic-surface drainage divide and lying upslope from a specified point on a stream. **D 4410,** D19

water softener—a vessel having a cation resin in the sodium form that removes cations such as calcium and magnesium from water and releases another ion such as sodium. The resin is usually regenerated. See **softener**. **D 6161,** D19

water solubility—the amount of material that is miscible or will dissolve in water at a given temperature. **D 4790,** D16

water soluble fraction (WSF), *n*—the filtrate or centrifugate of the water accommodated fraction which includes all parts of the WAF except the dispersed droplets of the poorly soluble material. **D 6384,** D02

water spotting—change in appearance of surface resulting solely from the action of cool water. **D 2825,** D21

water streak—a defect occurring in the bisque characterized by a washed-out pattern in the form of a streak. **C 286,** B08

water table—see **free water elevation**. **D 653,** D18

water table aquifer, *n*—see preferred term **unconfined aquifer**. **D 653,** D18

water table (ground-water table)—the surface of a ground-water body at which the water pressure equals atmospheric pressure. Earth material below the ground-water table is saturated with water. **D 653,** D18

water tolerance—the amount of water that a penetrant or emulsifier can absorb before its effectiveness is impaired. **E 1316,** E07

water vapor barrier—Use **water-vapor retarder,** the preferred term. **E 631,** E06

water vapor diffusion, *n*—the process by which water vapor spreads or moves through permeable materials caused by a difference in water vapor pressure. **C 168,** C16

water vapor permeability, *n*—the time rate of water vapor transmission through unit area of flat material of unit thickness induced by unit vapor pressure difference between two specific surfaces, under specified temperature and humidity conditions. **C 168,** C16

water vapor permeability—the rate of water vapor transmission per unit of thickness and per unit of vapor pressure differential. Test conditions must be specified. **D 1695,** D01

water vapor permeance, *n*—the time rate of water vapor transmission through unit area of flat material or construction induced by unit vapor pressure difference between two specific surfaces, under specified temperature and humidity conditions. **C 168,** C16

water-vapor permeance—time rate of water-vapor transmission through unit area of a flat material or construction induced by unit vapor-pressure difference between two specified surfaces, under specified temperature and humidity conditions. See **perm. C 168** **E 631,** E06

water vapor permeance (WVP), *n*—the steady state rate of water vapor movement through a free film induced by a vapor pressure difference ($\Delta\rho$) of one unit between the two surfaces of the film, where $\Delta\rho$ is expressed in inches or millimetres of mercury. Thus: WVP = WVT/$\Delta\rho$. See WVT. **D 16,** D01

water vapor pressure, *n*—the pressure of water vapor at a given temperature; also the component of atmospheric pressure contributed by the presence of water vapor. **C 168,** C16

water vapor pressure, *n*—the pressure of water vapor at a given temperature. **C 1145,** C28

water vapor resistance, *n*—the steady vapor pressure difference that induces unit time rate of vapor flow through unit area of a flat material (or construction that acts like a homogeneous body) for specific conditions of temperature and relative humidity at each surface. **C 168,** C16

water-vapor-resistance—measured ability to retard penetration and permeation by water-vapor. **D 996,** D10

water-vapor-resistant barrier—See **barrier material.** **D 996,** D10

water vapor resistivity, *n*—the steady vapor pressure difference that induces unit time rate of vapor flow through unit area and unit thickness of a flat material (or construction that acts like a homogeneous body), for specific conditions of temperature and relative humidity at each surface. **C 168,** C16

water-vapor retarder, *n*—material or system that impedes the transmission of water vapor under specified conditions. **E 631,** E06

water vapor retarder (barrier), *n*— a material or system that adequately impedes the transmission of water vapor under specified conditions. **C 168,** C16

water vapor transmission rate, *n*—the steady water vapor flow in unit time through unit area of a body, normal to specific parallel surfaces, under specific conditions of temperature and humidity at each surface. **C 168,** C16

water vapor transmission (WVT), *n*—the rate of water vapor flow, under steady specified conditions, through a unit area of a material, between its two parallel surfaces and normal to the surfaces. Metric unit of measurement is 1 g/24 h·m^2. See **permeability,permeance, perm. C 11,** C11

water vapor transmission (WVT) rate, *n*—the steady state rate of water vapor movement through a free film under specific conditions of temperature and humidity at each surface; customarily expressed in grains per square foot per hour (gr/ft^2·h) or grams per square metre per day (g/m^2·d). **D 16,** D01

watt, W—the unit of active power. One watt is energy, work, or quantity of heat expended at a rate of one joule per second. **A 340,** A06

wave—an optical effect due to uneven glass distribution or to striae. **C 162,** C14

wave—disturbance propagated in medium in such a manner that at any point in medium the amplitude is a function of time, while at any instant the displacement at point is function of position of point. **D 653,** D18

wave form—the shape of the peak-to-peak variation of a controlled mechanical test variable (for example, load, strain, displacement) as a function of time. **E 1823,** E08

wave front—moving surface in a medium at which a propagated disturbance first occurs. **D 653,** D18

wave front—(*1*) a continuous surface over which the phase of a wave that progresses in three dimensions is constant, or (*2*) a continuous line along which the phase of a surface wave is constant. (ISRM) **D 653,** D18

wave front—a continuous surface drawn through the most forward points in a wave disturbance which have the same phase. **E 1316,** E07

waveguide, acoustic emission—a device that couples elastic energy from a structure or other test object to a remotely mounted sensor during AE monitoring. An example of an acoustic emission waveguide would be a solid wire of rod that is coupled at one end to a monitored structure, and to a sensor at the other end. **E 1316,** E07

wave length—normal distance between two wave fronts with periodic characteristics in which amplitudes have phase difference of one complete cycle. **D 653,** D18

wavelength, λ— the distance, measured along the line of propagation, between two points that are in phase on adjacent waves. **E 131,** E13

wavelength, λ, *n—of an electromagnetic wave*, the distance in the direction of propagation between nearest points at which the electric vector has the same phase. (See also **complementary wavelength, dominant wavelength.**) **E 284,** E12

wavelength, *n*—distance in the direction of propagation of a periodic wave between two successive points at which the phase is the same (at the same time).

NOTE—The wavelength in a medium is equal to the wavelength in vacuo divided by the refractive index of the medium. Unless otherwise stated, values of wavelength are generally those in air. The refractive index of standard air (15°C, 101.325 N·m^{-2}) lies between 1.00027 and 1.00029 for visible radiations. **E 349,** E21

wavelength (X-rays)—the minimum distance between points at which the electric vector of an electromagnetic wave has the same value, measured along the direction of propagation of the wave. It is equal to the velocity divided by the frequency. (See also **electron wavelength.**) **E 7,** E04

wavenumber, $\bar{\nu}$—the number of waves per unit length. **E 131,** E13

wave number, *n*—the inverse of wavelength. **E 867,** E17

waves—uneven contours in the surface of a wafer visible to the unaided eye under large-area diffuse illumination. **F 1241,** F01

waves—*on semiconductor wafers*, uneven contours in the wafer surface. **F 1241,** F01

wave train—a succession of ultrasonic waves arising from the same source, having the same characteristics, and propagating along the same path. **E 1316,** E07

waviness—a long-order departure from flatness as opposed to sharp discontinuities. Amplitude is in excess of specified surface finish. In general, waviness will exhibit a number of inflection points that, if connected, would form a line whose path may be open or closed (Syn. *warp*).

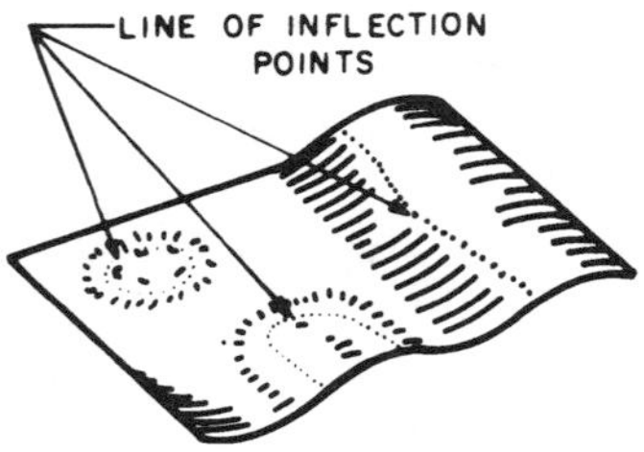

F 109, C21

waviness—the more widely spaced component of surface texture. **F 1241,** F01

wavy cloth, *n*—a cloth that will not lie flat on a cutting table. (Compare **tight pick.**) **D 123,** D13

wavy dressing—more pronounced imprints or markings made by machine knives in dressing than specified for knife marks. **D 9,** D07

wavy edges (waves—edge wave)—a continuous pattern of waves or wrinkles along the edges of the strip, with a relatively flat center portion. **B 846,** B05

wavy face, *n*—a surface condition characterized by a considerable variation in yarn diameter. **D 123,** D13

wax appearance point—the temperature at which wax or other solid substances first begin to separate from the liquid oil when it is cooled under prescribed conditions (Refer to D3117, Test Method for Wax Appearance Point of Distillate Fuels. **D 2864,** D27

wax content of insulating oil—the amount of solid material that separates from a mixture of liquid oil and a suitable wax antisolvent (methyl ethyl ketone, sec-butyl acetate, etc.) under prescribed test conditions. **D 2864,** D27

wax residue—film of wax that migrated onto the wafer surface from several possible sources. **F 1241,** F01

W-board—See **corrugated fiberboard.** **D 996,** D10

"W" configuration (aka "3" configuration)—boom positioned in a "W" shape. **F 818,** F20

wean-galvanized—See **hot-galvanized.** **F 547,** F16

wear—the removal of material or impairment of surface finish through friction or impact. **C 119,** C18

wear—attrition of polish film resulting from normal use. **D 2825,** D21

wear, *n*—the accumulative and integrative action of all the deleterious mechanical influences encountered in use which tend to impair a material's serviceability. Such influences include, but are not limited to abrasion, scratching, gouging and scuffing. **F 141,** F06

wear, *n*—damage to a solid surface, generally involving progressive loss of material, due to relative motion between that surface and a contacting substance or substances. **G 40,** G02

wear coefficient, *n*— *in tribology*, a wear parameter that relates sliding wear measurements to tribosystem parameters. Most commonly, but not invariably, it is defined as the dimensionless coefficient *k* in the equation

$$\text{wear volume} = k\left(\frac{\text{load} \times \text{sliding distance}}{\text{hardness of the softer material}}\right)$$

See also **wear factor.** **G 40,** G02

wear factor, *n*— *in tribology*, a wear parameter that relates sliding wear measurements to operating parameters. Most commonly, but not invariably, it is defined as the total wear volume divided by the normal force or load and also divided by the sliding distance; therefore, this term should always be fully defined in context. **G 40,** G02

wear in, v—See **run in.** **G 40,** G02

wear in, n—See **run in.** **G 40,** G02

wearing surface—See **traffic surface.** **C 717,** C24

wearlayer, *n*—the portion of a resilient floor covering that contains or protects the pattern and design exclusive of temporary finishes or maintenance coatings. **F 141,** F06

wear level, *n*—the number of wear refurbishing cycles to which an item has been subjected. **D 123,** D13

wear map, *n*—a calculated or experimentally determined diagram that identifies regions within each of which the wear mechanism or wear rate remains substantially the same, the regions being separated by transition lines or bands that are functions of two or more parameters. **G 40,** G02

wear mechanism map, n— see **wear map.** **G 40,** G02

wear-out, *n*—a tire condition where any point on the tread is reduced to a depth equal to the height of treadwear indicator. **F 538,** F09

wear performance index, *n*—a calculated value that relates the wear performance of a candidate tire to that of a control tire tested in the same test; it may be calculated on the basis of either percent loss or rate of tread wear. **F 538,** F09

wear rate, *n*—the rate of material removal or dimensional change due to wear per unit of exposure parameter, for example, quantity of material removed (mass, volume, thickness) in unit distance of sliding or unit time. **G 40,** G02

wear rate map, n— see **wear map.** **G 40,** G02

wear-refurbishing cycle, *n*—*for a specific wear testing program*, one complete series of events that may be terminated by laundering or dry cleaning. **D 123,** D13

wear resistance—ability of the carbon or ribbon to retain its image-producing capacity after repeated use. **F 221,** F05

wear-service conditions, *n*—the specific conditions under which a textile is used (for example, at school, at work, at leisure, or at home). **D 123,** D13

wear test, *n*—a test in which textiles are subjected to wear-service conditions and evaluated for performance. **D 123,** D13

weatherability—the ability of a polished surface to resist the effects of exposure. **D 2825,** D21

weather aging—placing materials in known environmental conditions to evaluate changes. **F 869,** F08

weather conditions, normal—the (actual or anticipated) range of environmental conditions (rain, snow, hail, wind, temperature, pollution) that will typically occur in a local climatic region over several years. **E 772,** E44

weather cracking, *n*—distinct surface cracks induced by action of ozone in those areas of sidewall that are under tension; the cracks usually form perpendicularly to the direction of stress. **F 538,** F09

weathering—natural alteration by either chemical or mechanical processes due to the action of constituents of the atmosphere, surface water or ground water, or to temperature change. **C 119,** C18

weathering—atmospheric attack on a glass surface. **C 162,** C14

weathering—the mechanical or chemical disintegration and discoloration of the surface of wood that is caused by exposure to light, the action of dust and sand carried by winds, and the alternate shrinking and swelling of the surface fibers with the continual variation in moisture content brought by changes in the weather. Weathering does not include decay. **D 9,** D07

weathering—the process of disintegration and decomposition as a consequence of exposure to the atmosphere, to chemical action, and to the action of frost, water, and heat. (ISRM) **D 653,** D18

weathering, *n*—the surface deterioration of a rubber article during outdoor exposure. **D 1566,** D11

weathering behind glass, *v*—a technique of weathering in which the test specimens are exposed in a glass covered frame constructed of wood, metal, or other satisfactory material which protects the specimen from the effects of rain and weather. The frame shall be open at the back or sides to allow ambient air to circulate over the specimens. **G 113,** G03

weathering of waterborne oil—the combined effects of evaporation, solution, chemical oxidation, and biological decomposition from environmental exposure. **D 1129,** D19

weathering reference material (WRM), ***n***—a reference material whose weathering degradation properties are well documented and repeatable when exposed to identical test conditions. A WRM differs from a SWRM in that its weathering history has not been certified by a recognized agency. **G 113,** G03

weatherproof—*as applied to plywood*, the term is synonymous with **exterior type plywood**. **D 1038,** D07

weather-resistance—measured ability of a material, assembly, or package to maintain designated physical properties and appearance under specified weathering conditions. **D 996,** D10

weather sealer—form of coating applied to the outer surface of a construction to augment its weather resistance. **E 631,** E06

weathertight, *adj*—*in building construction*, impermeable to the passage of air or water or both under certain conditions as determined by test. **C 717,** C24

weathertight—preventing passage of liquid water and gas. **C 717,** C24

weathertightness, *n*—*in building construction*, the degree of permeability to the passage of water or air or both under certain conditions as determined by test. **C 717,** C24

weave—a planar textile process that interlaces, in a specific pattern, strands or yarns oriented in two directions in a specific pattern. **D 3878,** D30

weaving wire, *n*—round wire intended especially for weaving. **B 899,** B02

web—that part of an H-type lock-strip gasket which extends between the flanges, forming two channels and that part of a reglet-type lock-strip gasket which extends between the flanges and spline. **C 717,** C24

webbing, *n*—*in textiles*, a stout narrow fabric with a mass per unit area of at least 0.5 kg/m^2(0.1 lb/ft^2) for each 25.4 mm (1 in.) of width. (Compare **narrow fabric, ribbon**, and **tape.**) **D 123,** D13

webbing, *n*—filaments or threads that may form when adhesive transfer surfaces are separated. (See also **legging** and **stringiness.**) (Compare **teeth.**) **D 907,** D14

Weber's law, *n*—the just-perceptible increment of a stimulus is an approximately constant fraction of the stimulus magnitude over a wide range. **E 284,** E12

weber, Wb—the unit of magnetic flux. The weber is the magnetic flux whose decrease to zero when linked with a single turn induces in the turn a voltage whose time integral is one volt-second. One weber equals 10^8 maxwells. See **magnetic flux.** **A 340,** A06

webs, *n*—the partitions dividing tile or hollow brick into cells. **C 43,** C15

webs—interior supports separating channels of vitrified clay filter block. **C 896,** C04

wedge, *n*—*in the field of adhesive technology*, a tool made from a rigid material, tapering to a thin edge, used to separate adherends by force. **D 907,** D14

wedge, *n*—*in mica*, a crystalline structure in a micablock, which upon splitting results in a yield of pieces thicker at one end than the other. **D 1711,** D09

wedge—in ultrasonic angle-beam examination by the contact method, a device used to direct ultrasonic energy into the material at an angle. **E 1316,** E07

wedge point—point with two convergent planes forming "V". **F 547,** F16

wedging of tile—the difference between two spaced measurements of the length or width of a tile, expressed as a percent of the distance between points of measurement. **C 242,** C21

week, *n*—*(1)* The week for workplace use is considered to be five workdays of approximately 8 h, within a period of seven consecutive days. *(2)* For reporting analysis of outdoor air on a weekly rate, results are calculated to a base of seven consecutive 24-h days. **D 1356,** D22

weekday traffic factor, *n*—the relationship of MAWDT to MADT and MAWET, from permanent counters, used to adjust short-term weekday counts to MADT and MAWET. **E 867,** E17

weep hole—a small diameter hole (usually ¼ in.) drilled into the protective casing above the ground surface that serves as a drain hole for water that may enter the protective casing annulus. **D 653,** D18

weephole, *n*—a small hole allowing drainage of fluid. Compare **drainage hole, vent hole.** **E 631,** E06

weeping, *n*—a condition where undesired ink on the faceplace which contaminates other ink colors. **F 1857,** F05

weft, *n*—see **filling.** **D 4439,** D35

weftless fabric, *n*—*as used in tire building*, a sheet of parallel cords surrounded by uncured rubber compound. **D 123,** D13

weigh—to measure the tire load on one or more tires by using a vehicle scale, an axle-load scale, a portable axle-load weigher, or a wheel-load weigher. These devices are usually subjected to field standard test weights at each locality of use and are adjusted to indicate units of mass. **E 867,** E17

weigh-in-motion, *n*—the process of estimating a moving vehicle's gross weight and the portion of that weight that is carried by each wheel, axle, and axle group, or both, by measurement and analysis of dynamic vehicle tire forces. (See E 1318, E 17.52 and E 1442, E 17.52). **E 867,** E17

weigh-in-motion (WIM)—the process of estimating a moving vehicle's gross weight and the portion of that weight that is carried by each wheel, axle, or axle group, or combination thereof, by measurement and analysis of dynamics forces applied by its tires to a measuring device. **E 867,** E17

weight, *v*—to determine the mass of a material. **D 123,** D13

weight, *n*—the force exerted on a body by gravity. (See also **mass** and**force.**) **D 123,** D13

weight, *n*—*in warp knitting*, the number of tex (yards per pound) of finished fabric. **D 123,** D13

weight, *n*—*as used with fabrics*, mass per unit area. **D 123,** D13

weight—synonymous with mass. The mass of a body is a measure of its inertia, or resistance to change in motion. **E 867,** E17

weighted bottom camber, h_B—the maximum height of the running surface measured from a plane horizontal surface, with the ski held in a plane horizontal orientation and thus subject to deflection due to its weight under the influence of the ski weight. X_{hB} is the location of h_B from the tail of the ski. **F 472,** F27

weight in-motion, (WIM), *n*—the process of estimating a moving vehicle's gross weight and the portion of that weight that is carried by each wheel, axle, or axle group, or combination thereof, by measurement and analysis of dynamic tire forces. **E 867,** E17

weight percent solids—the portion of a coating that remains as part of the dry film expressed as weight. **D 16,** D01

weight resistivity—the resistance in ohms at a specified temperature of a body of uniform cross section and of unit weight and unit length. **B 354,** B01

weir—a submerged refractory wall within a melter to direct glass flow. **C 162,** C14

weir boom (aka skimming boom/barrier)—boom that has a weir skimming device(s) built into its face. **F 818,** F20

Weissenberg camera—an X-ray diffraction camera used mainly for the determination of unknown crystal structures. A variation of the rotating crystal camera, it has a shield in front of the film which admits the diffracted beams of only one layer line. The film moves parallel to the camera axis and in synchronism with the rotation of the crystal. Overlapping spots are thus prevented and the film records not only the position of a diffraction spot but also the position of the crystal at the time the spot was produced. **E 7,** E04

weld decay—not a preferred term. Integranular corrosion, usually of stainless steels or certain nickel-base alloys, that occurs as the result of sensitization in the heat-affected zone during the welding operation. **G 15,** G01

welded pipe—product made from sheet, strip, or plate with a seam made by welding. **B 846,** B05

welded pipe, *n*—a round hollow product made by forming flat stock and joining the single longitudinal seam by welding; it is produced to the particular dimensions commercially known as standard pipe sizes. **B 899,** B02

welded tube—product made from sheet, strip, or plate with a seam made by welding. **B 846,** B05

welding—joining the ends of two wires, rods, metal whose length is great in comparison with the major axis of its cross section, or groups of wires (*1*) by fusing, using the application of heat or pressure or both, by means of a flame torch, electric arc, or electric current or (*2*) by cold pressure. **B 354,** B01

welding, *n—in tribology*, the bonding between metallic surfaces in direct contact, at any temperature. **G 40,** G02

welding glass—colored glass to protect a welder's eyes from injurious radiation. **C 162,** C14

weld-line, *n*—a discontinuity in a molded plastic part formed by the merging of two or more streams of plastic flowing together. **D 883,** D20

weld-mark, *n*—a visible weld-line. **D 883,** D20

weld mark—a deep groove or fissure formed by incomplete union of two or more particles or streams of material flowing together.

F 109, C21

weld-mark, *n*—a visible weld line. (D-20) **F 412,** F17

weld- or knit-line—a mark on, or weakness in, a molded plastic formed by the union of two or more streams of plastic flowing together. **D 883,** D20

weld reinforcement—the portion of the welded joint which extends beyond the inner and outer surface of the base metal of the welded pipe. **B 846,** B05

weld structure—the microstructure of a weld deposit and heat-affected base metal. **E 7,** E04

well, *n—in buttons*, the recess in the center of a sew through flange button that gives aesthetics and identifies the face side. **D 123,** D13

well completion diagram—a record that illustrates the details of a well installation. **D 653,** D18

well screen—a filtering device used to retain the primary or natural filter pack; usually a cylindrical pipe with openings of a uniform width, orientation, and spacing. **D 653,** D18

well screen jetting (hydraulic jetting)—when jetting is used for development, a jetting tool with nozzles and a high-pressure pump is used to force water outwardly through the screen, the filter pack, and sometimes into the adjacent geologic unit. **D 653,** D18

welt, *n*—the piping effect produced when welt cord and cover fabrics are sewn together for ornamental purposes to finish the edges between intersecting surfaces of upholstered furniture cushions, pillows, arms, or backs. **E 176,** E05

welt beating—operation of lightly beating the welt into a smooth position around the edge of a shoe. **F 869,** F08

welt cord, *n*—the continuous small-diameter cylindrical material that is wrapped in fabric and sewn as part of the cover to make a welt edge on upholstered furniture. **E 176,** E05

welted seam, *n—in upholstered furniture*, seam sewn with a strip of covered cord between the two fabric pieces, joined so that the welting shows on the exterior of the furniture unit. **D 123,** D13

welting, *n—in upholstered furniture*, a cord covered by strips of exterior fabric, used in welted seams of upholstery covering. **D 123,** D13

welting shoulder—the shoulder portion of the vegetable-tanned cattlehide leather, tanned with a low load to give the flexibility required for a welt. **D 1517,** D31

welt seam, *n—in home sewing*, a complex seam formed on the inside of the product with one trimmed cut edge enclosed and one stitching line visible on the face side. **D 4965,** D13

Westlake process—{archaic} an automatic process using vacuum gather for producing articles in paste molds. **C 162,** C14

wet, *adj*—covered or saturated with a liquid, generally water. **D 4538,** D33

wet adhesion—the ability of a coating film to adhere tightly to the substrate directly beneath it under wet conditions such as rain, dew, washing, etc. **D 16,** D01

wet-basis moisture content—the moisture content expressed as the ratio of the weight of water in the fuel to the total weight of the fuel. **E 1126,** E48

wet basis moisture content— see **total weight basis moisture content.** **E 1126,** E48

wet-basis moisture content—the moisture content expressed as the ratio of the weight of water in the fuel to the total weight of the fuel. **E 1705,** E48

wet basis moisture content—see **total weight basis moisture content.** **E 1705,** E48

wet blasting—a process for cleaning or finishing by means of a slurry of abrasive in water directed at high velocity against the work pieces. **B 374,** B08

wet-bulb depression ($t - t_w$), *n*—the difference between the dry-bulb temperature and the wet-bulb temperature. **D 1356,** D22

wet-bulb depression, $t - t_w$—the difference between the dry-bulb temperature and the wet-bulb temperature. **D 4023,** D22

wet-bulb temperature—the equilibrium temperature of a liquid vaporizing into a gas. With water and air, wet-bulb and dry-bulb temperatures give a measure of the relative humidity. **D 9,** D07

wet-bulb temperature—See **temperature.** **D 1356,** D22

wet-bulb temperature, *n*—the equilibrium temperature of a liquid vaporising into a gas. With water and air, wet-bulb and dry-bulb temperatures can be used to calculate the relative humidity. **G 113,** G03

wet-bulb temperature, t_w—the temperature indicated by the wet-bulb thermometer of a psychrometer. **D 4023,** D22

wet chop—fiber glass strand which has been chopped directly after application of sizing. See **chopped fiber** and **dry chop**. **C 162,** C14

wet cleaning—under **cleaning,** see *wet cleaning.* **D 459,** D12

wet cockle, *n*—the wave pattern on paper substrate, which may be created by the ink solvent during printing caused as paper fibers expand relieving stresses. **F 1857,** F05

wet crock—the degree of polish rub-off obtained during rubbing tests using water between a cloth (clothing) and the shoes. **D 2825,** D21

wet deposition—See **deposition.** **D 1356,** D22

wet edge—the edge of a wet, painted area that remains workable for continued painting. **D 16,** D01

wet edge extender—high boiling liquids such as propylene glycol added to latex or water-based paints to reduce the evaporation rate and thereby extend the (open) time that they are wet enough to brush into. **D 16,** D01

wet-felting—forming of a fibrous-felted board mat from a water suspension of fibers and fiber bundles by means of a deckle box, fourdrinier, or cylinder board machine. **D 1554,** D07

wet-film gage—a gage for measuring the thickness of a wet film. **C 717,** C24

wet-film thickness—the thickness of a liquid coating as it is applied. **C 717,** C24

wet gas—natural gas that contains water vapor in excess of sales or contractual specifications, or both. **D 4150,** D03

wet impingement—See **impingement.** **D 1356,** D22

wetland—land which has the water table at, near, or above the land surface, or which is saturated for long enough periods to promote hydrophylic vegetation and various kinds of biological activity which are adapted to the wet environment. **D 653,** D18

wet layup, *n*—a method of making a reinforced plastic in which the polymer compound is applied as a liquid and as the reinforcement is put in place. **D 883,** D20

wet milling—the grinding of porcelain enamel materials with sufficient liquid to form a slurry. **C 286,** B08

wet milling, *n*—the grinding of materials with sufficient liquid to form a slurry. **C 1145,** C28

wet pressing—See *wet pressing* under **pressing.** **C 242,** C21

wet process—See *wet process* under **process.** **C 242,** C21

wet process enameling—a method of porcelain enameling in which slip is applied to a metal article at ambienttemperature, dried and fired. **C 286,** B08

wet sieving—*for the purpose of Test Method D 4749*, the test method for the sieving of coal that uses water as a medium for facilitating the segregation of the sample into particle size; this is generally used when testing coal particles 600 μm (No. 30 U.S.A. Standard Series) or smaller. **D 121,** D05

wet slurry technique—a magnetic particle examination technique in which the magnetic particles are suspended in a high-viscosity vehicle. **E 1316,** E07

wet spots, *n*—local patches of high moisture content in raw rubber. **D 1566,** D11

wet storage stain (coatings)—See **white rust.** **D 16,** D01

wet strength, *n*—See *wet strength* under **bond strength.** **D 907,** D14

wet strength—See **strength, wet.** **E 1749,** E06

wet strength paper— See **paper.** **D 996,** D10

wettability—the rate at which particles can be made wet under specified conditions. **D 2652,** D28

wettage, *n*—See *static holdup.* **D 4175,** D02

wet technique—the examination technique in which the magnetic particles are suspended in a liquid vehicle. **E 1316,** E07

wet tensile strength, *n*— *of paper*, tensile strength of water-saturated paper or paperboard measured under specified conditions. **D 1968,** D06

wetting—the formation of a relatively uniform, smooth, unbroken, and adherent film of solder to a base material. **B 846,** B05

wetting—the property of a polish to uniformly and completely contact the solid surface to which it is applied. **D 2825,** D21

wetting action—the ability of a liquid to spread over and adhere to solid surfaces. **E 1316,** E07

wetting agent—a substance that reduces the surface tension of a liquid, thereby causing it to spread more readily on a solid surface. **B 374,** B08

wetting agent—a chemical additive that reduces the surface tension of a fluid, inducing it to spread readily on a surface to which it is applied, thus causing wetting of the surface of the solid with the fluids. **C 242,** C21

wetting agent, *n*—a chemical additive which reduces the surface tension of a fluid, inducing it to spread readily on a surface to which it is applied, thus causing wetting of the surface with the fluids. **C 1145,** C28

wetting agent—a material that increases the spreading of a liquid medium on a surface. **D 459,** D12

wetting agent—a substance capable of lowering the surface tension of liquids, facilitating the wetting of solid surfaces, and facilitating the penetration of liquids into the capillaries. **D 653,** D18

wetting agent—wetting agents can be considered synonymous with spreading agents in function. **E 1519,** E35

wetting agent (latex), *n*—a substance used to reduce the surface tension of latex and thereby facilitate spreading or impregnation of a surface by the latex. **D 1566,** D11

wetting and adhesion, surface, *n*—the mutual affinity of and bonding between finish and the surface to which it is applied. **C 168,** C16

wet unit weight—see **unit weight**. **D 653,** D18

wet winding, *n*—a method of making filament-wound reinforced plastics in which the fiber reinforcement is coated with a polymer compound as a liquid just prior to wrapping on a mandrel. **D 883,** D20

wheel, *n*—a rigid structure consisting of a rim connected to a central disk that permits rotationally centered attachment to an axle. **F 538,** F09

wheelbarrow sprayer—a sprayer apparatus mounted on a frame with wheelbarrow-type handles and one or two wheels. **E 1102,** E35

wheel center, *n*—the point at which the spin axis of a wheel intersects the wheel plane. **F 538,** F09

wheel force transducer system—a force-to-electrical signal converter system including transducer(s), associated signal condition, zeroing, amplifying, recording, and monitoring instrumentation.

E 867, E17

wheel load—the sum of the tire loads on all tires included in the wheel assembly which comprises a half-axle. **E 867,** E17

wheel plane, *n*—the central plane of a tire that is mounted on the wheel, normal to the spin axis. **F 538,** F09

wheel search unit—an ultrasonic device incorporating one or more piezoelectric elements mounted inside a liquid-filled flexible tire. The beam is coupled to the test surface through the rolling contact area of the tire. **E 1316,** E07

wheel torque, *n*—the external torque acting about the spin axis. **F 538,** F09

wheel track , *n*—a line or path followed by the tire of a road vehicle on a traveled surface. **E 867,** E17

whippers—a mechanical device used to beat air into a beverage so as to change its properties from a liquid drink to a frothy drink. **F 1827,** F26

whirling psychrometer—See **psychrometer, sling.** **D 1356,** D22

whisker, *n*—a short, single-crystal fiber. **D 883,** D20

whisker—two triangular pieces of metal formed during cutting of nail point. **F 547,** F16

whiskering, *n*—fine hair-like lines that are dragged from solid print areas to non-print areas. **D 6488,** D01

whiskers—metallic filamentary growths, often microscopic, sometimes formed during electrodeposition and sometimes spontaneously during storage or service, after finishing. **B 374,** B08

whiskers—See **striation**. **C 162,** C14

whiskers at trailing edge of solid areas, *n*—ragged edges around the printed image caused by improper roller setting or by an impression roll that is too hard (improper durometer). **D 6488,** D01

white, *adj*—*in color determination,* color description most usually applied to opaque, highly reflecting, highly diffusing, visually hueless specimens. **D 2946,** C17

white iron, *n*—cast iron in which substantially all of the carbon is in solution and in the combined form. The metal has a white fracture. **A 644,** A04

white light—see **visible light.** **E 1316,** E07

white light—light containing all wave lengths in the visible spectrum (in the range from 380 to 780 nm). **E 1316,** E07

whiteness, *n*—*in asbestos fiber and products*, the degree to which a near-white surface approaches *perfect white* defined as a 100 % reflectance over the whole visible spectrum. Other terms used for this property are **lightness** or **luminous apparent reflectance**. The concept of whiteness is not only applicable to near-white surfaces but also to dark and colored surfaces. Whiteness may be defined as the grading which an observer would assign to the surface, irrespective of its color or hue, when compared under daylight conditions against a scale of grays ranging from white to black. **C 460,** C17

whiteness—the degree of approach of the color of a substance to that of the ideal white, commonly represented by a standardized preparation of magnesium oxide. **D 1695,** D01

whiteness, *n*—*in asbestos fiber and products,* the degree to which a near-white surface approaches *perfect white* defined as a 100 % reflectance over the whole visible spectrum. Other terms used for this property are *lightness* or *luminous apparent reflectance.* The concept of whiteness is not only applicable to near-white surfaces but also to dark and colored surfaces. Whiteness may be defined as the grading which an observer would assign to the surface, irrespective of its color or hue, when compared under daylight conditions against a scale of grays ranging from white to black. **D 2946,** C17

whiteness, *n*—attribute of color perception by which an object color is judged to approach the preferred white. **E 284,** E12

whiteness index, *n*—a number, computed by a given procedure from colorimetric data, that indicates the degree of departure of an object color from a preferred white. **E 284,** E12

whiteness-retention—comparative whiteness of original and cleaned fabric. **D 459,** D12

white noise—noise with a continuous frequency spectrum and with equal power per unit bandwidth. For example, equal power in any band of 100-Hz width. **C 634,** E33

white rust—white corrosion products (zinc hydroxide and zinc oxide) on zinc-coated articles. They form when the parts are stored so close together that condensed moisture is entrapped between them and the air circulation is inadequate to assist drying. Also called wet storage stain. See **rust.** **D 16,** D01

white sidewall, *n*—a sidewall which contains a white (or light colored) compound as a part of the total sidewall. **F 538,** F09

white speck—in western softwoods, pockets of decay caused by *Fomes pini.* **D 9,** D07

white standard—a substance that reflects 100 % light and is used in calibrating test instruments. See **barium sulfate**. **F 149,** F05

whiteware—See **ceramic whiteware.** **C 242,** C21

white wash—a combination of hydrated lime (or slaked quicklime), water, and other materials to be used as a paintlike coating. **C 51,** C07

white weight—the weight of limed, unwashed stock. **D 1517,** D31

white wool, *n*—wool having shade variations from true white to creamy white but free of pigmented, dyed, or otherwise colored wools. **D 123,** D13

white write—a process in electrostatic printing where the photoconductive element is charged with a charge of the opposite sign as that of the toner. A light beam, acting like a "charge eraser" is used to discharge all areas of the photoconductor that are NOT to receive toner to form the image. The toner is attracted to the remaining charged areas of the photoconductor when the latent electrostatic image is developed. **F 909,** F05

white (X-rays)—containing a large number of wavelengths. **E 7,** E04

whiting—calcium carbonate powder of high purity. **C 242,** C21

whole air sampling, *n*—the collection of an air sample into a sealable container such as a canister, bottle, or bag for subsequent analysis of its contents. **D 1356,** D22

whole sediment—sediment and associated pore water that have had minimal manipulation. **E 943,** E47

whole tree chips (WTC)—wood chips produced by chipping whole trees, usually in the forest. Thus the chips contain both bark and wood. They are frequently produced from low quality trees or from tops, limbs, and other logging residues. **E 1126,** E48

wicking, *n*—transmission of a gas or liquid, due to a pressure differential or capillary action, along fibers incorporated in a rubber product. **D 1566,** D11

wicking—absorption of film by materials allowed to rest on polished substrate, for example, felt pads on bottoms of decorator pieces. **D 2825,** D21

wide crown—staple crown usually larger than 11⁄16 in. (17 mm) in width. **F 592,** F16

wide elastic fabric, *n*—an elastic fabric that is at least 150 mm or 6 in. in width. (Compare **narrow elastic fabric.**) **D 123,** D13

widefield eyepiece—an eyepiece that permits the observation of an extended field of view of the specimen. **E 7,** E04

wide ring—a rate of growth of less than four annual rings per inch, that is, a growth rate faster than that described by *medium grain.*

D 9, D07

wide strip tensile test, *n*—for geosynthetics, a tensile test in which the entire width of a 200 mm (8.0 in.) wide specimen is gripped in the clamps and the gage length is 100 mm (4.0 in.). **D 4439,** D35

wide-to-narrow-ratio, *n*—a calculated proportion comparing the width of wide elements to narrow elements in a bar code having two widths **F 1294,** F05

wide-width strip tensile test, *n—for geotextiles*, a uniaxial tensile test in which the entire width of a 200-mm (8.0-in.) wide specimen is gripped in the clamps and the gage length is 100 mm (4.0 in.). **D 4439,** D35

Widmannstätten structure—a precipitate structure, resulting from the precipitation of a new phase along certain crystallographic planes of the parent solid solution and characterized by a geometrical pattern appearance in the microstructure, originally observed in meteorites but readily produced in many other alloys with proper heat treatment. **E 7,** E04

width, *n*—the maximum dimension of the tile measured perpendicular to the length. **C 43,** C15

width, *n—of flat knit fabrics*, the perpendicular distance between the selvages when the fabric is under zero tension and free of folds or wrinkles. **D 123,** D13

width, *n—of circular knit fabrics*, twice the perpendicular distance between the enclosed edges of a flattened tube of fabric that is under zero tension and free of folds or wrinkles. **D 123,** D13

width, *n—of a raised-surface fabric*, the dimension included within the outer limits of the nap or pile, but excluding the selvages, or as otherwise agreed upon by the purchaser and seller. **D 123,** D13

width, *n—of a fabric*, the distance from the outer edge of one selvage to the outer edge of the other selvage, measured perpendicular to the selvages while the fabric is held under zero tension and is free of folds and wrinkles. **D 123,** D13

width, *n—of a fabric woven on a shuttleless loom*, the distance from the outer warp on one side to the outer warp on the other side, measured perpendicular to the warp yarns while the fabric is held under zero tension and is free of folds and wrinkles. **D 123,** D13

width—in the case of a beam, the cross-sectional dimension perpendicular to the direction in which the load is applied. **D 883,** D20

width—in the case of a beam, the shorter dimension perpendicular to the direction in which the load is applied. (D-20) **F 412,** F17

width—See **spacing.** **F 547,** F16

width—the width measurement is a straight line running across the bottom of the last at the ball or the widest part of the foot. **F 869,** F08

width, *b*—total distance measured perpendicular to the center line on the running surface. X_b indicates the location of ski width from the tail of the ski. **F 472,** F27

width code, *n—in bar code symbology*, a binary code that assigns a value of one to the wide element (bar or space) and a value of zero to the narrow element. **F 1294,** F05

width of metal connector plate—dimension of metal connector plate perpendicular to longitudinal axis of coiled metal strip from which plate was sheared during its fabrication. **E 1807,** E06

widthwise direction, *n—in textiles*, the direction in a machine-made fabric perpendicular to the direction of movement the fabric followed in the manufacturing machine. **D 123,** D13

widthwise direction, *n—in textiles*, the direction in machinemade fabric perpendicular to the warp. (Syn. *cross-machine direction, weftwise direction, fillgwise direction*). **D 5684,** D13

wilderness setting—situations in which the delivery of patient care by EMS providers is complicated by the remoteness with respect to logistics and access; a significant delay in the delivery of care to the patient; an environment that is physically stressful to both patients and rescuers; or the lack of equipment, supplies, and transporation. **F 1177,** F30

wildness, *n*—an obsolete term previously used to describe a number of cord properties including flare, straightness, and residual torsion. **D 123,** D13

wild type—the naturally occurring, original isolate. **E 1705,** E48

willow—(*1*) *willow grain*—refers to boarded leather.

(*2*) *willow tanned*—In the sporting goods industry, used to indicate flexible, well-oiled, chrome-tanned cattlehide or horsehide used for gloves. **D 1517,** D31

WIM system—a set of sensors and supporting instruments which measures the presence of a moving vehicle and the related dynamic tire forces at specified locations with respect to time; estimates tire loads, speed, axle spacing, vehicle class according to axle arrangement, and other parameters concerning the vehicle; and processes, displays, and stores this information. **E 867,** E17

winding—a process in which continuous material is applied under controlled tension to a form in a predetermined geometric relationship to make a structure. A matrix material to bind the fibers together may be added before, during or after winding. Filament winding is the most common type. **D 3878,** D30

winding loss, (copper loss), P_w—the power expended, as heat, in the conductors of an inductor or resistor, or both, as a result of the electric current in them. **A 340,** A06

winding system, *n—in open-end spinning machines*, a device which forms the yarn package. **D 123,** D13

winding wire—see **magnet wire.** **D 1711,** D09

window—in a chamois skin, a thin portion that transmits light when the skin is viewed against a window or light background. **D 1517,** D31

window—See **windows and doors.** **E 631,** E06

window, *n*—assembly consisting of a surrounding frame, and one or more window sashes, ventilators, or fixed lites (panes) of glass, or a combination of these, designed to be installed in a wall opening for the purpose of admitting light or air, or both. **E 1605,** E06

window frame, *n*—assembly of structural members that surrounds and supports the sash, ventilators, doors, panels, or glazing that is installed into an opening in a building envelope or wall. **E 1605,** E06

window glass—See **sheet glass.** **C 162,** C14

windows, *n—parts of zippers*, the openings in pin-lock and cam-lock sliders through which the locking pin and cams, respectively, may extend. **D 123,** D13

windows and doors:

back bedding—a **bead** of sealant, glazing compound, or putty, applied between the face of glass and the **frame** containing it.

back putty—Use **back bedding.**

balance, n—a mechanism used in hung window assemblies to provide mechanical assistance in raising the operable **sash** and providing a means of holding the **sash** in the open position.

bead, n—in glazing, (*1*) a strip of metal or wood used around the periphery of a **pane** of glass to secure it in place (also referred to as a "stop"). (*2*) a strip of sealant, glazing compound, or putty.

bite, n—the distance that the surround member (rail or stile) overlaps the **glazing.**

fixed, adj—describing a **sash, panel,** or **glazing** designed not to be opened (antonym: **operable**).

frame, n—an assembly of structural members that surrounds and supports the **sash, ventilators,** doors, **panels,** or **glazing** that is installed into an opening in a **building envelope** or **wall.**

glaze, v—to install **glazing.**

glazing, n—a material installed in a **sash, ventilator,** or **panel** such as glass, plastic, etc.

glazing bead, n—a **glazing material** used in a **sash, ventilator, panel, window** or door assembly that retains the glazing.

glazing material, n—the **components** used to install **glazing** into its surrounding edge members, such as gaskets, sealants, glazing retainers, etc.

head, n—an upper horizontal member of a **window** or door **frame.**

horizontal sliding window, n—a window assembly in which the operable **sash** (es) moves horizontally in the plane of the **window.**

hung window, n—a window assembly in which the operable **sash** (es) moves vertically in the plane of the **window** and having a balance(s) to aid in the operation of the sash. **Windows** may be single, double, or triple hung depending on the number of **operable sash** (es).

interlock, n—a set of **meeting rails** or meeting **stiles** that contains a provision for each of the **rails** or **stiles** to physically engage one another over their entire length.

jamb, n—a vertical member of a **window** or door **frame.**

light, n—Use **lite.**

lite, n—one piece of **glazing** (preferred term) (also spelled light) (synonym: pane).

meeting rail, n—a **rail** that overlaps another rail.

mull, v—to join or connect **frame** members of **windows** or doors; or a **frame** member to a **mullion.**

mullion, n—a member used between **windows** or doors as a means of connection, which may or may not be structural.

muntin, n—a member used between **lites** of **glazing** within a **sash, ventilator,** or **panel.**

"O"—in window and door design, a designation used to indicate a **fixed sash, panel,** or **lite.**

operable, adj—describing a **sash, ventilator,** or **panel** designed to be opened and closed (antonym: **fixed**).

pane, n—See **lite.**

panel, n—an assembly of one or more **lites** of **glazing,** encompassed by surrounding edge members, which when **operable,** slides horizontally in the plane of a sliding door.

rail, n—a horizontal surrounding edge member of a **sash, ventilator,** or **panel.**

sash, n—an assembly of one or more **lites** of **glazing,** encompassed by surrounding edge members, which when **operable,** slides in the plane of the **window.**

sealed insulating glass, n—an assembly of two or more **lites** separated by a dehydrated gaseous space(s), the entire assembly being sealed to resist passage of water vapor or gas.

sill, n—a lower horizontal member of a **window** or sliding door **frame.**

sliding glass door, n—a door assembly in which the **operable panel** (s) moves horizontally in the plane of the door.

stile, n—a vertical surrounding edge member of a **sash, ventilator,** or **panel.** **E 631,** E06

window sash—assembly of one or more lites (or panes) of glazing, encompassed by surrounding edge members, which when operable, slides in the plane of the window. **E 1605,** E06

window stool—flat, horizontal molding fitted over the sill, on the window interior, between jambs, that comes in contact with the bottom rail of the (lower) operating sash and the window sill. **E 1605,** E06

wind response—change in freeboard or draft due to wind force acting to displace the boom from rest. **F 818,** F20

wipe, *v—in upholstery cleaning*, to clean coated upholstery fabrics with a sponge or cloth wetted with a mild detergent, soap solution, or coated fabric cleaner formulated for the purpose. **D 123,** D13

wiper—see **slider**. The term "wiper" is sometimes used interchangeably with the more preferred term "slider." **B 542,** B02

wire, *n*—a single continuous length of metal, generally with a circular cross section, that is cold drawn from wire rod or bar. **A 902,** A05

wire—a rod or filament of drawn or rolled metal whose length is great in comparison with the major axis of its cross section. **B 354,** B01

wire—a rod or filament of drawn or rolled metal whose length is great in comparison with the major axis of its cross section. **B 713,** B01

wire, *n*—solid wrought product that is long in relation to its cross section, which is square or rectangular with sharp or rounded corners or edges, or is round, hexagonal, or octagonal, and whose diameter or greatest perpendicular distance between parallel faces is less than 0.375 in. [up through 10.00 mm].

Alclad wire, n—composite wire product comprised of an aluminum-alloy wire having on its surface a metallurgically bonded aluminum or aluminum-alloy coating that is anodic to the alloy to which it is bonded, thus electrolytically protecting the core against corrosion.

cold-heading wire, n—wire of a quality suitable for use in the manufacture of cold-headed products such as bolts and rivets.

drawn wire, n—wire brought to final dimension by drawing through a die.

extruded wire, n—wire produced by hot extruding.

flattened wire, n—wire having two parallel flat surfaces and rounded edges produced by roll-flattening round wire.

rivet wire, n—See *cold-heading wire.* **B 881,** B07

wire, *n*—a thin, flexible continuous length of metal, usually of uniform, round cross section. **B 899,** B02

wire—a single strand flexible anchor component with a circular cross section that connects the bony elements of the spine, pelvis, or ribs to each other or to other implant components. A series of wire components can be bound together to form a cable (see **cable**). **F 1582,** F04

wire—used extensively in fastener manufacturing. It is produced from hot rolled or annealed rods or bars by cold drawing. Steel sizes range from 0.062 to 1.375 in. **F 1789,** F16

wire bar—refinery shape used for rolling into rod or flat product for subsequent processing into wire, strip, or shape. Approximately 3½ to 5 in. (89 to 127 mm) square in cross section, usually 54 in. (1.56 m) in length, and ranging in weight from 200 to 420 lb (91 to 191 kg). Usually tapered on both ends. **B 846,** B05

wirebound box— See **box.** **D 996,** D10

wire brush surface—a roughness due to fibers protruding above the surface of the pultruded part.

NOTE—This condition usually occurs at the edge of a reinforcing mat strip and is typically located at the edges or corners of the profile. **D 3918,** D20

wired glass—flat glass with a layer of wire mesh completely embedded in the glass. **C 162,** C14

wire fabric—See **screen.** **E 1481,** E06

wire gage—measure used to describe, by system of arbitrary numbers, diameter of shank or wire from which nail is made. Because several systems of gage designations are in use, reference to particular gage system should be indicated. **F 547,** F16

wire mesh—See **railing systems.** **E 631,** E06

wire mesh—See **screen.** **E 1481,** E06

wire nail—nail manufactured from metal wire or rod. **F 547,** F16

wire rings (hog rings), *n*—wires formed to make attachments during splicing or under certain conditions to serve as an alternate to the tie wires (see *double loop ties*). **F 1379,** F14

wire rod, *n*—a hot-rolled, single continuous length of metal, generally of circular cross section, hot wound or laid into irregular coils, considered a semifinished product, primarily intended for wire drawing. **A 902,** A05

wire rope, *n*—a number of wire strands laid helically about an axis. **A 902,** A05

wire side, *n—of paper*, the side of paper that was in contact with the sheet-forming wire on a paper machine that utilized a single wire (versus a twin-wire) forming section (also called the "bottom side"). **D 1968,** D06

wire side—*in paper,* the side of a sheet next to the wire in paper manufacturing. **F 149,** F05

wires per unit length, *n—for woven pile yarn floor covering*, the number of binding sites per unit of floor covering length; wires in the widthwise direction being the usual means of forming the pile. **D 123,** D13

wire strand, *n*—a number of wires laid helically about an axis; may or may not contain a center wire. **A 902,** A05

withdrawal force—the force required to separate two mated connector halves. **B 542,** B02

withdrawal pressing, *n*—a powder consolidation method in which the die moves downward in relation to the lower punch(es) during compaction, it further descends over the fixed lower punch(es) for ejection, so that the compact may then be pushed off the tooling at this point. **B 243,** B09

with grain, *n—in manufactured carbon and graphite product technology*, the direction in a carbon or graphite body with preferred orientation that has the maximum a-axis alignment. **C 709,** D02

within-laboratory, between days variability—the precision of a method expressed as the agreement attainable between independent determinations (each the average of duplicates) performed by one analyst using the same apparatus and techniques on each of two days. **E 1547,** E15

within-laboratory standard deviation, *n*—the standard deviation of test results obtained within a laboratory for a single material under conditions that may include such elements as different operators, equipment, and longer time intervals.

Note—Because the training of operators, the agreement of different pieces of equipment in the same laboratory and the variation of environmental conditions with longer time intervals depend on the degree of within-laboratory control, the within-laboratory standard deviation is likely to vary appreciably from laboratory to laboratory. **E 456,** E11

witness tire, *n*—a reference tire with an extended period of stability for specified characteristic properties. **F 538,** F09

wobble—in electromagnetic testing, an effect that produces variations in coil spacing (operational lift-off) due to lateral motion of the test specimen in passing through an encircling coil. **E 1316,** E07

wollastonite—a calcium metasilicate mineral with the formula $CaSiO_3$ containing theoretically 48.3 % lime (CaO) and 51.7 % silica (SiO_2), occurring in acicular masses of elongated triclinic crystals, usually white or pale gray. **C 242,** C21

wood—the tissues of the stem, branches, and roots of a woody plant lying between the pith and cambium, serving for water conduction, mechanical strength, and food storage, and characterized by the presence of tracheids or vessels. **D 9,** D07

wood-base fiber and particle panel materials—a generic term applied to a group of board materials manufactured from wood or other lignocellulosic fibers or particles to which binding agents and other materials may be added during manufacture to obtain or improve certain properties. Composed of two broad types, fibrous-felted and particleboards. **D 1554,** D07

wood cellulose, purified—chemical cellulose from wood. **D 1695,** D01

wood-cement board—a panel material where wood usually in the form of excelsior is bonded with inorganic cement. **D 1554,** D07

wood-destroying organisms—principally the decay-producing fungi, beetles, termites, carpenter ants, marine borers, etc. **D 9,** D07

wood failure, *n*—the rupturing of wood fibers in strength tests on bonded specimens, usually expressed as the percentage of the total area involved which shows such failure. **D 907,** D14

wood failure—*(1) as applied to plywood glueline testing,* the area of wood fiber remaining at the glueline following completion of the specified shear test. Determination is by means of visual examination and expressed as a percent of the test area.

(2) as applied to failure in plywood not directly associated with the adhesive, a rupture, shelling, tearing or breaking of the wood itself. (See also **broken grain.**) **D 1038,** D07

wood fiber, *n*—insulation composed of wood fibers, with or without binders. **C 168,** C16

wood-fibered plaster, *n*—a calcined gypsum plaster containing shredded or ground wood fiber added during manufacture. **C 11,** C11

wood flour—very fine wood particles generated from wood reduced by a ball or similar mill until it resembles wheat flour in appearance, and of such a size that the particles usually will pass through a 40-mesh screen. **D 1554,** D07

wood fuel—fuel derived from biomass composed of woody trees or shrubs. **E 772,** E44

wood fuel—fuel derived from biomass composed of woody trees or shrubs. **E 1126,** E48

wood fuel—fuel derived from biomass composed of woody trees or shrubs. **E 1705,** E48

wood-heel attaching nail—bright, regular-stock-steel, annularly threaded or helically threaded, ⅝ by 0.054 to ⅞ by 0.072-in. nails with slightly countersunk flat head and medium or long diamond point. **F 547,** F16

wood-heel top-lift nail—bright, regular-stock-steel, ½ by 0.041 to ¾ by 0.048-in. nails with flat or brad head and long diamond point. **F 547,** F16

wood laminates, *n*—(See also **laminate,** *n*)

built-up laminated wood, n—an assembly made by joining layers of lumber with mechanical fastenings so that the grain of all laminations is essentially parallel.

glue-laminated wood, n—an assembly made by bonding layers of veneer or lumber with an adhesive so that the grain of all laminations is essentially parallel.

plywood, n— a panel generally flat built up of layers of veneer called plies, united under pressure by an adhesive to create a panel with the bond between the plies as strong as, or stronger than, the wood, and that has the following characteristics: (1) is constructed of an odd number of layers with grain of adjacent layers perpendicular, (2) with a layer consisting of either a single ply or two or more plies laminated with parallel grain direction, and (3) with outer layers and all odd numbered layers generally having the grain direction oriented parallel to the long dimension of the panel. **D 907,** D14

wood-lath nail—blued, regular-stock-steel, 1 and 1 ⅛ by 0.054 and 0.072-in. nails with flat ⅛ to 11⁄64-in. head and medium diamond point. **F 547,** F16

wood pellet—see **fuel pellet**. **E 1126,** E48

wood residue, *n*—the scrap and waste wood resulting from the harvest of trees. **D 1968,** D06

wood substance—the moisture-free material of which wood is composed.

Note—In laboratory tests, wood substance is usually on an extractive-free basis, but in common usage it usually includes all material remaining after oven-drying. There is little variation in the specific gravity of wood substance among temperate-zone species. Variation among species is caused primarily by the proportions of wood substance and air. **D 9,** D07

wood veneer, *n*—a thin sheet of wood, generally within the thickness range from 0.01 to 0.25 in. (0.3 to 6.3 mm) to be used in a laminate. **D 907,** D14

wood wool (excelsior)—long, curly, slender strands of wood used as an aggregate component for some particleboards. **D 1554,** D07

wool—a mass of staple fibers bonded into a three dimensional network, producing a low density material for thermal and/or acoustic insulation. **C 162,** C14

wool, *n*—*used in the generic sense in these tolerances*, the fiber from the fleece of the sheep or lamb, the hair of the Angora or Cashmere goat, rabbit hair, and the so-called specialty fibers from the hair of the camel, alpaca, llama, and vicuna. **D 123,** D13

wool, *n*—the fibrous covering of the sheep, *Ovis* species. **D 123,** D13

wool, *n*—*as defined in the Wool Products Labeling Act of 1939*, "the fiber from the fleece of the sheep or lamb, or hair of the Angora goat or Cashmere goat (and may include the so-called specialty fibers from the hair of the camel, alpaca, llama, and vicuna) which has never been reclaimed from any woven or felted wool product." **D 123,** D13

wool base, *n*—oven-dried scoured wool free of alcohol-extractable matter, mineral matter, vegetable matter, and all impurities. **D 123,** D13

wool content, *n*—the quantity of new and recycled wool, as defined in the Wool Products Labeling Act, which is determined by chemical analysis. **D 123,** D13

woolen run—an indirect yarn numbering system in the woolen system, equal to the number of 1600-yd lengths per pound. (Compare **woolen cut**, **worsted count**.) **D 123,** D13

woolen-spun, *adj*—of, or pertaining to, material produced by the woolen system of yarn spinning as distinct from materials made by the worsted system of spinning. (Compare **worsted-spun.**) **D 123,** D13

woolen system, *n*—a spinning system employing a minimum of drafting and producing yarns of low-bulk density. **D 123,** D13

woolen yarn, *n*—yarn spun from wool fibers which have been carded but not combed or gilled. **D 123,** D13

wool felt, *n*—a felt composed wholly of any one or a combination of new or recycled wool fibers. **D 123,** D13

woolskin—sheepskin tanned with the wool on. **D 1517,** D31

word processors—a combination of pieces of hardware consisting of keyboard, video display, and printer interfaced with a computer for the purpose of combining operator and computer logic control in preparing printed matter. **F 909,** F05

work, *n*—the energy expended in displacing a body; mathematically, force times distance. **D 123,** D13

workability—a subjective term used to describe the ease of fabrication of molten glass. **C 162,** C14

workability, *n*—the ability of mortar to be easily placed and spread. **C 1180,** C12

workability—the degree of ease and smoothness of cut obtainable on wood with hand or machine tools. **D 9,** D07

workability index, *n*—a measure of the moldability of plastic refractories, determined in accordance with Test Method C 181. **C 71,** C08

workability of concrete, *n*—that property determining the effort required to manipulate a freshly mixed quantity of concrete with minimum loss of homogeneity. **C 125,** C09

work area—assigned space within a building, or at an exterior location, that is designated as the region delineated by the scope of work. **E 1605,** E06

work function—the potential barrier that must be overcome to remove an electron from the Fermi level of a specimen to the vacuum level. **E 673,** E42

work hardening—a change in the hardness of a material as a result of plastic deformation. **E 7,** E04

working cotton standard, *n*—a reference standard developed primarily for use within a specific laboratory. (See **Precision**.) **D 123,** D13

working direction—*in manufactured carbon and graphite product technology*, direction of applied force used in forming a solid body; generally the direction of applied molding pressure for a uniaxially molded material and the extrusion direction for an extruded material. **C 709,** D02

working distance—the distance between the surface of the specimen being examined and the front surface of the objective lens. **E 7,** E04

working drawing—a detail drawing, usually produced by a draftsperson under direction of an architect, engineer, or other designer showing form, quantity, and relationship of construction elements and materials; indicating their location, identification, grades, dimensions, and connections. Compare **shop drawing.** **E 631,** E06

working drawing—See **drawing**. **E 1480,** E06

working electrode—the test or specimen electrode in an electrochemical cell. **G 15,** G01

working electrode (controlled-potential coulometry)—the electrode in an electrochemical cell at which the chemical species of interest is quantitatively electrolyzed by controlling the potential of the electrode at a selected constant value. **C 859,** C26

working elongation (WE), *n*—*of rope*, elongation which is immediately recoverable when tension is removed from the rope. **D 123,** D13

working end—the compartment of a melting furnace from which glass is delivered for forming. **C 162,** C14

working flow-rate standard—See **standard.** **D 1356,** D22

working life, *n*—*of a sealant*, the time interval after opening a container of a single component sealant, or after mixing the components of a multi-component sealant, during which application and tooling is possible. **C 717,** C24

working life, *n*—the period of time during which an adhesive, after mixing with catalyst, solvent, or other compounding ingredients, remains suitable for use. (Synonym **pot life**.) (Compare **storage life**.) **D 907,** D14

working life:

adhesive working life—the period of time during which an adhesive, after mixing with catalyst, solvent, or other compounding ingredients, remains suitable for use.

sealant working life—the amount of time faying surfaces can be left open once sealant has been applied and still squeeze out excess sealant on closure to a thickness of 0.005 in. (0.13 mm) or less. **E 1749,** E06

working pressure—the pressure adjudged best for any particular set of conditions encountered during grouting. **D 653,** D18

working range—the range of surface temperature in which glass is formed into ware in a specific process. The "upper end" refers to the temperature at which the glass is ready for working (generally corresponding to a viscosity of 10^2 to 10^3 Pa · s), while the "lower end" refers to the temperature at which it is sufficiently viscous to hold its formed shape (generally corresponding to a viscosity greater than 10^4 Pa · s). For comparative purposes, when no specific process is considered, the working range of glass is assumed to correspond to a viscosity range from 10^3 to $10^{6.6}$ Pa · s. **C 162,** C14

working reference material—a reference material usually prepared by a single laboratory for its own use as a calibration standard, as a measurement control standard, or for the qualification of a measurement method. **C 859,** C26

working standard—See **standard.** **D 1356,** D22

working standard, *n*—an instrument standard or laboratory standard in routine use. **E 284,** E12

working stresses—See **allowable properties.** Archaic. **D 9,** D07

working time, *n*—the elapsed time from the first addition of liquid during mixing of the refractory castable until the first indication that it can no longer be placed by the desired placement method; the total of the mixing time and placement time. **C 71,** C08

working time (chemical-resistant mortars)—the time interval in minutes after initial mixing of the component parts, at a specific temperature and in the absence of direct sunlight, in which the mortar may be applied to a brick or tile surface without curling behind the trowel. **C 904,** C03

workplace—See **space categories**. **E 1480,** E06

work (plating)—the material being plated or otherwise finished. **B 374,** B08

work-practice control—see **engineered control**. **E 1605,** E06

work recovery, *n*—the percent of recoverable work to the total work required to strain a fiber a specified amount under specified conditions. **D 123,** D13

workshop effort—the exclusive, intense concentration of the VA team on the project during the workshop period. **E 833,** E06

workspace—See **space categories**. **E 1480,** E06

workstation—See **building space.** **E 631,** E06

workstation—See **space categories**. **E 1480,** E06

work-to-break, *n*—the total energy required to rupture a specimen to the breaking force during a tensile test. **D 123,** D13

work to break, *n*—the energy expended up to the breaking force. (See **work-to-rupture**. Compare**toughness**.) **D 4848,** D13

work-to-break (W, LF)), *n*—*in tensile testing,* the total energy required to rupture a specimen. **D 4439,** D35

work-to-rupture, *n*—the energy expended to tear apart a material. (See **work-to-break.** Compare **toughness.**) **D 123,** D13

worst area difference—one of the measures of how easily an interpreter can identify the characters of a font set, it is the range of the surface areas enclosed by the nominal COL's of the characters; the greater this range, the more easily the characters can be identified. **F 149,** F05

worsted count, *n*—an indirect yarn numbering system in the worsted system equal to the number of 560-yd lengths per pound. (*Syn.* English worsted count. Compare **woolen run.**) **D 123,** D13

worsted-spun, *adj*—of, or pertaining to, materials produced by the worsted system of yarn spinning as distinct from materials made by the woolen system of spinning. **D 123,** D13

worsted system, *n*—a spinning system adapted to fibers 50 to 225 mm (2 to 9 in.) in length. **D 123,** D13

worsted yarn, *n*—yarn spun from wool fibers which have been carded, and either gilled or combed, or both. **D 123,** D13

worth—the value as defined in monetary terms of a specific function identified. **E 833,** E06

woven fabric, *n*—a structure produced when at least two sets of strands are interlaced, usually at right angles to each other, according to a predetermined pattern of interlacing, and such that at least one set is parallel to the axis along the lengthwise direction of the fabric. **D 123,** D13

woven fabric—See *woven fabric* under **fabric.** **D 3878,** D30

wppm, *abbr.*—an abbreviation for part per million by weight. **D 6384,** D02

WQA—water quality association. **D 6161,** D19

wrap, *n*—a flexible sheet material used to protect items in **packaging.** **D 996,** D10

wrap angle, *n*—*in yarn friction testing*, the cumulative angular contact of the test specimen against the friction-inducing device, expressed in radians. **D 123,** D13

wrap around—the display of misleading reflections from a previously transmitted pulse, caused by an excessively high pulse-repetition frequency. **E 1316,** E07

wrap-around bend, *n*—the bend obtained when a specimen is wrapped in a closed helix around a cylindrical mandrel. **E 6,** E28

wrap-in, *n*—*in vinyl-coated glass yarns*, a method of completing a package after a break by wrapping the two ends together on the package without splicing or tying a knot. **D 123,** D13

wrap seam—a depression or step in the surface finish caused by the lap of the flexible mold or carrier strip after it is removed from the cured pultrusion. **D 3918,** D20

Wratten filter—the trade name for a specific type of color filter. **E 7,** E04

wrinkle, *n*—an objectionable crease, generally short and irregular in shape. **D 123,** D13

wrinkle, *n*—an imperfection in reinforced plastics that has the appearance of a wave molded into one or more plies of fabric or other reinforcing material. **D 883,** D20

wrinkle—a permanent crease or furrow in the grain surface of a hide or leather, incapable of removal by rolling or plating. **D 1517,** D31

wrinkle depression—an undulation or series of undulations or waves on the surface of the pultruded part.

NOTE—This condition can occur in either the lengthwise or crosswise direction of the pultrusion and is caused by reinforcement shifting and crowding (see **folded reinforcement**). Wrinkles affect the flatness of the surface. **D 3918,** D20

wrinkle recovery, *n*—that property of a fabric which enables it to recover from folding deformations. **D 123,** D13

wrinkle resistance, *n*—that property of a fabric which enables it to resist the formation of wrinkles when subjected to a folding deformation. **D 123,** D13

wrinkles, *n*—a condition where ink being absorbed into a substrate coating causes it to swell and stretch to the point where it cannot recede back to its former dimensions, forming folds in the image area. **F 1857,** F05

wrinkling—*in protective coatings*, the formation of a surface appearance in a coating film resembling the skin of a prune. **D 4538,** D33

wrist, *n*—*in anatomy*, the joint which articulates between the end of the lower arm and the hand. **D 123,** D13

wrist, *n—in anatomy*, the joint which articulates between the end of the lower arm and the hand. **F 1494,** F23

wrist girth, *n—in body measurements*, the circumference over the prominence of the inner and the outer forearm bones. **D 123,** D13

wrist girth, *n— in body measurements*, the circumference over the prominence of the inner and outer forearm bones. **F 1494,** F23

write—the images produced by a ribbon or first carbon. **F 221,** F05

writing—the act of producing an image on a document, or other receptor media, one character or stroke at a time, such as by hand with a pen or pencil or by means of a typewriter or pen plotter. **F 335,** F05

writing—the act of producing an image on a document or other receptor media, one character or stroke at a time, such as by hand with a pen or pencil or by means of a typewriter or pen plotter. **F 909,** F05

wrong draw, *n—in woven fabric,* one or more incorrectly drawn warp ends in the harness or reed. **D 123,** D13

wrong side, *n*—See **back side**. **D 4965,** D13

"WS"—clean with either water-based shampoo or dry solvent. **D 5253,** D13

X

"X"—do not clean, use vacuum cleaner or light brush only. **D 5253,** D13

xanthate viscosity—see **viscosity, xanthate.** **D 1695,** D01

X_{bH}*, *X_{bM}*, and *X_{bV}—the *X* coordinates of the locations of these respective widths of the snowboard ski measured from the tail of the snowboard ski. **F 1107,** F27

X_{bH}, X_{bM}, X_{bV}—the *x* coordinates for the location of these respective widths of the ski measured from the tail of the ski. **F 472,** F27

X dimension, *n—in bar code symbology*, the intended width of the narrow elements as set by the application or specification, or both. **F 1294,** F05

xenon arc, *n*—a light source produced by a high intensity electrical current through a tube containing low pressure xenon gas. The spectral energy distribution of this light source is generated by the electrical current arcing through the xenon gas plasma between the electrodes, and is modified by filters. **G 113,** G03

XEPS—see **rigid cellular polystyrene thermal insulation board.** **E 631,** E06

xerographic copying—a form of plain paper copying which uses a dry toner. Synonym for *xerography.* **F 335,** F05

xerographic paper, *n*—a grade of paper suitable for copying by the electrostatic process. **D 1968,** D06

xerography—See **xerographic copying.** **F 335,** F05

xerography—the creation of an image by electro-photography. (See **electro-photography**.) **F 1156,** F12

X-radiation—electromagnetic radiation of the same nature as visible light but having a wavelength approximately 1/1000 that of visible light. **E 7,** E04

X-ray emission spectrometry, *n*—pertaining to emission spectrometry in the X-ray wavelength region of the electromagnetic spectrum. **E 135,** E01

X-ray ghost line—*XPS,* lines in a spectrum due to presence of contaminating X-ray photons from an impurity in the X-ray anode, from the X-ray window, or from certain elements present in the specimen. **E 673,** E42

X-ray image plane, *n—in tire testing*, a surface located at a specified distance from the X-ray tube "focal spot." **F 538,** F09

X-ray imaging system, *n—in tire testing*, a collection of the components and subsystems needed to produce a fluoroscopic or radiographic image of the tire. **F 538,** F09

X-ray linewidth—the energy width of the principal characteristic X-ray; in XPS it usually refers to that of the X-ray source. The X-ray linewidth contributes to the photo-electron peak widths. **E 673,** E42

X-ray monochromator—a device used to eliminate photons of energies other than those in a narrow band. **E 673,** E42

X-ray radiograph, *n*—an X-ray film, plate, or paper that is placed at the image plane and is used for recording an X-ray image of the object being examined. **F 538,** F09

X-rays—See **X-radiation**. **E 7,** E04

X-ray screen, *n*—a fluorescent screen, placed at the image plane, that produces an X-ray image of the object being examined. **F 538,** F09

X-ray standard field—an X-ray field that is well characterized as to the exposure rate, or absorbed dose rate produced in a specified material, at a specified location within the field for a given tube voltage, current, and filtration. **E 170,** E10

X-ray tube—a device for the production of X-rays by the impact of high-speed electrons on a metal target. **E 7,** E04

XRF analyzer—instrument that determines lead concentration, using the principle of X-ray fluorescence. **E 1605,** E06

XRF direct-reading analyzer—an analyzer that provides the operator with a display of a lead concentration calculated from the lead "K" x-ray intensity. Compare **XRF spectrum analyzer.** **E 1605,** E06

XRF reading cycle-XRF direct-reading analyzers—calculate the lead "K" x-ray intensity in a specific time interval (10 to 30 s) that is fixed by the manufacturer and related to the age of the source. The calculated result in this time interval is a *reading cycle.* **E 1605,** E06

XRF sample site—specific spot on a surface being tested for lead concentration by portable or laboratory XRF equipment. **E 1605,** E06

XRF-shoot, *v*—to obtain the results of an XRF single reading cycle. **E 1605,** E06

XRF spectrum analyzer—type of XRF analyzer that provides the operator with a plot of the energy and intensity of both" k" and "L" x-rays, as well as a calculated concentration of lead. Compare **XRF direct-reading analyzer**. **E 1605,** E06

XRF substrate effect—return of backscattered radiation from paint, substrate, or underlying material to the XRF analyzer. **E 1605,** E06

xylan—a pentosan giving almost exclusively xylose on hydrolysis. **D 1695,** D01

xylem—the portion of the tree trunk, branches, and roots that lies between the pith and the cambium. **D 9,** D07

xylene—a mixture of C_8 aromatic hydrocarbons. **D 4790,** D16

xylene, 10°C—xylene having a total distillation range of no more than 10°C. This range brackets the boiling points of the three individual isomers and ethylbenzene. Refer to Specification D 846 for complete specifications. **D 4790,** D16

xylene, nitration grade—xylene consisting principally of the meta isomer and having a total distillation range of no more than 5°C. Nonaromatic impurities is limited to a maximum of 4 volume %. Refer to Specification D 843 for complete specifications. **D 4790,** D16

xylenol(s)—dimethyl phenol(s), hydroxydimethylbenzene ($C_8H_{10}O$) mol weight 122.16; colorless crystalline powders. Five isomeric xylenols exist. **D 4790,** D16

Y

Y—conversion, recovery. **D 6161,** D19

yarn, *n*—a generic term for a continuous strand of textile fibers, filaments, or material in a form suitable for knitting, weaving, or otherwise intertwining to form a textile fabric. **D 123,** D13

yarn—*in fibrous composites,* a continuous, ordered assembly of essentially parallel, collimated filaments, normally with twist, and of either discontinuous or continuous filaments.

single yarn, n—an end in which each filament follows the same twist. **D 3878,** D30

yarn appearance, *n*—the visual effect obtained by viewing a sample of yarn wound with a designated traverse on a black board of designated size. **D 123,** D13

yarn break, *n*—*in sewn seams,* a mode of failure evidenced by yarns rupturing at the seam or at any other area in the test specimen. **D 123,** D13

yarn distortion, *n*—*in woven fabrics,* a condition in which the symmetrical surface appearance of a fabric is altered by the shifting or sliding of warp or filling yarns. **D 123,** D13

yarn number, *n*—a measure of the linear density of a yarn, expressed as "mass per unit length," or "length per unit mass," depending on the yarn numbering system used. (See also **yarn linear density**.) **D 123,** D13

yarn numbering system, *n*—a system expressing the size of a yarn as a relationship between its length and associated mass. (See also **yarn linear density, yarn number, direct yarn numbering system,** and **indirect yarn numbering system**.) **D 123,** D13

yarn number jute, *n*—mass per unit length of a yarn measured as the number of pounds per 13 167 m (14 400 yd), and expressed as pounds per spyndle. **D 123,** D13

yarn package, *n*—a length or parallel lengths of yarn in a form suitable for handling, storing, or shipping. **D 123,** D13

yarn severance, *n*—a numerical value expressed on a percentage basis from this test that is used as an index of the degree of cutting of fabric yarns by the sewing machine needle in making sewn seams. **D 123,** D13

yarn slippage, *n*—*in sewn seams in sewn fabrics,* the displacement of one or more yarns from the original position, causing differences in alignment, spacing, or both. **D 123,** D13

yarn take-up, *n*—*in fabric,* the additional length of yarn used to make a given length of fabric. **D 123,** D13

yaw, *n*—*in a vehicle,* the angular motion of a vehicle about its vertical axis through the center of gravity. **F 538,** F09

yaw velocity, *n*—the magnitude of the yaw (rotation or angular displacement); it may be measured by fore and aft, vehicle versus pavement, velocity sensors. **F 538,** F09

Y dimension, *n*—*in a bar code symbology,* the specified height of elements. **F 1294,** F05

yeast—single cell microorganisms (fungi) that produce alcohol and CO_2 under normal fermentation conditions. **E 1705,** E48

yellow, *adj*—hue description applied to light of wavelengths from 572 to 783 nm and to visually similar stimuli. **D 2946,** C17

yellowness, *n*—*in asbestos fiber and products,* the degree to which asbestos deviates from colorlessness or neutral grey toward yellowish or tan as determined by the yellowness index. **D 2946,** C17

yellowness, *n*—attribute of color perception by which an object color is judged to depart from colorless or a preferred white toward yellow. **E 284,** E12

yellowness index, *n*—*in asbestos fiber and products,* an index calculated arbitrarily from reflectance measurements using tristimulus filters. Use the following relationship for asbestos:

$$\text{yellowness index} = (A - B)/G$$

where:
A = reflectance with the amber tristimulus filter,
B = reflectance with the blue tristimulus filter, and
G = reflectance with the green tristimulus filter.

NOTE—Additional definitions are presented in Specification C 296, Specification C 428, Test Method C 458, Test Methods C 500, Specification C 508, Specification C 541, Specification C 663, Specification C 668, Specification C 875, Guide C 966, Terminology D 123, Test Method D 1918, Test Method D 2985, Test Method D 2590, Test Method D 3879, and STP 834. **D 2946,** C17

yellowness index, *n*—a number, computed by a given procedure from colorimetric or spectrophotometric data, that indicates the degree of departure of an object color from colorless, or from a preferred white, toward yellow. **E 284,** E12

yellow ware—a yellow semivitreous ware or an earthenware with a colorless, clear glaze. **C 242,** C21

yield, *n*—the weight percent of the feed that reports to the concentrate. **D 121,** D05

yield, *n*—*in knitted fabrics,* the number of finished square meter per kilogram (square yards per pound) of greige fabric. **D 123,** D13

yield, *n*—*of wool,* the percentage of a designated commercial composition obtained by processing a lot of raw wool. **D 123,** D13

yield, *n*—*in raw wool,* the combined weight of clean wool fiber present and vegetable matter present, as a percentage of the raw wool weight. **D 123,** D13

yield—*in grouting,* the volume of freshly mixed grout produced from a known quantity of ingredients. **D 653,** D18

yield bend angle—the angle at which the yield bend moment occurs. **F 1840,** F04

yield bend moment—the amount of moment required to initiate plastic deformation during a bend test. **F 1840,** F04

yield factor, *n*— *in papermaking,* either a proportion or percentage representing that portion of the input material that is retained after paper stock preparation for inclusion in the paper manufacturing process. **D 1968,** D06

yielding arch—type of support of arch shape, the joints of which deform plastically beyond a certain critical load, that is, continue to deform without increasing their resistance. (ISRM) **D 653,** D18

yield locus—plot of shear stress versus normal stress at failure. The yield locus (YL) is sometimes called the instantaneous yield locus to differentiate it from the time yield locus. **D 653,** D18

yield point, *n*—in a *stress strain curve,* the point beyond which work is not completely recoverable and permanent deformation takes place. **D 123,** D13

yield point, *n*—that point on the stress-strain curve, short of ultimate failure, where the rate of stress with respect to strain goes through a zero value and may become negative. **D 1566,** D11

yield point, *n*—*in geosynthetics,* the point on the force-elongation curve at which the first derivative equals zero (the first maximum). **D 4439,** D35

yield point elongation, *YPE, n*—in a uniaxial test of a material exhibiting discontinuous yielding at or near the onset of plastic deformation, the strain (expressed in percent) separating the stress-strain curve's first point of zero slope from the point of transition from discontinuous yielding to uniform strain hardening. If the transition occurs over a range of strain, the *YPE* end point is the intersection between (*a*) a horizontal line drawn tangent to the curve at the last zero slope and (*b*) a line drawn tangent to the strain hardening portion of the stress-strain curve at the point of inflection. If there is no point at or near the onset of yielding at which the slope reaches zero, the material has 0 % *YPE*. **E 6,** E28

yield point, *YP* [FL^{-2}], *n*—term previously used, by E 8 and E 8M, for the property which is now referred to as **upper yield strength**. **E 6,** E28

yield strain, *n*—the strain below which a material acts in an elastic manner, and above which it begins to exhibit permanent deformation or flow. **D 907,** D14

yield strain, *n*—the level of strain at the yield point. **D 1566,** D11

yield strength, *n*—the stress at which a material exhibits a specified limiting deviation from the proportionality of a stree strain. **D 123,** D13

yield strength—the tension-applied load at which the fastener experiences a specific amount of permanent deformation, that is, the bolt has been stressed beyond its elastic limit and is in the plastic zone. It is very difficult to test full size bolts for yield strength. Because of different strain rates in the threaded section, thread runout and unthreaded shank which together comprise the stressed length, a "proof load" concept was introduced. **F 1789,** F16

yield strength, σ_{YS} **[FL^{-2}]**—the stress at which a material exhibits a specific limiting deviation from the proportionality of stress to strain. This deviation is expressed in terms of strain. **E 1823,** E08

yield strength, *YS* or S_y [FL^{-2}], *n*—the engineering stress at which, by convention, it is considered that plastic elongation of the material has commenced. This stress may be specified in terms of (*a*) a specified deviation from a linear stress-strain relationship, (*b*) a specified total extension attained, or (*c*) maximum or minimum engineering stresses measured during discontinuous yielding. **E 6,** E28

yield stress—the stress beyond which the induced deformation is not fully annulled after complete destressing. (ISRM) **D 653,** D18

yield stress, *n*—the stress (either normal or shear) at which a marked increase in deformation occurs without an increase in load. (Synonym *yield value*)) **D 907,** D14

yield stress, *n*—the level of stress at the yield point. **D 1566,** D11

yield stress—the maximum shear stress that can be applied without causing permanent deformation. **D 2507,** F07

yield stress, *n— in solids*, the maximum stress that can be applied without causing permanent deformation. **D 4175,** D02

yield stress, *n— in fluids*, the shear stress required to initiate flow. **D 4175,** D02

yield stress, *n*—limit to internal force developed by application of external force or load or generation of internal strain to a material, member, connection component, or assembly beyond which a marked increase in the rate of deformation occurs without an increase in load; expressed in terms of units of force per unit area, psi, pounds force per square inch (Mpa, megapascals). **E 631,** E06

yield stress—limit to internal force developed by application of external force or load or generation of internal strain to a material, member, connection, component, or assembly beyond which a marked increase in deformation occurs without an appreciable increase in load; expressed in terms of units of force per unit of area, megapascals (pound-force per square inch). When the initial rate of force is nonlinear, an agreed upon convention shall apply. Sometimes incorrectly referred to as yield strength and as **ultimate strength**. **E 1807,** E06

yield value, n—Synonym **yield stress**. **D 907,** D14

yoke—a magnet that induces a magnetic field in the area of a part that lies between its poles. Yokes may be permanent magnets or either alternating-current or direct-current electromagnets. **E 1316,** E07

yoke magnetization—a longitudinal magnetic field induced in a part, or in an area of a part, by means of an external electromagnet shaped like a yoke. **E 1316,** E07

Youden square, *n*—a type of block design derived from certain Latin squares by deleting, or adding, rows (or columns) so that one block factor remains complete blocks and the second block factor constitutes balanced incomplete blocks. **E 456,** E11

Youden square, *n*—A type of block design derived from certain Latin squares by deleting, or adding, rows (or columns) so that one block factor remains complete blocks and the second block factor constitutes balanced incomplete blocks. **E 1325,** E11

Young's modulus, *n—in a stress-strain curve for an elastic material*, the ratio of change in stress to change in strain within the elastic region of the material. **D 123,** D13

Young's modulus—the ratio of the increase in stress on a test specimen to the resulting increase in strain under constant transverse stress limited to materials having a linear stress-strain relationship over the range of loading. Also called elastic modulus. **D 653,** D18

Z

Zahn cup—an apparatus for the measurement of liquid or slurry viscosity expressed as the number of seconds required for the liquid or slurry to drain from the cup through a hole of definite diameter. **C 242,** C21

Z-Axis, *n—in bar code reading*, a coordinate related to the orientation and movement of a symbol on a transport system. **F 1294,** F05

Z dimension, *n—in bar code reading*, a calculation of element width achieved by adding the average element width of bar and space and dividing by two. **F 1294,** F05

zeolite—any of various natural or synthetic hydrated aluminum silicates used as ion exchange substrates in water softening. **D 6161,** D19

zephyr yarn, *n*—a variety of soft worsted yarn characterized by a low twist and spun from wool which is as fine or finer in average diameter than U.S. Standard 64's grade tops. **D 123,** D13

zero air voids curve (saturation curve)—the curve showing the zero air voids unit weight as a function of water content. **D 653,** D18

zero air voids density (*zero air voids unit weight*)**—**see **unit weight.** **D 653,** D18

*zero carbon steel***—** see **decarburized enameling steel.** **C 286,** B08

zero crossings—*in fatigue loading*, the number of times that the load-time history crosses zero load level with a positive slope or negative slope, or both, as specified, during a given period. **E 1823,** E08

zero defects—a term which indicates that no deviation from the requirements of a specification are present within a statistically valid sample drawn from a given fastener lot. **F 1789,** F16

zero discharge—a condition whereby a facility discharges no process effluent. **D 6161,** D19

*zero drift***—**See **analyzer.** **D 1356,** D22

zero-filling—addition of zero-valued points to the end of a measured interferogram. **E 131,** E13

*zero path difference point***—** see **centerburst.** **E 131,** E13

*zero retardation point***—** see **centerburst.** **E 131,** E13

zero-span, *adj— in tensile testing*, having a grip separation of 0.00 mm (0.000 in.). **D 1968,** D06

zero–span tensile strength, *n—of paper*, the tensile strength of a sheet of fibrous material, measured with special jaws, at an apparent initial span of zero. **D 1968,** D06

zero suppression, *n—in bar code reading*, the technique used to shorten UPC codes by omitting zeros from the symbol. **F 1294,** F05

zero time, *n*—the time when the given loading or constraint conditions are initially obtained in creep and stress-relaxation tests, respectively. **E 6,** E28

zero time (T_o)—the travel time (correction factor), measured in seconds, associated with the electronic circuits in the pulse propagation system.

C 709, D02

zeta potential—colloidal stability measured in millivolts. High negative value (–10 to –30 mv) results in particulate stability. **D 6161,** D19

zigzagged seam finish, *n—in home sewing*, a seam finish in which machine zigzag stitching is placed 3 to 6 mm (⅛ to ¼ in.) from each of the cut edges of the seam allowances. (Compare **edge-stitched seam finish.**) **D 4965,** D13

zinc coated, galvanized—generic terms covering barrel-galvanized, dipped, electro-galvanized, electro-zinc-plated, flake-galvanized, hot-dipped, hot-dip galvanized, hot-dipped zinc coated, hot-galvanized, mechanically plated, peen-coated, peen galvanized, tumbler-galvanized, and wean-galvanized. **F 547,** F16

zinc-plated—surface provided with usually thin electrochemical deposit of zinc as a result of immersion in electrolytic bath or with mechanical deposit of zinc as a result of peen coating. **F 547,** F16

zinc-rich primer—a primer for ferrous metals, incorporating zinc dust at a concentration sufficient to make the dried film electrically conductive thus providing cathodic protection to the ferrous substrate. **D 16,** D01

zipper, *n*—a slide fastener consisting of interlockable elements each attached to one of the opposing edges of two tapes and a movable part called a "slider" that spans the interlockable elements, which when moved in one direction causes the elements on one tape to interlock with the elements on the other tape, and when moved in the opposite direction causes the elements to disengage. (Compare **continuous element zipper** and **separate-element zipper.**) **D 123,** D13

zipper gasket—See **lock-strip gasket**, the preferred term. **C 717,** C24

zirconium oxide refractory, *n*—refractory products consisting substantially of zirconium dioxide.

C 71, C08

*zircon porcelain***—**See *zircon porcelain* under **porcelain.** **C 242,** C21

zircon refractory, *n*—refractory products consisting substantially or entirely of crystalline zirconium orthosilicate ($ZrSiO_4$). **C 71,** C08

*zircon whiteware***—**See *zircon whiteware* under **ceramic whiteware.**

C 242, C21

Zn-5Al-MM—an abbreviation describing an alloy of zinc, 5 % aluminum, and mischmetal, employed as a metallic coating over steel wire or steel sheet used in the manufacture of chain link fabric, fence framework materials, and other products (see Specifications F 1234 and F 1345).

F 552, F14

Zn-5 Al-MM coated wire, *n*—steel wire-coated with either: (*1*) predominately Zn-5 Al-MM alloy as manufactured by the single-dip method, or (*2*) a Zn-Al-Fe intermetallic inner layer and a Zn-5 Al-MM alloy outer layer as manufactured by the double dip method. **A 902,** A05

zone—any group of crystal planes which are all parallel to one line, called the zone axis. **E 7,** E04

zone of mixing—the layer of the specimen surface within which the primary beam causes atomic mixing. **E 673,** E42

zone of saturation—a hydrologic zone in which all the interstices between particles of geologic material or all of the joints, fractures, or solution channels in a consolidated rock unit are filled with water under pressure greater than that of the atmosphere. **D 653,** D18